eGrade Plus

www.wiley.com/college/cummings

Based on the Activities You Do Every Day

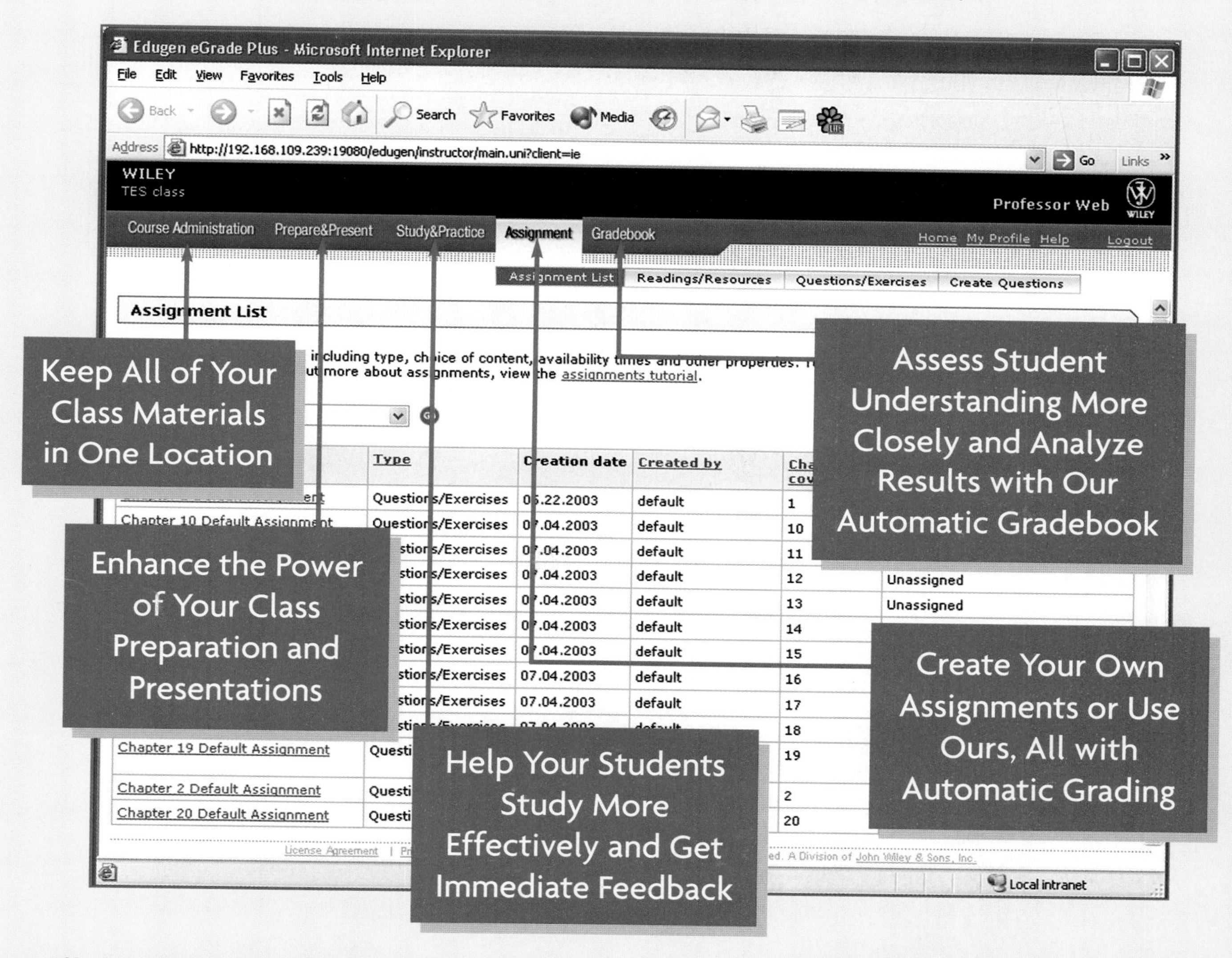

Keep All of Your Class Materials in One Location
Enhance the Power of Your Class Preparation and Presentations
Help Your Students Study More Effectively and Get Immediate Feedback
Assess Student Understanding More Closely and Analyze Results with Our Automatic Gradebook
Create Your Own Assignments or Use Ours, All with Automatic Grading

All the content and tools you need, all in one location, in an easy-to-use browser format.

Choose the resources you need, or rely on the arrangement supplied by us.

Now, many of Wiley's textbooks are available with eGrade Plus, a powerful online tool that provides a completely integrated suite of teaching and learning resources in one easy-to-use website. eGrade Plus integrates Wiley's world-renowned content with media, including a multimedia version of the text, PowerPoint slides, and more. Upon adoption of eGrade Plus, you can begin to customize your course with the resources shown here.

See for yourself!

Go to www.wiley.com/college/egradeplus for an online demonstration of this powerful new software.

Keep All of Your Class Materials in One Location

Course Administration tools allow you to manage your class and integrate your eGrade Plus resources with most Course Management Systems, allowing you to keep all of your class materials in one location.

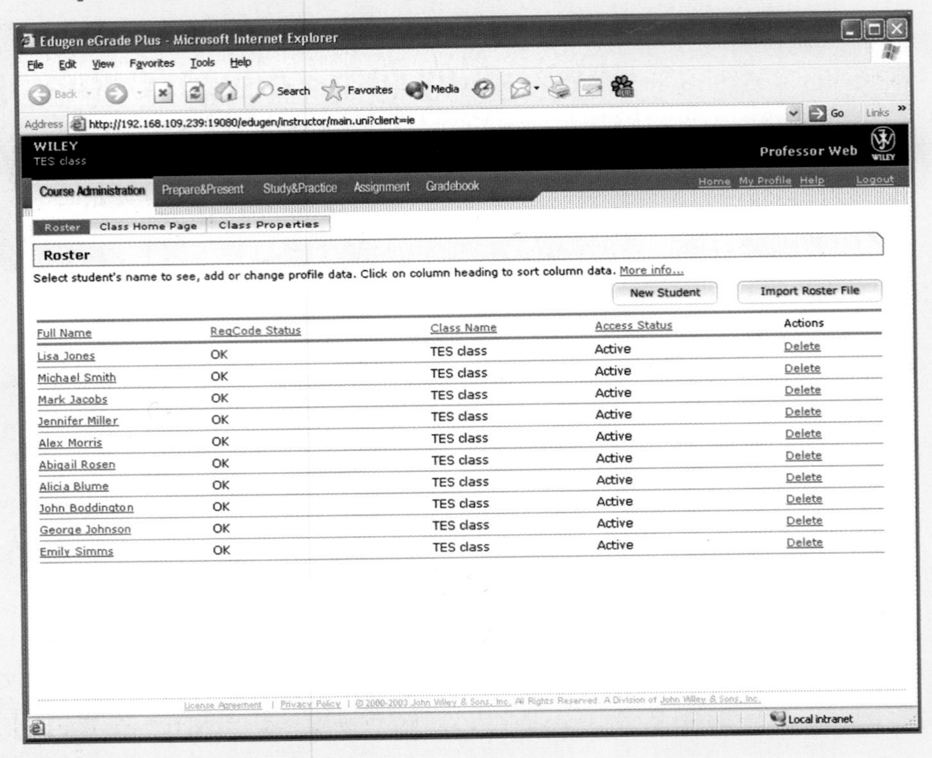

Enhance the Power of Your Class Preparation and Presentations

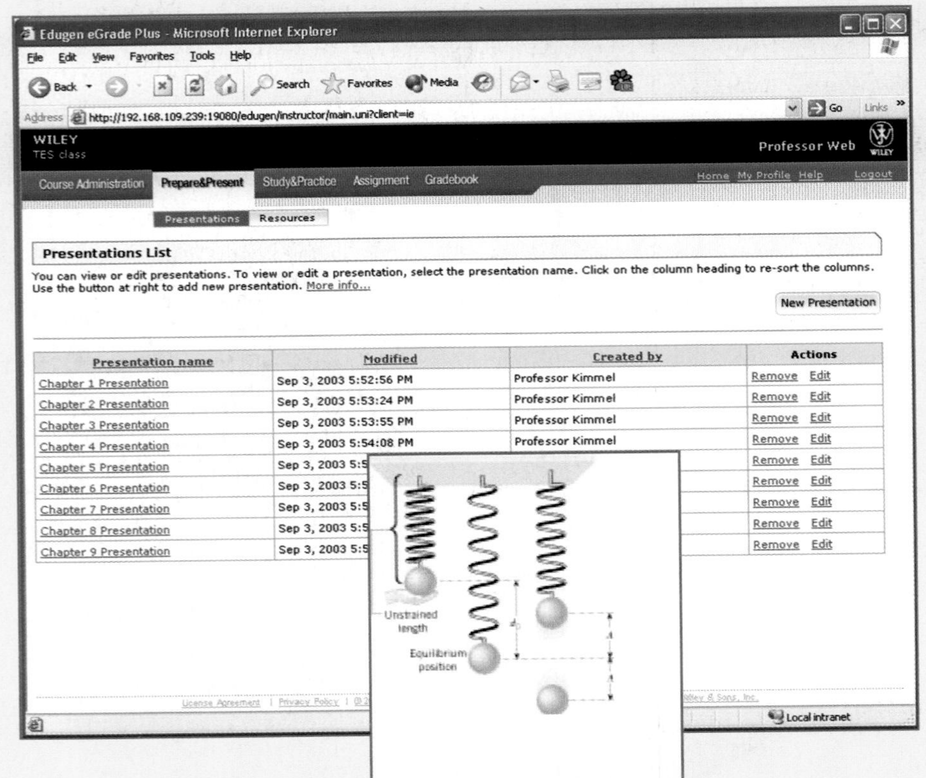

A **Prepare and Present tool** contains all of the Wiley-provided resources, such as **a multimedia version of the text, interactive chapter reviews,** and **PowerPoint slides,** making your preparation time more efficient. You may easily adapt, customize, and add to Wiley content to meet the needs of your course.

Create Your Own Assignments or Use Ours, All with Automatic Grading

An **Assignment** area allows you to create **student homework** and **quizzes** that utilize **Wiley-provided question banks,** and an **electronic version of the text.** One of the most powerful features of eGrade Plus is that student assignments will be automatically graded and recorded in your gradebook. This will not only save you time but will provide your students with immediate feedback on their work.

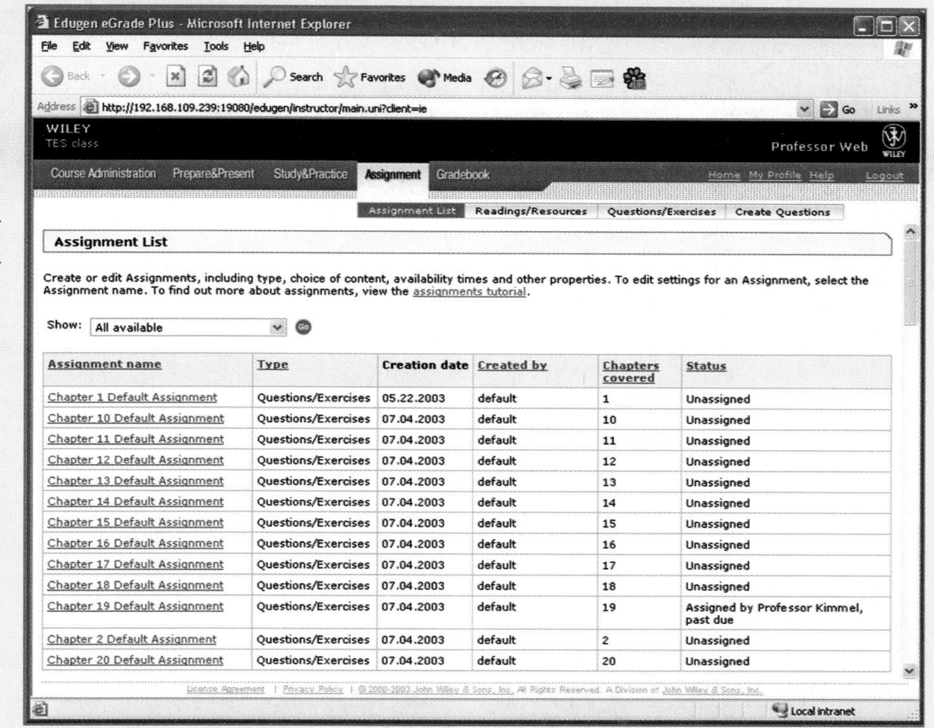

Assignment name	Type	Creation date	Created by	Chapters covered	Status
Chapter 1 Default Assignment	Questions/Exercises	05.22.2003	default	1	Unassigned
Chapter 10 Default Assignment	Questions/Exercises	07.04.2003	default	10	Unassigned
Chapter 11 Default Assignment	Questions/Exercises	07.04.2003	default	11	Unassigned
Chapter 12 Default Assignment	Questions/Exercises	07.04.2003	default	12	Unassigned
Chapter 13 Default Assignment	Questions/Exercises	07.04.2003	default	13	Unassigned
Chapter 14 Default Assignment	Questions/Exercises	07.04.2003	default	14	Unassigned
Chapter 15 Default Assignment	Questions/Exercises	07.04.2003	default	15	Unassigned
Chapter 16 Default Assignment	Questions/Exercises	07.04.2003	default	16	Unassigned
Chapter 17 Default Assignment	Questions/Exercises	07.04.2003	default	17	Unassigned
Chapter 18 Default Assignment	Questions/Exercises	07.04.2003	default	18	Unassigned
Chapter 19 Default Assignment	Questions/Exercises	07.04.2003	default	19	Assigned by Professor Kimmel, past due
Chapter 2 Default Assignment	Questions/Exercises	07.04.2003	default	2	Unassigned
Chapter 20 Default Assignment	Questions/Exercises	07.04.2003	default	20	Unassigned

Assess Student Understanding More Closely

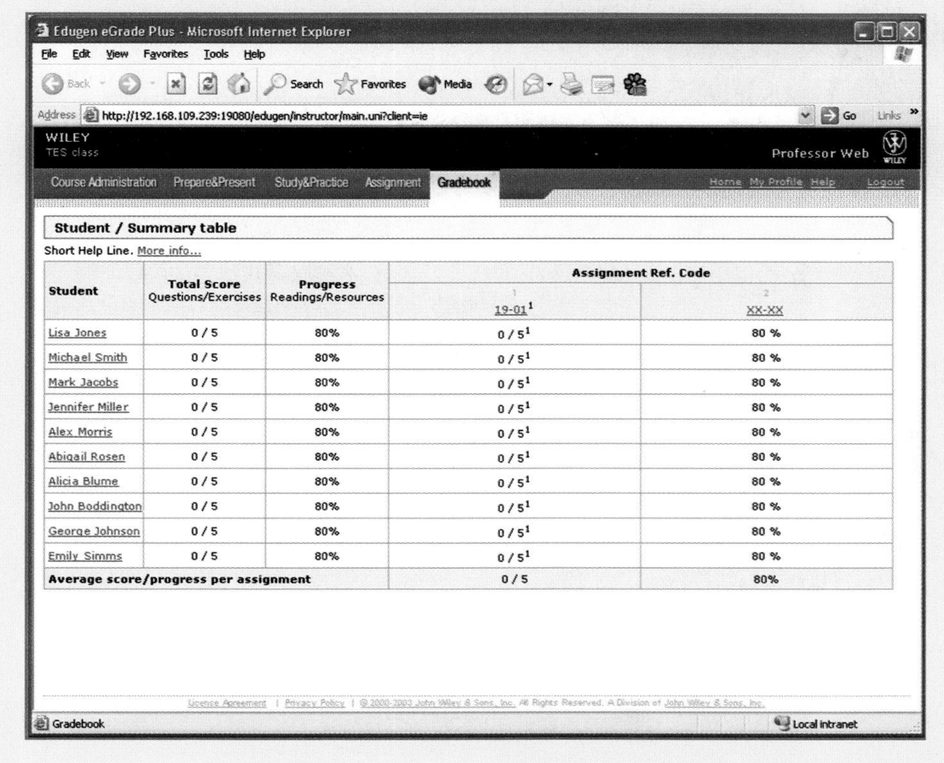

Student	Total Score Questions/Exercises	Progress Readings/Resources	Assignment Ref. Code 19-01[1]	XX-XX
Lisa Jones	0 / 5	80%	0 / 5[1]	80 %
Michael Smith	0 / 5	80%	0 / 5[1]	80 %
Mark Jacobs	0 / 5	80%	0 / 5[1]	80 %
Jennifer Miller	0 / 5	80%	0 / 5[1]	80 %
Alex Morris	0 / 5	80%	0 / 5[1]	80 %
Abigail Rosen	0 / 5	80%	0 / 5[1]	80 %
Alicia Blume	0 / 5	80%	0 / 5[1]	80 %
John Boddington	0 / 5	80%	0 / 5[1]	80 %
George Johnson	0 / 5	80%	0 / 5[1]	80 %
Emily Simms	0 / 5	80%	0 / 5[1]	80 %
Average score/progress per assignment			0 / 5	80%

An **Instructor's Gradebook** will keep track of your students' progress and allow you to analyze individual and overall class results to determine their progress and level of understanding.

Students,

eGrade Plus Allows You to:

Study More Effectively

Get Immediate Feedback When You Practice on Your Own

eGrade Plus problems links directly to the relevant sections of the **electronic book content,** so that you can review the text while you study and complete homework online. Additional resources include **self-assessment quizzing** with detailed feedback, **Interactive Learningware** with step-by-step problem solving tutorials, and **interactive simulations** to help you review key topics.

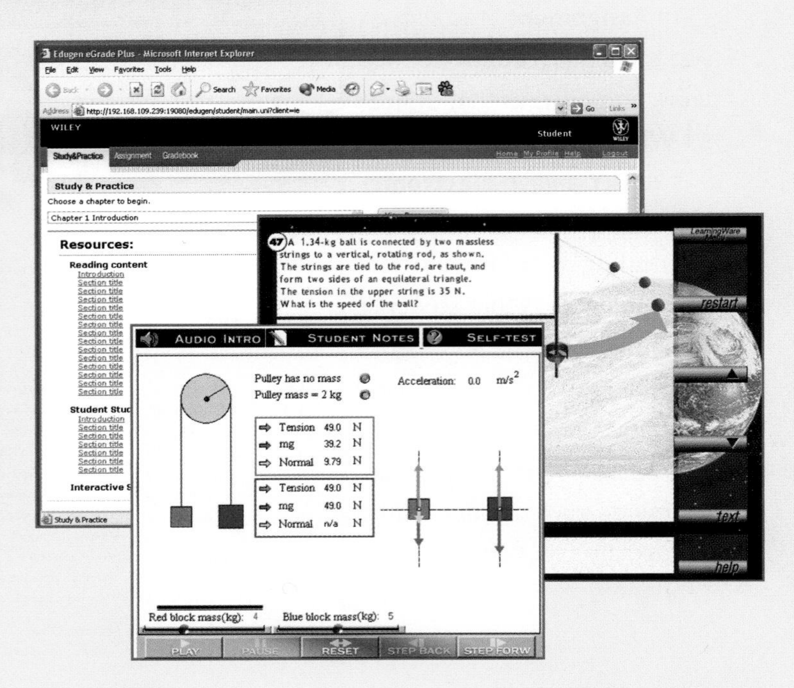

Complete Assignments / Get Help with Problem Solving

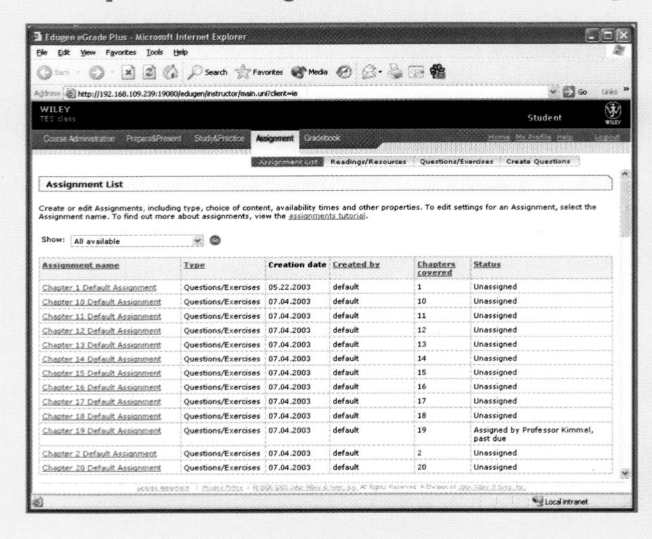

An **Assignment** area keeps all your assigned work in one location, making it easy for you to stay on task. In addition, many homework problems contain a **link** to the relevant section of the **electronic book,** providing you with a text explanation to help you conquer problem-solving obstacles as they arise.

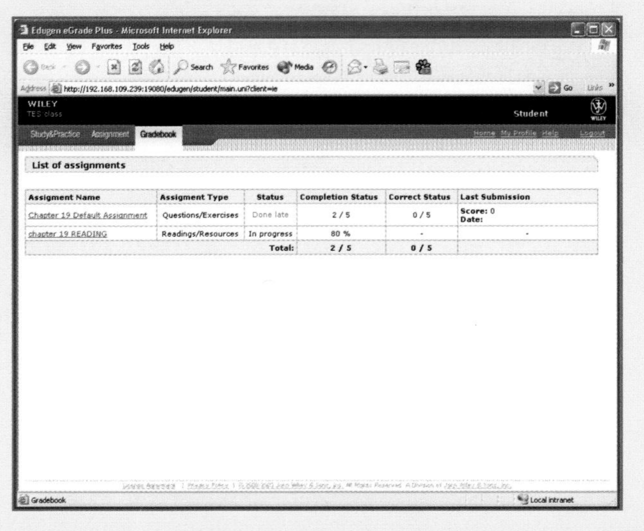

Keep Track of How You're Doing

A **Personal Gradebook** allows you to view your results from past assignments at any time.

UNDERSTANDING PHYSICS

Karen Cummings
Rensselaer Polytechnic Institute
Southern Connecticut State University

Priscilla W. Laws
Dickinson College

Edward F. Redish
University of Maryland

Patrick J. Cooney
Millersville University

GUEST AUTHOR

Edwin F. Taylor
Massachusetts Institute of Technology

ADDITIONAL MEMBERS OF ACTIVITY BASED PHYSICS GROUP

David R. Sokoloff
University of Oregon

Ronald K. Thornton
Tufts University

Understanding Physics is based on *Fundamentals of Physics* by David Halliday, Robert Resnick, and Jearl Walker.

WILEY
John Wiley & Sons, Inc.

This book is dedicated to Arnold Arons, whose pioneering work in physics education and reviews of early chapters have had a profound influence on our work.

SENIOR ACQUISITIONS EDITOR	Stuart Johnson
SENIOR DEVELOPMENT EDITOR	Ellen Ford
MARKETING MANAGER	Bob Smith
SENIOR PRODUCTION EDITOR	Elizabeth Swain
SENIOR DESIGNER	Kevin Murphy
INTERIOR DESIGN	Circa 86, Inc.
COVER DESIGN	David Levy
COVER PHOTO	© Antonio M. Rosario/The Image Bank/Getty Images
ILLUSTRATION EDITOR	Anna Melhorn
PHOTO EDITOR	Hilary Newman

This book was set in 10/12 Times Ten Roman by Progressive and printed and bound by Von Hoffmann Press. The cover was printed by Von Hoffmann Press.

This book is printed on acid free paper. ♾

Library of Congress Cataloging in Publication Data:

Understanding physics / Karen Cummings . . . [et al.]; with additional members of the Activity Based Physics Group.

p. cm.

Includes index.

ISBN 0-471-37099-1

1. Physics. I. Cummings, Karen. II. Activity Based Physics Group.

QC23.2.U54 2004
530—dc21 2003053481

L.C. Call no. Dewey Classification No. L.C. Card No.

Printed in the United States of America

10 9 8 7 6 5 4 3 2

Preface

Welcome to *Understanding Physics*. This book is built on the foundations of the 6th Edition of Halliday, Resnick, and Walker's *Fundamentals of Physics* which we often refer to as HRW 6th. The HRW 6th text and its ancestors, first written by David Halliday and Robert Resnick, have been best-selling introductory physics texts for the past 40 years. It sets the standard against which many other texts are judged. You are probably thinking, "Why mess with success?" Let us try to explain.

Why a Revised Text?

A physics major recently remarked that after struggling through the first half of his junior level mechanics course, he felt that the course was now going much better. What had changed? Did he have a better background in the material they were covering now? "No," he responded. "I started reading the book before every class. That helps me a lot. I wish I had done it in Physics One and Two." Clearly, this student learned something very important. It is something most physics instructors wish they could teach all of their students as soon as possible. Namely, no matter how smart your students are, no matter how well your introductory courses are designed and taught, your students will master more physics if they learn how to read an "understandable" textbook carefully.

We know from surveys that the vast majority of introductory physics students do not read their textbooks carefully. We think there are two major reasons why: (1) many students complain that physics textbooks are impossible to understand and too abstract, and (2) students are extremely busy juggling their academic work, jobs, personal obligations, social lives and interests. So they develop strategies for passing physics without spending time on careful reading. We address both of these reasons by making our revision to the sixth edition of *Fundamentals of Physics* easier for students to understand and by providing the instructor with more **Reading Exercises** (formerly known as Checkpoints) and additional strategies for encouraging students to read the text carefully. Fortunately, we are attempting to improve a fine textbook whose active author, Jearl Walker, has worked diligently to make each new edition more engaging and understandable.

In the next few sections we provide a summary of how we are building upon HRW 6th and shaping it into this new textbook.

A Narrative That Supports Student Learning

One of our primary goals is to help students make sense of the physics they are learning. We cannot achieve this goal if students see physics as a set of disconnected mathematical equations that each apply only to a small number of specific situations. We stress conceptual and qualitative understanding and continually make connections between mathematical equations and conceptual ideas. We also try to build on ideas that students can be expected to already understand, based on the resources they bring from everyday experiences.

In *Understanding Physics* we have tried to tell a story that flows from one chapter to the next. Each chapter begins with an introductory section that discusses why new topics introduced in the chapter are important, explains how the chapter builds on previous chapters, and prepares students for those that follow. We place explicit emphasis on basic concepts that recur throughout the book. We use extensive forward and backward referencing to reinforce connections between topics. For example, in the introduction of Chapter 16 on Oscillations we state: "Although your study of simple harmonic motion will enhance your understanding of mechanical systems it is also vital to understanding the topics in electricity and magnetism encountered in Chapters 30-37. Finally, a knowledge of SHM provides a basis for understanding the wave nature of light and how atoms and nuclei absorb and emit energy."

Emphasis on Observation and Experimentation

Observations and concrete everyday experiences are the starting points for development of mathematical expressions. Experiment-based theory building is a major feature of the book. We build ideas on experience that students either already have or can easily gain through careful observation.

Whenever possible, the physical concepts and theories developed in *Understanding Physics* grow out of simple observations or experimental data that can be obtained in typical introductory physics laboratories. We want our readers to develop the habit of asking themselves: What do our observations, experiences and data imply about the natural laws of physics? How do we know a given statement is true? Why do we believe we have developed correct models for the world?

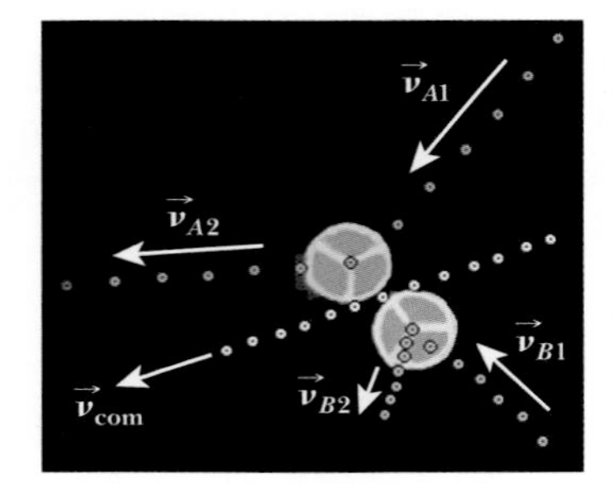

FIGURE P-1 ■ A video analysis shows that the center of mass of a two-puck system moves at a constant velocity.

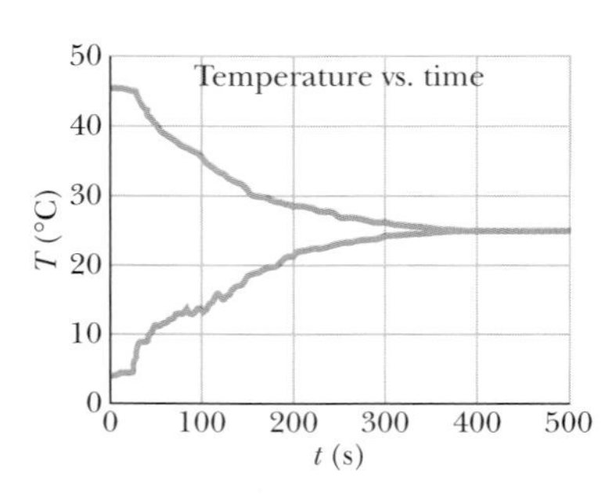

FIGURE P-2 ■ Electronic temperature sensors reveal that if equal amounts of hot and cold water mix the final temperature is the average of the initial temperatures.

Toward this end, the text often starts a chapter by describing everyday observations with which students are familiar. This makes *Understanding Physics* a text that is both relevant to students' everyday lives and draws on existing student knowledge. We try to follow Arnold Arons' principle "idea first, name after." That is, we make every attempt to begin a discussion by using everyday language to describe common experiences. Only then do we introduce formal physics terminology to represent the concepts being discussed. For example, everyday pushes, pulls, and their impact on the motion of an object are discussed before introducing the term "force" or Newton's Second Law. We discuss how a balloon shrivels when placed in a cold environment and how a pail of water cools to room temperature before introducing the ideal gas law or the concept of thermal energy transfer.

The "idea first, name after" philosophy helps build patterns of association between concepts students are trying to learn and knowledge they already have. It also helps students reinterpret their experiences in a way that is consistent with physical laws.

Examples and illustrations in *Understanding Physics* often present data from modern computer-based laboratory tools. These tools include computer-assisted data acquisition systems and digital video analysis software. We introduce students to these tools at the end of Chapter 1. Examples of these techniques are shown in Figs. P-1 and P-2 (on the left) and Fig. P-3 on the next page. Since many instructors use these computer tools in the laboratory or in lecture demonstrations, these tools are part of the introductory physics experience for more and more of our students. The use of real data has a number of advantages. It connects the text to the students' experience in other parts of the course and it connects the text directly to real world experience. Regardless of whether data acquisition and analysis tools are used in the student's own laboratory, our use of realistic rather that idealized data helps students develop an appreciation of the role that data evaluation and analysis plays in supporting theory.

FIGURE P-3 ■ A video analysis of human motion reveals that in free fall the center of mass of an extended body moves in a parabolic path under the influence of the Earth's gravitational force.

Using Physics Education Research

In re-writing the text we have taken advantage of two valuable findings of physics education research. One is the identification of concepts that are especially difficult for many students to learn. The other is the identification of active learning strategies to help students develop a more robust understanding of physics.

Addressing Learning Difficulties

Extensive scholarly research exists on the difficulties students have in learning physics.[1] We have made a concerted effort to address these difficulties. In *Understanding Physics,* issues that are known to confuse students are discussed with care. This is true even for topics like the nature of force and its effect on velocity and velocity changes that may seem trivial to professional physicists. We write about subtle, often counter-intuitive topics with carefully chosen language and examples designed to draw out and remediate common alternative student conceptions. For example, we know that students have trouble understanding passive forces such as normal and friction forces.[2] How can a rigid table exert a force on a book that rests on it? In Section 6-4 we present an idealized model of a solid that is analogous to an inner spring mattress with the repulsion forces between atoms acting as the springs. In addition, we invite our readers to push on a table with a finger and experience the fact that as they push harder on the table the table pushes harder on them in the opposite direction.

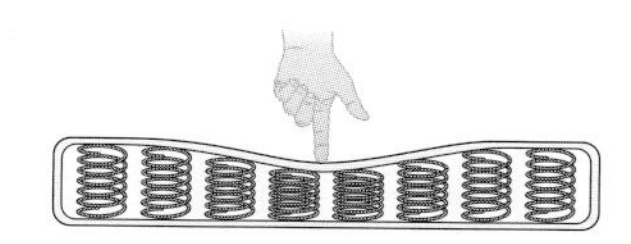

FIGURE P-4 ■ Compressing an innerspring mattress with a force. The mattress exerts an oppositely directed force, with the same magnitude, back on the finger.

Incorporating Active Learning Opportunities

We designed *Understanding Physics* to be more interactive and to foster thoughtful reading. We have retained a number of the excellent Checkpoint questions found at the end of HRW 6th chapter sections. We now call these questions **Reading Exercises.** We have created many new Reading Exercises that require students to reflect on the material in important chapter sections. For example, just after reading Section 6-2 that introduces the two-dimensional free-body diagram, students encounter Reading Exercise 6-1. This multiple-choice exercise requires students to identify the free-body diagram for a helicopter that experiences three non-collinear forces. The distractors were based on common problems students have with the construction of free-body diagrams. When used in "Just-In-Time Teaching" assignments or for in-class group discussion, this type of reading exercise can help students learn a vital problem solving skill as they read.

---

[1] L. C. McDermott and E. F. Redish, "Resource Letter PER-1: Physics Education Research," *Am. J. Phys.* **67**, 755-767 (1999)

[2] John J. Clement, "Expert novice similarities and instruction using analogies," *Int. J. Sci. Ed. 20*, 1271-1286 (1998)

We also created a set of **Touchstone Examples.** These are carefully chosen sample problems that illustrate key problem solving skills and help students learn how to use physical reasoning and concepts as an essential part of problem solving. We selected some of these touchstone examples from the outstanding collection of sample problems in HRW 6th and we created some new ones. In order to retain the flow of the narrative portions of each chapter, we have reduced the overall number of sample problems to those necessary to exemplify the application of fundamental principles. Also, we chose touchstone examples that require students to combine conceptual reasoning with mathematical problem-solving skills. Few, if any, of our touchstone examples are solvable using simple "plug-and-chug" or algorithmic pattern matching techniques.

Alternative problems have been added to the extensive, classroom tested end-of-chapter problem sets selected from HRW 6th. The design of these new problems are based on the authors' knowledge of research on student learning difficulties. Many of these new problems require careful qualitative reasoning. They explicitly connect conceptual understanding to quantitative problem solving. In addition, estimation problems, video analysis problems, and "real life" or "context rich" problems have been included.

The organization and style of *Understanding Physics* has been modified so that it can be easily used with other research-based curricular materials that make up what we call *The Physics Suite*. The *Suite* and its contents are explained at length at the end of this preface.

Reorganizing for Coherence and Clarity

For the most part we have retained the organization scheme inherited from HRW 6th. Instructors are familiar with the general organization of topics in a typical course sequence in calculus-based introductory physics texts. In fact, ordering of topics and their division into chapters is the same for 27 of the 38 chapters. The order of some topics has been modified to be more pedagogically coherent. Most of the reorganization was done in Chapters 3 through 10 where we adopted a sequence known as *New Mechanics*. In addition, we decided to move HRW 6th Chapter 25 on capacitors so it becomes the last chapter on DC circuits. Capacitors are now introduced in Chapter 28 in *Understanding Physics*.

The New Mechanics Sequence

HRW 6th and most other introductory textbooks use a familiar sequence in the treatment of classical mechanics. It starts with the development of the kinematic equations to describe constantly accelerated motion. Then two-dimensional vectors and the kinematics of projectile motion are treated. This is followed by the treatment of dynamics in which Newton's Laws are presented and used to help students understand both one- and two-dimensional motions. Finally energy, momentum conservation, and rotational motion are treated.

About 12 years ago when Priscilla Laws, Ron Thornton, and David Sokoloff were collaborating on the development of research-based curricular materials, they became concerned about the difficulties students had working with two-dimensional vectors and understanding projectile motion before studying dynamics.

At the same time Arnold Arons was advocating the introduction of the concept of momentum before energy.[3] Arons argued that (1) the momentum concept is simpler than the energy concept, in both historical and modern contexts and (2) the study

[3] Private Communication between Arnold Arons and Priscilla Laws by means of a document entitled "Preliminary Notes and Suggestions," August 19, 1990; and Arnold Arons, *Development of Concepts of Physics* (Addison-Wesley, Reading MA, 1965)

of momentum conservation entails development of the concept of center-of-mass which is needed for a proper development of energy concepts. Additionally, the impulse-momentum relationship is clearly an alternative statement of Newton's Second Law. Hence, its placement immediately after the coverage of Newton's laws is most natural.

In order to address these concerns about the traditional mechanics sequence, a small group of physics education researchers and curriculum developers convened in 1992 to discuss the introduction of a new order for mechanics.[4] One result of the conference was that Laws, Sokoloff, and Thornton have successfully incorporated a new sequence of topics in the mechanics portions of various curricular materials that are part of the Physics Suite discussed below.[5] These materials include *Workshop Physics*, the *RealTime Physics Laboratory Module in Mechanics*, and the *Interactive Lecture Demonstrations*. This sequence is incorporated in this book and has required a significant reorganization and revisions of HRW 6th Chapters 2 through 10.

The New Mechanics sequence incorporated into Chapters 2 through 10 of understanding physics includes:

- Chapter 2: One-dimensional kinematics using constant horizontal accelerations and vertical free fall as applications.
- Chapter 3: The study of one-dimensional dynamics begins with the application of Newton's laws of motion to systems with one or more forces acting along a single line. Readers consider observations that lead to the postulation of "gravity" as a constant invisible force acting vertically downward.
- Chapter 4: Two-dimensional vectors, vector displacements, unit vectors and the decomposition of vectors into components are treated.
- Chapter 5: The study of kinematics and dynamics is extended to two-dimensional motions with forces along only a single line. Examples include projectile motion and circular motion.
- Chapter 6: The study of kinematics and dynamics is extended to two-dimensional motions with two-dimensional forces.
- Chapters 7 & 8: Topics in these chapters deal with impulse and momentum change, momentum conservation, particle systems, center of mass, and the motion of the center-of-mass of an isolated system.
- Chapters 9 & 10: These chapters introduce kinetic energy, work, potential energy, and energy conservation.

Just-in-Time Mathematics

In general, we introduce mathematical topics in a "just-in-time" fashion. For example, we treat one-dimensional vector concepts in Chapter 2 along with the development of one-dimensional velocity and acceleration concepts. We hold the introduction of two- and three-dimensional vectors, vector addition and decomposition until Chapter 4, immediately before students are introduced to two-dimensional motion and forces in Chapters 5 and 6. We do not present vector products until they are needed. We wait to introduce the dot product until Chapter 9 when the concept of physical work is presented. Similarly, the cross product is first presented in Chapter 11 in association with the treatment of torque.

---

[4] The New Mechanics Conference was held August 6-7, 1992 at Tufts University. It was attended by Pat Cooney, Dewey Dykstra, David Hammer, David Hestenes, Priscilla Laws, Suzanne Lea, Lillian McDermott, Robert Morse, Hans Pfister, Edward F. Redish, David Sokoloff, and Ronald Thornton.

[5] Laws, P. W. "A New Order for Mechanics" pp. 125-136, *Proceedings of the Conference on the Introductory Physics Course*, Rensselaer Polytechnic Institute, Troy New York, May 20-23, Jack Wilson, Ed. 1993 (John Wiley & Sons, New York 1997)

Notation Changes

Mathematical notation is often confusing, and ambiguity in the meaning of a mathematical symbol can prevent a student from understanding an important relationship. It is also difficult to solve problems when the symbols used to represent different quantities are not distinctive. Some key features of the new notation include:

- We adhere to recent notation guidelines set by the U.S. National Institute of Standard and Technology Special Publication 811 (SP 811).
- We try to balance our desire to use familiar notation and our desire to avoid using the same symbol for different variables. For example, p is often used to denote momentum, pressure, and power. We have chosen to use lower case p for momentum and capital P for pressure since both variables appear in the kinetic theory derivation. But we stick with the convention of using capital P for power since it does not commonly appear side by side with pressure in equations.
- We denote vectors with an arrow instead of bolding so handwritten equations can be made to look like the printed equations.
- We label each vector component with a subscript that explicitly relates it to its coordinate axis. This eliminates the common ambiguity about whether a quantity represents a magnitude which is a scalar or a vector component which is not a scalar.
- We often use subscripts to spell out the names of objects that are associated with mathematical variables even though instructors and students will tend to use abbreviations. We also stress the fact that one object is exerting a force on another with an arrow in the subscript. For example, the force exerted by a rope on a block would be denoted as $\vec{F}_{\text{rope}\rightarrow\text{block}}$.

Our notation scheme is summarized in more detail in Appendix A4.

Encouraging Text Reading

We have described a number of changes that we feel will improve this textbook and its readability. But even the best textbook in the world is of no help to students who do not read it. So it is important that instructors make an effort to encourage busy students to develop effective reading habits. In our view the single most effective way to get students to read this textbook is to assign appropriate reading, reading exercises, and other reading questions after every class. Some effective ways to follow up on reading question assignments include:

1. Employ a method called "Just-In-Time-Teaching" (or JiTT) in which students submit their answers to questions about reading before class using just plain email or one of the many available computer based homework systems (Web Assign or E-Grade for example). You can often read enough answers before class to identify the difficult questions that need more discussion in class;
2. Ask students to bring the assigned questions to class and use the answers as a basis for small group discussions during the class period;
3. Assign multiple choice questions related to each section or chapter that can be graded automatically with a computer-based homework system; and
4. Require students to submit chapter summaries. Because this is a very effective assignment, we intentionally avoided doing chapter summaries for students.

Obviously, all of these approaches are more effective when students are given some credit for doing them. Thus you should arrange to grade all, or a random sample, of the submissions as incentives for students to read the text and think about the answers to Reading Exercises on a regular basis.

The Physics Suite

In 1997 and 1998, Wiley's physics editor, Stuart Johnson, and an informally constituted group of curriculum developers and educational reformers known as the *Activity Based Physics Group* began discussing the feasibility of integrating a broad array of curricular materials that are physics education research-based. This led to the assembly of an *Activity Based Physics Suite* that includes this textbook. The *Physics Suite* also includes materials that can be combined in different ways to meet the needs of instructors working in vastly different learning environments. The *Interactive Lecture Demonstration Series*[6] is designed primarily for use in lecture sessions. Other *Suite* materials can be used in laboratory settings including the *Workshop Physics Activity Guide,*[7] the *Real Time Physics Laboratory* modules,[8] and *Physics by Inquiry.*[9] Additional elements in the collection are suitable for use in recitation sessions such as the University of Washington *Tutorials in Introductory Physics* (available from Prentice Hall)[10] and a set of *Quantitative Tutorials*[11] developed at the University of Maryland. The *Activity Based Physics Suite* is rounded out with a collection of thinking problems developed at the University of Maryland. In addition to this **Understanding Physics** text, the Physics Suite elements include:

1. **Teaching Physics with the Physics Suite** by Edward F. Redish (University of Maryland). This book is not only the "Instructors Manual" for *Understanding Physics*, but it is also a book for anyone who is interested in learning about recent developments in physics education. It is a handbook with a variety of tools for improving both teaching and learning of physics—from new kinds of homework and exam problems, to surveys for figuring out what has happened in your class, to tools for taking and analyzing data using computers and video. The book comes with a Resource CD containing 14 conceptual and 3 attitude surveys, and more than 250 thinking problems covering all areas of introductory physics, resource materials from commercial vendors on the use of computerized data acquisition and video, and a variety of other useful reference materials. (Instructors can obtain a complimentary copy of the book and Resource CD, from John Wiley & Sons.)
2. **RealTime Physics** by David Sokoloff (University of Oregon), Priscilla Laws (Dickinson College), and Ronald Thornton (Tufts University). *RealTime Physics* is a set of laboratory materials that uses computer-assisted data acquisition to help students build concepts, learn representation translation, and develop an understanding of the empirical base of physics knowledge. There are three modules in the collection: Module 1: Mechanics (12 labs), Module 2: Heat and Thermodynamics (6 labs), and Module 3: Electric Circuits (8 labs). (Available both in print and in electronic form on *The Physics Suite CD.*)

---

[6]David R. Sokoloff and Ronald K. Thornton, "Using Interactive Lecture Demonstrations to Create an Active Learning Environment." *The Physics Teacher*, **35**, 340-347, September 1997.

[7]Priscilla W. Laws, *Workshop Physics Activity Guide*, Modules 1-4 w/ Appendices (John Wiley & Sons, New York, 1997).

[8]David R. Sokoloff, *RealTime Physics*, Modules 1-2, (John Wiley & Sons, New York, 1999).

[9]Lillian C. McDermott and the Physics Education Group at the University of Washington, *Physics by Inquiry* (John Wiley & Sons, New York, 1996).

[10]Lillian C. McDermott, Peter S. Shaffer, and the Physics Education Group at the University of Washington, *Tutorials in Introductory Physics*, First Edition (Prentice-Hall, Upper Saddle River, NJ, 2002).

[11]Richard N. Steinberg, Michael C. Wittmann, and Edward F. Redish, "Mathematical Tutorials in Introductory Physics," in, *The Changing Role Of Physics Departments In Modern Universities,* Edward F. Redish and John S. Rigden, editors, AIP Conference Proceedings **399**, (AIP, Woodbury NY, 1997), 1075-1092.

3. **Interactive Lecture Demonstrations** by David Sokoloff (University of Oregon) and Ronald Thornton (Tufts University). ILDs are worksheet-based guided demonstrations designed to focus on fundamental principles and address specific naïve conceptions. The demonstrations use computer-assisted data acquisition tools to collect and display high quality data in real time. Each ILD sequence is designed for delivery in a single lecture period. The demonstrations help students build concepts through a series of instructor led steps involving prediction, discussions with peers, viewing the demonstration and reflecting on its outcome. The ILD collection includes sequences in mechanics, thermodynamics, electricity, optics and more. (Available both in print and in electronic form on *The Physics Suite CD.*)

4. **Workshop Physics** by Priscilla Laws (Dickinson College). *Workshop Physics* consists of a four part activity guide designed for use in calculus-based introductory physics courses. Workshop Physics courses are designed to replace traditional lecture and laboratory sessions. Students use computer tools for data acquisition, visualization, analysis and modeling. The tools include computer-assisted data acquisition software and hardware, digital video capture and analysis software, and spreadsheet software for analytic mathematical modeling. Modules include classical mechanics (2 modules), thermodynamics & nuclear physics, and electricity & magnetism. (Available both in print and in electronic form on *The Physics Suite CD.*)

5. **Tutorials in Introductory Physics** by Lillian C. McDermott, Peter S. Shaffer and the Physics Education Group at the University of Washington. These tutorials consist of a set of worksheets designed to supplement instruction by lectures and textbook in standard introductory physics courses. Each tutorial is designed for use in a one-hour class session in a space where students can work in small groups using simple inexpensive apparatus. The emphasis in the tutorials is on helping students deepen their understanding of critical concepts and develop scientific reasoning skills. There are tutorials on mechanics, electricity and magnetism, waves, optics, and other selected topics. (Available in print from Prentice Hall, Upper Saddle River, New Jersey.)

6. **Physics by Inquiry** by Lillian C. McDermott and the Physics Education Group at the University of Washington. This self-contained curriculum consists of a set of laboratory-based modules that emphasize the development of fundamental concepts and scientific reasoning skills. Beginning with their observations, students construct a coherent conceptual framework through guided inquiry. Only simple inexpensive apparatus and supplies are required. Developed primarily for the preparation of precollege teachers, the modules have also proven effective in courses for liberal arts students and for underprepared students. The amount of material is sufficient for two years of academic study. (Available in print.)

7. **The Activity Based Physics Tutorials** by Edward F. Redish and the University of Maryland Physics Education Research Group. These tutorials, like those developed at the University of Washington, consist of a set of worksheets developed to supplement lectures and textbook work in standard introductory physics courses. But these tutorials integrate the computer software and hardware tools used in other Suite elements including computer data acquisition, digital video analysis, simulations, and spreadsheet analysis. Although these tutorials include a range of classical physics topics, they also include additional topics in modern physics. (Available only in electronic form on *The Physics Suite CD.*)

8. **The Understanding Physics Video CD for Students** by Priscilla Laws, et. al.: This CD contains a collection of the video clips that are introduced in *Understanding Physics* narrative and alternative problems. The CD includes a number of QuickTime movie segments of physical phenomena along with the QuickTime player

software. Students can view video clips as they read the text. If they have video analysis software available, they can reproduce data presented in text graphs or complete video analyses based on assignments designed by instructors.

9. **The Physics Suite CD.** This CD contains a variety of the Suite Elements in electronic format (Microsoft Word files). The electronic format allows instructors to modify and reprint materials to better fit into their individual course syllabi. The CD contains much useful material including complete electronic versions of the following: *RealTime Physics*, *Interactive Lecture Demonstrations, Workshop Physics, Activity Based Physics Tutorials.*

A Final Word to the Instructor

Over the past decade we have learned how valuable it is for us as teachers to focus on what most students actually need to do to learn physics, and how valuable it can be for students to work with research-based materials that promote active learning. We hope you and your students find this book and the other *Physics Suite* materials helpful in your quest to make physics both more exciting and understandable to your students.

Supplements for Use with Understanding Physics

Instructor Supplements

1. **Instructor's Solution Manual** prepared by Anand Batra (Howard University). This manual provides worked-out solutions for most of the end-of-chapter problems.
2. **Test Bank** by J. Richard Christman (U. S. Coast Guard Academy). This manual includes more than 2500 multiple-choice questions adapted from HRW 6th. These items are also available in the *Computerized Test Bank* (see below).
3. **Instructor's Resource CD.** This CD contains: The entire *Instructor's Solutions Manual* in both Microsoft Word© (IBM and Macintosh) and PDF files. A *Computerized Test Bank,* for use with both PCs and Macintosh computers with full editing features to help you customize tests. And all text illustrations, suitable for classroom projection, printing, and web posting.
4. **Online Homework and Quizzing:** *Understanding Physics* supports WebAssign and eGrade, two programs that give instructors the ability to deliver and grade homework and quizzes over the Internet.
5. **The Wiley Physics Demonstration Videos** by David Maiullo of Rutgers University consist of over a hundred classic physics demonstrations that will engage and instruct your students. Filmed, edited and produced by a professional film crew, the demonstrations include lying on a bed of nails, breaking glass with sound, and, in a show of atmospheric pressure, crushing a 55-gallon drum. Each demonstration is labeled according to the Physics Instructional Resource Association's demonstration classifying system. This system identifies the area, topic and concept presented in each demonstration. Go to www.pira.nu for more information about the Physics Instructional Resources Association and to download a spreadsheet of the demonstration classification systems.
6. **Wiley Physics Simulations CD-ROM** contains 50 interactive simulations (Java applets) that can be used for classroom demonstrations.

Student Supplements

1. **Student Study Guide** by J. Richard Christman (U. S. Coast Guard Academy). This student study guide provides chapter overviews, hints for solving selected end-of-chapter problems, and self-quizzes.

2. **Student Solutions Manual** by J. Richard Christman (U. S. Coast Guard Academy). This manual provides students with complete worked-out solutions for approximately 450 of the odd-numbered end-of-chapter problems.

Acknowledgements

Many individuals helped us create this book. The authors are grateful to the individuals who attended the weekend retreats at Airlie Center in 1997 and 1998 and to our editor, Stuart Johnson and to John Wiley & Sons for sponsoring the sessions. It was in these retreats that the ideas for *Understanding Physics* crystallized. We are grateful to Jearl Walker, David Halliday and Bob Resnick for graciously allowing us to attempt to make their already fine textbook better.

The authors owe special thanks to Sara Settlemyer who served as an informal project manager for the past few years. Her contributions included physics advice (based on her having completed Workshop Physics courses at Dickinson College), her use of Microsoft Word, Adobe Illustrator, Adobe Photoshop and Quark XPress to create the manuscript and visuals for this edition, and skillful attempts to keep our team on task—a job that has been rather like herding cats.

Karen Cummings: I would like to say "Thanks!" to: Bill Lanford (for endless advice, use of the kitchen table and convincing me that I really could keep the same address for more than a few years in a row), Ralph Kartel Jr. and Avery Murphy (for giving me an answer when people asked why I was working on a textbook), Susan and Lynda Cummings (for the comfort, love and support that only sisters can provide), Jeff Marx, Tim French and the poker crew (for their friendship and laughter), my colleagues at Southern Connecticut and Rensselaer, especially Leo Schowalter, Jim Napolitano and Jack Wilson (for the positive influence you have had on my professional life) and my students at Southern Connecticut and Rensselaer, Ron Thornton, Priscilla Laws, David Sokoloff, Pat Cooney, Joe Redish, Ken and Pat Heller and Lillian C. McDermott (for helping me learn how to teach).

Priscilla Laws: First of all I would like thank my husband and colleague Ken Laws for his quirky physical insights, for the Chapter 11 Kneecap puzzler, for the influence of his physics of dance work on this book, and for waiting for me countless times while I tried to finish "just one more thing" on this book. Thanks to my daughter Virginia Jackson and grandson Adam for all the fun times that keep me sane. My son Kevin Laws deserves special mention for sharing his creativity with us—best exemplified by his murder mystery problem, *A(dam)nable Man,* reprinted here as problem 5-68. I would like to thank Juliet Brosing of Pacific University who adapted many of the Workshop Physics problems developed at Dickinson for incorporation into the alternative problem collection in this book. Finally, I am grateful to my Dickinson College colleagues Robert Boyle, Kerry Browne, David Jackson, and Hans Pfister for advice they have given me on a number of topics.

Joe Redish: I would like to thank Ted Jacobsen for discussions of our chapter on relativity and Dan Lathrop for advice on the sources of the Earth's magnetic field, as well as many other of my colleagues at the University of Maryland for discussions on the teaching of introductory physics over many years.

Pat Cooney: I especially thank my wife Margaret for her patient support and constant encouragement and I am grateful to my colleagues at Millersville University: John Dooley, Bill Price, Mike Nolan, Joe Grosh, Tariq Gilani, Conrad Miziumski, Zenaida Uy, Ned Dixon, and Shawn Reinfried for many illuminating conversations.

We also appreciate the absolutely essential role many reviewers and classroom testers played. We took our reviewers very seriously. Several reviewers and testers deserve special mention. First and foremost is Arnold Arons who managed to review 29 of the 38 chapters either from the original HRW 6th material or from our early drafts before he passed away in February 2001. Vern Lindberg from the Rochester

Institute of Technology deserves special mention for his extensive and very insightful reviews of most of our first 18 chapters. Ed Adelson from Ohio State did a particularly good job reviewing most of our electricity chapters. Classroom tester Maxine Willis from Gettysburg Area High School deserves special recognition for compiling valuable comments that her advanced placement physics students made while class testing Chapters 1-12 of the preliminary version. Many other reviewers and class testers gave us useful comments in selected chapters.

Class Testers

Gary Adams
Rensselaer Polytechnic Institute

Marty Baumberger
Chestnut Hill Academy

Gary Bedrosian
Rensselaer Polytechnic Institute

Joseph Bellina,
Saint Mary's College

Juliet W. Brosing
Pacific University

Shao-Hsuan Chiu
Frostburg State

Chad Davies
Gordon College

Hang Deng-Luzader
Frostburg State

John Dooley
Millersville University

Diane Dutkevitch
Yavapai College

Timothy Hayes
Rensselaer Polytechnic Institute

Brant Hinrichs
Drury College

Kurt Hoffman
Whitman College

James Holliday
John Brown University

Michael Huster
Simpson College

Dennis Kuhl
Marietta College

John Lindberg
Seattle Pacific University

Vern Lindberg
Rochester Institute of Technology

Stephen Luzader
Frostburg State

Dawn Meredith
University of New Hampshire

Larry Robinson
Austin College

Michael Roth
University of Northern Iowa

John Schroeder
Rensselaer Polytechnic Institute

Cindy Schwarz
Vassar College

William Smith
Boise State University

Dan Sperber
Rensselaer Polytechnic Institute

Roger Stockbauer
Louisiana State University

Paul Stoler
Rensselaer Polytechnic Institute

Daniel F. Styer
Oberlin College

Rebecca Surman
Union College

Robert Teese
Muskingum College

Maxine Willis
Gettysburg Area High School

Gail Wyant
Cecil Community College

Anne Young
Rochester Institute of Technology

David Ziegler
Sedro-Woolley High School

Reviewers

Edward Adelson
Ohio State University

Arnold Arons
University of Washington

Arun Bansil
Northeastern University

Chadan Djalali
University of South Carolina

William Dawicke
Milwaukee School of Engineering

Robert Good
California State University-Hayware

Harold Hart
Western Illinois University

Harold Hastings
Hofstra University

Laurent Hodges
Iowa State University

Robert Hilborn
Amherst College

Theodore Jacobson
University of Maryland

Leonard Kahn
University of Rhode Island

Stephen Kanim
New Mexico State University

Hamed Kastro
Georgetown University

Debora Katz
U. S. Naval Academy

Todd Lief
Cloud Community College

Vern Lindberg
Rochester Institute of Technology

Mike Loverude
California State University-Fullerton

Robert Luke
Boise State University

Robert Marchini
Memphis State University

Tamar More
Portland State University

Gregor Novak
U. S. Air Force Academy

Jacques Richard
Chicago State University

Cindy Schwarz
Vassar College

Roger Sipson
Moorhead State University

George Spagna
Randolf-Macon College

Gay Stewart
University of Arkansas-Fayetteville

Sudha Swaminathan
Boise State University

We would like to thank our proof readers Georgia Mederer and Ernestine Franco, our copyeditor Helen Walden, and our illustrator Julie Horan.

Last but not least we would like to acknowledge the efforts of the Wiley staff; Senior Acquisitions Editor, Stuart Johnson, Ellen Ford (Senior Development Editor), Justin Bow (Program Assistant), Geraldine Osnato (Project Editor), Elizabeth Swain (Senior Production Editor), Hilary Newman (Senior Photo Editor), Anna Melhorn (Illustration Editor), Kevin Murphy (Senior Designer), and Bob Smith (Marketing Manager). Their dedication and attention to endless details was essential to the production of this book.

Brief Contents

Contents

CHAPTER 24 Gauss' Law 689

CHAPTER 25 Electric Potential 714

CHAPTER 26 Current and Resistance 744

CHAPTER 27 Circuits 772

CHAPTER 28 Capacitance 799

CHAPTER 29 Magnetic Fields 829

CHAPTER 30 Magnetic Fields Due to Currents 861

CHAPTER 31 Induction and Maxwell's Equations 888

Appendices

Introduction

The test of all knowledge is experiment. But what is the source of knowledge? Where do the laws that are to be tested come from? . . . Experiment, itself, helps to produce these laws, in the sense that it gives us hints. But also needed is imagination to create from these hints the great generalizations—to guess at the wonderful, simple, but very strange patterns beneath them all, and then to experiment to check again whether we have made the right guess.[1]

[1]R. P. Feynman, *The Feynman Lectures on Physics,* Ch. 1, (Addison-Wesley, Reading, MA, 1964).

The Nature of Physics and Learning Physics

Welcome to the study of physics. Physics is a process of learning about the physical world by finding ways to make sense of what we observe and measure. As the inspiring teacher Richard Feynman wrote, "Progress in all of the natural sciences depends on this interaction between experiment and theory."[2]

The point here is that to learn physics you must continually compare and contrast your observations to your intuitions and expectations. Sometimes your intuitions will be right, sometimes they'll be partially right, and sometimes they'll be dead wrong. Comparing observations to your intuitions will not only help you learn more physics, it will help you to understand how scientific knowledge is created.

Physics is supposed to help you make sense of the physical world. If a physical phenomenon doesn't make sense at first, keep thinking. Keep analyzing observations and experiments and considering what they mean. Einstein said, "Physics is the refinement of common sense." The key here is on the word "refinement." Physics is more than common sense. It's common sense made consistent by continued reference to both theory and experiment.

In some ways learning physics may seem much simpler than learning biology or chemistry. There are fewer things to consider and the systems we study are simpler. If you write down all the most basic equations you encounter in a physics course there are far fewer to remember than the number of organisms you encounter in a general biology course or the number of reactions you encounter in general chemistry. Also, many physical phenomena seem relatively simple. A system consisting of a ball rolling down an inclined plane or a battery connected to a bulb is a lot simpler than an octopus or the chemical cyclohexane. But many students complain that introductory physics is harder to learn than other sciences. What's going on? One problem is that it is easy to fall into the trap of thinking of physics as a jumble of separate equations to be memorized. *This is not so!* Most equations used in introductory physics courses can be derived from a relatively small number of fundamental relationships.

If you focus your efforts on trying to memorize the properties of hundreds of specific systems you will quickly get overwhelmed. Instead, you should focus on the nature of the scientific process by studying the behavior of a limited number of ideal systems. How can you tell whether a prediction you have made about the behavior of a physical system is correct? How do investigators discover or create "scientific laws?" How can we be sure a law or theory is valid? These questions are critical to solving real-world scientific problems such as how to create a new computer chip, diagnose an illness, or improve the performance of an athlete. Your efforts to *learn fundamental relationships* and to apply them to new scientific problems are the key to understanding physics.

The Art of Simplifying

In physics, we try to understand the rules that govern the way the natural world behaves. But the natural world is a very complex place. So, we start by considering the simplest system that allows us to observe and explain a type of behavior. For example, when studying motion we start with a small object whose structure and shape we can ignore. We pretend a football is just a tiny blob. We figure out how it moves after being thrown and under the influence of gravity only—pretending that it is in a vacuum and that it never rotates or deforms. These are clearly not good assumptions for a real football! But they provide an excellent starting point for making sense of its basic motion. Over small distances (a few feet), and for reasonably low speeds (below

[2]R. P. Feynman, *The Feynman Lectures on Physics,* Ch. 1, (Addison-Wesley, Reading, MA, 1964).

about 20 miles/hour) the idealized description works very well. As you get up to higher speeds and distances, the effects of the air grow in importance. However, this additional complication is manageable. Once you understand the basic principles of motion, you can add details to your "model" to account for the effects of air and thereby make the situations you understand more realistic and extend the number of cases you can treat.

A typical physicist's initial strategy is to understand simple systems as completely as possible by constructing physical laws that describe them. Once that is accomplished, the next step is to add more and more real-world complexity to the system one step at a time. This is the process investigators use to contribute to the powerful body of knowledge that is physics. This is the process we suggest you also use to construct and extend your knowledge of physics.

FIGURE I-1 ■ Jason being explicit about all the simplifications he is being asked to make.

Expect Surprises

You will probably find many surprises in your study of physics, and you don't need to wait until you study relativity or quantum mechanics to do so (though both topics are really interesting and lots of fun). Even the physics phenomena that we present in the early chapters of this book will reveal some facts about our everyday world that many people find surprising. For example, if you take a ball made of lead and a similar ball made of plastic, the lead ball may weigh 20 times as much as the plastic ball. Yet if you stand on a chair that is perched on a sturdy table and drop the two balls at the same time, they fall ten feet to the ground in almost exactly the same time. Why doesn't the lead ball go faster? Or, when an object is immersed in water, it seems to weigh less—and its weight reduction is equal to the weight of the water that it pushed out of the way. What could that water have to do with anything? That water is gone! When you connect two identical bulbs up to a battery, if you connect them in one way they'll both have the same brightness as a single bulb connected to the battery. But, if you connect them in another way, they both get much dimmer. Huh? Why does that happen? This book is full of such surprises.

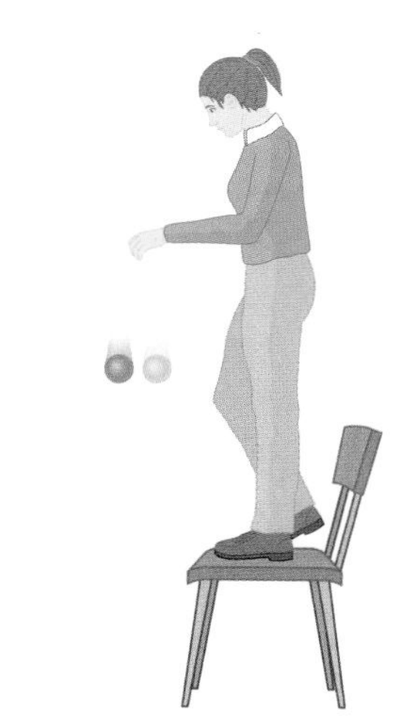

FIGURE I-2 ■ Two balls with different masses fall with the same acceleration whenever air drag is negligible.

Using This Book as a Learning Tool

This textbook is one of many resources that you will need to make use of in order to learn physics. It is very important that you read this textbook on a regular basis and do the *Reading Exercises* at the end of many sections in each chapter. We attempt to present both the experimental results that support theories and some of the reasoning that has gone into the development of theories. However, you will understand the physics only when you make your own observations and are actively engaged in reasoning. So, it is critical that you observe a physical phenomenon directly or ponder the outcome of an experiment that we describe. Then you need to *think* about whether the explanation of the phenomenon we present makes sense. In addition, you must test and refine your understanding of theoretical concepts by applying them to solving problems included at the end of each chapter. Solving problems requires you to use both the physical principles you have learned and the mathematical relationships that describe these principles. Finally, if possible, you will want to test your understanding of physical systems by predicting the outcomes of experiments that you can perform in a basic introductory physics laboratory.

We hope this book will help you enjoy the practice of physics as much as we do.

(Left to right): Priscilla W. Laws, Edward F. Redish, Karen Cummings, and Patrick J. Cooney. Photo by David Hildebrand.

1 Measurement

You can watch the Sun set and disappear over a calm ocean, once while lying on the beach, and then once again if you stand up. This is a surprising observation! Furthermore, if you measure the time between the two sunsets, you can approximate the Earth's radius by using an understanding of the shape and motion of the Earth relative to the Sun along with some basic high school mathematics.

How can such a simple observation be used to measure the Earth?

The answer is in this chapter.

1-1 Introduction

Physics is the study of the basic components of the universe and their interactions. The fact that you can use the time difference between sunsets while lying on the beach and then standing to estimate the size of the earth is indeed surprising to most people. It is one example of how the interplay between mathematics, theoretical principles, and observations allow us to develop a deeper understanding of the physical world. In fact, the ongoing quest of physics is to develop a unified set of ideas to explain apparently different phenomena. Scientific theories are only valid if they serve to explain and predict the outcomes of new observations and experiments. Many theories in physics are expressed in mathematical equations, and predictions usually involve quantities that can be measured.

Measurement is the process of associating numbers with physical quantities. In fact, *physical quantities are defined in terms of the procedures used to measure them.* But the numbers that result from measurements are not meaningful unless people who are using and interpreting them know what was measured and what units were used to obtain the numbers. For example, if you were asked to go to a store to buy 27, you would immediately ask 27 of what? If you were told 27 containers of milk, you might ask 27 of what size or unit—pints, quarts, or gallons? Unambiguous communication with others about the results of a scientific measurement requires agreement on (1) the definition of the physical quantity and (2) the basic units used for comparison when the measurements are made.

The focus in this chapter will be on the fundamental physical quantities and measurement processes used to study motion. Later on we introduce additional physical quantities defined for the study of thermal interactions, electricity, magnetism, and light. You will learn about common elements of physical measurements, reasons why precise measurements are highly valued, and the international system of standard basic units that allows scientists all over the world to communicate with each other.

1-2 Basic Measurements in the Study of Motion

A long jumper speeds up along a runway, leaps into the air, and then comes to a sudden stop in a sand pit. How can such a motion be described and studied scientifically?

In studying motion, at least three questions come to mind. How far has something moved and in what directions? How long did it take? How much stuff was moved? Let's consider length, time, and mass, the three basic physical quantities used in the study of motion. How are they usually defined? What procedures are used to measure them on an everyday basis?

Length: Our "How far?" question involves being able to measure the distance between two points. Suppose you had no measuring instrument. Is there any way you could meaningfully ask and answer the question, "What is the total distance that the jumper ran?" The only approach possible would be to compare this distance to the size of one of your body parts such as your hand or foot. It is not surprising that the hand and the foot have been used throughout history as basic units of measurement. The distance can then be described as a ratio between it and a convenient item chosen to be a length standard.

Time: To answer the question, "How long did it take?" you need to be able to measure a time interval. To do this, you define the time between repetitive events as a standard. Historically, repetitive events that have been used as time standards have included the day (the time it takes for the Sun to appear to revolve around the Earth), the year, and the time it takes for a pendulum of a certain length to swing back and

FIGURE 1-1 ■ A common method of determining mass assumes two objects have the same mass if they balance each other.

forth. A time interval, or time duration, is measured by determining how many years have passed or how many swings of a pendulum have occurred during the interval being measured.

Mass: Mass is a measure of "amount of stuff." Throughout recorded history, merchants and scientists have used balances to determine how many units of "standard mass" are needed to balance whatever is being measured. (See Fig. 1-1.) A standard of mass can be a certain object that everyone agrees should be used. Replicas of the standard mass that balance with it can be passed around and used by many people.

The everyday procedures outlined above for measuring length, time, and mass share common elements that characterize all physical quantities.

1. These quantities are defined by the procedures used to measure them.
2. Their measurement always involves the determination of a ratio between a unit, known as a base quantity, and the quantity being measured.
3. Such comparisons can only be made with limited precision.

As you will see, there are often many alternative procedures that can be used to measure the same quantity. Indeed, a major factor in the progress of science and technology has been the discovery of better, more **precise** methods of measurement.

READING EXERCISE 1-1: List one common base unit used for time, for length, and for mass not mentioned in the discussion in this section. ■

READING EXERCISE 1-2: What is a more precise base unit for length measurement that is reliable over a period of years—a 12-inch ruler or your foot? Explain the reason for your answer. ■

READING EXERCISE 1-3: What problems might arise when using the length of the day as a standard unit of time? ■

1-3 The Quest for Precision

Using a grocery store spring scale to find an apple's mass is fine for shopping purposes. But a mass can be determined to a far greater precision with a chemical microbalance. At best, the apple's mass can only be determined to the nearest gram, whereas the chemistry lab sample can be determined to the nearest hundred-thousandth of a gram.

Throughout history people have sought to measure physical quantities as precisely as possible, because reducing measurement uncertainties has been of tremendous importance in commerce, navigation, astronomical observation, engineering, and scientific research. For example, in 1707, the British navy lost almost 2000 men when four warships ran aground because navigators were unable to measure longitude with sufficient precision. In 1714, as a result of this mishap and others, the British government offered a prize of £20,000 (current value about $12 million) to anyone who could devise a scheme to measure longitude to within half a degree. John Harrison, a self-educated clockmaker, collected the prize in 1765 after designing a series of elaborate chronometers. His early models were driven by a combination of rust-proof brass and self-lubricating wooden gears that kept time to within 1 second per day.

HOW CAN TIME MEASUREMENTS BE USED TO DETERMINE LONGITUDE? Harrison's measurement technique is one of several examples of how a time standard and a knowledge of how fast something is moving are used to measure distance more precisely. In this case, since the Earth turns through 360° on its axis in 24 hours, a precise chronometer can be set so that it reads exactly noon when the Sun is at its highest point in a port with known longitude. Out at sea, the clock time that was set in port will differ from the local solar time by 4 minutes for each degree of longitude difference. Thus, the difference between the observed local noon and the clock reading can then be used to calculate longitude.

Of all the measured quantities, time and other measurements based on time are the most precise. By the end of the 20th century, many of us were wearing inexpensive digital watches driven by the oscillations of quartz crystals. These watches are 1000 times better than John Harrison's chronometer, since they are accurate to within 1 part in 10^8 or 1 thousandth of a second per day. Atomic clocks, precise to 3 billionths of a second per day, are now being used as time standards in many countries.

READING EXERCISE 1-4: A ship embarks from Southampton, England where its clock was set to 12:00:00 at local noon. After 14 days under sail its chronometer reads 12 h 20 min 13 s at the moment the Sun is highest in the sky (local noon). (a) By how many degrees has the ship's longitude changed? (b) Suppose the clock is not precise and has gained 2 minutes out of the 20 160 minutes that have elapsed since it set sail. How far off will the longitude measurement be? (c) The circumference of the Earth is 24 000 nautical miles. Suppose the ship was traveling along the equator. How many miles off course could the ship be if the uncertainty of longitude is 0.5°? ■

1-4 The International System of Units

In the past, communication between scientists was complicated by the fact that for every physical quantity there were a multitude of measurement procedures and basic units of comparison. In addition, there are so many physical quantities that it is a problem to organize them. Fortunately, these quantities are not all independent; for example, speed is the ratio of a length to a time. Thus, what we do is pick out—by international agreement—a small number of physical quantities, such as length and time, and assign standards to them alone. We then define all other physical quantities in terms of these *base quantities* and their standards (which we now call *base standards*). Speed, for example, is defined in terms of the base quantities length and time.

WHY IS IT IMPORTANT TO HAVE A STANDARD SYSTEM OF UNITS THAT IS USED BY ALL SCIENTISTS AND ENGINEERS? In December 1998, the National Aeronautics and Space Administration launched the *Mars Climate Orbiter* on a scientific mission to collect Martian climate data. Nine months later, on September 23, 1999, the *Orbiter* disappeared while approaching Mars at an unexpectedly low altitude. (See Fig. 1-2). An investigation revealed that the orbital calculations were incorrect due to an error in the transfer of information between the spacecraft's team in Colorado and the mission navigation team in California. One team was using English units such as feet and pounds for a critical calculation, while the other group assumed the result of the calculation was being reported in metric units such as meters and kilograms. This misunderstanding about the units being used cost U.S. taxpayers approximately 125 million dollars.

FIGURE 1-2 ■ The *Mars Climate Orbiter* failed to go into orbit around Mars and disappeared due to a miscalculation that resulted from confusion about what units were being used.

In 1971, the 14th General Conference on Weights and Measures recognized the need to use standard units for physical quantities. Conference attendees chose seven physical quantities as base quantities and defined a standard unit of measure for each one. Although other sets of physical quantities could be defined, the seven shown in Table 1-1 form the basis of the widely accepted International System of Units. The system is popularly known as the *metric system* or by its abbreviation, SI, which derives from its French name, *Système International.*

All other SI units are known as *derived units* because they can be expressed in terms of the base units. For example, the SI unit for power, called the **watt** (symbol: W), is defined in terms of the base units for mass, length, and time. As you will see in Chapter 9,

$$1 \text{ watt} = 1 \text{ W} = 1 \text{ kg} \cdot \text{m}^2/\text{s}^3. \tag{1-1}$$

The fact that the dozens of units used in different branches of physics can all be derived from a set of seven base units seems incredible and is a profound testimonial to the unity of physics.

To express the very large and very small quantities that we often run into in physics, we use *scientific notation*, which employs powers of 10. In this notation,

$$3\,560\,000\,000 \text{ m} = 3.56 \times 10^9 \text{ m} \tag{1-2}$$

and

$$0.000\,000\,492 \text{ s} = 4.92 \times 10^{-7} \text{ s}. \tag{1-3}$$

Scientific notation on computers sometimes takes on an even briefer look, as in 3.56 E9 and 4.92 E-7, where E stands for "exponent of ten." It is briefer still on some calculators, where E is replaced with an empty space.

TABLE 1-1
The SI Base Units

Quantity	Unit Name	Unit Symbol
Length	meter	m
Time	second	s
Mass	kilogram	kg
Amount of substance	mole	mol
Electric current	ampere	A
Thermodynamic temperature	kelvin	K
Luminous intensity	candela	cd

TABLE 1-2
Common Prefixes for SI Units

Factor	Prefix	Symbol	Factor	Prefix	Symbol
10^{12}	tera-	T	10^{-15}	femto-	f
10^{9}	giga-	G	10^{-12}	pico-	p
10^{6}	mega-	M	10^{-9}	nano-	n
10^{3}	kilo-	k	10^{-6}	micro-	μ
			10^{-3}	milli-	m
			10^{-2}	centi-	c
			10^{-1}	deci-	d

When reporting the results of very large or very small measurements, it is convenient to define prefixes that designate what power of ten a number has. For example, we can use the prefix kilo-, which represents 10^3, to express 1.0×10^3 grams as 1.0 kilogram. Some of the most common prefixes used in physics and engineering are listed in Table 1-2. A complete list of SI prefixes is included on the inside front cover. As you can see, each prefix represents a certain power of 10 as a factor. Attaching a prefix to an SI unit has the effect of multiplying it by the associated factor. Thus, we can express a particular electric power as

$$1.27 \times 10^9 \text{ watts} = 1.27 \text{ gigawatts} = 1.27 \text{ GW}, \tag{1-4}$$

or a particular length as

$$2.35 \times 10^{-9} \text{ m} = 2.35 \text{ nanometers} = 2.35 \text{ nm}. \tag{1-5}$$

FIGURE 1-3 ■ Two different surfaces of a crystal of pure silicon.

Some prefixes, as used in milliliter, centimeter, kilogram, and megabyte, may be familiar to you.

Once we have set up a standard unit—say, for length—we must work out procedures by which any length, be it the distance to a star or the radius of a hydrogen atom, can be expressed in terms of the standard. Rulers, which approximate our length standard, give us one such procedure for measuring length. We can use a ruler to measure another length by counting how many times the standard can be fit, laid end-to-end, to the other length. The count is our assigned length and is given in terms of the standard's unit. However, many of our comparisons must be indirect. You cannot use a ruler, for example, to measure the distance to a star or the radius of an atom. Figure 1-3 shows an image of the surface of a crystal of silicon obtained with a modern scanning probe microscope.

Base standards must be both accessible and invariable. If we define the length standard as the distance between one's nose and the index finger on an outstretched arm, we certainly have an accessible standard—but it will, of course, vary from person to person. The demand for precision in science and engineering pushes us to aim first for invariability. We then exert great effort to make duplicates of the base standards that are accessible to those who need them. In the United States, the National Institute of Standards and Technology (NIST) is responsible for maintaining base standards and researching issues related to measurement.

The topics that we will investigate first, those related to the physics of forces and motion, require that we make measurements of time, length, and mass. Therefore, we begin by discussing the formal SI definitions of these quantities.

1-5 The SI Standard of Time

Time has two separate aspects that are important in physics. We may want to note at what moment an event occurred or began, or we may want to know how long the event lasted. These are two very different aspects of the measurement of time. For example, the moment at which your physics teacher walks into the room for class on a given day will be measured differently by different students because their watches will not all be synchronized. However, the measured duration of the class will not be affected by the fact that the watches are not synchronized. Thus, "*When* did it happen?" and "What is its *duration*?" are two different questions.

Any phenomenon that regularly repeats itself is a possible time standard. The Earth's rotation, which determines the length of the day, has been used in this way for centuries. Originally the second was defined as the fraction 1/86 400 of a "mean solar day." Figure 1-4 shows a two-century-old example of a time-keeping instrument used to measure the Earth's rotation in terms of a 20-hour day. A quartz clock, in which a quartz ring is made to vibrate continuously, can be calibrated against Earth's rotation via astronomical observations and used to measure time intervals in the laboratory. However, even this calibration cannot be carried out with the accuracy called for by modern scientific and engineering technology.

FIGURE 1-4 ■ When the metric system was proposed in 1792, the hour was redefined to provide a 20-hour day. The idea did not catch on. The maker of this watch wisely provided a small dial that kept both 10-hour and conventional 12-hour time. Do the two dials indicate the same time?

To meet the need for more accuracy in the measurement of time, atomic clocks have been developed that replace the use of Earth's rotation in the definition of our time standard. In 1967, the 13th General Conference on Weights and Measures adopted a standard second based on the radiation absorption characteristics of the cesium-133 atom. Like other atoms, a cesium-133 atom can absorb electromagnetic radiation that has a very precise frequency when the atom makes a transition between two of its well-defined energy states known in technical jargon as "hyperfine levels." The fixed frequency of this external radiation is used to drive a cesium clock. Such a precisely repetitive event is just what is needed for a high-precision timekeeper. Although the technical details of how a cesium clock works is beyond the scope of this text, interested readers can consult the NIST web site at http://www.nist.gov for more information about how the cesium clock is used as a time standard. (See Fig. 1-5.) This new SI standard of time defines the second as follows:

> One second is the duration of 9 192 631 770 periods of the radiation corresponding to the transition between the two hyperfine levels of the ground state of the cesium-133 atom.

An atomic clock at NIST is the standard for Coordinated Universal Time (CUT) in the United States. Its time signals are available from NIST's Web site listed previously. You can also download a Java program from this site that will synchronize your computer's clock to Coordinated Universal Time so you can use your computer as a time standard by which to set other clocks.

Atomic clocks are so consistent that, in principle, two cesium clocks would have to run for 6000 years before their readings would differ by more than 1 second. This amounts to a precision better than 1 part in 10^{11}. Even such accuracy pales in comparison to that of clocks currently being developed; their precision may be as fine as 1 part in 10^{18}.

FIGURE 1-5 ■ The cesium fountain atomic frequency standard developed at the National Institute of Standards and Technology in Boulder, Colorado. It is the primary standard for the unit of time in the United States. To set your watch by it, call (303) 499-7111,
or call (900) 410-8463 or
http://tycho.usno.navy.mil/time.html for Naval Observatory time signals.

READING EXERCISE 1-5: (a) You and a friend are observing a storm. Each of you has your own watch. Describe under what conditions you will both measure the same time for a flash of lightning. Describe under what conditions you will both measure the same duration of time between the lightning flash and the clap of thunder. (b) Look at Fig. 1-4. Do the 10-hour and 12-hour clocks really show the same time? ■

TOUCHSTONE EXAMPLE 1-1*: Sunset

Suppose that while lying on a beach watching the Sun set over a calm ocean, you start a stopwatch just as the top of the Sun disappears. You then stand, elevating your eyes by a height $h = 1.70$ m, and stop the watch when the top of the Sun again disappears. If the elapsed time on the watch is $t = 11.1$ s, what is the radius r of Earth?

SOLUTION ■ A **Key Idea** here is that just as the Sun disappears, your line of sight to the top of the Sun is tangent to Earth's surface. Two such lines of sight are shown in Fig. 1-6. There your eyes are located at point A while you are lying, and at height h above point A while you are standing. For the latter situation, the line of sight is tangent to Earth's surface at point B. Let d represent the distance between point B and the location of your eyes when you are standing, and draw radii r as shown in Fig. 1-6. From the Pythagorean theorem, we then have

$$d^2 + r^2 = (r + h)^2 = r^2 + 2rh + h^2,$$

or

$$d^2 = 2rh + h^2. \tag{1-6}$$

Because the height h is so much smaller than Earth's radius r, the term h^2 is negligible compared to the term $2rh$, and we can rewrite Eq. 1-6 as

$$d^2 \approx 2rh. \tag{1-7}$$

In Fig. 1-6, the angle between the radii to the two tangent points A and B is θ, which is also the angle through which the Sun moves about Earth during the measured time $t = 11.1$ s. During a full day, which is approximately 24 h, the Sun moves through an angle of 360° about Earth. This allows us to write

$$\frac{\theta}{360°} = \frac{t}{24\text{ h}},$$

which, with $t = 11.1$ s, gives us

$$\theta = \frac{(360°)(11.1\text{ s})}{(24\text{ h})(60\text{ min/h})(60\text{ s/min})} = 0.04625°.$$

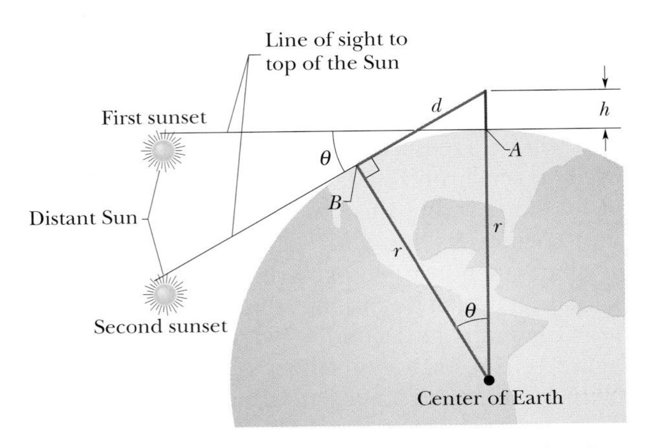

FIGURE 1-6 ■ Your line of sight to the top of the setting Sun rotates through the angle θ when you stand up at point A, and elevate your eyes by a distance h. (Angle θ and distance h are exaggerated here for clarity.)

Again in Fig. 1-6, we see that $d = r \tan \theta$. Substituting this for d in Eq. 1-7 gives us

$$r^2 \tan^2 \theta = 2rh,$$

or

$$r = \frac{2h}{\tan^2 \theta}.$$

Substituting $\theta = 0.04625°$ and $h = 1.70$ m, we find

$$r = \frac{(2)(1.70\text{ m})}{\tan^2 (0.04625°)} = 5.22 \times 10^6\text{ m}, \qquad \text{(Answer)}$$

which is within 20% of the accepted value (6.37×10^6 m) for the mean radius of Earth.

*Adapted from "Doubling Your Sunsets, or How Anyone Can Measure the Earth's Size with a Wristwatch and Meter Stick," by Dennis Rawlins, *American Journal of Physics*, Feb. 1979, Vol. 47, pp. 126–128. This technique works best at the equator.

1-6 The SI Standards of Length

In 1792, the newly born Republic of France established a new system of weights and measures. Its cornerstone was the meter, defined to be one ten-millionth of the distance from the North Pole to the equator. However, the first prototype of a 1-meter-long rod was short by 0.2 millimeter, because researchers miscalculated the flattening of the Earth due to its rotation. Nonetheless, this shortened length became the standard meter. For practical reasons, the meter came to be defined as the distance between two fine lines engraved near the ends of a special platinum-iridium bar, the **standard meter bar,** which was kept at the International Bureau of Weights and Measures near Paris. Accurate copies of the bar have been sent to standards laboratories throughout the world including NIST.

Eventually, modern science and technology required an even more precise standard. Today, the length standard is based on the speed of light. As you will learn in Chapter 38, one of the landmark discoveries of the 20th century was Einstein's recognition that the speed of light in a vacuum is the same for all observers. Since the speed of light can be measured to very high precision, it was adopted as a defined quantity in 1983. Time measurements with atomic clocks are also very precise, so it made sense to redefine the meter in terms of the time it takes light to travel 1 meter. By defining the speed of light c to be exactly

$$c = 299\,792\,458 \text{ m/s}, \tag{1-8}$$

light would travel 1 meter in a time period equal to 1/299 792 458 of a second. That is, if one takes this speed and multiplies by this time period, then the distance traveled by the light is exactly 1 meter. According to the 17th General Conference on Weights and Measures:

> The meter is the length of the path traveled by light in a vacuum during a time interval of 1/299 792 458 of a second.

This approach of measuring lengths in terms of a speed and time is similar to that taken by John Harrison in the 18th century when he proposed measuring longitude in terms of the angular speed of the Earth's rotation and time.

Defining the standard meter in terms of the time it takes light to travel a meter has not done away with the need for secondary standards like bars of metal with fine lines delineating the beginning and end points of a meter. We currently use the metal bar as a secondary standard against which we can easily compare other objects. Defining the meter in terms of the speed of light simply gives us a more precise way to verify that our secondary standard is correct.

1-7 SI Standards of Mass

Currently there are two accepted base units for mass—one suitable for determining large masses and the other for determining masses on an atomic scale.

The Standard Kilogram

The initial SI standard of mass is a platinum-iridium cylinder (Fig. 1-7) kept at the International Bureau of Weights and Measures near Paris. By international agreement, it is defined as a mass of 1 kilogram. Accurate replicas have been sent to standards laboratories in other countries, and the masses of other bodies can be determined by balancing them against a replica. The United States copy of the standard kilogram is housed in a vault at NIST. It is removed, no more than once a year, for the purpose of checking replicas used elsewhere. Since 1889, the U.S. replica of the standard kilogram has been taken to France twice for comparison with the primary standard.

FIGURE 1-7 ■ The international 1 kg standard of mass is a cylinder 39 mm in both height and diameter.

The Atomic Mass Unit

The mass of the known universe is estimated to be 1×10^{53} kg. In contrast, the electron, which plays a vital role in chemical bonding, has a mass of 9×10^{-31} kg. Obviously, the masses of electrons and atoms can be compared with each other more precisely than they can be compared with the standard kilogram. For this reason, we have a second mass standard. It is the carbon-12 atom, which, by international agreement, has been assigned a mass of 12 **atomic mass units** (u). The relation between the atomic mass unit and the kilogram is

$$1 \text{ u} = 1.660\,538\,73 \times 10^{-27} \text{ kg}, \tag{1-9}$$

with an uncertainty of ± 13 in the last two decimal places. Scientists can determine the masses of other atoms relative to the mass of carbon-12 with much better precision than they can using a standard kilogram.

We presently lack a reliable way to extend the precision of the atomic mass unit to more common units of mass, such as the kilogram. However, it is not hard to imagine how one might do this. If we had an object made up of carbon-12 atoms and knew the exact number of atoms in the object, than we could build a precise standard kilogram based on the atomic unit. Work on this is currently underway at NIST and other similar institutions.

READING EXERCISE 1-6: Describe a procedure for determining the mass of the object that has a mass much less than 1 kilogram. Assume that you have a balance, a replica of a standard kilogram, and a big blob of clay available to you. ■

1-8 Measurement Tools for Physics Labs

Institutions like NIST and the International Bureau of Weights and Measures in Paris have many exotic instruments for performing extremely precise measurements. Traditionally, physics students use more common measuring tools in the laboratory, such as meter sticks, vernier calipers (Fig. 1-8), mechanical and electronic balances, digital stopwatches, and multimeters. With careful use, these tools provide adequate precision for studying the time durations and distances investigated in introductory physics laboratories.

In the past few years, new computer tools have become popular in introductory laboratories and in interactive lecture demonstrations. These tools greatly enhance the speed and precision of measurements while allowing students to make many measurements easily and accurately. These tools include **computer data acquisition systems** (Fig. 1-9) **and video capture and analysis tools.** Data obtained using these new computer tools will be shown throughout this text. These data will be used to provide

FIGURE 1-8 ■ Vernier calipers are cleverly designed to make length measurements to within 1/10 of a millimeter.

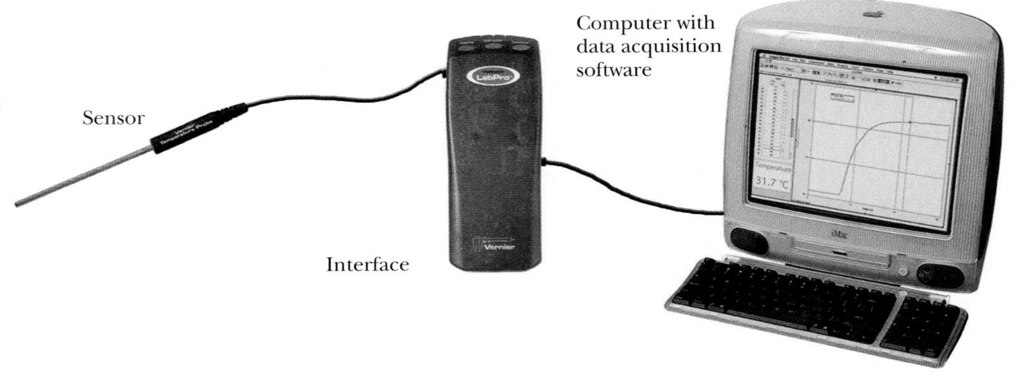

FIGURE 1-9 ■ The photo shows a computer data acquisition system consisting of a sensor, an interface, a computer, and software for real-time data collection.

experimental evidence to motivate and test various theories presented in this book. You may be replicating some of these experiments in laboratory or lecture sessions.

FIGURE 1-10 ■ An ultrasonic motion detector.

Computer Data Acquisition System

When a sensor is attached to a computer through an interface, a very powerful data collection, analysis, and display system is created.* Computers coupled with appropriate software packages are capable of analyzing signals and displaying them on the screen in easily understood formats. Using these capabilities, a graphical representation of data can be displayed in "real time."

A number of different sensors are used in contemporary introductory physics laboratories (Fig. 1-10). These include sensors for the detection of linear and rotational motion (Fig. 1-11), acceleration, force, temperature, pressure, voltage, current, and magnetic field. To determine distances, the most popular motion sensor emits pulses of ultra high frequency sound. Although these ultrasonic pulses are above the range of human hearing, the motion sensor can detect reflections of these pulses after they bounce off objects within the sensor's field of "view."

FIGURE 1-11 ■ Two electronic interfaces used in popular introductory physics computer data acquisition systems: The LabPro Interface (Vernier Software and Technology) and the Science Workshop 500 Interface (PASCO scientific).

Since the speed of ultrasound in room temperature air is known, the computer motion software can calculate the distance to an object by recording how long the pulse takes to reflect off the object and return to the sensor. This is similar to how a bat "sees," and how some auto-focus cameras determine the distance to an object. This approach to measuring a distance or length is not unlike that used by international standards organizations to define the meter in terms of the speed of light. Since ultrasonic motion detectors can send and receive short pulses up to 50 times a second, the computer software can also make rapid calculations of velocities and accelerations of slowly moving objects "on the fly," and graph them in real time. Sample graphs are shown in Fig. 1-12.

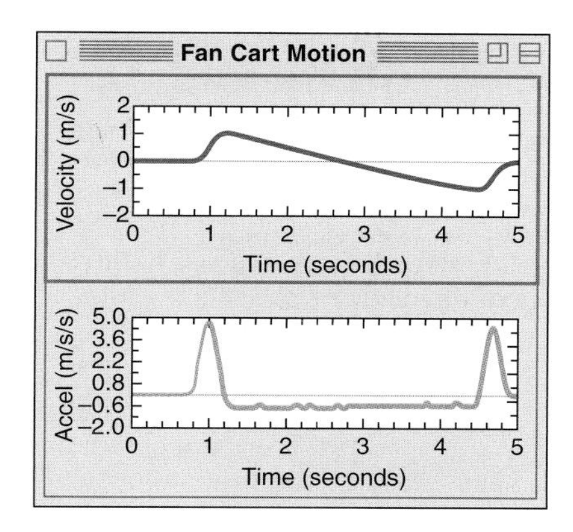

FIGURE 1-12 ■ Real-time graphs of position, velocity, or acceleration, as a function of time, can be generated by an ultrasonic motion detector.

Digital Video Capture and Analysis Tools

Software and hardware enable student investigators to digitize images from a video camera, VCR, or videodisc. Once a digital video movie is created, it can be analyzed using video analysis software. Video data are collected by locating items of interest in each frame of a movie as it is displayed on a computer screen. Video analysis is a useful tool for studying one- and two-dimensional motions, electrostatics, and digital simulations of molecular motions. Examples of digital video clips and their analysis will be presented in this text from time to time. (See Figs. 1-13 and 1-14).

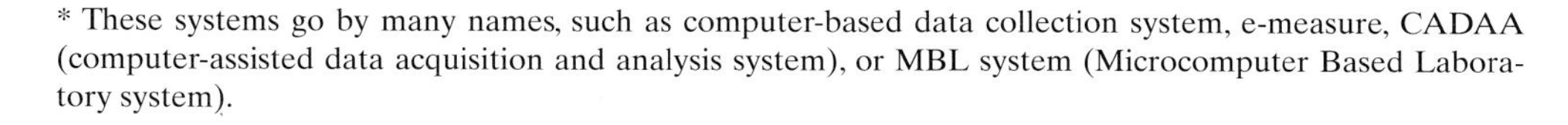

* These systems go by many names, such as computer-based data collection system, e-measure, CADAA (computer-assisted data acquisition and analysis system), or MBL system (Microcomputer Based Laboratory system).

FIGURE 1-13 ■ An overlay of five digital video frames showing a ballet dancer moving toward the left while performing a grand jeté.

1-9 Changing Units

An American traveling overseas notices a road sign indicating that the distance to the next town is 32 km. She wants to get a feel for how far away the town is, and needs to convert the kilometers to the more familiar units of miles. How would she go about doing that?

We often need to change the units in which a physical quantity is expressed. A good method is called *chain-link conversion*. In this method, we multiply the original measurement by one or more conversion factors. A **conversion factor** is defined as a ratio of units that is equal to 1. For example, because 1 mile and 1.61 kilometers are identical distances, we have

$$\frac{1 \text{ mi}}{1.61 \text{ km}} = 1 \quad \text{and also} \quad \frac{1.61 \text{ km}}{1 \text{ mi}} = 1. \tag{1-10}$$

Thus, the ratios (1 mi)/(1.61 km) and (1.61 km)/(1 mi) can be used as conversion factors. This is *not* the same as writing 1/1.61 = 1 or 1.61 = 1; each *number* and its *unit* must be treated together. Because multiplying any quantity by one leaves it unchanged, we can introduce such conversion factors wherever we find them useful. In chain-link conversion, we use the factors to cancel unwanted units. For example, to convert 32 kilometers to miles, we have

$$32 \text{ km} = (32 \cancel{\text{km}})\left(\frac{1 \text{ mi}}{1.61 \cancel{\text{km}}}\right) = 20 \text{ mi}. \tag{1-11}$$

Suppose instead that our traveler wanted to know how many feet there are in 32 kilometers. Then two conversion factors would be needed, so that

$$32 \text{ km} = (32 \cancel{\text{km}})\left(\frac{1 \cancel{\text{mi}}}{1.61 \cancel{\text{km}}}\right)\left(\frac{5280 \text{ ft}}{1 \cancel{\text{mi}}}\right) = 1.05 \times 10^5 \text{ ft}. \tag{1-12}$$

The number of feet is expressed in scientific notation so that the correct number of significant figures can be represented. See the next section and Appendix A for more details on how to represent significant figures properly.

Appendix D and the inside back cover give conversion factors between SI and other systems of units, including many of the non-SI units still used in the United States. However, the conversion factors are written in the style of "1 mi = 1.61 km" rather than the ratios we show here.

It is important to note that the **value of a physical quantity** is actually the product of a number and a unit. Thus, the number associated with a particular physical quantity depends on the unit in which it is expressed. For example, the distance to the trav-

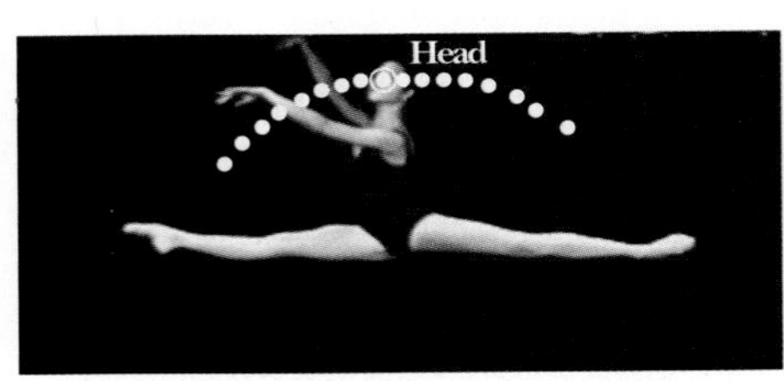

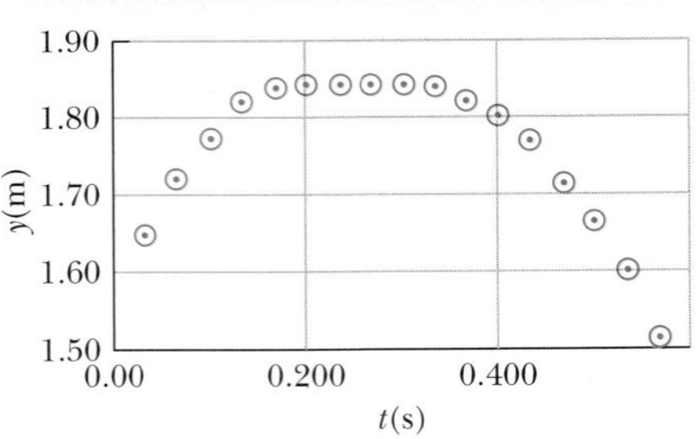

FIGURE 1-14 ■ A video analysis of the motion of the dancer reveals that while performing the grand jeté depicted in Fig. 1-13, her head is moving in a straight horizontal line between the times 0.180 s and 0.330 s. To observers following the motion of her head, the dancer appears to be floating for this short period of time. How does she accomplish this? In Chapter 8 we describe how video analysis helps us explore this question.

eler's town has a value of 32 km. The numerical component of its value expressed in the unit "kilometers" is 32. However, the value of the distance when expressed in miles is 20 mi, and the numerical component of its value when expressed in miles is 20. Since 20 miles is actually the *same distance* as 32 kilometers, it is meaningful to write 32 km = 20 mi. In this context the equal sign (=) signifies that 32 km is the *same distance* as 20 mi expressed in different units. However, it is totally meaningless to write 32 = 20. Thus it is extremely important to include appropriate units in all calculations.

TOUCHSTONE EXAMPLE 1-2: Marathon

When Pheidippides ran from Marathon to Athens in 490 B.C.E. to bring word of the Greek victory over the Persians, he probably ran at a speed of about 23 rides per hour (rides/h). The ride is an ancient Greek unit for length, as are the stadium and the plethron: 1 ride was defined to be 4 stadia, 1 stadium was defined to be 6 plethra, and, in terms of a modern unit, 1 plethron is 30.8 m. How fast did Pheidippides run in kilometers per second (km/s)?

SOLUTION ■ The **Key Idea** in chain-link conversions is to write the conversion factors as ratios that will eliminate unwanted units. Here we write

$$23\text{ rides/h} = \left(23\,\frac{\cancel{\text{rides}}}{\cancel{\text{h}}}\right)\left(\frac{4\ \cancel{\text{stadia}}}{1\ \cancel{\text{ride}}}\right)\left(\frac{6\ \cancel{\text{plethra}}}{1\ \cancel{\text{stadium}}}\right)\left(\frac{30.8\ \cancel{\text{m}}}{1\ \cancel{\text{plethron}}}\right)\left(\frac{1\text{ km}}{1000\ \cancel{\text{m}}}\right)\left(\frac{1\ \cancel{\text{h}}}{3600\text{ s}}\right)$$
$$= 4.7227 \times 10^{-3}\text{ km/s} \approx 4.7 \times 10^{-3}\text{ km/s.} \quad \text{(Answer)}$$

READING EXERCISE 1-7: (a) Explain why it is correct to write 1 min/60 s = 1, but it is not correct to write 1/60 = 1. (b) Use the relevant conversion factors and the method of chain-link conversions to calculate how many seconds there are in a day. ■

1-10 Calculations with Uncertain Quantities

Issue 1: Significant Figures and Decimal Places

In July 1988, in Indianapolis, Indiana, the U.S.'s Florence Griffith Joyner set a world record in the women's 100-meter dash with an official time of 10.49 seconds (Fig. 1-15). The timing in the race is considered good to the nearest 1/100 of a second. Suppose you had been asked to report the time in minutes instead of seconds. If you used a calculator to transform the 10.49 seconds into minutes by multiplying by (1 min)/(60 s), you might report the following by copying all the digits on your display:

$$10.49\text{ s} = (10.49\text{ s})\left(\frac{1\text{ min}}{60\text{ s}}\right) = 0.174\,833\,333\text{ min.} \tag{1-13}$$

No matter how precise a measuring instrument is, all measured quantities have uncertainties associated with them. The precision implied by the calculated time in minutes shown above is both meaningless and misleading! We should have rounded the answer to four significant digits, 0.1748 min, so as not to imply that it is more precise than the given data. The given time of 10.49 seconds consists of four digits, called **significant figures.** This tells us we should round the answer to four significant figures. In this text, final results of calculations are often rounded to match the least number of significant figures in the given data. *Significant figures* should not be confused with *decimal places*. Consider the lengths 35.6 mm, 3.56 cm, and 0.0356 m. They all have three significant figures, but they have one, two, and four decimal places, respectively.

FIGURE 1-15 ■ The late Florence Griffith Joyner set a world's record in the women's 100-meter dash in 1988.

As you work with scientific calculations in data analysis in the laboratory or complete the problems in this text, it is important to pay strict attention to reporting your answer to the same precision as the lowest precision in any of the factors used in your calculation. Information on how to keep track of significant figures and measurement uncertainties in calculations is included in Appendix A, and a table of fundamental constants that have been measured to high precision is in Appendix B.

Issue 2: Order of Magnitude

In order to make estimations, engineering and science professionals will sometimes round a number to be used in a calculation up or down to the nearest power of ten. This makes the number very easy to use in calculations. The result of this rounding procedure is known as the *order of magnitude* of a number. To determine an order of magnitude, we start by expressing the number of interest in scientific notation. Next, the mantissa is rounded up to 10 or down to 1 depending on which is closest. For example, if $A = 2.3 \times 10^4$, then the order of magnitude of A is 10^4 (ten to the fourth) since 2.3 is closer to 1 than it is to 10. On the other hand, if $B = 7.8 \times 10^4$, then the order of magnitude of B is 10^5 (ten to the fifth) since 7.8 is closer to 10 than it is to 1. Order of magnitude estimations are common when detailed or precise data are not required in a calculation or are not known.

READING EXERCISE 1-8: Using the method outlined in Appendix A, determine the number of significant figures in each of the following numbers: (a) 27 meters, (b) 27 cows, (c) 0.003 429 87 second, (d) $-1.970\,500 \times 10^{-11}$ coulombs, (e) 5280 ft/mi. (*Note:* By definition there are exactly 5280 feet in a mile.) ■

READING EXERCISE 1-9: A popular science book lists the radius of the Earth as 20 900 000 000 ft. (a) How many significant figures does this number have if you apply the method described in Appendix A for determining the number of significant figures? (b) How many significant figures did the author probably intend to report? (c) How could you rewrite this number so that it represents three significant figures? (d) What order of magnitude is the radius of the Earth in feet? ■

TOUCHSTONE EXAMPLE 1-3: Ball of String

The world's largest ball of string is about 2 m in radius. To the nearest order of magnitude, what is the total length L of the string in the ball?

SOLUTION ■ We could, of course, take the ball apart and measure the total length L, but that would take great effort and make the ball's builder most unhappy. A **Key Idea** here is that, because we want only the nearest order of magnitude, we can estimate any quantities required in the calculation.

Let us assume the ball is spherical with radius $R = 2$ m. The string in the ball is not closely packed (there are uncountable gaps between nearby sections of string). To allow for these gaps, let us somewhat overestimate the cross-sectional area of the string by assuming the cross section is square, with an edge length $d = 4$ mm. Then, with a cross-sectional area of d^2 and a length L, the string occupies a total volume of

$$V = (\text{cross-sectional area})(\text{length}) = d^2L.$$

This is approximately equal to the volume of the ball, given by $\frac{4}{3}\pi R^3$, which is about $4R^3$ because π is about 3. Thus, we have

$$d^2L = 4R^3,$$

or

$$L = \frac{4\,R^3}{d^2} = \frac{4(2\text{ m})^3}{(4 \times 10^{-3}\text{ m})^2} \quad \text{(Answer)}$$
$$= 2 \times 10^6\text{ m} \approx 10^6\text{ m} = 10^3\text{ km}.$$

(Note that you do not need a calculator for such a simplified calculation.) Thus, to the nearest order of magnitude, the ball contains about 1000 km of string!

READING EXERCISE 1-10: Suppose you are to calculate the volume of a cube that is $L = 1.4$ cm on a side and you start by calculating the area, A, of a face of the cube $A = L^2$ and then calculating $V = AL$. (a) What intermediate value for A should you use in the calculation for V? (b) What is the value of the volume to the correct number of significant figures? (c) What value do you get for V if you incorrectly retain only two significant figures after you calculate A? ■

READING EXERCISE 1-11: Perform the following calculations and express the answers to the correct number of significant figures. (a) Multiply 3.4 by 7.954. (b) Add 99.3 and 98.7. (c) Subtract 98.7 from 99.3. (d) Evaluate the cosine of 3°. (e) If five railroad track segments have an average length of 2.134 meters, what is the total length of these five rails when they lie end to end? ■

READING EXERCISE 1-12: Suppose you measure a time to the nearest 1/100 of a second and get a value of 1.78 s. (a) What is the absolute precision of your measurement? (b) What is the relative precision of your measurement? ■

Problems

SEC. 1-5 ■ THE SI STANDARD OF TIME

1. Speed of Light Express the speed of light, 3.0×10^8 m/s, in (a) feet per nanosecond and (b) millimeters per picosecond.

2. Fermi Physicist Enrico Fermi once pointed out that a standard lecture period (50 min) is close to 1 microcentury. (a) How long is a microcentury in minutes? (b) Using

$$\text{percentage difference} = \left(\frac{\text{actual} - \text{approximation}}{\text{actual}}\right) 100,$$

find the percentage difference from Fermi's approximation.

3. Five Clocks Five clocks are being tested in a laboratory. Exactly at noon, as determined by the WWV time signal, on successive days of a week the clocks read as in the following table. Rank the five clocks according to their relative value as good timekeepers, best to worst. Justify your choice.

Clock	Sun.	Mon.	Tues.	Wed.	Thurs.	Fri.	Sat.
A	12:36:40	12:36:56	12:37:12	12:37:27	12:37:44	12:37:59	12:38:14
B	11:59:59	12:00:02	11:59:57	12:00:07	12:00:02	11:59:56	12:00:03
C	15:50:45	15:51:43	15:52:41	15:53:39	15:54:37	15:55:35	15:56:33
D	12:03:59	12:02:52	12:01:45	12:00:38	11:59:31	11:58:24	11:57:17
E	12:03:59	12:02:49	12:01:54	12:01:52	12:01:32	12:01:22	12:01:12

4. The Shake A unit of time sometimes used in microscopic physics is the *shake*. One shake equals 10^{-8} s. (a) Are there more shakes in a second than there are seconds in a year? (b) Humans have existed for about 10^6 years, whereas the universe is about 10^{10} years old. If the age of the universe now is taken to be 1 "universe day," for how many "universe seconds" have humans existed?

5. Astronomical Units An astronomical unit (AU) is the average distance of Earth from the Sun, approximately 1.50×10^8 km. The speed of light is about 3.0×10^8 m/s. Express the speed of light in terms of astronomical units per minute.

6. Digital Clocks Three digital clocks A, B, and C run at different rates and do not have simultaneous readings of zero. Figure 1-16 shows simultaneous readings on pairs of the clocks for four occasions. (At the earliest occasion, for example, B reads 25.0 s and C reads 92.0 s.) If two events are 600 s apart on clock A, how far apart are they on (a) clock B and (b) clock C? (c) When clock A reads 400 s, what does clock B read? (d) When clock C reads 15.0 s, what does clock B read? (Assume negative readings for prezero times.)

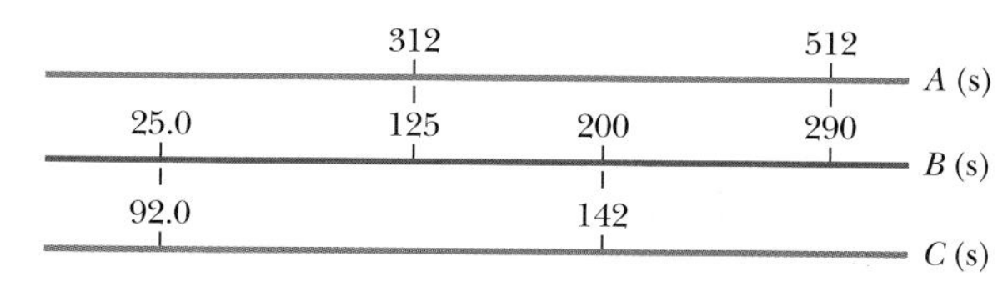

FIGURE 1-16 ■ Problem 6.

7. Length of Day Assuming the length of the day uniformly increases by 0.0010 s per century, calculate the cumulative effect on the measure of time over 20 centuries. (Such slowing of Earth's rotation is indicated by observations of the occurrences of solar eclipses during this period.)

8. Time Zones Until 1883, every city and town in the United States kept its own local time. Today, travelers reset their watches only when the time change equals 1.0 h. How far, on the average, must you travel in degrees of longitude until your watch must be reset by 1.0 h? (*Hint:* Earth rotates 360° in about 24 h.)

9. A Fortnight A fortnight is a charming English measure of time equal to 2.0 weeks (the word is a contraction of "fourteen nights"). That is a nice amount of time in pleasant company but perhaps a painful string of microseconds in unpleasant company. How many microseconds are in a fortnight?

10. Time Standards Time standards are now based on atomic clocks. A promising second standard is based on *pulsars*, which are rotating neutron stars (highly compact stars consisting only of neutrons). Some rotate at a rate that is highly stable, sending out a radio beacon that sweeps briefly across Earth once with each rotation, like a lighthouse beacon. Pulsar PSR 1937 + 21 is an example; it rotates once every 1.557 806 448 872 75 ± 3 ms, where the trailing ±3 indicates the uncertainty in the last decimal place (it does *not* mean ±3 ms). (a) How many times does PSR 1937 + 21 rotate in 7.00 days? (b) How much time does the pulsar take to rotate 1.0×10^6 times, and (c) what is the associated uncertainty?

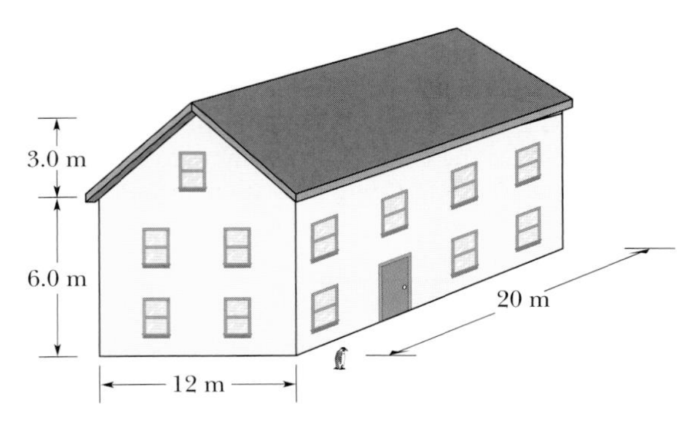

FIGURE 1-18 ■ Problem 18.

SEC. 1-6 ■ THE SI STANDARDS OF LENGTH

11. Furlongs Horses are to race over a certain English meadow for a distance of 4.0 furlongs. What is the race distance in units of (a) rods and (b) chains? (1 furlong = 201.168 m, 1 rod = 5.0292 m, and 1 chain = 20.117 m.)

12. Types of Barrels Two types of *barrel* units were in use in the 1920s in the United States. The apple barrel had a legally set volume of 7056 cubic inches; the cranberry barrel, 5826 cubic inches. If a merchant sells 20 cranberry barrels of goods to a customer who thinks he is receiving apple barrels, what is the discrepancy in the shipment volume in liters?

13. The Earth Earth is approximately a sphere of radius 6.37×10^6 m. What are (a) its circumference in kilometers, (b) its surface area in square kilometers, and (c) its volume in cubic kilometers?

14. Points and Picas Spacing in this book was generally done in units of points and picas: 12 points = 1 pica, and 6 picas = 1 inch. If a figure was misplaced in the page proofs by 0.80 cm, what was the misplacement in (a) points and (b) picas?

15. Antarctica Antarctica is roughly semicircular, with a radius of 2000 km (Fig. 1-17). The average thickness of its ice cover is 3000 m. How many cubic centimeters of ice does Antarctica contain? (Ignore the curvature of Earth.)

FIGURE 1-17 ■ Problem 15.

16. Roods and Perches An old manuscript reveals that a landowner in the time of King Arthur held 3.00 acres of plowed land plus a livestock area of 25.0 perches by 4.00 perches. What was the total area in (a) the old unit of roods and (b) the more modern unit of square meters? Here, 1 acre is an area of 40 perches by 4 perches, 1 rood is 40 perches by 1 perch, and 1 perch is 16.5 ft.

17. The Acre-Foot Hydraulic engineers in the United States often use, as a unit of volume of water, the *acre-foot*, defined as the volume of water that will cover 1 acre of land to a depth of 1 ft. A severe thunderstorm dumped 2.0 in. of rain in 30 min on a town of area 26 km^2. What volume of water, in acre-feet, fell on the town?

18. A Doll House In the United States, a doll house has the scale of 1:12 of a real house (that is, each length of the doll house is $\frac{1}{12}$ that of the real house) and a miniature house (a doll house to fit within a doll house) has the scale of 1:144 of a real house. Suppose a real house (Fig. 1-18) has a front length of 20 m, a depth of 12 m, a height of 6.0 m, and a standard sloped roof (vertical triangular faces on the ends) of height 3.0 m. In cubic meters, what are the volumes of the corresponding (a) doll house and (b) miniature house?

SEC. 1-7 ■ THE SI STANDARDS OF MASS

19. Earth's Mass Earth has a mass of 5.98×10^{24} kg. The average mass of the atoms that make up Earth is 40 u. How many atoms are there in Earth?

20. Gold Gold, which has a mass of 19.32 g for each cubic centimeter of volume, is the most ductile metal and can be pressed into a thin leaf or drawn out into a long fiber. (a) If 1.000 oz of gold, with a mass of 27.63 g, is pressed into a leaf of 1.000 μm thickness, what is the area of the leaf? (b) If, instead, the gold is drawn out into a cylindrical fiber of radius 2.500 μm, what is the length of the fiber?

21. Mass of Water (a) Assuming that each cubic centimeter of water has a mass of exactly 1 g, find the mass of one cubic meter of water in kilograms. (b) Suppose that it takes 10.0 h to drain a container of 5700 m^3 of water. What is the "mass flow rate," in kilograms per second, of water from the container?

22. The Thunderstorm What mass of water fell on the town in Problem 17 during the thunderstorm? One cubic meter of water has a mass of 10^3 kg.

23. Iron Iron has a mass of 7.87 g per cubic centimeter of volume, and the mass of an iron atom is 9.27×10^{-26} kg. If the atoms are spherical and tightly packed, (a) what is the volume of an iron atom and (b) what is the distance between the centers of adjacent atoms?

24. Grains of Sand Grains of fine California beach sand are approximately spheres with an average radius of 50 μm and are made of silicon dioxide. A solid cube of silicon dioxide with a volume of 1.00 m^3 has a mass of 2600 kg. What mass of sand grains would have a total surface area (the total area of all the individual spheres) equal to the surface area of a cube 1 m on an edge?

SEC. 1-9 ■ CHANGING UNITS

25. A Diet A person on a diet might lose 2.3 kg per week. Express the mass loss rate in milligrams per second, as if the dieter could sense the second-by-second loss.

26. Cats and Moles A mole of atoms is 6.02×10^{23} atoms. To the nearest order of magnitude, how many moles of atoms are in a large domestic cat? The masses of a hydrogen atom, an oxygen atom, and a carbon atom are 1.0 u, 16 u, and 12 u, respectively. (*Hint:* Cats are sometimes known to kill moles.)

27. Sugar Cube A typical sugar cube has an edge length of 1 cm. If you had a cubical box that contained a mole of sugar cubes, what would its edge length be? (One mole = 6.02×10^{23} units.)

28. Micrometer The micrometer (1 μm) is often called the *micron*. (a) How many microns make up 1.0 km? (b) What fraction of a centimeter equals 1.0 μm? (c) How many microns are in 1.0 yd?

29. Hydrogen Using conversions and data in the chapter, determine the number of hydrogen atoms required to obtain 1.0 kg of hydrogen. A hydrogen atom has a mass of 1.0 u.

30. A Gry A *gry* is an old English measure for length, defined as 1/10 of a line, where *line* is another old English measure for length, defined as 1/12 inch. A common measure for length in the publishing business is a *point*, defined as 1/72 inch. What is an area of 0.50 gry^2 in terms of points squared (points^2)?

Additional Problems

31. Harvard Bridge Harvard Bridge, which connects MIT with its fraternities across the Charles River, has a length of 364.4 Smoots plus one ear. The unit of one Smoot is based on the length of Oliver Reed Smoot, Jr., class of 1962, who was carried or dragged length by length across the bridge so that other pledge members of the Lambda Chi Alpha fraternity could mark off (with paint) 1-Smoot lengths along the bridge. The marks have been repainted biannually by fraternity pledges since the initial measurement, usually during times of traffic congestion so that the police could not easily interfere. (Presumably, the police were originally upset because a Smoot is not an SI base unit, but these days they seem to have accepted the unit.) Figure 1-19 shows three parallel paths, measured in Smoots (S), Willies (W), and Zeldas (Z). What is the length of 50.0 Smoots in (a) Willies and (b) Zeldas?

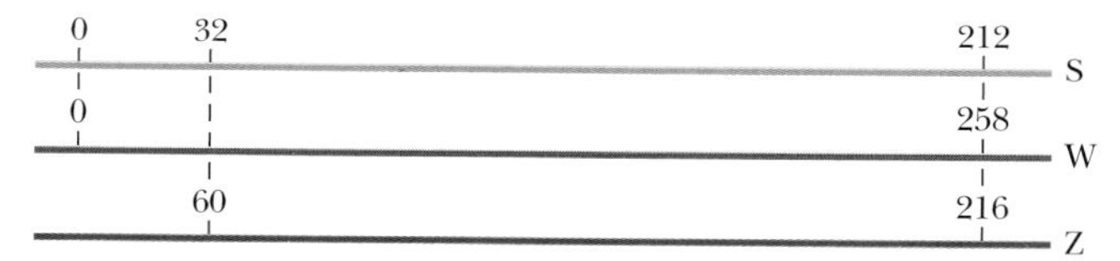

FIGURE 1-19 ■ Problem 31.

32. Little Miss Muffet An old English children's rhyme states, "Little Miss Muffet sat on her tuffet, eating her curds and whey, when along came a spider who sat down beside her. . . ." The spider sat down not because of the curds and whey but because Miss Muffet had a stash of 11 tuffets of dried flies. The volume measure of a tuffet is given by 1 tuffet = 2 pecks = 0.50 bushel, where 1 Imperial (British) bushel = 36.3687 liters (L). What was Miss Muffet's stash in (a) pecks, (b) bushels, and (c) liters?

33. Noctilucent Clouds During the summers at high latitudes, ghostly, silver-blue clouds occasionally appear after sunset when common clouds are in Earth's shadow and are no longer visible. The ghostly clouds have been called *noctilucent clouds* (NLC), which means "luminous night clouds," but now are often called *mesospheric clouds*, after the *mesosphere*, the name of the atmosphere at the altitude of the clouds.

These clouds were first seen in June 1885, after dust and water from the massive 1883 volcanic explosion of Krakatoa Island (near Java in the Southeast Pacific) reached the high altitudes in the Northern Hemisphere. In the low temperatures of the mesosphere, the water collected and froze on the volcanic dust (and perhaps on comet and meteor dust already present there) to form the particles that made up the first clouds. Since then, mesospheric clouds have generally increased in occurrence and brightness, probably because of the increased production of methane by industries, rice paddies, landfills, and livestock flatulence. The methane works its way into the upper atmosphere, undergoes chemical changes, and results in an increase of water molecules there, and also in bits of ice for the mesospheric clouds.

If mesospheric clouds are spotted 38 min after sunset and then quickly dim, what is their altitude if they are directly over the observer?

34. Staircase A standard interior staircase has steps each with a rise (height) of 19 cm and a run (horizontal depth) of 23 cm. Research suggests that the stairs would be safer for descent if the run were, instead, 28 cm. For a particular staircase of total height 4.57 m, how much farther would the staircase extend into the room at the foot of the stairs if this change in run were made?

35. Large and Small As a contrast between the old and the modern and between the large and the small, consider the following: In old rural England 1 hide (between 100 and 120 acres) was the area of land needed to sustain one family with a single plough for one year. (An area of 1 acre is equal to 4047 m^2.) Also, 1 wapentake was the area of land needed by 100 such families. In quantum physics, the cross-sectional area of a nucleus (defined in terms of the chance of a particle hitting and being absorbed by it) is measured in units of barns, where 1 barn is 1×10^{-28} m^2. (In nuclear physics jargon, if a nucleus is "large," then shooting a particle at it is like shooting a bullet at a barn door, which can hardly be missed.) What is the ratio of 25 wapentakes to 11 barns?

36. Cumulus Cloud A cubic centimeter in a typical cumulus cloud contains 50 to 500 water droplets, which have a typical radius of 10 μm. (a) How many cubic meters of water are in a cylindrical cumulus cloud of height 3.0 km and radius 1.0 km? (b) How many 1-liter pop bottles would that water fill? (c) Water has a mass per unit volume (or density) of 1000 kg/m^3. How much mass does the water in the cloud have?

37. Oysters In purchasing food for a political rally, you erroneously order shucked medium-size Pacific oysters (which come 8 to 12 per U.S. pint) instead of shucked medium-size Atlantic oysters (which come 26 to 38 per U.S. pint). The filled oyster container delivered to you has the interior measure of 1.0 m $\times$ 12 cm $\times$ 20 cm, and a U.S. pint is equivalent to 0.4732 liter. By how many oysters is the order short of your anticipated count?

38. U.K. Gallons A tourist purchases a car in England and ships it home to the United States. The car sticker advertised that the car's fuel consumption was at the rate of 40 miles per gallon on the open road. The tourist does not realize that the U.K. gallon differs from the U.S. gallon:

$$1 \text{ U.K. gallon} = 4.545\,963\,1 \text{ liters}$$
$$1 \text{ U.S. gallon} = 3.785\,306\,0 \text{ liters}.$$

For a trip of 750 miles (in the United States), how many gallons of fuel does (a) the mistaken tourist believe she needs and (b) the car actually require?

39. Types of Tons A ton is a measure of volume frequently used in shipping, but that use requires some care because there are at least three types of tons: A *displacement ton* is equal to 7 barrels bulk, a *freight ton* is equal to 8 barrels bulk, and a *register ton* is equal to 20 barrels bulk. A *barrel bulk* is another measure of volume: 1 barrel bulk = 0.1415 m^3. Suppose you spot a shipping order for "73 tons" of M&M candies, and you are certain that the client who sent the order intended "ton" to refer to volume (instead of weight or mass, as discussed in Chapter 6). If the client actually meant displacement tons, how many extra U.S. bushels of the candies will you erroneously ship to the client if you interpret the order as (a) 73 freight tons and (b) 73 register tons? One cubic meter is equivalent to 28.378 U.S bushels.

40. Wine Bottles The wine for a large European wedding reception is to be served in a stunning cut-glass receptacle with the interior dimensions of 40 cm × 40 cm × 30 cm (height). The receptacle is to be initially filled to the top. The wine can be purchased in bottles of the sizes given in the following table, where the volumes of the larger bottles are given in terms of the volume of a standard wine bottle. Purchasing a larger bottle instead of multiple smaller bottles decreases the overall cost of the wine. To minimize that overall cost, (a) which bottle sizes should be purchased and how many of each should be purchased, and (b) how much wine is left over once the receptacle is filled?

1 standard
1 magnum = 2 standard
1 jeroboam = 4 standard
1 rehoboam = 6 standard
1 methuselah = 8 standard
1 salmanazar = 12 standard
1 balthazar = 16 standard = 11.356 L
1 nebuchadnezzar = 20 standard

41. The Corn–Hog Ratio The *corn-hog ratio* is a financial term commonly used in the pig market and presumably is related to the cost of feeding a pig until it is large enough for market. It is defined as the ratio of the market price of a pig with a mass of 1460 slugs to the market price of a U.S. bushel of corn. The slug is the unit of mass in the English system. (The word "slug" is derived from an old German word that means "to hit"; we have the same meaning for "slug" as a verb in modern English.) A U.S. bushel is equal to 35.238 L. If the corn–hog ratio is listed as 5.7 on the market exchange, what is it in the metric units of

$$\frac{\text{price of 1 kilogram of pig}}{\text{price of 1 liter of corn}}?$$

(*Hint:* See the Mass table in Appendix D.)

42. Volume Measures in Spain You can easily convert common units and measures electronically, but you still should be able to use a conversion table, such as those in Appendix D. Table 1-3 is part of a conversion table for a system of volume measures once common in Spain; a volume of 1 fanega is equivalent to 55.501 dm^3 (cubic decimeters). (a) Complete the table, using three significant figures. Then express 7.00 almude in terms of (b) medio, (c) cahiz, and (d) cubic centimeters (cm^3).

TABLE 1-3
Problem 42

	cahiz	fanega	cuartilla	almude	medio
1 cahiz =	1	12	48	144	288
1 fanega =		1	4	12	24
1 cuartilla =			1	3	6
1 almude =				1	2
1 medio =					1

43. Pirate Ship You receive orders to sail due east for 24.5 mi to put your salvage ship directly over a sunken pirate ship. However, when your divers probe the ocean floor at that location and find no evidence of a ship, you radio back to your source of information, only to discover that the sailing distance was supposed to be 24.5 *nautical miles*, not regular miles. Use the Length table in Appendix D to calculate how far horizontally you are from the pirate ship in kilometers.

44. The French Revolution For about 10 years after the French revolution, the French government attempted to base measures of time on multiples of ten: One week consisted of 10 days, 1 day consisted of 10 hours, 1 hour consisted of 100 minutes, and 1 minute consisted of 100 seconds. What are the ratios of (a) the French decimal week to the standard week and (b) the French decimal second to the standard second?

45. Heavy Rain During heavy rain, a rectangular section of a mountainside measuring 2.5 km wide (horizontally), 0.80 km long (up along the slope), and 2.0 m deep suddenly slips into a valley in a mud slide. Assume that the mud ends up uniformly distributed over a valley section measuring 0.40 km × 0.40 km and that the mass of a cubic meter of mud is 1900 kg. What is the mass of the mud sitting above an area of 4.0 m^2 in that section?

46. Liquid Volume Prior to adopting metric systems of measurement, the United Kingdom employed some challenging measures of liquid volume. A few are shown in Table 1-4. (a) Complete the table, using three significant figures. (b) The volume of 1 bag is equivalent to a volume of 0.1091 m^3. If an old British story has a witch cooking up some vile liquid in a cauldron with a volume of 1.5 chaldrons, what is the volume in terms of cubic meters?

TABLE 1-4
Problem 46

	wey	chaldron	bag	pottle	gill
1 wey =	1	10/9	40/3	640	120 240
1 chaldron =					
1 bag =					
1 pottle =					
1 gill =					

47. The Dbug Traditional units of time have been based on astronomical measurements, such as the length of the day or year. However, one human-based measure of time can be found in Tibet, where the *dbug* is the average time between exhaled breaths. Estimate the number of dbugs in a day.

48. Tower of Pisa The following photograph of the Leaning Tower of Pisa was taken from an advertisement found in a 1994 airline magazine. Assume that the photo of the man talking on the telephone to the left has been dubbed in and is not part of the original photograph.

FIGURE 1-20 ■ Problem 48.

(a) Examine the photograph. Take the measurements in centimeters that are needed to find a scale factor that enables you to estimate the length of the tower in meters (i.e., its height if it were standing up straight.) Use only the evidence in the photograph—no other data are allowed. Then estimate the tower length in meters.
(b) According to data published in Sir Bannester Fletcher's *A History of Architecture* (U. of London Athlone Press, 1975, p. 470) the diameter of the lower part of the tower is 16.0 m. Using these data, find another scale factor for estimating the length of the tower, and then re-estimate the length of the tower using this new scale factor.
(c) Which of the scale factors (a) or (b) do you think will give the best estimate of the length of the tower? Explain the reasons for your answer.
(d) Using the scale factor you found in part (b), what is the length of the tower without the belfry or narrow top segment (i.e., just consider the bottom 7 stories)?

49. Mexican Food You are to fix dinners for 400 people at a convention of Mexican food fans. Your recipe calls for 2 jalapeño peppers per serving (one serving per person). However, you have only habanero peppers on hand. The spiciness of peppers is measured in terms of the *scoville heat unit* (SHU). On average, one jalapeño pepper has a spiciness of 4000 SHU and one habanero pepper has a spiciness of 300 000 SHU. To salvage the situation, how many (total) habanero peppers should you substitute for the jalapeño peppers in the recipe for the convention?

50. Big or Small? Discuss the question: "Is 500 feet big or small?" Before you do so, carry out the following estimates.

(a) You are on the top floor of a 500-ft-tall building. A fire breaks out in the building and the elevator stops working. You have to walk down to the ground floor. Estimate how long this would take you. (Your stairwell is on the other side of the building from the fire.)
(b) You are hiking the Appalachian Trail on a beautiful fall morning as part of a 10 mi hike with a group of friends. You are walking along a well-tended, level part of the trail. Estimate how long it would take you to walk 500 ft.
(c) You are driving on the New Jersey Turnpike at 65 mi/hr. You pass a sign that says "Lane ends 500 feet." How much time do you have in order to change lanes?

51. Doubling System Historically the English had a doubling system when measuring volumes; 2 mouthfuls equal 1 jigger, 2 jiggers equal 1 jack (also called a jackpot); 2 jacks equal 1 jill; 2 jills = 1 cup; 2 cups = 1 pint; 2 pints = 1 quart; 2 quarts = 1 pottle; 2 pottles = 1 gallon; 2 gallons = 1 pail. (The nursery rhyme "Jack and Jill" refers to these units and was a protest against King Charles I of England for his taxes on the jacks of liquor sold in the tavern. (See A. Kline, *The World of Measurement,* New York: Simon and Schuster, 1975, pp. 32–39.) American and British cooks today use teaspoons, tablespoons, and cups; 3 teaspoons = 1 tablespoon; 4 tablespoons = 1/4 cup. Assume that you find an old English recipe requiring 3 jiggers of milk. How many cups does this represent? How many tablespoons? You can assume that the cups in the two systems represent the same volume.

52. Fuel Efficiency In America, we measure fuel efficiency of our cars by citing the number of miles you can drive on 1 gallon of gas (miles/gallon). In Europe, the same information is given by quoting how many liters of gas it takes to go 100 kilometers (liter/100 kilometers).

(a) My current car gets 21 miles/gallon in highway travel. What number (in liter/100 kilometers) should I give to my Swedish friend so that he can compare it to the mileage for his Volvo?
(b) The car I drove in England last summer needed 6 liters of gas to go 100 kilometers. How many miles/gallon did it get?
(c) If my car has a fuel efficiency, f, in miles/gallon, what is its European efficiency, e, in liters/100 kilometers? (Write an equation that would permit an easy conversion.)

53. Pizza Sale Two terrapins decide to go to Jerry's for a pizza. When they get there they find that Jerry's is having a special:

SPECIAL TODAY:	one 20″ pizza	$15
REGULAR PRICE	one 10″ pizza	$5
	one 20″ pizza	$18

Raphael: "Great! Let's get a large one."

Donatello: "Don't be dumb. Let's get three of the small ones for the same price. That'll give us more pizza and be cheaper."

Raphael: "Why would it be a special if it's more than we could get for the regular price? Let's get the large."

Who's right? Which would you buy? What would the difference be if you were buying them at Ledo's (square pizzas)?

54. Dollar and Penny A student makes the following argument: "I can prove a dollar equals a penny. Since a dime (10 cents) is one-tenth of a dollar, I can write:

$$10\,¢ = \$0.1.$$

Square both sides of the equation. Since squares of equals are equal,

$$100\,¢ = \$0.1.$$

Since 100 ¢ = \$1 and \$0.01 = 1 ¢, it follows that \$1 = 1 ¢."
What's wrong with the argument?

55. Scaling Up Here are two related problems—one precise, one an estimation.

(a) A sculptor builds a model for a statue of a terrapin to replace Testudo.* She discovers that to cast her small scale model she needs 2 kg of bronze. When she is done, she finds that she can give it two coats of finishing polyurethane varnish using exactly one small can of varnish. The final statue is supposed to be 5 times as large as the model in each dimension. How much bronze will she need? How much varnish should she buy? (*Hint*: If this seems difficult, you might start by writing a simpler question that is easier to work on before tackling this one.)

FIGURE 1-21 ■ Problem 55.

(b) The human brain has 1000 times the surface area of a mouse's brain. The human brain is convoluted, the mouse's is not. How much of this factor is due just to size (the human brain is bigger)? How sensitive is your result to your estimations of the approximate dimensions of a human brain and a mouse brain?

56. Finding the Right Dose We know from our dimensional analysis that if an object maintains its shape but changes its size, its area changes as the square of its length and its volume changes as the cube of its length. Suppose you are a parent and your child is sick and has to take some medicine. You have taken this medicine previously and you know its dose for you. You are 5′10″ tall and weigh 180 lb, and your child is 2′11″ tall and weighs 30 lb. Estimate an appropriate dosage for your child's medicine in the following cases. Be sure to discuss your reasoning.

(a) The medicine is one that will enter the child's bloodstream and reach every cell in the body. Your dose is 250 mg.

(b) The medicine is one that is meant to coat the child's throat. Your dose is 15 ml.

57. Ping-Pong Ball Packing Estimate how many Ping-Pong balls it would take to fill your classroom (assuming all the doors and windows are closed).

58. Feeding the Cougar When visiting the Como Park Zoo in St. Paul, Minnesota, with my young grandson, we encountered the sign shown at the right on the cage of the mountain lion. The detailed numbers surprised me. The amount of food given to the cat was specified to the tenth of a gram and the average cat's weight was specified to within 10 grams—about 1/3 of an ounce. This seemed to be overly precise. Can you figure out what they were trying to say and what a plausible accuracy might be for those two numbers—the amount of food given and the average cat's weight?

* Testudo is the statue of a terrapin (the university mascot) in front of the main library on the University of Maryland campus.

COUGAR
North America

Natural Diet:	Hoofed animals, small animals
Zoo diet:	1.3608 kg. commercially prepared diet for large cats, six days a week
Average Weight:	90.72 kg.
Average Lifespan:	20 years

The cougar is also called mountain lion or puma.
It is the only large cat at Como Zoo that purrs.
Cougars are very solitary animals. They are seldom seen by humans.

FIGURE 1-22 ■ Problem 58.

59. Blowing Off the Units. Throughout your physics course, your instructor will expect you to be careful with the units in your calculations. Yet, some students tend to neglect them and just trust that they always work out properly. Maybe this real-world example will keep you from such a sloppy habit.

On July 23, 1983, Air Canada Flight 143 was being readied for its long trip from Montreal to Edmonton when the flight crew asked the ground crew to determine how much fuel was already onboard the airplane. The flight crew knew that they needed to begin the trip with 22 300 kg of fuel. They knew that amount in kilograms because Canada had recently switched to the metric system: previously fuel had been measured in pounds. The ground crew could measure the onboard fuel only in liters, which they reported as 7 682 L. Thus, to determine how much fuel was onboard and how much additional fuel must be added, the flight crew asked the ground crew for the conversion factor from liters to kilograms of fuel. The response was 1.77, which the flight crew used (1.77 kg corresponds to 1 L). (a) How many kilograms of fuel did the flight crew think they had? (In this problem, take all the given data as being exact.) (b) How many liters did they ask to be added to the airplane?

Unfortunately, the response from the ground crew was based on pre-metric habits—the number 1.77 was actually the conversion factor from liters to pounds of fuel (1.77 lb corresponds to 1 L). (c) How many kilograms of fuel were actually onboard? (Except for the given 1.77, use four significant figures for other conversion factors.) (d) How many liters of additional fuel were actually needed? (e) When the airplane left Montreal, what percentage of the required fuel did it actually have?

On route to Edmonton, at an altitude of 7.9 km, the airplane ran out of fuel and began to fall. Although the airplane then had no power, the pilot somehow managed to put it into a downward glide. However, the nearest working airport was too far to reach by only gliding, so the pilot somehow angled the glide toward an old nonworking airport.

Unfortunately, the runway at that airport had been converted to a track for race cars, and a steel barrier had been constructed across it. Fortunately, as the airplane hit the runway, the front landing gear collapsed, dropping the nose of the airplane onto the runway. The skidding slowed the airplane so that it stopped just short of the steel barrier, with stunned race drivers and fans looking on. All on board the airplane emerged safely. The point here is this: Take care of the units.

2 Motion Along a Straight Line

On September 26, 1993, Dave Munday, a diesel mechanic by trade, went over the Canadian edge of Niagara Falls for the second time, freely falling 48 m to the water (and rocks) below. On this attempt, he rode in a steel chamber with an airhole. Munday, keen on surviving this plunge that had killed other stuntmen, had done considerable research on the physics and engineering aspects of the plunge.

If he fell straight down, how could he predict the speed at which he would hit the water?

The answer is in this chapter.

2-1 Motion

The world, and everything in it, moves. Even a seemingly stationary thing, such as a roadway, moves because the Earth is moving. Not only is the Earth rotating and orbiting the Sun, but the Sun is also moving through space. The motion of objects can take many different forms. For example, a moving object's path might be a straight line, a curve, a circle, or something more complicated. The entity in motion might be something simple, like a ball, or something complex, like a human being or galaxy.

In physics, when we want to understand a phenomenon such as motion, we begin by exploring relatively simple motions. For this reason, in the study of motion we start with **kinematics,** which focuses on describing motion, rather than on **dynamics,** which deals with the causes of motion. Further, we begin our study of kinematics by developing the concepts required to measure motion and mathematical tools needed to describe them in one dimension (or in 1D). Only then do we extend our study to include a consideration of the causes of motion and motions in two and three dimensions. Further simplifications are helpful. Thus, in this chapter, our description of the motion of objects is restricted in two ways:

1. **The motion of the object is along a straight line.** The motion may be purely vertical (that of a falling stone), purely horizontal (that of a car on a level highway), or slanted (that of an airplane rising at an angle from a runway), but it must be a straight line.
2. **The object is effectively a particle** because its size and shape are not important to its motion. By "particle" we mean either: (a) a point-like object with dimensions that are small compared to the distance over which it moves (such as the size of the Earth relative to its orbit around the Sun), (b) an extended object in which all its parts move together (such as a falling basketball that is not spinning), or (c) that we are only interested in the path of a special point associated with the object (such as the belt buckle on a walking person).

We will start by introducing very precise definitions of words commonly used to describe motion like speed, velocity, and acceleration. These definitions may conflict with the way these terms are used in everyday speech. However, by using precise definitions rather than our casual definitions, we will be able to describe and predict the characteristics of common motions in graphical and mathematical terms. These mathematical descriptions of phenomena form the basic vocabulary of physics and engineering.

Although our treatment may seem ridiculously formal, we need to provide a foundation for the analysis of more complex and interesting motions.

READING EXERCISE 2-1: Which of the following motions are along a straight line: (a) a string of carts traveling up and down along a roller coaster, (b) a cannonball shot straight up, (c) a car traveling along a straight city street, (d) a ball rolling along a straight ramp tilted at a 45° angle. ■

READING EXERCISE 2-2: In reality there are no point particles. Rank the following everyday items from most particle-like to least particle-like: (a) a 2-m-tall long jumper relative to a 25 m distance covered in a jump, (b) a piece of lead shot from a shotgun shell relative to its range of 5 m, (c) the Earth of diameter 13×10^6 m relative to the approximate diameter of its orbit about the Sun of 3×10^{11} m. ■

2-2 Position and Displacement Along a Line

Defining a Coordinate System

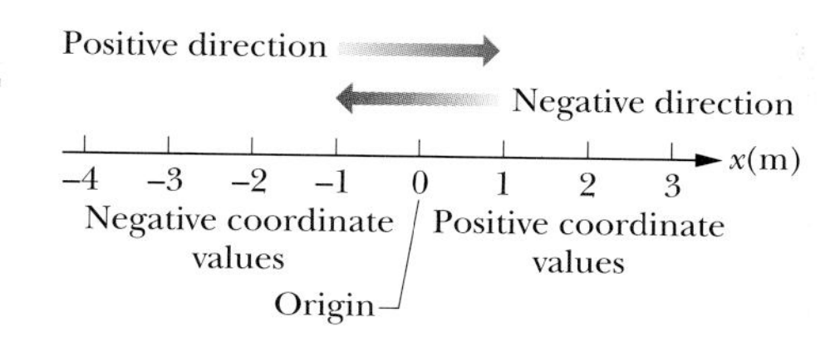

FIGURE 2-1 ■ Position is determined on an axis that is marked in units of meters and that extends indefinitely in opposite directions.

In order to study motion along a straight line, we must be able to specify the location of an object and how it changes over time. A convenient way to locate a point of interest or an object is to define a coordinate system. Houses in Costa Rican towns are commonly located with addresses such as "200 meters east of the Post Office." In order to locate a house, a distance scale must be agreed upon (meters are used in the example), and a reference point or origin (in this case the Post Office), and a direction (in this case east) must be specified. Thus, in locating an object that can move along a straight line, it is convenient to specify its position by choosing a one-dimensional **coordinate system.** The system consists of a *point of reference known as the origin (or zero point)*, a line that passes through the chosen origin called a *coordinate axis*, one direction along the coordinate axis, chosen as positive and the other direction as negative, and the units we use to measure a quantity. We have labeled the coordinate axis as the x axis, in Fig. 2-1, and placed an origin on it. The direction of increasing numbers (coordinates) is called the **positive direction,** which is toward the right in Fig. 2-1. The opposite direction is the **negative direction.**

Figure 2-1 is drawn in the traditional fashion, with negative coordinates to the left of the origin and positive coordinates to the right. It is also traditional in physics to use meters as the standard scale for distance. However, we have freedom to choose other units and to decide which side of the origin is labeled with negative coordinates and which is labeled with positive coordinates. Furthermore, we can choose to define an x axis that is vertical rather than horizontal, or inclined at some angle. In short, we are free to make choices about how we define our coordinate system.

Good choices make describing a situation much easier. For example, in our consideration of motion along a straight line, we would want to align the axis of our one-dimensional coordinate system along the line of motion. In Chapters 5 and 6, when we consider motions in two dimensions, we will be using more complex coordinate systems with a set of mutually perpendicular coordinate axes. Choosing a coordinate system that is appropriate to the physical situation being described can simplify your mathematical description of the situation. To describe a particle moving in a circle, you would probably choose a two-dimensional coordinate system in the plane of the circle with the origin placed at its center.

Defining Position as a Vector Quantity

The reason for choosing our standard one-dimensional coordinate axis and orienting it along the direction of motion is to be able to define the position of an object relative to our chosen origin, and then be able to keep track of how its position changes as the object moves. It turns out that the position of an object relative to a coordinate system can be described by a mathematical entity known as a **vector.** This is because, in order to find the position of an object, we must specify both how far and in which direction the object is from the origin of a coordinate system.

> A **VECTOR** is a mathematical entity that has both a magnitude and a direction. Vectors can be added, subtracted, multiplied, and transformed according to well-defined mathematical rules.

There are other physical quantities that also behave like vectors such as velocity, acceleration, force, momentum, and electric and magnetic fields.

However, not all physical quantities that have signs associated with them are vectors. For example, temperatures do not need to be described in terms of a coordinate

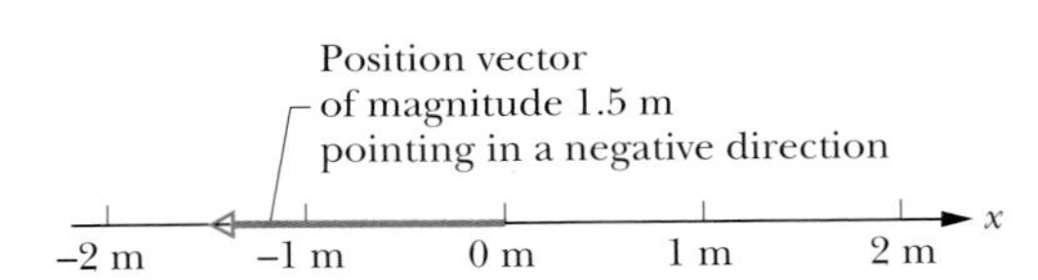

FIGURE 2-2 ■ A position vector can be represented by an arrow pointing from the origin of a chosen coordinate system to the location of the object.

system, and single numbers, such as $T = -5°C$ or $T = 12°C$, are sufficient to describe them. The minus sign, in this case, does not signify a direction. Mass, distance, length, area, and volume also have no directions associated with them and, although their values depend on the units used to measure them, their values do not depend on the orientation of a coordinate system. Such quantities are called **scalars.**

> A **SCALAR** is defined as a mathematical quantity whose value does not depend on the orientation of a coordinate system and has no direction associated with it.

In general, a one-dimensional vector can be represented by an arrow. The length of the arrow, which is inherently positive, represents the **magnitude** of the vector and the direction in which the arrow points represents the **direction** associated with the vector.

We begin this study of motion by introducing you to the properties of one-dimensional position and displacement vectors and some of the formal methods for representing and manipulating them. These formal methods for working with vectors will prove to be very useful later when working with two- and three-dimensional vectors.

A **one-dimensional position vector** is defined by the location of the origin of a chosen one-dimensional coordinate system and of the object of interest. The **magnitude** of the position vector is a scalar that denotes the distance between the object and the origin. For example, an object that has a position vector of magnitude 5 m could be located at the point +5 m or −5 m from the origin.

On a conventional x axis, the direction of the position vector is positive when the object is located to the right of the origin and negative when the object is located to the left of the origin. For example, in the system shown in Fig. 2-1, if a particle is located at a distance of 3 m to the left of the origin, its position vector has a magnitude of 3 m and a direction that is negative. One of many ways to represent a position vector is to draw an arrow from the origin to the object's location, as shown in Fig. 2-2, for an object that is 1.5 m to the left of the origin. Since the length of a vector arrow represents the magnitude of the vector, its length should be proportional to the distance from the origin to the object of interest. In addition, the direction of the arrow should be from the origin to the object.

Instead of using an arrow, a position vector can be represented mathematically. In order to develop a useful mathematical representation we need to define a **unit vector** associated with our x axis.

> A **UNIT VECTOR FOR A COORDINATE AXIS** is a dimensionless vector that points in the direction along a coordinate axis that is chosen to be positive.

It is customary to represent a unit vector that points along the positive x axis with the symbol $\hat{\text{i}}$ (i-hat), although some texts use the symbol $\hat{x}$ (x-hat) instead. When considering three-dimensional vectors, the unit vectors pointing along the designated positive y axis and z axis are denoted by $\hat{\text{j}}$ and $\hat{\text{k}}$, respectively.

These vectors are called "unit vectors" because they have a dimensionless value of one. However, you should not confuse the use of word "unit" with a physical unit. Unit vectors should be shown on coordinate axes as small pointers with no physical units, such as meters, associated with them. This is shown in Fig. 2-3 for the x axis unit vector. Since the scale used in the coordinate system has units, it is essential that the units always be associated with the number describing the location of an object along an axis. Figure 2-3 also shows how the unit vector is used to create a position vector corresponding to an object located at position −1.5 meters on our x axis. To do this we stretch or multiply the unit vector by the magnitude of the position vector, which

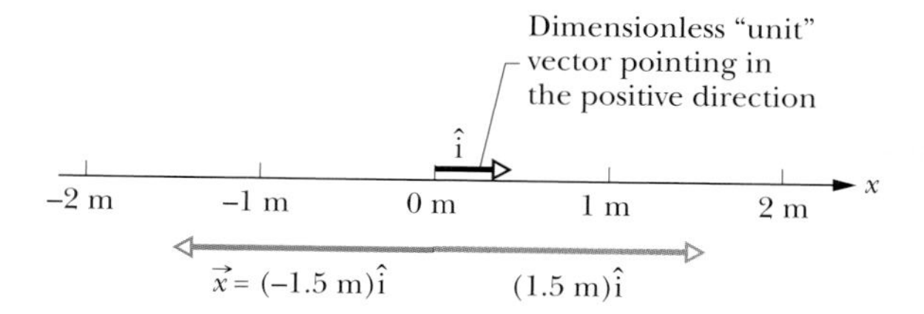

FIGURE 2-3 ■ Arrows representing: (1) a dimensionless unit vector, $\hat{i}$, pointing in the positive x direction; (2) a vector representing the unit vector multiplied by 1.5 meters; and (3) a vector multiplied by 1.5 meters and inverted by multiplication by −1 to create the position vector $\vec{x} = (-1.5 \text{ m})\hat{i}$. This position vector has a magnitude of 1.5 meters and points in a negative direction.

is 1.5 m. Note that we are using the coordinate axis to describe a position in meters relative to an origin, so it is essential to include the units with the number. This multiplication of the dimensionless unit vector by 1.5 m creates a 1.5-m-long vector that points in the same direction as the unit vector. It is denoted by $(1.5 \text{ m})\hat{i}$. However, the vector we want to create points in the negative direction, so the vector pointing in the positive direction must be inverted using a minus sign. The position vector we have created is denoted as $\vec{x}$. It can be divided into two parts—a vector component and a unit vector,

$$\vec{x} = (-1.5 \text{ m})\hat{i}.$$

In this example, the x-component of the position vector, denoted as x, is −1.5 m.

Here the quantity 1.5 m with no minus sign in front of it is known as the magnitude of this position vector. In general, the magnitude is denoted as $|\vec{x}|$. Thus, the one-dimensional position vector for the situation shown in Fig. 2-3 is denoted mathematically using the following symbols:

$$\vec{x} = x\hat{i} = (-1.5 \text{ m})\hat{i}.$$

The x-component of a position vector, denoted x, can be positive or negative depending on which side of the origin the particle is. Thus, in one dimension in terms of absolute values, the vector component x is either $+|x|$ or $-|x|$, depending on the object's location.

In general, a component of a vector along an axis, such as x in this case, is not a scalar since our x-component will change sign if we choose to reverse the orientation of our chosen coordinate system. In contrast, *the magnitude of a position vector is always positive, and it only tells us how far away the object is from the origin*, so the magnitude of a vector is always a scalar quantity. The sign of the component (+ or −) tells us in which direction the vector is pointing. The sign will be negative if the object is to the left of the origin and positive if it is to the right of the origin.

Defining Displacement as a Vector Quantity

The study of motion is primarily about how an object's location changes over time under the influence of forces. In physics the concept of **change** has an exact mathematical definition.

> **CHANGE** is defined as the difference between the state of a physical system (typically called the final state) and its state at an earlier time (typically called the initial state).

This definition of change is used to define displacement.

> **DISPLACEMENT** is defined as the change of an object's position that occurs during a period of time.

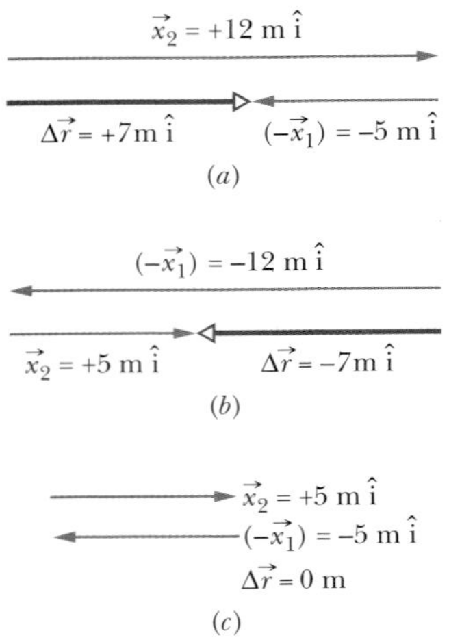

$\vec{x}_2 = +12$ m $\hat{\text{i}}$

$\Delta\vec{r} = +7$m $\hat{\text{i}}$ $(-\vec{x}_1) = -5$ m $\hat{\text{i}}$

(*a*)

$(-\vec{x}_1) = -12$ m $\hat{\text{i}}$

$\vec{x}_2 = +5$ m $\hat{\text{i}}$ $\Delta\vec{r} = -7$m $\hat{\text{i}}$

(*b*)

$\vec{x}_2 = +5$ m $\hat{\text{i}}$

$(-\vec{x}_1) = -5$ m $\hat{\text{i}}$

$\Delta\vec{r} = 0$ m

(*c*)

FIGURE 2-4 ■ The wide arrow shows the displacement vector $\Delta\vec{r}$ for three situations leading to: (*a*) a positive displacement, (*b*) a negative displacement, and (*c*) zero displacement.

Since position can be represented as a vector quantity, displacement is the difference between two vectors, and thus, is also a vector. So, in the case of motion along a line, an object moving from an "initial" position $\vec{x}_1$ to another "final" position $\vec{x}_2$ at a later time is said to undergo a **displacement** $\Delta\vec{r}$, given by the difference of two position vectors

$$\Delta\vec{r} \equiv \vec{x}_2 - \vec{x}_1 = \Delta x\hat{\text{i}} \quad \text{(displacement vector)}, \qquad (2\text{-}1)$$

where the symbol Δ is used to represent a change in a quantity, and the symbol "$\equiv$" signifies that the displacement $\Delta\vec{r}$ is given by $\vec{x}_2 - \vec{x}_1$ because we have *chosen* to define it that way.

As you will see when we begin to work with vectors in two and three dimensions, it is convenient to consider subtraction as the addition of one vector to another that has been inverted by multiplying the vector component by -1. We can use this idea of defining subtraction as the addition of an inverted vector to find displacements. Let's consider three situations:

(a) A particle moves along a line from $\vec{x}_1 = (5\text{ m})\hat{\text{i}}$ to $\vec{x}_2 = (12\text{ m})\hat{\text{i}}$. Since $\Delta\vec{r} = \vec{x}_2 - \vec{x}_1 = \vec{x}_2 + (-\vec{x}_1)$,

$$\Delta\vec{r} = (12\text{ m})\hat{\text{i}} - (5\text{ m})\hat{\text{i}} = (12\text{ m})\hat{\text{i}} + (-5\text{ m})\hat{\text{i}} = (7\text{ m})\hat{\text{i}}.$$

The *positive* result indicates that the motion is in the positive direction (toward the right in Fig. 2-4*a*).

(b) A particle moves from $\vec{x}_1 = (12\text{ m})\hat{\text{i}}$ to $\vec{x}_2 = (5\text{ m})\hat{\text{i}}$. Since $\Delta\vec{r} = \vec{x}_2 - \vec{x}_1 = \vec{x}_2 + (-\vec{x}_1)$,

$$\Delta\vec{r} = (5\text{ m})\hat{\text{i}} - (12\text{ m})\hat{\text{i}} = (5\text{ m})\hat{\text{i}} + (-12\text{ m})\hat{\text{i}} = (-7\text{ m})\hat{\text{i}}.$$

The negative result indicates that the displacement of the particle is in the negative direction (toward the left in Fig. 2-4*b*).

(c) A particle starts at 5 m, moves to 2 m, and then returns to 5 m. The displacement for the full trip is given by $\Delta\vec{r} = \vec{x}_2 - \vec{x}_1 = \vec{x}_2 + (-\vec{x}_1)$, where $\vec{x}_1 = (5\text{ m})\hat{\text{i}}$ and $\vec{x}_2 = (5\text{ m})\hat{\text{i}}$:

$$\Delta\vec{r} = (5\text{ m})\hat{\text{i}} + (-5\text{ m})\hat{\text{i}} = (0\text{ m})\hat{\text{i}}$$

and the particle's position hasn't changed, as in Fig. 2-4*c*. Since displacement involves only the original and final positions, the actual number of meters traced out by the particle while moving back and forth is immaterial.

If we ignore the sign of a particle's displacement (and thus its direction), we are left with the **magnitude** of the displacement. This is the distance between the original and final positions and is always positive. It is important to remember that displacement (or any other vector) has not been completely described until we state its direction.

We use the notation $\Delta\vec{r}$ for displacement because when we have motion in more than one dimension, the notation for the position vector is $\vec{r}$. For a one-dimensional motion along a straight line, we can also represent the displacement as $\Delta\vec{x}$. The magnitude of displacement is represented by surrounding the displacement vector symbol with absolute value signs:

$$\{\text{magnitude of displacement}\} = |\Delta\vec{r}| \quad \text{or} \quad |\Delta\vec{x}|$$

READING EXERCISE 2-3: Can a particle that moves from one position with a negative value, to another position with a negative value, undergo a positive displacement? ■

TOUCHSTONE EXAMPLE 2-1: Displacements

Three pairs of initial and final positions along an x axis represent the location of objects at two successive times: (pair 1) −3 m, +5 m; (pair 2) −3 m, −7 m; (pair 3) 7 m, −3 m.

(a) Which pairs give a negative displacement?

SOLUTION ▪ The **Key Idea** here is that the displacement is negative when the final position lies *to the left* of the initial position. As shown in Fig. 2-5, this happens when the final position is *more negative* than the initial position. Looking at pair 1, we see that the final position, +5 m, is positive while the initial position, −3 m, is negative. This means that the displacement is from left (more negative) to right (more positive) and so the displacement is positive for pair 1. (Answer)

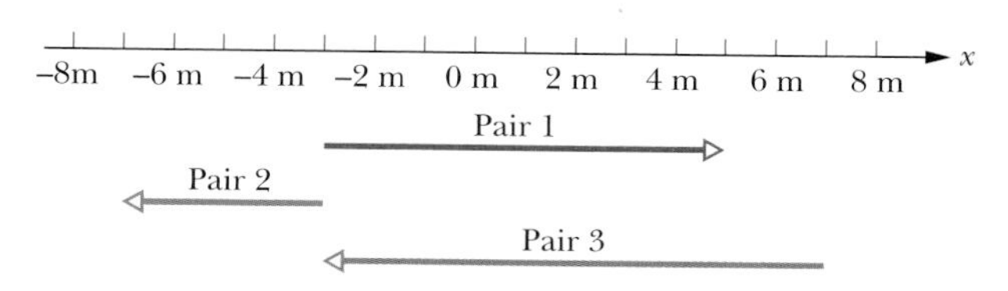

FIGURE 2-5 ▪ Displacement associated with three pairs of initial and final positions along an x axis.

For pair 2 the situation is different. The final position, −7 m, lies to the left of the initial position, −3 m, so the displacement is negative. (Answer)

For pair 3 the final position, −3 m, is to the left of the origin while the initial position, +7 m, is to the right of the origin. So the displacement is from the right of the origin to its left, a negative displacement. (Answer)

(b) Calculate the value of the displacement in each case using vector notation.

SOLUTION ▪ The **Key Idea** here is to use Eq. 2-1 to calculate the displacement for each pair of positions. It tells us the difference between the final position and the initial position, in that order,

$$\Delta\vec{x} = \vec{x}_2 - \vec{x}_1 \quad \text{(displacement).} \tag{2-2}$$

For pair 1 the final position is $\vec{x}_2 = (+5\text{ m})\hat{\text{i}}$ and the initial position is $\vec{x}_1 = (-3\text{ m})\hat{\text{i}}$, so the displacement between these two positions is just

$$\Delta\vec{x} = (+5\text{ m})\hat{\text{i}} - (-3\text{ m})\hat{\text{i}} = (+5\text{ m})\hat{\text{i}} + (3\text{ m})\hat{\text{i}} = (+8\text{ m})\hat{\text{i}}. \quad \text{(Answer)}$$

For pair 2 the same argument yields

$$\Delta\vec{x} = (-7\text{ m})\hat{\text{i}} - (-3\text{ m})\hat{\text{i}} = (-7\text{ m})\hat{\text{i}} + (3\text{ m})\hat{\text{i}} = (-4\text{ m})\hat{\text{i}}. \quad \text{(Answer)}$$

Finally, the displacement for pair 3 is

$$\Delta\vec{x} = (-3\text{ m})\hat{\text{i}} - (+7\text{ m})\hat{\text{i}} = (-3\text{ m})\hat{\text{i}} + (-7\text{ m})\hat{\text{i}} = (-10\text{ m})\hat{\text{i}}. \quad \text{(Answer)}$$

(c) What is the magnitude of each position vector?

SOLUTION ▪ Of the six position vectors given, one of them—namely $\vec{x}_1 = (-3\text{ m})\hat{\text{i}}$—appears in all three pairs. The remaining three positions are $\vec{x}_2 = (+5\text{ m})\hat{\text{i}}$, $\vec{x}_3 = (-7\text{ m})\hat{\text{i}}$, and $\vec{x}_4 = (+7\text{ m})\hat{\text{i}}$. The **Key Idea** here is that the magnitude of a position vector just tells us *how far* the point lies from the origin without regard to whether it lies to the left or to the right of the origin. Thus the magnitude of our first position vector is 3 m (Answer) since the position specified by $\vec{x} = (-3\text{ m})\hat{\text{i}}$ is 3 m to the left of the origin. It's *not* −3 m, because magnitudes only specify distance from the origin, not direction.

For the same reason, the magnitude of the second position vector is just 5 m (Answer) while the magnitude of the third and the fourth are *each* 7 m. (Answer) The fact that the third point lies 7 m to the left of the origin while the fourth lies 7 m to the right doesn't matter here.

(d) What is the value of the x-component of each of these position vectors?

SOLUTION ▪ To answer this question you need to remember what is meant by the component of a vector. The key equation relating a vector in one dimension to its component along its direction is $\vec{x} = x\,\hat{\text{i}}$, where $\vec{x}$ (with the arrow over it) is the vector itself and x (with no arrow over it) is the component of the vector in the direction specified by the unit vector $\hat{\text{i}}$. So the component of $\vec{x} = (-3\text{ m})\hat{\text{i}}$ is −3 m, while that of $\vec{x} = (+5\text{ m})\hat{\text{i}}$ is just +5 m, and $\vec{x} = (-7\text{ m})\hat{\text{i}}$ has as its component along the $\hat{\text{i}}$ direction (−7 m) while for $\vec{x} = (+7\text{ m})\hat{\text{i}}$ it's just (+7 m). In other words, the component of a vector in the direction of $\hat{\text{i}}$ is just the signed number (with its units) that multiplies $\hat{\text{i}}$. (Answer)

2-3 Velocity and Speed

Suppose a student stands still or speeds up and slows down along a straight line. How can we describe accurately and efficiently where she is and how fast she is moving? We will explore several ways to do this.

Representing Motion in Diagrams and Graphs

Motion Diagrams: Now that you have learned about position and displacement, it is quite easy to describe the motion of an object using pictures or sketches to chart how position changes over time. Such a representation is called a *motion diagram*. For example, Fig. 2-6 shows a student whom we treat as if she were concentrated into a particle located at the back of her belt. She is *standing still* at a position $\vec{x} = (-2.00 \text{ m})\hat{i}$ from a point on a sidewalk that we choose as our origin. Figure 2-7 shows a more complex diagram describing the student in motion. Suppose we see that just as we start timing her progress with a stopwatch (so $t = 0.0$ s), the back of her belt is 2.47 m to the left of our origin. The x-component of her position is then $x = -2.47$ m. The student then moves toward the origin, almost reaches the origin at $t = 1.5$ s, and then continues moving to the right so that her x-component of position has increasingly positive values. It is important to recognize that just as we chose an origin and direction for our coordinate axis, we also chose an origin in time. If we had chosen to start our timing 12 seconds earlier, then the new motion diagram would show the back of her belt as being at $x = -2.47$ m at $t = 12$ s.

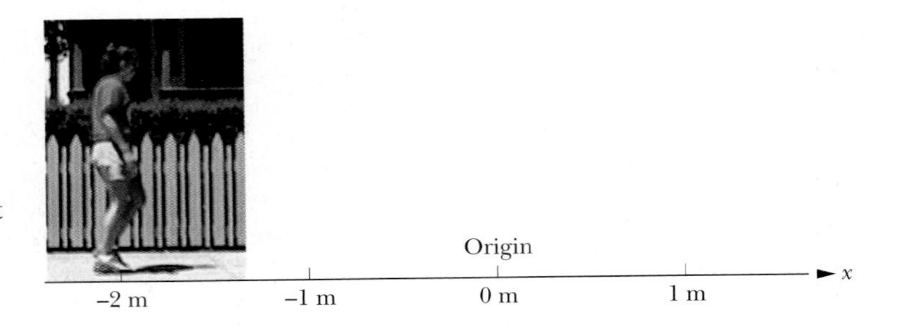

FIGURE 2-6 ■ A motion diagram of a student standing still with the back of her belt at a horizontal distance of 2.00 m to the left of a spot of the sidewalk designated as the origin.

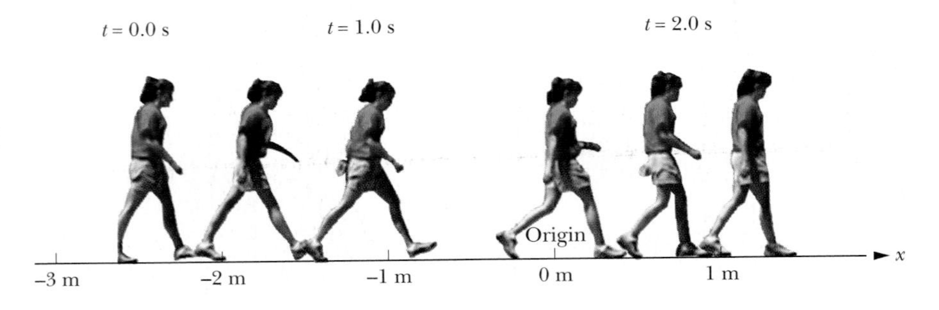

FIGURE 2-7 ■ A motion diagram of a student starting to walk slowly. The horizontal position of the back of her belt starts at a horizontal distance of 2.47 m to the left of a spot designated as the origin. She is speeding up for a few seconds and then slowing down.

Graphs: Another way to describe how the position of an object changes as time passes is with a graph. In such a graph, the x-component of the object's position, x, can be plotted as a function of time, t. This position–time graph has alternate names such as a graph of x as a function of t, $x(t)$, or x vs. t. For example, Fig. 2-8 shows a graph of the student *standing still* with the back of her belt located at a horizontal position of −2.00 m from a spot on the sidewalk that is chosen as the origin.

The graph of no motion shown in Fig. 2-8 is not more informative than the picture or a comment that the student is standing still for 3 seconds at a certain location. But it's another story when we consider the graph of a motion. Figure 2-9 is a graph of a student's x-component of position as a function of time. It represents the same information depicted in the motion diagram in Fig. 2-7. Data on the student's motion are first recorded at $t = 0.0$ s when the x-component of her position is $x = -2.47$ m. The student then moves toward $x = 0.00$ m, passes through that point at about $t = 1.5$ s, and then moves on to increasingly larger positive values of x while slowing down.

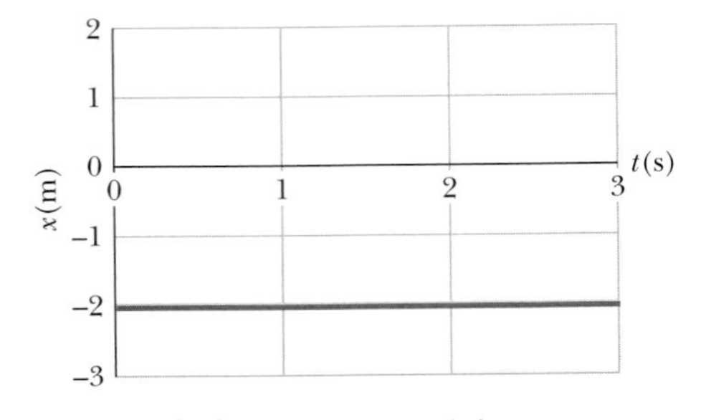

FIGURE 2-8 ■ The graph of the x-component of position for a student who is standing still at $x = -2.0$ m for at least 3 seconds.

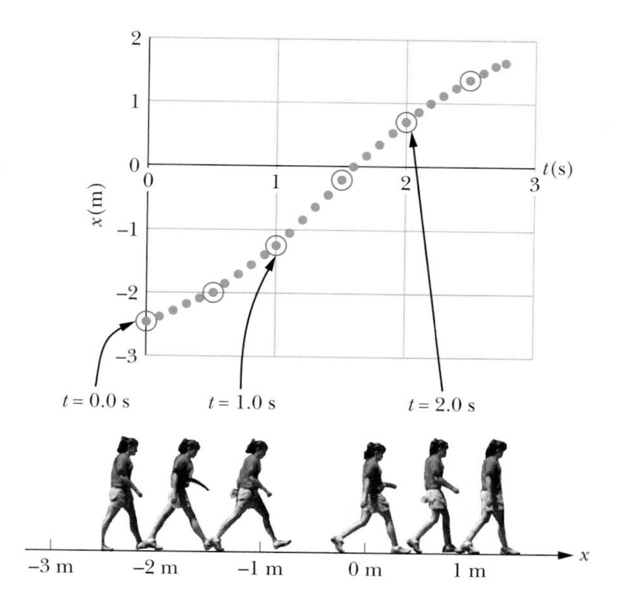

FIGURE 2-9 ■ A graph that represents how the position component, x, of the walking student shown in Fig. 2-7 changes over time. The motion diagram, shown below the graph, is associated with the graph at three points in time as indicated by the arrows.

Although the graph of the student's motion in Fig. 2-9 seems abstract and quite unlike a motion diagram, it is richer in information. For example, the graph allows us to estimate the motion of the student at times between those for which position measurements were made. Equally important, we can use the graph to tell us how fast the student moves at various times, and we deal with this aspect of motion graphs next.

What can motion diagrams and x vs. t graphs tell us about how fast and in what direction something moves along a line? It is clear from an examination of the motion diagram at the bottom of Fig. 2-9 that the student covers the most distance and so appears to be moving most rapidly between the two times $t_1 = 1.0$ s and $t_2 = 1.5$ s. But this time interval is also where the slope (or steepness) of the graph has the greatest magnitude. Recall from mathematics that the average slope of a curve between two points is defined as the ratio of the change in the variable plotted on the vertical axis (in this case the x-component of her position) to the change in the variable plotted on the horizontal axis (in this case the time). Hence, on position vs. time graphs (such as those shown in Fig. 2-8 and Fig. 2-9),

$$\text{average slope} \equiv \frac{\Delta x}{\Delta t} = \frac{x_2 - x_1}{t_2 - t_1} \quad \text{(definition of average slope).} \tag{2-3}$$

Since time moves forward, $t_2 > t_1$, so Δt always has a positive value. Thus, a slope will be positive whenever $x_2 > x_1$, so Δx is positive. In this case a straight line connecting the two points on the graph slants upward toward the right when the student is moving along the positive x-direction. On the other hand, if the student were to move "backwards" in the direction along the x axis we chose to call negative, then $x_2 < x_1$. In this case, the slope between the two times would be negative and the line connecting the points would slant downward to the right.

Average Velocity

For motion along a straight line, the steepness of the slope in an x vs. t graph over a time interval from t_1 to t_2 tells us "how fast" a particle moves. The direction of motion is indicated by the sign of the slope (positive or negative). Thus, this slope or ratio $\Delta x/\Delta t$ is a special quantity that tells us how fast and in what direction something moves. We haven't given the ratio $\Delta x/\Delta t$ a name yet. We do this to emphasize the fact that the ideas associated with figuring out how fast and in what direction something moves are more important than the names we assign to them. However, it is inconvenient not to have a name. The common name for this ratio is **average velocity,** which is defined as the ratio of displacement vector $\Delta\vec{x}$ for the motion of interest to the time interval Δt in which it occurs. This vector can be expressed in equation form as

$$\langle\vec{v}\rangle \equiv \frac{\Delta\vec{x}}{\Delta t} = \frac{\Delta x}{\Delta t}\hat{i} = \frac{x_2 - x_1}{t_2 - t_1}\hat{i} \quad \text{(definition of 1D average velocity)}, \tag{2-4}$$

where x_2 and x_1 are components of the position vectors at the final and initial times. Here we use angle brackets $\langle\ \rangle$ to denote the average of a quantity. Also, we use the special symbol "$\equiv$" for equality to emphasize that the term on the left is equal to the term on the right by definition. The time change is a positive scalar quantity because we never need to specify its direction explicitly. In defining $\langle\vec{v}\rangle$ we are basically multiplying the displacement vector, $\Delta\vec{x}$, by the scalar $(1/\Delta t)$. This action gives us a new vector that points in the same direction as the displacement vector.

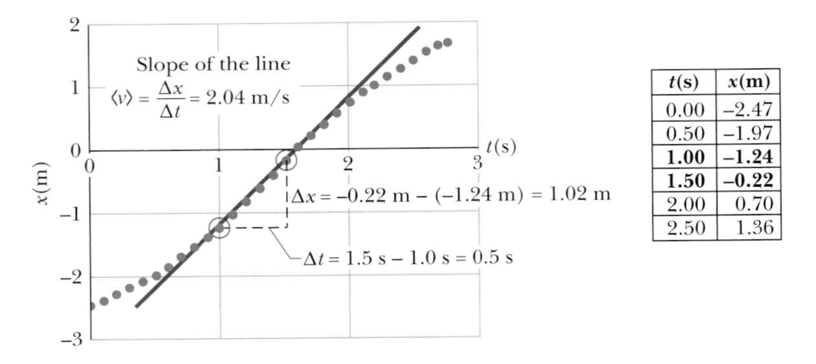

t(s)	x(m)
0.00	−2.47
0.50	−1.97
1.00	**−1.24**
1.50	**−0.22**
2.00	0.70
2.50	1.36

FIGURE 2-10 ■ Calculation of the slope of the line that connects the points on the curve at $t_1 = 1.0$ s and $t_2 = 1.5$ s. The x-component of the average velocity is given by this slope.

Figure 2-10 shows how to find the average velocity for the student motion represented by the graph shown in Fig. 2-9 between the times $t_1 = 1.0$ s and $t_2 = 1.5$ s. The average velocity during that time interval is

$$\langle\vec{v}\rangle \equiv \frac{\Delta x}{\Delta t}\hat{i} = \frac{x_2 - x_1}{t_2 - t_1}\hat{i} = \frac{-0.22\text{ m} - (-1.24\text{ m})}{(1.5\text{ s} - 1.0\text{ s})}\hat{i} = (2.04\text{ m/s})\hat{i}.$$

The x-component of the average velocity along the line of motion, $\langle v_x\rangle = 2.04$ m/s, is simply the slope of the straight line that connects the point on the curve at the beginning of our chosen interval and the point on the curve at the end of the interval. Since our student is speeding up and slowing down, the values of $\langle\vec{v}\rangle$ and $\langle v_x\rangle$ will in general be different when calculated using other time intervals.

Average Speed

Sometimes we don't care about the direction of an object's motion but simply want to keep track of the distance covered. For instance, we might want to know the total distance a student walks (number of steps times distance covered in each step). Our student could be pacing back and forth wearing out her shoes without having a vector displacement. Similarly, average speed, $\langle s\rangle$, is a different way of describing "how fast" an object moves. Whereas the average velocity involves the particle's displacement $\Delta\vec{x}$, which is a vector quantity, the average speed involves the total distance covered (for example, the product of the length of a step and the number of steps the student took), which is independent of direction. So **average speed** is defined as

$$\langle s \rangle \equiv \frac{\text{total distance}}{\Delta t} \quad \text{(definition of average speed).} \tag{2-5}$$

Since neither the total distance traveled nor the time interval over which the travel occurred has an associated direction, average speed does not include direction information. Both the total distance and the time period are always positive, so average speed is always positive too. Thus, an object that moves back and forth along a line can have no vector displacement, so it has zero velocity but a rather high average speed. At other times, while the object is moving in only one direction, the average speed $\langle s \rangle$ is the same as the magnitude of the average velocity $\langle \vec{v} \rangle$. However, as you can demonstrate in Reading Exercise 2-4, when an object doubles back on its path, the average speed is not simply the magnitude of the average velocity $|\langle \vec{v} \rangle|$.

Instantaneous Velocity and Speed

You have now seen two ways to describe how fast something moves: average velocity and average speed, both of which are measured over a time interval Δt. Clearly, however, something might speed up and slow down during that time interval. For example, in Fig. 2-9 we see that the student is moving more slowly at $t = 0.0$ s than she is at $t = 1.5$ s, so her velocity seems to be changing during the time interval between 0.0 s and 1.5 s. The average slope of the line seems to be increasing during this time interval. Can we refine our definition of velocity in such a way that we can determine the student's true velocity at any one "instant" in time? We envision something like the almost instantaneous speedometer readings we get as a car speeds up and slows down.

Defining an instant and instantaneous velocity is not a trivial task. As we noted in Chapter 1, the time interval of 1 second is defined by counting oscillations of radiation absorbed by a cesium atom. In general, even our everyday clocks work by counting oscillations in an electronic crystal, pendulum, and so on. We associate "instants in time" with positions on the hands of a clock, and "time intervals" with changes in the position of the hands.

For the purpose of finding a velocity at an instant, we can attempt to make the time interval we use in our calculation so small that it has almost zero duration. Of course the displacement we calculate also becomes very small. So **instantaneous velocity** along a line—like average velocity—is still defined in terms of the ratio of $\Delta\vec{x}/\Delta t$. But we have this ratio passing to a limit where Δt gets closer and closer to zero. Using standard calculus notation for this limit gives us the following definition:

$$\vec{v} \equiv \lim_{\Delta t \to 0} \frac{\Delta \vec{x}}{\Delta t} = \frac{d\vec{x}}{dt} \quad \text{(definition of 1D instantaneous velocity).} \tag{2-6}$$

> In the language of calculus, the **INSTANTANEOUS VELOCITY** is the rate at which a particle's position vector, $\vec{x}$, is changing with time at a given instant.

In passing to the limit the ratio $\Delta\vec{x}/\Delta t$ is not necessarily small, since both the numerator and denominator are getting small together. The first part of this expression,

$$\vec{v} = v_x \hat{\text{i}} = \lim_{\Delta t \to 0} \frac{\Delta \vec{x}}{\Delta t} \text{ or } \lim_{\Delta t \to 0} \frac{\Delta x}{\Delta t} \hat{\text{i}},$$

tells us that we can find the (instantaneous) velocity of an object by taking the slope of a graph of the position component vs. time at the point associated with that moment in time. If the graph is a curve rather than a straight line, the *slope at a point* is actually the tangent to the line at that point. Alternatively, the second part of the expression, shown in Eq. 2-6,

$$\vec{v} = \frac{d\vec{x}}{dt},$$

indicates that, if we can approximate the relationship between $\vec{x}$ and t as a continuous mathematical function such as $\vec{x} = (3.0 \text{ m/s}^2)t^2$, we can also find the object's instantaneous velocity by taking a derivative with respect to time of the object's position $\vec{x}$. When $\vec{x}$ varies continuously as time marches on, we often denote $\vec{x}$ as a position function $\vec{x}(t)$ to remind us that it varies with time.

Instantaneous speed, which is typically called simply **speed,** is just the magnitude of the instantaneous velocity vector, $|\vec{v}|$. Speed is a scalar quantity consisting of the velocity value that has been stripped of any indication of the direction the object is moving, either in words or via an algebraic sign. A velocity of $(+5 \text{ m/s})\hat{\text{i}}$ and one of $(-5 \text{ m/s})\hat{\text{i}}$ both have an associated speed of 5 m/s.

READING EXERCISE 2-4: Suppose that you drive 10 mi due east to a store. You suddenly realize that you forgot your money. You turn around and drive the 10 mi due west back to your home and then return to the store. The total trip took 30 min. (a) What is your average velocity for the entire trip? (Set up a coordinate system and express your result in vector notation.) (b) What was your average speed for the entire trip? (c) Discuss why you obtained different values for average velocity and average speed. ■

READING EXERCISE 2-5: Suppose that you are driving and look down at your speedometer. What does the speedometer tell you—average speed, instantaneous speed, average velocity, instantaneous velocity—or something else? Explain. ■

READING EXERCISE 2-6: The following equations give the position component, $x(t)$, along the x axis of a particle's motion in four situations (in each equation, x is in meters, t is in seconds, and $t > 0$): (1) $x = (3 \text{ m/s})t - (2 \text{ m})$; (2) $x = (-4 \text{ m/s}^2)t^2 - (2 \text{ m})$; (3) $x = (-4 \text{ m/s}^2)t^2$; and (4) $x = -2$ m.
(a) In which situations is the velocity $\vec{v}$ of the particle constant? (b) In which is the vector $\vec{v}$ pointing in the negative x direction? ■

READING EXERCISE 2-7: In Touchstone Example 2-2, suppose that right after refueling the truck you drive back to x_1 at 35 km/h. What is the average velocity for your entire trip? ■

TOUCHSTONE EXAMPLE 2-2: Out of Gas

You drive a beat-up pickup truck along a straight road for 8.4 km at 70 km/h, at which point the truck runs out of gasoline and stops. Over the next 30 min, you walk another 2.0 km farther along the road to a gasoline station.

(a) What is your overall displacement from the beginning of your drive to your arrival at the station?

SOLUTION ■ Assume, for convenience, that you move in the positive direction along an x axis, from a first position of $x_1 = 0$ to a second position of x_2 at the station. That second position must be at $x_2 = 8.4 \text{ km} + 2.0 \text{ km} = 10.4$ km. Then the **Key Idea** here is that your displacement Δx along the x axis is the second position minus the first position. From Eq. 2-1, we have

$$\Delta x = x_2 - x_1 = 10.4 \text{ km} - 0 = 10.4 \text{ km} \qquad \text{(Answer)}$$

Thus, your overall displacement is 10.4 km in the positive direction of the x axis.

(b) What is the time interval Δt from the beginning of your drive to your arrival at the station?

SOLUTION ■ We already know the time interval Δt_{wlk} (= 0.50 h) for the walk, but we lack the time interval Δt_{dr} for the drive. However, we know that for the drive the displacement Δx_{dr} is 8.4 km and the average velocity $\langle v_{dr\,x} \rangle$ is 70 km/h. A **Key Idea** to use here comes from Eq. 2-4: This average velocity is the ratio of the displacement for the drive to the time interval for the drive,

$$\langle v_{dr\,x} \rangle = \frac{\Delta x_{dr}}{\Delta t_{dr}}.$$

Rearranging and substituting data then give us

$$\Delta t_{dr} = \frac{\Delta x_{dr}}{\langle v_{dr\,x} \rangle} = \frac{8.4\text{ km}}{70\text{ km/h}} = 0.12\text{ h}.$$

Therefore,
$$\Delta t = \Delta t_{dr} + \Delta t_{wlk}$$
$$= 0.12\text{ h} + 0.50\text{ h} = 0.62\text{ h}.$$

(c) What is your average velocity $\langle v_x \rangle$ from the beginning of your drive to your arrival at the station? Find it both numerically and graphically.

SOLUTION ■ The **Key Idea** here again comes from Eq. 2-4: $\langle v_x \rangle$ for the entire trip is the ratio of the displacement of 10.4 km for the entire trip to the time interval of 0.62 h for the entire trip. With Eq. 2-4, we find it is

$$\langle v_x \rangle = \frac{\Delta x}{\Delta t} = \frac{10.4\text{ km}}{0.62\text{ h}} \qquad \text{(Answer)}$$
$$= 16.8\text{ km/h} \approx 17\text{ km/h}.$$

To find $\langle v_x \rangle$ graphically, first we graph $x(t)$ as shown in Fig. 2-11, where the beginning and arrival points on the graph are the origin and the point labeled "Station." The **Key Idea** here is that your average velocity in the x direction is the slope of the straight line connecting those points; that is, it is the ratio of the *rise* (Δx = 10.4 km) to the *run* (Δt = 0.62 h), which gives us $\langle v_x \rangle$ = 16.8 km/h.

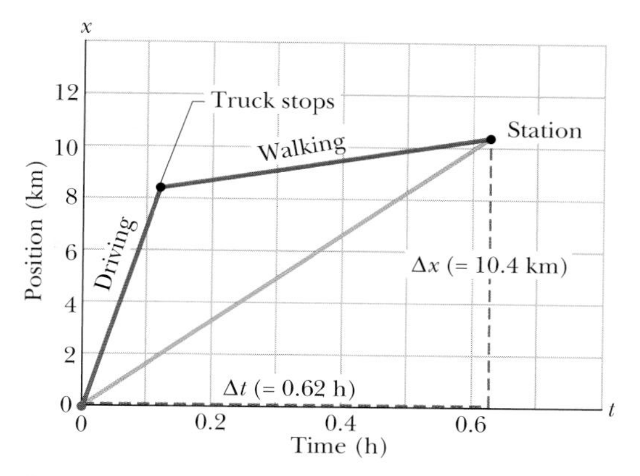

FIGURE 2-11 ■ The lines marked "Driving" and "Walking" are the position–time plots for the driving and walking stages. (The plot for the walking stage assumes a constant rate of walking.) The slope of the straight line joining the origin and the point labeled "Station" is the average velocity for the trip, from beginning to station.

(d) Suppose that to pump the gasoline, pay for it, and walk back to the truck takes you another 45 min. What is your average speed from the beginning of your drive to your return to the truck with the gasoline?

SOLUTION ■ The **Key Idea** here is that your average speed is the ratio of the total distance you move to the total time interval you take to make that move. The total distance is 8.4 km + 2.0 km + 2.0 km = 12.4 km. The total time interval is 0.12 h + 0.50 h + 0.75 h = 1.37 h. Thus, Eq. 2-5 gives us

$$\langle s \rangle = \frac{12.4\text{ km}}{1.37\text{ h}} = 9.1\text{ km/h}. \qquad \text{(Answer)}$$

2-4 Describing Velocity Change

The student shown in Fig. 2-9 is clearly speeding up and slowing down as she walks. We know that the slope of her position vs. time graph over small time intervals keeps *changing*. Now that we have defined velocity, it is meaningful to develop a mathematical description of how fast velocity changes. We see two approaches to describing velocity change. We could determine velocity change over an interval of displacement magnitude, $|\Delta x|$, and use $\Delta\vec{v}/|\Delta x|$ as our measure. Alternatively, we could use the ratio of velocity change to the interval of time, Δt, over which the change occurs or ($\Delta\vec{v}/\Delta t$). This is analogous to our definition of velocity.

Both of our proposals are possible ways of describing velocity change—neither is right or wrong. In the fourth century B.C.E., Aristotle believed that the ratio of velocity change to distance change was probably constant for any falling objects. Almost 2000 years later, the Italian scientist Galileo did experiments with ramps to slow down the motion of rolling objects. Instead he found that it was the second ratio, $\Delta\vec{v}/\Delta t$, that was constant.

Our modern definition of acceleration is based on Galileo's idea that $\Delta\vec{v}/\Delta t$ is the most useful concept in the description of velocity changes in falling objects.

Whenever a particle's velocity changes, we define it as having an **acceleration.** The **average acceleration,** $\langle\vec{a}\rangle$, over an interval Δt is defined as

$$\langle\vec{a}\rangle = \frac{\vec{v}_2 - \vec{v}_1}{t_2 - t_1} = \frac{\Delta\vec{v}}{\Delta t} \quad \text{(definition of 1D average acceleration).} \tag{2-7}$$

When the particle moves along a line (that is, an x axis in one-dimensional motion),

$$\langle\vec{a}\rangle = \frac{(v_{2x} - v_{1x})}{(t_2 - t_1)}\hat{\text{i}}.$$

It is important to note that an object is accelerated even if all that changes is only the *direction* of its velocity and not its speed. Directional changes are important as well.

Instantaneous Acceleration

If we want to determine how velocity changes during an instant of time, we need to define **instantaneous acceleration** (or simply **acceleration**) in a way that is similar to the way we defined instantaneous velocity:

$$\vec{a} \equiv \lim_{\Delta t \to 0} \frac{\Delta\vec{v}}{\Delta t} = \frac{d\vec{v}}{dt} \quad \text{(definition of 1D instantaneous acceleration).} \tag{2-8}$$

> In the language of calculus, the **ACCELERATION** of a particle at any instant is the rate at which its velocity is changing at that instant.

Using this definition, we can determine the acceleration by taking a time derivative of the velocity, $\vec{v}$. Furthermore, since velocity of an object moving along a line is the derivative of the position, $\vec{x}$, with respect to time, we can write

$$\vec{a} = \frac{d\vec{v}}{dt} = \frac{d}{dt}\left(\frac{d\vec{x}}{dt}\right) = \frac{d^2\vec{x}}{dt^2} \quad \text{(1D instantaneous acceleration).} \tag{2-9}$$

Equation 2-9 tells us that the instantaneous acceleration of a particle at any instant is equal to the second derivative of its position, $\vec{x}$, with respect to time. Note that if the object is moving along an x axis, then its acceleration can be expressed in terms of the x-component of its acceleration and the unit vector $\hat{\text{i}}$ along the x axis as

$$\vec{a} = a_x\hat{\text{i}} = \frac{dv_x}{dt}\hat{\text{i}} \quad \text{so} \quad a_x \equiv \frac{dv_x}{dt}.$$

Figure 2-12*c* shows a plot of the x-component of acceleration of an elevator cab. Compare the graph of the x-component of acceleration as a function of time (a_x vs. t) with the graph of the x-component of velocity as a function of time (v_x vs. t) in part *b*. Each point on the a_x vs. t graph is the derivative (slope or tangent) of the corresponding point on the v_x vs. t graph. When v_x is constant (at either 0 or 4 m/s), its time derivative is zero and hence so is the acceleration. When the cab first begins to move, the v_x vs. t graph has a positive derivative (the slope is positive), which means that a_x is positive. When the cab slows to a stop, the derivative or slope of the v_x vs. t graph is negative; that is, a_x is negative. Next compare the slopes of the v_x vs. t graphs during the

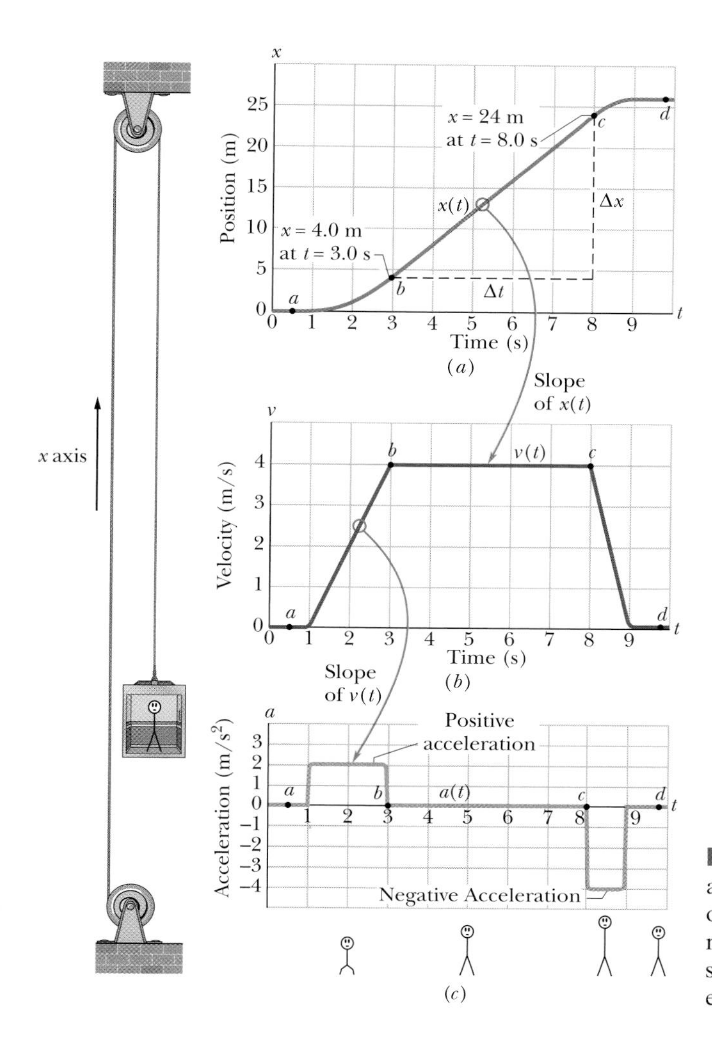

FIGURE 2-12 ■ (a) The x vs. t graph for an elevator cab that moves upward along an x axis. (b) The v_x vs. t graph for the cab. Note that it is the derivative of the x vs. t graph ($v_x = dx/dt$). (c) The a_x vs. t graph for the cab. It is the derivative of the v_x vs. t graph ($a_x = dv_x/dt$). The stick figures along the bottom suggest times that a passenger might feel light and long as the elevator accelerates downward or heavy and squashed as the elevator accelerates upward.

two acceleration periods. The slope associated with the cab's stopping is steeper, because the cab stops in half the time it took to get up to speed. The steeper slope means that the magnitude of the stopping acceleration is larger than that of the acceleration as the car is speeding up, as indicated in Fig. 2-12c.

Acceleration has both a magnitude and a direction and so it is a vector quantity. The algebraic sign of its component a_x represents the direction of velocity change along the chosen v_x axis. When acceleration and velocity are in the same direction (have the same sign) the object will speed up. If acceleration and velocity are in opposite directions (and have opposite signs) the object will slow down.

> It is important to realize that speeding up is not always associated with an acceleration that is positive. Likewise, slowing down is not always associated with an acceleration that is negative. The relative directions of an object's velocity and acceleration determine whether the object will speed up or slow down.

Since acceleration is defined as any change in velocity over time, whenever an object moving in a straight line has an acceleration it is either speeding up, slowing down, or turning around. Beware! In listening to common everyday language, you will probably hear the word acceleration used only to describe speeding up and the word deceleration to mean slowing down. It's best in studying physics to use the more formal definition of acceleration as a vector quantity that describes both the magnitude and direction of *any type of velocity change*. In short, an object is accelerating when it is slowing down as well as when it is speeding up. We suggest avoiding the use of the term deceleration while trying to learn the formal language of physics.

The fundamental unit of acceleration must be a velocity (displacement/time) divided by a time, which turns out to be displacement divided by time squared. Displacement is measured in meters and time in seconds in the SI system described in Chapter 1. Thus, the "official" unit of acceleration is m/s². You may encounter other units. For example, large accelerations are often expressed in terms of "g" units where g is directly related to the magnitude of the acceleration of a falling object near the Earth's surface. A g unit is given by

$$1\,g = 9.8\text{ m/s}^2. \tag{2-10}$$

On a roller coaster, you have brief accelerations up to $3g$, which, in standard SI units, is $(3)(9.8\text{ m/s}^2)$ or about 29 m/s^2. A more extreme example is shown in the photographs of Fig. 2-13, which were taken while a rocket sled was rapidly accelerated along a track and then rapidly braked to a stop.

FIGURE 2-13 ■ Colonel J.P. Stapp in a rocket sled as it is brought up to high speed (acceleration out of the page) and then very rapidly braked (acceleration into the page).

READING EXERCISE 2-8: A cat moves along an x axis. What is the sign of its acceleration if it is moving (a) in the positive direction with increasing speed, (b) in the positive direction with decreasing speed, (c) in the negative direction with increasing speed, and (d) in the negative direction with decreasing speed? ■

TOUCHSTONE EXAMPLE 2-3: Position and Motion

A particle's position on the x axis of Fig. 2-1 is given by

$$x = 4\text{ m} - (27\text{ m/s})\,t + (1\text{ m/s}^3)t^3,$$

with x in meters and t in seconds.

(a) Find the particle's velocity function $v_x(t)$ and acceleration function $a_x(t)$.

SOLUTION ■ One **Key Idea** is that to get the velocity function $v_x(t)$, we differentiate the position function $x(t)$ with respect to time. Here we find

$$v_x = -(27\text{ m/s}) + 3\cdot(1\text{ m/s}^3)t^2 = -(27\text{ m/s}) + (3\text{ m/s}^3)\,t^2 \quad \text{(Answer)}$$

with v_x in meters per second.

Another **Key Idea** is that to get the acceleration function $a_x(t)$, we differentiate the velocity function $v_x(t)$ with respect to time. This gives us

$$a_x = 2 \cdot 3 \cdot (1 \text{ m/s}^3)t = +(6 \text{ m/s}^3)t, \qquad \text{(Answer)}$$

with a_x in meters per second squared.

(b) Is there ever a time when $v_x = 0$?

SOLUTION ■ Setting $v_x(t) = 0$ yields

$$0 = -(27 \text{ m/s}) + (3 \text{ m/s}^3)t^2,$$

which has the solution

$$t = \pm 3 \text{ s}. \qquad \text{(Answer)}$$

Thus, the velocity is zero both 3 s before and 3 s after the clock reads 0.

(c) Describe the particle's motion for $t \geq 0$.

SOLUTION ■ The **Key Idea** is to examine the expressions for $x(t)$, $v_x(t)$, and $a_x(t)$.

At $t = 0$, the particle is at $x(0) = +4$ m and is moving with a velocity of $v_x(0) = -27$ m/s—that is, in the negative direction of the x axis. Its acceleration is $a_x(0) = 0$, because just then the particle's velocity is not changing.

For $0 < t < 3$ s, the particle still has a negative velocity, so it continues to move in the negative direction. However, its acceleration is no longer 0 but is increasing and positive. Because the signs of the velocity and the acceleration are opposite, the particle must be slowing.

Indeed, we already know that it turns around at $t = 3$ s. Just then the particle is as far to the left of the origin in Fig. 2-1 as it will ever get. Substituting $t = 3$ s into the expression for $x(t)$, we find that the particle's position just then is $x = -50$ m. Its acceleration is still positive.

For $t > 3$ s, the particle moves to the right on the axis. Its acceleration remains positive and grows progressively larger in magnitude. The velocity is now positive, and it too grows progressively larger in magnitude.

2-5 Constant Acceleration: A Special Case

If you watch a small steel ball bobbing up and down at the end of a spring, you will see the velocity changing continuously. But instead of either increasing or decreasing at a steady rate, we have a very nonuniform pattern of motion. First the ball speeds up and slows down moving in one direction, then it turns around and speeds up and then slows down in the other direction, and so on. This is an example of a nonconstant acceleration that keeps changing in time.

Although there are many examples of nonconstant accelerations, we also observe a surprising number of examples of constant or nearly constant acceleration. As we already discussed, Galileo discovered that if we choose to define acceleration in terms of the ratio $\Delta\vec{v}/\Delta t$, then a falling ball or a ball tossed into the air that slows down, turns around, and speeds up again is always increasing its velocity in a downward direction at the same rate—provided the ball is moving slowly enough that air drag is negligible.

There are many other common motions that involve constant accelerations. Suppose you measure the times and corresponding positions for an object that you suspect has a constant acceleration. If you then calculate the velocities and accelerations of the object and make graphs of them, the graphs will resemble those in Fig. 2-14. Some examples of motions that yield similar graphs to those shown in Fig. 2-14 include: a car that you accelerate as soon as a traffic light turns green; the same car when you apply its brakes steadily to bring it to a smooth stop; an airplane when first taking off or when completing a smooth landing; or a dolphin that speeds up suddenly after being startled.

Position component vs. time

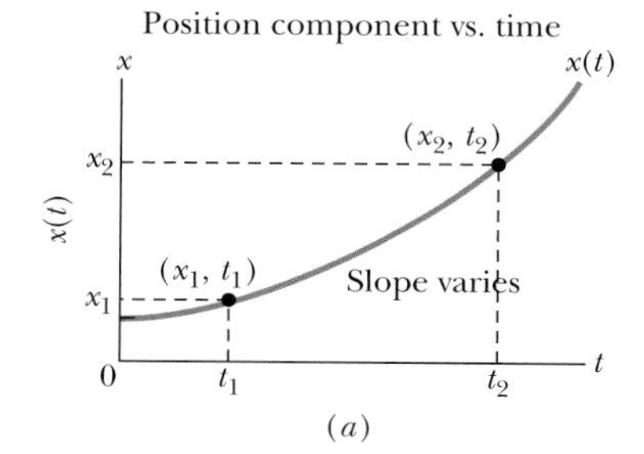

(a)

Velocity component vs. time

v
v(t)
(v_2, t_2)
v_2
$v_x(t)$
(v_1, t_1)
Slope = a_x
v_1
0
t_1
t_2
t

(b)

Acceleration component vs. time

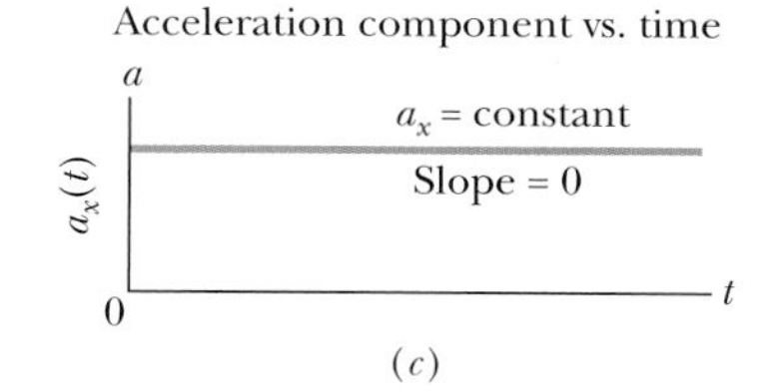

(c)

FIGURE 2-14 ■ (a) The position component $x(t)$ of a particle moving with constant acceleration. (b) Its velocity component $v_x(t)$, given at each point by the slope of the curve in (a). (c) Its (constant) component of acceleration, a_x, equal to the (constant) slope of $v_x(t)$.

Derivation of the Kinematic Equations

Because constant accelerations are common, it is useful to derive a special set of **kinematic equations** to describe the motion of any object that is moving along a line with a constant acceleration. We can use the definitions of acceleration and velocity and an assumption about average velocity to derive the kinematic equations. These equations allow us to use known values of the vector components describing positions, velocities, and accelerations, along with time intervals to predict the motions of constantly accelerated objects.

Let's start the derivation by noting that when the acceleration is constant, the average and instantaneous accelerations are equal. As usual we place our x axis along the line of the motion. We can now use vector notation to write

$$\vec{a} = a_x\hat{\text{i}} = \langle\vec{a}\rangle, \tag{2-11}$$

so that

$$\langle\vec{a}\rangle = \frac{(v_{2x} - v_{1x})}{t_2 - t_1}\hat{\text{i}},$$

where a_x is the component of acceleration along the line of motion of the object. We can use the definition of average acceleration (Eq. 2-7) to express the acceleration component a_x in terms of the object's velocity components along the line of motion, where v_{2x} and v_{1x} are the object's velocity components along the line of motion,

$$a_x = \frac{(v_{2x} - v_{1x})}{t_2 - t_1}. \tag{2-12}$$

This expression allows us to derive the kinematic equations in terms of the vector components needed to construct the actual one-dimensional velocity and acceleration vectors. The subscripts 1 and 2 in most of the equations in this chapter, including Eq. 2-12, refer to initial and final times, positions, and velocities.

If we solve Eq. 2-12 for v_{2x}, then the x-component of velocity at time t_2 is

$$v_{2x} = v_{1x} + a_x(t_2 - t_1) = v_{1x} + a_x\Delta t \quad \text{(primary kinematic } [a_x = \text{constant}] \text{ equation)},$$

$$\text{or} \quad \Delta v_x = a\Delta t. \tag{2-13}$$

This equation is the first of two primary equations that we will derive for use in analyzing motions involving constant acceleration. Before we move on, we should think carefully about what the expression $t_2 - t_1$ represents in this equation: *It represents the time interval in which we are tracking the motion.*

In a manner similar to what we have done above, we can rewrite Eq. 2-4, the expression for the average velocity along the x axis,

$$\langle\vec{v}\rangle = \langle v_x\rangle\hat{\text{i}} = \frac{\Delta x}{\Delta t}\hat{\text{i}} = \frac{(x_2 - x_1)}{t_2 - t_1}\hat{\text{i}}.$$

Hence, the x-component of the average velocity is given by

$$\langle v_x\rangle = \frac{(x_2 - x_1)}{(t_2 - t_1)}.$$

Solving for x_2 gives

$$x_2 = x_1 + \langle v_x\rangle(t_2 - t_1). \tag{2-14}$$

In this equation x_1 is the x-component of the position of the particle at $t = t_1$ and $\langle v_x\rangle$ is the component along the x axis of average velocity between $t = t_1$ and a later time $t = t_2$. Note that unless the velocity is constant, the average velocity component along the x axis, $\langle v_x\rangle$, is not equal to the instantaneous velocity component, v_x.

However, we do have a plausible alternative for expressing the average velocity component in the special case when the acceleration is constant. Figure 2-15 depicts the fact that velocity increases in a linear fashion over time for a constant acceleration. It seems reasonable to assume that the component along the x axis of the *average* velocity over any time interval is the average of the components for the in-

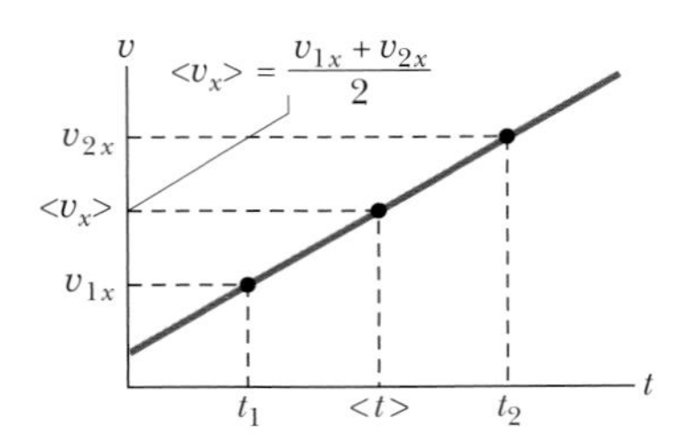

FIGURE 2-15 ■ When the acceleration is constant, then we assume (without rigorous proof) that the average velocity component in a time interval is the average of the velocity components at the beginning and end of the interval.

stantaneous velocity at the beginning of the interval, v_{1x}, and the instantaneous velocity component at the end of the interval, v_{2x}. So we expect that when a velocity increases linearly, the average velocity component over a given time interval will be

$$\langle v_x \rangle = \frac{v_{1x} + v_{2x}}{2}. \tag{2-15}$$

Using Eq. 2-13, we can substitute $v_{1x} + a_x(t_2 - t_1)$ for v_{2x} to get

$$\langle v_x \rangle = \frac{1}{2}\left[v_{1x} + v_{1x} + a_x(t_2 - t_1)\right] = v_{1x} + \tfrac{1}{2}a_x(t_2 - t_1). \tag{2-16}$$

Finally, substituting this equation into Eq. 2-14 yields

$$x_2 - x_1 = v_{1x}(t_2 - t_1) + \tfrac{1}{2}a_x(t_2 - t_1)^2 \quad \text{(primary kinematic } [a_x = \text{constant}] \text{ equation)}, \tag{2-17}$$

or

$$\Delta x = v_{1x}\Delta t + \tfrac{1}{2}a_x\Delta t^2$$

This is our second primary equation describing motion with constant acceleration. Figures 2-14*a* and 2-16 show plots of Eq. 2-17.

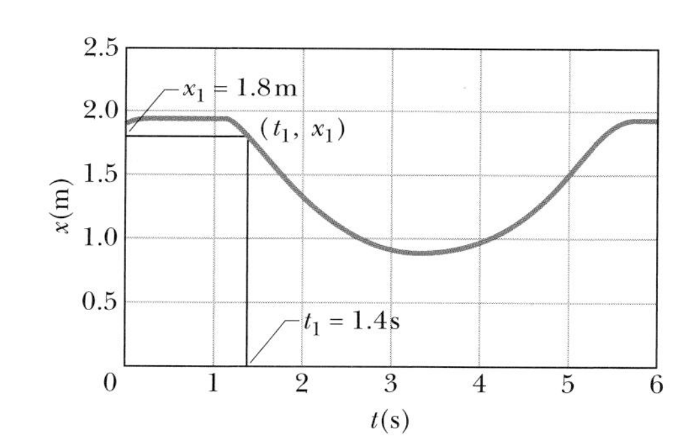

FIGURE 2-16 ■ A fan on a low-friction cart is being held in place about 1.2 s but isn't released fully until $t_1 = 1.4$ s. Data for the graph were collected with a computer data acquisition system outfitted with an ultrasonic motion detector. Between 1.4 s and about 5.4 s the cart appears to be undergoing a constant acceleration as it slows down, turns around, and speeds up again. Thus, the constant acceleration kinematic equations can be used to describe its motion but only during motion within that time interval. Thus, we can set t_1 to 1.4 s and x_1 to 1.8 m.

These two equations are very useful in the calculation of unknown quantities that can be used to characterize constantly accelerated motion. There are five or six quantities contained in our primary equations (Eqs. 2-13 and 2-17). The simplest kinematic calculations involve situations in which all but one of the quantities is known in one of the primary equations. In more complex situations, both equations are needed. Typically for a complex situation, we need to calculate more than one unknown. To do this, we find the first unknown using one of the primary equations and use the result in the other equations to find the second unknown. This method is illustrated in the next section and in Touchstone Examples 2-4 and 2-6.

The primary equations above, $v_{2x} = v_{1x} + a_x(t_2 - t_1) = v_{1x} + a_x\Delta t$ (Eq. 2-13), and $x_2 - x_1 = v_{1x}(t_2 - t_1) + \frac{1}{2}a_x(t_2 - t_1)^2$ (Eq. 2-17), are derived directly from the definitions of velocity and acceleration, with the condition that the acceleration is constant. These two equations can be combined in three ways to yield three additional equations. For example, solving for $t_2 - t_1$ in $v_{2x} = v_{1x} + a_x(t_2 - t_1)$ and substituting the result into $x_2 - x_1 = v_{1x}(t_2 - t_1) + \frac{1}{2}a_x(t_2 - t_1)^2$ gives us

$$v_{2x}^2 = v_{1x}^2 + 2a_x(x_2 - x_1).$$

We recommend that you learn the two primary equations and use them to derive other equations as needed. Then you will not need to remember so much. Table 2-1 lists our two primary equations. Note that a really nice alternative to using the two

TABLE 2-1
Equations of Motion with Constant Acceleration

Equation Number	Primary Vector Component Equation*
2-13	$v_{2x} = v_{1x} + a_x(t_2 - t_1)$
2-17	$x_2 - x_1 = v_{1x}(t_2 - t_1) + \frac{1}{2}a_x(t_2 - t_1)^2$

*A reminder: In cases where the initial time t_1 is chosen to be zero it is important to remember that whenever the term $(t_2 - t_1)$ is replaced by just t, then t actually represents a *time interval of* $\Delta t = t - 0$ over which the motion of interest takes place.

equations in Table 2-1 is to use the first of the equations (Eq. 2-13) along with the expression for the average velocity component in Eq. 2-15,

$$\langle v_x \rangle = \frac{\Delta x}{\Delta t} = \frac{v_{1x} + v_{2x}}{2} \quad \text{(an alternative "primary" equation),}$$

to derive all the other needed equations. The derivations of the kinematic equations that we present here are not rigorous mathematical proofs but rather what we call plausibility arguments. However, we know from the application of the kinematic equations to constantly accelerated motions that they do adequately describe these motions.

Analyzing the Niagara Falls Plunge

At the beginning of this chapter we asked questions about the motion of the steel chamber holding Dave Munday as he plunged into the water after falling 48 m from the top of Niagara Falls. How long did the fall take? That is, what is Δt? How fast was the chamber moving when it hit the water? (What is $\vec{v}$?) As you will learn in Chapter 3, if no significant air drag is present, objects near the surface of the Earth fall at a constant acceleration of magnitude $|a_x| = 9.8$ m/s^2. Thus, the kinematic equations can be used to calculate the time of fall and the impact speed.

Let's start by defining our coordinate system. We will take the x axis to be a vertical or up–down axis that is aligned with the downward path of the steel chamber. We place the origin at the bottom of the falls and define up to be positive as shown in Fig. 2-17. (Later when considering motions in two and three dimensions, we will often denote vertical axes as y axes and horizontal axes as x axes, but these changes in symbols will not affect the results of calculations.)

We know that the value of the vertical displacement is given by

$$x_2 - x_1 = (0 \text{ m}) - (+48 \text{ m}) = -48 \text{ m}$$

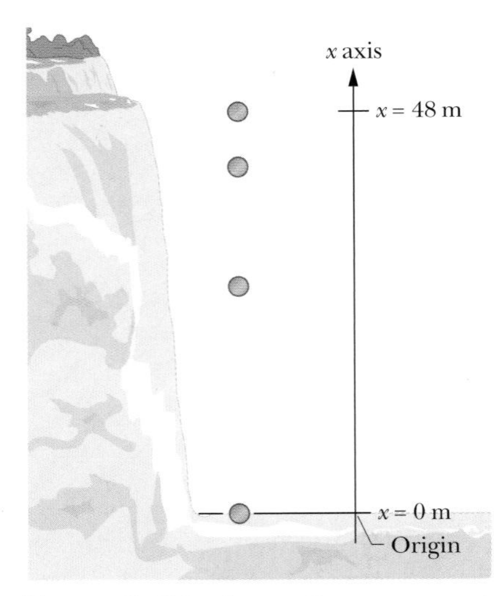

FIGURE 2-17 ■ A coordinate system chosen to analyze the fall of a steel chamber holding a man who falls 48 m from the top to the bottom of Niagara Falls.

and that the velocity is getting larger in magnitude in the downward (negative direction). Since the velocity is downward and the object is speeding up, the vertical acceleration is also downward (in the negative direction). Its component along the axis of motion is given by $a_x = -9.8$ m/s^2. Finally, we assume that Dave Munday's capsule dropped from rest, so $v_{1x} = 0$ m/s. Thus we can find the time of fall ($\Delta t = t_2 - t_1$) using Eq. 2-17. Solving this equation for the time elapsed during the fall ($t_2 - t_1$) when the initial velocity v_{1x} is zero gives

$$\Delta t = t_2 - t_1 = \sqrt{\frac{2(x_2 - x_1)}{a_x}} = \sqrt{\frac{2(-48 \text{ m})}{-9.8 \text{ m/s}^2}} = 3.13 \text{ s} = 3.1 \text{ s}.$$

This is a fast trip indeed!

Next we can use the time interval of the fall in the other primary kinematic equation, Eq. 2-13, to find the velocity at impact. This gives a component of impact velocity at the end of the fall of

$$v_{2x} = v_{1x} + a_x(t_2 - t_1) = 0 \text{ m/s} + (-9.8 \text{ m/s}^2)(3.13 \text{ s}) = -31 \text{ m/s}.$$

The minus sign indicates that the impact velocity component is negative and is, therefore, in the downward direction. In vector notation, the velocity $\vec{v} = v_x \hat{i}$ is thus $\vec{v} = (-31 \text{ m/s})\hat{i}$. Note that this is a speed of about 69 mi/hr. Since the time interval was put into the calculation of velocity of impact as an intermediate value, we retained an extra significant figure to use in the next calculation.

READING EXERCISE 2-9: The following equations give the x-component of position $x(t)$ of a particle in meters (denoted m) as a function of time in seconds for four situations: (1) $x = (3 \text{ m/s})t - 4$ m; (2) $x = (-5 \text{ m/s}^3)t^3 + (4 \text{ m/s})t + 6$ m; (3) $x = (2 \text{ m/s}^2)t^2 - (4 \text{ m/s})t$; (4) $x = (5 \text{ m/s}^2)t^2 - 3$ m. To which of these situations do the equations of Table 2-1 apply? Explain. ■

TOUCHSTONE EXAMPLE 2-4: Slowing Down

Spotting a police car, you brake your Porsche from a speed of 100 km/h to a speed of 80.0 km/h during a displacement of 88.0 m, at a constant acceleration.

(a) What is that acceleration?

SOLUTION ■ Assume that the motion is along the positive direction of an x axis. For simplicity, let us take the beginning of the braking to be at time $t_1 = 0$, at position x_1. The **Key Idea** here is that, with the acceleration constant, we can relate the car's acceleration to its velocity and displacement via the basic constant acceleration equations (Eqs. 2-13 and 2-17). The initial velocity is $v_{1x} = 100 \text{ km/h} = 27.78 \text{ m/s}$, the displacement is $x_2 - x_1 = 88.0$ m, and the velocity at the end of that displacement is $v_{2x} = 80.0 \text{ km/h} = 22.22 \text{ m/s}$. However, we do not know the acceleration a_x and time t_2, which appear in both basic equations, so we must solve those equations simultaneously.

To eliminate the unknown t_2, we use Eq. 2-13 to write

$$t_2 - t_1 = \frac{v_{2x} - v_{1x}}{a_x}, \tag{2-18}$$

and then we substitute this expression into Eq. 2-17 to write

$$x_2 - x_1 = v_{1x}\left(\frac{v_{2x} - v_{1x}}{a_x}\right) + \tfrac{1}{2}a_x\left(\frac{v_{2x} - v_{1x}}{a_x}\right)^2.$$

Solving for a_x and substituting known data then yields

$$a_x = \frac{v_{2x}^2 - v_{1x}^2}{2(x_2 - x_1)} = \frac{(22.22 \text{ m/s})^2 - (27.78 \text{ m/s})^2}{2(88.0 \text{ m})}$$
$$= -1.58 \text{ m/s}^2. \quad \text{(Answer)}$$

(b) How much time is required for the given decrease in speed?

SOLUTION ■ Now that we know a_x, we can use Eq. 2-18 to solve for t_2:

$$t_2 - t_1 = \frac{v_{2x} - v_{1x}}{a_x} = \frac{22.22 \text{ m/s} - 27.78 \text{ m/s}^2}{-1.58 \text{ m/s}^2} = 3.52 \text{ s}. \quad \text{(Answer)}$$

If you are initially speeding and trying to slow to the speed limit, there is plenty of time for the police officer to measure your excess speed.

You can use one of the alternate equations for motion with a constant acceleration, Eq. 2-15, to check this result. The **Key Idea** here is that the distance traveled is just the product of the average velocity and the elapsed time, when the acceleration is constant. The Porsche traveled 88.0 m while it slowed from 100 km/h down to 80 km/h. Thus its average velocity while it covered the 88.0 m was

$$\langle v_x \rangle = \frac{(100 \text{ km/h} + 80 \text{ km/h})}{2}$$
$$= 90\,\frac{\text{km}}{\text{h}} \cdot \left(\frac{1000 \text{ m}}{1 \text{ km}}\right) \cdot \left(\frac{1 \text{ h}}{3600 \text{ s}}\right) = 25.0 \text{ m/s},$$

so the time it took to slow down was just

$$t_2 - t_1 = \frac{x_2 - x_1}{\langle v_x \rangle} = \frac{88.0 \text{ m}}{25.0 \text{ m/s}} = 3.52 \text{ s}, \quad \text{(Answer)}$$

which still isn't enough time to avoid that speeding ticket!

TOUCHSTONE EXAMPLE 2-5: Motion Data

Suppose that you gave a box sitting on a carpeted floor a push and then recorded its position three times per second as it slid to a stop. The table gives the results of such a measurement. Let's analyze the position vs. time data for the box sliding on the carpet and use curve fitting and calculus to obtain the velocity measurements. We will use Excel spreadsheet software to perform our analysis, but other computer- or calculator-based fitting or modeling software can be used.

Box Sliding on Carpet

t[s]	x[m]
0.000	0.537
0.033	0.583
0.067	0.623
0.100	0.659
0.133	0.687
0.167	0.705
0.200	0.719
0.233	0.720

(a) Draw a graph of the x vs. t data and discuss whether the relationship appears to be linear or not.

SOLUTION ■ The **Key Idea** here is that the relationship between two variables is linear if the graph of the data points lie more or less along a straight line. There are many ways to graph the data for examination: by hand, with a graphing calculator, with a spreadsheet graphing routine, or with other graphing software such as Data Studio (available from PASCO scientific) or Graphical Analysis (available from Vernier Software and Technology). The graph in Fig. 2-18 that shows a curve and so the relationship between position, x, and time is not linear.

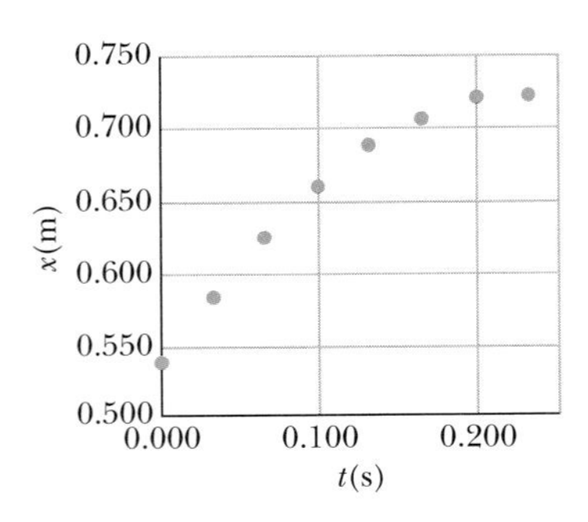

FIGURE 2-18 ■ Solution to Touchstone Example 2-5(a). A graph of position versus time for a box sliding across a carpet.

(b) Draw a motion diagram of the box as it comes to rest on the carpet.

SOLUTION ■ The **Key Idea** here is to use the data to sketch the position along a line at equal time intervals. In Fig. 2-19, the black circles represent the location of the rear of the box at intervals of 1/30 of a second.

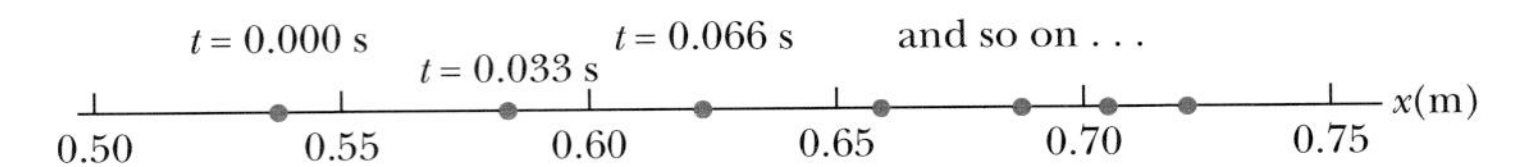

FIGURE 2-19 ■ Solution to Touchstone Example 2-5(b). A motion diagram for a box sliding across a carpet.

(c) Is the acceleration constant? If so, what is its component along the x axis?

SOLUTION ■ The **Key Idea** here is to explore whether or not the relationship between position and time of the box as it slides to a stop can be described with a quadratic (parabolic) function of time as described in Eq. 2-17. This can be done by entering the data that are given into a spreadsheet or graphing calculator and either doing a quadratic model or a fit to the data. The outcome of a quadratic model is shown in Fig. 2-20. The x-model column contains the results of calculating x using the equation $x_2 - x_1 = v_{1\,x}(t_2 - t_1) + \frac{1}{2}a_x(t_2 - t_1)^2$ for each of the times in the first column using the initial position, velocity and acceleration data shown in the boxes. The line shows the model data. If the kinematic equation fits the data, then we can conclude that the acceleration component is a constant given by $a_x = -6.6$ m/s². Thus the acceleration is in the negative y direction.

a_x	−6.7	(m/s²)
v_{1t}	1.6	(m/s)
x_1	0.537	(m)

Box Sliding on Carpet

t(s)	x-data (m)	x-model (m)
0.000	0.537	0.537
0.033	0.583	0.587
0.067	0.623	0.629
0.100	0.659	0.664
0.133	0.687	0.691
0.167	0.705	0.711
0.200	0.719	0.723
0.233	0.720	0.728

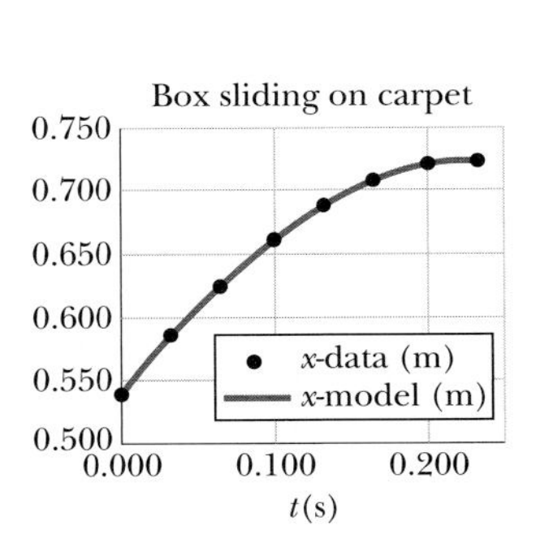

FIGURE 2-20 ■ Solution to Touchstone Example (c). Data and a graph of position as function of time for a box sliding over carpet. Actual data is compared to a model of what is expected from Eq. 2-17 (assumed constant acceleration). The value of acceleration which produced the best match between the model and actual data is −6.6 m/s².

TOUCHSTONE EXAMPLE 2-6: Distance Covered

Figure 2-21*b* shows a graph of a person riding on a low-friction cart being pulled along with a bungee cord as shown in Fig. 2-21*a*. Use information from the two graphs and the kinematic equations to determine approximately how far the student moved in the time interval between 1.1 s and 2.0 s.

SOLUTION ■ The **Key Idea** is that the initial velocity can be determined from the velocity vs. time graph on the left and the acceleration during the time interval can be determined from the acceleration vs. time graph on the right (or by finding the slope of the velocity vs. time graph on the left during the time interval). Note that the velocity at $t_1 = 1.1$ s is given by $v_{1x} \approx 0.4$ m/s. The acceleration during the time interval of interest is given by $a_x \approx 0.4$ m/s^2. Since the acceleration is constant over the time interval of interest, we can use the data in Eq. 2-17 to get

$$x_2 - x_1 = v_{1x}(t_2 - t_1) + \tfrac{1}{2}a_x(t_2 - t_1)^2$$

$$= (0.4 \text{ m/s})(2.0 \text{ s} - 1.1 \text{ s}) + \tfrac{1}{2}(0.4 \text{ m/s}^2)(2.0 \text{ s} - 1.1 \text{ s})^2$$

$$\approx 0.5 \text{ m.} \qquad \text{(Answer)}$$

Half a meter is not very far!

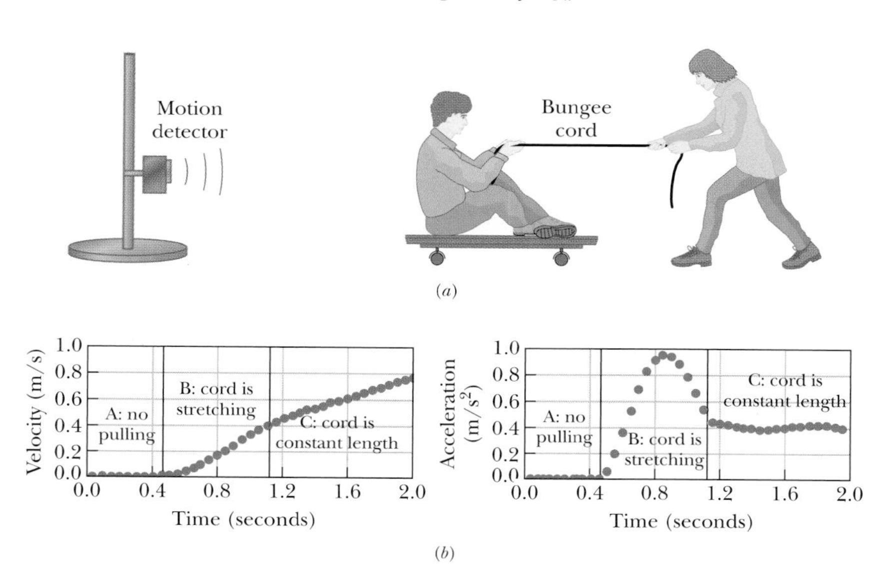

FIGURE 2-21 ■ (*a*) A person riding on a low-friction cart is pulled by another person who exerts a constant force along a straight line by keeping the length of a bungee cord constant. (*b*) These graphs show velocity and acceleration components vs. time for a rider on a cart. For the first 0.5 s (region A) the cart is at rest. Between 0.5 s and 1.1 s (region B) the cord is beginning to stretch. Between 1.1 s and 2.0 s (region C) a constant force is acting and the acceleration is also constant.

Problems

In several of the problems that follow you are asked to graph position, velocity, and acceleration versus time. Usually a sketch will suffice, appropriately labeled and with straight and curved portions apparent. If you have a computer or graphing calculator, you might use it to produce the graph.

SEC. 2-3 ■ VELOCITY AND SPEED

1. Fastball If a baseball pitcher throws a fastball at a horizontal speed of 160 km/h, how long does the ball take to reach home plate 18.4 m away?

2. Fastest Bicycle A world speed record for bicycles was set in 1992 by Chris Huber riding Cheetah, a high-tech bicycle built by three mechanical engineering graduates. The record (average) speed was 110.6 km/h through a measured length of 200.0 m on a desert road. At the end of the run, Huber commented, "Cogito ergo zoom!" (I think, therefore I go fast!) What was Huber's elapsed time through the 200.0 m?

3. Auto Trip An automobile travels on a straight road for 40 km at 30 km/h. It then continues in the same direction for another 40 km at 60 km/h. (a) What is the average velocity of the car during this 80 km trip? (Assume that it moves in the positive x direction.) (b) What is the average speed? (c) Graph x vs. t and indicate how the average velocity is found on the graph.

4. Radar Avoidance A top-gun pilot, practicing radar avoidance maneuvers, is manually flying horizontally at 1300 km/h, just 35 m above the level ground. Suddenly, the plane encounters terrain that slopes gently upward at 4.3°, an amount difficult to detect visually (Fig. 2-22). How much time does the pilot have to make a correction to avoid flying into the ground?

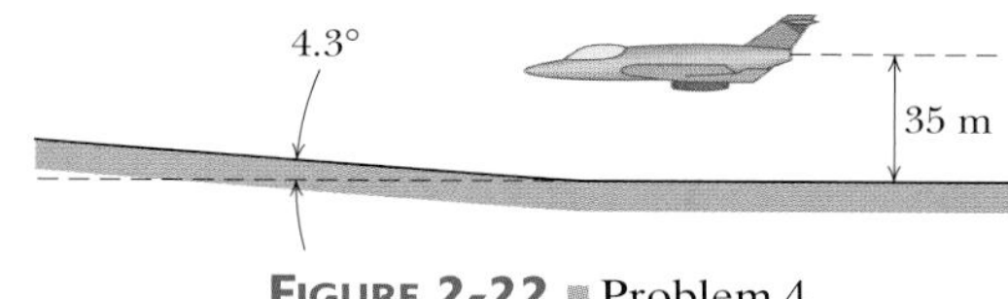

FIGURE 2-22 ■ Problem 4.

5. On Interstate 10 You drive on Interstate 10 from San Antonio to Houston, half the *time* at 55 km/h and the other half at 90 km/h. On the way back you travel half the *distance* at 55 km/h and the other half at 90 km/h. What is your average speed (a) from San Antonio to Houston, (b) from Houston back to San Antonio, and (c) for the entire trip? (d) What is your average velocity for the entire trip? (e) Sketch x vs. t for (a), assuming the motion is all in the positive x direction. Indicate how the average velocity can be found on the sketch.

6. Walk Then Run Compute your average velocity in the following two cases: (a) You walk 73.2 m at a speed of 1.22 m/s and then run 73.2 m at a speed of 3.05 m/s along a straight track. (b) You walk for 1.00 min at a speed of 1.22 m/s and then run for 1.00 min at 3.05 m/s along a straight track. (c) Graph x vs. t for both cases and indicate how the average velocity is found on the graph.

7. Position and Time The position of an object moving along an x axis is given by $x = (3\text{ m/s})t - (4\text{ m/s}^2)t^2 + (1\text{ m/s}^3)t^3$, where x is in meters and t in seconds. (a) What is the position of the object at $t =$ 1, 2, 3, and 4 s? (b) What is the object's displacement between $t_1 = 0$ s and $t_2 = 4$ s? (c) What is the average velocity between the time interval from $t_1 = 2$ s to $t_2 = 4$ s? (d) Graph x vs. t for $0 \leq t \leq 4$ s and indicate how the answer for (c) can be found on the graph.

8. Two Trains and a Bird Two trains, each having a speed of 30 km/h, are headed at each other on the same straight track. A bird that can fly 60 km/h flies off the front of one train when they are 60 km apart and heads directly for the other train. On reaching the other train it flies directly back to the first train, and so forth. (We have no idea *why* a bird would behave in this way.) What is the total distance the bird travels?

9. Two Winners On two *different* tracks, the winners of the 1 kilometer race ran their races in 2 min, 27.95 s and 2 min, 28.15 s. In order to conclude that the runner with the shorter time was indeed faster, how much longer can the other track be in *actual* length?

10. Scampering Armadillo The graph in Fig. 2-23 is for an armadillo that scampers left (negative direction of x) and right along an x axis. (a) When, if ever, is the animal to the left of the origin on the axis? When, if ever, is its velocity component (b) negative, (c) positive, or (d) zero?

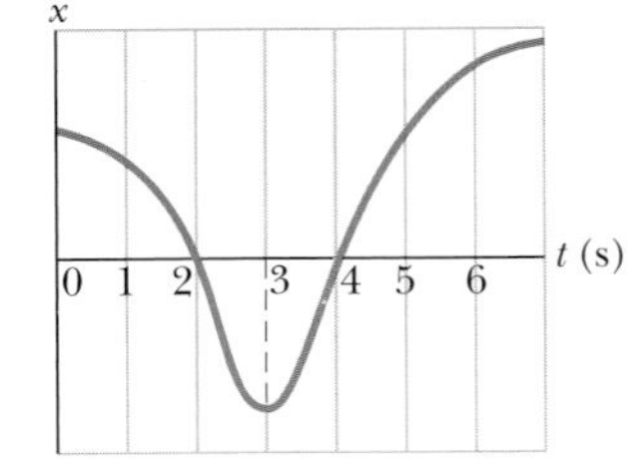

FIGURE 2-23 ■ Problem 10.

11. Position and Time (a) If a particle's position is given by $x = 4m - (12\text{ m/s})t + (3\text{ m/s}^2)t^2$ (where t is in seconds and x is in meters), what is its velocity at $t_1 = 1$ s? (b) Is it moving in the positive or negative direction of x just then? (c) What is its speed just then? (d) Is the speed larger or smaller at later times? (Try answering the next two questions without further calculation.) (e) Is there ever an instant when the velocity is zero? (f) Is there a time after $t_3 = 3$ s when the particle is moving in the negative direction of x?

12. Particle Position and Time The position of a particle moving along the x axis is given in meters by $x = 9.75\text{m} + (1.5\text{ m/s}^3)t^3$ where t is in seconds. Calculate (a) the average velocity during the time interval $t = 2.00$ s to $t = 3.00$ s; (b) the instantaneous velocity at $t = 2.00$ s; (c) the instantaneous velocity at $t = 3.00$ s; (d) the instantaneous velocity at $t = 2.50$ s; and (e) the instantaneous velocity when the particle is midway between its positions at $t = 2.00$ s and $t = 3.00$ s (f) Graph x vs. t and indicate your answers graphically.

13. Velocity–Time Graph How far does the runner whose velocity–time graph is shown in Fig. 2-24 travel in the time interval between (a) $t_2 = 2$ s and 10 s; (b) $t_{12} = 12$ s and $t_{16} = 16$ s?

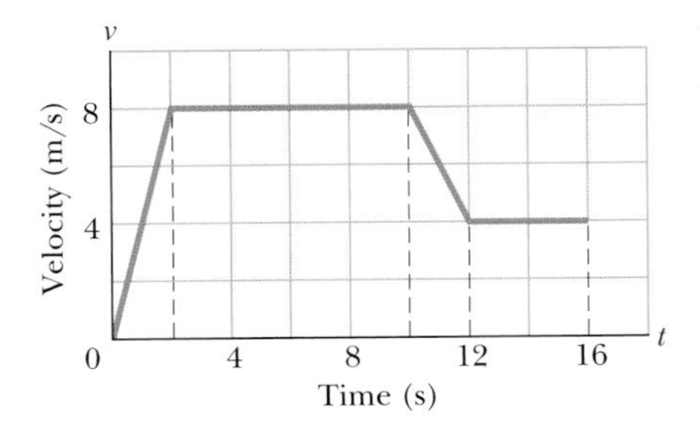

FIGURE 2-24 ■ Problem 13

SEC. 2-4 ■ DESCRIBING VELOCITY CHANGE

14. Various Motions Sketch a graph that is a possible description of position as a function of time for a particle that moves along the x axis and, at $t = 1$ s, has (a) zero velocity and positive acceleration; (b) zero velocity and negative acceleration; (c) negative velocity and positive acceleration; (d) negative velocity and negative acceleration. (e) For which of these situations is the speed of the particle increasing at $t = 1$ s?

15. Two Similar Expressions What do the quantities (a) $(dx/dt)^2$ and (b) d^2x/dt^2 represent? (c) What are their SI units?

16. Frightened Ostrich A frightened ostrich moves in a straight line with velocity described by the velocity–time graph of Fig. 2-25. Sketch acceleration vs. time.

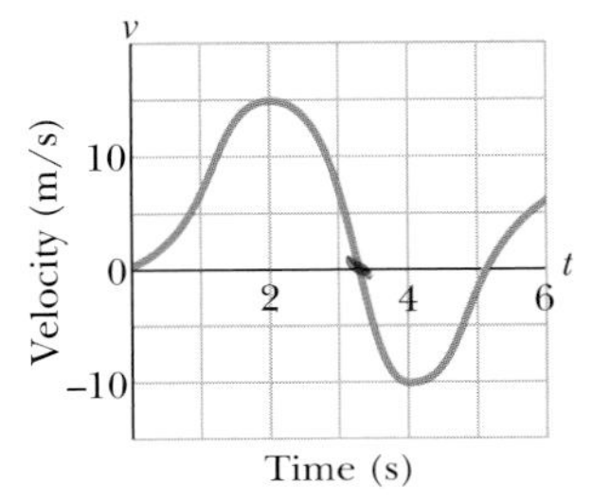

FIGURE 2-25 ■ Problem 16.

17. Speed Then and Now A particle had a speed of 18 m/s at a certain time, and 2.4 s later its speed was 30 m/s in the opposite direction. What were the magnitude and direction of the average acceleration of the particle during this 2.4 s interval?

18. Stand Then Walk From $t_0 = 0$ to $t_5 = 5.00$ min, a man stands still, and from $t_5 = 5.00$ min to $t_{10} = 10.0$ min, he walks briskly in a straight line at a constant speed of 2.20 m/s. What are (a) his average velocity $\langle \vec{v} \rangle$ and (b) his average acceleration $\langle \vec{a} \rangle$ in the time interval 2.00 min to 8.00 min? What are (c) $\langle \vec{v} \rangle$ and $\langle \vec{a} \rangle$ in the time interval 3.00 min to 9.00 min? (d) Sketch x vs. t and v vs. t, and indicate how the answers to (a) through (c) can be obtained from the graphs.

19. Particle Position and Time The position of a particle moving along the x axis depends on the time according to the equation $x = ct^2 - bt^3$, where x is in meters and t in seconds. (a) What units must c and b have? Let their numerical values be 3.0 and 2.0, respectively. (b) At what time does the particle reach its maximum positive x position? From $t_0 = 0.0$ s to $t_4 = 4.0$ s, (c) what distance does the particle move and (d) what is its displacement? At $t = 1.0$, 2.0, 3.0, and 4.0 s, what are (e) its velocities and (f) its accelerations?

SEC. 2-5 ■ CONSTANT ACCELERATION: A SPECIAL CASE

20. Driver and Rider An automobile driver on a straight road increases the speed at a constant rate from 25 km/h to 55 km/h in 0.50 min. A bicycle rider on a straight road speeds up at a constant rate from rest to 30 km/h in 0.50 min. Calculate their accelerations.

21. Stopping a Muon A muon (an elementary particle) moving in a straight line enters a region with a speed of 5.00×10^6 m/s and then is slowed at the rate of 1.25×10^{14} m/s^2. (a) How far does the muon take to stop? (b) Graph x vs. t and v vs. t for the muon.

22. Rattlesnake Striking The head of a rattlesnake can accelerate at 50 m/s^2 in striking a victim. If a car could do as well, how long would it take to reach a speed of 100 km/h from rest?

23. Accelerating an Electron An electron has a constant acceleration of $+3.2$ m/s$^2\hat{\text{i}}$. At a certain instant its velocity is $+9.6$ m/s$\hat{\text{i}}$. What is its velocity (a) 2.5 s earlier and (b) 2.5 s later?

24. Speeding Bullet The speed of a bullet is measured to be 640 m/s as the bullet emerges from a barrel of length 1.20 m. Assuming constant acceleration, find the time that the bullet spends in the barrel after it is fired.

25. Comfortable Acceleration Suppose a rocket ship in deep space moves with constant acceleration equal to 9.8 m/s^2, which gives the illusion of normal gravity during the flight. (a) If it starts from rest, how long will it take to acquire a speed one-tenth that of light, which travels at 3.0×10^8 m/s? (b) How far will it travel in so doing?

26. Taking Off A jumbo jet must reach a speed of 360 km/h on the runway for takeoff. What is the least constant acceleration needed for takeoff from a 1.80 km runway?

27. Even Faster Electrons An electron with initial velocity $v_1 = 1.50 \times 10^5$ m/s enters a region 1.0 cm long where it is electrically accelerated (Fig. 2-26). It emerges with velocity $v_2 = 5.70 \times 10^6$ m/s. What is its acceleration, assumed constant? (Such a process occurs in conventional television sets.)

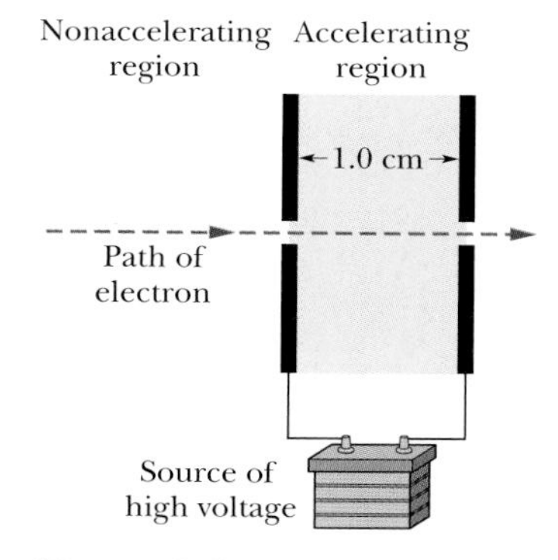

FIGURE 2-26 ■ Problem 27.

28. Stopping Col. Stapp A world's land speed record was set by Colonel John P. Stapp when in March 1954 he rode a rocket-propelled sled that moved along a track at 1020 km/h. He and the sled were brought to a stop in 1.4 s. (See Fig. 2-13) In g units, what acceleration did he experience while stopping?

29. Speed Trap The brakes on your automobile are capable of slowing down your car at a rate of 5.2 m/s^2. (a) If you are going 137 km/h and suddenly see a state trooper, what is the minimum time in which you can get your car under the 90 km/h speed limit? The answer reveals the futility of braking to keep your high speed from being detected with a radar or laser gun.) (b) Graph x vs. t and v vs. t for such a deceleration.

30. Judging Acceleration Figure 2-27 depicts the motion of a particle moving along an x axis with a constant acceleration. What are the magnitude and direction of the particle's acceleration?

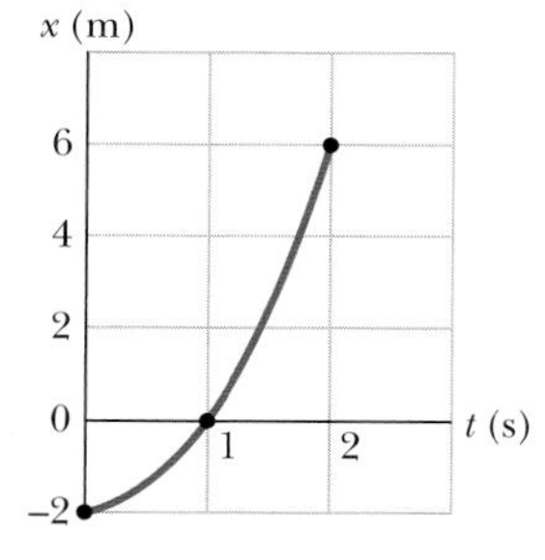

FIGURE 2-27 ■ Problem 30.

31. Hitting a Wall A car traveling 56.0 km/h is 24.0 m from a barrier when the driver slams on the brakes. The car hits the barrier 2.00 s later. (a) What is the car's constant acceleration before impact? (b) How fast is the car traveling at impact?

32. Red and Green Trains A red train traveling at 72 km/h and a green train traveling at 144 km/h are headed toward one another along a straight, level track. When they are 950 m apart, each engineer sees the other's train and applies the brakes. The brakes slow each train at the rate of 1.0 m/s^2. Is there a collision? If so, what is the speed of each train at impact? If not, what is the separation between the trains when they stop?

33. Between Two Points A car moving with constant acceleration covered the distance between two points 60.0 m apart in 6.00 s. Its speed as it passes the second point was 15.0 m/s. (a) What was the speed at the first point? (b) What was the acceleration? (c) At what prior distance from the first point was the car at rest? (d) Graph x vs. t and v vs. t for the car from rest ($t_1 = 0$ s).

34. Chasing a Truck At the instant the traffic light turns green, an automobile starts with a constant acceleration a of 2.2 m/s^2. At the same instant a truck, traveling with a constant speed of 9.5 m/s, overtakes and passes the automobile. (a) How far beyond the traffic signal will the automobile overtake the truck? (b) How fast will the car be traveling at that instant?

35. Reaction Time To stop a car, first you require a certain reaction time to begin braking; then the car slows under the constant braking. Suppose that the total distance moved by your car during these two phases is 56.7 m when its initial speed is 80.5 km/h, and 24.4 m when its initial speed is 48.3 km/h. What are (a) your reaction time and (b) the magnitude of the braking acceleration?

36. Avoiding a Collision When a high-speed passenger train traveling at 161 km/h rounds a bend, the engineer is shocked to see that a locomotive has improperly entered the track from a siding and is a distance $D = 676$ m ahead (Fig. 2-28). The locomotive is moving at 29.0 km/h. The engineer of the high-speed train immediately applies the brakes. (a) What must be the magnitude of the resulting constant acceleration if a collision is to be just avoided? (b) Assume that the engineer is at $x = 0$ when, at $t = 0$, he first spots the locomotive. Sketch the $x(t)$ curves representing the locomotive and, high-speed train for the situations in which a collision is just avoided and is not quite avoided.

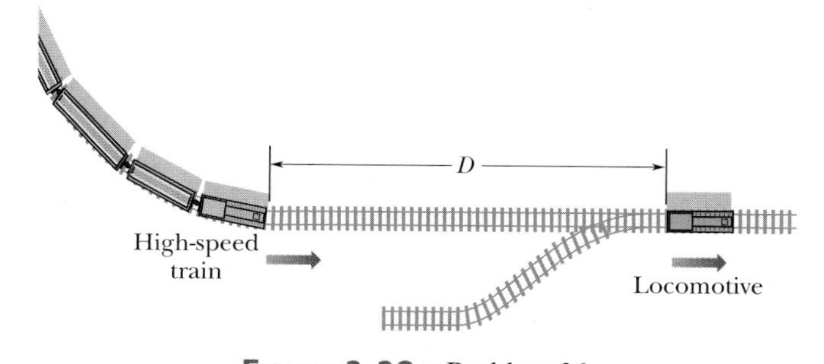

FIGURE 2-28 ■ Problem 36.

37. Going Up An elevator cab in the New York Marquis Marriott has a total run of 190 m. Its maximum speed is 305 m/min. Its acceleration (both speeding up and slowing) has a magnitude of 1.22 m/s^2. (a) How far does the cab move while accelerating to full speed from rest? (b) How long does it take to make the nonstop 190 m run, starting and ending at rest?

38. Shuffleboard Disk A shuffleboard disk is accelerated at a constant rate from rest to a speed of 6.0 m/s over a 1.8 m distance by a player using a cue. At this point the disk loses contact with the cue and slows at a constant rate of 2.5 m/s^2 until it stops. (a) How much time elapses from when the disk begins to accelerate until it stops? (b) What total distance does the disk travel?

39. Electric Vehicle An electric vehicle starts from rest and accelerates at a rate of 2.0 m/s^2 in a straight line until it reaches a speed of 20 m/s. The vehicle then slows at a constant rate of 1.0 m/s^2 until it stops. (a) How much time elapses from start to stop? (b) How far does the vehicle travel from start to stop?

40. Red Car–Green Car In Fig. 2-29 a red car and a green car, identical except for the color, move toward each other in adjacent lanes and parallel to an x axis. At time $t_1 = 0$ s, the red car is at $x_r = 0$ m and the green car is at $x_g = 220$ m. If the red car has a constant velocity of 20 km/h, the cars pass each other at $x = 44.5$ m, and if it has a constant velocity of 40 km/h, they pass each other at $x = 76.6$ m. If the green car has a constant acceleration, what are (a) its initial velocity and (b) its acceleration?

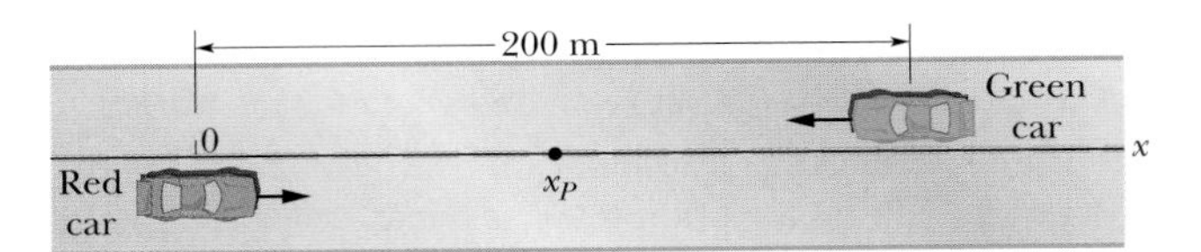

FIGURE 2-29 ■ Problem 40.

41. Position Function The position of a particle moving along an x axis is given by $x = (12 \text{ m/s}^2)t^2 - (2 \text{ m/s}^3)t^3$, where x is in meters and t is in seconds. (a) Determine the position, velocity, and acceleration of the particle at $t_3 = 3.0$ s. (b) What is the maximum positive coordinate reached by the particle and at what time is it reached? (c) What is the maximum positive velocity reached by the particle and at what time is it reached? (d) What is the acceleration of the particle at the instant the particle is not moving (other than at $t_0 = 0$)? (e) Determine the average velocity of the particle between $t_0 = 0$ and $t_3 = 3$ s.

Additional Problems

42. Kids in the Back! An unrestrained child is playing on the front seat of a car that is traveling in a residential neighborhood at 35 km/h. (How many mi/h is this? Is this car going too fast?) A small dog runs across the road and the driver applies the brakes, stopping the car quickly and missing the dog. Estimate the speed with which the child strikes the dashboard, presuming that the car stops before the child does so. Compare this speed with that of the world-record 100 m dash, which is run in about 10 s.

43. The Passat GLX Test results (*Car & Driver*, February 1993, p. 48) on a Volkswagen Passat GLX show that when the brakes are fully applied it has an average braking acceleration of *magnitude* 8.9 m/s^2. If a preoccupied driver who is moving at a speed of 42 mph looks up suddenly and sees a stop light 30 m in front of him, will he have sufficient time to stop? The weight of the Volkswagen is 3 152 lb.

44. Velocity and Pace When we drive a car we usually describe our motion in terms of speed or velocity. A speed limit, such as 60 mi/h, is a speed. When runners or joggers describe their motion, they often do so in terms of a *pace*—how long it takes to go a given distance. A 4-min mile (or better, "4 minutes/mile") is an example of a pace.

(a) Express the speed 60 mi/h as a pace in min/mi.
(b) I walk on my treadmill at a pace of 17 min/mi. What is my speed in mi/h?
(c) If I travel at a speed, v, given in mi/h, what is my pace, p, given in min/mi? (Write an equation that would permit easy conversion.)

45. Spirit of America The 9000 lb Spirit of America (designed to be the world's fastest car) accelerated from rest to a final velocity of 756 mph in a time of 45 s. What would the acceleration have been in meters per second? What distance would the driver, Craig Breedlove, have covered?

46. Driving to New York You and a friend decide to drive to New York from College Park, Maryland (near Washington, D.C.) on Saturday over the Thanksgiving break to go to a concert with some friends who live there. You figure you have to reach the vicinity of the city at 5 P.M. in order to meet your friends in time for dinner before the concert. It's about 220 mi from the entrance to Route 95 to the vicinity of New York City. You would like to get on the highway about noon and stop for a bite to eat along the way. What does your average velocity have to be? If you keep an approximately constant speed (not a realistic assumption!), what should your speedometer read while you are driving?

47. NASA Internship You are working as a student intern for the National Aeronautics and Space Administration (NASA) and your supervisor wants you to perform an indirect calculation of the upward velocity of the space shuttle relative to the Earth's surface just 5.5 s after it is launched when it has an altitude of 100 m. In order to obtain data, one of the engineers has wired a streamlined flare to the side of the shuttle that is gently released by remote control after 5.5 s. If the flare hits the ground 8.5 s after it is released, what is the upward velocity of the flare (and hence of the shuttle) at the time of its release? (Neglect any effects of air resistance on the flare.) *Note*: Although the flare idea is fictional, the data on a typical shuttle altitude and velocity at 5.5 s are straight from NASA!

48. Cell Phone Fight You are arguing over a cell phone while trailing an unmarked police car by 25 m; both your car and the police car are traveling at 110 km/h. Your argument diverts your attention from the police car for 2.0 s (long enough for you to look at the phone and yell, "I won't do that!"). At the beginning of that 2.0 s, the police officer begins emergency braking at 5.0 m/s^2. (a) What is the separation between the two cars when your attention finally returns? Suppose that you take another 0.40 s to realize your danger and begin braking. (b) If you too brake at 5.0 m/s^2, what is your speed when you hit the police car?

49. Reaction Distance When a driver brings a car to a stop by braking as hard as possible, the stopping distance can be regarded as the sum of a "reaction distance," which is initial speed multiplied by the driver's reaction time, and a "braking distance," which is the distance traveled during braking. The following table gives typical values. (a) What reaction time is the driver assumed to have? (b) What is the car's stopping distance if the initial speed is 25 m/s?

Initial Speed (m/s)	Reaction Distance (m)	Braking Distance (m)	Stopping Distance (m)
10	7.5	5.0	12.5
20	15	20	35
30	22.5	45	67.5

50. Tailgating In this problem we analyze the phenomenon of "tailgating" in a car on a highway at high speeds. This means traveling too close behind the car ahead of you. Tailgating leads to multiple car crashes when one of the cars in a line suddenly slows down. The question we want to answer is: "How close is too close?"

To answer this question, let's suppose you are driving on the highway at a speed of 100 km/h (a bit more than 60 mi/h). The driver ahead of you suddenly puts on his brakes. We need to calculate a number of things: how long it takes you to respond; how far you travel in that time, and how far the other car travels in that time.

(a) First let's estimate how long it takes you to respond. Two times are involved: how long it takes from the time you notice something happening till you start to move to the brake, and how long it takes to move your foot to the brake. You will need a ruler to do this. Take the ruler and have a friend hold it from the one end hanging straight down. Place your thumb and forefinger opposite the bottom of the ruler. As your friend releases the ruler suddenly, try to catch it with your thumb and forefinger. Measure how far it falls before you catch it. Do this three times and take the average distance. Assuming the ruler is falling freely without air resistance (not a bad assumption), calculate how much time it takes you to catch it, t_1. Now estimate the time, t_2, it takes you to move your foot from the gas pedal to the brake pedal. Your reaction time is $t_1 + t_2$.

(b) If you brake hard and fast, you can bring a typical car to rest from 100 km/h (about 60 mi/h) in 5 seconds.

1. Calculate your acceleration, $-a_0$, assuming that it is constant.
2. Suppose the driver ahead of you begins to brake with an acceleration $-a_0$. How far will he travel before he comes to a stop? (*Hint*: How much time will it take him to stop? What will be his average velocity over this time interval?)

(c) Now we can put these results together into a fairly realistic situation. You are driving on the highway at 100 km/hr and there is a driver in front of you going at the same speed.

1. You see him start to slow immediately (an unreasonable but simplifying assumption). If you are also traveling 100 km/h, how far (in meters) do you travel before you begin to brake? If you can also produce the acceleration $-a_0$ when you brake, what will be the total distance you travel before you come to a stop?
2. If you don't notice the driver ahead of you beginning to brake for 1 s, how much additional distance will you travel?
3. Discuss, on the basis of these calculations, what you think is a safe distance to stay behind a car at 60 mi/h. Express your distance in "car lengths" (about 15 ft). Would you include a safety factor beyond what you have calculated here? How much?

51. Testing the Motion Detector A motion detector that may be used in physics laboratories is shown in Fig. 2-30. It measures the distance to the nearest object by using a speaker and a microphone. The speaker clicks 30 times a second. The microphone detects the sound bouncing back from the nearest object in front of it. The computer calculates the time delay between making the sound and receiving the echo. It knows the speed of sound (about 343 m/s at room temperature), and from that it can calculate the distance to the object from the time delay.

FIGURE 2-30 ■ Problem 51.

(a) If the nearest object in front of the detector is too far away, the echo will not get back before a second click is emitted. Once that happens, the computer has no way of knowing that the echo isn't an echo from the second click and that the detector isn't giving correct results any more. How far away does the object have to be before that happens?

(b) The speed of sound changes a little bit with temperature. Let's try to get an idea of how important this is. At room temperature (72 °F) the speed of sound is about 343 m/s. At 62 °F it is about 1% smaller. Suppose we are measuring an object that is really 1.5 meters away at 72 °F. What is the time delay Δt that the computer detects before the echo returns? Now suppose the temperature is 62 °F. If the computer detects a time delay of Δt but (because it doesn't know the temperature) calculates the distance using the speed of sound appropriate for 72 °F, how far away does the computer report the object to be?

52. Hitting a Bowling Ball A bowling ball sits on a hard floor at a point that we take to be the origin. The ball is hit some number of times by a hammer. The ball moves along a line back and forth across the floor as a result of the hits. (See Fig. 2-31.) The region to the right of the origin is taken to be positive, but during its motion the ball is at times on both sides of the origin. After the ball has been moving for a while, a motion detector like the one discussed in Problem 51 is started and takes the following graph of the ball's velocity.

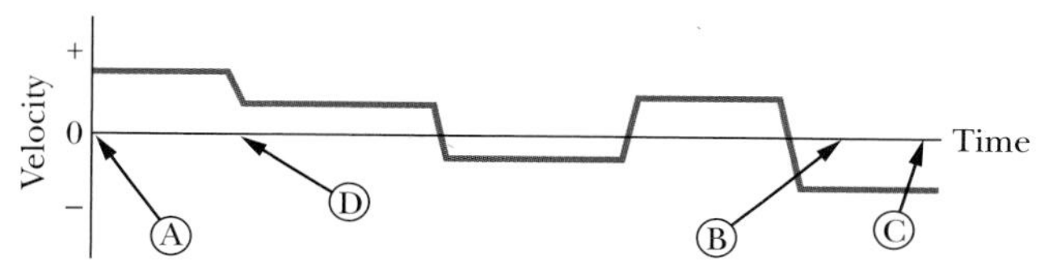

FIGURE 2-31 ■ Problem 52.

Answer the following questions with the symbols L (left), R (right), N (neither), or C (can't say which). Each question refers only to the time interval displayed by the computer.

(a) At which side of the origin is the ball for the time marked A?
(b) At the time marked B, in which direction is the ball moving?
(c) Between the times A and C, what is the direction of the ball's displacement?
(d) The ball receives a hit at the time marked D. In what direction is the ball moving after that hit?

53. Waking the Balrog In *The Fellowship of the Ring*, the hobbit Peregrine Took (Pippin for short) drops a rock into a well while the travelers are in the caves of Moria. This wakes a balrog (a bad thing) and causes all kinds of trouble. Pippin hears the rock hit the water 7.5 s after he drops it.

(a) Ignoring the time it takes the sound to get back up, how deep is the well?
(b) It is quite cool in the caves of Moria, and the speed of sound in air changes with temperature. Take the speed of sound to be 340 m/s (it is pretty cool in that part of Moria). Was it OK to ignore the time it takes sound to get back up? Discuss and support your answer with a calculation.

54. Two Balls, Passing in the Night* Figure 2-32 represents the position vs. clock reading of the motion of two balls, *A* and *B*,

*From A. Arons, *A Guide to Introductory Physics Teaching* (New York: John Wiley, 1990).

moving on parallel tracks. Carefully sketch the figure on your homework paper and answer the following questions:

(a) Along the t axis, mark with the symbol t_A any instant or instants at which one ball is passing the other.
(b) Which ball is moving faster at clock reading t_B?
(c) Mark with the symbol t_C any instant or instants at which the balls have the same velocity.
(d) Over the period of time shown in the diagram, which of the following is true of ball B? Explain your answer.

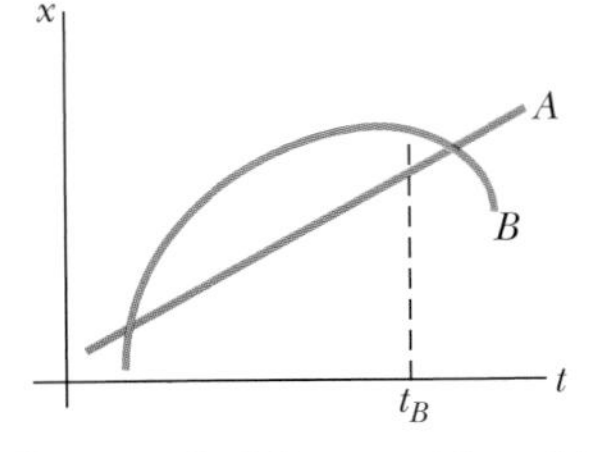

FIGURE 2-32 ■ Problem 54.

1. It is speeding up all the time.
2. It is slowing down all the time.
3. It is speeding up part of the time and slowing down part of the time.

55. Graph for a Cart on a Tilted Airtrack—with Spring The graph in Fig. 2-33 below shows the velocity graph of a cart moving on an air track. The track has a spring at one end and has its other end raised. The cart is started sliding up the track by pressing it against the spring and releasing it. The clock is started just as the cart leaves the spring. Take the direction the cart is moving in initially to be the positive x direction and take the bottom of the spring to be the origin.

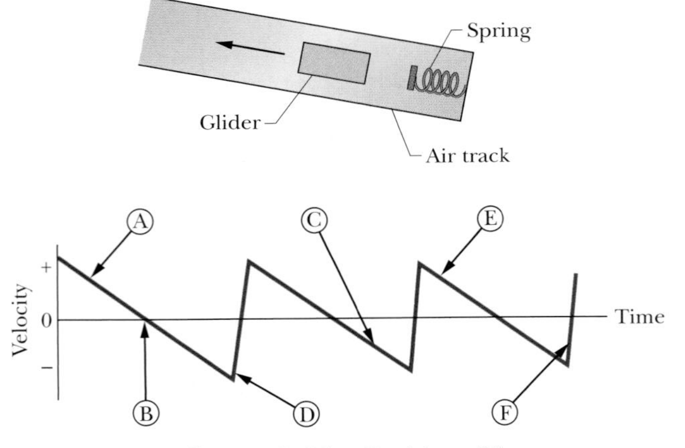

FIGURE 2-33 ■ Problem 55

Letters point to six points on the velocity curve. For the physical situations described below, identify which of the letters corresponds to the situation described. You may use each letter more than once, more than one letter may be used for each answer, or none may be appropriate. If none is appropriate, use the letter N.

(a) This point occurs when the cart is at its highest point on the track.
(b) At this point, the cart is instantaneously not moving.
(c) This is a point when the cart is in contact with the spring.
(d) At this point, the cart is moving down the track toward the origin.
(e) At this point, the cart has acceleration of zero.

56. Rolling Up and Down A ball is launched up a ramp by a spring as shown in Fig. 2-34. At the time when the clock starts, the ball is near the bottom of the ramp and is rolling up the ramp as shown. It goes to the top and then rolls back down. For the graphs shown in Fig. 2-34, the horizontal axis represents the time. The vertical axis is unspecified.

For each of the following quantities, select the letter of the graph that could provide a correct graph of the quantity for the ball in the situation shown (if the vertical axis were assigned the proper units). Use the x and y coordinates shown in the picture. If none of the graphs could work, write N.

(a) The x-component of the ball's position ________
(b) The y-component of the ball's velocity ________
(c) The x-component of the ball's acceleration ________
(d) The y-component of the normal force the ramp exerts on the ball ________
(e) The x-component of the ball's velocity ________
(f) The x-component of the force of gravity acting on the ball ________

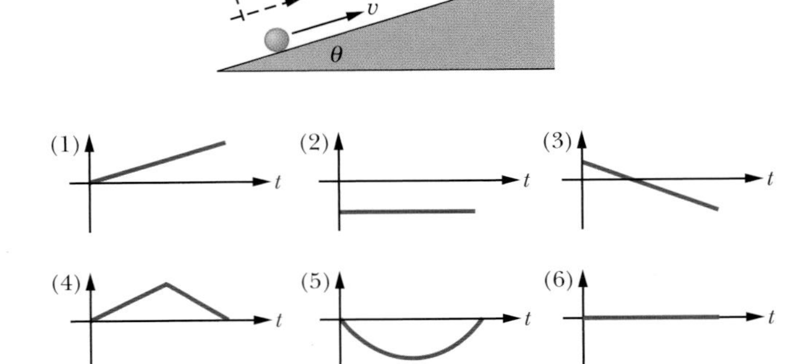

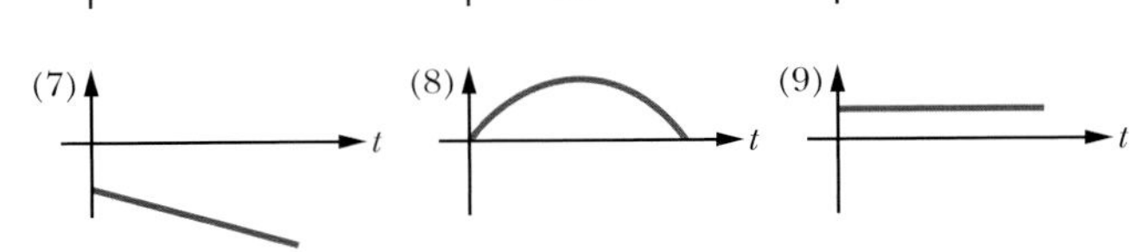

FIGURE 2-34 ■ Problem 56

57. Model Rocket A model rocket, propelled by burning fuel, takes off vertically. Plot qualitatively (numbers not required) graphs of y, v, and a versus t for the rocket's flight. Indicate when the fuel is exhausted, when the rocket reaches maximum height, and when it returns to the ground.

58. Rock Climber At time $t = 0$, a rock climber accidentally allows a piton to fall freely from a high point on the rock wall to the valley below him. Then, after a short delay, his climbing partner, who is 10 m higher on the wall, throws a piton downward. The positions y of the pitons versus t during the fall are given in Fig. 2-35. With what speed was the second piton thrown?

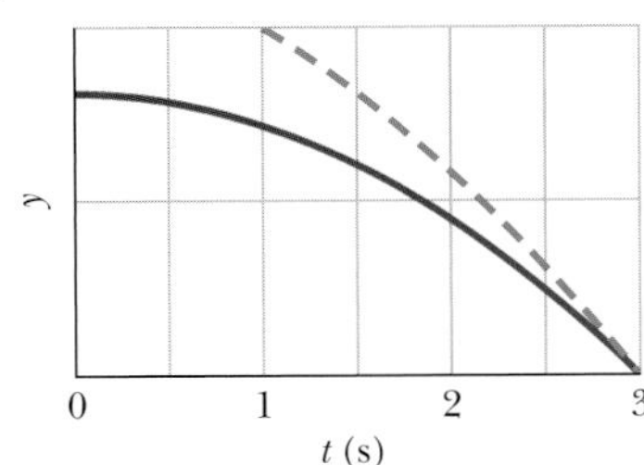

FIGURE 2-35 ■ Problem 58.

59. Two Trains As two trains move along a track, their conductors suddenly notice that they are headed toward each other. Figure 2-36 gives their velocities v as functions of time t as the conductors slow the trains.

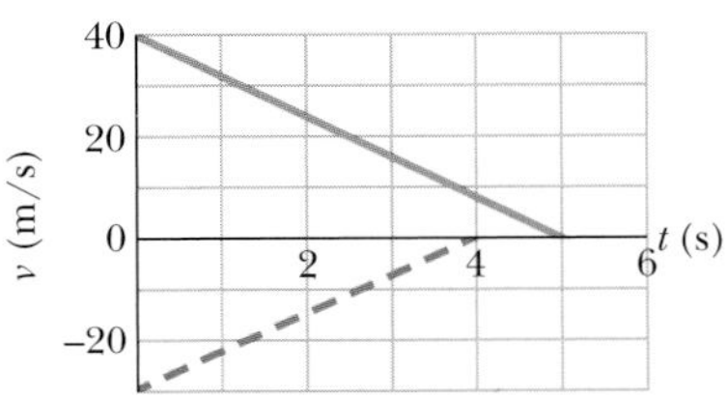

FIGURE 2-36 ■ Problem 59.

The slowing processes begin when the trains are 200 m apart. What is their separation when both trains have stopped?

60. Runaway Balloon As a runaway scientific balloon ascends at 19.6 m/s, one of its instrument packages breaks free of a harness and free-falls. Figure 2-37 gives the vertical velocity of the package versus time, from before it breaks free to when it reaches the ground. (a) What maximum height above the break-free point does it rise? (b) How high was the break-free point above the ground?

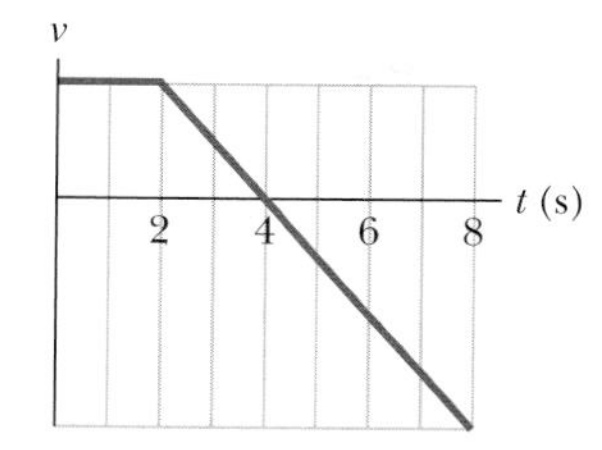

FIGURE 2-37 ■ Problem 60.

61. Position Function Two A particle moves along the x axis with position function $x(t)$ as shown in Fig. 2-38. Make rough sketches of the particle's velocity versus time and its acceleration versus time for this motion.

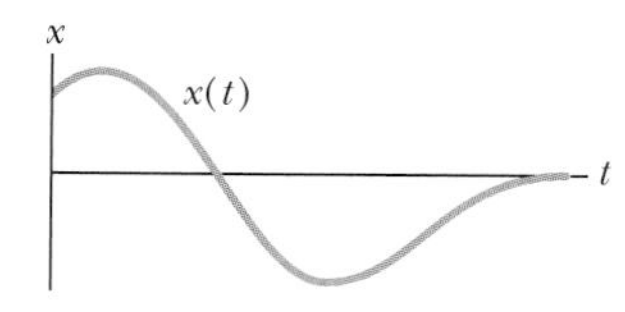

FIGURE 2-38 ■ Problem 61.

62. Velocity Curve Figure 2-39 gives the velocity v(m/s) versus time t (s) for a particle moving along an x axis. The area between the time axis and the plotted curve is given for the two portions of the graph. At $t = t_A$ (at one of the crossing points in the plotted figure), the particle's position is $x = 14$ m. What is its position at (a) $t = 0$ and (b) $t = t_B$?

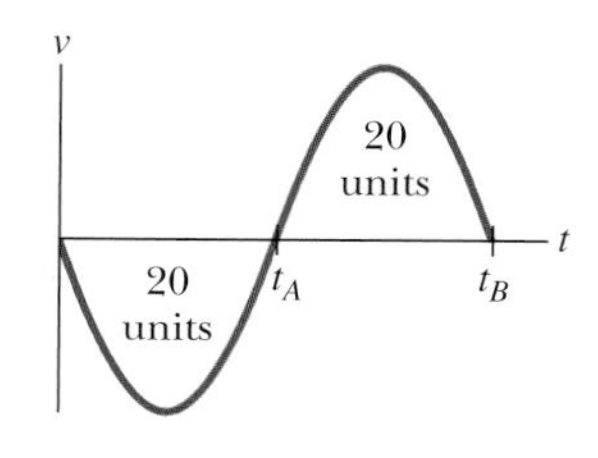

FIGURE 2-39 ■ Problem 62.

63. The Motion Detector Rag This assignment is based on the Physics Pholk Song CD distributed by Pasco scientific. The words to these songs are also available through the Dickinson College Web site at http://physics.dickinson.edu.

(a) Refer to the motion described in the first verse of the *Motion Detector Rag*; namely, you are moving for the same amount of time that you are standing. Sketch a position vs. time graph for this motion. Also, describe the shape of the graph in words.
(b) Refer to the motion described in the second verse of the *Motion Detector Rag*. In this verse, you are making a "steep down-slope," then a "gentle up-slope," and last a flat line. You spend the same amount of time engaged in each of these actions. Sketch a position vs. time graph of this motion. Also, describe what you are doing in words. That is, are you standing still, moving away from the origin (or motion detector), moving toward the origin (or motion detector)? Which motion is the most rapid, and so on?
(c) Refer to the motion described in the third verse of the *Motion Detector Rag*. You start from rest and move away from the motion detector at an acceleration of +1.0 m/s^2 for 5 seconds. Sketch the acceleration vs. time graph to this motion. Sketch the corresponding velocity vs. time graph. Sketch the shape of the corresponding position vs. time graph.

64. Hockey Puck At time $t = 0$, a hockey puck is sent sliding over a frozen lake, directly into a strong wind. Figure 2-40 gives the velocity v of the puck vs. time, as the puck moves along a single axis. At $t = 14$ s, what is its position relative to its position at $t = 0$?

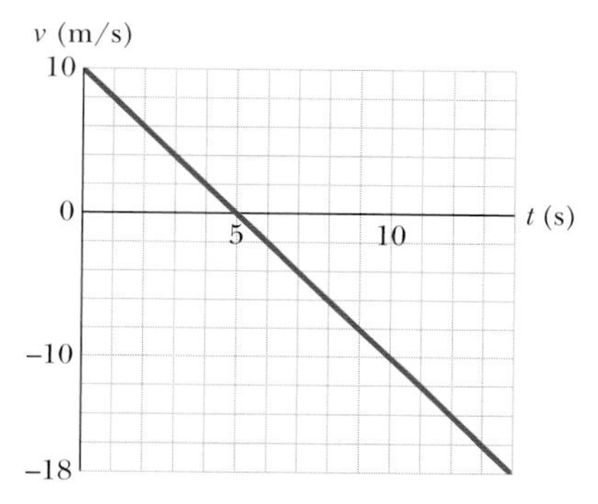

FIGURE 2-40 ■ Problem 64.

65. Describing One-Dimensional Velocity Changes In each of the following situations you will be asked to refer to the mathematical definitions and the concepts associated with the *number line*. Note that being more positive is the same as being less negative, and so on.

(a) Suppose an object undergoes a *change* in velocity from +1 m/s to +4 m/s. Is its velocity becoming *more positive* or *less positive*? What is meant by more positive? Less positive? Is the acceleration positive or negative?
(b) Suppose an object undergoes a *change* in velocity from −4 m/s to −1 m/s. Is its velocity becoming *more positive* or *less positive*? What is meant by more positive? Less positive? Is the acceleration positive or negative?
(c) Suppose an object is turning around so that it undergoes a *change* in velocity from −2 m/s to +2 m/s. Is its velocity becoming *more positive* or *less positive* than it was before? What is meant by more positive? Less positive? Is it undergoing an acceleration while it is turning around? Is the acceleration positive or negative?
(d) Another object is turning around so that it undergoes a *change* in velocity from +1 m/s to −1 m/s. Is its velocity becoming *more positive* or *less positive* than it was before? What is meant by more positive? Less positive? Is it undergoing an acceleration while it is turning around? Is the acceleration positive or negative?

66. Bowling Ball Graph A bowling ball was set into motion on a fairly smooth level surface, and data were collected for the total distance covered by the ball at each of four times. These data are shown in the table.

Average Time (s)	Distance (m)
0.00	0.0
0.92	2.0
1.85	4.0
2.87	6.0

(a) Plot the data points on a graph.
(b) Use a ruler to draw a straight line that passes as close as possible to the data points you have graphed.
(c) Using methods you were taught in algebra, calculate the value of the slope, m, and find the value of the intercept, b, of the line you have sketched through the data.

67. Modeling Bowling Ball Motion A bowling ball is set into motion on a smooth level surface, and data were collected for the total distance covered by the ball at each of four times. These data are shown in the table in Problem 66. Your job is to learn to use a spreadsheet program — for example, Microsoft Excel—to create a mathematical model of the bowling ball motion data shown. You are to find what you think is the best value for the slope, m, and the y-intercept, b. Practicing with a tutorial worksheet entitled MODTUT.XLS will help you to learn about the process of modeling for a linear relationship. Ask your instructor where to find this tutorial worksheet.

After using the tutorial, you can create a model for the bowling ball data given above. To do this:

(a) Open a new worksheet and enter a title for your bowling ball graph.
(b) Set the y-label to Distance (m) and the x-label to Time (s).
(c) Refer to the data table above. Enter the measured times for the bowling ball in the Time (s) column (formerly x-label).
(d) Set the y-exp column to D-data (m) and enter the measured distances for the bowling ball (probably something like 0.00 m, 2.00 m, 4.00 m, and 6.00 m.).
(e) Place the symbol m (for slope) in the cell B1. Place the symbol b (for y-intercept) in cell B2.
(f) Set the y-theory column to D-model (m) and then put the appropriate equation for a straight line of the form Distance = m*Time + b in cells C7 through C12. Be sure to refer to cells C1 for slope and C2 for y-intercept as absolutes; that is, use C1 and C2 when referring to them.
(g) Use the spreadsheet graphing feature to create a graph of the data in the D-exp and D-theory columns as a function of the data in the Time column.
(h) Change the values in cells Cl and C2 until your theoretical line matches as closely as possible your red experimental data points in the graph window.
(i) Discuss the meaning of the slope of a graph of distance vs. time. What does it tell you about the motion of the bowling ball?

68. A Strange Motion After doing a number of the exercises with carts and fans on ramps, it is easy to draw the conclusion that everything that moves is moving at either a constant velocity or a constant acceleration. Let's examine the horizontal motion of a triangular frame with a pendulum at its center that has been given a push. It undergoes an unusual motion. You should determine whether or not it is moving at either a constant velocity or constant acceleration. (*Note:* You may want to look at the motion of the triangular frame by viewing the digital movie entitled PASCO070. This movie is included on the VideoPoint compact disk. If you are not using VideoPoint, your instructor may make the movie available to you some other way.)

The images in Fig. 2-41 are taken from the 7th, 16th, and 25th frames of that movie.

Data for the position of the center of the horizontal bar of the triangle were taken every tenth of a second during its first second of motion. The origin was placed at the zero centimeter mark of a fixed meter stick. These data are in the table below.
(a) Examine the position vs. time graph of the data shown above. Does the triangle appear to have a constant velocity throughout the first second? A constant acceleration? Why or why not?
(b) Discuss the nature of the motion based on the shape of the graph. At approximately what time, if any, is the triangle changing direction? At approximately what time does it have the greatest negative velocity? The greatest positive velocity? Explain the reasons for your answers.
(c) Use the data table and the definition of average velocity to calculate the average velocity of the triangle at each of the times between 0.100 s and 0.900 s. In this case you should use the position just before the indicated time and the position just after the indicated time in your calculation. For example, to calculate the average velocity at $t_2 = 0.100$ seconds, use $x_3 = 44.5$ cm and $x_1 = 52.1$ cm along with the differences of the times at t_3 and t_1. *Hint:* Use only times and positions in the gray boxes to get a velocity in a gray box and use only times and positions in the white boxes to get a velocity in a white box.
(d) Since people usually refer to velocity as distance divided by time, maybe we can calculate the average velocities as simply x_1/t_1, x_2/t_2, x_3/t_3, and so on. This would be easier. Is this an equivalent method for

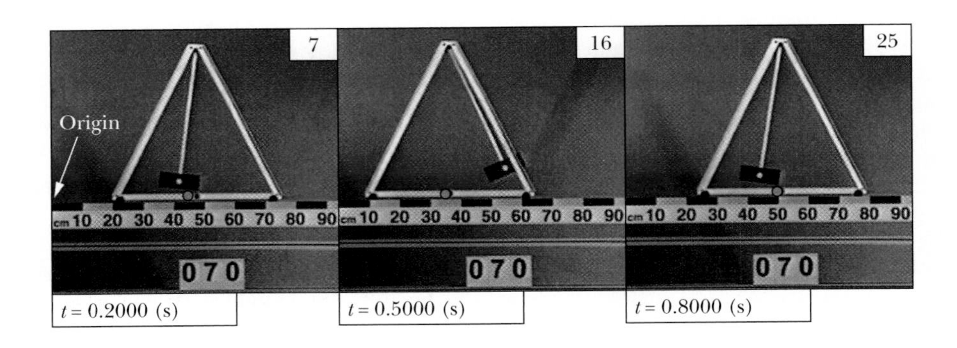

Fr#	Pr#	t(s)	x(cm)	$<v_x>$(cm/s)
1	1	0.000	52.1	no entry
4	2	0.100	49.9	−38.0
7	3	0.200	44.5	
10	4	0.300	39.1	
13	5	0.400	35.2	
16	6	0.500	34.8	
19	7	0.600	36.9	
22	8	0.700	43.0	
25	9	0.800	49.2	
28	10	0.900	53.6	
31	11	1.000	54.4	no entry

Triangle Motion
x(cm): 0, 10, 20, 30, 40, 50, 60
t(s): 0.000, 0.500, 1.000

FIGURE 2-41 ■ Problem 68.

finding the velocities at the different times? Try using this method of calculation if you are not sure. Give reasons for your answer.
(e) Often, when an oddly shaped but smooth graph is obtained from data it is possible to fit a polynomial to it. For example, a fourth-order polynomial that fits the data is

$$x = \{(-376 \text{ cm/s}^4)t^4 + (719 \text{ cm/s}^3)t^3 - (347 \text{ cm/s}^2)t^2 + (5.63 \text{ cm/s})t + 52.1 \text{ cm}\}$$

Using this polynomial approximation, find the *instantaneous* velocity at $t = 0.700$ s. Comment on how your answer compares to the average velocity you calculated at 0.700 s. Are the two values close? Is that what you expect?

69. Cedar Point At the Cedar Point Amusement Park in Ohio, a cage containing people is moving at a high initial velocity as the result of a previous free fall. It changes direction on a curved track and then coasts in a horizontal direction until the brakes are applied. This situation is depicted in a digital movie entitled DSON002. (*Note*: This movie is included on the VideoPoint compact disk. If you are not using VideoPoint, your instructor may make the movie available to you some other way.)

(a) Use video analysis software to gather data for the horizontal positions of the tail of the cage in meters as a function of time. Don't forget to use the scale on the title screen of the movie so your results are in meters rather than pixels. Summarize this data in a table or in a printout attached to your homework.
(b) Transfer your data to a spreadsheet and do a parabolic model to show that within 5% or better $x = (-7.5 \text{ m/s}^2)t^2 + (22.5 \text{ m/s})t + 2.38$ m. Please attach a printout of this model and graph with your name on it to your submission as "proof of completion."(*Note*: Your judgments about the location of the cage tail may lead to slightly different results.)
(c) Use the equation you found along with its interpretation as embodied in the first kinematic equation to determine the horizontal acceleration, a, of the cage as it slows down. What is its initial horizontal velocity, v_1, at time $t = 0$ s? What is the initial position, x_1, of the cage?
(d) The movie ends before the cage comes to a complete stop. Use your knowledge of a, v_1, and x_1 along with kinematic equations to determine the horizontal position of the cage when it comes to a *complete* stop so that the final velocity of the cage is given by $v_2 = v = 0.00$ m/s.

70. Three Digital Movies Three digital movies depicting the motions of four single objects have been selected for you to examine using a video-analysis program. They are as follows:

PASCO004: A cart moves on an upper track while another moves on a track just below.

PASCO153: A metal ball attached to a string swings gently.

HRSY003: A boat with people moves in a water trough at Hershey Amusement Park.

Please examine the horizontal motion of each object carefully by viewing the digital movies. In other words, just examine the motion in the x direction (and ignore any slight motions in the y direction). You may use LoggerPro 3, VideoPoint, VideoGraph, or World-in-Motion digital analysis software and a spreadsheet to analyze the motion in more detail if needed. Based on what you have learned so far, there is more than one analysis method that can be used to answer the questions that follow. *Note:* Since we are interested only in the nature of these motions (not exact values) you do not need to scale any of the movies. Working in pixel units is fine.

(a) Which of these four objects (upper cart, lower cart, metal ball, or boat), if any, move at a constant horizontal velocity? Cite the evidence for your conclusions.
(b) Which of these four objects, if any, move at a constant horizontal acceleration? Cite the evidence for your conclusions.
(c) Which of these four objects, if any, move at *neither* a constant horizontal velocity nor acceleration? Cite the evidence for your conclusions.
(d) The kinematic equations are very useful for describing motions. Which of the four motions, if any, cannot be described using the kinematic equations? Explain the reasons for your answer.

71. Speeding Up or Slowing Down Figure 2-42 shows the velocity vs. time graph for an object constrained to move in one dimension. The positive direction is to the right.

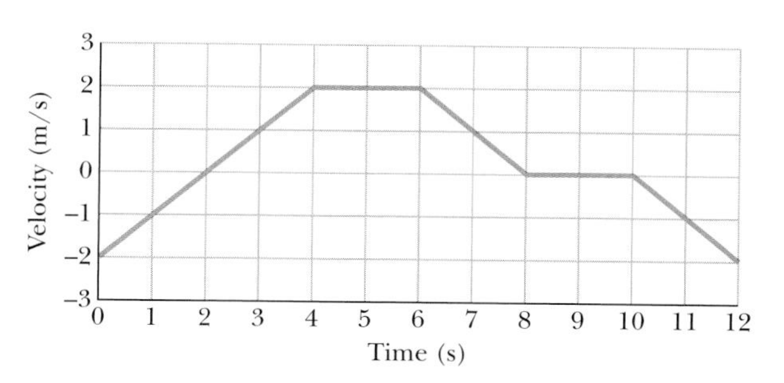

FIGURE 2-42 ■ Problems 71–74.

(a) At what times, or during what time periods, is the object speeding up?
(b) At what times, or during what time periods, is the object slowing down?
(c) At what times, or during what time periods, does the object have a constant velocity?
(d) At what times, or during what time periods, is the object at rest?

If there is no time or time period for which a given condition exists, state that explicitly.

72. Right or Left Figure 2-42 shows the velocity vs. time graph for an object constrained to move along a line. The positive direction is to the right.

(a) At what times, or during what time periods, is the object speeding up and moving to the right?
(b) At what times, or during what time periods, is the object slowing down and moving to the right?
(c) At what times, or during what time periods, does the object have a constant velocity to the right?
(d) At what times, or during what time periods, is the object speeding up and moving to the left?
(e) At what times, or during what time periods, is the object slowing down and moving to the left?
(f) At what times, or during what time periods, does the object have a constant velocity to the left?

If there is no time or time period for which a given condition exists, state that explicitly.

73. Constant Acceleration Figure 2-42 shows the velocity vs. time graph for an object constrained to move along a line. The positive direction is to the right.

(a) At what times, or during what time periods, is the object's acceleration zero?

(b) At what times, or during what time periods, is the object's acceleration constant?
(c) At what times, or during what time periods, is the object's acceleration changing?

If there is no time or time period for which a given condition exists, state that explicitly.

74. Acceleration to the Right or Left Figure 2-42 shows the velocity vs. time graph for an object constrained to move along a line. The positive direction is to the right.

(a) At what times, or during what time periods, is the object's acceleration increasing and directed to the right?
(b) At what times, or during what time periods, is the object's acceleration decreasing and directed to the right?
(c) At what times, or during what time periods, does the object have a constant acceleration to the right?
(d) At what times, or during what time periods, is the object's acceleration increasing and directed to the left?
(e) At what times, or during what time periods, is the object's acceleration decreasing and directed to the left?
(d) At what times, or during what time periods, does the object have a constant acceleration to the left?

If there is no time or time period for which a given condition exists, state that explicitly.

3 Forces and Motion Along a Line

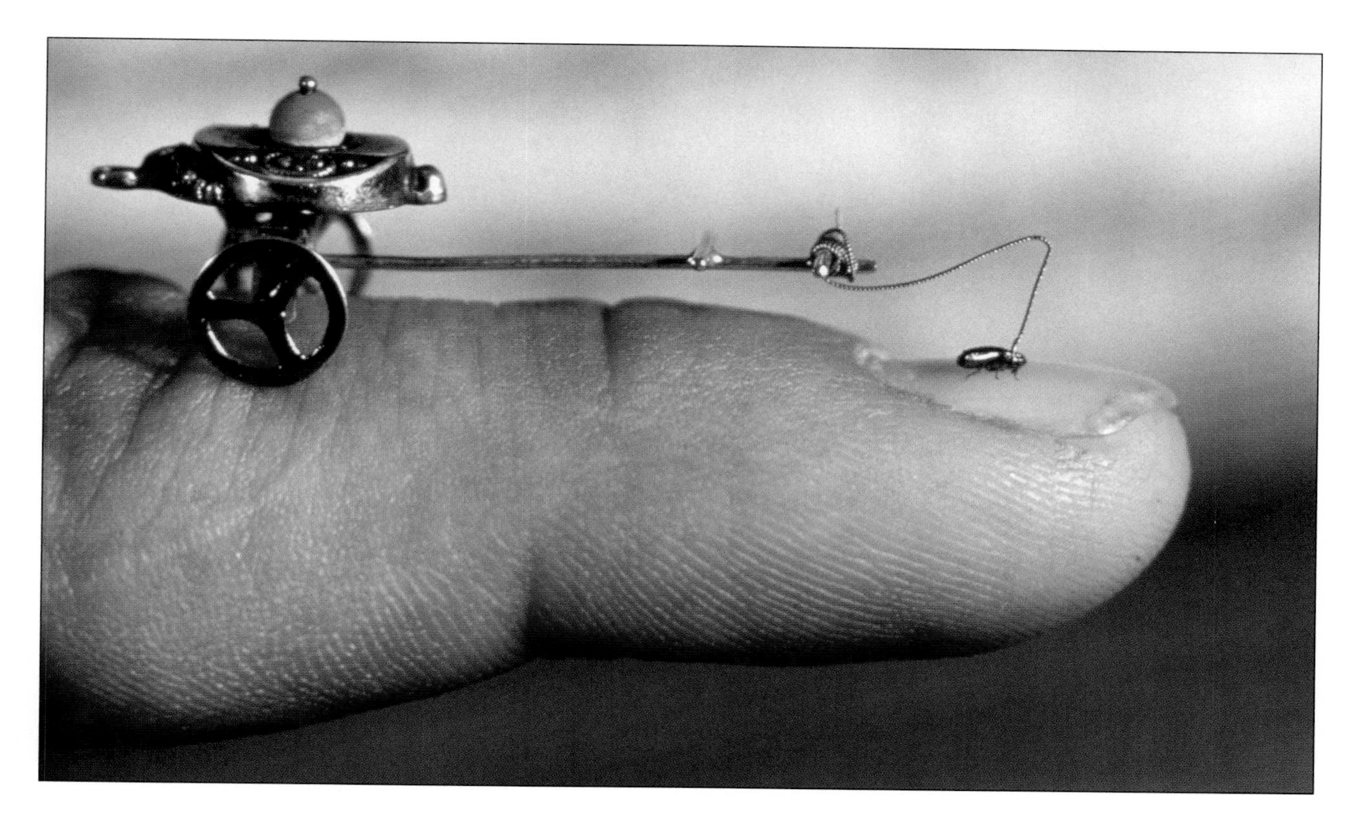

This photo shows a flea pulling a toy cart. In 1996 and 1997 Maria Fernanda Cardoso, a contemporary Colombian artist, created a circus of trained fleas and toured with them. Cardoso used a thin wire to attach Brutus, "the strongest flea on Earth," to a toy train car. She then used sound and carbon dioxide to induce Brutus to hop. Videos show that when Brutus hops, the train car jerks through a distance of about one centimeter. This is an amazing feat because the mass of the toy train car is 160,000 times greater than that of a flea.

How is it possible for a flea to pull 160 000 times its mass?

The answer is in this chapter.

FIGURE 3-1 ■ Isaac Newton (1642–1727) was the primary developer of the laws of classical mechanics.

3-1 What Causes Acceleration?

As part of our study of the kinematics of one-dimensional motion, we have introduced definitions of position, velocity, and acceleration. We have used these definitions to describe motion scientifically with graphs and equations. We now turn our attention to **dynamics**—the study of causes of motion. The central question in dynamics is: What causes a body to change its velocity or accelerate as it moves?

Everyday experience tells us that under certain circumstances an object can change its velocity when you interact with it with a push or pull of some sort. We call such a push or pull a **force.** For example, the velocity of a pitched baseball can suddenly change direction when a batter hits it, and a train can slow down when the engineer applies the brakes. However, at times an obvious interaction with an object does not cause a velocity change. Hitting or pushing on a massive object such as a brick wall does not cause it to move. To make matters more complex, many objects seem to undergo velocity changes even when no obvious interaction is present—a car rolls to a stop when you take your foot off the accelerator, and a falling object speeds up.

The laws of motion that relate external interactions between objects to their accelerations were first developed by Isaac Newton, pictured in Fig. 3-1. These laws lie at the heart of our modern interpretation of classical mechanics. Newton's laws are not absolute truths to be found in nature. Instead, they are part of a logically consistent conceptual framework that has emerged from the historical development of concepts, definitions, and measurement procedures.

Newton's laws have attained universal acceptance because they agree with countless observations made by scientists during the past 300 years. They have enabled us to learn about the fundamental nature of gravitational, electrical, and magnetic interactions. Engineers use the laws of motion and a knowledge of forces to predict precisely what motions will occur in the design of industrial-age devices such as engines, bridges, roadways, airplanes, and power plants.

In this chapter we begin our study of the causes of motion along a straight line. In chapters that follow we will extend this study to motions in two and eventually three dimensions.

3-2 Newton's First Law

In order to start thinking about what causes changes in an object's velocity, let's set up a thought experiment in which a small object sitting on a level surface is given a swift kick. How would you describe its motion in everyday language? Perhaps you might say something like, "The object speeds up quickly during the kick, but afterward, it begins to slow down as it slides or rolls along the surface, and eventually it comes to a stop." What caused the object to speed up (to change velocity) in the first place? The force of your kick did that. But after the kick is over, what caused the object to slow down? Before Newton's *Principia* was published in 1689, most scientists believed that the natural state of motion is rest and that a sliding object slows down and stops because there is no force to keep it moving.

Let's try to figure out whether this belief that a force is needed to keep an object in motion makes sense by looking at the outcome of an experiment. In the experiment, an object is given a kick and then its velocity is measured as a function of time as it slows down. In particular, the velocities of a plastic box and a small cart are measured as the object moves on different level surfaces—a rough carpet and a smooth track. In each case, the velocity of the slowing object is recorded by a motion detector attached to a microcomputer-based laboratory system. Figure 3-2 shows the experimental setup for two situations of interest—a cart rolling on a track and a plastic box sliding to a stop along a carpet.

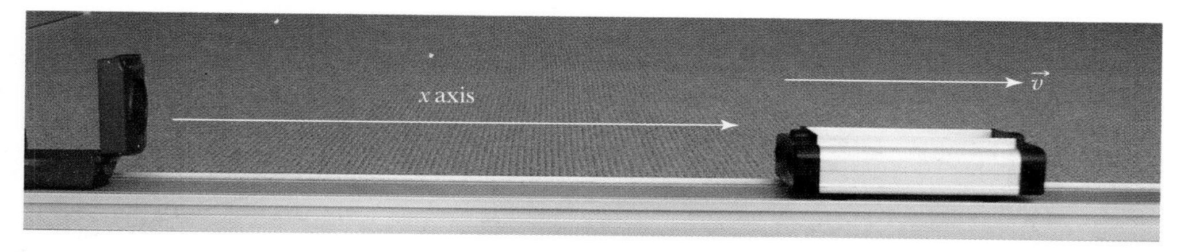

FIGURE 3-2 ■ Two objects are moving away from a motion detector. The cart on a level track is slowing down very little (top panel), and a plastic box sliding on a carpet is slowing to a stop much sooner than the cart on the track (bottom panel).

Figure 3-3 shows what happens to the x-components of velocity of the plastic box and cart in different situations. Each object is given roughly the same initial kick, but the objects slow down *differently*. The box sliding on the carpet comes to a stop in just over 0.2 s, but the cart rolling on the carpet takes 1.1 s to come to a full stop. Finally we see that the cart rolling on the smooth track still has 80% of its original speed. What enables the cart even after 1.2 s to move so much more freely on the track than the objects in the other situations?

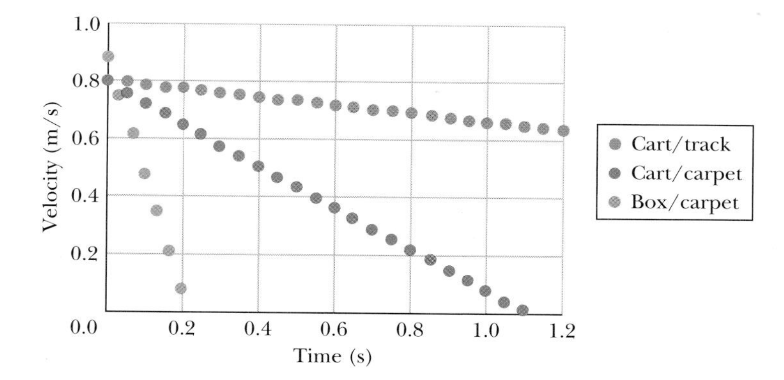

FIGURE 3-3 ■ An overlay graph of the x-component of velocity vs. time for objects slowing to a stop in three different situations. Although the rate of velocity decrease is linear in each case, the slowing rate is distinctly different for each object/surface combination. *Note*: Data for position vs. time were obtained using an ultrasonic motion detector. In each case, the position vs. time data were fit very accurately with a quadratic function and the first time derivative of each x vs. t fit equation was used to determine instantaneous velocity vs. time equations. Each of these v vs. t equations was plotted at the times that position data were recorded.

Let's return to the question that motivated the experiment: Is a force required to keep an object moving at a constant velocity? At first glance, the answer is yes, since the object of interest slows down after the kick in each case. But wait a minute! After the kick, the *rate of slowing* is different in each case. This suggests that the slowing is caused by different forces between the object and the surface over which it moves. We associate the longer slowing time with a smaller frictional force exerted on the object by the surface. A reasonable inference is that it doesn't require a force to keep an object moving at a constant velocity. Rather, forces are present that are causing it to slow down. So what is the natural state of motion in the absence of forces?

Imagine what would happen if we could make the surface that the cart and plastic box move on smoother and smoother or minimize the horizontal friction forces on an object by using an air track, hovercraft, or moving it in outer space. The object would move farther and farther. What if we could observe an object in motion that has no interactions with its surroundings and hence no forces on it? Our experiment suggests that it could move forever at a constant velocity. This was Newton's answer to this question and is embodied in his First Law of Motion, expressed here in contemporary English rather than 17th-century Latin:

NEWTON'S FIRST LAW: Consider a body on which no force acts. If the body is at rest, it will remain at rest. If the body is moving, it will continue to move with a constant velocity.

What is force? Clearly Newton is defining force here to be an agent acting on a body that changes its velocity. In the absence of force, a body's velocity will not change. We can state this definition of force more formally.

FORCE is that which causes the velocity of an object to change.

Newton's First Law and his definition of force seem sensible when applied to an object at rest or moving at a constant velocity in a typical physics laboratory. However, in order to measure the velocity of an object, we must choose a coordinate system or reference frame to measure the positions as a function of time. As you saw in Chapter 2, these measurements are needed to calculate velocities and accelerations.

Can we expect Newton's First Law to hold in any reference frame? It turns out that Newton's First Law doesn't hold in all frames of reference. For example, consider what happens to an object in a frame of reference that is accelerating. It is common to see pencils and other small objects that were at rest in a car's frame of reference spontaneously begin to roll around on a dashboard when a car suddenly speeds up or slows down. In this case, Newton's First Law doesn't appear to hold. For this reason, Newton's First Law is often called the law of inertia. Reference frames in which it holds are called inertial frames. Thus, any accelerating frame of reference, in which resting objects appear to start moving spontaneously such as those in a vehicle that is speeding up, slowing down, or turning, is a noninertial frame. Newton's First Law only holds in inertial reference frames. As we develop Newton's other laws of motion, we will restrict ourselves to working in inertial reference frames in which the first law is valid.

READING EXERCISE 3-1: Consider the graph shown in Fig. 3-3. (a) Roughly how many seconds does it take the cart rolling on the rough carpet to come to a complete stop? (b) Assuming the cart traveling on the smooth track has a speed of 0.8 m/s at $t = 0.0$ s, what percent of its initial speed does the cart rolling on the track still have just as the cart on the carpet has come to rest? ■

READING EXERCISE 3-2: (a) Describe a noninertial reference frame that you have been "at rest" in. (b) What observations did you make in that frame to lead you to conclude that it was noninertial? ■

3-3 A Single Force and Acceleration Along a Line

We will simplify our investigation of force and the changes in velocity that it produces by first considering situations in which a single force acts on an object in an inertial reference frame. After we study how a single force affects the motion of an object, we will investigate what happens to an object's motion when two or more forces are acting on an object along its line of motion.

Consider the motion of a person riding on a low-friction cart that can roll easily under the influence of a force. A steady pulling force is applied to the cart and rider. The force acts along the line of the cart's motion. The person who is pulling maintains a steady force on the cart and rider by keeping a short piece of bungee cord stretched to a constant length as shown in Fig. 3-4. By directing a motion detector toward the back of the cart rider, we can track the motion with a computer data acquisition system. If the pulling force is the only significant force on the rider in the direction of his motion, then the results displayed in Fig. 3-5 lead us to make the following observation.

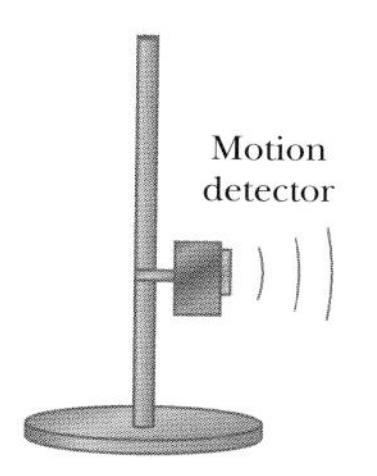

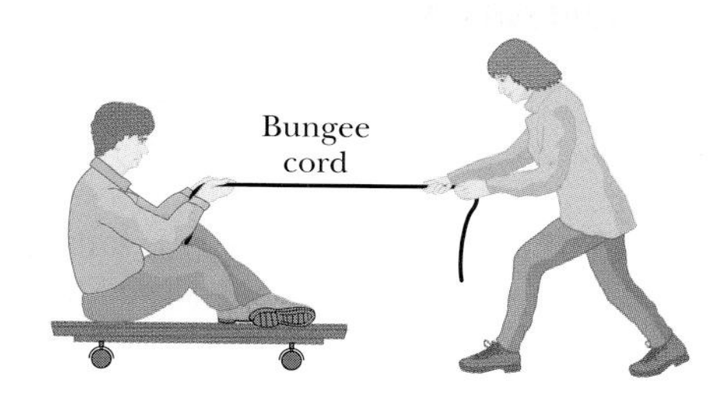

FIGURE 3-4 ■ A person riding on a low-friction cart is pulled by another person who exerts a constant force along a straight line by keeping the length of a bungee cord constant.

> **OBSERVATION:** A constant force acting on an object causes it to move along a straight line with a constant acceleration that is in the same direction as the force.

This observation has been verified many times for different objects moving under the influence of a constant push or pull when friction forces are small.

Many people believe that a constant force will cause a body to move at a constant velocity. This common belief stems from everyday experiences such as driving a car along a highway or sliding a heavy box along a floor. It takes a steady flow of gasoline to move the car at a constant velocity. Thus, the experimental result that a constant force causes a constant acceleration, as shown in Fig. 3-5, is surprising. Remember that we have designed our experiment to apply a single force to a low-friction cart so that there are no significant friction forces acting. Later in this chapter, we will discuss how contact forces, involving a direct push or pull or friction between surfaces like those experienced by a sliding box or a car moving along a highway, can cancel each other to yield zero *net force* on an object. This can then lead us to situations in which pushing or pulling forces, when counteracted by friction forces, do indeed cause bodies to move at a constant velocity.

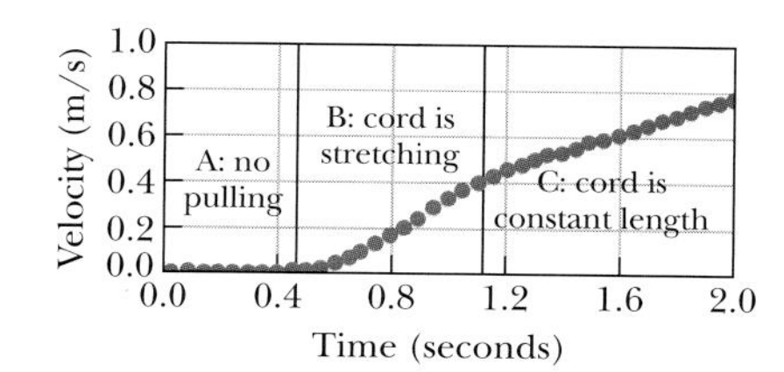

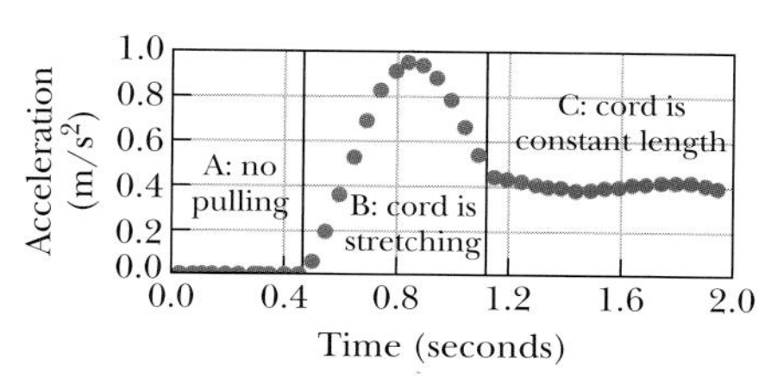

FIGURE 3-5 ■ These graphs show velocity and acceleration components vs. time for a rider on a cart. For the first 0.5 s (region A) the cart is at rest. Between 0.5 s and 1.1 s (region B) the cord is beginning to stretch. Between 1.1 s and 2.0 s (region C) a constant force is acting and the acceleration is observed to be constant as well.

READING EXERCISE 3-3: (a) Describe an experience you have had in which applying what seems like a steady force to an object did *not* cause it to accelerate. (b) Describe a situation in which an object accelerated when you applied what seemed to be a steady force to it. *Note:* You can experiment with applying a steady force to some objects readily available to you. ■

3-4 Measuring Forces

As we discussed in Chapter 1, in order to allow us to communicate with others precisely and unambiguously, we need to define a standard unit and a scale for force just as we did for distance, mass, and time. Since all physical quantities are defined by the procedures developed for measuring them, we must start by defining a procedure for measuring our standard unit of force. Our qualitative definition of force is that it is an interaction that causes acceleration, so our standard method for measuring force involves measuring how much acceleration a given force imparts to a standard object. We need to decide, as an international community of scientist and engineers, what the standard object we accelerate will be. It turns out that what we have chosen to use is the international standard kilogram discussed in Chapter 1. The SI unit of force is the newton.

> **DEFINITION OF THE STANDARD FORCE UNIT:** One newton of force is defined to be the force necessary to impart an acceleration of 1 m/s^2 to the international standard kilogram.

This definition of the newton assumes, of course, that all other forces experienced by the standard mass are small enough to be neglected. To measure any other force in newtons, we simply need to measure the acceleration of our standard object in a low-friction setting and compare its acceleration to 1 m/s^2.

Other units of force that are still used in the United States are summarized in Appendix D. These include the dyne, the pound, and the ton.

In order to measure a force in standard units, we must allow it to accelerate a 1 kg object that is free to move without experiencing significant friction forces. For practical reasons we have chosen to measure force by accelerating a low-friction cart on a smooth, level track instead of the actual standard kilogram. We start by adjusting the cart's mass so that it balances with a facsimile of the international standard kilogram.

Next we set up an ultrasonic motion detector with a computer data acquisition system to measure the change in position of the cart as a function of time as it accelerates. The computer data acquisition software can then be used to calculate velocity and acceleration values as a function of time from the position data.

Suppose that someone pulls our low friction cart along a track by means of a spring attached to one end of the cart. Assuming the spring is not yet stretched so far as to be permanently deformed, then the farther it is stretched the greater the size or magnitude of the pull force. The different strengths of pull impart different accelerations to our cart. For a certain strength of pull we find that we can impart an acceleration of 1 m/s^2 to the cart—measured by the computer data acquisition system. Of course, this length of the spring is by definition acting on the standard cart with a force of 1 N.

How could we exert a force on the cart of 2 N, 3 N, and so on? We can pull harder on the spring so it stretches enough to cause the cart to accelerate at 2 m/s^2. The process can be repeated to yield an acceleration of 3 m/s^2, and so on as illustrated in Fig. 3-6.

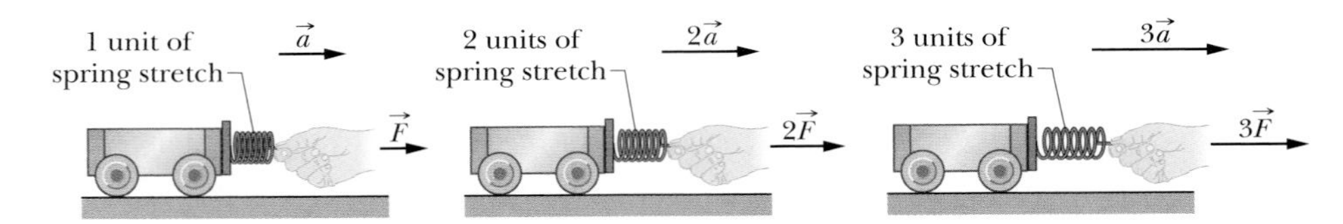

FIGURE 3-6 ■ An experiment in which a spring is used to apply steady forces to a 1 kg cart. First the spring is stretched enough to yield an acceleration of 1 m/s^2, so by definition the force applied to the 1 kg cart is 1 N. As the spring is stretched more and more, the forces on the cart become larger and accelerations of 2 m/s^2 and then 3 m/s^2 can be created.

Thus, a force can be measured by the acceleration it produces on a standard 1 kg object.

Acceleration is a vector quantity that has both a magnitude and direction. Is force also a vector quantity? Does it have a direction as well as a size associated with it? In order to answer the question of whether force is a vector quantity, consider the following question: Is a force of 1 N directed to the right different from a force of 1 N directed to the left? If so, how? The answer is "yes," these forces are different. A force directed to the right will cause an object to accelerate to the right, and a force directed to the left will cause an object to accelerate to the left. Thus, a force has both a magnitude and a direction associated with it. As we discussed in Section 2-2, to qualify as a vector, a force must also have certain other properties that we have not yet specified. However, it is reasonable for now to assume that force behaves like a vector.

Measuring force by setting up a system for measuring the acceleration of a standard object is very impractical. Most investigators take advantage of the fact that elastic devices such as springs, rubber bands, and electronic strain gauges (used in the electronic force sensor in Fig. 3-8) stretch more and more as greater forces are exerted on them. These devices can be calibrated "properly" by using the "official" method for measuring force. We can designate a 1 newton force as that which causes our standard mass to accelerate at 1 m/s^2 and record the amount of stretch or the electronic reading for the new device. Then we can designate a 2 newton force as that which leads to an acceleration of 2 m/s^2 and record the response of the new device and so on for other forces. More often, a secondary calibration can be performed by comparing the readings of a given force-measuring device to that of another force-

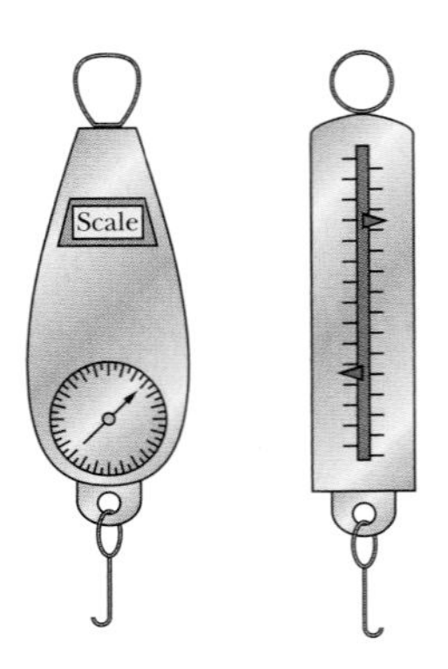

FIGURE 3-7 ■ Two types of spring scales that can be calibrated to measure forces in newtons by relating the gravitational force exerted by the Earth on a 1 kg weight to the amount of spring stretch.

measuring device that has already been properly calibrated. Spring scales, like those shown in Fig. 3-7, are very popular devices for measuring force. This popularity stems from the fortunate fact that the amount by which a spring stretches is directly proportional to the magnitude of the force acting on the spring—provided the spring is not overstretched. This proportionality was discovered in the 17th century by Robert Hooke, and will be discussed more formally in Chapter 9. The proportionality between spring stretch and force is a convenient property, but not necessary. We could just as well use a nonlinear device such as a piece of bungee cord.

FIGURE 3-8 ■ An electronic force sensor that can be used with a computer data acquisition system. When the hook at the bottom is pushed or pulled, a metal element is compressed or flexed. This is detected by an electronic strain gauge, which puts out a voltage proportional to force.

READING EXERCISE 3-4: A typical rubber band does not obey Hooke's law. However, it can be used as a force scale if not stretched to its limit. Describe how you might use a properly calibrated spring scale, like one of those shown in Fig. 3-7, to create a device that uses the elasticity of a rubber band to measure force. ■

3-5 Defining and Measuring Mass

We know from experience that if we push steadily on a wheelbarrow it is much harder to get it moving when it's full than when it's empty. We also know that it is much harder to lift a wheelbarrow when it's full. We can summarize these observations with the statement that a large amount of stuff is harder to move than a small amount of stuff. But how do we measure how much larger "an amount of stuff" on a loaded wheelbarrow is than on an unloaded one? Suppose we pile our wheelbarrow with a huge mound of hay and try to lift it or pull on it. What happens if we replace the hay with a relatively small lead brick? How much hay is the same amount of stuff as a small lead brick? How do we know?

In Section 1-2 we introduced the term mass as a measure of "amount of stuff" and stated that quantities are defined by the procedures used to measure them. In the last section we defined force in terms of basic procedures for measuring it. In this section we will do the same for mass. We introduce two quite different procedures for measuring mass based on two questions: How hard is it to lift a certain pile of stuff? And how hard is it to accelerate the pile of stuff with a standard force?

Measuring Gravitational Mass

As we mentioned in Chapter 1, the most common historical procedure for measuring mass is to compare the effect of the gravitational forces on two objects using a balance. As early as 5000 B.C.E., ancient Egyptians used the equal arm balance for comparing masses to a standard mass (Fig. 3-9).

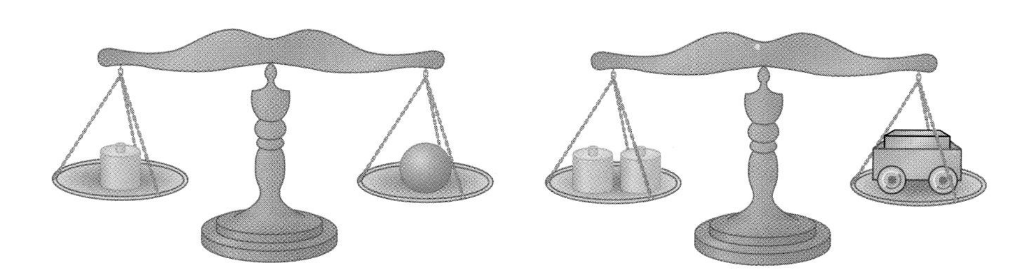

FIGURE 3-9 ■ An old fashioned balance is used to measure gravitational mass using 1 and then 2 replicas of a standard 1 kg mass. So the sphere has a gravitational mass of 1 kg. The cart loaded with extra mass has a gravitational mass of 2 kg.

We assume that two objects have the same mass if they balance with each other. If two masses balance, they are experiencing the same gravitational force. The mass of replicas of the standard 1 kg mass are adjusted using a balance. We can create a mass scale by assuming that masses add so that two replicas of the standard 1 kg mass have a combined mass of 2 kg, and so on. We can also create masses that are fractions of a

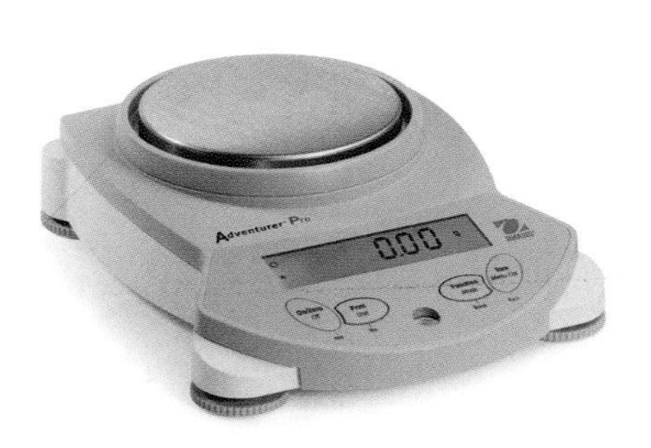

FIGURE 3-10 ■ A modern electronic balance uses an internal electronic strain gauge to measure gravitational mass. Although the principle on which it works is not obvious, it gives the same result as spring scales do.

kilogram. For example, we can create 1/2 kg masses by creating two less massive objects that balance with each other, but combine to balance with a standard 1 kg mass. Because this procedure for determining a mass involves balancing gravitational forces, we call this type of mass *gravitational mass*.

In modern laboratories, triple beam balances, spring scales, and electronic scales (Fig. 3-10) are used instead of the old-fashioned balance for measuring gravitational mass. As the Earth attracts a mass hanging from a spring, the spring will stretch. A mass on an electronic scale causes an electrical strain gauge to compress.

Measuring Inertial Mass

As we mentioned, another "measure" of how much stuff we have is to observe how hard it is to get an object moving, or accelerate it, with a known force. We know that by definition a 1 N force will cause a standard 1 kg mass to accelerate at 1 m/s^2. In general, when $m = 1$ kg, the magnitude of the acceleration is the same as that of the force. What happens to the relationship between a single force and acceleration when the mass is different from the standard mass?

If we set up a system to measure acceleration and force, such as the computer interface system shown in Fig. 3-11 with an accelerometer and an electronic force sensor attached firmly together, we can study how mass affects the relationship between force and acceleration. We do this by pushing and pulling in a horizontal direction on the force sensor–accelerometer system. We can then tape some additional mass on the system and repeat this procedure.

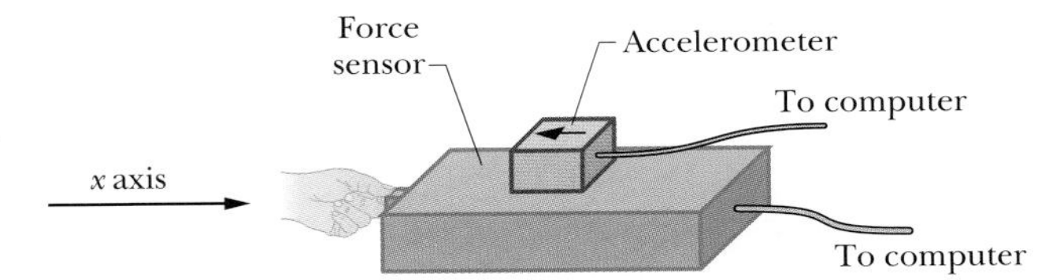

FIGURE 3-11 ■ Setup showing an electronic accelerometer tracking the acceleration as a function of the forces of a push or pull on a system consisting of itself, a force sensor, and additional mass. The system is held firmly by the hook that is attached to the sensitive area of the force sensor. It is then pushed and pulled horizontally in mid-air with gentle but rather erratic motions.

Figure 3-12 shows graphs of both the x-component of force vs. time and the x-component of acceleration vs. time for a system that has a *gravitational mass* of 150 g. We find that the force and acceleration components are directly proportional to each other on a moment-by-moment basis. The evidence for this is the fact that the graphs of force vs. time and acceleration vs. time have the same basic shape and are zero at the same times. By the same "graph shape" we mean that if the force is twice as large at one time than another, then so is the acceleration.

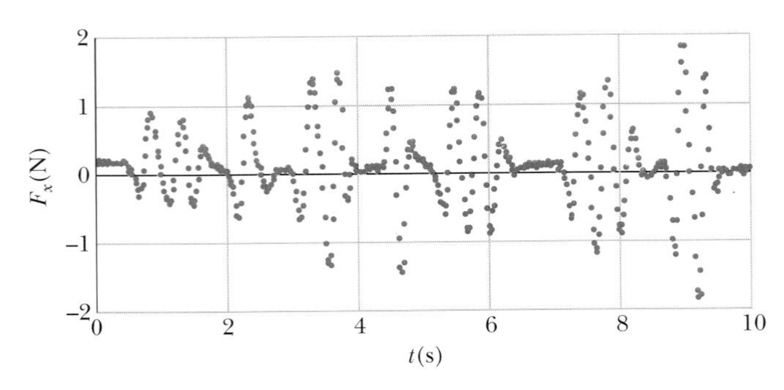

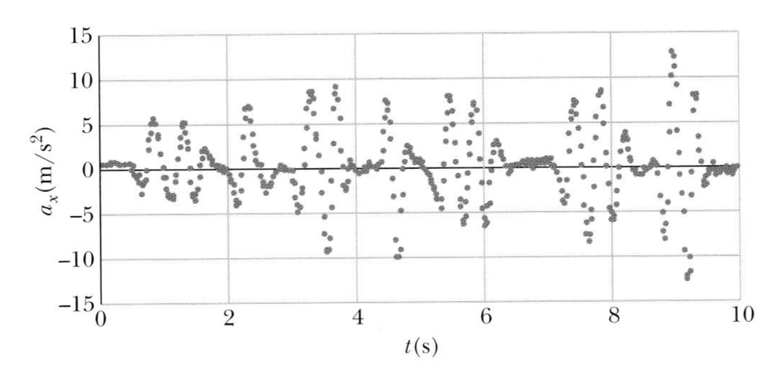

FIGURE 3-12 ■ Graphs of measured force and acceleration as a function of time. An accelerometer is attached to a force sensor as shown in Fig. 3-11. The combination is being pushed and pulled. Signals from these sensors were sent to a computer via a computer data acquisition interface. The similarity in the shapes of the real-time computer graphs reveal a moment-by-moment proportionality between force and acceleration.

In Figure 3-13 we use the same data displayed in Fig. 3-12 for the 150 g *gravitational mass* to graph the x-component of force as a function of the x-component of acceleration. The fact that this new graph is a straight line that passes through the origin is additional evidence that there is a direct proportionality between force and acceleration. The constant of proportionality is given by the slope of the graph.

We define the **INERTIAL MASS** of a system as the constant of proportionality between acceleration and the force that causes it.

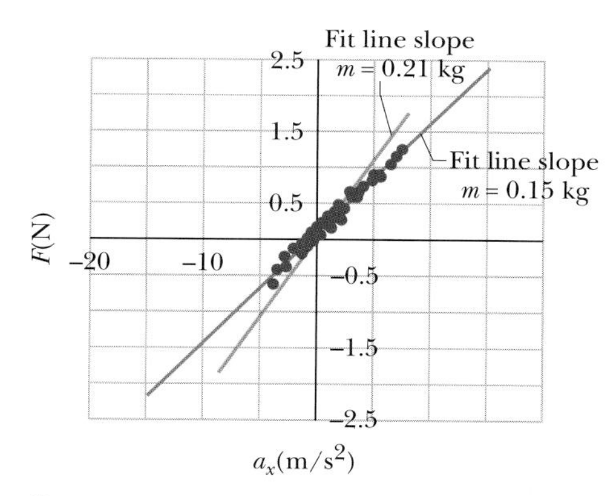

FIGURE 3-13 ■ Graphs of the horizontal force vs. acceleration components for the accelerometer–force sensor system (shown in Fig. 3-11) as it is being pushed and pulled in a horizontal direction. Two system masses were used, 0.150 kg and 0.200 kg. The resulting slopes for the F_x vs. a_x graphs show a proportionality between force and acceleration with the constant of proportionality being equal, within the limits of experimental uncertainty, to the gravitational mass of the system in each case.

Indeed, we see that if we now do the experiment shown in Fig. 3-13 with a 200 g gravitational mass we get a larger slope. This indicates that when there is more mass it takes more force to get the same acceleration. Perhaps the most interesting feature of Fig. 3-13 is that the *inertial masses* measured as the slopes of the F_x vs. a_x graphs are the *same* as the values of the *gravitational masses* measured with a balance—at least within the limits of experimental uncertainty.

The inertial mass of an object tells us how much it resists acceleration, whereas the gravitational mass is a measure of how hard the Earth pulls on an object. Sophisticated experiments involving precise measurements of the gravitational forces between two objects in a laboratory using a device known as a Cavendish balance have shown that there is no difference between the two types of mass to within less than one part in 10^{12}. Since the two types of mass seem to have the same values, we will drop the distinction between them and just refer to mass.

3-6 Newton's Second Law for a Single Force

The general relationship between force, mass, and acceleration discussed in Section 3-5 is known as Newton's Second Law. By pulling together conclusions we have reached so far, we will state this law for the case of a single force that acts alone along a line. We will then proceed, in this chapter and those that follow, to show that this law is also valid when more than one force acts and when forces act in two dimensions.

How is the acceleration of a body related to its mass and the force acting on it? The experimental evidence presented in Figs. 3-12 and 3-13 shows that the acceleration of a body is *directly proportional* to the force acting on it. The experimental results in Fig. 3-13 show acceleration to be *inversely proportional* to the mass of the body. That is, looking at the graph, we can see that for a given force acting on a body, the acceleration imparted to it is less when the body's mass is large than when it is small. Combining these two relations with our definition of the unit force as producing a unit acceleration of a unit mass, we can summarize what we now know in the single equation

$$\vec{a} = \frac{\vec{F}}{m}. \tag{3-1}$$

The arrows shown in Eq. 3-1 serve as a reminder that we believe that both force and acceleration are vector quantities that have magnitude and direction. The force on a body and acceleration caused by it are in the same direction. Mass is a scalar quantity that does not have a direction associated with it. Newton's Second Law can also be put in words:

NEWTON'S SECOND LAW FOR A SINGLE FORCE: When a single force acts on an object, it will cause the object to accelerate in the direction of the force. The amount of acceleration is given by the acting force divided by the object's mass.

Because it is easier to write, the most common way to refer to Newton's Second Law is in the form

$$\vec{F} = m\vec{a}. \tag{3-2}$$

If the force lies along the x axis, then $\vec{F} = F_x\,\hat{\imath}$ and $\vec{a} = a_x\,\hat{\imath}$. So we can also express the Second Law in terms of the force and acceleration components as $F_x = ma_x$.

Equations 3-1 and 3-2 represent an interesting combination of definitions and a law of nature. In both Eqs. 3-1 and 3-2, the equality sign does not mean that the two sides of an equation are the same physical quantities or that force is defined as the product of mass and acceleration. But, Eq. 3-1 provides a method for *predicting* the acceleration of an object when its mass and the force acting on it are known. Alternatively, Eq. 3-2 tells us that a measurement of acceleration and mass can be used to *determine* the force on a body that is causing it to accelerate.

For standard SI units, $\vec{F} = m\vec{a}$. tells us that

$$1\text{ N} = (1\text{ kg})(1\text{ m/s}^2) = 1\text{ kg}\cdot\text{m/s}^2. \tag{3-3}$$

Force units common in other systems of units are given in Appendix D.

So far we have been studying the relationship between motion and force under very limited circumstances. We have restricted our study to forces acting along a line in an inertial reference frame. We have also restricted ourselves to observations in which we think that the applied force acting on an object, such as a low-friction cart, is the only significant interaction the object is experiencing. By applying this rather unrealistic set of restrictions, we were able to formulate initial definitions of force and mass. We then combined these definitions with observations to develop two of Newton's three laws of motion.

As we already suggested, Newton's first two laws are not simply valid by definition in the way that $\vec{v} \equiv d\vec{x}/dt$ or $\vec{a} \equiv d\vec{v}/dt$ are. Rather, they represent a combination of definitions and natural laws. Can we refine these laws so they are valid in more complicated situations that describe forces and motion along a line? In particular, what happens when more than one force is acting at the same time? How do forces combine? What other forces besides the forces we apply can act on a body? What evidence is there that these forces are real? When forces are obviously due to interactions between two or more bodies, does a body acted upon also exert forces on the body acting on it? How are these related? The rest of this chapter will be devoted to dealing with these questions. Chapters 5 and 6 will deal with how to use Newton's laws to predict motions that result from forces that act in two and three dimensions.

The Flea Pulling a Train

Let us return to the question we asked at the beginning of the chapter. How can a jumping flea with a tiny mass pull an object that is 160,000 times more massive? When it comes to jumping, insects have a big advantage over larger animals. The strength of their legs increases as the square of the diameter of their legs while the mass that they push off with goes as the cube of their body dimensions. Thus the ratio of the mass they lift with their legs to their cross-sectional area is much smaller than it is for a large animal. While a world-class high jumper can barely jump his or her own height, a flea can jump up to 150 times its own height. So a 2-mm-tall flea can jump to a height of about 30 cm.

The flea's secret is that he can launch himself at a high speed. Suppose Brutus, whose mass is only about 2×10^{-3} g, starts a 30-cm high hop that takes him up and forward at the same time. Our flea will be moving at a pretty high horizontal speed. Using kinematic equations, we can estimate its initial hopping speed to be over 2 m/s.

Before the flea completes his hop, it will be rudely interrupted as he comes to the end of the wire. The wire begins to stretch and the wire then pulls Brutus to a sudden stop. But the force that Brutus exerts on his end of the wire while he is being stopped will be transmitted along the wire to the train. This causes the train, which has a mass of 32 g, to jerk forward. While Brutus is falling down, the friction in its wheels causes it to roll to a stop also.

Because Brutus is not pulling with a steady force, it is difficult to make detailed calculations of the motion of the train he is pulling on. Instead, in Touchstone Example 3-1 we calculate what happens to a man who pulls steadily on a pair of real passenger cars.

READING EXERCISE 3-5: A student sitting on a skateboard is pulled with a horizontal force to the left of magnitude 26 N and accelerates at 0.42 m/s^2. (a) Write the expressions for force and acceleration in vector notation using the unit vector. (b) What is the combined mass of the student and skateboard? (c) The mass of a student and her skateboard is measured using a European bathroom scale calibrated to read in kilograms. What is the scale reading? ■

READING EXERCISE 3-6: Consider your answers to Reading Exercise 3-5. (a) Which mass measurement is a determination of inertial mass, the one made in part (b) or part (c)? Explain. (b) What assumption did you make in determining your answer to part (c)? ■

TOUCHSTONE EXAMPLE 3-1: Pulling a Train

John Massis is shown in the photo pulling two passenger cars by applying a steady force to them at an angle of about 30° with respect to the horizontal. Assume instead that Massis had pulled the two cars of mass 8.0×10^4 kg with a horizontal force of 2.0×10^3 N. If there was no friction in the rails, what speed would the cars have after Massis moves them a distance of 1.0 m from their resting location?

In 1974, John Massis of Belgium managed to move two passenger cars belonging to New York's Long Island Railroad. He did so by clamping his teeth down on a bit that was attached to the cars with a rope and then leaning backward while pressing his feet against the railway ties. The cars together weighed about 80 tons, which is almost 1000 times more than the man's mass.

SOLUTION ■ **A Key Idea** here is that, from Newton's Second Law, the constant horizontal pulling force on the cars that Massis exerts causes a constant horizontal acceleration of the cars. Because the force is constant, and the motion is assumed to be one-dimensional, we can use the kinematic equations to find the horizontal velocity component v_{2x} at location x_2 (where $x_2 - x_1 = +1.0$ m).

Place an x axis along the direction of motion, as shown in Fig. 3-14. We know that the initial velocity component along the horizontal axis v_{1x} is 0, and that the displacement $x_2 - x_1$ is +1.0 m. However, we need to find the x-component of acceleration, a_x.

We can relate the x-component of the acceleration of the cars, a_x, to the pulling force on the cars from the rope by using Newton's Second Law. If we assume there are no friction forces, we can note that a single pulling force acting along the *horizontal axis* in Fig. 3-14 is

$$F_x^{\text{pull}} = Ma_x \tag{3-2}$$

where M is the mass of the cars and F_x^{pull} and a_x are the x-components of the force and acceleration vectors.

In Fig. 3-14, we see that Massis is pulling in the x-direction, so $F_x^{\text{pull}} = 2.0 \times 10^3$ N. Since the mass of the railroad cars, M, is 8.0×10^4 kg we can find a_x by rearranging Eq. 3-2 and substituting for F_x^{pull} and M. The acceleration component becomes

$$a_x = \frac{F_x^{\text{pull}}}{M} = \frac{2.0 \times 10^3\ \text{N}}{8.0 \times 10^4\ \text{kg}} = 0.025\ \text{m/s}^2.$$

Next we use Eq. 2-13

$$v_{2x} = v_{1x} + a_x(t_2 - t_1) = v_{1x} + a_x\Delta t \tag{3-4}$$

to find the velocity of the train after it has moved 1.0 m. Since $v_{1x} = 0$, we find that $v_{2x} = a_x\Delta t$ in this case.

To find Δt, we can use the fact that the train's average velocity is

$$\langle v_x \rangle = \frac{\Delta x}{\Delta t} = \frac{v_{1x} + v_{2x}}{2},$$

then solve this equation for Δt. Again using $v_{1x} = 0$, this yields $\Delta t = \Delta x/(v_{2x}/2)$. By substituting Δt back into Eq. 3-4, we find that $v_{2x} = a_x \Delta x/(v_{2x}/2)$ or, more simply,

$$v_{2x} = \sqrt{2a_x \Delta x} = \sqrt{(2)(0.025\text{m/s}^2)(1.0\text{ m})} = 0.22\text{m/s}. \quad \text{(Answer)}$$

We assumed in this calculation that the force Massis exerted on the railroad cars was horizontal. Actually his pull was not quite horizontal. This made his job harder. Can you see why?

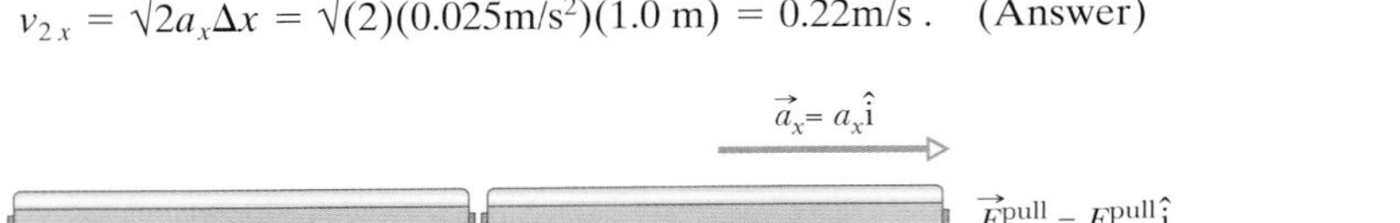

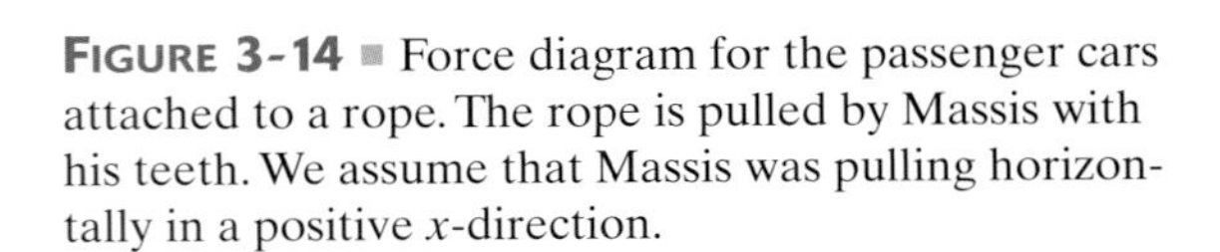

FIGURE 3-14 ■ Force diagram for the passenger cars attached to a rope. The rope is pulled by Massis with his teeth. We assume that Massis was pulling horizontally in a positive x-direction.

3-7 Combining Forces Along a Line

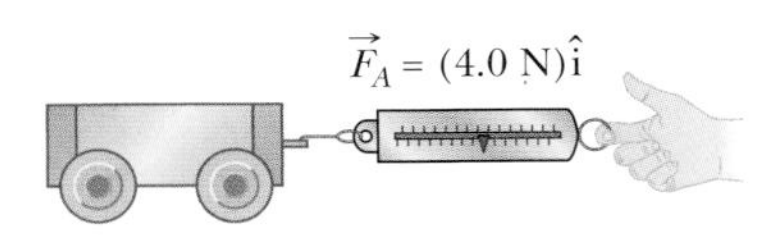

FIGURE 3-15 ■ Pulling to the right on a low-friction cart with a force of +4.0 N. If the cart has a mass of 0.50 kg, then it will accelerate to the right at $+8.0\text{ m/s}^2$.

We have discussed how a single applied force, such as a push or pull, affects the motion of an object. Now let's go one step further and think about what happens if a second applied force also acts on a body.

Suppose that you have a spring attached to a low friction cart like that shown in Fig. 3-15. You pull on the cart using the spring, keeping the spring constantly stretched to produce a constant force of magnitude 4.0 N to the right. Since you are applying a constant force to the cart, it will speed up with a constant acceleration. That is, the cart's velocity will increase at a constant rate.

What do you think would happen if a friend simultaneously pulled on the cart in the same manner, with the same magnitude of force, but in the opposite direction as shown in Fig. 3-16? Would the cart still accelerate? Clearly the answer is "no." How is the motion of the cart affected if you and your friend each apply a 2.0 N force to the cart in the *same direction*? Measurement reveals two things. First, the acceleration produced by a single 4.0 N force is twice that produced by a single 2.0 N force. Second, a single 4.0 N force produces the same acceleration as two 2.0 N forces applied in the same direction, as shown in Fig. 3-17.

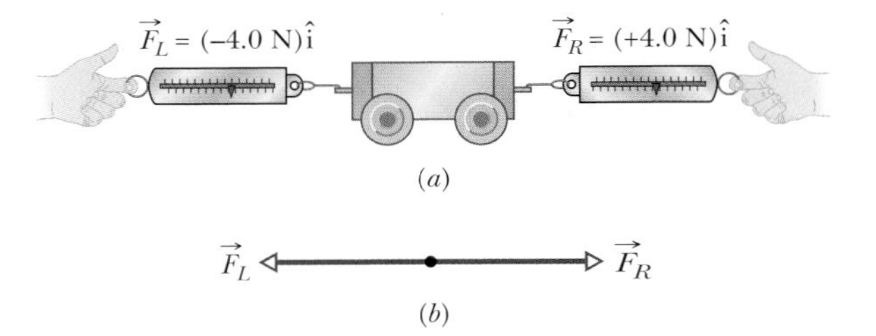

FIGURE 3-16 ■ (*a*) Pulling to the right on a low-friction cart with a force as someone pulls to the left with the same magnitude of force. Thus, $\vec{F}_R + \vec{F}_L = 0$, so the forces cancel and the cart doesn't move. (*b*) A simple diagram representing the forces acting on the cart. Such a diagram is called a free-body diagram.

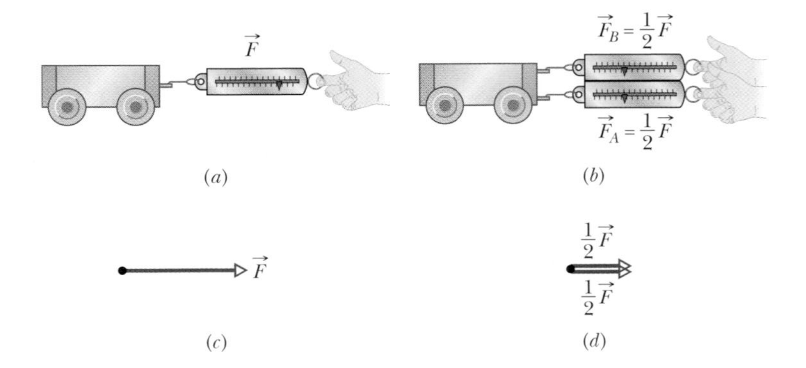

FIGURE 3-17 ■ (*a*) and (*b*): Pulling to the right on a low-friction cart with one force yields the same acceleration on it as two forces pulling to the right do when each has half the magnitude of the single force. (*c*) and (*d*): Free-body diagrams of situations (*a*) and (*b*).

These observations indicate that keeping track of the magnitudes and directions of all the forces acting on an object is very important if we want to be able to make predictions about the object's subsequent motion. A special type of diagram, called a **free-body diagram,** is an especially useful technique for doing this. Figures 3-16*b* and 3-17*c* and *d* show free-body diagrams that represent various situations.

We construct a free-body diagram by representing each object we are investigating as a point. For example, in Fig. 3-17, we are interested in the motion of the cart (not the hand) and so we represent the cart as a point. We then draw a force vector (as an arrow) for each force acting on the object. We place the tail of each force vector (arrow) on the point and draw the vector in the direction of the force. The relative magnitude of the forces is represented by the relative lengths of the arrows. Hence, the two equal force vectors in Fig. 3-17 are shown to have the same length. Finally, we label the force vectors so that we know which force each arrow represents.

Free-body diagrams help us to translate pictures or statements of a situation into mathematical expressions. That is, they help us to generate mathematical expressions in which we treat a force as a vector quantity with both a magnitude and a direction. As we discussed in Section 3-4, by normal convention, a horizontal force directed to the right has a positive x-component and one directed to the left has a negative x-component. Thus, each of the one-dimensional vectors we discussed can be represented as the combination of its magnitude and direction as follows:

Two 4.0 N forces acting in opposite directions:

$$\vec{F}_A = F_{Ax}\hat{\text{i}} = (+4.0\ \text{N})\hat{\text{i}} \qquad \vec{F}_B = F_{Bx}\hat{\text{i}} = (-4.0\ \text{N})\hat{\text{i}}$$

Two 2.0 N forces acting in the same direction:

$$\vec{F}_A = F_{Ax}\hat{\text{i}} = (+2.0\ \text{N})\hat{\text{i}} \qquad \vec{F}_B = F_{Bx}\hat{\text{i}} = (+2.0\ \text{N})\hat{\text{i}}$$

The plus or minus sign carried with the vector components to denote the direction of vectors makes it easy for us to remember in what direction the forces and acceleration point along a chosen x axis. The signs make it possible to combine forces mathematically using the rules of vector mathematics. As long as we denote direction with signs as we did above, we can determine the combined effect of multiple forces acting on an object simply by adding up the force components acting along a single line. For example, in the case of our two forces that are applied in opposite directions, we can determine the combined force, usually called the vector sum of the forces or **net force,** by calculating the vector sum, so that

$$\vec{F}^{\text{net}} = \vec{F}_A + \vec{F}_B = F_{Ax}\hat{\text{i}} + F_{Bx}\hat{\text{i}} = (F_{Ax} + F_{Bx})\hat{\text{i}} = (+4.0 - 4.0\ N)\hat{\text{i}} = 0.$$

The net force or vector sum of the forces for the situation depicted in Fig. 3-17*b* can be calculated as

$$\vec{F}^{\text{net}} = (F_{Ax} + F_{Bx})\hat{\text{i}} = [+2.0\ \text{N} + 2.0\ \text{N}]\hat{\text{i}} = (+4.0\ N)\hat{\text{i}}.$$

When the forces do not have the same magnitude or direction, we can still use vector sums. For instance, consider a 3 newton force to the right, denoted by its x-component of +3 N, and a 2 newton force to the left, denoted by its x-component of –2 N. These two forces combine to give a net force of

$$\vec{F}^{\text{net}} = \vec{F}_A + \vec{F}_B = (+3\ \text{N} - 2\ \text{N})\hat{\text{i}} = (+1\ \text{N})\hat{\text{i}}.$$

This means that part of the influence of the 3 N force to the right is counteracted by the application of a 2 N force to the left. In the end, an object with these two forces acting on it behaves as if only a 1 N force, directed to the right, is present. So,

> When two or more forces act on a body, we can find their net force or resultant force by adding the individual forces as vectors taking direction into account.

A single force with the magnitude and direction of the net force has the same effect on the body as all the individual forces together. This fact is called the **principle of superposition for forces.** The world would be quite strange if, for example, you and a friend were to pull on the cart in the same direction, each with a force of 5 N, and yet somehow the net pull was 20 N.

In this book a net force is represented with the vector symbol $\vec{F}^{\,\text{net}}$. Instead of what was previously given, the proper statement of Newton's First and Second Laws should now be rephrased in terms of net forces.

> **NEWTON'S FIRST LAW:** Consider a body on which no net force acts so that $\vec{F}^{\,\text{net}} = 0$. If the body is at rest, it will remain at rest. If the body is moving, it will continue to move with a constant velocity.

This statement means that there may be multiple forces acting on a body, but if the net force (the vector sum of the forces) is zero, then the body will not accelerate. Remember, this doesn't mean that the object is stationary. It simply means that the object will not speed up or slow down.

We can also rewrite Newton's Second Law in terms of net force.

> **NEWTON'S SECOND LAW FOR MULTIPLE FORCES:** The acceleration of a body is the *net* force acting on the body divided by the body's mass.

This statement can be expressed mathematically by replacing the force in Eq. 3-1 with net force, so that

$$\vec{a} = \frac{\vec{F}^{\,\text{net}}}{m} \quad \text{(Newton's Second Law).} \tag{3-5}$$

Once again, because it is easier to write down, a common way to write Newton's Second Law for multiple forces in vector form is

$$\vec{F}^{\,\text{net}} = m\vec{a} \quad \text{or} \quad F_x^{\text{net}} = ma_x,$$

where

$$\vec{F}^{\,\text{net}} = \vec{F}_A + \vec{F}_B + \vec{F}_C \cdots \vec{F}_N.$$

Hence, if we want to know the acceleration of an object on which more than one force acts, we can find it using the following procedure:

1. Draw a free-body diagram for the object of interest.
2. Determine the net force acting on the object.
3. Take the ratio of the net force to the mass of the object.

This procedure is used in the examples that follow. You will find it useful in completing many of the end-of-chapter problems as well.

READING EXERCISE 3-7: The figure shows two horizontal forces moving a cart along a frictionless track. Suppose a third horizontal force $\vec{F}_C$ could act on the cart. What are the magnitude and direction of $\vec{F}_C$ when the cart is (a) not moving and (b) moving to the left with a constant speed of 5 m/s? ■

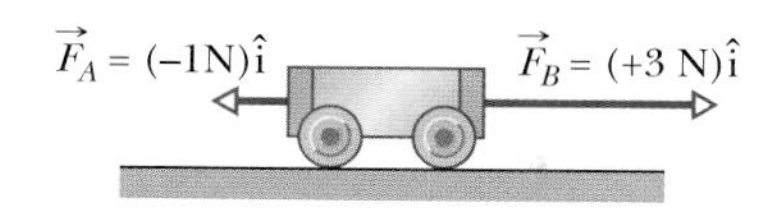

READING EXERCISE 3-8: The figures that follow show overhead views of four situations in which two forces accelerate the same cart along a frictionless track. Rank the situations according to the magnitudes of (a) the net force on the cart and (b) the acceleration of the cart, greatest first. ■

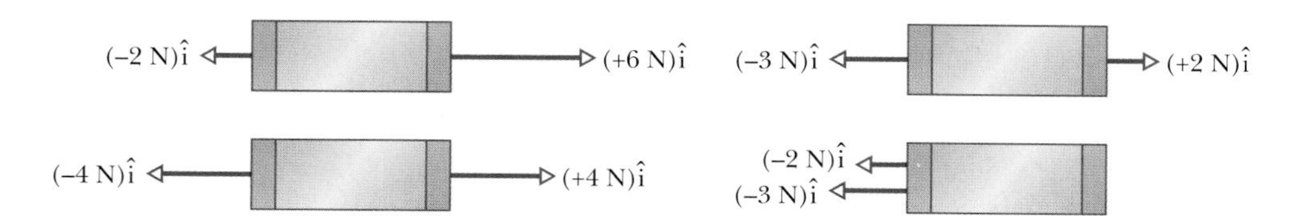

TOUCHSTONE EXAMPLE 3-2: Three Forces

In the overhead view of Fig. 3-18, a 2.0 kg cookie tin is accelerated at 3.0 m/s² in the direction shown by $\vec{a}$, over a frictionless horizontal surface. The acceleration is caused by three horizontal forces, only two of which are shown: $\vec{F}_A$ with a magnitude of 10 N and $\vec{F}_B$ with a magnitude of 20 N. Choose a coordinate system and then use it to express the third force $\vec{F}_C$ in unit-vector notation.

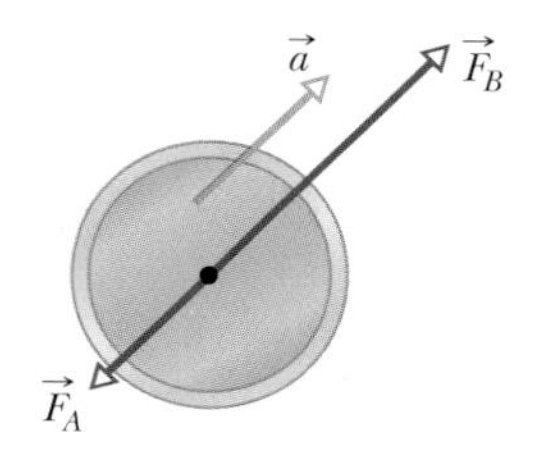

FIGURE 3-18 ■ Three forces act to produce an acceleration in the direction shown. Only two of the three forces causing this acceleration are included in this picture.

SOLUTION ■ The **Key Idea** here is that the net force $\vec{F}^{\text{net}}$ on the tin is the sum of the three forces and is related to the acceleration $\vec{a}$ of the tin via Newton's Second Law ($\vec{F}^{\text{net}} = m\vec{a}$). Thus,

$$\vec{F}_A + \vec{F}_B + \vec{F}_C = m\vec{a},$$

which gives us

$$\vec{F}_C = m\vec{a} - \vec{F}_A - \vec{F}_B. \qquad (3\text{-}6)$$

A second **Key Idea** is that this is a one-dimensional problem for which two of the forces and the acceleration are all along the same line. This means that the third force must also lie along the line of the acceleration. Thus we are able to choose a coordinate system in which the three forces lie along a single axis. If we choose our x axis to align with these forces, we have

$$\vec{F}_C = m\vec{a} - \vec{F}_A - \vec{F}_B = (ma_x)\hat{\text{i}} - F_{A\,x}\hat{\text{i}} - F_{B\,x}\hat{\text{i}}$$
$$= (ma_x - F_{A\,x} - F_{B\,x})\,\hat{\text{i}}.$$

Choosing the positive direction to be in the direction of the acceleration, components a_x and $F_{B\,x}$ are positive and the component $F_{A\,x}$ is negative. Thus $\vec{a} = (+3.0\text{ m/s}^2)\hat{\text{i}}$, $\vec{F}_B = (+20\text{ N})\hat{\text{i}}$, and $\vec{F}_A = (-10\text{ N})\hat{\text{i}}$.

Then, substituting known data and factoring $\hat{\text{i}}$ out of the equation, we find

$$\vec{F}_C = [(2.0\text{ kg})(3.0\text{ m/s}^2) - (-10\text{ N}) - (+20\text{ N})]\hat{\text{i}} = (-4\text{ N})\hat{\text{i}}. \qquad \text{(Answer)}$$

3-8 All Forces Result from Interaction

Careful observation of everyday motions should convince you that objects do not spontaneously speed up, slow down, or change direction. Clearly, pushes, pulls, bumps, winds, interactions with a surface during sliding motion, and so on, will influence an object's velocity by changing the object's speed, direction, or both. According to Newton's Second Law, changes in velocity (accelerations) occur only when the object experiences forces. Forces are always due to the presence of one or more other objects.

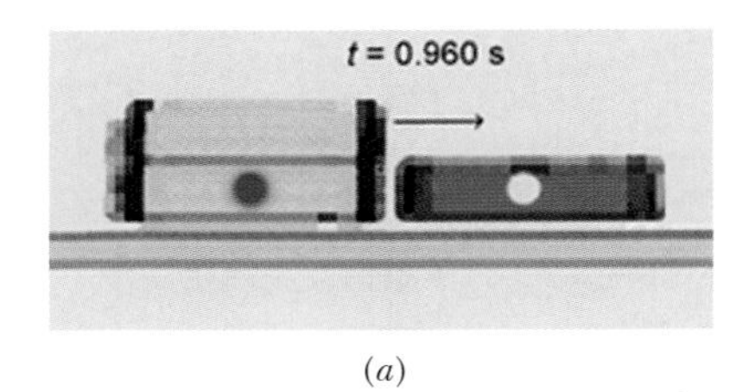

(a)

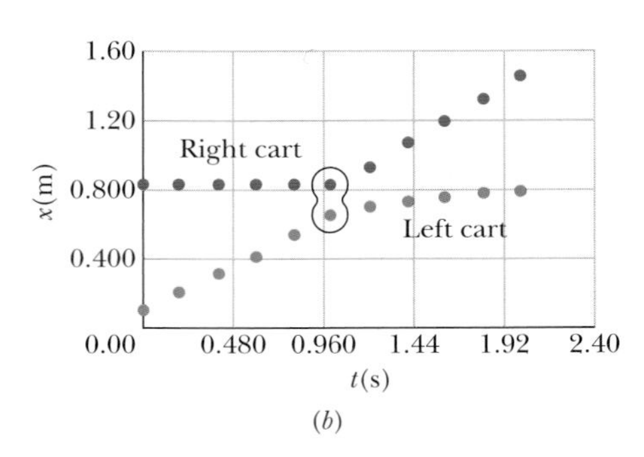

(b)

FIGURE 3-19 ■ Two low-friction carts are outfitted with neodymium magnets that repel each other. Initially the cart on the left bears down on the stationary cart on the right. (a) A video frame shows the carts interacting briefly at about 0.960 s but never touching. (b) Graphs of position vs. time for the two carts were obtained using video analysis. An examination of the changes in the slopes for each cart representing their velocity components enables us to deduce that the carts undergo velocity changes due to forces acting in opposite directions. A force to the left on the large cart slows it down. A force to the right on the small stationary cart starts it moving to the right.

In the course of your study of physics, you will be reading and hearing about dozens of forces. Adjectives such as net, combined, total, friction, contact, collision, normal, tension, spring, gravitational, electrostatic, magnetic, atomic, molecular, and so on, will be bandied about. It turns out that currently there are only four fundamental forces that are known: gravitational, electromagnetic, weak nuclear, and strong nuclear. However, essentially all of the types of forces introduced in this book (including "friction forces," "contact forces," and "collision forces") are actually fundamental forces (either electromagnetic or gravitational.) These other descriptive adjectives are used only to help us understand the physical situation in which various forces occur.

For example, in the first experiment we presented in this chapter, we tracked the motion of objects that roll or slide to a stop on different horizontal surfaces after a kick. Figure 3-3 showed that the rate of decrease of velocity for each of these objects was a constant. In other words, each object experienced a constant acceleration until it came to rest (or, in the case of the cart on the track, until data were no longer collected). What causes the objects to slow down? Consider Newton's Second Law and our definition of force (as an agent that causes an acceleration). We must conclude that each of the objects experienced a force. The only obvious interactions are interactions with the surface along which it was sliding or rolling. We call this type of contact interaction a friction force or, informally, friction. We found that in these cases the direction of the friction force on an object is opposite to the direction of the object's motion. We know this because the object slows down.

Forces like pushes, pulls, those experienced in a collision, and the friction force on a sliding or rolling object that moves over a surface are called **contact forces** because the objects involved appear to touch. The interactions that cause contact forces are ultimately due to a superposition of many small electromagnetic forces between the electrons and protons that the materials "in contact" are made of. Thus, contact forces are ultimately electromagnetic forces!

Another important contact force is a pull force exerted through a string, rope, cable, or rod attached to an object. This type of pulling force has a special name. It is called **tension.** Tension is always a pull force. Hence, the direction of a tension force is always the direction in which one would *pull* the object with a string or rope. The fundamental nature and origin of tension forces and frictional forces are discussed in more detail in Chapter 6.

Many other forces seem much less obvious than contact forces because they act at a distance. Electromagnetic and gravitational forces are capable of acting over large distances. But as you will learn in this chapter and later in this book, the source of these invisible or noncontact forces are not totally mysterious. If an everyday object experiences an "invisible force," we are always able to find it interacting with other objects that have some combination of electrical charges, magnets, electrical currents, or masses. An example of this is shown in Fig. 3-19, where two carts interact "at-a-distance" by means of magnetic forces that act in opposite directions.

READING EXERCISE 3-9: Consider Fig. 3-3 depicting the results of measurements on the motion of objects just after they have been given swift kicks along a positive x axis. Assume that the cart and the box both have the same mass of 0.5 kg. (a) What is the acceleration of the box on the carpet? Is it positive or negative? (b) What is the acceleration of the cart on the track? Is it positive or negative? ■

READING EXERCISE 3-10: Consider your answers to Reading Exercise 3-9. Assume that the cart and the box both have the same mass of 0.5 kg. (a) In each case is the friction force on the object constant or changing as the box or cart slows down? Cite evidence for your answer. (b) What is the magnitude of the friction force on the box due to its interaction with the carpet? Does it point to the right or the left? (c) What is the magnitude of the friction force on the cart due to the combined interaction of the cart wheels with the track and the cart axle? Does it point to the right or the left? ■

3-9 Gravitational Forces and Free Fall Motion

We now consider the forces the Earth exerts on objects near its surface. These forces are called "gravitational forces." Since we don't want to complicate our exploration with air resistance, we will limit ourselves to considering the motions of bodies that are relatively dense, small, and smooth like balls and coins. Also, assume that these objects are not moving at high speeds—say, in excess of about 5 m/s. In Chapter 6 we discuss situations where air resistance is a significant factor, and in Chapter 14, we explore the question of how masses such as galaxies and planets exert gravitational forces on each other in more general circumstances.

The Gravitational Acceleration Constant

We know that any object dropped near the Earth's surface falls, but the fall is so rapid that we can't easily describe it. Does the object suddenly speed up to a natural velocity and then fall at that rate or does it keep speeding up? A casual observation tells the story. Imagine lying on the floor while someone drops an apple on your forehead from different heights. The impact of the apple will feel harder when the apple is dropped from a greater height, so the apple must keep speeding up. The strobe photo in Fig. 3-20 confirms that an apple and a feather falling in a vacuum keep speeding up.

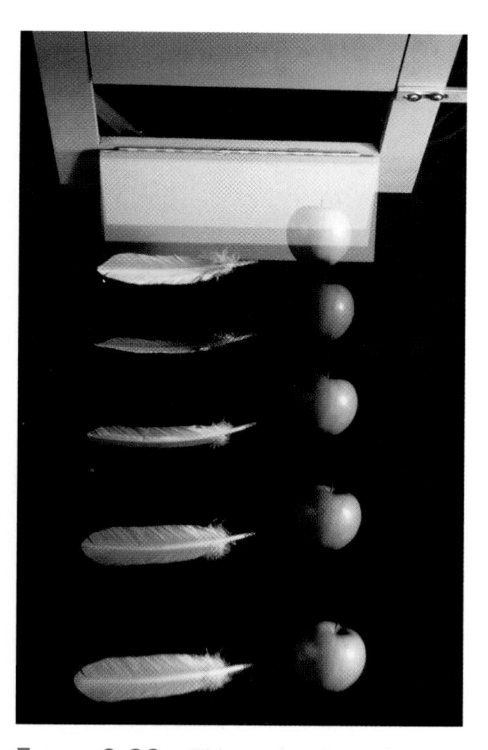

FIGURE 3-20 ■ This strobe photo shows a feather and an apple, undergoing free fall in a vacuum. The time interval between each exposure and the next is constant. The feather and the apple appear to be speeding up at the same rate, as evidenced by the increase in distance between the successive images.

At the end of Section 2-5 we asserted without any evidence that if there are no other forces on an object, it would move *downward* with a magnitude of acceleration $a = 9.8\ \text{m/s}^2$. But how do we know this? Back in the early 17th century Galileo rolled small balls of different masses down a ramp to slow their falling rates. He found that the velocities of all the balls increased at the same rate.

Today we can use modern technology such as ultrasonic motion detectors, video analysis, and strobe photos to make high-speed measurements of the position and time of an object falling straight down like the tossed ball in Fig. 3-21. For example, Figure 3-22 shows an analysis of a video clip of a small plastic ball shot vertically into the air with a spring-loaded launcher. The graph of the y-component of velocity vs. time was produced assuming that the y axis is pointing up. The graph shows that the ball is changing its velocity at the same constant rate when the ball is moving upward, turning around and moving downward. The measured acceleration component is the slope of the v_y vs. t graph, and a linear fit to the graph yields a vertical gravitational acceleration component of $-9.8\ \text{m/s}^2$. Within the limits of experimental uncertainty, we obtain the same result if we use a lead ball instead of a plastic one or for that matter any other object that doesn't experience much air resistance.

The magnitude of the acceleration we measured, denoted as $|\vec{a}|$ or a, is known as the **gravitational acceleration constant** given by $a = 9.8\ \text{m/s}^2$. If air resistance is significant, as is the case for a feather or sheet of paper falling through air, we will not obtain the gravitational acceleration constant from an experiment like the one we just described. However, the fact that the gravitational acceleration is independent of an object's mass, density, or shape can be verified by removing air in the vicinity of a falling object. In Fig. 3-20, a feather and an apple are shown accelerating downward at the same rate in a vacuum in spite of the fact that they have very different masses, shapes, and sizes.

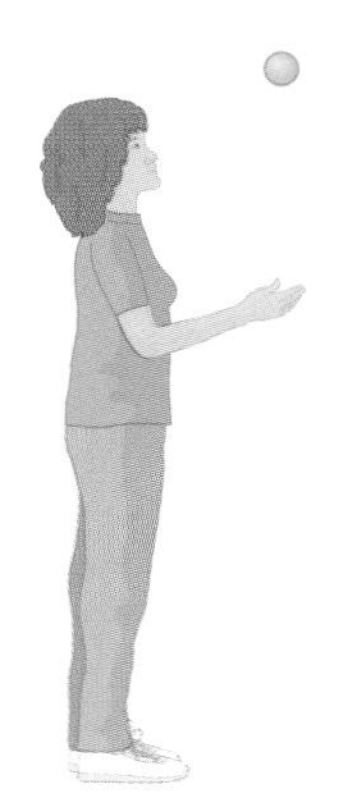

FIGURE 3-21 ■ A ball of arbitrary mass is tossed in the air near the surface of the Earth. What is its acceleration?

Gravitational Force and Mass Revisited

Since acceleration requires a net force, and the Earth doesn't need to touch an object to make it accelerate, we conclude that "gravity" is a *noncontact* force. Another piece of evidence that this force of attraction exists is that if you hang an object vertically from the spring force scale we developed in Section 3-4, the spring will be stretched. The stretch of the spring implies that there is something pulling down on the object and that the spring stretches just enough to pull the object up with the same magni-

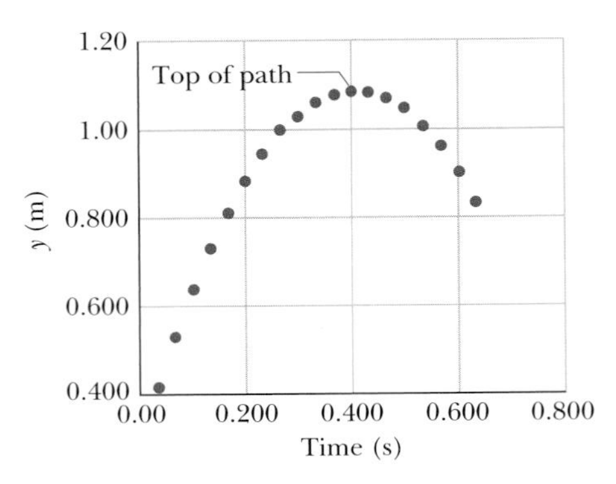

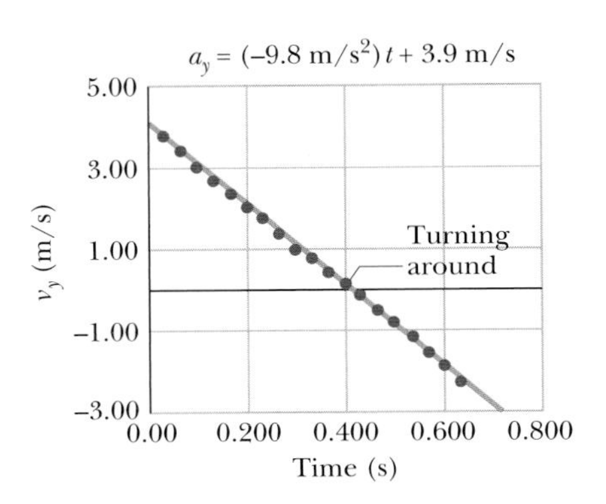

FIGURE 3-22 ■ Video analysis software is used to perform a frame-by-frame analysis of a digital movie depicting the motion of a small tossed ball. Graphs of position vs. time and the calculated average velocity vs. time are shown. A fit of the velocity vs. time graph reveals that the ball undergoes a constant acceleration in the downward direction of magnitude 9.8 m/s^2.

tude of force. If we use a spring scale like that shown in Fig. 3-23 to measure the gravitational force on an object, we find that it is directly proportional to the mass. This is not surprising since we know by experience that a bigger mass is harder to lift. We can express the proportionality between mass and gravitational force $\vec{F}^{\,\text{grav}}$ in terms of the force magnitude as

$$F^{\text{grav}} = mg, \qquad \text{so that } g = \frac{F^{\text{grav}}}{m} \tag{3-7}$$

where this constant of proportionality g is defined as the **local gravitational strength.** The magnitude of the gravitational force F^{grav} is commonly referred to as **weight.** Up to an altitude of 16 km or so, g can be expressed to two significant figures as

$$g = 9.8 \text{ N/kg} \qquad \text{(the Earth's local gravitational strength).}$$

Newton's Second Law predicts that, for an object of mass m that has no other forces on it except the gravitational force, the object will fall with an acceleration of magnitude

$$a = F^{\text{grav}}/m = 9.8 \text{ m/s}^2 \qquad \text{(gravitational acceleration constant).} \tag{3-8}$$

Thus we see that a and g have the same value and *different* but dimensionally equivalent units. We use m/s^2 when describing the gravitational acceleration a. We use the units N/kg when describing the local gravitational strength, g.

Equation 3-7 tells us that the Earth pulls harder on a larger mass, whereas Eq. 3-8 tells us the larger mass is harder to accelerate. These two mass-dependent effects cancel each other! Thus, near the Earth's surface,

> The *magnitude* of the acceleration of any falling object is that of the **GRAVITATIONAL ACCELERATION CONSTANT** 9.8 m/s^2, independent of the mass of the falling object.

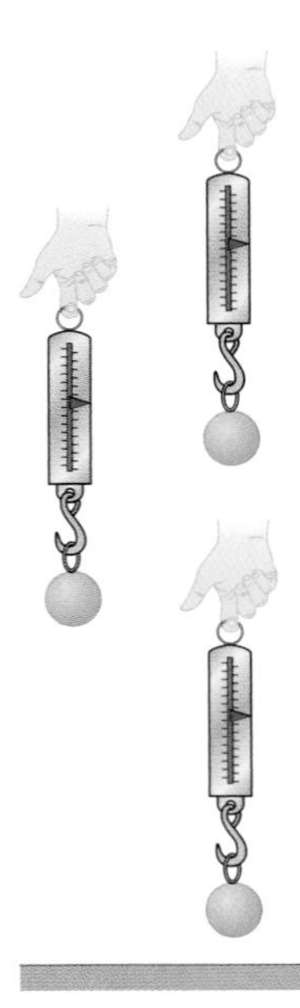

FIGURE 3-23 ■ Depiction of a spring scale used to determine the gravitational force on an object near the surface of the Earth. The scale reading is essentially the same at all heights reasonably near the Earth's surface (including those found in high flying passenger jets).

Other Properties of the Local Gravitational Force

So what are the characteristics of the gravitational force of attraction exerted by the Earth on objects near its surface? Does this force change as time passes or if the position of the object changes? The answers to these questions become clear if we consider an object hanging vertically from our spring force scale at different times and places.

Time Dependence: What we see when performing force measurements with a spring scale is that, for a given object, the amount that the spring stretches changes very little as time passes. Hence, we conclude that, at least over the span of a human life, *the force of gravitational attraction does not appear to be changing over time.*

Height Dependence: How does the gravitational force on a given object change with its height above the Earth's surface or its location? The stretch of a spring scale is approximately the same if we are standing at sea level, on top of a table, on top of a tall building, on top of a mountain, or inside of a high-flying passenger jet. This idea is pictured in Fig. 3-23. The gravitational force actually decreases with distance from the surface of the Earth, but the percentage change over the range of elevations that we have described is not measurable to the two significant figures that we have been using to describe g. There turns out to be a slight dependence on location and height. But for all heights and locations where people normally travel, the magnitude of the gravitational force of attraction the Earth exerts on another object is the same to two significant figures.

Direction: The direction of the gravitational force is apparently down. Since the Earth is approximately spherical, if we look at the Earth's gravitational force from the perspective of outer space, its direction changes from place to place. It will be different in Australia than in the United States.

Using the Kinematic Equations

Because the constant force of gravity near the surface of the Earth imparts a constant acceleration to objects on which it acts, the kinematic equations of motion derived in Chapter 2 (Table 2-1) can be used to describe free fall near the Earth's surface, but only as long as there are no other nonconstant forces present. The kinematic equations that you worked with in the last chapter describe motion along a line with constant acceleration.

Though the value of g does vary slightly with latitude and elevation, you may safely use a value of 9.8 m/s^2 (or 32 ft/s^2) in free fall calculations near the Earth's surface as long as air resistance is considered negligible. For many calculations, 10 m/s^2 is a convenient approximation, since it varies by only 2% from the more precise value.

When we introduced one-dimensional motion in Chapter 2, we noted that when thinking about the motion of objects, we have freedom to choose our coordinate system. However, to make communicating about these ideas easier, for now we will continue to use a *vertical* y axis that points up as shown in Fig. 3-24. In this coordinate system

$$\vec{F}^{\text{grav}} = -mg\,\hat{\text{j}} = ma_y\,\hat{\text{j}}, \tag{3-9}$$

where a_y is the y-component of the falling object's acceleration and $\hat{\text{j}}$ the dimensionless unit vector associated with the y axis.

We can easily construct kinematic equations to describe the relationships between vertical vector components for a freely falling object close to the Earth's surface. We simplify writing the equations in Table 2-1 in Chapter 2 by: (1) replacing position component x with the symbol y; (2) adding the subscript y to the velocity component to remind us that it is the component of velocity along the y axis; (3) replacing the component of acceleration along the vertical axis that was denoted a_x with $a_y = -g$. *Note:* We have chosen upward to be positive.

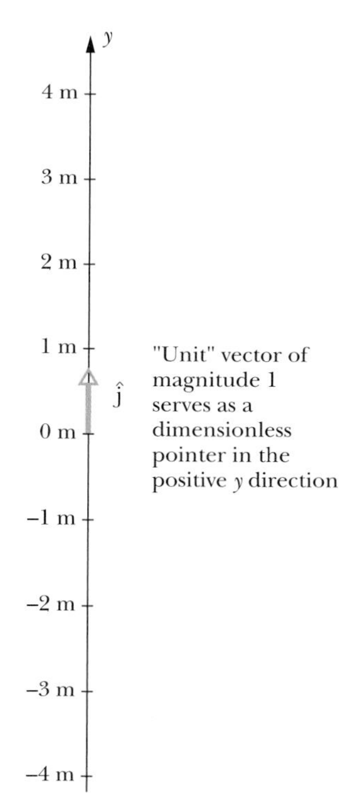

FIGURE 3-24 ■ It is customary to designate a vertical axis as the y axis, reserving the term x axis for the horizontal direction. The upward direction is typically given as positive. The unit vector is labeled $\hat{\text{j}}$ rather than $\hat{\text{i}}$, and it points upward in the positive y direction.

READING EXERCISE 3-11: Suppose that you throw an object upward and can ignore air resistance to the motion of the object. At the highest point in this motion, the object's velocity is instantaneously zero as it reverses direction. Does this mean that the object's acceleration is zero at that point? Explain how your answer is consistent with: (a) the definitions of instantaneous velocity and acceleration that you have learned; (b) the graphs in Fig. 3-22. ■

READING EXERCISE 3-12: Rewrite the equations in Table 2-1 so they describe the motion of an object in vertical free fall. Use a conventional coordinate system with the y axis pointing up. ■

TOUCHSTONE EXAMPLE 3-3: Model Rocket

A model rocket with a mass of 0.50 kg is fired vertically from the ground. Assume that it is streamlined enough that air resistance can be ignored. Suppose it ascends under the influence of a constant net force of 2.0 N acting in a vertical direction and travels for 6.0 s before its fuel is exhausted. Then it keeps moving as a particle-like object in free fall as it continues upward, turns around, and falls back down.

(a) How high is the rocket when it runs out of fuel? What is its velocity at that time?

SOLUTION ■ The net force on the rocket is a combination of the upward thrust of the rocket engine and the downward pull of the Earth. The **Key Idea** here is that we can use Newton's Second Law and our knowledge of the constant net force to find the rocket's constant acceleration and then use a kinematic equation to find out how high it will go in 6.0 seconds with that constant acceleration.

Using Eq. 3-5 for Newton's Second Law we get

$$\vec{a} = \frac{\vec{F}^{\text{net}}}{m} = \frac{(2.0\text{ N})\,\hat{\jmath}}{0.50\text{ kg}} = (+4.0\text{ m/s}^2)\,\hat{\jmath} = a_y\,\hat{\jmath}.$$

Thus the vertical component of acceleration is $a_y = +4.0\text{ m/s}^2$. The elapsed time since take-off is given by $t_2 - t_1 = 6.0$ s. Since the rocket is fired at the ground level, $y_1 = 0.0$ m and $v_{1y} = 0.0$ m/s. Thus we can put numbers in the primary kinematic equation (Eq. 2-17) to get the height of the rocket at the time the fuel has run out,

$$(y_2 - y_1) = v_{1y}(t_2 - t_1) + \tfrac{1}{2}a_y(t_2 - t_1)^2$$
$$= 0.0\text{ m} + \tfrac{1}{2}(4.0\text{ m/s}^2)(6.0\text{ s})^2 = 72\text{ m}. \quad \text{(Answer)}$$

We need to find the y-component of velocity just as the rocket's fuel runs out. This is given by the other primary kinematic equation (Eq. 2-13) with respect to time to get

$$v_{2y} = v_{1y} + a_y\Delta t = 0.0\text{ m/s} + (4.0\text{ m/s}^2)(6.0\text{ s}) = 24\text{ m/s}. \quad \text{(Answer)}$$

(b) What is the total height that the rocket rises?

SOLUTION ■ The rocket is now at 72 m above the ground, moving upward with a velocity component of 24 m/s. We need to know how much higher it will go when the only significant force acting on it is the gravitational pull of the Earth. Let's do this by using only the primary kinematic equations for free fall with $a_y = -g$ so that

$$v_{2y} = v_{1y} - g\Delta t \quad \text{(Eq. 2-13)}$$

$$\text{and} \quad (y_2 - y_1) = v_{1y}(t_2 - t_1) - \tfrac{1}{2}g(t_2 - t_1)^2. \quad \text{(Eq. 2-17)}$$

The **Key Idea** here is to use Eq. 2-13 to find the time it takes the rocket to go from its new initial velocity of 24 m/s to its "final" velocity of 0 m/s and then use Eq. 2-17 to find the additional distance moved in the upward direction. Solving $v_{2y} = v_{1y} - g\Delta t$ for the elapsed time Δt gives

$$\Delta t = \frac{v_{2y} - v_{1y}}{-g} = \frac{(0 - 24)\text{ m/s}}{-9.8\text{ m/s}^2} = 2.45\text{ s}.$$

Solving Eq. 2-17 for the additional rise of the rocket using $\Delta t = 2.45$ s gives

$$(y_2 - y_1) = v_{1y}\Delta t - \tfrac{1}{2}g(\Delta t)^2$$
$$= (24\text{ m/s})(2.45\text{ s}) - \tfrac{1}{2}9.8(2.45\text{ s})^2 = 29.4\text{ m}.$$

When added to the previous rise of the rocket under thrust we get

$$\text{maximum height} = 72\text{ m} + 29\text{ m} = 101\text{ m}. \quad \text{(Answer)}$$

(c) What is the net force on the rocket when it continues upward as a free fall particle? As it turns around? When it is traveling back toward the ground?

SOLUTION ■ A **Key Idea** here is that the only force on the rocket in free fall is the gravitational force. A second **Key Idea** is that this force is the same whether the rocket is moving up, turning around, or falling down. Its magnitude is given by the mass of the rocket times the gravitational constant g. In vector notation, the force is

$$\vec{F} = m\vec{a} = m(-g)\,\hat{\jmath} = (0.50\text{ kg})(-9.8\text{ N/kg})\,\hat{\jmath} = (-4.9\text{ N})\,\hat{\jmath}.$$

3-10 Newton's Third Law

Newton's first two laws of motion describe what happens to a *single* object that has forces acting on it. We made the claim in Section 3-8 that for every object that experiences a force there is another object causing that force. Further, we claimed that interactions between two objects always seem to go two ways. We begin this section with a discussion of observations Newton made of the two-way interaction between hanging magnets. We can then state Newton's Third Law, which deals with the relationship between the forces objects exert on each other. We end the section by presenting experimental evidence for the validity of Newton's Third Law using measurements of contact forces.

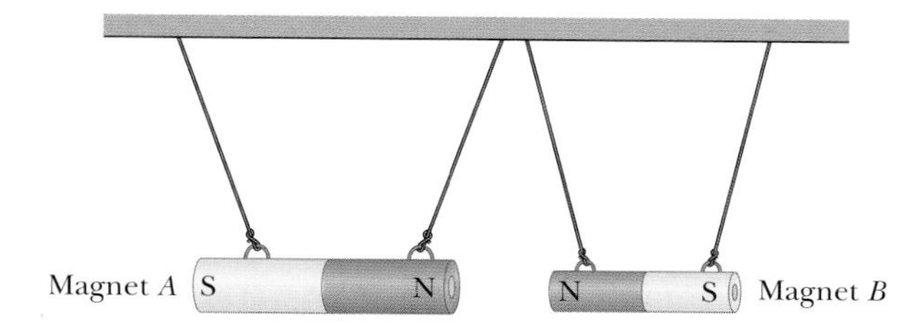

FIGURE 3-25 ■ Two magnets of different masses hang from long strings. They are pushed together and released. What happens to them as a result of magnetic repulsion forces?

Qualitative Considerations

Suppose that you hang two strong magnets side by side from long strings with their north poles facing each other as shown in Fig. 3-25. Many of us have observed that the north poles of magnets repel one another. If you were to hold the two north poles very close to each other and let go of the magnets, they would start to accelerate away from each other. The fact that *both* magnets are repelled and begin to accelerate implies that *each* magnet has a force acting on it. If you were to do this with magnets of the same mass, you would observe that the magnitudes of the two accelerations are identical. Observations of the accelerations of the magnets suggest that they are experiencing magnetic forces that have the same magnitude but are oppositely directed. (Actually, to get good measurements we should either mount our magnets on low-friction carts or hang them from long strings so the strings don't exert net horizontal forces on the magnets.)

This notion of equal and opposite forces is familiar to us in the case of contact forces. If you push on a wall it pushes back. This doesn't hurt if you push gently, but if you punch a wall hard it hurts very much. Newton hypothesized that any time two objects interact in such a way that a force is exerted on one of them, there is *always* a force that is equal in magnitude exerted in the opposite direction on the other object. This hypothesis is called Newton's Third Law, and we can state it simply in modern language.

> **NEWTON'S THIRD LAW:** If one object is exerting a force on a second object, then the second object is also exerting a force back on the first object. The two forces have exactly the same magnitude but act in opposite directions.

The most significant idea contained in Newton's Third Law is that *forces always exist in pairs*. It is very important that we realize we are talking about two *different forces* acting on two *different objects*.

In trying to visualize the application of this concept in the situation involving the magnets, it is helpful to draw a force vector at the center of each magnet showing the horizontal force it is experiencing from the other magnet. (Drawing the net force vector at the center of the object on which it acts is another of our many idealizations. The rod-shaped magnets are not really point particles, and each part of one magnet may be exerting forces on each part of the other and vice versa. However, in this situation, it turns out that assuming the rods are particle-like leads us to the same conclusions that treating them like rods would.)

Figure 3-26 shows the force diagrams for the two magnets discussed above, assuming that there are no other forces acting on them. The force exerted on object A by object B is denoted $\vec{F}_{B\to A}$ and the force exerted on object B by object A is denoted $\vec{F}_{A\to B}$. This notation allows us to write an equation that summarizes Newton's Third Law as follows:

$$\vec{F}_{B\to A} = -\vec{F}_{A\to B} \quad \text{(Newton's Third Law in equation form).} \tag{3-10}$$

The order of the letters in the subscripts on the force is very important because they tell us which object the force is acting on and the origin of the force. The first letter

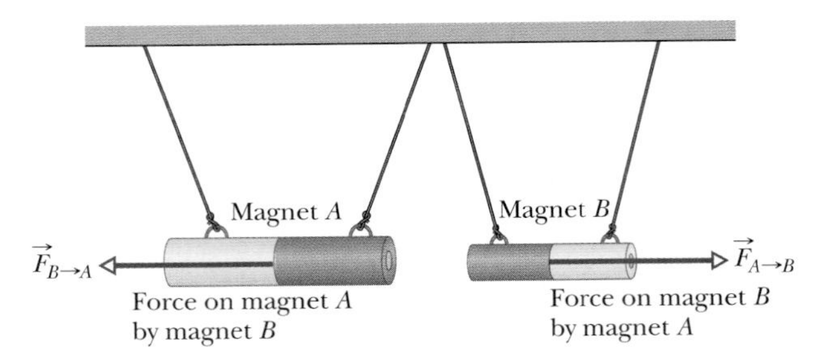

FIGURE 3-26 ■ We can draw an interaction force vector at the center of each magnet (assuming that Newton's Third Law describes the interactions between two magnets and that the magnets are particle-like in their mutual interaction).

denotes the object that exerts the force and the second letter denotes the object that feels the force. We call the forces shown in Eq. 3-10 between the two interacting magnets a **third-law force pair.** In situations where Newton's laws apply, we believe that if any two bodies are interacting, a third-law force pair is always present.

Experimental Verification for Contact Forces

We have developed Newton's Third Law in a qualitative fashion by doing a thought experiment. No measurements were taken to verify the law quantitatively. We have asserted that it holds whenever two bodies interact with each other. Now, let's consider whether the Third Law applies to objects that interact via contact (touching) forces. This time we will make measurements to verify the Third Law in a quantitative fashion.

Suppose two people hook the ends of two force sensors together as shown in Fig. 3-27 and have a back-and-forth tug of war. What happens?

If we interface these force sensors to a computer for data collection, the result would look something like what is shown in Fig. 3-28. This graph verifies that on a moment-by-moment basis the force ($\vec{F}_{B\to A}$) exerted on the person on the left by the person on the right is equal in magnitude but opposite in direction to the force ($\vec{F}_{A\to B}$) exerted on the person on the right by the person on the left.

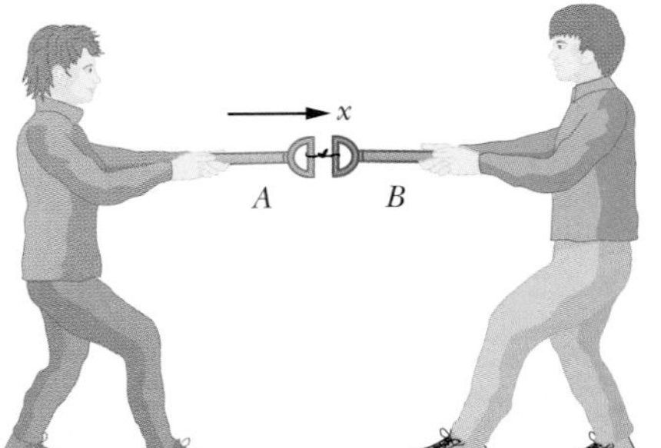

FIGURE 3-27 ■ Two people are playing tug-of-war with electronic force sensors hooked together.

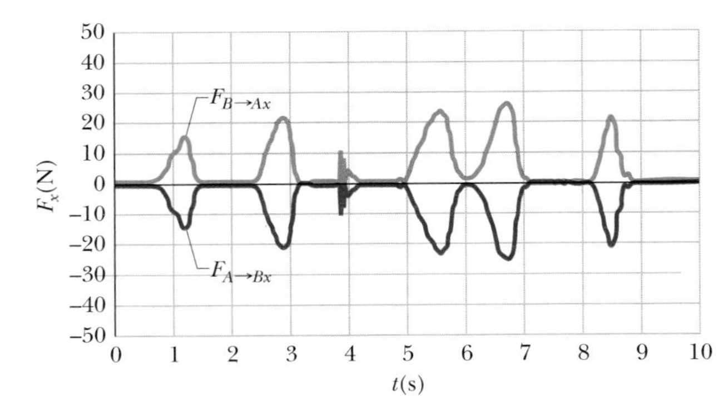

FIGURE 3-28 ■ A graph of the measured force vs. time for two people playing tug-of-war for 10 seconds. A computer data acquisition system was used to collect and display the data at a rate of 100 readings per second. $\vec{F}_{B\to A}$ is exerted on the person on the left and $\vec{F}_{A\to B}$ on the person on the right.

We have considered magnetic forces (one form of electromagnetic force) and contact forces (another form of electromagnetic force). Does Newton's Third Law also apply to cases where masses are very different and when gravitational forces are present? Is it true for high-speed collisions? Is it true when one object is stationary and the other is not moving at first? For example, is it true for a very heavy truck traveling at high speed that collides head-on with a small car that is at rest? Is it true for a baseball in free fall interacting with the Earth? The answer to all these questions is "yes." We will return to them in Chapter 7, where we discuss the experimental evidence that Newton's Third Law can help us predict the outcomes of collisions between objects.

READING EXERCISE 3-13: Suppose that the magnet on the left in Fig. 3-25 is replaced by a steel paper clip that is not magnetized. (a) What can you say about the force that the initially unmagnetized paper clip exerts on the magnet on the right compared to the force the magnet on the right exerts on the paper clip? (b) Do you think Newton's Third Law holds? Explain. ■

TOUCHSTONE EXAMPLE 3-4: Pushing Two Blocks

In Fig. 3-29a, a constant horizontal force $\vec{F}^{\,app}$ of magnitude 20 N is applied to block A of mass $m_A = 4.0$ kg, which pushes against block B of mass $m_B = 6.0$ kg. The blocks slide over a frictionless surface, along an x axis.

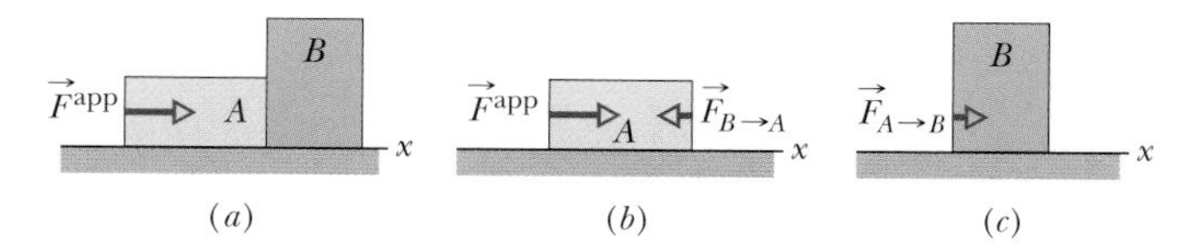

FIGURE 3-29 ■ (a) A constant horizontal force $\vec{F}^{\,app}$ is applied to block A, which pushes against block B. (b) Two horizontal forces act on block A: applied force $\vec{F}^{\,app}$ and force $\vec{F}_{B\to A}$ from block B. (c) Only one horizontal force acts on block B: force $\vec{F}_{A\to B}$ from block A.

(a) What is the acceleration of the blocks?

SOLUTION ■ We shall first examine a solution with a serious error, then a dead-end solution, and then a successful solution.

Serious Error: Because force $\vec{F}^{\,app}$ is applied directly to block A, we use Newton's Second Law to relate that force to the acceleration $\vec{a}$ of block A. Because the motion is along the x axis, we use that law for x-components ($F_x^{\,net} = ma_x$), writing it as

$$F_x^{\,net} = m_A\, a_x.$$

However, this is seriously wrong because $\vec{F}^{\,app}$ is not the only horizontal force acting on block A. There is also the force $\vec{F}_{B\to A}$ from block B (as shown in Fig. 3-29b).

Dead-End Solution: Let us now include force $\vec{F}_{B\to A}$ by writing, again for the x axis,

$$F_x^{\,app} + F_{B\to A\,x} = m_A\, a_x$$

where $F_x^{\,app}$ is positive, but $F_{B\to Ax}$ is negative. However, $F_{B\to Ax}$ is a second unknown, so we cannot solve this equation for the desired acceleration a_x.

Successful Solution: The **Key Idea** here is that, because of the direction in which force $\vec{F}^{\,app}$ is applied, the two blocks form a rigidly connected system. We can relate the net force *on the system* to the acceleration *of the system* with Newton's Second Law. Here, once again for the x axis, we can write that law as

$$F_x^{\,app} = (m_A + m_B)a_x,$$

where now we properly apply $\vec{F}^{\,app}$ to the system with total mass $m_A + m_B$. Solving for a_x and substituting known values, we find

$$a_x = \frac{F_x^{\,app}}{m_A + m_B} = \frac{20\text{ N}}{4.0\text{ kg} + 6.0\text{ kg}} = 2.0\text{ m/s}^2.$$

Thus, the acceleration of the system and of each block is in the positive direction of the x axis and has the magnitude 2.0 m/s^2.

(b) What is the force $\vec{F}_{A\to B}$ on block B from block A (Fig. 3-29c)?

SOLUTION ■ The **Key Idea** here is that we can relate the net force on block B to the block's acceleration with Newton's Second Law. Here we can write that law, still for components along the x axis, as

$$F_{A\to B\,x} = m_B\, a_x,$$

which, with known values, gives

$$F_{A\to B\,x} = (6.0\text{ kg})(2.0\text{ m/s}^2) = 12\text{ N}.$$

Thus, force $\vec{F}_{A\to B}$ is in the positive direction of the x axis and has a magnitude of 12 N.

TOUCHSTONE EXAMPLE 3-5: Pulling Two Blocks

Two blocks connected by a string are being pulled to the right across a horizontal frictional surface by another string, as shown in Fig. 3-30. The strings are horizontal and their masses are negligible compared to those of the blocks. The tension in the rightmost string is a constant 35 N. Find the tension in the other string if the mass of the left block is four times that of the right block.

SOLUTION ■ Note that we do not need to consider the tension in the rope between the objects when we treat them as a system, so as in the previous touchstone example, we choose to apply Newton's Second Law to the two-block system to find its acceleration. Although we don't know the masses of the individual blocks, we do know that $(m_{\text{left}}/m_{\text{right}}) = 4$. If we set $m_{\text{right}} = m$, then Newton's Second Law tells us that

$$F_x^{\,app} = (m + 4m)a_x.$$

Solving for a_x, we learn that

$$a_x = \frac{F_x^{\text{app}}}{5m}.$$

To find the tension, T, in the string between the two blocks, the **Key Idea** is to shift our attention from the two-block system to just the left-hand block. Its acceleration is the same as that of the two-block system, since they are joined by a string of constant length.

The second **Key Idea** is that the magnitude of the net force acting on the left block is equal to the tension, T_A, in the string joining the two blocks and that this force is directed to the right; that is, $\vec{F}_A = T_A \hat{\text{i}}$. Applying Newton's Second Law to the left block yields

$$F_{Ax} = T_A = 4ma_x.$$

But we've already seen that $a_x = F_x^{\text{app}}/5m$. Combining these two results tells us that

$$T_A = 4ma_x = \frac{4mF_x^{\text{app}}}{5m} = \tfrac{4}{5}F_x^{\text{app}}$$

$$= \tfrac{4}{5}(35\text{ N}) \qquad \text{(Answer)}$$

$$= 28\text{ N}.$$

We now see that T_A depends only on $|\vec{F}^{\text{app}}|$ and on the *ratio* of the two masses; we did not need to know the individual masses to solve the problem.

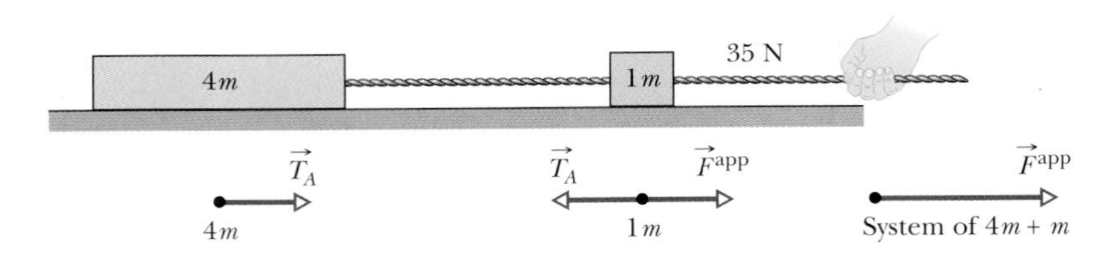

FIGURE 3-30

TOUCHSTONE EXAMPLE 3-6: Raising Bricks

Figure 3-31 shows a man raising a load of bricks from the ground to the first floor of a building using a rope hung over a pulley. Suppose the load of bricks weighs 900 N and the man weighs 1200 N. What is the maximum upward acceleration that the man can give to the load of bricks by pulling downward on his side of the rope?

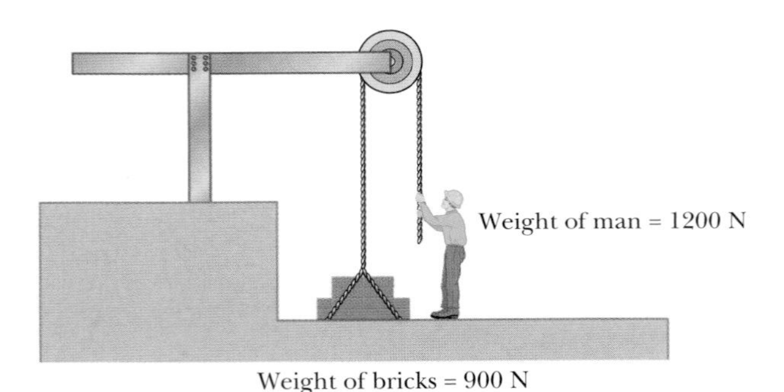

FIGURE 3-31

SOLUTION ■ One **Key Idea** here is that whatever force the man exerts downward on the rope, the rope in turn exerts upward on the man. Thus, if the man is not to accelerate upward, the maximum force he can exert downward on the rope is 1200 N. If he exceeds this, then he will experience a *net* upward force and accelerate upward. Thus the tension in the rope cannot exceed 1200 N.

Another **Key Idea** is that the tension in the rope is the same on each side of the pulley; ideal pulleys, such as this one, change the *direction* of the forces that ropes exert on objects but do not affect the *magnitude* of those forces. Thus the maximum *upward* force that the rope can exert on the load of bricks is 1200 N. Since gravity exerts a constant 900 N *downward* on the bricks, this limits the maximum vertical force on the bricks to

$$F_y^{\text{net max}} = F_{\text{rope}\rightarrow\text{bricks}\,y}^{\text{max}} + F_{\text{bricks}\,y}^{\text{grav}}$$

$$= 1200\text{ N} - 900\text{ N}$$

$$= 300\text{ N}.$$

Newton's Second Law then tells us that the maximum upward acceleration of the bricks is

$$a_y^{\text{max}} = \frac{F_y^{\text{net max}}}{m_{\text{bricks}}} = \frac{300\text{ N}}{(900\text{ N}/g)}$$

$$= \tfrac{1}{3}g$$

$$= 3.26\text{ m/s}^2. \qquad \text{(Answer)}$$

To find the mass of the bricks here, we have used the fact that the weight of the brick is equal to their mass times the local value of $g = 9.80\,\text{N/kg} = 9.80\,\text{m/s}^2$.

3-11 Comments on Classical Mechanics

A word of caution—classical mechanics does not apply to all situations. For instance, physicists know that if the speeds of the interacting bodies are very large—an appreciable fraction of the speed of light—we must replace Newtonian mechanics with *Einstein's special theory of relativity*. In addition, if the interacting bodies are molecules, atoms, or electrons within atoms, there are situations in which we must replace classical mechanics with *quantum mechanics*. Physicists now view Newtonian mechanics as a special case of these two more comprehensive theories. Still, classical mechanics is a very important special case of these other theories because it applies to the motion of objects ranging in size from that of large molecules to that of astronomical objects such as galaxies and galactic clusters. The domain of Newton's laws encompasses our everyday world including the translational, rotational and vibrational motions of cars, ships, airplanes, elevators, steam engines, our bodies, fluids, glaciers, the atmosphere, and oceans.

In the next chapter, we will introduce elements of vector mathematics that allow us to extend and apply Newtonian mechanics to more realistic situations involving motions in two dimensions. In spite of the limitations of Newton's laws, you will see throughout our study of basic classical physics that these laws are extraordinarily powerful in helping us describe, understand, and predict events involving motion in our everyday world and beyond.

Problems

SEC. 3-6 ■ NEWTON'S SECOND LAW FOR A SINGLE FORCE

1. Stopping a Neutron When a nucleus captures a stray neutron, it must bring the neutron to a stop within the diameter of the nucleus by means of the *strong force*. That force, which "glues" the nucleus together, is approximately zero outside the nucleus. Suppose that a stray neutron with an initial speed of 1.4×10^7 m/s is just barely captured by a nucleus with diameter $d = 1.0 \times 10^{-14}$ m. Assuming that the strong force on the neutron is constant, find the magnitude of that force. The neutron's mass is 1.67×10^{-27} kg.

2. Riding the Elevator A 50 kg passenger rides in an elevator that starts from rest on the ground floor of a building at $t = 0$ and rises to the top floor during a 10 s interval. The acceleration of the elevator as a function of the time is shown in Fig. 3-32, where positive values of the acceleration mean that it is directed upward. Give the magnitude and direction of the following forces: (a) the maximum force on the passenger from the floor, (b) the minimum force on the passenger from the floor, and (c) the maximum force on the floor from the passenger.

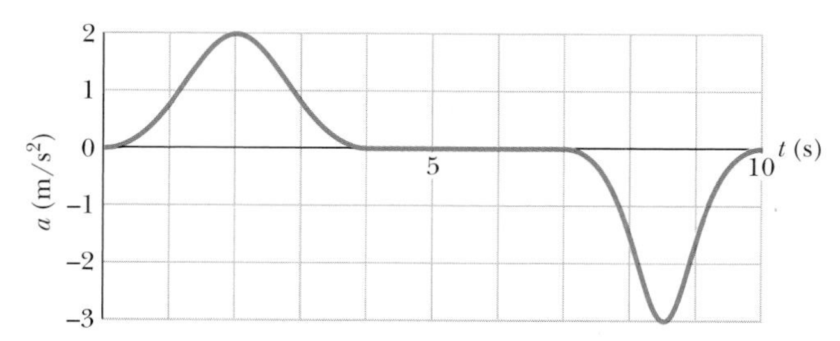

FIGURE 3-32 ■ Problem 2.

3. Sunjamming A "sun yacht" is a spacecraft with a large sail that is pushed by sunlight. Although such a push is tiny in everyday circumstances, it can be large enough to send the spacecraft outward from the Sun on a cost-free but slow trip. Suppose that the spacecraft has a mass of 900 kg and receives a push of 20 N. (a) What is the magnitude of the resulting acceleration? If the craft starts from rest, (b) how far will it travel in 1 day and (c) how fast will it then be moving?

4. Stopping a Salmon The tension at which a fishing line snaps is commonly called the line's "strength." What minimum strength is needed for a line that is to stop a salmon of weight 85 N in 11 cm if the fish is initially drifting at 2.8 m/s? Assume a constant acceleration.

5. Rocket Sled An experimental rocket sled can be accelerated at a constant rate from rest to 1600 km/h in 1.8 s. What is the magnitude of the required net force if the sled has a mass of 500 kg?

6. Stopping a Car A car with a mass of 1300 kg is initially moving at a speed of 40 km/h when the brakes are applied and the car is brought to a stop in 15 m. Assuming that the force that stops the car is constant, find (a) the magnitude of that force and (b) the time required for the change in speed. If the initial speed is doubled and the car experiences the same force during the braking, by what factors are (c) the stopping distance and (d) the stopping time multiplied? (There could be a lesson here about the danger of driving at high speeds.)

7. Rocket and Payload A rocket and its payload have a total mass of 5.0×10^4 kg. How large is the force produced by the engine (the thrust) when (a) the rocket is "hovering" over the launchpad just after ignition, and (b) the rocket is accelerating upward at 20 m/s^2?

8. Car Wreck A car traveling at 53 km/h hits a bridge abutment. A passenger in the car moves forward a distance of 65 cm (with respect to the road) while being brought to rest by an inflated air bag. What magnitude of force (assumed constant) acts on the passenger's upper torso, which has a mass of 41 kg?

9. The Fall An 80 kg man drops to a concrete patio from a window only 0.50 m above the patio. He neglects to bend his knees on landing, taking 2.0 cm to stop. (a) What is his average acceleration from when his feet first touch the patio to when he stops? (b) What is the magnitude of the average stopping force?

10. Starship An interstellar ship has a mass of 1.20×10^6 kg and is initially at rest relative to a star system. (a) What constant acceleration is needed to bring the ship up to a speed of $0.10c$ (where c is the speed of light, 3.0×10^8 m/s) relative to the star system in 3.0 days? (b) What is that acceleration in g units? (c) What force is required for the acceleration? (d) If the engines are shut down when $0.10c$ is reached (the speed then remains constant), how long does the ship take (start to finish) to journey 5.0 light-months, the distance that light travels in 5.0 months?

11. Force vs. Time Figure 3-33 gives, as a function of time t, the force component F_x that acts on a 3.00 kg ice block, which can move only along the x axis. At $t = 0$, the block is moving in the positive direction of the axis, with a speed of 3.0 m/s. What are its (a) speed and (b) direction of travel at $t = 11$ s?

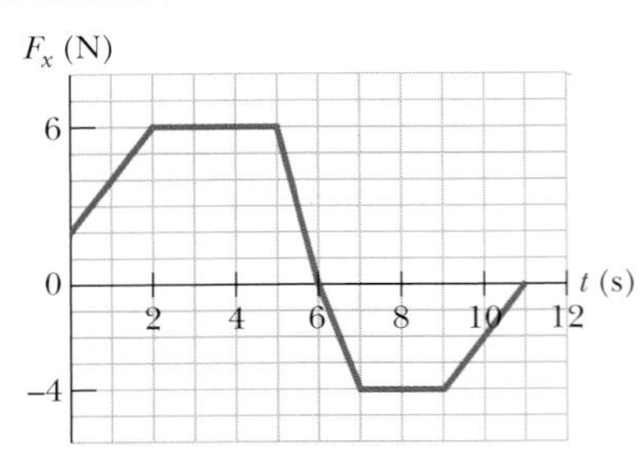

FIGURE 3-33 ■ Problem 11.

12. Variable Force A 2.0 kg particle moves along an x axis, being propelled by a variable force directed along that axis. Its position is given by

$$x = 3.0\text{ m} + (4.0\text{ m/s})t + ct^2 - (2.0\text{ m/s}^3)t^3,$$

with x in meters and t in seconds. The factor c is a constant. At $t = 3.0$ s, the force on the particle has a magnitude of 36 N and is in the negative direction of the axis. What is c? (Include units.)

SEC. 3-8 ■ ALL FORCES RESULT FROM INTERACTION

13. Two People Pull Two people pull with 90 N and 92 N in opposite directions on a 25 kg sled on frictionless ice. What is the sled's acceleration magnitude?

14. Take Off A Navy jet (Fig. 3-34) with a mass of 2.3×10^4 kg requires an airspeed of 85 m/s for liftoff. The engine develops a maximum force of 1.07×10^5 N, but that is insufficient for reaching takeoff speed in the 90 m runway available on an aircraft carrier. What minimum force (assumed constant) is needed from the catapult that is used to help launch the jet? Assume that the catapult and the jet's engine each exert a constant force over the 90 m distance used for takeoff.

FIGURE 3-34 ■ Problem 14.

15. Loaded Elevator An elevator and its load have a combined mass of 1600 kg. Find the pull (or tension) force supplied by in the supporting cable when the elevator, originally moving downward at 12 m/s, is brought to rest with constant acceleration in a distance of 42 m.

16. Four Penguins Figure 3-35 shows four penguins that are being playfully pulled along very slippery (frictionless) ice by a curator. The masses of three penguins and the tension in two of the cords are given. Find the penguin mass that is not given.

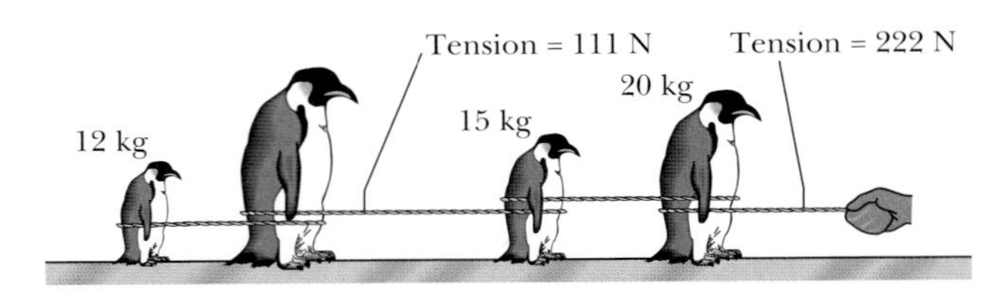

FIGURE 3-35 ■ Problem 16.

17. Elevator An elevator with a mass of 2840 kg is given an upward acceleration of 1.22 m/s² by a cable. (a) Calculate the tension in the cable. (b) What is the tension when the elevator is slowing at the rate of 1.22 m/s² but is still moving upward?

18. Three Blocks In Fig. 3-36 three blocks are connected and pulled to the right on a horizontal frictionless table by a force with a magnitude of $T_3 = 65.0$ N. If $m_A = 12.0$ kg, $m_B = 24.0$ kg, and $m_C = 31.0$ kg, calculate (a) the acceleration of the system and the magnitudes of the tensions (b) T_1 and (c) T_2 in the interconnecting cords.

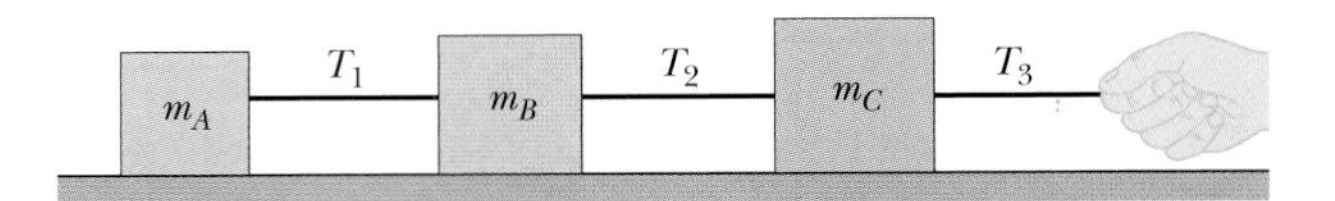

FIGURE 3-36 ■ Problem 18.

19. Hot-Air Balloon A hot-air balloon of mass M is descending vertically with downward acceleration of magnitude a. How much mass (ballast) must be thrown out to give the balloon an upward acceleration of magnitude a (same magnitude but opposite direction)? Assume that the upward force from the air (the lift) does not change because of the decrease in mass.

20. Lamp in Elevator A lamp hangs vertically from a cord in a descending elevator that slows down at 2.4 m/s². (a) If the tension in the cord is 89 N, what is the lamp's mass? (b) What is the cord's tension when the elevator ascends with an upward acceleration of 2.4 m/s²?

21. Two Forces, Two Blocks In Fig. 3-37 forces act on blocks A and B, which are connected by string. Force $\vec{F}_A = (12\text{ N})\hat{\text{i}}$ acts on block A, with mass 4.0 kg. Force $\vec{F}_B = (24\text{ N})\hat{\text{i}}$ acts on block B, with mass 6.0 kg. What is the tension in the string?

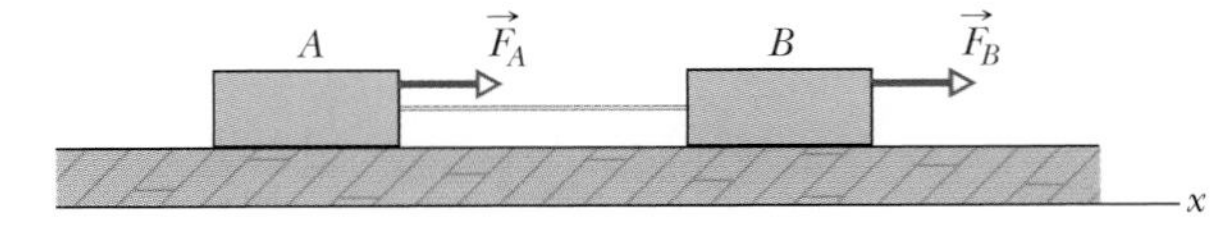

FIGURE 3-37 ■ Problem 21.

22. Coin Drop An elevator cab is pulled directly upward by a single cable. The elevator cab and its single occupant have a mass of 2000 kg. When that occupant drops a coin, its acceleration relative to the cab is 8.00 m/s² downward. What is the tension in the cable?

23. Links In Fig. 3-38, a chain consisting of five links, each of mass 0.100 kg, is lifted vertically with a constant acceleration of 2.50 m/s².

Find the magnitudes of (a) the force on link 1 from link 2, (b) the force on link 2 from link 3, (c) the force on link 3 from link 4, and (d) the force on link 4 from link 5. Then find the magnitudes of (e) the force $\vec{F}$ on the top link from the person lifting the chain and (f) the *net* force accelerating each link.

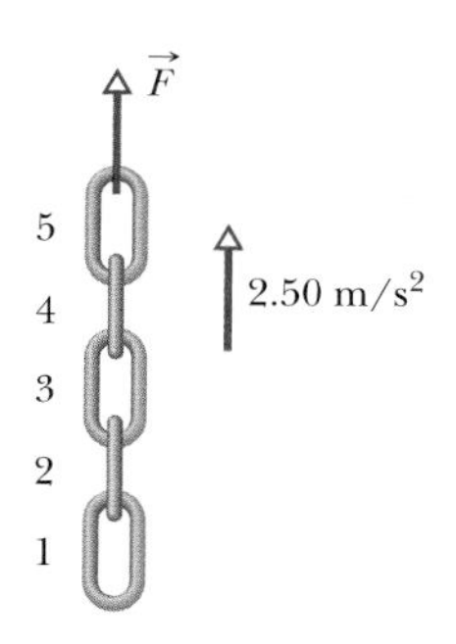

FIGURE 3-38 ■ Problem 23.

SEC. 3-9 ■ GRAVITATIONAL FORCES AND FREEFALL MOTION

24. Raindrops Raindrops fall 1700 m from a cloud to the ground. (a) If they were not slowed by air resistance, how fast would the drops be moving when they struck the ground? (b) Would it be safe to walk outside during a rainstorm?

25. Falling Rock A rock is dropped from a 100-m-high cliff. How long does it take to fall (a) the first 50 m and (b) the second 50 m?

26. Long Drop The Zero Gravity Research Facility at the NASA Lewis Research Center includes a 145 m drop tower. This is an evacuated vertical tower through which, among other possibilities, a 1 m diameter sphere containing an experimental package can be dropped. (a) How long is the sphere in free fall? (b) What is its speed just as it reaches a catching device at the bottom of the tower? (c) When caught, the sphere experiences an average acceleration of $25g$ as its speed is reduced to zero. Through what distance does it travel while stopping?

27. Leaping Armadillo A startled armadillo leaps upward, rising 0.544 m in the first 0.200 s. (a) What is its initial speed as it leaves the ground? (b) What is its speed at the height of 0.544 m? (c) How much higher does it go?

28. Ball Thrown Downward A ball is thrown *down* vertically with an initial *speed* of v_1 from a height of h. (a) What is its speed just before it strikes the ground? (b) How long does the ball take to reach the ground? What would be the answers to (c) part a and (d) part b if the ball were thrown *upward* from the same height and with the same initial speed? Before solving any equations, decide whether the answers to (c) and (d) should be greater than, less than, or the same as in (a) and (b).

29. Boat and Key A key falls from a bridge that is 45 m above the water. It falls directly into a model boat, moving with constant velocity, that is 12 m from the point of impact when the key is released. What is the speed of the boat?

30. Downward-Speeding Ball A ball is thrown vertically downward from the top of a 36.6-m-tall building. The ball passes the top of a window that is 12.2 m above the ground 2.00 s after being thrown. What is the speed of the ball as it passes the top of the window?

31. Drips Water drips from the nozzle of a shower onto the floor 200 cm below. The drops fall at regular (equal) intervals of time, the first drop striking the floor at the instant the fourth drop begins to fall. Find the locations of the second and third drops when the first strikes the floor.

32. Hang Time A basketball player, standing near the basket to grab a rebound, jumps 76.0 cm vertically. How much (total) time does the player spend (a) in the top 15.0 cm of this jump and (b) in the bottom 15.0 cm? Does this help explain why such players seem to hang in the air at the tops of their jumps?

33. Air Express A hot-air balloon is ascending at the rate of 12 m/s and is 80 m above the ground when a package is dropped over the side.

(a) How long does the package take to reach the ground?
(b) With what speed does it hit the ground?

34. Other-Worldly Pitch A ball is shot vertically upward from the surface of a planet in a distant solar system. A plot of y versus t for the ball is shown in Fig. 3-39, where y is the height of the ball above its starting point and $t = 0$ at the instant the ball is shot. What are the magnitudes of (a) the free-fall acceleration on the planet and (b) the initial velocity of the ball?

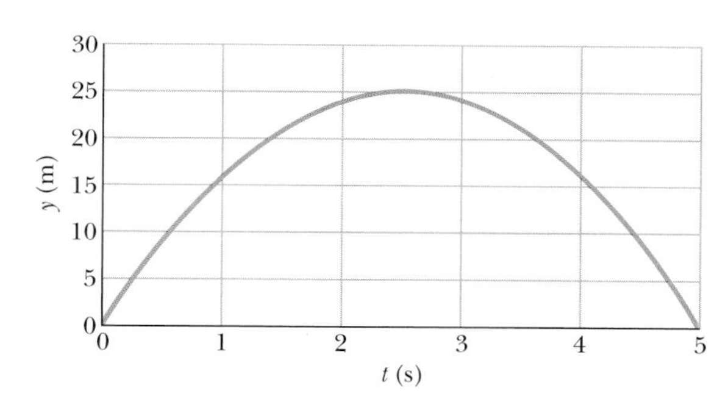

FIGURE 3-39 ■ Problem 34.

35. Reaction Time Figure 3-40 shows a simple device for measuring your reaction time. It consists of a cardboard strip marked with a scale and two large dots. A friend holds the strip *vertically*, with thumb and forefinger at the dot on the right in Fig. 3-40. You then position your thumb and forefinger at the other dot (on the left in Fig. 3-40), being careful not to touch the strip. Your friend releases the strip, and you try to pinch it as soon as possible after you see it begin to fall. The mark at the place where you pinch the strip gives your reaction time. (a) How far from the lower dot should you place the 50.0 ms mark? (b) How much higher should the marks for 100, 150, 200, and 250 ms be? (For example, should the 100 ms marker be two times as far from the dot as the 50 ms marker? Can you find any pattern in the answers?)

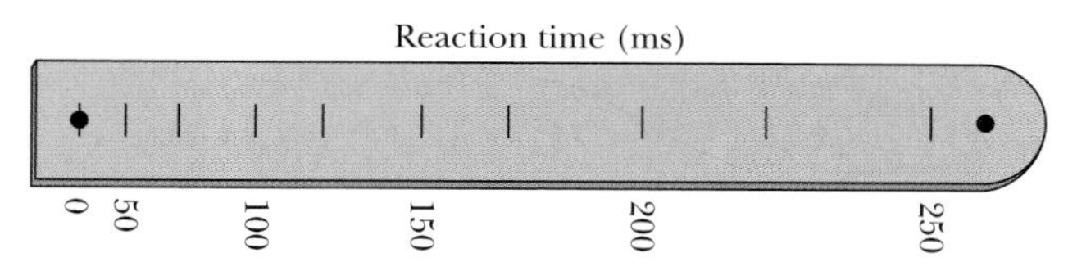

FIGURE 3-40 ■ Problem 35.

36. Juggling A certain juggler usually tosses balls vertically to a height H. To what height must they be tossed if they are to spend twice as much time in the air?

37. Dropping a Wrench At a construction site a pipe wrench struck the ground with a speed of 24 m/s. (a) From what height was it inadvertently dropped? (b) How long was it falling? (c) Sketch graphs of y, v_y, and a_y vs. t for the wrench.

38. Two Stones A stone is dropped into a river from a bridge 43.9 m above the water. Another stone is thrown vertically down 1.00 s after the first is dropped. Both stones strike the water at the same time. (a) What is the initial speed of the second stone?

(b) Plot velocity vs. time on a graph for each stone, taking zero time as the instant the first stone is released.

39. Callisto Imagine a landing craft approaching the surface of Callisto, one of Jupiter's moons. If the engine provides an upward force (thrust) of 3260 N, the craft descends at constant speed; if the engine provides only 2200 N, the craft accelerates downward at 0.39 m/s^2. (a) What is the weight of the landing craft in the vicinity of Callisto's surface? (b) What is the mass of the craft? (c) What is the magnitude of the free-fall acceleration near the surface of Callisto?

40. Rising Stone A stone is thrown vertically upward. On its way up it passes point A with speed v, and point B, 3.00 m higher than A, with speed $\frac{1}{2}v$. Calculate (a) the speed v and (b) the maximum height reached by the stone above point B.

41. Parachuting A parachutist bails out and freely falls 50 m. Then the parachute opens, and thereafter she slows at 2.0 m/s^2. She reaches the ground with a speed of 3.0 m/s. (a) How long is the parachutist in the air? (b) At what height does the fall begin?

42. Space Ranger's Weight Compute the weight of a 75 kg space ranger (a) on Earth, (b) on Mars, where $g = 3.8$ m/s^2, and (c) in interplanetary space, where $g = 0$. (d) What is the ranger's mass at each of these locations?

43. Different *g*'s A certain particle has a weight of 22 N at a point where $g = 9.8$ m/s^2. What are its (a) weight and (b) mass at a point where $g = 4.9$ m/s^2? What are its (c) weight and (d) mass if it is moved to a point in space where $g = 0$?

SEC. 3-10 ■ NEWTON'S THIRD LAW

44. A Child Stands Then Jumps A 29.0 kg child, with a 4.50 kg backpack on his back, first stands on a sidewalk and then jumps up into the air. Find the magnitude and direction of the force on the sidewalk from the child when the child is (a) standing still and (b) in the air. Now find the magnitude and direction of the *net* force on Earth due to the child when the child is (c) standing still and (d) in the air.

45. Sliding Down a Pole A firefighter with a weight of 712 N slides down a vertical pole with an acceleration of 3.00 m/s^2, directed downward. What are the magnitudes and directions of the vertical forces (a) on the firefighter from the pole and (b) on the pole from the firefighter?

46. Block A, Block B In Fig. 3-41a, a constant horizontal force $\vec{F}_a$ is applied to block A, which pushes against block B with a 20.0 N force horizontally to the right. In Fig. 3-41b, the same force $\vec{F}_a$ is applied to block B; now block A pushes on block B with a 10.0 N force horizontally to the left. The blocks have a total mass of 12.0 kg. What are the magnitudes of (a) their acceleration in Fig. 3-41a and (b) force $\vec{F}_a$?

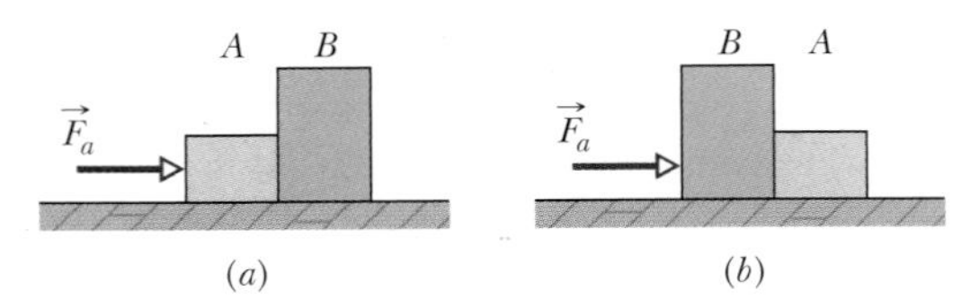

FIGURE 3-41 ■ Problem 46.

47. Two Blocks Two blocks are in contact on a frictionless table. A horizontal force is applied to the larger block, as shown in Fig. 3-42. (a) If $m_A = 2.3$ kg, $m_B = 1.2$ kg, and $F = 3.2$ N, find the magnitude of the force between the two blocks. (b) Show that if a force of the same magnitude F is applied to the smaller block but in the opposite direction, the magnitude of the force between the blocks is 2.1 N, which is not the same value calculated in (a). (c) Explain the difference.

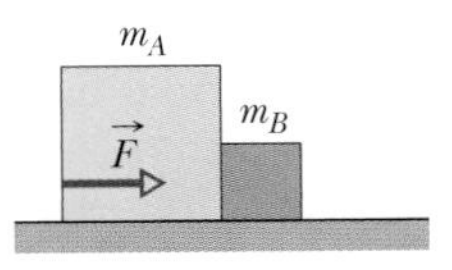

FIGURE 3-42 ■ Problem 47.

48. Parachuting Two An 80 kg person is parachuting and experiencing a downward acceleration of 2.5 m/s^2. The mass of the parachute is 5.0 kg. (a) What is the upward force on the open parachute from the air? (b) What is the downward force on the parachute from the person?

49. Getting Down An 85 kg man lowers himself to the ground from a height of 10.0 m by holding onto a rope that runs over a frictionless pulley to a 65 kg sandbag. With what speed does the man hit the ground if he started from rest?

50. Climbing a Rope A 10 kg monkey climbs up a massless rope that runs over a frictionless tree limb and back down to a 15 kg package on the ground (Fig. 3-43). (a) What is the magnitude of the least acceleration the monkey must have if it is to lift the package off the ground? If, after the package has been lifted, the monkey stops its climb and holds onto the rope, what are (b) the magnitude and (c) the direction of the monkey's acceleration, and (d) what is the tension in the rope?

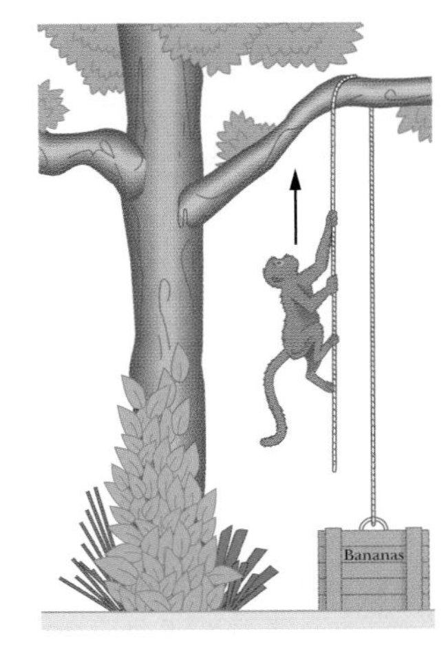

FIGURE 3-43 ■ Problem 50.

51. Bosun's Chair Figure 3-44 shows a man sitting in a bosun's chair that dangles from a massless rope, which runs over a massless, frictionless pulley and back down to the man's hand. The combined mass of man and chair is 95.0 kg. With what force magnitude must the man pull on the rope if he is to rise (a) with a constant velocity and (b) with an upward acceleration of 1.30 m/s^2? (*Hint:* A free-body diagram can really help.)

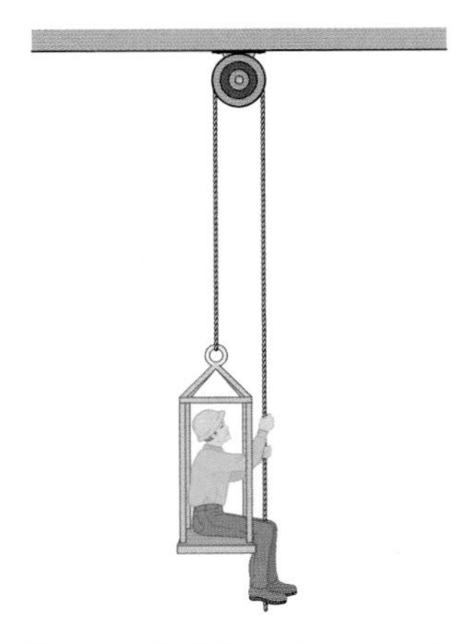

FIGURE 3-44 ■ Problem 51.

52. Girl and Sled A 40 kg girl and an 8.4 kg sled are on the frictionless ice of a frozen lake, 15 m apart but connected by a rope of negligible mass. The girl exerts a horizontal 5.2 N force on the rope. (a) What is the acceleration of the sled? (b) What is the acceleration of the girl? (c) How far from the girl's initial position do they meet?

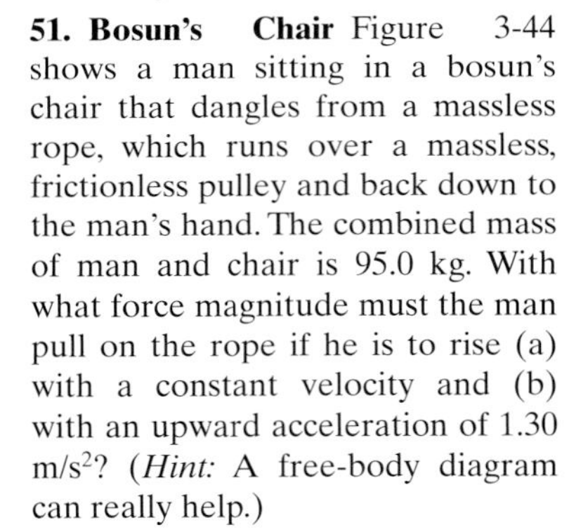

Additional Problems

53. Why Bother with N1? Newton's First Law states that an object will move with a constant velocity if nothing acts on it. This seems to contradict our everyday experience that a moving object comes to a rest unless something acts on it to keep it going. Does this everyday experience contradict Newton's First Law? If it does not, explain how this experience is consistent with Newton's First Law. If it does, explain why we bother to teach Newton's First Law anyway.

54. When Does N3 Hold? Newton's Third Law says that objects that touch each other exert forces on each other. These forces satisfy the rule:

If object A exerts a force on object B, then object B exerts a force back on object A and the two forces are equal in magnitude but opposite in direction.

Consider the following three situations concerning two identical cars and a much heavier truck.

(a) One car is parked and the other car crashes into it.
(b) One car is parked and the truck crashes into it.
(c) The truck is pushing the car, because the car's engine cannot start. The two are touching and the truck is speeding up.

For each situation, do you think Newton's Third Law holds or does not hold? Explain your reasons for saying so.

55. Why Bother with N2? Newton's Second Law written in equation form states

$$\vec{a} = \frac{\vec{F}^{\,\text{net}}}{m}.$$

Your roommate says "That's silly. Everyone knows it takes a force to keep something moving at a constant velocity, even when there's no acceleration." Do you agree with your roommate? If so, explain why physics classes bother to teach the law. If you disagree, how would you try to convince your roommate of the error of his/her ways?

56. Weight vs. Force A Frenchman, filling out a form, writes "78 kg" in the space marked poids (weight). However weight is a force and kg is a mass unit. What do the French (among others) have in mind when they use mass to report their weight? Why don't they report their weight in newtons? How many newtons does this Frenchman weigh? How many pounds?

57. Amy Is Pulled A student named Amy is being pulled across a smooth floor with a big rubber band that is stretched to a constant length. In one case she is riding on a low-friction cart and in the other case she is sliding along the floor. A motion detector is set up to track her motion in each case. The position–time graphs of her motion are shown in Fig. 3-45.

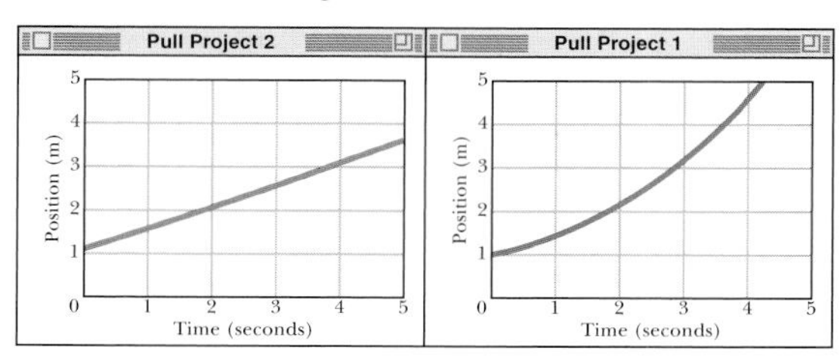

FIGURE 3-45 ■ Problem 57.

(a) Which graph depicts motion at a constant velocity? Pull Project 1 (on the right) or Pull Project 2 (on the left)? Explain.
(b) Which graph depicts motion at a roughly constant acceleration? Explain.
(c) Which graph demonstrates that something pulled with a constant force moves with a constant velocity? Explain.
(d) Which graph demonstrates that something pulled with a constant force moves with a constant acceleration? Explain.
(e) Which graph is most likely to show Amy's motion when she is rolling on the cart? Please justify your answer.
(f) Explain why it is possible to get two different types of motion even though Amy is being pulled with a constant force in both cases.

58. Inertial vs. Gravitational Mass Suppose you have the following equipment available: an electronic balance, a motion detector and an electronic force sensor attached to a computer-based laboratory system. You would like to determine the mass of a block of ice that can slide smoothly along a very level table top without noticeable friction.

(a) Describe how you would use some of the equipment to find the *gravitational* mass of the ice.
(b) Describe how you would use some of the equipment to find the *inertial* mass of the ice.
(c) Which of the two types of masses can be measured in outer space where gravitational forces are very small?

59. Free Fall Acceleration Your roommate peeks over your shoulder while you are reading a physics text and notices the following sentence: "In free fall the acceleration is always g and always straight downward regardless of the motion." Your roommate finds this peculiar and raises three objections:

(a) If I drop a balloon or a feather, it doesn't fall nearly as fast as a brick.
(b) Not everything falls straight down; if I throw a ball it can go sideways.
(c) If I hold a wooden ball in one hand and a steel ball in the other, I can tell that the steel ball is being pulled down much more strongly than the wooden one. It will probably fall faster.

How would you respond to these statements? Discuss the extent to which they invalidate the quoted statement. If they don't invalidate the statement, explain why.

60. Velocity and Force Graphs In the following situations friction is small and can be ignored. Consider whether the net or combined force on a small cart needs to be positive, negative, or zero to create the following motions. Sketch graphs that show the shapes of the velocity and force functions in each case. Use the format shown in Fig. 3-46. By convention, an object moving away from the origin has a positive velocity. (Draw a separate set of velocity vs. time and force vs. time graph for each of part (a) through (d).)

(a) The cart is moving away from the origin at a constant velocity.
(b) The cart moves toward the origin, speeding up at a steady rate until it reaches a constant velocity after 3 s.
(c) The cart moves toward the origin, slowing down at a steady rate, turns around after 2 s, and then moves away from the origin, speeding up at the *same* steady rate.

(d) The cart moves away from the origin, slows down for 3 s, and then speeds up for 3 seconds.

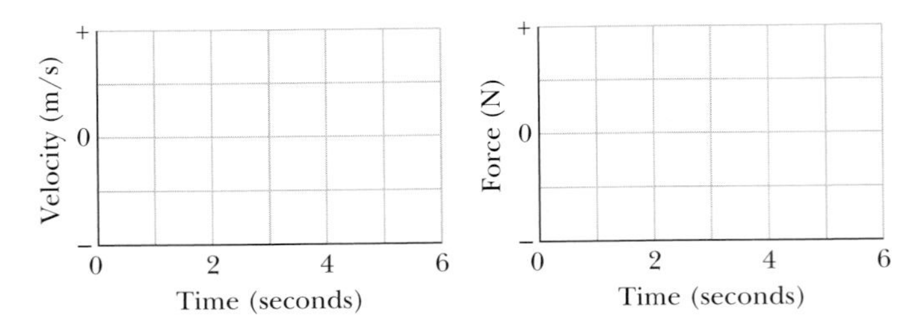

FIGURE 3-46 ■ Problem 60.

61. Toy Cars (a) Suppose a toy car moves along a horizontal line without friction and a constant force is applied to the car toward the left.

FIGURE 3-47 ■ Problem 61.

Sketch a set of axes like those shown in Fig. 3-48, and sketch the shape of the acceleration–time graph of the car using a solid line

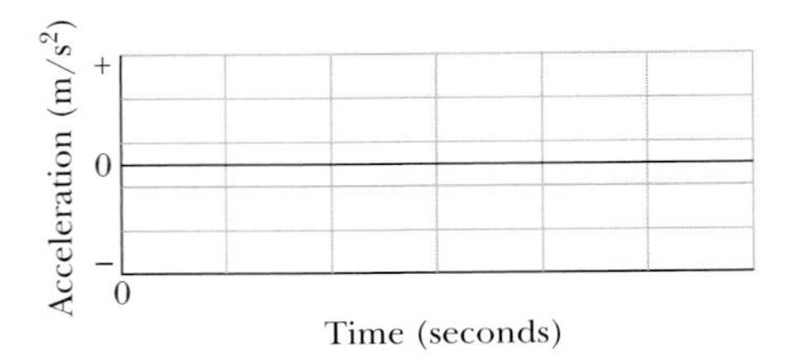

FIGURE 3-48 ■ Problem 61.

(b) What if two more identical cars are piled/glued on top of the first car and the same constant force is applied to the three cars? Use a dashed line to sketch the acceleration–time graph of the "triple-car." Explain any differences between this graph and the acceleration–time graph of the single car.

62. Rocket Thrust and Acceleration A wise being has placed a standard physics coordinate system in outer space far away from any massive bodies. A specially designed space cylinder that experiences no gravitational or frictional forces is moving along the x axis of this coordinate system. It has two identical rocket engines on each end. These engines can apply thrust forces that act in opposite directions but have equal magnitudes as shown in Fig. 3-49. Diagram A has engines on both ends on, diagram B has all engines off, diagram C has only the left engines on, and diagram D has only the right engines on.

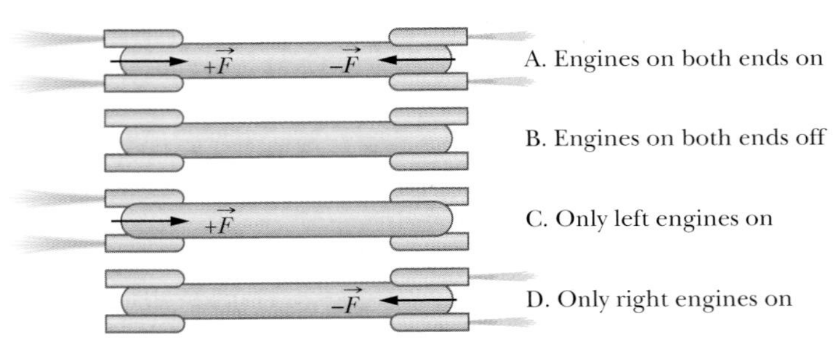

FIGURE 3-49 ■ Problem 62.

Choose all the force combinations (**A** through **D**) which could *keep the rocket moving* as described in each statement below. You may use a choice more than once or not at all. If you think that none is correct, answer choice **E.**

(a) Which force combinations could keep the rocket moving toward the right and speeding up at a steady rate (constant acceleration)?
(b) Which force combinations could keep the rocket moving toward the right at a steady (constant) velocity?
(c) The rocket is moving toward the right. Which force combinations could slow it down at a steady rate (constant acceleration)?
(d) Which force combinations could keep the rocket moving toward the left and speeding up at a steady rate (constant acceleration)?
(e) The rocket was started from rest and pushed until it reached a steady (constant) velocity toward the right. Which force combinations could keep the rocket moving at this velocity?
(f) The rocket is slowing down at a steady rate and has an acceleration to the right. Which force combinations could account for this motion?
(g) The rocket is moving toward the left. Which force combinations could slow it down at a steady rate (constant acceleration)?

63. Two Carts Two low-friction carts A and B have masses of 2.5 kg and 5.0 kg, respectively. Initially a student is pushing them with an applied force of $\vec{F}_B = -20.0$ N, which is exerted on cart B as shown in Fig. 3-50a.

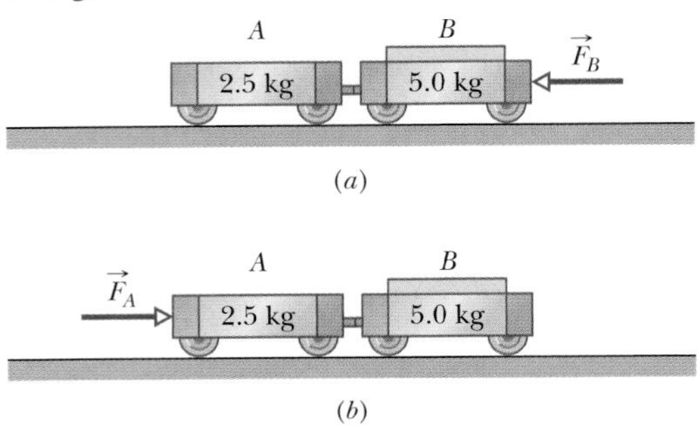

FIGURE 3-50 ■ Problem 63.

(a) Find the magnitude and direction of the interaction forces between the two carts $\vec{F}_{B\to A}$ and $\vec{F}_{A\to B}$ where $\vec{F}_{B\to A}$ represents the force on cart A due to cart B and $\vec{F}_{A\to B}$ represents the force on cart B due to cart A.
(b) If the student pushes on cart A with an applied force of $\vec{F}_A = +20.0$ N instead, as shown in part (b) of Fig. 3-50, determine the magnitude and direction of the interaction forces between the two carts $\vec{F}_{B\to A}$ and $\vec{F}_{A\to B}$ for this situation.
(c) Explain why the interaction forces are different in the two cases. *Hint:* If you consider the two carts together as a system with mass 7.5 kg, what is the acceleration of each of carts A and B? What does the *net* force on cart A have to be to result in this acceleration?

64. Spring Scale One The spring scale in Fig. 3-51 reads 10.5 N. The cart moves toward the right with an acceleration of 3.5 m/s^2.

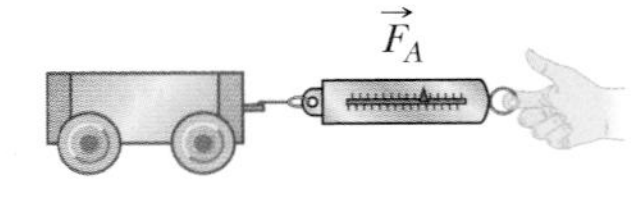

FIGURE 3-51 ■ Problem 64.

(a) Suppose a second spring scale is combined with the first and *acts in the same direction* as shown in Fig. 3-52. The spring scale $\vec{F}_A$ still reads 10.5 N.
The cart now moves toward the right with an acceleration of 4.50 m/s^2. What is the *net* force on the cart? What does spring scale $\vec{F}_B$ read? Show your calculations and explain.

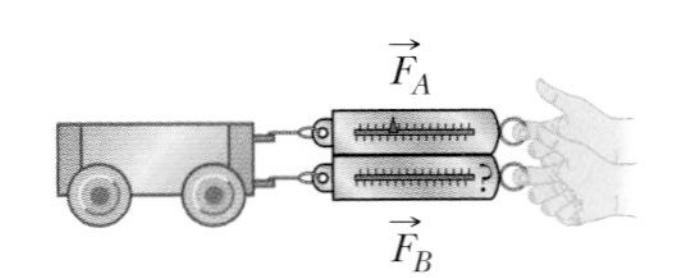

FIGURE 3-52 ■ Problem 64.

(b) Suppose a second spring scale is combined with the first and acts in the opposite direction as shown in Fig. 3-53. The spring scale $\vec{F}_A$ still reads 10.5 N.

FIGURE 3-53 ■ Problem 64.

The cart now moves toward the right with an acceleration of 2.50 m/s^2. What is the *net* force on the cart? What does spring scale $\vec{F}_B$ read? Show your calculations and explain.
(c) Which of Newton's first two laws apply to the situations in this problem?

65. Spring Scale Two Two forces are applied to a cart with two different spring scales as shown in Fig. 3-54. The spring scale $\vec{F}_A$ reads 15 N.

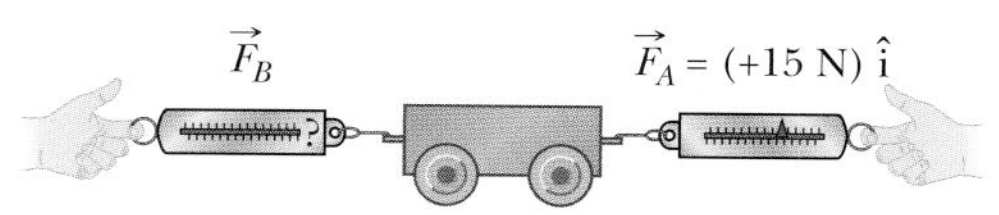

FIGURE 3-54 ■ Problem 65.

(a) The cart had an initial velocity of 0.00 m/s when the two forces were applied. It remains at rest after the combined forces are applied. What is the *net* force on the cart? What does spring scale $\vec{F}_B$ read? Show your calculations and explain.
(b) The cart had an initial velocity of +0.75 m/s and so it was moving to the right when the two forces were applied. It continues moving to the right at that same velocity after the combined forces are applied. What is the *net* force on the cart? What does spring scale $\vec{F}_B$ read? Show your calculations and explain.
(c) The cart had an initial velocity of −0.39 m/s and so it was moving to the left when the two forces were applied. It continues moving to the left at that same velocity after the combined forces are applied. What is the *net* force on the cart? What does spring scale $\vec{F}_B$ read? Show your calculations and explain.

66. Fire Ladder A physics student is standing on one of the steps of the fire ladder behind a building on campus doing a physics experiment. From there she drops a stone (without giving it any initial velocity) and notes that it takes approximately 2.45 s to hit the ground. The second time she throws the stone vertically upward and notes that it takes approximately 5.16 s for it to hit the ground.

(a) Calculate the height above the parking lot from which she releases the first stone.
(b) Calculate the initial velocity with which she has thrown the second stone upward.
(c) How high above the parking lot did the second stone rise before it started falling again?

Star Trek Problem Problems 67 and 68 both involve the following: "You are at the helm of the starship *Defiant* (*NCC-1764*), currently in orbit around the planet Iconia, near the Neutral Zone. Your mission: to rendezvous with a supply vessel at the other end of this solar system . . . You direct the impulse drive to be set at full power for leisurely half-light-speed travel . . . which should bring you to your destination in a few hours."* Assume that the diameter of the Iconian solar system is 100 Astronomical Units (an AU is the mean radius of the Earth's orbit about the Sun: 1AU = 1.49×10^{11} m).

*Krauss, Lawrence, *The Physics of Star Trek* (New York: Harper Perennial, 1996), p. 3.

67. Can You Stand the G-Forces? In order to minimize the *g*-forces on you, suppose you decide to accelerate with a constant acceleration such that you reach half the speed of light ($c/2 = 1.5 \times 10^8$ m/s) at the midpoint of your trip and then start slowing down so you are at rest just in time to dock with the supply vessel at the other end of this solar system.

(a) Draw a single motion diagram showing the speeding-up and slowing-down processes.
(b) In a coordinate system in which you move along the positive x axis, what is the direction and magnitude of your initial acceleration? In other words, is your acceleration positive or negative?
(c) In a coordinate system in which you move along the positive x axis, what is the direction and magnitude of your acceleration while you are slowing down for your rendezvous with the supply vessel? In other words, is your acceleration positive or negative? (*Hint*: The answer to part (b) and a symmetry argument can save you some effort.)
(d) How long will your overall trip take?
(e) If the *Defiant* has a mass of $M = 2.850 \times 10^8$ kg, what is the thrust force (in newtons) needed to accelerate your starship?
(f) The amount of force you feel being impressed on you by the back of your seat as the starship picks up speed is proportional to your acceleration. A common way to measure typical forces you might feel is to calculate *g*-forces. This is done by comparing the acceleration you experience to the acceleration you would experience while falling freely close to the surface of the Earth. Thus, you can find *g*-forces by dividing your acceleration by 9.8 m/s^2. What *g*-forces would you experience while accelerating in the *Defiant*?
(g) The maximum sustained *g*-force that a human can stand is about 3 g. What would happen to you during your leisurely acceleration to half the speed of light?

68. How Long Would a Trip Take If the Forces Were Bearable? Let's take the trip at a more reasonable acceleration of 3 g.

(a) What would your acceleration be in m/s^2?
(b) How long would it take you, starting from rest, to get halfway (i.e., d = 50 AU) across the Iconian solar system at this 3 g acceleration?
(c) What would your maximum speed be (i.e., the speed when you pass the d = 50 AU mark)?
(d) How long would it take you to slow down at a 3 g acceleration for docking with the supply vessel? What is the total trip time? Is this feasible?

69. The Demon Drop The Demon Drop is a popular ride at the Cedar Point Amusement Park in Ohio. It allows four people to get into a little cage and fall freely for a while. Physics professor Bob Speers of Firelands College in Huron, Ohio, took a video tape of the drop. It is called DSON001. Use VideoPoint and Excel to analyze and develop a mathematical model that describes the fall.

(a) Include a printout of your spreadsheet model along with the answers to questions (b) through (e).
(b) According to your model, what is the equation you think describes the vertical position of the bottom of the cage as a function of time?
(c) According to your model, what is the acceleration of the cage?
(d) Can you find values of initial position and velocity that allow you to obtain a good agreement between your model graph and the graph of the data using the accepted value of the free fall acceleration close to the surface of the Earth of $\vec{a} = -g = -9.8$ m/s^2?

(e) Suppose a group of four people with an average mass of 65 kg each are put in the Demon Drop cage of mass 2.0×10^3 lb. What is the force on the whole falling system consisting of the cage and the people? Be sure to indicate the direction of the force.

70. Force, Acceleration, and Velocity Graphs (a) A force is applied to an object that experiences very little friction. This force causes the object to move resulting in the acceleration vs. time graph shown in Fig. 3-55. Draw a set of graph axes with the same number of time units as that shown in the acceleration graph and carefully sketch the *shape* of a possible graph of the force vs. time for the object.

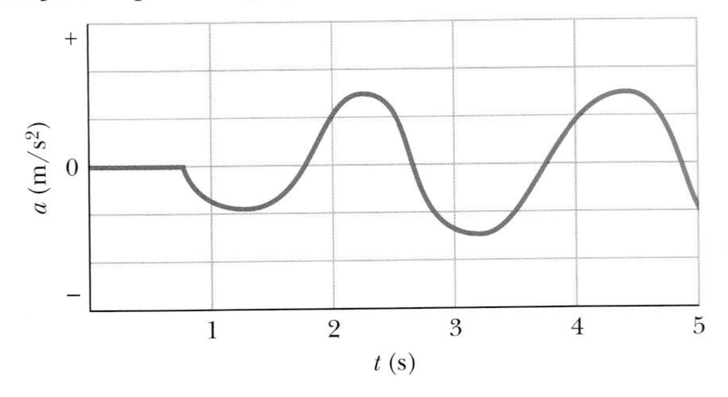

FIGURE 3-55 ■ Problem 70.

(b) A force is applied to an object that experiences very little friction. This force causes the object to move resulting in the velocity vs. time graph shown in Fig. 3-56. Draw a set of axes with the same number of time units as that shown in the velocity graph and carefully sketch the shape of a possible graph of acceleration vs. time for the object.

(c) Refer to the velocity vs. time graph shown in part (b) and the acceleration vs. time graph you sketched. Draw a set of graph axes with the same number of time units as that shown in the velocity graph and carefully sketch the shape of a possible graph of force vs. time for the object.

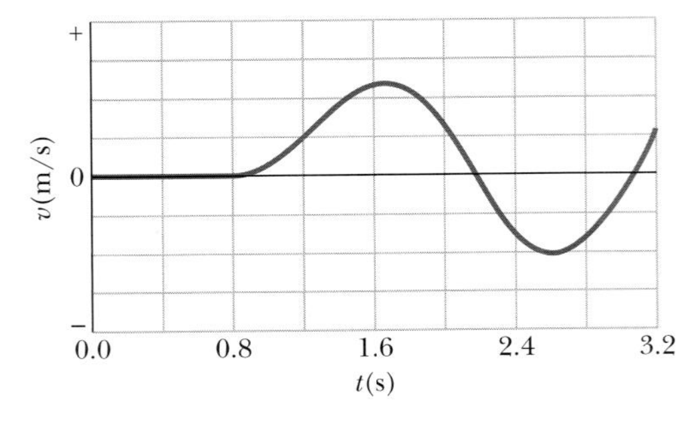

FIGURE 3-56 ■ Problem 70.

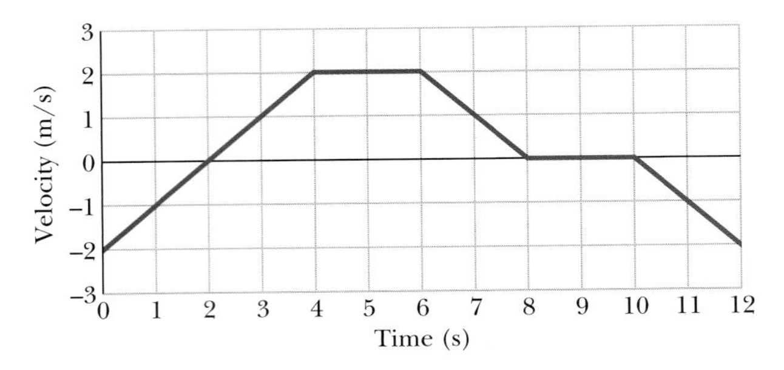

FIGURE 3-57 ■ Problems 71 and 72.

71. Force from Velocity One Figure 3-57 shows the velocity vs. time graph for an object constrained to move along a line. The positive direction is to the right.

(a) At what times, or during what time periods, is the net force acting on the object zero?
(b) At what times, or during what time periods, is the net force acting on the object constant and nonzero.
(c) At what times, or during what time periods, is the net force acting on the object changing?

In each case, explain your reasoning. Describe how your reasoning is consistent or inconsistent with Newton's Laws of Motion. If there is no time or time period for which a given condition exists, state that explicitly.

72. Force from Velocity Two Figure 3-57 shows the velocity vs. time graph for an object constrained to move along a line. The positive direction is to the right.

(a) At what times, or during what time periods, is the net force on the object increasing and directed to the right?
(b) At what times, or during what time periods, is the net force on the object decreasing and directed to the right?
(c) At what times, or during what time periods, is the net force on the object constant and directed to the right?
(d) At what times, or during what time periods, is the net force on the object increasing and directed to the left?
(e) At what times, or during what time periods, is the net force on the object decreasing and directed to the left?
(f) At what times, or during what time periods, is the net force on the object constant and directed to the left?

In each case, explain your reasoning. Describe how your reasoning is consistent or inconsistent with Newton's Laws of Motion. If there is no time or time period for which a given condition exists, state that explicitly.

4 Vectors

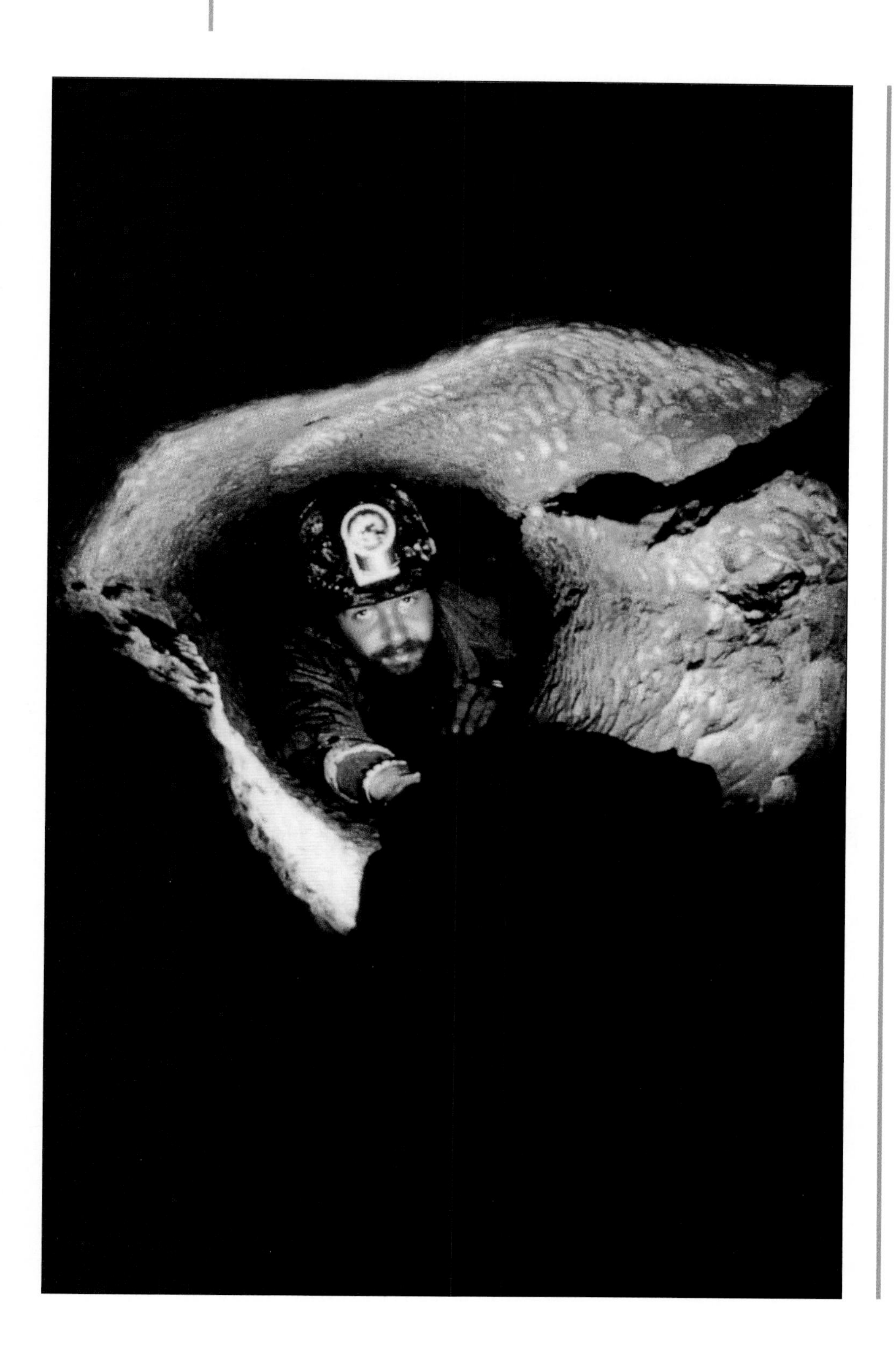

For two decades, spelunking teams crawled, climbed, and squirmed through 200 km of Mammoth Cave and the Flint Ridge cave system, seeking a connection. The photograph shows Richard Zopf pushing his pack through the Tight Tube, far inside the Flint Ridge system. After 12 hours of "caving" along a labyrinthine route, Zopf and six others waded through a stretch of chilling water and found themselves in Mammoth Cave. Their breakthrough established the Mammoth-Flint cave system as the longest cave in the world.

How can their final point be related to their initial point other than in terms of the actual route they covered?

The answer is found in this chapter.

4-1 Introduction

As you already learned in Chapters 2 and 3, it is useful to use vectors to represent several of the physical quantities that were used in our study of one-dimensional motion. These quantities include position, displacement, velocity, acceleration, and force. In Chapters 5 and 6, vector mathematics will be used in conjunction with Newton's Laws to study two-dimensional motions such as that of objects that move horizontally while falling (projectile motion), circular motion, motion when friction forces are present, and motions on inclined surfaces.

In order to study motion in two dimensions, you must learn to represent and add two-dimensional vectors both graphically and mathematically. This is not as simple as it is in one dimension. For example, you learned in Chapter 2 that when an x axis (or y axis) is assigned to describe a particle-like object moving along a line, the sign ($+$ or $-$) of the x-component of its velocity vector indicates the direction of motion. However, if the particle is not moving along a straight line, then keeping track of the changes in the direction of its velocity is not just a matter of using a single plus or minus sign.

We will start our general consideration of vectors and vector operations by extending the definitions of vectors developed in Chapter 2 to two dimensions. We only discuss three-dimensional vectors very briefly in this chapter. However, we will return to them later in the book.

4-2 Vector Displacements

In order to define velocity and acceleration in more than one dimension, we need to start with the general definition of displacement. As is true in one dimension, it is useful to represent displacement vectors in two or three dimensions by arrows. For example, if a particle changes its position by moving from point A point C, it turns out that its displacement, $\Delta\vec{r}$, can be represented by an arrow that points directly from A to C, as shown in Fig. 4-1a. Remember, a displacement vector tells us nothing about the actual path that the particle takes. Thus both the curved path and the two straight paths from A to B and B to C, shown in Fig. 4-1a, can lead a particle from point A to point C, so the displacement vector between A and C is the same in both cases.

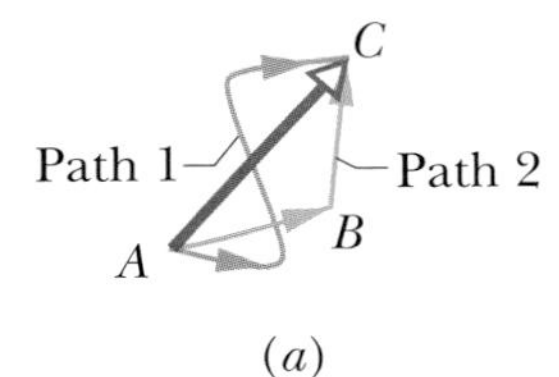

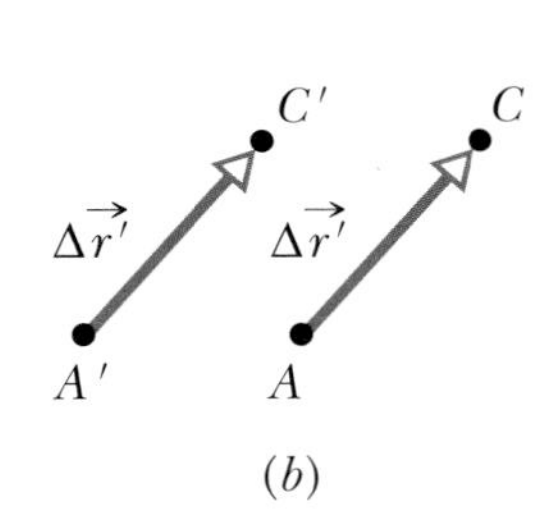

FIGURE 4-1 ■ (a) A displacement vector for a particle's motion between points A and C can be represented by an arrow pointing from A to C. Since displacement depends only on the relative locations of A and C, both paths shown result in the same displacement. (b) The vectors pointing from A to C and from A' to C' also represent the same displacement, $\Delta\vec{r}$, since displacement represents a change in position rather than the positions themselves, so $\Delta\vec{r} = \Delta\vec{r}'$.

In Figure 4-1b, the arrows pointing from A to C and from A' to C' have the same magnitude and direction. Thus, they represent identical displacement vectors because they signify the same *change of position* for the particle. Thus a displacement vector that is shifted in space without changing its magnitude (length) and direction is the same vector.

The fact that a displacement vector represents only the overall effect of a motion, and not its detailed path, can lead to miscommunication. The Foxtrot cartoon, Fig. 4-2, shows what can happen when one person assumes that getting from point A to point B is what counts, while another cares how one gets there. Figure 4-3 is a vector diagram of the path Jason takes vs. the path Peter wants him to take.

READING EXERCISE 4-1: A soccer field has goals on its north and south end. Consider the following displacement of a soccer ball.

The ball is initially sitting in the center of the field. It is kicked toward the west. After traveling 3 m, it is kicked toward the north. It travels 6 m before a player stops it.

Which of the following displacements, if any, are identical to the displacement described above?

(1) The ball is initially sitting directly in front of the south goal. It is kicked toward the east. After traveling 9 m it is kicked toward the north. It travels 6 m before a player kicks it due west. After it travels for 12 m, another player stops it.
(2) The ball is initially sitting in the center of the field. It is kicked toward the east. After traveling 3 m, it is kicked toward the south. It travels 6 m before a player stops it. ■

FIGURE 4-2 ■ Foxtrot cartoon in which Jason is using physics to focus on the path independence of displacement, while his brother Peter is interested in both Jason's displacement and his actual path.

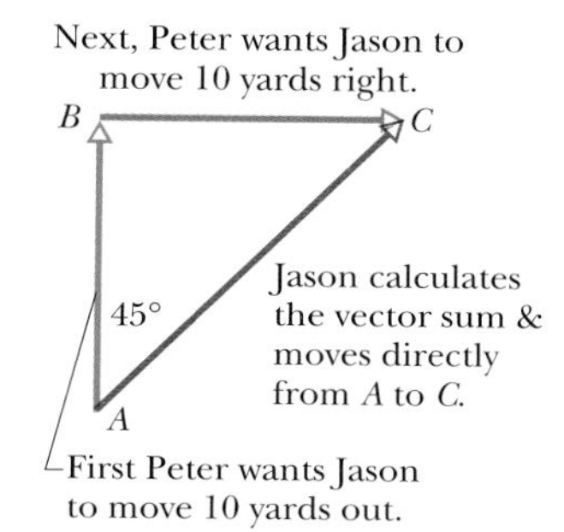

FIGURE 4-3 ■ Depiction of a vector sum of two successive displacements.

TOUCHSTONE EXAMPLE 4-1: Jogging in a Circle

Sara is running laps on a circular track. Each full lap is 400 m. If she starts at the northernmost point on the track, initially going east, find her displacement (magnitude and direction) after she has run (a) 100 m, (b) 200 m, (c) 300 m, and (d) 400 m.

SOLUTION ■ The **Key Idea** here is that Sara's displacement is a vector whose *magnitude* is the *straight-line* distance from her starting position to her current position. The *direction* of her displacement vector *points* straight from her starting point to her current location.

(a) When Sara has gone 100 m, she has gone one-quarter of the way around the track, as shown in Fig. 4-4*a*. As you can see there, her current position is the same as it would have been if she had gone a distance of one radius of the track due south and then the same distance due east. Since the angle between lines *AP* and *PB* in Fig. 4-4*a* is 90°, we can use the Pythagorean theorem to find the straight-line distance from *A* to *B*. This distance is just

$$\Delta \vec{r}_{AB} = \sqrt{R^2 + R^2} = \sqrt{2R^2} = \sqrt{2}R.$$

Since the track has a circumference of 400 m, its radius is $R = 400\text{ m}/(2\pi) = 63.6$ m. So the *magnitude* of Sara's displacement after she has run the 100 m from *A* to *B* is

$$|\Delta \vec{r}\,|_{AB} = (\sqrt{2})(400\text{ m})/(2\pi) = 90.0\text{ m}. \quad \text{(Answer)}$$

The *direction* of her displacement is seen from Fig. 4-4*a* to be due southeast. (Answer)

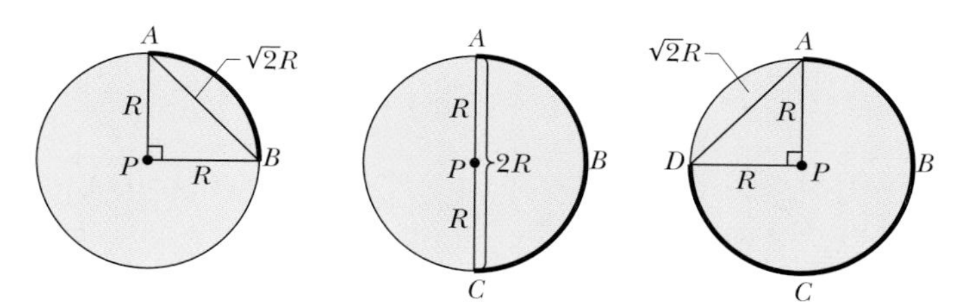

FIGURE 4-4 ■ (*a*) After Sara has run 100 m from *A* to *B*. (*b*) After she has run 200 m from *A* to *C*. (*c*) After she has run 300 m from *A* to *D*.

(b) When Sara has run 200 m from her starting point, we can see in Fig. 4-4*b* that she has covered exactly half the circumference of the track. This places her a distance $2R = 127$ m away from her starting point and due south of it. (Answer)

(c) Now Sara has covered three-quarters of the track's circumference, as shown in Fig. 4-4*c*. Her distance from her starting point is *the same* as it was when she had run only one-quarter of the way around the track, so now the *magnitude* of her displacement is once again $\sqrt{2}R = 90.0$ m. But the *direction* of her displacement from her starting point is due south*west*.

(d) Now that Sara has run 400 m, she has "come full circle" and returned to her starting point. Since her current position is the same as her starting position, her displacement from her starting position is now *zero*. (Answer)

4-3 Adding Vectors Graphically

The basic method for graphical addition of displacement vectors involves considering a single vector that describes the final outcome of two displacements. For example, in Fig. 4-2 big brother, coach Peter, wanted Jason to get from point A to point C by undergoing first one displacement by moving outward (forward along the field) for 10 yards from point A to B, and then moving to the right for 10 yards from point B to C. Instead Jason used the rules of vector addition to go directly from A to C by traveling a distance of $10\sqrt{2}$ yards at an angle of 45 degrees with respect to the outward direction as shown in Fig. 4-3.

Addition

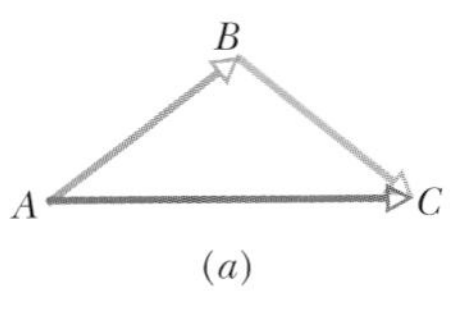

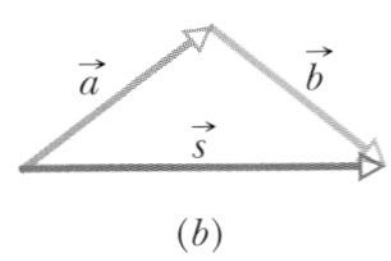

FIGURE 4-5 ■ (*a*) AC is the sum of the vectors AB and BC. (*b*) The same vectors with alternate labels $\vec{a}$, $\vec{b}$ and $\vec{s}$.

Suppose that, as in the vector diagram of Fig. 4-5*a*, a particle moves from A to B and then later from B to C. We can represent its overall displacement (no matter what its actual path) with two successive displacement vectors, AB and BC. The *net* displacement of these two displacements is a single displacement from A to C. We call AC the **vector sum** (or **resultant**) of the vectors AB and BC. This sum is not the usual algebraic sum.

In Figure 4-5*b*, we redraw the vectors of Figure 4-5*a* and relabel them in the way that we shall use from now on—namely, with an arrow over a symbol, as in $\vec{a}$. In adding two or more vectors, it's OK to move them to make the addition simpler, as long as the length of each vector and its orientation don't change. Recall from Chapter 2 that if we want to indicate only the magnitude or size of the vector (a quantity that lacks a sign or direction), we shall use the absolute value symbol, as in $|\vec{a}|$, or drop the arrow, as in a.

We can represent the relationship among the three vectors in Figure 4-5*b* with the vector equation

$$\vec{s} = \vec{a} + \vec{b}, \tag{4-1}$$

which says that the vector $\vec{s}$ is the vector sum or resultant of vectors $\vec{a}$ and $\vec{b}$. The symbol + in $\vec{s} = \vec{a} + \vec{b}$ and the words "sum" and "add" have different meanings for vectors than they do in algebra because they involve both magnitude and direction.

Figure 4-5 suggests a general procedure for adding two vectors $\vec{a}$ and $\vec{b}$ graphically: (1) On paper, sketch vector $\vec{a}$ to some convenient scale and at the proper angle. (2) Sketch vector $\vec{b}$ to the same scale, with its tail at the head of vector $\vec{a}$, again at the proper angle. (3) The vector sum $\vec{s}$ is the vector that extends from the tail of $\vec{a}$ to the head of $\vec{b}$.

Vector addition, defined in this way, has two important properties. First, the order of addition does not matter. That is,

$$\vec{a} + \vec{b} = \vec{b} + \vec{a} \qquad \text{(commutative law).} \tag{4-2}$$

Second, when there are more than two vectors, we can group them in any order as we add them, so

$$\left(\vec{a} + \vec{b}\right) + \vec{c} = \vec{a} + \left(\vec{b} + \vec{c}\right) \qquad \text{(associative law).} \tag{4-3}$$

A vector is not simply any entity that has both magnitude and direction. In fact, the rules for vector addition and the associative and commutative properties of vector addition are defining characteristics of vectors.

Subtraction

Subtracting one vector from another can be considered as the addition of one vector to the **additive inverse** of the other. The additive inverse (sometimes known as the "negative of a vector") is simply the vector we must add to the original vector to get zero. If we want to define the additive inverse of a vector $\vec{b}$, denoted as $-\vec{b}$, we can start with the understanding that $\vec{b} + (-\vec{b})$ should equal zero. Using the graphical method of adding vectors that we discussed above, this demands that the vector $-\vec{b}$ has the same magnitude as $\vec{b}$, but points in the opposite direction so that the two vectors cancel, as shown in Fig. 4-6. Thus, adding vector $\vec{b}$ to its additive inverse gives

$$\vec{b} + (-\vec{b}) = 0.$$

Finding an additive inverse is commonly referred to as **inverting** a vector. Adding $-\vec{b}$ has the effect of subtracting $\vec{b}$. We use this property to define the difference between any two vectors such as $\vec{d} = \vec{a} - \vec{b}$ as

$$\vec{d} = \vec{a} - \vec{b} = \vec{a} + (-\vec{b}) \quad \text{(vector subtraction as a form of addition).} \qquad (4\text{-}4)$$

That is, we find the difference vector $\vec{d}$ by adding the vector $-\vec{b}$ to the vector $\vec{a}$. Figure 4-7 shows how this is done geometrically.

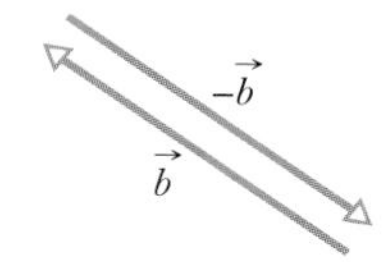

FIGURE 4-6 ■ The vectors $\vec{b}$ and $-\vec{b}$ have the same magnitude and opposite directions.

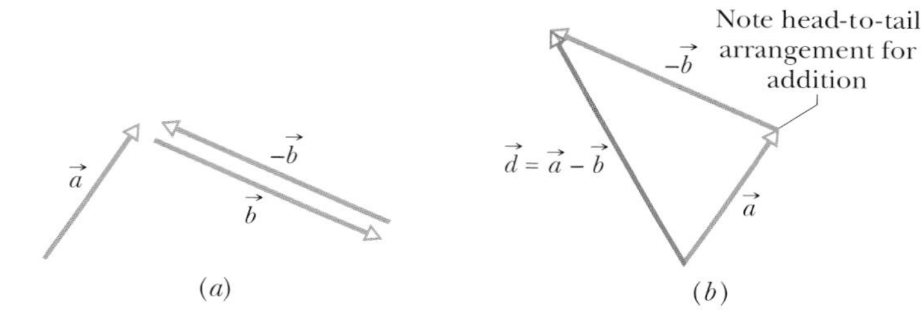

FIGURE 4-7 ■ Consider the vectors $\vec{a}$ and $\vec{b}$. To subtract vector $\vec{b}$ from vector $\vec{a}$, create vector $-\vec{b}$ from vector $\vec{b}$ as shown in Fig. 4-6. Then add vector $-\vec{b}$ to vector $\vec{a}$.

As another example, consider a car that speeds up (accelerates) along a straight road. The change in the car's velocity is given by $\Delta\vec{v} = \vec{v}_2 - \vec{v}_1$ and the car's average acceleration by $\langle\vec{a}\rangle = \Delta\vec{v}/\Delta t$. The use of vectors to depict how $\Delta\vec{v}$ can be found by vector subtraction is illustrated in Fig. 4-8.

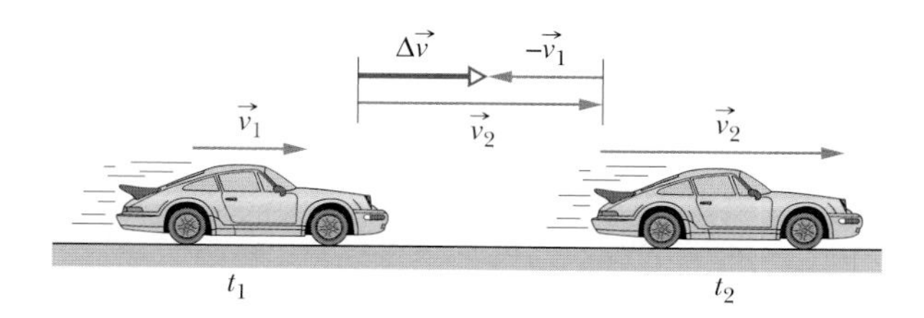

FIGURE 4-8 ■ Diagram showing how to find the change in velocity of a race car by taking a one-dimensional vector difference in which $\vec{v}_2$ is added to $-\vec{v}_1$ to get $\Delta\vec{v}$.

As in scalar algebra, we can move a term that includes a vector symbol from one side of a vector equation to the other, but we must change its sign. For example, if we are given $\vec{d} = \vec{a} - \vec{b}$ and need to solve for $\vec{a}$, we can rearrange the equation as $\vec{a} = \vec{d} + \vec{b}$. Remember, although we have used displacement vectors here, the rules for addition and subtraction hold for vectors of all kinds, whether they represent velocities, accelerations, forces, or any other vector quantity. However, we can add only vectors of the same kind. For example, we can add two displacements, or two velocities, but adding a displacement and a velocity makes no sense. In the arithmetic of scalars, that would be like trying to add 21 s and 12 m.

READING EXERCISE 4-2: The magnitudes of displacements $\vec{a}$ and $\vec{b}$ are 3 m and 4 m, respectively, and $\vec{c} = \vec{a} + \vec{b}$. Considering various orientations of $\vec{a}$ and $\vec{b}$, what are (a) the maximum possible magnitude for $\vec{c}$ and (b) the minimum possible magnitude? ■

4-4 Rectangular Vector Components

You have learned a method to find the vector sum or resultant of two vectors that do not point along the same line. Many times it is useful to do the opposite and **decompose** or **resolve** a vector into two or more vectors, which can be added to create the original vector. For example, consider what happens to the motion of a particle-like object when two forces that are not acting in the same direction are applied to it. It turns out that the object will accelerate as if a single force that is the vector sum of the forces is acting on it. In cases where a force vector can be resolved into two or more vectors, it is possible to break down even complex situations into simpler one-dimensional ones, so the skill of resolving vectors is very powerful. We will begin by considering how to describe a two-dimensional vector in a rectangular coordinate system as the sum of two one-dimensional vectors.

Resolving a Vector

It is typical to describe two dimensional vectors in a coordinate system in which the x and y axes are drawn in the plane of the page. We choose axes that are parallel to the edges of the paper as shown in Fig. 4-9*a*.

We already know how to represent, add, and subtract vectors that are parallel to an x axis or y axis. For this reason, it is convenient to decompose our vector into two component vectors—one parallel to the x axis and the other parallel to the y axis. In this case, the vector $\vec{a}$ is the sum of two component vectors $\vec{a}_x$ and $\vec{a}_y$ as shown in Fig. 4-9*a*. Therefore, $\vec{a} = \vec{a}_x + \vec{a}_y$.

FIGURE 4-9 ■ (*a*) The component vectors form the legs of a right triangle whose vector sum is the original vector. (*b*) The components a_x and a_y of vector $\vec{a}$ are determined by projections of the tail and tip of the vector on each axis. (*c*) The values of the components are unchanged if the vector is shifted, as long as its magnitude and orientation are the same.

As you can see in Fig. 4-9*a*, we have chosen the length of the component vectors $\vec{a}_x$ and $\vec{a}_y$ so that they conveniently add up to form vector $\vec{a}$. Since $\vec{a}_x = a_x\hat{\text{i}}$ and $\vec{a}_y = a_y\hat{\text{j}}$ (Section 3-2), we can say that we have expressed the component vectors in terms of their components a_x and a_y. Note that these components have *no arrows*. We define the **rectangular component** of a vector to be the projection of the vector on an axis. In Fig. 4-9*a*, for example, a_x is the component of vector $\vec{a}$ on (or along) the x axis and a_y is the component along the y axis. The practical way to get the projection or component of a vector along an axis is to draw lines from the two ends of the vector perpendicular to that axis, as shown in Fig. 4-9*b* or *c*.

In order to understand the idea of a vector component as the projection of the vector onto an axis, think about taking a distant spotlight and shining it onto the vector. The component (or projection) is the length of the shadow that is cast by the vector on one of the axes. For example, take a light and shine it straight down on the vector shown in Fig. 4-9*b* from the top of the page. The shadow of the vector will fall along the x axis. The length of the shadow is the x-component of this vector as shown in Fig. 4-9*b*. We would make the y-component of the vector visible as the length of a shadow by shining a light on the vector from the right side of the page. Then the shadow of the vector falls along the y axis and its length is the vector's y-component.

Figure 4-9 illustrates that the components (projections) of the vector do not change if we simply move the vector around within our coordinate system. In other words, when you shift a vector without changing its direction, its components, which are lengths, do not change. The length and direction of the projection on an axis tells what the vector component is. The projection of a vector on an x axis is called its **x-component** and is denoted as a_x. The projection on the y axis is called its **y-component** and is denoted a_y. We call the process of finding the components of a vector in a chosen coordinate system **resolving the vector.**

Positive and Negative Components

The components of a vector can be positive or negative depending on the overall orientation of the vector we are resolving relative to the coordinate system we have chosen. In a standard coordinate system, we indicate this by designating components that point up or to the right as positive; then those that point down or to the left are negative. Graphically, small arrowheads on each component can represent its direction. For example, in Fig. 4-9, a_x and a_y are both positive because $\vec{a}$ extends in the positive direction of both axes. (Note the small arrowheads on the components, to indicate their direction.) If we were to reverse vector $\vec{a}$, then both components would be negative and their arrowheads would point toward negative x and y. Resolving a different vector $\vec{b}$ shown in Fig. 4-10 yields a positive component b_x and a negative component b_y if we stick with the standard coordinate system.

Using Sines and Cosines to Find Components

In general, a two-dimensional vector has two components. As Figs. 4-9b and c imply, we can find the value of the components of $\vec{a}$ in Fig. 4-9b using the sine and cosine relations. Since

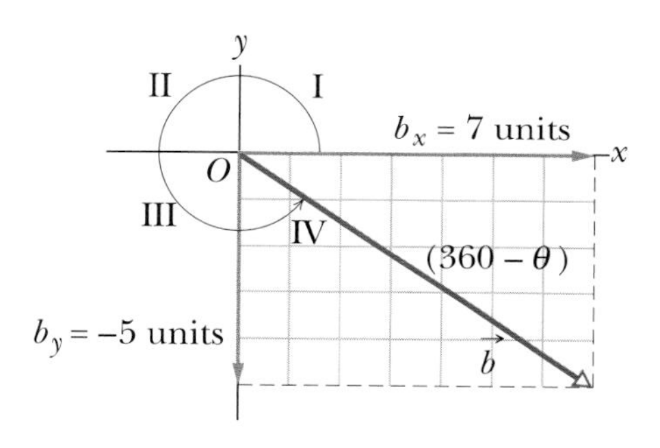

FIGURE 4-10 ■ The component of $\vec{b}$ on the x axis is positive, and the component on the y axis is negative.

$$\cos\theta = \frac{\text{adjacent side}}{\text{hypotenuse}} \quad \text{and} \quad \sin\theta = \frac{\text{opposite side}}{\text{hypotenuse}},$$

for the right triangle in Fig. 4-9a, the magnitude of the vector $\vec{a}$ is the hypotenuse, and

$$\cos\theta = \frac{a_x}{a} \quad \text{and} \quad \sin\theta = \frac{a_y}{a},$$

where θ is the angle that the vector $\vec{a}$ makes with the positive direction of the x axis. Remember, the symbols a and $|\vec{a}|$ provide alternate notations for the magnitude of $\vec{a}$, and a_x and a_y are the x- and y-components of $\vec{a}$, respectively. Rearranging these relationships, we find

$$a_x = a\cos\theta \quad \text{and} \quad a_y = a\sin\theta. \tag{4-5}$$

Reconstructing a Vector from Components

Look at Fig. 4-9a again. It shows that $\vec{a}$ and its x- and y-components form a right triangle. That means that we can reconstruct a vector $\vec{a}$ from its components. Graphically, we can arrange the components head to tail and then find $\vec{a}$ by completing a right triangle with the vector forming the hypotenuse, from the tail of one component to the head of the other component. We can also get the magnitude of $\vec{a}$ algebraically by using the Pythagorean theorem. That is,

$$a = |\vec{a}| = \sqrt{a_x^2 + a_y^2}.$$

Once a vector has been resolved into its components along a set of axes, the components themselves can be used in place of the vector. For example, $\vec{a}$ in Fig. 4-9*b* is represented (completely determined) by $|\vec{a}|$ and θ. It can also be completely determined by its components a_x and a_y. Both pairs of values contain the same information. If we know a vector in *component notation* (a_x and a_y) and want it in *magnitude-angle notation* (a and θ), we can use the equations

$$a = |\vec{a}| = \sqrt{a_x^2 + a_y^2} \quad \text{and} \quad \theta = \tan^{-1} = \left(\frac{a_y}{a_x}\right) \tag{4-6}$$

to transform the components into a magnitude and direction. Thus, it is common to represent a two-dimensional vector by ordered rectangular components such as a_x and a_y. However, in studying circular motion, it is more convenient to use polar coordinates (a, θ) to describe a vector. Finding θ using the inverse tangent must be done with care since the calculated value of θ must be replaced with $\theta + \pi$ if the $\vec{a}$ vector is in the second (II) or third (III) quadrants as shown in Fig. 4-10.

In the more general three-dimensional case, when using rectangular coordinates, we need to consider another axis, called the z axis, that is mutually perpendicular to the other two axes. In three dimensions the components a_x, a_y, and a_z can be used to represent a vector in a rectangular coordinate system. If a spherical coordinate system is used instead, then a magnitude and two angles (say, $|\vec{a}|$, θ, and ϕ) can be used to represent a vector. Three-dimensional vectors are used in the study of rotational motion. However, you will not be using three-dimensional vectors in the next few chapters.

READING EXERCISE 4-3: In the figures that follow, which of the indicated methods for combining the x- and y-components of vector $\vec{a}$ are correct?

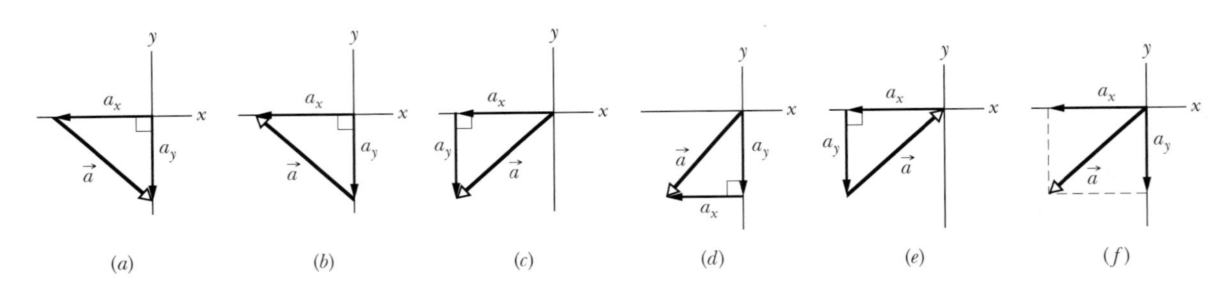

■

READING EXERCISE 4-4: Consider the following standard vector. Which vectors have been correctly repositioned so their components are the same as those of the original vector?

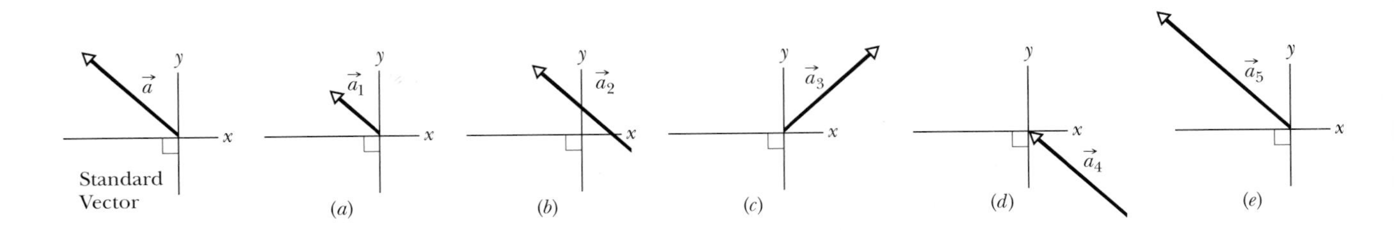

■

TOUCHSTONE EXAMPLE 4-2: Spelunking

The 1972 team that connected the Mammoth-Flint cave system went from Austin Entrance in the Flint Ridge system to Echo River in Mammoth Cave (Fig. 4-11*a*). Their horizontal travel (parallel to the Earth's surface) was a net 1.0 km westward and 4.2 km southward. What was their horizontal displacement vector from start to finish?

SOLUTION ■ The **Key Idea** here is that we have the components of a two-dimensional vector, and we need to find each vector's magnitude and direction to specify the displacement vector. We first choose a two-dimensional coordinate axis and then draw the x- and y-components of displacement as in Fig. 4-11*b*. The components ($\Delta x = 1.0$ km west and $\Delta y = 4.2$ km south) form the legs of a horizontal right triangle. The team's horizontal displacement forms the hypotenuse of the triangle, and its magnitude $|\Delta\vec{r}|$ is given by the Pythagorean theorem:

$$\Delta r = |\Delta\vec{r}| = \sqrt{(1.0\text{ km})^2 + (4.2\text{ km})^2} = 4.3\text{ km}.$$

Also from Fig. 4-11*b*, we see that this horizontal displacement is directed south of due west by an angle θ given by

$$\tan\theta = \frac{\Delta y}{\Delta x} = \frac{4.2\text{ km}}{1.0\text{ km}},$$

so

$$\theta = \tan^{-1}\frac{4.2\text{ km}}{1.0\text{ km}} = 77°. \quad \text{(Answer)}$$

In summary, the team's horizontal displacement vector had a magnitude of 4.3 km and was at an angle of 77° south of west. The team also traveled a net distance of 25 m upward. The net vertical motion was insignificant compared to the horizontal motion, so we ignored it. However, the relatively small net vertical displacement was of no comfort to the team. They had to climb up and down countless times to get through the cave. The route they actually covered was quite different from the horizontal displacement vector, which merely points in a straight line from start to finish.

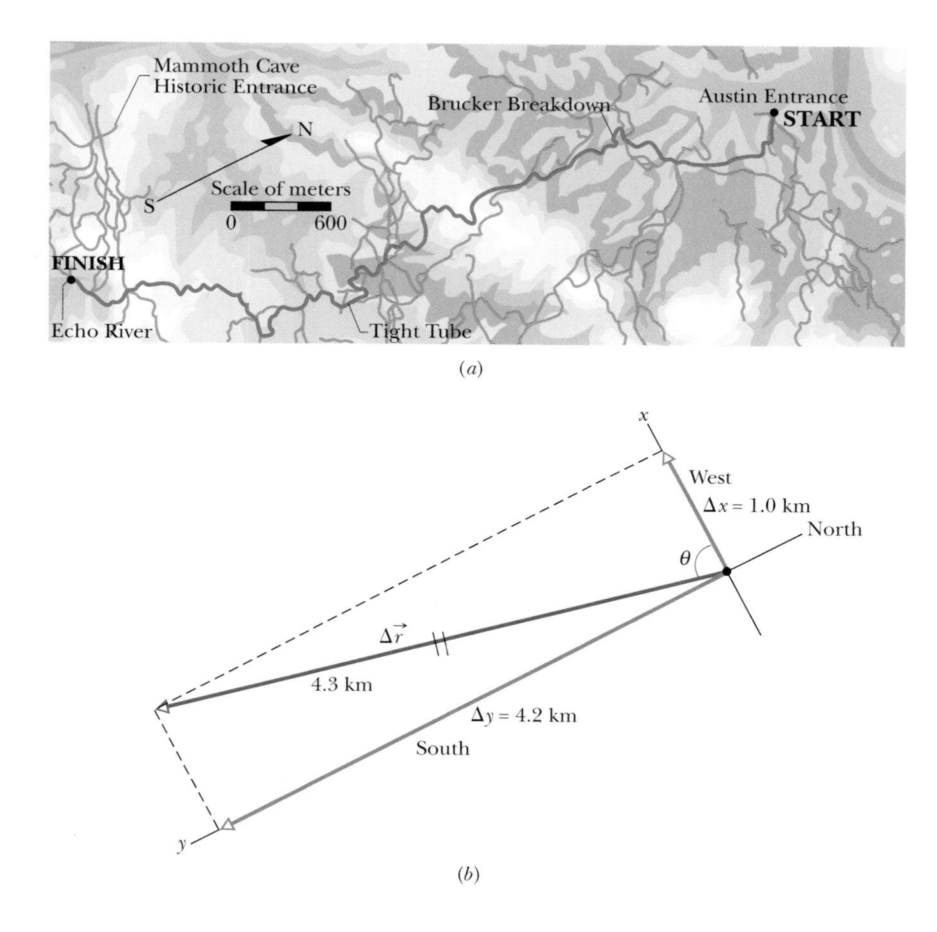

FIGURE 4-11 ■ (*a*) Part of the Mammoth-Flint cave system, with the spelunking team's route from Austin Entrance to Echo River indicated in red. (*b*) The components of the team's horizontal displacement Δx and Δy. They are to scale, but at a different scale than in part (*a*). (Adapted from a map by the Cave Research Foundation.)

4-5 Unit Vectors

In Section 2-2, we defined a unit vector as a dimensionless vector that points in the direction along a coordinate axis that is chosen to be positive. Its sole purpose is to point—that is, to specify a direction. The unit vectors that point in the positive directions of the x, y, and z axes are labeled $\hat{\text{i}}$, $\hat{\text{j}}$, and $\hat{\text{k}}$ (Fig. 4-12), where the hat ˆ (or caret) is used to note that these vectors are special.

The arrangement of axes in Fig. 4-12 is called a **right-handed coordinate system** because it can be constructed using the thumb and fingers of the right hand. There are several legitimate ways to construct a right-handed coordinate system using the right hand. One method is depicted in Fig. 4-12. The system remains right-handed if it is rotated rigidly to any new orientation. If we used the left hand to construct a coordinate system, $\hat{\text{i}}$ would point in the *opposite* direction than it does in Fig. 4-12 while the relative orientations of $\hat{\text{j}}$ and $\hat{\text{k}}$ would remain unchanged as in Fig. 4-13. Since the use of a right-handed system is standard in the scientific community, we use it exclusively in this book.

Unit vectors are very useful for expressing three-dimensional vectors; for example, we can express any vector in terms of the coordinate system in Fig. 4-12 as

$$\vec{a} = a_x\hat{\text{i}} + a_y\hat{\text{j}} + a_z\hat{\text{k}}. \tag{4-7}$$

y Right index finger
$\hat{\text{j}}$
Right thumb
x
$\hat{\text{k}}$
$\hat{\text{i}}$
z Right middle finger

FIGURE 4-12 ■ Unit vectors $\hat{\text{i}}$, $\hat{\text{j}}$, and $\hat{\text{k}}$ define the directions of a standard right-handed coordinate system. This system gets its name from the fact that the positive directions of the x, y, and z axes can be determined by the directions of fingers on a right hand. For example when the thumb, index finger, and middle finger of a right hand are arranged so they are at right angles to each other, they determine the directions of the positive x, y, and z axes, respectively.

The quantities $a_x\hat{\text{i}}$, $a_y\hat{\text{j}}$, and $a_z\hat{\text{k}}$ are vectors called the **component vectors** of $\vec{a}$. The quantities a_x, a_y, and a_z are called, respectively, the ***x*-component, *y*-component,** and ***z*-component** of $\vec{a}$ (or, as before, simply its components along the axes).

Note: The components a_x, a_y, and a_z are sometimes referred to in other books and articles by different names. They have been called "vector components" since the subscripts x, y, and z reveal what unit vectors can be used to construct the vectors that lie along each axis. Also they have incorrectly been called "scalar components." *However, they are not scalars*. Real scalars do not change when the coordinate axes are rotated, and the x-component, y-component, and z-component of a vector can change whenever a coordinate axis is rotated.

READING EXERCISE 4-5: (a) Using the procedure outlined in Fig. 4-12 and your left hand, sketch a left-handed coordinate system that depicts the positive x, y, and z axes and put the unit vectors, $\hat{\text{i}}$, $\hat{\text{j}}$, $\hat{\text{k}}$, in place. (b) Describe how the left-handed system differs from the right-handed one. ■

4-6 Adding Vectors Using Components

There are many different physical situations in which you will need to be able to add vectors in order to understand what is going on. For example, in Chapter 3 we found that we needed to add together force vectors to understand the motion of an object on which more than one force acts. Although we can add vectors geometrically using a sketch, or, if we have a vector-capable calculator, we can add them directly on the screen, perhaps the most practical method for finding the sum, $\vec{s}$, of two vectors is to combine their components, axis by axis.

To start, consider the mathematical statement for a vector sum in two dimensions

$$\vec{s} = \vec{a} + \vec{b}. \tag{4-8}$$

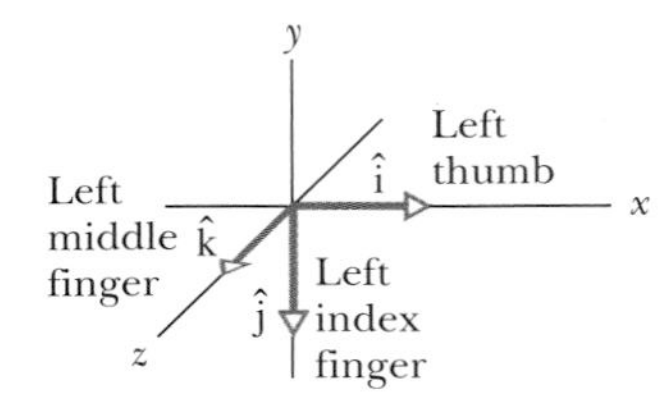

FIGURE 4-13 ■ A nonstandard left-handed coordinate system that is constructed using the same procedures shown in Fig. 4-12, using the left hand.

This statement implies that the vector $\vec{s}$ is the same as the vector $(\vec{a} + \vec{b})$. If this is so, then we can derive the relationships between the components of $\vec{s}$ and those of $\vec{a}$ and $\vec{b}$ mathematically as follows:

$$\vec{s} = (s_x\hat{\mathrm{i}} + s_y\hat{\mathrm{j}}) = \vec{a} + \vec{b} = (a_x\hat{\mathrm{i}} + a_y\hat{\mathrm{j}}) + (b_x\hat{\mathrm{i}} + b_y\hat{\mathrm{j}}) = (a_x + b_x)\hat{\mathrm{i}} + (a_y + b_y)\hat{\mathrm{j}}.$$

Thus, each component of $\vec{s}$ must be the same as the corresponding component of $(\vec{a} + \vec{b})$:

$$s_x = a_x + b_x, \tag{4-9}$$

$$s_y = a_y + b_y. \tag{4-10}$$

In other words, to add vectors $\vec{a}$ and $\vec{b}$, we must first resolve the vectors into their components. Next we must combine these components—taking direction (and thus sign) into account—axis by axis. This gives us the components of the sum vector $\vec{s}$. This is shown in Fig. 4-14. Once we get to this point, we have to make a choice about how to express the result. We can either:

(a) express $\vec{s}$ in unit-vector notation as $\vec{s} = s_x\hat{\mathrm{i}} + s_y\hat{\mathrm{j}}$, or

(b) combine the components of $\vec{s}$ to get $\vec{s}$ itself and express the vector in magnitude-angle notation, where $|\vec{s}| = \sqrt{s_x^2 + s_y^2}$ and $\tan\theta = s_y/s_x$.

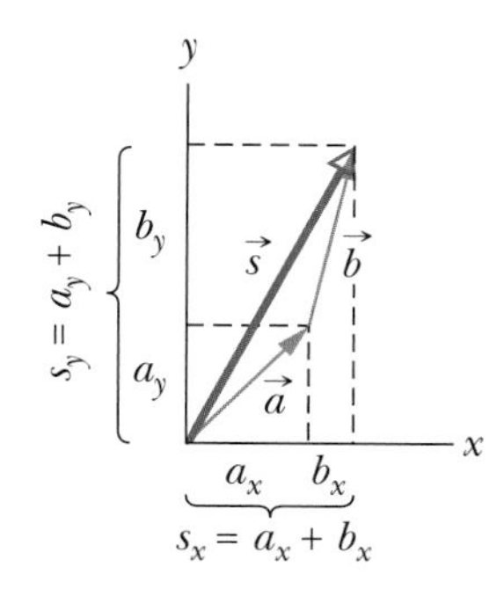

FIGURE 4-14 ■ Diagram showing how a vector sum can be constructed by adding component vectors.

This procedure for adding vectors by components also applies to vector subtractions. Recall that a subtraction such as $\vec{d} = \vec{a} - \vec{b}$ can be rewritten as an addition $\vec{d} = \vec{a} + (-\vec{b})$. To subtract, we simply add $\vec{a}$ and $-\vec{b}$ by components to get

$$d_x = a_x - b_x, \tag{4-11}$$

and

$$d_y = a_y - b_y, \tag{4-12}$$

where

$$\vec{d} = d_x\hat{\mathrm{i}} + d_y\hat{\mathrm{j}}.$$

READING EXERCISE 4-6: (a) In the figure, what are the signs of the x-components of $\vec{d}_1$ and $\vec{d}_2$? (b) What are the signs of the y-components of $\vec{d}_1$ and $\vec{d}_2$? (c) What are the signs of the x- and y-components of $\vec{d}_1 + \vec{d}_2$?

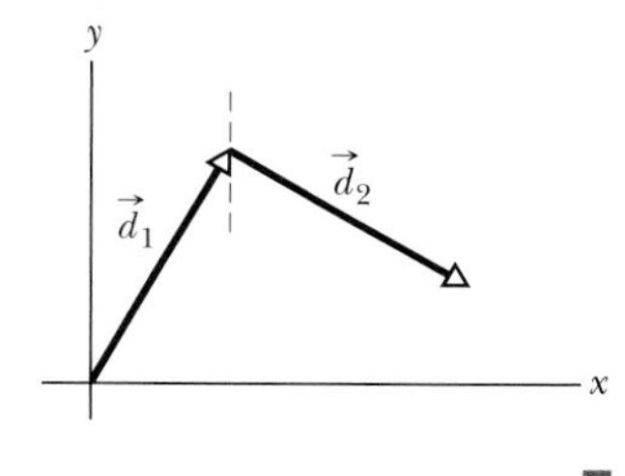

■

TOUCHSTONE EXAMPLE 4-3: Three Vectors

Figure 4-15a shows the following three vectors:

$$\vec{a} = (4.2\text{ m})\hat{\mathrm{i}} - (1.5\text{ m})\hat{\mathrm{j}},$$

$$\vec{b} = (-1.6\text{ m})\hat{\mathrm{i}} + (2.9\text{ m})\hat{\mathrm{j}},$$

and

$$\vec{c} = (-3.7\text{ m})\hat{\mathrm{j}}.$$

What is their vector sum $\vec{r}$, which is also shown?

SOLUTION ■ The **Key Idea** here is that we can add the three vectors by components, axis by axis. For the x axis, we add the x-components of $\vec{a}$, $\vec{b}$, and $\vec{c}$ to get the x-component of $\vec{r}$:

$$r_x = a_x + b_x + c_x$$
$$= 4.2\text{ m} - 1.6\text{ m} + 0 = 2.6\text{ m}.$$

Similarly, for the y axis,

$$r_y = a_y + b_y + c_y$$
$$= -1.5\text{ m} + 2.9\text{ m} - 3.7\text{ m} = -2.3\text{ m}.$$

Another **Key Idea** is that we can combine these components of $\vec{r}$ to write the vector in unit-vector notation:

$$\vec{r} = (2.6\text{ m})\hat{\mathrm{i}} - (2.3\text{ m})\hat{\mathrm{j}},$$

where $(2.6\ \text{m})\hat{\text{i}}$ is the vector component of $\vec{r}$ along the x axis and $-(2.3\ \text{m})\hat{\text{j}}$ is that along the y axis. Figure 4-15b shows one way to arrange these vector components to form $\vec{r}$. (Can you sketch the other way?)

A third **Key Idea** is that we can also answer the question by giving the magnitude and an angle for $\vec{r}$. From Eq. 4-6, the magnitude is

$$r = \sqrt{(2.6\ \text{m})^2 + (-2.3\ \text{m})^2} \approx 3.5\ \text{m}, \qquad \text{(Answer)}$$

and the angle (measured from the positive direction of x) is

$$\theta = \tan^{-1}\left(\frac{-2.3\ \text{m}}{2.6\ \text{m}}\right) = -41°, \qquad \text{(Answer)}$$

where the minus sign means that the angle is in the fourth quadrant.

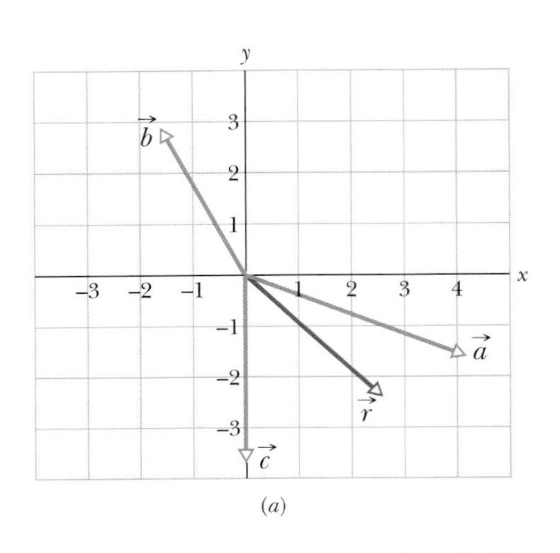

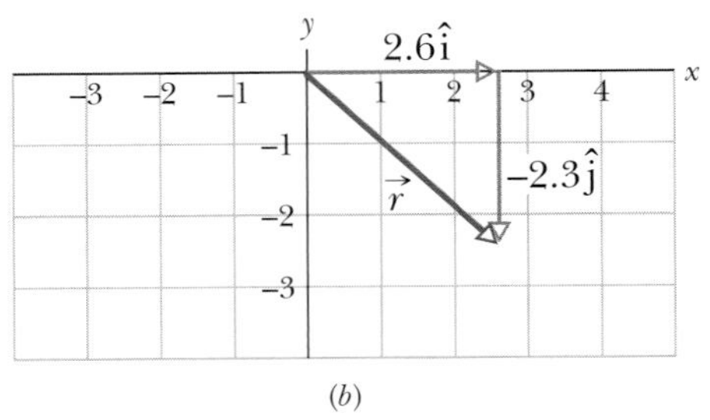

FIGURE 4-15 ■ Vector $\vec{r}$ is the vector sum of the other three vectors.

4-7 Multiplying and Dividing a Vector by a Scalar

We have encountered situations in which we need to multiply or divide a vector by a scalar. For example, we must divide a one-dimensional force vector by a scalar mass to predict the acceleration of an object. Conversely, if we measure an object's acceleration vector in one dimension, we need to multiply the vector by its mass to determine the net force acting on the object. The use of a time interval, Δt, to create a velocity vector from a displacement vector is another example of a scalar being divided into a vector. According to the rules of mathematics, if we multiply or divide a vector by a scalar we should get a new vector. Dividing a vector by a scalar s, can always be transformed into a multiplication. This is because dividing by s is the same as multiplying by $1/s$.

As shown in Fig. 4-16, multiplication of a vector by a scalar simply changes the magnitude of a vector without changing the "line" it lies along.

Although we have used familiar one-dimensional examples, it turns out that these rules for the multiplication of a vector by a scalar also work in two and three dimensions when using either graphical or component representations of vectors. The multiplication or division is distributive over addition so that the product of a scalar e and a vector $\vec{V}$, expressed in terms of its rectangular coordinates, can be expressed as

$$e\vec{V} = e(\vec{V}_x + \vec{V}_y) = e(V_x\hat{\text{i}} + V_y\hat{\text{j}}) = eV_x\hat{\text{i}} + eV_y\hat{\text{j}}.$$

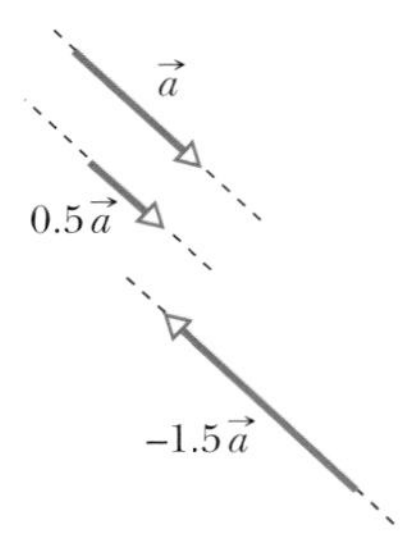

FIGURE 4-16 ■ The product of a scalar and a vector results in a new vector that is still pointing along the same line but has a new length.

There are two other types of vector multiplication that are commonly used in physics, but these both involve the product of two vectors. We call one the dot product, introduced in Section 9-8, and we call the other the cross product, introduced in Section

12-4.* These vector-vector products will be explained in later chapters when they will be needed for the study of work, energy, rotational, and magnetic phenomena.

READING EXERCISE 4-7: Use the rules governing multiplication and division of a vector by a scalar to sketch the indicated product vectors with proper magnitude and directions. Note that the average velocity $\langle \vec{v} \rangle$ and the force $\vec{F}$ are given by

$$\langle \vec{v} \rangle = \frac{\Delta \vec{r}}{\Delta t} \quad \text{and} \quad \vec{F} = m\vec{a}.$$

Using the vectors in the diagram: (a) Multiply the acceleration vector $\vec{a}$ by $m = 3$ kg and sketch the vector representing the force acting on the particle. (b) Divide the displacement vector $\Delta\vec{r}$ by the scalar time interval $\Delta t = 0.5$ s to sketch a vector describing a particle-like object's average velocity vector in cm/s.

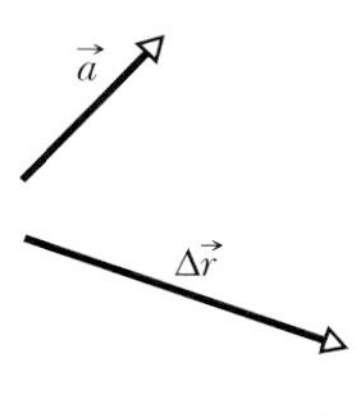

■

READING EXERCISE 4-8: Use the rules governing multiplication and division of a vector by a scalar to calculate the indicated product vectors in terms of its rectangular components and unit vectors. Don't forget to include units! Note that the average velocity $\langle \vec{v} \rangle$ and the force $\vec{F}$ are given by

$$\langle \vec{v} \rangle = \frac{\Delta \vec{r}}{\Delta t} \quad \text{and} \quad \vec{F} = m\vec{a}.$$

Multiply the acceleration vector $\vec{a} = a_x\hat{\text{i}} + a_y\hat{\text{j}} = (1.8\ \text{m/s}^2)\hat{\text{i}} + (1.0\ \text{m/s}^2)\hat{\text{j}}$ by $m = 3$ kg to calculate the vector representing the force acting on the particle. (b) Divide the displacement vector $\Delta\vec{r} = \Delta r_x\hat{\text{i}} + \Delta r_y\hat{\text{j}} = (3.2\ \text{m})\hat{\text{i}} + (-0.8\ \text{m})\hat{\text{j}}$ by the time interval $\Delta t = 0.5$ s to calculate the vector describing a particle-like object's average velocity.

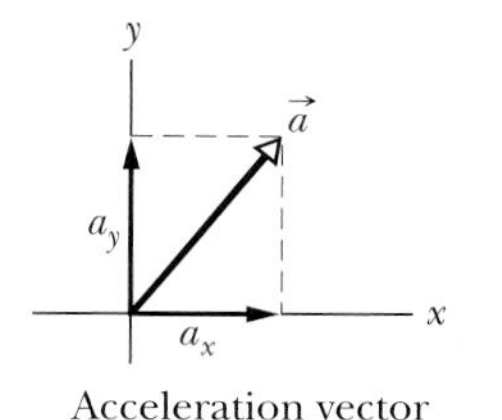

Acceleration vector

y
Δr_x
x
Δr_y
$\Delta\vec{r}$

Displacement vector

■

4-8 Vectors and the Laws of Physics

So far, in every figure that includes a coordinate system, the x and y axes are parallel to the edges of the book page. Thus, when a vector $\vec{a}$ is included, its components a_x and a_y are also parallel to the edges as in Fig. 4-17a. However, there are times when it is more convenient to choose a tilted coordinate system. For example, in studying the motion of a cart rolling down an inclined plane, it is easier to rotate the coordinate system so that one of the axes is aligned with the motion. If we choose to rotate the axes (but not the vector $\vec{a}$) through an angle ϕ in the x-y plane as in Fig. 4-17b, the components will have new values—call them a'_x and a'_y. Since there are an infinite number of choices of ϕ, there are an infinite number of different pairs of components for $\vec{a}$.

---

*The dot product is sometimes called the scalar product because even though it involves a multiplication of two vectors, this product is a scalar. This name can cause confusion since it does not represent the multiplication of a vector by a scalar, which we just discussed. See Section 9-8 for details.

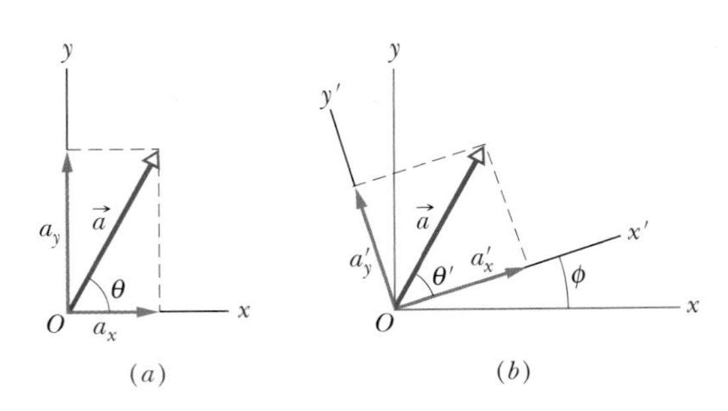

FIGURE 4-17 ■ (*a*) The vector $\vec{a}$ and its components. (*b*) The same vector, with the axes of the coordinate system rotated through an angle ϕ.

Which then is the "right" pair of components? The answer is that they are all equally valid mathematically. Each pair (with its axes) just gives us a different way of describing the same vector $\vec{a}$. All produce the same magnitude. In Fig. 4-17 we have

$$|\vec{a}| = \sqrt{a_x^2 + a_y^2} = \sqrt{a_x'^2 + a_y'^2}. \qquad (4\text{-}13)$$

Although the direction that the vector points in space does not change with coordinate rotation, the angle used to relate it to a new coordinate system is changed to

$$\theta' = \theta - \phi. \qquad (4\text{-}14)$$

In Section 2-2, we defined a scalar as a mathematical quantity whose value does not depend on the orientation of a coordinate system. The magnitude of a vector is a true scalar since it does not change when the coordinate axis is rotated. However, the angles θ and θ', as well as the rectangular components (a_x, a_y) and (a'_x, a'_y), *are not scalars*.

The point is that we really do have great freedom in choosing a coordinate system, because the mathematical relations among vectors (including, for example, vector addition) do not depend on the location of the origin or the orientation of the axes.

In Chapter 5, you will use the definitions of position, displacement, velocity, acceleration, and force vectors developed in Chapters 2 and 3 along with what we have learned in this chapter to study motion in two dimensions.

Problems

SEC. 4-3 ■ ADDING VECTORS GRAPHICALLY

1. Two Displacements Consider two displacements, one of magnitude 3 m and another of magnitude 4 m. Show at least one example of how the displacement vectors may be combined to get a resultant displacement of magnitude (a) 7 m, (b) 1 m, and (c) 5 m.

2. Bank Robbery A bank in downtown Boston is robbed (see the map in Fig. 4-18. To elude police, the robbers escape by helicopter, making three successive flights described by the following displacements: 32 km, 45° south of east; 53 km, 26° north of west; 26 km, 18° east of south. At the end of the third flight they are captured. In what town are they apprehended? (Use the geometrical method to add these displacements on the map.)

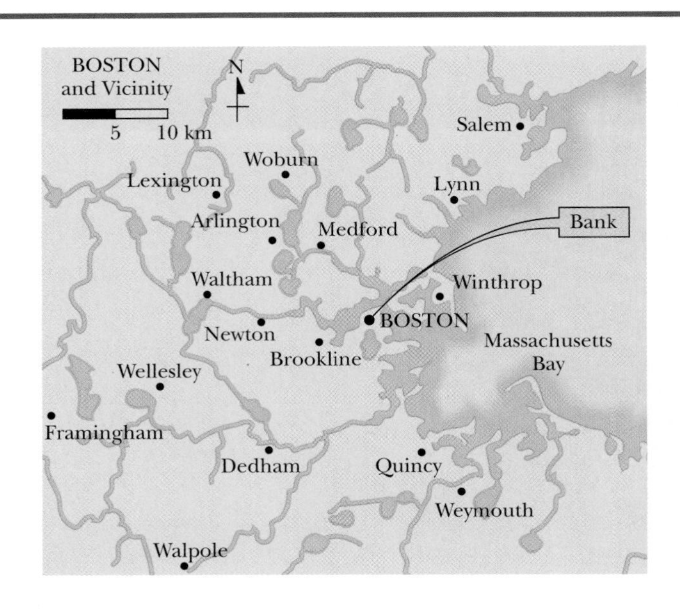

FIGURE 4-18 ■ Problem 2.

3. Velocity Vector Changes The motion of three objects is shown in the motion diagrams (*a*), (*b*), and (*c*) of Fig. 4-19. In each case the object is shown at three equally spaced times. A circle with no arrow indicates a velocity of zero magnitude, such as the final velocity in diagram (*a*). Indicate for each part which number is next to the arrow on the right side of the diagram that best shows the *direction* of the change in velocity. (*Hint*: Use the techniques developed in Section 4-3 to draw vectors representing the difference in velocity in each case.) *Note*: This exercise is adapted from a conceptual exercise developed by Dennis Albers of Columbia College.

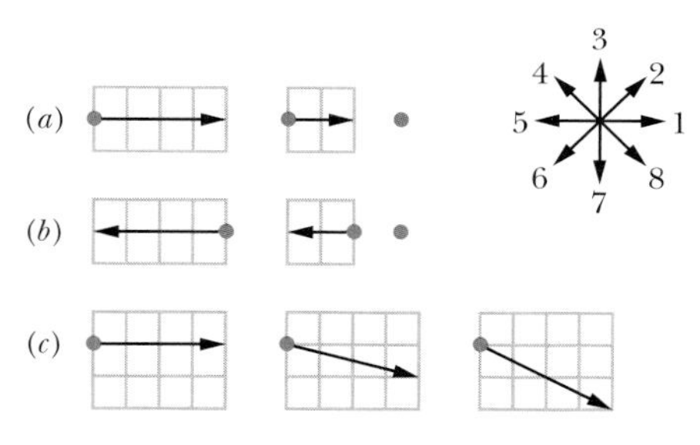

FIGURE 4-19 ■ Problem 3.

4. The Pea Shooter A pea leaves a pea shooter at a speed of 5.4 m/s. It makes an angle of $+30°$ with respect to the horizontal.

(a) Calculate the x-component of the pea's initial velocity.
(b) Calculate the y-component of the pea's initial velocity.
(c) Write an expression for the pea's velocity, $\vec{v}$, using unit vectors for the x direction and the y direction.

SEC. 4-4 ■ RECTANGULAR VECTOR COMPONENTS

5. Components What are (a) the x-component and (b) the y-component of a vector $\vec{a}$ in the xy plane if its direction is 250° counterclockwise from the positive direction of the x axis and its magnitude is 7.3 m?

6. Radians and Degrees Express the following angles in radians: (a) 20.0°, (b) 50.0°, (c) 100°. Convert the following angles to degrees: (d) 0.330 rad, (e) 2.10 rad, (f) 7.70 rad.

7. Magnitude and Angle The x-component of vector A is -25.0 m and the y-component is $+40.0$ m. (a) What is the magnitude of $\vec{A}$? (b) What is the angle between the direction of $\vec{A}$ and the positive direction of x?

8. Displacement Vector A displacement vector $\vec{r}$ in the xy plane is 15 m long and directed as shown in Fig. 4-20. Determine (a) the x-component and (b) the y-component of the vector.

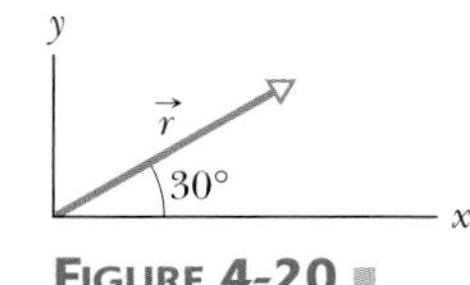

FIGURE 4-20 ■ Problem 8.

9. Rolling Wheel A wheel with a radius of 45.0 cm rolls without slipping along a horizontal floor (Fig. 4-21). At time t_1, the dot P painted on the rim of the wheel is at the point of contact between the wheel and the floor. At a later time t_2, the wheel has rolled through one-half of a revolution. What are (a) the magnitude and (b) the angle (relative to the floor) of the displacement of P during this interval?

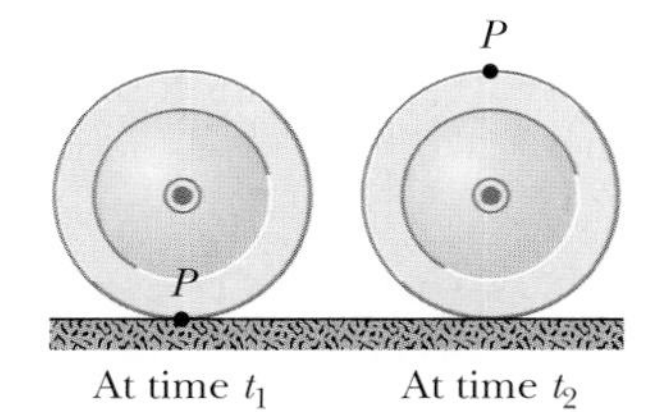

FIGURE 4-21 ■ Problem 9.

10. Rock Faults Rock *faults* are ruptures along which opposite faces of rock have slid past each other. In Fig. 4-22 points A and B coincided before the rock in the foreground slid down to the right. The net displacement $\overrightarrow{AB}$ is along the plane of the fault. The horizontal component of $\overrightarrow{AB}$ is the *strike-slip AC*. The component of $\overrightarrow{AB}$ that is directly down the plane of the fault is the *dip-slip AD*. (a) What is the magnitude of the net displacement $\overrightarrow{AB}$ if the strike-slip is 22.0 m and the dip-slip is 17.0 m? (b) If the plane of the fault is inclined 52.0° to the horizontal, what is the vertical component of $\overrightarrow{AB}$?

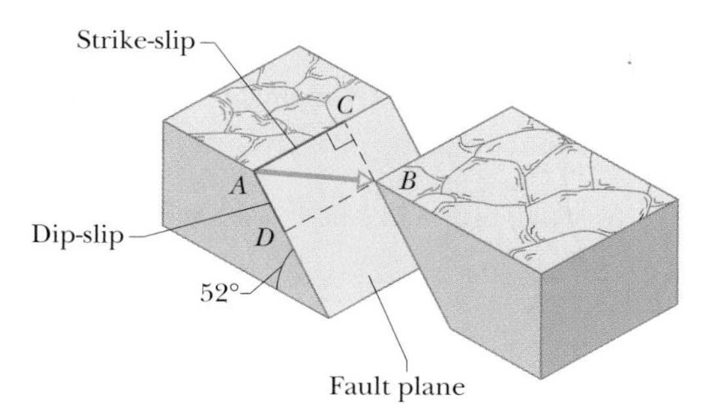

FIGURE 4-22 ■ Problem 10.

11. A Room A room has dimensions 3.00 m (height) × 3.70 m × 4.30 m. A fly starting at one corner flies around, ending up at the diagonally opposite corner. (a) What is the magnitude of its displacement? (b) Could the length of its path be less than this magnitude? (c) Greater than this magnitude? (d) Equal to this magnitude? (e) Choose a suitable coordinate system and find the components of the displacement vector in that system. (f) If the fly walks rather than flies, what is the length of the shortest path it can take? (*Hint*: This can be answered without calculus. The room is like a box. Unfold its walls to flatten them into a plane.)

SEC. 4-6 ■ ADDING VECTORS USING COMPONENTS

12. The Drive A car is driven east for a distance of 50 km, then north for 30 km, and then in a direction 30° east of north for 25 km. Sketch the vector diagram and determine (a) the magnitude and (b) the angle of the car's total displacement from its starting point.

13. A Walk A woman walks 250 m in the direction 30° east of north, then 175 m directly east. Find (a) the magnitude and (b) the angle of her final displacement from the starting point. (c) Find the distance she walks. (d) Which is greater, that distance or the magnitude of her displacement?

14. Another Walk A person walks in the following pattern: 3.1 km north, then 2.4 km west, and finally 5.2 km south. (a) Sketch the vector diagram that represents this motion. (b) How far and (c) in what direction would a bird fly in a straight line from the same starting point to the same final point?

15. Unit-vector (a) In unit-vector notation, what is the sum of

$$\vec{a} = (4.0\text{ m})\hat{\text{i}} + (3.0\text{ m})\hat{\text{j}} \quad \text{and} \quad \vec{b} = (-13.0\text{ m})\hat{\text{i}} + (7.0\text{ m})\hat{\text{j}}?$$

What are (b) the magnitude and (c) the direction of $\vec{a} + \vec{b}$ (relative to $\hat{\text{i}}$)?

16. Find the Components Find the (a) x- (b) y- and (c) z-components of the sum $\Delta\vec{r}$ of the displacements $\Delta\vec{c}$ and $\Delta\vec{d}$ whose components in meters along the three axes are $\Delta c_x = 7.4$, $\Delta c_y = -3.8$, $\Delta c_z = -6.1$; $\Delta d_x = 4.4$, $\Delta d_y = -2.0$, $\Delta d_z = 3.3$.

17. Two Vectors Vector $\vec{a}$ has a magnitude of 5.0 m and is directed east. Vector $\vec{b}$ has a magnitude of 4.0 m and is directed 35° west of north. What are (a) the magnitude and (b) the direction of $\vec{a} + \vec{b}$? What are (c) the magnitude and (d) the direction of $\vec{b} - \vec{a}$? (e) Draw a vector diagram for each combination.

18. For the Vectors For the vectors

$$\vec{a} = (3.0\text{ m})\hat{\text{i}} + (4.0\text{ m})\hat{\text{j}} \quad \text{and} \quad \vec{b} = (5.0\text{ m})\hat{\text{i}} + (-2.0\text{ m})\hat{\text{j}},$$

give $\vec{a} + \vec{b}$ in (a) unit-vector notation, and as (b) a magnitude and (c) an angle (relative to $\hat{\text{i}}$). Now give $\vec{b} - \vec{a}$ in (d) unit-vector notation, and as (e) a magnitude and (f) an angle.

19. Two Vectors Two Two vectors are given by

$$\vec{a} = (4.0\text{ m})\hat{\text{i}} - (3.0\text{ m})\hat{\text{j}} + (1.0\text{ m})\hat{\text{k}}$$

and

$$\vec{b} = (-1.0\text{ m})\hat{\text{i}} + (1.0\text{ m})\hat{\text{j}} + (4.0\text{ m})\hat{\text{k}}.$$

In unit-vector notation, find (a) $\vec{a} + \vec{b}$, (b) $\vec{a} - \vec{b}$, and (c) a third vector $\vec{c}$ such that $\vec{a} - \vec{b} + \vec{c} = 0$.

20. Two Vectors Three Here are two vectors:

$$\vec{a} = (4.0\text{ m})\hat{\text{i}} - (3.0\text{ m})\hat{\text{j}} \quad \text{and} \quad \vec{b} = (6.0\text{ m})\hat{\text{i}} + (8.0\text{ m})\hat{\text{j}}.$$

What are (a) the magnitude and (b) the angle (relative to $\hat{\text{i}}$) of $\vec{a}$? What are (c) the magnitude and (d) the angle of $\vec{b}$? What are (e) the magnitude and (f) the angle of $\vec{a} + \vec{b}$; (g) the magnitude and (h) the angle of $\vec{b} - \vec{a}$; and (i) the magnitude and (j) the angle of $\vec{a} - \vec{b}$? (k) What is the angle between the directions of $\vec{b} - \vec{a}$ and $\vec{a} - \vec{b}$?

21. Three Vectors Three vectors $\vec{a}$, and $\vec{b}$, and $\vec{c}$ each have a magnitude of 50 m and lie in an xy plane. Their directions relative to the positive direction of the x axis are 30°, 195°, and 315°, respectively. What are (a) the magnitude and (b) the angle of the vector $\vec{a} + \vec{b} + \vec{c}$, and (c) the magnitude and (d) the angle of $\vec{a} - \vec{b} + \vec{c}$? What are (e) the magnitude and (f) the angle of a fourth vector $\vec{d}$ such that $(\vec{a} + \vec{b}) - (\vec{c} + \vec{d}) = 0$.

22. Four Vectors What is the sum of the following four vectors in (a) unit-vector notation and (b) magnitude-angle notation? For the latter, give the angle in both degrees and radians. Positive angles are counterclockwise from the positive direction of the x axis; negative angles are clockwise.

$\vec{E}$: 6.00 m at +0.900 rad $\qquad$ $\vec{F}$: 5.00 m at −75.0°

$\vec{G}$: 4.00 m at +1.20 rad $\qquad$ $\vec{H}$: 6.00 m at −210°

23. Two Vectors Four The two vectors $\vec{a}$ and $\vec{b}$ in Fig. 4-23 have equal magnitudes of 10.0 m. Find (a) the x-component and (b) the y-component of their vector sum $\vec{r}$, (c) the magnitude of $\vec{r}$, and (d) the angle $\vec{r}$ makes with the positive direction of the x axis.

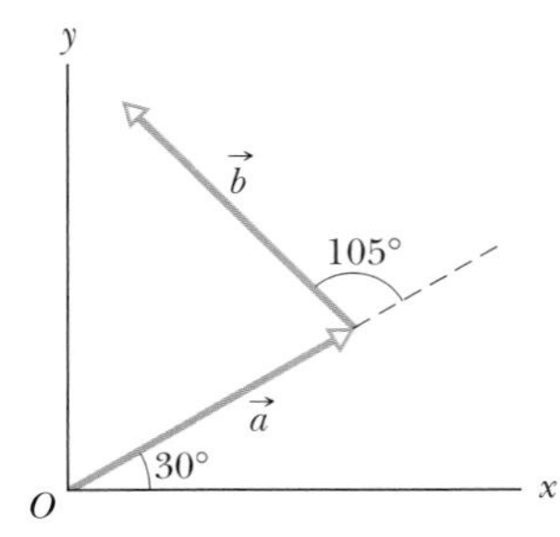

FIGURE 4-23 ■ Problem 23.

24. The Sum In the sum $\vec{A} + \vec{B} = \vec{C}$, vector $\vec{A}$ has a magnitude of 12.0 m and is angled 40.0° counterclockwise from the $+x$ direction, and vector $\vec{C}$ has a magnitude of 15.0 m and is angled 20.0° counterclockwise from the $-x$ direction. What are (a) the magnitude and (b) the angle (relative to $+x$) of $\vec{B}$?

25. Prove Prove that two vectors must have equal magnitudes if their sum is perpendicular to their difference.

26. The Sum of Four Find the sum of the following four vectors in (a) unit-vector notation, and as (b) a magnitude and (c) an angle relative to $+x$.

$\vec{P}$: 10.0 m, at 25.0° counterclockwise from $+x$

$\vec{Q}$: 12.0 m, at 10.0° counterclockwise from $+y$

$\vec{R}$: 8.00 m, at 20.0° clockwise from $-y$

$\vec{S}$: 9.00 m, at 40.0° counterclockwise from $-y$

27. Prove by Components Two vectors of magnitudes a and b make an angle θ with each other when placed tail to tail. Prove, by taking components along two perpendicular axes, that

$$r = \sqrt{a^2 + b^2 + 2ab\cos\theta}$$

gives the magnitude of the sum $\vec{r}$ of the two vectors.

28. The Sum of Four Again What is the sum of the following four vectors in (a) unit-vector notation, and as (b) a magnitude and (c) an angle? Positive angles are counterclockwise from the positive direction of the x axis; negative angles are clockwise.

$\vec{A} = (2.00\text{ mi})\hat{\text{i}} + (3.00\text{ mi})\hat{\text{j}}$ $\qquad$ $\vec{B}$: 4.00 m, at +65.0°

$\vec{C} = (-4.00\text{ m})\hat{\text{i}} - (6.00\text{ m})\hat{\text{j}}$ $\qquad$ $\vec{D}$: 5.00 m, at −235°

29. A Cube (a) Using unit vectors, write expressions for the four body diagonals (the straight lines from one corner to another through the center) of a cube in terms of its edges, which have length a. (b) Determine the angles that the body diagonals make with the adjacent edges. (c) Determine the length of the body diagonals in terms of a.

30. Oasis Oasis B is 25 km due east of oasis A. Starting from oasis A, a camel walks 24 km in a direction 15° south of east and then walks 8.0 km due north. How far is the camel then from oasis B?

31. A Plus B If $\vec{B}$ is added to $\vec{A}$, the result is $6.0\hat{\text{i}} + 1.0\hat{\text{j}}$. If $\vec{B}$ is subtracted from $\vec{A}$, the result is $-4.0\hat{\text{i}} + 7.0\hat{\text{j}}$. What is the magnitude of $\vec{A}$?

32. If-Then If $\vec{d}_1 + \vec{d}_2 = 5\vec{d}_3$, $\vec{d}_1 - \vec{d}_2 = 3\vec{d}_3$, and $\vec{d}_3 = 2\hat{\text{i}} + 4\hat{\text{j}}$, then what are (a) $\vec{d}_1$ and (b) $\vec{d}_2$?

33. Sailing A sailboat sets out from the U.S. side of Lake Erie for a point on the Canadian side, 90.0 km due north. The sailor, however, ends up 50.0 km due east of the starting point. (a) How far and (b) in what direction must the sailor now sail to reach the original destination?

SEC. 4-7 ■ MULTIPLYING AND DIVIDING A VECTOR BY A SCALAR

34. Vector by Scalar The three vectors in Fig. 4-24 have magnitudes $a = 3.00$ m, $b = 4.00$ m, and $c = 10.0$ m. What are (a) the

x-component and (b) the y-component of $\vec{a}$; (c) the x-component and (d) the y-component of $\vec{b}$; and (e) the x-component and (f) the y-component of $\vec{c}$? If $\vec{c} = p\vec{a} + q\vec{b}$, what are the values of (g) p and (h) q?

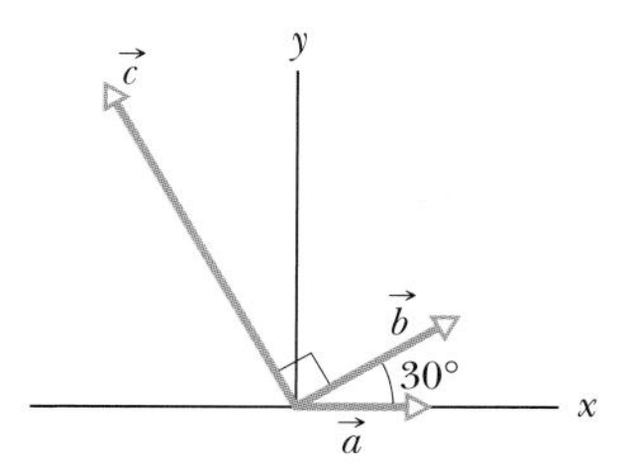

FIGURE 4-24 ■ Problem 34.

35. Five Times A vector $\vec{d}$ has a magnitude 3.0 m and is directed south. What are (a) the magnitude and (b) the direction of the vector $5.0\vec{d}$? What are (c) the magnitude and (d) the direction of the vector $-2.0\vec{d}$?

36. The Sum Is a Third Vector $\vec{A}$, which is directed along an x axis, is to be added to vector $\vec{B}$, which has a magnitude of 7.0 m. The sum is a third vector that is directed along the y axis, with a magnitude that is 3.0 times that of $\vec{A}$. What is that magnitude of $\vec{A}$?

Additional Problems

37. Explorer An explorer is caught in a whiteout (in which the snowfall is so thick that the ground cannot be distinguished from the sky) while returning to base camp. He was supposed to travel due north for 5.6 km, but when the snow clears, he discovers that he actually traveled 7.8 km at 50° north of due east. (a) How far and (b) in what direction must he now travel to reach base camp?

38. Bowling Balls In each case below, sketch the velocity vector. Find the magnitude and direction of motion with respect to the x axis of the coordinate system:

(a) $\vec{v} = (2.45\text{ m/s})\hat{\text{i}} + (3.67\text{ m/s})\hat{\text{j}}$
(b) $\vec{v} = (-2.45\text{ m/s})\hat{\text{i}} + (5.20\text{ m/s})\hat{\text{j}}$

39. Lawn Chess In a game of lawn chess, where pieces are moved between the centers of squares that are each 1.00 m on edge, a knight is moved in the following way: (1) two squares forward, one square rightward; (2) two squares leftward, one square forward; (3) two squares forward, one square leftward. What are (a) the magnitude and (b) the angle (relative to "forward") of the knight's overall displacement for the series of three moves?

40. Fire Ant A fire ant, searching for hot sauce in a picnic area, goes through three displacements along level ground: $\vec{d}_1$ for 0.40 m southwest (that is, at 45° from directly south and from directly west), $\vec{d}_2$ for 0.50 m due east (that is, directly east), $\vec{d}_3$ for 0.60 m at 60° north of east (that is 60.0° toward the north from due east). Let the positive x direction be east and the positive y direction be north. What are (a) the x-component and (b) the y-component of $\vec{d}_1$? What are (c) the x-component and (d) the y-component of $\vec{d}_2$? What are (e) the x-component and (f) the y-component of $\vec{d}_3$?
What are (g) the x-component, (h) the y-component, (i) the magnitude, and (j) the direction of the ant's net displacement? If the ant is to return directly to the starting point, (k) how far and (l) in what direction should it move?

41. A Heavy Object A heavy piece of machinery is raised by sliding it 12.5 m along a plank oriented at 20.0° to the horizontal, as shown in Fig. 4-25. (a) How high above its original position is it raised? (b) How far is it moved horizontally?

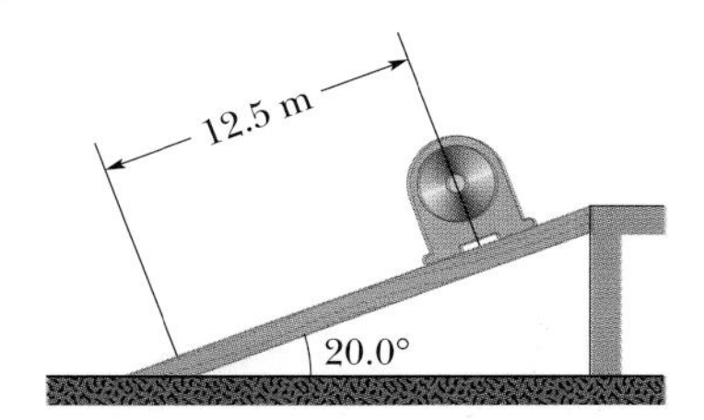

FIGURE 4-25 ■ Problem 41.

42. Two Beetles Two beetles run across flat sand, starting at the same point. Beetle 1 runs 0.50 m due east, then 0.80 m at 30° north of due east. Beetle 2 also makes two runs; the first is 1.6 m at 40° east of due north. What must be (a) the magnitude and (b) the direction of its second run if it is to end up at the new location of beetle 1?

43. Four Moves You are to make four straight-line moves over a flat desert floor, starting at the origin of an xy coordinate system and ending at the xy coordinates (−140 m, 30 m). The x-component and y-component of your moves are the following, respectively, in meters: (20 and 60), then (b_x and −70), then (−20 and c_y), then (−60 and −70). What are (a) component b_x and (b) component c_y? What are (c) the magnitude and (d) the angle (relative to the positive direction of the x axis) of the overall displacement?

44. Vector C The magnitude and angle of $\vec{A}$, which lies in an xy plane, are 4.00 and 130°, respectively. What are the components (a) A_x and (b) A_y? Vector $\vec{B}$ also lies in the xy plane, and it has components $B_x = -3.86$ and $B_y = -4.60$. What is $\vec{A} + \vec{B}$ in (c) magnitude-angle notation and (d) unit-vector notation? In (e) unit-vector notation and (f) magnitude-angle notation, find $\vec{C}$ such that $\vec{A} - \vec{C} = \vec{B}$. (g) Which of the vector diagrams in Fig. 4-26 correctly show the relationship between those three vectors?

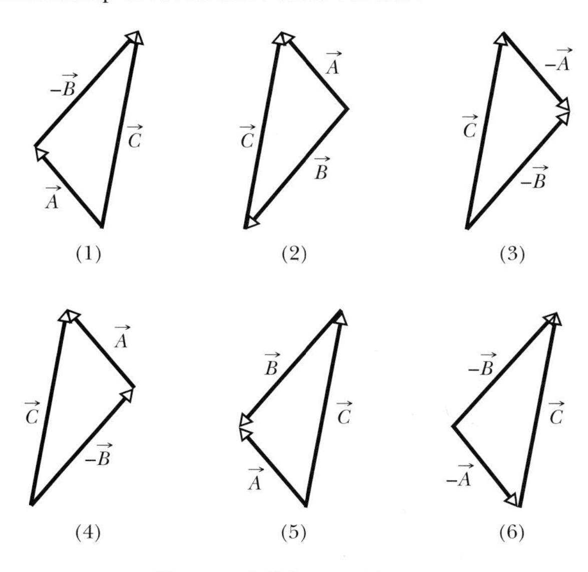

FIGURE 4-26 ■ Problem 44.

45. Twice the Magnitude A vector $\vec{B}$, with a magnitude of 8.0 m, is added to a vector $\vec{A}$, which lies along an x axis. The sum of these two vectors is a third vector that lies along the y axis and has a magnitude that is twice the magnitude of $\vec{A}$. What is the magnitude of $\vec{A}$?

46. To Reach a Point A person desires to reach a point that is 3.40 km from her present location and in a direction that is 35.0° north of east. However, she must travel along streets that are oriented either north–south or east–west. What is the minimum distance she could travel to reach her destination?

47. A Golfer A golfer takes three putts to get the ball into the hole. The first putt displaces the ball 3.66 m north, the second 1.83 m southeast, and the third 0.91 m southwest. What are (a) the magnitude and (b) the direction of the displacement needed to get the ball into the hole on the first putt?

48. Protestor's Sign A protester carries his sign of protest 40 m along a straight path, then 20 m along a perpendicular path to his left, and then 25 m up a water tower. (a) Choose and describe a coordinate system for this motion. In terms of that system and in unit-vector notation, what is the displacement of the sign from start to end? (b) The sign then falls to the foot of the tower. What is the magnitude of the displacement of the sign from start to this new end?

49. Rotated Coordinate System In Fig. 4.27, a vector $\vec{a}$ with a magnitude of 17.0 m is directed 56.0° counterclockwise from the $+x$ axis, as shown. What are the components (a) a_x and (b) a_y of the vector? A second coordinate system is inclined by 18.0° with respect to the first. What are the components (c) a'_x and (d) a'_y in this primed coordinate system?

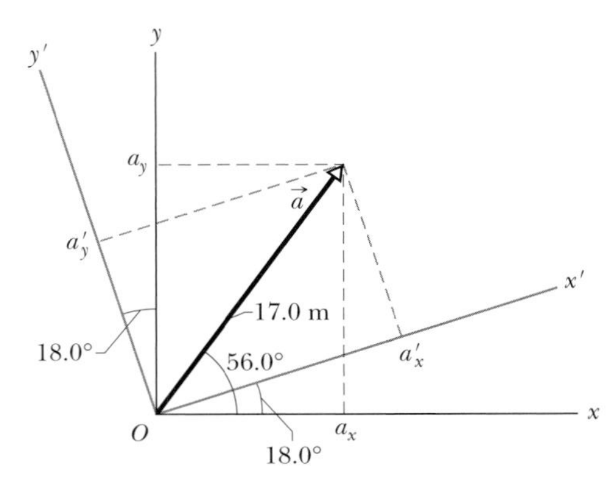

FIGURE 4-27 ■ Problem 49.

50. Shifted Coordinate System Consider how the components of a vector in the plane change if I change the reference point. Suppose I start with a coordinate system with an origin at O. An arbitrary vector $\vec{r} = x\hat{i} + y\hat{j}$ with coordinates (x, y) specifies a point in this system. Suppose also that I have another point O' specified in this coordinate system by a vector $\vec{A} = A_x\hat{i} + A_y\hat{j}$. If I change my origin to O' (without rotating the axes), what would the coordinates be for the point specified by $\vec{r}$?

51. A New System $\vec{A}$ has the magnitude 12.0 m and is angled 60.0° counterclockwise from the positive direction of the x axis of an xy coordinate system. Also, $\vec{B} = (12.0\text{ m})\hat{i} + (8.00\text{ m})\hat{j}$ on that same coordinate system. We now rotate the system, counterclockwise about the origin by 20.0°, to form an $x'y'$ system. On this new system, what are (a) $\vec{A}$ and (b) $\vec{B}$, both in unit-vector notation?

5 Net Force and Two-Dimensional Motion

In 1922, one of the Zacchinis, a famous family of circus performers, was the first human cannon ball to be shot across an arena into a net. To increase the excitement, the family gradually increased the height and distance of the flight until, in 1939 or 1940, Emanuel Zacchini soared over three Ferris wheels and through a horizontal distance of 69 m.

How could he know where to place the net, and how could he be certain he would clear the Ferris wheels?

The answer is in this chapter.

5-1 Introduction

In this chapter, we will apply Newton's Second Law of motion developed in Chapter 3 to the analysis of familiar two-dimensional motions. We start with an exploration of projectile motion. When a particle-like object is launched close to the surface of the Earth with a horizontal component of velocity and allowed to fall freely, we say it undergoes projectile motion. The second two-dimensional motion we consider is uniform circular motion in which a particle-like object moves in a circle at a constant speed. This motion can be produced by twirling a ball attached to a string in a circle or by watching a point on the edge of a spinning wheel.

You will use the vector algebra introduced in Chapter 4 and extend the concepts introduced in Chapters 2 and 3 to two dimensions. As you work with this chapter, you will want to review relevant sections in these chapters.

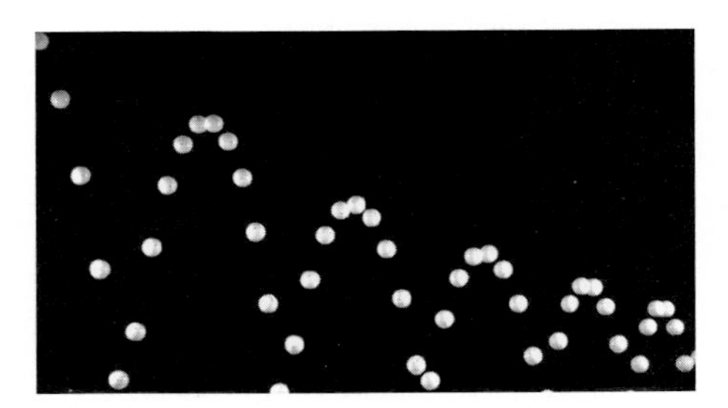

FIGURE 5-1 ■ A stroboscopic photograph of a golf ball bouncing off a hard surface. Between impacts, the ball undergoes projectile motion that is characterized by curved paths.

5-2 Projectile Motion

Projectile motion occurs near the Earth's surface whenever a ball rolls off a table, a basketball arcs toward a basket, a hailstone rolls off a steep roof, or a ball bounces (Fig. 5-1). All of these motions have curved paths. But why are the paths curved, and what sort of curves are they? In Section 3-2 we presented data on objects moving horizontally that provided evidence that in the absence of forces, moving objects tend to continue moving at a constant velocity (Newton's First Law). In Section 3-9 we presented data indicating that near the Earth's surface, objects fall freely in a vertical direction with an acceleration of magnitude 9.8 m/s^2. Projectile motions from basketballs to hailstones all involve a combination of horizontal and vertical motions. Galileo was the first to discover how to treat two-dimensional projectile motion as a combination of horizontal and vertical motions.

FIGURE 5-2 ■ Galileo (1564–1642) was the first scientist to deduce that projectile motion could be analyzed as a combination of two independent linear motions.

Galileo's Hypothesis

In his *Dialog Concerning Two New Sciences* published in 1632, Galileo (Fig. 5-2) observes:

> "... we have discussed the properties of uniform motion and of motion naturally accelerated along planes of all inclinations. I now propose to set forth those properties which belong to a body whose motion is compounded of two other motions, namely, one uniform and one naturally accelerated.... This is the kind of motion seen in a moving projectile; its origin I conceive to be as follows:
>
> Imagine any particle projected along a horizontal plane without friction; then we know ... that this particle will move along this same plane with a motion which is uniform and perpetual.... But if the plane is limited and elevated, then the moving particle ... will on passing over the edge of the plane acquire, in addition to its previous uniform and perpetual motion, a downward propensity due to its own weight; so that the resulting motion which I call projection is compounded of one which is uniform and horizontal and of another which is vertical and naturally accelerated."

Galileo went on to predict that the curve that describes projectile motion is the parabola. He deduces this using a construction (Fig. 5-3) which shows how uniform horizontal motion and uniformly accelerated vertical motion with displacements increasing in proportion to the square of time can be combined or superimposed to form a parabola. The superposition is not unlike that introduced in Chapter 3 with regard to the combination of forces. However, in this case the quantities must be at right angles to each other.

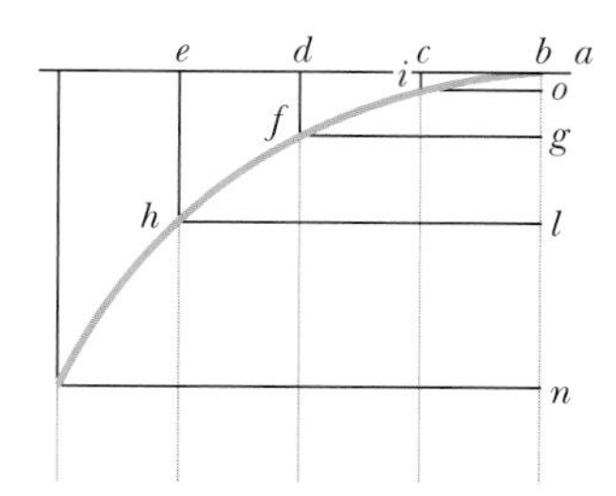

FIGURE 5-3 ■ Galileo's diagram showing how a parabola can be formed by a set of linearly increasing horizontal coordinates ($b \rightarrow e$) and a set of vertical coordinates that increase as the square of time ($o \rightarrow n$).

Experimental Evidence for Galileo's Hypothesis

Galileo's hypothesis was based on both observations and reasoning. By observing balls accelerating slowly on inclines and balls in free fall, he knew that their positions increased as the square of time. He determined the constant velocity of a ball rolling along a level ramp. Knowing the height of the ramp, he could predict how long the vertical falling motion should take and hence how far in a horizontal direction the ball should travel before it hit the floor, assuming the horizontal velocity was undisturbed by the introduction of the vertical falling motion. Since Galileo did not have contemporary technology such as strobe photography or video analysis at his disposal, his approach to understanding projectile motion was extraordinary.

It is instructive to confirm Galileo's hypothesis regarding a body moving off the edge of a ramp by using a digital video camera to record this motion, as in Fig. 5-4. What would Galileo have observed? If he had drawn the horizontal position component of the ball along the line of its original motion as shown in Fig. 5-4, he would see that the horizontal distance the ball traveled each time period remains constant. That is, the horizontal velocity is constant both before and after the ball reaches the edge of the table. Thus, the horizontal motion must be independent of the falling motion.

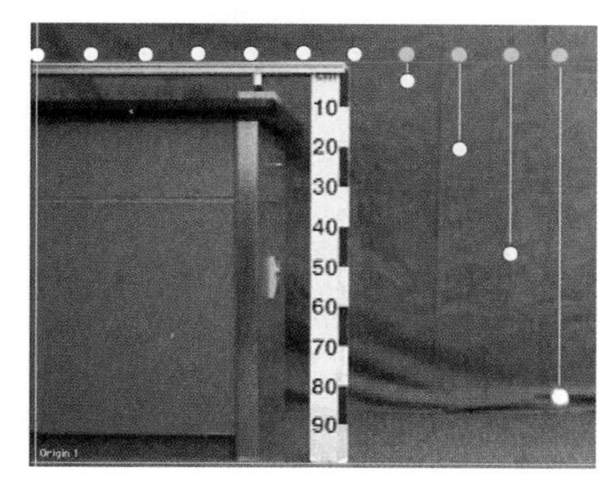

FIGURE 5-4 ■ Performing Galileo's thought experiment with modern equipment. This digital video image shows the location of a golf ball in the last frame, along with white markers left by the software showing the ball's location every tenth of a second as it rolls off the edge of a table and falls toward the floor.

Since our image shows that the ball has fallen a distance of approximately $(y_2 - y_1) \approx -0.85$ m in a time interval of $(t_2 - t_1) = 0.40$ s, we see that these quantities are consistent with the kinematic equation 2-17 given by $(y_2 - y_1) = v_{1y}(t_2 - t_1) + {}^1\!/_2\, a_y (t_2 - t_1)^2$. For this situation, the y-component of velocity equals zero at time t_1 (so that $v_{1y} = 0.0$ m/s) with the vertical acceleration component given by $a_y = -9.8$ m/s. This calculation suggests that the vertical motion is independent of the horizontal motion.

Perhaps the most compelling evidence for the independence of the vertical motion is a stroboscopic photograph of a ball that is released electronically just at the moment that a projectile is shot horizontally, as shown in Fig. 5-5. The vertical position components of these two balls are identical.

In summary, using new technology we can easily conclude, as Galileo did about 400 years ago, that

> The *horizontal and vertical motions of a projectile* (at right angles to each other) *are independent*, and the path of such a projectile can be found by combining its horizontal and vertical position components.

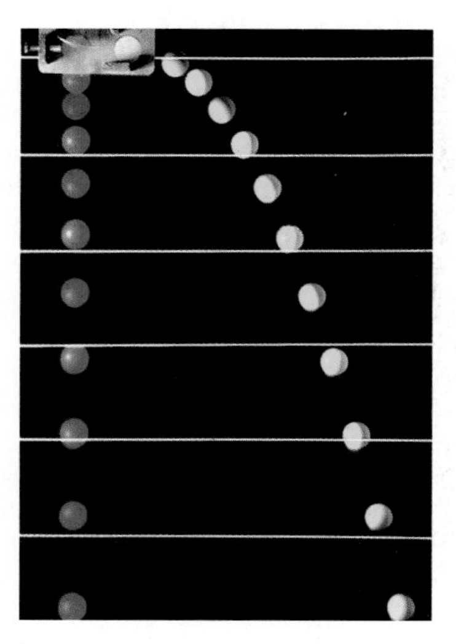

FIGURE 5-5 ■ One ball is released from rest at the same instant that another ball is shot horizontally to the right.

Ideal Projectile Motion

When Galileo wrote about projectiles in the *Two Sciences*, he made it quite clear that the moving object should be heavy and that friction should be avoided. In the next section, as we use vector mathematics to consider projectile motion, we will limit ourselves to "ideal" situations. We also assume the only significant force on an object is a constant gravitational force acting vertically downward. For example, the bouncing golf ball is undergoing ideal projectile motion between bounces because it is moving slowly enough that air resistance forces are negligible. On the other hand, flying airplanes and ducks are not ideal projectiles because their sustained flight depends on getting lift forces from air. Examples of ideal and nonideal projectile motion are shown in Fig. 5-6. Ideal projectile motion can be defined as follows:

> A particle-like object undergoes **IDEAL PROJECTILE MOTION** if the only significant force that acts on it is a constant gravitational force.

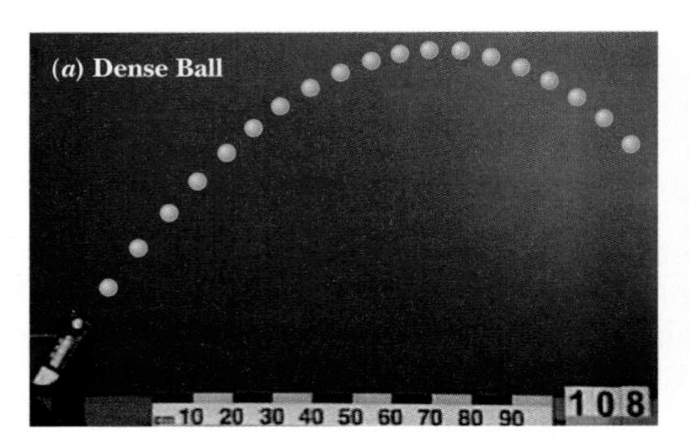

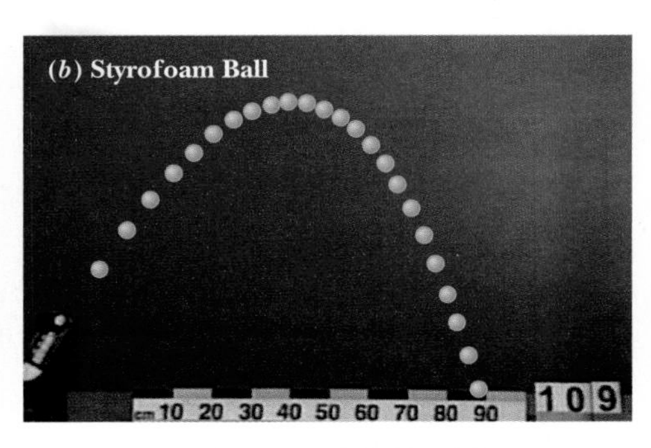

FIGURE 5-6 ■ Video analysis software is used to trace the paths of two small balls of the same size shot from a projectile launcher. The dots mark the location of the ball every 1/30 of a second. (*a*) The left video frame shows the path of a dense plastic ball. (*b*) The right video frame shows a Styrofoam ball path that is not as long or symmetric because it is influenced by air drag forces.

FIGURE 5-7 ■ The vertical component of this skateboarder's velocity is changing. However, during the entire time he is in the air, the skateboard stays underneath him, allowing him to land on it.

READING EXERCISE 5-1: (*a*) Consider the light-colored falling golf ball on the right in Fig. 5-5. Does its horizontal velocity change its vertical acceleration and the vertical velocities it normally acquires in free fall along a straight vertical line? Explain. (*b*) Does the fact that the light-colored golf ball on the right is falling have any effect on the rate that it is moving in the x direction? Explain. ■

READING EXERCISE 5-2: How does the fact that the skateboarder in Fig. 5-7 has a vertical acceleration and, therefore, vertical velocity affect his horizontal velocity while he is "flying" above his skateboard? ■

TOUCHSTONE EXAMPLE 5-1: Golf Ball

In Figure 5-4, a golf ball is rolling off a tabletop which is 1.0 meter above the floor. The golf ball has an initial horizontal velocity component of $v_{1x} = 1.3$ m/s. Its location has been marked at 0.100 s intervals.

(a) How long should the ball take to fall on the floor from the time it leaves the edge of the table?

SOLUTION ■ The **Key Idea** here is that the golf ball undergoes ideal projectile motion. Therefore, its horizontal and vertical motions are independent and can be considered separately (we need not consider the actual curved path of the ball). We choose a coordinate system in which the origin is at floor level and upward is the positive direction. Then, the first vertical position component of interest in our fall is $y_1 = 1.0$ m and our last vertical position of interest is $y_2 = 0.0$ m. We can use kinematic equation 2-17 to describe the descent along the y axis, so

$$(y_2 - y_1) = v_{1y}(t_2 - t_1) + \tfrac{1}{2}a_y(t_2 - t_1)^2.$$

Noting that at the table's edge the vertical velocity is zero, so $v_{1y} = 0.00$ m/s, we can solve our equation for $(t_2 - t_1)$ to get

$$t_2 - t_1 = \sqrt{\frac{2(y_2 - y_1)}{a_y}},$$

where in free fall the only force is the gravitational force, and so

$$a_y = -g = -9.8 \text{ m/s}^2$$

and $$(y_2 - y_1) = (0.0 \text{ m} - 1.0 \text{ m}) = -1.0 \text{ m}.$$

This gives $$(t_2 - t_1) = \sqrt{\frac{2(y_2 - y_1)}{-g}} = \sqrt{\frac{2(-1.0 \text{ m})}{-9.8 \text{ m/s}^2}} = 0.45 \text{ s}.$$

(b) How far will the ball travel in the horizontal direction from the edge of the table before it hits the floor?

SOLUTION ■ The **Key Idea** is that the ball will hit the floor after it has fallen for $(t_2 - t_1) = 0.45$ s. Since its horizontal velocity component doesn't change, its average velocity is the same as its instantaneous velocity. We can use the definition of average velocity to calculate the horizontal distance it travels, since

$$v_{1x} = \langle v_x \rangle = \frac{(x_2 - x_1)}{t_2 - t_1};$$

we can solve this equation for $x_2 - x_1$ to get our distance:

$$(x_2 - x_1) = v_{1x}(t_2 - t_1) = (1.3 \text{ m/s})(0.45 \text{ s}) = 0.59 \text{ m}.$$

An examination of Fig. 5-4 shows that this distance is reasonable.

5-3 Analyzing Ideal Projectile Motion

Now that we have established the independence of the horizontal and vertical components of the motion of an ideal projectile, we can analyze these motions mathematically. To do this we use the vector algebra we have developed to express the effect of the gravitational force on velocity components along horizontal and vertical axes.

Velocity Components and Launch Angle

A ball rolling off a level ramp has a launch angle of zero degrees with respect to the horizontal. In general, projectiles are launched at some angle θ with respect to the horizontal as shown in Fig 5-6(a). Suppose a projectile is launched at an angle θ_1 relative to the horizontal direction with an initial magnitude of velocity $|\vec{v}_1| = v_1$. If we use a standard rectangular coordinate system, then we can use the definitions of sine and cosine to find the initial x- and y-components of velocity at time t_1 (Fig. 5-8). These components are

$$v_{1x} = v_1 \cos\theta_1 \quad \text{and} \quad v_{1y} = v_1 \sin\theta_1. \tag{5-1}$$

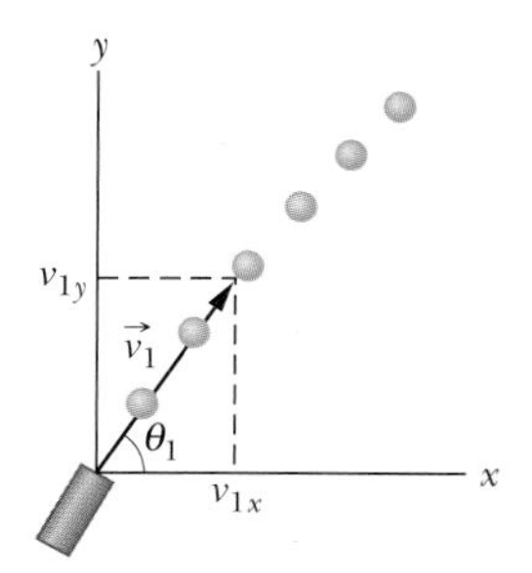

FIGURE 5-8 ■ If a projectile is launched at an initial angle θ_1 with respect to the horizontal with a magnitude v_1, its components can be calculated by using the trigonometric functions.

Alternatively, if the components of the initial velocity are known, we can rearrange Eq. 5-1 to determine the initial velocity vector, $\vec{v}_1$, and the angle of launch, θ_1,

$$\vec{v}_1 = v_{1x}\hat{\mathrm{i}} + v_{1y}\hat{\mathrm{j}} \quad \text{and} \quad \tan\theta_1 = \frac{v_1 \sin\theta_1}{v_1 \cos\theta_1} = \frac{v_{1y}}{v_{1x}}. \tag{5-2}$$

Solving for the launch angle at time t_1 gives

$$\theta_1 = \tan^{-1}\left(\frac{v_1 \sin\theta_1}{v_1 \cos\theta_1}\right) = \tan^{-1}\left(\frac{v_{1y}}{v_{1x}}\right). \tag{5-3}$$

During two-dimensional motion, an ideal projectile's position vector $\vec{r}$ and velocity vector $\vec{v}$ change continuously, but its acceleration vector $\vec{a}$ is constant—its value doesn't change and it is always directed vertically downward.

Position and Velocity Versus Time

In our standard coordinate system an ideal projectile experiences no net force in the x direction ($\vec{F}_x = (0)\hat{\mathrm{i}}$), so its horizontal acceleration stays constant at zero during its flight. The projectile experiences a constant gravitational force $\vec{F}_y = -mg\hat{\mathrm{j}}$ in the y direction, so the y-component of its acceleration is a constant with a value of $-g$. Since zero acceleration is a form of constant acceleration, the acceleration is constant in *both* directions. The object's projectile motion acceleration is given by

$$\vec{a}_x = 0\hat{\mathrm{i}} \quad \text{and} \quad \vec{a}_y = -g\hat{\mathrm{j}} \quad \text{(projectile motion accelerations).}$$

From Chapter 2 (Eq. 2-17) we know that (with $a_x = 0$) the changes in the object's position components are given by

$$x_2 - x_1 = v_{1x}(t_2 - t_1) \quad \text{and} \quad y_2 - y_1 = v_{1y}(t_2 - t_1) + \tfrac{1}{2}a_y(t_2 - t_1)^2 \quad \text{(position change),} \tag{2-17}$$

while the object's velocity component changes are given by Eq. 2-13 as

$$v_{2x} - v_{1x} = a_x(t_2 - t_1) \quad \text{and} \quad v_{2y} - v_{1y} = a_y(t_2 - t_1) \quad \text{(velocity change).} \tag{2-13}$$

If we call the time that we start measuring $t_1 = 0$, then the equations are somewhat simplified. However, independent of what time we start measuring, these two equations describe the motion of the object during a specified *interval* of time $t_2 - t_1$. They are our primary equations of motion and are valid for every situation involving constant acceleration, including cases where the acceleration is zero.

The Horizontal Motion

Suppose that at an initial time $t = t_1$ the projectile has a position component along the x axis of x_1 and a velocity component of v_{1x}. We must now use the notation $\vec{v}_1, v_{1x}, v_{1y}$ so we can distinguish the initial velocity vector $\vec{v}_1$ from its x- and y-components. We must also specify which component of acceleration, a_x or a_y, we are using in an equation. Using our new notation, we can find the x-component of the projectile's horizontal displacement x_2 at any later time t_2 using $x_2 - x_1 = v_{1x}(t_2 - t_1) + \frac{1}{2}a_x(t_2 - t_1)^2$. However, since there are *no forces* and hence *no acceleration* in the horizontal direction, the x-component of acceleration is zero. Noting that the only part of the initial velocity that affects the horizontal motion is its horizontal component v_{1x}, we can write this as:

$$x_2 - x_1 = v_{1x}(t_2 - t_1) \quad \text{(horizontal displacement).} \tag{5-4}$$

Because the horizontal component of the object's initial velocity is given by $v_{1x} = v_1 \cos \theta_1$, this equation for the displacement in the x direction can also be written as

$$x_2 - x_1 = (v_1 \cos \theta_1)(t_2 - t_1). \tag{5-5}$$

But the ratio of the displacement to the time interval over which it occurs, $(x_2 - x_1)/(t_2 - t_1)$, is just the x-component of average velocity $\langle v_x \rangle$. If the average velocity in the x direction is constant, this means that the instantaneous and average velocities in the x direction are the same. Thus Eq. 5-5 becomes

$$\langle v_x \rangle = \frac{x_2 - x_1}{t_2 - t_1} = v_{1x} = v_1 \cos \theta_1 = \text{a constant.} \tag{5-6}$$

Experimental verification of the constancy of the horizontal velocity component for ideal projectile motion is present in Figs. 5-4 and 5-5. As additional verification we can draw a graph of the x-component of position as a function of time for the projectile path depicted in Fig. 5-6a. This is shown in Fig. 5-9.

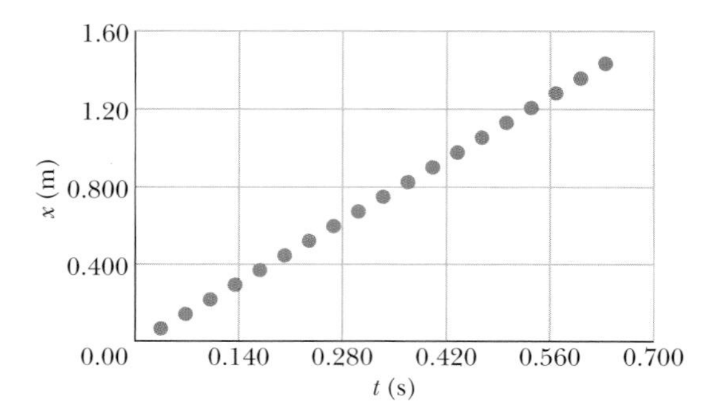

FIGURE 5-9 ■ A graph constructed from a video analysis of the motion of the projectile depicted in Fig. 5-6a. The x-component of position is plotted as a function of time. The coordinate system is chosen so that the initial value x_1 is zero at the launcher muzzle. The linearity of the graph confirms that the projectile's x-component of velocity is a constant given by the slope of the line so that $\langle v_x \rangle = v_{1x} = 2.3$ m/s.

The Vertical Motion

The vertical motion of an ideal projectile was discussed in Section 3-9 for a particle close to the surface of the Earth. Most important is that the acceleration resulting from the attractive gravitational force that the Earth exerts on the object is constant and directed downward. We denote its magnitude as g and recall that it has a value of 9.8 m/s^2. If, as usual, we take upward to be the positive y direction, then since the gravitational force points downward we can replace the y-component of acceleration, a_y, for position change along the y axis with $-g$. This allows us to rewrite our primary kinematic equation shown in Eq. 2-17 $[(x_2 - x_1) = v_{1\,x}(t_2 - t_1) + \frac{1}{2}a_x(t_2 - t_1)^2]$ for motion along a y axis as

$$y_2 - y_1 = v_{1\,y}(t_2 - t_1) + \tfrac{1}{2}(a_y)(t_2 - t_1)^2 \qquad \text{where } a_y = -g.$$

Note that we used only the y-component of the initial velocity in this equation. This is because the vertical acceleration affects only the vertical velocity and position components. Making the substitution $v_{1\,y} = v_1 \sin\theta_1$ from Eq. 5-1, we get

$$y_2 - y_1 = (v_1 \sin\theta_1)(t_2 - t_1) + \tfrac{1}{2}a_y(t_2 - t_1)^2 \qquad \text{where } a_y = -g. \tag{5-7}$$

Similarly, $v_2 - v_1 = a_x(t_2 - t_1)$ (primary equation 2-13) can be rewritten as

$$v_{2\,y} - v_{1\,y} = a_y(t_2 - t_1),$$

or

$$v_{2\,y} - (v_1 \sin\theta_1) = a_y(t_2 - t_1) \qquad \text{where } a_y = -g. \tag{5-8}$$

As is illustrated in Fig. 5-10, the vertical velocity component behaves just like that for a ball thrown vertically upward. At the instant the velocity is zero, the object must be at the highest point on its path, since the object then starts moving down. The magnitude of the velocity becomes larger with time as the projectile speeds up as it moves back down.

The velocity components for a projectile shot from a small vertical cannon mounted on a cart are shown in Fig. 5-11. Note that the horizontal velocity component does not change. The vertical component decreases in magnitude and becomes zero at the top of the path. This magnitude starts increasing again as the projectile descends and is finally "recaptured" by the cannon.

In the experimental results presented in this section, the effect of the air on the motion was negligible. Thus, we ignored air resistance and performed a mathematical analysis for ideal projectiles. Ignoring the effects of air works well for a compact, dense object such as a marble or a bowling ball, provided it is not launched at very high speeds. However, air resistance cannot be ignored for a less dense object like a crumpled piece of paper thrown rapidly or the styrofoam ball shown in Fig. 5-6b. Remember that the ideal projectile motion equations 5-4 through 5-8 have been derived assuming that resistance is negligible. These ideal projectile equations are summarized in Table 5-1. We shall discuss details of the effect of the air on motion in Chapter 6.

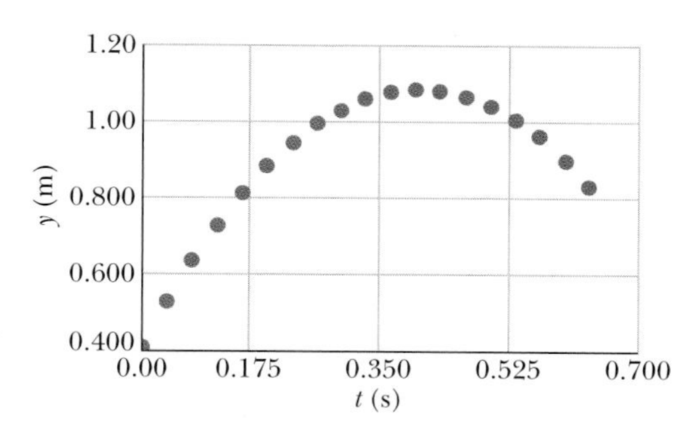

FIGURE 5-10 ■ A graph constructed from a video analysis of the motion of the projectile depicted in Fig. 5-6a. The y-component of position is plotted as a function of time. The coordinate system is chosen so that the initial value of the vertical position y_1 is zero at the launcher muzzle. A fit to the curve is parabolic and gives a y-component of acceleration of -9.7 m/s^2 and an initial vertical velocity component of $+3.5$ m/s.

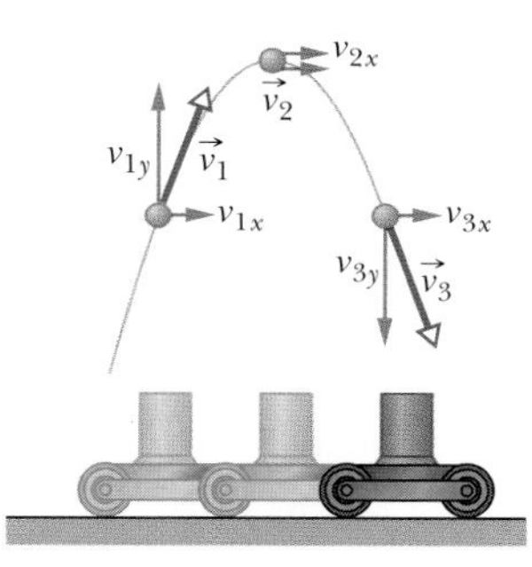

FIGURE 5-11 ■ Diagram showing sketches of three video frames. The actual movie recorded the motion of a cannon that shoots a projectile while moving. The projectile path and velocity vector components were calculated by modeling data acquired using digital video analysis software.

TABLE 5-1
Kinematic Equations for Ideal Projectile Motion

Quantity	Horizontal	Vertical
Forces	$\vec{F}_x = (0)\hat{\imath}$	$\vec{F}_y = (-mg)\hat{\jmath}$ (Eq. 3-7)
Acceleration Components	$a_x = 0$	$a_y = -g = -9.8$ m/s^2
Velocity components at t_1	$v_{1\,x} = v_1 \cos\theta_1$ (Eq. 5-1)	$v_{1\,y} = v_1 \sin\theta_1$ (Eq. 5-1)
Position component Change between t_1 and t_2	$x_2 - x_1 = v_{1\,x}(t_2 - t_1)$ (Eq. 5-4)	$y_2 - y_1 = v_{1y}(t_2 - t_1) + \frac{1}{2}a_y(t_2 - t_1)^2$ (Eq. 5-7)
Velocity component Change between t_1 and t_2	$v_{2\,x} - v_{1\,x} = 0$	$v_{2\,y} - v_{1\,y} = a_y(t_2 - t_1)$ (Eq. 5-8)

Note: Since g (the local gravitational field strength) is always positive, it is not a vector component. There fore, if we follow the standard practice of defining the negative y axis as down, we must put an explicit minus sign in front of it.

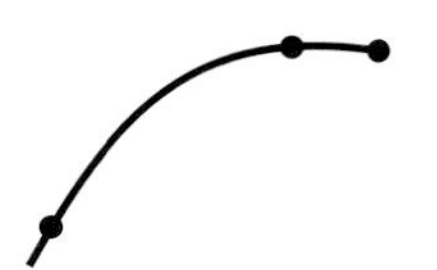

READING EXERCISE 5-3: Consider three points along an ideal projectile's path in space as shown in the figure and make three separate sketches of: (a) the force vectors showing the net force on the projectile at each point, (b) the acceleration vectors showing the acceleration of the projectile at each point, (c) the approximate horizontal and vertical components of velocity at each of the three points along with the velocity vector that is determined by these components. ■

READING EXERCISE 5-4: Consider the projectile launch shown in Fig. 5-6*a*. (a) According to the data in Figs. 5-9 and 5-10 the initial *x*- and *y*-components of the velocity are $v_{1x} = 2.3$ m/s and $v_{1y} = +3.5$ m/s . Use these values to find the launch angle of the projectile. (b) Use a protractor to measure the launch angle as indicated by the angle of the launcher shown in Fig. 5-6*a*. How do your calculated and measured launch angles compare? They should be approximately the same. ■

READING EXERCISE 5-5: A fly ball is hit to the outfield. During its flight (ignore the effects of the air), what happens to its (a) horizontal and (b) vertical components of velocity? What are the (c) horizontal and (d) vertical components of its acceleration during its ascent and its descent, and at the topmost point of its flight? ■

TOUCHSTONE EXAMPLE 5-2: Rescue Plane

In Fig. 5-12, a rescue plane flies at 198 km/h (= 55.0 m/s) and a constant elevation of 500 m toward a point directly over a boating accident victim struggling in the water. The pilot wants to release a rescue capsule so that it hits the water very close to the victim.

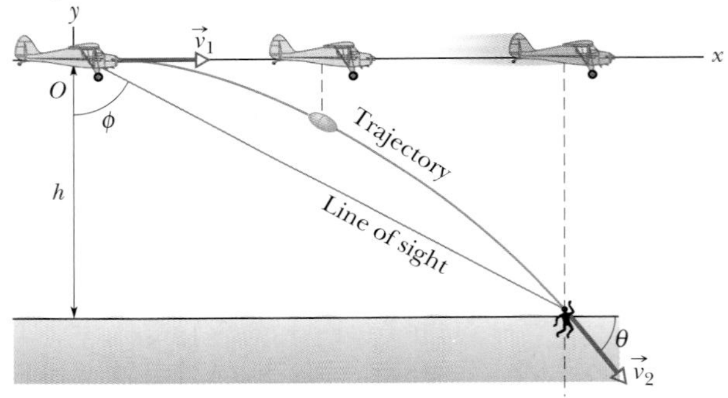

FIGURE 5-12 ■ A plane drops a rescue capsule while moving at constant velocity in level flight. While the capsule is falling, its horizontal velocity component remains equal to the velocity of the plane.

(a) What should be the angle ϕ of the pilot's line of sight to the victim when the release is made?

SOLUTION ■ The **Key Idea** here is that, once released, the capsule is a projectile, so its horizontal and vertical motions are independent and can be considered separately (we need not consider the actual curved path of the capsule). Figure 5-12 includes a coordinate system with its origin at the point of release, and we see there that ϕ is given by

$$\phi = \tan^{-1}\frac{(x_2 - x_1)}{h}, \tag{5-9}$$

where x_2 is the horizontal coordinate of the victim at release (and of the capsule when it hits the water), and h is the elevation of the plane. That elevation is 500 m, so we need only x_2 in order to find ϕ. We should be able to find x_2 with Eq. 5-4: $x_2 - x_1 = v_{1x}(t_2 - t_1)$. This can be written as

$$x_2 - x_1 = (v_1 \cos\theta_1)(t_2 - t_1), \tag{5-10}$$

where θ_1 is the angle between the initial velocity $\vec{v}_1$ and the positive x axis. For this problem, $\theta_1 = 0°$.

We know $x_1 = 0$ because the origin is placed at the point of release. Because the capsule is *released* and not shot from the plane, its initial velocity $\vec{v}_1$ is equal to the plane's velocity. Thus, we know also that the initial velocity has magnitude $v_1 = 55.0$ m/s and angle $\theta_1 = 0°$ (measured relative to the positive direction of the x axis). However, we do not know the elapsed time $t_2 - t_1$ the capsule takes to move from the plane to the victim.

To find $t_2 - t_1$, we next consider the vertical motion and specifically Eq. 5-7:

$$y_2 - y_1 = (v_1 \sin\theta_1)(t_2 - t_1) - \tfrac{1}{2}g(t_2 - t_1)^2. \tag{5-11}$$

Here the vertical displacement $y_2 - y_1$ of the capsule is -500 m (the negative value indicates that the capsule moves *downward*). Putting this and other known values into Eq. 5-7 gives us

$$-500 \text{ m} = (55.0 \text{ m/s})(\sin 0°)(t_2 - t_1) - \tfrac{1}{2}(9.8 \text{ m/s}^2)(t_2 - t_1)^2.$$

Solving for $t_2 - t_1$, we find $t_2 - t_1 = 10.1$ s. Using that value in Eq. 5-10 yields

$$x_2 - 0 \text{ m} = (55.0 \text{ m/s})(\cos 0°)(10.1 \text{ s}),$$

or

$$x_2 = 555.5 \text{ m}.$$

Then Eq. 5-9 gives us

$$\phi = \tan^{-1}\frac{555.5 \text{ m}}{500 \text{ m}} = 48°. \qquad \text{(Answer)}$$

(b) As the capsule reaches the water, what is its velocity $\vec{v}_2$ in unit-vector notation and as a magnitude and an angle?

SOLUTION ■ Again, we need the **Key Idea** that during the capsule's flight, the horizontal and vertical components of the capsule's velocity are independent of each other.

A second **Key Idea** is that the horizontal component of velocity v_x does not change from its initial value $v_{1x} = v_1 \cos\theta_1$ because there is no horizontal acceleration. Thus, when the capsule reaches the water,

$$v_{2x} = v_{1x} = v_1 \cos\theta_1 = (55.0 \text{ m/s})(\cos 0°) = 55.0 \text{ m/s}.$$

A third **Key Idea** is that the vertical component of velocity v_y changes from its initial value $v_{1y} = v_1 \sin\theta_1$ because there is a vertical acceleration. Using Eq. 5-8 and the capsule's time of fall $t_2 - t_1 = 10.1$ s, we find that when the capsule reaches the water,

$$\begin{aligned} v_{2y} &= v_1 \sin\theta_1 - g(t_2 - t_1) \\ &= (55.0 \text{ m/s})(\sin 0°) - (9.8 \text{ m/s}^2)(10.1 \text{ s}) \\ &= -99.0 \text{ m/s}. \end{aligned}$$

Thus, when the capsule reaches the water, it has the velocity

$$\vec{v}_2 = (55.0 \text{ m/s})\hat{\text{i}} - (99.0 \text{ m/s})\hat{\text{j}}. \qquad \text{(Answer)}$$

Using either the techniques developed in Section 4-4, or a vector-capable calculator, we find that the magnitude of the final velocity v_2 and the angle θ_2 are

$$v_2 = 113 \text{ m/s} \quad \text{and} \quad \theta_2 = -61°. \qquad \text{(Answer)}$$

TOUCHSTONE EXAMPLE 5-3: Ballistic Zacchini

Figure 5-13 illustrates the flight of Emanuel Zacchini over three Ferris wheels, located as shown, and each 18 m high. Zacchini is launched with speed $v_1 = 26.5$ m/s, at an angle $\theta_1 = 53°$ up from the horizontal and with an initial height of 3.0 m above the ground. The net in which he is to land is at the same height.

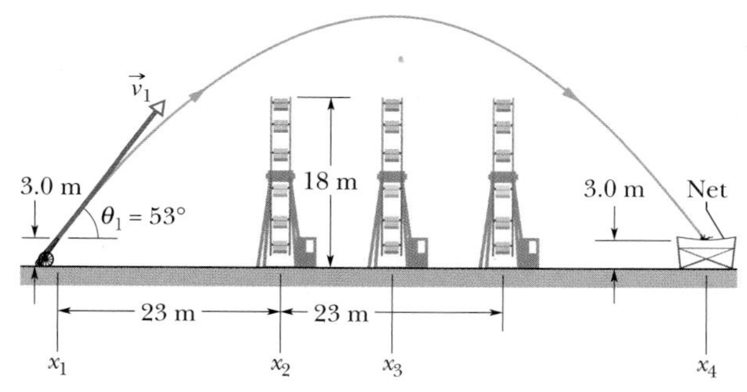

FIGURE 5-13 ■ The flight of a human cannonball over three Ferris wheels and into a net.

(a) Does he clear the first Ferris wheel?

SOLUTION ■ A **Key Idea** here is that Zacchini is a human projectile, so we can use the projectile equations. To do so, we place the origin of an xy coordinate system at the cannon muzzle. Then $x_1 = 0$ and $y_1 = 0$ and we want his height y_2 when $x_2 = 23$ m. However, we do not know the elapsed time $t_2 - t_1$ when he reaches that height. To relate y_2 to x_2 without $t_2 - t_1$, we can solve Eq. 5-5 for $t_2 - t_1$, which gives us

$$(t_2 - t_1) = \frac{x_2 - x_1}{(v_1 \cos\theta_1)}.$$

Then we can replace $t_2 - t_1$ everywhere that it appears in Eq. 5-7,

$$y_2 - y_1 = (v_1 \sin\theta_1)(t_2 - t_1) - \tfrac{1}{2} g(t_2 - t_1)^2.$$

We note that $(v_1 \sin\theta_1)/(v_1 \cos\theta_1) = \tan\theta_1$ to obtain:

$$\begin{aligned} y_2 - y_1 &= (\tan\theta_1)(x_2 - x_1) - \frac{g(x_2 - x_1)^2}{2(v_1 \cos\theta_1)^2} \\ &= (\tan 53°)(23 \text{ m}) - \frac{(9.8 \text{ m/s}^2)(23 \text{ m})^2}{2(26.5 \text{ m/s})^2 (\cos 53°)^2} \\ &= 20.3 \text{ m}. \end{aligned}$$

Since he begins 3.0 m off the ground, he clears the first Ferris wheel by about 5.3 m.

(b) If he reaches his maximum height when he is over the middle Ferris wheel, what is his clearance above it?

SOLUTION ■ A **Key Idea** here is that the vertical component v_y of his velocity is zero when he reaches his maximum height. We can combine Eqs. 5-7 and 5-8 to relate v_y and his height $y_3 - y_1$ to obtain:

$$v_{3y}^2 = v_{1y}^2 - 2g(y_3 - y_1) = (v_1 \sin\theta_1)^2 - 2g(y_3 - y_1) = 0.$$

Solving for $y_3 - y_1$ gives us

$$y_3 - y_1 = \frac{(v_1 \sin\theta_1)^2}{2g} = \frac{(26.5 \text{ m/s})^2(\sin 53°)^2}{(2)(9.8 \text{ m/s}^2)} = 22.9 \text{ m},$$

which means that he clears the middle Ferris wheel by 7.9 m.

(c) How far from the cannon should the center of the net be positioned?

SOLUTION ■ The additional **Key Idea** here is that, because Zacchini's initial and landing heights are the same, the horizontal distance from cannon muzzle to net is the value of $x_4 - x_1$ where time $t = t_4$ is when y_4 is once again zero. Then, since $y_1 = 0$ Eq. 5-7 becomes

$$0 = y_4 - y_1 = (v_1 \sin\theta_1)(t_4 - t_1) - \tfrac{1}{2}g(t_4 - t_1)^2.$$

Since $t_4 - t_1 \neq 0$ we can divide both sides of this equation by $t_4 - t_1$ and solve for the time he is airborne:

$$t_4 - t_1 = \frac{2v_1 \sin\theta_1}{g}.$$

Substituting this time interval in $x_4 - x_1 = (\vec{v}_1 \cos\theta_1)(t_4 - t_1)$ (Eq. 5-5) gives us the total horizontal distance he traveled:

$$x_4 - x_1 = \frac{2v_1^2}{g}\sin\theta_1\cos\theta_1 = \frac{2(26.5\text{ m/s})^2}{9.8\text{ m/s}^2}\sin(53°)\cos(53°)$$
$$= 69\text{ m}.$$

We can now answer the questions that opened this chapter: How could Zacchini know where to place the net, and how could he be certain he would clear the Ferris wheels? He (or someone) did the calculations as we have here. Although he could not take into account the complicated effects of the air on his flight, Zacchini knew that the air would slow him, and thus decrease his range from the calculated value. So, he used a wide net and biased it toward the cannon. He was then relatively safe whether the effects of the air in a particular flight happened to slow him considerably or very little. Still, the variability of this factor of air effects must have played on his imagination before each flight.

Zacchini still faced a subtle danger. Even for shorter flights, his propulsion through the cannon was so severe that he underwent a momentary blackout. If he landed during the blackout, he could break his neck. To avoid this, he had trained himself to awake quickly. Indeed, not waking up in time presents the only real danger to a human cannonball in the short flights today.

5-4 Displacement in Two Dimensions

We conclude this chapter with an exploration of objects that move in a circle at a constant speed. Before we begin this exploration, we need to learn more about finding displacement vectors in two dimensions—a task we began in Sections 2-2, 4-2, and 4-3.

How can we track the motion of a particle-like object that moves in a two-dimensional plane instead of being constrained to move along a line? As was the case for tracking motion along a line (treated in Chapter 2), it is useful to define a **position vector,** $\vec{r}$, that extends from the origin of a chosen coordinate system to the object. But this time, we have to choose a two-dimensional coordinate system. However, we can use the vector algebra introduced in the last chapter to resolve (decompose) the position vector into component vectors. This allows us to treat the two-dimensional motions using the techniques developed to describe one-dimensional motions.

Using Rectangular Coordinates

If we decide to use rectangular coordinates, then we denote the rectangular components of $\vec{r}$ as x and y. We can use the unit-vector notation of Section 4-5 to resolve the position vector into

$$\vec{r} = x\hat{\imath} + y\hat{\jmath}, \tag{5-12}$$

where $x\hat{\imath}$ and $y\hat{\jmath}$ are the rectangular component vectors. Note that for a rectangular coordinate system, the components x and y are the same as the coordinates of the object's location (x, y).

The coordinates x and y specify the particle's location along the coordinate axes relative to the origin. For instance, Fig. 5-14 shows a particle with rectangular coordinates (−3 m, 2 m) that has a position vector given by

$$\vec{r} = (-3\text{ m})\hat{\imath} + (2\text{ m})\hat{\jmath}.$$

The particle is located 3 meters from the y axis in the $-\hat{\imath}$ direction and 2 meters from the x axis in the $+\hat{\jmath}$ direction.

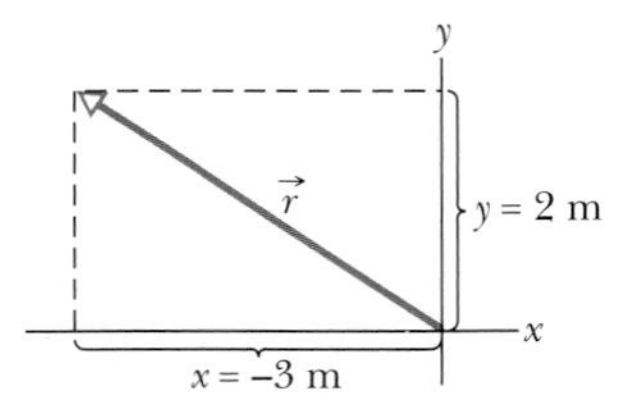

FIGURE 5-14 ■ The position vector $\vec{r}$ for a particle that has coordinates (−3 m, 2 m).

As a particle moves, its position vector changes in such a way that the vector always extends to the particle from the reference point (the origin). If the position vector changes—say, from $\vec{r}_1$ to $\vec{r}_2$ during a certain time interval—then the particle's **displacement** $\vec{r}$ during that time interval is

$$\Delta\vec{r} = \vec{r}_2 - \vec{r}_1. \tag{5-13}$$

Using unit-vector notation for $\vec{r}_2$ and $\vec{r}_1$, we can rewrite this displacement as

$$\Delta\vec{r} = (x_2\hat{\mathrm{i}} + y_2\hat{\mathrm{j}}) - (x_1\hat{\mathrm{i}} + y_1\hat{\mathrm{j}}).$$

By grouping terms having the same unit vector, we get

$$\Delta\vec{r} = (x_2 - x_1)\,\hat{\mathrm{i}} + (y_2 - y_1)\,\hat{\mathrm{j}}, \tag{5-14}$$

where the components (x_1, y_1) correspond to position vector $\vec{r}_1$ and components (x_2, y_2) correspond to position vector $\vec{r}_2$. We can also rewrite the displacement by substituting Δx for $(x_2 - x_1)$, and Δy for $(y_2 - y_1)$, so that

$$\Delta\vec{r} = \Delta x\,\hat{\mathrm{i}} + \Delta y\,\hat{\mathrm{j}}. \tag{5-15}$$

This expression is another example of a very important aspect of motion in more than one dimension. Notice that the coordinates of the displacement in each direction $(\Delta x, \Delta y)$ depend only on the change in the object's position in that one direction, and are independent of changes in position in the other directions. In other words, we don't have to simultaneously consider the change in the object's position in every direction. We can break the motion into two parts (motion in the x direction and motion in the y direction) and consider each direction separately. For motion in three dimensions we could add a z direction and do equivalent calculations.

Using Polar Coordinates

In dealing with circular motions or rotations it is often useful to describe positions and displacements in polar coordinates. In this case we locate a particle using r or $|\vec{r}|$ which represents the magnitude of its position vector $\vec{r}$ and its angle ϕ_1 which is measured in a counterclockwise direction from a chosen axis. The relationship between two-dimensional rectangular coordinates and polar coordinates is shown in Fig. 5-15. The transformation between these coordinate systems is based on the definitions of sine and cosine, so that

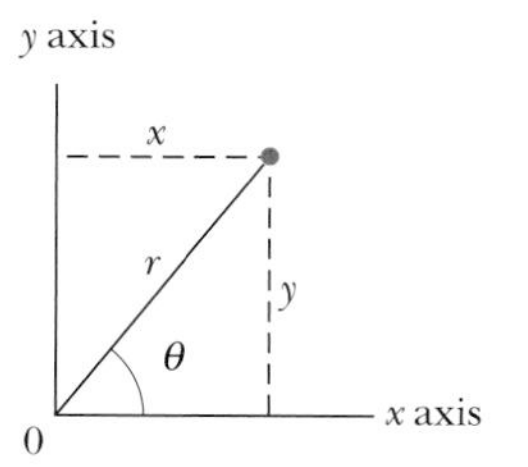

FIGURE 5-15 ■ Polar coordinates provide an alternative way to locate a particle that is confined to move in two dimensions. These coordinates are especially useful in the description of circular motion where the distance of a particle from the origin does not change.

$$x = r\cos(\theta) \quad \text{and} \quad y = r\sin(\theta). \tag{5-16}$$

Conversely,

$$r = \sqrt{x^2 + y^2} \quad \text{and} \quad \theta = \tan^{-1}\left(\frac{y}{x}\right). \tag{5-17}$$

In circular motion the distance of a particle from the center of the circle defining its motion does not change. So the magnitude of displacement Δr or $|\Delta\vec{r}|$ depends only on the change in angle and is given by

$$\Delta r = r(\theta_2 - \theta_1) \quad \text{(small angle circular displacement magnitude).} \tag{5-18}$$

for small angular differences. This expression should look familiar from mathematics classes. There you learned that the relationship between arc length, s, the radius, r, and angular displacement, $\Delta\theta$ is $s = r\Delta\theta$.

READING EXERCISE 5-6: (a) If a wily bat flies from x, y coordinates (-2 m, 4 m) to coordinates (6 m, -2 m), what is its displacement $\Delta\vec{r}$ in rectangular unit-vector notation? (b) Is $\Delta\vec{r}$ parallel to one of the two coordinate axes? If so, which axis? ■

TOUCHSTONE EXAMPLE 5-4: Displacement

In Fig. 5-16, the position vector for a particle is initially

$$\vec{r}_1 = (-3.0\text{ m})\hat{\text{i}} + (4.0\text{ m})\hat{\text{j}}$$

and then later is

$$\vec{r}_2 = (9.0\text{ m})\hat{\text{i}} + (-3.5\text{ m})\hat{\text{j}}.$$

What is the particle's displacement $\Delta\vec{r}$ from $\vec{r}_1$ to $\vec{r}_2$?

SOLUTION ■ The **Key Idea** is that the displacement $\Delta\vec{r}$ is obtained by subtracting the initial position vector $\vec{r}_1$ from the later position vector $\vec{r}_2$. That is most easily done by components:

$$\begin{aligned}\Delta\vec{r} &= \vec{r}_2 - \vec{r}_1\\ &= [9.0 - (-3.0)](\text{m})\hat{\text{i}} + [-3.5 - 4.0](\text{m})\hat{\text{j}}\\ &= (12\text{ m})\hat{\text{i}} + (-7.5\text{ m})\hat{\text{j}}.\end{aligned}\qquad\text{(Answer)}$$

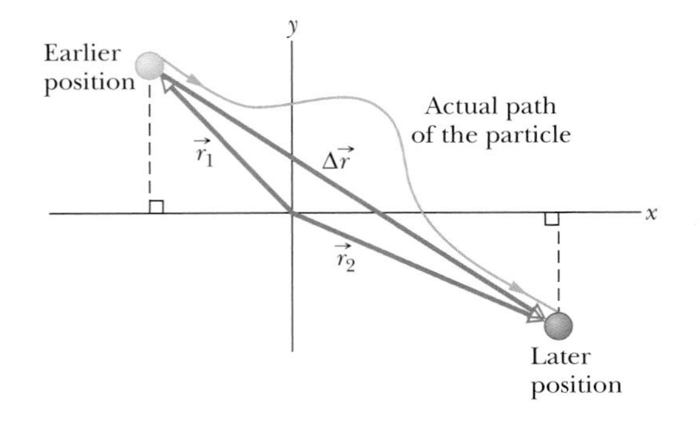

FIGURE 5-16 ■ The displacement $\Delta\vec{r} = \vec{r}_2 - \vec{r}_1$ extends from the head of the initial position vector $\vec{r}_1$ to the head of a later position vector $\vec{r}_2$ regardless of what path is actually taken.

TOUCHSTONE EXAMPLE 5-5: Rabbit's Trajectory

A rabbit runs across a parking lot on which a set of coordinate axes has, strangely enough, been drawn. The coordinates of the rabbit's position as functions of time t are given by

$$x = (-0.31\text{ m/s}^2)t^2 + (7.2\text{ m/s})t + 28\text{ m} \qquad (5\text{-}19)$$

and

$$y = (0.22\text{ m/s}^2)t^2 + (-9.1\text{ m/s})t + 30\text{ m}. \qquad (5\text{-}20)$$

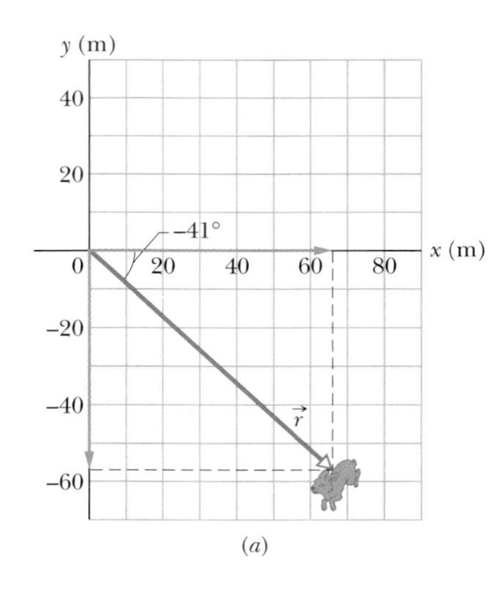

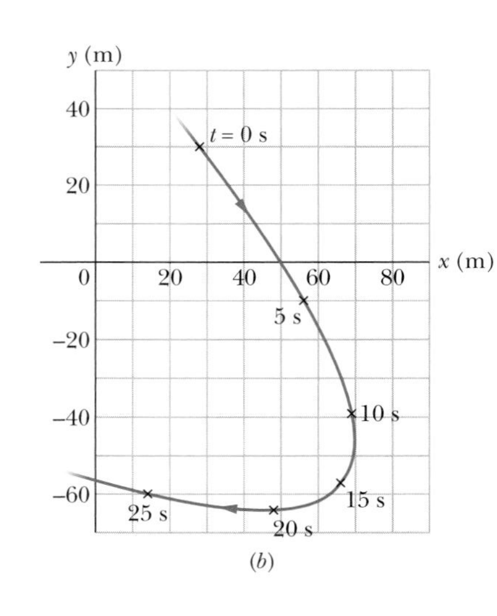

(a) At $t = 15$ s, what is the rabbit's position vector $\vec{r}$ in unit-vector notation and as a magnitude and an angle?

SOLUTION ■ The **Key Idea** here is that the x and y coordinates of the rabbit's position, as given by Eqs. 5-19 and 5-20, are the components of the rabbit's position vector $\vec{r}$. Thus, we can write

$$\vec{r}(t) = x(t)\hat{\text{i}} + y(t)\hat{\text{j}}. \qquad (5\text{-}21)$$

FIGURE 5-17 ■ (*a*) A rabbit's position vector $\vec{r}$ at time $t = 15$ s. The components of $\vec{r}$ are shown along the axes. (*b*) The rabbit's path and its position at five values of t.

(We write $\vec{r}(t)$ rather than $\vec{r}$ because the components are functions of t, and thus $\vec{r}$ is also.)

At $t = 15$ s, the components of the position vector are

$$x = (-0.31 \text{ m/s}^2)(15 \text{ s})^2 + (7.2 \text{ m/s})(15 \text{ s}) + 28 \text{ m} = 66 \text{ m},$$

and $$y = (0.22 \text{ m/s}^2)(15 \text{ s})^2 + (-9.1 \text{ m/s})(15 \text{ s}) + 30 \text{ m} = -57 \text{ m}.$$

Thus, at $t = 15$ s,

$$\vec{r} = (66 \text{ m})\hat{\text{i}} - (57 \text{ m})\hat{\text{j}},$$

which is drawn in Fig. 5-17a.

To get the magnitude and angle of $\vec{r}$, we can use a vector-capable calculator, or we can be guided by the Pythagorean theorem to write

$$r = \sqrt{x^2 + y^2} = \sqrt{(66 \text{ m})^2 + (-57 \text{ m})^2}$$
$$= 87 \text{ m},$$

and from trigonometric definition,

$$\theta = \tan^{-1}\frac{y}{x} = \tan^{-1}\left(\frac{-57 \text{ m}}{66 \text{ m}}\right) = -41°$$

(Although $\theta = 139°$ has the same tangent as $-41°$, study of the signs of the components of $\vec{r}$ rules out 139°.)

(b) Graph the rabbit's path for $t = 0$ to $t = 25$ s.

SOLUTION ■ We can repeat part (a) for several values of t and then plot the results. Figure 5-17b shows the plots for five values of t and the path connecting them. We can also use a graphing calculator to make a *parametric graph*; that is, we would have the calculator plot y versus x, where these coordinates are given by Eqs. 5-19 and 5-20 as functions of time t.

5-5 Average and Instantaneous Velocity

We have just shown that when tracking motions occurring in more than one dimension, position and displacement vectors can be resolved into rectangular component vectors. Can this also be done with velocity vectors? As is the case for motion in one dimension, if a particle moves through a displacement $\Delta\vec{r}$ in a time interval Δt (as shown in Fig. 5-18), then its **average velocity** $\langle\vec{v}\rangle$ is defined as

$$\text{average velocity} \equiv \frac{\text{displacement}}{\text{time interval}},$$

or using familiar symbols

$$\langle\vec{v}\rangle \equiv \frac{\Delta\vec{r}}{\Delta t}. \tag{5-22}$$

This tells us the direction of $\langle\vec{v}\rangle$ must be the same as the displacement $\Delta\vec{r}$. Using our new definition of displacement in two dimensions (Eq. 5-15), we can rewrite this as

$$\langle\vec{v}\rangle = \frac{\Delta x\,\hat{\text{i}} + \Delta y\,\hat{\text{j}}}{\Delta t} = \frac{\Delta x}{\Delta t}\hat{\text{i}} + \frac{\Delta y}{\Delta t}\hat{\text{j}}. \tag{5-23}$$

This equation can be simplified by noting that $\Delta x/\Delta t$ is defined in Chapter 2 as the component of average velocity in the x direction. If we use appropriate definitions for average velocity components in the y direction, then

$$\langle\vec{v}\rangle = \frac{\Delta x}{\Delta t}\hat{\text{i}} + \frac{\Delta y}{\Delta t}\hat{\text{j}} = \langle v_x\rangle\hat{\text{i}} + \langle v_y\rangle\hat{\text{j}}. \tag{5-24}$$

Here we are using the same notation introduced in Chapter 2 to denote the average velocity components. We use subscripts x and y to distinguish each of the average velocity components. Also the angle brackets $\langle\ \rangle$ are used to distinguish average velocity from instantaneous velocity.

As we mentioned in Chapter 2, when we speak of the **velocity** of a particle, we usually mean the particle's **instantaneous velocity** $\vec{v}$. $\vec{v}$ represents the limit the aver-

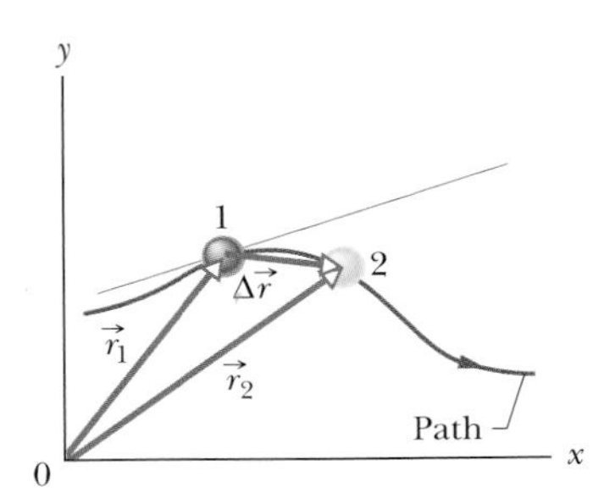

FIGURE 5-18 ■ The displacement $\Delta\vec{r}$ of a particle during a time interval Δt from position 1 with position vector $\vec{r}_1$ at time t_1 to position 2 with position vector $\vec{r}_2$ at time t_2.

age velocity $\langle\vec{v}\rangle$ approaches as we shrink the time interval Δt to zero. Using the language of calculus, we can also write $\vec{v}$ as the derivative

$$\vec{v} = \lim_{\Delta t \to 0} \frac{\Delta\vec{r}}{\Delta t} = \frac{d\vec{r}}{dt}. \tag{5-25}$$

If we substitute $\vec{r} = x\,\hat{\text{i}} + y\,\hat{\text{j}}$, then in unit-vector notation:

$$\vec{v} = \frac{d}{dt}(x\hat{\text{i}} + y\hat{\text{j}}) = \frac{dx}{dt}\hat{\text{i}} + \frac{dy}{dt}\hat{\text{j}}.$$

This equation can be simplified by recognizing that dx/dt is v_x and so on. Thus,

$$\vec{v} = v_x\hat{\text{i}} + v_y\hat{\text{j}}, \tag{5-26}$$

where v_x is the component of $\vec{v}$ along the x axis and v_y is the component of $\vec{v}$ along the y axis. The direction of $\vec{v}$ is tangent to the particle's path at the instant in question.

Figure 5-19 shows a velocity vector $\vec{v}$ of a moving particle and its x- and y-components. *Caution*: When a position vector is drawn as in Fig. 5-14, it is represented by an arrow that extends from one point (a "here") to another point (a "there"). However, when a velocity vector is drawn as in Fig. 5-19, it does *not* extend from one point to another. Rather, it shows the direction of travel of a particle at that instant, and the length of the arrow is proportional to the velocity magnitude. Since the unit for a velocity is a distance per unit time and is *not* a length, you are free to define a scale to use in depicting the relative magnitudes of a set of velocity vectors. For instance, each 2 cm of length on a velocity vector on a diagram could represent a velocity magnitude of 1 m/s.

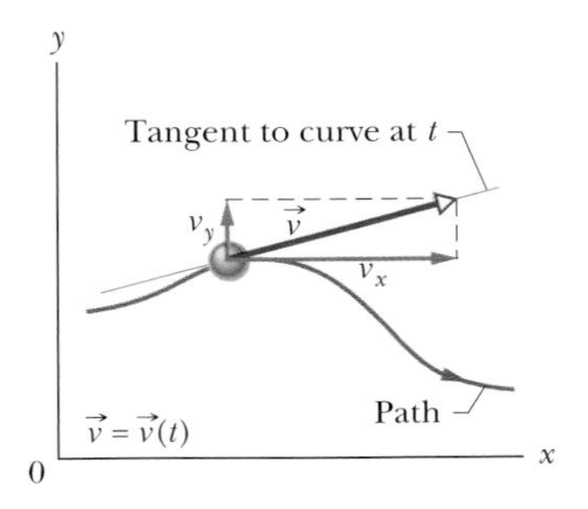

FIGURE 5-19 ■ A particle moves in a curved path. At a time t, the velocity $\vec{v}$ of a particle is shown along with its components v_x and v_y. The velocity vector points in the same direction as the tangent to the curve that traces the particle's motion.

Equations 5-24 and 5-26, developed in this section, show that the component of velocity of the object in one direction, such as the horizontal or x direction, can be considered completely separately from the component of velocity of the object in another direction, such as the vertical or y direction.

We assume that a curve that traces out a particle's motion is continuous. Mathematically, the tangent to the curve, its slope, and the instantaneous velocity are different names for the same quantity. Thus, we see that the velocity vector in Fig. 5-19 points along the tangent line that describes the slope of the graph at that point.

READING EXERCISE 5-7: The figure below shows a circular path taken by a particle about an origin. If the instantaneous velocity of the particle is $\vec{v} = (2\text{m/s})\hat{\text{i}} - (2\text{m/s})\hat{\text{j}}$, through which quadrant is the particle moving when it is traveling (a) clockwise and (b) counterclockwise around the circle? For both cases, draw $\vec{v}$ on the figure.

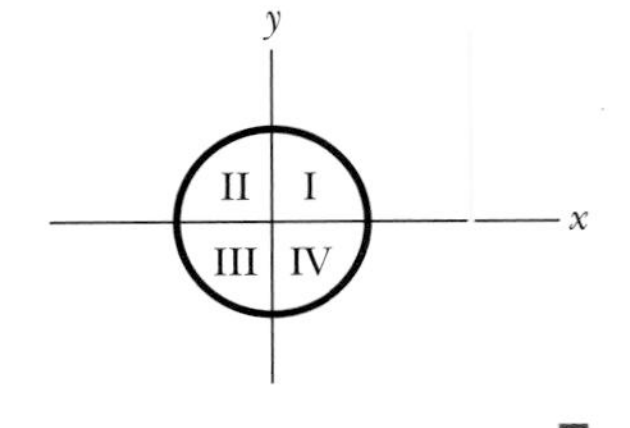

■

TOUCHSTONE EXAMPLE 5-6: A Rabbit's Velocity

For the rabbit in Touchstone Example 5-5, find the velocity $\vec{v}$ at time $t = 15$ s, in unit-vector notation and as a magnitude and an angle.

SOLUTION ■ There are two **Key Ideas** here: (1) We can find the rabbit's velocity $\vec{v}$ by first finding the velocity components. (2) We can find those components by taking derivatives of the components of the rabbit's position vector. Applying $\vec{v} = v_x\hat{\text{i}} + v_y\hat{\text{j}}$ (Eq. 5-26) with $v_x = dx/dt$ to the expression for the rabbit's x position from Touchstone Example 5-5 (Eq. 5-19), we find the x-component of $\vec{v}$ to be

$$v_x = \frac{dx}{dt} = \frac{d}{dt}[(-0.31\ \text{m/s}^2)t^2 + (7.2\ \text{m/s})t + 28\ \text{m})]$$
$$= (-0.62\ \text{m/s}^2)t + 7.2\ \text{m/s}. \quad (5\text{-}27)$$

At $t = 15$ s, this gives $v_x = -2.1$ m/s. Similarly, since $v_y = dy/dt$, using the expression for the rabbit's y position from Touchstone Example 5-5 (Eq. 5-20), we find that the y-component is

$$v_y = \frac{dy}{dt} = \frac{d}{dt}[(0.22\ \text{m/s}^2)t^2 + (-9.1\ \text{m/s})t + 30\ \text{m}]$$
$$= (0.44\ \text{m/s}^2)t - 9.1\ \text{m/s}. \quad (5\text{-}28)$$

At $t = 15$ s, this gives $v_y = -2.5$ m/s. Thus, by Equation 5-26,

$$\vec{v} = -(2.1\ \text{m/s})\hat{\text{i}} - (2.5\ \text{m/s})\hat{\text{j}}, \quad \text{(Answer)}$$

which is shown in Fig. 5-20, tangent to the rabbit's path and in the direction the rabbit is running at $t = 15$ s.

To get the magnitude and angle of $\vec{v}$, either we use a vector-capable calculator or we use the Pythagorean theorem and trigonometry to write

$$v = \sqrt{v_x^2 + v_y^2} = \sqrt{(-2.1\ \text{m/s})^2 + (-2.5\ \text{m/s})^2}$$
$$= 3.3\ \text{m/s}, \quad \text{(Answer)}$$

and

$$\theta = \tan^{-1}\frac{v_y}{v_x} = \tan^{-1}\left(\frac{-2.5\ \text{m/s}}{-2.1\ \text{m/s}}\right)$$
$$= \tan^{-1} 1.19 = -130°. \quad \text{(Answer)}$$

(Although 50° has the same tangent as −130°, inspection of the signs of the velocity components indicates that the desired angle is in the third quadrant, given by 50° − 180° = −130°.)

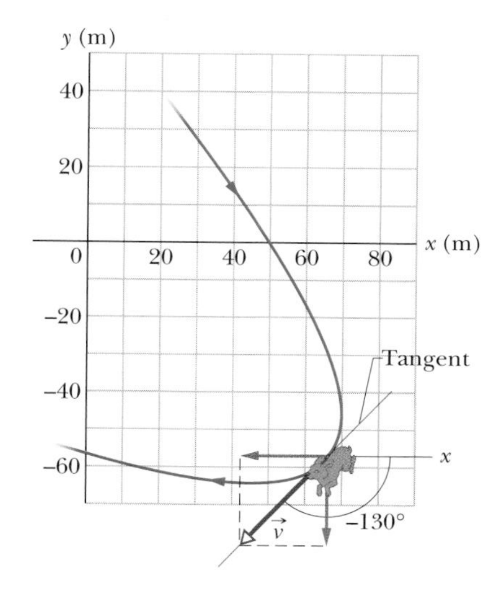

FIGURE 5-20 ■ The rabbit's velocity $\vec{v}$ at $t = 15$ s. The velocity vector is tangent to the path at the rabbit's position at that instant. The components of $\vec{v}$ are shown.

5-6 Average and Instantaneous Acceleration

We have just shown that velocity vectors can be resolved into rectangular component vectors. Can this also be done with acceleration vectors? As is the case for motion in one dimension, if a particle-like object undergoes a velocity change from $\vec{v}_1$ to $\vec{v}_2$ in a time interval Δt, its **average acceleration** $\langle\vec{a}\rangle$ during Δt is

$$\langle\vec{a}\rangle \equiv \frac{\text{change in velocity}}{\text{time interval}},$$

or

$$\langle\vec{a}\rangle = \frac{\vec{v}_2 - \vec{v}_1}{t_2 - t_1} = \frac{\Delta\vec{v}}{\Delta t}. \quad (5\text{-}29)$$

If we shrink Δt to zero about some instant, then in the limit the average acceleration $\langle\vec{a}\rangle$ approaches the **instantaneous acceleration** (or just **acceleration**) $\vec{a}$ at that instant. That is,

$$\vec{a} = \lim_{\Delta t \to 0} \frac{\Delta\vec{v}}{\Delta t} = \frac{d\vec{v}}{dt}. \quad (5\text{-}30)$$

If the velocity changes in *either* magnitude *or* direction (or both), the particle is accelerating. For example, a particle that moves in a circle at a constant speed (velocity magnitude) is always changing direction and hence accelerating. We can write this equation in unit-vector form by substituting for $\vec{v} = v_x\hat{\text{i}} + v_y\hat{\text{j}}$ to obtain

$$\vec{a} = \frac{d}{dt}(v_x\hat{\text{i}} + v_y\hat{\text{j}}) = \frac{dv_x}{dt}\hat{\text{i}} + \frac{dv_y}{dt}\hat{\text{j}}.$$

We can rewrite this as

$$\vec{a} = a_x\hat{\mathrm{i}} + a_y\hat{\mathrm{j}}, \tag{5-31}$$

where the components of $\vec{a}$ in two dimensions are given by

$$a_x = \frac{dv_x}{dt} \quad \text{and} \quad a_y = \frac{dv_y}{dt}. \tag{5-32}$$

Thus, we can find the components of $\vec{a}$ by differentiating the components of $\vec{v}$. As is the case for multidimensional position, displacement, and velocity vectors, an acceleration vector can be resolved mathematically into component vectors in a rectangular coordinate system.

As we saw in Chapter 2, the algebraic sign of an acceleration component (plus or minus) represents the direction of velocity change. Speeding up is *not* always associated with a positive acceleration component. Just as we discussed for one-dimensional motion, if the velocity and acceleration components along a given axis have the *same sign* then they are in the same direction. In this case, the object will *speed up*. If the acceleration and velocity components have *opposite signs*, then they are in opposite directions. Under these conditions, the object will *slow down*. So slowing down is not always associated with an acceleration that is negative. It is the relative directions of an object's velocity and acceleration that determine whether the object will speed up or slow down.

Figure 5-21 shows an acceleration vector $\vec{a}$ and its components for a particle moving in two dimensions. Again, when an acceleration vector is drawn as in Fig. 5-21, although its tail is located at the particle, the vector arrow does *not* extend from one position to another. Rather, it shows the direction of acceleration for the particle, and its length represents the acceleration magnitude. The length can be drawn to any convenient scale.

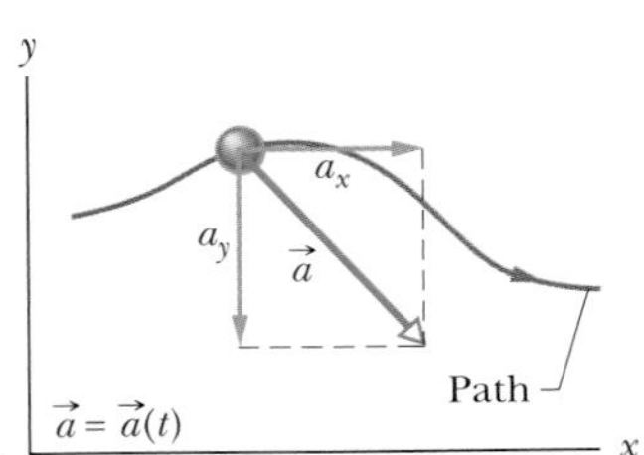

FIGURE 5-21 ■ A two-dimensional acceleration $\vec{a}$ of a particle at a time t is shown along with its x- and y-components.

READING EXERCISE 5-8: In Fig. 5-21 the particle is moving along a curved trajectory and its acceleration $\vec{a}$ is *not* tangent to the curve of the particle's trajectory. Under what circumstances, if any, would $\vec{a}$ be tangent to the trajectory? Under what circumstances if any, could $\vec{a}$ be perpendicular to a tangent to the trajectory? In order to answer these questions, you might want to examine the direction of the components of $\vec{a}$ and use what you learned in Chapter 2 about the relationship between velocity and acceleration. ■

TOUCHSTONE EXAMPLE 5-7: Rabbit's Acceleration

For the rabbit in Touchstone Examples 5-5 and 5-6, find the acceleration $\vec{a}$ at time $t = 15$ s, in unit-vector notation and as a magnitude and an angle.

SOLUTION ■ There are two **Key Ideas** here: (1) We can find the rabbit's acceleration $\vec{a}$ by first finding the acceleration components. (2) We can find those components by taking derivatives of the rabbit's velocity components. Applying $a_x = dv_x/dt$ (Eq. 5-32) to $v_x = (-0.62\ \text{m/s}^2)t + 7.2\ \text{m/s}$ (Eq. 5-27 giving the rabbit's x velocity in Touchstone Example 5-6), we find the x-component of $\vec{a}$ to be

$$a_x = \frac{dv_x}{dt} = \frac{d}{dt}[(-0.62\ \text{m/s}^2)t + 7.2\ \text{m/s})] = -0.62\ \text{m/s}^2.$$

Similarly, applying $a_y = dv_y/dt$ (Eq. 5-32) to the rabbit's y velocity from Touchstone Example 5-6 (Eq. 5-28) yields the y-component as

$$a_y = \frac{dv_y}{dt} = \frac{d}{dt}[(0.44\ \text{m/s}^2)t - 9.1\ \text{m/s}] = 0.44\ \text{m/s}^2.$$

We see that the acceleration does not vary with time (it is a constant) because the time variable t does not appear in the expression for either acceleration component. Therefore, by Eq. 5-31,

$$\vec{a} = (-0.62\ \text{m/s}^2)\hat{\mathrm{i}} + (0.44\ \text{m/s}^2)\hat{\mathrm{j}}, \quad \text{(Answer)}$$

which is shown superimposed on the rabbit's path in Fig. 5-22.

To get the magnitude and angle of $\vec{a}$, either we use a vector-capable calculator or we use the Pythagorean theorem and trigonometry. For the magnitude we have

$$a = \sqrt{a_x^2 + a_y^2} = \sqrt{(-0.62 \text{ m/s}^2)^2 + (0.44 \text{ m/s}^2)^2}$$
$$= 0.76 \text{ m/s}^2. \quad \text{(Answer)}$$

For the angle we have

$$\theta = \tan^{-1}\frac{a_y}{a_x} = \tan^{-1}\left(\frac{0.44 \text{ m/s}^2}{-0.62 \text{ m/s}^2}\right) = -35°.$$

However, this last result, which is what would be displayed on your calculator if you did the calculation, indicates that $\vec{a}$ is directed to the right and downward in Fig. 5-22. Yet, we know from the components above that $\vec{a}$ must be directed to the left and upward. To find the other angle that has the same tangent as $-35°$, but is not displayed on a calculator, we add 180°:

$$-35° + 180° = 145°. \quad \text{(Answer)}$$

This *is* consistent with the components of $\vec{a}$. Note that $\vec{a}$ has the same magnitude and direction throughout the rabbit's run because, as we noted previously, the acceleration is constant.

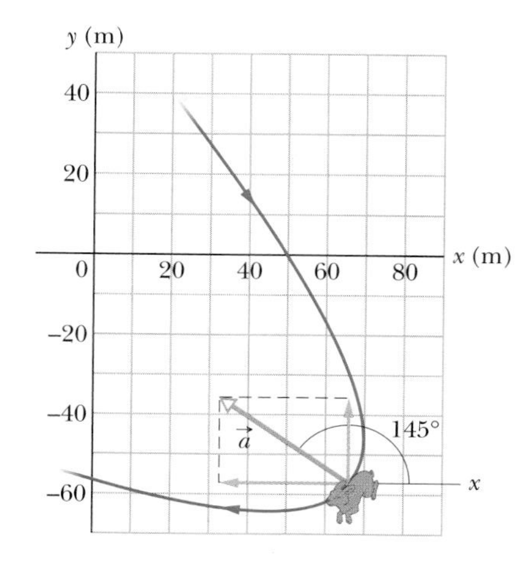

FIGURE 5-22 ■ The acceleration $\vec{a}$ of the rabbit at $t = 15$ s. The rabbit happens to have this same acceleration at all points along its path.

TOUCHSTONE EXAMPLE 5-8: Changing Velocity

A particle with velocity $\vec{v}_1 = (-2.0 \text{ m/s})\hat{\text{i}} + (4.0 \text{ m/s})\hat{\text{j}}$ at $t = 0$ undergoes a constant acceleration $\vec{a}$ of magnitude $a = 3.0 \text{ m/s}^2$ at an angle $\theta = 130°$ from the positive direction of the x axis. What is the particle's velocity $\vec{v}_2$ at $t = 5.0$ s, in unit-vector notation and as a magnitude and an angle?

SOLUTION ■ We first note that this is two-dimensional motion, in the xy plane. Then there are two **Key Ideas** here. One is that, because the acceleration is constant, Eq. 2-13 ($v_{2x} = v_{1x} + a_x\Delta t$) applies. The second is that, because Eq. 2-13 applies only to straight-line motion, we must apply it separately for motion parallel to the x axis and motion parallel to the y axis. That is, we must find the velocity components v_x and v_y from the equations

$$v_{2x} = v_{1x} + a_x\,\Delta t \quad \text{and} \quad v_{2y} = v_{1y} + a_y\,\Delta t.$$

In these equations, v_{1x} $(= -2.0 \text{ m/s})$ and v_{1y} $(= 4.0 \text{ m/s})$ are the x- and y-components of $\vec{v}_1$, and a_x and a_y are the x- and y-components of $\vec{a}$. To find a_x and a_y, we resolve $\vec{a}$ either with a vector-capable calculator or with trigonometry:

$$a_x = a\cos\theta = (3.0 \text{ m/s}^2)(\cos 130°) = -1.93 \text{ m/s}^2,$$
$$a_y = a\sin\theta = (3.0 \text{ m/s}^2)(\sin 130°) = +2.30 \text{ m/s}^2.$$

When these values are inserted into the equations for v_x and v_y, we find that, at time $t = 5.0$ s,

$$v_{2x} = -2.0 \text{ m/s} + (-1.93 \text{ m/s}^2)(5.0 \text{ s}) = -11.65 \text{ m/s},$$
$$v_{2y} = 4.0 \text{ m/s} + (2.30 \text{ m/s}^2)(5.0 \text{ s}) = 15.50 \text{ m/s}.$$

Thus, at $t = 5.0$ s, we have, after rounding,

$$\vec{v}_2 = (-12 \text{ m/s})\hat{\text{i}} + (16 \text{ m/s})\hat{\text{j}}. \quad \text{(Answer)}$$

We find that the magnitude and angle of $\vec{v}_2$ are

$$v_2 = \sqrt{v_{2x}^2 + v_{2y}^2} = 19.4 \text{ m/s} \approx 19 \text{ m/s}, \quad \text{(Answer)}$$

and

$$\theta_2 = \tan^{-1}\frac{v_{2y}}{v_{2x}} = 127° \approx 130°. \quad \text{(Answer)}$$

Check the last line with your calculator. Does 127° appear on the display, or does $-53°$ appear? Now sketch the vector v with its components to see which angle is reasonable.

5-7 Uniform Circular Motion

If a single Olympic event best captures the motions described in this chapter, it's the hammer throw. In this event an athlete spins a heavy steel ball attached to a wire rope with a handle in a circle. An athlete gains maximum distance by swinging a 16 lb ham-

FIGURE 5-23 ■ A Scotsman twirls a massive steel ball (called a "hammer") in a circle at a highland games competition. When he releases the hammer, it will travel in a direction that is tangent to its original circular path and undergo projectile motion.

mer repeatedly around his head while standing still to build up speed (Fig. 5-23). Finally, the athlete rotates quickly with the hammer before releasing it at the front of a throwing circle. Once the athlete releases the hammer, it begins to travel along a line tangent to the circle in which it was spinning and undergoes projectile motion.

Circular motion, like that of the spinning hammer, is another motion in two dimensions that can be analyzed using Newton's Second Law. Examples of motion that are approximately circular include the revolution of the Earth around the Sun, a race car zooming around a circular track, an electron moving near the center of a large electromagnet, and a stone tied to the end of a string that is twirled in a circle above one's head. In all of these cases if the object's speed is constant, we define its motion as **uniform circular motion.**

> A particle that travels around a circle or a circular arc at constant (*uniform*) speed is said to be undergoing **UNIFORM CIRCULAR MOTION**.

Not all circular motion is uniform. For example, a Hot Wheels® car or roller coaster cart doing a loop-the-loop slows down near the top of the loop and speeds up near the bottom. Analyzing loop-the-loop motions is more complex than analyzing uniform circular motion. For this reason, we start with the ideal case of uniform circular motion.

Centripetal Force

Consider an object, such as an ice hockey puck, that glides along a frictionless surface. Newton's First Law tells us that you cannot change either its direction or its speed without exerting a force on it. Giving the puck a kick along its line of motion will change its speed but not its direction. How can you have the opposite effect? How can you change the puck's direction without changing its speed? To do this you have to kick perpendicular to its direction of motion. This is an important statement regarding the accelerations that result when we apply a force.

> If a nonzero net force acts on an object, at any instant it can be decomposed into a component along the line of motion and a component perpendicular to the motion. The component of the net force that is *in line* with the object's motion produces only changes in the *magnitude* of the object's velocity (its speed). The component of the net force that is *perpendicular* to the line of motion produces only changes in the *direction* of the object's velocity.

What happens if you give the puck a series of short kicks but adjust their directions constantly so the kicks are always perpendicular to the current direction of motion? Does this lead to circular motion? We can use Newton's Second Law to answer this question. To help visualize the net force needed to maintain uniform circular motion, we consider a similar situation to that of the puck. Imagine twirling a ball at the end of a string in a perfectly horizontal circle of radius r. In order to keep the ball moving in a circle, you are constantly changing the direction of the force that the string exerts on the spinning ball. That is, there must constantly be a component of the net force that is perpendicular to the motion. The situation is complicated by the fact that there are actually two forces on the ball as shown in Fig. 5-24—the string force and the gravitational force.

To apply Newton's Second Law to the analysis of this motion, we must find the net force on the ball by taking the vector sum of the two forces acting on it. We start by resolving the string force into horizontal and vertical components, as shown in Fig. 5-24. Since the ball is rotating horizontally and so does not move up or down in a vertical direction, the net vertical force on it must be zero. Since the vertical force

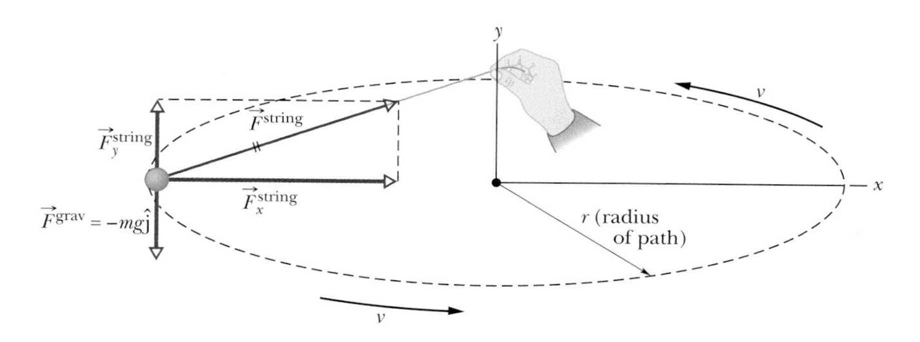

FIGURE 5-24 ■ If you spin a ball in a circle, the net force on the ball consists of a central or centripetal force $\vec{F}_x^{\,string}$ that lies in a horizontal plane. It points toward the center of the circular path.

components cancel, the net force is just the horizontal component of the string force vector, $\vec{F}_x^{\,string}$. If you carefully consider the situation depicted in Fig. 5-24, you should be convinced of two important points. First, the direction of the net force on the ball ($\vec{F}_x^{\,string}$) is always perpendicular to the line of motion of the ball. This means that the force results only in changes in the direction of the object's velocity. It does not cause changes in the object's speed. Second, the direction of the net force, $\vec{F}_x^{\,string}$, is constantly changing so that it always points toward the center of the circle in which the ball moves. We use the adjective **centripetal** to describe any force with this characteristic. The word *centripetal* comes from Latin and means "center-seeking."

> Centripetal is an adjective that *describes* any force or superposition of forces that is directed toward the center of curvature of the path of motion.

It is important to note that the horizontal component of the string force $\vec{F}_x^{\,string}$ *is* the centripetal force involved in the ball's motion. There is *not* another force, "the centripetal force," that must be added to the free-body diagram in Fig. 5-24.

If you suddenly let go of the string, the ball will fly off along a straight path that is tangent to the circle at the moment of release. This "linear flying off" phenomenon provides evidence that you *cannot maintain circular motion without a centripetal or center-seeking force.*

If we use a polar coordinate system with its origin at the center of the circular path, we can consider the centripetal force to be a kind of anti-radial force that points *inward* rather than outward in the direction of the circle's radius vector $\vec{r}$.

Centripetal Acceleration

A very simple example of uniform circular motion is shown in Fig. 5-25. There, an air hockey puck moves around in a circle at constant speed v while tied to a string looped around a central peg. Is this accelerated motion? We can predict that it is for two reasons:

- First, there is a net force on the puck due to the force exerted by the string. There is no vertical acceleration, so the upward force of the air jets and the downward gravitational force must cancel each other. According to Newton's Second Law, if there is a net force on an object there must be an acceleration.
- Second, although the puck moves with constant speed, the *direction* of the puck velocity is continuously changing. Recall that acceleration is related to the change in *velocity* (not speed), so we conclude that this motion is indeed accelerated motion.

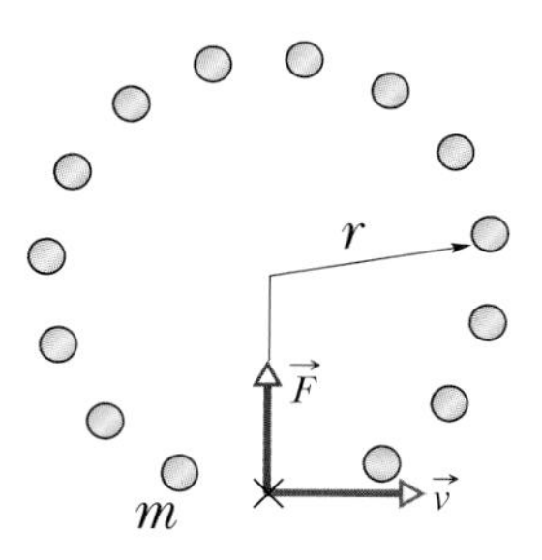

FIGURE 5-25 ■ A sketch of the locations of an air hockey puck of mass m moving with constant speed v in a circular path of radius r on a horizontal frictionless air table. The centripetal force on the puck, $\vec{F}^{cent}$, is the pull from the string directed inward toward the center of the circle traced out by the path of the puck.

What direction is this acceleration? Newton's Second Law tells us that the acceleration of an object of mass m is in the *same direction* as the force causing it and is

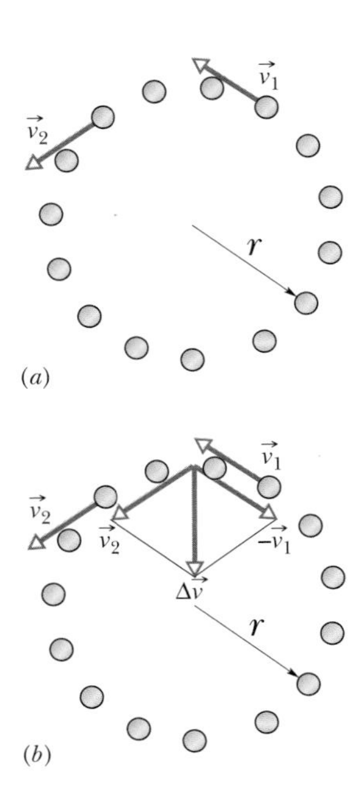

FIGURE 5-26 ■ (*a*) Sketch based on a video of an object moving counterclockwise on an airtable with constant speed in a circular path with the velocity vectors corresponding to its location at times t_1 and t_2. (*b*) Shows vectors $\vec{v}_2$ and $-\vec{v}_1$ and their sum $\Delta\vec{v}$ at the location of the object at an average time of $(t_1 + t_2)/2$.

given by Eq. 3-1, which is $\vec{a} = \vec{F}/m$. This suggests that, in uniform circular motion, the acceleration should also be directed radially inward. An acceleration that is directed radially inward is called a **centripetal acceleration.**

Next we will use the general definition of acceleration, a knowledge of geometry, and Newton's Second Law to show that uniform circular motion requires a centripetal acceleration of constant magnitude that depends on the radius of the path of a rotating object as well as its speed.

Proof that the acceleration is centripetal and has a constant magnitude: Let's start by considering the definition of average acceleration in Eq. 5-29,

$$\langle\vec{a}\rangle \equiv \frac{\vec{v}_2 - \vec{v}_1}{t_2 - t_1} = \frac{\Delta\vec{v}}{\Delta t} \quad \text{(average acceleration).}$$

Since Δt is a scalar, the acceleration must have the same direction as the difference between the two velocity vectors, $\vec{v}_2 - \vec{v}_1$. As usual the difference between the velocity vectors is actually the vector sum of $\vec{v}_2$ and the additive inverse of $\vec{v}_1$. Since the speed is constant (as indicated by the equally-spaced, "frame-by-frame" position markers in Fig. 5-26), the length of the velocity vectors and their additive inverses are the same at times t_1 and t_2. Another consequence of the speed being constant is that halfway between times t_1 and t_2, the puck is also midway between the two positions. If we place the tails of $\vec{v}_2$ and $-\vec{v}_1$ arrows at the midpoint between the two locations, we find that the vector sum points toward the center of the circular path taken by the object. This is shown in Fig. 5-26. So the acceleration is indeed centripetal (center-seeking). Furthermore, we could have created the same construction using any two points corresponding to other times that have the same *difference* Δt. It is obvious that the direction of the velocity change would be different, but it would still point toward the center. Furthermore, the magnitude of the Δv vector, and hence the acceleration magnitude, would be constant.

How does the centripetal acceleration depend on speed and path radius? We will prove that the magnitude of the acceleration of an object in uniform circular motion is given by

$$a = \frac{v^2}{R} \quad \text{(magnitude of centripetal acceleration),} \tag{5-33}$$

where R is the radius of the circular path of the object and $|\vec{v}|$ or v represents its speed. Here we start our proof by considering an object in Fig. 5-27, which happens to be moving in a circle in a counterclockwise direction at a constant speed. We choose to describe its motion in polar coordinates. We define θ_1 as zero at time t_1 and denote its location as $\vec{r}_1$. The object then moves at constant speed v through an angle $\theta_2 = \theta$ to a new location $\vec{r}_2$ at time t_2. Note that, in circular motion, velocity vectors are always perpendicular to their position vectors. This means that the angle between position vectors $\vec{r}_1$ and $\vec{r}_2$ is the same as the angle between velocity vectors $\vec{v}_1$ and $\vec{v}_2$. Furthermore, we note that if $\Delta\vec{v} = \vec{v}_2 - \vec{v}_1$ then $\vec{v}_2 = \Delta\vec{v} + \vec{v}_1$. Also, $\vec{r}_2 = \Delta\vec{r} + \vec{r}_1$. Thus the triangles shown in Fig. 5-27 are similar. According to Eq. 5-15 for small angles, $\Delta r = r(\theta_2 - \theta_1) = r\theta$. So we can write the ratio of magnitudes as

$$\frac{\Delta v}{v} = \frac{\Delta r}{R}.$$

We can solve the similar triangle ratios for the change in velocity and substitute into the expression that defines the magnitude of acceleration in terms of the magnitude of velocity change over a change in time to get

$$a = \frac{\Delta v}{\Delta t} = \frac{(\Delta r)(v)}{(\Delta t)(R)}.$$

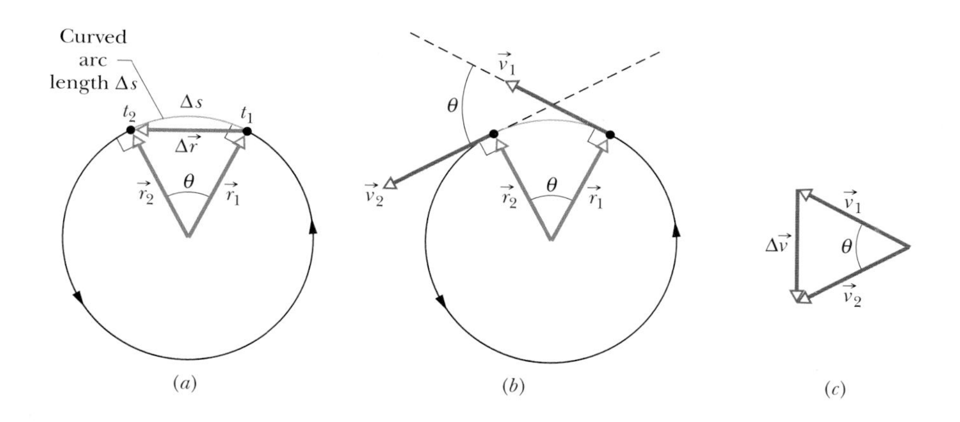

FIGURE 5-27 ■ Between times t_1 and t_2, a particle: (*a*) moves from location $\vec{r}_1$ to $\vec{r}_2$. The enclosed arc length Δs is curved and slightly longer than the magnitude of the vector displacement $|\Delta\vec{r}|$. (*b*) The velocity vector, which is always perpendicular to the position vector, changes direction but not magnitude. (*c*) Because the velocity vector is always perpendicular to the position vectors, the angle θ between them is the same as the angle between the position vectors.

However, the speed v is given by the arc length Δs along the circular path divided by the time interval, so that $v \approx \Delta s/\Delta t$. In the limit where the change in time is very small, the change in arc length Δs and the magnitude of the displacement Δr are essentially the same, so

$$v \approx \frac{\Delta s}{\Delta t} \approx \frac{\Delta r}{\Delta t} \quad \text{(when the time interval becomes small).}$$

So, we can replace $\Delta r/\Delta t$ in the expression for acceleration with the particle speed v to get

$$a = \frac{v^2}{R} \quad \text{(centripetal acceleration).} \tag{5-34}$$

Determining Average Speed, Period, and Frequency of Rotation

Often, we want to know how long it will take an object undergoing uniform circular motion to complete an entire revolution. For example, we might want to know how long it takes a race car on a circular track to complete one lap. This calculation is simplified because objects in uniform circular motion are moving at constant speeds. This means that the magnitude of velocity is constant, and therefore average and instantaneous values are the same. We can then use the relation for the speed of an object,

$$v = \langle v \rangle = \frac{\text{distance traveled}}{\text{time for the travel}}.$$

The distance traveled in one revolution is just the circumference of the circle ($2\pi r$). The time for a particle to go around a closed path exactly once has a special name. It is called the *period of revolution*, or simply the **period** of the motion. The period is represented with the symbol T, so,

$$v = \langle v \rangle = \frac{\text{distance traveled}}{\text{time for the travel}} = \frac{2\pi r}{T} \quad \text{(average speed).}$$

Recalling that the magnitudes of the average and instantaneous velocities are the same since the object is moving at constant speed, we can solve this expression for the period:

$$T = \frac{2\pi r}{v} \quad \text{(period of revolution).} \tag{5-35}$$

Another way to describe the motion of a particle moving in a circle is to cite the number of revolutions that the particle makes in a specific amount of time. This number of revolutions in a given time is known as the **frequency,** f, of revolution. From the definitions we have given for period and frequency, they are related by the expression

$$f = \frac{1}{T} \qquad \text{(frequency).}$$

Examples of Centripetal Forces

What are the forces involved in uniform circular motion? Suppose you were to undergo two different types of uniform circular motion—traveling in a tight circle while driving a car and orbiting the Earth in the space shuttle. You are experiencing centripetal forces that cause you to undergo a centripetal acceleration. In one case you feel that you are being rammed against the car door. In the other case you feel "weightless." Let us examine the forces involved in these two examples of uniform circular motion more closely.

Rounding a curve in a car: You are sitting in the center of the rear seat of a car moving at a constant high speed along a flat road. When the driver suddenly turns left, rounding a corner in a circular arc, you slide across the seat toward the right and then jam against the car door for the rest of the turn. What is going on?

While the car moves in the circular arc, it is in uniform circular motion. That is, it has an acceleration that is directed toward the center of the circle. By Newton's Second Law, $\vec{F} = m\vec{a}$, a force must cause this acceleration. Moreover, the force must also be directed toward the center of the circle. Thus, it is a centripetal force, where the adjective (centripetal) indicates the direction. In this example, the centripetal force is a frictional force on the tires from the road. Without it, the turn would not be possible. For example, imagine what would happen if you hit a patch of low-friction ice while trying to make such a turn.

If you are to move in uniform circular motion along with the car, there must also be a centripetal force on you as well. In our case, apparently the frictional force on you from the seat was not great enough to make you go in a circle with the car. Thus, the seat slid beneath you, until the right door of the car jammed into you. Then its push on you provided the needed centripetal force on you, and you joined the car's uniform circular motion. For you, the centripetal force is the push from the car door.

Orbiting the Earth: This time you are a passenger in the space shuttle *Atlantis*. As it (and you) orbit Earth, you float through your cabin. What is going on?

Both you and the shuttle are in uniform circular motion and have accelerations directed toward the center of the orbital circle. Again by Newton's Second Law, centripetal forces must cause these accelerations. This time the centripetal forces are gravitational pulls (the pull on you and the pull on the shuttle) by Earth, radially inward, toward the center of the Earth. You feel weightless, even though the Earth is pulling on you, because both you and the space shuttle are accelerating at the same rate. This is like feeling lighter when you descend in the elevator discussed in Section 2-4.

Differences between centripetal forces: In both car and shuttle, you are in uniform circular motion, acted on by a centripetal force—yet your sensations in the two situations are quite different. In the car, jammed up against the door, you are aware of being compressed by the door. In the orbiting shuttle, however, you are floating around with no sensation of any force acting on you. Why this difference? The difference is due to the nature of the two centripetal forces. In the car, the centripetal force is due to the push on the part of your body touching the car door. You can sense the compression on that part of your body. In the shuttle, the centripetal force is due to Earth's gravitational pull on every atom of your body. Thus, there is no compression (or pull) on any one part of your body and no sensation of force acting on you. (The sensation is said to be one of "weightlessness," but that description is tricky. The Earth's pull on you has certainly not disappeared and, in fact, is only a little less than it would be when you are on the ground.)

Recall also the example of a centripetal force shown in Fig. 5-25. There a hockey puck moves around a circle at constant speed v while tied to a string looped around a central peg. In this case the centripetal force is the radially inward pull on the puck from the string. Without that force, the puck would go off in a straight line instead of moving in a circle.

In the examples discussed above, the source of the centripetal force was different in each situation. The frictional force, the push of the right door of the car, the gravitational attraction of Earth, and the pull of a string, were all centripetal forces that we considered. This is an important point that was made earlier but is worth repeating. A centripetal force *is not* a new kind of force. It is not an additional force. The name merely indicates the direction in which the force acts. Under the right circumstances, a frictional force, a gravitational force, the force from a car door or a string, or any other kind of force can be centripetal. However, for any situation:

> A centripetal force accelerates a body by changing the direction of the body's velocity without changing the body's speed.

From Newton's Second Law with the centripetal acceleration given by $a = v^2/r$, we can determine what magnitude of the centripetal force, $\vec{F}^{\,\text{cent}}$, is needed to keep an object moving in a circle at a constant speed v. This is given by

$$F^{\text{cent}} = ma = \frac{mv^2}{r} \quad \text{(magnitude of centripetal force).} \qquad (5\text{-}36)$$

Because the speed v and radius r are constant, so are the magnitudes of the acceleration and the force. However, the directions of the centripetal acceleration and force change continuously so as to always point toward the center of a circle. Therefore, unlike the situation for ideal projectile motion, the motions in a chosen x and y direction cannot be treated as independent of each other.

Centripetal Versus Linear Forces and Accelerations

We find it interesting to contrast the forces involved in projectile motion with those of uniform circular motion. In both situations, a particle experiences a net force and accelerates. In projectile motion, the net force is linear with only a vertical component in rectangular coordinates. The acceleration results from a change in the magnitude of vertical velocity vectors and no change in direction.

In uniform circular motion, we can describe the forces in polar coordinates. In this case, we have a constant force pointing inward antiparallel to the r axis and no force in the direction of increasing θ. But since the r axis is changing direction at a constant rate, the magnitude of the velocity is the same and its acceleration is due to direction changes.

> In summary, linear accelerations are due purely to changes in the magnitude of the velocity, whereas uniform circular accelerations are due purely to *changes in the direction* of the velocity.

READING EXERCISE 5-9: Some people say that a centripetal force throws objects outward. For instance, in the example of the car rounding a turn in the discussion above, some might say that the passenger is thrown right when the car turns left. Is it true that centripetal forces throw objects outward? If so, explain how. If not, explain what is really going on. ■

TOUCHSTONE EXAMPLE 5-9: Little Casey Jones

Little Casey Jones is getting an electric train set for Christmas. In the box, he finds 8 pieces of track—4 pieces of straight track and 4 quarter-circle tracks. The straight tracks are 50 cm long, and the quarter-circle tracks will form a circle of radius 50 cm if they are all put together. Casey assembles them into the figure shown at the right in Fig. 5-28. Casey sets the engine on the track and brings it up to a constant speed. The engine has a mass M, and once it is up to speed, it takes 2.0 seconds to make one circuit of the track (at the speed Casey likes to use).

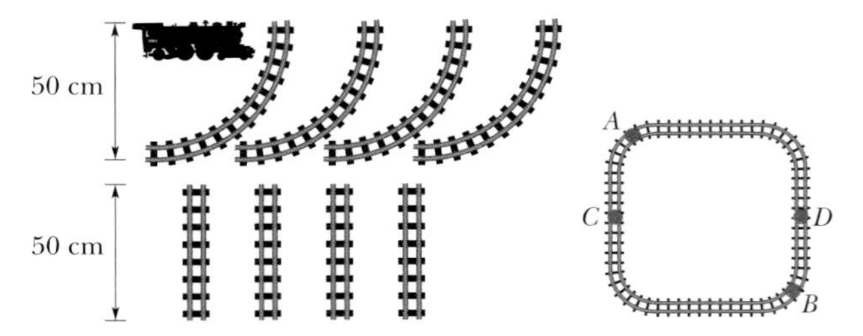

FIGURE 5-28 ■ Little Casey Jones' new train set.

(a) Without actually calculating values, compare the instantaneous velocities of the engine $\vec{v}_A$, $\vec{v}_B$, $\vec{v}_C$, and $\vec{v}_D$ when it is at points A, B, C, and D.

SOLUTION ■ Since the engine is traveling at a constant speed, this tells us that the magnitudes of these four velocity vectors are all the same; $|\vec{v}_A| = |\vec{v}_B| = |\vec{v}_C| = |\vec{v}_D| = v$, the speed at which the engine is moving.

However, each of these four velocity vectors has a different *direction* from each of the other three. A **Key Idea** here is that at each instant in time, the engine's velocity vector points in the direction that the engine is moving. We are not told whether the engine is going around the track in a clockwise or a counterclockwise sense. Let's say it is going around in a clockwise sense, as shown in Fig. 5-29.

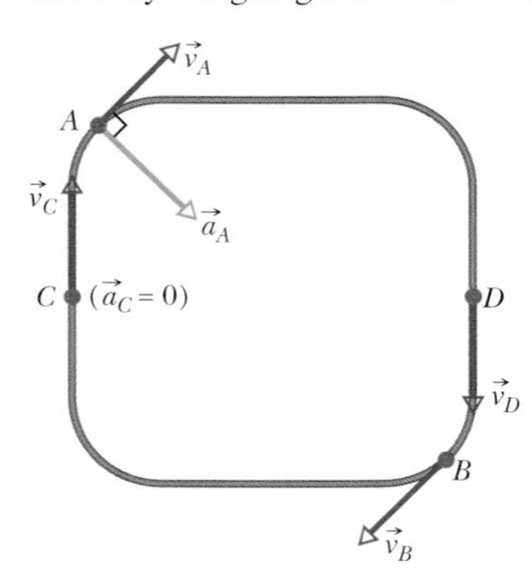

FIGURE 5-29 ■ The directions of the engine's velocity and acceleration at a number of points around the track.

A second **Key Idea** is that the velocity vector is always tangent to the path that the object is following. Thus, $\vec{v}_C$ would point upward along the track; $\vec{v}_A$ would point upward and to the right, tangent to the curved track; $\vec{v}_D$ would point downward, antiparallel to $\vec{v}_C$; and $\vec{v}_B$ would point downward and to the left, antiparallel to $\vec{v}_A$. The directions of these four vectors are shown in Fig. 5-29. (Answer)

(b) Calculate the average velocity of the engine, $\langle\vec{v}\rangle_{CD}$, for the time interval it takes to go from point C to point D.

SOLUTION ■ The **Key Idea** here is that the average velocity is given by Eq. 5-22:

$$\langle\vec{v}\rangle_{CD} = \frac{\Delta\vec{r}_{CD}}{\Delta t_{CD}} = \frac{\vec{r}_D - \vec{r}_C}{t_D - t_C}.$$

From the dimensions of the track given in Fig. 5-28, we see that point D is 50 cm (from the straight segment) + 100 cm (from the two curved segments added together) = 150 cm to the right of point C. We also see that the distance *along the track* from C to D is one-half of the length of one full circuit of the track. Since the engine is traveling at a constant speed, this means that it must take one-half of the time for one full circuit to go from C to D. Thus $\Delta t_{CD} = \frac{1}{2}(2.0\text{ s}) = 1.0\text{ s}$. So the magnitude of the engine's average velocity here is

$$|\langle\vec{v}\rangle_{CD}| = \frac{|\Delta\vec{r}_{CD}|}{\Delta t_{CD}} = \frac{150\text{ cm}}{1.0\text{ s}} = 150\text{ cm/s},$$

directed *to the right* in Fig. 5-29. (Answer)

(c) Calculate the instantaneous acceleration $\vec{a}_A$ and $\vec{a}_C$ of the engine when it is at points A and C.

SOLUTION ■ The **Key Idea** here is that the instantaneous acceleration is $\vec{a} = d\vec{v}/dt$. The acceleration will be zero only when both the magnitude and the direction of the velocity remain constant. If *either* the magnitude *or* the direction of the velocity are changing, then the acceleration will not be zero.

As the engine goes around the track, we are told that the *magnitude* of its velocity remains constant. On a straight track segment, such as at point C, the direction of travel of the engine is also constant, so $\vec{a}_C = 0$. (Answer)

On a curve, such as point A, the engine's direction of travel is changing, since it is turning right. In this case since the train is momentarily moving in a circle, we can use the expression for centripetal acceleration (Eq. 5-34) to determine that $|\vec{a}_A| = v_A^2/r_A$ and that $\vec{a}_A$ is directed down and to the right, as shown in Fig. 5-29.

To calculate the magnitude of $\vec{a}_C$, we note that the total length of track is $2\pi r + 4L = (2\pi \times 50\text{ cm}) + (4 \times 50\text{ cm}) = 514\text{ cm}$, so that

$$v_A = \frac{514\text{ cm}}{2\text{ s}} = 257\text{ cm/s},$$

and since the radius of the circle the train is momentarily moving in is 50 cm,

$$a_A = \frac{v_A^2}{r_A} = \frac{(257\text{ cm/s})^2}{50\text{ cm}} = 1322\text{ cm/s}^2,$$

directed inward toward the center of the curve. (Answer)

Problems

SEC. 5-3 ■ ANALYZING IDEAL PROJECTILE MOTION

In some of these problems, exclusion of the effects of the air is unwarranted but helps simplify the calculations.

1. Rifle and Bullet A rifle is aimed horizontally at a target 30 m away. The bullet hits the target 1.9 cm below the aiming point. What are (a) the bullet's time of flight and (b) its speed as it emerges from the rifle?

2. A Small Ball A small ball rolls horizontally off the edge of a tabletop that is 1.20 m high. It strikes the floor at a point 1.52 m horizontally away from the edge of the table. (a) How long is the ball in the air? (b) What is its speed at the instant it leaves the table?

3. Baseball A baseball leaves a pitcher's hand horizontally at a speed of 161 km/h. The distance to the batter is 18.3 m. (Ignore the effect of air resistance.) (a) How long does the ball take to travel the first half of that distance? (b) The second half? (c) How far does the ball fall freely during the first half? (d) During the second half? (e) Why aren't the quantities in (c) and (d) equal?

4. Dart A dart is thrown horizontally with an initial speed of 10 m/s toward point P, the bull's-eye on a dart board. It hits at point Q on the rim, vertically below P 0.19 s later. (a) What is the distance PQ? (b) How far away from the dart board is the dart released?

5. An Electron An electron, with an initial horizontal velocity of magnitude 1.00×10^9 cm/s, travels into the region between two horizontal metal plates that are electrically charged. In that region, it travels a horizontal distance of 2.00 cm and has a constant downward acceleration of magnitude 1.00×10^{17} cm/s^2 due to the charged plates. Find (a) the time required by the electron to travel the 2.00 cm and (b) the vertical distance it travels during that time. Also find the magnitudes of the (c) horizontal and (d) vertical velocity components of the electron as it emerges.

6. Mike Powell In the 1991 World Track and Field Championships in Tokyo, Mike Powell (Fig. 5-30) jumped 8.95 m, breaking the 23-year long-jump record set by Bob Beamon by a full 5 cm. Assume that Powell's speed on takeoff was 9.5 m/s (about equal to that of a sprinter) and that $g = 9.80$ m/s^2 in Tokyo. How much less was Powell's horizontal range than the maximum possible horizontal range (neglecting the effects of air) for a particle launched at the same speed of 9.5 m/s?

FIGURE 5-30 ■ Problem 6.

7. Catapulted A stone is catapulted at time $t_1 = 0$, with an initial velocity of magnitude 20.0 m/s and at an angle of 40.0° above the horizontal. What are the magnitudes of the (a) horizontal and (b) vertical components of its displacement from the catapult site at $t_2 = 1.10$ s? Repeat for the (c) horizontal and (d) vertical components at $t_3 = 1.80$ s, and for the (e) horizontal and (f) vertical components at $t_4 = 5.00$ s.

8. Golf Ball A golf ball is struck at ground level. The speed of the golf ball as a function of the time is shown in Fig. 5-31, where $t = 0$ at the instant the ball is struck. (a) How far does the golf ball travel horizontally before returning to ground level? (b) What is the maximum height above ground level attained by the ball?

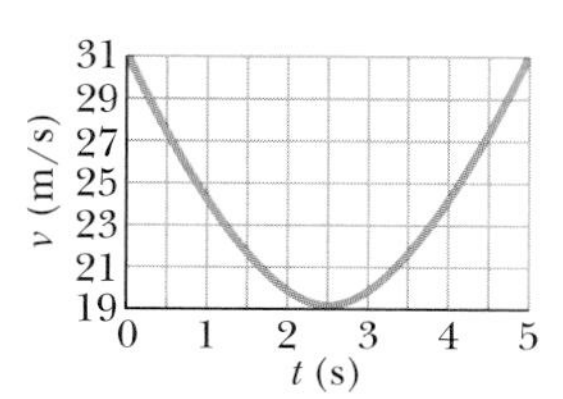

FIGURE 5-31 ■ Problem 8.

9. Fast Bullets A rifle that shoots bullets at 460 m/s is to be aimed at a target 45.7 m away and level with the rifle. How high above the target must the rifle barrel be pointed so that the bullet hits the target?

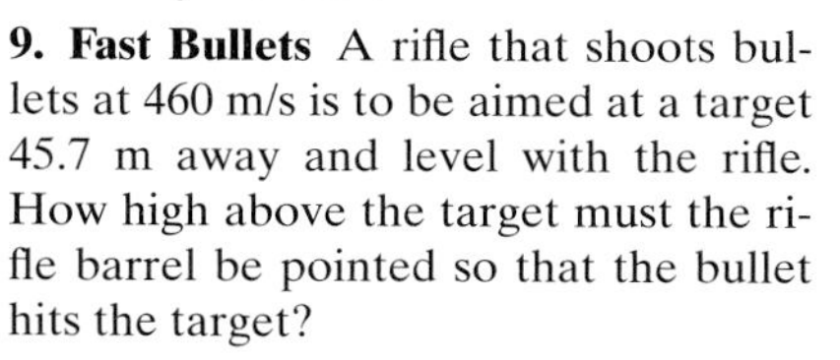

10. Slow-Pitch The pitcher in a slow-pitch softball game releases the ball at a point 3.0 ft above ground level. A stroboscopic plot of the position of the ball is shown in Fig. 5-32, where the readings are 0.25 s apart and the ball is released at $t = 0$. (a) What is the initial speed of the ball? (b) What is the speed of the ball at the instant it reaches its maximum height above ground level? (c) What is that maximum height?

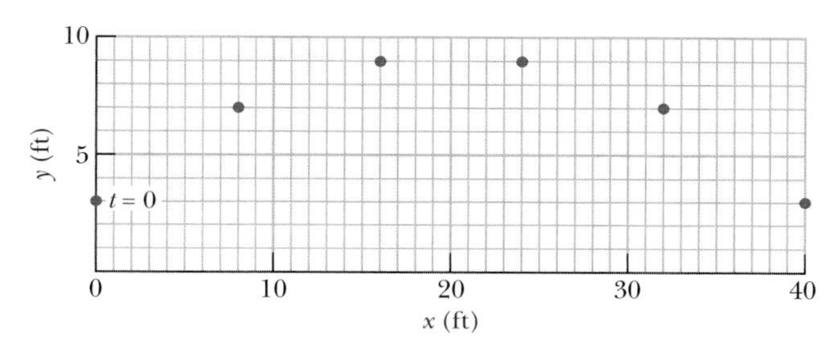

FIGURE 5-32 ■ Problem 10.

11. Maximum Height Show that the maximum height reached by a projectile is $y^{\max} = (v_1 \sin \theta_1)^2/2g$.

12. You Throw a Ball You throw a ball toward a wall with a speed of 25.0 m/s and at an angle of 40.0° above the horizontal (Fig. 5-33). The wall is 22.0 m from the release point of the ball. (a) How far above the release point does the ball hit the wall? (b) What are the horizontal and vertical components of its velocity as it hits the wall? (c) When it hits, has it passed the highest point on its trajectory?

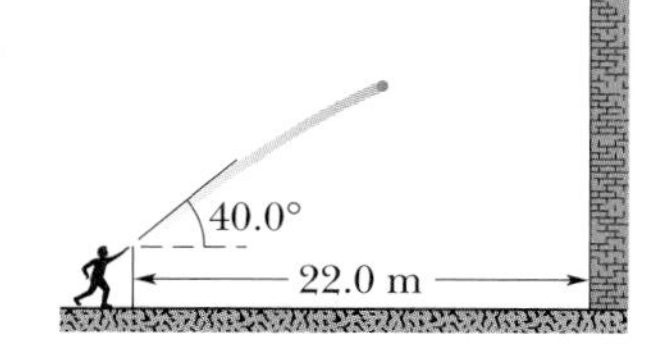

FIGURE 5-33 ■ Problem 12.

13. Shot into the Air A ball is shot from the ground into the air. At a height of 9.1 m. Its velocity is observed to be $\vec{v} = (7.6 \text{ m/s})\hat{i} + (6.1 \text{ m/s})\hat{j}$ ($\hat{i}$ horizontal, $\hat{j}$ upward). (a) To what maximum height does the ball rise? (b) What total horizontal distance does the ball travel? What are (c) the magnitude and (d) the direction of the ball's velocity just before it hits the ground?

14. Two Seconds Later Two seconds after being projected from ground level, a projectile is displaced 40 m horizontally and 53 m vertically above its point of projection. What are the (a) horizontal and (b) vertical components of the initial velocity of the projectile? (c) At the instant the projectile achieves its maximum height above ground level, how far is it displaced horizontally from its point of projection?

15. Football Player A football player punts the football so that it will have a "hang time" (time of flight) of 4.5 s and land 46 m away. If the ball leaves the player's foot 150 cm above the ground, what must be (a) the magnitude and (b) the direction of the ball's initial velocity?

16. Launching Speed The launching speed of a certain projectile is five times the speed it has at its maximum height. Calculate the elevation angle θ_1 at launching.

17. Airplane and Decoy A certain airplane has a speed of 290.0 km/h and is diving at an angle of 30.0° below the horizontal when the pilot releases a radar decoy (Fig. 5-34). The horizontal distance between the release point and the point where the decoy strikes the ground is 700 m. (a) How long is the decoy in the air? (b) How high was the released point?

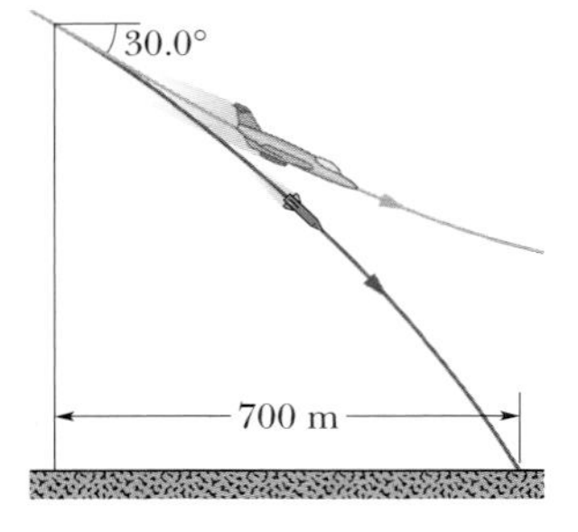

FIGURE 5-34 ■ Problem 17.

18. Soccer Ball A soccer ball is kicked from the ground with an initial speed of 19.5 m/s at an upward angle of 45°. A player 55 m away in the direction of the kick starts running to meet the ball at that instant. What must be his average speed if he is to meet the ball just before it hits the ground? Neglect air resistance.

19. Stairway A ball rolls horizontally off the top of a stairway with a speed of 1.52 m/s. The steps are 20.3 cm high and 20.3 cm wide. Which step does the ball hit first?

20. Volleyball For women's volleyball the top of the net is 2.24 m above the floor and the court measures 9.0 m by 9.0 m on each side of the net. Using a jump serve, a player strikes the ball at a point that is 3.0 m above the floor and a horizontal distance of 8.0 m from the net. If the initial velocity of the ball is horizontal, (a) what minimum magnitude must it have if the ball is to clear the net and (b) what maximum magnitude can it have if the ball is to strike the floor inside the back line on the other side of the net?

21. Airplane An airplane, diving at an angle of 53.0° with the vertical, releases a projectile at an altitude of 730 m. The projectile hits the ground 5.00 s after being released. (a) What is the speed of the aircraft? (b) How far did the projectile travel horizontally during its flight? What were the (c) horizontal and (d) vertical components of its velocity just before striking the ground?

22. Tennis Match During a tennis match, a player serves the ball at 23.6 m/s, with the center of the ball leaving the racquet horizontally 2.37 m above the court surface. The net is 12 m away and 0.90 m high. When the ball reaches the net, (a) does the ball clear it and (b) what is the distance between the center of the ball and the top of the net? Suppose that, instead, the ball is served as before but now it leaves the racquet at 5.00° below the horizontal. When the ball reaches the net, (c) does the ball clear it and (d) what now is the distance between the center of the ball and the top of the net?

23. The Batter A batter hits a pitched ball when the center of the ball is 1.22 m above the ground. The ball leaves the bat at an angle of 45° with the ground. With that launch, the ball should have a horizontal range (returning to the *launch* level) of 107 m. (a) Does the ball clear a 7.32-m-high fence that is 97.5 m horizontally from the launch point? (b) Either way, find the distance between the top of the fence and the center of the ball when the ball reaches the fence.

24. Detective Story In a detective story, a body is found 4.6 m from the base of a building and 24 m below an open window. (a) Assuming the victim left that window horizontally, what was the victim's speed just then? (b) Would you guess the death to be accidental? Explain your answer.

25. Football Kicker A football kicker can give the ball an initial speed of 25 m/s. Within what two elevation angles must he kick the ball to score a field goal from a point 50 m in front of goalposts whose horizontal bar is 3.44 m above the ground? (If you want to work this out algebraically, use $\sin^2\theta + \cos^2\theta = 1$ to get a relation between $\tan^2\theta$ and $1/\cos^2\theta$, substitute, and then solve the resulting quadratic equation.)

SEC. 5-4 ■ DISPLACEMENT IN TWO DIMENSIONS

26. Position Vector for an Electron The position vector for an electron is $\vec{r} = (5.0\text{ m})\hat{\imath} - (3.0\text{ m})\hat{\jmath}$. (a) Find the magnitude of $\vec{r}$. (b) Sketch the vector on a coordinate system.

27. Watermelon Seed A watermelon seed has the following coordinates: $x = -5.0$ m and $y = 8.0$ m. Find its position vector (a) in unit-vector notation and as (b) a magnitude and (c) an angle relative to the positive direction of the x axis. (d) Sketch the vector on a coordinate system. If the seed is moved to the coordinates (3.00 m, 0 m), what is its displacement (e) in unit-vector notation and as (f) a magnitude and (g) an angle relative to the positive direction of the x axis?

28. Radar Station A radar station detects an airplane approaching directly from the east. At first observation, the range to the plane is 360 m at 40° above the horizon. The airplane is tracked for another 123° in the vertical east–west plane, the range at final contact being 790 m. See Fig. 5-35. Find the displacement of the airplane during the period of observation.

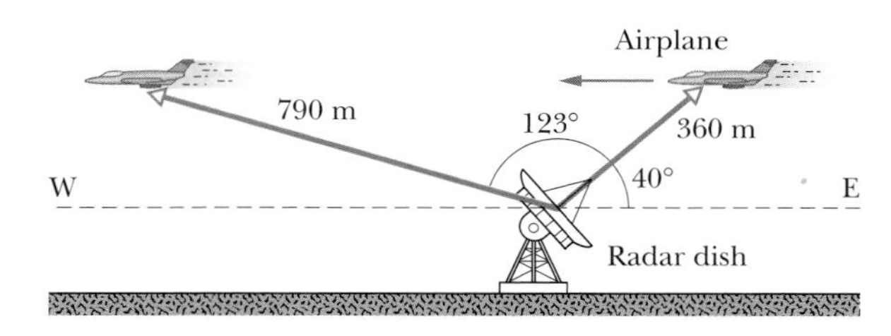

FIGURE 5-35 ■ Problem 28.

29. Position Vector for a Proton The position vector for a proton is initially $\vec{r}_1 = (5.0\text{ m})\hat{\imath} + (-6.0\text{ m})\hat{\jmath}$ and then later is $\vec{r}_2 = (-2.0\text{ m})\hat{\imath} + (6.0\text{ m})\hat{\jmath}$. (a) What is the proton's displacement vector, and (b) to what axis (if any) is that vector parallel?

30. Kidnapped You are kidnapped by armed political-science majors (who are upset because you told them that political science is not a real science). Although blindfolded, you can tell the speed of their car (by the whine of the engine), the time of travel (by mentally counting off seconds), and the direction of travel (by turns along the rectangular street system). From these clues, you know that you are taken along the following course: 50 km/h for 2.0 min, turn 90° to the right, 20 km/h for 4.0 min, turn 90° to the right, 20 km/h for 60 s, turn 90° to the left, 50 km/h for 60 s, turn 90° to the right, 20 km/h for 2.0 min, turn 90° to the left, 50 km/h for 30 s. At that point, (a) how far are you from your starting point and (b) in what direction relative to your initial direction of travel are you?

31. Drunk Skunk Figure 5-36 shows the path taken by my drunk skunk over level ground, from initial point i to final point f. The angles are $\theta_1 = 30.0°$, $\theta_2 = 50.0°$, and $\theta_3 = 80.0°$, and the distances are $d_1 = 5.00$ m, $d_2 = 8.00$ m, and $d_3 = 12.0$ m. In magnitude-angle notation, what is the skunk's displacement from i to f?

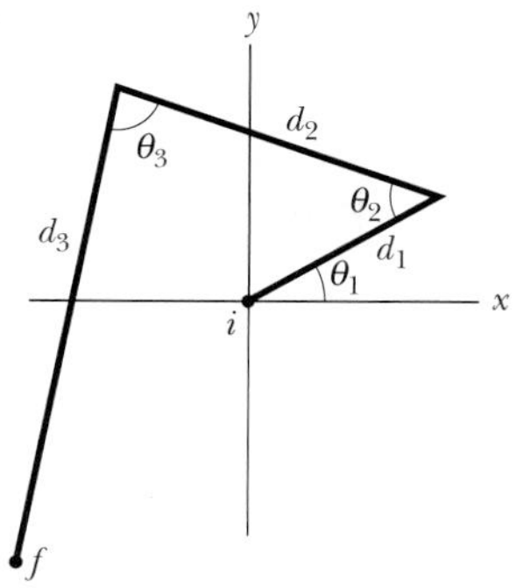

FIGURE 5-36 ■ Problem 31.

SEC. 5-5 ■ AVERAGE AND INSTANTANEOUS VELOCITY

32. Squirrel Path Figure 5-37 gives the path of a squirrel moving about on level ground, from point A (at time $t_1 = 0$), to points B (at $t_2 = 5.00$ min), C (at $t_3 = 10.0$ min), and finally D (at $t_4 = 15.0$ min). Consider the average velocities of the squirrel from point A to each of the other three points. (a) Of those three average velocities, which has the least magnitude, and what is the average velocity in magnitude-angle notation? (b) Which has the greatest magnitude, and what is the average velocity in magnitude-angle notation?

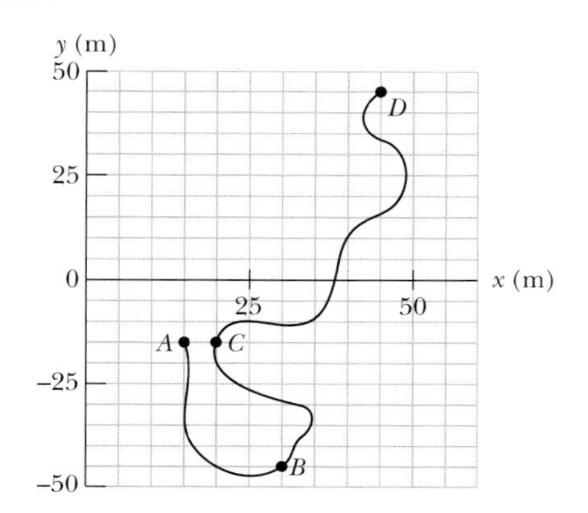

FIGURE 5-37 ■ Problem 32.

33. Train A train moving at a constant speed of 60.0 km/h moves east for 40.0 min. then in a direction 50.0° east of north for 20.0 min, and finally west for 50.0 min. What is the average velocity of the train during this trip?

34. Ion's Position An ion's position vector is initially $\vec{r}_1 = (5.0\text{ m})\hat{\text{i}} + (-6.0\text{ m})\hat{\text{j}}$, and 10 s later it is $\vec{r}_2 = (-2.0\text{ m})\hat{\text{i}} + (8.0\text{ m})\hat{\text{j}}$. What is its average velocity during the 10 s?

35. Electron's Position The position of an electron is given by $\vec{r}(t) = [(3.00\text{ m/s})\,t]\,\hat{\text{i}} + [(-4.00\text{ m/s}^2)\,t^2]\hat{\text{j}}$. (a) What is the electron's velocity $\vec{v}(t)$? At $t = 2.00$ s, what is $\vec{v}$ (b) in unit-vector notation and as (c) a magnitude and (d) an angle relative to the positive direction of the x axis?

36. Oasis Oasis A is 90 km west of oasis B. A camel leaves oasis A and during a 50 h period walks 75 km in a direction 37° north of east. The camel then walks toward the south a distance of 65 km in a 35 h period after which it rests for 5.0 h. (a) What is the camel's displacement with respect to oasis A after resting? (b) What is the camel's average velocity from the time it leaves oasis A until it finishes resting? (c) What is the camel's average speed from the time it leaves oasis A until it finishes resting? (d) If the camel is able to go without water for five days (120 h), what must its average velocity be after resting if it is to reach oasis B just in time?

37. Jet Ski You are to ride a jet-cycle over a lake, starting from rest at point 1: First, moving at 30° north of due east:

1. Increase your speed at 0.400 m/s^2 for 6.00 s.
2. With whatever speed you then have, move for 8.00 s.
3. Then slow at 0.400 m/s^2 for 6.00 s.

Immediately next, moving due west:

4. Increase your speed at 0.400 m/s^2 for 5.00 s.
5. With whatever speed you then have, move for 10.0 s.
6. Then slow at 0.400 m/s^2 until you stop.

In magnitude-angle notation, what then is your average velocity for the trip from point 1?

SEC. 5-6 ■ AVERAGE AND INSTANTANEOUS ACCELERATION

38. A Proton A proton initially has $\vec{v}_1 = (4.0\text{ m/s})\hat{\text{i}} + (-2.0\text{ m/s})\hat{\text{j}}$ and then 4.0 s later has $\vec{v}_2 = (-2.0\text{ m/s})\hat{\text{i}} + (-2.0\text{ m/s})\hat{\text{j}}$. For that 4.0 s. what is the proton's average acceleration $\langle\vec{a}\rangle$ (a) in unit-vector notation and (b) as a magnitude and a direction?

39. Particle in *xy* Plane The position $\vec{r}$ of a particle moving in an xy plane is given by $\vec{r}(t) = [(2.00\text{ m/s}^3)t^3 - (5.00\text{ m/s})\,t]\hat{\text{i}} + [(6.00\text{ m}) - (7.00\text{ m/s}^4)\,t^4]\hat{\text{j}}$. Calculate (a) $\vec{r}$, (b) $\vec{v}$, and (c) $\vec{a}$ for $t = 2.00$ s.

40. Iceboat An iceboat sails across the surface of a frozen lake with constant acceleration produced by the wind. At a certain instant the boat's velocity is $\vec{v}_1 = (6.30\text{ m/s})\hat{\text{i}} + (-8.42\text{ m/s})\hat{\text{j}}$. Three seconds later, because of a wind shift, the boat is instantaneously at rest. What is its average acceleration for this 3 s interval?

41. Particle Leaves Origin A particle leaves the origin with an initial velocity $\vec{v}_1 = (3.00\text{ m/s})\hat{\text{i}}$ and a constant acceleration $\vec{a} = (-1.00\text{ m/s}^2)\hat{\text{i}} + (-0.500\text{ m/s}^2)\hat{\text{j}}$. When the particle reaches its maximum x coordinate, what are (a) its velocity and (b) its position vector?

42. Particle *A* Particle *B* Particle A moves along the line $y = 30$ m with a constant velocity $\vec{v}$ of magnitude 3.0 m/s and directed parallel to the positive x axis (Fig. 5-38). Particle B starts at the origin with zero speed and constant acceleration $\vec{a}$ (of magnitude 0.40 m/s^2) at the same instant that particle A passes the y axis. What angle θ between $\vec{a}$ and the positive y axis would result in a collision between these two particles? (If your computation involves an equation with a term such as t^4, substitute $u = t^2$ and then consider solving the resulting quadratic equation to get u.)

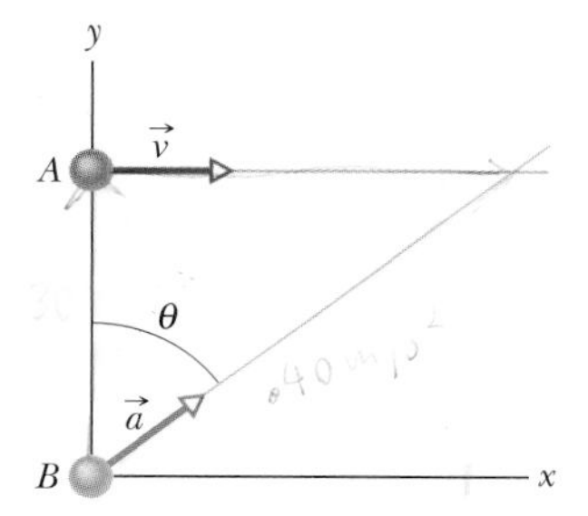

FIGURE 5-38 ■ Problem 42.

43. Particle Starts from Origin A particle starts from the origin at $t = 0$ with a velocity of $\vec{v}_1 = (8.0\text{ m/s})\hat{\text{j}}$ and moves in the xy plane with a constant acceleration of $\vec{a} = (4.0\text{ m/s}^2)\hat{\text{i}} + (2.0\text{ m/s}^2)\hat{\text{j}}$. At the instant the particle's x coordinate is 29 m, what are (a) its y coordinate and (b) its speed?

44. The Wind and a Pebble A moderate wind accelerates a smooth pebble over a horizontal xy plane with a constant acceleration

$$\vec{a} = (5.00\text{ m/s}^2)\hat{\text{i}} + (7.00\text{ m/s}^2)\hat{\text{j}}.$$

At time $t = 0$, its velocity is $(4.00\text{ m/s})\hat{\text{i}}$. In magnitude-angle notation, what is its velocity when it has been displaced by 12.0 m parallel to the x axis?

45. Particle Acceleration A particle moves so that its position as a function of time is $\vec{r}(t) = (1\text{ m})\hat{i} + [(4\text{ m/s}^2)\,t^2]\,\hat{j}$. Write expressions for (a) its velocity and (b) its acceleration as functions of time.

SEC. 5-7 ■ UNIFORM CIRCULAR MOTION

46. Sprinter What is the magnitude of the acceleration of a sprinter running at 10 m/s when rounding a turn with a radius of 25 m?

47. Sprinter on Circular Path A sprinter runs at 9.2 m/s around a circular track with a centripetal acceleration of magnitude 3.8 m/s^2. (a) What is the track radius? (b) What is the period of the motion?

48. Rotating Fan A rotating fan completes 1200 revolutions every minute. Consider the tip of a blade, at a radius of 0.15 m. (a) Through what distance does the tip move in one revolution? What are (b) the tip's speed and (c) the magnitude of its acceleration? (d) What is the period of the motion?

49. An Earth Satellite An Earth satellite moves in a circular orbit 640 km above Earth's surface with a period of 98.0 min. What are (a) the speed and (b) the magnitude of the centripetal acceleration of the satellite?

50. Merry-Go-Round A carnival merry-go-round rotates about a vertical axis at a constant rate. A passenger standing on the edge of the merry-go-round has a constant speed of 3.66 m/s. For each of the following instantaneous situations, state how far the passenger is from the center of the merry-go-round, and in which direction. (a) The passenger has an acceleration of 1.83 m/s^2, east. (b) The passenger has an acceleration of 1.83 m/s^2, south.

51. Astronaut An astronaut is rotated in a horizontal centrifuge at a radius of 5.0 m. (a) What is the astronaut's speed if the centripetal acceleration has a magnitude of 7.0g? (b) How many revolutions per minute are required to produce this acceleration? (c) What is the period of the motion?

52. TGV The fast French train known as the TGV (Train à Grande Vitesse) has a scheduled average speed of 216 km/h. (a) If the train goes around a curve at that speed and the magnitude of the acceleration experienced by the passengers is to be limited to 0.050g, what is the smallest radius of curvature for the track that can be tolerated? (b) At what speed must the train go around a curve with a 1.00 km radius to be at the acceleration limit?

53. Object on the Equator (a) What is the magnitude of the centripetal acceleration of an object on Earth's equator due to the rotation of Earth? (b) What would the period of rotation of Earth have to be for objects on the equator to have a centripetal acceleration with a magnitude of 9.8 m/s^2?

54. Supernova When a large star becomes a *supernova*, its core may be compressed so tightly that it becomes a *neutron star*, with a radius of about 20 km (about the size of the San Francisco area). If a neutron star rotates once every second, (a) what is the speed of a particle on the star's equator and (b) what is the magnitude of the particle's centripetal acceleration? (c) If the neutron star rotates faster, do the answers to (a) and (b) increase, decrease, or remain the same?

55. Ferris Wheel A carnival Ferris wheel has a 15 m radius and completes five turns about its horizontal axis every minute. (a) What is the period of the motion? What is the centripetal acceleration of a passenger at (b) the highest point and (c) the lowest point, assuming the passenger is at a 15 m radius?

56. A Particle at Constant Speed A particle P travels with constant speed on a circle of radius $r = 3.00$ m (Fig. 5-39) and completes one revolution in 20.0 s. The particle passes through O at time $t = 0$. State the following vectors in magnitude-angle notation (angle relative to the positive direction of x). With respect to O, find the particle's position vector at the times t of (a) 5.00 s, (b) 7.50 s, and (c) 10.0 s. (d) For the 5.00 s interval from the end of the fifth second to the end of the tenth second, find the particle's displacement. (e) For the same interval, find its average velocity. Find its velocity at (f) the beginning and (g) the end of that 5.00 s interval. Next, find the acceleration at (h) the beginning and (i) the end of that interval.

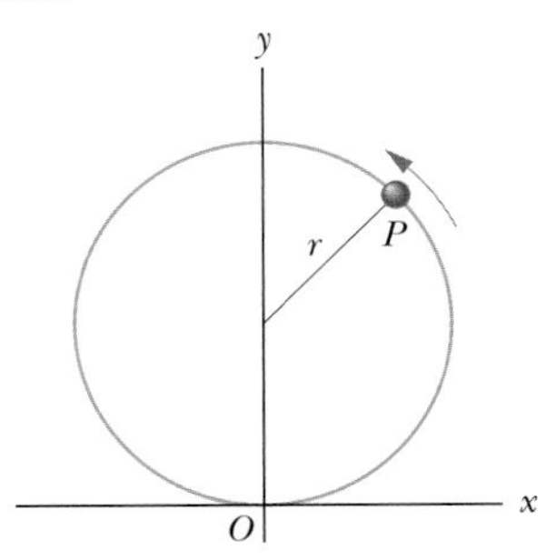

FIGURE 5-39 ■ Problem 56.

57. Stone on a String A boy whirls a stone in a horizontal circle of radius 1.5 m and at height 2.0 m above level ground. The string breaks, and the stone flies off horizontally and strikes the ground after traveling a horizontal distance of 10 m. What is the magnitude of the centripetal acceleration of the stone while in circular motion?

58. Cat on a Merry-Go-Round A cat rides a merry-go-round while turning with uniform circular motion. At time $t_1 = 2.00$ s, the cat's velocity is

$$\vec{v}_1 = (3.00\text{ m/s})\hat{i} + (4.00\text{ m/s})\hat{j},$$

measured on a horizontal xy coordinate system. At time $t_2 = 5.00$ s, its velocity is

$$\vec{v}_2 = (-3.00\text{ m/s})\hat{i} + (-4.00\text{ m/s})\hat{j}.$$

What are (a) the magnitude of the cat's centripetal acceleration and (b) the cat's average acceleration during the time interval $t_2 - t_1$?

59. Center of Circular Path A particle moves horizontally in uniform circular motion, over a horizontal xy plane. At one instant, it moves through the point at coordinates (4.00 m, 4.00 m) with a velocity of $(-5.00\text{ m/s})\hat{i}$ and an acceleration of $(12.5\text{ m/s}^2)\hat{j}$. What are the coordinates of the center of the circular path?

Additional Problems

60. Keeping Mars in Orbit Although the planet Mars orbits the Sun in a Kepler ellipse with an eccentricity of 0.09, we can approximate its orbit by a circle. If you have faith in Newton's laws then you must conclude that there is an invisible centripetal force holding Mars in orbit. The data on the orbit of Mars around the sun are shown in the Fig. 5-40.

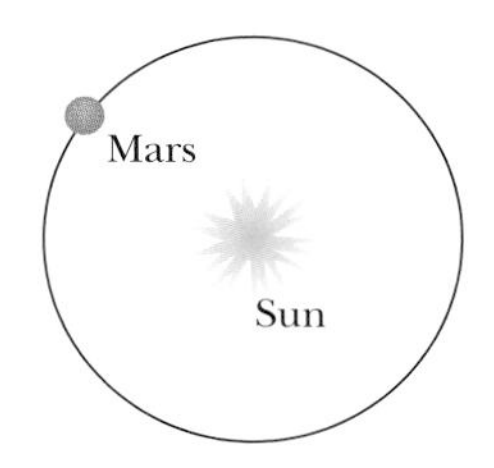

$m_{\text{Sun}} = 2.00 \times 10^{30}$ kg

$m_{\text{Mars}} = 0.107$ Earth masses or
$= 6.42 \times 10^{23}$ kg

$d_{\text{Mars}} = 1.523$ AU $= 2.28 \times 10^{11}$ m $=$
distance from the sun (= radius of circular orbit)

$\langle v \rangle = 24.13$ km/s (mean orbital speed)

FIGURE 5-40 ■ Problem 60.

(a) Calculate the magnitude of the centripetal force needed to hold Mars in its circular orbit. Please use the proper number of significant figures. (b) What is the direction of the force as Mars orbits around the Sun? (c) What is the most likely source of this force? (d) Could this force have anything in common with the force that attracts objects to the Earth?

61. Playing Catch A boy and a girl are tossing an apple back and forth between them. Figure 5-41 shows the path the apple followed when watched by an observer looking on from the side. The apple is moving from left to right. Five points are marked on the path. Ignore air resistance. (a) Make a copy of this figure. At each of the marked points, draw an arrow that indicates the magnitude and direction of the force on the apple when it passes through that point. (b) Make a second copy of the figure. This time, at each marked point, place an arrow indicating the magnitude and direction of the apple's velocity at the instant it passes that point. (c) Did you change your answer to the first question after solving the second? If so, explain what you were thinking at first and why you changed it.

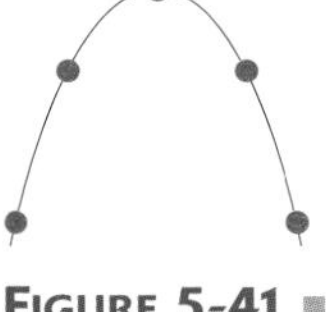

FIGURE 5-41 ■ Problem 61.

62. The Cut Pendulum A pendulum (i.e., a string with a ball at the end) is set swinging by holding it at the point marked A in Fig. 5-42*a* and releasing it. The x and y coordinates are shown with the origin at the crossing point of the axes and the positive directions indicated by the arrowheads. (a) During one swing, the string breaks exactly at the bottom-most point of the swing (the point labeled B in the figure) as the ball is moving from A to B toward C. Make a copy of this figure. Using solid lines, sketch on the figure the path of the ball after the string has broken. Sketch qualitatively the x and y coordinates of the ball and the x- and y-components of its velocity on graphs like those shown in Fig. 5-42*b*. Take $t = 0$ to be the instant the string breaks. (b) During a second trial, the string breaks again, but this time at the top-most point of the swing (the point labeled C in the figure). Using dashed lines, sketch on the figure the path of the ball after the string has broken. Sketch qualitatively the x and y coordinates of the ball and the x- and y-components of its velocity on graphs like those shown in Fig. 5-42*b*. Take $t = 0$ to be the instant the string breaks.

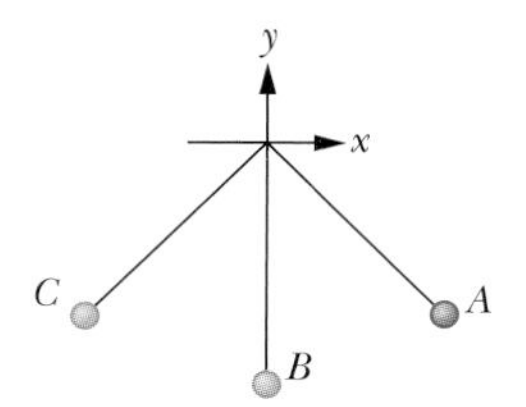

FIGURE 5-42*a* ■ Problem 62.

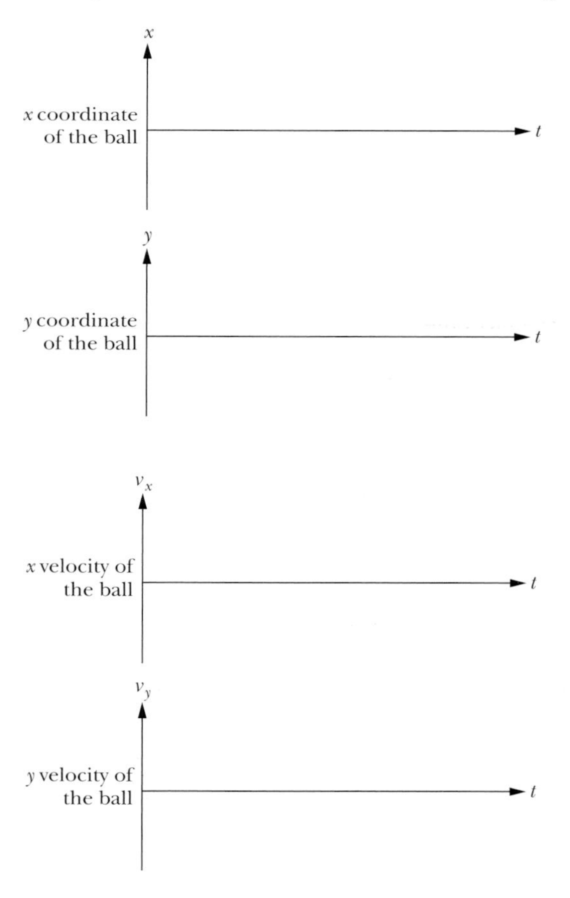

FIGURE 5-42*b* ■ Problem 62.

63. Projectile Graphs A pop-gun is angled so that it shoots a small dense ball through the air as shown in Fig. 5-43*a*.

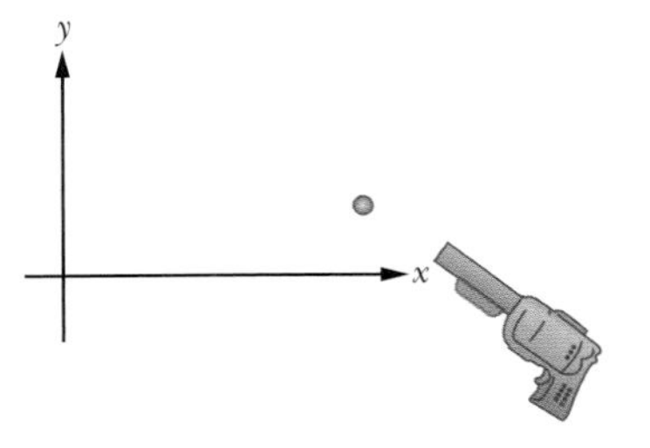

FIGURE 5-43*a* ■ Problem 63.

(a) Sketch the path that the ball will follow on the figure. For the graphs shown in Fig. 5-43*b*, the horizontal axis represents the time. The vertical axis is unspecified. For each of the following quantities, select the letter of the graph that could provide a correct graph of the quantity for the ball in the situation shown (if the vertical axis were assigned the proper units). Use the x and y coordinates shown in the picture. The arrow heads point in the positive direction. If none of the graphs could work, write N. The time graphs begin just after the ball leaves the gun.

(b) y coordinate
(c) x-component of the velocity
(d) y-component of the net force
(e) y-component of the velocity
(f) x coordinate
(g) y-component of the acceleration
(h) x-component of the net force

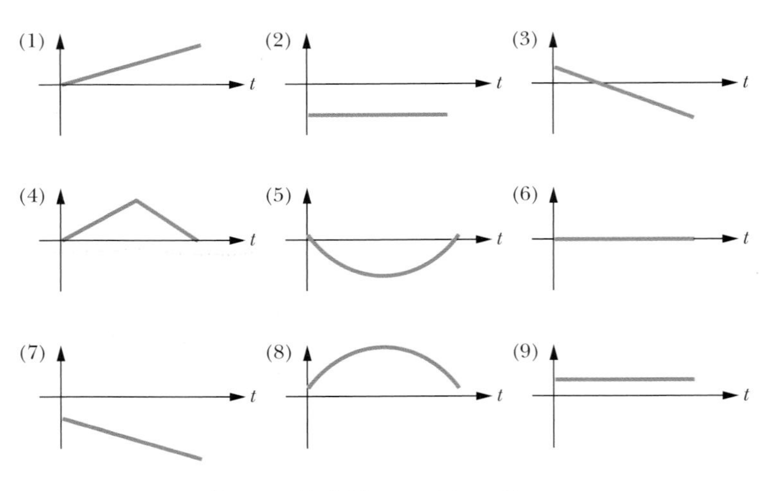

FIGURE 5-43*b* ■ Problem 63.

64. Shoot and Drop In the demonstration discussed in Section 5-2, two identical objects were dropped, one straight down and the other shot off to the side by a spring. Both objects seemed to hit the ground at about the same time. Explain why this happens in terms of the physics we have learned. Does it matter how fast we shoot the one launched sideways? How would the outcome of this experiment change if the objects had different masses? (*Hint*: See Fig. 5-5.)

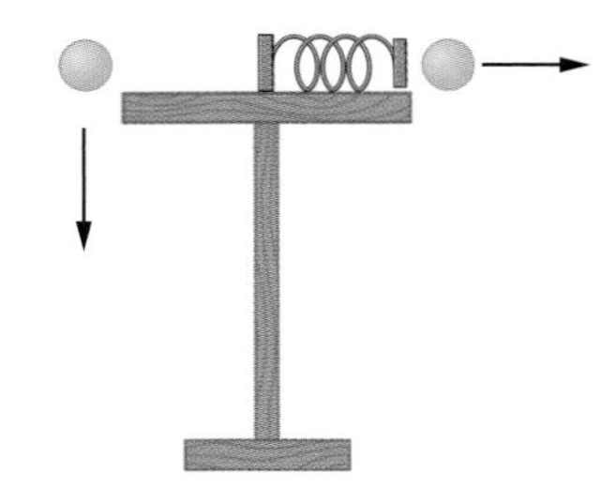
FIGURE 5-44 ■ Problem 64.

65. Billiards over the Edge Two identical billiard balls are labeled A and B. Maryland Fats places ball A at the very edge of the table. He places ball B at the other side. He strikes ball B with his cue so that it flies across the table and off the edge. As it passes A, it just touches ball A lightly, knocking it off. Figure 5-45*a* shows the balls just at the instant they have left the table. Ball B is moving with a speed v_1, and ball A is essentially at rest.

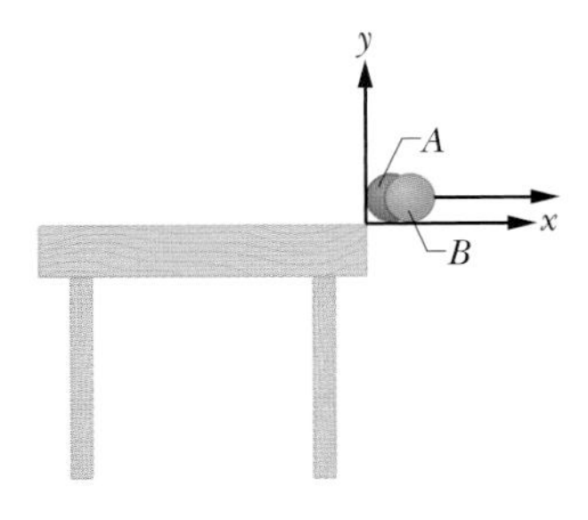

FIGURE 5-45*a* ■ Problem 65.

(a) Which ball do you think will hit the ground first? Explain your reasons for thinking so.

Figure 5-45*b* shows a number of graphs of a quantity vs. time. In each case, the horizontal axis is the time axis. The vertical axis is unspecified. For each of the items below, select which graph could be a plot of that quantity vs. time. If none of the graphs are possible, write N. The time axes are taken to have $t = 0$ at the instant both balls leave the table. Use the x and y axes shown in the figure.
(b) the x-component of the velocity of ball B? (c) the y-component of the velocity of ball A? (d) the y-component of the acceleration of ball A? (e) the y-component of the force on ball B? (f) the y-component of the force on ball A? (g) the x-component of the velocity of ball A? (h) the y-component of the acceleration of ball B?

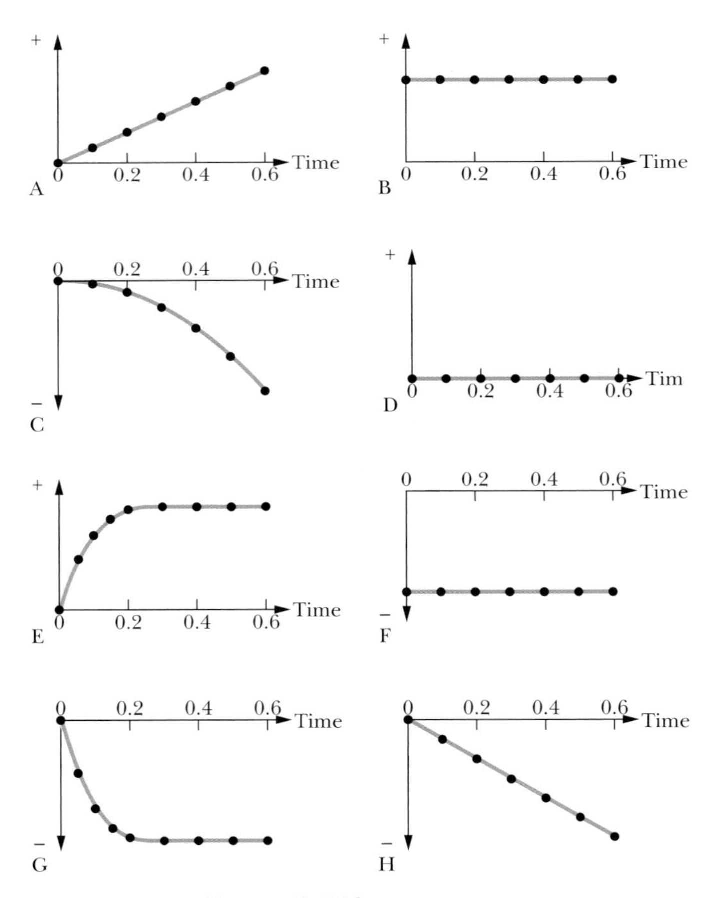

FIGURE 5-45*b* ■ Problem 65.

66. Properties of a Projectile A heavy projectile is thrown and follows a path something like the one shown in Fig. 5-46. For each of the quantities in the list (a)–(d) below, select a direction from the list (A–G) that describes it. If you think that none of the choices apply, write N.

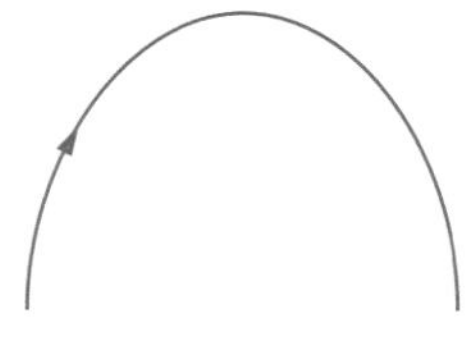

FIGURE 5-46 ■ Problem 66.

Quantities:

(a) The projectile's velocity when it is at the highest point
(b) The force on the projectile when it is part way up
(c) The force on the projectile when it is at the highest point
(d) The projectile's acceleration when it is part way down

Choices:

A. Points straight up
B. Points straight down
C. Points directly to the left
D. Points directly to the right
E. Is equal to zero
F. Points somewhat upward and to the right
G. Points somewhat upward and to the left
N. None of the above

67. Passing by the Spanish Guns In C. S. Forster's novel *Lieutenant Hornblower* (set in the early 1800s), a British naval vessel tries to sneak by a Spanish garrison. The ship passes as far away from the Spanish guns as it can—a distance s. The Spanish gunner knows that his gun has a muzzle velocity whose magnitude is equal to v_1. (a) Once the gun is fired, what controls the motion of the cannonball? Write the equations that determine the vector position of the cannonball after it leaves the cannon. You may ignore air resistance. (b) Suppose the gunner inclines his gun upward at an angle θ to the horizontal. Solve the equations you have written in part (a) to obtain expressions that can be evaluated to give the position of the cannonball at any time, t. (c) If the gunner wants the cannonball to hit the ship, he must choose his angle correctly. Explain how he can calculate the correct angle. (Again, you may ignore air resistance.) (d) If the muzzle velocity of the cannonball has a magnitude of 100 m/s and the ship is a distance of half a kilometer away, find the angle the gunner should use. (Take g to be 10 m/s^2.)

FIGURE 5-47 ■ Problem 67.

You may need one or more of the following trigonometric identities (i.e., these are true for all angles, θ):

$\sin^2\theta + \cos^2\theta = 1$ $\qquad \cos^2\theta - \sin^2\theta = \cos 2\theta$

$\tan\theta = (\sin\theta)/(\cos\theta)$ $\qquad 2\sin\theta\cos\theta = \sin 2\theta$

68. Who Killed Adam Able? A person shoved out of a window makes just as good a projectile as a golf ball rolling off a table. Read the murder mystery entitled A Damnable Man that follows. In order to solve the crime, read the section on projectile motion in your text carefully and reason out for yourself what variables might be important in solving the crime. In fact, not all of the information given in the mystery is relevant and some information, which you can find for yourself by observation and experiment, is missing. Solve the crime by presenting a clear explanation of the equations and calculations you used. (*Hint*: If you are in the physics lab and shove your lab partner fairly hard, you will likely find that your partner ends up with a speed of about 2 m/s.)

A Damnable Man
by Kevin Laws

It is a warm, quiet, humid night in the city—the traffic has died away and there isn't even a cooling breeze. There is a busy hotel that is so well built that sounds don't carry through the windows. The hotel has impressively large rooms. This is obvious from the outside because there is more space between floors and rooms than normal. The rooms appear to have 14-foot-high ceilings, nice plate glass windows that slide open, and fully two-foot-thick floors for ducting and sound insulation. This is the type of hotel that people like to stay at when someone else is paying the bill.

Outside the hotel, a man is speaking quietly with the doorman, then begins to measure the plush runway carpet for replacement. He is reeling out the tape measure between the hotel and the curb when a scream breaks the quiet. Looking up, he sees a man falling toward him. Stunned, he drops the tape measure and runs for the safety of the hotel. The doorman stands, horrified, as the man completes his fall with a sickening sound, ensuring that the carpet must be replaced. At intense times, people can think of the strangest things, and the carpet-man finds this to be true . . . all he can think of are the bloodstains left on his tape measure. Even if they are cleaned off, he doesn't think he can use it again without thinking of tonight. Even measuring with another will be hard, and 18 feet will be indelibly marked in his memory—that's where the blood stains are.

The police arrive and quickly conclude that it is not a suicide—among the victim's personal effects, they find pictures and records that indicate he has been blackmailing four other occupants of the hotel. He also has bruises on his shins where the ledge at the bottom of the tall hotel window would have hit them; he must have been pushed pretty hard. Adam Able is the dead man's name, as it appears on the driver's license in his wallet. His license indicates that Adam was 5' 11" tall and weighed 160 lb. He has been blackmailing Adrianna Myers, a frail widow in Room 356; Steven Caine, a newspaper reporter in Room 852; Mark Johnson, a body builder in Room 1956; and Stanley Michaels, an actor in Room 2754. All of the suspects admit they were in their rooms at the time of the murder. **WHO KILLED ADAM ABLE?**

69. Digital Projectile One In this problem and the one that follows you will be asked to use VideoPoint, VideoGraph, or some other video analysis program and a spreadsheet to explore and analyze the nature of a projectile launch depicted in a digital movie. If you use VideoPoint, one appropriate movie has filename PASCO106. In this movie a small ball of mass 9.5 g is launched at an angle, θ, with respect to the horizontal. Your instructor may suggest an alternative file for your use.

Open the movie PASCO106. For simplicity you might want to set the origin in the video analysis at the location of the ball at time $t = 0$. Also, for immediate visual feedback on your results you should use the *View Window* to set up graphs of x vs. t and y vs. t before you begin the analysis.

(a) What is the approximate launch angle θ? Measure this angle with respect to the horizontal. Explain how you found the angle.
(b) Explain in which direction, x or y, the ball has a constant velocity and cite the real evidence (not just theoretical) for this constant velocity. (*Hint*: Use markers of various sorts on the digital movie to demonstrate that the ball is moving at a constant velocity in one of the directions and not in the other.)
(c) Explain in which direction, x or y, the ball is accelerating. Cite real evidence (not just theoretical) for this acceleration. (*Hint*: Use markers of various sorts on the digital movie.)
(d) Theoretically, what is the net vertical force on the 9.5 g ball when it is rising? Falling? Turning around? What is the observational basis for this theoretical assumption?
(e) Theoretically, what is the net horizontal force on the 9.5 g ball when it is rising? Falling? Turning around? What is the observational basis for this theoretical assumption?
(f) What do you predict will happen to the shapes of the x vs. t and y vs. t graphs if you rotate your coordinate system so that the x axis points in the vertical direction and the y axis points in the horizontal direction?
(g) Rotate your coordinate system so that the x axis points in the vertical direction and the y axis points in the horizontal direction. What happens to the shapes of the graphs? Is this what you predicted?

70. Digital Projectile Two In this problem you will use VideoPoint, VideoGraph, or some other video analysis program and a spreadsheet to explore and analyze the nature of a projectile launch depicted in a digital movie. If you use VideoPoint, one appropriate movie has filename PASCO106. In this movie a small ball of mass 9.5 g is launched at an angle, θ, with respect to the horizontal. Your instructor may suggest an alternative file for your use.

Open the movie PASCO106. Use the VideoPoint software and spreadsheet modeling to find the equation that describes: the horizontal motion x vs. t and the equations that describe the vertical motion y vs. t.

(a) Hand in the printout of your two models. Place your name, date and section # on it, and answer questions (b) through (d) at the bottom of the page.
(b) According to your horizontal model, what is the equation that describes the horizontal position of the ball, x, as a function of time? What is its horizontal acceleration, a_x? What is its initial horizontal velocity, v_{1x}?
(c) According to your vertical model, what is the equation that describes the vertical position, y, of the ball as a function of time? What is the value of the ball's vertical acceleration, a_y? What is its initial vertical velocity, v_{1y}?
(d) Use the components v_{1x} and v_{1y} to compute the initial speed of the ball. What is the launch angle with respect to the horizontal?
(e) Compare your answer to part (d) to your approximation from part (a) of the previous problem.

71. Curtain of Death A large metallic asteroid strikes Earth and quickly digs a crater into the rocky material below ground level by launching rocks upward and outward. The following table gives five pairs of launch speeds and angles (from the horizontal) for such rocks, based on a model of crater formation. (Other rocks, with intermediate speeds and angles, are also launched.) Suppose that you are at $x = 20$ km when the asteroid strikes the ground at time $t_1 = 0$ and position $x = 0$ (Fig. 5-48). (a) At $t_2 = 20$ s, what are the x and y coordinates of the rocks headed in your direction from launches A through E? (b) Plot these coordinates and then sketch a curve through the points to include rocks with intermediate launch speeds and angles. The curve should give you an idea of what you would see as you look up into the approaching rocks and what dinosaurs must have seen during asteroid strikes long ago.

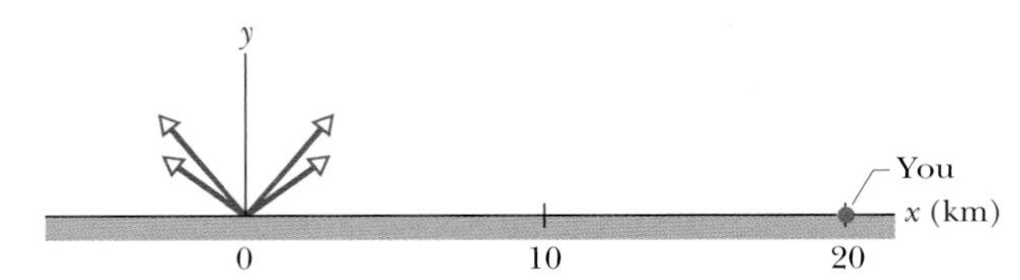

FIGURE 5-48 ■ Problem 71.

Launch	Speed (m/s)	Angle (degrees)
A	520	14.0
B	630	16.0
C	750	18.0
D	870	20.0
E	1000	22.0

6 Identifying and Using Forces

Cats, who enjoy sleeping on window sills, are often kept in apartment buildings. When a cat accidentally falls out of a window and onto a sidewalk, the extent of injury (such as the number of fractured bones or the certainty of death) *decreases* with height if the fall is more than seven or eight floors. (There is even a record of a cat who fell 32 floors and suffered only slight damage to its thorax and one tooth.)

How can the damage possibly decrease with height?

The answer is in this chapter.

6-1 Combining Everyday Forces

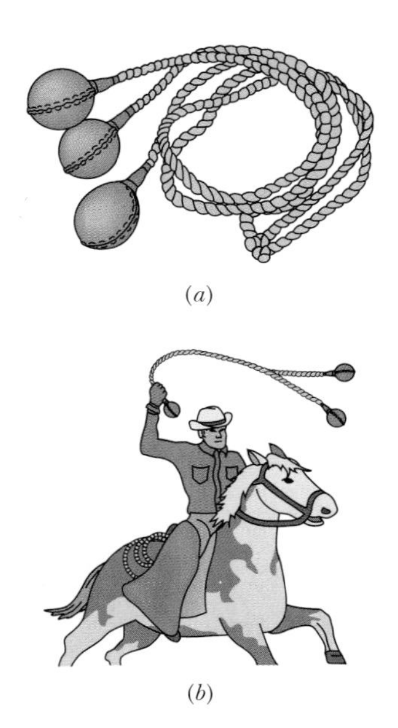

FIGURE 6-1 ■ (*a*) A modern Inuit bola. (*b*) A sketch of a gaucho using a bola.

It is common for objects to experience multiple forces that do not act along the same line. We saw examples of this in Section 5-2 in our brief consideration of the motion of a ball falling under the influence of both a gravitational force and air drag forces. In Section 5-7 we discussed the motion involved in the hammer throw and that of a rock rotating on a string. The hammer and the rock experience both a vertical gravitational force and changing centripetal forces that are almost horizontal.

The bola shown in Fig. 6-1 is another example of a system that experiences multiple forces acting in more than one dimension. The bola is a prehistoric weapon devised for capturing relatively large animals. The analysis of the bola's motion as it is whirled about, released, and encounters an animal is very complex. At any given moment the spherical end of a flying bola experiences a gravitational force, the pull of the rope, and an air drag force.

In this chapter, you will learn more about the characteristics of these everyday forces and how they can be superimposed using vector addition to find net forces. In addition, we will consider how to apply Newton's laws to predict motion and to identify hidden forces. As you will see, the ability to identify forces and use them along with Newton's laws to predict motion is extremely useful for two reasons. First, engineers can use their knowledge of the forces on a system to predict the motion of system components. This ability is vital in the design of a range of devices from bridges to aircraft. Second, the belief physicists have in the validity of Newton's laws of motion leads them to combine acceleration measurements with Newtonian analysis techniques to identify and characterize invisible forces. This approach to the discovery of forces was introduced in Section 3-9.

6-2 Net Force as a Vector Sum

In Chapter 3 we presented experiments that demonstrate that when two or more forces act on an object that moves in a straight line, it is the *net* force that determines how the object's motion will change. For one-dimensional motion the net force turns out to be the vector sum of the forces acting on the object. We call this the **principle of superposition for forces.** If we use the rules of two-dimensional vector addition that we learned about in Chapter 4, can we apply the principle of superposition in cases where the forces do not lie along a single line?

Countless experiments have demonstrated that the principle of superposition also works in two (and three) dimensions. For example, consider the rotating rock discussed in Fig. 5-24. As the rock rotates, it experiences both a gravitational and a string force as shown in Fig. 6-2. We already know that $\vec{F}^{\text{grav}} = -mg\hat{j}$ where m is the mass of the rock. If we attach a spring scale between the rock and the string, we can measure the string force $\vec{F}^{\text{string}}$. If the rock is rotating in a circle in a horizontal plane and we measure its centripetal acceleration, we find that it is related to the forces on the rock by

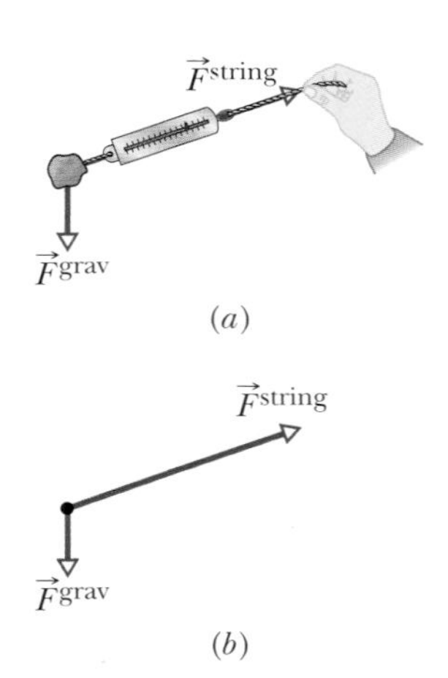

FIGURE 6-2 ■ (*a*) At any particular moment there are two forces on a rock twirled on the end of a string—a gravitational force and a string force. Here the directions of the forces are indicated for the case where the rock rotates in a horizontal plane. (*b*) A free-body diagram showing the tails of the two force vectors at a point that represents the rock on which they act.

$$\vec{F}^{\text{net}} = \vec{F}^{\text{grav}} + \vec{F}^{\text{string}} = m\vec{a}. \tag{6-1}$$

Here the net force that leads to the measured acceleration turns out indeed to be the two-dimensional vector sum (or superposition) of the two forces acting on the rock. We can find the vector sum of two or more force vectors by using the graphical method explained in Section 4-3, or we can resolve the vectors into components using the method presented in Section 4-4.

Another way to verify experimentally that the superposition of force vectors in two dimensions is a vector sum is to set up a situation in which the net force in a plane is zero. For example, we can pull on a ring with three spring scales in such a way that the ring is stationary. In this case, we know the acceleration of the ring, and hence the net force on the ring is zero. Every time we do this, we find that the vector sum of three forces is zero. An example is shown in Fig. 6-3. Here the sum of $F_{Ax}\hat{\text{i}}$ and $F_{Bx}\hat{\text{i}}$ gives us a vector component that has the same magnitude as $F_{Cx}\hat{\text{i}}$ but has the opposite sign so as to cancel it. Thus, the net force is zero as shown in Fig. 6-3*d*. After many such experiments, we become convinced that the net force on an object is the vector sum of the individual forces acting on the object, even if those forces do not act along a single line.

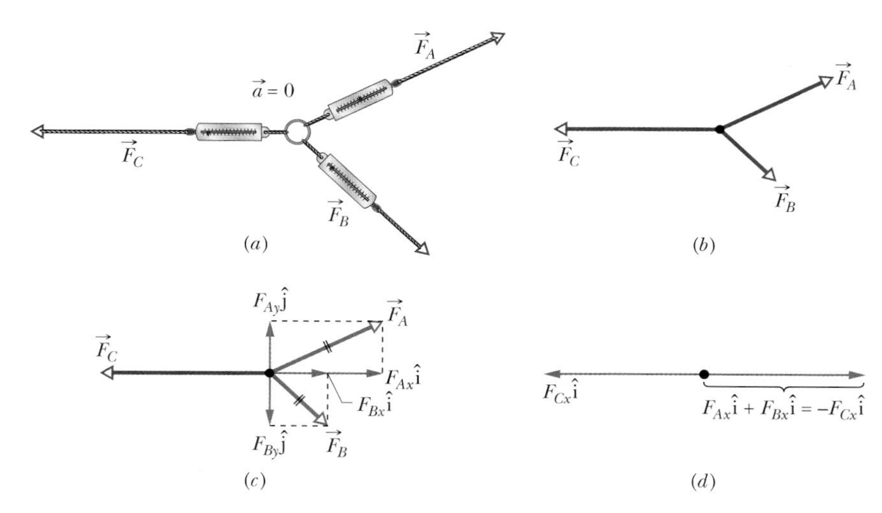

FIGURE 6-3 ■ (*a*) If the ring does not accelerate under the influence of the three forces, we conclude that the net force on it is zero, hence the vector sum of $\vec{F}_A$, $\vec{F}_B$, and $\vec{F}_C$ is zero. (*b*) A free-body diagram showing the tails of the three vectors at a point that represents the center of the ring on which they act. (*c*) Using the component method of resolving the vectors $\vec{F}_A$, $\vec{F}_B$, and $\vec{F}_C$ verifies that their sum is zero. (*d*) The sum of the *x*-components of F_{Ax} and F_{Bx} is $-F_{Cx}$.

Free-Body Diagrams in Two Dimensions

In Chapter 3 we found that it was important to keep track of the magnitudes and directions of the forces acting on an object if we wanted to use Newton's Second Law ($\vec{F}^{\text{net}} = m\vec{a}$) to determine the object's acceleration. The same is true for cases in which the forces do not lie along a single line. We introduced the idea of using a *free-body diagram* for this purpose in Section 3-7. The procedures for drawing free-body diagrams for two- and three-dimensional forces are similar to those used for one-dimensional forces: (1) Identify the object for which the motion is to be analyzed and represent it as a point. (2) Identify all the forces acting on the object and represent each force vector with an arrow. The tail of each force vector should be on the point. Draw the arrow in the direction of the force. Represent the relative magnitudes of the forces through the relative lengths of the arrows. (3) Label each force vector so that it is clear which force it represents.

Figures 6-2*b* and 6-3*b* are free-body diagrams for the situations depicted in the first part of those figures.

Newton's Second Law in Multiple Dimensions

The preceding example hints at another important point regarding multiple forces acting along different lines. Namely, forces (or components of forces) in perpendicular dimensions are independent and separable. That is, Newton's Second Law $\vec{F}^{\text{net}} = m\vec{a}$ can be written as two (or three) component equations:

$$F_x^{\text{net}} = ma_x, \quad F_y^{\text{net}} = ma_y, \quad \text{and} \quad F_z^{\text{net}} = ma_z.$$

We will focus on two-dimensional examples in this chapter.

This statement regarding the separable nature of forces and components of forces should not be especially surprising. Recall from Chapter 5 that horizontal and vertical motions are independent and separable. That is, an acceleration in one dimension only affects the motion in that dimension. Therefore, we could treat two-dimensional motions as two separate one-dimensional cases. Since net force and acceleration are directly related, the independent and separable nature of acceleration is a direct hint that forces behave this way.

If three forces act on an object, then we can expand $F_x^{net} = ma_x$ to get

$$F_{A\,x} + F_{B\,x} + F_{C\,x} = ma_x,$$

where $F_{A\,x}$ is the x-component of force A, $F_{B\,x}$ is the x-component of force B, and so on. The x-component of the acceleration is a_x. These components are signed quantities. This means that although the components are not vectors, they can still be either positive or negative. So, we need to be careful when we begin substituting in actual values for the components that we include the correct sign.

We can use a similar expansion to find ma_y and so on.

A Word about Notation

Recall from earlier chapters that $\vec{F}$ represents a vector. The magnitude (that is, size) of the vector is represented by $|\vec{F}|$ when we want to stress that the value is always positive. More commonly, the magnitude is simply represented as F. That is, a vector quantity represented without the arrow over it is the magnitude of the vector, which is always positive. F_x and F_y represent vector components and may be positive or negative depending on what direction $\vec{F}$ points in relation to the chosen coordinate system.

We have already introduced several different forces including gravitational, tension, and friction forces. These are important, everyday forces. In the rest of this chapter, we will add to our list of common forces and discuss those we have already introduced in more detail.

READING EXERCISE 6-1: A helicopter is moving to the right at a constant horizontal velocity due to the force on it caused by its rotor. It also experiences a downward gravitational force and a horizontal drag force as shown in the diagram below. Which of the following diagrams is a correct free-body diagram representing the forces on the helicopter?

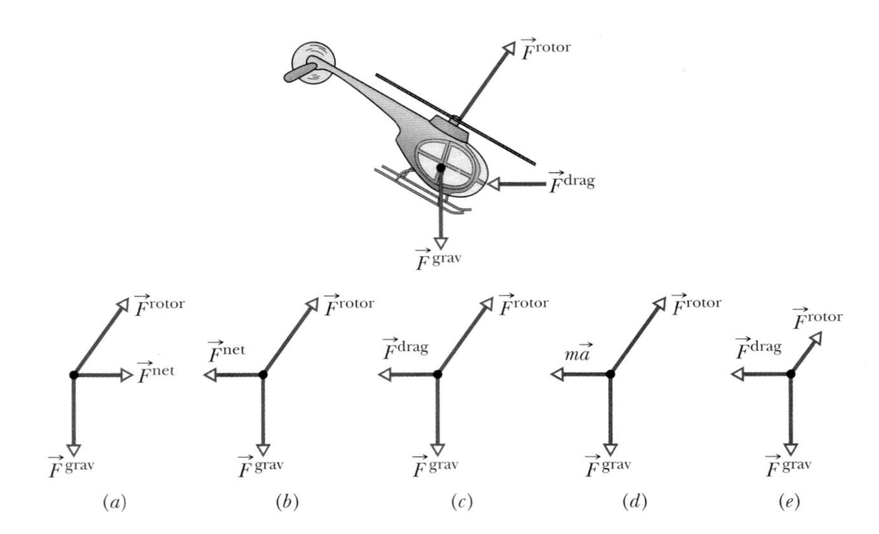

■

TOUCHSTONE EXAMPLE 6-1: Tug-of-War

In a two-dimensional tug-of-war, Alex, Betty, and Charles pull horizontally on an automobile tire at the angles shown in the overhead view of Fig. 6-4*a*. The tire remains stationary in spite of the three pulls. Alex pulls with a force $\vec{F}_A$ of magnitude 220 N, and Charles pulls with force $\vec{F}_C$ of magnitude 170 N. The direction of $\vec{F}_C$ is not given. What is the magnitude of Betty's force $\vec{F}_B$?

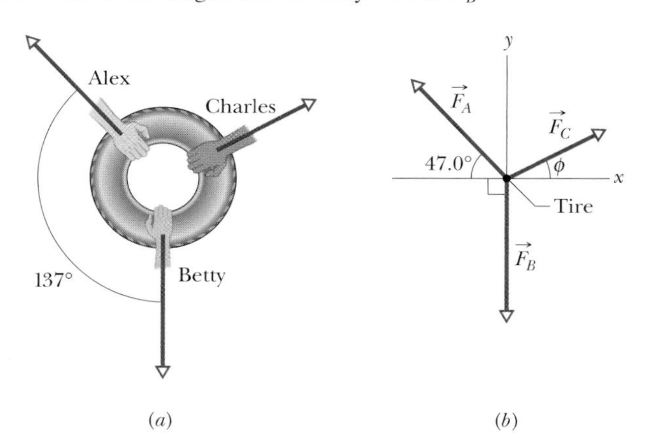

FIGURE 6-4 ■ (*a*) An overhead view of three people pulling on a tire. (*b*) A free-body diagram for the tire.

SOLUTION ■ Because the three forces pulling on the tire do not accelerate the tire, the tire's acceleration is $\vec{a} = 0$ (that is, the forces are in equilibrium). The **Key Idea** here is that we can relate that acceleration to the net force $\vec{F}^{\text{net}}$ on the tire with Newton's Second Law ($\vec{F}^{\text{net}} = m\vec{a}$), which we can write as

$$\vec{F}_A + \vec{F}_B + \vec{F}_C = m(0) = 0,$$

or

$$\vec{F}_B = -\vec{F}_A - \vec{F}_C. \tag{6-2}$$

The free-body diagram for the tire is shown in Fig. 6-4*b*, where we have conveniently centered a coordinate system on the tire and assigned ϕ to the angle between the x axis and $\vec{F}_C$.

We want to solve for the magnitude of $\vec{F}_B$. Although we know both magnitude and direction for $\vec{F}_A$, we know only the magnitude of $\vec{F}_C$ and not its direction. Thus, with unknowns on both sides of Eq. 6-2, we cannot directly solve it on a vector-capable calculator. Instead we must rewrite Eq. 6-2 in terms of components for either the x or the y axis. If the sum of the forces is zero, it must also be that the sum of the x-components of the forces is zero *and* the sum of the y-components is zero. Since $\vec{F}_B$ is directed along the y axis, we choose that axis and write

$$F_{By} = -F_{Ay} - F_{Cy}.$$

Note that we have dropped the arrows over our symbols and added a subscript "y" here. We did this because we are now dealing with components of the vectors as opposed to the vectors themselves. Evaluating these components with their angles and using the angle 133° (= 180° − 47.0°) for $\vec{F}_A$, we obtain

$$F_B \sin(-90°) = -F_A \sin 133° - F_C \sin \phi,$$

where F_A, F_B, and F_C denote vector magnitudes (not components). Using the given data for the magnitudes, yields

$$-F_B = -(220 \text{ N})(\sin 133°) - (170 \text{ N}) \sin \phi. \tag{6-3}$$

However, we do not know ϕ.

We can find ϕ by rewriting Eq. 6-2 for the x axis as

$$F_{Bx} = -F_{Ax} - F_{Cx}$$

and then as

$$F_B \cos(-90°) = -F_A \cos 133° - F_C \cos \phi,$$

which gives us

$$0 = -(220 \text{ N})(\cos 133°) - (170 \text{ N}) \cos \phi$$

and

$$\phi = \cos^{-1} -\frac{(220 \text{ N})(\cos 133°)}{170 \text{ N}} = 28.04°.$$

Inserting this into Eq. 6-3, we find

$$F_B = 241 \text{ N}. \qquad \text{(Answer)}$$

6-3 Gravitational Force and Weight

Gravitational Force

As we discussed in Section 3-9, gravitational forces result from interactions between masses and can act over long distances. Although gravitational interactions between any two masses are always present, they are only noticeable when at least one of the masses is very large. We have already presented experimental evidence in Section 3-9 that the gravitational pull of the Earth on an object is directly proportional to the object's mass. We use the constant of proportionality, denoted g, to relate the gravitational force to mass from Eq. 3-9:

$$\vec{F}^{\text{grav}} = -mg\hat{j}, \tag{6-4}$$

where the constant g, known as the **local gravitational strength,** is a positive scalar and $\hat{j}$ is a unit vector that points up. The minus sign tells us that the gravitational force points down. Close to the Earth's surface, the value of g is 9.8 N/kg.

Weight

Weight is a commonly used synonym for the magnitude of the gravitational force acting on an object.

> The weight W of a body is a scalar quantity that equals the magnitude $|\vec{F}^{\text{grav}}|$ of the local gravitational force exerted by the Earth or some other massive astronomical object (such as the moon) on the body.

$$W = |\vec{F}^{\text{grav}}| = mg \qquad \text{(weight)} \tag{6-5}$$

To *weigh* a body means to measure its weight. As we mentioned in Section 3-9, we can measure gravitational force and hence weight, using a balance, a spring scale, or an electronic scale. Sometimes scales are marked in mass units. Since the value of g changes as we move away from the Earth, scales are only accurate for measuring mass when the value of g is the same as it is where the scale was calibrated.

Weight must be measured when the body is not accelerating vertically relative to the astronomical object attracting it. For example, you can accurately measure your weight on a scale in your bathroom or on a fast train moving horizontally. However, if you repeat the measurement with the scale in an accelerating elevator, the reading on the scale differs from your weight because of the vertical acceleration. This was first discussed in Section 2-4.

Note that the weight W, which has SI units of newtons, and the local gravitational strength g, which has SI units of newtons per kilogram, are not components of vectors, which can be positive or negative. Instead they are both magnitudes and *are always positive*.

Mass Versus Weight

Unfortunately, everyday speech sometimes leads us to believe that the terms "weight" and "mass" are interchangeable. Although the weight of a body (given by $W = mg$) is proportional to its mass, *weight and mass are not the same thing*. Mass has a standard unit of kilograms whereas weight is the magnitude of a force, with a standard unit of newtons. If you move a body to a location such as the surface of the Moon where the value of the local gravitational strength g is different, the body's mass (how much "stuff" the object is made up of) is *not* different, but its weight is. For example, the weight of a bowling ball with a mass of 7.2 kg is 71 N on Earth. On the Moon, this same bowling ball would have the same mass, but a weight of only 12 N. This is because the local gravitational strength is only about one-sixth of its value on Earth.

READING EXERCISE 6-2: Suppose you are given two different objects, a balance like the one shown in Fig. 3-9, and a spring scale like the one shown in Fig. 3-23. Describe how you could determine whether the two objects have the same mass. What might you do to determine the weight of one of the objects? Is the weight of each object the same as the mass of the object? Is the ratio of the masses the same as the ratio of the weights? ■

READING EXERCISE 6-3: Comment on the accuracy of the statement the patient is making in the *Frank & Ernest* cartoon.

Frank and Ernest
M.D.
LOOK, THE ENTIRE PLANET IS YANKING DOWN ON ME WITH AN ACCELERATION OF THIRTY-TWO FEET PER SECOND SQUARED, AND IT'S MY FAULT THAT I WEIGH SO MUCH?!
THAVES

6-4 Contact Forces

As we have mentioned, the gravitational force can act over large distances and exists even if the two interacting objects are not touching. Hence, we sometimes refer to the gravitational force as an "action at a distance" force. In contrast, forces such as tension and frictional forces only exist when there is contact between interacting objects. We call forces of this kind "contact" forces. In order to understand the nature of contact forces between solid objects, it is helpful to learn more about the atomic nature of solids.

An Idealized Model of a Solid

Modern scientists have strong evidence that solids in our everyday world are made of atoms. It is very hard to compress a solid object or pull it apart. The forces between atoms seem to behave like springs. When you push on a spring that is at its natural or equilibrium length, it resists compression by pushing back on you. But when you pull on a spring, it also resists stretching by pulling back on you. This has led physicists to create an idealized model for a solid as an array of atoms held together by forces that behave like very stiff springs, each having an equilibrium length of about 10^{-10} m. A three-dimensional model of a possible array of atoms in a simple solid is shown in Fig. 6-5*a*. This model is explained in more detail in Section 13-5. (As we will see in Chapter 22, the force between atoms in a solid can be understood in terms of the electromagnetic forces between the charged particles in atoms.)

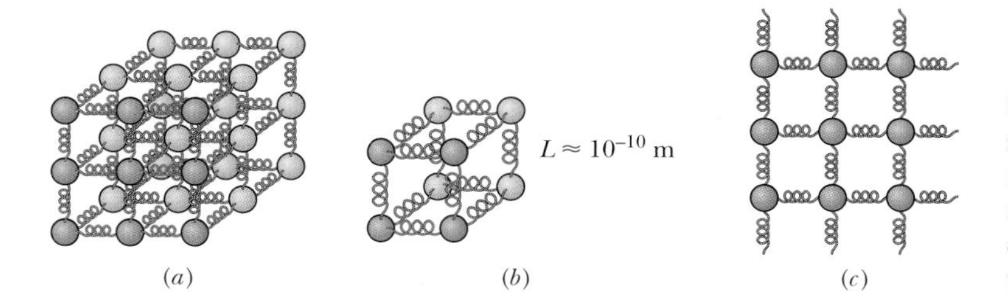

FIGURE 6-5 ■ An idealized model of a solid consisting of atoms separated by tiny springs. (*a*) A model consisting of stiff springs ("atomic bonds") holding balls ("atoms") together. (*b*) Eight atoms at the corner of a cube show the three-dimensional nature of a small hunk of the idealized solid. (*c*) A depiction of a few of the atoms that lie in the plane of the paper.

Using the Model to Understand Contact Forces

How can we use this simplified model to help us understand contact forces? Let's consider what happens when you push on an innerspring mattress (Fig. 6-6). As you push, the springs in the mattress become compressed under your finger and push back

FIGURE 6-6 ■ This physical model of a solid as a matrix of atoms separated by tiny springs behaves rather like an innerspring mattress. Our "solid" is compressed just slightly by the force exerted on it by a finger. According to Newton's Third Law the "solid" then exerts an equal and opposite upward force back on the finger.

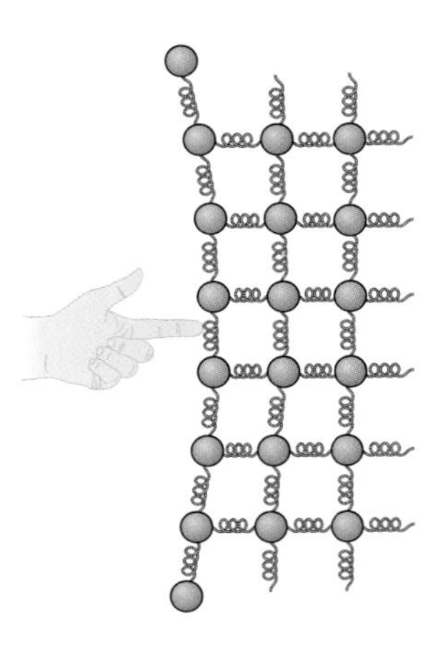

FIGURE 6-7 ■ Compressing an idealized solid wall with a force exerted by a finger. The deformation of the wall is exaggerated. The wall exerts an oppositely directed force with the same magnitude back on the finger.

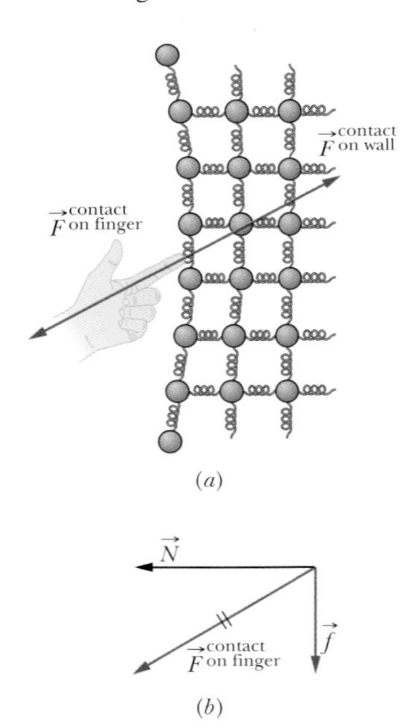

FIGURE 6-8 ■ (*a*) Compressing an idealized solid surface with a contact force that is neither purely perpendicular nor purely parallel to the surface. (*b*) This force exerted by the wall on the finger can be decomposed into parallel and perpendicular components.

on your finger. According to Newton's Third Law, the force you exert on the mattress springs is equal in magnitude and opposite in direction to the force the mattress springs exert on your finger.

Similarly, if you push on a wall it compresses (it is deformed, bent, or buckled ever so slightly), and it pushes back on you (Fig. 6-7). The compression of the wall is hard to see because its billions and billions of tiny atomic springs are much stiffer than the mattress springs. But the harder you push, the more compressed the wall becomes and the larger the force the wall exerts on your finger. When you push harder, it hurts, because in accord with Newton's Third Law, the surface is also pushing back on your finger with a larger force. You can feel (and see) your finger becoming more and more compressed due to the force exerted on it by the wall. Try it.

We call a force exerted perpendicular to a surface a **normal force** and denote it as $\vec{N}$. Note that in this context *normal* is a technical term that derives from a Latin term *norma* meaning "carpenter's square." It is a synonym for perpendicular and does not mean "ordinary."

When one object, such as a wall, exerts a contact force on another object, such as your finger, the force is not necessarily perpendicular to the surfaces in contact. However, you can decompose the force vector into a parallel component and a perpendicular component as shown in Fig. 6-8*b*. We call the component vector perpendicular to the surfaces in contact the **normal force.** We call the component vector parallel to the surfaces the **friction force** and denote it as $\vec{f}$.

In mathematical terms the decomposition of the contact force on your finger, $\vec{F}^{\text{contact}}_{\text{on finger}}$, is given by the sum of the two perpendicular force vectors,

$$\vec{F}^{\text{contact}}_{\text{on finger}} = \vec{N} + \vec{f}. \tag{6-6}$$

The Normal Force

Let's consider a couple of situations in which normal forces are exerted on stationary blocks as shown in Fig. 6-9. The normal force exerted on one object by another object is always directed perpendicular to the surfaces that are in contact and away from the surface of the object exerting the force. We can use our idealized atomic model to explain this. The atoms at the surfaces of the objects that are in close contact interact so as to oppose being pushed closer together. As a result of Newton's Third Law, we can see that:

> When one body exerts a force with a component that is perpendicular to the surface of another body, the other body (even one with a seemingly rigid surface) deforms and pushes back on the first body with an opposing normal force $\vec{N}$ that is also perpendicular to the surfaces that are in contact.

1. A Vertical Wall: A block that is pushed against a wall experiences a normal force from the wall. An example of this is shown in Fig. 6-9*a*. Since the block is not moving,

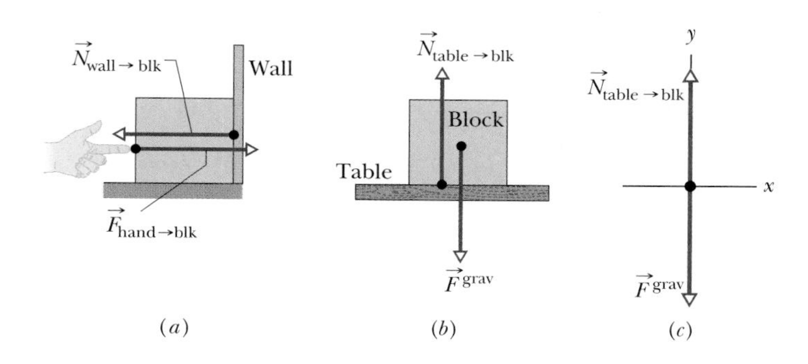

FIGURE 6-9 ■ (*a*) A hand pushes a block into a wall with a force $\vec{F}_{hand\to blk}$. Since the block can't move, it compresses the wall, which pushes back on it with a normal force $\vec{N}_{wall\to blk}$. (*b*) A block resting on a tabletop experiences a normal force $\vec{N}$ perpendicular to the tabletop. (*c*) The corresponding free-body diagram for the block.

the net force on it must be zero. For now we will just consider the horizontal forces on the block given by

$$\vec{F}_{hand\to blk} - \vec{N}_{wall\to blk} = 0 \quad \text{or} \quad \vec{N}_{wall\to blk} = -\vec{F}_{hand\to blk} \quad \text{(special case 1).} \quad (6\text{-}7)$$

2. A Horizontal Table: Likewise, any object that rests on a table, shelf, or the ground near the Earth's surface experiences a normal force. Figure 6-9*b* shows an example. A block of mass m lies on a table's horizontal surface. It is not moving in spite of the fact that it has a gravitational force $\vec{F}^{grav}$ on it due to the Earth. In other words, the block should fall but the table is in the way! We must conclude that if the block does not accelerate, the net force on the block must be zero,

$$\vec{F}^{net} = \vec{F}^{grav} + \vec{N}_{table\to blk} = 0 \quad \text{or} \quad \vec{N}_{table\to blk} = -\vec{F}^{grav}_{blk} \quad \text{(special case 2).}$$

So the table must be pushing up on the block with normal force $\vec{N}_{table\to blk}$ that is equal to $-\vec{F}^{grav}_{blk}$. A free-body diagram for the block is shown in Fig. 6-9*c*. Forces $\vec{F}^{grav}$ and $\vec{N}_{table\to blk}$ are the only two forces on the block, and they are both vertical. We can write Newton's Second Law in terms of components along a positive upward y axis.

The component, F_y^{grav}, of the gravitational force is $-mg$. So, if there is no vertical acceleration and no other vertical forces act on the object, the magnitude of the normal force on an object resting on a horizontal surface is mg. Since its direction is up, if we use the coordinates shown in Fig. 6-9*c*, then

$$\vec{N}_{table\to blk} = +mg\,\hat{j} \quad \text{(special case 2).} \quad (6\text{-}8)$$

Single Normal Force as an Idealization: The normal force exerted by the surface of the table on the block is actually the sum of billions of contact interactions between surface atoms in the table and block. However, the use of a single force vector to summarize external forces that act in the same direction as shown in Fig. 6-9 is a useful simplification. It is conventional to draw a single upward arrow at the point where the middle of the bottom surface of the block touches the table, as shown in Fig. 6-10.

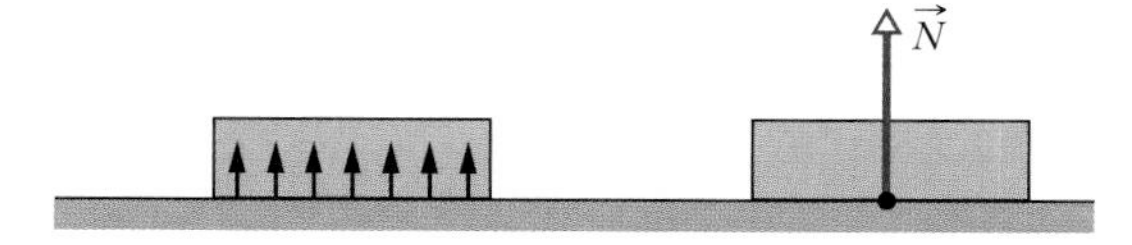

FIGURE 6-10 ■ For simplification, many small force vectors supporting the bottom of the block are replaced by a single large force vector acting through the center of the block.

Normal Force in an Elevator: Suppose a block is placed in an elevator that is accelerating in an upward direction. How would that change the normal force it experiences? In Chapter 2, we discussed how a person riding in such an elevator would feel heavy while accelerating upward and feel light while accelerating downward (see Fig. 2-12). This brings us to the idea of *apparent weight*. A common bathroom scale reading is a measurement of the normal force exerted by the scale on your feet. In normal usage of the scale (in other words, you are standing still on the scale in a space that is not accelerating vertically), the scale will measure your weight. This is because the scale reading (normal force from the scale on your feet) is related to your weight through Newton's $F_y^{net} = ma_y$ relation.

READING EXERCISE 6-4: In Figure 6-9*b*, is the magnitude of the normal force $\vec{N}$ greater than, less than, or equal to mg if the body and table are in an elevator that is moving upward (a) at constant speed, (b) at increasing speed, and (c) at decreasing speed? ■

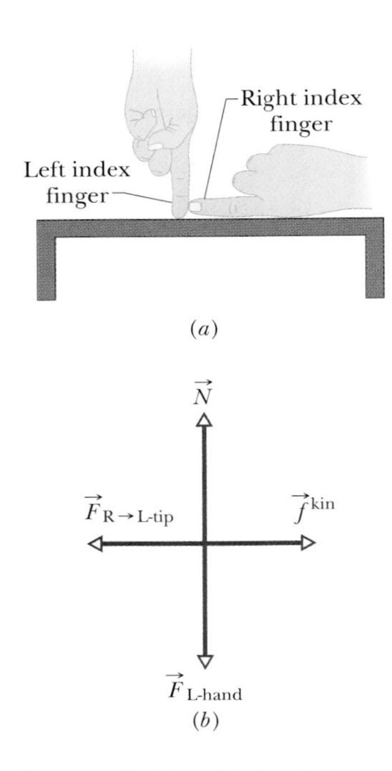

FIGURE 6-11 ■ (*a*) Applying a horizontal force toward the left to an object (such as a fingertip) that is in contact with a surface. (*b*) A free-body diagram showing the forces on the left fingertip. If the object is not accelerating, there must be a friction force on it toward the right that is equal in magnitude and opposite in direction to the force applied on the left fingertip by the right index finger.

The Friction Force Component

Let's consider the friction component of a general contact force. As we discussed earlier, this is the component of the contact force that is parallel to the surface. Suppose the tip of your finger is the object of interest. You would like to study the friction component of the contact force that a fairly smooth table can exert on your fingertip and how it might be related to the normal force. Tilt your left finger so it is vertical (is at an angle of about 90° with the horizontal). Try the following activities while maintaining the 90° angle with respect to the surface of the table:

Activity 1: Press on the table with your left index finger, first with a small force and then a larger force, and feel the increase in the normal force the table exerts on your fingertip.

Activity 2: Now take the index finger of your right hand and apply enough horizontal force to your left index finger so that it glides along at a constant velocity. (See Figure 6-11.) Is there a horizontal friction force acting? If so, why?

NOTE: In order to make your fingertip slide across the table at constant velocity, you must continually push on it in a horizontal direction. Can your applied force be the only horizontal force on your fingertip? No, because if it were, then your fingertip would accelerate. Thus, if we are not willing to give up on Newton's Second Law, we must assume that there is a second force, directed opposite to the applied force but with the same magnitude, so that the two forces balance out. This idea that a second force exists is represented in both Fig. 6-11 and by the following x-component equations:

$$F_x^{net} = F_{r\rightarrow L\text{-tip}\,x} + f_x^{kin} = ma_x = 0\,\text{N} \quad \text{so} \quad F_{R\rightarrow L\text{-tip}\,x} = -f_x^{kin}.$$

Since both forces are purely horizontal, this gives us

$$\vec{F}_{R\rightarrow L\text{-tip}} = -\vec{f}_x^{\,kin}. \tag{6-9}$$

Activity 3: What happens to the friction force when the normal force on your fingertip increases? Once again, adjust your constant applied force so that your left fingertip is moving at a constant velocity. Next, increase the normal force on your left fingertip just enough so your fingertip stops moving. Then get your left fingertip moving at a constant velocity again by applying more horizontal force with your right finger.

If you do Activity 3 carefully, you should conclude that the friction force on an object opposes the direction of its slipping over the surface and that it is greater when the normal force on the object becomes larger.

Contact friction forces are unavoidable in our daily lives. They are literally everywhere. If we were not able to counteract them, they would stop every moving object and bring to a halt every rotating shaft. On the other hand, if friction were totally absent, we could not walk, travel in a car, or ride a bicycle. In some cases, the effects of friction are very small compared to other forces and can be ignored. In other cases, to simplify a situation, friction is assumed to be negligible even though it may not really be. In either case, if the intention is to ignore the effects of friction, the interface between the object and the surface is called *frictionless.*

Contact friction depends on many factors, and it turns out that friction forces can behave very differently depending on the normal force, the nature of the surfaces that are in contact, and other factors. It is not always obvious when looking at surfaces whether the friction forces will be large or small. Sometimes smooth surfaces have greater friction forces than rough ones.

Understanding the relationship between friction forces and atomic and molecular interactions is a very active field of research in both physics and the engineering sciences. These relationships are not completely understood. Unlike Newton's laws of motion, which scientists believe hold to a high degree of accuracy when applied to everyday objects in our surroundings, some of the characteristics of friction that we describe here are only valid for certain common types of interacting surfaces. *Thus, the friction equations that we present here are sometimes useful approximations, but they do not always apply.*

In the next two subsections, we will explore some common characteristics of *kinetic friction*, in which one surface moves relative to another, and of *static friction*, in which the surfaces in contact are stationary relative to one another.

Kinetic Friction Forces

Imagine that you give a book a quick push and send it sliding across a long horizontal countertop. As you expect, the book slows and then stops. We showed data on this behavior in Section 3-2. What does this observation tell us about the nature of the interaction between the book and the countertop? Based on our definition of velocity and on the data shown in Fig. 3-3, we suspect that the book has a *constant* acceleration. This acceleration is parallel to the surface, and in the direction opposite the book's velocity. Once again, we have no reason to believe that Newton's Second Law is not valid in this situation. Hence, from $F_x^{\text{net}} = ma_x$, we must assume that a contact friction force that is constant acts on the book in the same direction as the acceleration (parallel to the counter surface, in the direction opposite the book's velocity relative to the table) as is shown in Fig. 6-12.

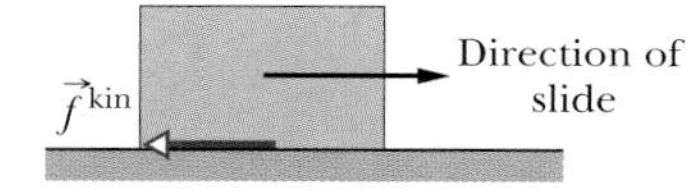

FIGURE 6-12 ■ A friction force $\vec{f}^{\text{kin}}$ opposes the slide of a body over a surface.

In both the example of keeping your fingertip moving at a constant velocity and the example of watching a book with an initial velocity slide to a stop with a constant acceleration, an object is experiencing a **kinetic friction force** $\vec{f}^{\text{kin}}$. The word "kinetic" indicates that the object is moving relative to a surface. The phenomenon of "contact friction" can be explained by assuming that there is an attractive force between the atoms at the surfaces of the two objects. The attraction between two very smooth surfaces such as glass panes is consistent with this assumption and is known as **adhesion.**

What might the kinetic friction force depend on? Imagine sending an object sliding across a countertop as we discussed above. Would the book slow down more or less quickly if we slide the book across a carpeted floor instead of the smooth countertop? Would it slow down more or less quickly if we slide it across ice instead? Does the rate at which an object slows down seem to depend on its velocity? Would the book slow down more or less quickly if it has more mass or an additional applied downward force on it so the normal force between the surfaces is larger?

We can answer some of these questions for several situations by looking at the graph presented in Fig. 3-3. This graph shows the velocity as a function of time for three situations where objects slide to a stop on surfaces. You will likely find it helpful to refer back to that figure now (page 59). We can also draw inferences from the fingertip motions earlier in this section. Here are some observations and conclusions about kinetic friction.

KINETIC FRICTION—SOME OBSERVATIONS AND CONCLUSIONS

OBSERVATION 1 ON THE INFLUENCE OF THE RELATIVE VELOCITY BETWEEN SURFACES: The graphs in Fig. 3-3 tell us that in three situations involving different combinations of objects and surfaces, the objects all slow to a stop with constant acceleration and hence experience a constant kinetic friction force. *Conclusion: Kinetic friction forces appear to be independent of the magnitude of the velocity of the object relative to the surface over which the object is sliding, but act in a direction opposite to the direction of the velocity.*

OBSERVATION 2 ON THE NATURE OF THE SLIDING SURFACES: The graphs in Fig. 3-3 tell us that in three situations involving different combinations of object and surfaces, the rate of the stopping acceleration is different. *Conclusion: Kinetic friction forces appear to depend on the nature of the surfaces that are in contact with one another.*

OBSERVATION 3 ON THE INFLUENCE OF THE NORMAL FORCE: When you completed Activity 3 earlier in this section, you observed that the applied force needed to keep your fingertip moving at a constant velocity increases when the normal force on your fingertip becomes larger. *Conclusion: Kinetic friction forces appear to increase when the normal force on a sliding object increases and thus depend on how hard the objects are being pushed together.*

Is there a mathematical relationship between the magnitude of the kinetic friction force on an object and the magnitude of the normal force the object experiences? A plausible relationship would be that these two force magnitudes are proportional to each other. Let's look at the results of a simple experiment in which we can measure the kinetic friction force as a function of the normal force on a sliding block. In this experiment, we use a spring scale to measure how much horizontal force we need to apply to pull a wooden block along at a constant velocity. (See Fig. 6-13.) We can determine the magnitude of the friction force by using the fact that it must be equal to the magnitude of the applied force if the moving block doesn't accelerate (Eq. 6-9). If the table surface is horizontal, and the only other vertical force on the block is the gravitational force, then the normal force is given by $\vec{N} = mg\hat{j}$ (Eq. 6-8). That is, $N_y = mg$. The normal force can be changed by piling more mass on the block. We can then measure the kinetic friction force again.

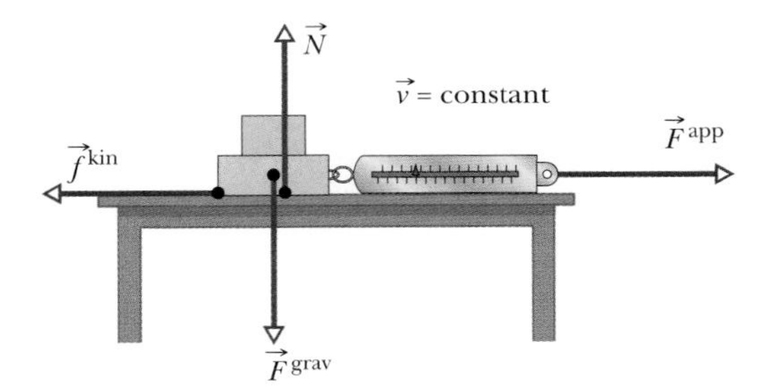

FIGURE 6-13 ■ A block is pulled along at a constant velocity with a horizontal applied force, measured by a spring scale. This force is countered by a kinetic friction force of the same magnitude. Since the tabletop is horizontal, the magnitude of the normal force on the block is equal to the product of its mass m and the gravitational acceleration constant g.

The data shown in Fig. 6-14 reveal that for a Velcro-covered wood block sliding on a Formica table surface, the magnitude of the friction force is proportional to that of the normal force with a constant of proportionality given by $\mu^{kin} = 0.21$. Turning the block on its side to reduce the area in contact does not affect this constant of proportionality.

Results similar to those shown in Fig. 6-14 for many situations reveal that the magnitude of the friction force for dry sliding is usually proportional to the magnitude of the normal forces pressing surfaces together and does not depend on other factors. Thus, for the purposes of the systems we will deal with in this book, the magnitude of the kinetic friction force, $\vec{f}^{kin}$, can be expressed as

$$f^{kin} = \mu^{kin} N, \tag{6-10}$$

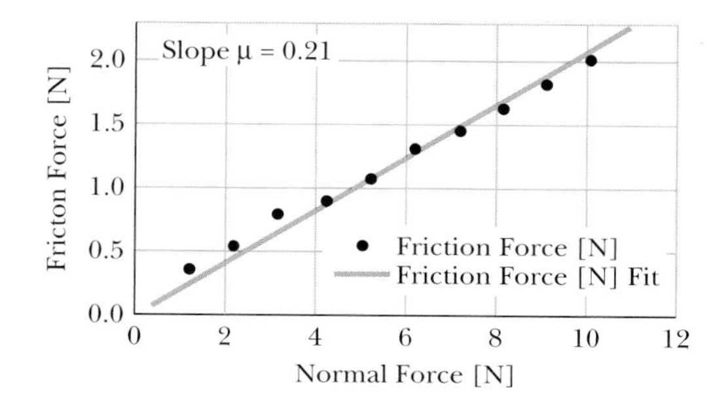

FIGURE 6-14 ■ A graph of data showing that when a block is pulled along at a constant velocity, the magnitude of the kinetic friction force is directly proportional to the normal force exerted by the surface it slides over.

where μ^{kin} is the slope of the linear graph that relates the magnitude of the kinetic friction force f^{kin} and the magnitude of the normal force N. The slope μ^{kin} is called the **coefficient of kinetic friction.**

The coefficient μ^{kin} is a dimensionless scalar that must be determined experimentally. Its value depends on certain properties of both the body and the surface. Hence, the coefficients are usually referred to with the preposition "between," as in "the value of μ^{kin} *between* a book and countertop is 0.04, but the value *between* rock-climbing shoes and rock is as much as 0.9." Based on our observations, we assume that the value of μ^{kin} does not depend on the speed at which the body slides along the surface. Note that $f^{\text{kin}} = \mu^{\text{kin}}N$ is *not* a vector equation. The direction of $\vec{f}^{\text{kin}}$ is always parallel to the surface and *opposes* the sliding motion.

Static Friction Forces

Do friction forces continue to act on an object once it stops sliding? The answer to this question is more complicated than simply "yes" or "no." Start out by imagining a large, heavy box sitting on a horizontal, carpeted floor. You push on the box, but the box does not move. Unless we are to believe that Newton's Second Law ($\vec{F}^{\text{net}} = m\vec{a}$) is not valid in this situation, we must assume that there is some other force acting on the box that is counteracting the application of the push force. That is, there must be a force acting in the opposite direction that is exactly equal in magnitude to the push force. We will call this opposing force a **static friction force.** The word "static" is used to signify that the object is not moving relative to the surface as shown in Fig. 6-15*b-d*.

Now imagine that you push even harder on the box as shown in Fig. 6-15*c* and *d*. The box still does not move. Apparently the friction force can change in magnitude, otherwise it would no longer balance your applied force. In other words, if you push on an object in an attempt to slide it across a surface and the object does not slide, then we know that there is a static friction force. This force acts in the direction opposite the push with a matching magnitude, regardless of how hard you push. If you stop pushing on the box, that oppositely directed force must disappear. How do we know this? Because if you removed the push force, and the static friction force *did not* disappear as well, then the box would accelerate in the direction of the friction force. We know from everyday observation that this does not happen. So the static friction force appears to be a very strange force that changes magnitude in response to other forces.

This situation is in no way specific to the example of the box on carpet. At the interface between any two solids prior to slipping, the static friction force starts at zero when no applied force is present and increases as the force that tends to produce slipping increases. The static friction force adjusts in magnitude to exactly counteract the applied force (usually a push or pull) at every instant. The static friction force mirrors the applied force. If the applied force is zero, then the static friction force is zero. If the applied force has a horizontal component that is 10 N, the static friction force has a horizontal component that is 10 N. We call forces that behave like the static friction force **passive forces.** Passive forces are forces that change in magnitude in response to other forces.

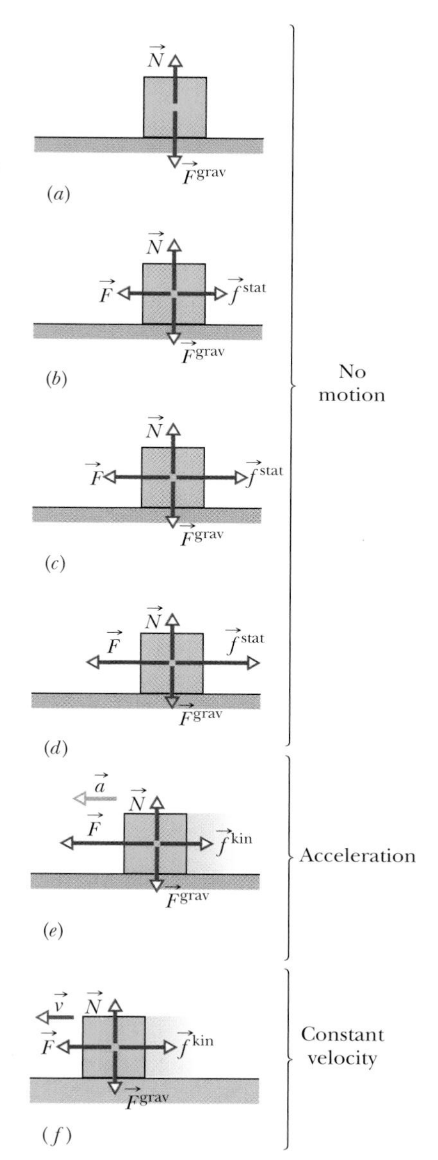

FIGURE 6-15 ■ (*a*) There are no horizontal forces on a stationary block. (*b–d*) An external force $\vec{F}$ applied to the block is balanced by a static friction force $\vec{f}^{\text{stat}}$. As $\vec{F}$ is increased, $\vec{f}^{\text{stat}}$ also increases, until $\vec{f}^{\text{stat}}$ reaches a certain maximum value. (*e*) The block then "breaks away," accelerating suddenly in the direction of $\vec{F}$. (*f*) If the block is now to move with constant velocity, the magnitude $|\vec{F}|$ of the applied force must be reduced from the maximum value it had just before the block broke away.

Now imagine that you push on the box with all your strength. Finally, the box begins to slide. Evidently, there is a maximum magnitude of the static friction force. When you exceed that maximum magnitude, your push force is larger than the opposing static friction force and the box accelerates in the direction of your push. A typical sequence of static frictional force responses to applied forces is shown in Fig. 6-15. This sequence is consistent with the experimental results obtained when an electronic force sensor is used to monitor the force on a block as a function of time. The experimental setup is shown in Fig. 6-16 and a graph of the results is shown in Fig. 6-17.

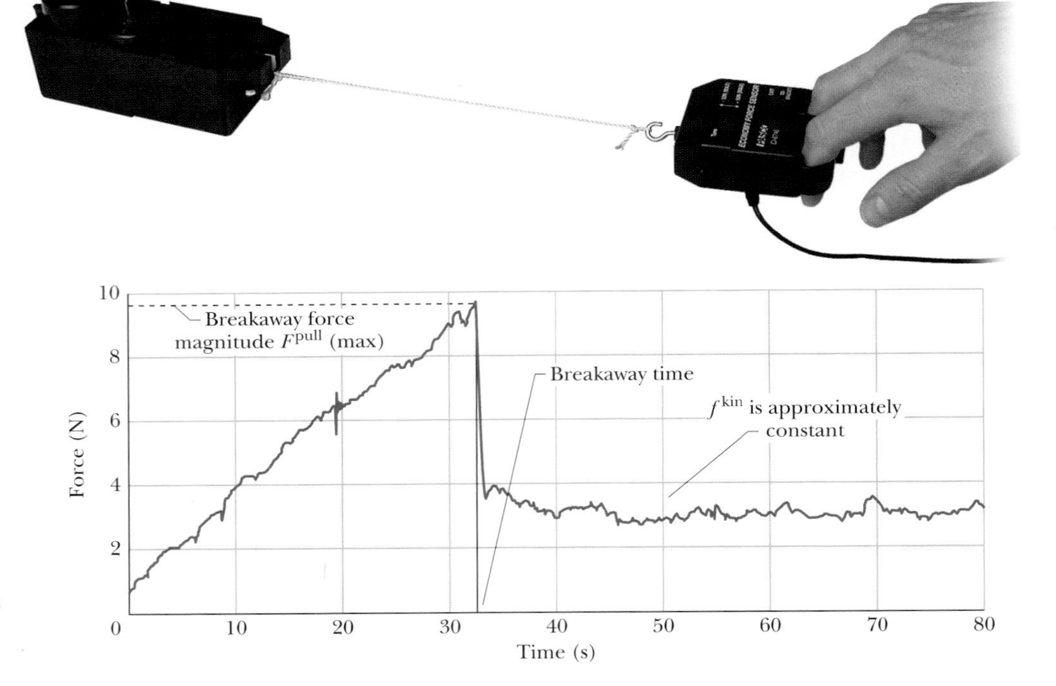

FIGURE 6-16 ■ The apparatus used for the static friction experiment includes an electronic force sensor attached to a computer data acquisition system (not pictured).

FIGURE 6-17 ■ Graph of the magnitude of the static friction force on a wooden block as a function of time. This force opposes a steadily increasing applied force between 0.0 s and 32 s. At 32 s the block suddenly "breaks away" and starts moving. At about 40 s, it starts moving at a steady velocity as a kinetic friction force with a magnitude that is less than the static force starts acting.

As the pulling force, $\vec{F}^{pull}$, increases, the block remains at rest. Then, when a "breakaway" force is reached, it moves very suddenly. That is, the magnitude of the friction force, $\vec{f}^{stat}$, keeps increasing to oppose the pulling force in accordance with Newton's Second Law until the object "breaks free" and starts to move. Hence, we express the magnitude of the static friction force as

$$f^{stat} = |\vec{f}^{stat}| \leq \mu^{stat} N, \tag{6-11}$$

where μ^{stat} is known as the **coefficient of static friction** and N is the magnitude of the normal force on the body from the surface. Just as for kinetic friction, the coefficient μ^{stat} is dimensionless and determined experimentally. Its value depends on certain properties of both the body and the surface, and so is referred to with the preposition "between."

Usually, the magnitude of the kinetic friction force, which acts when there is motion, is less than the maximum magnitude of the static friction force, which acts when there is no motion. We see this in the data shown in Fig. 6-17. Thus, if you wish the block to move across the surface with a constant speed, you must usually decrease the magnitude of the applied force once the block begins to move, as in Fig. 6-15*f*. Another common behavior for a certain range of applied forces is to see slip-and-stick behavior in which an object breaks away, slides to a stop, breaks away again, and so on. We will not deal with the slip-stick phenomenon in this book.

READING EXERCISE 6-5: Figure 6-17 shows the result of the experiment in which a 295.6 g block with a 500 g mass on it is pulled along a table with a steadily increasing force until it breaks away at $t = 32$ s. (a) What is the coefficient of static friction, μ^{stat}, between the table and the mass? (b) What is the coefficient of kinetic friction μ^{kin}? ■

READING EXERCISE 6-6: A block lies on a floor. (a) What is the magnitude of the friction force exerted on it by the floor if the block is not being pushed? (b) If a horizontal force of 5 N is now applied to the block, but the block does not move, what is the magnitude of the friction force on it? (c) If the maximum value $\vec{f}^{max}$ of the static friction force on the block is 10 N, will the block move if the magnitude of the horizontally applied force is 8 N? (d) If the magnitude is 12 N? (e) What is the magnitude of the friction force in part (c)? ■

READING EXERCISE 6-7: Discuss and explain the following statement using the terms related to friction forces that are presented above: "If we were not able to counteract them, frictional forces would stop every moving object and bring to a halt every rotating shaft. On the other hand, if friction were totally absent, we could not walk or ride a bicycle." ■

Tension

So far we considered contact forces between objects that are not attached and that can be pulled apart fairly easily. Let's consider one more type of contact force—a force that occurs when a long thin object such as a rod or string is attached to other objects at each of its ends. For example, consider a leash with a dog straining at one end and the dog's owner pulling the other end, a handle bolted to a pot that is too massive to move and is being pulled by a cook, or a string with one end attached to a ceiling and the other end attached to a hanging mass. In all three cases, a long narrow object that is stretched is transmitting forces from an object at one of its ends to an object at its other end. We say that a long narrow object that is being pulled taut by opposing forces is under **tension.** In order to use Newton's laws of motion to analyze the forces and motions of the objects that are attached to the ends of strings or rods, we need to understand more about the phenomenon of tension.

What do we observe about tension? Let's consider a stationary rubber band that connects two force probes like that shown in Fig. 6-18. We observe that the forces the rubber band exerts on the force probes at each end have the same magnitude but act in opposite directions.

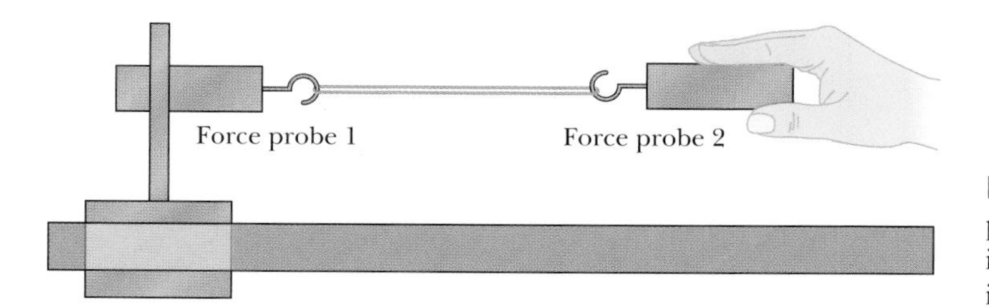

FIGURE 6-18 ■ A rubber band is connecting two force probes. Each probe detects the same magnitude of force, but the force on probe 1 is in the opposite direction of the force on probe 2. This observation is not surprising as it is entirely consistent with Newton's laws.

We also observe that the tension force is present everywhere along the rubber band. Although it is not readily observable, the tension everywhere along the rubber band is in fact equal in magnitude to the applied forces at the ends that caused the rubber band or string to stretch. Thus, when a taut rubber band (that is not accelerating) is attached to an object, it exerts a tension force on the object that is directed along the rubber band and away from the object. This tension expresses itself as a pulling force, but only at the ends of the rubber band. These same observations hold for most long thin connectors including strings, cords, and ropes.

An Atomic Model for Tension Our simple model of solid matter, as consisting of atoms connected by springs, is very helpful in understanding how objects that are under tension can transmit forces. Suppose a very, very thin string, having only one strand of atoms, is connected by small interatomic springs. Figure 6-19*a* shows the natural length of the string. Figure 6-19*b* shows the string when it is extended by equal and opposite forces applied to its ends so that it is not accelerating.

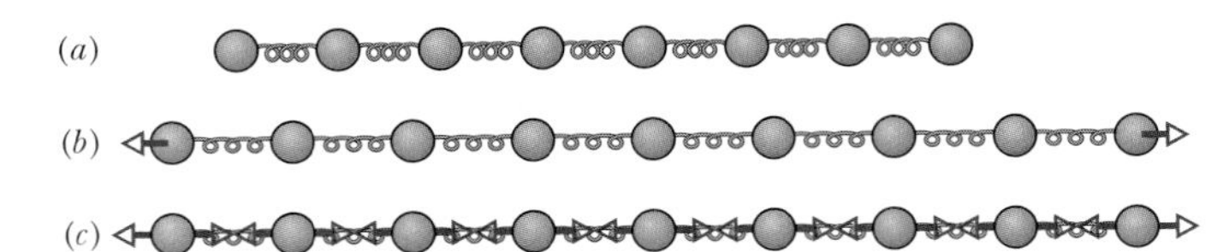

FIGURE 6-19 ■ A string is idealized as a line of atoms with springs representing the mutual interaction forces between them.

In our idealization, we have assumed only one strand of atoms. Obviously, real strings, cords, and ropes have many strands of molecules consisting of complex arrays of atoms. Although many strands will make a string or rope stronger, it will not change the ideas presented in our simple model.

Assuming that our ideal string is not accelerating, each atom must have zero net force on it. The atom on the left end of the string must be experiencing an attractive force from the neighboring atom to its right that is equal in magnitude and opposite in direction to the applied force on the end. However, each stretched spring represents a force of interaction between neighboring atoms that must obey Newton's Third Law. Thus, the leftmost atom must be exerting an attractive force on its neighboring atom that is "equal and opposite" to the force that atom exerts on it. These pairs of mutual interaction forces exist throughout the string, as shown in Fig. 6-19*c*. The magnitude of these interaction forces each atom experiences has been given a special name. It is called the *tension in the string*, which we denote as T. In contrast to the tension force, tension, which we often denote with a T, is a scalar quantity that is always positive with no inherent direction associated with it. Hence, we will often denote a tension force $\vec{T}$ that points (for example) in the positive y direction as $\vec{T} = +T\hat{j}$ and one that points in a negative direction as $\vec{T} = -T\hat{j}$.

Next, let's use Newton's laws to examine the effect of tension associated with the motion of a skier being towed by a snowmobile by means of a nylon cord as depicted in Fig. 6-20. We consider two situations—one in which the system is not accelerating and the other in which it is. The snowmobile moves forward when its treads, which are turning, dig into the snow and push against it. However, assume for now that the runners on the skis and those at the front of the snowmobile experience no friction forces.

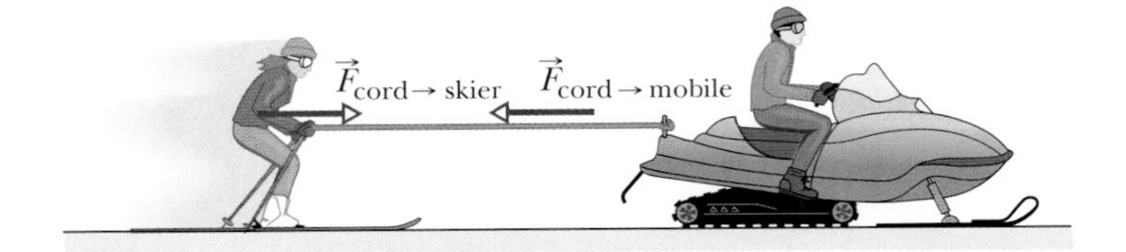

FIGURE 6-20 ■ The cord connecting a skier and a snowmobile exerts oppositely directed forces on the skier and the snowmobile.

Tension for a Nonaccelerating System Remember that if the system is not accelerating, then it moves at a constant velocity. Furthermore, Newton's Third Law tells us that the force between any two objects in the system that are in contact is equal and opposite. For example, at the left end of the cord, the skier feels a pulling force acting along the direction of the cord, which we denote as $\vec{F}_{\text{cord}\rightarrow\text{skier}}$, and the cord experiences an oppositely directed force from the skier, denoted by $\vec{F}_{\text{skier}\rightarrow\text{cord}}$. A similar situation applies to the interaction forces at the right end of the cord, so that

$$\vec{F}_{\text{cord}\rightarrow\text{skier}} = -\vec{F}_{\text{skier}\rightarrow\text{cord}} \quad \text{and} \quad \vec{F}_{\text{cord}\rightarrow\text{mobile}} = -\vec{F}_{\text{mobile}\rightarrow\text{cord}}. \tag{6-12}$$

But we already know from Newton's Second Law for $\vec{a} = 0$ that the net force on the cord must be zero. Since the net force is zero, F_x^{net} (the sum of the x-components of the forces) must be zero. Hence,

$$F_{\text{cord } x}^{\text{net}} = F_{\text{skier}\to\text{cord } x} + F_{\text{mobile}\to\text{cord } x} = 0 \quad \text{or} \quad F_{\text{skier}\to\text{cord } x} = -F_{\text{mobile}\to\text{cord } x}.$$

The forces the skier and snowmobile exert on the cord are purely horizontal. So we have

$$\vec{F}_{\text{skier}\to\text{cord}} = -\vec{F}_{\text{mobile}\to\text{cord}}. \tag{6-13}$$

This result agrees with the observation reported in Fig. 6-18. Namely, forces exerted by the ends of a taut cord have the same magnitude. Even more significantly, we can combine Eqs. 6-12 and 6-13 to show that

$$\vec{F}_{\text{skier}\to\text{mobile}} = -\vec{F}_{\text{mobile}\to\text{skier}}.$$

When nonaccelerating objects are connected by a string, cord, or rope, they interact in accordance with Newton's Third Law *through the connector* as if they are in direct contact.

An Accelerating System Suppose the snowmobile driver pushes in his throttle and increases his velocity at a constant rate. Now the system has an acceleration $\vec{a}$, and the cord connecting the skier to the snowmobile must experience the same acceleration. In terms of the x-components we get

$$F_{\text{cord } x}^{\text{net}} = F_{\text{skier}\to\text{cord } x} + F_{\text{mobile}\to\text{cord } x} = m_{\text{cord}}\, a_x. \tag{6-14}$$

This tells us that if the cord has a nonzero mass, then the force of the snowmobile on the right end of the cord must be greater than the force on the left end to maintain an acceleration. However, in many situations, including this one showing the snowmobile pulling a skier, the mass of the cord is so much less than the mass of the entire system that it can be taken to be zero.

Taking the direction of motion to be along the positive x axis, we can write the tension forces in terms of the positive scalar T representing the tension,

$$\vec{F}_{\text{skier}\to\text{cord}} = -T_L\hat{\text{i}} \quad \text{and} \quad \vec{F}_{\text{mobile}\to\text{cord}} = +T_R\hat{\text{i}}, \tag{6-15}$$

where T_L is the tension on the left side of the cord and T_R is the tension on the right side of the cord. Then we can rewrite Eq. 6-14 in terms of the tension difference and the x-component of acceleration to get $T_R - T_L = m_{\text{cord}}\, a_x$. But the snowmobile force, which serves to accelerate the entire system, is given by $F_{\text{mobile}\to\text{sys } x} = m^{\text{tot}} a_x$. Solving the last two equations for a_x and rearranging terms gives us the ratio

$$\frac{T_L - T_R}{F_{\text{mobile}\to\text{sys } x}} = \frac{m_{\text{cord}}}{m^{\text{tot}}}. \tag{6-16}$$

Let's consider the implications of this equation. In most situations, the mass of the cord is much less than the mass of the system. Whenever that is true, the difference in tension at the ends of the cord is much smaller than the force that accelerates the system. For example, assume the skier's mass and the snowmobile's mass together total 200 kg, and the mass of the cord she grips is 1 kg. The ratio of these masses gives us only a 0.5% difference in tension forces at the ends of the cord. For most everyday purposes, this force difference at the ends of an accelerating cord is negligible. In laboratory experiments, masses of between 100 g and 2 kg are typically connected by

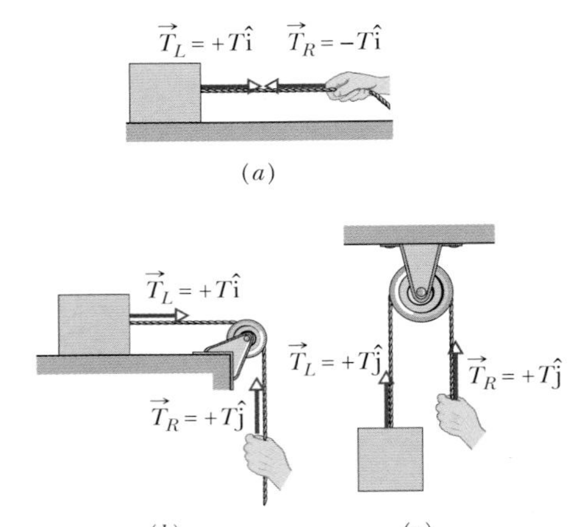

FIGURE 6-21 ■ (*a*) The cord, pulled taut, is under tension. If its mass is negligible, it pulls on the body and the hand with force of magnitude $T = |\vec{T}|$, even if it runs around a massless, frictionless pulley as in (*b*) and (*c*).

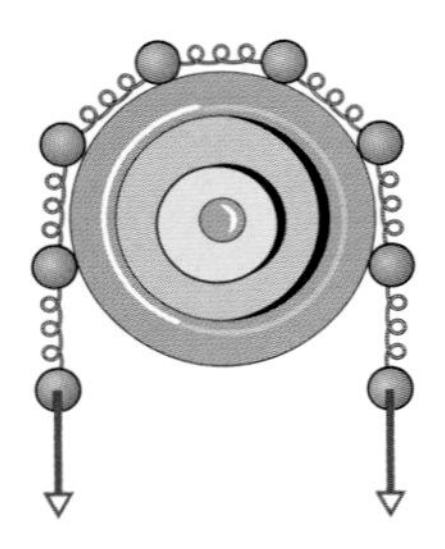

FIGURE 6-22 ■ Tension in a taut string still exists even when it undergoes direction changes.

fishing line capable of sustaining tensions of well over 100 N. A 1 m length of this type of fishing line has a mass of about 0.25 g, so the force differences are usually less than 1%. For cases where the mass of a connecting cord is very small compared to the masses of the objects attached to its ends, we can assume that the tension is *essentially* the same at all points along the cord. When we can legitimately make this simplifying approximation, we say that we have a **massless string.**

Pulleys and Direction Change What happens if a "massless" cord stretched over a "massless" pulley changes direction as shown in Fig. 6-21*b* and *c*? Is the tension still the same everywhere in the cord? Let's examine our atomic model. If a string is wrapped around a pulley, its direction is different at one end than at the other. However, each tiny segment of the string only changes direction ever so slightly. The direction change is less than it is in Fig. 6-22 where we have only placed eight atoms in the chain. We conclude that the magnitude of the tension forces that are spread throughout the string also do not change significantly when the string bends around other objects. This conclusion is supported by experiments in which spring scales are inserted in various places along a string that bends while it is under tension.

Any solid object that is attached at two ends and pulled can transmit tension forces from one end to another. Some objects are quite elastic, such as rubber bands or weak springs, others are more rigid, such as strings and rods. Small rubber bands, light-duty springs, and strings cannot stand compressive forces. They are so long and narrow that they buckle under compression. Alternatively, rods and heavy springs do not buckle under compression. In some of the analyses that follow, you will be dealing with "massless" strings and springs that buckle under compression forces.

READING EXERCISE 6-8: Consider Figure 6-21*c* and assume that the pulley is massless but the cord is *not*. Is the magnitude of the pull force on the cord exerted by the hand equal to (=), less than (<), or greater than (>) the magnitude of the pull force exerted by the block when the block is moving upward (a) at constant speed, (b) at increasing speed, and (c) at decreasing speed? Explain. ■

TOUCHSTONE EXAMPLE 6-2: Einstein's Elevator

In Fig. 6-23*a*, a passenger of mass $m = 72.2$ kg stands on a platform scale in an elevator cab. We are concerned with the scale readings when the cab is stationary and when it is moving up or down.

(a) Find a general solution for the scale reading, whatever the vertical motion of the cab happens to be.

SOLUTION ■ One **Key Idea** here is that the scale reading is equal to the magnitude of the normal force $\vec{N}$ the scale exerts on the passenger. The only other force acting on the passenger is the gravitational force $\vec{F}^{\,\text{grav}}$, as shown in the free-body diagram of the passenger in Fig. 6-23*b*.

A second **Key Idea** is that we can relate the forces on the passenger to the acceleration $\vec{a}$ of the passenger with Newton's Second Law ($\vec{F}^{\,\text{net}} = m\vec{a}$). However, recall that we can use this law only in an inertial frame. If the cab accelerates, then it is *not* an inertial frame. So we choose the ground to be our inertial frame and make any measure of the passenger's acceleration relative to it.

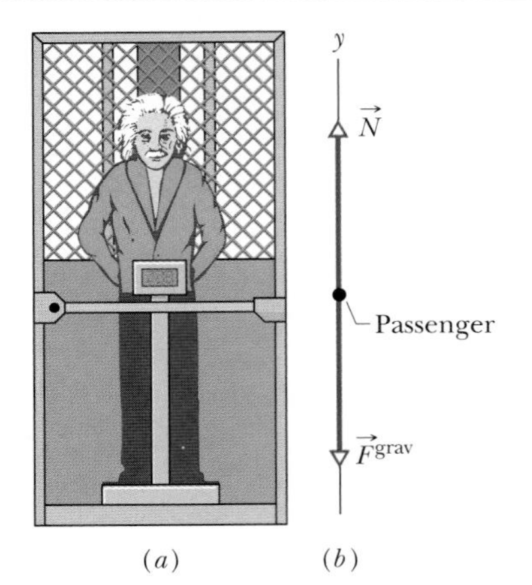

FIGURE 6-23 ■ (*a*) A passenger stands on a platform scale that indicates his weight or apparent weight. (*b*) The free-body diagram for the passenger, showing the normal force $\vec{N}$ on him from the scale and the gravitational force $\vec{F}^{\text{grav}}$.

Because the two forces on the passenger and the passenger's acceleration are all directed vertically, along the *y* axis shown in Fig.

6-23*b*, we can use Newton's Second Law written for *y*-components ($F_y^{net} = ma_y$) to get

$$N_y + F_y^{grav} = ma_y$$

or

$$N_y = -F_y^{grav} + ma_y. \tag{6-17}$$

This tells us that the scale reading, which is equal to N_y (provided $N_y \geq 0$), depends on the vertical acceleration a_y of the cab. Since $\vec{F}^{grav} = -mg\,\hat{j}$, the *y*-component of the gravitational force, $F_y^{grav} = -mg$. This gives us

$$N_y = m(g + a_y). \quad \text{(Answer)} \tag{6-18}$$

This tells us that the scale reading is larger than the passenger's static weight, mg, when the elevator accelerates upward, since then $a_y > 0$. But if the elevator is accelerating *downward*, then a_y is negative and the scale reads less than the passenger's static weight. This is true, as long as the downward acceleration is smaller than g. If the downward acceleration is greater than g, $(g + a_y)$ in Eq. 6-18 is a negative value. In that case, $N_y = 0$, since N_y can never be negative. (Why not?)

(b) What does the scale read if the cab is stationary or moving upward at a constant 0.50 m/s?

SOLUTION ■ The **Key Idea** here is that for any constant velocity (zero or otherwise), the acceleration a_y of the passenger is zero. Substituting this and other known values into Eq. 6-18, we find

$$N_y = (72.2\text{ kg})(9.8\text{ m/s}^2 + 0) = 708\text{ N}. \quad \text{(Answer)}$$

This is just the weight of the passenger and is equal to the magnitude F^{grav} of the gravitational force on him.

(c) What does the scale read if the cab accelerates upward at 3.20 m/s² and downward at 3.20 m/s²?

SOLUTION ■ For $a_y = +3.20\text{ m/s}^2$, Eq. 6-18 gives

$$N_y = (72.2\text{ kg})(9.8\text{ m/s}^2 + 3.20\text{ m/s}^2)$$
$$= 939\text{ N}, \quad \text{(Answer)}$$

and for $a_y = -3.20\text{ m/s}^2$, it gives

$$N_y = (72.2\text{ kg})(9.8\text{ m/s}^2 - 3.20\text{ m/s}^2)$$
$$= 477\text{ N}. \quad \text{(Answer)}$$

So for an upward acceleration (either the cab's upward speed is increasing or its downward speed is decreasing), the scale reading is greater than the passenger's weight. Similarly, for a downward acceleration (either the cab's upward speed is decreasing or its downward speed is increasing), the scale reading is less than the passenger's weight.

(d) During the upward acceleration in part (c), what is the magnitude F^{net} of the net force on the passenger, and what is the magnitude $a_{p,\,cab}$ of the passenger's acceleration as measured in the frame of the cab? Does $\vec{F}^{net} = m\vec{a}_{p,\,cab}$?

SOLUTION ■ One **Key Idea** here is that the magnitude F^{grav} of the gravitational force on the passenger does not depend on the motion of the passenger or the cab, so from part (b), F^{grav} is 708 N. From part (c), the magnitude N of the normal force on the passenger during the upward acceleration is the 939 N reading on the scale. Thus, the net force on the passenger is

$$F_y^{net} = N_y + F_y^{grav} = N - F^{grav} = 939\text{ N} - 708\text{ N} = 231\text{ N}, \quad \text{(Answer)}$$

during the upward acceleration. However, the acceleration $\vec{a}_{p,\,cab}$ of the passenger relative to the frame of the cab is zero. Thus, in the noninertial frame of the accelerating cab, $\vec{F}^{net}$ is not equal to $m\vec{a}_{p,\,cab}$. This is an example of the fact that Newton's Second Law does not hold in noninertial (that is, accelerating) frames of reference.

TOUCHSTONE EXAMPLE 6-3: Pulling a Block

In Fig. 6-24*a*, a hand H pulls on a taut horizontal rope R (of mass $m = 0.200$ kg) that is attached to a block B (of mass $M = 5.00$ kg). The resulting acceleration $\vec{a}$ of the rope and block across the frictionless surface has constant magnitude 0.300 m/s² and is directed to the right. We will call this the positive direction for the x axis. Note that this rope is not "massless;" we return to this feature in part (d).

(a) Identify all the third-law force pairs for the horizontal forces in Fig. 6-24*a* and show how the vectors in each pair are related.

SOLUTION ■ The **Key Idea** here is that a third-law force pair arises when two bodies interact; the forces of the pair are equal in magnitude and opposite in direction, and the force on each body is due to the other body. The "exploded view" of Fig. 6-24*b* shows

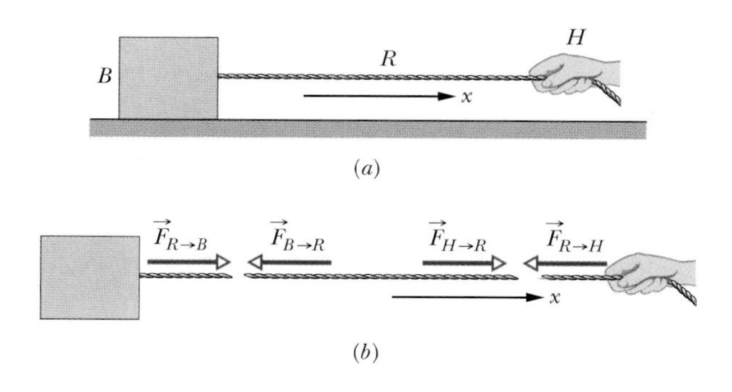

FIGURE 6-24 ■ (*a*) Hand H pulls on rope R, which is attached to block B. (b) An exploded view of block, rope, and hand, with the forces between block and rope and between rope and hand.

that here there are two such force pairs for the horizontal forces. At the hand-rope boundary, we have the force $\vec{F}_{H\to R}$ exerted by the hand on the rope and the force $\vec{F}_{R\to H}$ exerted by the rope on the hand. These forces are a Newton's Third Law force pair and so are equal in magnitude and opposite in direction. They are related by

$$\vec{F}_{H\to R} = -\vec{F}_{R\to H}. \quad \text{(Answer)}$$

Similarly, at the rope–block boundary we have

$$\vec{F}_{R\to B} = -\vec{F}_{B\to R}. \quad \text{(Answer)}$$

(b) What is the magnitude of the force $\vec{F}_{R\to B}$ that the rope exerts on the block?

SOLUTION We know that the block has an acceleration $\vec{a}$ in the positive direction of the x axis. The only force acting on the block along that axis is $\vec{F}_{R\to B}$. The **Key Idea** here is that we can relate force $\vec{F}_{R\to B}$ to acceleration $\vec{a}$ by Newton's Second Law. Because both vectors are along the x axis, we use the x component version of the law ($F_x^{net} = ma_x$), writing

$$F_{R\to B\,x} = Ma_x.$$

Substituting known values, we find that the magnitude of $\vec{F}_{R\to B}$, which we denote $F_{R\to B}$ and is equal to $F_{R\to B\,x}$, is

$$F_{R\to B} = (5.00\text{ kg})(0.300\text{ m/s}^2) = 1.50\text{ N}. \quad \text{(Answer)}$$

(c) What is the magnitude of the force $\vec{F}_{B\to R}$ that the block exerts on the rope?

SOLUTION From (a), we know that $\vec{F}_{B\to R} = -\vec{F}_{R\to B}$, so $\vec{F}_{B\to R}$ has the magnitude

$$F_{B\to R} = F_{R\to B} = 1.50\text{ N}. \quad \text{(Answer)}$$

(d) What is the magnitude of the force $\vec{F}_{H\to R}$ that the hand exerts on the rope?

SOLUTION A **Key Idea** here is that, with the rope taut, the rope and block form a system on which $\vec{F}_{H\to R}$ acts. The mass of the system is $m + M$. For this system, Newton's Second Law for x-components gives us

$$\begin{aligned} F_{H\to R\,x} &= (m + M)a_x \\ &= (0.200\text{ kg} + 5.00\text{ kg})(0.300\text{ m/s}^2) \\ &= 1.56\text{ N} \end{aligned} \quad \text{(Answer)} \quad (6\text{-}19)$$

Now note that the magnitude of the force $\vec{F}_{H\to R}$ on the rope from the hand (1.56 N) is greater than the magnitude of the force $\vec{F}_{R\to B}$ on the block from the rope [1.50 N, from part (b) above]. The reason is that $\vec{F}_{R\to B}$ must accelerate only the block but $\vec{F}_{H\to R}$ must accelerate both the block and the rope, and the rope's mass m is not negligible. If we let $m \to 0$ in Eq. 6-19, then we find 1.50 N, the same magnitude as at the other end. We often assume that an interconnecting rope is massless so that we can approximate the forces at its two ends as having the same magnitude.

TOUCHSTONE EXAMPLE 6-4: Three Cords

In Fig. 6-25*a*, a block B of mass M = 15 kg hangs by a cord from a knot K of mass m_K, which hangs from a ceiling by means of two other cords. The cords have negligible mass, and the magnitude of the gravitational force on the knot is negligible compared to the gravitational force on the block. What are the tensions in the three cords?

SOLUTION Let's start with the block because it has only one attached cord. The free-body diagram in Fig. 6-25*b* shows the forces on the block: gravitational force $\vec{F}^{grav}$ (with a magnitude of Mg) and force $\vec{T}_C$ from the attached cord. A **Key Idea** is that we can relate these forces to the acceleration of the block via Newton's Second Law ($\vec{F}^{net} = m\vec{a}$). Because the forces are both vertical, we choose the vertical component version of the law, $F_y^{net} = ma_y$, and write

$$F_y^{net} = T_{Cy} + F_y^{grav} = T_C - Mg = Ma_y.$$

Substituting 0 for the block's acceleration a_y, we find

$$T_{Cy} - Mg = M(0) = 0.$$

This means that the two forces on the block are equal in magnitude. Substituting for M (= 15.0 kg) and g and solving for T_{Cy} yields

$$T_{Cy} = 147\text{ N}. \quad \text{(Answer)}$$

Note: Although $\vec{T}_C$ and $\vec{F}^{grav}$ are equal in magnitude and opposite in direction, they are *not* a Newton's Third Law force pair. Why?

We next consider the knot in the free-body diagram of Fig. 6-25*c*, where the negligible gravitational force on the knot is not included. The **Key Idea** here is that we can relate the three other forces acting on the knot to the acceleration of the knot via Newton's Second Law ($\vec{F}^{net} = m\vec{a}$) by writing

$$\vec{T}_A + \vec{T}_B + \vec{T}_C = m_K\vec{a}_K.$$

Substituting 0 for the knot's acceleration $\vec{a}_K$ yields

$$\vec{T}_A + \vec{T}_B + \vec{T}_C = 0, \quad (6\text{-}20)$$

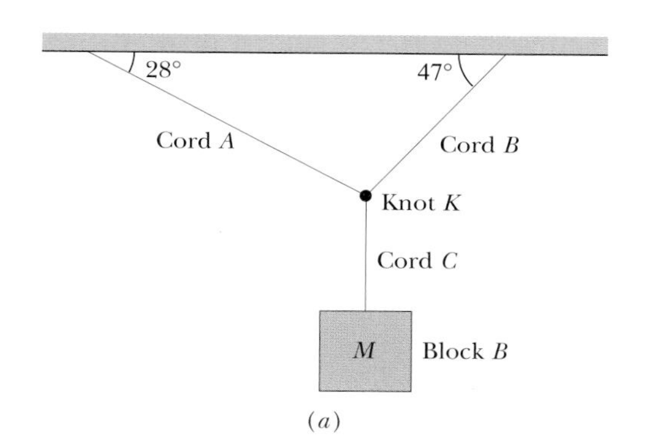

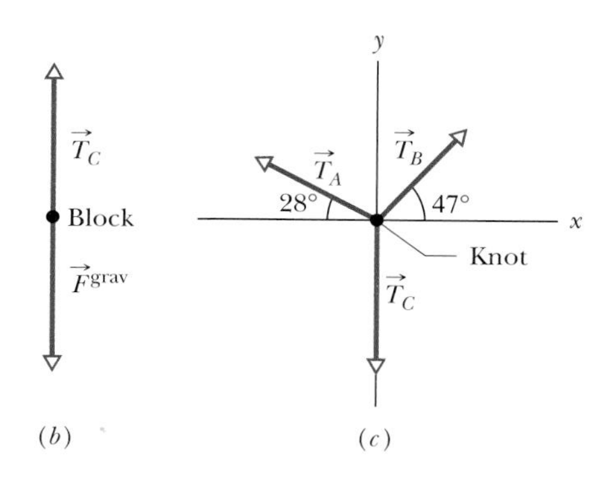

FIGURE 6-25 ■ (*a*) A block of mass *m* hangs from three cords. (*b*) A free-body diagram for the block. (*c*) A free-body diagram for the knot at the intersection of the three cords.

which means that the three forces on the knot are in equilibrium. Although we know both magnitude and angle for $\vec{T}_C$, we know only the angles and not the magnitudes for $\vec{T}_A$ and $\vec{T}_B$. With unknowns in two vectors, we cannot solve Eq. 6-20 for $\vec{T}_A$ or $\vec{T}_B$ directly on a vector-capable calculator. Instead, we rewrite Eq. 6-20 in terms of components along the x and y axes. For the x axis, we have

$$T_{Ax} + T_{Bx} + T_{Cx} = 0,$$

which, using the given data, yields

$$-|\vec{T}_A|\cos 28° + |\vec{T}_B|\cos 47° + 0 = 0, \qquad (6\text{-}21)$$

or alternatively

$$|\vec{T}_A|\cos 152° + |\vec{T}_B|\cos 47° + 0 = 0.$$

Similarly, for the y axis we rewrite Eq. 6-20 as

$$T_{Ay} + T_{By} + T_{Cy} = 0$$

or

$$|\vec{T}_A|\sin 28° + |\vec{T}_B|\sin 47° - |\vec{T}_C| = 0.$$

Substituting our previous result for T_C then gives us

$$|\vec{T}_A|\sin 28° + |\vec{T}_B|\sin 47° - 147\text{ N} = 0. \qquad (6\text{-}22)$$

We cannot solve Eq. 6-21 or Eq. 6-22 separately because each contains two unknowns, but we can solve them simultaneously because they contain the same two unknowns. Doing so (either by substitution, by adding or subtracting the equations appropriately, or by using the equation-solving capability of a calculator), we discover

$$|\vec{T}_A| = 104\text{ N} \quad \text{and} \quad |\vec{T}_B| = 134\text{ N}. \qquad \text{(Answer)}$$

Thus, the magnitudes of the tensions in the cords are 104 N in cord *A*, 134 N in cord *B*, and 147 N in cord *C*.

6-5 Drag Force and Terminal Speed

If you are riding in a car and put your hand out the window, you feel nothing when the car is first starting up. But as you speed up, the forces on your hand become larger and larger. The force you feel on your hand is called **air drag.** The magnitude of the air drag increases as the velocity of your hand relative to the air increases. Air drag is another common force, but it is only important when an object is moving relatively rapidly.

Air is a fluid. A **fluid** is anything that can flow—generally either a gas or a liquid. When there is a relative velocity between a fluid and a body (either because the body moves through the fluid or because the fluid moves past the body), the body experiences a **drag force** $\vec{D}$ that opposes the relative motion and points in the direction in which the fluid flows relative to the body. Like contact forces, air drag forces are ultimately the result of billions of tiny electromagnetic forces between air molecules and another object.

Here we examine only cases in which air is the fluid, the body is blunt (like your hand or a baseball) rather than slender (like a javelin), and the relative motion is fast enough so that the air becomes turbulent (breaks up into swirls) behind the body. In such cases, experiments reveal that the magnitude $D = |\vec{D}|$ of the drag force is re-

FIGURE 6-26 ■ This skier crouches in an "egg position" to minimize her effective cross-sectional area and thus the air drag acting on her.

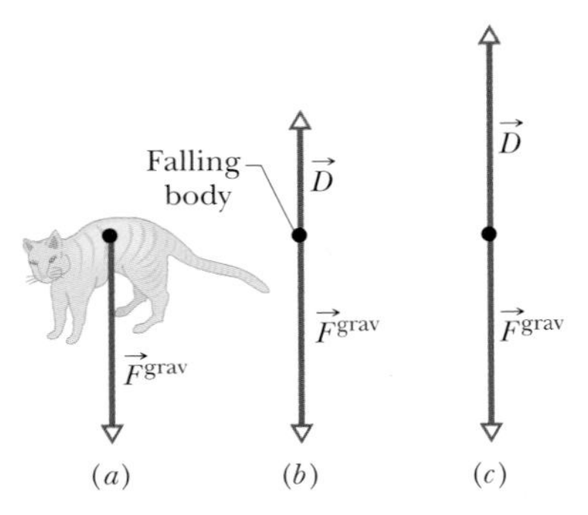

FIGURE 6-27 ■ The forces that act on a body falling through air: (*a*) the body when it has just begun to fall and (*b*) the free-body diagram a little later, after a drag force has developed. (*c*) The drag force has increased until it balances the gravitational force on the body. The body now falls at its constant terminal speed.

lated to the relative speed $v = |\vec{v}|$ by an experimentally determined **drag coefficient** C according to

$$D = \tfrac{1}{2}C\rho A v^2, \tag{6-23}$$

where ρ is the air density (mass per volume) and A is the **effective cross-sectional area** of the body (the area of a cross section taken perpendicular to the velocity $\vec{v}$). The drag coefficient C (typical values range from 0.4 to 1.0) is not truly a constant for a given body, because if v varies significantly, the value of C can vary as well. Here, we ignore such complications.

Downhill speed skiers know well that drag depends on the cross-sectional area (A) and speed squared (v^2). To reach high speeds a skier must reduce the drag force as much as possible by, for example, riding the skis in the "egg position" (Fig. 6-26) to minimize cross-sectional area A.

When a blunt body falls from rest through air, the drag force $\vec{D}$ is directed upward; its magnitude gradually increases from zero as the speed of the body increases. This upward force $\vec{D}$ opposes the downward gravitational force, $\vec{F}^{grav} = -mg\,\hat{j}$, on the body. We can relate these forces to the body's acceleration by writing Newton's Second Law in terms of vector components for a vertical y axis ($F_y^{net} = ma_y$),

$$F_y^{net} = (D_y + F_y^{grav}) = (+D - mg) = ma_y, \tag{6-24}$$

where m is the mass of the body. Experience tells us that D increases as the velocity of the falling object relative to the air increases. As suggested in Fig. 6-27, if the body falls long enough the force magnitudes, D and F^{grav}, eventually equal each other as shown in Fig. 6-27*c*. According to Eq. 6-24, when this happens $a_y = 0$, and the body's speed no longer increases. The body then falls at a constant speed, called the *terminal speed* v_t. To find the terminal speed, we set $a_y = 0$ in Eq. 6-24 and use that relation for the magnitude of the drag force given by $D = \tfrac{1}{2}C\rho A v^2$ (Eq. 6-23). Then the terminal speed is given by

$$v_t = \sqrt{\frac{2mg}{C\rho A}}. \tag{6-25}$$

Table 6-1 gives values of the terminal speed for some common objects.

According to calculations* based on the assumption that $D = \tfrac{1}{2}CA\rho v^2$, a cat must fall about six floors to reach terminal speed. Until it does so, $mg > D_y$ and the cat ac-

TABLE 6-1
Some Terminal Speeds in Air

Object	Terminal Speed (m/s)	95% Distance[a] (m)
Shot (from shot put)	145	2500
Sky diver (typical)	60	430
Baseball	42	210
Tennis ball	31	115
Basketball	20	47
Ping-Pong ball	9	10
Raindrop (radius = 1.5 mm)	7	6
Parachutist (typical)	5	3

[a]This is the distance through which the body must fall from rest to reach 95% of its terminal speed.
Source: Adapted from Peter J. Brancazio, *Sport Science* New York: Simon & Schuster (1984).

celerates downward because of the net downward force. Recall from Chapter 2 that your body is an accelerometer, not a speedometer. Because the cat also senses the acceleration, it is frightened and keeps its feet underneath its body, its head tucked in, and its spine bent upward, making its cross-sectional area (A) small, so its terminal speed v_t becomes relatively large. If the cat maintains this position, it could be injured on landing. However, if the cat shown at the top of the chapter opening photo reaches v_t, its acceleration vanishes so it relaxes, stretching its legs and neck horizontally outward and straightening its spine (it then resembles a flying squirrel). These actions increase its area A and hence the magnitude of the drag force D_y acting on it. The cat begins to slow its descent because now the magnitude of its upward drag force is greater than the downward gravitational force. Eventually, a new, smaller terminal velocity is reached. The decrease in terminal velocity reduces the possibility of serious injury on landing. Just before hitting the ground, the cat pulls its legs back beneath its body to prepare for the landing.

Humans often fall from great heights for fun when sky diving. However, in April 1987, during a jump, sky diver Gregory Robertson noticed that fellow sky diver Debbie Williams had been knocked unconscious in a collision with a third sky diver and was unable to open her parachute. Robertson, who was well above Williams at the time and who had not yet opened his parachute for the 4 km plunge, reoriented his body head-down to minimize his cross-sectional area and maximize his downward speed. Reaching an estimated terminal velocity of 320 km/h, he caught up with Williams and then went into a horizontal "spread eagle" (as shown in Fig. 6-28) to increase his drag force. He could then grab her. He opened her parachute and then, after releasing her, his own, a scant 10 s before impact. Williams received extensive internal injuries due to her lack of control on landing but survived.

FIGURE 6-28 ■ A sky diver in a horizontal "spread eagle" maximizes the air drag.

READING EXERCISE 6-9: Near the ground, is the speed of large raindrops greater than (>), less than (<), or equal to (=) the speed of small raindrops? Assume that all raindrops are spherical and have the same drag coefficient C. **Beware!** More than one factor is involved. ■

6-6 Applying Newton's Laws

Now that you have learned about several types of forces that can act on an object, you have the basic knowledge needed to analyze the accelerations and forces experienced by bodies in an interacting system. However, you will need to use your knowledge in an organized fashion to predict how a system will move or to identify unknown forces based on observations of system motions.

There are several key steps that we suggest you use in performing an analysis. These steps are an extension of those presented in Sections 3-7 and 6-2. The steps are outlined in more detail in Touchstone Example 6-5:

1. Construct a diagram of the system you wish to analyze.
2. Isolate the bodies of interest in the system on your diagram. Identify the types, directions, and approximate magnitudes of the forces acting on each body. Label the forces to indicate the type of force ($\vec{F}^{\text{grav}}$, $\vec{N}$, $\vec{f}$, $\vec{T}$).
3. Construct a free-body diagram representing each body as a point. Place the tails of the labeled force vectors for that body at its point. If possible, show the angles these vectors make with respect to each other as well as the relative magnitudes of the vectors.

*W. O. Whitney and C. J. Mehlhaff, "High-Rise Syndrome in Cats," *The Journal of the American Veterinary Medical Association*, 1987, Vol. 191, pp. 1399–1403.

4. Predict the direction of the acceleration and draw a special acceleration vector in that direction and label it with $\vec{a}$. Then choose a coordinate system so that one axis lies parallel to the direction of the predicted acceleration.
5. Write down Newton's Second Law in vector form for each body in the system. Then decompose the vectors into a pair of one-dimensional equations for each body,

$$\vec{a} = \frac{1}{m}\vec{F}^{\text{net}} \Rightarrow \quad a_x = \frac{1}{m}F_x^{\text{net}} \quad \text{and} \quad a_y = \frac{1}{m}F_y^{\text{net}}.$$

Remember that we drop the vector notation (arrows) when we write the one-dimensional equations. These equations associate the components of vectors.

6. Solve the set of equations for each dimension (x and y) separately to find the unknown vector components.

In Touchstone Example 6-5 we show how these six steps can be used to find the forces that act on a block of known mass as it slides up an incline.

TOUCHSTONE EXAMPLE 6-5: Sliding Up a Ramp

Figure 6-29 shows a 300 g block on a 30° incline. The block is moving up the incline at a constant velocity because a string that passes over a pulley is attached to a falling mass. We assume that the mass of the string and pulley are negligible and that there is no friction in the pulley and no friction between the incline and the block. (a) What force is the string exerting on the block during the time that the block is moving up the plane at constant velocity? (b) What is the normal force that the incline is exerting on the block?

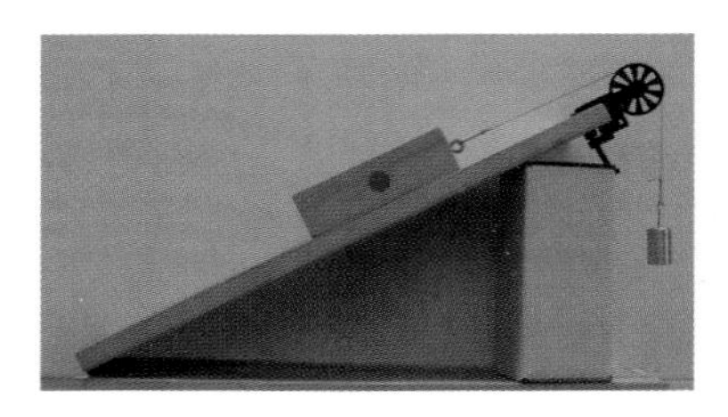

FIGURE 6-29 ■ Photograph of a block on an inclined plane that is moving at a constant velocity.

SOLUTION ■ The **Key Idea** is that because the block is not accelerating, the net force on it must be zero (according to Newton's Second Law). If we follow the steps outlined in Section 6-6 we can identify the forces on the block, choose a coordinate system, and decompose the vectors into components. Since the components along each axis must add up to zero, we can solve our equations for the magnitude and direction of the force of the string on the block.

Step One: Construct a Diagram of the System Figure 6-30*a* shows the essential features of the system of interest needed to answer question (a) including the incline, the block, and the string pulling on the block. The figure is more abstract than the photograph of Fig. 6-29.

Step Two: Isolate the Objects of Interest and Identify the Forces There is only one object of interest in this problem—the block.

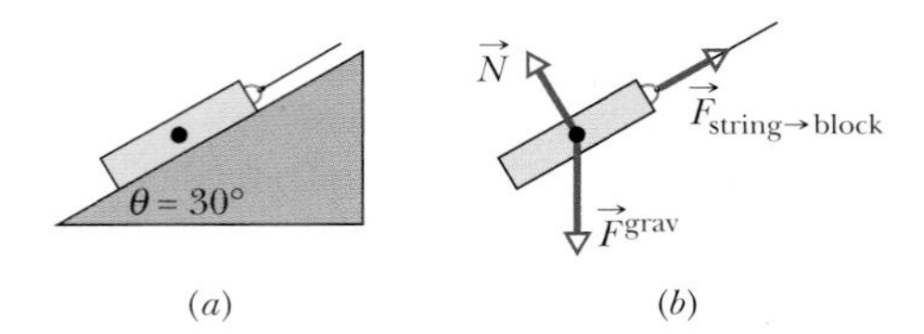

FIGURE 6-30 ■ (*a*) Step one sketch of just those parts of the system of interest for solving the problem. (*b*) Step two sketch of just the block and the forces acting on it with labels.

Thus, we only need to diagram and identify the forces on it. There are three forces acting on the block. First, there is the gravitational force that the Earth exerts on the block that acts vertically downward. Next, there is the normal force that is at right angles (normal) to the surface of the incline. Finally, there is the tension force along the direction of the string that is exerted on the upper end of the block. These forces are shown in Fig. 6-30*b*. *Note*: Although each bit of mass on the block is being pulled downward by the Earth, we can idealize this force and assume it acts at the center of the block. Likewise we assume that the normal force exerted on the block by the inclined plane surface acts like a single force at the middle of the surface of the block that is in contact with the incline. We realize it is the vector sum of billions of smaller normal vectors acting at all points along the surface of contact of the block.

Step Three: Construct a Free-Body Diagram To analyze a system using Newton's Second Law, we draw a free-body diagram for each object in our system. Usually the object experiencing forces is represented by a dot. Then, a vector representing each force that acts on that object is drawn with its tail on the dot. Each vector should be pointing in the direction of the particular force being represented. Also, if the relative magnitudes of the forces are known, the

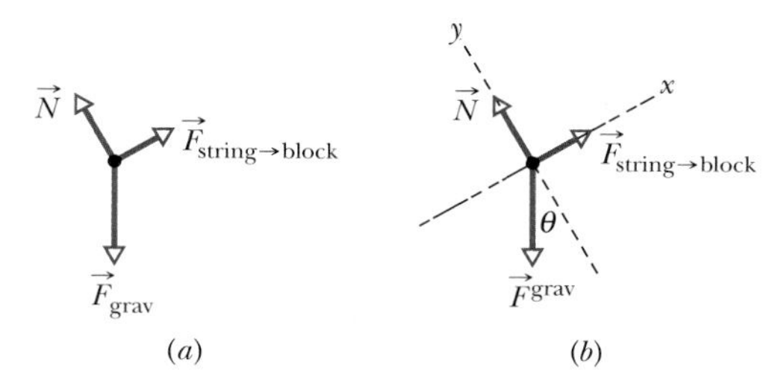

FIGURE 6-31 ■ Steps three and four free-body diagrams for the forces on the block: (*a*) without a coordinate system and (*b*) with a coordinate system.

lengths of the vectors should represent those magnitudes. In this example, we only need a free-body diagram for one object—the block. A clearly labeled arrow showing the predicted direction of the acceleration of the object should also be included. We have no acceleration in this case, so no acceleration vector is included. The free-body diagram for the block is shown in Fig. 6-31*a*.

Step Four: Predict the Direction of the Acceleration and Choose a Coordinate System In choosing the coordinate system for this particular situation, it is useful to break away from our standard practice of having the y axis be a vertical axis and the x axis be horizontal. In general, it is helpful to have one of the axes chosen so it is in the direction of either the acceleration of the object of interest or the forces we are trying to find. One force on the block points up the incline (the string force). Another force is perpendicular to the incline (the normal force). Let's choose "up the incline" as the direction of the positive x axis and a y axis that is perpendicular to the incline (shown in Fig. 6-31*b*). In this coordinate system, only the gravitational force vector will need to be decomposed. Note that using a standard coordinate system would not be incorrect, just less convenient.

We can use some basic geometry to convince ourselves that the gravitational force vector makes an angle of 30° with respect to the negative y axis.

Step Five: Apply Newton's Second Law and Decompose the Force Vectors Recall that the block is moving with constant velocity so the vector sum of the forces acting on it must be zero. Thus we can write

$$\vec{F}^{\text{net}} = m\vec{a} = 0 \quad \text{so} \quad \vec{F}^{\text{grav}} + \vec{N} + \vec{F}_{\text{string}\to\text{block}} = 0.$$

But in order for $\vec{F}^{\text{net}} = 0$ we must have $F_x^{\text{net}} = 0$ and $F_y^{\text{net}} = 0$. Therefore,

$$F_x^{\text{grav}} + N_x + F_{\text{string}\to\text{block}\,x} = 0 \quad \text{and}$$

$$F_y^{\text{grav}} + N_y + F_{\text{string}\to\text{block}\,y} = 0.$$

Recall that here F_x^{grav} denotes the x-component of the gravitational force, N_x denotes the x-component of the normal force, and so on. These components are not vectors and so do not have arrows above them. The only vector that needs decomposition is the gravitational force vector. This decomposition is shown in Fig. 6-32. Since we know by inspection that the gravitational force components are negative, they are expressed with explicit signs as

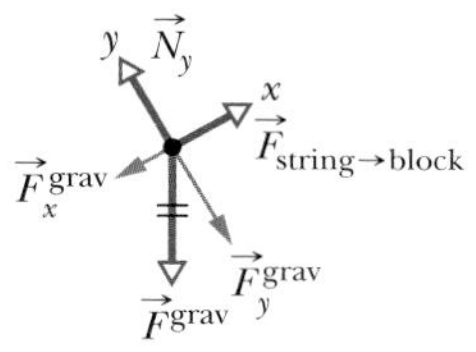

FIGURE 6-32 ■ Decomposition of the gravitational force vector into components along the chosen x and y axes.

$$F_x^{\text{grav}} = -F^{\text{grav}} \sin\theta \quad \text{and} \quad F_y^{\text{grav}} = -F^{\text{grav}} \cos\theta.$$

The angle θ between the downward-pointing force vector and the y axis is 30°. By inspecting the diagram, we see that the normal force vector points along the positive y axis and the tension force vector points along the positive x axis. So, these vectors can be written as

$$\vec{N} = +N\hat{\text{j}} \quad \text{and} \quad \vec{F}_{\text{string}\to\text{block}} = +T\hat{\text{i}},$$

where T is a positive scalar representing the tension in the string and N is a positive scalar representing the magnitude of the normal force. Our expression for $F_x^{\text{net}} = 0$ then becomes $+T - F^{\text{grav}} \sin\theta = 0$. Our expression for $F_y^{\text{net}} = 0$ then becomes $N - F^{\text{grav}} \cos\theta = 0$. Thus,

$$T = mg \sin\theta,$$

and

$$N = mg \cos\theta.$$

We know how to find the values of the gravitational force components in terms of the mass of the block m, the local gravitational strength constant g, and the angle θ:

$$T = mg \sin\theta = 0.300 \text{ kg} \times 9.8 \text{ m/s}^2 \times \sin 30° = 1.47 \text{ N}$$

$$N = mg \cos\theta = 0.300 \text{ kg} \times 9.8 \text{ m/s}^2 \times \sin 30° = 2.55 \text{ N}.$$

Finally, rounding to 2 significant figures gives us

$$\vec{F}_{\text{string}\to\text{block}} = +T\hat{\text{i}} = +(1.5 \text{ N})\hat{\text{i}} \qquad \text{(Answer)}$$

$$\vec{N} = +(2.5 \text{ N})\hat{\text{j}}. \qquad \text{(Answer)}$$

A final note: This example shows the basics for a relatively simple analysis. If we had taken friction into account and picked a part of the motion that is accelerated, the problem would have been more complicated. However, the basic steps would be exactly the same. To master the techniques of analysis for more complex situations, you will also need to study the rest of the of touchstone examples in this chapter.

TOUCHSTONE EXAMPLE 6-6: Breaking Loose

Figure 6-33*a* shows a coin of mass m at rest on a book that has been tilted at an angle θ with the horizontal. By experimenting, you find that when θ is increased to 13°, the coin is on the *verge* of sliding down the book, which means that even a slight increase beyond 13° produces sliding. What is the coefficient of static friction μ^{stat} between the coin and the book?

SOLUTION ■ If the book were frictionless, the coin would surely slide down it for any tilt of the book because of the gravitational force on the coin. Thus, one **Key Idea** here is that a frictional force f^{stat} must be holding the coin in place. A second **Key Idea** is that, because the coin is *on the verge* of sliding *down* the book, that force is at its *maximum* magnitude $f^{\max}$ and is directed *up* the book. Also, from Eq. 6-11, we know that $f^{\max} = \mu^{\text{stat}} N$, where N is the magnitude of the normal force $\vec{N}$ on the coin from the book. Thus,

$$f^{\max} = \mu^{\text{stat}} N,$$

from which

$$\mu^{\text{stat}} = \frac{f^{\text{stat}}}{N}. \tag{6-26}$$

To evaluate this equation, we need to find the force magnitudes f^{stat} and N. To do that, we use another **Key Idea:** When the coin is on the verge of sliding, it is stationary and thus its acceleration $\vec{a}$ is zero. We can relate this acceleration to the forces on the coin with Newton's Second Law ($\vec{F}^{\text{net}} = m\vec{a}$). As shown in the free-body diagram of the coin in Fig. 6-33*b*, these forces are (1) the frictional force $\vec{f}^{\text{stat}}$, (2) the normal force $\vec{N}$, and (3) the gravitational force $\vec{F}^{\text{grav}}$ on the coin, with magnitude equal to mg. Then, from Newton's Second Law with $\vec{a} = 0$, we have

$$\vec{f}^{\text{stat}} + \vec{N} + \vec{F}^{\text{grav}} = 0. \tag{6-27}$$

To find f^{stat} and N, we rewrite Eq. 6-27 for components along the x and y axes of the tilted coordinate system in Fig. 6-33*b*. For the x axis and with mg substituted for $|\vec{F}^{\text{grav}}|$, we have

$$f_x^{\text{stat}} + N_x + F_x^{\text{grav}} = f_x^{\text{stat}} + 0 - mg\sin\theta = 0,$$

so

$$f_x^{\text{stat}} = f^{\text{stat}} = +mg\sin\theta. \tag{6-28}$$

Similarly, for the y axis we have

$$f_y^{\text{stat}} + N_y + F_y^{\text{grav}} = 0 + N - mg\cos\theta = 0,$$

so

$$N = +mg\cos\theta. \tag{6-29}$$

Substituting Eqs. 6-28 and 6-29 into Eq. 6-26 produces

$$\mu^{\text{stat}} = \frac{mg\sin\theta}{mg\cos\theta} = \tan\theta, \tag{6-30}$$

which here means

$$\mu^{\text{stat}} = \tan 13° = 0.23. \quad \text{(Answer)}$$

Actually, you do not need to measure θ to get μ^{stat}. Instead, measure the two lengths shown in Fig. 6-33*a* and then substitute h/d for $\tan\theta$ in Eq. 6-30.

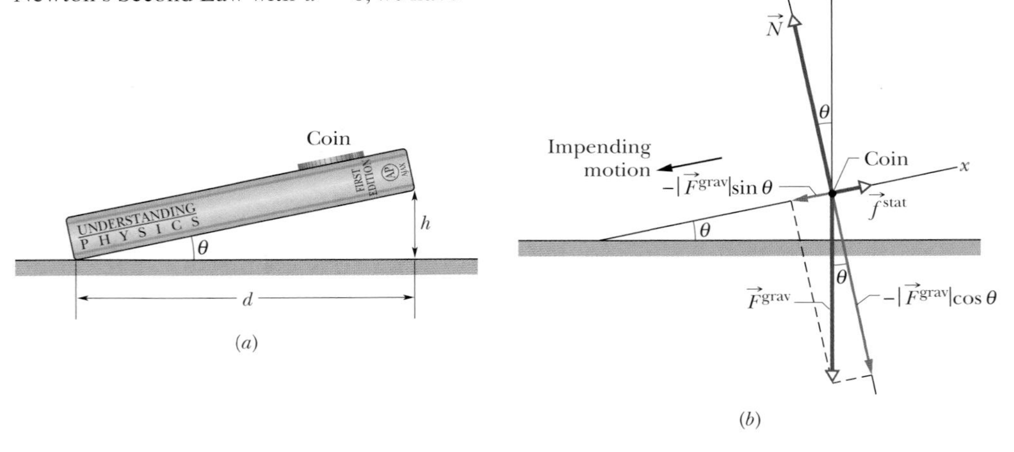

FIGURE 6-33 ■ (*a*) A coin on the verge of sliding down a book. (*b*) A free-body diagram for the coin, showing the three forces (drawn to scale) that act on it. The gravitational force $\vec{F}^{\text{grav}}$ is shown resolved into its components along the x and the y axes, whose orientations are chosen to simplify the problem. Component $F_x^{\text{grav}} = -|\vec{F}^{\text{grav}}|\sin\theta$ tends to slide the coin down the book. Component $F_y^{\text{grav}} = -|\vec{F}^{\text{grav}}|\cos\theta$ presses the coin onto the book.

TOUCHSTONE EXAMPLE 6-7: Accelerated by Friction

A 40 kg slab rests on a frictionless floor. A 10 kg block rests on top of the slab (Fig. 6-34). The coefficient of static friction μ^{stat} between the block and the slab is 0.60, whereas their kinetic friction coefficient μ^{kin} is 0.40. The 10 kg block is pulled by a horizontal force with a magnitude of 100 N. What are the resulting accelerations of (a) the slab and (b) the block?

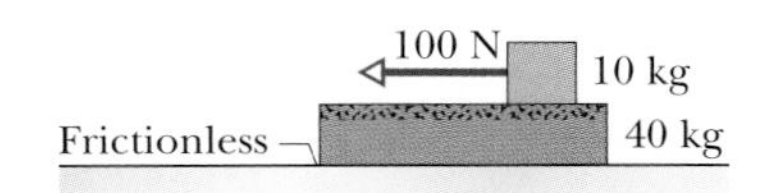

FIGURE 6-34

SOLUTION ■ The first **Key Idea** here is that we should apply Newton's Second Law *separately* to the slab (m_{slab} = 40 kg) and to the block (m_{block} = 10 kg) to obtain the acceleration of each:

$$\vec{a}_{slab} = \frac{\vec{F}^{net}_{slab}}{m_{slab}} \quad \text{and} \quad \vec{a}_{block} = \frac{\vec{F}^{net}_{block}}{m_{block}}.$$

To find the net force on each of these objects, we can draw a free-body diagram for each, as shown in Fig. 6-35. The direction of the frictional force from the slab on the block, $\vec{f}_{slab \to block}$, is determined by considering the direction of the block's impending motion. The direction of the frictional force from the block on the slab, $\vec{f}_{block \to slab}$, is inferred from Newton's Third Law.

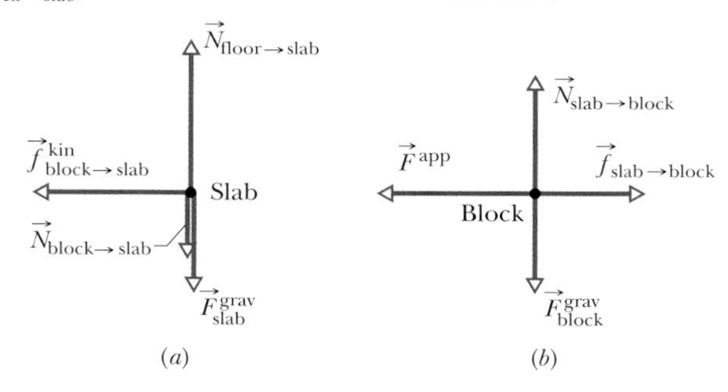

FIGURE 6-35 ■ The free-body diagram showing all the forces acting on (a) the 40 kg slab and (b) the 10 kg block. These forces are not drawn to scale.

Since we expect the block and the slab to accelerate to the left, let's decide that the positive x axis is pointing to the left and the y axis is pointing straight up. Then

$$a_{slab\,y} = 0 \quad \text{so} \quad F^{net}_{slab\,y} = 0, \tag{6-31}$$

and

$$a_{block\,y} = 0 \quad \text{so} \quad F^{net}_{block\,y} = 0. \tag{6-32}$$

To calculate the horizontal accelerations of the block and the slab, we must first determine whether the frictional force of interaction between them is static or kinetic. The **Key Idea** here is that the maximum static frictional force's magnitude is limited to be no larger than $f^{max} = \mu^{stat} N$. Applying this to the slab, we find that

$$f^{max}_{block \to slab} = \mu^{stat} N_{block \to slab}. \tag{6-33}$$

Newton's Third Law tells us that

$$N_{block \to slab} = N_{slab \to block}, \tag{6-34}$$

and Fig. 6-35b and Eq. 6-31 tell us that

$$N_{slab \to block} = F^{grav}_{block} = m_{block}\, g. \tag{6-35}$$

Combining these ideas yields

$$f^{max}_{block \to slab} = \mu^{stat} m_{block}\, g$$
$$= (0.60)(10\text{ kg})(9.8\text{ N/kg}) = 59\text{ N}.$$

If this limit is not exceeded, then static friction would keep the block and the slab locked together, accelerating with a common acceleration

$$a_x = \frac{F_x^{net}}{(m_{slab} + m_{block})}$$
$$= 100\text{ N}/(40\text{ kg} + 10\text{ kg}) = +2.00\text{ m/s}^2.$$

Newton's Second Law (written in terms of x-components) tells us that this acceleration requires $F^{net}_{slab\,x} = m_{slab}\, a_x = (40\text{ kg})(2.00\text{ m/s}^2) =$ 80.0 N. But Fig. 6-35a shows that the only horizontal force acting on the slab is the frictional force from the block. Since 80 N are required to accelerate the slab but we found the static frictional force is limited to 59 N, we can conclude that the block and the slab *cannot* be locked together by static friction. The block must be sliding to the left on top of the slab. This means that

$$F^{net}_{block\,x} = F_x^{app} + f^{kin}_{slab \to block}, \tag{6-36}$$

and

$$F^{net}_{slab\,x} = f^{kin}_{block \to slab} = \mu^{kin} N_{block \to slab}. \tag{6-37}$$

Combining Eqs. 6-34, Eq. 6-35, and Eq. 6-37 yields $F^{net}_{slab\,x} = \mu^{kin} m_{block}\, g$ or

$$\vec{F}^{net}_{slab} = (\mu^{kin} m_{block}\, g)\,\hat{\imath}, \tag{6-38}$$

so

$$\vec{a}_{slab} = \frac{\vec{F}^{net}_{slab}}{m_{slab}}$$
$$= (0.40)(10\text{ kg})(9.8\text{ m/s}^2)/(40\text{ kg})\hat{\imath}$$
$$= (0.98\text{ m/s}^2)\hat{\imath}. \qquad \text{(Answer)}$$

It's interesting to note that a frictional force causes the slab to speed up, not slow down. The same is true when you start running from rest. To accelerate, you push backwards on the ground with your shoes. The ground, courtesy of Newton's Third Law, pushes forward on you, accelerating you forward. It is actually the static frictional force that the ground exerts on you that accelerates you.

Finally, to calculate the acceleration of the block, we note that the net force on the block in the y direction is zero and that (by Newton's Third Law) $\vec{f}^{kin}_{slab \to block} = -\vec{f}^{kin}_{block \to slab}$. Therefore,

$$\vec{F}^{net}_{block} = \vec{F}^{app} + \vec{f}^{kin}_{slab \to block}$$
$$= \vec{F}^{app} - \vec{f}^{kin}_{block \to slab}$$
$$= \vec{F}^{app} - \mu^{kin} m_{block}\, g\,\hat{\imath}.$$

So

$$\vec{a}_{block} = \frac{\vec{F}^{net}_{block}}{m_{block}}$$
$$= \frac{(100\text{ N})\hat{\imath} - (0.40)(10\text{ kg})(9.8\text{ m/s}^2)\hat{\imath}}{10\text{ kg}}$$
$$= [10\text{ m/s}^2 - (0.4)(9.8\text{ m/s}^2)]\hat{\imath}$$
$$= (+6.1\text{ m/s}^2)\hat{\imath}. \qquad \text{(Answer)}$$

TOUCHSTONE EXAMPLE 6-8: Banked Curve

You cannot always count on friction to get your car around a curve, especially if the road is icy or wet. That is why highway curves are banked. Suppose that a car of mass m moves at a constant speed v of 20 m/s around a curve, now banked, whose radius R is 190 m (Fig. 6-36a). What bank angle θ makes reliance on friction unnecessary?

SOLUTION ■ A centripetal force must act on the car if the car is to move along the circular path. A **Key Idea** is that the track is banked so as to tilt the normal force $\vec{N}$ on the car toward the center of the circle (Fig. 6-36b). Thus, $\vec{N}$ now has a centripetal component N_r, directed inward along a radial axis r. We want to find the value of the bank angle θ such that this centripetal component keeps the car on the circular track without need of friction.

A second **Key Idea** is to keep the y axis vertical and the x axis horizontal rather than in the direction of the incline. This enables us to find the radial component of the normal force more easily.

As Fig. 6-36b shows (and as you should verify), the angle that $\vec{N}$ makes with the vertical is equal to the bank angle θ of the track. Thus, the radial component N_r is equal to $+N \sin\theta$ where N is the magnitude of the normal force. We can now write Newton's Second Law for components along the r axis ($F_r^{net} = ma_r$) as

$$+N \sin\theta = m\left(+\frac{v^2}{R}\right). \tag{6-39}$$

We cannot solve this equation for the value of θ because it also contains the unknowns N and m.

We next consider the forces and acceleration along the y axis in Fig. 6-36b. The vertical component of the normal force is $N_y = N\cos\theta$, the gravitational force $\vec{F}^{grav}$ on the car is $(-mg)\hat{j}$, and the acceleration of the car along the y axis is zero. Thus, we can write Newton's Second Law for components along the y axis ($F_y^{net} = ma_y$) as

$$+N\cos\theta - mg = m(0),$$

from which

$$N\cos\theta = mg. \tag{6-40}$$

This too contains the unknowns N and m, but note that dividing Eq. 6-39 by Eq. 6-40 neatly eliminates both those unknowns. Doing so, replacing $\sin\theta/\cos\theta$ with $\tan\theta$ and solving for θ, then yield

$$\theta = \tan^{-1}\left(\frac{v^2}{gR}\right)$$

$$= \tan^{-1}\left(\frac{(20 \text{ m/s})^2}{(9.8 \text{ m/s}^2)(190 \text{ m})}\right) = 12°. \quad \text{(Answer)}$$

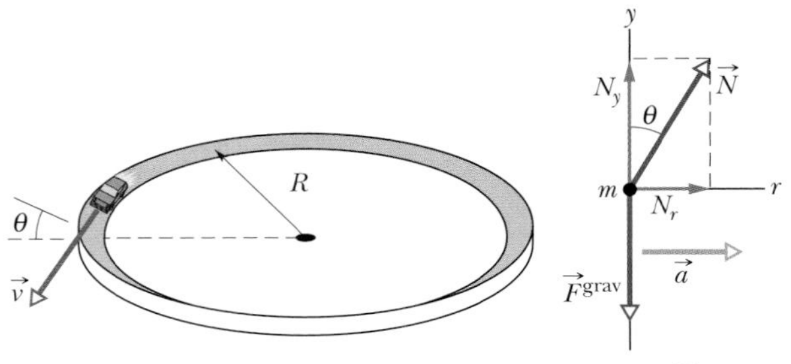

FIGURE 6-36 ■ (a) A car moves around a curved banked road at constant speed. The bank angle is exaggerated for clarity. (b) A free-body diagram for the car, assuming that friction between tires and road is zero. The radially inward component of the normal force provides the necessary centripetal force. The resulting acceleration is also radially inward.

6-7 The Fundamental Forces of Nature

According to Newton's Third Law, forces between two objects always act in pairs. In the study of the structure of matter, physicists have used a belief in mutual interactions to study the nature of forces. As we learn more about matter and how it behaves, we explore the nature of forces by observing changes in the motion of objects that interact. Using these observations, scientists have identified only four types of forces.

The most familiar of these forces are the **gravitational force,** of which falling and weight are our most familiar examples, and the **electromagnetic force,** which, at a fundamental level, is the basis of all the other forces we considered in this chapter. The electromagnetic force is the combination of electrical forces and magnetic forces. Electromagnetic forces enable an electrically charged balloon to stick to a wall and a magnet to pick up an iron nail. In fact, aside from the gravitational force, *any* force that we can experience directly as a push or pull is electromagnetic in nature. That is, all such forces, including friction forces, normal forces, contact forces, and tension forces arise from electromagnetic forces exerted by one atom on another. For example, the tension in a taut cord exists only because its atoms attract one another. When pulled apart a bit, while normal forces result from atoms repelling each other when being pushed together.

Only two other fundamental forces are known, and they both act over such short distances that we cannot experience them directly through our senses. They are the **weak force,** which is involved in certain kinds of radioactive decay, and the **strong force,** which binds together the quarks that make up protons and neutrons and is the "glue" that holds together an atomic nucleus.

Physicists have long believed that nature has an underlying simplicity and that the number of fundamental forces can be reduced. Einstein spent most of his working life trying to interpret these forces as different aspects of a single *superforce.* He failed, but in the 1960s and 1970s, other physicists showed that the weak force and the electromagnetic force are different aspects of a single **electroweak force.** The quest for further reduction continues today, at the very forefront of physics. Table 6-2 lists the progress that has been made toward **unification** (as the goal is called) and gives some hints about the future.

TABLE 6-2
The Quest for the Superforce—A Progress Report

Date	Researcher	Achievement
1687	Newton	Showed that the same laws apply to astronomical bodies and to objects on Earth. Unified celestial and terrestrial mechanics.
1820 1830s	Oersted Faraday	Showed, by brilliant experiments, that the then separate sciences of electricity and magnetism are intimately linked.
1873	Maxwell	Unified the sciences of electricity, magnetism, and optics into the single subject of electromagnetism.
1979	Glashow, Salam, Weinberg	Received the Nobel Prize for showing that the weak force and the electromagnetic force could be different aspects of a single *electroweak force.* This combination of forces reduced the number of forces viewed as fundamental forces from four to three.
1984	Rubbia, van der Meer	Received the Nobel Prize for verifying experimentally the predictions of the theory of the electroweak force.
		Work in Progress
Grand unification theories (*GUTs*): Seek to unify the electroweak force and the strong force.		
Supersymmetry theories: Seek to unify all forces, including the gravitational force, within a single framework.		
Superstring theories: Interpret point-like particles, such as electrons, as being unimaginably tiny, closed loops. Strangely, extra dimensions beyond the familiar four dimensions of space-time appear to be required.		

Problems

SEC. 6-2 ■ NET FORCE AS A VECTOR SUM

1. Standard Body If the 1 kg standard body has an acceleration of 2.00 m/s^2 at 20° to the positive direction of the x axis, then what are (a) the x-component and (b) the y-component of the net force on it, and (c) what is the net force in unit-vector notation?

2. Chopping Block Two horizontal forces act on a 2.0 kg chopping block that can slide over a frictionless kitchen counter, which lies in an xy plane. One force is $\vec{F}_A = (3.0\text{ N})\hat{i} + (4.0\text{ N})\hat{j}$. Find the acceleration of the chopping block in unit-vector notation when the other force is (a) $\vec{F}_B = (-3.0\text{ N})\hat{i} + (-4.0\text{ N})\hat{j}$, (b) $\vec{F}_B = (-3.0\text{ N})\hat{i} + (4.0\text{ N})\hat{j}$, and (c) $\vec{F}_B = (3.0\text{ N})\hat{i} + (-4.0\text{ N})\hat{j}$.

3. Two Horizontal Forces Only two horizontal forces act on a 3.0 kg body. One force is 9.0 N, acting due east, and the other is 8.0 N, acting 62° north of west. What is the magnitude of the body's acceleration?

4. Two Forces While two forces act on it, a particle is to move at the constant velocity $\vec{v} = (3 \text{ m/s})\hat{i} - (4 \text{ m/s})\hat{j}$. One of the forces is $\vec{F}_A = (2 \text{ N})\hat{i} + (-6 \text{ N})\hat{j}$. What is the other force?

5. Three Forces Three forces act on a particle that moves with unchanging velocity $\vec{v} = (2 \text{ m/s})\hat{i} - (7 \text{ m/s})\hat{j}$. Two of the forces are $\vec{F}_A = (2 \text{ N})\hat{i} + (3 \text{ N})\hat{j}$ and $\vec{F}_B = (-5 \text{ N})\hat{i} + (8 \text{ N})\hat{j}$. What is the third force?

6. Three Astronauts Three astronauts, propelled by jet backpacks, push and guide a 120 kg asteroid toward a processing dock, exerting the forces shown in Fig. 6-37. What is the asteroid's acceleration (a) in unit-vector notation and as (b) a magnitude and (c) a direction?

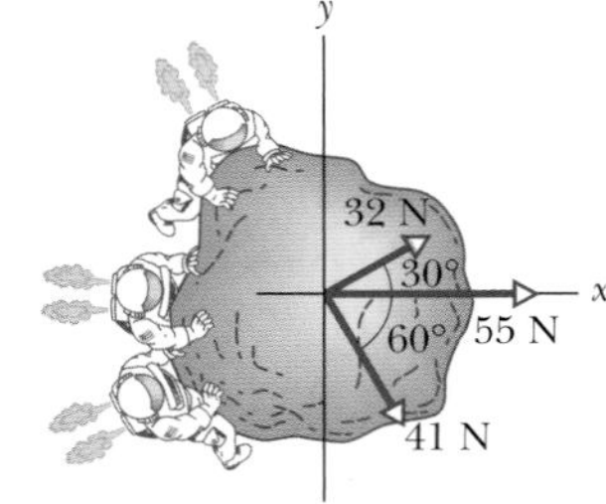

FIGURE 6-37 ■ Problem 6.

7. The Box There are two forces on the 2.0 kg box in the overhead view of Fig. 6-38 but only one is shown. The figure also shows the acceleration of the box. Find the second force (a) in unit-vector notation and as (b) a magnitude and (c) a direction.

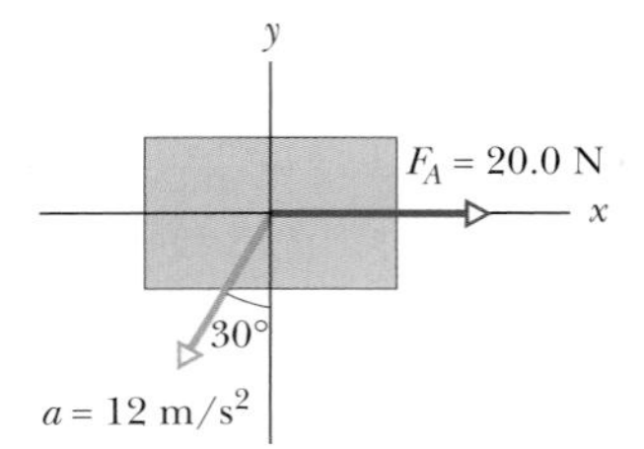

FIGURE 6-38 ■ Problem 7.

8. A Tire Figure 6-39 is an overhead view of a 12 kg tire that is to be pulled by three ropes. One force ($\vec{F}_A$, with magnitude 50 N) is indicated. Orient the other two forces $\vec{F}_B$ and $\vec{F}_C$ so that the magnitude of the resulting acceleration of the tire is least, and find that magnitude if (a) $F_B = 30$ N, $F_C = 20$ N; (b) $F_B = 30$ N, $F_C = 10$ N; and (c) $F_B = F_C = 30$ N.

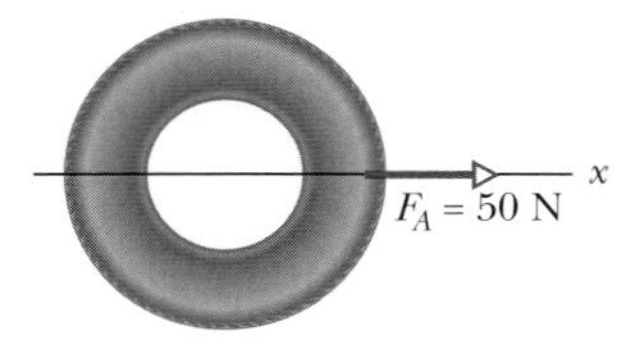

FIGURE 6-39 ■ Problem 8.

SEC. 6-3 ■ GRAVITATIONAL FORCE AND WEIGHT

9. Salami on a Cord (a) An 11.0 kg salami is supported by a cord that runs to a spring scale, which is supported by another cord from the ceiling (Fig. 6-40*a*). What is the reading on the scale, which is marked in weight units? (b) In Fig. 6-40*b* the salami is supported by a cord that runs around a pulley and to a scale. The opposite end of the scale is attached by a cord to a wall. What is the reading on the scale? (c) In Fig. 6-40*c* the wall has been replaced with a second 11.0 kg salami on the left, and the assembly is stationary. What is the reading on the scale now?

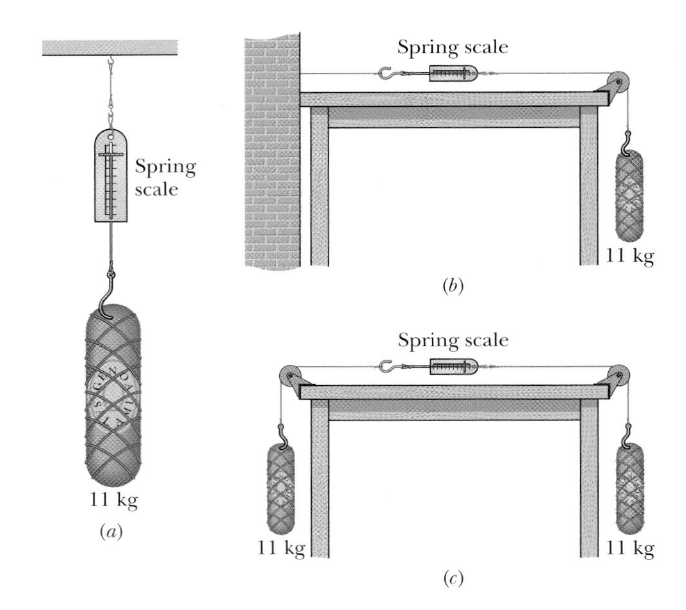

FIGURE 6-40 ■ Problem 9.

10. Spaceship on the Moon A spaceship lifts off vertically from the Moon, where the freefall acceleration is 1.6 m/s². If the spaceship has an upward acceleration of 1.0 m/s² as it lifts off, what is the magnitude of the force of the spaceship on its pilot, who weighs 735 N on Earth?

SEC. ■ 6-4 CONTACT FORCES

11. A Bureau A bedroom bureau with a mass of 45 kg. including drawers and clothing, rests on the floor. (a) If the coefficient of static friction between the bureau and the floor is 0.45, what is the magnitude of the minimum horizontal force that a person must apply to start the bureau moving? (b) If the drawers and clothing, with 17 kg mass, are removed before the bureau is pushed, what is the new minimum magnitude?

12. Scrambled Eggs The coefficient of static friction between Teflon and scrambled eggs is about 0.04. What is the smallest angle from the horizontal that will cause the eggs to slide across the bottom of a Teflon-coated skillet?

13. Baseball Player A baseball player with mass $m = 79$ kg, sliding into second base, is retarded by a frictional force of magnitude 470 N. What is the coefficient of kinetic friction μ^{kin} between the player and the ground?

14. The Mysterious Sliding Stones Along the remote Racetrack Playa in Death Valley. California, stones sometimes gouge out prominent trails in the desert floor, as if they had been migrating (Fig 6-41). For years curiosity mounted about why the stones moved. One explanation was that strong winds during the occasional rainstorms would drag the rough stones over ground softened by rain. When the desert dried out, the trails behind the stones were hard-baked in place. According to measurements, the coefficient of kinetic friction between the stones and the wet playa ground is about 0.80. What horizontal force is needed on a stone of typical mass 20 kg to maintain the stone's motion once a gust has started it moving? (Story continues with Problem 42.)

FIGURE 6-41 ■ Problem 14.

15. A Crate A person pushes horizontally with a force of 220 N on a 55 kg crate to move it across a level floor. The coefficient of kinetic friction is 0.35. (a) What is the magnitude of the frictional force? (b) What is the magnitude of the crate's acceleration?

16. A House on a Hill A house is built on the top of a hill with a nearby 45° slope (Fig. 6-42). An engineering study indicates that the slope angle should be reduced because the top layers of soil along the slope might slip past the lower layers. If the static coefficient of friction between two such layers is 0.5, what is the least angle ϕ through which the present slope should be reduced to prevent slippage?

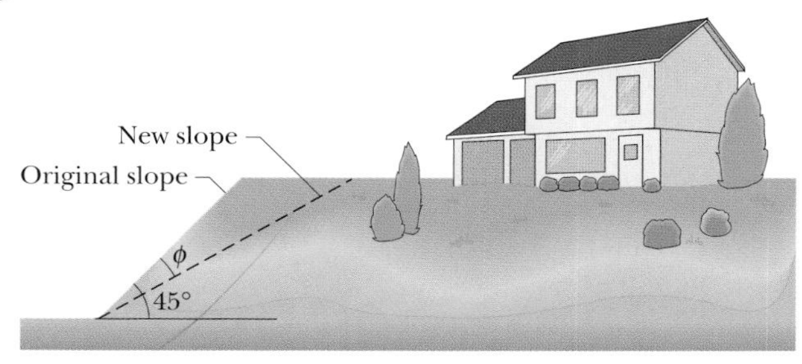

FIGURE 6-42 ■ Problem 16.

17. Hockey Puck A 110 g hockey puck sent sliding over ice is stopped in 15 m by the frictional force on it from the ice. (a) If its initial speed is 6.0 m/s, what is the magnitude of the frictional force? (b) What is the coefficient of friction between the puck and the ice?

18. Rock Climber In Fig. 6-43 a 49 kg rock climber is climbing a "chimney" between two rock slabs. The static coefficient of friction between her shoes and the rock is 1.2; between her back and the rock it is 0.80. She has reduced her push against the rock until her back and her shoes are on the verge of slipping. (a) Draw a free-body diagram of the climber. (b) What is her push against the rock? (c) What fraction of her weight is supported by the frictional force on her shoes?

FIGURE 6-43 ■ Problem 18.

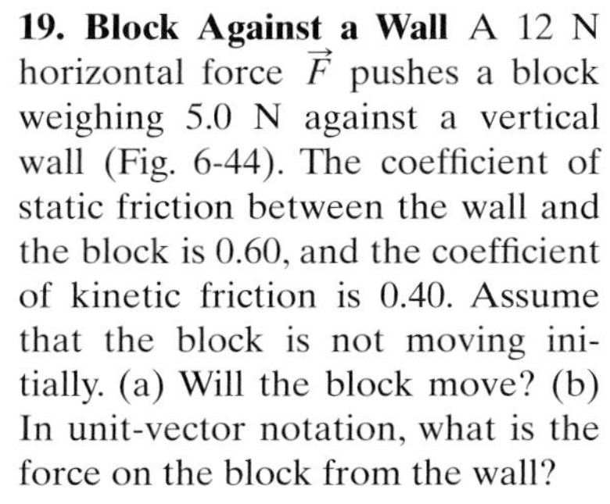

19. Block Against a Wall A 12 N horizontal force $\vec{F}$ pushes a block weighing 5.0 N against a vertical wall (Fig. 6-44). The coefficient of static friction between the wall and the block is 0.60, and the coefficient of kinetic friction is 0.40. Assume that the block is not moving initially. (a) Will the block move? (b) In unit-vector notation, what is the force on the block from the wall?

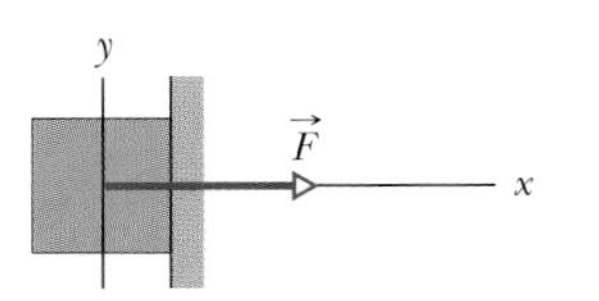

FIGURE 6-44 ■ Problem 19.

20. Block on a Horizontal Surface A 2.5 kg block is initially at rest on a horizontal surface. A 6.0 N horizontal force and a vertical force $\vec{P}$ are applied to the block as shown in Fig. 6-45. The coefficients of friction for the block and surface are $\mu^{\text{stat}} = 0.40$ and $\mu^{\text{kin}} = 0.25$. Determine the magnitude and direction of the frictional force acting on the block if the magnitude of $\vec{P}$ is (a) 8.0 N, (b) 10 N, and (c) 12 N.

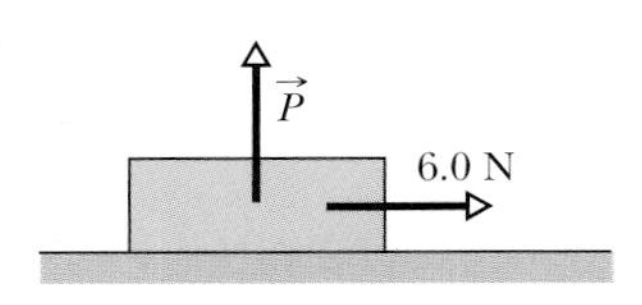

FIGURE 6-45 ■ Problem 20.

21. Pile of Sand A worker wishes to pile a cone of sand onto a circular area in his yard. The radius of the circle is R, and no sand is to spill onto the surrounding area (Fig. 6-46). If μ^{stat} is the static coefficient of friction between each layer of sand along the slope and the sand beneath it (along which is might slip), show that the greatest volume of sand that can be stored in this manner is $\pi\mu^{\text{stat}}R^3/3$. (The volume of a cone is $Ah/3$, where A is the base area and h is the cone's height.)

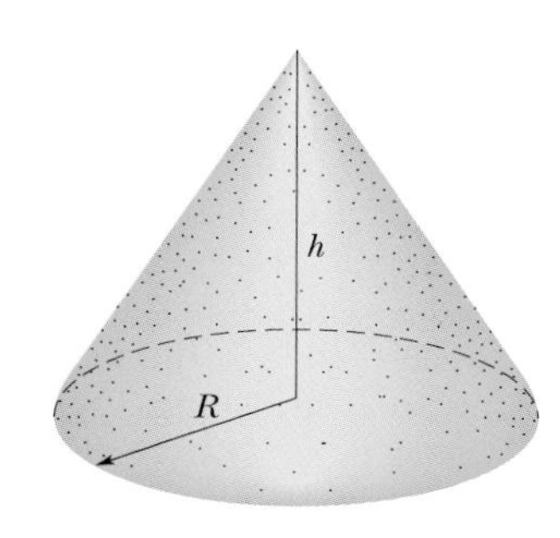

FIGURE 6-46 ■ Problem 21.

22. Worker and Crate A worker pushes horizontally on a 35 kg crate with a force of magnitude 110 N. The coefficient of static friction between the crate and the floor is 0.37. (a) What is the frictional force on the crate from the floor? (b) What is the maximum magnitude $f^{\text{stat}}_{\text{max}}$ of the static frictional force under the circumstances? (c) Does the crate move? (d) Suppose, next, that a second worker pulls directly upward on the crate to help out. What is the least vertical pull that will allow the first worker's 110 N push to move the crate? (e) If, instead, the second worker pulls horizontally to help out, what is the least pull that will get the crate moving?

23. A Crate is Dragged A 68 kg crate is dragged across a floor by pulling on a rope attached to the crate and inclined 15° above the horizontal. (a) If the coefficient of static friction is 0.50, what minimum force magnitude is required from the rope to start the crate moving? (b) If $\mu^{\text{kin}} = 0.35$, what is the magnitude of the initial acceleration of the crate?

FIGURE 6-47 ■ Problem 24.

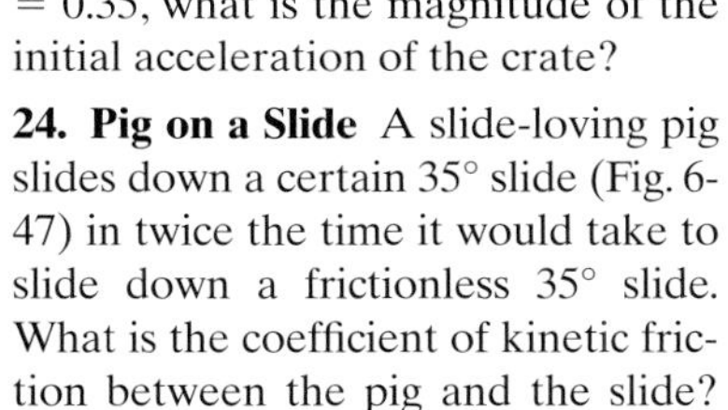

24. Pig on a Slide A slide-loving pig slides down a certain 35° slide (Fig. 6-47) in twice the time it would take to slide down a frictionless 35° slide. What is the coefficient of kinetic friction between the pig and the slide?

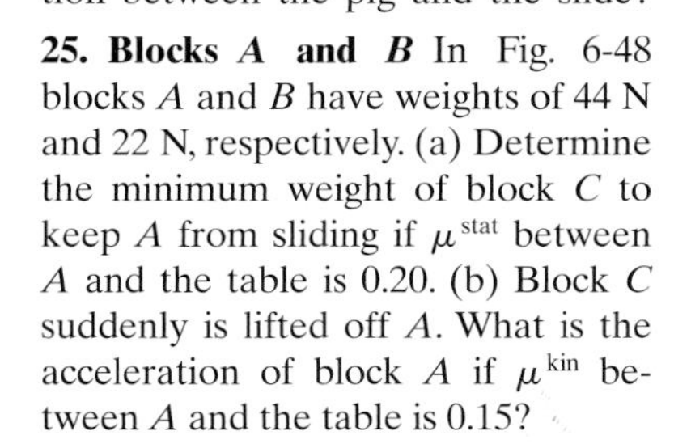

25. Blocks *A* and *B* In Fig. 6-48 blocks A and B have weights of 44 N and 22 N, respectively. (a) Determine the minimum weight of block C to keep A from sliding if μ^{stat} between A and the table is 0.20. (b) Block C suddenly is lifted off A. What is the acceleration of block A if μ^{kin} between A and the table is 0.15?

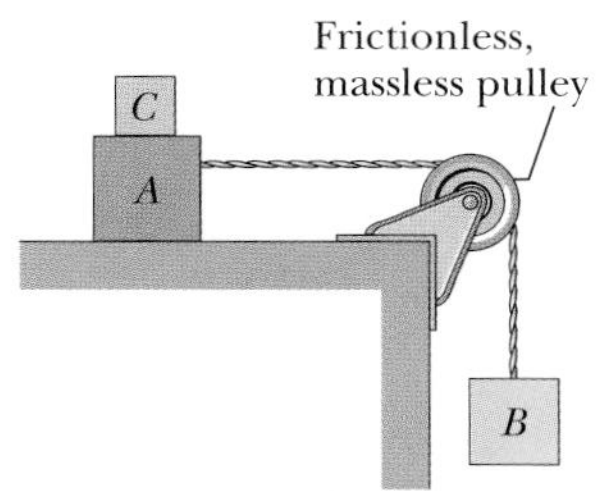

FIGURE 6-48 ■ Problem 25.

26. Block Pushed at an Angle A 3.5 kg block is pushed along a horizontal floor by a force $\vec{F}$ of magnitude 15 N at an angle $\theta = 40°$ with the horizontal (Fig. 6-49). The coefficient of kinetic friction between the block and the floor is 0.25. Calculate

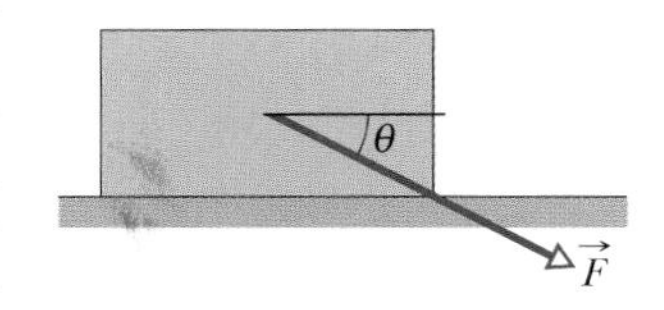

FIGURE 6-49 ■ Problem 26.

the magnitudes of (a) the frictional force on the block from the floor and (b) the acceleration of the block.

27. Mountain Side Figure 6-50 shows the cross section of a road cut into the side of a mountain. The solid line AA' represents a weak bedding plane along which sliding is possible. Block B directly above the highway is separated from uphill rock by a large crack (called a *joint*), so that only friction between the block and the bedding plane prevents sliding. The mass of the block is 1.8×10^7 kg, the *dip angle* θ of the bedding plane is 24°, and the coefficient of static friction between block and plane is 0.63. (a) Show that the block will not slide. (b) Water seeps into the joint and expands upon freezing, exerting on the block a force $\vec{F}$ parallel to AA'. What minimum value of F will trigger a slide?

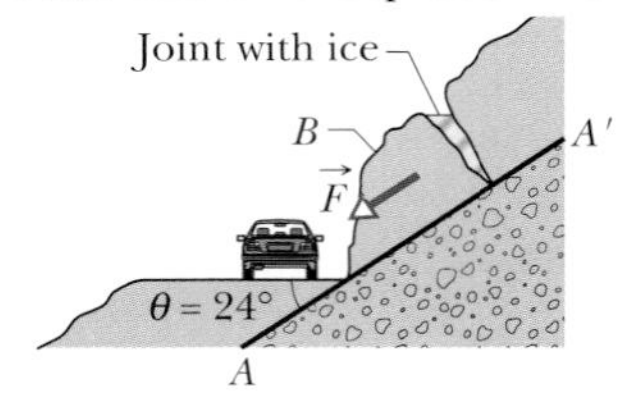

FIGURE 6-50 ▪ Problem 27.

28. Penguin Sled A loaded penguin sled weighing 80 N rests on a plane inclined at 20° to the horizontal (Fig. 6-51). Between the sled and the plane, the coefficient of static friction is 0.25, and the coefficient of kinetic friction is 0.15. (a) What is the minimum magnitude of the force $\vec{F}$, parallel to the plane, that will prevent the sled from slipping down the plane? (b) What is the minimum magnitude F that will start the sled moving up the plane? (c) What value of F is required to move the sled up the plane at constant velocity?

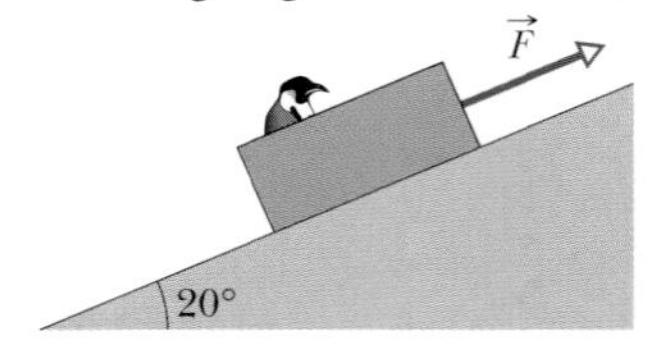

FIGURE 6-51 ▪ Problem 28.

29. Block on a Table Block B in Fig. 6-52 weighs 711 N. The coefficient of static friction between block and table is 0.25; assume that the cord between B and the knot is horizontal. Find the maximum weight of block A for which the system will be stationary.

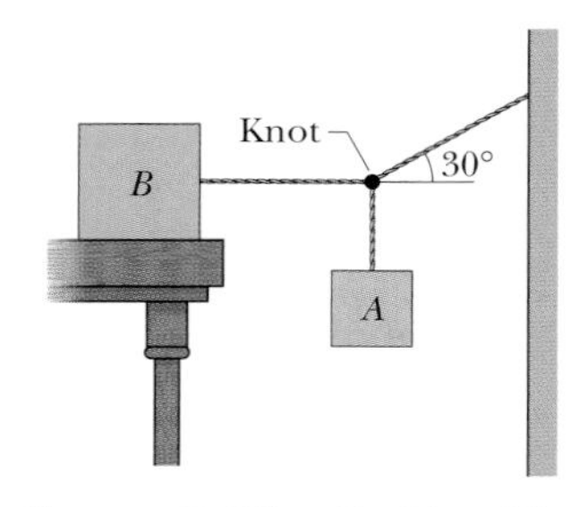

FIGURE 6-52 ▪ Problem 29.

30. Force Parallel to a Surface A force $\vec{P}$, parallel to a surface inclined 15° above the horizontal, acts on a 45 N block, as shown in Fig. 6-53. The coefficients of friction for the block and surface are $\mu^{\text{stat}} = 0.50$ and $\mu^{\text{kin}} = 0.34$. If the block is initially at rest, determine the magnitude and direction of the frictional force acting on the block for magnitudes of $\vec{P}$ of (a) 5.0 N, (b) 8.0 N, and (c) 15 N.

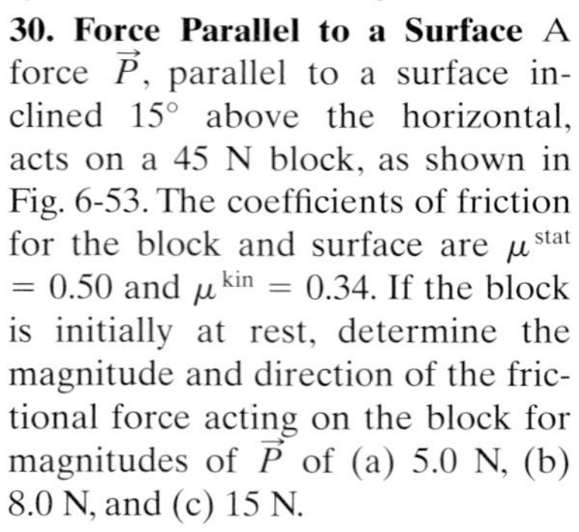

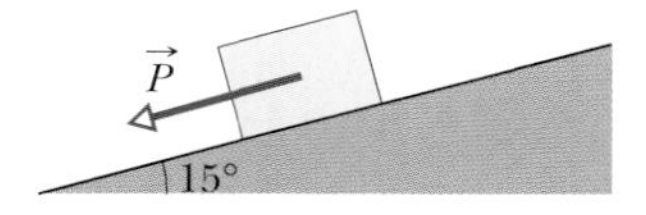

FIGURE 6-53 ▪ Problem 30.

31. Body A–Body B Body A in Fig. 6-54 weighs 102 N, and body B weighs 32 N. The coefficients of friction between A and the incline are $\mu^{\text{stat}} = 0.56$ and $\mu^{\text{kin}} = 0.25$. Angle θ is 40°. Find the acceleration of A if (a) A is initially at rest, (b) A is initially moving up the incline, and (c) A is initially moving down the incline.

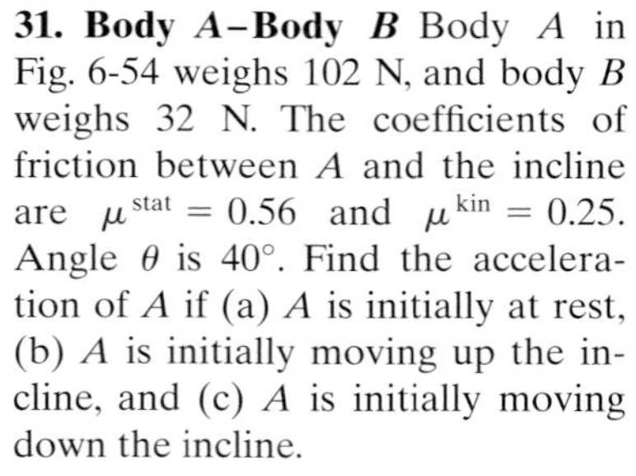

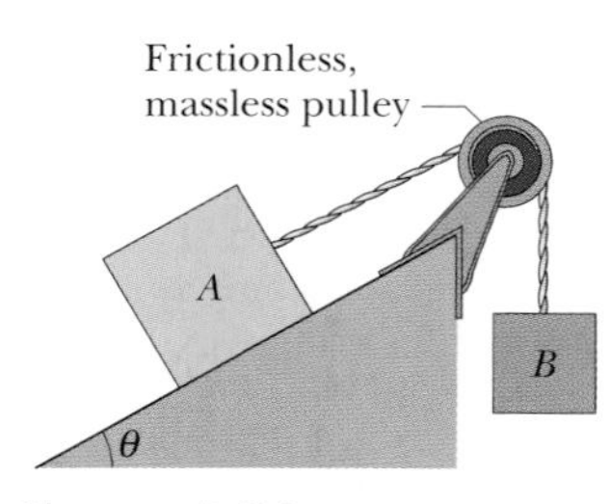

FIGURE 6-54 ▪ Problem 31 and 32.

32. Two Blocks and a Pulley In Fig. 6-54, two blocks are connected over a pulley. The mass of block A is 10 kg and the coefficient of kinetic friction between A and the incline is 0.20. Angle θ of the incline is 30°. Block A slides down the incline at constant speed. What is the mass of block B?

33. Two Blocks Massless String Two blocks of weights 3.6 N and 7.2 N are connected by a massless string and slide down a 30° inclined plane. The coefficient of kinetic friction between the lighter block and the plane is 0.10; that between the heavier block and the plane is 0.20. Assuming that the lighter block leads, find (a) the magnitude of the acceleration of the blocks and (b) the tension in the string. (c) Describe the motion if, instead, the heavier block leads.

34. Box of Cheerios® In Fig. 6-55, a box of Cheerios® and a box of Wheaties® are accelerated across a horizontal surface by a horizontal force $\vec{F}$ applied to the Cheerios® box. The magnitude of the frictional force on the Cheerios® box is 2.0 N, and the magnitude of the frictional force on the Wheaties® box is 4.0 N. If the magnitude of $\vec{F}$ is 12 N, what is the magnitude of the force on the Wheaties® box from the Cheerios® box?

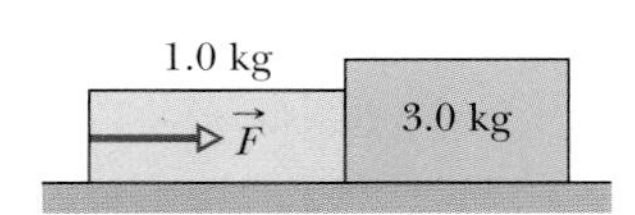

FIGURE 6-55 ▪ Problem 34.

35. Blocks Not Attached The two blocks (with $m = 16$ kg and $M = 88$ kg) shown in Fig. 6-56 are not attached. The coefficient of static friction between the blocks is $\mu^{\text{stat}} = 0.38$, but the surface beneath the larger block is frictionless. What is the minimum magnitude of the horizontal force $\vec{F}$ required to keep the smaller block from slipping down the larger block?

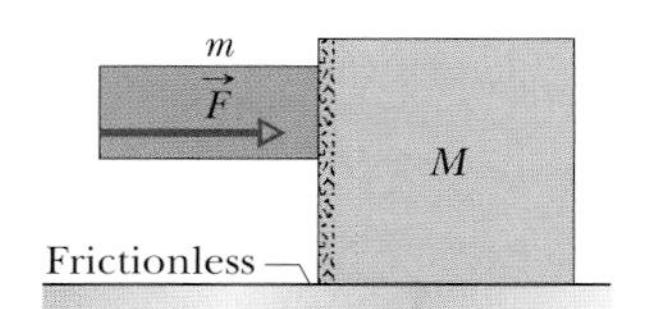

FIGURE 6-56 ▪ Problem 35.

36. Aunts and Uncles In Fig. 6-57, a box of ant aunts (total mass $m_A = 1.65$ kg) and a box of ant uncles (total mass $m_B = 3.30$ kg) slide down an inclined plane while attached by a massless rod parallel to the plane. The angle of incline is $\theta = 30°$. The coefficient of kinetic friction between the aunt box and the incline is $\mu_A^{\text{kin}} = 0.226$; that between the uncle box and the incline is $\mu_B^{\text{kin}} = 0.113$. Compute (a) the tension in the rod and (b) the common acceleration of the two boxes. (c) How would the answers to (a) and (b) change if the uncles trailed the aunts?

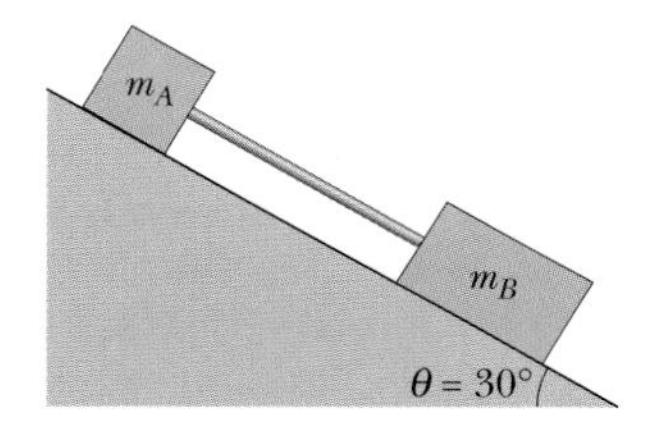

FIGURE 6-57 ▪ Problem 36.

37. Block on a Slab A 40 kg slab rests on a frictionless floor. A 10 kg block rests on top of the slab (Fig. 6-58). The coefficient of static friction μ^{stat} between the block and the slab is 0.60, whereas their kinetic friction coefficient μ^{kin} is 0.40. The 10 kg block is pulled by a horizontal force with a magnitude of 100 N. What are the resulting accelerations of (a) the block and (b) the slab?

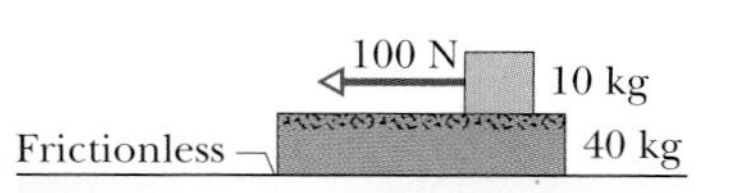

FIGURE 6-58 ▪ Problem 37.

38. A Locomotive A locomotive accelerates a 25-car train along a level track. Every car has a mass of 5.0×10^4 kg and is subject to a frictional force $f^{kin} = (250 \text{ N} \cdot \text{s/m})\vec{v}$. At the instant when the speed of the train is 30 km/h, the magnitude of its acceleration is 0.20 m/s^2. (a) What is the tension in the coupling between the first car and the locomotive? (b) If this tension is equal to the maximum force the locomotive can exert on the train, what is the steepest grade up which the locomotive can pull the train at 30 km/h?

39. Crate in a Trough In Fig. 6-59, a crate slides down an inclined right-angled trough. The coefficient of kinetic friction between the crate and the trough is μ^{kin}. What is the acceleration of the crate in terms of μ^{kin}, θ, and g?

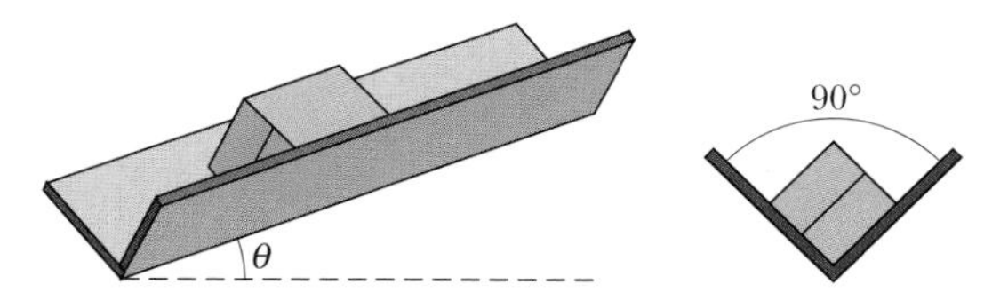

FIGURE 6-59 ■ Problem 39.

40. Box of Sand An initially stationary box of sand is to be pulled across a floor by means of a cable in which the tension should not exceed 1100 N. The coefficient of static friction between the box and the floor is 0.35. (a) What should be the angle between the cable and the horizontal in order to pull the greatest possible amount of sand, and (b) what is the weight of the sand and box in that situation?

41. Boat with Engine Off A 1000 kg boat is traveling at 90 km/h when its engine is shut off. The magnitude of the frictional force $\vec{f}^{kin}$ between boat and water is proportional to the speed v of the boat; $\vec{f}^{kin} = (70 \text{ N} \cdot \text{s/m})\vec{v}$. Find the time required for the boat to slow to 45 km/h.

SEC. 6-5 ■ DRAG FORCE AND TERMINAL SPEED

42. Continuation of Problem 14 First reread the explanation of how the wind might drag desert stones across the playa. Now assume that Eq. 6-23 gives the magnitude of the air drag force on the typical 20 kg stone, which presents a vertical cross-sectional area to the wind of 0.040 m^2 and has a drag coefficient C of 0.80. Take the air density to be 1.21 kg/m^3, and the coefficient of kinetic friction to be 0.80. (a) In kilometers per hour, what wind speed V along the ground is needed to maintain the stone's motion once it has started moving? Because winds along the ground are retarded by the ground, the wind speeds reported for storms are often measured at a height of 10 m. Assume wind speeds are 2.00 times those along the ground, (b) For your answer to (a), what wind speed would be reported for the storm and is that value reasonable for a high-speed wind in a storm?

43. Missile Calculate the drag force on a missile 53 cm in diameter cruising with a speed of 250 m/s at low altitude, where the density of air is 1.2 kg/m^3. Assume $C = 0.75$.

44. Sky Diver The terminal speed of a sky diver is 160 km/h in the spread-eagle position and 310 km/h in the nosedive position. Assuming that the diver's drag coefficient C does not change from one position to the other, find the ratio of the effective cross-sectional area A in the slower position to that in the faster position.

45. Jet Vs. Prop-Driven Transport Calculate the ratio of the drag force on a passenger jet flying with a speed of 1000 km/h at an altitude of 10 km to the drag force on a prop-driven transport flying at half the speed and half the altitude of the jet. At 10 km the density of air is 0.38 kg/m^3, and at 5.0 km it is 0.67 kg/m^3. Assume that the airplanes have the same effective cross-sectional area and the same drag coefficient C.

SEC. 6-6 ■ APPLYING NEWTON'S LAWS

46. Block on an Incline Refer to Fig. 6-29. Let the mass of the block be 8.5 kg and the angle θ be 30°. The block moves at constant velocity. Find (a) the tension in the cord and (b) the normal force acting on the block. (c) If the cord is cut, find the magnitude of the block's acceleration.

47. Electron Moving Horizontally An electron with a speed of 1.2×10^7 m/s moves horizontally into a region where a constant vertical force of 4.5×10^{-16} N acts on it. The mass of the electron is 9.11×10^{-31} kg. Determine the vertical distance the electron is deflected during the time it has moved 30 mm horizontally.

48. Tarzan Tarzan, who weighs 820 N, swings from a cliff at the end of a 20 m vine that hangs from a high tree limb and initially makes an angle of 22° with the vertical. Immediately after Tarzan steps off the cliff, the tension in the vine is 760 N. Choose a coordinate system for which the x axis points horizontally away from the edge of the cliff and the y axis points upward. (a) What is the force of the vine on Tarzan in unit-vector notation? (b) What is the net force acting on Tarzan in unit-vector notation? What are the (c) magnitude and (d) direction of the net force acting on Tarzan? What are the (e) magnitude and (f) direction of Tarzan's acceleration?

49. Skier on a Rope Tow A 50 kg skier is pulled up a frictionless ski slope that makes an angle of 8.0° with the horizontal by holding onto a tow rope that moves parallel to the slope. Determine the magnitude of the force of the rope on the skier at an instant when (a) the rope is moving with a constant speed of 2.0 m/s and (b) the rope is moving with a speed of 2.0 m/s but that speed is increasing at a rate of 0.10 m/s^2.

50. Running Armadillo For sport, a 12 kg armadillo runs onto a large pond of level, frictionless ice with an initial velocity of 5.0 m/s along the positive direction of an x axis. Take its initial position on the ice as being the origin. It slips over the ice while being pushed by a wind with a force of 17 N in the positive direction of the y axis. In unit-vector notation, what are the animal's (a) velocity and (b) position vector when it has slid for 3.0 s?

51. Sphere Suspended from a Cord A sphere of mass 3.0×10^{-4} kg is suspended from a cord. A steady horizontal breeze pushes the sphere so that the cord makes a constant angle of 37° with the vertical. Find (a) the magnitude of that push and (b) the tension in the cord.

52. Skier in the Wind A 40 kg skier comes directly down a frictionless ski slope that is inclined at an angle of 10° with the horizontal while a strong wind blows parallel to the slope. Determine the magnitude and direction of the force of the wind on the skier if (a) the magnitude of the skier's velocity is constant, (b) the magnitude of the skier's velocity is increasing at a rate of 1.0 m/s^2. and (c) the magnitude of the skier's velocity is increasing at a rate of 2.0 m/s^2.

53. Jet Engine A 1400 kg jet engine is fastened to the fuselage of a passenger jet by just three bolts (this is the usual practice). Assume that each bolt supports one-third of the load. (a) Calculate the force on each bolt as the plane waits in line for clearance to take off. (b) During flight, the plane encounters turbulence, which suddenly imparts an upward vertical acceleration of 2.6 m/s^2 to the plane. Calculate the force on each bolt now.

54. Pulling a Crate A worker drags a crate across a factory floor by pulling on a rope tied to the crate (Fig. 6-60). The worker exerts a force of 450 N on the rope, which is inclined at 38° to the horizontal, and the floor exerts a horizontal force of 125 N that opposes the motion. Calculate the magnitude of the acceleration of the crate if (a) its mass is 310 kg or (b) its weight is 310 N.

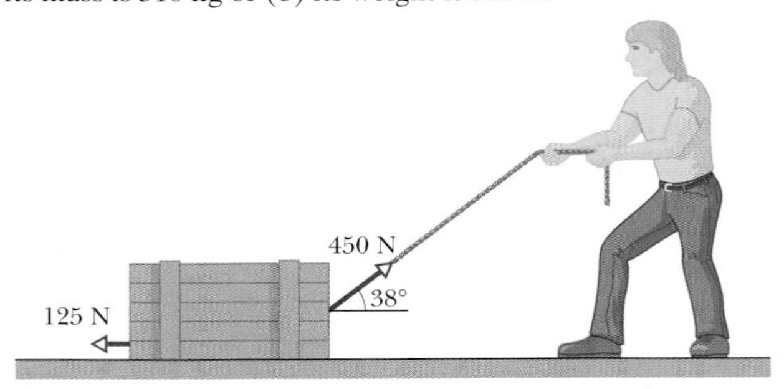

FIGURE 6-60 ■ Problem 54.

55. Motorcycle Rider A motorcycle and 60.0 kg rider accelerate at 3.0 m/s² up a ramp inclined 10° above the horizontal. (a) What is the magnitude of the net force acting on the rider? (b) What is the magnitude of the force on the rider from the motorcycle?

56. One on an Incline—One Hanging A block of mass m_A = 3.70 kg on a frictionless inclined plane of angle 30.0° is connected by a cord over a massless, frictionless pulley to a second block of mass m_B = 2.30 kg hanging vertically (Fig. 6-61). What are (a) the magnitude of the acceleration of each block and (b) the direction of the acceleration of the hanging block? (c) What is the tension in the cord?

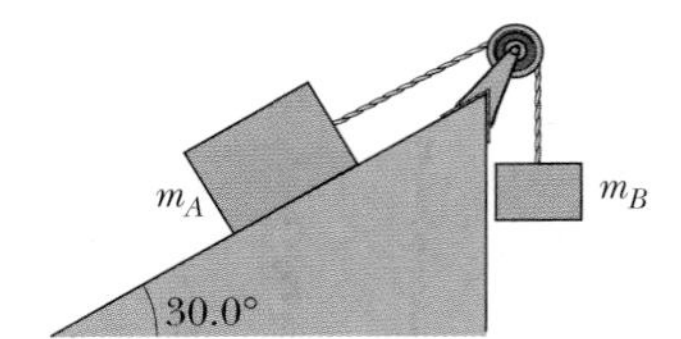

FIGURE 6-61 ■ Problem 56.

57. Pencil Box In Fig. 6-62, a 1.0 kg pencil box on a 30° frictionless incline is connected to a 3.0 kg pen box on a horizontal frictionless surface. The pulley is frictionless and massless. (a) If the magnitude of the applied force $\vec{F}$ is 2.3 N, what is the tension in the connecting cord? (b) What is the largest value that the magnitude of $\vec{F}$ may have without the connecting cord becoming slack?

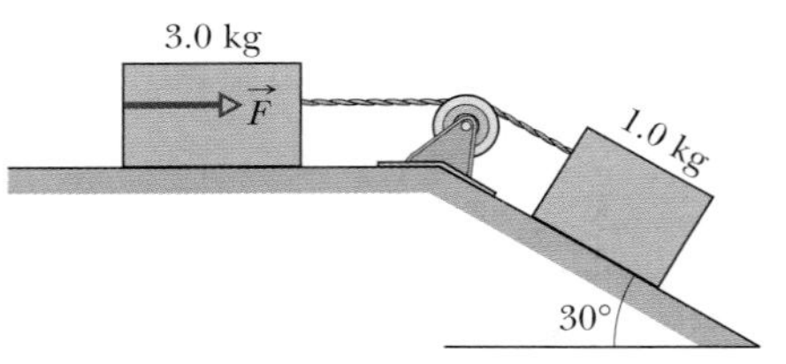

FIGURE 6-62 ■ Problem 57.

58. Projected Up an Incline A block is projected up a frictionless inclined plane with initial speed v_1 = 3.50 m/s. The angle of incline is θ = 32.0°. (a) How far up the plane does it go? (b) How long does it take to get there? (c) What is its speed when it gets back to the bottom?

59. Horse-Drawn Barge In earlier days, horses pulled barges down canals in the manner shown in Fig. 6-63. Suppose the horse pulls on the rope with a force of 7900 N at an angle of 18° to the direction of motion of the barge, which is headed straight along the canal. The mass of the barge is 9500 kg, and its acceleration is 0.12 m/s². What are the (a) magnitude and (b) direction of the force on the barge from the water?

FIGURE 6-63 ■ Problem 59.

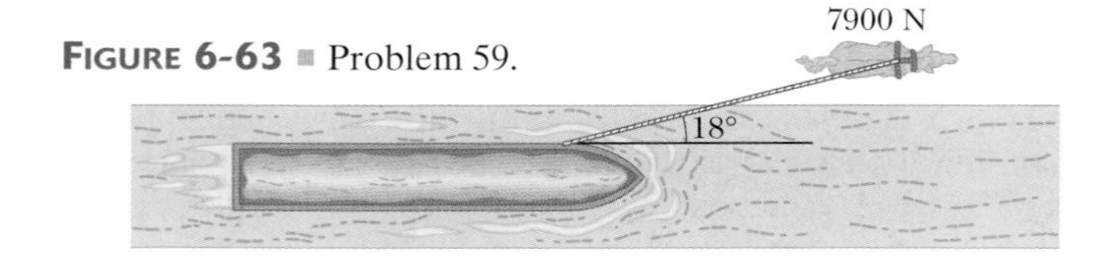

60. Lifting a Block In Fig. 6-64, a 5.00 kg block is pulled along a horizontal frictionless floor by a cord that exerts a force of magnitude F = 12.0 N at an angle θ = 25.0° above the horizontal. (a) What is the magnitude of the block's acceleration? (b) The force magnitude F is slowly increased. What is its value just before the block is lifted (completely) off the floor? (c) What is the magnitude of the block's acceleration just before it is lifted (completely) off the floor?

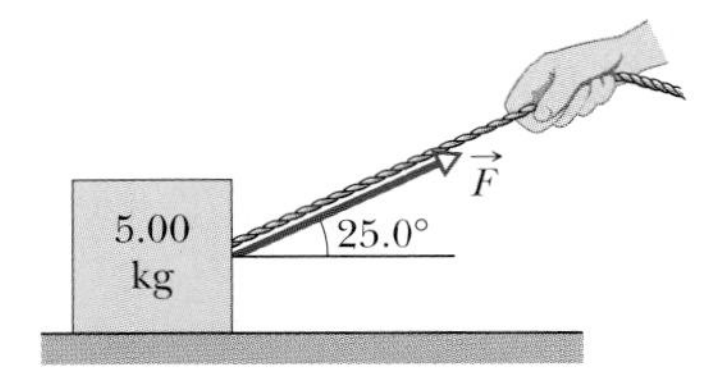

FIGURE 6-64 ■ Problem 60.

61. A Rope Must Sag A block of mass M is pulled along a horizontal frictionless surface by a rope of mass m, as shown in Fig. 6-65. A horizontal force $\vec{F}$ is applied to one end of the rope. (a) Show that the rope *must* sag, even if only by an imperceptible amount. Then, assuming the sag is negligible, find (b) the acceleration of rope and block, (c) the force on the block from the rope, and (d) the tension in the rope at its midpoint.

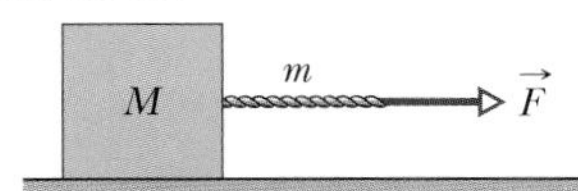

FIGURE 6-65 ■ Problem 61.

62. Crate at Constant Speed In Fig. 6-66, a 100 kg crate is pushed at constant speed up the frictionless 30.0° ramp by a horizontal force $\vec{F}$. What are the magnitudes of (a) $\vec{F}$ and (b) the force on the crate from the ramp?

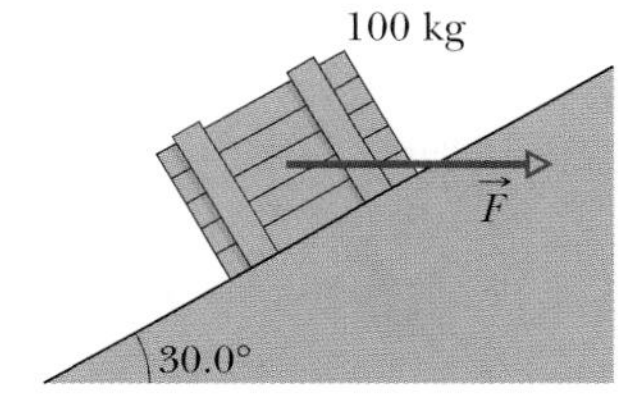

FIGURE 6-66 ■ Problem 62.

63. Alpine Cable Car Figure 6-67 shows a section of an alpine cable-car system. The maximum permissible mass of each car with occupants is 2800 kg. The cars, riding on a support cable, are pulled by a second cable attached to each pylon (support tower); assume the cables are straight. What is the difference in tension between adjacent sections of pull cable if the cars are at the maximum permissible mass and are being accelerated up the 35° incline at 0.81 m/s²?

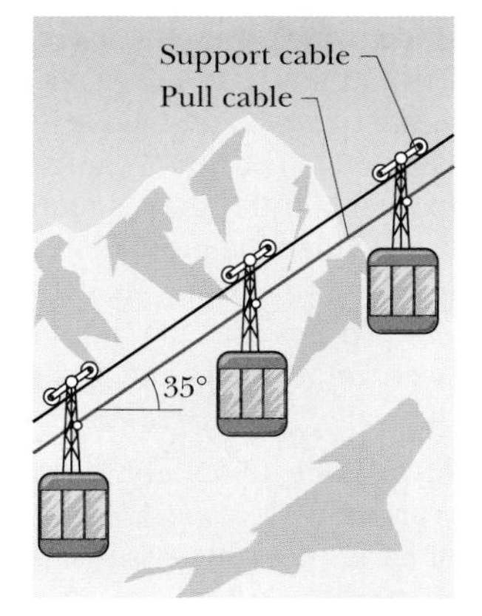

FIGURE 6-67 ■ Problem 63.

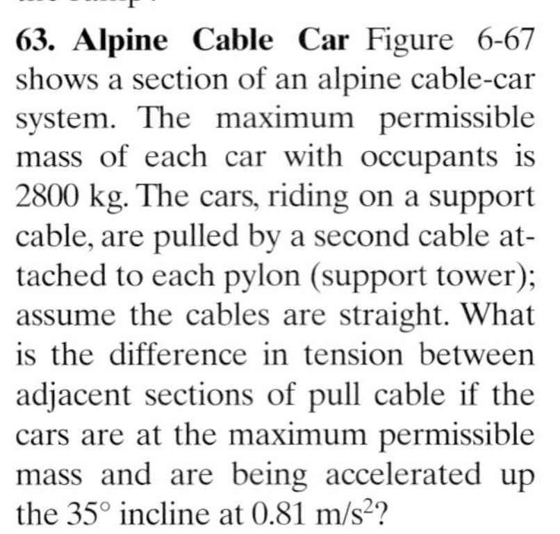

64. Bobsled Run During an Olympic bobsled run, the Jamaican team makes a turn of radius 7.6 m at a speed of 96.6 km/h. What is their acceleration in g-units? (1 g-unit = 9.8 m/s².)

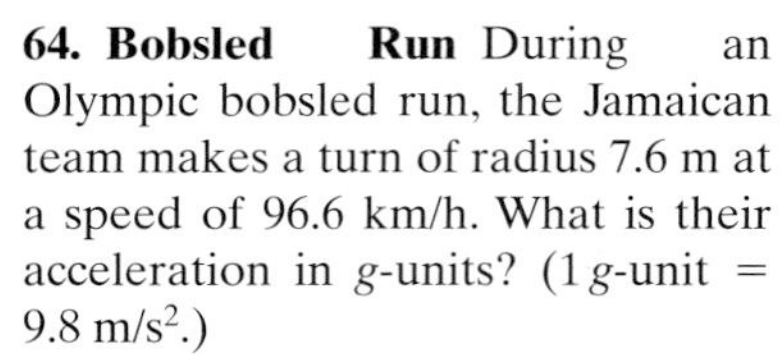

65. Grand Prix Suppose the coefficient of static friction between the road and the tires on a Formula One car is 0.6 during a Grand Prix auto race. What speed will put the car on the verge of sliding as it rounds a level curve of 30.5 m radius?

66. Roller Coaster A roller-coaster car has a mass of 1200 kg when fully loaded with passengers. As the car passes over the top of a circular hill of radius 18 m, its speed is not changing. What are the magnitude and direction of the force of the track on the car at the top of the hill if the car's speed is (a) 11 m/s and (b) 14 m/s?

67. Flat Track What is the smallest radius of an unbanked (flat) track around which a bicyclist can travel if her speed is 29 km/h and the coefficient of static friction between tires and track is 0.32?

68. Amusement Park Ride An amusement park ride consists of a car moving in a vertical circle on the end of a rigid boom of negligible mass. The combined weight of the car and riders is 5.0 kN, and the radius of the circle is 10 m. What are the magnitude and direction of the force of the boom on the car at the top of the circle if the car's speed there is (a) 5.0 m/s and (b) 12 m/s?

69. Puck on a Table A puck of mass m slides on a frictionless table while attached to a hanging cylinder of mass M by a cord through a hole in the table (Fig. 6-68). What speed keeps the cylinder at rest?

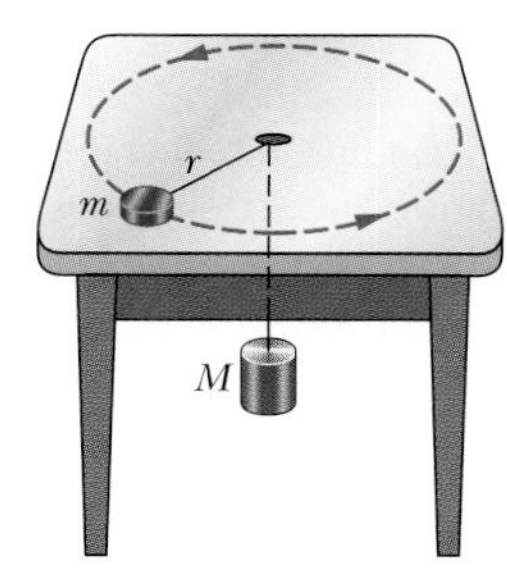

FIGURE 6-68 ■ Problem 69.

70. Bicyclist A bicyclist travels in a circle of radius 25.0 m at a constant speed of 9.00 m/s. The bicycle–rider mass is 85.0 kg. Calculate the magnitudes of (a) the force of friction on the bicycle from the road and (b) the *total* force on the bicycle from the road.

71. Student on Ferris Wheel A student of weight 667 N rides a steadily rotating Ferris wheel (the student sits upright). At the highest point, the magnitude of the normal force $\vec{N}$ on the student from the seat is 556 N. (a) Does the student feel "light" or "heavy" there? (b) What is the magnitude of $\vec{N}$ at the lowest point? (c) What is the magnitude N if the wheel's speed is doubled?

72. Old Streetcar An old streetcar rounds a flat corner of radius 9.1 m, at 16 km/h. What angle with the vertical will be made by the loosely hanging hand straps?

73. Flying in a Circle An airplane is flying in a horizontal circle at a speed of 480 km/h. If its wings are tilted 40° to the horizontal, what is the radius of the circle in which the plane is flying? (See Fig. 6-69.) Assume that the required force is provided entirely by an "aerodynamic lift" that is perpendicular to the wing surface.

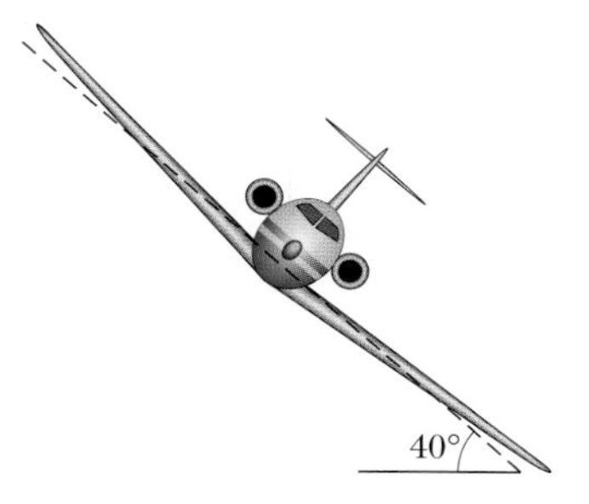

FIGURE 6-69 ■ Problem 73.

74. High-Speed Railway A high-speed railway car goes around a flat, horizontal circle of radius 470 m at a constant speed. The magnitudes of the horizontal and vertical components of the force of the car on a 51.0 kg passenger are 210 N and 500 N, respectively. (a) What is the magnitude of the net force (of *all* the forces) on the passenger? (b) What is the speed of the car?

75. Ball Connected to a Rod As shown in Fig. 6-70, a 1.34 kg ball is connected by means of two massless strings to a vertical, rotating rod. The strings are tied to the rod and are taut. The tension in the upper string is 35 N. (a) Draw the free-body diagram for the ball. What are (b) the tension in the lower string, (c) the net force on the ball, and (d) the speed of the ball?

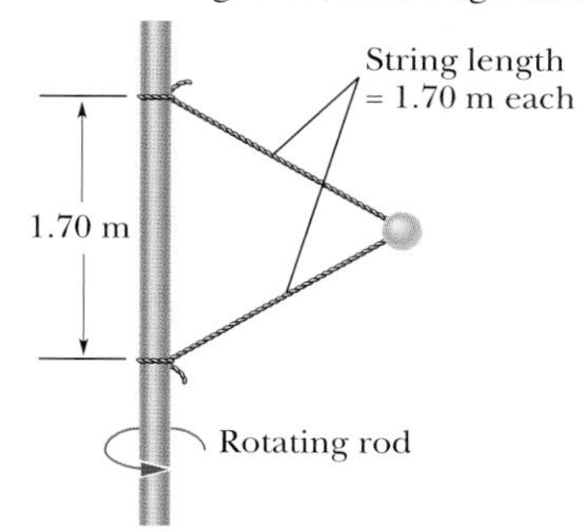

FIGURE 6-70 ■ Problem 75.

76. Pushing the Second Block A 2.0 kg block and a 1.0 kg block are connected by a string and are pushed across a horizontal surface by a force applied to the 1.0 kg block as shown in Fig. 6-71. The coefficient of kinetic friction between the blocks and the horizontal surface is 0.20. If the magnitude of $\vec{F}$ is 20 N, what is the tension in the string that connects the blocks?

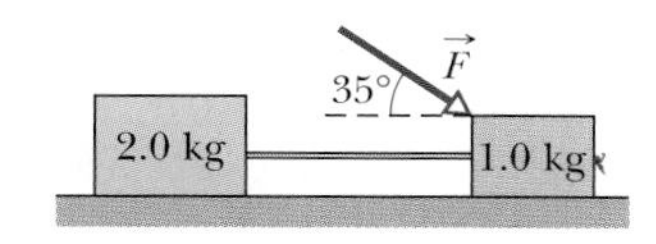

FIGURE 6-71 ■ Problem 76.

Additional Problems

77. Engineering a Highway Curve If a car goes through a curve too fast, the car tends to slide out of the curve, as discussed in Touchstone Example 6-8. For a banked curve with friction, a frictional force acts on a fast car to oppose the tendency to slide out of the curve; the force is directed down the bank (in the direction in which water would drain). Consider a circular curve of radius $R = 200$ m and bank angle θ, where the coefficient of static friction between tires and pavement is μ^{stat}. A car is driven around the curve as shown in Fig. 6-72. (a) Find an expression for the car speed v^{max} that puts the car on the verge of sliding out. (b) On the same graph, plot v^{max} versus angle θ for the range 0° to 50°, first for $\mu^{\text{stat}} = 0.60$ (dry pavement) and then for $\mu^{\text{stat}} = 0.050$ (wet or icy pavement). In kilometers per hour, evaluate v^{max} for a bank angle of $\theta = 10°$ and for (c) $\mu^{\text{stat}} = 0.60$ and (d) $\mu^{\text{stat}} = 0.050$. (Now you can see why accidents occur in highway curves when wet or icy conditions are not obvious to drivers, who tend to drive at normal speeds.)

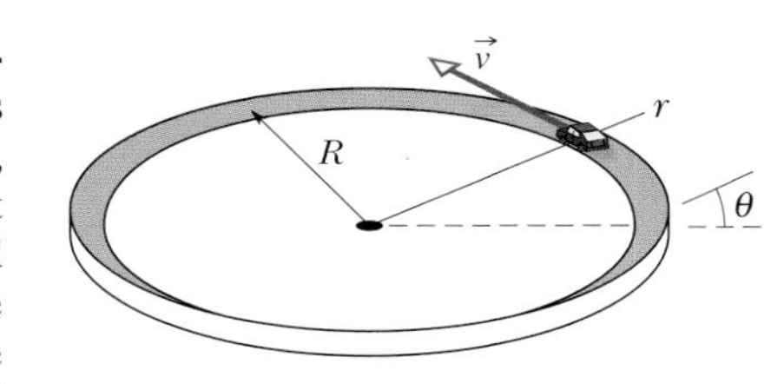

FIGURE 6-72 ■ Problem 77.

78. Change in Conditions In the early afternoon, a car is parked on a street that runs down a steep hill, at an angle of 35.0° relative to the horizontal. Just then the coefficient of static friction between the tires and the street surface is 0.725. Later, after nightfall, a sleet storm hits the area, and the coefficient decreases due to both the ice and a chemical change in the road surface because of the temperature decrease. By what percentage must the coefficient decrease if the car is to be in danger of sliding down the street?

79. Moving People at the Airport While traveling, I passed through Charles de Gaulle Airport in Paris, France. The airport has some interesting devices, including a "people mover"—a moving strip of rubber like a horizontal escalator without steps. It became interesting when the mover entered a plastic tube bent up at an angle to take me to the next terminal. I managed to get a photograph of it (Fig. 6-73). If you were building this people mover for the architect, what material would you choose for the surface of the moving strip? (*Hint*: You want to be sure that people standing on the strip do not tend to slide down it. Figure out what coefficient of friction you need to keep from sliding down and then look up coefficients of friction in tables in reference books to get a material appropriate for the slipperiest shoes.)

FIGURE 6-73 ■ Problem 79.

80. Expert Witness You testify as an *expert witness* in a case involving an accident in which car A slid into the rear of car B, which was stopped at a red light along a road headed down a hill (Fig. 6-74). You find that the slope of the hill is $\theta = 12.0°$, that the cars were separated by distance $d = 24.0$ m when the driver of car A put the car into a slide (it lacked any automatic anti-brake-lock system), and that the speed of car A at the onset of braking was $v_1 = 18.0$ m/s. With what speed did car A hit car B if the coefficient of kinetic friction was (a) 0.60 (dry road surface) and (b) 0.10 (road surface covered with wet leaves)?

FIGURE 6-74 ■ Problem 80.

81. Luggage Transport Luggage is transported from one location to another in an airport by a conveyor belt. At a certain location, the belt moves down an incline that makes an angle of 2.5° with the horizontal. Assume that with such a slight angle there is no slipping of the luggage. Determine the magnitude and direction of the frictional force by the belt on a box weighing 69 N when the box is on the inclined portion of the belt for the following situations: (a) The belt is stationary. (b) The belt has a speed of 0.65 m/s that is constant. (c) The belt has a speed of 0.65 m/s that is increasing at a rate of 0.20 m/s^2. (d) The belt has a speed of 0.65 m/s that is decreasing at a rate of 0.20 m/s^2. (e) The belt has a speed of 0.65 m/s that is increasing at a rate of 0.57 m/s^2.

82. Bolt on a Rod A bolt is threaded onto one end of a thin horizontal rod, and the rod is then rotated horizontally about its other end. An engineer monitors the motion by flashing a strobe lamp onto the rod and bolt, adjusting the strobe rate until the bolt appears to be in the same eight places during each full rotation of the rod (Fig. 6-75). The strobe rate is 2000 flashes per second; the bolt has mass 30 g and is at radius 3.5 cm. What is the magnitude of the force on the bolt from the rod?

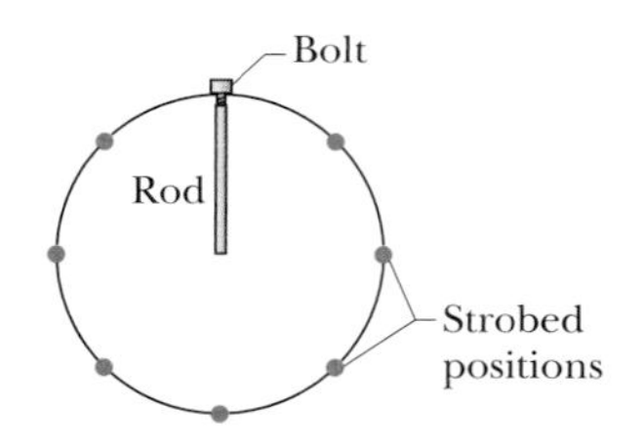

FIGURE 6-75 ■ Problem 82.

83. From the Graph A 4.10 kg block is pushed along a floor by a constant applied force that is horizontal and has a magnitude of 40.0 N. Figure 6-76 gives the block's speed v versus time t as the block moves along an x axis on the floor. What is the coefficient of kinetic friction between the block and the floor?

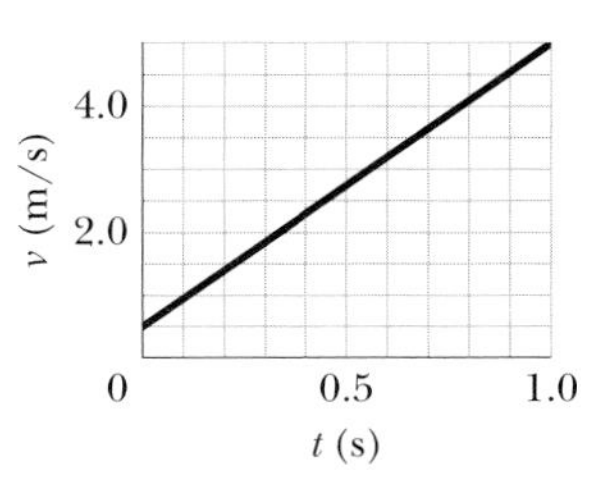

FIGURE 6-76 ■ Problem 83.

84. Tapping a Rolling Ball Figure 6-77 shows a multiple exposure strobe photograph of a ball rolling on a horizontal table. The image marked with a heavy arrow occurs at time $t = 0$ and the ball moves to the right at that instant. Each image of the ball occurs 1/30 s later than the one immediately to its left. Using the coordinate system shown in Fig. 6-77, sketch qualitatively accurate (i.e., we don't care about the values but we do care about the shape) graphs of each of the following variables as a function of time: x coordinate, y coordinate, x-component of velocity, y-component of velocity, x-component of the net force on the ball, and y-component of the net force on the ball. The time at which the "kink" in the path occurs is $t = t_1$. Be sure to note this important time on your graphs.

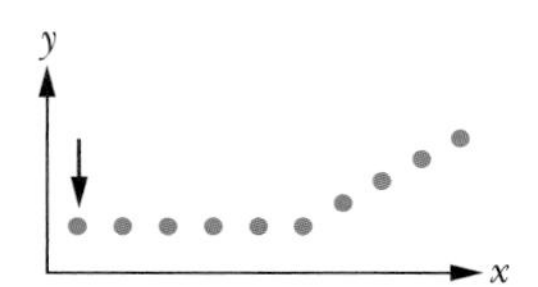

FIGURE 6-77 ■ Problem 84.

85. Ball on a Ramp Figure 6-78 shows a multiple-exposure photograph of a ball rolling up an inclined plane. (The ball is rolling in the dark, the camera lens is held open, and a brief flash occurs every 3/4 sec, four times in total.) The leftmost ball corresponds to an instant just after the ball was released. The rightmost ball is at the highest point the ball reaches.

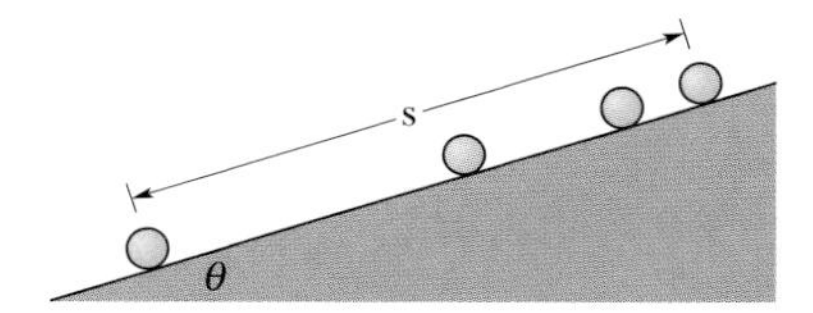

FIGURE 6-78 ■ Problem 85.

(a) Copy this picture on your paper. Draw an arrow at each of the four ball locations to indicate the velocity of the ball at that instant. Make the relative lengths of the arrows indicate the relative magnitudes of the velocities. Explain what is happening ("tell the story" of the picture).

(b) For the instant of time when the ball is at the second position shown from the left, draw a free-body diagram for the ball and indicate all forces acting on it.

(c) If your force diagram doesn't include an arrow pointing up the ramp, explain why the ball keeps rolling up the ramp.

(d) If the mass of the ball is m, what is its acceleration?

(e) If the angle θ is equal to 30°, how long is the distance s?

86. Motion Graphs (a) Suppose you were to push on a bowling ball on a smooth floor at a 45° angle as shown in Fig. 6-79*a* and then leave it alone to roll. Sketch a graph frame like that shown in Fig. 6-79*a*, and then sketch a prediction of the ball's motion both before and after you stop pushing. Note on your graph the point at which you stop pushing and explain the basis for your prediction.

(b) If the initial speed of the ball is 3.5 m/s, what is the magnitude of the x-component of velocity, v_{1x}? Is it positive or negative? What is v_{1y}? Is it positive or negative?

(0, 0) Ball starts at 45 deg & rolls freely on a floor +x −y (a)

Direction of second tapper (0, 0) Direction of first tapper Ball starts at rest on a floor +x −y (b)

(0, 0) Rocket is thrust in the x-direction and falls freely in the y-direction +x −y (c)

FIGURE 6-79 ■ Problem 86.

(c) Suppose you and your partner were to tap the ball *very* rapidly. Each set of taps is at right angles to the other as shown in Fig. 6-79*b*. Sketch a graph frame like that shown in Fig. 6-79*b*, and sketch a prediction of the ball's motion on your graph. Explain the basis for your prediction.

(d) Suppose a rocket ship is thrust from a tower at a constant acceleration that has a magnitude of about 9.8 m/s^2 in the x direction and is allowed to fall freely toward Earth in the y direction. Sketch a graph frame like that shown in Fig. 6-79*c*, and sketch a prediction of the rocket's motion on your graph. Explain the basis for your prediction.

87. Wanda Lifts Weights Wanda is working out with weights and manages to lift a light rope with a 10 kg mass hanging from it. When she is through lifting the right side of the rope and the left side of the rope each make an angle of $\theta = 15°$ with respect to the horizontal. See Fig. 6-80.

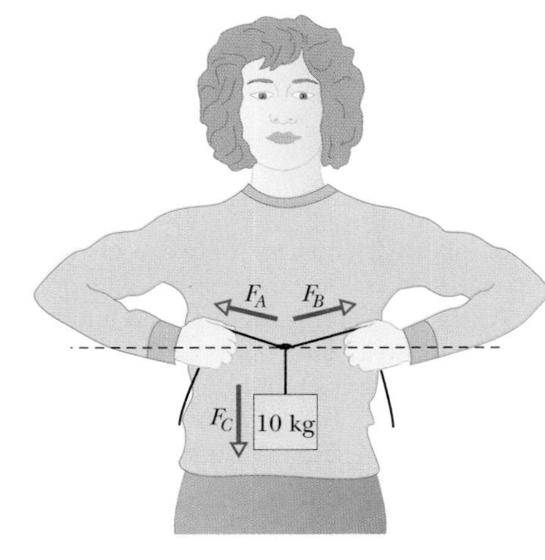

FIGURE 6-80 ■ Problem 87.

(a) Draw a free-body diagram showing the forces on the midpoint of the rope (where it is the lowest).

(b) What are the magnitudes of each of her pulling forces $\vec{F}_A$ and $\vec{F}_B$?

(c) How hard would Wanda have to pull with each hand to raise the 10 kg mass so that the rope becomes perfectly horizontal?

88. Constant Speed on a Race Track The race track shown in Fig. 6-81 has two straight sections connected by semicircular ends. A car is traveling in a clockwise direction around the track at a constant speed. Assume that air resistance is negligible. Draw three sketches of the race track.

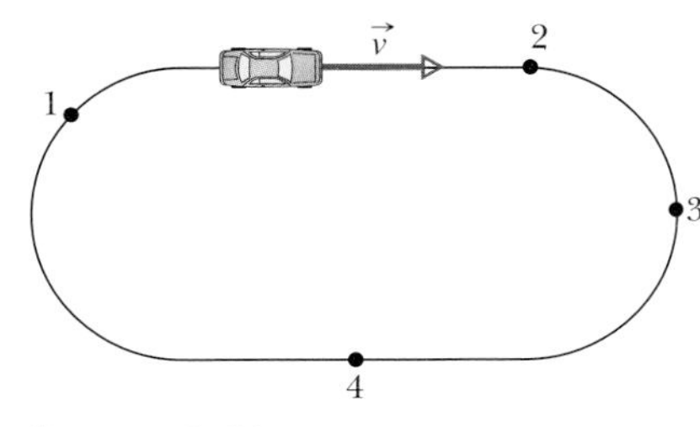

FIGURE 6-81 ■ Problem 88.

(a) On the first sketch show the velocity vector at each of the numbered points 1–4. Make the relative lengths of the vectors consistent with the relative magnitudes of the velocity at the four points.

(b) On the second sketch show the acceleration vectors at each of the numbered points 1–4. Make the relative lengths of the vectors consistent with the relative magnitudes of the acceleration at the four points. *Hint*: Use the techniques developed in Chapter 5 to draw vectors representing the acceleration or change in velocity.

(c) Horizontal forces are needed to maintain the car's motion around the track. These are provided by road friction and by road forces where the track is banked at the curves. On the third sketch show the vectors representing the required horizontal forces at each of the numbered points 1–4. Make the relative lengths of the vectors consistent with the relative magnitudes of the force at the four points.

Note: This exercise is adapted from A. Arons, *Homework and Test Questions for Introductory Physics Teaching* (New York: Wiley, 1994), Chapter 3.

89. Pulling on the Ceiling Suppose a person exerts a force of 50 N on one end of a rope as shown in Fig. 6-82.

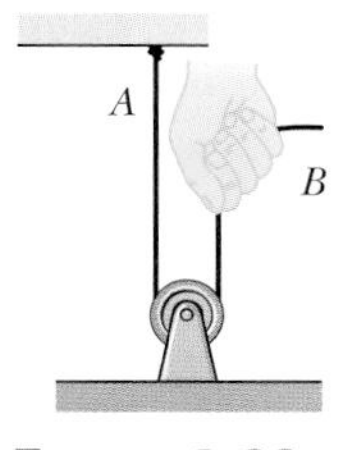

FIGURE 6-82 ■ Problem 89.

(a) What are the magnitude and direction of the force at point A exerted on the rope by the ceiling?

(b) What are the magnitude and direction of the force exerted on the ceiling by the rope? How does the force get transmitted from one end of the rope to the other? What does the stretching of the rope have to do with this?

(c) What are the magnitude and direction of the force the rope exerts on the person's hand at point B?

(d) Draw a diagram with vector arrows indicating the *relative magnitudes* and *directions* of the forces the rope exerts on the ceiling at point A and the force the rope exerts on the person's hand at point B.

90. Thinking About Normal Forces Suppose you push on a flexible piece of stretched fabric with a force of 5.0 N as shown in Fig. 6-83*a*. The fabric assembly is fixed and does not move.

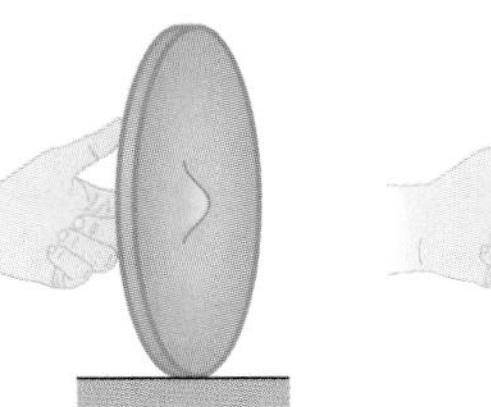

FIGURE 6-83*a* ■ Problem 90.

(a) What are the direction and magnitude of the normal force exerted back on the finger by the sheet? Is this normal force zero? If not, is it larger, smaller, or the same as the normal force would be if the fabric did not stretch?
(b) Discuss the role the stretching of the fabric plays in regard to this normal force.
(c) Suppose you push in the same way on a wall as shown in Fig. 6-83*b*. What are the direction and magnitude of the normal force exerted back on the finger by the wall?
(d) Does the wall stretch noticeably? What causes the wall to be able to exert a force on the finger? How does the wall "know" what force to exert back on the hand?

91. Forces in a Car Suppose you are sitting in a car that is speeding up. Assume the car has rear-wheel drive.

(a) Draw free-body diagrams for your own body, the seat in which you are sitting (apart from the car), the car (apart from the seat), and the road surface where the tires and the road interact.
(b) Describe each force in words; show larger forces with longer arrows.
(c) Identify the third-law pairs of forces.
(d) Explain carefully in your own words the origin of the force imparting acceleration to the car.

92. The Sliding Pizza One day I was coming home late from work and stopped to pick up a pizza for dinner. I put the pizza box on the dashboard of my car and pushed it forward against the windshield and left against the steering wheel to prevent it from falling. (See Fig. 6-84.) Before I started driving, I realized that the box could still slide to the right or back toward the seat. When driving, do I have to worry more about it sliding when I turn left or when I turn right? Do I have to worry more when I speed up or when I slow down? Explain your answer in terms of the physics you have learned.

FIGURE 6-84 ■ Problem 92.

93. The Farmer and the Donkey An old Yiddish joke is told about a farmer in Chelm, a town famous for the lack of wisdom of its inhabitants. One day the farmer was going to the mill to have a bag of wheat ground into flour. He was riding to the mill on his donkey, with the sack of wheat thrown over the donkey's back behind him. On his way, he met a friend. His friend chastised him. "Look at you! You must weigh 200 pounds and that sack of flour must weigh 100. That's a very small donkey! Together, you're too much weight for him to carry!" On his way to the mill the farmer thought about what his friend had said. On his way home, he passed his friend again, confident that this time the friend would be satisfied. The farmer still rode the donkey, but this time he carried the 100 pound bag of flour on his own shoulder!

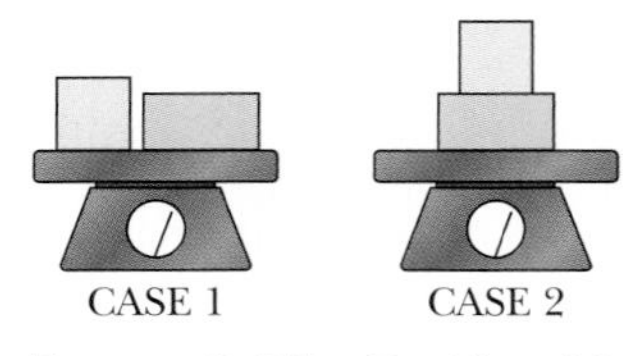

FIGURE 6-85 ■ Problem 93.

Our common sense and intuitions seem to suggest that it doesn't matter how you arrange things; they'll weigh the same. Let's be certain that the Newtonian framework we are developing yields our intuitive result. Analyze the problem by considering the simplified picture shown in Fig. 6-85. Two blocks rest on a scale. One block weighs 10 N, the other 25 N. In case 1 the blocks are arranged on the scale as shown in the figure on the left. In case 2 the blocks are arranged as shown on the right. Each system has come to rest. Analyze the forces on the blocks and on the scale in the two cases by isolating the objects—each block and the scale—and show that according to the principles of Newton's laws, the total force exerted on the scale by both blocks together must be the same in both cases. (*Note*: It's not enough to say: "They have to be the same." That's just restating your intuition. We need to see that *reasoning using only the principles of our Newtonian framework* leads to the same conclusion.)

94. Pulling Two Boxes (a) A worker is trying to pull a pair of heavy crates along the floor with a rope. The rope is attached to the lower crate, which has a mass M. The upper crate has a mass m and the coefficient of static friction between the crate and the floor is μ^{stat}. If the rope is held at an angle θ as shown in Fig. 6-86, what is the magnitude of the maximum force the worker can exert without the lower crate beginning to slide?

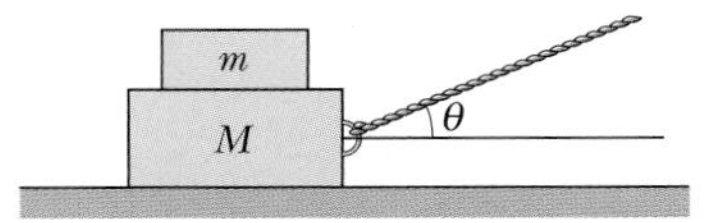

FIGURE 6-86 ■ Problem 94.

(b) The worker knows that the lower crate has a mass of 50 kg and the upper crate has a mass of 10 kg. She finds that if she pulls with a force of 120 N at an angle of 60° she can keep the crates sliding at a constant speed. Can you use this information to find the coefficient of kinetic friction μ^{kin} between the lower crate and the floor? If you can, do it. If you can't, explain why not.
(c) In a different situation, she finds that she can pull a lower crate of mass 30 kg and an upper crate of mass 7.5 kg with a constant velocity of 50 cm/s pulling at an angle of 45°. Can you use this information to find the coefficient of kinetic friction μ^{kin} between the lower crate and the floor? If you can, do it. If you can't, explain why not.

95. Tricking Bill A student, whom we will call Bill, was about to go out on a date when his roommate, Bob, asked him to hold a pail against the ceiling with a broom for a moment. After Bill complied, the roommate mentioned that the pail was filled with water and left. See Fig. 6-87.

FIGURE 6-87 ■ Problem 95.

(a) Draw a free-body diagram showing all the forces acting on the pail. For each force, be sure you identify the kind of force and the object whose interaction with the pail is responsible for the force.
(b) Suppose Bill wants to slide the pail a few feet to one side so he can get to a chair in the room. Are there any other forces not specified in your answer to part (a) that become relevant?

(c) Suppose the pail weighs 1 pound, it has 6 pounds of water in it, the maximum coefficient of static friction $\mu_{\text{broom}}^{\text{stat}}$ between the broom and pail is 0.3, and the maximum coefficient of static friction $\mu_{\text{ceiling}}^{\text{stat}}$ between the pail and the ceiling is 0.5. Can Bill slide the pail? Explain.

96. Friction is Doing *What?* A large block is resting on the table. On top of that block rests another, smaller block, as shown in Fig. 6-88. You press on the larger block to start it moving. After about 0.25 s, it is moving at a constant speed and the block on the top is not slipping.

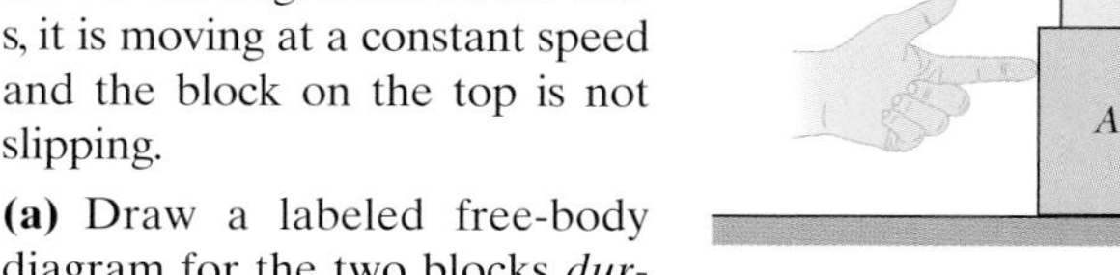

FIGURE 6-88 ■ Problem 96.

(a) Draw a labeled free-body diagram for the two blocks *during the time when they are accelerating*, specifying all the forces acting on the blocks. (Be sure to specify the type of force and the object causing each force.) Wherever you can, compare the magnitudes of forces.

(b) Draw a labeled free-body diagram for the two blocks *during the time when they are moving at a constant speed*, specifying all the forces acting on the blocks. (Be sure to specify the type of force and the object causing each force.) Wherever you can, compare the magnitudes of forces.

(c) Suppose the bottom block has a mass of 0.4 kg and the coefficient of friction between the block and the table is 0.3. The top block has a mass of 0.1 kg and the coefficient of friction between the two blocks is 0.2. What force do you need to exert to keep the blocks moving at a constant speed of 10 cm/s? (You may use $g =$ 10 N/kg and you may treat kinetic and static friction as the same.)

97. Al and George Pushing the Truck George left the lights on in his truck while at a truck stop in Kansas and his battery went dead. Fortunately, his friend Al is there, although Al is driving his Geo Metro. Since the road is very flat, George is able to convince Al to give his truck a long, slow push to get it up to 20 miles/hour. At this speed, George can engage the truck's clutch and the truck's engine should start up. (See Fig. 6-89.)

FIGURE 6-89 ■ Problem 97.

(a) Al begins to push the truck. It takes him 5 minutes to get the truck up to a speed of 20 miles/hour. Draw separate free-body diagrams for the Geo and for the truck during the time that Al's Geo is pushing the truck. List all the horizontal forces in order by magnitude from largest to smallest. If any are equal, state that explicitly. Explain your reasoning.

(b) If the truck is accelerating uniformly over the 5 minutes, how far does Al have to push the truck before George can engage the clutch?

(c) Suppose the mass of the truck is 4000 kg, the mass of the car is 800 kg, and the coefficient of static friction between the vehicles and the road is 0.1. At one instant when they are trying to get the truck moving, the car is pushing the truck and exerting a force of 1000 N, but neither vehicle moves. What is the static frictional force between the truck and the road? Explain your reasoning.

98. Pushing a Carriage A young man is pushing a baby carriage at a constant velocity along a level street. A friend comes by to chat and the young man lets go of the carriage. It rolls on for a bit, slows, and comes to a stop. At time $t = 0$ the young man is walking with a constant velocity. At time t_1 he releases the carriage. At time t_2 the carriage comes to rest. Sketch qualitatively accurate (i.e., we don't care about the values but we do care about the shape) graphs of each of the following variables versus time:

(a) position of the carriage, (b) velocity of the carriage, (c) acceleration of the carriage, (d) net force on the carriage, (e) force the man exerts on the carriage, (f) force of friction on the carriage. Be sure to note the important times $t = 0, t_1$, and t_2 on the time axes of your graphs. Take the positive direction to be the direction in which the man was initially walking.

99. A Two-Stage Rocket Students in a school rocketry club have prepared a two-stage rocket. The rocket has two small engines. The first will fire for a time, getting the rocket up partway. Then the first-stage engine drops off, revealing a second engine. After a little time, that engine will fire and take the rocket up even higher.

The rocket starts firing its engines at a time $t = 0$. From that instant, it begins to move upward with a constant acceleration. This continues until time t_1. The rocket drops the first stage and continues upward briefly until time t_2, at which point the second stage begins to fire and the rocket again accelerates upward, this time with a larger (but again constant) acceleration. Sometime during this second period of acceleration, our recording apparatus stops.

Sketch qualitatively accurate (i.e., we don't care about the values but we do care about the shape) graphs of the height of the rocket, y, its velocity, v_y, its acceleration, a_y, the force on the rocket that results from the firing of the engine, $\vec{F}_y$, and the net force on the rocket, $\vec{F}_y^{\text{net}}$. Take the positive direction as upward. Be sure to note times $t = 0, t_1$, and t_2 on the time axes of your graphs.

100. Pushing a Cart A worker is pushing a cart along the floor. At first, the worker has to push hard in order to get the cart moving. After a while, it is easier to push. Finally, the worker has to pull back on the cart in order to bring it to a stop before it hits the wall. The force exerted by the worker on the cart is purely horizontal. Take the direction the worker is going as positive.

Figure 6-90 shows graphs of some of the physical variables of the problem. Match the graphs with the variables in the list at the

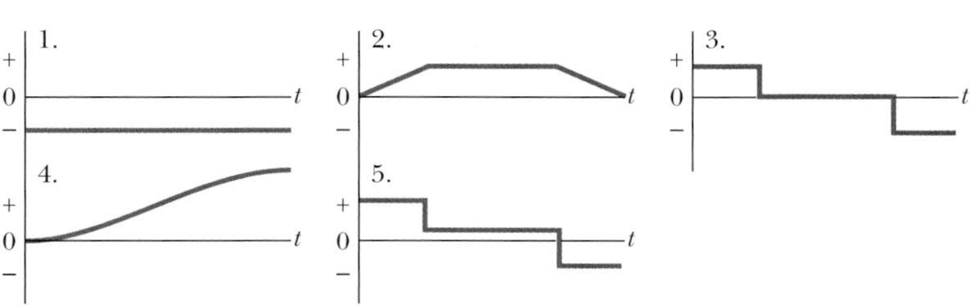

FIGURE 6-90 ■ Problem 100.

left below. You may use a graph more than once or not at all. *Note*: The time axes are to the same scale, but the ordinates y axes are not.

(a) Friction force
(b) Force exerted by the worker
(c) Net force
(d) Acceleration
(e) Velocity

101. Comparing a Light and Heavy Object Consider a metal sphere two inches in diameter and a feather. For each quantity in the list below, indicate the relation between the quantity for the sphere and feather. Is it the same, greater, or lesser? Explain in each case why you gave the answer you did.

(a) The gravitational force
(b) The time it will take to fall a given distance in air
(c) The time it will take to fall a given distance in vacuum
(d) The total force on the object when falling in vacuum
(e) The total force on the object when falling in air

102. Hitting the Green A golfer is trying to hit a golf ball onto the green as shown in Fig. 6-91. The green is a horizontal distance s from his tee and it is up on the side of a hill a height h above his tee. When he strikes the ball it leaves the tee at an angle θ to the horizontal. He wants to know with what speed, v_1, the ball must leave the tee in order to reach the height h at the distance s.

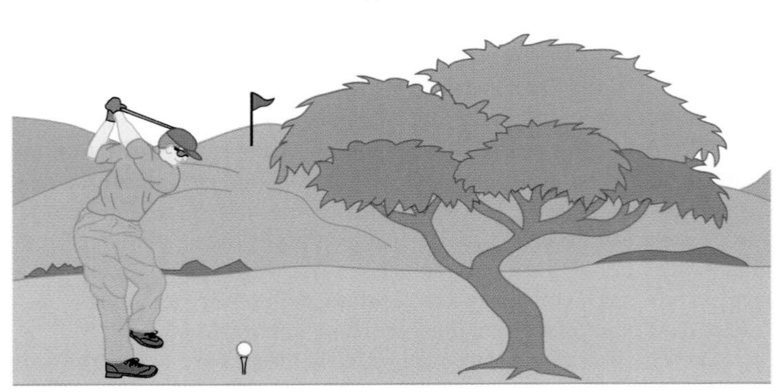

FIGURE 6-91 ■ Problem 102.

(a) Once he has struck the ball, what controls its motion? Write the equations that determine the vector acceleration of the golf ball afler it leaves the tee. Be sure to specify your coordinate system. For this part of the problem you may ignore air resistance.
(b) Solve the equations you have written in (a) to obtain expressions that can be evaluated to give the position of the ball at any time, t.
(c) If the golfer wants his ball to land in the right place, he must hit it so that it leaves the tee with the right speed. Explain how he can calculate it. (Again, you may ignore air resistance.) Find an equation for the initial speed in terms of the problem's givens.
(d) If the ball leaves the tee at an angle of 30°, $s = 100$ m, and $h = 10$ m, find the speed with which the ball leaves the tee.
(e) Now consider the effect of air resistance. Suppose that a good model for the force of air resistance is Newton's drag law,

$$\vec{F} = -b\,|\vec{v}|\,\vec{v}$$

where $|\vec{v}|$ is the speed and b is a constant. Consider three points on the ball's trajectory: halfway up, at its highest point, and halfway down. Discuss the direction of the resistance force at each point. Qualitatively (do not attempt a calculation!), what will the effect of air resistance be on the ball's motion?

103. Air Resistance 1: Dimensional Analysis We know that as an object passes through the air, the air exerts a resistive force on it. Suppose we have a spherical object of radius R and mass m. What might the force plausibly depend on?

- It might depend on the properties of the object. The only ones that seem relevant are m and R.
- It might depend on the object's coordinate and its derivatives: $\vec{r}, \vec{v}, \vec{a}, \ldots$.
- It might depend on the properties of the air, such as the density, ρ.

(a) Explain why it is plausible that the force the air exerts on a sphere depends on R but implausible that it depends on m.
(b) Explain why it is plausible that the force the air exerts depends on the object's speed through it, $|\vec{v}|$, but not on its position, $\vec{r}$, or acceleration, $\vec{a}$.
(c) Dimensional analysis is the use of units (e.g., meters, seconds, or newtons) associated with quantities to reason about the relationship between the quantities. Using dimensional analysis, construct a plausible form for the force that air exerts on a spherical body moving through it.

104. Counterweights The use of counterweights to help devices move up and down with a minimum of effort is common in engineering. For example, counterweights are used to help people open and close old-fashioned windows and to move up and down in elevators. Imagine that an engineer working for the Disney Epcot Center is asked to design a ride that allows people to travel up and down a sloped hill to get a view of the entire Epcot Center while other tourists move straight up and down an artificial cliff on the other side of the incline. Our engineer builds a small prototype of his device using a low-friction cart on an inclined track attached to a falling mass. His goal is to see whether he can actually apply Newton's laws to this situation and if it is okay to neglect the effects of friction.

In this exercise you will analyze data collected from a digital movie of the situation discussed above and shown in Fig. 6-92a. If you have access to VideoPoint you can view the digital movie yourself. It is entitled PASCO098. Your instructor may provide you with a different but similar movie. The cart in PASCO098 has a mass $m_c = .510$ kg and is accelerated up a ramp that has a 21° incline. A string attached to the cart exerts a force on it. The string transmits a force to the cart because its other end is attached by means of a pulley to a falling mass of $m_f = .184$ kg.

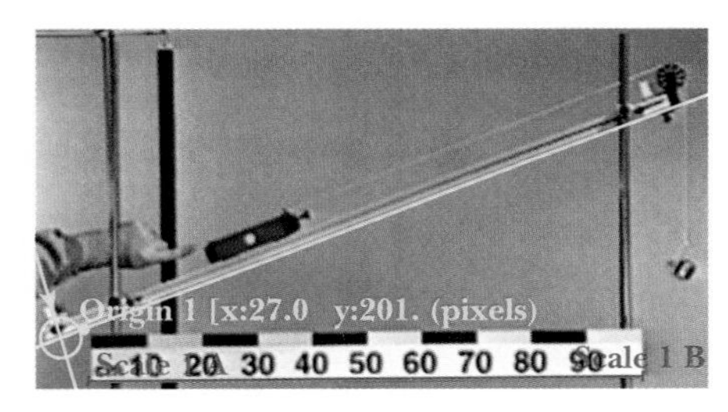

FIGURE 6-92a ■ Problem 104.

Table 6-3 contains position vs. time measurements for the cart in PASCO098 along an x axis. The x axis is rotated from the horizontal direction so that it lies along the ramp. Using these data you can determine the acceleration, if any, of the cart. (It is best to enter the data into a spreadsheet for analysis.) Finally, you will use Newton's laws along with the information on the angle of the incline and the masses of the cart and the falling mass to determine (theoretically) what the acceleration of the cart is. Our goal is to determine whether the theoretically calculated motion and the actual motion (as described by the data in Table 6-3) agree.

(a) Enter the data in Table 6-3 into a spreadsheet program. Determine what kind of motion the cart experiences. Is it a constant velocity? If so what are the magnitude and direction of the velocity? Is the motion a constant acceleration? If so, what are the magnitude and direction of the acceleration? (You may want to use equation-fitting software in answering this question). Cite the evidence that leads you to give the answers you did.

(b) What is the value of the net force on the cart in the x direction (along the incline)?

(c) Sketch a diagram of the cart like that shown in Fig. 6-92*b*. Draw a free-body diagram showing the directions of *all* the forces on the cart including the gravitational force, $\vec{F}^{\text{grav}}$, the normal force, $\vec{N}$, and the string force due to its tension, T.

TABLE 6-3
Problem 104

Time (sec)	x(m)
0.000	0.002929
0.2050	0.03956
0.4100	0.08465
0.6150	0.1221
0.8200	0.1659
1.025	0.2038
1.230	0.2463
1.435	0.2885
1.640	0.3301
1.845	0.3676
2.050	0.4114
2.255	0.4472
2.460	0.4931
2.665	0.5297
2.870	0.5748
3.075	0.6165
3.280	0.6624

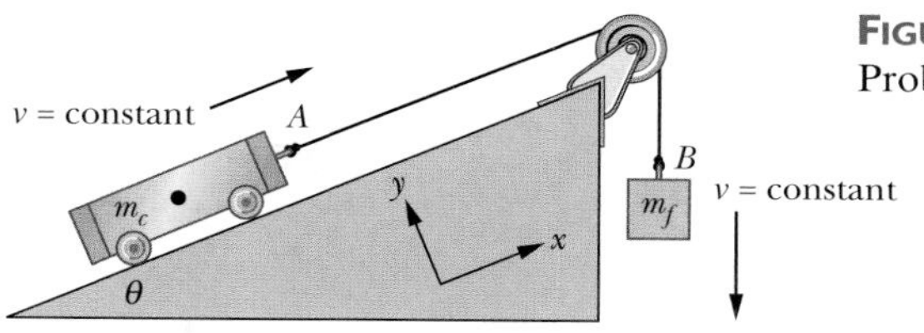

(d) Consider the situation in which the cart and falling mass move with a constant velocity. Choose a coordinate system in which the positive x axis is directed up along the ramp (rotated from the horizontal). *Assume that there is no friction in the pulley or cart bearings!* Show that by taking components of these forces along the x axis the magnitude, F_x^{net}, of the net force on the cart in the x direction, can be calculated using the equation

$$F_x^{\text{net}} = T - m_c g \sin\theta = 0,$$

where the gravitational constant, g, is $+9.8$ N/kg.

FIGURE 6-92*b* Problem 104.

(e) Assume that since the cart and falling mass are connected by the string they have the same magnitude of velocity. Also assume that the tension in the string is the same at all points along the string so that the magnitude of the string force at point A on the cart is the same as the magnitude of the string force at point B on the falling mass. Show that if the net force on the falling mass is zero, then $T - F^{\text{grav}} = 0$, where $F^{\text{grav}} = m_f g$.

(f) Use the equations you derived in parts (d) and (e) to show that if the velocity of the cart and falling mass system are constant, then theoretically $m_f g$ ought to equal $m_c g \sin\theta$.

(g) Use the given values of m_c and m_f (also available on the title screen of the PASCO098 movie) along with the angle of the incline to verify that $m_f g$ and $m_c g \sin\theta$ have the same values to two significant digits. This equality, if it exists, confirms the agreement between theory and experiment.

(h) Also discuss why the answers should only be good to two significant figures.

7 Translational Momentum

A karate master undergoes extensive training to thicken the bones and strengthen the muscles in his or her hands. This enables him or her to break stacks of concrete patio blocks with a single blow. Although novices cannot perform this feat, they are able to break 3/4-inch-thick pine boards quite easily. For example, Tom Casiani, an introductory physics student at Dickinson College, broke a stack of nine pine boards in spite of the fact that he had never done any karate before taking physics.

How can novices break pine boards, but not concrete slabs, without sustaining injuries?

The answer is in this chapter.

7-1 Collisions and Explosions

In the last few chapters, we have focused on understanding how forces affect the motion of an object. We have specifically discussed several common forces and practiced using Newton's Second Law, $\vec{F}^{\text{net}} = m\vec{a}$, to determine the acceleration of an object that experiences steady forces. Although there are many situations in which one can determine the acceleration of an object by summing the forces acting on it, there are other situations in which using the equation $\vec{F}^{\text{net}} = m\vec{a}$ is not possible. For example, when objects collide or a large object explodes into smaller fragments, the event can happen so rapidly that it is impossible to keep track of the interaction forces.

Collisions and explosions range from the microscopic scale of subatomic particles (Fig. 7-1*b*) to the astronomic scale of colliding stars and galaxies, so an understanding of these processes is of great interest to physicists. In fact, many physicists today spend their time playing "the collision game." The goal of this game is to find out as much as possible about the forces that act during rapid interactions between particles or during the explosion of a particle into fragments. We have developed techniques for learning about rapid interactions by determining the state of the particles before and after they interact. Indeed, most of our understanding of the subatomic world—electrons, protons, neutrons, muons, quarks, and the like—comes from experiments involving collisions and explosions.

In this chapter we will define a new quantity known as linear or translational momentum to help us study collision processes. Since explosions are actually collision processes in reverse, it turns out that we can use the same methods in studying both phenomena. We shall use the following formal definition of a collision.

> A **COLLISION** or **EXPLOSION** is an isolated event in which two or more bodies exert relatively strong forces on each other over a short time compared to the period over which their motions take place.

By *relatively strong forces* we mean that the collision or explosion forces are considerably larger than other forces that might be acting on the system. Similarly, a *relatively short time* means that the weaker forces (other than the collision or explosion forces) have not had enough time to accelerate the system elements noticeably.

In order to analyze collisions or explosions, we distinguish between times that are *before*, *during*, and *after* an event, as suggested in Fig. 7-2. Figure 7-2 shows two colliding bodies and indicates that the forces associated with the collision are forces that the bodies exert *on each other*.

READING EXERCISE 7-1: According to our definition of collision, which, if any, of the following events qualify as collisions? Explain. (*a*) Suppose it took the ocean liner *Titanic* 60 seconds to plow into an iceberg and come to a stop. (*b*) During a volley, a tennis ball usually is in the air for less than a second. Suppose a tennis racket is in contact with a ball for 2 seconds. ■

(*a*)

(*b*)

(*c*)

FIGURE 7-1 ■ Collisions range widely in scale. (*a*) Meteor Crater in Arizona is about 1200 m wide and 200 m deep. (*b*) An alpha particle coming in from the left bounces off a nitrogen nucleus that had been stationary and that now moves toward the bottom right. (*c*) In a tennis match, the ball is in contact with the racquet for about 4 ms in each collision (for a cumulative time of approximately 1 s in a set).

7-2 Translational Momentum of a Particle

Consider a winter accident on a narrow icy road in which a compact car skids into a loaded pickup truck that is moving toward it. If you want to predict the motions of the vehicles after that crash, what do you need to know about the vehicles? Many people would guess that both the mass and the velocity of each vehicle make a difference. It turns out that the product of these two quantities, which we will soon begin calling by the name momentum, is a very useful concept in predicting the outcome of collisions.

In fact, Newton did not use the ideas of acceleration and velocity in his original descriptions of the laws of motion. He developed his laws, in part, by studying colli-

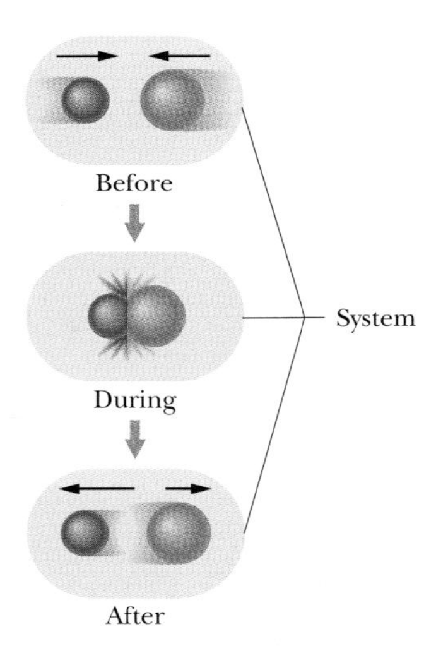

FIGURE 7-2 ■ Stages in a collision between two bodies.

sions, and this led him to introduce the concept of momentum. As is the case for "acceleration," momentum is a word that has several meanings in everyday language but only a single precise meaning in physics. The **translational momentum** of a particle is a vector $\vec{p}$, defined as

$$\vec{p} \equiv m\vec{v} \quad \text{(definition of translational momentum),} \tag{7-1}$$

where m is the mass of the particle and $\vec{v}$ is its instantaneous velocity. Since m is always a positive scalar quantity, the momentum vector and the velocity vector are always in the same direction. This relation also tells us that the SI unit for momentum is the kilogram-meter per second (the unit for mass multiplied by the unit for velocity).

Some people use the phrase "linear momentum" rather than the phase "translational momentum" when discussing the product $m\vec{v}$. However, this momentum is associated with the movement of an object from one position to another, regardless of whether the overall motion of the object occurs along a line. For example, the equation $\vec{p} = m\vec{v}$ serves to define the momentum of a projectile following a parabolic path or a small rock rotating in a circle. Hence, the term "translational" is a better adjective than "linear."

The adjective "translational" is often dropped, leaving us with just the term "momentum." However, it serves to distinguish this type of momentum from *rotational momentum*, which is introduced in Chapter 12.

Newton expressed his second law of motion in terms of momentum as follows:

> The rate of change of the momentum of a particle is proportional to the net force acting on the particle and is in the direction of that force.

In equation form this statement is

$$\vec{F}^{\text{net}} = \frac{d\vec{p}}{dt} \quad \text{(single particle).} \tag{7-2}$$

We can relate this statement of the second law to the familiar $\vec{F}^{\text{net}} = m\vec{a}$ by substituting $m\vec{v}$ for $\vec{p}$ and pulling the mass, which is a constant, out of the derivative so that

$$\vec{F}^{\text{net}} = \frac{d\vec{p}}{dt} = \frac{d}{dt}(m\vec{v}) = m\frac{d\vec{v}}{dt} = m\vec{a}.$$

Thus, the equations $\vec{F}^{\text{net}} = d\vec{p}/dt$ and $\vec{F}^{\text{net}} = m\vec{a}$ are equivalent expressions of Newton's Second Law of Motion as it applies to the motion of a particle whose mass remains constant.

What these relations are telling us is that a nonzero net force on a body causes it to undergo a momentum change. This should not come as a surprise. A nonzero net force results in an acceleration of the object on which the force acts. That acceleration produces a change in velocity, and the change in velocity is associated with a change in momentum. Another way to think of this is that a nonzero net force—for example, the push on a cart—is what "gives" the cart its change in momentum.

READING EXERCISE 7-2: The figure to the right gives the x-component of translational momentum versus time for a particle moving along an x-axis. A force directed along the axis acts on the particle, causing its momentum to change. (a) Rank the four regions indicated according to the magnitude of the force, greatest first. (b) In which region is the particle slowing down? ■

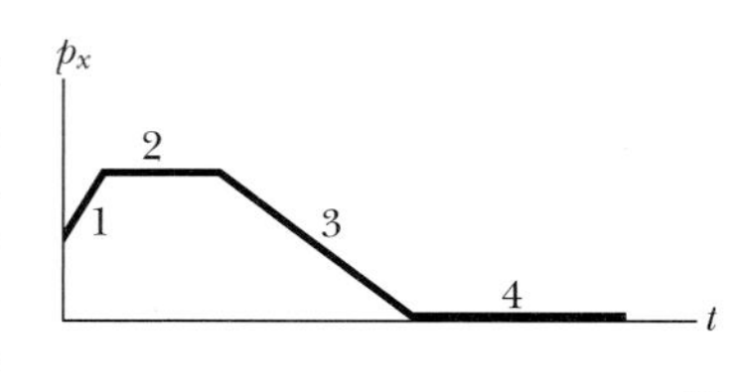

7-3 Isolated Systems of Particles

We are often interested in the behavior of a collection of particles that interact only with each other. We can draw an imaginary boundary around the particles, but complications can arise if some of the particles experience net forces that originate outside the "system" boundary. In order to effectively study interactions between particles, we must limit our focus to **isolated systems** of particles. (See Fig. 7-3.)

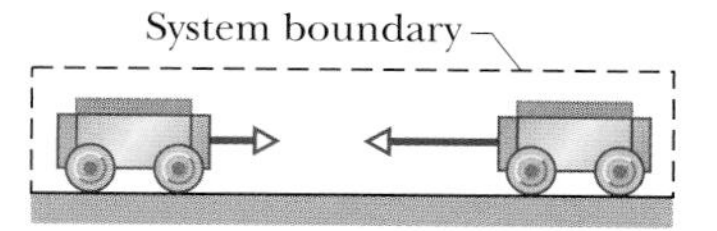

FIGURE 7-3 ■ The vector sum of the normal force on each cart and the gravitational force on it is zero. If the track is also essentially frictionless, the carts form an *isolated system*. The collision forces they exert on each other are inside the system and can be studied fairly easily.

> An **ISOLATED SYSTEM** is defined as a collection of particles that can interact with each other but whose interactions with the environment outside the collection have a negligible effect on their motions.

Basically the particles in an isolated system experience no significant net external forces. Two carts that are about to collide on a frictionless track form an isolated system. Why? Because even though each cart experiences a normal force from the track and a gravitational force from the Earth, the net force on each cart that originates outside the system boundary is zero.

Let's consider another example. Suppose two galaxies collide in outer space and exert a complex series of gravitational and electromagnetic forces on each other. Let's take these two galaxies to be our system. If these galaxies are far from other astronomical bodies, their gravitational interactions with entities outside the system will be small—especially in comparison to the internal forces they exert on each other. These galaxies can be considered to be an isolated system. If the particles in our two galaxies are strongly attracted to neighboring galaxies, they do not form an isolated system. Additional examples of isolated systems are shown in Fig. 7-4.

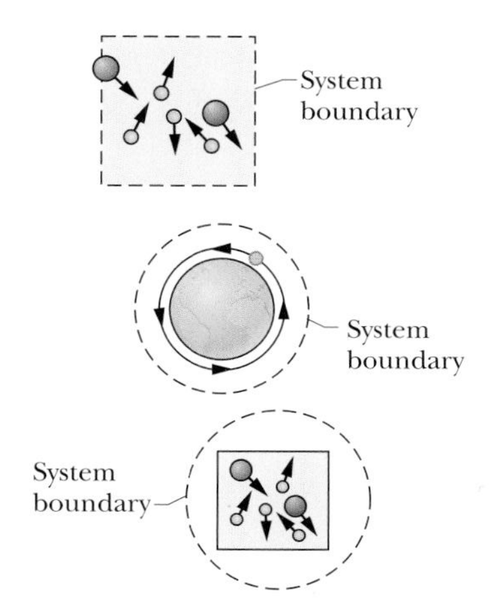

Pucks riding on a cushion of air on an air table interact with each other before hitting the walls of the table. Friction forces with the surface of the table are negligible. The system is temporarily isolated—until a puck hits an air table wall.

An orbiting satellite and the Earth interact. Forces between these objects and others such as the sun and moon are considered to have negligible effect on their motions.

Gas molecules interact with each other and with the walls of their container. Other forces, such as those of the table holding up the container and the gravitational force, are considered to have a negligible effect on the motions of the molecules and container.

FIGURE 7-4 ■ Examples of isolated systems.

Momentum for a System of Particles

If we apply Newton's Second and Third Laws to an isolated system of particles, we can learn about changes in the total momentum of the system. Let's consider a system of n particles, each with its own mass, velocity, and translational momentum. The particles in the system may interact with each other. The system as a whole has a total translational momentum $\vec{p}_{sys}$, which is defined to be the vector sum of the individual particles' translational momenta. Thus,

$$\vec{p}_{sys} = \vec{p}_A + \vec{p}_B + \vec{p}_C + \cdots + \vec{p}_n$$
$$= m_A\vec{v}_A + m_B\vec{v}_B + m_C\vec{v}_C + \cdots + m_n\vec{v}_n. \quad (7\text{-}3)$$

> The translational momentum of a system is the vector sum of the momenta of the individual particles.

If the system is not isolated, then external forces are also acting on the system. Recall that for a single particle such as particle A,

$$\vec{F}_A^{\text{net}} = \frac{d\vec{p}_A}{dt}.$$

Hence, we have a separate Newton's Second Law equation for each of the n particles, telling how that particle will respond to the forces it feels:

$$\vec{F}_A^{\text{net}} = \frac{d\vec{p}_A}{dt}, \quad \vec{F}_B^{\text{net}} = \frac{d\vec{p}_B}{dt}, \quad \vec{F}_C^{\text{net}} = \frac{d\vec{p}_C}{dt}, \ldots .$$

But the total momentum of the system, $\vec{p}_{\text{sys}}$, is given by the sum of the momenta of the particles in the system, so that

$$\vec{p}_{\text{sys}} = \vec{p}_A + \vec{p}_B + \vec{p}_C + \cdots .$$

Since the derivative of a sum is the same as the sum of the derivatives, the rate of change of the system momentum is given by

$$\frac{d\vec{p}_{\text{sys}}}{dt} = \frac{d(\vec{p}_A + \vec{p}_B + \vec{p}_C \ldots)}{dt} = \frac{d\vec{p}_A}{dt} + \frac{d\vec{p}_B}{dt} + \frac{d\vec{p}_C}{dt} + \cdots . \qquad (7\text{-}4)$$

However, according to Eq. 7-2, the net force on any one of the particles is given by $\vec{F}^{\text{net}} = d\vec{p}/dt$, so the rate of change of the total momentum of the system is equal to the sum of the forces felt by each of the n particles:

$$\frac{d\vec{p}_{\text{sys}}}{dt} = \vec{F}_A^{\text{net}} + \vec{F}_B^{\text{net}} + \vec{F}_C^{\text{net}} + \cdots = \vec{F}_{\text{sys}}^{\text{net}}.$$

In words, the sum of all forces acting on all the particles in the system is equal to the time rate of change of the total momentum of the system. That leaves us with the general statement:

$$\vec{F}_{\text{sys}}^{\text{net}} = \frac{d\vec{p}_{\text{sys}}}{dt} \quad \text{(system of particles).} \qquad (7\text{-}5)$$

In principle, $\vec{F}^{\text{net}}$ is the sum of all forces on particles in the system. This includes forces from particles within the system acting on other particles within the system (called **internal forces**). It also includes forces from objects outside the system acting on objects within the system (called **external forces**). In practice, all the internal forces occur as third-law pairs that cancel. Thus, the contribution of the internal forces to the overall net force is zero. Hence, $\vec{F}^{\text{net}}$ is always just the sum of all *external* forces acting on the system. This equation is the generalization of the single-particle equation $\vec{F}^{\text{net}} = d\vec{p}/dt$ (Eq. 7-2) to a system of many particles.

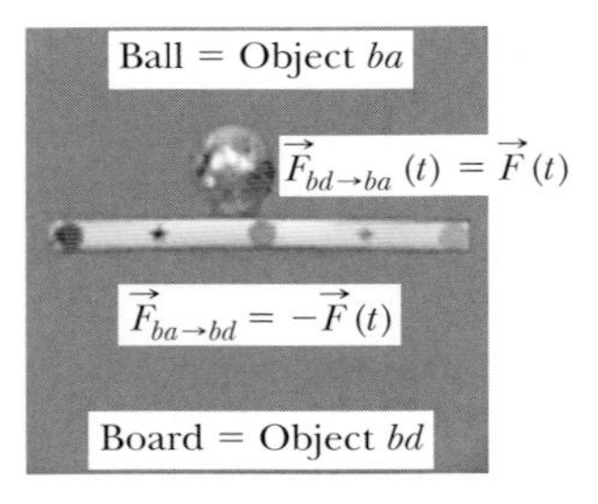

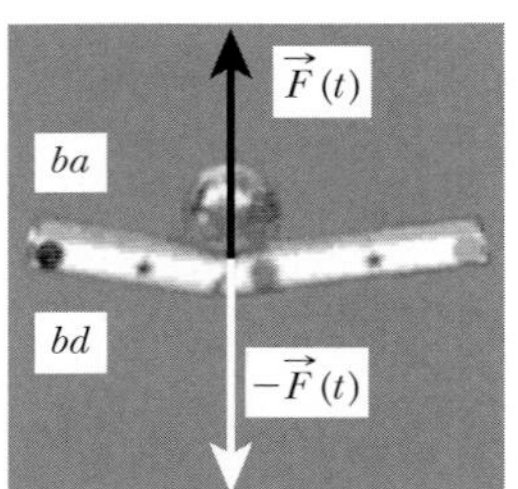

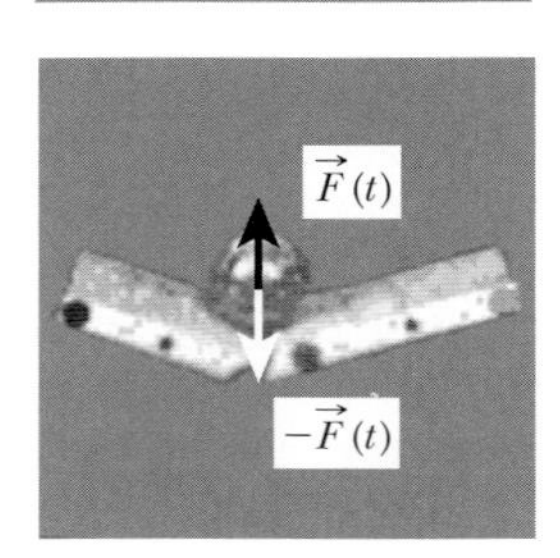

FIGURE 7-5 ■ A lead ball (ba) of mass 0.850 kg collides with a pine board (bd). During the collision, the lead ball exerts a force of $\vec{F}_{ba\to bd} = -\vec{F}(t)$ on the board and the board exerts force $\vec{F}_{bd\to ba} = +F(t)$ on the ball. Forces $\vec{F}(t)$ and $-\vec{F}(t)$ are a third-law force pair. Their magnitudes vary with time during the collision, but at any given instant those magnitudes are equal. A digital video clip of the collision was recorded at 250 frames/second. The ball and the board are in contact for only about 0.012 s. (Courtesy of Robert Teese.)

7-4 Impulse and Momentum Change

Although a pine board is not really very particle-like, it is instructive to consider a two-body system consisting of a falling lead ball (ba) that collides with a pine board (bd) (shown in Fig. 7-5). At any given moment, the force that the board exerts on the ball can be denoted as $\vec{F}_{bd\to ba}$. It is obvious that before the falling ball makes contact with the board, the interaction forces are negligible. When the ball first makes contact with the board, the magnitudes of the interaction forces are relatively small. As the force magnitudes reach a maximum, the ball causes the board to flex and begin to break. After the board breaks, the forces are zero again. Thus, we expect a graph of the magnitude of the force exerted on the ball by the board to look something like that shown in Fig. 7-6*a*.

A video analysis of the acceleration of the ball shows that the peak force exerted on it by the board is about 800 N. This is also the peak force that a karate expert breaking a board would experience. However, the fact that it takes less than one-hundredth of a second (two frames in Fig. 7-5) helps prevent injury. Note that the gravitational force is less than 10 N, so we can neglect it.

As the forces act over time they change the translational momentum of both objects. The amount of change will depend on how the forces vary over time. To see this quantitatively, let us apply Newton's Second Law in the form $\vec{F}_{bd\to ba} = d\vec{p}_{ba}/dt$ to the lead ball depicted in Fig. 7-5. If we denote the net force on the ball as $\vec{F}^{\,\text{net}}(t) = \vec{F}_{bd\to ba} + \vec{F}^{\text{grav}}_{ba} \approx \vec{F}_{bd\to ba}$ and $\vec{p}_{ba}$ as $\vec{p}$, then

$$d\vec{p} = \vec{F}^{\,\text{net}}(t)\,dt, \tag{7-6}$$

in which $\vec{F}^{\,\text{net}}(t)$ is a time-varying force on the ball with magnitude given by the curve in Fig. 7-6*a*. Let us integrate $d\vec{p} = \vec{F}^{\,\text{net}}(t)\,dt$ over the collision interval Δt from an initial time t_1 (just before the collision) to a final time t_2 (just after the collision). We obtain

$$\int_{\vec{p}_1}^{\vec{p}_2} d\vec{p} = \int_{t_1}^{t_2} \vec{F}^{\text{net}}(t)\,dt, \tag{7-7}$$

where $\vec{p}_1$ represents the momentum of the ball at time t_1 just before the collision, and $\vec{p}_2$ represents the momentum at time t_2 just after the collision. The left side of this equation is $\vec{p}_2 - \vec{p}_1$, which is the change in translational momentum of the lead ball. The right side of the equation is a measure of both the strength and the duration of the collision force exerted on the ball by the board. It is defined as a vector quantity called the **impulse** $\vec{J}$. In general, the impulse an object experiences due to a collision force $\vec{F}^{\text{net}}(t)$ is defined as

$$\vec{J} \equiv \int_{t_1}^{t_2} \vec{F}^{\text{net}}(t)\,dt \quad \text{(impulse defined).} \tag{7-8}$$

When $\vec{F}^{\,\text{net}}(t)$ does not change direction during the collision, this relation tells us that the magnitude of the impulse is equal to the area under the $|\vec{F}^{\text{net}}(t)|$ curve of Fig. 7-6*a*.

When an object undergoes a collision it is considerably easier to measure its momentum change than it is to determine its impulse curve. Thus it is quite common to consider the time interval over which the colliding objects are in contact with each other. Then, if we assume that the force is constant during that time interval, we have a feel for the magnitude of the average force that an object experiences. If we denote the average net force during a collision as $\langle \vec{F}^{\text{net}} \rangle$, then we can relate it to momentum change using the expression

$$\langle \vec{F}^{\,\text{net}} \rangle = \frac{\vec{J}}{\Delta t} = \frac{\vec{p}_2 - \vec{p}_1}{\Delta t} = \frac{\Delta \vec{p}}{\Delta t} \quad \text{(average net force during a collision).} \tag{7-9}$$

Graphically we can represent the magnitude of the average force on an object $|\vec{F}^{\,\text{net}}|$ as the area within the rectangle of Fig. 7-6*b* that is equal to the area under the $|\vec{F}^{\,\text{net}}(t)|$ curve of Fig. 7-6*a* over the same time interval.

In the case of the example in Fig. 7-5 of the ball (object *ba*) breaking the board (object *bd*), the time period in which the ball is in contact with the board turns out to

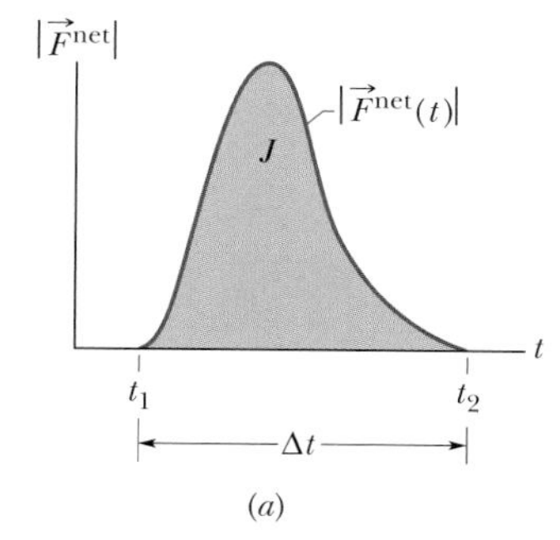

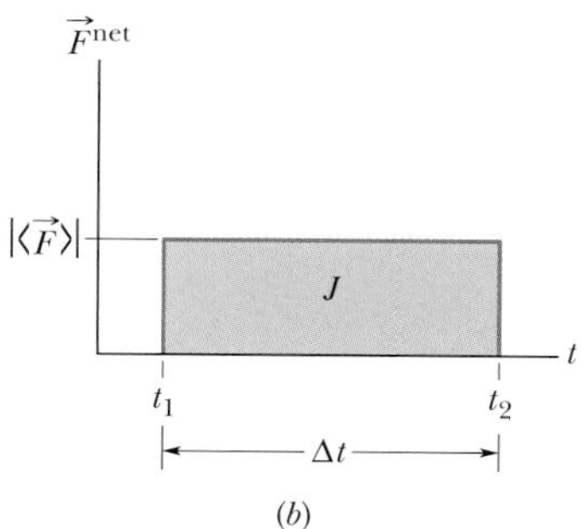

FIGURE 7-6 ■ (*a*) The sketched graph is an idealization of what the magnitude of the time-varying net force that acts on an object during a collision shown in Fig. 7-5 might look like. The time interval Δt for the collision is only about 1/100th of a second for the ball falling on a board as shown in Fig. 7-5. (*b*) The height of the rectangle represents the magnitude of the average net force acting on an object during the same time interval Δt. The area within the rectangle is equal to the area under the curve (or integral) in (*a*).

be only $\Delta t \approx 0.01$ s while the magnitude of the momentum change is $|\Delta \vec{p}_{ba}| =$ 0.53 kg · m/s. This gives us an average net collision force on the ball of magnitude

$$|\langle \vec{F}^{\text{net}}_{ba} \rangle| = \frac{|\Delta \vec{p}_{ba}|}{\Delta t} = \frac{.53 \text{ kg} \cdot \text{m/s}}{0.01 \text{ s}} \approx 50 \text{ N}.$$

The general relation

$$\int_{\vec{p}_1}^{\vec{p}_2} d\vec{p} = \int_{t_1}^{t_2} \vec{F}^{\text{net}}(t)\, dt$$

tells us that the change in the translational momentum of any object is equal to the impulse that acts on that object. Thus, for an object in a collision:

$$\vec{J} = \vec{p}_2 - \vec{p}_1 \quad \text{(impulse-momentum theorem).} \tag{7-10}$$

This relation is called the **impulse-momentum theorem;** it tells us that impulse and translational momentum are both vectors and have the same units and dimensions. The impulse-momentum theorem can also be written in component form as

$$p_{2x} - p_{1x} = \Delta p_x = J_x, \tag{7-11}$$

$$p_{2y} - p_{1y} = \Delta p_y = J_y, \tag{7-12}$$

and

$$p_{2z} - p_{1z} = \Delta p_z = J_z. \tag{7-13}$$

In an isolated, two-body system, forces exerted between body A and body B form third-law force pairs. That is, the force of body B on body A is equal and opposite to the force of body A on body B. So the impulses on the two objects have the same magnitudes but opposite directions. This can be represented by the expression $\vec{F}_{B\to A} = -\vec{F}_{A\to B}$,

so that

$$\vec{J}_{B\to A} = -\vec{J}_{A\to B} \quad \text{(isolated 2-body system).}$$

READING EXERCISE 7-3: Have you ever been in an egg-tossing contest? The idea of this adventure is to work as a team of two people tossing a raw egg back and forth. After each successful toss (success = unbroken egg), each team member must take a step back. Pretty soon, you have to throw the egg quite hard to get it across to your partner. If you catch an egg of mass m that is coming toward you with velocity $\vec{v}$, what is the magnitude of the change in momentum that the egg undergoes? Would this value change if you catch the egg more quickly or more slowly? Explain. Suppose that the time it takes you to bring the egg to a stop in your hand is Δt. Are you more likely to have a "successful" catch if Δt is large or small? Why? How do you physically react in order to make Δt larger? ■

READING EXERCISE 7-4: The figure to the right shows an overhead view of a ball bouncing from a vertical wall without any change in its speed. Consider the change $\Delta \vec{p}$ in the ball's translational momentum. (a) Is Δp_x positive, negative, or zero? (b) Is Δp_y positive, negative, or zero? (c) What is the direction of $\Delta \vec{p}$?

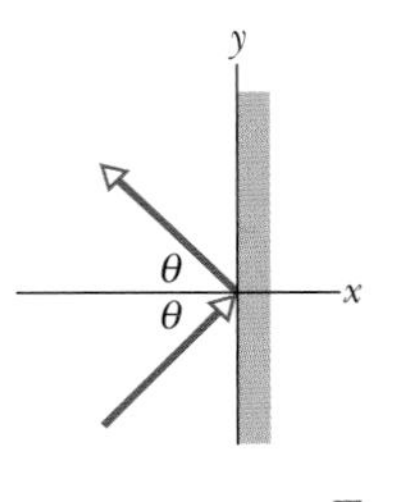

■

TOUCHSTONE EXAMPLE 7-1: Ball and Bat

A pitched 140 g baseball, in horizontal flight with a speed v_1 of 39.0 m/s, is struck by a bat. After leaving the bat, the ball travels in the opposite direction with speed v_2, also 39.0 m/s.

(a) What impulse $\vec{J}$ acts on the ball while it is in contact with the bat during the collision?

SOLUTION ■ The **Key Idea** here is that momentum is a vector quantity, so even though the magnitude of the momentum does not change, there is a significant change in momentum due to the direction change of the ball. We must calculate the impulse from the change in the ball's translational momentum, using Eq. 7-10 for one-dimensional motion. Let us choose the direction in which the ball is initially moving to be the negative direction. From Eq. 7-10 we have

$$\begin{aligned} J_x &= p_{2x} - p_{1x} = mv_{2x} - mv_{1x} \\ &= (0.140 \text{ kg})(39.0 \text{ m/s}) - (0.140 \text{ kg})(-39.0 \text{ m/s}) \\ &= 10.9 \text{ kg}\cdot\text{m/s}. \qquad \text{(Answer)} \end{aligned}$$

With our sign convention, the initial velocity of the ball is negative and the final velocity is positive. The impulse turns out to be positive, which tells us that the direction of the impulse vector acting on the ball is the direction in which the bat is swinging.

(b) The impact time Δt for the baseballl–bat collision is 1.20 ms. What average net force acts on the baseball?

SOLUTION ■ The **Key Idea** here is that the average net force is the ratio of the impulse $\vec{J}$ to the duration Δt of the collision (see Eq. 7-9). Thus,

$$\begin{aligned} \langle F_x^{\text{net}} \rangle &= \frac{J_x}{\Delta t} = \frac{10.9 \text{ kg}\cdot\text{m/s}}{0.00120 \text{ s}} \\ &= 9080 \text{ N}. \qquad \text{(Answer)} \end{aligned}$$

Note that this is the *average* net force. The *maximum* net force is larger. The sign of the average force on the ball from the bat is positive, which means that the direction of the force vector is the same as that of the impulse vector.

In defining a collision, we assumed that no significant external force acts on the colliding bodies. The gravitational force always acts on the ball, whether the ball is in flight or in contact with the bat. However, this force, with a magnitude of mg = 1.37 N, is negligible compared to the average force exerted by the bat, which has a magnitude of 9080 N. We are quite safe in treating the collision as "isolated during the short collision time period."

(c) Now suppose the collision is not head-on, and the ball leaves the bat with a speed v_2 of 45.0 m/s at an upward angle of 30.0° (Fig. 7-7). What now is the impulse on the ball?

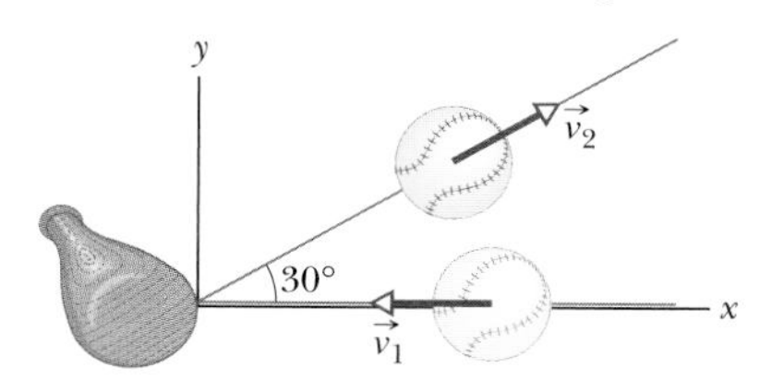

FIGURE 7-7 ■ A bat collides with a pitched baseball, sending the ball off at an angle of 30° from the horizontal.

SOLUTION ■ The **Key Idea** here is that now the collision is two-dimensional because the ball's outward path is not along the same axis as its incoming path. Thus, we must use vectors to find the impulse $\vec{J}$. From Eq. 7-10, we can write

$$\vec{J} = \vec{p}_2 - \vec{p}_1 = m\vec{v}_2 - m\vec{v}_1.$$

Thus,

$$\vec{J} = m(\vec{v}_2 - \vec{v}_1). \qquad (7\text{-}14)$$

We can evaluate the right side of this equation directly on a vector-capable calculator, since we know that the mass m is 0.140 kg, the final velocity $\vec{v}_2$ is 45.0 m/s at 30.0°, and the initial velocity $\vec{v}_1$ is 39.0 m/s at 180°.

Instead, we can evaluate Eq. 7-14 in component form. To do so, we first place an xy coordinate system as shown in Fig. 7-7. Then along the x axis we have

$$\begin{aligned} J_x &= p_{2x} - p_{1x} = m(v_{2x} - v_{1x}) \\ &= (0.140 \text{ kg})[(45.0 \text{ m/s})(\cos 30.0°) - (-39.0 \text{ m/s})] \\ &= 10.92 \text{ kg}\cdot\text{m/s}. \end{aligned}$$

Along the y axis,

$$\begin{aligned} J_y &= p_{2y} - p_{1y} = m(v_{2y} - v_{1y}) \\ &= (0.140 \text{ kg})[(45.0 \text{ m/s})(\sin 30.0°) - 0] \\ &= 3.150 \text{ kg}\cdot\text{m/s}. \end{aligned}$$

The impulse is then

$$\vec{J} = (10.9\hat{\text{i}} + 3.15\hat{\text{j}}) \text{ kg}\cdot\text{m/s}, \qquad \text{(Answer)}$$

and the magnitude and direction of $\vec{J}$ are

$$J = |\vec{J}| = \sqrt{J_x^2 + J_y^2} = 11.4 \text{ kg}\cdot\text{m/s}$$

and

$$\theta = \tan^{-1}\frac{J_y}{J_x} = 16°. \qquad \text{(Answer)}$$

TOUCHSTONE EXAMPLE 7-2: Carts Colliding

A moving cart coming from the left has a mass of 1.8 kg and an initial velocity component of +0.3 m/s. It then collides with a stationary cart to its right with a mass of 0.8 kg. After the collision the 1.8 kg cart slows down and the 0.8 kg cart moves away from it at a brisk velocity as shown in Fig. 7-8.

Let's consider two collisions for which the right cart is given the *same* momentum after the collision. In one case the right cart has a deformable rubber stopper attached to its force sensor. In the other case the rubber stopper is replaced with a more rigid metal hook. *What effect does the deformability of the surfaces in contact during the collision have on the collision process? In particular, what information do measured impulse curves give us about the duration of the collision and the maximum force experienced by the right cart?* How can we use the impulse curve to estimate the momentum transferred to the right cart during the collision?

(a) Use the measured impulse curves shown in Fig. 7-9 to find the approximate collision times when the collision involves contact between a metal hook and a deformable rubber stopper (shown in case a). Compare that to the collision time when the contact is between two metal hooks (shown in case b).

SOLUTION The **Key Idea** here is that during the time that the collision force is significantly above zero, the two colliding objects are in contact. It is clear from the graph (case a) that the collision time when the deformable rubber stopper is the point of contact is about 22 ms or 22×10^{-3} s. When the rubber stopper is replaced with a more rigid metal hook, the collision time, as shown on the graph (case b), is reduced to about 15 ms or 15×10^{-3} s. Another **Key Idea** is that the collision times for highly deformable objects are greater than they are for less deformable objects.

(b) Also use the measured impulse curves to compare the maximum forces experienced by the initially stationary cart for the two types of collisions (metal–rubber and metal–metal). Use the impulse-momentum theorem to explain why one maximum force is greater than the other.

SOLUTION It is clear from the graph (case a) that the peak force when the deformable rubber stopper is the point of contact is about 40 N (case a), while the peak force when the rubber stopper is replaced with a more rigid metal hook is greater at approximately 49 N (case b). The **Key Idea** here is that if an object experiences a certain momentum change, the impulse-momentum theorem can be used to relate the momentum change to the impulse curve by the equation

$$p_{2x} - p_{1x} = J_x = \int_{t_1}^{t_2} F_x\, dt = \langle F_x \rangle \Delta t.$$

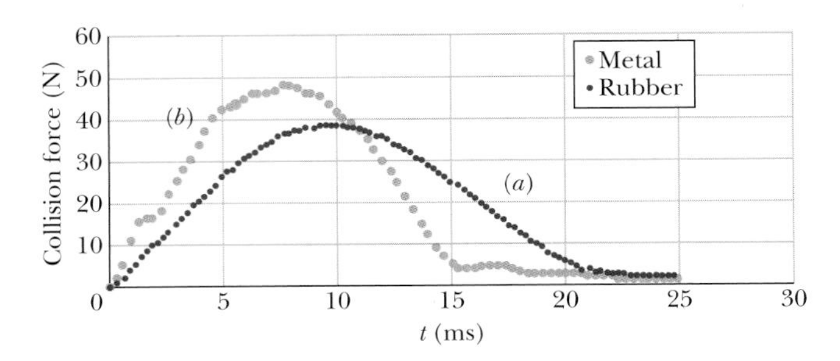

FIGURE 7-9 ■ Impulse curves when (a) the point of contact on the force sensor is a deformable rubber stopper, and (b) the point of contact on the force sensor is a piece of hard metal.

Thus, since the duration of the contact, Δt, is longer in a slow collision than in a fast one, the average force and hence the peak force must be smaller in a slow collision. Conversely the peak force during a rapid collision is greater than it would be in a slow collision.

(c) Use the measured impulse curves to estimate the magnitude of the momentum transferred to the right cart during each type of collision. You can approximate the impulse "curves" as triangles with the base being the contact time and the height equal to the peak force. Verify that both curves predict that approximately (in this case, to one significant figure) the same momentum was imparted to the cart in each case in spite of the fact that the collision times and peak forces are different.

SOLUTION The **Key Ideas** here are that the momentum change of the right cart is equal to the impulse imparted to it and that this impulse is an integral that can be calculated by finding the area under the impulse or force vs. time curve. In the special case where the right cart is initially at rest, this momentum change is also the final momentum of the right cart. If we approximate this curve as a triangle, this area can be computed using the familiar equation Area = $(\frac{1}{2})$ base × height.

Rubber stopper:

$$J_x = \text{Area} = \tfrac{1}{2}bh = \tfrac{1}{2}\Delta t F_x^{\text{peak}}$$
$$= \tfrac{1}{2}(22 \text{ ms} \times 40 \text{ N})$$
$$= \tfrac{1}{2}(22 \times 10^{-3} \text{ s} \times 40 \text{ N})$$
$$= 0.4 \text{ N}\cdot\text{s}. \quad \text{(Answer)}$$

$$p_{2x} - p_{1x} = p_{2x} - 0 = \Delta p_x = J_x$$
$$p_{2x} = 0.4 \text{ N}\cdot\text{s} = 0.4 \text{ kg}\cdot\text{m/s}.$$

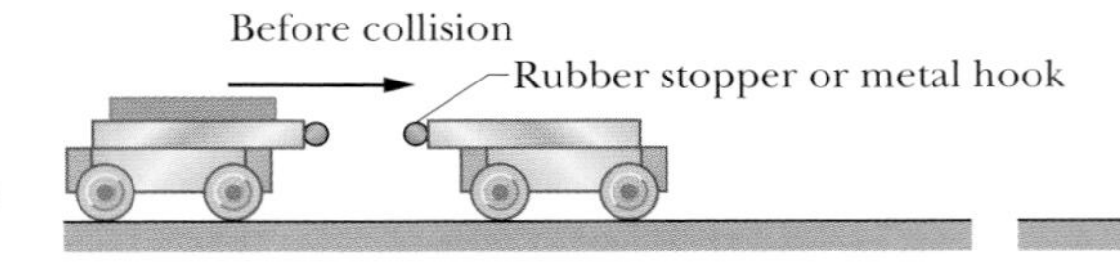

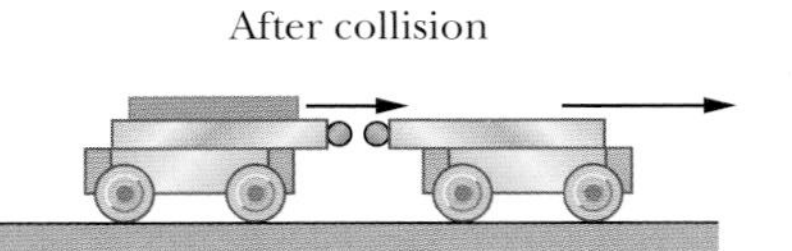

FIGURE 7-8 ■ A depiction of the motion of two carts before and after a collision.

Metal hook:

$$J_x = \text{Area} = \tfrac{1}{2}bh = \tfrac{1}{2}\Delta t F_x^{\text{peak}}$$
$$= \tfrac{1}{2}(15 \text{ ms} \times 49 \text{ N})$$
$$= \tfrac{1}{2}(15 \times 10^{-3} \text{ s} \times 49 \text{ N})$$
$$= 0.4 \text{ N}\cdot\text{s}. \qquad \text{(Answer)}$$

$$p_{2x} - p_{1x} = p_{2x} - 0 = \Delta p_x = J_x$$
$$p_{2x} = 0.4 \text{ N}\cdot\text{s} = 0.4 \text{ kg}\cdot\text{m/s}.$$

Note on Impulse and Karate Injuries: We can use the differences in the shapes of the two impulse curves to explain how it is possible for beginners who are not trained in the art of karate to break pine boards, but not patio blocks, without sustaining injuries. Breaking a board is a complex process that must obey the law of conservation of momentum that will be introduced in Section 7-5, and work and energy relationships that will be introduced in Chapter 9. It turns out that the concept of impulse is one of the critical factors in karate, so that in order to break a concrete block or a board, a certain impulse must be imparted to it.

When struck, the board or block bends, storing energy like a stretched spring does, until a critical deformation needed to break it is reached. In fact, a clean, knot-free pine board that is hit along its grain is relatively easy to break. One of several factors that make breaking a pine board less injurious is the fact that a board deforms much more than a concrete block before breaking. Thus, for a given impulse, the duration of the collision is significantly longer when a pine board breaks than when a concrete block breaks. As we saw in part (b) of this touchstone example, this means that for a given impulse much less peak force will be exerted on the board by the hand. Since Newton's Third Law holds, then it also means that the hand experiences a much lower peak force than it would striking a concrete block hard enough to break it. A lower peak force on the hand reduces the chance that the fifth metacarpal bone in the hand will break.

7-5 Newton's Laws and Momentum Conservation

What happens to the momentum, $\vec{p}_{\text{sys}}$, of a system of particles that is *isolated* so there is no net force acting? Assume that the particles are interacting with each other and undergoing all sorts of collisions that obey Newton's Third Law. What happens to the momentum of the overall system if $\vec{F}^{\text{net}} = 0$ from all sources both external and internal? We know from Newton's Second Law that

$$\vec{F}^{\text{net}} = \frac{d\vec{p}_{\text{sys}}}{dt} = 0,$$

and so

$$\vec{p}_{\text{sys}} = \text{constant} \qquad \text{(for an isolated system).} \qquad (7\text{-}15)$$

If no net external force acts on a system of particles, the total translational momentum $\vec{p}_{\text{sys}}$ of the system cannot change.

This result is called the **law of conservation of translational momentum.** It is a natural consequence of Newton's laws. This law can also be written in equation form as

$$\vec{p}_{\text{sys}\,1} = \vec{p}_{\text{sys}\,2} \qquad \text{(isolated system),} \qquad (7\text{-}16)$$

where $\vec{p}_{\text{sys}1}$ is the total momentum of all the particles in a system at time t_1 and $\vec{p}_{\text{sys}2}$ is the system momentum at time t_2. In words, this equation says that, for an isolated system, the total translational momentum at any initial time t_1 is equal to the total translational momentum at any later time t_2. This is not to say that the momenta of individual particles within the system do not change. Particles inside a system can undergo changes in momentum. However, they must do so by exchanging momentum with other particles in the system so that the total system momentum remains constant.

In the next section we will consider two colliding carts that form an isolated system and look at how they exchange their momenta in a way that conserves the total momentum of the system.

7-6 Simple Collisions and Conservation of Momentum

Suppose that two very low friction carts roll along a smooth, level track. What happens to them before, during, and after they collide? We know by analysis with Newton's Second Law that the external forces on the carts (the gravitational force pulling downward and the normal forces of the track holding them up) cancel each other out, so $\vec{F}^{\text{net}} = 0$. Thus, the system is isolated, so we predict that the total momentum of the two-cart system will be conserved. In other words, each cart should change its momentum in such a way that the total change in system momentum is zero.

In this section, we will examine two different types of collisions for simple systems that are isolated: (1) a collision in which the hard rubber end of a more massive cart hits the hard rubber end of a less massive cart and the two carts bounce off each other, and (2) a collision in which the rubber ends are replaced with Velcro or clay so that the carts stick together after the collision. Is it possible for momentum to be conserved in these two very different situations?

A Bouncy Collision

Our first case, the bouncy collision, is depicted in Fig. 7-10. Two bodies having almost the same speed but different masses are just about to have a *one-dimensional collision* (meaning that the motions before and after the collision are along the same straight line). Imagine that these two objects bounce off one another immediately following their collision. What happens during the collision? Does the cart on the right with more mass on it exert more force on the cart on the left? Less force? The same force?

FIGURE 7-10 ■ Two carts of different masses undergo a "bouncy" collision. The collision forces can be measured 4000 times a second using electronic force sensors attached to a computer data acquisition system.

These carts are outfitted with electronic force sensors, so we can measure the collision forces. The impulse curves indicating the changes in forces on each of the carts during the time of impact are shown in Fig. 7-11.

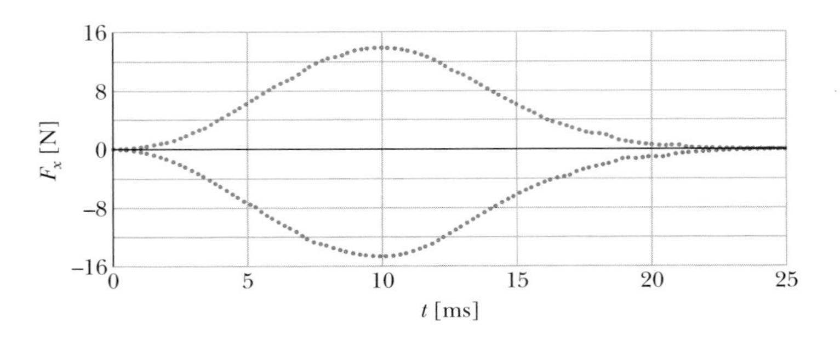

FIGURE 7-11 ■ The top graph displays the x-component of force the left cart exerts on the right cart. The bottom graph shows the force the right cart exerts on the left cart. The collision forces for the two carts are equal and opposite on a moment-by-moment basis. The time of contact is less 25 ms or about 1/40th of a second.

The fact that the interaction forces have equal magnitudes and are oppositely directed at every moment of contact is yet another experimental verification of Newton's Third Law. It shows that there is no net internal force in this two "particle" system. If the total momentum of the system is to be considered, we expect that the change in momentum of the left cart will be equal and opposite to the change in momentum of the right cart. However, since the mass of the right cart is greater and momentum is the product of mass and velocity, the right cart must have a smaller change in velocity than the less massive cart on the left. You are familiar with this fact. When a massive bowling ball hits a bowling pin, the magnitude of the pin's velocity is much larger than that of the ball. An observation of the two carts bouncing off each other confirms the prediction that the more massive cart on the right undergoes less velocity change than the cart on the left.

This conclusion can be expressed mathematically. Using Eq. 7-16,

Total momentum $\vec{p}_{\text{sys }1}$ (before the collision) = total momentum $\vec{p}_{\text{sys }2}$ (after the collision).

We can also express this mathematically in terms of the momentum of each cart as

$$\vec{p}_{A1} + \vec{p}_{B1} = \vec{p}_{A2} + \vec{p}_{B2} \quad \text{(conservation of translational momentum).} \tag{7-17}$$

Because the motion is one-dimensional, we can drop the vector arrows and use only components along the direction of the motion. Thus, from $\vec{p} = m\vec{v}$, we can rewrite this expression in terms of the masses and velocity components of the particles. For example, if we choose an x axis along the line of motion, then

$$m_A v_{A\,x}(t_1) + m_B v_{B\,x}(t_1) = m_A v_{A\,x}(t_2) + m_B v_{B\,x}(t_2) \quad (x\text{-component}), \tag{7-18}$$

where $v_{A\,x}(t_1)$ is the x-component of object A's velocity at time t_1. As we discussed while treating one-dimensional motions in previous chapters, it is essential when substituting actual values for the components into an equation that we use the correct sign (+ or −) to denote the direction of motion of each object along the chosen axis.

Here we have used an experimental verification of Newton's Third Law and a belief that Newton's Second Law is valid to assert that momentum ought to be conserved for an isolated system. Are we correct? Indeed, if we measure masses and use a computer data acquisition system or video analysis software to find velocity components before and after a collision, it is possible to verify momentum conservation experimentally for bouncy collisions. In the next subsection we will describe the details of this type of experimental verification for a sticky collision.

A Sticky Collision

To discuss a collision in which the particles stick together, we can replace the rubber cart bumpers with Velcro or gooey clay blobs. Another way to explore a sticky collision is to drop a stationary mass onto our low-friction cart. We can gently place the stationary mass on top of the moving cart and record what happens to the cart velocity with a video camera. We will describe how a video analysis of the cart position on video frames (1) enables us to confirm that momentum is conserved and (2) enables us to use our knowledge of momentum conservation to predict the final velocity of any sticky collision between two particle-like objects that form an isolated system. (See Fig. 7-12.)

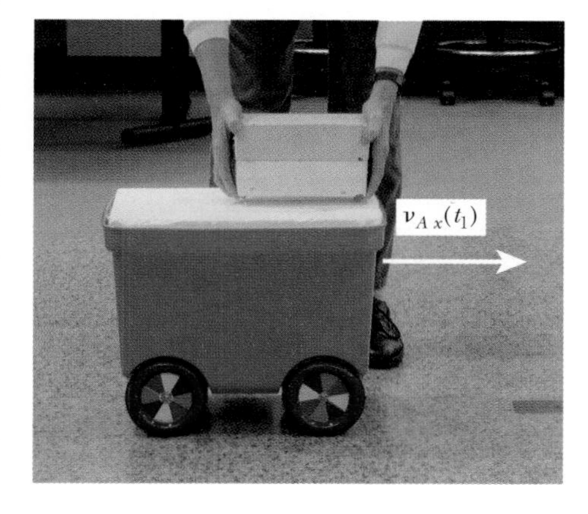

FIGURE 7-12 ■ A single frame of a video clip shows two bricks being placed on top of a cart as it moves toward the right with an initial velocity component of 1.78 m/s. The cart slows down noticeably once the bricks are placed on top of it.

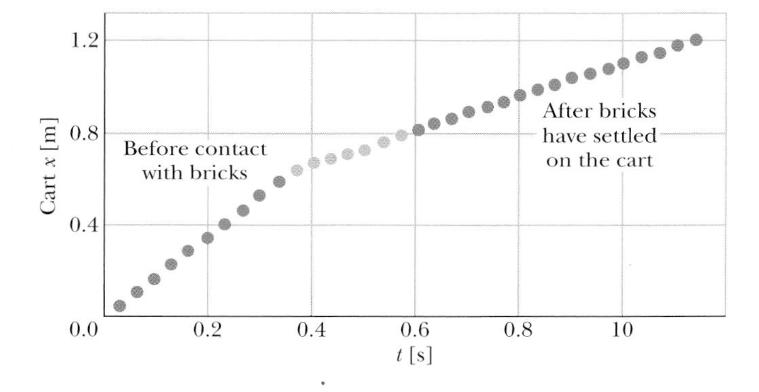

FIGURE 7-13 ■ A graph based on video analysis shows how the position of the moving cart in Fig. 7-12 changes before, during, and after bricks are placed gently on top of it. The slope of the graph before the mass touches the cart (0.00 to 0.33 s) gives the cart's initial velocity. The slope of the graph after the mass has fully settled on the cart (0.60 to 1.13 s) gives the final velocity of the cart–mass system.

The Video Analysis: As we view the video frames and locate the cart position in each frame, we see that in this case the two objects "stick" together following their gentle "collision." By determining the slope of the position vs. time graph for the first few frames we find that the cart of mass $m_A = 2.84$ kg is moving from left to right with an initial x-component of velocity of $v_{A\,x}(t_1) = +1.78$ m/s. (See Fig. 7-13.) After two bricks of total mass $m_B = 4.26$ kg are placed gently on the cart, the combined masses continue to move more slowly from left to right with a system velocity component given by $v_{\text{sys}\,x}(t_2) = 0.712$ m/s.

(1) Confirmation of Momentum Conservation: Let's check to see that $\vec{p}_{\text{sys}\,2} = \vec{p}_{\text{sys}\,1}$ (that is, Eq. 7-16 holds). We use our data to find the initial momentum of the system. The bricks (denoted as B) have no initial velocity, so

$$\begin{aligned}\vec{p}_{\text{sys}\,x}(t_1) &= p_{A\,x}(t_1)\hat{\imath} + p_{B\,x}(t_1)\hat{\imath} = m_A v_{A\,x}(t_1)\hat{\imath} + m_B v_{B\,x}(t_1)\hat{\imath}\\ &= (2.84\text{ kg})(1.78\text{ m/s})\hat{\imath} + 0\,\hat{\imath} \qquad (7\text{-}19)\\ \vec{p}_{\text{sys}\,x}(t_1) &= \vec{p}_{\text{sys}\,1} = (5.06\text{ kg}\cdot\text{m/s})\hat{\imath}.\end{aligned}$$

To find the final momentum of the system we note that after their collision, the cart and the bricks move together with the *same* velocity. Thus

$$\vec{p}_{\text{sys}\,x}(t_2) = p_{A\,x}(t_2)\hat{\imath} + p_{B\,x}(t_2)\hat{\imath} = m_A v_{A\,x}(t_2)\hat{\imath} + m_B v_{B\,x}(t_2)\hat{\imath} \qquad (7\text{-}20)$$

with

$$v_{A\,x}(t_2) = v_{B\,x}(t_2) = v_{\text{sys}\,x}(t_2),$$

so

$$\begin{aligned}\vec{p}_{\text{sys}\,x}(t_2) &= \vec{p}_{\text{sys}\,2} = (m_A + m_B)\vec{v}_{\text{sys}\,x}(t_2)\\ &= (2.84\text{ kg} + 4.26\text{ kg})(0.712\text{ m/s})\hat{\imath}\\ &= (5.06\text{ kg}\cdot\text{m/s})\hat{\imath}.\end{aligned}$$

There is uncertainty associated with any experimental measurements. Even though video analysis is a very fine tool for motion analysis, we were quite fortunate to have our initial and final momentum values agree to three significant figures. That doesn't usually happen in momentum conservation experiments.

(2) Predicting the Final Velocity: If you can correctly identify an isolated system and apply momentum conservation, a knowledge of the initial velocities of a two-particle system can enable you to predict the velocities after a sticky collision. We merely need to equate the last terms in Eqs. 7-19 and 7-20 and solve for the final velocity. For example, with $\vec{v}_{B\,x}(t_1) = 0$, this gives

$$v_{\text{sys}\,x}(t_2)\hat{\imath} = \frac{m_A}{m_A + m_B} v_{A\,x}(t_1)\hat{\imath}. \qquad (7\text{-}21)$$

For our cart–brick situation this would give us a predicted final velocity of

$$v_{\text{sys}\,x}(t_2) = \frac{2.84\text{ kg}}{2.84\text{ kg} + 4.26\text{ kg}}(1.78\text{ m/s}) = 0.712\text{ m/s} \qquad \text{(predicted final speed).}$$

Note that the speed $|\vec{v}_{\text{sys}\,x}(t_2)|$ of the combined masses after the collision must be less than the speed $|\vec{v}_{A\,x}(t_1)|$ of the mass that was moving before the collision, because the mass ratio $m_A/(m_A + m_B)$ is always less than one.

Remember that regardless of whether the objects involved in the collision bounce off one another or stick together, the total translational momentum of a system is conserved so long as there is no net external force acting on it. Friction is an external force that often renders a system nonisolated and hence interferes with momentum conservation.

Our consideration of bouncy and sticky collisions is enough to get us started analyzing collisions. However, many collisions are not completely bouncy or completely sticky. In Chapter 10, we will use the concept of mechanical energy conservation to refine our understanding of collisions.

READING EXERCISE 7-5: Consider two small frictionless carts of equal mass that are resting on a level track with a firecracker wedged between them. When the firecracker explodes, the carts fly apart. Is translational momentum conserved in this case? (State any assumptions you made in formulating your answer.) Explain in detail why momentum is conserved or why it isn't. ■

TOUCHSTONE EXAMPLE 7-3: Exploding Box

A fireworks box with mass $m = 6.0$ kg slides with speed $v = 4.0$ m/s across a frictionless floor in the positive direction along an x axis. It suddenly explodes into two pieces. One piece, with mass $m_A = 2.0$ kg, moves in the positive direction along the x axis with speed $v_A = 8.0$ m/s. What is the velocity of the second piece, with mass m_B?

SOLUTION ■ There are two **Key Ideas** here. First, we could get the velocity of the second piece if we knew its momentum, because we already know its mass is $m_B = m - m_A = 4.0$ kg. Second, we can relate the momenta of the two pieces to the original momentum of the box if momentum is conserved. Let's check.

Our reference frame will be that of the floor. Our system consists initially of the box and then of the two pieces. The box and pieces each experience a normal force from the floor and a gravitational force. However, those forces are both vertical and cancel out (sum to zero). The forces produced by the explosion are internal to the system. Thus, the horizontal component of the momentum of the system is conserved, and we can apply momentum conservation (Eq. 7-16) along the x axis.

The initial momentum of the system is that of the box:

$$\vec{p}_{\text{sys }1} = m\vec{v}.$$

Similarly, we can write the final momenta of the two pieces as

$$\vec{p}_{A\,2} = m_A\vec{v}_A \quad \text{and} \quad \vec{p}_{B\,2} = m_B\vec{v}_B.$$

The final total momentum $\vec{p}_{\text{sys }2}$ of the system is the vector sum of the momenta of the two pieces:

$$\vec{p}_{\text{sys }2} = \vec{p}_{A\,2} + \vec{p}_{B\,2} = m_A\vec{v}_A + m_B\vec{v}_B.$$

Since all the velocities and momenta in this problem are vectors along the x axis, we can write them in terms of their x-components. Doing so, we now obtain

$$p_{\text{sys }x}(t_1) = p_{\text{sys }x}(t_2)$$

or

$$mv_x(t_1) = m_A v_{A\,x}(t_2) + m_B v_{B\,x}(t_2).$$

Inserting known data, we find

$$(6.0\text{ kg})(4.0\text{ m/s}) = (2.0\text{ kg})(8.0\text{ m/s}) + (4.0\text{ kg})v_{B\,x}(t_2)$$

and thus

$$v_{B\,x}(t_2) = 2.0\text{ m/s}.$$

Since all the momenta and velocities in the vertical direction are zero, our final result is

$$\vec{v}_B = v_{B\,x}\hat{\text{i}} = (2.0\text{ m/s})\hat{\text{i}}, \qquad \text{(Answer)}$$

and the second piece also moves in the positive direction along the x axis.

7-7 Conservation of Momentum in Two Dimensions

What happens when one object strikes another with a glancing blow? As shown in Fig. 7-14, the objects can come off at an angle with respect to each other. Can we still apply the law of conservation of momentum?

The principle of conservation of momentum is applicable to collisions in two or three dimensions just as it is in one dimension, as long as the net force on the system is zero in each of the dimensions. If the net force is not zero in one of the dimensions, momentum is not conserved in that dimension in accordance with Eqs. 7-11, 7-12, and 7-13. For convenience, we choose a two-dimensional coordinate system. Then we can

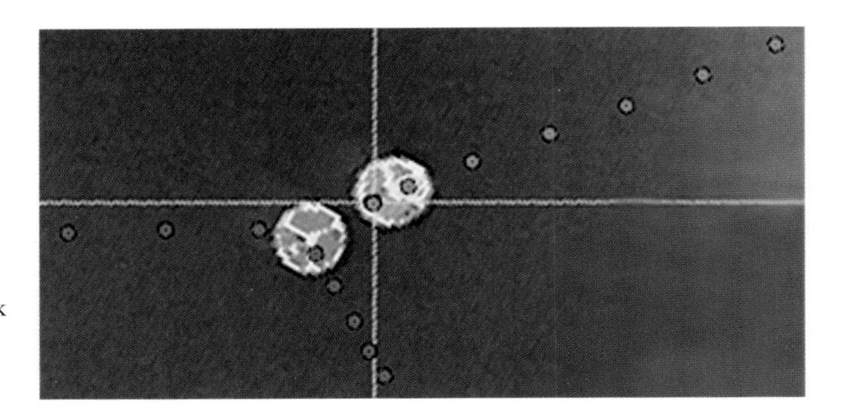

FIGURE 7-14 ■ A video analysis of a collision between two pucks. A puck traveling on an air table hits another stationary puck with a glancing blow. They both travel off in different velocities in such a way that momentum is conserved.

decompose the momentum conservation equation $\vec{p}_{A1} + \vec{p}_{B1} = \vec{p}_{A2} + \vec{p}_{B2}$ (Eq. 7-17) into components. When conservation of momentum is applied to multidimensional motion, it is applied in each direction separately. In other words, the single expression $\vec{p}_{A1} + \vec{p}_{B1} = \vec{p}_{A2} + \vec{p}_{B2}$ can be replaced with up to three expressions that involve unit vectors associated with three orthogonal coordinate axes directions. In the two-dimensional case, these are

$$p_{Ax}(t_1)\hat{\mathrm{i}} + p_{Bx}(t_1)\hat{\mathrm{i}} = p_{Ax}(t_2)\hat{\mathrm{i}} + p_{Bx}(t_2)\hat{\mathrm{i}}, \tag{7-22}$$

and

$$p_{Ay}(t_1)\hat{\mathrm{j}} + p_{By}(t_1)\hat{\mathrm{j}} = p_{Ay}(t_2)\hat{\mathrm{j}} + p_{By}(t_2)\hat{\mathrm{j}}, \tag{7-23}$$

where the subscripts denote the momenta for particles A and B along each of the coordinate axes x and y.

This set of equations describes the relationships that have to be satisfied by the initial and final momenta of the particles as a result of momentum conservation in two dimensions. In terms of the object's masses and velocities, the equations above can be expressed in terms of components:

x-components: $$m_A v_{Ax}(t_1) + m_B v_{Bx}(t_1) = m_A v_{Ax}(t_2) + m_B v_{Bx}(t_2), \tag{7-24}$$

and y-components: $$m_A v_{Ay}(t_1) = +m_B v_{By}(t_1) = m_A v_{Ay}(t_2) + m_B v_{By}(t_2). \tag{7-25}$$

We can use either set of two equations above (7-22 and 7-23 or 7-24 and 7-25) to analyze a collision. We will choose which set to use based on the information available to us.

If we determine the angles that the objects make with respect to various axes before and after a collision, we can often calculate the x- and y-components of the momenta or velocities using trigonometry. This is shown in Fig. 7-15 for two pucks that have different masses. *We also must take special care to associate the correct sign (to denote direction) with each term in the expressions above.* For example, Fig. 7-15 shows a collision between a projectile body and a target body initially at rest. The impulses between the bodies have sent the bodies off at angles θ_A and θ_B measured relative to the x axis, along which object A initially traveled. In this situation, we would rewrite $\vec{p}_{A1} + \vec{p}_{B1} = \vec{p}_{A2} + \vec{p}_{B2}$ for components along the x axis as

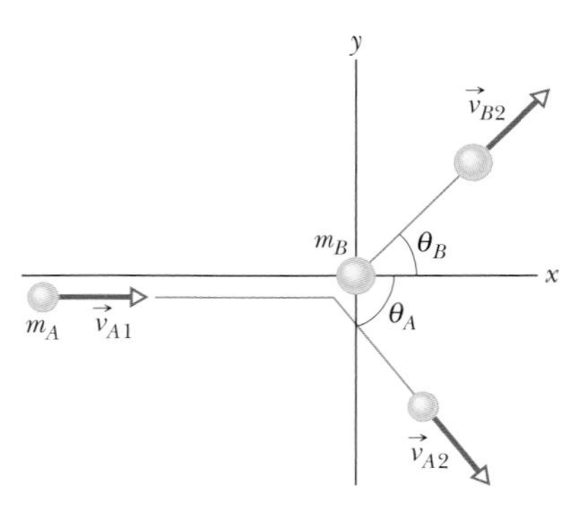

FIGURE 7-15 ■ An object of mass m_A hits a second object of mass m_B at a glancing blow, and each object moves off at an angle with respect to the original line of motion (defined here as the positive x axis).

$$m_A v_{A\,x}(t_1) + m_B v_{B\,x}(t_1) = m_A v_{A\,x}(t_2) + m_B v_{B\,x}(t_2),$$

or

$$m_A|\vec{v}_{A1}| + 0 = m_A|\vec{v}_{A2}|\cos\theta_A + m_B|\vec{v}_{B2}|\cos\theta_B,$$

and along the y axis as

$$m_A v_{A\,y}(t_1) + m_B v_{B\,y}(t_1) = m_A v_{A\,y}(t_2) + m_B v_{B\,y}(t_2),$$

or $$0 + 0 = -m_A|\vec{v}_{A\,2}|\sin\theta_A + m_B|\vec{v}_{B\,2}|\sin\theta_B.$$

The minus sign in the first term to the right of the equal sign above is very important. It indicates that the y-component of velocity for m_A is downward.

READING EXERCISE 7-6: An initially stationary device lying on a frictionless floor explodes into two pieces, which then slide across the floor. One piece slides in the positive direction along an x axis. (a) What is the sum of the momenta of the two pieces after the explosion? (b) Can the second piece move at an angle to the x axis? Why or why not? (c) What is the direction of the momentum of the second piece? ■

READING EXERCISE 7-7: Consider a system that contains the Earth and a grapefruit. The grapefruit starts off at rest and falls a certain distance, at which point its velocity has increased to 2 m/s. What is the change in momentum of the grapefruit? What is the change in momentum of the Earth? What is the approximate change in speed of the Earth associated with this change in momentum? State any estimates you made in answering the question. ■

TOUCHSTONE EXAMPLE 7-4: Skaters Embrace

Two skaters collide and embrace, "sticking" together after impact, as suggested by Fig. 7-16, where the origin is placed at the point of collision. Alfred, whose mass m_A is 83 kg, is originally moving east with speed $v_A = 6.2$ km/h. Barbara, whose mass m_B is 55 kg, is originally moving north with speed $v_B = 7.8$ km/h.

(a) What is the velocity $\vec{v}_{\text{sys}\,2}$ of the couple after they collide?

SOLUTION ■ One **Key Idea** here is the assumption that the two skaters form an isolated system. That is, during the collision we assume no *net* external force acts on them. In particular, we neglect any frictional force on their skates from the ice because the peak collision forces are much larger than the friction forces. With that assumption, we can apply conservation of the total translational momentum $\vec{p}_{\text{sys}}$ by writing $\vec{p}_{\text{sys}\,1} = \vec{p}_{\text{sys}\,2}$ as

$$m_A\vec{v}_{A\,1} + m_B\vec{v}_{B\,1} = (m_A + m_B)\vec{v}_{\text{sys}\,2}. \tag{7-26}$$

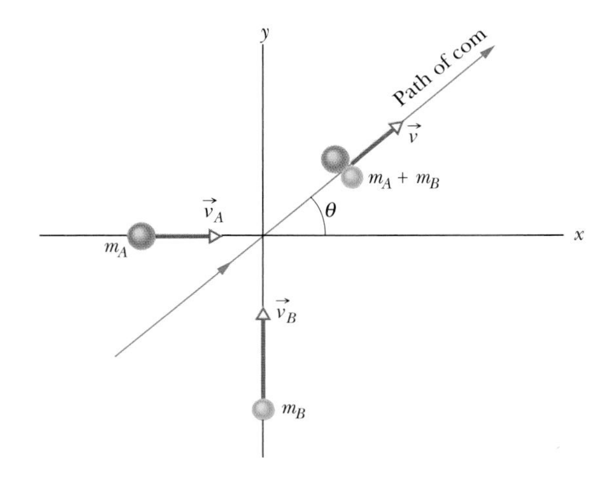

FIGURE 7-16 ■ Two skaters, Alfred (A) and Barbara (B), represented by spheres in this simplified overhead view, have a "sticky" collision. Afterward, they move off together at angle θ, with speed v.

Solving for the system velocity $\vec{v}_{\text{sys}\,2} = \vec{v}$ after collision gives us

$$\vec{v} = \frac{m_A\vec{v}_A + m_B\vec{v}_B}{m_A + m_B}.$$

We can solve this directly on a vector-capable calculator by substituting given data for the symbols on the right side. We can also solve it by applying a second **Key Idea** (one we have used before) and then some algebra: The idea is that the total translational momentum of the system is conserved separately for components along the x axis and y axis shown in Fig. 7-16. Writing Eq. 7-26 in component form for the x axis and noting that $\vec{v}_A = v_A\hat{\text{i}} + 0\hat{\text{j}}$ yields

$$m_Av_A + m_B(0) = (m_A + m_B)|\vec{v}|\cos\theta, \tag{7-27}$$

and for the y axis, since $\vec{v}_B = 0\hat{\text{i}} + v_B\hat{\text{j}}$,

$$m_A(0) + m_Bv_B = (m_A + m_B)|\vec{v}|\sin\theta. \tag{7-28}$$

We cannot solve either of these equations separately because they both contain two unknowns ($|\vec{v}|$ and θ), but we can solve them simultaneously by dividing Eq. 7-28 by Eq. 7-27. We get

$$\tan\theta = \frac{m_Bv_B}{m_Av_A} = \frac{(55\text{ kg})(7.8\text{ km/h})}{(83\text{ kg})(6.2\text{ km/h})} = 0.834.$$

Thus,

$$\theta = \tan^{-1}0.834 = 39.8° \approx 40°. \qquad \text{(Answer)}$$

From Eq. 7-28, with $m_A + m_B = 138$ kg, we then have a final system speed of

$$v = |\vec{v}| = \frac{m_Bv_B}{(m_A + m_B)\sin\theta} = \frac{(55\text{ kg})(7.8\text{ km/h})}{(138\text{ kg})(\sin 39.8°)}$$
$$= 4.86\text{ km/h} \approx 4.9\text{ km/h}. \qquad \text{(Answer)}$$

FIGURE 7-17 ■ Liftoff of Project Mercury spacecraft.

7-8 A System with Mass Exchange—A Rocket and Its Ejected Fuel

In the systems we have dealt with so far, we have assumed that the total mass of the system remains constant; no mass is added or removed from the system. Such systems are called **closed.** Sometimes, as in a rocket (Fig. 7-17), the mass does not stay constant. Most of the mass of a rocket on its launching pad is fuel, all of which will eventually be burned and ejected from the nozzle of the rocket engine. A rocket accelerates by ejecting some of its own mass in the form of exhaust gases. It turns out that both the rate at which the fuel burns and the velocity of the ejected fuel particles relative to the rocket are constant.

We handle the variation of the mass of the rocket as the rocket accelerates by applying Newton's Second Law, not to the rocket alone but to the rocket and its ejected combustion products taken together. The mass of *this* system does *not* change as the rocket accelerates.

Finding the Acceleration

Let's consider the acceleration of this rocket in deep space with no gravitational or atmospheric drag forces acting on it. To simplify our observation of what happens, suppose that at an arbitrary time t_1 when the rocket has a total mass M, we happen to be in an inertial reference frame that moves at a constant velocity that is exactly the same as the rocket's velocity. What do we observe in a short time interval dt?

At time t_1 the rocket is not moving relative to us (see Fig. 7-18*a*). After a time interval dt, the rocket has ejected a small amount of burned fuel of mass dm at a velocity relative to the rocket, which we call $\vec{v}^{\,\text{rel}}$.

Our system consists of the rocket and the exhaust products released during interval dt. The system is closed and isolated, so the translational momentum of the system must be conserved during dt; that is,

$$\vec{p}_{\text{sys}\,1} = \vec{p}_{\text{sys}\,2}. \tag{7-29}$$

However, at time t_1 when the rocket is not moving relative to us, we observe that the initial momentum of the system is zero. Thus, at a later time dt the total momentum of the system must still be zero. As the mass dm of burned fuel flies off at a velocity $\vec{v}^{\,\text{rel}}$ the rocket that now has a very slightly smaller mass of $M - dm$ must recoil in the opposite direction with a small increase in its velocity of $d\vec{v}$ as shown in Fig. 7-18*b*. In order to keep the total momentum of the rocket–fuel system zero we must have

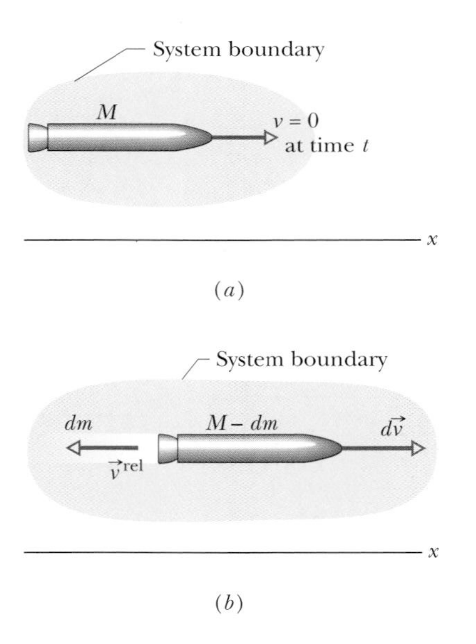

FIGURE 7-18 ■ (*a*) An accelerating rocket of mass M at time t_1, as seen from an inertial reference frame. (*b*) The same rocket, but at time $t_1 + dt$. The exhaust products released during interval dt are shown.

$$\vec{p}_{\text{sys}\,1} = 0 = \vec{p}_{\text{sys}\,2} = dm(\vec{v}^{\,\text{rel}}) + (M - dm)\,d\vec{v}. \tag{7-30}$$

Since the rocket mass $M \gg dm$, the total rocket mass M is always much greater than the mass of fuel ejected in a short time, so we can rewrite the momentum conservation equation as

$$dm(\vec{v}^{\,\text{rel}}) + M\,d\vec{v} \approx 0. \tag{7-31}$$

Dividing each term by dt and rearranging terms gives us

$$-\frac{dm}{dt}\,\vec{v}^{\,\text{rel}} = M\frac{d\vec{v}}{dt}. \tag{7-32}$$

If we note that the change in the rocket mass due to the loss of the ejected fuel during the time interval dt is given by $dM = -dm$, we can replace $-dm/dt$ with dM/dt. Since

dv/dt is the acceleration of the rocket relative to the inertial reference, the expression above becomes

$$\frac{dM}{dt}\vec{v}^{\,\text{rel}} = M\vec{a} \qquad \text{(first rocket equation).} \tag{7-33}$$

This equation holds at any instant, with the mass M, the *fuel consumption rate* $R = -dM/dt$, and the acceleration $\vec{a}$ evaluated at that instant. Note that $\vec{v}^{\,\text{rel}}$ and $\vec{a}$ point in opposite directions because we chose $\vec{v}^{\,\text{rel}}$ to be the velocity of the *ejected gas relative to the rocket* rather than the other way around. This is not at first apparent in Eq. 7-33 until you remember that dM/dt is negative. The left side of this equation has the dimensions of a force ($\text{kg}\cdot\text{m/s}^2 = \text{N}$) and depends only on design characteristics of the rocket engine—namely, the rate R at which it consumes fuel mass and the speed $\vec{v}^{\,\text{rel}}$ with which that mass is ejected relative to the rocket.

We call the term $-R\vec{v}^{\,\text{rel}}$ the **thrust** of the rocket engine and represent it with $\vec{F}^{\text{thrust}}$. Newton's Second Law emerges clearly if we write $-R\vec{v}^{\,\text{rel}} = M\vec{a}$ as $\vec{F}^{\text{thrust}} = M\vec{a}$, in which $\vec{a}$ is the acceleration of the rocket at the time that its mass is M. Notice that $\vec{F}^{\text{thrust}}$ points in the same direction that the rocket is accelerating, even though $\vec{v}^{\,\text{rel}}$ points in the opposite direction. Since dM/dt is intrinsically negative, $R = -dM/dt$ is positive.

Finding the Velocity Change

How will the velocity of a rocket change as it consumes its fuel? Recall that the change in the rocket mass due to the loss of the ejected fuel during the time interval dt is given by $dM = -dm$. Then we can rewrite Eq. 7-30, which is $Md\vec{v} = -(dm)\vec{v}^{\,\text{rel}}$, and rearrange the terms to get

$$d\vec{v} = \vec{v}^{\,\text{rel}}\frac{dM}{M}$$

where integrating gives us

$$\int_{\vec{v}_1}^{\vec{v}_2} d\vec{v} = \vec{v}^{\,\text{rel}}\int_{M_1}^{M_2}\frac{dM}{M},$$

in which $M_1 = M(t_1)$ represents the initial mass of the rocket at time t_1 and $M_2 = M(t_2)$ is the mass of the rocket at some later ("final") time t_2. Evaluating the integrals then gives

$$\vec{v}_2 - \vec{v}_1 = \vec{v}^{\,\text{rel}}\ln\frac{M_2}{M_1} = -\vec{v}^{\,\text{rel}}\ln\frac{M_1}{M_2} \qquad \text{(second rocket equation),} \tag{7-34}$$

for the increase in the speed of the rocket during the change in mass from M_1 to M_2. (The symbol "ln" in this equation means the *natural logarithm.*) The final mass is always less than the initial mass so the natural log will always be positive. But the velocity of the ejected fuel relative to the rocket is also in the opposite direction as the velocity change of the rocket. This always gives us a velocity change in a direction opposite that of mass ejection.

We see here the advantage of multistage rockets, in which M_2 is reduced by discarding successive stages when their fuel is depleted. Discarding rocket stages means there is less mass to accelerate. An ideal rocket would reach its destination with only its payload remaining.

FIGURE 7-19a ■ Liftoff of the Mercury-Redstone rocket that sent the first American astronaut, Alan Shepard, into space in 1961.

Thrust Forces at Liftoff

In the first few seconds of liftoff, the fuel consumption rate is not large enough to change the overall mass M of a typical modern rocket by a noticeable amount. Thus, its mass M is approximately constant. We can use this fact along with Eq. 7-33 in the analysis of video images of a NASA rocket to find the thrust forces of the rocket. As an example, we will do an analysis of the Mercury-Redstone rocket that lifted Alan Shepard into space in 1961. An image of the rocket during liftoff is shown in Fig. 7-19*a*. However, at liftoff we are not in deep space, so the net force on the rocket is the vector sum of the thrust force of the rocket acting in an upward direction and the downward force of the gravitational attraction of the Earth. Therefore,

$$\vec{F}^{\text{thrust}} + \vec{F}^{\text{grav}} = M\vec{a} \qquad \text{(at liftoff from the Earth's surface).}$$

Taking the positive y direction to be vertically upward, this simplifies to

$$F_y^{\text{thrust}}\,\hat{\jmath} - Mg\,\hat{\jmath} = Ma_y\,\hat{\jmath}.$$

The y position as a function of time of the Mercury-Redstone rocket liftoff is shown in Fig. 7-19*b*.

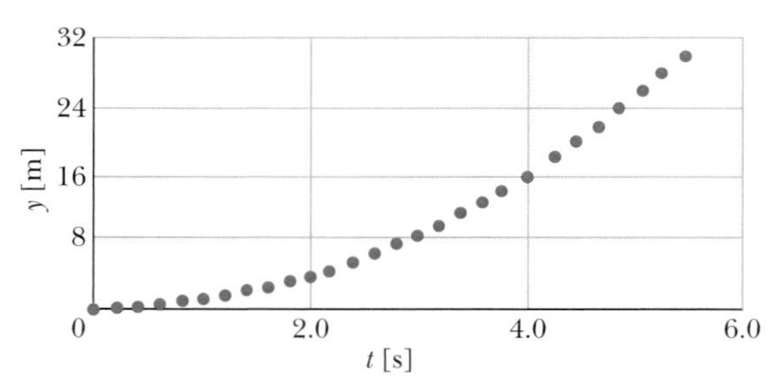

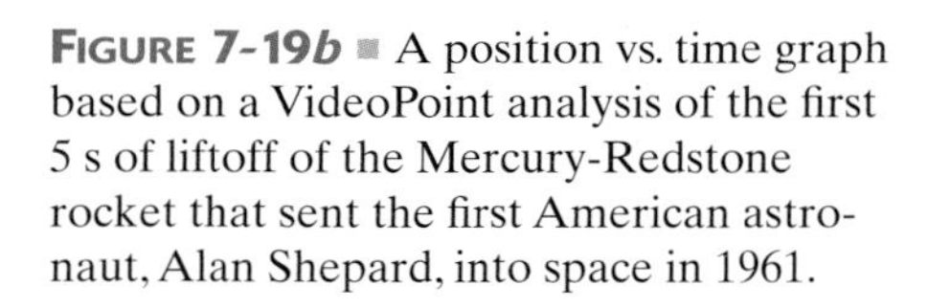

FIGURE 7-19b ■ A position vs. time graph based on a VideoPoint analysis of the first 5 s of liftoff of the Mercury-Redstone rocket that sent the first American astronaut, Alan Shepard, into space in 1961.

Fitting the curve with a quadratic function gives an upward acceleration of magnitude 1.1 m/s^2. The mass of the Mercury-Redstone rocket with full fuel and payload is $M = 3.0 \times 10^4$ kg. Thus, the y-component of the thrust force is given by

$$F_y^{\text{thrust}} = M(a_y + g) = (3.0 \times 10^4\,\text{kg})(1.1 + 9.8)\,\text{m/s}^2 = 33 \times 10^5\,\text{N}.$$

If we know the fuel consumption rate we can also find the relative velocity with which fuel is ejected from the rocket using the first rocket equation (Eq. 7-33) given by $R\vec{v}^{\text{ rel}} = M\vec{a}$.

TOUCHSTONE EXAMPLE 7-5: Rocket Thrust

A rocket whose initial mass M_1 is 850 kg consumes fuel at the rate $R = 2.3$ kg/s. The speed v^{rel} of the exhaust gases relative to the rocket engine is 2800 m/s.

(a) What thrust does the rocket engine provide?

SOLUTION ■ The **Key Idea** here is that the magnitude of the thrust F^{thrust} is equal to the product of the fuel consumption rate R and the relative speed v_{rel} at which exhaust gases are expelled:

$$F^{\text{thrust}} = Rv^{\text{rel}} = (2.3 \text{ kg/s})(2800 \text{ m/s})$$
$$= 6440 \text{ N} \approx 6400 \text{ N}.$$

(b) What is the initial acceleration of the rocket launched from a spacecraft?

SOLUTION ■ We can relate the thrust $\vec{F}^{\text{thrust}}$ of a rocket to the resulting acceleration $\vec{a}$ with $\vec{F}^{\text{thrust}} = M\vec{a}$, where M is the rocket's mass. The **Key Idea**, however, is that M decreases and the magnitude of the acceleration a increases as fuel is consumed. Because we want the initial value of the acceleration here, we must use the initial value M_1 of the mass, finding that

$$\vec{a} = \frac{\vec{F}^{\text{thrust}}}{M_1} = \frac{6440 \text{ N } \hat{\text{i}}}{850 \text{ kg}} = (7.6 \text{ m/s}^2)\, \hat{\text{i}}. \quad \text{(Answer)}$$

(c) Suppose that the mass M_2 of the rocket when its fuel is exhausted is 180 kg. What is its speed relative to the spacecraft at that time? Assume that the spacecraft is so massive that the launch does not alter its speed.

SOLUTION ■ The **Key Idea** here is that the rocket's final speed v_2 (when the fuel is exhausted) depends on the ratio M_1/M_2 of its initial mass to its final mass, as given by Eq. 7-34. With the initial speed $v_1 = 0$, we have

$$\vec{v}_2 = -\vec{v}^{\text{ rel}} \ln\left(\frac{M_1}{M_2}\right)$$
$$= -(-2800 \text{ m/s}\, \hat{\text{i}}) \ln\left(\frac{850 \text{ kg}}{180 \text{ kg}}\right)$$
$$= (2800 \text{ m/s}) \ln(4.72)\, \hat{\text{i}} \approx 4300 \text{ m/s}\, \hat{\text{i}}. \quad \text{(Answer)}$$

Note that the ultimate speed of the rocket can exceed the exhaust speed v^{rel}.

Problems

SEC. 7-2 ■ TRANSLATIONAL MOMENTUM OF A PARTICLE

1. Same Momentum Suppose that your mass is 80 kg. How fast would you have to run to have the same translational momentum as a 1600 kg car moving at 1.2 km/h?

2. VW Beetle How fast must an 816 kg VW Beetle travel to have the same translational momentum as a 2650 kg Cadillac going 16 km/h?

3. Radar An object is tracked by a radar station and found to have a position vector given by $\vec{r} = [(3500 \text{ m}) - (160 \text{ m/s})t]\hat{\text{i}} + (2700 \text{ m})\hat{\text{j}}$ with $\vec{r}$ in meters and t in seconds. The radar station's x axis points east, its y axis north, and its z axis vertically up. If the object is a 250 kg meteorological missile, what are (a) its translational momentum and (b) its direction of motion?

SEC. 7-4 ■ IMPULSE AND MOMENTUM CHANGE

4. Ball Moving Horizontally A 0.70 kg ball is moving horizontally with a speed of 5.0 m/s when it strikes a vertical wall. The ball rebounds with a speed of 2.0 m/s. What is the magnitude of the change in translational momentum of the ball?

5. Cue Ball A 0.165 kg cue ball with an initial speed of 2.00 m/s bounces off the rail in a game of pool, as shown from an overhead view in Fig. 7-20. For x and y axes located as shown, the bounce reverses the y-component of the ball's velocity but does not alter the x-component. (a) What is θ in Fig 7-20? (b) What is the change in the ball's momentum in unit-vector notation? (The fact that the ball rolls is not relevant to either question.)

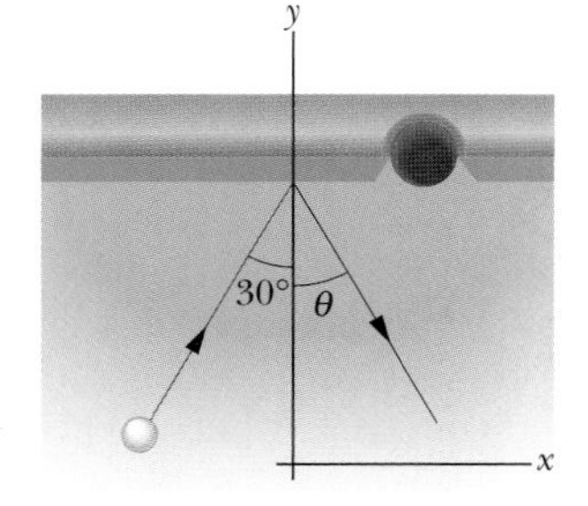

FIGURE 7-20 ■ Problem 5.

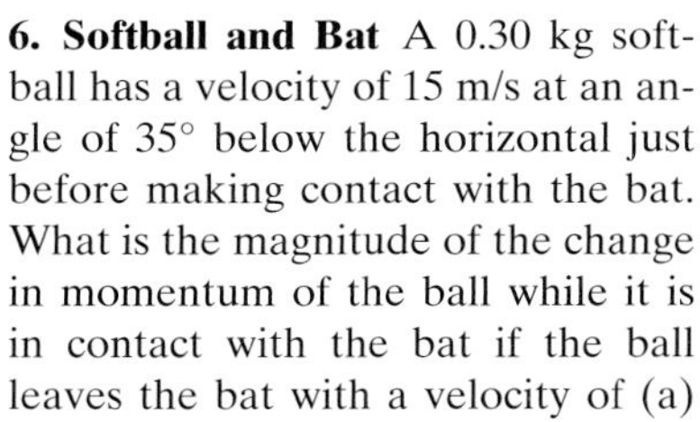

6. Softball and Bat A 0.30 kg softball has a velocity of 15 m/s at an angle of 35° below the horizontal just before making contact with the bat. What is the magnitude of the change in momentum of the ball while it is in contact with the bat if the ball leaves the bat with a velocity of (a) 20 m/s, vertically downward and (b) 20 m/s, horizontally away from the batter and back toward the pitcher?

7. Stationary Ball-Impulse A cue stick strikes a stationary pool ball, with an average force of 50 N over a time of 10 ms. If the ball has mass 0.20 kg, what speed does it have just after impact?

8. Average Force During Crash The National Transportation Safety Board is testing the crash-worthiness of a new car. The 2300 kg vehicle, moving at 15 m/s, is allowed to collide with a bridge abutment, which stops it in 0.56 s. What is the magnitude of the average force that acts on the car during the impact?

9. Average Force of Bat A 150 g baseball pitched at a speed of 40 m/s is hit straight back to the pitcher at a speed of 60 m/s. What is the magnitude of the average force on the ball from the bat if the bat is in contact with the ball for 5.0 ms?

10. Henri LaMothe Until he was in his seventies, Henri LaMothe excited audiences by belly-flopping from a height of 12 m into 30 cm of water (Fig. 7-21). Assuming that he stops just as he reaches the bottom of the water and estimating his mass, find the magnitudes of (a) the average force and (b) the average impulse on him from the water.

FIGURE 7-21 ■ Problem 10.

11. Steel Ball A force magnitude that averages 1200 N is applied to a 0.40 kg steel ball moving at 14 m/s in a collision lasting 27 ms. If the force is in a direction opposite the initial velocity of the ball, find the final speed and direction of the ball.

12. Chute Failure In February 1955, a paratrooper fell 370 m from an airplane without being able to open his chute but happened to land in snow, suffering only minor injuries. Assume that his speed at impact was 56 m/s (terminal speed), that his mass (including gear) was 85 kg, and that the magnitude of the force on him from the snow was at the survivable limit of 1.2×10^5 N. What are (a) the minimum depth of snow that would have stopped him safely and (b) the magnitude of the impulse on him from the snow?

13. Rebounding Ball A 1.2 kg ball drops vertically onto a floor, hitting with a speed of 25 m/s. It rebounds with a speed of 10 m/s. (a) What impulse acts on the ball during the contact? (b) If the ball is in contact with the floor for 0.020 s, what is the magnitude of the average force on the floor from the ball?

14. Superman It is well known that bullets and other missiles fired at Superman simply bounce off his chest (Fig.7-22). Suppose that a gangster sprays Superman's chest with 3 g bullets at the rate of 100 bullets/min, and the speed of each bullet is 500 m/s. Suppose too that the bullets rebound straight back with no change in speed. What is the magnitude of the average force on Superman's chest from the stream of bullets?

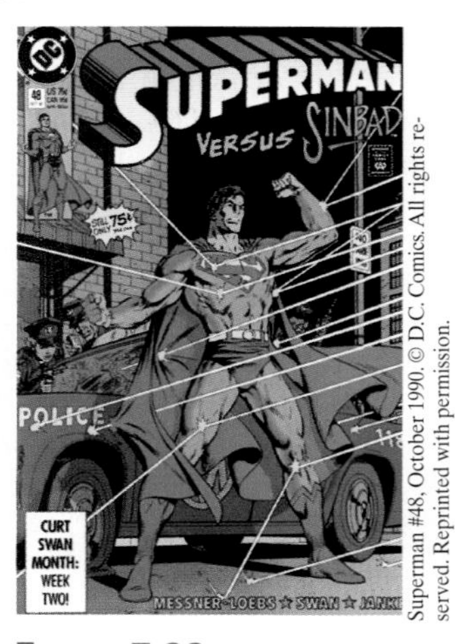

FIGURE 7-22 ■ Problem 14.

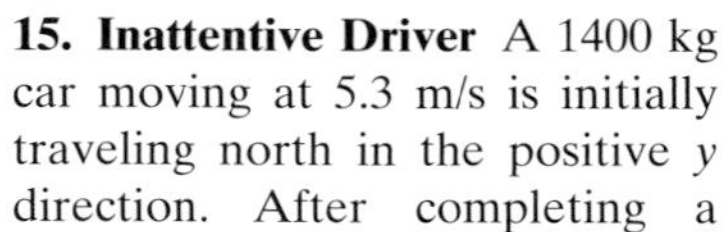

15. Inattentive Driver A 1400 kg car moving at 5.3 m/s is initially traveling north in the positive y direction. After completing a 90° right-hand turn to the positive x direction in 4.6 s, the inattentive operator drives into a tree, which stops the car in 350 ms. In unit-vector notation, what is the impulse on the car (a) due to the turn and (b) due to the collision? What is the magnitude of the average force that acts on the car (c) during the turn and (d) during the collision? (e) What is the angle between the average force in (c) and the positive x direction?

16. Softball A 0.30 kg softball has a speed of 12 m/s at an angle of 35° below the horizontal just before making contact with a bat. The ball leaves the bat 2.0 ms later with a vertical velocity of magnitude 10 m/s as shown in Fig. 7-23. What is the magnitude of the average force of the bat on the ball during the ball–bat contact?

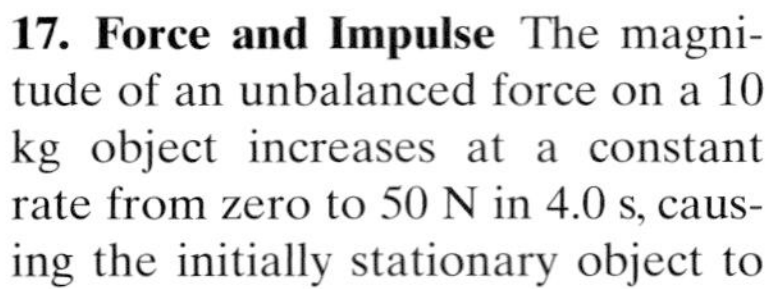
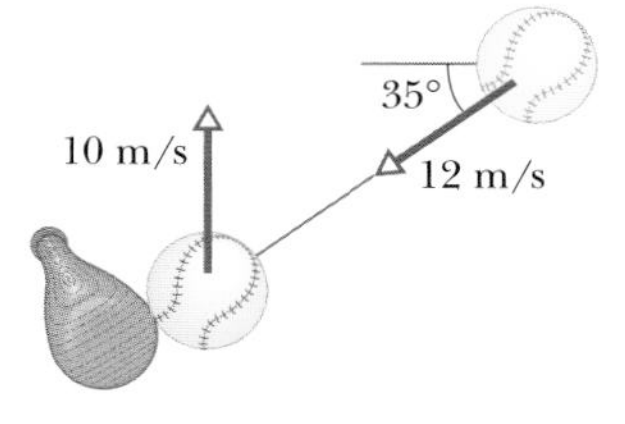

FIGURE 7-23 ■ Problem 16.

17. Force and Impulse The magnitude of an unbalanced force on a 10 kg object increases at a constant rate from zero to 50 N in 4.0 s, causing the initially stationary object to move. What is the object's speed at end of the 4.0 s?

18. Thunderstorm During a violent thunderstorm, hail of diameter 1.0 cm falls directly downward at a speed of 25 m/s. There are estimated to be 120 hailstones per cubic meter of air. (a) What is the mass of each hailstone (density = 0.92 g/cm^3)? (b) Assuming that the hail does not bounce, find the magnitude of the average force on a flat roof measuring 10 m $\times$ 20 m due to the impact of the hail. (*Hint*: During impact, the force on a hailstone from the roof is approximately equal to the net force on the hailstone, because the gravitational force on it is small.)

19. Pellet Gun A pellet gun fires ten 2.0 g pellets per second with a speed of 500 m/s. The pellets are stopped by a rigid wall. What are (a) the momentum of each pellet and (b) the magnitude of the average force on the wall from the stream of pellets? (c) If each pellet is in contact with the wall for 0.6 ms, what is the magnitude of the average force on the wall from each pellet during contact? (d) Why is this average force so different from the average force calculated in (b)?

20. Superball Hits Wall Figure 7-24 shows an approximate plot of force magnitude versus time during the collision of a 58 g Superball with a wall. The initial velocity of the ball is 34 m/s perpendicular to the wall; it rebounds directly back with approximately the same speed, also perpendicular to the wall. What is F^{max}, the maximum magnitude of the force on the ball from the wall during the collision?

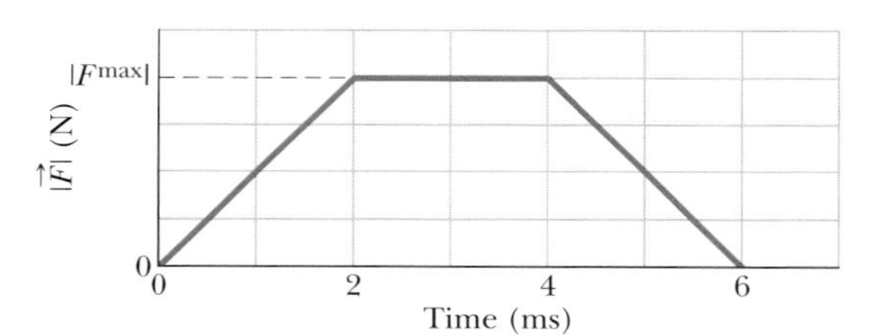

FIGURE 7-24 ■ Problem 20.

21. Spacecraft A spacecraft is separated into two parts by detonating the explosive bolts that hold them together. The masses of the parts are 1200 kg and 1800 kg; the magnitude of the impulse on each part from the bolts is 300 N · s. With what relative speed do the two parts separate because of the detonation?

22. Ball Strikes Wall In the overhead of Fig. 7-25, a 300 g ball with a speed v of 6.0 m/s strikes a wall at an angle θ of 30° and then rebounds with the same speed and angle. It is in contact with the wall for 10 ms. (a) What is the impulse on the ball from the wall? (b) What is the average force on the wall from the ball?

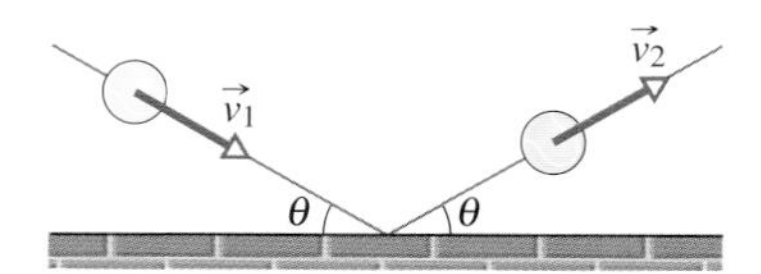

FIGURE 7-25 ■ Problem 22.

23. Two Barges In Fig. 7-26, two long barges are moving in the same direction in still water, one with a speed of 10 km/h and the other with a speed of 20 km/h. While they are passing each other, coal is shoveled from the slower to the faster one at a rate of 1000 kg/min. How much additional force must be provided by the driving engines of (a) the fast barge and (b) the slow barge if neither is to change speed? Assume that the shoveling is always perfectly sideways and that the frictional forces between the barges and the water do not depend on the mass of the barges.

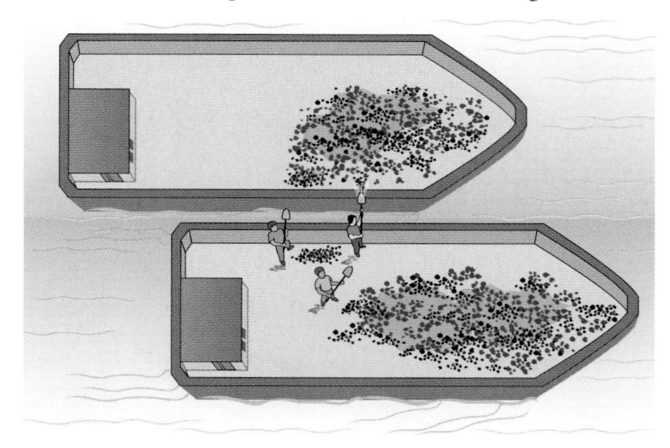

FIGURE 7-26 ■ Problem 23.

SEC. 7-6 ■ SIMPLE COLLISIONS AND CONSERVATION OF MOMENTUM

24. Two Blocks Two blocks of masses 1.0 kg and 3.0 kg on a frictionless surface are connected by a stretched spring and initially are held at rest. Then the two blocks are simultaneously released from rest. Shortly after the spring starts contracting we find that the 1.0 kg block is traveling toward the other at 1.7 m/s. What is the velocity of the other block at that moment?

25. Meteor Impact Meteor Crater in Arizona (Fig 7-1*a*) is thought to have been formed by the impact of a meteor with Earth some 20,000 years ago. The mass of the meteor is estimated at 5×10^{10} kg, and its speed at 7200 m/s. What speed would such a meteor give Earth in a head-on collision?

26. Bullet Strikes Wooden Block A 5.20 g bullet moving at 672 m/s strikes a 700 g wooden block at rest on a frictionless surface. The bullet emerges, traveling in the same direction with its speed reduced to 428 m/s. What is the resulting speed of the block?

27. Man Throws Stone A 91 kg man lying on a surface of negligible friction shoves a 68 g stone away from him, giving it a speed of 4.0 m/s. What velocity does the man acquire as a result?

28. Mechanical Toys A mechanical toy slides along an x axis on a frictionless surface with a velocity of $(-0.40\ \text{m/s})\hat{\text{i}}$ when two internal springs separate the toy into three parts, as given in the table. What is the velocity of part A?

Part	Mass (kg)	Velocity (m/s)
A	0.50	?
B	0.60	$0.20\hat{\text{i}}$
C	0.20	$0.30\hat{\text{i}}$

29. Icy Road Two cars A and B slide on an icy road as they attempt to stop at a traffic light. The mass of A is 1100 kg, and the mass of B is 1400 kg. The coefficient of kinetic friction between the locked wheels of either car and the road is 0.13. Car A succeeds in stopping at the light, but car B cannot stop and rear-ends car A. After the collision, A stops 8.2 m ahead of its position at impact, and B 6.1 m ahead; see Fig. 7-27. Both drivers had their brakes locked throughout the incident. Using the material in Chapters 2 and 6, find the speed of (a) car A and (b) car B immediately after impact. (c) Use conservation of translational momentum to find the speed at which car B struck car A. On what grounds can the use of momentum conservation be criticized here?

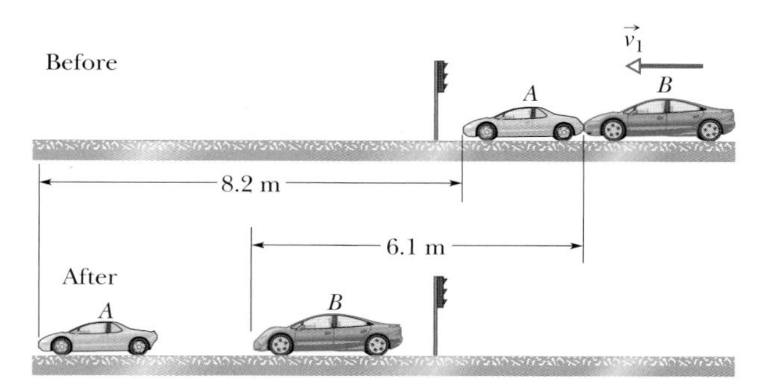

FIGURE 7-27 ■ Problem 29.

30. Bullet and Two Blocks In Fig. 7-28*a*, a 3.50 g bullet is fired horizontally at two blocks at rest on a frictionless tabletop. The bullet passes through the first block, with mass 1.20 kg, and embeds itself in the second, with mass 1.80 kg. Speeds of 0.630 m/s and 1.40 m/s, respectively, are thereby given to the blocks (Fig. 7-28*b*). Neglecting the mass removed from the first block by the bullet, find (a) the speed of the bullet immediately after it emerges from the first block and (b) the bullet's original speed.

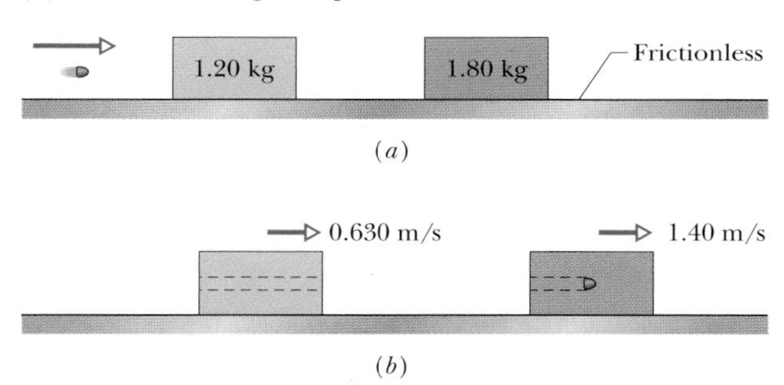

FIGURE 7-28 ■ Problem 30.

31. Man on a Cart A 75 kg man is riding on a 39 kg cart traveling at a speed of 2.3 m/s. He jumps off with zero horizontal speed relative to the ground. What is the resulting change in the speed of the cart?

32. Block and Bullet A bullet of mass 4.5 g is fired horizontally into a 2.4 kg wooden block at rest on a horizontal surface. The bullet is embedded in the block. The speed of the block immediately after the bullet stops relative to it is 2.7 m/s. At what speed is the bullet fired?

33. Water in a Rocket Sled A rocket sled with a mass of 2900 kg moves at 250 m/s on a set of rails. At a certain point, a scoop on the sled dips into a trough of water located between the tracks and scoops water into an empty tank on the sled. By applying the principle of conservation of translational momentum, determine the speed of the sled after 920 kg of water has been scooped up. Ignore any retarding force on the scoop.

34. Bullet Fired Upward A 10 g bullet moving directly upward at 1000 m/s strikes and passes through the center of a 5.0 kg block initially at rest (Fig. 7-29). The bullet emerges from the block moving directly upward at 400 m/s. To what maximum height does the block then rise above its initial position? (*Hint*: Use free-fall equations from Chapter 3.)

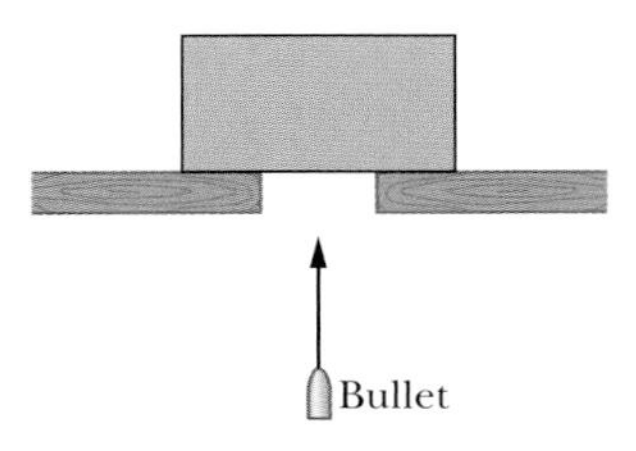

FIGURE 7-29 ■ Problem 34.

35. Projectile Body A projectile body of mass m_A and initial velocity $\vec{v}_{A1}$ collides with an initially stationary target body of mass m_B in a one-dimensional collision. What are the velocities of the bodies after the collision if they stick together?

36. Two Blocks Collide A 5.0 kg block with a speed of 3.0 m/s collides with a 10 kg block that has a speed of 2.0 m/s in the same direction. After the collision, the 10 kg block is observed to be traveling in the original direction with a speed of 2.5 m/s. What is the velocity of the 5.0 kg block immediately after the collision?

37. Last Stage of a Rocket The last stage of a rocket, which is traveling at a speed of 7600 m/s, consists of two parts that are clamped together: a rocket case with a mass of 290.0 kg and a payload capsule with a mass of 150.0 kg. When the clamp is released, a compressed spring causes the two parts to separate with a relative speed of 910.0 m/s. What are the speeds of (a) the rocket case and (b) the payload after they have separated? Assume that all velocities are along the same line.

38. Man on a Flatcar A railroad flatcar of weight W can roll without friction along a straight horizontal track. Initially, a man of weight w is standing on the car, which is moving to the right with speed v_{c1} (see Fig. 7-30). What is the change in velocity of the car if the man runs to the left (in the figure) so that his speed relative to the car is v^{rel}?

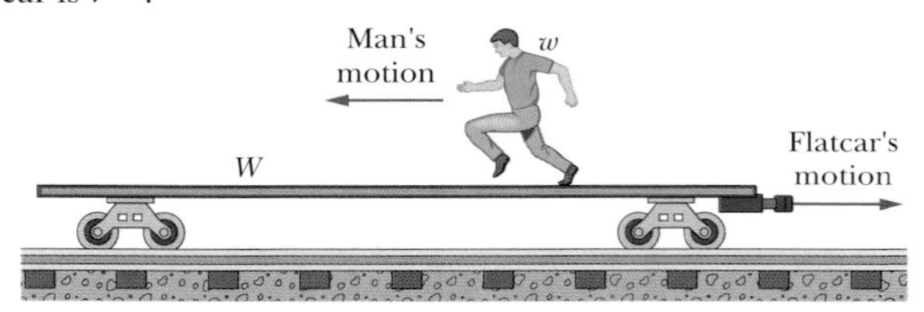

FIGURE 7-30 ■ Problem 38.

39. Space Vehicle A space vehicle is traveling at 4300 km/h relative to Earth when the exhausted rocket motor is disengaged and sent backward with a speed of 82 km/h relative to the command module. The mass of the motor is four times the mass of the module. What is the speed of the command module relative to Earth just after the separation?

40. Projectile Body Two A projectile body of mass m_A and initial x-component velocity $v_{Ax}(t_1) = 10.0$ m/s collides with an initially stationary target body of mass $m_B = 2.00\, m_A$ in a one-dimensional collision. What is the velocity of m_B following the collision if the two masses stick together?

SEC. 7-7 ■ CONSERVATION OF MOMENTUM IN TWO DIMENSIONS

41. Ice-Skating Man A 60 kg man is ice-skating due north with a velocity of 6.0 m/s when he collides with a 38 kg child. The man and child stay together and have a velocity of 3.0 m/s at an angle of 35° north of east immediately after the collision. What are the magnitude and direction of the velocity of the child just before the collision?

42. Barge Collision A barge with mass 1.50×10^5 kg is proceeding downriver at 6.2 m/s in heavy fog when it collides with a barge heading directly across the river (see Fig. 7-31). The second barge has mass 2.78×10^5 kg and before the collision is moving at 4.3 m/s. Immediately after impact, the second barge finds its course deflected by 18° in the downriver direction and its speed increased to 5.1 m/s. The river current is approximately zero at the time of the accident. What are the speed and direction of motion of the first barge immediately after the collision?

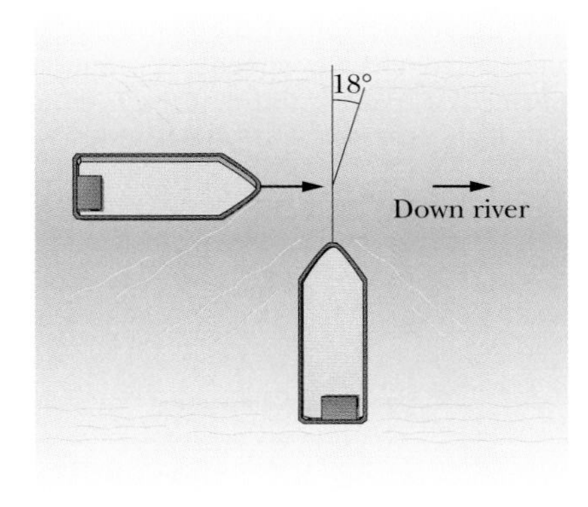

FIGURE 7-31 ■ Problem 42.

43. Package Explodes A 2.65 kg stationary package explodes into three parts that then slide across a frictionless floor. The package had been at the origin of a coordinate system. Part A has mass $m_A = 0.500$ kg and velocity $(10.0 \text{ m/s}\, \hat{i} + 12.0 \text{ m/s}\, \hat{j})$. Part B has mass $m_B = 0.750$ kg, a speed of 14.0 m/s, and travels at an angle 110° counterclockwise from the positive direction of the x axis. (a) What is the speed of part C? (b) In what direction does it travel?

44. Particle Collision A 2.00 kg "particle" traveling with velocity $\vec{v}_{A1} = (4.0 \text{ m/s})\hat{i}$ collides with a 4.00 kg "particle" traveling with velocity $\vec{v}_{B1} = (2.0 \text{ m/s})\hat{j}$. The collision connects the two particles. What then is their velocity in (a) unit-vector notation and (b) magnitude-angle notation?

45. Two Vehicles Two vehicles A and B are traveling west and south, respectively, toward the same intersection, where they collide and lock together. Before the collision, A (total weight 12.0 kN) has a speed of 64.4 km/h, and B (total weight 16.0 kN) has a speed of 96.6 km/h. Find the (a) magnitude and (b) direction of the velocity of the (interlocked) vehicles immediately after the collision, assuming the collision is isolated.

46. Tin Cookie A 2.0 kg tin cookie, with an initial velocity of 8.0 m/s to the east, collides with a stationary 4.0 kg cookie tin. Just after the collision, the cookie has a velocity of 4.0 m/s at an angle of 37° north of east. Just then, what are (a) the magnitude and (b) the direction of the velocity of the cookie tin?

47. Colliding Balls A 5.0 kg ball moving due east at 4.0 m/s collides with a 4.0 kg ball moving due west at 3.0 m/s. Just after the collision, the 5.0 kg ball has a velocity of 1.2 m/s, due south. What is the magnitude of the velocity of the 4.0 kg ball just after the collision?

48. Particle Collision Two A collision occurs between a 2.00 kg particle traveling with velocity $\vec{v}_{A1} = (-4.00 \text{ m/s})\hat{i} + (-5.00 \text{ m/s})\hat{j}$ and a 4.00 kg particle traveling with velocity $\vec{v}_{B1} = (6.00 \text{ m/s})\hat{i} + (-2.00 \text{ m/s})\hat{j}$. The collision connects the two particles. What then is their velocity in (a) unit-vector notation and (b) magnitude-angle notation?

49. Suspicious Package A suspicious package is sliding on frictionless surface when it explodes into three pieces of equal masses and with the velocities (1) 7.0 m/s, north, (2) 4.0 m/s, 30° south of west, and (3) 4.0 m/s, 30° south of east. (a) What is the velocity (magnitude and direction) of the package before it exploded?

50. Mess Kit A 4.0 kg mess kit sliding on a frictionless surface explodes into two 2.0 kg parts, one moving at 3.0 m/s, due north, and the other at 5.0 m/s, 30° north of east. What is the original speed of the mess kit?

51. Radioactive Nucleus A certain radioactive nucleus can transform to another nucleus by emitting an electron and a neutrino. (The *neutrino* is one of the fundamental particles of physics.) Suppose that in such a transformation, the initial nucleus is stationary, the electron and neutrino are emitted along perpendicular paths, and the magnitudes of the translational momenta are 1.2×10^{-22} kg · m/s for the electron and 6.4×10^{-23} kg · m/s for the neutrino. As a result of the emissions, the new nucleus moves (recoils). (a) What is the magnitude of its translational momentum? What is the angle between its path and the path of (b) the electron (c) the neutrino?

52. Internal Explosion A 20.0 kg body is moving in the positive x direction with a speed of 200 m/s when, due to an internal explosion, it breaks into three parts. One part, with a mass of 10.0 kg, moves away from the point of explosion with a speed of 100 m/s in the positive y direction. A second fragment, with a mass of 4.00 kg, moves in the negative x direction with a speed of 500 m/s. What is the velocity of the third (6.00 kg) fragment?

53. Vessel at Rest Explodes A vessel at rest explodes, breaking into three pieces. Two pieces, having equal mass, fly off perpendicular to one another with the same speed of 30 m/s. The third piece has three times the mass of each other piece. What are the magnitude and direction of its velocity immediately after the explosion?

54. Proton–Proton Collision A proton with a speed of 500 m/s collides with another proton initially at rest. The projectile and target protons then move along perpendicular paths, with the projectile path at 60° from the original direction. After the collision, what are the speeds of (a) the target proton and (b) the projectile proton?

55. Box Sled A 6.0 kg box sled is coasting across frictionless ice at a speed of 9.0 m/s when a 12 kg package is dropped into it from above. What is the new speed of the sled?

56. Two Balls Two balls A and B, having different but unknown masses, collide. Initially, A is at rest and B has speed v_B. After the collision, B has speed $v_B/2$ and moves perpendicularly to its original motion. (a) Find the direction in which ball A moves after the collision. (b) Show that you cannot determine the speed of A from the information given.

57. Two Objects, Same Mass After a collision, two objects of the same mass and same initial speed are found to move away together at $\frac{1}{2}$ their initial speed. Find the angle between the initial velocities of the objects.

58. Sliding on Ice Two 30 kg children, each with a speed of 4.0 m/s, are sliding on a frictionless frozen pond when they collide and stick together because they have Velcro straps on their jackets. The two children then collide and stick to a 75 kg man who was sliding at 2.0 m/s. After this collision, the three-person composite is stationary. What is the angle between the initial velocity vectors of the two children?

59. Alpha Particle and Oxygen An alpha particle collides with an oxygen nucleus that is initially at rest. The alpha particle is scattered at an angle of 64.0° from its initial direction of motion, and the oxygen nucleus recoils at an angle of 51.0° on the opposite side of that initial direction. The final speed of the nucleus is 1.20×10^5 m/s. Find (a) the final speed and (b) the initial speed of the alpha particle. (In atomic mass units, the mass of an alpha particle is 4.0 u, and the mass of an oxygen nucleus is 16 u.)

60. Two Bodies Collide Two 2.0 kg bodies, A and B, collide. The velocities before the collision are $\vec{v}_{A1} = (15\text{ m/s})\hat{\text{i}} + (30\text{ m/s})\hat{\text{j}}$ and $\vec{v}_{B1} = (-10\text{ m/s})\hat{\text{i}} + (5.0\text{ m/s})\hat{\text{j}}$. After the collision, $\vec{v}_{A2} = (-5.0\text{ m/s})\hat{\text{i}} + (20\text{ m/s})\hat{\text{j}}$. What is the final velocity of B?

61. Game of Pool In a game of pool, the cue ball strikes another ball of the same mass and initially at rest. After the collision, the cue ball moves at 3.50 m/s along a line making an angle of 22.0° with its original direction of motion, and the second ball has a speed of 2.00 m/s. Find (a) the angle between the direction of motion of the second ball and the original direction of motion of the cue ball and (b) the original speed of the cue ball.

62. Billiard Ball A billiard ball moving at a speed of 2.2 m/s strikes an identical stationary ball with a glancing blow. After the collision, one ball is found to be moving at a speed of 1.1 m/s in a direction making a 60° angle with the original line of motion. Find the velocity of the other ball.

63. Three Balls In Fig. 7-32, ball A with an initial speed of 10 m/s collides with stationary balls B and C, whose centers are on a line perpendicular to the initial velocity of ball A and that are initially in contact with each other. The three balls are identical. Ball A is aimed directly at the contact point, and all motion is frictionless. After the collision, balls B and C have the same speed 6.93 m/s, but ball B moves at an angle of 30° above the horizontal and ball C moves at an angle of 30° below the horizontal. What is the velocity of ball A after the collision?

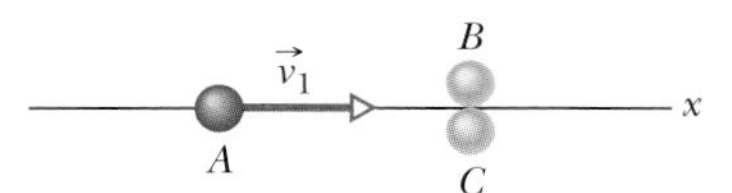

FIGURE 7-32 ■ Problem 63.

SEC. 7-8 ■ A SYSTEM WITH MASS EXCHANGE—A ROCKET AND ITS EJECTED FUEL

64. Railroad Car with Grain A railroad car moves at a constant speed of 3.20 m/s under a grain elevator. Grain drops into it at the rate of 540 kg/min. What is the magnitude of the force needed to keep the car moving at constant speed if friction is negligible?

65. Space Probe A 6090 kg space probe, moving nose-first toward Jupiter at 105 m/s relative to the Sun, fires its rocket engine, ejecting 80.0 kg of exhaust at a speed of 253 m/s relative to the space probe. What is the final velocity of the probe?

66. Moving Away From Solar System A rocket is moving away from the solar system at a speed of 6.0×10^3 m/s. It fires its engine, which ejects exhaust with a speed of 3.0×10^3 m/s relative to the rocket. The mass of the rocket at this time is 4.0×10^4 kg, and its acceleration is 2.0 m/s². (a) What is the thrust of the engine? (b) At what rate, in kilograms per second is exhaust ejected during the firing?

67. Deep Space A rocket, which is in deep space and initially at rest relative to an inertial reference frame, has a mass of 2.55×10^5 kg, of which 1.81×10^5 kg is fuel. The rocket engine is then fired for 250 s, during which fuel is consumed at the rate of 480 kg/s. The speed of the exhaust products relative to the rocket is 3.27 km/s. (a) What is the rocket's thrust? After the 250 s firing, what are the (b) mass and (c) speed of the rocket?

68. Mass Ratio Consider a rocket that is in deep space and at rest relative to an inertial reference frame. The rocket's engine is to be fired for a certain interval. What must be the rocket's *mass ratio* (ratio of initial to final mass) over that interval if the rocket's original speed relative to the inertial frame is to be equal to (a) the exhaust speed (speed of the exhaust products relative to the rocket) and (b) 2.0 times the exhaust speed?

69. Lunar Mission During a lunar mission, it is necessary to increase the speed of a spacecraft by 2.2 m/s when it is moving at 400 m/s relative to the Moon. The speed of the exhaust products from the rocket engine is 1000 m/s relative to the spacecraft. What fraction of the initial mass of the spacecraft must be burned and ejected to accomplish the speed increase?

70. Set for Vertical Firing A 6100 kg rocket is set for vertical firing from the ground. If the exhaust speed is 1200 m/s, how much gas must be ejected each second if the thrust (a) is to equal the magnitude of the gravitational force on the rocket and (b) is to give the rocket an initial upward acceleration of 21 m/s^2?

Additional Problems

71. Break a Leg (Not!) When jumping straight down, you can be seriously injured if you land stiff-legged. One way to avoid injury is to bend your knees upon landing to reduce the force of the impact. Suppose you have a mass m and you jump off a wall of height h.

(a) Use what you learned about constant acceleration motion to find the speed with which you hit the ground. Assume you simply step off the wall, so your initial y velocity is zero. Ignore air resistance. (Express your answer in terms of the symbols given.)
(b) Suppose that the time interval starting when your feet first touch the ground until you stop is Δt. Calculate the (average) net force acting on you during that interval. (Again, express your answer in terms of the symbols given.)
(c) Suppose $h = 1$ m. If you land stiff-legged, the time it takes you to stop may be as short as 2 ms, whereas if you bend your knees, it might be as long as 0.1 s. Calculate the average net force that would act on you in the two cases.
(d) The net force on you while you are stopping includes both the force of gravity and the force of the ground pushing up. Which of these forces do you think does you the injury? Explain your reasoning.
(e) For the two cases in part (c), calculate the upward force the ground exerts on you.

72. Finding Momentum Change and Impulse Consider the graphs shown in Fig. 7-33. These graphs depict two force magnitude vs. time curves and several related momentum vs. time graphs. They describe a low-friction cart traveling along an x axis with a force sensor attached to it. The cart–force sensor system has a mass of 0.50 kg. The cart undergoes a series of collisions. It collides with a hard wall and with a wall that is padded with soft foam. Sometimes there is a small clay blob on the wall causing the cart–force sensor system to stick to the wall after the collision.

(a) What is the approximate momentum change associated with graph a? With graph d? Determine this change by taking approximate readings from the graphs. Show your calculations!
(b) Which of the two impulse curves, A or B, might lead to the momentum change depicted in graph a? In graph d? Explain the reasons for your answer.
(c) Suppose the forces on the cart–force sensor system were described by graph A. What would its velocity change be?

73. Relating Impulse Curves to Collisions Suppose you collected F_x vs. t and p_x vs. t data for a series of collisions for an important project report and then you lost your notes. Fortunately you still

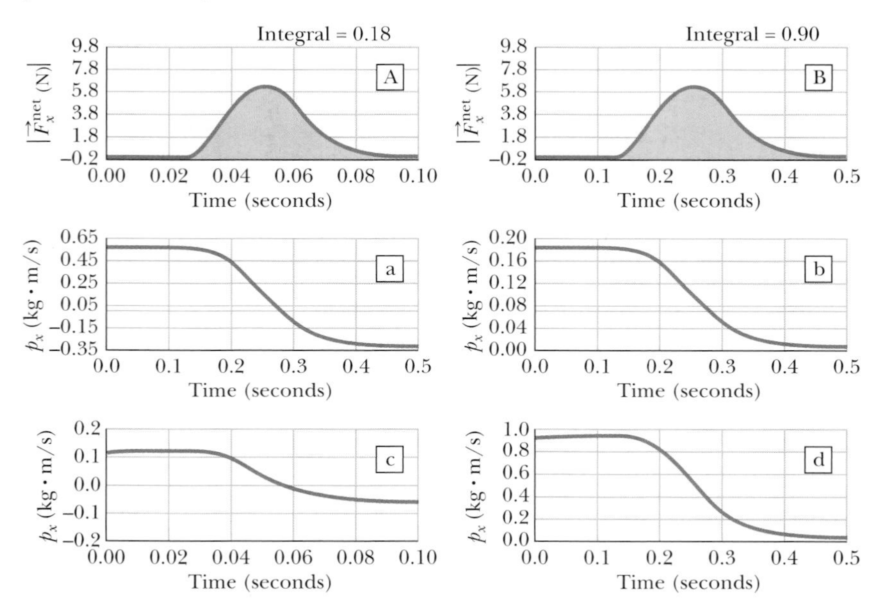

FIGURE 7-33 ■ Problems 72 and 73.

have your data on a computer disk. You open up the files and find the graphs shown in Fig. 7-33. You don't know which graph corresponds to which collision, but you are able to reconstruct some of your work by asking and answering the following questions:

(a) Which F_x^{net} vs. t graph, A or B, probably resulted from collisions between the cart–force sensor system and a soft, padded wall? Which one probably resulted from collisions between the force sensor and a hard wall? Explain in words the reasons for your answer.
(b) Which p_x vs. t graphs probably resulted from collisions between the cart–force sensor system and a padded wall? Which ones probably resulted from collisions between the cart–force sensor system and a hard wall? Explain the reasons for your answers. (*Hint*: There may be more than one graph for each type of collision.)
(c) Which p_x vs. t graphs correspond to a situation in which the cart bounces back? Which p_x vs. t graphs correspond to a situation in which you placed a small clay blob on the force sensor hook so the cart sticks to the wall that it collides with? Explain the reasons for your answers. (*Hint*: There may be more than one graph for each type of collision.)

74. Carts and Graphs Two carts on an air track are pushed toward each other. Initially, cart A moves in the positive x direction and cart B moves in the negative x direction. The carts bounce off each other. The graphs in Fig. 7-34 describe some of the variables associated with the motion as a function of time. For each item in the list below, identify which graph is a possible display of that variable as a function of time. If none apply, write N (for none).

(a) the momentum of cart A
(b) the force on cart B
(c) the force on cart A
(d) the position of cart A
(e) the position of cart B

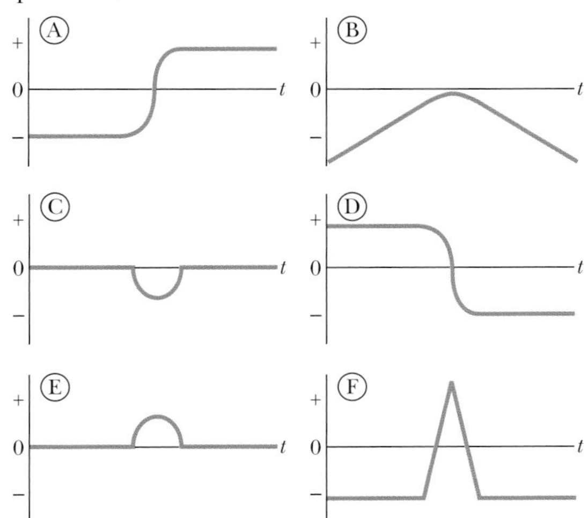

FIGURE 7-34 ■ Problem 74.

75. Colliding Carts Two carts are riding on an air track as shown in Fig. 7-35a. At clock time $t = 0$, cart B is at the origin traveling in the negative x direction with a velocity $\vec{v}_{B1}$. At that time, cart A is at the position shown and is at rest. Cart B has twice the mass of cart A. The carts "bump" each other, but don't stick.

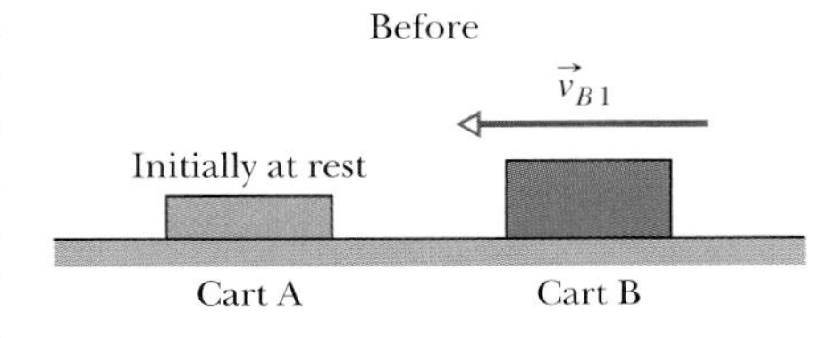

FIGURE 7-35a ■ Problem 75.

The graphs shown in Fig. 7-35b are a number of possible plots for the various physical parameters associated with the two carts. Each graph has two curves, one for each cart and labeled with the cart's letter. For each property (a)–(e), select the number 1, 2, etc., of the graphs that could be a plot of the property.

(a) The forces *exerted by* the carts
(b) The position of the carts
(c) The velocity of the carts
(d) The acceleration of the carts
(e) The momentum of the carts

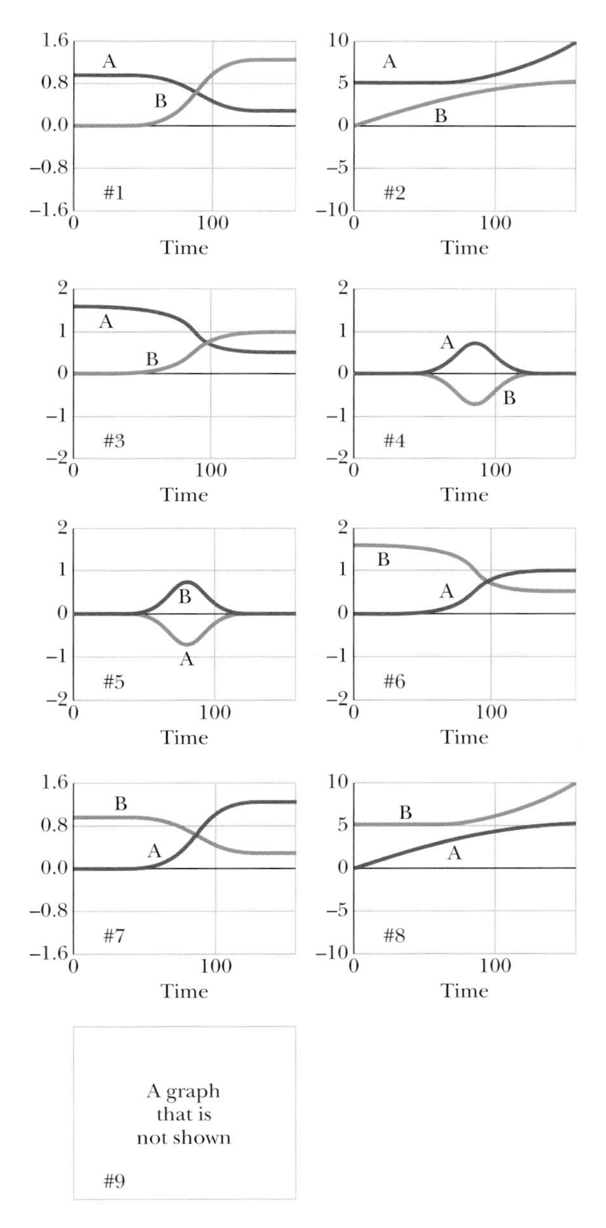

FIGURE 7-35b ■ Problem 75.

76. Could Newton Predict the "Third Law"? Isaac Newton studied many types of collisions and invented the definition of momentum about twenty years before he developed his three laws of motion. As a result of his observations of collision processes, he formulated the law of conservation of momentum as a statement of experimental fact.

Let's assume for the sake of argument that Newton had already defined the concepts of force and momentum but had not yet formulated his laws of motion. Also assume that he had an electronic force sensor and was able to verify the impulse-momentum theorem. Explain in words how Newton could use the impulse-momentum theorem and the law of conservation of momentum to predict the existence of the third law of motion and to explain the nature of the interaction forces between two colliding objects.

77. Taking Cyrano to the Moon In Edmund Rostand's famous play, *Cyrano de Bergerac*, Cyrano, in an attempt to distract a suitor from visiting Roxanne, claims to have descended to Earth from the Moon and proclaims to have invented six novel and fantastical methods for traveling to the Moon. One is as follows.

Sitting on an iron platform—thence
To throw a magnet in the air. This is
A method well conceived—the magnet flown,
Infallibly the iron will pursue:
Then quick! relaunch your magnet, and you thus
*Can mount and mount unmeasured distances!**

In an old cartoon, there is another version of this method. A character in the old West is on a hand-pumped, two-person rail car. After getting tired of pumping the handle up and down to make the car move along the rails, he takes out a magnet, hangs it from a fishing pole, and holds it in front of the cart. The magnet pulls the cart toward it, which pushes the magnet forward, and so on, so the cart moves forward continually. What do you think of these methods? Can some version of them work? Discuss in terms of the physics you have learned.

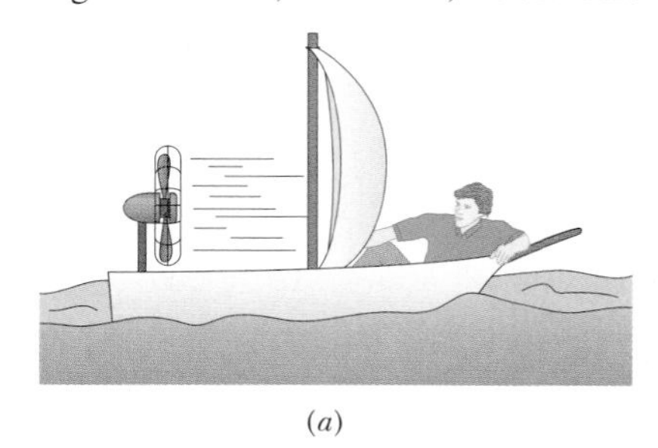
(a)

FIGURE 7-36a ■ Problem 78.

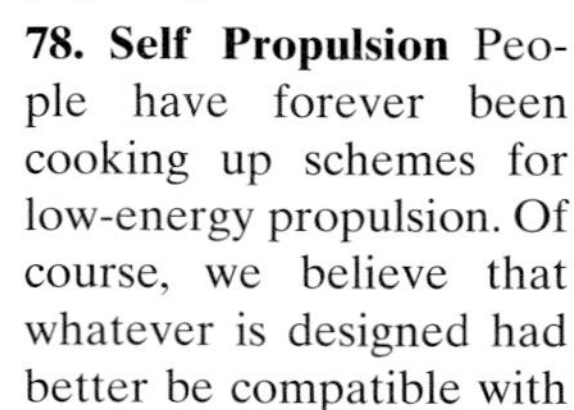

78. Self Propulsion People have forever been cooking up schemes for low-energy propulsion. Of course, we believe that whatever is designed had better be compatible with the laws of physics. Several schemes are shown below. Which ones do you think will work? Answer the questions detailed in (a) through (d) by referring to Fig. 7-36.

(a) In Fig. 7-36*a*, a lazy fisherman turns on a battery-operated fan and blows air onto the sail of his boat. Will he go anywhere? If he moves, what will his direction be? Explain.
(b) In Fig. 7-36*b*, a clever child is dangling a large magnet out in front of her wagon. It attracts a smaller magnet that she has attached to the front of her cart. Will she go anywhere? If she moves, what will her direction be? Explain.

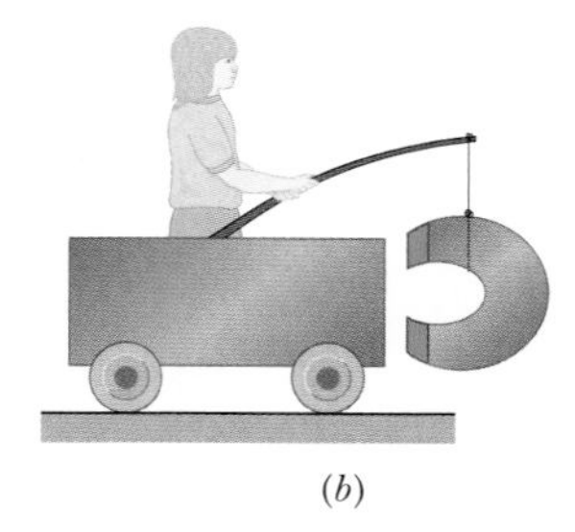
(b)

FIGURE 7-36b ■ Problem 78.

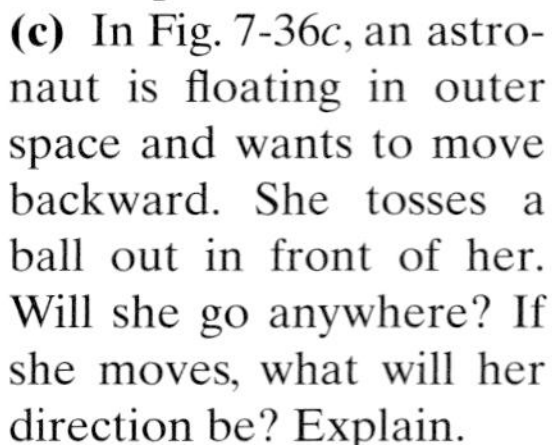

(c) In Fig. 7-36*c*, an astronaut is floating in outer space and wants to move backward. She tosses a ball out in front of her. Will she go anywhere? If she moves, what will her direction be? Explain.

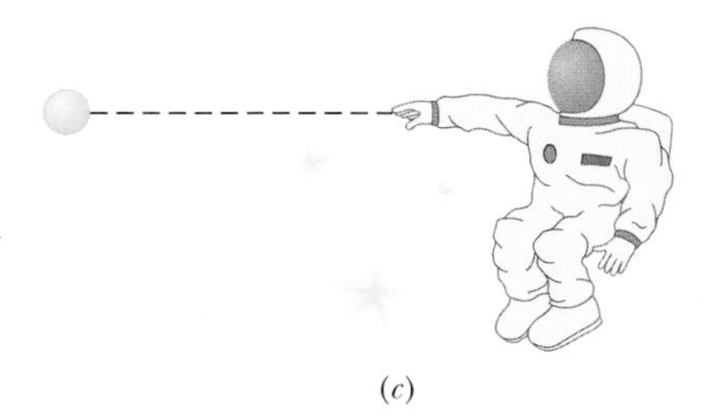
(c)

FIGURE 7-36c ■ Problem 78.

(d) In Fig. 7-36*d*, a college student on roller blades has a carbon dioxide container strapped to her back. The carbon dioxide jets out behind her as shown. Will she go anywhere? If she moves, what will her direction be? Explain.

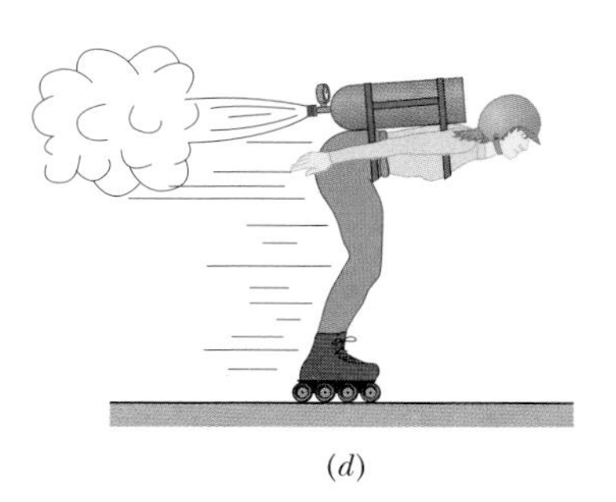
(d)

FIGURE 7-36d ■ Problem 78.

79. The Ice-Skating Professor A professor of physics is going ice skating for the first time. He has gotten himself into the middle of an ice rink and cannot figure out how to make the skates work. Every motion he makes simply causes his feet to slip on the ice and leaves him in the same place he started. He decides that he can get off the ice by throwing his gloves in the opposite direction.

(a) Suppose he has a mass M and his gloves have a mass m. If he throws the gloves as hard as he can away from him, they leave his hand with a velocity $\vec{v}_{\text{glove}}$. Explain whether or not he will move. If he does move, calculate his velocity, $\vec{v}_{\text{prof}}$.
(b) Discuss his motion from the point of view of the forces acting on him.
(c) If the ice rink is 10 m in diameter and the skater starts in the center, estimate how long it will take him to reach the edge, assuming there is no friction at all.

80. When Can You Conserve Momentum? The principle of conservation of momentum is useful in some situations and not in others. Describe how you obtain the impulse-momentum theorem from Newton's Second Law and what situations lead to momentum conservation. How would you decide whether conservation of momentum could be useful in a particular problem?

81. Momentum Conservation in Subsystems Can a system whose momentum is conserved be made up of smaller systems whose individual momenta are not conserved? Explain why or why not and give an example.

82. The Rabbit and the Eagle You are working for the Defenders of Wildlife on the protection of the bald eagle, an endangered species. Walt Disney Productions, Inc. has agreed to help your cause by producing an animated movie about the bald eagle. You have set up a dramatic scene in which a young rabbit is frightened by the shadow of the eagle and starts bounding toward the east at 30 m/s as the eagle swoops down vertically at a speed of 15 m/s. A moment before the eagle contacts it, the rabbit bounds off a cliff and is captured in mid-air. (See Fig. 7-37.) The animators want to know how

*Translated from the French by Gladys Thomas and Mary F. Guillemard, e-text prepared by Sue Asscher, distributed by Project Gutenberg.

to portray what happens just after the capture. If the eagle has a mass of 2.5 kg and the rabbit has a mass of 0.8 kg, what is the *velocity* of the eagle with the rabbit in its talons just after the capture? (Include a diagram of the situation before and after capture with vectors showing the initial and final velocities.)

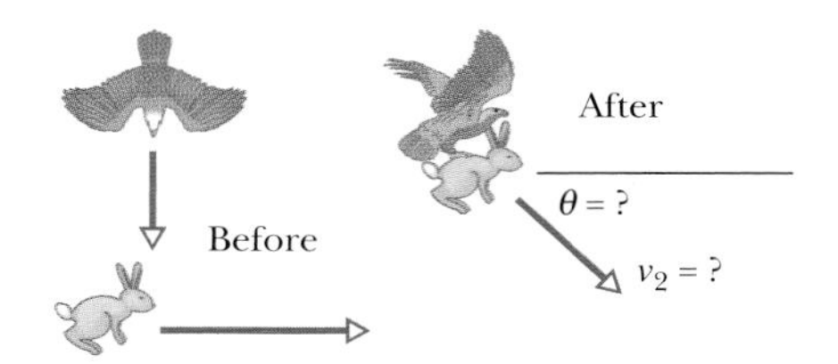

FIGURE 7-37 ■ Problem 82.

83. Air Resistance 1: Estimating the Effect The force of air resistance on a sphere of radius R can plausibly be argued to have the form

$$\vec{F}^{\text{ drag}} = -\tfrac{1}{2}C\rho R^2|v|\vec{v} = -b|v|\vec{v},$$

where $\vec{v}$ is the vector velocity and $|\vec{v}|$ is its magnitude (the speed). The density of the air, ρ, is about 1 kg/m^3—1/1000 that of water. The parameter C is a dimensionless constant.

If we drop a steel ball and a styrofoam ball from a height of s, the steel ball reaches the ground when the styrofoam ball is still a bit above the ground. Call this distance h. Estimate the air resistance coefficient C as follows:

(a) Assume the effect of air resistance on the steel sphere is negligible. Calculate approximately how long the steel sphere takes to fall to the ground (Δt_{ste}) and how fast it is traveling just before it hits (v_{ste}). Express your answers in terms of s, g, and m.
(b) Since the steel and styrofoam were not very different, use $\langle\vec{v}_{\text{ste}}\rangle$, the average velocity of the steel ball during its fall to calculate an average air resistance force, $\langle\vec{F}^{\text{ drag}}\rangle = -b\langle\vec{v}\rangle^2$ acting on the styrofoam sphere during its fall. Express this force in terms of b, m (the mass of the styrofoam sphere), g, s, and h.
(c) The average velocity of the steel ball is $\langle\vec{v}_{\text{ste}}\rangle = s/\Delta t_{\text{ste}}$. The average velocity of the styrofoam sphere was $\langle\vec{v}_{\text{sty}}\rangle = (s-h)/\Delta t_{\text{ste}}$. Assume this difference, $\Delta\langle\vec{v}\rangle$, is caused by the average air resistance force acting over the time Δt_{ste} with our basic Newton's law formula:

$$\langle\vec{F}^{\text{drag}}\rangle\Delta t_{\text{ste}} = m\,\Delta\langle\vec{v}\rangle.$$

Use this to show that

$$b \cong \frac{mh}{s^2}.$$

(d) A styrofoam ball of radius $R = 5$ cm and mass $m = 50$ g is dropped with a steel ball from a height of $s = 2$ m. When the steel ball hits, the styrofoam is about $h = 10$ cm above the ground. Calculate b (for the styrofoam sphere) and C (for any sphere).

84. Air Resistance 2: Deriving the Equation In this problem, you will derive an explicit form of Newton's drag law for air resistance, whose structure we derived by dimensional analysis in Problem 6-103. The derivation below will provide the dimensionless coefficient that we were unable to find by dimensional analysis.

(a) Consider a small particle of mass m that is initially at rest. (Ignore gravity.) The particle is approached by a very massive wall moving toward it along an x axis with a speed v. After the wall hits it, what speed will the small particle have? (*Hint*: Consider first the case of the small particle moving toward a stationary wall with a velocity $-v$. Analyze what happens.)
(b) Suppose the moving wall is a disk of radius R moving at a speed v in a direction perpendicular to the plane of the disk. If there are N small particles per unit volume in the region of space the disk is sweeping through, how many of them will the disk encounter in a small time Δt?
(c) Calculate the total momentum transferred to the air in the time Δt by the disk, assuming that there are N air particles per unit volume and they each have mass m.
(d) Find the force the disk exerts on the air and the force the air exerts on the disk. How do you know?
(e) Show that the force you calculated has the form

$$\vec{F}^{\text{ drag}} = -\tfrac{1}{2}C\rho R^2|\vec{v}|\vec{v}$$

and find the dimensionless constant, C.

85. Juggler This problem is based on the analysis of a digital movie depicting a juggler. If you are using VideoPoint, view the movie entitled DSON007. Your instructor may provide you with a different movie to analyze or ask you to use the data presented in Fig. 7-38*b*. We track the motion of the white baseball of mass 0.138 kg in Fig. 7-38*a*, which is being caught and thrown in a smooth motion. The figure shows alternate frames depicting the catch and throw from just before to just after the juggler's hand is in contact with the ball. The data presented in Fig. 7-38*b* include a least-squares fit for frames 33–39 of the digital video shown in Fig. 7-38*a*. During all of these frames the ball is in contact with the juggler's hand. (Although the time codes are correct, the digital capture system missed recording a few frames between $t = 1.567$ s and $t = 1.700$ s.)

The goal of this problem is to consider the catch–throw process as a slow collision between the juggler's hand and the ball. In particular we would like you to verify that the impulse-momentum theorem holds for this situation. You should assume that the data and analysis presented here are correct and that Newton's Second Law is valid.

(a) Examine the y position of the ball as a function of time for a time period during which the ball is in the juggler's hand (frames 33–39 in Fig. 7-38*a*). Express each fit coefficient and its uncertainty (that is, the standard deviation of the mean) to the correct number of significant figures. Write down the equation that allows you to calculate y as a function of t.
(b) What is the nature of the vertical motion of the ball during the time it is being caught and thrown? Is its vertical velocity component zero, a constant, constantly changing, or is something else going on? Cite the reasons for your answer. What are the magnitude and direction of the vertical acceleration, a_y, of the ball?
(c) Calculate the *instantaneous* vertical velocity of the ball just as it's being caught (frame 33). Calculate the *instantaneous* vertical velocity of the ball just as it's being released (frame 39). (*Hints*: Use three significant figures in your coefficients. You can either interpret the physical meaning of the fit coefficient a_1 and then use the kinematic equation relating velocity to acceleration, initial velocity (at $t = 0.000$ s), and time, or you can take the derivative with respect to time of the y vs. t equation you just wrote down in part (a).)
(d) Assuming the vertical acceleration of the ball is constant while it is in the juggler's hand, what is the *net* vertical force on the ball during the entire catch–throw process? Draw a free-body diagram showing the magnitudes and directions of the forces on the ball. What are the magnitude and direction of the gravitational force on the ball? What are the magnitude and direction of the vertical force the juggler exerts on the ball?

FIGURE 7-38a ■ Problem 85.

DSON007: Juggling Data

Frame	t(s)	y(m)
33	1.500	0.416
34	1.533	0.330
35	1.567	0.249
36	1.700	0.152
37	1.733	0.213
38	1.767	0.305
39	1.800	0.421

$a_0 = 34.7656$ m

$a_1 = -41.9051$ m/s

$a_2 = 12.6780$ m/s^2

The fit graph is given by $y_{\text{fit}}(m) = a_0 + a_1 t + a_2 t^2$.

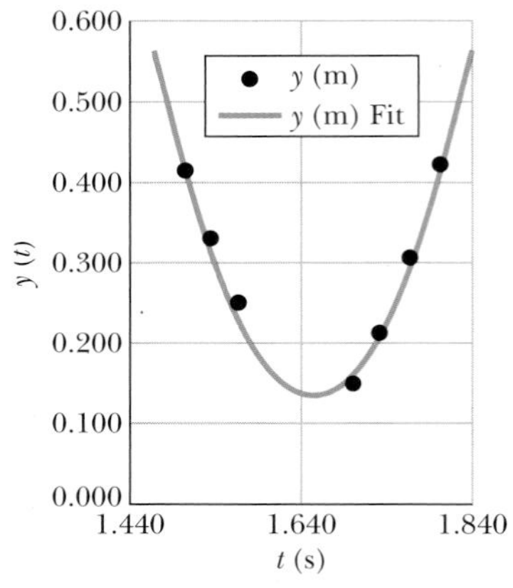

FIGURE 7-38b ■ Problem 85.

(e) Identify any Newton's Third Law pairs for this situation. Identify what object is exerting the gravitational force on the ball. According to Newton's Third Law, how is the ball interacting with that object?
(f) Find the vertical momentum of the ball when it first falls into the juggler's hand (as in frame 33). Also find the vertical momentum of the ball when it is just about to leave the juggler's hand (as in frame 39). What are the magnitude and direction of the *momentum change*, Δp_y, in the vertical direction that the ball undergoes during this time period? *Beware*: Momentum is a vector quantity. Do not fall into the trap of simply subtracting the *magnitudes* of the two momenta.
(g) How much time, Δt, does the ball spend in the hand of the juggler? Calculate the *impulse* transmitted to the ball by the net force on it during the catch–throw "collision."
(h) Compare the change in momentum to the impulse imparted to the ball. Does the impulse-momentum theorem seem to hold to the appropriate number of significant figures?

86. Momentum Before and After a Sticky Collision This problem is based on the analysis of a digital movie depicting a collision between two carts. Before the collision, one cart is moving and one cart is stationary. Following the collision, the two carts stick together. If you are using VideoPoint, view the movie entitled PASCO028. It depicts a cart of mass 2 kg colliding with a stationary cart of mass 1 kg. Your instructor may provide you with a different movie to analyze.

(a) Use video analysis software and a spreadsheet to find the initial momentum of the two-cart system before collision. Explain the method you used and show all your data and calculations.
(b) Use video analysis software and a spreadsheet to find the final momentum of the two-cart system after collision. Explain the method you used and show all your data and calculations.
(c) What is the percent difference between the momentum of the system before the collision and after the collision? Within the limits of experimental uncertainty, is the total momentum of the two-cart system conserved? Why or why not?
(d) If you found that the total momentum after collision is less than that before the collision, you can either conclude that: (1) momentum is still conserved but some of it is transferred to the track (that is the whole Earth) or (2) the law of conservation of momentum has failed. Assuming that the law of conservation of momentum still holds, how much momentum is transferred to the track and Earth? Remember that momentum is a vector quantity, and you must specify both the magnitude and direction of this momentum.
(e) Why don't you see the track move just after the collision?

8 Extended Systems

If you leap forward, chances are that your head and torso will follow a parabolic path, like a baseball thrown in from the outfield. However, when a skilled ballet dancer leaps across the stage in a *grand jeté*, the path taken by her head and torso is nearly horizontal during much of the jump. She seems to be floating across the stage. The audience may not know much about projectile motion, but they still sense that something unusual has happened.

How does the ballerina seemingly "turn off" the gravitational force?

The answer is in this chapter.

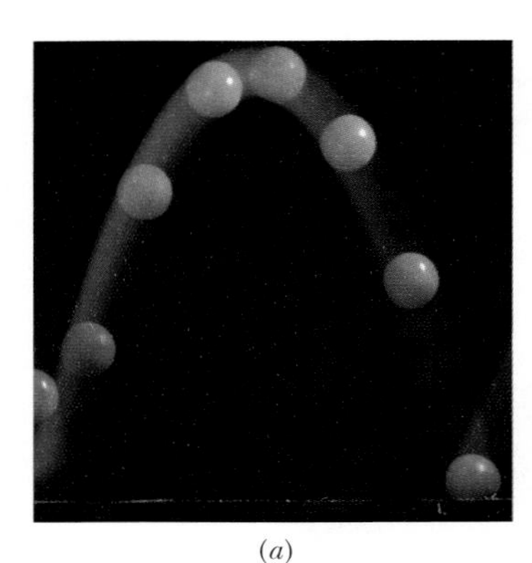

(a)

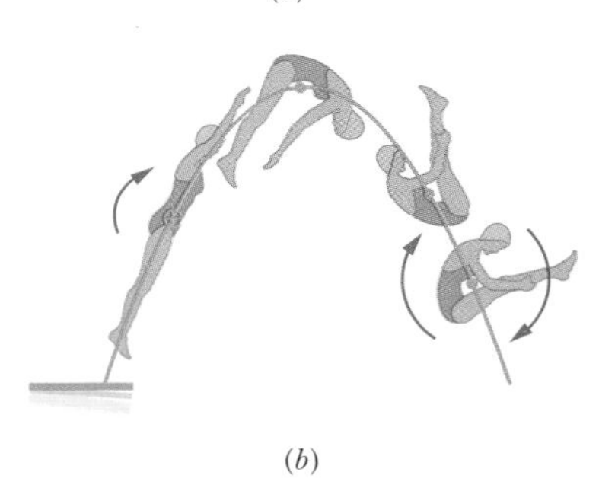

(b)

FIGURE 8-1 ■ (a) A bouncing ball follows a parabolic path. (b) A diver bounces off a board. Even though many points on her body that are not marked follow complex paths, a special point that also follows a parabolic path (shown by the dots) can be calculated based on the positions of the diver's body parts.

8-1 The Motion of Complex Objects

Up to this point we have focused our discussions on objects that can be considered to move as particles. In order to treat an object as a particle, every point on the object must be moving with the same velocity and acceleration. Although this requirement simplifies the analysis of motion, it is not commonly the case with everyday objects.

Here is an example. The motion of a rotating diver shown in Fig. 8-1*b* is clearly more complicated than that of the bouncing ball in Fig. 8-1*a*. Every part of the diver moves in a different manner than every other part, so we cannot describe her as a tossed particle. Instead, we must consider her as a system of particles. In large, complicated systems, it is often difficult to keep track of all the parts, and we cannot make predictions about the motion of the parts using the physics we have learned for particles. In fact, even a baseball, which seems to move as a particle, is usually spinning as it moves through the air.

So why is it that we have been able to treat objects like baseballs as particles in the previous chapters? And how do we handle the analysis of more complex systems, like divers and rotating baseball bats? We answer these questions in this chapter.

8-2 Defining the Position of a Complex Object

Even if we only have two objects in a system, their motions can be quite complex. Suppose two stars attract each other gravitationally so they are moving relative to one another. At the same time that the stars are exerting forces on each other, external forces could cause this two-star system to accelerate. But what is it that accelerates? Where is this two-star system actually located? At the location of the first star? The second star? Somewhere else? In this section we will show that we can define a position that can be used to describe accurately where a system is located and how the system accelerates.

Let's start by considering two particles, A and B, that attract one another as shown in Fig. 8-2. Suppose they have external forces $\vec{F}_A^{\text{ext}}$ and $\vec{F}_B^{\text{ext}}$ acting on them. Applying Newton's Second Law ($\vec{F}^{\text{net}} = m\vec{a}$) to each particle in this system gives us

$$\vec{F}_A^{\text{ext}} + \vec{F}_{B\to A} = m_A\vec{a}_A, \tag{8-1}$$

and

$$\vec{F}_B^{\text{ext}} + \vec{F}_{A\to B} = m_B\vec{a}_B, \tag{8-2}$$

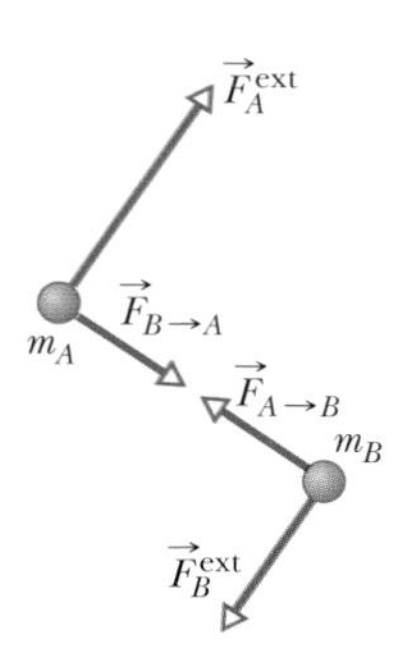

FIGURE 8-2 ■ Two particles connected by a "massless" rod (not shown) can experience different external forces while exerting equal and opposite internal forces on each other.

where $\vec{F}_{B\to A}$ and $\vec{F}_{A\to B}$ are the internal forces that the two particles in the system exert on each other. In order to get the net force acting on the "system," we must add Eqs. 8-1 and 8-2 together. Since $\vec{F}_{B\to A}$ and $\vec{F}_{A\to B}$ are equal and opposite forces (by Newton's Third Law), they cancel each other and we are left with an expression for the net force on the system of

$$\vec{F}_{\text{sys}}^{\text{net}} = \vec{F}_A^{\text{ext}} + \vec{F}_B^{\text{ext}} = m_A\vec{a}_A + m_B\vec{a}_B. \tag{8-3}$$

However, applying Newton's Second Law directly to the entire system also gives us

$$\vec{F}_{\text{sys}}^{\text{net}} = M_{\text{sys}}\vec{a}_{\text{sys}}, \tag{8-4}$$

where $M_{\text{sys}} = m_A + m_B$ is the total mass and $\vec{a}_{\text{sys}}$ is the acceleration of the system taken as a whole. A look at Fig. 8-2 tells us that the particles could be in orbit about each other or moving together while the system is rotating and accelerating along a line. So what do we mean by the "acceleration of the system as a whole?" If we combine Eqs. 8-3 and 8-4 we can use the result as a basis for defining a point in space that represents the system's acceleration. This result is given by

$$\vec{F}^{\,\text{net}}_{\text{sys}} = M_{\text{sys}}\vec{a}_{\text{sys}} = m_A\vec{a}_A + m_B\vec{a}_B. \tag{8-5}$$

Solving for the system acceleration gives us

$$\vec{a}_{\text{sys}} = \frac{1}{M_{\text{sys}}}(m_A\vec{a}_A + m_B\vec{a}_B). \tag{8-6}$$

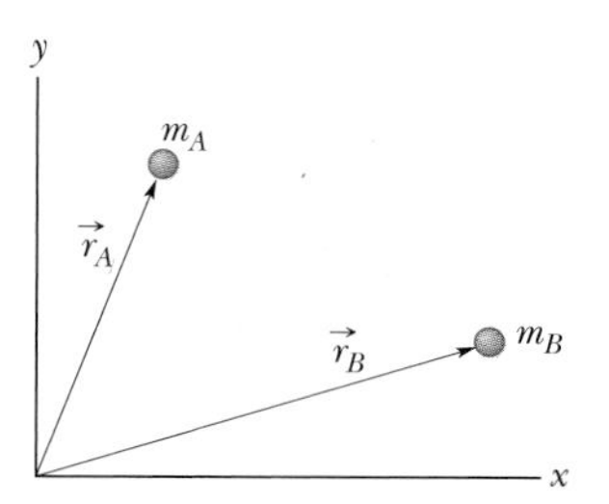

FIGURE 8-3 ■ If we choose a coordinate system to describe our two particles mathematically, the position vectors describing the location of each of the particles are $\vec{r}_A$ and $\vec{r}_B$, respectively.

This expression indicates that the acceleration of the system can be viewed as a "weighted average" of the particle accelerations.

If we choose a coordinate system, we can locate the particles in the system in terms of their position vectors $\vec{r}_A$ and $\vec{r}_B$ as shown in Fig. 8-3. Remember that acceleration is related to position by $\vec{a} = d^2\vec{r}/dt^2$—our expression for the acceleration suggests that we can define the effective "position" of the system as

$$\vec{R}^{\,\text{eff}} = \frac{1}{M_{\text{sys}}}(m_A\vec{r}_A + m_B\vec{r}_B) \quad \text{(two particle system)}. \tag{8-7}$$

We can verify that this expression for the position of the object makes sense by taking its derivative with respect to time twice. When we do that, we find that we get back the equation for the system acceleration that we derived in Eq. 8-6.

If we had considered a more complex system of N particles we would have come to a similar expression for the effective position, $\vec{R}^{\,\text{eff}}$, of the system in terms of the system mass and the sum of the products of the individual masses and position vectors,

$$\vec{R}^{\,\text{eff}} = \frac{1}{M_{\text{sys}}}(m_A\vec{r}_A + m_B\vec{r}_B + m_C\vec{r}_C + \cdots + m_N\vec{r}_N) \quad (N \text{ particle system}). \tag{8-8}$$

In the next section we explore the properties of this expression and compare $\vec{R}^{\,\text{eff}}$ to the location of the balancing point for a system of objects.

8-3 The Effective Position—Center of Mass

Consider the system shown in Fig. 8-4. If the two masses are equal, $m_A = m_B$, then from

$$\vec{R}^{\,\text{eff}} = \frac{1}{M_{\text{sys}}}(m_A\vec{r}_A + m_B\vec{r}_B),$$

we get

$$\vec{R}^{\,\text{eff}} = \tfrac{1}{2}(\vec{r}_A + \vec{r}_B) \quad \text{(equal masses)}.$$

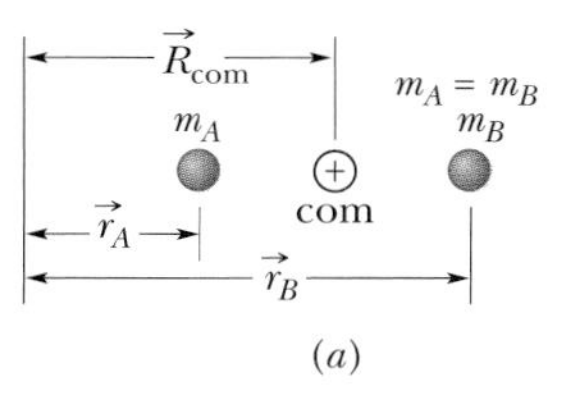

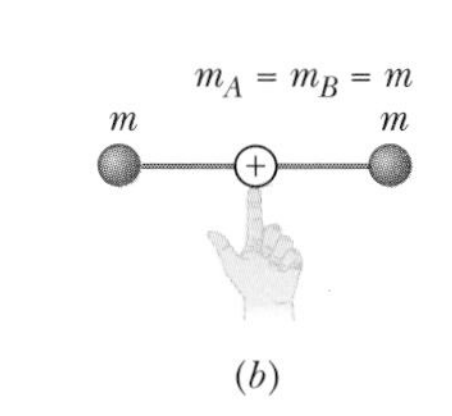

FIGURE 8-4 ■ (*a*) If the masses in a two-particle system are equal, then the center of mass of the system is always on a line halfway between the two particles. (*b*) If we connect the two particles having identical masses with a massless rod, the balance point of the system also turns out to be halfway between the particles.

If we want to consider the two-particle system's effective position quantitatively, then we must pick a coordinate system to determine $\vec{r}_A$ and $\vec{r}_B$. If we choose one of the axes of the coordinate system to lie on a line connecting the particles, it is easy to see that when the masses are equal, then $\vec{R}^{\,\text{eff}}$, the effective position of the system, is midway between the two objects on the line connecting them (Fig. 8-4*a*). If we imagined that the system particles are connected by a massless rod and tried to balance such an object on our finger, we would find that the balancing point is also halfway between the two masses whenever m_A is equal to m_B (Fig. 8-4*b*). For this reason we define the effective position of a system that is calculated using Eq. 8-8 as the **center**

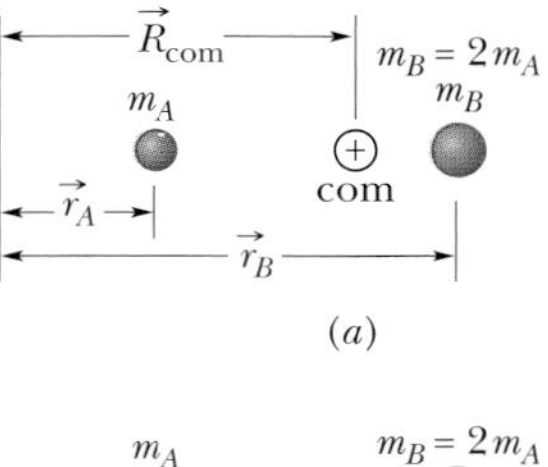

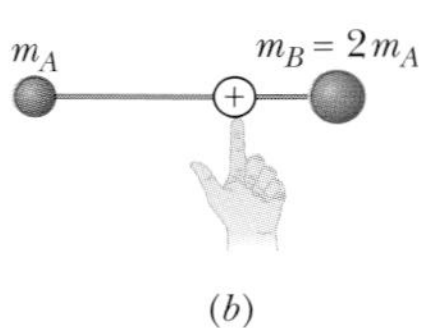

FIGURE 8-5 ■ (*a*) If the masses in a two-particle system are not equal, then the center of mass is always on a line between the two particles but is closer to the more massive particle. (*b*) If we connect the two particles having unequal masses with a massless rod, the balance point of the system turns out to be at the same location as the calculated center of mass of the system.

of mass (com) of the system. We denote the location of the center of mass as $\vec{R}_{com} \equiv \vec{R}^{eff}$. With experiment and careful observation, we can determine that special balancing point or the center of mass of almost any system. In general:

> The center of mass (com) of a body or a system of bodies is its balancing point. It is the point that moves as though all of the mass were concentrated there and the system behaves as if all the external forces are applied there.

What happens to the center of mass of a two-particle system if the masses are not equal? If we let particle B be twice the mass of particle A so $m_B = 2m_A$, we find that

$$\vec{R}_{com} = \frac{1}{M_{sys}}(m_A\vec{r}_A + m_B\vec{r}_B) = \frac{1}{m_A + m_B}(m_A\vec{r}_A + 2m_A\vec{r}_B)$$

$$= \frac{m_A}{m_A + 2m_A}(\vec{r}_A + 2\vec{r}_B),$$

or

$$\vec{R}_{com} = \tfrac{1}{3}(\vec{r}_A + 2\vec{r}_B) \qquad \text{(special case for } m_B = 2m_A).$$

In words, the center of mass or "effective position" of this system is located along the line joining the centers of the two masses, two-thirds of the way from the less massive object m_A and one-third of the way from the more massive object m_B (Fig. 8-5*a*). If we connected the two particles in the system with a massless rod and tried to balance it on our finger, we would find that the center of mass is also the same as the balancing point (Fig. 8-5*b*).

Physicists love to look at something complicated and find something simple and familiar in it. Fortunately, this turns out to be the case with the complicated motions of particle systems. For example, recall the diver who is rotating as she falls through the air as shown in Fig. 8-1*b*. We can consider her body to be a system made up of many individual particles that can exert internal forces on each other. If we neglect air drag, then the only significant external force on her is a constant gravitational force that acts downward. If we were to calculate the location of her center of mass at each moment during her fall, we would find that the calculated center of mass of the diver moves in a very simple parabolic path.

Actually calculating the center of mass of an athlete or dancer who is constantly changing her configuration seems like an impossible task. However, we can think of an athlete as a series of particles connected by massless rods. We locate each particle near the center of a linear body segment (such as ankle to toe, knee to ankle, hip to knee, and so on) and assign it the mass of the body part it represents. We find that we can use the techniques described in the next section to perform computer-aided calculations of an athlete's center of mass. Such an analysis performed on a series of video frames always gives a parabolic path when the athlete or dancer is jumping. An example of this is presented at the end of Section 8-5 for a ballerina performing a *grand jeté*. This analysis and countless others provide us with experimental verification that the concept of center of mass is useful when tracking the motion of complex systems that experience external forces.

8-4 Locating a System's Center of Mass

Let's consider how to calculate the center of mass (com) for a system consisting of a few particles that lie along a chosen x axis. Figure 8-5 shows two particles of masses m_A and m_B separated by a distance d. Here, x_A is the x coordinate of m_A's position and

x_B is the x coordinate of m_B's position. We can write the expression for the x coordinate of the center of mass of this system as

$$x_{\text{com}} = \frac{m_A x_A + m_B x_B}{M_{\text{sys}}}, \tag{8-9}$$

in which M_{sys} is the total mass of the system. (Here, $M_{\text{sys}} = m_A + m_B$.) We can extend this equation to a more general situation in which N particles are strung out along the x axis. Then the total mass is $M_{\text{sys}} = m_A + m_B + \cdots + m_N$, and the location of the center of mass is

$$x_{\text{com}} = \frac{m_A x_A + m_B x_B + m_C x_C + \cdots + m_N x_N}{M_{\text{sys}}}. \tag{8-10}$$

If the particles are distributed in three dimensions, then we can start with our expression for $\vec{R}_{\text{com}}$ (Eq. 8-8) and express each position vector in terms of its x-, y-, and z-components. For example, the ith position vector is given by

$$\vec{r}_i = x_i\hat{\text{i}} + y_i\hat{\text{j}} + z_i\hat{\text{k}}.$$

It is not difficult to show that when all the position vectors in a system of N particles are expressed in their rectangular coordinates using the equation above, then

$$\vec{R}_{\text{com}} = X_{\text{com}}\hat{\text{i}} + Y_{\text{com}}\hat{\text{j}} + Z_{\text{com}}\hat{\text{k}}.$$

The components of the center of mass of a system of particles are

$$X_{\text{com}} = \frac{1}{M_{\text{sys}}}(m_A x_A + m_B x_B + m_C x_C + \cdots),$$

$$Y_{\text{com}} = \frac{1}{M_{\text{sys}}}(m_A y_A + m_B y_B + m_C y_C + \cdots), \quad \text{(center of mass vector components-particle system)}. \tag{8-11}$$

$$Z_{\text{com}} = \frac{1}{M_{\text{sys}}}(m_A z_A + m_B z_B + m_C z_C + \cdots)$$

We can use the equations for X_{com} and Y_{com} to calculate the center of mass of a system of three pucks on an air table that have masses of 100, 200, and 300 g. The location of the center of each puck is shown in the data table in Fig. 8-6. The diagram shows the locations of the pucks in a rectangular coordinate system along with the calculated location of the center of mass of the system.

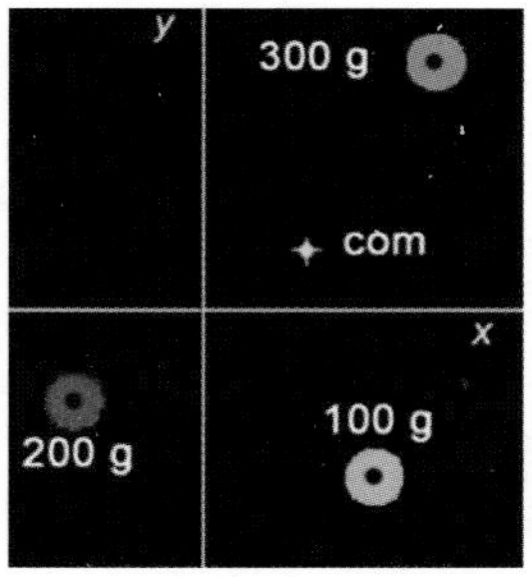

m(g)	x(cm)	y(cm)
100	16.8	–16.8
200	–12.2	–9.4
300	22.7	24.1
M_{sys}(g)	X_{com}(cm)	Y_{com}(cm)
600	10.1	6.1

FIGURE 8-6 ■ Three pucks gliding on an air table form a system. Equation 8-11 can be used to locate the center of mass of the system at $x = 10.1$ cm and $y = 6.1$ cm.

Solid Bodies

Some systems have too many particles to keep track of individually. It would be an enormous task to calculate the location of the center of mass using the summation technique described above. A solid object can be treated as a "continuous distribution" of matter. The term *continuous* implies that the "particles" that make up the object are no longer clearly separable. The particles then become differential mass elements, dm, the sums (shown in Eq. 8-11) become integrals, and the coordinates of the center of mass vector components are defined as

$$X_{\text{com}} = \frac{1}{M_{\text{sys}}} \int x\,dm, \quad Y_{\text{com}} = \frac{1}{M_{\text{sys}}} \int y\,dm, \quad Z_{\text{com}} = \frac{1}{M_{\text{sys}}} \int z\,dm \tag{8-12}$$

(center of mass vector components-continuous system),

where M_{sys} is the mass of the system.

If you are clever and don't enjoy doing unnecessary integrations, you can bypass one or more of the integrals above if an object has a point, a line, or a plane of symmetry. In these cases, the center of mass of such an object then lies at that point, on that line, or in that plane. For example, the center of mass of a uniform sphere (which has a point of symmetry) is at the center of the sphere (which is the point of symmetry). The center of mass of a uniform cone (whose axis is a line of symmetry) lies on the axis of the cone. The task required to determine the location of the center of mass of the cone is then reduced to determining where along this axis the center of mass is located. For example, the center of mass of a banana (which has a plane of symmetry that splits it into two equal parts) lies somewhere in that plane.

The center of mass of an object need not lie within the object. There is no dough at the center of mass of a doughnut, and no iron at the center of mass of a horseshoe.

Evaluating these integrals for most common objects (like a television set) would be difficult, so here we shall consider only *uniform* solid objects. Such an object has *uniform density,* or mass per unit volume. That is, the density ρ (Greek letter rho) is the same for any given segment of the object as for the whole object:

$$\rho = \frac{dm}{dV} = \frac{M_{\text{sys}}}{V} \quad \text{(uniform object density)}, \tag{8-13}$$

where dV is the volume occupied by a mass element dm, and V is the total volume of the object. If we substitute $dm = (M_{\text{sys}}/V)\,dV$ into Eq. 8-12 we find that

$$X_{\text{com}} = \frac{1}{V} \int x\,dV, \quad Y_{\text{com}} = \frac{1}{V} \int y\,dV, \quad Z_{\text{com}} = \frac{1}{V} \int z\,dV. \tag{8-14}$$

READING EXERCISE 8-1: The figure shows a uniform square plate from which four identical squares at the corners will be removed. (a) Where is the center of mass of the plate originally? Where is it after the removal of (b) square 1, (c) squares 1 and 2, (d) squares 1 and 3, (e) squares 1, 2, and 3, (f) all four squares? Answer in terms of quadrants, axes, or points (without calculation, of course).

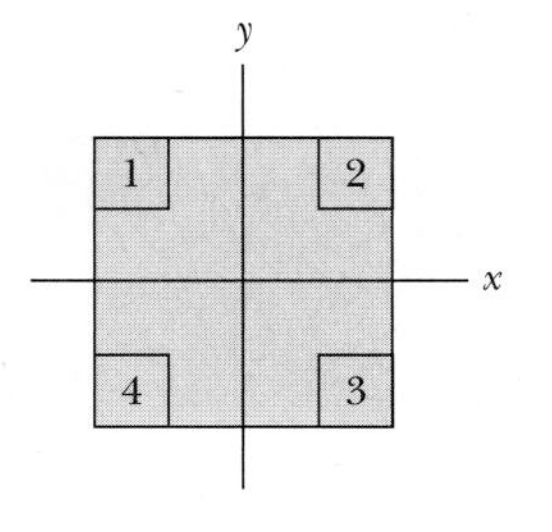

■

TOUCHSTONE EXAMPLE 8-1: Three Masses

Three particles of masses $m_A = 1.2$ kg, $m_B = 2.5$ kg, and $m_C = 3.4$ kg form an equilateral triangle of edge length $a = 140$ cm. Where is the center of mass of this three-particle system?

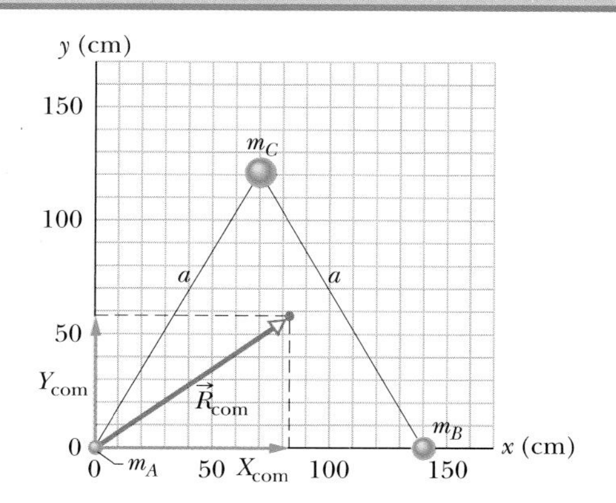

Figure 8-7 ■ Three particles form an equilateral triangle of edge length a. The center of mass is located by the position vector $\vec{R}_{com}$.

SOLUTION ■ A **Key Idea** to get us started is that we are dealing with particles instead of an extended solid body, so we can use Eq. 8-11 to locate their center of mass. The particles are in the plane of the equilateral triangle, so we need only the first two equations. A second **Key Idea** is that we can simplify the calculations by choosing the x and y axes so that one of the particles is located at the origin and the x axis coincides with one of the triangle's sides (Fig. 8-7). The three particles then have the following coordinates:

Particle	Mass (kg)	X (cm)	Y (cm)
A	1.2	0	0
B	2.5	140	0
C	3.4	70	121

The total mass M_{sys} of the system is 7.1 kg.

From Eq. 8-11, the coordinates of the center of mass are

$$X_{com} = \frac{m_A X_A + m_B X_B + m_C X_C}{M_{sys}}$$

$$= \frac{(1.2\text{ kg})(0) + (2.5\text{ kg})(140\text{ cm}) + (3.4\text{ kg})(70\text{ cm})}{7.1\text{ kg}}$$

$$= 83\text{ cm}, \qquad \text{(Answer)}$$

and

$$Y_{com} = \frac{m_A Y_A + m_B Y_B + m_C Y_C}{M_{sys}}$$

$$= \frac{(1.2\text{ kg})(0) + (2.5\text{ kg})(0) + (3.4\text{ kg})(121\text{ cm})}{7.1\text{ kg}}$$

$$= 58\text{ cm}. \qquad \text{(Answer)}$$

In Fig. 8-7, the center of mass is located by the position vector $\vec{R}_{com}$, which has components X_{com} and Y_{com}.

TOUCHSTONE EXAMPLE 8-2: U-Shaped Object

The U-shaped object pictured in Fig. 8-8 has outside dimensions of 100 mm on each side, and each of its three sides is 20 mm wide. It was cut from a uniform sheet of plastic 6.0 mm thick. Locate the center of mass of this object.

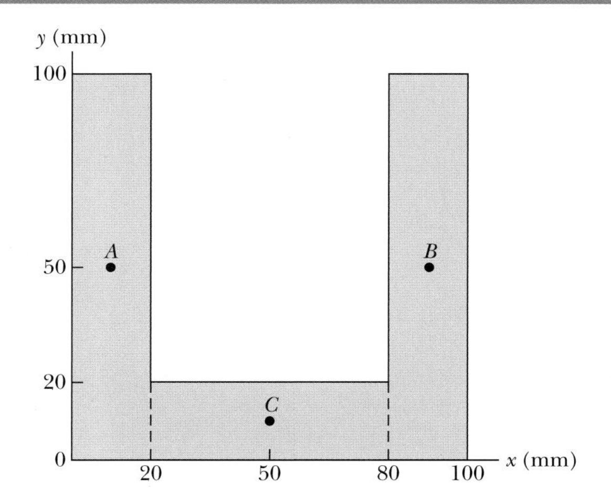

Figure 8-8 ■ The U-shaped object can be broken up into three uniform rectangular bars, A, B, and C, as shown in the figure. For convenience we'll take the corners of the U to be square and not rounded. The dot on each bar shows the location of its center of mass. We will use the coordinate system shown to locate the center of mass of each bar and then to locate the center of mass of the whole object.

SOLUTION ■ A **Key Idea** here is to break the U-shaped object up into pieces, each having an easily located center of mass. We can then replace each piece by a point mass located at the center of mass of that piece, and then use Eq. 8-11 to locate the center of mass of the whole object.

As shown in Fig. 8-8, we can think of the U-shaped object as made up of two vertical bars, each 100 mm long by 20 mm wide, joined together by one horizontal bar 60 mm long and 20 mm wide. Let's place the origin of our coordinate system at the lower-left rear corner of the U, with the x axis across its base and the y axis along its left edge.

To locate the center of mass of each of the bars, we will use the **Key Idea** that the center of mass of a symmetric object of uniform density is located at its geometric center. This means that the center of mass of each bar is exactly halfway from either end, halfway from either side, and halfway between the top and bottom surfaces of the plastic sheet. Putting a dot at the center of each of the bars, A, B, and C in Fig. 8-8, we can now write down their locations in the coordinate system pictured there:

Object	Mass	X	Y	Z
Left bar	M_A	10 mm	50 mm	3 mm
Right bar	M_B	90 mm	50 mm	3 mm
Bottom bar	M_C	50 mm	10 mm	3 mm

To learn more about the relative masses of the three bars, we can use the **Key Idea** that the mass of each bar is proportional to its volume since the bars have a uniform common density. In this case the relationship is even simpler: since the three bars are each the same width and thickness, each one's mass is directly proportional to its length, so

$$M_A = M_B = M \quad \text{and} \quad M_C = M(60\ \text{mm})/(100\ \text{mm}) = 0.6M,$$

and $\quad M_{\text{sys}} = M_A + M_B + M_C = M + M + 0.6M = 2.6M.$

Replacing each bar by a point mass at its center of mass, Eq. 8-11 gives us the location of the center of mass of the entire U-shaped object:

$$X_{\text{com}} = \frac{M_A X_A + M_B X_B + M_C X_C}{M_{\text{sys}}} = \frac{\cancel{M}(10\ \text{mm}) + \cancel{M}(90\ \text{mm}) + 0.6\cancel{M}(50\ \text{mm})}{2.6\cancel{M}}$$

$$= 50\ \text{mm}, \qquad \text{(Answer)}$$

$$Y_{\text{com}} = \frac{M_A Y_A + M_B Y_B + M_C Y_C}{M_{\text{sys}}} = \frac{\cancel{M}(50\ \text{mm}) + \cancel{M}(50\ \text{mm}) + 0.6\cancel{M}(10\ \text{mm})}{2.6\cancel{M}}$$

$$= 40.769\ \text{mm} \cong 41\ \text{mm}, \qquad \text{(Answer)}$$

and

$$Z_{\text{com}} = \frac{M_A Z_A + M_B Z_B + M_C Z_C}{M_{\text{sys}}} = \frac{\cancel{M}(3\ \text{mm}) + \cancel{M}(3\ \text{mm}) + 0.6\cancel{M}(3\ \text{mm})}{2.6\cancel{M}}$$

$$= 3\ \text{mm}, \qquad \text{(Answer)}$$

so

$$\vec{R}_{\text{com}} = (50\ \text{mm})\hat{\text{i}} + (41\ \text{mm})\hat{\text{j}} + (3\ \text{mm})\hat{\text{k}}. \qquad \text{(Answer)}$$

TOUCHSTONE EXAMPLE 8-3: Crescent-Shaped Object

Figure 8-9a shows a uniform metal plate P of radius $2R$ from which a disk of radius R has been stamped out (removed) in an assembly line. Using the xy coordinate system shown, locate the center of mass, com_P, of the plate.

SOLUTION ■ First, let us roughly locate the center of mass of plate P by using the **Key Idea** of symmetry. We note that the plate is symmetric about the x axis (we get the portion below that axis by rotating the upper portion about the axis). Thus, com_P must be on the x axis. The plate is not symmetric about the y axis. However, because there is somewhat more mass on the right of the y axis, com_P must be somewhat to the right of that axis. Thus, the location of com_P should be roughly as indicated in Fig. 8-9a.

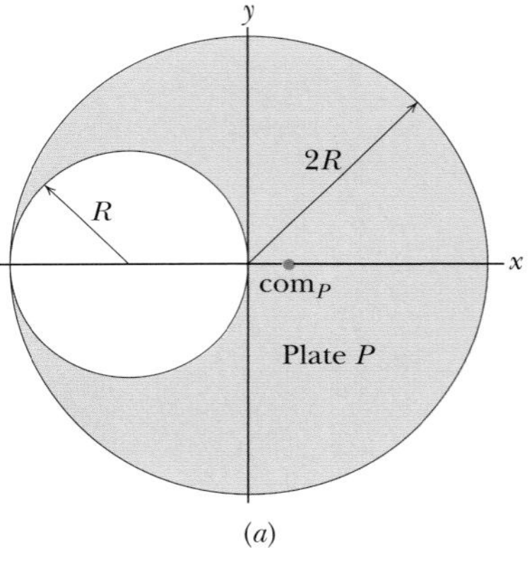

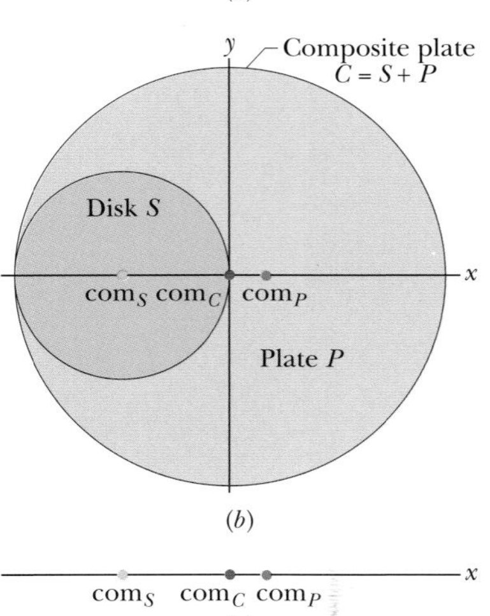

FIGURE 8-9 ■ (a) Plate P is a metal plate of radius $2R$, with a circular hole of radius R. The center of mass of P is at point com_P. (b) Disk S has been put back into place to form a composite plate C. The center of mass, com_S, of disk S and the center of mass, com_C, of plate C are shown. (c) The center of mass, com_{S+P}, of the combination of S and P coincides with com_C, which is at $x = 0$.

Another **Key Idea** here is that plate P is an extended solid body, so we can use Eq. 8-14 to find the actual coordinates of com_P. However, that procedure is difficult. A much easier way is to use this **Key Idea**: In working with centers of mass, we can take the mass of any *uniform* object to be concentrated in a particle located at the object's center of mass. Here is how we do so:

First, put the stamped-out disk (call it disk S) back into place (Fig. 8-9b) to form the original composite plate (call it plate C). Because of its circular symmetry, the center of mass, com_S, for disk S is at the center of S, at $x = -R$ (as shown). Similarly, the center of mass, com_C, for composite plate C is at the center of C, at the origin (as shown). We then have the following:

Plate	Center of Mass	Location of of com	Mass
P	com_P	$X_P = ?$	m_P
S	com_S	$X_S = -R$	m_S
C	com_C	$X_C = 0$	$m_C = m_S + m_P$

Now we use the **Key Idea** of concentrated mass: Assume that mass m_S of disk S is concentrated in a particle at $X_S = -R$, and mass m_P is concentrated in a particle at X_P (Fig. 8-9c). Next treat these two particles as a two-particle system, using Eq. 8-9 to find their center of mass X_{S+P}. We get

$$X_{S+P} = \frac{m_S X_S + m_P X_P}{m_S + m_P}. \quad (8\text{-}15)$$

Next note that the combination of disk S and plate P is composite plate C. Thus, the position X_{S+P} of com_{S+P} must coincide with the position X_C of com_C, which is at the origin, so $X_{S+P} = X_C = 0$. Substituting this into Eq. 8-15 and solving for X_P, we get

$$X_P = -X_S \frac{m_S}{m_P}. \quad (8\text{-}16)$$

Now we seem to have a problem, because we do not know the masses in Eq. 8-16. However, we can relate the masses to the face areas of S and P by noting that

$$\begin{aligned}\text{Mass} &= \text{density} \times \text{volume}\\ &= \text{density} \times \text{thickness} \times \text{area}.\end{aligned}$$

Then $$\frac{m_S}{m_P} = \frac{\text{density}_S}{\text{density}_P} \times \frac{\text{thickness}_S}{\text{thickness}_P} \times \frac{\text{area}_S}{\text{area}_P}.$$

Because the plate is uniform, the densities and thicknesses are equal; we are left with

$$\frac{m_S}{m_P} = \frac{\text{area}_S}{\text{area}_P} = \frac{\text{area}_S}{\text{area}_C - \text{area}_S} = \frac{\pi R^2}{\pi(2R)^2 - \pi R^2} = \frac{1}{3}.$$

Substituting this and $X_S = -R$ into Eq. 8-16, we have

$$X_P = \tfrac{1}{3}R. \quad \text{(Answer)}$$

8-5 Newton's Laws for a System of Particles

If you roll a billiard ball at a second billiard ball that is at rest, you expect that the two-ball system will continue to have some forward motion after impact. You would be surprised to see both balls come back toward you. But what do we actually observe when one billiard ball rolling at a constant velocity hits another resting ball that has the same mass?

What we observe is that the center of mass of the two-ball system continues to move forward, its motion completely unaffected by the collision. If you focus on the center of mass (which is always halfway between two particles that have the same mass) you can easily convince yourself that this is so. No matter whether the collision is glancing, head-on, or somewhere in between, the center of mass continues to move forward, just as if the collision had never occurred. This is depicted in Fig. 8-10 for a head-on collision.

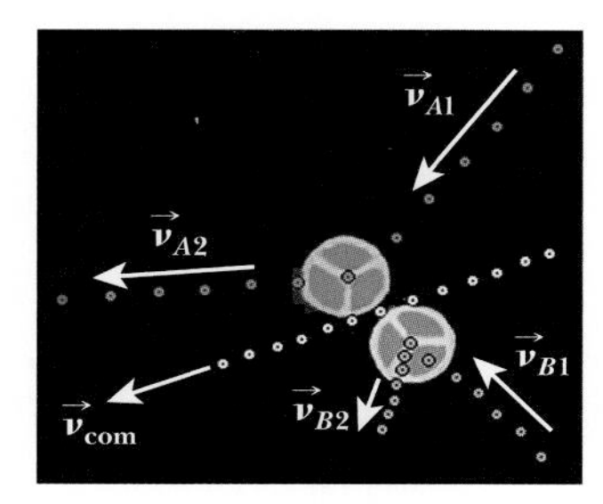

FIGURE 8-10 ■ Two pucks of equal mass glide along an air table. They strike each other a glancing blow. A video analysis shows a point halfway between them in each frame moving at a constant velocity.

Let's consider another simple situation. Two pucks with the same mass are moving and collide with a glancing blow. Using a digital video clip of this collision, we can track the locations of the two pucks frame by frame and mark a point halfway between the two puck centers. These halfway points, shown as white dots in Fig. 8-10, represent the center of mass of the two-puck system. It is clear that the center of mass of the system is moving in a straight line at constant speed.

Let us look into this center of mass motion theoretically. Why should we expect the center of mass of the billiard balls or a collection of pucks on an air table to move with a constant velocity? Let's start with an assemblage of N particles of different masses and shapes like those shown in Fig. 8-11. These objects are floating just above a level air table. We are interested not in the individual motions of these particles, but *only* in the motion of the center of mass of the system. We use balancing points to find the center of mass of each of the oddly shaped objects in the system. We can then use these locations for each object to calculate the center of mass of the system as a whole. As we discussed in Section 8-2,

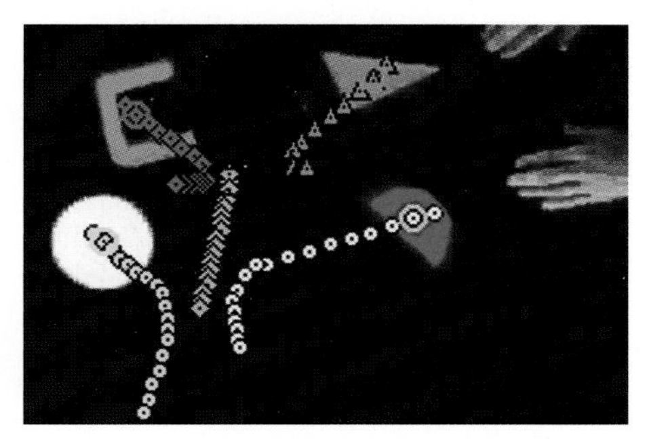

FIGURE 8-11 ■ Four shapes, each having a different mass, collide in the center of an air table. The path of the center of mass of each shape is found using video analysis. The calculated path of the center of mass of the system is shown as diamonds. Note that this calculated system center of mass moves along a straight line at a constant velocity before, during, and after the collisions that take place between the various objects.

$$\vec{F}^{\,\text{net}}_{\text{sys}} = M_{\text{sys}}\vec{a}_{\text{sys}} \quad \text{(system of particles)}, \quad (8\text{-}17)$$

where M_{sys} is the total mass of the system and as we now know, $\vec{a}_{\text{sys}} = \vec{a}_{\text{com}}$ is the acceleration of the system's center of mass. This equation is Newton's Second Law for the motion of the center of mass of a system of particles. However, the meaning of the three quantities that appear in $\vec{F}^{\,\text{net}}_{\text{sys}} = M_{\text{sys}}\vec{a}_{\text{sys}}$ must be carefully interpreted.

1. $\vec{F}^{\,\text{net}}_{\text{sys}}$ is the sum of *all external forces* that act on the system. Forces on one part of the system from another (*internal forces*) do not matter. (By Newton's Third Law, we know that the internal forces cancel each other out when the system is considered as a whole).
2. M_{sys} is the *total mass* of the objects in the system. We assume that no mass enters or leaves the system as it moves, so that M_{sys} remains constant. Such a system is said to be **closed.**
3. Although Newton's Second Law allows us to determine the acceleration of the *center of mass*, $\vec{a}_{\text{com}}$ of the system from the net force on it, in some situations we may have no information about the acceleration of any other point in the system.

$\vec{F}^{\,\text{net}}_{\text{sys}} = M_{\text{sys}}\vec{a}_{\text{com}}$ is equivalent to three equations involving the components of $\vec{F}^{\text{net}}$ and $\vec{a}_{\text{com}}$ along three coordinate axes that can be chosen. These equations are

$$F^{\text{net}}_{\text{sys }x} = M_{\text{sys}}a_{\text{com }x}, \quad F^{\text{net}}_{\text{sys }y} = M_{\text{sys}}a_{\text{com }y}, \quad F^{\text{net}}_{\text{sys }z} = M_{\text{sys}}a_{\text{com }z}. \tag{8-18}$$

Application to the Air Table Objects Once the objects on the air table are set into motion, no net external force acts on the system. This is because the external forces consist of a downward gravitational force and the upward normal force on each object. These forces cancel each other out. Thus there is no net force on the system. Since $\vec{F}^{\,\text{net}}_{\text{sys}} = 0$, we know that $\vec{a}_{\text{com}} = 0$ also. Because acceleration is the rate of change of velocity, we conclude that the velocity of the center of mass of the system of four objects does not change. When various objects collide, the forces that come into play are *internal* forces on one object from another. Such forces do not contribute to the net force, which remains zero. Thus, even though the velocities of the four objects change individually as a result of the forces the objects feel from within the system, the center of mass of the system continues to move with unchanged velocity (Fig. 8-11).

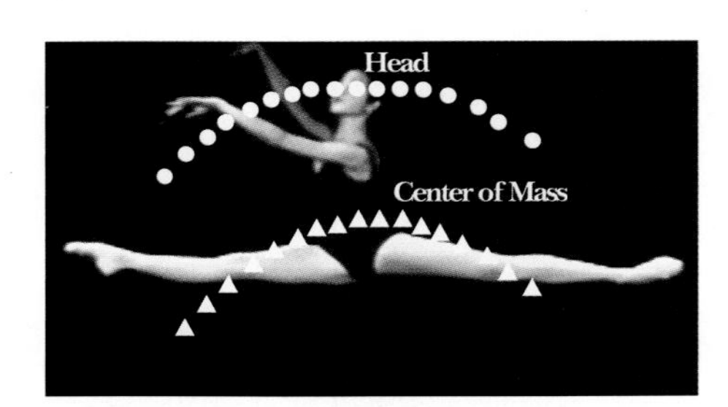

FIGURE 8-12 ■ A video analysis of the *grand jeté* shows that the center of mass of the dancer moves in a parabolic path while her head moves horizontally at the peak of her jump. (See Fig. 1-14 in Ch. 1 for more details).

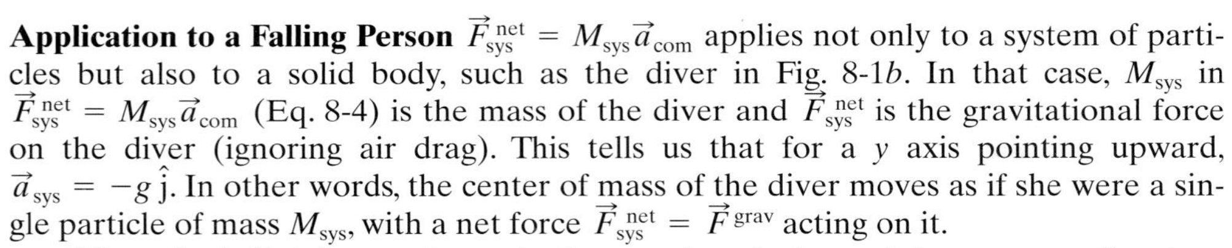

Application to a Falling Person $\vec{F}^{\,\text{net}}_{\text{sys}} = M_{\text{sys}}\vec{a}_{\text{com}}$ applies not only to a system of particles but also to a solid body, such as the diver in Fig. 8-1*b*. In that case, M_{sys} in $\vec{F}^{\,\text{net}}_{\text{sys}} = M_{\text{sys}}\vec{a}_{\text{com}}$ (Eq. 8-4) is the mass of the diver and $\vec{F}^{\,\text{net}}_{\text{sys}}$ is the gravitational force on the diver (ignoring air drag). This tells us that for a y axis pointing upward, $\vec{a}_{\text{sys}} = -g\,\hat{\text{j}}$. In other words, the center of mass of the diver moves as if she were a single particle of mass M_{sys}, with a net force $\vec{F}^{\,\text{net}}_{\text{sys}} = \vec{F}^{\,\text{grav}}$ acting on it.

When the ballet dancer shown in the opening photograph leaps across the stage in a *grand jeté*, she raises her arms and stretches her legs out horizontally as soon as her feet leave the stage (Fig. 8-12). These actions shift her center of mass upward through her body. Although the shifting center of mass faithfully follows a parabolic path across the stage, its movement, relative to the body, decreases the height that is attained by her head and torso, relative to that of a normal jump. The result is that the head and torso follow a nearly horizontal path, giving an illusion that the dancer is floating as shown in the Fig. 8-12 video analysis.

READING EXERCISE 8-2: The halfway point between the two pucks in Fig. 8-10 is moving in a straight line. If each frame is exactly 1/15th of a second later than the previous frame, (a) what evidence is there that the speed is constant? (b) If the distance between the first location of the center of mass and the last location is 0.41 m, what is the speed of the center of mass? ■

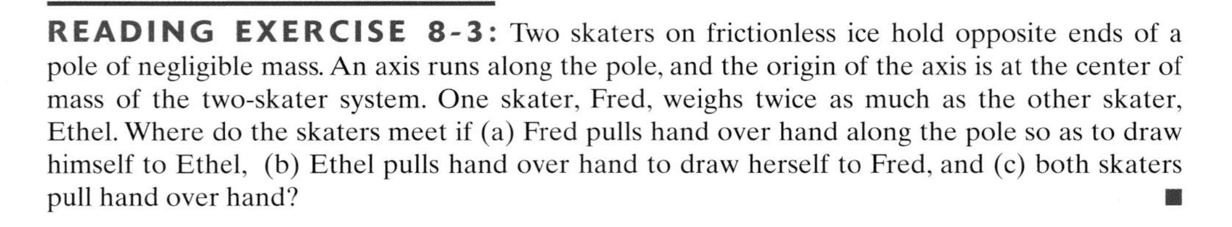

READING EXERCISE 8-3: Two skaters on frictionless ice hold opposite ends of a pole of negligible mass. An axis runs along the pole, and the origin of the axis is at the center of mass of the two-skater system. One skater, Fred, weighs twice as much as the other skater, Ethel. Where do the skaters meet if (a) Fred pulls hand over hand along the pole so as to draw himself to Ethel, (b) Ethel pulls hand over hand to draw herself to Fred, and (c) both skaters pull hand over hand? ■

TOUCHSTONE EXAMPLE 8-4: Center-of-Mass Acceleration

The three particles in Fig. 8-13*a* are initially at rest. Each experiences an *external* force due to bodies outside the three-particle system. The directions are indicated, and the magnitudes are $F_A = 6.0$ N, $F_B = 12$ N, and $F_C = 14$ N. What is the magnitude of the acceleration of the center of mass of the system, and in what direction does it move?

SOLUTION ■ The position of the center of mass, calculated by the method of Touchstone Example 8-1, is marked by a dot in Fig. 8-13. One **Key Idea** here is that we can treat the center of mass as if it were a real particle, with a mass equal to the system's total mass $M_{sys} = 16$ kg. We can also treat the three external forces as if they act at the center of mass (Fig. 8-13*b*).

A second **Key Idea** is that we can now apply Newton's Second Law ($\vec{F}^{net} = m\vec{a}$) to the center of mass, writing

$$\vec{F}_{sys}^{net} = M_{sys}\vec{a}_{com}, \tag{8-19}$$

or

$$\vec{F}_A + \vec{F}_B + \vec{F}_C = M_{sys}\vec{a}_{com},$$

so

$$\vec{a}_{com} = \frac{\vec{F}_A + \vec{F}_B + \vec{F}_C}{M_{sys}}. \tag{8-20}$$

Equation 8-19 tells us that the acceleration $\vec{a}_{com}$ of the center of mass is in the same direction as the net external force $\vec{F}_{sys}^{net}$ on the system (Fig. 8-13*b*). Because the particles are initially at rest, the center of mass must also be at rest. As the center of mass then begins to accelerate, it must move off in the common direction of $\vec{a}_{com}$ and $\vec{F}_{sys}^{net}$.

We can evaluate the right side of Eq. 8-20 directly on a vector-capable calculator, or we can rewrite Eq. 8-20 in component form, find the components of $\vec{a}_{com}$, and then find $\vec{a}_{com}$. Along the x axis, we have

$$a_{com\,x} = \frac{F_{Ax} + F_{Bx} + F_{Cx}}{M_{sys}}$$

$$= \frac{-6.0\text{ N} + (12\text{ N})\cos 45° + 14\text{ N}}{16\text{ kg}} = 1.03\text{ m/s}^2.$$

Along the y axis, we have

$$a_{com\,y} = \frac{F_{Ay} + F_{By} + F_{Cy}}{M_{sys}}$$

$$= \frac{0 + (12\text{ N})\sin 45° + 0}{16\text{ kg}} = 0.530\text{ m/s}^2.$$

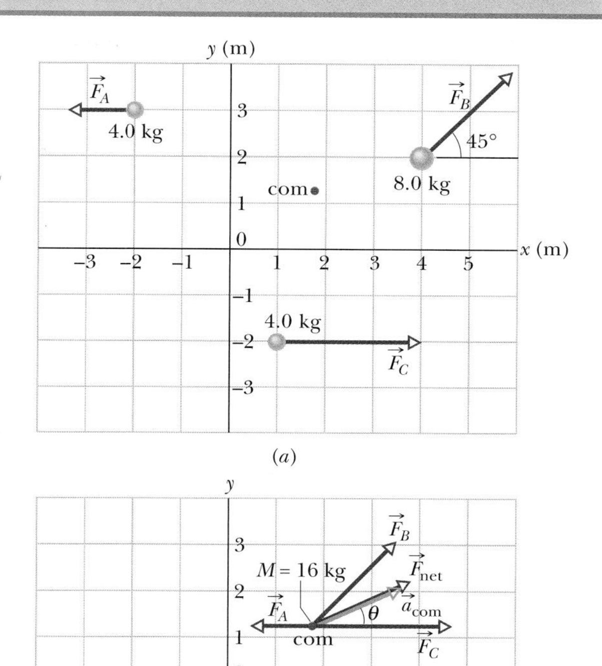

FIGURE 8-13 ■ (*a*) Three particles, initially at rest in the positions shown, are acted on by the external forces shown. The center of mass, com, of the system is marked. (*b*) The forces are now transferred to the center of mass of the system, which behaves like a particle with a mass M equal to the total mass of the system. The net external force $\vec{F}^{net}$ and the acceleration $\vec{a}_{com}$ of the center of mass are shown.

From these components, we find that $\vec{a}_{com}$ has the magnitude

$$a_{com} = \sqrt{(a_{com\,x})^2 + (a_{com\,y})^2}$$

$$= 1.16\text{ m/s}^2 \approx 1.2\text{ m/s}^2, \quad \text{(Answer)}$$

and the angle (from the positive direction of the x axis)

$$\theta = \tan^{-1}\frac{a_{com\,y}}{a_{com\,x}} = 27°. \quad \text{(Answer)}$$

8-6 The Momentum of a Particle System

Since the effective position vector describing a system of N particles is the same as its center of mass, we can express Eq. 8-8 as

$$\vec{R}_{com} = \vec{R}^{eff} = \frac{1}{M_{sys}}(m_A\vec{r}_A + m_B\vec{r}_B + m_C\vec{r}_C + \cdots), \tag{8-21}$$

in which M_{sys} is the system's total mass and $m_A\vec{r}_A$, $m_B\vec{r}_B$, and so on represent the product of the mass and position vector of each of the particles in the system. This expression can be rewritten as

$$M_{sys}\vec{R}_{com} = m_A\vec{r}_A + m_B\vec{r}_B + m_C\vec{r}_C + \cdots. \tag{8-22}$$

Differentiating the expression above with respect to time gives

$$M_{sys}\vec{v}_{com} = m_A\vec{v}_A + m_B\vec{v}_B + m_C\vec{v}_C + \cdots. \tag{8-23}$$

Here $\vec{v}_A(=d\vec{r}_A/dt)$ is the velocity of particle A and $\vec{v}_{com}(=d\vec{R}_{com}/dt)$ is the velocity of the center of mass.

Now consider the translational momentum of this same system. The system as a whole has a total translational momentum $\vec{p}_{sys}$, which is defined to be the vector sum of translational momenta of the particles in the system. Thus,

$$\begin{aligned}\vec{p}_{sys} &= \vec{p}_A + \vec{p}_B + \vec{p}_C + \cdots \\ &= m_A\vec{v}_A + m_B\vec{v}_B + m_C\vec{v}_C + \cdots.\end{aligned} \tag{8-24}$$

If we compare this equation with $M_{sys}\vec{v}_{com} = m_A\vec{v}_A + m_B\vec{v}_B + m_C\vec{v}_C + \cdots$ (Eq. 8-23), we see that

$$\vec{p}_{sys} = M_{sys}\vec{v}_{com} \quad \text{(translational momentum, system of particles)}, \tag{8-25}$$

which gives us another way to define the translational momentum of a system of particles:

> The translational momentum of a system of particles is equal to the product of the total mass M_{sys} of the system and the velocity $\vec{v}_{com}$ of the center of mass.

So we can determine the total momentum of a system either by determining the vector sum of the individual momenta of parts or by taking the total mass of the system and multiplying by the velocity of the center of mass of the system. Either path leads us to the same value.

If we take the time derivative of $\vec{p}_{sys} = M_{sys}\vec{v}_{com}$, we find

$$\frac{d\vec{p}_{sys}}{dt} = M_{sys}\frac{d\vec{v}_{com}}{dt}, \tag{8-26}$$

or

$$\frac{d\vec{p}_{sys}}{dt} = M_{sys}\vec{a}_{com}. \tag{8-27}$$

Comparing $\vec{F}^{net}_{sys} = M\vec{a}_{com}$ (Eq. 8-17) with Eq. 8-27 allows us to write Newton's Second Law for a system of particles in the equivalent form

$$\vec{F}^{net}_{sys} = \frac{d\vec{p}_{sys}}{dt} \quad \text{(system of particles)}, \tag{8-28}$$

where $\vec{F}^{net}_{sys}$ is the net external force acting on the particles in the system. As we discussed in Chapter 7, this equation is the generalization of the single-particle equation $\vec{F}^{net} = d\vec{p}/dt$ to a system of many particles. In its new form, $d\vec{p}_{sys}/dt = M_{sys}\vec{a}_{com}$, the introduction of the concept of the center of mass of a system gives us an additional technique for determining the rate at which the total momentum of a system changes.

Problems

SEC. 8-4 ■ LOCATING A SYSTEM'S CENTER OF MASS

1. Particle-Like Object A 4.0 kg particle-like object is located at $x = 0$, $y = 2.0$ m; a 3.0 kg particle-like object is located at $x = 3.0$ m, $y = 1.0$ m. At what (a) x and (b) y coordinates must a 2.0 kg particle-like object be placed for the center of mass of the three-particle system to be located at the origin?

2. 2D Center of Mass of Three Objects Consider Fig. 8-14. Three masses located in the x-y plane have the following coordinates; a 5 kg mass has coordinates given by (2, −3) m; a 4 kg mass has coordinates (−4, 2) m; a 2 kg mass has coordinates (3, 3) m. Find the coordinates of the center of mass to two significant figures.

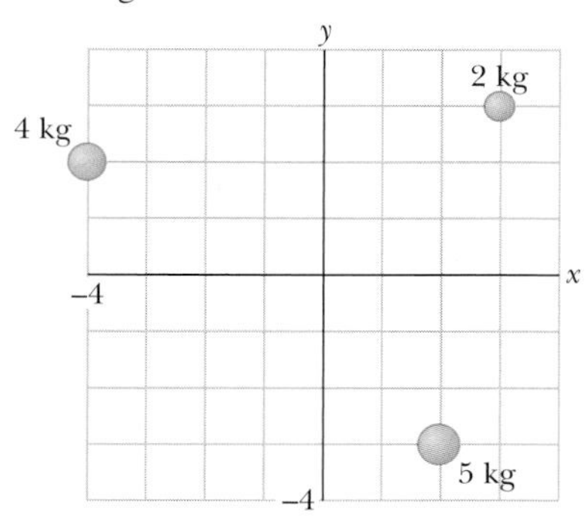

FIGURE 8-14 ■ Problem 2.

3. Earth–Moon System (a) How far is the center of mass of the Earth–Moon system from the center of Earth? (Appendix C gives the masses of Earth and the Moon and the distance between the two.) (b) Express the answer to (a) as a fraction of Earth's radius R_e.

4. Carbon Monoxide A distance of 1.131×10^{-10} m lies between the centers of the carbon and oxygen atoms in a carbon monoxide (CO) gas molecule. Locate the center of mass of a CO molecule relative to the carbon atom. (Find the masses of C and O in Appendix F.)

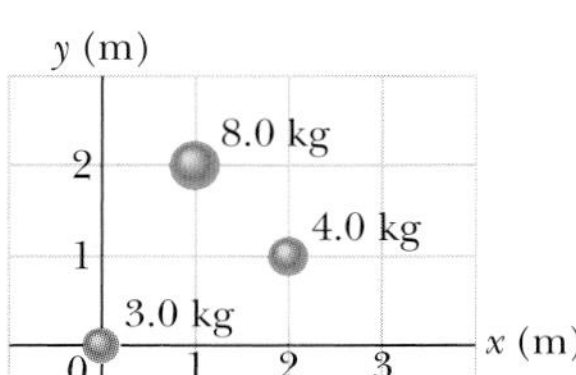

FIGURE 8-15 ■ Problem 5.

5. Three-Particle System What are (a) the x coordinate and (b) the y coordinate of the center of mass of the three-particle system shown in Fig. 8-15? (c) What happens to the center of mass as the mass of the topmost particle is gradually increased?

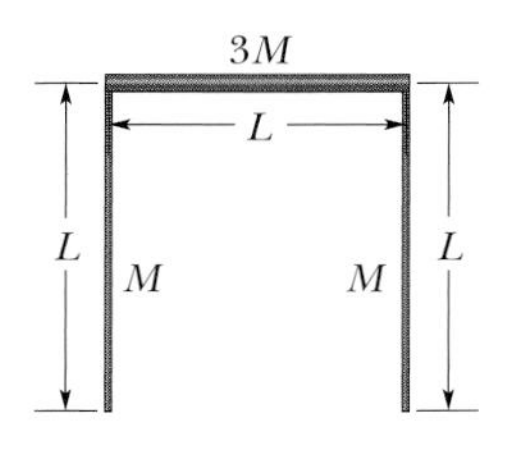

FIGURE 8-16 ■ Problem 6.

6. Three Thin Rods Three thin rods, each of length L, are arranged in an inverted U, as shown in Fig. 8-16. The two rods on the arms of the U each have mass M; the third rod has mass $3M$. Where is the center of mass of the assembly?

7. Uniform Square A uniform square plate 6 m on a side has had a square piece 2 m on a side cut out of it (Fig. 8-17). The center of that piece is at $x = 2$ m, $y = 0$. The center of the square plate is at $x = y = 0$. Find (a) the x coordinate and (b) the y coordinate of the center of mass of the remaining piece.

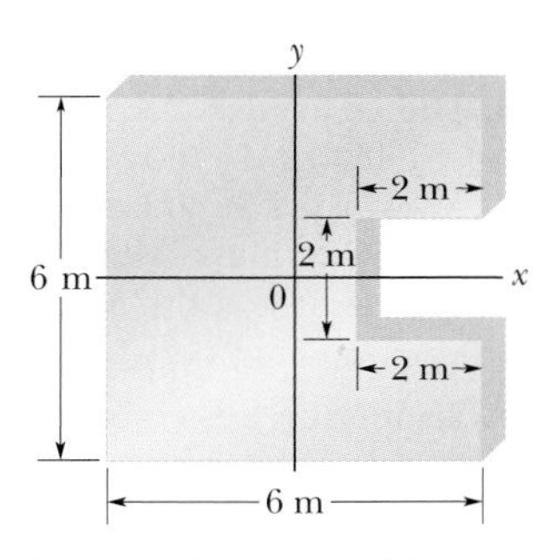

FIGURE 8-17 ■ Problem 7.

8. Composite Slab Figure 8-18 shows the dimensions of a composite slab; half the slab is made of aluminum (density = 2.70 g/cm³) and half is made of iron (density = 7.85 g/cm³). Where is the center of mass of the slab?

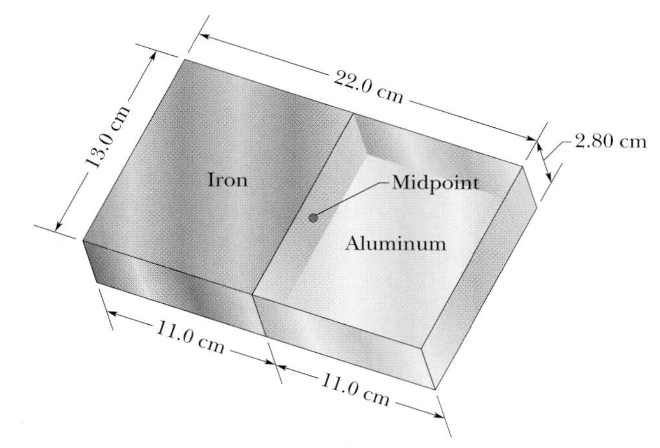

FIGURE 8-18 ■ Problem 8.

9. Ammonia In the ammonia (NH_3) molecule (see Fig. 8-19), the three hydrogen (H) atoms form an equilateral triangle; the center of the triangle is 9.40×10^{-11} m from each hydrogen atom. The nitrogen (N) atom is at the apex of a pyramid, with the three hydrogen atoms forming the base. The nitrogen-to-hydrogen atomic mass ratio is 13.9, and the nitrogen-to-hydrogen distance is 10.14×10^{-11} m. Locate the center of mass of the molecule relative to the nitrogen atom.

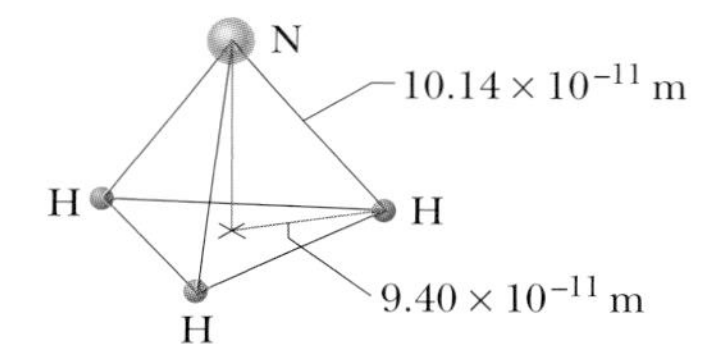

FIGURE 8-19 ■ Problem 9.

10. Metal Cube Figure 8-20 shows a cubical box that has been constructed from a metal plate of uniform density and negligible thickness. The box is open at the top and has edge length 40 cm. Find (a) the x coordinate, (b) the y coordinate, and (c) the z coordinate of the center of mass of the box.

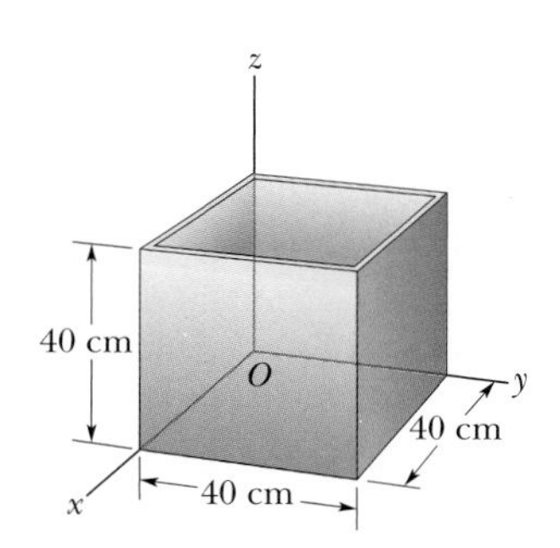

FIGURE 8-20 ■ Problem 10.

11. Cylindrical Can A right cylindrical can with mass M, height H, and uniform density is initially filled with soda of mass m (Fig. 8-21). We punch small holes in the top and bottom to drain the soda; we then consider the height h of the center of mass of the can and any soda within it. What is h (a) initially and (b) when all the soda has drained? (c) What happens to h during the

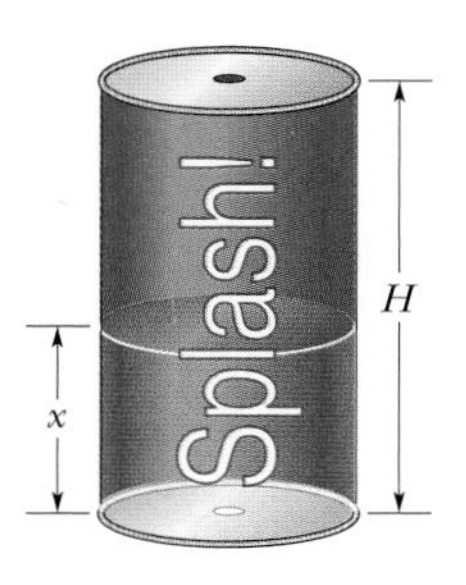

FIGURE 8-21 ■ Problem 11.

draining of the soda? (d) If x is the height of the remaining soda at any given instant, find x (in terms of M, H, and m) when the center of mass reaches its lowest point.

12. Clustered Problem 1 In Fig. 8-22*a*, a uniform wire forms an isosceles triangle of base B and height H. (a) Find the x and y coordinates of the figure's center of mass by assuming that each side can be replaced with a particle of the same mass as that side and positioned at the center of the side. (*Be careful*: Note that the base and, say, the left-hand side do not have the same mass.) (b) Use Eq. 8-12 to find the x and y coordinates of the center of mass of the left-hand side.

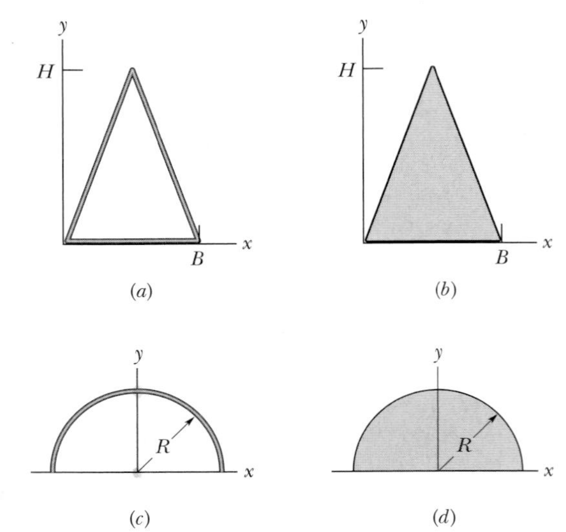

FIGURE 8-22 ■ Problems 12 through 15.

13. Clustered Problem 2 Figure 8-22*b* shows a uniform, solid plate in the shape of an isosceles triangle with base B and height H. What are the x and y coordinates of the plate's center of mass?

14. Clustered Problem 3 In Fig. 8-22*c*, a uniform wire forms a semicircle of radius R. What are the x and y coordinates of the figure's center of mass?

15. Clustered Problem 4 Figure 8-22*d* shows a uniform, solid plate in the shape of a semicircle with radius R. What are the x and y coordinates of the plate's center of mass?

16. Great Pyramid The Great Pyramid of Cheops at El Gizeh, Egypt (Fig. 8-23*a*), had height $H = 147$ m before its topmost stone fell. Its base is a square with edge length $L = 230$ m (see Fig. 8-23*b*). Its volume V is equal to $L^2H/3$. Assuming $\rho = 1.8 \times 10^3$ kg/m^3 is its uniform density, find the original height of its center of mass above the base.

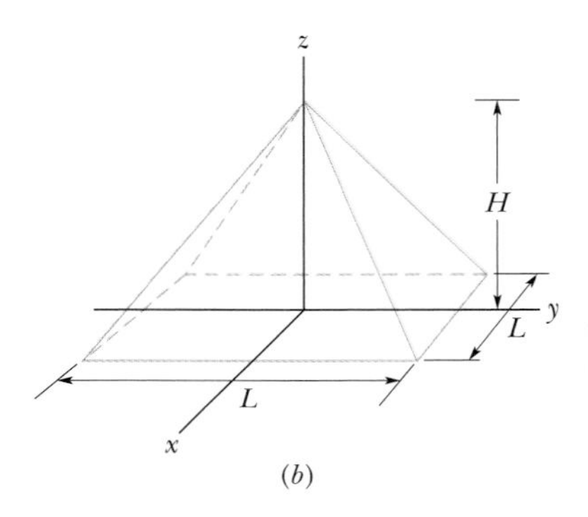

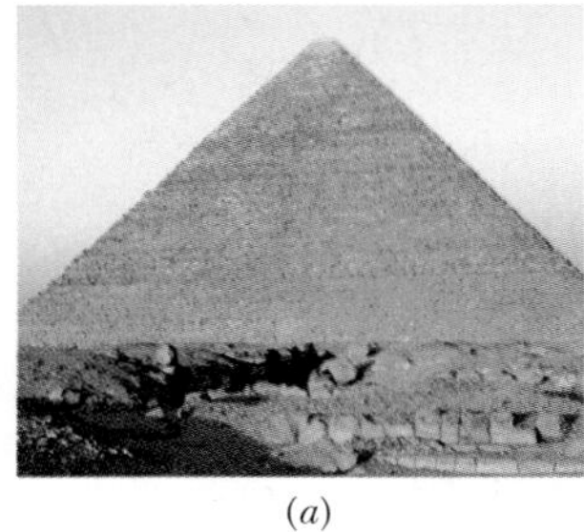

FIGURE 8-23 ■ Problem 16.

17. Four Particles At a certain instant, four particles have the xy coordinates and velocities given in the following table. At that instant, what are (a) the coordinates of their center of mass and (b) the velocity of their center of mass?

Particle	Mass (kg)	Position (m)	Velocity (m/s)
1	2.0	0, 3.0	-9.0 m/s $\hat{j}$
2	4.0	3.0, 0	6.0 m/s $\hat{i}$
3	3.0	0, -2.0	6.0 m/s $\hat{j}$
4	12	-1.0, 0	-2.0 m/s $\hat{i}$

18. Inverse Ratios Show that the ratio of the distances of two particles from their center of mass is the inverse ratio of their masses.

19. *xy* Coordinates A 2.00 kg particle has the xy coordinates (-1.20 m, 0.500 m) and a 4.00 kg particle has the xy coordinates (0.600 m, -0.750 m). Both lie on a horizontal plane. At what xy coordinates must you place a 3.00 kg particle such that the center of mass of the three-particle system has the coordinates (-0.500 m, -0.700 m)?

20. Uniform Plate What are (a) the x coordinate and (b) the y coordinate of the center of mass for the uniform plate shown in Fig. 8-24?

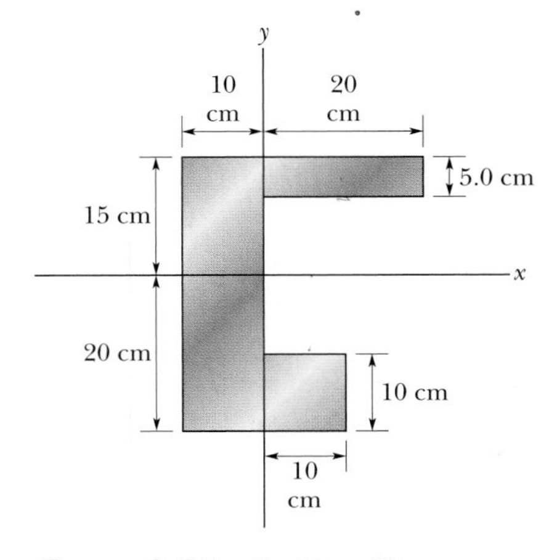

FIGURE 8-24 ■ Problem 20.

SEC. 8-6 ■ MOMENTUM OF A PARTICLE SYSTEM

21. Peanut Butter and Jelly At $t_1 = 0$, a 1.0 kg jelly jar is projected vertically upward from the base of a 50-m-tall building with an initial velocity of 40 m/s. At the same instant and directly overhead, a 2.0 kg peanut butter jar is dropped from rest from the top of the building. How far above ground level is the center of mass of the two-jar system at $t_2 = 3.0$ s?

22. Two Skaters With Pole Two skaters, one with mass 65 kg and the other with mass 40 kg, stand on an ice rink holding a pole of length 10 m and negligible mass. Starting from the ends of the pole, the skaters pull themselves along the pole until they meet. How far does the 40 kg skater move?

23. Old Chrysler An old Chrysler with mass 2400 kg is moving along a straight stretch of road at 80 km/h. It is followed by a Ford with mass 1600 kg moving at 60 km/h. How fast is the center of mass of the two cars moving?

FIGURE 8-25 ■ Problem 24.

24. Ladder on a Balloon A man of mass m clings to a rope ladder suspended below a balloon of mass M; see Fig. 8-25. The balloon is stationary with respect to the ground. (a) If the man begins to climb the ladder at speed v (with respect to the ladder), in what direction and with what speed (with respect to the ground) will the balloon move? (b) What is the state of the motion after the man stops climbing?

25. A Stone is Dropped A stone is dropped at $t_1 = 0$. A second stone, with twice the mass of the first, is dropped from the same point at $t_2 = 100$ ms. (a) How far below the release point is the center of mass of the two stones at $t_3 = 300$ ms? (Neither stone has yet reached the ground.) (b) How fast is the center of mass of the two-stone system moving at that time?

26. Traffic Signal A 1000 kg automobile is at rest at a traffic signal. At the instant the light turns green, the automobile starts to move with a constant acceleration of 4.0 m/s^2. At the same instant a 2000 kg truck, traveling at a constant speed of 8.0 m/s, overtakes and passes the automobile. (a) How far is the center of mass of the automobile–truck system from the traffic light at $t_2 = 3.0$ s? (b) What is the speed of the center of mass of the automobile–truck system then?

27. Shell Explodes A shell is shot with an initial velocity $\vec{v}_1$ of 20 m/s, at an angle of 60° with the horizontal. At the top of the trajectory, the shell explodes into two fragments of equal mass (Fig. 8-26). One fragment, whose speed immediately after the explosion is zero, falls vertically. How far from the gun does the other fragment land, assuming that the terrain is level and that air drag is negligible?

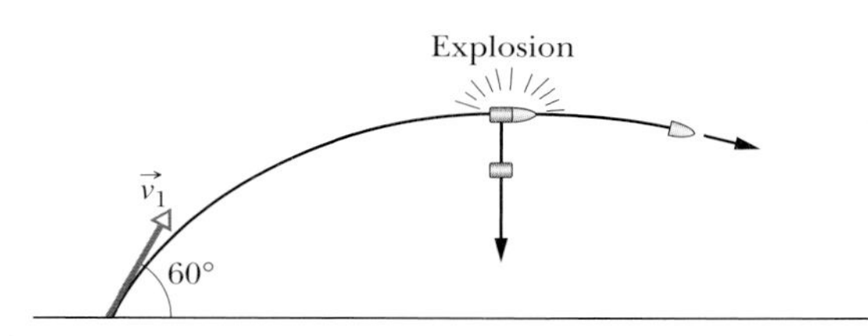

FIGURE 8-26 ■ Problem 27.

28. Big Olive A big olive ($m = 0.50$ kg) lies at the origin and a big Brazil nut ($M = 1.5$ kg) lies at the point (1.0, 2.0) m in an xy plane. At $t_1 = 0$, a force $\vec{F}_o = (2.0\text{ N})\hat{\text{i}} + (3.0\text{ N})\hat{\text{j}}$ begins to act on the olive, and a force $\vec{F}_n = (-3.0\text{ N})\hat{\text{i}} + (-2.0\text{ N})\hat{\text{j}}$ begins to act on the nut. In unit-vector notation, what is the displacement of the center of mass of the olive–nut system at $t_2 = 4.0$ s, with respect to its position at $t_1 = 0$?

29. Sugar Containers Two identical containers of sugar are connected by a massless cord that passes over a massless, frictionless pulley with a diameter of 50 mm (Fig. 8-27). The two containers are at the same level. Each originally has a mass of 500 g. (a) What is the horizontal position of their center of mass? (b) Now 20 g of sugar is transferred from one container to the other, but the containers are prevented from moving. What is the new horizontal position of their center of mass, relative to the central axis through the lighter container? (c) The two containers are now released. In what direction does the center of mass move? (d) What is its acceleration?

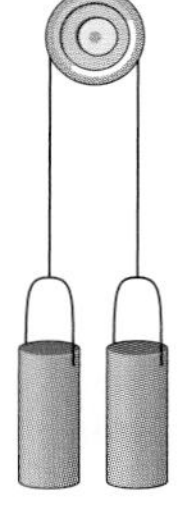

FIGURE 8-27 ■ Problem 29.

30. Ricardo and Carmelita Ricardo, of mass 80 kg, and Carmelita, who is lighter, are enjoying Lake Merced at dusk in a 30 kg canoe. When the canoe is at rest in the placid water, they exchange seats, which are 3.0 m apart and symmetrically located with respect to the canoe's center. Ricardo notices that the canoe moves 40 cm relative to a submerged log during the exchange and calculates Carmelita's mass, which she has not told him. What is it?

31. Dog in a Boat In Fig. 8-28a, a 4.5 kg dog stands on an 18 kg flatboat and is 6.1 m from the shore. He walks 2.4 m along the boat toward shore and then stops. Assuming there is no friction between the boat and the water, find how far the dog is then from the shore. (*Hint*: See Fig. 8-28b. The dog moves leftward and the boat moves rightward, but does the center of mass of the *boat* + *dog* system move?)

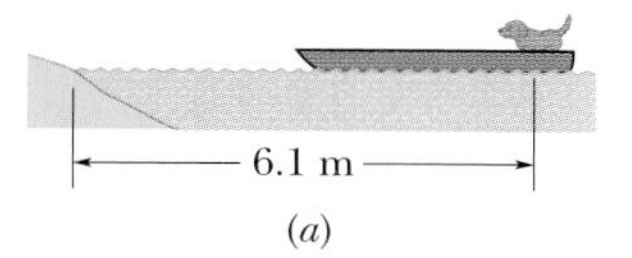

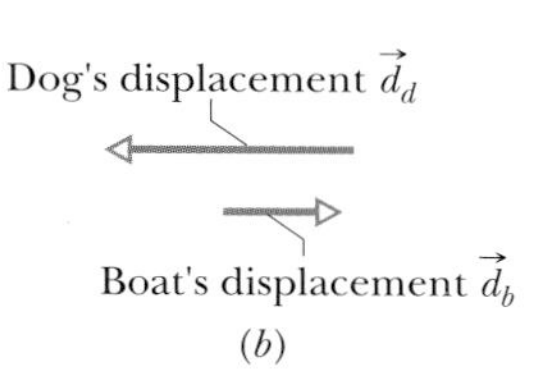

FIGURE 8-28 ■ Problem 31.

32. A Certain Nucleus A certain nucleus, at rest, transforms into three particles. Two of them are detected; their masses and velocities are as shown in Fig. 8-29. In unit-vector notation, what is the translational momentum of the third particle, with a mass of 11.7×10^{-27} kg?

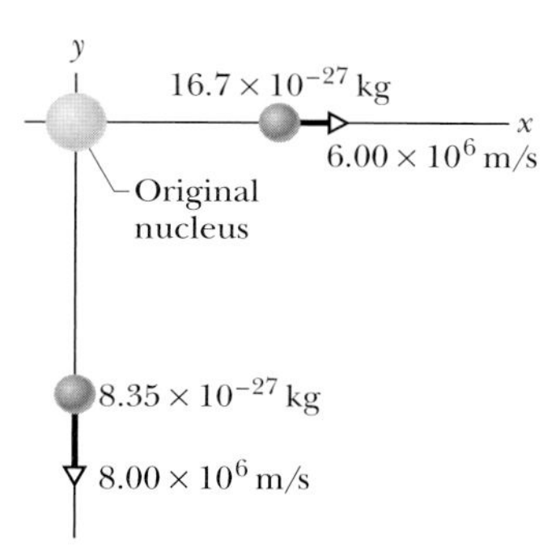

FIGURE 8-29 ■ Problem 32.

33. Father and Child A 40 kg child and her 75 kg father simultaneously dive from a 100 kg boat that is initially motionless. The child dives horizontally toward the east with a speed of 2.0 m/s, and the father dives toward the south with a speed of 1.5 m/s at an angle of 37° above the horizontal. (Assume the boat's vertical motion due to the father's dive does not alter its horizontal motion.) Determine the magnitude and direction of the velocity of the boat along the water's surface immediately after their dives.

34. Sumo Wrestler A 2140 kg railroad flatcar, which can move with negligible friction, is motionless next to a platform. A 242 kg sumo wrestler runs at 5.3 m/s along the platform (parallel to the track) and then jumps onto the flatcar. What is the speed of the flatcar if he then (a) stands on it, (b) runs at 5.3 m/s relative to the flatcar in his original direction, and (c) turns and runs at 5.3 m/s relative to the flatcar opposite his original direction?

35. Block Released from Rest A 2.00 kg block is released from rest over the side of a very tall building at time $t_1 = 0$. At time $t_2 = 1.00$ s, a 3.00 kg block is released from rest at the same point. The first block hits the ground at $t_3 = 5.00$ s. Plot, for the time interval $t_1 = 0$ to $t_4 = 6.00$ s, (a) the position and (b) the speed of the center of mass of the two-block system. Take $y = 0$ at the release point.

36. Speed of COM At the instant a 3.0 kg particle has a velocity of 6.0 m/s in the negative y direction, a 4.0 kg particle has a velocity of 7.0 m/s in the positive x direction. What is the speed of the center of mass of the two-particle system?

37. Car and Truck A 1500 kg car and a 4000 kg truck are moving north and east, respectively, with constant velocities. The center of mass of the car–truck system has a velocity of 11 m/s in a direction 55° north of east. (a) What is the magnitude of the car's velocity? (b) What is the magnitude of the truck's velocity?

38. Cannon in a Flatcar A cannon and a supply of cannonballs are inside a sealed railroad car of length L, as in Fig.8-30. The cannon fires to the right, the car recoils to the left. Fired cannonballs travel a horizontal distance L and remain in the car after hitting the far wall and landing on the floor there. (a) After all the cannonballs have been fired, what is the greatest distance the car could have moved from its original position? (b) What is the speed of the car just after the last cannonball has completed its motion?

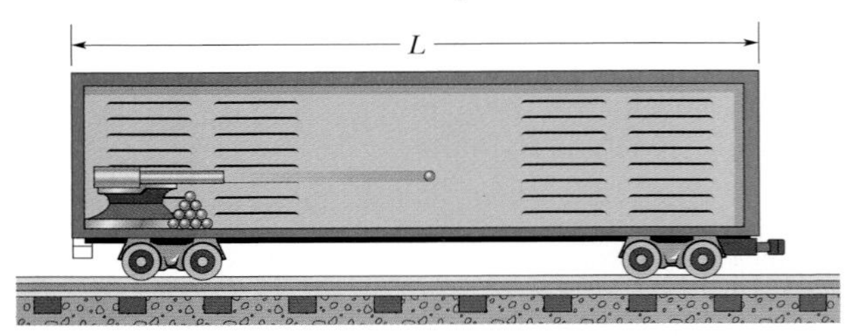

FIGURE 8-30 ■ Problem 38.

39. Cannon Tilted Up A 1400 kg cannon, which fires a 70.0 kg shell with a speed of 556 m/s relative to the muzzle, is set at an elevation angle of 39.0° above the horizontal. The cannon is mounted on frictionless rails so that it can recoil freely. (a) At what speed relative to the ground is the shell fired? (b) At what angle with the ground is the shell fired? (*Hint*: The horizontal component of the momentum of the system remains unchanged as the cannon is fired.)

40. Table of Three The following table gives the masses of three objects and, at a certain instant, the coordinates (x, y) and the velocities of the objects. At that instant, what are the (a) position and (b) velocity of the center of mass of the three-particle system, and (c) what is the net translational momentum of the system?

Object	Mass (kg)	Coordinates (m)	Velocity (m/s)
1	4.00	(0.00, 0.00)	$(1.50\ \text{m/s})\hat{\text{i}} - (2.50\text{m/s})\hat{\text{j}}$
2	3.00	(7.00, 3.00)	0.00
3	5.00	(3.00, 2.00)	$(2.00\ \text{m/s})\hat{\text{i}} - (1.00\text{m/s})\hat{\text{j}}$

Additional Problems

41. Iceboat You are on an iceboat on frictionless, flat ice; you and the boat have a combined mass M. Along with you are two stones with masses m_A and m_B such that $M = 6.00m_A = 12.0m_B$. To get the boat moving, you throw the stones rearward, either in succession or together, but in each case with a certain speed v^{rel} relative to the boat after the stone is thrown. What is the resulting speed of the boat if you throw the stones (a) simultaneously, (b) in the order m_A and then m_B, and (c) in the order m_B and then m_A?

42. *P* and *Q* Two particles P and Q are initially at rest 1.0 m apart. P has a mass of 0.10 kg and Q a mass of 0.30 kg. P and Q attract each other with a constant force of 1.0×10^{-2} N. No external forces act on the system. (a) Describe the motion of the center of mass. (b) At what distance from P's original position do the particles collide?

43. Suspicious Package A suspicious package is sliding on a frictionless surface when it explodes into three pieces of equal masses and with the velocities (1) 7.0 m/s, north, (2) 4.0 m/s, 30° south of west, and (3) 4.0 m/s, 30° south of east. (a) What is the velocity (magnitude and direction) of the package before it explodes? (b) What is the displacement of the center of mass of the three-piece system (with respect to the point where the explosion occurs) 3.0 s after the explosion?

44. Mass on an Air Track Figure 8-31 shows an arrangement with an air track, in which a cart is connected by a cord to a hanging block. The cart has mass $m_A = 0.600$ kg and its center is initially at xy coordinates $(-0.500\ \text{m}, 0.000\ \text{m})$; the block has mass $m_B = 0.400$ kg and its center is initially at xy coordinates $(0, -0.100\ \text{m})$. The mass of the cord and pulley are negligible. The cart is released from rest, and both cart and block move until the cart hits the pulley. The friction between the cart and the air track and between the pulley and its axle is negligible. (a) In unit-vector notation, what is the acceleration of the center of mass of the cart–block system? (b) What is the velocity of the center of mass as a function of time t? (c) Sketch the path taken by the system's center of mass. (d) If the path is curved, does it bulge upward to the right or downward to the left? If, instead, it is straight, give the angle between it and the x axis.

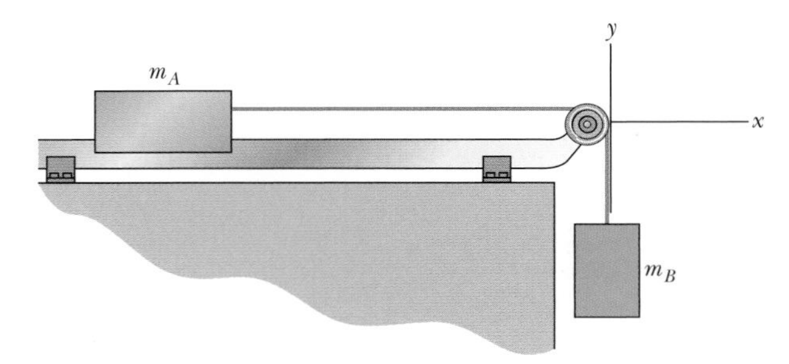

FIGURE 8-31 ■ Problem 44.

45. *Left Alone, Write Your Own* For one or more of the following situations, write a problem involving physics in this chapter, using the style of the Touchstone Examples and providing realistic data, graphs of the variables, and explained solutions: (a) determining the center of mass of a large object, (b) a system separated into parts by an internal explosion, (c) someone climbing or descending a structure, (d) track and field events.

46. Car on a Boat The script for an action movie calls for a small race car (of mass 1500 kg and length 3.0 m) to accelerate along a flat-top boat (of mass 4000 kg and length 14 m), from one end to the other. The car will then jump the gap between the boat and a somewhat lower dock. You are the technical advisor for the movie. The boat will initially touch the dock as shown in Fig. 8-32. Assume the boat can slide through the water without significant resistance, and that both the car and the boat can be approximated as uniform in their mass distribution. Determine what the width of the gap will be just as the car is about to make the jump.

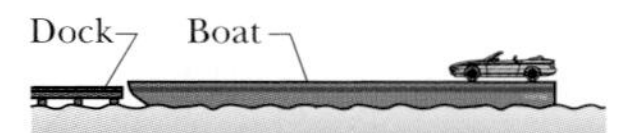

FIGURE 8-32 ■ Problem 46.

47. Two Carts with Unequal Masses Suppose you examine a digital movie of two carts with different masses that undergo a collision (for example, PASCO020 in VideoPoint). You will find there is a point between the two carts that moves at the same constant velocity both before and after the collision. We call this special point the center of mass of the two-cart system. In the PASCO020 movie, where one cart has twice the mass of the other, analysis of the video indicates that the center of mass is one-third of the distance between the two carts (measured relative to the more massive cart).

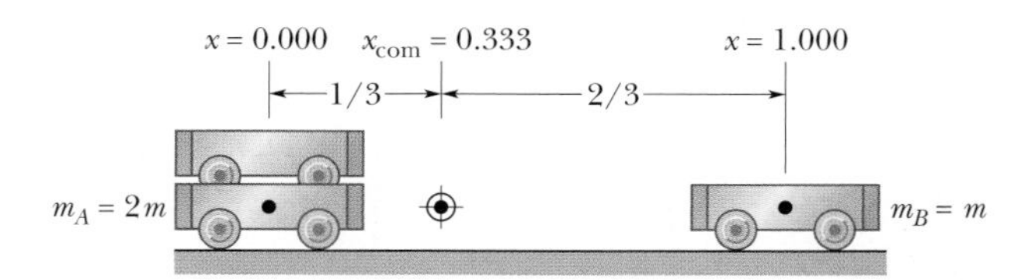

FIGURE 8-33 ■ Problem 47.

A similar situation is depicted in Fig. 8-33. The figure shows a moment in time when the cart *centers* just happen to be 1.000 m apart. For the situation in Fig. 8-33, show that the equation

$$x_{com} = \frac{m_A x_A + m_B x_B}{m_A + m_B}$$

gives a center of mass for these two carts that is one-third of the distance between them (measured from the more massive cart).

9 Kinetic Energy and Work

In the weight-lifting competition of the 1996 Olympics, Andrey Chemerkin lifted a record-breaking 260.0 kg from the floor to over his head (about 2 m). In 1957, Paul Anderson stooped beneath a reinforced wood platform, placed his hands on a short stool to brace himself, and then pushed upward on the platform with his back, lifting the platform and its load about a centimeter. On the platform were auto parts and a safe filled with lead. The composite weight of the load was 27 900 N (6270 lb)!

Who did more work on the objects he lifted—Chemerkin or Anderson?

The answer is in this chapter.

9-1 Introduction

A ski jumper who wants to understand her motion along a curved track is presented with a special challenge. If she tries to use Newton's laws of motion to predict her speed at each location along that track, she has to account for the fact that the net force and her acceleration keep changing as the slope of the track changes. The goal of this chapter is to devise a way to simplify the analysis of motions like those of our ski jumper shown in Fig. 9-1.

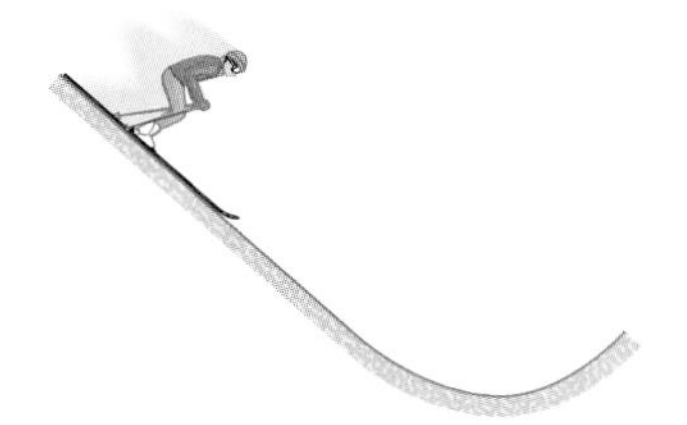

FIGURE 9-1 ■ It is difficult to use Newton's Second Law to analyze the motion of a ski jumper traveling down a curved ramp and predict her velocity at the bottom of the ramp.

We can begin by drawing on ideas presented in Chapter 7. There we introduce two new concepts—*momentum* and *impulse*—and use them to derive an alternate form of Newton's Second Law known as the *impulse–momentum theorem*. This theorem, expressed in Eq. 7-10, tells us that the impulse $\vec{J}$ on a moving particle is equal to its momentum change.

$$\vec{J} = \int_{t_1}^{t_2} \vec{F}^{\,\text{net}}(t)\,dt = m\vec{v}_2 - m\vec{v}_1 \qquad \text{(impulse–momentum theorem).} \qquad (7\text{-}10)$$

One of the most useful aspects of the impulse–momentum theorem is that we can use it without having to keep track of the particle's position.

Can we derive another alternate form of Newton's law to relate a particle's velocity and position changes without keeping track of time? In this chapter we simplify the analysis of complex motions like that of the skier by proving an analogous theorem called the *net work–kinetic energy theorem.* But, in order to "derive" our new theorem we introduce two additional concepts—*work* and *kinetic energy*.

We will start our development of the new theorem by introducing the concept of work, W, in analogy to the impulse represented by the integral in Eq. 7-10. Initially we consider a very simple situation in which a net force acts along the line of motion of a particle. In this case the concept of work as a one-dimensional analogy to impulse involving position changes rather than time changes would be

$$W = \int_{x_1}^{x_2} F_x^{\text{net}}\,dx \qquad \text{(one-dimensional position analogy to impulse).} \qquad (9\text{-}1)$$

Here F_x^{net} and dx are components of force and infinitesimal position change vectors along an x axis, x_1 is the initial position of the particle and x_2 is its position at a later time. In order for the integral to be unique and well defined we will also add the requirement that the component of the force is either constant or only varies with x. That is, we will consider the integral

$$\int_{x_1}^{x_2} F_x^{\text{net}}(x)\,dx$$

where the x in parentheses signifies the force on the particle varies with location along the x axis. After developing the concept of work we introduce the concept of kinetic energy and then derive the *net work–kinetic energy theorem* for one-dimensional motions. Next we apply the concept of work and the new theorem to the analysis of some motions that result from the actions of common one-dimensional forces.

Although we begin with one-dimensional situations, we will extend the equations we derive to two (and three) dimensions. As part of this process we will also introduce a method for finding a scalar product of two vectors. Then in Section 9-9 toward the end of the chapter, we demonstrate how the two-dimensional form of our new net work–kinetic energy theorem enables us to determine the speed of the skier as a function of her location along a frictionless ramp in a very simple manner.

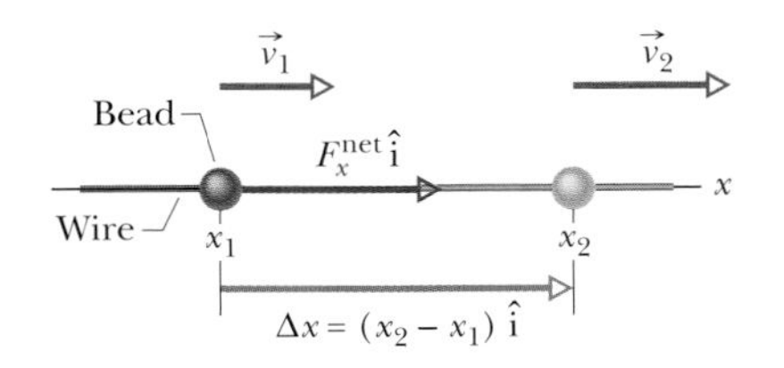

FIGURE 9-2 ■ A simple situation showing a bead on a wire that experiences a net force that is directed along the wire. The bead moves in the same direction as the force.

9-2 Introduction to Work and Kinetic Energy

One-Dimensional Relationship for a Net Force

In order to find an alternative form of Newton's Second Law that relates position and velocity, suppose our particle is a bead moving along a wire, as in Fig. 9-2. Suppose the net force on the bead is a combination of an applied force and a friction force that acts as the bead slides along. We know that if a net force acts on the bead along its direction of motion, its position will change in the direction of the force and its speed will increase. If the direction of the net force is opposite to that of the bead's motion, the bead's speed will decrease.

In our example, the force, $\vec{F}^{\text{net}} = F_x^{\text{net}}\hat{\mathrm{i}}$, is directed along the wire as shown in Fig. 9-2. This force causes the bead to accelerate in the same direction as the force. We can use Newton's Second Law to relate the force and acceleration components as

$$F_x^{\text{net}}(x) = ma_x, \tag{9-2}$$

where m is the bead's mass. As the bead moves through a displacement $\Delta\vec{x} = (x_2 - x_1)\hat{\mathrm{i}}$, the force changes the bead's velocity from an initial value $\vec{v}_1$ to another value we will call $\vec{v}_2$. Using the definition of acceleration as the rate of velocity change over a short time interval dt, this gives us

$$F_x^{\text{net}}(x) = ma_x = m\frac{dv_x}{dt}.$$

In order to relate the velocity and position, we perform two mathematical operations: First we multiply each term in the equation above by the x-component of velocity, v_x. Second, we use the definition of v_x as dx/dt to substitute for v_x on the left side of our new equation. This gives us

$$F_x^{\text{net}}(x)\,\frac{dx}{dt} = mv_x\frac{dv_x}{dt}. \tag{9-3}$$

Since the bead's mass m is constant, we can use the chain rule of differentiation to see that the term on the right can be rewritten as

$$mv_x\frac{dv_x}{dt} = \frac{d(\frac{1}{2}mv_x^2)}{dt}. \tag{9-4}$$

If we substitute the expression on the right side of Eq. 9-4 for the term $mv_x\,dv_x/dt$ in Eq. 9-3, we get

$$F_x^{\text{net}}(x)\frac{dx}{dt} = \frac{d(\frac{1}{2}mv_x^2)}{dt}.$$

Now we can eliminate dt from Eq. 9-3 by realizing that during the same infinitesimal time interval dt the x-component of force times the infinitesimal change in x is equal to the change in the expression $\frac{1}{2}mv_x^2$. This gives us the following equality between differentials

$$F_x^{\text{net}}(x)\,dx = d(\tfrac{1}{2}mv_x^2). \tag{9-5}$$

Because the net force on the bead and its velocity are not necessarily constant over the full displacement shown in Fig. 9-2, we must integrate both sides of Eq. 9-5 to determine the relationship between position change and velocity change due to our variable force.

$$\int_{x_1}^{x_2} F_x^{\text{net}}(x)\,dx = \int_{v_{1x}}^{v_{2x}} d(\tfrac{1}{2}mv_x^2) = \tfrac{1}{2}mv_{2x}^2 - \tfrac{1}{2}mv_{1x}^2. \tag{9-6}$$

In summary, by using Newton's Second Law, the definitions of velocity and acceleration, and the rules of calculus, we have derived a new form of the second law that relates how a variable force acting over a distance will change the speed of a particle of mass m.

Note that the expression on the left side of this equation,

$$\int_{x_1}^{x_2} F_x^{\text{net}}(x)\,dx,$$

is identical to the expression we developed (in Eq. 9-1) as the one-dimensional (1D) integral of force over position that is analogous to the integral of force over time used in the impulse-momentum theorem.

Initial Definitions of Work and Kinetic Energy

The left and right sides of Eq. 9-6 are the basis for two new and very important definitions. We define the integral as the **net work,** W^{net}, done on a particle as it moves from an initial to a final position, so that

$$W^{\text{net}} \equiv \int_{x_1}^{x_2} F_x^{\text{net}}(x)\,dx \quad \text{(**net work** definition—1D net force and displacement).} \tag{9-7}$$

If the net force component $F_x^{\text{net}}(x)$ along a line is made up of the sum of several force components $F_{Ax}(x) + F_{Bx}(x) + F_{Cx}(x) + \cdots$, we see that the contribution of each force to the net work is

$$W^{\text{net}} \equiv \int_{x_1}^{x_2} F_x^{\text{net}}(x)\,dx = \int_{x_1}^{x_2} (F_{Ax}(x) + F_{Bx}(x) + F_{Cx}(x) + \cdots)\,dx$$
$$= \int_{x_1}^{x_2} F_{Ax}(x)\,dx + \int_{x_1}^{x_2} F_{Bx}(x)\,dx + \int_{x_1}^{x_2} F_{Cx}(x)\,dx + \cdots.$$

If we define the work associated with a single force component $F_{Ax}(x)$ as

$$W_A \equiv \int_{x_1}^{x_2} F_{Ax}(x)\,dx \quad \text{(definition of work—1D **single** force and displacement),} \tag{9-8}$$

we see that the net work is given by

$$W^{\text{net}} = W_A + W_B + W_C + \cdots. \tag{9-9}$$

So there are two ways to calculate the net work. One is to sum the force components before the net work is calculated. The other is to calculate the work associated with each of the components separately. The net work is then determined by adding up the work done by each force component.

The right side of Eq. 9-6 tells us how the speed of a particle of mass m is changed by the net work done on it. This $\frac{1}{2}mv_x^2$ is the factor that is changed by the work done on the bead. Because we are talking about translational motion (as distinct from rotational motion), we call this factor **translational kinetic energy** (or often just "kinetic energy"). In the most general sense, kinetic energy is a quantity associated with motion. However, in this chapter, we limit our discussion to the motion of a single particle, or the center of mass of systems of particles and extended objects. When we turn our attention to the study of thermodynamics later in the book, we will have to revisit the concept of kinetic energy to take into account the fact that particles within objects may well be moving, even if the center of mass is not. So we define the translational kinetic energy, K, of a particle-like object in terms of its mass and the square of the speed of its center of mass as

$$K \equiv \tfrac{1}{2}mv_x^2 \qquad \text{(definition of kinetic energy for 1D motion).} \tag{9-10}$$

According to this definition the term on the right side of Eq. 9-6 given by $\frac{1}{2}mv_{2x}^2 - \frac{1}{2}mv_{1x}^2$ represents the *change* in the object's translational kinetic energy.

Using our new definitions for net work and translational kinetic energy, Eq. 9-6 can be rewritten in streamlined form as

$$W^{\text{net}} = K_2 - K_1 = \Delta K \qquad \text{(the net work-energy theorem).} \tag{9-11}$$

Equation 9-11 is known as the **net work-kinetic energy theorem.** This theorem tells us that when a net force acts along the direction of motion of a particle that moves from one position to another, the particle's kinetic energy changes by $\frac{1}{2}mv_{2x}^2 - \frac{1}{2}mv_{1x}^2$. Equation 9-11 is known as a theorem because Eqs. 9-6 and 9-9 were derived mathematically from Newton's Second Law. It is analogous in many ways to the *impulse-momentum theorem* (also derived from Newton's Second Law) that relates the action of a net force over time to the change in momentum of a particle.

Units of Work and Energy

The SI unit for both work and kinetic energy (and every other type of energy) is the **joule** (J). It is defined directly from $K = \frac{1}{2}mv^2$ (Eq. 9-10) in terms of the units for mass and velocity as $\text{kg} \cdot \text{m}^2/\text{s}^2$. It is easy to show that the units for work, which is the product of a force in newtons (N) and a distance in meters (m), are $\text{N} \cdot \text{m}$, which also reduce to $\text{kg} \cdot \text{m}^2/\text{s}^2$. In summary,

$$1 \text{ joule} = 1 \text{ J} = 1 \text{ kg} \cdot \text{m}^2/\text{s}^2 = 1 \text{ N} \cdot \text{m} \qquad \text{(SI unit for energy).}$$

Other units of energy which you may encounter are the erg (or $\text{g} \cdot \text{cm}^2/\text{s}^2$) and the foot-pound.

Generalizing Work and Kinetic Energy Concepts

So far we have only related position and velocity for the very special case of a net force acting along a line of motion. What if the forces on a particle that make up the net force have components that do not lie along the direction of motion of our body of interest? If forces do not act parallel to a particle's displacement, how can we multiply and then integrate two vectors such as force and displacement to calculate work?

The remainder of this chapter is devoted to understanding how to calculate the work done by forces in some common situations. This will enable us to apply the net work-kinetic energy theorem to relate how changes in a particle's position due to a net force are related to changes in its speed.

READING EXERCISE 9-1: A particle moves along an x axis. Does the kinetic energy of the particle increase, decrease, or remain the same if the particle's velocity changes (a) from -3 m/s to -2 m/s and (b) from -2 m/s to 2 m/s? (c) In each situation, is the net work done on the particle positive, negative, or zero? ■

9-3 The Concept of Physical Work

So far we have only discussed the work done on small particles with no internal structure. Now we would like to apply the concept of work to changes in motion of familiar extended objects. If we push on an object that deforms and changes its shape as it moves, then it's impossible to describe its motion in terms of a single displacement. For this reason, when we apply the concept of work to extended objects, we are assuming that these objects are particle-like as defined in Section 2-1. Thus, when we refer to doing work on an object, we assume the object is rigid enough that the work done distorting it is negligible compared to the work that displaces its center of mass.

Now let's get back to work. In casual conversation, most of us think of work as an expenditure of effort. It takes effort to push a rigid box down the hallway or to lift it. But you also expend effort to hold a heavy object steady in midair or to shove against a massive object that won't budge. If we examine our expression for work (Eq. 9-7), we see that at least for one-dimensional motion, work is given by the product of the components of a force along the line of motion of a particle and the particle's displacement along that line. This means that even though shoving really hard on a massive object that is at rest takes a lot of effort, *no physical work is done on it* unless it starts to move in the direction of the force. As we examine ways to define work in more general situations, we will find that the definition of work in physics requires that

> No work is done on a rigid object by a force unless there is a component of the force along the object's line of motion.

We have defined work in such a way that it requires a force and a displacement. How do we know how much work is done? Let's consider how much effort it takes to push a heavy, very rigid box down a hallway with a steady force. In this special case the force and the displacement of the box are in the same direction. In the next section we will consider what happens when the force acting on an object is in the opposite direction as its displacement.

Suppose you push the box to the right so the x-component of its displacement is $\Delta x = x_2 - x_1$. Now imagine that you use the same steady force to push the same box through twice the distance so its displacement component is $2\Delta x$. How much more effort did this take? How much effort would it take to just watch the box? The answers to these questions give us important insights into the nature of work. Namely, for a given force the magnitude of physical work *should* be proportional to the distance that an object is moved. This is consistent with the way we have defined work—so that it is proportional to the distance an object moves under the influence of forces.

Imagine pushing another larger box through a rightward displacement of Δx using twice the force. How much more effort would you guess it takes to push the larger box than it takes to push the smaller box? You can get a feel for this by pushing one and then two larger textbooks of identical mass across a tabletop as shown in Fig. 9-3.

If you took a moment to do this experiment, you found that it takes about twice the effort to push two books through a displacement of Δx using a force of $2\vec{F}_x$ as it did to push one book through the same displacement with a force of $\vec{F}_x$. We can conclude that the amount (that is, the absolute value) of work done is not just proportional to the distance an object is moved—it is also proportional to the magnitude of the force acting on the object. The concepts of proportionality between displacement force and the amount of physical work done can be summarized by the equation

$$|W| = |F_x \Delta x| \quad \text{(amount of work done by a steady force along a line of motion).}$$

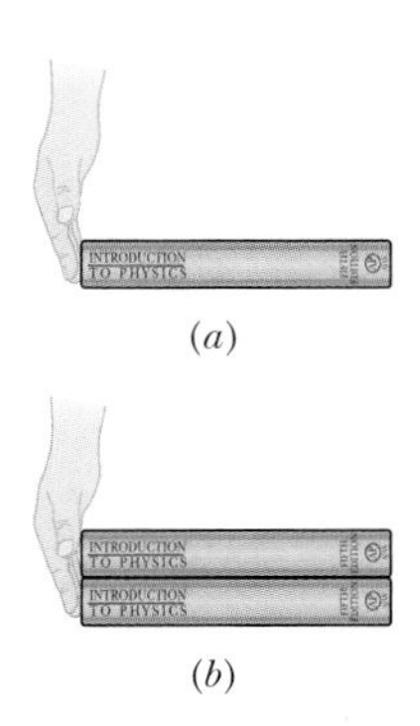

FIGURE 9-3 ■ Pushing one and then two textbooks across a tabletop with a small but steady force.

In the next section we will use the mathematical definition of work to consider how much work is done under the influence of a steady force and also the circumstances under which work is positive or negative.

READING EXERCISE 9-2: The figure shows four situations in which a force acts on a box while the box either slides to the right with a displacement $\Delta\vec{x}$ or doesn't budge as indicated. The force on each box is shown. Rank the situations according to the amount of the work done by the force on the box during its displacement from greatest to least.

$\vec{F}_x = (1\text{ N})\,\hat{\text{i}}$, $\Delta\vec{x} = (1\text{ m})\,\hat{\text{i}}$ (*a*)
$\vec{F}_x = (3\text{ N})\,\hat{\text{i}}$, $\Delta\vec{x} = (0\text{ m})\,\hat{\text{i}}$ (*b*)
$\vec{F}_x = (2\text{ N})\,\hat{\text{i}}$, $\Delta\vec{x} = (1\text{ m})\,\hat{\text{i}}$ (*c*)
$\vec{F}_y = (2\text{ N})\,\hat{\text{j}}$, $\Delta\vec{x} = (2\text{ m})\,\hat{\text{i}}$ (*d*)

■

9-4 Calculating Work for Constant Forces

One-Dimensional Forces and Motions Along the Same Line

Let's start by reconsidering the formal definition of the work associated with one-dimensional motion (presented in Eq. 9-8),

$$W \equiv \int_{x_1}^{x_2} F_x(x)\,dx \quad \text{(definition of work for a 1D motion and force).} \tag{9-12}$$

As we established in Section 9-1, $F_x(x)$ denotes the x-component of a force that can vary with x. However, if the force does not vary with x, then we can simply denote $F_x(x)$ as F_x and take it out of the integral. This allows us to write

$$W \equiv \int_{x_1}^{x_2} F_x\,dx = F_x(x_2 - x_1) = F_x\Delta x,$$

so that

$$W = F_x\,\Delta x \quad \text{(work done by a constant force along a line of motion).} \tag{9-13}$$

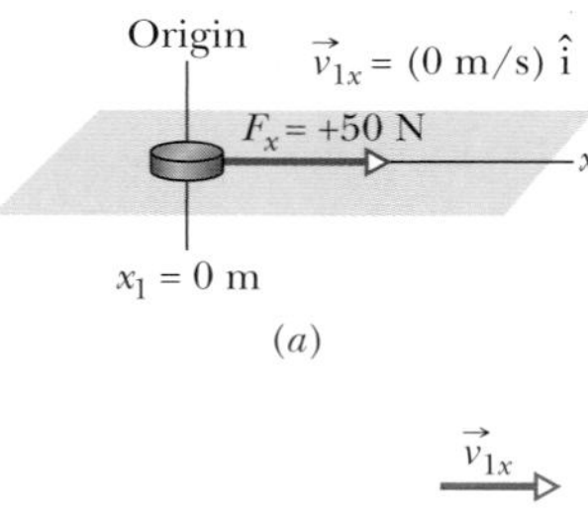

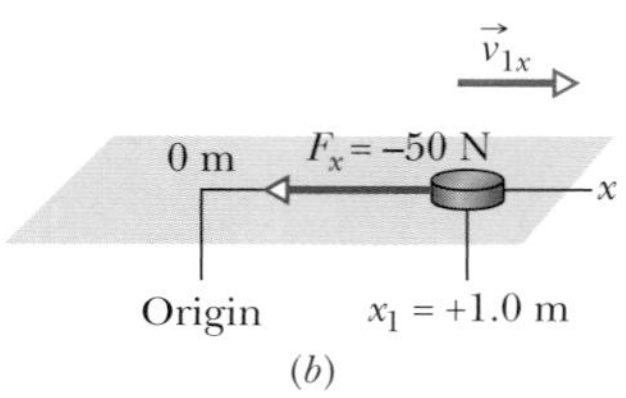

FIGURE 9-4 ■ (*a*) A puck is pushed from rest with a positive force component. (*b*) A puck that is already moving in a positive direction is pushed with a force that has a negative force component along a chosen x axis.

Positive vs. Negative Work At the end of the last section we presented the equation $|W| = |F_x\Delta x|$ to represent the amount of work done on a rigid object. This is an informal expression we developed by imagining the effort needed to slide rigid objects on a table or down a hall. Equation 9-13 that we just derived is very similar except it has no absolute value signs. Since both F_x and Δx represent components of vectors along an axis, either component can be positive or negative. If this is the case, then the sign of the work, W, calculated as the products of these components can also be positive and negative. This raises some questions. How can we tell when the work done by a force will be positive? Negative? Does this mean that work is a vector component? To answer these questions let's consider the work done on a puck that is free to move along a line on a sheet of ice with no friction forces on it.

Imagine that the puck is initially at rest at your chosen origin. When you push it to the right with a steady horizontal force component of $F_x = +50$ N, it speeds up until its x position is $+1$ m (Fig. 9-4*a*). The work you do on the puck is positive since it is given by

$$W = F_x\Delta x = F_x(x_2 - x_1) = (+50\text{ N})(+1\text{ m} - 0\text{ m}) = +50\text{ J} \quad \text{(speeding up).}$$

Suppose that now as the puck is moving away from the origin along your chosen positive x axis, you suddenly take your other hand and push on it in the opposite direction with a steady horizontal force component of $F_x = -50$ N. Since you are pushing the puck in a direction opposite to its motion, it starts slowing down and reaches a zero velocity at a distance of 2.0 m from the origin (Fig. 9-4*b*). In this case the work you do on the puck while slowing it down is negative since it is given by

$$W = F_x \Delta x = F_x(x_2 - x_1) = (-50\text{ N})(+2\text{ m} - 1\text{ m}) = -50\text{ J} \quad \text{(slowing down)}.$$

If we consider many similar situations with different types of forces acting, we can make the following general statement about the sign of the work done by a single force or net force acting on the center of mass of a rigid object:

POSITIVE VS. NEGATIVE WORK: If a single (or net) force has a component that acts in the direction of the object's displacement, the work that the force does is *positive*. If a single (or net) force has a component that acts in a direction opposite to the object's displacement, the work it does is *negative*.

Work Is a Scalar Quantity Recall that in Section 2-2 we defined a scalar (unlike the component of a vector) to be a quantity that is independent of coordinate systems. Based on this definition, we can see that work is a scalar quantity even though it can be positive or negative. If we rotate our x axis by 180° so that all the components of force and displacement in our puck example above change sign, the sign of the work (which is the product of components) would not change sign. So even though there are directions associated with both force and displacement, there is no direction associated with the positive or negative work done by a force on an object. Therefore, *work is a scalar quantity.*

Work Done by a Gravitational Force

We next examine the work done on an object by a particular type of constant force—namely, the gravitational force. Suppose a particle-like object of mass m, such as a tomato, is thrown upward with initial speed of $v_{1\,y}$ as in Fig. 9–5*a*. As it rises, it is slowed by a gravitational force $\vec{F}^{\text{grav}}$ that acts downward in the direction opposite the tomato's motion. We expect that $\vec{F}^{\text{grav}}$ does negative work on the tomato as it rises because the force is in the direction opposite the motion.

To verify this, let's choose our y axis to be positive in the upward direction, so that the y-component of the gravitational force (acts downward) is given by $F_y^{\text{grav}} = -mg$. We calculate the work done on the tomato as $W^{\text{grav}} = F_y^{\text{grav}} \Delta y$ where the y-component of displacement is given by $\Delta y = y_2 - y_1$.

Since during the rise y_2 is greater than y_1, Δy is positive. The gravitational force component is negative and so we can write

$$W^{\text{grav}} = -mg(y_2 - y_1) = -mg\,|\Delta y| \quad \text{(rising object)}. \tag{9-14}$$

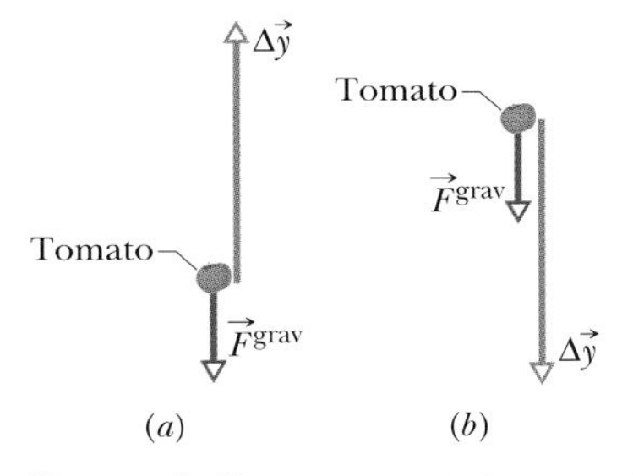

FIGURE 9-5 ■ If the only force acting on a tomato is gravitational: (*a*) As the tomato rises, the gravitational force does negative work on the object. (*b*) As the tomato falls downward, the gravitational force does positive work on it.

After the object has reached its maximum height it begins falling back down. We expect the work done by the gravitational force to be positive in this case because the force and motion are in the same direction. Here y_2 is less than y_1 (Fig. 9-5*b*). Hence both the y-component of the gravitational force F_y^{grav} and displacement $\Delta y = y_2 - y_1$ are negative. This gives us an expression for positive work of

$$W^{\text{grav}} = mg\,|\Delta y| \quad \text{(falling object)}. \tag{9-15}$$

Thus, as we saw for the puck on the ice, the work done by the gravitational force is positive when the force and the displacement of the tomato are in the same direction and the work done by the force is negative when the force and displacement are in opposite directions.

TOUCHSTONE EXAMPLE 9-1: Crepe Crate

During a storm, a crate of crepe is sliding across a slick, oily parking lot through a displacement $\Delta\vec{x} = (-3.0\text{ m})\hat{\imath}$ while a steady wind pushes against the crate with a force $\vec{F} = (2.0\text{ N})\hat{\imath}$. The situation and coordinate axes are shown in Fig. 9-6.

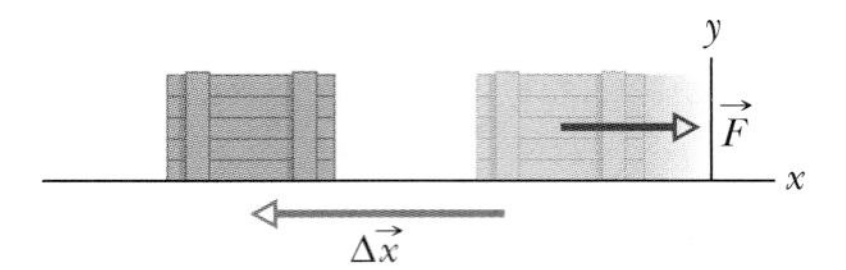

FIGURE 9-6 ■ A constant force $\vec{F}$ created by the wind slows a crate down as it undergoes a displacement $\Delta\vec{x}$.

(a) How much work does this force from the wind do on the crate during the displacement?

SOLUTION ■ The **Key Idea** here is that, because we can treat the crate as a particle and because the wind force is constant ("steady") in both magnitude and direction during the displacement, we can use Eq. 9-13 ($W = F_x\,\Delta x$) to calculate the work,

$$\begin{aligned} W &= F_x\,\Delta x \\ &= (2.0\text{ N})(-3.0\text{ m}) \qquad \text{(Answer)} \\ &= -6.0\text{ J}. \end{aligned}$$

So, the wind's force does negative 6.0 J of work on the crate.

(b) If the crate has a kinetic energy of 10 J at the beginning of displacement $\Delta\vec{x}$, what is its kinetic energy at the end of $\Delta\vec{x}$ assuming $\vec{F} = \vec{F}^{\text{net}}$?

SOLUTION ■ The **Key Idea** here is that, because the force does negative work on the crate, it reduces the crate's kinetic energy. Using the work-kinetic energy theorem in the form of Eq. 9-11, we have

$$K_2 = K_1 + W^{\text{net}} = 10\text{ J} + (-6.0\text{ J}) = 4.0\text{ J}. \qquad \text{(Answer)}$$

Because the kinetic energy is decreased to 4.0 J, the crate has been slowed.

9-5 Work Done by a Spring Force

So far, we have limited our discussion to the work done by *constant* forces that do not change with position or time. Our goal in this section and the next is to explore how to calculate the work done by variable forces. A very common one-dimensional variable force is the spring force. The spring force is of great interest because many forces in our natural and man-made surroundings have the same mathematical form as the spring force. Examples include the interaction between atoms bound in a solid, the flexing of a bridge under the weight of vehicles, and the sway of a building during an earthquake—as long as displacements remain small. Thus, by examining this one idealized force, you can gain an understanding of many phenomena.

As you may have experienced, the magnitude of the force exerted by a spring increases when it is stretched or compressed more. Figure 9-7*a* shows a spring in its **relaxed state**—that is, neither compressed nor extended. One end is fixed, and a rigid block is attached to the other, free end. If we stretch the spring by pulling the block to the right, as in Fig. 9-7*b*, the spring pulls back on the block toward the left. (Because a spring's force acts to restore the relaxed state, it is sometimes said to be a *restoring force*.) If we compress the spring by pushing the block to the left, as in Fig. 9-7*c*, the spring now pushes on the block back toward the right.

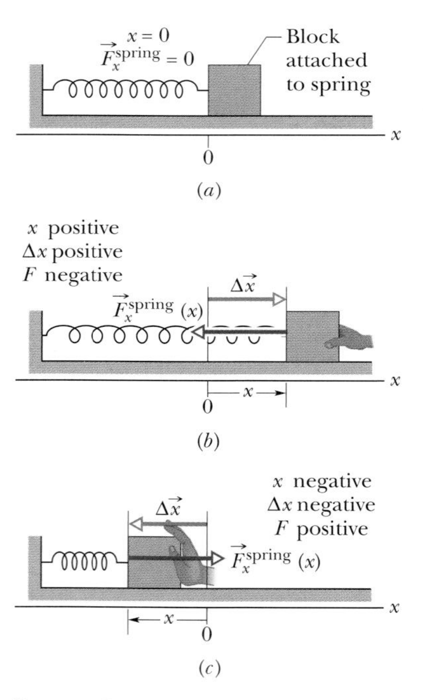

FIGURE 9-7 ■ One end of a spring is attached to a fixed wall and the other end to a block that is free to slide. (*a*) The origin of an x axis is located at the point where the relaxed spring is connected to the block. (*b*) The block and spring are given a positive displacement. Note the direction of the restoring force $\vec{F}_x^{\text{spring}}(x)$ exerted by the spring. (*c*) The spring is compressed by a negative amount x. Again, note the direction of the restoring force.

To a good approximation for many springs, the force $\vec{F}_x^{\text{spring}}$ exerted by it is proportional to the displacement $\Delta\vec{x}$ of the free end from its relaxed position. As usual, the fact that $\vec{F}_x^{\text{spring}}$ depends on x is symbolized by writing it as a function of x, $\vec{F}_x^{\text{spring}}(x)$. If its displacement is not too large, many spring-like objects have a *spring force* given by

$$\vec{F}_x^{\text{spring}}(x) = -k\,\Delta\vec{x} \qquad \text{(Hooke's law for a 1D ideal spring).} \qquad (9\text{-}16)$$

This "law" is named after Robert Hooke, an English scientist of the late 1600s. Since Hooke's law is based on the measured behavior of specific objects, it does not have the same status as Newton's laws.

The minus sign in $\vec{F}_x^{\text{spring}}(x) = -k\,\Delta\vec{x}$ (Eq. 9-16) indicates that the spring force is always opposite in direction from the displacement of the free end so the force is "restoring." The constant of proportionality k is called the **spring constant.** It is always positive and is a measure of the stiffness of the spring. The larger k is, the stiffer the spring—that is, the stronger will be its pull or push for a given displacement. The SI unit for k is the N/m.

In Fig. 9-7, an x axis has been placed parallel to the length of a spring, with the origin ($x = 0$) at the position of the free end when the spring is in its relaxed state. For this coordinate system and arrangement, we can write $\vec{F}_x^{\text{spring}}(x) = -k\Delta\vec{x}$ in component form as

$$F_x^{\text{spring}}(x) = -kx \qquad \text{(Hooke's law for } x = 0 \text{ in the relaxed state).} \qquad (9\text{-}17)$$

The equation correctly describes ideal spring behavior. It tells us that if x is positive (the spring is stretched toward the right on the x axis), then the component $F_x^{\text{spring}}(x)$ is negative (it is a pull toward the left). If x is negative (the spring is compressed toward the left), then the component $F_x^{\text{spring}}(x)$ is positive (it is a push toward the right). Also note that Hooke's law gives us a *linear* relationship between F_x and x.

Work Done by a Spring Force

In the situation shown in Fig. 9-7, the spring force components and displacements lie along the same line, so we can substitute the spring force in the more general expression presented in Eq. 9-8 to determine the work done by a one-dimensional variable force. We get

$$W^{\text{spring}} \equiv \int_{x_1}^{x_2} F_x^{\text{spring}}(x)\,dx. \qquad \text{(Eq. 9-8)}$$

To apply this equation to the work done by the spring force as the block in Fig. 9-7*a* moves, let us make two simplifying assumptions about the spring and block. (1) The spring is *massless*; that is, its mass is negligible compared to the block's mass. (2) The spring is *ideal* so it obeys Hooke's law exactly. Making these simplifying assumptions might seem to make the results unreal. But for many interesting cases, these simplifications give us results that agree fairly well with experimental findings.

Back to the integral. We use Hooke's law (Eq. 9-16) to substitute $-kx$ for the component $F_x^{\text{spring}}(x)$. We also pull k out of the integral since it is constant. Thus, we get

$$\begin{aligned} W^{\text{spring}} &= \int_{x_1}^{x_2} (-kx)\,dx = -k\int_{x_1}^{x_2} x\,dx \\ &= (-\tfrac{1}{2}k)[x^2]_{x_1}^{x_2} = -\tfrac{1}{2}k(x_2^2 - x_1^2), \end{aligned} \qquad (9\text{-}18)$$

so the work done on the block by the spring force as the block moves is

$$W^{\text{spring}} = +\tfrac{1}{2}kx_1^2 - \tfrac{1}{2}kx_2^2 \qquad \text{(work by a spring force).} \qquad (9\text{-}19)$$

This work W^{spring}, done by the spring force, can have a positive or negative value, depending on whether the block is moving toward or away from its zero position. This is quite similar to the way the gravitational work done on the tomato in the previous section changes as the tomato rises and falls.

Note that the final position x_2 appears in the *second* term on the right side of Eq. 9-19. Therefore,

The work done by the spring force on the block W^{spring} is positive if the block moves closer to the relaxed position ($x = 0$). The work done by the spring force on the block is negative if the block moves farther away from $x = 0$. It is zero if the block ends up at the same distance from $x = 0$.

READING EXERCISE 9-3: For three situations, the initial and final positions, along the x axis for the block in Fig. 9-7 are, respectively, (a) −3 cm, 2 cm; (b) 2 cm, 3 cm; and (c) −2 cm, 2 cm. In each situation, is the work done by the spring force on the block positive, negative, or zero? ■

TOUCHSTONE EXAMPLE 9-2: Pralines and a Spring

A package of spicy Cajun pralines lies on a frictionless floor, attached to the free end of a spring in the arrangement of Fig. 9-7*a*. An applied force of magnitude $F^{\text{app}} = 4.9$ N would be needed to hold the package stationary at $x_2 = 12$ mm (Fig. 9-7*b*).

(a) How much work does the spring force do on the package if the package is pulled rightward from $x_1 = 0$ to $x_3 = 17$ mm?

SOLUTION ■ A **Key Idea** here is that as the package moves from one position to another, the spring force does work on it as given by Eq. 9-19. We know that the initial position x_1 is 0 and the final position x_3 is 17 mm, but we do not know the spring constant k.

We can probably find k with Eq. 9-16 (Hooke's law), but we need a second **Key Idea** to use it: if the package were held stationary at $x_2 = 12$ mm, the spring force would have to balance the applied force (by Newton's Second Law). Thus, the x-component of the spring force F_x^{spring} would have to be −4.9 N (toward the left in Fig. 9-7*b*), so Eq. 9-16 gives us

$$k = -\left(\frac{F_x^{\text{spring}}}{x_2}\right) = -\left(\frac{-4.9\ \text{N}}{12 \times 10^{-3}\ \text{m}}\right) = +408\ \text{N/m}.$$

Now, with the package at $x_3 = 17$ mm, Eq. 9-19 yields

$$W^{\text{spring}} = -\tfrac{1}{2}kx_3^2 = -\tfrac{1}{2}(408\ \text{N/m})(17 \times 10^{-3}\ \text{m})^2$$
$$= -0.059\ \text{J} = -59\ \text{mJ}. \qquad \text{(Answer)}$$

(b) Next, the package is moved leftward from $x_3 = 17$ mm to $x_4 = -12$ mm. How much work does the spring force do on the package during this displacement? Explain the sign of this work.

SOLUTION ■ The **Key Idea** here is the first one we noted in part (a). Now $x_3 = +17$ mm and $x_4 = -12$ mm where x_3 and x_4 are two positions of the spring relative to its equilibrium position. They do not represent displacements. Eq. 9-19 yields

$$W^{\text{spring}} = \tfrac{1}{2}kx_3^2 - \tfrac{1}{2}kx_4^2 = \tfrac{1}{2}k(x_3^2 - x_4^2)$$
$$= \tfrac{1}{2}(408\ \text{N/m})[(17 \times 10^{-3}\ \text{m})^2 - (-12 \times 10^{-3}\ \text{m})^2]$$
$$= 0.030\ \text{J} = 30\ \text{mJ}. \qquad \text{(Answer)}$$

This work done on the block by the spring force is positive because the block ends up closer to the spring's relaxed position.

TOUCHSTONE EXAMPLE 9-3: Cumin Canister

In Fig. 9-8, a cumin canister of mass $m = 0.40$ kg slides across a horizontal frictionless counter with velocity $\vec{v} = v_x\hat{\text{i}}(-0.50\ \text{m/s})\hat{\text{i}}$. It then runs into and compresses a spring of spring constant $k = 750$ N/m. When the canister is momentarily stopped by the spring, by what amount Δx is the spring compressed?

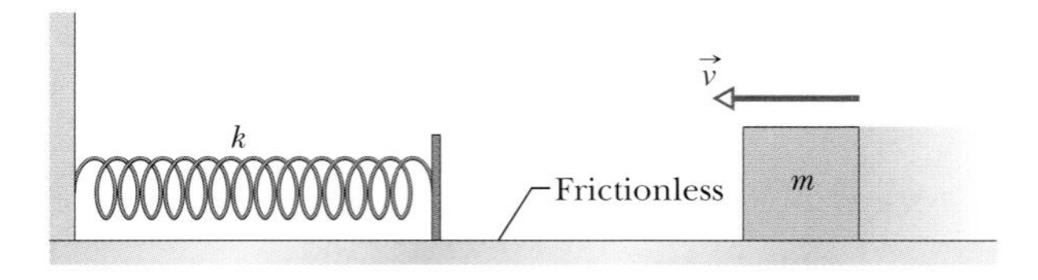

FIGURE 9-8 ■ A canister of mass m moves at velocity $\vec{v}$ toward a spring with spring constant k.

SOLUTION ■ There are three **Key Ideas** here:

1. The work W^{spring} done on the canister by the spring force is related to the requested displacement $\Delta x = x_2 - x_1$ by Eq. 9-19 ($W^{\text{spring}} = \frac{1}{2}kx_1^2 - \frac{1}{2}kx_2^2$).
2. Since $\vec{F}^{\text{spring}} = \vec{F}^{\text{net}}$, the work W^{spring} is also related to the kinetic energy of the canister by Eq. 9-11 ($W^{\text{net}} = K_2 - K_1$).
3. The canister's kinetic energy has an initial value of $K_1 = \frac{1}{2}mv_x^2$ and a value of zero when the canister is momentarily at rest.

Putting the first two of these ideas together, and noting that $x_1 = 0$ here since the spring is initially uncompressed, we write the net work-kinetic energy theorem for the canister as

$$K_2 - K_1 = -\tfrac{1}{2}kx_2^2.$$

Substituting according to the third idea makes this

$$0 - \tfrac{1}{2}mv_x^2 = -\tfrac{1}{2}kx_2^2.$$

Simplifying, solving for x, and substituting known data then give us

$$x_2 = \pm\sqrt{\frac{mv_x^2}{k}} = \pm\sqrt{\frac{(0.40\text{ kg})(-0.50\text{ m/s})^2}{750\text{ N/m}}}$$
$$= \pm 1.2 \times 10^{-2}\text{ m}$$
$$= \pm 1.2\text{ cm}.$$

We reject $x_2 = +1.2$ cm as a solution, since clearly the mass moves to the left as it compresses the spring. So

$$\Delta x = x_2 - x_1$$
$$= -1.2\text{ cm} - 0\text{ cm}$$
$$= -1.2\text{ cm}. \qquad \text{(Answer)}$$

9-6 Work for a One-Dimensional Variable Force—General Considerations

Calculating Work for Well-Behaved Forces

In the previous section, we were able to find the work done by our spring force using calculus to perform the integration called for in Eq. 9-8. This is because the spring force is a "well-behaved," continuous mathematical function that can be integrated. If you know the function $\vec{F}_x(x)$, you can substitute it into Eq. 9-8, introduce the proper limits of integration, carry out the integration, and thus find the work. (Appendix E contains a list of common integrals.) In summary, whenever a one-dimensional variable force is a function that can be integrated using the rules of calculus, the use of Eq. 9-8 is the preferred way to find the work done by the force on an object that moves along that same line.

Calculating Work Using Numerical Integration

Suppose that instead of a spring force, our variable force on an object is caused by someone pushing and pulling erratically on the bead sliding along a wire depicted in Fig. 9-2. In that case, the force will probably not vary with x the way a familiar mathematical function does, so we cannot use the rules of calculus to perform our integration. Whenever this is the case, we can use numerical methods to examine the variable force during small displacements where the force is approximately constant. We can then calculate the work done during each small displacement, and we can add each contribution to the work together to determine the total work. In this situation we are doing a **numerical integration.**

Let's start our exploration of numerical integration by considering the x-component of a one-dimensional force that varies as a particle moves. A general plot of such a *one-dimensional variable force* is shown in Fig. 9-9*a*. One method for finding the work done on the particle is to divide the distance between the initial location of a particle, x_1, and its final location, x_N, into N small steps of width Δx. We can choose a large N so that the values of Δx are small enough so the force component along the x axis $\vec{F}_x(x)$ is reasonably constant over that interval. Let $\langle F_{x\,n}(x)\rangle$ be the component representing the average value of $F_x(x)$ within the nth interval. As shown in Fig. 9-9*b* or *c*, $\langle F_{x\,n}(x)\rangle$ is the height of the nth strip. The value of x for the nth strip is given by $x_n = (n - \frac{1}{2})\,\Delta x$, where $\Delta x = (x_N - x_0)/N$.

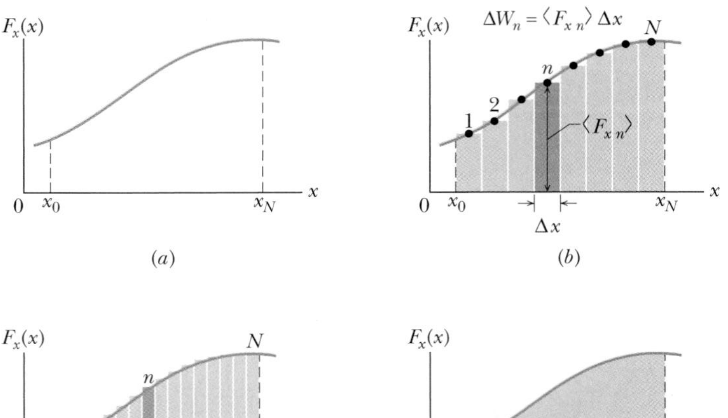

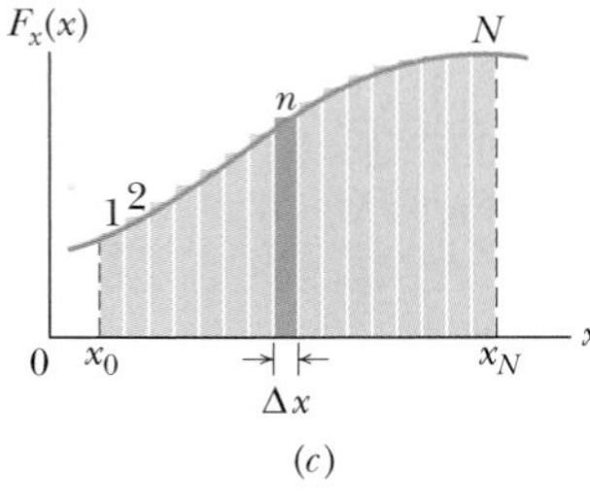

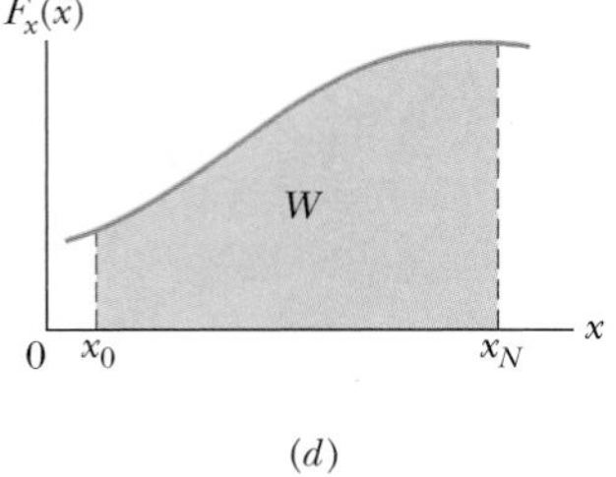

FIGURE 9-9 ■ A particle only moves in one dimension. (*a*) A one-dimensional force component $F_x(x)$ is plotted against the displacement $x_N - x_0$ of the particle. (*b*) Same as (*a*) but with the area under the curve divided into narrow strips. (*c*) Same as (*b*) but with the area divided into narrower strips. (*d*) The limiting case. The work done by the force is given in Eq. 9-8 and is represented by the shaded area between the curve and the x axis and between x_0 and x_N.

With $\langle F_{x\,n} \rangle$ taken to be constant, the small increment of work ΔW_n done by the force in the nth interval is approximately given by Eq. 9-13 as

$$\Delta W_n \cong \langle F_{x\,n} \rangle \Delta x. \tag{9-20}$$

Referring to the most darkly shaded region in Fig. 9-9*b* or *c*, we see that ΔW_n is then equal to the area of the nth rectangular strip.

To approximate the total work W done by the force as the particle moves from x_0 to x_N, we add the areas of all the strips between x_0 and x_N in Fig. 9-9*c*,

$$W \cong \sum_{n=1}^{N} \Delta W_n = \sum_{n=1}^{N} \langle F_{x\,n}(x) \rangle \Delta x. \tag{9-21}$$

This is not an exact calculation of the actual work done because the broken "skyline" formed by the tops of the rectangular strips in Fig. 9-9*b* (representing the values of $\langle F_{x\,n} \rangle$ as constants) only approximates the actual curve of $\vec{F}_x(x)$.

If needed in a particular situation we can make the approximation better by reducing the strip width Δx and using more strips, as in Fig. 9-9*c*. Once the strip width is sufficiently small, Eq. 9-21 can be used to compute the total work done by the variable force.

Defining the Integral

It is interesting to note that in the limit where the strip width approaches zero, the number of strips then becomes infinitely large and we approach an exact result,

$$W = \lim_{\Delta x \to 0} \sum_{n=1}^{N} \langle F_{n\,x}(x) \rangle \Delta x. \tag{9-22}$$

This limit is precisely what we mean by the integral of the function $F_x(x)$ between the limits x_0 and x_N. Thus, Eq. 9-22 becomes

$$W = \int_{x_0}^{x_N} F_x(x)\, dx \quad \text{(work done by a variable force in one dimension).} \qquad \text{(Eq. 9-8)}$$

Geometrically, the work is equal to the area between the $\vec{F}_x(x)$ curve and the x axis, taken between the limits x_0 and x_N (shaded in Fig. 9-9*d*). Remember that whenever F_x is negative, the area between the graph of F_x and the x axis is also negative.

TOUCHSTONE EXAMPLE 9-4: Work on a Stone

A 2.0 kg stone moves along an x axis on a horizontal frictionless surface, acted on by only a force $F_x(x)$ that varies with the stone's position as shown in Fig. 9-10.

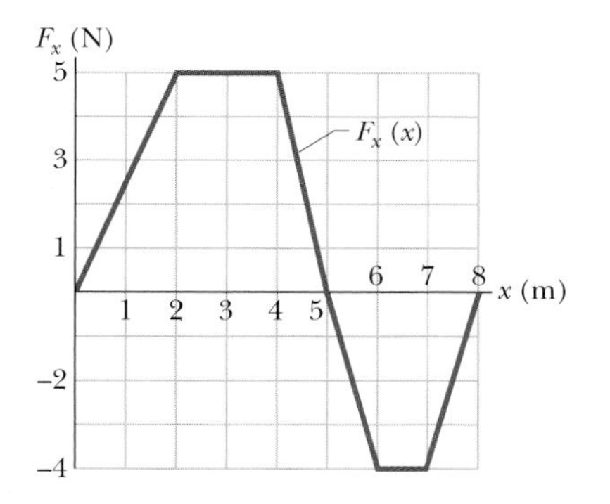

FIGURE 9-10 ■ A graph showing the variation of a one-dimensional force component with a stone's position.

(a) How much work is done on the stone by the force as the stone moves from its initial point at $x_1 = 0$ to $x_2 = 5$ m?

SOLUTION ■ A **Key Idea** is that the work done by a single one-dimensional force is given by Eq. 9-8:

$$W = \int_{x_1}^{x_2} F_x(x)\, dx.$$

Here the limits are $x_1 = 0$ m and $x_2 = 5$ m, and $F_x(x)$ is given by Fig. 9-10. A second **Key Idea** is that we can easily evaluate the integral graphically from Fig. 9-10. To do so, we find the area between the plot of $F_x(x)$ and the x axis, between the limits $x_1 = 0$ m and $x_2 = 5$ m. Note that we can split that area into three parts: a right triangle at the left (from $x = 0$ m to $x = 2$ m), a central rectangle (from $x = 2$ m to $x = 4$ m), and a triangle at the right (from $x = 4$ m to $x = 5$ m).

Recall that the area of a triangle is $\frac{1}{2}$(base)(height). The work $W_{0\to 5}$ that was done on the stone from $x_1 = 0$ and $x_2 = 5$ m is then

$$W_{0\to 5} = \tfrac{1}{2}(2\text{ m})(5\text{ N}) + (2\text{ m})(5\text{ N}) + \tfrac{1}{2}(1\text{ m})(5\text{ N})$$
$$= 17.5\text{ J}. \qquad \text{(Answer)}$$

(b) The stone starts from rest at $x_1 = 0$ m. What is its speed at $x = 8$ m?

SOLUTION ■ A **Key Idea** here is that the stone's speed is related to its kinetic energy, and its kinetic energy is changed because of the net work done on the stone by the force. Because the stone is initially at rest, its initial kinetic energy K_1 is 0. If we write its final kinetic energy at $x_3 = 8$ m as $K_3 = \frac{1}{2}mv_3^2$, then we can write the work-kinetic energy theorem of Eq. 9-11 ($K_3 = K_1 + W^{\text{net}}$) as

$$\tfrac{1}{2}mv_3^2 = 0 + W_{0\to 8}, \qquad (9\text{-}23)$$

where $W_{0\to 8}$ is the work done on the stone from $x_1 = 0$ m to $x_3 = 8$ m.

A second **Key Idea** is that, as in part (a), we can find the work graphically from Fig. 9-10 by finding the area between the plotted curve and the x axis. However, we must be careful about signs. We must take an area to be positive when the plotted curve is above the x axis and negative when it is below the x axis. We already know that work $W_{0\to 5} = 17.5$ J, so completing the calculation of the area gives us

$$W_{0\to 8} = W_{0\to 5} + W_{5\to 8}$$
$$= 17.5\text{ J} - \tfrac{1}{2}(1\text{ m})(4\text{ N}) - (1\text{ m})(4\text{ N}) - \tfrac{1}{2}(1\text{ m})(4\text{ N})$$
$$= 9.5\text{ J}.$$

Substituting this and $m = 2.0$ kg into Eq. 9-23 and solving for v_3, we find

$$v_3 = 3.1\text{ m/s}. \qquad \text{(Answer)}$$

9-7 Force and Displacement in More Than One Dimension

In this section we will explore a quite general situation in which a particle moves in a curved three-dimensional path while acted upon by a three-dimensional force that could vary with the position of the particle and might not act in the same direction as the particle's displacement. How can we calculate the work done on the particle by the force in this much more complex situation?

Before undertaking this more general treatment of work, we will actually start our exploration with a simple example of the work done by a constant two-dimensional force acting on a sled that is moving in only one dimension. Our simple example will lead us to conclude that we need to devise a general method for finding work as the product of two vectors.

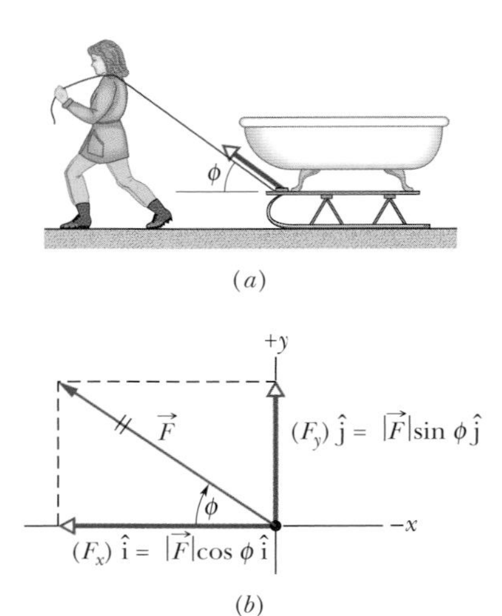

FIGURE 9-11 ■ (*a*) A sled is pulled by a rope that makes an angle ϕ with the ground as it moves toward the left. (*b*) The components of the pulling force $\vec{F}$ along a positive x axis and perpendicular to it.

Work Done by a Force Applied at an Angle

Imagine that you are pulling a loaded sled with no friction present (Fig. 9-11*a*). You hold the rope handle of the sled at some angle ϕ relative to the ground. You pull as hard as you can and the sled starts to move. However, you find that you are getting tired quickly and still have a significant distance to go. What would you do? Is the situation hopeless? One thing that you could try is to change the angle at which you pull on the handle of the sled. Should you make the angle ϕ larger or smaller?

As you probably know from your everyday experiences, you must pull more or less horizontally on a heavy object to pull it along. If you make the angle ϕ smaller, then you will pull the sled more efficiently. As discussed in Chapter 6, this is because the perpendicular force component can only change the direction of the motion (and in this situation we assume the sled glides on top of packed snow that prevents it from moving down). Only the component of a force along the line of motion is effective in changing an object's speed. So, by the work-kinetic energy theorem, it must be that only the component of a force along the line of motion contributes to the work done by the force. Saying this more formally:

> To calculate the work done on an object by a force during a displacement, we use only the component of force along the line of the object's displacement. The component of force perpendicular to the displacement does zero work.

From Fig. 9-11*b*, we see that we can write the x-component of the force F_x in terms of the magnitude of the force and the angle ϕ between the force and the positive x axis. That is,

$$F_x = |\vec{F}|\cos\phi. \tag{9-24}$$

To find the work done, we can use Eq. 9-13 ($W = F_x\Delta x$) for a constant force to get

$$W = F_x\,\Delta x = (|\vec{F}|\cos\phi)\,\Delta x \quad \text{(work for displacement parallel to an } x \text{ axis)}, \tag{9-25}$$

where Δx is the sled's displacement. Since the sled is moving to the left in the direction of the pull, both F_x and Δx are positive in our coordinate system and so the work done by the force is positive.

We can derive a similar but more general expression for the work done by a two-dimensional force along any line of displacement $\Delta\vec{r}$ (that does not necessarily lie along a chosen axis). To do this we must always take the angle ϕ *between the force and the direction of the displacement* (rather than the direction of a positive axis). In this case the expression for the work becomes

$$W = |\vec{F}||\Delta\vec{r}|\cos\phi \quad \text{(work in terms of angle between } \Delta\vec{r} \text{ and } \vec{F}\text{)}. \tag{9-26}$$

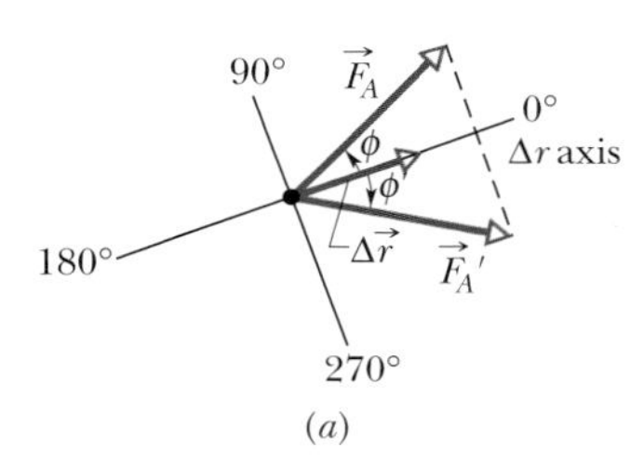

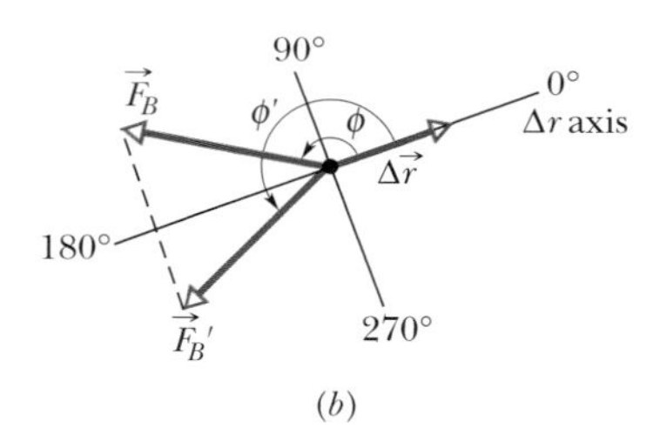

FIGURE 9-12 ■ (*a*) If force $\vec{F}_A$ or $\vec{F}_A'$ has an angle $\phi < 90°$ or $\phi > 270°$ with respect to the displacement vector $\Delta\vec{r}$ of an object, its component relative to the displacement will be positive. (*b*) If a force $\vec{F}_B$ or $\vec{F}'_B$ has an angle $90° < \phi < 270°$ with respect to the displacement $\Delta\vec{r}$ of an object, its components relative to the displacement will be negative.

As shown in Fig. 9-12, using the angle between the force and displacement and *absolute values* for both, the sign of the work comes out correctly.

Why the Sign of the Work Is Correct in Eq. 9-26 As shown in Fig. 9-12, if we set the angle ϕ in $W = |\vec{F}||\Delta\vec{r}|\cos\phi$ (Eq. 9-26) to any value less than 90°, then $\cos\phi$ is positive and so is the work. If ϕ is greater than 90° (up to 180°), then $\cos\phi$ is negative and so is the work. Referring to Fig. 9-11, we see that this way of determining the sign of the work done by an applied force is equivalent to determining the sign based on whether there is a component of force in the same or opposite direction as the motion. (Can you see why the work is zero when $\phi = 90°$?)

You can use similar considerations to determine the sign of work for $180° < \phi < 270°$ and for $270° < \phi < 360°$. Notice that once again it is the *relative* directions of the force and displacement vectors that determine the work done. As we already stated, no matter which way you choose to have the coordinate system pointing, the work for a particular process (including its sign) stays the same. Thus, work is indeed a scalar quantity.

Cautions: There are two restrictions to using the equations above to calculate work done on an object by a force. First, the force must be a *constant force*; that is, it must not change in magnitude or direction as the object moves through its displacement $\Delta\vec{r}$. Second, the object must be *particle-like*. This means that the object must be *rigid* and not change shape as its center of mass moves.

Net Work Done by Several Forces Suppose the net force on a rigid object is given by $\vec{F}^{\text{net}} = \vec{F}_A + \vec{F}_B + \vec{F}_C \cdots$, and we want to calculate the net work done by these forces. As we discussed earlier, it is simple to prove mathematically that the **net work** done on the object is the sum of the work done by the individual forces. We can calculate the net work in two ways: (1) We can use $W_A = |\vec{F}_A||\Delta\vec{r}|\cos\phi$ (Eq. 9-26) where ϕ is the angle between the direction of $\vec{F}_A$ and the object's displacement to find the work done by each force and then sum those works. Work is a scalar quantity, so summing the work done by the forces is as simple as adding up positive and negative numbers. (2) Alternatively, we can first find the net force $\vec{F}^{\text{net}}$ by finding the vector sum of the individual forces. Then we can use $W = |\vec{F}||\Delta\vec{r}|\cos\phi$ (Eq. 9-26), substituting the magnitude of $\vec{F}^{\text{net}}$ for the magnitude of $\vec{F}$, and the angle between the directions of the net force and the displacement for ϕ.

Work Done by a Three-Dimensional Variable Force

In general, even if a force varies with position, a particle can move through an infinitesimal displacement $d\vec{r}$ while being acted on by a three-dimensional force $\vec{F}(\vec{r})$. The displacement can be expressed in rectangular coordinates as

$$d\vec{r} = dx\hat{\text{i}} + dy\hat{\text{j}} + dz\hat{\text{k}}. \tag{9-27}$$

If we restrict ourselves to considering forces with rectangular components that depend only on the position component of the particle along a given axis, then

$$\vec{F}(\vec{r}) = F_x(x)\hat{\text{i}} + F_y(y)\hat{\text{j}} + F_z(z)\hat{\text{k}}. \tag{9-28}$$

Given the fact that no work is done unless there is a force component along the line of displacement, we can write the infinitesimal amount of work dW done on the particle by the force $\vec{F}(\vec{r})$ as

$$dW = F_x(x)\,dx + F_y(y)\,dy + F_z(z)\,dz. \tag{9-29}$$

The work W done by $\vec{F}$ while the particle moves from an initial position $\vec{r}_1$ with coordinates (x_1, y_1, z_1) to a final position $\vec{r}_2$ with coordinates (x_2, y_2, z_2) is then

$$W = \int_{\vec{r}_1}^{\vec{r}_2} dW = \int_{x_1}^{x_2} F_x(x)\,dx + \int_{y_1}^{y_2} F_y(y)\,dy + \int_{z_1}^{z_2} F_z(z)\,dz. \tag{9-30}$$

Note that if $\vec{F}(\vec{r})$ has only an x-component, then the y and z terms in the equation above are zero, so this equation reduces to

$$W = \int_{x_1}^{x_2} F_x(x)\,dx. \qquad \text{(Eq. 9-8)}$$

READING EXERCISE 9-4: The figure shows four situations in which a force acts on a box while the box slides rightward a distance $|\Delta\vec{x}|$ across a frictionless floor. The magnitudes of the forces are identical; their orientations are as shown. Rank the situations according to the work done on the box by the force during the displacement, from most positive to most negative.

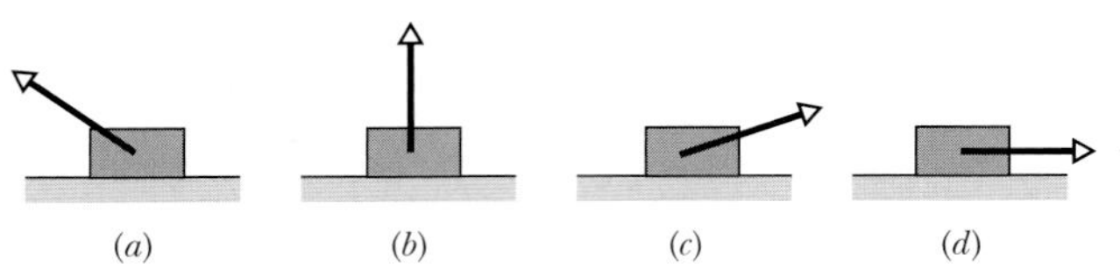

TOUCHSTONE EXAMPLE 9-5: Sliding a Safe

Figure 9-13*a* shows two industrial spies sliding an initially stationary 225 kg floor safe a displacement $\Delta\vec{r}$ of magnitude 8.50 m, along a straight line toward their truck. The push $\vec{F}_1$ of Spy 001 is 12.0 N, directed at an angle of 30° downward from the horizontal; the pull $\vec{F}_2$ of Spy 002 is 10.0 N, directed at 40° above the horizontal. The magnitudes and directions of these forces do not change as the safe moves, and the floor and safe make frictionless contact.

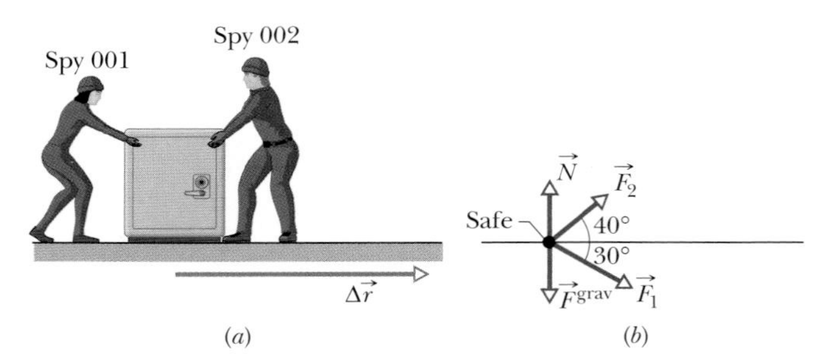

FIGURE 9-13 ■ (*a*) Two spies move a floor safe through displacement $\Delta\vec{r}$. (*b*) A free-body diagram showing the forces on the safe.

(a) What is the work done on the safe by forces $\vec{F}_1$ and $\vec{F}_2$ during the displacement $\Delta\vec{r}$?

SOLUTION ■ We use two **Key Ideas** here. First, the work W done on the safe by the two forces is the sum of the works they do individually. Second, because we can treat the safe as a particle and the forces are constant in both magnitude and direction, we can use Eq. 9-26,

$$(W = |\vec{F}||\Delta\vec{r}|\cos\phi),$$

to calculate those works. Note: $\cos(+\phi) = \cos(-\phi)$. From this and the free-body diagram for the safe in Fig. 9-13*b*, the work done by $\vec{F}_1$ is

$$W_1 = |\vec{F}_1||\Delta\vec{r}|\cos\phi_1 = (12.0\text{ N})(8.50\text{ m})(\cos 30°)$$
$$= 88.33\text{ J},$$

and the work done by $\vec{F}_2$ is

$$W_2 = |\vec{F}_2||\Delta\vec{r}|\cos\phi_2 = (10.0\text{ N})(8.50\text{ m})(\cos 40°)$$
$$= 65.11\text{ J}.$$

Thus, the work done by both forces is

$$W = W_1 + W_2 = 88.33\text{ J} + 65.11\text{ J}$$
$$= 153.4\text{ J} \approx 153\text{ J}. \qquad \text{(Answer)}$$

During the 8.50 m displacement, therefore, the spies transfer 153 J of energy to the kinetic energy of the safe.

(b) During the displacement, what is the work W^{grav} done on the safe by the gravitational force $\vec{F}^{\text{grav}}$ and what is the work W^{Normal} done on the safe by the normal force $\vec{N}$ from the floor?

SOLUTION ■ The **Key Idea** is that, because these forces are constant in both magnitude and direction, we can find the work they do with Eq. 9-26. Thus, with mg as the magnitude of the gravitational force, we write

$$W^{\text{grav}} = mg|\Delta\vec{r}|\cos 90° = mg|\Delta\vec{r}|(0) = 0, \qquad \text{(Answer)}$$

and

$$W^{\text{Normal}} = N|\Delta\vec{r}|\cos 90° = N|\Delta\vec{r}|(0) = 0. \qquad \text{(Answer)}$$

We should have known this result. Because these forces are perpendicular to the displacement of the safe, they do zero work on the safe and do not transfer any energy to or from it.

9-8 Multiplying a Vector by a Vector: The Dot Product

In the previous section, we discussed how to calculate the work done by a force that acts at some angle to the direction of an object's motion. We saw that in one dimension work is defined as the scalar product of two vector components (force and displacement). This may seem strange, but it is because only the *component* of the force along a line relative to the direction of displacement contributes to the work done by the force. This type of relationship between two vector quantities is so common that mathematicians have defined an operation to represent it. That operation is called the *dot* (or *scalar*) *product*. Learning about how to represent and calculate this product will lead us to a more general mathematical definition of work for three-dimensional situations. Application of the dot product will make the key equations we derived in the previous section easier to represent.

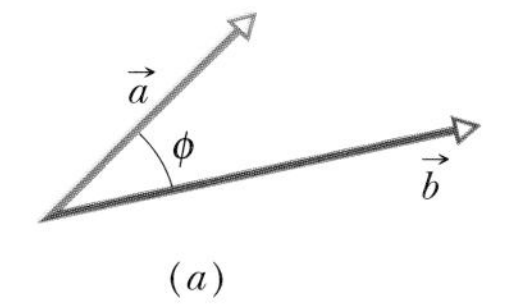

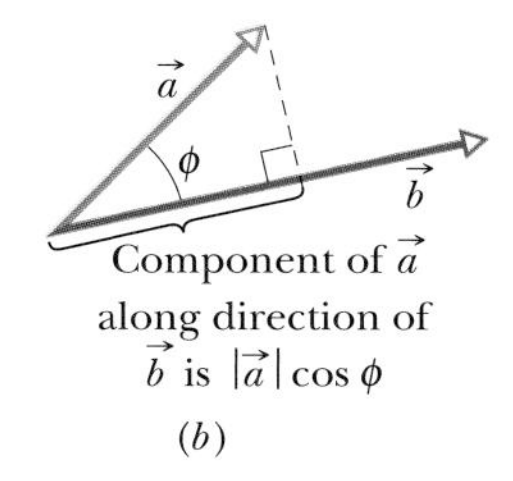

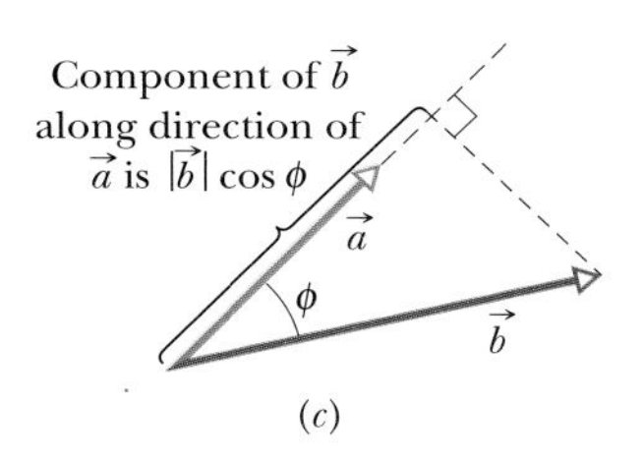

FIGURE 9-14 (*a*) Two vectors $\vec{a}$ and $\vec{b}$ with an angle ϕ between them. Since each vector has a component along the direction of the other vector, the same dot product results from: (*b*) multiplying the component of $\vec{a}$ on $\vec{b}$ by $|\vec{b}|$ or (*c*) multiplying the component of $\vec{b}$ on $\vec{a}$ by $|\vec{a}|$.

The Dot Product of Two Vectors

The **scalar or dot product** of the vectors $\vec{a}$ and $\vec{b}$ in Fig. 9-14*a* is written as $\vec{a}\cdot\vec{b}$ and defined to be

$$\vec{a}\cdot\vec{b} \equiv |\vec{a}||\vec{b}|\cos\phi \quad \text{(definition of scalar product)}, \tag{9-31}$$

where $|\vec{a}|$ is the magnitude of $\vec{a}$, $|\vec{b}|$ is the magnitude of $\vec{b}$, and ϕ is the angle between $\vec{a}$ and $\vec{b}$ (or, more properly, between the directions of $\vec{a}$ and $\vec{b}$). There are actually two such angles, ϕ and $360° - \phi$. Either can be used in $\vec{a}\cdot\vec{b} = |\vec{a}||\vec{b}|\cos\phi$, because their cosines are the same.

Note that there are only scalars on the right side of $\vec{a}\cdot\vec{b} = |\vec{a}||\vec{b}|\cos\phi$ (including the value of $\cos\phi$). Thus $\vec{a}\cdot\vec{b}$ on the left side represents a scalar quantity. Being scalars, the values of these quantities do not change, no matter how we choose to define our coordinate system. Because of the dot placed between the two vectors to denote this product, the name usually used for it is "dot product" and $\vec{a}\cdot\vec{b}$ is spoken as "a dot b."

As in the case of work, the dot product can be regarded as the product of two quantities: (1) the magnitude of one of the vectors and (2) the component of the second vector along the direction of the first vector. For example, in Fig. 9-14*b*, $\vec{a}$ has a component ($|\vec{a}|\cos\phi$) along the direction of $\vec{b}$. Note that a perpendicular dropped from the head of $\vec{a}$ to $\vec{b}$ determines that component. Alternatively, $\vec{b}$ has a component $|\vec{b}|\cos\phi$ along the direction of $\vec{a}$.

> If the angle ϕ between two vectors is 0°, the component of one vector along the other is maximum, and so also is the dot product of the vectors. If the angle ϕ between two vectors is 180°, the component of one vector along the other is a minimum. If, instead, ϕ is 90°or 270°, the component of one vector along the other is zero, and so is the dot product.

Equation 9-31 ($\vec{a}\cdot\vec{b} = |\vec{a}||\vec{b}|\cos\phi$) is sometimes rewritten as follows to emphasize the components:

$$\vec{a}\cdot\vec{b} = |\vec{a}|(|\vec{b}|\cos\phi) = (|\vec{a}|\cos\phi)|\vec{b}|. \tag{9-32}$$

Here, ($|\vec{b}|\cos\phi$) is the component of $\vec{b}$ along $\vec{a}$, and ($|\vec{a}|\cos\phi$) is the component of $\vec{a}$ along $\vec{b}$. The commutative law applies to a scalar product, so we can write

$$\vec{a}\cdot\vec{b} = \vec{b}\cdot\vec{a}.$$

When two vectors are in unit-vector notation in one, two, or three dimensions, it can be shown mathematically that we will get the same result shown in Eq. 9-32 by writing the dot product as

$$\vec{a} \cdot \vec{b} = (a_x\hat{\mathrm{i}} + a_y\hat{\mathrm{j}} + a_z\hat{\mathrm{k}}) \cdot (b_x\hat{\mathrm{i}} + b_y\hat{\mathrm{j}} + b_z\hat{\mathrm{k}}), \tag{9-33}$$

which we can expand according to the distributive law: Each component of the first vector is to be "dotted" with each component of the second vector. For example, the first step is

$$a_x\hat{\mathrm{i}} \cdot (b_x\hat{\mathrm{i}} + b_y\hat{\mathrm{j}} + b_z\hat{\mathrm{k}}) = a_xb_x(\hat{\mathrm{i}} \cdot \hat{\mathrm{i}}) + a_xb_y(\hat{\mathrm{i}} \cdot \hat{\mathrm{j}}) + a_xb_z(\hat{\mathrm{i}} \cdot \hat{\mathrm{k}}).$$

Since $\hat{\mathrm{i}}$ is perpendicular to both $\hat{\mathrm{j}}$ and $\hat{\mathrm{k}}$, there is no component of $\hat{\mathrm{i}}$ along the other two unit vectors, the angle between them is 90°, and so $\hat{\mathrm{i}} \cdot \hat{\mathrm{j}} = \hat{\mathrm{i}} \cdot \hat{\mathrm{k}} = 0$. On the other hand, $\hat{\mathrm{i}}$ is completely along $\hat{\mathrm{i}}$, the angle here is 0°, and so $\hat{\mathrm{i}} \cdot \hat{\mathrm{i}} = 1$. Therefore,

$$a_x\hat{\mathrm{i}} \cdot (b_x\hat{\mathrm{i}} + b_y\hat{\mathrm{j}} + b_z\hat{\mathrm{k}}) = a_xb_x.$$

If we continue along these lines, we find that

$$\vec{a} \cdot \vec{b} = a_xb_x + a_yb_y + a_zb_z. \tag{9-34}$$

Defining the Work Done as a Dot Product

If the force is constant over a displacement $\Delta\vec{r}$, we can use the definition of a dot product above and the relationship $W = |\vec{F}||\Delta\vec{r}| \cos\phi$ in Eq. 9-26 to produce an alternative mathematical definition for work,

$$W \equiv \vec{F} \cdot \Delta\vec{r} \quad \text{(definition of work done by a constant force).} \tag{9-35}$$

If the force is variable we can still use the definition of a dot product above along with the relationship presented in Eq. 9-30, where we integrated over infinitesimal displacements:

$$W = \int_{\vec{r}_1}^{\vec{r}_2} dW = \int_{x_1}^{x_2} F_x(x)\,dx + \int_{y_1}^{y_2} F_y(y)\,dy + \int_{z_1}^{z_2} F_z(z)\,dz,$$

to produce a more general alternative mathematical definition for work:

$$W \equiv \int_{\vec{r}_1}^{\vec{r}_2} \vec{F}(\vec{r}) \cdot d\vec{r} \quad \text{(definition of work done by a variable force).} \tag{9-36}$$

This dot product representation of work has some advantages. For one, the notation is more compact. It is also especially useful for calculating work when $\vec{F}$ and $d\vec{r}$ or $\Delta\vec{r}$ are given in unit-vector notation because we can exploit the fact that $\vec{a} \cdot \vec{b} = a_xb_x + a_yb_y + a_zb_z$ (Eq. 9-34).

9-9 Net Work and Translational Kinetic Energy

Generalizing the Net Work-Kinetic Energy Theorem

We know from Newton's laws that if you apply a force to an object in the same direction as the object's motion, the object's speed will increase. From our discussion of work, we also know that the force does positive work on the object. If you apply the force in the direction opposite the direction of the object's motion, the object's speed will decrease, and we know that in that case the force will do negative work on the object. This suggests that work done by forces correlates with changes in speed. We used these considerations to relate work and kinetic energy for the special case of a bead on a wire that experiences a single force in the direction of the wire. In doing so, we developed a net work-kinetic energy theorem for one-dimensional forces and motions given by

$$W^{net} = K_2 - K_1 \quad \text{(the 1D net work-kinetic energy theorem)}, \qquad \text{(Eq. 9-11)}$$

where $$W^{net} \equiv \int_{x_1}^{x_2} F_x^{net}(x)\,dx \text{ (Eq. 9-7)} \quad \text{and} \quad K = \tfrac{1}{2}mv_x^2. \qquad \text{(Eq. 9-10)}$$

Can we extend this to our more general three-dimensional situation? Fortunately we can combine

$$\int_{x_1}^{x_2} F_x^{net}(x)\,dx = \tfrac{1}{2}mv_{2x}^2 - \tfrac{1}{2}mv_{1x}^2, \qquad \text{(Eq. 9-6)}$$

and our general expression for work in three dimensions (Eq. 9-30) and rearrange terms to get

$$\begin{aligned} W^{net} &= \int_{x_1}^{x_2} F_x^{net}(x)\,dx + \int_{y_1}^{y_2} F_y^{net}(y)\,dy + \int_{z_1}^{z_2} F_z^{net}(z)\,dz \\ &= \tfrac{1}{2}m(v_{2x}^2 + v_{2y}^2 + v_{2z}^2) - \tfrac{1}{2}m(v_{1x}^2 + v_{1y}^2 + v_{1z}^2). \end{aligned} \qquad (9\text{-}37)$$

Since the speed of a particle moving in three dimensions is given by $v^2 = v_x^2 + v_y^2 + v_z^2$ and

$$W \equiv \int_{\vec{r}_1}^{\vec{r}_2} \vec{F}(\vec{r})\cdot d\vec{r} = \int_{x_1}^{x_2} F_x(x)\,dx + \int_{y_1}^{y_2} F_y(y)\,dy + \int_{z_1}^{z_2} F_z(z)\,dz,$$

this reduces to

$$W^{net} = K_2 - K_1 \quad \text{(3D net work-kinetic energy theorem for variable force)}, \qquad (9\text{-}38)$$

where $K \equiv \frac{1}{2}mv^2$ represents a more general definition of the kinetic energy of a particle of mass m with its center of mass moving with speed v. In words, Eq. 9-38 tells us that

Net work done on the particle = Change in its translational kinetic energy.

This relationship is valid in one, two, or three dimensions.

Experimental Verification of the Net Work-Kinetic Energy Theorem

Experimental verification of the one-dimensional net work-kinetic energy theorem is shown in Figs. 9-15 and 9-16. A low-friction cart with a force sensor attached to it is pulled along a smooth track from $x_1 = 0.6$ m to $x_2 = 1.2$ m with a variable applied force. The applied force is measured with a force sensor. The distance along the track is measured with a motion detector. Both measurements are fed to a computer for

FIGURE 9-15 ■ A variable force is applied to a force sensor attached to a cart on a horizontal track. The cart's position and velocity are recorded using a motion sensor. The friction force is negligible.

display. If we ignore friction, then $W^{\text{net}} = W^{\text{app}}$. The net work is given by the area under the curve obtained when data for the net force vs. distance is graphed. This area (determined by numerical integration as in Fig. 9-9) gives us

$$W^{\text{net}} = \int_{x_1}^{x_2} F_x^{\text{net}}(x)\, dx = 1.3 \text{ J}.$$

The distance and time data are used to determine the velocity of the cart at each location. The cart mass ($m = 1.5$ kg) and velocity are then used to determine the translational kinetic energy of the cart as a function of its location along the track. The change in kinetic energy between $x_1 = 0.6$ m and $x_2 = 1.6$ m is

$$\Delta K = K_2 - K_1 = 1.4 \text{ J} - 0.1 \text{ J} = 1.3 \text{ J},$$

as expected according to the net work-kinetic energy theorem.

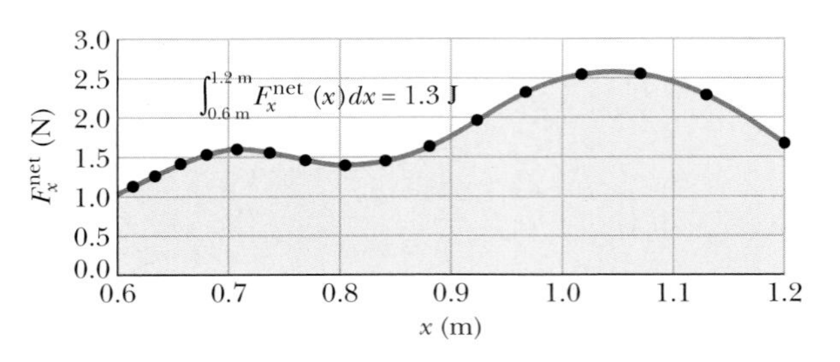

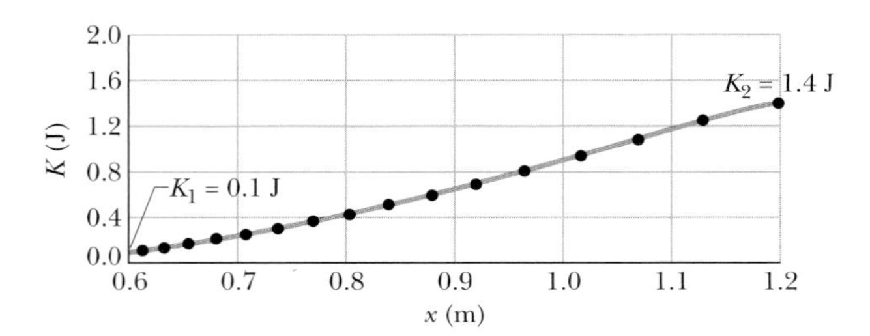

FIGURE 9-16 ■ Experimental verification of the net work-kinetic energy theorem for a cart undergoing one-dimensional horizontal motion under the influence of a variable applied force and a negligible friction force.

Lifting and Lowering—Net Work and Kinetic Energy

Suppose we *lift* a particle-like object by applying a vertical force $\vec{F}_y$ to it as shown in Fig. 9-17. During the upward displacement, our applied force does *positive* work W^{app} on the object while the gravitational force does *negative* work W^{grav} on it. Our force adds energy to (or transfers energy *to*) the object while the gravitational force removes energy from (or transfers energy *from*) it. By $\Delta K = K_2 - K_1$ (Eq. 9-38), the change ΔK in the translational kinetic energy of the object due to these two energy transfers is

$$\Delta K = K_2 - K_1 = W^{\text{net}} = W^{\text{app}} + W^{\text{grav}}. \qquad \text{(9-39)}$$

(a) (b)

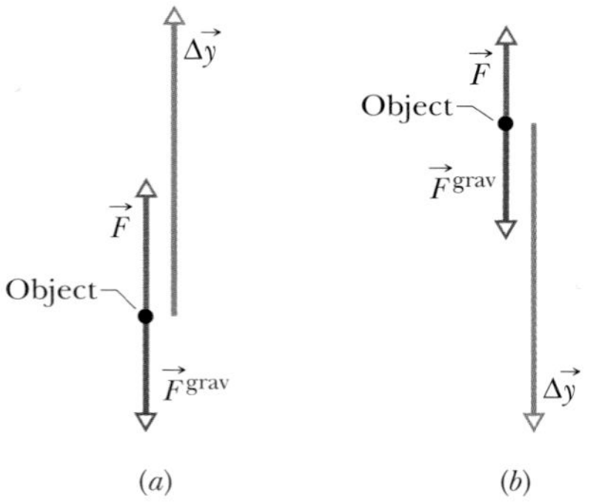

FIGURE 9-17 ■ An upward force is applied to an object in the presence of a downward gravitational force: (*a*) As the object rises, the applied force does positive work while the gravitational force does negative work. (*b*) As the object is lowered the applied force does negative work on the object while the gravitational force does positive work.

This equation also applies if we lower the object. However, then the gravitational force tends to transfer energy *to* the object whereas our force tends to transfer energy *from* it.

A common situation involves an object that is stationary before and after being lifted. For example, suppose you lift a book from the floor to a shelf. Then K_2 and K_1 are both zero, and $\Delta K = K_2 - K_1 = W^{\text{net}} = W^{\text{app}} + W^{\text{grav}}$ reduces to

$$W^{\text{net}} = W^{\text{app}} + W^{\text{grav}} = 0 \text{ N},$$

or

$$W^{\text{app}} = -W^{\text{grav}} \quad \text{(if object starts and ends at rest).} \qquad \text{(9-40)}$$

Note that we get the same result if K_2 and K_1 are not zero but are still equal. This result means that the work done by the applied force is the negative of the work done by the gravitational force. That is, the applied force transfers the same amount of energy to the object as the gravitational force takes away from it (whenever the initial and final speeds of an object are the same).

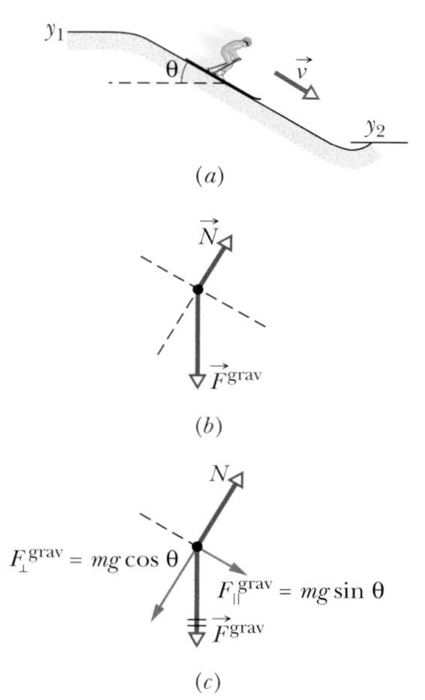

FIGURE 9-18 ■ (*a*) A curved ramp makes an angle θ with respect to the horizontal at the location of a skier. (*b*) A free-body diagram showing the forces on the skier. (*c*) A diagram showing the resolution of $\vec{F}^{grav}$ into the components parallel and perpendicular to the ramp.

Falling on an Incline—The Skier on a Curved Ramp

Finally, we are ready to return to the question of how to find the speed of a skier (shown in Fig. 9-1) as a function of how far she has descended on a frictionless ramp. Suppose we would like to know the speed of the skier shown in Fig. 9-1 as she leaves the end of a long curved ramp in order to predict how far she can jump. Let us revisit our initial claim that the net work-kinetic energy theorem is much more useful than Newton's Second Law for this calculation. The net work-kinetic energy theorem is only useful in this particular case if we make the simplifying assumptions that: (1) we can neglect frictional forces; (2) the skier doesn't push with her poles as she slides down the ramp; and (3) she holds her body rigid.

Given these assumptions, we can determine the net force on the skier's center of mass when she is at an arbitrary location on the ramp (Fig. 9-18*a*). This net force is the sum of forces shown in the free-body diagram in Fig. 9-18*b*. Figure 9-18*c* shows the components of the normal force and the gravitational force parallel and perpendicular to the ramp. Since there is no motion perpendicular to the ramp at a given location, the force components perpendicular to the ramp cancel out. So the net force acts parallel to the ramp. Its component down the ramp is given by $F_{\parallel}^{net} = mg\sin\theta$ where θ is the angle between the horizontal and the ramp.

Our problem now is to take into account the fact that θ keeps changing along the curved ramp. To do this we can divide the ramp into a whole series of tiny ramps having sides dx, dy, and a length dr as shown in Fig. 9-19*a*. A greatly enlarged picture of one of these infinitesimal ramps is shown in Fig. 9-19*b*. The infinitesimal work done in traveling a distance dr down any one of the tiny ramps is given by

$$dW = F_{\parallel}dr = (-mg\sin\theta\, dr), \tag{9-41}$$

but since $\sin\theta = dy/dr$, we see that dW becomes simply $(-mg)\,dy$. This is a very profound result because it tells us that the work done by a rigid object as it falls down a frictionless ramp does not depend on the angle of the ramp but only on the constant factor $-mg$ and the vertical distance through which the object's center of mass falls. We will return to this idea in Chapter 10.

If we integrate the net force over the collection of tiny ramps we get

$$\begin{aligned} W^{net} &= \int_{\vec{r}_1}^{\vec{r}_2} \vec{F}^{net}(\vec{r}) \cdot d\vec{r} = \int_{r_1}^{r_2} F_{\parallel}\, dr \\ &= \int_{y_1}^{y_2} (-mg)\, dy = -mg(y_2 - y_1) = -mg\,\Delta y = F_y^{grav}\,\Delta y, \end{aligned} \tag{9-42}$$

where Δy (like F_y^{grav}) is a negative quantity because $y_2 < y_1$.

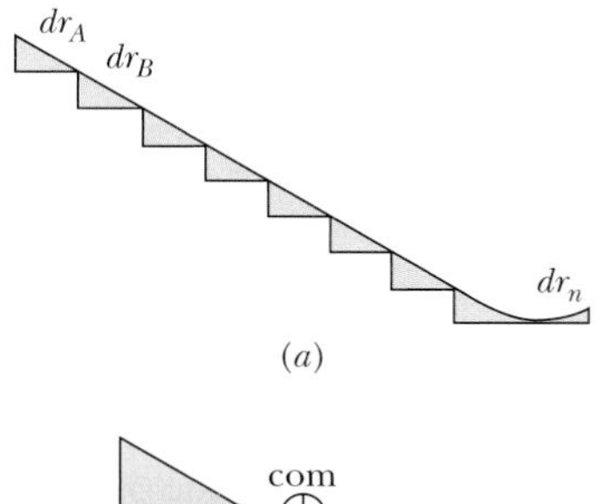

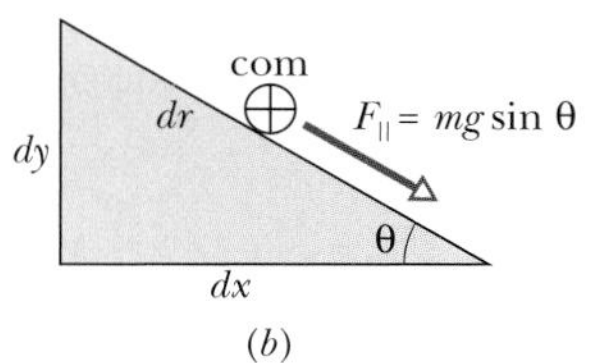

FIGURE 9-19 ■ (*a*) The ramp can be divided into many smaller ramps, each possibly having a different θ. (*b*) A ramp of infinitesimal length dr with a vertical component dy and a horizontal component dx.

Now that we have obtained a simple expression for the net work done by the gravitational force as the skier goes down the ramp, we can use the net work-kinetic energy theorem (Eq. 9-38) to find the skier's speed at the bottom of the ramp. If the skier starts from rest so that her initial speed is $v_1 = 0$ m/s, then

$$W^{net} = K_2 - K_1 \text{ so that } W^{net} = mg\Delta y = \tfrac{1}{2}m(v_2^2). \tag{9-43}$$

Solving for the final speed gives us

$$v_2 = \sqrt{2g\,\Delta y}. \tag{9-44}$$

Note that using Newton's Second Law to find this speed would be extremely difficult because it requires us to keep track of the angle of the ramp at each location. We will further explore the advantages of the net work-kinetic energy theorem for other situations in the next chapter.

TOUCHSTONE EXAMPLE 9-6: Weight Lifting

Let us return to the lifting feats of Andrey Chemerkin shown on the opening page of this chapter.

(a) Chemerkin made his record-breaking lift with rigidly connected objects (a barbell and disk weights) having a total mass $m = 260.0$ kg. He lifted them a distance of 2.0 m. During the lift, how much work was done on the objects by the gravitational force $\vec{F}^{grav}$ acting on them?

SOLUTION ■ The **Key Idea** here is that we can treat the rigidly connected objects as a single particle and thus use Eq. 9-14,

$$W^{grav} = -mg\Delta y,$$

to find the work W^{grav} done on them by $\vec{F}^{grav}$. The total weight mg was 2548 N, and $\Delta y = +2.0$ m. Thus,

$$W^{grav} = -mg\Delta y = -(2548 \text{ N})(2.0 \text{ m}) = -5100 \text{ J}. \quad \text{(Answer)}$$

(b) How much work was done on the objects by Chemerkin's force during the lift?

SOLUTION ■ We do not have an expression for Chemerkin's force on the object, and even if we did, his force was certainly not constant. Thus, one **Key Idea** here is that we *cannot* just substitute his force into Eq. 9-12 to find his work. However, we know that the objects were stationary at the start and end of the lift, so that $K_2 - K_1 = 0$. Therefore, as a second **Key Idea**, we know by the net work-kinetic energy theorem that the work W^{app} done by Chemerkin's applied force was the negative of the work W^{grav} done by the gravitational force $\vec{F}^{grav}$. Equation 9-40 expresses this fact and gives us

$$W^{app} = -W^{grav} = +5100 \text{ J}. \quad \text{(Answer)}$$

(c) While Chemerkin held the objects stationary above his head, how much work was done on them by his force?

SOLUTION ■ The **Key Idea** is that when he supported the objects, they were stationary. Thus, their displacement $\Delta\vec{r} = 0$ and, by Eq. 9-36, the work done on them was zero (even though supporting them was a very tiring task).

(d) How much work was done by the force Paul Anderson applied to lift objects with a total weight of 27 900 N a distance of 1.0 cm?

SOLUTION ■ Following the argument of parts (a) and (b) but now with $mg = 27\,900$ N and $\Delta y = 1.0$ cm, we find

$$W^{app} = -W^{grav} = -(-mg\,\Delta y) = +mg\,\Delta y = (27\,900 \text{ N})(0.010 \text{ m}) = 280 \text{ J}. \quad \text{(Answer)}$$

Anderson's lift required a tremendous upward force but only a small energy transfer of 280 J because of the short displacement involved.

TOUCHSTONE EXAMPLE 9-7: Crate on a Ramp

An initially stationary 15.0 kg crate of cheese wheels is pulled, via a cable, a distance $d = 5.70$ m up a frictionless ramp to a height h of 2.50 m, where it stops (Fig. 9-20*a*).

(a) How much work W^{grav} is done on the crate by the gravitational force $\vec{F}^{grav}$ during the lift?

SOLUTION ■ A **Key Idea** is that we can treat the crate as a particle and thus use Eq. 9-26 ($W = |\vec{F}||\Delta\vec{r}|\cos\phi$) to find the work W^{grav} done by $\vec{F}^{grav}$. However, we do not know the angle ϕ between the directions of $\vec{F}^{grav}$ and displacement $\Delta\vec{r}$. From the crate's free-body diagram in Fig. 9-20*b*, we find that ϕ is $\theta + 90°$, where θ is the (unknown) angle of the ramp. Equation 9-26 then gives us

$$W^{grav} = mgd\cos(\theta + 90°) = -mgd\sin\theta, \quad (9\text{-}45)$$

where we have used a trigonometric identity to simplify the expression. The result seems to be useless because θ is unknown. But (continuing with physics courage) we see from Fig. 9-20*a* that $\sin\theta = h/d$, where h is a known quantity. With this substitution, Eq. 9-45 becomes

$$W^{grav} = -mgh = -(15.0 \text{ kg})(9.8 \text{ N/kg})(2.50 \text{ m}) = -368 \text{ J}. \quad (9\text{-}46) \quad \text{(Answer)}$$

Note that Eq. 9-46 tells us that the work W^{grav} done by the gravitational force depends on the vertical displacement but (perhaps surprisingly) not on the horizontal displacement. (Again, we return to this point in Chapter 10.)

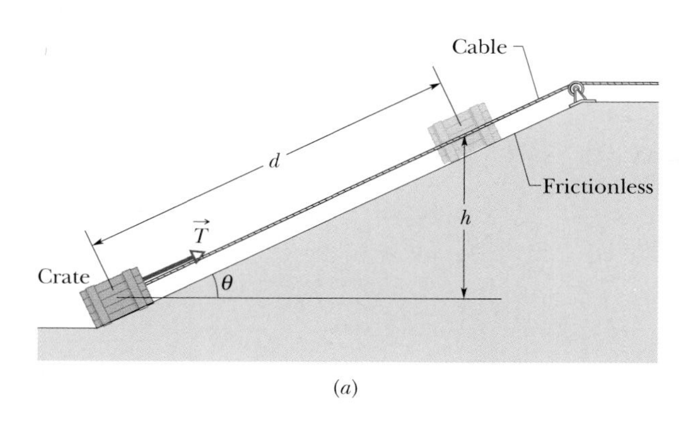

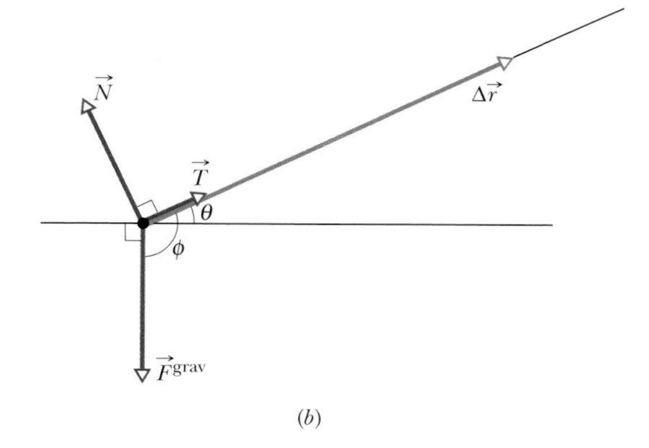

FIGURE 9-20 ▪ (*a*) A crate is pulled up a frictionless ramp by a force $\vec{T}$ parallel to the ramp. (*b*) A free-body diagram for the crate, showing all the forces on it. Its displacement $\Delta\vec{r}$ is also shown.

(b) How much work W^{rope} is done on the crate by the force $\vec{T}$ from the cable during the lift?

SOLUTION ▪ We cannot just substitute the force magnitude T for $|\vec{F}|$ in Eq. 9-26 ($W = |\vec{F}||\Delta\vec{r}|\cos\phi$) because we do not know the value of T. However, a **Key Idea** to get us going is that we can treat the crate as a particle and then apply the work-kinetic energy theorem ($W^{net} = \Delta K$) to it. Because the crate is stationary before and after the lift, the change ΔK in its kinetic energy is zero. For the net work W^{net} done on the crate, we must sum the works done by all three forces acting on the crate. From (a), the work W^{grav} done by the gravitational force $\vec{F}^{grav}$ is -368 J. The work W^{Normal} done by the normal force $\vec{N}$ on the crate from the ramp is zero because $\vec{N}$ is perpendicular to the displacement. We want the work W^{rope} done by $\vec{T}$. Thus, the work-kinetic energy theorem gives us

$$\Delta K = W^{rope} + W^{grav} + W^{Normal},$$

or

$$0 = W^{rope} - 368\text{ J} + 0,$$

and so

$$W^{rope} = 368\text{ J}. \qquad \text{(Answer)}$$

9-10 Power

A contractor wishes to lift a load of bricks from the sidewalk to the top of a building using a winch. We can now calculate how much work the force applied by the winch must do on the load to make the lift. The contractor, however, is much more interested in the *rate* at which that work is done. Will the job take 5 minutes (acceptable) or a week (unacceptable)?

The rate at which work is done by a force is called the **power.** If an amount of work W is done in an amount of time Δt by a force, we define the **average power** due to the work done by a force during that time interval as

$$\langle P \rangle \equiv \frac{W}{\Delta t} \quad \text{(definition of average power).} \qquad (9\text{-}47)$$

We define the **instantaneous power** P as the instantaneous rate of doing work, so that

$$P \equiv \frac{dW}{dt} \quad \text{(definition of instantaneous power),} \qquad (9\text{-}48)$$

where dW is the infinitesimal amount of work done in an infinitesimal time interval dt. Suppose we know the work $W(t)$ done by a force as a continuous well-behaved function of time. Then to get the instantaneous power P at, say, time $t = 3.0$ s during the work, we would first take the time derivative of $W(t)$, and then evaluate the result for $t = 3.0$ s.

The SI unit of power is the joule per second. This unit is used so often that it has a special name, the **watt** (W), after James Watt (who greatly improved the rate at which steam engines could do work). In the British system, the unit of power is the foot-pound per second. Often the horsepower is used. Some relations among these units are

$$1 \text{ watt} = 1 \text{ W} = 1 \text{ J/s} = 0.738 \text{ ft} \cdot \text{lb/s} \tag{9-49}$$

and

$$1 \text{ horsepower} = 1 \text{ hp} = 550 \text{ ft} \cdot \text{lb/s} = 746 \text{ W}. \tag{9-50}$$

Inspection of Eq. 9-47 shows that we can express work as power multiplied by time, $W = \langle P \rangle \Delta t$. When we do this, we commonly use the unit of kilowatt-hour. Thus,

$$\begin{aligned} 1 \text{ kilowatt-hour} &= 1 \text{ kW} \cdot \text{h} = (10^3 \text{ W})(3600 \text{ s}) \\ &= 3.6 \times 10^6 \text{ J} = 3.6 \text{ MJ}. \end{aligned} \tag{9-51}$$

Perhaps because the unit of kilowatt-hour appears on our utility bills, it has become identified as an electrical unit. However, the kilowatt-hour can be used equally well as a unit for other examples of work (or energy). Thus, if you pick up this book from the floor and put it on a tabletop, you are free to report the work that you have done as 4×10^{-6} kW · h (or alternatively converting to milliwatts to get as 4 mW · h).

We can also express the rate at which a force does work on a particle (or particle-like object) in terms of that force and the body's velocity. For a particle that is moving along a straight line (say, the x axis) and acted on by a constant force $\vec{F}$ directed at some angle ϕ to that line, $P = dW/dt$ (Eq. 9-48) becomes

$$P = \frac{dW}{dt} = \frac{(|\vec{F}|\cos\phi)|dx|}{dt} = |\vec{F}|\cos\phi\left|\frac{dx}{dt}\right|,$$

but since $v_x = dx/dt$, we get

$$P = |\vec{F}||v_x|\cos\phi. \tag{9-52}$$

Reorganizing the right side of this equation as the dot product $\vec{F} \cdot \vec{v}$ we may rewrite Eq. 9-52 as

$$P = \vec{F} \cdot \vec{v} \quad \text{(instantaneous power)}. \tag{9-53}$$

For example, the truck in Fig. 9-21 exerts a force $\vec{F}$ on the trailing load, which has velocity $\vec{v}$ at some instant. The instantaneous power due to $\vec{F}$ is the rate at which $\vec{F}$ does work on the load at that instant and is given by Eq. 9-52 and $P = \vec{F} \cdot \vec{v}$ (Eq. 9-53). Saying that this power is "the power of the truck" is often acceptable, but we should keep in mind what is meant: Power is the rate at which the applied *force* does work.

FIGURE 9-21 ■ The power due to the truck's applied force on the trailing load is the rate at which that force does work on the load.

READING EXERCISE 9-5: A block moves with uniform circular motion because a cord tied to the block is anchored at the center of a circle. Is the power due to the force on the block from the cord positive, negative, or zero? ■

TOUCHSTONE EXAMPLE 9-8: Average and Instantaneous Power

A horizontal cable accelerates a suspicious package across a frictionless horizontal floor. The amount of work that has been done by the cable's force on the package is given by $W(t) = (0.20\ \text{J/s}^2)t^2$.

(a) What is the average power $\langle P \rangle$ due to the cable's force in the time interval $t_1 = 0$ s to $t_2 = 10$ s?

SOLUTION ■ The **Key Idea** here is that the average power $\langle P \rangle$ is the ratio of the amount of work W done in the given time interval to that time interval (Eq. 9-47). To find the work W, we evaluate the amount of work that has been done, $W(t)$, at $t = 0$ s and $t = 10$ s. At those times, the cable has done work W_1 and W_2, respectively:

$$W_1 = (0.20\ \text{J/s}^2)(0\ \text{s})^2 = 0\ \text{J} \quad \text{and} \quad W_2 = (0.20\ \text{J/s}^2)(10\ \text{s})^2 = 20\ \text{J}.$$

Therefore, in the 10 s interval, the work done is $W_2 - W_1 = 20$ J. Equation 9-47 then gives us

$$\langle P \rangle = \frac{W}{\Delta t} = \frac{20\ \text{J}}{10\ \text{s}} = 2.0\ \text{W}. \quad \text{(Answer)}$$

Thus, during the 10 s interval, the cable does work at the average rate of 2.0 joules per second.

(b) What is the instantaneous power P due to the cable's force at $t = 3.0$ s, and is P then increasing or decreasing?

SOLUTION ■ The **Key Idea** here is that the instantaneous power P at $t = 3.0$ s is the time derivative of the work dW/dt evaluated at $t = 3.0$ s (Eq. 9-48). Taking the derivative of $W(t)$ gives us

$$P = \frac{dW}{dt} = \frac{d}{dt}[(0.20\ \text{J/s}^2)t^2] = (0.40\ \text{J/s}^2)t.$$

This result tells us that as time t increases, so does P. Evaluating P for $t = 3.0$ s, we find

$$P = (0.40\ \text{J/s}^2)(3.0\ \text{s}) = 1.20\ \text{W}. \quad \text{(Answer)}$$

Thus, at $t = 3.0$s, the cable is doing work at the rate of 1.20 joules per second, and that rate is increasing.

Problems

SEC. 9-2 ■ INTRODUCTION TO WORK AND KINETIC ENERGY

1. Electron in Copper If an electron (mass $m = 9.11 \times 10^{-31}$ kg) in copper near the lowest possible temperature has a kinetic energy of 6.7×10^{-19} J, what is the speed of the electron?

2. Large Meteorite vs. TNT On August 10, 1972, a large meteorite skipped across the atmosphere above western United States and Canada, much like a stone skipped across water. The accompanying fireball was so bright that it could be seen in the daytime sky (see Fig. 9-22 for a similar event). The meteorite's mass was about 4×10^6 kg; its speed was about 15 km/s. Had it entered the atmosphere vertically, it would have hit Earth's surface with about the same speed. (a) Calculate the meteorite's loss of kinetic energy (in joules) that would have been associated with the vertical impact. (b) Express the energy as a multiple of the explosive energy of 1 megaton of TNT, which is 4.2×10^{15} J. (c) The energy associated with the atomic bomb explosion over Hiroshima was equivalent to 13 kilotons of TNT. To how many Hiroshima bombs would the meteorite impact have been equivalent?

FIGURE 9-22 ■ Problem 2. A large meteorite skips across the atmosphere in the sky above the Ottawa region.

3. Calculate Kinetic Energy Calculate the kinetic energies of the following objects moving at the given speeds: (a) a 110 kg football linebacker running at 8.1 m/s; (b) a 4.2 g bullet at 950 m/s; (c) the aircraft carrier *Nimitz*, 40.2×10^8 kg at 32 knots.

4. Father Racing Son A father racing his son has half the kinetic energy of the son, who has half the mass of the father. The father speeds up by 1.0 m/s and then has the same kinetic energy as the son. What are the original speeds of (a) the father and (b) the son?

5. A Proton is Accelerated A proton (mass $m = 1.67 \times 10^{-27}$ kg) is being accelerated along a straight line at 3.6×10^{13} m/s² in a machine. If the proton has an initial speed of 2.4×10^7 m/s and travels 3.5 cm, what then is (a) its speed and (b) the increase in its kinetic energy?

6. Vehicle's Kinetic Energy If a vehicle with a mass of 1200 kg has a speed of 120 km/h, what is the vehicle's kinetic energy as determined by someone at rest alongside the vehicle's road?

7. Truck Traveling North A 2100 kg truck traveling north at 41 km/h turns east and accelerates to 51 km/h. (a) What is the change in the kinetic energy of the truck? What are the (b) magnitude and (c) direction of the change in the translational momentum of the truck?

8. Two Pieces from One An object, with mass m and speed v relative to an observer, explodes into two pieces, one three times as massive as the other; the explosion takes place in deep space. The less massive piece stops relative to the observer. How much kinetic energy is added to the system in the explosion, as measured in the observer's reference frame? *Hint*: Translational momentum is conserved.

9. Freight Car A railroad freight car of mass 3.18×10^4 kg collides with a stationary caboose car. They couple together, and 27.0% of the initial kinetic energy is transferred to nonconservative forms of energy (thermal, sound, vibrational, and so on). Find the mass of the caboose. *Hint:* Translational momentum is conserved.

10. Two Chunks An 8.0 kg body is traveling at 2.0 m/s with no external force acting on it. At a certain instant an internal explosion occurs, splitting the body into two chunks of 4.0 kg mass each. The explosion gives the chunks an additional 16 J of kinetic energy. Neither chunk leaves the line of original motion. Determine the speed and direction of motion of each of the chunks after the explosion. *Hint:* Translational momentum is conserved.

11. Kinetic Energy and Impulse A ball having a mass of 150 g strikes a wall with a speed of 5.2 m/s and rebounds straight back with only 50% of its initial kinetic energy. (a) What is the speed of the ball immediately after rebounding? (b) What is the magnitude of the impulse on the wall from the ball? (c) If the ball was in contact with the wall for 7.6 ms, what was the magnitude of the average force on the ball from the wall during this time interval?

12. Unmanned Space Probe A 2500 kg unmanned space probe is moving in a straight line at a constant speed of 300 m/s. Control rockets on the space probe execute a burn in which a thrust of 3000 N acts for 65.0 s. (a) What is the change in the magnitude of the probe's translational momentum if the thrust is backward, forward, or directly sideways? (b) What is the change in kinetic energy under the same three conditions? Assume that the mass of the ejected burn products is negligible compared to the mass of the space probe.

SEC. 9-6 ■ WORK FOR A ONE-DIMENSIONAL VARIABLE FORCE

13. Graph of Acceleration Figure 9-23 gives the acceleration of a 2.00 kg particle as it moves from rest along an x axis while an applied force $\vec{F}^{\text{app}}$ acts on it from $x = 0$ m to $x = 9$ m. How much work has the force done on the particle when the particle reaches (a) $x = 4$ m, (b) $x = 7$ m, and (c) $x = 9$ m? What is the particle's speed and direction of travel when it reaches (d) $x = 4$ m, (e) $x = 7$ m, and (f) $x = 9$ m?

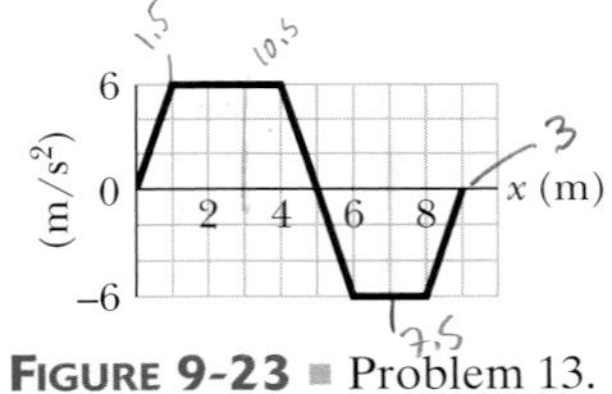

FIGURE 9-23 ■ Problem 13.

14. Can of Nuts and Bolts A can of nuts and bolts is pushed 2.00 m along an x axis by a broom along the greasy (frictionless) floor of a car repair shop in a version of shuffleboard. Figure 9-24 gives the work W done on the can by the constant horizontal force from the broom, versus the can's position x. (a) What is the magnitude of that force? (b) If the can had an initial kinetic energy of 3.00 J, moving in the positive direction of the x axis, what is its kinetic energy at the end of the 2.00 m displacement?

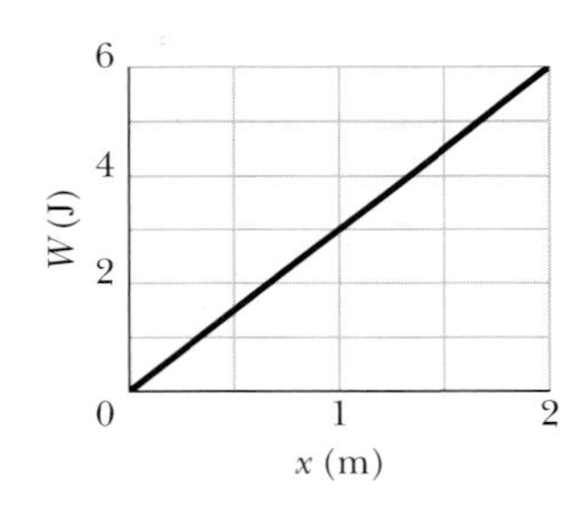

FIGURE 9-24 ■ Problem 14.

15. Single Force A single force acts on a body that moves along an x axis. Figure 9-25 shows the velocity component v_x versus time t for the body. For each of the intervals AB, BC, CD, and DE, give the sign (plus or minus) of the work done by the force on the body or state that the work is zero.

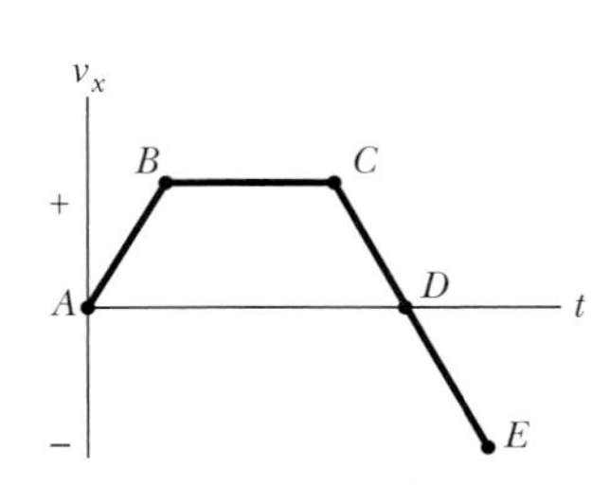

FIGURE 9-25 ■ Problem 15.

16. Block Attached to a Spring The block in Fig. 9-7 lies on a horizontal frictionless surface and is attached to the free end of the spring, with a spring constant of 50 N/m. Initially, the spring is at its relaxed length and the block is stationary at position $x = 0$ m. Then an applied force with a constant magnitude of 3.0 N pulls the block in the positive direction of the x axis, stretching the spring until the block stops. When that stopping point is reached, what are (a) the position of the block, (b) the work that has been done on the block by the applied force, and (c) the work that has been done on the block by the spring force? During the block's displacement, what are (d) the block's position when its kinetic energy is maximum and (e) the value of that maximum kinetic energy?

17. Luge Rider A luge and rider, with a total mass of 85 kg, emerge from a downhill track onto a horizontal straight track with an initial speed of 37 m/s. If they slow at a constant rate of 2.0 m/s², (a) what magnitude F is required for the slowing force, (b) what distance d do they travel while slowing, and (c) what work W is done on them by the slowing force? What are (d) F, (e) d, and (f) W if the luge and the rider slow at a rate of 4.0 m/s²?

18. Work from Graph A 5.0 kg block moves in a straight line on a horizontal frictionless surface under the influence of a force that varies with position as shown in Fig. 9-26. How much work is done by the force as the block moves from the origin to $x = 8.0$ m?

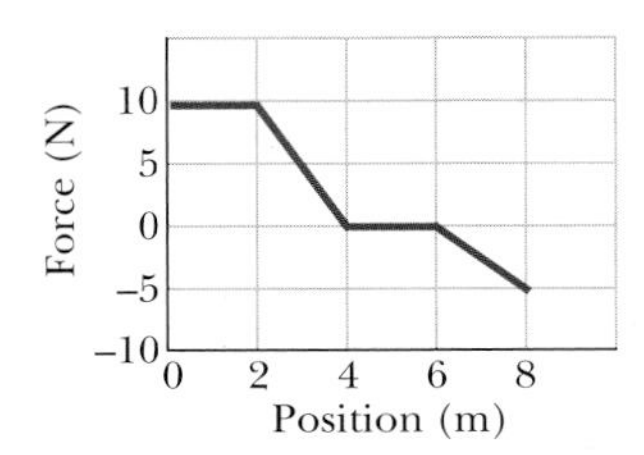

FIGURE 9-26 ■ Problem 18.

19. Brick A 10 kg brick moves along an x axis. Its acceleration as a function of its position is shown in Fig. 9-27. What is the net work done on the brick by the force causing the acceleration as the brick moves from $x = 0$ m to $x = 8.0$ m?

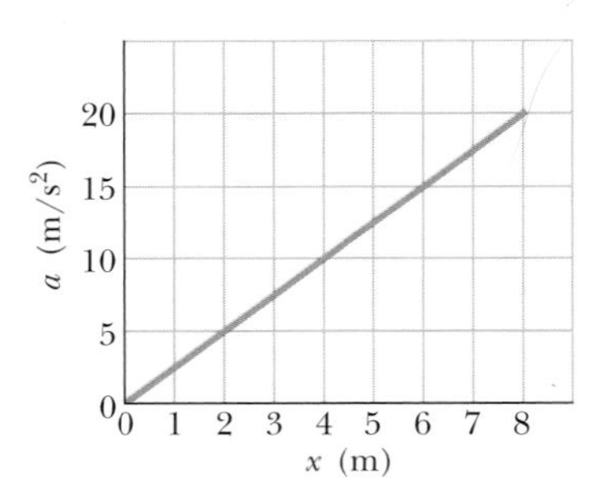

FIGURE 9-27 ■ Problem 19.

20. Velodrome (a) In 1975 the roof of Montreal's Velodrome, with a weight of 360 kN, was lifted by 10 cm so that it could be centered. How much work was done on the roof by the forces making the lift? (b) In 1960, Mrs. Maxwell Rogers of Tampa, Florida, reportedly raised one end of a car that had fallen onto her son when a jack failed. If her panic lift effectively raised 4000 N (about $\frac{1}{4}$ of the car's weight) by 5.0 cm, how much work did her force do on the car?

21. Two Pulleys and a Canister In Fig. 9-28, a cord runs around two massless, frictionless pulleys; a canister with mass $m = 20$ kg hangs from one pulley; and you exert a force $\vec{F}$ on the free end of the cord. (a) What must be the magnitude of $\vec{F}$ if you are to lift the canister at a constant speed? (b) To lift the canister by 2.0 cm, how far must you pull the free end of the cord? During that lift, what is the work done on the canister by (c) your force (via the cord) and (d) the gravitational force on the canister? (*Hint*: When a cord loops around a pulley as shown, it pulls on the pulley with a net force that is twice the tension in the cord.)

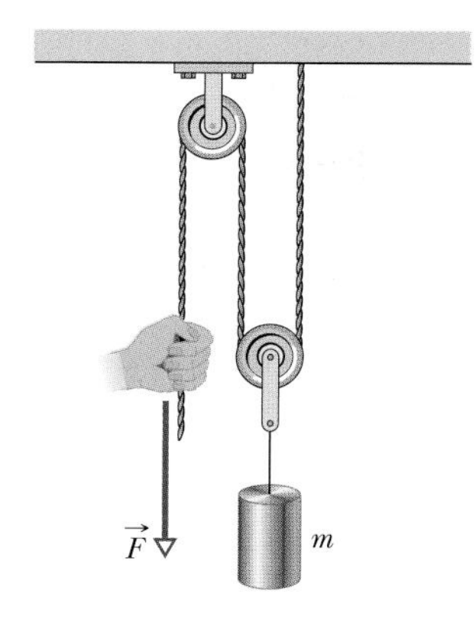

FIGURE 9-28 ■ Problem 21.

22. Spring at MIT During spring semester at MIT, residents of the parallel buildings of the East Campus dorms battle one another with large catapults that are made with surgical hose mounted on a window frame. A balloon filled with dyed water is placed in a pouch attached to the hose, which is then stretched through the width of the room. Assume that the stretching of the hose obeys Hooke's law with a spring constant of 100 N/m. If the hose is stretched by 5.00 m and then released, how much work does the force from the hose do on the balloon in the pouch by the time the hose reaches its relaxed length?

23. Plot $F(x)$ The force on a particle is directed along an x axis and given by $F_x = F_0(x/x_0 - 1)$. Find the work done by the force in moving the particle from $x = 0$ to $x = 2x_0$ by (a) plotting $F_x(x)$ and measuring the work from the graph and (b) integrating $F_x(x)$.

24. Block Dropped on a Spring A 250 g block is dropped onto a relaxed vertical spring that has a spring constant of k = 2.5 N/cm (Fig. 9-29). The block becomes attached to the spring and compresses the spring 12 cm before turning around. While the spring is being compressed, what work is done on the block by (a) the gravitational force on it and (b) the spring force? (c) What is the speed of the block just before it hits the spring? (Assume that friction is negligible.) (d) If the speed at impact is doubled, what is the maximum compression of the spring?

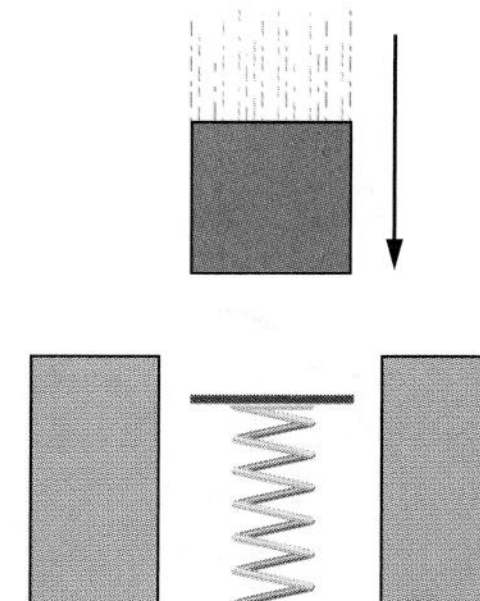

FIGURE 9-29 ■ Problem 24.

25. Bird Cage A spring with a spring constant of 15 N/cm has a cage attached to one end (Fig. 9-30). (a) How much work does the spring force do on the cage when the spring is stretched from its relaxed length by 7.6 mm? (b) How much additional work is done by the spring force when the spring is stretched by an additional 7.6 mm?

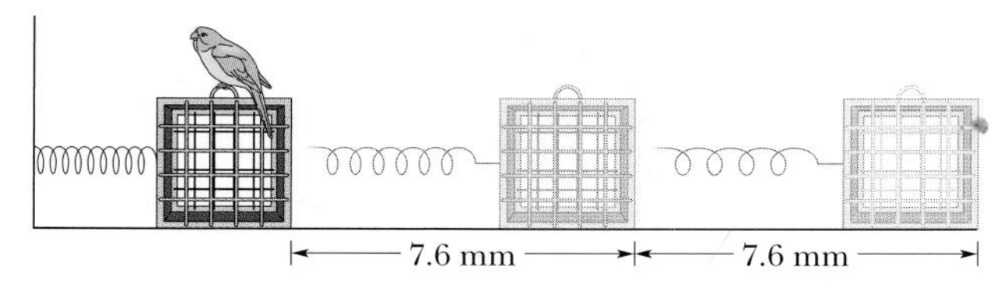

FIGURE 9-30 ■ Problem 25.

SEC. 9-7 ■ FORCE AND DISPLACEMENT IN MORE THAN ONE DIMENSION

26. Constant Force A constant force of magnitude 10 N makes an angle of 150° (measured counterclockwise) with the positive x direction as it acts on a 2.0 kg object moving in the xy plane. How much work is done on the object by the force as the object moves from the origin to the point with position vector $(2.0\text{ m})\hat{i} - (4.0\text{ m})\hat{j}$?

27. Force on a Particle A force $\vec{F} = (4.0\text{ N})\hat{i} + (c\text{ N})\hat{j}$ acts on a particle as the particle goes through displacement $\vec{d} = (3.0\text{ m})\hat{i} - (2.0\text{ m})\hat{j}$. (Other forces also act on the particle.) What is the value of c if the work done on the particle by force $\vec{F}$ is (a) zero, (b) 17 J, and (c) −18 J?

28. Crate on an Incline To push a 25.0 kg crate up a frictionless incline, angled at 25.0° to the horizontal, a worker exerts a force of magnitude 209 N parallel to the incline. As the crate slides 1.50 m, how much work is done on the crate by (a) the worker's applied force, (b) the gravitational force on the crate, and (c) the normal force exerted by the incline on the crate? (d) What is the total work done on the crate?

29. Cargo Canister Figure 9-31 shows an overhead view of three horizontal forces acting on a cargo canister that was initially stationary but that now moves across a frictionless floor. The force magnitudes are $F_A = 3.00$ N, $F_B = 4.00$ N, and $F_C = 10.0$ N. What is the net work done on the canister by the three forces during the first 4.00 m of displacement?

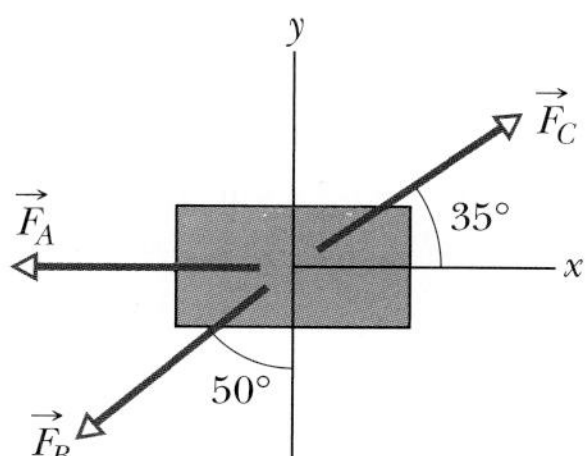

FIGURE 9-31 ■ Problem 29.

30. A Particle Moves A particle moves along a straight path through displacement $\vec{d} = (8\text{ m})\hat{i} - (c\text{ m})\hat{j}$ while force $\vec{F} = (2\text{ N})\hat{i} - (4\text{ N})\hat{j}$ acts on it. (Other forces also act on the particle.) What is the value of c if the work done by $\vec{F}$ on the particle is (a) zero, (b) positive, and (c) negative?

31. Worker Pulling Crate To pull a 50 kg crate across a horizontal frictionless floor, a worker applies a force of 210 N, directed 20° above the horizontal. As the crate moves 3.0 m, what work is done on the crate by (a) the worker's force, (b) the gravitational force on the crate, and (c) the normal force on the crate from the floor? (d) What is the total work done on the crate?

32. Floating Ice Block A floating ice block is pushed through a displacement $\vec{d} = (15\text{ m})\hat{i} - (12\text{ m})\hat{j}$ along a straight embankment by rushing water, which exerts a force $\vec{F} = (210\text{ N})\hat{i} - (150\text{ N})\hat{j}$ on the block. How much work does the force do on the block during the displacement?

33. Coin on a Frictionless Plane A coin slides over a frictionless plane and across an xy coordinate system from the origin to a point with xy coordinates (3.0 m, 4.0 m) while a constant force acts on it. The force has magnitude 2.0 N and is directed at a counterclockwise angle of 100° from the positive direction of the x axis. How much work is done by the force on the coin during the displacement?

34. Work Done by 2-D Force What work is done by a force $\vec{F} = ((2\text{ N/m})x)\hat{i} + (3\text{ N})\hat{j}$, with x in meters, that moves a particle from a position $\vec{r}_1 = (2\text{ m})\hat{i} + (3\text{ m})\hat{j}$ to a position $\vec{r}_2 = -(4\text{ m})\hat{i} - (3\text{ m})\hat{j}$?

SEC. 9-9 ■ NET WORK AND TRANSLATIONAL KINETIC ENERGY

35. Cold Hot Dogs Figure 9-32 shows a cold package of hot dogs sliding rightward across a frictionless floor through a distance d = 20.0 cm while three forces are applied to it. Two of the forces are horizontal and have the magnitudes $F_A = 5.00$ N and $F_B = 1.00$ N;

the third force is angled down by $\theta = -60.0°$ and has the magnitude $F_C = 4.00$ N. (a) For the 20.0 cm displacement, what is the *net* work done on the package by the three applied forces, the gravitational force on the package, and the normal force on the package? (b) If the package has a mass of 2.0 kg and an initial kinetic energy of 0 J, what is its speed at the end of the displacement?

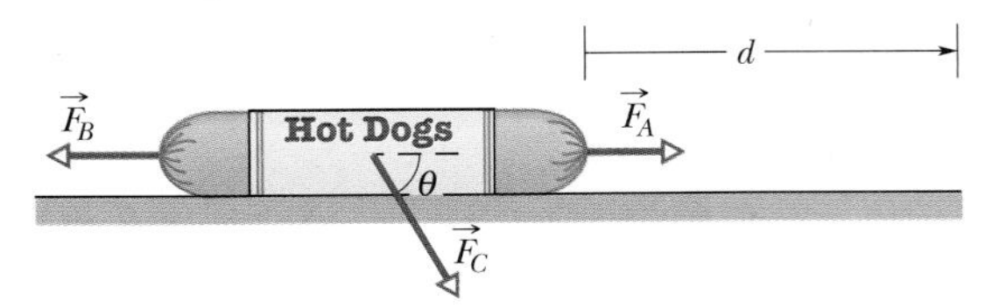

FIGURE 9-32 ■ Problem 35.

36. Air Track A 1.0 kg standard body is at rest on a frictionless horizontal air track when a constant horizontal force $\vec{F}$ acting in the positive direction of an x axis along the track is applied to the body. A stroboscopic graph of the position of the body as it slides to the right is shown in Fig. 9-33. The force $\vec{F}$ is applied to the body at $t_1 = 0.0$ s, and the graph records the position of the body at 0.50 s intervals. How much work is done on the body by the applied force $\vec{F}$ between $t_1 = 0.0$ s and $t_2 = 2.0$ s?

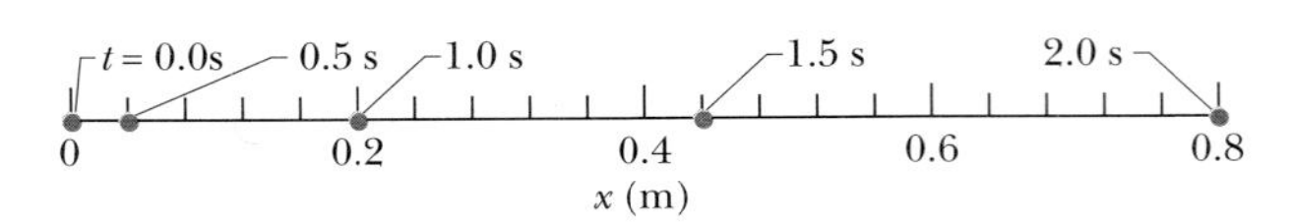

FIGURE 9-33 ■ Problem 36.

37. Three Forces Figure 9-34 shows three forces applied to a trunk that moves leftward by 3.00 m over a frictionless floor. The force magnitudes are $F_A = 5.00$ N, $F_B = 9.00$ N, and $F_C = 3.00$ N. During the displacement, (a) what is the net work done on the trunk by the three forces and (b) does the kinetic energy of the trunk increase or decrease?

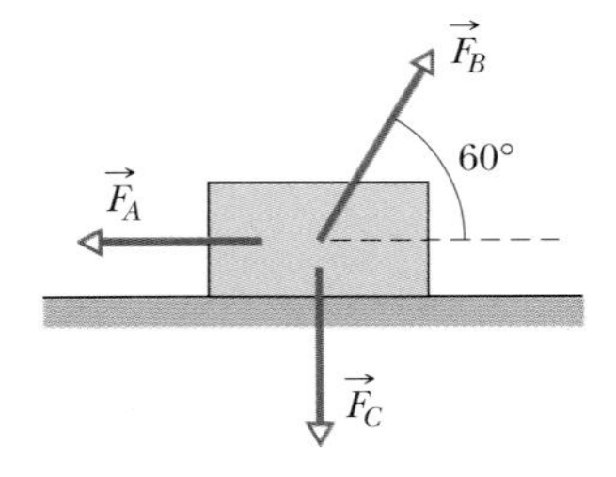

FIGURE 9-34 ■ Problem 37.

38. Block of Ice In Fig. 9-35, a block of ice slides down a frictionless ramp at angle $\theta = 50°$, while an ice worker pulls up the ramp (via a rope) with a force of magnitude F_r = 50 N. As the block slides through distance $d = 0.50$ m along the ramp, its kinetic energy increases by 80 J. How much greater would its kinetic energy have been if the rope had not been attached to the block?

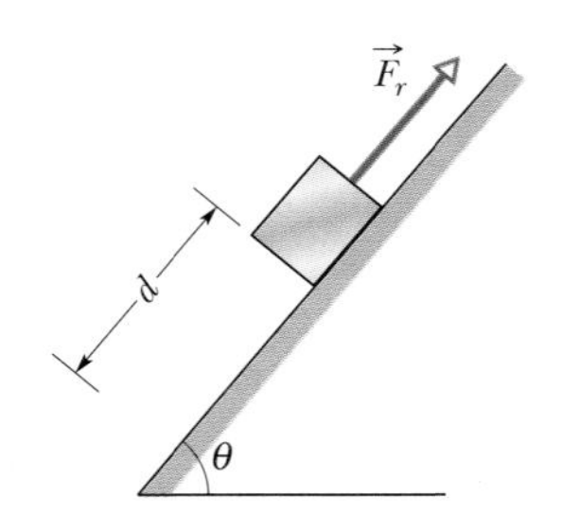

FIGURE 9-35 ■ Problem 38.

39. Helicopter A helicopter hoists a 72 kg astronaut 15 m vertically from the ocean by means of a cable. The acceleration of the astronaut is $g/10$. How much work is done on the astronaut by (a) the force from the helicopter and (b) the gravitational force on her? What are the (c) kinetic energy and (d) speed of the astronaut just before she reaches the helicopter?

40. Given *x*(*t*) A force acts on a 3.0 kg particle-like object in such a way that the position of the object as a function of time is given by $x = (3 \text{ m/s})t - (4 \text{ m/s}^2)t^2 + (1 \text{ m/s}^3)t^3$ with x in meters and t in seconds. Find the work done on the object by the force from $t_1 = 0.0$ s to $t_2 = 4.0$ s. (*Hint*: What are the speeds at those times?)

41. Lowering a Block A cord is used to vertically lower an initially stationary block of mass M at a constant downward acceleration of $g/4$. When the block has fallen a distance d, find (a) the work done by the cord's force on the block, (b) the work done by the gravitational force on the block, (c) the kinetic energy of the block, and (d) the speed of the block.

42. Force Applied Downward In Fig. 9-36*a*, a 2.0 N force is applied to a 4.0 kg block at a downward angle θ as the block moves rightward through 1.0 m across a frictionless floor. Find an expression for the speed v_2 of the block at the end of that distance if the block's initial velocity is (a) 0.0 m/s and (b) 1.0 m/s to the right. (c) The situation in Fig. 9-36*b* is similar in that the block is initially moving at 1.0 m/s to the right, but now the 2.0 N force is directed downward to the left. Find an expression for the speed v_2 of the block at the end of the 1.0 m distance. (d) Graph all three expressions for v_2 versus downward angle θ, for $\theta = 0°$ to $\theta = -90°$. Interpret the graphs.

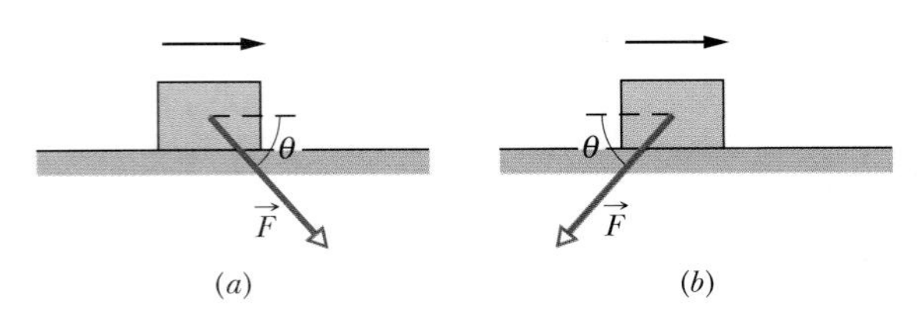

FIGURE 9-36 ■ Problem 42.

43. Canister and One Force The only force acting on a 2.0 kg canister that is moving in an xy plane has a magnitude of 5.0 N. The canister initially has a velocity of 4.0 m/s in the positive x direction, and some time later has a velocity of 6.0 m/s in the positive y direction. How much work is done on the canister by the 5.0 N force during this time?

44. Block of Ice Slides A 45 kg block of ice slides down a frictionless incline 1.5 m long and 0.91 m high. A worker pushes up against the ice, parallel to the incline, so that the block slides down at constant speed. (a) Find the magnitude of the worker's force. How much work is done on the block by (b) the worker's force, (c) the gravitational force on the block, (d) the normal force on the block from the surface of the incline, and (e) the net force on the block?

45. Cave Rescue A cave rescue team lifts an injured spelunker directly upward and out of a sinkhole by means of a motor-driven cable. The lift is performed in three stages, each requiring a vertical distance of 10.0 m: (a) the initially stationary spelunker is accelerated to a speed of 5.00 m/s; (b) he is then lifted at the constant speed of 5.00 m/s; (c) finally he is slowed to zero speed. How much work is done on the 80.0 kg rescuee by the force lifting him during each stage?

46. Work-Kinetic Energy The only force acting on a 2.0 kg body as the body moves along the x axis varies as shown in Fig. 9-37. The velocity of the body at $x = 0.0$ m is 4.0 m/s. (a) What is the kinetic energy of the body at $x = 3.0$ m? (b) At what value of x will the body have a ki-

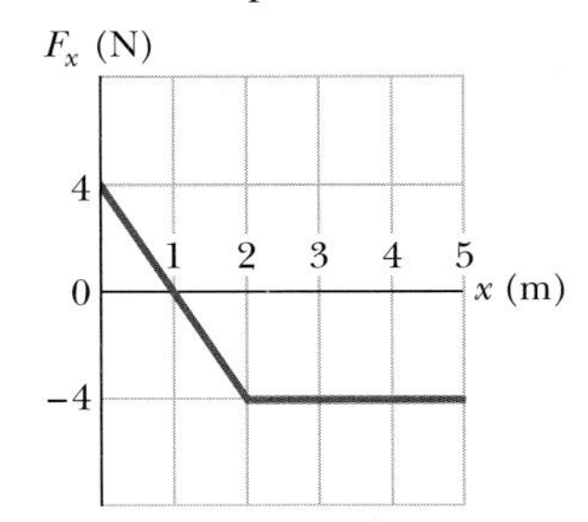

FIGURE 9-37 ■ Problem 46.

netic energy of 8.0 J? (c) What is the maximum kinetic energy attained by the body between $x = 0.0$ m and $x = 5.0$ m?

47. Block at Rest A 1.5 kg block is initially at rest on a horizontal frictionless surface when a horizontal force in the positive direction of an x axis is applied to the block. The force is given by $\vec{F}(x) = (2.5\text{ N} - x^2\text{ N/m}^2)\hat{\text{i}}$, where x is in meters and the initial position of the block is $x = 0.0$ m. (a) What is the kinetic energy of the block as it passes through $x = 2.0$ m? (b) What is the maximum kinetic energy of the block between $x = 0.0$ m and $x = 2.0$ m?

SEC. 9-10 ■ POWER

48. Average Rate of Work The loaded cab of an elevator has a mass of 3.0×10^3 kg and moves 210 m up the shaft in 23 s at constant speed. At what average rate does the force from the cable do work on the cab?

49. Block Pulled at Constant Speed A 100 kg block is pulled at a constant speed of 5.0 m/s across a horizontal floor by an applied force of 122 N directed 37° above the horizontal. What is the rate at which the force does work on the block?

50. Resistance to Motion Resistance to the motion of an automobile consists of road friction, which is almost independent of speed, and air drag, which is proportional to speed-squared. For a certain car with a weight of 12,000 N, the net resistant force $\vec{F}$ is given by $\vec{F} = [300\text{ N} + (1.8\text{ N}\cdot\text{s}^2/\text{m}^2)v_x^2]\hat{\text{i}}$, where $\vec{F}$ is in newtons and v_x is in meters per second. Calculate the power (in horsepower) required to accelerate the car at 0.92 m/s^2 when the speed is 80 km/h.

51. A Force Acts on a Body A force of 5.0 N acts on a 15 kg body initially at rest. Compute the work done by the force in (a) the first, (b) the second, and (c) the third seconds and (d) the instantaneous power due to the force at the end of the third second.

52. Rope Tow A skier is pulled by a tow rope up a frictionless ski slope that makes an angle of 12° with the horizontal. The rope moves parallel to the slope with a constant speed of 1.0 m/s. The force of the rope does 900 J of work on the skier as the skier moves a distance of 8.0 m up the incline. (a) If the rope moved with a constant speed of 2.0 m/s, how much work would the force of the rope do on the skier as the skier moved a distance of 8.0 m up the incline? At what rate is the force of the rope doing work on the skier when the rope moves with a speed of (b) 1.0 m/s and (c) 2.0 m/s?

53. Freight Elevator A fully loaded, slow-moving freight elevator has a cab with a total mass of 1200 kg, which is required to travel upward 54 m in 3.0 min, starting and ending at rest. The elevator's counterweight has a mass of only 950 kg, so the elevator motor must help pull the cab upward. What average power is required of the force the motor exerts on the cab via the cable?

54. Ladle Attached to Spring A 0.30 kg ladle sliding on a horizontal frictionless surface is attached to one end of a horizontal spring (with $k = 500$ N/m) whose other end is fixed. The ladle has a kinetic energy of 10 J as it passes through its equilibrium position (the point at which the spring force is zero). (a) At what rate is the spring doing work on the ladle as the ladle passes through its equilibrium position? (b) At what rate is the spring doing work on the ladle when the spring is compressed 0.10 m and the ladle is moving away from the equilibrium position?

55. Towing a Boat The force (but not the power) required to tow a boat at constant velocity is proportional to the speed. If a speed of 4.0 km/h requires 7.5 kW, how much power does a speed of 12 km/h require?

56. Transporting Boxes Boxes are transported from one location to another in a warehouse by means of a conveyor belt that moves with a constant speed of 0.50 m/s. At a certain location the conveyor belt moves for 2.0 m up an incline that makes an angle of 10° with the horizontal, then for 2.0 m horizontally, and finally for 2.0 m down an incline that makes an angle of 10° with the horizontal. Assume that a 2.0 kg box rides on the belt without slipping. At what rate is the force of the conveyor belt doing work on the box (a) as the box moves up the 10° incline, (b) as the box moves horizontally, and (c) as the box moves down the 10° incline?

57. Horse Pulls Cart A horse pulls a cart with a force of 40 lb at an angle of 30° above the horizontal and moves along at a speed of 6.0 mi/h. (a) How much work does the force do in 10 min? (b) What is the average power (in horsepower) of the force?

58. Object Accelerates Horizontally An initially stationary 2.0 kg object accelerates horizontally and uniformly to a speed of 10 m/s in 3.0 s. (a) In that 3.0 s interval, how much work is done on the object by the force accelerating it? What is the instantaneous power due to that force (b) at the end of the interval and (c) at the end of the first half of the interval?

59. A Sprinter A sprinter who weighs 670 N runs the first 7.0 m of a race in 1.6 s, starting from rest and accelerating uniformly. What are the sprinter's (a) speed and (b) kinetic energy at the end of the 1.6 s? (c) What average power does the sprinter generate during the 1.6 s interval?

60. The *Queen Elizabeth 2* The luxury liner *Queen Elizabeth 2* has a diesel-electric powerplant with a maximum power of 92 MW at a cruising speed of 32.5 knots. What forward force is exerted on the ship at this speed? (1 knot = 1.852 km/h.)

61. Swimmer A swimmer moves through the water at a constant speed of 0.22 m/s. The average drag force opposing this motion is 110 N. What average power is required of the swimmer?

62. Auto Starts from Rest A 1500 kg automobile starts from rest on a horizontal road and gains a speed of 72 km/h in 30 s. (a) What is the kinetic energy of the auto at the end of the 30 s? (b) What is the average power required of the car during the 30 s interval? (c) What is the instantaneous power at the end of the 30 s interval, assuming that the acceleration is constant?

63. A Locomotive A locomotive with a power capability of 1.5 MW can accelerate a train from a speed of 10 m/s to 25 m/s in 6.0 min. (a) Calculate the mass of the train. Find (b) the speed of the train and (c) the force accelerating the train as functions of time (in seconds) during the 6.0 min interval. (d) Find the distance moved by the train during the interval.

Additional Problems

64. Estimate, Then Integrate (a) Estimate the work done by the force represented by the graph of Fig. 9-38 in displacing a particle from $x_1 = 1$ m to $x_2 = 3$ m. (b) The curve is given by $F_x = a/x^2$, with $a = 9\ \text{N}\cdot\text{m}^2$. Calculate the work using integration.

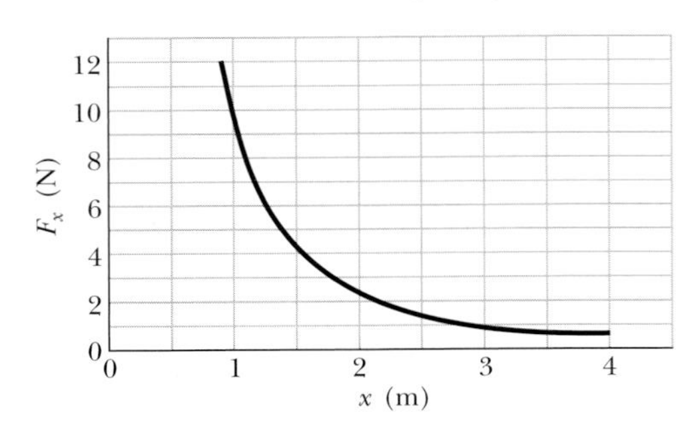

FIGURE 9-38 ■ Problem 64.

65. Explosion at Ground Level An explosion at ground level leaves a crater with a diameter that is proportional to the energy of the explosion raised to the $\frac{1}{3}$ power; an explosion of 1 megaton of TNT leaves a crater with a 1 km diameter. Below Lake Huron in Michigan there appears to be an ancient impact crater with a 50 km diameter. What was the kinetic energy associated with that impact, in terms of (a) megatons of TNT (1 megaton yields 4.2×10^{15} J) and (b) Hiroshima bomb equivalents (13 kilotons of TNT each)? (Ancient meteorite or comet impacts may have significantly altered Earth's climate and contributed to the extinction of the dinosaurs and other life-forms.)

66. Pushing a Block A hand pushes a 3 kg block along a table from point A to point C as shown in Fig. 9-39. The table has been prepared so that the left half of the table (from A to B) is frictionless. The right half (from B to C) has a nonzero coefficient of friction equal to μ^{kin}. The hand pushes the block from A to C using a constant force of 5 N. The block starts off at rest at point A and comes to a stop when it reaches point C. The distance from A to B is $\frac{1}{2}$ meter and the distance from B to C is also $\frac{1}{2}$ meter.

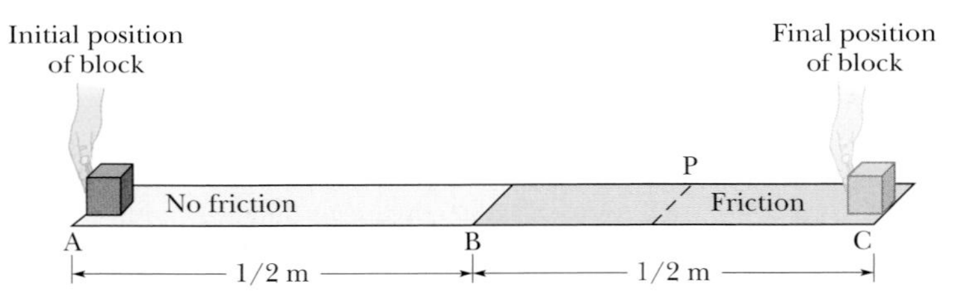

FIGURE 9-39 ■ Problem 66.

(a) Describe in words the motion of the block as it moves from A to C.
(b) Draw a free-body diagram for the block when it is at point P.
(c) What is the direction of the acceleration of the block at point P? If it is 0, state that explicitly. Explain your reasoning.
(d) Does the magnitude of the acceleration increase, decrease, or remain the same as the block moves from B to C? Explain your reasoning.
(e) What is the net work done on the object as it moves from A to B? From B to C?
(f) Calculate the coefficient of friction μ^{kin}.

67. Continental Drift According to some recent highly accurate measurements made from satellites, the continent of North America is drifting at a rate of about 1 cm per year. Assuming a continent is about 50 km thick, estimate the kinetic energy the continental United States has as a result of this motion.

68. Fan Carts P&E Two fan carts labeled A and B are placed on opposite sides of a table with their fans pointed in the same direction as shown in Fig. 9-40. Cart A is weighted with iron bars so it is twice as massive as cart B. When each fan is turned on, it provides the same constant force on the cart independent of its mass. Assume that friction is small enough to be neglected. The fans are set with a timer so that after they are switched on, *they stay on for a fixed length of time*, Δt, *and then turn off.*

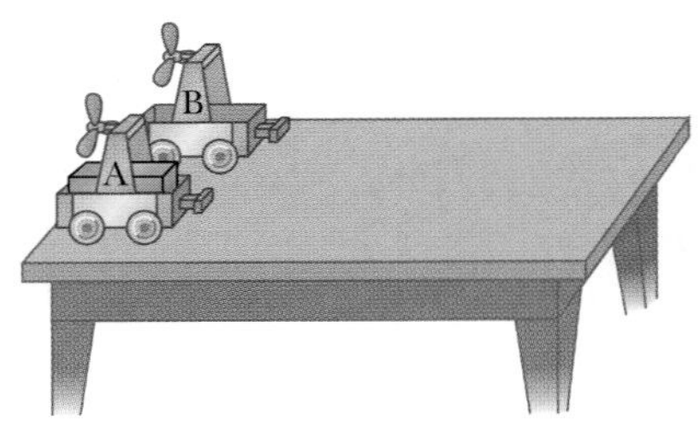

FIGURE 9-40 ■ Problem 68.

(a) *Just after the fans turn off*, which of the following statements is true about the magnitude of the momenta of the two carts?

(i) $p_A > p_B$
(ii) $p_A = p_B$
(iii) $p_A < p_B$

(b) *Just after the fans turn off*, which of the following statements is true about the kinetic energies of the two carts?

(i) $K_A > K_B$
(ii) $K_A = K_B$
(iii) $K_A < K_B$

(c) Which of the following statements are true? You may choose as many as you like, or none. If you choose none, write N.

(i) After the fans are turned on, each cart moves at a constant velocity, but the two velocities are different from each other.
(ii) The kinetic energy of each cart is conserved.
(iii) The momentum of each cart is conserved.

69. Sticky Carts Two identical carts labeled A and B are initially resting on a smooth track. The coordinate system is shown in Fig. 9-41a. The cart on the right, cart B, is given a push to the left and is released. The clock is then started. At $t_1 = 0$, cart B moves in the direction shown with a speed v_1. The carts hit and stick to each other. The graphs in Fig. 9-41b describe some of the variables associated with the motion as a function of time, but without labels on the vertical axis. For the experiment described and for each item in the list below, identify which graph (or graphs) is a possible display of that variable as a function of time, assuming a proper scale and units. "The system" refers to carts A and B together. Friction is so small that it can be ignored. If none apply, write N.

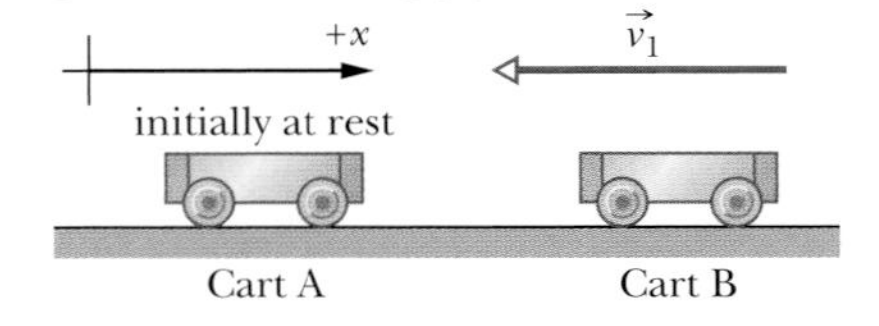

FIGURE 9-41a ■ Problem 69.

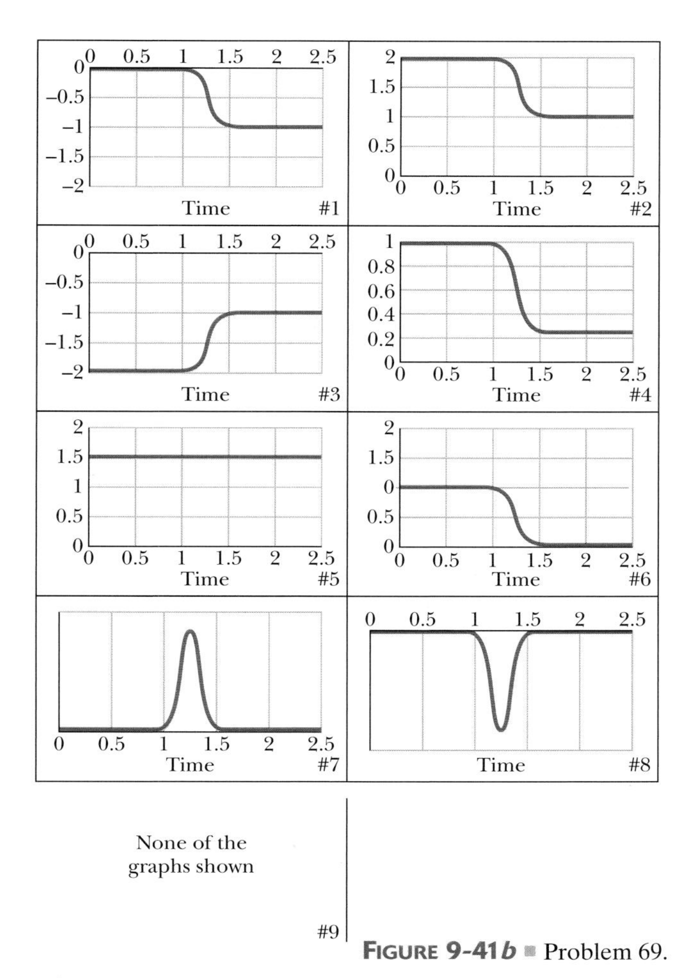

FIGURE 9-41*b* ■ Problem 69.

(a) The x-component of momentum of cart B
(b) The x-component of force on cart A
(c) The x-component of total momentum of the system
(d) The kinetic energy of cart B
(e) The total kinetic energy of the system

70. Graphs and Carts Two *identical* carts are riding on an air track. Cart A is given a quick push in the positive x direction toward cart B. When the carts hit, they stick to each other. The graphs shown in Fig. 9-42 describe some of the variables associated with the motion as a function of time beginning just *after* the push is completed. For the experiment described and for each item in the list below, identify which graph (or graphs) is a possible display of that variable as a function of time.

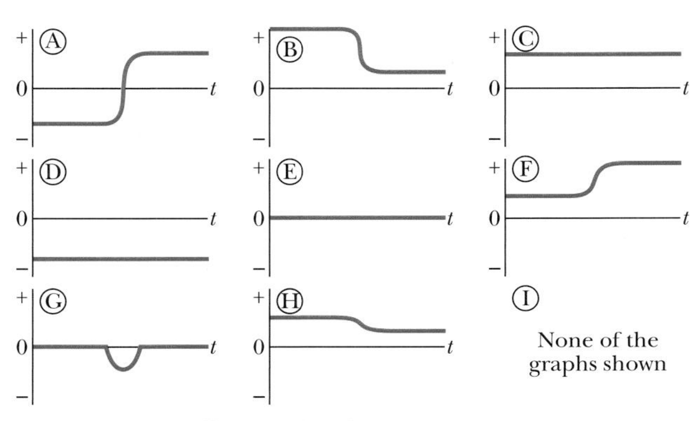

FIGURE 9-42 ■ Problem 70.

(a) The momentum of cart A
(b) The total momentum of the two carts
(c) The kinetic energy of cart A
(d) The force on cart A
(e) The force on cart B

71. Rebound to the Left A 5.0 kg block travels to the right on a rough, horizontal surface and collides with a spring. The speed of the block *just before* the collision is 3.0 m/s. The block continues to move to the right, compressing the spring to some maximum extent. The spring then forces the block to begin moving to the left. As the block rebounds to the left, it leaves the now uncompressed spring at 2.2 m/s. If the coefficient of kinetic friction between the block and surface is 0.30, determine (a) the work done by friction while the block is in contact with the spring and (b) the maximum distance the spring is compressed.

72. Rescue A helicopter lifts a stretcher with a 74 kg accident victim in it out of a canyon by applying a vertical force on the stretcher. The stretcher is attached to a guide rope, which is 50 meters long and makes an angle of 37° with respect to the horizontal. See Fig. 9-43. What is the work done by the helicopter on the injured person and stretcher?

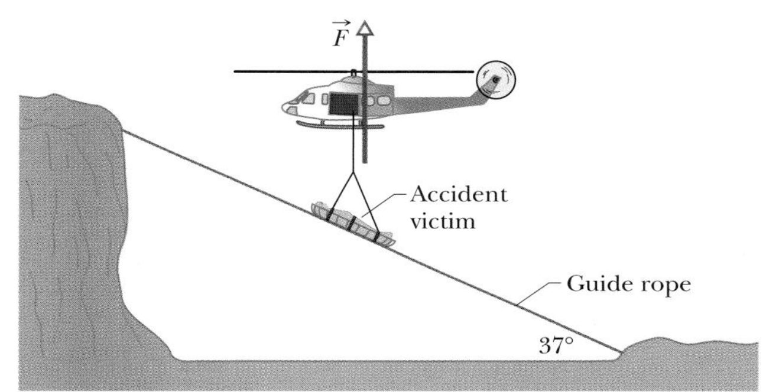

FIGURE 9-43 ■ Problem 72.

73. A Spring Idealized data for a spring's displacement $\Delta\vec{x}$ from its equilibrium position as a function of an external force, $\vec{F}^{\text{ext}}$, are shown in Fig. 9-44.

$\vec{F}^{\text{ext}}$ [N]	$\Delta\vec{x}$ [cm]
0.0	0
1.0	5
2.0	10
3.0	15
4.0	20

FIGURE 9-44 ■ Problem 73.

(a) Draw a properly scaled and carefully labeled graph of $\vec{F}^{\text{ext}}$ vs. $\Delta\vec{x}$ for these data.
(b) Does this spring obey Hooke's law? Why or why not?
(c) What is the value of its spring constant k?
(d) Shade the area on your graph that represents the amount of work done in stretching the spring from a displacement or extension of 0 cm to one of 5 cm. Also shade the area on the graph that represents the amount of work done in stretching the spring from a displacement or extension of 15 cm to one of 20 cm. Are the shaded areas approximately the same size? What does the size of the shaded area indicate about the work done in these two cases?
(e) Explain why the amount of work done in the second case is different from the amount done in the first case, even though the change in length of the spring is the same in both cases.

74. Variable Force The center of mass of a cart having a mass of 0.62 kg starts with a velocity of −2.5 m/s along an x axis. The cart starts from a position of 5.0 m and moves without any noticeable friction acting on it to a position of 0.0 meters. During this motion, a fan assembly exerts a force on the cart in a positive x direction.

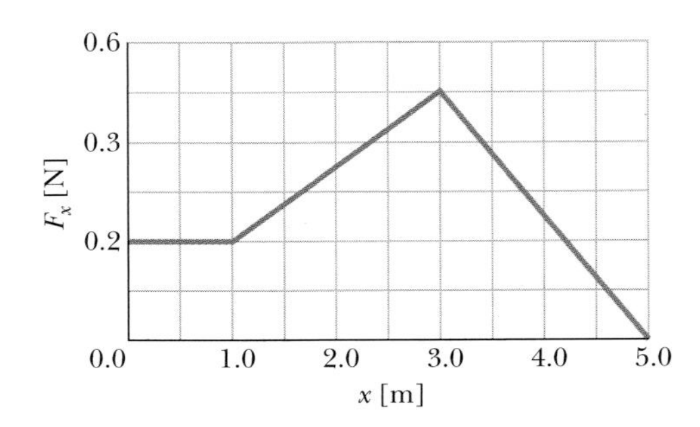

FIGURE 9-45 ■ Problem 74.

However, instead of being powered by batteries, the fan is driven by a voltage source that is programmed to change with its distance from a motion detector. This program leads to a variable force as shown in Fig. 9-45.

(a) What is the work done on the cart by the fan as the cart moves from 5.0 m to 0.0 m? (b) What is the change in kinetic energy of the cart between 5.0 m and 0.0 m? (c) What is the final velocity of the cart when it is at 0.0 m?

75. Given a Shove An ice skater of mass m is given a shove on a frozen pond. After the shove she has a speed of $v_1 = 2$ m/s. Assume that the only horizontal force that acts on her is a slight frictional force between the blades of the skates and the ice.

(a) Draw a free-body diagram showing the horizontal force and the two vertical forces that act on the skater. Identify these forces. **(b)** Use the net work-kinetic energy theorem to find the distance the skater moves before coming to rest. Assume that the coefficient of kinetic friction between the blades of the skates and the ice is $\mu^{\text{kin}} = 0.12$.

76. Karate Board Tester The karate board tester shown in Fig. 9-46*a* is a destructive testing device that allows one to determine the deformation of the center of a pine karate board as a function of the forces applied to it.

The displacement component, Δx, of the center of a pine board from its equilibrium position increases as a function of the x-component of an external force, F_x^{ext}, applied to it as shown in the data table of Fig. 9-46*b*.

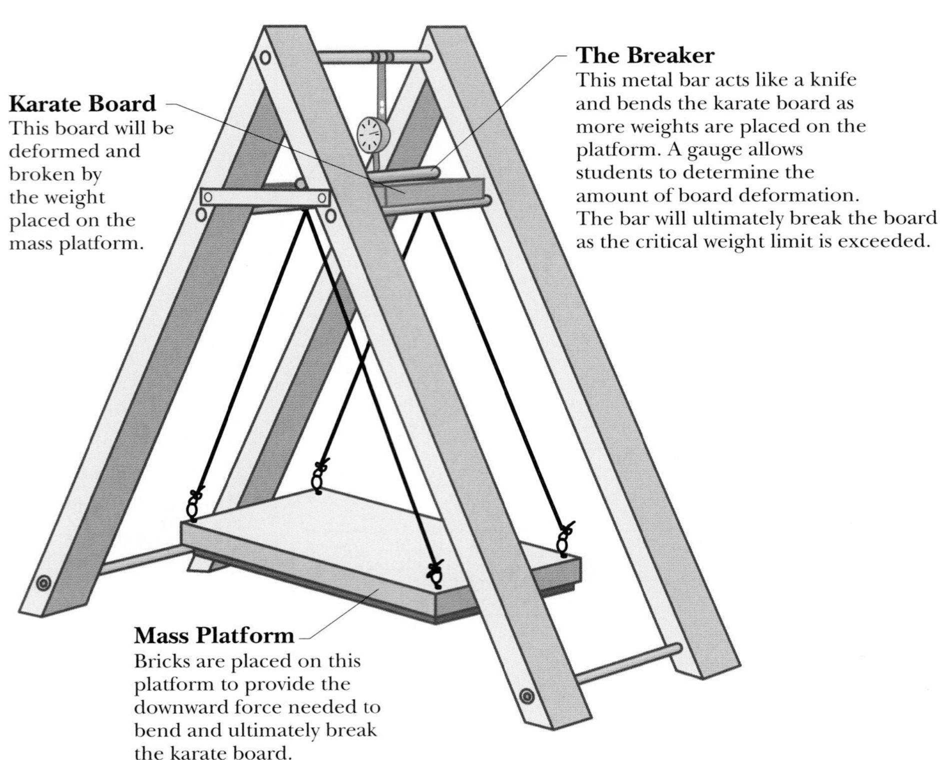

Karate Board
This board will be deformed and broken by the weight placed on the mass platform.
The Breaker
This metal bar acts like a knife and bends the karate board as more weights are placed on the platform. A gauge allows students to determine the amount of board deformation. The bar will ultimately break the board as the critical weight limit is exceeded.
Mass Platform
Bricks are placed on this platform to provide the downward force needed to bend and ultimately break the karate board.

FIGURE 9-46*a* ■ Problem 76.

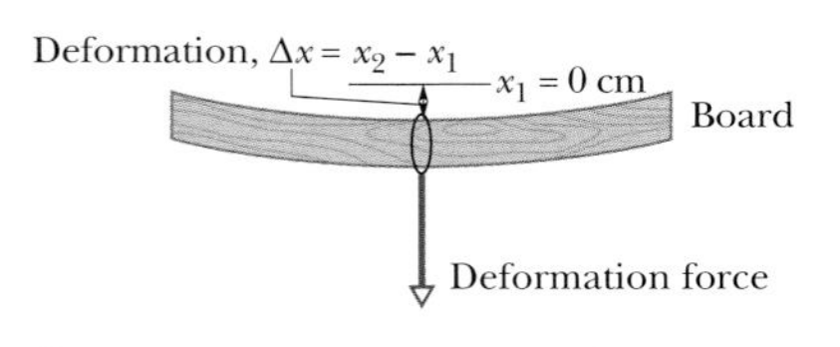

F_x^{ext} [N]	Δx[cm]
0	0.000
167	0.156
326	0.312
461	0.446
567	0.602

FIGURE 9-46*b* ■ Problem 76.

(a) Draw a properly scaled and carefully labeled graph of F^{ext} vs. Δx for these data.
(b) Does this pine board obey Hooke's law? Why or why not?
(c) What is the value of the effective spring constant k for the board?
(d) Shade the area on the graph that represents the amount of work done in stretching the center of the board from a displacement or extension of 0.000 cm to one of 0.156 cm. Also shade the area on the graph that represents the amount of work done in stretching the board from a displacement or extension of 0.446 cm to one of 0.602 cm. Are the shaded areas approximately the same size? What does the size of the shaded area indicate about the work done in these two cases?
(e) Explain why the amount of work done in the second case is different from the amount done in the first case, even though the change in displacement of the board is the same in both cases.

77. Karate Chop Movie In the movie DSON012 (available in VideoPoint or from your instructor) a physics student breaks a stack of eight pine boards. The thickness of the stack of boards with spacers is 0.34 m. In answering the following questions, treat any work done by gravitational forces on the student's hand as negligible.

(a) Use video analysis software to analyze the motion of the student's hand in the vertical or y direction. By using data from frames 3–5, find the velocity of the student's hand just before he hits the boards. By using data from frames 7–9, find the velocity of the student's hand just after he breaks all the boards.
(b) Assume that the effective mass of the student's hand is 1.0 kg. Use the net work-kinetic energy theorem to find the work done on the student's hand by the boards.

10 Potential Energy and Energy Conservation

The prehistoric people of Easter Island carved hundreds of giant stone statues in their quarry, then moved them to sites all over the island. How they managed to move them by as much as 10 km without the use of sophisticated machines has been a hotly debated subject, with many fanciful theories about the source of the required energy.

How could this have been accomplished using only primitive means?

The answer is in this chapter.

FIGURE 10-1 ■ While lifting a massive barbell, a powerlifter increases the separation between the barbell and Earth and rearranges the Earth–barbell system.

10-1 Introduction

In Chapter 9 we introduced the concepts of work and kinetic energy. We then derived a net work-kinetic energy theorem to describe what happens to the kinetic energy of a single rigid object when work is done on it. In this chapter we will consider systems composed of several objects that interact with one another. We are interested in what happens when forces from objects outside the system (*external* forces) change the arrangement of the interacting parts.

Let's consider two systems that can be reconfigured by external forces. The first system consists of an Earth–barbell system that has its arrangement changed when a weight lifter (outside of the system) pulls the barbell and the Earth apart by pulling up on the barbell with his arms and pushing down on the Earth with his feet (Fig. 10-1). The second system consists of two crates and a floor. This system is rearranged by a person (again, outside the system) who pushes the crates apart by pushing on one crate with her back and the other with her feet (Fig. 10-2). Although the external forces changing each system's configuration are exerted by a person pushing in opposite directions on two objects, there is an obvious difference between these two situations. Namely, as soon as the weight lifter stops pushing in both directions, the system's parts (barbell and Earth) fall back together. When the person stops pushing in opposite directions on the crates, the crates do not snap back together.

The internal interaction forces between the parts of these two systems differ. (Recall that we can call forces between objects within a system *internal forces*.) The lifter's forces are opposed by gravitational forces, but the crate-separator's forces are opposed by sliding friction forces. The lifter has to do a considerable amount of work to raise the barbell. However, when the barbell is dropped, we know that it picks up speed and gains kinetic energy as the Earth and the barbell move toward each other. In what sense can we say that the work the weight lifter did has been stored in the new configuration of the Earth–barbell system? And why does the work done by the woman separating the crates seem to be lost rather than stored away?

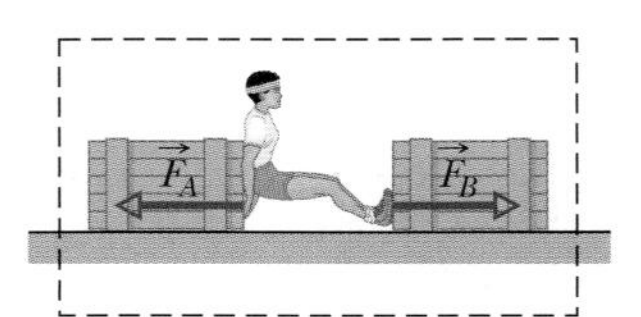

FIGURE 10-2 ■ A woman exerts equal and opposite forces on two crates. The work she does on the crates causes the crate–floor system to be rearranged.

10-2 Work and Path Dependence

There are many types of internal forces that can do work on a system of interacting objects. Examples include gravitational forces, sliding friction forces, spring forces, and air drag forces. How can we tell whether the work done by a certain type of internal force is "stored" or "used up" when the arrangement of a system changes?

A test has been devised for determining whether the work done by a particular type of force is "stored" or "used up." This test involves considering the work done by an internal interaction force when one part of a system moves. Consider a preliminary description of this test, which we will refine later:

> **TEST OF A SYSTEM'S ABILITY TO "STORE" WORK DONE BY INTERNAL FORCES (PRELIMINARY STATEMENT):** If the work done by a force between two objects within a system as some object in the system moves does not depend on the path taken, then the work done by this (internal) force can be stored in the system.

This test doesn't seem so strange when we apply it to some simple situations we have already discussed involving gravitational and friction forces.

The Path Independence Test for a Gravitational Force

Consider the skier traveling down a curved frictionless ramp as shown in Fig. 10-3. We showed in Section 9-9 that the net work done on the skier as she travels down the ramp is given by

$$W^{\text{net}} = -mg(y_2 - y_1) = F_y^{\text{grav}} \Delta y \qquad \text{(Eq. 9-42)}$$

and so does not depend on the shape of the ramp but only on the vertical component of the gravitational force and the vertical displacement of her center of mass. This result derives from that fact that whenever the skier has a component of motion in a horizontal direction the horizontal displacement is perpendicular to the Earth's gravitational force. These horizontal "detours" do not contribute to the work done by the Earth's gravitational force on the skier's center of mass.

Thus the work done on a particle by a gravitational force seems to be independent of the path taken to get from y_1 to y_2 as shown for the skier in Fig. 10-3. The gravitational force passes the test! If we do work on the system of skier and Earth to raise the skier to the top of the ramp, she can fall down again gaining kinetic energy, just as the barbell a weight lifter raises can fall down again. In both cases, we have to overcome the opposing internal gravitational force. In both cases, our work seems to be stored within the system.

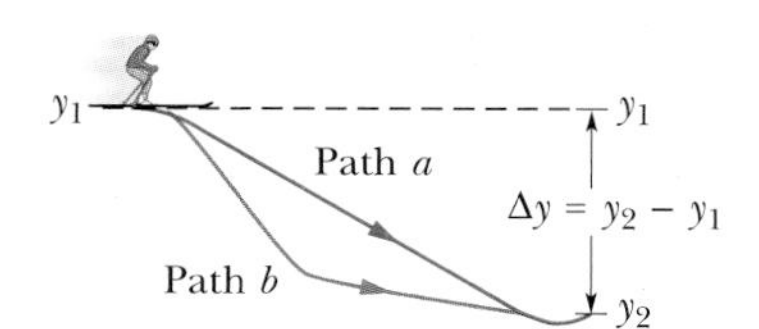

FIGURE 10-3 ■ The gravitational work done on a skier descending on a frictionless ramp depends only on the gravitational force and her vertical displacement Δy and not on the shape of the ramp.

Path Dependence of Work Done by a Friction Force

Our consideration of the skier indicates that the gravitational force does work that is path independent. But what about an object such as a crate that is displaced in the presence of a friction force? Is the work that the opposing friction force does on the crate path independent?

Consider pushing one of the crates shown in Fig. 10-2 along a level floor. Suppose the surface of the floor is quite uniform so that when you push with a constant magnitude of force, the crate moves at a constant speed. According to Newton's Second Law, if the acceleration of the crate is zero the net force on it is zero. So the external force you apply and the internal friction force must be equal and opposite. In this special case the friction force is steady. It always acts in a direction that opposes the displacement. If you push the crate directly from point 1 to point 2, you are taking it along path a as shown in Fig. 10-4. The work done by friction along that path is always negative (since the force and displacement are in opposite directions) and is given by

$$W_{1\to 2}^{\text{fric}} = -|f^{\text{kin}}|d,$$

where d is the distance between points 1 and 2. Suppose instead we push the crate along path b from points 1 to 4, then points 4 to 3 and then points 3 to 2, where the distance on each leg of the path is also d. The work done by the friction force is still negative and is given by

$$W_{1\to 4\to 3\to 2}^{\text{fric}} = (-|f^{\text{kin}}|d) + (-|f^{\text{kin}}|d) + (-|f^{\text{kin}}|d) = -3|f^{\text{kin}}|d.$$

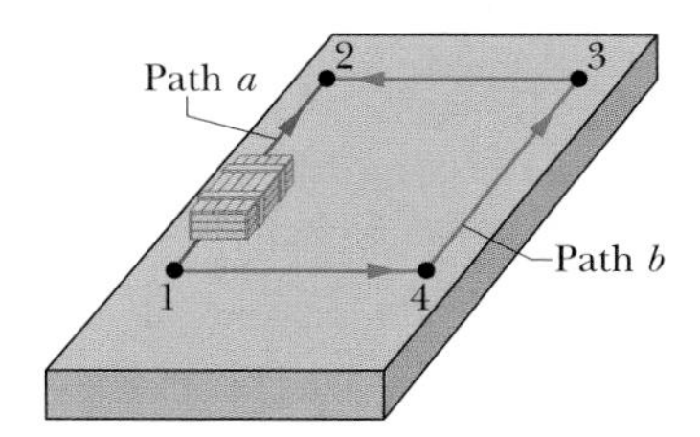

FIGURE 10-4 ■ The work done on a crate by the friction forces acting on it is much greater when it is pushed from point 1 to 2 along path b than along path a.

We see that the kinetic friction force does not pass the path independence test! The negative work done on the crate by the friction force is three times greater for path b than for path a. In general, if a friction force of constant magnitude is the only internal force in a system, then the work needed to get from point 1 to point 2 is proportional to the length of the path taken. Thus kinetic friction is a path-dependent force. This suggests that path dependence is an indicator of whether or not external work done on a system can be stored. When you do external work on the crate that is part of a crate–floor system, the system cannot then use the external work you do on it to rearrange itself after you stop pushing.

Conservative Forces and Path Independence

So far we have seen that the work done on a system that has gravitational forces acting between its parts seems to be path independent and seems to "store external work." Alternatively, if friction forces act between system parts, the opposite seems to hold. The system cannot store external work, and the friction forces do work that is not path independent. It is customary to define gravitational and other forces that do path independent work as **conservative forces** and forces that do not as **nonconservative forces.** The term "conservative" implies that something related to work is "stored" or conserved when the parts of a system are rearranged. In the next few sections of this chapter we will explore the concept of "conserved" in more detail.

General Statements about Conservative Forces

There is an alternative way to apply the path independence test to the work done by a force. It has to do with the net work associated with motion along a closed path—that is, motion in which an object makes a round trip through space, returning to its original location. Figure 10-5*b* shows an arbitrary round trip for a particle that has work done on it by a single force during its trip. The particle moves from an initial point 1 to point 2 along path *a* and then back to point 1 along path *b*. The internal force does work on the particle as the particle moves along each path. Without worrying about where positive work is done and where negative work is done, let us just represent the work between points 1 and 2 as the particle moves along path *a* as $W_{1\to 2}^{\text{path } a}$. Then we can denote the work done between points 1 and 2 if the particle moves along path *b* as $W_{1\to 2}^{\text{path } b}$ (Fig 10-5*a*). If the force is conservative, then the net work done is the same for either path,

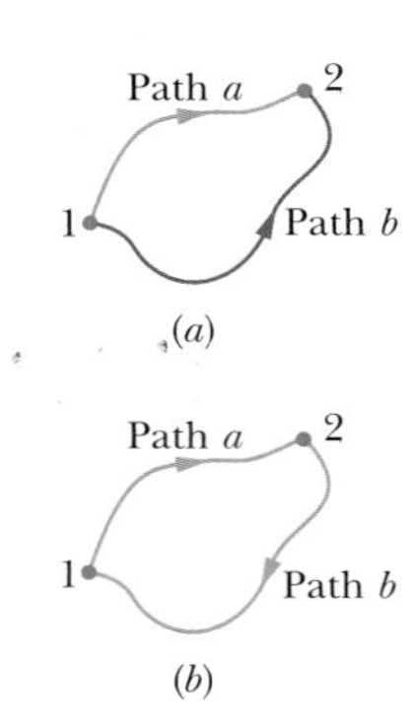

FIGURE 10-5 ■ When a particle is acted on by a conservative force, the work done by the force is: (*a*) *independent* of whether the particle moves from point 1 to point 2 by following either path *a* or path *b*; or (*b*) *zero* if the particle makes any possible round trip from 1 back to point 1. One possible round trip includes moving to point 2 along path *a* and then back to point 1 along path *b*.

$$W_{1\to 2}^{\text{path } a} = W_{1\to 2}^{\text{path } b} \quad \text{so that} \quad W_{1\to 2}^{\text{path } a} - W_{1\to 2}^{\text{path } b} = 0.$$

However, if we move in the opposite direction and go along path *b* from point 2 to point 1, then all the increments of displacement change sign, and work done in one direction is the negative of work done in the other direction. This is given by

$$W_{1\to 2}^{\text{path } b} = -W_{2\to 1}^{\text{path } b}.$$

Thus we get the following expression for the work done on a particle as it makes a round trip along a closed path traveling from point 1 to point 2 along path *a* and then back from point 2 to point 1 along path *b*,

$$W_{1\to 2}^{\text{path } a} + W_{2\to 1}^{\text{path } b} = 0 \qquad \text{(conservative force only).} \tag{10-1}$$

This equation tells us that the work done by a conservative force along any closed path is zero.

> **CONSERVATIVE FORCE TEST:** The work done by a conservative force on a particle moving between two points does not depend on the path taken by the particle. An alternative statement of this test is that the net work done by a conservative force on a particle moving around any closed path is zero.

The path independence of conservative forces has another useful aspect. If you need to calculate the work done by a conservative force along a given path between two points and the calculation is difficult, you can find the work by using another path between those two points for which the calculation is easier.

The Conservative Force Test for a Spring Force

So far the only systems we have introduced that have conservative internal forces are those in which the gravitational force acts alone. Let's consider another system consisting of a wall, a rigid block, and an ideal table that does not exert friction forces on the block. We assume that the wall and the block interact because they are connected by an ideal spring (see Fig. 10-6). The end of the spring that is free to move exerts a force on the block in a direction opposite to the displacement from the spring's relaxed position. According to Hooke's law, the component of the spring force on the block is given by $F_x^{\text{spring}}(x) = -kx$, where x is the displacement from a relaxed state at $x = 0$ (Eq. 9-17). We used this force and the definition of work to show that the work done on the block by the spring force is

$$W^{\text{spring}} = +\tfrac{1}{2}kx_1^2 - \tfrac{1}{2}kx_2^2, \qquad \text{(Eq. 9-19)}$$

whenever the spring is stretched or compressed from position x_1 to position x_2, along one-dimensional paths.

Before discussing the application of the conservative force test, we need to point out that a single spring exerts a force that is inherently one-dimensional. So thinking about paths that the block might take under the influence of net external forces does not make sense unless we restrict ourselves to situations for which the net external force acts along the line of the spring. For the sake of discussion we choose the line of the spring to be the x axis.

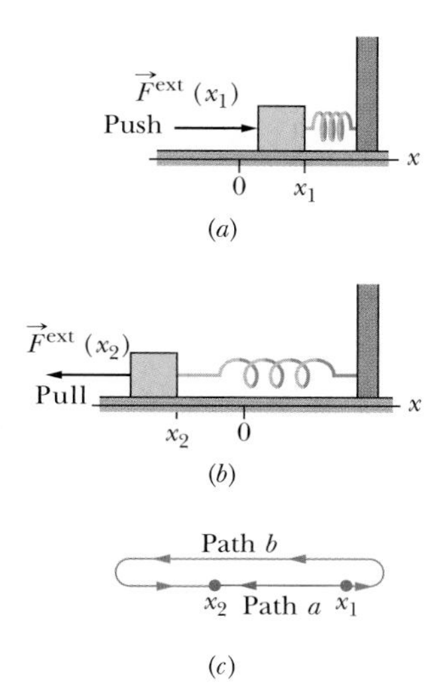

FIGURE 10-6 ■ A block is attached to a spring that is anchored to a wall. (a) An external force is used to push the block so the position component of the end of the spring is x_1. (b) Then, the block is pulled out so the position component of the end of the spring is x_2. (c) Two of many possible paths the spring end can take to get from x_1 to x_2.

Let's apply the test that says that if the spring force is conservative, then the work done along any one-dimensional path is the same. Figure 10-6 shows a block that is pushed inward and then pulled outward by an external force. Eq. 9-19 indicates the work done by the spring. The equation describing the spring's work depends only on the two locations x_1 and x_2 and not on how the spring got from one location to the other. For example, you could start the spring end at location x_1 and push it in further, then pull it out past x_2 and finally back to x_2. The work will be the same no matter what one-dimensional path you take—that is, as long as you don't impose very large displacements on the spring that cause Hooke's law to break down.

Since the work done by ideal spring forces is path independent, we can conclude that

> The ideal spring force is a conservative force, as is the gravitational force.

The alternate test that requires the work done by the spring force to be zero on a round trip is also true, since for a round trip $x_2 = x_1$ so $W^{\text{spring}} = \frac{1}{2}kx_1^2 - \frac{1}{2}kx_2^2 = 0$. The fact that the spring force is zero in a round trip makes sense. Suppose the spring starts out in a compressed position. When you push on a spring and compress it further, the spring force opposes its displacement and the work done by the spring is negative. If you then pull the spring back to its original but still compressed position, the spring force and the displacement are in the same direction and the work done by the spring is positive. So the negative work done by the spring while it is being pushed in and the positive work it does while being pulled out add up to zero.

When a spring attached to a wall is stretched and released it naturally heads toward its equilibrium position. This is not unlike the weight lifter's mass naturally falling back toward the Earth. The external work done on the block in opposition to the spring force seems to be stored in the wall–spring–block system.

The Conservative Force Test for a Car on a Hot Wheels® Track

Let's see how our two conservative force tests are applied to a fairly complex system consisting of a low-friction toy car, a Hot Wheels® track, and the Earth. Are the internal forces the system exerts on the car conservative?

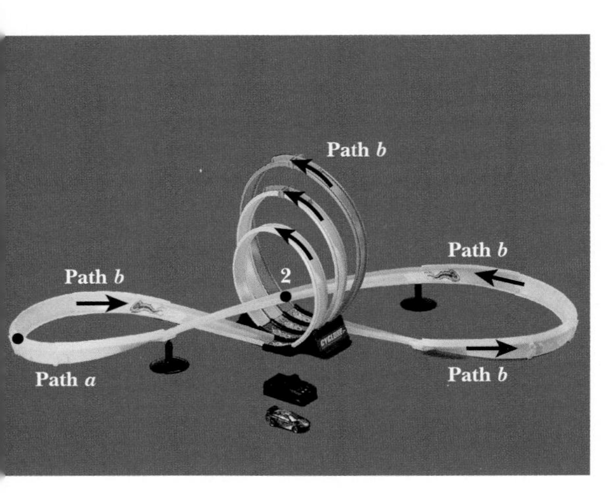

FIGURE 10-7 ■ This Hot Wheels® race track allows toy cars to travel around a closed path. The point labeled 2 is an overpass.

Using the Path Independence Test There are two possible paths that a "low-friction" toy car could take traveling between points 1 and 2 on the track shown in Fig. 10-7. The car could travel uphill directly from point 1 to point 2 along path a. Alternatively it could travel many times further by taking path b. On that path it proceeds downhill, passes under the top ramp, goes through three loop-the-loops, and traverses the figure-eight loop on the left before returning to point 2.

We can use the net work-kinetic energy theorem ($W^{\text{net}} = \Delta K$) developed in Chapter 9 to determine whether or not the internal forces do work on the car along certain paths. The net work is given by the sum of the works done by the internal forces that the rest of the system exerts on the car. In equation form this is $W^{\text{net}} = W^{\text{int}} = W^{\text{norm}} + W^{\text{grav}} + W^{\text{fric}}$.

The normal forces do not contribute to the work done on the car. This is because these forces are always perpendicular to the direction of the car's displacement at any point along the track. So the internal work is done by only a combination of the gravitational and friction forces so that $W^{\text{net}} = W^{\text{grav}} + W^{\text{fric}}$.

If we start the car along path a at point 1 with a certain amount of kinetic energy, then we can measure its kinetic energy at point 2 to determine the net internal work done on it by the rest of the system. We can make the same observation for the car as it goes from point 1 to point 2 along path b. If the kinetic energy change was the same along both paths, then the net work done along each path would be the same and we would conclude that the combination of the gravitational and friction forces is conservative. However, measurements tell us that a different amount of kinetic energy is lost along path b than along the more direct path a. We conclude that the combination of friction and gravitational forces acting on the car is not conservative. Since we believe that gravitational forces are conservative, we suspect that friction is the problem.

Closed-Path Test It turns out that the closed-path test is a lot easier to apply in this case. All we have to do is ask the question: Is the net work done on the car in going around a closed loop (say, from point 1 to point 1) zero? If the answer is yes, then according to the net work-kinetic energy theorem given by $W^{\text{net}} = W^{\text{grav}} + W^{\text{fric}} = \Delta K = 0$, the car would lose no kinetic energy in making a complete loop. However, for this Hot Wheels® track we observe that the car does slow down, so there must be a loss in kinetic energy. As expected, therefore, the combination of gravitational and friction forces does not pass this logically equivalent conservative force test.

What if friction were not present? If we could devise a magic car with no friction in its wheel bearings, then the only type of force capable of doing work on the car as it traveled would be the conservative gravitational force on the car due to the Earth. In this case, the net work done on the car would be zero around a closed loop, and the car would lose no kinetic energy.

READING EXERCISE 10-1: The figure shows three paths connecting points 1 and 2. A single force $\vec{F}$ does the indicated work on a particle moving along each path in the indicated direction. On the basis of this information, is force $\vec{F}$ conservative?

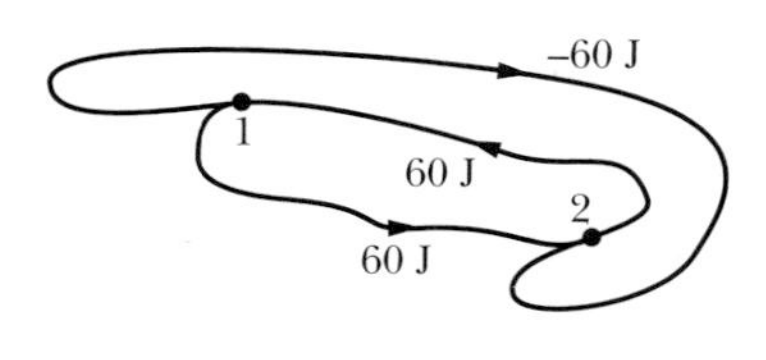

■

READING EXERCISE 10-2: In applying the path independence test for conservative forces to the car traveling on the Hot Wheels® track, we made the statement: "Measurements tell us that a different amount of kinetic energy is lost along path b than along the more direct path a." Which path do you think will have the most kinetic energy loss associated with it? Explain the reasons for your answer. ■

TOUCHSTONE EXAMPLE 10-1: Cheese on a Track

Figure 10-8a shows a 2.0 kg block of slippery cheese that slides along a frictionless track from point 1 to point 2. The cheese travels through a total distance of 2.0 m along the track, and a net vertical distance of 0.80 m. How much work is done on the cheese by the gravitational force during the slide?

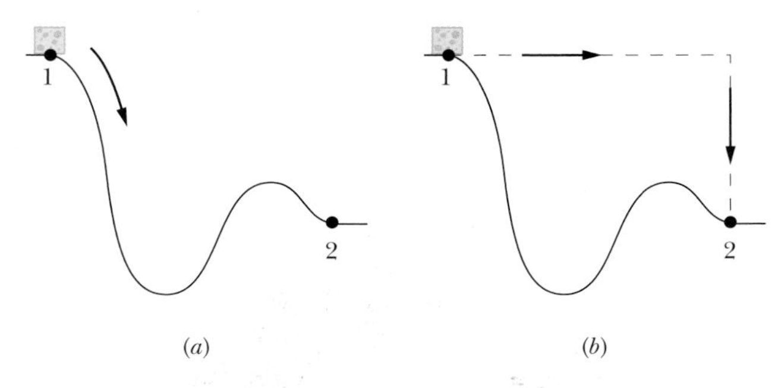

FIGURE 10-8 ■ (a) A block of cheese slides along a frictionless track from point 1 to point 2. (b) Finding the work done on the cheese by the gravitational force is easier along the dashed path than along the actual path taken by the cheese; the result is the same for both paths.

SOLUTION ■ A **Key Idea** here is that we *cannot* use Eq. 9-26 ($W^{\text{grav}} = |\vec{F}^{\text{grav}}||\Delta\vec{r}|\cos\phi$) to calculate the work done by the gravitational force $\vec{F}^{\text{grav}}$ as the cheese moves along the track. The reason is that the angle ϕ between the directions of $\vec{F}^{\text{grav}}$ and the displacement $\Delta\vec{r}$ varies along the track in an unknown way. (Even if we did know the shape of the track and could calculate ϕ along it, the calculation could be very difficult.)

A second **Key Idea** is that because $\vec{F}^{\text{grav}}$ is a conservative force, we can find the work by choosing some other path between 1 and 2—one that makes the calculation easy. Let us choose the dashed path in Fig. 10-8b; it consists of two straight segments. Along the horizontal segment, the angle ϕ is a constant 90°. Even though we do not know the displacement along that horizontal segment, Eq. 9-26 tells us that the work W^{horiz} done there is

$$W^{\text{horiz}} = mg|\Delta\vec{r}|\cos 90^\circ = 0.$$

Along the vertical segment, the magnitude of the displacement $|\Delta\vec{r}|$ is 0.80 m and, with $\vec{F}^{\text{grav}}$ and $\Delta\vec{r}$ both downward, the angle ϕ is a constant 0°. Thus, Eq. 9-26 gives us, for the work W^{vert} done along the vertical part of the dashed path,

$$\begin{aligned} W^{\text{vert}} &= mg|\Delta\vec{r}|\cos 0^\circ \\ &= (2.0\text{ kg})(9.8\text{ m/s}^2)(0.80\text{ m})(1) = 15.7\text{ J}. \end{aligned}$$

The total work done on the cheese by $\vec{F}^{\text{grav}}$ as the cheese moves from point a to point b along the dashed path is then

$$W = W^{\text{horiz}} + W^{\text{vert}} = 0 + 15.7\text{ J} \approx 16\text{ J}. \qquad \text{(Answer)}$$

This is also the work done as the cheese moves along the track from 1 to 2.

10-3 Potential Energy as "Stored Work"

If external work can be stored when a system of objects is rearranged, we refer to the system as a "conservative system." In this section we will define a new quantity called potential energy as a measure of stored work in a conservative system.

Rearranging a Gravitational System

Consider the external work that weightlifter Sun Ruiping does separating the Earth and the 118.5 kg barbell shown in Fig 10-9. Ruiping acts as an external agent that does work on *both* the Earth and the barbell in opposing the gravitational force as she pushes up on the barbell with her hands and down on the Earth with her feet. The *net* work done on the Earth–barbell system during the time the barbell is raised is the sum of the external work, $W_{\text{sys}}^{\text{ext}}$, done on the system by our weightlifter and the internal gravitational work, $W_{\text{sys}}^{\text{int}} = W_{\text{sys}}^{\text{grav}}$, that the two objects in the system exert on each other. This can be summarized by the equation

$$W_{\text{sys}}^{\text{net}} = W_{\text{sys}}^{\text{ext}} + W_{\text{sys}}^{\text{int}} \quad \text{(net work on a system).} \qquad (10\text{-}2)$$

Since the velocities of both the barbell and the Earth are zero before and after the lift, there is no change in the kinetic energy of the Earth–barbell system as a

(a)

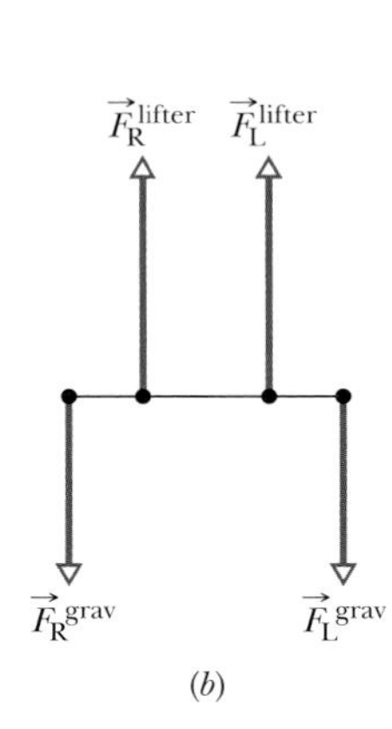

(b)

FIGURE 10-9 ■ (*a*) China's Sun Ruiping broke the world record for the snatch lift in October 2002. In the middle part of her lift the barbell has an upward acceleration. Thus the positive force she exerted on the barbell is greater than the negative force exerted by the Earth. The lifter does positive external work on the Earth–barbell system while the system does negative internal work on itself. (*b*) A modified free-body diagram of forces on the barbell. The net force of the lifter exceeds the net gravitational force.

result of the lift. So, the net work-kinetic energy theorem developed in Chapter 9 tells us that

$$W_{\text{barbell}}^{\text{net}} = W_{\text{barbell}}^{\text{ext}} + W_{\text{barbell}}^{\text{grav}} = \Delta K_{\text{barbell}} = 0 \quad \text{(net work on the system).} \quad (10\text{-}3)$$

Since the net work on the system is zero,

$$W_{\text{barbell}}^{\text{ext}} = -W_{\text{barbell}}^{\text{grav}}. \quad (10\text{-}4)$$

As shown in Fig 10-10, an analysis of another lift, the work the lifter has to do to separate the barbell from the Earth is experimentally confirmed to be equal in magnitude to the gravitational work done by the Earth on the barbell.

In general, when an object is lifted near the surface of the Earth we often carelessly think that the work done on the lifted object is different from the work done on the system. However, as long as we calculate the work using the change in separation of the Earth and object, there is no difference.

FIGURE 10-10 ■ Results of video analysis of the lifting of a 372.5 kg barbell by about a half-meter. (*a*) y vs. t of the barbell during the lift. This graph was fitted with a polynomial to allow determination of the acceleration and hence the variation over time of net force on the barbell during the lift. (*b*) A graph of net force on the barbell during the lift vs. the y-component of the height shows that the area under the F_y^{net} vs y curve that defines the net work on the system is zero.

Defining Potential Energy Change

Alas, our weightlifter's labor did not lead to a change in kinetic energy! However, suppose the lifter dropped her barbell. The barbell would *gain* an amount of kinetic energy while falling that is just equal to the work the lifter had to do on the system to raise it. We call this increased potential for kinetic energy gain a *potential energy* change, ΔU. Basically this *change* in potential energy is "stored work." The term "change" is used to allow for the possibility that the system already had some potential energy stored in it before the external work was done.

However, according to $W_{\text{sys}}^{\text{net}} = W_{\text{sys}}^{\text{ext}} + W_{\text{sys}}^{\text{int}}$ (Eq. 10-2) when the kinetic energy of the system does not change, the external work is equal to the *negative* of the internal work done by interaction forces. This leads us to a general definition of potential energy change for a conservative system (one with only conservative internal forces) in terms of the work done by internal forces on parts of the system.

POTENTIAL ENERGY CHANGE FOR A CONSERVATIVE SYSTEM is defined as the negative of the internal work the system does on itself when it undergoes a reconfiguration.

Even though we used an example of a two-object system to motivate this definition, we can also apply it to many-body systems. This is discussed in more detail in Chapter 25, where we deal with the potential energy associated with electrostatic forces. Symbolically, the general definition of potential energy change for a single conservative force is

$$\Delta U \equiv -W^{\text{cons}} \quad \text{(definition of potential energy change).} \quad (10\text{-}5)$$

Here W^{cons} is the work done by a specific conservative force and ΔU is the change in potential energy associated with that force.

Gravitational Potential Energy When we are near the surface of the Earth, we can use the expression $W^{grav} = -mg(y_2 - y_1)$, (Eq. 9-14), to derive an expression for the change in gravitational potential energy (ΔU^{grav}) of an object that is lifted from one height y_1 to another height y_2. W^{grav} represents the internal work done by the system, and

$$\Delta U^{grav} = -W^{grav} = +mg(y_2 - y_1) \quad \text{(gravitational PE change near the Earth's surface).} \quad (10\text{-}6)$$

Only *changes* ΔU in gravitational potential energy (or any other type of potential energy) are physically meaningful. In an object–Earth system, there is no special separation between the center of the Earth and an object that obviously has zero potential energy. However, to simplify a calculation or a discussion, we often choose to set the gravitational potential energy value U^{grav} to zero when the object is at a certain height. To do so, we rewrite Eq. 10-6 as

$$\Delta U^{grav} = U_2^{grav} - U_1^{grav} = mg(y_2 - y_1). \quad (10\text{-}7)$$

Then we take U_1 to be the gravitational potential energy (GPE) of the system when it is in a **reference configuration** (in which the object is at a **reference point** y_1). Usually we take the reference point to be $y_1 = 0$ so $U_1^{grav} = U^{grav}(y_1) = 0$. If we do this, and replace the specific point y_2 with the more general y, Eq. 10-7 becomes

$$U^{grav}(y) = mgy \quad \text{(GPE } \textit{relative} \text{ to a chosen origin).} \quad (10\text{-}8)$$

This equation tells us that:

Near the Earth's surface, the gravitational potential energy associated with an object–Earth system depends only on the vertical position y (or height) of the object relative to the reference height $y_1 = 0$, not on its horizontal location.

Elastic (or Spring) Potential Energy The same definition of change in potential energy (in Eq. 10-5) applies equally well to a block–spring–wall system like that shown in Fig. 10-6. So $\Delta U^{spring} = -W^{spring}$; that is,

$$W^{spring} = +\tfrac{1}{2}kx_1^2 - \tfrac{1}{2}kx_2^2, \quad \text{(Eq. 9-19)}$$

and so

$$\Delta U^{spring} = -W^{spring} = \tfrac{1}{2}kx_2^2 - \tfrac{1}{2}kx_1^2 \quad \text{(ideal spring PE change).} \quad (10\text{-}9)$$

A spring–block system has a natural zero point for potential energy when the spring is unstretched. So, to associate an elastic potential energy (EPE) value U^{spring} with the block at position x_2, we choose the reference point to be the block's location when the spring is at its relaxed length. If we let $x_1 = 0$ at that point, then the elastic potential energy U^{spring} is 0 there, and Eq. 10-9 becomes

$$\Delta U^{spring} = U_2^{spring} - U_1^{spring} = \tfrac{1}{2}kx_2^2 - 0,$$

which gives us the general expression

$$U^{spring}(x) = \tfrac{1}{2}kx^2 \quad \text{(ideal EPE relative to block location with spring relaxed).} \quad (10\text{-}10)$$

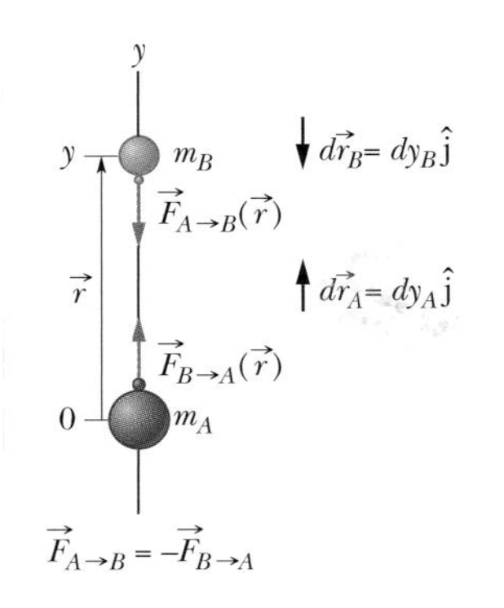

FIGURE 10-11 ■ Two bodies interact by means of a conservative force. According to Newton's Third Law, they exert "equal and opposite" forces on each other. In general, the magnitude of these forces depends only on their separation, $|\vec{r}|$.

Potential Energy Change for any Conservative Two-Body System

Now that we have considered two systems that can undergo potential energy change, we are ready to derive a general expression for the PE changes for any conservative two-body system. Let's start by defining a y axis that passes through two interacting bodies, body A and body B, with its origin located at body A as shown in Fig. 10-11. For this choice of coordinate system, the internal interaction forces all point along the y axis.

Since the change in the system's potential energy is $\Delta U = -W^{\text{cons}} - W^{\text{int}}$, the key to finding the potential energy change is to determine the internal work the particles do on each other as a result of a change in their separation. For an infinitesimally small change in separation, dr, the increment of internal work the system bodies do on each other, dW^{int}, is given by

$$dW^{\text{int}} = \vec{F}_{B\to A}(\vec{r}) \cdot d\vec{r}_A + \vec{F}_{A\to B}(\vec{r}) \cdot d\vec{r}_B.$$

Because we chose our y axis along the displacement direction, we can represent the forces in terms of their y-components as $\vec{F}_{B\to A} = F_{A\,y}\,\hat{j}$ and $\vec{F}_{A\to B} = F_{B\,y}\,\hat{j}$. Here we shorten $F_{B\to A\,y}$ to $F_{A\,y}$ and $F_{A\to B\,y}$ to $F_{B\,y}$ and recall that these components can be positive or negative. This allows us to eliminate the dot products to get

$$dW^{\text{int}} = F_{A\,y}(y)\,dy_A + F_{B\,y}(y)\,dy_B.$$

Since Newton's Third Law tells us that $\vec{F}_{A\to B} = -\vec{F}_{B\to A}$ we know that the y-components of these vectors are related by $F_{A\,y} = -F_{B\,y}$, and we can rewrite dW^{int} as

$$dW^{\text{int}} = F_{A\,y}(y)\,dy_A + F_{B\,y}(y)\,dy_B = F_{B\,y}(y)d(y_B - y_A).$$

The interaction forces always point along an axis through the two particles, so that in vector notation $\vec{r} = y\,\hat{j}$. The changes in the interaction forces due to changes in separation depend only on y. The expression for dW^{int} can be expressed in terms of only the force that particle A exerts on particle B and the variable y. This simplifies dW^{int} to

$$dW^{\text{int}} = F_{B\,y}\,(y)\,dy.$$

For the most general case in which particle B moves relative to particle A from an initial location y_1 to a final location y_2, the internal work is given by the integral of dW^{int} with respect to y,

$$W^{\text{int}} = \int_{y_1}^{y_2} dW^{\text{int}} = \int_{y_1}^{y_2} F_{B\,y}(y)\,dy. \tag{10-11}$$

Substituting this into $\Delta U = -W^{\text{int}}$, we find that the change in the system's potential energy due to the change in configuration along our chosen y axis is

$$\Delta U = -W^{\text{int}} = -\int_{y_1}^{y_2} F_{B\,y}(y)\,dy. \tag{10-12}$$

We can equally well decide to place an y' axis instead of a y axis through the two points of interest or to develop the equation for ΔU in terms of the force of particle B on particle A. In the absence of choosing a specific coordinate system we can substitute

$$W^{\text{int}} \equiv \int_{\vec{r}_1}^{\vec{r}_2} \vec{F}(\vec{r}) \cdot d\vec{r} \qquad \text{(Eq. 9-36)}$$

for W^{int} to get a general expression for the change in a conservative system's potential energy in terms of the dot product, where $\vec{r}$ is the radius vector pointing from particle A to particle B,

$$\Delta U = -W^{\text{int}} = -\int_{r_1}^{r_2} \vec{F}_B(\vec{r}) \cdot d\vec{r}. \tag{10-13}$$

The valuable conclusion we have reached is that

> The potential energy change of a two-body system with only conservative internal forces that depend only on the separation between the particles can be determined by considering the internal force on only *one* of the bodies.

Since W^{int} is path independent, we can write ΔU as $\Delta U = U(\vec{r}_2) - U(\vec{r}_1)$. As we did earlier, we can choose a reference separation point $\vec{r}_1$ such that $U(\vec{r}_1) = 0$. Then we can express the potential energy of a particle relative to this reference for any $\vec{r}_2$ (or more generally $\vec{r}$) as $U(\vec{r})$.

READING EXERCISE 10-3: The net work-kinetic energy theorem predicts that the net work should be zero when the barbell is raised from its low point to its high point. Are the data in Fig. 10-10*b* consistent with this prediction? Explain. ■

READING EXERCISE 10-4: A particle is to move along the x axis from $x_1 = 0$ to x_2 while a conservative internal force from a second particle force, directed along the x axis, acts on the first particle. The figure shows three situations in which the x-component of that force varies with x. The force has the same maximum magnitude F_1 in all three situations. Rank the situations according to the change in the associated potential energy during the particle's motion, most positive first.

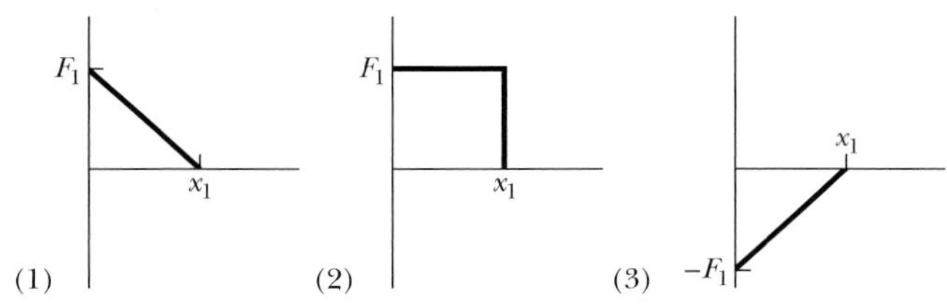

■

TOUCHSTONE EXAMPLE 10-2: Sloth's Energy

A 2.0 kg sloth clings to a limb that is 5.0 m above the ground (Fig. 10-12).

(a) What is the gravitational potential energy U^{grav} of the sloth–Earth system if we take the reference point $y_1 = 0$ to be (1) at the ground, (2) at a balcony floor that is 3.0 m above the ground, (3) at the limb, and (4) 1.0 m above the limb? Take the gravitational potential energy to be zero at $y_1 = 0$ and denote y_2 as y.

SOLUTION ■ The **Key Idea** here is that once we have chosen the reference point for $y_1 = 0$, we can calculate the gravitational potential energy U^{grav} of the system *relative to that reference point* with Eq. 10-8. For example, for choice (1) the sloth is at $y = 5.0$ m, and

$$U^{\text{grav}} = mgy = (2.0\text{ kg})(9.8\text{ m/s}^2)(5.0\text{ m})$$
$$= 98\text{ J}. \qquad \text{(Answer)}$$

For the other choices, the values of U^{grav} are

(2) $U^{\text{grav}} = mgy = mg(2.0\text{ m}) = 39\text{ J}$,

(3) $U^{\text{grav}} = mgy = mg(0) = 0\text{ J}$,

(4) $U^{\text{grav}} = mgy = mg(-1.0\text{ m}) = -19.6\text{ J} \approx -20\text{ J}$. (Answer)

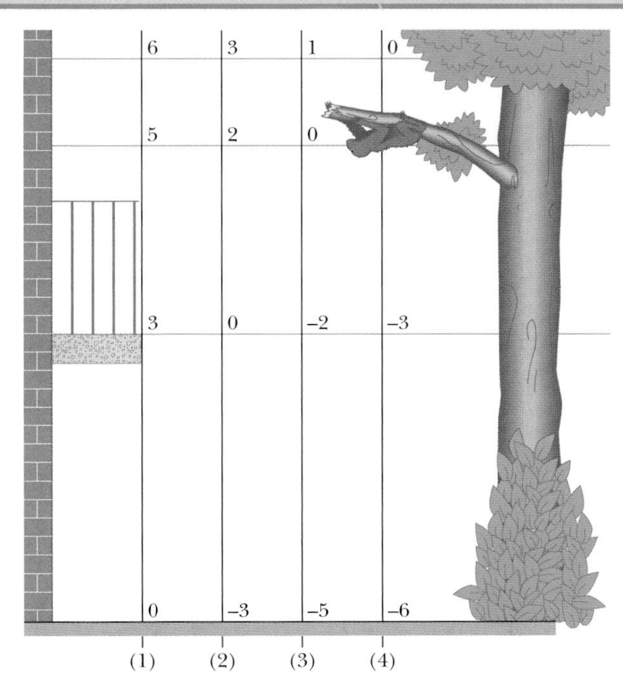

FIGURE 10-12 ■ Four choices of reference point $y_1 = 0$. Each y axis is marked in units of meters.

(b) The sloth drops to the ground. For each choice of reference point, what is the change ΔU^{grav} in the potential energy of the sloth–Earth system due to the fall?

SOLUTION ■ The **Key Idea** here is that the *change* in potential energy does not depend on the choice of the reference point for $y_1 = 0$; instead, it depends on the change in height Δy. For all four situations, we have the same $\Delta y = -5.0$ m. Thus, for (1) to (4), Eq. 10-6 tells us that

$$\Delta U^{\text{grav}} = mg\,\Delta y = (2.0\text{ kg})(9.8\text{ m/s}^2)(-5.0\text{ m})$$
$$= -98\text{ J}. \qquad \text{(Answer)}$$

10-4 Mechanical Energy Conservation

Let's consider a collection of rigid objects or particles in a system that interact only by means of conservative internal forces. Basically we are assuming that the system is *isolated* from its environment, so that no *external force* is present to do work on the system.

The system parts can have kinetic energy. For example, when a barbell is dropped the Earth–barbell system acquires kinetic energy as its gravitational potential energy decreases. This decrease in potential energy with increase in kinetic energy leads us to suspect that for isolated conservative systems, the sum of these two energies might be constant. As we saw in Chapter 7 on momentum, when a quantity is constant over time, physicists say that quantity is **conserved.** We will explore this possibility by defining a new quantity we call mechanical energy, or E^{mec}, that is the sum of the kinetic energy K and the potential energy U of a system. Symbolically we get

$$E^{\text{mec}} \equiv K + U \quad \text{(definition of mechanical energy).} \qquad (10\text{-}14)$$

In this section, we examine what happens to the mechanical energy of an isolated system when all of its internal forces are conservative.

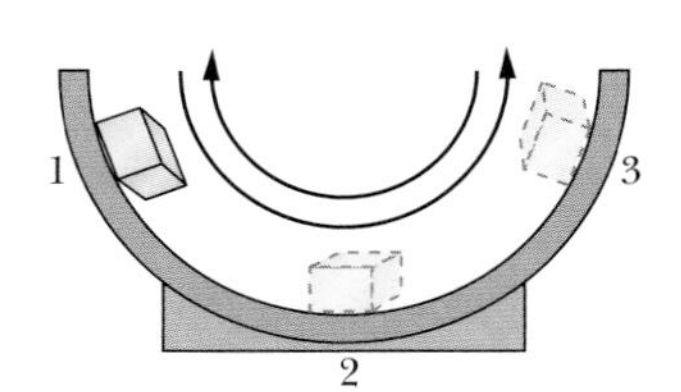

FIGURE 10-13 ■ A small ice cube oscillates back and forth on a curved frictionless ramp constantly trading energy between potential energy and kinetic energy.

Imagine a small ice cube that is placed on a curved, frictionless ramp bolted to a table, as in Fig. 10-13. If the ice cube is released from point 1 it will oscillate back and forth. When first released it falls toward point 2 under the influence of the conservative gravitational force component parallel to the surface of the ramp. The kinetic energy of the ice cube will increase and it loses potential energy. The system's potential energy will be a minimum at point 2 when its kinetic energy is maximum. As it rises toward point 3 it loses kinetic energy and it gains potential energy.

In general, when an internal conservative force does work W^{int} on an object within the system and no other objects in the system move appreciably, the system transfers energy between kinetic energy K of the object and potential energy U of the system. According to the net work-kinetic energy theorem, if the only work done on an object in a system is the internal work, then the change ΔK in kinetic energy is

$$\Delta K = W^{\text{net}} = W^{\text{int}}. \qquad (10\text{-}15)$$

The change ΔU in the potential energy of the system is

$$\Delta U = -W^{\text{int}} \qquad (10\text{-}16)$$

where W^{int} is the sum of all works done by all the conservative internal forces. Combining these two equations, we find that

$$\Delta K = -\Delta U. \qquad (10\text{-}17)$$

This shows that one of these energies increases exactly as much as the other decreases.

We can rewrite $\Delta K = -\Delta U$ as

$$K_2 - K_1 = -(U_2 - U_1), \qquad (10\text{-}18)$$

where the subscripts refer to two different instants and thus to two different arrangements of the objects in the system. Rearranging $K_2 - K_1 = -(U_2 - U_1)$ yields

$$E^{\text{mec}} = K_1 + U_1 = K_2 + U_2 \quad \text{(conservation of mechanical energy).} \qquad (10\text{-}19)$$

In words, this equation says that

> In a system where (1) no work is done on it by external forces and (2) only conservative internal forces act on the system elements, then the internal forces in the system can cause energy to be transferred between kinetic energy and potential energy, but their sum, the mechanical energy E^{mec} of the system, cannot change.

This result is called the **conservation of mechanical energy.** Beware! Conservation of mechanical energy only holds under the special conditions we just outlined. Now you can see where *conservative* forces got their name. With the aid of $\Delta K = -\Delta U$ (Eq. 10-17), we can write this principle in one more form, as

$$\Delta E^{\text{mec}} = \Delta K + \Delta U = 0. \tag{10-20}$$

In cases where it holds, conservation of mechanical energy allows us to solve problems that would be quite difficult to solve using only Newton's laws:

> When the mechanical energy of a system is conserved, we can relate the sum of kinetic energy and potential energy at one instant to that at any other instant *without considering the intermediate motion* and *without finding the work done by the forces involved.*

Figure 10-14 shows an example in which the principle of conservation of mechanical energy can be applied. As a pendulum swings, the energy of the pendulum–Earth system is transferred back and forth between kinetic energy K and gravitational potential energy U, with the sum $K + U$ being constant. If we are given the gravitational potential energy when the pendulum bob is at its highest point (Fig 10-14, stage 1), we can find the kinetic energy of the bob at the lowest point (Fig. 10-14, stage 3) using $K_2 + U_2 = K_1 + U_1$ (Eq. 10-19). The continual exchange back and forth between potential energy and kinetic energy is shown in the graph in Fig. 10-14.

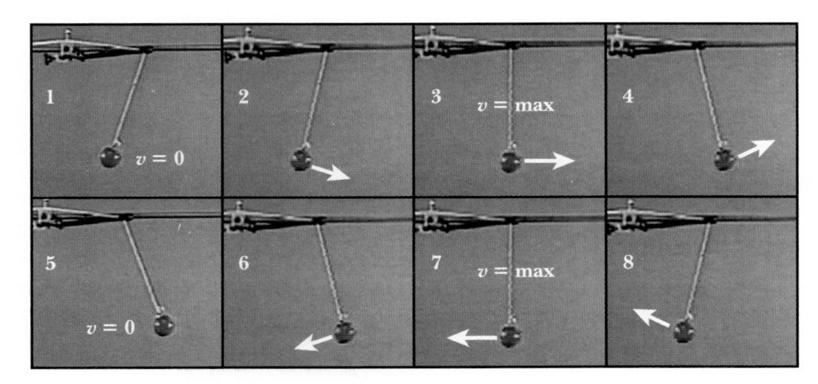

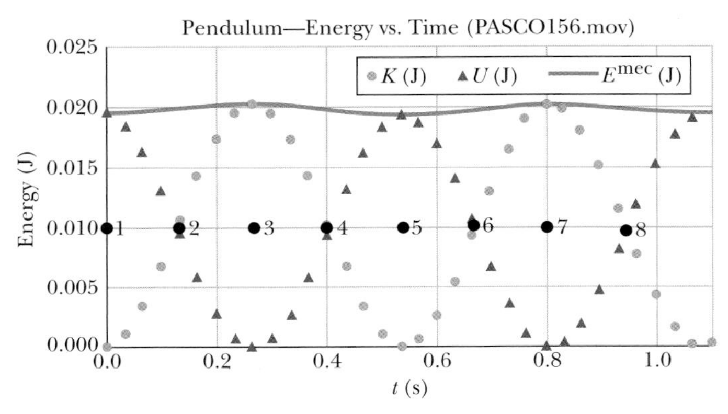

FIGURE 10-14 ■ A pendulum with its mass of 0.10 kg concentrated in a bob at the lower end. A selection of video frames that capture its motion shows the potential and kinetic energy of the bob as it swings back and forth for one full cycle of motion. A local fit of data for angular position vs time and its first derivative was used to calculate gravitational potential energy, U, and kinetic energy, K, of the bob on a moment-by-moment basis. During the cycle, the values of the potential and kinetic energies of the pendulum–Earth system vary as the bob rises and falls. But, as shown in the graph, the total mechanical energy, E^{mec}, of the system remains constant within the limits of experimental uncertainty. In stages 3 and 7, all the energy is kinetic. The bob has its greatest speed while passing rapidly through its lowest point. In stages 1 and 5, all the energy is potential energy. In stages 2, 4, 6, and 8, the energy is split between potential and kinetic. The forces on the pendulum appear to be conservative when only one cycle is observed. However, the friction at the point of attachment and the presence of drag forces due to the air will cause the total mechanical energy of the system to decrease slowly with time.

For example, let us choose the lowest point of the pendulum as the reference point and set the corresponding gravitational potential energy $U_2 = 0.00$ J. Note then that the potential energy at the highest point is approximately given by $U_1 = 0.20$ J relative to the reference point. Because the bob momentarily has speed $v = 0$ at its highest point, the kinetic energy there is $K_1 = 0.00$ J. Substituting these values into $K_2 + U_2 = K_1 + U_1$ gives us the kinetic energy K_2 at the lowest point,

$$K_2 + 0.00\text{ J} = 0.00\text{ J} + 0.20\text{ J} \quad \text{or} \quad K_2 = 0.20\text{ J}.$$

Note that we get this result without considering the motion between the highest and lowest points (such as in Fig. 10-14, stage 7) and without finding the work done by any forces involved in the motion.

READING EXERCISE 10-5: The figure shows four situations—one in which an initially stationary block is dropped and three in which the block is allowed to slide down frictionless ramps. (a) Rank the situations according to the kinetic energy of the block at point B, greatest first. (b) Rank them according to the speed of the block at point B, greatest first.

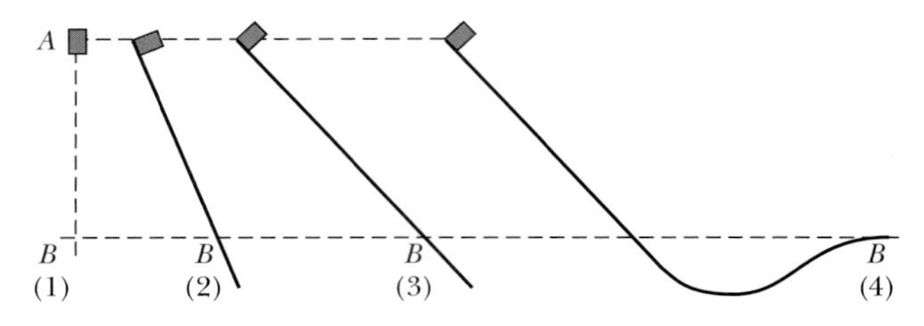

TOUCHSTONE EXAMPLE 10-3: Bungee Jumper

A 61.0 kg bungee-cord jumper is on a bridge 45.0 m above a river. The elastic bungee cord has a relaxed length of $L =$ 25.0 m. Assume that the cord obeys Hooke's law, with a spring constant of 160 N/m. If the jumper stops before reaching the water, what is the height h of her feet above the water at her lowest point?

SOLUTION ■ Figure 10-15 shows the jumper at the lowest point, with her feet at height h and with the cord stretched by distance d from its relaxed length. If we knew d, we could find h. One **Key Idea** is that perhaps we can solve for d by applying the principle of conservation of mechanical energy, between her initial point (on the bridge) and her lowest point. In that case, a second **Key Idea** is that mechanical energy is conserved in an isolated system when only conservative forces cause energy transfers. Let's check.

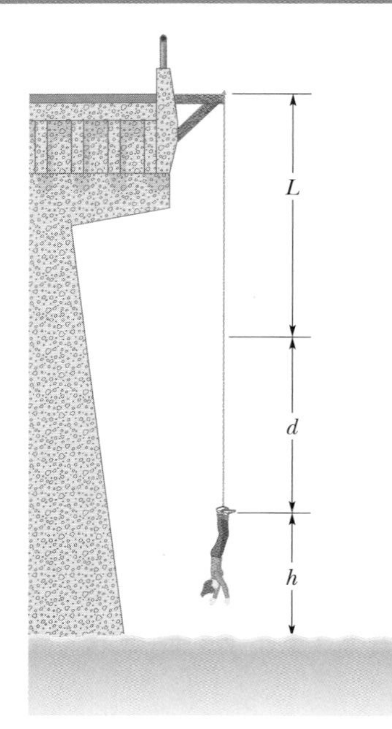

FIGURE 10-15 ■ A bungee-cord jumper at the lowest point of the jump.

Forces: The gravitational force does work on the jumper throughout her fall. Once the bungee cord becomes taut, the spring-like force from it does work on her, transferring energy to elastic potential energy of the cord. The force from the cord also pulls on the bridge, which is attached to Earth. The gravitational force and the spring-like force are conservative.

System: The jumper–Earth–cord system includes all these forces and energy transfers, and we can take it to be isolated (no work done by external forces). Thus, we *can* apply the principle of conservation of mechanical energy to the system. From Eq. 10-20, we can write the principle as

$$\Delta K + \Delta U^{\text{elas}} + \Delta U^{\text{grav}} = 0, \quad (10\text{-}21)$$

where ΔK is the change in the jumper's kinetic energy, ΔU^{elas} is the change in the elastic potential energy of the bungee cord, and ΔU^{grav} is the change in gravitational potential energy. All these changes must be computed between her initial point and her lowest point. Because she is stationary (at least momentarily) both initially and at her lowest point, $\Delta K = 0$. From Fig. 10-15 (with the bridge as origin and downward the negative y direction), we see that the change Δy in her height is $-(L + d)$, so we have

$$\Delta U^{\text{grav}} = mg\,\Delta y = -mg(L + d),$$

where m is her mass. Also from Fig. 10-15, we see that the bungee cord is stretched by distance d. Thus, we also have

$$\Delta U^{\text{elas}} = \tfrac{1}{2}kd^2.$$

Inserting these expressions and the given data into Eq. 10-21, we obtain

$$0 + \tfrac{1}{2}kd^2 - mg(L + d) = 0,$$

or

$$\tfrac{1}{2}kd^2 - mgL - mgd = 0,$$

and then

$$\tfrac{1}{2}(160\ \text{N/m})d^2 - (61.0\ \text{kg})(9.8\ \text{m/s}^2)(25.0\ \text{m}) - (61.0\ \text{kg})(9.8\ \text{m/s}^2)d = 0.$$

Solving this quadratic equation yields

$$d = 17.9\ \text{m}.$$

The jumper's feet are then a distance of $(L + d) = 42.9$ m below their initial height. Thus,

$$h = 45.0\ \text{m} - 42.9\ \text{m} = 2.1\ \text{m}. \qquad \text{(Answer)}$$

10-5 Reading a Potential Energy Curve

Once again we consider a particle that is part of a system in which a conservative force acts. This time suppose that the particle is constrained to move along an x axis while the conservative force does work on it. We can learn a lot about the motion of the particle from a plot of the system's potential energy $U(x)$. However, before we discuss such plots, we need one more relationship.

Finding the Force Analytically for a Two-Body System

According to Eq. 10-12, if we choose a vertical x axis passing from particle A through particle B (like that shown in Fig. 10-11) with its origin at particle A, then the change in potential energy of the system can be expressed as

$$\Delta U = -\int_{x_1}^{x_2} F_{Bx}(x)\,dx.$$

This is the potential energy change that occurs when one of the particles, chosen as B, moves between x_1 and x_2 along a (vertical) x axis.

Suppose we have the reverse situation. That is, suppose we happen to know ΔU and we would like to know the internal force acting on particle B denoted as $F_{Bx}(x)$. If the force on particle B does not vary rapidly with x, the potential energy change in the system as particle B moves through a distance Δx is approximately

$$\Delta U(x) \approx -F_{Bx}(x)\,\Delta x.$$

If we solve for $F_{Bx}(x)$, pass to the differential limit, and drop the label B (so the x-component of force denotes the internal force on whichever particle in the system is displaced relative to the other) we have

$$F_x^{\text{cons}}(x) = -\frac{dU(x)}{dx} \quad \text{(one-dimensional internal force).} \qquad (10\text{-}22)$$

We can check our result with $U(x) = \tfrac{1}{2}kx^2$, which is the elastic potential energy function for a spring force. Equation 10-22 ($F_x^{\text{cons}}(x) = -dU(x)/dx$) then yields, as expected, $F_x^{\text{cons}}(x) = F_x^{\text{spring}}(x) = -kx$, which is Hooke's law. Similarly, we can substitute $U(y) = mgy$, which is the gravitational potential energy function for a particle–Earth system, with a particle of mass m at height y above Earth's surface. $F_y^{\text{cons}}(y) = F_y^{\text{grav}}(y) = -dU(y)/dy$ then yields $F_y^{\text{grav}}(y) = -mg$, which is the y-component of gravitational force on the particle.

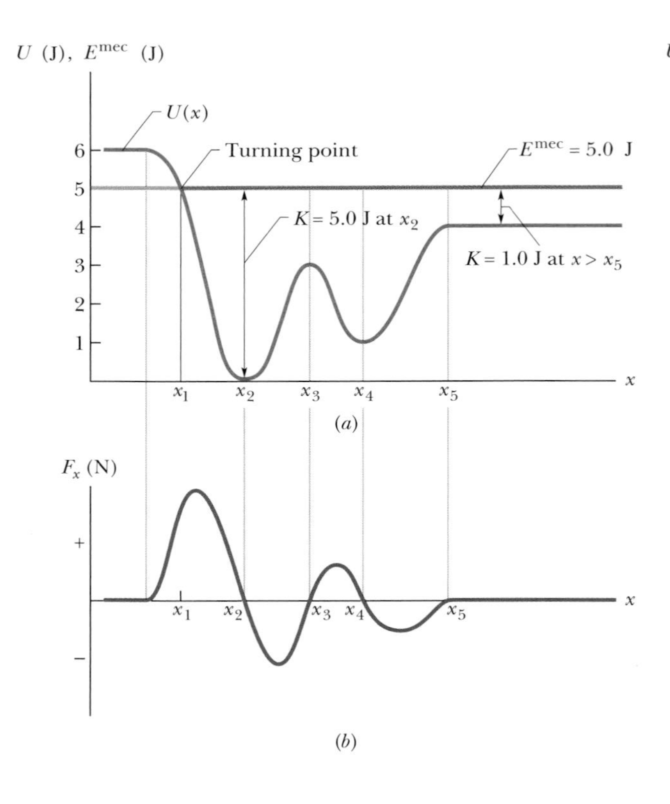

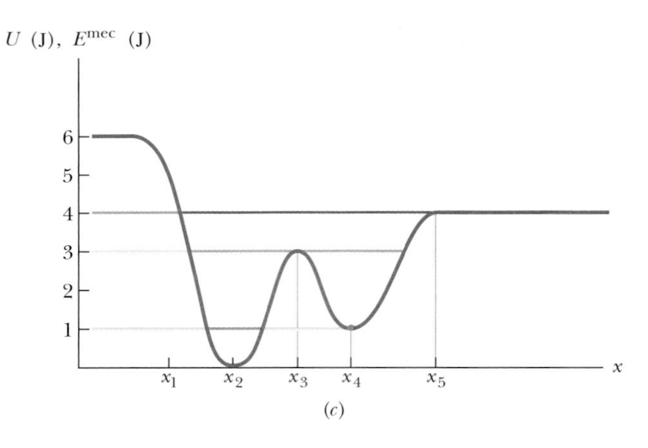

FIGURE 10-16 ■ (*a*) A plot of $U(x)$, the potential energy function of a system containing a particle confined to move along the x axis. There is no friction, so mechanical energy is conserved. (*b*) A plot of the force $F(x)$ acting on the particle, derived from the potential energy plot by taking its slope at various points. (*c*) The $U(x)$ plot of (*a*) with three different possible values of E^{mec} shown.

The Potential Energy Curve

Figure 10-16*a* is a plot of a potential energy function $U(x)$ for a system in which a particle is in one-dimensional motion while a conservative internal force $\vec{F}_x^{int}(x)$ does work on it. Since $dU(x)/dx$ in Eq. 10-22 is the slope of the $U(x)$ vs. x curve, we can easily find $F_x^{int}(x)$ by (graphically) taking the slope of the $U(x)$ curve at various points and negating it. Figure 10-16*b* is a plot of $F_x^{int}(x)$ found in this way.

Turning Points

As we discussed in Section 10-4, in the absence of a nonconservative force, the mechanical energy E^{mec} of the system has a constant value given by

$$U(x) + K(x) = E^{mec}. \tag{10-23}$$

Here $K(x)$ is the *kinetic energy function* of the particle (this $K(x)$ gives the kinetic energy as a function of the particle's location x). We may rewrite this expression as

$$K(x) = E^{mec} - U(x). \tag{10-24}$$

Suppose that E^{mec} (which has a constant value for a conservative isolated system) happens to be 5.0 J. It would be represented in Fig. 10-16*a* by a horizontal line that runs through the value 5.0 J on the energy axis. (It is, in fact, shown there.)

Equation 10-24 ($K(x) = E^{mec} - U(x)$) tells us how to determine the kinetic energy K for any location x of the particle: On the $U(x)$ curve, find U for that location x and then subtract U from E^{mec}. For example, if the particle is at any point to the right of x_5, then $K = 1.0$ J. The value of K is greatest (5.0 J) when the particle is at x_2, and least (0 J) when the particle is at x_1.

Since K can never be negative (because v^2 is always positive), the particle can never move to the left of x_1, where $E^{mec} - U$ is negative. Instead, as the particle moves toward x_1 from x_2, K decreases (the particle slows) until $K = 0$ at x_1 (the particle stops there).

Note that when the particle reaches x_1, the x-component of the internal force on the particle due to the rest of the system, given by

$$F_x^{\text{int}}(x) = -\frac{dU(x)}{dx},$$

is positive (because the slope dU/dx is negative). This means that the particle does not remain at x_1 but instead begins to move to the right, opposite its earlier motion. Hence x_1 is a **turning point,** a place where $K = 0$ (because $U = E$) and the particle changes direction. There is no turning point (where $K = 0$) on the right side of the graph. When the particle heads to the right and $x > x_5$, there is no force on it, and it will continue indefinitely.

Equilibrium Points

Figure 10-16*c* shows three different values for E^{mec} superimposed on the plot of the same potential energy function $U(x)$. Let us see how they would change the situation. If $E^{\text{mec}} = 3.0$ J (line running through the value 3.0 J on the energy axis), there are two turning points: one is between x_1 and x_2 and the other is between x_4 and x_5. In addition, x_3 is a point at which $K = 0$. If the particle is located exactly there, the force on it is also zero (the slope of the curve is zero), and the particle remains stationary. However, if it is displaced even slightly in either direction, a nonzero force pushes it further in the same direction, and the particle continues to move. A particle at such a position is said to be in **unstable equilibrium.** (A marble balanced on top of a bowling ball is an example.)

Next consider the particle's behavior if $E^{\text{mec}} = 1.0$ J (line running through the value 1.0 J on the energy axis). If we place the particle at x_4, it is stuck there. It cannot move left or right on its own because to do so would require a negative kinetic energy. If we push it slightly left or right, a restoring force appears that moves it back to x_4. A particle at such a position is said to be in **stable equilibrium.** (A marble placed at the bottom of a hemispherical bowl is an example.) If we place the particle in the cup-like *potential well* centered at x_2, it is between two turning points. It can still move left and right somewhat, but only partway to x_1 or to x_3.

If $E^{\text{mec}} = 4.0$ J (line running through the value 4.0 J on the energy axis), the turning point shifts from x_1 to a point between x_1 and x_2. Also, at any point to the right of x_5, the system's mechanical energy is equal to its potential energy; thus, the particle has no kinetic energy and (by $F_x^{\text{int}}(x) = -dU(x)/dx$) no force acts on it. So it must be stationary. A particle at such a position is said to be in **neutral equilibrium.** (A marble placed on a horizontal tabletop is in that state.)

READING EXERCISE 10-6: The figure gives the potential energy function $U(x)$ for a system in which a particle is in one-dimensional motion. (a) Rank regions AB, BC, and CD according to the magnitude of the force on the particle, greatest first. (b) What is the direction of the force when the particle is in region AB?

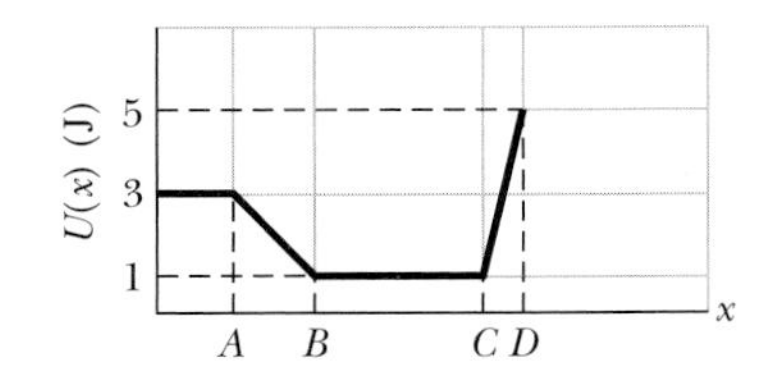

■

TOUCHSTONE EXAMPLE 10-4: Shifting the Zero

Suppose that you shifted the origin of the graph shown in Fig. 10-16*a* up by 6.0 J so that the potential energy reference point from which all the values of $U(x)$ were measured was located at $x = 0$ rather than $x = x_2$.

(a) What effect would this have on the values of $U(x)$?

SOLUTION ■ While the plot of $U(x)$ would still have the same shape as in Fig. 10-16*a*, all of its values would be reduced by 6.0 J and so would now be negative rather than positive. For example, now $U(0) = 0$, $U(x_1) = -1$ J, $U(x_2) = -6$ J, and so on. (Answer)

(b) What effect would this have on the values of the kinetic energy?

SOLUTION ■ The particle's kinetic energy, $K = \frac{1}{2}mv^2$, depends only on the particle's mass and speed. Since neither of these depend on our choice of reference point from which we measure the particle's potential energy, the values of the particle's kinetic energy at each location will be the same as before. (Answer)

(c) What effect would this have on the values of E^{mec}?

SOLUTION ■ The **Key Idea** here is that $E^{mec} = U + K$. Since each value of U is reduced by 6.0 J by this shift of reference point, and the values of K remain the same, then E^{mec} is also reduced by 6.0 J in each case. (Answer)

This change in E^{mec} does *not* mean that E^{mec} no longer has a constant value. It's just that it now has a *different* constant value than it had before. For example, in Fig. 10-16*a*, $E^{mec} = +5.0$ J everywhere to the right of $x = x_1$ before we shifted the potential energy's reference point. After the shift, $E^{mec} = -1.0$ J, still constant and independent of location, but now with a different value.

(d) What effect would this have on the values of $F_x(x)$, the force experienced by the particle, as pictured in Fig. 10-16*b*?

SOLUTION ■ The **Key Idea** here is that $F_x(x) = -dU/dx$. Subtracting a constant from $U(x)$ has no effect on its derivative since

$$\frac{d(U(x) - \text{constant})}{dx} = \frac{dU(x)}{dx} - \frac{d(\text{constant})}{dx} = \frac{dU(x)}{dx} - 0 = \frac{dU(x)}{dx}.$$

Therefore, $F_x(x)$ will be unchanged. (Answer)

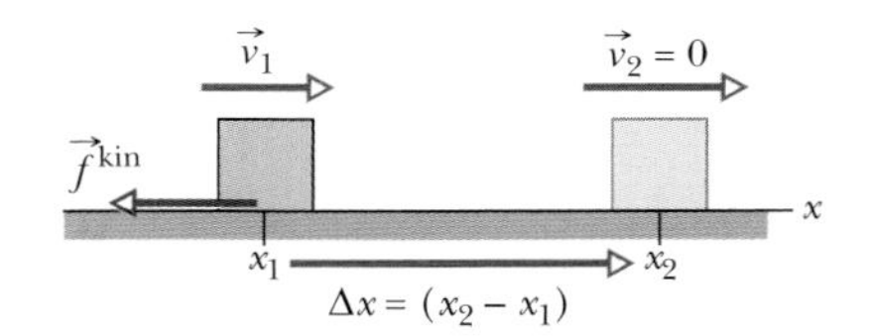

FIGURE 10-17 ■ (*a*) A block slides across a floor while a kinetic frictional force $\vec{f}^{kin}$ opposes the motion. The block has velocity $\vec{v}_1$ at the start of a displacement $\Delta\vec{x}$ and a velocity $\vec{v}_2 = 0$ at the end of the displacement.

10-6 Nonconservative Forces and Energy

We have made the claim that mechanical energy is conserved in an isolated system (no net work is done on the system by external forces) whose internal forces are conservative. Let's now consider an isolated system whose elements interact by means of nonconservative kinetic friction forces. Our example is an isolated system consisting of a sliding block and a floor (Fig. 10-17). Assume the block has an initial velocity $\vec{v}_1$. What happens to its initial kinetic energy as it slides along the floor and comes to rest so $\vec{v}_2 = 0$? According to the net work-kinetic energy theorem, the net work done on the block from the sum of all the forces acting on it will result in a kinetic energy change of the block given by

$$W^{net} = \Delta K = \tfrac{1}{2}mv_2^2 - \tfrac{1}{2}mv_1^2.$$

Since the net work is calculated from the net force, we need to write down an expression for the net force on the block. The block has a friction force, a downward gravitational force, and an upward normal force exerted on it. Since there is no motion in the vertical direction, we know that the gravitational and normal forces cancel each other out. The net force on the block is just the horizontal kinetic friction force, so $\vec{F}^{net} = \vec{f}^{kin}$. Since there are no external forces acting on the system, all the work done on the system is done by internal forces. So, $W^{net} = W^{int}$. If the block has a displacement $\Delta\vec{x}$ as it slides to rest, then

$$W^{net} = W^{int} = f_x^{kin}\Delta x = -\tfrac{1}{2}mv_1^2. \quad (10\text{-}25)$$

The product of $f_x^{kin}\Delta x$, which represents the internal work done on the system, is negative since friction forces always act in a direction opposite to an object's displacement. Since the friction force is nonconservative (the amount of work it does depends on the path taken), we cannot associate a potential energy change with it. Instead, as Eq. 10-25 indicates, the internal work done on the system has caused a *loss of kinetic energy*. This represents a loss of the only form of mechanical energy such a system can have. We can conclude that

> If the internal forces in an isolated system include nonconservative forces, then mechanical energy is not conserved.

By experimenting, we find that the block and the portion of the floor along which it slides become warmer as the block slides to a stop. If we associate the temperature of an object with a new kind of energy, thermal energy E^{thermal}, we may be able to continue to make use of energy conservation methods. In fact, it turns out that the kinetic energy lost in Eq. 10-25 does cause a gain in thermal energy where

$$\Delta E^{\text{thermal}} = -f_x^{\text{kin}} \Delta x \quad \text{(increase in thermal energy due to kinetic friction).} \qquad (10\text{-}26)$$

As we shall discuss in Chapter 19, the thermal energy of an object is related to temperatures and can be associated with the random motions of atoms and molecules in objects.

We define the *total energy* of the system to be the sum of its mechanical energy and other forms of energy including thermal energy, chemical energy, light, sound, and so on. Doing so, we see that we have a new principle of energy conservation for isolated systems even in the presence of nonconservative forces given by

$$\Delta E^{\text{total}} = \Delta E^{\text{mec}} + \Delta E^{\text{noncons}} = 0. \qquad (10\text{-}27)$$

Here, $\Delta E^{\text{noncons}} = \Delta E^{\text{thermal}} + \Delta E^{\text{other}}$ where ΔE^{other} includes light, sound, and so on.

READING EXERCISE 10-7: In three trials, a block starts with the same kinetic energy and slides across a floor that is not frictionless, as in Fig. 10-17. In all three trials, the block is allowed to slide through the same distance Δx but has not yet come to rest. Rank the three trials according to the change in the thermal energy of the block and floor that occurs, greatest first.

Trial	f_x^{kin}	Block's Displacement Δx
a	5.0 N	0.20 m
b	7.0 N	0.30 m
c	8.0 N	0.10 m

■

TOUCHSTONE EXAMPLE 10-5: Tamale Stops Here

In Fig. 10-18, a 2.0 kg package of tamales slides along a floor with speed $v_1 = 4.0$ m/s. It then runs into and compresses a spring, until the package momentarily stops. Its path to the initially relaxed spring is frictionless, but as it compresses the spring, a kinetic frictional force from the floor, of magnitude 15 N, acts on it. The spring constant is 10 000 N/m. By what distance d is the spring compressed when the package stops?

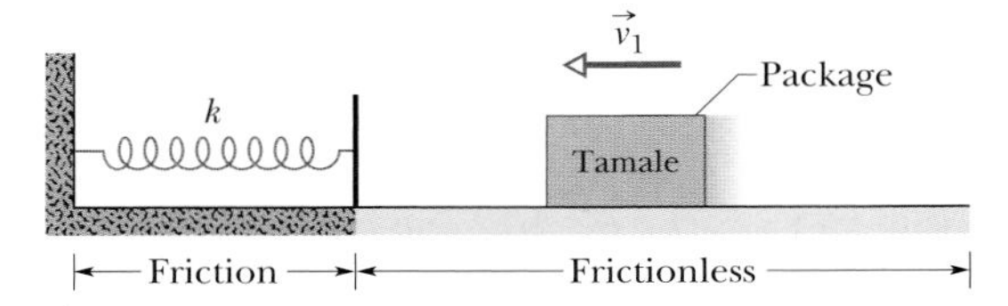

FIGURE 10-18 ■ A package slides across a frictionless floor with velocity $\vec{v}_1$ toward a spring of spring constant k. When the package reaches the spring, a frictional force from the floor acts on it.

SOLUTION ■ A starting **Key Idea** is to examine all the forces acting on the package, and then to determine whether we have an isolated system or a system on which an external force is doing work.

Forces: The normal force on the package from the floor does no work on the package, because its direction is always perpendicular to that of the package's displacement. For the same reason, the gravitational force on the package does no work. As the spring is compressed, however, a spring force does work on the package, transferring energy to elastic potential energy of the spring. The spring force also pushes against a rigid wall. Because there is friction between the package and the floor, the sliding of the package across the floor increases their thermal energies.

System: The package–spring–floor–wall system includes all these forces and energy transfers in one isolated system. Therefore, a second **Key Idea** is that, because the system is isolated, its total energy cannot change. We can then apply the law of conservation of energy in the form of Eq. 10-27 to the system:

$$\Delta E^{\text{mec}} + \Delta E^{\text{noncons}} = 0.$$

But $\quad \Delta E^{\text{mec}} = E_2^{\text{mec}} - E_1^{\text{mec}}$ and $\Delta E^{\text{noncons}} = \Delta E^{\text{thermal}}$,

so

$$E_2^{\text{mec}} = E_1^{\text{mec}} - \Delta E^{\text{thermal}}. \tag{10-29}$$

Let subscript 1 correspond to the initial state of the sliding package and subscript 2 correspond to the state in which the package is momentarily stopped and the spring is compressed by distance d. For both states the mechanical energy of the system is the sum of the package's kinetic energy ($K = \frac{1}{2}mv^2$) and the spring's potential energy ($U = \frac{1}{2}kx^2$). For state 1, $U = 0$ (because the spring is not compressed), and the package's speed is v_1. Thus, we have

$$E_1^{\text{mec}} = K_1 + U_1 = \tfrac{1}{2}mv_1^2 + 0.$$

For state 2, $K = 0$ (because the package is stopped), and the compression distance is d. Therefore, we have

$$E_2^{\text{mec}} = K_2 + U_2 = 0 + \tfrac{1}{2}kd^2.$$

Finally, by Eq. 10-26, we can substitute $(-f_x^{\text{kin}}\Delta x) = f_x^{\text{kin}}d$ for the change $\Delta E^{\text{thermal}}$ in the thermal energy of the package and the floor. We can now rewrite Eq. 10-29 as

$$\tfrac{1}{2}kd^2 = \tfrac{1}{2}mv_1^2 - f_d^{\text{kin}}d.$$

Rearranging and substituting known data give us

$$(5000\text{ N/m})d^2 + (15\text{ N})d - (16\text{ J}) = 0.$$

Solving this quadratic equation yields

$$d = 0.055\text{ m} = 5.5\text{ cm}. \qquad \text{(Answer)}$$

10-7 Conservation of Energy

We now have discussed several situations in which energy is transferred between objects within systems. In each situation, we assume that the energy that was involved could always be accounted for. That is, energy could not appear or disappear. In more formal language, we assumed that energy obeys a law called the **law of conservation of energy,** which is concerned with the **total energy** E^{total} of a system. There are many complex situations in which it is difficult to account for all the energy. But physicists have always found that if a change in a system takes place and some energy seems to be missing, it simply has taken on a new form. This is the case with the thermal energy we talked about in the previous section. It can be accounted for by developing methods for keeping track of the kinetic energy stored in the random motions of atoms and molecules in the sliding block and floor and in the potential energy associated with the chemical bonds that hold them together.

We define total energy of a system as the sum of the system's mechanical energy, thermal energy, and other forms of energy we will not touch on here that are associated with things like sound and light. The law of conservation of energy states that:

> The total energy E^{total} of a system can change only by amounts of energy that are transferred to or from the system.

When is energy transferred to or from a system? This occurs when an external force does work W^{ext} on the system. The external work W^{ext} done on a system is not merely a calculation procedure. It is an energy transfer process. Thus, the **law of conservation of energy** can be stated in very general terms as

$$W^{\text{ext}} = \Delta E^{\text{total}} = \Delta E^{\text{mec}} + \Delta E^{\text{noncons}}, \tag{10-28}$$

where $\Delta E^{\text{noncons}}$ is any change in thermal energy or the many other forms of energy that we have not discussed here. Included in ΔE^{mec} are changes ΔK in kinetic energy and changes ΔU in potential energy due to conservative forces such as elastic, gravitational, and electrostatic forces (which we discuss in Chapter 25).

As you may have noticed, this law of conservation of energy is *not* something we have derived from basic physics principles. It is more speculative. But in the past, whenever it has appeared to fail, scientists and engineers have been able to identify new forms of energy that allow us to hold on to the law of conservation of energy. Furthermore, each time a new form of energy has been identified, we have been able to understand whole new classes of phenomena, such as how stars shine or how radioactive atoms decay.

TOUCHSTONE EXAMPLE 10-6: Easter Island

The giant stone statues of Easter Island were most likely moved by the prehistoric islanders by cradling each statue in a wooden sled and then pulling the sled over a "runway" consisting of almost identical logs acting as rollers. In a modern reenactment of this technique, 25 men were able to move a 9000 kg Easter Island-type statue 45 m over level ground in 2 min.

(a) Estimate the work the external force $\vec{F}^{\,ext}$ from the men did during the 45 m displacement of the statue, and determine the system on which that force did the work.

SOLUTION ■ One **Key Idea** is that we can calculate the work done with Eq. 9-26 ($W = |\vec{F}||\Delta\vec{r}|\cos\phi$). Here $|\Delta\vec{r}|$ is the distance 45 m, $|\vec{F}^{\,ext}|$ is the magnitude of the external force on the statue from the 25 men, and $\phi = 0°$. Let us estimate that each man pulled with a force magnitude equal to twice his weight, which we take to be the same value mg for all the men. Thus, the magnitude of the external force was $|\vec{F}^{\,ext}| = (25)(2)(mg) = 50mg$. Estimating a man's mass as 80 kg, we can then write Eq. 9-26 as

$$\begin{aligned} W^{ext} &= |\vec{F}^{\,ext}||\Delta\vec{r}|\cos\phi = 50mgd\cos\phi \\ &= (50)(80\text{ kg})(9.8\text{ N/kg})(45\text{ m})\cos 0° \qquad \text{(Answer)} \\ &= 1.8 \times 10^6\text{ J} \approx 2\text{ MJ}. \end{aligned}$$

The **Key Idea** in determining the system on which the work is done is to see which energies change. Because the statue moved, there was certainly a change ΔK in its kinetic energy during the motion. We can easily guess that there must have been considerable kinetic friction between the sled, logs, and ground, resulting in a change $\Delta E^{thermal}$ in their thermal energies. Thus, the system on which the work was done consisted of the statue, sled, logs, and ground.

(b) What was the increase $\Delta E^{thermal}$ in the thermal energy of the system during the 45 m displacement?

SOLUTION ■ The **Key Idea** here is that we can relate $\Delta E^{noncons} = \Delta E^{thermal}$ to the work W^{ext} done by $\vec{F}^{ext}$ with the energy statement of Eq. 10-28,

$$W^{ext} = \Delta E^{mec} + \Delta E^{thermal}.$$

We know the value of W^{ext} from (a). The change ΔE^{mec} in the statue's mechanical energy was zero because the statue was stationary at the beginning and end of the move and did not change in elevation. Thus, we find

$$\Delta E^{thermal} = W^{ext} = 1.8 \times 10^6\text{ J} \approx 2\text{ MJ}. \qquad \text{(Answer)}$$

(c) Estimate the work that would have been done by the 25 men if they had moved the statue 10 km across level ground on Easter Island. Also estimate the total change $\Delta E^{thermal}$ that would have occurred in the statue–sled–logs–ground system.

SOLUTION ■ The **Key Ideas** here are the same as in (a) and (b). Thus we calculate W^{ext} as in (a), but with 1×10^4 m now substituted for $|\Delta\vec{r}|$. Also, we again equate $\Delta E^{thermal}$ to W^{ext}. We get

$$W^{ext} = \Delta E^{thermal} = 3.9 \times 10^8\text{ J} \approx 400\text{ MJ}. \qquad \text{(Answer)}$$

This would have been a significant amount of energy for the men to have transferred during the movement of a statue. Still, the 25 men *could* have moved the statue 10 km, and the required energy does not suggest some mysterious source.

10-8 One-Dimensional Energy and Momentum Conservation

Recall that in some situations, the conservation of translational momentum allowed us to figure out what was going to happen, even when we didn't know what the forces were. (For example, when two objects collide and stick together.) Now that we have identified a second conservation law—the conservation of energy—we can figure out what is going to happen in a larger class of situations.

Consider a system of two colliding bodies. If there is to be a collision, then at least one of the bodies must be moving, so the system has a certain kinetic energy and a certain translational momentum before the collision. During the collision, the kinetic energy and translational momentum of each body are changed by the impulse from the other body. We can discuss these changes—and also the changes in the kinetic energy and translational momentum of the system as a whole—without knowing the details of the impulses that determine the changes. As was the case in Chapter 7 where we first discussed collisions, the discussion here will be limited to collisions in systems that are **closed** (no mass enters or leaves them) and **isolated** (no net external forces act on the bodies within the system).

FIGURE 10-19 ■ Two cars after an almost head-on, almost completely inelastic collision.

Elastic versus Inelastic Collisions

Collisions that we casually called bouncy and sticky in Chapter 7 can be classified in terms of whether or not mechanical energy is conserved. Except for a brief period during a collision, typically no potential energy is stored in a system of objects before and after the collision. So most of the time the mechanical energy in the system is equal to the total kinetic energy of the colliding objects.

Elastic Collisions: If the total kinetic energy of the system of two colliding bodies is unchanged by the collision, the collision is called a completely **elastic collision.** This happens if the forces between the objects during the collision are approximately conservative and spring-like. Some of the "bouncy" collisions we discussed in Chapter 7 may have been elastic collisions. However, most "bouncy" collisions are in fact not completely elastic collisions.

Inelastic Collisions: In everyday collisions of common bodies, such as between two cars or a ball and a bat, some energy is always transferred from kinetic energy to other forms of energy, such as thermal energy or energy of sound. Thus, the kinetic energy of the system is *not* conserved. Such a collision is defined as an **inelastic collision.** Figure 10-19 shows a dramatic example of a **completely inelastic collision.** In such collisions, the bodies always stick together and lose all their kinetic energy. Most real collisions are partially elastic and partially inelastic.

Almost Elastic Collisions: In some situations, we can *approximate* a collision of common bodies as elastic. Suppose that you drop a Superball onto a hard floor. If the collision between the ball and floor (or Earth) were elastic, the ball would lose no kinetic energy because of the collision and would rebound to its original height. However, the actual rebound height is somewhat short of the starting point, showing that at least some kinetic energy is lost in the collision and thus that the collision is somewhat inelastic. Still, we might choose to neglect that small loss of kinetic energy to approximate the collision as elastic.

Distinguishing Energy and Momentum Conservation

It is easy to confuse momentum conservation with energy conservation. However, *they are not the same*. Momentum is a vector quantity defined as the product of mass and velocity. Energy is a scalar quantity that has no direction associated with it. As far as we know, *momentum is always conserved* as a result of interactions between objects in an isolated system. This is not the case for mechanical energy. *Mechanical energy is only conserved when the internal forces that do work on the system are conservative.*

Let's perform three thought experiments that illustrate some of the differences between the two conservation laws. To do this, imagine three types of collision processes described in Chapter 7 on collisions and momentum. One is a completely *inelastic collision* in which the colliding objects stick together. Another is a completely *elastic collision* in which the objects bounce off one another and the system consisting of the colliding objects loses no mechanical energy. The third is a *superelastic collision*, or explosion, in which some energy is released so the system has more mechanical energy than it did before.

In all three thought experiments, two identical carts with negligible friction are moving toward each other at the same speed (Fig. 10-20). Since they are moving on a horizontal ramp they have no change in gravitational potential energy as they move. In this special circumstance, the mechanical energy of each of the two-cart systems is the same as its kinetic energy.

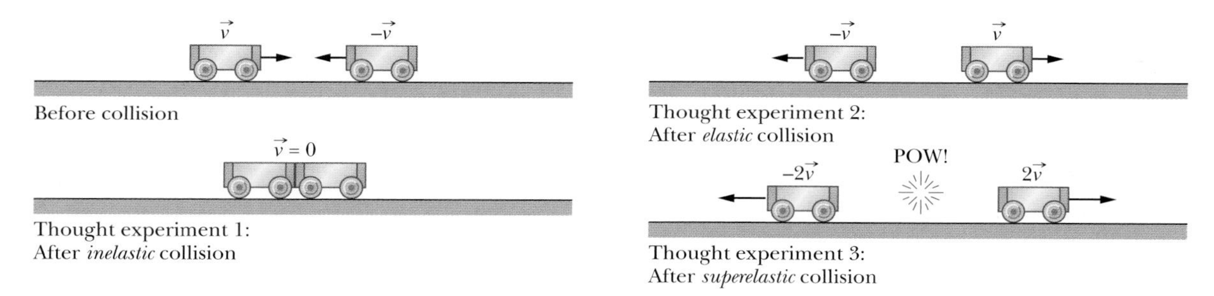

FIGURE 10-20 ■ Three possible outcomes of a head-on collision between two identical carts. Momentum is conserved in all three cases but not necessarily mechanical energy.

From the information in the figure, it is apparent that the momentum before and after the collision is the same in all three cases and is zero. However, in the inelastic collision shown in experiment 1, there is Velcro on the ends of the carts so they come to a dead halt when they stick together. In Section 7-6, we referred to this type of collision as "sticky." Although the temperature of the Velcro rises, there is no kinetic energy left after the collision and, hence, mechanical energy is not conserved. In experiment 2, the carts have magnets embedded in the ends that repel, causing the carts to rebound with the same speed but not the same velocity as before. In this case kinetic energy, and hence mechanical energy, is conserved. Finally, in experiment 3, wads of gunpowder glued to the cart ends ignite. Chemical potential energy is released in an explosion that causes the carts to rebound with a greater kinetic energy than they had before. Once again mechanical energy is not conserved, but translational momentum still is.

Translational Momentum

Regardless of the details of the impulses in a collision and regardless of what happens to the total kinetic energy of the system, the total translational momentum $\vec{p}_{sys}$ of a closed, isolated system *cannot* change. The reason is that $\vec{p}_{sys}$ can be changed only by external forces (from outside the system), but the forces in the collision are internal forces (inside the system). Thus, we have this important rule:

> In a closed, isolated system in which a collision occurs, the translational momentum of each colliding body may change but the total translational momentum of the system $\vec{p}_{sys}$ cannot change, whether the collision is elastic or inelastic.

This is actually another statement of the **law of conservation of translational momentum** that we first discussed in Section 7-6. In Section 7-7, we explored translational momentum conservation for inelastic collisions—that is, "sticky collisions." In the next two sections we apply this law to elastic collisions.

TOUCHSTONE EXAMPLE 10-7: Ballistic Pendulum

The *ballistic pendulum* was used to measure the speeds of bullets before electronic timing devices were developed. The version shown in Fig. 10-21 consists of a large block of wood of mass $M = 5.4$ kg, hanging from two long cords. A bullet of mass $m = 9.5$ g is fired into the block, coming quickly to rest. The *block + bullet* then swing upward, their centers of mass rising a vertical distance $h = 6.3$ cm before the pendulum comes momentarily to rest at the end of its arc. What is the speed of the bullet just prior to the collision?

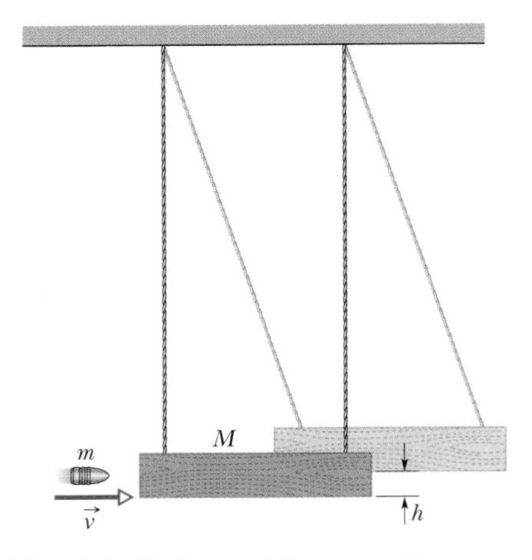

FIGURE 10-21 ■ A ballistic pendulum, used to measure the speeds of bullets.

SOLUTION ■ We can see that the bullet's speed v must determine the rise height h. However, a **Key Idea** is that we cannot use the conservation of mechanical energy to relate these two quantities because surely energy is transferred from mechanical energy to other forms as the bullet penetrates the block. Another **Key Idea** helps—we can split this complicated motion into two steps that we can separately analyze: (1) the bullet–block collision and (2) the bullet–block rise, during which mechanical energy *is* conserved.

Step 1. Because the collision within the bullet–block system is so brief, we can make two important assumptions: (1) During the collision, the gravitational force on the block and the force on the block from the cords are still balanced. Thus, during the collision, the net external force on the bullet–block system is zero. Therefore, the system is isolated and its total translational momentum is conserved. (2) The collision is one-dimensional in the sense that the direction of the bullet and block *just after the collision* is in the bullet's original direction of motion.

Because the collision is one-dimensional, the block is initially at rest, and the bullet sticks in the block, we use Eq. 7-21 to express the conservation of linear momentum. If the speed of the block just after the collision is V, we have $mv + 0 = mV + MV$ or

$$V = \frac{m}{m + M} v. \tag{10-30}$$

Step 2. After the "collision" between the bullet and the block is over, the bullet and block now swing up together, and the mechanical energy of the bullet–block–Earth system is conserved. (This mechanical energy is not changed by the force of the cords on the block, because that force is always directed perpendicular to the block's direction of travel.) Let's take the block's initial level as our reference level of zero gravitational potential energy. Then conservation of mechanical energy means that the system's kinetic energy at the start of the swing must equal its gravitational potential energy at the highest point of the swing. Because the speed of the bullet and block at the start of the swing is the speed V immediately after the collision, we may write this conservation as

$$\tfrac{1}{2}(m + M)V^2 + 0 = 0 + (m + M)gh.$$

Substituting this result for V in Eq. 10-30 leads to

$$\begin{aligned} v &= \frac{m + M}{m}\sqrt{2gh} \\ &= \left(\frac{0.0095 \text{ kg} + 5.4 \text{ kg}}{0.0095 \text{ kg}}\right)\sqrt{(2)(9.8 \text{ m/s}^2)(0.063 \text{ m})} \\ &= 630 \text{ m/s}. \end{aligned} \quad \text{(Answer)}$$

The ballistic pendulum is a kind of "transformer," exchanging the high speed of a light object (the bullet) for the low—and thus more easily measurable—speed of a massive object (the block).

10-9 One-Dimensional Elastic Collisions

Stationary Target

As we discussed in Section 10-8, everyday collisions are inelastic but we can approximate some of them as being elastic. That is, we can assume that the total kinetic energy of the colliding bodies is approximately conserved and is not transferred to other forms of energy:

(total kinetic energy before the collision) $\approx$ (total kinetic energy after the collision).

This does not mean that the kinetic energy of each colliding body cannot change. Rather, it means this:

> In an elastic collision, the kinetic energy of each colliding body may change, but the total kinetic energy of the system is the same before the collision as it is after.

For example, consider the collision of a cue ball with an object ball of approximately the same mass in a game of pool. If the collision is head-on (the cue ball heads directly toward the object ball), the kinetic energy of the cue ball can be transferred almost entirely to the object ball. (Still, the fact that the collision makes a sound means that at least a little of the kinetic energy is transferred to the energy of the sound.)

Figure 10-22 shows two bodies whose masses are not necessarily different before and after they have a one-dimensional collision, like a head-on collision between pool balls. A projectile body of mass m_A and initial velocity $\vec{v}_{A1}$ moves toward a target body of mass m_B that is initially at rest with velocity $\vec{v}_{B1} = 0$. Let's assume that this two-body system is closed and isolated. Then the net linear momentum of the system is conserved, and from Eq. 7-18 we can write

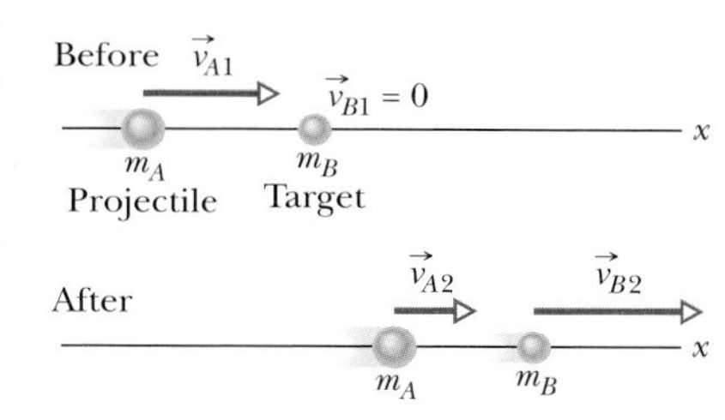

FIGURE 10-22 ■ Body 1 moves along an x axis before having an elastic collision with body 2, initially at rest. Both bodies move along that axis after the collision.

$$m_A \vec{v}_{A1} = m_A \vec{v}_{A2} + m_B \vec{v}_{B2} \quad \text{(linear momentum conservation)}, \tag{10-31}$$

where in general $\vec{v}_A(t) = v_{Ax}(t)\hat{\text{i}}$ and $\vec{v}_B(t) = v_{Bx}(t)\hat{\text{i}}$.

If the collision is also completely elastic, then the total kinetic energy is conserved and we can write

$$\tfrac{1}{2} m_A v_{A1}^2 = \tfrac{1}{2} m_A v_{A2}^2 + \tfrac{1}{2} m_B v_{B2}^2 \quad \text{(kinetic energy conservation)}. \tag{10-32}$$

In each of these equations, "1" signifies a time before the collision and "2" signifies a time after the collision. If we know the masses of the bodies and if we also know $\vec{v}_{A1}$, the initial velocity of body A, the only unknown quantities are $\vec{v}_{A2}$ and $\vec{v}_{B2}$, the final velocities of the two bodies. With two equations at our disposal, we should be able to find these two unknowns.

To do so, we express the velocities in terms of their x-components and rewrite Eq. 10-31 as

$$m_A[v_{Ax}(t_1) - v_{Ax}(t_2)] = m_B v_{Bx}(t_2), \tag{10-33}$$

and Eq. 10-32 as*

$$m_A[v_{Ax}(t_1) - v_{Ax}(t_2)][v_{Ax}(t_1) + v_{Ax}(t_2)] = m_B[v_{Bx}(t_2)]^2. \tag{10-34}$$

After dividing Eq. 10-34 by Eq. 10-33 and doing some more algebra, we obtain

$$v_{Ax}(t_2) = \frac{m_A - m_B}{m_A + m_B} v_{Ax}(t_1), \tag{10-35}$$

and

$$v_{Bx}(t_2) = \frac{2m_A}{m_A + m_B} v_{Ax}(t_1). \tag{10-36}$$

* In this step, we use the identity $a^2 - b^2 = (a - b)(a + b)$. It reduces the amount of algebra needed to solve the simultaneous equations, Eqs. 10-33 and 10-34.

We note from Eq. 10-36 that $v_{Bx}(t_2)$ is always positive (the target body with mass m_B always moves forward). From Eq. 10-35 we see that $v_{Ax}(t_2)$ may be of either sign (the projectile body with mass m_A moves forward if $m_A > m_B$ but rebounds if $m_A < m_B$).

Let us look at a few special situations.

1. **Equal masses.** If $m_A = m_B$, Eqs. 10-35 and 10-36 reduce to

$$v_{Ax}(t_2) = 0 \quad \text{and} \quad v_{Bx}(t_2) = v_{Ax}(t_1),$$

which we might call a pool player's result. It predicts that after a head-on collision of bodies with equal masses, body *A* (initially moving) stops dead in its tracks and body *B* (initially at rest) takes off with the initial speed of body *A*. In head-on collisions, bodies of equal mass simply exchange velocities. This is true even if the target particle (body *B*) is not initially at rest.

2. **A massive target.** In terms of Fig. 10-22, a massive target means that $m_B >> m_A$. For example, we might fire a golf ball at a cannonball. Equations 10-35 and 10-36 then reduce to

$$v_{Ax}(t_2) \approx -v_{Ax}(t_1) \quad \text{and} \quad v_{Bx}(t_2) \approx \left(\frac{2m_A}{m_B}\right)v_{Ax}(t_1). \tag{10-37}$$

This tells us that body *A* (the golf ball) simply bounces back in the same direction from which it came, its speed essentially unchanged. Body *B* (the cannonball) moves forward at a very low speed, because the quantity in parentheses in Eq. 10-37 is much less than unity. All this is what we should expect.

3. **A massive projectile.** This is the opposite case; that is, $m_A >> m_B$. This time, we fire a cannonball at a golf ball. Equations 10-35 and 10-36 reduce to

$$v_{Ax}(t_2) \approx v_{Ax}(t_1) \quad \text{and} \quad v_{Bx}(t_2) \approx 2v_{Ax}(t_1). \tag{10-38}$$

Equation 10-38 tells us that body *A* (the cannonball) simply keeps on going, scarcely slowed by the collision. Body *B* (the golf ball) charges ahead at twice the speed of the cannonball.

You may wonder: Why twice the speed? As a starting point in thinking about the matter, recall the collision described by Eq. 10-37, in which the velocity of the incident light body (the golf ball) changed from $v_{Ax}(t_1)$ to $-v_{Ax}(t_1)$, a velocity *change* of magnitude $|2v_{Ax}(t_1)|$. The same magnitude of *change* in velocity (from 0 to $|2v_{Ax}(t_1)|$) occurs in this example also.

Moving Target

Now that we have examined the elastic collision of a projectile and a stationary target, let us examine the situation in which both bodies are moving before they undergo an elastic collision.

For the situation of Fig. 10-23, the conservation of linear momentum is written as

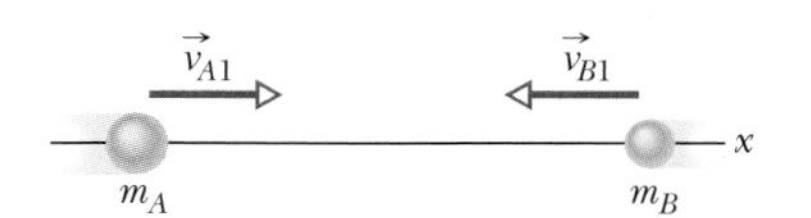

FIGURE 10-23 ■ Two bodies headed for a one-dimensional elastic collision.

$$m_A\vec{v}_{A1} + m_B\vec{v}_{B1} = m_A\vec{v}_{A2} + m_B\vec{v}_{B2}, \tag{10-39}$$

and the conservation of kinetic energy is written as

$$\tfrac{1}{2}m_Av_{A1}^2 + \tfrac{1}{2}m_Bv_{B1}^2 = \tfrac{1}{2}m_Av_{A2}^2 + \tfrac{1}{2}m_Bv_{B2}^2. \tag{10-40}$$

If we use similar procedures to those used in deriving Eqs. 10-35 and 10-36, we get

$$v_{Ax}(t_2) = \frac{m_A - m_B}{m_A + m_B} v_{Ax}(t_1) + \frac{2m_B}{m_A + m_B} v_{Bx}(t_1), \tag{10-41}$$

and

$$v_{Bx}(t_2) = \frac{2m_A}{m_A + m_B} v_{Ax}(t_1) + \frac{m_B - m_A}{m_A + m_B} v_{Bx}(t_1). \tag{10-42}$$

Note that the assignment of subscripts A and B to the bodies is arbitrary. If we exchange those subscripts in Fig. 10-23 and in Eqs. 10-41 and 10-42, we end up with the same set of equations. Note also that if we set $v_{Bx}(t_1) = 0$, body B becomes a stationary target, and Eqs. 10-41 and 10-42 reduce to Eqs. 10-35 and 10-36, respectively.

READING EXERCISE 10-8: What is the final translational momentum of the target in Fig. 10-22 if the initial translational momentum of the projectile is 6 kg·m/s and the final translational momentum of the projectile is (a) 2 kg·m/s and (b) −2 kg·m/s? (c) If the collision is elastic, what is the final kinetic energy of the target if the initial and final kinetic energies of the projectile are, respectively, 5 J and 2 J? ■

TOUCHSTONE EXAMPLE 10-8: Colliding Pendula

Two metal spheres, suspended by vertical cords, initially just touch, as shown in Fig. 10-24. Sphere A, with mass $m_A = 30$ g, is pulled to the left to height $h_0 = 8.0$ cm and then released from rest. After swinging down, it undergoes an elastic collision with sphere B, whose mass $m_B = 75$ g. What is the velocity $\vec{v}_{A2}$ of sphere A just after the collision?

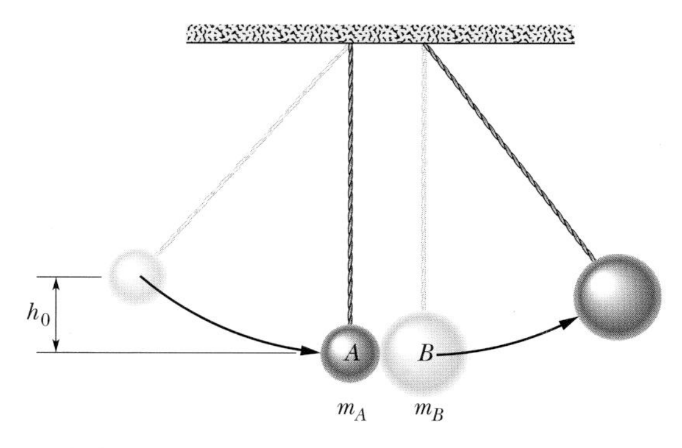

FIGURE 10-24 ■ Two metal spheres suspended by cords just touch when they are at rest. Sphere A, with mass m_A, is pulled to the left to height h_0 and then released.

SOLUTION ■ A first **Key Idea** is that we can split this complicated motion into two steps that we can separately analyze: (1) the descent of sphere A and (2) the two-sphere collision.

Step 1. The **Key Idea** here is that as sphere A swings down, the mechanical energy of the sphere–Earth system is conserved. (The mechanical energy is not changed by the force of the cord on sphere A because that force is always directed perpendicular to the sphere's direction of travel.) Let's take the lowest level as our reference level of zero gravitational potential energy. Then the kinetic energy of sphere A at the lowest level must equal the gravitational potential energy of the system when sphere A is at the initial height. Thus,

$$\tfrac{1}{2} m_A v_{A1}^2 = m_A g h_0,$$

which we solve for the speed v_{A1} of sphere A just before the collision:

$$|\vec{v}_{A1}| = \sqrt{2gh_0} = \sqrt{(2)(9.8\text{ m/s}^2)(0.080\text{ m})} = 1.252\text{ m/s}.$$

Step 2. Here we can make two assumptions in addition to the assumption that the collision is elastic. First, we can assume that the collision is one-dimensional because the motions of the spheres are approximately horizontal from just before the collision ($\vec{v}_{A1} = v_{Ax}(t_1)\hat{\text{i}}$) to just after it ($\vec{v}_{A2} = v_{Ax}(t_2)\hat{\text{i}}$). Second, because the collision is so brief, we can assume that the two-sphere system is closed and isolated. This gives the **Key Idea** that the total translational momentum of the system is conserved. Thus, we can use Eq. 10-35 to find the velocity of sphere A just after the collision:

$$v_{Ax}(t_2) = \frac{m_A - m_B}{m_A + m_B} v_{Ax}(t_1) = \frac{0.030 - 0.075\text{ kg}}{0.030 + 0.075\text{ kg}}(1.252\text{ m/s})$$

$$= -0.537\text{ m/s} \approx -0.54\text{ m/s}. \qquad \text{(Answer)}$$

The minus sign tells us that sphere A moves to the left just after the collision.

10-10 Two-Dimensional Energy and Momentum Conservation

When two bodies collide, the impulses of one on the other determine the directions in which they then travel. In particular, when the collision is not head-on, the bodies do not end up traveling along their initial axis. For such two-dimensional collisions in a closed, isolated system, the total translational momentum must still be conserved:

$$\vec{p}_{A1} + \vec{p}_{B1} = \vec{p}_{A2} + \vec{p}_{B2}. \qquad (10\text{-}43)$$

If the collision is also elastic (a special case), then the total kinetic energy is also conserved

$$K_{A1} + K_{B1} = K_{A2} + K_{B2}. \qquad (10\text{-}44)$$

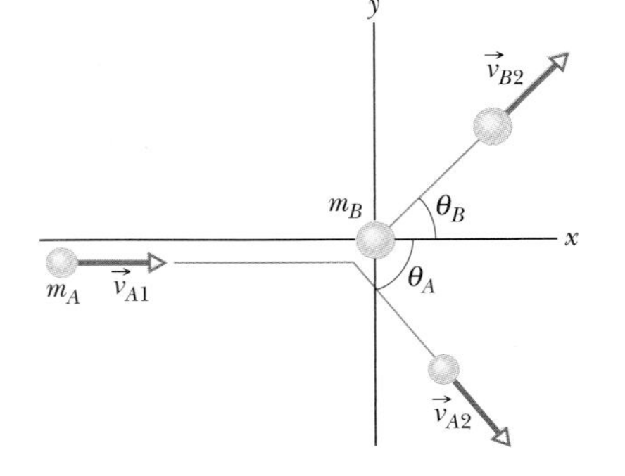

FIGURE 10-25 ■ An elastic collision between two bodies in which the collision is not head-on. The body with mass m_B (the target) is initially at rest.

Equation 10-43 is often more useful for analyzing a two-dimensional collision if we write it in terms of components on an *xy*-coordinate system. For example, let's revisit the momentum conservation analysis we did in Section 7-7 for a glancing two-dimensional collision. This time we will add the requirement that the collision be elastic so kinetic energy is conserved. Figure 10-25 shows a *glancing collision* (it is not head-on) between a projectile body and a target body initially at rest. The impulses between the bodies have sent the bodies off at angles θ_A and θ_B with respect to the x axis, along which the projectile traveled initially. In this situation, we would rewrite the momentum conservation equation initially presented in Section 7-7 (Eqs. 7-24 and 7-25) in terms of components along the x axis as

$$m_A|\vec{v}_{A1}|\cos 0° = m_A|\vec{v}_{A2}|\cos\theta_A + m_B|\vec{v}_{B2}|\cos\theta_B, \qquad (10\text{-}45)$$

and along the y axis as

$$0 = -m_A|\vec{v}_{A2}|\sin\theta_A + m_B|\vec{v}_{B2}|\sin\theta_B. \qquad (10\text{-}46)$$

We can also write Eq. 10-44 for this situation as

$$\tfrac{1}{2}m_A v_{A1}^2 = \tfrac{1}{2}m_A v_{A2}^2 + \tfrac{1}{2}m_B v_{B2}^2 \quad \text{(kinetic energy target initially at rest).} \qquad (10\text{-}47)$$

Equations 10-45 to 10-47 contain seven variables: two masses, m_A and m_B; three velocity magnitudes, v_{A1}, v_{A2}, and v_{B2}; and two angles, θ_A and θ_B. If we know any four of these quantities, we can solve the three equations for the remaining three quantities.

READING EXERCISE 10-9: In Fig. 10-25, suppose that the projectile has an initial x-component of momentum of 6 kg·m/s, a final x-component of momentum of 4 kg·m/s, and a final y-component of momentum of -3 kg·m/s. For the target, what then are (a) the final x-component of momentum and (b) the final y-component of momentum? ■

Problems

SEC. 10-3 ■ POTENTIAL ENERGY AS STORED WORK

1. Spring Constant What is the spring constant of a spring that stores 25 J of elastic potential energy when compressed by 7.5 cm from its relaxed length?

2. Dropping a Textbook You drop a 2.00 kg textbook to a friend who stands on the ground 10.0 m below the textbook with outstretched hands 1.50 m above the ground (Fig. 10-26). (a) How much work W^{grav} is done on the textbook by the gravitational force as it drops to your friend's hands? (b) What is the change ΔU in the gravitational potential energy of the textbook–Earth system during the drop? If the gravitational potential energy U of that system is taken to be zero at ground level, what is U when the textbook (c) is released and (d) reaches the hands? Now take U to be 100 J at ground level and again find (e) W^{grav} (f) ΔU, (g) U at the release point, and (h) U at the hands.

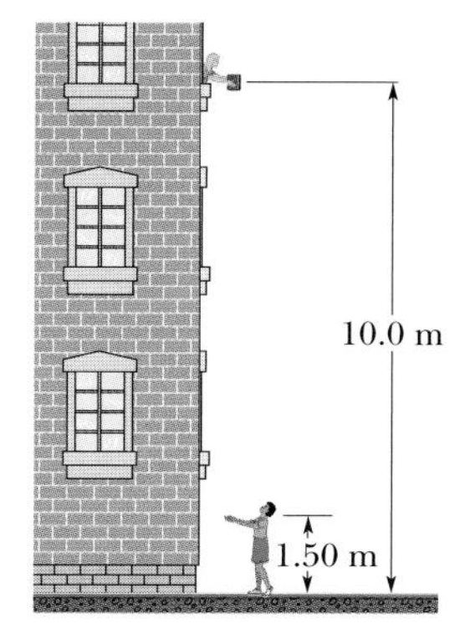

FIGURE 10-26 ■ Problems 2 and 12.

3. Ice Flake In Fig. 10-27, a 2.00 g ice flake is released from the edge of a hemispherical bowl whose radius r is 22.0 cm. The flake–bowl contact is frictionless. (a) How much work is done on the flake by the gravitational force during the flake's descent to the bottom of the bowl? (b) What is the change in the potential energy of the flake–Earth system during that descent? (c) If that potential energy is taken to be zero at the bottom of the bowl, what is its value when the flake is released? (d) If, instead, the potential energy is taken to be zero at the release point, what is its value when the flake reaches the bottom of the bowl? (e) If the mass of the flake were doubled, would the magnitudes of the answers to (a) through (d) increase, decrease, or remain the same?

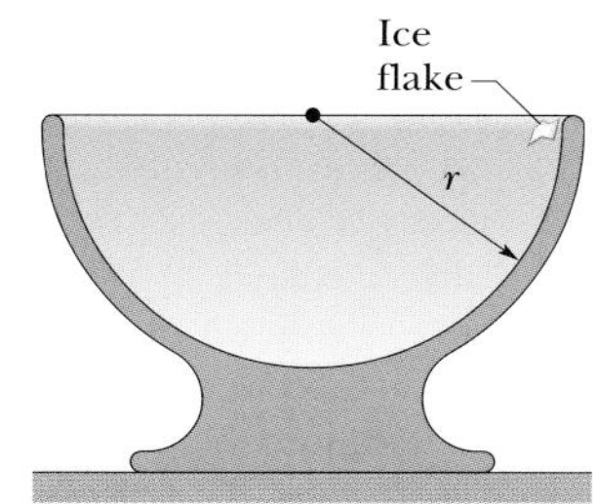

FIGURE 10-27 ■ Problems 3 and 11.

4. Roller Coaster In Fig. 10-28, a frictionless roller coaster of mass m tops the first hill with speed v_1. How much work does the gravitational force do on it from that point to (a) point A, (b) point B, and (c) point C? If the gravitational potential energy of the coaster–Earth system is taken to be zero at point C, what is its value when the coaster is at (d) point B and (e) point A? (f) If mass m were doubled, would the change in the gravitational potential energy of the system between points A and B increase, decrease, or remain the same?

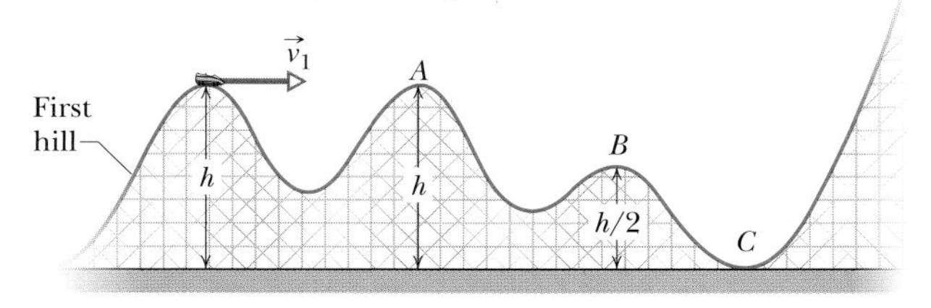

FIGURE 10-28 ■ Problems 4 and 14.

5. Ball Attached to a Rod Figure 10-29 shows a ball with mass m attached to the end of a thin rod with length L and negligible mass. The other end of the rod is pivoted so that the ball can move in a vertical circle. The rod is held in the horizontal position as shown and then given enough of a downward push to cause the ball to swing down and around and just reach the vertically upward position, with zero speed there. How much work is done on the ball by the gravitational force from the initial point to (a) the lowest point, (b) the highest point, and (c) the point on the right at which the ball is level with the initial point? If the gravitational potential energy of the ball–Earth system is taken to be zero at the initial point, what is its value when the ball reaches (d) the lowest point, (e) the highest point, and (f) the point on the right that is level with the initial point? (g) Suppose the rod were pushed harder so that the ball passed through the highest point with a nonzero speed. Would the change in the gravitational potential energy from the lowest point to the highest point then be greater, less, or the same?

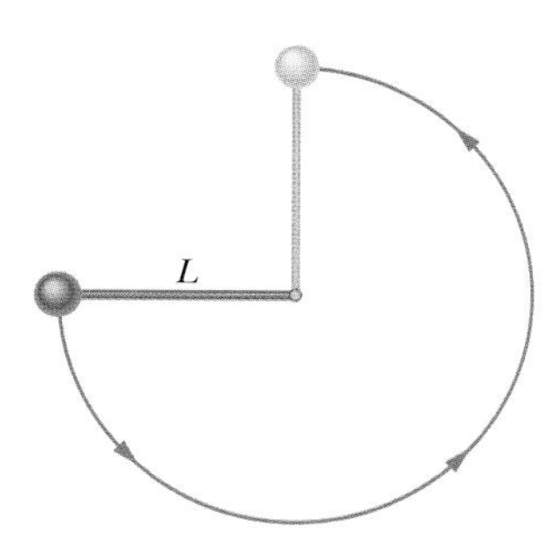

FIGURE 10-29 ■ Problems 5 and 13.

6. Loop-the-Loop In Fig. 10-30, a small block of mass m can slide along the frictionless loop-the-loop. The block is released from rest at point P, at height $h = 5R$ above the bottom of the loop. How much work does the gravitational force do on the block as the block travels from point P to (a) point Q and (b) the top of the loop? If the gravitational potential energy of the block–Earth system is taken to be zero at the bottom of the loop, what is that potential energy when the block is (c) at point P, (d) at point Q, and (e) at the top of the loop? (f) If, instead of being released, the block is given some initial speed downward along the track, do the answers to (a) through (e) increase, decrease, or remain the same?

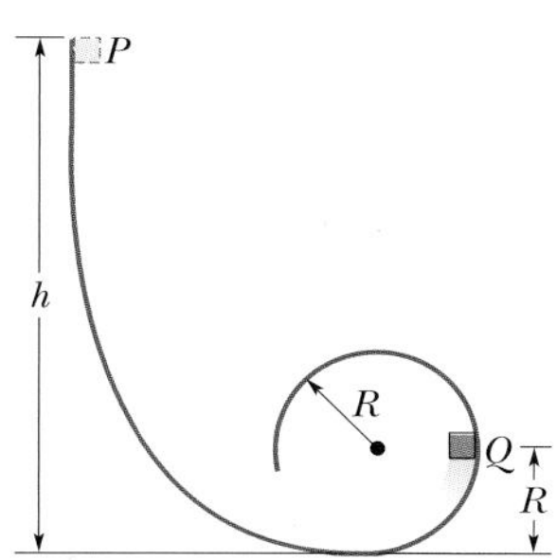

FIGURE 10-30 ■ Problems 6 and 22.

7. Snowball A 1.50 kg snowball is fired from a cliff 12.5 m high with an initial velocity of 14.0 m/s, directed 41.0° above the horizontal. (a) How much work is done on the snowball by the gravitational force during its flight to the flat ground below the cliff? (b) What is the change in the gravitational potential energy of the snowball–Earth system during the flight? (c) If that gravitational potential energy is taken to be zero at the height of the cliff, what is its value when the snowball reaches the ground?

8. Thin Rod Figure 10-31 shows a thin rod, of length L and negligible mass, that can pivot about one end to rotate in a vertical circle. A heavy ball of mass m is attached to the other end. The rod is pulled aside through an angle θ and released. As the ball descends to its lowest point, (a) how much work does the gravitational force do on it and (b) what is the change in the gravitational potential energy of the ball–Earth system? (c) If the gravitational potential energy is taken to be zero at the lowest point, what is its value just as the ball is released? (d) Do the magnitudes of the answers to (a) through (c) increase, decrease, or remain the same if angle θ is increased?

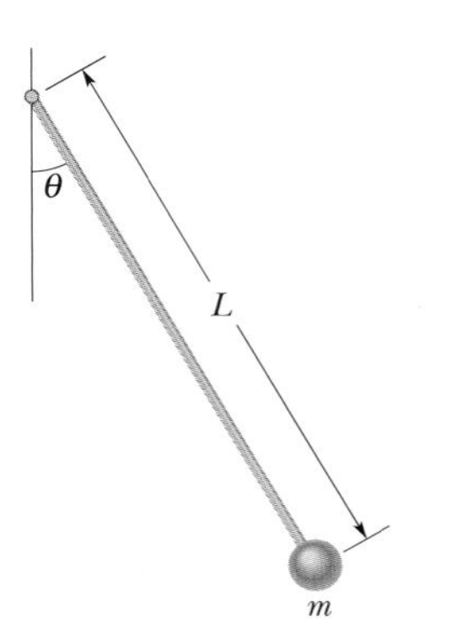

FIGURE 10-31 ■ Problems 8 and 16.

9. Ball Thrown from Tower At $t_1 = 0$ a 1.0 kg ball is thrown from the top of a tall tower with velocity $\vec{v}_1 = (18 \text{ m/s})\hat{i} + (24 \text{ m/s})\hat{j}$. What is the change in the potential energy of the ball–Earth system between $t_1 = 0$ and $t_2 = 6.0$ s?

SEC. 10-4 ■ MECHANICAL ENERGY CONSERVATION

10. Block Dropped on a Spring A 250 g block is dropped onto a relaxed vertical spring that has a spring constant of $k = 2.5$ N/cm (Fig. 10-32). The block becomes attached to the spring and compresses the spring 12 cm before momentarily stopping. While the spring is being compressed, what work is done on the block by (a) the gravitational force on it and (b) the spring force? (c) What is the speed of the block just before it hits the spring? (Assume that friction is negligible.) (d) If the speed at impact is doubled, what is the maximum compression of the spring?

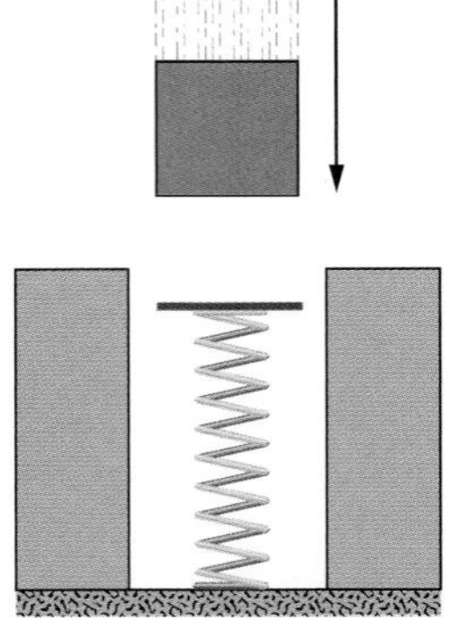

FIGURE 10-32 ■ Problem 10.

11. Speed of Flake (a) In Problem 3, what is the speed of the flake when it reaches the bottom of the bowl? (b) If we substituted a second flake with twice the mass, what would its speed be? (c) If, instead, we gave the flake an initial downward speed along the bowl, would the answer to (a) increase, decrease, or remain the same?

12. Speed of Textbook (a) In Problem 2, what is the speed of the textbook when it reaches the hands? (b) If we substituted a second textbook with twice the mass, what would its speed be? (c) If, instead, the textbook were thrown down, would the answer to (a) increase, decrease, or remain the same?

13. Zero Speed at Vertical (a) In Problem 5, what initial speed must be given the ball so that it reaches the vertically upward position with zero speed? What then is its speed at (b) the lowest point and (c) the point on the right at which the ball is level with the initial point? (d) If the ball's mass were doubled, would the answers to (a) through (c) increase, decrease, or remain the same?

14. Speed of Coaster In Problem 4, what is the speed of the coaster at (a) point A, (b) point B, and (c) point C? (d) How high will it go on the last hill, which is too high for it to cross? (e) If we substitute a second coaster with twice the mass, what then are the answers to (a) through (d)?

15. Runaway Truck In Fig. 10-33, a runaway truck with failed brakes is moving downgrade at 130 km/h just before the driver steers the truck up a frictionless emergency escape ramp with an inclination of 15°. The truck's mass is 5000 kg. (a) What minimum length L must the ramp have if the truck is to stop (momentarily) along it? (Assume the truck is a particle, and justify that assumption.) Does the minimum length L increase, decrease, or remain the same if (b) the truck's mass is decreased and (c) its speed is decreased?

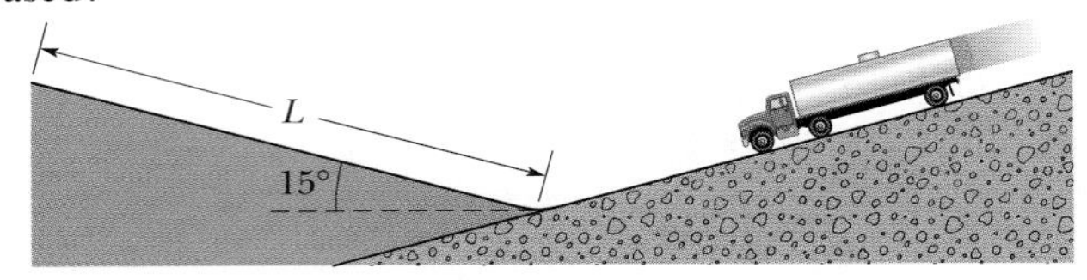

FIGURE 10-33 ■ Problem 15.

16. Speed at Lowest Point (a) In Problem 8, what is the speed of the ball at the lowest point if $L = 2.00$ m, $\theta = 30.0°$, and $m = 5.00$ kg? (b) Does the speed increase, decrease, or remain the same if the mass is increased?

17. Snowball Reaches Ground (a) In Problem 7, using energy techniques rather than the techniques of Chapter 5, find the speed of the snowball as it reaches the ground below the cliff. What is that speed (b) if the launch angle is changed to 41.0° *below* the horizontal and (c) if the mass is changed to 2.50 kg?

18. Stone Rests on Spring Figure 10-34 shows an 8.00 kg stone at rest on a spring. The spring is compressed 10.0 cm by the stone. (a) What is the spring constant? (b) The stone is pushed down an additional 30.0 cm and released. What is the elastic potential energy of the compressed spring just before that release? (c) What is the change in the gravitational potential energy of the stone–Earth system when the stone moves from the release point to its maximum height? (d) What is that maximum height, measured from the release point?

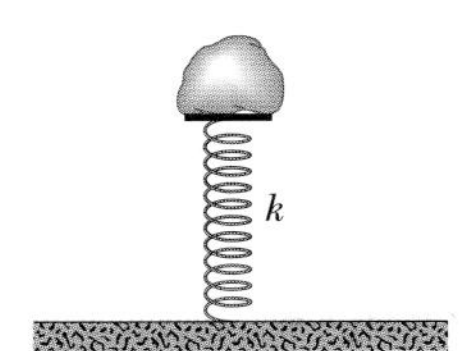

FIGURE 10-34 ■ Problem 18.

19. Marble Fired Vertically A 5.0 g marble is fired vertically upward using a spring gun. The spring must be compressed 8.0 cm if the marble is to just reach a target 20 m above the marble's position on the compressed spring. (a) What is the change ΔU^{grav} in the gravitational potential energy of the marble–Earth system during the 20 m ascent? (b) What is the change ΔU^{elas} in the elastic potential energy of the spring during its launch of the marble? (c) What is the spring constant of the spring?

20. Pendulum Figure 10-35 shows a pendulum of length L. Its bob (which effectively has all the mass) has speed v_1 when the cord makes an angle θ_1 with the vertical. (a) Derive an expression for the speed of the bob when it is in its lowest position. What is the least value that v_1 can have if the pendulum is to swing down and then up (b) to a horizontal position, and (c) to a vertical position with the cord remaining straight? (d) Do the answers to (b) and (c) increase, decrease, or remain the same if θ_1 is increased by a few degrees?

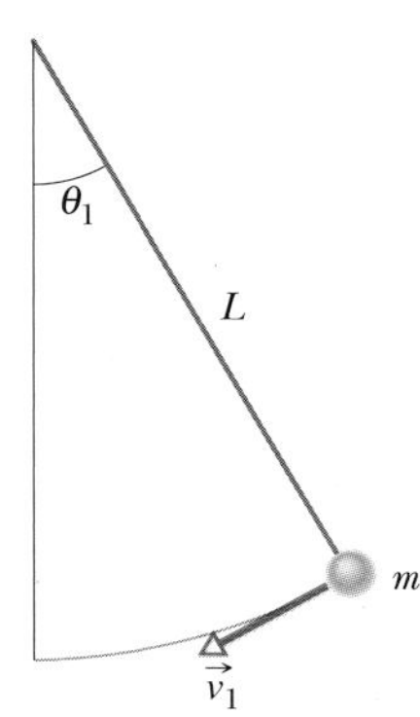

FIGURE 10-35 ■ Problem 20.

21. Block–Spring–Incline A 2.00 kg block is placed against a spring on a frictionless 30.0° incline (Fig. 10-36). (The block is not attached to the spring.) The spring, whose spring constant is 19.6 N/cm, is compressed 20.0 cm and then released. (a) What is the elastic potential energy of the compressed spring? (b) What is the change in the gravitational potential energy of the block–Earth system as the bock moves from the release point to its highest point on the incline? (c) How far along the incline is the highest point from the release point?

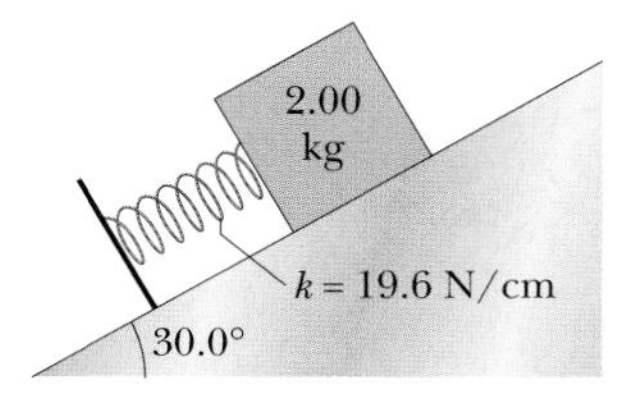

FIGURE 10-36 ■ Problem 21.

22. Horizontal and Vertical Components In Problem 6, what are (a) the horizontal component and (b) the vertical component of the *net* force acting on the block at point Q? (c) At what height h should the block be released from rest so that it is on the verge of losing contact with the track at the top of the loop? (*On the verge of losing contact* means that the normal force on the block from the track has just then become zero.) (d) Graph the magnitude of the normal force on the block at the top of the loop versus initial height h, for the range $h = 0$ to $h = 6R$.

23. Block on Incline Collides with Spring In Fig. 10-37, a 12 kg block is released from rest on a 30° frictionless incline. Below the block is a spring that can be compressed 2.0 cm by a force of 270 N. The block momentarily stops when it compresses the spring by 5.5 cm. (a) How far does the block move down the incline from its rest position to this stopping point? (b) What is the speed of the block just as it touches spring?

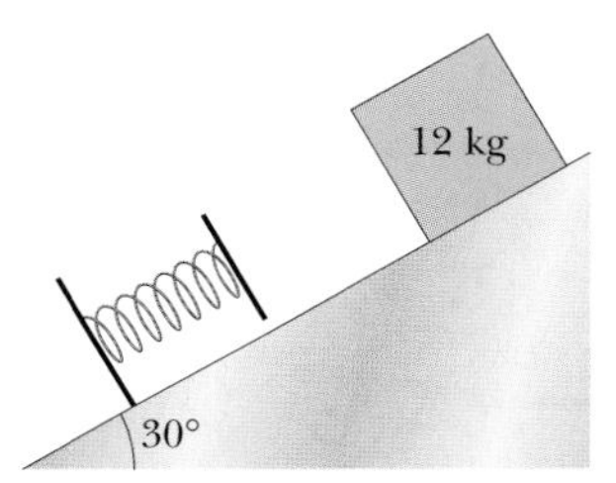

FIGURE 10-37 ■ Problem 23.

24. Ski-Jump Ramp A 60 kg skier starts from rest at a height of 20 m above the end of a ski-jump ramp as shown in Fig. 10-38. As the skier leaves the ramp, his velocity makes an angle of 28° with the horizontal. Neglect the effects of air resistance and assume the ramp is frictionless. (a) What is the maximum height h of his jump above the end of the ramp? (b) If he increased his weight by putting on a backpack, would h then be greater, less, or the same?

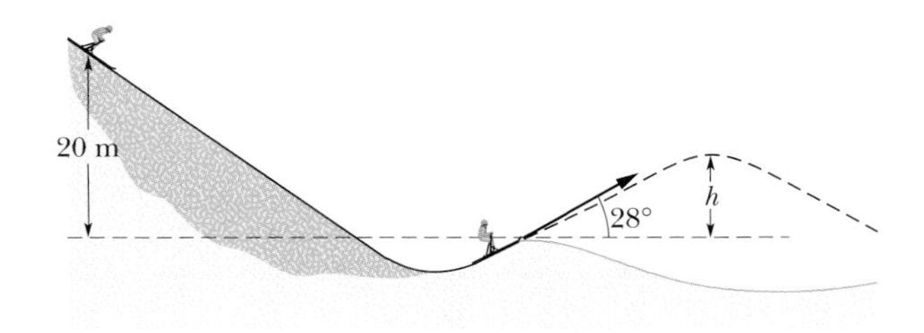

FIGURE 10-38 ■ Problem 24.

25. Block Dropped on a Spring Two A 2.0 kg block is dropped from a height of 40 cm onto a spring of spring constant $k = 1960$ N/m (Fig. 10-39). Find the maximum distance the spring is compressed.

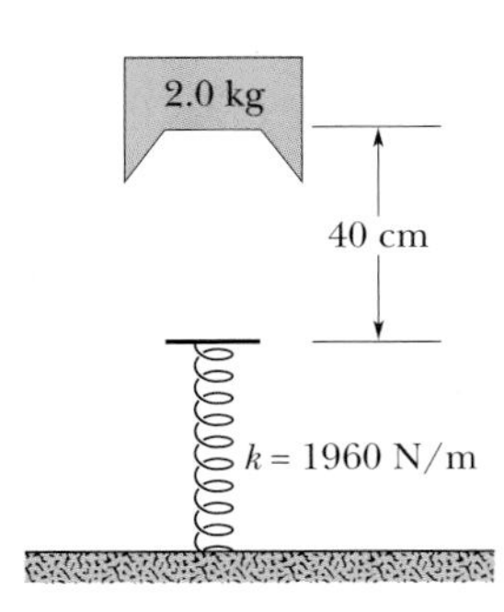

FIGURE 10-39 ■ Problem 25.

26. Tarzan Tarzan, who weighs 688 N, swings from a cliff at the end of a convenient vine that is 18 m long (Fig. 10-40). From the top of the cliff to the bottom of the swing, he descends by 3.2 m. The vine will break if the force on it exceeds 950 N. (a) Does the vine break? (b) If no, what is the greatest force on it during the swing? If yes, at what angle with the vertical does it break?

FIGURE 10-40 ■ Problem 26.

27. Two Children Play Two children are playing a game in which they try to hit a small box on the floor with a marble fired from a spring loaded gun that is mounted on a table. The target box is 2.20 m horizontally from the edge of the table; see Fig. 10-41. Bobby compresses the spring 1.10 cm, but the center of the marble falls 27.0 cm short of the center of the box. How far should Rhoda compress the spring to score a direct hit? Assume that neither the spring nor the ball encounters friction in the gun.

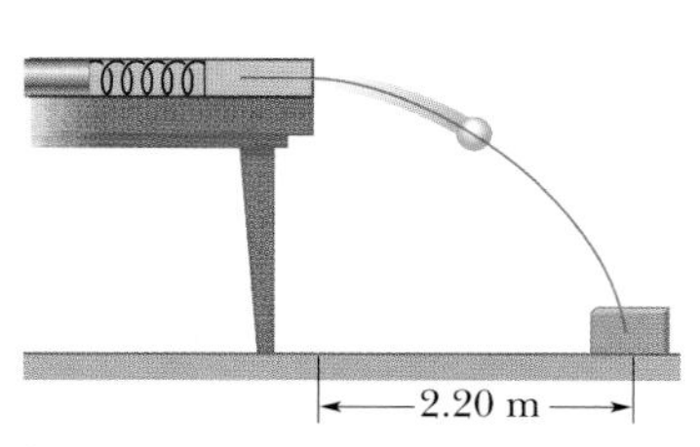

FIGURE 10-41 ■ Problem 27.

28. Block Sticks to Spring A 700 g block is released from rest at height h_1 above a vertical spring with spring constant $k = 400$ N/m and negligible mass. The block sticks to the spring and momentarily stops after compressing the spring 19.0 cm. How much work is done (a) by the block on the spring and (b) by the spring on the block? (c) What is the value of h_1? (d) If the block were released from height $2h_1$ above the spring, what would be the maximum compression of the spring?

29. Complete Swing In Fig. 10-42 show that, if the ball is to swing completely around the fixed peg, then $d > 3L/5$. (*Hint*: The ball must still be moving at the top of its swing. Do you see why?)

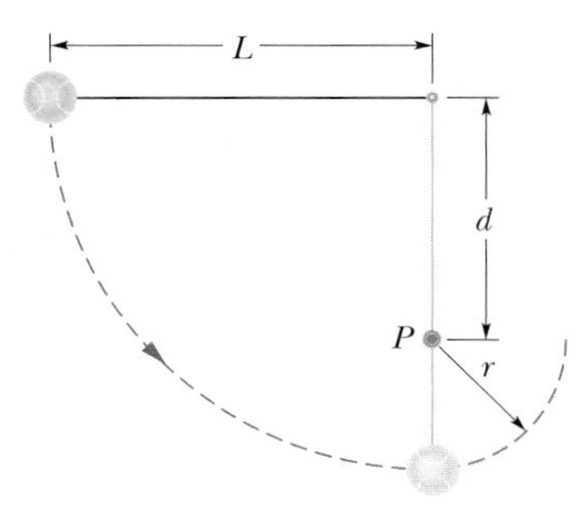

FIGURE 10-42 ■ Problems 29 and 36.

30. To Make a Pendulum To make a pendulum, a 300 g ball is attached to one end of a string that has a length of 1.4 m and negligible mass. (The other end of the string is fixed.) The ball is pulled to one side until the string makes an angle of 30.0° with the vertical; then (with the string taut) the ball is released from rest. Find (a) the speed of the ball when the string makes an angle of 20.0° with the vertical and (b) the maximum speed of the ball. (c) What is the angle between the string and the vertical when the speed of the ball is one-third its maximum value?

31. Rigid Rod A rigid rod of length L and negligible mass has a ball with mass m attached to one end and its other end fixed, to form a pendulum. The pendulum is inverted, with the rod straight

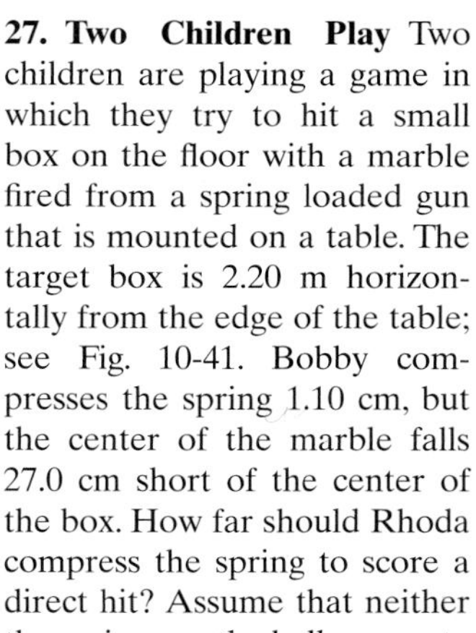

up, and then released. At the lowest point, what are (a) the ball's speed and (b) the tension in the rod? (c) The pendulum is next released at rest from a horizontal position. At what angle from the vertical does the tension in the rod equal the weight of the ball?

32. Spring at the Top of an Incline In Fig. 10-43, a spring with spring constant $k = 170$ N/m is at the top of a 37.0° frictionless incline. The lower end of the incline is 1.00 m from the end of the spring, which is at its relaxed length. A 2.00 kg canister is pushed against the spring until the spring is compressed 0.200 m and released from rest. (a) What is the speed of the canister at the instant the spring returns to its relaxed length (which is when the canister loses contact with the spring)? (b) What is the speed of the canister when it reaches the lower end of the incline?

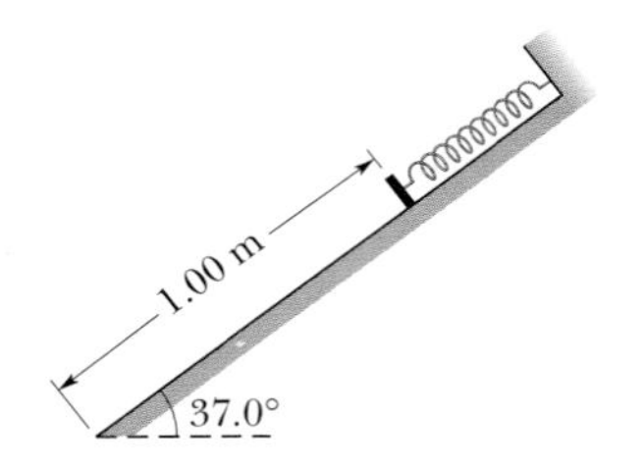

FIGURE 10-43 ■ Problem 32.

33. Chain on Table In Fig. 10-44, a chain is held on a frictionless table with one-fourth of its length hanging over the edge. If the chain has length L and mass m, how much work is required to pull the hanging part back onto the table?

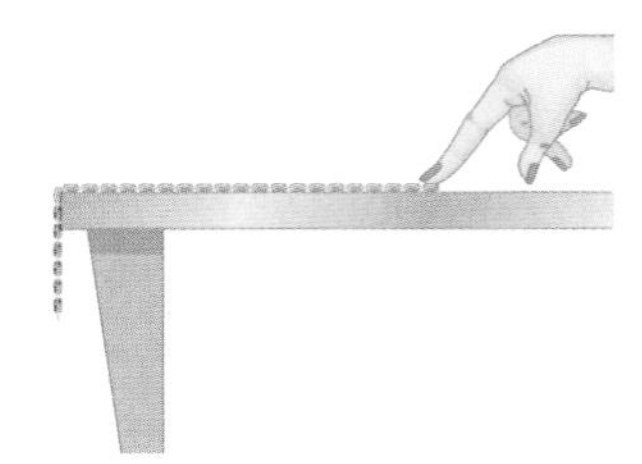

FIGURE 10-44 ■ Problem 33.

34. Vertical Spring A spring with spring constant $k = 400$ N/m is placed in a vertical orientation with its lower end supported by a horizontal surface. The upper end is depressed 25.0 cm, and a block with a weight of 40.0 N is placed (unattached) on the depressed spring. The system is then released from rest. Assume the gravitational potential energy U^{grav} of the block is zero at the release point ($y_1 = 0$) and calculate the gravitational potential energy, the elastic potential energy U^{elas}, and the kinetic energy K of the block for y_2 equal to (a) 0, (b) 5.00 cm, (c) 10.0 cm, (d) 15.0 cm, (e) 20.0 cm, (f) 25.0 cm, and (g) 30.0 cm, Also, (h) how far above its point of release does the block rise?

35. Ice Mound A boy is seated on the top of a hemispherical mound of ice (Fig. 10-45). He is given a very small push and starts sliding down the ice. Show that he leaves the ice at a point whose height is $2R/3$ if the ice is frictionless. (*Hint*: The normal force vanishes as he leaves the ice.)

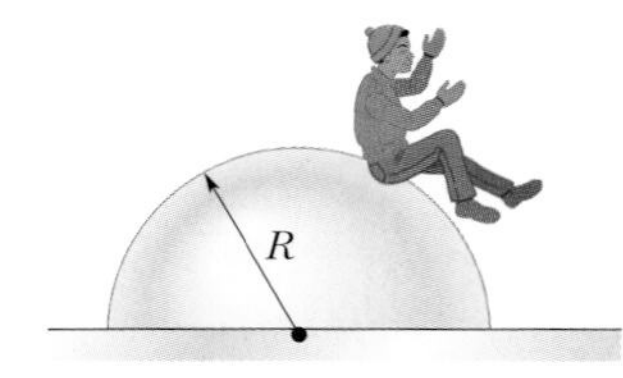

FIGURE 10-45 ■ Problem 35.

36. Ball on a String The string in Fig. 10-42 is $L = 120$ cm long, has a ball attached to one end, and is fixed at its other end. The distance d to the fixed peg at point P is 75.0 cm. When the initially stationary ball is released with the string horizontal as shown, it will swing along the dashed arc. What is its speed when it reaches (a) its lowest point and (b) its highest point after the string catches on the peg?

SEC. 10-5 ■ READING A POTENTIAL ENERGY CURVE

37. Diatomic Molecule The potential energy of a diatomic molecule (a two-atom system like H_2 or O_2) is given by

$$U = \frac{A}{r^{12}} - \frac{B}{r^6},$$

where r is the separation of the two atoms of the molecule and A and B are positive constants. This potential energy is associated with the force that binds the two atoms together. (a) Find the *equilibrium separation*—that is, the distance between the atoms at which the force on each atom is zero. Is the force repulsive (the atoms are pushed apart) or attractive (they are pulled together) if their separation is (b) smaller and (c) larger than the equilibrium separation?

38. Potential Energy Graph A conservative force $F(x)$ acts on a 2.0 kg particle that moves along the x axis. The potential energy $U(x)$ associated with $F(x)$ is graphed in Fig. 10-46. When the particle is at $x = 2.0$ m, its velocity is -1.5 m/s. (a) What are the magnitude and direction of $F(x)$ at this position? (b) Between what limits of x does the particle move? (c) What is its speed at $x = 7.0$ m?

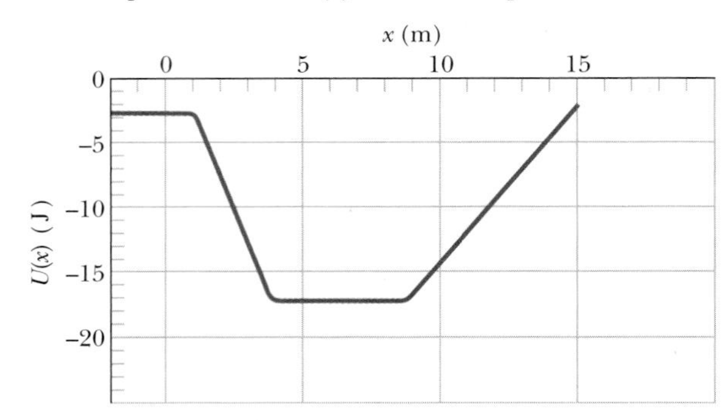

FIGURE 10-46 ■ Problem 38.

39. Potential Energy Function A single conservative force $F(x)$ acts on a 1.0 kg particle that moves along an x axis. The potential energy $U(x)$ associated with $F(x)$ is given by

$$U(x) = (-4.00 \text{ J/m})\, e^{(-x/(4.00 \text{ m}))}.$$

At $x = 5.0$ m the particle has a kinetic energy of 2.0 J. (a) What is the mechanical energy of the system? (b) Make a plot of $U(x)$ as a function of x for $0 \leq x \leq 10$ m, and on the same graph draw the line that represents the mechanical energy of the system. Use part (b) to determine (c) the least value of x and (d) the greatest value of x between which the particle can move. Use part (b) to determine (e) the maximum kinetic energy of the particle and (f) the value of x at which it occurs. (g) Determine the equation for $F(x)$ as a function of x. (h) For what (finite) value of x does $F(x) = 0$?

SEC. 10-7 ■ CONSERVATION OF ENERGY

40. Plastic Cube The temperature of a plastic cube is monitored while the cube is pushed 3.0 m across a floor at constant speed by a horizontal force of 15 N. The monitoring reveals that the thermal energy of the cube increases by 20 J. What is the increase in the thermal energy of the floor along which the cube slides?

41. Block Drawn by Rope A 3.57 kg block is drawn at constant speed 4.06 m along a horizontal floor by a rope. The force on the block from the rope has a magnitude of 7.68 N and is directed 15.0° above the horizontal. What are (a) the work done by the rope's force, (b) the increase in thermal energy of the block–floor system, and (c) the coefficient of kinetic friction between the block and floor?

42. Worker Pushes Block A worker pushed a 27 kg block 9.2 m along a level floor at constant speed with a force directed 32° below the horizontal. If the coefficient of kinetic friction between block and floor was 0.20, what were (a) the work done by the worker's force and (b) the increase in thermal energy of the block–floor system?

43. The Collie A Collie drags its bed box across a floor by applying a horizontal force of 8.0 N. The kinetic frictional force acting on the box has magnitude 5.0 N. As the box is dragged through 0.70 m along the way, what are (a) the work done by the collie's applied force and (b) the increase in thermal energy of the bed and floor?

44. Bullet Hits Wall A 30 g bullet, with a horizontal velocity of 500 m/s, comes to a stop 12 cm within a solid wall. (a) What is the change in its mechanical energy? (b) What is the magnitude of the average force from the wall stopping it?

45. Ski Jumper A 60 kg skier leaves the end of a ski-jump ramp with a velocity of 24 m/s directed 25° above the horizontal. Suppose that as a result of air drag the skier returns to the ground with a speed of 22 m/s, landing 14 m vertically below the end of the ramp. From the launch to the return to the ground, by how much is the mechanical energy of the skier–Earth system reduced because of air drag?

46. Frisbee A 75 g Frisbee is thrown from a point 1.1 m above the ground with a speed of 12 m/s. When it has reached a height of 2.1 m, its speed is 10.5 m/s. What was the reduction in the mechanical energy of the Frisbee–Earth system because of air drag?

47. Outfielder Throws An outfielder throws a baseball with an initial speed of 81.8 mi/h. Just before an infielder catches the ball at the same level, the ball's speed is 110 ft/s. In foot-pounds, by how much is the mechanical energy of the ball–Earth system reduced because of air drag? (The weight of a baseball is 9.0 oz.)

48. Niagara Falls Approximately 5.5×10^6 kg of water fall 50 m over Niagara Falls each second. (a) What is the decrease in the gravitational potential energy of the water–Earth system each second? (b) If all this energy could be converted to electrical energy (it cannot be), at what rate would electrical energy be supplied? (The mass of 1 m^3 of water is 1000 kg.) (c) If the electrical energy were sold at 1 cent/kW · h. what would be the yearly cost?

49. Rock Slide During a rockslide, a 520 kg rock slides from rest down a hillside that is 500 m long and 300 m high. The coefficient of kinetic friction between the rock and the hill surface is 0.25. (a) If the gravitational potential energy U of the rock–Earth system is zero at the bottom of the hill, what is the value of U just before the slide? (b) How much energy is transferred to thermal energy during the slide? (c) What is the kinetic energy of the rock as it reaches the bottom of the hill? (d) What is its speed then?

50. Block Against Horizontal Spring You push a 2.0 kg block against a horizontal spring, compressing the spring by 15 cm. Then you release the block, and the spring sends it sliding across a tabletop. It stops 75 cm from where you released it. The spring constant is 200 N/m. What is the coefficient of kinetic friction between the block and the table?

51. Horizontal Spring As Fig. 10-47 shows, a 3.5 kg block is accelerated by a compressed spring whose spring constant is 640 N/m. After leaving the spring at the spring's relaxed length, the block travels over a horizontal surface, with a coefficient of kinetic friction of 0.25, for a distance of 7.8 m before stopping. (a) What is the increase in the thermal energy of the block–floor system? (b) What is the maximum kinetic energy of the block? (c) Through what distance is the spring compressed before the block begins to move?

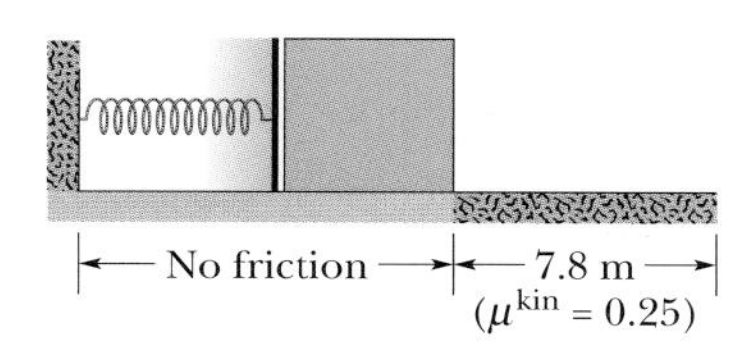

FIGURE 10-47 ■ Problem 51.

52. Block Slides Down an Incline In Fig. 10-48, a block is moved down an incline a distance of 5.0 m from point A to point B by a force $\vec{F}$ that is parallel to the incline and has magnitude 2.0 N. The magnitude of the frictional force acting on the block is 10 N. If the kinetic energy of the block increases by 35 J between A and B, how much work is done on the block by the gravitational force as the block moves from A to B?

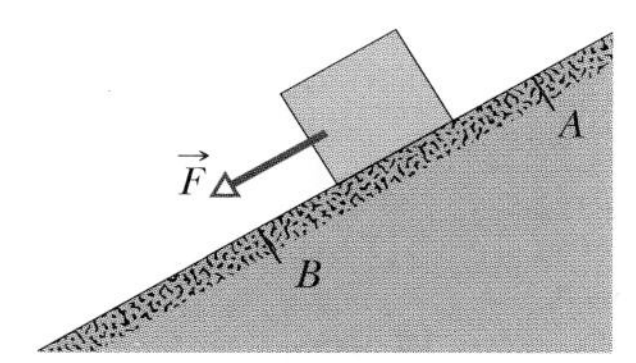

FIGURE 10-48 ■ Problem 52.

53. Nonconforming Spring A certain spring is found *not* to conform to Hooke's law. The force (in newtons) it exerts when stretched a distance x (in meters) is found to have magnitude $(52.8 \text{ N/m})x + (38.4 \text{ N/m}^2)x^2$ in the direction opposing the stretch. (a) Compute the work required to stretch the spring from $x_1 = 0.500$ m to $x_2 = 1.00$ m. (b) With one end of the spring fixed, a particle of mass 2.17 kg is attached to the other end of the spring when it is extended by an amount $x_2 = 1.00$ m. If the particle is then released from rest, what is its speed at the instant the spring has returned to the configuration in which the extension is $x_1 = 0.500$ m? (c) Is the force exerted by the spring conservative or nonconservative? Explain.

54. Bundle A 4.0 kg bundle starts up a 30° incline with 128 J of kinetic energy. How far will it slide up the incline if the coefficient of kinetic friction between bundle and incline is 0.30?

55. Two Snowy Peaks Two snowy peaks are 850 m and 750 m above the valley between them. A ski run extends down from the top of the higher peak and then back up to the top of the lower one, with a total length of 3.2 km and an average slope of 30° (Fig. 10-49). (a) A skier starts from rest at the top of the higher peak. At what speed will he arrive at the top of the lower peak if he coasts without using ski poles? Ignore friction. (b) Approximately what coefficient of kinetic friction between snow and skis would make him stop just at the top of the lower peak?

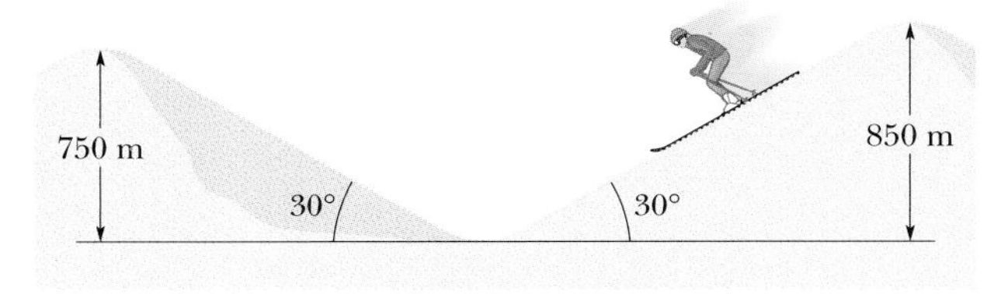

FIGURE 10-49 ■ Problem 55.

56. Playground Slide A girl whose weight is 267 N slides down a 6.1 m playground slide that makes an angle of 20° with the horizontal. The coefficient of kinetic friction between slide and child is 0.10.

(a) How much energy is transferred to thermal energy? (b) If the girl starts at the top with a speed of 0.457 m/s, what is her speed at the bottom?

57. Block and Horizontal Spring In Fig. 10-50, a 2.5 kg block slides head on into a spring with a spring constant of 320 N/m. When the block stops, it has compressed the spring by 7.5 cm. The coefficient of kinetic friction between the block and the horizontal surface is 0.25. While the block is in contact with the spring and being brought to rest, what are (a) the work done by the spring force and (b) the increase in thermal energy of the block–floor system? (c) What is the block's speed just as the block reaches the spring?

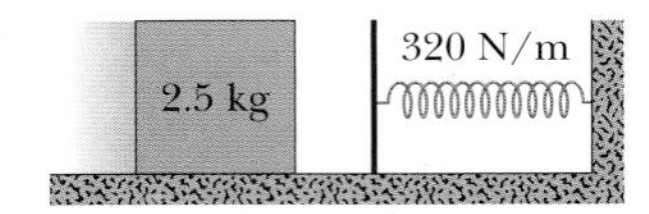

FIGURE 10-50 ■ Problem 57.

58. Factory Worker A factory worker accidentally releases a 180 kg crate that was being held at rest at the top of a 3.7 m-long-ramp inclined at 39° to the horizontal. The coefficient of kinetic friction between the crate and the ramp, and between the crate and the horizontal factory floor, is 0.28. (a) How fast is the crate moving as it reaches the bottom of the ramp? (b) How far will it subsequently slide across the factory floor? (Assume that the crate's kinetic energy does not change as it moves from the ramp onto the floor.) (c) Do the answers to (a) and (b) increase, decrease, or remain the same if we halve the mass of the crate?

59. Block on a Track In Fig. 10-51, a block slides along a track from one level to a higher level, by moving through an intermediate valley. The track is frictionless until the block reaches the higher level. There a frictional force stops the block in a distance d. The block's initial speed v_1 is 6.0 m/s; the height difference h is 1.1 m; and the coefficient of kinetic friction μ^{kin} is 0.60. Find d.

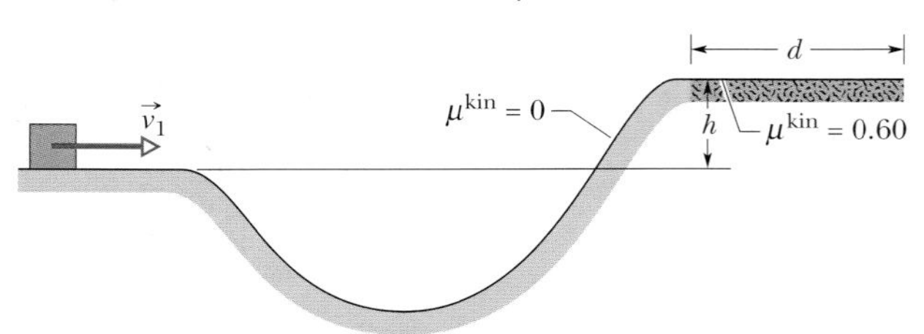

FIGURE 10-51 ■ Problem 59.

60. Cookie Jar A cookie jar is moving up a 40° incline. At a point 55 cm from the bottom of the incline (measured along the incline), it has a speed of 1.4 m/s. The coefficient of kinetic friction between jar and incline is 0.15. (a) How much farther up the incline will the jar move? (b) How fast will it be going when it has slid back to the bottom of the incline? (c) Do the answers to (a) and (b) increase, decrease, or remain the same if we decrease the coefficient of kinetic friction (but do not change the given speed or location)?

61. Stone Thrown Vertically A stone with weight w is thrown vertically upward into the air from ground level with initial speed v_1. If a constant force f due to air drag acts on the stone throughout its flight, (a) show that the maximum height reached by the stone is

$$h = \frac{v_1^2}{2g(1 + f/w)}.$$

(b) Show that the stone's speed is

$$v_2 = v_1\left(\frac{w - f}{w + f}\right)^{1/2}$$

just before impact with the ground.

62. Playground Slide Two A playground slide is in the form of an arc of a circle with a maximum height of 4.0 m, with a radius of 12 m, and with the ground tangent to the circle (Fig. 10-52). A 25 kg child starts from rest at the top of the slide and has a speed of 6.2 m/s at the bottom. (a) What is the length of the slide? (b) What average frictional force acts on the child over this distance? If, instead of the ground, a vertical line through the *top of the slide* is tangent to the circle, what are (c) the length of the slide and (d) the average frictional force on the child?

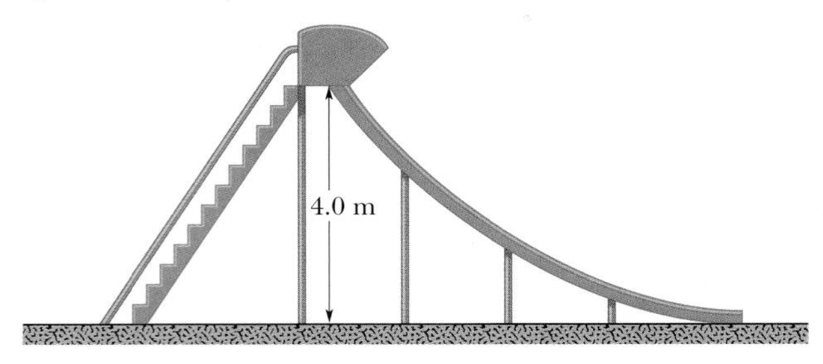

FIGURE 10-52 ■ Problem 62.

63. Particle on a Slide A particle can slide along a track with elevated ends and a flat central part, as shown in Fig. 10-53. The flat part has length L. The curved portions of the track are frictionless, but for the flat part the coefficient of kinetic friction is $\mu^{kin} = 0.20$. The particle is released from rest at point A, which is a height $h = L/2$ above the flat part of the track. Where does the particle finally stop?

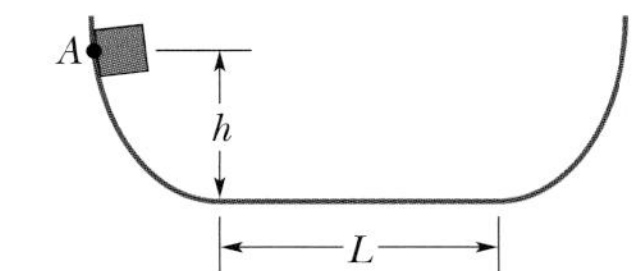

FIGURE 10-53 ■ Problem 63.

64. Cable Breaks The cable of the 1800 kg elevator cab in Fig. 10-54 snaps when the cab is at rest at the first floor, where the cab bottom is a distance $d = 3.7$ m above a cushioning spring whose spring constant is $k = 0.15$ MN/m. A safety device clamps the cab against guide rails so that a constant frictional force of 4.4 kN opposes the cab's motion. (a) Find the speed of the cab just before it hits the spring. (b) Find the maximum distance x that the spring is compressed (the frictional force still acts during this compression). (c) Find the distance that the cab will bounce back up the shaft. (d) Using conservation of energy, find the approximate total distance that the cab will move before coming to rest. (Assume that the frictional force on the cab is negligible when the cab is stationary.)

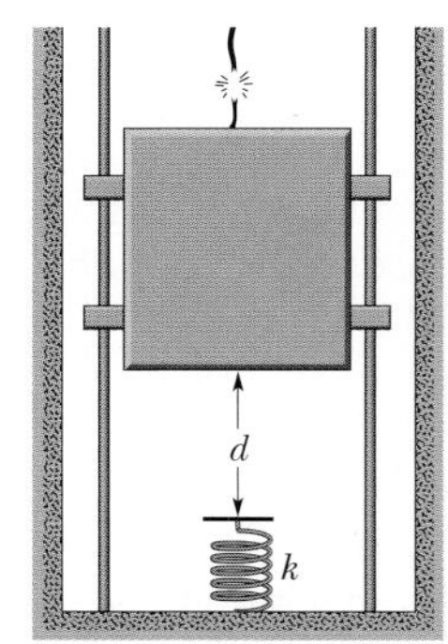

FIGURE 10-54 ■ Problem 64.

65. At a Factory At a certain factory, 300 kg crates are dropped vertically from a packing machine onto a conveyor belt moving at 1.20 m/s (Fig 10-55). (A motor maintains the belt's constant speed.) The coefficient of kinetic friction between the belt and each crate is 0.400. After a short time, slipping between the belt and the crate ceases, and the crate then moves along with the belt. For the period of time during which the crate is being brought to rest relative to the belt, calculate, for a coordinate system at rest in the factory, (a) the kinetic energy supplied to the crate, (b) the magnitude of the kinetic frictional force acting on the crate, and (c) the energy supplied by the motor. (d) Explain why the answers to (a) and (c) are different.

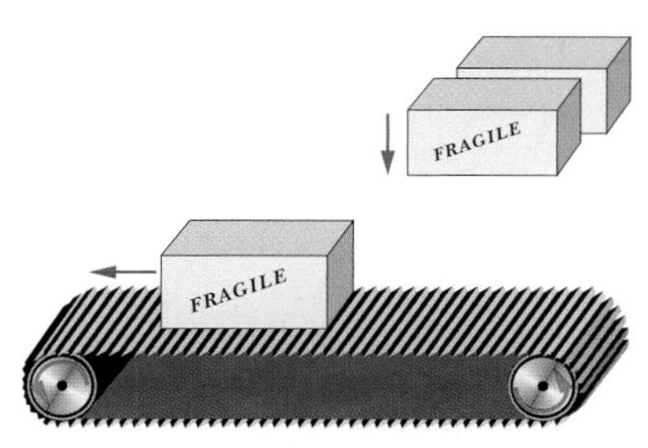

FIGURE 10-55 ■ Problem 65.

66. A Bear Slides A 25 kg bear slides, from rest, 12 m down a lodgepole pine tree, moving with a speed of 5.6 m/s just before hitting the ground. (a) What change occurs in the gravitational potential energy of the bear–Earth system during the slide? (b) What is the kinetic energy of the bear just before hitting the ground? (c) What is the average frictional force that acts on the sliding bear?

67. Daniel Goodwin In 1981, Daniel Goodwin climbed 443 m up the *exterior* of the Sears Building in Chicago using suction cups and metal clips. (a) Approximate his mass and then compute how much energy he had to transfer from biomechanical (internal) energy to the gravitational potential energy of the Earth–Goodwin system to lift his center of mass to that height. (b) How much energy would he have had to transfer if he had, instead, taken the stairs inside the building (to the same height)?

68. Mount Everest The summit of Mount Everest is 8850 m above sea level. (a) How much energy would a 90 kg climber expend against the gravitational force on him in climbing to the summit from sea level? (b) How many candy bars, at 1.25 MJ per bar, would supply an energy equivalent to this? Your answer should suggest that work done against the gravitational force is a very small part of the energy expended in climbing a mountain.

69. A Woman Leaps Vertically A 55 kg woman leaps vertically from a crouching position in which her center of mass is 40 cm above the ground. As her feet leave the floor, her center of mass is 90 cm above the ground; it rises to 120 cm at the top of her leap. (a) As she is pressing down on the ground during the leap, what is the average magnitude of the force on her from the ground? (b) What maximum speed does she attain?

70. An Automobile with Passengers An automobile with passengers has weight 16,400 N and is moving at 113 km/h when the driver brakes to a stop. The frictional force on the wheels from the road has a magnitude of 8230 N. Find the stopping distance.

SECS. 10-8 TO 10-10 ■ CONSERVATION OF ENERGY AND MOMENTUM

71. Box of Marbles A box is put on a scale that is marked in units of mass and adjusted to read zero when the box is empty. A stream of marbles is then poured into the box from a height h above its bottom at a rate of R (marbles per second). Each marble has mass m. (a) If the collisions between the marbles and the box are completely inelastic, find the scale reading at time t after the marbles begin to fill the box. (b) Determine a numerical answer when $R = 100\ \text{s}^{-1}$, $h = 7.60$ m, $m = 4.50$ g, and $t = 10.0$ s.

72. Particle *A* and Particle *B* Particle A and particle B are held together with a compressed spring between them. When they are released, the spring pushes them apart and they then fly off in opposite directions, free of the spring. The mass of A is 2.00 times the mass of B, and the energy stored in the spring was 60 J. Assume that the spring has negligible mass and that all its stored energy is transferred to the particles. Once that transfer is complete, what are the kinetic energies of (a) particle A and (b) particle B?

73. Ball and Spring Gun In Fig. 10-56, a ball of mass m is shot with speed v_1 into the barrel of a spring gun of mass M initially at rest on a frictionless surface. The ball sticks in the barrel at the point of maximum compression of the spring. Assume that the increase in thermal energy due to friction between the ball and the barrel is negligible. (a) What is the speed of the spring gun after the ball stops in the barrel? (b) What fraction of the initial kinetic energy of the ball is stored in the spring?

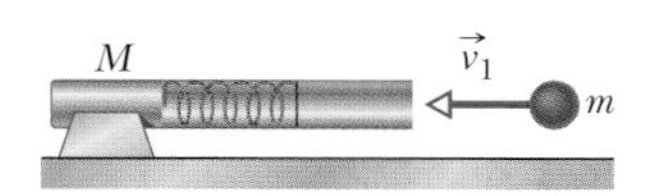

FIGURE 10-56 ■ Problem 73.

74. Ballistic Pendulum A bullet of mass 10 g strikes a ballistic pendulum of mass 2.0 kg. The center of mass of the pendulum rises a vertical distance of 12 cm. Assuming that the bullet remains embedded in the pendulum, calculate the bullet's initial speed.

75. Two Blocks and a Spring A block of mass $m_A = 2.0$ kg slides along a frictionless table with a speed of 10 m/s. Directly in front of it, and moving in the same direction, is a block of mass $m_B = 5.0$ kg moving at 3.0 m/s. A massless spring with spring constant $k = 1120$ N/m is attached to the near side of m_B, as shown in Fig. 10-57. When the blocks collide, what is the maximum compression of the spring? (*Hint*: At the moment of maximum compression of the spring, the two blocks move as one. Find the velocity by noting that the collision is completely inelastic at this point.)

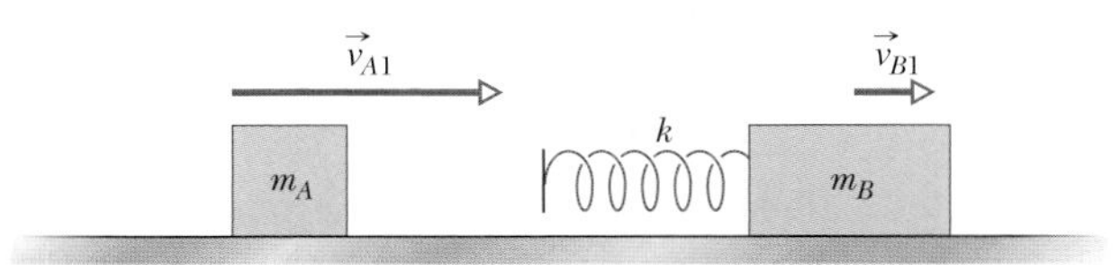

FIGURE 10-57 ■ Problem 75.

76. Physics Book A 4.0 kg physics book and a 6.0 kg calculus book, connected by a spring, are stationary on a horizontal frictionless surface. The spring constant is 8000 N/m. The books are pushed together, compressing the spring, and then they are released from rest. When the spring has returned to its unstretched length, the speed of the calculus book is 4.0 m/s. How much energy is stored in the spring at the instant the books are released?

77. Neutron Scattering Show that if a neutron is scattered through 90° in an elastic collision with an initially stationary deuteron, the neutron loses $\frac{2}{3}$ of its initial kinetic energy to the deuteron. (In atomic mass units, the mass of a neutron is 1.0 u and the mass of a deuteron is 2.0 u.)

78. Spring Attached to Wall A 1.0 kg block at rest on a horizontal frictionless surface is connected to an unstretched spring ($k = 200$ N/m) whose other end is fixed (Fig. 10-58). A 2.0 kg block moving at 4.0 m/s collides with the 1.0 kg block. If the two blocks stick together after the one-dimensional collision, what maximum compression of the spring occurs when the blocks momentarily stop?

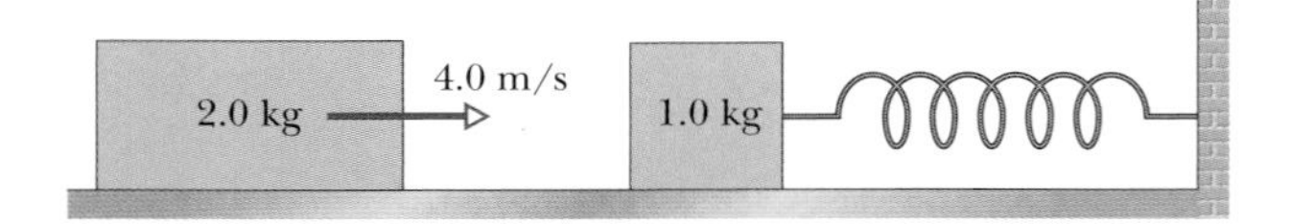

FIGURE 10-58 ■ Problem 78.

79. Game of Pool In a game of pool, the cue ball strikes another ball of the same mass and initially at rest. After the collision, the cue ball moves at 3.50 m/s along a line making an angle of 22.0° with its original direction of motion, and the second ball has a speed of 2.00 m/s. Find (a) the angle between the direction of motion of the second ball and the original direction of motion of the cue ball and (b) the original speed of the cue ball. (c) Is kinetic energy (of the centers of mass, don't consider the rotation) conserved?

80. Billiard Ball A billiard ball moving at a speed of 2.2 m/s strikes an identical stationary ball a glancing blow. After the collision, one ball is found to be moving at a speed of 1.1 m/s in a direction making a 60° angle with the original line of motion. (a) Find the velocity of the other ball. (b) Can the collision be inelastic, given these data?

81. Three Balls In Fig. 10-59, ball A with an initial speed of 10 m/s collides elastically with stationary balls B and C, whose centers are on a line perpendicular to the initial velocity of ball A and that are initially in contact with each other. The three balls are identical. Ball A is aimed directly at the contact point, and all motion is frictionless. After the collision, what are the velocities of (a) ball B, (b) ball C, and (c) ball A? (*Hint*: With friction absent, each impulse is directed along the line connecting the centers of the colliding balls, normal to the colliding surfaces.)

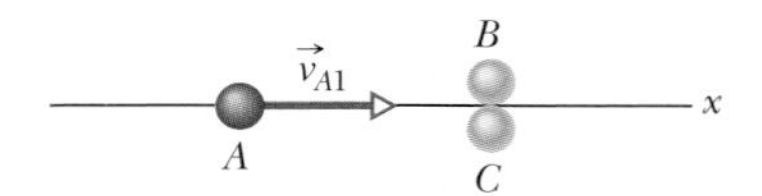

FIGURE 10-59 ■ Problem 81.

82. Two Bodies Collide Two 2.0 kg bodies, A and B, collide. The velocities before the collision are $\vec{v}_{A1} = (15\text{ m/s})\hat{\text{i}} + (30\text{ m/s})\hat{\text{j}}$ and $\vec{v}_{B1} = (-10\text{ m/s})\hat{\text{i}} + (5.0\text{ m/s})\hat{\text{j}}$. After the collision, $\vec{v}_{A2} = (-5.0\text{ m/s})\hat{\text{i}} + (20\text{ m/s})\hat{\text{j}}$. (a) What is the final velocity of B? (b) How much kinetic energy is gained or lost in the collision?

83. Elastic Collision of Cart A cart with mass 340 g moving on a frictionless linear air track at an initial speed of 1.2 m/s undergoes an elastic collision with an initially stationary cart of unknown mass. After the collision, the first cart continues in its original direction at 0.66 m/s. (a) What is the mass of the second cart? (b) What is its speed after impact? (c) What is the speed of the two-cart center of mass?

84. Electron Collision An electron undergoes a one-dimensional elastic collision with an initially stationary hydrogen atom. What percentage of the electron's initial kinetic energy is transferred to kinetic energy of the hydrogen atom? (The mass of the hydrogen atom is 1840 times the mass of the electron.)

85. Alpha Particle An alpha particle (mass 4 u) experiences an elastic head-on collision with a gold nucleus (mass 197 u) that is originally at rest. (The symbol u represents the atomic mass unit.) What percentage of its original kinetic energy does the alpha particle lose?

86. Voyager 2 Spacecraft *Voyager 2* (of mass m and speed v relative to the Sun) approaches the planet Jupiter (of mass M and speed V_J relative to the Sun) as shown in Fig. 10-60. The spacecraft rounds the planet and departs in the opposite direction. What is its speed, relative to the Sun, after this slingshot encounter, which can be analyzed as a collision? Assume $v = 12$ km/s and $V_J = 13$ km/s (the orbital speed of Jupiter). The mass of Jupiter is very much greater than the mass of the spacecraft ($M \gg m$).

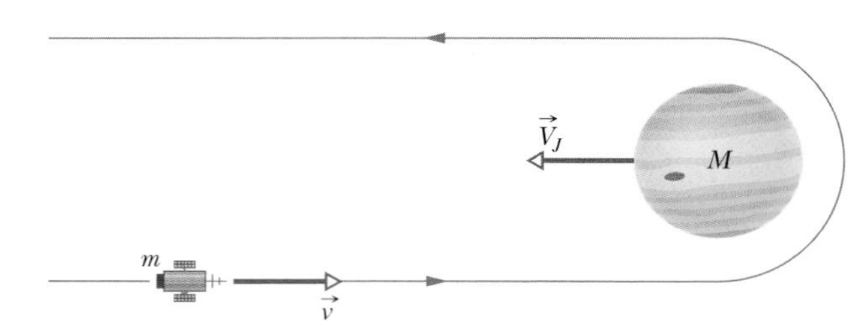

FIGURE 10-60 ■ Problem 86.

87. Elastic Collision A body of mass 2.0 kg makes an elastic collision with another body at rest and continues to move in the original direction but with one-fourth of its original speed. (a) What is the mass of the other body? (b) What is the speed of the two-body center of mass if the initial speed of the 2.0 kg body was 4.0 m/s?

88. Steel Ball and Block A steel ball of mass 0.500 kg is fastened to a cord that is 70.0 cm long and fixed at the far end. The ball is then released when the cord is horizontal (Fig. 10-61). At the bottom of its path, the ball strikes a 2.50 kg steel block initially at rest on a frictionless surface. The collision is elastic. Find (a) the speed of the ball and (b) the speed of the block, both just after the collision.

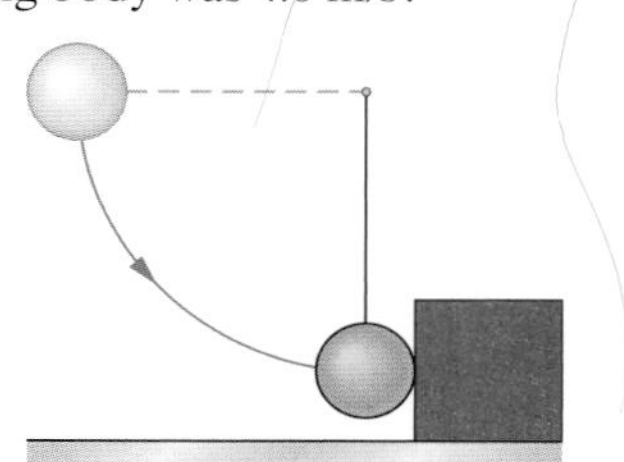

FIGURE 10-61 ■ Problem 88.

89. Two Titanium Spheres Two titanium spheres approach each other head-on with the same speed and collide elastically. After the collision, one of the spheres, whose mass is 300 g, remains at rest. (a) What is the mass of the other sphere? (b) What is the speed of the two-sphere center of mass if the initial speed of each sphere is 2.0 m/s?

90. Two-Sphere Arrangement In the two-sphere arrangement of Touchstone Example 10-8, assume that sphere A has a mass of 50 g and an initial height of 9.0 cm and that sphere B has a mass of 85 g. After the collision, what height is reached by (a) sphere A and (b) sphere B? After the next (elastic) collision, what height is reached by (c) sphere A and (d) sphere B? (*Hint*: Do not use rounded-off values.)

91. Blocks without Friction The blocks in Fig. 10-62 slide without friction. (a) What is the velocity $\vec{v}$ of the 1.6 kg block after the collision? (b) Is the collision elastic? (c) Suppose the initial velocity of the 2.4 kg block is the reverse of what is shown. Can the velocity $\vec{v}$ of the 1.6 kg block after the collision be in the direction shown?

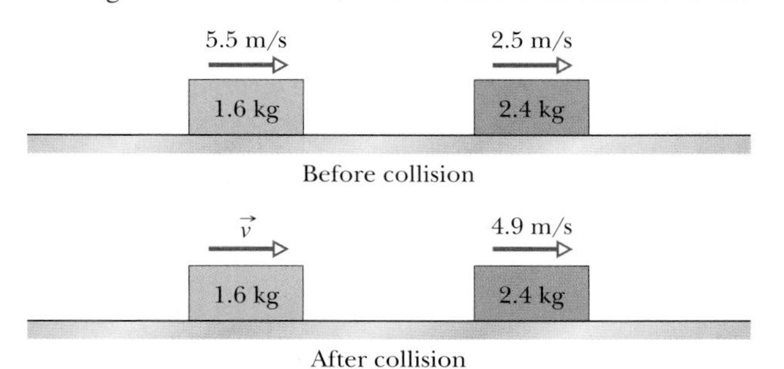

FIGURE 10-62 ■ Problem 91.

92. Two Blocks on Frictionless Table In Fig. 10-63, block A of mass m_A, is at rest on a long frictionless table that is up against a wall. Block B of mass m_B is placed between block A and the wall and sent sliding to the left, toward block A, with constant speed v_{B1}. Assuming that all collisions are elastic, find the value of m_B (in terms of m_A) for which both blocks move with the same velocity after block B has collided once with block A and once with the wall. Assume the wall to have infinite mass..

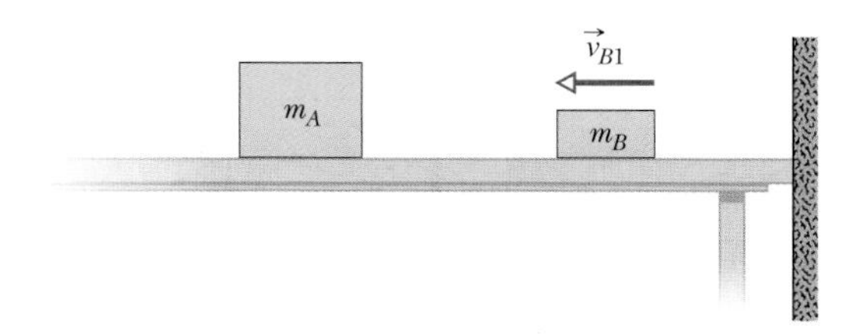

FIGURE 10-63 ■ Problem 92.

93. Small Ball Above Larger A small ball of mass m is aligned above a larger ball of mass M (with a slight separation, and the two are dropped simultaneously from h. (Assume the radius of each ball is negligible compared to h.) (a) If the larger ball rebounds elastically from the floor and then the small ball rebounds elastically from the larger ball, what ratio m/M results in the larger ball stopping upon its collision with the small ball? (The answer is approximately the mass ratio of a baseball to a basketball.) (b) What height does the small ball then reach?

Additional Problems

94. Frictionless Ramp In Fig. 10-64, block A of mass m_A slides from rest along a frictionless ramp from a height of 2.50 m and then collides with stationary block B, which has mass $m_B = 2.00m_A$. After the collision, block B slides into a region where the coefficient of kinetic friction is 0.500 and comes to a stop in distance d within that region. What is the value of distance d if the collision is (a) elastic and (b) completely inelastic?

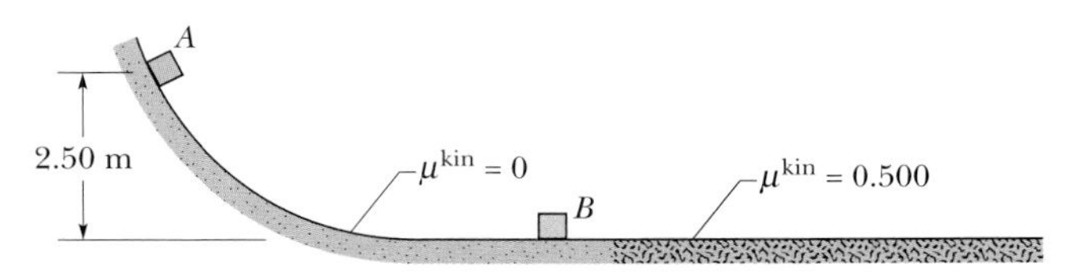

FIGURE 10-64 ■ Problem 94.

95. Pucks on Table In Fig. 10-65, puck A of mass $m_A = 0.20$ kg is sent sliding across a frictionless lab bench, to undergo a one-dimensional elastic collision with stationary puck B. Puck B then slides off the bench and lands a distance d from the base of the bench. Puck A rebounds from the collision and slides off the opposite edge of the bench, landing a distance $2d$ from the base of the bench. What is the mass of puck B? (*Hint*: Be careful with signs.)

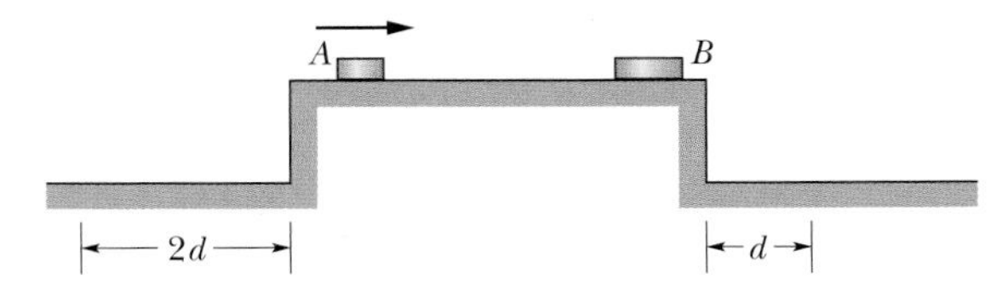

FIGURE 10-65 ■ Problem 95.

96. Speed Amplifier In Fig. 10-66, block A of mass m_A slides along an x axis on a frictionless floor with a speed of $v_{A1} = 1.00$ m/s. Then it undergoes a one-dimensional elastic collision with stationary block B of mass $m_B = 0.500m_A$. Next, block B undergoes a one-dimensional elastic collision with stationary block C of mass $m_C = 0.500m_B$. (a) What then is the speed of block C? Are (b) the speed, (c) the kinetic energy, and (d) the momentum of block C greater than, less than, or the same as the initial values for block A?

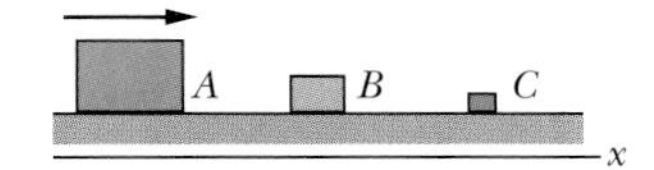

FIGURE 10-66 ■ Problem 96.

97. Speed Amplifier Graphs For the two-collision sequence of Problem 96, Figure 10-67a shows the speed V_A of block A plotted versus time t. The times for the first collision (t_1) and the second collision (t_2) are indicated. (a) On the same graph, plot the speeds of blocks B and C. Figure 10-67b shows a plot of the kinetic energy of block A versus time, where kinetic energy is given in terms of the initial kinetic energy $K_{A1} = 1.00$ J. (b) On the same graph, plot the kinetic energies of blocks B and C, all in terms of K_{A1}. After the second collision, what percentage of the total kinetic energy do (c) block A, (d) block B, and (e) block C have?

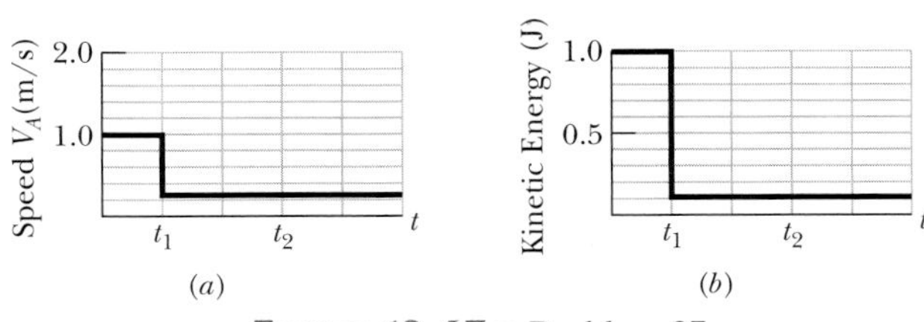

FIGURE 10-67 ■ Problem 97.

98. The Janitor Suppose a janitor wants to slide a trash barrel across the floor to a large trash bin. If the coefficient of kinetic friction is 0.123, determine the work done by a kinetic friction force on a 25 kg trash barrel that is pushed horizontally at a constant speed: (a) around a semicircle of diameter 2.3 m and (b) straight across the diameter.

99. Loading Dock Loading docks often have spring-loaded bumpers on them so that big trucks don't accidentally ruin the docks when backing up. See Fig. 10-68. Suppose a 6.45×10^3 kg truck backs into a spring-loaded dock at a speed of 2.51 m/s. If the truck compresses the dock bumper springs by 0.15 m when it slows down to zero speed, what is the effective spring constant of the bumper system? Use the correct number of significant figures.

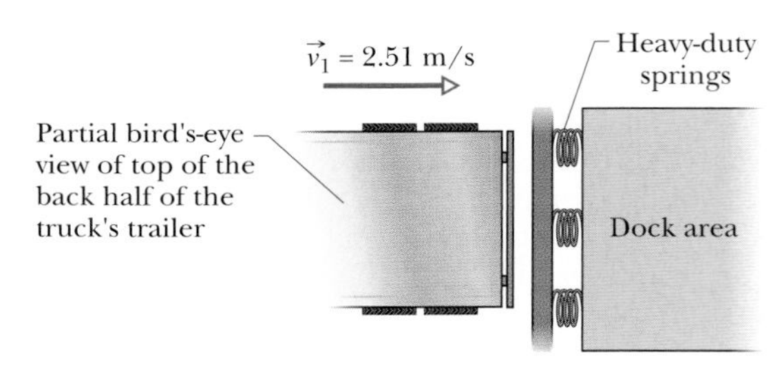

FIGURE 10-68 ■ Problem 99.

100. Jumping into a Haystack Tom Sawyer wanders out to the barn one fine summer's day. He notices that a haystack has recently been built just outside the barn. The barn has a second-story door into which the hay will be hauled into the barn by a crane. Tom decides it would be a neat idea to jump out of the second-story door onto the haystack. However, he knows from sad experience that if he jumps out of the second-story door onto the ground, that he is likely to break his leg. Knowing lots of physics, Tom decides to estimate whether the haystack will be able to break his fall.

He estimates the height of the haystack to be 3 meters. He presses down on top of the stack and discovers that to compress the stack by 25 cm, he has to exert a force of about 50 N. The barn door is 6 meters above the ground. Solve the problem by breaking it into pieces as follows:

1. Model the haystack by a spring. What is its spring constant?

2. Is the haystack tall enough to bring his speed to zero? (Estimate using conservation of energy.)

3. If he does come to a stop before he hits the ground, what will the average force exerted on him be?

101. Closing the Door A student is in her dorm room, sitting on her bed doing her physics homework. The door to her room is open. All of a sudden, she hears the voice of her ex-boyfriend talking to the girl in the room next door. She wants to shut the door quickly, so she throws a superball (which she keeps next to her bed for this purpose) against the door. The ball follows the path shown in Fig. 10-69. It hits the door squarely and bounces straight back.

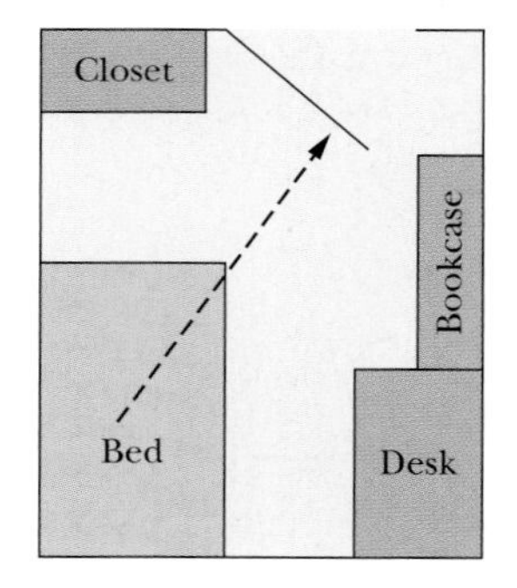

FIGURE 10-69 ■ Problem 101.

(a) If the ball has a mass m, hits the door with a speed v, and bounces back

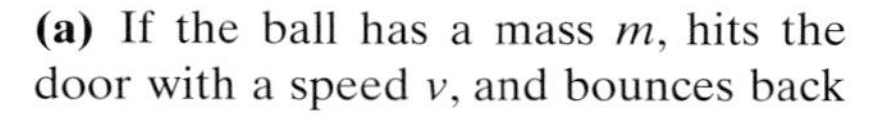

with a speed equal to v, what is the change in the ball's momentum? **(b)** If the ball was in contact with the door for a time Δt, what was the average force that the door exerted on the ball?
(c) Would she have been better off with a clay ball of the same mass that stuck to the door? Explain your reasoning.

102. The Astronaut and the Cream Pie* A 77 kg astronaut, freely floating at 6 m/s, is hit by a large 36 kg lemon cream pie moving oppositely at 9 m/s. See Fig. 10-70. How much thermal energy is generated by the collision?

FIGURE 10-70 ■ Problem 102.

103. Various Slopes A skier wants to try different slopes of the same overall vertical height, h, to see which one would give him the most speed when he reaches the end of the hill (points 1, 2, and 3 in Fig. 10-71). His options are shown in the figure.

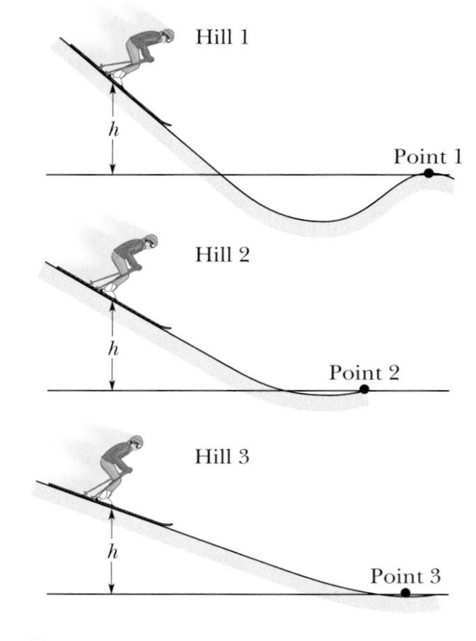

FIGURE 10-71 ■ Problem 103.

(a) Assuming there is no friction force between the skis and the snow, which hill would leave him with the most speed? Which would leave him with the least speed? Explain the basis for your answer.
(b) Assuming there is a noticeable friction force between the skis and the snow, which hill would leave him with the most speed? Which would leave him with the least speed? Explain the basis for your answer.

104. Rolling Carts Down Hill** Two carts A and B are identical in all respects. They roll down a hill and collide as shown in Fig. 10-72.

Figure 10-72a: (Case 1) cart A starts from rest on a hill at a height h above the ground. It rolls down and collides head-on with cart B, which is initially at rest on the ground. The two carts stick together.

Figure 10-72b: (Case 2) carts A and B are at rest on opposite hills at heights $h/2$ above the ground. They roll down, collide head-on with each other on the ground, and stick together.

* From Patrick H. Canan, *A Beginner's Guide to Classical Physics,* Corvallis, OR, School District (1982).
** Adapted from "Energy Concepts Survey" to be published in the *American Journal of Physics* by Chandralekha Singh.

Which of the following statements are true about the two-cart system *just before the carts collide* in the two cases? Give all the statements that are true. If none are true, write N.

(a) The kinetic energy of the system is zero in case 2.
(b) The kinetic energy of the system is greater in case 1 than in case 2.
(c) The kinetic energy of the system is the same in both cases.
(d) The total momentum of the system is greater in case 2 than in case 1.
(e) The total momentum of the system is the same in both cases.

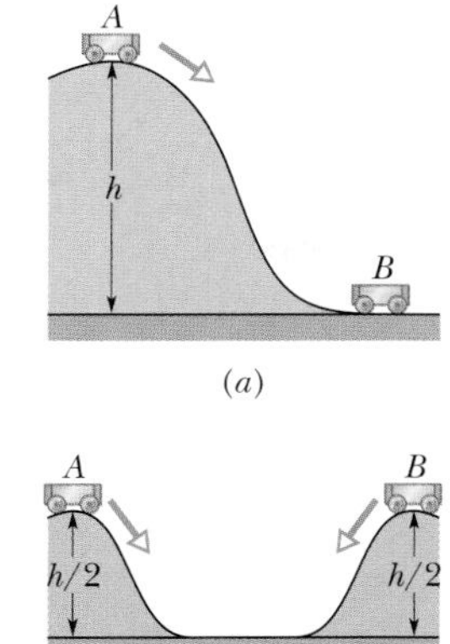

FIGURE 10-72 ■ Problem 104.

Which of the following statements are true about the two-cart system *just after the carts collide* in the two cases? Give all the statements that are true. If none are true, write N.

(f) The kinetic energy of the system is greater in case 2 than in case 1.
(g) The kinetic energy of the system is the same in both cases.
(h) The momentum of the system is greater in case 2 than in case 1.
(i) The total momentum of the system is nonzero in case 1 whereas it is zero in case 2.
(j) The total momentum of the system is the same in both cases.

105. Billiards Over the Edge Two identical billiard balls are labeled A and B as shown in Fig. 10-73*a*. Maryland Fats places ball A at the very edge of the table and ball B at the other side. He strikes ball B with his cue so that it flies across the table and off the edge. As it passes A, it just touches ball A lightly, knocking it off. The balls are shown just at the instant they have left the table. Ball B is moving with a speed v_{B1}, and ball A is essentially at rest.

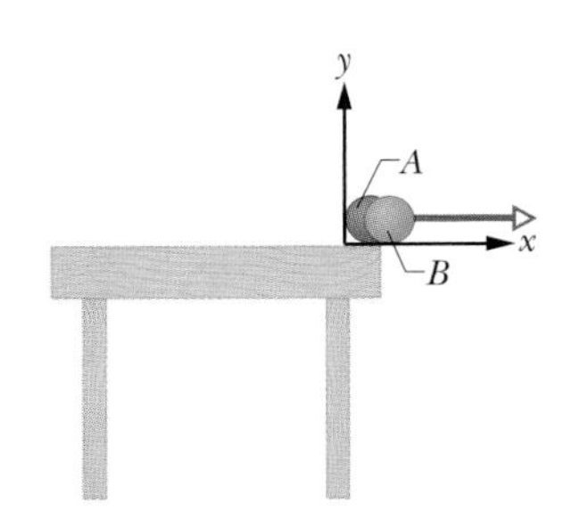

FIGURE 10-73*a* ■ Problem 105.

(a) Which ball do you think will hit the ground first? Explain your reasons for thinking so.

Fig. 10-73*b* shows a number of graphs of a quantity versus time. For each of the items below, select which graph could be a plot of that quantity vs. time. If none of the graphs are possible, write N. The time axes are taken to have $t = 0$ at the instant both balls leave the table. Use the x and y axes shown in Fig. 10-73*a*. For each of the following, which graph could represent:

(b) The x-component of the velocity of ball B?
(c) The y-component of the velocity of ball A?
(d) The y-component of the acceleration of ball A?
(e) The y-component of the force on ball B?
(f) The y-component of the force on ball A?
(g) The x-component of the velocity of ball A?
(h) The y-component of the acceleration of ball B?

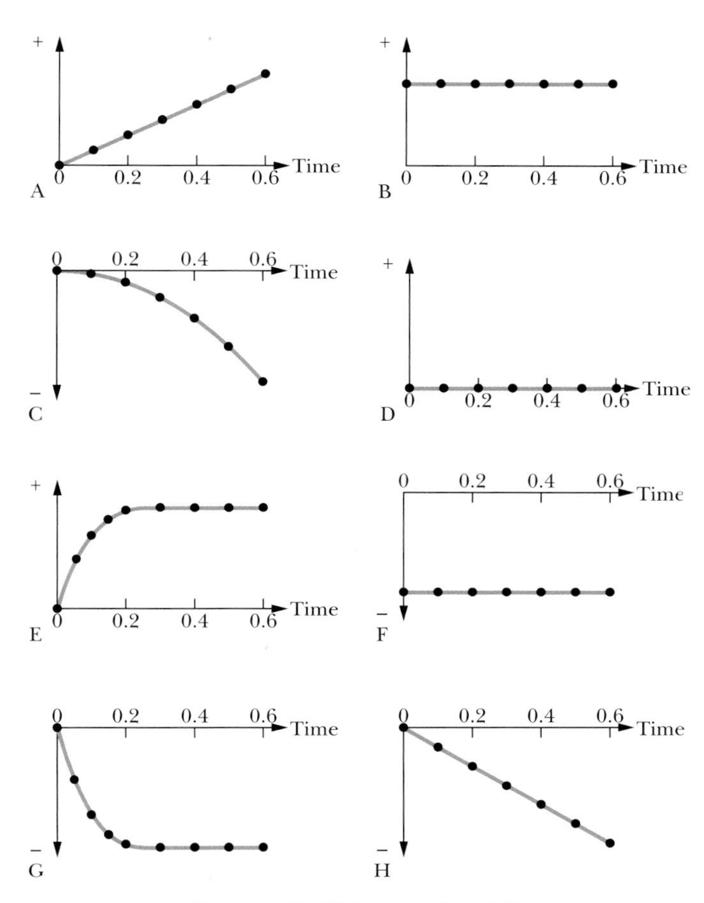

FIGURE 10-73*b* ■ Problem 105.

106. When Can You Conserve Energy? Mechanical energy conservation is sometimes a useful principle in helping us solve problems concerning the motion of objects. Suppose a single object is moving subject to a number of forces. Describe how you would know whether energy conservation would hold for the given example and in what kinds of problems you might find it appropriate to use this principle.

107. Conserving Momentum but Not Energy? Is it possible for a system of interacting objects to conserve momentum but not mechanical energy (kinetic plus potential)? Discuss and defend your answer, then given an example that illustrates the case you are trying to make.

108. Momentum *and* Energy? Is it possible for a system of interacting objects to conserve momentum and also mechanical energy (kinetic plus potential)? Discuss and defend your answer, then give an example that illustrates the case you are trying to make.

109. Frames of Reference Different observers can choose to use different coordinate systems. A frictionless roller coaster has been invented in which a single rider in a little cart can roll from the highest point to the lowest point, picking up kinetic energy as the cart goes downhill. The support struts for the roller coaster (shown as the grid in Fig. 10-74) are 4.00 meters apart, and the cart and rider have a combined mass of 195 kg. (a) What is the total mechanical energy of the cart-rider–Earth system according to Consuelo (an observer at the highest point on the track)? (b) What is the total mechanical energy of the cart-rider–Earth system according to

Mike (an observer at the ground level)? (c) Do Consuelo and Mike agree on the *value* of the total mechanical energy? Why or why not? (d) Do Consuelo and Mike agree that mechanical energy is conserved? Explain. (e) Assuming that mechanical energy is conserved, what is the kinetic energy of the cart and rider when it rolls over the top of the second smaller hill?

Hint: If you have access to the VideoPoint movie collection you may want to look at some of the roller coaster movies in the Hershey Park collection. For example, HRSY018 and HRSY019 provide similar scenarios. Although real roller coasters are not frictionless, using the VideoPoint software to find the location of a car at the top of a hill from the perspectives of two coordinate systems might be helpful.

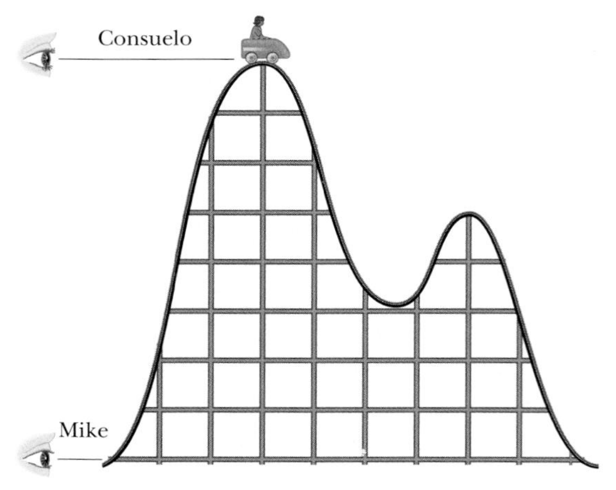

FIGURE 10-74 ■ Problem 109.

110. Largest and Smallest A ball is thrown from ground level with initial horizontal velocity component $v_{x\,1}$ and the initial vertical velocity component $v_{y\,1}$ and returns to ground level. Neglecting air resistance and explaining your reasoning in each instance, write expressions in terms of these two velocities for:

(a) The largest kinetic energy of the ball during its flight
(b) The smallest kinetic energy of the ball during its flight
(c) The maximum potential energy of the ball–earth system during the flight.
Hint: If you have access to the VideoPoint movie collection you may want to view the movies entitled PASCO104 and PASCO106 to remind you of the nature of a ball's path.

111. Coffee Filter Drop If a flat-bottomed coffee filter is dropped from rest near the surface of the Earth, it will fall more slowly than a small dense object of the same mass. You are to investigate whether or not mechanical energy is conserved during the fall of a small coffee filter using video analysis software. If you have access to the VideoPoint movie collection, you can use the movie entitled PASCO121 for this analysis. Your instructor may provide access to the movie some other way.

(a) If the coffee filter is dropped from rest, what is its initial velocity and kinetic energy?
(b) What are the final velocity and kinetic energy of the coffee filter (at the time of the last frame)? Explain how you arrived at the final velocity.
(c) What are the initial and final potential energies of the coffee filter?
(d) Is mechanical energy conserved as the coffee filter falls? Cite the evidence based on your measurements and calculations.
(e) How much mechanical energy, if any, is lost?
(f) What is the most likely source of a nonconservative force on the coffee filter? Where would missing mechanical energy probably go? Use conservation of energy concepts to explain why a paper coffee filter fall more slowly than a small dense object?

11 Rotation

These volleyball players leap high to block a spike. The height of their jumps would be far less if kneecaps were not a part of the human leg structure.

How do kneecaps help people jump more effectively?

The answer is in this chapter.

(a)

(b)

FIGURE 11-1 ■ Figure skater Sarah Hughes in motion of (a) pure translation in a fixed direction and (b) pure rotation about a vertical axis.

11-1 Translation and Rotation

The graceful movement of figure skaters can be used to illustrate two kinds of pure motion. Figure 11-1*a* shows a skater gliding across the ice in a straight line with constant speed. Her motion is one of pure **translation.** Figure 11-1*b* shows her spinning at a constant rate about a vertical axis in a motion of pure **rotation.**

Translation is motion along a line, which we have considered in previous chapters. Rotation is turning motions, like those of wheels, gears, motors, planets, clock hands, jet engine rotors, and helicopter blades. It is our focus in this chapter. Rotational motion is everywhere around us, because most everyday objects are extended (rather than point masses) and can rotate about their centers of mass when moving freely. The characteristics of rotational motion are quite analogous to those of translational motion, and so the study of rotations will help you obtain a deeper understanding of both kinematics and the laws of translational motion. Examples of translational and rotational motion are shown in Fig. 11-2.

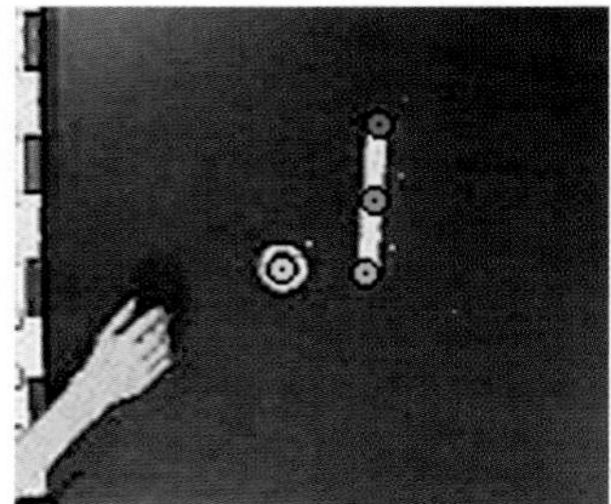

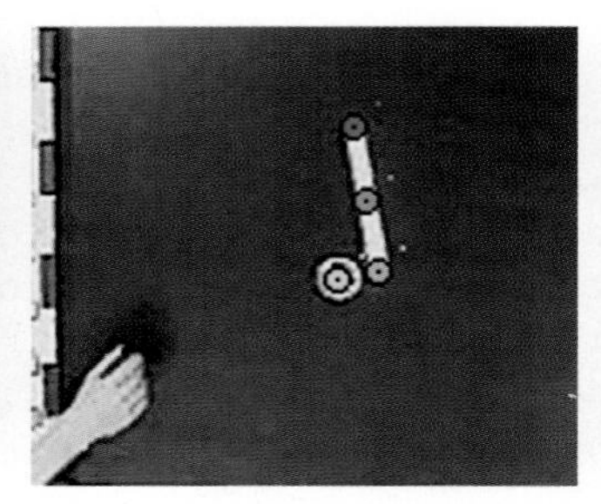

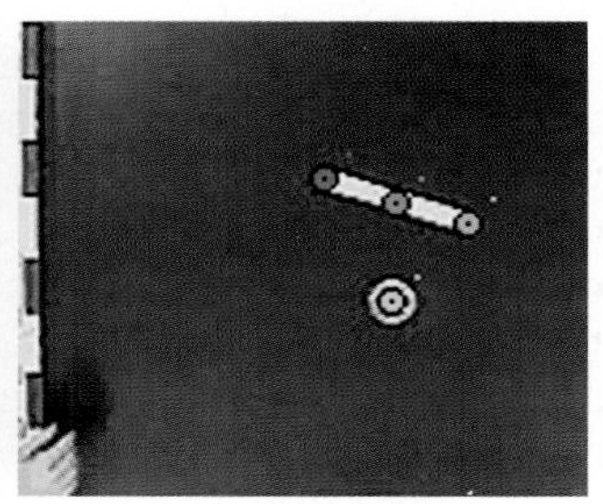

FIGURE 11-2 ■ These video clips show a small puck colliding with a stationary rod on an air table. The puck's motion before and after the collision is purely translational. After the objects collide, the rod has a combination of rotational motion about its center of mass and translational motion of its center of mass. In Chapters 11 and 12, you will learn to use the laws of translational and rotational motion to predict the detailed outcome of collisions like this.

In this chapter we consider simple examples of rotational motion. It is the rotational analogy of motion along a line, so that we will not have to deal with rotational variables as two- or three-dimensional vectors. For example, we will limit our considerations to motions for which the axis of rotation is fixed, or at least does not accelerate, so we can always pick a frame of reference in which it doesn't move.

In Chapter 12, we will consider more complex motions involving axes of rotation that are not fixed, such as yo-yo motion. There we will also learn more about the advantages of treating rotational variables as vectors, even when the rotations are about a fixed axis. This will allow us to extend our understanding of the types of forces that can cause rotational accelerations.

11-2 The Rotational Variables

As is usual in physics, we like to start with a simple case so that we can make sense of the basic ideas. The big difference between what we've done before and what we are going to do now is that now we are going to consider the rotation of extended objects. This can get quite complicated if we allow the object to deform or to twist in an arbitrary way. Let's simplify by considering an object that is solid enough that we can treat it as if it has a fixed shape throughout its motion—that is, it is **perfectly rigid.** Many of these objects will have an **axis of symmetry,** a line about which the object may be turned and still look the same, like the line through the center of a cylinder or a ball. A rigid object and a nonrigid object are shown in Fig. 11-3.

FIGURE 11-3 ■ A coffee cup serves as an example of a rigid object (upper), whereas a cloud is an example of a nonrigid object (lower).

In summary, in this chapter we wish to examine the rotations of rigid bodies about fixed axes. Figure 11-4 shows a rigid body in rotation about a fixed axis. The axis is called the **axis of rotation** or the **rotation axis.** This is defined as a **rigid body** because it can rotate with all its parts locked together, without any change in its shape. For example, in Fig. 11-1*b*, if the skater holds her shape while spinning, she is temporarily acting as a rigid body. But when she is moving her arms and legs relative to her body to change from one pose to another, she is not rigid. Therefore, we will not analyze the rotations of dancers and athletes except during those parts of their motions that are approximately rigid. Similarly, we will not examine the rotational motion of the Sun, because it is a ball of gas whose parts are not locked together.

A **fixed axis** means the rotation occurs about an axis that does not move. We also will not yet examine an object like a bowling ball rolling along a bowling alley, because the ball rotates about an axis that moves (the ball's motion is a mixture of rotation and translation).

As we know from our previous study of linear motions, in pure translational motion, every point on a body moves in a straight line. In other words, every point moves through the same *linear distance* during a particular time interval. In pure rotational motion, every point on the body moves in a circle whose center lies on the body's axis of rotation. Since the parts of a rigid body are locked together, every point moves through the same angle during a particular time interval. Hence, there are similarities and differences between translational and rotational motions. Comparisons between rotational and translational motion will appear throughout this chapter.

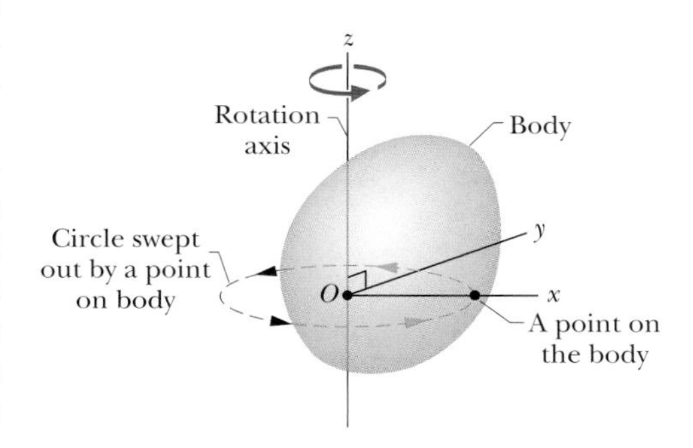

FIGURE 11-4 ■ A rigid body of arbitrary shape in pure rotation about an axis. It is customary to choose a coordinate system in which the z axis is aligned with the axis of rotation. The position of the *reference line* with respect to the rigid body is arbitrary, but it is perpendicular to the rotation axis. It is fixed in the body and rotates with the body. In this case, it must lie in the x-y plane.

We deal now—one at a time—with specifying how an object is placed and moves rotationally. We will point out the rotational (or angular) equivalents of the translational (or linear) quantities position, displacement, velocity, and acceleration. The first step in introducing rotational quantities is to specify a coordinate system and reference line to aid in the description of motion.

Although there are many ways to specify a system for the analysis of rotational motion, the one shown in Fig. 11-4 is the most conventional. We start by choosing a rectangular coordinate system that is fixed in space. It is customary to orient the z axis along the rotation axis. Next we choose a **reference line** that is perpendicular to the axis of rotation so it lies in the x-y plane. The reference line is fixed with respect to the rotating body so that it rotates around the z axis as the body rotates.

Rotational Position

We define the rotational position θ of the body as the angle between the reference line at a given moment and the positive x axis, as shown in Fig. 11-5.

For a rigid object rotating around a fixed axis, each point within the object moves in a circle around the axis of rotation. Consider a point along the reference line that is a distance r from the axis. From geometry, we know that the magnitude of θ is given by

$$|\theta| = \frac{s}{r} \quad \text{(radian measure).} \tag{11-1}$$

Here s is a scalar quantity that represents the length of arc (or the arc distance) between the x axis (the zero rotational position) and the reference line; r is the radius of that circle.

For the equation

$$|\theta| = \frac{s}{r}$$

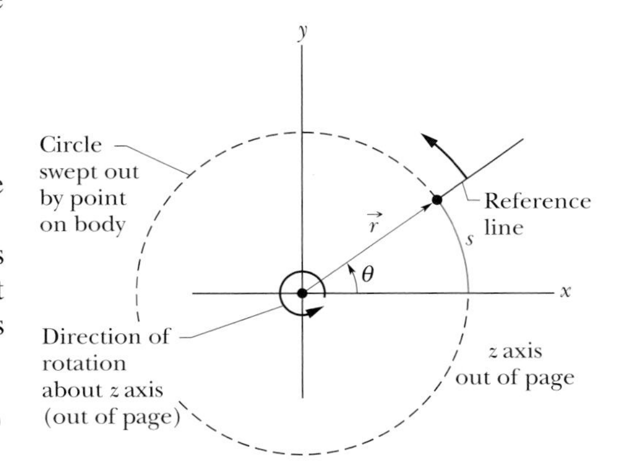

FIGURE 11-5 ■ The rotating rigid body of Fig. 11-4 in cross section, viewed from above. The plane of the cross section is perpendicular to the rotation axis, which now extends out of the page, toward you. In this position of the body, the reference line makes an angle θ with the x axis.

to be valid, the angle must be measured in radians (rad) rather than in revolutions (rev) or degrees. An angle of one radian is defined as the angle for which the length of the arc is equal to the radius of the circle. The radian, being the ratio of two lengths, is a pure number and thus has no dimensions. Although angles have no dimensions, they do have units, and it is vital to keep track of them. Because the circumference of a circle of radius r is $2\pi r$, there are 2π radians in a complete circle. There are three common units used to measure angles. They are related by the equation

$$1 \text{ revolution} = 360° = \frac{2\pi r}{r} \text{ radians} = 2\pi \text{ radians}. \tag{11-2}$$

By rearranging terms algebraically, we find that

$$1 \text{ rad} = 57.3° = 0.159 \text{ rev}. \tag{11-3}$$

We do *not* reset θ to zero with each complete rotation of the reference line about the rotation axis. If we did, a smoothly rotating object would be described by a variable that jumps discontinuously. We have to keep in mind the physical, as well as the mathematical, meaning of the rotational variable. Although θ, $\theta + 2\pi$, $\theta + 4\pi$, and so on, all represent the same physical position, they represent different total displacements. For example, if the reference line completes two revolutions from the zero rotational position, it is back at its starting point, but it has traveled through an angle of $\theta = 4\pi$ rad.

For pure translational motion along the x direction, we can know all there is to know about a moving body if we are given $x(t)$, which is its position as a function of time. Similarly, for pure rotation, we can know all there is to know about the motion of a rigid rotating body about a fixed axis of rotation if we are given $\theta(t)$, the rotational position of the body's reference line as a function of time.

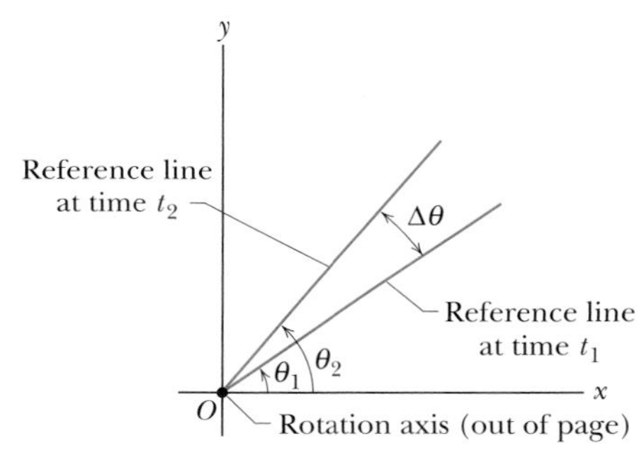

FIGURE 11-6 ■ Bird's eye view of Fig. 11-4. The point on the reference line is at rotational position θ_1 at time t_1, and at rotational position θ_2 at a later time t_2. The quantity $\Delta\theta (= \theta_2 - \theta_1)$ is the rotational displacement that occurs during the interval $\Delta t\ (= t_2 - t_1)$. The body itself is not shown.

Rotational Displacement

If the body of Fig. 11-5 rotates about the rotation axis as in Fig. 11-6, changing the rotational position of the reference line from θ_1 to θ_2, the body undergoes a rotational (or angular) displacement $\Delta\theta$ given by

$$\Delta\theta = \theta_2 - \theta_1. \tag{11-4}$$

This definition of rotational displacement holds not only for the rigid body as a whole, but also for *every particle within that body*, because the particles are all locked together.

If a body is in translational motion along an x axis, its displacement Δx is either positive or negative, depending on whether the body is moving in the positive or negative direction (as we have assigned them). Similarly, the rotational displacement $\Delta\theta$ of a rotating body can be either positive or negative.

Just as was the case for translational motion, the terms "positive" and "negative" are only meaningful once we have defined a coordinate system. For any situation that involves rotation about a fixed axis—for example, the rotation of the record shown in Fig. 11-7—the rotational displacement θ has a direction that is tied to the axis of rotation. Consequently, it makes sense to define a coordinate system that has one of its axes along the axis of rotation. It is standard practice to align the axis of rotation of a body along the z axis of the rectangular right-handed coordinate system introduced in Section 4-5. Thus, by convention, if our right-handed rectangular coordinate system happens to be drawn so that the positive z axis is out of the page along the axis of rotation of the body we are describing, then we define upward along the "vertical axis" as the positive y direction and rightward along the "horizontal axis" as the positive x direction.

Once we have established a coordinate system to describe a rotational motion, we can establish whether the rotational quantities of position, velocity, and acceleration are positive or negative by using a **right-hand rule,** as shown in Fig. 11-7*c*. Curl the fingers of your right hand in the direction of the rotation. If your extended thumb then points in the negative direction along the chosen axis of rotation (as is the case for the record in Fig. 11-7), we call the rotational displacement negative. If the record were to rotate in the opposite sense, the right-hand rule would tell you that the rotational displacement was positive, because your thumb would point in the opposite (positive) direction along the axis of rotation.

By using the right-hand rule, we can consider a rotational displacement to be a one-dimensional vector, where $\Delta\theta$ is its component along the axis of rotation. This assignment makes sense, since rotational displacements are meaningless unless we know what axis to relate them to. When the rotational displacement $\Delta\theta$ is positive, the object is rotating one way and when it is negative, the object is rotating the opposite way.

For the basic types of motion that we will treat in this book, the axis of rotation will not change orientation over time. In such cases, rotational displacements are said to commute. That is, the order in which you make the rotations doesn't matter. However, in more complex motions where the orientation of the axis of rotation changes direction over time, rotational displacements *do not* commute. In those cases, rotational displacements do not behave as vectors.

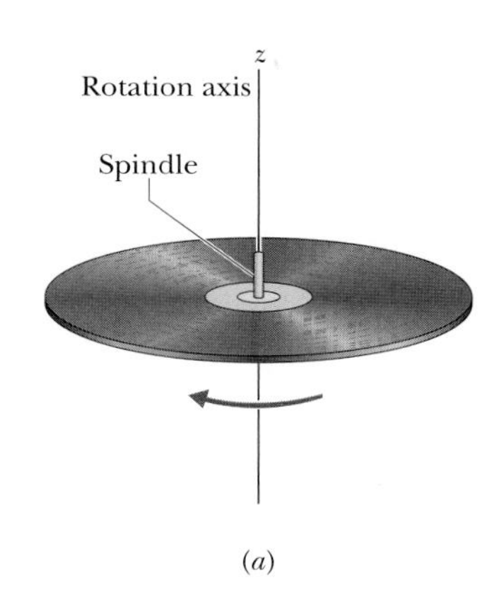

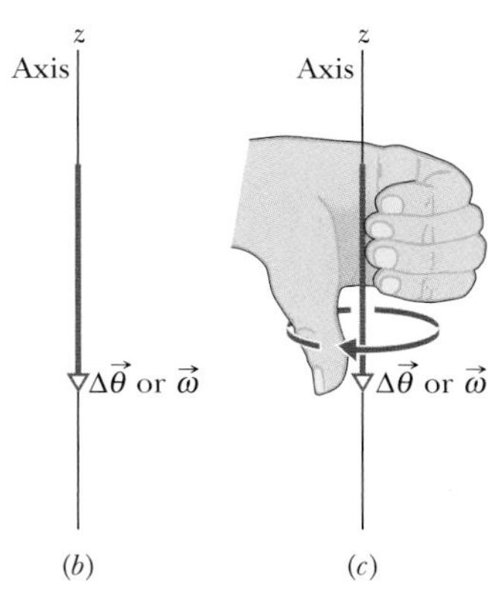

FIGURE 11-7 ■ (*a*) A record rotating about a vertical axis that coincides with the axis of the spindle. (*b*) We establish that the rotational displacement component and the rotational velocity component are negative because our thumb points downward when using the right-hand rule. (*c*) We establish the direction of the rotational velocity vector as downward by using the right-hand rule. When the fingers of the right hand curl around the record and point the way it is moving, the extended thumb points in the direction of $\vec{\omega}$.

The Rotational Velocity Component

Suppose (see Fig. 11-6) that our rotating body is at rotational position θ_1 at time t_1 and at rotational position θ_2 at time t_2. A body's **average rotational velocity** component along its axis of rotation is defined as

$$\frac{\theta_2 - \theta_1}{t_2 - t_1} = \frac{\Delta\theta}{\Delta t}, \quad \text{(average rotational velocity component),} \tag{11-5}$$

where $\Delta\theta$ is the rotational displacement that occurs during the time interval $\Delta t = t_2 - t_1$ and ω is the lowercase Greek letter omega. Rotational velocity is often referred to as angular velocity. Note that when z is the axis of rotation $\langle\omega\rangle = \langle\omega_z\rangle$.

The component ω_z of the **(instantaneous) rotational velocity** with which we shall be most concerned, is the limit of the ratio in the equation above as Δt approaches zero. Thus,

$$\omega \equiv \lim_{\Delta t \to 0} \frac{\Delta\theta}{\Delta t} = \frac{d\theta}{dt} \quad \text{(instantaneous rotational velocity component).} \tag{11-6}$$

Once again when z is the rotation axis $\omega_z = \omega$. If we know $\theta(t)$ and it is continuous, we can find the rotational velocity component ω_z by differentiation. As is the case for the rotational displacement, the rotational velocity ω_z in this context represents the component of a one-dimensional vector along the axis of rotation relative to the coordinate system chosen to describe the motion. As a component, ω_z can be positive or negative and we do not use a vector arrow over it. Whenever the rotational position θ is becoming more positive ω_z is positive and, conversely, whenever the rotational position θ is becoming more negative ω_z is negative. Happily we will get the same result using the right-hand rule to determine whether ω_z is positive or negative.

Strictly speaking, we should always define an axis of rotation as z and call ω_z the rotational velocity component. But, for the simple rotations considered in this chapter, we often refer to it casually as the rotational velocity ω. Also, because the particles in a rigid body are all linked to each other, the rotational velocity is the same for *every particle in a rotating rigid body.*

Rotational Speed The magnitude (or absolute value) of rotational velocity is called the rotational speed. Since we have designated ω (or more correctly ω_z) to be a component along the axis of rotation that can be either positive or negative, the rotational speed must be represented using an absolute value sign. Thus, to avoid confusion we always denote rotational speed as $|\omega|$ or $|\vec{\omega}|$.

Units for Rotational Velocity The preferred scientific unit of rotational velocity is the radian per second (rad/s). In some cases the unit revolution per second (rev/s) is used instead. Another popular measure of rotational velocity is rpm or revolutions per minute, used in automobile tachometers that measure the turning rate of engine crankshafts. The rpm is also used in conjunction with turntables used to play vinyl phonograph records.

The Rotational Acceleration Component

If the rotational velocity of a rotating body is not constant, then the body has rotational acceleration. Let ω_2 and ω_1 be its rotational velocity components at times t_2 and t_1, respectively. The component of the **average rotational acceleration** along the axis of rotation of the body in the interval from t_1 to t_2 is defined as

$$\langle\alpha\rangle = \langle\alpha_z\rangle \equiv \frac{\omega_2 - \omega_1}{t_2 - t_1} = \frac{\Delta\omega}{\Delta t} \quad \text{(average rotational acceleration component)}, \qquad (11\text{-}7)$$

in which $\Delta\omega$ is the component of the change in rotational velocity that occurs during the time interval Δt. The **(instantaneous) rotational acceleration** component α, with which we shall be most concerned, is the limit of this quantity as Δt approaches zero. Thus,

$$\alpha = \alpha_z \equiv \lim_{\Delta t \to 0} \frac{\Delta\omega}{\Delta t} = \frac{d\omega}{dt} \quad \text{(instantaneous rotational acceleration component)}. \qquad (11\text{-}8)$$

Just as was the case for the rotational velocity ω, these expressions for the rotational acceleration hold not only for the rotating rigid body as a whole, but also for *every particle of that body.*

Whenever the rotational velocity component ω is becoming more positive α is positive and, conversely, whenever the rotational velocity component ω is becoming more negative α is negative. Thus, the relationship between the directions of velocity and acceleration that we hold to be true for translational motion are analogous to those that apply to rotational motion.

Note that rotational acceleration, as introduced in this simple context, like rotational displacement and rotational velocity, is a component of a one-dimensional vector relative to the chosen coordinate axis aligned with the axis of rotation of the body that is rotating. Rotational acceleration is often referred to as angular acceleration.

Once again we have followed the convention of not designating what axis of rotation the acceleration component refers to. For this reason the rotational acceleration component is casually called the rotational acceleration. So for its magnitude we always denote the magnitude of rotation acceleration as $|\vec{\alpha}|$ or $|\alpha|$.

Units for Rotational Acceleration The preferred scientific unit of rotational acceleration is commonly the radian per second-squared (rad/s^2). Another common unit is the revolution per second-squared (rev/s^2).

READING EXERCISE 11-1: The Sign of Rotational Velocity: An off-center egg like the one shown in Fig. 11-4 is rotating about a z axis. What is the sign of its rotational velocity component if it is rotating so (a) the angle between its reference line and the x axis is increasing; (b) the angle between its reference line and the x axis is decreasing? ■

READING EXERCISE 11-2: The Sign of Rotational Acceleration—An off-center egg like the one shown in Fig. 11-4 is rotating about a z axis. What is the sign of its rotational acceleration if it is rotating so (a) the angle between its reference line and the x axis is increasing and so is its speed; (b) the angle between its reference line and the x axis is increasing and its speed is decreasing; (c) the angle between its reference line and the x axis is decreasing but its speed is increasing; and (d) the angle between its reference line and the x axis is decreasing and its speed is decreasing? ■

TOUCHSTONE EXAMPLE 11-1: Rotating Disk

The disk in Fig. 11-8a is rotating about its central axis like a merry-go-round. The rotational position $\theta(t)$ of a reference line on the disk is given by

$$\theta(t) = -(1.00 \text{ rad}) - (0.600 \text{ rad/s})t + (0.250 \text{ rad/s}^2)t^2, \quad (11\text{-}9)$$

with the zero rotational position as indicated in the figure.

(a) Graph the rotational position of the disk versus time from $t = -3.0$ s to $t = 6.0$ s. Sketch the disk and its rotational position reference line at $t = -2.0$ s, 0 s, and 4.0 s, and when the curve crosses the t axis.

SOLUTION ■ The **Key Idea** here is that the rotational position of the disk is the rotational position $\theta(t)$ of its reference line, which is given by Eq. 11-9 as a function of time. So we graph Eq. 11-9; the result is shown in Fig. 11-8b.

To sketch the disk and its reference line at a particular time, we need to determine θ for that time. To do so, we substitute the time into Eq. 11-9. For $t = -2.0$ s, we get

$$\theta = -(1.00 \text{ rad}) - (0.600 \text{ rad/ s})(-2.0 \text{ s}) + (0.250 \text{ rad/s}^2)(-2.0 \text{ s})^2$$

$$= 1.2 \text{ rad} = 1.2 \text{ rad}\,\frac{360°}{2\pi \text{ rad}} = 69°.$$

This means that at $t = -2.0$ s the reference line on the disk is rotated counterclockwise from the zero rotational position by 1.2 rad or 69° (counterclockwise because θ is positive). Sketch A in Fig. 11-8b shows this rotational position of the reference line. Similarly, for $t = 0$, we find $\theta = -1.00$ rad $= -57°$, which means that the reference line is rotated clockwise from the zero rotational position by 1.0 rad or 57°, as shown in sketch C. For $t = 4.0$ s, we find $\theta = 0.60$ rad $= 34°$ (sketch E). Drawing sketches for when the curve crosses the t axis is easy, because then $\theta = 0$ and the reference line is momentarily aligned with the zero rotational position (sketches B and D).

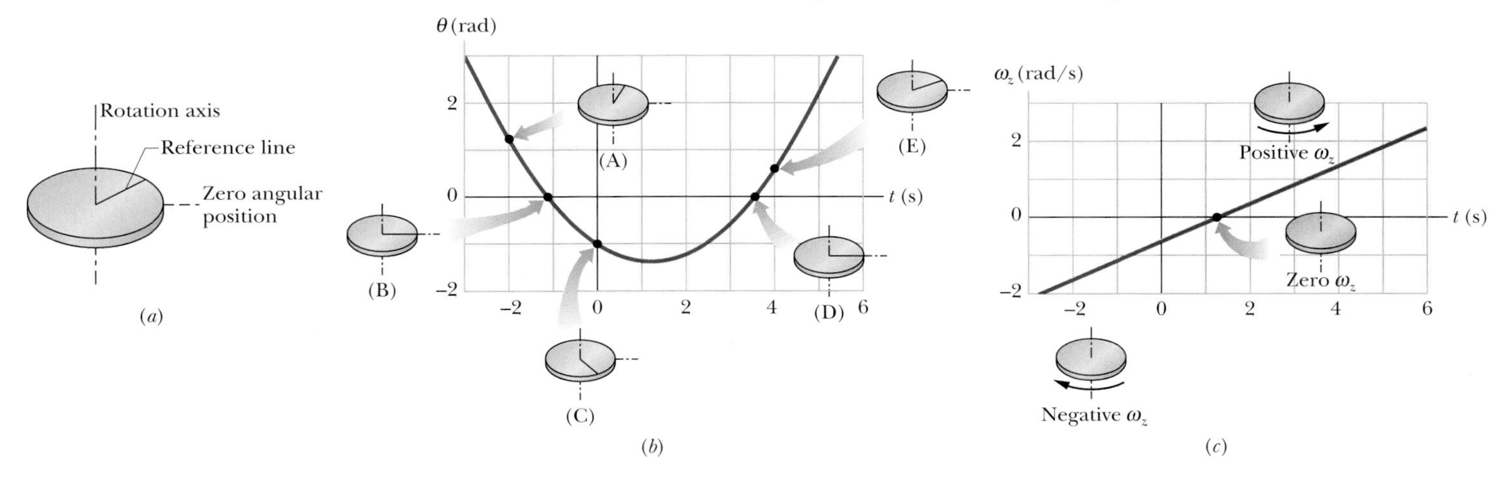

FIGURE 11-8 ■ (a) A rotating disk. (b) A plot of the disk's rotational position $\theta(t)$. Five sketches indicate the rotational position of the reference line on the disk for five points on the curve A-E. (c) A plot of the z-component of the disk's rotational velocity $\omega_z(t)$. Positive values of ω_z correspond to counterclockwise rotation, and negative values to clockwise rotation.

(b) At what time $t^{\min}$ does $\theta(t)$ reach the minimum value shown in Fig. 11-8*b*? What is that minimum value?

SOLUTION ■ The **Key Idea** here is that to find the extreme value (here the minimum) of a function, we take the first derivative of the function and set the result to zero. The first derivative of $\theta(t)$ is

$$\frac{d\theta}{dt} = -(.600 \text{ rad/s}) + (0.500 \text{ rad/s}^2)t. \tag{11-10}$$

Setting this to zero and solving for t give us the time at which $\theta(t)$ is minimum:

$$t^{\min} = 1.20 \text{ s.} \qquad \text{(Answer)}$$

To get the minimum value of θ, we next substitute $t^{\min}$ into Eq. 11-9, finding

$$\theta^{\min} = -1.36 \text{ rad} = -77.9°. \qquad \text{(Answer)}$$

This *minimum* of $\theta(t)$ (the bottom of the curve in Fig. 11-8*b*) corresponds to the *extreme clockwise* rotation of the disk from the zero rotational position, somewhat more than is shown in sketch C.

(c) Graph the rotational velocity ω of the disk versus time from $t = -3.0$ s to $t = 6.0$ s. Sketch the disk and indicate the direction of turning and the sign of ω at $t = -2.0$ s and 4.0 s, and also at $t^{\min}$.

SOLUTION ■ The **Key Idea** here is that, from Eq. 11-6, the rotational velocity ω is equal to $d\theta/dt$ as given in Eq. 11-10. So, we have

$$\omega = -(.600 \text{ rad/s}) + (0.500 \text{ rad/s}^2)t. \tag{11-11}$$

The graph of this function $\omega(t)$ is shown in Fig. 11-8*c*.

To sketch the disk at $t = -2.0$ s, we substitute that value into Eq. 11-11, obtaining

$$\omega = -1.6 \text{ rad/s} \qquad \text{(Answer)}$$

The minus sign tells us that at $t = -2.0$ s, the disk is turning clockwise as suggested by the lowest sketch in Fig. 11-8*c*.

Substituting $t = 4.0$ s into Eq. 11-11 gives us

$$\omega = 1.4 \text{ rad/s.} \qquad \text{(Answer)}$$

The implied plus sign tells us that at $t = 4.0$ s, the disk is turning counterclockwise (the highest sketch in Fig. 11-8*c*).

For $t^{\min}$, we already know that $d\theta/dt = 0$. So, we must also have $\omega = 0$. That is, the disk is changing its direction of rotation when the reference line reaches the minimum value of θ in Fig. 11-8*b* as suggested by the center sketch in Fig. 11-8*c*.

(d) Use the results in parts (a) through (c) to describe the motion of the disk from $t = -3.0$ s to $t = 6.0$ s.

SOLUTION ■ When we first observe the disk at $t = -3.0$ s, it has a positive rotational position and is turning clockwise but slowing. It reverses its direction of rotation at rotational position $\theta = -1.36$ rad and then begins to turn counterclockwise, with its rotational position eventually becoming positive again.

11-3 Rotation with Constant Rotational Acceleration

In pure translation, motion with a *constant translational acceleration* (for example, that of a falling body) is an important special case. In Table 2-1, we displayed a series of equations that hold for such motion.

Recall that in Chapter 2 we derived two primary equations $v_{2x} = v_{1x} + a_x(t_2 - t_1)$ (Eq. 2-13) and $x_2 - x_1 = v_{1x}(t_2 - t_1) + \frac{1}{2}a_x(t_2 - t_1)^2$ (Eq. 2-17) that describe velocity and position changes of an object that undergoes a constant translational acceleration. In pure rotation, the case of *constant rotational acceleration* is also important, and a parallel set of equations holds for this case also. Since the logic used to derive the analogous rotational equations is identical, we shall not derive them here. We can simply write them from the corresponding translational equations, substituting equivalent rotational quantities for the translational ones. This is done in Table 11-1. Figure 11-9 shows a situation that you can analyze using these equations. The equations for constant rotational acceleration are

$$\omega_2 = \omega_1 + \alpha(t_2 - t_1), \tag{11-12}$$

and

$$\theta_2 - \theta_1 = \omega_1(t_2 - t_1) + \tfrac{1}{2}\alpha(t_2 - t_1)^2. \tag{11-13}$$

Note that it is possible to derive other useful secondary equations from these two primary equations.

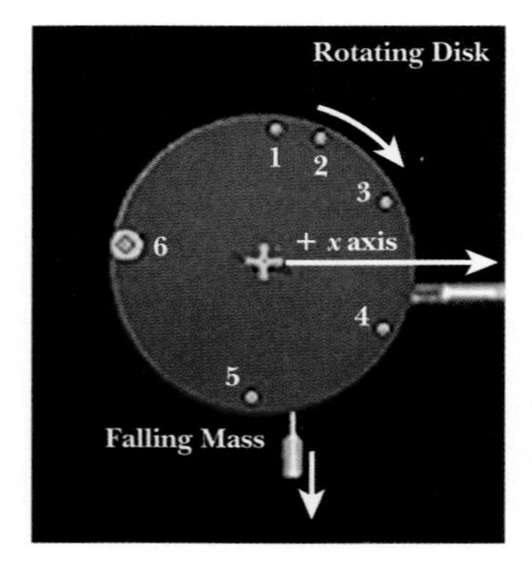

FIGURE 11-9 ■ A falling mass is attached to the axle of a rotating disk. The mass falls with a constant translational acceleration. Video analysis shows that as the mass falls, the disk rotates from position 1 to position 6 with a constant rotational acceleration. By defining a coordinate system, both the translational and rotational accelerations can be determined using the equations in Table 11-1.

TABLE 11-1
Equations of Motion with Constant Translational Acceleration and with Constant Rotational Acceleration

Equation Number	Translational Equation	Rotational Equation	Equation Number
Primary Vector Component Equations:*		Primary Rotational Vector Component Equations	
(2-13)	$v_{2x} = v_{1x} + a_x(t_2 - t_1)$	$\omega_2 = \omega_1 + \alpha(t_2 - t_1)$	(11-12)
(2-17)	$x_2 - x_1 = v_{1x}(t_2 - t_1) + \frac{1}{2}a_x(t_2 - t_1)^2$	$\theta_2 - \theta_1 = \omega_1(t_2 - t_1) + \frac{1}{2}\alpha(t_2 - t_1)^2$	(11-13)

**A reminder*: In cases where the initial time t_1 is chosen to be zero and t_2 is denoted as t, it is important to remember that whenever the term $(t_2 - t_1)$ is replaced by just t, then t actually represents a *time period of* $\Delta t = t_2 - t_1 = t - 0$ over which the motion of interest takes place.

READING EXERCISE 11-3: In four situations, a rotating body has rotational position $\theta(t)$ of:

(a) $\theta(t) = 3\left[\frac{\text{rad}}{\text{s}}\right]t - 4[\text{rad}]$,

(b) $\theta(t) = -5\left[\frac{\text{rad}}{\text{s}^3}\right]t^3 + 4\left[\frac{\text{rad}}{\text{s}^2}\right]t^2 + 6[\text{rad}]$,

(c) $\theta(t) = \frac{2[\text{rad}\cdot\text{s}^2]}{t^2} - \frac{4[\text{rad}\cdot\text{s}]}{t}$, and

(d) $\theta(t) = 5\left[\frac{\text{rad}}{\text{s}^2}\right]t^2 - 3[\text{rad}]$.

To which situations do the rotational equations of Table 11-1 apply? ■

TOUCHSTONE EXAMPLE 11-2: Grindstone

A grindstone (Fig. 11-10) rotates at constant rotational acceleration $\alpha = 0.35$ rad/s². At time $t_1 = 0$ s, it has a rotational velocity of $\omega_1 = -4.6$ rad/s and a reference line on it is horizontal, at the rotational position $\theta_1 = 0.0$ rad.

(a) At what time after $t = 0.0$ s is the reference line at the rotational position $\theta = 5.0$ rev?

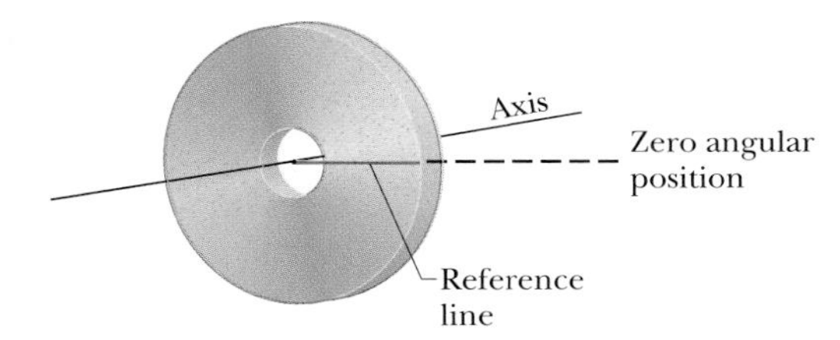

FIGURE 11-10 ■ A grindstone. At $t_1 = 0$ the reference line (which we imagine to be marked on the stone) is horizontal.

SOLUTION ■ The **Key Idea** here is that the rotational acceleration is constant, so we can use the rotation equations of Table 11-1. We choose Eq. 11-13,

$$\theta_2 - \theta_1 = \omega_1(t_2 - t_1) + \tfrac{1}{2}\alpha(t_2 - t_1)^2,$$

because the only unknown variable it contains is the desired time $(t_2 - t_1)$. Substituting known values and setting $\theta_1 = 0.0$ rad and $\theta_2 = 5.0 \text{ rev} = 10\pi$ rad give us

$$10\pi\,\text{rad} = (-4.6\,\text{rad/s})(t_2 - t_1) + \tfrac{1}{2}(0.35\,\text{rad/s}^2)(t_2 - t_1)^2.$$

(We converted 5.0 rev to 10π rad to keep the units consistent.) Solving this quadratic equation for $t_2 - t_1$, we find

$$t_2 - t_1 = 32\text{ s}. \quad \text{(Answer)}$$

(b) Describe the grindstone's rotation between $t_1 = 0$ and $t_2 = 32$ s.

SOLUTION ■ The wheel is initially rotating in the negative direction with rotational velocity $\omega_1 = -4.6$ rad/s, but its rotational acceleration α is positive. This initial opposition of the signs of rotational velocity and rotational acceleration means that the wheel slows in its rotation in the negative direction and then reverses to rotate in the positive direction. After the reference line comes back through its initial orientation of $\theta_1 = 0.0$ rad, the wheel turns an additional 5.0 rev by time $t_2 = 32$ s. (Answer)

(c) At what time t_3 does the grindstone change its direction of rotation?

SOLUTION ■ We again go to the table of equations for constant rotational acceleration, and again we need an equation that contains only the desired unknown variable $(t_3 - t_1)$. However, now we use another **Key Idea**. The equation must also contain the variable ω, so that we can set it to 0 and then solve for the corresponding time t_3. We choose Eq. 11-12, which yields

$$t_3 - t_1 = \frac{\omega_3 - \omega_1}{\alpha} = \frac{0.0\,\text{rad/s} - (-4.6\,\text{rad/s})}{0.35\,\text{rad/s}^2} = 13\text{ s}. \quad \text{(Answer)}$$

11-4 Relating Translational and Rotational Variables

In Section 5-7, we discussed uniform circular motion, in which a particle travels at constant translational speed v along a circle and around an axis of rotation. When a rigid body, such as a merry-go-round, rotates around an axis, each particle in the body moves in its own circle around that axis. Since the body is rigid, all the particles make one revolution in the same amount of time; that is, they all have the same rotational speed $|\omega|$.

However, try the following experiment. Pretend that you are a dancer or a skater. Although people are not rigid objects, they often configure their bodies into poses that are temporarily rigid. Hold out your arms like a dancer or skater and spin around in place (Fig. 11-11). What point on your body is moving fastest? Your shoulder, elbow, or fingertip?

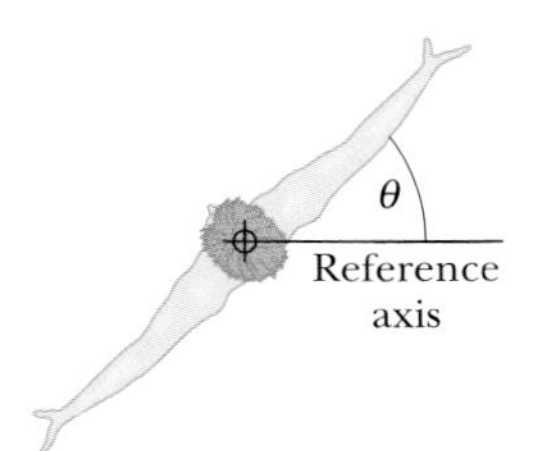

FIGURE 11-11 ■ If you rotate your body about a fixed vertical axis with your arms extended, what moves fastest: your shoulder, your elbow, or your fingers?

As you may have gathered from the observation described above, a particle far from the axis of rotation moves at a greater translational speed v than a particle close to the axis of rotation. This is because the farther the object is from the axis, the greater the circumference of the circular path the object takes in rotating about the axis. Since all the points on the object complete a revolution in the same time interval (they all have the same rotational speed $|\omega|$), those points that must travel a larger circumference must move at a higher translational speed. Hence, all points on a rotating object have the same rotational speed $|\omega|$, but not the same translational speed $|\vec{v}|$. You can also notice this on a merry-go-round. You turn with the same rotational speed $|\omega|$ regardless of your distance from the center, but your translational speed $|\vec{v}|$ increases noticeably as you move from the center to the outside edge of the merry-go-round. This is the reason we describe rotation using rotational, rather than translational, variables.

We often need to relate the translational variables s, $|\vec{v}|$, and $|\vec{a}|$ for a particular point in a rotating body to the rotational variables $|\theta|$, $|\omega|$, and $|\alpha|$ for that body. For example, we may know the rotational velocity and need to know the associated translational velocity. The two sets of variables (translational and rotational) are related by r, the *perpendicular distance* of the point from the rotation axis. This perpendicular distance is the distance between the point and the rotation axis, measured along a perpendicular to the axis. It is also the radius r of the circle traveled by the point around the axis of rotation.

Rotational Position and Distance Moved

If a reference line on a rigid body rotates through an angle θ, a point within the body at a distance r from the rotation axis moves a distance s along a circular arc, where s is given by

$$s = |\theta|\, r \quad \text{(for radian measure only).} \tag{11-14}$$

This is the first of our translational-rotational relations. *Caution*: The angle θ here *must be measured in radians* because $s = |\theta| r$ is derived from the definition of the radian.

Relating Rotational and Translational Speed

How can we compare the translational speed v of a rotating particle to its rotational speed $|\omega|$? Any small element of a rotating object that is rigid stays a fixed distance, r, from the axis of rotation throughout its rotation around the axis. In Section 5-7, we showed that if a rotating particle moves from one point on a circle to another, then the magnitude of its translational displacement, $|\Delta\vec{r}|$, between those points and the distance it moves along the arc of the circle, Δs, are essentially the same when the displacement is infinitesimally small. For this reason, we can find the magnitude of the instantaneous translational velocity by taking the time derivative of $s = |\theta| r$ (Eq. 11-14). In other words,

$$|\vec{v}| = v = \frac{|d\vec{r}|}{dt} = \frac{ds}{dt} = \frac{d|\theta|}{dt} r.$$

However, $d|\theta|/dt$ is the rotational speed $|\omega|$ of the rotating body, so the translational speed is given by

$$|\vec{v}| = v = |\omega| r \quad \text{(for radian measure only).} \tag{11-15}$$

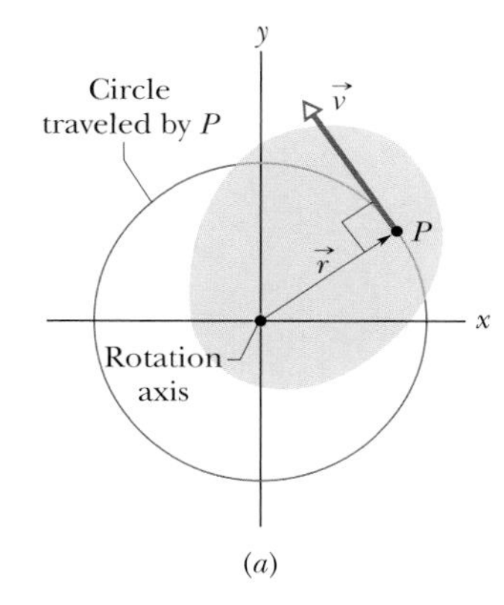

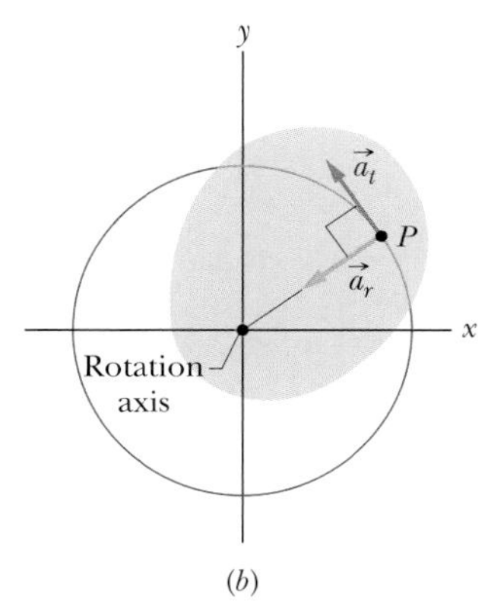

FIGURE 11-12 ■ The rotating rigid body of Fig. 11-4, shown in cross section viewed from above. Every point of the body (such as P) moves in a circle around the rotation axis. (a) The translational velocity v of every point is tangent to the circle in which the point moves. (b) The translational acceleration $\vec{a}$ of the point has (in general) two components, a tangential component a_t and a radial component a_r.

Caution: Rotational speed $|\omega|$ *must be expressed in radian measure* and be denoted with an absolute value sign since ω represents a vector component.

Equation 11-15 ($v = |\omega| r$) tells us that all points within the rigid body have the same rotational speed, $|\omega|$, while points with greater radius r, have greater translational speed $|\vec{v}|$. This equation verifies the conclusion we already reached. Namely, that when all points on an object complete a revolution in the same time interval they all have the same rotational speed $|\omega|$. But those points that are a larger distance from the axis of rotation must travel a larger circumference and must move at a higher translational velocity. Figure 11-12a reminds us that the translational velocity is always tangent to the circular path of the point in question.

If the rotational speed $|\omega|$ of the rigid body is constant, then $v = |\omega| r$ (Eq. 11-15) tells us that the translational speed v of any point within it is also constant. Thus, each point within the body undergoes uniform circular motion. We can find the period of revolution T by recalling that this is the time for one revolution (which is a linear distance $2\pi r$). The rate at which that distance is traveled is equal to the circumference divided by the time needed to make one revolution. Hence,

$$v = \frac{2\pi r}{T},$$

and the period of revolution T, for the motion of each point and for the rigid body itself is given by

$$T = \frac{2\pi r}{v}. \tag{11-16}$$

Substituting for v from $v = |\omega| r$ (Eq. 11-15) and canceling r, we find also that

$$T = \frac{2\pi}{|\omega|} \quad \text{(radian measure).} \tag{11-17}$$

The Acceleration

Differentiating Eq. 11-15 with respect to time—again, with r held constant—leads to

$$\frac{dv}{dt} = \frac{d|\omega|}{dt}r. \quad (11\text{-}18)$$

Here we run up against a complication. In this equation, dv/dt represents only the part of the magnitude of translational acceleration that is responsible for changes in the *magnitude* $|\vec{v}|$ of the translational velocity $\vec{v}$. Like $\vec{v}$, that part of the translational acceleration is tangent to the path of the point in question. We call it the *tangential component* a_t of the translational acceleration of the point, and we express its magnitude

$$|\vec{a}_t| = |\alpha|r \quad \text{(radian measure)}, \quad (11\text{-}19)$$

where the component of rotational acceleration is given by $\alpha = d\omega/dt$. *Caution:* Once again the rotational acceleration α in the expression $|\vec{a}_t| = |\alpha|\,r$ (Eq. 11-19) must be expressed in radian measure.

In addition, we know from our previous work that a particle (or point) moving in a circular path (even at constant velocity) has a *radial component vector* of translational acceleration, which we called the centripetal acceleration, $|\vec{a}_r| = v^2/r$ (directed radially inward), that is responsible for changes in the *direction* of the translational velocity $\vec{v}$. By substituting for v from $v = |\omega|r$ (Eq. 11-15), we can write this component as

$$|\vec{a}_r| = \frac{v^2}{r} = \omega^2 r \quad \text{(radian measure)}. \quad (11\text{-}20)$$

Thus, as Fig. 11-12*b* shows, the translational acceleration of a point on a rotating rigid body has, in general, two components. The radially inward component

$$|\vec{a}_r| = \frac{v^2}{r} = \omega^2 r,$$

is present whenever the rotational velocity of the body is not zero. That is, this component is nonzero whenever an object undergoes rotational motion. In addition, there is a tangential component $|\vec{a}_t| = |\alpha|\,r$ (Eq. 11-19) which is present whenever the rotational acceleration is nonzero. That is, this component is nonzero only if the object's rotation rate is increasing or decreasing. The total translational acceleration of a rotating rigid object is found using $|\vec{a}^{\text{tot}}|^2 = |\vec{a}_r|^2 + |\vec{a}_t|^2$.

READING EXERCISE 11-4: In Eq. 11-20 we did not bother to represent the squares of the magnitude of the translational and rotational speeds as $|\vec{v}|^2$ and $|\omega|^2$. Rather, we just use v^2 and ω^2. Why is this legitimate? ■

READING EXERCISE 11-5: A beetle rides the rim of a rotating merry-go-round. If the rotational speed of this system (merry-go-round + beetle) is constant, does the beetle have (a) radial acceleration and (b) tangential acceleration? If the rotational speed is decreasing, does the beetle have (c) radial acceleration and (d) tangential acceleration? ■

TOUCHSTONE EXAMPLE 11-3: Human Centrifuge

Figure 11-13 shows a centrifuge used to accustom astronaut trainees to high accelerations. The radius r of the circle traveled by an astronaut is 15 m.

(a) At what constant rotational speed must the centrifuge rotate if the astronaut is to have a translational acceleration of magnitude $11g$?

SOLUTION ■ The **Key Idea** is this: Because the rotational speed is constant, the rotational acceleration $\alpha(= d\omega/dt)$ is zero and so is the tangential component vector of the translational acceleration ($|\vec{a}_t| = |\alpha| r$). This leaves only the radial component vector. From Eq. 11-20 ($|\vec{a}_r| = \omega^2 r$), with $|\vec{a}_r| = 11g$, we have

$$\omega = \sqrt{\frac{|\vec{a}_r|}{r}} = \sqrt{\frac{(11)\,(9.8\text{ m/s}^2)}{15\text{ m}}}$$

$$= 2.68\text{ rad/s} \approx 26\text{ rev/min.} \qquad \text{(Answer)}$$

(b) What is the tangential acceleration of the astronaut if the centrifuge accelerates at a constant rate from rest to the rotational speed found in part (a) in 120 s?

SOLUTION ■ The **Key Idea** here is that the magnitude of the tangential acceleration $|\vec{a}_t|$ is related to the rotational acceleration α by Eq. 11-19 ($|\vec{a}_t| = |\alpha| r$). Also, because the rotational acceleration is constant, we can use Eq. 11-12 ($\omega_2 = \omega_1 + \alpha(t_2 - t_1)$) from Table 11-1 to find α from the given rotational speeds. Putting these two equations together, we find

FIGURE 11-13 ■ A centrifuge in Cologne, Germany, is used to accustom astronauts to the large acceleration experienced during a liftoff.

$$|\vec{a}_t| = |\alpha| r = \left|\frac{\omega_2 - \omega_1}{t_2 - t_1}\right| r$$

$$= \left|\frac{2.68\text{ rad/s} - 0}{120\text{ s}}\right| (15\text{ m}) = 0.34\text{ m/s}^2$$

$$= 0.034\,g. \qquad \text{(Answer)}$$

Although the magnitude of the final radial acceleration $|\vec{a}_r| = 11g$ is large (and alarming), the astronaut's tangential acceleration a_t during the speed-up is not.

11-5 Kinetic Energy of Rotation

The rapidly rotating blade of a table saw certainly has kinetic energy (KE) due to that rotation. How can we express the energy? We need to treat the table saw (and any other rotating rigid body) as a collection of particles with different speeds. We can then add up the kinetic energies of all the particles $A, B, C \ldots$ to find the kinetic energy of the body as a whole. In this way we obtain, for the kinetic energy of a rigid rotating body,

$$K = \tfrac{1}{2}m_A v_A^2 + \tfrac{1}{2}m_B v_B^2 + \tfrac{1}{2}m_C v_C^2 + \cdots. \qquad (11\text{-}21)$$

The sum is taken over all the particles in the body. The problem with this sum is that the values of translational velocity are not the same for all particles. We can solve this problem by substituting for v using Eq. 11-15 ($v = \omega r$) so that we have

$$\begin{aligned} K &= \tfrac{1}{2}m_A r_A^2 \omega_A^2 + \tfrac{1}{2}m_B r_B^2 \omega_B^2 + \tfrac{1}{2}m_C r_C^2 \omega_C^2 + \cdots \\ &= \tfrac{1}{2}\omega^2\{m_A r_A^2 + m_B r_B^2 + m_C r_C^2 + \cdots\}, \end{aligned} \qquad (11\text{-}22)$$

since ω is the same for all particles.

The quantity in brackets on the right side of this equation, $\{m_A r_A^2 + m_B r_B^2 + m_C r_C^2 + \cdots\}$, tells us how the mass of the rotating body is distributed about its axis of rotation. We call that quantity the **rotational inertia** (or *moment of inertia*) I of the body with respect to the axis of the rotation. It is a constant for a particular rigid body and a particular rotation axis. (That axis must always be specified if the value of I is to be meaningful.)

We may write an expression defining the rotational inertia for a collection of particles as

$$I \equiv \sum m_i r_i^2 \quad \text{(rotational inertia)}, \tag{11-23}$$

where Σ is a summation sign signifying that we sum over all the particles in the rigid rotating system. We can substitute into Eq. 11-22 ($K = \frac{1}{2}\omega^2\{m_A r_A^2 + m_B r_B^2 + m_C r_C^2 + \cdots\}$), obtaining

$$K = \tfrac{1}{2} I \omega^2 \quad \text{(rotational KE, radian measure)}, \tag{11-24}$$

as the expression we seek for the rotational kinetic energy. Because we have used the relation $v = \omega r$ in deriving $K = \frac{1}{2}I\omega^2$, ω must be expressed in radian measure. The SI unit for I is the kilogram-meter-squared ($\text{kg} \cdot \text{m}^2$).

Equation 11-24 ($K = \frac{1}{2}I\omega^2$), which gives the kinetic energy of a rigid body in pure rotation, is the rotational equivalent of the formula $K = \frac{1}{2}Mv_{\text{com}}^2$, which gives the kinetic energy of a rigid body in pure translation. In both formulas, there is a factor of $\frac{1}{2}$. Where mass M appears in one equation, I (which involves both mass and distribution) appears in the other. Finally, each equation contains a factor of the square of a speed—translational or rotational as appropriate. The kinetic energies of translation and rotation are not different kinds of energy. They are both kinetic energy, expressed in ways that are appropriate to the motion at hand.

We noted previously that the rotational inertia of a rotating body involves not only its mass but also how that mass is distributed. Here is an example that you can literally feel. Rotate a long rod (a pole, a length of lumber, a twirling baton, or something similar), first around its central (longitudinal) axis (Fig. 11-14*a*) and then around an axis perpendicular to the rod and through the center (Fig. 11-14*b*). Both rotations involve the very same mass, but the mass of the object in the first rotation is much closer to the rotation axis. As a result, the rotational inertia of the rod is much smaller in Fig. 11-14*a* than in Fig. 11-14*b*. In general, smaller rotational inertia means easier rotation.

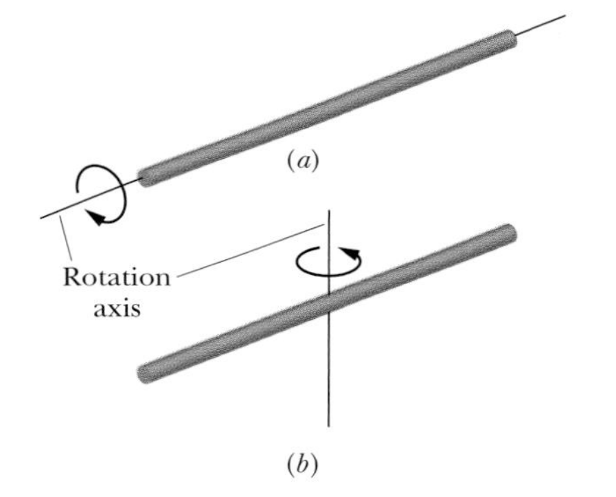

FIGURE 11-14 ■ A long rod is much easier to rotate about (*a*) its central (longitudinal) axis through its center and perpendicular to its length because the mass is distributed closer to the rotation axis in (*a*) than in (*b*).

READING EXERCISE 11-6: The figure shows three small spheres that rotate about a vertical axis. The perpendicular distance between the axis and the center of each sphere is given. Rank the three spheres according to their rotational inertia about the axis, greatest first.

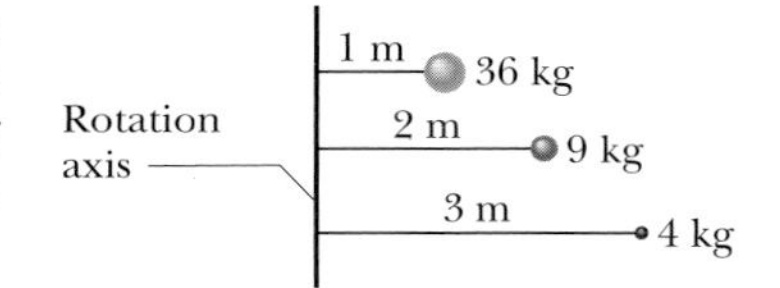

■

11-6 Calculating Rotational Inertia

If a rigid body consists of a few particles, we can calculate its rotational inertia about a given rotational axis with Eq. 11-23 ($I = \Sigma\, m_i r_i^2$). For example, consider the rotational inertia of a lump of clay (considered to be a point mass) with mass M at a distance r from the axis of rotation. The rotational inertia of such an object is simply Mr^2. Consider the rotational inertia of the object if the clay is now split into two pieces of equal mass, or eight pieces of equal mass, or even a very large number of point masses. As shown in Fig. 11-15, these pieces can be made to fashion a hoop of mass m and radius r.

Since the total mass M is divided into n equal masses, we can write the rotational inertia of this hoop as the sum of the rotational inertias of each of its elements:

$$\sum_{i=1}^{i=n} m_i r_i^2 = \frac{M}{n} r_1^2 + \frac{M}{n} r_2^2 + \frac{M}{n} r_3^2 + \cdots.$$

Further, since all the point masses that make up the hoop are located the same distance r away from the axis of rotation,

$$\begin{aligned} \sum_{i=1}^{i=n} m_i r_i^2 &= \left(\frac{M}{n} r^2 + \frac{M}{n} r^2 + \frac{M}{n} r^2 + \cdots\right) \\ &= r^2 \left(\frac{M}{n} + \frac{M}{n} + \frac{M}{n} + \cdots\right) \qquad (11\text{-}25) \\ I_{\text{hoop}} &= Mr^2. \end{aligned}$$

If a rigid body consists of a great many adjacent particles (it is *continuous*, like a Frisbee), using $I = \Sigma m_i r_i^2$ would require a tedious computer calculation. Instead, for a body that has a simple geometric form, we can replace the sum $\Sigma m_i r_i^2$ with an integral, and define the rotational inertia of the body as

$$I = \int r^2\, dm \quad \text{(rotational inertia, continuous body).} \qquad (11\text{-}26)$$

Table 11-2 gives the results of such integration for nine common body shapes and the indicated axes of rotation.

Studying the equations in Table 11-2 is helpful. *For example, for objects all having the same radius and mass, an object with its mass distributed very close to the axis of rotation has a smaller rotational inertia than an object with mass distributed farther out.* A case in point is the rotation of a cylinder (or equivalently a disk) about a central diameter. Table 11-2 shows that

$$I_{\text{disk}} = I_{\text{cylinder}} = \int r^2\, dm = \tfrac{1}{2} MR^2. \qquad (11\text{-}27)$$

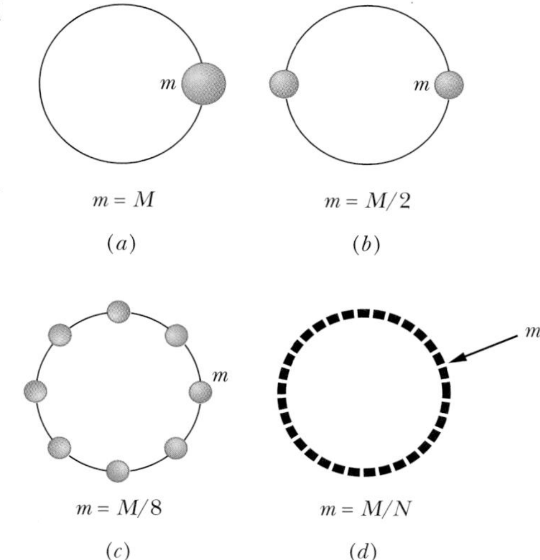

FIGURE 11-15 ■ Imagine a thin wire that provides a "massless" circular frame for clay blobs. As a clay blob of mass m is divided into more and more parts of equal mass, the distance of each smaller blob from the center of the circle is still the same. How does I compare for each of the objects, (*a*), (*b*), (*c*), and (*d*)?

TABLE 11-2
Some Rotational Inertias

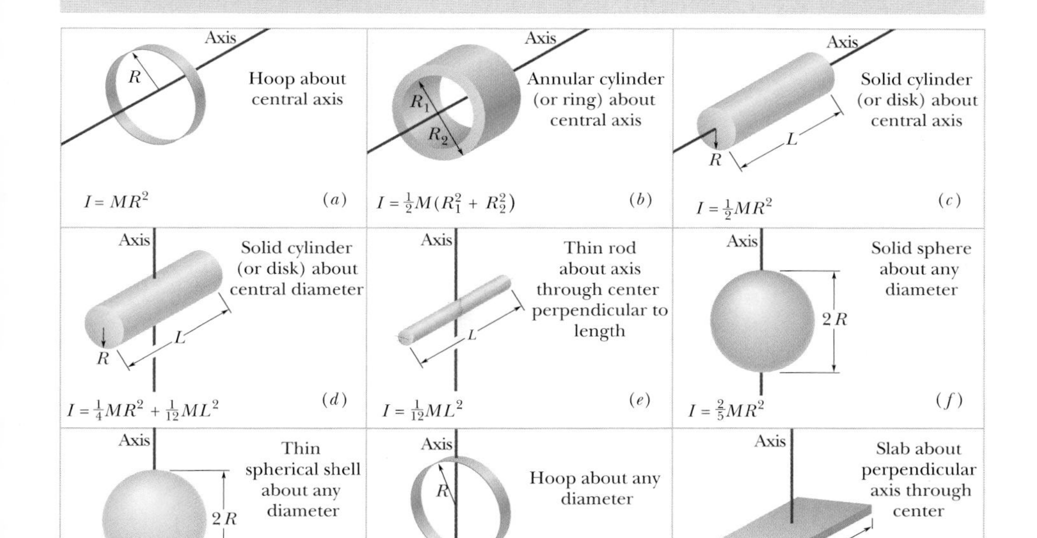

Thus, the rotational inertia of the hoop is twice that of a disk with the same mass and radius. This is because a hoop of the same radius as the disk has *all* its mass distributed as far away from the axis of rotation as possible.

Note that an object can have more than one axis of rotation. For example, you can roll a cylinder (or disk) along a table so its axis of rotation is perpendicular to the flat face of the cylinder (as in Table 11-2(*c*)). Alternatively Table 11-2(*d*) shows a different rotational inertia equation for an axis parallel to its face. In the next subsection we present a parallel-axis theorem that will allow us to determine the rotational inertia about any rotational axis once we know its rotational inertia about another axis parallel to it.

Parallel-Axis Theorem

Suppose we want to find the rotational inertia I of a body of mass M about a given axis. In principle, we can always find I using integration of

$$I = \int r^2\, dm.$$

However, it is easier mathematically to find the rotational inertia of an object about an axis of symmetry that passes through the object's center of mass. Fortunately, in certain circumstances, there is a shortcut. If we know the rotational inertia of a symmetric object rotating about an axis passing through its center of mass (for example, from Table 11-2), then the rotational inertia I about another parallel axis is

$$I = I^{\text{com}} + Mh^2 \qquad \text{(parallel-axis theorem).} \qquad (11\text{-}28)$$

Here h is the perpendicular distance between the given axis and the axis through the center of mass (remember that these two axes must be parallel). The proof of this equation, known as the **parallel-axis theorem,** is fairly straightforward, because it takes advantage of the fact that the object is symmetric about its center-of-mass axis of rotation.

READING EXERCISE 11-7: The figure shows a book-like object (one side is longer than the other) and four choices of rotation axes, all perpendicular to the face of the object. Rank the choices according to the rotational inertia of the object about the axis, greatest first.

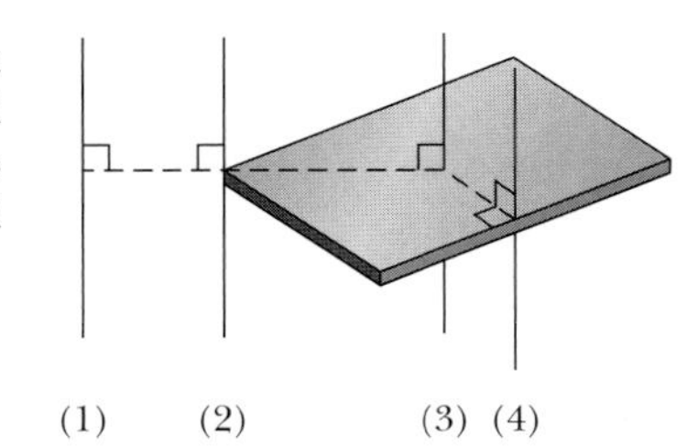

READING EXERCISE 11-8: Four objects having the same "radius" and mass are shown in the figure that follows. Rank the objects according to the rotational inertia about the axis shown, greatest first.

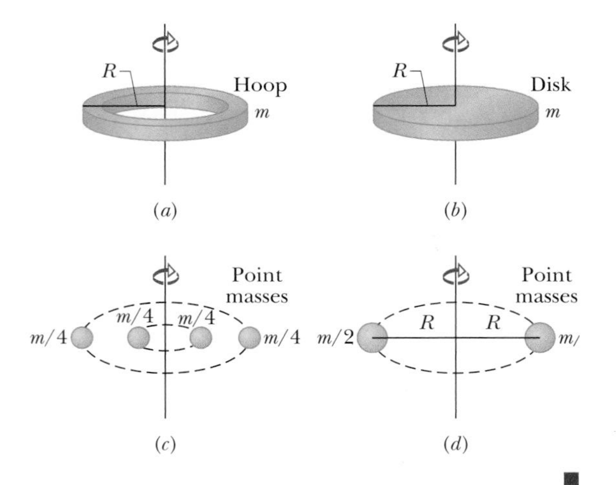

TOUCHSTONE EXAMPLE 11-4: Rotor Failure

Large machine components that undergo prolonged, high-speed rotation are first examined for the possibility of failure in a *spin test system*. In this system, a component is *spun up* (brought up to high speed) while inside a cylindrical arrangement of lead bricks and containment liner, all within a steel shell that is closed by a lid clamped into place. If the rotation causes the component to shatter, the soft lead bricks are supposed to catch the pieces so that the failure can then be analyzed.

In early 1985, Test Devices Inc. (www.testdevices.com) was spin-testing a sample of a solid steel rotor (a disk) of mass $M = 272$ kg and radius $R = 38.0$ cm. When the sample reached a rotational speed ω of 14 000 rev/min, the test engineers heard a dull thump from the test system, which was located one floor down and one room over from them. Investigating, they found that lead bricks had been thrown out in the hallway leading to the test room, a door to the room had been hurled into the adjacent parking lot, one lead brick had shot from the test site through the wall of a neighbor's kitchen, the structural beams of the test building had been damaged, the concrete floor beneath the spin chamber had been shoved downward by about 0.5 cm, and the 900 kg lid had been blown upward through the ceiling and had then crashed back onto the test equipment (Fig. 11-16). The exploding pieces had not penetrated the room of the test engineers only by luck.

How much energy was released in the explosion of the rotor?

SOLUTION ■ The **Key Idea** here is that this released energy was equal to the rotational kinetic energy K of the rotor just as it reached the rotational speed of 14 000 rev/min. We can find K with Eq. 11-24 ($K = \frac{1}{2}I\omega^2$), but first we need an expression for the rotational inertia I. Because the rotor was a disk that rotated like a merry-go-round, I is given by the expression in Table 11-2(*c*) ($I = \frac{1}{2}MR^2$). Thus we have

$$I = \tfrac{1}{2}MR^2 = \tfrac{1}{2}(272\text{ kg})(0.38\text{ m})^2 = 19.64\text{ kg}\cdot\text{m}^2.$$

FIGURE 11-16 ■ Some of the destruction caused by the explosion of a rapidly rotating steel disk.

The rotational speed of the rotor was

$$\omega = (14\,000\text{ rev/min})(2\pi\text{ rad/rev})\left(\frac{1\text{ min}}{60\text{ s}}\right)$$
$$= 1.466\times10^3\text{ rad/s}.$$

Now we can use Eq. 11-24 to write

$$K = \tfrac{1}{2}I\omega^2 = \tfrac{1}{2}(19.64\text{ kg}\cdot\text{m}^2)(1.466\times10^3\text{ rad/s})^2$$
$$= 2.1\times10^7\text{ J.} \qquad \text{(Answer)}$$

Being near this explosion was like being near an exploding bomb.

11-7 Torque

Now that we have defined the variables needed to describe the rotation of an object, we need to determine how forces can affect rotational motion. Because we are talking about more complex objects than point particles, we need to consider not only the forces that act on a rotating body but also the locations of those forces.

For instance, a doorknob is located as far as possible from the door's hinge line for a good reason. If you want to open a heavy door, you must certainly apply a force; that alone, however, is not enough. Where you apply that force and in what direction you push are also important. If you apply your force nearer to the hinge line than to the knob, or at any angle other than 90° to the plane of the door, you must use a greater force to move the door than if you apply the force at the knob and perpendicular to the door's plane. If you have never noticed this phenomenon, compare the force you need to open a heavy door near the hinge to that at the handle.

Figure 11-17*a* shows a cross section of a body that is free to rotate about an axis passing through O and perpendicular to the cross section. A force $\vec{F}$ is applied at point P, whose position relative to O is defined by a position vector $\vec{r}$. The directions

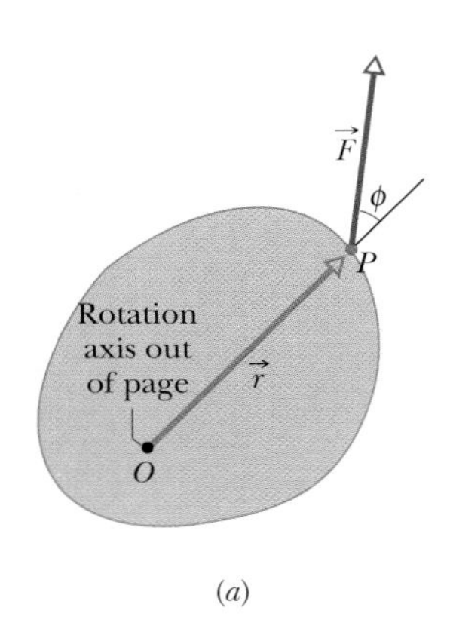

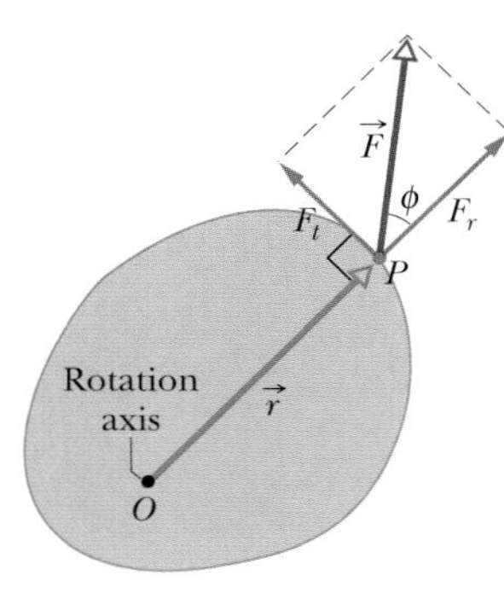

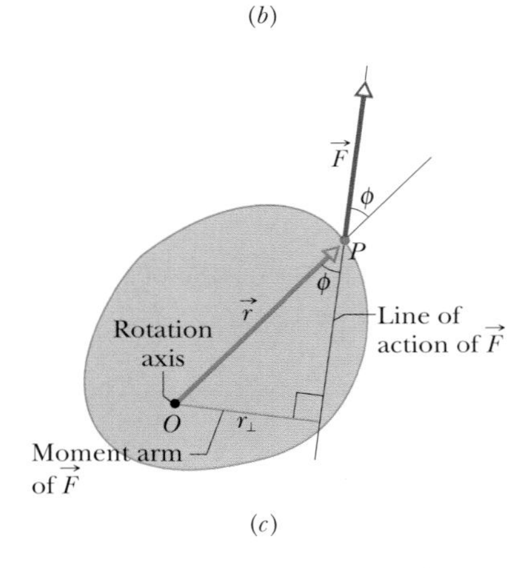

FIGURE 11-17 ■ (*a*) A force $\vec{F}$ acts at point P on a rigid body that is free to rotate about an axis through O. The axis is perpendicular to the plane of the cross section shown here. (*b*) The torque due to this force is $|\vec{r}||\vec{F}|\sin\phi$. We can also write it as $|\vec{r}||\vec{F}_t|$, where $\vec{F}_t$ is the tangential component vector of $\vec{F}$. (*c*) The torque magnitude can also be written as $r_\perp|\vec{F}|$, where $r_\perp$ is the moment arm of $\vec{F}$.

of vectors $\vec{F}$ and $\vec{r}$ make an angle ϕ ($0 \le \phi \le 180°$) with each other. For simplicity, we consider only forces that have no component parallel to the rotation axis; in other words, $\vec{F}$ is in the plane of the page.

To determine how $\vec{F}$ results in a rotation of the body around the rotation axis, we resolve $\vec{F}$ into two components (Fig. 11-17*b*). One component, called the *radial component vector* $\vec{F}_r$, points along $\vec{r}$. This component does not cause rotation, because it acts along a line that extends through O. (If you pull on a door parallel to the plane of the door, you are stretching and compressing the door, but you do not cause the door to rotate.) The other vector component of $\vec{F}$, called the *tangential component vector* $\vec{F}_t$, is perpendicular to $\vec{r}$ and has magnitude $|\vec{F}_t| = |\vec{F}|\sin\phi$. This component *does* cause rotation. (If you pull on a door perpendicular to its plane, you can rotate the door.)

The ability of $\vec{F}$ to rotate the body depends not only on the magnitude of its tangential component $|\vec{F}_t|$, but also on just how far from O the force is applied. To include both these factors, we define a new quantity called **torque.** In general, torque is a three-dimensional vector whose direction depends on the location and direction of a net force that acts on a rigid object that can rotate. Since we are only considering fixed rotation axes in this chapter, we can represent torque here as a one-dimensional vector (as we have with the other rotational variables). For now, we will describe torque in terms of its component τ_z along a z axis of rotation of the body experiencing a net force.

The torque component τ_z, often denoted as simply τ, has either a positive or negative value, depending on the direction of rotation it would give a body initially at rest. If a body rotates so the thumb of the right hand points along the positive direction assigned to the axis of rotation, the torque component is positive. If the object rotates in the opposite way, the torque component is negative.

The magnitude of the torque can be written as the product of the magnitude of a moment arm $|\vec{r}|$ and the magnitude of the tangential component of the force $|\vec{F}_t|$. As you can see in Fig. 11-17*b*, $|\vec{F}_t| = |\vec{F}|\sin\phi$.

$$|\vec{\tau}| = |\vec{r}||\vec{F}_t| = |\vec{r}||\vec{F}|\sin\phi. \tag{11-29}$$

Two equivalent ways of computing the magnitude of torque are

$$|\vec{\tau}| = |\vec{r}|(|\vec{F}|\sin\phi) = |\vec{r}||\vec{F}_t|, \tag{11-30}$$

and

$$|\vec{\tau}| = (r\sin\phi)|\vec{F}| = r_\perp|\vec{F}|, \tag{11-31}$$

where $r_\perp$ is the perpendicular distance between the rotation axis at O and an extended line running through the vector $\vec{F}$ (Fig. 11-17*c*). This extended line is called the **line of action** of $\vec{F}$, and $r_\perp$ is called the **moment arm** of $\vec{F}$. Figure 11-17*c* shows that we can describe r, the magnitude of $\vec{r}$, as being the moment arm of the force component F_t.

Torque, which comes from the Latin word meaning "to twist," may be loosely identified as the turning or twisting action of the force $\vec{F}$. When you apply a force to an object—such as a screwdriver or torque wrench—with the purpose of turning that object, you are applying a torque. The SI unit of torque is the newton-meter (N · m). *Caution*: The newton-meter is also the unit of work. Torque and work, however, are quite different quantities and must not be confused. Work is often expressed in joules (1 J = 1 N · m), but torque never is.

In the next chapter, we shall discuss cases in which torque must be represented by a vector that changes direction over time.

Torques obey the superposition principle that we discussed in Chapter 3 for forces: When several torques act on a body, the **net torque** (or **resultant torque**) component is the vector sum of the individual torques. The symbol for net torque component along the axis of rotation is τ_z^{net}.

Using Torque to Jump

So how do kneecaps allow us to jump higher? When we jump, we create a large torque in the knee joint in order to straighten the leg. If the force exerted by the strong thigh muscle (the quadriceps) is exerted along a line that is close to the pivot in the knee joint (represented by the dashed arrow in Fig. 11-18), the torque is not very great. A kneecap allows that force to be exerted farther from the pivot (represented by the solid arrow in Fig. 11-18). Recall that it is more effective to open a door by pushing far from the hinge. Similarly, the leg with a kneecap achieves more leg-straightening torque, thereby allowing for a higher jump!

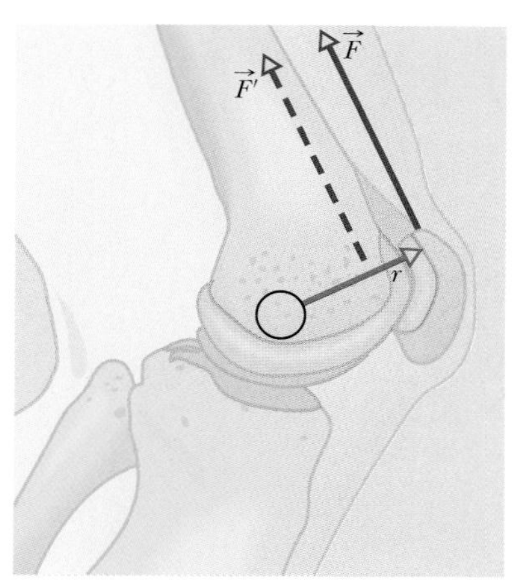

FIGURE 11-18 ■ The structure of the human knee. Note the force exerted by the quadriceps muscle, shown as $\vec{F}$, acting on the kneecap at a distance, r, from the pivot axis. If the kneecap were not there, the force, shown as $\vec{F}'$, would be acting along the dashed line a smaller distance r from the pivot axis, which is located approximately at the small circle.

READING EXERCISE 11-9: The figure shows an overhead view of a meter stick that can pivot about the dot at the position marked 20 (for 20 cm). All five forces on the stick have the same magnitude. Rank those forces according to the magnitude of the torque that they produce, greatest first.

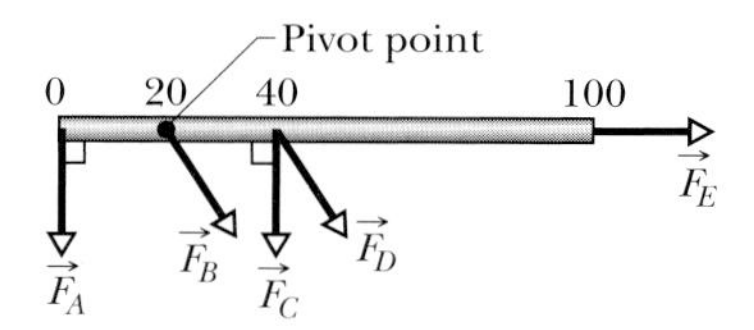

■

11-8 Newton's Second Law for Rotation

A torque can cause rotation of a rigid body, such as when you open a door about its hinge. Here we want to consider a special case in which a rigid body is symmetric about its axis of rotation. For this case we can relate the net torque component τ^{net} that acts on the body to the rotational acceleration component α the torque causes about a rotation axis. A good guess is to do so by analogy to the one-dimensional form of Newton's Second Law. If a one-dimensional net force F_x^{net} is acting along the x axis, then $F_x^{\text{net}} = ma_x$, where a_x is the acceleration component of a body of mass m, due to the net force acting along the x axis. For a rotation about a z axis we replace F_x^{net} with τ_z^{net}, m with I, and a_x with α_z, writing

$$\tau_z^{\text{net}} = I\alpha_z \quad \text{(Newton's Second Law for rotation).} \tag{11-32}$$

Remember in this context that τ_z^{net} and α_z are vector components, that we have chosen to represent as τ^{net} and α respectively. We can then rewrite Eq. 11-32 as

$$\tau^{\text{net}} = I\alpha \quad \text{(Newton's Second Law for symmetric rotations),} \tag{11-33}$$

where α must be in radian measure. This rotational analog to one-dimensional translational motion only holds when the axis of rotation does not change direction and when the body is symmetric about its axis of rotation.

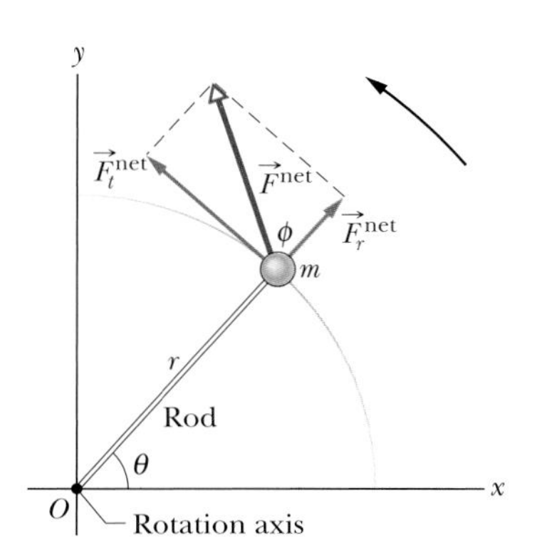

FIGURE 11-19 ■ A simple rigid body, free to rotate about an axis through O, consists of a particle of mass m fastened to the end of a rod of length r and negligible mass. An applied force $\vec{F}^{\text{net}}$ causes the body to rotate.

Proof of Equation 11-33

To see that Eq. 11-33 is, in fact, valid, let us see whether we can prove mathematically that $\tau^{\text{net}} = I\alpha$ by first considering the simple situation shown in Fig. 11-19. The rigid body there consists of a particle of mass m on one end of a massless rod of length r. The rod can move only by rotating about its other end, around a rotation axis (an axle) that is perpendicular to the plane of the page. Thus, the particle can move only in a circular path that has the rotation axis at its center.

A force $\vec{F}^{\text{net}}$ acts on the particle. However, because the particle can move only along the circular path, only the tangential component $\vec{F}_t^{\text{net}}$ of the force (the component that is tangent to the circular path) can accelerate the particle along the path. We can relate $\vec{F}_t^{\text{net}}$ to the particle's tangential acceleration component a_t along the path with Newton's Second Law, writing

$$\vec{F}_t^{\text{net}} = m\vec{a}_t.$$

So now, the magnitude of the torque acting on the particle is given by Eq. 11-30 as

$$|\vec{\tau}^{\text{net}}| = |\vec{r}|(|\vec{F}^{\text{net}}|\sin\phi) = |\vec{r}||\vec{F}_t^{\text{net}}|.$$

Note that we define a net tangential force component $\vec{F}_t^{\text{net}}$ as positive if it causes rotational and tangential accelerations that have positive components according to the right-hand rule. Conversely, $\vec{F}_t^{\text{net}}$ is negative if it leads to negative acceleration components along the axis of rotation. Since the distance $r = |\vec{r}|$ is the magnitude of a vector perpendicular to the rotation axis that points to the rotating particle, it is always positive. So the torque component can be expressed in terms of the net tangential force and acceleration components associated with a rotating body of mass m as

$$\tau^{\text{net}} = F_t^{\text{net}}\, r = ma_t\, r.$$

From Eq. 11-19 ($a_t = \alpha r$), we can write this as

$$\tau^{\text{net}} = m(\alpha r)r = (mr^2)\alpha. \tag{11-34}$$

Since the quantity in parentheses on the right side of this equation is the rotational inertia, mr^2, of the particle about the rotation axis, Eq. 11-34 reduces to

$$\tau^{\text{net}} = I\alpha \quad \text{(radian measure)}, \qquad \text{(Eq. 11-33)}$$

which is the expression we set out to prove. We can extend this equation to any rigid body rotating about an axis of symmetry, because any such body can always be analyzed as an assembly of single particles. Both α and τ^{net} are vector components along the rotation axis. Since I is inherently positive, α and τ^{net} must always have the same sign.

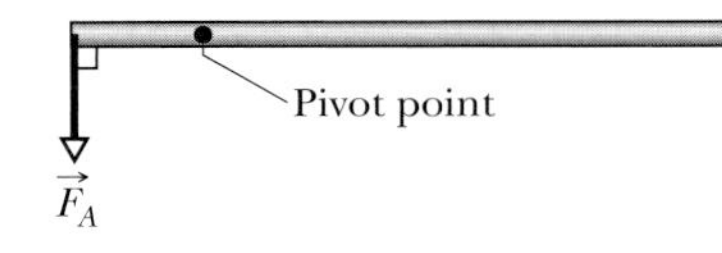

READING EXERCISE 11-10: The figure shows an overhead view of a meter stick that can pivot about a vertical axis at the point indicated, which is to the left of the stick's midpoint. Two horizontal forces, $\vec{F}_A$ and $\vec{F}_B$, are applied to the stick. Only $\vec{F}_A$ is shown. Force $\vec{F}_B$ is perpendicular to the stick and is applied at the right end. If the stick does not turn, (a) Is $\vec{F}_B$ in the same or opposite direction as $\vec{F}_A$ and (b) should the magnitude of $\vec{F}_B$ be greater than, less than, or equal to $\vec{F}_A$? ■

TOUCHSTONE EXAMPLE 11-5: Accelerating a Wheel

Figure 11-20*a* shows a uniform disk, with mass $M = 2.5$ kg and radius $R = 20$ cm, mounted on a fixed horizontal axle. A block with mass $m = 1.2$ kg hangs from a massless cord that is wrapped around the rim of the disk. Find the acceleration of the falling block, the rotational acceleration of the disk, and the tension in the cord. The cord does not slip, and there is no friction at the axle.

SOLUTION ■ One **Key Idea** here is that, taking the block as a system, we can relate its acceleration a to the forces acting on it with Newton's Second Law ($\vec{F}^{\text{net}} = m\vec{a}$). Those forces are shown in the block's free-body diagram in Fig. 11-20*b*: The force from the cord is $\vec{F}^{\text{cord}}$ and the gravitational force is $\vec{F}^{\text{grav}}$, of magnitude mg. We can now write Newton's Second Law for components along a vertical y axis $F_y^{\text{net}} = ma_y$ as

$$|\vec{F}^{\text{cord}}| - mg = ma_y. \tag{11-35}$$

However, we cannot solve this equation for a_y because it also contains the unknown $|\vec{F}^{\text{cord}}|$.

Previously, when we got stuck on the y axis, we would switch to the x axis. Here, we switch to the rotation of the disk and use this **Key Idea**: Taking the disk as a system, we can relate its rotational

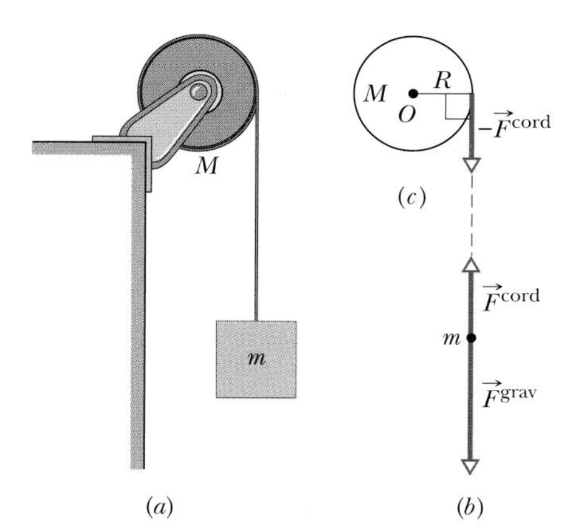

FIGURE 11-20 ■ (*a*) The falling block causes the disk to rotate. (*b*) A free-body diagram for the block. (*c*) An incomplete free-body diagram for the disk.

acceleration α to the torque acting on it with Newton's Second Law for rotation ($\tau^{\text{net}} = I\alpha$). To calculate the torques and the rotational inertia I, we take the rotation axis to be perpendicular to the disk and through its center, at point O in Fig. 11-20*c*.

The torques are then given by Eq. 11-29 ($|\vec{\tau}| = |\vec{r}||\vec{F}_t|$). The gravitational force on the disk and the force on the disk from the axle both act at the center of the disk and thus at distance $r = 0$, so their torques are zero. The force $\vec{F}^{\text{cord}}$ on the disk due to the cord acts at distance $r = R$ and is tangent to the rim of the disk. Therefore, the magnitude of its torque is $|\vec{\tau}| = R|\vec{F}^{\text{cord}}|$. From Table 11-2(*c*), the rotational inertia I of the disk is $\frac{1}{2}MR^2$. Thus we can write ($\tau_z^{\text{net}} = I\alpha_z$) as

$$|\vec{\tau}| = R|\vec{F}^{\text{cord}}| = \tfrac{1}{2}MR^2|\alpha|. \tag{11-36}$$

This equation seems equally useless because it has two unknowns, α and $|\vec{F}^{\text{cord}}|$, neither of which is the desired acceleration a. However, mustering physics courage, we can make it useful with a third **Key Idea**: Because the cord does not slip, the magnitudes of the translational acceleration $|a_y|$ of the block and of the (tangential) translational acceleration $|\vec{a}_t|$ of the rim of the disk are equal. Then, by Eq. 11-19 ($|\vec{a}_t| = |\alpha|r$) we see that here $|\alpha| = |a_y|/R$. Substituting this in Eq. 11-36 yields

$$R|\vec{F}^{\text{cord}}| = \frac{\frac{1}{2}MR^2|a_y|}{R} \quad \text{or} \quad |\vec{F}^{\text{cord}}| = \tfrac{1}{2}M|a_y|. \tag{11-37}$$

From Fig. 11-20*a* it's apparent that a_y is negative, which along with Eq. 11-37 tells us that

$$a_y = -\frac{2|\vec{F}^{\text{cord}}|}{M}. \tag{11-38}$$

Now combining Eqs. 11-35 and 11-38 leads to

$$a_y = -g\frac{2m}{M + 2m} = -(9.8\ \text{m/s}^2)\frac{(2)(1.2\ \text{kg})}{2.5\ \text{kg} + (2)(1.2\ \text{kg})}$$
$$= -4.8\ \text{m/s}^2. \qquad \text{(Answer)}$$

We then use Eq. 11-37 to find $|\vec{F}^{\text{cord}}|$:

$$|\vec{F}^{\text{cord}}| = \tfrac{1}{2}M|a_y| = \tfrac{1}{2}(2.5\ \text{kg})(4.8\ \text{m/s}^2) = 6.0\ \text{N}. \qquad \text{(Answer)}$$

As we should expect, the magnitude of the acceleration of the falling block is less than g, and the tension in the cord ($= 6.0$ N) is less than the gravitational force on the hanging block ($= mg = 11.8$ N). We see also that the acceleration of the block and the tension depend on the mass of the disk but not on its radius. As a check, we note that the formulas derived above predict $a_y = -g$ and $T = 0$ for the case of a massless disk ($M = 0$). This is what we would expect; the block simply falls as a free body, trailing the string behind it.

From Eq. 11-19, the magnitude of the rotational acceleration of the disk is

$$|\alpha| = \frac{|a_y|}{R} = \frac{4.8\ \text{m/s}^2}{0.20\ \text{m}} = 24\ \text{rad/s}^2. \qquad \text{(Answer)}$$

TOUCHSTONE EXAMPLE 11-6: Judo

To throw an 80 kg opponent with a basic judo hip throw, you intend to pull his uniform with a force $\vec{F}$ and a moment arm $d_1 = 0.30$ m from a pivot point (rotation axis) on your right hip (Fig. 11-21). You wish to rotate him about the pivot point with an rotational acceleration α of -6.0 rad/s²—that is, with an rotational acceleration that is *clockwise* in the figure. Assume that his rotational inertia I relative to the pivot point is 15 kg · m².

(a) What must the magnitude of $\vec{F}$ be if, before you throw him, you bend your opponent forward to bring his center of mass to your hip (Fig. 11-21*a*)?

SOLUTION ■ One **Key Idea** here is that we can relate your pull $\vec{F}$ on him to the given rotational acceleration α via Newton's Second Law for rotation ($\tau^{\text{net}} = I\alpha$). As his feet leave the floor, we can assume that only three forces act on him: your pull $\vec{F}$, a force $\vec{N}$ on him from you at the pivot point (this force is not indicated in Fig. 11-21), and the gravitational force $\vec{F}^{\text{grav}}$. To use $\tau^{\text{net}} = I\alpha$, we need the corresponding three torques, each about the pivot point.

From Eq. 11-31 ($|\tau| = r_\perp|\vec{F}|$), the torque due to your pull $\vec{F}$ is equal to $-d_1F$, where d_1 is the moment arm $r_\perp$ and the sign indicates the clockwise rotation this torque tends to cause. The torque due to $\vec{N}$ is zero, because $\vec{N}$ acts at the pivot point and thus has moment arm $r_\perp = 0.0$ m.

To evaluate the torque due to $\vec{F}^{\text{grav}}$, we need a **Key Idea** from Chapter 8: We can assume that $\vec{F}^{\text{grav}}$ acts at your opponent's center of mass. With the center of mass at the pivot point, $\vec{F}^{\text{grav}}$ has moment arm $r_\perp = 0.0$ m and thus the torque due to $\vec{F}^{\text{grav}}$ is zero. Thus, the only torque on your opponent is due to your pull $\vec{F}$, and we can write $\tau^{\text{net}} = I\alpha$ as

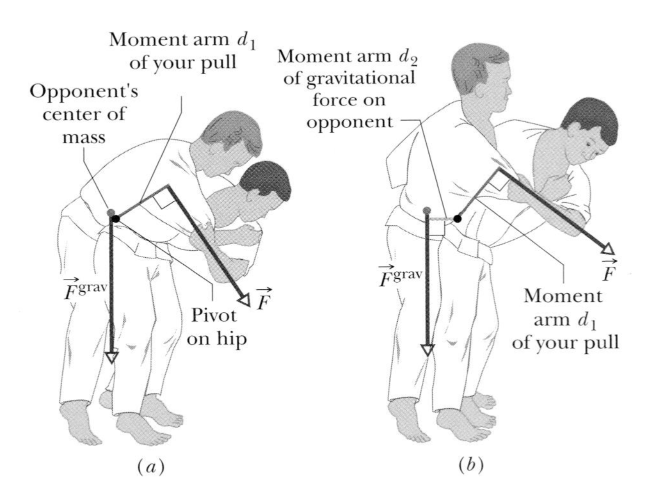

FIGURE 11-21 ■ A judo hip throw (*a*) correctly executed and (*b*) incorrectly executed.

$$-d_1|\vec{F}| = I\alpha.$$

We then find

$$|\vec{F}| = \frac{-I\alpha}{d_1} = \frac{-(15\ \text{kg}\cdot\text{m}^2)(-6.0\ \text{rad/s}^2)}{0.30\ \text{m}}$$
$$= 300\ \text{N}. \qquad \text{(Answer)}$$

(b) What must the magnitude of $\vec{F}$ be if your opponent remains upright before you throw him, so that $\vec{F}^{grav}$ has a moment arm $d_2 = 0.12$ m from the pivot point (Fig. 11-21*b*)?

SOLUTION ■ The **Key Ideas** we need here are similar to those in (a) with one exception: Because the moment arm for $\vec{F}^{grav}$ is no longer zero, the torque due to $\vec{F}^{grav}$ is now equal to d_2mg, and is positive because the torque attempts counterclockwise rotation. Now we write $\tau^{net} = I\alpha$ as

$$-d_1|\vec{F}| + d_2mg = I\alpha,$$

which gives

$$|\vec{F}| = -\frac{I\alpha}{d_1} + \frac{d_2mg}{d_1}.$$

From (a), we know that the first term on the right is equal to 300 N. Substituting this and the given data, we have

$$|\vec{F}| = 300\ \text{N} + \frac{(0.12\ \text{m})(80\ \text{kg})(9.8\ \text{m/s}^2)}{0.30\ \text{m}}$$
$$= 613.6\ \text{N} \approx 610\ \text{N}. \qquad \text{(Answer)}$$

The results indicate that you will have to pull much harder if you do not initially bend your opponent to bring his center of mass to your hip. A good judo fighter knows this lesson from physics. (An analysis of the physics of judo and aikido is given in "The Amateur Scientist" by J. Walker, *Scientific American*, July 1980, Vol. 243, pp. 150–161.)

11-9 Work and Rotational Kinetic Energy

Net Work-Kinetic Energy Theorem for Translational Motion in One Dimension

As we discussed in Chapter 9, when a net force causes the center of mass of a rigid body of mass m to accelerate along a coordinate axis, it does net work, W^{net}, on the body. Thus, the body's translational kinetic energy ($K = \frac{1}{2}mv^2$) can change. We can use the net work-kinetic energy theorem to relate these two quantities:

$$W^{net} = K_2 - K_1 = \Delta K \qquad \text{(work-kinetic energy theorem)}, \qquad \text{(Eq. 9-11)}$$

where K_1 is the kinetic energy of the object when it is located at an initial position and K_2 is its kinetic energy when it is displaced to a new position.

For translational motion confined to a single axis we choose to be the x axis, we can calculate the net work using the expression

$$W^{net} = \int_{x_1}^{x_2} F_x^{net}(x)dx \qquad \text{(work, one-dimensional motion)}. \qquad (11\text{-}39)$$

This reduces to $W^{net} = F_x^{net}\Delta x$ when the net force is constant and the body's displacement is $\Delta x = x_2 - x_1$. The rate at which the work is done is the power, which we can find with

$$P = \frac{dW}{dt} = Fv \quad \text{(power, one-dimensional motion).} \tag{11-40}$$

Net Work-Kinetic Energy Theorem for Rotational Motion—Fixed Axis

A similar situation exists for rotational motion. When a net torque accelerates a rigid body in rotation about a fixed axis, it also does work on the body—rotational work. Therefore, the body's rotational kinetic energy as derived in Section 11-5 as $K^{\text{rot}} = \frac{1}{2}I\omega^2$ (Eq. 11-24) can change. We can show that it is also possible to relate the change in rotational kinetic energy to the net rotational *work using the work-kinetic energy theorem* where we use rotational quantities to determine the net work and kinetic energy.

$$W^{\text{net-rot}} = \Delta K = K_2 - K_1 = \tfrac{1}{2}I\omega_2^2 - \tfrac{1}{2}I\omega_1^2 \quad \text{(rotational net work-kinetic energy theorem).} \tag{11-41}$$

Here I is the rotational inertia of the body about the fixed axis and ω_1 and ω_2 are the rotational speeds of the body before and after the rotational work is done, respectively.

Derivation of Rotational Work-Energy Theorem

We have already derived an expression for rotational kinetic energy as shown in Eq. 11-24. In order to derive Eq. 11-41, we need to use the definition of work to find an expression for the net rotational work $W^{\text{net-rot}}$. Then we can relate the work W done on the body in Fig. 11-19 to the net torque τ^{net} (which is due to a net force $\vec{F}^{\text{net}}$ that produces it). To do this we use the relationships between rotational and translational variables. We start by considering how the net force affects a single particle located at a distance r from the axis of rotation.

When a single particle moves a distance ds along its circular path, only the tangential component $\vec{F}_t$ of the force accelerates the particle along the path. Therefore only F_t does work on the particle. We write that infinitesimal increment of work dW as $F_t\, ds$. However, we can replace ds with $r\, d\theta$, where $d\theta$ is the angle through which the particle moves with respect to the x axis. Thus we have

$$dW^{\text{net-rot}} = F_t^{\text{net}} r\, d\theta. \tag{11-42}$$

However, the product $\vec{F}_t^{\text{net}} r$ is equal to the net torque τ^{net}, so we can rewrite Eq. 11-42 as

$$dW^{\text{net-rot}} = \tau^{\text{net}}\, d\theta. \tag{11-43}$$

The work done on a single rotating particle during a finite rotational displacement from θ_1 to θ_2 is then

$$W^{\text{net-rot}} = \int_{\theta_1}^{\theta_2} \tau^{\text{net}}\, d\theta \quad \text{(rotational work, fixed axis).} \tag{11-44}$$

If all the particles in a body rotate together, this equation for rotational work also applies to the extended body that is rigid. So we now have expressions for determining both the net rotational work and the change in rotational kinetic energy in terms of rotational variables and the same basic definitions of work and kinetic energy. This verifies that we can use the work-kinetic energy theorem to relate net work and kinetic energy change when a rigid body rotates about a fixed axis.

As is the case for work done by translational forces, rotational work is a scalar quantity that can be either positive or negative, depending on whether work is done on the rotating body or by it. The work is calculated using the product of the signed quantities torque and rotational displacement.

Power for a Rotating Body

In addition, we can find the power P associated with the rotational motion of a rigid object about a fixed axis using the equation $dW = \tau\, d\theta$ (Eq. 11-43):

$$P = \frac{dW}{dt} = \tau \frac{d\theta}{dt} = \tau\omega. \tag{11-45}$$

The signs of both torque and rotational velocity depend on the sign of the rotation as determined by the right-hand rule.

Table 11-3 summarizes the equations that apply to the rotation of a rigid body about a fixed axis and the corresponding equations for translational motion.

TABLE 11-3
Corresponding Relations for Translational and Rotational Motion

Pure Translation (x axis)		Pure Rotation (Symmetry about a Fixed Rotation Axis)	
Position component	x	Rotational position component	θ
Velocity component	$v_x = dx/dt$	Rotational velocity component	$\omega = d\theta/dt$
Acceleration	$a_x = dv_x/dt$	Rotational acceleration component	$\alpha = d\omega/dt$
Mass	m	Rotational inertia	I
Newton's Second Law	$F_x^{\text{net}} = ma_x$	Newton's Second Law	$\tau^{\text{net}} = I\alpha$
Work	$W = \int F_x\, dx$	Work	$W = \int \tau\, d\theta$
Kinetic energy	$K = \frac{1}{2}mv_x^2$	Kinetic energy	$K = \frac{1}{2}I\omega^2$
Power	$P = F_x v_x$	Power	$P = \tau\omega$
Work-kinetic energy theorem	$W = \Delta K$	Work-kinetic energy theorem	$W^{\text{rot}} = \Delta K^{\text{rot}}$

TOUCHSTONE EXAMPLE 11-7: Rotating Sculpture

A rigid sculpture consists of a thin hoop (of mass m and radius $R = 0.15$ m) and a thin radial rod (of mass m and length $L = 2.0\,R$), arranged as shown in Fig. 11-22. The sculpture can pivot around a horizontal axis in the plane of the hoop, passing through its center.

(a) In terms of m and R, what is the sculpture's rotational inertia I about the rotation axis?

SOLUTION ■ A **Key Idea** here is that we can separately find the rotational inertias of the hoop and the rod and then add the results to get the sculpture's total rotational inertia I. From Table 11-2(h), the hoop has rotational inertia $I_{\text{hoop}} = \frac{1}{2}mR^2$ about its diameter. From Table 11-2(e), the rod has rotational inertia $I_{\text{com}} = mL^2/12$ about an axis through its center of mass and parallel to the sculpture's rotation axis. To find its rotational

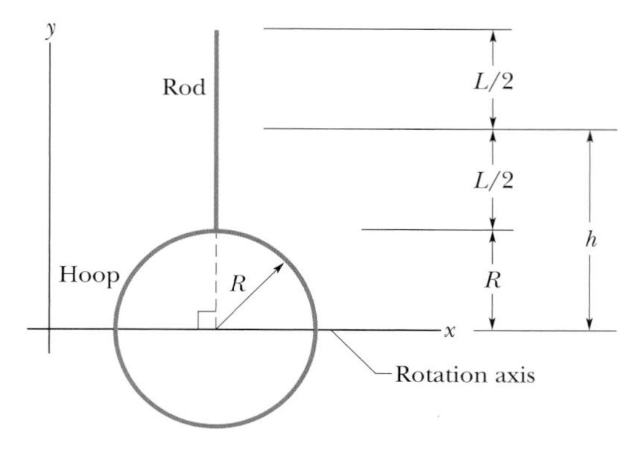

FIGURE 11-22 ■ A rigid sculpture consisting of a hoop and two rods can rotate around a horizontal axis.

inertia I_{rod} about that rotation axis, we use Eq. 11-28, the parallel-axis theorem:

$$I_{rod} = I_{com} + mh_{com}^2 = \frac{mL^2}{12} + m\left(R + \frac{L}{2}\right)^2$$
$$= 4.33mR^2,$$

where we have used the fact that $L = 2.0R$ and where the perpendicular distance between the rod's center of mass and the rotation axis $h = R + L/2$. Thus, the rotational inertia I of the sculpture about the rotation axis is

$$I = I_{hoop} + I_{rod} = \tfrac{1}{2}mR^2 + 4.33mR^2$$
$$= 4.83mR^2 \approx 4.8mR^2. \qquad \text{(Answer)}$$

(b) Starting from rest, the sculpture rotates around the rotation axis from the initial upright orientation of Fig. 11-22. What is its rotational speed ω about the axis when it is inverted?

SOLUTION ■ Three **Key Ideas** are required here:

1. We can relate the sculpture's speed ω to its rotational kinetic energy K with Eq. 11-24 ($K = \frac{1}{2}I\omega^2$).
2. We can relate K to the gravitational potential energy U^{grav} of the sculpture via the conservation of the sculpture's mechanical energy E^{mec} during the rotation. Thus, during the rotation, E^{mec} does not change ($\Delta E^{mec} = 0$) as energy is transferred from U^{grav} to K.
3. For the gravitational potential energy we can treat the rigid sculpture as a particle located at the center of mass, with the total mass $2m$ concentrated there.

We can write the conservation of mechanical energy ($\Delta E^{mec} = 0$) as

$$\Delta K + \Delta U^{grav} = 0. \qquad (11\text{-}46)$$

As the sculpture rotates from its initial position at rest to its inverted position, when the rotational speed is ω, the change ΔK in its kinetic energy is

$$\Delta K = K_2 - K_1 = \tfrac{1}{2}I\omega^2 - 0 = \tfrac{1}{2}I\omega^2. \qquad (11\text{-}47)$$

From Eq. 10-6 ($\Delta U^{grav} = mg\Delta y$), the corresponding change ΔU^{grav} in the gravitational potential energy is

$$\Delta U^{grav} = (2m)g\,\Delta y_{com}, \qquad (11\text{-}48)$$

where $2m$ is the sculpture's total mass, and Δy_{com} is the vertical displacement of its center of mass during the rotation.

To find Δy_{com}, we first find the initial location y_{com} of the center of mass in Fig. 11-22. The hoop (with mass m) is centered at $y = 0$. The rod (with mass m) is centered at $y = R + L/2$. Thus, from Eq. 8-11, the sculpture's center of mass is at

$$y_{com} = \frac{m(0) + m(R + L/2)}{2m} = \frac{0 + m(R + 2R/2)}{2m} = R.$$

When the sculpture is inverted, the center of mass is this same distance R from the rotation axis but *below* it. Therefore, the vertical displacement of the center of mass from the initial position to the inverted position is $\Delta y_{com} = -2R$.

Now let's pull these results together. Substituting Eqs. 11-47 and 11-48 into 11-46 gives us

$$\tfrac{1}{2}I\omega^2 + (2m)g\,\Delta y_{com} = 0.$$

Substituting $I = 4.83mR^2$ from (a) and $\Delta y_{com} = -2R$ from above and solving for ω, we find

$$\omega = \sqrt{\frac{8g}{4.83\,R}} = \sqrt{\frac{(8)(9.8\text{ m/s}^2)}{(4.83)(0.15\text{ m})}}$$
$$= 10\text{ rad/s}. \qquad \text{(Answer)}$$

Problems

SEC. 11-2 ■ THE ROTATIONAL VARIABLES

1. Flywheel The rotational position of a flywheel on a generator is given by $\theta = (a\text{ rad/s})t + (b\text{ rad/s}^3)t^3 - (c\text{ rad/s}^4)t^4$, where a, b, and c are constants. Write expressions for the wheel's (a) rotational velocity and (b) rotational acceleration.

2. Hands of a Clock What is the rotational speed of (a) the second hand, (b) the minute hand, and (c) the hour hand of a smoothly running analog watch? Answer in radians per second.

3. Milky Way Our Sun is 2.3×10^4 ly (light-years) from the center of our Milky Way galaxy and is moving in a circle around the center at a speed of 250 km/s. (a) How long does it take the Sun to make one revolution about the galactic center? (b) How many revolutions has the Sun completed since it was formed about 4.5×10^9 years ago?

4. Rotating Wheel The rotational position of a point on the rim of a rotating wheel is given by $\theta = (4.0\text{ rad/s})t + (3.0\text{ rad/s}^2)t^2 + (1\text{ rad/s}^3)t^3$, where θ is in radians and t is in seconds. What are the rotational velocities at (a) $t_1 = 2.0$ s and (b) $t_2 = 4.0$ s? (c) What is the average rotational acceleration for the time interval that begins at $t_1 = 2.0$ s and ends at $t_2 = 4.0$ s? What are the instantaneous rotational accelerations at (d) the beginning and (e) the end of this time interval?

5. Rotational Position The rotational position of a point on a rotating wheel is given by $\theta = 2.0\text{ rad} + (4.0\text{ rad/s}^2)t^2 + (2.0\text{ rad/s}^3)t^3$, where θ is in radians and t is in seconds. At $t_1 = 0$, what are (a) the point's rotational position and (b) its rotational velocity? (c) What is its rotational velocity at $t_3 = 4.0$ s? (d) Calculate its rotational acceleration at $t_2 = 2.0$ s. (e) Is its rotational acceleration constant?

6. The Wheel The wheel in Fig.11-23 has eight equally spaced spokes and a radius of 30 cm. It is mounted on a fixed axle and is spinning at 2.5 rev/s. You want to shoot a 20-cm-long arrow parallel to this axle and through the wheel without hitting any of the spokes. Assume that the arrow and the spokes are very thin. (a) What minimum speed must the arrow have? (b) Does it matter where between the axle and rim of the wheel you aim? If so, what is the best location?

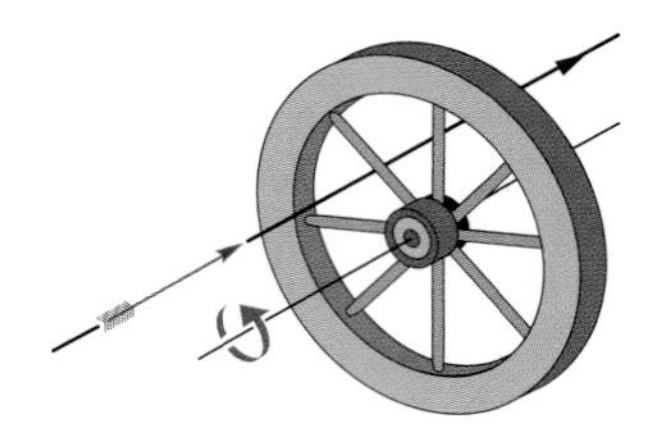

FIGURE 11-23 ■ Problem 6.

7. A Diver A diver makes 2.5 revolutions on the way from a 10-m-high platform to the water. Assuming zero initial vertical velocity, find the diver's average rotational velocity during a dive.

SEC. 11-3 ■ ROTATION WITH CONSTANT ROTATIONAL ACCELERATION

8. Automobile Engine The rotational speed of an automobile engine is increased at a constant rate from 1200 rev/min to 3000 rev/min in 12 s. (a) What is its rotational acceleration in revolutions per minute-squared? (b) How many revolutions does the engine make during this 12 s interval?

9. Turntable A record turntable rotating at $33\frac{1}{3}$ rev/min slows down and stops in 30 s after the motor is turned off. (a) Find its (constant) rotational acceleration in revolutions per minute-squared. (b) How many revolutions does it make in this time?

10. A Disk A disk, initially rotating at 120 rad/s, is slowed down with a constant rotational acceleration of magnitude 4.0 rad/s^2. (a) How much time does the disk take to stop? (b) Through what angle does the disk rotate during that time?

11. Heavy Flywheel A heavy flywheel rotating on its central axis is slowing down because of friction in its bearings. At the end of the first minute of slowing, its rotational speed is 0.90 of its initial rotational speed of 250 rev/min. Assuming a constant rotational acceleration, find its rotational speed at the end of the second minute.

12. A Disk Rotates Starting from rest, a disk rotates about its central axis with constant rotational acceleration. In 5.0 s, it rotates 25 rad. During that time, what are the magnitudes of (a) the rotational acceleration and (b) the average rotational velocity? (c) What is the instantaneous rotational velocity of the disk at the end of the 5.0 s? (d) With the rotational acceleration unchanged, through what additional angle will the disk turn during the next 5.0 s?

13. Constant Rotational Acceleration A wheel has a constant rotational acceleration of 3.0 rad/s^2. During a certain 4.0 s interval, it turns through an angle of 120 rad. Assuming that the wheel starts from rest, how long is it in motion at the start of this 4.0 s interval?

14. Starting from Rest A wheel, starting from rest, rotates with a constant rotational acceleration of 2.00 rad/s^2. During a certain 3.00 s interval, it turns through 90.0 rad. (a) How long is the wheel turning before the start of the 3.00 s interval? (b) What is the rotational velocity of the wheel at the start of the 3.00 s interval?

15. A Flywheel Has a Rotational Velocity At $t_1 = 0$, a flywheel has a rotational velocity of 4.7 rad/s, a rotational acceleration of -0.25 rad/s^2, and a reference line at $\theta_1 = 0$. (a) Through what maximum angle $\theta^{\max}$ will the reference line turn in the positive direction? For what length of time will the reference line turn in the positive direction? At what times will the reference line be at (b) $\theta = \frac{1}{2}\theta^{\max}$ and (c) $\theta = -10.5$ rad (consider both positive and negative values of t)? (d) Graph θ versus t, and indicate the answers to (a), (b), and (c) on the graph.

16. A Disk Rotates A disk rotates about its central axis starting from rest and accelerates with constant rotational acceleration. At one time it is rotating at 10 rev/s; 60 revolutions later, its rotational speed is 15 rev/s. Calculate (a) the rotational acceleration, (b) the time required to complete the 60 revolutions, (c) the time required to reach the 10 rev/s rotational speed, and (d) the number of revolutions from rest until the time the disk reaches the 10 rev/s rotational speed.

17. A Flywheel Turns A flywheel turns through 40 rev as it slows from an rotational speed of 1.5 rad/s to a stop. (a) Assuming a constant rotational acceleration, find the time for it to come to rest. (b) What is its rotational acceleration? (c) How much time is required for it to complete the first 20 of the 40 revolutions?

18. A Wheel Rotating A wheel rotating about a fixed axis through its center has a constant rotational acceleration of 4.0 rad/s^2. In a certain 4.0 s interval the wheel turns through an angle of 80 rad. (a) What is the rotational velocity of the wheel at the start of the 4.0 s interval? (b) Assuming that the wheel starts from rest, how long is it in motion at the start of the 4.0 s interval?

SEC. 11-4 ■ RELATING THE TRANSLATIONAL AND ROTATIONAL VARIABLES

19. Record What is the translational acceleration of a point on the rim of a 30-cm-diameter record rotating at a constant rotational speed of $33\frac{1}{3}$ rev/min?

20. Vinyl Record A vinyl record on a turntable rotates at $33\frac{1}{3}$ rev/min. (a) What is its rotational speed in radians per second? What is the translational speed of a point on the record at the needle when the needle is (b) 15 cm and (c) 7.4 cm from the turntable axis?

21. Rotational Speed of Car What is the rotational speed of car traveling at 50 km/h and rounding a circular turn of radius 110 m?

22. Flywheel Rotating A flywheel with a diameter of 1.20 m has a rotational speed of 200 rev/min. (a) What is the rotational speed of the flywheel in radians per second? (b) What is the translational speed of a point on the rim of the flywheel? (c) What constant rotational acceleration (in revolutions per minute-squared) will increase the wheel's rotational speed to 1000 rev/min in 60 s? (d) How many revolutions does the wheel make during that 60 s?

23. Astronaut in Centrifuge An astronaut is being tested in a centrifuge. The centrifuge has a radius of 10 m and, in starting, rotates according to $\theta = (0.30 \text{ rad/s}^2)t^2$, where t is in seconds and θ is in radians. When $t = 5.0$ s, what are the magnitudes of the astronaut's (a) rotational velocity, (b) translational velocity, (c) tangential acceleration, and (d) radial acceleration?

24. Spaceship What are the magnitudes of (a) the rotational velocity, (b) the radial acceleration, and (c) the tangential acceleration of a spaceship taking a circular turn of radius 3220 km at a speed of 29 000 km/h?

25. Speed of Light An early method of measuring the speed of light makes use of a rotating slotted wheel. A beam of light passes through a slot at the outside edge of the wheel, as in Fig. 11-24, travels to a distant mirror, and returns to the wheel just in time to pass through the next slot in the wheel. One such slotted wheel has a radius of 5.0 cm and 500 slots at its edge. Measurements taken when the mirror is $L = 500$ m from the wheel indicate a speed of light of 3.0×10^5 km/s. (a) What is the (constant) rotational speed of the wheel? (b) What is the translational speed of a point on the edge of the wheel?

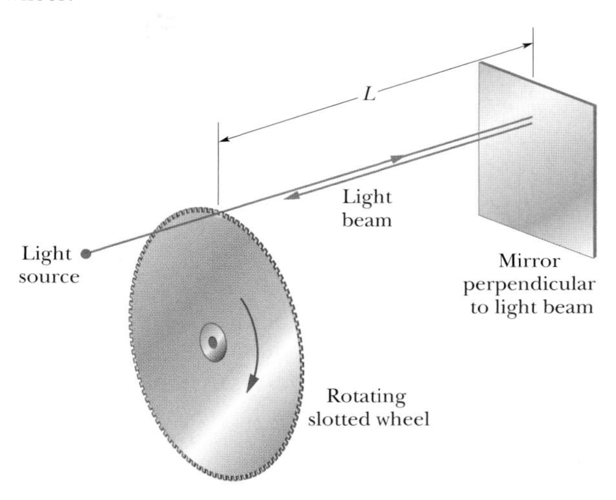

FIGURE 11-24 ■ Problem 25.

26. Steam Engine The flywheel of a steam engine runs with a constant rotational velocity of 150 rev/min. When steam is shut off, the friction of the bearings and of the air stops the wheel in 2.2 h. (a) What is the constant rotational acceleration, in revolutions per minute-squared, of the wheel during the slowdown? (b) How many rotations does the wheel make during the slowdown? (c) At the instant the flywheel is turning at 75 rev/min, what is the tangential component of the translational acceleration of a flywheel particle that is 50 cm from the axis of rotation? (d) What is the magnitude of the net translational acceleration of the particle in (c)?

27. Polar Axis of Earth (a) What is the rotational speed ω about the polar axis of a point on Earth's surface at a latitude of 40° N? (Earth rotates about that axis.) (b) What is the translational speed v of the point? What are (c) ω and (d) v for a point at the equator?

28. Gyroscope A gyroscope flywheel of radius 2.83 cm is accelerated from rest at 14.2 rad/s^2 until its rotational speed is 2760 rev/min. (a) What is the tangential acceleration of a point on the rim of the flywheel during this spin-up process? (b) What is the radial acceleration of this point when the flywheel is spinning at full speed? (c) Through what distance does a point on the rim move during the spin-up?

29. Coupled Wheels In Fig. 11-25, wheel A of radius $r_A = 10$ cm is coupled by belt B to wheel C of radius $r_C = 25$ cm. The rotational speed of wheel A is increased from rest at a constant rate of 1.6 rad/s^2. Find the time for wheel C to reach a rotational speed of 100 rev/min, assuming the belt does not slip. (*Hint*: If the belt does not slip, the translational speeds at the rims of the two wheels must be equal.)

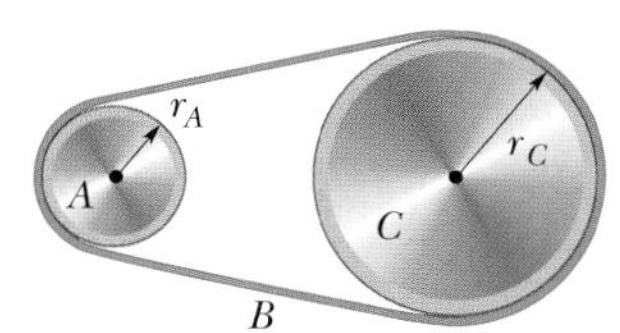

FIGURE 11-25 ■ Problem 29.

30. Fixed Axis An object rotates about a fixed axis, and the rotational position of a reference line on the object is given by $\theta = (0.40 \text{ rad})\, e^{(2.0\text{s}^{-1})t}$. Consider a point on the object that is 4.0 cm from the axis of rotation. At $t = 0$, what are the magnitudes of the point's (a) tangential component of acceleration and (b) radial component of acceleration?

31. Pulsar A pulsar is a rapidly rotating neutron star that emits a radio beam like a lighthouse emits a light beam. We receive a radio pulse for each rotation of the star. The period T of rotation is found by measuring the time between pulses. The pulsar in the Crab nebula (Fig. 11-26) has a period of rotation of $T = 0.033$ s that is increasing at the rate of 1.26×10^{-5} s/y. (a) What is the pulsar's rotational acceleration? (b) If its rotational acceleration is constant, how many years from now will the pulsar stop rotating? (c) The pulsar originated in a supernova explosion seen in the year 1054. What was the intial T of the pulsar? (Assume constant rotational acceleration since the pulsar originated.)

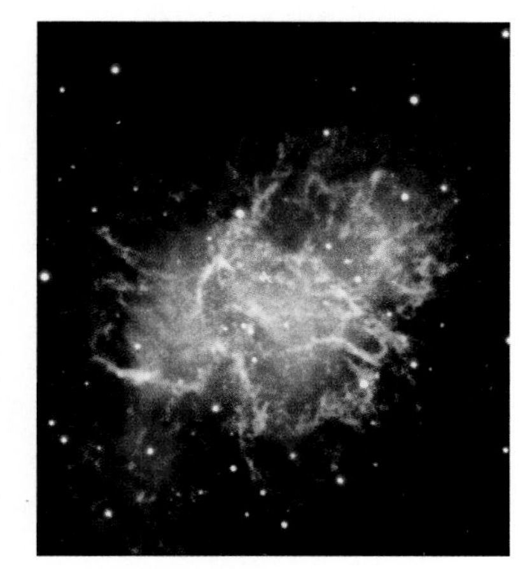

FIGURE 11-26 ■ Problem 31. The Crab nebula resulted from a star whose explosion was seen in 1054. In addition to the gaseous debris seen here, the explosion left a spinning neutron star at its center. The star has a diameter of only 30 km.

32. Turntable Two A record turntable is rotating at $33\frac{1}{3}$ rev/min. A watermelon seed is on the turntable 6.0 cm from the axis of rotation. (a) Calculate the translational acceleration of the seed, assuming that it does not slip. (b) What is the minimum value of the coefficient of static friction, μ^{stat}, between the seed and the turntable if the seed is not to slip? (c) Suppose that the turntable achieves its rotational speed by starting from rest and undergoing a constant rotational acceleration for 0.25 s. Calculate the minimum μ^{stat} required for the seed not to slip during the acceleration period.

SEC. 11-5 ■ KINETIC ENERGY OF ROTATION

33. Rotational Inertia of Wheel Calculate the rotational inertia of a wheel that has a kinetic energy of 24 400 J when rotating at 602 rev/min.

34. Oxygen Molecule The oxygen molecule O_2 has a mass of 5.30×10^{-26} kg and a rotational inertia of 1.94×10^{-46} kg · m^2 about an axis through the center of the line joining the atoms and perpendicular to that line. Suppose the center of mass of an O_2 molecule in a gas has a translational speed of 500 m/s and the molecule has a rotational kinetic energy that is $\frac{2}{3}$ of the translational kinetic energy of its center of mass. What then is the molecule's rotational speed about the center of mass?

SEC. 11-6 ■ CALCULATING ROTATIONAL INERTIA

35. Two Solid Cylinders Two uniform solid cylinders, each rotating about its central (longitudinal) axis, have the same mass of 1.25 kg and rotate with the same rotational speed of 235 rad/s, but they differ in radius. What is the rotational kinetic energy of (a) the smaller cylinder, of radius 0.25 m, and (b) the larger cylinder, of radius 0.75 m?

36. Communications Satellite A communications satellite is a solid cylinder with mass 1210 kg, diameter 1.21 m, and length 1.75 m. Prior to launching from the shuttle cargo bay, it is set spinning at 1.52 rev/s about the cylinder axis (Fig. 11-27). Calculate the satellite's (a) rotational inertia about the rotation axis and (b) rotational kinetic energy.

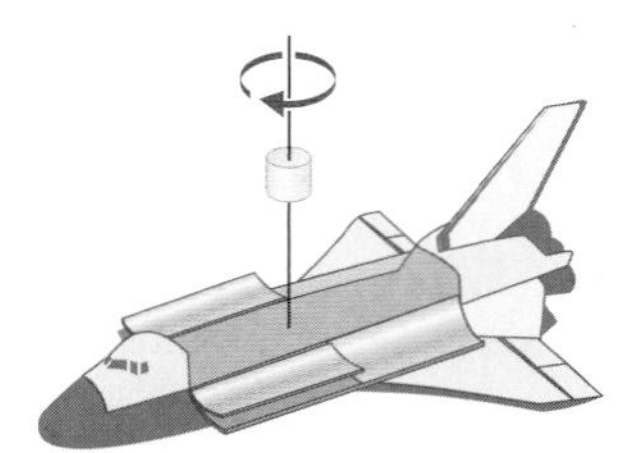

FIGURE 11-27 ■ Problem 36.

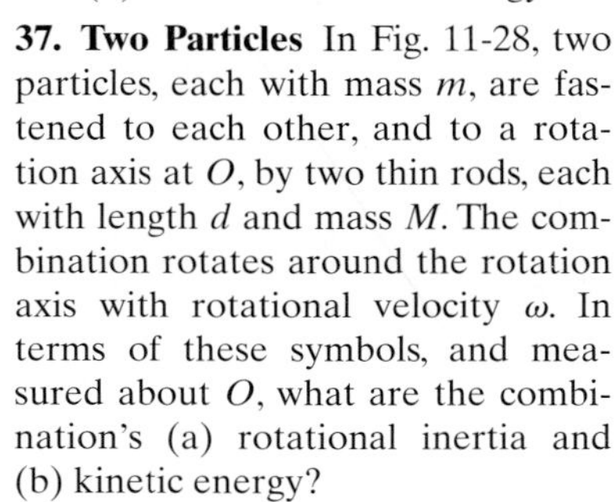

37. Two Particles In Fig. 11-28, two particles, each with mass m, are fastened to each other, and to a rotation axis at O, by two thin rods, each with length d and mass M. The combination rotates around the rotation axis with rotational velocity ω. In terms of these symbols, and measured about O, what are the combination's (a) rotational inertia and (b) kinetic energy?

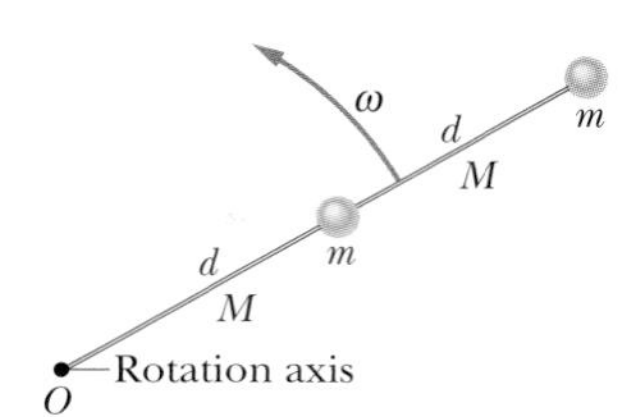

FIGURE 11-28 ■ Problem 37.

38. Helicopter Blades Each of the three helicopter rotor blades shown in Fig. 11-29 is 5.20 m long and has a mass of 240 kg. The rotor is rotating at 350 rev/min. (a) What is the rotational inertia of the rotor assembly about the axis of rotation? (Each blade can be considered to be a thin rod rotated about one end.) (b) What is the total kinetic energy of rotation?

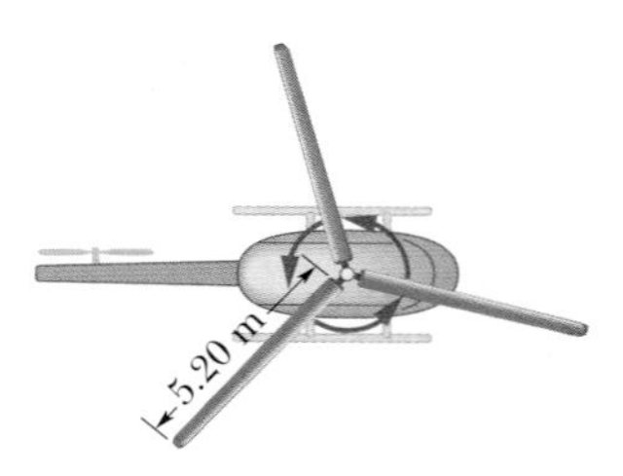

FIGURE 11-29 ■ Problem 38.

39. Meter Stick Calculate the rotational inertia of a meter stick, with mass 0.56 kg, about an axis perpendicular to the stick and located at the 20 cm mark. (Treat the stick as a thin rod.)

40. Four Identical Particles Four identical particles of mass 0.50 kg each are placed at the vertices of a 2.0 m × 2.0 m square and held there by four massless rods, which form the sides of the square. What is the rotational inertia of this rigid body about an axis that (a) passes through the midpoints of opposite sides and lies in the plane of the square, (b) passes through the midpoint of one of the sides and is perpendicular to the plane of the square, and (c) lies in the plane of the square and passes through two diagonally opposite particles?

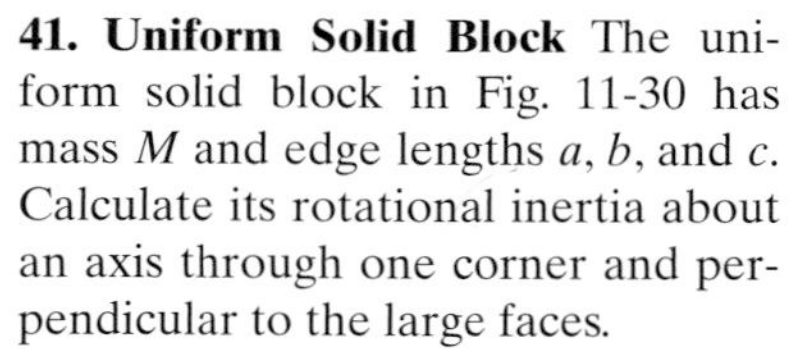

41. Uniform Solid Block The uniform solid block in Fig. 11-30 has mass M and edge lengths a, b, and c. Calculate its rotational inertia about an axis through one corner and perpendicular to the large faces.

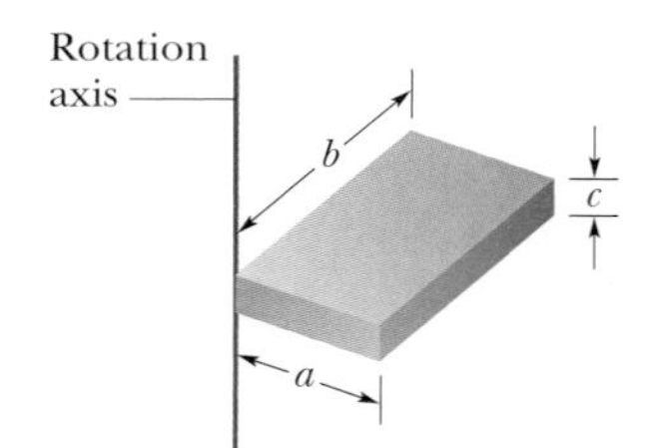

FIGURE 11-30 ■ Problem 41.

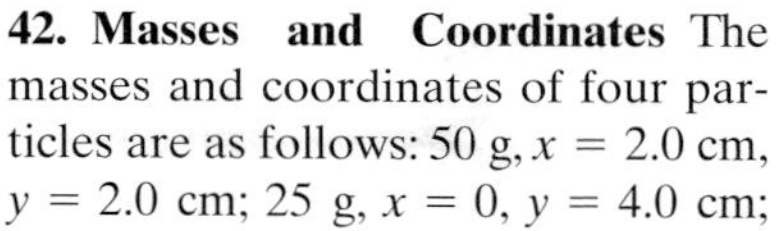

42. Masses and Coordinates The masses and coordinates of four particles are as follows: 50 g, $x = 2.0$ cm, $y = 2.0$ cm; 25 g, $x = 0$, $y = 4.0$ cm; 25 g, $x = -3.0$ cm, $y = -3.0$ cm; 30 g, $x = -2.0$ cm, $y = 4.0$ cm. What are the rotational inertias of this collection about the (a) x, (b) y, and (c) z axes? (d) Suppose the answers to (a) and (b) are A and B, respectively. Then what is the answer to *(c)* in terms of A and B?

43. Solid Cylinder—Thin Hoop (a) Show that the rotational inertia of a solid cylinder of mass M and radius R about its central axis is equal to the rotational inertia of a thin hoop of mass M and radius $R/\sqrt{2}$ about its central axis. (b) Show that the rotational inertia I of any given body of mass M about any given axis is equal to the rotational inertia of an *equivalent hoop* about that axis, if the hoop has the same mass M and a radius k given by

$$k = \sqrt{\frac{I}{M}}.$$

The radius k of the equivalent hoop is called the *radius of gyration* of the given body.

44. Delivery Trucks Delivery trucks that operate by making use of energy stored in a rotating flywheel have been used in Europe. The trucks are charged by using an electric motor to get the flywheel up to its top speed of 200π rad/s. One such flywheel is a solid, uniform cylinder with a mass of 500 kg and a radius of 1.0 m. (a) What is the kinetic energy of the flywheel after charging? (b) If the truck operates with an average power requirement of 8.0 kW, for how many minutes can it operate between chargings?

SEC. 11-7 ■ TORQUE

45. Small Ball A small ball of mass 0.75 kg is attached to one end of a 1.25-m-long massless rod, and the other end of the rod is hung from a pivot. When the resulting pendulum is 30° from the vertical, what is the magnitude of the torque about the pivot?

46. Bicycle Pedal Arm The length of a bicycle pedal arm is 0.152 m, and a downward force of 111 N is applied to the pedal by the rider's foot. What is the magnitude of the torque about the pedal arm's pivot point when the arm makes an angle of (a) 30°, (b) 90°, and (c) 180° with the vertical?

47. Pivoted at *O* The body in Fig. 11-31 is pivoted at O, and two forces act on it as shown. (a) Find an expression for the net torque on the body about the pivot. (b) If $r_A = 1.30$ m, $r_B = 2.15$ m, $F_A = 4.20$ N, $F_B = 4.90$ N, $\theta_A = 75.0°$, and $\theta_B = 60.0°$, what is the net torque about the pivot?

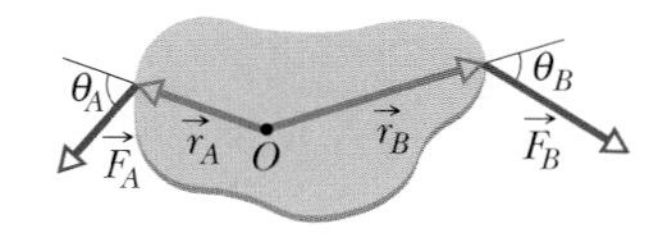

FIGURE 11-31 ■ Problem 47.

48. Three Force The body in Fig. 11-32 is pivoted at O. Three forces act on it in the directions shown: $F_A = 10$ N at point A, 8.0 m from O; $F_B = 16$ N at point B, 4.0 m from O; and $F_C = 19$ N at point C, 3.0 m from O. What is the net torque about O?

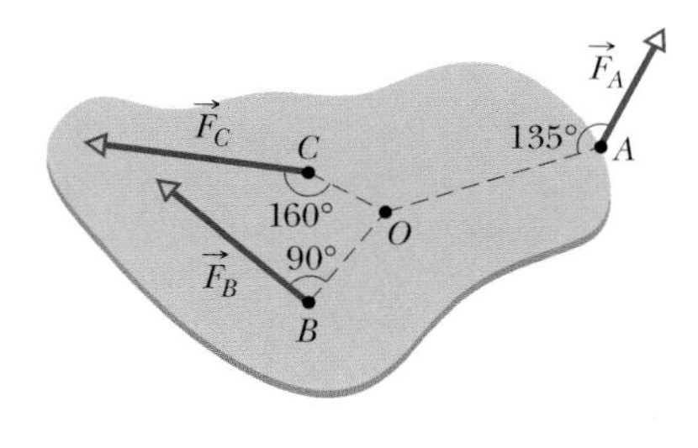

FIGURE 11-32 ■ Problem 48.

SEC. 11-8 ■ NEWTON'S SECOND LAW FOR ROTATION

49. Diver's Launch During the launch from a board, a diver's rotational speed about her center of mass changes from zero to 6.20 rad/s in 220 ms. Her rotational inertia about her center of mass is 12.0 kg · m². During the launch, what are the magnitudes of (a) her average rotational acceleration and (b) the average external torque on her from the board?

50. Torque on a Certain Wheel A torque of 32.0 N · m on a certain wheel causes a rotational acceleration of 25.0 rad/s². What is the wheel's rotational inertia?

51. Thin Spherical Shell A thin spherical shell has a radius of 1.90 m. An applied torque of 960 N · m gives the shell a rotational acceleration of 6.20 rad/s² about an axis through the center of the shell. What are (a) the rotational inertia of the shell about that axis and (b) the mass of the shell?

52. Cylinder Having Mass In Fig. 11-33, a cylinder having a mass of 2.0 kg can rotate about its central axis through point O. Forces are applied as shown: $F_A = 6.0$ N, $F_B = 4.0$ N, $F_C = 2.0$ N, and $F_D = 5.0$ N. Also, $R_1 = 5.0$ cm and $R_2 = 12$ cm. Find the magnitude and direction of the rotational acceleration of the cylinder. (During the rotation, the forces maintain their same angles relative to the cylinder.)

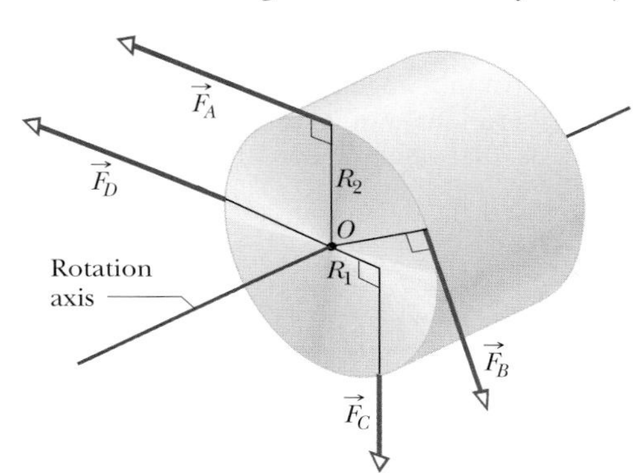

FIGURE 11-33 ■ Problem 52.

53. Lawrence Livermore Door Figure 11-34 shows the massive shield door at a neutron test facility at Lawrence Livermore Laboratory; this is the world's heaviest hinged door. The door has a mass of 44,000 kg, a rotational inertia about a vertical axis through its huge hinges of 8.7×10^4 kg · m², and a (front) face width of 2.4 m. Neglecting friction, what steady force, applied at its outer edge and perpendicular to the plane of the door can move it from rest through an angle of 90° in 30 s?

FIGURE 11-34 ■ Problem 53.

54. Wheel on a Frictionless Axis A wheel of radius 0.20 m is mounted on a frictionless horizontal axis. The rotational inertia of the wheel about the axis is 0.050 kg · m². A massless cord wrapped around the wheel is attached to a 2.0 kg block that slides on a horizontal frictionless surface. If a horizontal force of magnitude $P =$ 3.0 N is applied to the block as shown in Fig. 11-35, what is the magnitude of the rotational acceleration of the wheel? Assume that the string does not slip on the wheel.

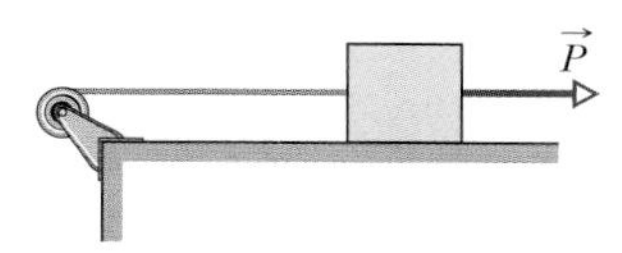

FIGURE 11-35 ■ Problem 54.

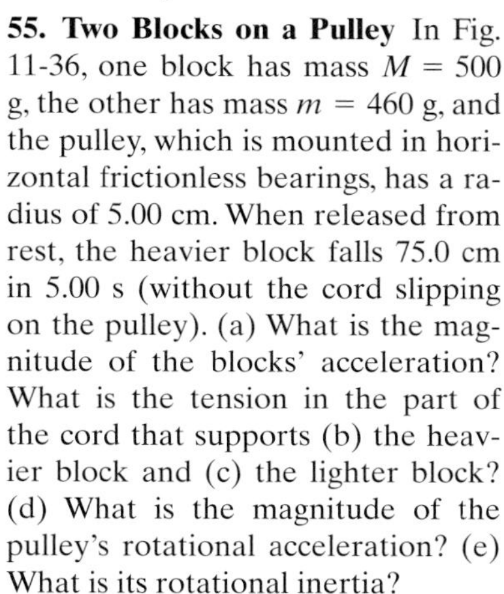

55. Two Blocks on a Pulley In Fig. 11-36, one block has mass $M = 500$ g, the other has mass $m = 460$ g, and the pulley, which is mounted in horizontal frictionless bearings, has a radius of 5.00 cm. When released from rest, the heavier block falls 75.0 cm in 5.00 s (without the cord slipping on the pulley). (a) What is the magnitude of the blocks' acceleration? What is the tension in the part of the cord that supports (b) the heavier block and (c) the lighter block? (d) What is the magnitude of the pulley's rotational acceleration? (e) What is its rotational inertia?

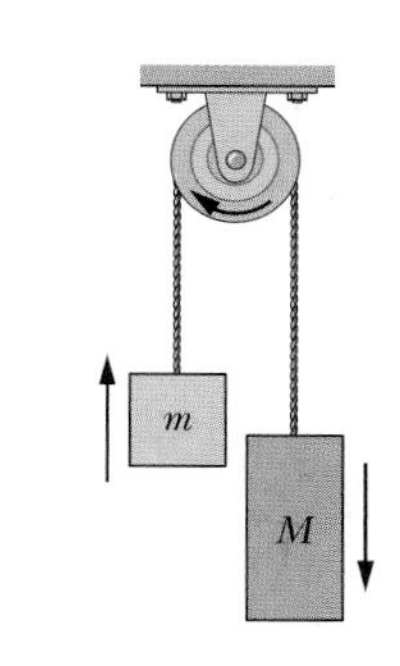

FIGURE 11-36 ■ Problem 55.

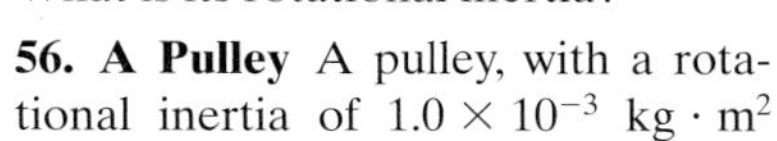

56. A Pulley A pulley, with a rotational inertia of 1.0×10^{-3} kg · m² about its axle and a radius of 10 cm, is acted on by a force applied tangentially at its rim. The force magnitude varies in time as $F = (0.50 \text{ N/s})t + (0.30 \text{ N/s}^2)t^2$, with F in newtons and t in seconds. The pulley is initially at rest. At $t = 3.0$ s what are (a) its rotational acceleration and (b) its rotational speed?

57. Two Blocks on a Rod Figure 11-37 shows two blocks, each of mass m, suspended from the ends of a rigid massless rod of length $L_1 + L_2$, with $L_1 = 20$ cm and $L_2 = 80$ cm. The rod is held horizontally on the fulcrum and then released. What are the magnitudes of the initial accelerations of (a) the block closer to the fulcrum and (b) the other block?

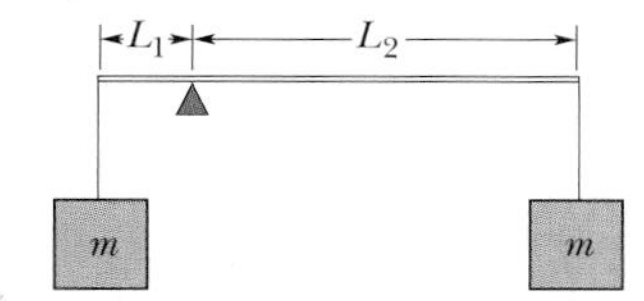

FIGURE 11-37 ■ Problem 57.

SEC. 11-9 ■ WORK AND ROTATIONAL KINETIC ENERGY

58. Speed of the Block (a) If $R = 12$ cm, $M = 400$ g, and $m = 50$ g in Fig. 11-20, find the speed of the block after it has descended 50 cm starting from rest. Solve the problem using energy conservation principles. (b) Repeat (a) with $R = 5.0$ cm.

59. Crankshaft An automobile crankshaft transfers energy from the engine to the axle at the rate of 100 hp (=74.6 kW) when rotating at a speed of 1800 rev/min. What torque (in newton-meters) does the crankshaft deliver?

60. Thin Hoop A 32.0 kg wheel, essentially a thin hoop with radius 1.20 m, is rotating at 280 rev/min. It must be brought to a stop in 15.0 s. (a) How much work must be done to stop it? (b) What is the required average power?

61. Thin Rod of Length L A thin rod of length L and mass m is suspended freely from one end. It is pulled to one side and then allowed to swing like a pendulum, passing through its lowest position with rotational speed ω. In terms of these symbols and g, and neglecting friction and air resistance, find (a) the rod's kinetic energy at its lowest position and (b) how far above that position the center of mass rises.

62. Accelerating the Earth Calculate (a) the torque, (b) the energy, and (c) the average power required to accelerate Earth in 1 day from rest to its present rotational speed about its axis.

63. Meter Stick Held Vertically A meter stick is held vertically with one end on the floor and is then allowed to fall. Find the speed of the other end when it hits the floor, assuming that the end on the floor does not slip. (*Hint*: Consider the stick to be a thin rod and use the conservation of energy principle.)

64. Cylinder Rotates about Horizontal A uniform cylinder of radius 10 cm and mass 20 kg is mounted so as to rotate freely about a horizontal axis that is parallel to and 5.0 cm from the central longitudinal axis of the cylinder. (a) What is the rotational inertia of the cylinder about the axis of rotation? (b) If the cylinder is released from rest with its central longitudinal axis at the same height as the axis about which the cylinder rotates, what is the rotational speed of the cylinder as it passes through its lowest position?

65. The Letter *H* A rigid body is made of three identical thin rods, each with length L, fastened together in the form of a letter H (Fig. 11-38). The body is free to rotate about a horizontal axis that runs along the length of one of the legs of the H. The body is allowed to fall from rest from a position in which the plane of the H is horizontal. What is the rotational speed of the body when the plane of the H is vertical?

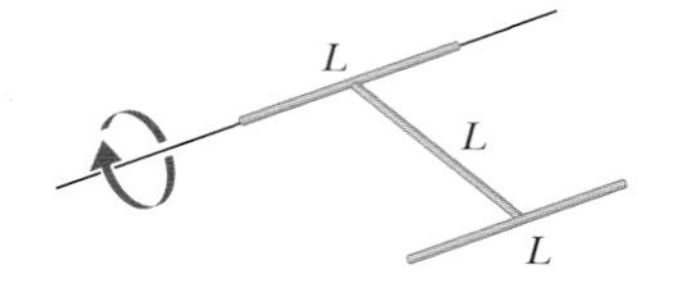

FIGURE 11-38 ■ Problem 65.

66. Uniform Spherical Shell A uniform spherical shell of mass M and radius R rotates about a vertical axis on frictionless bearings (Fig. 11-39). A massless cord passes around the equator of the shell, over a pulley of rotational inertia I and radius r, and is attached to a small object of mass m. There is no friction on the pulley's axle; the cord does not slip on the pulley. What is the speed of the object after it falls a distance h from rest? Use energy considerations.

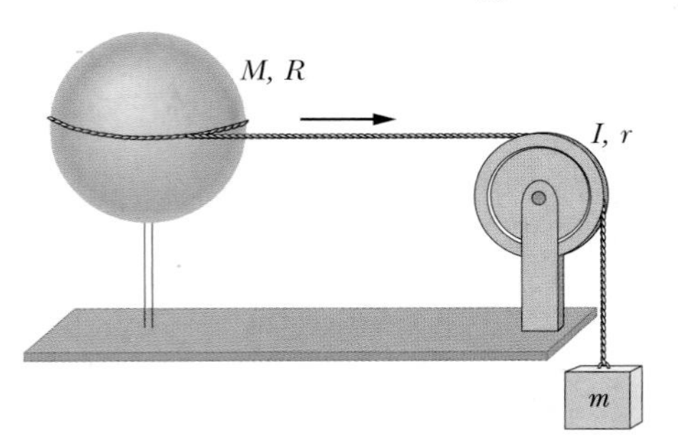

FIGURE 11-39 ■ Problem 66.

67. Tall Cylinder-Shaped Chimney A tall, cylinder-shaped chimney falls over when its base is ruptured. Treat the chimney as a thin rod of length H, and let θ be the angle the chimney makes with the vertical. In terms of these symbols and g, express the following: (a) the rotational speed of the chimney, (b) the radial acceleration of the chimney's top, and (c) the tangential acceleration of the top. (*Hint*: Use energy considerations, not a torque. In part (c) recall that $\alpha = d\omega/dt$.) (d) At what angle θ does the tangential acceleration equal g?

Additional Problems

68. Judo In a judo foot-sweep move, you sweep your opponent's left foot out from under him while pulling on his gi (uniform) toward that side. As a result, your opponent rotates around his right foot and onto the mat. Figure 11-40 shows a simplified diagram of your opponent as you face him, with his left foot swept out. The rotational axis is through point O. The gravitational force $\vec{F}^{\text{grav}}$ on him effectively acts at his center of mass, which is a horizontal distance of $d = 28$ cm from point O. His mass is 70 kg, and his rotational inertia about point O is 65 kg · m². What is the magnitude of his initial rotational acceleration about point O if your pull $\vec{F}^{\text{app}}$ on his gi is (a) negligible and (b) horizontal with a magnitude of 300 N and applied at height $h = 1.4$ m?

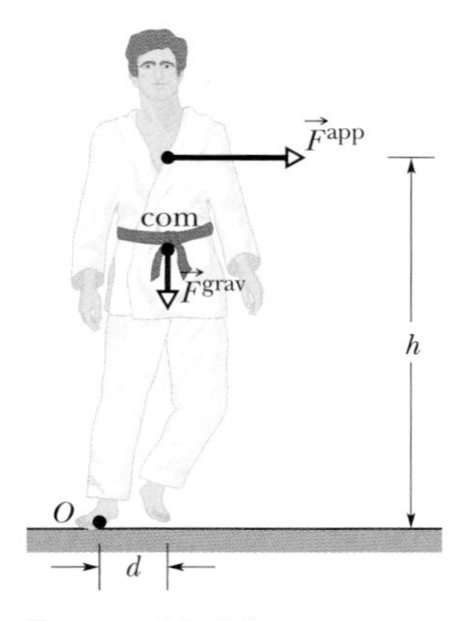

FIGURE 11-40 ■ Problem 68.

69. Disk Rod Figure 11-41 shows an arrangement of 15 identical disks that have been glued together in a rod-like shape of length L and (total) mass M. The arrangement can rotate about a perpendicular axis through its central disk at point O. (a) What is the rotational inertia of the arrangement about that axis? (b) If we approximated the arrangement as being a uniform rod of mass M and length L, what percentage error would we make in using the formula in Table 11-2*e* to calculate the rotational inertia?

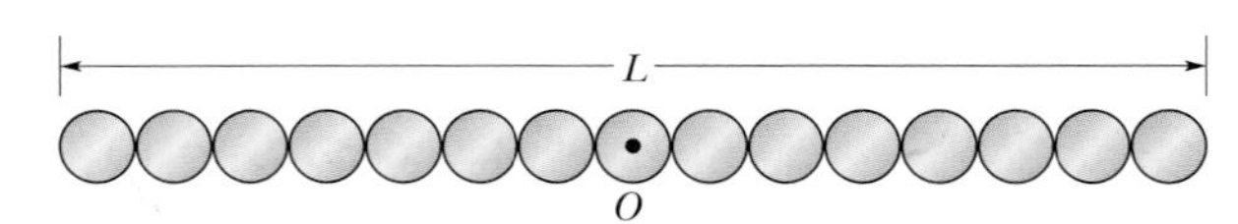

FIGURE 11-41 ■ Problem 69.

70. Summing Up to Estimate Rotational Inertia. By performing an integration it can be shown that the general equation for the rotational inertia of a thin rod of length L and mass M about an axis through one end of the rod that is perpendicular to its length is given by

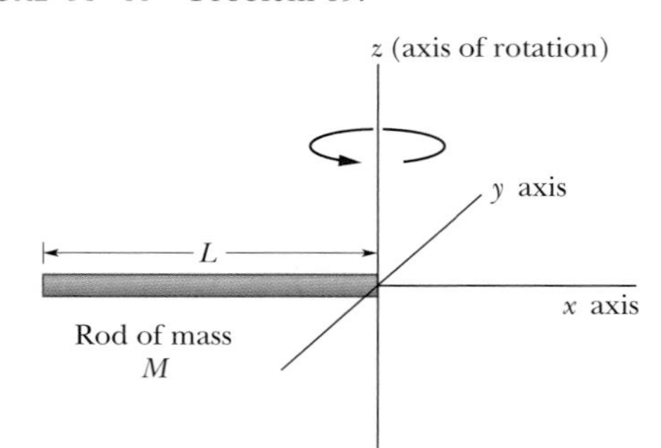

FIGURE 11-42 ■ Problem 70.

$$I = \tfrac{1}{3}ML^2.$$

Consider a rod of length $L = 0.50$ m that has a mass of $M = 1.2$ kg rotating as shown in Fig. 11-42.

(a) Calculate the theoretical value of the rotational inertia.

(b) Estimate the rotational inertia of the rod by breaking it into 50 small point masses each having a mass of $M/50$, with the first point mass being 0.01 m from the axis of rotation, the second mass being 0.02 m from the axis of rotation, and so on. Use a spreadsheet to do your estimated calculations of the rotational inertia of the rod. **(c)** Compare the theoretically calculated value with the estimated value. Are they similar?

71. Calculation of Torque (Angle Method) Before the finger holes are drilled, a uniform bowling ball of radius 0.120 m has a net gravitational force of 65 N exerted on it by the Earth. Assume that this force acts through the center of mass of the bowling ball. Determine the magnitude and direction of this net force and the resulting torques on the bowling ball about four axes that are *perpendicular* to the plane of the paper passing through points A, B, C, and D as shown in Fig. 11-43. *Hint*: You can use the $|\vec{\tau}| = |\vec{r}||\vec{F}|\sin\theta$ form of the torque equation.

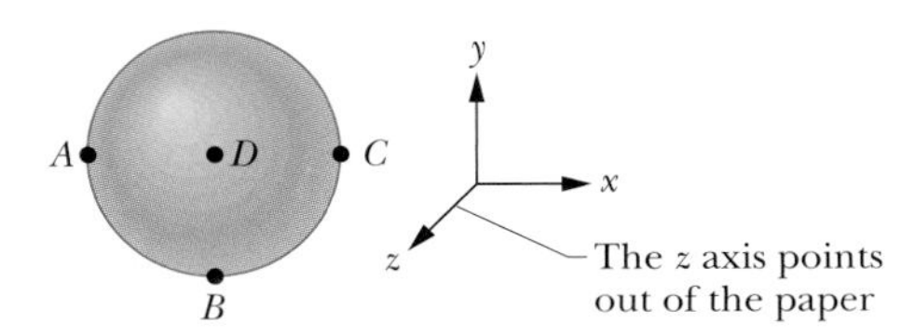

FIGURE 11-43 ■ Problem 71.

72. Simple Yo-Yo Consider a "yo-yo" consisting of a disk fixed to an axle that has two strings wrapped around it. As the axle rolls off the strings, the disk and the axle fall as shown in Fig. 11-44.

(a) If the disk has fallen through a vertical distance of $d = 30$ cm and the radius of the string and axle is given by $r = 50$ mm, how many revolutions has the disk gone through?
(b) If the disk is rotating faster and faster with a constant rotational acceleration and takes 25 s to fall through the distance d from rest, what is the magnitude of its rotational acceleration α?
(c) What is the magnitude of its rotational velocity ω after the 25 seconds have elapsed? *Hint*: Use the rotational kinematic equations.

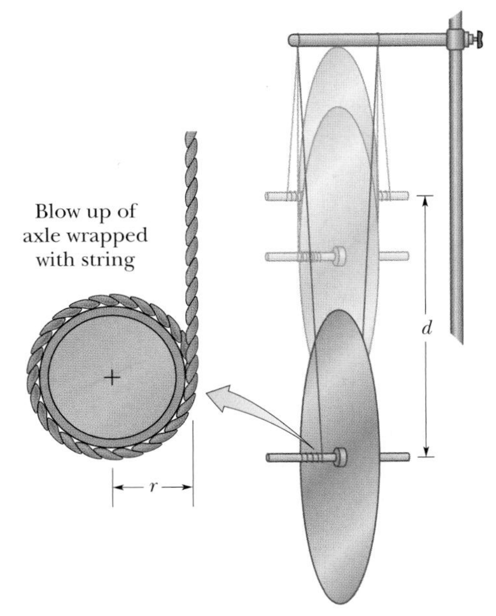

FIGURE 11-44 ■ Problem 72.

73. Buying Wire Wire is often delivered wrapped on a large cylindrical spool. Suppose such a spool is supported by resting on a horizontal metal rod pushed through a hole that runs through the center of the spool. A worker is pulling some wire off the spool by exerting a force on it as shown in Fig. 11-45. (If you have ever bought wire in a hardware store, this is the way they usually store and dispense it.)

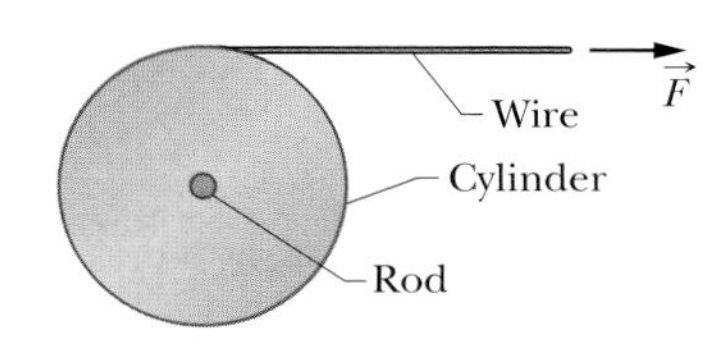

FIGURE 11-45 ■ Problem 73.

Suppose the spool rotates on the rod essentially without friction. The spool is approximately a uniform cylinder with a mass of 50 kg and a radius of 30 cm. The worker pulls on the wire for 2 s with a force of 30 N. At the end of the 2 s he immediately clamps on a brake that very quickly stops the spool's rotation. Just before he puts the brake on, how fast is the spool rotating? How much wire does he pull off the spool?

74. Fly on an LP An old-fashioned record player spins a disk at approximately a constant angular velocity, ω. A fly of mass m is sitting on the disk as it turns, at a point a distance R from the center.

(a) What force keeps the fly from sliding off the rotating disk? What direction does the force point? How big is it? For the last question, express your answer in terms of the symbols given in the description above.

(b) If the fly has a mass of 0.5 grams, is sitting 10 cm from the center of the disk, and the disk is turning at a rate of 33 rev/min, what is the coefficient of friction?

75. Rotational Inertia and Rotational Acceleration A small spool of radius r_s and a large Lucite disk of radius r_d are connected by an axle that is free to rotate in an almost frictionless manner inside of a bearing as shown in Fig. 11-46. A string is wrapped around the spool and a mass, m, which is attached to the string, is allowed to fall.

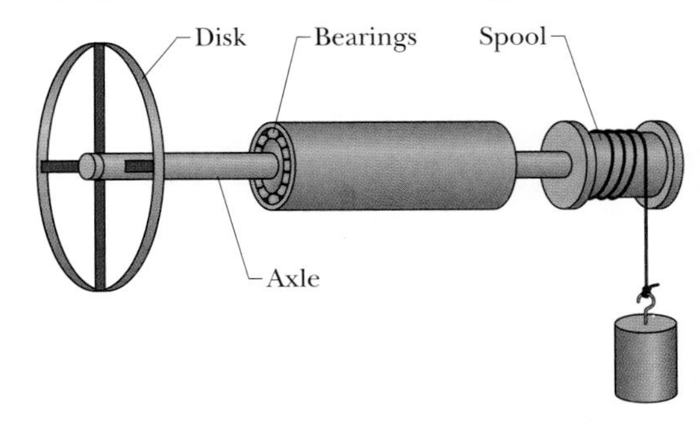

FIGURE 11-46 ■ Problem 75.

(a) Draw a free-body diagram showing the forces on the falling mass, m, in terms of m, g, and $\vec{F}^{\text{tension}}$.
(b) If the magnitude of the translational acceleration of the mass is measured to be a, what is the equation that should be used to calculate, $|\vec{F}^{\text{tension}}|$ in the string? In other words what equation relates m, g, $|\vec{F}^{\text{tension}}|$ and a? *Note*: In a system where $|\vec{F}^{\text{tension}}| - mg = ma$, if $a \ll g$ then $|\vec{F}^{\text{tension}}| \approx mg$.
(c) What is the magnitude of torque, τ, on the spool–axle–disk system as a result of the tension in the string, $|\vec{F}^{\text{tension}}|$, acting on the spool?
(d) What is the magnitude of the rotational acceleration, α, of the rotating system as a function of the translational acceleration, a, of the falling mass, and the radius, r_s, of the spool?

(e) The rotational inertias of the axle and the spool are so small compared to the rotational inertia of the disk the they can be neglected. If only the rotational inertia, I_d, of the large disk of radius r_d is considered, what is the equation that can be used to predict the value of I_d as a function of the torque on the system, τ, and the magnitude of the rotational acceleration, α, of the disk?

(f) What is the theoretical value of the rotational inertia, I_d, of a disk of mass M and radius r_d in terms of M_d and r_d?

76. Round and Round Little Jay is enjoying his first ride on a merry-go-round. (He is riding a stationary horse rather than one that goes up and down.) A schematic view of the merry-go-round as seen from above is shown in Fig. 11-47*a* with a convenient coordinate system. A bit after the merry-go-round has started and is going around uniformly, we start our clock. Little Jay's position and velocity at time $t_1 = 0$ are shown as a dot and arrow. At $t_1 = 0$ is the net force acting on Jay equal to zero? If it is, write "Yes" and give a reason why you think so. If it isn't, write "No" and specify the type of force and the object responsible for exerting it.

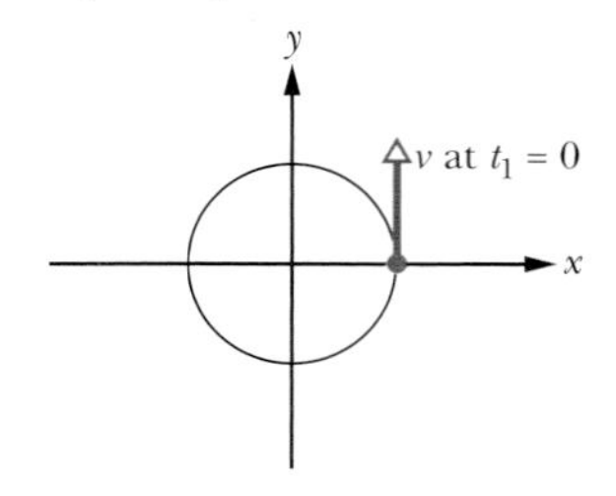

FIGURE 11-47*a* ■ Problem 76.

For the next six parts, specify which of the graphs shown in Fig. 11-47*b* could represent the indicated variable for Jay's motion. If none of the graphs work, write "N."

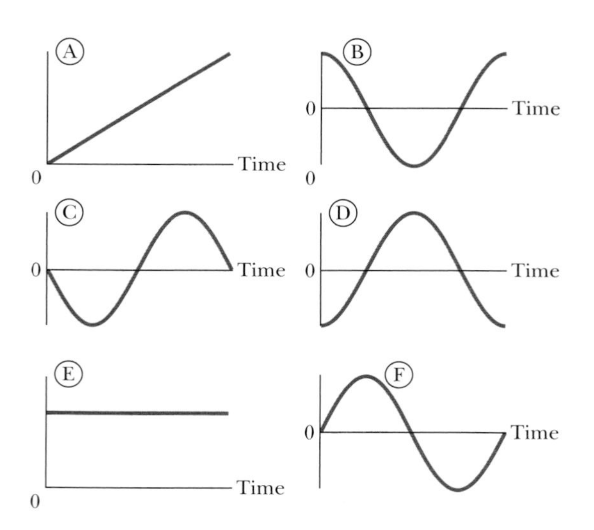

FIGURE 11-47*b* ■ Problem 76.

(a) The x-component of Jay's velocity
(b) The angle Jay's position vector makes with the x axis
(c) The y-component of the force keeping Jay moving in a circle
(d) Jay's rotational velocity
(e) Jay's translational speed
(f) The x-component of Jay's position

77. Comparing Rotational Inertias If all three of the objects shown in Fig. 11-48 have the same radius and mass, which one has the most rotational inertia about its indicated axis of rotation? Which one has the least rotational inertia? Explain the reasons for your answer. *Hint*: Consider which one has its mass distributed farthest from the axis of rotation.

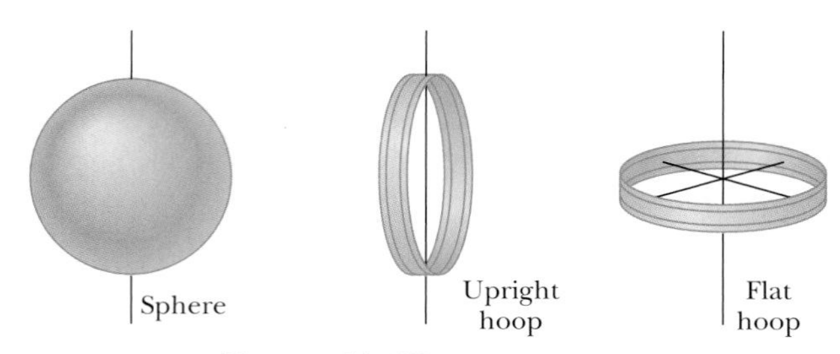

FIGURE 11-48 ■ Problem 77.

78. Rotational Vs. Translational Energy of Motion

(a) Describe how a solid ball can move so that

i. Its total kinetic energy is just the energy of motion of its center of mass

ii. Its total kinetic energy is the energy of its motion relative to its center of mass

(b) Two bowling balls are moving down a bowling alley so that their centers of mass have the same velocity, but one just slides down the alley, while the other rolls down the alley. Which ball has more energy? Explain your reasoning.

79. Closing the Door A student is in her dorm room, sitting on her bed doing her physics homework. The door to her room is open. Suddenly she hears the voice of her ex-boyfriend down the hall, talking to the girl in the room next door. She wants to shut the door quickly, so she throws a superball (which she keeps next to her bed for this purpose) against the door. The ball follows the path shown in Fig. 11-49. It hits the door squarely and bounces straight back. Does the ball's effectiveness in closing the door depend on where on the door the ball hits? If it does, where should it hit to be most effective? Explain your reasoning.

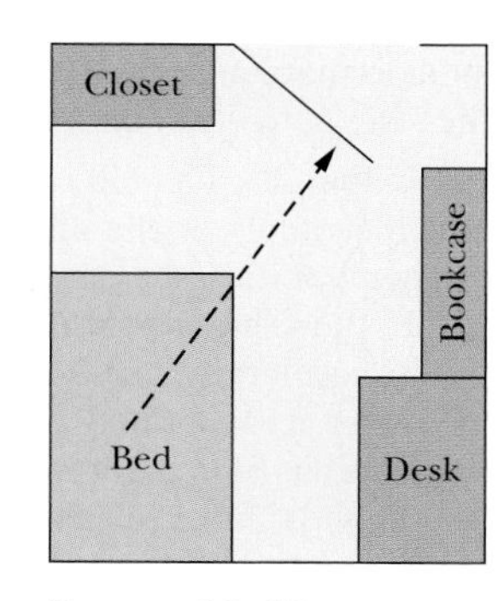

FIGURE 11-49 ■ Problem 79.

80. Cleaning Up with Flywheels One proposal for reducing air pollution is the flywheel-driven automobile. Instead of an engine, the car contains a large steel disk, or flywheel, which is mounted to rotate about a vertical axis. It is set spinning at a high rotational velocity in the early morning using electric power (from plugging it into the wall). If the car is to be about the same size as a typical car today, estimate the amount of energy that could be stored in a rotating steel flywheel that fits under the car's hood. You may find some of the following numbers useful:

- density of steel = 6 g/cm^3
- mass of a typical car = 1000 kg
- maximum speed of flywheel = 1000 revolutions/minute
- fraction of carbon monoxide pollution produced by vehicles = 60%

81. Spinning with the Earth Because the earth is spinning about its axis once a day, you are also spinning about the earth's axis once a day. Estimate the rotational kinetic energy you have as a result of this motion.

82. Keep the Dust Off Your Hard Drive! A current-generation hard drive in a computer spins at a rate of 7000 rpm. The disk in the drive is about the same size as a floppy disk. Estimate the coefficient of friction that would permit the frictional force to keep a speck of dust sitting on the disk from sliding off. Assume that the speck has a mass of 50 mg. Discuss the implications of your result.

83. Kinetic Energy of a Bicycle Wheel Estimate the rotational kinetic energy of a bicycle wheel as the bicycle it is a part of is being ridden down the street.

84. Ferris Wheel Use a video analysis software program to analyze the motion of a Ferris wheel. If you have access to the VideoPoint movie collection, use the movie with filename HRSY001. This is a movie of the Cyclops Ferris wheel at Hershey Park. A sample frame is shown in Fig. 11-50.

FIGURE 11-50 ■ Problem 84.

(a) What is the nature of the rotational speed of a Ferris wheel as a function of time? Is it increasing, decreasing, or remaining constant? Cite the evidence for your answer.
(b) At $t = 0.1000$ s, what is the translational speed of a point on the inner circle?
(c) At $t = 0.1000$ s, what is the translational speed of a point on the outer circle?

85. Falling Mass Turns Disk Use a video analysis software program to analyze the motion of a disk that is attached to a spool. A falling mass attached by a string to the spool causes the spool and disk to undergo a rotational acceleration. If you have access to the VideoPoint movie collection, use the movie with filename DSON014 and analyze the first 12 frames.

(a) Is the acceleration of the disk constant? Explain what you did and cite the evidence for your conclusions.
(b) Describe the nature of the rotational acceleration. Does it increase, decrease, or stay the same? If you concluded that the rotational acceleration is constant, then determine what its value is in rad/s^2. Explain how you arrived at your conclusions. Show relevant data and graphs.
(c) What is the equation that describes the angle through which the disk has moved as a function of time? Explain how you determined this equation.
(d) What is the equation that describes the rotational velocity of the disk as a function of time? Explain how you derived this equation.

12 Complex Rotations

Image courtesy Ringling Brothers and Barnum & Bailey® THE GREATEST SHOW ON EARTH

In 1897, a European "aerialist" made the first triple somersault during the flight from a swinging trapeze to the hands of a partner. For the next 85 years aerialists attempted to complete a *quadruple* somersault, but not until 1982 was it done before an audience. Miguel Vazquez of the Ringling Bros. and Barnum & Bailey Circus rotated his body in four complete circles in midair before his brother Juan caught him. Both were stunned by their success.

Why was the feat so difficult, and what feature of physics made it (finally) possible?

The answer is in this chapter.

12-1 About Complex Rotations

This chapter presents an extension of the concepts in rotational motion that we began discussing in the last chapter. In Chapter 11, we studied the relationships between translational and rotational quantities like position and angle, translational velocity and rotational velocity, translational acceleration and rotational acceleration, and force and torque. We limited our discussion to the rotation of rigid bodies with a constant rotational inertia I (constant mass distribution) about a fixed axis. Furthermore, because the rotation took place about a fixed axis, we were able to treat rotational quantities as one-dimensional components along the axis of rotation. In this chapter, we extend our study to more complex motion in which either the axis of rotation does not stay fixed in space, or the rotational inertia of the rotating body changes over time.

There are several distinct types of complex motion of interest to scientists and engineers that we will consider:

1. Rotations about an axis of rotation that moves but does not change direction. This type of motion is a combination of rotational motion and translational motion. Examples include the motion of yo-yos, wheels, and bowling balls (Fig. 12-1).
2. Rotations about an axis of rotation which changes direction. The axes of rotation for Frisbees and boomerangs change direction as a result of interactions both with air molecules and the Earth's gravity. The axis of a spinning top changes direction as it loses energy. A simple example of this type of motion is someone flipping the axis of a spinning wheel (Fig. 12-2).
3. Rotating objects that have fixed axes of rotation but undergo changes in rotational inertia while spinning. For example, skaters who pull in their arms are reducing their rotational inertia (Fig. 12-3). Stellar matter does the same thing when collapsing into a neutron star.

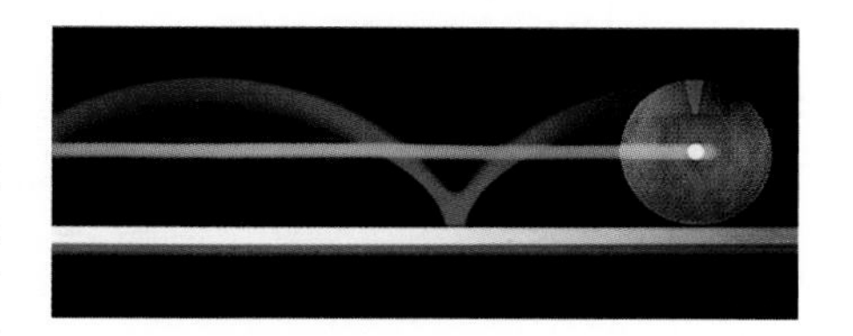

FIGURE 12-1 ■ A time exposure photograph of a rolling disk. Small lights have been attached to the disk, one at its center and one at its edge. The latter traces out a curve called a *cycloid*.

FIGURE 12-2 ■ A student applies a torque to the axis of a rotating bicycle wheel in order to change the direction of the wheel's axis of rotation. (Photo courtesy of PASCO scientific.)

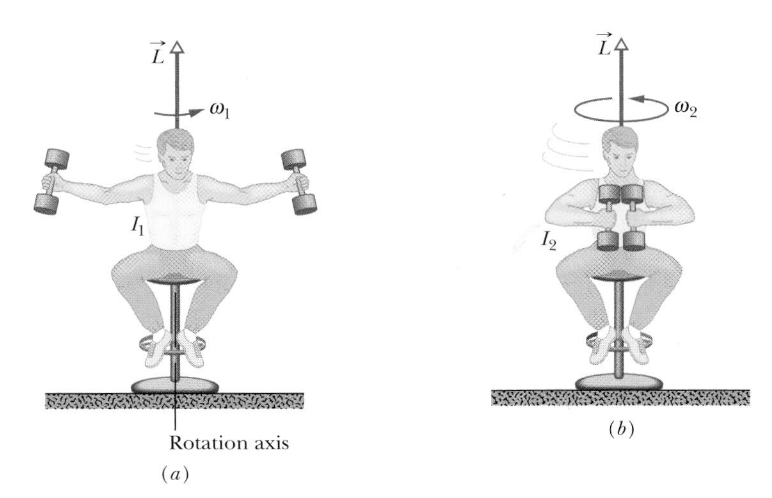

FIGURE 12-3 ■ A student reduces his rotational inertia while rotating by pulling in his arms.

We start our study of complex rotations by considering the kinetic energy associated with combined translational and rotational motions. We will then apply Newton's Second Law in both a translational form and a rotational form to the motion of a yo-yo traveling up and down a string. This will provide us with a model of how one might find an expression for the translational acceleration in a complex motion. Before moving on to the task of analyzing motions involving changes in axis direction and rotational inertia, we will develop the mathematical tools we will require in order to treat torque and other rotational quantities as three-dimensional vectors.

Finally, because the law of conservation of translational momentum in cases of zero net force is such a powerful analysis tool, we will explore the concept of rotational momentum as a rotational corollary of translational momentum. We can then recast Newton's Second Law of rotation so it relates changes in rotational momentum to the net applied torque, showing that rotational momentum is conserved when the net torque acting on the system is zero. In the last sections of the chapter, we bring all of this together, using Newton's laws in their rotational form, the new concept of conservation of rotational momentum, and the vector mathematics we developed to explain the complex motions of aerialists, divers, spacecraft navigation, and neutron star rotation.

12-2 Combining Translations with Simple Rotations

We begin our discussion of complex rotations by considering motions that are combinations of translation and rotation. For example, when a bicycle moves along a straight track, the center of each wheel moves forward with a translational speed v_{com}. At any given instant if the wheel is rolling without slipping, the top point on the wheel is moving forward at twice v_{com} relative to the track, and the bottom point on the wheel is not moving. However, every point on the wheel also rotates about the center with rotational speed ω. Hence, the rolling motion of a wheel is a combination of purely translational and purely rotational motions.

A yo-yo is another example of this type of motion. As a yo-yo rolls down a string, it undergoes rotational motion. However, it also undergoes translational motion as it falls. One way to view such motion is as rotation about an axis that is moving (translating) downward. We will more carefully consider this type of motion by analyzing the forces and torques at work in the case of the falling yo-yo. But first, let's consider energy issues involved in motions that combine rotation with translation.

Energy Considerations

If a yo-yo rolls down its string for a distance h, the yo-yo-Earth system loses potential energy in the amount of mgh but gains kinetic energy in both translational ($\frac{1}{2}mv_{com}^2$) and rotational ($\frac{1}{2}I_{com}\omega^2$) forms. As the yo-yo climbs back up, the system loses kinetic energy and regains potential energy.

An object that undergoes combined rotational and translation motion has two types of kinetic energy: a rotational kinetic energy ($\frac{1}{2}I_{com}\omega^2$) due to its rotation about its center of mass and a translational kinetic energy ($\frac{1}{2}Mv_{com}^2$) due to translation of its center of mass. The total kinetic energy of the object is the sum of these two.

In a modern yo-yo, the string is not tied to the axle but is looped around it. When the yo-yo "hits" the bottom of its string, an upward force on the axle from the string stops the descent. The yo-yo then spins about its axle inside the loop and has only rotational kinetic energy. The yo-yo keeps spinning ("sleeping") until you "wake it" by jerking on the string, causing the string to catch on the axle and the yo-yo to climb back up. The rotational kinetic energy of the yo-yo at the bottom of its string (and thus the sleeping time) can be considerably increased by throwing the yo-yo downward so it starts down the string with initial speeds v_{com} and ω instead of rolling down from rest.

The Forces of Rolling

The simultaneous application of Newton's Second Law in both its translational and rotational forms allows us to calculate the acceleration of an object in situations where the motion combines rotation and translation. As an example of this technique, let's attempt to find an expression for the translational acceleration a_{com} of a yo-yo rolling down a string. We will use Newton's Second Law, noting the following points:

1. The yo-yo rolls down a string that makes angle $\theta = 90°$ with the horizontal.
2. The yo-yo rolls on an axle of radius R_0 (Fig. 12-4*a*).
3. The yo-yo is slowed by the tension force, $\vec{T}$, exerted on it by the string (Fig. 12-4*b*).

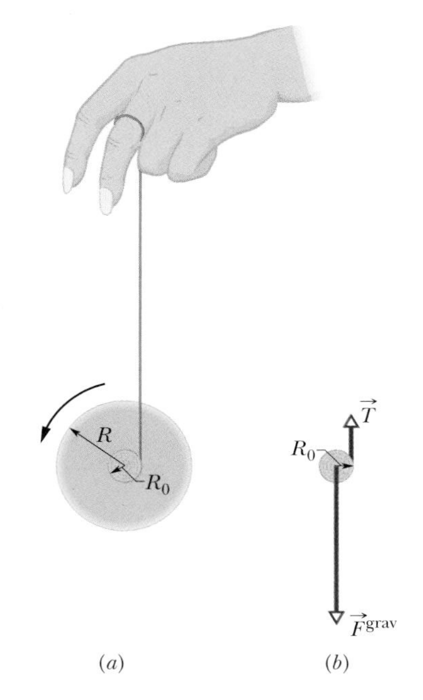

FIGURE 12-4 ■ (*a*) A yo-yo, shown in cross section. The string, of negligible thickness, is wound around an axle of radius R_0. (*b*) A free-body diagram for the falling yo-yo. Only the axle is shown.

The net force acting on the yo-yo is the vector sum of the gravitational force of the Earth on the yo-yo and the tension in the string. This net force causes the yo-yo to speed up or slow down. That is, the net force causes a translational acceleration $\vec{a}_{com}$ of the center of mass along the direction of travel. The net force also causes the yo-yo to rotate faster or slower, which means it causes a rotational acceleration α about the center of mass. From Chapter 11, we know that we can relate the magnitudes of the translational acceleration $\vec{a}_{com}$ and the rotational acceleration α by

$$a_{com} = \alpha R_0 \quad \text{(smooth rolling motion).} \tag{12-1}$$

If we want to find an expression for the yo-yo's acceleration $a_{com\,y}$ down the string, we can do this by using Newton's Second Law in the component form of both its translational version ($F_y^{net} = Ma_y$) and its rotational version ($\tau^{net} = I\alpha$).

We start by drawing the forces on the body as shown in Fig. 12-4:

1. The gravitational force $\vec{F}^{grav}$ on the body is directed downward. It acts at the center of mass of the yo-yo.
2. The tension in the string is directed upward. It acts at the point of contact outside of the yo-yo's central axis.

We can write Newton's Second Law for components along the y axis in Fig. 12-4 ($F_y^{net} = ma_y$) as

$$T_y - Mg = -Ma_{com\,y}. \tag{12-2}$$

Here M is the mass of the yo-yo. This equation contains two unknowns: the positive tension force component ($T_y = +|\vec{T}|$) and the component describing the vertical acceleration of the center of mass ($a_{com\,y}$).

Now we can use Newton's Second Law in rotational form to analyze the yo-yo's rotation about its center of mass (which coincides with its central axis). First, we shall use $|\vec{\tau}| = r_\perp|\vec{F}|$ (Eq. 11-31) to determine the magnitude of torque on the yo-yo about that point. The perpendicular distance from the rotation axis to the tension force (or moment arm) is R_0. So, the magnitude of torque that causes the yo-yo to rotate is given by $|\vec{T}||\vec{R}_0| = T_y R_0$. By the right-hand rule that we learned in Chapter 11, the resulting rotational acceleration would be positive (out of the page). Since the rotational acceleration is positive, we know the torque that produced the rotational acceleration is also positive.

The other force acting on the yo-yo, the gravitational force $\vec{F}^{grav}$, acts at the center of mass of the yo-yo. That is the center of the object itself, and so the gravitational force has a zero moment arm ($r_\perp = 0$) about the center of mass. Thus, the gravitational force produces zero torque. So we can write the rotational version of Newton's Second Law in component form ($\tau_z^{net} = I\alpha_z$) about an axis through the body's center of mass as

$$T_y R_0 = I_{com}\alpha_z. \tag{12-3}$$

As was the case for the equation resulting from the application of Newton's Second Law in its translational form, this equation contains two unknowns, T_y and α_z.

However, there is a relationship between $a_{\text{com } y}$ and α. We can use that relationship ($a_{\text{com}} = \alpha_z R_0$) to tie together the rotational and translational expressions of Newton's Second Law. Thus, we substitute $a_{\text{com } y}/R_0$ for α_z in the expression above (Eq. 12-3) and solve for the magnitude of the tension force, $T = T_y$, to obtain

$$T_y = I_{\text{com}} \frac{a_{\text{com } y}}{R_0^2}. \tag{12-4}$$

Substituting the right side of the equation above for T_y in the relationship we derived based on the translational motion of the yo-yo,

$$T_y - Mg = -Ma_{\text{com } y}, \tag{12-5}$$

we then find

$$a_{\text{com } y} = -\frac{g}{1 + I_{\text{com}}/MR_0^2}, \tag{12-6}$$

where I_{com} is the yo-yo's rotational inertia about its center, R_0 is its axle radius, and M is its mass. A yo-yo has the same downward acceleration when it is climbing back up the string, because the forces on it are still those shown in Fig. 12-4*b*.

TOUCHSTONE EXAMPLE 12-1: Hoop, Disk, Sphere

Consider a hoop, a disk, and a sphere, each of mass M and radius R, that roll smoothly along a horizontal table. For each, what fraction of its kinetic energy is associated with the translation of its center of mass?

SOLUTION The **Key Idea** is that the kinetic energy of a smoothly rolling body is the sum of its translational kinetic energy ($\frac{1}{2}Mv_{\text{com}}^2$) and its rotational kinetic energy ($\frac{1}{2}I_{\text{com}}\omega^2$). Therefore, the fraction of the kinetic energy associated with translation is

$$\text{frac} = \frac{\frac{1}{2}Mv_{\text{com}}^2}{\frac{1}{2}Mv_{\text{com}}^2 + \frac{1}{2}I_{\text{com}}\omega^2}. \tag{12-7}$$

We can greatly simplify the right side of Eq. 12-7 by substituting v_{com}/R for ω (Eq. 11-15) and realizing that the expressions for rotational inertia in Table 11-2 are all of the form βMR^2, where β is a numerical coefficient (the "front number"). Here β is 1 for a hoop, $\frac{1}{2}$ for a disk, and $\frac{2}{5}$ for a sphere. Thus, we can substitute βMR^2 for I_{com} in Eq. 12-7.

After these substitutions and some cancellations, Eq. 12-7 becomes

$$\text{frac} = \frac{1}{1 + \beta}. \tag{12-8}$$

Now, substituting the β values for the hoop, disk, and sphere, we can generate Table 12-1 to show the fractional splits of translational and rotational kinetic energy. For example, 0.67 of the kinetic energy of the disk is associated with the translation.

The relative split between translational and rotational energy depends on the relative size of the rotational inertia of the rolling object. As Table 12-1 shows, the rolling object (the hoop) that has its mass farthest from the central axis of rotation (and so has the largest rotational inertia) has the largest share of its kinetic energy in rotational motion. The object (the sphere) that has its mass closest to the central axis of rotation (and so has the smallest rotational inertia) has the smallest share in rotational motion.

TABLE 12-1
The Relative Splits between Rotational and Translational Energy for Rolling Objects

Object	Rotational Inertia I_{com}	Fraction of Energy in: Translation	Rotation
Hoop	$1MR^2$	0.50	0.50
Disk	$\frac{1}{2}MR^2$	0.67	0.33
Sphere	$\frac{2}{5}MR^2$	0.71	0.29
General[a]	βMR^2	$\frac{1}{1+\beta}$	$\frac{\beta}{1+\beta}$

[a] β may be computed for any rolling object as I_{com}/MR^2.

TOUCHSTONE EXAMPLE 12-2: Racing Down a Ramp

A uniform hoop, disk, and sphere, with the same mass M and same radius R, are released simultaneously from rest at the top of a ramp of length $L = 2.5$ m and angle $\theta = 12°$ with the horizontal. The objects roll without slipping down the ramp. No appreciable energy is lost to friction.

(a) Which object wins the race down to the bottom of the ramp?

SOLUTION ■ Two **Key Ideas** are these: First, the objects begin with the same mechanical energy E^{mec}, because they start from rest and the same height. Second, E^{mec} is conserved during the race to the bottom, because the only force doing work on the object-ramp-Earth system is the gravitational force. (The normal force on them from the ramp and the frictional force at their point of contact with the ramp do not cause energy transfers). Further, at any given point along the ramp, the objects must have the same kinetic energy K because the same amount of energy has been transferred from gravitational potential energy to kinetic energy.

If the objects were sliding down the ramp, this means they would have the same speed. However, another **Key Idea** is that they do not have the same speed v_{com} because each object shares its kinetic energy between its translational motion down the ramp and its rotational motion around its center of mass. As we saw in Touchstone Example 12-1 and Table 12-1, the sphere has the greatest fraction (0.71) as translational energy, so it has the greatest v_{com} and wins the race. Figure 12-5 shows the order of the objects during the race.

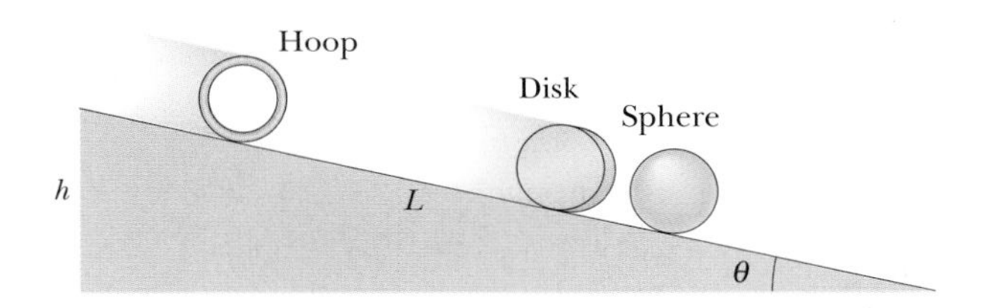

FIGURE 12-5 ■ A hoop, a disk, and a sphere roll smoothly from rest down the last segment of a very long ramp of angle θ.

(b) What is v_{com} for each object at the bottom of the ramp?

SOLUTION ■ Again, the **Key Idea** here is that mechanical energy is conserved. Let us choose the bottom of the ramp as our reference height for zero gravitational potential energy, so at the finish each object-ramp-Earth system has $U_2 = 0$. The initial kinetic energy for all three objects is $K_1 = 0$. The initial potential energy is $U_1 = Mgh = Mg(L\sin\theta)$. Now we can write the conservation of mechanical energy $E_2^{\text{mec}} = E_1^{\text{mec}}$ as

$$K_2 + U_2 = K_1 + U_1$$

or

$$(\tfrac{1}{2}I_{\text{com}}\omega^2 + \tfrac{1}{2}Mv_{\text{com}}^2) + 0 = 0 + Mg(L\sin\theta).$$

Substituting $\omega = v_{\text{com}}/R$ and solving for v_{com} give us

$$v_{\text{com}} = \sqrt{\frac{2gL\sin\theta}{1 + I_{\text{com}}/MR^2}}, \quad \text{(Answer)} \qquad (12\text{-}9)$$

which is the symbolic answer to the question.

Note that the speed depends not on the mass or the radius of the rolling object but only on the distribution of its mass about its central axis, which enters through the term I_{com}/MR^2. A marble and a bowling ball will have the same speed at the bottom of the ramp and will thus roll down the ramp in the same time. A bowling ball will beat a disk of any mass or radius, and almost anything that rolls will beat a hoop.

For the rolling hoop (see the hoop listing in Table 12-1) we have $I_{\text{com}}/MR^2 = 1$, so Eq. 12-9 yields

$$v_{\text{com}} = \sqrt{\frac{2gL\sin\theta}{1 + I_{\text{com}}/MR^2}} = \sqrt{\frac{(2)(9.8\ \text{m/s}^2)(2.5\ \text{m})(\sin 12°)}{1+1}} = 2.3\ \text{m/s}. \quad \text{(Answer)}$$

From a similar calculation, we obtain $v_{\text{com}} = 2.6$ m/s for the disk ($I_{\text{com}}/MR^2 = \frac{1}{2}$) and 2.7 m/s for the sphere ($I_{\text{com}}/MR^2 = \frac{2}{5}$).

12-3 Rotational Variables as Vectors

In the previous chapter, we considered only rotations that are about a fixed axis. We used the right-hand rule to determine whether the alignments for rotational displacement and velocity, representing the direction of rotation, are positive or negative. By assigning a standard coordinate system with the z axis along the axis of rotation, we treated the variables $\Delta\theta$ and ω as components along the z axis. Since rotational acceleration is defined in terms of changes of rotational velocity over time, the variable α could also be treated as a component along the axis of rotation. Thus, we developed a useful foundation for treating rotational quantities as vectors.

How can we work with rotational variables mathematically in cases where the axis of rotation is changing direction? For example, as a spinning top loses energy, its

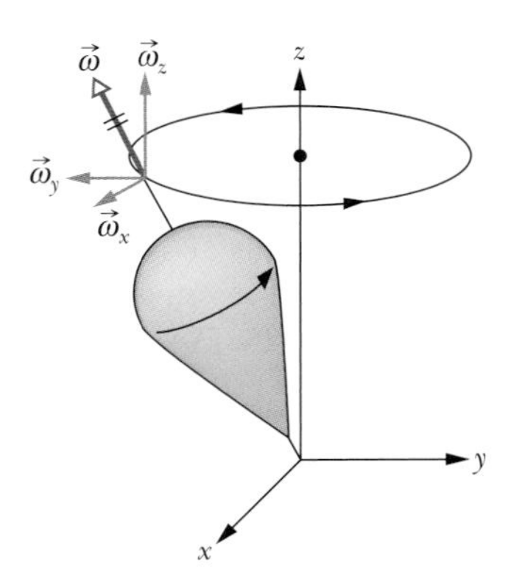

FIGURE 12-6 ■ The rotational velocity of a top rotating about an axis of symmetry always points along its axis of rotation. In the case where the rotational velocity changes direction, it must be described as a three-dimensional vector. Its components at one moment in time are shown relative to a right-handed coordinate system in which the z axis points up in the vertical direction.

axis of rotation begins turning (in technical terms, "precessing") around a vertical axis as shown in Fig. 12-6. In this example, and many others like it, it seems logical to explore the feasibility of expressing rotational variables as three-dimensional vectors. In Fig. 12-6, we can then choose to define a right-hand coordinate system with the z axis vertical. At any particular moment, the rotational displacement or velocity can be thought of as a vector pointing along the axis of rotation of the top.

In such a system, the rotational displacement must be described as a three-dimensional vector. Although we have not worked with three-dimensional vectors very much, we did introduce the decomposition of vectors into rectangular components in Section 4-4. In order to decompose a vector into components, we use unit vectors $\hat{\text{i}}$, $\hat{\text{j}}$, and $\hat{\text{k}}$ (discussed in Section 4-5) that point, respectively, in the positive directions of the x, y, and z axes shown in Fig. 12-6.

This method of using unit vectors enables us to decompose a rotational variable in terms of vector components in the familiar way. Using the rotational velocity vector as an example, we get

$$\vec{\omega} = \vec{\omega}_x + \vec{\omega}_y + \vec{\omega}_z = \omega_x\hat{\text{i}} + \omega_y\hat{\text{j}} + \omega_z\hat{\text{k}}. \tag{12-10}$$

Do Rotational Displacements and Velocities Behave Like Vectors?

It is not easy to get used to the way in which rotational quantities are represented as vectors. We instinctively expect that something should be moving *along* the direction of a vector. That is not the case when we attempt to use vectors to describe rotations. In the world of pure rotation, a vector defines an axis of rotation, not a direction in which something moves. Instead, a single particle or the many particles that make up a rigid body rotate *around* the direction of the vector. Nonetheless, a vector can be used to describe a rotational motion if it obeys the rules for vector manipulation discussed in Chapters 2 and 4. In particular, we stated in Chapter 2 that a vector is a mathematical entity that has both magnitude and direction, and that can be added, subtracted, multiplied, and transformed according to well-accepted mathematical rules. We have established that rotational variables seem to have both magnitude and direction. But we were vague about what the "well-accepted mathematical rules" for vector operations really are. One of these rules, used when vector addition was defined in Chapter 4, requires that the order of vector addition not matter, so that, for instance,

$$\vec{a} + \vec{b} = \vec{b} + \vec{a}.$$

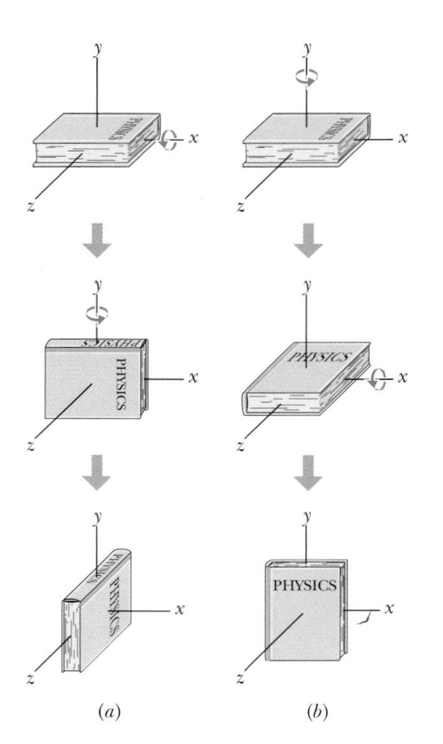

FIGURE 12-7 ■ (*a*) From its initial position, at the top in the figure, the book is given two successive 90° rotations, first about the (horizontal) x axis and then about the (vertical) y axis. (*b*) The book is given the same rotations, but in the reverse order.

Now for a caution: It turns out that *large rotational displacements cannot* be treated as vectors. Why not? We can certainly give them both magnitude and direction, as we did for the rotational velocity vector in Fig. 12-6. However, to be represented as a vector, a quantity must *also* obey the rules of vector addition. Rotational displacements fail this test.

Figure 12-7 shows an example of how large rotational displacements can fail the test. A book that is horizontal is given two 90° rotational displacements, first in the order shown in Fig. 12-7*a* and then in the order shown in Fig. 12-7*b*. Although each of the two rotational displacements are identical, the order in which they are applied is not. The book ends up with different orientations. Thus, the addition of the two *large* rotational displacements depends on their order and they cannot be vectors.

Fortunately, it can be shown mathematically that for small displacements, the order of the rotations does not matter. Since instantaneous rotational velocity is defined as

$$\vec{\omega} = \lim_{\Delta t \to 0} \frac{\Delta\vec{\theta}}{\Delta t} = \frac{d\vec{\theta}}{dt}, \tag{Eq. 11-6}$$

it is made up of infinitesimally small displacements. Thus, it appears that any series of small rotational displacements, as well as instantaneous rotational velocities, behave like vectors. Since rotational acceleration is constructed as a vector difference between rotational velocity vectors, it should behave like a vector also. Thus, we conclude that the basic rotational velocity and acceleration variables behave like vectors so long as they are determined by using small rotational displacements $\Delta\vec{\theta}$.

Can Torque Be Described as a Vector?

Recall that torque is a kind of "turning force" that can cause rotational accelerations about an axis. It was constructed mathematically for a very simple situation by combining the force acting on a single particle and the distance between that force and the particle's rotational axis. To see whether it is feasible to define torque as a three-dimensional vector, let's revisit the simple situation presented in Chapter 11.

In Section 11-8, we considered a single force acting on a particle that is attached to a "massless" rigid rod, which is, in turn, connected to a point (we'll call that point the origin of a coordinate system). We find that this force can cause the particle to rotate in a circle, but only when the force has a component that is tangent to the circle. As shown in Fig. 11-19, *this circle lies in the same plane as the force*. This means that the direction of the particle's rotational velocity and acceleration will be along its axis of rotation.

In Sections 11-7 and 11-8 we explored the relationship between torque (τ) and rotational acceleration (α) for special cases where the axis of rotation of a symmetric body is aligned with an axis of symmetry. We used the definition of rotational inertia (I) and Newton's Second Law for translational motion to show that for a single force acting on a particle,

$$\tau = I\alpha, \qquad \text{(Eq. 11-33)}$$

provided that we define the magnitude of torque (τ) in this situation to be given by

$$|\tau| = |\vec{r}||\vec{F}|\sin\phi, \qquad \text{(Eq. 11-29)}$$

where ϕ is the *smaller* of the two angles between the vectors $\vec{r}$ and $\vec{F}$.

In fact, by regarding the rotational inertia I as a scalar, and α and τ as components of a vector along the axis of rotation of the particle, we presented $\tau_z = I\alpha_z$ as the one-dimensional rotational analog to the expression $F_x = ma_x$ that describes motion along a straight line. We assume that for both expressions the acceleration that results from the application of a torque (or force) is in the same direction as the force (or torque). If we generalize this analogy between the translational and rotational laws of motion to three dimensions, then we expect that if $\vec{F}^{\text{net}} = m\vec{a}$, then

$$\vec{\tau}^{\text{net}} = I\vec{\alpha}, \qquad (12\text{-}11)$$

where $\vec{\alpha}$ and $\vec{\tau}^{\text{net}}$ are three-dimensional vectors that point in the *same direction*. If this is the case, then $\vec{\tau}$ is a vector that must be *perpendicular* to both the applied force $\vec{F}$ and the position vector $\vec{r}$ that extends from the axis of rotation to the particle experiencing the force. The torque vector must also have a magnitude given by $|\tau| = |\vec{r}||\vec{F}|\sin\phi$ (Eq. 11-29).

In Section 9-8, we discussed the fact that there are two different methods defined by mathematicians for multiplying vectors. One, known as the scalar (or dot) product, is used to define the amount of work, W, done on an object that undergoes a translational displacement $\vec{d}$ under the influence of a constant force $\vec{F}$. Work is a scalar quantity that is invariant to coordinate rotations and is given by

$$W = \vec{F}\cdot\vec{d} = |\vec{F}||\vec{d}|\cos\theta.$$

The other type of vector multiplication is known as the **vector** (or **cross**) **product.** The vector product $\vec{c}$ of two vectors $\vec{a}$ and $\vec{b}$ is given by

$$\vec{c} = \vec{a} \times \vec{b}.$$

As its name suggests, the **vector product** of two vectors is itself a vector. It is not hard to convince yourself that any two vectors determine a plane. We define the vector that results when a vector product is calculated *to be perpendicular to the plane determined by the vectors being multiplied.*

Recall that the plane a particle rotates in is perpendicular to the axis of rotation along which we expect the torque vector to point. This suggests that we may be able to express torque as a vector product. It also turns out that the magnitude of a vector product is equal to the product of the magnitudes of the two vectors being multiplied times the sine of the angle between them. This is also how the magnitude of a torque about a fixed axis is determined.

It appears it may be valid to define torque as the vector product of the position vector $\vec{r}$ and the force vector $\vec{F}$ so that

$$\vec{\tau} = \vec{r} \times \vec{F} \quad \text{(tentative definition of torque).}$$

In the next section, we discuss the mathematical properties of the vector product.

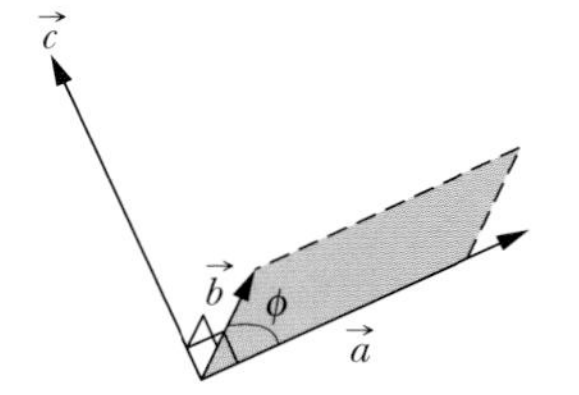

FIGURE 12-8 ■ A parallelogram of area $A = |\vec{a}||\vec{b}|\sin\phi$ with $\vec{c}$ being perpendicular to $\vec{a}$ and $\vec{b}$. $\vec{a}$ and $\vec{b}$ lie in the same plane.

12-4 The Vector or Cross Product

Is there a natural way to associate a vector with the product of a pair of vectors? If we think about a pair of vectors in three-dimensional space, we see that they have two directions (unless they point in the same direction). There are only three mutually perpendicular directions, so we could choose the direction not used—the one perpendicular to the plane determined by the two vectors we are trying to multiply—as the direction of the vector product. Here's one way to think about it. Consider two vectors of lengths $|\vec{a}|$ and $|\vec{b}|$ pointing in different directions with ϕ being the *smaller angle* between them. The two vectors can be considered to be two sides of a parallelogram of area $A = |\vec{a}||\vec{b}|\sin\phi$.

This area has a direction, though we don't often think of area that way. The same area can be turned and oriented in different ways in space. We can choose to describe its orientation by an arrow perpendicular to the area. This suggests that we create a vector product $\vec{a} \times \vec{b} = \vec{c}$ that has the magnitude equal to the size of the area, $A = |\vec{a}||\vec{b}|\sin\phi$, and a direction perpendicular to the two vectors $\vec{a}$ and $\vec{b}$ (Fig. 12-8).

We should point out, though, that the area could actually have two different directions associated with it, with one direction pointing perpendicular to one side of the area and one direction pointing perpendicular to the other side of the area. These two vectors point in opposite directions, so they are just the negative of each other. Since the choice between these two directions is arbitrary, we will use the right-hand rule to choose which direction to associate with the product. Applying the right-hand rule to this vector product means if we point our straightened fingers on our right hand in the direction of the first vector so we can curl them to the direction of the second vector, then the direction of our extended thumb will be the direction associated with the vector product as shown in Fig. 12-9.

This discussion also implies that the vector (or cross) product of two vectors that point in the same direction must be zero. We know this because the area created by two such vectors is zero. Furthermore, we can't know in what direction a zero area would point!

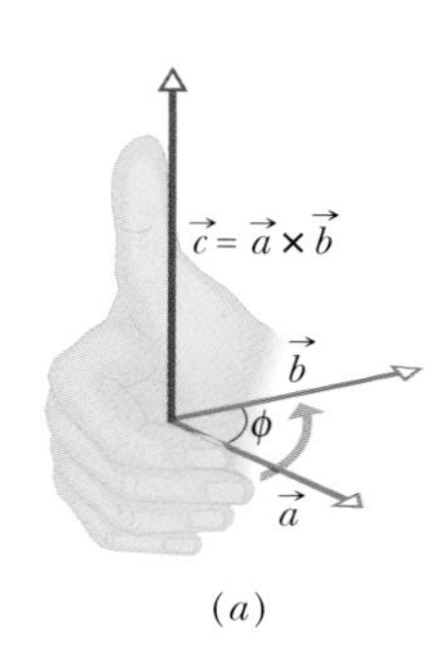

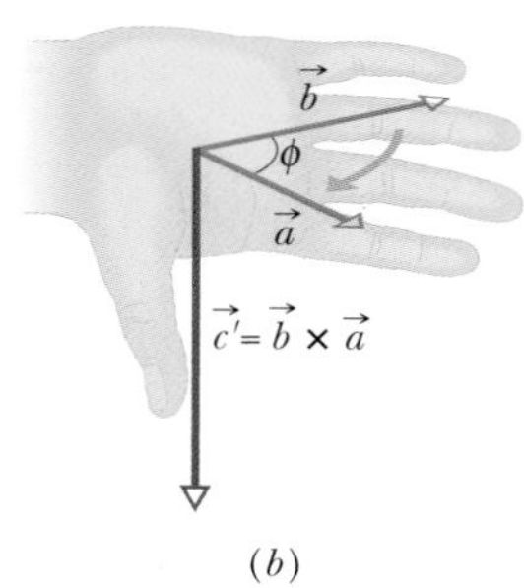

FIGURE 12-9 ■ Illustration of the right-hand rule for vector products. (*a*) Sweep vector $\vec{a}$ into vector $\vec{b}$ with the fingers of your right hand. Your outstretched thumb shows the direction of vector $\vec{c} = \vec{a} \times \vec{b}$. (*b*) Showing that $\vec{a} \times \vec{b}$ is the reverse of $\vec{b} \times \vec{a}$.

If the vectors $\vec{a}$ and $\vec{b}$ are not actually length vectors (but have some other unit), we can just generalize the discussion above as if we were working with an area. So, in general, the vector product of any two vectors $\vec{a}$ and $\vec{b}$, written $\vec{a} \times \vec{b}$, produces a third vector $\vec{c}$ whose magnitude is

$$|\vec{c}| = |\vec{a}||\vec{b}|\sin\phi, \tag{12-12}$$

where ϕ (phi) is the *smaller* of the two angles between $\vec{a}$ and $\vec{b}$. (You must use the smaller of the two angles between the vectors because $\sin\phi$ and $\sin(360° - \phi)$ differ in algebraic sign.) Because of the notation $\vec{a} \times \vec{b}$, the vector product is known as the **cross product** of $\vec{a}$ and $\vec{b}$ or, more simply, "a cross b."

> If $\vec{a}$ and $\vec{b}$ are parallel or antiparallel, $\vec{a} \times \vec{b} = 0$. The magnitude of $\vec{a} \times \vec{b}$, which can be written as $|\vec{a} \times \vec{b}|$, is maximum when $\vec{a}$ and $\vec{b}$ are perpendicular to each other.

Remember that the order of the vector multiplication is important. In Fig. 12-9*b*, we are determining the direction of $\vec{c}' = \vec{b} \times \vec{a}$, so the fingers are placed to sweep $\vec{b}$ into $\vec{a}$ through the smaller angle. The thumb ends up in the opposite direction from before, and so it must be that $\vec{c}' = -\vec{c}$ or $\vec{a} \times \vec{b} = -\vec{b} \times \vec{a}$.

In unit-vector notation, we can write

$$\vec{a} \times \vec{b} = (a_x\hat{\text{i}} + a_y\hat{\text{j}} + a_z\hat{\text{k}}) \times (b_x\hat{\text{i}} + b_y\hat{\text{j}} + b_z\hat{\text{k}}), \tag{12-13}$$

which can be expanded according to the distributive law. That is, each component of the first vector is to be crossed with each component of the second vector. The cross products of unit vectors are given in Appendix E (see Products of Vectors). For example, in the expansion of the equation above, we have

$$a_x\hat{\text{i}} \times b_x\hat{\text{i}} = a_xb_x(\hat{\text{i}} \times \hat{\text{i}}) = 0, \tag{12-14}$$

because the two unit vectors $\hat{\text{i}}$ and $\hat{\text{i}}$ are parallel and thus have a zero cross product. Similarly, we have

$$a_x\hat{\text{i}} \times b_y\hat{\text{j}} = a_xb_y(\hat{\text{i}} \times \hat{\text{j}}) = a_xb_y\hat{\text{k}}. \tag{12-15}$$

In the last step, we used Eq. 12-12 to evaluate the magnitude of $\hat{\text{i}} \times \hat{\text{j}}$ as unity (one). (The vectors $\hat{\text{i}}$ and $\hat{\text{j}}$ each have a dimensionless magnitude of unity, and the angle between them is 90°.) Also, we used the right-hand rule to get the direction of $\hat{\text{i}} \times \hat{\text{j}}$ as being in the positive direction of the z axis (thus in the direction of $\hat{\text{k}}$).

Continuing to expand Eq. 12-13, we can show that

$$\vec{a} \times \vec{b} = (a_yb_z - b_ya_z)\hat{\text{i}} + (a_zb_x - b_za_x)\hat{\text{j}} + (a_xb_y - b_xa_y)\hat{\text{k}}. \tag{12-16}$$

We can also evaluate a cross product by setting up and evaluating a determinant (as shown in Appendix E) or by using a vector-capable calculator.

To check whether any xyz coordinate system is a right-handed coordinate system, use the right-hand rule shown in Fig. 12-9 for the cross product $\hat{\text{i}} \times \hat{\text{j}} = \hat{\text{k}}$ with that system. If your fingers sweep $\hat{\text{i}}$ (positive direction of x) into $\hat{\text{j}}$ (positive direction of y) with the outstretched thumb pointing in the positive direction of z, then the system is right-handed.

READING EXERCISE 12-1: Vectors $\vec{c}$ and $\vec{d}$ have magnitudes of 3 units and 4 units, respectively. What is the angle between the directions of $\vec{c}$ and $\vec{d}$ if the magnitude of the vector product $\vec{c} \times \vec{d}$ is (a) zero, (b) 12 units, (c) 6 units? ■

TOUCHSTONE EXAMPLE 12-3: Vector Product

In Fig. 12-10, vector $\vec{a}$ lies in the xy plane, has a magnitude of 18 units, and points in a direction 250° from the positive direction of x. Also, vector $\vec{b}$ has a magnitude of 12 units and points along the positive direction of z. What is the vector product $\vec{c} = \vec{a} \times \vec{b}$?

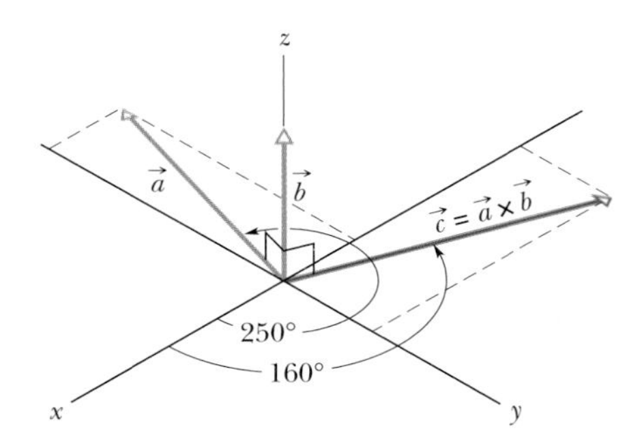

FIGURE 12-10 ■ Vector $\vec{c}$ (in xy plane) is the vector (or cross) product of vectors $\vec{a}$ and $\vec{b}$.

SOLUTION ■ One **Key Idea** is that when we have two vectors in magnitude-angle notation, we find the magnitude of their cross product (that is, the vector that results from taking their cross product) with Eq. 12-12. Here that means the magnitude of $\vec{c}$ is

$$|\vec{c}| = |\vec{a}||\vec{b}|\sin\phi = (18)(12)(\sin 90°) = 216. \quad \text{(Answer)}$$

A second **Key Idea** is that with two vectors in magnitude-angle notation, we find the direction of their cross product with the right-hand rule of Fig. 12-9. In Fig. 12-10, imagine placing the fingers of your right hand around a line perpendicular to the plane of $\vec{a}$ and $\vec{b}$ (the line on which $\vec{c}$ is shown) such that your fingers sweep $\vec{a}$ into $\vec{b}$. Your outstretched thumb then gives the direction of $\vec{c}$. Thus, as shown in Fig. 12-10, $\vec{c}$ lies in the xy plane. Because its direction is perpendicular to the direction of $\vec{a}$, it is at an angle of

$$250° - 90° = 160° \quad \text{(Answer)}$$

from the positive direction of x.

12-5 Torque as a Vector Product

In Chapter 11, we defined the torque component, τ_z, for a rigid body that can rotate around a fixed axis. In that case, each particle in the body was forced to move in a path that is a circle about that axis. We now use the vector product to expand the definition of torque to apply it to an individual particle that moves along *any* path relative to a fixed *point* (rather than a fixed axis). The path need no longer be a circle, and we must write the torque as a vector $\vec{\tau}$ that may have any direction.

Figure 12-11*a* shows a particle at point A in the xy plane. A single force $\vec{F}$ in that plane acts on the particle. The particle's position relative to the origin O is given by position vector $\vec{r}$. The torque $\vec{\tau}$ acting on the particle relative to the fixed point O is a vector quantity defined as the vector product of $\vec{r}$ and $\vec{F}$ so that

$$\vec{\tau} = \vec{r} \times \vec{F} \quad \text{(torque defined).} \tag{12-17}$$

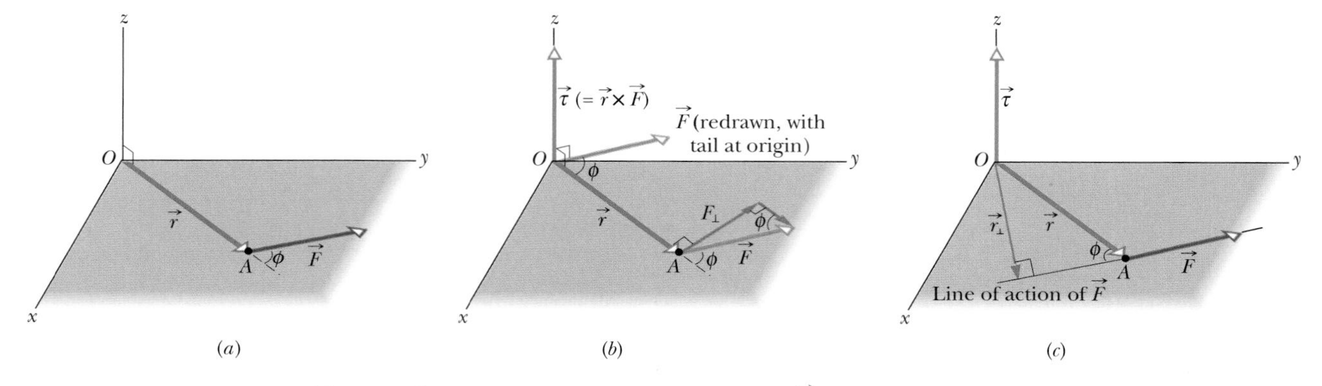

FIGURE 12-11 ■ Defining torque. (*a*) A force $\vec{F}$, lying in the xy plane, acts on a particle at point A. (*b*) This force produces a torque $\vec{\tau}(= \vec{r} \times \vec{F})$ on the particle with respect to the origin O. By the right-hand rule for vector (cross) products, the torque vector points in the positive direction of z. Its magnitude is equivalently given by $rF_\perp$ in (*b*) and by $r_\perp F$ in (*c*).

We can evaluate the vector (or cross) product in this definition of $\vec{\tau}$ by using the rules for such products given above. To find the direction of $\vec{\tau}$, we slide the vector $\vec{F}$ (without changing its direction) until its tail is at the origin O, so that the two vectors in the vector product are tail to tail as in Fig. 12-11*b*. We then use the right-hand rule for vector products in Fig. 12-9, sweeping the fingers of the right hand from $\vec{r}$ (the first vector in the product) into $\vec{F}$ (the second vector). The outstretched right thumb then gives the direction of $\vec{\tau}$. In Fig. 12-11*c*, the direction of $\vec{\tau}$ is again shown to be in the positive direction of the z axis.

When drawing diagrams of three-dimensional vectors, we often need a way to show that a vector points into or out of the plane of the page.

> The symbol ⊗ is used to denote a vector that points into the plane of the page. The symbol ⊙ denotes a vector pointing out of the page.

TOUCHSTONE EXAMPLE 12-4: Three Torques

In Fig. 12-12*a*, three forces, each of magnitude 2.0 N, act on a particle. The particle is in the xz plane at point a given by position vector $\vec{r}$, where $r = 3.0$ m and $\theta = 30°$. Force $\vec{F}_A$ is antiparallel to the x axis, force $\vec{F}_B$ is antiparallel to the z axis, and force $\vec{F}_C$ is antiparallel to the y axis. What is the torque, with respect to the origin O, due to each force?

SOLUTION ■ The **Key Idea** here is that, because the three force vectors do not lie in a plane, we cannot evaluate their torques as in Chapter 11. Instead, we must use vector (or cross) products, given by Eq. 12-17 ($\vec{\tau} = \vec{r} \times \vec{F}$) with their directions given by the right-hand rule for vector products.

Because we want the torques with respect to the origin O, the vector $\vec{r}$ required for each cross product is the given position vector. To determine the angle ϕ between the direction of $\vec{r}$ and the direction of each force, we shift the force vectors of Fig. 12-12*a*, each in turn, so that their tails are at the origin. Figures 12-12*b*, *c*, and *d*, which are direct views of the xz plane, show the shifted force vectors $\vec{F}_A$, $\vec{F}_B$, and $\vec{F}_C$, respectively. (Note how much easier the angles are to see.) In Fig. 12-12*d*, the angle between the directions of $\vec{r}$ and $\vec{F}_C$ is 90° and the symbol ⊗ means $\vec{F}_C$ is directed into the page.

Now, applying Eq. 12-17 for each force, we find the magnitudes of the torques to be

$$\tau_A = rF_A \sin\phi_A = (3.0 \text{ m})(2.0 \text{ N})(\sin 150°) = 3.0 \text{ N}\cdot\text{m},$$

$$\tau_B = rF_B \sin\phi_B = (3.0 \text{ m})(2.0 \text{ N})(\sin 120°) = 5.2 \text{ N}\cdot\text{m},$$

and $$\tau_C = rF_C \sin\phi_C = (3.0 \text{ m})(2.0 \text{ N})(\sin 90°) = 6.0 \text{ N}\cdot\text{m}.$$

(Answer)

To find the directions of these torques, we use the right-hand rule, placing the fingers of the right hand so as to rotate $\vec{r}$ into $\vec{F}$ through the *smaller* of the two angles between their directions. The thumb points in the direction of the torque. Thus $\vec{\tau}_A$ is directed into the page in Fig. 12-12*b*, $\vec{\tau}_B$ is directed out of the page in Fig. 12-12*c*, and $\vec{\tau}_C$ is directed as shown in Fig. 12-12*d*. All three torque vectors are shown in Fig. 12-12*e*

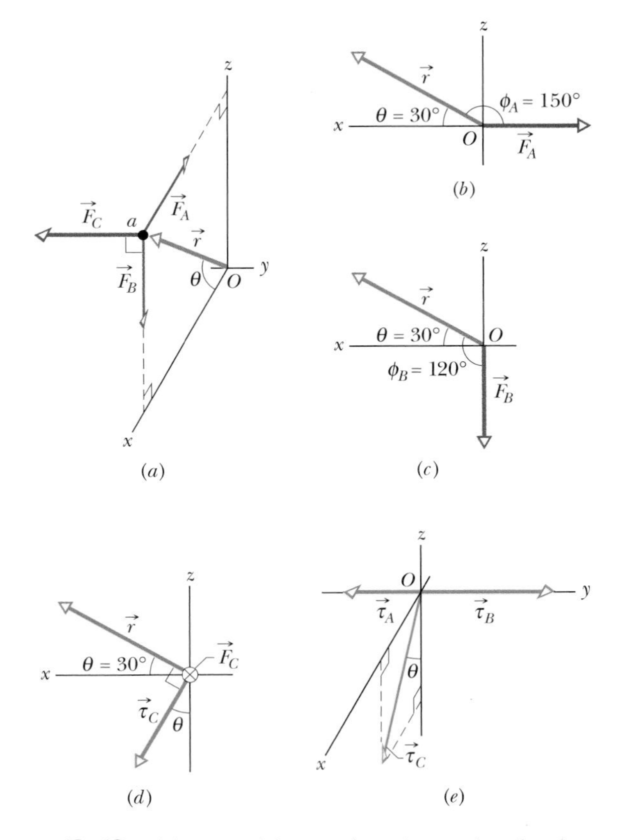

FIGURE 12-12 ■ (*a*) A particle at point a is acted on by three forces, each antiparallel to a coordinate axis. The angle ϕ (used in finding torque) is shown (*b*) for $\vec{F}_A$ and (*c*) for $\vec{F}_B$. (*d*) Torque $\vec{\tau}_C$ is perpendicular to both $\vec{r}$ and $\vec{F}_C$ (force $\vec{F}_C$ is directed into the plane of the figure). (*e*) The torques (relative to the origin O) acting on the particle.

12-6 Rotational Form of Newton's Second Law

Recall that the concept of translational momentum $\vec{p}$ and the principle of conservation of momentum are extremely powerful tools. They allow us to predict the outcome of, say, a collision between two cars without knowing the details of what goes on during the collision. Here we begin a discussion of the rotational counterpart of $\vec{p}$. In Chapter 7, we found that we could write Newton's Second Law in the form

$$\vec{F}^{\text{net}} = m\vec{a} = \frac{d\vec{p}}{dt} \quad \text{(single particle).} \tag{12-18}$$

This relationship expresses the close relation between force and translational momentum for a single particle. It can be generalized to extended bodies. It also leads directly to the powerful idea that translational momentum is conserved in the absence of a net external force.

We have seen enough of the parallelism between translational and rotational quantities to be hopeful that there is a rotational corollary to $\vec{F}^{\text{net}} = d\vec{p}/dt$. In search of the equivalent expression, we start with

$$\vec{\tau}^{\text{net}} = \vec{r} \times \vec{F}^{\text{net}}, \tag{12-19}$$

and replace the force vector with $m\vec{a}$. This gives us

$$\vec{\tau}^{\text{net}} = \vec{r} \times m\vec{a} \quad \text{or} \quad \vec{\tau}^{\text{net}} = \vec{r} \times m\frac{d\vec{v}}{dt}. \tag{12-20}$$

For a constant mass, the expression $\vec{r} \times m\,d\vec{v}/dt$ above can be replaced with $d(\vec{r} \times m\vec{v})/dt$.

The equality of these two expressions is more clearly seen in reverse. Namely,

$$\frac{d(\vec{r} \times m\vec{v})}{dt} = \frac{md(\vec{r} \times \vec{v})}{dt} \quad \text{(for constant mass).}$$

Then applying the product rule of derivatives

$$\frac{d(\vec{r} \times m\vec{v})}{dt} = m\left(\vec{r} \times \frac{d\vec{v}}{dt} + \frac{d\vec{r}}{dt} \times \vec{v}\right). \tag{12-21}$$

However, $d\vec{r}/dt$ is the object's velocity $\vec{v}$, and $\vec{v} \times \vec{v} = 0$. Thus, we can rewrite the equation above as

$$\frac{d(\vec{r} \times m\vec{v})}{dt} = m\left(\vec{r} \times \frac{d\vec{v}}{dt}\right),$$

or

$$\frac{d(\vec{r} \times m\vec{v})}{dt} = \vec{r} \times m\frac{d\vec{v}}{dt}. \tag{12-22}$$

So, from Eq. 12-20 above,

$$\vec{\tau}^{\text{net}} = \vec{r} \times m\frac{d\vec{v}}{dt} = \frac{d(\vec{r} \times m\vec{v})}{dt},$$

or

$$\vec{\tau}^{\text{net}} = \frac{d(\vec{r} \times \vec{p})}{dt}. \tag{12-23}$$

Comparing this expression to $\vec{F}^{\text{net}} = d\vec{p}/dt$, we see that if we choose to define the rotational momentum, $\vec{\ell}$, as the rotational corollary of translational momentum, then

$\vec{\ell} = \vec{r} \times \vec{p}$. We now have an equivalent expression for rotations as we do for translations. Namely,

$$\vec{\tau}^{\,\text{net}} = \frac{d\vec{\ell}}{dt} \quad \text{(single particle).} \tag{12-24}$$

In words,

> The (vector) sum of all the torques acting on a particle is equal to the time rate of change of the rotational momentum of that particle.

Be careful, though: $\vec{\tau}^{\,\text{net}} = d\vec{\ell}/dt$ has no meaning unless the net torque $\vec{\tau}^{\,\text{net}}$, and the rotational momentum $\vec{\ell}$, are defined with respect to the same origin. Many texts refer to rotational momentum as angular momentum.

READING EXERCISE 12-2: The figure shows the position vector $\vec{r}$ of a particle at a certain instant, and four choices for the direction of a force that is to accelerate the particle. All four choices lie in the xy plane. Rank the choices according to the magnitude of the time rate of change ($d\vec{\ell}/dt$) they produce in the rotational momentum of the particle about point O, greatest first.

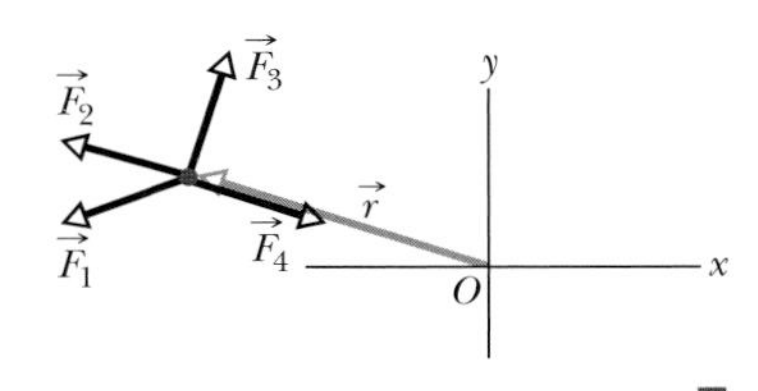

12-7 Rotational Momentum

Figure 12-13 shows a particle of mass m with translational momentum $\vec{p} = m\vec{v}$ as it passes through point A in the xy plane. The **rotational momentum** $\vec{\ell}$ of this particle with respect to the origin O is a vector quantity defined as

$$\vec{\ell} \equiv \vec{r} \times \vec{p} = m(\vec{r} \times \vec{v}) \quad \text{(rotational momentum defined),} \tag{12-25}$$

where $\vec{r}$ is the position vector of the particle with respect to O. Note carefully that to have rotational momentum about O, the particle does *not* have to rotate around O. Comparison of $\vec{\tau} = \vec{r} \times \vec{F}$ (Eq. 12-17) and $\vec{\ell} = \vec{r} \times \vec{p}$ (Eq. 12-25) shows that rotational momentum bears the same relation to translational momentum as torque does to force. The SI unit of rotational momentum is the kilogram-meter-squared per second ($\text{kg} \cdot \text{m}^2/\text{s}$), equivalent to the joule-second ($\text{J} \cdot \text{s}$).

To find the direction of the rotational momentum vector $\vec{\ell}$ in Fig. 12-13, we slide the vector $\vec{p}$ until its tail is at the origin O. Then we use the right-hand rule for vector products, sweeping our right-hand fingers from $\vec{r}$ into $\vec{p}$. The outstretched thumb then shows that the direction of $\vec{\ell}$ is in the positive (upward) direction of the z axis in Fig. 12-13. To find the magnitude of $\vec{\ell}$, we use the general definition of a cross product to write

$$|\vec{\ell}| = |\vec{r}||mv|\sin\phi \quad \text{or} \quad \ell = rmv\sin\phi, \tag{12-26}$$

where ϕ is the smaller angle between $\vec{r}$ and $\vec{p}$. From Fig. 12-13a, we see that $\ell = rmv\sin\phi$ can be rewritten as

$$\ell = rp_\perp = rmv_\perp, \tag{12-27}$$

where $p_\perp$ is the component of $\vec{p}$ perpendicular to $\vec{r}$, $v_\perp$ is the component of $\vec{v}$ perpendicular to $\vec{r}$ and r is the magnitude of $\vec{r}$. From Fig. 12-13b, we see that $\ell = rmv\sin\phi$. So, $\vec{\ell} = \vec{r} \times \vec{p}$ (Eq. 12-25) can also be rewritten as

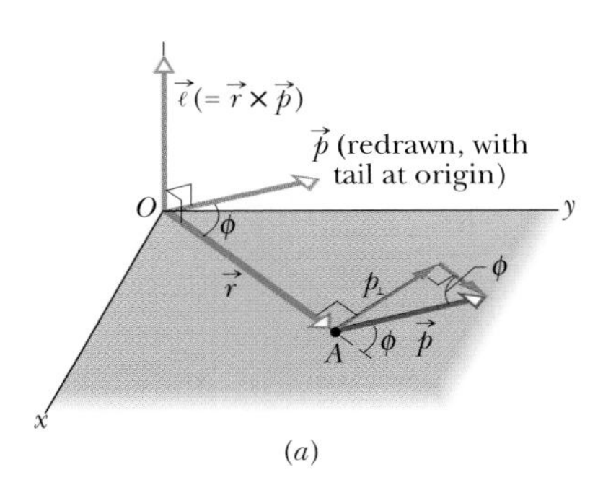

(a)

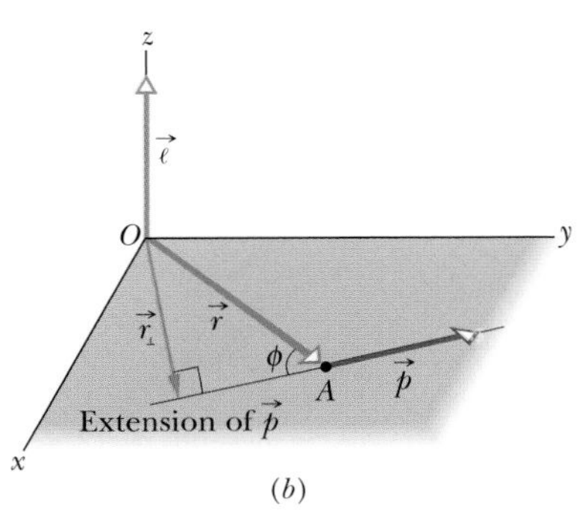

(b)

FIGURE 12-13 Defining rotational momentum. A particle passing through point A has translational momentum $\vec{p} = m\vec{v}$, with the vector $\vec{p}$ lying in the xy plane. The particle has rotational momentum $\vec{\ell} = \vec{r} \times \vec{p}$ with respect to the origin O. By the right-hand rule, the rotational momentum vector points in the positive direction of z. (a) The magnitude of $\vec{\ell}$ is given by $\ell = rp_\perp = rmv_\perp$. ($b$) The magnitude of $\vec{\ell}$ is also given by $\ell = r_\perp p = r_\perp mv$.

$$\ell = r_\perp p = r_\perp mv, \tag{12-28}$$

where $r_\perp$ is the perpendicular distance between O and the extension of $\vec{p}$.

Just as is true for torque, rotational momentum has meaning only with respect to a specified origin. Moreover, if the particle in Fig. 12-13 did not lie in the xy plane, or if the translational momentum $\vec{p}$ of the particle did not also lie in that plane, the rotational momentum $\vec{\ell}$ would not be parallel to the z axis. The direction of the rotational momentum vector is always perpendicular to the plane formed by the position and translational momentum vectors $\vec{r}$ and $\vec{p}$.

READING EXERCISE 12-3: In the diagrams below there is an axis of rotation perpendicular to the page that intersects the page at point O. Figure (*a*) shows particles 1 and 2 moving around point O in opposite rotational directions, in circles with radii 2 m and 4 m. Figure (*b*) shows particles 3 and 4 traveling in the same direction, along straight lines at perpendicular distances of 2 m and 4 m from point O. Particle 5 moves directly away from O. All five particles have the same mass and the same constant speed. (a) Rank the particles according to the magnitudes of their rotational momentum about point O, greatest first. (b) Which particles have rotational momentum about point O that is directed into the page?

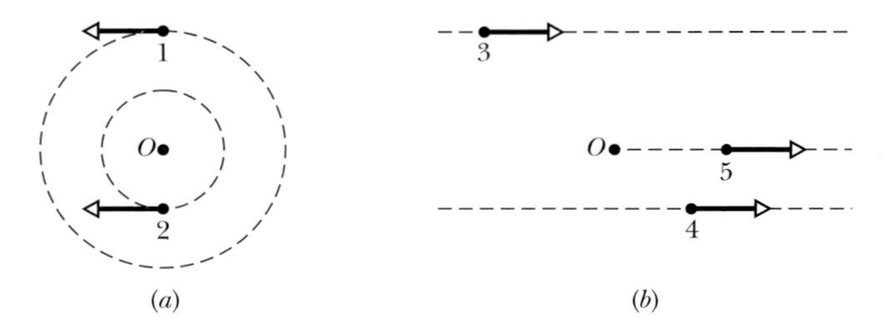

12-8 The Rotational Momentum of a System of Particles

Having a rotational equivalent of translational momentum is interesting, but what we would really like to do with such a quantity is to use it to understand the rotational motion of complex objects in dynamic situations. This is what made translational momentum so useful. For example, why is it that a skater spins faster when she pulls in her arms? How do we steer spaceships? Why do neutron stars spin so much faster than other stars? To understand these and other real-world situations, we must develop an expression for the rotational momentum of a system of particles.

Just as we did for translational momentum, we can use a principle of superposition for rotational momentum. We define the total rotational momentum $\vec{L}$ of a system of particles to be the vector sum of the rotational momenta $\vec{\ell}$ of the individual particles

$$\vec{L} = \vec{\ell}_A + \vec{\ell}_B + \vec{\ell}_C + \cdots + \vec{\ell}_n = \sum_{i=A}^{n} \vec{\ell}_i, \tag{12-29}$$

in which i ($A,B,C,\ldots$) labels the particles. With time, the rotational momenta of individual particles may change, either because of interactions within the system (between the individual particles) or because of influences that may act on the system from the outside.

We can find the change in $\vec{L}$ as these changes take place by taking the time derivative of

$$\vec{L} = \sum_{i=A}^{n} \vec{\ell}_i.$$

Thus,

$$\frac{d\vec{L}}{dt} = \sum_{i=A}^{n} \frac{d\vec{\ell}_i}{dt}. \qquad (12\text{-}30)$$

From $\vec{\tau}^{\,\text{net}} = d\vec{\ell}/dt$ (Eq. 12-24), we see that

$$\frac{d\vec{L}}{dt} = \sum_{i=1}^{n} \vec{\tau}_i^{\,\text{net}}. \qquad (12\text{-}31)$$

In the equation above, the right side is the sum of the torques acting on the particles that make up the system. This sum includes torques that result from all the forces acting on the system, whether they originate from within the system (internal forces) or outside of it (external forces). However, the internal torques sum to zero, as did the internal forces in the analogous expression $\vec{F}^{\,\text{net}} = d\vec{P}/dt$.

In general,

$$\vec{\tau}^{\,\text{net}} = \frac{d\vec{L}}{dt} \quad \text{(system of particles)}, \qquad (12\text{-}32)$$

where $\vec{\tau}^{\,\text{net}}$ is the net torque acting on the system. In practice, this is just the vector sum of all external torques on all particles in the system, since the internal torques sum to zero.

This equation is Newton's Second Law in rotational form, for a system of particles. It says:

> The net (external) torque $\vec{\tau}^{\,\text{net}}$ acting on a system of particles is equal to the time rate of change of the system's total rotational momentum $\vec{L}$.

$\vec{\tau}^{\,\text{net}} = d\vec{L}/dt$ (Eq. 12-32) is analogous to $\vec{F}^{\,\text{net}} = d\vec{P}/dt$. However, it requires extra caution: Torques and the system's rotational momentum must be measured relative to the same origin.

12-9 The Rotational Momentum of a Rigid Body Rotating About a Fixed Axis

We next evaluate the rotational momentum of an extended system of particles that form a rigid body that rotates about a fixed axis. Figure 12-14 shows such a body. In Chapter 8, when we discussed the translational motion of extended systems, we derived an expression for the translational momentum of the object in terms of the velocity of its center of mass,

$$\vec{p}_{\text{sys}} = M_{\text{sys}}\vec{v}_{\text{com}} \quad \text{(translational momentum, system of particles)}. \qquad (12\text{-}33)$$

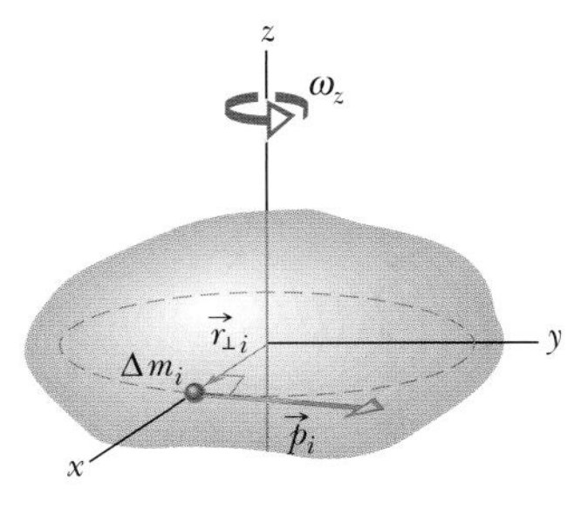

FIGURE 12-14 ■ A rigid body rotates about the z axis with rotational speed ω. A mass element of mass Δm_i within the body moves about the z axis in a circle with radius r. The translational momentum $\vec{p}_i$ and the position vector $\vec{r}_i$ of the mass element relative to the axis of rotation change constantly as the mass element rotates.

We can develop an analogous expression for rotational motion. Let's start our development with a single mass element Δm_i that rotates with a rotational velocity whose component along the axis of rotation is ω. In Fig. 12-14 we see that the mass element has a translational momentum $\vec{p}_i$ and a position vector $\vec{r}_i$ relative to the axis of rotation. These vectors change constantly as the mass element rotates in a circle about its axis of rotation. However, the translational momentum and position vectors are always perpendicular to each other and lie in the x-y plane. This means that the rotational momentum vector only has a component in the z direction. Since rotational momentum is defined as $\vec{\ell} \equiv \vec{r} \times \vec{p} = m(\vec{r} \times \vec{v})$ (Eq. 12-25) we see that for our simple situation

$$\vec{\ell}_i = \vec{r}_i \times \vec{p}_i = \ell_{iz}\hat{\text{k}} = r_i p_i \hat{\text{k}} = r_i(\Delta m_i v_i)\hat{\text{k}}.$$

If we replace the translational speed v_i with ωr_i, where the rotational velocity component $\omega = \omega_z$ does not depend on which mass element we are considering (that is, the entire object moves as one), this equation reduces to

$$\ell_{iz} = r_i(\Delta m_i \omega_i r_i) = \Delta m_i r_i^2 \omega.$$

Aha! The term $\Delta m_i r_i^2$ is just the rotational inertia, ΔI_i, of the ith mass element. So we can sum over all the mass elements to get a total rotational momentum of the rotating body given by

$$L_z = \sum \ell_{iz} = \sum \Delta m_i r_i^2 \omega_z = \sum \Delta I_i \omega = I\omega_z.$$

Recalling that the rotational analogy of mass is rotational inertia I, and that all points in a rigid rotating body move with the same rotational velocity $\omega = \omega_z$, we write the analogous expression for the rotational momentum of an extended object for an arbitrary choice of coordinate axes as

$$\vec{L} = I\vec{\omega} \quad \text{(rigid symmetric body, fixed axis through com).} \tag{12-34}$$

As you will see in the next section, this expression is very useful in situations where rotational momentum is conserved. It allows us to explain why rotating objects that change from one shape to another (such as a spinning ice skater) can speed up or slow down the rate of turn. However, you must remember that the rotational momentum $\vec{L}$, can only be expressed as $I\vec{\omega}$ when the rotational momentum and the rotational inertia, I, are taken about the same axis.

If an extended body is not symmetric with respect to its axis of rotation and its rotation axis does not pass through its center of mass, calculation of rotational inertia and momentum can become quite complex. For example, you can get different values of I when the object rotates about different axes. (Compare, for example, a long rod rotating about its central axis and about one end.) Furthermore, in some cases, the rotational momentum is not aligned along the axis of rotation. These more complicated cases require the mathematics of "tensors" to handle them correctly; which is beyond the scope of this book.

READING EXERCISE 12-4: In the figure, a disk, a hoop, and a solid sphere are made to spin about fixed central axes (like a top) by means of strings wrapped around them, with the strings producing the same constant tangential force $\vec{F}$ on all three objects. The three objects have the same mass and radius, and they are initially stationary. Rank the objects according to (a) their rotational momentum about their central axes and (b) their rotational speed, greatest first, when the strings have been pulled for a certain time t.

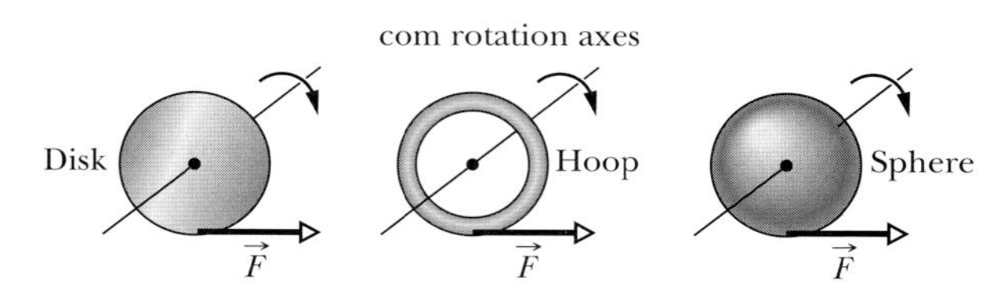

TOUCHSTONE EXAMPLE 12-5: First Ferris Wheel

George Washington Gale Ferris, Jr., a civil engineering graduate from Rensselaer Polytechnic Institute, built the original Ferris wheel (Fig. 12-15) for the 1893 World's Columbian Exposition in Chicago. The wheel, an astounding engineering construction at the time, carried 36 wooden cars, each holding as many as 60 passengers, around a circle of radius $R = 38$ m. The mass of each car was about 1.1×10^4 kg. The mass of the wheel's structure was about 6×10^5 kg, which was mostly in the circular grid at the rim of the wheel from which the cars were suspended. The cars were loaded 6 at a time, and once all 36 cars were full, the wheel made a complete rotation at a constant rotational speed ω_F in about 2 min.

FIGURE 12-15 ■ The original Ferris wheel, built in 1893 near the University of Chicago, towered over the surrounding buildings.

(a) Estimate the magnitude L of the rotational momentum of the wheel and its passengers while the wheel rotated at ω_F.

SOLUTION ■ The **Key Idea** here is that we can treat the wheel, cars, and passengers as a rigid object rotating about a fixed axis, at the wheel's axle. Then Eq. 12-34 ($\vec{L} = I\vec{\omega}$) gives the magnitude of the rotational momentum of that object. We need to find the rotational inertia I of this object and the rotational speed ω_F.

To find I, let us start with the loaded cars. Because we can treat them as particles, at distance R from the axis of rotation, we know from Eq. 11-23 that their rotational inertia is $I_{pc} = M_{pc}R^2$, where M_{pc} is their total mass. Let us assume that the 36 cars are each filled with 60 passengers, each of mass 70 kg. Then their total mass is

$$M_{pc} = 36[1.1 \times 10^4 \text{ kg} + 60(70 \text{ kg})] = 5.47 \times 10^5 \text{ kg}$$

and their rotational inertia is

$$I_{pc} = M_{pc}R^2 = (5.47 \times 10^5 \text{ kg})(38 \text{ m})^2$$
$$= 7.90 \times 10^8 \text{ kg} \cdot \text{m}^2.$$

Next we consider the structure of the wheel. Let us assume that the rotational inertia of the structure is due mainly to the circular grid suspending the cars. Further, let us assume that the grid forms a hoop of radius R, with a mass M_{hoop} of 3×10^5 kg (half the wheel's mass). From Table 11-2(*a*), the rotational inertia of the hoop is

$$I_{hoop} = M_{hoop}R^2 = (3.0 \times 10^5 \text{ kg})(38 \text{ m})^2$$
$$= 4.33 \times 10^8 \text{ kg} \cdot \text{m}^2.$$

The combined rotational inertia I of the cars, passengers, and hoop is then

$$I = I_{pc} + I_{hoop} = 7.90 \times 10^8 \text{ kg} \cdot \text{m}^2 + 4.33 \times 10^8 \text{ kg} \cdot \text{m}^2$$
$$= 1.22 \times 10^9 \text{ kg} \cdot \text{m}^2.$$

To find the rotational speed ω_F, we use Eq. 11-5 ($\langle\omega_z\rangle = \Delta\theta/\Delta t$). Here the wheel goes through a rotational displacement of $\Delta\theta = 2\pi$ rad in a time period $\Delta t = 2$ min. Thus, we have

$$|\langle\omega_F\rangle| = |\omega_F| = \frac{2\pi \text{ rad}}{(2 \text{ min})(60 \text{ s/min})} = 0.0524 \text{ rad/s},$$

since at constant rotational speed $\langle\omega_F\rangle = \omega_F$. Now we can find the magnitude L of the rotational momentum with Eq. 12-34:

$$|\vec{L}| = I|\vec{\omega}_F| = (1.22 \times 10^9 \text{ kg} \cdot \text{m}^2)(0.0524 \text{ rad/s})$$
$$= 6.39 \times 10^7 \text{ kg} \cdot \text{m}^2/\text{s} \approx 6.4 \times 10^7 \text{ kg} \cdot \text{m}^2/\text{s}. \quad \text{(Answer)}$$

(b) Assume that the fully loaded wheel is rotated from rest to ω_F in a time period $\Delta t = 5.0$ s. What is the magnitude $|\langle\tau\rangle|$ of the average net external torque acting on it during Δt?

SOLUTION ■ The **Key Idea** here is that the average net external torque is related to the rate of change in the rotational momentum of the loaded wheel by Eq. 12-32 ($\vec{\tau}^{\,net} = d\vec{L}/dt$). The wheel rotates about a fixed axis to reach rotational speed ω_F in time period Δt and the change ΔL is from zero to the answer for part (a). Thus, we have

$$|\langle\vec{\tau}^{\,net}\rangle| = \left|\frac{\Delta\vec{L}}{\Delta t_1}\right| = \frac{6.39 \times 10^7 \text{ kg} \cdot \text{m}^2/\text{s} - 0}{5.0 \text{ s}}$$
$$\approx 1.3 \times 10^7 \text{ N} \cdot \text{m}. \quad \text{(Answer)}$$

12-10 Conservation of Rotational Momentum

So far we have discussed two powerful conservation laws, the conservation of energy and the conservation of translational momentum. Now we meet a third law of this type, involving the conservation of rotational momentum. We start from Eq. 12-32 ($\vec{\tau}^{\,net} = d\vec{L}/dt$), which is Newton's Second Law in rotational form. If no net external torque acts on the system, this equation becomes $d\vec{L}/dt = 0$, or

$$\vec{L} = \text{a constant} \qquad (\vec{\tau}^{\,net} = 0). \tag{12-35}$$

This result, called the **law of conservation of rotational momentum,** can also be written as

$$\begin{Bmatrix}\text{Net rotational momentum}\\ \text{at some initial time } t_1\end{Bmatrix} = \begin{Bmatrix}\text{Net rotational momentum}\\ \text{at some later time } t_2\end{Bmatrix}$$

or

$$\vec{L}_1 = \vec{L}_2 \qquad (\vec{\tau}^{\,net} = 0). \tag{12-36}$$

Equation 12-35 ($\vec{L} =$ a constant) and Eq. 12-36 ($\vec{L}_1 = \vec{L}_2$) tell us:

> If the net (external) torque acting on a system is zero, the rotational momentum $\vec{L}$ of the system remains constant, no matter what changes take place within the system.

Equations 12-32 ($\vec{\tau}^{\,net} = d\vec{L}/dt$) and 12-36 ($\vec{L}_1 = \vec{L}_2$) are vector equations. As such, they are equivalent to three component equations corresponding to the conservation of rotational momentum in three mutually perpendicular directions. Depending on the torques acting on a system, the rotational momentum of the system might be conserved in only one or two directions but not in all three:

> If the component of the net *external* torque on a system along a certain axis is zero, then the component of the rotational momentum of the system along that axis cannot change, no matter what changes take place within the system.

We can apply this law to the isolated body in Fig. 12-14, which rotates around the z axis. Suppose that the initially rigid body somehow redistributes its mass relative to that rotation axis, changing its rotational inertia about that axis. Equation 12-35 ($\vec{L} =$ a constant) and Eq. 12-36 ($\vec{L}_1 = \vec{L}_2$) state that the rotational momentum of the body cannot change in the absence of a net external torque. Substituting $\vec{L} = I\vec{\omega}$ (Eq. 12-34) for the rotational momentum along the rotational axis into Eq. 12-36, we write this conservation law as

$$I_1\vec{\omega}_1 = I_2\vec{\omega}_2 \qquad (\vec{\tau}^{\,net} = 0). \tag{12-37}$$

Here the subscripts refer to the values of the rotational inertia I and rotational speed ω before and after the redistribution of mass.

Like the other two conservation laws that we have discussed, $\vec{L} =$ a constant and $\vec{L}_1 = \vec{L}_2$ hold beyond the limitations of Newtonian mechanics. They hold for particles whose speeds approach that of light (where the theory of special relativity reigns), and they remain true in the world of subatomic particles (where quantum physics reigns). No exceptions to the law of conservation of rotational momentum have ever been found.

We now discuss four examples involving this law.

1. ***The spinning volunteer.*** Figure 12-16 shows a student seated on a stool that can rotate freely about a vertical axis. The student, who has been set into rotation at a modest initial rotational speed ω_1, holds two dumbbells in his outstretched hands. His rotational momentum vector $\vec{L}$ lies along the vertical rotation axis, pointing upward.

   The instructor now asks the student to pull in his arms. This action reduces his rotational inertia from its initial value I_1 to a smaller value I_2 because he moves mass closer to the rotation axis. His rate of rotation increases markedly, from $\vec{\omega}_1$ to $\vec{\omega}_2$. The student can then slow down by extending his arms once more.

   No net external torque acts along the vertical axis of the system consisting of the student, stool, and dumbbells. Thus, the rotational momentum of that system about the rotation axis must remain constant. In Fig. 12-16*a*, the student's rotational speed $|\vec{\omega}_1|$ is relatively low and his rotational inertia I_1 is relatively high. According to Eq. 12-37, ($I_1\vec{\omega}_1 = I_2\vec{\omega}_2$) his rotational speed in Fig. 12-16*b* must be greater to compensate for the decreased rotational inertia.

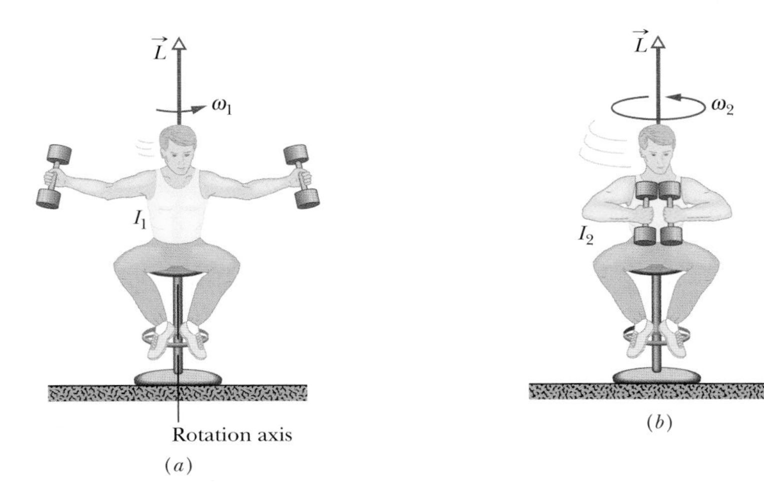

FIGURE 12-16 ■ (*a*) The student has a relatively large rotational inertia and a relatively small rotational speed. (*b*) By decreasing his rotational inertia, the student automatically increases his rotational speed. The rotational momentum $\vec{L}$ of the rotating system remains unchanged.

2. ***The springboard diver.*** Figure 12-17 shows a diver doing a forward one-and-a-half-somersault dive. As you should expect from our discussion in Chapter 8, her center of mass follows a parabolic path. She leaves the springboard with a definite rotational momentum $\vec{L}$ about an axis through her center of mass, represented by a vector pointing into the plane of Fig. 12-17, perpendicular to the page. When she is in the air, no net external torque acts on her about her center of mass (assuming air drag is negligible). So, her rotational momentum about her center of mass cannot change. By pulling her arms and legs into the *closed pike* position (in the fourth image), she reduces her rotational inertia about the same axis and thus, according to $I_1\vec{\omega}_1 = I_2\vec{\omega}_2$ (Eq. 12-37), increases her rotational speed. Pulling out of the closed pike position (and back into the *open layout position*) at the end of the dive increases her rotational inertia. This slows her rotation rate so she can enter the water with little splash. Even in a more complicated dive involving both twisting and somersaulting, the rotational momentum of the diver must be conserved, in both magnitude *and* direction, throughout the dive.

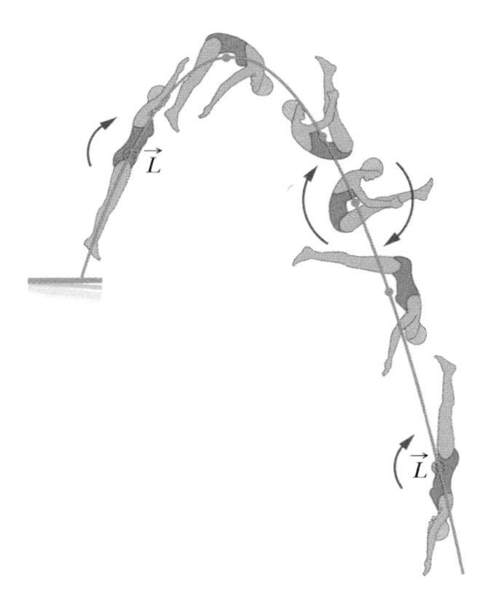

FIGURE 12-17 ■ A diver rotates about her center of mass as she falls. Since she has no net torque relative to her center of mass, her rotational momentum is constant throughout the dive. Note also that her center of mass (see the dots) follows a parabolic path as she falls.

3. ***Spacecraft orientation.*** Figure 12-18, which represents a spacecraft with a rigidly mounted flywheel, suggests a scheme (albeit crude) for orientation control. The *spacecraft + flywheel* form a system on which no net torque acts. Therefore, if the system's total rotational momentum $\vec{L}$ is zero because neither spacecraft nor flywheel is turning, it must remain zero (as long as the system remains isolated).

   To change the orientation of the spacecraft, the flywheel is made to rotate (Fig. 12-18*a*). The spacecraft will start to rotate in the opposite direction to maintain the system's rotational momentum at zero. When the flywheel is then brought to rest, the spacecraft will also stop rotating but will have changed its orientation (Fig. 12-18*b*). Throughout, the rotational momentum of the system *spacecraft + flywheel* never differs from zero.

Interestingly, the spacecraft *Voyager 2*, on its 1986 flyby of the planet Uranus, was set into unwanted rotation by this flywheel effect every time its tape recorder was turned on at high speed. The ground staff at the Jet Propulsion Laboratory had to program the on-board computer to turn on counteracting thruster jets every time the tape recorder was turned on.

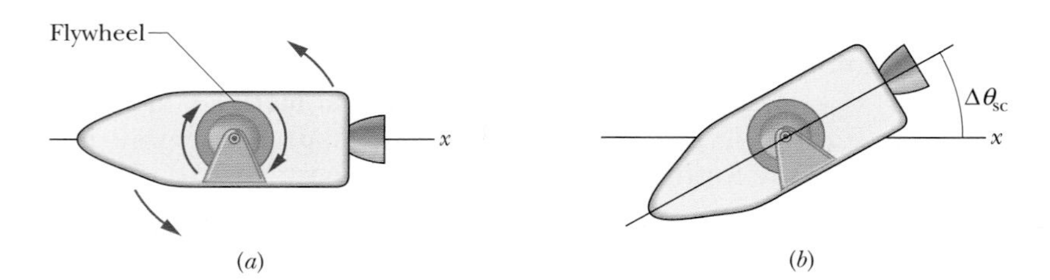

FIGURE 12-18 ■ (*a*) An idealized spacecraft containing a flywheel. If the flywheel is made to rotate clockwise as shown, the spacecraft itself will rotate counterclockwise because the total rotational momentum must remain zero. (*b*) When the flywheel is braked to a stop, the spacecraft will also stop rotating but will have reoriented its axis by the angle $\Delta\theta_{sc}$.

4. ***The incredible shrinking star.*** When the nuclear fire in the core of a star burns low, the star may eventually begin to collapse, building up pressure in its interior. The collapse may go so far as to reduce the radius of the star from something like that of the Sun to the incredibly small value of a few kilometers. The star then becomes a *neutron star*—its material has been compressed to an extremely dense gas of neutrons.

   During this shrinking process, the star is an isolated system and its rotational momentum $\vec{L}$ cannot change. Because its rotational inertia is greatly reduced, its rotational speed is correspondingly greatly increased, to as much as 600 to 800 revolutions per *second.* For comparison, the Sun, a typical star, rotates at about one revolution per month.

Summary of Rotational vs. Translational Equations

Table 12-2 supplements Table 11-3 with some of the new equations developed in this chapter. It extends our list of corresponding translational and rotational relations.

TABLE 12-2
More Corresponding Relations for Translational and Rotational Motion[a]

Translational		Rotational	
Force	$\vec{F}$	Torque	$\vec{\tau} = \vec{r} \times \vec{F}$
Translational momentum	$\vec{p}^{\,\text{sys}}$	Rotational momentum	$\vec{\ell} = \vec{r} \times \vec{p}$
Translational momentum[b]	$\vec{p}^{\,\text{sys}} = \Sigma\vec{p}_i$	Rotational momentum[b]	$\vec{L} = \Sigma\vec{\ell}_i$
Translational momentum[b]	$\vec{p}^{\,\text{sys}} = M\vec{v}_{\text{com}}$	Rotational momentum[c]	$\vec{L} = I\vec{\omega}$
Newton's Second Law[b]	$\Sigma\vec{F}^{\text{ext}} = \dfrac{d\vec{p}^{\,\text{sys}}}{dt}$	Newton's Second Law[b]	$\Sigma\vec{\tau}^{\text{ext}} = \dfrac{d\vec{L}}{dt}$
Conservation law[d]	$\vec{p}^{\,\text{sys}} =$ a constant	Conservation law[d]	$\vec{L} =$ a constant

[a] See also Table 11-3.
[b] For systems of particles, including rigid bodies.
[c] For a rigid body about a fixed axis, with L being the component along that axis.
[d] For a closed, isolated system ($\vec{F}^{\text{net}} = 0$, $\vec{\tau}^{\text{net}} = 0$).

READING EXERCISE 12-5: A rhinoceros beetle rides the rim of a small disk that rotates like a merry-go-round. If the beetle crawls toward the center of the disk, do the following (each relative to the central axis) increase, decrease, or remain the same: (a) the rotational inertia of the beetle–disk system, (b) the rotational momentum of the system, and (c) the rotational speed of the beetle and disk? ■

TOUCHSTONE EXAMPLE 12-6: Student with a Wheel

Figure 12-19*a* shows a student sitting on a stool that can rotate freely about a vertical axis. The student, initially at rest, is holding a bicycle wheel whose rim is loaded with lead and whose rotational inertia I_{wh} about its central axis is 1.2 kg · m². The wheel is rotating at a rotational speed ω_{wh} of 3.9 rev/s; as seen from overhead, the rotation is counterclockwise. The axis of the wheel is vertical, and the rotational momentum $\vec{L}_{wh}$ of the wheel points vertically upward. The student now inverts the wheel (Fig. 12-19*b*) so that, as seen from overhead, it is rotating clockwise. Its rotational momentum is then $-\vec{L}_{wh}$. The inversion results in the student, the stool, and the wheel's center rotating together as a composite rigid body about the stool's rotation axis, with rotational inertia $I_b = 6.8$ kg · m². (The fact that the wheel is also rotating about its center does not affect the mass distribution of this composite body; thus, I_b has the same value whether or not the wheel rotates.) With what rotational speed ω_b and in what direction does the composite body rotate after the inversion of the wheel?

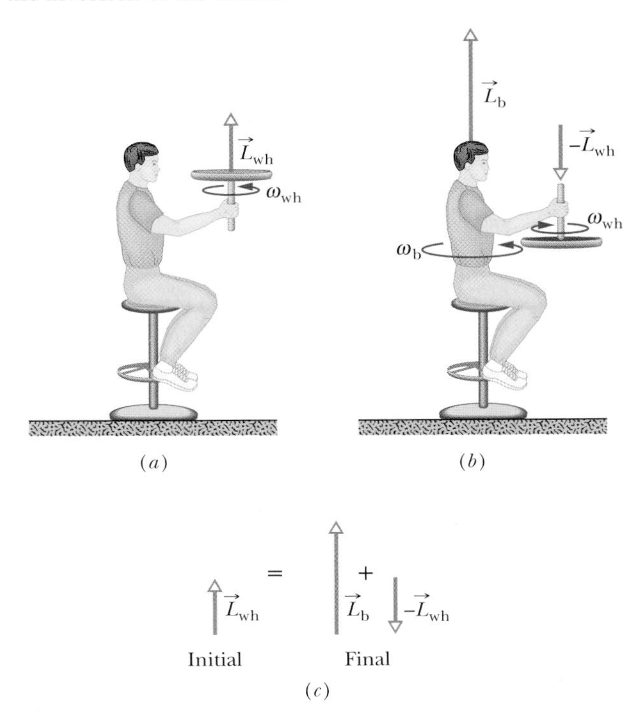

FIGURE 12-19 ■ (*a*) A student holds a bicycle wheel rotating around the vertical. (*b*) The student inverts the wheel, setting himself into rotation. (*c*) The net angular momentum of the system must remain the same in spite of the inversion.

SOLUTION ■ The **Key Ideas** here are these:

1. The rotational speed ω_b we seek is related to the final rotational momentum $\vec{L}_b$ of the composite body about the stool's rotation axis by Eq. 12-34 ($\vec{L} = I\vec{\omega}$).
2. The initial rotational speed ω_{wh} of the wheel is related to the rotational momentum $\vec{L}_{wh}$ of the wheel's rotation about its center by the same equation.
3. The vector addition of $\vec{L}_b$ and $\vec{L}_{wh}$ gives the total rotational momentum $\vec{L}^{tot}$ of the system of student, stool, and wheel.
4. As the wheel is inverted, no net *external* torque acts on that system to change $\vec{L}^{tot}$ about any vertical axis. (Torques due to forces between the student and the wheel as the student inverts the wheel are *internal* to the system.) So, the system's total rotational momentum is conserved about any vertical axis.

The conservation of $\vec{L}^{tot}$ is represented with vectors in Fig. 12-19*c*. We can also write it in terms of components along a vertical axis as

$$L_{b\,y}(t_2) + L_{wh\,y}(t_2) = L_{b\,y}(t_1) + L_{wh\,y}(t_1), \tag{12-38}$$

where t_1 and t_2 refer to the initial state (before inversion of the wheel) and the final state (after inversion). Because inversion of the wheel inverted the wheel's rotational momentum vector, we substitute $-L_{wh\,y}(t_1)$ for $L_{wh\,y}(t_2)$. Then, if we set $L_{b\,y}(t_1) = 0$ (because the student, the stool, and the wheel's center were initially at rest), Eq. 12-38 yields

$$L_{b\,y}(t_2) = 2L_{wh\,y}(t_1).$$

We next substitute $I_b\omega_{b\,y}$ for $L_{b\,y}$ and $I_{wh}\omega_{wh\,y}$ for $L_{wh\,y}$ and solve for ω_b, finding

$$\omega_{b\,y}\hat{j} = \frac{2I_{wh}}{I_b}\omega_{wh\,y}\hat{j}$$

$$= \frac{(2)(1.2\text{ kg}\cdot\text{m}^2)(3.9\text{ rev/s})}{6.8\text{ kg}\cdot\text{m}^2}\hat{j} = (1.4\text{ rev/s})\,\hat{j}.$$

(Answer)

The fact that this final rotational velocity points upward tells us that the student rotates counterclockwise about the stool axis as seen from overhead. If the student wishes to stop rotating, he has only to invert the wheel once more.

TOUCHSTONE EXAMPLE 12-7: Quadruple Somersault

During a jump to his partner, an aerialist is to make a quadruple somersault lasting a time $t = 1.87$ s. For the first and last quarter-revolution, he is in the extended orientation shown in Fig. 12-20, with rotational inertia $I_1 = 19.9 \text{ kg} \cdot \text{m}^2$ around his center of mass (the dot). During the rest of the flight he is in a tight tuck, with rotational inertia $I_2 = 3.93 \text{ kg} \cdot \text{m}^2$. What must be his rotational speed ω_2 around his center of mass during the tuck?

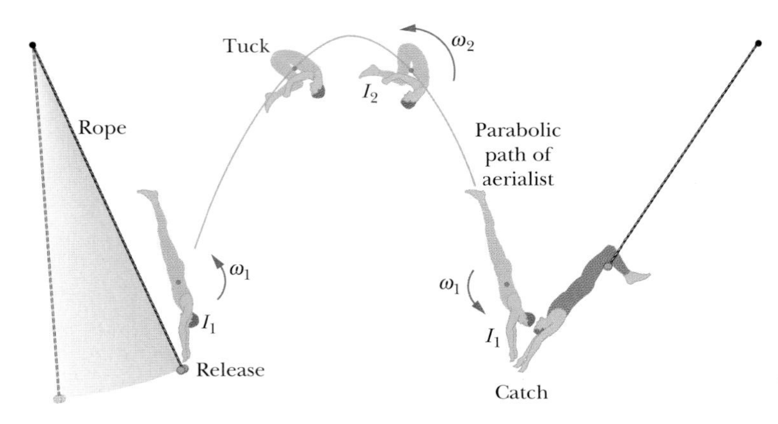

FIGURE 12-20 ■ An aerialist performing a multiple somersault.

SOLUTION ■ Obviously he must turn fast enough to complete the 4.0 rev required for a quadruple somersault in the given 1.87 s. To do so, he increases his rotational speed to ω_2 by tucking. We can relate ω_2 to his initial rotational speed ω_1 with this **Key Idea:** His rotational momentum about his center of mass is conserved throughout the free flight because there is no net external torque about his center of mass to change it. From Eq. 12-37, we can write the conservation of rotational momentum ($\vec{L}_1 = \vec{L}_2$) as

$$I_1\vec{\omega}_1 = I_2\vec{\omega}_2,$$

or

$$\vec{\omega}_1 = \frac{I_2}{I_1}\vec{\omega}_2. \tag{12-39}$$

A second **Key Idea** is that these rotational speeds are related to the angles through which he must rotate and the time available to do so. At the start and at the end, he must rotate in the extended orientation for a total angle of $\Delta\theta_1 = 0.500$ rev (two quarter-turns) in a time we shall call Δt_1. In the tuck, he must rotate through an angle of $\Delta\theta_2 = 3.50$ rev in a time Δt_2. From Eq. 11-5 ($\langle\omega\rangle = \Delta\theta/\Delta t$), we can write

$$\Delta t_1 = \frac{\Delta\theta_1}{\omega_1} \quad \text{and} \quad \Delta t_2 = \frac{\Delta\theta_2}{\omega_2}.$$

Thus, his total flight time is

$$\Delta t = \Delta t_1 + \Delta t_2 = \frac{\Delta\theta_1}{\omega_1} + \frac{\Delta\theta_2}{\omega_2},$$

which we know to be 1.87 s. Now substituting from Eq. 12-39 yields for ω_1

$$\Delta t = \frac{(\Delta\theta_1)I_1}{\omega_2 I_2} + \frac{\Delta\theta_2}{\omega_2} = \frac{1}{\omega_2}\left(\Delta\theta_1\frac{I_1}{I_2} + \Delta\theta_2\right).$$

Inserting the known data, we obtain

$$1.87 \text{ s} = \frac{1}{\omega_2}\left((0.500 \text{ rev})\frac{19.9 \text{ kg}\cdot\text{m}^2}{3.93 \text{ kg}\cdot\text{m}^2} + 3.50 \text{ rev}\right),$$

which gives us

$$\omega_2 = 3.23 \text{ rev/s}. \quad \text{(Answer)}$$

This rotational speed is so fast that the aerialist cannot clearly see his surroundings or fine-tune his rotation by adjusting his tuck. The possibility of an aerialist making a four-and-a-half-somersault flight, which would require a greater value of ω_2 and thus a smaller I_2 via a tighter tuck, seems very small.

TOUCHSTONE EXAMPLE 12-8: Turnstile Takes a Hit

(This touchstone example is long and challenging, but it is helpful because it pulls together many ideas of Chapters 11 and 12.) In the overhead view of Fig. 12-21, four thin, uniform rods, each of mass M and length $d = 0.50$ m, are rigidly connected to a vertical axle to form a turnstile. The turnstile rotates clockwise about the axle, which is attached to a floor, with initial rotational velocity $\vec{\omega}_1 = (-2.0 \text{ rad/s})\hat{j}$. A mud ball of mass $m = \frac{1}{3}M$ and initial speed $v_1 = 12$ m/s is thrown along the path shown and sticks to the end of one rod. What is the final rotational velocity $\vec{\omega}_2$ of the ball–turnstile system?

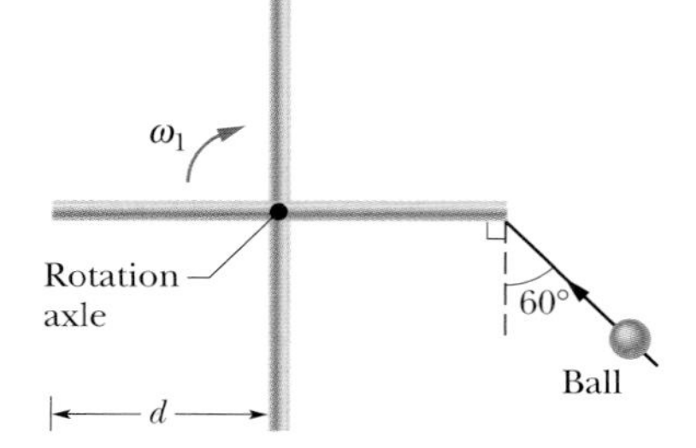

FIGURE 12-21 ■ An overhead view of four rigidly connected rods rotating freely around a central axle, and the path a mud ball takes to stick onto one of the rods.

SOLUTION ■ A **Key Idea** here can be stated in a question-and-answer format. The question is this: Does the system have a quantity that is conserved during the collision and that involves rotational velocity, so that we can solve for $\vec{\omega}_2$? To answer, let us check the conservation possibilities:

1. The total kinetic energy K is *not* conserved, because the collision between ball and rod is completely inelastic (the ball sticks). So, some energy must be transferred from kinetic energy to other types of energy (such as thermal energy). For the same reason, total mechanical energy is not conserved.
2. The total translational momentum $\vec{P}$ is also *not* conserved, because during the collision an external force acts on the turnstile at the attachment of the axle to the floor. (This is the force that keeps the turnstile from moving across the floor when it is hit by the mud ball.)
3. The total rotational momentum $\vec{L} = L_y\hat{\text{j}}$ of the system about the axle *is* conserved because there is no net external torque to change $\vec{L}$. (The forces in the collision produce only internal torques; the external force on the turnstile acts at the axle, has zero moment arm, and thus does not produce an external torque.)

We can write the conservation of the system's total rotational momentum ($\vec{L}_2 = \vec{L}_1$) about the axle as

$$L_{\text{ts}\,y}(t_2) + L_{\text{ball}\,y}(t_2) = L_{\text{ts}\,y}(t_1) + L_{\text{ball}\,y}(t_1), \qquad (12\text{-}40)$$

where ts stands for turnstile. The final rotational velocity $\vec{\omega}_2$ is contained in the terms $L_{\text{ts}\,y}(t_2)$ and $L_{\text{ball}\,y}(t_2)$ because those final rotational momenta depend on how fast the turnstile and ball are rotating. To find $\vec{\omega}_2$, we consider first the turnstile and then the ball, and then we return to Eq. 12-40.

Turnstile: The **Key Idea** here is that, because the turnstile is a rotating rigid object, Eq. 12-34 ($\vec{L} = I\vec{\omega}$) gives its rotational momentum. Thus we can write its final and initial rotational momenta about the axle as

$$L_{\text{ts}\,y}(t_2) = I_{\text{ts}}\omega_{2\,y} \quad \text{and} \quad L_{\text{ts}\,y}(t_1) = I_{\text{ts}}\omega_{1\,y}. \qquad (12\text{-}41)$$

Because the turnstile consists of four rods, each rotating around an end, the rotational inertia I_{ts} of the turnstile is four times the rotational inertia I_{rod} of each rod about its end. From Table 11-2(e), we know that the rotational inertia I_{com} of a rod about its center is $\frac{1}{12}Md^2$, where M is its mass and d is its length. To get I_{rod}, we use the parallel-axis theorem of Eq. 11-28 ($I = I_{\text{com}} + Mh^2$). Here perpendicular distance h is $d/2$. Thus, we find

$$I_{\text{rod}} = \tfrac{1}{12}Md^2 + M\left(\frac{d}{2}\right)^2 = \tfrac{1}{3}Md^2.$$

With four rods in the turnstile, we then have

$$I_{\text{ts}} = \tfrac{4}{3}Md^2. \qquad (12\text{-}42)$$

Ball: Before the collision, the ball is like a particle moving along a straight line, as in Fig. 12-13. So, to find the ball's initial rotational momentum $L_{\text{ball}\,y}(t_1)$ about the axle, we can use any of Eqs. 12-25 through 12-28, but Eq. 12-27 ($\ell = rmv_\perp$) is easiest. Here ℓ is $L_{\text{ball}\,y}(t_1)$. Just before the ball hits, its radial distance r from the axle is d, and the component $v_\perp$ of the ball's velocity perpendicular to r is $v_1 \cos 60°$.

To give a sign to this rotational momentum, we mentally draw a position vector from the turnstile's axle to the ball. As the ball approaches the turnstile, this position vector rotates counterclockwise about the axle, so the ball's rotational momentum is a positive quantity. We can now rewrite $\ell = rmv_\perp$ as

$$L_{\text{ball}\,y}(t_1) = mdv_1 \cos 60°. \qquad (12\text{-}43)$$

After the collision, the ball is like a particle rotating in a circle of radius d. So, from Eq. 11-23 ($I = \Sigma m_i r_i{}^2$), we have $I_{\text{ball}} = md^2$ about the axle. Then from Eq. 12-34 ($\vec{L} = I\vec{\omega}$), we can write the final rotational momentum of the ball about the axle as

$$L_{\text{ball}\,y}(t_2) = I_{\text{ball}}\omega_{2y} = md^2\omega_{2y}. \qquad (12\text{-}44)$$

Return to Eq. 12-40: Substituting from Eqs. 12-41 through 12-44 into Eq. 12-40, we have

$$\tfrac{4}{3}Md^2\omega_{2\,y} + md^2\omega_{2\,y} = \tfrac{4}{3}Md^2\omega_{1\,y} + mdv_1 \cos 60°.$$

Substituting $M = 3m$ and solving for $\omega_{2\,y}$, we find

$$\begin{aligned}\omega_{2\,y} &= \frac{1}{5d}(4d\omega_{1\,y} + v_1 \cos 60°)\\ &= \frac{1}{5(0.50\text{ m})}[4(0.50\text{ m})(-2.0\text{ rad/s}) + (12\text{ m/s})(\cos 60°)]\\ &= 0.80\text{ rad/s}. \qquad \text{(Answer)}\end{aligned}$$

Thus, the turnstile is now turning counterclockwise.

Problems

SEC. 12-2 ■ COMBINING TRANSLATIONS WITH SIMPLE ROTATIONS

Unless otherwise noted, rolling occurs without slipping.

1. An Automobile Traveling An automobile traveling 80.0 km/h has tires of 75.0 cm diameter. (a) What is the rotational speed of the tires about their axles? (b) If the car is brought to a stop uniformly in 30.0 complete turns of the tires (without skidding), what is the magnitude of the rotational acceleration of the wheels? (c) How far does the car move during the braking?

2. Car's Tire Consider a 66-cm-diameter tire on a car traveling at 80 km/h on a level road in the positive direction of an x axis. Relative to a woman in the car, what are (a) the translational velocity $\vec{v}_{\text{center}}$ and (b) the magnitude a_{center} of the translational acceleration of the center of the wheel? What are (c) $\vec{v}_{\text{top}}$ and (d) a_{top} for a point at the top of the tire? What are (e) $\vec{v}_{\text{bot}}$ and (f) a_{bot} for a point at the bottom of the tire?

Now repeat the questions relative to a hitchhiker sitting near the road: What are (g) $\vec{v}$ at the wheel's center, (h) a at the wheel's center, (i) $\vec{v}$ at the tire top, (j) a at the tire top, (k) $\vec{v}$ at the tire bottom, and (l) a at the tire bottom?

3. A Hoop Rolls A 140 kg hoop rolls along a horizontal floor so that its center of mass has a speed of 0.150 m/s. How much work must be done on the hoop to stop it?

4. Thin-Walled Pipe A thin-walled pipe rolls along the floor. What is the ratio of its translational kinetic energy to its rotational kinetic energy about an axis parallel to its length and through its center of mass?

5. Car Has Four Wheels A 1000 kg car has four 10 kg wheels. When the car is moving, what fraction of the total kinetic energy of the car is due to rotation of the wheels about their axles? Assume that the wheels have the same rotational inertia as uniform disks of the same mass and size. Why do you not need the radius of the wheels?

6. A Body of Radius *R* A body of radius R and mass m is rolling smoothly with speed v on a horizontal surface. It then rolls up a hill to a maximum height h. (a) If $h = 3v^2/4g$, what is the body's rotational inertia about the rotational axis through its center of mass? (b) What might the body be?

7. A Uniform Solid Sphere A uniform solid sphere rolls down an incline. (a) What must be the incline angle if the translational acceleration of the center of the sphere is to have a magnitude of $0.10g$? (b) If a frictionless block were to slide down the incline at that angle, would its acceleration magnitude be more than, less than, or equal to $0.10g$? Why?

8. A Hollow Sphere A hollow sphere of radius 0.15 m, with rotational inertia $I = 0.040\ \text{kg}\cdot\text{m}^2$ about a line through its center of mass, rolls without slipping up a surface inclined at 30° to the horizontal. At a certain initial position, the sphere's total kinetic energy is 20 J. (a) How much of this initial kinetic energy is rotational? (b) What is the speed of the center of mass of the sphere at the initial position? What are (c) the total kinetic energy of the sphere and (d) the speed of its center of mass after it has moved 1.0 m up along the incline from its initial position?

9. Yo-Yo's Inertia A yo-yo has a rotational inertia of $950\ \text{g}\cdot\text{cm}^2$ and a mass of 120 g. Its axle radius is 3.2 mm, and its string is 120 cm long. The yo-yo rolls from rest down to the end of the string. (a) What is the magnitude of its translational acceleration? (b) How long does it take to reach the end of the string? As it reaches the end of the string, what are its (c) translational speed, (d) translational kinetic energy, (e) rotational kinetic energy, and (f) rotational speed?

10. Instead of Rolling Suppose that the yo-yo in Problem 9, instead of rolling from rest, is thrown so that its initial speed down the string is 1.3 m/s. (a) How long does the yo-yo take to reach the end of the string? As it reaches the end of the string, what are its (b) total kinetic energy, (c) translational speed, (d) translational kinetic energy, (e) rotational speed, and (f) rotational kinetic energy?

SEC. 12-4 ■ THE VECTOR OR CROSS PRODUCT

11. Area of Triangle Show that the area of the triangle contained between $\vec{a}$ and $\vec{b}$ and the solid line connecting their tips in Fig. 12-22 is $\frac{1}{2}|\vec{a} \times \vec{b}|$.

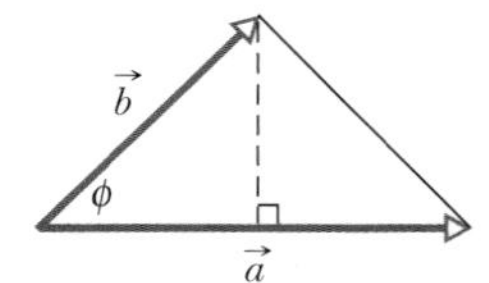

FIGURE 12-22 ■ Problem 11.

12. The Product In the product $\vec{F} = q\vec{v} \times \vec{B}$, take $q = 2$,

$$\vec{v} = 2.0\,\hat{\text{i}} + 4.0\,\hat{\text{j}} + 6.0\,\hat{\text{k}}$$

and

$$\vec{F} = 4.0\,\hat{\text{i}} - 20\,\hat{\text{j}} + 12\,\hat{\text{k}}.$$

What then is $\vec{B}$ in unit-vector notation if $B_x = B_y$?

13. Show That (a) Show that $\vec{a}\cdot(\vec{b} \times \vec{a})$ is zero for all vectors $\vec{a}$ and $\vec{b}$. (b) What is the magnitude of $\vec{a} \times (\vec{b} \times \vec{a})$ if there is an angle ϕ between the directions of $\vec{a}$ and $\vec{b}$?

14. For the Following For the following three vectors, what is $3\vec{C}\cdot(2\vec{A} \times \vec{B})$?

$$\vec{A} = 2.00\hat{\text{i}} + 3.00\hat{\text{j}} - 4.00\hat{\text{k}}$$

$$\vec{B} = -3.00\hat{\text{i}} + 4.00\hat{\text{j}} + 2.00\hat{\text{k}}$$

$$\vec{C} = 7.00\hat{\text{i}} - 8.00\hat{\text{j}}$$

SEC. 12-5 ■ TORQUE AS A VECTOR PRODUCT

15. In a Given Plane Show that, if $\vec{r}$ and $\vec{F}$ lie in a given plane, the torque $\vec{\tau} = \vec{r} \times \vec{F}$ has no component in that plane.

16. A Plum What are the magnitude and direction of the torque about the origin on a plum located at coordinates (−2.0, 0.0, 4.0) m due to force $\vec{F}$ whose only component is (a) $F_x = 6.0$ N, (b) $F_x = -6.0$ N, (c) $F_z = 6.0$ N, and (d) $F_z = -6.0$ N?

17. Particle Located at What are the magnitude and direction of the torque about the origin on a particle located at coordinates (0.0, −4.0, 3.0) m due to (a) force $\vec{F}_A$ with components $F_{A\,x} = 2.0$ N and $F_{A\,y} = F_{A\,z} = 0$, and (b) force $\vec{F}_B$ with components $F_{B\,x} = 0$, $F_{B\,y} = 2.0$ N, and $F_{B\,z} = 4.0$ N?

18. Pebble Force $\vec{F} = (2.0\ \text{N})\hat{\text{i}} - (3.0\ \text{N})\hat{\text{k}}$ acts on a pebble with position vector $\vec{r} = (0.50\ \text{m})\hat{\text{j}} - (2.0\ \text{m})\hat{\text{k}}$, relative to the origin. What is the resulting torque acting on the pebble about (a) the origin and (b) a point with coordinates (2.0, 0.0, −3.0) m?

19. Particle at Origin Force $\vec{F} = (-8.0\ \text{N})\hat{\text{i}} + (6.0\ \text{N})\hat{\text{j}}$ acts on a particle with position vector $\vec{r} = (3.0\ \text{m})\hat{\text{i}} + (4.0\ \text{m})\hat{\text{j}}$. What are (a) the torque on the particle about the origin and (b) the angle between the directions of $\vec{r}$ and $\vec{F}$?

20. Jar of Jalapeños What is the torque about the origin on a jar of jalapeño peppers located at coordinates (3.0 m, −2.0 m, 4.0 m) due to (a) force $\vec{F}_A = (3.0\ \text{N})\hat{\text{i}} - (4.0\ \text{N})\hat{\text{j}} + (5.0\ \text{N})\hat{\text{k}}$, (b) force $\vec{F}_B = (-3.0\ \text{N})\hat{\text{i}} - (4.0\ \text{N})\hat{\text{j}} - (5.0\ \text{N})\hat{\text{k}}$, and (c) the vector sum of $\vec{F}_A$ and $\vec{F}_B$? (d) Repeat part (c) about a point with coordinates (3.0 m, 2.0 m, 4.0 m) instead of about the origin.

SEC. 12-6 ■ ROTATIONAL FORM OF NEWTON'S SECOND LAW

21. A Particle with Velocity A 3.0 kg particle with velocity $\vec{v} = (5.0\ \text{m/s})\hat{\text{i}} - (6.0\ \text{m/s})\hat{\text{j}}$ is at $x = 3.0$ m, $y = 8.0$ m. It is pulled by a 7.0 N force in the negative x direction. (a) What is the rotational momentum of the particle about the origin? (b) What torque about the origin acts on the particle? (c) At what rate is the rotational momentum of the particle changing with time?

22. Acted on by Two Torques A particle is acted on by two torques about the origin: $\vec{\tau}_1$ has a magnitude of $2.0\ \text{N}\cdot\text{m}$ and is directed in the positive direction of the x axis, and $\vec{\tau}_2$ has a magnitude of

4.0 N · m and is directed in the negative direction of the y axis. What are the magnitude and direction of $d\vec{\ell}/dt$, where $\vec{\ell}$ is the rotational momentum of the particle about the origin?

23. Torque About the Origin What torque about the origin acts on a particle moving in the xy plane, clockwise about the origin, if the particle has the following magnitudes of rotational momentum about the origin:

(a) 4.0 kg · m²/s,
(b) $(4.0\,\frac{1}{s^2})t^2$ kg · m²/s,
(c) $(4.0\,\frac{1}{s^{1/2}})\sqrt{t}$ kg · m²/s,
(d) $(4.0\text{ s}^2)/t^2$ kg · m²/s?

24. At Time t At time $t = 0$, a 2.0 kg particle has position vector $\vec{r} = (4.0\text{ m})\hat{i} - (2.0\text{ m})\hat{j}$ relative to the origin. Its velocity just then is given by $\vec{v} = (-6.0\text{ m/s}^3)t^2\ \hat{i}$. About the origin and for $t > 0$, what are (a) the particle's rotational momentum and (b) the torque acting on the particle? (c) Repeat (a) and (b) about a point with coordinates $(-2.0, -3.0, 0.0)$ m instead of about the origin.

SEC. 12-7 ■ ROTATIONAL MOMENTUM

25. Two Objects Two objects are moving as shown in Fig. 12-23. What is their total rotational momentum about point O?

FIGURE 12-23 ■ Problem 25.

26. A Particle P In Fig. 12-24, a particle P with mass 2.0 kg has position vector $\vec{r}$ of magnitude 3.0 m and velocity $\vec{v}$ of magnitude 4.0 m/s. A force $\vec{F}$ of magnitude 2.0 N acts on the particle. All three vectors lie in the xy plane oriented as shown. About the origin, what are (a) the rotational momentum of the particle and (b) the torque acting on the particle?

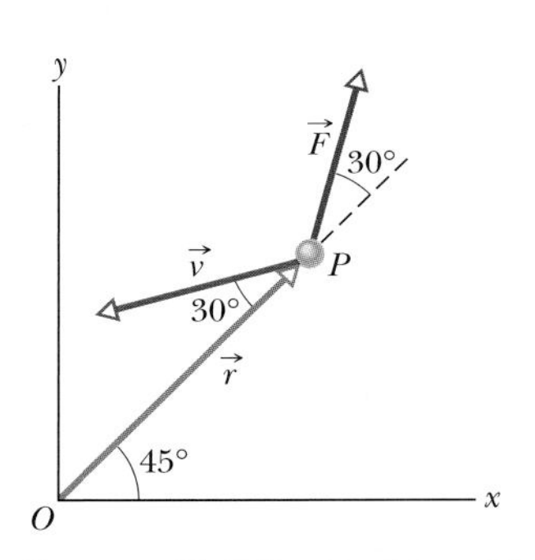

FIGURE 12-24 ■ Problem 26.

27. At a Certain Time At a certain time, a 0.25 kg object has a position vector $\vec{r} = (2.0\text{ m})\hat{i} + (-2.0\text{ m})\hat{y}$ in meters. At that instant, its velocity in meters per second is $\vec{v} = (-5.0\text{ m/s})\hat{i} + (5.0\text{ m/s})\hat{j}$ and the force in newtons acting on it is $\vec{F} = (4.0\text{ N})\hat{j}$. (a) What is the rotational momentum of the object about the origin? (b) What torque acts on it?

28. Particle-Like Object A 2.0 kg particle-like object moves in a plane with velocity components $v_x = 30$ m/s and $v_y = 60$ m/s as it passes through the point with (x, y) coordinates of $(3.0, -4.0)$ m. Just then, what is its rotational momentum relative to (a) the origin and (b) the point $(-2.0, -2.0)$ m?

29. Two Particles of Mass m Two particles, each of mass m and speed v, travel in opposite directions along parallel lines separated by a distance d. (a) In terms of m, v, and d, find an expression for the magnitude L of the rotational momentum of the two-particle system around a point midway between the two lines. (b) Does the expression change if the point about which L is calculated is not midway between the lines? (c) Now reverse the direction of travel for one of the particles and repeat (a) and (b).

30. At the Instant A 4.0 kg particle moves in an xy plane. At the instant when the particle's position and velocity are $\vec{r} = (2.0\text{ m})\hat{i} + (4.0\text{ m})\hat{j}$ and $\vec{v} = (-4.0\text{ m/s})\hat{j}$, the force on the particle is $\vec{F} = (-3.0\text{ N})\hat{i}$. At this instant, determine (a) the particle's rotational momentum about the origin, (b) the particle's rotational momentum about the point $x = 0$, $y = 4.0$ m, (c) the torque acting on the particle about the origin, and (d) the torque acting on the particle about the point $x = 0.0$ m, $y = 4.0$ m.

SEC. 12-9 ■ THE ROTATIONAL MOMENTUM OF A RIGID BODY ROTATING ABOUT A FIXED AXIS

31. Flywheel The rotational momentum of a flywheel having a rotational inertia of 0.140 kg · m² about its central axis decreases from 3.00 to 0.800 kg · m²/s in 1.50 s. (a) What is the magnitude of the average torque acting on the flywheel about its central axis during this period? (b) Assuming a constant rotational acceleration, through what angle does the flywheel turn? (c) How much work is done on the wheel? (d) What is the average power of the flywheel?

32. Sanding Disk A sanding disk with rotational inertia 1.2×10^{-3} kg · m² is attached to an electric drill whose motor delivers a torque of 16 N · m. Find (a) the rotational momentum of the disk about its central axis and (b) the rotational speed of the disk 33 ms after the motor is turned on.

33. d Apart Three particles, each of mass m, are fastened to each other and to a rotation axis at O by three massless strings, each with length d as shown in Fig. 12-25. The combination rotates around the rotational axis with rotational velocity ω in such a way that the particles remain in a straight line. In terms of m, d, and ω, and relative to point O, what are (a) the rotational inertia of the combination, (b) the rotational momentum of the middle particle, and (c) the total rotational momentum of the three particles?

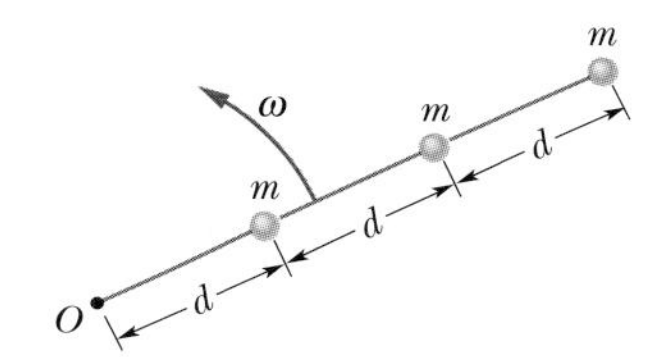

FIGURE 12-25 ■ Problem 33.

34. Impulsive Force An impulsive force $\vec{F}(t) = F_x(t)\hat{i}$ acts for a short time Δt on a rotating rigid body constrained to rotate about the z axis with rotational inertia I. Show that

$$\left(\int \tau_z\, dt\right)\hat{k} = (|\langle\vec{F}\rangle| R\, \Delta t)\hat{k} = I(\omega_{2\,z} - \omega_{1\,z})\hat{k},$$

where $\tau_z\hat{k}$ is the torque due to the force, R is the moment arm of the force, $\langle\vec{F}\rangle$ is the average value of the force during the time it acts on the body, and $\omega_{1\,z}\hat{k}$ and $\omega_{2\,z}\hat{k}$ are the rotational velocities of the body just before and just after the force acts. (The quantity $(\int \tau_z\, dt)\hat{k} = (|\langle\vec{F}\rangle| R\, \Delta t)\hat{k}$ is called the rotational *impulse,* in analogy with $\langle\vec{F}\rangle\Delta t$, the translational impulse.)

35. Two Cylinders Two cylinders having radii R_A and R_B and rotational inertias I_A and I_B about their central axes are supported by axles perpendicular to the plane of Fig. 12-26. The large cylinder is initially rotating clockwise with rotational velocity $\vec{\omega}_1$.

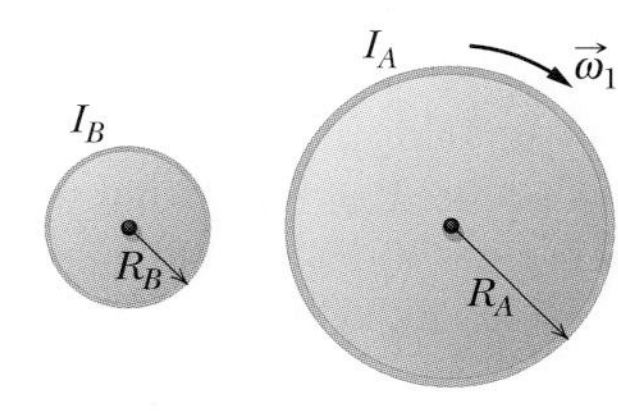

FIGURE 12-26 ■ Problem 35.

The small cylinder is moved to the right until it touches the large cylinder and is caused to rotate by the frictional force between the two. Eventually, slipping ceases, and the two cylinders rotate at constant rates in opposite directions. Find the final rotational velocity $\vec{\omega}_2$ of the small cylinder in terms of I_A, I_B, R_A, R_B, and $\vec{\omega}_1$. (*Hint*: Neither rotational momentum nor kinetic energy is conserved. Apply the rotational impulse equation of Problem 34.)

36. Rigid Structure Figure 12-27 shows a rigid structure consisting of a circular hoop of radius R and mass m, and a square made of four thin bars, each of length R and mass m. The rigid structure rotates at a constant speed about a vertical axis, with a period of rotation of 2.5 s. Assuming $R = 0.50$ m and $m = 2.0$ kg, calculate (a) the structure's rotational inertia about the axis of rotation and (b) its rotational momentum about that axis.

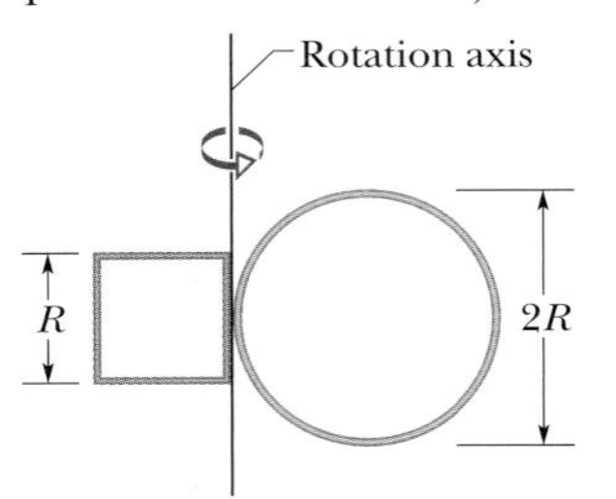

FIGURE 12-27 ■ Problem 36.

SEC. 12-10 ■ CONSERVATION OF ROTATIONAL MOMENTUM

37. A Man Stands on a Platform A man stands on a platform that is rotating (without friction) with a rotational speed of 1.2 rev/s; his arms are outstretched and he holds a brick in each hand. The rotational inertia of the system consisting of the man, bricks, and platform about the central axis is 6.0 kg · m². If by moving the bricks the man decreases the rotational inertia of the system to 2.0 kg · m², (a) what is the resulting rotational speed of the platform and (b) what is the ratio of the new kinetic energy of the system to the original kinetic energy? (c) What provided the added kinetic energy?

38. Rotor The rotor of an electric motor has rotational inertia $I_m = 2.0 \times 10^{-3}$ kg · m² about its central axis. The motor is used to change the orientation of the space probe in which it is mounted. The motor axis is mounted parallel to the axis of the probe, which has rotational inertia $I_p = 12$ kg · m² about its axis. Calculate the number of revolutions of the rotor required to turn the probe through 30° about its axis.

39. Wheel is Rotating A wheel is rotating freely at rotational speed 800 rev/min on a shaft whose rotational inertia is negligible. A second wheel, initially at rest and with twice the rotational inertia of the first, is suddenly coupled to the same shaft. (a) What is the rotational speed of the resultant combination of the shaft and two wheels? (b) What fraction of the original rotational kinetic energy is lost?

40. Two Disks Two disks are mounted on low-friction bearings on the same axle and can be brought together so that they couple and rotate as one unit. (a) The first disk, with rotational inertia 3.3 kg · m² about its central axis, is set spinning at 450 rev/min. The second disk, with rotational inertia 6.6 kg · m² about its central axis, is set spinning at 900 rev/min in the same direction as the first. They then couple together. What is their rotational speed after coupling? (b) If instead the second disk is set spinning at 900 rev/min in the direction opposite the first disk's rotation, what is their rotational speed and direction of rotation after coupling?

41. Playground In a playground, there is a small merry-go-round of radius 1.20 m and mass 180 kg. Its radius of gyration (see Problem 43 of Chapter 11) is 91.0 cm. A child of mass 44.0 kg runs at a speed of 3.00 m/s along a path that is tangent to the rim of the initially stationary merry-go-round and then jumps on. Neglect friction between the bearings and the shaft of the merry-go-round. Calculate (a) the rotational inertia of the merry-go-round about its axis of rotation, (b) the magnitude of the rotational momentum of the running child about the axis of rotation of the merry-go-round, and (c) the rotational speed of the merry-go-round and child after the child has jumped on.

42. Collapsing Spinning Star The rotational inertia of a collapsing spinning star changes to $\frac{1}{3}$ its initial value. What is the ratio of the new rotational kinetic energy to the initial rotational kinetic energy?

43. Track on a Wheel A track is mounted on a large wheel that is free to turn with negligible friction about a vertical axis (Fig. 12-28). A toy train of mass m is placed on the track and, with the system initially at rest, the electrical power is turned on. The train reaches a steady speed v with respect to the track. What is the rotational speed of the wheel if its mass is M and its radius is R? (Treat the wheel as a hoop, and neglect the mass of the spokes and hub.)

FIGURE 12-28 ■ Problem 43.

44. Two Skaters In Fig. 12-29, two skaters, each of mass 50 kg, approach each other along parallel paths separated by 3.0 m. They have opposite velocities of 1.4 m/s each. One skater carries one end of a long pole with negligible mass, and the other skater grabs the other end of it as she passes. Assume frictionless ice. (a) Describe quantitatively the motion of the skaters after they have become connected by the pole. (b) What is the kinetic energy of the two-skater system?

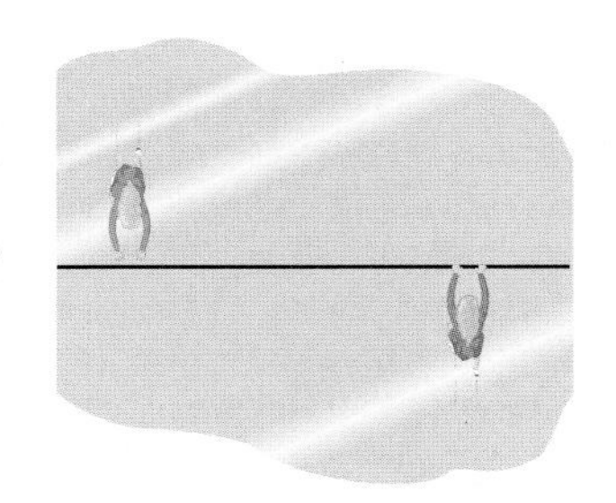
FIGURE 12-29 ■ Problem 44.

Next, the skaters each pull along the pole so as to reduce their separation to 1.0 m. What then are (c) their rotational speed and (d) the kinetic energy of the system? (e) Explain the source of the increased kinetic energy.

45. A Cockroach A cockroach of mass m runs counterclockwise around the rim of a lazy Susan (a circular dish mounted on a vertical axle) of radius R and rotational inertia I and having frictionless bearings. The cockroach's speed (relative to the ground) is v, whereas the lazy Susan turns clockwise with rotational speed ω_1. The cockroach finds a bread crumb on the rim and, of course, stops. (a) What is the rotational speed of the lazy Susan after the cockroach stops? (b) Is mechanical energy conserved?

46. Girl on a Merry-go-Round A girl of mass M stands on the rim of a frictionless merry-go-round of radius R and rotational inertia I that is not moving. She throws a rock of mass m horizontally in a direction that is tangent to the outer edge of the merry-go-round. The speed of the rock, relative to the ground, is v. Afterward, what are (a) the rotational speed of the merry-go-round and (b) the translational speed of the girl?

47. Vinyl Record A horizontal vinyl record of mass 0.10 kg and radius 0.10 m rotates freely about a vertical axis through its center with a rotational speed of 4.7 rad/s. The rotational inertia of the record about its axis of rotation is 5.0×10^{-4} kg · m^2. A wad of wet putty of mass 0.020 kg drops vertically onto the record from above and sticks to the edge of the record. What is the rotational speed of the record immediately after the putty sticks to it?

48. Uniform Thin Rod A uniform thin rod of length 0.50 m and mass 4.0 kg can rotate in a horizontal plane about a vertical axis through its center. The rod is at rest when a 3.0 g bullet traveling in the horizontal plane of the rod is fired into one end of the rod. As viewed from above, the direction of the bullet's velocity makes an angle of 60° with the rod (Fig. 12-30). If the bullet lodges in the rod and the rotational velocity of the rod is 10 rad/s immediately after the collision, what is the bullet's speed just before impact?

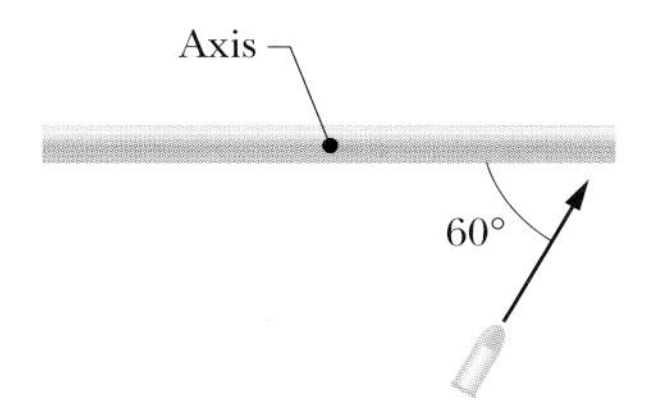

FIGURE 12-30 ■ Problem 48.

49. Putty Wad Two 2.00 kg balls are attached to the ends of a thin rod of negligible mass, 50.0 cm long. The rod is free to rotate in a vertical plane without friction about a horizontal axis through its center. With the rod initially horizontal (Fig. 12-31), a 50.0 g wad of wet putty drops onto one of the balls, hitting it with a speed of 3.00 m/s and then sticking to it. (a) What is the rotational speed of the system just after the putty wad hits? (b) What is the ratio of the kinetic energy of the entire system after the collision to that of the putty wad just before? (c) Through what angle will the system rotate until it momentarily stops?

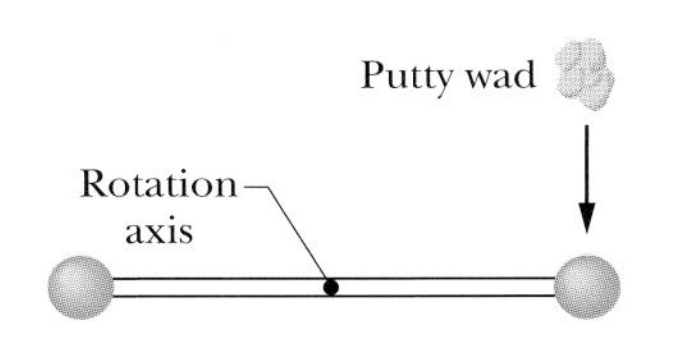

FIGURE 12-31 ■ Problem 49.

50. Cockroach on a Disk A cockroach of mass m lies on the rim of a uniform disk of mass $10.0m$ that can rotate freely about its center like a merry-go-round. Initially the cockroach and disk rotate together with a rotational velocity of ω_1. Then the cockroach walks halfway to the center of the disk. (a) What is the change $\Delta\omega$ in the rotational velocity of the cockroach–disk system? (b) What is the ratio K_2/K_1 of the new kinetic energy of the system to its initial kinetic energy? (c) What accounts for the change in the kinetic energy?

51. Earth's Polar Ice Caps If Earth's polar ice caps fully melted and the water returned to the oceans, the oceans would be deeper by about 30 m. What effect would this have on Earth's rotation? Make an estimate of the resulting change in the length of the day. (Concern has been expressed that warming of the atmosphere resulting from industrial pollution could cause the ice caps to melt.)

52. Horizontal Platform A horizontal platform in the shape of a circular disk rotates on a frictionless bearing about a vertical axle through the center of the disk. The platform has a mass of 150 kg, a radius of 2.0 m, and a rotational inertia of 300 kg · m^2 about the axis of rotation. A 60 kg student walks slowly from the rim of the platform toward the center. If the rotational speed of the system is 1.5 rad/s when the student starts at the rim, what is the rotational speed when she is 0.50 m from the center?

53. Uniform Disk A uniform disk of mass $10m$ and radius $3.0r$ can rotate freely about its fixed center like a merry-go-round. A smaller uniform disk of mass m and radius r lies on top of the larger disk, concentric with it. Initially the two disks rotate together with a rotational velocity of 20 rad/s. Then a slight disturbance causes the smaller disk to slide outward across the larger disk, until the outer edge of the smaller disk catches on the outer edge of the larger disk. Afterward, the two disks again rotate together (without further sliding). (a) What then is their rotational velocity about the center of the larger disk? (b) What is the ratio K_2/K_1 of the new kinetic energy of the two-disk system to the system's initial kinetic energy?

54. A Child Stands A 30 kg child stands on the edge of a stationary merry-go-round of mass 100 kg and radius 2.0 m. The rotational inertia of the merry-go-round about its axis of rotation is 150 kg · m^2. The child catches a ball of mass 1.0 kg thrown by a friend. Just before the ball is caught, it has a horizontal velocity of 12 m/s that makes an angle of 37° with a line tangent to the outer edge of the merry-go-round, as shown in the overhead view of Fig. 12-32. What is the rotational speed of the merry-go-round just after the ball is caught?

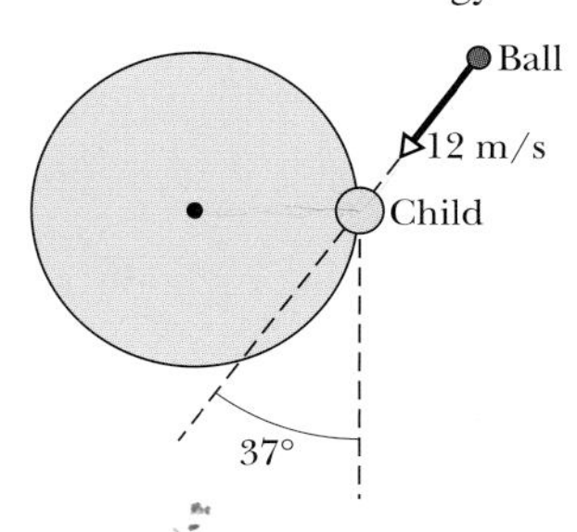

FIGURE 12-32 ■ Problem 54.

55. Bullet Hits Block In Fig. 12-33, a 1.0 g bullet is fired into a 0.50 kg block that is mounted on the end of a 0.60 m nonuniform rod of mass 0.50 kg. The block–rod–bullet system then rotates about a fixed axis at point A. The rotational inertia of the rod alone about A is 0.060 kg · m^2. Assume the block is small enough to treat as a particle on the end of the rod. (a) What is the rotational inertia of the block–rod–bullet system about point A? (b) If the rotational speed of the system about A just after the bullet's impact is 4.5 rad/s, what is the speed of the bullet just before the impact?

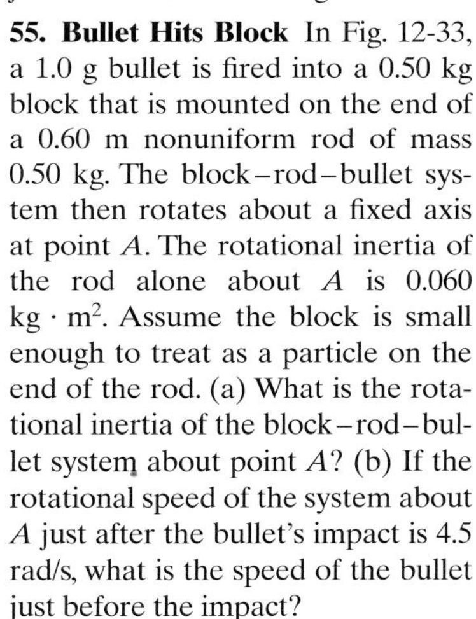

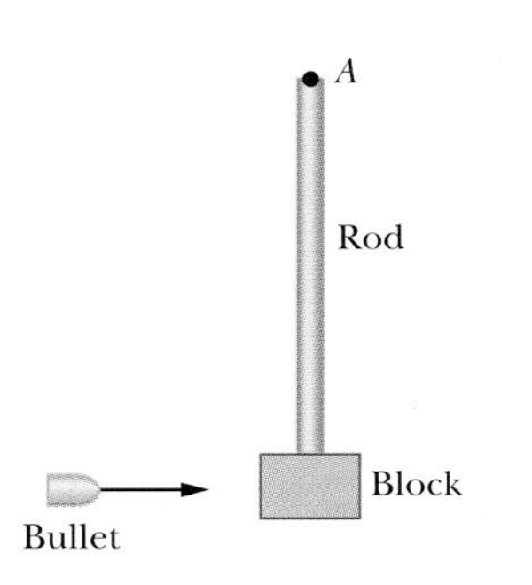

FIGURE 12-33 ■ Problem 55.

56. Uniform Rod In Fig. 12-34, a uniform rod (length = 0.60 m, mass 1.0 kg) rotates about an axis through one end, with a rotational inertia of 0.12 kg · m^2. As the rod swings through its lowest position, the end of the rod collides with a small 0.20 kg putty wad that sticks to the end of the rod. If the rotational speed of the rod just before the collision is 2.4 rad/s, what is the rotational speed of the rod–putty system immediately after the collision?

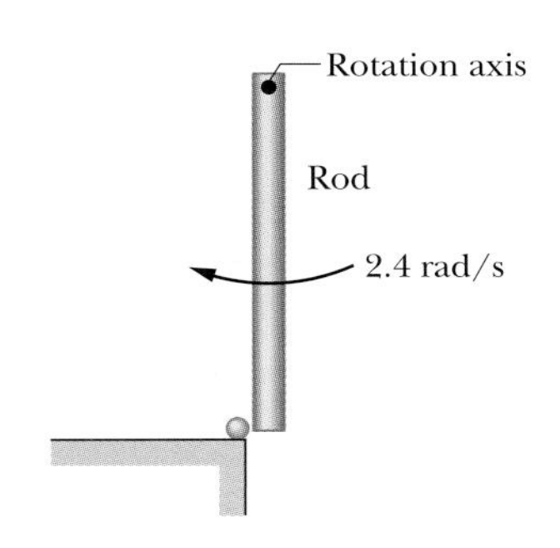

FIGURE 12-34 ■ Problem 56.

57. Particle on a Slide The particle of mass m in Fig. 12-35 slides down the frictionless surface through height h and collides with the uniform vertical rod (of mass M and length d), sticking to it. The rod pivots about point O through the angle θ before momentarily stopping. Find θ.

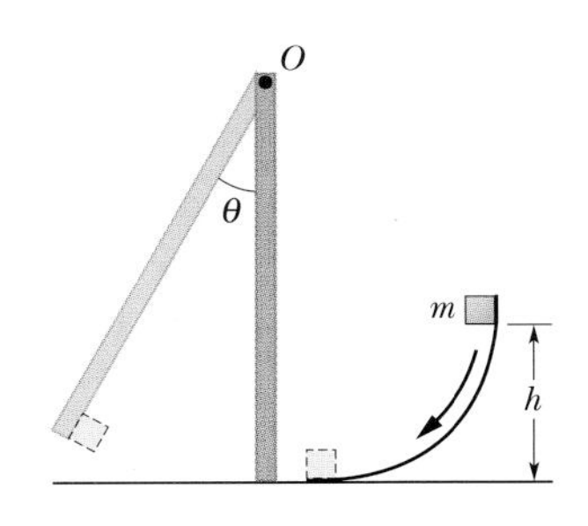

FIGURE 12-35 ■ Problem 57.

Additional Problems

58. Finding a Mistake Using Dimensional Analysis As part of an examination a few years ago, a student went through the algebraic manipulations on an exam shown in Fig. 12-36. At this point you don't know what the symbols mean, but given the information about the dimensions associated with each symbol, decide the following:

(a) Is it possible that the final equation in Fig. 12-36 is correct? Justify your answer.
(b) If the final equation is not correct, does that mean that the starting equation is necessarily wrong? Explain.
(c) If the final equation is not correct and the starting equation is not wrong, can you find the error using dimensional analysis? If so, do so. If not, explain why.

$[M] = \mathrm{M}$

$[g] = \mathrm{L/T^2}$

$[h] = \mathrm{L}$

$[\omega] = \mathrm{1/T}$

$[v] = \mathrm{L/T}$

$[R] = \mathrm{L}$

$[I] = \mathrm{ML^2}$

$$Mgh = \tfrac{1}{2}Mv^2 + \tfrac{1}{2}I\omega^2$$

$$Mgh = \tfrac{1}{2}Mv^2 + \tfrac{1}{2}(MR^2)\omega^2$$

$$Mgh = \tfrac{1}{2}Mv^2 + \tfrac{1}{2}(MR^2)\left(\frac{v^2}{R}\right)^2$$

$$gh = \tfrac{1}{2}v^2 + \tfrac{1}{2}v^4$$

FIGURE 12-36 ■ Problem 58.

Note: M stands for a mass unit, L is for length unit, and T is for a time unit.

59. Comparing Conserved Quantities The four objects in Fig. 12-37 are moving as indicated by the arrows. A curved arrow indicates rolling without slipping in the direction. For object (a), use the coordinates shown. For the others, take the origin at the center of the circle. Use the directions associated with the coordinate axes shown for object (a). Construct a table with the values of the magnitudes total translational momentum, total rotational momentum, and total energy of motion at the instant shown for each case. Express your answers in terms of m, v, and R. (Include an indicator of the direction where appropriate.) Which system has the largest and smallest of each of the quantities? Explain your reasoning.

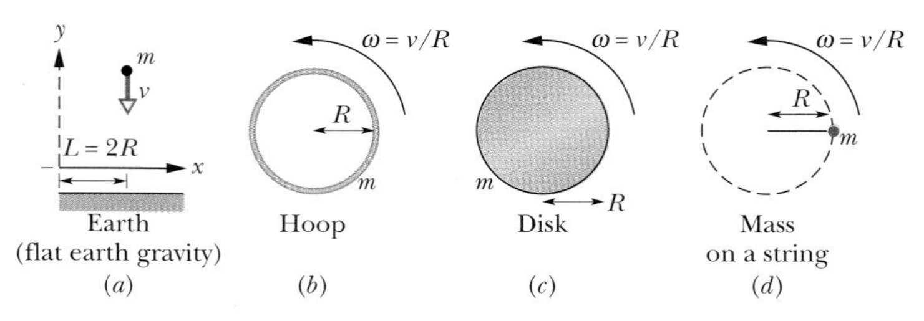

FIGURE 12-37 ■ Problem 59.

60. Designing a Yo-Yo In testing a design for a yo-yo, an engineer begins by constructing a simple prototype—a string wound about the rim of a wooden disk. She puts an axle riding on nearly frictionless ball bearings through the axis of the wooden disk and fixes the ends of the axle. See Fig. 12-38. In order to measure the moment of inertia of the disk, she attaches a weight of mass m to the string and measures how long it takes to fall a given distance. (a) Assuming the rotational inertia of the disk is given by I, and the radius of the disk is R, find the time for the mass to fall a distance h starting from rest. (b) The engineer doesn't have a very accurate stopwatch but wants to get a measurement good to a few percent. She decides that a fall time of 2 seconds would work. How big a mass should she use? Imagine you were setting up this experiment, and make reasonable estimates of the parameters you need.

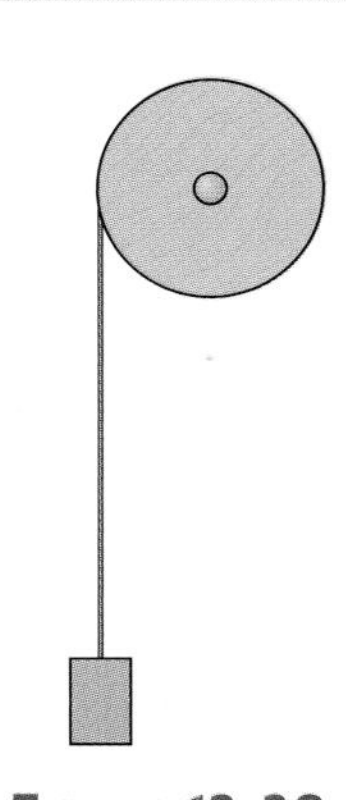

FIGURE 12-38 ■ Problem 60.

61. Approximating Atwood Figure 12-39 shows an Atwood's machine with two unequal masses attached by a massless string. The pulley has a mass of 20 g and a radius of 2 cm. (a) State three approximations that you can make to simplify your calculation of the motion of the blocks. ("Making an approximation" is the process of ignoring a physical effect because you expect it to be small and have little effect on your result if you only care about a few significant figures. If you want more significant figures, you may have to include those effects.) (b) Using your approximations, find the acceleration of block A. (c) What happens to your result if the two masses are equal? Is the result what you expect? Explain. (d) If you have ignored the rotational inertia of the pulley in your calculation in part (a) of this problem, set up the equations that would allow you to solve for the acceleration when it is included (but don't solve them).

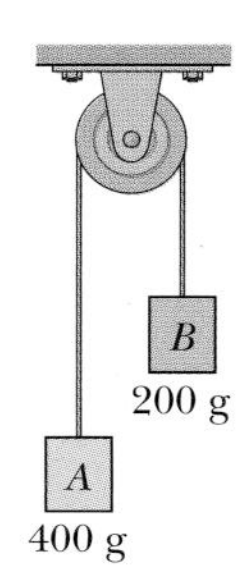

FIGURE 12-39 ■ Problem 61.

62. The Refrigerator Door A refrigerator has separate shelves on the door for storing bottles. Thin plastic straps keep the bottles from falling off the door. Someone in the house slams the door with a bit too much vigor and a heavy bottle breaks the strap. Do you think the bottle would be more likely to break the plastic strap if it is close to the hinge? Close to the handle? Or doesn't it matter? Explain your answer in terms of the physics we have learned.

13 Equilibrium and Elasticity

Rock climbing may be the ultimate physics exam. Failure can mean death, and even "partial credit" can mean severe injury. For example, in a long chimney climb, in which your torso is pressed against one wall of a wide vertical fissure and your feet are pressed against the opposite wall, you need to rest occasionally or you will fall due to exhaustion. Here the exam consists of a single question: What can you do to relax your push on the walls in order to rest? If you relax without considering the physics, the walls will not hold you up.

What is the answer to this life-and-death, one-question exam?

The answer is in this chapter.

13-1 Introduction

In this chapter we consider objects that remain motionless in the presence of external forces and torques. In particular we address the questions: (1) Under what conditions can objects that experience external forces and torques remain motionless (a state we will come to call static equilibrium)? (2) Under what conditions can objects that experience external forces and torques deform in order to remain in static equilibrium? The answers to these questions are of vital importance to engineers in the design of structures such as buildings, dams, roads, and bridges. The design engineer must identify all the external forces and torques that may act on a structure and, by good design and wise choice of materials, ensure that the structure will remain stable under these loads.

We begin this chapter by defining static equilibrium. We then use Newton's laws to summarize the conditions needed to keep structures in static equilibrium. For simplicity, the conditions for static equilibrium are developed assuming that objects do not change shape in the presence of external forces.

In Section 6-4 we discussed how the contact forces that an object can exert, such as normal and tension forces, result from the deformation of tiny spring-like bonds that separate atoms. For this reason no material is perfectly rigid. "Perfect rigidity" is an idealization, just like the assumptions that air resistance is negligible or that a surface is frictionless. Even a table sags under the load of a sheet of paper. Often, the change in shape is so small that we cannot observe it directly. However, sometimes the change in shape is easily noticeable. Engineers need to predict how an object of a given composition, shape, and size will deform as a function of the external forces on it for two reasons. First, its change of shape may, in turn, change the nature of the external forces it experiences and thereby force it out of static equilibrium. Second, the object could be deformed to the breaking point. For these reasons, the remainder of the chapter deals with an introduction to how structures deform in the presence of external forces.

13-2 Equilibrium

Consider these objects: (1) a book resting on a table, (2) a hockey puck sliding across a frictionless surface with constant velocity, (3) the rotating blades of a ceiling fan, and (4) the wheel of a bicycle that is traveling along a straight path at constant speed. For each of these four objects:

1. The translational momentum $\vec{p}_{\text{com}}$ of its center of mass is constant.
2. The rotational momentum $\vec{L}_{\text{com}}$ about its center of mass, or about any other point, is also constant.

Even though all four of these objects are moving, we say that they are in **equilibrium** because in each case both the translational momentum and rotational momentum of the object's center of mass are constant. Thus, the two requirements for equilibrium are

$$\vec{p}_{\text{com}} = \text{a constant} \quad \text{and} \quad \vec{L}_{\text{com}} = \text{a constant}. \tag{13-1}$$

Static Equilibrium and Stability

Our primary concern in this chapter is with objects that are not moving in any way—either in translation or in rotation—in the reference frame from which we observe them. Such objects are defined as being in **static equilibrium** whenever both the translational momentum and rotational momentum of the center of mass of the system is zero. In other words, if an object is in static equilibrium the constants in Eq. 13-1 must be zero. Of the four objects mentioned at the beginning of this section, only one—the book resting on the table—is in static equilibrium.

The balancing rock of Fig. 13-1 is another example of an object that, for the present at least, is in static equilibrium. It shares this property with countless other structures, such as cathedrals, houses, filing cabinets, and taco stands, that remain stationary over time.

As we discussed in Chapter 10, if a body returns to a state of static equilibrium after having been displaced from it by a force, the body is said to be in *stable* static equilibrium. A marble placed at the bottom of a hemispherical bowl is an example. However, if a small force can displace the body and end the equilibrium, the body is in *unstable* static equilibrium.

As an example of unstable static equilibrium, suppose we balance a domino with the domino's center of mass vertically above the supporting edge as in Fig. 13-2*a*. The torque about the supporting edge due to the gravitational force $\vec{F}^{grav}$ on the domino is zero, because the line of action of $\vec{F}^{grav}$ is through that edge. Thus, the domino is in equilibrium. Of course, even a slight force on it due to some chance disturbance ends the equilibrium. As the line of action of $\vec{F}^{grav}$ moves to one side of the supporting edge (as in Fig. 13-2*b*), the torque due to $\vec{F}^{grav}$ will cause the domino's rotation. Thus, the domino in Fig. 13-2*a* is in unstable static equilibrium.

FIGURE 13-1 ■ A balanced rock in the Arches National Park, Utah. Although its perch seems precarious, the rock is in static equilibrium.

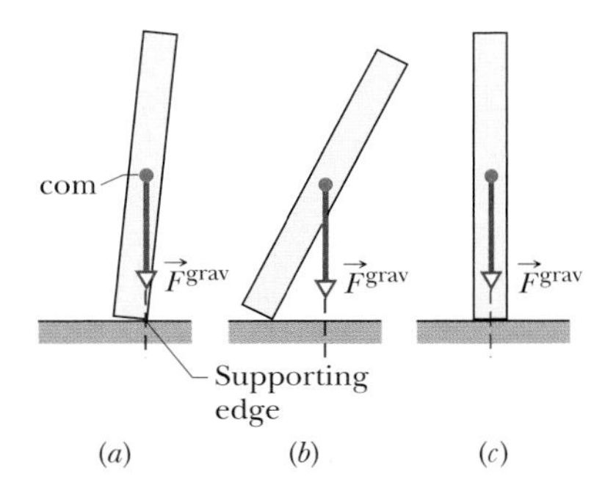

FIGURE 13-2 ■ (*a*) A domino balanced on one edge, with its center of mass vertically above that edge. The gravitational force $\vec{F}^{grav}$ on the domino is directed through the supporting edge. (*b*) If the domino is rotated even slightly clockwise from the balanced orientation, then $\vec{F}^{grav}$ causes a torque that increases the rotation. (*c*) A domino upright on a narrow side is somewhat more stable than the domino in (*a*). (*d*) A cubical block is even more stable.

The domino in Fig. 13-2*c* is slightly more stable. To topple this domino, a force would have to rotate it through and then beyond the balance position of Fig. 13-2*a*, in which the center of mass is above a supporting edge. A slight force will not topple this domino, but a vigorous flick of the finger against the domino certainly will. (If we arrange a chain of such upright dominos, a finger flick against the first can cause the whole chain to fall.)

The child's cubical block in Fig. 13-2*d* is even more stable because its center of mass would have to be moved even farther to get it to pass above a supporting edge. A flick of the finger may not topple the block. (This is why you never see a chain of toppling blocks.) The worker in Fig. 13-3 is like both the domino and the square block. Parallel to the beam, his stance is wide and he is stable. Perpendicular to the beam, his stance is narrow and he is unstable (and at the mercy of a chance gust of wind).

FIGURE 13-3 ■ A construction worker balanced above New York City is in static equilibrium but is more stable parallel to the beam than perpendicular to it.

The Conditions for Static Equilibrium

The translational motion of a body is governed by Newton's Second Law. In its translational momentum form, this relation is given as

$$\vec{F}^{\text{net}} = \frac{d\vec{p}}{dt}. \tag{13-2}$$

If the body is in translational equilibrium—that is, if $\vec{p}$ is a constant—then $d\vec{p}/dt = 0$ and we must have

$$\vec{F}^{\text{net}} = 0 \quad \text{(balance of forces).} \tag{13-3}$$

The rotational motion of a body is governed by Newton's Second Law in its rotational momentum form, given by Eq. 12-32 as

$$\vec{\tau}^{\text{net}} = \frac{d\vec{L}}{dt}. \tag{13-4}$$

If the body is in rotational equilibrium—that is, if $\vec{L}$ is a constant—then $d\vec{L}/dt = 0$ and we must have

$$\vec{\tau}^{\text{net}} = 0 \quad \text{(balance of torques).} \tag{13-5}$$

Thus, two requirements for a body to be in equilibrium are as follows:

> **If a body is in equilibrium:** (1) The vector sum of all the external forces that act on it must be zero; and (2) the vector sum of all the external torques that act on it, measured about *any* possible point, must also be zero.

Although these requirements obviously hold for *static* equilibrium, they also hold for the more general equilibrium in which $\vec{p}$ and $\vec{L}$ are constant but not zero.

We can express the vector form of equilibrium represented by Eqs. 13-3 and 13-5 in terms of three independent component equations, one for each axis in the chosen coordinate system:

Balance of Force Components	Balance of Torque Components
$F_x^{\text{net}} = \Sigma F_x = 0$	$\tau_x^{\text{net}} = \Sigma \tau_x = 0$
$F_y^{\text{net}} = \Sigma F_y = 0$	$\tau_y^{\text{net}} = \Sigma \tau_y = 0$
$F_z^{\text{net}} = \Sigma F_z = 0$	$\tau_z^{\text{net}} = \Sigma \tau_z = 0$

(13-6)

We shall simplify matters by considering only situations in which the forces that act on the body lie in the x-y plane. This means that the only torques that can act on the body must tend to cause rotation around an axis parallel to the z axis. With this assumption, we eliminate one force equation and two torque equations leaving

$$F_x^{\text{net}} = 0 \quad \text{(balance of forces),} \tag{13-7}$$

$$F_y^{\text{net}} = 0 \quad \text{(balance of forces),} \tag{13-8}$$

$$\tau_z^{\text{net}} = 0 \quad \text{(balance of torques).} \tag{13-9}$$

Here, τ_z^{net} is the net torque that the external forces produce either about the z axis or about *any* axis parallel to the z axis.

The conditions for static equilibrium are more stringent than those for general equilibrium. For example, a hockey puck that is sliding at constant velocity over ice while spinning about its center of mass with a constant rotational velocity satisfies the conditions for general equilibrium. But the puck is *not in static* equilibrium. The requirements for static equilibrium are that:

> **In static equilibrium** all parts of a body must be at rest in an inertial (nonaccelerating) frame of reference with no net force and no net torque acting on it.

READING EXERCISE 13-1: The figure gives six overhead views of a uniform rod on which two or more forces act perpendicular to the rod. If the magnitudes of the forces are adjusted properly (but kept nonzero), in which situations can the rod be in static equilibrium?

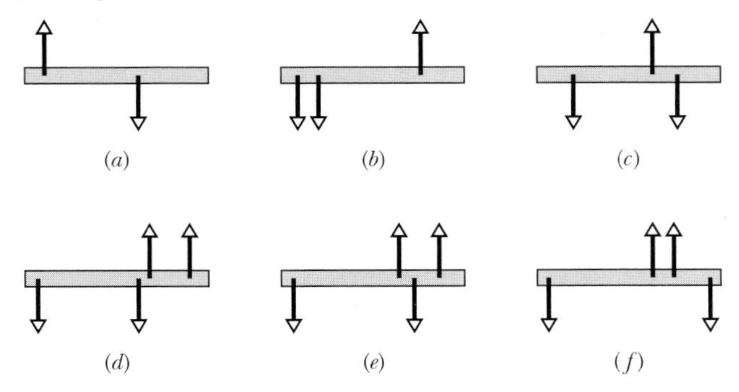

13-3 The Center of Gravity

Consider an extended body that is close to the surface of the Earth. The gravitational force on this body is the vector sum of the gravitational forces acting on the individual elements (the atoms) of the body. Instead of considering all those individual elements, we can say:

> The gravitational force $\vec{F}^{grav}$ on a body effectively acts at a single point, called the **center of gravity** (cog) of the body.

Here the word "effectively" means that if the gravitational forces on the individual elements were somehow turned off and force $\vec{F}^{grav}$ at the center of gravity were turned on, the net force and the net torque (about any point) acting on the body would not change.

Until now, we have assumed that the gravitational force $\vec{F}^{grav}$ acts at the center of mass (com) of the body. This is equivalent to assuming that the center of gravity is at the center of mass. Considering Fig. 13-4, it can be shown mathematically that

> If the local gravitational strength, g, is the same for all elements of a body, then the body's center of gravity (cog) is coincident with the body's center of mass (com).

This constancy is a very good approximation for everyday objects because, as we explained in Section 3-9, g varies only slightly along Earth's surface and with altitude. Thus, for objects like a mouse or a skyscraper, we can assume that the gravitational force acts at the center of mass.

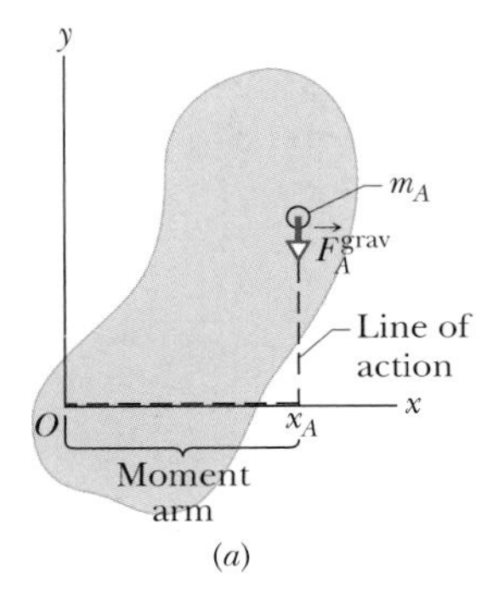

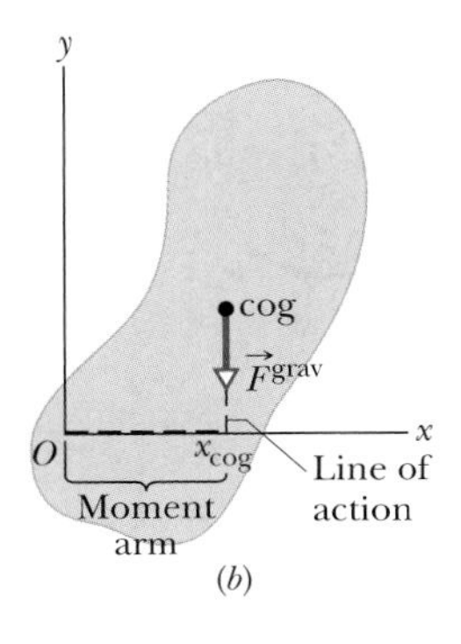

FIGURE 13-4 (*a*) An element of mass m_A in an extended body. The y-component of the gravitational force F_{Ay}^{grav} on it has moment arm x_A about the origin O of the coordinate system. (*b*) The gravitational force $\vec{F}^{grav}$ on a body is said to act at the center of gravity (cog) of the body. Here it has moment arm x_{cog} about origin O.

READING EXERCISE 13-2: Suppose that you skewer an apple with a thin rod, missing the apple's center of gravity. When you hold the rod horizontally and allow the apple to rotate freely, where does the center of gravity end up and why? ■

READING EXERCISE 13-3: A ladder is leaning against a wall as shown in the figure. If you were going to climb the ladder, are you better off if there is no friction between the ladder and the wall or no friction between the ladder and the floor? Explain your answer in terms of the physics you have learned. *Hint:* It will be helpful to draw an extended free-body diagram like that shown in Touchstone Example 13-2 and then consider the conditions that must be met to keep the ladder from moving.

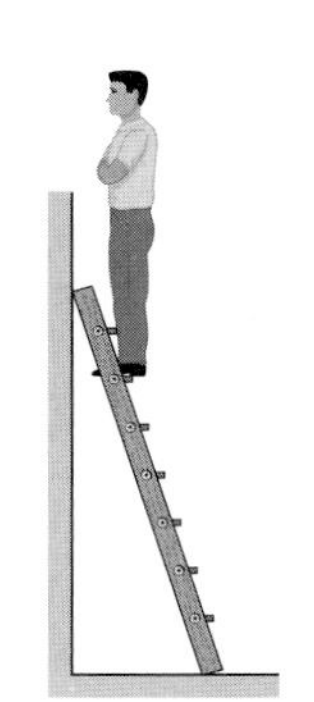

■

TOUCHSTONE EXAMPLE 13-1: Cat on a Plank

Two workmen are carrying a 6.0-m-long plank as shown in Fig. 13-5. The plank has a mass of 15 kg. A cat, with a mass of 5.0 kg, jumps on the plank and hangs on, 1.0 m from the end of the plank. Assuming that the workers are walking at a constant velocity, how much force does each workman have to exert to hold the plank up?

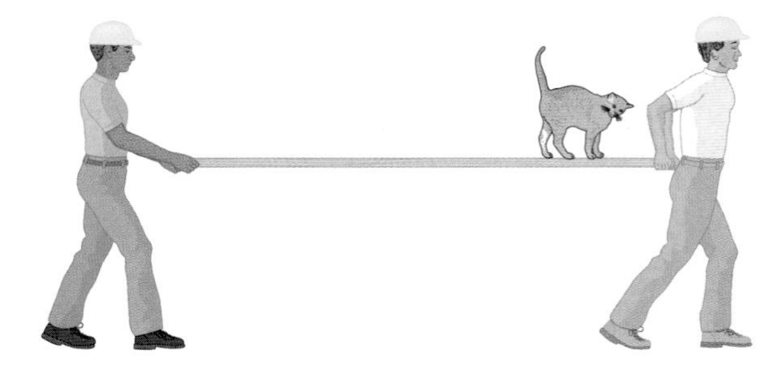

FIGURE 13-5 ■ Workmen holding up a cat and a plank.

SOLUTION ■ The first **Key Idea** here is that for an extended object such as the plank and cat to move at a constant velocity without rotating, it must be in equilibrium. Two conditions must be satisfied for equilibrium: (1) The net force on the plank must be zero to ensure that its center of mass is not accelerating; and (2) the net torque on it must be zero to ensure that it is not rotating. The second **Key Idea** here is that we can treat the downward gravitational forces on each of the mass elements that make up the plank as a single force acting at the plank's center of gravity.

The first step of our solution is to identify and locate the forces acting on the plank and display them as an extended free-body diagram as shown in Fig. 13-6. There are two downward forces on the plank: the gravitational force assumed to be acting as its center of gravity and the force exerted by the cat due to its weight. These must be counterbalanced by the upward forces exerted by worker A on the left and worker B on the right.

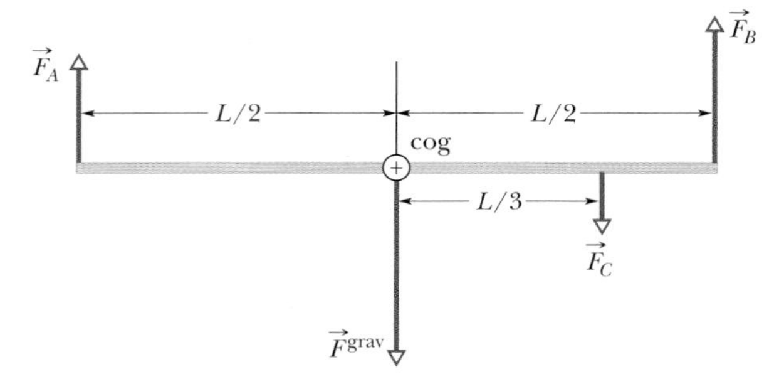

FIGURE 13-6 ■ An extended free-body diagram.

Next we can set up equations to balance the force and torque components shown in Eq. 13-6. Let us choose a conventional rectangular coordinate system with the origin at the center of gravity of the plank. The y axis is up, the x axis is in the plane of the page, and the z axis points out of the page. In this case we find that all the forces are parallel to the y axis, so there is only one force component balance equation,

$$F_y^{\text{net}} = F_{A\,y} + F_y^{\text{grav}} + F_{B\,y} + F_{C\,y} = 0 \quad \text{(force component balance).} \tag{13-10}$$

In this case, the equation for balancing torque components must be taken about the z axis (or any axis parallel to it). Let's stick with the z axis since it rather conveniently passes through the center of gravity of the plank. The torque balance equation that follows has a positive (counterclockwise) z-component of torque due to worker B's force and two negative (clockwise) z-components of torque due to worker A's force and the force exerted by the cat. We can express this as

$$\tau_z^{\text{net}} = +\tfrac{L}{2}|F_{B\,y}| - \tfrac{L}{2}|F_{A\,y}| - \tfrac{L}{3}|F_{C\,y}| = 0.$$

Since we know that the y-components of the workers' forces are both positive and the cat's force component is negative, we can rewrite the torque balance equation as

$$\tau_z^{net} = +\tfrac{L}{2}F_{By} - \tfrac{L}{2}F_{Ay} + \tfrac{L}{3}F_{Cy} = 0 \quad \text{(torque component balance).}$$

If we then eliminate the length of the plank by dividing each term by $L/2$, we get

$$F_{By} - F_{Ay} + \tfrac{2}{3}F_{Cy} = 0. \qquad (13\text{-}11)$$

Recall that we are trying to use our balance equations to determine $\vec{F}_A = F_{Ay}\hat{j}$ and $\vec{F}_B = F_{By}\hat{j}$. We have been given the information needed to find the gravitational force of the center of gravity of the plank and the force the cat exerts on the plank. In particular,

$$F_y^{grav} = -Mg = -(15\text{ kg})(9.8\text{ N/kg}) = -147\text{ N}, \qquad (13\text{-}12)$$

and $$F_{Cy} = -m_{cat}g = -(5.0\text{ kg})(9.8\text{ N/kg}) = -49.0\text{ N}. \qquad (13\text{-}13)$$

Thus we have two equations (Eqs. 13-10 and 13-11) with two unknowns, so we should be able to find our unknowns F_{Ay} and F_{By}. If we add Eq. 13-10 to Eq. 13-11 and solve for F_{By}, we get

$$F_{By} = \tfrac{1}{2}(-F_y^{grav} - \tfrac{5}{3}F_{Cy}) = \tfrac{1}{2}(147\text{ N} + (\tfrac{5}{3})49\text{ N}) = 114.333\text{ N} \cong 114\text{ N}. \quad \text{(Answer)}$$

Finally, we can solve Eq. 13-11 for F_{Ay} to get

$$F_{Ay} = F_{By} + \tfrac{2}{3}F_{Cy} = 114\text{ N} + \tfrac{2}{3}(-49\text{ N}) = 81.666\text{ N} \cong 82\text{ N}. \quad \text{(Answer)}$$

As you might have predicted, worker B, who is closer to the cat, has to exert a larger force than worker A does. However, note that we were able to factor the length L out of the torque balance equation, so the forces exerted by the workers do not depend on the length of the plank but only on the fraction of the distance that the cat is from the ends of the plank.

TOUCHSTONE EXAMPLE 13-2: Fireman on a Ladder

In Fig. 13-7a, a ladder of length $L = 12$ m and mass $m = 45$ kg leans against a slick (frictionless) wall. Its upper end is at height $h = 9.3$ m above the pavement on which the lower end rests (the pavement is not frictionless). The ladder's center of mass is $L/3$ from the lower end. A firefighter of mass $M = 72$ kg climbs the ladder until her center of mass is $L/2$ from the lower end. What then are the magnitudes of the forces on the ladder from the wall and the pavement?

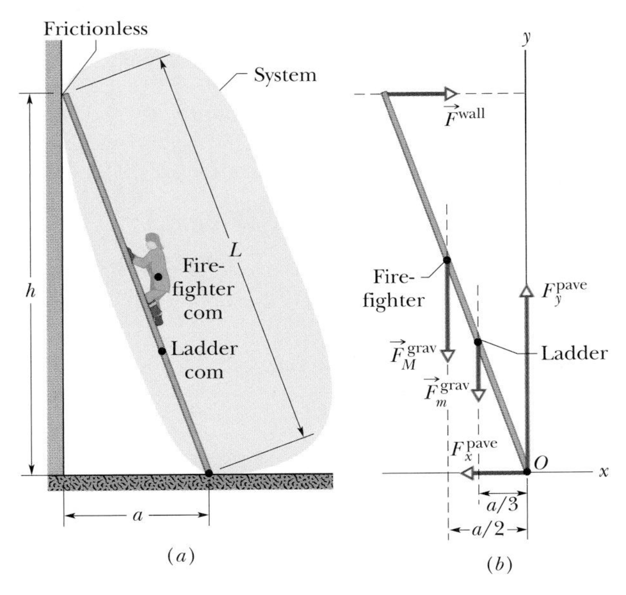

FIGURE 13-7 ■ (a) A firefighter climbs halfway up a ladder that is leaning against a frictionless wall. The pavement beneath the ladder is not frictionless. (b) A free-body diagram, showing the forces that act on the firefighter–ladder system. The origin O of a coordinate system is placed at the point of application of the unknown force $\vec{F}^{pave}$ (whose components F_x^{pave} and F_y^{pave} are shown).

SOLUTION ■ First, we choose our system as being the firefighter and ladder together, and then we draw the free-body diagram shown in Fig. 13-7b. The firefighter is represented with a dot within the boundary of the ladder. The gravitational force on her, $\vec{F}_M^{grav}$, has been shifted along its line of action, so that its tail is on the dot. (The shift does not alter a torque due to $\vec{F}_M^{grav}$ about any axis perpendicular to the figure.)

The only force on the ladder from the wall is the horizontal force $\vec{F}^{wall}$ (there cannot be frictional force along a frictionless wall). The force $\vec{F}^{pave}$ on the ladder from the pavement has a horizontal component F_x^{pave} that is a static frictional force and a vertical component F_y^{pave} that is a normal force.

A **Key Idea** here is that the system is in static equilibrium, so the balancing equations (Eqs. 13-7 through 13-9) apply to it. Let us start with Eq. 13-9 ($\tau_z^{net} = 0$). To choose an axis about which to calculate the torques, note that we have unknown forces ($\vec{F}^{wall}$ and $\vec{F}^{pave}$) at the two ends of the ladder. To eliminate, say, $\vec{F}^{pave}$ from the calculation, we place the axis at point O, perpendicular to the figure. We also place the origin of an xy coordinate system at O. We can find torques about O with any of Eqs. 11-29 through 11-31, but Eq. 11-31 ($|\vec{\tau}| = r_\perp|\vec{F}|$) is easiest to use here.

To find the moment arm $r_\perp$ of $\vec{F}^{wall}$, we draw a line of action through that vector (Fig. 13-7b). Then $r_\perp$ is the perpendicular distance between O and the line of action. In Fig. 13-7b, it extends along the y axis and is equal to the height h. Similarly, we draw lines of action for $\vec{F}_M^{grav}$ and $\vec{F}_m^{grav}$ and see that their moment arms extend along the x axis. For the distance a shown in Fig. 13-7a, the moment arms are $a/2$ (the firefighter is halfway up the ladder) and $a/3$ (the ladder's center of mass is one-third of the way up the ladder), respectively. The moment arms for F_x^{pave} and F_y^{pave} are zero.

Now, the torques can be written in the form $r_\perp F$. The balancing equation $\tau_z^{net} = 0$ becomes

$$-(h)(|\vec{F}^{\text{wall}}|) + (a/2)(Mg) + (a/3)(mg) + (0)(|F_x^{\text{pave}}|) + (0)(|F_y^{\text{pave}}|) = 0. \quad (13\text{-}14)$$

(Recall our rule: A positive torque corresponds to counterclockwise rotation and a negative torque corresponds to clockwise rotation.)

Using the Pythagorean theorem, we find that

$$a = \sqrt{L^2 - h^2} = 7.58 \text{ m}.$$

Then Eq. 13-14 gives us

$$|\vec{F}^{\text{wall}}| = \frac{ga(M/2 + m/3)}{h} = \frac{(9.8 \text{ m/s}^2)(7.58 \text{ m})(72/2 \text{ kg} + 45/3 \text{ kg})}{9.3 \text{ m}} = 4.1 \times 10^2 \text{ N}. \quad \text{(Answer)}$$

But since $\vec{F}^{\text{wall}}$ points to the right, its component F_x^{wall} is also +410 N.

Now we need to use the force-balancing equations. The equation $F_x^{\text{net}} = 0$ gives us

$$F_x^{\text{wall}} + F_x^{\text{pave}} = 0,$$

so

$$F_x^{\text{pave}} = -F_x^{\text{wall}} = -4.1 \times 10^2 \text{ N}, \quad \text{(Answer)}$$

where the minus sign tells us F_x^{pave} points to the left.

Since gravitational forces are negative whereas the local gravitational strength g and the force component F_y^{pave} are positive, the equation $F_y^{\text{net}} = 0$ gives us

$$F_y^{\text{pave}} - Mg - mg = 0.0 \text{ N} \quad \text{(force component balance)},$$

so

$$F_y^{\text{pave}} = (M + m)g = (72 \text{ kg} + 45 \text{ kg})(9.8 \text{ m/s}^2) = 1146.6 \text{ N} \approx 1.15 \times 10^3 \text{ N}. \quad \text{(Answer)}$$

TOUCHSTONE EXAMPLE 13-3: Safe on a Boom

Figure 13-8a shows a safe, of mass $M = 430$ kg, hanging by a rope from a boom with dimensions $a = 1.9$ m and $b = 2.5$ m. The boom consists of a hinged beam and a horizontal cable that connects the beam to a wall. The uniform beam has a mass m of 85 kg; the mass of the cable and rope are negligible.

(a) What is the tension T^{cable} in the cable? In other words, what is the magnitude of the force $\vec{T}^{\text{cable}}$ on the beam from the cable?

SOLUTION ■ The system here is the beam alone, and the forces on it are shown in the free-body diagram of Fig. 13-8b. The force from the cable is $\vec{T}^{\text{cable}}$. The gravitational force on the beam acts at the beam's center of mass (at the beam's center) and is represented by its equivalent $\vec{F}_m^{\text{grav}}$. The vertical component of the force on the beam from the hinge is F_y^{hinge} and the horizontal component of the force from the hinge is F_x^{hinge}. The force from the rope supporting the safe is $\vec{T}^{\text{rope}}$. Because beam, rope, and safe are stationary, the magnitude of $\vec{T}^{\text{rope}}$ is equal to the weight of the safe: $|\vec{T}^{\text{rope}}| = Mg$. We place the origin O of an xy coordinate system at the hinge.

One **Key Idea** here is that our system is in static equilibrium, so the balancing equations apply to it. Let us start with Eq. 13-9 $\tau_z^{\text{net}} = 0$. Note that we are asked for the magnitude of force $\vec{T}^{\text{cable}}$ and not of force components F_x^{hinge} and F_y^{hinge} acting at the hinge, at point O. Thus, a second **Key Idea** is that, to eliminate F_x^{hinge} and F_y^{hinge} from the torque calculation, we should calculate torques about an axis that is perpendicular to the figure at point O. Then F_x^{hinge} and F_y^{hinge} will have moment arms of zero. The lines of action for $\vec{T}^{\text{cable}}$, $\vec{T}^{\text{rope}}$, and $\vec{F}_m^{\text{grav}}$ are dashed in Fig. 13-8b. The corresponding moment arms are a, b, and $b/2$.

Writing torques in the form of $r_\perp|\vec{F}^{\text{hinge}}|$ and using our rule about signs for torques, the balancing equation $\tau_z^{\text{net}} = 0$ becomes

$$(a)(|\vec{T}^{\text{cable}}|) - (b)(|\vec{T}^{\text{rope}}|) - (\tfrac{1}{2}b)(mg) = 0.$$

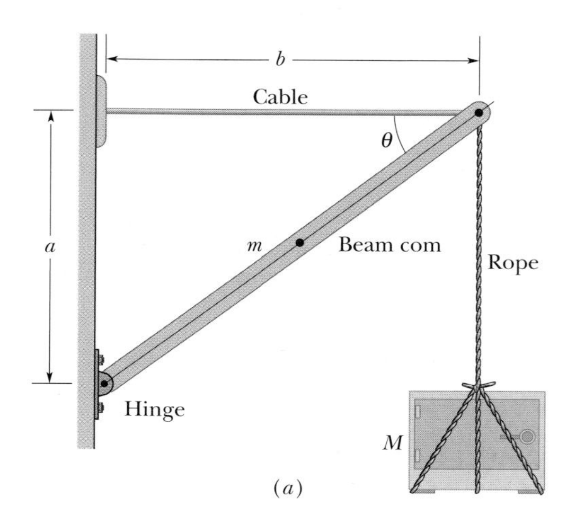

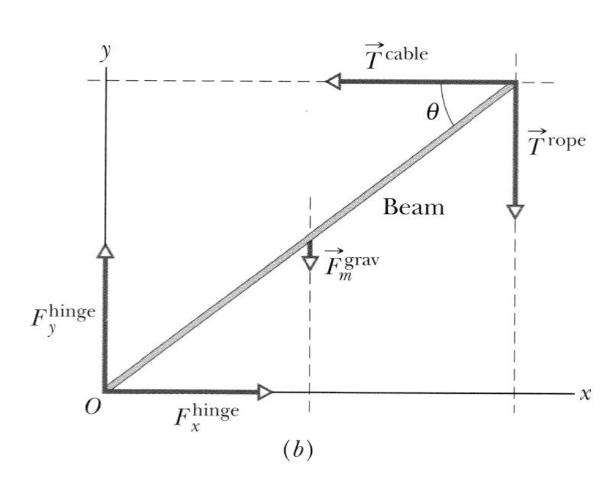

FIGURE 13-8 ■ (a) A heavy safe is hung from a boom consisting of a horizontal steel cable and a uniform beam. (b) A free-body diagram for the beam.

Substituting $|\vec{F}_M^{grav}|$ for $|\vec{T}^{rope}|$ and solving for $|\vec{T}^{cable}|$, we find that

$$|\vec{T}^{cable}| = \frac{gb(M + \frac{1}{2}m)}{a}$$

$$= \frac{(9.8 \text{ m/s}^2)(2.5 \text{ m})(430 \text{ kg} + 85/2 \text{ kg})}{1.9 \text{ m}}$$

$$= 6093 \text{ N} \approx 6100 \text{ N}. \qquad \text{(Answer)}$$

Since $\vec{T}^{cable}$ points along the negative x axis, its component T_x^{cable} is negative, so

$$T_x^{cable} = -|\vec{T}^{cable}| = -6.1 \times 10^3 \text{ N}.$$

(b) Find the magnitude $|\vec{F}^{hinge}|$ of the net force on the beam from the hinge.

SOLUTION ■ Now we want to know the values of the force components F_x^{hinge} and F_y^{hinge} so we can combine them to get $|\vec{F}^{hinge}|$. Because we know T_x^{cable}, our **Key Idea** here is to apply the force-balancing equations to the beam. For the horizontal balance, we write $F_x^{net} = 0$ as

$$F_x^{hinge} + T_x^{cable} = 0,$$

and so

$$F_x^{hinge} = -T_x^{cable} = +6093 \text{ N}.$$

For the vertical balance, we write $F_y^{net} = 0$ as

$$F_y^{hinge} - mg + T_y^{rope} = 0.$$

Substituting $-Mg$ for T_y^{rope} and solving for F_y^{hinge}, we find that

$$F_y^{hinge} = (m + M)g = (85 \text{ kg} + 430 \text{ kg})(9.8 \text{ m/s}^2)$$

$$= +5047 \text{ N} \approx 5.0 \times 10^3 \text{ N}.$$

The net force vector is then

$$\vec{F}^{hinge} = F_x^{hinge}\,\hat{i} + F_y^{hinge}\,\hat{j}$$

$$= (6093 \text{ N})\hat{i} + (5047 \text{ N})\hat{j}.$$

From the Pythagorean theorem, we now have

$$|\vec{F}^{hinge}| = \sqrt{(F_x^{hinge})^2 + (F_y^{hinge})^2}$$

$$= \sqrt{(6093 \text{ N})^2 + (5047 \text{ N})^2} \approx 7.9 \times 10^3 \text{ N}. \qquad \text{(Answer)}$$

Note that $|\vec{F}^{hinge}|$ is substantially greater than either the combined weights of the safe and the beam, which is 5.0×10^3 N, or the tension in the horizontal cable, which is 6.1×10^3 N.

TOUCHSTONE EXAMPLE 13-4: Chimney Climbing

In Fig. 13-9, a rock climber with mass $m = 55$ kg rests during a "chimney climb," pressing only with her shoulders and feet against the walls of a fissure of width $w = 1.0$ m. Her center of mass is a horizontal distance $d = 0.20$ m from the wall against which her shoulders are pressed. The coefficient of static friction between her shoes and the wall is $\mu_{shoes}^{stat} = 1.1$, and between her shoulders and the wall it is $\mu_{shoulders}^{stat} = 0.70$. To rest, the climber wants to minimize her horizontal push on the walls. The minimum occurs when her feet and her shoulders are both on the verge of sliding.

(a) What is that minimum horizontal push on the walls?

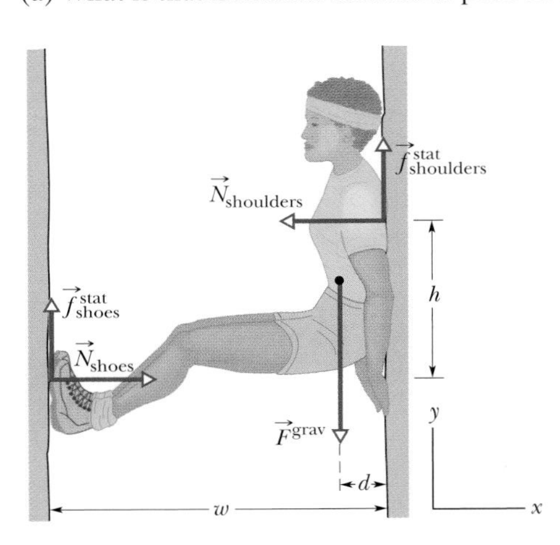

FIGURE 13-9 ■ The forces on a climber resting in a rock chimney. The push of the climber on the chimney walls results in a rise of the normal forces $\vec{N}_{shoes}$ and $\vec{N}_{shoulders}$ and the static frictional forces $\vec{f}_{shoes}^{stat}$ and $\vec{f}_{shoulders}^{stat}$.

SOLUTION ■ Our system is the climber, and Fig. 13-9 shows the forces that act on her. The only horizontal forces are the normal forces $\vec{N}_{shoes}$ and $\vec{N}_{shoulders}$ on her from the walls, at her feet and shoulders. The static frictional forces on her are $\vec{f}_{shoes}^{stat}$ and $\vec{f}_{shoulders}^{stat}$, directed upward. The gravitational force $\vec{F}^{grav}$ acts downward at her center of gravity.

A **Key Idea** is that, because the system is in static equilibrium, we can apply the force-balancing equations (Eqs. 13-7 and 13-8) to it. The equation $F_x^{net} = 0$ tells us that the two normal forces on her must be equal in magnitude and opposite in direction. We seek the magnitude $|\vec{N}|$ of these two forces, which is also the magnitude of her push against either wall.

The balancing equation $F_y^{net} = 0$ gives us

$$F_y^{net}\hat{j} = \vec{f}_{shoes}^{stat} + \vec{f}_{shoulders}^{stat} + \vec{F}^{grav} = 0, \quad \text{where } \vec{F}^{grav} = -mg\hat{j}. \qquad (13\text{-}15)$$

We want the climber to be on the verge of sliding at both her feet and her shoulders. That means we want the static frictional forces there to be at their maximum values. Those maximum magnitudes are, from Eq. 6-11, ($|\vec{f}_{max}^{stat}| = \mu^{stat}|\vec{N}|$),

$$|\vec{f}_{shoes}^{stat}| = \mu_{shoes}^{stat}|\vec{N}_{shoes}| \quad \text{and} \quad |\vec{f}_{shoulders}^{stat}| = \mu_{shoulders}^{stat}|\vec{N}_{shoulders}|, \qquad (13\text{-}16)$$

where $|\vec{N}| = |\vec{N}_{shoes}| = |\vec{N}_{shoulders}|$. Substituting these expressions into Eq. 13-15 and solving for the magnitude of $|\vec{N}|$ gives us

$$|\vec{N}| = \frac{mg}{\mu^{\text{stat}}_{\text{shoes}} + \mu^{\text{stat}}_{\text{shoulders}}} = \frac{(55\text{ kg})(9.8\text{ m/s}^2)}{1.1 + 0.70}$$
$$= 299\text{ N} \approx 3.0 \times 10^2\text{ N}.$$

Thus, her minimum horizontal push must be about 300 N.

(b) For that push, what must be the vertical distance h between her feet and her shoulder if she is to be stable?

SOLUTION ■ A **Key Idea** here is that the climber will be stable if the torque-balancing equation giving the z-component of torque ($\tau^{\text{net}}_z = 0$) applies to her. This means that the forces on her must not produce a net torque about any rotation axis. Another **Key Idea** is that we are free to choose a rotation axis that helps simplify the calculation. We shall write the torques in the form $r_\perp|\vec{F}|$, where $r_\perp$ is the moment arm of force $\vec{F}$. In Fig. 13-9, we choose a rotation axis at her shoulders, perpendicular to the figure's plane. Then the moment arms of the forces acting there ($\vec{N}_{\text{shoulders}}$ and $\vec{f}^{\text{stat}}_{\text{shoulders}}$) are zero. Frictional force $\vec{f}^{\text{stat}}_{\text{shoes}}$, the normal force $\vec{N}_{\text{shoes}}$ at her feet, and the gravitational force $\vec{F}^{\text{grav}} = -mg\hat{\text{j}}$ have the corresponding moment arms w, h, and d.

Recalling our rule about the signs of torques and the corresponding directions, we can now write the torque component $\tau^{\text{net}}_z = 0$ as

$$-(w)(|\vec{f}^{\text{stat}}_{\text{shoes}}|) + (h)(\vec{N}_{\text{shoes}}) + (d)(mg) + (0)(|\vec{f}^{\text{stat}}_{\text{shoulders}}|) + (0)(|\vec{N}_{\text{shoulders}}|) = 0. \quad (13\text{-}17)$$

(Note how the choice of rotation axis neatly eliminates $|\vec{f}^{\text{stat}}_{\text{shoulders}}|$ from the calculation.) Next, solving Eq. 13-17 for h, setting $|\vec{f}^{\text{stat}}_{\text{shoes}}| = \mu^{\text{stat}}_{\text{shoes}}|\vec{N}_{\text{shoes}}|$, and substituting $|\vec{N}| = |\vec{N}_{\text{shoes}}| = |\vec{N}_{\text{shoulders}}| = 299$ N and other known values, we find that

$$h = \frac{|\vec{f}^{\text{stat}}_{\text{shoes}}|w - mgd}{|\vec{N}|} = \frac{\mu^{\text{stat}}_{\text{shoes}}|\vec{N}|w - mgd}{|\vec{N}|} = \mu^{\text{stat}}_{\text{shoes}}w - \frac{mgd}{|\vec{N}|}$$
$$= (1.1)(1.0\text{ m}) - \frac{(55\text{ kg})(9.8\text{ m/s}^2)(0.20\text{ m})}{299\text{ N}} \quad \text{(Answer)}$$
$$= 0.739\text{ m} \approx 0.74\text{ m}.$$

We would find the same required value of h if we wrote the torques about any other rotation axis perpendicular to the page, such as one at her feet.

If h is more than *or* less than 0.74 m, she must exert a force greater than 299 N on the walls to be stable. Here, then, is the advantage of knowing the physics before you climb a chimney. When you need to rest, you will avoid the (dire) error of novice climbers who place their feet too high or too low. Instead, you will know that there is a "best" distance between shoulders and feet, requiring the least push, and giving you a good chance to rest.

13-4 Indeterminate Equilibrium Problems

We decided earlier in the chapter to reduce the complication of equilibrium calculations by only working with situations in which forces that act on a body all lie in the x-y plane. In these cases we have only three independent equations at our disposal. These are the two balance of force components equations (typically for the x and y axis force components) and the one balance of torque components equation about a rotation axis (typically the z axis). If a problem has more than three unknowns, we cannot solve it.

It is easy to find such problems. For example, the seesaw shown in Fig. 13-10 with its board weighing 100 N will remain in static equilibrium if the magnitudes of the three forces $\vec{F}_B$, $\vec{F}_C$, and $\vec{F}_D$ are 150 N, 100 N, and 50 N, respectively. But it will also remain in equilibrium for a force combination of 160 N, 80 N, 60 N, and so on. A variant on Touchstone Example 13-2 provides another example. We could have assumed that there is friction between the wall and the top of the ladder. Then there would have been a vertical frictional force acting where the ladder touches the wall, making a total of four unknown forces. With only three equations, we could not have solved this problem. Problems like these, in which there are more unknowns than equations, are called **indeterminate.**

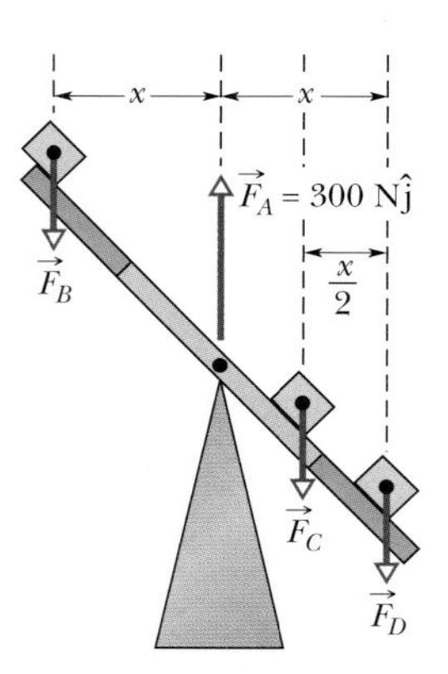

FIGURE 13-10 ■ The seesaw with three small boxes of unknown weight placed on its rough board is in static equilibrium. The board weighing 100 N is pivoted at its center of mass. However the system forms an indeterminate structure even when the total downward force on the seesaw pivot is known to be $(-400\text{ N})\hat{\text{j}}$. This is because the weights of the three boxes cannot be found from the conditions for static equilibrium alone.

Yet solutions to indeterminate problems exist in the real world even in cases where the forces on an object do not necessarily lie in one plane. If you rest the tires of the car on four platform scales, each scale will register a definite reading, the sum of the readings being the weight of the car. What is eluding us in our efforts to find the individual forces by solving equations?

The problem is that we have assumed—without making a great point of it—that the bodies to which we apply the equations of static equilibrium are perfectly rigid. By this we mean that they do not deform when forces are applied to them. Strictly, there are no such bodies. The tires of the car, for example, deform easily under this load until the car settles into a position of static equilibrium.

We have all had experience with a wobbly restaurant table, which we usually level by putting folded paper under one of the legs. If a big enough elephant sat on such a table, however, you may be sure that if the table did not collapse, it would deform just like the tires of a car. Its legs would all touch the floor, the forces acting upward on the table legs would all assume definite (and different) values, and the table would no longer wobble. How do we find the values of those forces acting on the legs?

To solve such indeterminate equilibrium problems, we must supplement equilibrium equations with some knowledge of deformation or *elasticity,* the branch of physics and engineering that describes how real bodies deform when forces are applied to them. The next section provides an introduction to this subject.

READING EXERCISE 13-4: A horizontal uniform bar of weight 10 N is to hang from a ceiling by three wires that exert upward forces $\vec{F}_A$, $\vec{F}_B$, and $\vec{F}_C$ on the bar. The figure shows three arrangements for the wires. Which arrangements, if any, are indeterminate (so that we cannot solve for numerical values of $\vec{F}_A$, $\vec{F}_B$, and $\vec{F}_C$)?

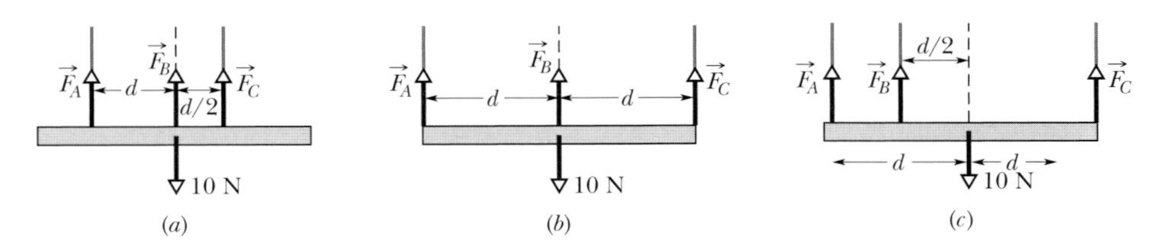

■

13-5 Elasticity

In Section 6-4 we introduced an idealized model based on atomic physics to help us explain the behavior of contact forces. This model can also be used to help us understand the elastic properties of solids. When a large number of atoms come together to form a metallic solid, such as an iron nail, they settle into equilibrium positions in a three-dimensional *lattice,* a repetitive arrangement in which each atom has a well-defined equilibrium distance from its nearest neighbors. The atoms are held together by interatomic forces that act like tiny springs. A two-dimensional drawing of this model is shown in Fig. 6-7 and a three-dimensional picture of this model is shown in Figs. 6-5 and Fig. 13-11. As shown in Fig. 6-19*c*, when an object that is in equilibrium is stretched or compressed by forces acting at opposite ends, each atom or molecule within the material feels oppositely directed forces on it that balance.

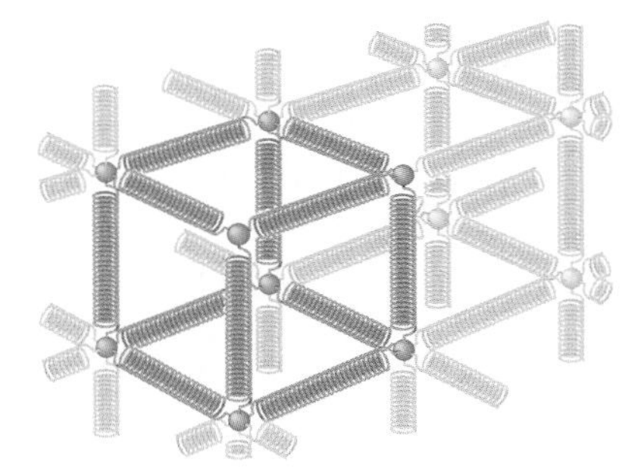

FIGURE 13-11 ■ The atoms of a metallic solid are distributed on a repetitive three-dimensional lattice. The interatomic forces behave like tiny springs.

The lattice of a metallic solid is remarkably rigid. This is another way of saying that the "interatomic springs" are extremely stiff. It is for this reason that we perceive many ordinary objects such as metal ladders, tables, and spoons as perfectly rigid. Of course, some ordinary objects, such as garden hoses or rubber gloves, do not strike us as rigid at all. The atoms that make up these objects *do not* form a rigid lattice like that of Fig. 13-11 but are aligned in long, flexible molecular chains, each chain being only loosely bound to its neighbors.

Given the atomic-molecular picture presented above, we can visualize all objects as being made up of discrete particles. Can we feel or observe the spring-like interactions between particles that make up an object? Consider what happens when we pull on an object. For example, suppose that you pull on your finger. Your finger does not immediately come apart, but you feel stretching forces along the entire length of your finger. These forces along the finger feel greater as you increase the magnitude of your pull. What you are feeling is an opposing, balancing force that arises within your finger to oppose your pull. This opposing force works to keep the particles that make up your finger from moving away from one another (which ultimately would result in your finger coming apart). This experience can be generalized by saying that the particles that make up your finger (or other object) are held together by forces that are *attractive* forces when a stretching force is applied. This attractive force increases as stretching occurs and the particles that make up the object move apart. Stretching forces like your pull are called **tensile forces.**

If instead, you push on your finger (in trying to shorten it), you will again sense balancing, opposing forces along the length of your finger. Push harder on your finger. The feeling will change and likely convince you that these opposing forces also increase with increasing applied force. From this experience, we gather that the particles that make up our finger (or other objects) will encounter *repulsive* forces if we try to push them closer together. Furthermore, the repulsive forces increase with decreasing separation. Squeezing forces like the one you applied to your finger (in an attempt to shorten it) are called **compressive forces.** As it turns out, the attractive and repulsive forces between the particles that make up an object are associated with the electrical nature of the particles that comprise atoms as well as their interactions on a microscopic scale. However, the nature of the forces between the particles that make up an object is very much like the nature of a spring force. Hence, we can model these forces quite well as spring forces. This is why Fig. 13-11 shows the particles connected with springs.

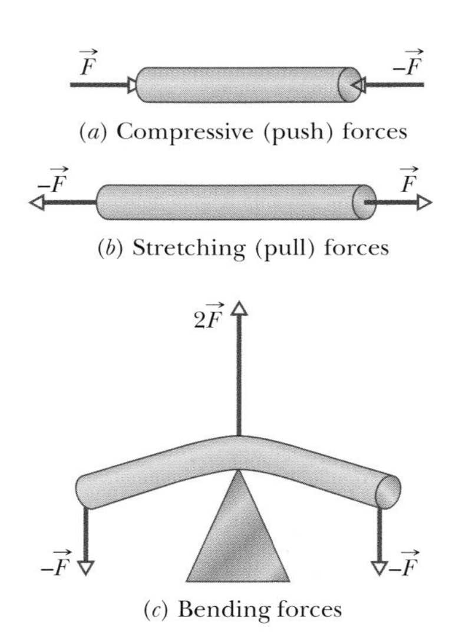

FIGURE 13-12 ■ Forces leading to (*a*) compression, (*b*) stretching, and (*c*) bending.

As you surely have experienced, neither tensile nor compressive forces can be increased indefinitely. Eventually the object simply breaks. In general, solid materials tend to be strongest under compression forces and weakest under stretching (tensile) forces. There is another way to break a long thin object—by *bending* it. If you have ever tried to break a dry stick, you probably observed that it breaks more easily when it is bent than when it is stretched or compressed. This makes sense when you visualize what is happening to the atoms or molecules that makeup an object. When bending occurs, these tiny particles are compressed on one side while they are simultaneously stretched on the opposite side. Figure 13-12 shows combinations of forces that can lead to compression, stretching, and bending.

Tensile and Compressive Forces

Let's consider what can happen when we pull on each end of a long, thin, apparently rigid object with forces of greater and greater magnitude. If you fix one end of a steel rod that is 1 m long and 1 cm in diameter and hang a subcompact car from the end, the rod will stretch. However, it stretches only about 0.5 mm, or 0.05% as shown in Fig. 13-13. In this case the rod acts like an *elastic* spring that obeys Hooke's law, $\vec{F}_x^{\text{spring}}(x) = -k\Delta\vec{x}$(Eq. 9-16), so that the magnitude of the applied force on it is proportional to the displacement of the spring.

> If the deformation of an object is proportional to the magnitude of the applied forces on its ends and if the object returns to its original length when the forces are removed, then the deformation is **elastic.**

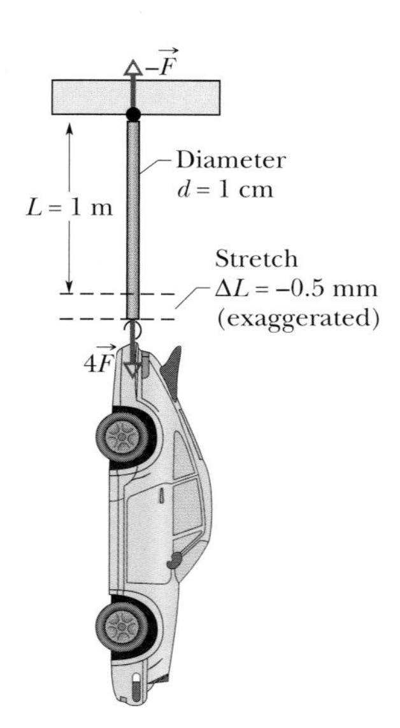

FIGURE 13-13 ■ A subcompact car is hanging from a 1-m-long steel rod that is 1 cm in diameter. Even though the thin rod supports the entire weight of the car, the rod will only stretch about 0.5 mm.

If you hang two cars from the rod, the rod will be permanently stretched and will not recover its original length when you remove the load. The stretching is now **inelastic.** If you hang three cars from the rod, the rod will break or **rupture.** Just before breaking, the elongation of the rod will be less than 0.2%. Although deformations of this size seem small, they are important in engineering practice. (Whether a wing under load will stay on an airplane is obviously important.)

We want to quantify the relationship between applied tensile or compressive forces and deformations. Let's think about elastic deformations (those that are proportional to the applied forces that cause them). How do these deformations depend on the particular geometry of the object and nature of the material being placed under tension or compression forces? Let's see.

If we pull on the ends of a rod of length L parallel to it and with equal but opposite forces of magnitude F, we can measure the change in the rod's length, ΔL. In practice we can do this by pulling on one end while the other end is attached to an immovable object. What happens if we now apply the same force to the end of a rod of the same material and diameter that is twice as long? By how much will it stretch? We can figure this out by imagining that the double-length rod is just two rods connected in the middle as shown in Figure 13-14.

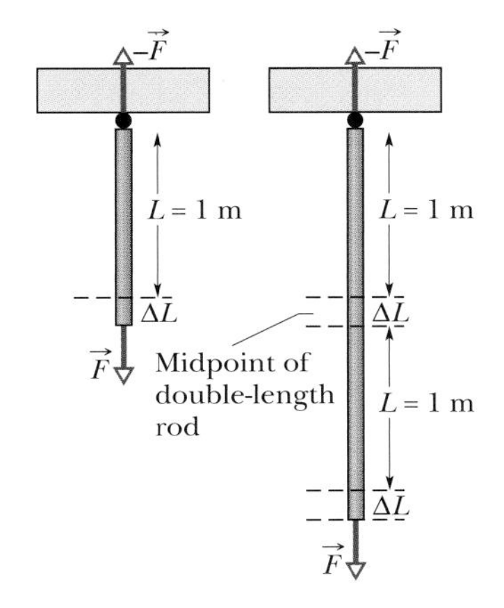

FIGURE 13-14 ■ If the forces applied to the ends of a double-length rod are the same as those applied to a single-length rod the double-length rod will experience twice the deformation. The extent of the deformation is exaggerated.

Let's assume that the magnitude of pulling force on the ends of our double-length rod is much greater than the rod's weight. In static equilibrium, the support point at the other end of the rod exerts a force of the same magnitude in the opposite direction. What happens to a small cross section of the rod at its midpoint? As we saw in our discussion of tension in Chapter 6, every cross-sectional element along the rod will experience the same pair of opposite pull forces. So the pull force from the lower half of the rod must be equal in magnitude and opposite in direction to the pull force from the upper half of the rod. Therefore, each half of the rod will stretch by an amount ΔL as shown in Fig. 13-14. As a result, the total stretch for the double length rod will be $2\,\Delta L$. We can easily generalize this argument to show that for a rod of arbitrary length L, the amount of stretch ΔL produced by a given force will be proportional to the length of the rod—assuming that we keep everything else about the rod and situation the same.

Similarly, we can consider a rod of the same length as our original rod but with double the cross-sectional area. We can picture this as two rods with our original diameter placed right next to each other but not touching as shown in Fig. 13-15. Now we see that our two rods would have to *share* the force $\vec{F}$ between them. This means that each effectively feels only half of the total force. As a result, the double diameter rods will only stretch half the amount of the single rod. This implies that the change in length should be *inversely* proportional to the cross-sectional area of the rod. That is, the deformation ΔL is proportional to $1/A$ where A is the cross-sectional area of the rod.

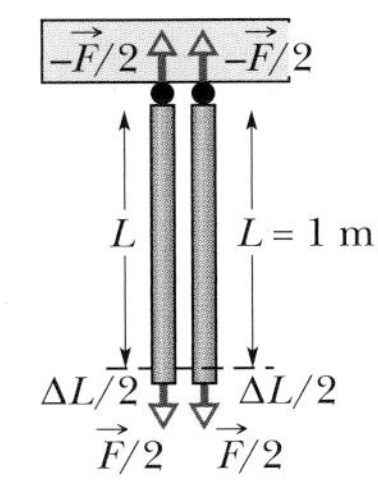

FIGURE 13-15 ■ A rod that has twice the cross-sectional area but has the same forces applied to its ends will experience half the deformation. The deformation is exaggerated.

Recapping what we discussed in the paragraphs above: The deformation of the rod ΔL is proportional to the magnitude of the applied force F, proportional to the length of the rod L, and inversely proportional to the cross-sectional area of the rod A. If we combine these results, we get

$$\Delta L \propto \frac{LF}{A} \quad \text{or} \quad \Delta L = c\frac{LF}{A},$$

where c is a constant of proportionality that may well be present. It is customary to express this proportionality so that we can relate the force magnitude to the length change for a given rod. Thus, we can write

$$F = \left(\frac{1}{c}\right)\left(\frac{A}{L}\right)\Delta L. \tag{13-18}$$

Let's examine the factors in parentheses in the equation shown above. First, the $(1/c)$ factor is a constant, which turns out to depend on the material an object is made of but not on its shape. This factor is a kind of stiffness constant that characterizes the tensile and compressive strength of a material. This factor $(1/c)$ is represented in engineering practice by the symbol E and is called the **Young's modulus.** So, we can write

$$F = E\left(\frac{A}{L}\right)\Delta L. \tag{13-19}$$

The second factor given by (A/L) depends only on the geometry of the rod.

As we mentioned in the beginning of this section, whenever the magnitude of the force applied to an object is proportional to its deformation in the direction of the force, the object is behaving like a Hooke's law spring. Note that if we represent $E(A/L)$ for a given rod by k, then we can write Eq. 13-19 in the form $F = k\Delta L$. In this form, Eq. 13-19 and Eq. 9-16 describing Hooke's law for a spring in one dimension are essentially the same. However, a length of rod or wire in general is much stiffer (that is, more resistant to deformation) than the same rod or wire would be if it were coiled into a spring.

It is customary to write Eq. 13-19 in a form known as the stress–strain equation:

$$\frac{F}{A} = E\left(\frac{\Delta L}{L}\right) \quad \text{(stress–strain equation).} \tag{13-20}$$

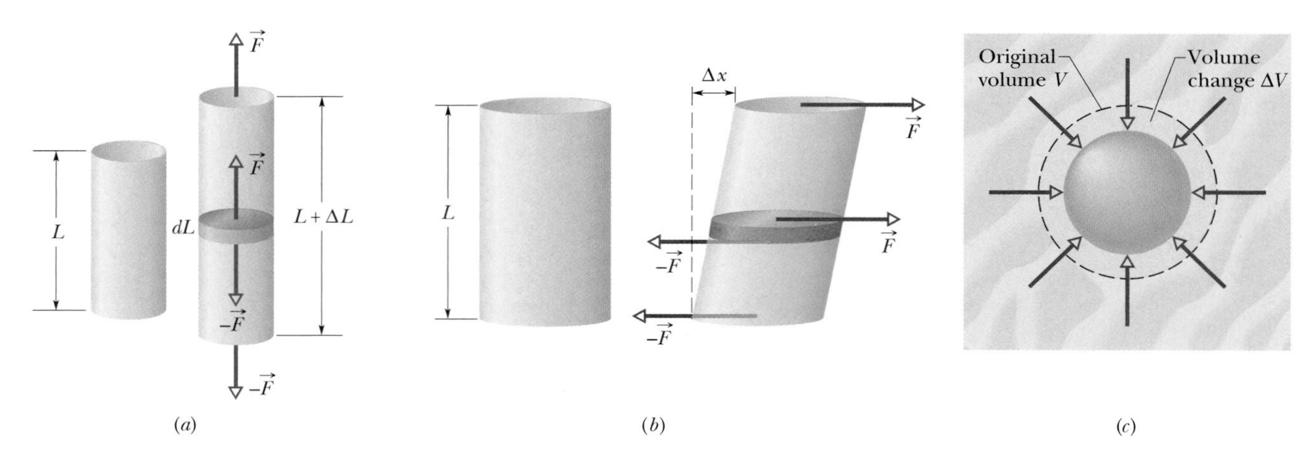

FIGURE 13-16 ■ When solid matter is in static equilibrium, it can experience oppositely directed tensile, compressive, or shearing forces that deform it. These deformation forces are experienced by every tiny element that the matter is composed of, such as the cross-sectional disks of infinitesimal thickness dl depicted here. (*a*) A cylinder subject to *tensile stress* stretches by an amount ΔL. (*b*) A cylinder subject to *shearing stress* deforms by an amount Δx, somewhat like a pack of playing cards would. (*c*) A solid sphere that is immersed in a fluid is subject to compressive, uniform *hydraulic stress*. It shrinks in volume by an amount ΔV. All the deformations shown are greatly exaggerated.

Figure 13-16 shows three ways in which a solid might change its dimensions when forces act on it. In Fig. 13-16*a*, a cylinder is stretched. In Fig. 13-16*b*, a cylinder is deformed by a force perpendicular to its axis, much as we might deform a pack of cards or a book. In Fig. 13-16*c*, a solid object, placed in a fluid under high pressure, is compressed uniformly on all sides. What the three deformation types have in common is that a deforming force per unit area F/A produces a unit deformation $\Delta L/L$. The deforming force per unit area F/A is called a **stress** and the unit deformation $\Delta L/L$ is called a **strain.** So, the stresses and strains take different forms in the three situations of Fig. 13-16. When engineers design structures, they usually work with strains that are small enough that the materials are elastic (so that stress and strain are proportional to each other). That is,

$$\text{stress} = \text{modulus} \times \text{strain}. \tag{13-21}$$

As we already mentioned, the modulus is a stiffness factor. Obviously Eq. 13-21 tells us that when the modulus is very large, it takes a lot of force per unit of cross-sectional area (stress) to produce a small deformation (strain). In Fig. 13-16, *tensile stress* (associated with stretching) is illustrated in (*a*), *shearing stress* in (*b*), and *hydraulic stress* in (*c*).

To get a better feel for the relationship between the applied force and the force per unit area (or stress), consider the normal force exerted by the table on a chunk of clay. Suppose that the clay starts as a tall cylinder. It is then squashed down into a flat pancake of clay, without any change in mass. Comparing these two geometries for the clay, we note that the total normal force does not change (the weight has not changed). However, the area of the clay does change. Therefore, the force per unit area or stress must also change. As Eq. 13-21 shows, thinking about deformations is more direct when we consider stress rather than force.

In a typical test of tensile (stretching) properties, the stress on a test specimen (like that in Figs. 13-17*b* and 13-18) is slowly increased until the cylinder fractures, and the stress vs. strain are carefully measured and plotted. For metals, the result is a graph like those shown in Fig. 13-17 or Fig. 13-19. For a substantial range of applied stresses, the stress–strain relation is linear, and the specimen recovers its original dimensions when the stress is removed. Here stress = modulus × strain (Eq. 13-21) applies. If the stress is increased beyond the **yield strength** S^{yield} of the specimen, the specimen becomes permanently deformed. If the stress continues to increase, the specimen eventually ruptures, at a stress called the **ultimate strength** S^{ultimate}.

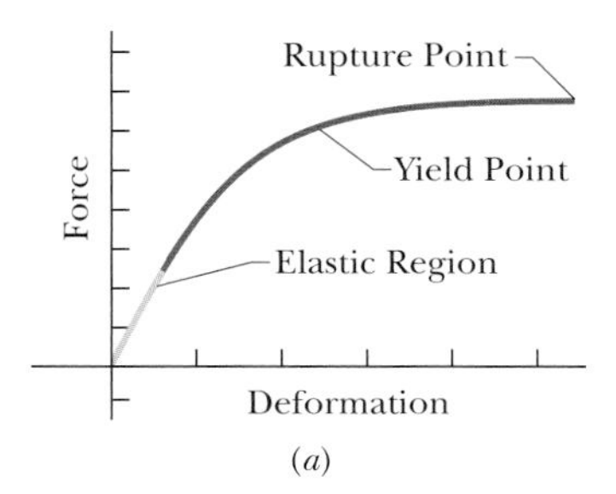

(*a*)

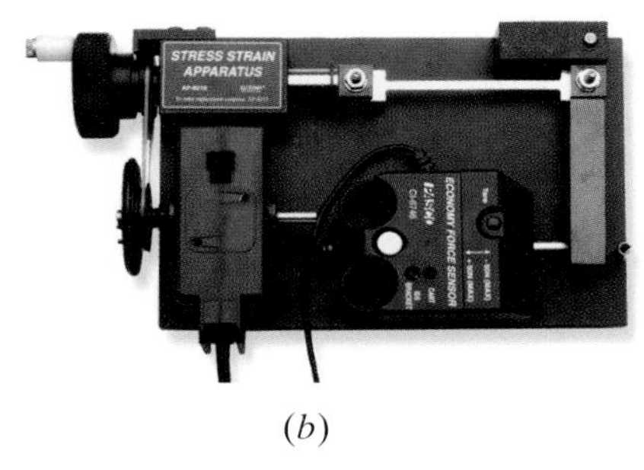

(*b*)

FIGURE 13-17 ■ (*a*) A graph showing the deformation of a hard steel bar as a function of the magnitude of the forces applied to its ends. The applied forces are increased until the bar ruptures. (*b*) The measurements are made using an electronic force sensor and a rotary motion sensor to gauge the deformation of the bar. (Courtesy of PASCO scientific.)

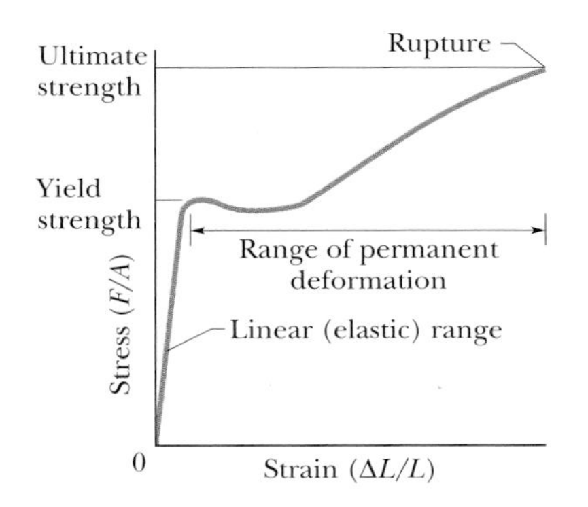

FIGURE 13-19 ■ A stress–strain curve for a soft steel test specimen such as that of Fig. 13-18. Young's modulus is the slope of the first linear portion of the graph. The specimen deforms permanently when the stress is equal to the *yield strength* of the material. It ruptures when the stress is equal to the *ultimate strength* of the material.

For simple tension or compression, the stress on an object is defined as F/A, where F represents the magnitude of the force applied perpendicular to the cross-sectional area A of the object. The strain, or unit deformation, is then the dimensionless quantity $\Delta L/L$. This is the fractional (or sometimes percentage) change in the length of the specimen. Because the strain is dimensionless, and stress = modulus × strain (Eq. 13-21), the modulus has the same dimensions as the stress—namely, force per unit area.

The strain $\Delta L/L$ in a specimen can often be measured conveniently with a *strain gage* (Fig. 13-20). This simple and useful device, which can be attached directly to operating machinery with an adhesive, is based on the principle that its electrical properties are dependent on the strain it undergoes.

FIGURE 13-18 ■ A test specimen, used to determine a stress–strain curve such as that of Fig. 13-19. The change ΔL that occurs in a certain length L is measured in a tensile stress–strain test.

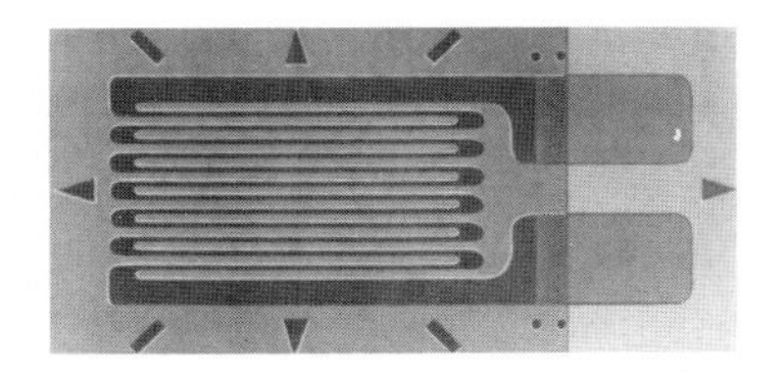

FIGURE 13-20 ■ A strain gage of overall dimensions 9.8 mm by 4.6 mm. The gage is fastened with adhesive to the object whose strain is to be measured; it experiences the same strain as the object. The electrical resistance of the gage varies with the strain, permitting deformations up to about 3 mm to be measured.

Although the Young's modulus for an object is typically almost the same for tension and compression, the object's ultimate strength or yield strength may well be different for the two types of stress. Concrete, for example, is very strong in compression but is so weak in tension that it is almost never used in that manner. In fact, steel bars are often embedded in concrete to enhance the tensile strength of a concrete structure. Table 13-1 shows the Young's modulus and other elastic properties for some materials of engineering interest.

TABLE 13-1
Some Elastic Properties of Selected Materials of Engineering Interest

Material	Density ρ (kg/m^3)	Young's Modulus E (10^9 N/m^2)	Ultimate Strength S^{ultimate} (10^6 N/m^2)	Yield Strength S^{yield} (10^6 N/m^2)
Steel[a]	7860	200	400	250
Aluminum	2710	70	110	95
Glass	2190	65	50[b]	—
Concrete[c]	2320	30	40[b]	—
Wood[d]	525	13	50[b]	—
Bone	1900	9[b]	170[b]	—
Polystyrene	1050	3	48	—

[a]Structural steel (ASTM-A36). [b]In compression. [c]High strength. [d]Douglas fir.

Shearing

Consider Fig. 13-16*b*. Here the force is applied parallel to the cross-sectional area. We call this force orientation *shearing stress.* In the case of shearing, the stress is still a force per unit area, but the force vector lies in the plane of the area rather than perpendicular to it. The strain is effectively the angle of the shear in radians as shown in Fig. 13-16*b*. It is given by the dimensionless ratio $\Delta x/L$, with the quantities defined as shown in Fig. 13-16*b*. The corresponding modulus, which is given the symbol G in engineering practice, is called the **shear modulus.** For shearing, the stress = modulus × strain equation (Eq. 13-21) is written as

$$\frac{F^{\text{perp}}}{A} = G\frac{\Delta x}{L}, \tag{13-22}$$

where F^{perp} is the magnitude of the component of an applied force that is applied perpendicular to the length of the rod. Shearing stresses play a critical role in the buckling of shafts that rotate under load and in bone fractures caused by bending.

Hydraulic Stress

In Fig. 13-16*c*, the stress is the fluid pressure P on the object. We will discuss fluid pressure in more detail in Chapter 15. For example, as you will see in Eq. 15-1, if the forces that act on an area A are uniform, then pressure is the ratio of the force component perpendicular to the area and to the area itself. The strain is then $|\Delta V|/V$, where V is the original volume of the specimen and ΔV is the absolute value of the change in volume. The corresponding modulus, with symbol B, is called the **bulk modulus** of the material. The object is said to be under *hydraulic compression,* and the pressure can be called the *hydraulic stress.* For this situation, we write stress = modulus × strain (Eq. 13-21) as

$$P = B\frac{|\Delta V|}{V}. \tag{13-23}$$

The bulk modulus is 2.2×10^9 N/m^2 for water and 16×10^{10} N/m^2 for steel. The pressure at the bottom of the Pacific Ocean, at its average depth of about 4000 m, is 4.0×10^7 N/m^2. The fractional compression $\Delta V/V$ of a volume of water due to this pressure is 1.8%; that for a steel object is only about 0.025%. In general, solids—with their rigid atomic lattices—are less compressible than liquids, in which the atoms or molecules are less tightly coupled to their neighbors.

READING EXERCISE 13-5: Four cylindrical rods are stretched as in Fig. 13-16*a*. The force magnitudes, the cross-sectional areas, the initial lengths, and the changes in length are shown in the table below. Rank the rods from largest to smallest Young's modulus.

Rod	Force Magnitude	Area	Length Change	Initial Length
1	F	A	ΔL	L
2	$2F$	$2A$	$2\Delta L$	L
3	F	$2A$	$2\Delta L$	$2L$
4	$2F$	A	ΔL	$2L$

■

READING EXERCISE 13-6: Visualizing the microscopic particles that make up objects as shown in Fig. 13-11, discuss the difference between the bending shown in Fig. 13-12*c* and the shearing shown in Fig. 13-16*b*. ■

TOUCHSTONE EXAMPLE 13-5: Stretching a Rod

A structural steel rod has a radius R of 9.5 mm and a length L of 81 cm. A 6.2×10^4 N force of magnitude F stretches it along its length. What are the stress on the rod and the elongation and strain of the rod?

SOLUTION ■ The first **Key Idea** here has to do with what is meant by the second sentence in the problem statement. We assume the rod is held stationary by, say, a clamp or vise at one end. Then force $\vec{F}$ is applied at the other end, parallel to the length of the rod and thus perpendicular to the end face there. Therefore, the situation is like that in Fig. 13-16*a*.

The next **Key Idea** is that we assume the force is applied uniformly across the end face and thus over an area $A = \pi R^2$. Then the stress on the rod is given by the left side of Eq. 13-20,

$$\text{stress} = \frac{F}{A} = \frac{\text{F}}{\pi R^2} = \frac{6.2 \times 10^4\,\text{N}}{(\pi)(9.5 \times 10^{-3}\,\text{m})^2} \quad \text{(Answer)}$$
$$= 2.2 \times 10^8\,\text{N/m}^2.$$

The yield strength for structural steel is 2.5×10^8 N/m², so this rod is dangerously close to its yield strength.

Another **Key Idea** is that the elongation of the rod depends on the stress, the original length L, and the type of material in the rod. The last determines which value we use for Young's modulus E (from Table 13-1). Using the value for steel, Eq. 13-20 gives us

$$\Delta L = \frac{(F/A)L}{E} = \frac{(2.2 \times 10^8\,\text{N/m}^2)(0.81\,\text{m})}{2.0 \times 10^{11}\,\text{N/m}^2} \quad \text{(Answer)}$$
$$= 8.9 \times 10^{-4}\,\text{m} = 0.89\,\text{mm}.$$

The last **Key Idea** we need here is that strain, which is the dimensionless ratio of the change in length to the original length, is

$$\frac{\Delta L}{L} = \frac{8.9 \times 10^{-4}\,\text{m}}{0.81\,\text{m}} \quad \text{(Answer)}$$
$$= 1.1 \times 10^{-3} = 0.11\%.$$

Problems

SEC. 13-3 ■ THE CENTER OF GRAVITY

1. Brady Bunch A physics Brady Bunch, whose weights in newtons are indicated in Fig. 13-21, is balanced on a seesaw. What is the number of the person who causes the largest torque, about the rotation axis at *fulcrum f*, directed (a) out of the page and (b) into the page?

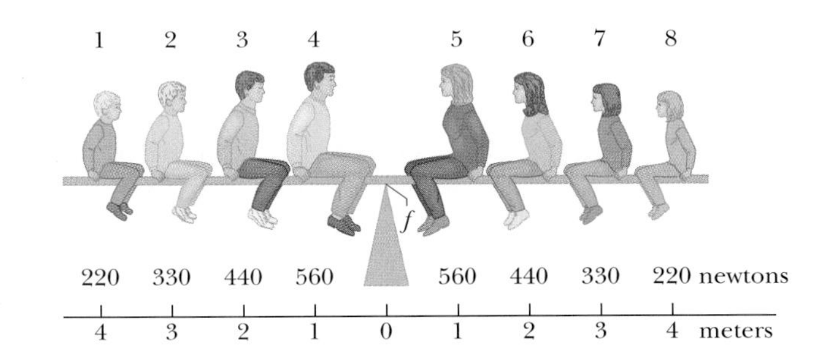

FIGURE 13-21 ■ Problem 1.

2. Tower of Pisa The leaning Tower of Pisa (Fig. 13-22) is 55 m high and 7.0 m in diameter. The top of the tower is displaced 4.5 m from the vertical. Treat the tower as a uniform, circular cylinder. (a) What additional displacement, measured at the top, would bring the tower to the verge of toppling? (b) What angle would the tower then make with the vertical?

FIGURE 13-22 ■ Problem 2.

3. Particle Acted on A particle is acted on by forces given by $\vec{F}_A = (10\text{ N})\hat{\text{i}} + (-4\text{ N})\hat{\text{j}}$ and $\vec{F}_B = (17\text{ N})\hat{\text{i}} + (2\text{ N})\hat{\text{j}}$. (a) What force $\vec{F}_C$ balances these forces? (b) What direction does $\vec{F}_C$ have relative to the x axis?

4. A Bow is Drawn A bow is drawn at its midpoint until the tension in the string is equal to the force exerted by the archer. What is the angle between the two halves of the string?

5. Rope of Negligible Mass A rope of negligible mass is stretched horizontally between two supports that are 3.44 m apart. When an object of weight 3160 N is hung at the center of the rope, the rope is observed to sag by 35.0 cm. What is the tension in the rope?

6. Scaffold A scaffold of mass 60 kg and length 5.0 m is supported in a horizontal position by a vertical cable at each end. A window washer of mass 80 kg stands at a point 1.5 m from one end. What is the tension in (a) the nearer cable and (b) the farther cable?

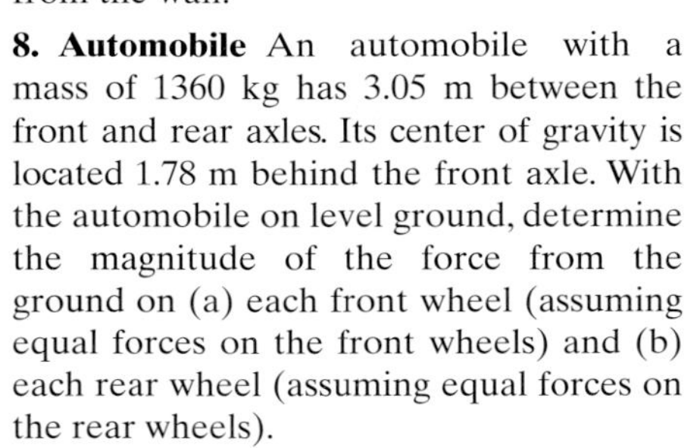

7. Uniform Sphere In Fig. 13-23 a uniform sphere of mass m and radius r is held in place by a massless rope attached to a frictionless wall a distance L above the center of the sphere. Find (a) the tension in the rope and (b) the force on the sphere from the wall.

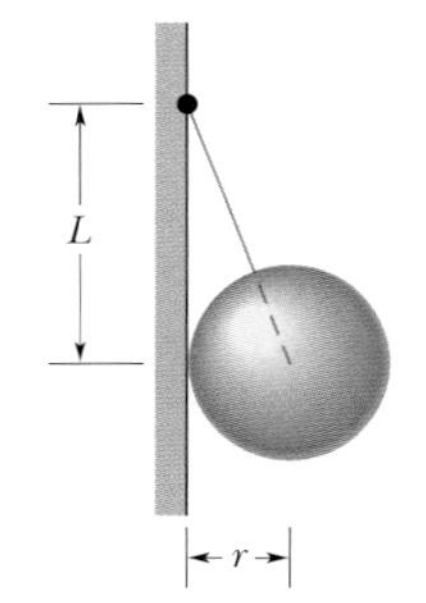

FIGURE 13-23 ■ Problem 7.

8. Automobile An automobile with a mass of 1360 kg has 3.05 m between the front and rear axles. Its center of gravity is located 1.78 m behind the front axle. With the automobile on level ground, determine the magnitude of the force from the ground on (a) each front wheel (assuming equal forces on the front wheels) and (b) each rear wheel (assuming equal forces on the rear wheels).

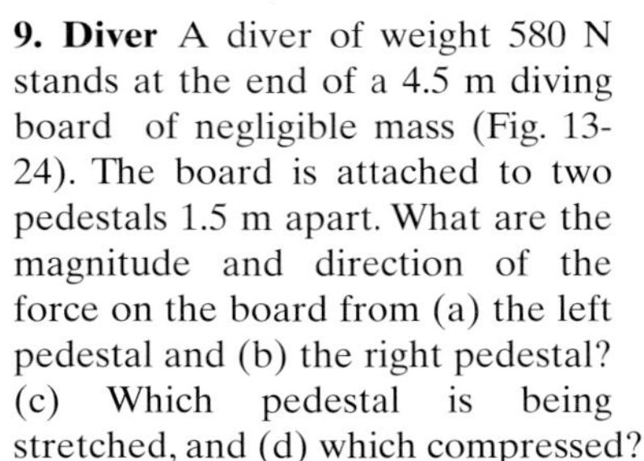

9. Diver A diver of weight 580 N stands at the end of a 4.5 m diving board of negligible mass (Fig. 13-24). The board is attached to two pedestals 1.5 m apart. What are the magnitude and direction of the force on the board from (a) the left pedestal and (b) the right pedestal? (c) Which pedestal is being stretched, and (d) which compressed?

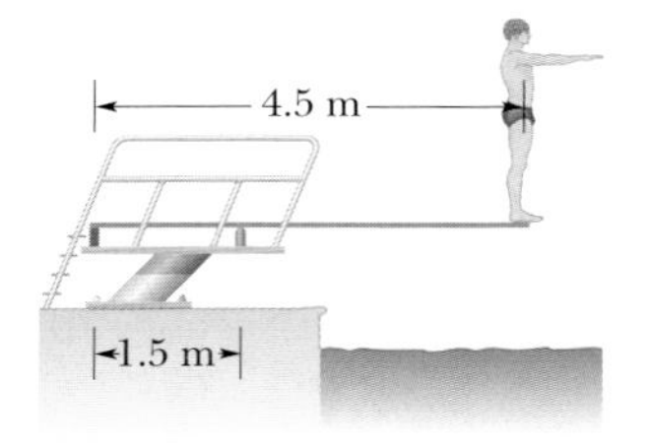

FIGURE 13-24 ■ Problem 9.

10. Car in the Mud In Fig. 13-25, a man is trying to get his car out of mud on the shoulder of a road. He ties one end of a rope tightly around the front bumper and the other end tightly around a utility pole 18 m away. He then pushes sideways on the rope at its midpoint with a force of 550 N, displacing the center of the rope 0.30 m from its previous position, and the car barely moves. What is the magnitude of the force on the car from the rope? (The rope stretches somewhat.)

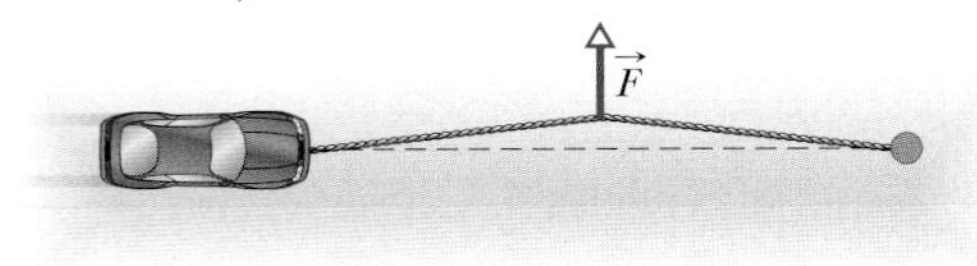

FIGURE 13-25 ■ Problem 10.

11. Meter Stick A meter stick balances horizontally on a knife-edge at the 50.0 cm mark. With two 5.0 g coins stacked over the 12.0 cm mark, the stick is found to balance at the 45.5 cm mark. What is the mass of the meter stick?

12. Uniform Cubical A uniform cubical crate is 0.750 m on each side and weighs 500 N. It rests on a floor with one edge against a very small, fixed obstruction. At what least height above the floor must a horizontal force of magnitude 350 N be applied to the crate to tip it?

13. Window Cleaner A 75 kg window cleaner uses a 10 kg ladder that is 5.0 m long. He places one end on the ground 2.5 m from a wall, rests the upper end against a cracked window, and climbs the ladder. He is 3.0 m up along the ladder when the window breaks. Neglecting friction between the ladder and window and assuming that the base of the ladder does not slip, find (a) the magnitude of the force on the window from the ladder just before the window breaks and (b) the magnitude and direction of the force on the ladder from the ground just before the window breaks.

14. Lower Leg Figure 13-26 shows the anatomical structures in the lower leg and foot that are involved in standing tiptoe with the heel raised off the floor so the foot effectively contacts the floor at only one point, shown as P in the figure. Calculate, in terms of a person's weight W, the forces on the foot from (a) the calf muscle (at A) and (b) the lower-leg bones (at B) when the person stands tiptoe on one foot. Assume that $a = 5.0$ cm and $b = 15$ cm.

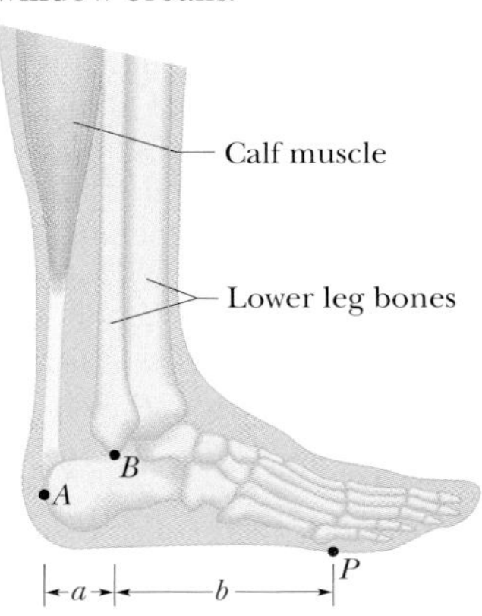

FIGURE 13-26 ■ Problem 14.

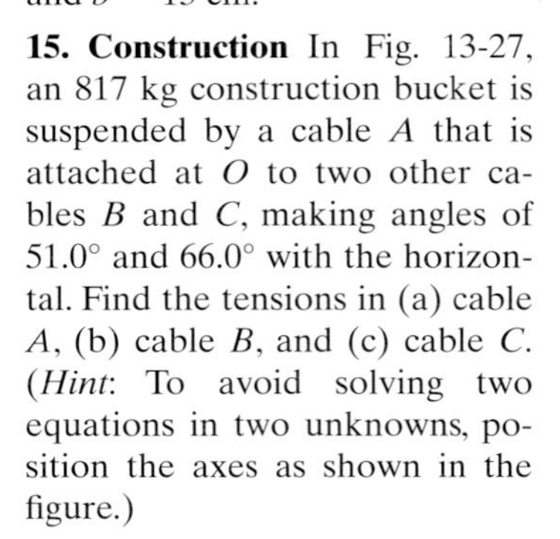

15. Construction In Fig. 13-27, an 817 kg construction bucket is suspended by a cable A that is attached at O to two other cables B and C, making angles of 51.0° and 66.0° with the horizontal. Find the tensions in (a) cable A, (b) cable B, and (c) cable C. (*Hint*: To avoid solving two equations in two unknowns, position the axes as shown in the figure.)

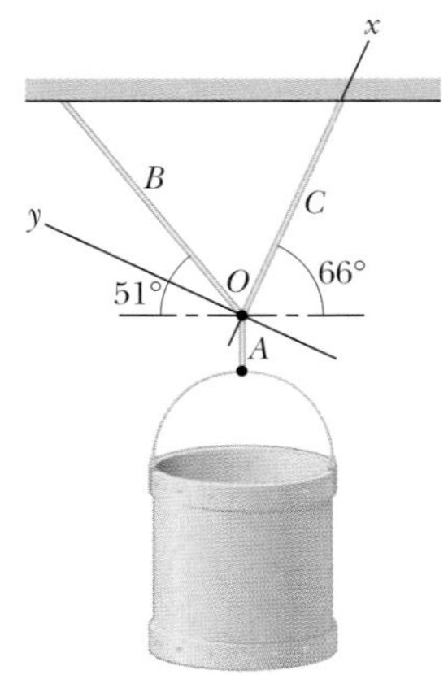

FIGURE 13-27 ■ Problem 15.

16. System in Equilibrium The system in Fig. 13-28 is in equilibrium, with the string in the center exactly horizontal. Find (a) tension T_A, (b) tension T_B, (c) tension T_C, and (d) angle θ.

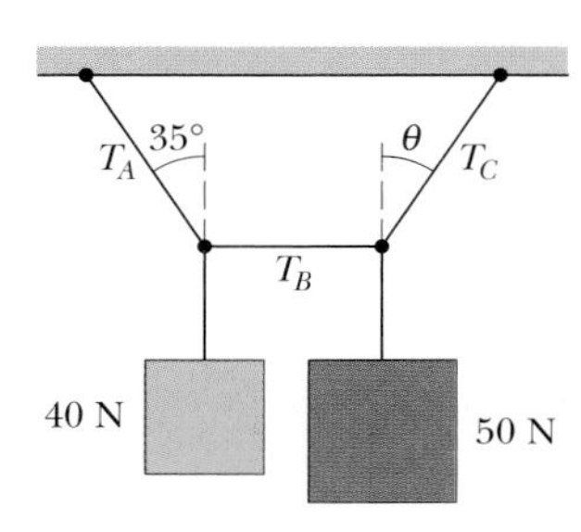

FIGURE 13-28 ■ Problem 16.

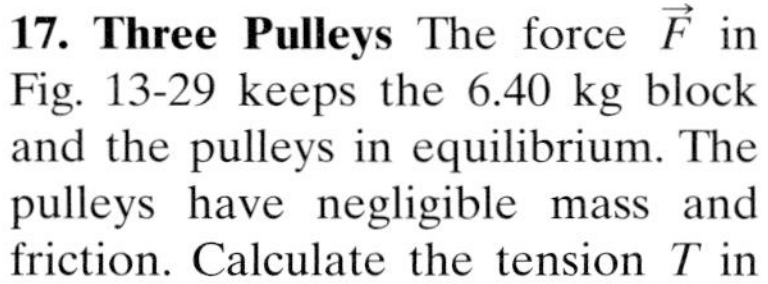

17. Three Pulleys The force $\vec{F}$ in Fig. 13-29 keeps the 6.40 kg block and the pulleys in equilibrium. The pulleys have negligible mass and friction. Calculate the tension T in the upper cable. (*Hint:* When a cable wraps halfway around a pulley as here, the magnitude of its net force on the pulley is twice the tension in the cable.)

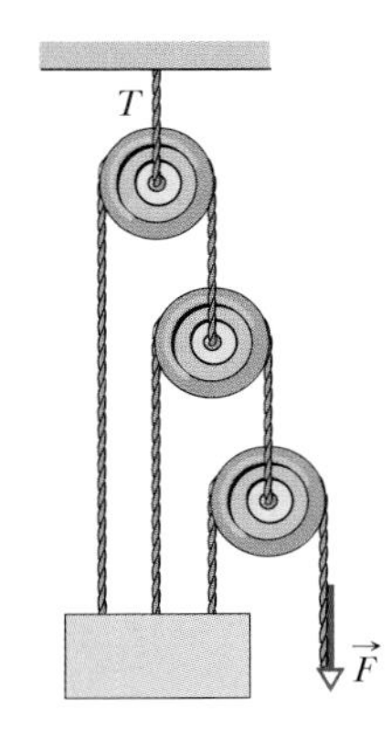

FIGURE 13-29 ■ Problem 17.

18. Triceps A 15 kg block is being lifted by the pulley system shown in Fig. 13-30. The upper arm is vertical, whereas the forearm makes an angle of 30° with the horizontal. What are the forces on the forearm from (a) the triceps muscle and (b) the upper-arm bone (the humerus)? The forearm and hand together have a mass of 2.0 kg with a center of mass 15 cm (measured along the arm) from the point where the forearm and upper-arm bones are in contact. The triceps muscle pulls vertically upward at a point 2.5 cm behind that contact point.

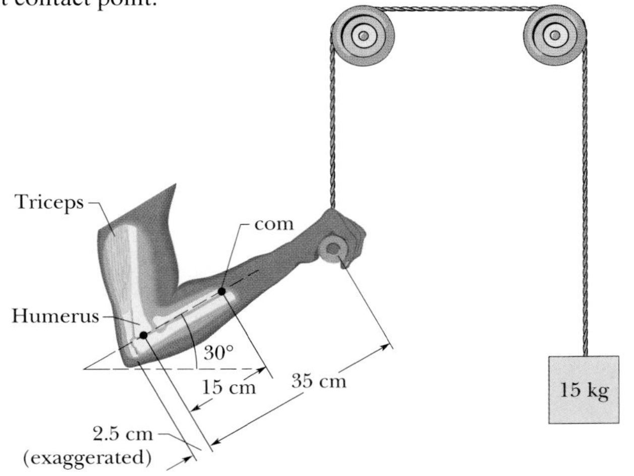

FIGURE 13-30 ■ Problem 18.

19. Forces on Structure Forces $\vec{F}_A$, $\vec{F}_B$, and $\vec{F}_C$ act on the structure of Fig. 13-31 shown in an overhead view. We wish to put the structure in equilibrium by applying a fourth force, at a point such as P. The fourth force has vector components $\vec{F}_h$ and $\vec{F}_v$. We are given that $a = 2.0$ m, $b = 3.0$ m, $c = 1.0$ m, $|\vec{F}_A| = 20$ N, $|\vec{F}_B| = 10$ N, and $|\vec{F}_C| = 5.0$ N. Find (a) $|\vec{F}_h|$, (b) $|\vec{F}_v|$, and (c) d.

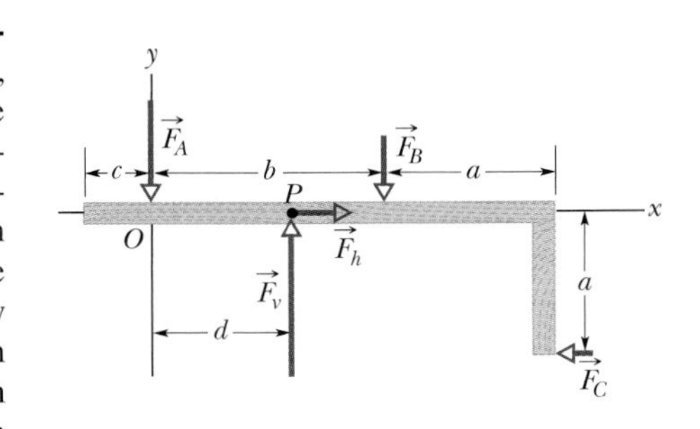

FIGURE 13-31 ■ Problem 19.

20. Square Sign In Fig. 13-32, a 50.0 kg uniform square sign, 2.00 m on a side, is hung from a 3.00 m horizontal rod of negligible mass. A cable is attached to the end of the rod and to a point on the wall 4.00 m above the point where the rod is hinged to the wall. (a) What is the tension in the cable? What are the magnitudes and directions of the (b) horizontal and (c) vertical components of the force on the rod from the wall?

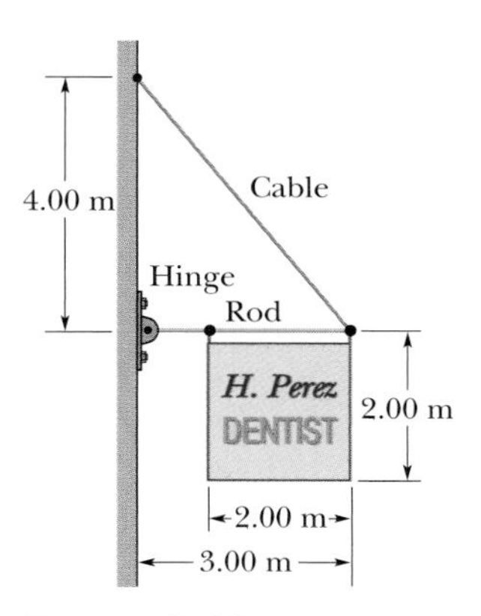

FIGURE 13-32 ■ Problem 20.

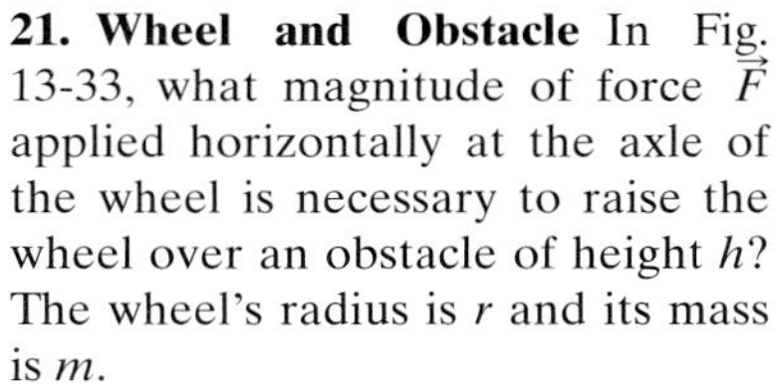

21. Wheel and Obstacle In Fig. 13-33, what magnitude of force $\vec{F}$ applied horizontally at the axle of the wheel is necessary to raise the wheel over an obstacle of height h? The wheel's radius is r and its mass is m.

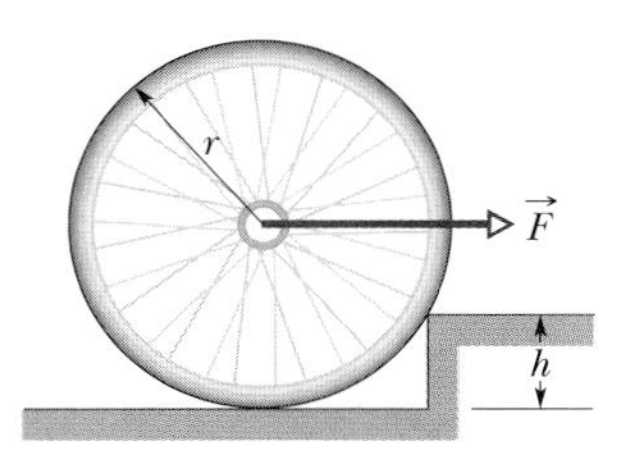

FIGURE 13-33 ■ Problem 21.

22. Rock Climber In Fig. 13-34, a 55 kg rock climber is in a lie-back climb along a fissure, with hands pulling on one side of the fissure and feet pressed against the opposite side. The fissure has width $w = 0.20$ m, and the center of mass of the climber is a horizontal distance $d = 0.40$ m from the fissure. The coefficient of static friction between hands and rock is $\mu_{\text{hands}}^{\text{stat}} = 0.40$, and between boots and rock it is $\mu_{\text{boots}}^{\text{stat}} = 1.2$. (a) What is the least horizontal pull by the hands and push by the feet that will keep the climber stable? (b) For the horizontal pull of (a), what must be the vertical distance h between hands and feet? (c) If the climber encounters wet rock, so that $\mu_{\text{hands}}^{\text{stat}}$ and $\mu_{\text{boots}}^{\text{stat}}$ are reduced, what happens to the answers to (a) and (b), respectively?

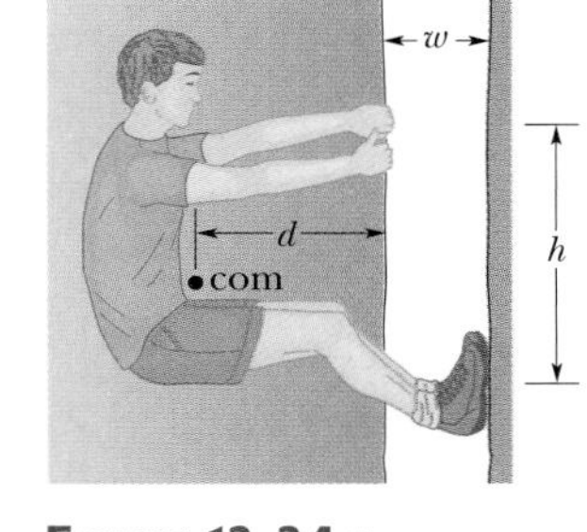

FIGURE 13-34 ■ Problem 22.

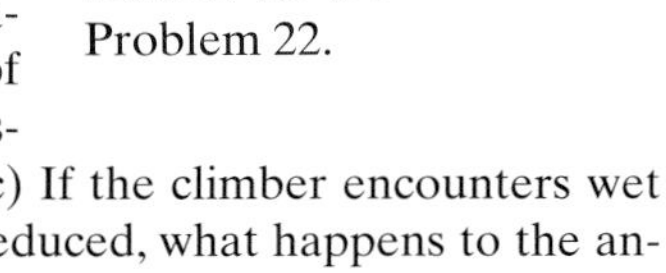

23. Beam and Hinge In Fig. 13-35, one end of a uniform beam that weighs 222 N is attached to a wall with a hinge. The other end is supported by a wire. (a) Find the tension in the wire. What are the (b) horizontal and (c) vertical components of the force of the hinge on the beam?

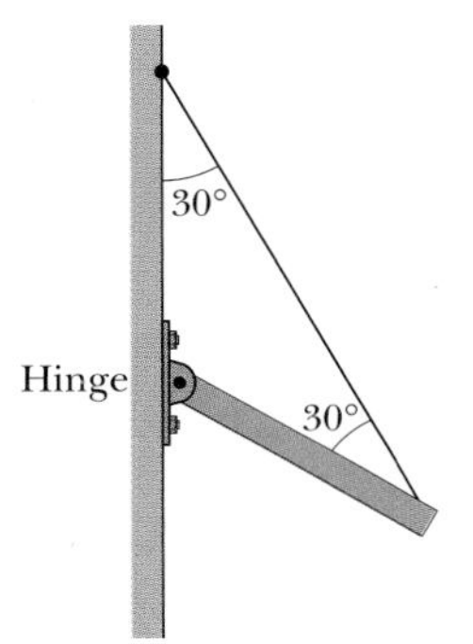

FIGURE 13-35 ■ Problem 23.

24. Four Bricks Four bricks of length L, identical and uniform, are stacked on top of one another (Fig. 13-36) in such a way that part of each extends beyond the one beneath.

Find, in terms of L, the maximum values of (a) a_A, (b) a_B, (c) a_C, (d) a_D, and (e) h, such that the stack is in equilibrium.

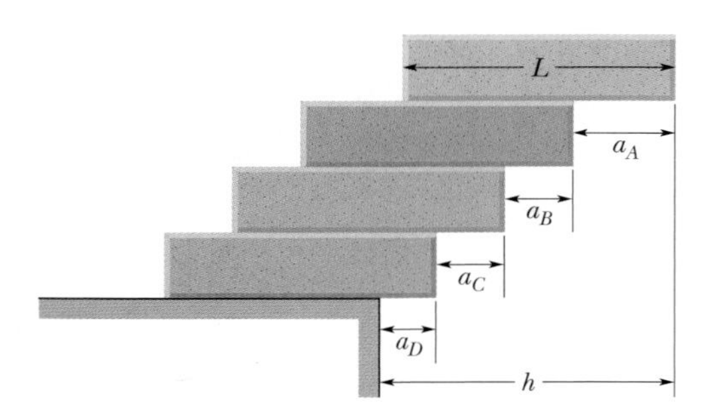

FIGURE 13-36 ■ Problem 24.

25. Concrete Block The system in Fig. 13-37 is in equilibrium. A concrete block of mass 225 kg hangs from the end of the uniform strut whose mass is 45.0 kg. Find (a) the tension T in the cable and the (b) horizontal and (c) vertical force components on the strut from the hinge.

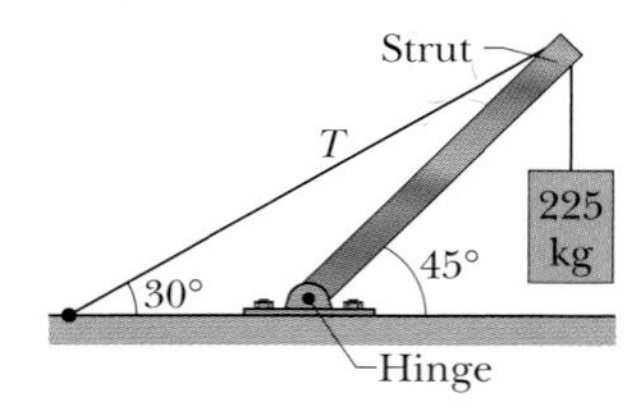

FIGURE 13-37 ■ Problem 25.

26. A Door A door 2.1 m high and 0.91 m wide has a mass of 27 kg. A hinge 0.30 m from the top and another 0.30 m from the bottom each support half the door's mass. Assume that the center of gravity is at the geometrical center of the door, and determine the (a) vertical and (b) horizontal components of the force from each hinge on the door.

27. Nonuniform Bar A nonuniform bar is suspended at rest in a horizontal position by two massless cords as shown in Fig. 13-38. One cord makes the angle $\theta = 36.9°$ with the vertical; the other makes the angle $\phi = 53.1°$ with the vertical. If the length L of the bar is 6.10 m, compute the distance x from the left-hand end of the bar to its center of mass.

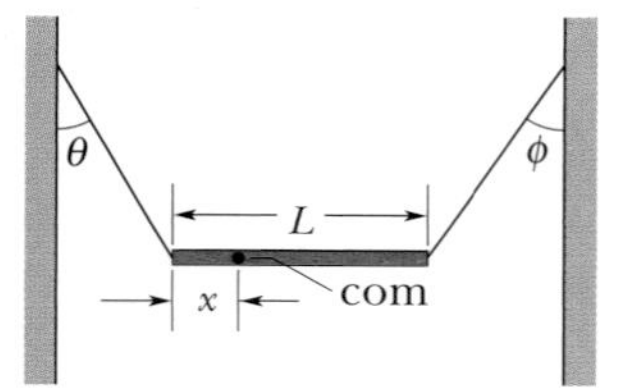

FIGURE 13-38 ■ Problem 27.

28. Thin Horizontal Bar In Fig. 13-39 a thin horizontal bar AB of negligible weight and length L is hinged to a vertical wall at A and supported at B by a thin wire BC that makes an angle θ with the horizontal. A load of weight W can be moved anywhere along the bar; its position is defined by the distance x from the wall to its center of mass. As a function of x, find (a) the tension in the wire, and the (b) horizontal and (c) vertical components of the force on the bar from the hinge at A.

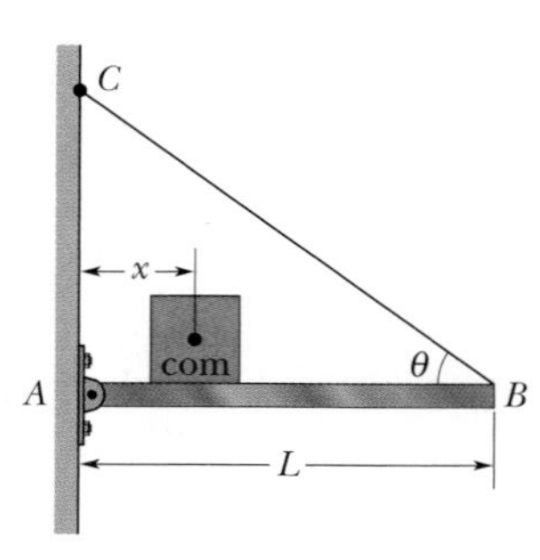

FIGURE 13-39 ■ Problems 28 and 30.

29. Uniform Plank In Fig. 13-40, a uniform plank, with a length L of 6.10 m and a weight of 445 N, rests on the ground and against a frictionless roller at the top of a wall of height $h = 3.05$ m. The plank remains in equilibrium for any value of $\theta \geq 70°$ but slips if $\theta < 70°$. Find the coefficient of static friction between the plank and the ground.

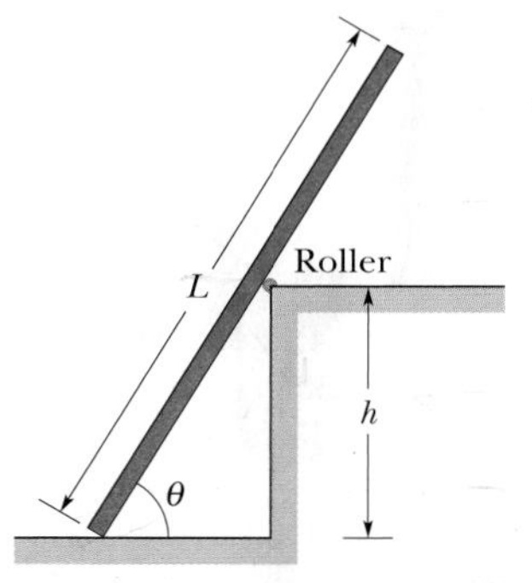

FIGURE 13-40 ■ Problem 29.

30. Max Tension In. Fig. 13-39, suppose the length L of the uniform bar is 3.0 m and its weight is 200 N. Also, let the load's weight $W = 300$ N and the angle $\theta = 30°$. The wire can withstand a maximum tension of 500 N. (a) What is the maximum possible distance x before the wire breaks? With the load placed at this maximum x, what are the (b) horizontal and (c) vertical components of the force on the bar from the hinge at A?

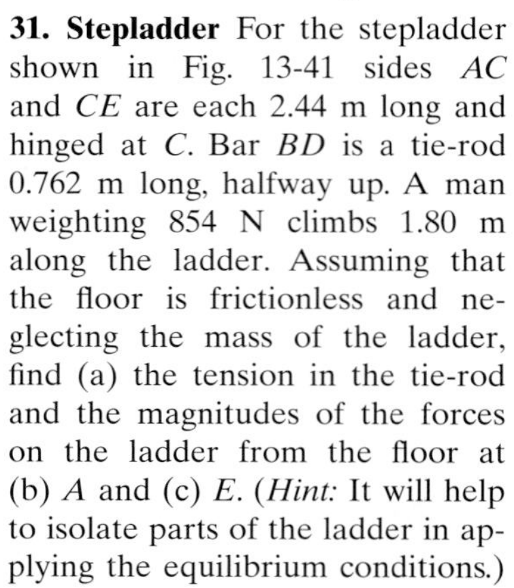

31. Stepladder For the stepladder shown in Fig. 13-41 sides AC and CE are each 2.44 m long and hinged at C. Bar BD is a tie-rod 0.762 m long, halfway up. A man weighting 854 N climbs 1.80 m along the ladder. Assuming that the floor is frictionless and neglecting the mass of the ladder, find (a) the tension in the tie-rod and the magnitudes of the forces on the ladder from the floor at (b) A and (c) E. (*Hint:* It will help to isolate parts of the ladder in applying the equilibrium conditions.)

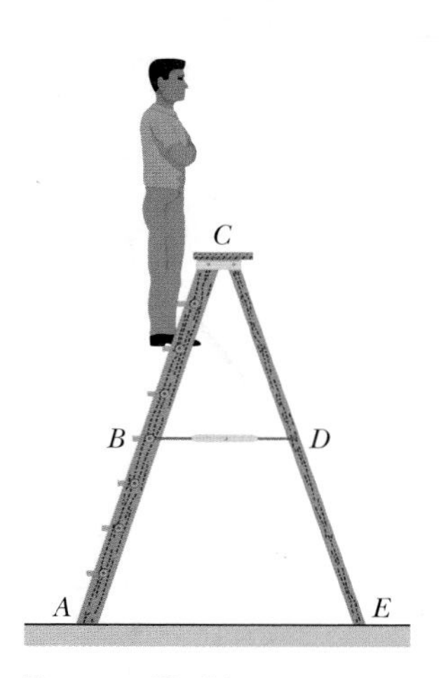

FIGURE 13-41 ■ Problem 31.

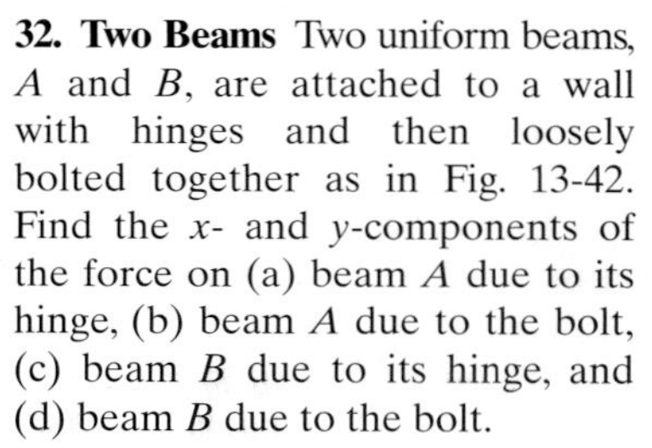

32. Two Beams Two uniform beams, A and B, are attached to a wall with hinges and then loosely bolted together as in Fig. 13-42. Find the x- and y-components of the force on (a) beam A due to its hinge, (b) beam A due to the bolt, (c) beam B due to its hinge, and (d) beam B due to the bolt.

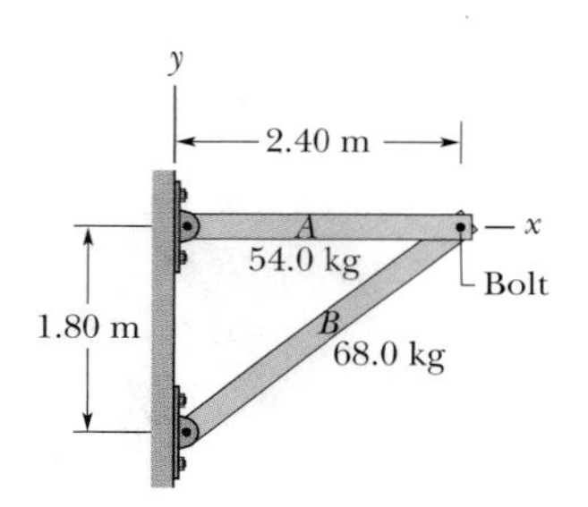

FIGURE 13-42 ■ Problem 32.

33. Box of Sand A cubical box is filled with sand and weighs 890 N. We wish to tip the box by pushing horizontally on one of the upper edges. (a) What minimum force is required? (b) What minimum coefficient of static friction between box and floor is required? (c) Is there a more efficient way to tip the box? If so, find the smallest possible force that would have to be applied directly to the box to tip it. (*Hint:* At the onset of tipping, where is the normal force located?)

34. Two Arrangements Four bricks of length L, identical and uniform, are stacked on a table in two ways, as shown in Fig. 13-43 (compare

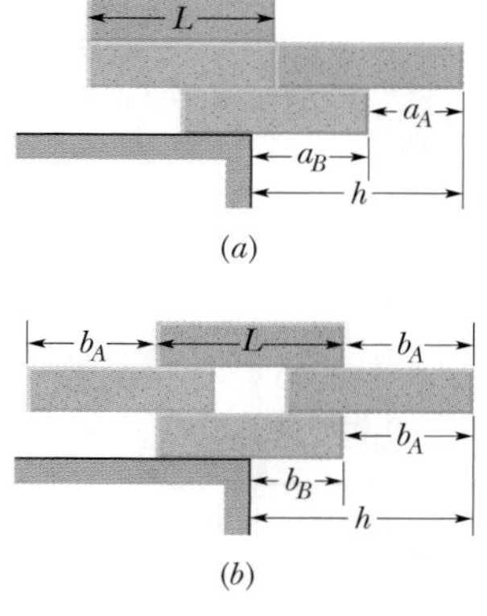

FIGURE 13-43 ■ Problem 34.

with Problem 24). We seek to maximize the overhang distance h in both arrangements. Find the optimum distances a_A, a_B, b_A, b_B, and calculate for the two arrangements. [See "The Amateur Scientist," *Scientific American*, June 1985, pp. 133–134, for a discussion and an even better version of arrangement (*b*).]

35. Cubical Crate A crate, in the form of a cube with edge lengths of 1.2 m, contains a piece of machinery; the center of mass of the crate and its contents is located 0.30 m above the crate's geometrical center. The crate rests on a ramp that makes an angle θ with the horizontal. As θ is increased from zero, an angle will be reached at which the crate will either start to slide down the ramp or tip over. Which event will occur (a) when the coefficient of static friction between ramp and crate is 0.60 and (b) when it is 0.70? In each case, give the angle at which the event occurs. (*Hint:* At the onset of tipping, where is the normal force located?)

SEC. 13-5 ■ ELASTICITY

36. Young's Modulus Figure 13-44 shows the stress–strain curve for quartzite. What are (a) the Young's modulus and (b) the approximate yield strength for this material?

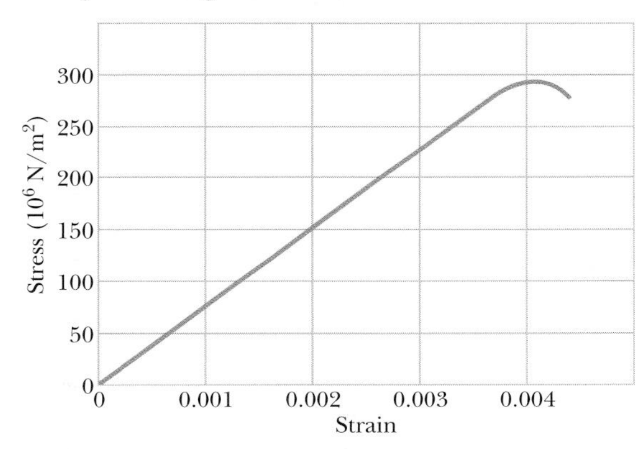

FIGURE 13-44 ■ Problem 36.

37. Aluminum Rod A horizontal aluminum rod 4.8 cm in diameter projects 5.3 cm from a wall. A 1200 kg object is suspended from the end of the rod. The shear modulus of aluminum is 3.0×10^{10} N/m^2. Neglecting the rod's mass, find (a) the shear stress on the rod and (b) the vertical deflection of the end of the rod.

38. Lead Brick In Fig. 13-45, a lead brick rests horizontally on cylinders A and B. The areas of the top faces of the cylinders are related by $A_A = 2A_B$; the Young's moduli of the cylinders are related by $E_A = 2E_B$. The cylinders had identical lengths before the brick was placed on them. What fraction of the brick's mass is supported (a) by cylinder A and (b) by cylinder B? The horizontal distances between the center of mass of the brick and the centerlines of the cylinders are d_A for cylinder A and d_B for cylinder B. (c) What is the ratio d_A/d_B?

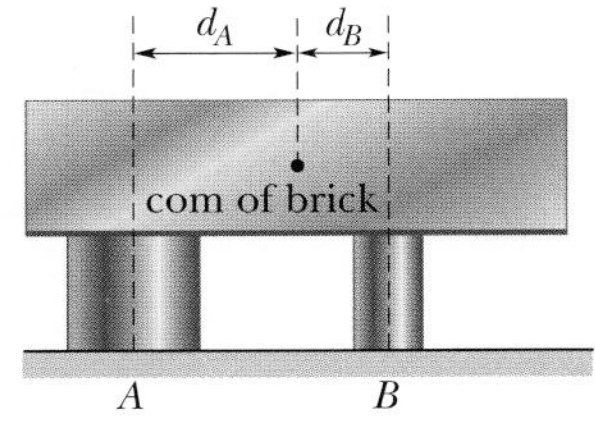

FIGURE 13-45 ■ Problem 38.

39. Uniform Log In Fig. 13-46, 103 kg uniform log hangs by two steel wires, A and B, both of radius 1.20 mm. Initially, wire A was 2.50 m long and 2.00 mm shorter than wire B. The log is now horizontal. What are the magnitudes of the forces on it from (a) wire A and (b) wire B? (c) What is the ratio d_A/d_B?

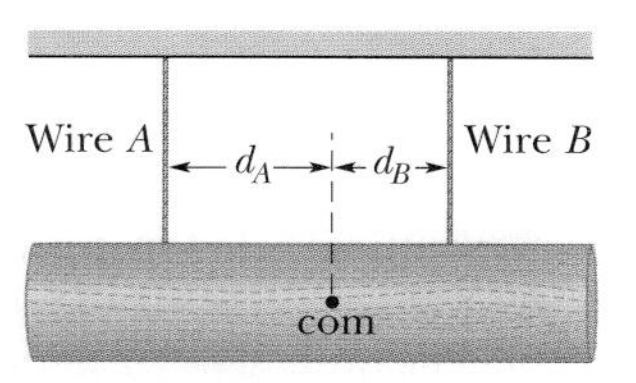

FIGURE 13-46 ■ Problem 39.

40. Tunnel A tunnel 150 m long, 7.2 m high, and 5.8 m wide (with a flat roof) is to be constructed 60 m beneath the ground. (See Fig. 13-47.) The tunnel roof is to be supported entirely by square steel columns, each with a cross-sectional area of 960 cm^2. The density of the ground material is 2.8 g/cm^3. (a) What is the total mass of the material that the columns must support? (b) How many columns are needed to keep the compressive stress on each column at one-half its ultimate strength?

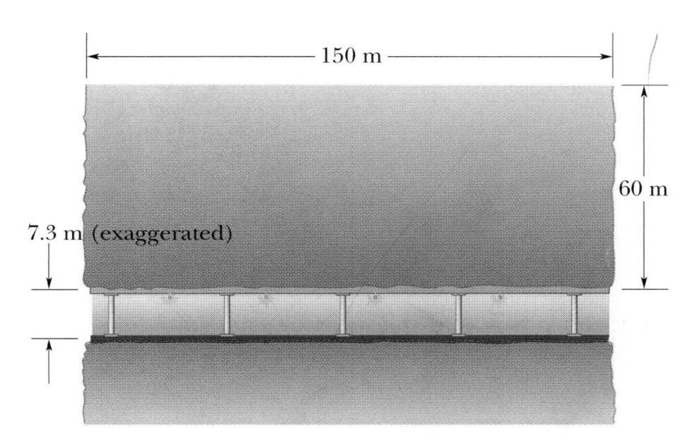

FIGURE 13-47 ■ Problem 40.

41. Cylindrical Aluminum Rod A cylindrical aluminum rod, with an initial length of 0.8000 m and radius 1000.0 μm, is clamped in place at one end and then stretched by a machine pulling parallel to its length at its other end. Assuming that the rod's density (mass per unit volume) does not change, find the force magnitude that is required of the machine to decrease the radius to 999.9 μm. (The yield strength is not exceeded.)

42. Stress Versus Strain Figure 13-48 shows the stress versus strain plot for an aluminum wire that is stretched by a machine pulling in opposite directions at the two ends of the wire. The wire has an initial length of 0.800 m and an initial cross-sectional area of 2.00×10^{-6} m^2. How much work does the force from the machine do on the wire to produce a strain of 1.00×10^{-3}?

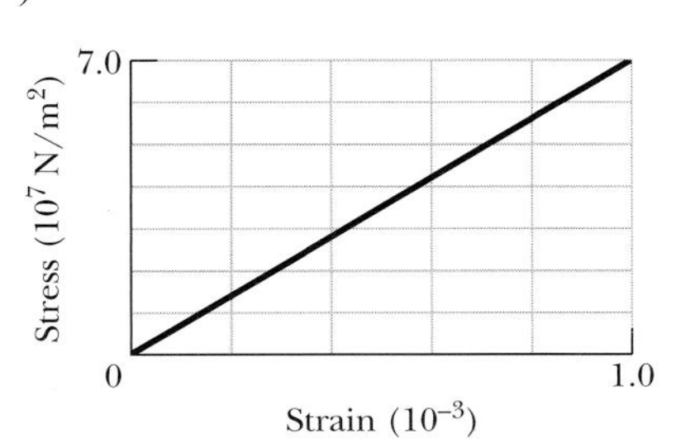

FIGURE 13-48 ■ Problem 42.

Additional Problems

43. Preventing Velociraptors In the movie Jurassic Park, there is a scene in which some members of the visiting group are trapped in the kitchen with dinosaurs outside the door. The paleontologist is pressing his shoulder near the center of the door, trying to keep out the dinosaurs who are on the other side. The botanist throws herself against the door at the edge right next to the hinge. A pivotal point in the film is that she cannot reach a gun on the floor because she is trying to help hold the door closed. Would they improve or worsen their situation if the paleontologist moved to the outer edge of the door and the botanist went for the gun? Estimate the change in the torque they are exerting on the door due to the change in their positions.

44. Rollerboards In the past few years, luggage carts that are rolling suitcases with handles, called "rollerboards," have become commonplace in airports around the country. Often, you will see people with a briefcase or additional small bag hung on the cart in one of the two ways shown, either hanging over the front of the cart (Fig. 13-49*a)* or resting on the handle (Fig. 13-49*b*). In this problem, we will figure out which way is easier for the traveler.

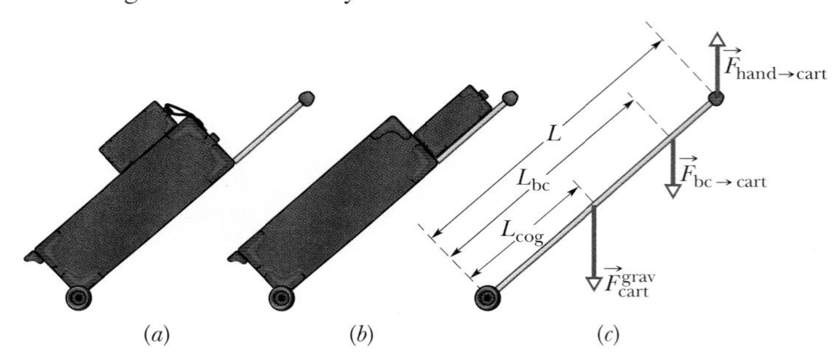

FIGURE 13-49 ■ Problem 44.

In Fig. 13-49*c* we have sketched a simplified idealization of the cart as a thin rod with forces acting on it. Three of the forces are shown, the gravitational attraction of the earth on the cart ($\vec{F}^{grav}_{cart}$), the force of the briefcase on the cart ($\vec{F}_{bc}$), and the force of the traveler's hand holding the cart up ($\vec{F}_{hand \to cart}$). Take the angle the cart makes with the ground to be θ, the mass of the cart to be M, and the mass of the briefcase to be m. Assume that the total length of the cart, handle and all, is L, the center of gravity of the cart is a distance L_{cog} from the wheel, and the center of gravity of the briefcase is a distance L_{bc} from the wheel.

(a) Find an expression for the force the hand has to exert in order to hold up the luggage cart. Express your answer in terms of the symbols given above.
(b) Does the force the cart exerts on the floor depend on the positioning of the briefcase? Explain.
(c) Estimate how different the force the hand has to exert would be in the two cases shown in Figs. 13-49*a* and 13-49*b*.

45. TV on a Handtruck You are working as a staff person on the Internet chat program "Ask Dr. Science." The following e-mail message comes in and needs a quick answer.

My wife just called me at the office and asked the following question. We had a large computer monitor delivered to her home office this morning. The delivery person was kind enough to put the box on our hand truck (see attached picture) but he put it on while the truck was lying flat. My wife has some back problems and doesn't want to have to exert more than 50 pounds of force. The monitor in its box weighs about 85 pounds. She's having a business meeting at the house later and would like to get the box out of the front room. What I want to know is, can she stand the truck upright safely without hurting her back?

A schematic diagram of the hand truck with the box on it is shown in Fig. 13-50. Is it safe for the man's wife to pull the hand truck upright so she can roll the box into the back room? Be sure to explain why you think so.

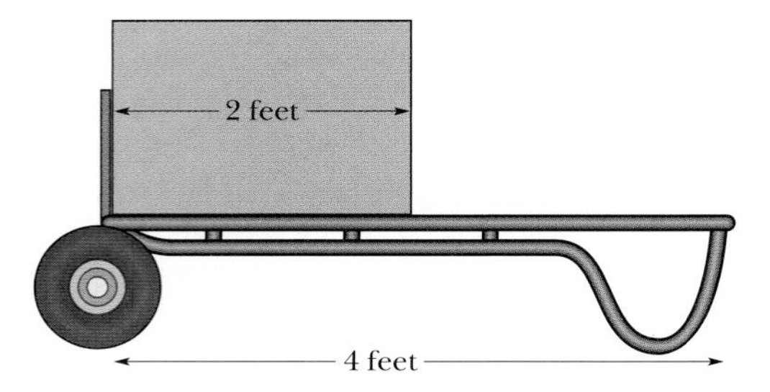

FIGURE 13-50 ■ Problem 45.

46. Refrigerator Shelf The shelves in your refrigerator are metal lattices that are held up by being slipped into two small (about 1 inch long) hollow boxes or "pockets" attached to the interior back wall of the refrigerator. See Fig. 13-51. If you put a full gallon of milk on the shelf, is it more likely to break the pocket if you place it near the back of the refrigerator or near the front? Explain your answer in terms of the physics you have learned. If the milk is the only thing on the shelf, estimate the downward force that the shelf exerts on the front of the pocket when the milk is placed at the front of the shelf.

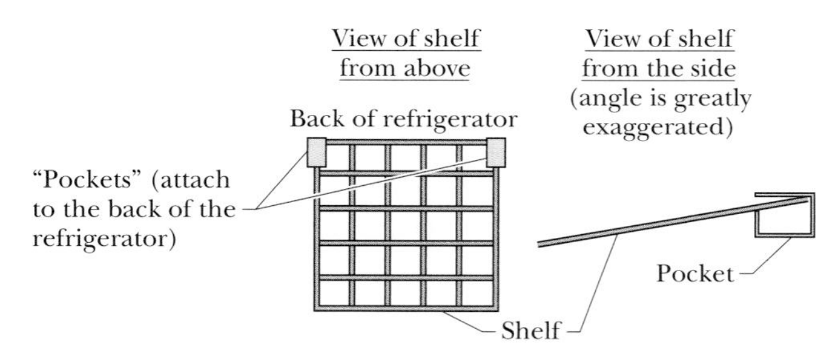

FIGURE 13-51 ■ Problem 46.

47. Weighing a Big Suitcase When preparing to travel to Australia last summer, a friend was concerned that her suitcase was too heavy. (There is a 20 kg limit on suitcases for international travel). Unfortunately, she only had a small bathroom scale. When she placed the suitcase directly on the scale, it covered the dial. She tried standing on the scale, measuring her weight, and then standing on the scale holding the suitcase. Unfortunately, when she was holding the suitcase, she couldn't see the markings on the scale, and it was too heavy to hold behind her. Design a way for her to measure the weight of the suitcase.

48. Indoor Playground You have been hired to build a large sand mound in an indoor playground and must be careful about the stress that the sand will put on the floor. Consulting research literature, you are surprised to find that the greatest stress occurs, not directly beneath the apex (top) of the mound, but at points that are a distance r^{max} from that central point (Fig. 13-52a). This outward displacement of the maximum stress is presumably due to the sand grains forming arches within the mound. For a mound of height $H = 3.00$ m and angle $\theta = 33°$, and with sand of density $\rho = 1800$ kg/m^3, Fig. 13-52b gives the stress σ as a function of radius r from the central point of the mound's base. In that figure, $\sigma^{center} = 40\ 000$ N/m^2, $\sigma^{max} = 40\ 024$ N/m^2, and $r^{max} = 1.82$ m.

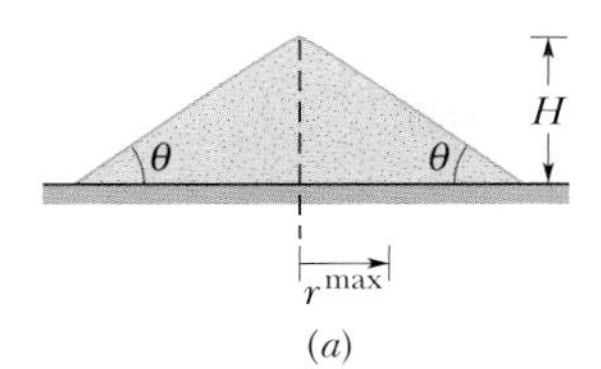

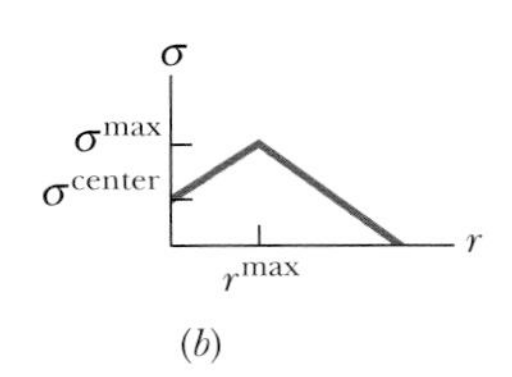

FIGURE 13-52 ■ Problem 48.

(a) What is the volume of sand contained in the mound for $r \leq r^{max}/2$? (*Hint:* The volume is that of a vertical cylinder plus a cone on top of the cylinder. The volume of the cone is $\pi R^2 h/3$, where R is the cone's radius and h is the cone's height.) (b) What is the weight W of that volume of sand? (c) Use Fig. 13-52b to write an expression for the stress σ on the floor as a function of radius r, for $r \leq r^{max}$. (d) On the floor, what is the area dA of a thin ring of radius r centered on the mound's central axis and with radial width dr? (e) What then is the magnitude dF of the downward force on the ring due to the sand? (f) What is the magnitude F of the net downward force on the floor due to all the sand contained in the mound for $r \leq r^{max}/2$? [*Hint:* Integrate the expression of (e) from $r = 0$ to $r = r^{max}/2$.] Now note the surprise: This force magnitude F on the floor is less than the weight W of the sand above the floor, as found in (b). (g) By what fraction is F reduced from W; that is, what is $(F - W)/W$?

49. Moving a Heavy Log Here is a way to move a heavy log through a tropical forest. Find a young tree in the general direction of travel; find a vine that hangs from the top of the tree down to ground level; pull the vine over to the log; wrap the vine around a limb on the log: pull hard enough on the vine to bend the tree over; and then tie off the vine on the limb. Repeat this procedure with several trees; eventually the net force of the vines on the log moves the log forward. Although tedious, this technique allowed workers to move heavy logs long before modern machinery was available. Figure 13-53 shows the essentials of the technique. There, a single vine is shown attached to a branch at one end of a uniform log of mass M. The coefficient of static friction between the log and the ground is 0.80. If the log is on the verge of sliding, with the left end raised slightly by the vine, what are (a) the angle θ and (b) the magnitude T of the force on the log from the vine?

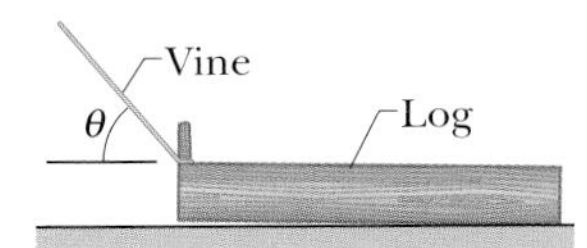

FIGURE 13-53 ■ Problem 49.

50. Uniform Ramp Figure 13-54a shows a uniform ramp between two buildings that allows for motion between the buildings due to strong winds. At its left end, it is hinged to the building wall; at its right end, it has a roller that can roll along the building wall. There is no vertical force on the roller from the building, only a horizontal force with magnitude F^{horiz}. The horizontal distance between the buildings is $D = 4.00$ m. The rise of the ramp is $h = 0.490$ m. A man walks across the ramp from the left. Figure 13-54b gives F^{horiz} as a function of the horizontal distance x of the man from the building at the left. What are the masses of (a) the ramp and (b) the man?

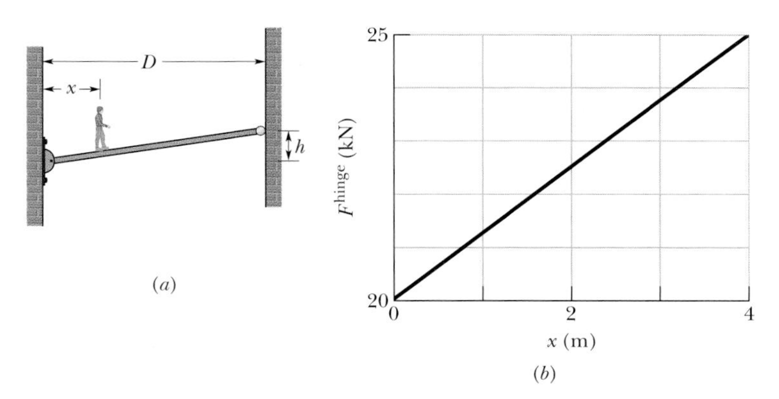

FIGURE 13-54 ■ Problem 50.

51. Diving Board In Fig. 13-55, a uniform diving board (mass = 40 kg) is 3.5 m long and is attached to two supports. When a diver stands on the end of the board, the support on the other end exerts a downward force of 1200 N on the board. Where on the board should the diver stand in order to reduce that force to zero?

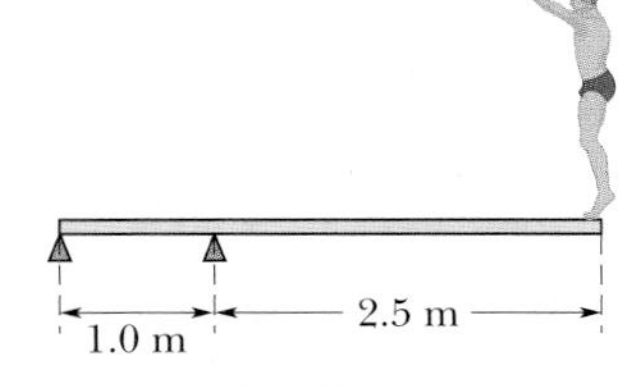

FIGURE 13-55 ■ Problem 51.

52. Rollers In Fig. 13-56a, a uniform 40 kg beam is centered over two rollers. Vertical lines across the beam mark off equal lengths. Two of the lines are centered over the rollers; a 10 kg package of tamale is centered over roller B. What are the magnitudes of the forces on the beam from (a) roller A and (b) roller B? The beam is then rolled to the left until the right-hand end is centered over roller B (Fig. 13-56b). What now are the magnitudes of the forces on the beam from (c) roller A and (d) roller B? Next, the beam is rolled to the right. Assume that it has a length of 0.800 m. (e) What horizontal distance between the package and roller B puts the beam on the verge of losing contact with roller A?

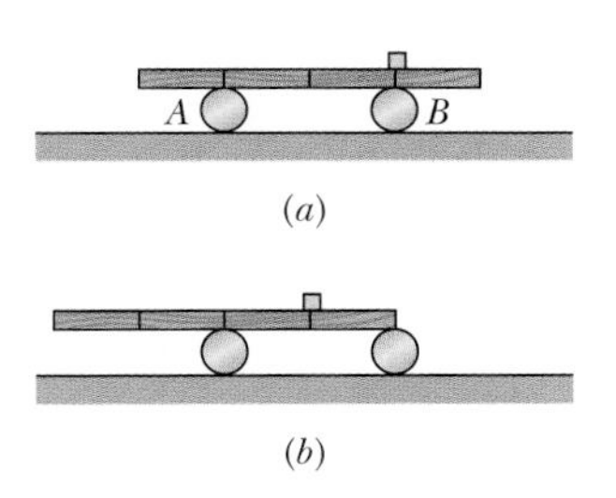

FIGURE 13-56 ■ Problem 52.

53. Horizontal Uniform Beam Figure 13-57a shows a horizontal uniform beam of mass m_{beam} and length L that is supported on the left by a hinge with a wall and on the right by a cable at angle θ with the horizontal. A package of mass m_{pack} is positioned on the beam at a distance x from the left end. The total mass is m_{beam} +

$m_{pack} = 61.22$ kg. Figure 13-57*b* gives the tension T in the cable as a function of the package's position given as a fraction x/L of the beam length. Evaluate (a) angle θ, (b) mass m_{beam}, and (c) mass in m_{pack}.

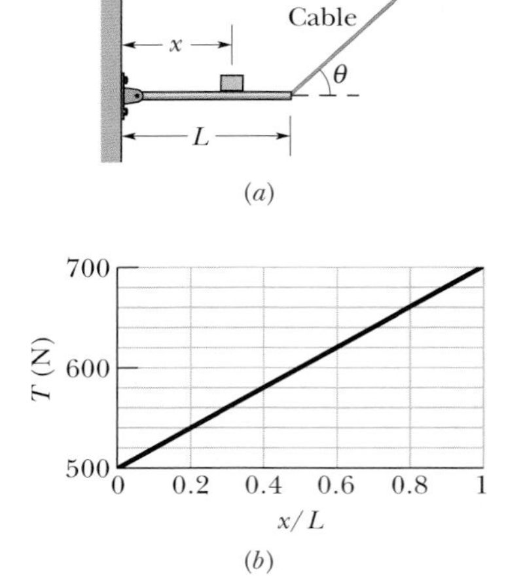

FIGURE 13-57 ■ Problem 53.

54. Vertical Uniform Beam Figure 13-58*a* shows a vertical uniform beam of length L that is hinged at its lower end. A horizontal force $\vec{F}^{app}$ is applied to the beam at a distance y from the lower end. The beam remains vertical because of a cable attached at the upper end, at angle θ with the horizontal. Figure 13-58*b* gives the tension T in the cable as a function of the position of the applied force given as a fraction y/L of the beam length. Figure 13-58*c* gives the magnitude F^{hinge} of the horizontal force on the beam from the hinge, also as a function of y/L. Evaluate (a) angle θ and (b) the magnitude of $\vec{F}^{app}$.

55. Makeshift Swing A makeshift swing is constructed by making a loop in one end of a rope and tying the other end to a tree limb. A child is sitting in the loop with the rope hanging vertically when an adult pulls on the child with a horizontal force and displaces the child to one side. Just before the child is released from rest, the rope makes an angle of 15° with the vertical and the tension in the rope is 280 N. (a) How much does the child weigh? (b) What is the magnitude of the (horizontal) force of the adult on the child just before the child is released? (c) If the maximum horizontal force that the adult can exert on the child is 93 N, what is the maximum angle with the vertical that the rope can make while the adult is pulling horizontally?

56. Emergency Stop A car on a horizontal road makes an emergency stop by applying the brakes so that all four wheels lock and skid along the road. The coefficient of kinetic friction between tires and road is 0.40. The separation between the front and rear axles is 4.2 m, and the center of mass of the car is located 1.8 m behind the front axle and 0.75 m above the road; see Fig. 13-59. The car weighs 11 kN. Calculate (a) the braking acceleration of the car, (b) the normal force on each wheel, and (c) the braking force on each wheel. (*Hint:* Although the car is not in translational equilibrium, it *is* in rotational equilibrium.)

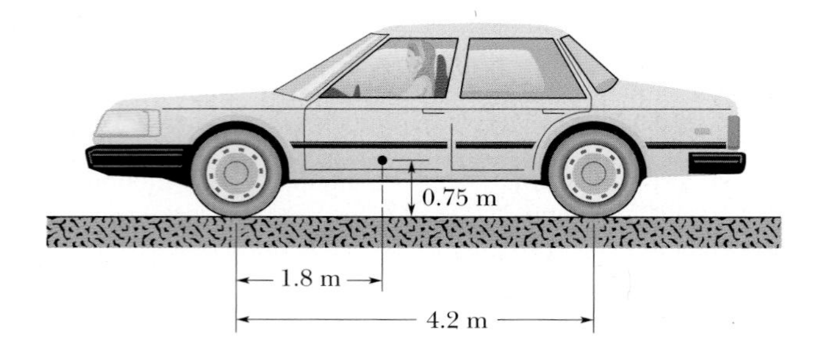

FIGURE 13-59 ■ Problem 56.

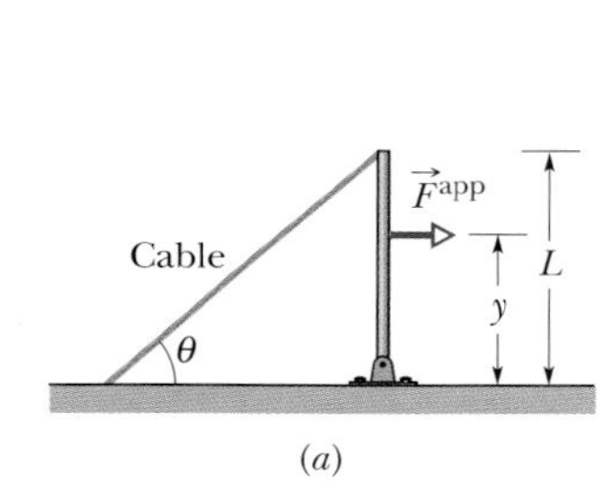

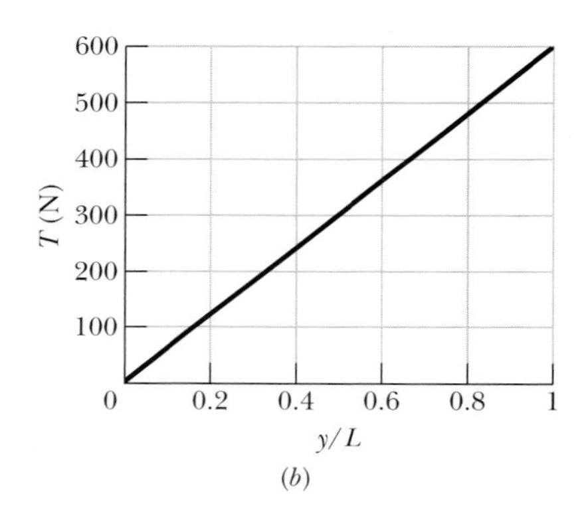

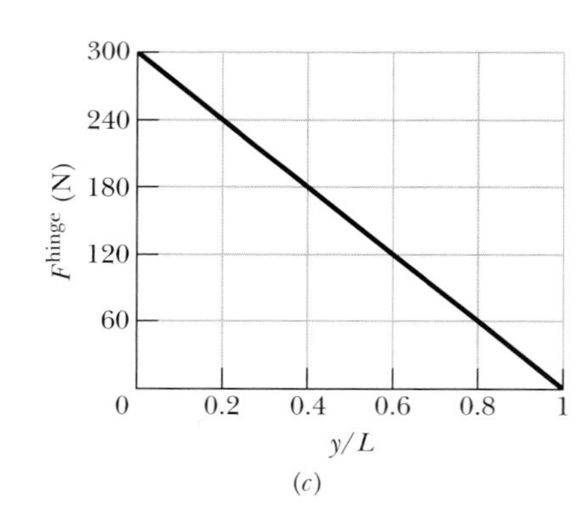

FIGURE 13-58 ■ Problem 54.

14 Gravitation

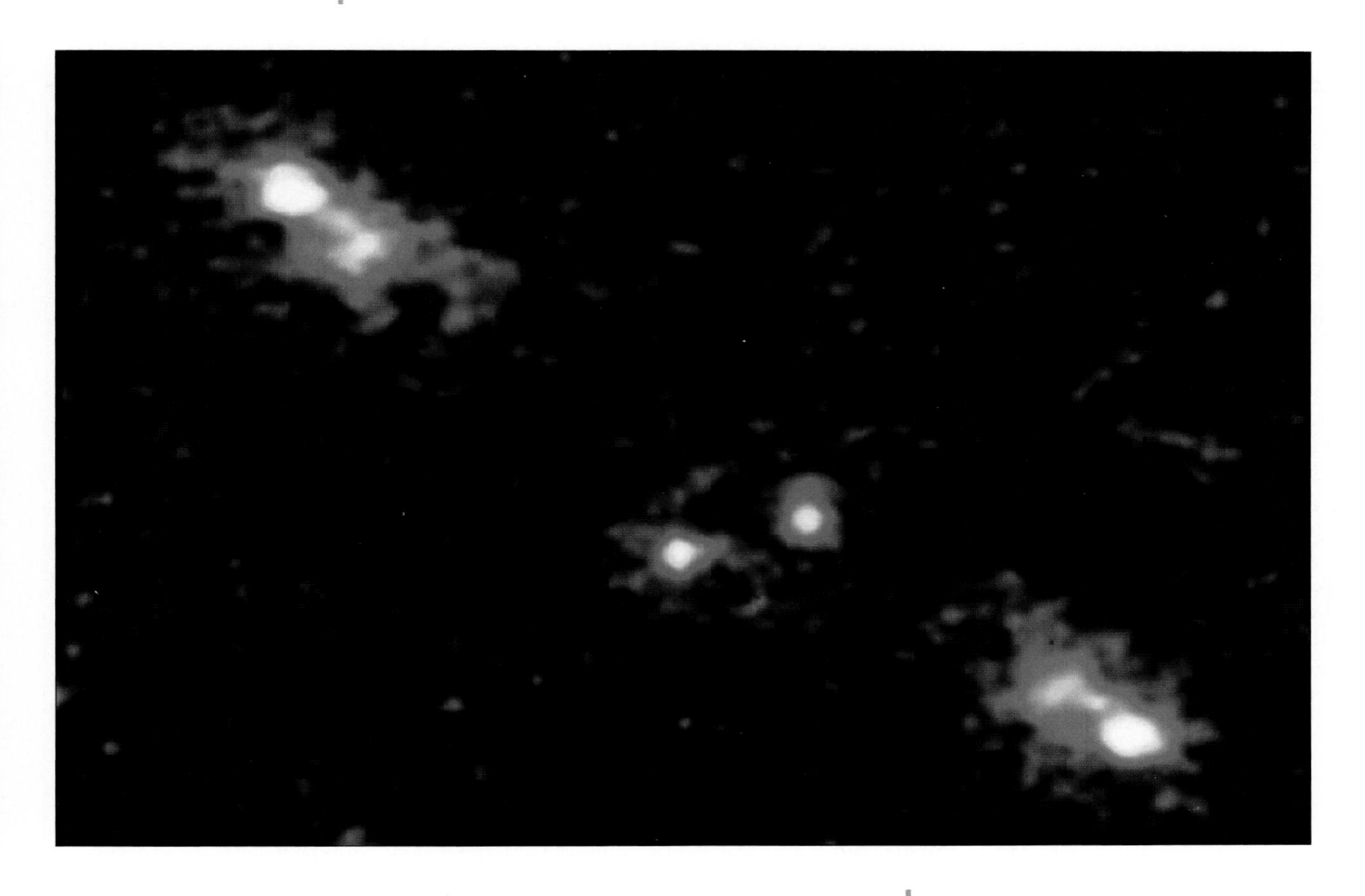

Quasars (or quasi-stellar objects) are highly luminous compact objects that are larger than stars and smaller than whole galaxies. Although some quasars emit energy at a trillion times the rate of our Sun and are bigger than our entire solar system, they cannot be detected with the naked eye. This is because quasars are the most distant objects yet detected in our universe. In 1979, astronomers using a powerful telescope were astonished to discover two similar-looking quasars that have the same spectrum of light coming from them. These are shown in the upper left and lower right corners of the photograph. Is this a rare coincidence or is there another explanation for the similarity of the two images?

How can Einstein's theory of gravitation be used to explain this coincidence?

The answer is in this chapter.

14-1 Our Galaxy and the Gravitational Force

Our Milky Way galaxy is a disk-shaped collection of gas, dust, and billions of stars, including our Sun and solar system. Figure 14-1 shows how our galaxy would look if we could view it from outside. Earth is near the edge of the disk of the galaxy, about 26 000 light-years (2.5×10^{20} m) from its central bulge. Our galaxy is a member of the Local Group of galaxies, which includes the Andromeda galaxy at a distance of 2.5×10^6 light-years, and several closer dwarf galaxies, such as the Large Magellanic Cloud.

FIGURE 14-1 ■ A scientifically constructed image of our Milky Way galaxy from the perspective of an observer outside the galaxy. Painting by Jon Lomberg, taken from a mural in the *Where Next, Columbus*? exhibit at the National Air and Space Museum.

The Local Group is part of the Local Supercluster of galaxies. Measurements taken during and since the 1980s suggest that the Local Supercluster and the supercluster consisting of the clusters Hydra and Centaurus are all moving toward an exceptionally massive region called the Great Attractor. This region appears to be about 300 million light-years away, on the opposite side of the Milky Way from us, past the clusters Hydra and Centaurus.

The force that binds together these progressively larger structures, from star to galaxy to supercluster, and may be drawing them all toward the Great Attractor, is known as the **gravitational force.** This force not only holds you on Earth but also reaches out across intergalactic space and acts between galaxies. It is our focus in this chapter.

14-2 Newton's Law of Gravitation

One of the strengths of physics as a scientific discipline is that physicists can often find connections between seemingly unrelated phenomena. The physicist's search for unification has been going on for centuries, and we continue the tradition in this chapter. We will search for connections between what we have already learned about gravitation close to the Earth's surface and a more general theory of gravitation.

In order to get started on this, we now turn our attention to the universe beyond Earth and see what Newton's laws reveal about the motion of a heavenly body like the Moon. Based on careful astronomical observation, we know the Moon orbits the Earth in an approximately circular path. Newton's laws of motion say that since the Moon is not going in a straight line there must be a force acting on it. This is because the direction of the speed (and so the velocity) is changing.

What force keeps the Moon in its circular orbit? Well, we know everything is pulled to Earth by a gravitational force, so this is a logical place to start. Suppose we assume that the magnitude of the force exerted on the moon by the Earth is given by the mass of the Moon, m, times the local gravitational strength at the Earth's surface given by $g = 9.8$ N/kg. If we also assume that this magnitude of the gravitational force on the Moon results in a centripetal acceleration, then the following equation should hold:

$$mg = \frac{mv^2}{r} \quad \text{(proposed relationship)},$$

where v is the Moon's orbital speed, and r is its distance from the Earth. However, when we calculate the Moon's speed based on its orbit around the Earth, we discover that the actual magnitude of the Moon's centripetal acceleration, v^2/r, is thousands of times smaller than this equation would suggest. The gravitational force that we know and love (after all, it keeps our feet on the ground and our air from leaking away) is much too big to keep the Moon in its orbit. In fact, the gravitational force is about 3600 times too big. Such a gravitational force would cause the Moon to spiral in toward the Earth very rapidly.

Rather than abandon the idea that the gravitational force holds objects (including the Moon) near the Earth's surface, Newton suggested that perhaps gravity got weaker as you got farther from the center of the Earth. Since the Moon is about 60 Earth radii away, a gravitational force that decreased as the inverse square of the distance r between the Earth and the Moon would be just strong enough to be the centripetal force holding the Moon in its orbit. Furthermore, in our everyday life we only observe objects falling extremely short distances compared with the radius of the Earth. For this reason, a gravitational force that decreases as $1/r^2$ would appear constant to us. So, such a form for the gravitational force would be consistent with both astronomical observations and those made of objects close to the Earth's surface.

Suppose we pull together what we can observe and infer in an effort to construct a general statement regarding the gravitational force. We know from $|\vec{F}^{\text{grav}}| = mg$ that the gravitational force from the Earth on a mass m is proportional to the mass. Therefore we expect for two interacting masses m_A and m_B that the gravitational force on m_A is proportional to m_A and the force on m_B is proportional to m_B. However, Newton's Third Law tells us that these two forces must have equal magnitudes. So, the force magnitudes must be proportional to both masses. If the masses are separated by a distance r, our discussion of a $1/r^2$ dependence above implies that the force magnitudes should be given by

$$F^{\text{grav}}_{B \to A} = F^{\text{grav}}_{A \to B} \propto \frac{m_A m_B}{r^2}. \tag{14-1}$$

In 1665, the 23-year-old Isaac Newton figured this out and made a historic contribution to physics. He showed that the force that holds the Moon in its orbit is the same force that makes an apple fall. We take this so much for granted now that it is not easy for us to comprehend the ancient belief that the motions of Earth-bound and heavenly bodies were governed by different laws. Furthermore, Newton determined that not only does Earth attract an apple and the Moon, but every body in the universe attracts every other body; this generalized tendency of bodies to move toward each other is called **gravitation.**

Newton's conclusion takes a little getting used to, because the familiar attraction of Earth for Earth-bound bodies is so great that it overwhelms the attraction that

Earth-bound bodies have for each other. For example, Earth attracts an apple with a force magnitude of about 0.8 N. You also attract a nearby apple (and it attracts you), but this force of attraction has less magnitude than the weight of a speck of dust, so we don't notice it.

However, if you consider the expression above (Eq. 14-1), you may notice that the units don't match. We have a unit of Newtons on the left, but not on the right. Newton suggested using a constant of proportionality, G, in the expression to create the equation describing the magnitude of the gravitational force between two particles:

$$F^{grav}_{B\to A} = F^{grav}_{A\to B} = G\frac{m_A m_B}{r^2} \quad \text{(Newton's law of gravitation for particles).} \qquad (14\text{-}2)$$

This equation is the symbolic form of Newton's law of gravitation, which is expressed in words as follows:

Newton's law of gravitation: Every particle attracts any other particle with a *gravitational force*. This force has (1) a magnitude that is directly proportional to the product of the masses of the two particles and inversely proportional to the square of the distance between them; and (2) a direction that points along a line connecting the centers of the interacting particles.

The constant of proportionality G is known as the **gravitational constant.** Careful measurements show that in SI units G has a value of

$$\begin{aligned} G &= 6.67 \times 10^{-11}\ \text{N} \cdot \text{m}^2/\text{kg}^2 \\ &= 6.67 \times 10^{-11}\ \text{m}^3/\text{kg} \cdot \text{s}^2. \end{aligned} \qquad (14\text{-}3)$$

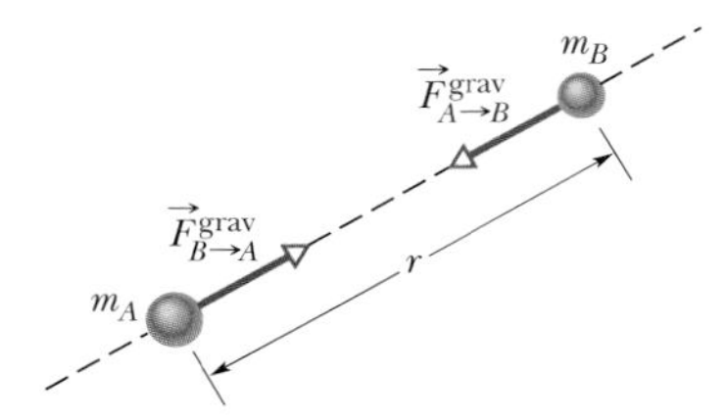

FIGURE 14-2 ■ Two particles, of masses m_A and m_B and with separation r, attract each other according to Newton's law of gravitation described in Eq. 14-2. The mutual forces of attraction are equal in magnitude and opposite in direction so that $\vec{F}^{grav}_{B\to A} = -\vec{F}^{grav}_{A\to B}$.

As Fig. 14-2 shows, a particle m_B attracts a particle m_A with a gravitational force $\vec{F}^{grav}_{B\to A}$ that is directed toward particle m_B. Particle m_A attracts particle m_B with a gravitational force $\vec{F}^{grav}_{A\to B}$ that is directed toward m_A. The forces $\vec{F}^{grav}_{B\to A}$ and $\vec{F}^{grav}_{A\to B}$ form a third-law force pair. So, we know that they must be opposite in direction but equal in magnitude. Thus,

$$\vec{F}^{grav}_{B\to A} = -\vec{F}^{grav}_{A\to B}.$$

These interaction forces depend on the separation of the two particles, but not on their location: the particles could be in a deep cave or in deep space. Also forces $\vec{F}^{grav}_{B\to A}$ and $\vec{F}^{grav}_{A\to B}$ are not altered by the presence of other bodies, even if those bodies lie between the two particles we are considering. We know this because we observe that interposing an object (like a table) between the Earth and a book does not affect the gravitational force that the Earth exerts on the book.

Applying the Law of Gravitation to Spherical Objects

Although Newton's law of gravitation applies strictly to particles, we can also apply it to real objects as long as the sizes of the objects are small compared to the distance between them. The Moon and Earth are far enough apart so that, to a good approximation, we can treat them both as particles. But, what about an apple and Earth? From the point of view of the apple, the broad and level Earth, stretching out to the horizon beneath the apple, certainly does not look like a particle.

Newton solved the apple–Earth problem by proving an important theorem called the *shell theorem:*

> A uniform spherical shell of matter attracts a particle that is outside the shell as if all the shell's mass were concentrated at its center.*

Earth can be thought of as a nest of such shells, one within another, with the outer shells having less density. The shell theorem tells us that each shell will attract a particle outside Earth's surface as if the mass of that shell were at the center of the shell. If we also invoke the principle of superposition, then from the apple's point of view, the Earth *does* behave like a particle, one that is located at the center of Earth and has a total mass equal to that of Earth. So for spherical objects that don't overlap each other,

> The value of r in the expression $Gm_A m_B/r^2$ is always the center-to-center separation of two objects provided they do not overlap.

Suppose, as shown in Fig. 14-3, the Earth pulls down on an apple with a force of magnitude 0.80 N. The apple must then pull up on Earth with a force of magnitude 0.80 N, which we take to act at the center of Earth. Although the forces are matched in magnitude, they produce different accelerations due to the difference in the masses of the two objects. For the apple, the magnitude of the acceleration is about 9.8 m/s^2, the familiar acceleration of a falling body near Earth's surface. For Earth, the acceleration magnitude measured in a reference frame attached to the center of mass of the apple–Earth system is only about 1×10^{-25} m/s^2.

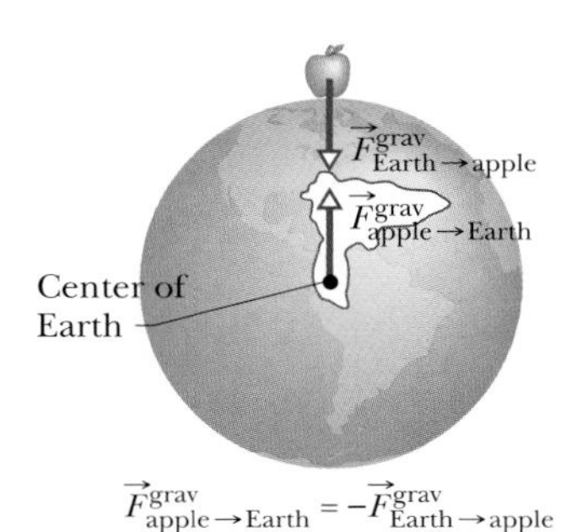

FIGURE 14-3 ■ The apple pulls up on Earth just as hard as Earth pulls down on the apple.

It is important to remember that Newton's theory of gravitation is an "action-at-a-distance" theory. That is, two objects exert gravitational forces on one another even if they do not touch one another. Newton's Third Law holds, so the forces between the two interacting bodies are equal and opposite *instant by instant*. This is interesting in that changes in the forces occur instantaneously at two different locations. So, variations cannot be propagated through intervening space at finite velocity or there would be an elapsed time interval. Massive objects do manage to interact instantaneously over huge intervening spaces. In the "General Scholium" section at the end of Book III of his *Principia*, Newton acknowledged this difficulty:

> *"But hitherto I have not been able to discover the cause of those properties of gravity from phenomena, and I frame no hypotheses . . . . to us it is enough that gravity does really exist, and acts according to the laws which we have explained, and abundantly serves to account for all the motions of the celestial bodies and of our sea."*†

Newton was clearly pleased that his theory could unify known astronomical observations. Ultimately, it was Einstein who successfully undertook a deeper inquiry into the very troubling and fundamental problem of how action-at-a-distance forces "travel" across space instantaneously. The answer he found lies in the fact that forces are not instantaneously transmitted. Rather, they are just transmitted very fast—at the speed of light. Although we introduce Einstein's theory of gravitation in Section 14-7, his explanation of action-at-a-distance phenomena is beyond the scope of this text.

READING EXERCISE 14-1: In order to calculate the relative amount of force the Earth exerts on the Moon as compared to a mass close to the Earth's surface having the same mass as the Moon, you do not need to know the mass of the Moon. Why? ■

* This is an example of Gauss' law applied to gravitational forces. We introduce Gauss' law in more detail in Chapter 24.

† Isaac Newton. *Principia,* Vol 2: The System of the World. Translated by Andrew Motte and revised by Florian Cajori. (University of California Press Berkeley: 1962), p. 547.

READING EXERCISE 14-2: In 1666, the following facts stood by themselves without additional ramifications or supporting evidence: (1) The centripetal acceleration of the Moon is 3600 times smaller than the gravitational acceleration near the Earth's surface and (2) the square of the ratio of the Earth's radius to the mean radius of the Moon's orbit is 1/3600. How would you interpret the meaning of these facts? Do they "prove" that the Moon is held in its orbit by gravity? ■

14-3 Gravitation and Superposition

Suppose that we are given a group of particles and we want to find the net (or resultant) gravitational force on any one of them due to the others. How would we go about doing this? Previously, we found by observation that we could get the net force on an object by finding the vector sum of all of the forces acting on the object. This straightforward vector addition procedure is called "superposition." However, we should keep in mind that there are instances in which a simple linear superposition does *not* work. For example, superposition doesn't work on the atomic level. If we bring a neutron and proton together to form a heavy hydrogen nucleus, the mass of our nucleus is less than the sum of the neutron and proton masses.

When we test this idea of using the **principle of superposition** for gravitational forces, we find that it does work. Thus, we can compute the gravitational force that acts on our selected particle due to each of the other particles, in turn, by adding these forces vectorially. For n interacting particles, the force on the first particle is given by the vector addition

$$\vec{F}_A^{\,\text{net}} = \vec{F}_{B\to A}^{\,\text{grav}} + \vec{F}_{C\to A}^{\,\text{grav}} + \vec{F}_{D\to A}^{\,\text{grav}} + \vec{F}_{E\to A}^{\,\text{grav}} + \cdots + \vec{F}_{n\to A}^{\,\text{grav}}. \tag{14-4}$$

Here $\vec{F}_A^{\,\text{net}}$ is the net gravitational force on particle A, $\vec{F}_{B\to A}^{\,\text{grav}}$ is the gravitational force on particle A from particle B, and so on. We can express this equation more compactly as a vector sum:

$$\vec{F}_A^{\,\text{net}} = \sum_{i=B}^{n} \vec{F}_{i\to A}^{\,\text{grav}}. \tag{14-5}$$

What about the gravitational force on a particle from a real extended object? The force can be found by dividing the object into small particle-like parts, and then calculating the vector sum of the forces on the particle from all the parts. In the limiting case, we can divide the extended object into differential parts of mass dm, each of which produces a differential gravitational force $d\vec{F}^{\,\text{grav}}$ on particle A. In this limit, the net gravitational force on the particle is given by an integral

$$\vec{F}_A^{\,\text{net}} = \int d\vec{F}^{\,\text{grav}}, \tag{14-6}$$

taken over the entire extended object. If the extended object has spherical symmetry and if particle A lies outside of the sphere, we can avoid the integration by assuming that the extended object's mass is concentrated at its center. In this case we can use Newton's law of gravitation for particles described in Eq. 14-2,

$$F_{B\to A}^{\text{grav}} = F_{A\to B}^{\text{grav}} = G\frac{m_A m_B}{r^2}.$$

READING EXERCISE 14-3: A particle is to be placed, in turn, at the same distance, r, from the center of the four objects, shown in the figure, each of mass m: (1) a small uniform solid sphere, (2) a small uniform spherical shell (3) a large uniform spherical shell and (4) a large uniform solid sphere. In each situation, the distance between the particle and the center of the object is r. Rank the objects according to the magnitude of the gravitational force they exert on the particle, greatest first.

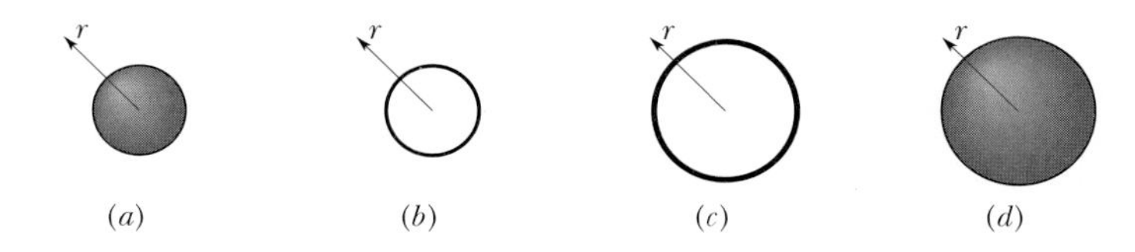

TOUCHSTONE EXAMPLE 14-1: Three Particles

Figure 14-4 shows an arrangement of three particles, particle A having mass $m_A = 6.0$ kg and particles B and C having mass $m_B = m_C = 4.0$ kg, and with distance $a = 2.0$ cm. What is the net gravitational force $\vec{F}_A^{\text{net}}$ that acts on particle A due to the other particles?

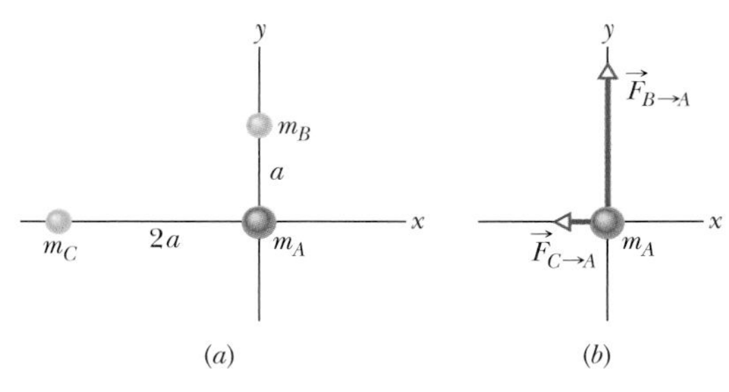

FIGURE 14-4 (a) An arrangement of three particles. (b) The forces acting on the particle of mass m_A due to the other particles.

SOLUTION One **Key Idea** here is that, because we have particles, the magnitude of the gravitational force on particle A due to particle B is given by Eq. 14-2 ($F_{B\to A} = Gm_Am_B/r^2$). Thus, the magnitude of the force $\vec{F}_{B\to A}$ on particle A from particle B is

$$F_{B\to A} = \frac{Gm_Am_B}{a^2} = \frac{(6.67\times 10^{-11}\ \text{m}^3/\text{kg}\cdot\text{s}^2)(6.0\ \text{kg})(4.0\ \text{kg})}{(0.020\ \text{m})^2} = 4.00\times 10^{-6}\ \text{N}.$$

Similarly, the magnitude of force $\vec{F}_{C\to A}$ on particle A from particle C is

$$F_{C\to A} = \frac{Gm_Am_C}{(2a)^2} = \frac{(6.67\times 10^{-11}\ \text{m}^3/\text{kg}\cdot\text{s}^2)(6.0\ \text{kg})(4.0\ \text{kg})}{(0.040\ \text{m})^2} = 1.00\times 10^{-6}\ \text{N}.$$

To determine the directions of $\vec{F}_{B\to A}$ and $\vec{F}_{C\to A}$ we use this **Key Idea**: Each force on particle A is directed toward the particle responsible for that force. Thus, $\vec{F}_{B\to A}$ is directed in the positive direction of y (Fig. 14-4b) and has only the y-component $F_{A\,y} = F_{B\to A}$. Similarly, $\vec{F}_{C\to A}$ is directed in the negative direction of x and has only the x-component $F_{A\,x} = -F_{C\to A}$.

To find the net force $\vec{F}_A^{\text{net}}$ on particle A, we first use this very important **Key Idea:** Because the forces are not directed along the same line, we *cannot* simply add or subtract their magnitudes or their components to get their net force. Instead, we must add them as vectors.

We can do so on a vector-capable calculator. However, here we note that $-F_{C\to A}$ and $F_{B\to A}$ are actually the x- and y-components of $\vec{F}_A^{\text{net}}$. Therefore, we shall follow the guide of Eq. 4-6 to find first the magnitude and then the direction of $\vec{F}_A^{\text{net}}$. The magnitude is

$$F_A^{\text{net}} = \sqrt{(F_{B\to A})^2 + (-F_{C\to A})^2} = \sqrt{(4.00\times 10^{-6}\ \text{N})^2 + (-1.00\times 10^{-6}\ \text{N})^2} = 4.1\times 10^{-6}\ \text{N}. \quad \text{(Answer)}$$

Relative to the positive direction of the x axis, Eq. 4-6 gives the direction of $\vec{F}_A^{\text{net}}$ as

$$\theta = \tan^{-1}\frac{F_{B\to A}}{-F_{C\to A}} = \tan^{-1}\frac{4.00\times 10^{-6}\ \text{N}}{-1.00\times 10^{-6}\ \text{N}} = -76°.$$

Is this a reasonable direction? No, the direction of $\vec{F}_A^{\text{net}}$ must be between the directions of $\vec{F}_{B\to A}$ and $\vec{F}_{C\to A}$. A calculator displays only one of the two possible answers to a $\tan^{-1}$ function. We find the other answer by adding 180°. That gives us

$$-76° + 180° = 104°, \quad \text{(Answer)}$$

which is a reasonable direction for $\vec{F}_A^{\text{net}}$.

14-4 Gravitation in the Earth's Vicinity

The Earth has a mean radius of just over 6000 km. In this section we will consider the gravitational acceleration constants of objects located at various altitudes between 0 km (at the Earth's surface) and about 36 000 km (at the greatest altitude communications satellites achieve). We will also consider how the Earth's rotation and nonuniformity can cause relatively small changes in the measured weight for objects at the Earth's surface.

Gravitational Forces in the Vicinity of a Spherical Earth

Let's assume that Earth is spherically symmetric and has a total mass M and radius R. The magnitude of the gravitational force from the Earth on a particle of mass m, located outside Earth a distance $r > R$ from Earth's center can be expressed by modifying Eq. 14-2 as

$$F^{\text{grav}} = G\frac{Mm}{r^2}. \tag{14-7}$$

Let's focus on the particle of mass m. In Chapter 3 we introduced the local gravitational strength, g, as the ratio of the particle's gravitational force magnitude and its mass. In symbols this ratio is expressed as $g = F^{\text{grav}}/m$ (Eq. 3-7). If we combine Eq. 3-7 with Eq. 14-7 we see that the local gravitational strength can be expressed as

$$g = G\frac{M}{r^2}, \tag{14-8}$$

where the units for g are N/kg.

If our mass m experiences no other forces when it is released, it will fall toward the center of Earth under the influence of the only gravitational force $\vec{F}^{\text{grav}}$. As we saw in Chapter 3, according to Newton's Second Law the particle's **gravitational acceleration constant,** $\vec{a}$, is given by

$$\frac{\vec{F}^{\text{grav}}}{m} = \vec{a}.$$

By combining $F^{\text{grav}} = G\,Mm/r^2$ (Eq. 14-7) and the equation immediately above, the magnitude of the gravitational acceleration constant can be expressed in terms of the gravitational constant, G, the mass of the Earth, M, and the distance of the particle from the Earth's center, r, as

$$a = G\frac{M}{r^2} \quad \text{(gravitational acceleration constant)}, \tag{14-9}$$

where the units for a are m/s^2. The similarity of Eqs. 14-8 and 14-9 remind us that the *gravitational acceleration constant* and the Earth's *local gravitational strength* have the same value. However, as we observed in Section 3-9, we use different but dimensionally equivalent units to describe these two quantities.

Table 14-1 shows the calculated values of the magnitude of the gravitational acceleration constant of an object as a function of altitude. The calculations are made using $a = G\,(M/r^2)$ (Eq. 14-9) along with the known value for the mass of the Earth of $M = 5.98 \times 10^{24}$ kg. We note from the calculations in the table that anywhere on the Earth's land surface, from the bottom of the Dead Sea to the top of Mt. Everest, the gravitational acceleration constant calculated from the expression GM/r^2 is the same to two significant figures. In other words, the principles developed in this chapter

TABLE 14-1
Calculated Values of Gravitational Acceleration Constant with Altitude

Altitude* (km)	a (m/s²)	Altitude Example
0.0	*9.8*	*Mean Earth radius*
8.8	9.8	Mt. Everest
36.6	9.7	Highest manned balloon
400	8.7	Space shuttle orbit
35 700	0.2	Communications satellite

*Altitude = $r - R$, where the radius of the Earth R = 6370 km.

reduce to the same familiar gravitational acceleration constant 9.8 m/s² that has been measured countless times in physics laboratories throughout the world.

The *reduction* of our "new" theory of gravitation discussed above to what we already had found to be true for the specific case of gravitation close to the surface of the Earth is an example of a general requirement for any "new" scientific model. When a model is developed to explain new, more general or more complicated phenomena, the "new" model must provide correct predictive information for the set of phenomena it describes and it must also be consistent with any simpler, more specific, or previously investigated phenomena. For example, in future chapters we will see that relativistic physics (for very high velocities) reduces to the nonrelativistic physics we have been studying so far when velocities are much less than the speed of light. Quantum physics reduces to classical physics for large, low-energy objects.

It is interesting to note that we can use our measurement of the gravitational acceleration constant at the Earth's surface (Eq. 14-9) along with our knowledge of G and r to calculate the mass of Earth!

Variations of Gravitational Forces over the Earth's Surface

In Section 6-3 we made two assumptions. First, we ignored the variations of the gravitational force and gravitational acceleration constant for an object at different locations on the Earth's surface. Second, we assumed that weight of an object as measured on a scale and the gravitational force on it, given by Eq. 14-7, are the same. Although these two assumptions are approximately true, geophysicists have measured slight variations or anomalies in the Earth's local gravitational field strength at different locations. There are many reasons for these anomalies. Some of the most significant are:

1. ***The Earth has an uneven surface.*** The Earth is covered with hills and mountains that rise above sea level and some valleys that are below sea level. The gravitational acceleration constant depends on altitude. When an object is closer to the dense core at a low altitude, it experiences a greater gravitational acceleration than it would at a high altitude. The distance between an object on dry land at the Earth's surface and the Earth's mean sea level varies from −0.414 km at the Dead Sea to +8.85 km at the summit of Mt. Everest. There is a difference of 0.03 m/s² in gravitational acceleration constant between sea level and the top of Everest.
2. ***The Earth bulges at its equator.*** Even if we were to "sand down" the bumps that represent hills and mountains rising above sea level and fill in the valleys below sea level, the Earth has the shape of an oblate spheroid. In other words, its shape is that of a sphere flattened at the poles and bulging at the equator. The equatorial radius is greater than its polar radius by 21 km. Thus, an object at the poles is

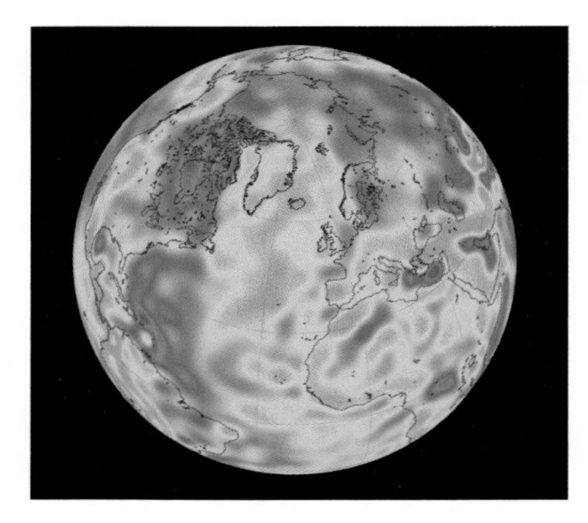

FIGURE 14-5 ■ Colored free-air gravity anomaly map of the Earth, centered on the Atlantic Ocean. These anomalies are the differences between the theoretical value for the gravity at the surface and the measured value. The colors range from purple (low gravity), through blue, green (normal), yellow, red, to white (high). The anomalies are only a tiny fraction of the gravitational field strength, but they can provide information on the Earth's internal structure. The gravity low in Hudson Bay, Canada (upper left), occurs partly because the area's rocks are recovering from being compressed during the ice age.

closer to the dense core of Earth than an object at the equator is. This is one of the reasons why the gravitational acceleration constant increases when we move it from the equator (where the latitude is 0°) toward either pole (where the latitude is 90°). This difference is about one-half of one percent or about 0.05 m/s^2.

3. ***The Earth is rotating.*** The rotation axis runs through the north and south poles of Earth. An object located on Earth's surface anywhere except at those poles must rotate in a circle about the rotation axis and thus must have a centripetal acceleration directed toward the center of the circle. This centripetal acceleration is caused by a centripetal force that is zero at the poles and most pronounced at the equator. This centripetal acceleration causes the apparent weight and measured gravitational acceleration of an object at the equator to be about 0.35% smaller than it would be at a pole. This latitude-dependent reduction in gravitational acceleration constant is about 0.03 m/s^2. (See below for more details.)
4. ***The Earth has a crust of uneven thickness and density.*** Even after geophysicists make corrections for the effects of altitude and latitude, measurements show that the gravitational force the Earth exerts on an object varies from location to location for other reasons. This is attributed to variations in: (1) the thickness of the Earth's crust (or outer section) and (2) the density of the rocks at the surface. Measurements in gravitational variations are useful in locating oil and mineral deposits. An example of regional variations is shown in Fig. 14-5.

Calculating the Effects of the Earth's Rotation

Recall from Chapter 6 that the weight we perceive for a mass, the object's apparent weight, is associated with the normal force on the object and can vary from the value mg if other forces act on the object. To see how Earth's rotation causes the apparent weight of an object at the Earth's surface to differ from the magnitude of the gravitational force on it, let us analyze a simple situation in which a crate of mass m is on a scale at the equator. Figure 14-6a shows this situation as viewed from a point in space above the north pole.

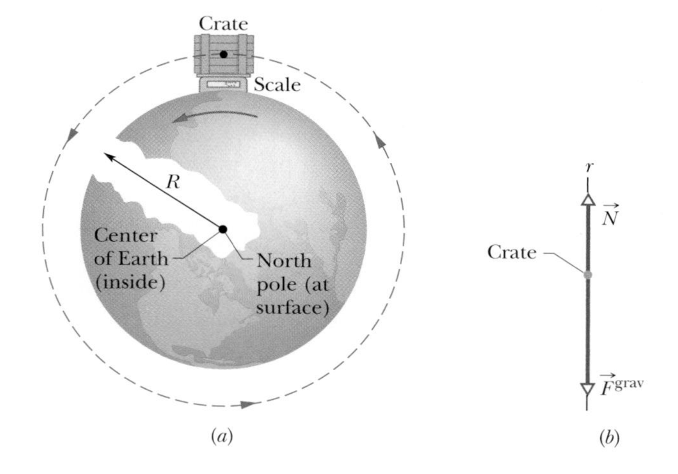

FIGURE 14-6 ■ (a) A crate lies on a scale at Earth's equator, as seen along Earth's rotation axis from above the north pole. (b) A free-body diagram for the crate, with a radially outward r axis. The gravitational force on the crate is represented by $\vec{F}^{\,grav}$. The normal force (or apparent weight) on the crate as read on a scale is represented by $\vec{N}$. Because of Earth's rotation, the crate also has a centripetal acceleration and hence a net centripetal force directed toward Earth's center.

Figure 14-6b, a free-body diagram for the crate, shows the two forces on the crate, both acting along a radial axis r that extends from Earth's center. The normal force $\vec{N}$ on the crate exerted on it by the scale is directed outward, in the positive direction of axis r. The gravitational force, $\vec{F}^{\,grav}$, acts inward on the object of mass m. Because the crate travels in a circle about the center of Earth as the Earth turns, the crate also experiences a centripetal acceleration directed inward with its radial component given by a_r. From Eq. 11-20, we know the magnitude of this acceleration is equal to $\omega^2 R$, where ω is Earth's angular speed and R is the circle's radius (approximately Earth's radius). Thus, we can write Newton's Second Law in component form along the r axis ($F_r^{\text{net}} = ma_r$) as

$$N_r + F_r^{\text{grav}} = ma_r \quad \text{or} \quad |\vec{N}| - |\vec{F}^{\text{grav}}| = -m(\omega^2 R). \tag{14-10}$$

As we determined in Chapter 6, the magnitude N of the normal force is equal to the weight read on the scale. Solving Eq. 14-10 for the apparent (or measured) weight as the magnitude of the normal force gives us

$$|\vec{N}| = |\vec{F}^{\text{grav}}| - m(\omega^2 R) \quad \text{(apparent weight at the equator).} \tag{14-11}$$

We see that the measured weight of the crate is actually slightly less than the magnitude of the gravitational force on the crate, because of Earth's rotation,

$$\text{(the measured weight)} = \begin{pmatrix}\text{the magnitude of} \\ \text{the gravitational force}\end{pmatrix} - \begin{pmatrix}\text{the object's mass times} \\ \text{its centripetal acceleration}\end{pmatrix}.$$

To find the difference at the equator, we can use Eq. 11-5 ($\langle\omega\rangle = \Delta\theta/\Delta t$) and Earth's radius $R = 6.37 \times 10^6$ m. For one Earth rotation ($\theta = 2\pi$ rad) the period is $\Delta t = 24$ h. Using these values (and converting hours to seconds), we find that $|\vec{N}|$ differs from $|\vec{F}^{\text{grav}}|$ by less than four-tenths of a percent (0.35%). Therefore, neglecting the difference between the apparent or measured weight and gravitational force magnitude is usually justified.

As we already mentioned, the difference between measured weight and the gravitational force magnitude is greatest on the equator (for one reason, the radius of the circle traveled by the crate is greatest there). At latitudes other than 0° it can be shown that Eq. 14-11 can be modified, to a very good approximation, to take the more general form,

$$|\vec{N}| = |\vec{F}^{\text{grav}}| - m(\omega^2 r) \quad \text{(apparent weight at any latitude),} \tag{14-12}$$

where r is the perpendicular distance from a location on the Earth's surface to the axis of rotation and varies from R at the equator to zero at the poles.

READING EXERCISE 14-4: In the discussion above, we talked about the impact of the Earth's rotation on an object's measured weight as compared to the magnitude of the gravitational force on it. Do the factors that we discussed affect the direction of $\vec{F}^{\text{grav}}$ due to the Earth? Is $\vec{F}^{\text{grav}}$ directed toward the center of the Earth at all points on Earth? To answer this question, consider the arguments made above in regard to the effect of the Earth's rotation on the object's apparent weight. Then, without full algebraic analysis, do some visualization based on a relevant force diagram. ■

TOUCHSTONE EXAMPLE 14-2: Floor to Ceiling

How much less will a mass of 1.000 kg weigh at a ceiling of height $h = 3.00$ m compared to its weight on the floor? Assume that the local gravitational strength at floor level has the internationally adopted standard value of 9.80665 N/kg.

SOLUTION ■ The **Key Idea** here is that the mass is slightly further from the center of the Earth on the ceiling than on the floor so its local gravitational strength is slightly smaller.

The first step in our solution is to find the ratio of the local gravitational strength on the ceiling to that on the floor. We can start by using the equation $g = GM/R^2$ (Eq. 14-8) where R is the distance from the center of the Earth to the location of an object near its surface. If we represent the distance from the Earth's center to the ceiling as R_{ceiling} then we can represent the distance to the floor as

$$R_{\text{floor}} = R_{\text{ceiling}} - h.$$

Next we substitute the symbols representing distances to the floor and to ceiling from the center of the Earth into Eq. 14-8 and take a ratio. This gives

$$\frac{g_{\text{ceiling}}}{g_{\text{floor}}} = \frac{GM/R_{\text{ceiling}}^2}{GM/(R_{\text{ceiling}} - h)^2} = \frac{(R_{\text{ceiling}} - h)^2}{R_{\text{ceiling}}^2} = \frac{R_{\text{ceiling}}^2 - 2hR_{\text{ceiling}} + h^2}{R_{\text{ceiling}}^2}.$$

We can then note in Appendix B that the mean radius of the Earth, which we can take as the Earth center to floor distance, is millions of meters. More precisely, it is 6.4×10^6 m. The ceiling height is so much less than R_{ceiling} that we can ignore h^2. So we can express the ratio to very good approximation as

$$\frac{g_{\text{ceiling}}}{g_{\text{floor}}} = \frac{R_{\text{ceiling}}^2 - 2hR_{\text{ceiling}} + h^2}{R_{\text{ceiling}}^2} \approx 1 - \frac{2h}{R_{\text{ceiling}}}.$$

Since the weight of our mass m is just $F^{\text{grav}} = mg$ in either location, the ratio of its weight at the ceiling to its weight at the floor is just

$$\frac{F_{\text{ceiling}}^{\text{grav}}}{F_{\text{floor}}^{\text{grav}}} = \frac{\cancel{m}g_{\text{ceiling}}}{\cancel{m}g_{\text{floor}}} \approx 1 - \frac{2h}{R_{\text{ceiling}}}.$$

So the difference in the weight of the mass between the ceiling and the floor is

$$F_{\text{ceiling}}^{\text{grav}} - F_{\text{floor}}^{\text{grav}} = -\frac{2hF_{\text{floor}}^{\text{grav}}}{R_{\text{ceiling}}}$$

$$= \frac{-(2)(3.00\text{ m})(1.000\text{ kg})(9.80665\text{ m/s}^2)}{6.37 \times 10^6\text{ m}}$$

$$= -9.24 \times 10^{-6}\text{ N}.$$

Hanging out on the ceiling is certainly not a viable way to lose a measurable amount of weight!

14-5 Gravitation Inside Earth

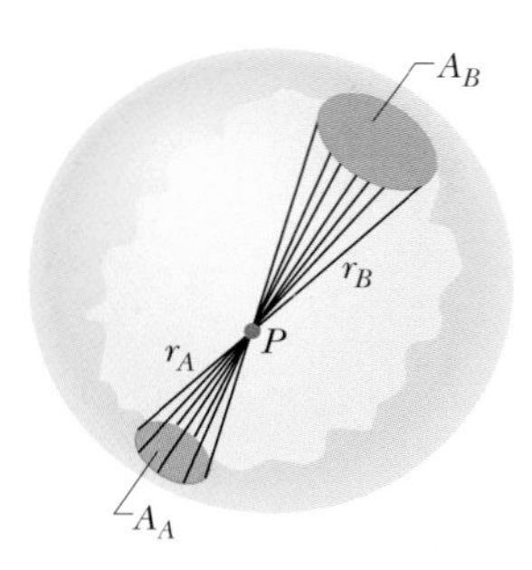

FIGURE 14-7 ■ The gravitational force at a point P due to an element of mass of area A is directly proportional to the area and inversely proportional to the square of the distance between the area and point P. The area of intersection increases with the square of the distance from it to point P. The gravitational forces of areas A_A and A_B on either side of any cone pair cancel each other out. This proves that there is no gravitational force acting anywhere inside a sphere of uniform mass.

The *shell theorem*, discussed in Section 14-2, states that "a uniform spherical shell of matter attracts a particle that is *outside* the shell as if all the shell's mass were concentrated at its center." However, we are still left with the question of what the force is on a particle that is *inside* the spherical shell of mass. In order to answer this question, let us refer to Fig. 14-7 and develop a geometric argument to show that a spherical shell with uniformly distributed mass exerts no net gravitational force on a particle inside itself.

Consider a mass m placed at a point P somewhere in the interior of this shell. Let's construct two cones that come together at P and have equal vertex angles. The cones will intercept patches of mass with areas A_A and A_B on opposite sides of the shell. The area of patch A_A and hence its mass m_A will be proportional to r_A^2, so that $m_A = Cr_A^2$ where C is a constant. Similarly the area of patch A_B and its mass m_B will be proportional to r_B^2 and $m_B = Cr_B^2$. Let's consider the magnitude of the gravitational force on an object of mass m at point P due to a patch having an area A_A and a mass m_A. This force magnitude is given by

$$F_{m_A \to m}^{\text{grav}} = G\frac{mm_A}{r_A^2} = G\frac{mCr_A^2}{r_A^2} = GmC.$$

We end up with a force that is independent of *both* the distance to the point (r) and the area of the patch (A). The same can obviously be said for the force on m from the mass m_B in the patch of area A_B. So, the forces exerted at P by patches A_A and A_B will be equal in magnitude. As we can see from Fig. 14-7, these forces are also opposite in direction. So, the forces from the two patches sum to zero and cancel out. We can cover the entire shell with such opposing patches, so overall the net force at P must be zero. Using this geometric argument, we can state a shell theorem as follows:

> A uniform shell of matter exerts no *net* gravitational force on a particle located anywhere inside it.

Be careful though: This statement does *not* mean that the gravitational forces on the particle from the various elements of the shell magically disappear. Rather, it means that the *sum* of the force vectors on the particle from all the elements is zero. In addition, if the gravitational force did not obey the inverse square law, the areas would not be canceled patch by patch by the radial distances, and the net force at P would not be zero. Therefore, this theorem is valid *only* for a force that obeys an inverse square law.

If we think of Earth as a series of concentric shells of uniform density, this shell theorem implies that the gravitational force acting on a particle would be a maximum at Earth's surface. If the particle were to move inward, perhaps down a deep mine shaft, the gravitational force would change for two reasons. (1) It would tend to increase because the particle would be moving closer to the center of Earth. (2) It would tend to decrease because the thickening shell of material lying outside the particle's radial position would not exert any net force on the particle. For a uniform Earth, the second influence would prevail and the force on the particle would steadily decrease to zero as the particle approached the center of Earth. However, for the real (nonuniform) Earth, the force on the particle actually increases as the particle begins to descend. The force reaches a maximum at a certain depth: Only then does it begin to decrease as the particle descends farther. This is because the Earth's crust is low density as compared to the average density of the Earth.

READING EXERCISE 14-5: How would the force of gravity from the Earth on a particle change in the following three cases? Case A: The particle starts at the surface of the Earth and moves outward from its center. Case B: The particle starts at the surface of the Earth and moves in toward the center of the Earth (assumed to have a uniform density). Case C: The particle starts at the surface of the Earth and moves in toward the center of the Earth (with the Earth's real density distribution). ■

TOUCHSTONE EXAMPLE 14-3: Pole to Pole

In *Pole to Pole*, an early science fiction story by George Griffith, three explorers attempt to travel by capsule through a naturally formed tunnel between the south pole and the north pole (Fig. 14-8). According to the story, as the capsule approaches Earth's center, the gravitational force on the explorers becomes alarmingly large and then, exactly at the center, it suddenly but only momentarily disappears. Then the capsule travels through the second half of the tunnel, to the north pole.

Check Griffith's description by finding the gravitational force on the capsule of mass m when it reaches a distance r from Earth's center. Assume that Earth is a sphere of uniform density ρ (mass per unit volume).

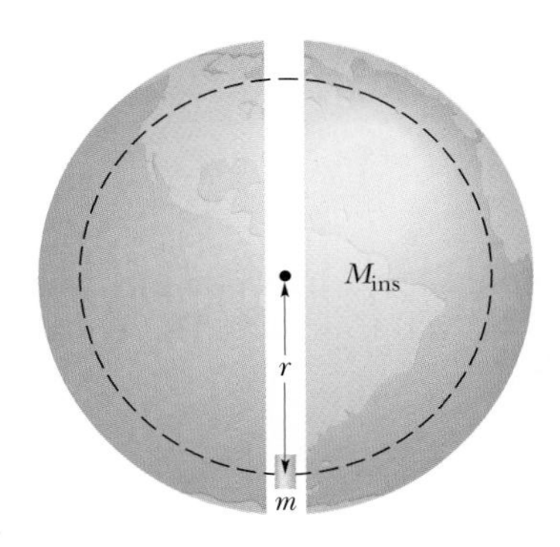

FIGURE 14-8 ■ A capsule of mass m falls from rest through a tunnel that connects Earth's south and north poles. When the capsule is at distance r from Earth's center, the portion of Earth's mass that is contained in a sphere of that radius is M_{ins}.

SOLUTION ■ Newton's shell theorem gives us three **Key Ideas** here:

1. When the capsule is at a radius r from Earth's center, the portion of Earth that lies outside a sphere of radius r does *not* produce a net gravitational force on the capsule.
2. The portion that lies inside that sphere *does* produce a net gravitational force on the capsule.
3. At a given location inside the Earth, we can treat the mass M_{ins} of the inside portion of Earth at that location as being the mass of a particle located at Earth's center.

All three ideas tell us that we can write Eq. 14-2, for the magnitude of the gravitational force on the capsule, as

$$F^{grav} = \frac{GmM_{ins}}{r^2}. \tag{14-13}$$

To write the mass M_{ins} in terms of the radius r, we note that the volume V_{ins} containing this mass is $\frac{4}{3}\pi r^3$. Also, its density is Earth's density ρ. Thus, we have

$$M_{ins} = \rho V_{ins} = \rho \frac{4\pi r^3}{3}. \tag{14-14}$$

Then, after substituting this expression into Eq. 14-13 and canceling, we have

$$F^{\text{grav}} = \frac{4\pi G m \rho}{3} r. \qquad \text{(Answer) (14-15)}$$

This equation tells us that the gravitational force magnitude F^{grav} depends linearly on the capsule's distance r from Earth's center. Thus, as r decreases, F^{grav} also decreases (the opposite of Griffith's description), until it is zero at Earth's center. At least Griffith got zero-at-the-center correct. However, forces near the Earth's center are not large, but "alarmingly" small instead.

Equation 14-15 can also be written in terms of the force vector $\vec{F}^{\text{grav}}$ and the capsule's position vector $\vec{r}$ along a radial axis extending from Earth's center. Let K represent the collection of constants $4\pi G m\rho/3$. Then Eq. 14-15 becomes

$$\vec{F}^{\text{grav}} = -K\vec{r}, \qquad (14\text{-}16)$$

in which we have inserted a minus sign to indicate that $\vec{F}^{\text{grav}}$ and $\vec{r}$ have opposite directions, since $\vec{r}$ represents the displacement of an object from the Earth's center. Equation 14-15 has the form of Hooke's law (Eq. 9-16). Thus, under the idealized conditions of the story, the capsule would oscillate like a block on a spring, with the center of the oscillation at Earth's center. After the capsule had fallen from the south pole to Earth's center, it would travel from the center to the north pole (as Griffith said) and then back again.

14-6 Gravitational Potential Energy

In Chapter 10, we defined potential energy and derived an expression for the change in potential energy, ΔU, associated with any conservative force when a system of two objects is reconfigured as shown in Fig. 10-11. We did this by finding the internal work done when one of the objects exerts a force on the other during the reconfiguration. Our expression for the gravitational force was

$$\Delta U = -W^{\text{int}} = -\int_{r_1}^{r_2} \vec{F}^{\text{grav}}_{A\to B}(r)\cdot d\vec{r}, \qquad \text{(Eq. 10-13)}$$

where r_1 is the original separation between objects in the system and r_2 is the separation of the objects in the system at some later time. We then used this equation to find changes in gravitational potential energy (GPE) for an Earth–object system. Because we did not yet have a general expression for the gravitational forces between two objects, we only considered the special case in which the object is close to the Earth's surface and found $\Delta U = mg\Delta y$ (Eq. 10-6). However, we cannot use this expression to determine how much energy it would take to launch a rocket that escapes the gravitational pull of the Earth.

Gravitational Potential Energy Changes for Any Two-Particle System

A key objective of this section is to find an equation for the gravitational potential energy for a two-particle system and use it to describe an Earth–object system for objects at any distance from the Earth's surface. In Section 10-3 we determined the potential energy change for a two-particle system that interacts by means of a conservative force. This is where we will start (with Eq. 10-13) and use the gravitational force as our conservative force. So,

$$\Delta U = -\int_{r_1}^{r_2} \vec{F}^{\text{grav}}_{A\to B}(r)\cdot d\vec{r},$$

for a general equation of the magnitude of the gravitational force between any two particles of masses m_A and m_B (represented in both Fig. 14-2 and Eq. 14-2). This equation, which is simply Newton's law of gravitation, is given by

$$F^{\text{grav}}_{B\to A} = F^{\text{grav}}_{A\to B} = G\frac{m_A m_B}{r^2}. \qquad \text{(Eq. 14-2)}$$

If we increase the separation of the two masses from an initial separation r_1 to a final separation r_2, the direction of $\vec{F}^{\text{grav}}_{A\to B}$ is opposite to the direction of dr. If we take the direction of dr to be positive, then the r-component of the gravitational force exerted on particle B by particle A, $(F^{\text{grav}}_{A\to B})_r$, must be negative. Thus,

$$(F^{\text{grav}}_{A\to B})_r = -G\frac{m_A m_B}{r^2}.$$

We can substitute this expression for the gravitational force component along r into the integral and evaluate as we separate the particles from r_1 to r_2. This gives

$$\int_{r_1}^{r_2}(F^{\text{grav}}_{A\to B})_r\,dr = \int_{r_1}^{r_2}\left(\frac{-Gm_A m_B}{r^2}\right)dr = \left[\frac{+\,Gm_A m_B}{r}\right]_{r_1}^{r_2} = \frac{Gm_A m_B}{r_2} - \frac{Gm_A m_B}{r_1}.$$

We now substitute the value of this integral into Eq. 10-13 to find the gravitational potential energy change,

$$\Delta U = U(r_2) - U(r_1) = -\int_{r_1}^{r_2}(F^{\text{grav}}_{A\to B})_r\,dr = -\frac{Gm_A m_B}{r_2} + \frac{Gm_A m_B}{r_1}. \quad (14\text{-}17)$$

Defining an Absolute Gravitational Potential Energy for a Two-Particle System

In Chapter 10, we discussed the gravitational potential energy of a particle–Earth system where the particle is close to the Earth's surface. For that special case we found it useful to choose a "zero potential energy" configuration in which the particle was located at the surface of the Earth (or some other convenient height near the Earth's surface). In this more general situation in which the particles can be very far apart, we find it more useful to define a different reference configuration for which the potential energy is equal to zero. Since gravitational forces decrease rapidly to zero with separation, (in fact as $1/r^2$), it is very convenient to define our potential energy to be zero when separation distance r between particles is *infinite.* We can then define the gravitational potential energy as minus the internal work done on particle B by particle A as the separation of the two particle changes from an initial separation of infinity (denoted by ∞) to a final separation of r.

For this situation, Eq. 14-17 tell us

$$\Delta U = U(r) - U(\infty) = -\frac{Gm_A m_B}{r} + \frac{Gm_A m_B}{\infty} = -\frac{Gm_A m_B}{r} + 0.$$

Since we have defined our reference potential to be zero at infinity so $U(\infty) = 0$, the equation above reduces further to

$$U(r) = -\frac{Gm_A m_B}{r} \quad \text{(gravitational PE relative to infinite separation).} \quad (14\text{-}18)$$

Here $G = 6.67 \times 10^{-11}\ \text{N}\cdot\text{m}^2/\text{kg}^2$ is the gravitational constant, m_A is the mass of one object, m_B is the mass of the other object, and r is the center-to-center separation of the two particle-like masses. Note that $U(r)$ approaches zero as r approaches infinity and that for any finite value of r, the value of $U(r)$ is negative (Fig. 14-9).

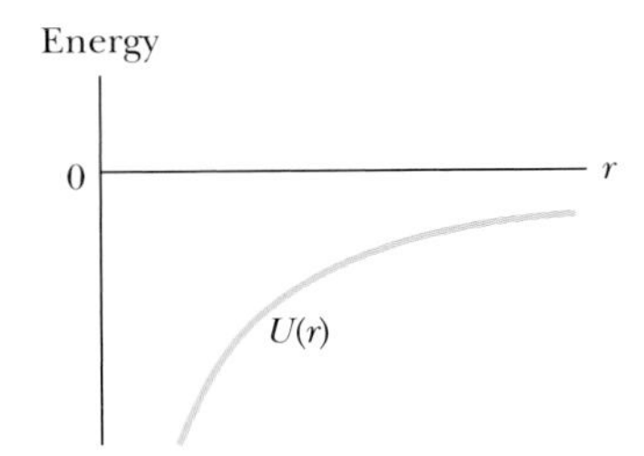

FIGURE 14-9 ■ The gravitational potential energy of a two-mass system. Note that the PE is negative everywhere. It has a very large magnitude as the distance r between the masses approaches 0, but it approaches 0 as r approaches infinity.

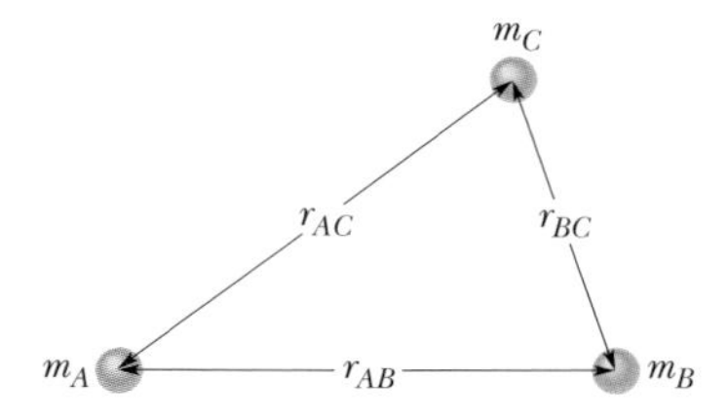

FIGURE 14-10 ■ Three particles form a system. (The separation for each pair of particles is labeled with a double subscript to indicate the particles.) The gravitational potential energy *of the system* is the sum of the gravitational potential energies of all three pairs of particles.

Gravitational Potential Energy for a Many-Particle System

The potential energy given by this expression is a property of the system of two particles rather than of either particle alone. If our system contains more than two particles, we consider each pair of particles separately. We calculate the gravitational potential energy of that pair with this equation as if the other particles were not there. We then algebraically sum the results (energy is a scalar). Applying Eq. 14-18 to each of the three pairs of charges in Fig. 14-10, for example, gives the potential energy of this system as

$$U = -\left(\frac{Gm_A m_B}{r_{AB}} + \frac{Gm_A m_C}{r_{AC}} + \frac{Gm_B m_C}{r_{BC}}\right) \quad \text{(3-particle system).} \quad (14\text{-}19)$$

Gravitational Potential Energy for an Earth–Object System

Suppose we bring a baseball of mass m from infinity along the y axis to a point P, a distance r from the center of the Earth, which has mass M as shown in Fig. 14-11. If $r \geq R_E$, where R_E is the Earth's radius, what is the *general* expression for the gravitational potential energy of the Earth–baseball system? We simply substitute our new symbols into Eq. 14-18 shown above. This gives us

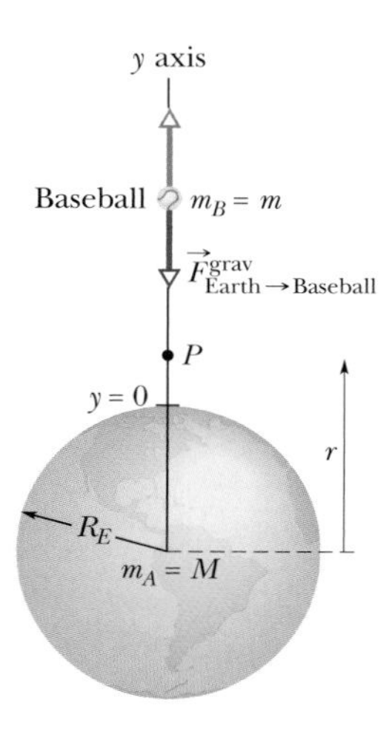

FIGURE 14-11 ■ A baseball is moved along a y axis to a point P which is a distance $r > R_E$ from the center of the Earth. Since the gravitational force is attractive, the Earth–baseball system loses gravitational PE as the particles get closer together.

$$U(r) = -\frac{GMm}{r} \quad \text{(Earth–object system gravitational PE relative to } r = \infty\text{).} \quad (14\text{-}20)$$

However, this general expression must be consistent with what we derived in Chapter 10 for the special case of an object *close* to the Earth's surface. When we chose to define $y = 0$ at some convenient height near the Earth's surface, we found that

$$U(y) = mgy \quad \text{(near Earth gravitational PE relative to } y = 0\text{).} \quad \text{(Eq. 10-8)}$$

Although the two expressions look quite different at first glance, we see that in both cases the potential energy decreases (becomes progressively more negative) as the Earth and the baseball move closer together. Our two expressions for gravitational potential energy are consistent in this regard.

However, if our general expression for gravitational potential energy in Eq. 14-20 is valid for all separations, it must be consistent with the more specific expression $U(y) = mgy$. To see that this is the case, suppose the object starts at the surface of the Earth. Its potential energy at this location is

$$U(R_E) = -\frac{GMm}{R_E}.$$

The object then moves upward to a height Δy above the Earth's surface. The potential energy of the object at this location is

$$U(R_E + \Delta y) = -\frac{GMm}{R_E + \Delta y}.$$

So, the change in potential energy between these two configurations is

$$\begin{aligned}\Delta U = U_2 - U_1 &= U(R_E + \Delta y) - U(R_E) \\ &= -\frac{GMm}{R_E + \Delta y} - \left(-\frac{GMm}{R_E}\right) \\ &= -\frac{GMm}{R_E + \Delta y} + \frac{GMm}{R_E}.\end{aligned}$$

Simplifying this expression by finding a common denominator and factoring gives us

$$\Delta U = \frac{GMm\Delta y}{R_E(R_E + \Delta y)}.$$

However, the radius of the Earth, R_E, is *orders of magnitude* (much, much) larger than the additional height Δy for situations in which the object is close to the surface of the Earth. So,

$$R_E + \Delta y \approx R_E,$$

and

$$\Delta U = \frac{GMm\Delta y}{R_E(R_E + \Delta y)} \approx \frac{GMm\Delta y}{R_E^2}.$$

Equation 14-9 tells us that the magnitude of the gravitational acceleration constant at the surface of the Earth (with radius R_E and mass M) is

$$a = \frac{GM}{R_E^2} \quad \text{(gravitational acceleration constant).}$$

If we ignore small variations in the gravitational acceleration constant due to the Earth's rotation and its small deviations from a spherical shape, then the gravitational acceleration constant and the local gravitational strength have the same magnitude so that $a = g$.

Substituting these last two equations into the equation for the change in the gravitational potential energy expression gives

$$\Delta U \approx \frac{GMm\,\Delta y}{R_E^2} = mg\,\Delta y.$$

In other words, if the height Δy above the surface of the Earth is small compared to the radius of the Earth, the gravitational acceleration constant and the local gravitational strength g are essentially constant, and our general expression for gravitational potential energy allows us to predict the same changes in gravitational potential energy as the more specific one we used in Chapter 10. This is true even though we have chosen very different zero points for our general and near-Earth potential energies. The two expressions are consistent because they allow us to calculate the same changes in gravitational potential energy as long as we are near the Earth's surface.

Path Independence

In the equations derived in this section, we have made the simplifying assumption that our particles move apart or come together along a line connecting their centers. But, because the gravitational force is conservative, potential energy changes of the system are path independent as discussed in Section 10-2. Thus, our equations hold even when we allow the interacting particles to separate along any crazy path.

An example of this path independence is shown in Fig. 14-12. We imagine moving a baseball from point A to point G along a path consisting of three radial lengths and three circular arcs (centered on Earth). We are interested in the total work W done by Earth's gravitational force on the ball as it moves from A to G. The work done along each circular arc is zero, because the direction of the force is perpendicular to the arc at every point. Thus, the only work done by the force is along the three radial lengths, and the total work W is the sum of the work done along the radial lengths.

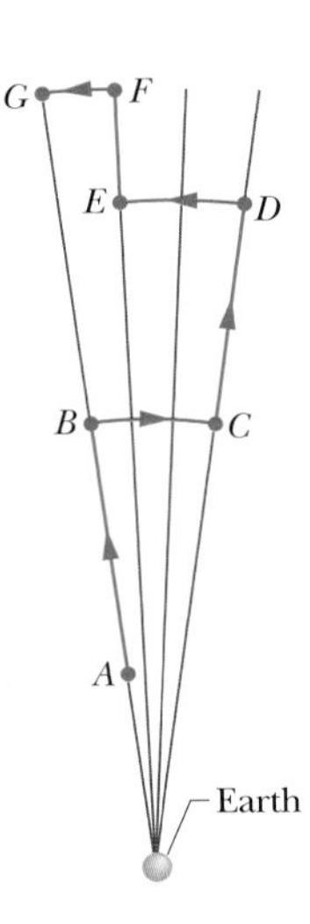

FIGURE 14-12 ■ Near Earth, a baseball is moved from point A to point G along a path consisting of radial lengths and circular arcs.

Now suppose we mentally shrink the arcs to zero. We would then be moving the ball directly from A to G along a single radial length. Does that change the total work done? No. Because no work was done along the arcs, eliminating them does not change the work. The path taken from A to G now is clearly different, but the work done by the force on the baseball is the same. This example reminds us that the internal work done by a system can be independent of the actual path taken. In that case, the change ΔU in the system's gravitational potential energy is path independent as well. This example of path independence for the gravitational force is consistent with what we have shown in Section 10-2.

Escape Speed

When launching a rocket, how fast does it need to be moving to escape the gravitational pull of the Earth? If you throw an object upward, it will usually slow, turn around, and then speed up as it travels back toward the Earth. There is, however, a minimum initial speed that will allow an object to move upward forever. In this case, the rocket or any other object launched upward slows and approaches zero speed as the object gets farther and farther away from the Earth, but never reverses direction and returns. This special initial speed is called the (Earth) **escape speed.**

In order to calculate the escape speed from Earth (or some other spherical astronomical body), consider a projectile of mass m, leaving the surface of a planet of radius R with escape speed v. Assume the rocket has an initial kinetic energy K given by $\frac{1}{2}mv^2$ and a potential energy U given by Eq. 14-20,

$$U = -\frac{GMm}{R},$$

where M is the mass of the planet, and R is its radius.

As the rocket gets far away, its speed approaches zero and the kinetic energy of the Earth–rocket system approaches zero. In addition, the system potential energy approaches zero because the planet–rocket separation is very large. So, total energy of the system at an "infinite" separation is zero. If we ignore the relatively small amount of energy lost to air drag, then mechanical energy is approximately conserved. In this case the rocket's total energy at the planet's surface must also have been zero, so we can use Eq. 10-19 to get

$$E^{\text{mec}} = K + U = \tfrac{1}{2}mv^2 + \left(-\frac{GMm}{R}\right) = 0.$$

This yields

$$v = \sqrt{\frac{2GM}{R}}. \tag{14-21}$$

The escape speed v does not depend on the direction in which a projectile is fired from a planet. However, attaining that speed is easier if the projectile is fired in the direction the launch site is moving as the planet rotates about its axis. For example, rockets are launched eastward at Cape Canaveral to take advantage of the Cape's eastward speed of 0.5 km/s due to Earth's rotation.

Equation 14-21 can be applied to find the escape speed of a projectile from any astronomical body, provided we substitute the mass of the body for M and the radius of the body for R. Table 14-2 shows escape speeds from some astronomical bodies. It is interesting to note that objects of any size can escape from an astronomical body. For example, gas molecules in planetary atmospheres sometimes reach escape speeds.

TABLE 14-2
Some Escape Speeds

Body	Mass (kg)	Radius (m)	Escape Speed (km/s)
Ceres[a]	1.17×10^{21}	3.8×10^5	0.64
Earth's Moon	7.36×10^{22}	1.74×10^6	2.38
Earth	5.98×10^{24}	6.37×10^6	11.2
Jupiter	1.90×10^{27}	7.15×10^7	59.5
Sun	1.99×10^{30}	6.96×10^8	618
Sirius B[b]	2×10^{30}	1×10^7	5200
Neutron star[c]	6×10^{30}	1×10^4	3.4×10^5

[a]The most massive of the asteroids.
[b]A *white dwarf* (a very compact old star that has burned its nuclear fuel) that is in orbit around the bright star Sirius. Sirius B has roughly the mass of our Sun and a radius close to that of the Earth.
[c]The collapsed core of a massive star that remains after that star has exploded in a *supernova* event and is more compact than a white dwarf (with 3 times the mass of our Sun and a diameter of only a few kilometers).

READING EXERCISE 14-6: You move a ball of mass m away from a sphere of mass M. (a) Does the gravitational potential energy of the ball-sphere system increase or decrease? (b) Is positive or negative work done by the gravitational force between the ball and the sphere? ■

TOUCHSTONE EXAMPLE 14-4: Asteroid

An asteroid, headed directly toward Earth, has a speed of 12 km/s relative to the planet when it is at a distance of 10 Earth radii from Earth's center. Neglecting the effects of Earth's atmosphere on the asteroid, find the asteroid's speed v_2 when it reaches Earth's surface.

SOLUTION ■ One **Key Idea** is that, because we are to neglect the effects of the atmosphere on the asteroid, the mechanical energy of the asteroid-Earth system is conserved during the fall. Thus, the final mechanical energy (when the asteroid reaches Earth's surface) is equal to the initial mechanical energy. We can write this as

$$E^{\text{mec}} = K_1 + U_1 = K_2 + U_2, \qquad \text{(Eq. 10-19)}$$

where K is kinetic energy and U is gravitational potential energy.

A second **Key Idea** is that, if we assume the system is isolated, the system's linear momentum must be conserved during the fall. Therefore, the momentum change of the asteroid and that of Earth must be equal in magnitude and opposite in sign. However, because Earth's mass is so great relative to the asteroid's mass, the change in Earth's speed is negligible relative to the change in the asteroid's speed. So, the change in Earth's kinetic energy is also negligible. Thus, we can assume that the kinetic energies in Eq. 10-19 are those of the asteroid alone.

Let m represent the asteroid's mass and M represent Earth's mass (5.98×10^{24} kg). The asteroid is initially at the distance $10R_E$ and finally at the distance R_E, where R_E is Earth's radius (6.37×10^6 m). Substituting Eq. 14-20 for U and $\frac{1}{2}mv^2$ for K, we rewrite Eq. 10-19 as

$$\tfrac{1}{2}mv_2^2 - \frac{GMm}{R_E} = \tfrac{1}{2}mv_1^2 - \frac{GMm}{10R_E}.$$

Rearranging and substituting known values, we find

$$\begin{aligned} v_2^2 &= v_1^2 + \frac{2GM}{R_E}\left(1 - \frac{1}{10}\right) \\ &= (12 \times 10^3 \text{ m/s})^2 \\ &\quad + \frac{2(6.67 \times 10^{-11} \text{ m}^3/\text{kg}\cdot\text{s}^2)(5.98 \times 10^{24} \text{ kg})}{6.37 \times 10^6 \text{ m}} 0.9 \\ &= 2.567 \times 10^8 \text{ m}^2/\text{s}^2, \end{aligned}$$

and thus the magnitude of the impact velocity is

$$v_2 = 1.60 \times 10^4 \text{ m/s} = 16 \text{ km/s}. \qquad \text{(Answer)}$$

At this speed, the asteroid would not have to be particularly large to do considerable damage at impact. As an example, if it were only 5 m across, the impact could release about as much energy as the nuclear explosion at Hiroshima. Alarmingly, about 500 million asteroids of this size are near Earth's orbit, and in 1994 one of them apparently penetrated Earth's atmosphere and exploded at an altitude of 20 km near a remote South Pacific island (setting off nuclear-explosion warnings on six military satellites). The impact of an asteroid 500 m across (there may be a million of them near Earth's orbit) could end modern civilization and almost eliminate humans worldwide.

TOUCHSTONE EXAMPLE 14-5: Escape Speeds

(a) Suppose a particle in space is the same distance from the Sun as the Earth is. At this distance, what is the particle's escape speed from the Sun?

SOLUTION ■ We can use Eq. 14-21 to find the escape speed relative to the Sun. The **Key Idea** here is that the particle is not on the surface of the Sun but at a distance equivalent to the mean distance between the Earth and the Sun given by $R = 1.5 \times 10^{11}$ m. Since the mass of the Sun is $M = 1.99 \times 10^{30}$ kg, we get

$$v = \sqrt{\frac{2GM}{R}} = \sqrt{\frac{2(6.67 \times 10^{-11}\,\text{N}\cdot\text{m}^2/\text{kg}^2)(1.99 \times 10^{30}\,\text{kg})}{1.5 \times 10^{11}\text{m}}}$$

$$= v = 4.2 \times 10^4\,\text{m/s}. \qquad \text{(Answer)}$$

(b) How does the escape speed you just calculated compare to the particle's escape speed from the surface of the Earth?

SOLUTION ■ We can look up the escape speed from the Earth's surface in Table 14-2. The value is given by

$$v = 11.2\,\text{km/s} = 1.12 \times 10^4\,\text{m/s}.$$

The escape speed from the Earth's surface is about one-fourth (=1.12/4.2) of that needed to escape the Sun at an "Earth orbit distance" from it. The **Key Idea** here is that though the particle at the surface of the Earth is much closer to the center of the Earth than it is to the center of the Sun in situation *a*, the Sun is much more massive than the Earth is. (Answer)

(c) How does the escape speed calculated in part (a) compare to the particle's escape speed from the surface of the Sun?

SOLUTION ■ Once again we can look up the escape speed in Table 14-2. This time we need to list the speed needed to escape from the surface of the Sun,

$$v = 618\,\text{km/s} = 61.8 \times 10^4\,\text{km/s}.$$

The escape speed from the Sun's surface is about 15 times larger (= 61.8/4.2) that needed to escape the Sun at an "Earth orbit distance" from it. The **Key Idea** here is that the particle in this case is much closer to the Sun. (Answer)

14-7 Einstein and Gravitation

Principle of Equivalence

When we casually discuss gravitational forces, we often say things like "we can feel the pull of gravity" or "we can feel the pull of the Earth." However, careful observation will convince you that what we actually "feel" is the upward push of the floor or a chair. If we hang from a rope, we feel the upward pull of the rope. We do not feel any push or pull if the floor, chair, or rope is taken away.

In contrast, if we jump off a ladder or cliff, we are in free fall, and we feel no forces at all even though we are subject to the uncomfortable sensation that is (unfortunately) called "weightlessness." This is what Albert Einstein was referring to when he said: "I was . . . in the patent office at Bern when all of a sudden a thought occurred to me: 'If a person falls freely, he will not feel his own weight.' I was startled. This simple thought made a deep impression on me. It impelled me toward a theory of gravitation."

Thus Einstein tells us how he began to form his **general theory of relativity.** The fundamental postulate of this theory about gravitation (the gravitating of objects toward each other) is called the **principle of equivalence,** which says that gravitation and acceleration are equivalent. If a physicist were locked up in a small box as in Fig. 14-13, he would not be able to tell whether the box was at rest on Earth (and subject only to Earth's gravitational force), as in Fig. 14-13*a*, or accelerating through interstellar space at 9.8 m/s^2 (and subject only to the force producing that acceleration), as in Fig. 14-13*b*. In both situations he would feel the same and would read the same value for his weight on a scale. Moreover, if he watched an object fall past him, the object would have the same acceleration relative to him in both situations.

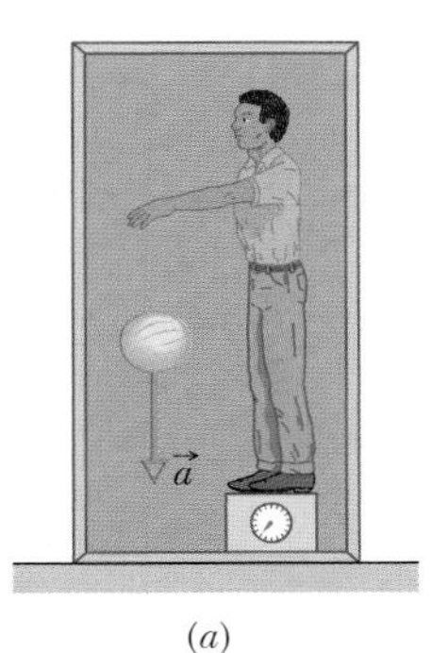

(*a*)

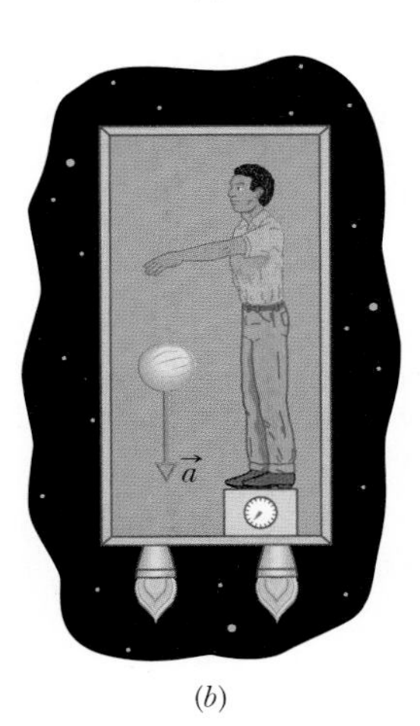

(*b*)

FIGURE 14-13 ▪ (*a*) A physicist in a box resting on Earth sees a cantaloupe falling with acceleration $a = 9.8$ m/ s^2. (*b*) If he and the box accelerate in deep space at 9.8 m/s^2, the cantaloupe has the same acceleration relative to him. It is not possible, by doing experiments within the box, for the physicist to tell which situation he is in. For example, the platform scale on which he stands reads the same weight in both situations.

Curvature of Space

We have introduced the concept of gravitation to explain the interaction forces between masses. Einstein introduced an alternative explanation of gravitation as a curvature (or shape) of space that is caused by masses. (As we will discuss in Chapter 38, space and time are entangled so the curvature of which Einstein spoke is really a curvature of *spacetime,* the combined four dimensions of our universe.)

Picturing how space (such as vacuum) can have curvature is difficult, but an analogy might help. Suppose that from orbit we watch a race in which two boats begin on the equator with a separation of 20 km and head due south (Fig. 14-14*a*). To the sailors, the boats travel along flat, parallel paths. However, with time the boats draw together until, nearer the south pole, they touch. The sailors in the boats can interpret this drawing together in terms of a force acting on the boats. However, we can see that the boats draw together simply because of the curvature of Earth's surface. We can see this because we are viewing the race from "outside" that surface.

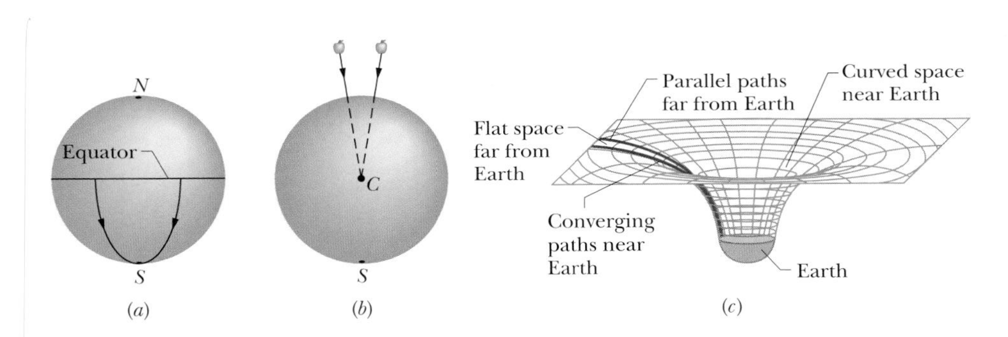

FIGURE 14-14 ▪ (*a*) Two objects moving along lines of longitude toward the south pole converge because of the curvature of Earth's surface. (*b*) Two objects falling freely near Earth move along lines that converge toward the center of Earth because of the curvature of space near Earth. (*c*) Far from Earth (and other masses), space is flat and parallel paths remain parallel. Close to Earth, the parallel paths begin to converge because space is curved by Earth's mass.

Figure 14-14*b* shows a similar race: Two horizontally separated apples are dropped from the same height above Earth. Although the apples may appear to travel along parallel paths, they actually move toward each other because they both fall toward Earth's center. We can interpret the motion of the apples in terms of the gravitational force on the apples from Earth. We can also interpret the motion in terms of a curvature of the space near Earth, due to the presence of Earth's mass. This time we cannot see the curvature because we cannot get "outside" the curved space, as we got "outside" the curved Earth in the boat example. However, we can depict the curvature with a drawing like Fig. 14-14*c*. There the apples would move along a surface that curves toward Earth because of Earth's mass.

When light passes near Earth, its path bends slightly because of the curvature of space there, an effect called *gravitational lensing.* When it passes a more massive

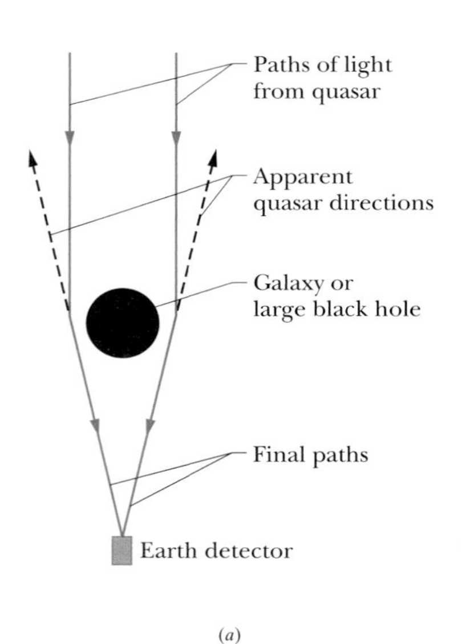

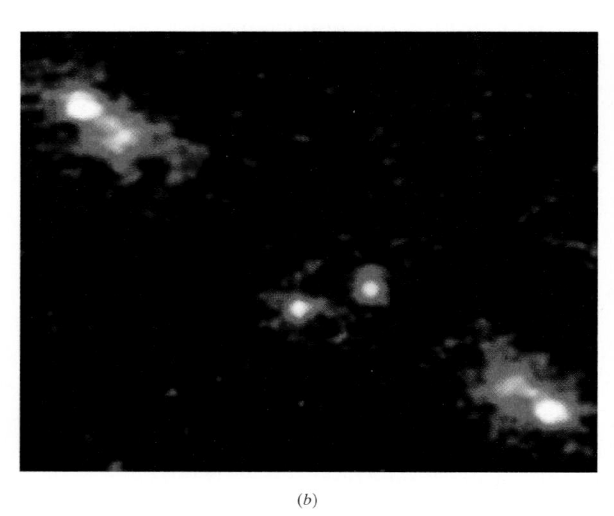

FIGURE 14-15 ■ (*a*) Light from a distant quasar named AC 114 follows curved paths around a galaxy because the mass of the galaxy has curved the adjacent space. If the light is detected, it appears to have originated along the backward extensions of the final paths (dashed lines). (*b*) An image showing identical quasars. The source of the light is far behind a large, unseen "lensing" galaxy that has just the right shape and orientation to produce two images of the quasar. The two objects near the center of the image are believed to be unrelated galaxies in front of the lensing galaxy.

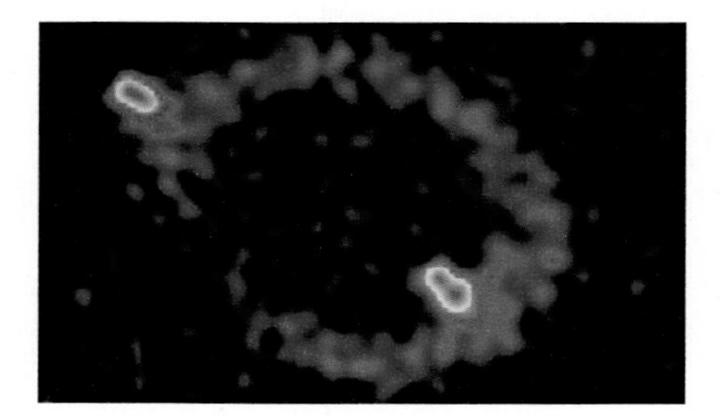

FIGURE 14-16 ■ The Einstein ring known as MG1131+0456 as it appeared on the computer screen of a radio telescope.

structure like a galaxy, its path can be bent more. If such a massive structure is between us and a quasar (an extremely bright, distant source of light), the light from the quasar can bend around the massive structure and toward us (Fig. 14-15*a*). Then, because the light seems to be coming to us from a number of slightly different directions in the sky, *we see the same quasar appearing to be located in two different directions* (Fig. 14-15*b*). In situations where the light from a distant quasar lies precisely behind the center of the lensing galaxy, the images of a single quasar can blend together to form a full ring of light known as an *Einstein ring* (Fig. 14-16).

Should we attribute gravitation to the curvature of spacetime due to the presence of masses or to a force between masses? Or should we attribute it to the actions of a type of fundamental particle called a *graviton*, as conjectured in some modern physics theories? We do not know.

Problems

SEC. 14-2 ■ NEWTON'S LAW OF GRAVITATION

1. What Separation? What must the separation be between a 5.2 kg particle and a 2.4 kg particle for their gravitational attraction to have a magnitude of 2.3×10^{-12} N?

2. Horoscopes Some believe that the positions of the planets at the time of birth influence the newborn. Others deride this belief and claim that the gravitational force exerted on a baby by the obstetrician is greater than that exerted by the planets. To check this claim, calculate and compare the magnitude of the gravitational force exerted on a 3 kg baby (a) by a 70 kg obstetrician who is 1 m away and roughly approximated as a point mass, (b) by the massive planet Jupiter ($m = 2 \times 10^{27}$ kg) at its closest approach to Earth ($= 6 \times 10^{11}$ m), and (c) by Jupiter at its greatest distance from Earth ($= 9 \times 10^{11}$ m). (d) Is the claim correct?

3. Echo Satellites One of the *Echo* satellites consisted of an inflated spherical aluminum balloon 30 m in diameter and of mass 20 kg. Suppose a meteor having a mass of 7.0 kg passes within 3.0 m of the surface of the satellite. What is the magnitude of the gravitational force on the meteor from the satellite at the closest approach?

4. Sun and Earth The Sun and Earth each exert a gravitational force on the Moon. What is the ratio $F_{\text{Sun}\rightarrow\text{Moon}}/F_{\text{Earth}\rightarrow\text{Moon}}$ of the magnitudes of these two forces? (The average Sun–Moon distance is equal to the Sun–Earth distance.)

5. Split into Two A mass M is split into two parts, m and $M - m$, which are then separated by a certain distance. What ratio m/M maximizes the magnitude of the gravitational force between the parts?

SEC 14-3 ■ GRAVITATION AND SUPERPOSITION

6. Zero Net Force A spaceship is on a straight-line path between Earth and its moon. At what distance from Earth is the net gravitational force (due to the Earth and the Moon only) on the spaceship zero?

7. Space Probe How far from Earth must a space probe be along a line toward the Sun so that the Sun's gravitational pull on the probe balances Earth's pull?

8. Three Spheres Three 5.0 kg spheres are located in the xy plane as shown in Fig. 14-17. What is the magnitude of the net gravitational force on the sphere at the origin due to the other two spheres?

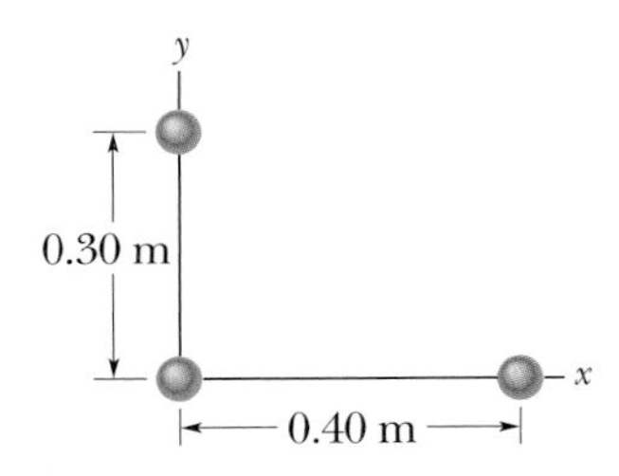

FIGURE 14-17 ■ Problem 8.

9. Four Spheres In Fig. 14-18a, four spheres form the corners of a square whose side is 2.0 cm long. What are the magnitude and direction of the net gravitational force from them on a central sphere with mass m_A = 250 kg?

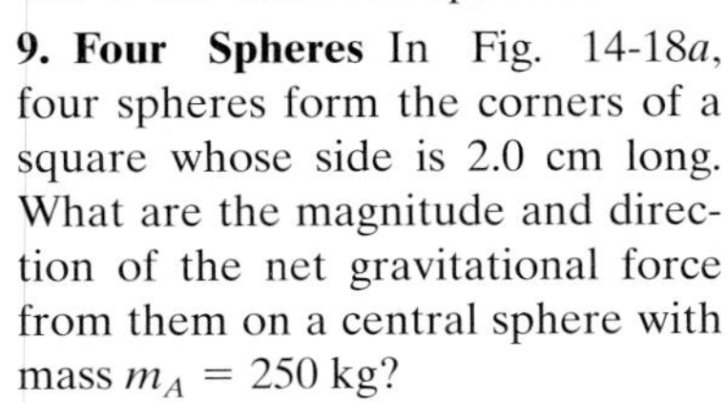

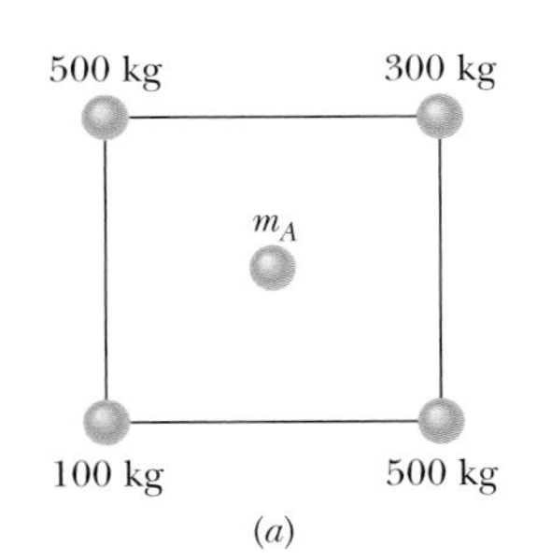

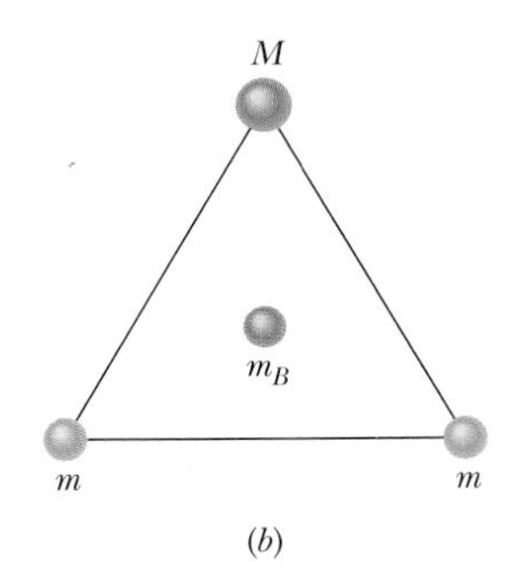

FIGURE 14-18 ■ Problems 9 and 10.

10. Two Spheres In Fig. 14-18b, two spheres of mass m and a third sphere of mass M form an equilateral triangle, and a fourth sphere of mass m_B is at the center of the triangle. The net gravitational force on that central sphere from the three other spheres is zero. (a) What is M in terms of m? (b) If we double the value of m_B, what then is the magnitude of the net gravitational force on the central sphere?

11. Masses and Coordinates Given The masses and coordinates of three spheres are as follows: 20 kg, x = 0.50 m, y = 1.0 m; 40 kg, x = −1.0 m, y = −1.0 m; 60 kg, x = 0 m, y = −0.50 m. What is the magnitude of the gravitational force on a 20 kg sphere located at the origin due to the other spheres?

12. Four Uniform Spheres Four uniform spheres, with masses m_A = 400 kg, m_B = 350 kg, m_C = 2000 kg, and m_D = 500 kg, have (x, y) coordinates of (0, 50) cm, (0, 0) cm, (−80, 0) cm, and (40, 0) cm, respectively. What is the net gravitational force on sphere B due to the other spheres?

13. Spherical Hollow Figure 14-19 shows a spherical hollow inside a lead sphere of radius R; the surface of the hollow passes through the center of the sphere and "touches" the right side of the sphere. The mass of the sphere before hollowing was M. With what gravitational force does the hollowed-out lead sphere attract a small sphere of mass m that lies at a distance d from the center of the lead sphere, on the straight line connecting the centers of the spheres and of the hollow?

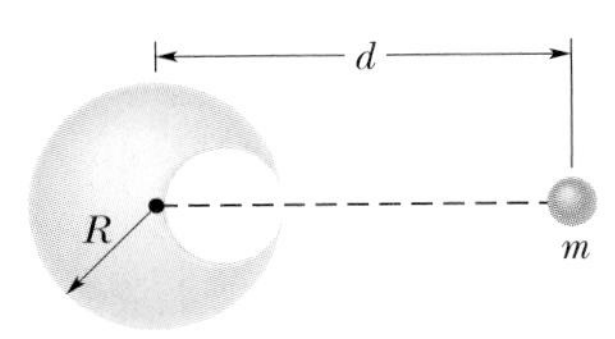

FIGURE 14-19 ■ Problem 13.

SEC. 14-4 ■ GRAVITATION IN THE EARTH'S VICINITY.

14. Empire State Building You weigh 530 N at sidewalk level outside the Empire State Building in New York City. Suppose that you ride from this level to the 102nd floor tower, a height of 373 m. Ignoring Earth's rotation, how much less would you weigh there (because you are slightly farther from the center of Earth)?

15. g = 4.9 m/s² At which altitude above Earth's surface would the gravitational acceleration be 4.9 m/s²?

16. Moon's Surface (a) What will an object weigh on the Moon's surface if it weighs 100 N on Earth's surface? (b) How many Earth radii must this same object be from the center of Earth if it is to weigh the same as it does on the Moon?

17. Rate of Rotation The fastest possible rate of rotation of a planet is that for which the gravitational force on material at the equator just barely provides the centripetal force needed for the rotation. (Why?) (a) Show that the corresponding shortest period of rotation is

$$T = \sqrt{\frac{3\pi}{G\rho}},$$

where ρ is the uniform density of the spherical planet. (b) Calculate the rotation period assuming a density of 3.0 g/cm³, typical of many planets, satellites, and asteroids. No astronomical object has ever been found to be spinning with a period shorter than that determined by this analysis.

18. Model of a Planet One model for a certain planet has a core of radius R and mass M surrounded by an outer shell of inner radius R, outer radius $2R$, and mass $4M$. If $M = 4.1 \times 10^{24}$ kg and $R = 6.0 \times 10^6$ m, what is the gravitational acceleration of a particle at points (a) R and (b)$3R$ from the center of the planet?

19. Spring Scale A body is suspended from a spring scale in a ship sailing along the equator with speed v. (a) Show that the scale reading will be very close to $W_0\,(1 \pm 2\,\omega v/g)$, where ω is the rotational speed of Earth and W_0 is the scale reading when the ship is at rest. (b) Explain the ± sign.

20. Neutron Stars Certain neutron stars (extremely dense stars) are believed to be rotating at about 1 rev/s. If such a star has a radius of 20 km, what must be its minimum mass so that material on its surface remains in place during the rapid rotation?

SEC. 14-5 ■ GRAVITATION INSIDE EARTH

21. Apple and Tunnel Assume that a planet is a sphere of radius R with a uniform density and (somehow) has a narrow radial tunnel through its center. Also assume that we can position an apple anywhere along the tunnel or outside the sphere. Let F_R be the magnitude of the gravitational force on the apple when it is located at the planet's surface. How far from the surface is a point where the magnitude of the gravitational force on the apple is $\frac{1}{2}F_R$ if we move the apple (a) away from the planet and (b) into the tunnel?

22. Two Concentric Shells Two concentric shells of uniform density having masses M_1 and M_2 are situated as shown in Fig. 14-20. Find the magnitude of the net gravitational force on a particle of mass m, due to the shells, when the particle is located at (a) point A, at distance $r = a$ from the center, (b) point B at $r = b$, and (c) point C at $r = c$. The distance r is measured from the center of the shells.

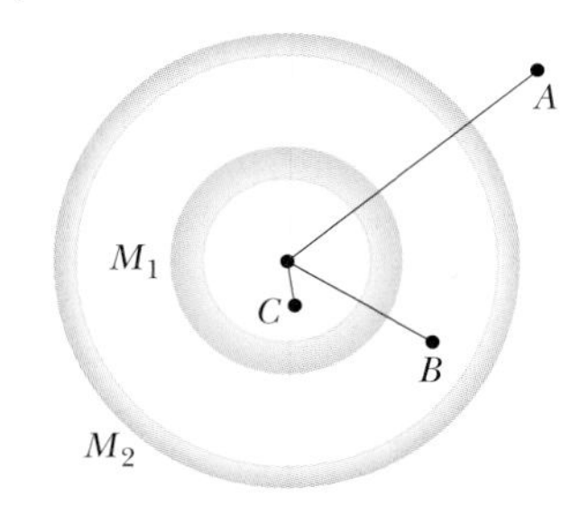

FIGURE 14-20 ■ Problem 22.

23. Solid Sphere A solid sphere of uniform density has a mass of 1.0×10^4 kg and a radius of 1.0 m. What is the magnitude of the gravitational force due to the sphere on a particle of mass m located at a distance of (a) 1.5 m and (b) 0.50 m from the center of the sphere? (c) Write a general expression for the magnitude of the gravitational force on the particle at a distance $r \leq 1.0$ m from the center of the sphere.

24. Uniform Solid Sphere A uniform solid sphere of radius R has a gravitational strength g_{local} at its surface. At what two distances from the center of the sphere is the gravitational strength $g_{\text{local}}/3$? (*Hint:* Consider distances both inside and outside the sphere.)

25. Crust, Mantle, Core Figure 14-21 shows, not to scale, a cross section through the interior of Earth. Rather than being uniform throughout, Earth is divided into three zones: an outer *crust*, a *mantle*, and an inner *core*. The dimensions of these zones and the masses contained within them are shown on the figure. Earth has a total mass of 5.98×10^{24} kg and a radius of 6370 km. Ignore rotation and assume that Earth is spherical. (a) Calculate the local gravitational strength g at the surface. (b) Suppose that a bore hole (the *Mohole*) is driven to the crust–mantle interface at a depth of 25 km. What would be the value of g at the bottom of the hole? (c) Suppose that Earth were a uniform sphere with the same total mass and size. What would be the value of g at a depth of 25 km? (Precise measurements of g are sensitive probes of the interior structure of Earth, although results can be clouded by local density variations.)

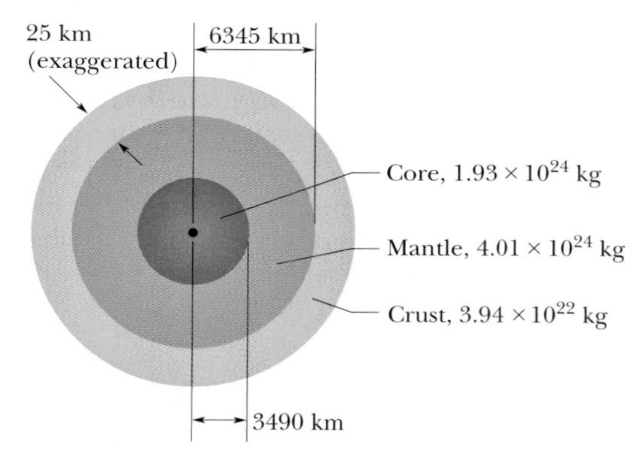

FIGURE 14-21 ■ Problem 25.

SEC. 14-6 ■ GRAVITATIONAL POTENTIAL ENERGY

26. Potential Energy (a) What is the gravitational potential energy of the two-particle system in Problem 1? If you triple the separation between the particles, how much work is done (b) by the gravitational force between the particles and (c) by you?

27. Remove Sphere *A* (a) In Problem 12, remove sphere A and calculate the gravitational potential energy of the remaining three-particle system. (b) If A is then put back in place, is the potential energy of the four-particle system more or less than that of the system in (a)? (c) In (a), is the work done by you to remove A positive or negative? (d) In (b), is the work done by you to replace A positive or negative?

28. Ratio *m/M* In Problem 5, what ratio m/M gives the least gravitational potential energy for the system?

29. Mars and Earth The mean diameters of Mars and Earth are 6.9×10^3 km and 1.3×10^4 km, respectively. The mass of Mars is 0.11 times Earth's mass. (a) What is the ratio of the mean density of Mars to that of Earth? (b) What is the value of the gravitational acceleration on Mars? (c) What is the escape speed on Mars?

30. Escape Calculate the amount of energy required to escape from (a) Earth's moon and (b) Jupiter relative to that required to escape from Earth.

31. Three Other Spheres The three spheres in Fig. 14-22, with masses $m_A = 800$ g, $m_B = 100$ g, and $m_C = 200$ g, have their centers on a common line, with $L = 12$ cm and $d = 4.0$ cm. You move sphere B along the line until its center-to-center separation from C is $d = 4.0$ cm. How much work is done on sphere B (a) by you and (b) by the net gravitational force on B due to spheres A and C?

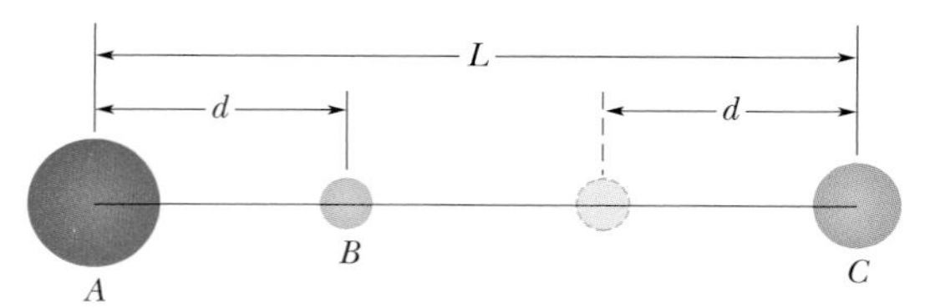

FIGURE 14-22 ■ Problem 31.

32. Zero Zero, a hypothetical planet, has a mass of 5.0×10^{23} kg, a radius of 3.0×10^6 m, and no atmosphere. A 10 kg space probe is to be launched vertically from its surface. (a) If the probe is launched with an initial energy of 5.0×10^7 J, what will be its kinetic energy when it is 4.0×10^6 m from the center of Zero? (b) If the probe is to achieve a maximum distance of 8.0×10^6 m from the center of Zero, with what initial kinetic energy must it be launched from the surface of Zero?

33. Rocket Accelerated A rocket is accelerated to speed $v = 2\sqrt{gR_E}$ near Earth's surface (where Earth's radius is R_E), and it then coasts upward. (a) Show that it will escape from Earth. (b) Show that very far from Earth its speed will be $v = \sqrt{2gR_E}$.

34. Roton Planet Roton, with a mass of 7.0×10^{24} kg and a radius of 1600 km, gravitationally attracts a meteorite that is initially at rest relative to the planet, at a great enough distance to take as infinite. The meteorite falls toward the planet. Assuming the planet is airless, find the speed of the meteorite when it reaches the planet's surface.

35. Escape Speed (a) What is the escape speed on a spherical asteroid whose radius is 500 km and whose gravitational acceleration at the surface is 3.0 m/s²? (b) How far from the surface will a particle go if it leaves the asteroid's surface with a radial speed of 1000 m/s? (c) With what speed will an object hit the asteroid if it is dropped from 1000 km above the surface?

36. Rocket Moving Radially A 150.0 kg rocket moving radially outward from Earth has a speed of 3.70 km/s when its engine shuts off 200 km above Earth's surface. (a) Assuming negligible air drag, find the rocket's kinetic energy when the rocket is 1000 km above Earth's surface. (b) What maximum height above the surface is reached by the rocket?

37. Two Neutron Stars Two neutron stars are separated by a distance of 10^{10} m. They each have a mass of 10^{30} kg and a radius of 10^5 m. They are initially at rest with respect to each other. As measured from that rest frame, how fast are they moving when (a) their separation has decreased to one-half its initial value and (b) they are about to collide?

38. Deep Space In deep space, sphere A of mass 20 kg is located at the origin of an x axis and sphere B of mass 10 kg is located on the axis at $x = 0.80$ m. Sphere B is released from rest while sphere A is held at the origin. (a) What is the gravitational potential energy of the two-sphere system as B is released? (b) What is the kinetic energy of B when it has moved 0.20 m toward A?

39. Projectile A projectile is fired vertically from Earth's surface with an initial speed of 10 km/s. Neglecting air drag, how far above the surface of Earth will it go?

SEC. 14-7 ■ EINSTEIN AND GRAVITATION

40. Cantaloupe In Fig. 14-13*b*, the scale on which the 60 kg physicist stands reads 220 N. How long will the cantaloupe take to reach the floor if the physicist drops it from rest (relative to himself), 2.1 m from the floor?

Additional Problems

41. Frames of Reference Figure 14-23 shows two identical spheres, each with mass 2.00 kg and radius $R =$ 0.0200 m, that initially touch, somewhere in deep space. Suppose the spheres are blown apart such that they initially separate at the relative speed 1.05×10^{-4} m/s. They then slow due to the gravitational force between them.

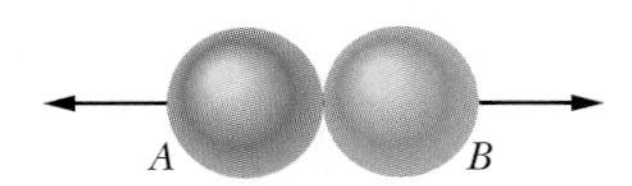

FIGURE 14-23 ■ Problem 41.

Center-of-mass frame: Assume that we are in an inertial reference frame that is stationary with respect to the center of mass of the two-sphere system. Use the principle of conservation of mechanical energy ($K_2 + U_2 = K_1 + U_1$) to find the following when the center-to-center separation is $10R$: (a) the kinetic energy of each sphere and (b) the speed of sphere B relative to sphere A.

Sphere frame: Next assume that we are in a reference frame attached to sphere A (we ride on the body). Now we see sphere B move away from us. From this reference frame, again use $K_2 + U_2 = K_1 + U_1$ to find the following when the center-to-center separation is $10R$: (c) the kinetic energy of sphere B and (d) the speed of sphere B relative to sphere A. (e) Why are the answers to (b) and (d) different? Which answer is correct?

42. Black Hole The radius R_h of a black hole is the radius of a mathematical sphere, called the event horizon, that is centered on the black hole. Information from events inside the event horizon cannot reach the outside world. According to Einstein's general theory of relativity, $R_h = 2GM/c^2$, where M is the mass of the black hole and c is the speed of light.

Suppose that you wish to study black holes near them, at a radial distance of $50R_h$. However, you do not want the difference in gravitational acceleration between your feet and your head to exceed 10 m/s² when you are feet down (or head down) toward the black hole. (a) As a multiple of our sun's mass, what is the limit to the mass of the black hole you can tolerate at the given radial distance? (You need to estimate your height.) (b) Is the limit an upper limit (you can tolerate smaller masses) or a lower limit (you can tolerate larger masses)?

43. Romeo and Juliet Two schoolmates, Romeo and Juliet, catch each other's eye across a crowded dance floor at a school dance. Estimate the gravitational attraction they exert on each other.

44. The Alignment of the Planets Some authors seeking public attention have suggested that when many planets are "aligned" (i.e., are close together in the sky) their gravitational pull on the Earth all acting together might produce earthquakes and other disasters. To get an idea of whether this is plausible, set up the following calculation: (a) Draw a sketch of the solar system and arrange the planets so that Mars, Jupiter, and Saturn are on the same side of the Sun as the Earth. Look up (there is a table in the back of *Understanding Physics*) the radii of the planetary orbits and their masses. (b) Infer the distances these planets would be from Earth in this arrangement. (c) Without doing all the calculations, decide which of the three planets would exert the strongest gravitational force on the Earth. (*Hint:* Use the dependence of Newton's universal gravitation law on mass and distance.) (d) Calculate the gravitational force of the most important planet on the Earth. (e) Calculate how this compares to the gravitational force the Moon exerts on the Earth.

Note: In fact, it is not the gravitational force itself that produces the possibly dangerous effects, but the tidal forces—the derivative of the gravitational force. This reduces the effect by another factor of the distance. That is, the tidal force goes like $1/r^3$ instead of like $1/r^2$. This weakens the planet's gravitational effect compared to the Moon's by an additional factor of $r_{\text{Earth–moon}}/r_{\text{Earth–planet}}$, a number much less than 1.

45. Is Newton's Law of Gravity Wrong? A professional scientist (not a physicist) stops you in the hall and says: "I can prove Newton's theory of gravity is wrong. The Sun is 320,000 times as massive as the Earth, but only 400 times as far from the Moon as is the Earth. Therefore, the force of the Sun's gravity on the Moon should be twice as big as the Earth's and the Moon should go around the Sun instead of around the Earth. Since it doesn't, Newton's theory of gravity must be wrong!" What's the matter with this reasoning?

46. In the Shuttle When we see the astronauts in orbit in the space shuttle on TV, they seem to float. If they let go of something, it just stays where they put it. It doesn't fall. What happens to gravity for objects in orbit? Does gravity stop at the Earth's atmosphere? Explain what's happening in terms of the physics you have learned.

15 Fluids

The force exerted by water on the body of a descending diver increases noticeably, even for a relatively shallow descent to the bottom of a swimming pool. However, in 1975, using scuba gear with a special gas mixture for breathing, William Rhodes emerged from a chamber that had been lowered 300 m into the Gulf of Mexico, and he then swam to a record depth of 350 m. Strangely, a novice scuba diver practicing in a swimming pool might be in more danger from the force exerted by the water than was Rhodes. Occasionally, novice scuba divers die because they have neglected that danger.

What is this potentially lethal risk?

The answer is in this chapter.

15-1 Fluids and the World Around Us

Fluids—which include both liquids and gases—play a central role in our daily lives. We breathe and drink them, and a rather vital fluid circulates in the human cardiovascular system. The Earth's oceans and atmosphere consist of fluids.

Cars and jet planes need many different fluids to operate, including fluids in their tires, fuel tanks, and engine combustion chambers. They also need fluids for their air conditioning, lubrication, and hydraulic systems. Windmills transform the kinetic energy in air to electrical energy, and hydroelectric plants convert the gravitational potential energy in water to electrical energy. Over long time periods, air and water carve out and reshape the Earth's landscape.

In our study of fluids, we will start by examining simple physical situations that we encounter every day. First, we will study the forces acting on fluids that are in static equilibrium and consider the forces on objects in fluids. Then we will examine how a hydraulic system can be used as a lever. Later in the chapter, we will study the motions of fluids as they flow through pipes and around objects.

15-2 What Is a Fluid?

To understand what we mean by the term "fluid," let us compare solids, liquids, and gases. A solid vertical column that rests on a table can retain its shape without external support. Since a gravitational force is acting on each of the columns shown in Fig. 15-1, each exerts a downward normal force on the table that is equal in magnitude to its weight. What happens if we try to make a column out of a liquid? Without external support, the gravitational forces on the liquid will cause it to collapse and flow into a puddle. (In the more formal terms introduced in Section 13-5, liquids cannot withstand shear stresses.)

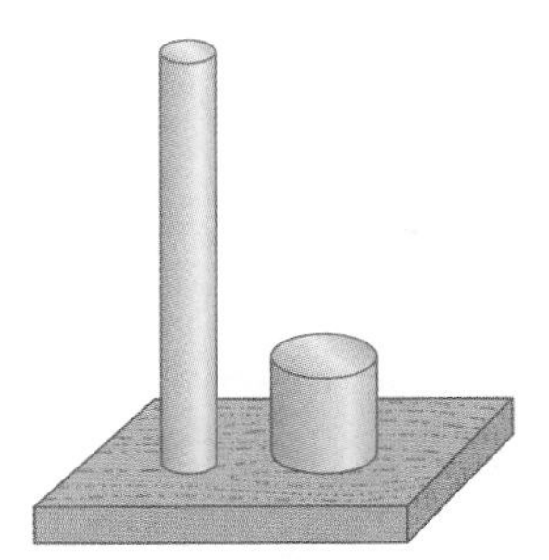

FIGURE 15-1 ■ Two columns resting on a table have the same mass, so they each exert the same downward normal force on the table.

However, we can maintain a vertical column of liquid if we provide it with solid walls. In this case, the liquid presses sideways against the walls and the walls press back against the liquid. Thus, the vertical columns of the liquid and solid differ in that the column of liquid needs external forces acting on it to maintain its shape whereas the solid does not. However, both a solid column and a container full of liquid will exert normal forces on a table.

When external forces are present, a **fluid,** unlike a solid, can flow until it conforms to the boundaries of its container. Obviously gases, such as the air that surrounds us, are also fluids, because they can conform to the shape of a container quite rapidly. Some gooey materials, such as heavy syrup and silly putty, take a longer time to conform to the boundaries of a container. But since they can do so eventually, we also classify them as fluids.

15-3 Pressure and Density

Defining Pressure for Uniform Forces

Let us consider the properties of the two solid columns shown in Fig. 15-1. Since they both have the same weight, they exert the same downward forces on the table. However, if you placed your hand under each of the columns, you would *feel* a difference. Why? Because the forces are spread out over different areas. It is this difference in *pressure* you feel when placing each column on your hand. If a force is evenly distributed over every point of an area (as is the case for the normal forces exerted by the cylinders), we say that it is *uniform* over the area. For a force that is both uniformly distributed over an area and perpendicular or normal to it, the **pressure** P on a surface is defined as the magnitude of the net force acting on the surface divided by its area. Thus, it can be expressed by the equation

$$P \equiv \frac{|\vec{F}_{\perp}|}{A} \quad \text{(uniform forces normal to area } A\text{)}, \tag{15-1}$$

where, as usual, the symbol $\equiv$ is used to signify that the equation holds "by definition." The column on the left in Fig. 15-1 has one-fourth the area that the right-hand column does. Even though the left and right columns exert the same net force on the table, the left column exerts four times more pressure on the table. Later on in this section we will refine our discussion of pressure to handle situations in which the normal forces acting on a surface are not uniform.

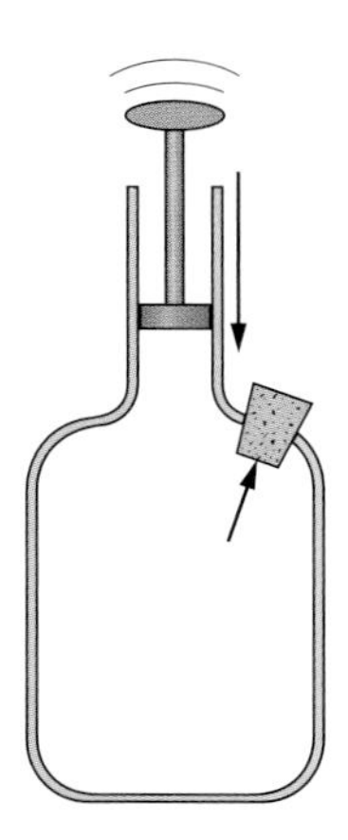

FIGURE 15-2 ■ Suppose a hole is drilled at some random place on a bottle and plugged with a cork. If an airtight plunger is thrust down the bottle's neck, the increased pressure in the bottle can cause the cork to pop out no matter what direction it faces. This can happen whenever the bottle contains either a gas (such as air) or a liquid (such as water).

Is Pressure a Vector or Scalar?

In order to think about the idea of pressure exerted by a fluid, consider a bottle full of a fluid that has a piston on top. Although there is a hole in the bottle, it is plugged with a cork as shown in Fig. 15-2. If we press on the piston, the cork will pop out. This indicates that the fluid exerts a perpendicular force on the face of the cork that is sticking into the bottle. What's remarkable is that no matter where the hole and cork are located on the bottle, the cork would still pop out! Somehow, the downward force we apply with the piston to one part of the fluid is translated into "internal forces" that act in all directions. Thus, *the fluid pressure acting at the surface of a container appears to have no preferred direction.*

Let us consider a fluid that is not moving so that we can define it as being in a state of static equilibrium. What is the pressure like inside the fluid? We can consider this question both experimentally and theoretically.

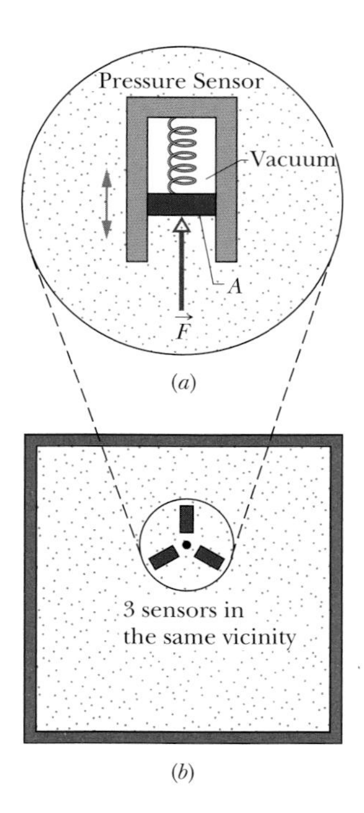

FIGURE 15-3 ■ (a) A tiny pressure sensor that uses spring compression to measure the net force normal to the area, A, of a piston. (b) When an array of pressure sensors pointing in different directions are placed in the vicinity of a single point in a fluid, their pressure measurements are identical.

Experimental Results: If we want to measure the pressure exerted by a fluid at a point inside a container of fluid, we can design a small pressure sensor like that shown in Fig. 15-3a. The sensor consists of a piston with a small cross-sectional area A. The piston fits snugly in an evacuated cylinder, so the cylinder contains no matter other than a coiled spring that is lodged behind the piston. By measuring the spring compression we can determine the normal force the fluid exerts on the piston.

Suppose we place an array of three tiny pistons at the point of interest inside the container as shown in Fig. 15-3. We find that the magnitude of force on each of the pistons is the same independent of the directions the pistons are facing. Thus, we only need to place a single pressure sensor at a point of interest and measure the force on its piston to calculate the pressure using Eq. 15-1.

> Experiments reveal that at a given point in a fluid that is in static equilibrium, the pressure P has the same value in all directions. In other words, pressure is a scalar, having no directional properties.

Agreement between Experiment and Theory: We can use the fact that we have chosen to examine a fluid that is in static equilibrium to see why we should indeed expect the pressure near a point in the fluid to be nondirectional. Let us simplify the situation by assuming that a container of fluid is located where there are no gravitational forces on it. In Section 15-4, we will revisit this idea for the more common case of nonzero gravitational forces. Next we can draw an imaginary cubical boundary around a tiny parcel of fluid (Fig. 15-4) centered on some point in the container. Since the parcel of fluid is in equilibrium it cannot be accelerating, and we must conclude that the net force on its boundaries is zero as shown in Fig. 15-4. Since the force vectors on opposite faces of the cubical parcel must be equal in magnitude and opposite in direction, the pressure on opposite faces must be the same. Furthermore, there are no gravitational forces acting in our special case and thus there is no preferred direction. These facts allow us to conclude that if no part of the fluid is accelerating, the pressure must be the same in all directions throughout the entire container. So far we have considered a very special shape for our parcel in the absence of gravitational forces. In Section 15-4 we consider what happens to the pressure in fluids in static equilibrium close to the Earth's surface or at other locations where gravitational forces must be taken into account.

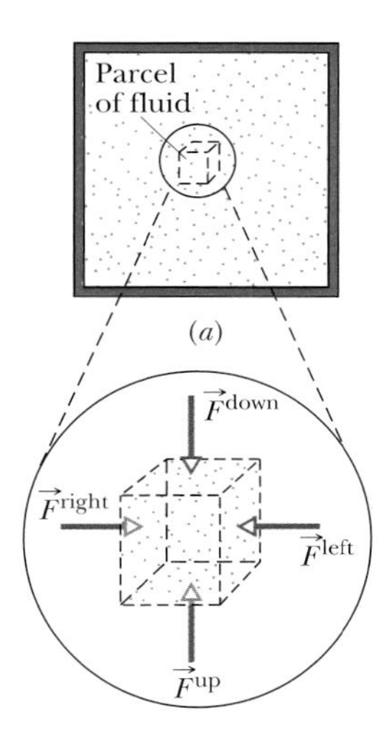

FIGURE 15-4 ■ (*a*) A tiny parcel of fluid with no gravitational forces acting on it in static equilibrium. (*b*) Since the parcel does not accelerate, the net force on it due to the surrounding fluid must be zero, so the pressure on the parcel is the same in all directions. For clarity, force vectors on the front and back parcel faces are not shown.

Defining Pressure for Nonuniform Forces and Surfaces

If the forces on an area are not uniform or if the area is curved, we can still use our basic definition of pressure by breaking area A into segments. The area segments must be small enough so that the normal forces acting on each segment are uniform and each area segment is essentially flat (Fig. 15-5). If we do this, the pressure at the location of the ith segment of the area can be defined as

$$P_i \equiv \lim_{A_i \to 0} \frac{|\vec{F}_i|}{A_i} \quad \text{(pressure at a point, nonuniform forces),} \qquad (15\text{-}2)$$

where $|\vec{F}_{i\perp}|$ is the magnitude of the net force normal to the ith area. That is, the pressure at any point is the limit of this ratio as the area A_i centered on that point is made smaller and smaller. Obviously, the net force acting on smaller areas will be smaller so the ratio is still physically meaningful.

The SI unit of pressure is the newton per square meter, which is given a special name, the **pascal** (Pa). In countries using the metric system, tire pressure gauges are calibrated in kilopascals (kPa). The pascal is related to some other common (non-SI) pressure units as follows:

$$1.00 \text{ atm} \equiv 101\,325 \text{ Pa} \equiv 760 \text{ Torr} \approx 14.7 \text{ lb/in}^2.$$

The *atmosphere* (atm) is, as the name suggests, the approximate average pressure of the atmosphere at sea level. The *torr* (named for Evangelista Torricelli, who invented the mercury barometer in 1674) was formerly called the *millimeter of mercury* (mm Hg). The pound per square inch is often abbreviated psi. Table 15-1 shows some pressures.

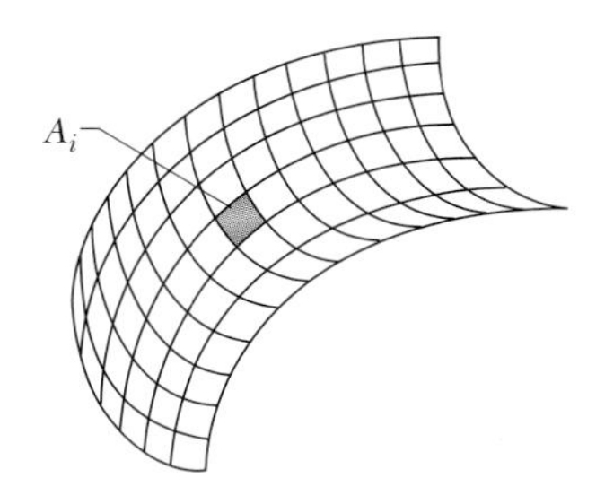

FIGURE 15-5 ■ To calculate the pressure exerted by fluids on curved surfaces or surfaces that have nonuniform forces acting on them, a surface area must be divided into a large number of small area elements, A_1, A_2, and so on. The ith area is shown in the diagram.

Density

Let us return one more time to the column of solid we discussed above. Clearly, the weight of a column of a given size and height depends on what the column is made of. If a certain column was constructed of a material like Styrofoam® it would be much lighter than if it was made of lead. Hence, it would be convenient to have a way to predict the weight of an object that has a certain size and shape.

For this purpose, we invent a new quantity that is a measure of the mass of one cubic meter of a material. To determine this value for a given substance, we measure the total mass M in a measured volume V of the material and calculate M/V. This quantity is called *density*. In general, density is a measure of mass per unit of volume. The standard symbol for density is ρ.

TABLE 15-1
Some Pressures

	Pressures (Pa)
Center of the Sun	2×10^{16}
Center of Earth	4×10^{11}
Highest sustained laboratory pressure	1.5×10^{10}
Deepest ocean trench (bottom)	1.1×10^{8}
Spike heels on a dance floor	1×10^{6}
Automobile tire[a]	2×10^{5}
Atmosphere at sea level	1.0×10^{5}
Normal blood pressure[a,b]	1.6×10^{4}
Best laboratory vacuum	10^{-12}

[a] Pressure in excess of atmospheric.

[b] The systolic pressure (120 torr on a physician's pressure gauge).

TABLE 15-2
Some Densities

Material or Object	Density (kg/m³)
Stray atoms in interstellar space	10^{-20}
Air remaining in the best laboratory vacuum	10^{-17}
Air: 20°C and 1 atm pressure	1.21
20°C and 50 atm pressure	60.5
Styrofoam	1×10^{2}
Ice	0.917×10^{3}
Water: 20°C and 1 atm	0.998×10^{3}
20°C and 50 atm	1.000×10^{3}
Seawater: 20°C and 1 atm	1.024×10^{3}
Whole blood	1.060×10^{3}
Iron	7.9×10^{3}
Mercury (the metal)	13.6×10^{3}
Earth: average	5.5×10^{3}
core	9.5×10^{3}
crust	2.8×10^{3}
Sun: average	1.4×10^{3}
core	1.6×10^{5}
White dwarf star (core)	10^{10}
Uranium nucleus	3×10^{17}
Neutron star (core)	10^{18}
Black hole (1 solar mass)	10^{19}

Table 15-2 shows the densities of several substances and the average densities of some objects. Notice that the density of a gas (see Air in the table) varies considerably with pressure, but the density of a liquid (see Water) does not. That is, gases are readily *compressible* but liquids are not.

The density of a fluid is not always uniform. For example, the density of the gas molecules and other particles that make up the Earth's atmosphere is much greater close to the surface of the Earth than in the stratosphere. As is the case for pressure, we can find the density ρ of any fluid at point i if we isolate a small volume element V_i around that point and measure the mass m_i of the fluid contained within that element. The **density** is then

$$\rho_i = \frac{m_i}{V_i} \quad \text{(density in the vicinity of a point } i\text{).} \qquad (15\text{-}3)$$

In theory, the density at any point in a fluid is the limit of this ratio as the volume element V at that point is made smaller and smaller. In practice, many fluid samples are large compared to atomic dimensions and are thus "smooth" rather than "lumpy" with atoms. If it is reasonable to assume further that the sample has a uniform density, we can simplify Eq. 15-3 to

$$\rho = \frac{m}{V} \quad \text{(uniform density),} \qquad (15\text{-}4)$$

where m and V are the total mass and volume of the sample. Density is a scalar property; its SI unit is the kilogram per cubic meter.

Density and pressure are fundamental concepts in regard to fluids. When we discuss solids, we are concerned with particular lumps of matter, such as wooden blocks, baseballs, or metal rods. Physical quantities that we find useful, and in whose terms we express Newton's laws, are *mass* and *force.* We might speak, for example, of a 3.6 kg block acted on by a 25 N force. With fluids, we are more interested in the extended substance, and in properties that can vary from point to point in that substance. In these cases, it is more useful to speak of *density* and *pressure* than of mass and force.

READING EXERCISE 15-1: Estimate the pressure in pascals exerted on a dance floor by just the spike heels worn by a 125 lb woman who is standing on both feet. Assume that half of her weight is on the spike heels and half is on the front soles of her shoes. How does your estimate compare with the number given in Table 15-1? Discuss why it is possible for the spike heels to exert more pressure on the floor than the pressure exerted on the road by a single tire holding up its share of a 2500 lb automobile. ■

READING EXERCISE 15-2: Examine the densities listed in Table 15-2. Use the fact that air and water have significantly different densities to develop a plausible explanation for the fact that air is a much more compressible fluid than is water. ■

READING EXERCISE 15-3: Consider a book of dimensions 8 in. by 10 in. Show that the downward force on the book by the atmosphere is about 1200 lb. The downward force on a smooth thick rubber mat of the same dimensions is also 1200 lb. You find that you cannot lift the rubber mat when it is placed on a smooth Formica table so no air can get under it. However, you can easily lift the book. Can you explain this phenomenon? ■

TOUCHSTONE EXAMPLE 15-1: Force Due to Air Pressure

A living room has floor dimensions of 3.5 m and 4.2 m and a height of 2.4 m.

(a) What does the air in the room weigh when the air pressure is 1.0 atm?

SOLUTION ■ The **Key Ideas** here are these: (1) The air's weight is equal to mg, where m is its mass. (2) Mass m is related to the air density ρ and the air's volume V by Eq. 15-4 ($\rho = m/V$). Putting these two ideas together and taking the density of air at 1.0 atm from Table 15-1, we find

$$\begin{aligned} mg &= (\rho V)g \\ &= (1.21 \text{ kg/m}^3)(3.5 \text{ m} \times 4.2 \text{ m} \times 2.4 \text{ m})(9.8 \text{ m/s}^2) \\ &= 418 \text{ N} \approx 420 \text{ N}. \end{aligned}$$

This is the weight of about 126 cans of soda.

(b) What is the magnitude of the atmosphere's force on the ceiling of the room?

SOLUTION ■ The **Key Idea** here is that the atmosphere pushes up on the ceiling with a force of magnitude $F_\perp = |\vec{F}_\perp|$ that is uniform over the ceiling. Thus, it produces a pressure that is related to F and the flat area A of the ceiling by Eq. 15-1 ($P = F_\perp/A$), which gives us

$$\begin{aligned} F_\perp = PA &= (1.0 \text{ atm})\left(\frac{1.01 \times 10^5 \text{ N/m}^2}{1.0 \text{ atm}}\right)(3.5 \text{ m})(4.2 \text{ m}) \\ &= 1.5 \times 10^6 \text{ N}. \qquad \text{(Answer)} \end{aligned}$$

This enormous force is equal to the weight of the column of air that has as its base the horizontal area of the room and extends all the way to the top of the atmosphere.

15-4 Gravitational Forces and Fluids at Rest

In the last section we considered the forces on a parcel of fluid that was "at rest" and does not experience gravitational forces. We found that the pressure in any fluid in a container was the same in every direction at every location in the container. This is not so in fluids that experience gravitational forces. In this section we will consider how the presence of gravitational forces leads to pressure differences at different levels in a container of fluid. As every diver knows, the pressure *increases* with depth below the air–water interface. The diver's depth gauge, in fact, is a pressure sensor much like that of Fig. 15-3*a*. As every mountaineer knows, the pressure *decreases* with altitude as one ascends into the atmosphere. The pressures encountered by the diver and the mountaineer are usually called **hydrostatic pressures** because they are pressures due to fluids that are static (at rest). Here we want to find an expression for hydrostatic pressure as a function of depth or altitude.

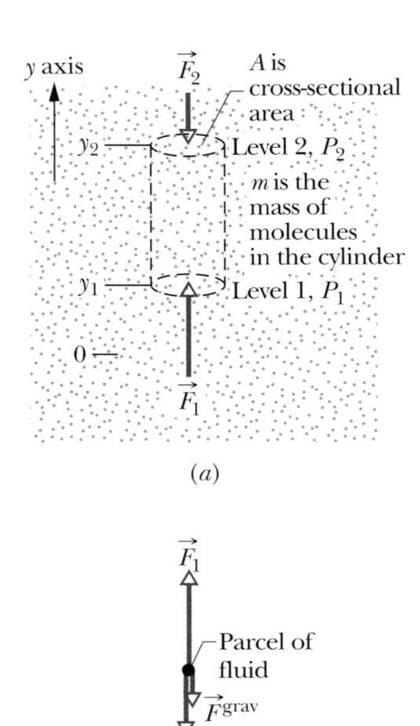

FIGURE 15-6 ■ (*a*) A parcel of fluid such as air or water is contained in an imaginary cylinder of cross-sectional area A. Forces $\vec{F}_1$ and $\vec{F}_2$ act, respectively, on the bottom and top of the cylinder. The gravitational force of the parcel of fluid is $\vec{F}^{grav} = -mg\,\hat{j}$. (*b*) A free-body diagram of the forces that act on the parcel of fluid in the cylinder.

Let us consider a fluid such as air or water near the surface of the Earth. What happens to its pressure when one changes from an initial level y_1 to a final level y_2? Since a fluid is made up of lots of molecules, we can pick any subset of them as our "object" and apply Newton's laws to that parcel of fluid. So let us imagine a parcel of the fluid consisting of all the molecules contained in a cylindrical column of cross-sectional area A that extends between the two levels y_1 and y_2, as shown in Fig. 15-6*a*. The total mass of the molecules in the cylinder is m.

If the parcel of fluid is at rest, in *static equilibrium*, the horizontal forces on it from the sides must add up to zero. Similarly, static equilibrium requires that the vector sum of the vertical forces on the parcel of fluid must be zero too. There are three vertical forces that act on the parcel of fluid in the cylinder. Figure 15-6*b* shows a free-body diagram of these forces. Force $\vec{F}_1$ acts at the bottom surface of the cylinder and is due to the water below the cylinder. Similarly, force $\vec{F}_2$ acts at the top surface of the cylinder and is due to the fluid above the cylinder. The gravitational force on the water in the cylinder can be represented by $\vec{F}^{grav} = -mg\,\hat{j}$. The net force comprised of these forces must be zero, so that

$$\vec{F}^{net} = \vec{F}_1 + \vec{F}_2 + \vec{F}^{grav} = 0.$$

Since we know that all the forces act in the vertical direction, we can rewrite this equation in terms of the y-components of the vectors as

$$F^{net}_y = F_{1\,y} + F_{2\,y} + F^{grav}_y = F_{1\,y} + F_{2\,y} + (-mg) = 0. \tag{15-5}$$

We know that force $\vec{F}_1$ acts in an upward direction and is inherently positive while force $\vec{F}_2$ acts in a downward direction and is inherently negative. So we can replace the force *components*, with the corresponding pressures and areas, with

$$F_{1\,y} = +P_1 A \quad \text{and} \quad F_{2\,y} = -P_2 A. \tag{15-6}$$

We use the explicit minus sign in front of the inherently positive P_2A term to signify the fact that the force component $F_{2\,y}$ must be negative. The mass m of the fluid in the cylinder is $m = \rho V$, where ρ represents the density of the fluid and where V represents the volume of the cylinder. Since the volume is the product of its face area A and its height $\Delta y = y_2 - y_1$, the mass m is equal to $\rho A(y_2 - y_1)$. Using these facts and substituting Eq. 15-6 into Eq. 15-5, we get

$$P_2 = P_1 - \rho g(y_2 - y_1) = P_1 - \rho g\,\Delta y \quad \text{(only if } \rho \text{ is uniform)}, \tag{15-7}$$

or since the pressure difference $\Delta P = P_2 - P_1$, we can also write

$$\Delta P = -\rho g\,\Delta y. \tag{15-8}$$

If SI units are used for ρ, g, and Δy in calculations, then the pressure will be in pascals.

Equation 15-7 is a general expression that can be used to find pressure changes in either a liquid (as a function of depth) or in the atmosphere (as a function of altitude or height) (Fig. 15-7). However, it is only valid when the density of the fluid and the local gravitational strength factors are essentially constant between the levels under consideration.

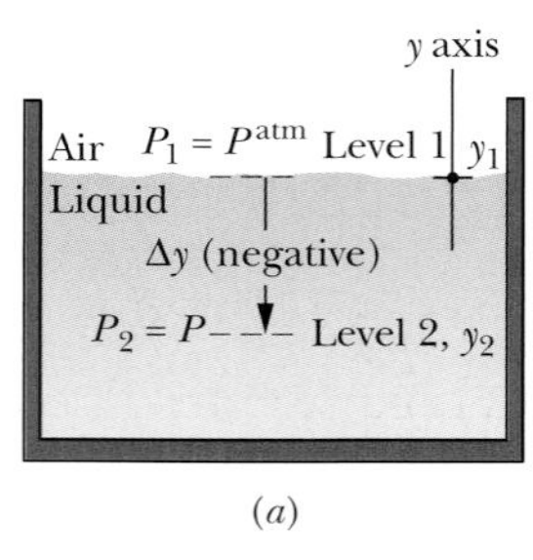

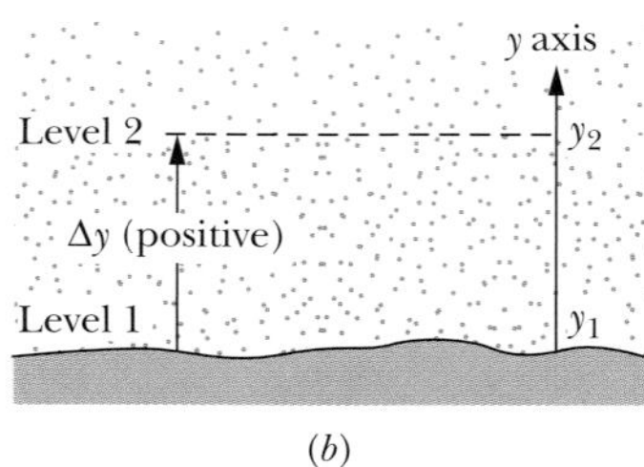

FIGURE 15-7 ■ Equation 15-7 can be used to determine (*a*) either the pressure underwater or (*b*) the atmospheric pressure above the surface of the Earth.

Special Case 1: Pressure in the Earth's Atmosphere

Equation 15-7 can be used to determine the pressure of the atmosphere at a given distance above a reference point in terms of the atmospheric pressure P_1 at that level. If we denote the density as $\rho = \rho_{air}$, we can write Eq. 15-7 as

$$P_2 = P_1 - \rho_{air}\, g\, \Delta y \quad \text{(only if } \rho_{air} \text{ and } g \approx \text{const).} \tag{15-9}$$

Since we are *above* the reference level, we know that $y_2 > y_1$ so that Δy is *positive.* Thus, we obtain an expression for pressure that predicts that pressure decreases in a linear manner with altitude.

Observations like those shown in Fig. 15-8 verifies that pressure does indeed decrease with altitude. Data in Table 14-1 indicate that the local gravitational strength is essentially constant at any altitude found on Earth, including Mt. Everest at 8.8 km. However, the density of air decreases with altitude. Fortunately the density of air ρ_{air} is reasonably constant up to about 5000 ft (or 1500 m). This means that Eq. 15-9 can be used to calculate pressures for the range of altitudes encountered in the trip across the cascades described in Fig. 15-8. However, Eq. 15-9 would not be accurate when considering higher elevations.

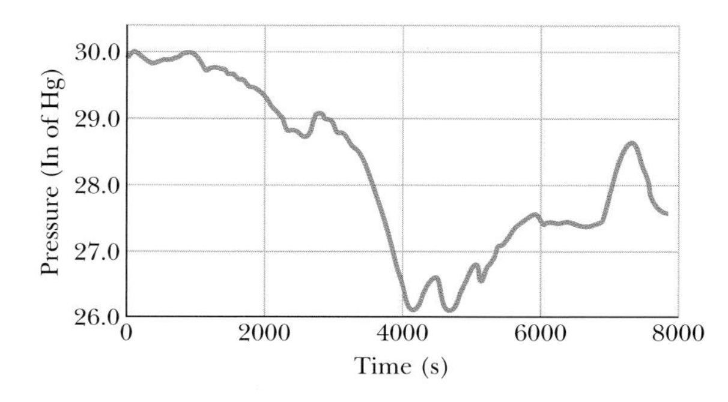

FIGURE 15-8 ■ Here is a graph of pressure data collected while driving a car from Portland, Oregon (altitude of about 3 m above sea level) over the Cascade Mountain Range to Madras, Oregon. The first pressure minimum came at about 4200 s when going over the pass near Government Camp. The second minimum came at about 4700 s while going over Blue Box pass, which is almost as high. Both passes are just over 1200 m above sea level. (Data courtesy of David Vernier.)

Special Case 2: Underwater Pressure

Equation 15-7 can also be used to determine the pressure underwater at a given distance below a reference point in terms of the pressure P_1 at the reference level. If we denote the density as $\rho = \rho_{water}$, we can write Eq. 15-7 as

$$P_2 = P_1 - \rho_{water}\, g\, \Delta y \quad \text{(only if } \rho_{water} \text{ and } g \approx \text{const).} \tag{15-10}$$

Since we are *below* the reference level, we know that $y_2 < y_1$ so that Δy is *negative*. Thus, we obtain an expression for pressure that predicts that pressure increases in a linear manner with depth.

As you can see in Fig. 15-9, our theory, which predicts a linear increase in pressure as a function of depth (with a slope given by the factor $\rho_{water}\, g$), compares nicely with experiment. Although we observe that the pressure in a liquid increases with depth, it does not depend on the horizontal location of the parcel of liquid. If we have a liquid other than water, obviously we need to use the density of that liquid in place of the density of water.

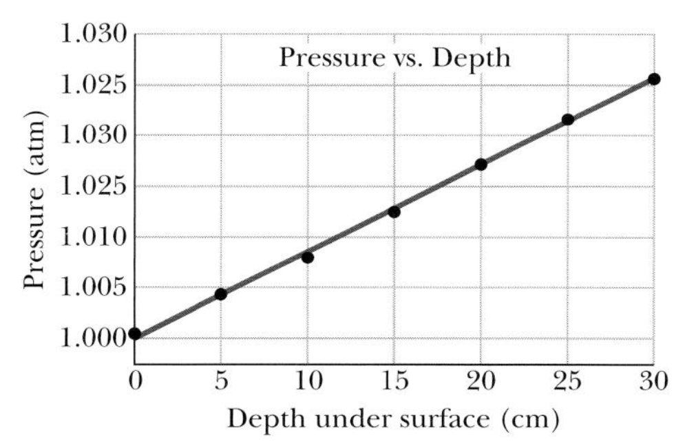

FIGURE 15-9 ■ Pressure as a function of depth for distances between 0 cm and 30 cm below the surface of a container of water. Measurements were made using a computer data acquisition system outfitted with a gas pressure sensor like that shown in the next section. The slope of the graph when expressed in SI units should be equal to the factor $\rho_{water}\, g$ shown in Eq. 15-10. Within the limits of experimental uncertainty, the data shown are compatible with theory.

> When gravitational forces are present, the pressure at a point in a fluid in static equilibrium depends on the depth of that point but not on any horizontal dimension of the fluid or the shape of its container.

READING EXERCISE 15-4: Scuba divers know that if they descend to a depth of 10 m the pressure they experience doubles. However, alpine mountain climbers must ascend to about 5.5 km to cut the atmospheric pressure in half. What factor in Eq. 15-7 accounts for the fact that the pressure change with distance is so much smaller in air than in water? ■

READING EXERCISE 15-5: The figure shows four containers of olive oil. Rank them according to the pressure at y_2, greatest first.

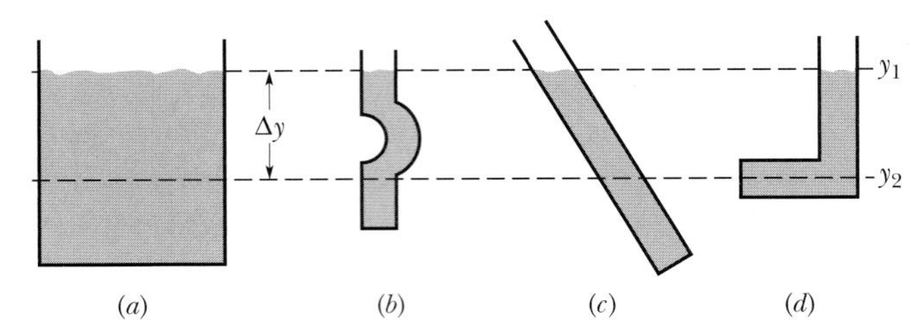

TOUCHSTONE EXAMPLE 15-2: Exhale!

A novice scuba diver practicing in a swimming pool takes enough air from his tank to fully expand his lungs before abandoning the tank at depth L and swimming to the surface. He ignores instructions and fails to exhale during his ascent. When he reaches the surface, the difference between the external pressure on him and the air pressure in his lungs is 9.3 kPa. From what depth does he start? What potentially lethal danger does he face?

SOLUTION ■ The **Key Idea** here is that when he fills his lungs at depth Δy, the external pressure on him (and thus the air pressure within his lungs) is greater than normal.

Assuming that P_1 equals the atmospheric pressure at the water's surface, we can rewrite Eq. 15-10 as

$$P_2 = P_1 - \rho_{\text{water}} g \,\Delta y = P^{\text{atm}} - \rho_{\text{water}} g \,\Delta y,$$

ρ_{water} is the water's density (998 kg/m³), given in Table 15-2. As the diver ascends, the external pressure on him decreases, until it is atmospheric pressure P^{atm} at the surface. His blood pressure also decreases, until it is normal. However, because he does not exhale, the air pressure in his lungs remains at the value it had at depth Δy. At the surface, the pressure difference between the higher pressure in his lungs and the lower pressure on his chest is

$$\Delta P = P_2 - P^{\text{atm}} = -\rho_{\text{water}} g \,\Delta y,$$

from which we find

$$\Delta y = \frac{\Delta P}{-\rho g} = -\frac{9300 \text{ Pa}}{(998 \text{ kg/m}^3)(9.8 \text{ m/s}^2)} = -0.95 \text{ m}. \quad \text{(Answer)}$$

This is not deep! Yet, the pressure difference of 9.3 kPa (about 9% of atmospheric pressure) is sufficient to rupture the diver's lungs and force air from them into the depressurized blood, which then carries the air to the heart, killing the diver. If the diver follows instructions and gradually exhales as he ascends, he allows the pressure in his lungs to equalize with the external pressure, and then there is no danger.

TOUCHSTONE EXAMPLE 15-3: U-Tube

The U-tube in Fig. 15-10 contains two liquids in static equilibrium: Water of density ρ_{water} (= 998 kg/m³) is in the right arm, and oil of unknown density ρ_x is in the left. Measurement gives l = 135 mm and d = 12.3 mm. What is the density of the oil?

SOLUTION ■ One **Key Idea** here is that the pressure P^{int} at the oil–water interface in the left arm depends on the density ρ_x and height of the oil above the interface. A second **Key Idea** is

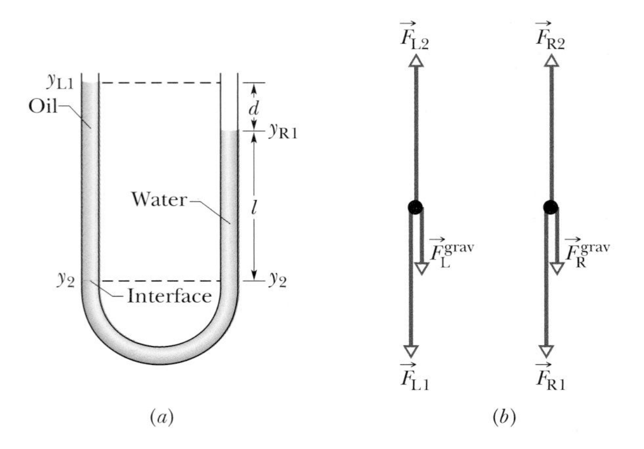

FIGURE 15-10 ■ (*a*) The oil in the left arm stands higher than the water in the right arm because the oil is less dense than the water. Both fluid columns produce the same pressure P^{int} at the level of the interface. (*b*) Free-body diagrams showing the forces on the parcel of oil on the left and the forces on the parcel of water above the interface on the right.

that the water in the right arm *at the same level* must be at the same pressure P^{int}. The reason is that, because the water is in static equilibrium, pressures at points in the water at the same level must be the same even if the points are separated horizontally.

In the right arm, the interface is a distance l below the free surface of the *water* and we have, from Eq. 15-10, $P_2 = P_1 - \rho_{\text{water}}\, g\, \Delta y$ where $P_2 = P^{\text{int}}$ and $\Delta y = y_2 - y_1 = -l$, so

$$P^{\text{int}} = P^{\text{atm}} - \rho_{\text{water}}\, g(-l) \quad \text{(right arm)}.$$

In the left arm, the interface is a distance $l + d$ below the free surface of the *oil* and we have, again from Eq. 15-10,

$$P^{\text{int}} = P^{\text{atm}} - \rho_x g[-(l + d)] \quad \text{(left arm)}.$$

Equating these two expressions and solving for the unknown density yield

$$\rho_x = \rho_{\text{water}}\left(\frac{l}{l+d}\right) = (998\ \text{kg/m}^3)\,\frac{135\ \text{mm}}{135\ \text{mm} + 12.3\ \text{mm}}$$

$$= 915\ \text{kg/m}^3 \quad \text{(Answer)}$$

Note that the answer does not depend on the atmospheric pressure P^{atm} or the local gravitational strength g.

15-5 Measuring Pressure

Electronic Pressure Sensors

One of the most popular methods for measuring absolute pressure in a gas is to use an electronic sensor. Many electronic sensors work in much the same way as the test sensor shown in Fig. 15-3. A common electronic pressure sensor has a flexible membrane with a vacuum chamber on one side of it, examples of which are shown in Fig. 15-11. The flexing of the membrane under pressure is sensed electronically. These devices are used in recording barometers found in weather stations and in physics laboratories.

FIGURE 15-11 ■ Two popular gas pressure sensors used in contemporary physics laboratories can record between 0 atm and about 7 atm of pressure. (*a*) Vernier pressure sensor. (*b*) PASCO pressure sensor. (Photos used with permission of Vernier Software and Technology and PASCO scientific.)

Typically an electronic gas sensor will be damaged when immersed in a liquid. However, when air-filled tubing connected to a gas sensor is immersed in a liquid, the pressure at various depths in the liquid can also be measured.

The Mercury Barometer

For historical and practical reasons, the *mercury barometer* and the open tube *manometer* are still popular methods for measuring atmospheric pressure and pressures near atmospheric pressure.

Figure 15-12*a* shows a very basic *mercury barometer,* a device used to measure the pressure of the atmosphere. The long glass tube is filled with mercury and inverted with its open end in a dish of mercury, as the figure shows. The space above the mercury column contains only mercury vapor, whose pressure is so small at ordinary temperatures that it can be neglected.

We can use Eq. 15-10 to find the local atmospheric pressure P^{atm} in terms of the height $\Delta y = y_2 - y_1$ of the mercury column. Since the chemical symbol for mercury is

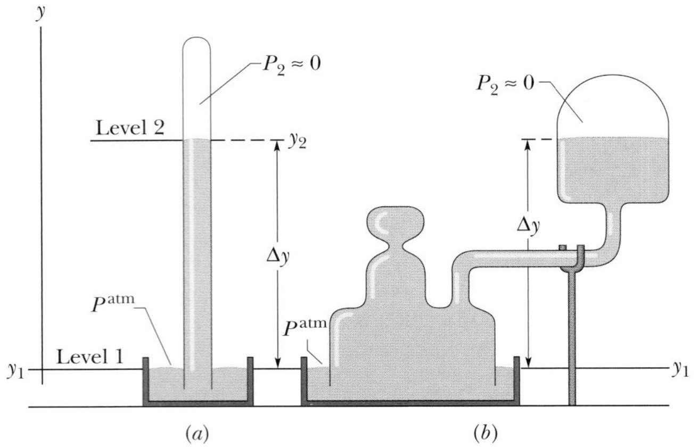

FIGURE 15-12 ■ (*a*) A mercury barometer. (*b*) Another mercury barometer. The difference Δy between liquid levels is the same in both cases.

Hg, we denote the density of the mercury by ρ_{Hg}. If we choose level 1 of Fig. 15-12*a* to be that of the air–mercury interface and level 2 to be that of the top of the mercury column, as labeled in Fig. 15-12*a*, we can substitute

$$P_1 = P^{atm}, \quad P_2 = 0, \quad \text{and} \quad \rho = \rho_{Hg},$$

into Eq. 15-10 to get

$$P^{atm} = \rho_{Hg} g \, \Delta y \tag{15-11}$$

where $\Delta y = y_2 - y_1$ is positive.

For a given pressure, the height Δy of the mercury column does not depend on the cross-sectional area of the vertical tube. The fanciful mercury barometer of Fig. 15-12*b* gives the same reading as that of Fig. 15-12*a*; all that counts is the vertical distance Δy between the mercury levels.

Equation 15-11 shows that, for a given pressure, the height of the column of mercury depends on the value of the local gravitational constant g at the location of the barometer and on the density of mercury, which varies only slightly with temperature. The column height (in millimeters) is numerically equal to the pressure (in torr) *only* if the barometer is at a place where g has its accepted average value of $+9.80665$ N/kg *and* the temperature of the mercury is 0°C. If these conditions do not prevail (and they rarely do), small corrections must be made before the height of the mercury column can be transformed into a pressure.

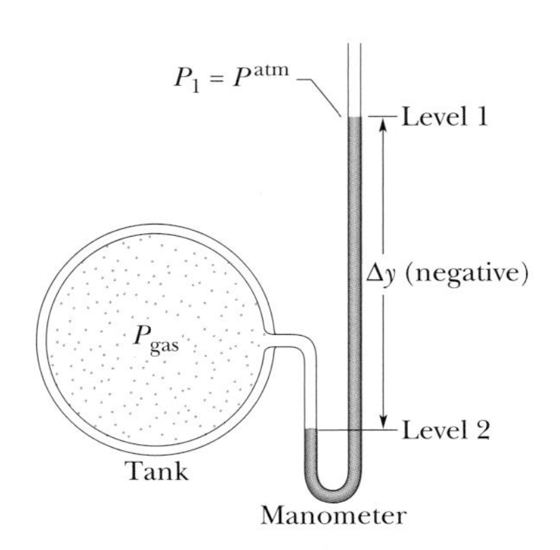

FIGURE 15-13 ■ An open-tube manometer, connected to measure the pressure of the gas in the tank on the left. The right arm of the U-tube is open to the atmosphere.

The Open-Tube Manometer

An *open-tube manometer* (Fig. 15-13) measures the pressure P_{gas} of a gas. It consists of a U-tube containing a liquid, with one end of the tube connected to the vessel whose pressure we wish to measure and the other end open to the atmosphere. Looking at the figure, we see that the "U" of fluid below the line marked "Level 2" is being pushed on the left by the force from the pressure of the gas in the tank and is being pushed on the right by the force arising from the pressure built up by everything above it, including the weight of the column between levels 1 and 2 and the pressure of the atmosphere. When the column is in equilibrium (no rising or falling), these forces must balance. Keeping this in mind, we can use Eq. 15-10 to find the pressure in terms of the distance from level 1 to level 2, Δy, shown in Fig. 15-13. Let us choose levels 1 and 2 as shown in Fig. 15-13. We then substitute

$$P_1 = P^{atm} \quad \text{and} \quad P_2 = P_{gas}$$

into Eq. 15-9, finding that

$$P_{gas} = P^{atm} - \rho g \, \Delta y, \tag{15-12}$$

where ρ is the density of the liquid in the tube.

In Fig. 15-13, the Δy factor is negative since level 2 is lower than level 1. So this figure depicts a gas pressure that is greater than the atmospheric pressure. Inflated tires and the human circulatory network are examples of systems with pressures that are greater than atmospheric. As you can see from Fig. 15-13 if level 2 were above level 1, Δy would be positive. In this case the gas pressure calculated using Eq. 15-12 would be less than the atmospheric pressure. For example, when you suck on a straw to pull fluid up the straw, the (absolute) pressure in your lungs is actually less than atmospheric pressure.

Gauge Pressure

Often when measuring pressure we are interested in knowing the difference between the pressure at some point such as level 2 in Fig. 15-13, which we call the **absolute pressure,** and a reference pressure P_1, which is taken to be the atmospheric pressure. In general, the difference between an absolute pressure and an atmospheric pressure is called the **gauge pressure.** (The name comes from the use of a gauge to measure this difference in pressures.) For example, most tire pressure devices measure gauge pressure.

The gauge pressure for water or another liquid used in a barometer or open-tube manometer is given by Eq. 15-8, which is $\Delta P = P_2 - P_1 = -\rho g \, \Delta y$. Since the gauge pressure is simply ΔP in the case where $P_1 = P^{atm}$, we can write

$$P^{gauge} = P_2 - P_1 = P_2 - P^{atm} = -\rho g \, \Delta y. \tag{15-13}$$

15-6 Pascal's Principle

When you push down the plunger in the bottle shown in Fig. 15-2 and the cork pops out, you are watching **Pascal's principle** in action. This principle is also the reason why toothpaste comes out of the open end of a tube when you squeeze on the other end, and it is the basis of the Heimlich maneuver. In that maneuver a sharp pressure increase properly applied to the abdomen is transmitted to the throat, forcefully ejecting food lodged there. Pascal's principle was first stated clearly in 1652 by Blaise Pascal (for whom the unit of pressure is named):

> A change in the pressure applied to an enclosed fluid is transmitted undiminished to every portion of the fluid and to the walls of its container.

This is just what we explained at the start of our discussion of pressure when we applied a force to the piston in Fig. 15-2. A cork inserted in the side of a bottle pops out no matter where it is located.

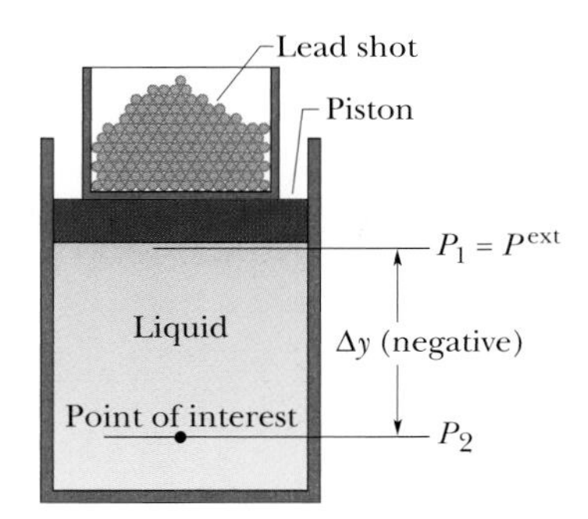

FIGURE 15-14 ■ The atmosphere and container with lead shot (small balls of lead) combine to create a pressure P^{ext} at the top of the enclosed (incompressible) liquid. If P^{ext} is increased, by adding more lead shot, the pressure increases by the same amount at all points within the liquid.

A Mathematical "Proof" of Pascal's Principle

Consider the case in which the incompressible fluid is a liquid contained in a tall cylinder, as in Fig. 15-14. The cylinder is fitted with a piston on which a container of lead shot rests. The atmosphere, container, and shot put pressure P^{ext} on the piston and thus on the liquid. The pressure P at any point in the liquid a distance Δy below the piston is then

$$P_2 = P_1 - \rho g \, \Delta y = P^{ext} - \rho g \, \Delta y, \tag{15-14}$$

where, as usual, when level 2 is below level 1, Δy is negative. Let us add a little more lead shot to the container to increase P^{ext} by an amount ΔP^{ext}. Since the fluid is assumed to be incompressible, the quantities ρ, g, and Δy in Eq. 15-14 are unchanged. Thus the pressure change at any point is

$$\Delta P = \Delta P^{\text{ext}}. \tag{15-15}$$

This pressure *change* is independent of Δy, so it must hold for all points within the liquid, as Pascal's principle states.

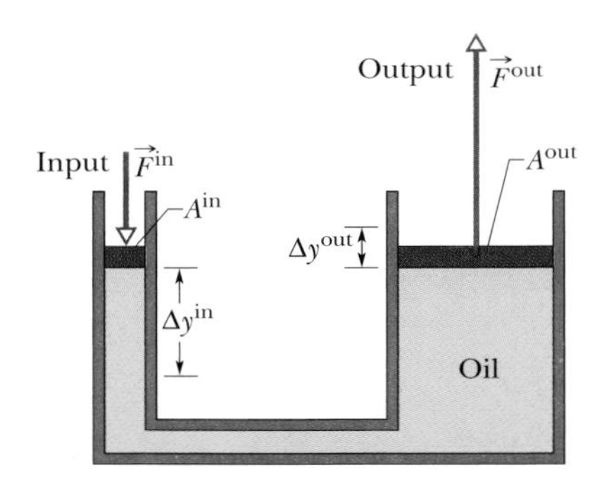

FIGURE 15-15 ■ A hydraulic arrangement that can be used to magnify a force $\vec{F}^{\text{in}}$. The work done is, however, not magnified and is the same for both the input and output forces.

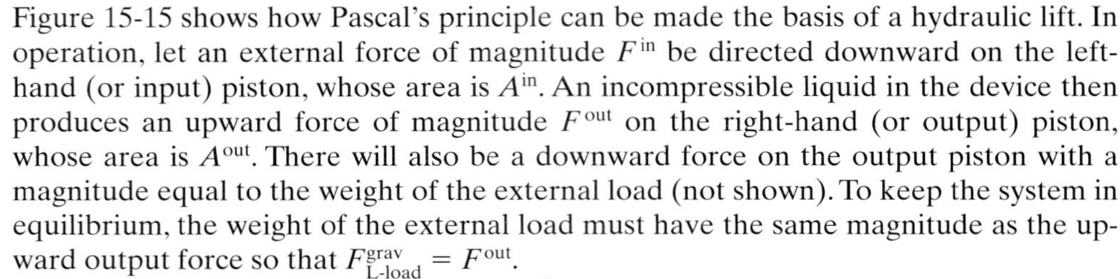

Pascal's Principle and the Hydraulic Lift

Figure 15-15 shows how Pascal's principle can be made the basis of a hydraulic lift. In operation, let an external force of magnitude F^{in} be directed downward on the left-hand (or input) piston, whose area is A^{in}. An incompressible liquid in the device then produces an upward force of magnitude F^{out} on the right-hand (or output) piston, whose area is A^{out}. There will also be a downward force on the output piston with a magnitude equal to the weight of the external load (not shown). To keep the system in equilibrium, the weight of the external load must have the same magnitude as the upward output force so that $F^{\text{grav}}_{\text{L-load}} = F^{\text{out}}$.

The magnitude of the input force F^{in} applied on the left and the magnitude of the downward force from the load on the right, $F^{\text{grav}}_{\text{R-load}}$, both serve to produce a change ΔP in the pressure of the liquid. Since $F^{\text{grav}}_{\text{R-load}} = F^{\text{out}}$, this pressure change is given by

$$\Delta P = \frac{F^{\text{in}}}{A^{\text{in}}} = \frac{F^{\text{out}}}{A^{\text{out}}},$$

so

$$F^{\text{out}} = F^{\text{in}}\frac{A^{\text{out}}}{A^{\text{in}}}. \tag{15-16}$$

Equation 15-16 shows that the magnitude of the output force F^{out} on the load must be greater than the magnitude of the input force F^{in} if $A^{\text{out}} > A^{\text{in}}$, as is the case in Fig. 15-15.

If we move the input piston downward a distance equal to the magnitude of Δy^{in}, the output piston moves upward a distance equal to the magnitude of Δy^{out}, such that the same volume V of the incompressible liquid is displaced at both pistons. Then

$$V = A^{\text{in}}|\Delta y^{\text{in}}| = A^{\text{out}}|\Delta y^{\text{out}}|,$$

which we can write as

$$A^{\text{out}} = A^{\text{out}}\frac{|\Delta y^{\text{in}}|}{|\Delta y^{\text{out}}|}. \tag{15-17}$$

This shows that, if $A^{\text{out}} > A^{\text{in}}$ (as in Fig. 15-15), the output piston moves a smaller distance than the input piston moves.

By combining Eqs. 15-16 and 15-17 and noting that the displacements on both the input and output sides of the lift are in the *same* direction as the forces that cause them, we get the following expression relating work out to work in,

$$W^{\text{out}} = F^{\text{out}}|\Delta y^{\text{out}}| = \left(F^{\text{out}}\frac{A^{\text{out}}}{A^{\text{in}}}\right)\left(|\Delta y^{\text{in}}|\frac{A^{\text{in}}}{A^{\text{out}}}\right) = F^{\text{in}}|\Delta y^{\text{in}}| = W^{\text{in}}. \tag{15-18}$$

This equation shows that in a hydraulic lift, the work W^{in} done *on* the input piston by the applied force should be equal to the work W^{out} done on the load by the output piston in lifting it.

The advantage of a hydraulic lever is this:

> With a hydraulic lift, a given force applied over a given distance can be transformed to a greater force applied over a smaller distance.

The product of force and distance remains unchanged so that the same work is done. A small version of the hydraulic lift we have described is the jack used to change automobile tires. Most of us, for example, cannot lift an automobile directly but can with a hydraulic jack, even though we have to pump the handle farther than the automobile rises. In this device, the displacement is accomplished not in a single stroke but over a series of small strokes.

READING EXERCISE 15-6: Consider the cylinder of a real hydraulic jack filled with oil. The oil is slightly compressible. Discuss how the relations presented in this section are affected by the compressibility of the oil. For example, is the work put in still equal to the work out? If not, which is larger? What (if any) are the energy transformations that take place? ■

READING EXERCISE 15-7: The pressure on the bottom of a container with sloping walls is determined by the height Δy of the central column. Relate this observation to the concepts addressed in this section.

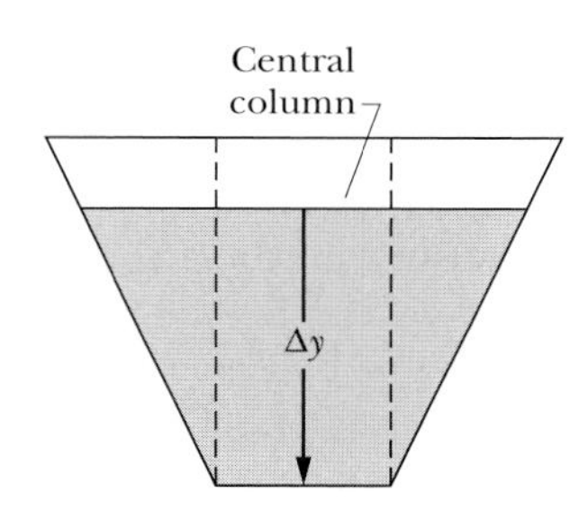

■

TOUCHSTONE EXAMPLE 15-4: Car Lift

You are a new employee at an auto repair shop. The exhaust system on a car you are repairing is leaking. Since you don't want to crawl under the car to work on the problem, you tell your boss that you'll use Pascal's principle to design a hydraulic lift that anyone can use to lift cars by hand that weigh up to 2000 kg using a lifting post with a 50 cm diameter. Describe your design.

SOLUTION ■ The **Key Ideas** here are these: (1) From Pascal's principle you know that the pressure is constant at the same level everywhere in a container of fluid, and (2) since $F = PA$ if two ends of a container have different cross-sectional areas, then the force exerted at the end with the small area can be much less than the force exerted at the end with the much larger area. Basically a hydraulic lift like that shown in Fig. 15-16 is a device that enables you to exert a small force over a large distance on one side of a container of fluid and to transmit a large force acting over a small distance on the other side.

Assume the lift is in equilibrium and the level of the liquid on the input and output of the hydraulic lift shown in Fig. 15-16 is the same. Then $P^{\text{in}} = P^{\text{out}}$ where $P^{\text{in}} = F^{\text{in}}/A^{\text{in}}$ and $P^{\text{out}} = F^{\text{out}}/A^{\text{out}}$, so

$$F^{\text{out}} = F^{\text{in}}\frac{A^{\text{out}}}{A^{\text{in}}}, \tag{15-19}$$

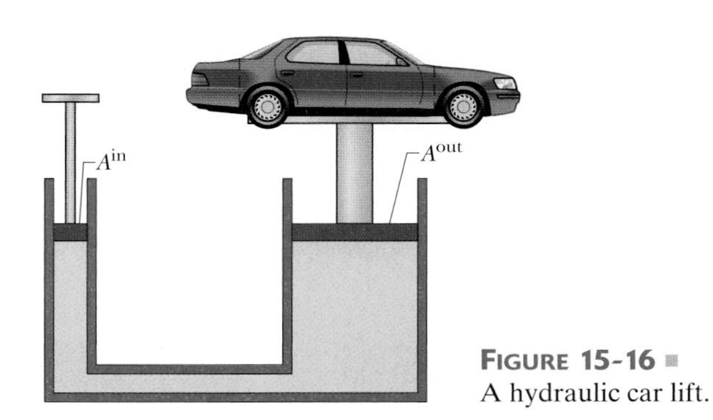

FIGURE 15-16 ■ A hydraulic car lift.

where F^{in} and F^{out} are the magnitudes of the forces pressing down on the pistons at the two ends of the hydraulic lift. You know that the magnitude of the force F^{out} needed to hold a car up is given by

$$F^{\text{out}} = F^{\text{grav}}_{\text{car}} = mg = 2000\text{ kg} \times 9.8\text{ m/s} = 2.0 \times 10^4\text{ N}.$$

Most people can manage to lift a 20 kg mass without too much difficulty, so you assume that the input force you exert is given by

$$F^{\text{in}} = 20\text{ kg} \times 9.8\text{ m/s} = 2.0 \times 10^2\text{ N},$$

which is 100 times less than the force you need to exert on a car to hold it up. So we can rewrite Eq. 15-19 to get

$$\frac{A^{out}}{A^{in}} = \frac{F^{out}}{F^{in}} = \frac{2.0 \times 10^4 \text{ N}}{2.0 \times 10^2 \text{ N}} = 100.$$

Using the fact that $A = \pi r^2$ and that $A^{out} = \pi r_{out}^2 = \pi(50 \text{ cm}/2)^2$, we can rearrange Eq. 15-19 to calculate the size of the plunger you can use in the input side to exert the force needed to hold the car up,

$$\frac{A^{out}}{A^{in}} = \frac{\pi r_{out}^2}{\pi r_{in}^2} = 100$$

so that

$$r_{in} = \sqrt{\frac{r_{out}^2}{100}} = \frac{r_{out}}{10} = \frac{50 \text{ cm}/2}{10} = 2.5 \text{ cm}.$$

Thus, in your design, you construct a column of diameter 2×2.5 cm $= 5.0$ cm at the input end and push a piston into it with a force of just over

$$F^{in} = 2.0 \times 10^2 \text{ N}. \quad \text{(Answer)}$$

15-7 Archimedes' Principle

FIGURE 15-17 ■ A thin-walled plastic sack of water is in static equilibrium in the pool. Its weight must be balanced by a net upward force on the water in the sack from the surrounding water.

Let us think about what happens when we immerse an object in a fluid. Your first guess might be that the weight of the water above the object would push it to the bottom. But think about what happens when we try to push a beach ball down under the surface in a swimming pool. It will be hard to push it down. If we do get it under the surface and release it, it will go shooting up into the air. What's going on here? It turns out that the critical issue is that pressure is a scalar. You can get forces in all directions—up as well as down. Let us see how this works.

Figure 15-17 shows a student in a swimming pool, manipulating a very thin plastic sack (of negligible mass) that is filled with water. She finds that the sack and its contained water are in static equilibrium, tending neither to rise nor to sink. The downward gravitational force $\vec{F}^{grav}$ on the contained water must be balanced by a net upward force from the water surrounding the sack.

This net upward force on an object in a fluid is called a **buoyant force** $\vec{F}^{buoy}$. It exists because the pressure in the surrounding water increases with depth below the surface. Thus, the pressure near the bottom of the sack is greater than the pressure near the top. Then the forces on the sack due to this pressure are greater in magnitude near the bottom of the sack than near the top. Some of the forces are represented in Fig. 15-18*a*, where the space occupied by the sack has been left empty. As you can see, the force vectors drawn near the bottom of that space (with upward components) have longer lengths than those drawn near the top of the sack (with downward components). If we vectorially add all the forces on the sack from the water, the horizontal components cancel and the vertical components add to yield the upward buoyant force $\vec{F}^{buoy}$ on the sack. (Force $\vec{F}^{buoy}$ is shown to the right of the pool in Fig. 15-18*a*.)

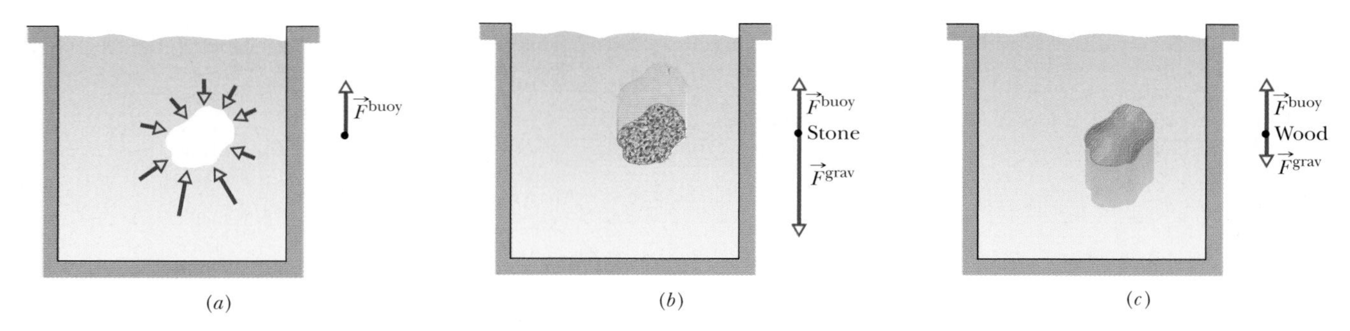

FIGURE 15-18 ■ (*a*) The water surrounding the hole in the water produces a net upward buoyant force on whatever fills the hole. (*b*) For a stone of the same volume as the hole, the gravitational force exceeds the buoyant force in magnitude. (*c*) For a lump of wood of the same volume, the gravitational force is less than the buoyant force in magnitude.

Let's denote the mass of the fluid in the sack of water as m_f. In static equilibrium, the buoyant force magnitude F^{buoy} is equal to the gravitational force magnitude $F^{\text{grav}} = m_f g$ experienced by the sack of water so that

$$F^{\text{buoy}} = m_f g. \tag{15-20}$$

In short, the magnitude of the buoyant force is equal to the weight of the water in the sack.

In Fig. 15-18*b*, we have replaced the sack of water with a stone that exactly fills the hole in Fig. 15-18*a*. The stone is said to *displace* the water, meaning that it occupies space that would otherwise be occupied by water. The object pushes the water out of the way. We have changed nothing about the shape of the hole, so the forces at the hole's surface must be the same as when the water-filled sack was in place. Thus, the same upward buoyant force that acted on the water-filled sack now acts on the stone. That is, the magnitude F^{buoy} of the buoyant force is still equal to $m_f g$, the weight of the water displaced by the stone.

Unlike the water-filled sack, the stone is not in static equilibrium. Because the stone has a higher density than water, the magnitude of the downward gravitational force F^{grav} on the stone is greater in magnitude than that of the upward buoyant force. This is shown in the free-body diagram to the right of the pool in Fig. 15-18*b*. The stone thus accelerates downward, sinking to the bottom of the pool. Once on the bottom, the normal force from the floor balances $\vec{F}^{\text{grav}}$ and the stone stops moving.

Let us next exactly fill the hole in Fig. 15-18*a* with a block of low-density wood, as in Fig. 15-18*c*. Again, nothing has changed about the forces at the hole's surface, so the magnitude F^{buoy} of the buoyant force is still equal to $m_f g$, the weight of the displaced water. Like the stone, the block is not in static equilibrium. However, this time the magnitude of the gravitational force F^{grav} is less than the buoyant force (as shown to the right of the pool), and so the block accelerates upward, rising to the top surface of the water.

Our results with the sack, stone, and block apply to all fluids and are summarized in **Archimedes' principle:**

> When a body is fully or partially submerged in a fluid, a buoyant force $\vec{F}^{\text{buoy}}$ from the surrounding fluid acts on the body. The buoyant force is directed upward and has a magnitude equal to the weight $m_f g$ of the fluid that has been displaced by the body.

The buoyant force on a body in a fluid has the magnitude

$$F^{\text{buoy}} = m_f g \quad \text{(buoyant force magnitude)}, \tag{15-21}$$

where m_f is the mass of the fluid that is displaced by the body.

In the late evening of August 21, 1986, something (possibly a volcanic tremor) disturbed Cameroon's Lake Nyos, which has a high concentration of dissolved carbon dioxide. The disturbance caused that gas to form bubbles. Being less dense than the surrounding fluid (the water), those bubbles were buoyed to the surface, where they released the carbon dioxide. The gas, being more dense than the surrounding fluid (now the air), rushed down the mountainside like a river, asphyxiating 1700 persons and the scores of animals seen here.

Floating

When we release a block of lightweight wood just above the water in a pool, it moves into the water because the gravitational force on it pulls it downward. As the block displaces more and more water, the magnitude F^{buoy} of the upward buoyant force acting on it increases. Eventually, F^{buoy} is large enough to equal the magnitude F^{grav} of the downward gravitational force on the block, and the block comes to rest. The block is then in static equilibrium and is said to be *floating* in the water. In general,

> When a body floats in a fluid, the magnitude F^{buoy} of the buoyant force on the body is equal to the magnitude F^{grav} of the gravitational force on the body.

We can write this statement in terms of the magnitude of the forces as

$$F^{\text{buoy}} = F^{\text{grav}}. \tag{15-22}$$

From Eq. 15-21, we know that $F^{\text{buoy}} = m_f g$. Thus,

> When a body floats in a fluid, the magnitude F^{grav} of the gravitational force on the body is equal to the weight $m_f g$ of the fluid that has been displaced by the body.

We can write this statement

$$F^{\text{grav}} = m_f g \quad \text{(floating condition).} \tag{15-23}$$

In thinking about whether an object will sink or float, density is the key consideration, *not* the total mass of the object. If the density is less than that of water, the object can displace a mass of water equal to its own weight by being only partially submerged. If the density is greater than that of water, even completely immersing the object produces a buoyant force that is less than the object's weight. In that case, part of the object's weight will be unbalanced; there will be a net downward force and the object will sink to the bottom.

Apparent Weight in a Fluid

If we place a stone on a scale that is calibrated to measure weight, then the reading on the scale is the stone's weight. However, if we do this underwater, the upward buoyant force on the stone from the water decreases the reading. That reading is then an apparent weight. In general, an apparent weight or net downward force on the body is related to the actual weight of a body and the buoyant force on the body by

$$\text{(apparent weight)} = \text{(actual weight)} - \text{(magnitude of buoyant force)},$$

which we can write as

$$F^{\text{app}} = F^{\text{grav}} - F^{\text{buoy}} \quad \text{(apparent weight).} \tag{15-24}$$

If, in some strange test of strength, you had to lift a heavy stone, you could do it more easily with the stone underwater. Then your applied force would need to exceed only the stone's apparent weight, not its larger actual weight, because the upward buoyant force would help you lift the stone.

The magnitude of the buoyant force on a floating body is equal to the body's weight. Equation 15-24 thus tells us that a floating body has an apparent weight of zero—the body would produce a reading of zero on a scale. (When astronauts prepare to perform a complex task in space, they practice the task floating underwater, where their apparent weight is zero as it is in space.)

READING EXERCISE 15-8: A penguin floats first in fluid A of density ρ_A, then in fluid B of density $\rho_B = 0.95\rho_A$, and then in fluid C of density $\rho_C = 1.10\rho_A$. (a) Rank the fluids according to the magnitude of the buoyant force on the penguin, greatest first. (b) Rank the fluids according to the amount of fluid displaced by the penguin, greatest first. ■

READING EXERCISE 15-9: Each year student teams and colleges and universities design, build, and race full-size canoes made entirely of reinforced concrete. In the figure, a group of University of Maryland boat designers show off their award-winning concrete canoe. How can you get a concrete canoe to float? ■

TOUCHSTONE EXAMPLE 15-5: Iceberg

What fraction of the volume of an iceberg floating in seawater is visible?

SOLUTION ■ Let V_{iceberg} be the total volume of the iceberg. The nonvisible portion is below water and thus is equal to the volume V_{fluid} of the fluid (the seawater) displaced by the iceberg. We seek the fraction (call it frac)

$$\text{frac} = \frac{V_{\text{iceberg}} - V_{\text{fluid}}}{V_{\text{iceberg}}} = 1 - \frac{V_{\text{fluid}}}{V_{\text{iceberg}}}, \tag{15-25}$$

but we know neither volume. A **Key Idea** here is that, because the iceberg is floating, Eq. 15-23 ($F^{\text{grav}} = m_{\text{fluid}}\, g$) applies. We can write that equation as

$$m_{\text{iceberg}}\, g = m_{\text{fluid}}\, g,$$

from which we see that $m_{\text{iceberg}} = m_{\text{fluid}}$. Thus, the mass of the iceberg is equal to the mass of the displaced fluid (seawater). Although we know neither mass, we can relate them to the densities of ice and seawater given in Table 15-2 by using Eq. 15-1 ($\rho = m/V$). Because $m_{\text{iceberg}} = m_{\text{fluid}}$, we can write

$$\rho_{\text{iceberg}} V_{\text{iceberg}} = \rho_{\text{fluid}} V_{\text{fluid}},$$

or

$$\frac{V_{\text{fluid}}}{V_{\text{iceberg}}} = \frac{\rho_{\text{iceberg}}}{\rho_{\text{fluid}}}.$$

Substituting this into Eq. 15-25 and then using the known densities, we find

$$\text{frac} = 1 - \frac{\rho_{\text{iceberg}}}{\rho_{\text{fluid}}} = 1 - \frac{917\ \text{kg/m}^3}{1024\ \text{kg/m}^3}$$

$$= 0.10 \text{ or } 10\%. \qquad \text{(Answer)}$$

The fact that only 10% of an iceberg can be seen above water is the source of a common expression: "That's only the tip of the iceberg."

TOUCHSTONE EXAMPLE 15-6: Water-Filled Sack

Consider the thin-walled sack full of water suspended in a swimming pool shown in Fig. 15-17. The atmospheric pressure at the surface of the pool is the mean sea level pressure. Assume that the bottom of the sack is 60 cm below the top of the sack.

(a) If the top of the sack is 40 cm below the surface of the pool, what is the pressure at the center of the sack?

SOLUTION ■ To determine the pressure at the center of the sack we need to find how far it is below the water surface. The middle of a sack with a top to bottom distance of 60 cm is 60 cm/2 = 30 cm below the top of the sack. If the top of the sack is 40 cm below the surface, then the middle of the sack is at a depth given by $\Delta y = -40\text{ cm} - 30\text{ cm} = -70\text{ cm}$. Assuming that the water is incompressible, we can use Eq. 15-10 to find the pressure, P_2, at the center of the sack,

$$P_2 = P_1 - \rho_{\text{water}}\, g\, \Delta y = P^{\text{atm}} - \rho_{\text{water}}\, g\, \Delta y$$

$$= 101\,325\ \text{Pa} - (998\ \text{kg/m}^3)(9.8\ \text{N/kg})(-0.70\ \text{m})$$

$$= 108.2\ \text{Pa}. \qquad \text{(Answer)}$$

(b) If the water inside the sack has a mass of 100 kg, what is the buoyant force on the sack?

SOLUTION ■ The **Key Idea** here is that according to Archimedes' principle, the buoyant force has the same magnitude as the weight of the pool water *displaced* by the sack of water. Since the plastic sack is thin with essentially no mass, the weight of the sack with the water in it is the same as the weight of the water the sack displaced. We can use Eq. 6-5 to find the weight of the displaced water,

$$\text{Weight} = |\vec{F}^{\,\text{grav}}| = mg = (100\,\text{kg})(9.80\,\text{m/s}^2) = 980\,\text{N}.$$

Since the buoyant force acts upward, using Archimedes' principle (Eq. 15-21),

$$\vec{F}^{\,\text{buoy}} = +|\vec{F}^{\,\text{grav}}|\hat{\text{j}} = (980\,\text{N})\hat{\text{j}}. \qquad \text{(Answer)}$$

(c) If the swimmer gently pulls the sack down deeper into the pool, what happens to the pressure at the center of the sack?

SOLUTION ■ The **Key Idea** here is that according to Eq. 15-10, the pressure at the center of the sack (and in fact at all locations in the sack) increases as the depth of the sack increases.

(d) What happens to the buoyant force on the sack as it is pulled down further?

SOLUTION ■ The **Key Idea** here is that according to Archimedes' principle, the buoyant force must remain the same as the weight of the water displaced. But the displaced weight doesn't change with depth (to the extent that the local gravitational field strength g does not change with depth).

(e) If the buoyant force is caused by the pressure of water on the sack and the pressure increases with depth, why doesn't the buoyant force on the water in the sack *increase*?

SOLUTION ■ The **Key Idea** here is that the buoyant force on the water in the sack is a *net force* that depends on the vector sum of the upward forces on the bottom surface elements of the sack due to the pool's water pressure, and the downward forces on the top surface elements of the sack also due to the pool's water pressure. Ultimately the net force is related to *pressure differences* between the bottom and top sack elements. Since pressure increases linearly with depth, the *pressure differences do not* depend on depth.

15-8 Ideal Fluids in Motion

FIGURE 15-19 ■ At a certain point, the rising flow of smoke and heated gas changes from steady to turbulent.

The motion of *real fluids* is very complicated and not yet fully understood. Instead, we shall discuss the motion of an **ideal fluid,** which is simpler to handle mathematically and yet provides useful results. Here are four assumptions that we make about our ideal fluid, all concerned with flow:

1. ***Steady flow*** In *steady* (or *laminar*) *flow,* the velocity of the moving fluid at any fixed point does not change with time, either in magnitude or in direction. The gentle flow of water near the center of a quiet stream is steady; that in a chain of rapids is not. Figure 15-19 shows a transition from steady flow to *nonsteady* (or *turbulent*) flow for a rising stream of smoke. The speed of the smoke particles increases as they rise and, at a certain critical speed, the flow changes from steady to nonsteady (that is, from laminar to *nonlaminar* flow).
2. ***Incompressible flow*** We assume, as we have already done for fluids at rest, that our ideal fluid is incompressible; that is, its density has a constant, uniform value.
3. ***Nonviscous flow*** Roughly speaking, the viscosity of a fluid is a measure of how resistive the fluid is to flow. For example, thick honey is more resistive to flow than water, and so honey is said to be more viscous than water. Viscosity is the fluid analog of friction between solids; both are mechanisms by which the kinetic energy of moving objects can be transferred to thermal energy. In the absence of friction, a block could glide at constant speed along a horizontal surface. In the same way, an object moving through a nonviscous fluid would experience no *viscous drag force*—that is, no resistive force due to viscosity; it could move at constant speed through the fluid. The British scientist Lord Rayleigh noted that in an ideal fluid a ship's propeller would not work but, on the other hand, a ship (once set into motion) would not need a propeller!

We can make the flow of a fluid visible by adding a *tracer.* This might be a dye injected into many points across a liquid stream (Fig. 15-20) or smoke particles added to

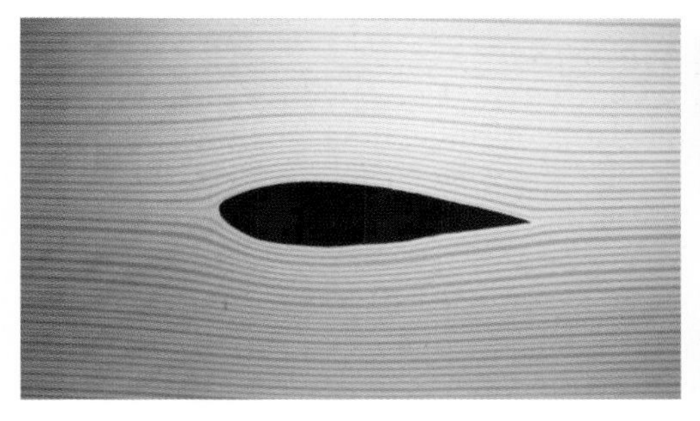

FIGURE 15-20 ■ Streamlined flow around an airfoil.

FIGURE 15-21 ■ Smoke reveals streamlines in airflow past a car in a wind-tunnel test.

a gas flow (Figs. 15-19 and 15-21). Each bit of a tracer follows a *streamline,* which is the path that a tiny element of the fluid would take as the fluid flows. Recall from Chapter 2 that the velocity of a particle is always tangent to the path taken by the particle. Here the particle is the fluid element, and its velocity $\vec{v}$ is always tangent to a streamline (Fig. 15-22). For this reason, two streamlines cannot intersect with a finite fluid velocity, since a fluid element cannot flow in two directions simultaneously. However, at a point of zero velocity, a stagnation point, there is no direction defined and two or more streamlines may intersect at a stagnation point.

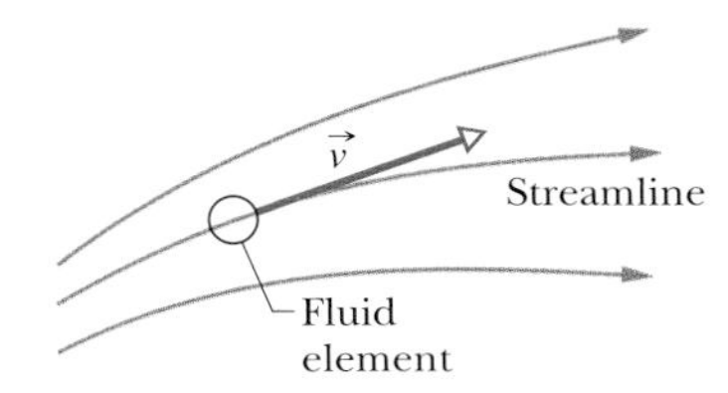

FIGURE 15-22 ■ A fluid element traces out a streamline as it moves. The velocity vector of the element is tangent to the streamline at every point.

15-9 The Equation of Continuity

You may have noticed that you can increase the speed of the water emerging from a garden hose by partially closing the hose opening with your thumb. Apparently the speed $|\vec{v}|$ (often denoted simply as v) of the water depends on the cross-sectional area A through which the water flows. This makes sense when we realize that the faucet is putting water out at a certain rate. If that much water has to get through a smaller hole, it has to go faster.

Here we wish to derive an expression that relates the speed of a fluid v to the cross-sectional area A of the pipe through which it flows. Let's consider the steady flow of an ideal fluid through a pipe with varying cross section, like that in Fig. 15-23. The flow is toward the right, and the pipe segment shown (part of a longer pipe) has length L. The fluid has speeds v_1 at the left end of the segment and v_2 at the right end. The pipe has cross-sectional areas A_1 at the left end and A_2 at the right end. Suppose that in a time interval Δt a volume ΔV of fluid enters the pipe segment at its left end (that volume is colored purple in Fig. 15-23*a*). Then, because the fluid is incompressible, an identical volume ΔV must emerge from the right end of the segment (it is colored green in Fig. 15-23*b*).

We can use this common volume ΔV to relate the speeds and areas. To do so, we first consider Fig. 15-24, which shows a side view of a pipe of *uniform* cross-sectional area A. In Fig. 15-24*a*, a fluid element e is about to pass through the dashed line drawn across the pipe width. The element's speed is v, so during a time interval Δt, the element moves along the pipe a distance $\Delta x = v\,\Delta t$. The volume ΔV of fluid that has passed through the dashed line in that time interval Δt is

$$\Delta V = A\,\Delta x = Av\,\Delta t. \tag{15-26}$$

Applying these ideas to both the left and right ends of the pipe segment in Fig. 15-23, we have

$$\Delta V = A_1 v_1\,\Delta t = A_2 v_2\,\Delta t$$

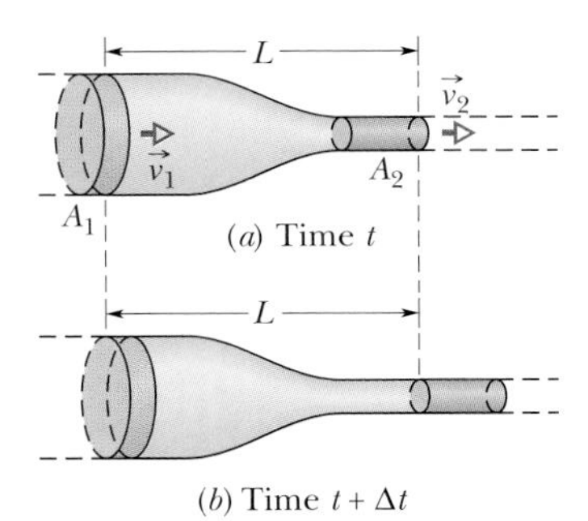

FIGURE 15-23 ■ Fluid flows from left to right at a steady rate through a tube segment of length L. The fluid's velocity is $\vec{v}_1$ at the left side and $\vec{v}_2$ at the right side. The tube's cross-sectional area is A_1 at the left side and A_2 at the right side. From time t in (*a*) to time $t + \Delta t$ in (*b*), the amount of fluid in purple enters at the left side and the equal amount of fluid shown in green emerges at the right side.

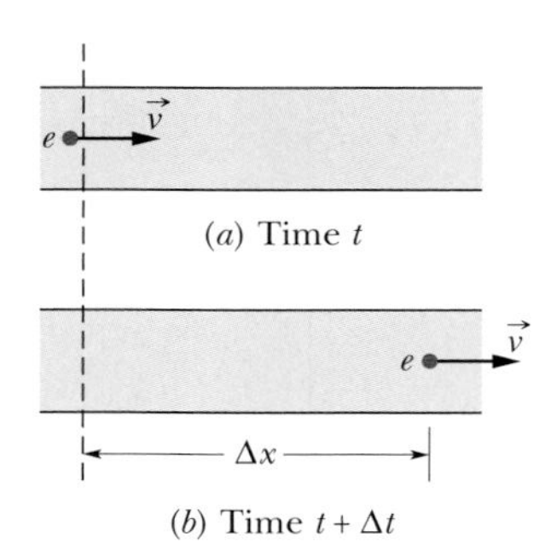

FIGURE 15-24 ■ Fluid flows at a constant velocity $\vec{v}$ through a tube. (*a*) At time *t*, fluid element *e* is about to pass the dashed line. (*b*) At time $t + \Delta t$, element *e* is a distance $\Delta x = v_x \Delta t$ from the dashed line where v_x is the *x*-component of $\vec{v}$.

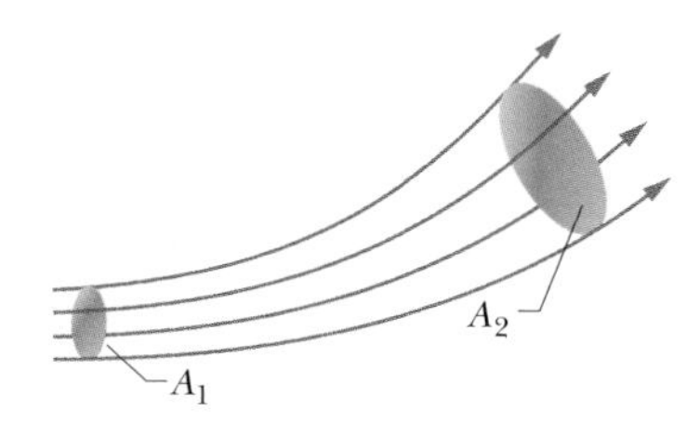

FIGURE 15-25 ■ A tube of flow is defined by the streamlines that form the boundary. The volume flow rate must be the same for all cross sections of the tube of flow.

or $$A_1v_1 = A_2v_2 \quad \text{(equation of continuity).} \tag{15-27}$$

This relation between speed and cross-sectional area is called the **equation of continuity** for the flow of an ideal fluid. It tells us that the flow speed increases when we decrease the cross-sectional area through which the fluid flows (as when we partially close off a garden hose with a thumb).

The equation of continuity (Eq. 15-27) applies not only to an actual pipe but also to any so-called *tube of flow,* or imaginary tube whose boundary consists of streamlines. Such a tube acts like a real pipe because no fluid element can cross a streamline. Thus, all the fluid within a tube of flow must remain within its boundary. Figure 15-25 shows a tube of flow in which the cross-sectional area increases from area A_1 to area A_2 along the flow direction. From Eq. 15-27 we know that, with the increase in area, the speed must decrease, as is indicated by the greater spacing between streamlines at the right in Fig. 15-25. Similarly, you can see that in Fig. 15-20 the speed of the flow is greatest just above and just below the cylinder.

We can rewrite the equation of continuity (Eq. 15-27) as

$$R_V = Av = \text{a constant} \quad \text{(volume flow rate, equation of continuity),} \tag{15-28}$$

in which R_V is the **volume flow rate** of the fluid (volume per unit time). Its SI unit is the cubic meter per second (m^3/s). If the density ρ of the fluid is uniform, we can multiply this expression (Eq. 15-28) by that density to get the **mass flow rate** R_m (mass per unit time):

$$R_m = \rho R_V = \rho Av = \text{a constant} \quad \text{(mass flow rate, the equation of continuity).} \tag{15-29}$$

The SI unit of mass flow rate is the kilogram per second (kg/s). Equation 15-29 says that the mass that flows into the pipe segment of Fig. 15-23 each second must be equal to the mass that flows out of that segment each second.

Engineers can use velocity measurements and the equation of continuity as a tool for determining volume flow rates of incompressible fluids such as oil flowing in a pipe of variable cross section or water flowing in a stream of variable cross section. But the equation of continuity (Eqs. 15-27 to 15-29) can only be used under certain conditions. In deriving it we assumed that fluid isn't being added to or subtracted from the system. For example, we assume that our oil pipe doesn't leak or that new water is not being added to a stream by a tributary as it flows along. In cases where fluid is added or subtracted from a tube of flow, we would find that $v_1A_1 \neq v_2A_2$.

Even when the continuity condition holds, whenever we use the equation of continuity we assume that we have a uniform flow of fluid over a cross-sectional area. In other words, we assume that the velocity of each of the elements of the fluid is moving at the same speed in a direction that is perpendicular to a cross-sectional area. In real pipes, fluid flows more slowly near the surface of the pipe than in the middle and it is not obvious how to calculate the product vA at a given location along our pipe.

In the next section we will introduce the mathematical concept of *flux* to help us deal with more realistic situations in which the simplified application of the equation of continuity cannot be used either because we do not have true continuity or because we do not have a nice uniform flow of fluid over a cross-sectional area.

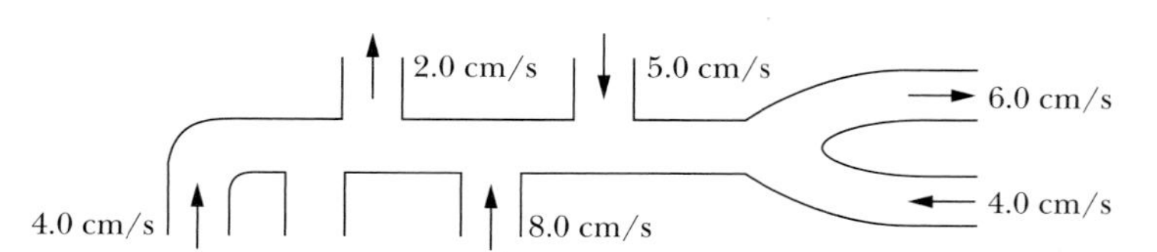

READING EXERCISE 15-10: The figure shows a pipe and gives the volume flow rate (in cm^3/s) and the direction of flow for all but one section. What are the volume flow rate and the direction of flow for that section?

■

TOUCHSTONE EXAMPLE 15-7: Blood Flow

The cross-sectional area A_1 of the aorta (the major blood vessel emerging from the heart) of a normal resting person is 3 cm^2, and the average speed v_1 of the blood is 30 cm/s. A typical capillary (diameter $\approx$ 6 μm) has a cross-sectional area A_2 of 3×10^{-7} cm^2 and an average flow speed v_2 of 0.05 cm/s. How many capillaries does such a person have?

SOLUTION ■ The **Key Idea** here is that all the blood that passes through the capillaries must have passed through the aorta. Therefore, the volume flow rate through the aorta must equal the total volume flow rate through the capillaries. Let us assume that the capillaries are identical, with the given cross-sectional area A_2 and average flow speed v_2. Then, from Eq. 15-27 we have

$$A_1 v_1 = n A_2 v_2,$$

where n is the number of capillaries. Solving for n yields

$$n = \frac{A_1 v_1}{A_2 v_2} = \frac{(3\ \text{cm}^2)(30\ \text{cm/s})}{(3 \times 10^{-7}\ \text{cm}^2)(0.05\ \text{cm/s})}$$

$$= 6 \times 10^9 \text{ or 6 billion.} \qquad \text{(Answer)}$$

You can easily show that the combined cross-sectional area of the capillaries is about 600 times the cross-sectional area of the aorta.

TOUCHSTONE EXAMPLE 15-8: Necking Down

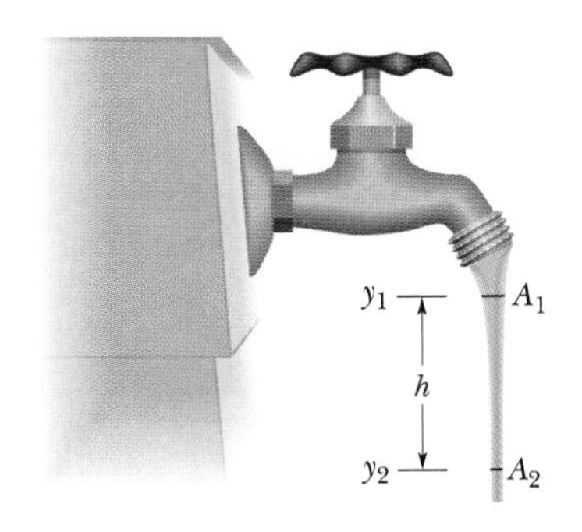

FIGURE 15-26 ■ As water falls from a tap, its speed increases. Because the flow rate must be the same at all cross-sections, the stream must "neck down."

Figure 15-26 shows how the stream of water emerging from a faucet "necks down" as it falls. The indicated cross-sectional areas are A_1 = 1.2 cm^2 and A_2 = 0.35 cm^2. The two levels are separated by a vertical distance h = 45 mm. What is the volume flow rate from the tap?

SOLUTION ■ The **Key Idea** here is simply that the volume flow rate through the higher cross section must be the same as that through the lower cross section. Thus, from Eq. 15-27, we have

$$A_1 v_1 = A_2 v_2, \qquad (15\text{-}30)$$

where v_1 and v_2 are the water speeds at the levels corresponding to A_1 and A_2. Adapting the last equation on page 43 to this situation we get

$$v_{2y}^2 = v_{1y}^2 + 2(-g)(y_2 - y_1), \qquad (15\text{-}31)$$

where $h = y_1 - y_2$. Eliminating v_2 between Eqs. 15-30 and 15-31 and solving for v_1, we obtain

$$v_1 = \sqrt{\frac{2ghA_2^2}{A_1^2 - A_2^2}}$$

$$= \sqrt{\frac{(2)(9.8\ \text{m/s}^2)(0.045\ \text{m})(0.35\ \text{cm}^2)^2}{(1.2\ \text{cm}^2)^2 - (0.35\ \text{cm}^2)^2}}$$

$$= 0.286\ \text{m/s} = 28.6\ \text{cm/s}.$$

From Eq. 15-28, the volume flow rate R_V is then

$$R_V = A_1 v_1 = (1.2\ \text{cm}^2)(28.6\ \text{cm/s})$$

$$= 34\ \text{cm}^3/\text{s}. \qquad \text{(Answer)}$$

15-10 Volume Flux

The term *volume flux* is a synonym for volume flow rate, R_V. In this section we consider a more careful mathematical definition of volume flux that will allow us to apply the equation of continuity to more realistic situations. The word "flux" comes from the Latin word meaning "to flow." Often the word "flux" is used in science and engineering to describe the rate of flow or penetration of matter or energy through a surface. However, later in this text we will introduce definitions of electric and magnetic flux. Even though the mathematical definitions of electric and magnetic flux are analogous to that of volume flow rate, it is surprising to find that nothing at all is flowing.

Volume Flux in a Stream

In this section we develop the concept of flux to calculate the rate at which water flows though a cross-sectional area of a stream. Suppose we have a stream of water undergoing steady laminar flow (as described in Sections 15-8 and 15-9). Assume all the water is traveling in the same direction through a wide shallow channel as shown in Fig. 15-27 but that the elements of water closer to the stream bottom and sides are moving more slowly than the water in the top central part of the stream. We cannot simply find a single velocity, $\vec{v}$, at all points along our cross-sectional area.

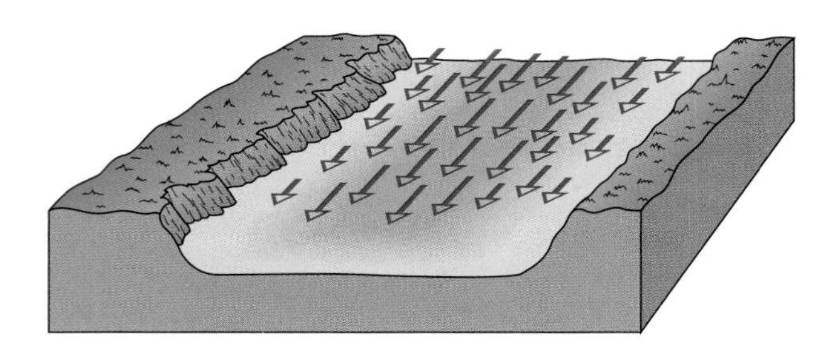

FIGURE 15-27 ■ Water moving in laminar flow through a wide shallow channel. The darker regions in the diagram with longer velocity vectors indicate areas farther from the sides and bottom of the stream where the water encounters less drag and hence flows faster. The flow can be described by a vector field consisting of the velocity vectors at each location in the stream.

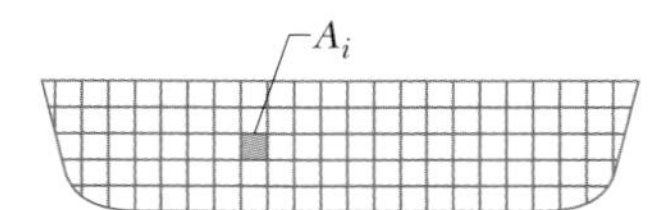

FIGURE 15-28 ■ The ith of many, many small area elements that make up the cross-sectional area of a stream channel. The size of the elements is chosen to be so small that the stream velocity vector is constant over an element of area.

Let us consider stream water passing through many small elements. We can denote the ith area element as ΔA_i, where we use the delta notation (Δ) to signify that the area is part of a larger area, as shown in Fig. 15-28. In fact, we typically choose the area elements ΔA_i to be small enough that the velocity of the part of the stream flowing through it is essentially constant. Suppose that the volume flux, Φ_i, represents the rate at which water flows through the ith element of area.

VOLUME FLUX is defined as the rate at which something passes through an area.

Obviously if the water velocity $\vec{v}$ is parallel to the plane of the area element ΔA_i, water passes by the area without going through it. In that case the flux element, Φ_i, is zero (Fig. 15-29*a*). On the other hand, if $\vec{v}$ is perpendicular to the plane of the area, the flux element, Φ_i, is a maximum (Fig. 15-29*b*). If the area is oriented so it is between perpendicular and parallel the volume flux is in between zero and its maximum value (Fig. 15-29*c*). Thus, the volume flux depends on the *angle* between the velocity vector $\vec{v}$ representing the flow of water and the orientation of the areas shown in Fig. 15-29.

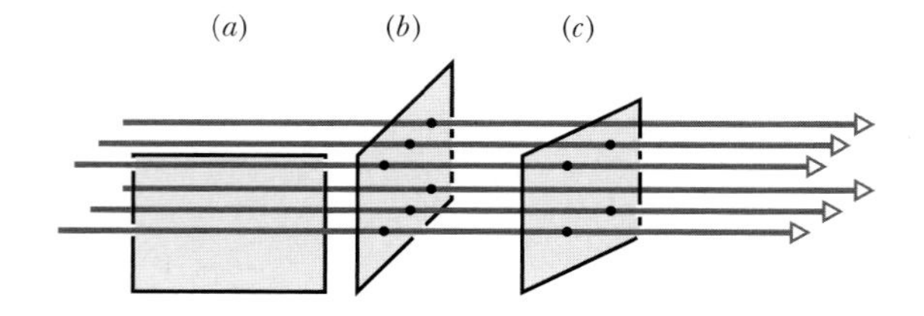

FIGURE 15-29 ■ The velocity vector field of a stream is represented here as imaginary streamlines. The amount of water that passes through a small imaginary area depends on the area's orientation. Three orientations are shown for the same area element A_i. (*a*) No water passes through the area when it is parallel to the velocity vector. (*b*) The largest volume of water passes through an area that is perpendicular to the velocity vector. (*c*) Less water passes through when the orientation of the plane of the area is between perpendicular and parallel.

As shown in Eq. 15-26, the volume of water, ΔV_i, that passes in a time Δt through a small area element, A_i, perpendicular to the stream's direction of flow is given by $\Delta V_i = \Delta x\ A_i = (v_{ix}\Delta t)\ A_i$ (Fig. 15-29). Thus, when an area is perpendicular to the x-component of the stream velocity v_x, the flux element, $\Phi_\perp$, is given by the volume rate of flow through the area

$$\Phi_{i\perp} = \frac{\Delta V_i}{\Delta t} = \frac{\Delta x\ A_i}{\Delta t} = v_{ix}A_i. \tag{15-32}$$

Defining the Normal Vector for an Area

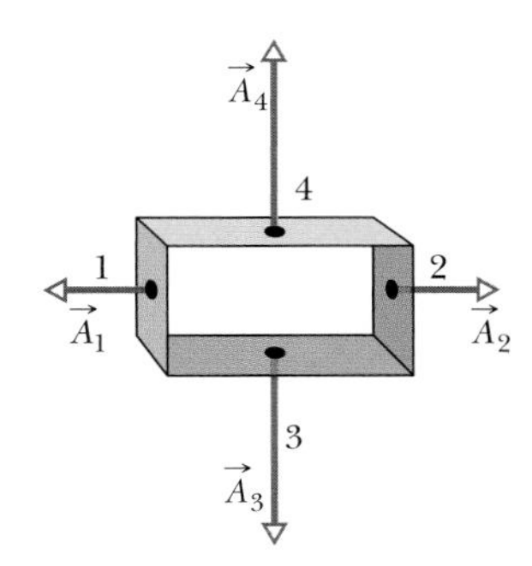

FIGURE 15-30 ■ The directions of the normal vectors are shown for four of the six faces that make up a container. Each normal vector points out of the container. Since area elements 3 and 4 have twice the area as 1 and 2, their normal vectors have twice the length.

In order to find the flux at intermediate angles, it will be convenient to represent the orientation of our small area mathematically. An area uses two of the three dimensions of space, but a direction perpendicular to the plane of the area lies along a single line. Thus it is easier to define the orientation of the ith area element mathematically by a vector having a magnitude of A_i that points in a direction *perpendicular* to the plane of the area. Since the term normal is a synonym for perpendicular, a vector used to describe the direction and magnitude of an area is known as a **normal vector** and can be denoted as $\vec{A}_i$. But in what direction should our normal vector point? If the element of area is part of an imaginary closed container, it is conventional to choose our normal vector for a flat surface pointing *out* of a surface rather than into a surface. The box in Fig. 15-30 has an inside and outside even if it has no front or back. Thus, the normal vectors representing each element of area point outward. If we put a front and back on the box, we would define it as a *closed surface* because if it were a real box, we could trap or enclose something inside of it.

How does the angle between the normal to an area and a stream velocity vector affect the flux? Suppose water in a straight stream channel is moving from left to right as it flows through a small imaginary box shown in Fig. 15-31. What will the flux be through each face of the box as a function of the angle θ between the normal vector of a box face and the stream velocity vector $\vec{v}_i$? Since the flux is a measure of the volume of water that flows through a surface per unit time, it will depend on the component of $\vec{v}_i$ normal to the plane of the area element and the magnitude of the area $|\Delta\vec{A}_i|$ (Fig. 15-30). But as seen in Fig. 15-31, the component of $\vec{v}_i$ along the direction of the normal vector is $|\vec{v}_i|\cos\theta_i$, and so the flux element Φ_i is given by

$$\Phi_i = (|\vec{v}_i|\cos\theta_i)|\vec{A}_i| = \vec{v}_i \cdot \vec{A}_i$$

where A_i represents the ith area element.

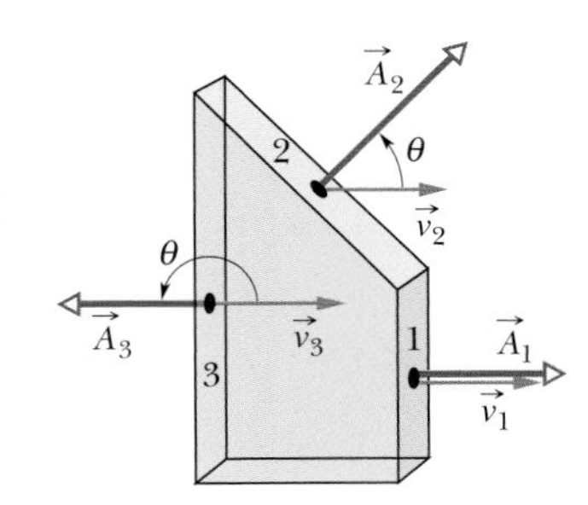

FIGURE 15-31 ■ A small imaginary box with six faces (or surfaces) is placed in an area of the stream where the velocity of the water is uniform over all the faces of the box. The angle between the velocity vector and the normal vectors for three of the six faces of the box are shown.

To make our definition of flux more generally useful, we will develop an equation for flux in terms of the velocity and area vectors $\vec{v}$ and $\vec{A}$. We define the flux of water through the ith element of area mathematically using the scalar product (introduced in Section 9-4) as

$$\Phi_i \equiv \vec{v}_i \cdot \vec{A}_i \quad \text{(volume flux definition for a small area element).} \qquad (15\text{-}33)$$

Let us consider how to calculate the net flow of water through an imaginary surface placed in a stream such as the box shown in Fig. 15-31. We can assign a velocity vector to each point in the stream of water. If our area elements are small enough, the velocity vector is the same everywhere on the surface of a given area element. Thus, if we know the area and orientation of each face and the magnitude and direction of the stream velocity vectors, we can use Eq. 15-33 to calculate the net flux through the cross-sectional area of the stream by adding up the products of the normal velocity vector components and the area elements through which water flows. Mathematically this is given by

$$\Phi^{\text{net}} = \Phi_1 + \Phi_2 + \cdots + \Phi_N = \vec{v}_1\cdot\vec{A}_1 + \vec{v}_2\cdot\vec{A}_2 + \cdots + \vec{v}_N\cdot\vec{A}_N = \sum_{n=1}^{N} \vec{v}_n\cdot\vec{A}_n \qquad (15\text{-}34)$$

where $\vec{v}_1, \vec{v}_2, \vec{v}_N$, and so on represent the velocity vectors at the location of each of the N area elements.

Net Fluid Volume Flux through a Closed Container

Let us consider how fluid flows into face 3 and out of faces 1 and 2 of a small imaginary box we have placed in our stream (Fig. 15-31). If our fluid, in this case water, is incompressible and is not created or destroyed inside our imaginary closed surface, we expect that the rate at which it flows into the box should be the same as the rate at which it flows out of the box. The total flux through all the faces of a closed container is known as the **net flux.** Thus, we expect the net flux of an incompressible fluid through the box to be zero. Does the mathematics of our method for finding net flux tell us this?

We have chosen our area elements to be small enough that the flow velocity vector at the location of any particular area element is uniform. However, that doesn't mean that magnitudes of the velocity vectors and relative angle between the velocity vectors and the normal vectors are necessarily the same at the location of each flat surface area. Thus, in general the net flux through a surface is the sum of the flux through each surface area and is still given by the application of Eq. 15-34 to our new situation. Our definition of net flux allows us to deal with a general case for which we have different velocity vectors and normal vector orientations at each face. This might be the case if the stream is very turbulent.

When the velocity and area vectors point in the same general direction as they do in faces 1 and 2, the scalar product of $\vec{v}_i$ and $\vec{A}_i$ is positive. This tells us that water *flowing out of a surface* is defined as a *positive* flux. But the flux at face 3 is different. The velocity and area vectors are in opposite directions and their scalar product is negative. Thus, when water flows *into a surface* the flux is *negative*. Another feature of our box is that the areas associated with the bottom face and the front and back faces all have normal vectors that are perpendicular to the stream velocity vectors. The scalar product rules give us no flux or volume flow through these additional faces. It can be shown that the sum of the negative flux of water into face 3, the positive flux of water out of face 1, and the positive flux out of face 2 add up to zero.

If we refine the equation of continuity to take the direction of flow of fluid into a closed surface as negative volume flux and the flow out of a closed surface as positive volume flux, we indeed expect the net flux through our imaginary box to be zero. This makes sense physically as long as water is incompressible and as long as we can't spontaneously create new water inside the box or remove water from the box. Later when we define electric and magnetic flux, we will see that it may be possible to have a net flux at the boundaries of a closed surface.

READING EXERCISE 15-11: Consider the imaginary box in Fig. 15-31 and assume that it has a width of 4.0 cm, a depth of 1.0 cm, a height on the left side of 8.0 cm, and a height on the right side of 4.0 cm. (a) Find the magnitude of the area of faces 1, 2, and 3, respectively. Report your answer in square meters. (b) What is the total surface area of the box? [Be sure to include the areas of all six area elements (faces) in your calculation.] ■

READING EXERCISE 15-12: Consider the imaginary box in Fig. 15-31 and assume that it is placed in a stream that moves from left to right through the box with a uniform velocity. In other words, we make the simplifying assumption that the velocity vector is the same at every point and on the surface of the small box. Suppose the box has a width of 4.0 cm, a depth of 1.0 cm, a height on the left side of 8.0 cm, and a height on the right side of 4.0 cm. If the magnitude of the stream velocity at the location of the box is 0.50 m/s, (a) find the flux through faces 1, 2, and 3 respectively in m^3/s; (b) explain why only faces 1, 2, and 3 have nonzero flux, and (c) show that the net flux through the closed surface defined by the box is zero. ■

15-11 Bernoulli's Equation

As we discussed in Section 15-8, ideal fluids are incompressible and flow in a streamlined fashion without experiencing friction forces. Water that flows in slow-moving rivers or large pipes acts like an ideal fluid. In this section we present a very useful equation that relates the pressure, velocity, and vertical location of an ideal fluid as it flows.

Consider an ideal fluid flowing through a pipe like that shown in Fig. 15-32. In a time interval Δt, a parcel of fluid of volume ΔV (colored purple in Fig. 15-32*a*) enters the pipe at the left input end and an identical volume (colored green in Fig. 15-32*b*) emerges at the right output end. Since we are assuming that the fluid is incompressible, the emerging volume must be the same as the entering volume. In addition we assume that the fluid has a constant density ρ.

Let y_1, v_1, and P_1 be the elevation, speed, and pressure of the fluid entering at the left, and y_2, v_2, and P_2 be the corresponding quantities for the fluid emerging at the right. By applying the principle of conservation of energy to the fluid, we can show that these quantities are related by

$$P_1 + \tfrac{1}{2}\rho v_1^2 + \rho g y_1 = P_2 + \tfrac{1}{2}\rho v_2^2 + \rho g y_2. \qquad (15\text{-}35)$$

We can also write this equation as

$$P + \tfrac{1}{2}\rho v^2 + \rho g y = \text{a constant} \quad \text{(ideal fluid-flow equation).} \qquad (15\text{-}36)$$

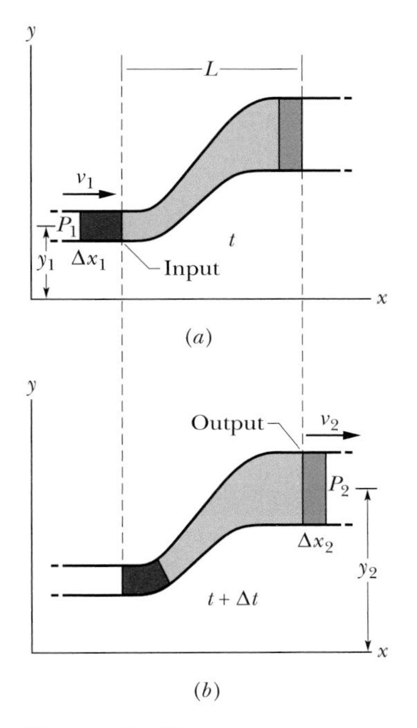

FIGURE 15-32 ■ Fluid flows at a steady rate through a length L of a tube from the input end at the left to the output end at the right. From time t in (*a*) to time $t + \Delta t$ in (*b*), the parcel of fluid of mass m shown in purple enters the input end and the equal amount shown in green emerges from the output end.

Equations 15-35 and 15-36 are equivalent forms of **Bernoulli's equation,** named after Daniel Bernoulli, who studied fluid flow in the 1700s.* Like the equation of continuity (Eq. 15-28), Bernoulli's equation is not a new principle but simply the reformulation of a familiar principle in a form more suitable to fluid mechanics. As a check let us consider Bernoulli's equation for two special cases.

Case 1, No flow: Here we set the speeds at the two ends of the pipe equal to zero in Eq. 15-35 so that $v_1 = v_2$. This gives us

$$P_2 = P_1 - \rho g(y_2 - y_1) \quad \text{(elevation change but no flow).}$$

In this case, Bernoulli's equation simply reduces to Eq. 15-7.

Case 2, No Elevation Change: A major prediction of Bernoulli's equation emerges if we assume the fluid does not change elevation as it flows so that $y_1 = y_2$. Equation 15-35 then becomes

$$P_1 + \tfrac{1}{2}\rho v_1^2 = P_2 + \tfrac{1}{2}\rho v_2^2, \qquad (15\text{-}37)$$

which tells us that:

> If the speed of a fluid element increases as it travels along a horizontal streamline, the pressure of the fluid must decrease.

Put another way, where the speed of the fluid is relatively high, its pressure is relatively low. Conversely, where the speed of the fluid is relatively low, its pressure is relatively high.

The link between a change in pressure and a change in speed makes sense if you consider what happens to the parcel of fluid as it crosses a boundary between high and low pressure. As the parcel of fluid that is moving from left to right nears a narrow region, the higher pressure behind it exerts a force on it toward the right of

*For irrotational flow (which we assume), the constant in Eq. 15-36 has the same value for all points within the tube of flow; the points do not have to lie along the same streamline. Similarly, the points 1 and 2 in Eq. 15-35 can lie anywhere within the tube of flow.

magnitude $F_1 = P_1\Delta A$ where ΔA is the cross-sectional area of the parcel. Likewise the lower pressure in front of it exerts a force on it toward the left of magnitude $F_2 = P_2\Delta A$. But since $P_1 > P_2$, then $F_1 > F_2$ and the parcel experiences a net force to the right. It then accelerates over a short distance and speeds up. In passing from a region of lower pressure to one of higher pressure $F_2 > F_1$, so the parcel slows down.

Bernoulli's equation is strictly valid only to the extent that the fluid is ideal. If viscous forces are present, thermal energy will be involved. We take no account of this in the derivation that follows.

Proof of Bernoulli's Equation

Since our ideal fluid system experiences no friction, we can assume that mechanical energy is conserved by any parcel of fluid lying between the two vertical planes separated by a distance L in Fig. 15-32 as it moves from its initial state (Fig. 15-32*a*) to its final state (Fig. 15-32*b*). We also assume that the fluid does not change its properties during this process and that the flow is steady. Thus, we need be concerned only with changes that take place at the input and output ends of the pipe.

The work-kinetic energy theorem tells us that if energy is conserved,

$$W^{\text{net}} = \Delta K. \tag{15-38}$$

In other words, the change in the kinetic energy of our parcel must equal the net work done on it. But the change in kinetic energy results from the change in speed of the parcel of fluid at the ends of the pipe. Thus, if a parcel of fluid flows into end 1 and out end 2 of the pipe, then

$$\begin{aligned}\Delta K &= \tfrac{1}{2}mv_2^2 - \tfrac{1}{2}mv_1^2 \\ &= \tfrac{1}{2}\rho V(v_2^2 - v_1^2),\end{aligned} \tag{15-39}$$

where $m = \rho V$ is the mass of the parcel of fluid of volume V.

The work done on the parcel arises from two sources. One is the negative gravitational work done on the parcel of fluid during the vertical lift of the mass from the input to the output level. A second source of work done on the parcel of fluid results from the forces exerted on it by the fluid behind it and the fluid in front of it as it flows due to pressure differences at various locations in the pipe.

Gravitational Work: The work W^{grav} done by the gravitational force $(mg\hat{\mathrm{j}})$ on the parcel of fluid of mass m during the vertical lift of the mass from the input to the output level is given by Eq. 9-14

$$\begin{aligned}W^{\text{grav}} &= -mg(y_2 - y_1) \\ &= -\rho g V(y_2 - y_1).\end{aligned} \tag{15-40}$$

This work is negative because the upward displacement and the downward gravitational force have opposite directions. (Note that in this context the notation W^{grav} has nothing to do with the weight of the parcel.)

Pressure Difference Work: Work is done on the parcel of moving fluid as a result of the pressure difference between its input and output ends. We start by finding the equation for the work needed to move a parcel of fluid through a distance Δx in a pipe of cross section A. If the parcel of fluid is under pressure P, it is given by

$$W = F_x\Delta x = (PA)\Delta x = P(A\Delta x) = PV,$$

where $V = A\,\Delta x$ is the volume of the parcel.

We now consider the parcel of fluid shown at the input end in Fig. 15-32. It is at pressure P_1 and has a volume V. Since the displacement and the force due to the pressure at the input end are in the *same* direction, the work done on it by the fluid behind it is positive and given by $W_1 = +P_1V$. Now consider the work done on the parcel of fluid shown at the output end in Fig. 15-32. It is at pressure P_2 but also has a volume V. Since the displacement and the force exerted due to the pressure at the output end are in *opposite* directions, the work done on it by the fluid behind it is negative and given by $W_2 = +P_2V$. Thus, the total work W^{pd} done on the incoming and emerging parcels due to pressure difference is given by

$$\begin{aligned} W^{\text{pd}} &= W_1 - W_2 = P_1V - P_2V \\ &= -(P_2 - P_1)V. \end{aligned} \tag{15-41}$$

The work-kinetic energy theorem of Eq. 15-38 now becomes

$$W^{\text{net}} = W^{\text{grav}} + W^{\text{pd}} = \Delta K.$$

Substituting from Eqs. 15-39, 15-40, and 15-41 yields

$$-\rho gV(y_2 - y_1) - V(P_2 - P_1) = \tfrac{1}{2}\rho V(v_2^2 - v_1^2).$$

This, after a slight rearrangement, matches Eq. 15-35, which we set out to prove.

READING EXERCISE 15-13: Water flows smoothly through the pipe shown in the figure, descending in the process. Rank the four numbered sections of pipe according to (a) the volume flow rate R_V through them, (b) the flow speed v through them, and (c) the water pressure P within them, greatest first.

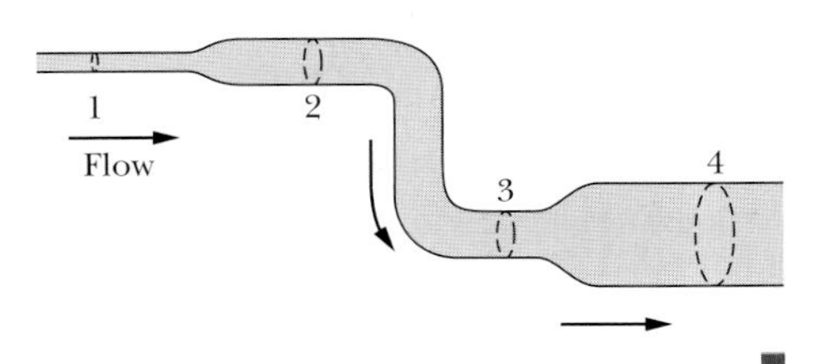

TOUCHSTONE EXAMPLE 15-9: Ethanol Flow

Ethanol of density $\rho = 791$ kg/m^3 flows smoothly through a horizontal pipe that tapers in cross-sectional area from $A_1 = 1.20 \times 10^{-3}$ m^2 to $A_2 = A_1/2$. The pressure difference between the wide and narrow sections of pipe is 4120 Pa. What is the volume flow rate R_V of the ethanol?

SOLUTION ■ One **Key Idea** here is that, because the fluid flowing through the wide section of pipe must entirely pass through the narrow section, the volume flow rate R_V must be the same in the two sections. Thus, from Eq. 15-28,

$$R_V = v_1A_1 = v_2A_2. \tag{15-42}$$

However, with two unknown speeds, we cannot evaluate this equation for R_V.

A second **Key Idea** is that, because the flow is smooth, we can apply Bernoulli's equation. From Eq. 15-35, we can write

$$P_1 + \tfrac{1}{2}\rho v_1^2 + \rho gy = P_2 + \tfrac{1}{2}\rho v_2^2 + \rho gy, \tag{15-43}$$

where subscripts 1 and 2 refer to the wide and narrow sections of pipe, respectively, and y is their common elevation. This equation hardly seems to help because it does not contain the desired volume flow R_V and it contains the unknown speeds v_1 and v_2.

However, there is a neat way to make it work for us: First, we can use Eq. 15-42 and the fact that $A_2 = A_1/2$ to write

$$v_1 = \frac{R_V}{A_1} \quad \text{and} \quad v_2 = \frac{R_V}{A_2} = \frac{2R_V}{A_1}. \tag{15-44}$$

Then we can substitute these expressions into Eq. 15-43 to eliminate the unknown speeds and introduce the desired volume flow rate. Doing this and solving for R_V yield

$$R_V = A_1\sqrt{\frac{2(P_1 - P_2)}{3\rho}}. \tag{15-45}$$

We still have a decision to make: We know that the pressure difference between the two sections is 4120 Pa, but does that mean that $P_1 - P_2$ is 4120 Pa or -4120 Pa? We could guess the former is true, or otherwise the square root in Eq. 15-45 would give us an imaginary number. Instead of guessing, however, let's try some reasoning. From Eq. 15-42 we see that speed v_2 in the narrow section (small A_2) must be greater than speed v_1 in the wider section (larger A_1). Recall that if the speed of a fluid increases as it travels along a horizontal path (as here), the pressure of the fluid must decrease. Thus, P_1 is greater than P_2, and $P_1 - P_2 = +4120$ Pa. Inserting this and known data into Eq. 15-45 gives

$$R_V = 1.20 \times 10^{-3}\ \text{m}^2 \sqrt{\frac{(2)(4120\ \text{Pa})}{(3)(791\ \text{kg/m}^3)}}$$

$$= 2.24 \times 10^{-3}\ \text{m}^3/\text{s}. \qquad \text{(Answer)}$$

Problems

SEC. 15-3 ■ PRESSURE AND DENSITY

1. Syringe Find the pressure increase in the fluid in a syringe when a nurse applies a force of 42 N to the syringe's circular piston, which has a radius of 1.1 cm.

2. Three Liquids Three liquids that will not mix are poured into a cylindrical container. The volumes and densities of the liquids are 0.50 L, 2.6 g/cm^3; 0.25 L, 1.0 g/cm^3; and 0.40 L, 0.80 g/cm^3. What is the force on the bottom of the container due to these liquids? One liter = 1 L = 1000 cm^3. (Ignore the contribution due to the atmosphere.)

3. Office Window An office window has dimensions 3.4 m by 2.1 m. As a result of the passage of a storm, the outside air pressure drops to 0.96 atm, but inside the pressure is held at 1.0 atm. What net force pushes out on the window?

4. Front Tires You inflate the front tires on your car to 28 psi. Later, you measure your blood pressure, obtaining a reading of 120/80, the readings being in mm Hg. In countries using the metric system (which is to say, most of the world), these pressures are customarily reported in kilopascals (kPa). In kilopascals, what are (a) your tire pressure and (b) your blood pressure?

5. Fish A fish maintains its depth in fresh water by adjusting the air content of porous bone or air sacs to make its average density the same as that of the water. Suppose that with its air sacs collapsed, a fish has a density of 1.08 g/cm^3. To what fraction of its expanded body volume must the fish inflate the air sacs to reduce its density to that of water?

6. Airtight Container An airtight container having a lid with negligible mass and an area of 77 cm^2 is partially evacuated. If a 480 N force is required to pull the lid off the container and the atmospheric pressure is 1.0×10^5 Pa, what is the air pressure in the container before it is opened?

7. Otto Von Guericke In 1654 Otto von Guericke, inventor of the air pump, gave a demonstration before the noblemen of the Holy Roman Empire in which two teams of eight horses could not pull apart two evacuated brass hemispheres. (a) Assuming that the hemispheres have thin walls, so that R in Fig. 15-33 may be considered both the inside and outside radius, show that the force $\vec{F}$ required to pull apart the hemispheres has magnitude $|\vec{F}| = \pi R^2 \Delta P$, where ΔP is the difference between the pressures outside and inside the sphere. (b) Taking R as 30 cm, the inside pressure as 0.10 atm, and the outside pressure as 1.00 atm, find the force magnitude the teams of horses would have had to exert to pull apart the hemispheres. (c) Explain why one team of horses could have proved the point just as well if the hemispheres were attached to a sturdy wall.

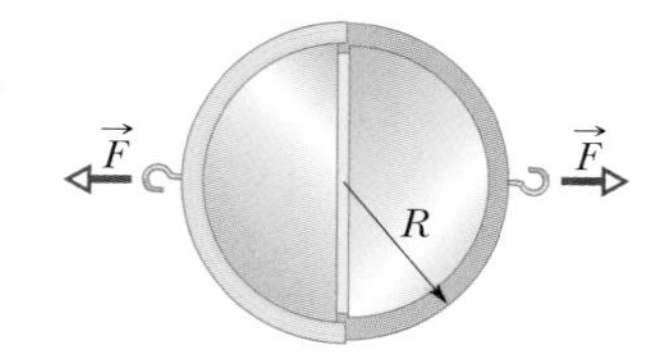

FIGURE 15-33 ■ Problem 7.

SEC. 15-4 ■ GRAVITATIONAL FORCES AND FLUIDS AT REST

8. Hydrostatic Difference Calculate the hydrostatic difference in blood pressure between the brain and the foot in a person of height 1.83 m. The density of blood is 1.06×10^3 kg/m^3.

9. Sewage Outlet The sewage outlet of a house constructed on a slope is 8.2 m below street level. If the sewer is 2.1 m below street level, find the minimum pressure difference that must be created by the sewage pump to transfer waste of average density 900 kg/m^3 from outlet to sewer.

10. Phase Diagram Figure 15-34 displays the *phase diagram* of carbon, showing the ranges of temperature and pressure in which carbon will crystallize either as diamond or graphite. What is the minimum depth at which diamonds can form if the temperature at that depth is 1000°C and the rocks there have density 3.1 g/cm^3? Assume that, as in a fluid, the pressure at any level is due to the gravitational force on the material lying above that level, and neglect variation of g with depth.

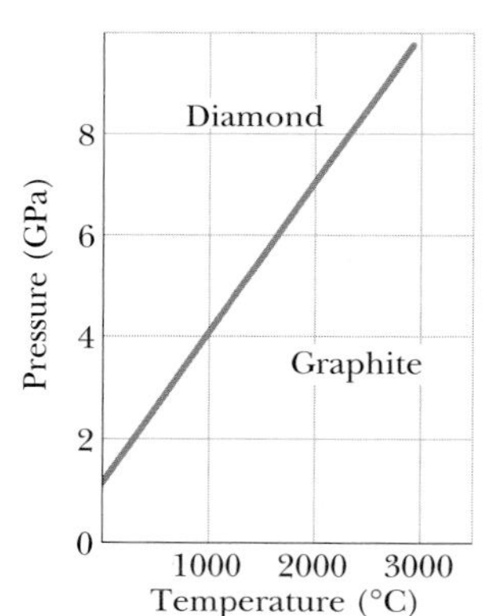

FIGURE 15-34 ■ Problem 10.

11. Swimming Pool A swimming pool has the dimensions 24 m × 9.0 m × 2.5 m. When it is filled with water, what is the force (resulting from the water alone) on (a) the bottom, (b) each short side, and (c) each long side? (d) If you are concerned with the possibility that the concrete walls and floor will collapse, is it appropriate to take the atmospheric pressure into account? Why?

12. Seawater (a) Assuming the density of seawater is 1.03 g/cm^3, find the total weight of water on top of a nuclear submarine at a depth of 200 m if its (horizontal cross-sectional) hull area is 3000 m^2. (b) In atmospheres, what water pressure would a diver experience at this depth? Do you think that occupants of a damaged submarine at this depth could escape without special equipment?

13. Crew Members Crew members attempt to escape from a damaged submarine 100 m below the surface. What force must be applied to a pop-out hatch, which is 1.2 m by 0.60 m, to push it out at that depth? Assume that the density of the ocean water is 1025 kg/m^3.

14. Barrel A cylindrical barrel has a narrow tube fixed to the top, as shown (with dimensions) in Fig. 15-35. The vessel is filled with water to the top of the tube. Calculate the ratio of the hydrostatic force on the bottom of the barrel to the gravitational force on the water contained inside the barrel. Why is that ratio not equal to one? (You need not consider the atmospheric pressure.)

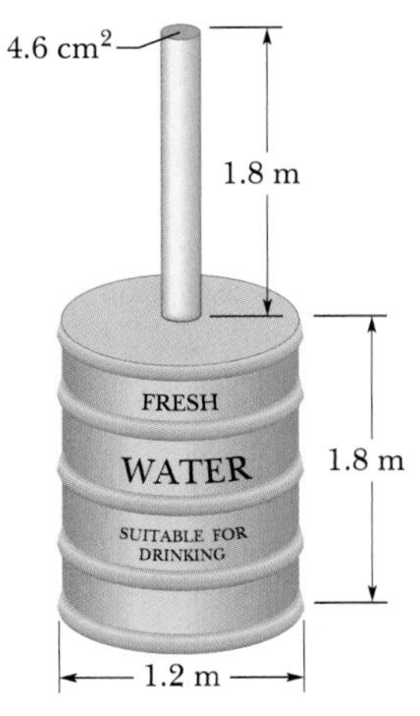

FIGURE 15-35 ■ Problem 14.

15. Cylindrical Vessels Two identical cylindrical vessels with their bases at the same level each contain a liquid of density ρ. The area of each base is A, but in one vessel the liquid height is h_A, and in the other it is h_B. Find the work done by the gravitational force in equalizing the levels when the two vessels are connected.

16. Geological Features in analyzing certain geological features, it is often appropriate to assume that the pressure at some horizontal *level of compensation,* deep inside Earth, is the same over a large region and is equal to the pressure due to the gravitational force on the overlying material. Thus, the pressure on the level of compensation is given by the fluid pressure formula. This model requires, for one thing, that mountains have roots of continental rock extending into the denser mantle (Fig. 15-36). Consider a mountain 6.0 km high. The continental rocks have a density of 2.9 g/cm^3, and beneath the continent the mantle has a density of 3.3 g/cm^3. Calculate the depth D of the root. (*Hint:* Set the pressure at points a and b equal; the depth y of the level of compensation will cancel out.)

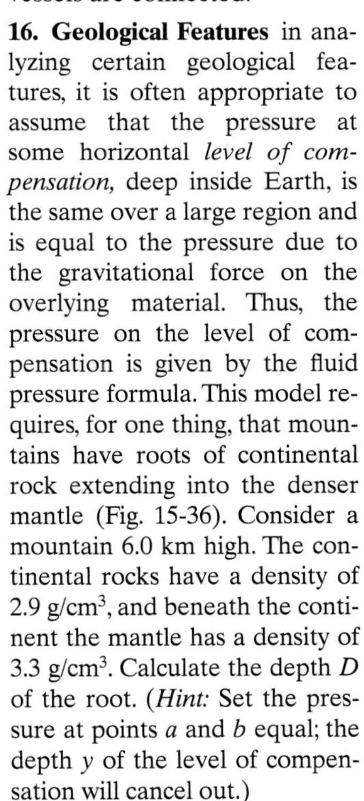

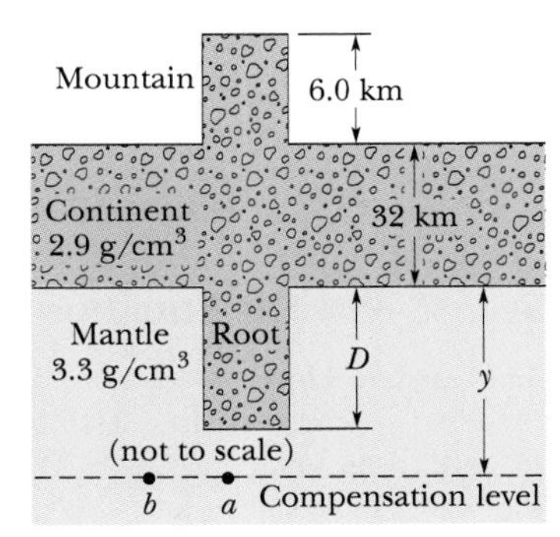

FIGURE 15-36 ■ Problem 16.

17. Ocean Figure 15-37 shows the juncture of ocean and continent. Find the depth h of the ocean using the level-of-compensation technique presented in Problem 16.

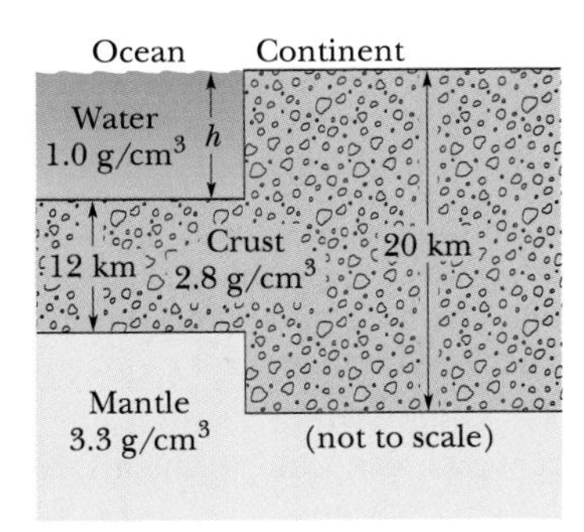

FIGURE 15-37 ■ Problem 17.

18. L-shaped Tank The L-shaped tank shown in Fig. 15-38 is filled with water and is open at the top. If $d = 5.0$ m, what are (a) the force on face A and (b) the force on face B due to the water?

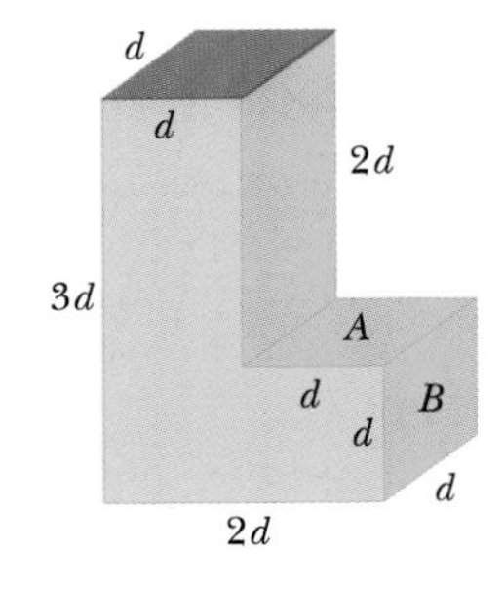

FIGURE 15-38 ■ Problem 18.

19. Water Stands Water stands at a depth D behind the vertical upstream face of a dam, as shown in Fig. 15-39. Let W be the width of the dam. Find (a) the net horizontal force on the dam from the gauge pressure of the water and (b) the net torque due to that force (and thus gauge pressure) about a line through O parallel to the width of the dam. (c) Find the moment arm of the net horizontal force about the line through O.

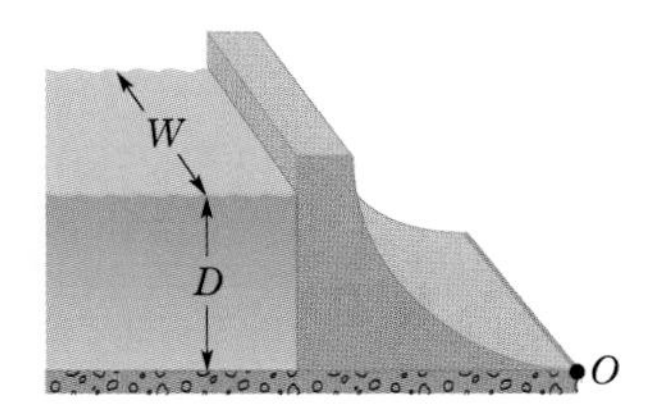

FIGURE 15-39 ■ Problem 19.

SEC. 15-5 ■ MEASURING PRESSURE

20. Lemonade To suck lemonade of density 1000 kg/m^3 up a straw to a maximum height of 4.0 cm, what minimum gauge pressure (in atmospheres) must you produce in your mouth?

21. Atmosphere What would be the height of the atmosphere if the air density (a) were uniform and (b) decreased linearly to zero with height? Assume that at sea level the air pressure is 1.0 atm and the air density is 1.3 kg/m^3.

SEC. 15-6 ■ PASCAL'S PRINCIPLE

22. Piston A piston of small cross-sectional area a is used in a hydraulic press to exert a small force $\vec{f}$ on the enclosed liquid. A connecting pipe leads to a larger piston of cross-sectional area A (Fig. 15-40). (a) What force magnitude $|\vec{F}|$ will the larger piston sustain without moving? (b) If the small piston has a diameter of 3.80 cm and the large piston one of 53.0 cm, what force magnitude on the small piston will balance a 20.0 kN force on the large piston?

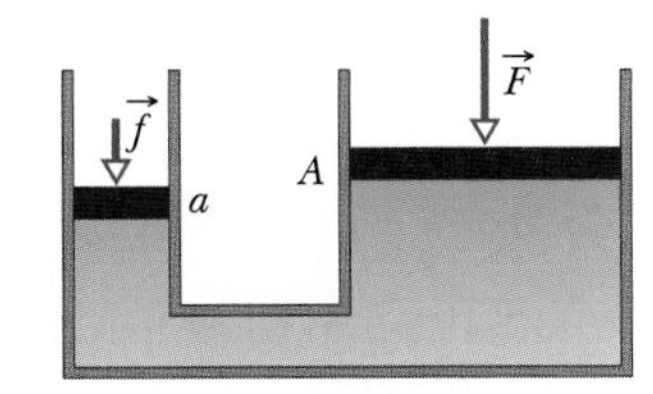

FIGURE 15-40 ■ Problems 22 and 23.

23. Hydraulic Press In the hydraulic press of Problem 22, through what distance must the large piston be moved to raise the small piston a distance of 0.85 m?

SEC. 15-7 ■ ARCHIMEDES' PRINCIPLE

24. A Boat Floats A boat floating in fresh water displaces water weighing 35.6 kN. (a) What is the weight of the water that this boat would displace if it were floating in salt water with a density of 1.10×10^3 kg/m^3? (b) Would the volume of the displaced water change? If so, by how much?

25. Iron Anchor An iron anchor of density 7870 kg/m^3 appears 200 N lighter in water than in air. (a) What is the volume of the anchor? (b) How much does it weigh in air?

26. Cubical Object In Fig. 15-41 a cubical object of dimensions $L = 0.600$ m on a side and with a mass of 450 kg is suspended by a rope in an open tank of liquid of

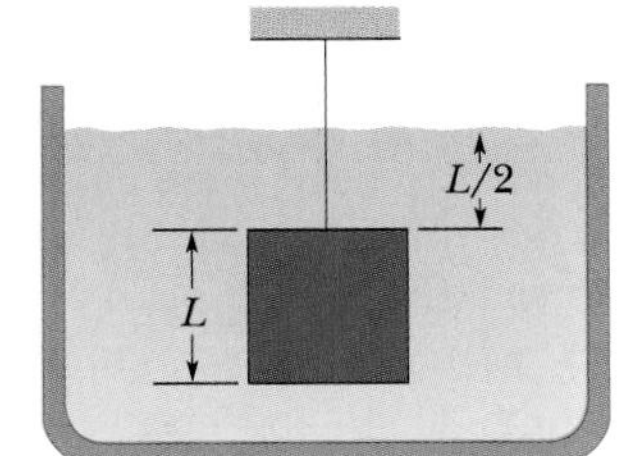

FIGURE 15-41 ■ Problem 26.

density 1030 kg/m^3. (a) Find the magnitude of the total downward force on the top of the object from the liquid and the atmosphere, assuming that atmospheric pressure is 1.00 atm. (b) Find the magnitude of the total upward force on the bottom of the object. (c) Find the tension in the rope. (d) Calculate the magnitude of the buoyant force on the object using Archimedes' principle. What relation exists among all these quantities?

27. Block of Wood A block of wood floats in fresh water with two-thirds of its volume submerged. In oil the block floats with 0.90 of its volume submerged. Find the density of (a) the wood and (b) the oil.

28. Blimp A blimp is cruising slowly at low altitude, filled as usual with helium gas. Its maximum useful payload, including crew and cargo, is 1280 kg. The volume of the helium-filled interior space is 5000 m^3. The density of helium gas is 0.16 kg/m^3, and the density of hydrogen is 0.081 kg/m^3. How much more payload could the blimp carry if you replaced the helium with hydrogen? (Why not do it?)

29. Hollow Sphere A hollow sphere of inner radius 8.0 cm and outer radius 9.0 cm floats half-submerged in a liquid of density 800 kg/m^3. (a) What is the mass of the sphere? (b) Calculate the density of the material of which the sphere is made.

30. Dead Sea About one-third of the body of a person floating in the Dead Sea will be above the water line. Assuming that the human body density is 0.98 g/cm^3, find the density of the water in the Dead Sea. (Why is it so much greater than 1.0 g/cm^3?)

31. Iron Shell A hollow spherical iron shell floats almost completely submerged in water. The outer diameter is 60.0 cm, and the density of iron is 7.87 g/cm^3. Find the inner diameter.

32. Wood with Lead A block of wood has a mass of 3.67 kg and a density of 600 kg/m^3. It is to be loaded with lead so that it will float in water with 0.90 of its volume submerged. What mass of lead is needed (a) if the lead is attached to the top of the wood and (b) if the lead is attached to the bottom of the wood? The density of lead is 1.13×10^4 kg/m^3.

33. Iron Casting An iron casting containing a number of cavities weighs 6000 N in air and 4000 N in water. What is the total volume of all the cavities in the casting? The density of iron (that is, a sample with no cavities) is 7.87 g/cm^3.

34. Density of Brass Assume the density of brass weights to be 8.0 g/cm^3 and that of air to be 0.0012 g/cm^3. What percent error arises from neglecting the buoyancy of air in weighing an object of mass m and density ρ on a beam balance?

35. Slab of Ice (a) What is the minimum area of the top surface of a slab of ice 0.30 m thick floating on fresh water that will hold up an automobile of mass 1100 kg? (b) Does it matter where the car is placed on the block of ice?

36. Three Children Three children, each of weight 356 N, make a log raft by lashing together logs of diameter 0.30 m and length 1.80 m. How many logs will be needed to keep them afloat in fresh water? Take the density of the logs to be 800 kg/m^3.

37. Metal Rod A metal rod of length 80 cm and mass 1.6 kg has a uniform cross-sectional area of 6.0 cm^2. Due to a nonuniform density, the center of mass of the rod is 20 cm from one end of the rod. The rod is suspended in a horizontal position in water by ropes attached to both ends (Fig. 15-42). (a) What is the tension in the rope closer to the center of mass? (b) What is the tension in the rope farther from the center of mass? (*Hint:* The buoyancy force on the rod effectively acts at the rod's geometric center.)

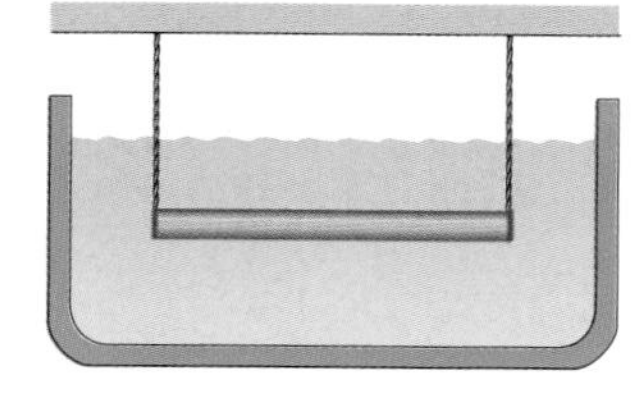

FIGURE 15-42 ■ Problem 37.

38. Floating Car A car has a total mass of 1800 kg. The volume of air space in the passenger compartment is 5.00 m^3. The volume of the motor and front wheels is 0.750 m^3, and the volume of the rear wheels, gas tank, and trunk is 0.800 m^3; water cannot enter these areas. The car is parked on a hill; the handbrake cable snaps and the car rolls down the hill into a lake (Fig. 15-43). (a) At first, no water enters the passenger compartment. How much of the car, in cubic meters, is below the water surface with the car floating as shown? (b) As water slowly enters, the car sinks. How many cubic meters of water are in the car as it disappears below the water surface? (The car, with a heavy load in the trunk, remains horizontal.)

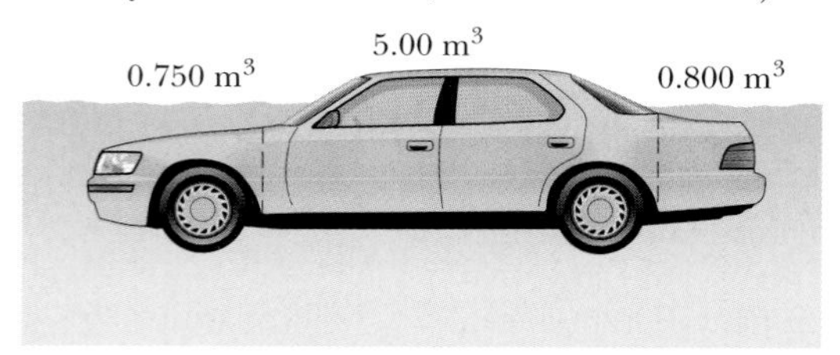

FIGURE 15-43 ■ Problem 38.

SEC. 15-9 ■ THE EQUATION OF CONTINUITY

39. Garden Hose A garden hose with an internal diameter of 1.9 cm is connected to a (stationary) lawn sprinkler that consists merely of an enclosure with 24 holes, each 0.13 cm in diameter. If the water in the hose has a speed of 0.91 m/s, at what speed does it leave the sprinkler holes?

40. Two Streams Two streams merge to form a river. One stream has a width of 8.2 m, depth of 3.4 m, and current speed of 2.3 m/s. The other stream is 6.8 m wide and 3.2 m deep, and flows at 2.6 m/s. The width of the river is 10.5 m, and the current speed is 2.9 m/s. What is its depth?

41. Flooded Basement Water is pumped steadily out of a flooded basement at a speed of 5.0 m/s through a uniform hose of radius 1.0 cm. The hose passes out through a window 3.0 m above the waterline. What is the power of the pump?

42. Water Pipe The water flowing through a 1.9 cm (inside diameter) pipe flows out through three 1.3 cm pipes. (a) If the flow rates in the three smaller pipes are 26, 19, and 11 L/min, what is the flow rate in the 1.9 cm pipe? (b) What is the ratio of the speed of water in the 1.9 cm pipe to that in the pipe carrying 26 L/min?

SEC. 15-11 ■ BERNOULLI'S EQUATION

43. Pipe Increases in Area Water is moving with a speed of 5.0 m/s through a pipe with a cross-sectional area of 4.0 cm^2. The water gradually descends 10 m as the pipe increases in area to 8.0 cm^2. (a) What is the speed at the lower level? (b) If the pressure at the upper level is 1.5×10^5 Pa, what is the pressure at the lower level?

44. Torpedoes Models of torpedoes are sometimes tested in a horizontal pipe of flowing water, much as a wind tunnel is used to test model airplanes. Consider a circular pipe of internal diameter 25.0 cm and a torpedo model, aligned along the axis of the pipe, with a diameter of 5.00 cm. The model is to be tested with water flowing past it at 2.50 m/s. (a) With what speed must the water flow in the part of the pipe that is unconstricted by the model? (b) What will the pressure difference be between the constricted and unconstricted parts of the pipe?

45. Basement Pipe A water pipe having a 2.5 cm inside diameter carries water into the basement of a house at a speed of 0.90 m/s and a pressure of 170 kPa. If the pipe tapers to 1.2 cm and rises to the second floor 7.6 m above the input point, what are (a) the speed and (b) the water pressure at the second floor?

46. Water Intake A water intake at a pump storage reservoir (Fig. 15-44) has a cross-sectional area of 0.74 m². The water flows in at a speed of 0.40 m/s. At the generator building 180 m below the intake point, the cross-sectional area is smaller than at the intake and the water flows out at 9.5 m/s. What is the difference in pressure, in megapascals, between inlet and outlet?

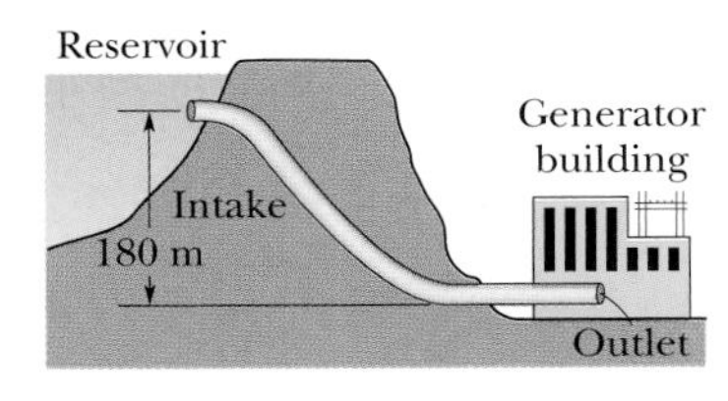

FIGURE 15-44 ■ Problem 46.

47. Large Area A tank of large area is filled with water to a depth $D = 0.30$ m. A hole of cross-sectional area $A = 6.5$ cm² in the bottom of the tank allows water to drain out. (a) What is the rate at which water flows out, in cubic meters per second? (b) At what distance below the bottom of the tank is the cross-sectional area of the stream equal to one-half the area of the hole?

48. Air Flows Air flows over the top of an airplane wing of area A with speed $|\vec{v}_{top}|$ and past the underside of the wing (also of area A) with speed $|\vec{v}_{under}|$. Show that in this simplified situation Bernoulli's equation predicts that the magnitude $|\vec{L}|$ of the upward lift force on the wing will be

$$|\vec{L}| = \tfrac{1}{2}\rho A(v_{top}^2 - v_{under}^2),$$

where ρ is the density of the air.

49. Airplane Wing If the speed of flow past the lower surface of an airplane wing is 110 m/s, what speed of flow over the upper surface will give a pressure difference of 900 Pa between upper and lower surfaces? Take the density of air to be 1.30×10^{-3} g/cm³, and see Problem 48.

50. Two Tanks Suppose that two tanks, A and B, each with a large opening at the top, contain different liquids. A small hole is made in the side of each tank at the same depth d below the liquid surface, but the hole in tank A has half the cross-sectional area of the hole in tank B. (a) What is the ratio ρ_A/ρ_B of the densities of the liquids if the mass flow rate is the same for the two holes? (b) What is the ratio of the volume flow rates from the two tanks? (c) To what height above the hole in tank B should liquid be added or drained to equalize the volume flow rates?

51. Water in the Horizontal Pipe In Fig, 15-45, water flows through a horizontal pipe, and then out into the atmosphere at a speed of 15 m/s. The diameters of the left and right sections of the pipe are 5.0 cm and 3.0 cm, respectively. (a) What volume of water flows into the atmosphere during a 10 min period? In the left section of the pipe, what are (b) the speed v_B, and (c) the gauge pressure?

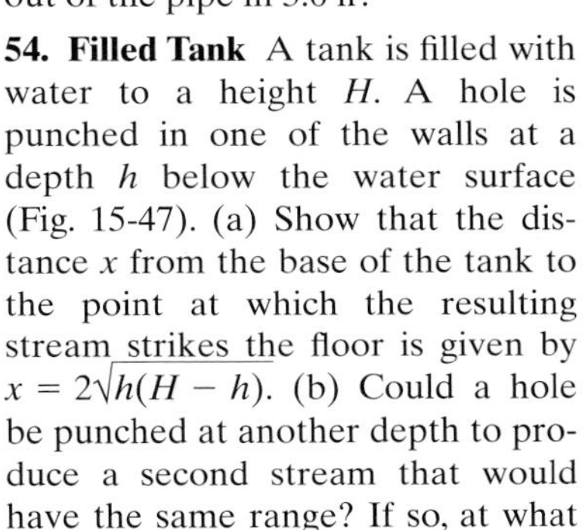

FIGURE 15-45 ■ Problem 51.

52. Beverage Keg An opening of area 0.25 cm² in an otherwise closed beverage keg is 50 cm below the level of the liquid (of density 1.0 g/cm³) in the keg. What is the speed of the liquid flowing through the opening if the gauge pressure in the air space above the liquid is (a) zero and (b) 0.40 atm?

53. Dam The fresh water behind a reservoir dam is 15 m deep. A horizontal pipe 4.0 cm in diameter passes through the dam 6.0 m below the water surface, as shown in Fig. 15-46. A plug secures the pipe opening. (a) Find the magnitude of the frictional force between plug and pipe wall. (b) The plug is removed. What volume of water flows out of the pipe in 3.0 h?

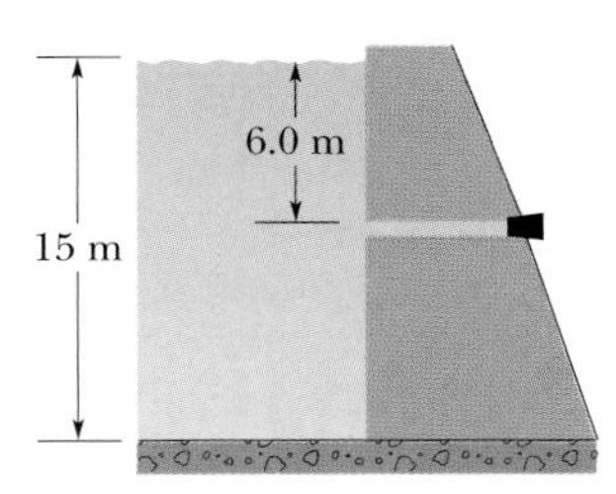

FIGURE 15-46 ■ Problem 53.

54. Filled Tank A tank is filled with water to a height H. A hole is punched in one of the walls at a depth h below the water surface (Fig. 15-47). (a) Show that the distance x from the base of the tank to the point at which the resulting stream strikes the floor is given by $x = 2\sqrt{h(H - h)}$. (b) Could a hole be punched at another depth to produce a second stream that would have the same range? If so, at what depth? (c) At what depth should the hole be placed to make the emerging stream strike the ground at the maximum distance from the base of the tank?

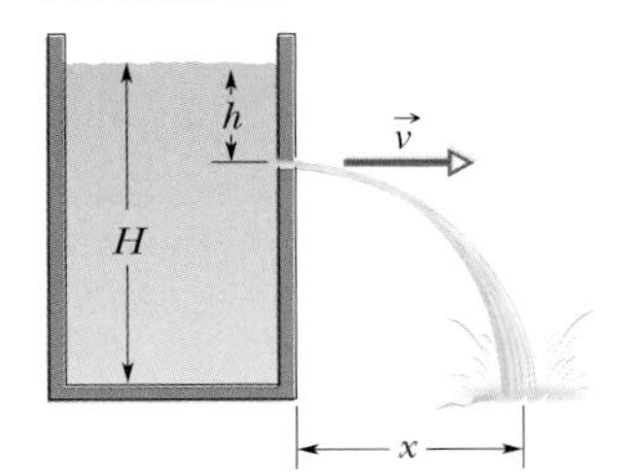

FIGURE 15-47 ■ Problem 54.

55. Venturi Meter A *venturi meter* is used to measure the flow speed of a fluid in a pipe. The meter is connected between two sections of the pipe (Fig. 15-48); the cross-sectional area A of the entrance and exit of the meter matches the pipe's cross-sectional area. At the entrance and exit, the fluid flows through the pipe with speed $v_A = |\vec{v}_A|$. But it flows through a narrow "throat" of cross-sectional area B with speed $v_B = |\vec{v}_B|$. A manometer connects the wider portion of the meter to the narrower portion. The change in the fluid's speed is accompanied by a change ΔP in the fluid's pressure, which

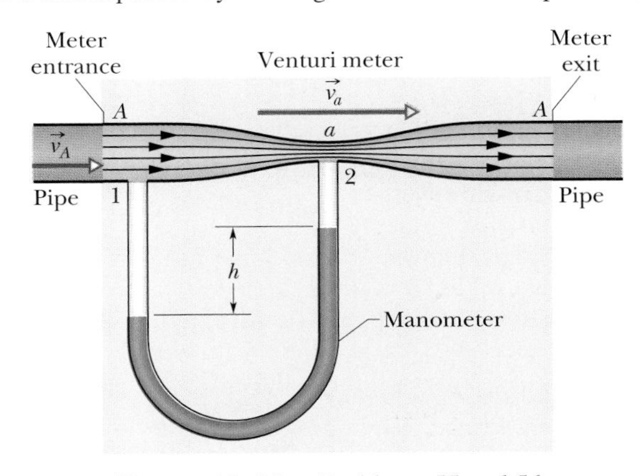

FIGURE 15-48 ■ Problems 55 and 56.

causes a height difference h of the liquid in the two arms of the manometer. (Here ΔP means pressure in the throat minus pressure in the pipe.) (a) By applying Bernoulli's equation and the equation of continuity to points 1 and 2 in Fig. 15-48, show that

$$\vec{v}_A = \sqrt{\frac{2B^2\,\Delta P}{\rho(B^2 - A^2)}},$$

where ρ is the density of the fluid. (b) Suppose that the fluid is fresh water, that the cross-sectional areas are 64 cm^2 in the pipe and 32 cm^2 in the throat, and that the pressure is 55 kPa in the pipe and 41 kPa in the throat. What is the rate of water flow in cubic meters per second?

56. Venturi Tube Consider the venturi tube of Problem 55 and Fig. 15-48 without the manometer. Let A equal $5a$. Suppose that the pressure P_1 at A is 2.0 atm. Compute the values of (a) $|\vec{V}_A|$ at A and (b) $|\vec{v}_a|$ at a that would make the pressure P_2 at a equal to zero. (c) Compute the corresponding volume flow rate if the diameter at A is 5.0 cm. The phenomenon that occurs at a when P_2 falls to nearly zero is known as cavitation. The water vaporizes into small bubbles.

57. Pitot Tube A pitot tube (Fig. 15-49) is used to determine the airspeed of an airplane. It consists of an outer tube with a number of small holes B (four are shown) that allow air into the tube; that tube is connected to one arm of a U-tube. The other arm of the U-tube is connected to hole A at the front end of the device, which points in the direction the plane is headed. At A the air becomes stagnant so that $v_A = 0$. At B, however, the speed of the air presumably equals the airspeed v of the aircraft. (a) Use Bernoulli's equation to show that

$$v = \sqrt{\frac{2\rho g h}{\rho_{\text{air}}}},$$

where ρ is the density of the liquid in the U-tube and h is the difference in the fluid levels in that tube. (b) Suppose that the tube contains alcohol and indicates a level difference h of 26.0 cm. What is the plane's speed relative to the air? The density of the air is 1.03 kg/m^3 and that of alcohol is 810 kg/m^3.

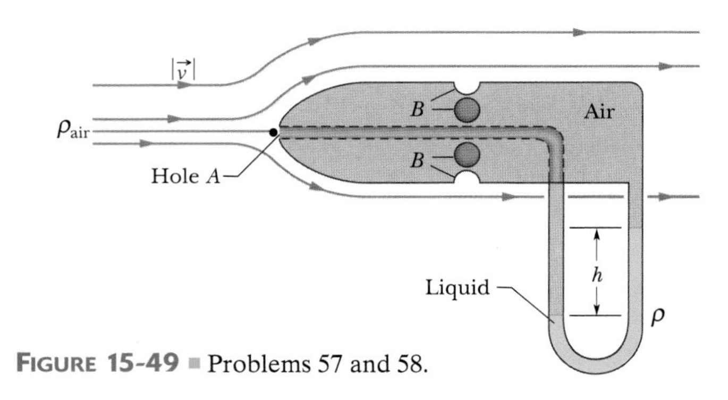

FIGURE 15-49 ■ Problems 57 and 58.

58. High-Altitude Aircraft A pitot tube (see Problem 57) on a high-altitude aircraft measures a differential pressure of 180 Pa. What is the airspeed if the density of the air is 0.031 kg/m^3?

Additional Problems

59. Pool Filling You have been asked to review plans for a swimming pool in a new hotel. The water is to be supplied to the hotel by a horizontal main pipe of radius $R_1 = 6.00$ cm, with water under pressure of 2.00 atm. A vertical pipe of radius $R_2 = 1.00$ cm is to carry the water to a height of 9.40 m, where the water is to pour out freely into a square pool of width 10.0 m and (proposed) water depth of 2.00 m. (a) How much time will be required to fill the pool? (b) If more than a few days is considered unacceptable and less than a few hours is considered dangerous, is the filling time acceptable and safe?

60. Hydraulic Engineers Figure 15-50 shows two sections of an old pipe system that runs through a hill. On each side of the hill, the pipe radius is 2.00 cm. However, the radius of the pipe inside the hill is no longer known. To determine it, hydraulic engineers first establish that water flows through the left-hand and right-hand sections at 2.50 m/s. Then they release a dye in the water at point A and find that it takes 88.8 s to reach point B. What is the radius (or average radius) of the pipe within the hill?

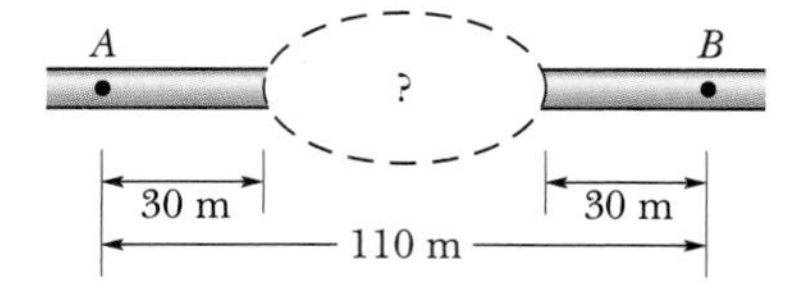

FIGURE 15-50 ■ Problem 60.

61. Floating and Sinking Suppose you have the following collection of objects: a pencil, a coin, an empty plastic box for CDs with its edges taped shut, the same box opened up, a needle, an unopened can of soda pop, and an empty can of soda pop. Which of these objects do you expect will float on water and which will sink? Will it make a difference if you carefully place the object with its largest surface on the surface of the water? In which cases? Discuss the criteria you come up with, explaining carefully why you decided on each one and why it plays a role. After you have written your answer, perform the experiments and compare your results with your predictions.

62. Balloon in a Car Explain why a helium balloon in a closed automobile moves to the front of the car when the car accelerates, whereas the passengers feel pushed backwards. Discuss this in terms of the physics you have learned.

63. At the Pool If an inflated beach ball is placed beneath the surface of a pool and released, it shoots upward, out of the water. Explain why.

64. The Meteor and the Dolphin The curator of a science museum is transporting a chunk of meteor iron (i.e., a piece of iron that fell from the sky—see Fig. 15-51a) from one part of the museum to another. Since the chunk of iron weighs 250 lb and is too big for her to lift by herself, she is using a handtruck (see Fig. 15-51b). While passing through the marine mammals section of the museum, she accidentally hits a bump and the meteorite tips off the handtruck and falls into the dolphin pool. Fortunately, the iron doesn't hit a dolphin, but it quickly sinks to the bottom. "Rats!" she cries. Unfortunately, the meteorite has many sharp edges and she is worried that the dolphins, curious creatures that they are, will come to inspect it and be cut when they rub against it. She wants to get it up out of the pool as quickly as possible. Fortunately, the meteorite has lots of holes in it and there are ropes with hooks on one end lying around. If she could get a hook into one of the holes, she might be able to pull it up to the top, tie the rope around a post, and lever it out with the handtruck. Unfortunately, she remembers that the meteorite is too heavy for her to lift. (a) Will the fact that the meteorite is in the pool under water make it harder or easier for her to lift with the rope? Explain. (b) The meteorite is sitting on the concrete bottom of the pool. Is the force the meteorite exerts on the bottom bigger or smaller than the force it would exert if the pool had no water in it? Explain. (c) Can she lift the meteorite? Calculate how much force she would have to exert on a rope hooked to the meteorite to pull it up from the bottom of the pool. She can lift about 100 pounds, the pool is 12 feet deep, and the density of iron is about 8000 kg/m^3.

FIGURE 15-51(a) ■ Problem 64..

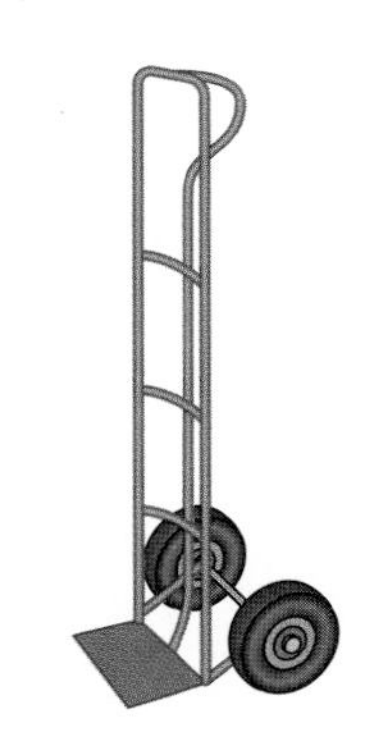

FIGURE 15-51(b) ■ Problem 64.

65. Pushing Iron For each of the following partial sentences, indicate whether they are correctly completed by the symbol corresponding to the phrase *greater than* (>), *less than* (<), or *the same as* (=). (a) A chunk of iron is sitting on a table. It is then moved from the table into a bucket of water sitting on the table. The iron now rests on the bottom of the bucket. The force the bucket exerts on the block when the block is sitting on the bottom of the bucket is _____ the force that the table exerted on the block when the block was sitting on the table. (b) A chunk of iron is sitting on a table. It is then moved from the table into a bucket of water sitting on the table. The iron now rests on the bottom of the bucket. The total force on the block when it is sitting on the bottom of the bucket is _____ it was on the table. (c) A chunk of iron is sitting on a table. It is then covered by a bell jar, which has a nozzle connected to a vacuum pump. The air is extracted from the bell jar. The force the table exerts on the block when the block is sitting in a vacuum is _____ the force that the table exerted on the block when the block was sitting in the air. (d) A chunk of iron is sitting on a scale. The iron and the scale are then both immersed in a large vat of water. After being immersed in the water, the scale reading will be _____ the scale reading when they were simply sitting in the air. (Assume the scale would read zero if nothing were sitting on it, even when it is underwater.)

66. The Three-Vase Puzzle* Water is poured to the same level in each of the three vessels shown in Fig. 15-52. Each vessel has the same base area. Since the water is to the same depth in each vessel, each will have the same pressure at the bottom. Since the area and pressure are the same, each liquid should exert the same force on the base of the vessel. Yet, if the vessels are weighed, three different values are obtained. (The one in the center clearly holds less liquid than the one at the left, so it must weigh less.) How can you justify this apparent contradiction?

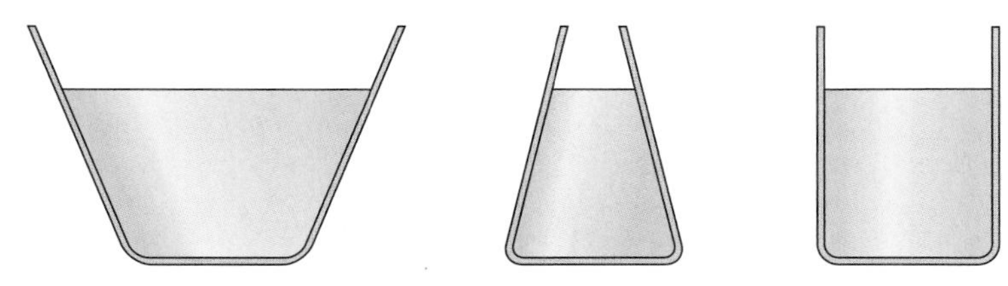

FIGURE 15-52 ■ Problem 66.

67. Hanging Blocks† Three cubical blocks of equal volume are suspended from strings. Blocks A and B have the same mass and block C has less mass. Each block is lowered into a fish tank and they hang at rest as shown in Fig. 15-53. (a) Is the force exerted by the water on the top surface of block A greater than, less than, or equal to the force exerted by the water on the top surface of block B? Explain. (b) Is the force exerted by the water on the top surface of block A greater than, less than, or equal to the force exerted by the water on the top surface of block C? Explain. (c) Is the force exerted on the water by block C greater than, less than or equal to the force exerted on the water by block A? Explain. (d) Rank the buoyant forces acting on the three blocks from largest to smallest. If any buoyant forces are equal, indicate that explicitly. Explain.

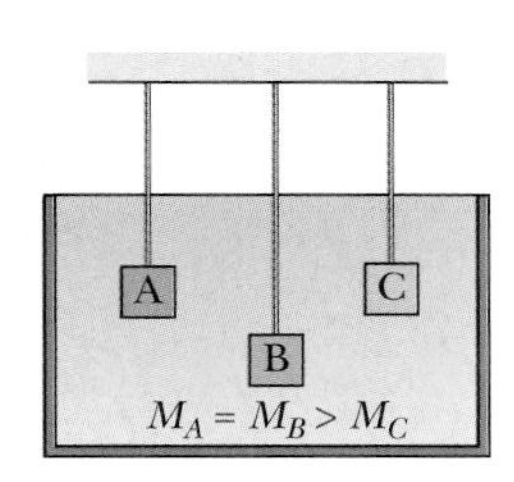

FIGURE 15-53 ■ Problem 67.

68. Floating Blocks‡ Figure 15-54 shows five blocks increasing in mass from block A to block E as indicated. The blocks have equal volumes but different masses. The blocks are placed in an aquarium tank filled with water and blocks B and E come to rest as shown in Fig. 15-54. Sketch on the figure where you would expect blocks A, C, and D to come to rest. (The differences in mass between successive blocks is significant—not just a tiny amount.)

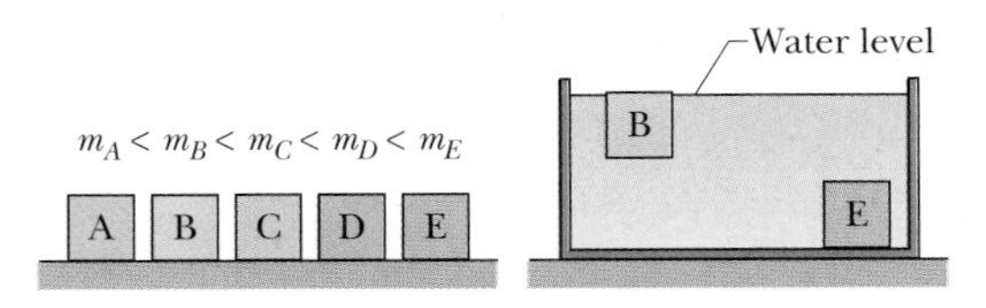

FIGURE 15-54 ■ Problem 68.

*From A. Arons, *A Guide to Introductory Physics Teaching* (New York: John Wiley, 1990).

†From M. E. Loverude, "Investigation of Student Understanding of Hydrostatics and Thermal Physics and the Underlying Concepts from Mechanics," Ph.D. thesis, University of Washington, 1999.

‡From M. E. Loverude, "Investigation of Student Understanding of Hydrostatics and Thermal Physics and the Underlying Concepts from Mechanics," Ph.D. thesis, University of Washington, 1999.

16 Oscillations

On September 19, 1985, seismic waves from an earthquake that originated along the west coast of Mexico caused terrible and widespread damage in Mexico City, about 400 km from the origin.

Why did the seismic waves cause such extensive damage in Mexico City but almost none on the way there?

The answer is in this chapter.

16-1 Periodic Motion: An Overview

Any measurable quantity that repeats itself at regular time intervals is defined as undergoing **periodic** behavior. We are surrounded by systems with quantities that vary periodically. The systems with periodic behavior that are most familiar involve obvious mechanical oscillations or motions. There are swinging chandeliers, boats bobbing at anchor, and the surging pistons in the engines of cars.

The motions associated with some periodic behavior are not obvious. For example, we cannot see the oscillations of the air molecules that transmit the sensation of sound, the oscillations of the atoms in a solid that convey the sensation of temperature, and the oscillations of the electrons in the antennas of radio and TV transmitters that convey information. Some examples of periodic changes are shown in Figs. 16-1 and 16-2. They include electrical signals associated with human heart beats and the air pressure changes that occur when musical instruments are played.

It is obvious from looking at Figs. 16-1 and 16-2 that the variations of electrical signals from the heart and the sound pressure from the trumpet are both periodic but quite complex. On the other hand, the oscillation of air pressure caused by the flute is much simpler. In fact the flute pattern looks like the graph of a sine or cosine function. If the periodic variation of a physical quantity over time has the shape of a sine (or cosine) function, we call it a **sinusoidal oscillation.**

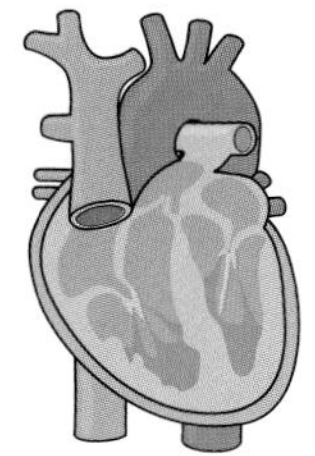

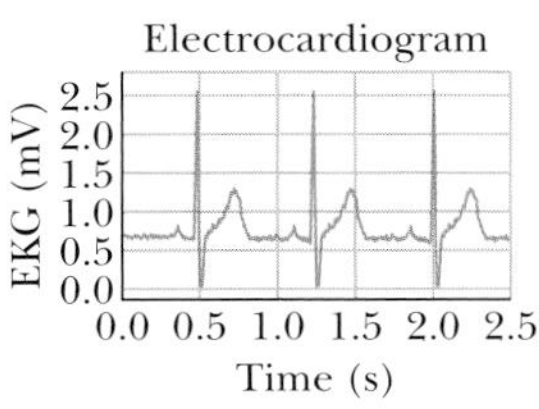

FIGURE 16-1 ■ An electrocardiogram showing the periodic pattern of electrical signals that drive human heart beats. Data recorded with a computer data acquisiton EKG sensor. (Courtesy of Vernier Software and Technology.)

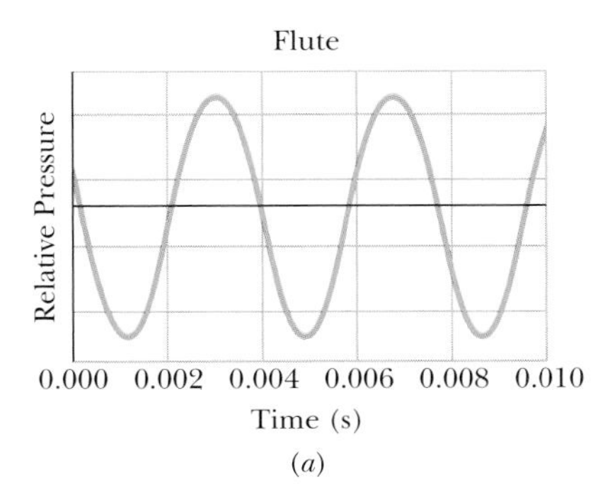

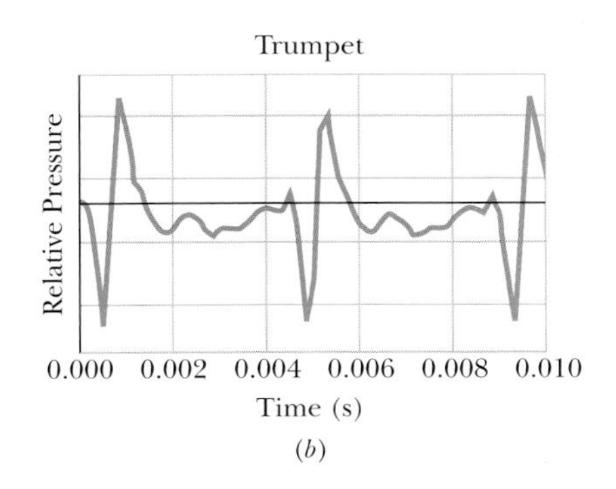

FIGURE 16-2 ■ The disturbance of air molecules causes variations in air pressure near musical instruments. The pattern of these pressure variations repeats itself at regular time intervals so that the sound is periodic. The pressure variations that are proportional to the voltage output of a small microphone can be recorded with a computer data acquisition system. (*a*) A sustained note from a flute and (*b*) from a trumpet. (Data courtesy of Vernier Software and Technology.)

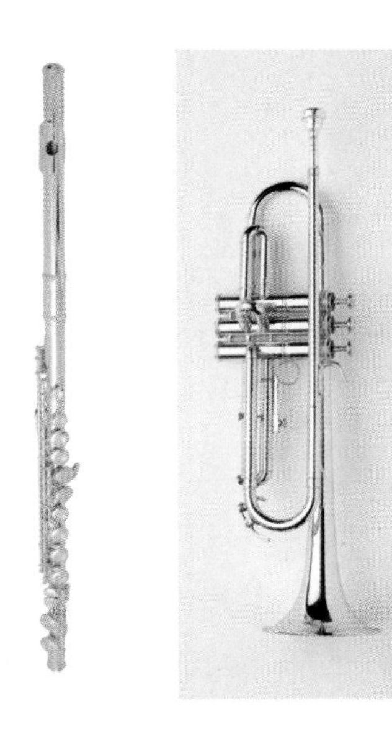

Sinusoidal oscillations are surprisingly common and learning about them helps us understand more complex oscillations. For this reason we begin this chapter by exploring the mathematics of sinusoidal oscillations and how oscillations can be related to the uniform circular motion we studied in Chapters 5, 11, and 12. Mastering the mathematical description of sinusoidal motion is critical to acquiring a full understanding of periodic physical systems. It is also vital to obtaining a full appreciation of the transmission of both mechanical and sound waves treated in Chapters 17 and 18.

As you will see, physicists and engineers refer to the sinusoidal motions of particles in mechanical systems as **simple harmonic motion (or SHM).** In fact, most of the chapter is devoted to understanding how certain forces found in our everyday surroundings cause the sinusoidal oscillations that we call SHM.

Although your study of simple harmonic motion will enhance your understanding of mechanical systems, it is also vital to understanding the topics in waves, electricity, magnetism, and light encountered in Chapters 30–37. Finally, a knowledge of SHM provides a basis for understanding modern physics, including the wave nature of the light and how atoms and nuclei absorb and emit energy.

16-2 The Mathematics of Sinusoidal Oscillations

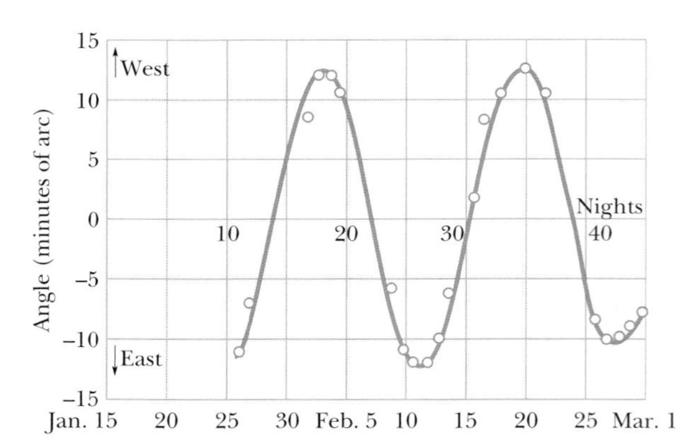

FIGURE 16-3 ■ The angle between Jupiter and its moon Callisto as seen from Earth. The circles are based on Galileo's 1610 measurements. The curve is a best fit, strongly suggesting sinusoidal motion. At Jupiter's mean distance, 10 minutes of arc corresponds to about 2×10^6 km. (Adapted from A. P. French, *Newtonian Mechanics*, W. W. Norton, New York, 1971, p. 288.)

Sinusoidal Oscillations and Uniform Circular Motion

In 1610, Galileo used his newly constructed telescope to discover the four principal moons of Jupiter. Over weeks of observation, each moon seemed to him to be moving back and forth relative to the planet in what today we would call sinusoidal motion; the disk of the planet was the midpoint of the motion. The record of Galileo's observations, written in his own hand, is still available. A. P. French of MIT used Galileo's data to work out the position of the moon Callisto relative to Jupiter. In the results shown in Fig. 16-3, the circles are Galileo's data which looks sinusoidal. The curve shows the best fit of a sinusoidal function to the data. A full oscillation takes about 16.8 days, as can be seen on the plot.

Actually, Callisto moves with essentially constant speed in an essentially circular orbit around Jupiter. The moon's true motion—far from being sinusoidal—is uniform circular motion. What Galileo saw—and what you can see with a good pair of binoculars and a little patience—is the projection of this uniform circular motion on a line in the plane of the motion. We are led by Galileo's remarkable observations to the conclusion that the sinusoidal motion he observed is actually uniform circular motion viewed edge-on. In more formal language:

> The projection of uniform circular motion on a diameter of the circle in which this motion occurs is sinusoidal.

We can explore the relationship between the sinusoidal oscillations and uniform circular motion that we studied in Chapter 5 more carefully using an everyday object instead of a moon that orbits around a distant planet. Consider a spot on a disk that is rotating about an axis at a constant rotational velocity shown in Fig. 16-4. A graph of the spot's vertical displacement x versus time is shown in Fig. 16-5. For this case we have chosen to point the x axis up and the y axis to the right. If we take a series of side views of the disk (so you only see one of the dimensions it is moving in), the projection of the spot on the x axis as a function of time gives us a sinusoidal graph as shown in Fig. 16-4.

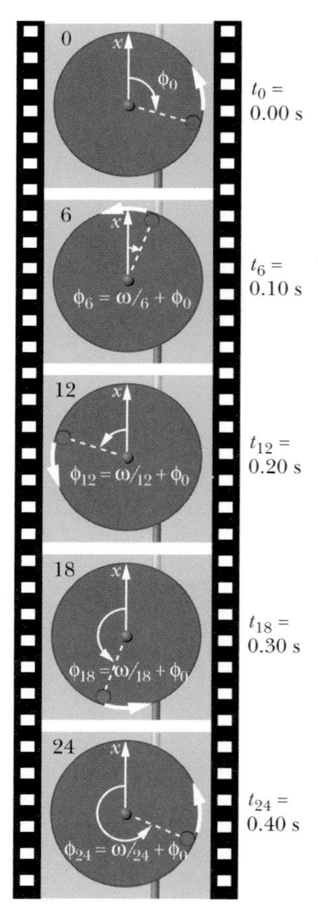

FIGURE 16-4 ■ This selection of frames shows the positions of a spot on a rotating disk every 1/10th of a second. They represent every 6th frame of a video sequence recorded at 60 frames/second. The angle the spot makes with respect to the chosen x axis starts out negative and then increases at a constant rate. *Note*: In order to tie in with Fig. 16-4 we have chosen to orient the x axis vertically.

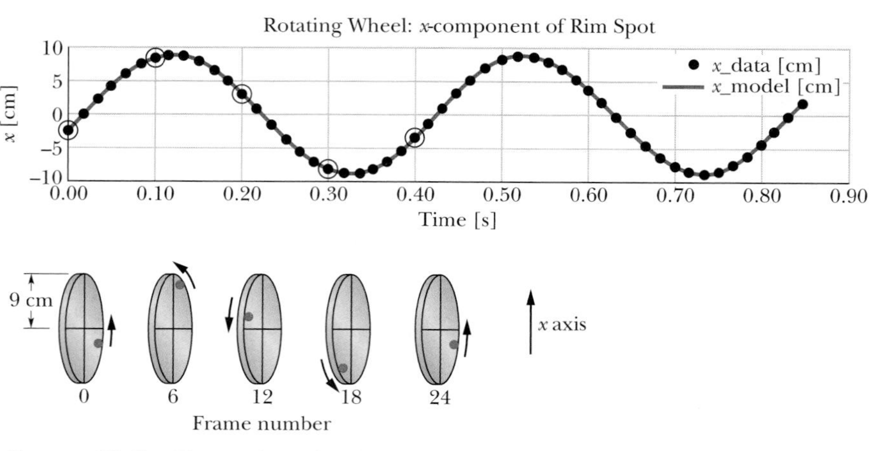

FIGURE 16-5 ■ The motion of a disk with a constant rotational velocity was recorded at 60 frames a second. A plot of the x-component of a spot on the disk has a sinusoidal shape. Thus it oscillates with SHM. The location of the spot on our sideways depictions of the disk is shown below the graph for every 6th frame. *Note:* In order to use conventional polar coordinates to relate the spot on the disk with the x-component of its position, we have chosen the x axis to be vertical.

Period and Frequency

As part of our study of uniform circular motion in Section 5-7 we introduced the idea of a **period** as the time it takes an object rotating about an axis at a regular rate to complete a single revolution. For example, the video frames of a rotating disk in Fig. 16-5 show that it has a period of 0.40 s. In Fig.16-4, we see that the projection of the spot on our chosen x axis appears to oscillate up and down with a period of $T =$ 0.40 s. Thus, the time for one oscillation of the spot's x-component is the same as the period of the rotation of the disk.

Whereas the period tells us the time for one rotation or oscillation, **frequency** is a related quantity that tells us how many oscillations or cycles there are in a given time. For example, we can see from the Fig. 16-4 graph that the spot has a frequency of 2.5 oscillations each second because that's the number of complete oscillations that would occur if we had taken data for 1 s rather than for only 0.85 s. The symbol for frequency is f, and its SI unit is the hertz (abbreviated Hz). Alternative names and symbols for the hertz include

$$1 \text{ hertz} = 1 \text{ Hz} = 1 \text{ cycle/second} = 1 \text{ oscillation/second}. \tag{16-1}$$

Clearly, when the period of oscillation is very short there are many more oscillations in a second so the frequency goes up. The converse is true also; when the period is long, the frequency goes down. In fact, we see that the period and frequency of the oscillations shown in Fig. 16-4 are inversely related to each other. This inverse relationship holds in general, so that

$$T = \frac{1}{f}. \tag{16-2}$$

The Equation Describing Sinusoidal Motion

We have rather glibly described the graphs shown in Figs. 16-2a and 16-4 as representing sinusoidal functions. We do this because the graphs look like that of a sine or cosine as a function of angle. Recall that the cosine of an angle, ϕ, is defined as the ratio of the distance of a point of interest from the y axis, denoted as x, and the magnitude of the distance of the point from the origin denoted as r as shown in Fig. 16-6. The sine function is similarly defined in terms of a ratio involving the distance from the x axis:

$$\cos \phi \equiv \frac{x}{r} \tag{16-3}$$

and

$$\sin \phi \equiv \frac{y}{r}. \tag{16-4}$$

In considering rotational positions, we continue the convention of describing angles in radians (or rads) used in Chapters 11 and 12. Although it would be possible to use degrees, radian measure is required if we want to take derivatives of mathematical functions that involve angles. Recall that in Eq. 11-1 the magnitude of the radian is defined as the magnitude of the ratio of the arc length s of a rotating object and the perpendicular distance from its axis of rotation ($|\phi| = s/r$) to the object (or spot). As the spot moves through a complete cycle, it sweeps out an arc length of $s = 2\pi r$ and thus an angle of $(2\pi r)/r = 2\pi$ radians. Since 2π radians $= 360°$, we can convert from radians to degrees by multiplying by the factor

$$\phi\,(\text{deg}) = \left(\frac{180°}{\pi \text{ rad}}\right) \phi\,(\text{rad}) \quad \text{(conversion from radians to degrees).}$$

Obviously we divide by the factor to convert to radians from degrees.

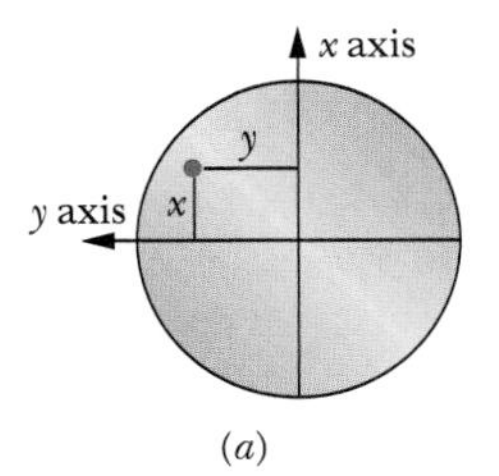

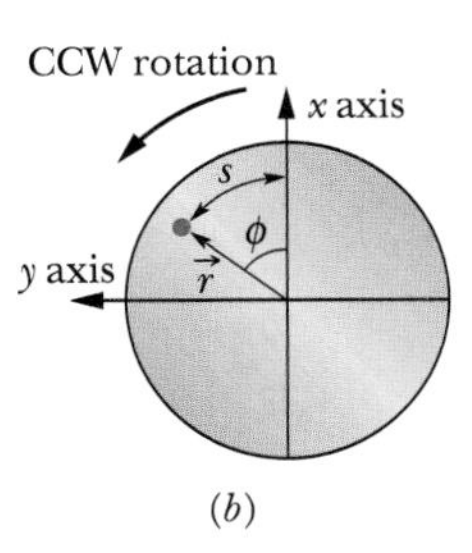

FIGURE 16-6 ■ The location of any spot on a disk can be described in either (a) Cartesian or (b) polar coordinates. Rotational position is denoted as ϕ. A counterclockwise arc from the axis is defined as a positive ϕ. *Note*: In order to coordinate with Fig. 16-4, we have also chosen the x axis to be vertical here.

How do our definitions of the sine and cosine functions in Eq. 16-3 and Eq. 16-4 lead to the graph shapes shown in Figs. 16-2*a* and 16-4? Let's use the cosine function as an example. In Fig. 16-5, a spot on a disk is turning with a constant rotational velocity ω. If we denote the initial angular position of the disk spot when $t = 0$ s as ϕ_0, then the angular position increases at a constant rate according to the equation $\phi(t) = \omega t + \phi_0$. This is shown is Fig. 16-5, in which the initial angular position ϕ_0 is taken to be negative since it is marked as being in a clockwise direction relative to the x axis. As time goes on the angular position $\phi(t)$ increases, passes through zero, and becomes positive.

Let us think about what happens in Cartesian coordinates. How does the x-component of the vector, $\vec{r}$, shown in Fig 16-6 vary over time for a counterclockwise rotation? In the time period where $\phi(t)$ is near zero the value of the x-component, denoted as x, does not change very rapidly but after a quarter-turn near $\phi(t) = \pi/2$ the value of x is changing very rapidly. The rate of change of x slows down again near $\phi(t) = \pi$. Between π and 2π, x is negative but its rate of change speeds up and slows down again. This clearly leads to a sinusoidal graph shape that goes on and on as the disk spot turns round and round.

In general for a sinusoidal motion, the value of the x-component of the spot as a function of time *can be described by either a sine or cosine function,* depending on which axis the angle is measured from. The values of x taken relative to the origin are typically called the displacement. Using the cosine function we find that the sinusoidal variation over time of the **displacement,** x, can be represented by the equation

$$x(t) = X \cos(\omega t + \phi_0) \quad \text{(sinusoidal displacement)}, \tag{16-5}$$

where X, ω, and ϕ_0 are constants.

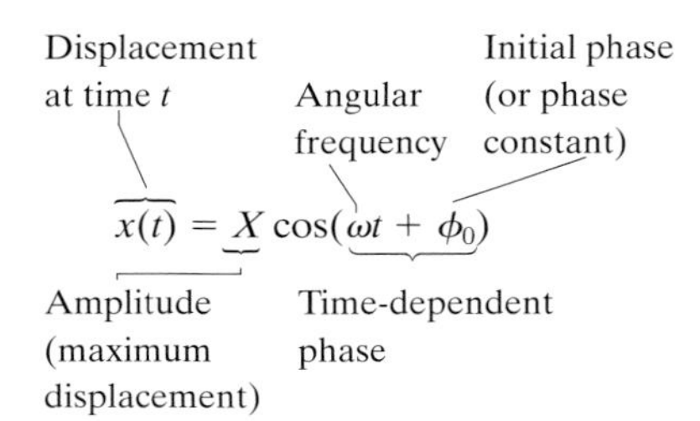

FIGURE 16-7 ■ A handy reference to the quantities in Eq. 16-5 for simple harmonic motion.

When a sine function is shifted left by 90° or $+\frac{\pi}{2}$rad it looks exactly like a cosine function. So an alternate, equally viable equation for displacement $x(t)$ would be $x(t) = X \sin(\omega t + \phi_0')$, where $\phi_0' = (\phi_0 + \frac{\pi}{2})$.

For convenience the quantities that determine the shape of the graph of Eq. 16-5 are named and displayed in Fig. 16-7. X is defined as the **amplitude** of the x-component motion. It is a positive constant whose value represents the magnitude of the maximum displacement of the particle in either direction from its so-called equilibrium value. The cosine function in Eq. 16-5 varies between the limits ± 1, so the displacement $x(t)$ varies between the limits $\pm X$. For example, in Fig. 16-4 the amplitude of the cosine curve is obviously 9 cm. This is also the maximum distance from the axis of rotation of the disk to the spot.

The constant ϕ_0 is called the **initial phase.** It is also sometimes called the *phase constant* or *phase angle.* The value of the initial phase ϕ_0 allows us to calculate the magnitude of the displacement of the x-component, denoted as $x(0)$, at the starting time $t = 0$ s. Since the expression $\omega t = 0$ rad when $t = 0$ s, we get $x(0) = X\cos(\phi_0)$ or $\phi_0 = \pm\cos^{-1}(x(0)/X)$. Knowing the initial phase and displacement allows us to determine the initial angular velocity. The initial phase plays the same role as the x_1 or y_1 terms in the kinematic equations because it determines the initial value of the function $x(t)$.

$\phi(t) = \omega t + \phi_0$ is defined as the **time-dependent phase** of $x(t)$. Some time-dependent phases for the rotating disk are shown in Fig. 16-4. Each "frame" shown consists of every 6th frame of a more complete set of video images. The phases in the selection of frames shown in the figure are denoted as ϕ_0, $\omega t_7 + \phi_0$, $\omega t_{13} + \phi_0$, and so on, where the corresponding times are $t_1 = 0.00$ s, $t_6 = 0.10$ s, $t_{12} = 0.30$ s.

We find that if X is a maximum when $t = 0$ s the initial phase is either $\phi_0 = 0$ rad, $\pm 2\pi$ rad, $\pm 4\pi$ rad, . . . , and so on. This is because in this case $x(0) = X\cos(0) = X$. In other words, whenever the initial phase is a multiple of 2π radians, the displacement is a maximum at $t = 0$ s. For simplicity, in the $x(t)$ plots in Fig. 16-8*a* the initial phase (or phase constant) ϕ_0 has been set to zero radians.

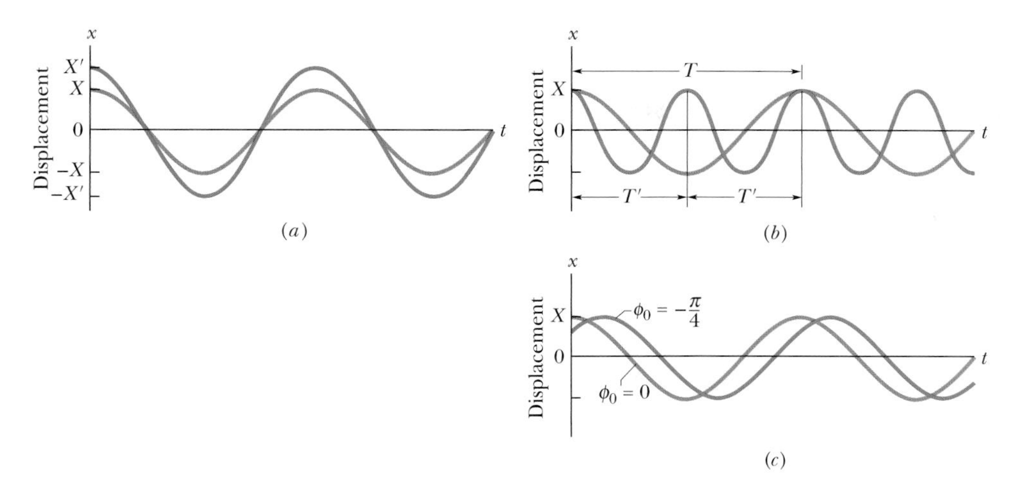

FIGURE 16-8 ■ Graphs of two sinusoidal motions. In each case, the blue curve is obtained from Eq. 16-5 with $\phi_0 = 0$ rad. (*a*) The red curve differs from the blue curve *only* in that its amplitude X' is greater (the red curve extremes of displacement are higher and lower). (*b*) The red curve differs from the blue curve *only* in that its period is $T' = T/2$ (the red curve is compressed horizontally). (*c*) The red curve differs from the blue curve *only* in that $\phi_0 = -\pi/4$ rad rather than zero (the negative value of ϕ_0 shifts the red curve to the right).

In describing sinusoidal motion, the constant ω is known as the **angular frequency** of the motion. When the sinusoidal equation represents the projection of a spot on a steadily rotating object along an axis, *the rotational velocity of the object and the angular frequency of the projection are identical in value*. This is certainly the case in Fig. 16-4.

The SI unit of angular frequency is the radian per second. Figure 16-8 compares $x(t)$ for two sinusoidal motions that differ either in amplitude, in period (and thus in frequency and angular frequency), or in initial phase.

We can use the fact that the rotational velocity of a spinning object and the angular frequency of a corresponding oscillation are identical to derive an important relationship. In particular, we can relate the angular frequency, ω, of an oscillating object to its oscillation frequency f. The derivation goes as follows: A rotating object undergoing uniform circular motion sweeps through an angle of 2π radians in a single period T so that its rotational velocity is given by $\omega = 2\pi/T$. But since the frequency is inversely proportional to the period (that is, since $f = 1/T$), it is obvious that

$$\omega = \frac{2\pi}{T} = 2\pi f. \tag{16-6}$$

READING EXERCISE 16-1: Although it is not conventional to do so, the equation $x(t) = X' \sin[\omega' t + \phi'_0]$ can also be used to describe sinusoidal motion. Suppose the same motion has been described by both the cosine function in Eq. 16-5 and by the sine function shown here. Consider the amplitude, angular frequency, and initial phase associated with the motion. Which factors stay the same? Which will change? Explain. *Hint*: You may want to think about the spot on a disk as it undergoes uniform circular motion. ■

READING EXERCISE 16-2: A particle undergoes sinusoidal oscillations of period T (such as the curve with a value of $+X$ at $t = 0$ s in Fig. 16-8*a*). Assume the particle is at $-X$ at time $t = 0$. (a) When $t = 2.00T$, where is the particle? At $-X$? $+X$? Zero? Between $-X$ and 0 m? Or between 0 m and $+X$? Answer the same questions for the following times: (b) $t = 3.50T$, and (c) $t = 5.25T$. ■

16-3 Simple Harmonic Motion: The Mass–Spring System

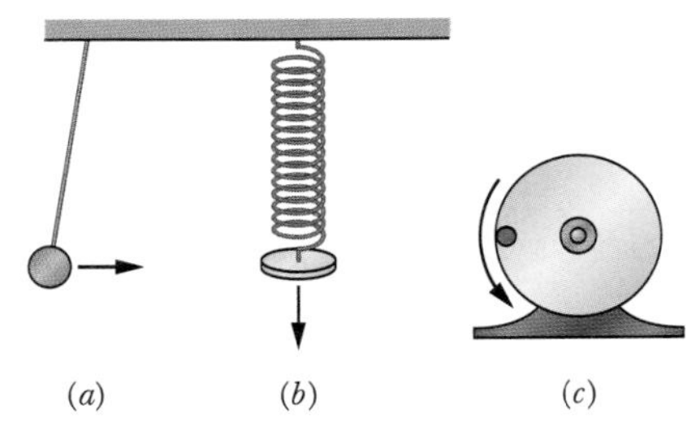

FIGURE 16-9 ■ We know that the (a) pendulum oscillating at small angles, the (b) mass on the spring, and the (c) Cartesian components of a spot on a disk undergo sinusoidal motion. These can be made to move with the same period and phase. The spring and the spot on the rotating disk can also have the same amplitude. What are the mathematical characteristics of forces needed to induce sinusoidal motion in a mass–spring system? In a pendulum? Are they the same?

So far we have described sinusoidal motion mathematically as in Eq. 16-5. We considered a particle on a rotating disk undergoing uniform circular motion, as shown in Fig. 16-9(c), and we observed something special about the component along an axis in the plane of motion passing through the center of rotation: It varies sinusoidally. In this section we explore the behavior of particles that move back and forth sinusoidally along a straight line. We will use a mass–spring system as a model, as shown in the middle of Fig. 16-9. In particular, we are interested in: (1) determining the behavior of a particular spring force experimentally and showing that it causes sinusoidal motion of a mass attached to that spring; (2) using Newton's Second Law to predict theoretically that a spring should cause a mass to oscillate with one-dimensional sinusoidal motion; and (3) discussing how our theory predicts some surprising characteristics of that motion that leads us to define it as simple harmonic motion.

In Section 9-5 we explored Hooke's law for an ideal spring. We imagined a horizontally oriented spring attached to a wall at one end and to a mass at the other end (like that shown in Fig. 16-10). We noted that in its relaxed state, the spring exerts no forces. However, if the spring is displaced from its relaxed state so it is stretched or compressed, it exerts a force on anything attached to its ends. The direction of the spring force always acts in a direction *opposite* to its displacement. The force tries to bring the spring to its relaxed state. For this reason, we describe a spring force as a **restoring force.** If we are careful to put the origin of our chosen x axis at the relaxed position, then Hooke's law can be expressed using Eq. 9-17,

$$F_x^{\text{spring}} = -kx,$$

where F_x^{spring} is the x-component of the spring force, x is the displacement from its equilibrium position, and k is the spring constant (or stiffness factor).

Experimental Findings: Forces, Displacement, and Time

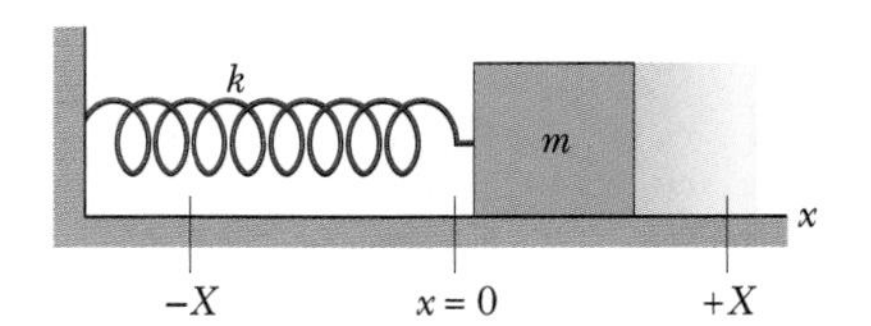

FIGURE 16-10 ■ We can imagine a *horizontal oscillator* that consists of a mass–spring system. In theory, the mass can oscillate back and forth on a frictionless surface when it is displaced from equilibrium. In practice, this system is very difficult to set up.

In physics we usually analyze a simplified model system before considering more complex "real-world" systems. Unfortunately, in picking a model mass–spring system we are presented with a dilemma. The simplest system to model mathematically is a *horizontal oscillator* with a partially extended spring attached to a block that slides on a perfectly frictionless surface as shown in Fig. 16-10. However, it is not easy without special equipment (such as an air track) to set up a friction-free experiment using such a system. The simplest system to set up experimentally is a *vertical oscillator* in which a mass descends as it stretches until there is no net force on the mass. This vertical location of a mass hanging on a spring is called its **equilibrium position.** However, both theory and experiments show that the corresponding components of net forces caused by displacements from equilibrium on masses hanging vertically and horizontal masses are the *same.** Thus, for simplicity, we will consider our model system to be a mass hanging down vertically from a spring.

We start by presenting the results of measurements made on our vertical mass–spring system. Next we show how our experimental knowledge of forces on the system can be used in conjunction with Newton's Second Law to derive its position vs. time equation theoretically. At the same time we can determine theoretically how factors such as mass, spring stiffness, and amplitude influence the period of oscillation. In a later section we will show how the forces experienced by a pendulum bob have the same mathematical characteristics as those that drive a mass on a spring.

*The theoretical equivalency of the horizontal and vertical mass-spring system is developed in Touchstone Example 16-3.

A 100 g mass is attached to a light 10.1 g spring and suspended from an electronic force sensor, shown in Fig. 16-11. An ultrasonic motion detector is placed underneath it to record its displacement from equilibrium. The mass is pulled down to a maximum displacement of about 4.0 cm and released. About a half second later, a computer data acquisition system begins to measure the forces exerted by the spring on the mass as a function of the displacement of the mass (Fig. 16-12). At the same time, the displacement of the mass is tracked with the motion detector. The experimental results for the net force on the mass as a function of time are shown in Fig. 16-13*a*. Similarly, the experimental results for the displacement from equilibrium as a function of time are shown in Fig. 16-13*b*.

FIGURE 16-11 ■ A common way to measure forces and displacements for a vertical mass–spring oscillator is to hang a light spring with an attached mass from an electronic force sensor. An ultrasonic motion detector can be placed underneath the mass to measure displacements. If we displace the mass vertically from its equilibrium point, it will oscillate between a maximum value of displacement from equilibrium $+Y$ and a minimum value $-Y$.

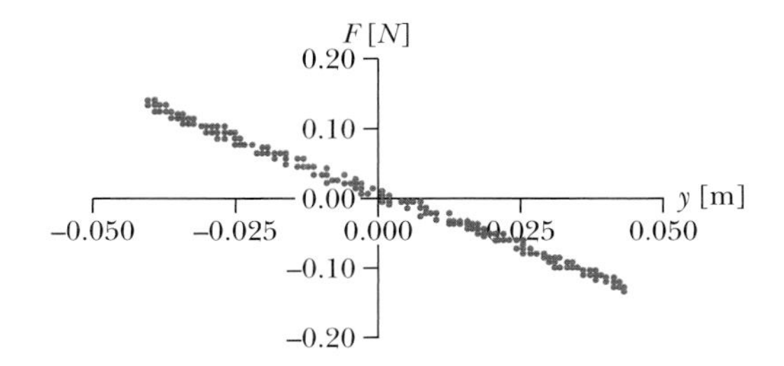

FIGURE 16-12 ■ A graph of the data showing net force on the mass as a function of its displacement from equilibrium. It is obvious that Hooke's law holds with $F_y^{net} = -ky$. A fit to the data shows that the magnitude of the slope of the graph (which is the value of k) is given by $k = 3.23$ N/m.

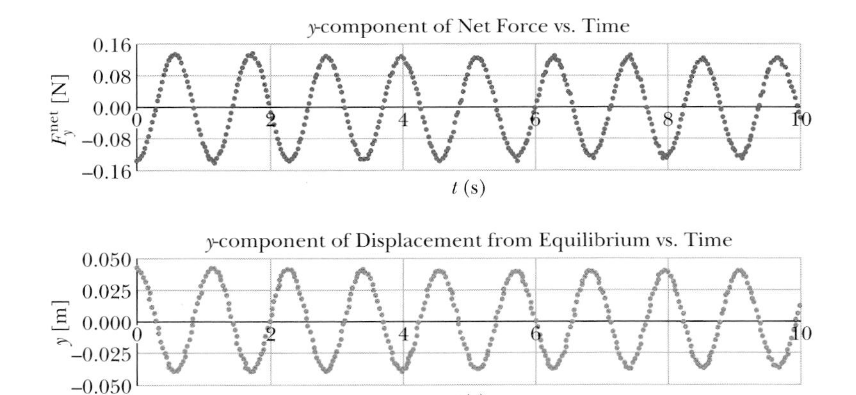

FIGURE 16-13 ■ These graphs of experimental data show that the net force exerted on a 100 g mass by a spring causes it to oscillate with what appears to be sinusoidal motion. Examination of the two graphs shows that on a moment-by-moment basis the net force is proportional to the displacement but oppositely directed.

Careful examination of the graph in Fig. 16-13 gives a period of oscillation for our 100 g mass that is $T^{exp} = 1.1$ s. A model of the plotted data shows that the equation that describes the displacement versus time has exactly the same form as Eq. 16-5. In particular,

$$x(t) = X\cos[\omega t + \phi_0] = (0.040 \text{ m})\cos[(5.5 \text{ rad/s})\,t + 0.0 \text{ rad}]. \quad (16\text{-}7)$$

The value of the maximum displacement or amplitude X is 0.040 m, or only 4.0 cm. The angular frequency, ω, is 5.5 rad/s, and the initial phase (or phase constant), ϕ_0, is 0.0 rad. Equation 16-7 provides experimental verification that our vertical spring–mass system oscillates *sinusoidally*.

We see from Fig. 16-13*a* that the restoring force component in the y direction is proportional to the displacement component but opposite in sign—the familiar Hooke's law as described earlier in Eq. 9-17 (with y playing the role of x). We can use the experimental fact that the spring force causes our mass to oscillate sinusoidally to develop a definition of simple harmonic motion:

> **Simple harmonic motion (or SHM)** is the sinusoidal motion executed by a particle of mass m subject to a one-dimensional net force that is proportional to the displacement of the particle from equilibrium but opposite in sign.

As we show later in this section, the displacement can be either linear or rotational.

Theoretical Prediction: Spring Forces Cause SHM

We can combine Newton's Second Law with our experimental verification of Eq. 9-17 for a spring–mass system that undergoes SHM to get

$$F^{\text{net}} = -ma_x = -kx, \tag{16-8}$$

where F^{net}, a_x, and x are the respective x-components of force, acceleration, and displacement. This equation can be used to explain why Eq. 16-5 does indeed describe the motion of our mass–spring system. To do this, we need to express the acceleration of the mass as the second derivative of the displacement. By doing this we can rewrite Eq. 16-8 as

$$F^{\text{net}} = ma_x = m\frac{d^2x}{dt^2} = -kx, \quad \text{so that} \quad \frac{d^2x}{dt^2} = -\frac{k}{m}x. \tag{16-9}$$

What happens when we actually take the first and then the second derivative of Eq. 16-5? Do we get a minus sign and a positive constant that can be associated with the ratio k/m? The first derivative is

$$\frac{dx}{dt} = \frac{d[X\cos(\omega t + \phi_0)]}{dt} = -\omega X\sin(\omega t + \phi_0),$$

and the second derivative is

$$\frac{d^2x}{dt^2} = \frac{d[-\omega X\sin(\omega t + \phi_0)]}{dt} = -\omega^2 X\cos(\omega t + \phi_0) = -\omega^2 x. \tag{16-10}$$

We do indeed get a negative sign times a positive constant in front of the x term, but for the theoretical equation derived from Newton's Second Law to match our experimentally determined equation for displacement versus time, we must have the angular frequency be equal to

$$\omega = \sqrt{\frac{k}{m}} \quad \text{(angular frequency).} \tag{16-11}$$

By combining Eqs. 16-6 and 16-11, we can write, for the **period** of the linear oscillator shown in Fig. 16-11,

$$T = 2\pi\sqrt{\frac{m}{k}} \quad \text{(period).} \tag{16-12}$$

Equations 16-11 and 16-12 tell us that a large angular frequency (and thus a small period) goes with a stiff spring (large k) and a low-mass object (small m). This seems like a very reasonable pair of relationships. Let us verify whether our data for the mass–spring system satisfies Eq. 16-12. In order to simplify our theoretical model *we have chosen to ignore the relatively small 10.1 g mass of the spring* and assume that the mass of the system can be adequately represented by only the hanging mass. Using this assumption our theoretical model tells us that a spring with k of 3.23 N/m and hanging mass, m, of 0.10 kg will have a period of

$$T^{\text{theory}} = 2\pi\sqrt{\frac{m}{k}} = 2\pi\sqrt{\frac{0.10\text{ kg}}{3.23\text{ N/m}}} = 1.1\text{ s.}$$

This predicted period is the same as our experimental value of $T^{\text{exp}} = 1.1$ s to 2 significant figures.

Describing How Spring Forces Cause Oscillations

One quite surprising result of our theoretical predictions is that Eq. 16-12 tells us that *the period of the oscillations do not depend on amplitude*. You might guess that if you stretch or compress the spring more, the distance the mass travels in an oscillation will be greater. But additional experiments show that period does not depend on amplitude as long as we don't distort the spring so that Hooke's law fails. This is because if we start with a larger initial displacement (amplitude) the forces and accelerations are greater and our mass moves faster.

We can now see why an object that experiences a spring force oscillates. If we hold the object out at some displacement and release it, the net force acts opposite to the displacement so the object will start to accelerate toward its equilibrium position. It moves faster and faster toward the equilibrium position but as it gets there, the force is becoming smaller, so when it gets to its equilibrium position it is moving with some velocity. By Newton's First Law the object keeps going and overshoots. The force now acts in the opposite direction to slow it down, but by the time the force has brought the object to its turn-around point the object has another displacement, this time on the other side. The process repeats, and if there is no friction or damping, the oscillations will go on forever.

Not All Sinusoidal Motions Are SHM

Every oscillating system such as a diving board or a violin string has an element of "springiness" and an element of "inertia" or mass, and thus behaves like the linear oscillator of Fig. 16-11. As long as the forces that act on objects in an oscillating system are linear restoring forces, we will get simple harmonic motion that is sinusoidal. Almost all sinusoidal motions qualify as simple harmonic motion. One notable exception is the sinusoidal motion of the moon Callisto about Jupiter that Galileo observed (Fig. 16-3). The motion Galileo observed is the projection of a nearly circular orbit seen edge-on. The gravitational force law that keeps the moon in orbit is not proportional to its displacement from the center of the orbit. Therefore, Callisto is not undergoing simple harmonic motion. One consequence of this is that the period of the moon's motion is not amplitude-independent—it depends of the orbital radius. Although the motion Galileo observed is sinusoidal, it does not qualify as SHM.

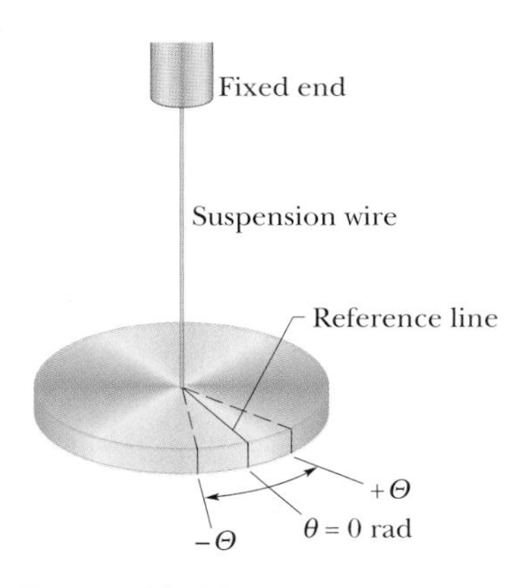

FIGURE 16-14 ■ An angular simple harmonic oscillator, or torsion oscillator, is an angular version of the linear simple harmonic oscillator of Fig. 16-10. The disk oscillates in a horizontal plane; the reference line oscillates with angular amplitude Θ. The twist in the suspension wire stores potential energy as a spring does and provides the restoring torque.

A Rotational Simple Harmonic Oscillator

Figure 16-14 shows a rotational version of a simple harmonic oscillator; the element of springiness or elasticity is associated with the restoring torque the suspension wire can exert when it's twisted. This is rather like the forces a spring can exert when it is stretched or compressed. A device that can exert a restoring torque on an object is called a **torsion oscillator** (or torsion pendulum), with *torsion* referring to the twisting.

If we rotate the disk in Fig. 16-14 by some rotational displacement of magnitude $|\Theta|$ from its equilibrium position (where the reference line is at $\theta = 0$) and release it, it will oscillate with a rotational amplitude of Θ about that position in **rotational simple harmonic motion.** Displacing the disk through any angle θ in either direction from its equilibrium orientation results in the suspension wire of a restoring torque given by

$$\tau^{\text{net}} = I\alpha = -\kappa\theta. \tag{16-13}$$

Here κ (Greek *kappa*) is a constant, called the **torsion constant,** that depends on the length, diameter, and the suspension wire's shear modulus (as defined in Section 13-5).

Comparison of Eq. 16-13 with Eq. 16-8 leads us to suspect that Eq. 16-13 is the rotational form of Hooke's law, and that we can transform Eq. 16-12, which gives the period of linear SHM, into an equation for the period of rotational SHM; we replace the spring constant k in Eq. 16-12 with its equivalent, the constant κ of Eq. 16-13, and we replace the mass m in Eq. 16-12 with its equivalent, the rotational inertia I of the oscillating disk. These replacements lead to

$$T = 2\pi\sqrt{\frac{I}{\kappa}} \quad \text{(torsion oscillator),} \tag{16-14}$$

which is the correct equation for the period of a rotational simple harmonic oscillator, or torsion pendulum.

READING EXERCISE 16-3: The experimental period of the 100 g vertical mass oscillating on the spring shown in Fig. 16-11 is about 3% larger than the period calculated using a simplified theoretical model in which the spring is assumed to be massless, but we know the spring actually has a mass of 10.1 g. Consider the nature of Eq. 16-12 and explain why the measured period should be a bit longer than the theoretical value we reported. No calculation is needed. ■

READING EXERCISE 16-4: Which of the following relationships between the x-component force F_x on a particle and the particle's position x implies simple harmonic oscillation: (a) $F_x = (-5\text{ N/m})x$, (b) $F_x = (-400\text{ N/m}^2)x^2$, (c) $F_x = (+10\text{ N/m})x$, (d) $F_x = (3\text{ N/m}^2)x^2$? ■

16-4 Velocity and Acceleration for SHM

The Velocity for Simple Harmonic Motion

Let us imagine how the velocity of the mass on the spring in Fig. 16-11 changes as it moves through a complete oscillation cycle. Is the magnitude of the velocity a maximum, a minimum or zero when the magnitude of its displacement is the greatest? Obviously the mass has a velocity of zero when it is turning around. But the mass turns around when the magnitude of the displacement is a maximum. This means that the velocity of the mass is out of phase with its displacement in the same way that the cosine and sine functions are out of phase with each other. By differentiating Eq. 16-5, we find that the expression for the velocity of a particle moving with simple harmonic motion is indeed a sine function whenever the displacement is a cosine function; that is,

$$v(t) = \frac{dx(t)}{dt} = \frac{d}{dt}[X\cos(\omega t + \phi_0)],$$

or $$v(t) = -\omega X \sin[\omega t + \phi_0] \quad \text{(velocity)}. \tag{16-15}$$

Figure 16-15a is a plot of the x-component of displacement given by Eq. 16-5 with $\phi_0 = 0$ rad. Figure 16-15b shows Eq. 16-15, also with $\phi_0 = 0$ rad. Analogous to the amplitude X in Eq. 16-5, the positive quantity ωX in Eq. 16-15 is the maximum velocity V and is called the **velocity amplitude.** As you can see in Fig. 16-15b, the velocity of the oscillating particle varies between the limits $\pm\omega X$. Note also in that figure that the curve of $v(t)$ is *shifted* (to the left) from the curve of $x(t)$ by one-quarter period; when the magnitude of the displacement is greatest (that is, $x(t) = X$), then the magnitude of the velocity is least [that is, $v(t) = 0$ m/s]. When the magnitude of the displacement is least (that is, zero), the magnitude of the velocity is greatest (that is, $V = \omega X$).

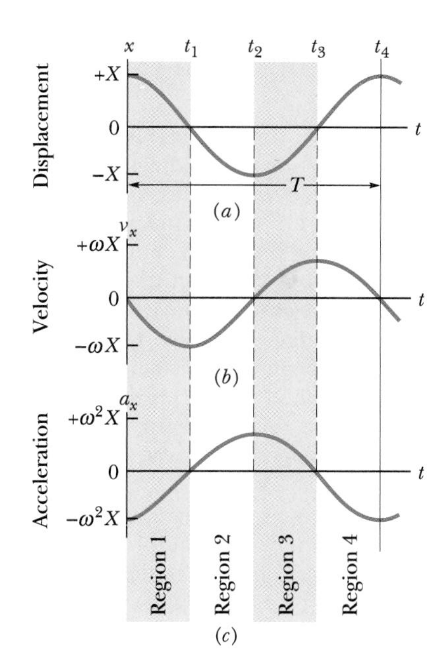

FIGURE 16-15 ■ Assume $t_1 = 0$ s. (a) The displacement $x(t)$ of a particle undergoing SHM with an initial phase of $\phi_0 = 0$ rad. The period T marks one complete oscillation. (b) The velocity $v_x(t)$ of the particle. (c) The acceleration $a_x(t)$ of the particle.

The Acceleration of SHM

Knowing the velocity $v(t)$ for simple harmonic motion, we can find an expression for the acceleration of the oscillating particle by differentiating once more. Thus, we have, from Eq. 16-15,

$$a(t) = \frac{dx(t)}{dt} = \frac{d}{dt}[-\omega X \sin(\omega t + \phi_0)],$$

or $$a(t) = -\omega^2 X \cos(\omega t + \phi_0) \quad \text{(acceleration)}. \tag{16-16}$$

Figure 16-15c is a plot of Eq. 16-16 for the case where the initial phase is zero (or $\phi_0 = 0$ rad). The positive quantity $\omega^2 X$ in Eq. 16-16 is equal to the maximum acceleration called the **acceleration amplitude** A; that is, the acceleration of the particle varies between the limits $\pm A = \pm\omega^2 X$, as Fig. 16-15c shows. Note also that the curve of $a(t)$ is shifted (to the left) by one-quarter period relative to the curve of $v(t)$.

We can combine Eqs. 16-5 and 16-16 to yield

$$a(t) = -\omega^2 x(t), \tag{16-17}$$

which is the hallmark of simple harmonic motion:

> In SHM, the acceleration is proportional to the displacement but opposite in sign, and the two quantities are related by the square of the angular frequency.

Thus, as Fig. 16-15 shows, when the displacement has its greatest positive value, the acceleration has its greatest negative value, and conversely, when the displacement is zero, the acceleration is also zero.

READING EXERCISE 16-5: Consider Fig. 16-15b, which shows the velocities of a mass on spring as a function of time. Identify at what time or times (t_1, t_2, t_3, and t_4) the vertical component of velocity of the mass has a maximum value, a minimum value, and is zero. ■

READING EXERCISE 16-6: Consider Fig. 16-15c, which shows the acceleration of a mass on spring as a function of time. In which region or regions (1, 2, 3, or 4) is the vertical component of acceleration of the mass increasing? Decreasing? ■

TOUCHSTONE EXAMPLE 16-1: Spider Oscillations

Orb web spiders are commonly found outdoors near buildings. An orb web spider with a mass of about 2 g drops off a tree branch onto the center of her horizontal web (Fig. 16-16). As a result, the web undergoes a vertical displacement from its original location to the equilibrium point of the spider-web system (as determined by the vertical location of the spider when the oscillations damp out).

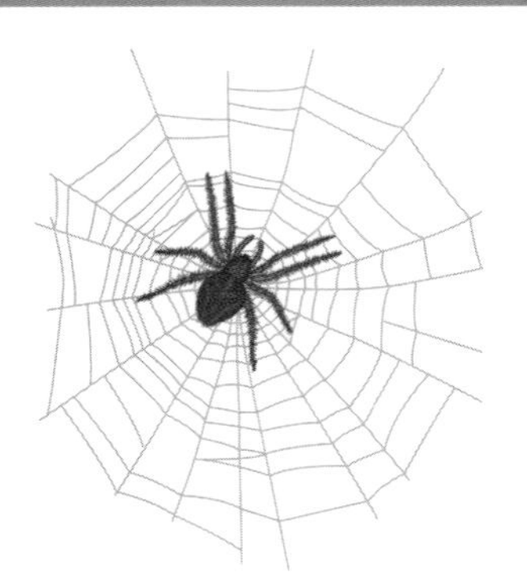

FIGURE 16-16 ■ Top view of a spider dropping onto her web.

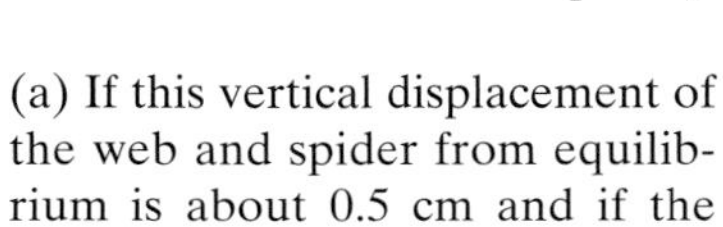

(a) If this vertical displacement of the web and spider from equilibrium is about 0.5 cm and if the vertical restoring force the web exerts on the spider is proportional to her displacement, estimate the frequency and period of the vertical oscillation of the spider-web system.

SOLUTION ■ The **Key Idea** here is that as the spider steps on her web the degree of sag caused by the force of her weight enables us to find the web's spring constant k. Once we know the spring constant we can use that along with her estimated mass to determine the frequency and period of the spider's oscillation.

We choose a y axis that points vertically upward and use Eq. 9-16 $[F_y^{spring} = -k(y_2 - y_1)]$. By noting that the vertical component of the web's net upward restoring force is $+mg$ and that $y_2 < y_1$, we get

$$k = \frac{F_y^{spring}}{-(y_2 - y_1)} = \frac{+(2 \times 10^{-3}\,\text{kg})(9.8\ \text{N/kg})}{0.5 \times 10^{-2}\,\text{m}} = 3.9\,\text{N/m}.$$

The angular frequency is then given by Eq. 16-11 as

$$\omega = \sqrt{\frac{k}{m}} = \sqrt{\frac{3.9\ \text{N/m}}{2 \times 10^{-3}\,\text{kg}}} = 44\ \text{rad/s}.$$

But we know that $\omega = 2\pi f$, so

$$f = \frac{\omega}{2\pi} = \frac{44\ \text{rad/s}}{2\pi} = 7.0\ \text{Hz} \quad \text{and} \quad T = \frac{1}{f} = \frac{1}{7\,\text{Hz}} = 0.14\ \text{s}. \qquad \text{(Answer)}$$

(b) If we start tracking the spider's oscillation when she is at her lowest vertical position relative to the spider-web system equilibrium point, what is her initial phase?

SOLUTION ■ The **Key Idea** here is that we must take the equilibrium point to be at $y = 0$ m and that $y(0) = -Y$ when $t = 0$ s. Using Eq. 16-5 with y instead of x denoting displacement, we get

$$-Y = Y\cos(\omega t + \phi_0)$$

so that

$$\cos(\phi_0) = -1,$$

which gives
$$\phi_0 = \pm\pi\ \text{rad}. \qquad \text{(Answer)}$$

(c) If the maximum displacement of the spider from the equilibrium point of the spider-web system is given by $Y = 1.0$ cm, what is the maximum acceleration of the spider as she oscillates?

SOLUTION ■ The **Key Idea** here is that the magnitude A of the maximum acceleration is the acceleration amplitude $\omega^2 Y$ as shown in Eq. 16-16. Thus,

$$A = \omega^2 Y = (44\ \text{rad/s})^2(10 \times 10^{-3}\,\text{m}) = 20\ \text{m/s}^2. \qquad \text{(Answer)}$$

The maximum magnitude of acceleration occurs when the spider is turning around at the ends of her path. This is when the force on her is a maximum. The graphs in Fig. 16-15 also show the magnitude of displacement and acceleration to be maximum at the same time.

16-5 Gravitational Pendula

We turn now to a class of simple harmonic oscillators in which the restoring force is associated with the gravitational force rather than with the elastic properties of a twisted wire or a spring. Oscillators that depend on gravitational restoring forces or torques hang, and so they are considered to be types of pendulums.

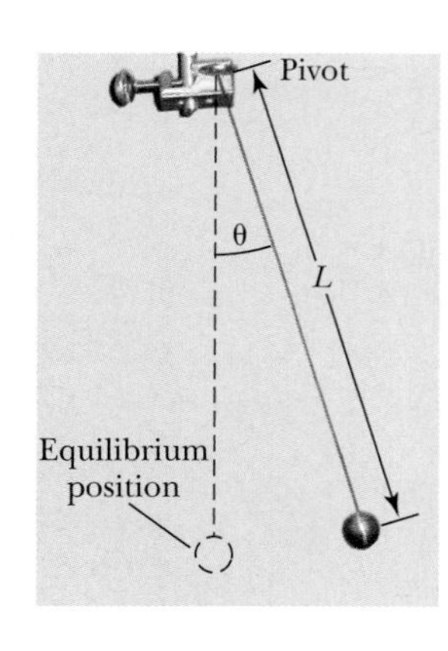

FIGURE 16-17 ■ A pendulum bob swings back and forth at a small angle. Its angular displacement θ from its vertical equilibrium is measured using a rotary motion sensor attached to a computer data acquisition system. The length L of this pendulum measured from the pivot to the center of the bob is 32 cm.

The Simple Pendulum Oscillating at a Small Angle

Consider a small particle of mass m (called a *bob*) that hangs from the end of a wire or string. Assume that the wire has a small mass compared to the mass of the particle and that it can't stretch noticeably. If you fix the wire at its upper end, you have constructed a **simple pendulum.** An example of a simple pendulum is shown in Fig. 16-17. The bob is in its equilibrium position when it hangs vertically. But suppose you pull the

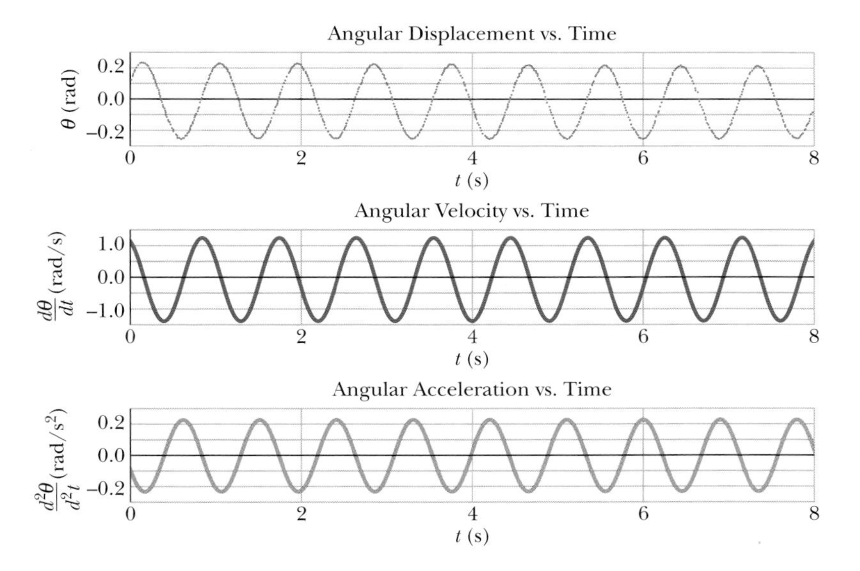

FIGURE 16-18 ■ Measurements for the *angular displacement* (θ) of the simple pendulum shown in Fig. 16-17 were obtained using a computer data acquisition system outfitted with a rotary motion sensor. A graph of the *angular velocity* ($d\theta/dt$) was also constructed using an algorithm to find a smoothed first derivative of the angular data. This process is repeated using the velocity data to find how the angular acceleration (α) varies with time.

bob up so the initial angular displacement θ between the wire and the vertical is small. What happens to the angle θ between the vertical and the wire as the pendulum bob swings back and forth? You easily see that the bob's motion is periodic. Is it, in fact, simple harmonic motion? If so, what factors does the period T depend on?

We can verify that the simple pendulum shown in Fig. 16-17 undergoes SHM by examining the graphs of the data shown in Fig. 16-18.

Careful examination of the θ vs. t graph in Fig. 16-18 gives a period of oscillation for our pendulum bob of $T^{\text{exp}} = 1.1$ s. A model of the plotted data shows that the equation that describes the angular displacement versus time has exactly the *same form* as Eq. 16-7. In particular,

$$\theta(t) = \Theta\cos(\omega t + \phi_0) = (0.24\text{ rad})\cos([7.0\text{ rad/s}]t - 0.90\text{ rad}). \tag{16-18}$$

The value of the maximum angular displacement or amplitude is $\Theta = 0.24$ rad, the angular frequency, ω, is 7.0 rad/s, and the initial phase (or phase constant), ϕ_0, is -0.90 rad. The maximum angular displacement of 0.24 rad can be expressed in degrees as

$$\Theta = 0.24\text{ rad} = \left(\frac{180°}{\pi\text{ rad}}\right)0.24\text{ rad} \approx 15°.$$

If we repeat the experiment for smaller amplitudes, we find that the angular frequency ω stays the same to two significant figures. This fact when combined with the sinusoidal oscillation shown in Eq. 16-18 strongly suggests that the pendulum is undergoing simple harmonic motion. If this is the case then the net horizontal force on the pendulum bob is proportional to its horizontal displacement—at least when angles are small.

Let us see if our experimental results could have been predicted theoretically.

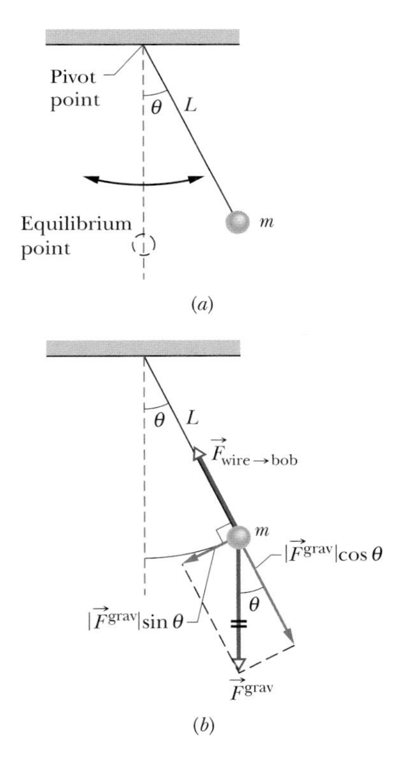

FIGURE 16-19 ■ (*a*) A simple pendulum displaced from its equilibrium position by an angle θ. (*b*) The forces acting on the bob are the gravitational force $\vec{F}^{\text{grav}}$ and the tension force $\vec{F}_{\text{wire}\to\text{bob}}$ from the wire. The tangential component $|\vec{F}^{\text{grav}}|\sin\theta$ of the gravitational force is a restoring force that tends to bring the pendulum back to its central position.

Theoretical Derivation of Simple Pendulum Forces

To find the net horizontal forces on the pendulum, we can set up a free-body diagram using methods introduced in Chapter 6. The forces acting on the bob are the tension force $\vec{F}_{\text{wire}\to\text{bob}}$ that the wire exerts on the bob and the gravitational force $\vec{F}^{\text{grav}}$, as shown in Fig. 16-19*b* where the string makes an angle θ with the vertical. We resolve $\vec{F}^{\text{grav}}$ into a radial component $|\vec{F}^{\text{grav}}|\cos\theta$ and a component $|\vec{F}^{\text{grav}}|\sin\theta$ that is tangent to the path taken by the bob. This tangential component produces a restoring torque about the pendulum's pivot point, because it always acts opposite the displacement of the bob so as to bring the bob back toward its central location. That location is called the *equilibrium position* ($\theta = 0$), because if the pendulum were not swinging it would be at rest at that position.

From Eq. 11-30 ($|\vec{\tau}| = |\vec{r}||\vec{F}_t|$), we can write the magnitude of the restoring torque as the product of the magnitude of a moment arm ($|\vec{r}| = L$) about the pivot arm and the tangential component of the gravitational force ($|F^{\text{grav}}|\sin\theta$). The tension force in the wire $\vec{F}_{\text{wire}\to\text{bob}}$ does not contribute to the restoring torque because it always acts parallel to the moment arm. Thus the z-component of the torque can be expressed as

$$\tau_z = -L(|F^{\text{grav}}|\sin\theta). \tag{16-19}$$

The minus sign indicates that the torque acts to reduce θ. Substituting Eq. 16-19 into Eq. 11-32 ($\tau_z^{\text{net}} = I\alpha_z$) and then substituting mg for the magnitude of $\vec{F}^{\text{grav}}$, we obtain

$$-L(mg\sin\theta) = I\alpha_z, \tag{16-20}$$

where I is the pendulum's rotational inertia about the pivot point and α_z is the z-component of its angular acceleration about that point.

We want to focus on the nature of the motion when the maximum (and minimum) angle of displacement is small. Note that whenever an angle θ is small, then the arc length s *that the pendulum bob sweeps through* (with respect to its equilibrium) and the value of x have essentially the same magnitude as shown in Fig. 16-20, so that

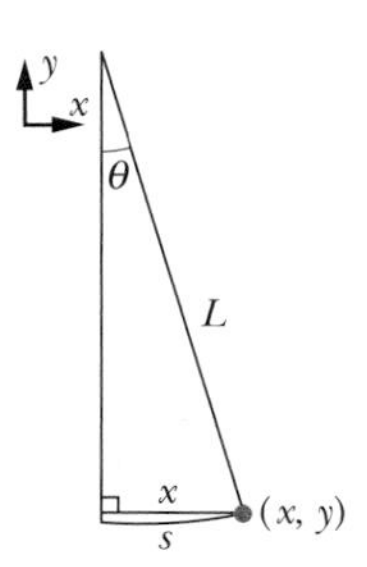

FIGURE 16-20 ■ This diagram illustrates that the arc length s and the value of y are approximately the same when the angle θ is small.

$$\sin\theta = \frac{x}{L} \approx \frac{s}{L} = \theta \quad \text{(approximation for small } \theta). \tag{16-21}$$

Thus when the angle θ is small, we can replace $\sin\theta$ with θ (expressed in radian measure). (As an example, if $\theta = 15.0° = 0.262$ rad, then $\sin\theta = 0.265$, a difference of only about 1%.) Using this approximation and rearranging terms, we can then express the z-component of angular acceleration as

$$\alpha_z \approx -\frac{mgL}{I}\,\theta. \tag{16-22}$$

Note that the z axis passes through pivot point and is perpendicular to the plane of oscillation. This equation is the angular equivalent of Eq. 16-5, the hallmark of SHM. It tells us that the angular acceleration $\vec{\alpha} = \alpha_z\hat{k}$ of the pendulum is proportional to the angular displacement $\theta\hat{k}$ but opposite in sign.

Thus, as the pendulum bob moves to, say, the right as in Fig. 16-19*a*, its acceleration *to the left* increases until the bob turns around and begins moving to the left. Then, when it is on the left, its acceleration to the right tends to return it to the right, and so on, as it swings back and forth in SHM. More precisely, we have verified both experimentally and theoretically that the motion of a *simple pendulum swinging through small angles* is approximately SHM. We can state this restriction to small angles another way: to be correct on predicting the period of a motion to within about 1%, the **angular amplitude** Θ of the motion (the maximum angle of swing) must be about 15° or less.

Comparing Eq. 16-22 and Eq. 16-17, we see that the angular frequency of the pendulum is $\omega = \sqrt{mgL/I}$. Next, if we substitute this expression for ω into Eq. 16-6 ($\omega = 2\pi/T$), we see that the period of the pendulum may be written as

$$T = 2\pi\sqrt{\frac{I}{mgL}}. \tag{16-23}$$

All the mass of a simple pendulum is concentrated in the mass m of the particle-like bob, which is at radius L from the pivot point. Thus, we can use Eq. 11-23 ($I = mr^2$) to write ($I = mL^2$) for the rotational inertia of the pendulum. Substituting this into Eq. 16-23 and simplifying yields

$$T = 2\pi\sqrt{\frac{L}{g}} \quad \text{(simple pendulum, small amplitude)}, \tag{16-24}$$

as a simpler expression for the period of a simple pendulum swinging through only small angles. (We also assume small-angle swinging in the problems of this chapter.) Of course we can use Eq. 16-24 to predict the period of the pendulum of length L = 32 cm described in Figs. 16-17 and 16-18. We get

$$T^{\text{theory}} = 2\pi\sqrt{\frac{L}{g}} = 2\pi\sqrt{\frac{0.32\text{ m}}{9.80\text{ N/kg}}} = 1.1\text{ s}.$$

This result matches our experimental result to at least two significant figures. This period is also identical to that of our spring-mass system described by Eq. 16-7. This is because just for fun we chose the pendulum length L so that we would get the same period for the two systems!

A very surprising outcome of this theoretical derivation is that, *for small displacement angles, the period of the simple pendulum does not depend on its bob mass*. If you reflect on this, you should be able to see that the motion of the simple pendulum bob is mass-independent for the same reason that the motion of a falling object does not depend on its mass.

Measuring g with a Simple Pendulum

Geologists often use a pendulum to determine the local gravitational strength, g, at a particular location on Earth's surface. If a simple pendulum oscillating at small angles is used, we can solve Eq. 16-24 for g to get

$$g = \frac{4\pi^2 L}{T^2}. \tag{16-25}$$

Thus, by measuring L and the period T, we can find the value of g. In order to make more precise measurements, a number of refinements are needed. Geophysicists often use a physical pendulum consisting of a solid rod in conjunction with a more sophisticated equation than Eq. 16-25. They can also place the pendulum in an evacuated chamber.

The Physical Pendulum

A "real" pendulum that isn't just a point mass suspended from a massless string is usually called a **physical pendulum,** and it can have a complicated distribution of mass, much different from that of a simple pendulum. Does a physical pendulum also undergo SHM? If so, what is its period?

Figure 16-21 shows an arbitrary physical pendulum displaced to one side by angle θ. The gravitational force $\vec{F}^{\text{grav}}$ acts at its center of mass C, at a distance h from the pivot point O. In spite of their shapes, comparison of Figs. 16-21 and 16-19b reveals only one important difference between an arbitrary physical pendulum and a simple pendulum. For a physical pendulum, the restoring component $|\vec{F}^{\text{grav}}|\sin\theta$ of the gravitational force has a moment arm of distance h about the pivot point rather than of wire length L. In all other respects, an analysis of the physical pendulum would duplicate our analysis of the simple pendulum up through Eq. 16-23. Again, for a small angular amplitude Θ, we would find that the motion is approximately SHM.

If we replace L with h in Eq. 16-23, we can write the period of a physical pendulum as

$$T = 2\pi\sqrt{\frac{I}{mgh}} \quad \text{(physical pendulum, small amplitude).} \tag{16-26}$$

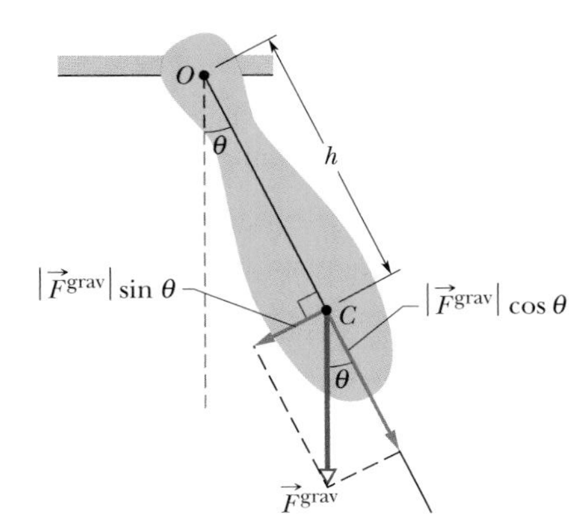

FIGURE 16-21 ■ A physical pendulum. The magnitude of the restoring torque is $h|\vec{F}^{\text{grav}}|\sin\theta$. When $\theta = 0$, the center of mass C hangs directly below pivot point O.

As with the simple pendulum, I is the rotational inertia of the pendulum about O. However, now I is not simply mL^2. It depends both on the shape of the physical pendulum and on the axis about which it rotates. Table 11-2 shows the rotational inertia equations for common shapes, but these equations describe the rotational inertia about the center of mass (I_{com}). In essentially all physical pendula, the axis of rotation is parallel to an axis through the center of mass but offset by a distance h. In these cases, the parallel axis theorem $I = I_{com} + mh^2$ (Eq. 11-28) can be used to find the required equation for I.

A physical pendulum will not swing if it pivots at its center of mass. Formally, this corresponds to putting $h = 0$ in Eq. 16-26. That equation then predicts $T \to \infty$, which implies that such a pendulum will never complete one swing.

Corresponding to any physical pendulum that oscillates about a given pivot point O with period T is a simple pendulum of length L_0 with the same period T. We can find L_0 with Eq. 16-24. The point along the physical pendulum at distance L_0 from point O is defined as the *center of oscillation* of the physical pendulum for the given suspension point.

READING EXERCISE 16-7: The vertical acceleration of a falling object is independent of its mass. Likewise the period of a simple pendulum oscillating at small angles is independent of bob mass. Can you explain why in each case? What is similar about the two situations? ■

READING EXERCISE 16-8: Three physical pendula, of masses m, $2m$, and $3m$, have the same shape and size and are suspended at the same point. Rank the masses according to the periods of the pendulum, greatest period first. ■

TOUCHSTONE EXAMPLE 16-2: T-Shaped Pendulum

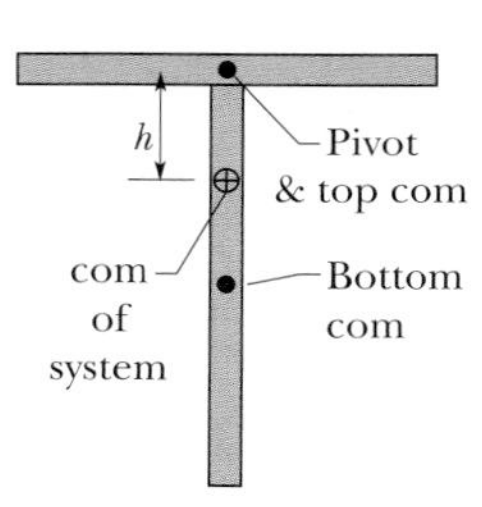

FIGURE 16-22 ■ A T-shaped physical pendulum pivoted at the top of the "T."

A physics student has devised a physical pendulum from two meter sticks of negligible width that are joined together as shown in Fig. 16-22. Assuming the oscillations about the designated pivot occur at small angles, what is the frequency of oscillation?

SOLUTION ■ We need to use Eq. 16-26, which relates the period of a physical pendulum to its rotational inertia (I), mass (m), and moment arm distance (h),

$$T = 2\pi\sqrt{\frac{I}{mgh}}.$$

The **Key Idea** here is to figure out what the rotational inertia of this oddly shaped pendulum is about the chosen pivot and to find the moment arm distance.

The *moment arm h* is the distance from the pivot to the center of mass of the pendulum. To find the com of the T-pendulum we need to clarify our notation. If the total mass of the meter stick system is denoted as m, then the top stick has a mass of $m_{top} = m/2$ and the bottom stick also has a mass of $m_{bottom} = m/2$. If we set $y = 0$ m at the pivot point, then the center of mass of the bottom stick is given by $y_{bottom} = -L/2$. We can now use Eq. 8-11 to find the center of mass of the pendulum (and hence h),

$$Y_{com} = \frac{1}{M_{sys}}(m_{top}y_{top} + m_{bottom}y_{bottom})$$

$$= \frac{1}{m}\left[\left(\frac{m}{2}\right)(0) + \left(\frac{m}{2}\right)\left(-\frac{L}{2}\right)\right] = -\frac{L}{4}.$$

Thus

$$h = |Y_{com}| = \frac{L}{4}. \qquad (16\text{-}27)$$

The *rotational inertia* of the T-shaped system is the sum of the rotational inertias of the top and bottom sticks about the pivot point. The top rod is rotating about its center of mass. The rotational inertia of a stick or rod rotating about its center of mass is shown in Table 11-2 as $I_{com} = \frac{1}{12}ML^2$ so the rotational inertia of the top rod is given by $I_{top} = \frac{1}{12}(m/2)L^2$. The bottom rod is rotating about its end, and we must use the parallel axis theorem (Eq. 11-28) to find its rotational inertia. Since the distance between the axis of rotation of the bottom rod and the pivot point is $L/2$, the rotational inertia of the bottom rod turns out to be $I_{end} = \frac{1}{12}ML^2 + M(L/2)^2 = \frac{1}{3}ML^2$. Thus, the pendulum's total rotational inertia I is given by

$$I = I_{top} + I_{bottom} = \tfrac{1}{12}(m/2)L^2 + \tfrac{1}{3}(m/2)L^2 = \tfrac{5}{24}(mL^2). \qquad (16\text{-}28)$$

Finally we can substitute Eqs. 16-27 and 16-28 into Eq. 16-26 to get

$$T = 2\pi\sqrt{\frac{I}{mgh}} = 2\pi\sqrt{\frac{\frac{5}{24}mL^2}{mg(L/4)}} = 2\pi\sqrt{\frac{20\,L}{24g}}.$$

Since $L = 1.00$ m,

$$T = 2\pi\sqrt{\frac{20\,L}{24g}} = 2\pi\sqrt{\frac{20(1.00\text{ m})}{24(9.8\text{ m/s}^2)}} = 1.83\text{ s}. \qquad \text{(Answer)}$$

16-6 Energy in Simple Harmonic Motion

In Chapter 10 we saw that a simple pendulum swinging at a small angle transfers energy back and forth between kinetic energy and potential energy, while the sum of the two—the mechanical energy E of the oscillating Earth–pendulum system—remains constant. What about the linear oscillator made up of the mass–spring system that was considered in Section 16-3? Does it trade energy back and forth as it oscillates? Let us try to answer this question for the linear oscillator using theoretical considerations.

The potential energy of a horizontal linear oscillator like that of Fig. 16-10 is associated entirely with the mass–spring system. Its value depends on how much the spring is stretched or compressed—that is, on $x(t)$. We can use Eqs. 10-14 and 16-5 to find

$$U(t) = \tfrac{1}{2}kx^2 = \tfrac{1}{2}kX^2\cos^2(\omega t + \phi). \qquad (16\text{-}29)$$

Note carefully that for any angle α, a function written in the form $\cos^2\alpha$ (as here) means $(\cos\alpha)^2$. This is *not* the same as a function written as $\cos\alpha^2$, which means $\cos(\alpha^2)$.

The kinetic energy of the system of Fig. 16-11 is associated entirely with the hanging mass. Its value depends on how fast the mass is moving—that is, on $v(t)$. We can use Eq. 16-15 to find

$$K(t) = \tfrac{1}{2}mv^2 = \tfrac{1}{2}m\omega^2X^2\sin^2(\omega t + \phi_0). \qquad (16\text{-}30)$$

If we use Eq. 16-11 to substitute k/m for ω^2, we can write Eq. 16-30 as

$$K(t) = \tfrac{1}{2}mv^2 = \tfrac{1}{2}kX^2\sin^2(\omega t + \phi_0). \qquad (16\text{-}31)$$

The mechanical energy follows from Eqs. 16-29 and 16-31 and is

$$\begin{aligned} E &= U + K \\ &= \tfrac{1}{2}kX^2\cos^2(\omega t + \phi_0) + \tfrac{1}{2}kX^2\sin^2(\omega t + \phi_0) \\ &= \tfrac{1}{2}kX^2[\cos^2(\omega t + \phi_0) + \sin^2(\omega t + \phi_0)]. \end{aligned}$$

For any angle α,

$$\cos^2\alpha + \sin^2\alpha = 1.$$

Thus, the quantity in the square brackets above is unity and we have

$$E = U + K = \tfrac{1}{2}kX^2. \qquad (16\text{-}32)$$

The mechanical energy of a horizontal mass–spring system oscillator is indeed constant and independent of time. The potential energy and kinetic energy of this oscillator are shown as functions of time t in Fig. 16-23*a*, and as functions of displacement x in Fig. 16-23*b*.

Since a linear oscillation trades energy back and forth in a symmetric fashion, it turns out that the average kinetic energy is the same as the average potential energy. Each average energy is 1/2 of the total. In equation form this can be expressed as

$$\langle K\rangle = \langle U\rangle = \frac{E^{\text{tot}}}{2}. \qquad (16\text{-}33)$$

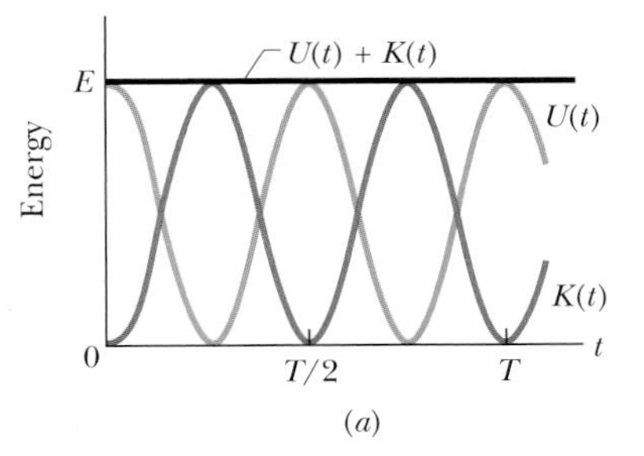

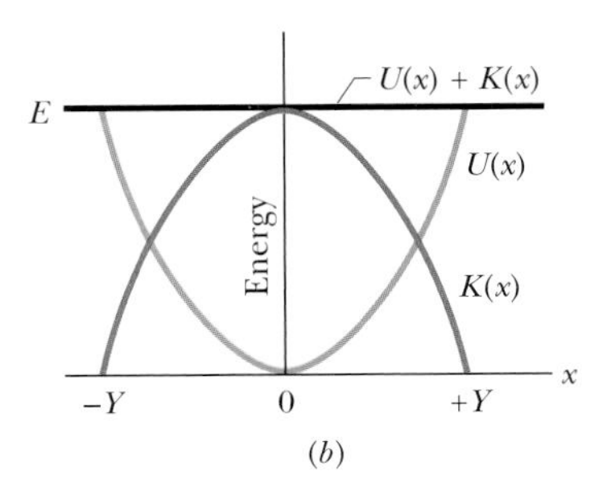

FIGURE 16-23 (*a*) Potential energy $U(t)$, kinetic energy $K(t)$, and mechanical energy E as functions of time t for a linear harmonic oscillator. Note that all energies are positive and that the potential energy and the kinetic energy peak twice during every period. (*b*) Potential energy $U(t)$, kinetic energy $K(x)$, and mechanical energy E as functions of position x for a linear harmonic oscillator with amplitude Y. For $x = 0$ the energy is all kinetic, and for $x = \pm Y$ it is all potential.

We also examined the behavior of a vertical mass–spring–Earth system (Fig. 16.11). To consider the mechanical energy exchange between potential and kinetic energies in that system, we need to take gravitational energy into account. This complication is treated in Touchstone Example 16-3.

We presented actual data in Fig. 10-14 showing very similar trading of kinetic and gravitational potential energy for a simple pendulum.

READING EXERCISE 16-9: Assume that the spring–block system shown in Fig. 16-10 has a kinetic energy of 3 J and an elastic potential energy of 2 J when the block is at $x = +20$ cm. (a) What is the kinetic energy when the block is at $x = 0$ cm? What is the elastic potential energy when the block is at (b) $x = -X$? ■

TOUCHSTONE EXAMPLE 16-3: Oscillation Energy

Consider the 100 g mass hanging vertically from the spring as shown in Fig. 16-11. It oscillates with a period of $T = 1.1$ s trading between kinetic and potential energy. Its displacement from equilibrium as a function of time is given by Eq. 16-7 and the motion of the mass is shown graphically in Fig. 16-13. Since the 100 g mass is much greater than the mass of the spring, we will ignore the mass of the spring as we analyze the energy transformations during a single oscillation.

(a) What is the total energy of the oscillating system?

SOLUTION ■ Equation 16-32 can be used to express total energy of a SHM oscillator in terms of the spring constant k and the amplitude of the displacement Y. Since the angular frequency is related to k and the oscillating mass m by Eq. 16-11 ($\omega = \sqrt{k/m}$), we can write Eq. 16-32 as

$$E = U + K = \tfrac{1}{2}kY^2 = \tfrac{1}{2}m\omega^2Y^2. \qquad \text{(Eq. 16-32)}$$

We know that $m = 0.10$ kg. By examining Eq. 16-7 we see that $Y = 0.040$ m and $\omega = 5.5$ rad/s. This gives us a total system energy of

$$E = \tfrac{1}{2}m\omega^2Y^2 = \tfrac{1}{2}(0.10\text{ kg})(5.5\text{ rad/s})^2(0.040\text{ m})^2$$
$$= 2.4 \times 10^{-3}\text{ J}. \qquad \text{(Answer)}$$

(b) At what time or times during the first oscillation is the total mechanical energy of the mass–spring system equal to its potential energy?

SOLUTION ■ When the system's total and potential energy are the same, its kinetic energy must be zero. That occurs whenever the magnitude of velocity of the mass is zero, which happens when the magnitude of the displacement is a maximum and the mass is turning around. Since the period $T = 1.1$ s, an examination of Fig. 16-13 shows this occurring at $t = 0$ s and 1.1 s. It is also turning around halfway between these times, so

$$E = U \quad \text{at 0.0 s, 0.55 s, and 1.1 s.} \qquad \text{(Answer)}$$

An alternate way to arrive at the same answer is to examine Fig. 16-23 and note that the potential energy peaks at 0, $T/2$, and T.

(c) At what time or times during the first oscillation is the total mechanical energy of the mass–spring system equal to its kinetic energy?

SOLUTION ■ When the system's total and kinetic energy are the same, the speed of the mass is a maximum. That occurs whenever the mass passes through its equilibrium point. According to Eq. 16-7, if we set $t_1 = 0$ then this happens when

$$\cos(\omega t) = \cos([5.5\text{ rad}]t) = 0,$$

which occurs when

$$\omega t = (5.5\text{ rad})t = \pi/2 \text{ or } 3\pi/2 \text{ at about 0.28 s and 0.83 s.} \qquad \text{(Answer)}$$

An alternate way to arrive at the same answer is to examine Fig. 16-23 and note that the kinetic energy peaks at $T/4$ and at $3T/4$.

(d) At what time or times are the potential and kinetic energy the same so $K = U = \frac{1}{2}E$?

SOLUTION ■ An examination of Fig. 16-23 shows that the system's potential and kinetic energy are the same at

$$\tfrac{1}{8}T,\ \tfrac{3}{8}T,\ \tfrac{5}{8}T, \text{ and } \tfrac{7}{8}T, \quad \text{or} \quad \text{0.138 s, 0.413 s, 0.688 s, and 0.963 s.} \qquad \text{(Answer)}$$

(e) What types of potential energy is stored in the system as it oscillates? Is it really legitimate to use Eqs. 16-29 and 16-32, which measure spring potential energy, when both gravitational and spring potential energy are present?

SOLUTION ■ If our mass–spring system were oscillating horizontally the potential energy would consist entirely of the elastic energy stored in the spring. The situation for the hanging mass is not so simple since we have both spring energy and gravitational potential energy involved. In this case $U = U^{\text{grav}} + U^{\text{spring}}$. Actually the energy relationships presented in Section 16-6 are still valid.

To explain why the energy equations in Section 16-6 are valid, let us choose our origin so that $y = 0$ m at the hanging equilibrium point of the system shown in Fig. 16-24. At any point during an

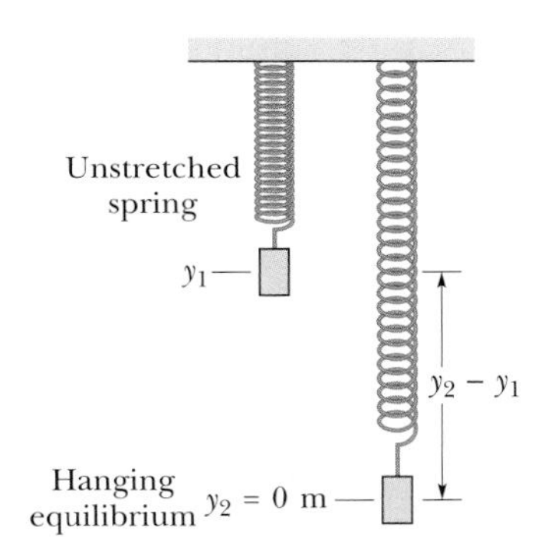

FIGURE 16-24 ■ A hanging mass-spring system.

oscillation, the net force on the hanging mass is the sum of an upward (positive) spring force and a downward (negative gravitational force). This can be expressed in terms of the vertical components as

$$F_y^{\text{net}} = F_y^{\text{spring}} + F_y^{\text{grav}}.$$

When the mass is placed on the spring, the weight of the mass causes the spring to descend from its natural equilibrium point at y_1 to its hanging equilibrium point at y_2. But we chose $y_2 = 0$ m. At this hanging equilibrium point, the net force on the mass is zero, so that

$$F_y^{\text{net}} = F_y^{\text{spring}} + F_y^{\text{grav}} = -k(y_2 - y_1) - mg$$
$$= -k(0 - y_1) - mg = 0\text{ N}.$$

A **Key Idea** here is that at the hanging equilibrium location $ky_1 = mg$.

Since we are free to set an arbitrary reference level for gravitational potential, let's set the gravitational potential energy at $y = 0$ m to $U^{\text{grav}}(0) = -\frac{1}{2}mgy_1$ where y_1 is the location of the center of the mass when the spring is unstretched (as shown in Fig. 16-24). Next let's calculate the total potential energy of the spring–mass system when the mass has a displacement of y from its hanging equilibrium position,

$$U = U^{\text{grav}} + U^{\text{spring}} = (mgy - \tfrac{1}{2}mgy_1) + \tfrac{1}{2}k(y - y_1)^2,$$

but $ky_1 = mg$, so that

$$U = (ky_1y - \tfrac{1}{2}ky_1^2) + \tfrac{1}{2}k(y - y_1)^2$$
$$= (ky_1y - \tfrac{1}{2}ky_1^2) + (\tfrac{1}{2}ky^2 - ky_1y + \tfrac{1}{2}ky_1^2)$$
$$= \tfrac{1}{2}ky^2.$$

This is a very important result because it tells us that with regard to energy considerations, the potential energy of a hanging mass when oscillating about its equilibrium position can be treated as if the spring were unstretched at the hanging equilibrium. This depends on choosing our gravitational potential energy reference point carefully (which we are free to do). This result provides theoretical verification for the claim made in Section 16-3, that a mass oscillating from a hanging spring and one oscillating horizontally have the same type of mathematical behavior.

16-7 Damped Simple Harmonic Motion

The equations we have developed to describe harmonic motion predict that it goes on forever with the same amplitude. Whatever starting value of t you put into Eq. 16-5, it will oscillate from then on with the same amplitude it started with. Oscillations in the real world usually die out gradually, transferring mechanical energy to thermal energy by the action of frictional forces. A pendulum will swing only briefly under water, because the water exerts a drag force on the pendulum that quickly eliminates the motion. A pendulum swinging in air does better, but still the motion dies out eventually, because the air exerts a drag force on the pendulum and friction acts at its support. These forces reduce the mechanical energy of the pendulum's motion.

When the motion of an oscillator is reduced by external friction or drag forces, the oscillator and its motion are said to be **damped.** An idealized example of a damped oscillator is shown in Fig. 16-25, where a block with mass m oscillates vertically on a spring with spring constant k. From the block, a rod extends to a vane (both assumed to have negligible mass) that is submerged in a liquid. As the vane moves up and down, the liquid exerts an inhibiting drag force on it and thus on the entire oscillating system. With time, the mechanical energy of the block–spring system decreases, as energy is transferred to thermal energy of the liquid and vane. If the liquid is alcohol that is not very viscous, the drag forces will be small. But a liquid like honey or molasses could exert much larger drag forces.

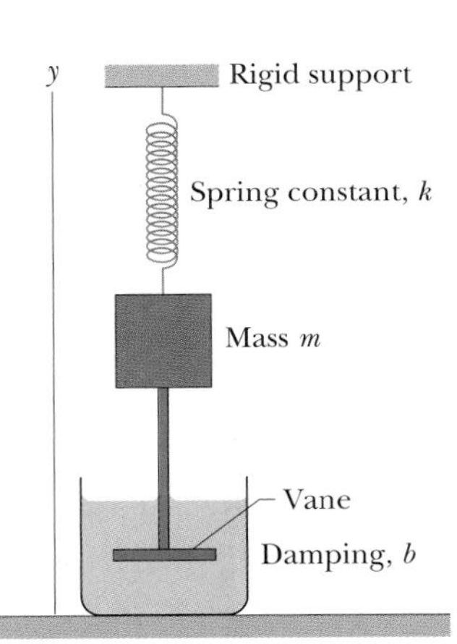

FIGURE 16-25 ■ An idealized damped simple harmonic oscillator. A vane immersed in a liquid exerts a damping force on the block as the block oscillates parallel to the x axis.

Theoretical Analysis

Let us assume the liquid in the system shown in Fig. 16-25 exerts a **damping force** $\vec{F}^{\text{drag}}$ that is *proportional in magnitude to the velocity* $\vec{v}$ *of the vane and block* (an assumption that is accurate if the vane moves slowly). Then, for forces along the y axis in Fig. 16-25, we have

$$\vec{F}^{\text{drag}} = -b\vec{v}, \tag{16-34}$$

where b is a **damping constant** that depends on the characteristics of both the vane and the liquid and has the SI unit of kilogram per second. The minus sign indicates that $\vec{F}^{\text{drag}}$ opposes the motion.

The y-component of the force on the block from the spring is $F_y^{\text{spring}} = -ky$. Let us assume that the gravitational force on the block is negligible compared to $\vec{F}^{\text{drag}}$ and $\vec{F}^{\text{spring}}$. Then we can write Newton's Second Law for components along the y axis ($F_y^{\text{net}} = ma_y$) as

$$-bv_y - ky = ma_y. \tag{16-35}$$

Substituting dy/dt for v_y and d^2x/dt^2 for a_y and rearranging give us the differential equation

$$m\frac{d^2y}{dt^2} + b\frac{dy}{dt} + ky = 0. \tag{16-36}$$

The solution of this equation is

$$y(t) = Ye^{-bt/2m}\cos(\omega' t + \phi_0), \tag{16-37}$$

where Y is the initial amplitude, ϕ_0 is the initial phase, and ω' is the angular frequency of the damped oscillator. This angular frequency is given by

$$\omega' = \sqrt{\frac{k}{m} - \frac{b^2}{4m^2}}. \tag{16-38}$$

If $b = 0$ (there is no damping), then Eq. 16-38 reduces to Eq. 16-11 ($\omega = \sqrt{k/m}$) for the angular frequency of an undamped oscillator, and Eq. 16-37 reduces to Eq. 16-5 for the displacement of an undamped oscillator. If the damping constant is small but not zero (so that $b \ll \sqrt{km}$), then $\omega' \approx \omega$. We define the pendulum as **underdamped** whenever $k/m > b^2/4m^2$ so that $\omega' < \omega$.

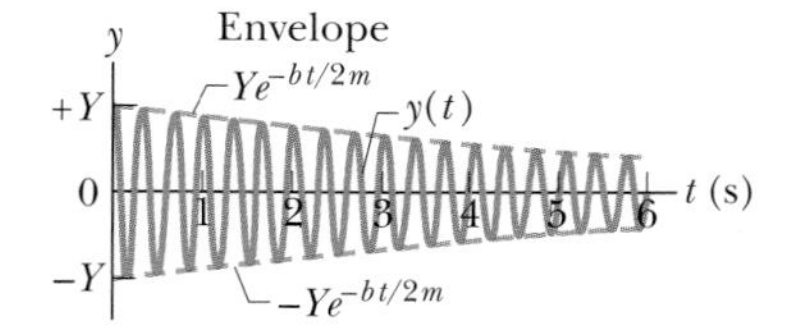

FIGURE 16-26 ■ The displacement function $y(t)$ for the damped oscillator of Fig. 16-25, with $m = 250$ g, $k = 85$ N/m, and $b = 70$ g/s. The amplitude, which is $Ye^{-bt/2m}$, decreases exponentially, though this exponential decrease does not show up well since the damping coefficient b is not large. The exponential drop-off is more pronounced in Fig. 16-28.

We can regard Eq. 16-37 as a cosine function whose amplitude, which is $Ye^{-bt/2m}$, gradually decreases with time, as Fig. 16-26 suggests. For an undamped oscillator, the mechanical energy is constant and is given by Eq. 16-32 ($E = \frac{1}{2}kY^2$). If the oscillator is damped, the mechanical energy is not constant but decreases with time. If the damping is small, we can find $E(t)$ by replacing the amplitude, Y, in Eq. 16-32 with $Ye^{-bt/2m}$, the amplitude of the damped oscillations. By doing so, we find that

$$E(t) \approx \tfrac{1}{2}kY^2 e^{-bt/m}, \tag{16-39}$$

which tells us that, like the amplitude, the mechanical energy decreases exponentially with time.

Experimental Results

Here we present actual data for another damped oscillator that has identical mathematical behavior to that of the block oscillating in a viscous liquid. It consists of an

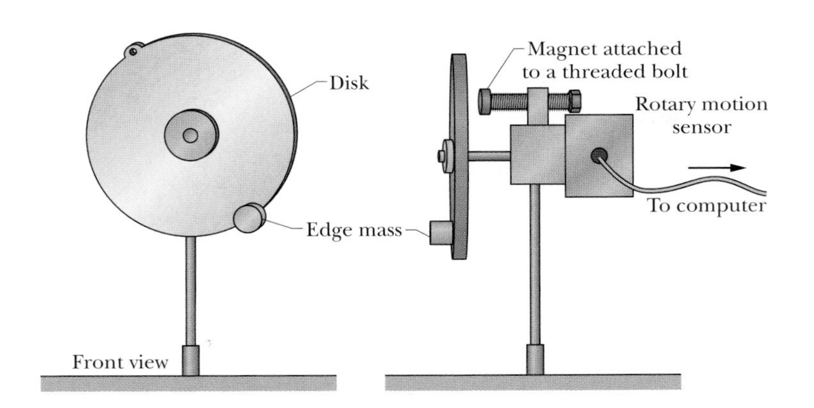

FIGURE 16-27 ■ A physical pendulum consisting of an aluminum disk with an edge mass attached to it. Magnetic damping is used to provide a drag torque that reduces the amplitude of the pendulum over time.

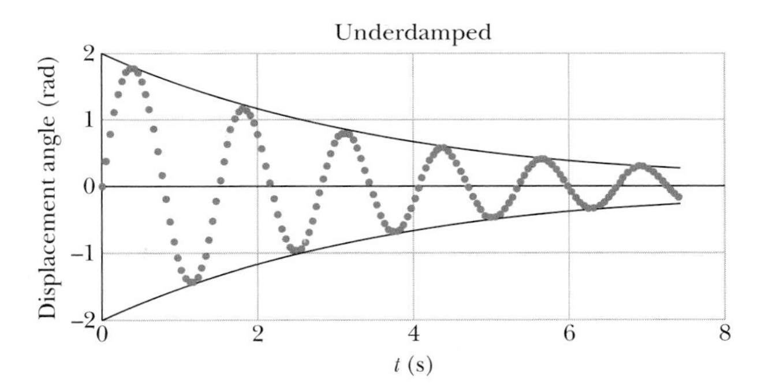

FIGURE 16-28 ■ Actual data for the angular displacement from the vertical equilibrium of the damped physical pendulum (shown in Fig. 16-27). Of most interest here is that the magnetic damping force causes the amplitude of oscillation to decrease exponentially in time. This is shown by the shape of the "envelope function" that best conforms to the displacement maxima or minima, found to be $\Theta e^{-(b/2I)t} = (2\text{ rad})e^{-(0.27s^{-1})t}$.

edge mass attached to an aluminum disk to create the physical pendulum shown in Fig. 16-27. A strong magnet placed in the vicinity of the disk exerts a drag force on the disk that is proportional to the angular velocity of the disk $d\theta/dt$. (This kind of damping is a result of eddy currents induced in the disk. You can refer to Section 31-5 for more details.)

Both the pendulum and the mass–spring systems have linear restoring forces and velocity-dependent drag forces. Thus, we can modify Eqs. 16-37, 16-38, and 16-39 to describe the pendulum motion. We need to replace the linear displacement y with an angular displacement θ, the spring mass m with I (where I is the total rotational inertia of the disk and its edge mass), and the spring constant k with a torque strength mgL (where in our case L is the radius of the disk and m is the edge mass). This gives us a new motion equation for small angles of pendulum oscillation

$$\theta(t) = \Theta e^{-bt/2I}\cos(\omega' t + \phi_0), \tag{16-40}$$

where the angular amplitude (or "envelope" function) $\Theta e^{-bt/2I}$ varies in time.

The feature of the data that we are most interested in here is the way the angular amplitude of the pendulum decreases with time when the pendulum is underdamped. This is shown in Fig. 16-28. The exponential behavior is obvious for these data.

Critical Damping and Overdamping

In Figs. 16-27 and 16-28 we have shown underdamped motions for two different but mathematically similar systems. When the damping coefficient in our systems are so large that

$$\omega' = \sqrt{\frac{mgL}{I} - \frac{b^2}{4I^2}} = 0 \quad \text{(physical pendulum)}$$

$$\text{or} \quad \omega' = \sqrt{\frac{k}{m} - \frac{b^2}{4m^2}} = 0 \quad \text{(mass–spring)},$$

the system undergoes **critical damping.** Under this circumstance, a system that is displaced will settle back exponentially to its equilibrium point in the *minimum possible time* without oscillating. We have not presented the equations that describe what happens when the damping coefficient b in Eq. 16-38 gets so large that the square root becomes negative, a condition known as **overdamping.** An overdamped system also settles back to its equilibrium exponentially, but it takes more time (Fig. 16-29).

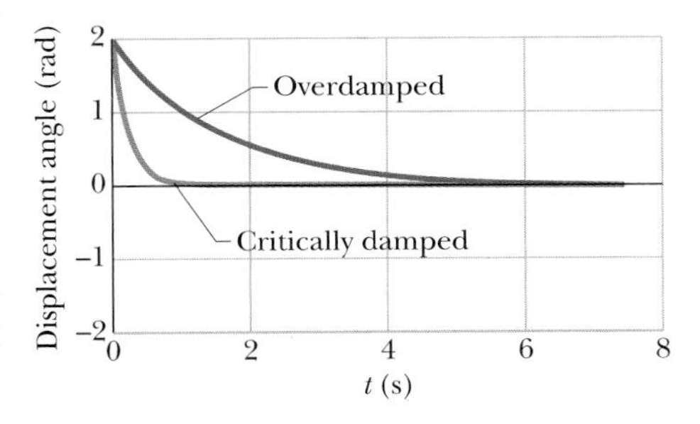

FIGURE 16-29 ■ The theoretically calculated angular displacement vs. time values for the damped pendulum shown in Fig. 16-27. The bottom exponential curve shows what happens when the pendulum edge mass is critically damped and settles back to equilibrium without oscillating in a minimum time of about 1 s. The top exponential curve shows an overdamped condition when the edge mass takes about 6 s to reach equilibrium.

Engineers make use of their understanding of critical damping and overdamping in the design of automobile shock absorbers. A shock absorber is designed so that when a car hits a typical bump, the springs that connect the chassis to the wheels return to equilibrium slowly without oscillating too much. Too little damping causes the passengers to bounce up and down, and too much damping gives a rough ride because the car cannot respond quickly to a bump. Note what happens when you push the front of a car down suddenly and let go. If it returns to equilibrium without oscillating, the car is either critically damped or overdamped. If is oscillates a bit, it is underdamped.

READING EXERCISE 16-10: Here are three sets of values for the damping constant and mass for the damped oscillator of Fig. 16-25. Using Eq. 16-39, rank the sets according to the time required for the mechanical energy to decrease to one-fourth of its initial value, greatest first. No calculations are needed.

Set 1:	b_0	m_0
Set 2:	$6b_0$	$4m_0$
Set 3:	$3b_0$	m_0

■

TOUCHSTONE EXAMPLE 16-4: Damped Mass–Spring

For the damped oscillator of Fig. 16-25, $m = 250$ g, $k = 85$ N/m, and $b = 70$ g/s.

(a) What is the period of the motion?

SOLUTION ■ The **Key Idea** here is that because $b \ll \sqrt{km} = 4.6$ kg/s, the period is approximately that of the undamped oscillator. From Eq. 16-12, we then have

$$T = 2\pi\sqrt{\frac{m}{k}} = 2\pi\sqrt{\frac{0.25 \text{ kg}}{85 \text{ N/m}}} = 0.34 \text{ s.} \qquad \text{(Answer)}$$

(b) How long does it take for the amplitude of the damped oscillations to drop to half its initial value?

SOLUTION ■ Now the **Key Idea** is that the amplitude at time t is displayed in Eq. 16-37 as $Ye^{-bt/2m}$. It has the value Y at $t = 0$. Thus, we must find the value of t for which

$$Ye^{-bt/2m} = \tfrac{1}{2}Y.$$

Canceling Y and taking the natural logarithm of the equation that remains, we have $\ln\frac{1}{2}$ on the right side and

$$\ln(e^{-bt/2m}) = -bt/2m$$

on the left side. Thus,

$$t = \frac{-2m\ln\frac{1}{2}}{b} = \frac{-(2)(0.25 \text{ kg})(\ln\frac{1}{2})}{0.070 \text{ kg/s}}$$
$$= 5.0 \text{ s.} \qquad \text{(Answer)}$$

Because $T = 0.34$ s, this is about 15 periods of oscillation.

(c) How long does it take for the mechanical energy to drop to one-half its initial value?

SOLUTION ■ Here the **Key Idea** is that, from Eq. 16-39, the mechanical energy at time t is $\frac{1}{2}kY^2e^{-bt/m}$. It has the value $\frac{1}{2}kY^2$ at $t = 0$. Thus, we must find the value of t for which

$$\tfrac{1}{2}kY^2e^{-bt/m} = \tfrac{1}{2}(\tfrac{1}{2}kY^2).$$

If we divide both sides of this equation by $\frac{1}{2}kY^2$ and solve for t as we did above, we find

$$t = \frac{-m\ln\frac{1}{2}}{b} = \frac{-(0.25 \text{ kg})(\ln\frac{1}{2})}{0.070 \text{ kg/s}} = 2.5 \text{ s.} \qquad \text{(Answer)}$$

This is exactly half the time we calculated in (b), or about 7.5 periods of oscillation. Figure 16-30 was drawn to illustrate this touchstone example. Since the system's mechanical energy depends on the square of the amplitude, it decreases more rapidly than the amplitude does.

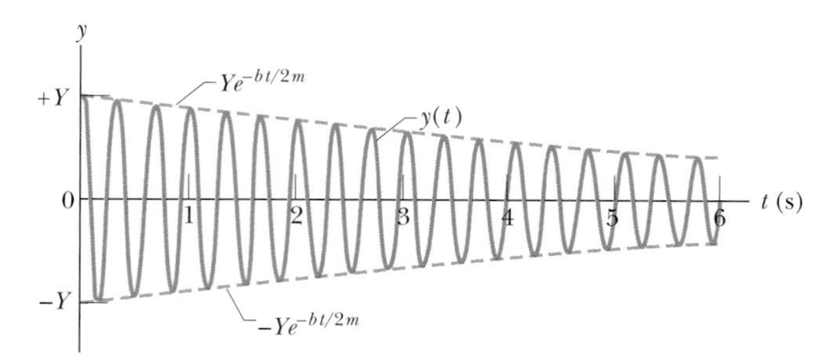

FIGURE 16-30 ■ (a) This diagram shows a period of about 0.34 s. (b) Since amplitude has dropped to about $0.7Y$ after 2.5 s, the diagram shows a system mechanical energy of $(0.7Y)^2 = 0.5Y^2$ at that time.

16-8 Forced Oscillations and Resonance

Sometimes we would like to maintain oscillations longer than they would naturally continue because of the damping forces. For example, if you were on a swing that was given only one big push, you would go up and back a few times before the mechanical energy was completely lost and you came to a stop. Although we cannot totally eliminate such loss of mechanical energy, we can replenish the energy from some source. As an example, you know that by swinging your legs or torso you can "pump" a swing to maintain or enhance the oscillations. In doing this, you transfer biochemical energy to mechanical energy of the oscillating system.

A person swinging without pumping their legs or without anyone pushing undergoes *free oscillation.* However, if someone pushes the swing periodically, the swing has *forced,* or *driven, oscillations. Two* angular frequencies are associated with a system undergoing driven oscillations: (1) the *natural* angular frequency ω of the system while oscillating freely with the inevitable damping forces present, and (2) the angular frequency ω_d of the external driving force.

We can use Fig. 16-25 to represent an idealized forced simple harmonic oscillator if we cause the structure marked "rigid support" to move up and down at an angular frequency ω_d that we can adjust. This forced oscillator will settle down to oscillate at our chosen angular frequency ω_d of the driving force, and its displacement $y(t)$ is given by

$$y(t) = Y\cos(\omega_d t + \phi_1), \tag{16-41}$$

where Y is the amplitude of the oscillations.

How large the displacement amplitude Y is depends on a complicated function of ω_d and ω. The maximum velocity $V = \omega' Y$ of the oscillations is easier to describe: it is greatest when

$$\omega_d = \omega' \quad \text{(resonance)}, \tag{16-42}$$

which is a condition called **resonance.** Equation 16-42 is also *approximately* the condition at which the displacement amplitude Y of the oscillations is greatest. Thus, if you push a swing at its natural angular frequency, the displacement and maximum velocities will increase to large values, a fact that children learn quickly by trial and error. If you push at other angular frequencies, either higher or lower, the displacement and maximum velocities will be smaller.

Figure 16-31 shows how the displacement amplitude of a damped, driven oscillator depends on the angular frequency ω_d of the driving force, for three values of the damping coefficient b. Note that for all three the amplitude is approximately greatest when $\omega_d/\omega' = 1$—that is, when the resonance condition of Eq. 16-42 is satisfied. The curves of Fig. 16-31 show that less damping gives a taller and narrower *resonance peak.* A system with a tall, narrow resonance curve is referred to as having a "high-Q."

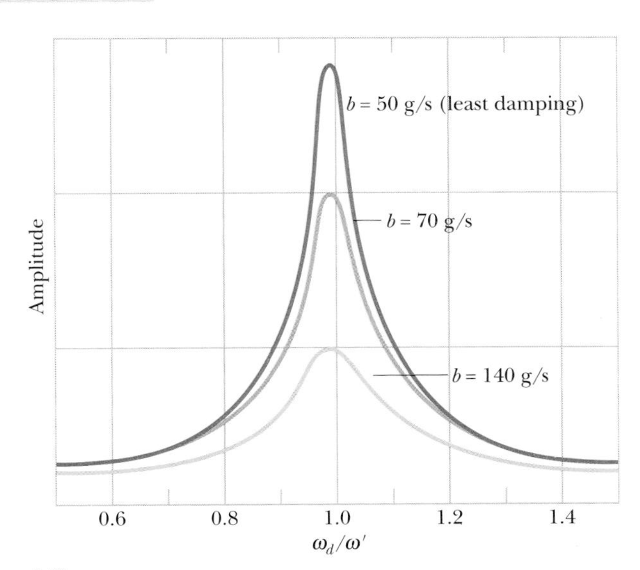

FIGURE 16-31 ■ The displacement amplitude of a forced oscillator varies as the angular frequency ω_d of the driving force is varied. The amplitude is greatest *approximately* at $\omega_d/\omega' = 1$, the resonance condition. The curves here correspond to three values of the damping constant b.

Resonance and Earthquake Damage

All mechanical structures have one or more natural angular frequencies, and if a structure is subjected to a strong external driving force that matches one of these angular frequencies, the resulting oscillations of the structure may rupture it. Thus, for example, aircraft designers must make sure that none of the natural angular frequencies at which a wing can oscillate matches the angular frequency of the engines in flight. A wing that flaps violently at certain engine speeds would obviously be dangerous.

Mexico's earthquake in September 1985 was a major earthquake (8.1 on the Richter scale), but the seismic waves from it should have been too weak to cause extensive damage when they reached Mexico City about 400 km away. However, Mexico City is largely built on an ancient lake bed, where the soil is still soft with water. Although the amplitude of the seismic waves was weak in the firmer ground en route to Mexico City, their amplitude substantially increased in the loose soil of the city. Maximum accelerations of the waves were as much as $0.20g$, and the angular frequency was (surprisingly) concentrated around 3 rad/s. Not only did the ground begin oscillating with large amplitudes, but also many of the buildings with intermediate height had resonant angular frequencies of about 3 rad/s. Most of those buildings collapsed during the violent shaking, while shorter buildings (higher resonant angular frequencies) and taller buildings (with lower resonant angular frequencies) remained standing.

Problems

SEC. 16-4 ■ VELOCITY AND ACCELERATION FOR SHM

1. Object Undergoing SHM An object undergoing simple harmonic motion takes 0.25 s to travel from one point of zero velocity to the next such point. The distance between those points is 36 cm. Calculate the object's (a) period, (b) frequency, and (c) amplitude.

2. Oscillating Block An oscillating block–spring system takes 0.75 s to begin repeating its motion. Find its (a) period, (b) frequency in hertz, and (c) angular frequency in radians per second.

3. Oscillator An oscillator consists of a block of mass 0.500 kg connected to a spring. When set into oscillation with amplitude 35.0 cm, the oscillator repeats its motion every 0.500 s. Find (a) the period, (b) the frequency, (c) the angular frequency, (d) the spring constant, (e) the maximum speed, and (f) the magnitude of the maximum force on the block from the spring.

4. Maximum Acceleration What is the maximum acceleration of a platform that oscillates with an amplitude of 2.20 cm at a frequency of 6.60 Hz?

5. Loudspeaker A loudspeaker produces a musical sound by means of the oscillation of a diaphragm. If the amplitude of oscillation is limited to 1.0×10^{-3} mm, what frequencies will result in the magnitude of the diaphragm's acceleration exceeding g?

6. Spring Balance The scale of a spring balance that reads from 0 to 15.0 kg is 12.0 cm long. A package suspended from the balance is found to oscillate vertically with a frequency of 2.00 Hz. (a) What is the spring constant? (b) How much does the package weigh?

7. A Particle of Mass A particle with a mass of 1.00×10^{-20} kg is oscillating with simple harmonic motion with a period of 1.00×10^{-5} s and a maximum speed of 1.00×10^{3} m/s. Calculate (a) the angular frequency and (b) the maximum displacement of the particle.

8. A Small Body A small body of mass 0.12 kg is undergoing simple harmonic motion of amplitude 8.5 cm and period 0.20 s. (a) What is the magnitude of the maximum force acting on it? (b) If the oscillations are produced by a spring, what is the spring constant?

9. Electric Shaver In an electric shaver, the blade moves back and forth over a distance of 2.0 mm in simple harmonic motion, with frequency 120 Hz. Find (a) the amplitude, (b) the maximum blade speed, and (c) the magnitude of the maximum blade acceleration.

10. Speaker Diaphragm A loudspeaker diaphragm is oscillating in simple harmonic motion with a frequency of 440 Hz and a maximum displacement of 0.75 mm. What are (a) the angular frequency, (b) the maximum speed and (c) the magnitude of the maximum acceleration?

11. Automobile Spring An automobile can be considered to be mounted on four identical springs as far as vertical oscillations are concerned. The springs of a certain car are adjusted so that the oscillations have a frequency of 3.00 Hz. (a) What is the spring constant of each spring if the mass of the car is 1450 kg and the mass is evenly distributed over the springs? (b) What will be the oscillation frequency if five passengers, averaging 73.0 kg each, ride in the car? (Again, consider an even distribution of mass.)

12. A Body Oscillates A body oscillates with simple harmonic motion according to the equation

$$x = (6.0 \text{ m}) \cos[(3\pi \text{ rad/s})t + \pi/3 \text{ rad}].$$

At $t = 2.0$ s, what are (a) the displacement, (b) the velocity, (c) the acceleration, and (d) the phase of the motion? Also, what are (e) the frequency and (f) the period of the motion?

13. Piston in Cylinder The piston in the cylinder head of a locomotive has a stroke (twice the amplitude) of 0.76 m. If the piston moves with simple harmonic motion with a frequency of 180 rev/min, what is its maximum speed?

14. BMMD Astronauts sometimes use a device called a body-mass measuring device (BMMD). Designed for use on orbiting space vehicles, its purpose is to allow astronauts to measure their mass in the "weightless" conditions in Earth orbit. The BMMD is a spring-mounted chair; an astronaut measures his or her period of oscillation in the chair; the mass follows from the formula for the period of an oscillating block–spring system. (a) If M is the mass of the astronaut and m the effective mass of that part of the BMMD that also oscillates, show that

$$M = (k/4\pi^2)T^2 - m,$$

where T is the period of oscillation and k is the spring constant. (b) The spring constant was $k = 605.6$ N/m for the BMMD on Skylab Mission Two; the period of oscillation of the empty chair was 0.90149 s. Calculate the effective mass of the chair. (c) With an astronaut in the chair, the period of oscillation became 2.08832 s. Calculate the mass of the astronaut.

15. Harbor At a certain harbor, the tides cause the ocean surface to rise and fall a distance d (from highest level to lowest level) in simple harmonic motion, with a period of 12.5 h. How long does it take for the water to fall a distance $d/4$ from its highest level?

16. Two Blocks In Fig. 16-32 two blocks (m = 1.0 kg and M = 10 kg) and a spring (k = 200 N/m) are arranged on a horizontal, frictionless surface. The coefficient of static friction between the two blocks is 0.40. What amplitude of simple harmonic motion of the spring–blocks system puts the smaller block on the verge of slipping over the larger block?

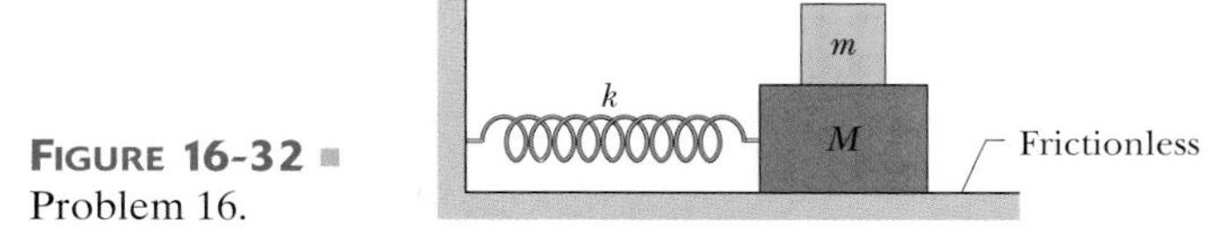

FIGURE 16-32 ■ Problem 16.

17. Shake Table A block is on a horizontal surface (a shake table) that is moving back and forth horizontally with simple harmonic motion of frequency 2.0 Hz. The coefficient of static friction between block and surface is 0.50. How great can the amplitude of the SHM be if the block is not to slip along the surface?

18. Block and Piston A block rides on a piston that is moving vertically with simple harmonic motion. (a) If the SHM has period 1.0 s, at what amplitude of motion will the block and piston separate? (b) If the piston has an amplitude of 5.0 cm, what is the maximum frequency for which the block and piston will be in contact continuously?

19. Oscillator An oscillator consists of a block attached to a spring (k = 400 N/m). At some time t, the position (measured from the system's equilibrium location), velocity, and acceleration of the block are x = 0.100 m, v = −13.6 m/s, and a = −123 m/s^2. Calculate (a) the frequency of oscillation, (b) the mass of the block, and (c) the amplitude of the motion.

20. Simple Harmonic Oscillator A simple harmonic oscillator consists of a block of mass 2.00 kg attached to a spring of spring constant 100 N/m. When t = 1.00 s, the position and velocity of the block are x = 0.129 m and v = 3.415 m/s. (a) What is the amplitude of the oscillations? What were the (b) position and (c) velocity of the block at t = 0 s?

21. Massless Spring A massless spring hangs from the ceiling with a small object attached to its lower end. The object is initially held at rest in a position y_1 such that the spring is at its rest length. The object is then released from y_1 and oscillates up and down, with its lowest position being 10 cm below y_1. (a) What is the frequency of the oscillation? (b) What is the speed of the object when it is 8.0 cm below the initial position? (c) An object of mass 300 g is attached to the first object, after which the system oscillates with half the original frequency. What is the mass of the first object? (d) Relative to y_1 where is the new equilibrium (rest) position with both objects attached to the spring?

22. Two Particles Two particles execute simple harmonic motion of the same amplitude and frequency along close parallel lines. They pass each other moving in opposite directions each time their displacement is half their amplitude. What is their phase difference?

23. Two Particles Oscillate Two particles oscillate in simple harmonic motion along a common straight-line segment of length A. Each particle has a period of 1.5 s, but they differ in phase by $\pi/6$ rad. (a) How far apart are they (in terms of A) 0.50 s after the lagging particle leaves one end of the path? (b) Are they then moving in the same direction, toward each other, or away from each other?

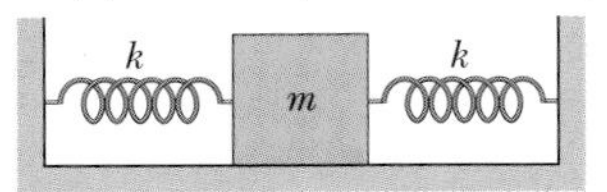

FIGURE 16-33 ■ Problems 24 and 25.

24. Two Identical Springs In Fig. 16-33, two identical springs of spring constant k are attached to a block of mass m and to fixed supports. Show that the block's frequency of oscillation on the frictionless surface is

$$f = \frac{1}{2\pi}\sqrt{\frac{2k}{m}}.$$

25. Block and Two Springs Suppose that the two springs in Fig. 16-33 have different spring constants k_1 and k_2. Show that the frequency f of oscillation of the block is then given by

$$f = \sqrt{f_1^2 + f_2^2},$$

where f_1 and f_2 are the frequencies at which the block would oscillate if connected only to spring 1 or only to spring 2.

26. Tuning Fork The end of one of the prongs of a tuning fork that executes simple harmonic motion of frequency 1000 Hz has an amplitude of 0.40 mm. Find (a) the magnitude of the maximum acceleration and (b) the maximum speed of the end of the prong. Find (c) the magnitude of the acceleration and (d) the speed of the end of the prong when the end has a displacement of 0.20 mm.

27. Two Springs Are Joined In Fig. 16-34, two springs are joined and connected to a block of mass m. The surface is frictionless. If the springs both have spring constant k, show that

$$f = \frac{1}{2\pi}\sqrt{\frac{k}{2m}}$$

gives the block's frequency of oscillation.

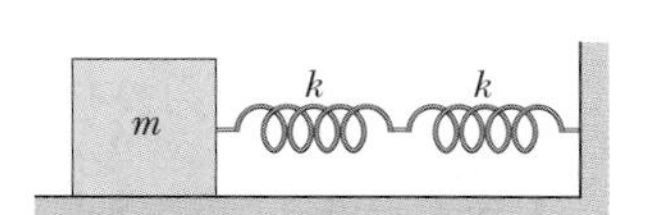

FIGURE 16-34 ■ Problem 27.

28. Block on Incline In Fig. 16-35, a block weighing 14.0 N, which slides without friction on a 40.0° incline, is connected to the top of the incline by a massless spring of unstretched length 0.450 m and spring constant 120 N/m. (a) How far from the top of the incline does the block stop? (b) If the block is pulled slightly down the incline and released, what is the period of the resulting oscillations?

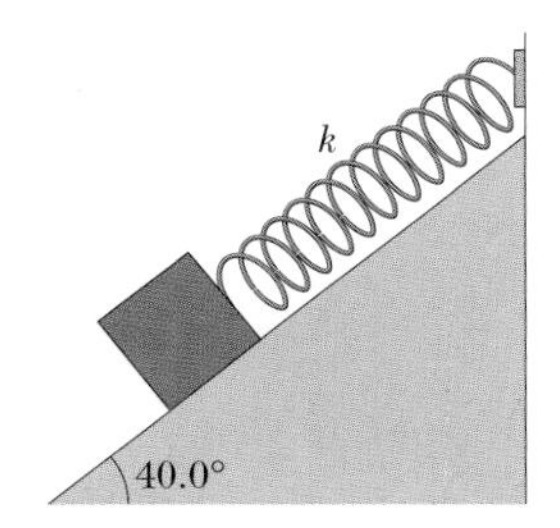

FIGURE 16-35 ■ Problem 28.

29. Unstretched Length A uniform spring with unstretched length L and spring constant k is cut into two pieces of unstretched lengths L_1 and L_2, with $L_1 = nL_2$. What are the corresponding spring constants (a) k_1 and (b) k_2 in terms of n and k? If a block is attached to the original spring, as in Fig. 16-10, it oscillates with frequency f. If the spring is replaced with the piece L_1 or L_2, the corresponding frequency is f_1 or f_2. Find (c) f_1 and (d) f_2 in terms of f.

30. Ore Cars In Fig. 16-36, three 10 000 kg ore cars are held at rest on a 30° incline on a mine railway using a cable that is parallel to the incline. The cable stretches 15 cm just before the coupling between the two lower cars breaks, detaching the lowest car. Assuming that the cable obeys Hooke's law, find (a) the frequency and (b) the amplitude of the resulting oscillations of the remaining two cars.

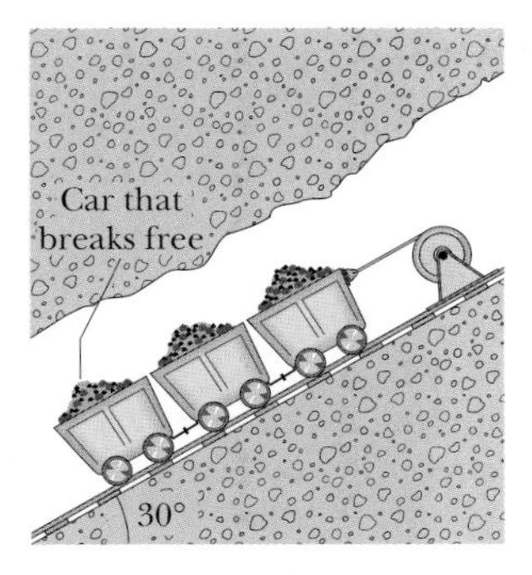

FIGURE 16-36 ■ Problem 30.

31. Balance Wheel The balance wheel of a watch oscillates with a rotational amplitude of π rad and a period of 0.500 s. Find (a) the maximum rotational speed of the wheel, (b) the rotational speed of the wheel when its displacement is $\pi/2$ rad, and (c) the magnitude of the rotational acceleration of the wheel when its displacement is $\pi/4$ rad.

32. Flat Disk A flat uniform circular disk has a mass of 3.00 kg and a radius of 70.0 cm. It is suspended in a horizontal plane by a vertical wire attached to its center. If the disk is rotated 2.50 rad about the wire, a torque of 0.0600 N · m is required to maintain that orientation. Calculate (a) the rotational inertia of the disk about the wire, (b) the torsion constant, and (c) the angular frequency of this torsion pendulum when it is set oscillating.

SEC. 16-5 ■ GRAVITATIONAL PENDULA

33. Simple Pendulum What is the length of a simple pendulum that marks seconds by completing a full swing from left to right and then back again every 2.0 s?

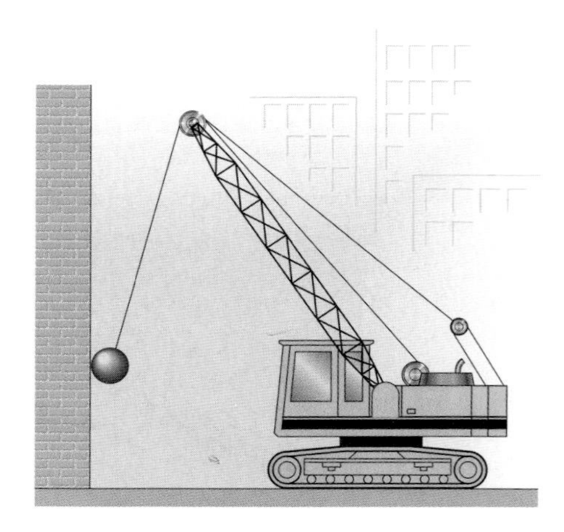

FIGURE 16-37 ■ Problem 34.

34. Demolition Ball In Fig. 16-37, a 2500 kg demolition ball swings from the end of a crane. The length of the swinging segment of cable is 17 m. (a) Find the period of the swinging, assuming that the system can be treated as a simple pendulum. (b) Does the period depend on the ball's mass?

35. Physical Pendulum A physical pendulum consists of a meter stick that is pivoted at a small hole drilled through the stick a distance d from the 50 cm mark. The period of oscillation is 2.5 s. Find d.

36. Trapeze A performer seated on a trapeze is swinging back and forth with a period of 8.85 s. If she stands up, thus raising the center of mass of the *trapeze + performer* system by 35.0 cm, what will be the new period of the system? Treat *trapeze + performer* as a simple pendulum.

37. Pivoting Long Rod A pendulum is formed by pivoting a long thin rod of length L and mass m about a point on the rod that is a distance d above the center of the rod. (a) Find the period of this pendulum in terms of d, L, m, and g, assuming small-amplitude swinging. What happens to the period if (b) d is decreased, (c) L is increased, or (d) m is increased?

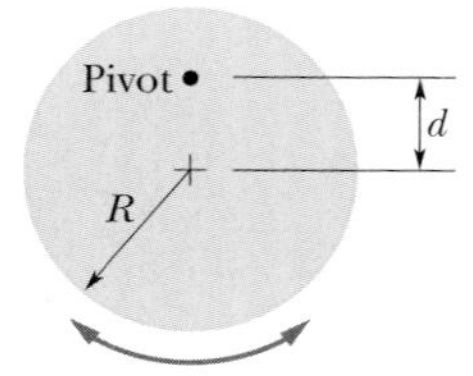

FIGURE 16-38 ■ Problem 38.

38. Solid Disk In Fig. 16-38, a physical pendulum consists of a uniform solid disk (of mass M and radius R) supported in a vertical plane by a pivot located a distance d from the center of the disk. The disk is displaced by a small angle and released. Find an expression for the period of the resulting simple harmonic motion.

39. Oscillating Physical Pendulum The pendulum in Fig. 16-39, consists of a uniform disk with radius 10.0 cm and mass 500 g attached to a uniform rod with length 500 mm and mass 270 g. (a) Calculate the rotational inertia of the pendulum about the pivot point. (b) What is the distance between the pivot point and the center of mass of the pendulum? (c) Calculate the period of oscillation.

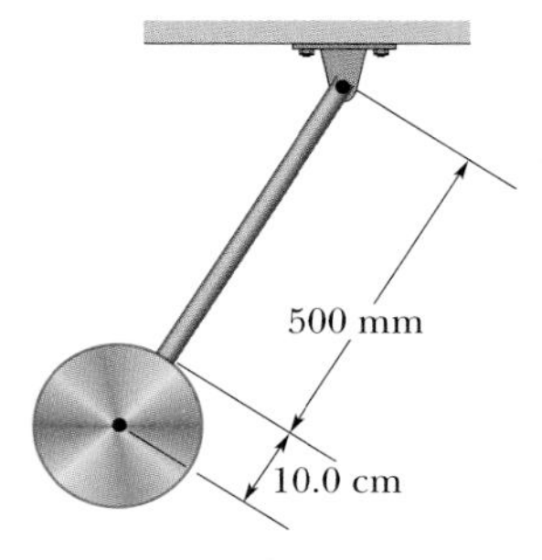

FIGURE 16-39 ■ Problem 39.

40. Pendulum with Disk A uniform circular disk whose radius R is 12.5 cm is suspended as a physical pendulum from a point on its rim. (a) What is its period? (b) At what radial distance $r < R$ is there a pivot point that gives the same period?

41. Long Uniform Rod In the overhead view of Fig. 16-40, a long uniform rod of length L and mass m is free to rotate in a horizontal plane about a vertical axis through its center. A spring with force constant k is connected horizontally between one end of the rod and a fixed wall. When the rod is in equilibrium, it is parallel to the wall. What is the period of the small oscillations that result when the rod is rotated slightly and released?

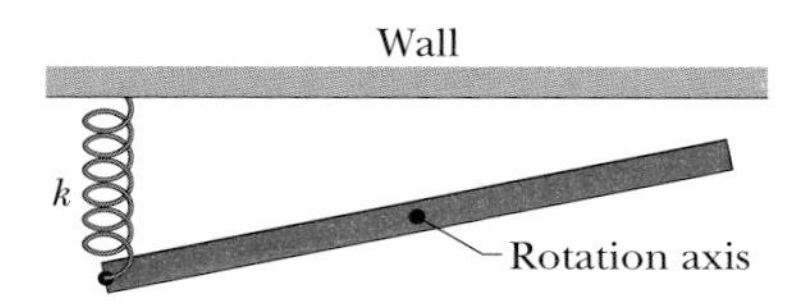

FIGURE 16-40 ■ Problem 41.

42. A Stick A stick with length L oscillates as a physical pendulum, pivoted about point O in Fig. 16-41. (a) Derive an expression for the period of the pendulum in terms of L and x, the distance from the pivot point to the center of mass of the pendulum. (b) For what value of x/L is the period a minimum? (c) Show that if L = 1.00 m and g = 9.80 m/s^2, this minimum period is 1.53 s.

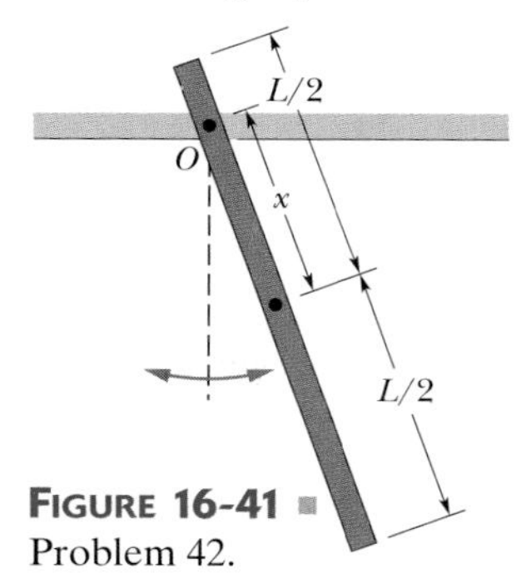

FIGURE 16-41 ■ Problem 42.

43. Frequency What is the frequency of a simple pendulum 2.0 m long (a) in a room, (b) in an elevator accelerating upward at a rate of 2.0 m/s^2, and (c) in free fall?

44. In a Car A simple pendulum of length L and mass m is suspended in a car that is traveling with constant speed $|\vec{v}|$ around a circle of radius R. If the pendulum undergoes small oscillations in a radial direction about its equilibrium position, what will be its frequency of oscillation?

45. The Bob The bob on a simple pendulum of length R moves in an arc of a circle. (a) By considering that the radial acceleration of the bob as it moves through its equilibrium position is that for uniform circular motion (v^2/R), show that the tension in the string at that position is $mg\,(1 + \Theta^2)$ if the angular amplitude Θ is small. (See "Trigonometric Expansions" in Appendix E.)

(b) Is the tension at other positions of the bob greater, smaller, or the same?

46. Angular Amplitude For a simple pendulum, find the angular amplitude Θ at which the restoring torque required for simple harmonic motion deviates from the actual restoring torque by 1.0%. (See "Trigonometric Expansions" in Appendix E.)

47. Wheel Rotates A wheel is free to rotate about its fixed axle. A spring is attached to one of its spokes a distance r from the axle, as shown in Fig. 16-42. (a) Assuming that the wheel is a hoop of mass m and radius R, obtain the angular frequency of small oscillations of this system in terms of m, R, r, and the spring constant k. How does the result change if (b) $r = R$ and (c) $r = 0$?

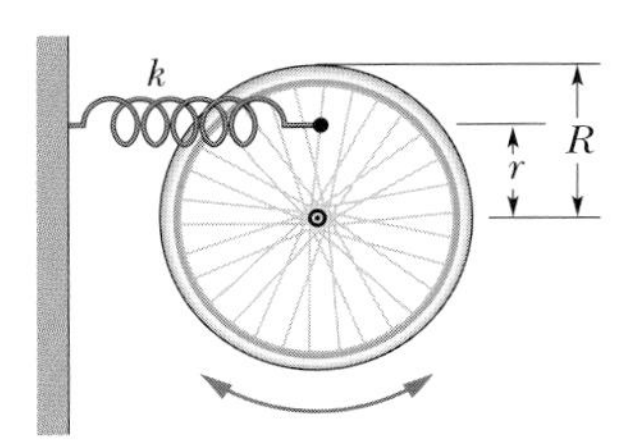

FIGURE 16-42 ■ Problem 47.

SEC. 16-6 ■ ENERGY IN SIMPLE HARMONIC MOTION

48. Large Slingshot A (hypothetical) large slingshot is stretched 1.50 m to launch a 130 g projectile with speed sufficient to escape from Earth (11.2 km/s). Assume the elastic bands of the slingshot obey Hooke's law. (a) What is the spring constant of the device, if all the elastic potential energy is converted to kinetic energy? (b) Assume that an average person can exert a force of 220 N. How many people are required to stretch the elastic bands?

49. Mechanical Energy Find the mechanical energy of a block–spring system having a spring constant of 1.3 N/cm and an oscillation amplitude of 2.4 cm.

50. Block–Spring An oscillating block–spring system has a mechanical energy of 1.00 J, an amplitude of 10.0 cm, and a maximum speed of 1.20 m/s. Find (a) the spring constant, (b) the mass of the block, and (c) the frequency of oscillation.

51. Horizontal Frictionless A 5.00 kg object on a horizontal frictionless surface is attached to a spring with spring constant 1000 N/m. The object is displaced from equilibrium 50.0 cm horizontally and given an initial velocity of 10.0 m/s back toward the equilibrium position. (a) What is the frequency of the motion? What are (b) the initial potential energy of the block–spring system, (c) the initial kinetic energy, and (d) the amplitude of the oscillation?

52. Block of Mass M A block of mass M, at rest on a horizontal frictionless table, is attached to a rigid support by a spring of constant k. A bullet of mass m and velocity $\vec{v}$ strikes the block as shown in Fig. 16-43. The bullet is embedded in the block. Determine (a) the speed of the block immediately after the collision and (b) the amplitude of the resulting simple harmonic motion.

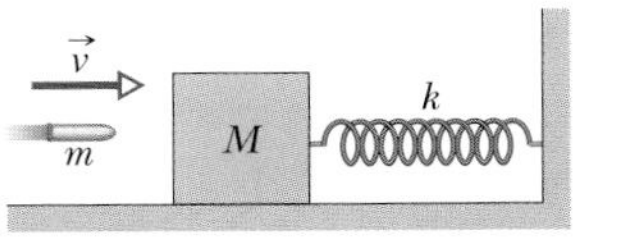

FIGURE 16-43 ■ Problem 52.

53. Displacement in SHM When the displacement in SHM is one-half the amplitude X, what fraction of the total energy is (a) kinetic energy and (b) potential energy? (c) At what displacement, in terms of the amplitude, is the energy of the system half kinetic energy and half potential energy?

54. Particle Undergoing SHM A 10 g particle is undergoing simple harmonic motion with an amplitude of 2.0×10^{-3} m and a maximum acceleration of magnitude 8.0×10^{-3} m/s^2. The phase constant is $-\pi/3$ rad. (a) Write an equation for the force on the particle as a function of time. (b) What is the period of the motion? (c) What is the maximum speed of the particle? (d) What is the total mechanical energy of this simple harmonic oscillator?

55. Block Suspended from Spring A 4.0 kg block is suspended from a spring with a spring constant of 500 N/m. A 50 g bullet is fired into the block from directly below with a speed of 150 m/s and becomes embedded in the block. (a) Find the amplitude of the resulting simple harmonic motion. (b) What fraction of the original kinetic energy of the bullet is transferred to mechanical energy of the harmonic oscillator?

56. Vertical Spring A vertical spring stretches 9.6 cm when a 1.3 kg block is hung from its end. (a) Calculate the spring constant. This block is then displaced an additional 5.0 cm downward and released from rest. Find (b) the period, (c) the frequency, (d) the amplitude, and (e) the maximum speed of the resulting SHM.

SEC. 16-7 ■ DAMPED SIMPLE HARMONIC MOTION

57. Amplitude Ratio In Touchstone Example 16-4, what is the ratio of the amplitude of the damped oscillations to the initial amplitude when 20 full oscillations have elapsed?

58. Lightly Damped The amplitude of a lightly damped oscillator decreases by 3.0% during each cycle. What fraction of the mechanical energy of the oscillator is lost in each full oscillation?

59. System Shown For the system shown in Fig. 16-25, the block has a mass of 1.50 kg and the spring constant is 8.00 N/m. The damping force is given by $-b(dy/dt)$, where $b = 230$ g/s. Suppose that the block is initially pulled down a distance 12.0 cm and released. (a) Calculate the time required for the amplitude of the resulting oscillations to fall to one-third of its initial value. (b) How many oscillations are made by the block in this time?

60. Suspension System Assume that you are examining the oscillation characteristics of the suspension system of a 2000 kg automobile. The suspension "sags" 10 cm when the entire automobile is placed on it. Also, the amplitude of oscillation decreases by 50% during one complete oscillation. Estimate the values of (a) the spring constant k and (b) the damping constant b for the spring and shock absorber system of one wheel, assuming each wheel supports 500 kg.

SEC. 16-8 ■ FORCED OSCILLATIONS AND RESONANCE

61. Rigid Support For Eq. 16-41, suppose the amplitude Y is given by

$$Y = \frac{F^{\max}}{[m^2(\omega_d^2 - \omega^2)^2 + b^2\omega_d^2]^{1/2}},$$

where $F^{\max}$ is the (constant) amplitude of the external oscillating force exerted on the spring by the rigid support in Fig. 16-25. At resonance, what are (a) the amplitude and (b) the velocity amplitude of the oscillating object?

62. Washboard Road A 1000 kg car carrying four 82 kg people travels over a rough "washboard" dirt road with corrugations 4.0 m apart, which cause the car to bounce on its spring suspension. The car bounces with maximum amplitude when its speed is 16 km/h. The car now stops, and the four people get out. By how much does the car body rise on its suspension due to this decrease in mass?

Additional Problems

63. Where's the Force? A 50 gram mass is hanging from a spring whose unstretched length is 10 cm and whose spring constant is 2.5 N/m, as shown in Fig. 16-44. In the list below are described five situations. In some of the situations, the mass is at rest and remains at rest. In other situations, at the instant described, the mass is in the middle of an oscillation initiated by a person pulling the mass downward 5 cm from its equilibrium position and releasing it. Ignore both air resistance and internal damping in the spring. At the time the situation occurs, indicate whether the force vector requested points up (U), down (D), or has a magnitude of zero newtons (0). (a) The force on the mass exerted by the spring when the mass is at its equilibrium position and is at rest. (b) The force on the mass exerted by the spring when the mass is at its equilibrium position and is moving downward. (c) The net force on the mass when the mass is at its equilibrium position and is moving upward. (d) The force on the mass exerted by the spring when it is at the top of its oscillation. (e) The net force on the mass when it is at the top of its oscillation.

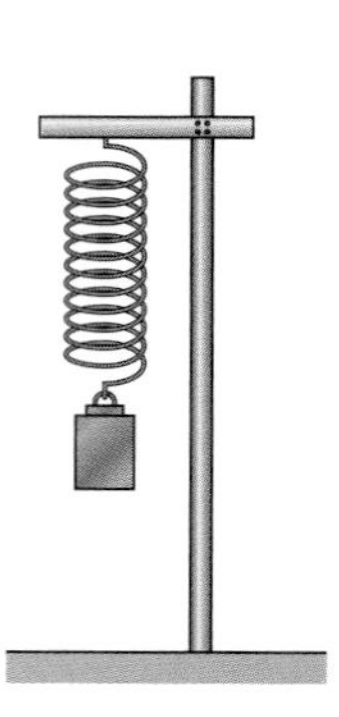

FIGURE 16-44 ■ Problem 63.

64. Swinging Ball A pendulum consisting of a massive ball on a light but nearly rigid rod is shown at two successive times in Fig. 16-45. The maximum angle of displacement of the pendulum from its equilibrium point is 25°. (a) If the length of the rod is R and the mass of the ball is m, find an expression for the speed of the ball at the point shown in Fig. 16-45*b*. (b) If the ball in Fig 16-45*b* is moving to the left at the instant of the snapshot, in what direction does its acceleration point at this instant in time? (c) In what direction does the acceleration of the ball in Fig. 16-45*a* point?

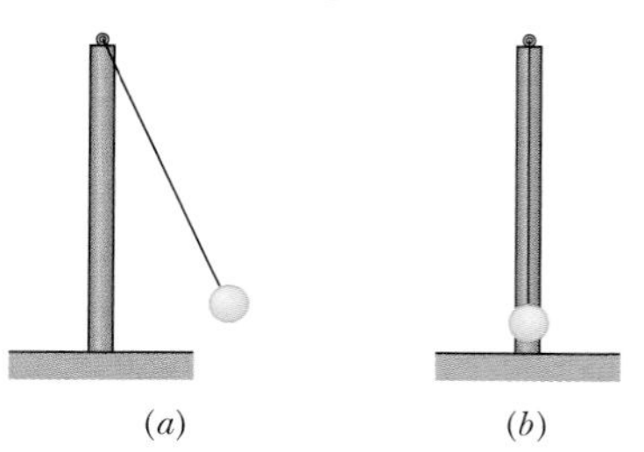

FIGURE 16-45 ■ Problem 64.

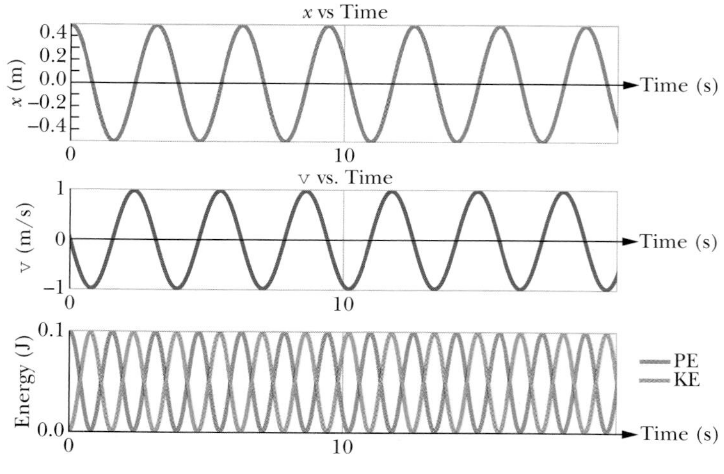

FIGURE 16-46 ■ Problem 65.

65. Oscillating Energies The graphs in Fig. 16-46 represent a computer simulation of a mass on a spring. The kinetic and potential energies are plotted in the bottom graph, in addition to the position and velocity of the oscillator (upper two graphs). The potential and kinetic energy curves oscillate, but not about zero. They also seem to oscillate twice as fast as the position and velocity curves. Is this correct? Explain.

66. Oscillating Graphs A mass is hanging from a spring off the edge of a table. The position of the mass is measured by a sonic ranger sitting on the floor 25 cm below the mass's equilibrium position. At some time, the mass is started oscillating. At a later time, the sonic ranger begins to take data.

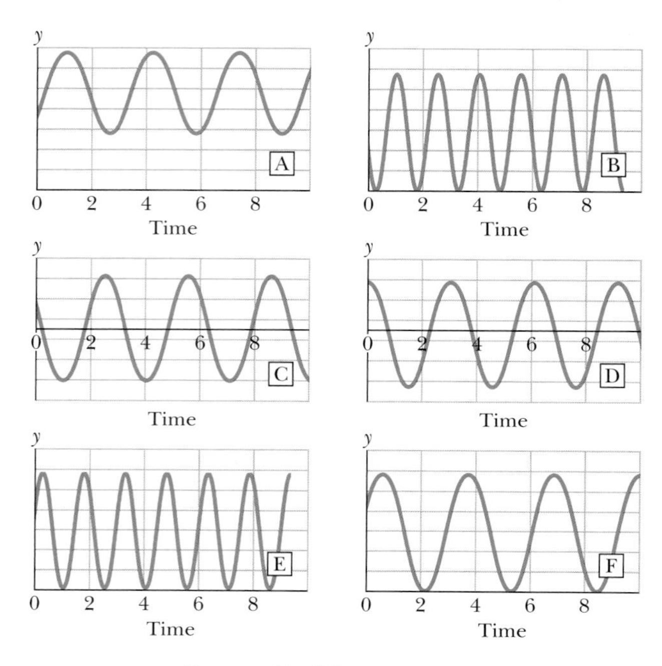

FIGURE 16-47 ■ Problem 66.

Figure 16-47 shows a series of graphs associated with the motion of the mass and a series of physical quantities. The graph labeled (A) is a graph of the mass's position as measured by the ranger. For each physical quantity, identify which graph could represent that quantity for this situation. If none are possible, answer N.

(a) Velocity of the mass
(b) Net force on the mass
(c) Force exerted by the spring on the mass
(d) Kinetic energy of the spring–mass system
(e) Potential energy of the spring–mass–Earth system
(f) Gravitational potential energy of the spring–mass–Earth system

67. Pendulum Graphs Some of the graphs shown in Fig. 16-48 represent the motion of a pendulum—a massive ball attached to a rigid, nearly massless rod, which in turn is attached to a rigid, nearly frictionless pivot. Four graphs of the pendulum's angle as a function of time are shown. Below are a set of four initial conditions and a denial. Match each graph with its most likely initial conditions (or with the denial). Note that the scales on the y axes are not necessarily the same. (There is not necessarily a one-to-one match.)

(1) $\theta_0 = 120°, (d\theta/dt)_0 = 0°/\text{s}$
(2) $\theta_0 = 173°, (d\theta/dt)_0 = 30°/\text{s}$
(3) $\theta_0 = 6°, (d\theta/dt)_0 = 0°/\text{s}$
(4) $\theta_0 = 173°, (d\theta/dt)_0 = 0°/\text{s}$
(5) Not a possible pendulum graph.

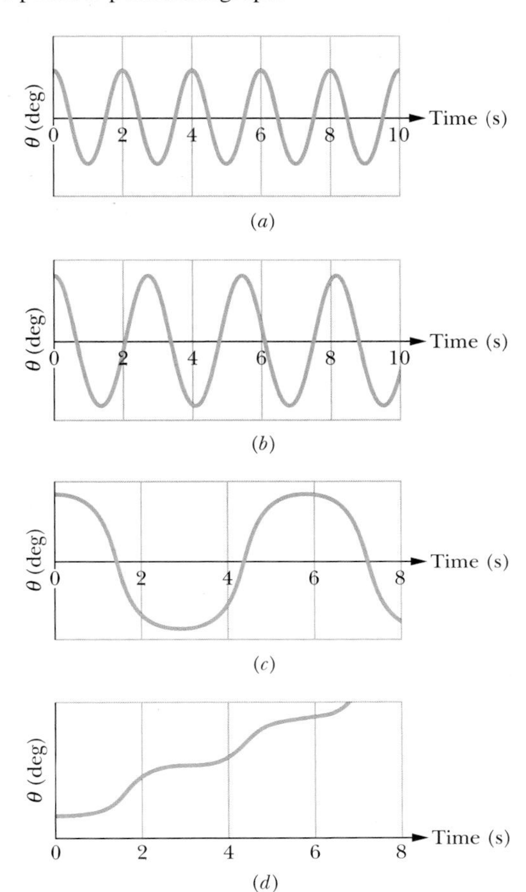

FIGURE 16-48 ■ Problem 67.

68. Swingin' in the Rain There is a forest in Italy that has many waterfalls. At one of the waterfalls, a long rope hangs down from the top of the cliff near the waterfall and has a seat on the bottom. Adventurous visitors could hop onto the seat and swing down into the waterfall. Their starting angle seems to be about 20°. It takes one of these adventurous people 8 seconds to swing out and back. Estimate the length of the rope and the speed with which they pass through the waterfall.

69. To What Angle? A small metal ball of mass m hangs from a pivot by a rigid, light metal rod of length R. The ball is swinging back and forth with an amplitude that remains small throughout its motion, $\theta^{max} \leq 5°$. Ignore all damping. (a) The equation of motion of this ideal pendulum can be derived in a variety of ways and is

$$\frac{d^2\theta}{dt^2} = -\frac{g}{R}\sin\theta.$$

For small angles, show how this can be replaced by an approximate equation of motion that can be solved more easily than the one given. (b) Write a general solution for the approximate equation of motion you obtained in (a) that works for any starting angle and angular velocity (as long as the angles stay in the range where the approximation is OK). Demonstrate that what you have written *is* a solution and show that at a time $t = 0$ your solution can have any given starting position and velocity. (c) If the length of the rod is 0.3 m, the mass of the ball is 0.2 kg, and the clock is started at a time when the ball is passing through the center ($\theta = 0$) and is moving with an angular speed of 0.1 rad/s, find the maximum angle your solution says the ball will reach. Can you use the approximate equation of motion for this motion? If the starting angle is not small, you cannot easily solve the equation of motion without a computer. But there are still things you can do. (d) Derive the energy conservation equation for the motion of the pendulum. (Do *not* use the small-amplitude approximation.) (e) If the pendulum is released from a starting angle of θ_1, what will be the maximum speed it travels at any point on its swing?

70. What's Wrong with cos? Observation of the oscillation of a mass on the end of a spring reveals that the detailed structure of the position as a function of time is fit very well by a function of the form

$$x(t) = X\cos(\omega t + \phi_0).$$

Yet subsequent observations give convincing evidence that this cannot be a good representation of the motion for long time periods. Explain what observation leads to this conclusion and resolve the apparent contradiction.

71. Where Is the Energy? A block of mass m is attached to a spring of spring constant k that is attached to a wall as shown in Fig. 16-49. (a) If the block starts at time $t = 0$ with the spring being at its rest length but the block having a velocity v_1, find a solution for the mass's position at all subsequent times. Make any assumptions you like in order to have a plausible but solvable model, but state your assumptions explicitly. (b) Are the energies at the times

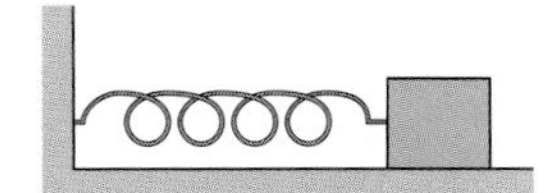

FIGURE 16-49 ■ Problem 71.

$$t = 0, \quad t = \frac{\pi}{2}\sqrt{\frac{m}{k}}, \quad t = 2\pi\sqrt{\frac{m}{k}}$$

kinetic, potential, or a mixture of the two? (This is *not* a short-answer question. Show how you know.)

72. Damped Oscillator A class looked at the oscillation of a mass on a spring. They observed for 10 seconds and found its oscillation was well fit by assuming that the mass's motion was governed by Newton's Second Law with the spring force $F_x^{\text{spring}} = -k\Delta x$, where Δx represents the stretch or squeeze of the spring and k is the spring constant. However, it was also clear that this was not an adequate representation for times on the order of 10 minutes, since by that time the mass had stopped oscillating. (a) Suppose the mass was started at an initial position x_1 with a velocity 0. Write down the solution, $x(t)$ and $v_x(t)$, for the equations of motion of the mass using only the spring force. What is the total energy of the oscillating mass? (b) Assume that there is also a velocity-dependent force (the *damping* force) that the spring exerts on an object when it is moving. Let's make the simplest assumption that the dynamic-spring force is linear in the velocity, $F_x^{\text{spring-dyn}} = -\gamma v_x$. Assume further that over one period of oscillation the velocity-dependent piece is small. Therefore, take $x(t)$ and $v_x(t)$ to be given by the oscillation without damping. Calculate the work done by the damping force over one period in terms of the parameters k, m, and γ and the fraction of the energy lost in one period. (c) The mass used was 1 kg. From the description of the experiment, estimate the spring constant and the number of periods it took to lose half the energy. Use this and your result from (a) to estimate the approximate size of the damping constant γ.

73. Catching a Pellet and Oscillating The following problem is a standard problem found in this text (and in many others). A block of mass M is at rest on a horizontal frictionless table. It is attached to a rigid support by a spring of constant k. A clay pellet having mass m and velocity v strikes the block as shown in the figure and sticks to it. See Fig. 16-50. (a) Determine the velocity of the block immediately after the collision. (b) Determine the amplitude of the resulting simple harmonic motion. In order to solve this problem you must make a number of simplifying assumption; some are stated in the problem and some are not. First, solve the problem as stated. (c) Discuss the approximations you had to make in order to solve the problem. (There are at least five.)

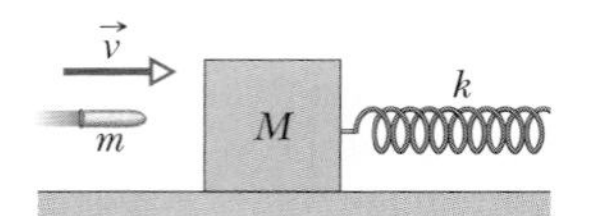

FIGURE 16-50 ■ Problem 73.

74. Bungee Jump As part of an open house, a physics department sets up a bungee jump from the top of a crane. See Fig. 16-51. Assume that one end of an elastic band will be firmly attached to the top of the crane and the other to the waist of a courageous participant. The participant will step off the edge of the crane house to be slowed and brought back up by the elastic band before hitting the ground. Assume the elastic band behaves like a Hooke's law spring. Estimate the length and spring constant of the elastic you would recommend be used.

FIGURE 16-51 ■ Problem 74.

17 Transverse Mechanical Waves

When a beetle moves along the sand within a few tens of centimeters of this sand scorpion, the scorpion immediately turns toward the beetle and dashes to it (for lunch). The scorpion can do this without seeing (it is nocturnal) or hearing the beetle.

How can the scorpion so precisely locate its prey?

The answer is in this chapter.

17-1 Waves and Particles

Two ways to get in touch with a friend in a distant city are to write a letter and to use the telephone.

The first choice (the letter) involves the concept of "particles": A material object moves from one point to another, carrying information with it. Most of the preceding chapters deal with particles or with systems of particles.

The second choice (the telephone) involves the concept of "waves," the subject of this chapter and the next. In your telephone call, a sound wave carries your message from your vocal cords to the telephone. There, an electromagnetic wave takes over, passing along a copper wire or an optical fiber or through the atmosphere, possibly by way of a communications satellite. At the receiving end there is another sound wave, from a telephone to your friend's ear. Although the message is passed, nothing that you have touched reaches your friend. For example, if you have the flu, she can't catch it by talking to you on the phone since no matter (and therefore no virus) is passed between the two of you. In a wave, momentum and energy move from one point to another but no material object makes that journey.

Leonardo da Vinci understood waves when he wrote of water waves: "It often happens that the wave flees the place of its creation, while the water does not; like the waves made in a field of grain by the wind, where we see the waves running across the field while the grain remains in place."

Particles and *waves* are the two important entities described in classical physics. Both entities have positions and velocities associated with them. But there are ways in which these two entities are very different. The word *particle* suggests a tiny concentration of matter capable of transmitting energy and momentum through its movement from one place to another. For example, a baseball is a particle that can transmit the energy and momentum imparted to it by a pitcher to the catcher by moving from one to the other. Alternatively the energy and momentum associated with a *wave* is spread out in space and is transmitted without any matter moving from one place to another. For example, when the pitcher shouts at a catcher, she can disturb the air particles in her vicinity, but the sound wave that transmits her voice depends on a chain of temporary disturbances of the air and not on molecules moving from the pitcher to the catcher.

In this chapter and the next we will put particles aside for a while and learn about mechanical waves and how they travel. Then in Chapters 34, 36, and 37, we will explore the behavior of electromagnetic waves including radio and light waves.

17-2 Types of Waves

Waves are of three main types:

1. ***Mechanical waves.*** These waves are most familiar because we encounter them almost constantly; common examples include water waves, sound waves, and seismic waves. All these waves have certain central features: They are governed by Newton's laws, and they can only move through a material medium, such as water, air, and rock.
2. ***Electromagnetic waves.*** These waves are less familiar, but you use them constantly; common examples include visible and ultraviolet light, radio and television waves, microwaves, x-rays, and radar waves. These waves can transmit energy and momentum without a material medium. Light waves from stars, for example, travel through the vacuum of space to reach us. Electromagnetic waves are treated in Chapters 32, 34, 36 and 37.

3. ***Matter waves.*** Although these waves are commonly used in modern technology, they are probably very unfamiliar to you. These waves are associated with electrons, protons, and other fundamental particles, and even atoms and molecules. Because we commonly think of these things as constituting matter, such waves are called matter waves. Their behavior is described by the laws of quantum mechanics.

Much of what we discuss in this chapter applies to waves of all kinds. However, for specific examples we shall refer to mechanical waves because they are the simplest and most familiar. In this chapter we consider how to best describe ideal mechanical waves mathematically. An **ideal** mechanical wave does not lose mechanical energy or change its shape as it travels through a medium.

17-3 Pulses and Waves

Consider Figure 17-1, which shows a "Slinky wave demonstrator" consisting of a very long Slinky that hangs from long, evenly spaced strings. If we give a quick pull outward (to the left) on the left end of the Slinky and then give it a quick push back inward (to the right), a compression-expansion disturbance like that shown in Fig. 17-1 will propagate along the length of the Slinky. (You need to look carefully at Fig. 17-1, rather than Fig. 17-2, to see the effect). Since the back-and-forth oscillations of the Slinky coils are parallel to the direction in which the disturbance travels, the motion is said to be **longitudinal.**

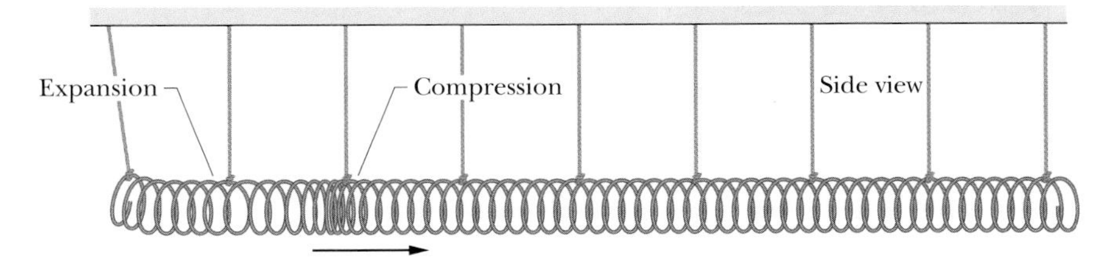

FIGURE 17-1 ■ A large Slinky demonstrator hanging from long, evenly spaced strings. Note the compression disturbance moving from left to right near the second and third strings with an expansion just behind it.

On the other hand, if we give the end of the Slinky a quick jerk back and forth (into and out of the page) at right angles to the line of the Slinky, a disturbance like that shown in Fig. 17-2 results and moves down the length of the Slinky. In this case, the displacement of the coils in the Slinky is *perpendicular* to the direction in which the disturbance travels (i.e., along the length of the Slinky). Such a disturbance is called **transverse.**

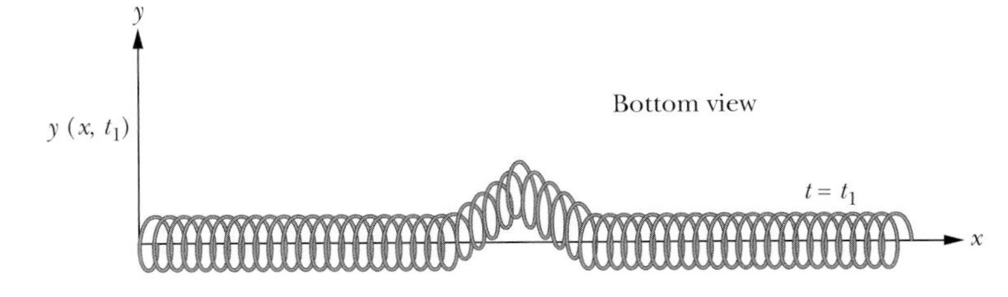

FIGURE 17-2 ■ A transverse or sideways disturbance or pulse moving from left to right along a Slinky wave demonstrator. In this case the viewer is lying on the floor underneath the Slinky looking up at time $t = t_1$.

In both cases discussed above, the singular disturbance is referred to as a **wave pulse** or **short wave.** If we repeat the motion that causes the disturbance (either the back-forth jerk or the push-pull motion) at regular time intervals, the result is a traveling and repeating disturbance that is referred to as a **continuous wave.** Just as with single pulses, waves can be longitudinal or transverse.

For example, if you give one end of a stretched string a single up-and-down jerk, a single *pulse* travels along the string as in Fig. 17-3*a*. This pulse and its motion can occur because the string is under tension. During the upward part of the jerk, when you pull your end of the string upward, it begins to pull upward on the

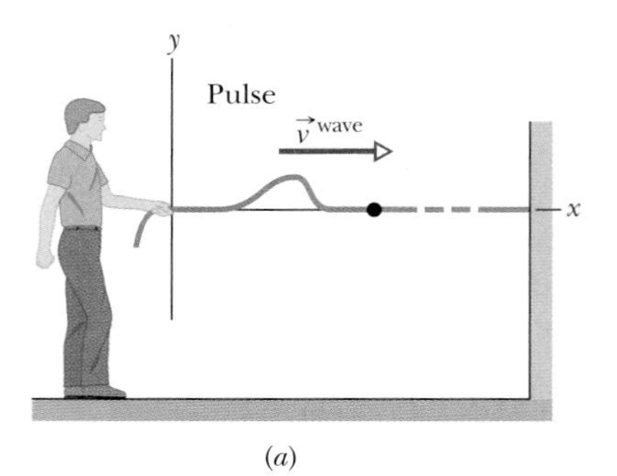

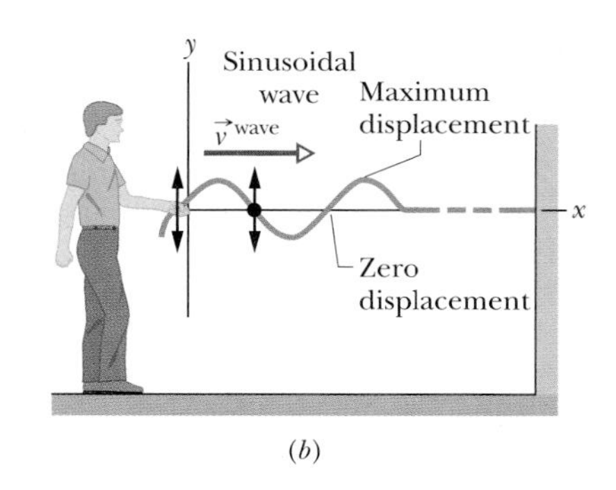

(a) (b)

FIGURE 17-3 ■ (*a*) A single pulse travels along a stretched string. A typical string piece (marked with a dot) moves up and back only once as the pulse passes. Since the string piece's displacement is perpendicular to the wave direction, the pulse is called a *transverse wave.* (*b*) A continuous sinusoidal displacement at the left end also causes the piece of string to move up and down in a transverse direction except now the wave is continuous.

adjacent section of the string via tension between the two sections. As the adjacent section moves upward, it begins to pull the next section upward, and so on. Meanwhile, suppose that you have now pulled down on your end of the string, completing the up-down stroke. As each section is moving upward in turn, it begins to be pulled back downward by neighboring sections that are already on the way down. The net result is that a distortion in the string's shape (the pulse) moves along the string at some velocity $\vec{v}^{\,\text{wave}}$. Note that although the disturbance moves down the string to the right, the parts of the string itself just move up and down. This is an extremely important, but subtle, point regarding pulses and waves. The wave or pulse travels, but the particles that make up the medium (string or otherwise) do not.

Sinusoidal Transverse Waves

If, instead of a single up-down stroke, you move your hand up and down repeatedly with simple harmonic motion as in Fig. 17-3*b*, a continuous *wave* travels along the string with a wave velocity denoted by $\vec{v}^{\,\text{wave}} = v_x^{\text{wave}}\hat{\text{i}}$. The vertical displacement of the string is perpendicular to the direction that the wave propagates (along the string), and so this wave is called a **transverse wave.** The larger the vertical displacement of your hand during the up-down stroke, the higher the peak and the lower the valley of the wave will be. This characteristic, measured as the magnitude of the maximum displacement of a small bit of string from its equilibrium position as the wave passes through it, is called the **amplitude** Y of the wave. This definition of amplitude is very similar to the one developed in the last chapter in regard to oscillations.

If we take a photograph of a transverse wave (perhaps the wave in the string of Fig. 17-3*b*) at some time $t = t_1$, we can see the wave shape and try to find a mathematical function that describes that shape or **wave form.** For example, if the motion of your hand is a sinusoidal function of time, the wave has a sinusoidal shape at any given instant, as in Fig. 17-3*b*. That is, the wave has the shape of a continuously repeating sine curve or cosine curve. (We consider here only an "undamped" string, in which no friction-like forces within the string cause the wave to die out as it travels along. In addition, we assume that the string is so long that we need not consider a wave rebounding from the far end and that the wave amplitude is small.)

The wave in Fig. 17-3*b* travels along a single line, so we call it a one-dimensional wave. For example, sound waves traveling in a pipe can only move in one dimension (but they can travel in two or three dimensions in other circumstances). For the wave in Fig. 17-3*b*, or any other one-dimensional wave or pulse, the convention is to define the line of travel as the x axis. As we can see from Fig. 17-3*b*, the vertical displacement of a bit of string at a given time depends on how far down along the string we make the measurement. Hence, we say that the vertical displacement of the string is a function of the horizontal position, x. The distance (in this example the horizontal distance) the wave travels in one cycle (for example, the peak-to-peak or valley-to-valley distance) is defined as the **wavelength** λ.

How a Sand Scorpion Catches Its Prey

The sand scorpion shown in the photograph opening this chapter uses waves of both transverse and longitudinal motion to locate its prey. When a beetle even slightly disturbs the sand, it sends pulses along the sand's surface (Fig. 17-4). One set of pulses is longitudinal, traveling with speed $|\vec{v}^{\text{ L-wave}}| = v^{\text{L-wave}} = 150$ m/s. A second set is transverse, traveling with speed $|\vec{v}^{\text{ T-wave}}| = v^{\text{T-wave}} = 50$ m/s.

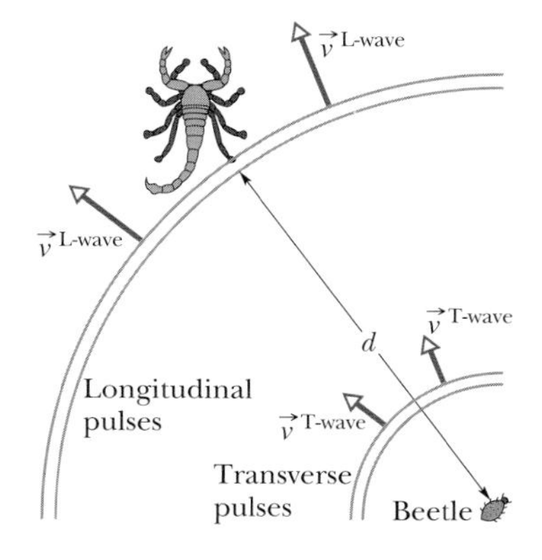

FIGURE 17-4 ■ A beetle's motion sends fast longitudinal pulses and slower transverse pulses along the sand's surface. The sand scorpion first intercepts the longitudinal pulses; here, it is the rear-most right leg that senses the pulses earliest.

The scorpion, with its eight legs spread roughly in a circle about 5 cm in diameter, intercepts the faster longitudinal pulses first and learns the direction of the beetle; it is in the direction of whichever leg is disturbed earliest by the pulses. The scorpion then senses the time interval Δt between that first interception and the interception of the slower transverse waves and uses it to determine the distance d to the beetle. The time interval is given by

$$\Delta t = \frac{d}{v^{\text{T-wave}}} - \frac{d}{v^{\text{L-wave}}}.$$

Solving for the distance gives us

$$d = (75 \text{ m/s})\,\Delta t.$$

For example, if $\Delta t = 4.0$ ms, then $d = 30$ cm, which gives the scorpion a perfect fix on the beetle.

Of course, the scorpion no more does these calculations in its head as it hunts a beetle than you do a conscious recalculation of the location of your center of mass as you decide in what way to throw out your arms when you slip. Instead, instinct and habits based on successful personal experiences rule the scorpion (and you) in such situations. But there is physics behind such "instincts"—they are not arbitrarily successful.

Waves and Oscillations

As we discussed above, if we take a picture of a wave, we can freeze the wave in time and investigate the spatial characteristics of the wave, like the wave's amplitude and its length. Alternatively, we could pick a special place, x_P, along the wave and peek at it through a slit, as in Fig. 17-5. As we peek though the slit, we would see the bits of string or Slinky rise and fall as the wave travels through space. If we made measurements while peeking, we could plot the displacement as a function of time for this location $x = x_P$. Doing so allows us to focus in on the characteristics of the wave that are associated with the passage of time. When we do this, we see several aspects of a wave that remind us of the oscillations we learned about in the last chapter.

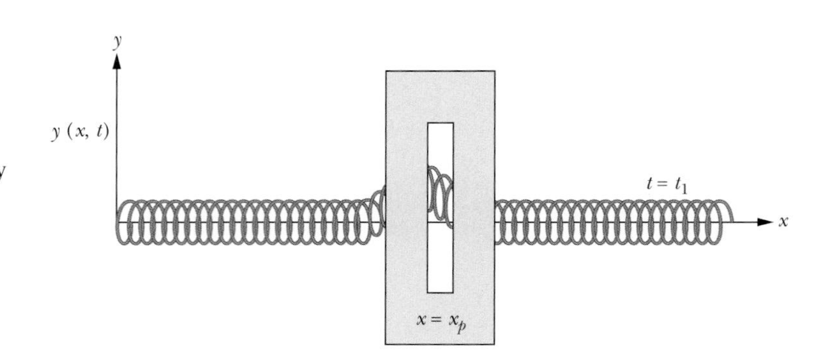

FIGURE 17-5 ■ A transverse wave traveling along a hanging Slinky as seen from the perspective of someone looking up through a slit while lying on the floor. The x axis is horizontal and chosen to be along the axis of the Slinky. The y axis is also in the horizontal plane but it lies along the direction of transverse displacement of Slinky coils. The viewer watches displacement y from the Slinky's equilibrium point as a function of time at a fixed position x_P.

For example, let us examine the small piece of string boxed in Fig. 17-6. This figure shows five "snapshots" of a sinusoidal wave traveling in the positive direction of an x axis. The time between snapshots is constant. The movement of the wave is indicated by the rightward progress of the short arrow pointing to a high point of the wave. From snapshot to snapshot, the short arrow moves to the right with the wave shape, but the pieces of the string move *only* parallel to the y axis. Let us follow the motion of the boxed string segment at $x = 0$. In the first snapshot (Fig. 17-6a), it is at displacement $y = 0$. In the next snapshot, it is at its extreme downward displacement because a *valley* (or extreme low point) of the wave is passing through it. It then moves back up through $y = 0$. In the fourth snapshot, it is at its extreme upward displacement because a *peak* (or extreme high point) of the wave is passing through it. In the fifth snapshot, it is again at $y = 0$, having completed one full oscillation. Notice that the wave in Fig. 17-6 moves to the right by $\frac{1}{4}\lambda$ from one snapshot to the next. Thus, by the fifth snapshot, it has moved to the right by 1λ. The time required for the cycle of rising and falling to repeat is called the **period** T of the wave.

FIGURE 17-6 ■ Five "snapshots" of a traveling string wave taken at times $0, \frac{1}{4}T, \frac{1}{2}T, \frac{3}{4}T$, and T, where T is the period of oscillation of a fixed piece of string. A particular bit of string (the piece around $x = 0$) has been boxed so you can see how it moves back and forth as the continuous wave disturbance travels horizontally to the right along the string. The amplitude Y is indicated. A wavelength λ is also indicated. The circle in each snapshot shows the location of the wave crest as it moves from left to right.

READING EXERCISE 17-1: The figure just below shows an asymmetric pulse traveling from left to right along a stretched string. If this snapshot was taken at time $t = 0$ s: (a) which figure represents the graph of the displacement y versus position x at $t = 0$ s? (b) Which figure represents the graph of the displacement y versus time t at $x = x_1$ (as marked with a dot on the snapshot)?

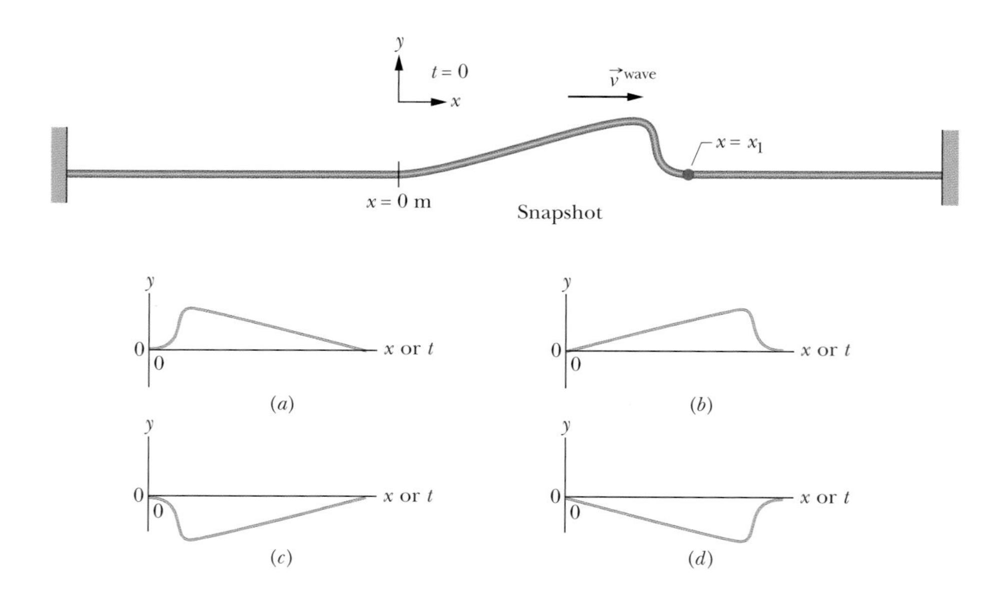

■

TOUCHSTONE EXAMPLE 17-1: Propagating Pulse

Consider a pulse propagating along a long stretched string in the x direction as shown in Fig. 17-7. Note that the scale in the y direction has been exaggerated for visibility.

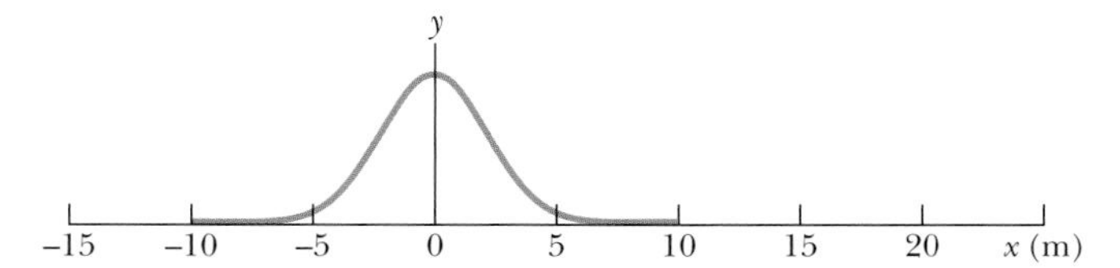

FIGURE 17-7 ■ A pulse propagates along a string in the x direction.

(a) Sketch the shape of the string after the wave pulse has traveled a distance of 10 m to the right. Explain why you sketched the shape you did.

SOLUTION ■ The **Key Idea** here is that the crest of the wave, which was at $x = 0$ m, has moved so that it is now at $x = 10$ m, but the wave shape does not change. This is like the continuous sinusoidal wave shown in Fig. 17-6 that is depicted as moving without shape change. Figure 17-8 shows the sketch.

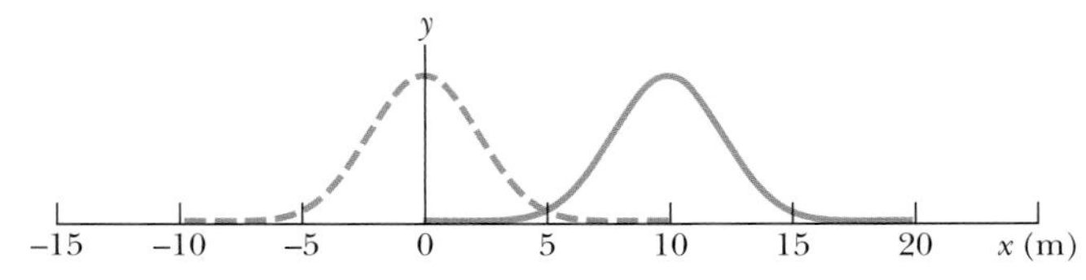

FIGURE 17-8 ■ The pulse after it has moved a distance of 10 m to the right. The original position of the pulse is the dashed line.

(b) If the wave takes 2.0 s to move the 10 m to the right, what are its velocity and wave speed?

SOLUTION ■ The **Key Idea** is that we assume that the wave shape does not change and that it moves at a constant rate determined by the motion of its crest. If the wave moves to the right 10 m in 2 s, then

$$\vec{v}^{\text{ wave}} = v_x^{\text{wave}}\hat{\mathrm{i}} = \left(\frac{10\text{ m}}{2\text{ s}}\right)\hat{\mathrm{i}} = (5\text{ m/s})\hat{\mathrm{i}}$$

(wave velocity L → R movement),

and

$$v^{\text{wave}} = |\vec{v}^{\text{ wave}}| = 5\text{ m/s}$$

(wave speed—or velocity magnitude).

(c) Imagine that a small piece of essentially massless tape is placed on the string at $x = 0$ m when $t = 0$ s. Describe the motion of the tape for the next 2 s.

SOLUTION ■ The **Key Idea** here is that as the wave pulse moves along the piece of string with the tape on it, the piece of tape does not move in the horizontal direction. Only its displacement in the y direction changes. At first the tape is at the location of the wave crest and so is at its maximum value of y. It then moves down toward a zero value of y, slowly at first and then more rapidly. In a little more than a second, the tape comes to rest at $y = 0$. The wave pulse has passed and so the tape remains there.

(d) Suppose the pulse shown in Fig. 17-7 is moving to the left instead. Sketch the shape of the string after the wave pulse has traveled a distance of 5 m to the left. Explain why you sketched the shape you did.

SOLUTION ■ The **Key Idea** here is that the crest of the wave which was at $x = 0$ m, has moved without changing shape so that it is now at $x = -5$ m. Figure 17-9 shows the sketch.

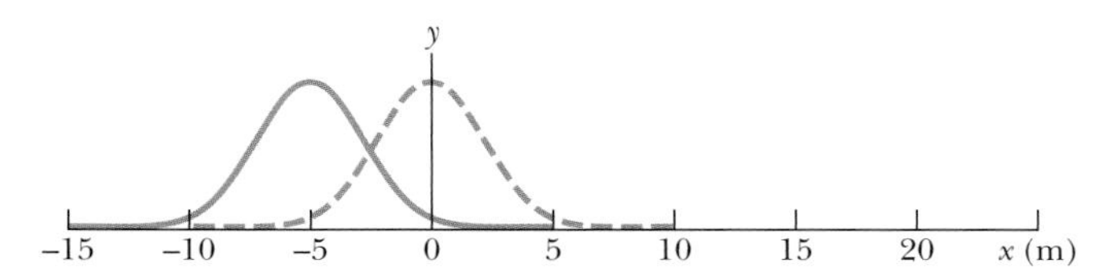

FIGURE 17-9 ■ The pulse after it has moved a distance of 5 m to the left. The original position of the pulse is the dashed line.

(e) If the wave takes 1 s to move the 5 m to the left, what are its velocity and wave speed?

SOLUTION ■ Once again we assume that the wave shape does not change and that it moves at a constant rate determined by the motion of its crest. If the wave moves 5 m to the left in 1 s, then

$$\vec{v}^{\text{ wave}} = v_x^{\text{wave}}\hat{\mathrm{i}} = \left(\frac{-5\text{ m}}{1\text{ s}}\right)\hat{\mathrm{i}} = (-5\text{m/s})\hat{\mathrm{i}}$$

(wave velocity R → L movement),

and

$$v^{\text{ wave}} = |\vec{v}^{\text{ wave}}| = 5\text{ m/s}$$

(wave speed—or velocity magnitude).

17-4 The Mathematical Expression for a Sinusoidal Wave

We found that the disturbance (whether pulse or wave, transverse or longitudinal) depends on both position x and time t. If we call the displacement y, we can write $y = f(x,t)$ or $y(x,t)$ to represent this functional dependence on time and position. In the example of the transverse pulse traveling along a Slinky as pictured in Fig. 17-2,

$y(x,t)$ represents the transverse (vertical) displacement of the Slinky rings from their equilibrium position at given position x and time t. (Alternatively, in the longitudinal wave on the Slinky shown in Fig. 17-1, $y(x,t)$ could represent the number of Slinky coils per centimeter at a given x and t.)

We can completely describe any wave or pulse that does not change shape over time and travels at a constant velocity using the relation $y = f(x,t)$, in which y is the displacement as a function f of the time t and the position x. In general, a wave can have any shape so long as it is not too sharp. The trick then is to find the correct expression for the function, $f(x,t)$.

Fortunately, it turns out that any shape pulse or wave can be constructed by adding up different sinusoidal oscillations. This makes the description of sinusoidal waves especially useful. So, for the rest of this section we'll discuss the properties and descriptions of continuous waves produced by displacing a stretched string using a sinusoidal motion like that shown in Fig. 17-3*b*. We will start by using the equation we developed in Chapter 16 to describe for sinusoidal motion at the location of a single piece of string. As we did in looking through the slit in Fig. 17-5, we will only let time vary. Next we can consider how to describe a snapshot that records the displacement of many pieces of the string at a single time. Finally, we can combine our snapshot with the results of peeking through a slit to get a single equation that ought to describe the propagation of a single sinusoidal wave. Basically we are trying to describe the displacement y of every piece of the string from its equilibrium point at every time. We are looking for $y(x,t)$.

Looking Through a Slit: Sinusoidal Wave Displacement at $x = 0$

If we choose a coordinate system so that $x = 0$ m at the left end of the string in Fig. 17-3*b*, then the motion at the left end of the string can be described using Eq. 16-5 with the string displacement from equilibrium represented by $y(x,t) = y(0,t)$ rather than by simply $y(t)$. To simplify our consideration we assume that the initial phase of the string oscillation at $x = 0$ m and $t = 0$ s is zero. This gives us

$$y(0,t) = Y\cos(\omega t), \qquad \text{(Eq. 16-5)}$$

where the angular frequency can be related to the period of oscillation by $\omega = 2\pi/T$. Although we use the cosine function in Chapter 16 to describe simple harmonic motion, *it is customary to use the sine function to describe wave motion*. As we mentioned in Chapter 16, when a sine function is shifted to the left by $\pi/2$ it looks like a cosine function. So we can also describe the same string displacement as a function of time at $x = 0$ m as

$$y(0,t) = Y\cos(\omega t) = Y\sin(\omega t + \pi/2). \qquad (17\text{-}1)$$

Note that using the sine function requires a different, nonzero initial phase angle given by $\pi/2$. If we locate our slit at another nonzero value of x as shown in Fig. 17-5, then the initial phase (at $t = 0$ s) will often turn out to be different from $\pi/2$. In fact this initial phase is a function of the location x of the piece of string we are considering.

A Snapshot: Sinusoidal Wave Displacement at $t = 0$

Imagine that the man has been moving the end of the string up and down as shown in Fig. 17-3*b* for a long time using a sinusoidal motion. Instead of looking through a slit as time varies, we take a snapshot of the string at a time $t = 0$s similar to that shown in Fig. 17-3*b*. Then we expect our snapshot to be described by the equation

$$y(x,0) = Y\sin(kx + \pi/2) \qquad (17\text{-}2)$$

where k is a constant and the "initial" phase when x is zero must also be $\pi/2$. Note that if the snapshot of the string were taken at another time, the initial phase would probably be different.

Combining Expressions for *x* and *t*

Equation 17-1 describes the displacement at all times for just the piece of string located at $x = 0$ m. Equation 17-2 describes the displacement of all the pieces of string at $t = 0$ s. We can make an intelligent guess that the equation describing $y(x,t)$ is some combination of these two expressions given by

$$y(x,t) = Y\sin[(kx \pm \omega t) + \pi/2)], \tag{17-3}$$

where $\pi/2$ represents the initial phase when $x = 0$ m and $t = 0$ s for the special case we considered. In general we can describe the motion of our sinusoidal wave with an arbitrary initial phase by modifying Eq. 17-3 to get

$$y(x,t) = Y\sin[(kx \pm \omega t) + \phi_0)] \quad \text{(sinusoidal wave motion, arbitrary initial phase)}, \tag{17-4}$$

where ϕ_0 is the initial phase (or phase constant) when both $x = 0$m and $t = 0$s. The $\pm$ sign refers to the direction of motion of the wave as we shall see in Section 17-5. In cases where the initial phase is not important, we can simplify Eq. 17-3 by choosing an initial time and origin of the x axis that lies along the line of motion of the wave so that $\phi_0 = 0$ rad.

Amplitude and Phase

The argument $\phi(x,t) = (kx \pm \omega t) + \phi_0$ of the sine function is called the **time- and space-dependent phase** of the wave where k and ω are constants that we can determine for a particular wave. Although the phase $\phi(x,t)$ is a function of both time and position, it is *neither* a time *nor* a position. Rather, the phase $\phi(x,t)$ *must* be an angle because we can only take the sine of angles.

As the wave sweeps through a string segment at a particular position x, the phase changes linearly with time t. This means that the sine also changes, oscillating between $+1$ and -1. Its extreme positive value $(+1)$ corresponds to a peak of the wave moving through the segment; then, the value of y at position x is Y. Its extreme negative value (-1) corresponds to a valley of the wave moving through the segment; then, the value of y at position x is $-Y$. Thus, the sine function and the time-dependent phase of a wave correspond to the oscillation of a string segment, and the amplitude of the wave determines the extremes of the segment's displacement.

Wavelength and Wave Number

In order to come up with a more useful mathematical expression for the wave, we must more carefully investigate the nature of the phase of the wave. We know that the sine function repeats itself every 2π radians. We also know that the wavelength λ of a wave is the distance (parallel to the direction of the wave's travel) between repetitions of the shape of the wave (or *wave form*). A typical wavelength is marked in Fig. 17-6*a*, which is a snapshot of the wave at time $t = 0$ s. If we take the simple case for which the initial phase is zero, at that time Eq. 17-4, $y(x,t) = Y\sin[(kx \pm \omega t) + \phi_0)]$ becomes

$$y(x,0) = Y\sin(kx). \tag{17-5}$$

By definition, the displacement y is the same at both ends of one wavelength—that is, y is the same at $x = x_1$ and at $x = x_1 + \lambda$. (Again, x does not represent a displacement here but just the horizontal position of a string element. By the equation above,

$$Y\sin kx_1 = Y\sin k(x_1 + \lambda) = Y\sin(kx_1 + k\lambda). \quad (17\text{-}6)$$

A sine function begins to repeat itself when its angle (or argument) is increased by 2π rad, so in the equation above, we must have $k\lambda = 2\pi$, or

$$k = \frac{2\pi}{\lambda} \quad \text{(wave number)}. \quad (17\text{-}7)$$

We call k the **wave number.** The wave number is inversely proportional to the wavelength. Its SI unit is the radian per meter, or the inverse meter. (Note that the symbol k here does *not* represent a spring constant as previously.)

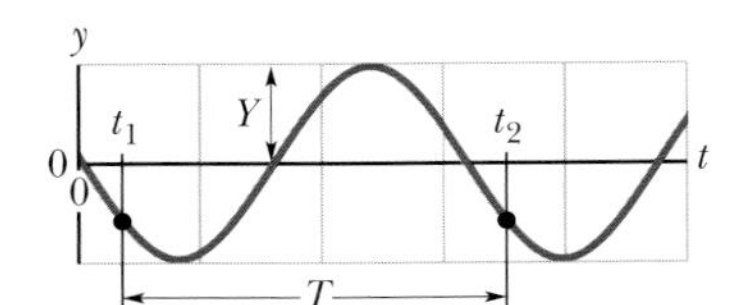

FIGURE 17-10 ■ A graph with unspecified units of the displacement of the string segment at $x = 0$ as a function of time, as the sinusoidal wave of Fig. 17-6 passes through it. The amplitude Y is indicated. A typical period T, measured from an arbitrary time t_1, is also indicated.

Period, Angular Frequency, and Frequency

In contrast to the graphs in Fig. 17-6, which represent pictures of the string at a particular instant, we will now focus on a particular bit of string and consider how it moves as a function of time. As we noted earlier, if you were to peek at the string through a slit, you would see that the single segment of the string at that position moves up and down in simple harmonic motion. Figure 17-10 shows a graph of the displacement y of a bit of string versus time t at a certain position along the string, taken to be $x = 0$. This motion is described by $y(x,t) = Y\sin(kx \pm \omega t + \phi_0)$. Setting $x = 0$ and arbitrarily choosing a zero phase constant and a negative time-dependent term, we get

$$y(0,t) = Y\sin(-\omega t) = -Y\sin(\omega t). \quad (17\text{-}8)$$

Here we have made use of the fact that $\sin(-\alpha) = -\sin\alpha$. Figure 17-10 is a graph of this equation. Be careful though: This figure *does not* show the shape of the wave. It shows a graph of the time-dependent variation in the displacement of a small bit of the string.

Since we have defined the period of oscillation T of a wave to be the time any string segment takes to move through one full oscillation, we can apply this equation to both ends of this time interval. Equating the results yields

$$Y\sin\omega t_1 = Y\sin\omega(t_1 + T)$$
$$Y\sin\omega t_1 = Y\sin(\omega t_1 + \omega T) \quad (17\text{-}9)$$

which can be true only if $\omega T = 2\pi$. In other words,

$$\omega = \frac{2\pi}{T} \quad \text{(angular frequency)}. \quad (17\text{-}10)$$

We call ω the **angular frequency** of the wave; its SI unit is the radian per second.

Recall from Chapter 16 that the **frequency** f of an oscillation is defined as $1/T$. Hence, its relationship to angular frequency is

$$f = \frac{1}{T} = \frac{\omega}{2\pi} \quad \text{(frequency)}. \quad (17\text{-}11)$$

Like the frequency of simple harmonic motion in Chapter 16, this frequency f is a number of oscillations per unit time—here, the number made by a string segment as the wave moves through it. As in Chapter 16, f is usually measured in hertz or its multiples, such as kilohertz.

To summarize our discussion of Eq. 17-4 given by $y(x,t) = Y\sin[(kx \pm \omega t) + \phi_0]$, the names of the quantities are displayed in Fig. 17-11 for your reference.

$$y(x,t) = Y\sin[(kx \pm \omega t) + \phi(0,0)]$$

Time and space-dependent phase
Amplitude
Displacement
Oscillating term
Wave number
Position
Angular frequency
Time
Initial phase—the phase constant used to determine displacement at $t = 0$ and $x = 0$

FIGURE 17-11 ■ The names of the quantities in Eq. 17-4, for a transverse sinusoidal wave.

READING EXERCISE 17-2: The figure is a composite of three snapshots, each of a wave traveling along a particular string. The time-dependent phases for the waves are given by (a) $(2\text{ rad/m})x - (4\text{ rad/s})t$, (b) $(4\text{ rad/m})x - (8\text{ rad/s})t$, and (c) $(8\text{ rad/m})x - (16\text{ rad/s})t$. Which phase corresponds to each snapshot?

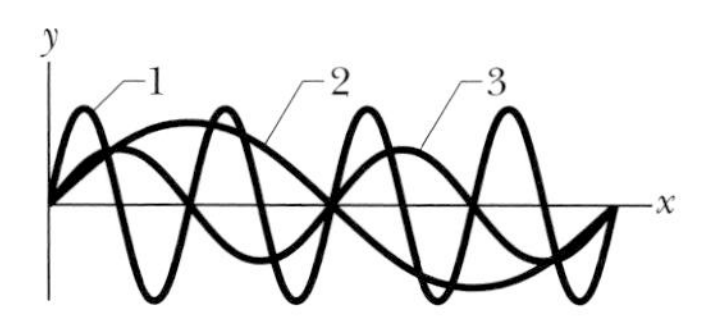

■

TOUCHSTONE EXAMPLE 17-2: Traveling Wave Equations

A traveling sinusoidal wave train can be described in terms of a wave number k and an angular frequency ω using Eq. 17-4. If the wave train is moving to the right and has an initial phase of $\phi_0 = 0$ rad then we can write our wave equation as

$$y(x,t) = Y\sin(kx - \omega t).$$

Use the definition of various terms along with Eqs. 17-7, 17-10, and 17-11 to:

(a) Explain what T and λ mean physically.

SOLUTION ■ The symbols λ and T stand for the wavelength and the period of the wave, respectively.

The wavelength, λ, is the distance one has to move along the string at a fixed instant in time (for instance, when looking at a snapshot of the wave) in order to pass by a full oscillation of the wave.

The period, T, is the time one has to wait at a particular point in space (along an axis that is lined up with the undisturbed string) for a small piece of the string to go through one complete oscillation.

(b) Show that displacement of the string from its undisturbed condition given by $y(x,t)$ can also be described by the equation

$$y(x,t) = Y\sin\left[2\pi\left(\frac{x}{\lambda} - ft\right)\right].$$

SOLUTION ■ The **Key Idea** here is that the most important symbols in the sinusoidal wave equation are x and t. They represent the variables. All the other symbols represent constants that describe a particular wave. As a result, every sinusoidal wave with a zero initial phase will look like $\sin[(\text{mess})x \pm (\text{another mess})t]$. By identifying how the "mess" in each situation is related to the constants in another situation, the transformation of Eq. 17-4 into different forms is straightforward. Also we only need to show that the argument given by terms inside the square brackets are the same as the argument $[kx - \omega t]$. For the situation at hand we can use the relationships in Eqs. 17-7, 17-10, and 17-11 ($k = 2\pi/\lambda$, $\omega = 2\pi/T$, and $f = 1/T = \omega/2\pi$) to verify that

$$2\pi\left(\frac{x}{\lambda} - ft\right) = \frac{2\pi}{\lambda}x - 2\pi ft = kx - \frac{2\pi}{T}t = kx - \omega t. \quad \text{(Answer)}$$

(c) Show that displacement of the string from its undisturbed condition given by $y(x, t)$ can also be described by the equation

$$y(x,t) = Y\sin\left[2\pi\left(\frac{x}{\lambda} - \frac{t}{T}\right)\right].$$

SOLUTION ■ Using the same approach as we used in part (b), we see that

$$2\pi\left(\frac{x}{\lambda} - \frac{t}{T}\right) = 2\pi\left(\frac{x}{\lambda} - ft\right),$$

but we showed in part (a) that

$$2\pi\left(\frac{x}{\lambda} - ft\right) = kx - \omega t. \quad \text{(Answer)}$$

17-5 Wave Velocity

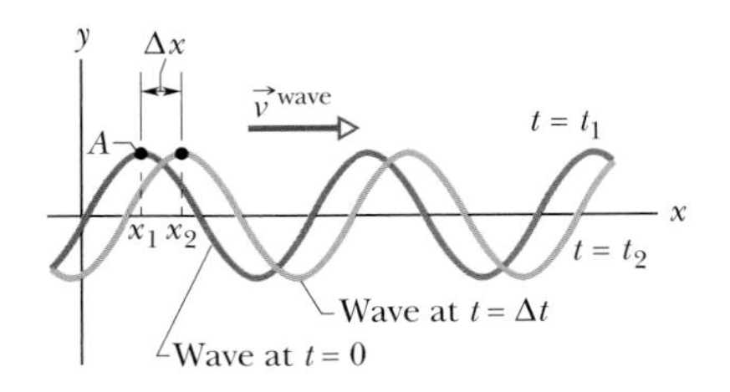

FIGURE 17-12 ■ One end of a string that is under tension is moved up and down continuously in a sinusoidal fashion as shown in Fig. 17-3*b*. Two snapshots of the sinusoidal wave that results are shown, at time $t = t_1$ and then at time t_2. As the wave moves to the right with an x-component of velocity v_x^{wave}, the entire sine curve shifts by Δx during Δt. The string segment at x_1 moves down during $\Delta t = t_2 - t_1$, while the piece of string at x_2 moves up as the local wave crest moves from left to right.

Figure 17-12 shows two snapshots of a sinusoidal wave taken a small time interval Δt apart. The wave is traveling in the positive x direction (to the right in Fig. 17-12*b*), and the entire wave pattern is moving a distance Δx in that direction during the interval Δt. The ratio $\Delta x/\Delta t$ (or, in the differential limit, dx/dt) is the **wave velocity** component along the x axis, which we denote as v_x^{wave}. How can we find the wave velocity?

As the wave in Fig. 17-12 moves, each point of the moving wave form, such as point A marked on a peak, retains its displacement y. (Points on the string do not retain their displacement, but points on the wave *form* do.) That is, time passes and the location of point A changes (so x and t are both changing), but the value of $y(x,t)$ associated with point A does not change; it remains the maximum value. If point A retains its displacement as it moves, the phase in $y(x,t) = Y\sin(kx \pm \omega t)$ must remain a constant. So,

$$kx \pm \omega t = \text{a constant.} \tag{17-12}$$

Note that although this argument is constant, in this case, both x and t are increasing. [The time t always increases and x increases because the wave is moving in the (rightward) positive direction]. In order for the phase $kx \pm \omega t$ to remain constant with both t and x increasing, the negative sign must be chosen. In other words,

> The equation for a sinusoidal wave traveling right is $y(x,t) = Y\sin(kx - \omega t)$ and the equation for a sinusoidal wave moving left is $y(x,t) = Y\sin(kx + \omega t)$.

By assuming that the wave moves at a constant velocity $\vec{v}^{\text{wave}}$, we can express its velocity in terms of how far an imaginary point on the wave form has moved in a short time interval. Since the wave form only moves parallel to the x axis, we can write

$$\vec{v}^{\text{wave}} \equiv \frac{\Delta\vec{x}}{\Delta t} \quad \text{(definition of wave velocity),}$$

or in component form

$$v_x^{\text{wave}} = \frac{\Delta x}{\Delta t}.$$

If we take the time interval under consideration to be the period of the wave T, then we know that the wave travels one wavelength λ. So the magnitude of the wave velocity, or wave speed, is

$$|v_x^{\text{wave}}| = v^{\text{wave}} = \frac{\lambda}{T} = \lambda f = \frac{\omega}{k} \quad \text{(speed related to wave length and period).} \tag{17-13}$$

Are all traveling waves sinusoidal? Not necessarily. Consider a wave of arbitrary shape, given by

$$y(x,t) = h(kx \pm \omega t), \tag{17-14}$$

where h represents *any* function, the sine or cosine function being one possibility. Since the variables x and t enter in the combination $kx \pm \omega t$, it is possible for waves to have other shapes associated with them. For example, a wave pulse traveling from left to right could be represented by

$$y(x,t) = \frac{Y}{(kx - \omega t)^2 + 1} \quad \text{(a possible traveling wave pulse).}$$

Particle Velocity and Acceleration in Mechanical Waves

In discussing the velocity of a wave as we did above, it is important to remember that the magnitude of this velocity is the speed at which the wave disturbance moves, *not* the speed of the bits or pieces of string through which the wave travels. To repeat the quote of Leonardo da Vinci, consider " . . . waves made in a field of grain by the wind, where we see the waves running across the field while the grain remains in place."

We have already begun to develop an understanding of the velocity of the wave itself. We now consider the velocity of the particles that are displaced as a transverse wave moves through. A piece of string can only move in the y direction as a disturbance passes by. Thus, at a given time t we can denote the velocity of the piece of string that is located at x in terms of its y-component $v_y^{\text{string}} = v_y(x,t)$. Since only the string pieces can move in the y direction, we often shorten this y-component notation to simplify $v_y(x,t)$. To find the velocity component $v_y(x,t)$ as a sinusoidal disturbance passes through, we simply need to differentiate the expression for the displacement of the string segment, $y(x,t) = Y\sin[(kx \pm \omega t) + \phi_0)]$ with respect to time while holding x constant. This is similar to what we did to find the velocity of a mass oscillating at the end of a spring in Section 16-4,

$$v_y = \frac{\partial y(x,t)}{\partial t} = \pm\omega Y\cos[(kx \pm \omega t) + \phi_0], \tag{17-15}$$

where the use of the symbol $\partial/\partial t$ signifies that we are taking a partial derivative in which the location along the string, x, is considered to be a constant. If we find a second partial derivative using Eq. 17-15, we have the following expression for the acceleration of the piece of string at location x at a time t,

$$a_y = \frac{\partial v_y}{\partial t} = \frac{\partial^2 y(x,t)}{\partial t^2} = \mp\,\omega^2 Y\sin[(kx \pm \omega t) + \phi_0]. \tag{17-16}$$

TOUCHSTONE EXAMPLE 17-3: Wave on a String

A wave traveling along a string is described by

$$y(x,t) = (0.00327\text{ m})\sin[(72.1\text{ rad/m})x - (2.72\text{ rad/s})t], \tag{17-17}$$

where $y(x,t)$ represents the transverse displacement from equilibrium of a small piece of string with a horizontal location of x at time t.

(a) What is the amplitude of this wave?

SOLUTION ■ The **Key Idea** is that Eq. 17-17 is of the same form as Eq. 17-4, with $\phi_0 = 0$ rad,

$$y = Y\sin(kx - \omega t + \phi_0), \tag{17-18}$$

so we have a sinusoidal wave. By comparing the two equations, we see that the amplitude is

$$Y = 0.00327\text{ m} = 3.27\text{ mm}. \quad \text{(Answer)}$$

(b) What are the wavelength, period, and frequency of this wave?

SOLUTION ■ By comparing Eqs. 17-17 and 17-18, we see that the wave number and angular frequency are

$$k = 72.1\text{ rad/m} \quad \text{and} \quad \omega = 2.72\text{ rad/s}.$$

We then relate wavelength λ to k via Eq. 17-7:

$$\lambda = \frac{2\pi}{k} = \frac{2\pi\text{ rad}}{72.1\text{ rad/m}} = 0.0871\text{ m} = 8.71\text{ cm}. \quad \text{(Answer)}$$

Next, we relate T to ω using Eq. 17-10:

$$T = \frac{2\pi}{\omega} = \frac{2\pi\text{ rad}}{2.72\text{ rad/s}} = 2.31\text{ s}, \quad \text{(Answer)}$$

and from Eq. 17-11 we have

$$f = \frac{1}{T} = \frac{1}{2.31\text{ s}} = 0.433\text{ Hz}. \quad \text{(Answer)}$$

(c) What is the velocity with which the wave moves along the string?

SOLUTION ■ The speed of the wave is given by Eq. 17-13:

$$v^{\text{wave}} = \frac{\omega}{k} = \frac{2.72 \text{ rad/s}}{72.1 \text{ rad/m}} = 0.0377 \text{ m/s} = 3.77 \text{ cm/s}. \quad \text{(Answer)}$$

Because the phase in Eq. 17-17 contains the variable x, the wave is moving along the x axis. According to Eq. 17-4, the *minus* sign in front of the ωt term indicates that the wave is moving in the *positive* x direction. (Note that the (b) and (c) answers are independent of the wave amplitude.)

(d) What is the displacement y of the small piece of string located at $x = 22.5$ cm when $t = 18.9$ s?

SOLUTION ■ The **Key Idea** here is that Eq. 17-17 gives the displacement as a function of position x and time t. Substituting the given values into the equation yields

$$\begin{aligned} y &= (0.00327 \text{ m}) \sin[(72.1 \text{ rad/m}) \times (0.225 \text{ m}) - (2.72 \text{ rad/s}) \times (18.9 \text{ s})] \\ &= (0.00327 \text{ m}) \sin[-35.1855 \text{ rad}] = (0.00327 \text{ m})(0.588) \\ &= 0.00192 \text{ m} = 1.92 \text{ mm}. \quad \text{(Answer)} \end{aligned}$$

Thus, the transverse displacement from the string's equilibrium position is positive.

TOUCHSTONE EXAMPLE 17-4: Transverse String Motion

In Touchstone Example 17-3d, we showed that at $t = 18.9$ s the transverse displacement component y of the piece of string at $x = 0.255$ m due to the wave of Eq. 17-17 is 1.92 mm.

(a) What is v_y, the transverse velocity component of the same element of the string, at that time? (This speed, which is associated with the transverse oscillation of an element of the string, is in the y direction. Do not confuse it with $\vec{v}^{\text{ wave}}$, the constant velocity at which the *wave form* travels along the x axis.)

SOLUTION ■ The **Key Idea** here is that as the wave travels, the displacement of each piece of string is transverse, where x represents the horizontal position of a piece of string. Another **Key Idea** is that the horizontal position, x, of a piece of string never changes. In general, the displacement of a piece of string is given by

$$y(x,t) = Y \sin(kx - \omega t + \phi_0). \quad (17\text{-}19)$$

For a piece of string at a certain location (x, y) we find the rate of change of y by taking the partial derivative of Eq. 17-19 with respect to t while treating x as a constant. Here we have $\phi_0 = 0$ rad, so according to Eq. 17-15,

$$v_y = \frac{\partial y}{\partial t} = -\omega Y \cos(kx - \omega t), \quad (17\text{-}20)$$

where the amplitude Y is always taken as positive. Next, substituting numerical values from Touchstone Example 17-3 we obtain

$$\begin{aligned} v_y &= (-2.72 \text{ rad/s})(3.27 \text{ mm}) \cos(-35.1855 \text{ rad}) \\ &= 7.20 \text{ mm/s}. \quad \text{(Answer)} \end{aligned}$$

Thus, at $t = 18.9$ s, the piece of string at the horizontal location $x = 22.5$ cm is moving in the y direction, with a velocity of 7.20 mm/s.

(b) What is the transverse acceleration component a_y of the same element at that time?

SOLUTION ■ The **Key Idea** here is that the transverse acceleration component a_y is the rate at which the transverse velocity of a given piece of string is changing. From Eq. 17-20, again treating x as constant but allowing t to vary, we find

$$a_y = \frac{\partial v_y}{\partial t} = -\omega^2 Y \sin(kx - \omega t).$$

Comparison with Eq. 17-19 shows that we can write this as

$$a_y = -\omega^2 y.$$

We see that the transverse acceleration component of a piece of string that is oscillating is proportional to its transverse component of displacement but opposite in sign. This is completely consistent with the action of the string piece—namely, that it is moving transversely in simple harmonic motion. Substituting numerical values yields

$$\begin{aligned} a_y &= -(2.72 \text{ rad/s})^2(1.92 \text{ mm}) \\ &= -14.2 \text{ mm/s}^2. \end{aligned} \quad \text{(Answer)}$$

Thus, at $t = 18.9$ s, the piece of string at $x = 22.5$ cm is displaced from its equilibrium position by 1.92 mm in the positive direction and has an acceleration of magnitude 14.2 mm/s^2 in the negative y direction.

17-6 Wave Speed on a Stretched String

If we distort a string that is under tension, we observe that this distortion (or wave) will travel rapidly along the string (Fig. 17-1). This is true whether the distortion is a single pulse (Figs. 17-2 or 17-3*a*) or continuous sinusoidal pattern (Fig. 17-3*b*). In Section 17-3 we asserted that a net force on a piece of string could result from forces exerted on its ends by the pieces of string just adjacent to it. If these end forces act in different directions on a string segment, the string piece can move in a vertical direction. As it moves vertically, it can exert a force on adjacent string pieces. This could cause a set of disturbances to propagate along the string as a wave.

In general, the speed of a wave is determined by the properties of the medium through which it travels. For instance, how do the properties of the stretched string affect the speed of a traveling wave? Intuitively we expect that the wave will travel faster when the string is stretched to a higher tension and that the wave will travel more slowly on a thicker string where each segment of string has a greater mass. In this section we use the impulse-momentum theorem—a form of Newton's Second Law—to derive an equation that relates the speed of the traveling wave to the tension of the string and its massiveness. In our derivation we imagine that the string consists of many tiny segments or pieces that are connected together.

Consider a horizontal stretched string as shown in Fig. 17-13. Suppose the string is under tension—a nonvector quantity we denote as F^{tension}*. What happens if the tiny segment at one end of the string is jerked abruptly upward and then returned to its equilibrium position? The pulse that results propagates along the pieces of string as a wave like the one shown in Fig. 17-13. Although the amplitude in Fig. 17-13 is exaggerated for clarity, we assume the wave amplitude is small enough that: (1) the tension in the string does not change significantly except at the boundary between a displaced and undisplaced segment of string (as shown in Fig. 17-13*a*) and (2) the angle θ between the horizontal and any segment (or small piece) of string that is displaced is so small that $\sin\theta \approx \tan\theta \approx \theta$. We also assume that the pulse propagates along the string at a *constant* wave velocity $\vec{v}^{\text{wave}} = v_x^{\text{wave}}\hat{\text{i}}$. As is often the case in physics, we can test the validity of our assumptions by seeing whether the expression we derive for the wave speed is consistent with observation.

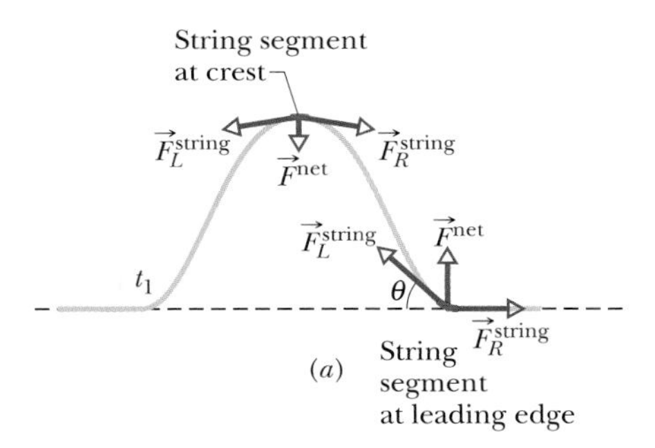

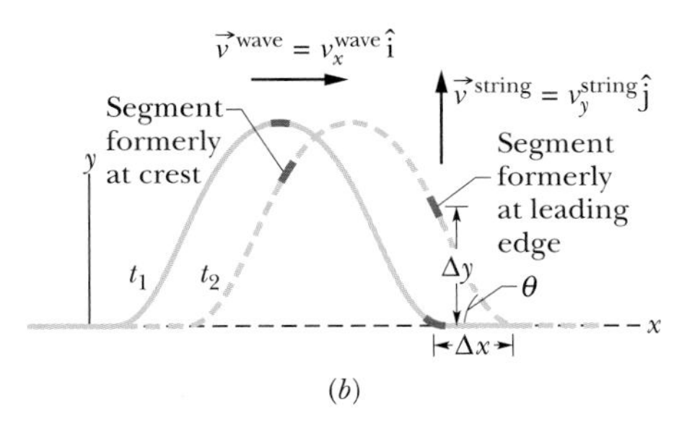

FIGURE 17-13 ■ Depictions of a traveling wave pulse. The amplitude is exaggerated for clarity. (*a*) A single pulse travels to the right along a string under constant tension. The piece of string at the wave crest experiences a net force downward while the string segment at the leading edge experiences a net upward force. (*b*) Two snapshots of the wave at times t_1 and t_2 show the wave form traveling with velocity $\vec{v}^{\text{ wave}}$, so it shifts right a distance Δx during a time interval $\Delta t = t_2 - t_1$. As a result of the net forces on the piece of strings, the string segment at the wave crest moves down during Δt while the piece of string at the leading edge moves up.

Let us consider how a segment of string moves as a wave pulse travels by it in a short time interval. If we define the mass per unit length of the string as its linear mass density, μ, then the mass of the segment of string that the wave pulse affects in a time interval Δt is given by

$$m = \mu v^{\text{wave}}\Delta t, \qquad (17\text{-}21)$$

where v^{wave} is the speed of propagation of the wave pulse. We start by examining the tension forces that act at each end of our small string segment. For example, in Fig. 17-13*a* the segment at the crest of the wave experiences a net downward force, but the segment at the leading edge of the wave pulse experiences a net upward force. Thus a short time Δt later, the front segment has moved up while the segment at the crest has moved down. If we take all the string segments into account, we see that the entire pulse shape "moves" to the right even though each of the string segments has only moved vertically—either up or down. This is shown in Fig. 17-13*b*.

What happens when the leading edge of the pulse encounters a segment of the string? A vertical net force $\vec{F}^{\text{net}} = F_y^{\text{net}}\hat{\text{j}}$ due to the unbalanced tension forces at

*To avoid confusing wave period, denoted as T, with tension, we use the notation F^{tension} for tension here and in the next chapter.

the string segment's ends shown in Fig. 17-13*a* causes the segment to accelerate upward. In fact, the piece of string just to the left of the leading edge is stretched so its tension is greater than in the undisturbed string piece just to the right. As a result, the string piece at the leading edge undergoes a change in vertical position of Δy and a momentum change, $\Delta \vec{p}_y$ during the time Δt. We can use the impulse-momentum theorem to describe the motion of the piece of string in terms of the vertical components of net force and momentum as

$$F_y^{\text{net}} \Delta t = \Delta p_y. \tag{17-22}$$

The segment of string has a mass m, and we assume the segment has no vertical velocity at time t_1. If we denote the vertical velocity change imparted to the segment of string in the short time interval Δt as $\Delta \vec{v}_y^{\text{ string}}$, then

$$F_y^{\text{net}} \Delta t = m \Delta v_y^{\text{string}} = m(v_y^{\text{string}} - 0) = m v_y^{\text{string}}.$$

But we can see from Fig. 17-13*a* that the magnitude of the net force on the string segment is approximately $F^{\text{tension}} \sin\theta$, where F^{tension} is a scalar quantity denoting the tension in the string. The impulse delivered to the string segment by the traveling wave can be related to the newly acquired vertical velocity of the string segment by

$$(F^{\text{tension}} \sin\theta)\Delta t \approx m|v_y^{\text{string}}| = m|\vec{v}^{\text{ string}}|.$$

From Fig. 17-13*b* we also see that the ratio of the magnitude of the vertical velocity of the leading edge mass segment and the magnitude of the horizontal wave velocity of the temporary disturbance is given by

$$\tan\theta = \frac{\Delta y}{\Delta x} = \frac{|\vec{v}^{\text{ string}}|}{|\vec{v}^{\text{ wave}}|} = \frac{v^{\text{string}}}{v^{\text{wave}}}. \tag{17-23}$$

Combining the last two expressions we derived gives us

$$(F^{\text{tension}} \sin\theta)\Delta t = m v^{\text{wave}} \tan\theta \quad \text{or} \quad F^{\text{tension}} = \frac{m}{\Delta t}\left(\frac{v^{\text{wave}}}{\cos\theta}\right). \tag{17-24}$$

At this point we need to express the mass of the segment in terms of the length of the string segment and the linear density of the string. Thus the mass of the segment in the string can be determined as the density multiplied by the length of the segment so that $m = \mu \Delta x$. However, during the time interval Δt the waveform has moved a distance $\Delta x = |\vec{v}^{\text{ wave}}|\Delta t$. We can now rewrite Eq. 17-24 as

$$F^{\text{tension}} = \mu (v^{\text{wave}})^2 \frac{1}{\cos\theta},$$

but for small-amplitude waves the angle of string displacement relative to the horizontal, $\cos\theta$ is approximately 1, so

$$(v^{\text{wave}})^2 = \frac{F^{\text{tension}}}{\mu}.$$

Therefore, $$v^{\text{wave}} = \sqrt{\frac{F^{\text{tension}}}{\mu}} \quad \text{(speed)}. \tag{17-25}$$

Equation 17-25 gives the speed of the pulse in Fig. 17-13 and the speed of *any* other wave pulse or continuous wave on the same string under the same tension. If we take the square root of the ratio of F^{tension} (dimension MLT^{-2}) and μ (dimension ML^{-1}) we get a dimension of velocity. Thus Eq. 17-25 is dimensionally correct.

Equation 17-25 tells us that

> The speed of a small-amplitude pulse traveling along a stretched string depends only on the tension and linear density of the string and not on the amplitude or frequency of the disturbance.

The amplitude independence is much like that for systems oscillating with simple harmonic motion. It indicates a proportionality between the net forces exerted on a piece of string and its displacement from equilibrium.

If the simplifying assumptions we used in the derivation of Eq. 17-25 are reasonable, then we expect that measurements of wave speed, tension, and linear mass density can be used to verify Eq. 17-25. Indeed, this theoretical prediction has been experimentally confirmed many times. In fact, it turns out that Eq. 17-25 is just as valid for continuous waves as it is for a single wave pulse, and so the speed of a continuous wave is not a function of frequency. The frequency of the wave is fixed entirely by whatever generates the wave (for example, the person jerking the string up and down in Fig. 17-3*b*). However, once the frequency of a continuous wave is set, the wave speed determines the relationship between frequency and wavelength since $\lambda = v^{\text{wave}}/f$ (Eq. 17-13).

READING EXERCISE 17-3: In the derivation of the wave speed in a string shown above, there are references to two different velocities; one is referred to as $\vec{v}_y^{\text{ string}}$ and one is referred to as $\vec{v}_x^{\text{ wave}}$. As completely as possible, describe the differences between these two velocities. Explain which velocity is used in deriving the expression for the mass, m, of a string piece being displaced and why. Explain which velocity is used to find the y-component of momentum change, Δp_y, of a piece of string and why. ■

READING EXERCISE 17-4: You send a continuous traveling wave along a string by moving one end up and down sinusoidally. If you increase the frequency of the oscillations, do (a) the speed of the wave and (b) the wavelength of the wave increase, decrease, or remain the same? If, instead, you increase the tension in the string, do (c) the speed of the wave and (d) the wavelength of the wave increase, decrease, or remain the same? ■

17-7 Energy and Power Transported by a Traveling Wave in a String

When we start up a wave or pulse in a stretched string, *we* provide the energy for the motion of the string. We impart a momentum to a segment of the string. As the wave or pulse moves away, the momentum is transferred from one segment of the string to the next. In addition, the wave transports energy as both kinetic energy and elastic potential energy. Let us consider each of these forms of energy in turn.

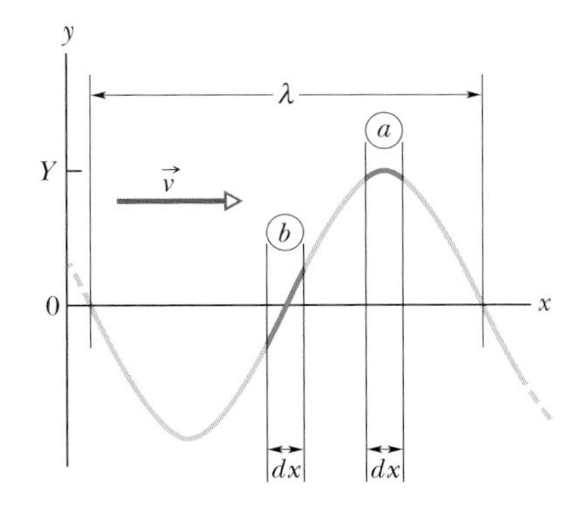

FIGURE 17-14 ■ A snapshot of a traveling wave on a string at time $t = 0$. String segment a is at displacement $y = Y$, and string segment b is at displacement $y = 0$. The kinetic energy of the string segment at each position depends on the transverse velocity of the segment. The potential energy depends on the amount by which the string is displaced from equilibrium as the wave passes through it.

Kinetic Energy

A segment of the string of mass dm, oscillating transversely in simple harmonic motion as the wave passes through it, has kinetic energy associated with its transverse velocity v_y. When the segment is rushing through its $y = 0$ position (segment b in Fig. 17-14), its transverse velocity—and thus its kinetic energy—is a maximum. When the segment is at its extreme position $y = Y$ (as is segment a), its transverse velocity—and thus its kinetic energy—is zero.

Elastic Potential Energy

To send a sinusoidal wave along a previously straight string, the wave must necessarily stretch the string. As a string segment of length dx oscillates transversely, its length must increase and decrease in a periodic way if the string segment is to fit the sinusoidal wave form. Elastic potential energy is associated with these length changes, just as for a spring.

When the string segment is at its $y = Y$ position (segment a in Fig. 17-14), its length has its normal undisturbed value dx, so its elastic potential energy is zero. However, when the segment is rushing through its $y = 0$ position, it is stretched to its maximum extent, and its elastic potential energy then is a maximum.

Energy Transport

The oscillating string segment thus has both its maximum kinetic energy and its maximum elastic potential energy at $y = 0$. In the snapshot of Fig. 17-14, the regions of the string at maximum displacement have no energy, and the regions at zero displacement have maximum energy. As the wave travels along the string, forces due to the tension in the string continuously do work to transfer energy from regions with energy to regions with no energy.

Suppose we set up a wave on a string stretched along a horizontal x axis so that Eq. 17-4 describes the string's displacement. We might send a wave along the string by oscillating one end of the string, as in Fig. 17-3b. In doing so, we provide energy for the motion and stretching of the string—as the string sections oscillate perpendicularly to the x axis, they have kinetic energy and elastic potential energy. As the wave moves into sections that were previously at rest, energy is transferred into those new sections. Thus, we say that the wave *transports* the energy along the string.

The Rate of Energy Transmission

The kinetic energy dK associated with a string segment of mass dm is given by

$$dK = \tfrac{1}{2}\, dm\, (v_y^{\text{string}})^2, \qquad (17\text{-}26)$$

where v_y^{string} is the transverse component of velocity of the oscillating string segment. If we assume the initial phase ϕ_0 is zero, then the y-component of the string element velocity is given by Eq. 17-15 as

$$v_y^{\text{string}} = -\omega Y \cos(kx - \omega t).$$

Using this relation and putting $dm = \mu\, dx$, we rewrite Eq. 17-26 as

$$dK = \tfrac{1}{2}(\mu\, dx)[(-\omega Y)\cos^2(kx - \omega t)]^2. \qquad (17\text{-}27)$$

Dividing both sides of Eq. 17-27 by dt, we get the rate at which the kinetic energy of a string segment changes and thus the rate at which kinetic energy is carried along by the wave. The result is given by

$$\frac{dK}{dt} = \tfrac{1}{2}\mu \frac{dx}{dt} \omega^2 Y^2 \cos^2(kx - \omega t) = \tfrac{1}{2}\mu v^{\text{wave}} \omega^2 Y^2 \cos^2(kx - \omega t), \qquad (17\text{-}28)$$

where the ratio dx/dt, which is positive for a wave moving from left to right, has been replaced by the wave speed v^{wave}. The *average* rate at which kinetic energy is transported is

$$\frac{dK}{dt} = \langle \tfrac{1}{2}\mu v^{\text{wave}}\omega^2 Y^2[\cos^2(kx - \omega t)]\rangle = \tfrac{1}{4}\mu v^{\text{wave}}\omega^2 Y^2. \tag{17-29}$$

Here we have taken the average over an integer number of wavelengths and have used the fact that the average value of the square of a cosine function over an integer number of periods is $\frac{1}{2}$.

Elastic potential energy is also carried along with the wave, and at the same average rate given by Eq. 17-29. Although we shall not examine the proof, you should recall that, in an oscillating system such as a pendulum or a spring–block system, the average kinetic energy and the average potential energy are indeed equal.

The **average power,** which is the average rate at which energy of both kinds is transmitted by the wave, is then

$$\langle P\rangle = 2\langle\frac{dK}{dt}\rangle, \tag{17-30}$$

or, from Eq. 17-29,

$$\langle P\rangle = \tfrac{1}{2}\mu v^{\text{wave}}\omega^2 Y^2 \quad \text{(average power).} \tag{17-31}$$

The factors μ and v^{wave} in this equation depend on the material and tension of the string. The factors ω and Y depend on the process that generates the wave.

> The dependence of the average power of a wave on the square of its amplitude and also on the square of its angular frequency is a general result, true for sinusoidal waves of all types.

17-8 The Principle of Superposition for Waves

It often happens that two or more waves pass simultaneously through the same region. When we listen to a concert, for example, sound waves from many instruments fall simultaneously on our eardrums. The electrons in the antennas of our radio and television receivers are set in motion by the net effect of many electromagnetic waves from many different broadcasting centers. The water of a lake or harbor may be churned up by waves in the wakes of many boats.

In many real-world cases, we find the interaction between two overlapping waves to be quite complex. However, we also observe that when wave disturbances move along the same straight line and their amplitudes are small, like those shown in Fig. 17-15, their interaction is well behaved. We observe that when well behaved waves interact, they produce a resultant wave with an amplitude equal to the sum of the amplitudes of the two original waves. This effect is seen in Fig. 17-15, where a sequence of snapshots of two pulses traveling in opposite directions on the same stretched string is shown.

> When two well behaved linear waves overlap, the displacement of each point on the string is the sum of the two displacements it would have had from each wave independently.

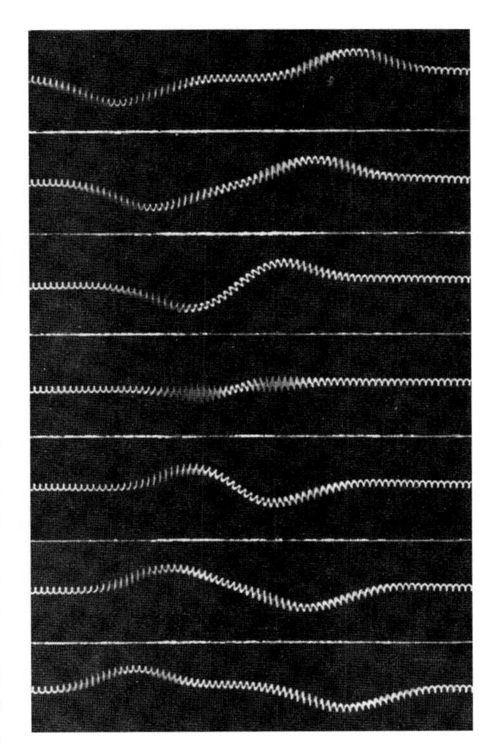

FIGURE 17-15 ■ An experimental demonstration of wave pulse superposition. In this sequence of movie frames two pulses are set in motion in opposite directions along a taut spring. The pulses have almost the same shape except that the one starting on the right has positive displacements while the other starting on the left has negative displacements. In the fourth frame the two pulses almost cancel each other.

Moreover, we observe that each pulse moves through the other, as if the other were not present:

> When well behaved linear waves overlap, one wave does not in any way alter the travel of the other.

This superposition is one of the many ways that waves and particles differ. When particles overlap by colliding they alter each other's motions.

In order to make these two important observations quantitative, let $y_1(x,t)$ and $y_2(x,t)$ be the displacements associated with two waves traveling simultaneously along the same stretched string. Since the result of the overlapping waves is a resultant wave that is the sum of the two, the displacement of the string when the waves overlap is given by the algebraic sum

$$y'(x,t) = y_1(x,t) + y_2(x,t). \tag{17-32}$$

This is an example of the **principle of superposition,** which says that when several effects occur simultaneously, their net effect is the sum of the individual effects.

We can look in more detail at how the sum of two waves passing through each other can look rather odd while the waves overlap, but after they pass each other they look more normal again. This situation is shown in the visualization of the superposition of two waves having different amplitudes in Figs. 17-16 and 17-17.

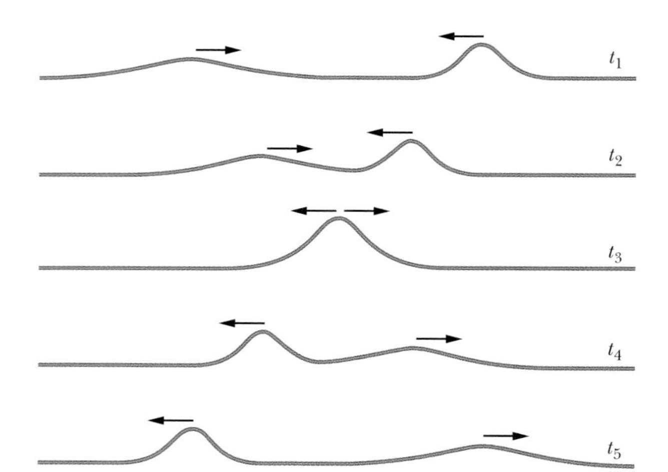

FIGURE 17-16 ■ A visualization of two pulses on a taut string passing through each other. A short broad pulse moves from left to right while a sharp tall pulse moves from right to left. When they overlap at times t_2, t_3, and t_4, the wave form is the sum or superposition of the displacement of each pulse at each location along the string. A close-up of this type of superposition is shown in Fig. 17-17.

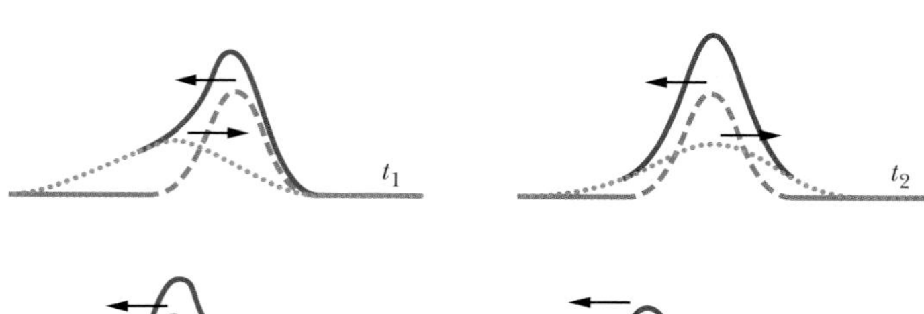

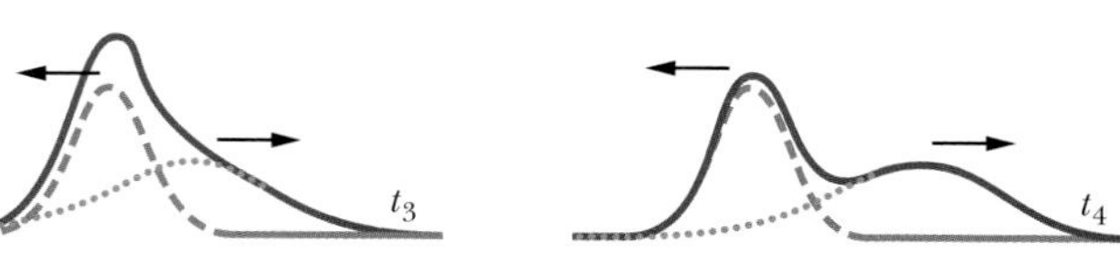

FIGURE 17-17 ■ Closeup view of a short, broad pulse and a tall sharp pulse passing through each other. The vertical scale relative to the horizontal scale has been enlarged to show more details. *Note that the resultant wave is the sum of the displacements contributed by each of the two pulses at each location along the string.*

TOUCHSTONE EXAMPLE 17-5: Overlapping Pulses

At the time $t = 0$ s, the string has the shape shown in Fig. 17-18. The pulse on the left is moving toward the right and the pulse on the right is moving toward the left. Assume that each box in Fig. 17-18 is 1 cm square.

(a) The leading edges of the pulses will just touch in a time of 0.05 s. What is the speed with which each pulse is traveling?

SOLUTION ■ The **Key Ideas** here are that (1) since these waves travel on the same string they must have the same speed so their leading edges will meet in the middle, and (2) if the wave pulses don't change shape, the motion of any location on the wave shape tells us about the velocity and speed of the wave as a whole. At first the leading edges of the waves are 4.0 cm apart, so each wave has time to travel 2.0 cm in the 0.05 s. This gives us a wave speed of

$$v^{\text{wave}} = |\vec{v}^{\text{wave}}| = \left(\frac{2.0\text{ cm}}{0.05\text{ s}}\right) = 40\text{ cm/s} \quad \text{(speed of each wave).} \quad \text{(Answer)}$$

(b) Two points on the string are marked with red dots and with the letters A and B. At the instant shown, what are the velocities of the dots? Give magnitude and direction (up, down, left, right, or some combination of them).

SOLUTION ■ There are three **Key Ideas** here. (1) The piece of string located at a given point can only move up or down but NOT along the string. (2) If the displacement of the string is less just behind a point on a wave (relative to its direction of motion), then the velocity of the piece of string is downward in the negative y direction. The opposite is true if the displacement behind the string is greater, then the velocity is upward. (3) The change Δy of a point on a string as a wave disturbance passes depends on both the wave velocity component $v_x^{\text{wave}} = \Delta x/\Delta t$ and the current slope of the wave shape $\Delta y/\Delta x$ at the point of interest along the string.

At point A: Wave slope is $\Delta y/\Delta x = -2$ so that $\Delta y = -2\Delta x$. But the x-component of the wave speed is given by $v_x^{\text{wave}} = \Delta x/\Delta t = 40$ cm/s, so

$$v_{Ay} = \frac{\Delta y}{\Delta t} = -\frac{2\Delta x}{\Delta t} = -80\text{ cm/s} \quad \text{(downward motion).} \quad \text{(Answer)}$$

At point B: Wave slope is $\Delta y/\Delta x = +\frac{2}{3}$ so that $\Delta y = +(\frac{2}{3})\Delta x$. Therefore,

$$v_{By} = \frac{\Delta y}{\Delta t} = \frac{+(\frac{2}{3})\Delta x}{\Delta t} = 27\text{ cm} \quad \text{(upward motion).} \quad \text{(Answer)}$$

(c) In Fig. 17-19 are dashed lines indicating where the pulses would be at a time $t = 0.075$ s. Draw a heavy line to show what the shape of the string would look like at this instant. Explain why you think it would look like your sketch.

SOLUTION ■ The **Key Idea** here is that we can use the principle of superposition to sum the contribution of each wave. Right in the middle where the waves overlap, they add up to a constant because the two overlapping slopes are equal in magnitude and opposite in sign (Fig. 17-20).

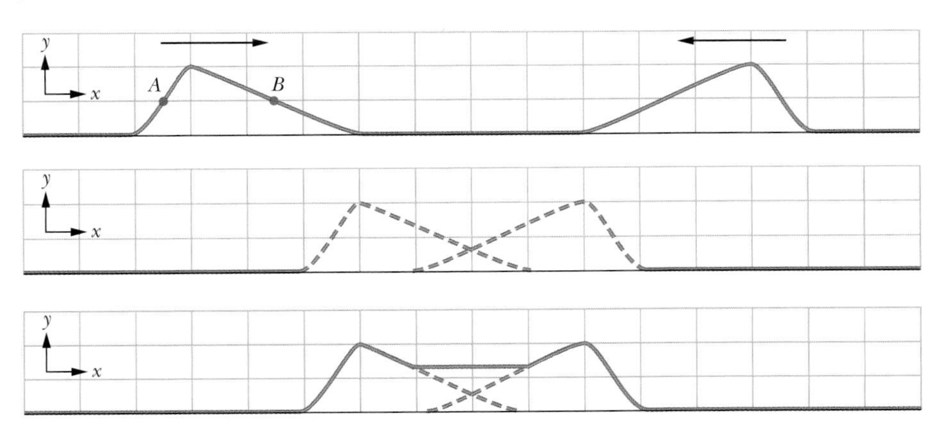

FIGURE 17-18 ■ Two pulses move toward each other on a string.

FIGURE 17-19 ■ Overlapping pulses at time $t = 0.075$ s.

FIGURE 17-20 ■ Answer to the question posed in part (c).

17-9 Interference of Waves

Suppose we send two sinusoidal waves of the same wavelength and amplitude in the same direction along a stretched string. The superposition principle applies. What resultant wave does it predict for the string?

The resultant wave depends on the extent to which the waves are *in phase* (in step) with respect to each other—that is, how much one wave form is shifted from the other wave form. If the waves are exactly in phase (so that the peaks and valleys of one are exactly aligned with those of the other), they combine to double the displacement of either wave acting alone. If they are exactly out of phase (the peaks of one are exactly aligned with the valleys of the other), they combine to cancel everywhere, and the string remains straight. We call this phenomenon of combining waves **interference,** and the waves are said to **interfere.** (These terms refer only to the displacements of the waves; the travel of the waves is unaffected.)

Let one wave traveling along a stretched string be given by

$$y_1(x,t) = Y\sin(kx - \omega t), \tag{17-33}$$

and another, shifted from the first, by an initial phase of ϕ_0 so that

$$y_2(x,t) = Y\sin(kx - \omega t + \phi_0). \tag{17-34}$$

The waves in question have the same angular frequency ω (that is, the same frequency f), the same wave number k (that is, the same wavelength λ), and the same amplitude Y. They both travel in the positive direction of the x axis, with the same speed, given by $|v^{\text{wave}}| = \sqrt{F^{\text{tension}}/\mu}$ (Eq. 17-25). They differ only by the nonzero initial phase ϕ_0 of wave 2, which we call the **phase difference.** These waves are said to be *out of phase* by ϕ_0 or to have a *phase difference* of $\Delta\phi = (\phi_0)_2 - (\phi_0)_1 = \phi_0$, or one wave is said to be *phase-shifted* from the other by ϕ_0.

From the principle of superposition, the resultant wave is the algebraic sum of the two interfering waves and has displacement

$$\begin{aligned} y'(x,t) &= y_1(x,t) + y_2(x,t) \\ &= Y\sin(kx - \omega t) + Y\sin(kx - \omega t + \phi_0). \end{aligned} \tag{17-35}$$

In Appendix E we see that we can write the sum of the sines of two angles α and β as

$$\sin\alpha + \sin\beta = 2\sin\tfrac{1}{2}(\alpha + \beta)\cos\tfrac{1}{2}(\alpha - \beta). \tag{17-36}$$

Applying this relation to Eq. 17-35 leads to

$$y'(x,t) = [2Y\cos\tfrac{1}{2}\phi_0]\sin(kx - \omega t + \tfrac{1}{2}\phi_0), \tag{17-37}$$

where ϕ_0 represents the phase difference $\Delta\phi$ between the two waves. As Fig. 17-21 shows, the resultant wave is also a sinusoidal wave traveling in the direction of increasing x. It is the only wave you would actually see on the string (you would *not* see the two interfering waves of Eqs. 17-33 and 17-34).

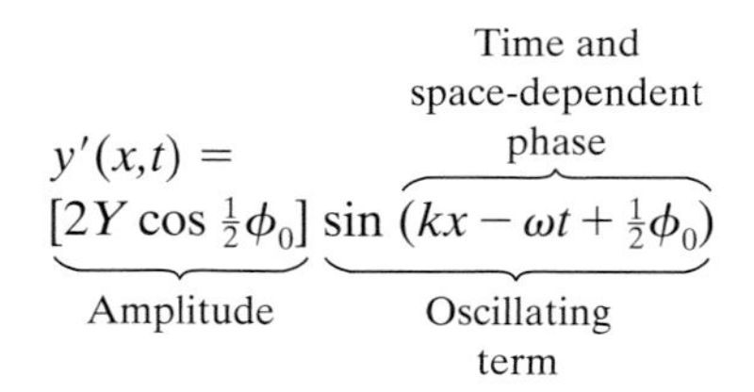

FIGURE 17-21 ■ The resultant wave of Eq. 17-37, due to the interference of two sinusoidal transverse waves, is also a sinusoidal wave, with an amplitude and an oscillating term.

> If two sinusoidal waves of the same amplitude and wavelength travel in the *same* direction along a stretched string, they interfere to produce a resultant sinusoidal wave traveling in that direction.

The resultant wave differs from the interfering waves in two respects: (1) its initial *phase* is $\frac{1}{2}\phi_0$, and (2) its amplitude Y' is the quantity in the brackets in Eq. 17-37:

$$Y' = 2Y\cos\tfrac{1}{2}\phi_0 \quad \text{(amplitude)}. \tag{17-38}$$

The resultant wave of Eq. 17-37, due to the interference of two sinusoidal transverse waves, is also a sinusoidal transverse wave, with an amplitude and an oscillating term.

Let's consider a couple of special and very important cases. If $\phi_0 = 0$ rad (or 0°), the two interfering waves are exactly in phase, as in Fig. 17-22*a*. Then Eq. 17-37 reduces to

$$y'(x,t) = 2Y\sin(kx - \omega t) \qquad (\phi_0 = 0 \text{ rad}). \tag{17-39}$$

This resultant wave is plotted in Fig. 17-22*d*. Note from both that figure and $y'(x,t) = 2Y\sin(kx - \omega t)$ that the amplitude of the resultant wave is twice the amplitude of either interfering wave. That is the greatest amplitude the resultant wave can have, because the cosine term in $y'(x,t) = [2Y\cos\frac{1}{2}\phi_0]\sin(kx - \omega t + \frac{1}{2}\phi_0)$ has its greatest value (unity) when $\phi_0 = 0$. Interference that produces the greatest possible amplitude is called *fully constructive interference*.

If $\Delta\phi = \phi_0 \pi$ rad (or 180°), the interfering waves are exactly out of phase as in Fig. 17-22*b*. Then $\cos\frac{1}{2}\phi_0$ becomes $\cos \pi/2 = 0$, and the amplitude of the resultant wave as given by Eq. 17-38 is zero. We then have, for all values of x and t,

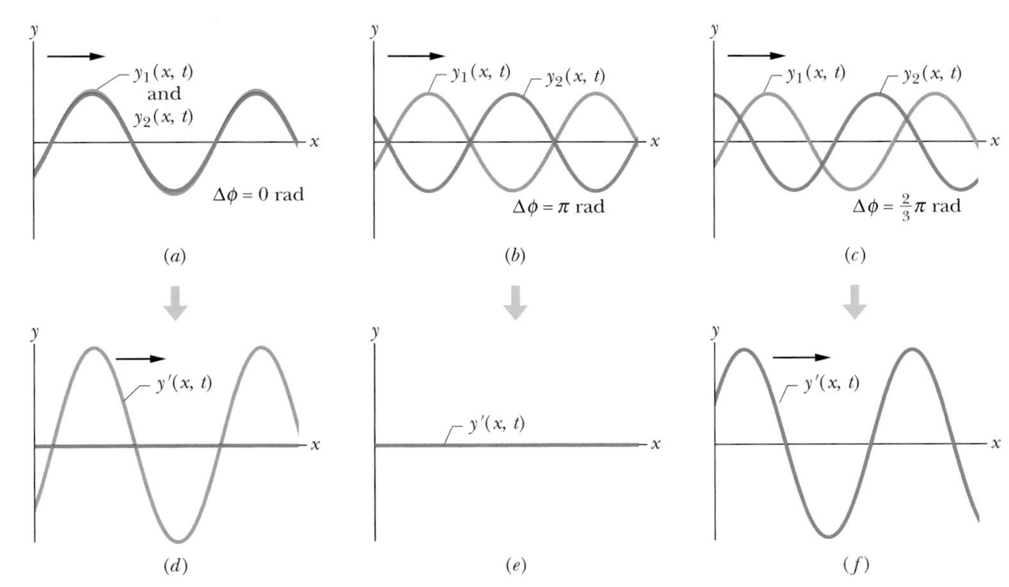

FIGURE 17-22 ■ Two sinusoidal waves with the same k, ω, and y, $y_1(x,t)$ and $y_2(x,t)$, travel along a string in the positive x direction. Here the units for ϕ_0 and t are unspecified. They interfere to give a resultant wave $y'(x,t)$. The resultant wave is what is actually seen on the string. The phase difference $\Delta\phi$ between the two interfering waves is (*a*) 0 rad or 0°, (*b*) π rad or 180°, and (*c*) $\frac{2}{3}\pi$ rad or 120°. The corresponding resultant waves are shown in (*d*), (*e*), and (*f*).

$$y'(x,t) = 0 \qquad (\phi = \pi \text{ rad}). \tag{17-40}$$

The resultant wave is plotted in Fig. 17-22*e*. Although we sent two waves along the string, we see no motion of the string. This type of interference is called *fully destructive interference.*

Because a sinusoidal wave repeats its shape every 2π rad, a phase difference $\Delta\phi_0 = 2\pi$ rad (or 360°) corresponds to a shift of one wave relative to the other wave by a distance equivalent to one wavelength. Thus, phase differences can be described in terms of wavelengths as well as angles. For example, in Fig. 17-22*b* the waves may be said to be 0.50 wavelength out of phase. Table 17-1 shows some other examples of phase differences and the interference they produce. Note that when interference is neither fully constructive nor fully destructive, it is called *intermediate interference.* The amplitude of the resultant wave is then intermediate between 0 and $2Y$. For example, from Table 17-1, if the interfering waves have a phase difference of 120° ($\Delta\phi = \frac{2}{3}\pi$ rad = 0.33 wavelength), then the resultant wave has an amplitude of Y, the same as the interfering waves (see Figs. 17-22*c* and *f*).

TABLE 17-1
Phase Differences and Resulting Interference Types[a]

Phase Difference, ($\Delta\phi$), in			Amplitude of Resultant Wave	Type of Interference
Degrees	**Radians**	**Wavelengths**		
0	0	0	$2Y$	Fully constructive
120	$\frac{2}{3}\pi$	0.33	Y	Intermediate
180	π	0.50	0	Fully destructive
240	$\frac{4}{3}\pi$	0.67	Y	Intermediate
360	2π	1.00	$2Y$	Fully constructive
865	15.1	2.40	$0.60Y$	Intermediate

[a] The phase difference is between two otherwise identical waves, with amplitude Y, moving in the same direction.

TOUCHSTONE EXAMPLE 17-6: Two Sine Waves

Two identical sinusoidal waves, moving in the same direction along a stretched string, interfere with each other. The amplitude Y of each wave is 9.8 mm, and the phase difference $\Delta\phi = \phi_0$ between them is $100°$.

(a) What is the amplitude Y of the resultant wave due to the interference of these two waves, and what type of interference occurs?

SOLUTION ■ The **Key Idea** here is that these are identical sinusoidal waves traveling in the *same direction* along a string, so they interfere to produce a sinusoidal traveling wave. Because they are identical except for their initial phase, they have the *same amplitude.* Thus, the amplitude Y' of the resultant wave is given by Eq. 17-38:

$$Y' = 2Y' \cos\tfrac{1}{2}\phi_0 = (2)(9.8 \text{ mm}) \cos(100°/2)$$

$$= 12.6 \text{ mm}. \qquad \text{(Answer)}$$

Here we have assumed that wave 2 has an initial phase of $100°$ relative to wave 1. We can tell that the interference is *intermediate* in two ways. The phase difference is between 0 and $180°$ and, correspondingly, amplitude Y' is between 0 and $2Y$ (= 19.6 mm).

(b) What phase difference, in radians and wavelengths, will give the resultant wave an amplitude of 4.9 mm?

SOLUTION ■ The same **Key Idea** applies here as in part (a), but now we are given Y' and seek ϕ_0. From Eq. 17-38,

$$Y' = 2Y \cos \tfrac{1}{2}\phi_0.$$

We now have

$$4.9 \text{ mm} = (2)(9.8 \text{ mm})\cos \tfrac{1}{2}\phi_0,$$

which gives us (with a calculator in the radian mode)

$$\phi_0 = 2 \cos^{-1}\frac{4.9 \text{ mm}}{(2)(9.8 \text{ mm})}$$

$$= \pm 2.636 \text{ rad} \approx \pm 2.6 \text{ rad}. \qquad \text{(Answer)}$$

There are two solutions because we can obtain the same resultant wave by letting the first wave *lead* (travel ahead of) or lag (travel behind) the second wave by 2.6 rad. In wavelengths, the phase difference is

$$\frac{\phi_0}{2\pi \text{ rad/wavelength}} = \frac{\pm 2.636 \text{ rad}}{2\pi \text{ rad/wavelength}}$$

$$= \pm 0.42 \text{ wavelength}. \qquad \text{(Answer)}$$

17-10 Reflections at a Boundary and Standing Waves

In the preceding two sections, we discussed two sinusoidal waves of the same wavelength and amplitude traveling *in the same direction* along a stretched string. What if they travel in opposite directions? We can again find the resultant wave by applying the superposition principle. Figure 17-23 suggests the situation graphically. It shows the two combining waves, one traveling to the left in Fig. 17-23*a*, the other to the right

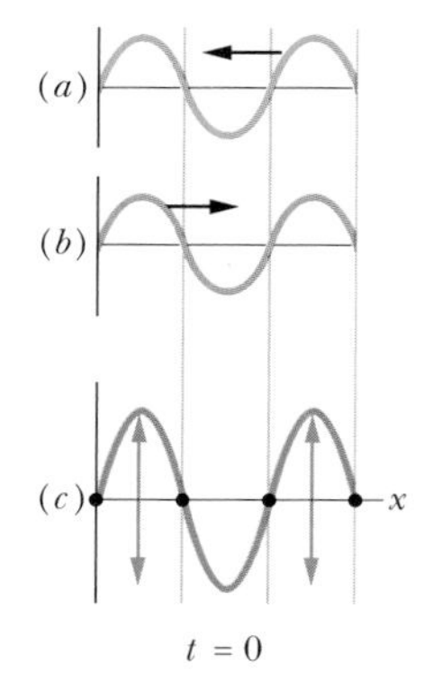

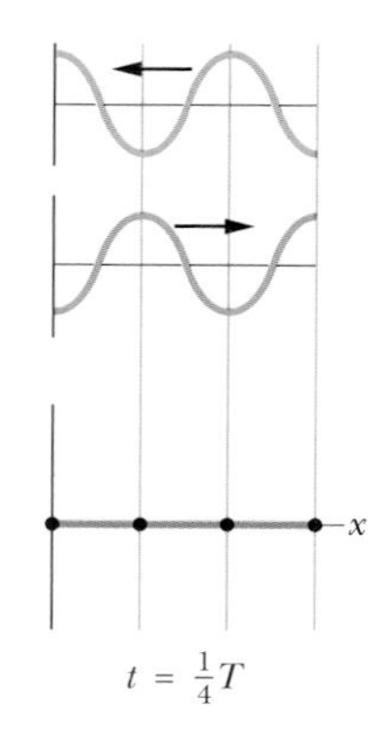

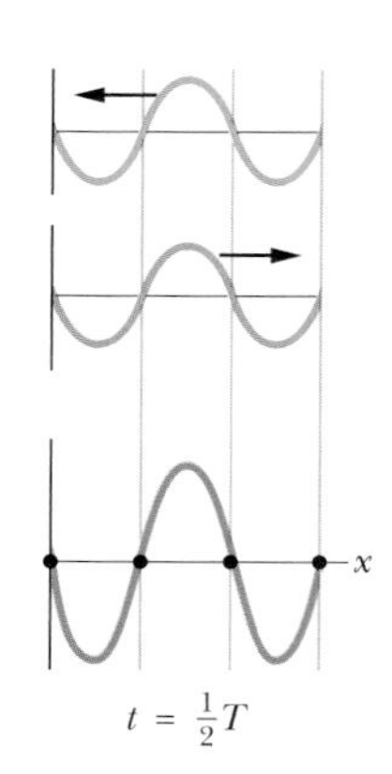

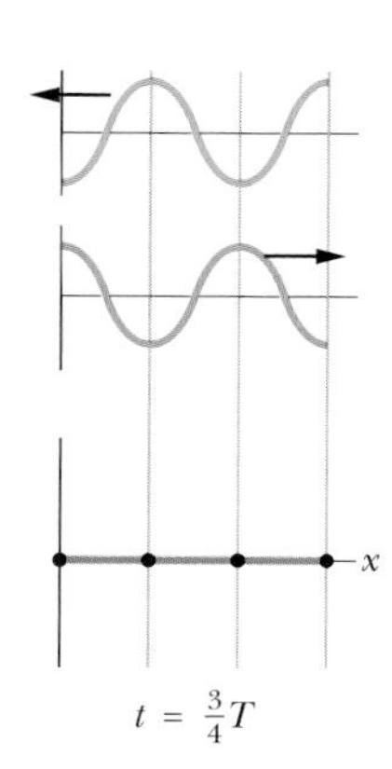

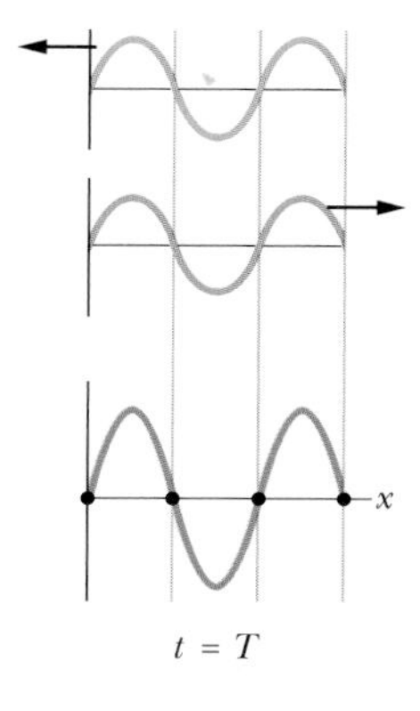

FIGURE 17-23 ■ (*a*) Five snapshots of a wave traveling to the left, at the times t indicated below part (*c*) (T is the period of oscillation). (*b*) Five snapshots of a wave identical to that in (*a*) but traveling to the right, at the same times t. (*c*) Corresponding snapshots for the superposition of the two waves on the same string. At $t = 0, \frac{1}{2}T$, and T, fully constructive interference occurs because of the alignment of peaks with peaks and valleys with valleys. At $t = \frac{1}{4}T$ and $\frac{3}{4}T$, fully destructive interference occurs because of the alignment of peaks with valleys. Some points (the nodes, marked with dots) never oscillate; some points (the antinodes) oscillate the most.

in Fig. 17-23*b*. Figure 17-23*c* shows their sum, obtained by applying the superposition principle graphically.

The outstanding feature of the resultant wave is that there are places along the string, called **nodes,** where the string never moves. Four such nodes are marked by dots in Fig. 17-23*c*. Halfway between adjacent nodes are **antinodes,** where the amplitude of the resultant wave is a maximum. Wave patterns such as that of Fig. 17-23*c* are called **standing waves** because the wave patterns do not move left or right; the locations of the maxima and minima do not change.

> If two sinusoidal waves of the same amplitude and wavelength travel in *opposite* directions along a stretched string, their interference with each other produces a standing wave.

To analyze a standing wave, we represent the two combining waves with the equations

$$y_1(x,t) = Y\sin(kx - \omega t) \tag{17-41}$$

and

$$y_2(x,t) = Y\sin(kx + \omega t). \tag{17-42}$$

The principle of superposition gives, for the combined wave,

$$y'(x,t) = y_1(x,t) + y_2(x,t) = Y\sin(kx - \omega t) + Y\sin(kx + \omega t).$$

Applying the trigonometric relation of $\sin\alpha + \sin\beta = 2\sin\frac{1}{2}(\alpha + \beta)\cos\frac{1}{2}(\alpha - \beta)$ leads to

$$y'(x,t) = [2Y\sin kx]\cos\omega t, \tag{17-43}$$

which is displayed in Fig. 17-24. This equation does not describe a traveling wave because it is not of the form of $y(x,t) = h(kx \pm \omega t)$. Instead, it describes a standing wave.

$$\overbrace{y'(x,t)}^{\text{Displacement}} = \underbrace{[2Y\sin kx]}_{\substack{\text{Amplitude}\\ \text{at position } x}}\underbrace{\cos\omega t}_{\substack{\text{Oscillating}\\ \text{term}}}$$

FIGURE 17-24 ■ The resultant wave of Eq. 17-43 is a standing wave and is due to the interference of two sinusoidal waves of the same amplitude and wavelength that travel in opposite directions with the same initial phase.

The quantity $2Y\sin kx$ in the brackets of $y'(x,t) = [2Y\sin kx]\cos\omega t$ can be viewed as the amplitude of oscillation of the string segment that is located at position x. However, since an amplitude is always positive and $\sin kx$ can be negative, we take the absolute value of the quantity $2Y\sin kx$ to be the amplitude at x.

In a traveling sinusoidal wave, the amplitude of the wave is the same for all string segments. That is not true for a standing wave, in which the amplitude *varies with position.* In the standing wave of $y'(x,t) = [2Y\sin kx]\cos\omega t$, for example, the amplitude is zero for values of kx that give $\sin kx = 0$. Those values are

$$kx = n\pi, \quad \text{for } n = 0, 1, 2, \ldots. \tag{17-44}$$

Substituting $k = 2\pi/\lambda$ in this equation and rearranging, we get

$$x = n\frac{\lambda}{2}, \quad \text{for } n = 0, 1, 2, \ldots \quad \text{(nodes)}, \tag{17-45}$$

as the positions of zero amplitude—the nodes—for the standing wave of Eq. 17-43. Note that adjacent nodes are separated by $\lambda/2$, half a wavelength.

The amplitude of the standing wave of $y'(x,t) = [2Y\sin kx]\cos\omega t$ has a maximum value of $2Y$, which occurs for values of kx that give $|\sin kx| = 1$. Those values are

$$\begin{aligned} kx &= \tfrac{1}{2}\pi, \tfrac{3}{2}\pi, \tfrac{5}{2}\pi, \ldots \\ &= (n + \tfrac{1}{2})\pi, \quad \text{for } n = 0, 1, 2, \ldots. \end{aligned} \tag{17-46}$$

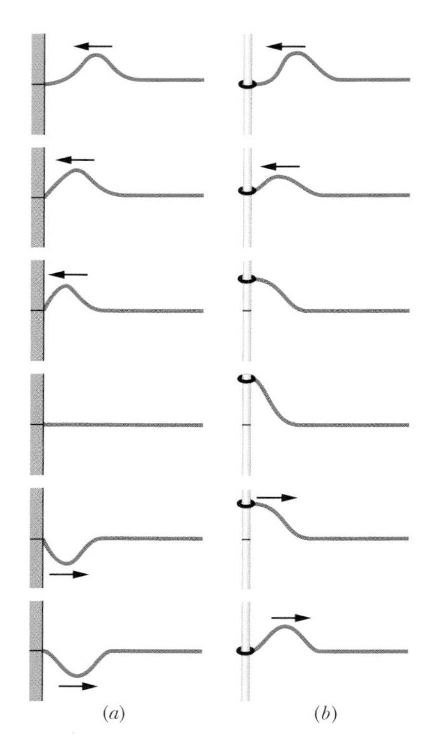

FIGURE 17-25 ■ (*a*) A pulse incident from the right is reflected at the left end of the string, which is tied to a wall. Note that the reflected pulse is inverted from the incident pulse. (*b*) Here the left end of the string is tied to a ring that can slide without friction up and down the rod. Now the pulse is not inverted by the reflection.

Substituting $k = 2\pi/\lambda$ in Eq. 17-43 and rearranging, we get

$$x = (n + \tfrac{1}{2})\frac{\lambda}{2} \quad \text{for } n = 0, 1, 2, \ldots \quad \text{(antinodes)}, \tag{17-47}$$

as the positions of maximum amplitude—the antinodes—of the standing wave of Eq. 17-43. The antinodes are separated by $\lambda/2$ and are located halfway between pairs of nodes.

Reflections at a Boundary

We can set up a standing wave in a stretched string by allowing a traveling wave to be reflected from the far end of the string so that it travels back through itself. The incident (original) wave and the reflected wave can then be described by Eqs. 17-41 and 17-42, respectively, and they can combine to form a pattern of standing waves.

In Fig. 17-25, we use a single pulse to show how such reflections take place. In Fig. 17-25*a*, the string is fixed at its left end. When the pulse arrives at that end, it exerts an upward force on the support (the wall). By Newton's Third Law, the support exerts an opposite force of equal magnitude on the string. This force generates a pulse at the support. This pulse travels back along the string in the direction opposite that of the incident pulse causing transverse string displacements that are inverted. In a "hard" reflection of this kind, there must be a node at the support because the string is fixed there. The reflected and incident pulses must have opposite signs, so as to cancel each other at that point.

In Fig. 17-25*b*, the left end of the string is fastened to a light ring that is free to slide without friction along a rod. When the incident pulse arrives, the ring moves up the rod. As the ring moves, it pulls on the string, stretching the string and producing a reflected pulse with the same sign and amplitude as the incident pulse. This reflected pulse is not inverted. Thus, in such a "soft" reflection, the incident and reflected pulses reinforce each other, creating an antinode at the end of the string; the maximum displacement of the ring is twice the amplitude of either of these pulses. The same types of reflections occur for continuous sinusoidal waves.

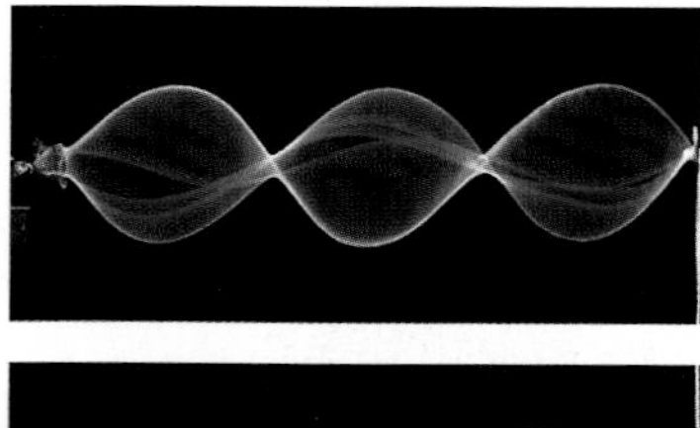

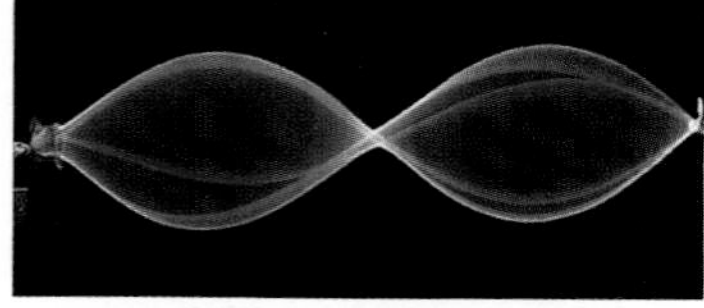

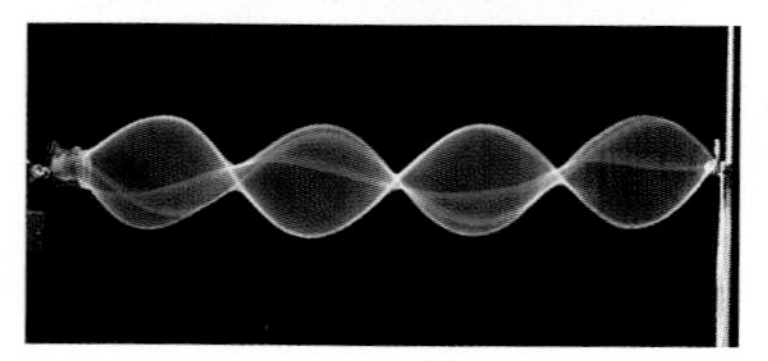

FIGURE 17-26 ■ Stroboscopic photographs reveal (imperfect) standing wave patterns on a string being made to oscillate by a vibrator at the left end. The patterns occur at certain frequencies of oscillation.

READING EXERCISE 17-5: Two waves with the same amplitude and wavelength interfere in three different situations to produce resultant waves with the following equations:
(1) $y'(x,t) = (0.004\text{ m})\sin((5\text{ rad/m})x - (4 \times 10^3\text{ rad/s})t)$,
(2) $y'(x,t) = [(0.004\text{ m})\sin((5\text{ rad/m})x)]\cos((4 \times 10^3\text{ rad/s})t)$, and
(3) $y'(x,t) = (0.004\text{ m})\sin((5\text{ rad/m})x + (4 \times 10^3\text{ rad/s})t)$.
In which situation are the two combining waves traveling (a) toward positive x, (b) toward negative x, and (c) in opposite directions? ■

17-11 Standing Waves and Resonance

Consider a string, such as a guitar string, that is stretched between two clamps. Suppose we send a sinusoidal wave of a certain frequency along the string, say, toward the right. When the wave reaches the right end, it reflects and begins to travel back to the left. That left-going wave then overlaps the wave that is still traveling to the right. When the left-going wave reaches the left end, it reflects again and the newly reflected wave begins to travel to the right, overlapping the left-going and right-going waves. In short, we very soon have many overlapping traveling waves, which interfere with one another.

For certain frequencies, the interference produces a standing wave pattern (or normal **oscillation mode**) with nodes and large antinodes like those in Fig. 17-26. Such a standing wave is said to be produced at **resonance,** and the string is said to *resonate* at these certain frequencies, called **resonant frequencies.** This standing wave resonance is not unlike the harmonic oscillator resonance discussed in Section 16-8. If the string is oscillated at some frequency other than a resonant frequency, a standing wave is not set up. Then the interference of the right-going and left-going traveling waves results in only small (perhaps imperceptible) oscillations of the string.

Let a string be stretched between two clamps separated by a fixed distance L. To find expressions for the resonant frequencies of the string, we note that a node must exist at each of its ends, because each end is fixed and cannot oscillate. The simplest pattern that meets this key requirement is that in Fig. 17-27*a*, which shows the string at both its extreme displacements (one solid and one dashed, together forming a single "loop"). There is only one antinode, which is at the center of the string. Note that half a wavelength spans the length L, which we take to be the string's length. Thus, for this pattern, $\lambda/2 = L$. This condition tells us that if the left-going and right-going traveling waves are to set up this pattern by their interference, they must have the wavelength $\lambda = 2L$.

A second simple pattern meeting the requirement of nodes at the fixed ends is shown in Fig. 17-27*b*. This pattern has three nodes and two antinodes and is said to be a two-loop pattern. For the left-going and right-going waves to set it up, they must have a wavelength $\lambda = L$. A third pattern is shown in Fig. 17-27*c*. It has four nodes, three antinodes, and three loops, and the wavelength is $\lambda = \frac{2}{3}L$. We could continue this progression by drawing increasingly more complicated patterns. In each step of the progression, the pattern would have one more node and one more antinode than the preceding step, and an additional $\lambda/2$ would be fitted into the distance L.

Thus, a standing wave can be set up on a string of length L by a wave with a wavelength equal to one of the values

$$\lambda = \frac{2L}{n}, \quad \text{for } n = 1, 2, 3 \ldots. \tag{17-48}$$

The resonant frequencies that correspond to these wavelengths follow from Eq. 17-16:

$$f = \frac{v^{\text{wave}}}{\lambda} = n\frac{v^{\text{wave}}}{2L}, \quad \text{for } n = 1, 2, 3, \ldots \quad \text{(string fixed at both ends)}, \tag{17-49}$$

where v^{wave} is the speed of traveling waves on the string.

Equation 17-49 tells us that the resonant frequencies are integer multiples of the lowest resonant frequency, $f = v^{\text{wave}}/2L$, which corresponds to $n = 1$. The oscillation mode with that lowest frequency is called the *fundamental mode* or the *first harmonic.* The *second harmonic* is the oscillation mode with $n = 2$, the *third harmonic* is that with $n = 3$, and so on. The frequencies associated with these modes are often labeled f_1, f_2, f_3 and so on. The collection of all possible oscillation modes is called the **harmonic series,** and n is called the **harmonic number** of the nth harmonic.

The phenomenon of resonance is common to all oscillating systems and can occur in two and three dimensions. For example, Fig. 17-28 shows a two-dimensional standing wave pattern on the oscillating head of a kettledrum.

READING EXERCISE 17-6: In the following series of resonant frequencies, one frequency (lower than 400 Hz) is missing: 150, 225, 300, 375 Hz. (a) What is the missing frequency? (b) What is the frequency of the seventh harmonic? ■

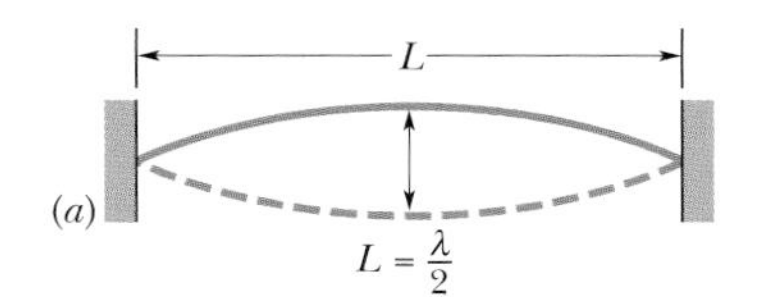

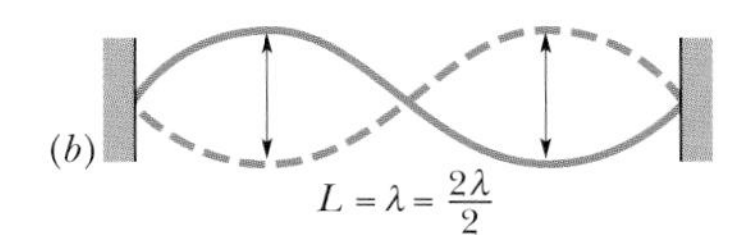

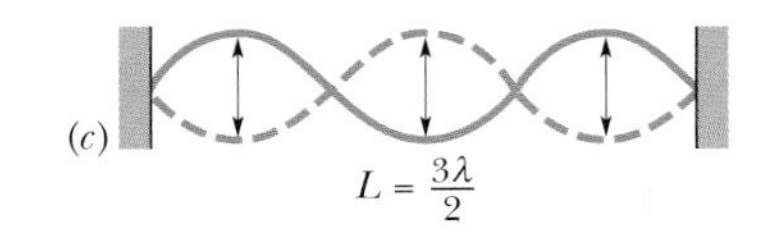

FIGURE 17-27 ■ A string, stretched between two clamps, is made to oscillate in standing-wave patterns. (*a*) The simplest possible pattern consists of one *loop,* which refers to the composite shape formed by the string in its extreme displacements (the solid and dashed lines). (*b*) The next simplest pattern has two loops. (*c*) The next has three loops.

FIGURE 17-28 ■ One of many possible standing wave patterns for a kettledrum head, made visible by dark powder sprinkled on the drumhead. As the head is set into oscillation at a single frequency by a mechanical vibrator at the upper left of the photograph, the powder collects at the nodes, which are circles and straight lines (rather than points) in this two-dimensional example.

TOUCHSTONE EXAMPLE 17-7: String Harmonics

In Fig. 17-29, a string, tied to a sinusoidal vibrator at P and running over a fixed pulley at Q, is stretched by a block of mass m. The separation L between P and Q is 1.2 m, the linear density of the string is 1.6 g/m, and the frequency f of the vibrator is fixed at 120 Hz. The amplitude of the motion at P is small enough for that point to be considered a node. A node also exists at Q.

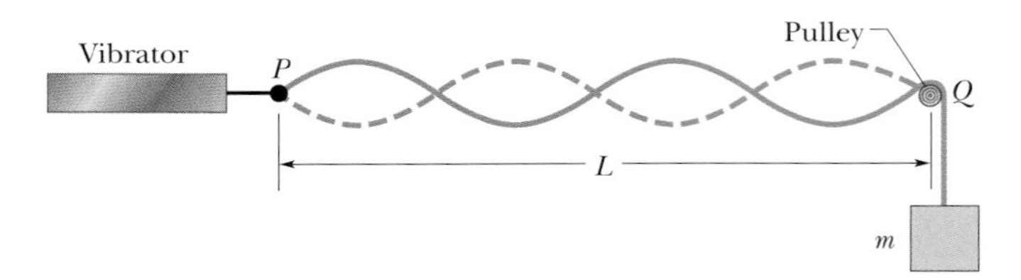

FIGURE 17-29 ■ A string under tension connected to a vibrator. For a fixed vibrator frequency, standing wave patterns will occur for certain values of the string tension.

(a) What mass m allows the vibrator to set up the fourth harmonic on the string?

SOLUTION ■ One **Key Idea** here is that the string will resonate at only certain frequencies, determined by the wave speed $|\vec{v}^{\text{wave}}|$ on the string and the length L of the string. From Eq. 17-49, these resonant frequencies are

$$f = n\frac{v^{\text{wave}}}{2L}, \quad \text{for } n = 1, 2, 3, \ldots. \tag{17-50}$$

To set up the fourth harmonic (for which $n = 4$), we need to adjust the right side of this equation, with $n = 4$, so that the left side equals the frequency of the vibrator (120 Hz).

We cannot adjust L in Eq. 17-50; it is set. However, a second **Key Idea** is that we *can* adjust v^{wave} because it depends on how much mass m we hang on the string. According to Eq. 17-25, wave speed $v^{\text{wave}} = \sqrt{F^{\text{tension}}/\mu}$. Here the tension F^{tension} in the string is equal to the magnitude of the weight mg of the block. Thus,

$$v^{\text{wave}} = \sqrt{\frac{F^{\text{tension}}}{\mu}} = \sqrt{\frac{mg}{\mu}}. \tag{17-51}$$

Substituting v^{wave} from Eq. 17-51 into Eq. 17-50, setting $n = 4$ for the fourth harmonic, and solving for m give us

$$m = \frac{4L^2f^2\mu}{n^2g} = \frac{(4)(1.2\text{ m})^2(120\text{ Hz})^2(0.0016\text{ kg/m})}{(4)^2(9.8\text{ m/s}^2)}$$

$$= 0.846\text{ kg} \approx 0.85\text{ kg}. \quad \text{(Answer)} \tag{17-52}$$

(b) What standing wave mode is set up if $m = 1.00$ kg?

SOLUTION ■ If we insert this value of m into Eq. 17-52 and solve for n, we find that $n = 3.7$. A **Key Idea** here is that n must be an integer, so $n = 3.7$ is impossible. Thus, with $m = 1.00$ kg, the vibrator cannot set up a standing wave on the string, and any oscillation of the string will be small, perhaps even imperceptible.

17-12 Phasors

We can represent a wave on a string (or any other type of wave) with a **phasor.** In essence, a phasor is an arrow that has a magnitude equal to the amplitude of the wave and that rotates around an origin; the angular speed of the phasor is equal to the angular frequency ω of the wave. For example, the wave

$$y_1(x,t) = Y_1\sin(kx - \omega t), \tag{17-53}$$

which has an initial phase of $(\phi_0)_1 = 0$ rad. It is represented at a time $t > 0$ s, and location x along a string by the phasor shown in Fig. 17-30*a*. The magnitude of the phasor is the amplitude Y_1 of the wave. As the phasor rotates around the origin at angular speed ω, its projection y_1 on the vertical axis varies sinusoidally, from a maximum of Y_1 through zero to a minimum of $-Y_1$ and then back to Y_1. This variation corresponds to the sinusoidal variation in the displacement $y_1(x,t)$ of any chosen point x along the string as the wave passes through it.

When two waves travel along the same string in the same direction, we can represent them and their resultant wave in a *phasor diagram.* The phasors in Fig. 17-30*b* represent the wave of Eq. 17-53 and a second wave given by

$$y_2(x,t) = Y_2\sin(kx - \omega t + \phi_0), \tag{17-54}$$

and has an initial phase of $(\phi_0)_2$. So, this second wave is phase-shifted from the first wave with a phase difference of $\Delta\phi = (\phi_0)_2 - (\phi_0)_1 = \phi_0 - 0 = \phi_0$. Because

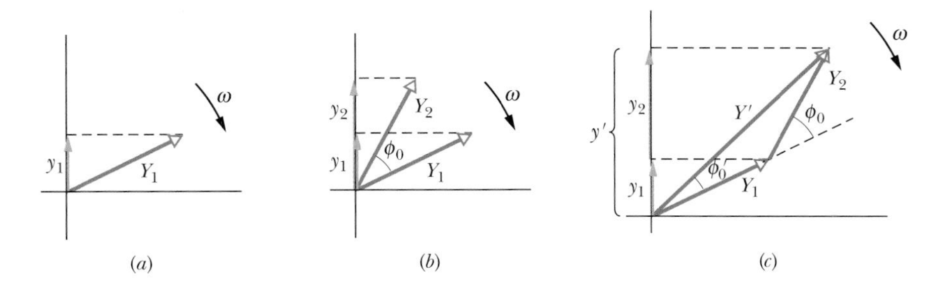

FIGURE 17-30 ■ (*a*) A phasor of magnitude Y_1 rotating about an origin at angular speed ω represents a sinusoidal wave with an initial phase of zero. The phasor's projection y_1 on the vertical axis represents the displacement at a point X at a time t through which the wave passes. (*b*) A second phasor, also of angular speed ω but of magnitude Y_2 and rotating at a constant angle ϕ_0 from the first phasor, represents a second wave, with an initial phase of ϕ_0. (*c*) The resultant wave of the two waves is represented by the vector-like phasor sum Y' of the two phasors. The projection y' on the vertical axis represents the displacement of the some point x and time t as that resultant wave passes through it.

the phasors rotate at the same angular speed ω, the angle between the two phasors is always $\Delta\phi = \phi_0$. If $\Delta\phi$ is a *positive* quantity, then the phasor for wave 2 *lags* the phasor for wave 1 as they rotate, as drawn in Fig. 17-30*b*. If $\Delta\phi$ is a negative quantity, then the phasor for wave 2 *leads* the phasor for wave 1.

Since waves y_1 and y_2 have the same wave number k and angular frequency ω, we know from Eq. 17-37 that if the two waves have the same amplitude ($Y = Y_1 = Y_2$) then their resultant is of the form

$$y'(x,t) = Y'\sin(kx - \omega t + \phi'_0), \tag{17-55}$$

where $Y' = 2Y\cos(\frac{1}{2}\phi_0)$ is the amplitude of the resultant wave and $\phi'_0 = \frac{1}{2}\phi_0$ is its phase. We used Eq. 17-37 to determine the equations for Y' and ϕ'_0 by superimposing the two combining waves.

Finding the superimposed wave is much easier using a phasor diagram. This is because even though phasors are not really vectors that have defined vector dot and cross products, they add like vectors. So even if the amplitudes Y_1 and Y_2 and frequencies ω_1 and ω_2 of two waves are not the same, we can use a vector-like phasor sum to find an equation for the resultant wave at any instant during their rotation. For example, we simply use the rules of vector addition to sum of the two phasors at any instant during their rotation. Fig. 17-30*c* shows how phasor Y_2 can be shifted to the head of phasor Y_1. The magnitude of the phasor sum equals the amplitude Y' in Eq. 17-55. The angle between the phasor sum and the phasor for y_1 equals the initial phase of the combined wave given in Eq. 17-55 as $\phi'_0 = \frac{1}{2}\phi_0$.

Although we have shown how phasors can be combined for a situation for which the amplitudes and frequencies are the same, it is important to note that:

> We can use phasors to combine waves *even if their amplitudes and frequencies are different.*

TOUCHSTONE EXAMPLE 17-8: Combining Two Waves

Two sinusoidal waves $y_1(x, t)$ and $y_2(x, t)$ have the same wavelength and travel together in the same direction along a string. Their amplitudes are $Y_1 = 4.0$ mm and $Y_2 = 3.0$ mm, and their initial phases are 0 and $\pi/3$ rad, respectively. What are the amplitude Y' and initial phase ϕ'_0 of the resultant wave (in the form of Eq. 17-55)?

SOLUTION ■ One **Key Idea** here is that the two waves have a number of properties in common: they travel along the same string and so have the same speed v^{wave}, angular wave number $k(= 2\pi/\lambda)$, and the same angular frequency $\omega(= kv^{\text{wave}})$.

A second **Key Idea** is that the waves can be represented by phasors rotating at the same angular speed ω about an origin. Because the initial phase for wave 2 is *greater* than that for wave 1 by $\pi/3$, phasor 2 must *lag* phasor 1 by $\pi/3$ rad in their clockwise rotation, as shown in Fig. 17-31. The resultant wave due to the interference of waves 1 and 2 can then be represented by a phasor that is a vector–like sum of phasors 1 and 2.

To simplify the phasor summation, we drew phasors 1 and 2 in Fig. 17-31*a* at the instant when phasor 1 lies along the horizontal axis and the lagging phasor 2 at positive angle $\pi/3$ rad. In Fig. 17-31*b* we shifted phasor 2 so its tail is at the head of phasor 1. Then we draw the phasor Y' of the resultant wave from the tail of phasor 1 to the head of phasor 2. The initial phase ϕ'_0 of the combined is the angle phasor 2 makes with respect to phasor 1.

Though phasors are not really vectors, they can be added using vector addition rules. To find values for Y' and ϕ'_0 we can sum phasors 1 and 2 directly on a vector-capable calculator, by

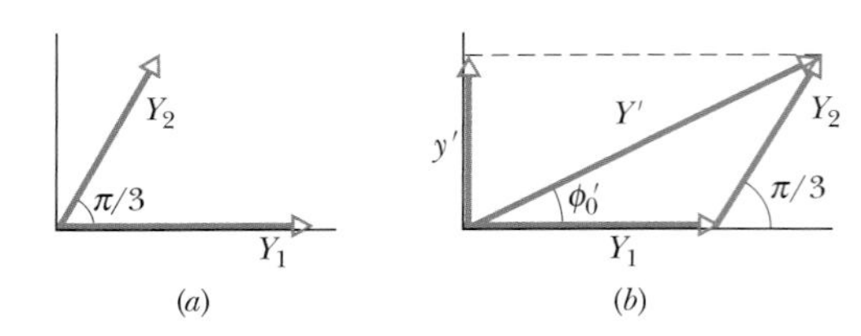

FIGURE 17-31 ■ (*a*) Two phasors of magnitudes Y_1 and Y_2 with phase difference $\pi/3$. (*b*) Phasor addition at any instant during their rotation gives the magnitude Y' of the phasor for the resultant wave.

adding a vector of magnitude 4.0 mm and angle 0 rad to a vector of magnitude 3.0 mm and angle $\pi/3$ rad, or we can add the phasors using vector components. For the horizontal components we have

$$Y_h' = Y_1 \cos 0 + Y_2 \cos \pi/3$$
$$= 4.0 \text{ mm} + (3.0 \text{ mm}) \cos \pi/3 = 5.50 \text{ mm}.$$

For the vertical components we have

$$Y_v' = Y_1 \sin 0 + Y_2 \sin \pi/3$$
$$= 0 + (3.0 \text{ mm}) \sin \pi/3 = 2.60 \text{ mm}.$$

Thus, the resultant wave has an amplitude of

$$Y' = \sqrt{(5.50 \text{ mm})^2 + (2.60 \text{ mm})^2}$$
$$= 6.1 \text{ mm} \qquad \text{(Answer)}$$

and an initial resultant wave phase of

$$\phi_0' = \tan^{-1}\frac{2.60 \text{ mm}}{5.50 \text{ mm}} = 0.44 \text{ rad.} \qquad \text{(Answer)}$$

From Fig. 17-31*b*, the initial phase for wave 2, given by ϕ_0, is a *positive* angle relative to the initial phase of wave 1. Thus, the resultant wave *lags* wave 1 by an initial phase of $\phi'_0 = 0.44$ rad. From Eq. 17-55, we can write the resultant wave as

$$y'(x,t) = (6.1 \text{ mm}) \sin(kx - \omega t + 0.44 \text{ rad}). \qquad \text{(Answer)}$$

Problems

SEC. 17-5 ■ WAVE VELOCITY

1. Angular Frequency A wave has an angular frequency of 110 rad/s and a wavelength of 1.80 m. Calculate (a) the angular wave number and (b) the speed of the wave.

2. Electromagnetic Waves The speed of electromagnetic waves (which include visible light, radio, and x-rays) in vacuum is 3.0×10^8 m/s. (a) Wavelengths of visible light waves range from about 400 nm in the violet to about 700 nm in the red. What is the range of frequencies of these waves? (b) The range of frequencies for shortwave radio (for example, FM radio and VHF television) is 1.5 to 300 MHz. What is the corresponding wavelength range? (c) X-ray wavelengths range from about 5.0 nm to about 1.0×10^{-2} nm. What is the frequency range for x-rays?

3. Sinusoidal Wave A sinusoidal wave travels along a string. The time for a particular point to move from maximum displacement to zero is 0.170 s. What are the (a) period and (b) frequency? (c) The wave length is 1.40 m; what is the wave speed?

4. Write the Equation Write the equation for a sinusoidal wave traveling in the negative direction along an x axis and having an amplitude of 0.010 m, a frequency of 550 Hz, and a speed of 330 m/s.

5. Show That Show that

$$y(x,t) = Y \sin k(x - vt), \qquad y(x,t) = Y \sin 2\pi\left(\frac{x}{\lambda} - ft\right),$$

$$y(x,t) = Y \sin \omega\left(\frac{x}{v} - t\right), \qquad y(x,t) = Y \sin 2\pi\left(\frac{x}{\lambda} - \frac{t}{T}\right)$$

are all equivalent to $y(x,t) = Y \sin(kx - \omega t)$.

6. Equation of a Transverse The equation of a transverse wave traveling along a very long string is $y(x,t) = (6.0 \text{ cm}) \sin\{(0.020\ \pi \text{ rad/cm})\, x + (4.0\ \pi \text{ rad/s})t\}$ where x and y are expressed in centimeters and t is in seconds. Determine (a) the amplitude, (b) the wavelength, (c) the frequency, (d) the speed, (e) the direction of propagation of the wave, and (f) the maximum transverse speed of a particle in the string. (g) What is the transverse displacement at $x = 3.5$ cm when $t = 0.26$ s?

7. Write an Equation (a) Write an equation describing a sinusoidal transverse wave traveling on a cord in the $+x$ direction with a wavelength of 10 cm, a frequency of 400 Hz, and an amplitude of 2.0 cm. (b) What is the maximum speed of a point on the cord? (c) What is the speed of the wave?

8. Transverse Sinusoidal A transverse sinusoidal wave of wavelength 20 cm is moving along a string in the positive x direction. The transverse displacement of the string particle at $x = 0$ cm as a function of time is shown in Fig. 17-32. (a) Make a rough sketch of one wavelength of the wave (the portion between $x = 0$ cm and $x = 20$ cm) at time $t = 0$ s. (b) What is the speed of the wave? (c) Write the equation for the wave with all the constants evaluated. (d) What is the transverse velocity of the particle at $x = 0$ m at $t = 5.0$ s?

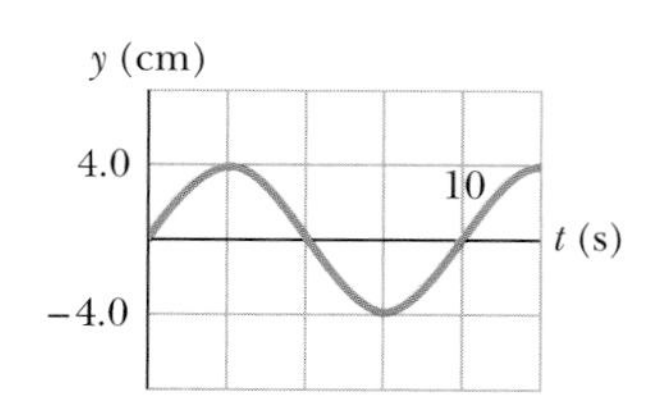

FIGURE 17-32 ■ Problem 8.

9. Sinusoidal Wave Two A sinusoidal wave of frequency 500 Hz has a speed of 350 m/s. (a) How far apart are two points that differ in phase by $\pi/3$ rad? (b) What is the phase difference between two displacements at a certain point at times 1.00 ms apart?

SEC. 17-6 ■ WAVE SPEED ON A STRETCHED STRING

10. Violin Strings The heaviest and lightest strings on a certain violin have linear densities of 3.0 and 0.29 g/m. (a) What is the ratio of the diameter of the heaviest string to that of the lightest string, assuming that the strings are of the same material? (b) What is the ratio of speeds if the strings have the same tension?

11. What Is the Speed What is the speed of a transverse wave in a rope of length 2.00 m and mass 60.0 g under a tension of 500 N?

12. Wire Clamped The tension in a wire clamped at both ends is doubled without appreciably changing the wire's length between the clamps. What is the ratio of the new to the old wave speed for transverse waves traveling along this wire?

13. Linear Density The linear density of a string is 1.6×10^{-4} kg/m. A transverse wave on the string is described by the equation

$$y(x,t) = (0.021 \text{ m}) \sin[(2.0 \text{ rad/m})x + (30 \text{ rad/s})t].$$

What is (a) the wave speed and (b) the tension in the string?

14. The Equation of The equation of a transverse wave on a string is

$$y(x,t) = (2.0 \text{ mm}) \sin[(20 \text{ rad/m})x - (600 \text{ rad/s})t].$$

The tension in the string is 15 N. (a) What is the wave speed? (b) Find the linear density of the string in grams per meter.

15. Stretched String A stretched string has a mass per unit length of 5.0 g/cm and a tension of 10 N. A sinusoidal wave on this string has an amplitude of 0.12 mm and a frequency of 100 Hz and is traveling in the negative direction of x. Write an equation for this wave.

16. The Fastest Wave What is the fastest transverse wave that can be sent along a steel wire? For safety reasons, the maximum tensile stress to which steel wires should be subjected is 7.0×10^8 N/m^2. The density of steel is 7800 kg/m^3. Show that your answer does not depend on the diameter of the wire.

17. Single Particle A sinusoidal transverse wave of amplitude Y and wavelength λ travels on a stretched cord. (a) Find the ratio of the maximum particle speed (the speed with which a single particle in the cord moves transverse to the wave) to the wave speed. (b) If a wave having a certain wavelength and amplitude is sent along a cord, would this speed ratio depend on the material of which the cord is made such as wire or nylon?

18. Displacement of Particles A sinusoidal wave is traveling on a string with speed 40 cm/s. The displacement of the particles of the string at $x = 10$ cm is found to vary with time according to the equation $y(x,t) = (5.0 \text{ cm}) \sin [1.0 \text{ rad/cm} - (4.0 \text{ rad/s})t]$. The linear density of the string is 4.0 g/cm. What are the (a) frequency and (b) wavelength of the wave? (c) Write the general equation giving the transverse displacement of the particles of the string as a function of position and time. (d) Calculate the tension in the string.

19. As a Function of Position A sinusoidal transverse wave is traveling along a string in the negative direction of an x axis. Figure 17-33 shows a plot of the displacement as a function of position at time $t = 0$ s; the y intercept is 4.0 cm. The string tension is 3.6 N, and its linear density is 25 g/m. Find the (a) amplitude, (b) wavelength, (c) wave speed, and (d) period of the wave. (e) Find the maximum transverse speed of a particle in the string. (f) Write an equation describing the traveling wave.

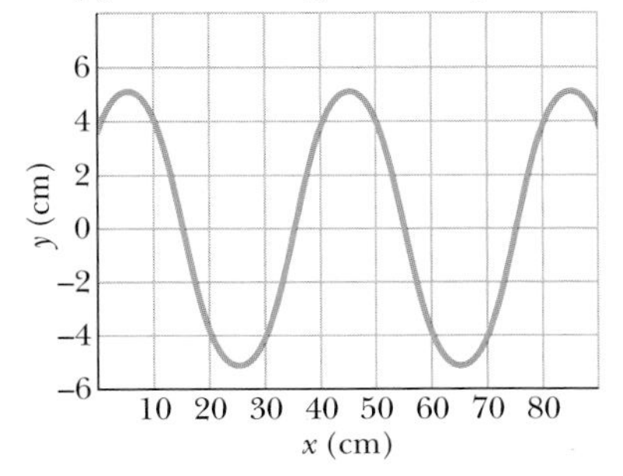

FIGURE 17-33 ■ Problem 19.

20. Three Pulleys, One Mass In Fig. 17-34a string 1 has a linear density of 3.00 g/m, and string 2 has a linear density of 5.00 g/m. They are under tension owing to the hanging block of mass $M = 500$ g. Calculate the wave speed on (a) string 1 and (b) string 2. (*Hint*: When a string loops halfway around a pulley, it pulls on the pulley with a net force that is twice the tension in the string.) Next the block is divided into two blocks (with $M_1 + M_2 = M$) and the apparatus is rearranged as shown in Fig. 17-34b. Find (c) M_1 and (d) M_2 such that the wave speeds in the two strings are equal.

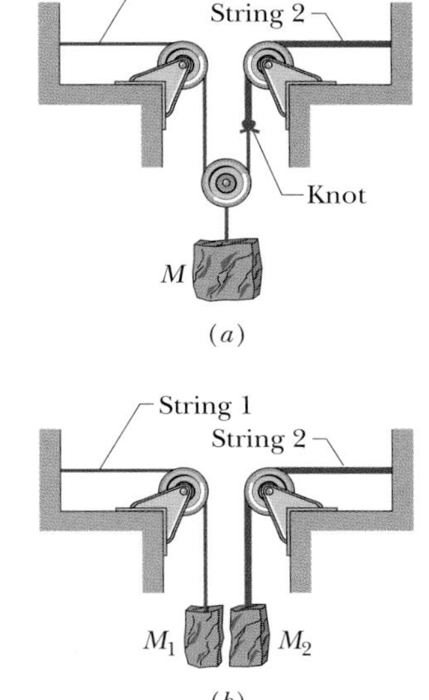

FIGURE 17-34 ■ Problem 20.

21. Two Pulses Two A wire 10.0 m long and having a mass of 100 g is stretched under a tension of 250 N. If two pulses, separated in time by 30.0 ms, are generated, one at each end of the wire, where will the pulses first meet?

22. Baseball Rubber Band The type of rubber band used inside some baseballs and golf balls obeys Hooke's law over a wide range of elongation of the band. A segment of this material has an unstretched length ℓ and a mass m. When a force $\vec{F}$ is applied, the band stretches an additional length $\Delta\ell$. (a) What is the speed (in terms of m, $\Delta\ell$, and the spring constant k) of transverse waves on this stretched rubber band? (b) Using your answer to (a), show that the time required for a transverse pulse to travel the length of the rubber band is proportional to $1/\sqrt{\Delta\ell}$ if $\Delta\ell \ll \ell$ and is constant if $\Delta\ell \gg \ell$.

23. Uniform Rope A uniform rope of mass m and length L hangs from a ceiling. (a) Show that the speed of a transverse wave on the rope is a function of y, the distance from the lower end, and is given by $v^{\text{wave}} = \sqrt{gy}$. (b) Show that the time a transverse wave takes to travel the length of the rope is given by $t = 2\sqrt{L/g}$.

SEC. 17-7 ■ ENERGY AND POWER TRANSPORTED BY A TRAVELING WAVE IN A STRING

24. Average Power A string along which waves can travel is 2.70 m long and has a mass of 260 g. The tension in the string is 36.0 N. What must be the frequency of traveling waves of amplitude 7.70 mm for the average power to be 85.0 W?

SEC. 17-9 ■ INTERFERENCE OF WAVES

25. Two Identical Traveling Waves Two identical traveling waves, moving in the same direction, are out of phase by $\pi/2$ rad. What is the amplitude of the resultant wave in terms of the common amplitude Y of the two combining waves?

26. What Phase Difference What phase difference between two otherwise identical traveling waves, moving in the same direction along a stretched string, will result in the combined wave having an amplitude 1.50 times that of the common amplitude of the two combining waves? Express your answer in (a) degrees, (b) radians, and (c) wavelengths.

27. Identical Except for Phase Two sinusoidal waves, identical except for phase, travel in the same direction along a string and interfere to produce a resultant wave given by $y'(x,t) = (3.0 \text{ mm}) \sin[(20 \text{ rad/m})x - (4.0 \text{ rad/s})t + 0.820 \text{ rad}]$, with x in meters and t in seconds. What are (a) the wavelength λ of the two waves, (b) the phase difference between them, and (c) their amplitude Y?

SEC. 17-11 ■ STANDING WAVES AND RESONANCE

28. A String Under Tension A string under tension F^{tension} oscillates in the third harmonic at frequency f_3, and the waves on the string have wavelength λ_3. If the tension is increased to $\tau_f = 4\tau_i$ and the string is again made to oscillate in the third harmonic, what then are (a) the frequency of oscillation in terms of f_3 and (b) the wavelength of the waves in terms of λ_3?

29. Nylon Guitar String A nylon guitar string has a linear density of 7.2 g/m and is under a tension of 150 N. The fixed supports are 90 cm apart. The string is oscillating in the standing wave pattern shown in Fig. 17-35. Calculate the (a) speed, (b) wavelength, and (c) frequency of the traveling waves whose superposition gives this standing wave.

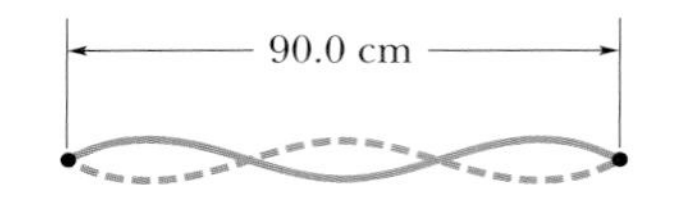

FIGURE 17-35 ■ Problem 29.

30. Two Sinusoidal Waves Two sinusoidal waves with identical wavelengths and amplitudes travel in opposite directions along a string with a speed of 10 cm/s. If the time interval between instants when the string is flat is 0.50 s, what is the wavelength of the waves?

31. String Fixed at Both Ends A string fixed at both ends is 8.40 m long and has a mass of 0.120 kg. It is subjected to a tension of 96.0 N and set oscillating. (a) What is the speed of the waves on the string? (b) What is the longest possible wavelength for standing wave? (c) Give the frequency of that wave.

32. Between Fixed Supports A 125 cm length of string has a mass of 2.00 g. It is stretched with a tension of 7.00 N between fixed supports. (a) What is the wave speed for this string? (b) What is the lowest resonant frequency of this string?

33. Three Lowest Frequencies What are the three lowest frequencies for standing waves on a wire 10.0 m long having a mass of 100 g, which is stretched under a tension of 250 N?

34. String A, String B String A is stretched between two clamps separated by distance L. String B, with the same linear density and under the same tension as string A, is stretched between two clamps separated by distance $4L$. Consider the first eight harmonics of string B. Which, if any, has a resonant frequency that matches a resonant frequency of string A?

35. Resonant Frequencies A string that is stretched between fixed supports separated by 75.0 cm has resonant frequencies of 420 and 315 Hz, with no intermediate resonant frequencies. What are (a) the lowest resonant frequency and (b) the wave speed?

36. Two Pulses In Fig. 17-36, two pulses travel along a string in opposite directions. The wave speed v^{wave} is 2.0 m/s and the pulses are 6.0 cm apart at $t = 0$. (a) Sketch the wave patterns when t is equal to 5.0, 10, 15, 20, and 25 ms. (b) In what form (or type) is the energy of the pulses at $t = 15$ ms?

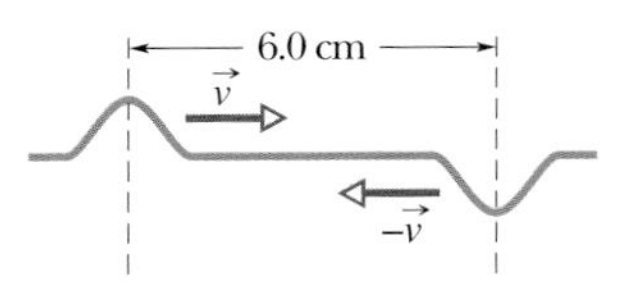

FIGURE 17-36 ■ Problem 36.

37. A String Oscillates A string oscillates according to the equation

$$y'(x,t) = (0.50 \text{ cm}) \sin\left[\left(\frac{\pi}{3} \text{ rad/cm}\right)x\right] \cos[(40\pi \text{ rad/s})t].$$

What are the (a) amplitude and (b) speed of the two waves (identical except for direction of travel) whose superposition gives this oscillation? (c) What is the distance between nodes? (d) What is the speed of a particle of the string at the position $x = 1.5$ cm when $t = \frac{9}{8}$ s?

38. Standing Wave A standing wave results from the sum of two transverse waves traveling in opposite directions given by

$$y_1 = (0.05 \text{ m}) \cos((\pi \text{ rad/m})x - (4\pi \text{ rad/s})t)$$

and

$$y_2 = (0.05 \text{ m}) \cos((\pi \text{ rad/m})x + (4\pi \text{ rad/s})t).$$

(a) What is the smallest positive value of x that corresponds to a node? (b) At what times during the interval $0 \le t \le 0.50$ s will the particle at $x = 0.00$ m have zero velocity?

39. Three-Loop Standing Wave A string 3.0 m long is oscillating as a three-loop standing wave with an amplitude of 1.0 cm. The wave speed is 100 m/s. (a) What is the frequency? (b) Write equations for two waves that, when combined, will result in this standing wave.

40. In an Experiment In an experiment on standing waves, a string 90 cm long is attached to the prong of an electrically driven tuning fork that oscillates perpendicular to the length of the string at a frequency of 60 Hz. The mass of the string is 0.044 kg. What tension must the string be under (weights are attached to the other end) if it is to oscillate in four loops?

41. Tuning Fork Oscillation of a 600 Hz tuning fork sets up standing waves in a string clamped at both ends. The wave speed for the string is 400 m/s. The standing wave has four loops and an amplitude of 2.0 mm. (a) What is the length of the string? (b) Write an equation for the displacement of the string as a function of position and time.

42. Second Harmonic A rope, under a tension of 200 N and fixed at both ends, oscillates in a second-harmonic standing wave pattern. The displacement of the rope is given by

$$y(x,t) = (0.10 \text{ m})\left(\sin\left(\left(\frac{\pi}{2} \text{ rad/m}\right)x\right) \sin((12\pi \text{ rad/s})t)\right),$$

where $x = 0.00$ m at one end of the rope, x is in meters, and t is in seconds. What are (a) the length of the rope, (b) the speed of the waves on the rope, and (c) the mass of the rope? (d) If the rope oscillates in a third-harmonic standing wave pattern, what will be the period of oscillation?

43. A Generator A generator at one end of a very long string creates a wave given by

$$y(x,t) = (6.0\text{ cm})\cos\frac{\pi}{2}[(2.0\text{ rad/m})x + (8.0\text{ rad/s})t],$$

and one at the other end creates the wave

$$y(x,t) = (6.0\text{ cm})\cos\frac{\pi}{2}[(2.0\text{ rad/m})x - (8.0\text{ rad/s})t].$$

Calculate the (a) frequency, (b) wavelength, and (c) speed of each wave. At what x values are the (d) nodes and (e) antinodes?

44. Standing Wave Pattern A standing wave pattern on a string is described by

$$y(x,t) = (0.040\text{ m})\sin((5\pi\text{ rad/m})x)\cos((40\pi\text{ rad/s})t).$$

(a) Determine the location of all nodes for $0 \leq x \leq 0.40$ m. (b) What is the period of the oscillatory motion of any (nonnode) point on the string? What are (c) the speed and (d) the amplitude of the two traveling waves that interfere to produce this wave? (e) At what times for $0.000\text{ s} \leq t \leq 0.050$ s will all the points on the string have zero transverse velocity?

45. Maximum Kinetic Energy Show that the maximum kinetic energy in each loop of a standing wave produced by two traveling waves of identical amplitudes is $2\pi^2\mu Y^2 f v^{\text{wave}}$.

46. Antinode For a certain transverse standing wave on a long string, an antinode is at $x = 0.00$ m and a node is at $x = 0.10$ m. The displacement $y(t)$ of the string particle at $x = 0.00$ m is shown in Fig. 17-37. When $t = 0.50$ s, what are the displacements of the string particles at (a) $x = 0.20$ m and (b) $x = 0.30$ m? At $x = 0.20$ m, what are the transverse velocities of the string particles at (c) $t = 0.50$ s and (d) $t = 1.0$ s? (e) Sketch the standing wave at $t = 0.50$ s for the range $x = 0.00$ m to $x = 0.40$ m.

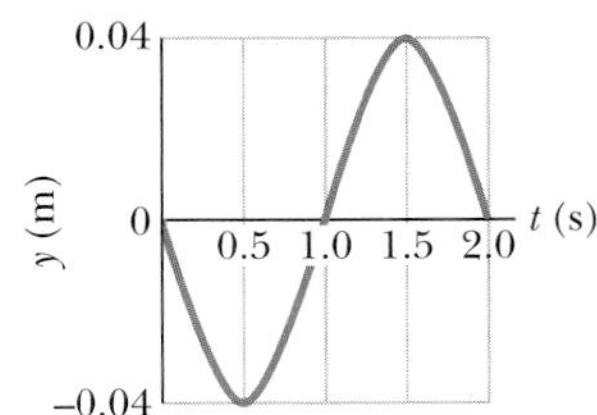

FIGURE 17-37 ■ Problem 46.

47. Aluminum Wire In Fig. 17-38, an aluminum wire, of length $L_1 = 60.0$ cm, cross-sectional area 1.00×10^{-2} cm^2, and density 2.60 g/cm^3, is joined to a steel wire, of density 7.80 g/cm^3 and the same cross-sectional area. The compound wire, loaded with a block of mass $m = 10.0$ kg, is arranged so that the distance L_2 from the joint to the supporting pulley is 86.6 cm. Transverse waves are set up in the wire by using an external source of variable frequency; a node is located at the pulley. (a) Find the lowest frequency of excitation for which standing waves are observed such that the joint in the wire is one of the nodes. (b) How many nodes are observed at this frequency?

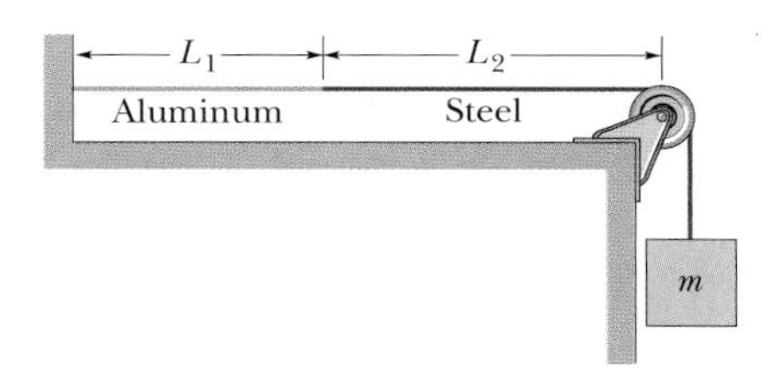

FIGURE 17-38 ■ Problem 47.

SEC. 17-12 ■ PHASORS

48. Of the Same Period Two sinusoidal waves of the same period, with amplitudes of 5.0 and 7.0 mm, travel in the same direction along a stretched string; they produce a resultant wave with an amplitude of 9.0 mm. The initial phase of the 5.0 mm wave is 0.0 rad. What is the initial phase of the 7.0 mm wave?

49. Amplitude of the Resultant Determine the amplitude of the resultant wave when two sinusoidal string waves having the same frequency and traveling in the same direction on the same string are combined, if their amplitudes are 3.0 cm and 4.0 cm and they have initial phases of 0.0 and $\pi/2$ rad, respectively.

50. Three Sinusoidal Waves Three sinusoidal waves of the same frequency travel along a string in the positive direction of an x axis. Their amplitudes are y_1, $y_1/2$, and $y_1/3$, and their initial phases are 0, $\pi/2$, and π rad, respectively. What are (a) the amplitude and (b) the phase constant of the resultant wave? (c) Plot the wave form of the resultant wave at $t = 0.00$ s, and discuss its behavior as t increases.

Additional Problems

51. Graphing a Pulse on a String Consider the motion of a pulse on a long taut string. We will choose our coordinate system so that when the string is at rest, the string lies along the x axis of the coordinate system. We will take the positive direction of the x axis to be to the right on this page and the positive direction of the y axis to be up. Ignore gravity. A pulse is started on the string moving to the right. At a time $t_1 = 0$ s a photograph of the string would look like Fig. 17-39*A*. A point on the string to the right of the pulse is marked by a spot of red paint.

For each of the items that follow, identify which figure (*B-F*) would look most like the graph of the indicated quantity. (Take the positive axis as up.) If none of the figures look like you expect the graph to look, write N. Assume that the axes would be labeled with correct units.

(a) The graph of the y displacement of the spot of red paint as a function of time.

(b) The graph of the x velocity of the spot of red paint as a function of time.

(c) The graph of the y velocity of the spot of red paint as a function of time.

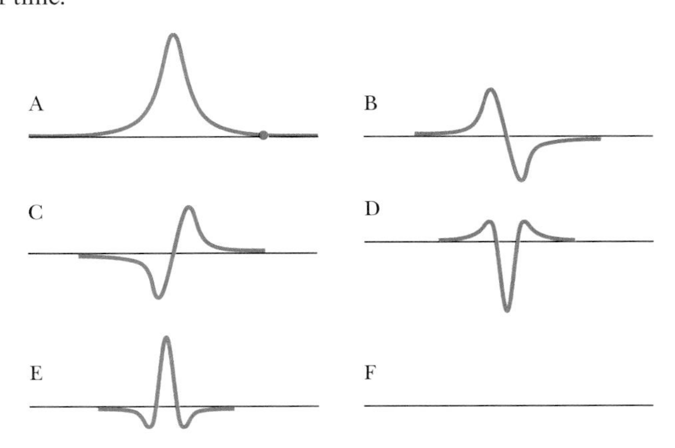

FIGURE 17-39 ■ Problem 51.

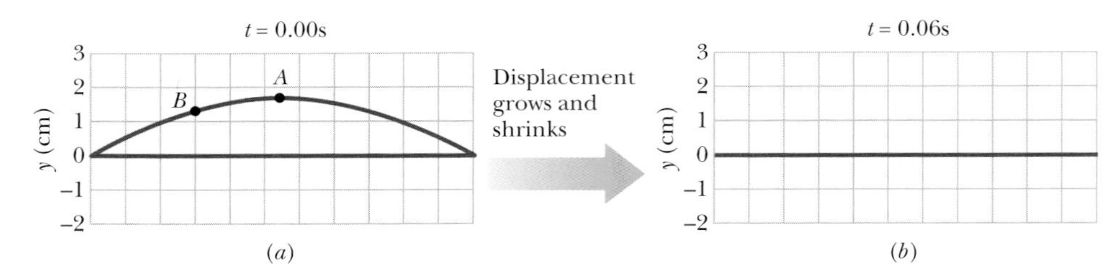

FIGURE 17-40 ■ Problem 52.

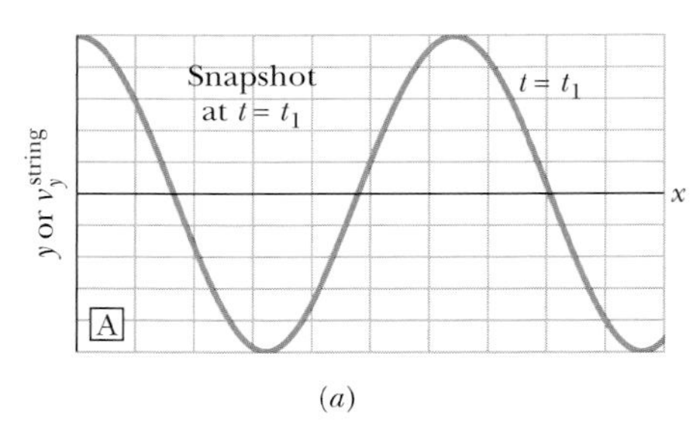

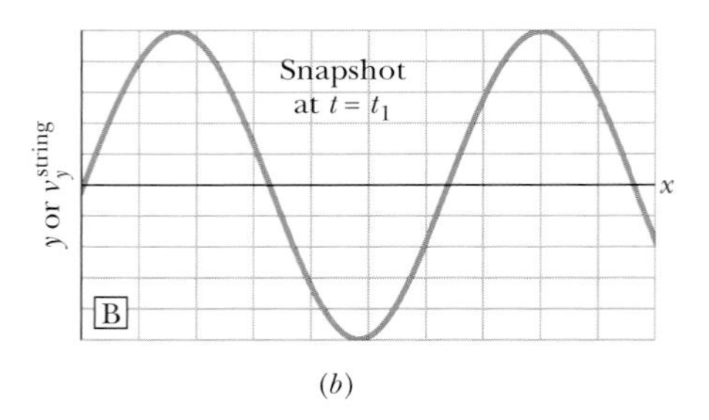

FIGURE 17-41 ■ Problem 53.

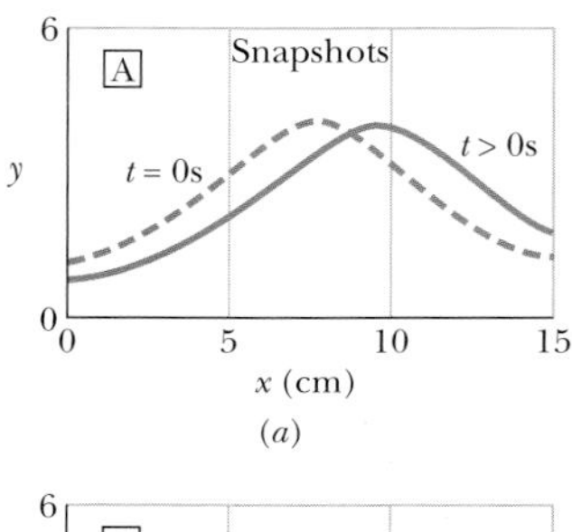

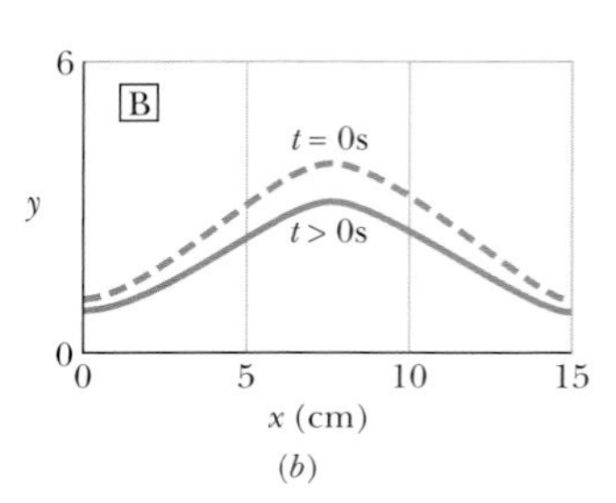

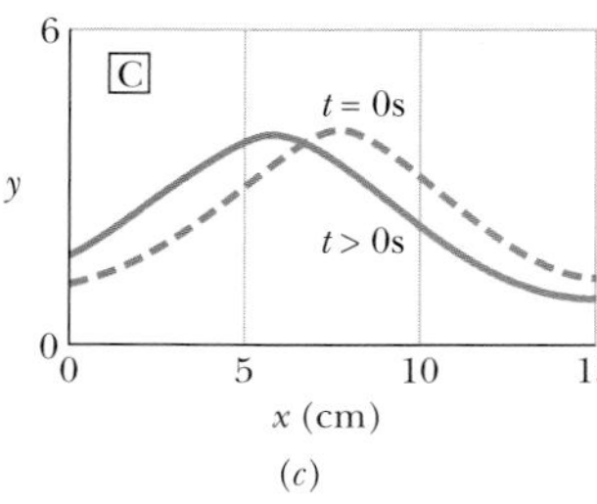

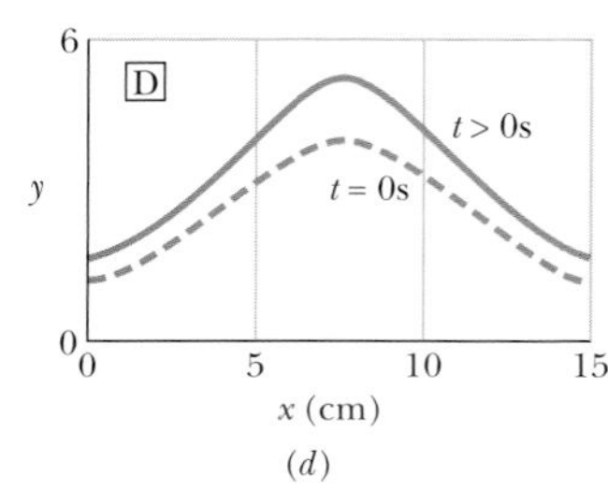

FIGURE 17-42(a) ■ Problem 55.

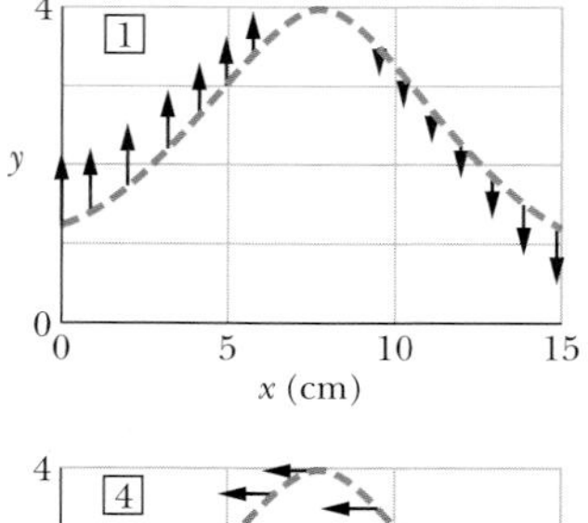

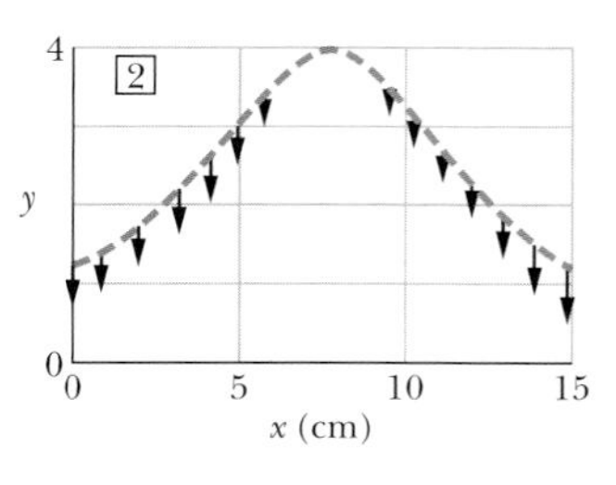

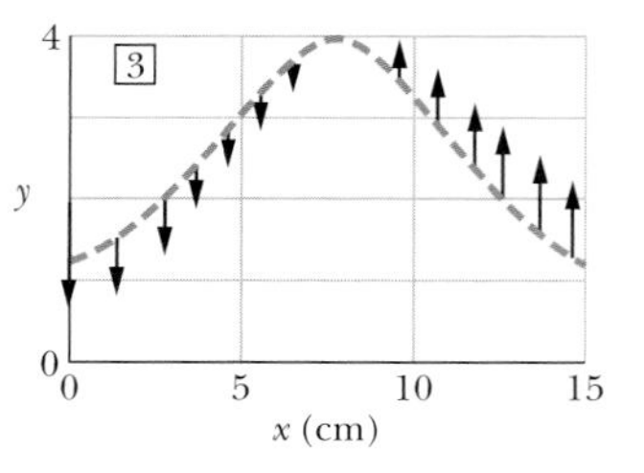

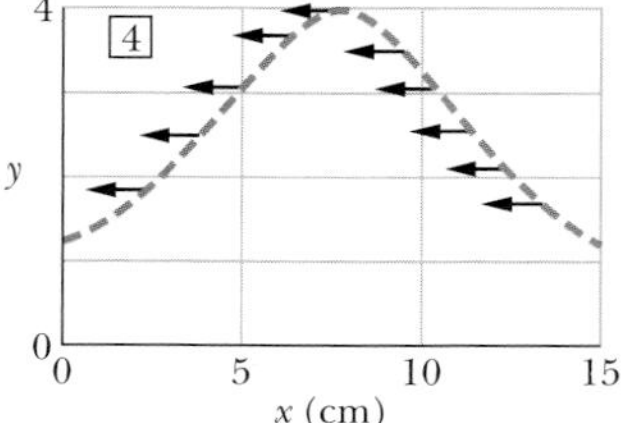

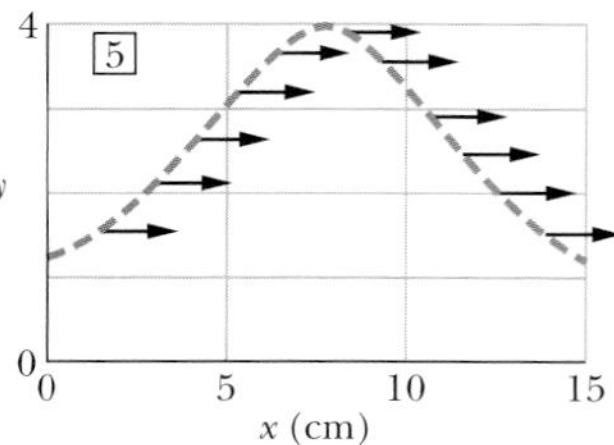

6

None of the graphs shown

FIGURE 17-42(b) ■ Problem 55.

(d) The graph of the y-component of the force on the piece of string marked by the red spot as a function of time.

52. Motion in a Standing Wave A string is connected at one end to a vibrating reed and on the other to a weight pulling the string tight over a pulley. At the time $t = 0.00$ s a photograph of the string looks like Fig. 17-40*a*. The displacement of the wave is then observed to grow, reaching a maximum of 3 cm at $t = 0.02$ s. The displacement of the wave then decreases with time and at time $t = 0.06$ s appears flat, as shown in Fig. 17-40*b*.

Each box in the grid has a side of 1 cm. (a) Two points on the string are marked with heavy black dots and with the letters A and B. At $t = 0.00$ s, in what direction is each of them moving and which one (if any) is moving faster? (b) At $t = 0.06$ s, mark the position of the two points on the string in the figure (b). In what direction is each of them moving and which one (if any) is moving faster? (c) What is the period of the oscillation? (d) Explain what could be changed to make the velocity of point A zero throughout the oscillation. Give two different ways to do this, and explain why such a change would work.

53. Displacement and Velocity Patterns in Waves Each graph shown in Fig. 17-41 may represent either a picture of the shape of a wave on a string at a particular instant in time, t_1, or the transverse (up and down) velocity of the mass points of that string at that time. Depending on which it is, the wave on the string may be:

(a) A right-moving traveling wave, (b) A left-moving traveling wave, (c) A part of a standing wave with a displacement that increases in time and is shown at time t_1, (d) A part of a standing wave with a displacement that decreases in time and is shown at time t_1. For each of the following cases, decide what the wave is doing and choose one of the four letters (a)–(d).

i. Graph A is a graph of the string's shape and graph B is a graph of the string's velocity.

ii. Graph A is a graph of both the string's shape and the string's velocity.

iii. Graph B is a graph of the string's shape and graph A is a graph of the string's velocity.

54. Interpreting an Oscillatory Equation Consider the equation

$$y(x,t) = Y\cos(at)\sin(bx)$$

Let x represent some position and t represent time. (a) Describe a physical situation represented by this equation. As part of your description include a sketch and a written description. Indicate what y and x correspond to in the situation you describe. (b) How, if at all, would the physical situation you described in part (a) be different if a were twice as large? Explain how you determined your answer. (c) How, if at all, would the physical situation you described in part (a) be different if b were twice as large? Explain how you determined your answer.

55. Waves and Velocities The four graphs labeled (A)–(D) in Fig. 17-42*a* show snapshots of waves on a long, taut string. The dashed line shows a picture of a part of the string at the time $t = 0$ s. The solid line is a picture at a time a little bit later. Each of these pictures looks the same at $t = 0$ s, but the results differ because the parts of the string have different velocities at $t = 0$ in each case.

For each of the four cases, select one of the six patterns shown in Fig. 17-42*b* as the correct velocity pattern for the motion of adjacent string elements to lead to the solid line in Fig. 17-42*a*, showing the displacement y. (Note the arrows indicate mainly direction. Their lengths are scaled somewhat in proportion to their magnitude, but not strictly so.)

56. Making a Pulse Move A transverse wave pulse is traveling in the $+x$ direction along a long stretched string. The origin is taken at a point on the string that is far from the ends. The speed of the wave is v (a positive number). At time $t = 0$ s, the displacement of the string is described by the function

$$y = f(x,0) = \begin{cases} A\left(1 - \left|\frac{x}{\ell}\right|\right) & |x| \le \ell \\ 0 & |x| \ge \ell \end{cases}$$

(a) Construct a graph that portrays the actual shape of the string at $t = 0$ s for the case $A = \ell/2$. (b) Sketch a graph of the wave form at the following times: 1. $t = \ell/v$, 2. $t = 2\ell/v$, 3. $t = -3\ell/2v$.

57. Comparing Waves Figure 17-43 shows a snapshot of a piece of a wave at a time $t =$ 0 s. Make four sketches of this picture and use a dotted line to sketch what each pulse would look like at a slightly later time (a time that is small compared to the time it would take the pulse to move a distance equal to its own width but large enough to see a change in the shape of the string) for the following four cases:

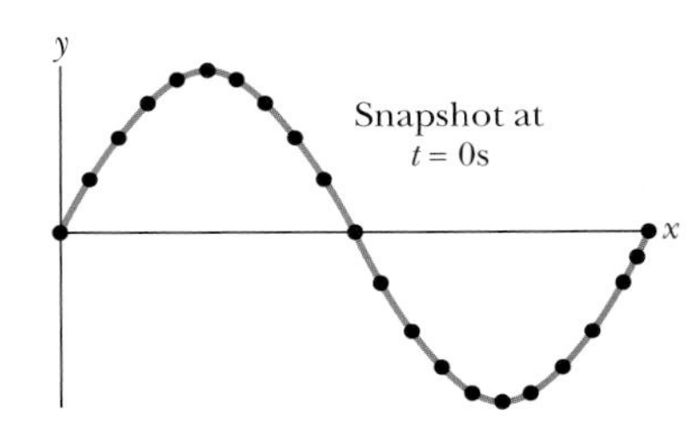

FIGURE 17-43 ■ Problem 57.

A. The pulse is a traveling wave moving to the right.
B. The pulse is a traveling wave moving to the left.
C. The pulse is a standing wave with a displacement that increases in time.
D. The pulse is a standing wave with a displacement that decreases in time.

On each picture, draw arrows to show the velocity of the marked points at time $t = 0$ s.

58. Combining Pulses Figure 17-44 shows graphs that could represent properties of pulses on a stretched string. For the situation and the properties (a)–(e), select which graph provides the best representation of the given property. If none of the graphs are correct, write "none."

Two pulses are started on a stretched string. At time $t = t_1$, an upward pulse is started on the right that moves to the left. At the same instant, a downward pulse is started on the left that moves to the right. At t_1 their peaks are separated by a distance $2s$. The distance between the pulses is much larger than their individual widths. The pulses move on the string with a speed v_1. The scales in the graphs are arbitrary and not necessarily the same. (a) Which graph best represents the appearance of the string at time t_1? (b) Which graph best represents the appearance of the string at a time $t_1 + s/v_1$? (c) Which graph best represents the appearance of the string at a time $t_1 + 2s/v_1$? (d) Which graph best represents the velocity of the string at a time $t_1 + s/v_1$? (e) Which graph best represents the appearance of the string at a time $t_1 + s/v_1 + \varepsilon$ where ε is small compared to s/v_1?

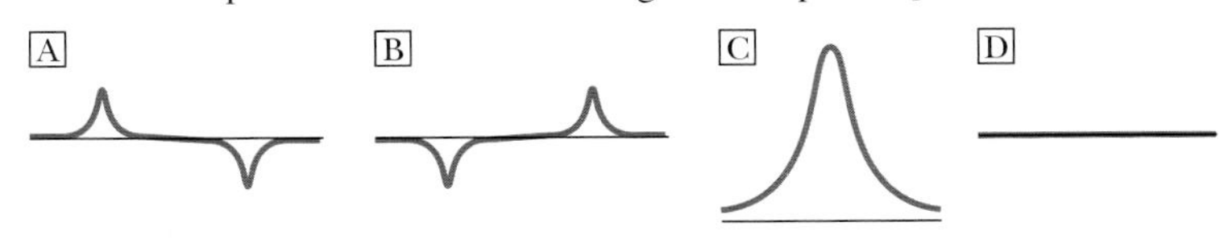

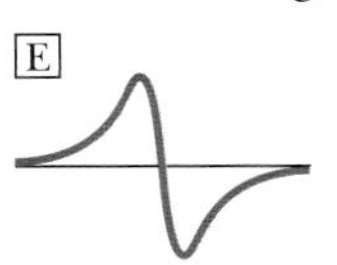

FIGURE 17-44 ■ Problem 58.

59. Moving a Nonsymmetric Triangular Pulse A long taut spring is started at a time $t = 0$ with a pulse moving in the $+x$ direction in the shape given by the function $f(x)$ with

$$f(x) = \begin{cases} \frac{1}{4}x + 1 & -4 < x < 0 \\ -x + 1 & 0 < x < 1 \\ 0 & \text{otherwise} \end{cases}$$

(The units of x and f are in centimeters.) (a) Draw a labeled graph showing the shape of the string at $t = 0$. (b) If the mass density of the string is 50 g/m and it is under a tension of 5 N, draw a labeled graph that shows the shape and position of the string at a time $t =$ 0.001 s. (c) Write the solution of the wave equation that explicitly gives the displacement of any piece of the string at any time. (d) What is the speed of a piece of string that is moving up after the pulse has reached it but before it has risen to its maximum displacement? What is its speed while it is returning to its original position after the pulse's peak has passed it?

60. Which Wave Is Which? Figure 17-45*a* shows a picture of a string at a time t_1. The pieces of the string are each moving with velocities that are indicated by arrows in the picture. (The vertical displacements are small and don't show up in the picture.) Figure 17-45*b* shows five graphs that could give the shape of the string at the instant for which the velocities are displayed above. (*Note*: The vertical scale magnifies the displacement by a factor of 100.)

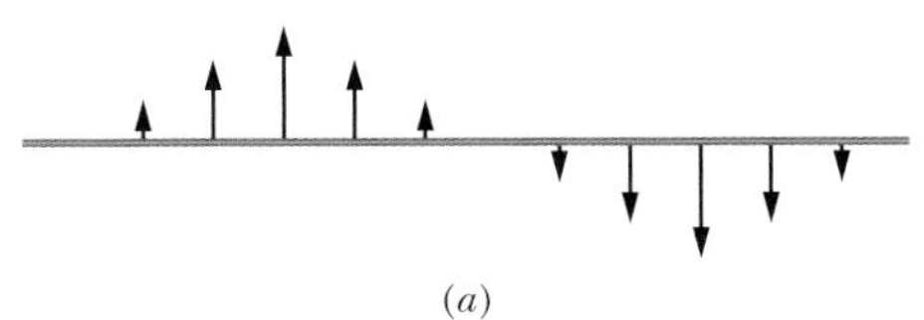

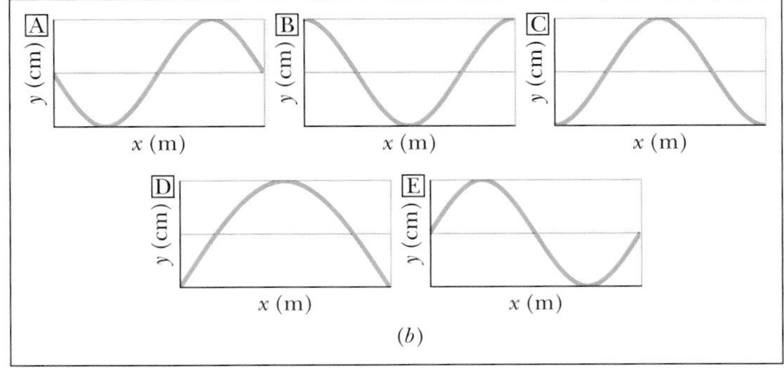

FIGURE 17-45 ■ Problem 60.

On your paper, place the letters A–E. Next to these letters, indicate for the graphs labeled by those letters, whether the string is moving as a: (L) left-traveling wave, (R) right-traveling wave, (S+) standing wave with a displacement that increases in time, (S−) standing wave with a displacement that decreases in time, or (N) none of the above.

61. Spring vs. String In this course, we analyzed the motion of a mass on a spring and the oscillations of a taut string. Discuss these two systems, explaining similarities and differences, and give an equation of motion for each.

62. Parsing a Pulse An instructor is demonstrating the motion of waves on a long, taut spring. He is holding the spring at one end and will move it so the spring will move back and forth on the floor. The spring is rigidly connected to a metal rod at its other end. The spring is under a tension T and it has a mass density μ. The instructor starts a pulse moving toward the right as shown in Fig. 17-46. The pulse is triangular and is *not* symmetric. The figure is shown at a time t_1. (a) Calculate the time τ it will take the peak of the pulse to reach the wall (to travel a distance s). (b) What will the spring look like at the time $t_1 + \tau$? Draw a carefully constructed and labeled diagram to show what it looks like and how you got your result. (c) What will the spring look like a bit later—say, at a time $t_1 + 2\tau$? What is responsible for this result? (d) The width of the pulse is 0.5 m. If the tension in the spring is 5 N and it has a mass density of 0.1 kg/m, how much time did the professor take to generate the pulse?

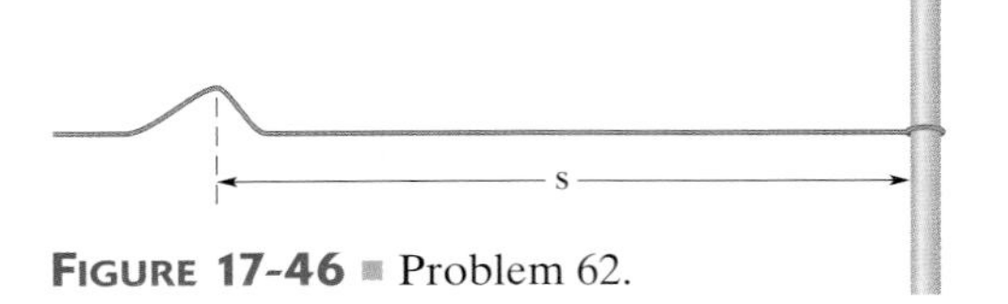

FIGURE 17-46 ■ Problem 62.

63. Modified Harmonics A taut string is tied down at both ends. Assume that the fundamental mode of oscillation of the string has a period T_0. For each of the changes described below give the factor by which the period changes. For example, if the change described resulted in a period twice as long, you would put the number "2." Do not accumulate changes. That is, before each change, assume you are back at the original starting situation. (a) The string is replaced by one of twice the mass but of the same length. (b) The wave length of the starting shape is divided by three. (c) The amplitude of the oscillation is doubled. (d) The tension of the string is halved.

64. Spring vs. String: Graphs Consider two physical systems: System A is a mass hanging from a light spring fixed at one end to a ringstand on a table above the floor. System B is a long spring of uniform density held under tension and able to move transversely in a horizontal plane.

Figure 17-47 shows three graphs labeled #1, #2, and #3, with unmarked axes. They could represent many quantities in the physical systems A and B. For the five physical quantities below, indicate which of the graphs could be obtained for the system indicated. If more than one graph applies, give *all* the possible choices. If none applies, write N.

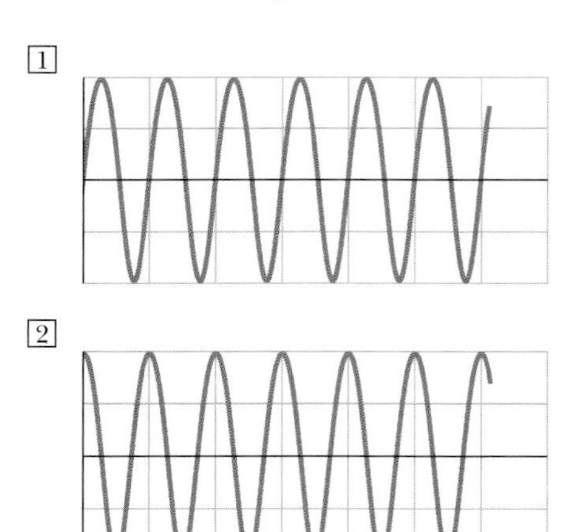

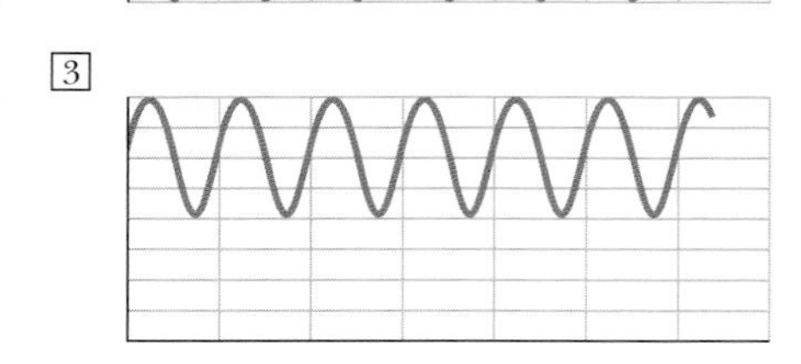

FIGURE 17-47 ■ Problem 64.

(a) The height of the mass hanging from the spring (system A) above the ground as a function of time for some time interval after the mass has been set into oscillation
(b) The transverse displacement from equilibrium of some portion of the long spring (system B) as a function of position at a particular instant of time while it is carrying a harmonic wave
(c) The velocity of the mass hanging from the spring (system A) as a function of time given that its displacement from equilibrium as a function of time is given by graph #1
(d) The transverse velocity of a small piece of the long spring (system B) as a function of time as a single pulse moves down the spring
(e) The velocity of a small piece of the long spring (system B) as a function of time given that its displacement from equilibrium as a function of time is given by graph #2

65. Explaining the Wave Equation The wave equation

$$\frac{\partial^2 y}{\partial x^2} = \frac{1}{v_1^2}\frac{\partial^2 y}{\partial t^2}$$

is often used to describe the transverse displacement of waves on a stretched spring. Explain the meaning of each of the elements of this equation with reference to the physical spring and discuss under what circumstances you expect it to be a good description.

66. Propagating a Gaussian Pulse Figure 17-48 shows the shape of a pulse on a stretched spring at the time $t = 0$ s. The dispacement of the spring from its equilibrium position at that time is given by

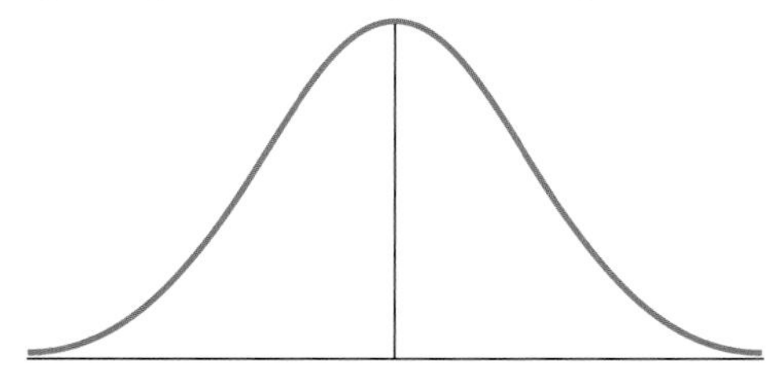
FIGURE 17-48 ■ Problem 66.

$$F(x) = Ae^{-(x/b)^2}$$

The pulse is moving in the positive x direction with a velocity v_1. (a) Sketch a graph showing the shape of the spring at a later time, $t = t_2$. Specify the height and position of the peak in terms of the symbols given. (b) Write an equation for the displacement of any portion of the spring at any time $y(x,t)$. (c) Sketch a graph of the velocity of the piece of the spring at the position $x = 2b$ as a function of time.

67. Varying a Pulse* A long, taut string is attached to a distant wall as shown in Fig. 17-49. A demonstrator moves her hand and creates a very small amplitude pulse that reaches the wall in a time t_1. A small red dot is painted on the string halfway between the demonstrator's hand and the wall. For each situation below, state which of the actions 1–10 listed (taken by itself) will produce the desired result. For each question, *more than one answer may be correct*. If so, give them all.

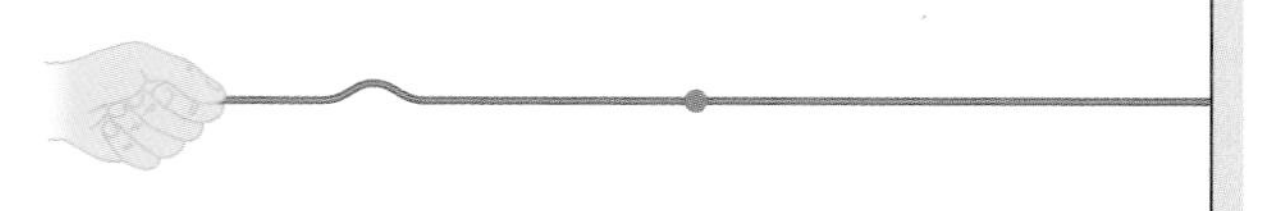
FIGURE 17-49 ■ Problem 67.

*From the *Wave Test* by M. Wittmann.

Tell how, if at all, the demonstrator can repeat the original experiment to produce: (a) A pulse that takes a longer time to reach the wall (b) A pulse that is wider than the original pulse (c) A pulse that makes the red dot travel a further distance than in the original experiment

1. Move her hand more quickly (but still only up and down once and still by the same amount)
2. Move her hand more slowly (but still only up and down once and still by the same amount)
3. Move her hand a larger distance but up and down in the same amount of time
4. Move her hand a smaller distance but up and down in the same amount of time
5. Use a heavier string of the same length under the same tension
6. Use a lighter string of the same length under the same tension
7. Use a string of the same density, but decrease the tension
8. Use a string of the same density, but increase the tension
9. Put more force into the wave
10. Put less force into the wave

68. Reflecting on a Textbook Error Figure 17-50 is taken from the first edition of a popular standard textbook. Column (a) shows a pulse approaching and reflecting from a fixed end. Column (b) a pulse approaching and reflecting from a free end (ring sliding on a frictionless rod). At least six (6!) of the figures are incorrect. State which and explain why. (*Hint*: The second edition of the book fixed most of the problems by making the pulses symmetric.)

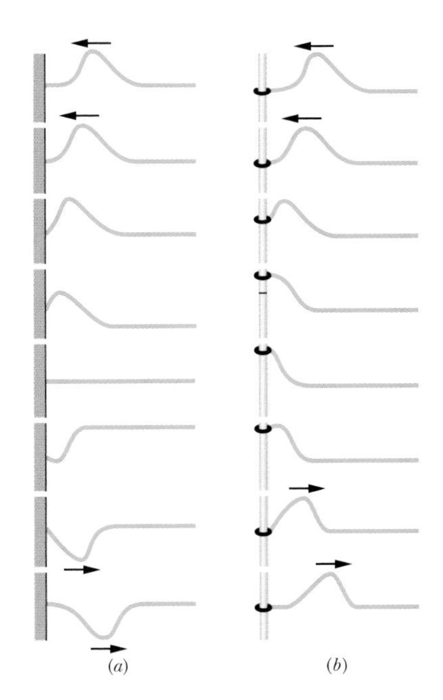

FIGURE 17-50 ▪ Problem 68.

69. Graphing Another Pulse on a String Figure 17-51*a* shows a photograph of a pulse on a taut string moving to the right. The red dot the right of the figure is a small bead of negligible mass attached to the string.

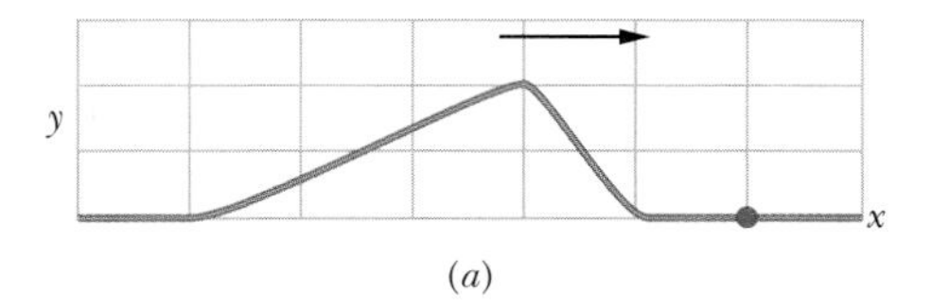

FIGURE 17-51(*a*) ▪ Problem 69.

For each of the following quantities, select the letter of the graph in Fig. 17-51*b* that could provide a correct graph of the quantity (if the vertical axis were assigned the proper units) and write it on your answer sheet. If none of the graphs could work, write N.

(a) The vertical (up-down) displacement of the bead
(b) The vertical velocity of the bead
(c) The horizontal (left-right) displacement of the bead
(d) The horizontal velocity of the bead

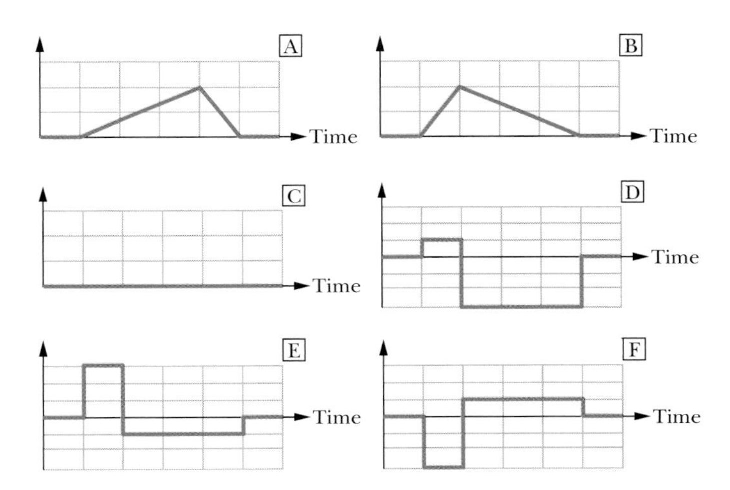

FIGURE 17-51(*b*) ▪ Problem 69.

18 Sound Waves

This horseshoe bat not only can locate a moth flying in total darkness, but can also determine the moth's relative speed, to home in on the insect.

How does the bat's detection system work, and how can a moth "jam" the system or otherwise reduce its effectiveness?

The answer is in this chapter.

18-1 Sound Waves

What Is a Sound Wave?

As we saw in Chapter 17, there are two types of mechanical waves that require a material medium to exist: *transverse waves* that involve displacements of a medium perpendicular to the direction in which the wave travels, and *longitudinal waves* that have displacements of a medium parallel to the direction of wave travel. Examples of longitudinal waves include the expansion/compression waves in the Slinky described in the last chapter (Fig. 17-1) and the waves in an air-filled pipe shown in Fig. 18-1.

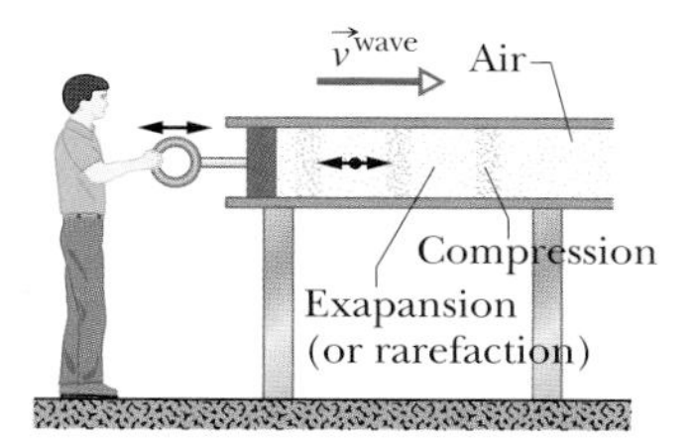

FIGURE 18-1 ■ A sound wave is set up in an air-filled pipe by moving a piston back and forth. Because the oscillations of a segment of the air (represented by the black dot) are parallel to the direction in which the wave travels, the wave is a *longitudinal wave*.

When we hear sounds, we are detecting longitudinal waves passing through air that have frequencies within the range of human hearing. A **sound wave** can be defined more broadly as a longitudinal wave of any frequency passing through a medium. The medium can be a solid, liquid, or gas. For example, geological prospecting teams use sound waves to probe the Earth's crust for oil. Ships carry sound-emitting gear (sonar) to detect underwater obstacles. Medical personnel use high-frequency sound waves (ultrasound) to create computer-processed images of soft tissues (as shown in Fig. 18-2). Physics students use ultrasound pulses to track the motion of objects in the laboratory.

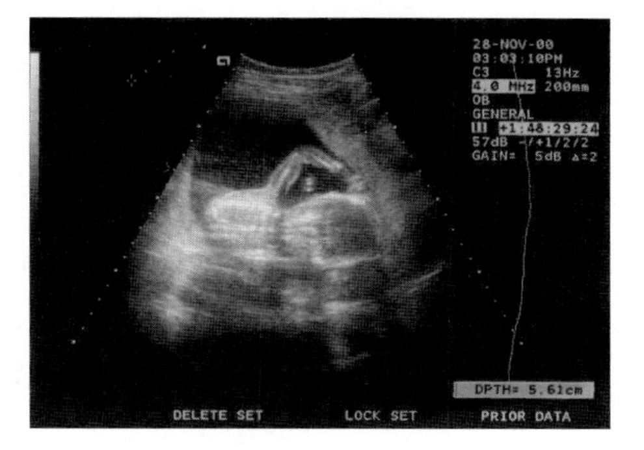

FIGURE 18-2 ■ An image of a fetus flexing an arm. This image is made with ultrasound waves that have a frequency of 4 MHz—two hundred times higher than the threshold of human hearing.

In this chapter we focus on the characteristics of audible sound waves that travel through air. We start by considering some differences in how waves propagate in one-, two-, and three-dimensional spaces.

Wave Dimensions

Unlike the wave pulses that travel in a straight line along a string, a sound wave from a small "point-like" source is usually not constrained to travel in only one direction. The same is true for the electromagnetic waves that we will study in Chapters 34–38. In order to understand how sound waves and electromagnetic waves travel in more than one dimension, we need to consider how the dimensionality of the space through which a wave propagates affects it.

In the previous chapter we observed that the crest of a wave passing along a one-dimensional string moves along a line. If we constrain a sound wave to travel in

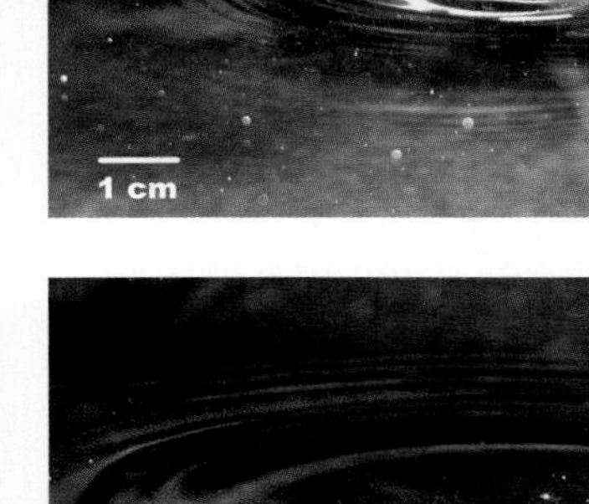

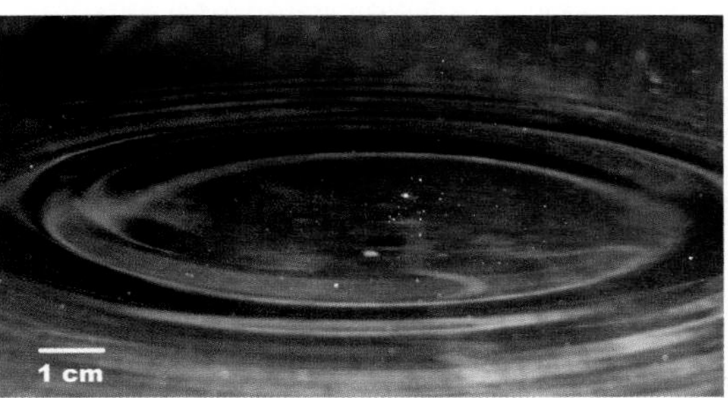

FIGURE 18-3 ■ Water on the surface of a tank of water is disturbed as a steel ball of diameter 0.79 cm falls into the water. The wave crest propagates outward from the source of the disturbance in a widening circle. The wave crest moves at a constant speed of approximately 30 cm/s.

a long tube, its compression wave crests would lie in planes perpendicular to the axis of the tube as shown in Fig. 18-1. Such a one-dimensional wave is known as a **plane wave.** For example, the relatively narrow beam of sound waves that are emitted from an ultrasonic motion detector described in Section 1-8 behave like one-dimensional plane waves.

On the other hand, if you put a small sound source between two large sheets of plywood, the sound wave crests will propagate outward in expanding circles. Although we cannot see sound waves as they propagate in two dimensions, we have all seen two-dimensional surface waves on water. For example, when a pebble is dropped onto the surface of a pond, a wave crest propagates out from the pebble in an expanding circle as shown in Fig. 18-3. This two-dimensional wave is known as a **circular wave.**

If you snap your fingers or ring a tiny sleigh bell, a compression wave is created. At distances that are large relative to the size of the source, compression wave crests travel out in three dimensions as expanding spheres (Fig. 18-4). A three-dimensional wave is defined as a **spherical wave,** provided there is no preferred direction for the propagation of the wave energy.

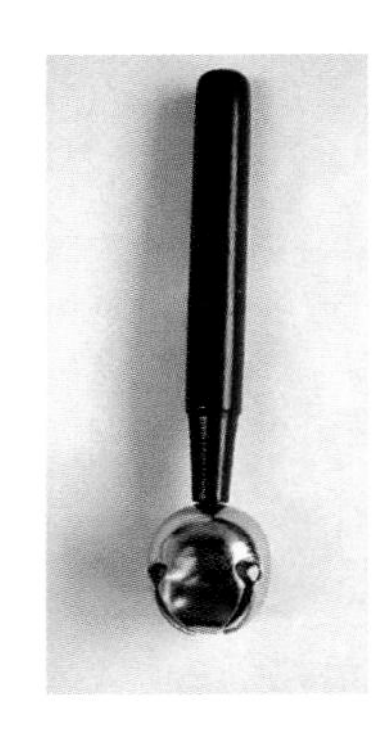

(*a*)

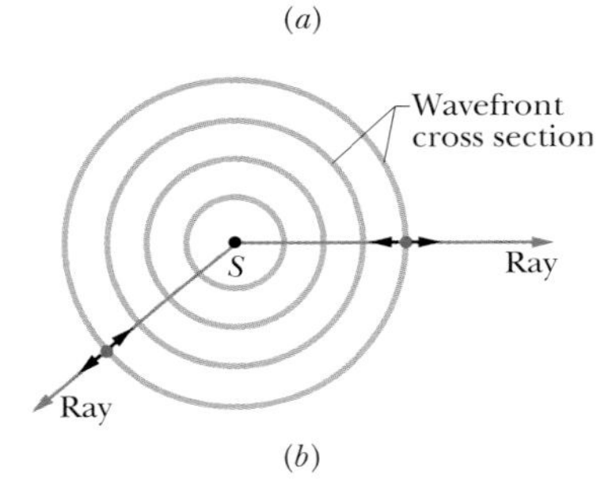

(*b*)

FIGURE 18-4 ■ (*a*) A small sleigh bell that is rung at a steady rate. (*b*) If the distances from the bell to its listeners are large compared to the size of the bell, then the bell acts as a three-dimensional point source of sound. Each compression wave crest moves out in an expanding sphere. The two-dimensional cross-section of each of four wave crest spheres is shown. The short double arrows signify that the air particles oscillate parallel to the direction of motion of the wave crests. The lines drawn perpendicular to the wave crests are *rays.*

It is important to understand that the dimensionality of a sound source (such as a one-dimensional wave in a guitar string) and the dimensionality of a sound wave produced by the source are not necessarily the same. Sound waves often travel through a medium with uniform density such as air or water. In this case, we can associate the dimensionality of a wave with the curvature of its wave crest as it spreads, rather than with the dimensionality of the source. In general, we can define the dimensionality of a wave in terms of any point on a propagating waveform. This is because for a wave of a given dimension, the *shape* of the wave crests (maxima) and wave troughs (minima) and points of no deflection (nodes) are the same. For example, in a two-dimensional wave the crests, troughs, and points of zero deflection are all circular or at least circular arcs.

It is useful for us to define a **wavefront** as the collection of all adjacent points on an expanding wave that have the same phase. For example, a wavefront can be the collection of all adjacent crest points. Or it can be taken to be a collection of all adjacent nodes, or all adjacent troughs. **Rays** are defined as lines directed perpendicular to the wavefronts that indicate the direction of travel of the wavefronts. The short double arrows superimposed on the rays of Fig. 18-4 indicate that the longitudinal oscillations of the air, which transmits the wave, are parallel to the rays.

A one-dimensional or plane wave is defined as any wave that has a wavefront that lies in a plane. The wavefront of a two-dimensional or circular wave lies along an expanding circle. The wavefront of a three-dimensional or spherical wave lies along an expanding sphere. These definitions are often still useful in situations where we do not have ideal point sources. For instance, a flat speaker mounted in the door of a car only emits waves in a "forward" direction. In this case we would have half-spherical three-dimensional wavefronts.

As the wavefronts move outward and their radii become larger, their curvatures decrease. Far from the source, these spheres associated with a three-dimensional wave are so large that we lose track of their curved nature all together. For example, when you listen to a singer in a concert hall, your ear is so far away from her that you cannot detect any curvature in the wavefronts she sends out. In such cases we can treat the *local* portion of a wavefront as if it lies in a plane. Thus, there are times in our study of sound when we will treat sound waves as if they are one-dimensional plane waves. Similarly, in Chapter 36 when we study how light waves interfere when passing through slits and traveling parallel to a two-dimensional surface, we can treat light waves as if they were two-dimensional rather than three-dimensional.

At this point let us return to our primary task in this chapter, which is to learn more about the nature of sound waves.

18-2 The Speed of Sound

The ability to calculate the speed of a sound wave as a function of the properties of the medium through which it travels is of great practical importance. For example, we can use our knowledge of the speed of sound in air to estimate how far away a lightning bolt is. Since sound travels approximately 0.2 mi/s in air, it will take about 5 s to travel one mile. Thus a 5 s interval between a lightning flash and a thunderclap tells us that the electrical storm is about a mile away. The ultrasonic motion detectors used in many physics laboratories bounce sound pulses off objects in their surroundings. The time of travel of the sound pulses emitted and then reflected from the motion detector is used to measure the distance between it and the reflecting object.

What properties of a medium does the speed of sound depend on? Let's draw an analogy between the speed of sound and the speed of a wave traveling along a string (Section 17-6). We found that the wave speed increases with the tension, T, in the string. However, the tension determines the magnitude of the restoring force that brings a displaced section of string back toward its equilibrium position. We also found, as expected, that the wave speed decreases with the mass of each disturbed length of the string. This makes sense since the linear mass density of the string, μ, is an inertial property that determines how rapidly, or slowly, the string can respond to the restoring forces acting on it. Thus, we can generalize (Eq. 17-25), which we derived for the wave speed along a stretched string, to

$$v^{\text{wave}} = |\vec{v}^{\text{wave}}| = \sqrt{\frac{F^{\text{tension}}}{\mu}} = \sqrt{\frac{\text{restoring property}}{\text{inertial property}}}. \qquad \text{(Eq. 17-25)}$$

If the medium is a fluid (such as air or water) and the wave is longitudinal, we can guess that the inertial property, corresponding to the linear density of the string μ, is the volume density ρ of the fluid. What shall we define as its restoring property?

As a sound wave passes through a fluid, elements of mass in the fluid undergo compressions and expansions due to pressure differences within the fluid. When a mass element in the fluid is compressed it has a higher pressure and pushes on adjacent fluid. This new mass element of fluid becomes compressed into a smaller volume. Then it pushes in turn on another mass element, and so on. This is how the compression wave travels through a fluid.

As shown in Fig. 18-5, the restoring property of the fluid is determined by the extent to which an element of mass changes volume when it experiences a difference in pressure (force per unit area). As a compression wave travels, the density of each mass element increases temporarily. When a fluid mass element becomes more dense its pressure rises, causing a difference in pressure between it and the mass element down the line. Just as the factor k tells us how stiff a spring is, we would like to define a factor B that tells us how "stiff" a three-dimensional medium is. The ratio of the pressure difference to the relative volume change is a property of the fluid we will call the **bulk modulus,** and define as

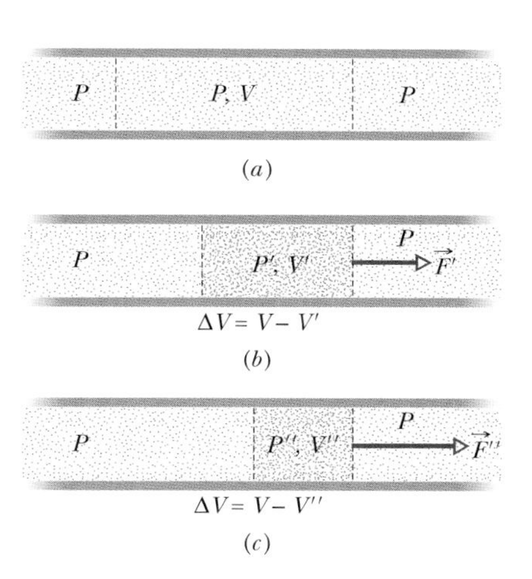

FIGURE 18-5 ■ (a) An undisturbed tube of fluid at pressure P. (b) A compression wave passing through a fluid with a small bulk modulus causes a mass element to be compressed to a smaller volume, V'. (c) The same type of compression wave with a larger bulk modulus. It causes a mass element to undergo a greater volume reduction.

$$B \equiv -\frac{\Delta P}{\Delta V/V} \quad \text{(definition of bulk modulus)}, \qquad (18\text{-}1)$$

where $\Delta V/V$ is the fractional change in volume produced by a change in pressure ΔP. [As explained in Section 15-3, the SI unit for pressure is the newton per square meter, which is given a special name, the *pascal* (Pa).] The minus sign signifies that a rise in pressure on a medium causes its volume to decrease.

For the pressure changes associated with small amplitude sound waves, the bulk modulus in a given material is approximately constant and is a restoring factor just as tension is in a string or k is in a spring. So we can substitute B for F^{tension} and ρ for μ in Eq. 17-25 ($v^{\text{wave}} = \sqrt{F^{\text{tension}}/\mu}$) to get

$$v^{\text{wave}} = |\vec{v}^{\text{wave}}| = \sqrt{\frac{B}{\rho}} \quad \text{(speed of sound in a fluid).} \tag{18-2}$$

Notice that as is the case for waves on a string, we expect the wave velocity to be independent of frequency and amplitude.

It can be easily shown that the dimensions of $\sqrt{B/\rho}$ are those of a velocity. It is possible to derive Eq. 18-2 mathematically using methods similar to those used in Section 17-6 to find the expression for the wave speed along a stretched string. Once again experimental results confirm the validity of Eq. 18-2.

Table 18-1 lists the speed of sound in various media.

TABLE 18-1
The Speed of Sound[a]

Medium	Speed (m/s)	Density (kg/m^3)	Bulk Modulus (N/m^2)
Gases			
Air (0°C)	331		
Air (20°C)	343	1.21	1.4×10^5
Helium	965		
Hydrogen	1284		
Liquids			
Water (0°C)	1402		
Water (20°C)	1482	990	2.2×10^9
Seawater[b]	1522		
Solids			
Aluminum	6420		
Steel	5941	7850	1.6×10^{11}
Granite	6000		

[a]At 0°C and 1 atm pressure, except where noted.
[b]At 20°C and 3.5% salinity.

Note that the density of water is almost 1000 times greater than the density of air. If this were the only relevant factor, we would expect from Eq. 18-2 that the speed of sound in water would be considerably less than the speed of sound in air. However, Table 18-1 shows us that the reverse is true. We conclude (again from Eq. 18-2) that the bulk modulus of water must be more than 1000 times greater than that of air. This is indeed the case. Water is less compressible than air, which (see Eq. 18-1) is another way of saying that its bulk modulus is much greater.

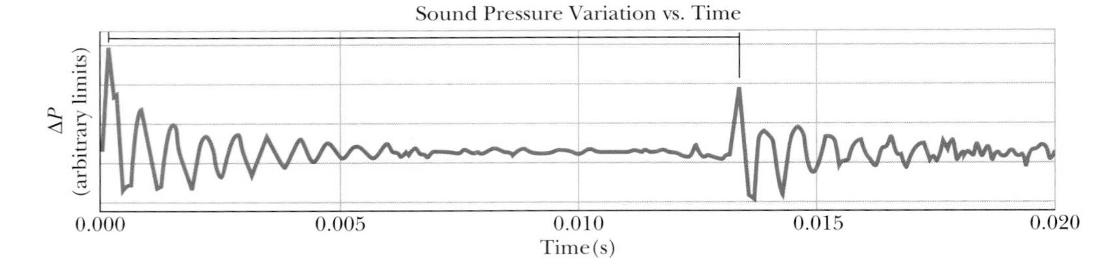

FIGURE 18-6 ■ Measurement of the speed of sound. The sound pressure variations from a finger snap travel down a 2.4 m tube and back. A microphone sensor connected to a computer data acquisition system is used to measure the time between the initial sound pulses and the reflected pulses.

A graph that shows the outcome of a simple measurement of the speed of sound in air is shown in Fig. 18-6. Analysis of the data shows a speed, which is temperature-dependent, of about 345 m/s.

More About Traveling Sound Waves in Air

If the hanging Slinky discussed in Chapter 17 is pushed back and forth at one end with a continuous sinusoidal motion, we can set up a series of compressions and rarefactions like those shown in Fig. 18-7*a*. We can do the same thing when pushing a piston back and forth in a column of air as shown in Fig. 18-1. Figure 18-7*b* displays such a wave traveling rightward through a long air-filled tube. The piston's rightward motion compresses the air next to it; the piston's leftward motion allows the element of air to move back to the left and the pressure to decrease. As each element of air pushes on the next element in turn, the right-left motion of the air and the change in its pressure is passed to the next bit of air along the tube.

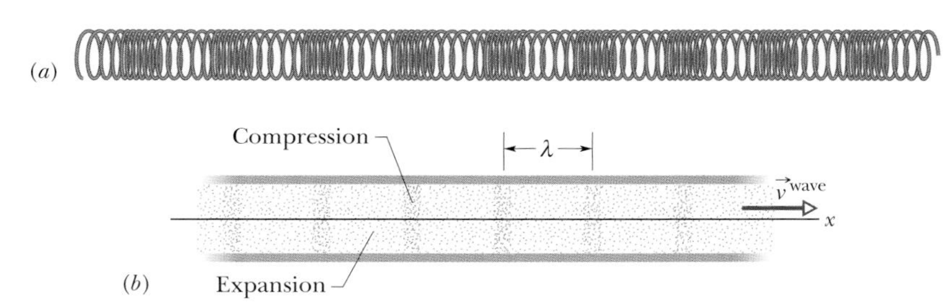

FIGURE 18-7 ■ (*a*) A continuous longitudinal wave traveling along a hanging Slinky. (*b*) A sound wave, traveling through a long air-filled tube with speed $\vec{v}^{\,\text{wave}}$, consists of a moving, periodic pattern of expansions and compressions of the air. The wave is shown at an arbitrary instant. As the wave passes, a fluid element of thickness Δx (not shown) oscillates left and right in simple harmonic motion about its equilibrium position.

The alternating compressions and rarefactions (reductions in pressure) propagate as a sound wave. As a reminder of what we learned in Chapter 17, note that it is the oscillations of pressure that propagate, not the air molecules. However, the air molecules (or molecules of another fluid) are disturbed and oscillate back and forth about their initial positions.

As the wave moves, the air pressure changes at any position x in Fig. 18-7*b* in a sinusoidal fashion, like the displacement of a string element in a transverse wave, however the pressure variations are not transverse, they are longitudinal. To describe this pressure change from the local atmospheric pressure as a function of position and time for purely sinusoidal oscillations, consider Fig. 18-8. We can model our equation on Eq. 17-4, which describes the propagation of a sinusoidal transverse wave along a string. However, instead of the vertical displacement of a bit of string, we are interested in how pressure varies from an equilibrium value. Modifying Eq. 17-4 gives us

$$\Delta P(x, t) = \Delta P^{\max} \sin[(kx \pm \omega t) + \phi_0], \tag{18-3}$$

where $\Delta P^{\max}$ is called the **pressure amplitude,** which is the maximum change from local atmospheric pressure due to the wave. It turns out that for typical sound waves that

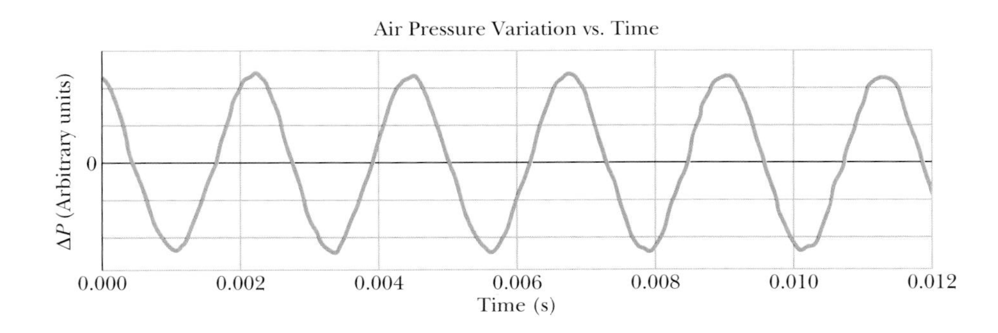

FIGURE 18-8 ■ A computer data acquisition system is used to measure air pressure variations from equilibrium using a small microphone. The sampling rate was 25 000 data points/second. This pressure variation was caused by a speaker cone that oscillated sinusoidally.

we hear, ΔP^{max} is extremely small compared to the equilibrium pressure P^{atm} present when no wave is present.

The wave number k, angular frequency ω, frequency f, wavelength λ, wave speed v^{wave}, period T, and time- and space-dependent phase $\phi(x, t)$ for a longitudinal sound wave are defined and interrelated exactly as for a transverse wave. However, note that the wavelength λ is now the distance (again along the direction of travel) in which the pattern of compression and expansion due to the wave begins to repeat itself (see Fig. 18-7*b*). Also note that Eq. 17-13 still holds so that the wave speed is given by $v^{wave} = \lambda f = \omega/k$.

A negative value of ΔP in Fig. 18-8 and Eq. 18-3 corresponds to an expansion of a packet of air, and a positive value corresponds to a compression.

READING EXERCISE 18-1: Verify that Eq. 18-2 is dimensionally correct. In other words, show that the term $\sqrt{B/\rho}$ has the dimensions of velocity in SI units. ■

READING EXERCISE 18-2: Examine Fig. 18-6. The maximum of the finger snap pulse set is at 0.0002 s and the maximum of the reflected pulse is at 0.0133 s. Use these times to calculate the measured value of the wave speed for the set of sound pulses as they travel back and forth through the air inside the 2.26-m-long tube. Is your calculated speed approximately the same as the speed of sound of air at room temperature as reported in Table 18-1? *Note:* The air temperature in the room was not recorded. ■

TOUCHSTONE EXAMPLE 18-1: Sound Arrival Delay

One clue used by your brain to determine the direction of a source of sound is the time delay Δt between the arrival of the sound at the ear closer to the source and the arrival at the farther ear. Assume that the source is distant so that a wavefront from it is approximately planar when it reaches you, and let D represent the separation between your ears.

(a) Find an expression that gives Δt in terms of D and the angle θ between the direction of the source and the forward direction.

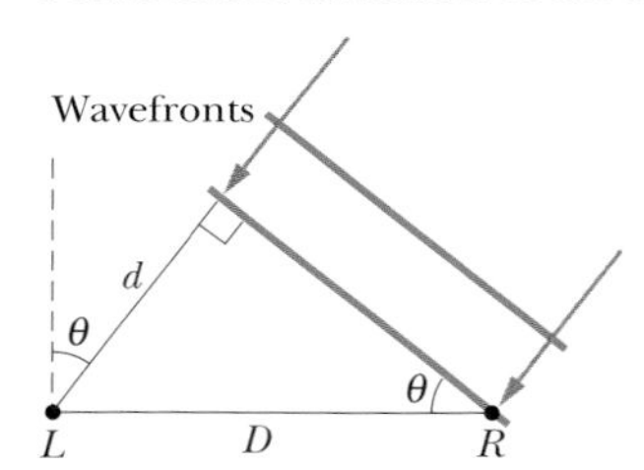

FIGURE 18-9 ■ A wavefront travels a distance $d(= D \sin\theta)$ farther to reach the left ear (L) than to reach the right ear (R).

SOLUTION ■ The situation is shown (from an overhead view) in Fig. 18-9, where wavefronts approach you from a source that is located in front of you and to your right. The **Key Idea** here is that the time delay Δt is due to the distance d that each wavefront must travel to reach your left ear (L) after it reaches your right ear (R). From Fig. 18-9, we find

$$\Delta t = \frac{d}{v^{air}} = \frac{D\sin\theta}{v^{air}}, \qquad \text{(Answer)} \quad (18\text{-}4)$$

where v^{air} is the speed of the sound wave in air. Based on a lifetime of experience, your brain correlates each detected value of Δt (from zero to the maximum value) with a value of θ (from zero to 90°) for the direction of the sound source.

(b) Suppose that you are submerged in water at 20°C when a wavefront arrives from directly to your right (θ = 90°). Based on the time-delay clue, at what angle θ from the forward direction does the source seem to be?

SOLUTION ■ The **Key Idea** here is that the speed is now the speed of the sound wave in water, v^{water}, so in Eq. 18-4 we substitute v^{water} for v^{air} and 90° for θ, finding that

$$\Delta t_w = \frac{D\sin\theta}{v^{water}} = \frac{D}{v^{water}}, \qquad (18\text{-}5)$$

Since v^{water} is about four times v^{air}, delay Δt_w is about one-fourth the maximum time delay in air. Based on experience, your brain will process the water time delay as if it occurred in air. Thus, the sound source appears to be at an angle θ smaller than 90°. To find that apparent angle we substitute the time delay D/v^{water} from Eq. 18-5 for Δt in Eq. 18-4, obtaining

$$\frac{D}{v^{water}} = \frac{D\sin\theta}{v^{air}}. \qquad (18\text{-}6)$$

Then, to solve for θ we substitute v^{air} = 343 m/s and v^{water} = 1482 m/s (from Table 18-1) into Eq. 18-5, finding

$$\sin\theta = \frac{v^{air}}{v^{water}} = \frac{343 \text{ m/s}}{1482 \text{ m/s}} = 0.231,$$

and thus

$$\theta = 13°. \qquad \text{(Answer)}$$

18-3 Interference

Like transverse waves, sound waves undergo interference when two waves pass through the same point at the same time. Let us consider, in particular, the interference between two sound waves with the same frequency, wavelength, amplitude, and phase. The only difference between these waves is that they are traveling in slightly different directions. Figure 18-10 shows how we can set up such a situation: Two point sources S_1 and S_2 emit sound waves that are in phase. Thus, the sources themselves are said to be in phase; that is, as the waves emerge from the sources, their displacements are always identical. We are interested in the waves that then travel through point P in Fig. 18-10. We assume that the distance to P is much greater than the distance between the sources so that the waves are traveling in almost the same direction at P.

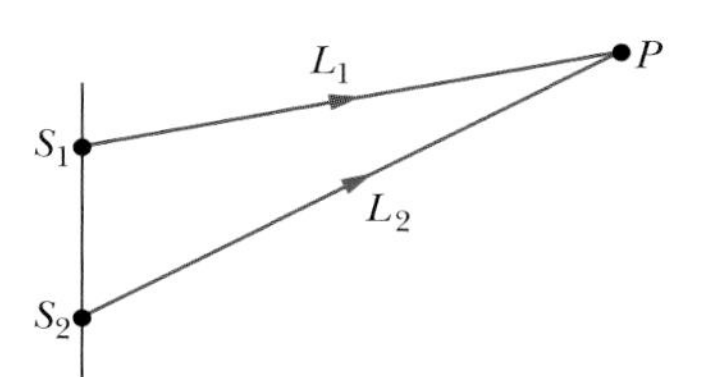

FIGURE 18-10 ■ Two point sources S_1 and S_2 emit spherical sound waves in phase. The rays indicate that the waves pass through a common point P. (Recall that rays are lines that run perpendicular to the wavefront and indicate direction of travel.)

If the waves (which both start out at the same point in their pressure oscillation, i.e., *in phase*) traveled along paths with identical lengths to reach point P, they would still be in phase there. In this case, the displacements of the two waves would add. As with transverse waves, this means that they would undergo fully constructive interference there. However, in Fig. 18-10, path L_2 traveled by the wave from S_2 is longer than path L_1 traveled by the wave from S_1. The difference in path lengths means that the waves may not be in phase at point P and so might be at different points in their oscillations.

We can use the definition of phase (from Section 17-4) to determine the phase difference. We specified that the waves had the same phase at locations S_1 and S_2. So we can set $(\phi_0)_1 = (\phi_0)_2$. Thus, the phase difference when the two waves arrive at point P at a common time t is given by

$$\Delta\phi = \phi_2(x,t) - \phi_1(x,t) = [kL_2 - \omega t + (\phi_0)_2] - [kL_1 - \omega t + (\phi_0)_1] = k(L_2 - L_1). \qquad (18\text{-}7)$$

Indeed the magnitude of the phase difference $\Delta\phi$ at P depends on the **path length difference** $\Delta L = |L_2 - L_1|$ of the two waves. Since the wave number $k = 2\pi/\lambda$, we can rewrite $\Delta\phi = k\Delta L$ as

$$|\Delta\phi| = \frac{\Delta L}{\lambda}2\pi \quad \text{(path length–phase difference relation).} \qquad (18\text{-}8)$$

Fully constructive interference occurs when the phase difference, $\Delta\phi$, is zero, 2π, or any integer multiple of 2π.

We can write this condition as

$$|\Delta\phi| = m(2\pi), \quad \text{for } m = 0, 1, 2, \ldots \quad \text{(fully constructive interference),} \qquad (18\text{-}9)$$

or from Eq. 18-8, this also occurs when the ratio $\Delta L/\lambda$ is

$$\frac{\Delta L}{\lambda} = m \quad \text{(fully constructive interference),} \qquad (18\text{-}10)$$

where m is zero or any positive integer.

For example, if the magnitude of the path length difference $\Delta L = |L_2 - L_1|$ in Fig. 18-10 is equal to 2λ, then $\Delta L/\lambda = 2$ and the waves undergo fully constructive interference at point P. The interference is fully constructive because the wave from S_2 is phase-shifted relative to the wave from S_1 by 2λ, putting the two waves *exactly in phase* at P.

Fully destructive interference occurs when the magnitude of the phase difference between the two waves $\Delta\phi$ is an odd multiple of π.

We can write this as

$$|\Delta\phi| = (2m + 1)\pi \qquad \text{(fully destructive interference)}, \tag{18-11}$$

where once again m is zero or any positive integer. From $|\Delta\phi| = (\Delta L/\lambda)2\pi$ (Eq. 18-8), this we see occurs when the ratio $\Delta L/\lambda$ is

$$\frac{\Delta L}{\lambda} = \frac{2m + 1}{2} = m + \tfrac{1}{2} \qquad \text{(fully destructive interference)}. \tag{18-12}$$

For example, if the path length difference $\Delta L = |L_2 - L_1|$ in Fig. 18-10 is equal to 2.5λ, then $\Delta L/\lambda = 2.5$ and the waves undergo fully destructive interference at point P. The interference is fully destructive because the wave from S_2 is phase-shifted relative to the wave from S_1 by 2.5 wavelengths, which puts the two waves *exactly out of phase* at P.

Of course, two waves could produce intermediate interference as, say, when $\Delta L/\lambda = 1.2$. This would be closer to fully constructive interference ($\Delta L/\lambda = 1.0$) than to fully destructive interference ($\Delta L/\lambda = 1.5$).

If the initial phase of one of the interfering waves is different than that of the other, we would have to derive a different set of conditions for constructive and destructive interference.

TOUCHSTONE EXAMPLE 18-2: Constructive Interference

In Fig. 18-11*a*, two point sources S_1 and S_2, which are in phase and separated by distance $D = 1.5\lambda$, emit identical sound waves of wavelength λ.

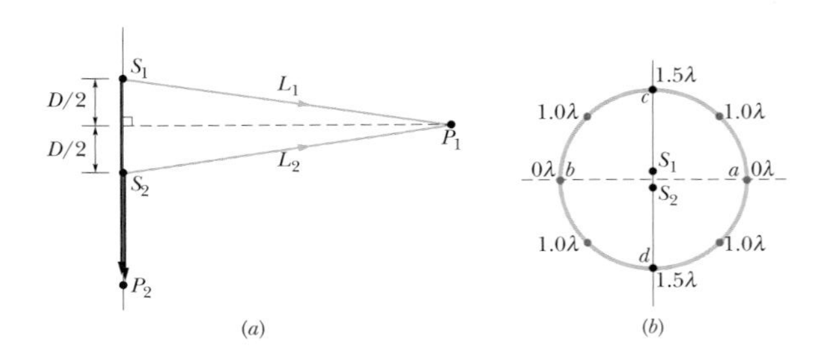

FIGURE 18-11 ■ (*a*) Two point sources S_1 and S_2, separated by distance D, emit spherical sound waves in phase. The waves travel equal distances to reach point P_1. Point P_2 is on the line extending through S_1 and S_2. (*b*) The path length difference (in terms of wavelength) between the wave from S_1 and S_2, at eight points on a large circle around the sources.

(a) What is the path length difference of the waves from S_1 and S_2 at point P_1, which lies on the perpendicular bisector of distance D, at a distance greater than D from the sources? What type of interference occurs at P_1?

SOLUTION ■ The **Key Idea** here is that, because the waves travel identical distances to reach P_1, their path length difference is

$$\Delta L = 0.0\lambda. \qquad \text{(Answer)}$$

From Eq. 18-10, this means that the waves undergo fully constructive interference at P_1.

(b) What are the path length difference and type of interference at point P_2 in Fig. 18-11*a*?

SOLUTION ■ Now the **Key Idea** is that the wave from S_1 travels the extra distance D ($= 1.5\lambda$) to reach P_2. Thus, the path length difference is

$$\Delta L = 1.5\lambda. \qquad \text{(Answer)}$$

From Eq. 18-12, this means that the waves are exactly out of phase at P_2 and undergo fully destructive interference there.

(c) Figure 18-11*b* shows a circle with a radius much greater than D, centered on the midpoint between sources S_1 and S_2. What is the number of points N around this circle at which the interference is fully constructive?

SOLUTION ■ Imagine that, starting at point a, we move clockwise along the circle to point d. One **Key Idea** here is that as we move to point d, the path length difference ΔL increases and so the type of interference changes. From (a), we know that the path length difference is $\Delta L = 0.0\lambda$ at point a. From (b), we know that $\Delta L = 1.5\lambda$ at point d. Thus, there must be one point along the circle between a and d at which $\Delta L = \lambda$, as indicated in Fig. 18-11*b*. From Eq. 18-10, fully constructive interference occurs at that point. Also, there can be no other point along the way from point a to point d at which fully constructive interference occurs, because there is no other integer than 1 between 0.0 and 1.5.

Another **Key Idea** here is to use symmetry to locate the other points of fully constructive interference along the rest of the circle. Symmetry about line cd gives us point b, at which $\Delta L = 0\lambda$. Also, there are three more points at which $\Delta L = \lambda$. In all we have constructive interference at

$$N = 6 \text{ points}. \qquad \text{(Answer)}$$

18-4 Intensity and Sound Level

In Section 17-7 we derived the fact that the average power in a one-dimensional continuous sinusoidal wave that propagates along a string depends on the squares of both the frequency and amplitude as shown in Eq. 17-31 ($\langle\text{Power}\rangle = \frac{1}{2}\mu v^{\text{wave}}\omega^2 Y^2$).* Let's compare Eq. 17-31 to the equation for the power transmission of a sinusoidal one-dimensional sound wave. Suppose our sound is produced by an oscillating piston in an air-filled pipe of cross-sectional area A shown in Fig. 18-1. The relationship for the average power transmission of the sound waves is given by

$$\langle\text{Power}\rangle = \tfrac{1}{2}(\rho A)v^{\text{wave}}\omega^2 S^2, \tag{18-13}$$

where S represents a maximum displacement. In this case S denotes the maximum longitudinal displacement about an equilibrium point that particles in the medium carrying the sound undergo. This power relationship can be derived by calculating the average kinetic and potential energy of a given volume of air that undergoes simple harmonic motion. But without using a formal derivation we can see that Eq. 17-31 and Eq. 18-13 are essentially the same. Both S and Y are displacements and the term ρA is equal to the linear density μ of the air in the tube.

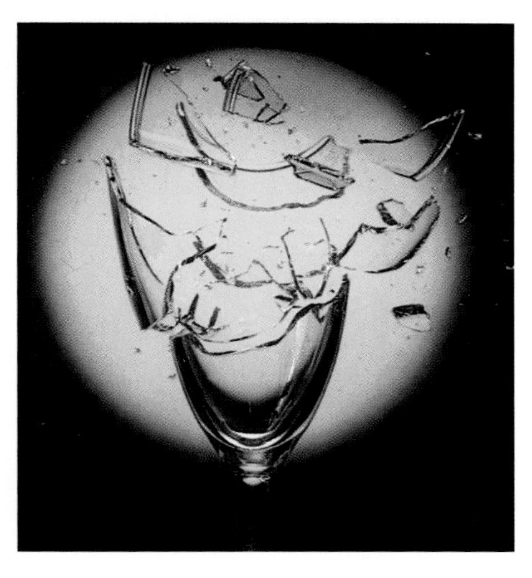

Sound can cause the wall of a drinking glass to oscillate. If the sound produces a standing wave of oscillations and if the intensity of the sound is large enough, the glass will shatter.

In previous sections we have been using maximum pressure difference, ΔP^{max}, rather than maximum particle displacement, S, as an indicator of how strong a sinusoidal sound disturbance is. It can be shown that these two measures of wave strength are related, by the expression

$$\Delta P^{\text{max}} = v^{\text{wave}}\rho\omega S.$$

If you have ever tried to sleep while someone played loud music nearby, you are well aware that there is more to sound than frequency, wavelength, and speed. Humans also detect how *loud* a sound is. Although the human ear does detect pressure amplitudes, it turns out that we are more sensitive to the *energy fluctuations* in a propagating wave than we are to the pressure alone. Hence, we will define a new energy related quantity associated with waves. The **intensity** I of a sound wave at a surface is the average rate per unit area, A, at which energy is transferred by the wave through or onto the surface. This is what we commonly refer to as loudness. By definition, the intensity of a wave is

$$I \equiv \frac{\langle\text{Power}\rangle}{A}, \tag{18-14}$$

where "Power" is the time rate of energy transfer (the power) of the sound wave, and A is the area of the surface intercepting the sound.

We can use Eq. 18-13 to show the relationship between the energy-related quantity *intensity* and the maximum particle displacement associated with a sound wave,

$$I = \tfrac{1}{2}\rho v^{\text{wave}}\omega^2 S^2. \tag{18-15}$$

The equation $\Delta P^{\text{max}} = v^{\text{wave}}\rho\omega S$ shown above can be used to express the intensity as a function of pressure change, so that

> The intensity of a continuous sinusoidal wave is proportional to both the square of its displacement amplitude and the square of maximum pressure change in the medium caused by the sound waves.

*In order to avoid confusion between power, usually denoted as P, and pressure, also denoted as P, we choose to spell out the word Power in this section.

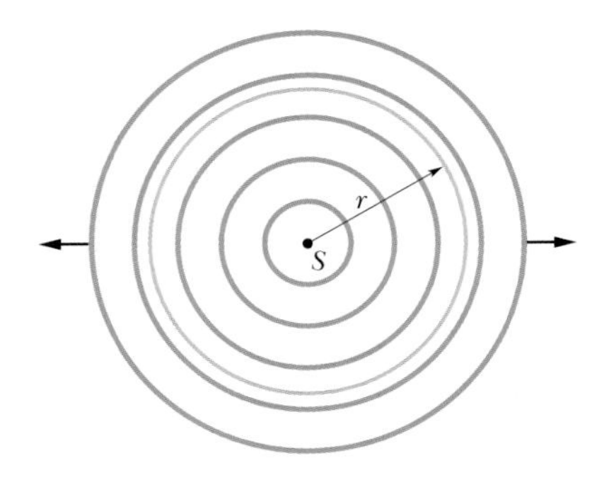

FIGURE 18-12 ■ A point source S emits sound waves uniformly in all directions. The waves pass through an imaginary sphere of radius r that is centered on S.

Variation of Intensity with Distance

How intensity varies with distance from a real sound source is often complex. Some real sources (like loudspeakers) may transmit sound only in particular directions, and objects in the surroundings usually produce echoes (reflected sound waves) that overlap the direct sound waves. In some situations, however, we can ignore echoes and assume that the sound source is a point source that emits the sound *isotropically*—that is, with equal intensity in all directions. The wavefronts spreading from such an isotropic point source S at a particular instant are shown in Fig. 18-12.

Let us assume that the mechanical energy of the sound waves is conserved as they spread from this source. Let us also center an imaginary sphere of radius r on the source, as shown in Fig. 18-12. All the energy emitted by the source must pass through the surface of the sphere. Thus, the time rate at which energy is transferred through the surface by the sound waves must equal the time rate at which energy is emitted by the source (that is, the power, Power_s, of the source). From the fact that intensity is equal to the ratio of power to area (Eq. 18-14), the intensity I at the sphere must then be

$$I = \frac{\text{Power}_s}{4\pi r^2}, \tag{18-16}$$

where $4\pi r^2$ is the area of the sphere. Equation 18-16 tells us that the intensity of sound from an isotropic point source decreases with the square of the distance r from the source.

The Decibel Scale

The displacement amplitude at the human ear ranges from about 10^{-5} m for the loudest tolerable sound to about 10^{-11} m for the faintest detectable sound, a ratio of 10^6. From our discussions above, we know that the intensity of a sound varies as the *square* of its amplitudes, so the ratio of intensities at these two limits of the human auditory system is 10^{12}. Humans can hear over an enormous range of intensities.

We deal with such an enormous range of values by using base 10 logarithms. Consider the relation

$$y = \log x,$$

in which x and y are variables. It is a property of this equation that if we *multiply* x by 10, then y increases by 1. To see this, we write

$$y' = \log(10x) = \log 10 + \log x = 1 + y.$$

Similarly, if we multiply x by 10^{12}, y increases by only 12. An important characteristic of human hearing is that we have logarithmic ears. Within the normal range of hearing when the measured sound intensity increases by a factor of 10, the sound only seems twice as loud to us.

Thus, instead of speaking of the intensity I of a sound wave, it is much more convenient to speak of its **sound level** β, defined as

$$\beta = (10\text{ dB})\log\frac{I}{I_0}. \tag{18-17}$$

Here dB is the abbreviation for **decibel,** the unit of sound level, a name that was chosen to recognize the work of Alexander Graham Bell. I_0 in Eq. 18-17 is a standard reference intensity ($= 10^{-12}$ W/m^2), chosen because it is near the lower limit of the human range of hearing at a frequency of 1 kHz. For $I = I_0$, Eq. 18-17 gives $\beta = 10 \log 1 = 0$, so our standard reference level, which we can barely hear, corresponds to zero decibels. Then β increases by 10 dB every time the sound intensity increases by an order of magnitude (a factor of 10). But as we mentioned, a 10 dB increase seems to be twice the loudness. And the sound of a typical conversation, which has a β of 60 dB, corresponds to an intensity that is 10^6 times our normal hearing threshold! Table 18-2 lists the sound levels for a variety of environments.

TABLE 18-2
Some Sound Levels (dB)

Hearing threshold	0	Rock concert	110
Rustle of leaves	10	Pain threshold	120
Conversation	60	Jet engine	130

The decibel scale is one of many logarithmic scales used by scientists. Others include the pH scale, star magnitudes, and the Richter scale used to determine the severity of earthquakes.

READING EXERCISE 18-3: The figure indicates the near edge of three small patches 1, 2, and 3 that lie on the surfaces of two imaginary spheres; the spheres are centered on an isotropic point source S of sound. The rates at which energy is transmitted through the three patches by the sound waves are equal. Rank the patches according to (a) the intensity of the sound on them and (b) their area, greatest first. ■

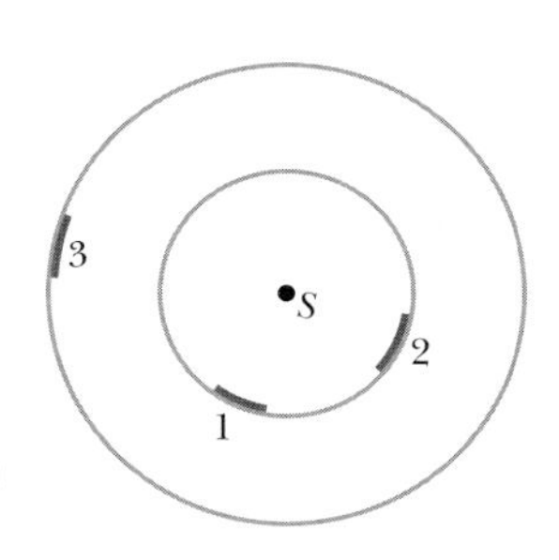

TOUCHSTONE EXAMPLE 18-3: Spark Sound

An electric spark jumps along a straight line of length $L = 10$ m, emitting a pulse of sound that travels radially outward from the spark. (The spark is said to be a line source of sound.) The power of the emission is $\text{Power}_s = 1.6 \times 10^4$ W.

(a) What is the intensity I of the sound when it reaches a distance $r = 6$ m from the spark?

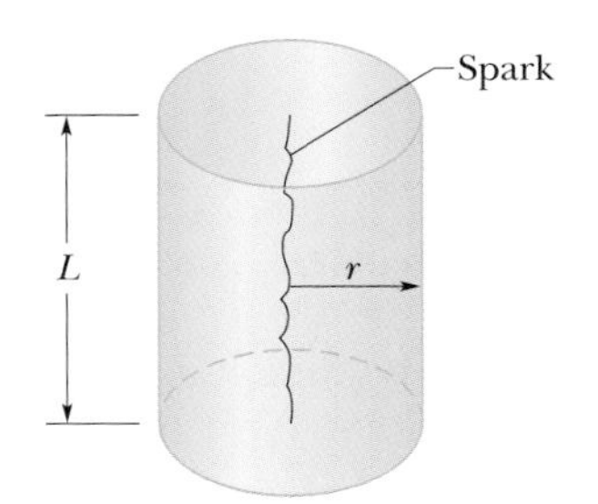

FIGURE 18-13 ■ A spark along a straight line of length L emits sound waves radially outward. The waves pass through an imaginary cylinder of radius r and length L that is centered on the spark.

SOLUTION ■ Let us center an imaginary cylinder of radius $r = 6$ m and length $L = 10$ m (open at both ends) on the spark, as shown in Fig. 18-13. One **Key Idea** here is that the intensity I at the cylindrical surface is the ratio $\langle\text{Power}\rangle/A$ of the average rate per unit area at which sound energy passes through a surface of area A on the cylinder. Another **Key Idea** is to assume that the principle of conservation of energy applies to the sound energy. This means that the Power or rate at which energy is transferred through the cylinder must equal the rate at which energy is emitted by the source. Putting these ideas together and noting that the area of a length L of the cylindrical surface is $A = 2\pi rL$, and neglecting the relatively small amount of sound energy that flows out through the ends of the cylinder, we have, from Eq. 18-14,

$$I = \frac{\langle\text{Power}\rangle}{A} = \frac{\langle\text{Power}_s\rangle}{2\pi rL}. \tag{18-18}$$

This tells us that the intensity of the sound from a line source decreases with distance r (rather than with the square of distance r as for the point source described in Eq. 18-16). Substituting the given data, we find

$$I = \frac{1.6 \times 10^4 \text{ W}}{2\pi(6 \text{ m})(10 \text{ m})} = 42.4 \text{ W/m}^2 \approx 42 \text{ W/m}^2. \quad \text{(Answer)}$$

(b) At what average $\langle\text{Power}_d\rangle$ is sound energy intercepted by an acoustic detector of area $A_d = 2.0\text{ cm}^2$? Assume that the detector is aimed at the spark and located a distance $r = 6$ m from the spark.

SOLUTION ■ Applying the first **Key Idea** of part (a), we know that the intensity of sound at the detector is the ratio of the energy transfer rate $\langle\text{Power}_d\rangle$ there to the detector's area A_d:

$$I = \frac{\langle\text{Power}_d\rangle}{A_d}. \tag{18-19}$$

We can imagine that the detector lies on the cylindrical surface of (a). Then the sound intensity at the detector is the intensity I ($= 42.4$ W/m^2) at the cylindrical surface. Solving Eq. 18-19 for $\langle\text{Power}_d\rangle$ gives us

$$\langle\text{Power}_d\rangle = (42.4 \text{ W/m}^2)(2.0 \times 10^{-4}\text{m}^2) = 8.5 \text{ mW}. \quad \text{(Answer)}$$

TOUCHSTONE EXAMPLE 18-4: Concert Sounds

In 1976, the Who set a record for the loudest concert—the sound level 46 m in front of the speaker systems was $\beta_2 = 120$ dB. What is the ratio of the intensity I_2 of the band at that spot to the intensity I_1 of a jackhammer operating at sound level $\beta_1 = 92$ dB?

SOLUTION ■ The **Key Idea** here is that for both the Who and the jackhammer, the sound level β is related to the intensity by the definition of sound level in Eq. 18-17. For the Who, we have

$$\beta_2 = (10 \text{ dB}) \log \frac{I_2}{I_0},$$

and for the jackhammer, we have

$$\beta_1 = (10 \text{ dB}) \log \frac{I_1}{I_0}.$$

The difference in the sound levels is

$$\beta_2 - \beta_1 = (10 \text{ dB}) \left(\log \frac{I_2}{I_0} - \log \frac{I_1}{I_0} \right). \tag{18-20}$$

Using the identity

$$\log \frac{a}{b} - \log \frac{c}{d} = \log \frac{ad}{bc},$$

we can rewrite Eq. 18-20 as

$$\beta_2 - \beta_1 = (10 \text{ dB}) \log \frac{I_2}{I_1}. \tag{18-21}$$

Rearranging and substituting the known sound levels now yield

$$\log \frac{I_2}{I_1} = \frac{\beta_2 - \beta_1}{10 \text{ dB}} = \frac{120 \text{ dB} - 92 \text{ dB}}{10 \text{ dB}} = 2.8.$$

Taking the antilog of the far left and far right sides of this equation (the antilog key on your calculator is probably marked as 10^x), we find

$$\frac{I_2}{I_1} = \log^{-1} 2.8 = 630. \quad \text{(Answer)}$$

Thus, the Who was *very* loud.

Temporary exposure to sound intensities as great as those of a jackhammer and the 1976 Who concert results in a temporary reduction of hearing. Repeated or prolonged exposure can result in permanent reduction of hearing (Fig. 18-14). Loss of hearing is a clear risk for any one continually listening to, say, heavy metal at high volume, especially on headphones.

FIGURE 18-14 ■ Pete Townshend of the Who, playing in front of a speaker system. He suffered a permanent reduction in his hearing ability due to his exposure to high-intensity sound, not so much during on-stage performances as from wearing headphones in recording studios and at home.

FIGURE 18-15 ■ The air column within a fujara oscillates when that traditional Slovakian instrument is played.

18-5 Sources of Musical Sound

Musical sounds can be set up by oscillating strings (guitar, piano, violin), membranes (kettledrum, snare drum), air columns (flute, oboe, pipe organ, and the fujara of Fig. 18-15), wooden blocks or steel bars (marimba, xylophone), and many other oscillating bodies. Most instruments involve more than a single oscillating part. In the violin, for example, both the strings and the body of the instrument participate in producing the music.

Standing Waves and Musical Instruments

Recall from Chapter 17 that standing waves can be set up on a stretched string that is fixed at both ends. They arise because waves traveling along the string are reflected back onto the string at each end. If the wavelength of the waves is suitably matched to the length of the string, the superposition of waves traveling in opposite directions produces a standing wave pattern (or oscillation mode). The wavelength required of the waves for such a match is one that corresponds to a *resonant frequency* of the string. The advantage of setting up standing waves is that the string then oscillates with a large, sustained amplitude, pushing back and forth against the surrounding air and thus generating a noticeable sound wave with the same frequency as the oscillations of the string. This production of sound is of obvious importance to, say, a guitarist.

We can set up standing waves of sound in an air-filled pipe in a similar way. As sound waves travel through the air in the pipe, they are reflected at each end and travel back through the pipe. (The reflection occurs even if an end is open, but the reflection is not as complete as when the end is closed.) If the wavelength of the sound waves is suitably matched to the length of the pipe, the superposition of waves traveling in opposite directions through the pipe sets up a standing wave pattern. The wavelength required of the sound waves for such a match is one that corresponds to a resonant frequency of the pipe. The advantage of such a standing wave is that the air in the pipe oscillates with a large, sustained amplitude, emitting at any open end a sound wave that has the same frequency as the oscillations in the pipe. This emission of sound is of obvious importance to, say, an organist.

Many other aspects of standing sound wave patterns are similar to those of string waves: The closed end of a pipe is like the fixed end of a string in that there must be a displacement node (zero displacement in a string, zero motion of molecules in a fluid like air) located there. A zero particle velocity immediately against the closed end of the pipe leads to a large change in pressure (a pressure antinode). The open end of a pipe is like the end of a string attached to a freely moving ring, as in Fig. 17-19*b*, in that there must be a displacement antinode (maximum displacement in a string, maximum motion of molecules in a fluid like air) located there. This too makes sense, as the molecules of air (or other fluid) are completely unconstrained once they leave the end of the pipe. The large particle velocities here lead to very small changes in pressure (a pressure antinode). To be perfectly precise, the antinode for the open end of a pipe is located slightly beyond the end. We will ignore this difference in our discussions.

So, the simplest standing wave pattern that can be set up in a pipe with two open ends is one with a particle displacement antinode (maximum particle velocities and zero change in pressure) at both ends and no other antinodes between them. Under these conditions, the only way that there can be a standing wave pattern at all is for there to be a displacement node (zero particle velocity and maximum change in pressure) in the middle of the pipe. This is shown in Fig. 18-16*a*. An alternate representation of the standing wave pattern as a graph of maximum molecular displacement at each position along the length of the page is shown in Fig. 18-16*b*.

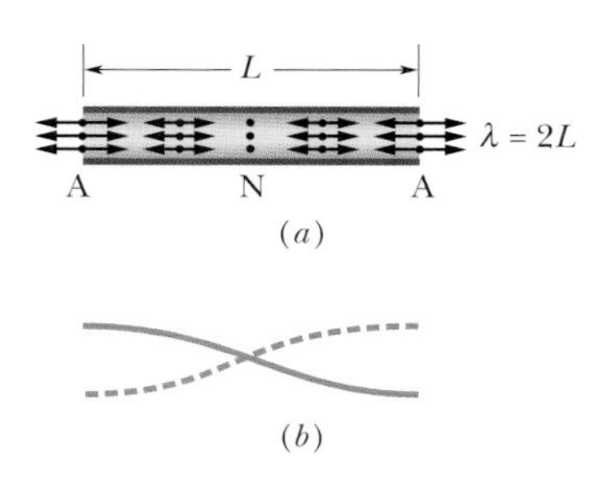

FIGURE 18-16 ■ (*a*) The simplest standing wave pattern of air molecule displacements from equilibrium for (longitudinal) sound waves in a pipe with both ends open. It shows an antinode (A) across each end and a node (N) across the middle of the pipe. (The longitudinal displacements represented by the double arrows are greatly exaggerated.) (*b*) A graph of maximum molecular displacements versus position.

Harmonics

The standing wave pattern of Fig. 18-16*a* is called the *fundamental mode* or *first harmonic*. For it to be set up, the sound waves in a pipe of length L must have a wavelength given by $L = \lambda/2$, so that $\lambda = 2L$. Several more standing sound wave patterns for a pipe with two open ends are shown in Fig. 18-17*a* with graphs of air molecule displacements from equilibrium shown next to the pipes. The *second harmonic* requires sound waves of wavelength $\lambda = L$, the *third harmonic* requires wavelength $\lambda = 2L/3$, and so on.

More generally, the resonant frequencies for a pipe of length L with two open ends correspond to the wavelengths

$$\lambda = \frac{2L}{n} \quad \text{(pipe, two open ends).} \tag{18-22}$$

Here n is a positive integer (1, 2, 3, . . . etc.) called the *harmonic number*. The resonant frequencies for a pipe with two open ends are then given by

$$f = \frac{v^{\text{air}}}{\lambda} = n\left(\frac{v^{\text{air}}}{2L}\right) \quad \text{(pipe, two open ends),} \tag{18-23}$$

where v^{air} is the speed of the sound wave in air.

Figure 18-17*b* shows (using pressure variation graphs) some of the standing sound wave patterns that can be set up in a pipe with only one open end. As required, across

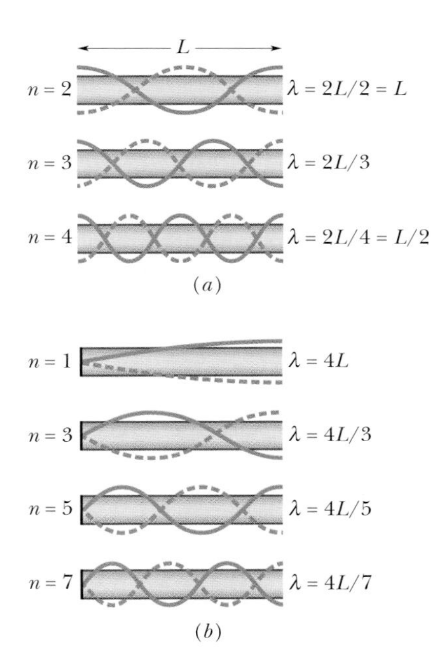

FIGURE 18-17 ■ Standing wave patterns are shown as graphs of the extremes of maximum and minimum air molecule displacements versus location along the pipe. (*a*) With *both* ends of the pipe open, any harmonic can be set up in the pipe. (*b*) With only *one* end open, only odd harmonics can be set up.

the open end there is a displacement antinode (maximum particle velocities and zero pressure change) and across the closed end there is a displacement node (zero particle velocities and maximum pressure change). The simplest pattern requires sound waves having a wavelength given by $L = \lambda/4$, so that $\lambda = 4L$. The next simplest pattern requires a wavelength given by $L = 3\lambda/4$, so that $\lambda = 4L/3$, and so on.

More generally, the resonant frequencies for a pipe of length L with only one open end correspond to the wavelengths

$$\lambda = \frac{4L}{2n+1} \qquad \text{for } n = 0, 1, 2, \ldots, \tag{18-24}$$

where n still represents a positive integer with $2n + 1$ giving us odd positive integers. The resonant frequencies are then given by

$$f = \frac{v^{\text{air}}}{\lambda} = (2n+1)\frac{v^{\text{air}}}{\lambda}n \qquad \text{(pipe, one open end).} \tag{18-25}$$

Note again that only odd harmonics can exist in a pipe with one open end. For example, the second harmonic, with $2n + 1 = 2$, cannot be set up in such a pipe. Note also that for such a pipe the numeric adjective (e.g., first, second, third, . . .) before the word harmonic in a phrase such as "the third harmonic" always refers to the harmonic number n and not to the nth *possible* harmonic.

The length of a musical instrument is related to the range of frequencies over which the instrument is designed to function, and smaller length implies higher frequencies. Figure 18-18, for example, shows the saxophone and violin families, with their frequency ranges suggested by the piano keyboard. Note that, for every instrument, there is overlap with its higher- and lower-frequency neighbors.

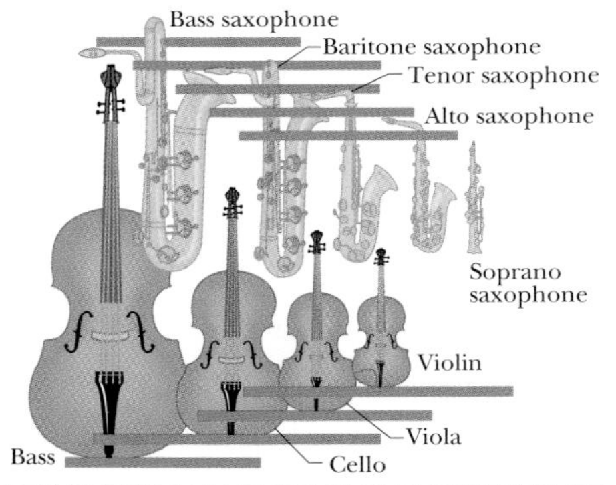

FIGURE 18-18 ■ The saxophone and violin families, showing the relations between instrument length and frequency range. The frequency range of each instrument is indicated by a horizontal bar along a frequency scale suggested by the piano keyboard at the bottom; the frequency increases toward the right.

What Makes Instruments Distinctive?

In any oscillating system that gives rise to a musical sound, whether it is a violin string or the air in an organ pipe, the fundamental and one or more of the higher harmonics are usually generated simultaneously. Thus, you hear them together—that is, superimposed into a combined wave. When different instruments are playing the same note, they produce the same fundamental frequency but different intensities for the higher harmonics. For example, the fourth harmonic of middle C might be relatively loud on one instrument

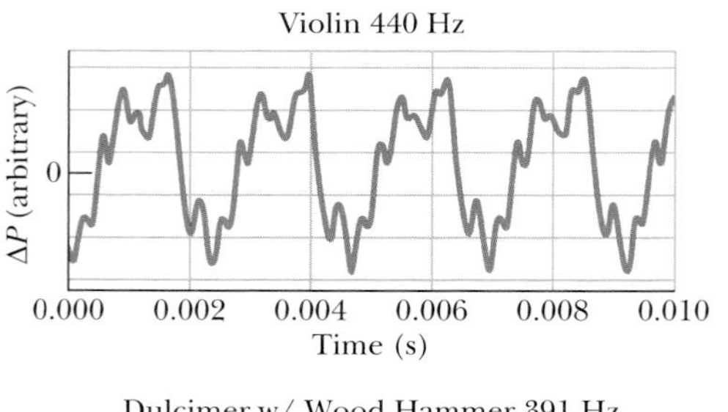

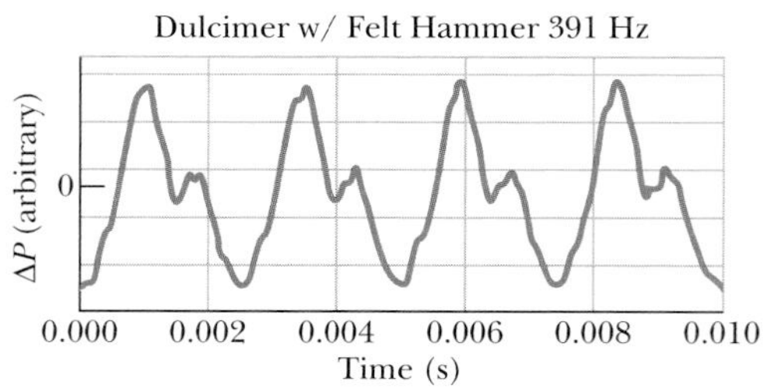

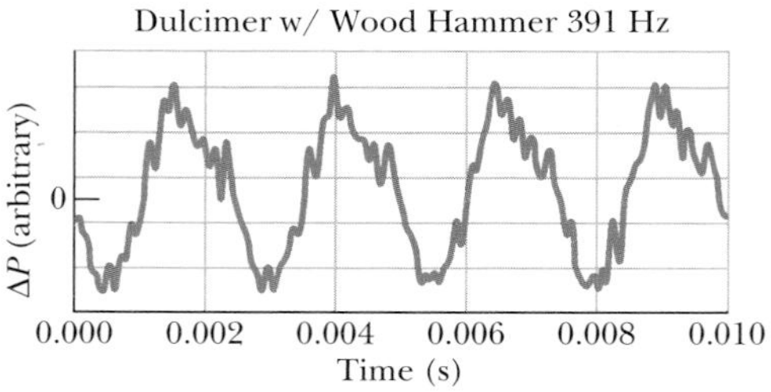

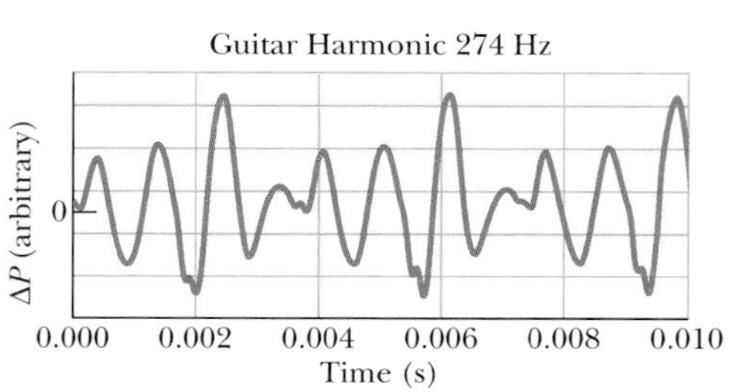

FIGURE 18-19 ■ The wave forms produced by a violin, a hammer dulcimer struck with a felt and a wood hammer, and a guitar. The fundamental frequency is listed for each wave. These graphs of the relative sound pressure variation versus time are produced using a computer data acquisition system with a microphone sensor attached.

and relatively quiet or even missing on another. Thus, different instruments sound different to you even when they are playing the same note. That would be the case for the combined waves shown for the violin, hammer dulcimer and guitar in Fig. 18-19 even if these three instruments were playing the same note (which they are not).

In Fig. 18-19 we can also see that the pressure variation versus time for the hammer dulcimer hit with a wood hammer has more high frequency harmonics than the corresponding graph for the dulcimer hit with a felt hammer. But, in both cases the dulcimer has the same fundamental frequency of 391 Hz. There is a mathematical technique know as **Fourier analysis** that can be used to determine the amplitudes of each of the harmonic frequencies that combines to make a full sound. A fast Fourier transform (FFT) of the two hammer dulcimer sounds is shown in Fig. 18-20. As you can see, the relative amplitudes for harmonic frequencies between 1500 Hz and 2500 Hz are much higher when the dulcimer is hit with a wood hammer than when it is hit with a felt hammer.

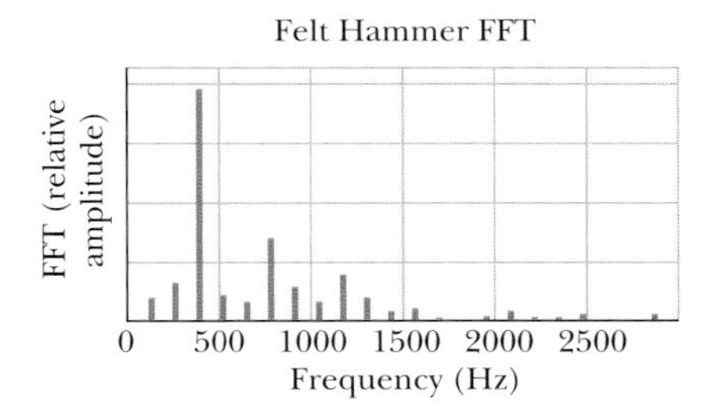

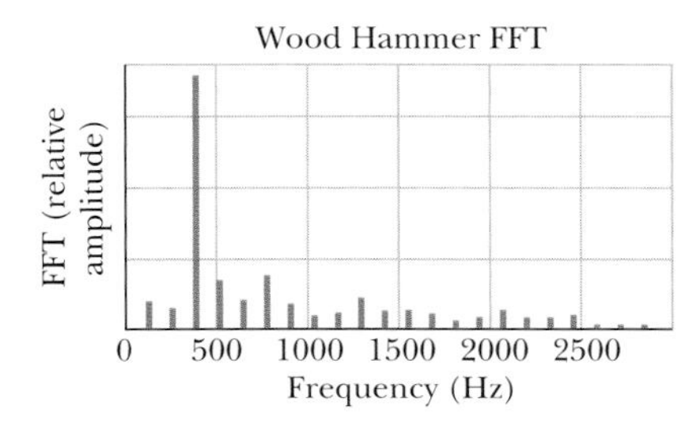

FIGURE 18-20 ■ A fast Fourier transform showing the amplitudes of the harmonics for a hammered dulcimer when it is (a) struck with a felt hammer and (b) struck with a wood hammer. The wood hammer stimulates more very high harmonics.

READING EXERCISE 18-4: Pipe A, with length L, and pipe B, with length $2L$, both have two open ends. Which harmonic of pipe B has the same frequency as the fundamental of pipe A? ■

TOUCHSTONE EXAMPLE 18-5: Sound in a Tube

Weak background noises from a room set up the fundamental standing wave in a cardboard tube of length $L = 67.0$ cm with two open ends. Assume that the speed of sound in the air within the tube is 343 m/s.

(a) What frequency do you hear from the tube?

SOLUTION ■ The **Key Idea** here is that, with both pipe ends open, we have a symmetric situation in which the standing wave has a displacement antinode at each end of the tube. The standing wave displacement variation pattern is that of Fig. 18-16b. The frequency is given by Eq. 18-23 with $n = 1$ for the fundamental mode:

$$f = \frac{nv^{\text{air}}}{2L} = \frac{(1)(343 \text{ m/s})}{(2)(0.670 \text{ m})} = 256 \text{ Hz}. \quad \text{(Answer)}$$

If the background noises set up any higher harmonics, such as the second harmonic, you must also hear frequencies that are *integer* multiples of 256 Hz.

(b) If you jam your ear against one end of the tube, what fundamental frequency do you hear from the tube?

SOLUTION ■ The **Key Idea** now is that, with your ear effectively closing one end of the tube we have an asymmetric situation—a displacement antinode still exists at the open end but a displacement node is now at the other (closed) end. The standing wave pattern is the top one in Fig. 18-17b. The frequency is given by Eq. 18-25 with $n = 1$ for the fundamental mode:

$$f = \frac{nv^{\text{air}}}{4L} = \frac{(1)(343 \text{ m/s})}{4(0.670 \text{ m})} = 128 \text{ Hz}. \quad \text{(Answer)}$$

If the background noises set up any higher harmonics, they will be *odd* multiples of 128 Hz. That means that the frequency of 256 Hz (which is an even multiple) cannot now occur.

18-6 Beats

If you listen separately to two sounds whose frequencies are, say, 50 and 52 Hz, most of us cannot tell which one has the higher frequency. However, if the sounds reach our ears simultaneously, what we hear is a sound whose frequency turns out to be 51 Hz, the *average* of the two combining frequencies. We also hear a striking variation in the intensity of this sound—it increases and decreases in slow, wavering **beats** that repeat at a frequency of 2 Hz, the *difference* between the two combining frequencies. Figure 18-21 shows this beat phenomenon (for sounds of frequencies of 8 Hz and 10 Hz).

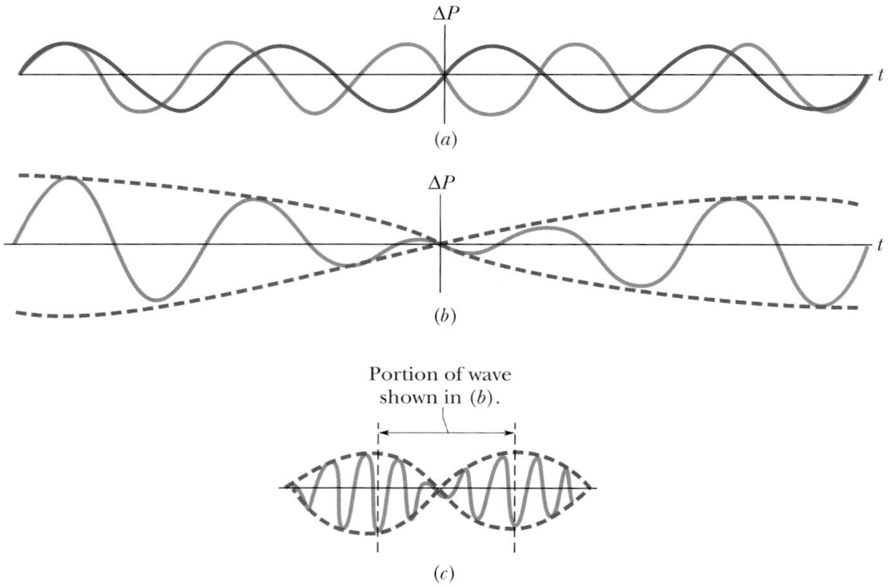

FIGURE 18-21 ■ (a) The pressure variations ΔP of two sound waves as they would be detected separately are plotted as a function of time. The frequencies of the waves are nearly equal. (b) If the two waves are detected simultaneously the resultant pressure variation is the superposition of the two waves. Notice how the waves cancel each other at the center of the plot and reinforce each other at the two ends of the plot. (c) When plotted over a longer time period the two waves shown in (a) and (b) show a beat pattern of 2 Hz if the frequencies of the original waves are 10 and 12 Hz.

Let the time-dependent variations of pressure due to two sound waves at a particular location be

$$\Delta P_1 = \Delta P^{\max}\cos\omega_1 t \quad \text{and} \quad \Delta P_2 = \Delta P^{\max}\cos\omega_2 t, \tag{18-26}$$

where $\omega_1 > \omega_2$. We have assumed, for simplicity, that the waves have the same amplitude and initial phase. According to the superposition principle, the resultant pressure variation is

$$\Delta P = \Delta P_1 + \Delta P_2 = \Delta P^{\max}(\cos\omega_1 t + \cos\omega_2 t).$$

Using the trigonometric identity (see Appendix E)

$$\cos\alpha + \cos\beta = 2\cos\tfrac{1}{2}(\alpha - \beta)\cos\tfrac{1}{2}(\alpha + \beta)$$

allows us to write the resultant pressure variation as

$$\Delta P = 2\Delta P^{\max}\cos\tfrac{1}{2}(\omega_1 - \omega_2)t\,\cos\tfrac{1}{2}(\omega_1 + \omega_2)t. \tag{18-27}$$

If we write

$$\omega' = \tfrac{1}{2}(\omega_1 - \omega_2) \quad \text{and} \quad \omega = \tfrac{1}{2}(\omega_1 + \omega_2) \tag{18-28}$$

we can then write Eq. 18-27 as

$$\Delta P(t) = [2\Delta P^{\max}\cos\omega' t]\cos\omega t. \tag{18-29}$$

We now assume that the angular frequencies ω_1 and ω_2 of the combining waves are almost equal, which means that $\omega \gg \omega'$ in Eq. 18-28. We can then regard Eq. 18-29 as a cosine function whose angular frequency is ω and whose amplitude (which is not constant but varies with angular frequency ω') is the quantity in the square brackets.

A maximum amplitude will occur whenever $\cos \omega' t$ in Eq. 18-29 has the value $+1$ or -1, which happens twice in each repetition of the cosine function. Because $\cos \omega' t$ has angular frequency ω', the angular frequency ω_{beat} at which beats occur is $\omega_{\text{beat}} = 2\omega'$. Then, with the aid of Eq. 18-28, we can write

$$\omega_{\text{beat}} = 2\omega' = (2)(\tfrac{1}{2})(\omega_1 - \omega_2) = \omega_1 - \omega_2.$$

Because $\omega = 2\pi f$, we can recast this as

$$f_{\text{beat}} = f_1 - f_2 \quad \text{(beat frequency).} \tag{18-30}$$

Musicians use the beat phenomenon in tuning their instruments. If an instrument is sounded against a standard frequency (for example, the lead oboe's reference A) and tuned until the beat disappears, then the instrument is in tune with that standard. In musical Vienna, concert A (440 Hz) is available as a telephone service for the benefit of the city's many professional and amateur musicians.

Types of Superposition

So far the three simple cases we have discussed involving the superposition of sound waves are:

1. *Standing waves* created by the interference of two identical waves moving the opposite directions with the same speed
2. *Wave interference* caused by waves that have the same frequency, wavelength, and amplitude that are moving in almost the same direction and overlap at some point
3. *Beats* created by two waves with the same amplitude moving in the same direction with slightly different frequencies

There are many other more complex types of interference of interest to scientists and engineers. The mathematical techniques used to analyze these situations are similar to those we introduced in this section.

18-7 The Doppler Effect

An ambulance is parked by the side of the highway, sounding its 1000 Hz siren. If you are also parked by the highway, you will hear that same frequency. However, if there is relative motion between you and the ambulance, either toward or away from each other, you will hear a different frequency. For example, if you are driving *toward* the ambulance at 120 km/h (about 75 mi/h), you will hear a *higher* frequency (1096 Hz, an *increase* of 96 Hz). If you are driving *away from* the ambulance at that same speed, you will hear a *lower* frequency (904 Hz, a *decrease* of 96 Hz).

These motion-related frequency changes are examples of the **Doppler effect.** The effect was proposed (although not fully worked out) in 1842 by Austrian physicist Johann Christian Doppler. It was tested experimentally in 1845 by Buys Ballot in Holland, "using a locomotive drawing an open car with several trumpeters."

The Doppler effect holds not only for sound waves but also for electromagnetic waves, including microwaves, radio waves, and visible light. Here, we shall consider only sound waves for the special case where no wind is present. This means that we shall measure the speeds of a source S of sound waves and a detector D of those waves *relative to a body of air that is not moving.* We shall assume that S and D move either directly toward or directly away from each other, at speeds less than the speed of sound.

If either the detector or the source is moving, or both are moving, the emitted frequency f and the detected frequency f' are related by

$$f' = f\frac{v^{air} \pm v_D}{v^{air} \pm v_S} \quad \text{(1D Doppler effect for sound in still air),} \tag{18-31}$$

where v^{air} is the speed of sound through the air, v_D is the detector's speed relative to the air, and v_S is the source's speed relative to the air. The choice of plus or minus signs is set by this rule:

> When a sound source and a detector are moving toward each other, the sign on the sound's speed and the detector's speed must give upward shifts in frequency. When a sound source and a detector are moving away from each other, the signs on the speeds must give downward shifts in frequency.

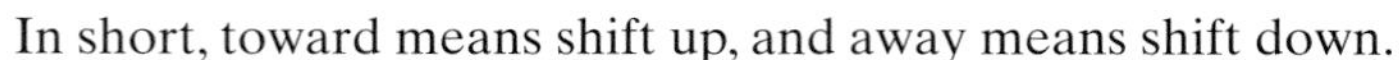

In short, toward means shift up, and away means shift down.

Here are some examples of the rule. If the detector moves toward the source, use the plus sign in the numerator of Eq. 18-31 to get a shift up in the frequency. If it moves away, use the minus sign in the numerator to get a shift down. If it is stationary, substitute 0 for v_D. If the source moves toward the detector, use the minus sign in the denominator of Eq. 18-31 to get a shift up in the frequency. If it moves away, use the plus sign in the denominator to get a shift down. If the source is stationary, substitute 0 for v_S.

Next, we derive equations for the Doppler effect for two specific situations and then derive Eq. 18-31 for the general situation.

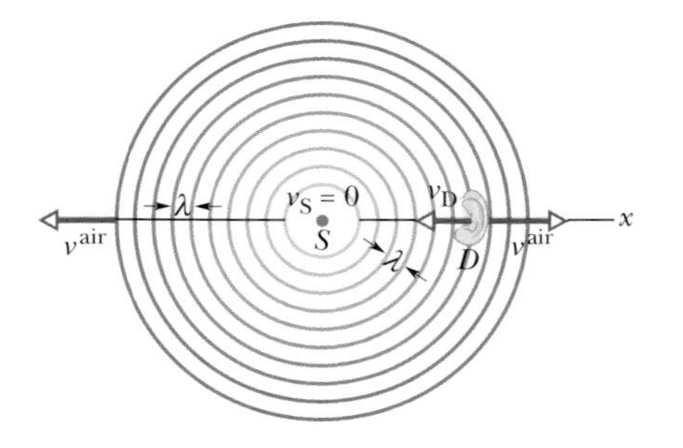

FIGURE 18-22 ■ A stationary source of sound S emits spherical wavefronts, shown one wavelength apart, that expand outward at speed v^{air}. A sound detector D, represented by an ear, moves with velocity $\vec{v}_D$ toward the source. The detector senses a higher frequency because of its motion.

1. When the detector moves relative to the air and the source is stationary relative to the air, the motion changes the frequency at which the detector intercepts wavefronts and thus the detected frequency of the sound wave.
2. When the source moves relative to the air and the detector is stationary relative to the air, the motion changes the wavelength of the sound wave and thus the detected frequency (recall that frequency is related to wavelength).

Detector Moving; Source Stationary

In Fig. 18-22, a detector D (represented by an ear) is moving at speed v_D toward a stationary source S that emits spherical wavefronts, of wavelength λ and frequency f, moving at the speed v^{air} of sound in air. A cross section of the wavefronts are drawn one wavelength apart. The frequency detected by detector D is the rate at which D intercepts wavefronts (or individual wavelengths). If D were stationary, that rate would be f, but since D is moving into the wavefronts, the rate of interception is greater, and thus the detected frequency f' is greater than f.

Let us for the moment consider the situation in which D is stationary (Fig. 18-23). In time t, the wavefronts move to the right a distance $v^{air}t$. The number of wavelengths in that distance $v^{air}t$ is the number of wavelengths intercepted by D in time t, and that number is $v^{air}t/\lambda$. The rate at which D intercepts wavelengths, which is the frequency f detected by D, is

$$f = \frac{v^{air}t/\lambda}{t} = \frac{v^{air}}{\lambda}. \tag{18-32}$$

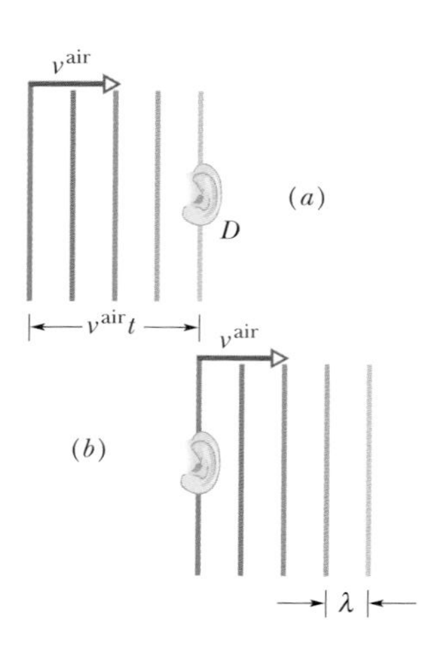

FIGURE 18-23 ■ Wavefronts of Fig. 18-22, assumed planar, (a) reach and (b) pass a stationary detector D; they move a distance $v^{air}t$ to the right in time t.

In this situation, with D stationary, there is no Doppler effect—the frequency detected by D is the frequency emitted by S.

Now let us again consider the situation in which D moves opposite the wavefronts (Fig. 18-24). In time t, the wavefronts move to the right a distance $v^{air}t$ as previously, but now D moves to the left a distance $v_D t$. Thus, in this time t, the distance moved by the wavefronts relative to D is $v^{air}t + v_D t$. The number of wavelengths in this relative distance $v^{air}t + v_D t$ is the number of wavelengths intercepted by D in time t, and is $(v^{air}t + v_D t)/\lambda$. The *rate* at which D intercepts wavelengths in this situation is the frequency f', given by

$$f' = \frac{(v^{air}t + v_D t)/\lambda}{t} = \frac{v^{air} + v_D}{\lambda}. \tag{18-33}$$

From Eq. 18-32, we have $\lambda = v^{air}/f$. Then Eq. 18-33 becomes

$$f' = \frac{v^{air} + v_D}{v^{air}/f} = f\left(\frac{v^{air} + v_D}{v^{air}}\right). \tag{18-34}$$

Note that in Eq. 18-34 f' must be greater than f unless the detector is stationary so that $v_D = 0$.

Similarly, we can find the frequency detected by D if D moves away from the source. In this situation, the wavefronts move a distance $v^{air}t - \vec{v}_D t$ relative to D in time t, and f' is given by

$$f' = f\left(\frac{v^{air} - v_D}{v^{air}}\right). \tag{18-35}$$

In Eq. 18-35, f' must be less than f unless $v_D = 0$.

We can summarize Eqs. 18-34 and 18-35 with

$$f' = f\left(\frac{v^{air} \pm v_D}{v^{air}}\right) \quad \text{(detector moving; source stationary).} \tag{18-36}$$

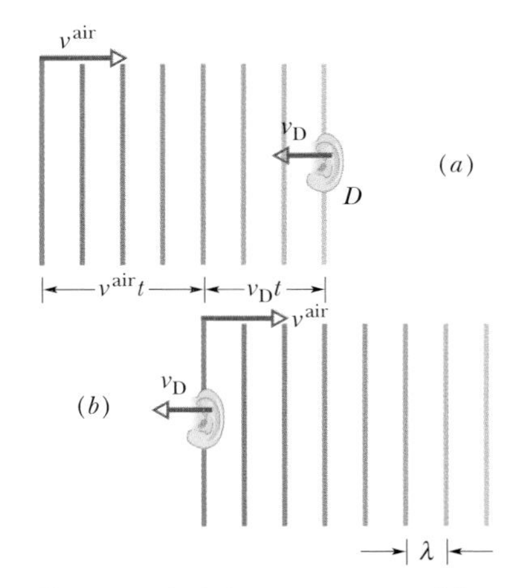

FIGURE 18-24 ■ Wavefronts (*a*) reach and (*b*) pass detector D, which moves opposite the wavefronts. In time t, the wavefronts move a distance $v^{air}t$ to the right and D moves a distance $v_D t$ to the left.

Source Moving; Detector Stationary

Let detector D be stationary with respect to the body of air, and let source S move toward D at speed $|\vec{v}_S|$ (Fig. 18-25). The motion of S changes the wavelength of the sound waves it emits, and thus the frequency detected by D.

To see this change, let $T(= 1/f)$ be the time between the emission of any pair of successive wavefronts W_1 and W_2. During T, wavefront W_1 moves a distance $v^{air}T$ and the source moves a distance $v_S T$. At the end of T, wavefront W_2 is emitted. In the direction in which S moves, the distance between W_1 and W_2, which is the wavelength λ' of the waves moving in that direction, is $v^{air}T - v_S T$. If D detects those waves, it detects frequency f' given by

$$\begin{aligned} f' &= \frac{v^{air}}{\lambda'} = \frac{v^{air}}{v^{air}T - v_S T} = \frac{v^{air}}{v^{air}/f - v_S/f} \\ &= f\frac{v^{air}}{v^{air} - v_S}. \end{aligned} \tag{18-37}$$

Note that f' must be greater than f unless $|\vec{v}_S| = 0$.

In the direction opposite that taken by S, the wavelength λ' of the waves is $v^{air}T - v_S T$. If D detects those waves, it detects frequency f', given by

$$f' = f\frac{v^{air}}{v^{air} + v_S}. \tag{18-38}$$

Now f' must be less than f unless $v_S = 0$.

We can summarize Eqs. 18-37 and 18-38 with

$$f' = f\frac{v^{air}}{v^{air} \pm v_S} \quad \text{(source moving; detector stationary).} \tag{18-39}$$

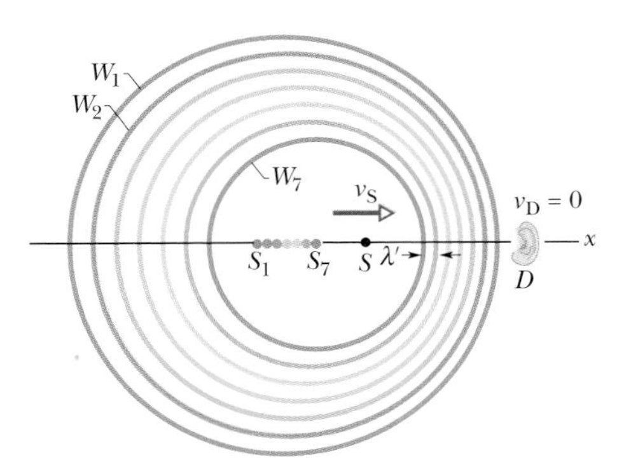

FIGURE 18-25 ■ A detector D is stationary, and a source S is moving toward it at speed v_S. Wavefront W_1 was emitted when the source was at S_1, wavefront W_7 when the source was at S_7. At the moment depicted, the source is at S. The detector perceives a higher frequency because the moving source, chasing its own wavefronts, emits a reduced wavelength λ' in the direction of its motion.

General Doppler Effect Equation for Sound

For the one-dimensional case, we can now derive the general Doppler effect equation by replacing f in Eq. 18-39 (the frequency of the source) with f' of Eq. 18-36 (the frequency associated with motion of the detector). The result is Eq. 18-31 for the general Doppler effect.

That general equation holds not only when both detector and source are moving but also in the two specific situations we just discussed. For the situation in which the detector is moving and the source is stationary, substitution of $v_S = 0$ into Eq. 18-31 gives us Eq. 18-36, which we previously found. For the situation in which the source is moving and the detector is stationary, substitution of $v_D = 0$ into Eq. 18-30 gives us Eq. 18-39, which we previously found. Thus, Eq. 18-31 is the equation to remember.

Similar Doppler frequency shifts occur with electromagnetic waves treated in Chapters 34 and 38, but the equations are slightly different (see Section 38-14).

The Doppler Effect in More Than One Dimension

We can even use Eqs. 18-36 and 18-39 if the source and/or the detector are moving relative to each other at constant velocities that *don't lie along the same line.* Simply draw a line from one object to the other at a particular time when the sound of interest is being emitted and call it the x axis. Start by finding the x-components of the velocities. You can then use the speeds along the x axis in Eq. 18-36 or Eq. 18-37 to determine the frequency shifts. For example, the Doppler shift for light waves is used by astronomers to help them determine how fast a distant object is moving relative to Earth. However, they can only use Doppler shift measurements to find the object's velocity *component* along a line between Earth and the object.

Bat Navigation and Feeding

An example of the use of the three-dimensional Doppler effect in nature is provided by bats. Bats navigate and search out prey by emitting and then detecting reflections of ultrasonic waves. These are sound waves with frequencies greater than can be heard by a human.* For example, a horseshoe bat emits ultrasonic waves at 83 kHz, well above the 20 kHz limit of human hearing.

After the sound is emitted through the bat's nostrils, it might reflect (echo) from a moth, and then return to the bat's ears. The motions of the bat and the moth relative to the air cause the frequency heard by the bat to differ by a few kilohertz from the frequency it emitted. The bat automatically translates this difference into a relative speed between itself and the moth, so it can zero in on the moth.

Some moths evade capture by flying away from the direction in which they hear ultrasonic waves. That choice of flight path reduces the frequency difference between what the bat emits and what it hears, and then the bat may not notice the echo. Some moths avoid capture by clicking to produce their own ultrasonic waves, thus "jamming" the detection system and confusing the bat. (Surprisingly, moths and bats do all this without first studying physics.)

READING EXERCISE 18-5: The figure indicates the directions of motion of a sound source and a detector for six situations in stationary air. In general, $v_0 \neq v_S$. For each situation, is the detected frequency greater than or less than the emitted frequency, or can't we tell without more information about the actual speeds?

	Source	Detector		Source	Detector
(*a*)	⟶	• 0 speed	(*d*)	⟵	⟵
(*b*)	⟵	• 0 speed	(*e*)	⟶	⟵
(*c*)	⟶	⟶	(*f*)	⟵	⟶

■

*Ultrasonic motion detectors use the same technique for locating objects.

TOUCHSTONE EXAMPLE 18-6: Rocket Sounds

A rocket moves at a speed of 242 m/s directly toward a stationary pole (through stationary air) while emitting sound waves at frequency $f = 1250$ Hz.

(a) What frequency f' is measured by a detector that is attached to the pole?

SOLUTION ■ We can find f' with Eq. 18-31 for the general Doppler effect. The **Key Idea** here is that, because the sound source (the rocket) moves through the air *toward* the stationary detector on the pole, we need to choose the sign on v_S that gives a *shift up* in the frequency of the sound. Thus, in Eq. 18-31 we use the minus sign in the denominator. We then substitute 0 for the detector speed v_D, 242 m/s for the source speed v_S, 343 m/s for the speed of sound v^{air} (from Table 18-1), and 1250 Hz for the emitted frequency f. We find

$$f' = f\frac{v^{air} \pm v_D}{v^{air} \pm v_S} = (1250\text{ Hz})\frac{343\text{ m/s} \pm 0}{343\text{ m/s} - 242\text{m/s}},$$

$$= 4245\text{ Hz} \approx 4250\text{ Hz}, \qquad \text{(Answer)}$$

which, indeed, is a greater frequency than the emitted frequency.

(b) Some of the sound reaching the pole reflects back to the rocket as an echo. What frequency f'' does a detector on the rocket detect for the echo?

SOLUTION ■ Two **Key Ideas** here are the following:

1. The pole is now the source of sound (because it is the source of the echo), and the rocket's detector is now the detector (because it detects the echo).
2. The frequency of the sound emitted by the source (the pole) is equal to f', the frequency of the sound the pole intercepts and reflects.

We can rewrite Eq. 18-31 in terms of the source frequency f' and the detected frequency f'' as

$$f'' = f'\frac{v^{air} \pm v_D}{v^{air} \pm v_S}. \qquad (18\text{-}40)$$

A third **Key Idea** here is that, because the detector (on the rocket) moves through the air *toward* the stationary source, we need to use the *sign* on v_D that gives a *shift up* in the frequency of the sound. Thus, we use the plus sign in the numerator of Eq. 18-40. Also, we substitute $v_D = 242$ m/s, $v_S = 0$, $v^{air} = 343$ m/s, and $f' = 4245$ Hz. We find

$$f'' = (4245\text{ Hz})\frac{343\text{ m/s} + 242\text{ m/s}}{343\text{ m/s} \pm 0}$$

$$= 7240\text{ Hz}, \qquad \text{(Answer)}$$

which, indeed, is greater than the frequency of the sound reflected by the pole.

18-8 Supersonic Speeds; Shock Waves

If a source is moving toward a stationary detector at a speed equal to the speed of sound in a medium—that is, if $v_S = v^{air}$ or $v_S = v^{water}$ and so on—Eqs. 18-31 and 18-39 predict that the detected frequency f' will be infinitely great. This means that the

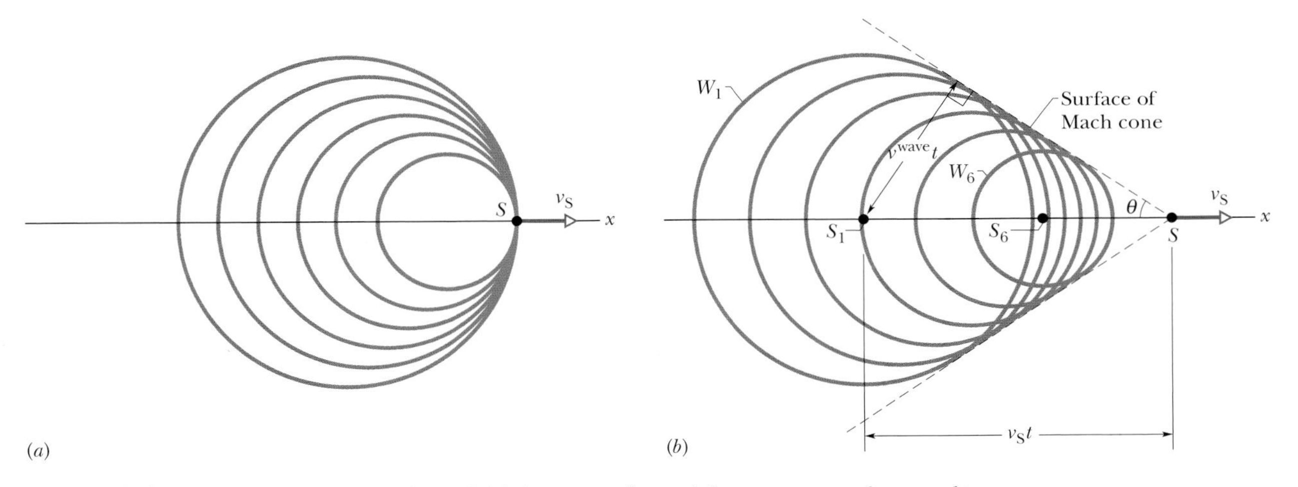

FIGURE 18-26 ■ Cross-sectional drawings of: (*a*) A source of sound S moves at speed v_S equal to the speed of sound and thus as fast as the wavefronts it generates. (*b*) A source S moves at speed v_S faster than the speed of sound and thus faster than the wavefronts. When the source was at position S_1 it generated wavefront W_1, and at position S_6 it generated W_6. All the spherical wavefronts expand at the speed of sound v^{wave} and bunch along the surface of a cone called the Mach cone, forming a shock wave. The surface of the cone has half-angle θ and is tangent to all the wavefronts.

source is moving so fast that it keeps pace with its own spherical wavefronts, as Fig. 18-26*a* suggests. What happens when the speed of the source exceeds the speed of sound?

For such supersonic speeds, Eqs. 18-31 and 18-39 no longer apply. Figure 18-26*b* depicts the spherical wavefronts that originated at various positions of the source. The radius of any wavefront in this figure is $v^{wave}t$, where v^{wave} is the speed of sound in the medium (for example, v^{air} or v^{water}) and t is the time that has elapsed since the source emitted that wavefront. Note that all the wavefronts bunch along a V-shaped envelope in the two-dimensional drawing of Fig. 18-26*b*. The wavefronts actually extend in three dimensions, and the bunching forms a cone called the *Mach cone*. A *shock wave* is said to exist along the surface of this cone, because the bunching of wavefronts causes an abrupt rise and fall of air pressure as the surface passes by any point. An observer hears the wavefront bunching as a sharp loud sound known as a **sonic boom.** From Fig. 18-26*b*, we see that the half-angle θ of the cone, called the *Mach cone angle,* is given by

$$\sin\theta = \frac{v^{wave}t}{v_S t} = \frac{v^{wave}}{v_S} \quad \text{(Mach cone angle).} \tag{18-41}$$

FIGURE 18-27 ■ A cloud formation is produced off the wings of this Navy jet. One plausible explanation is that the cloud forms because of the sudden decrease in air pressure that results as a supersonic shock wave propagates. This causes water vapor in the air to condense.

The ratio v_S/v^{wave} is called the *Mach number.* When you hear that a particular plane has flown at Mach 2.3, it means that its speed was 2.3 times the speed of sound in the air through which the plane was flying. There is a common misconception that a sonic boom is a single burst of sound that is generated at the moment a plane breaks the sound barrier. However, the shock wave is generated continuously as long as the speed of the plane is greater than the speed of sound. When a shock wavefront generated by a supersonic aircraft (Fig. 18-27) passes by an observer she hears the sonic boom.

Part of the sound that is heard when a rifle is fired is the sonic boom produced by the bullet. A sonic boom can also be heard from a long bullwhip when it is snapped quickly: Near the end of the whip's motion, its tip is moving faster than sound and produces a small sonic boom—the *crack* of the whip.

Problems

Where needed in the problems, use

speed of sound in air = 343 m/s *and*

density of air = 1.21 kg/m^3

unless otherwise specified.

SEC. 18-2 ■ THE SPEED OF SOUND

1. Devise a Rule Devise a rule for finding your distance in kilometers from a lightning flash by counting the seconds from the time you see the flash until you hear the thunder. Assume that the sound travels to you along a straight line.

2. Outdoor Concert You are at a large outdoor concert, seated 300 m from the speaker system. The concert is also being broadcast live via satellite (at the speed of light, 3.0×10^8 m/s). Consider a listener 5000 km away who receives the broadcast. Who hears the music first, you or the listener and by what time difference?

3. Two Spectators Two spectators at a soccer game in Montjuic Stadium see, and a moment later hear, the ball being kicked on the playing field. The time delay for one spectator is 0.23 s and for the other 0.12 s. Sight lines from the two spectators to the player kicking the ball meet at an angle of 90°. (a) How far is each spectator from the player? (b) How far are the spectators from each other?

4. Column of Soldiers A column of soldiers, marching at 120 paces per minute, keep in step with the beat of a drummer at the head of the column. It is observed that the soldiers in the rear end of the column are striding forward with the left foot when the drummer is advancing with the right. What is the approximate length of the column?

5. Earthquakes Earthquakes generate sound waves inside Earth. Unlike a gas, Earth can experience both transverse (S) and longitudinal (P) sound waves. Typically, the speed of S waves is about 4.5 km/s, and that of P waves 8.0 km/s. A seismograph records P and S waves from an earthquake. The first P waves arrive 3.0 min before the first S waves (Fig. 18-28). Assuming the waves travel in a straight line, how far away does the earthquake occur?

6. Speed of Sound The speed of sound in a certain metal is v^{metal}. One end of a long pipe of that metal of length L is struck a hard blow. A listener at the other end hears two sounds, one from the wave that travels along the pipe and the other from the wave that travels through the air. (a) If v^{air} is the speed of sound in air, what time interval Δt elapses between the arrivals of the two sounds? (b) Suppose that $\Delta t = 1.00$ s and the metal is steel. Find the length L.

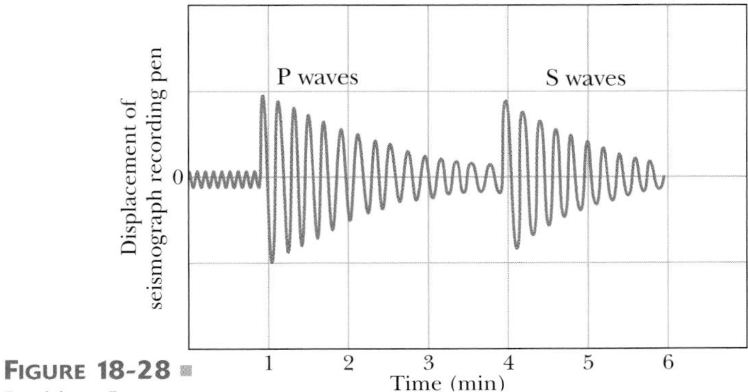

FIGURE 18-28 ■ Problem 5.

7. Stone Is Dropped A stone is dropped into a well. The sound of the splash is heard 3.00 s later. What is the depth of the well?

8. Audible Frequency The audible frequency range for normal hearing is from about 20 Hz to 20 kHz. What are the wavelengths of sound waves at these frequencies?

9. Diagnostic Ultrasound Diagnostic ultrasound of frequency 4.50 MHz is used to examine tumors in soft tissue. (a) What is the wavelength in air of such a sound wave? (b) If the speed of sound in tissue is 1500 m/s, what is the wavelength of this wave in tissue?

10. Pressure in Traveling Wave The pressure in a traveling sound wave is given by the equation

$$\Delta P(x, t) = (1.50 \text{ Pa}) \sin \pi [(0.900 \text{ rad/m})x - (315 \text{ rad/s})t].$$

Find the (a) pressure amplitude, (b) frequency, (c) wavelength, and (d) speed of the wave.

SEC. 18-3 ■ INTERFERENCE

11. Two Loudspeakers In Fig. 18-29, two loudspeakers, separated by a distance of 2.00 m, are in phase. Assume the amplitudes of the sound from the speakers are approximately the same at the position of a listener, who is 3.75 m directly in front of one of the speakers. (a) For what frequencies in the audible range (20 Hz to 20 kHz) does the listener hear a minimum signal? (b) For what frequencies is the signal a maximum?

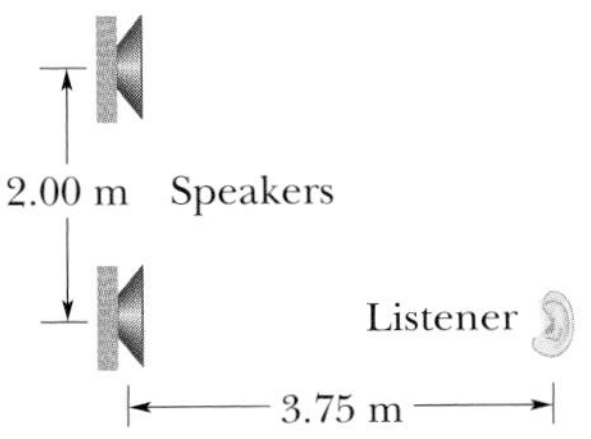

FIGURE 18-29 ■ Problem 11

12. Two Point Sources Two point sources of sound waves of identical wavelength λ and amplitude are separated by distance $D = 2.0\lambda$. The sources are in phase. (a) How many points of maximum signal (that is, maximum constructive interference) lie along a large circle around the sources? (b) How many points of minimum signal (destructive interference) lie around the circle?

13. Loudspeakers on Outdoor Stage Two loudspeakers are located 3.55 m apart on an outdoor stage. A listener is 18.3 m from one and 19.5 m from the other. During the sound check, a signal generator drives the two speakers in phase with the same amplitude and frequency. The transmitted frequency is swept through the audible range (20 Hz to 20 kHz). (a) What are the three lowest frequencies at which the listener will hear a minimum signal because of destructive interference? (b) What are the three lowest frequencies at which the listener will hear a maximum signal?

14. Two Sound Waves Two sound waves, from two different sources with the same frequency, 540 Hz, travel in the same direction at 330 m/s. The sources are in phase. What is the phase difference of the waves at a point that is 4.40 m from one source and 4.00 m from the other?

15. Half-Circle In Fig. 18-30, sound with a 40.0 cm wavelength travels rightward from a source and through a tube that consists of a straight portion and a half-circle. Part of the sound wave travels through the half-circle and then rejoins the rest of the wave, which goes directly through the straight portion. This rejoining results in interference. What is the smallest radius r that results in an intensity minimum at the detector?

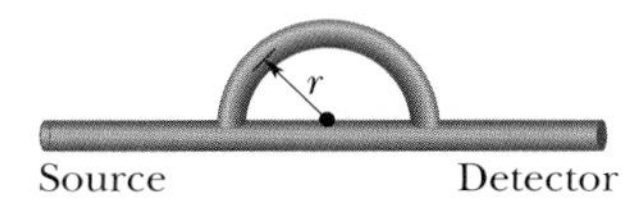

FIGURE 18-30 ■ Problem 15.

SEC. 18-4 ■ INTENSITY AND SOUND LEVEL

16. Point Source A 1.0 W point source emits sound waves isotropically. Assuming that the energy of the waves is conserved, find the intensity (a) 10 m from the source and (b) 2.5 m from the source.

17. A Source Emits A source emits sound waves isotropically. The intensity of the waves 2.50 m from the source is 1.91×10^{-4} W/m^2. Assuming that the energy of the waves is conserved, find the power of the source.

18. Differ in Level Two sounds differ in sound level by 1.00 dB. What is the ratio of the greater intensity to the smaller intensity?

19. Increased in Level A certain sound source is increased in sound level by 30 dB. By what multiple is (a) its intensity increased and (b) its pressure amplitude increased?

20. The Source of a Sound The source of a sound wave has a power of 1.00 μW. If it is a point source, (a) what is the intensity 3.00 m away and (b) what is the sound level in decibels at that distance?

21. One in Air, One in Water (a) If two sound waves, one in air and one in (fresh) water, are equal in intensity, what is the ratio of the pressure amplitude of the wave in water to that of the wave in air? Assume the water and the air are at 20°C. (See Table 15-2.) (b) If the pressure amplitudes are equal instead, what is the ratio of the intensities of the waves?

22. Noisy Freight Train Assume that a noisy freight train on a straight track emits a cylindrical, expanding sound wave, and that the air absorbs no energy. How does the amplitude $\Delta P^{\max}$ of the wave depend on the perpendicular distance r from the source?

23. Ratios Find the ratios (greater to smaller) of (a) the intensities, and (b) the pressure amplitudes for two sounds whose sound levels differ by 37 dB.

24. Point Source Two A point source emits 30.0 W of sound isotropically. A small microphone intercepts the sound in an area of 0.750 cm^2, 200 m from the source. Calculate (a) the sound intensity there and (b) the power intercepted by the microphone.

25. Acoustic Interferometer Figure 18-31 shows an air-filled, acoustic interferometer, used to demonstrate

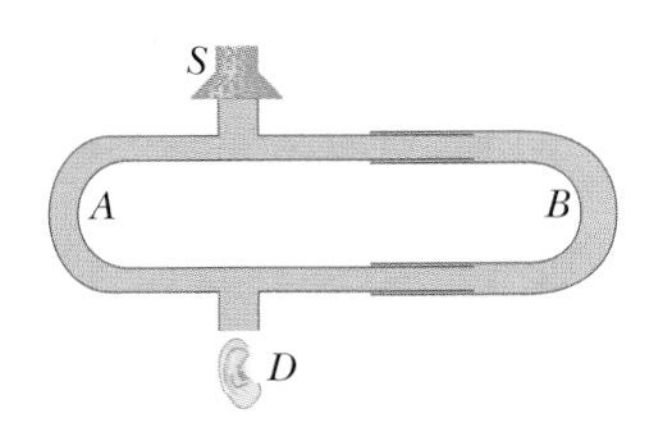

FIGURE 18-31 ■ Problem 25.

the interference of sound waves. Sound source S is an oscillating diaphragm; D is a sound detector, such as the ear or a microphone. Path SBD can be varied in length, but path SAD is fixed. At D, the sound wave coming along path SBD interferes with that coming along path SAD. In one demonstration, the sound intensity at D has a minimum value of 100 units at one position of the movable arm and continuously climbs to a maximum value of 900 units when that arm is shifted by 1.65 cm. Find (a) the frequency of the sound emitted by the source and (b) the ratio of the amplitude at D of the SAD wave to that of the SBD wave. (c) How can it happen that these waves have different amplitudes, considering that they originate at the same source?

SEC. 18-5 ■ SOURCES OF MUSICAL SOUND

26. A Violin String A violin string 15.0 cm long and fixed at both ends oscillates in its $n = 1$ mode. The speed of waves on the string is 250 m/s, and the speed of sound in air is 348 m/s. What are (a) the frequency and (b) the wavelength of the emitted sound wave?

27. Organ Pipe Organ pipe A, with both ends open, had a fundamental frequency of 300 Hz. The third harmonic of organ pipe B, with one end open, has the same frequency as the second harmonic of pipe A. How long are (a) pipe A and (b) pipe B?

28. Glass Tube The water level in a vertical glass tube 1.00 m long can be adjusted to any position in the tube. A tuning fork vibrating at 686 Hz is held just over the open top end of the tube, to set up a standing wave of sound in the air-filled top portion of the tube. (That air-filled top portion acts as a tube with one end closed and the other end open.) At what positions of the water level is there resonance?

29. Speed of Waves (a) Find the speed of waves on a violin string of mass 800 mg and length 22.0 cm if the fundamental frequency is 920 Hz. (b) What is the tension in the string? For the fundamental, what is the wavelength of (c) the waves on the string and (d) the sound waves emitted by the string?

30. Violin String A certain violin string is 30 cm long between its fixed ends and has a mass of 2.0 g. The "open" string (no applied finger) sounds an A note (440 Hz). (a) To play a C note (523 Hz), how far down the string must one place a finger? (b) What is the ratio of the wavelength of the string waves required for an A note to that required for a C note? (c) What is the ratio of the wavelength of the sound wave for an A note to that for a C note?

31. Small Loudspeaker In Fig. 18-32, S is a small loudspeaker driven by an audio oscillator and amplifier, adjustable in frequency from 1000 to 2000 Hz only. Tube D is a piece of cylindrical sheet-metal pipe 45.7 cm long and open at both ends. (a) If the speed of sound in air is 344 m/s at the existing temperature, at what frequencies will resonance occur in the pipe when the frequency emitted by the speaker is varied from 1000 Hz to 2000 Hz? (b) Sketch the standing wave (using the style of Fig. 18-16b) for each resonant frequency.

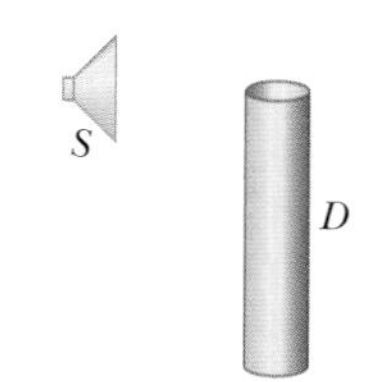

FIGURE 18-32 ■ Problem 31.

32. Cello String A string on a cello has length L, for which the fundamental frequency is f. (a) By what length l must the string be shortened by fingering to change the fundamental frequency to rf? (b) What is l if $L = 0.80$ m and $r = 1.2$? (c) For $r = 1.2$, what is the ratio of the wavelength of the new sound wave emitted by the string to that of the wave emitted before fingering?

33. Well A well with vertical sides and water at the bottom resonates at 7.00 Hz and at no lower frequency. (The air-filled portion of the well acts as a tube with one closed end and one open end.) The air in the well has a density of 1.10 kg/m^3 and a bulk modulus of 1.33×10^5 Pa. How far down in the well is the water surface?

34. A Tube A tube 1.20 m long is closed at one end. A stretched wire is placed near the open end. The wire is 0.330 m long and has a mass of 9.60 g. It is fixed at both ends and oscillates in its fundamental mode. By resonance, it sets the air column in the tube into oscillation at that column's fundamental frequency. Find (a) that frequency and (b) the tension in the wire.

35. Pulsating Variable Star The period of a pulsating variable star may be estimated by considering the star to be executing *radial* longitudinal pulsations in the fundamental standing wave mode. That is, the star's radius varies periodically with time, with a displacement antinode at the star's surface. (a) Would you expect the center of the star to be a displacement node or antinode? (b) By analogy with a pipe with one open end, show that the period of pulsation T is given by

$$T = \frac{4R}{\langle v \rangle},$$

where R is the equilibrium radius of the star and $\langle v \rangle$ is the average sound speed in the material of the star. (c) Typical white dwarf stars are composed of material with a bulk modulus of 1.33×10^{22} Pa and a density of 10^{10} kg/m^3. They have radii equal to 9.0×10^{-3} solar radius. What is the approximate pulsation period of a white dwarf?

36. Pipes A and B Pipe A, which is 1.2 m long and open at both ends, oscillates at its third lowest harmonic frequency. It is filled with air for which the speed of sound is 343 m/s. Pipe B, which is closed at one end, oscillates at its second lowest harmonic frequency. These frequencies of pipes A and B happen to match. (a) If an x axis extends along the interior of pipe A, with $x = 0$ at one end, where along the axis are the displacement nodes? (b) How long is pipe B? (c) What is the lowest harmonic frequency of pipe A?

37. Violin String and Loudspeaker A violin string 30.0 cm long with linear density 0.650 g/m is placed near a loudspeaker that is fed by an audio oscillator of variable frequency. It is found that the string is set into oscillation only at the frequencies 880 and 1320 Hz as the frequency of the oscillator is varied over the range 500–1500 Hz. What is the tension in the string?

SEC. 18-6 ■ BEATS

38. Too Tightly Stretched The A string of a violin is a little too tightly stretched. Four beats per second are heard when the string is sounded together with a tuning fork that is oscillating accurately at concert A (440 Hz). What is the period of the violin string oscillation?

39. Unknown Tuning Fork A tuning fork of unknown frequency makes three beats per second with a standard fork of frequency

384 Hz. The beat frequency decreases when a small piece of wax is put on a prong of the first fork. What is the freauency of this fork?

40. Five Tuning Forks You have five tuning forks that oscillate at close but different frequencies. What are the (a) maximum and (b) minimum number of different beat frequencies you can produce by sounding the forks two at a time depending on how the frequencies differ?

41. Two Piano Wires Two identical piano wires have a fundamental frequency of 600 Hz when kept under the same tension. What fractional increase in the tension of one wire will lead to the occurrence of 6 beats/s when both wires oscillate simultaneously?

SEC. 18-7 ■ THE DOPPLER EFFECT

42. Trooper *B* Trooper B is chasing speeder A along a straight stretch of road. Both are moving at a speed of 160 km/h. Trooper B, failing to catch up, sounds his siren again. Take the speed of sound in air to be 343 m/s and the frequency of the source to be 500 Hz. What is the Doppler shift in the frequency heard by speeder A?

43. Turbine Whine The 16 000 Hz whine of the turbines in the jet engines of an aircraft moving with speed 200 m/s is heard at what frequency by the pilot of a second craft trying to overtake the first at a speed of 250 m/s?

44. Ambulance Siren An ambulance with a siren emitting a whine at 1600 Hz overtakes and passes a cyclist pedaling a bike at 2.44 m/s. After being passed, the cyclist hears a frequency of 1590 Hz. How fast is the ambulance moving?

45. A Whistle A whistle of frequency 540 Hz moves in a circle of radius 60.0 cm at a rotational speed of 15.0 rad/s. What are (a) the lowest and (b) the highest frequencies heard by a listener a long distance away, at rest with respect to the center of the circle?

46. Motion Detector A stationary motion detector sends sound waves of frequency 0.150 MHz toward a truck approaching at a speed of 45.0 m/s. What is the frequency of the waves reflected back to the detector?

47. French Submarine A French submarine and a U.S. submarine move toward each other during maneuvers in motionless water in the North Atlantic (Fig. 18-33). The French sub moves at 50.0 km/h, and the U.S. sub at 70.0 km/h. The French sub sends out a sonar signal (sound wave in water) at 1000 Hz. Sonar waves travel at 5470 km/h. (a) What is the signal's frequency as detected by the U.S. sub? (b) What frequency is detected by the French sub in the signal reflected back to it by the U.S. sub?

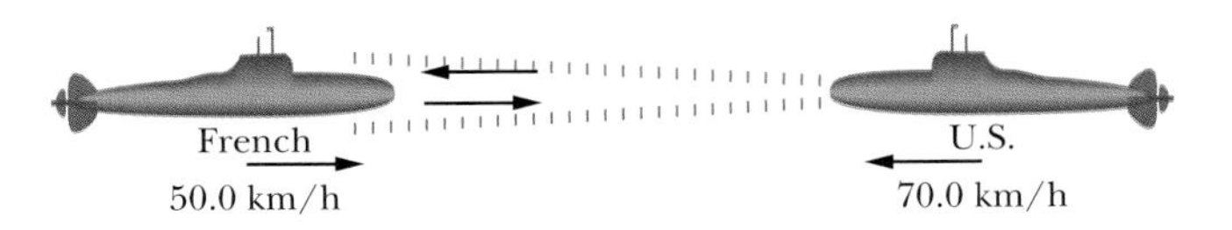

FIGURE 18-33 ■ Problem 47.

48. Sound Source A sound source A and a reflecting surface B move directly toward each other. Relative to the air, the speed of source A is 29.9 m/s, the speed of surface B is 65.8 m/s, and the speed of sound is 329 m/s. The source emits waves at frequency 1200 Hz as measured in the source frame. In the reflector frame, what are (a) the frequency and (b) the wavelength of the arriving sound waves? In the source frame, what are (c) the frequency and (d) the wavelength of the sound waves reflected back to the source?

49. Burglar Alarm An acoustic burglar alarm consists of a source emitting waves of frequency 28.0 kHz. What is the beat frequency between the source waves and the waves reflected from an intruder walking at an average speed of 0.950 m/s directly away from the alarm?

50. A Bat A bat is flitting about in a cave, navigating via ultrasonic bleeps. Assume that the sound emission frequency of the bat is 39 000 Hz. During one fast swoop directly toward a flat wall surface, the bat is moving at 0.025 times the speed of sound in air. What frequency does the bat hear reflected off the wall?

51. Girl in Window A girl is sitting near the open window of a train that is moving at a velocity of 10.00 m/s to the east. The girl's uncle stands near the tracks and watches the train move away. The locomotive whistle emits sound at frequency 500.0 Hz. The air is still. (a) What frequency does the uncle hear? (b) What frequency does the girl hear? A wind begins to blow from the east at 10.00 m/s. (c) What frequency does the uncle now hear? (d) What frequency does the girl now hear?

52. Civil Defense Official A 2000 Hz siren and a civil defense official are both at rest with respect to the ground. What frequency does the official hear if the wind is blowing at 12 m/s (a) from source to official and (b) from official to source?

53. Two Trains Two trains are traveling toward each other at 30.5 m/s relative to the ground. One train is blowing a whistle at 500 Hz. (a) What frequency is heard on the other train in still air? (b) What frequency is heard on the other train if the wind is blowing at 30.5 m/s toward the whistle and away from the listener? (c) What frequency is heard if the wind direction is reversed?

SEC. 18-8 ■ SUPERSONIC SPEEDS; SHOCK WAVES

54. Bullet Fired A bullet is fired with a speed of 685 m/s. Find the half angle made by the shock cone with the line of motion of the bullet.

55. Jet Plane A jet plane passes over you at a height of 5000 m and a speed of Mach 1.5. (a) Find the Mach cone half angle. (b) How long after the jet passes directly overhead does the shock wave reach you? Use 331 m/s for the speed of sound.

56. Plane Flies A plane flies at 1.25 times the speed of sound. Its sonic boom reaches a man on the ground 1.00 min after the plane passes directly overhead. What is the altitude of the plane? Assume the speed of sound to be 330 m/s.

Additional Problems

57. Building a Pipe Organ You decide to build a pipe organ in your dormitory room using PVC pipe. Estimate whether you could build an organ that would cover the entire range of human hearing without bending any pipes.

58. Arranging the Patio Speakers You have set up two stereo speakers on your back patio railing as shown in the top view diagram in Fig. 18-34. You are worried that at certain positions you will lose frequencies as a result of interference. The coordinate grid on the edge of the picture has its large tick marks separated by 1 meter. For ease of calculation, make the following assumptions:

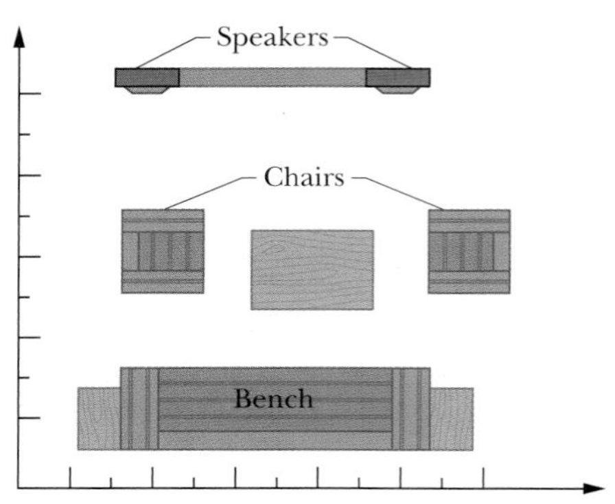

FIGURE 18-34 ■ Problem 58.

- Assume that the relevant objects lie on integer or half-integer grid points of the coordinate system.
- Take the speed of sound to be 343 m/s.
- Ignore the reflection of sound from the house, trees, and so on.
- The speakers are in phase.

(a) What will happen if you are sitting in the middle of the bench? (b) If you are sitting in the lawn chair on the left, what will be the lowest frequency you will lose to destructive interference? (c) Can you restore the frequency lost in part (a) by switching the leads to one of the speakers, thereby reversing the phase of that source? (d) With the leads reversed, what will happen to the sound for a person sitting at the center of the bench?

59. Truthful Salesman? A salesperson claimed that a stereo system had a maximum audio power of 120 W. Testing the system with several speakers set up so as to simulate a point source, the consumer noted that she could get as close as 1.2 m with the volume full on before the sound hurt her ears. Was the salesperson truthful? Explain your answer with a calculation.

60. Experimenter An experimenter wishes to measure the speed of sound in an aluminum rod 10 cm long by measuring the time it takes for a sound pulse to travel the length of the rod. If results good to four significant figures are desired, how precisely must the length of the rod be known and how closely must the experimenter be able to resolve time intervals?

19 The First Law of Thermodynamics

The giant hornet *Vespa mandarinia japonica* preys on Japanese bees. However, if one of the hornets attempts to invade a bee hive, several hundred of the bees quickly form a compact ball around the hornet to stop it. After about 20 minutes the hornet is dead, although the bees do not sting, bite, crush, or suffocate it.

Why, then, does the hornet die?

The answer is in this chapter.

19-1 Thermodynamics

In the next three chapters we focus on a new subject—thermodynamics. The development of thermodynamic principles is one of humankind's most profound intellectual achievements. Why? The steam engine that powered the industrial revolution operates according to thermodynamic principles, as do many modern power plants. Thermodynamics has enriched our fundamental understanding of phenomena ranging from the metabolism of a lizard to the evolution of the universe.

Recall from Chapter 10 that the total mechanical energy of a system is the sum of the *macroscopic* kinetic and potential energies associated with the motion and configuration of the objects within the system. By macroscopic we mean that the particles in a system were large enough so that we could observe how fast they were moving (associated with kinetic energy) and also see their configuration (associated with potential energy). We found that when conservative forces act on the particles in an isolated system, the system's mechanical energy is conserved even if its potential energy is converted to kinetic energy or vice versa. Also recall that we can add mechanical energy to a system by doing work on it.

Furthermore, in Chapter 10 we observed that when a block slides along a surface and comes to a stop due to a nonconservative friction force, the temperatures of the block and surface rise. A decrease in the total mechanical energy of the block-surface system was accompanied by an increase in the temperature of the parts of the system. In response to this observation, we defined a new kind of energy, **thermal energy,** E^{thermal}, associated with the temperature of an object. Then, the overall system energy would still be conserved.

Our study of thermodynamics begins with learning how to quantify the hidden *internal* energy stored in ordinary matter on a microscopic (and hence invisible) scale. It also involves an examination of the role that temperature plays in determining whether a system's internal energy will increase or decrease when it comes into contact with another system. But, when hot steam is injected into a cylinder with a piston on top of it, the piston can be raised. So we believe that under other circumstances the internal energy in a system can be transformed back into mechanical energy.

In this chapter we consider temperature and various ways it can be measured. We then quantify the invisible transfer of thermal energy between objects of different temperatures. We also introduce the first law of thermodynamics, a statement of energy conservation, which relates internal energy change to both the thermal energy transferred to or from a system and the mechanical work done on or by it.

In Chapter 20 we introduce **kinetic theory** as an idealized model of how microscopic scale kinetic and potential energies associated with the motions and configurations of atoms and molecules can be added together to explain internal energy. The 19th century development of kinetic theory to explain internal energy is similar in character and importance to Einstein's 20th century discovery that matter also contains hidden energy by virtue of its mass. We then begin a study of heat engines by considering how the internal energy in hot expanding gases can be transformed into mechanical work.

In Chapter 21 we explain the principles that govern the operation of heat engines. We than introduce the concept of entropy that helps us understand why no one has ever invented a heat engine that can transform internal energy into mechanical work with anything close to 100% efficiency.

19-2 Thermometers and Temperature Scales

Temperature and its measurement are central to understanding the behavior of macroscopic systems that are heated and cooled. Although we have a natural ability to sense hot and cold, we can only use our sense of touch to tell whether an object is hot or cold over a relatively narrow range of temperatures. But, we will now need to *quantify* our intuitive sense of hotness. Recall that any characteristic of a material or object that is measurable can be referred to as a **quantity** or **measurable property.** In other words, a measurable property is one that can be quantified through physical comparison with a reference (Sections 1-1 and 1-2). Careful observation of our everyday world tells us that some objects such as a balloon full of air or a metal rod have characteristics that change as the object gets hotter or colder. For example, put a balloon full of air into the freezer and you can observe for yourself that it gets smaller. A metal rod may grow a little longer when heated. Volume, length, electrical resistance, and pressure are examples of measurable properties of a material object that can change with temperature. We can use any one of the properties of materials that change as an object gets hotter or colder to design crude (or not so crude) devices that quantify the hotness of an object. Such a device is called a **thermometer.**

Designing an accurate thermometer is not a trivial task. Nevertheless, thermometers are common devices and so many of us have a basic, "common sense" understanding of what a thermometer is and how to use one. We begin our study of thermodynamics with temperature. We will also reconsider this topic later in the chapter when we will refine and expand our understanding of the concept of temperature and its measurement.

Most, but not all, substances expand when heated. You can loosen a tight metal lid on a jar by holding it under a stream of hot water. Both the metal of the lid and the glass of the jar expand as the hot water transfers some of its hidden (or *internal*) thermal energy to both the jar and lid. This happens because with the added energy, the atoms in the lid can move a bit farther from each other than usual, pulling against the spring-like interatomic forces that hold every solid together. (See Chapter 6 if you need to remind yourself of our spring-atom model for solids.) However, the metal lid expands more than the glass, and so the lid is loosened. The familiar sealed liquid-in-glass thermometer (Fig. 19-1*a*) works in a similar way. The mercury or colored alcohol contained in the hollow glass bulb and tube expands more than the glass that surrounds it.

We call the transfer of energy from a "hotter" system to a "colder" system by invisible atomic and molecular collisions and chemical reactions **thermal energy transfer.** It is important to note that thermal energy transfers often involve no exchange of matter. For example, in the situation described above, no water penetrates the jar or its lid. How is this possible? Imagine a high kinetic energy cue ball colliding with a ball at rest in the middle of a billiard table. The cue ball loses all its energy and the other ball gains the amount that was lost—a massless transfer! So we can imagine that if the molecules in the hot water are vibrating more vigorously than the glass molecules in the jar, energy is transferred from the water molecules to the jar molecules.

Suppose you dip an unmarked liquid thermometer filled with red colored alcohol into a cup of cold water. Since the water feels cold to the touch you can make a scratch in the tube where the liquid stands and define that height as a cold temperature. Then you can transfer your unmarked thermometer to a cup of water that feels hot to the touch. You make another scratch where the liquid now stands and define that height as a hot temperature. What if you dip the thermometer into a cup of water that feels neither hot nor cold to your touch and the height of the alcohol is halfway between? Is it reasonable to assume that the new temperature is halfway between the two original temperatures? The assumption that this is true has guided the development of the historical Fahrenheit and Celsius temperature scales, and it is not too far from the truth. Thus, it is useful to start our study of thermodynamics with a very crude definition of temperature change as *a quantity that is proportional to changes of the height of the liquid inside a thermometer.* In order to quantify temperature, we need to assign numbers to various heights of the liquid in our glass tube. That is, we must set up a *temperature scale*.

(*a*)

(*b*)

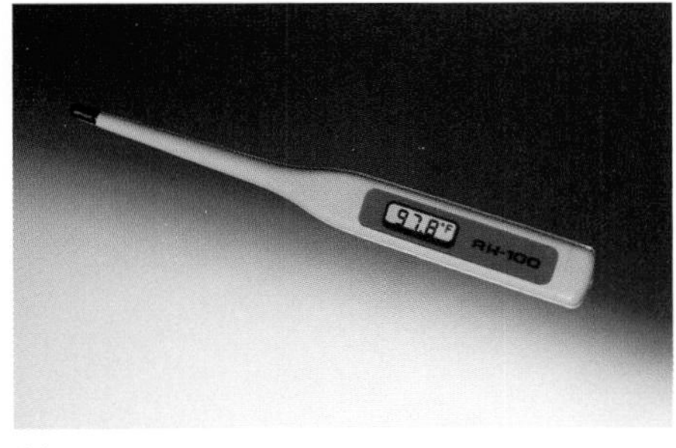

(*c*)

FIGURE 19-1 ■ Two types of thermometers based on changes in measurable properties of materials with hotness. (*a*) A liquid thermometer in which the liquid in a tube expands more than the glass that contains it when placed in hotter surroundings. (*b*) and (*c*) Electronic thermometers in which the electrical resistance of a sensing element embedded at the end of a thin rod changes when placed in hotter surroundings.

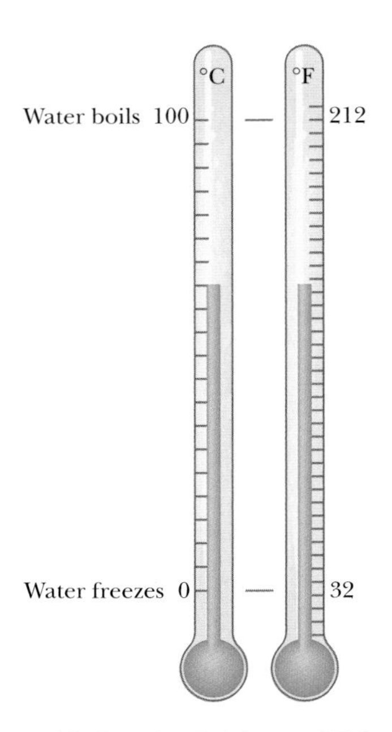

FIGURE 19-2 ■ The Celsius and Fahrenheit temperature scales compared for the freezing and boiling points of water at standard atmospheric pressure at sea level. Each tic mark represents 5°.

The Fahrenheit and Celsius scales are the two temperature scales in common use today. You are probably familiar with the Fahrenheit scale from U.S. weather forecasts and you may have worked with the Celsius scale in other science courses. Each scale is set up using a reproducible low temperature "fixed point" and another at a higher temperature. A temperature is assigned to each of the two fixed points on the thermometer column. Then the distance between the two fixed points is divided into equally spaced **degrees.**

The Fahrenheit Scale

The scale set up by Gabriel Fahrenheit in 1714 starts with a zero point (0°F) defined to be the lowest temperature attainable by a mixture of ice and salt. His upper fixed point was set at human body temperature and defined as 96°F. So Fahrenheit put 96 divisions along the glass tube between his two fixed points. The Fahrenheit scale has proven to be quite awkward. For example, the freezing point of water at sea level turns out to be about 32°F and the boiling point of water at sea level turns out to be about 212°F as shown in Fig. 19-2.

The Celsius Scale

In 1742 a Swedish investigator named Celsius devised a more sensible scale that he called the centigrade scale. Celsius defined the freezing point of water at sea level as 0°C and the boiling point as 100°C. The modern Celsius scale was developed based on a degree that is almost the same "size" as the centigrade scale. However, it has been adjusted so one of its fixed points is the triple point of water. The triple point of water is the temperature and pressure at which solid ice, liquid water, and water vapor coexist. The triple point temperature has been defined as 0.01°C. So, a modern Celsius thermometer and the historical centigrade thermometer will give essentially the same reading. We discuss the triple point in more detail later in this chapter. Another refinement has been to define a standard value for the atmospheric pressure at sea level to help make the boiling point temperature more stable.

The Kelvin Scale

A 19th-century British physicist, Lord Kelvin, discovered that there is a natural limit to how cold any object can get. Kelvin defined an important temperature scale used in thermodynamics that is based on this natural zero point for temperature. We discuss the Kelvin scale used by scientists and engineers in more detail in Section 19-9.

Temperature Conversions

The Fahrenheit scale employs a smaller degree than the Celsius scale and a different zero of temperature. You can easily verify both these differences by examining an ordinary room thermometer on which both scales are marked as shown in Fig. 19-2. The equation for converting between these two scales can be derived quite easily by remembering a few corresponding fixed-point temperatures for each scale (see Table 19-1 or Fig. 19-2). The equation is

$$T_F = \tfrac{9}{5}T_C + 32°, \tag{19-1}$$

where T_F is Fahrenheit temperature and T_C is Celsius temperature.

TABLE 19-1
Some Corresponding Temperatures

Temperature	°C	°F
Boiling point of water[a]	100	212
Normal body temperature	37.0	98.6
Accepted comfort level	20	68
Freezing point of water[a]	0	32
Zero of Fahrenheit scale	≈−18	0
Scales coincide	−40	−40

[a]Strictly, the boiling point of water on the Celsius scale is 99.975°C, and the freezing point is 0.00°C. Thus there is slightly less than 100°C between those two points.

In general, people use the letters C and F to distinguish measurements and degrees on the two scales. Thus,

$$0°\text{C} = 32°\text{F} \tag{19-2}$$

means that 0° on the Celsius scale measures the same temperature as 32° on the Fahrenheit scale. We might also say

$$\Delta T = 5\ \text{C}° = 9\ \text{F}°$$

which means that a temperature difference of 5 Celsius degrees (note the degree symbol appears *after* C) is equivalent to a temperature difference of 9 Fahrenheit degrees.

Problems with the Initial Definition of Temperature

There are several problems with basing our definition of temperature on the height of liquid in a scaled liquid thermometer: (1) the historically chosen fixed points are not highly reproducible since we will find that freezing and boiling points depend on pressure, (2) the height range over which liquids can vary in a glass tube is quite limited (liquids freeze when very cold or vaporize when very hot), (3) the assumption that liquids expand in proportion to temperature may not always be accurate and, the glass bulb thermometer must be small enough compared to the system that it doesn't affect it. We will return to our discussion of how to define temperature and design a better thermometer in Section 19-9. Meanwhile, we use our initial definition of temperature as the height of a liquid in a hollow glass thermometer as the starting point for our study of thermal interactions.

READING EXERCISE 19-1: List several additional measurable properties of an object. List several properties of an object that you believe are not measurable. ■

19-3 Thermal Interactions

Now that we have a means to measure temperature we can explore what happens to temperatures when two systems come into contact with each other. Consider the following two observations:

1. When a metal bucket of hot water is placed in a room, the reading of a thermometer placed in the water always decreases until, eventually, it matches the reading of a (assumed identical) thermometer in the room.

2. When a bucket of cold water is placed in a room, the reading of a thermometer in the bucket always increases until, eventually, it matches the reading of a thermometer in the room.

Our common sense tells us these observations indicate that there has been an interaction between the bucket of water and its surroundings. These interactions are examples of **thermal interactions.** We will call the bucket of water our *system*. We first introduced the term "system" in Chapter 8 in regard to the concept of conservation of momentum. Just as we did there, we will define a **system** to be the object or objects that are the primary focus of our interest. That, of course, means that we can make choices about what objects we include in our system. Just as in our study of conservation of energy and momentum, we will often be considering the interaction between two or more systems that can exchange energy with each other without exchanging matter. At other times we will consider the interaction between a single system and its surroundings.

The **environment** is defined as a system's surroundings or everything outside of it. Many thermal interactions of interest are interactions between a system (for example, the bucket of water) and its environment (for example, the air in the room). For practical purposes, the environment can often be taken as a system's immediate surroundings (without extending our consideration to the entire universe). All we must do is to be sure that the surroundings are much bigger than the system. For example, the environment for a bucket of water could be the room full of air that surrounds it, rather than the entire planet.

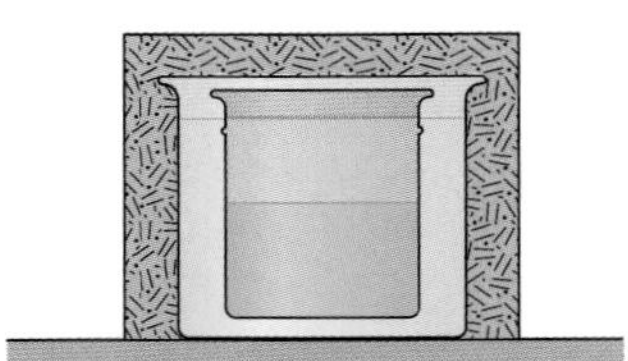

FIGURE 19-3 ■ Two vessels of water that have different temperatures at first are placed in thermal contact. They are surrounded by an insulating Styrofoam cup with a lid so that they don't interact significantly with their surroundings (room air). What happens to their temperatures?

Let's now consider two additional observations for objects that are insulated from the surrounding environment:

3. When two similar objects at different temperatures are brought into contact with one another, the thermometer reading for the hotter object decreases and the thermometer reading of the colder object increases until the two readings are the same. (See Figs. 19-3 and 19-4).

4. When two similar objects at the same temperature are brought together, no changes in the thermometer readings occur.

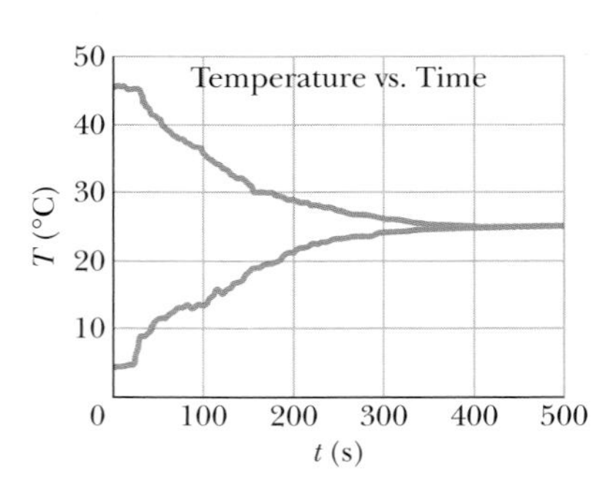

FIGURE 19-4 ■ A system consisting of $m_1 =$ 200 g of cool water at about 5°C in a thin plastic cup is brought into thermal contact with another system consisting of $m_2 = 200$ g of warm water in a larger insulated container at about 45°C. A computer-based data acquisition system is used to monitor their temperatures. Even though no matter is exchanged between the systems (as shown in Fig. 19-3), the temperature of the cool water rises while that of the warm water falls until they reach thermal equilibrium at about 25°C, 400 seconds later.

All four of the observations discussed above include examples of a condition we will call *thermal equilibrium*. If two objects (for example, our system and the surrounding environment) produce the same thermometer readings (assuming identical thermometers) then the two objects are said to be in **thermal equilibrium.** That is, two objects are in thermal equilibrium if they have the same temperature. Observing the tendency of systems in contact to reach thermal equilibrium suggests that the hotter system is transferring thermal energy to the colder one until they both become warm.

The concept of thermal equilibrium is important in understanding temperature measurements. For example, as you likely know from common experience, it takes some time for a thermometer reading to stabilize. If you place a glass bulb thermometer under your tongue, it does not immediately measure your correct body temperature. This is (at least in part) because it takes some time for a thermometer to reach thermal equilibrium with another object like your body. When a thermometer comes to thermal equilibrium with an object that is not heating up or cooling down, the thermometer reading will reach a constant value. Only then do we have an accurate measurement of the object's temperature.

Thermal equilibrium is important in the measurement of temperature in another way. Consider three objects: object A, object B and object T. (Object T may be a thermometer, but it doesn't need to be.) Object A and object B are placed in separate, well-insulated environments as shown in Fig. 19-5. We measure the temperatures of the three objects, two at a time. Suppose we find that object A and object T are in thermal equilibrium. We then compare the temperatures of object B and object T and find that they are in thermal equilibrium with each other as well (at the same thermometer reading as object A). Are object A and object B then necessarily in thermal equilibrium too? Experimentation provides the answer, which is referred to as the **zeroth law of thermodynamics:**

> If bodies A and B are each in thermal equilibrium with a third body T, then they are in thermal equilibrium with each other.

If we choose object T to be a thermometer, we see that we use the zeroth law constantly in science laboratories. It is the basis of our acceptance of the use of thermometers to compare temperatures. For example, if we want to know whether the liquids in two different students' beakers are at the same temperature, we can measure the two temperatures separately with a single thermometer and compare them. Often when measuring temperature, the thermometer starts at room temperature and its temperature must rise or fall when it is placed in contact with body A or body B. Again, we must be careful not to have the thermometer system be so large that either body loses or gains enough thermal energy to change its temperature noticeably.

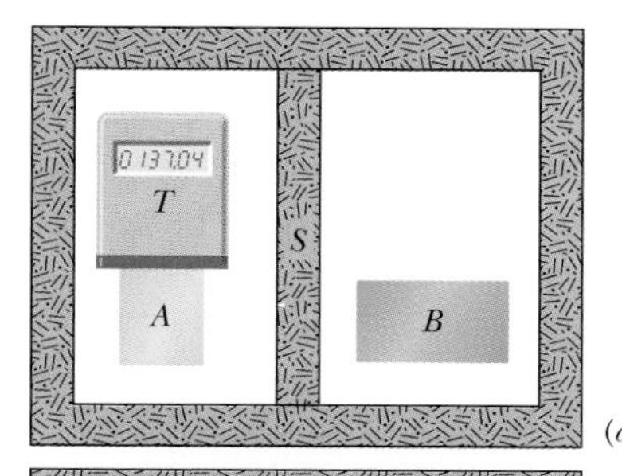

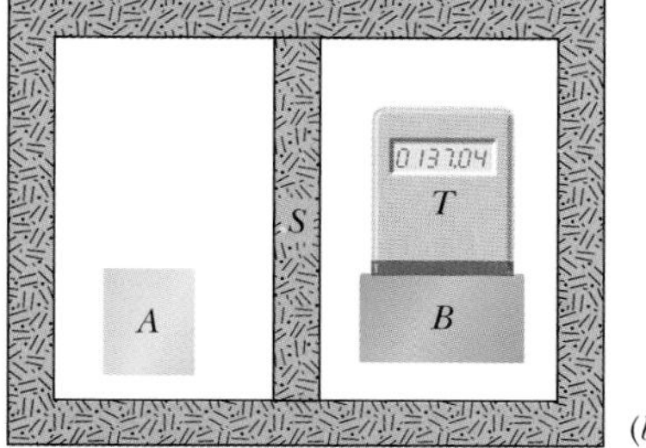

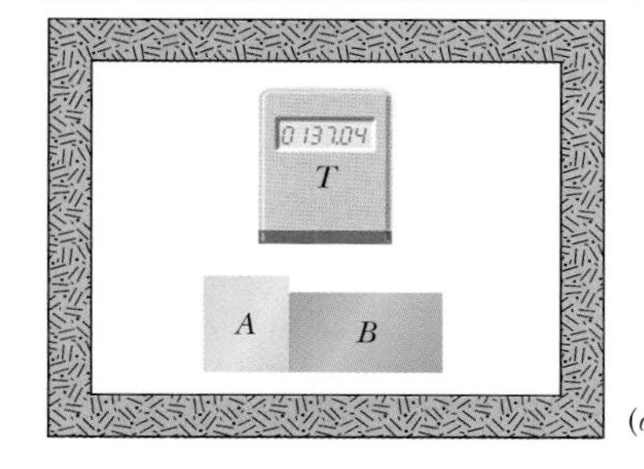

FIGURE 19-5 ■ (*a*) Body T (perhaps a thermometer) and body A are in thermal equilibrium. (Body S is a thermally insulating screen.) (*b*) Body T and body B are also in thermal equilibrium (*c*) If (*a*) and (*b*) are true, the zeroth law of thermodynamics states that body A and body B are also in thermal equilibrium.

19-4 Heating, Cooling, and Temperature

Is just measuring the initial and final temperature of an object that is changing temperature a good way to learn about how objects heat or cool? In order to answer this question, let us consider several examples of how water cools:

1. A small cup of very hot water is put in a room that is being maintained at a comfortable air temperature of 20°C. The water will cool down until it reaches thermal equilibrium with the surrounding air.
2. We place a large bucket of hot water in the same room, it will also cool down, but it will take longer than the cup of water does to cool.
3. We first place a small cup of hot water in a sealed thermos bottle and then place the bottle in the room, the water will still reach thermal equilibrium with the room but it will take much longer to cool down to room temperature than it did before.
4. We put a cup containing water at 30°C in a very cold freezer. We find the water freezes and becomes ice. During the freezing process we keep the water-ice mixture well-stirred and measure its temperature as a function of time. As shown in Fig. 19-6, once the mixture of ice and water are cooled to 0°C, the mixture undergoes no change in temperature until all the ice is frozen.

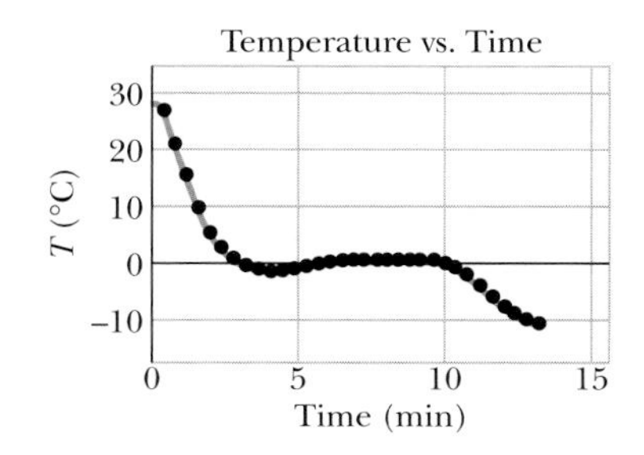

FIGURE 19-6 ■ A temperature vs. time graph recorded by a data acquisition system shows what happens to a cup of water after it is placed in a cold freezer. When the water temperature decreases to 0°C ice begins to form. While ice is forming the temperature does not change. Once all the water is changed to ice the temperature starts decreasing again.

The examples above are evidence that our ability to analyze thermal interactions is significantly limited if we rely solely on initial and final temperature measurements. Hence, we are motivated to invent a new, broader concept that we can use in discussing thermal interactions—even in cases where temperature changes are not the significant feature. The name of the process we will introduce is "heating," which we can use to describe the interaction between a hotter body and colder body, *even if no temperature change occurs*. For example, if you surround ice water with air or water which has a higher temperature, the ice will melt. Or, heating cold water takes longer if it is insulated from its environment than if it is not. Observations such as those above form the basis of our understanding of the *process* called heating.

In order to gain some additional insight into the process of heating, consider two more observations.

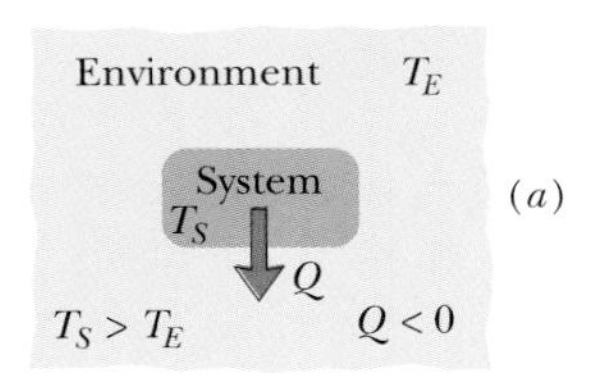

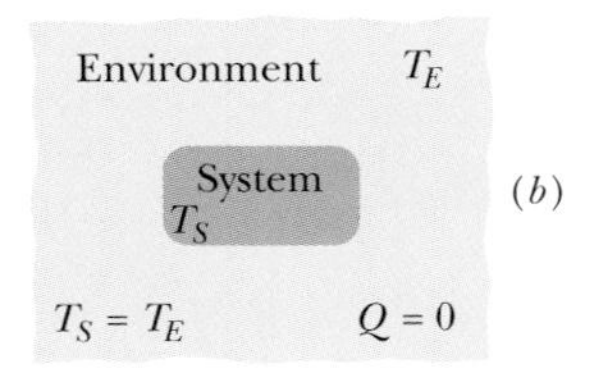

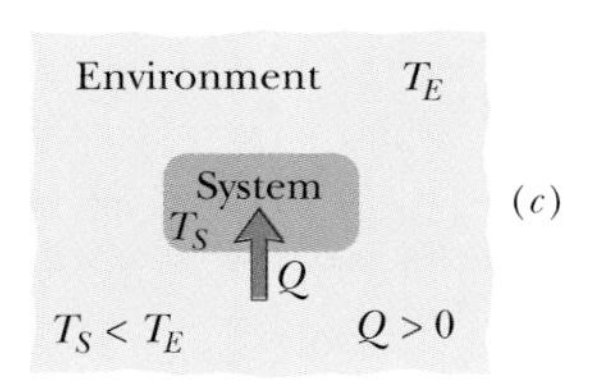

FIGURE 19-7 ■ If the temperature of a system exceeds that of its environment as in (*a*), an amount of thermal energy Q is transferred by the system to the environment until thermal equilibrium as shown in part (*b*) is established. (*c*) If the temperature of the system is below that of the environment, thermal energy is absorbed by the system until thermal equilibrium is established.

1. A container of water is placed on a burner to raise its temperature. It takes fuel (an energy source) for this process to occur.
2. If we start with two containers of water, one large and one small, at the same temperature and place them over identical burners, it takes a longer time (and so more fuel) to elevate the temperature of the larger amount of water to the same final value as the smaller.

Both of these observations are direct indications that there is energy involved in heating, just as we assumed in Chapter 10 when discussing energy conservation. Furthermore, the idea that the heating process is a *transfer of energy* from a hotter object to a colder object is consistent with all the other observations we have discussed in this chapter. Hence, we are led then to this important statement:

> **Heating** is the transfer of energy from a system with a higher temperature to one with lower temperature that is in contact with it that occurs simply because a temperature difference exists between the systems.

The net amount of transferred thermal energy as the result of microscopic energy exchanges between systems (or parts of a system) is denoted by the letter Q. The thermal energy transferred between systems is often called "heat." Because the term *heat* is sometimes used casually to mean the total thermal energy in a system, people often confuse heat with internal energy when describing thermal interactions. For this reason we try to avoid passive terms like *heat* or *stored heat* when describing thermal interactions. Since the phrase "heating a system" suggests an active process we will use terms like "heating," "cooling," or "thermal energy transfer" to refer to the additional energy that is added to (or subtracted from) a system through microscopic energy exchanges.

> The amount of transferred thermal energy Q is taken to be *positive* when it is transferred *to a system* (then we say that thermal energy is absorbed). The transferred thermal energy Q is *negative* when it is transferred *from the system* (we then say that thermal energy is released or lost by the system).

This transfer of energy is shown in Fig. 19-7. In the situation of Fig. 19-7*a*, in which the temperature of the system T_S is greater than the temperature of the environment T_E ($T_S > T_E$), energy is transferred from the system to the environment, so Q is negative. In Fig. 19-7*b*, in which $T_S = T_E$, no thermal energy transfer takes place, Q is zero, and energy is neither released nor absorbed. In Fig. 19-7*c*, in which $T_S < T_E$, the transfer is to the system from the environment, so Q is positive.

It took scientists a while to realize that "heating" was associated with the transfer of energy from one system to another. Hence, thermal energy ended up with its own unit. Since heating was initially considered strictly in terms of temperature change, the **calorie** (cal) was defined as the amount of energy that would raise the temperature of 1 g of water from 14.5°C to 15.5°C. In the British system, the corresponding unit of thermal energy was the **British thermal unit** (Btu), defined as the amount of energy that would raise the temperature of 1 lb of water from 63°F to 64°F.

FIGURE 19-8 ■ In the United States we have not adopted the international system of units when referring to food energy, so the **food calorie** (denoted Cal rather than cal) is used in diet books and in all government-regulated food labels is actually a kilocalorie. Most other countries use *joules* (the accepted SI unit for energy) on food labels. The can in the photo above was purchased in Australia.

In 1948, the scientific community decided that since heating (like work) is an energy transfer process, the SI unit for thermal energy transferred (or "heat added") should be the one we use for all other energy—namely, the **joule.** Joules are used instead of calories in most countries, as seen on the can of soda from Australia in Fig. 19-8. The calorie is now defined to be 4.1860 J (exactly), with no reference to the heating of water. (The "calorie" used in nutrition, sometimes called the Calorie with a capital C or Cal for short, is really a kilocalorie or 1000 calories.) The relations among the various thermal energy units are

$$1 \text{ cal} = 3.969 \times 10^{-3} \text{ Btu} = 4.1860 \text{ J} = 0.001 \text{ Cal (or food calorie)}. \quad (19\text{-}3)$$

Mechanisms for Transfer of Thermal Energy

We have discussed heating, which is the transfer of energy between a system and its environment that takes place simply because a temperature difference exists between them, but we have not yet described how that transfer takes place. We will briefly describe the three heating mechanisms here and return to discuss them more fully at the end of this chapter.

If you leave the end of a metal poker in a fire for enough time, its handle will get hot. There are large vibrations of the atoms and electrons of the metal at the fire end of the poker because of the high temperature of their environment. These increased vibrational amplitudes, and thus the associated energy, are passed along the poker, from atom to atom, during collisions between adjacent atoms. In this way, a region of increasing temperature extends itself along the poker to the handle. It is important to note though that there has been no flow of matter—only energy that is transmitted along the poker. This type of heating process is called (thermal) **conduction.**

Conduction is, by definition, a transfer mechanism that requires direct contact between two objects at different temperatures. Consider our poker to be a series of systems that can have different temperatures. If there is not direct contact between the colder object and hotter object (one atom and the next in our poker example), there cannot be a transfer of thermal energy by conduction. As we mentioned above, *conduction does not involve any mass transfer*. However, in some real situations it is sometimes impossible to avoid the transfer of material from one object to the other during heating. In such cases, the process is no longer one of "pure" conduction.

When you open a low and a high window in a heated house, you can feel cold outside air rushing into the room through the low window and warm room air rushing outside through the high window. This is because the cold air is more dense than the warm air so it displaces the warm air at the bottom of the room. The warm air rises as a result of buoyant forces and flows out the top window. The room gets colder because of the exchange of cold and warm air. In this situation, thermal energy is being transported by the flow of matter (air currents in this case). The transfer by the exchange of hotter and cooler fluids is known as **convection.**

Examples of heating by convection are everywhere. This is how air circulation helps spread the warmth through a room when a radiator or heater gets hot. It is why *all* the water in a tea kettle gets hot (as opposed to only the water in contact with the hot kettle surface) when the kettle is placed on a hot stove. Convection is part of many other natural processes. Atmospheric convection plays a fundamental role in determining global climate patterns and daily weather variations. Glider pilots and birds alike seek rising thermals (convection currents of warm air) that keep them aloft. Huge energy transfers take place within the oceans by the same process. Finally, energy is transported to the surface of the sun from the nuclear furnace at its core by enormous cells of convection, in which hot gas rises to the surface along the cell core and cooler gas around the core descends below the surface. In all examples of heating by convection, gravitational forces play a vital role. Without gravitational forces, hotter materials would not rise above cooler materials. Hence, there is no heating by convection on a space station.

When you sit in the sun you can feel your skin getting warmer. If you put a shield between you and the sun, the sensation of warmth immediately disappears. How is this energy transfer taking place? We know that the light from the sun has to pass through over a hundred million kilometers of almost empty space. So, this thermal energy transfer can't be attributed to either convection or conduction. Solar energy transfer is attributed to a third transfer process—the absorption of **electromagnetic radiation.** Although visible light is one kind of electromagnetic radiation, the sun and a hot fire that can also warm you emit both visible light and invisible infrared radiation that has a longer characteristic wavelength than light. (See Chapter 34 for more details on electromagnetic waves.) No medium is required for energy transfer via electromagnetic radiation. Thermal energy transferred by infrared electromagnetic waves is often called **thermal radiation** or **radiant energy.**

READING EXERCISE 19-2: Explain the function of insulation in homes. ■

(a)

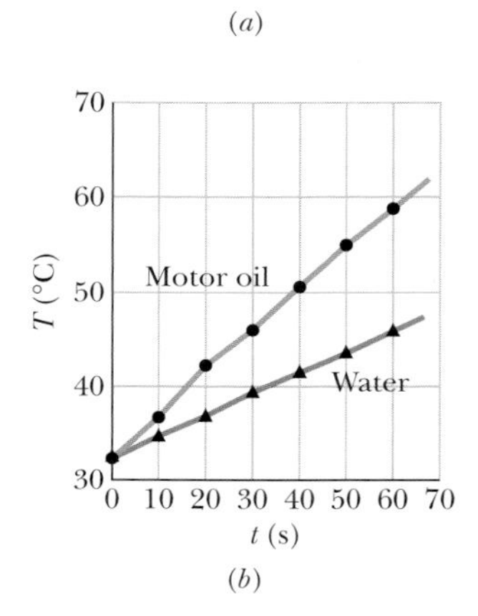

(b)

FIGURE 19-9 ■ (a) A computer-based data collection system with a digital temperature sensor can be used to monitor the temperature rise in a liquid while thermal energy is transferred to it at a constant rate by an immersion heater. The liquid in this photo is motor oil. (b) If thermal energy is transferred to two different liquids that have the same mass, the temperature does not usually rise at the same rate. So we define a different specific heat for each liquid. For example, a small immersion heater shown in Fig. 19-9a "heats up" 175 g of motor oil faster than 175 g of water. Data were taken once each second for 70 seconds with a computer-based data collection system.

19-5 Thermal Energy Transfer to Solids and Liquids

If we start with two containers of water, one large and one small, at the same temperature and place them to heat over identical burners, it takes a longer time (and so more fuel and thus more thermal energy) to elevate the temperature of the larger amount of water to the same final value as the smaller. It takes almost twice as much energy to heat a cup of water as it does to heat a cup of motor oil to the same temperature. This is shown in Fig. 19-9*a* and *b*. A common electric immersion heater is placed in a container of motor oil and plugged in. It puts out a bit less than 200 W of power at a constant rate. Since energy is power × time the total amount of thermal energy transferred to the oil is directly proportional to the time the heater has been on. The graph in Fig. 19-9*b* shows that after 70 seconds, the change in water temperature, $T_f - T_i$, is about 15°C while the oil temperature has risen about 30°C. We can say the "water holds its heat" twice as well as motor oil. It would be more correct to say that water absorbs the thermal energy transferred to it without showing as much temperature change.

If we look at heating and cooling curves for a large number of different objects and different substances (see Fig. 19-9*b* for an example), we find that the relationship between thermal energy transfer and temperature is linear (as long as the material doesn't melt, freeze, or vaporize). However, the scaling factor (proportionality constant) changes from material to material. The **heat capacity** C of an object is the name that we give to the proportionality constant between the thermal energy transferred Q and the resulting temperature change ΔT of the object; that is,

$$Q = C\,\Delta T = C(T_f - T_i), \tag{19-4}$$

in which T_i and T_f are the initial and final temperatures of the object. Heat capacity C has the unit of energy per degree Celsius. The heat capacity C of, say, a marble slab used in a bun warmer might be 179 cal/C°.

The word "capacity" in this context is really misleading in that it suggests analogy with the capacity of a bucket to hold water. Since heat is not a substance, *that analogy is misleading,* and you should not think of the object as "containing" thermal energy or being limited in its ability to absorb thermal energy. Thermal energy transfer can proceed without limit as long as the necessary temperature difference between the object and its surroundings is maintained. The object may, of course, melt or vaporize during the process.

Specific Heat

Two objects made of the same material—say, marble—will have heat capacities proportional to their masses. It is therefore convenient to define a "heat capacity per unit mass" or **specific heat** c that refers not to an object but to a unit mass of the material of which the object is made. Equation 19-4 then becomes

$$Q = cm\,\Delta T = cm(T_f - T_i), \tag{19-5}$$

where $C = mc$.

Through experiment we would find that although the heat capacity of a particular marble slab might be 179 cal/C°, the specific heat of the marble (in that slab or in any other marble object) is 0.21 cal/g · C°.

From the way the calorie and the British thermal unit were initially defined, measurements show that the specific heat of water is

$$c = 1\ \text{cal/g}\cdot\text{C}^\circ = 1\ \text{Btu/lb}\cdot\text{F}^\circ = 4190\ \text{J/kg}\cdot\text{C}^\circ. \tag{19-6}$$

Table 19-2 shows the specific heats of some substances at room temperature. Note that the value for water is relatively high. The specific heat of any substance actually depends somewhat on temperature, but the values in Table 19-2 apply reasonably well in a range of temperatures near room temperature.

TABLE 19-2
Specific Heats of Some Substances at Room Temperature and Constant Pressure

Substance	Specific Heat $\frac{\text{cal}}{\text{g}\cdot\text{C}^\circ}$	Specific Heat $\frac{\text{J}}{\text{kg}\cdot\text{C}^\circ}$	Molar Specific Heat $\frac{\text{J}}{\text{mol}\cdot\text{C}^\circ}$
Elemental Solids			
Lead	0.0305	128	26.5
Tungsten	0.0321	134	24.8
Silver	0.0564	236	25.5
Copper	0.0923	386	24.5
Aluminum	0.215	900	24.4
Other Solids			
Brass	0.092	380	
Granite	0.19	790	
Glass	0.20	840	
Ice (−10°C)	0.530	2220	
Liquids			
Mercury	0.033	140	
Ethyl alcohol	0.58	2430	
Seawater	0.93	3900	
Water	1.00	4190	

Molar Specific Heat

In many instances the most convenient unit for specifying the amount of a substance is the mole (mol), where

$$1\ \text{mol} = 6.02 \times 10^{23}\ \text{basic units}$$

of *any* substance. Thus 1 mol of aluminum means 6.02×10^{23} atoms (the atom being the elementary unit), and 1 mol of aluminum oxide means 6.02×10^{23} molecules of the oxide (because the molecule is the basic unit of a compound).

When quantities are expressed in moles, specific heats must also involve moles (rather than a mass unit); they are then called **molar specific heats.** Table 19-2 shows the values for some elemental solids (each consisting of a single element) at room temperature.

Specific Heats at Constant Pressure and Constant Volume

In determining and then using the specific heat of any substance, we need to know the conditions under which thermal energy is transferred. For solids and liquids, we usually assume that the sample is under constant pressure (usually atmospheric) during the transfer. It is also conceivable that the sample is held at constant volume while the thermal energy is absorbed. This means that thermal expansion of the sample is prevented by applying external pressure. For solids and liquids, this is very hard to arrange experimentally but the effect can be calculated, and it turns out that the specific heats under constant pressure and constant volume for any solid or liquid differ usually by no more than a few percent. Gases, as you will see, have quite different values for their specific heats under constant-pressure conditions and under constant-volume conditions.

Heats of Transformation

Recall our example above of the ice (solid water). "Solid" is a description of the water that we will call its **phase.** There are, in general, three phases of matter: solid, liquid, and vapor. As a *solid,* the molecules of a sample are locked into a fairly rigid structure by their mutual attraction. As a *liquid,* the molecules have more energy and move about more. They may form brief clusters, but the sample does not have a rigid structure and can flow or settle into a container. As a *gas* or *vapor,* the molecules have even more energy, are free of one another, and can fill up the full volume of a container. To fully describe a material for thermodynamic purposes we must specify not only the phase (solid, liquid, or vapor) but also the temperature, pressure, and volume. The phase of a material along with the temperature, pressure, and volume of the material specify the **state** of the material.

As we all know, we find that ice melts when exposed to a warm room and becomes liquid water. The melting is called a **change of phase.** When thermal energy is absorbed or lost by a solid, liquid, or vapor, the temperature of the sample does not necessarily change. Instead, the sample may change from one phase to another. Through experiments we have found that while a material is undergoing a change in phase additional transfers of thermal energy do not change the temperature of the material. We saw one example of this in Fig. 19-6.

To *melt* a solid means to change it from the solid phase to the liquid phase. The process requires energy because the molecules of the solid must be freed from their rigid structure. Melting an ice cube to form liquid water is a common example. To *freeze* a liquid to form a solid is the reverse of melting and requires that energy be removed from the liquid, so that the molecules can settle into a rigid structure. To *vaporize* a liquid means to change it from the liquid phase to the vapor or gas phase. This process, like melting, requires energy because the molecules must be freed from their clusters. Boiling liquid water transforms it to water vapor (or steam—a gas of individual water molecules) is a common example. *Condensing* a gas to form a liquid is the reverse of vaporizing; it requires that energy be removed from the gas, so that the molecules can cluster instead of flying away from one another.

The amount of energy per unit mass that must be transferred as thermal energy when a sample completely undergoes a phase change is called the heat of transformation L. Thus, when a sample of mass m completely undergoes a phase change, the total energy transferred is

$$Q = Lm. \tag{19-7}$$

When the phase change is from liquid to gas (then the sample must absorb thermal energy) or from gas to liquid (then the sample must release thermal energy), the thermal energy required for this transformation is called the **heat of vaporization** L_V.* For water at its normal boiling or condensation temperature,

$$L_V = 539 \text{ cal/g} = 40.7 \text{ kJ/mol} = 2256 \text{ kJ/kg} \quad \text{(water} \rightarrow \text{steam).} \tag{19-8}$$

*Chemists often denote entropy as ΔH and call it enthalpy of vaporization.

When the phase change is from solid to liquid (then the sample must absorb thermal energy) or from liquid to solid (then the sample must release thermal energy), the heat of transformation is called the **heat of fusion** L_F. For water at its normal freezing or melting temperature,

$$L_F = 79.5 \text{ cal/g} = 6.01 \text{ kJ/mol} = 333 \text{ kJ/kg} \quad \text{(ice} \rightarrow \text{water).} \tag{19-9}$$

Table 19-3 shows the heat of transformation for some substances.

TABLE 19-3
Some Heats of Transformation

	Melting		Boiling	
Substance	**Melting Point (°C)**	**Heat of Fusion L_F (kJ/kg)**	**Boiling Point (°C)**	**Heat of Vaporization L_V (kJ/kg)**
Hydrogen	−259.2	58.0	−252.9	455
Oxygen	−218.4	13.9	−183.0	213
Mercury	−39.2	11.4	356.9	296
Water	−0.1	333	99.9	2256
Lead	327.9	23.2	1743.9	858
Silver	961.9	105	2049.9	2336
Copper	1082.9	207	2594.9	4730

Heating and Internal Energy

Consider melting a 0°C block of ice so that it melts and then becomes hot water. This system undergoes *both* a phase change and a temperature increase. What happens to the system's hidden internal energy that we mentioned in the introduction? Heating via thermal energy transfer increases its internal energy in two ways:

1. We change the system's microscopic configuration by allowing its atoms and molecules to move further away from each other (without a temperature change).
2. We increase its microscopic kinetic energy (or thermal energy) by increasing the motions of its atoms and molecules (which can be measured as a rise in temperature).

READING EXERCISE 19-3: If an amount of thermal energy Q is transferred to object A, it will cause each gram of A to rise in temperature by 3 C°. If the same amount of energy is transferred to object B, then the temperature of each gram of B will rise by 4 C°. If object A and B have the same mass, which one has the greater specific heat? ■

READING EXERCISE 19-4: Notice in Table 19-3 that water has a very large heat of vaporization. What is the significance of this to a firefighter who finds that the water sprayed on a very hot fire is converted to steam? Suppose that 1 g of steam comes in contact with 1 g of a firefighter's flesh and condenses. Assuming that flesh has the same heat capacity as water, what would be the temperature rise in 1 g of the firefighter's flesh? ■

TOUCHSTONE EXAMPLE 19-1: Melting Ice

(a) How much thermal energy must be absorbed by ice of mass $m = 720$ g at −10°C to take it to a liquid state at 15°C?

SOLUTION ■ The first **Key Idea** is that the heating process is accomplished in three steps.

Step 1. The **Key Idea** here is that the ice cannot melt at a temperature below the freezing point—so initially, any thermal energy transferred to the ice can only increase the temperature of the ice. The energy Q_1 needed to increase that temperature from the initial value $T_i = -10$°C to a final value $T_f = 0$°C (so that the ice can then melt) is given by Eq. 19-5 ($Q = cm\,\Delta T$). Using the specific heat of ice c_{ice} in Table 19-2 gives us

$$\begin{aligned} Q_1 &= c_{ice} m(T_f - T_i) \\ &= (2220\ \text{J/kg}\cdot{}^\circ\text{C})(0.720\ \text{kg})[0^\circ\text{C}-(-10^\circ\text{C})] \\ &= 15\,984\ \text{J} = 15.98\ \text{kJ}. \end{aligned}$$

Step 2. The next **Key Idea** is that the temperature cannot increase from 0°C until all the ice melts—so any energy transferred to the ice due to heating now can only change ice to liquid water. The thermal energy Q_2 needed to melt all the ice is given by Eq. 19-7 ($Q = Lm$). Here L is the heat of fusion L_F, with the value given in Eq. 19-9 and Table 19-3. We find

$$Q_2 = L_F m = (333\ \text{kJ/kg})(0.720\ \text{kg}) = 239.8\ \text{kJ}.$$

Step 3. Now we have liquid water at 0°C. The next **Key Idea** is that the energy transferred to the liquid water during heating now can only increase the temperature of the liquid water. The Q_3 needed to increase the temperature of the water from the initial value $T_i = 0^\circ\text{C}$ to the final value $T_f = 15^\circ\text{C}$ is given by Eq. 19-5 (with the specific heat of liquid water c_{liq}):

$$\begin{aligned} Q_3 &= c_{liq} m(T_f - T_i) \\ &= (4190\ \text{J/kg}\cdot{}^\circ\text{C})(0.720\ \text{kg})(15^\circ\text{C} - 0^\circ\text{C}) \\ &= 45\,252\ \text{J} \approx 45.25\ \text{kJ}. \end{aligned}$$

The total required thermal energy transfer Q^{tot} is the sum of the amounts required in the three steps:

$$\begin{aligned} Q^{tot} &= Q_1 + Q_2 + Q_3 \\ &= 15.98\ \text{kJ} + 239.8\ \text{kJ} + 45.25\ \text{kJ} \\ &\approx 300\ \text{kJ}. \qquad \text{(Answer)} \end{aligned}$$

Note that the energy transfer required to melt the ice is much greater than the energy transfer required to raise the temperature of either the ice or the liquid water.

(b) If we supply the ice with a total energy of only 210 kJ (as heat), what then are the final state and temperature of the water?

SOLUTION ■ From step 1, we know that 15.98 kJ is needed to raise the temperature of the ice to the melting point. The remaining energy required Q^{rem} is then 210 kJ − 15.98 kJ, or about 194 kJ. From step 2, we can see that this amount of energy is insufficient to melt all the ice. Then this **Key Idea** becomes important: Because the melting of the ice is incomplete, we must end up with a mixture of ice and liquid; the temperature of the mixture must be the freezing point, 0°C.

We can find the mass m of ice that is melted by the available energy Q^{rem} by using Eq. 19-7 with L_F:

$$m = \frac{Q^{rem}}{L_F} = \frac{194\ \text{kJ}}{333\ \text{kJ/kg}} = 0.583\ \text{kg} \approx 580\ \text{g}.$$

Thus, the mass of the ice that remains is 720 g − 580 g, or 140 g, and we have

580 g water and 140 g ice at 0°C. (Answer)

TOUCHSTONE EXAMPLE 19-2: Copper Slug

A copper slug whose mass m_c is 75 g is heated in a laboratory oven to a temperature T of 312°C. The slug is then dropped into a glass beaker containing a mass $m_w = 220$ g of water. The heat capacity C_b of the beaker is 45 cal/K. The initial temperature T_i of the water and the beaker is 12°C. Assuming that the slug, beaker, and water are an isolated system and the water does not vaporize, find the final temperature T_f of the system at thermal equilibrium.

SOLUTION ■ One **Key Idea** here is that, with the system isolated, only transfers of thermal energy can occur. There are three such transfers, all as thermal energy. The slug loses energy, the water gains energy, and the beaker gains energy. Another **Key Idea** is that, because these transfers do not involve a phase change, the energy transfers can only change the temperatures. To relate the transfers to the temperature changes, we can use Eqs. 19-4 and 19-5 to write

$$\text{for the water:} \quad Q_w = c_w m_w (T_f - T_i); \qquad (19\text{-}10)$$

$$\text{for the beaker:} \quad Q_b = C_b (T_f - T_i); \qquad (19\text{-}11)$$

$$\text{for the copper:} \quad Q_c = c_c m_c (T_f - T). \qquad (19\text{-}12)$$

A third **Key Idea** is that, with the system thermally isolated, the total energy of the system cannot change. This means that the sum of these three thermal energy transfers is zero:

$$Q_w + Q_b + Q_c = 0. \qquad (19\text{-}13)$$

Substituting Eqs. 19-10 through 19-12 into Eq. 19-13 yields

$$c_w m_w (T_f - T_i) + C_b (T_f - T_i) + c_c m_c (T_f - T) = 0. \qquad (19\text{-}14)$$

Temperatures are contained in Eq. 19-14 only as differences. Thus, because the differences on the Celsius and Kelvin scales are identical, we can use either of those scales in this equation. Solving it for T_f, we obtain

$$T_f = \frac{c_c m_c T + C_b T_i + c_w m_w T_i}{c_w m_w + C_b + c_c m_c}.$$

Using Celsius temperatures and taking values for c_c and c_w from Table 19-2, we find the numerator to be

$$(0.0923 \text{ cal/g}\cdot°\text{C})(75 \text{ g})(312°\text{C}) + (45 \text{ cal/°C})(12°\text{C})$$
$$+ (1.00 \text{ cal/g}\cdot°\text{C})(220 \text{ g})(12°\text{C}) = 5339.8 \text{ cal},$$

and the denominator to be

$$(1.00 \text{ cal/g}\cdot°\text{C})(220 \text{ g}) + 45 \text{ cal/°C} + (0.0923 \text{ cal/g}\cdot°\text{C})(75 \text{ g})$$
$$= 271.9 \text{ cal/°C}.$$

We then have

$$T_f = \frac{5339.8 \text{ cal}}{271.9 \text{ cal/°C}} = 19.6°\text{C} \approx 20°\text{C}.$$

From the given data you can show that

$$Q_w \approx 1670 \text{ cal}, \qquad Q_b \approx 342 \text{ cal}, \qquad Q_c \approx -2020 \text{ cal}.$$

Apart from rounding errors, the algebraic sum of these three thermal energy transfers is indeed zero, as Eq. 19-13 requires.

19-6 Thermal Energy and Work

Here we look in some detail at how internal energy can be transferred into or out of both as thermal energy and as macroscopic physical work (involving the displacement of the system in the presence of net forces). Let us take as our system a gas confined to a cylinder with a movable piston, as in Fig. 19-10. This is the kind of device that is used to drive a steam engine, the engine of an automobile, and many other tools—all of which convert thermal energy into work.

In our piston–cylinder system, the upward force on the piston due to the pressure of the confined gas is equal to the weight of lead shot loaded onto the top of the piston. The walls of the cylinder are made of insulating material that does not allow any transfer of thermal energy. The bottom of the cylinder, however, rests on a reservoir of thermal energy—a *thermal reservoir* (perhaps a hot plate) whose temperature T can be held constant.

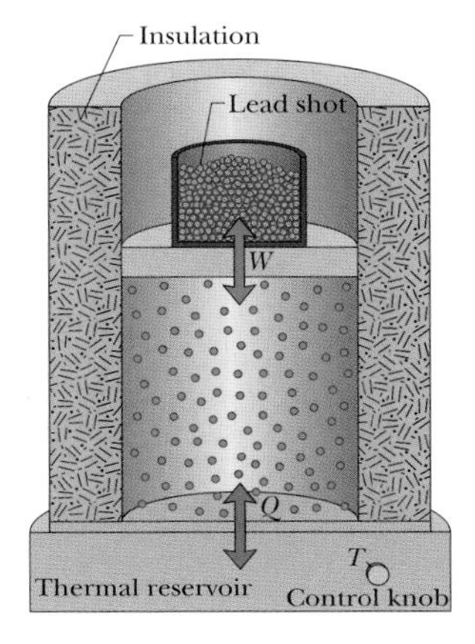

FIGURE 19-10 ■ Cross section of a system that confines a gas to a cylinder with a movable piston. (Its insulating lid is not shown.) Thermal energy Q can be transferred to or from the gas by regulating the temperature T of the adjustable thermal reservoir. Work W is done on the gas when the piston rises or falls.

The system (the gas) starts from an *initial state i*, described by a pressure P_i, a volume V_i, and a temperature T_i. You want to change the system to a final state f, described by a pressure P_f, a volume V_f, and a temperature T_f. The procedure by which you change the system from its initial state to its final state is called a *thermodynamic process*. During such a process, an amount of energy Q may be transferred into or out of the system from the thermal reservoir. Also, work can be done by the system to raise the loaded piston (positive work) or lower it (negative work). *We assume that all such changes occur slowly, so that all parts of the system are always in (approximate) thermal equilibrium.*

Suppose that you remove a few of the lead shot from the piston of Fig. 19-10, allowing the gas to push the piston and remaining shot upward through a differential displacement $d\vec{s}$ with an upward force $\vec{F}$. Since the displacement is tiny, we can assume that $\vec{F}$ is constant during the displacement. Then $\vec{F}$ has a magnitude that is equal to PA, where P is the pressure of the gas and A is the face area of the piston. The differential work dW done by the gas during the displacement is

$$\begin{aligned} dW &= \vec{F}\cdot d\vec{s} = (PA)(|d\vec{s}\,|) = P(A|d\vec{s}\,|) \\ &= PdV, \end{aligned} \tag{19-15}$$

in which dV is the differential change in the volume of the gas due to the movement of the piston. When you have removed enough shot to allow the gas to change its volume from V_i to V_f, the total work done by the gas is

$$W = \int dW = \int_{V_i}^{V_f} PdV. \tag{19-16}$$

During the change in volume, the pressure and temperature of the gas may also change. To evaluate the integral in this expression directly, we need to know how pressure varies with volume for the actual process by which the system changes from state i to state f.

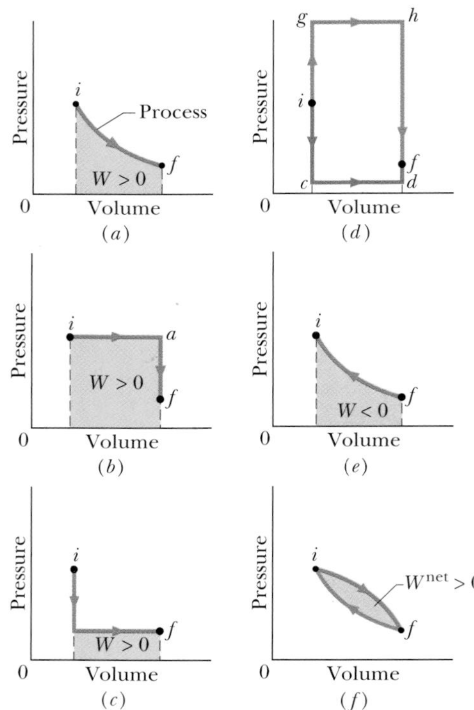

FIGURE 19-11 ■ (*a*) The shaded area represents the work W done by a system as it goes from an initial state i to a final state f. Work W is positive because the system's volume increases. (*b*) W is still positive, but now greater. (*c*) W is still positive, but now smaller. (*d*) W can be even smaller (path *icdf*) or larger (path *ighf*). (*e*) Here the system goes from state i to state f, as the gas is compressed to less volume by an external force. The work W done by the system is now negative. (*f*) The net work W^{net} done by the system during a complete cycle is represented by the shaded area.

There are actually many ways to take the gas from state i to state f. One way is shown in Fig. 19-11*a*, which is a plot of the pressure of the gas versus its volume and is called a *P-V* diagram. In Fig. 19-11*a*, the curve indicates that the pressure decreases as the volume increases. The integral in Eq. 19-16 (and thus the work W done by the gas) is represented by the shaded area under the curve between points i and f. Regardless of exactly what we do to take the gas along the curve, that work by the gas or system is positive, due to the fact that the gas increases its volume by forcing the piston upward.

Another way to get from state i to state f is shown in Fig. 19-11*b*: the change takes place in two steps—the first from state i to state a and the second from state a to state f.

Step *ia* of this process is carried out at constant pressure, which means that you leave undisturbed the lead shot that ride on top of the piston in Fig. 19-10. You cause the volume to increase (from V_i to V_f) by slowly turning up the temperature control knob, raising the temperature of the gas to some higher value T_a. (Increasing the temperature increases the force from the gas on the piston, moving it upward.) During this step, positive work is done by the expanding gas (to lift the loaded piston) and thermal energy is absorbed by the system from the thermal reservoir (in response to the arbitrarily small temperature differences that you create as you turn up the temperature). The thermal energy transferred (Q) is positive because it is added to the system.

Step *af* of the process of Fig. 19-11*b* is carried out at constant volume, so you must wedge the piston, preventing it from moving. Then as you use a control knob to decrease the reservoir temperature, you find that the pressure drops from P_a to its final value P_f. During this step, thermal energy is lost by the system to the thermal reservoir.

For the overall process *iaf*, the work W, which is positive and is carried out only during step *ia*, is represented by the shaded area under the curve. Energy is transferred as thermal energy during both steps *ia* and *af*, with a net thermal energy transfer Q.

Figure 19-11*c* shows a process in which the previous two steps are carried out in reverse order. The work W in this case is smaller than for Fig. 19-11*b*, as is the net thermal energy absorbed. Figure 19-11*d* suggests that you can make the work done by the gas as small as you want (by following a path like *icdf*) or as large as you want (by following a path like *ighf*).

To sum up: A system can be taken from a given initial state to a given final state by an infinite number of processes. Thermal energy transfers may or may not be involved, and in general, the work W and the thermal energy transfer Q will have different values for different processes. We say that thermal energy and work are *path-dependent* quantities.

Figure 19-11*e* shows an example in which negative work is done by a system as some external force compresses the system, reducing its volume. The absolute value of the work done is still equal to the area beneath the *P-V* curve, but because the gas is compressed, the work done by the gas is negative.

Figure 19-11*f* shows a thermodynamic cycle in which the system is taken from some initial state i to some other state f and then back to i. The net work done by the system during the cycle is the sum of the positive work done during the expansion and the negative work done during the compression as shown by the shaded area inside the path. In Fig. 19-11*f*, the net work is positive because the area under the expansion curve (i to f) is greater than the area under the compression curve (f to i).

READING EXERCISE 19-5: The *P-V* diagram here shows six curved paths (connected by vertical paths) that can be followed by a gas. Which two of them should be part of a closed cycle if the net work done by the gas is to be at its maximum positive value?

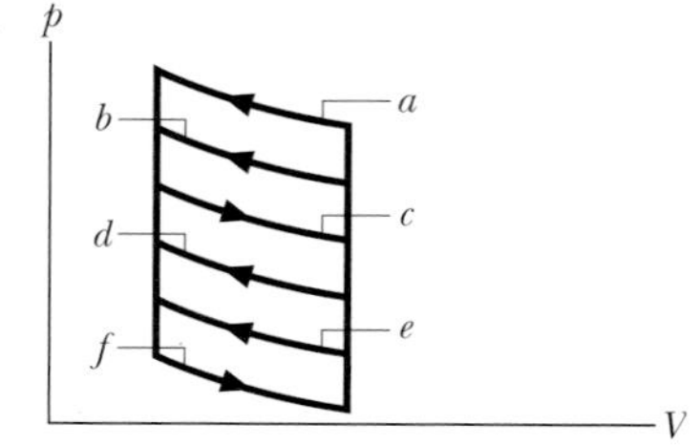

■

19-7 The First Law of Thermodynamics

You have just seen that when a system changes from a given initial state to a given final state, both the work W done by the system and the thermal energy transferred to the system Q depend on the nature of the process.* Experimentally, however, we find a surprising thing. The quantity $Q - W$ *is the same for all processes*. It depends only on the initial and final states and does not depend at all on how the system gets from one to the other. All other combinations of Q and W, including Q alone, W alone, $Q + W$, and $Q - 2W$, are path dependent; only the quantity $Q - W$ is not.

So, we come to believe that the quantity $Q - W$ must represent a change in some intrinsic property of the system. Since we know that both heating and work are energy transfer processes, we assume that the quantity $Q - W$ results in a change in energy. But a change in the energy of what? As we pointed out in the introduction to this chapter, matter has hidden within it energy associated with the motions and configurations of the microscopic atoms and molecules of which it is composed. We call this energy associated with the microscopic kinetic and potential energies of the system the **internal energy** E^{int}. So, we can then write

$$\Delta E^{\text{int}} = E_f^{\text{int}} - E_i^{\text{int}} = Q - W \quad \text{(first law for } W = \text{work by system)} \tag{19-17}$$

where Q represents the thermal or heat energy transferred *to the system* and W represents the work done on the surroundings *by the system*. This expression represents a very important relationship and is called the **first law of thermodynamics.** If the thermodynamic system undergoes only a differential change, we can write the first law as†

$$dE^{\text{int}} = dQ - dW \quad \text{(first law).} \tag{19-18}$$

> The internal energy E^{int} of a system increases if thermal energy Q is transferred to the system and decreases if thermal energy ($Q < 0$) is transferred from the system.
>
> The internal energy E^{int} of a system decreases if it does an amount of work $W > 0$ on its surroundings and increases when it has an amount of work $W < 0$ done on it by the surroundings. (Work done on a system is negative. Work done by a system is positive.)

The first law of thermodynamics should remind you of issues we first raised in Chapter 10. However, in Chapter 10, we discussed energy conservation as it applies to isolated systems—that is, to systems in which no energy enters or leaves the system. The first law of thermodynamics is an extension of that principle to systems that are *not* isolated. In such cases, energy may be transferred into or out of the system as some combination of macroscopic physical work W done by a system and transfer of the microscopic thermal energy Q into a system. In our statement of the first law of thermodynamics above, we assume that there are no changes in the *macroscopic* kinetic energy or the potential energy of the system as a whole; that is, $\Delta K = \Delta U = 0$.

Before this chapter, the term *work* and the symbol W always meant the work done on a system. However, starting with Eq. 19-15 above and continuing through the next two chapters about thermodynamics, we focus on the work done *by* a system, such as the gas in Fig. 19-10. This confusing reversal of focus is left over from the 19th

---

*Recall that a system's state is specified by its pressure, volume, temperature, and phase.

†Here dQ and dW, unlike dE^{int}, are not true differentials; that is, there are no such functions as $Q(P, V)$ and $W(P, V)$ that depend only on the system state. Both dQ and dW are inexact differentials and often represented by the symbols $đQ$ and $đW$. Here we can treat them simply as infinitesimally small energy transfers.

century when scientists and engineers were most interested in how much physical work could be done *by* an engine on its surroundings.

The work done *on* a system is always the negative of the work done by the system, so if we rewrite $\Delta E^{int} = E_f^{int} - E_i^{int} = Q - W$ in terms of the work W^{on} done *on* the system, we have $\Delta E^{int} = Q + W^{on}$. This tells us the following: the internal energy of a system tends to increase if thermal energy is absorbed by the system or if positive work is done *on* the system. Conversely, the internal energy tends to decrease if thermal energy is lost by the system or if negative work is done *on* the system. This can be a bit confusing, so it is a good idea to spend a little time now to make sure that this distinction is clear.

READING EXERCISE 19-6: The figure here shows four paths on a P-V diagram along which a gas can be taken from state i to state f. Rank the paths according to (a) the change ΔE^{int}, (b) the work W done by the gas, and (c) the magnitude of the thermal energy Q transferred to the gas, greatest first.

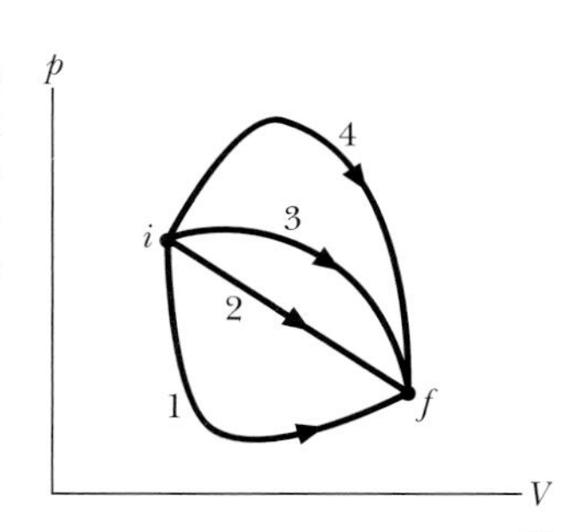

■

19-8 Some Special Cases of the First Law of Thermodynamics

Here we look at four different thermodynamic processes, in each of which a certain restriction is imposed on the system. We then see what consequences follow when we apply the first law of thermodynamics to the process. The results are summarized in Table 19-4.

TABLE 19-4
The First Laws of Thermodynamics: Four Special Cases

The First Law: $\Delta E^{int} = Q - W$ (Eq. 19-17)[a]

Process	Restriction	Consequence
Adiabatic	$Q = 0$	$\Delta E^{int} = -W$
Constant volume	$W = 0$	$\Delta E^{int} = Q$
Closed cycle	$\Delta E^{int} = 0$	$Q = W$
Free expansion	$Q = W = 0$	$\Delta E^{int} = 0$

[a] $Q > 0$ represents thermal or heat energy added to the system; $W > 0$ represents work done by system.

1. ***Adiabatic processes.*** An adiabatic process is one in which *no transfer of thermal energy* occurs between the system and its environment. Adiabatic processes can occur if a system that is so well insulated that no thermal energy transfers can occur. A process can also be made adiabatic by having it occur so quickly that there is no time for the energy transfer. In either case, putting $Q = 0$ in the first law ($\Delta E^{int} = E_f^{int} - E_i^{int} = Q - W$) yields

$$\Delta E^{int} = -W \qquad \text{(adiabatic process).} \tag{19-19}$$

This tells us that if the work done *by* the system on its surroundings is positive, the internal energy of the system decreases by an amount equal to the work it did. Conversely, if work is done *on* the system (that is, if W is negative), the internal energy of the system increases by that amount.

Figure 19-12 shows an idealized adiabatic process. Thermal energy cannot be transferred to or from a system because it is insulated from its surroundings. Thus, the only way energy can be transferred between the system and its environment is if work is either done by the system or on the system. If we remove shot from the piston and allow the gas to expand, work is done by the system (the gas) and thus is positive. Thus the internal energy of the gas must decrease. If, instead, we add shot and compress the gas, the work done by the system is negative and the internal energy of the gas increases.

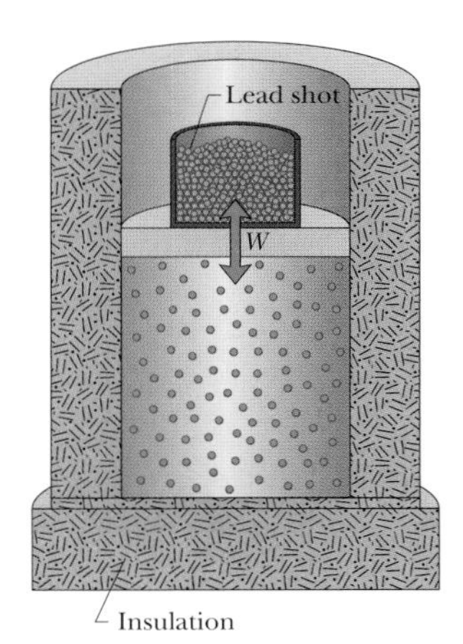

FIGURE 19-12 ■ An adiabatic expansion can be carried out by removing lead shot from the top of the piston. Adding lead shot reverses the process at any stage. (The insulating lid is not shown.)

2. ***Constant-volume processes.*** If the volume of a system (such as a gas in a container) is held constant, that system can do no work. Putting $W = 0$ in the first law ($\Delta E^{\text{int}} = \Delta E_f^{\text{int}} - E_i^{\text{int}} = Q - W$) yields

$$\Delta E^{\text{int}} = Q \qquad \text{(constant-volume process).} \qquad (19\text{-}20)$$

Thus, all the thermal energy transferred to a system (so that Q is positive), contributes to an increase in the system's internal energy. Conversely, if thermal energy is transferred from the system to its surroundings so that ΔQ is negative, the internal energy of the system must decrease.

3. ***Cyclical processes.*** There are processes in which, after certain interchanges of thermal energy and work, the system is restored to its initial state. Engines undergo this type of process. Since the system is restored to its initial state, no intrinsic property of the system—including its internal energy—can possibly change. Putting $\Delta E^{\text{int}} = 0$ in the first law ($\Delta E^{\text{int}} = E_f^{\text{int}} - E_i^{\text{int}} = Q - W$) yields

$$Q(\text{in}) = W(\text{by}) \qquad \text{(cyclical process).} \qquad (19\text{-}21)$$

Thus, the net work done by the system on its surroundings during the process must exactly equal the net amount of thermal energy transferred to the system. So the store of internal energy hidden in the system remains unchanged. Cyclical processes form a closed loop on a P-V plot, as shown in Fig. 19-11*f*.

4. ***Free expansions.*** These are adiabatic processes (that is, ones in which no transfer of thermal energy occurs between the system and its environment) and no work is done on or by the system. Thus, $Q = W = 0$ and the first law requires that

$$\Delta E^{\text{int}} = 0 \qquad \text{(free expansion).} \qquad (19\text{-}22)$$

Figure 19-13 shows how such an expansion can be carried out. A gas, which is in thermal equilibrium within itself, is initially confined by a closed stopcock to one half of an insulated double chamber; the other half is evacuated. The stopcock is opened, and the gas expands freely to fill both halves of the chamber. No thermal energy is transferred to or from the gas because of the insulation. No work is done by the gas because it rushes into a vacuum; the motion of the gas atoms or molecules is not opposed by any pressure.

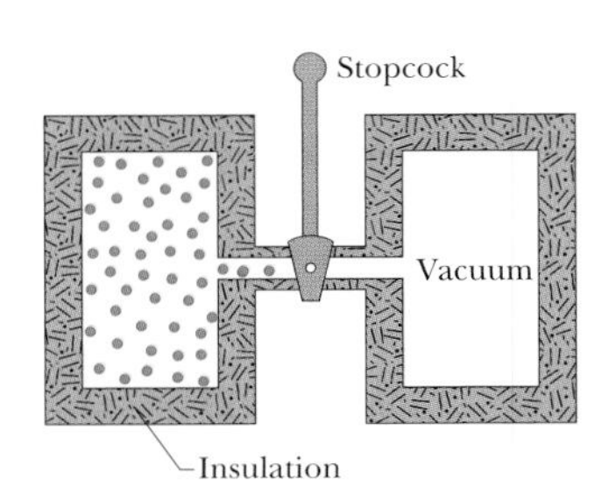

FIGURE 19-13 ■ The initial stage of a free-expansion process. After the stopcock is opened, the gas fills both chambers and eventually reaches an equilibrium state.

A free expansion differs from all other processes we have considered because it cannot be done slowly and in a controlled way. As a result, at any given instant during the sudden expansion, the gas is not in thermal equilibrium and its pressure is not the same everywhere. Therefore, although we can plot the initial and final states on a P-V diagram, we cannot plot the expansion itself.

READING EXERCISE 19-7: For one complete cycle as shown in the P-V diagram here, (a) is the internal energy change ΔE^{int} of the gas positive, negative, or zero? and (b) what about the net thermal energy, Q, transferred to the gas?

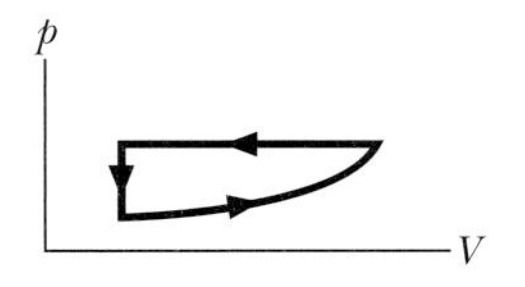

TOUCHSTONE EXAMPLE 19-3: Boiling Water

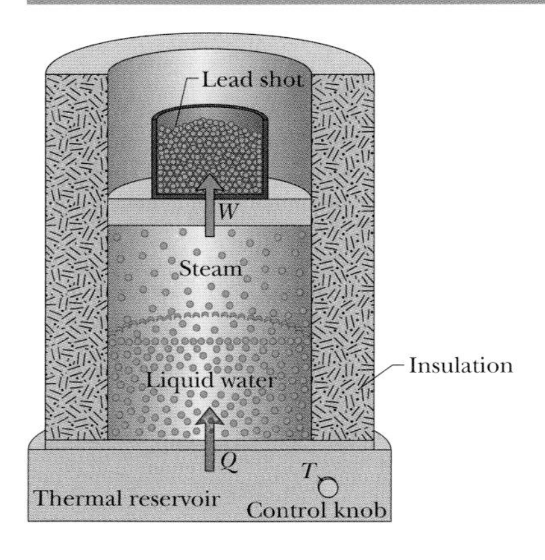

FIGURE 19-14 ■ Water boiling at constant pressure. Thermal energy is transferred from the thermal reservoir until the liquid water has changed completely into steam. Work is done by the expanding gas as it lifts the loaded piston. The insulating lid is not shown.

Let 1.00 kg of liquid water at 100°C be converted to steam at 100°C by boiling at standard atmospheric pressure (which is 1.00 atm or 1.01×10^5 Pa) in the arrangement of Fig. 19-14. The volume of that water changes from an initial value of 1.00×10^{-3} m^3 as a liquid to 1.671 m^3 as steam.

(a) How much work is done by the system during this process?

SOLUTION ■ The **Key Idea** here is that the system must do positive work because the volume increases. In the general case we would calculate the work W done by integrating the pressure with respect to the volume (Eq. 19-16). However, here the pressure is constant at 1.01×10^5 Pa, so we can take P outside the integral. We then have

$$W = \int_{V_i}^{V_f} P\,dV = P\int_{V_i}^{V_f} dV = P(V_f - V_i)$$

$$= (1.01 \times 10^5 \text{ Pa})(1.671 \text{ m}^3 - 1.00 \times 10^{-3} \text{ m}^3) \quad \text{(Answer)}$$

$$= 1.69 \times 10^5 \text{ J} = 169 \text{ kJ}.$$

(b) How much thermal energy is transferred as heat during the process?

SOLUTION ■ The **Key Idea** here is that the thermal energy causes only a phase change and not a change in temperature, so it is given fully by Eq. 19-7 ($Q = Lm$). Because the change is from a liquid to a gaseous phase, L is the heat of vaporization L_V, with the value given in Eq. 19-8 and Table 19-3. We find

$$Q = L_V m = (2256 \text{ kJ/kg})(1.00 \text{ kg})$$

$$= 2256 \text{ kJ} \approx 2260 \text{ kJ}. \quad \text{(Answer)}$$

(c) What is the change in the system's internal energy during the process?

SOLUTION ■ The **Key Idea** here is that the change in the system's internal energy is related to the thermal energy transferred into the system and the work done on the surroundings which transfers out of the system) by the first law of thermodynamics (Eq. 19-17). Thus, we can write

$$\Delta E^{\text{int}} = Q - W = 2256 \text{ kJ} - 169 \text{ kJ}$$

$$\approx 2090 \text{ kJ} = 2.09 \text{ MJ}. \quad \text{(Answer)}$$

This quantity is positive, indicating that the internal energy of the system has increased during the boiling process. This energy goes into separating the H_2O molecules, which strongly attract each other in the liquid state. We see that, when water is boiled, about 7.5% (= 169 kJ/2260 kJ) of the thermal energy added goes into the work of pushing back the atmosphere. The rest of the thermal energy added goes into the system's internal energy.

19-9 More on Temperature Measurement

In Section 1-4 we discussed the fact that there are only seven fundamental quantities in physics that serve as base units for the entire international system (or SI) of units. Temperature in kelvins is one of them. How is the kelvin unit defined? Why has a gas thermometer operating under constant volume become the standard thermometer? What are its fixed points? Why is it superior to the liquid thermometers we discussed in Section 19-2?

As we discussed in Section 19-2, there are several difficulties with defining temperature in terms of the height of a liquid column in a liquid thermometer. First, the historical fixed points (such as freezing and boiling points at sea level, body temperature, and lowest ice/salt mixture temperature) are not reproducible to a high accuracy. Second, liquid thermometers are only usable in a narrow range of temperatures between the freezing and boiling points of the liquid. Third, no two liquids expand and contract in exactly the same way as their temperatures change.

These difficulties prompted a search for highly reproducible fixed points and a way to measure temperature that is independent of the behavior of any one particular substance. Gases are more promising substances for temperature measurements, since they are already "boiling" and have no upper limit except at the melting point of their container. Also, gases do not tend to liquefy until they reach very low temperatures. For example, air liquefies at about −200°C. Generally a gas volume (at constant pressure) or a gas pressure (at constant volume) can be measured between two fixed points. The scales can be determined in the same way as they are for liquids. The various gas scales have been found to agree among themselves better than liquid scales do. Other thermometers are based on changes of the electrical properties or materials or changes in the light given off by glowing substances and so on. Although we have dozens of thermometric scales based on the behaviors of various substances, none of them can be proven exactly true. Here we present some methods for identifying better fixed points and designing more accurate thermometers.

Defining Standard Fixed Points

Using a low-density gas instead of liquid to measure low temperatures gives very interesting results. As we suggested in Section 19-2, when thermal energy is extracted from a gas held at constant pressure, its temperature and volume drop. A simple apparatus showing these results for a limited range of temperatures is shown in Fig. 19-15.*

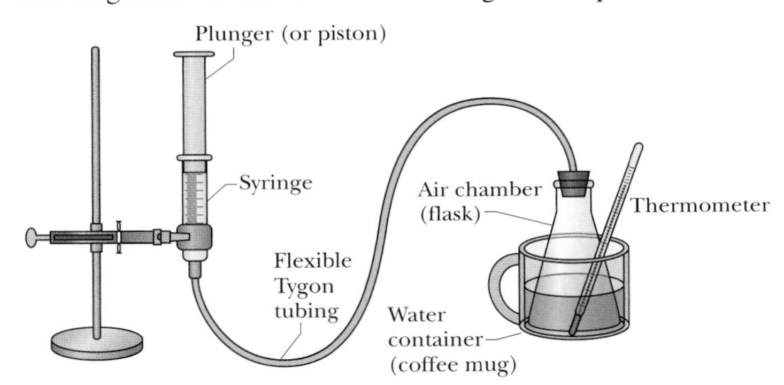

FIGURE 19-15 ■ Diagram of apparatus that can be used to measure volume changes as a function of the temperature of air or some other gas trapped at constant pressure. The pressure is a combination of atmospheric pressure plus the pressure exerted by the weight of the plunger of cross-sectional area A, so that $P = P^{atm} + mg/A$.

If you plot data for the volume of any low-density gas as a function of temperature, the graph is linear. By extrapolating the graph to zero volume, you can predict that the volume will go to zero at a temperature of approximately −273°C. In reality any gas will liquefy before its temperature gets that low. An alternative approach is to hold the volume of a gas constant and observe that its pressure will also approach zero at a temperature of about −273°C. A simple apparatus for doing this experiment is shown in Fig. 19-16.

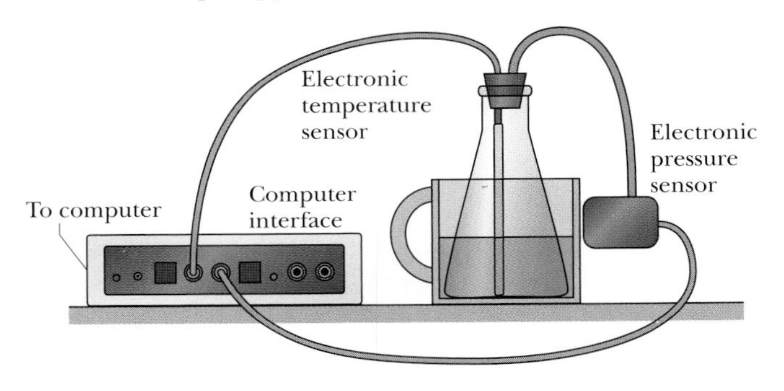

FIGURE 19-16 ■ Diagram of apparatus that uses a computer data acquisition system to measure pressure changes as a function of the temperature of air or some other gas trapped at constant volume.

*The temperature sensor should be placed inside the flask in Fig. 19-15.

Actually the extrapolation to zero pressure gives a slightly different result depending on how much gas is placed in the flask that holds it to a constant volume. However, as we place smaller and smaller masses of gas in the flask and retake the data we find that the temperature at zero pressure converges to a lower limit of −273.16°C. Student data for this experiment using different apparatus are shown in Fig. 19-17.

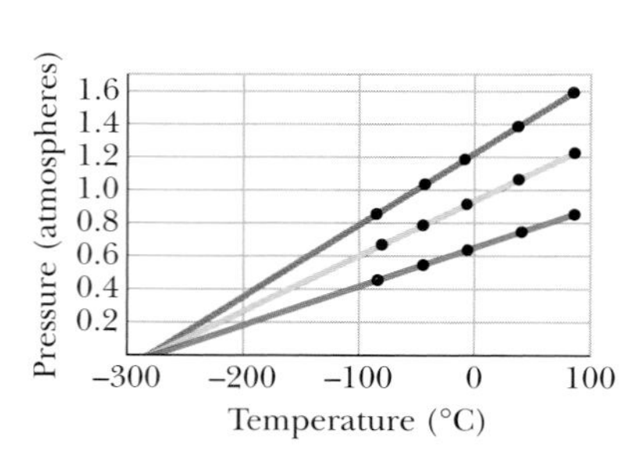

FIGURE 19-17 ■ Data from a student experiment measuring the pressure of a fixed volume of gas at various temperatures. The experimental data are represented by the points in the graphs and the lines are fits to these data. The three data sets are for three different masses of gas (air) in a container. Regardless of the amount of gas, the pressure is a linear function of temperature that extrapolates to zero at approximately −280°C. (More precise measurements show that the zero point does depend slightly on the amount of gas, but has a well-defined limit of −273.16°C as the density of the gas goes to zero.)

Since vanishingly small samples of all gases appear to approach the same minimum temperature regardless of their chemical composition, this temperature seems to be a fundamental property of nature. We call this minimum temperature **absolute zero.** We define absolute zero as the temperature of a body when it has the minimum possible internal energy. *Because of the universality of this minimum temperature of −273.16°C it has become our standard low temperature fixed point.*

But what should we use if we want a second standard fixed point? We could, for example, select the freezing point or the boiling point of water. However, water's change of phase, for example from liquid to vapor (boiling), depends not just on temperature, but also on pressure. This introduces some technical difficulties and reduces our confidence in using boiling or freezing as fixed, reproducible thermal phenomena. As we suggested in Section 19-2 a state known as the triple point of water is a good candidate for a fixed point.

We have found through extensive experimentation that three phases—liquid water, solid ice, and water vapor (gaseous water)—can coexist, in thermal equilibrium, at one and only one set of values of pressure and temperature. Thus, this temperature is called the **triple point of water.** Since both the temperature and pressure are well defined at this point, *the triple point of water is often used as a standard high temperature fixed point in designing a thermometer.*

Figure 19-18 shows a triple-point cell, in which the triple point of water can be set up in a laboratory. For reasons that we will explain when introducing the thermodynamic temperature scale, the triple point of water has been assigned a value of 0.01°C.

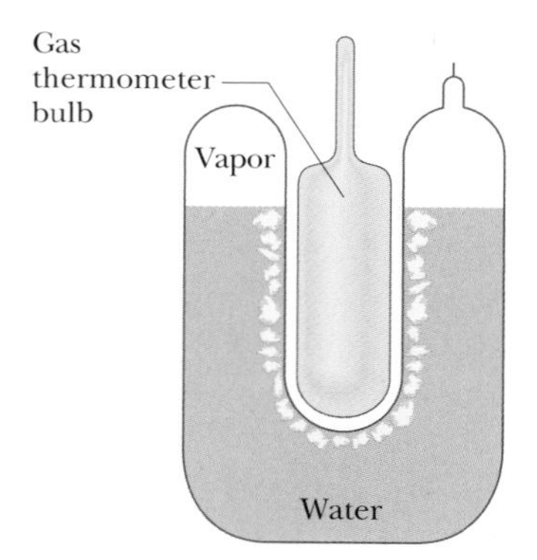

FIGURE 19-18 ■ A triple-point cell, in which solid ice, liquid water, and water vapor coexist in thermal equilibrium. By international agreement, the temperature of this mixture has been defined to be 273.16 K. The bulb of a constant-volume gas thermometer is shown inserted into the well of the cell.

The Thermodynamic Temperature Scale

The accepted SI unit of temperature is the kelvin. The kelvin is named after Lord Kelvin who first proposed that there is a natural limit to how cold any object can get. If we assume that temperature is an indicator of the amount of internal energy in a system, then the natural limit occurs when $E^{\text{int}} = 0$. Kelvin proposed that when $E^{\text{int}} = 0$ then $T = 0$. This zero is known as **absolute zero.** It is defined by using absolute zero and the triple point of water as its two fixed points. In order to tie in with the popular Celsius scale, the thermodynamic scale was set to have the same "size" degree as the Celsius scale. This yields a rather odd definition. Basically the minimum possible temperature is defined as zero kelvin or 0 K, and the triple point of water is defined as exactly +273.16 K.

Since the triple point of water is 0.01°C, the conversion between a Celsius scale temperature and the thermodynamic scale temperature is very simple since one merely needs to add 273.15 to the Celsius temperature, T_C, to get the thermodynamic temperature T. In other words,

$$T = T_C + 273.15. \tag{19-23}$$

Figure 19-19 shows a wide range of temperatures in kelvin, either measured or conjectured. When expressing temperatures on the Fahrenheit or Celsius scales, we commonly say degrees Fahrenheit (°F) or degrees Celsius (°C), but thermodynamic temperatures are simply called "kelvin" (abbreviated K). Thus, we do not say "degrees kelvin" or write °K. Figure 19-20 compares the thermodynamic, Celsius, and Fahrenheit scales.

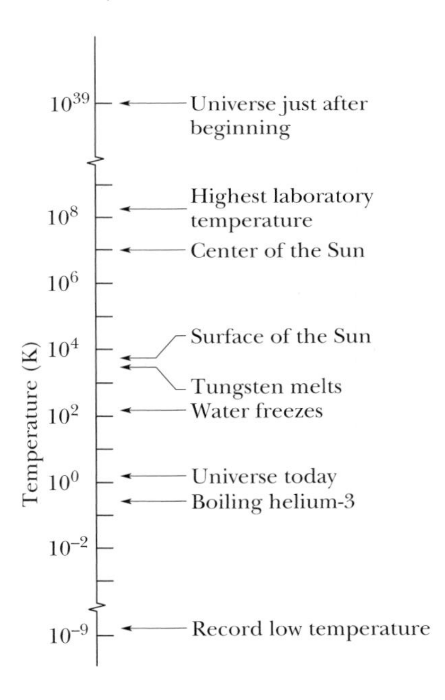

FIGURE 19-19 ■ Some temperatures on the thermodynamic scale. Temperature $T = 0$ corresponds to $10^{-\infty}$ and cannot be plotted on this thermodynamic scale.

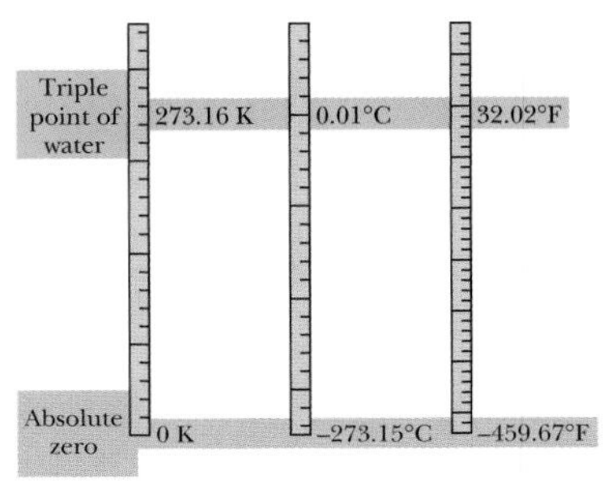

FIGURE 19-20 ■ A comparison of the thermodynamic, Celsius, and Fahrenheit temperature scales.

When the universe began, some 10 to 20 billion years ago, its temperature was about 10^{39} K. As the universe expanded it cooled, and it has now reached an average temperature of about 3 K. We on Earth are a little warmer than that because we happen to live near a star. Without our sun, we too would be at 3 K (and we could not exist).

The Constant-Volume Gas Thermometer

Now that we have better fixed points we can use them as part of a standard thermometer that uses gas rather than a liquid as its medium. The standard thermometer, against which all other thermometers are calibrated, is based on the effect of temperature changes on the pressure of a gas occupying a fixed volume. Figure 19-21 shows such a **constant-volume gas thermometer;** it consists of a gas-filled bulb connected by a tube to a mercury manometer. By raising and lowering reservoir R, the mercury level on the left can always be brought to the zero of the scale to keep the gas volume constant (variations in the gas volume can affect temperature measurements).

The temperature of any body in thermal contact with the bulb (like the liquid in Fig. 19-21) is then defined to be

$$T = CP, \tag{19-24}$$

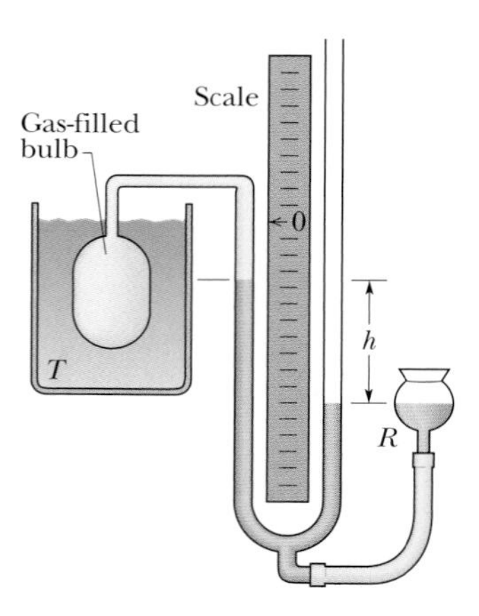

FIGURE 19-21 ■ A constant-volume gas thermometer, the gas-filled bulb on the left is immersed in a liquid whose temperature T is being measured. The mercury-filled bulb on the right is raised or lowered as P changes to keep the left-hand column of mercury at the zero point on the scale so the gas volume stays constant.

in which P is the pressure within the gas and C is a constant. From Eq. 15-10, the pressure P is

$$P = P^{\text{atm}} - \rho_{\text{Hg}} g h, \tag{19-25}$$

in which P^{atm} is the atmospheric pressure, ρ_{Hg} is the density of the mercury in the open-tube manometer (like that described in Section 15-5), and h is the measured difference between the mercury levels in the two arms of the tube.*

If we next put the bulb in a triple-point cell (Fig. 19-18), the temperature being measured is

$$T_3 = CP_3 \tag{19-26}$$

*For pressure units, we shall use units introduced in Section 15-3. The SI unit for pressure is the newton per square meter, which is called the pascal (Pa). The pascal is related to other common pressure units by

$$1.00 \text{ atm} = 1.01 \times 10^5 \text{ Pa} = 760 \text{ torr} \approx 14.7 \text{ lb/in.}^2$$

in which P_3 is the gas pressure now. Eliminating C between Eqs. 19-24 and 19-26 gives us the temperature as

$$T = T_3\left(\frac{P}{P_3}\right) = (273.16\text{ K})\left(\frac{P}{P_3}\right) \quad \text{(provisional)}. \tag{19-27}$$

We still have a problem with this thermometer. If we use it to measure a given temperature, we find that different gases in the bulb give slightly different results. However, as we use smaller and smaller masses of gas to fill the bulb, the readings converge nicely to a single temperature, no matter what gas we use.

Thus the recipe for measuring a temperature with a gas thermometer is

$$T = (273.16\text{ K})\left(\lim_{\text{gas}\to 0}\frac{P}{P_3}\right). \tag{19-28}$$

The recipe instructs us to measure an unknown temperature T as follows: Fill the thermometer bulb with an arbitrary amount of *any* gas (for example, nitrogen) and measure P_3 (using a triple-point cell) and P, the gas pressure at the temperature being measured. (Keep the gas volume the same.) Calculate the ratio P/P_3. Then repeat both measurements with a smaller amount of gas in the bulb, and again calculate this ratio. Continue this way, using smaller and smaller amounts of gas, until you can extrapolate to the ratio P/P_3 that you would find if there were approximately no gas in the bulb. Calculate the temperature T by substituting that extrapolated ratio into Eq. 19-28. (The temperature is called the *ideal gas temperature*.)

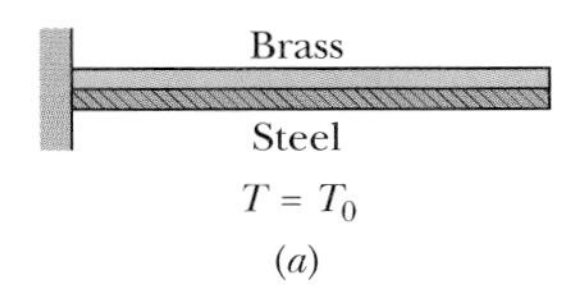

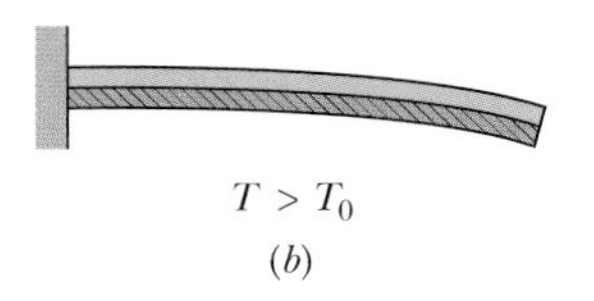

FIGURE 19-22 ■ (*a*) A bimetal strip, consisting of a strip of brass and a strip of steel welded together, at temperature T_0. (*b*) The strip bends as shown at temperatures above this reference temperature. Below the reference temperature the strip bends the other way. Many thermostats operate on this principle, making and breaking an electrical contact as the temperature rises and falls.

19-10 Thermal Expansion

We already know that most materials expand when heated and contract when cooled. Indeed, we used this principle in our first definition of temperature. The design of thermometers and thermostats is often based on the differences in expansion between the components of a *bimetal strip* (Fig. 19-22). In aircraft manufacture, rivets and other fasteners are often cooled in dry ice before insertion and then allowed to expand to a tight fit. However, such **thermal expansion** is not always desirable, as Fig. 19-23 suggests. To prevent buckling, expansion slots must be placed in bridges to accommodate roadway expansion on hot days. Dental materials used for fillings must be matched in their thermal expansion properties to those of tooth enamel (otherwise consuming hot coffee or cold ice cream would be quite painful). Regardless of whether we might wish to exploit or avoid thermal expansion, this property has many important implications and we need to consider how it works in detail.

FIGURE 19-23 ■ Railroad tracks in Asbury Park, New Jersey, distorted because of thermal expansion on a very hot July day.

Linear Expansion

If the temperature of a metal rod of length L is raised by a small amount ΔT, its length is found to increase by an amount ΔL according to

$$\frac{\Delta L}{L} = \alpha\,\Delta T, \tag{19-29}$$

in which α is a constant called the coefficient of linear expansion. For small changes in temperature, the fractional length change ($\Delta L/L$) is proportional to the change in temperature. The coefficient α should be thought of as the ratio of how much the length changes per unit length given a certain change in temperature. It has the unit "per degree" or "per kelvin" and depends on the material. Although α varies somewhat with temperature, for most practical purposes it can be taken as constant for a particular material. Table 19-5 shows some coefficients of linear expansion. Note that the unit C° there could be replaced with the unit K.

TABLE 19-5
Some Coefficients of Linear Expansion[a]

Substance	α (10^{-6}/C°)	Substance	α (10^{-6}/C°)
Ice (at 0°C)	51	Steel	11
Lead	29	Glass (ordinary)	9
Aluminum	23	Glass (Pyrex)	3.2
Brass	19	Diamond	1.2
Copper	17	Invar[b]	0.7
Concrete	12	Fused quartz	0.5

[a]Room temperature values except for the listing for ice.

[b]This alloy was designed to have a low coefficient of expansion. The word is a shortened form of "invariable."

The thermal expansion of a solid is like a (three-dimensional) photographic enlargement. Figure 19-24*b* shows the (exaggerated) expansion of a steel ruler after its temperature is increased from that of Fig. 19-24*a*. Equation 19-29 applies to every linear dimension of the ruler, including its edge, thickness, diagonals, and the diameters of the circle etched on it and the circular hole cut in it. If the disk cut from that hole originally fits snugly in the hole, it will continue to fit snugly if it undergoes the same temperature increase as the ruler.

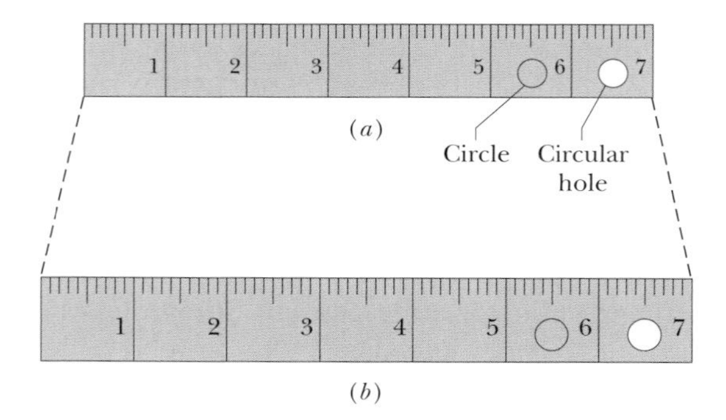

FIGURE 19-24 ■ The same steel ruler at two different temperatures. When the ruler expands, the scale, the numbers, the thickness, and the diameters of the circle and circular hole are all increased by the same factor. (The expansion has been exaggerated for clarity.)

Volume Expansion

If all dimensions of a solid expand with temperature, the volume of that solid must also expand. For liquids, volume expansion is the only meaningful expansion parameter. If the temperature of a solid or liquid whose volume is V is increased by a small amount ΔT, the increase in volume is found to be ΔV, where

$$\frac{\Delta V}{V} = \beta\,\Delta T \qquad (19\text{-}30)$$

and β is the **coefficient of volume expansion** of the solid or liquid.

The coefficients of volume expansion β and linear expansion α for a solid are related. To see how, consider a rectangular solid of height h, width w, and length l. The volume of the rectangular solid is hwl. If the solid is heated, each dimension expands linearly so that the height becomes $h + \Delta h$, the width becomes $w + \Delta w$, and the length becomes $l + \Delta l$. Hence, the new volume is $(h + \Delta h)(w + \Delta w)(l + \Delta l)$. If we multiply this out, ignoring all terms with two or more deltas (Δ) because those terms will be very small, we get $hwl + hw\,\Delta l + h\,\Delta wl + \Delta h\,wl$. Since each dimension expands linearly, this means that our new volume is equal to $hwl + hwl(3\alpha\,\Delta T)$. Hence,

$$\beta = 3\alpha. \tag{19-31}$$

The most common liquid, water, does not behave like other liquids. Above about 4°C, water expands as the temperature rises, as we would expect. Between 0 and about 4°C, however, water *contracts* with increasing temperature. Thus, at about 4°C, the density of water passes through a maximum. At all other temperatures, the density of water is less than this maximum value.

This behavior of water is the reason why lakes freeze from the top down rather than from the bottom up. As water on the surface is cooled from, say, 10°C toward the freezing point, it becomes denser ("heavier") than lower water and sinks to the bottom. Below 4°C, however, further cooling makes the water then on the surface *less* dense ("lighter") than the lower water, so it stays on the surface until it freezes. Thus the surface freezes while the lower water is still liquid. If lakes froze from the bottom up, the ice so formed would tend not to melt completely during the summer, because it would be insulated by the water above. After a few years, many bodies of open water in the temperate zones of Earth would be frozen solid all year round—and aquatic life as we know it could not exist.

READING EXERCISE 19-8: The figure below shows four rectangular metal plates, with sides of L, $2L$, or $3L$. They are all made of the same material, and their temperature is to be increased by the same amount. Rank the plates according to the expected increase in (a) their vertical heights and (b) their areas, greatest first.

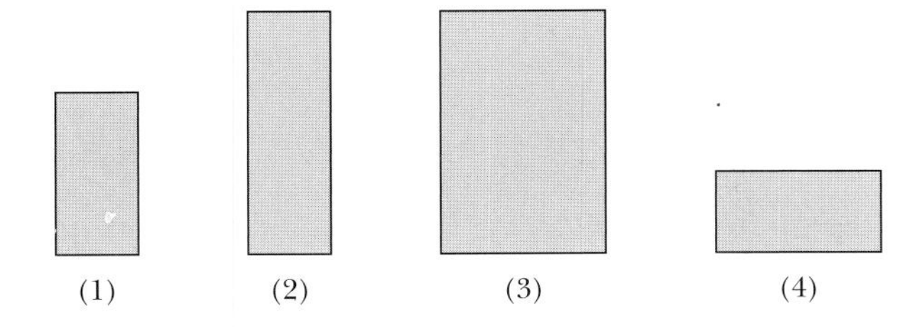

■

READING EXERCISE 19-9: Suppose that one of the plates shown above has a round hole cut out of its center. The temperature is increased. Does the hole in the center get larger, smaller, or remain unchanged in size? Explain your reasoning. ■

READING EXERCISE 19-10: Consider a cylindrical metal rod that stands on its base in a vertical orientation. The temperature increases uniformly. What happens to the pressure at the base of the rod? Does it increase, decrease, or remain unchanged? Explain your reasoning. ■

TOUCHSTONE EXAMPLE 19-4: Hot Day in Vegas

On a hot day in Las Vegas, an oil trucker loaded 37 000 L of diesel fuel. He encountered cold weather on the way to Payson, Utah, where the temperature was 23.0 K lower than in Las Vegas, and where he delivered his entire load. How many liters did he deliver? The coefficient of volume expansion for diesel fuel is 9.50×10^{-4}/°C, and the coefficient of linear expansion for his steel truck tank is 11×10^{-6}/°C.

SOLUTION ■ The **Key Idea** here is that the volume of the diesel fuel depends directly on the temperature. Thus, because the temperature decreased, the volume of the fuel did also. From Eq. 19-25, the volume change is

$$\Delta V = V\beta\,\Delta T$$
$$= (37\,000\text{ L})(9.50 \times 10^{-4}/°\text{C})(-23.0\text{ °C}) = -808\text{ L}.$$

Thus, the amount delivered was

$$V_{\text{del}} = V + \Delta V = 37\,000\text{ L} - 808\text{ L}$$
$$= 36\,192\text{ L}. \qquad \text{(Answer)}$$

Note that the thermal expansion of the steel tank has nothing to do with the problem. Question: Who paid for the "missing" diesel fuel?

19-11 More on Thermal Energy Transfer Mechanisms

In Section 19-4, during our initial discussion of the transfer of thermal energy between a system and its environment, we qualitatively discussed the three transfer mechanisms (conduction, convection, and radiation). In order to enhance our understanding of transfer mechanisms, we expand upon our discussion of conduction and radiation here.

Conduction

We all have a natural ability to sense hot and cold. But unfortunately, our "temperature sense" is in fact not always reliable. On a cold winter day, for example, why does an iron railing seem much colder to the touch than a wooden fence post when both are at the same temperature? Why are frying pans made out of metal while pot holders are made out of cloth and other fibers? The answer is that some materials are much more effective than others at transferring thermal energy via conduction.

Consider a slab of face area A and thickness L, whose faces are maintained at temperatures T_H and T_C by a hot reservoir and a cold reservoir, as in Fig. 19-25. Let Q be the energy that is transferred as thermal energy through the slab, from its hot face to its cold face, in a time interval Δt. Experiment shows that the thermal energy *conduction rate* P^{cond} (the power or thermal energy transferred per unit time—not a pressure) is

$$P^{\text{cond}} = \frac{Q}{\Delta t} = kA\frac{T_H - T_C}{L}, \qquad (19\text{-}32)$$

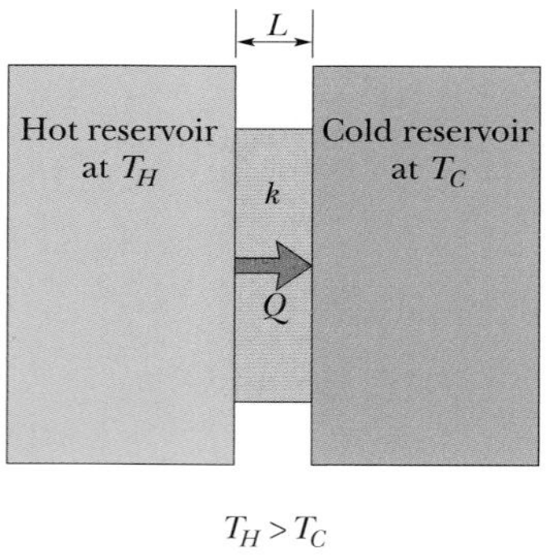

Hot reservoir at T_H | Cold reservoir at T_C | L | k | Q | $T_H > T_C$

FIGURE 19-25 ■ Thermal conduction. Thermal energy Q is transferred from a reservoir at temperature T_H to a cooler reservoir at temperature T_C through a conducting slab of thickness L and thermal conductivity k.

in which k, called the *thermal conductivity,* is a constant that depends on the material of which the slab is made. That is, the thermal conductivity k is a *property* of the material. A material that readily transfers energy by conduction is a *good thermal conductor* and has a high value of k. Table 19-6 gives the thermal conductivities of some common metals, gases, and building materials.

Thermal Resistance to Conduction (R-Value)

If you are interested in insulating your house or in keeping cola cans cold on a picnic, you are more concerned with poor conductors of thermal energy than with good ones. For this reason, the concept of *thermal resistance* R has been introduced into engineering practice. The R-value of a slab of thickness L is defined as

$$R = \frac{L}{k}. \qquad (19\text{-}33)$$

The lower the thermal conductivity of the material of which a slab is made, the higher is the R-value of the slab, so something that has a high R-value is a *poor thermal conductor* and thus a *good thermal insulator*.

Note that R is a property attributed to a slab of a specified thickness, not to a material. That is, R is *not* a property of a material alone. The commonly used unit for R (which, in the United States at least, is almost never stated) is the square foot-fahrenheit degree-hour per British thermal unit ($ft^2 \cdot F° \cdot h/Btu$).

TABLE 19-6
Some Thermal Conductivities[a]

Substance	k(W/m · K)
Metals	
Stainless steel	14
Lead	35
Aluminum	235
Copper	401
Silver	428
Gases	
Air (dry)	0.026
Helium	0.15
Hydrogen	0.18
Building Materials	
Polyurethane foam	0.024
Rock wool	0.043
Fiberglass	0.048
White pine	0.11
Window glass	1.0

[a]Conductivities change somewhat with temperature. The given values are at room temperature.

Conduction Through a Composite Slab

Figure 19-26 shows a composite slab, consisting of two materials having different thicknesses L_1 and L_2 and different thermal conductivities k_1 and k_2. The temperatures of the outer surfaces of the slab are T_H and T_C. Each face of the slab has area A. Let us derive an expression for the conduction rate through the slab under the assumption that the transfer is a *steady-state* process; that is, the temperatures everywhere in the slab and the rate of energy transfer do not change with time.

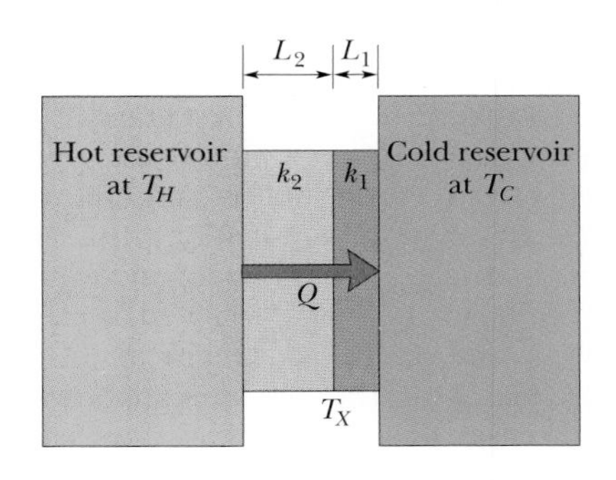

FIGURE 19-26 ■ Thermal energy Q is transferred from a hot reservoir to a cold reservoir at a steady rate through a composite slab made up of two different materials with different thicknesses and different thermal conductivities. The steady-state temperature at the interface of the two materials is denoted T_X.

In the steady state, the conduction rates through the two materials must be equal. This is the same as saying that the energy transferred through one material in a certain time must be equal to that transferred through the other material in the same time. If this were not true, temperatures in the slab would be changing and we would not have a steady-state situation. Letting T_X be the temperature of the interface between the two materials, we can now use Eq. 19-33 to express the rate of thermal energy transfer as

$$P^{\text{cond}} = \frac{k_2 A(T_H - T_X)}{L_2} = \frac{k_1 A(T_X - T_C)}{L_1}. \tag{19-34}$$

Solving for T_X yields, after a little algebra,

$$T_X = \frac{k_1 L_2 T_C + k_2 L_1 T_H}{k_1 L_2 + k_2 L_1}. \tag{19-35}$$

Substituting this expression for T_X into either equality of Eq. 19-34 yields

$$P^{\text{cond}} = \frac{A(T_H - T_C)}{L_1/k_1 + L_2/k_2}. \tag{19-36}$$

We can extend Eq. 19-36 to apply to any number n of materials making up a slab:

$$P^{\text{cond}} = \frac{A(T_H - T_C)}{\sum_{i=1}^{n}(L_i/k_i)}. \tag{19-37}$$

The summation sign in the denominator tells us to add the values of L/k for all the materials.

Radiation

The rate P^{rad} at which an object emits energy via electromagnetic radiation depends on the object's surface area A and the temperature T of that area in kelvins and is given by

$$P^{\text{rad}} = \sigma \varepsilon A T^4. \tag{19-38}$$

Here $\sigma = 5.6703 \times 10^{-8}\ \text{W/m}^2 \cdot \text{K}^4$ is called the *Stefan-Boltzmann constant* after Josef Stefan (who discovered Eq. 19-33 experimentally in 1879) and Ludwig Boltzmann (who derived it theoretically soon after). The symbol ε represents the *emissivity* of the object's surface, which has a value between 0 and 1, depending on the composition of the surface. A surface with the maximum emissivity of 1.0 is said to be a *blackbody radiator,* but such a surface is an ideal limit and does not occur in nature. Note again that the temperature in Eq. 19-38 must be in kelvins so that a temperature of absolute zero corresponds to no radiation. Note also that every object whose temperature is above 0 K—including you—emits thermal radiation. (See Fig. 19-27.)

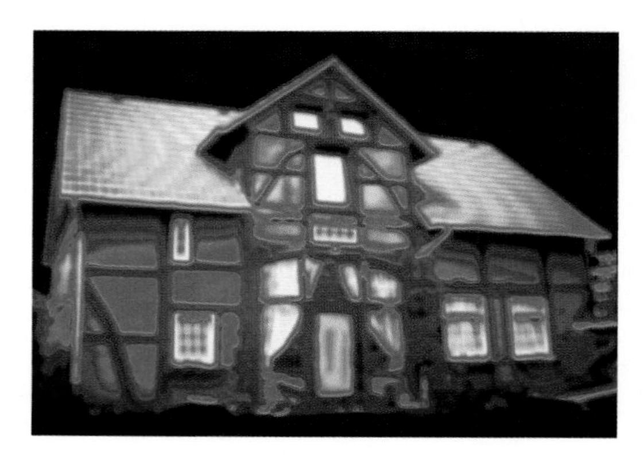

FIGURE 19-27 ■ Thermogram of a house showing the distribution of heat over its surface. The color coding ranges from white to yellow for the warmest areas (greatest heat loss from windows, etc) through red to purple and green for the coolest areas (greatest insulation). This thermogram shows that the roof and windows (yellow) are poorly insulated, while the walls (red, purple, and green) are losing the least heat. Thermograms are often used to check houses for heat loss, so that they can be made more energy efficient through improved insulation.

The rate P^{abs} at which an object absorbs energy via thermal radiation from its environment, which we take to be at uniform temperature T_{env} (in kelvins), is

$$P^{\text{abs}} = \sigma \varepsilon A T_{\text{env}}^4. \tag{19-39}$$

The emissivity ε in Eq. 19-39 is the same as that in Eq. 19-38. An idealized blackbody radiator, with $\varepsilon = 1$, will absorb all the radiated energy it intercepts (rather than sending a portion back away from itself through reflection or scattering).

Because an object will radiate energy to the environment while it absorbs energy from the environment, the object's net rate P^{net} of energy exchange due to thermal radiation is

$$P^{\text{net}} = P^{\text{abs}} - P^{\text{rad}} = \sigma \varepsilon A (T_{\text{env}}^4 - T^4). \tag{19-40}$$

P^{net} is positive if net energy is being absorbed via radiation, and negative if it is being lost via radiation.

READING EXERCISE 19-11: The figure shows the face and interface temperatures of a composite slab consisting of four materials, of identical thicknesses, through which the thermal energy transfer is steady. Rank the materials according to their thermal conductivities, greatest first.

25°C 15°C 10°C −5.0°C −10°C

a *b* *c* *d*

■

TOUCHSTONE EXAMPLE 19-5: Thermal Conduction

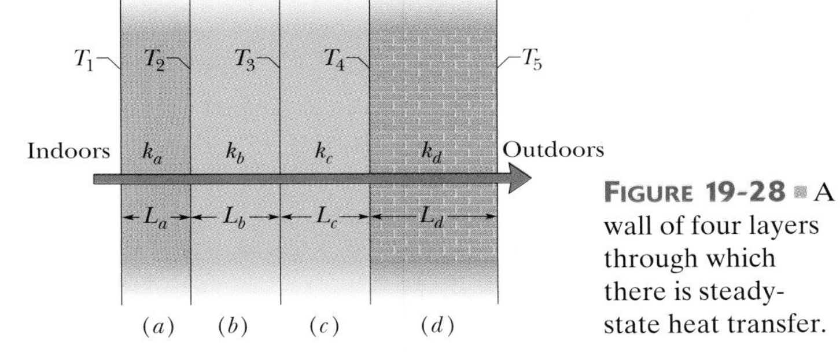

FIGURE 19-28 A wall of four layers through which there is steady-state heat transfer.

Figure 19-28 shows the cross section of a wall made of white pine of thickness L_a and brick of thickness L_d ($= 2.0L_a$), sandwiching two layers of unknown material with identical thicknesses and thermal conductivities. The thermal conductivity of the pine is k_a and that of the brick is k_d ($= 5.0k_a$). The face area A of the wall is unknown. Thermal conduction through the wall has reached the steady state; the only known interface temperatures are $T_1 = 25°C$, $T_2 = 20°C$, and $T_5 = -10°C$. What is interface temperature T_4?

SOLUTION One **Key Idea** here is that temperature T_4 helps determine the rate P_d^{cond} at which energy is conducted through the brick, as given by Eq. 19-32. However, we lack enough data to solve Eq. 19-32 for T_4. A second **Key Idea** is that because the conduction is steady, the conduction rate P_d^{cond} through the brick must equal the conduction rate P_a^{cond} through the pine. From Eq. 19-32 and Fig. 19-28, we can write

$$P_a^{\text{cond}} = k_a A \frac{T_1 - T_2}{L_a} \quad \text{and} \quad P_d^{\text{cond}} = k_d A \frac{T_4 - T_5}{L_d}.$$

Setting $P_a^{\text{cond}} = P_d^{\text{cond}}$ and solving for T_4 yield

$$T_4 = \frac{k_a L_d}{k_d L_a}(T_1 - T_2) + T_5.$$

Letting $L_d = 2.0L_a$ and $k_d = 5.0k_a$, and inserting the known temperatures, we find

$$T_4 = \frac{k_a(2.0L_a)}{(5.0k_a)L_a}(25°C - 20°C) + (-10°C)$$

$$= -8.0°C. \quad \text{(Answer)}$$

TOUCHSTONE EXAMPLE 19-6: Japanese Bees

FIGURE 19-29 The bees were unharmed by their increased body temperature, which the hornet could not withstand.

When hundreds of Japanese bees form a compact ball around a giant hornet that attempts to invade their hive, they can quickly raise their body temperature from the normal 35°C to 47°C or 48°C. That higher temperature is lethal to the hornet but not to the bees (Fig. 19-29). Assume the following: 500 bees form a ball of radius $R = 2.0$ cm for a time $\Delta t = 20$ min, the primary loss of energy by the ball is by thermal radiation, the ball's surface has emissivity $\varepsilon = 0.80$, and the ball has a uniform temperature. On average, how much additional energy must each bee produce during the 20 min to maintain 47°C?

SOLUTION The **Key Idea** here is that, because the surface temperature of the bee ball increases after the ball forms, the rate at which energy is radiated by the ball also increases. Thus, the bees lose an additional amount of energy to thermal radiation. We can relate the surface temperature to the rate of radiation (energy per unit time) with Eq. 19-38 ($P^{\text{rad}} = \sigma\varepsilon AT^4$), in which A is the ball's surface area and T is the ball's surface temperature in kelvins. This rate is an energy per unit time; that is,

$$P^{\text{rad}} = \frac{\Delta E}{\Delta t}.$$

Thus, the amount of energy ΔE radiated in time t is $\Delta E = P^{\text{rad}}\Delta t$.

At the normal temperature $T_1 = 35°C$, the radiation rate is P_1^{rad} and the amount of energy radiated in time Δt is $\Delta E_1 = P_1^{\text{rad}}\,\Delta t$. At the increased temperature $T_2 = 47°C$, the (greater) radiation rate is P_2^{rad} and the (greater) amount of energy radiated in time Δt is $\Delta E_2 = P_2^{\text{rad}}\,\Delta t$. Thus, in maintaining the ball at T_2 for time Δt, the bees must (together) provide an additional amount of energy of $E = \Delta E_2 - \Delta E_1$.

We can now write

$$E = \Delta E_2 - \Delta E_1 = P_2^{\text{rad}}\,\Delta t - P_1^{\text{rad}}\,\Delta t$$

$$= (\sigma\varepsilon A T_2^4)\,\Delta t - (\sigma\varepsilon A T_1^4)\,\Delta t = \sigma\varepsilon A\,\Delta t(T_2^4 - T_1^4). \quad (19\text{-}41)$$

The temperatures here *must* be in kelvins; thus, we write them as

$$T_2 = 47°C + 273°C = 320 \text{ K}$$

and

$$T_1 = 35°C + 273°C = 308 \text{ K}.$$

The surface area A of the ball is

$$A = 4\pi R^2 = (4\pi)(0.020 \text{ m})^2 = 5.027 \times 10^{-3} \text{ m}^2,$$

and the time Δt is 20 min = 1200 s. Substituting these and other known values into Eq. 19-41, we find

$$E = (5.6703 \times 10^{-8} \text{ W/m}^2 \cdot \text{K}^4)(0.80)(5.027 \times 10^{-3} \text{ m}^2)$$

$$\times (1200 \text{ s})[(320 \text{ K})^4 - (308 \text{ K})^4] = 406.8 \text{ J}.$$

Thus, with 500 bees in the ball, each bee must produce an additional energy of

$$\frac{E}{500} = \frac{406.8 \text{ J}}{500} = 0.81 \text{ J}. \quad \text{(Answer)}$$

Problems

SEC. 19-2 ■ THERMOMETERS AND TEMPERATURE SCALES

1. Fahrenheit and Celsius At what temperature is the Fahrenheit scale reading equal to (a) twice that of the Celsius and (b) half that of the Celsius?

2. Oymyakon (a) In 1964, the temperature in the Siberian village of Oymyakon reached $-71°C$. What temperature is this on the Fahrenheit scale? (b) The highest officially recorded temperature in the continental United States was 134°F in Death Valley, California. What is this temperature on the Celsius scale?

SEC. 19-4 ■ HEATING, COOLING, AND TEMPERATURE

3. Hot and Cold It is an everyday observation that hot and cold objects cool down or warm up to the temperature of their surroundings. If the temperature difference ΔT between an object and its surroundings ($\Delta T = T_{obj} - T_{sur}$) is not too great, the rate of cooling or warming of the object is proportional, approximately, to this temperature difference; that is,

$$\frac{d\,(\Delta T)}{dt} = -A(\Delta T),$$

where A is a constant. (The minus sign appears because ΔT decreases with time if ΔT is positive and increases if ΔT is negative.) This is known as *Newton's law of cooling*. (a) On what factors does A depend? What are its dimensions? (b) If at some instant $t_1 = 0$ the temperature difference is ΔT_1, show that it is

$$\Delta T = \Delta T_1 e^{-At_2}$$

at a later time t_2.

4. House Heater The heater of a house breaks down one day when the outside temperature is 7.0°C. As a result, the inside temperature drops from 22°C to 18°C in 1.0 h. The owner fixes the heater and adds insulation to the house. Now she finds that, on a similar day, the house takes twice as long to drop from 22°C to 18°C when the heater is not operating. What is the ratio of the new value of constant A in Newton's law of cooling (see Problem 3) to the previous value?

SEC. 19-5 ■ THERMAL ENERGY TRANSFER TO SOLIDS AND LIQUIDS

5. A Certain Substance A certain substance has a mass per mole of 50 g/mol. When 314 J of thermal energy is transferred to a 30.0 g sample, the sample's temperature rises from 25.0°C to 45.0°C. What are (a) the specific heat and (b) the molar specific heat of this substance? (c) How many moles are present?

6. Diet Doctor A certain diet doctor encourages people to diet by drinking ice water. His theory is that the body must burn off enough fat to raise the temperature of the water from 0.00°C to the body temperature of 37.0°C. How many liters of ice water would have to be consumed to burn off 454 g (about 1 lb) of fat, assuming that this much fat burning requires 3500 Cal be transferred to the ice water? Why is it not advisable to follow this diet? (One liter = 10^3 cm^3. The density of water is 1.00 g/cm^3.)

7. Minimum Energy Calculate the minimum amount of energy, in joules required to completely melt 130 g of silver initially at 15.0°C.

8. How Much Unfrozen How much water remains unfrozen after 50.2 kJ of thermal energy is transferred from 260 g of liquid water initially at its freezing point?

9. Energetic Athlete An energetic athlete can use up all the energy from a diet of 4000 food calories/day where a food calorie = 1000 cal. If he were to use up this energy at a steady rate, how would his rate of energy use compare with the power of a 100 W bulb? (The power of 100 W is the rate at which the bulb converts electrical energy to thermal energy and the energy of visible light.)

10. Four Lightbulbs A room is lighted by four 100 W incandescent lightbulbs. (The power of 100 W is the rate at which a bulb converts electrical energy to thermal energy and the energy of visible light.) Assuming that 90% of the energy is converted to thermal energy, how much thermal energy is transferred to the room in 1.00 h?

11. Drilling a Hole A power of 0.400 hp is required for 2.00 min to drill a hole in a 1.60 lb copper block. (a) If the full power is the rate at which thermal energy is generated, how much is generated in Btu? (b) What is the rise in temperature of the copper if the copper absorbs 75.0% of this energy? (Use the energy conversion 1 ft·lb = 1.285×10^{-3} Btu.)

12. How Much Butter How many grams of butter, which has a usable energy content of 6.0 Cal/g (= 6000 cal/g), would be equivalent to the change in gravitational potential energy of a 73.0 kg man who ascends from sea level to the top of Mt. Everest, at elevation 8.84 km? Assume that the average value of g is 9.80 m/s^2.

13. Immersion Heater A small electric immersion heater is used to heat 100 g of water for a cup of instant coffee. The heater is labeled "200 watts," so it converts electrical energy to thermal energy that is transferred to the water at this rate. Calculate the time required to bring the water from 23°C to 100°C ignoring any thermal energy that transfers out of the cup.

14. Tub of Water One way to keep the contents of a garage from becoming too cold on a night when a severe subfreezing temperature is forecast is to put a tub of water in the garage. If the mass of the water is 125 kg and its initial temperature is 20°C, (a) how much thermal energy must the water transfer to its surroundings in order to freeze completely and (b) what is the lowest possible temperature of the water and its surroundings until that happens?

15. A Chef A chef, on finding his stove out of order, decides to boil the water for his wife's coffee by shaking it in a thermos flask. Suppose that he uses tap water at 15°C and that the water falls 30 cm each shake, the chef making 30 shakes each minute. Neglecting any transfer of thermal energy out of the flask, how long must he shake the flask for the water to reach 100°C?

16. Copper Bowl A 150 g copper bowl contains 220 g of water, both at 20.0°C. A very hot 300 g copper cylinder is dropped into the water, causing the water to boil, with 5.00 g being converted to steam. The final temperature of the system is 100°C. Neglect energy transfers with the environment. (a) How much energy (in calories) is transferred to the water? (b) How much to the bowl? (d) What is the original temperature of the cylinder?

17. Ethyl Alcohol Ethyl alcohol has a boiling point of 78°C, a freezing point of −114°C, a heat of vaporization of 879 kJ/kg, a heat of fusion of 109 kJ/kg, and a specific heat of 2.43 kJ/kg · C°. How much thermal energy must be transferred out of 0.510 kg of ethyl alcohol that is initially a gas at 78°C so that it becomes a solid at −114°C?

18. Metric–Nonmetric *Nonmetric version:* How long does a 2.0×10^5 Btu/h water heater take to raise the temperature of 40 gal of water from 70°F to 100°F? *Metric version:* How long does a 59 kW water heater take to raise the temperature of 150 L of water from 21°C to 38°C?

19. Buick A 1500 kg Buick moving at 90 km/h brakes to a stop, at a uniform rate and without skidding, over a distance of 80 m. At what average rate is mechanical energy transformed into thermal energy in the brake system?

20. Solar Water Heater In a solar water heater, radiant energy from the Sun is transferred to water that circulates through tubes in a rooftop collector. The solar radiation enters the collector through a transparent cover and warms the water in the tubes; this water is pumped into a holding tank. Assume that the efficiency of the overall system is 20% (that is, 80% of the incident solar energy is lost from the system). What collector area is necessary to raise the temperature of 200 L of water in the tank from 20°C to 40°C in 1.0 h when the intensity of incident sunlight is 700 W/m²?

21. Steam What mass of steam at 100°C must be mixed with 150 g of ice at its melting point, in a thermally insulated container, to produce liquid water at 50°C?

22. Iced Tea A person makes a quantity of iced tea by mixing 500 g of hot tea (essentially water) with an equal mass of ice at its melting point. If the initial hot tea is at a temperature of (a) 90°C and (b) 70°C, what are the temperature and mass of the remaining ice when the tea and ice reach a common temperature? Neglect energy transfers with the environment.

23. Ice Cubes (a) Two 50 g ice cubes are dropped into 200 g of water in a thermally insulated container. If the water is initially at 25°C, and the ice comes directly from a freezer at −15°C, what is the final temperature of the drink when the drink reaches thermal equilibrium? (b) What is the final temperature if only one ice cube is used?

24. Thermos of Coffee An insulated Thermos contains 130 cm³ of hot coffee, at a temperature of 80.0°C. You put in a 12.0 g ice cube at its melting point to cool the coffee. By how many degrees has your coffee cooled once the ice has melted? Treat the coffee as though it were pure water and neglect energy transfers with the environment.

SEC. 19-8 ■ SOME SPECIAL CASES OF THE FIRST LAW OF THERMODYNAMICS

25. Gas Expands A sample of gas expands from 1.0 m³ to 4.0 m³ while its pressure decreases from 40 Pa to 10 Pa. How much work is done by the gas if its pressure changes with volume via each of the three paths shown in the *P-V* diagram in Fig. 19-30?

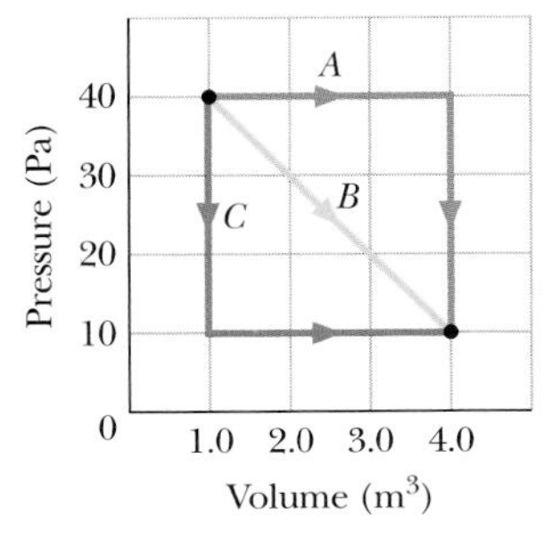

FIGURE 19-30 ■ Problem 25.

26. Work Done Consider that 200 J of work is done on a system and 70.0 cal of thermal energy is transferred out of the system. In the sense of the first law of thermodynamics, what are the values (including algebraic signs) of (a) W, (b) Q, and (c) ΔE^{int}?

27. Closed Chamber Gas within a closed chamber undergoes the cycle shown in the *P-V* diagram of Fig. 19-31. Calculate the net thermal energy added to the system during one complete cycle.

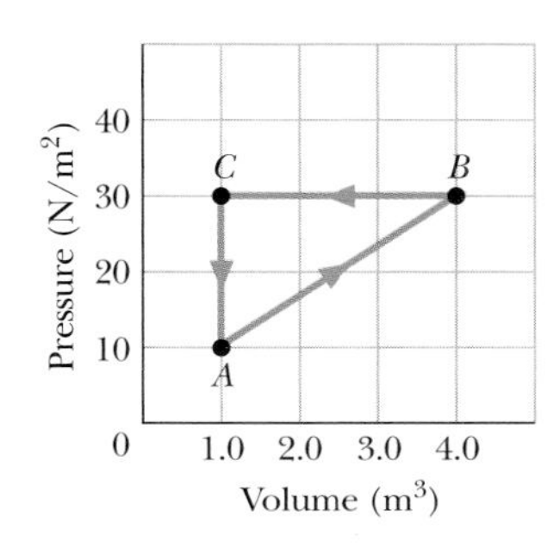

FIGURE 19-31 ■ Problem 27.

28. Thermodynamic A thermodynamic system is taken from an initial state A to another state B and back again to A, via state C, as shown by path $ABCA$ in the *P-V* diagram of Fig. 19-31*a*. (a) Complete the table in Fig. 19-32*b* by filling in either + or − for the sign of each thermodynamic quantity associated with each step of the cycle. (b) Calculate the numerical value of the work done by the system for the complete cycle $ABCA$.

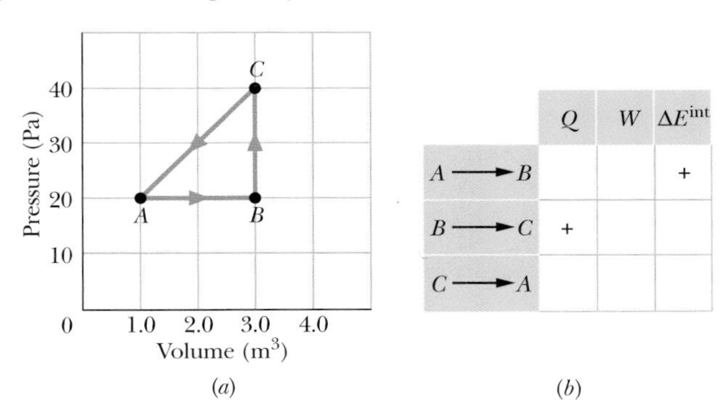

	Q	W	ΔE^{int}
$A \rightarrow B$			+
$B \rightarrow C$	+		
$C \rightarrow A$			

FIGURE 19-32 ■ Problem 28.

29. From *i* to *f* When a system is taken from state i to state f along path iaf in Fig. 19-33. $Q = 50$ cal and $W = 20$ cal. Along path ibf, $Q = 36$ cal. (a) What is W along path ibf? (b) If $W = -13$ cal for the return path fi, what is Q for this path? (c) Take $E_i^{\text{int}} = 10$ cal. What is E_f^{int}? (d) If $E_b^{\text{int}} = 22$ cal, what are the values of Q for path ib and path bf?

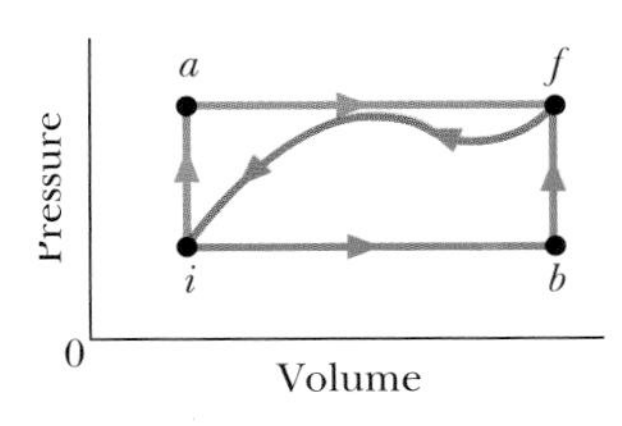

FIGURE 19-33 ■ Problem 29.

30. Gas Within Gas within a chamber passes through the cycle shown in Fig. 19-34. Determine the thermal energy transferred by the system during process CA if the thermal energy added Q_{AB}, during process AB is 20.0 J, no thermal energy is transferred during process BC, and the net work done during the cycle is 15.0 J.

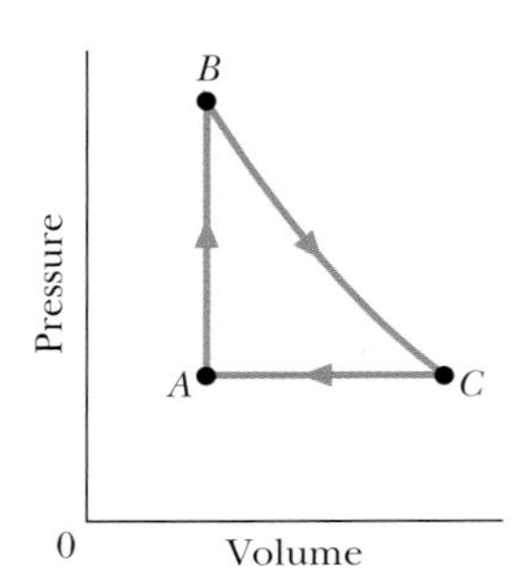

FIGURE 19-34 ■ Problem 30.

SEC. 19-9 ■ MORE ON TEMPERATURE MEASUREMENT

31. Gas Thermometer A particular gas thermometer is constructed of two gas-containing bulbs, each of which is put into a water bath, as shown in Fig. 19-35. The pressure difference between the two bulbs is measured by a mercury manometer as shown. Appropriate reservoirs, not shown in the diagram, maintain constant gas volume in the two bulbs. There is no difference in pressure when both baths are at the triple point of water. The pressure difference is 120 torr when one bath is at the triple point and the other is at the boiling point of water. It is 90.0 torr when one bath is at the triple point and the other is at an unknown temperature to be measured. What is the unknown temperature?

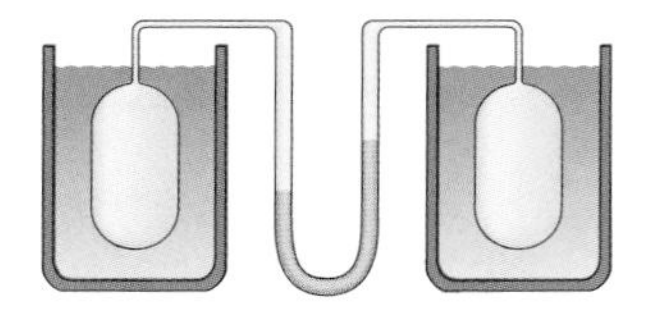

FIGURE 19-35 ■ Problem 31.

32. A Gas at Boiling Suppose the temperature of a gas at the boiling point of water is 373.15 K. What then is the limiting value of the ratio of the pressure of the gas at that boiling point to its pressure at the triple point of water? (Assume the volume of the gas is the same at both temperatures.)

33. Pairs of Scales At what temperature do the following pairs of scales read the same, if ever: (a) Fahrenheit and Celsius (verify the listing in Table 19-1), (b) Fahrenheit and Kelvin, and (c) Celsius and Kelvin?

SEC. 19-10 ■ THERMAL EXPANSION

34. Aluminum Flagpole An aluminum flagpole is 33 m high. By how much does its length increase as the temperature increases by 15 C°?

35. Pyrex Glass The Pyrex glass mirror in the telescope at the Mt. Palomar Observatory has a diameter of 200 in. The temperature ranges from −10°C to 50°C on Mt. Palomar. In micrometers, what is the maximum change in the diameter of the mirror, assuming that the glass can freely expand and contract?

36. Aluminum Alloy An aluminum-alloy rod has a length of 10.000 cm at 20.000°C and a length of 10.015 cm at the boiling point of water. (a) What is the length of the rod at the freezing point of water? (b) What is the temperature if the length of the rod is 10.009 cm?

37. Circular Hole A circular hole in an aluminum plate is 2.725 cm in diameter at 0.000°C. What is its diameter when the temperature of the plate is raised to 100.0°C?

38. Lead Ball What is the volume of a lead ball at 30°C if the ball's volume at 60°C is 50 cm^3?

39. Change in Volume Find the change in volume of an aluminum sphere with an initial radius of 10 cm when the sphere is heated from 0.0°C to 100°C.

40. Area Rectangular The area A of a rectangular plate is ab. Its coefficient of linear expansion is α. After a temperature rise ΔT, side a is longer by Δa and side b is longer by Δb (Fig. 19-36). Show that if the small quantity $(\Delta a\, \Delta b)/ab$ is neglected, then $\Delta A = 2\alpha A\, \Delta T$.

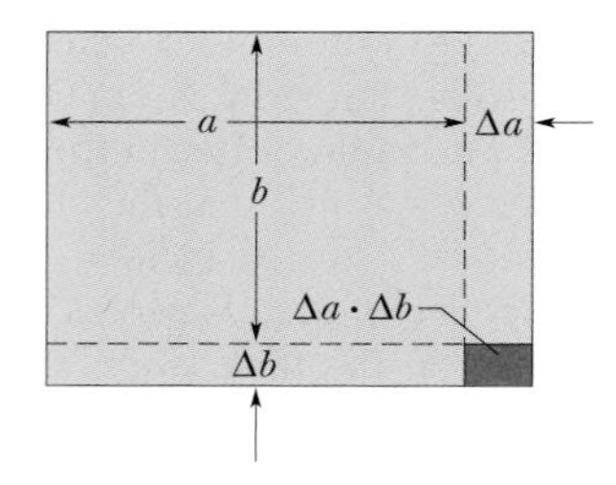

FIGURE 19-36 ■ Problem 40.

41. Aluminum Cup An aluminum cup of 100 cm^3 capacity is completely filled with glycerin at 22°C. How much glycerin, if any, will spill out of the cup if the temperature of both the cup and glycerin is increased to 28°C? (The coefficient of volume expansion of glycerin is 5.1×10^{-4}/C°.)

42. Rod At 20°C, a rod is exactly 20.05 cm long on a steel ruler. Both the rod and the ruler are placed in an oven at 270°C, where the rod now measures 20.11 cm on the same ruler. What is the coefficient of thermal expansion for the material of which the rod is made?

43. Steel Rod A steel rod is 3.000 cm in diameter at 25°C. A brass ring has an interior diameter of 2.992 cm at 25°C. At what common temperature will the ring just slide onto the rod?

44. Metal Cylinder When the temperature of a metal cylinder is raised from 0.0°C to 100°C, its length increases by 0.23%. (a) Find the percent change in density. (b) What is the metal? Use Table 19-5.

45. Barometer Show that when the temperature of a liquid in a barometer changes by ΔT and the pressure is constant, the liquid's height h changes by $\Delta h = \beta h\, \Delta T$, where β is the coefficient of volume expansion. Neglect the expansion of the glass tube.

46. Copper Coin When the temperature of a copper coin is raised by 100 C°, its diameter increases by 0.18%. To two significant figures, give the percent increase in (a) the area of a face, (b) the thickness, (c) the volume, and (d) the mass of the coin. (e) Calculate the coefficient of linear expansion of the coin.

47. Pendulum Clock A pendulum clock with a pendulum made of brass is designed to keep accurate time at 20°C. If the clock operates at 0.0°C, what is the magnitude of its error, in seconds per hour, and does the clock run fast or slow?

48. Radioactive Source In a certain experiment, a small radioactive source must move at selected, extremely slow speeds. This motion is accomplished by fastening the source to one end of an aluminum rod and heating the central section of the rod in a controlled way. If the effective heated section of the rod in Fig. 19-37 is 2.00 cm, at what constant rate must the temperature of the rod be changed if the source is to move at a constant speed of 100 nm/s?

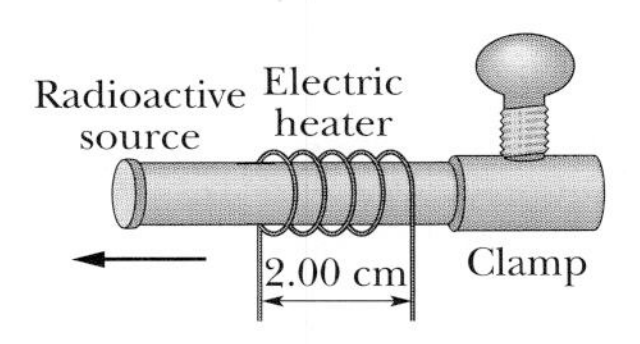

FIGURE 19-37 ■ Problem 48.

49. Temperature Rise As a result of a temperature rise of 32°C, a bar with a crack at its center buckles upward (Fig. 19-38). If the fixed distance L_0 is 3.77 m and the coefficient of linear expansion of the bar is 25×10^{-6}/C°, find the rise x of the center.

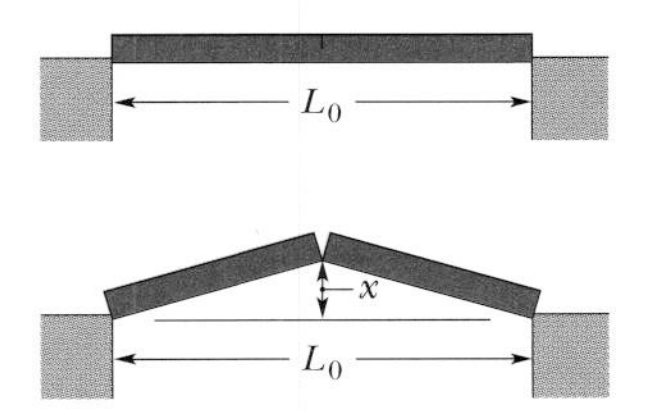

FIGURE 19-38 ■ Problem 49.

50. Copper Ring A 20.0 g copper ring has a diameter of 2.54000 cm at its temperature of 0.000°C. An aluminum sphere has a diameter of 2.54508 cm at its temperature of 100.0°C. The sphere is placed on top of the ring (Fig. 19-39), and the two are allowed to come to thermal equilibrium, with no thermal energy transferred to the surroundings. The sphere just passes through the ring at the equilibrium temperature. What is the mass of the sphere?

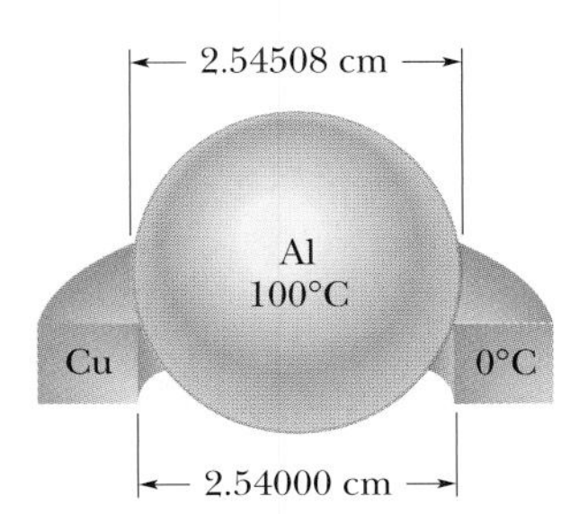

FIGURE 19-39 ■ Problem 50.

SEC. 19-11 ■ MORE ON THERMAL ENERGY TRANSFER MECHANISMS

51. Single-Family Dwelling The ceiling of a single-family dwelling in a cold climate should have an R-value of 30. To give such insulation, how thick would a layer of (a) polyurethane foam and (b) silver have to be?

52. North America The average rate at which energy is conducted outward through the ground surface in North America is 54.0 mW/m^2, and the average thermal conductivity of the near-surface rocks is 2.50 W/m · K. Assuming a surface temperature of 10.0°C, find the temperature at a depth of 35.0 km (near the base of the crust). Ignore the thermal energy transferred from the radioactive elements.

53. Slab Consider the slab shown in Fig. 19-25. Suppose that $L =$ 25.0 cm, $A =$ 90.0 cm^2, and the material is copper. If $T_H =$ 125°C, $T_C =$ 10.0°C, and a steady state is reached, find the conduction rate through the slab.

54. Body Heat (a) Calculate the rate at which body heat is conducted through the clothing of a skier in a steady-state process, given the following data: the body surface area is 1.8 m^2 and the clothing is 1.0 cm thick; the skin surface temperature is 33°C and the outer surface of the clothing is at 1.0°C; the thermal conductivity of the clothing is 0.040 W/m · K. (b) How would the answer to (a) change if, after a fall, the skier's clothes became soaked with water of thermal conductivity 0.60 W/m · K?

55. Copper Rod A cylindrical copper rod of length 1.2 m and cross-sectional area 4.8 cm^2 is insulated to prevent thermal energy from being transferred through its surface. The ends are maintained at a temperature difference of 100°C by having one end in a water–ice mixture and the other in boiling water and steam. (a) Find the rate at which thermal energy is conducted along the rod. (b) Find the rate at which ice melts at the cold end.

56. Without a Spacesuit If you were to walk briefly in space without a spacesuit while far from the Sun (as an astronaut does in the movie *2001*), you would feel the cold of space—while you radiated thermal energy, you would absorb almost none from your environment. (a) At what rate would you lose thermal energy? (b) How much thermal energy would you lose in 30 s? Assume that your emissivity is 0.90, and estimate other data needed in the calculations.

57. Rectangular Rods Two identical rectangular rods of metal are welded end to end as shown in Fig. 19-40*a*, and 10 J of thermal energy is conducted (in a steady-state process) through the rods in 2.0 min. How long would it take for 10 J to be conducted through the rods if they were welded together as shown in Fig. 19-40*b*?

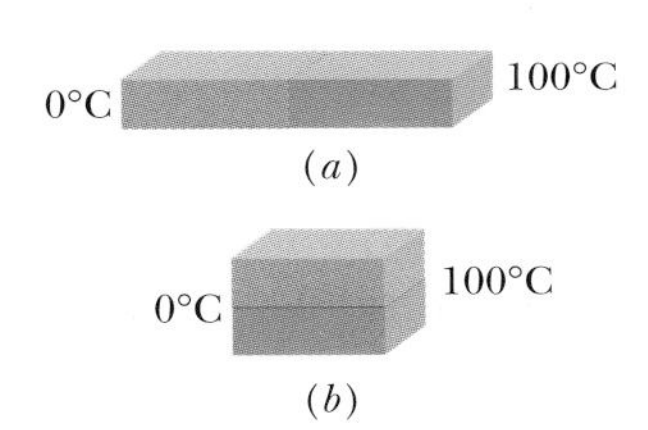

FIGURE 19-40 ■ Problem 57.

58. Four Squares Four square pieces of insulation of two different materials, all with the same thickness and area A, are available to cover an opening of area $2A$. This can be done in either of the two ways shown in Fig. 19-41. Which arrangement, (*a*) or (*b*), gives the lower thermal energy flow if $k_2 \neq k_1$?

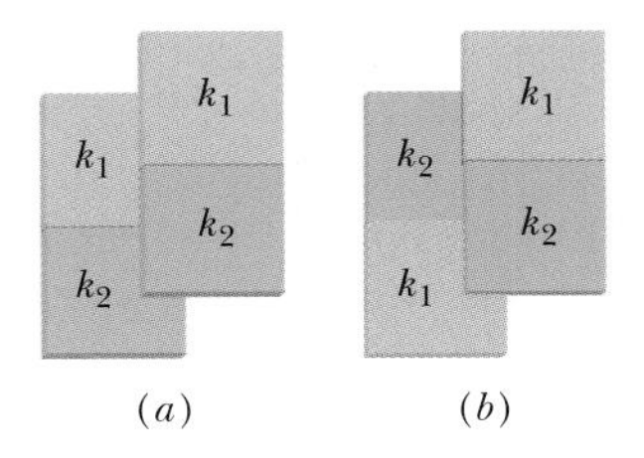

FIGURE 19-41 ■ Problem 58.

59. Glass Window (a) What is the rate of thermal energy transfer in watts per square meter through a glass window 3.0 mm thick if the outside temperature is $-20°F$ and the inside temperature is $+72°F$? (b) A storm window having the same thickness of glass is installed parallel to the first window, with an air gap of 7.5 cm between the two windows. What now is the rate of energy loss if conduction is the only important energy-transfer mechanism?

60. A Sphere A sphere of radius 0.500 m, temperature 27.0°C, and emissivity 0.850 is located in an environment of temperature 77.0°C. At what rate does the sphere (a) emit and (b) absorb thermal radiation? (c) What is the sphere's net rate of energy exchange?

61. Tank of Water A tank of water has been outdoors in cold weather, and a slab of ice 5.0 cm thick has formed on its surface (Fig. 19-42). The air above the ice is at $-10°C$. Calculate the rate of formation of ice (in centimeters per hour) on the ice slab. Take the thermal conductivity and density of ice to be $0.0040\ \text{cal/s}\cdot\text{cm}\cdot\text{C}°$ and $0.92\ \text{g/cm}^3$. Assume that energy is not transferred through the walls or bottom of the tank.

FIGURE 19-42 ■ Problem 61.

62. A Wall Figure 19-43 shows (in cross section) a wall that consists of four layers. The thermal conductivities are $k_1 = 0.060$ W/m·K, $k_3 = 0.040$ W/m·K, and $k_4 = 0.12$ W/m·K (k_2 is not known). The layer thicknesses are $L_1 = 1.5$ cm, $L_3 = 2.8$ cm, and $L_4 = 3.5$ cm (L_2 is not known). Energy transfer through the wall is steady. What is the temperature of the interface indicated?

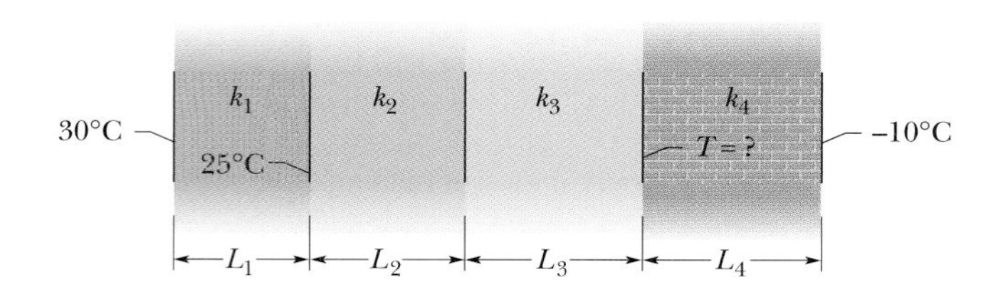

FIGURE 19-43 ■ Problem 62.

Additional Problems

63. 300 F Club You can join the semi-secret "300 F" club at the Amundsen–Scott South Pole Station only when the outside temperature is below $-70°C$. On such a day, you first bask in a hot sauna and then run outside wearing only your shoes. (This is, of course, extremely dangerous, but the rite is effectively a protest against the constant danger of the winter cold at the south pole.)

Assume that when you step out of the sauna, your skin temperature is 102°F and the walls, ceiling, and floor of the sauna room have a temperature of 30°C. Estimate your surface area, and take your skin emissivity to be 0.80. (a) What is the approximate net rate P^{net} at which you lose energy via thermal radiation transfer to the room? Next, assume that when you are outside half your surface area transfers thermal radiation to the sky at a temperature of $-25°C$ and the other half transfers thermal radiation to the snow and ground at a temperature of $-80°C$. What is the approximate net rate at which you lose energy via thermal radiation exchanges with (b) the sky and (c) the snow and ground?

64. Shallow Pond Ice has formed on a shallow pond and a steady state has been reached, with the air above the ice at $-5.0°C$ and the bottom of the pond at 4.0°C. If the total depth of *ice + water* is 1.4 m, how thick is the ice? (Assume that the thermal conductivities of ice and water are 0.40 and 0.12 cal/m·C°·s, respectively.)

65. Emperor Penguins Emperor penguins, those large penguins that resemble stuffy English butlers, breed and hatch their young even during severe Antarctic winters. Once an egg is laid, the father balances the egg on his feet to prevent the egg from freezing. He must do this for the full incubation period of 105 to 115 days, during which he cannot eat because his food is in the water. He can survive this long without food only if he can reduce his loss of internal food energy significantly. If he is alone, he loses that energy too quickly to stay warm, and eventually abandons the egg in order to eat. To protect themselves and each other from the cold so as to reduce the loss of internal energy, penguin fathers huddle closely together, in groups of perhaps several thousand. In addition to providing other benefits, the huddling reduces the rate at which the penguins thermally radiate energy to their surroundings.

Assume that a penguin father is a circular cylinder with top surface area a, height h, surface temperature T, and emissivity ε. (a) Find an expression for the rate P_i at which an individual father would radiate energy to the environment from his top surface and his side surface were he alone with his egg.

If N identical fathers were well apart from one another, the total rate of energy loss via radiation would be NP_i. Suppose, instead, that they huddle closely to form a *huddled cylinder* with top surface area Na and height h. (b) Find an expression for the rate P_h at which energy is radiated by the top surface and the side surface of the huddled cylinder.

(c) Assuming $a = 0.34\ \text{m}^2$ and $h = 1.1$ m and using the expressions you obtained for P_i and P_h, graph the ratio P_h/NP_i versus N_h. Of course, the penguins know nothing about algebra or graphing, but their instinctive huddling reduces this ratio so that more of their eggs survive to the hatching stage. From the graphs (as you will see, you probably need more than one version), approximate how many penguins must huddle so that P_h/NP_i is reduced to (d) 0.5, (e) 0.4, (f) 0.3, (g) 0.2, and (h) 0.15. (i) For the assumed data, what is the lower limiting value for P_h/NP_i?

66. The Penny and the Jelly Donut You see a penny lying on the ground. A penny won't buy much these days, so you think: "If I bend down to pick it up I will do work. To do that work I will have to burn some energy. It probably costs me more to buy the fuel (food) to provide that energy than I would gain by picking up the penny. It's not cost effective." You pass it by. Is the argument correct? Estimate the energy cost for picking up a penny. You may find the following information useful: A jelly donut contains about 250 Calories (1 Calorie = 1 Kcal).

67. Considering Changes For each of the situations described below, the object considered is undergoing some changes. Among the possible changes you should consider are: (Q) The object is absorbing or giving off thermal energy. (T) The object's temperature is changing. (E^{int}) The object's internal energy is changing. (W) The object is doing mechanical work or having work done on it. For each of the situations described below, identify which of the four changes are taking place and write as many of the letters Q, T, E^{int}, W, (or none) as are appropriate. (a) A cylinder with a piston on top contains a compressed gas and is sitting on a thermal reservoir (a large iron block). After everything has come to thermal equilibrium, the piston is moved upward somewhat (very slowly). The object to be considered is the gas in the cylinder. (b) Consider the same cylinder as in part (a), but it is wrapped in styrofoam, a very good thermal insulator, instead of sitting on a thermal reservoir. The piston is pressed downward (again, very slowly), compressing the gas. The object to be considered is the gas in the cylinder. (c) An ice cube that is sitting in the open air and is melting.

68. KE and Temperature Converting kinetic energy into thermal energy produces small rises in temperature. This was in part responsible for the difficulty in discovering the law of conservation of energy. It also implies that hot objects contain a lot of energy. (This latter comment is largely responsible for the industrial revolution in the 19th century.) To get some feel for these numbers, assume all mechanical energy is converted to thermal energy and carry out three estimates:

(a) A steel ball is dropped from a height of 3 m onto a concrete floor. It bounces a large number of times but eventually comes to rest. Estimate the ball's rise in temperature.

(b) Suppose the steel ball you used in part (a) is at room temperature. If you converted all its thermal energy to translational kinetic energy, how fast would it be moving? (Give your answer in units of miles per hour. Also, ignore the fact that you would have to create momentum.)

(c) Suppose a nickel–iron meteor falls to Earth from deep space. Estimate how much its temperature would rise on impact.

20 The Kinetic Theory of Gases

When a container of cold champagne, soda pop, or any other carbonated drink is opened, a slight fog forms around the opening and some of the liquid sprays outward. (In the photograph, the fog is the white cloud that surrounds the stopper, and the spray has formed streaks within the cloud.)

What causes the fog?

The answer is in this chapter.

20-1 Molecules and Thermal Gas Behavior

In our studies of mechanics and thermodynamics we have found a number of strange and interesting results. In mechanics, we saw that moving objects tend to run down and come to a stop. We attributed this to the inevitable presence of friction and drag forces. Without these nonconservative forces mechanical energy would be conserved and perpetual motion would be possible. In thermodynamics, we discovered that ordinary objects, by virtue of their temperature, contain huge quantities of internal energy. This is where the "lost" energy resulting from friction forces is hidden. In this chapter, we will learn about some ways that matter can store internal energy. What you are about to learn may be counterintuitive. Instead of finding that the "natural state" of a system is to lose energy, you will find considerable evidence that the "natural state" of a system is quite the opposite. It is one in which its fundamental parts (atoms and molecules) are traveling every which way—in a state of perpetual motion.

Classical thermodynamics—the subject of the previous chapter—has nothing to say about atoms or molecules. Its laws are concerned only with such macroscopic variables as pressure, volume, and temperature. In this chapter we begin an exploration of the atomic and molecular basis of thermodynamics. As is usual in the development of new theories in physics, we start with a simple model. The fact that gases are fluid and compressible is evidence that their molecules are quite small relative to the average spacing between them. If so, we expect that gas molecules are relatively free and independent of one another. For this reason, we believe that the thermal behavior of gases will be easier to understand than that of liquids and solids. Thus, we begin an exploration of the atomic and molecular basis of thermodynamics by developing the kinetic theory of gases—a simplified model of gas behavior based on the laws of classical mechanics.

We start with a discussion of how the ideal gas law characterizes the macroscopic behavior of simple gases. This macroscopic law relates the amount of gas and its pressure, temperature, and volume to each other. Next we consider how kinetic theory, which provides us with a molecular (or microscopic) model of gas behavior, can be used to explain observed macroscopic relationships between gas pressure, volume, and temperature. We then move on to using kinetic theory as an underlying model of the characteristics of an ideal gas. The basic ideas of kinetic theory are that: (1) an ideal gas at a given temperature consists of a collection of tiny particles (atoms or molecules) that are in perpetual motion—colliding with each other and the walls of their container; and (2) the hidden internal energy of an ideal gas is directly proportional to the kinetic energy of its particles.

20-2 The Macroscopic Behavior of Gases

Any gas can be described by its macroscopic variables volume V, pressure P, and temperature T. Simple experiments were performed on low density gases in the 17th and 18th centuries to relate these variables. Robert Boyle (b. 1627) determined that at a constant temperature the product of pressure and volume remains constant. (See Fig. 20-1.)

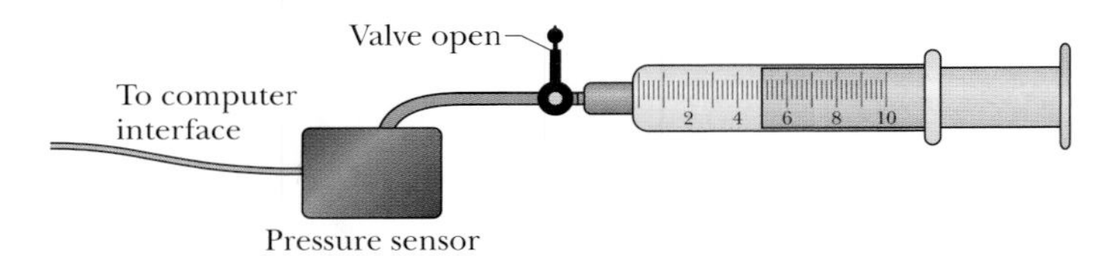

FIGURE 20-1 ■ A contemporary setup for determining the relationship between gas pressure and volume using an inexpensive medical syringe and an electronic pressure sensor attached to a computer data acquisition system. The volume is changed when the plunger is pushed or pulled. When temperature is held constant, P and V turn out to be inversely proportional to each other so that PV is constant.

French scientists Jacques Charles (b. 1746) and Joseph Gay-Lussac (b. 1778) found that as the Kelvin temperature of a fixed volume of gas is raised its pressure increases proportionally. (See Fig. 19-16.) Similarly, Charles, who was a hot-air balloonist, discovered that for a constant pressure (such as atmospheric pressure) the volume of a gas is proportional to its temperature. (See Fig. 19-15.) By combining the results of all three of these experiments we must conclude that there is a proportionality between PV and T:

$$PV \propto T.$$

The Molecular Form of the Ideal Gas Law

If we can find a constant of proportionality between PV and T for a relatively low density gas, then we will have formulated a gas law. An examination of the student-generated P vs. T data shown in Fig. 19-17 indicates that the constant of proportionality between the product PV and the variable T (determined by the slopes) of the graphs decreases as the mass of gas confined to the same volume decreases. Similar experiments have shown that the slope of a P vs. T graph will change if the same volume and mass of a different kind of gas is used. This suggests that the constant of proportionality we are looking for must be a function of *both* the mass and type of gas. It was puzzling to early investigators that the slopes of their P vs. T and V vs. T graphs were not just proportional to the mass of the different gases used in the experiments.

The key to finding a constant of proportionality that embodies both gas type and mass was a hypothesis developed in the early 19th century by the Italian scientist Amadeo Avogadro (1776–1856). In 1811, Avogadro proposed that equal volumes of any kind of gas at the same pressure would have the same number of molecules and occupy the same volume. Eventually it was discovered that the constant of proportionality needed for the fledgling gas law was one that is directly proportional to the number of molecules of a gas rather than its mass, so that

$$PV = Nk_BT \quad \text{(molecular ideal gas law)}, \tag{20-1}$$

where N is the number of molecules of confined gas and k_B is a proportionality constant needed to shift from kelvins to joules, the SI units for the product PV. The experimentally determined value of k_B is known historically as the Boltzmann constant. Its measured value is

$$k_B = 1.38 \times 10^{-23}\ \text{J/K} \quad \text{(Boltzmann constant)}. \tag{20-2}$$

It turns out that common gases such as O_2, N_2, and Ar behave like ideal gases at relatively low pressure (< 10 atm) when their temperatures are well above their boiling points. For example, air near room temperature and 1 atm of pressure behaves like an ideal gas.

Avogadro's Number and the Mole

The problem with the molecular form of the ideal gas law we just presented is that it is hard to count molecules. It is much easier to measure the mass of a sample of gas or its volume at a standard pressure. In this subsection we will define two new quantities—*mole* and *molar mass*. Although these quantities are related to the number of molecules in a gas, they can be measured macroscopically, so it is useful to reformulate the ideal gas law in terms of moles.

Let's start our reformulation of the ideal gas law with definitions of mole and molar mass. In Section 1-7, we presented the SI definition of the *atomic mass unit* in terms of the mass of a carbon-12 atom. In particular, carbon-12 is assigned an atomic mass of exactly 12 u. Here the atomic mass unit u represents grams per mole (g/mol). In a related fashion, the SI definition of the *mole* (or *mol* for short) relates the number of particles in a substance to its macroscopic mass.

A **mole** is defined as the amount of any substance that contains the same number of atoms or molecules as there are in *exactly* 12 g of carbon-12.

The results of many different types of experiments, including x-ray diffraction studies in crystals, have revealed that there are a very large number of atoms in 12 g of carbon-12. The number of atoms is known as Avogadro's number and is denoted as N_A.

$$N_A = 6.022137 \times 10^{23}\ \text{mol}^{-1} \quad \text{(Avogadro's number)}. \tag{20-3}$$

Here the symbol mol^{-1} represents the inverse mole or "per mole." Usually we round off the value to three significant figures so that $N_A = 6.02 \times 10^{23}\ \text{mol}^{-1}$.

The number of moles n contained in a sample of any substance is equal to the ratio of the number of atoms or molecules N in the sample to the number of atoms or molecules N_A in 1 mole of the same substance:

$$n = \frac{N}{N_A}. \tag{20-4}$$

(*Caution:* The three symbols in this equation can easily be confused with one another, so you should sort them with their meanings now, before you end in "N-confusion.")

We can easily calculate the mass of one mole of atoms or molecules in any sample, defined as the **molar mass** (denoted as M), by looking in a table of atomic or molecular masses.

Note that if we refer to Appendix F to find the molar mass, M, in grams of a sample of matter, we can determine the number of moles in the sample by determining its mass M_{sam} and using the equation

$$n = \frac{M_{sam}}{M}. \tag{20-5}$$

For atoms, the molar mass is just the atomic mass so that molar mass also has the unit g/mol, which is often denoted as u.

It is puzzling to note that the atomic mass of carbon that is listed in Appendix F is given as 12.01115 u rather than 12.00000 u. This is because a natural sample of carbon does not consist of only carbon-12. Instead it contains a relatively small percentage of carbon-13, which has an extra neutron in its nucleus. Nevertheless, by definition, a mole of pure carbon-12 and a mole of a naturally occurring mixture of carbon-12 and carbon-13 both contain Avogadro's number of atoms.

The Molar Form of the Ideal Gas Law

We can rewrite the molecular ideal gas law expressed in Eq. 20-1 in an alternative form by using Eq. 20-4, so that

$$PV = Nk_BT = nN_Ak_BT.$$

Since both Avogadro's number and the Boltzmann constant are constants, we can replace their product with a new constant R, which is called the **universal gas constant** because it has the same value for all ideal gases—namely,

$$R = N_A k_B = 6.02 \times 10^{23}\ \text{mol}^{-1}(1.38 \times 10^{-23}\ \text{J/K}) = 8.31\ \text{J/mol} \cdot \text{K}. \quad (20\text{-}6)$$

This allows us to write

$$nR = Nk_B. \quad (20\text{-}7)$$

Substituting this into Eq. 20-1 gives a second expression for the **ideal gas law:**

$$PV = nRT \quad \text{(molar ideal gas law)}, \quad (20\text{-}8)$$

in which P is the absolute (not gauge) pressure, V is the volume, n is the number of moles of gas present, and T is the temperature in Kelvin. Provided the gas density is low, the ideal gas law as represented in either Eq. 20-1 or Eq. 20-8 holds for any single gas or for any mixture of different gases. (For a mixture, n is the total number of moles in the mixture.)

Note the difference between the two expressions for the ideal gas law—Eq. 20-8 involves the number of moles n and Eq. 20-1 involves the number of atoms N. That is, the Boltzmann constant k_B tells us about individual atomic particles, whereas the gas constant R tells us about moles of particles. Recall that moles are defined via macroscopic measurements that are easily done in the lab—such as 1 mol of carbon has a mass of 12 g. As a result, R is easily measured in the lab. On the other hand, since k_B is about individual atoms, to get to it from a lab measurement we have to count the number of molecules in a mole. This is a decidedly nontrivial task.

You may well ask, "What is an *ideal gas* and what is so 'ideal' about it?" The answer lies in the simplicity of the law (Eqs. 20-1 and 20-8) that describes the macroscopic properties of a gas. Using this law—as you will see—we can deduce many properties of the ideal gas in a simple way. There is no such thing in nature as a truly ideal gas. But *all* gases approach the ideal state at low enough densities—that is, under conditions in which their molecules are far enough apart that they do not interact with one another as much as they do with the walls of their containers. Thus, the two equivalent ideal gas equations allow us to gain useful insights into the behavior of most real gases at low densities.

TOUCHSTONE EXAMPLE 20-1: Final Pressure

A cylinder contains 12 L of oxygen at 20°C and 15 atm. The temperature is raised to 35°C, and the volume is reduced to 8.5 L. What is the final pressure of the gas in atmospheres? Assume that the gas is ideal.

SOLUTION ■ The **Key Idea** here is that, because the gas is ideal, its pressure, volume, temperature, and number of moles are related by the ideal gas law, both in the initial state i and in the final state f (after the changes). Thus, from Eq. 20-8 we can write $P_i V_i = nRT_i$ and $P_f V_f = nRT_f$. Dividing the second equation by the first equation and solving for P_f yields

$$P_f = \frac{P_i T_f V_i}{T_i V_f}. \quad (20\text{-}9)$$

Note here that if we converted the given initial and final volumes from liters to SI units of cubic meters, the multiplying conversion factors would cancel out of Eq. 20-9. The same would be true for conversion factors that convert the pressures from atmospheres to the more accepted SI unit of pascals. However, to convert the given temperatures to kelvins requires the addition of an amount that would not cancel and thus must be included. Hence, we must write

$$T_i = (273 + 20)\ \text{K} = 293\ \text{K}$$

and

$$T_f = (273 + 35)\ \text{K} = 308\ \text{K}.$$

Inserting the given data into Eq. 20-9 then yields

$$P_f = \frac{(15\ \text{atm})(308\ \text{K})(12\ \text{L})}{(293\ \text{K})(8.5\ \text{L})} = 22\ \text{atm}. \quad \text{(Answer)}$$

20-3 Work Done by Ideal Gases

Heat engines are devices that can absorb thermal energy and do useful work on their surroundings. As you will see in the next chapter, air, which is typically used as a working medium in heat engines, behaves like an ideal gas in some circumstances. For this reason engineers are interested in knowing how to calculate the work done by ideal gases. Before we turn our attention to how the action of molecules that make up an ideal gas can be used to explain the ideal gas law, we first consider how to calculate the work done by ideal gases under various conditions. We restrict ourselves to expansions that occur slowly enough that the gas is very close to thermal equilibrium throughout its volume.

Work Done by an Ideal Gas at Constant Temperature

Suppose we put an ideal gas in a piston–cylinder arrangement like those in Chapter 19. Suppose also that we allow the gas to expand from an initial volume V_i to a final volume V_f while we keep the temperature T of the gas constant. Such a process, at *constant temperature,* is called an **isothermal expansion** (and the reverse is called an **isothermal compression**).

On a P-V diagram, an *isotherm* is a curve that connects points that have the same temperature. Thus, it is a graph of pressure versus volume for a gas whose temperature T is held constant. For n moles of an ideal gas, it is a graph of the equation

$$P = nRT\frac{1}{V} = (\text{a constant})\,\frac{1}{V}. \tag{20-10}$$

Figure 20-2 shows three isotherms, each corresponding to a different (constant) value of T. (Note that the values of T for the isotherms increase upward to the right.) Superimposed on the middle isotherm is the path followed by a gas during an isothermal expansion from state i to state f at a constant temperature of 310 K.

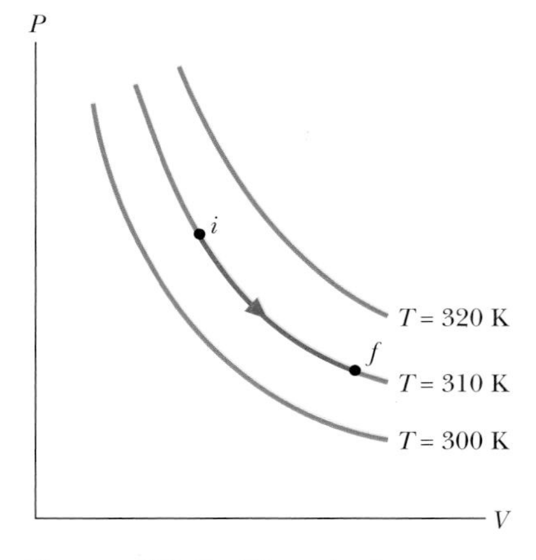

FIGURE 20-2 ■ Three isotherms on a P-V diagram. The path shown along the middle isotherm represents an isothermal expansion of a gas from an initial state i to a final state f. The path from f to i along the isotherm would represent the reverse process, an isothermal compression.

To find the work done by an ideal gas during an isothermal expansion, we start with Eq. 19-16,

$$W = \int_{V_i}^{V_f} P\,dV. \tag{20-11}$$

This is a general expression for the work done during any change in volume of any gas. For an ideal gas, we can use Eq. 20-8 to substitute for P, obtaining

$$W = \int_{V_i}^{V_f} \frac{nRT}{V}\,dV. \tag{20-12}$$

Because we are considering an isothermal expansion, T is constant and we can move it in front of the integral sign to write

$$W = nRT\int_{V_i}^{V_f} \frac{dV}{V} = nRT[\ln V]_{V_i}^{V_f}. \tag{20-13}$$

By evaluating the expression in brackets at the limits and then using the relationship $\ln a - \ln b = \ln(a/b)$, we find that

$$W = nRT\ln\frac{V_f}{V_i} = Nk_B\ln\frac{V_f}{V_i} \quad \text{(ideal gas, isothermal process).} \tag{20-14}$$

Recall that the symbol ln specifies a *natural* logarithm, which has base e.

As we often do in science and engineering we have derived a mathematical relationship. Before using this relationship it's a good idea to check our equation to see whether it makes sense. Unless a gas is undergoing a free expansion into a vacuum, we know that an expanding gas does work on its surroundings. If the gas contracts we expect that the surroundings have done work on the gas instead. Is this what Eq. 20-14 tells us? For an expansion, V_f is greater than V_i, so the ratio V_f/V_i in Eq. 20-14 is greater than unity. The natural logarithm of a quantity greater than unity is positive, and so the work W done by an ideal gas during an isothermal expansion is positive, as we expect. For a compression, V_f is less than V_i, so the ratio of volumes in Eq. 20-14 is less than unity. The natural logarithm in that equation—hence the work W—is negative, again as we expect.

Work Done at Constant Volume and at Constant Pressure

Equation 20-14 does not give the work W done by an ideal gas during *every* thermodynamic process. Instead, it gives the work only for a process in which the temperature is held constant. If the temperature varies, then the symbol T in Eq. 20-12 cannot be moved in front of the integral symbol as in Eq. 20-13, and thus we do not end up with Eq. 20-14.

However, we can go back to Eq. 20-11 to find the work W done by an ideal gas (or any other gas) during two more processes—a constant-volume process and a constant-pressure process. If the volume of the gas is constant, then Eq. 20-11 yields

$$W = 0 \quad \text{(constant-volume process).} \tag{20-15}$$

If, instead, the volume changes while the pressure P of the gas is held constant, then Eq. 20-11 becomes

$$W = P(V_f - V_i) = P\Delta V \quad \text{(constant-pressure process).} \tag{20-16}$$

READING EXERCISE 20-1: An ideal gas has an initial pressure of 3 pressure units and an initial volume of 4 volume units. The table gives the final pressure and volume of the gas (in those same units) in five processes. Which processes start and end on the same isotherm?

	a	*b*	*c*	*d*	*e*
P	12	6	5	4	1
V	1	2	7	3	12

■

TOUCHSTONE EXAMPLE 20-2: Work Done by Expansion

One mole of oxygen (assume it to be an ideal gas) expands at a constant temperature T of 310 K from an initial volume V_i of 12 L to a final volume V_f of 19 L. How much work is done by the gas during the expansion?

SOLUTION ■ The **Key Idea** is this: Generally we find the work by integrating the gas pressure with respect to the gas volume, using Eq. 20-11. However, because the gas here is ideal and the expansion is isothermal, that integration leads to Eq. 20-14. Therefore, we can write

$$W = nRT\ln\frac{V_f}{V_i}$$

$$= (1\text{ mol})(8.31\text{ J/mol}\cdot\text{K})(310\text{ K})\ln\left(\frac{19\text{ L}}{12\text{ L}}\right)$$

$$= 1180\text{ J}. \quad \text{(Answer)}$$

The expansion is graphed in the P-V diagram of Fig. 20-3. The work done by the gas during the expansion is represented by the area beneath the curve between i and f.

You can show that if the expansion is now reversed, with the gas undergoing an isothermal compression from 19 L to 12 L, the work done by the gas will be -1180 J. Thus, an external force would have to do 1180 J of work on the gas to compress it.

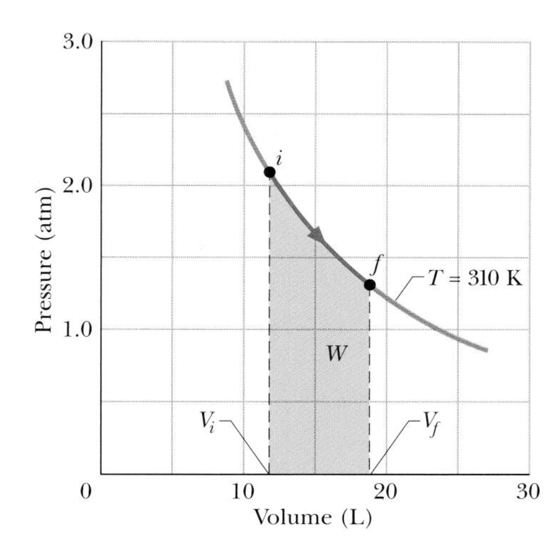

FIGURE 20-3 ■ The shaded area represents the work done by 1 mol of oxygen in expanding from V_i to V_f at a constant temperature T of 310 K.

20-4 Pressure, Temperature, and Molecular Kinetic Energy

In terms of our everyday experiences, molecules and atoms are invisible. Only in the past 40 years or so have scientists been able to "see" molecules using electron microscopes and field ion microscopes. But long before atoms and molecules could be "seen," 19th-century scientists such as James Clerk Maxwell and Ludwig Boltzmann in Europe and Josiah Willard Gibbs in the United States constructed models that made the description and prediction of the *macroscopic* (visible to the naked eye) behavior of thermodynamic systems possible. Their models were based on the yet unseen *microscopic* atoms and molecules.

Is it possible to describe the behavior of an ideal gas that obeys the first law of thermodynamics microscopically as a collection of moving molecules? To answer this question, let's observe the pressure exerted by a hypothetical molecule undergoing perfectly elastic collisions with the walls of a cubical box. By using the laws of mechanics we can derive a mathematical expression for the pressure exerted by just one of the molecules as a function of the volume of the box. Next we can extend our "ideal gas" so it is a low-density collection of molecules all having the same mass. By low density we mean that the volume occupied by the molecules is negligible compared to the volume of their container. This means that the molecules are far enough apart on the average that attractive interactions between molecules are also negligible. For this reason an ideal gas has internal energy related to its configuration. If we then define temperature as being related to the average kinetic energy of the molecules in an ideal gas, we can show that kinetic theory is a powerful construct for explaining both the ideal gas law and the first law of thermodynamics.

We start developing our idealized kinetic theory model by considering N molecules of an ideal gas that are confined in a cubical box of volume V, as in Fig. 20-4. The walls of the box are held at temperature T. How is the pressure P exerted by the gas on the walls related to the speeds of the molecules? Remember from our discussions of fluids in Chapter 15 that pressure is a scalar defined as the ratio of the magnitude of force (exerted normal to a surface) and the area of the surface. In the example at hand, a gas confined to a box, the pressure results from the motion of molecules in all directions resulting in elastic collisions between gas molecules and the walls of the box. We ignore (for the time being) collisions of the molecules with one another and consider only elastic collisions with the walls.

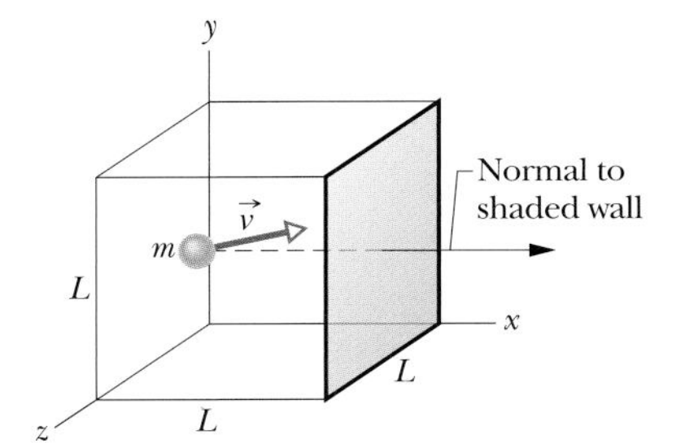

FIGURE 20-4 ■ We assume a cubical box of edge L contains N ideal gas molecules (not shown) that move around perpetually without losing energy. One of the molecules of mass m and velocity $\vec{v}$ is shown heading for a collision with the shaded wall of area L^2. The normal to the shaded wall points in the positive x direction.

Figure 20-4 shows a typical gas molecule, of mass m and velocity $\vec{v}$, which is about to collide with the shaded wall. Because we assume that any collision of a molecule with a wall is elastic, when this molecule collides with the shaded wall, the only component of its velocity that is changed by the collision is its x-component. That x-component has the same magnitude after collision but its sign is reversed. This means that the only change in the particle's momentum is along the x axis, so

$$(\Delta p_x)_{\text{molecule}} = p_{fx} - p_{ix} = (-m|v_x|) - (+m|v_x|) = -2m|v_x|.$$

But the law of conservation of momentum tells us that the momentum change $(\Delta p_x)_{\text{wall}}$ that the wall experiences after a molecule collides with it is $+2m|v_x|$. Remember that in this book $\vec{p}$, p_x, p_y, and p_z denote momentum vectors or vector components and capital P represents pressure. *Be careful not to confuse them.*

The molecule of Fig. 20-4 will hit the shaded wall repeatedly. The time Δt between collisions is the time the molecule takes to travel to the opposite wall and back again (a distance of $2L$) at speed $|v_x|$. Thus, Δt is equal to $2L/|v_x|$. (Note that this result holds even if the molecule bounces off any of the other walls along the way, because those walls are parallel to x and so cannot change $|v_x|$.) Therefore, the average rate at which momentum is delivered to the shaded wall by this single molecule is

$$\frac{(\Delta p_x)_{\text{wall}}}{\Delta t} = \frac{+2m|v_x|}{2L/|v_x|} = \frac{mv_x^2}{L}.$$

From Newton's Second Law ($\vec{F} = d\vec{p}/dt$), the rate at which momentum is delivered to the wall is the force acting on that wall. To find the total force, we must add up the contributions of all of the N molecules that strike the wall during a short time interval Δt. We will allow for the possibility that all the molecules have different velocities. Then we can divide the magnitude of the total force acting normal to the shaded wall $|F_x|$ by the area of the wall (L^2) to determine the pressure P on that wall. Thus,

$$P = \frac{|F_x|}{L^2} = \frac{mv_{x\,1}^2/L + mv_{x\,2}^2/L + \cdots + mv_{x\,N}^2/L}{L^2} \tag{20-17}$$

$$= \left(\frac{m}{L^3}\right)(v_{x\,1}^2 + v_{x\,2}^2 + \cdots + v_{x\,N}^2).$$

Since by definition $\langle v_x^2 \rangle = (v_{x\,1}^2 + v_{x\,2}^2 + \cdots + v_{x\,N}^2)/N$ we can replace the sum of squares of the velocities in the second parentheses of Eq. 20-17 by $N\langle v_x^2 \rangle$, where $\langle v_x^2 \rangle$ is the average value of the square of the x-components of all the speeds. Equation 20-17 for the pressure on the container wall then reduces to

$$P = \frac{Nm}{L^3}\langle v_x^2 \rangle = \frac{Nm}{V}\langle v_x^2 \rangle, \tag{20-18}$$

since the volume V of the cubical box is just L^3.

It is reasonable to assume that molecules are moving at random in three dimensions rather than just in the x direction that we considered initially, so that $\langle v_x^2 \rangle = \langle v_y^2 \rangle = \langle v_z^2 \rangle$ and

$$\langle v^2 \rangle = \langle v_x^2 + v_y^2 + v_z^2 \rangle = \langle v_x^2 \rangle + \langle v_y^2 \rangle + \langle v_z^2 \rangle = 3\langle v_x^2 \rangle,$$

or

$$\langle v_x^2 \rangle = \langle v^2 \rangle/3.$$

Thus, we can rewrite the expression above as

$$P = \frac{Nm\langle v^2 \rangle}{3V}. \tag{20-19}$$

The square root of $\langle v^2 \rangle$ is a kind of average speed, called the **root-mean-square speed** of the molecules and symbolized by v^{rms}. Its name describes it rather well: You *square* each speed, you find the *mean* (that is, the average) of all these squared speeds, and then you take the square *root* of that mean. With $\sqrt{\langle v^2 \rangle} = v^{\mathrm{rms}}$, we can then write Eq. 20-19 as

$$P = \frac{Nm(v^{\mathrm{rms}})^2}{3V}. \tag{20-20}$$

Equation 20-20 is very much in the spirit of kinetic theory. It tells us how the pressure of the gas (a purely macroscopic quantity) depends on the speed of the molecules (a purely microscopic quantity). We can turn Eq. 20-20 around and use it to calculate v^{rms} as

$$v^{\mathrm{rms}} = \sqrt{\frac{3PV}{Nm}}.$$

Combining this with the molecular form of the ideal gas law in Eq. 20-1 ($PV = Nk_BT$) gives us

$$v^{\mathrm{rms}} = \sqrt{\frac{3k_BT}{m}} \quad \text{(ideal gas)}, \tag{20-21}$$

where m is the mass of a single molecule in kilograms.

Table 20-1 shows some rms speeds calculated from Eq. 20-21. The speeds are surprisingly high. For hydrogen molecules at room temperature (300 K), the rms speed is 1920 m/s or 4300 mi/h—faster than a speeding bullet! Remember too that the rms speed is only a kind of average speed; some molecules move much faster than this, and some much slower.

TABLE 20-1
Some Molecular Speeds at Room Temperature (T = 300 K)[a]

Gas	Molar Mass $M = mN_A$ (10^{-3} kg/mol)	v^{rms} (m/s)
Hydrogen (H_2)	2.02	1920
Helium (He)	4.0	1370
Water vapor (H_2O)	18.0	645
Nitrogen (N_2)	28.0	517
Oxygen (O_2)	32.0	483
Carbon dioxide (CO_2)	44.0	412
Sulfur dioxide (SO_2)	64.1	342

[a] For convenience, we often set room temperature at 300 K even though (at 27°C or 81°F) that represents a fairly warm room.

The speed of sound in a gas is closely related to the rms speed of the molecules of that gas. In a sound wave, the disturbance is passed on from molecule to molecule by means of collisions. The wave cannot move any faster than the "average" speed of the molecules. In fact, the speed of sound must be somewhat less than this "average" molecular speed because not all molecules are moving in exactly the same direction as the wave. As examples, at room temperature, the rms speeds of hydrogen and nitrogen molecules are 1920 m/s and 517 m/s, respectively. The speeds of sound in these two gases at this temperature are 1350 m/s and 350 m/s, respectively.

Translational Kinetic Energy

Let's again consider a single molecule of an ideal gas as it moves around in the box of Fig. 20-4, but we now assume that its speed changes when it collides with other molecules. Its translational kinetic energy at any instant is $\frac{1}{2}mv^2$. Its *average* translational kinetic energy over the time that we watch it is

$$\langle K\rangle = \tfrac{1}{2}\langle mv^2\rangle = \tfrac{1}{2}m\langle v^2\rangle = \tfrac{1}{2}m(v^{\text{rms}})^2, \tag{20-22}$$

in which we make the assumption that the average speed of the molecule during our observation is the same as the average speed of all the molecules at any given instant. (Provided the total energy of the gas is not changing and we observe our molecule for long enough, this assumption is appropriate.) Substituting for v^{rms} from Eq. 20-21 leads to

$$\langle K\rangle = (\tfrac{1}{2}m)\frac{3k_BT}{m}$$

so that
$$\langle K\rangle = \tfrac{3}{2}k_BT \quad \text{(one ideal gas molecule).} \tag{20-23}$$

This equation tells us something unexpected:

> At a given temperature T, all ideal gas molecules—no matter what their mass—have the same average translational kinetic energy—namely, $(\frac{3}{2})k_BT$. When we measure the temperature of a gas, we are also measuring the average translational kinetic energy of its molecules.

READING EXERCISE 20-2: What happens to the average translational kinetic energy of each molecule in a gas when its temperature in kelvin: (a) doubles and (b) is reduced to zero? ■

READING EXERCISE 20-3: A gas mixture consists of molecules of types 1, 2, and 3, with molecular masses $m_1 > m_2 > m_3$. Rank the three types according to (a) average kinetic energy and (b) rms speed, greatest first. ■

20-5 Mean Free Path

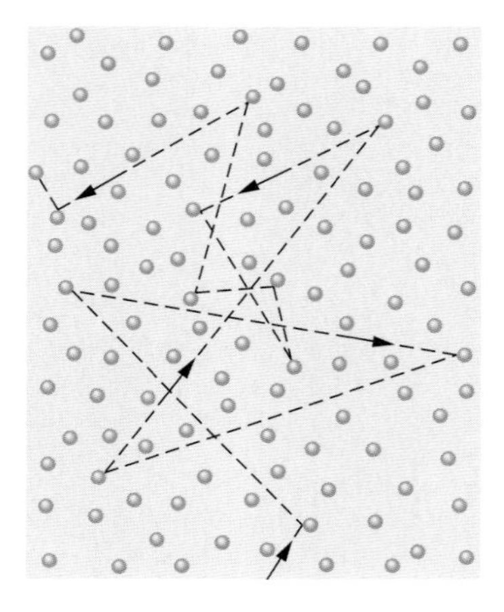

FIGURE 20-5 ■ A molecule traveling through a gas, colliding with other gas molecules in its path. Although the other molecules are shown as stationary, we believe they are also moving in a similar fashion.

In considering the motion of molecules, a question often arises: If molecules move so fast (hundreds of meters per second), why does it take as long as a minute or so before you can smell perfume when someone opens a bottle across a room (only a few meters away)? To answer this question, we continue to examine the motion of molecules in an ideal gas. Figure 20-5 shows the path of a typical molecule as it moves through the gas, changing both speed and direction abruptly as it collides elastically with other molecules. Between collisions, our typical molecule moves in a straight line at constant speed. Although the figure shows all the other molecules as stationary, they too are moving similarly.

One useful parameter to describe this random motion is the **mean free path** λ of the molecules. As its name implies, λ is the average distance traversed by a molecule between collisions. We expect λ to vary inversely with N/V, the number of molecules per unit volume (or "number density" of molecules). The larger N/V is, the more collisions there should be and the smaller the mean free path. We also expect λ to vary inversely with the size of the molecules, say, with their diameter d. (If the molecules were points, as we have assumed them to be, they would never collide and the mean free path would be infinite.) Thus, the larger the molecules are, the smaller the mean free path. We can even predict that λ should vary (inversely) as the *square* of the molecular diameter because the cross section of a molecule—not its diameter—determines its effective target area.

The expression for the mean free path does, in fact, turn out to be

$$\lambda = \frac{1}{\sqrt{2}\pi d^2\, N/V} \quad \text{(ideal gas mean free path).} \tag{20-24}$$

To justify Eq. 20-24, we focus attention on a single molecule and assume—as Fig. 20-5 suggests—that our molecule is traveling with a constant speed v and that all the other molecules are at rest. Later, we shall relax this assumption.

We assume further that the molecules are spheres of diameter d. A collision will then take place if the centers of the molecules come within a distance d of each other, as in Fig. 20-6*a*. Another, more helpful way to look at the situation is to consider our single molecule to have a *radius* of d and all the other molecules to be *points*, as in Fig. 20-6*b*. This does not change our criterion for a collision.

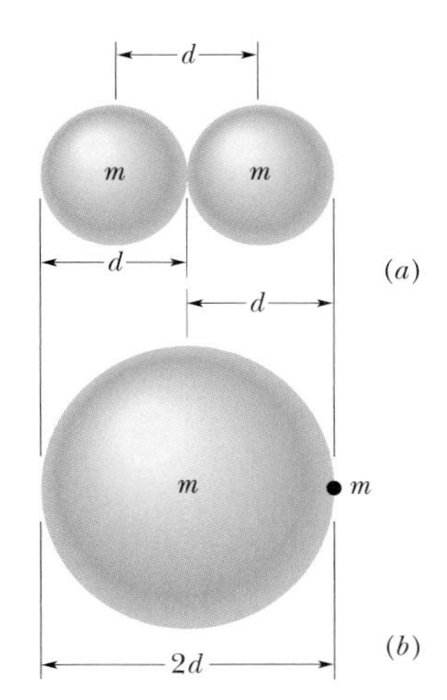

FIGURE 20-6 ■ (*a*) A collision occurs when the centers of two molecules come within a distance d of each other, d being the molecular diameter. (*b*) An equivalent but more convenient representation is to think of the moving molecule of interest as having a *radius* d and all other molecules as being points. The condition for a collision is unchanged.

As our single molecule zigzags through the gas, it sweeps out a short cylinder of cross-sectional area πd^2 between successive collisions. If we watch this molecule for a time interval Δt, it moves a distance $v\Delta t$, where v is its assumed speed. Thus, if we align all the short cylinders swept out in Δt, we form a composite cylinder (Fig. 20-7) of length $v\Delta t$ and volume $(\pi d^2)(v\Delta t)$. The number of collisions that occur in time Δt is then equal to the number of (point) molecules that lie within this cylinder.

Since N/V is the number of molecules per unit volume, the number of molecules in the cylinder is N/V times the volume of the cylinder, or $(N/V)(\pi d^2 v\Delta t)$. This is also the number of collisions in time Δt. The mean free path is the length of the path (and of the cylinder) divided by this number:

$$\lambda = \frac{\text{length of path}}{\text{number of collisions}} \approx \frac{v\Delta t}{\pi d^2 v\Delta t\, N/V} = \frac{1}{\pi d^2\, N/V}. \tag{20-25}$$

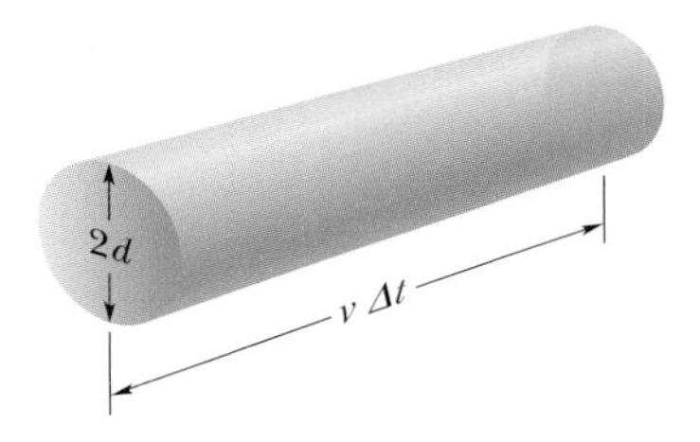

FIGURE 20-7 ■ In time Δt the moving molecule effectively sweeps out a cylinder of length $v\,\Delta t$ and radius d.

This equation is only approximate because it is based on the assumption that all the molecules except one are at rest. In fact, *all* the molecules are moving; when this is taken properly into account, Eq. 20-24 results. Note that it differs from the (approximate) Eq. 20-25 only by a factor of $1/\sqrt{2}$.

We can even get a glimpse of what is "approximate" about Eq. 20-25. The v in the numerator and that in the denominator are—strictly—not the same. The v in the numerator is $\langle v\rangle$, the mean speed of the molecule *relative to the container*. The v in the denominator is $\langle v^{\text{rel}}\rangle$, the mean speed of our single molecule *relative to the other molecules*, which are moving. It is this latter average speed that determines the number of collisions. A detailed calculation, taking into account the actual speed distribution of the molecules, gives $\langle v^{\text{rel}}\rangle = \sqrt{2}\,\langle v\rangle$ and thus the factor $\sqrt{2}$.

The mean free path of air molecules at sea level is about 0.1 μm. At an altitude of 100 km, the density of air has dropped to such an extent that the mean free path rises to about 16 cm. At 300 km, the mean free path is about 20 km. A problem faced by those who would study the physics and chemistry of the upper atmosphere in the laboratory is the unavailability of containers large enough to hold gas samples that simulate upper atmospheric conditions. Yet studies of the concentrations of freon, carbon dioxide, and ozone in the upper atmosphere are of vital public concern.

Recall the question that began this section: If molecules move so fast, why does it take as long as a minute or so before you can smell perfume when someone opens a bottle across a room? We now know part of the answer. In still air, each perfume molecule moves away from the bottle only very slowly because its repeated collisions with other molecules prevent it from moving directly across the room to you.

TOUCHSTONE EXAMPLE 20-3: Mean Free Path

(a) What is the mean free path λ for oxygen molecules at temperature $T = 300$ K and pressure $P = 1.00$ atm? Assume that the molecular diameter is $d = 290$ pm and the gas is ideal.

SOLUTION ■ The **Key Idea** here is that each oxygen molecule moves among other *moving* oxygen molecules in a zigzag path due to the resulting collisions. Thus, we use Eq. 20-24 for the mean free path, for which we need the number of molecules per unit volume, N/V. Because we assume the gas is ideal, we can use the ideal gas law of Eq. 20-1 ($PV = Nk_BT$) to write $N/V = P/k_BT$. Substituting this into Eq. 20-24, we find

$$\lambda = \frac{1}{\sqrt{2}\pi d^2\, N/V} = \frac{k_BT}{\sqrt{2}\pi d^2 P}$$

$$= \frac{(1.38 \times 10^{-23}\ \text{J/K})(300\ \text{K})}{\sqrt{2}\pi(2.9 \times 10^{-10}\ \text{m})^2\,(1.01 \times 10^5\ \text{Pa})} \qquad \text{(Answer)}$$

$$= 1.1 \times 10^{-7}\ \text{m}.$$

This is about 380 molecular diameters.

(b) Assume the average speed of the oxygen molecules is $\langle v \rangle = 450$ m/s. What is the average time interval Δt between successive collisions for any given molecule? At what rate does the molecule collide; that is, what is the frequency f of its collisions?

SOLUTION ■ To find the time interval Δt between collisions, we use this **Key Idea**: Between collisions, the molecule travels, on average, the mean free path λ at average speed $\langle v \rangle$. Thus, the average time between collisions is

$$\langle \Delta t \rangle = \frac{\text{(distance)}}{\text{(average speed)}} = \frac{\lambda}{\langle v \rangle} = \frac{1.1 \times 10^{-7}\ \text{m}}{450\ \text{m/s}} \qquad \text{(Answer)}$$

$$= 2.44 \times 10^{-10}\ \text{s} \approx 0.24\ \text{ns}.$$

This tells us that, on average, any given oxygen molecule has less than a nanosecond between collisions.

To find the frequency f of the collisions, we use this **Key Idea**: The average rate or frequency at which the collisions occur is the inverse of the average time $\langle \Delta t \rangle$ between collisions. Thus,

$$f = \frac{1}{2.44 \times 10^{-10}\ \text{s}} = 4.1 \times 10^9\ \text{s}^{-1}. \qquad \text{(Answer)}$$

This tells us that, on average, any given oxygen molecule makes about 4 billion collisions per second.

20-6 The Distribution of Molecular Speeds

The root-mean-square speed v^{rms} gives us a general idea of molecular speeds in a gas at a given temperature. We often want to know more. For example, what fraction of the molecules have speeds greater than the rms value? Greater than twice the rms value? To answer such questions, we need to know how the possible values of speed are distributed among the molecules. Figure 20-8*a* shows this distribution for oxygen molecules at room temperature ($T = 300$ K); Fig. 20-8*b* compares it with the distribution at $T = 80$ K.

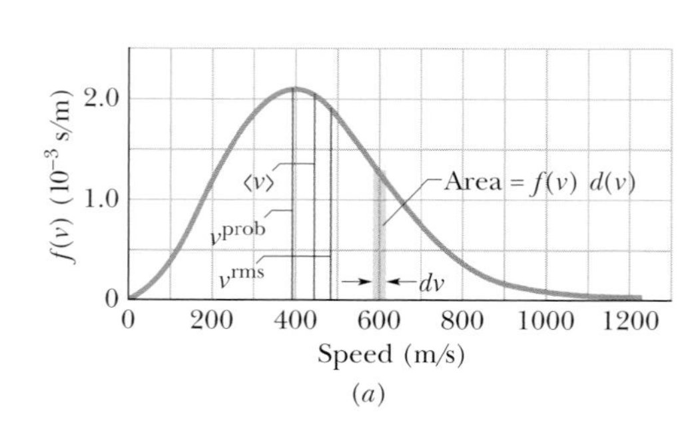

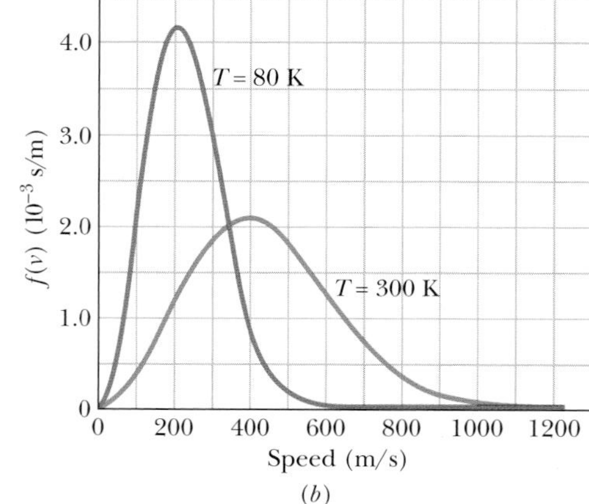

FIGURE 20-8 ■ (*a*) The Maxwell speed distribution for oxygen molecules at $T = 300$ K. The three characteristic speeds are marked. (*b*) The curves for 300 K and 80 K. Note that the molecules move more slowly at the lower temperature. Because these are probability distributions, the area under each curve has a numerical value of unity.

In 1852, Scottish physicist James Clerk Maxwell first solved the problem of finding the speed distribution of gas molecules. His result, known as **Maxwell's speed distribution law,** is

$$f(v) = 4\pi\left(\frac{m}{2\pi k_B T}\right)^{3/2} v^2 e^{-mv^2/2k_B T}. \tag{20-26}$$

Here v is the molecular speed, T is the gas temperature, m is the mass of a single gas molecule, and k_B is Boltzmann's constant. It is this equation that is plotted in Fig. 20-8*a*, *b*. The quantity $f(v)$ in Eq. 20-26 and Fig. 20-8 is a *probability distribution function:* For any speed v, the product $f(v)dv$ (a dimensionless quantity) is the fraction of molecules whose speeds lie in the interval of width dv centered on speed v.

As Fig. 20-8*a* shows, this fraction is equal to the area of a strip with height $f(v)$ and width dv. The total area under the distribution curve corresponds to the fraction of the molecules whose speeds lie between zero and infinity. All molecules fall into this category, so the value of this total area is unity; that is,

$$\int_0^\infty f(v)\, dv = 1. \tag{20-27}$$

The fraction of molecules with speeds in an interval of, say, v_1 to v_2 is then

$$\text{fraction} = \int_{v_1}^{v_2} f(v)\, dv. \tag{20-28}$$

Average, RMS, and Most Probable Speeds

In principle, we can find the **average speed** $\langle v \rangle$ of the molecules in a gas with the following procedure: We *weight* each value of v in the distribution; that is, we multiply it by the fraction $f(v)dv$ of molecules with speeds in a differential interval dv centered on v. Then we add up all these values of $vf(v)dv$. The result is $\langle \vec{v} \rangle$. In practice, we do all this by evaluating

$$\langle v \rangle = \int_0^\infty v\, f(v)\, dv. \tag{20-29}$$

Substituting for $f(v)$ from Eq. 20-26 and using definite integral 20 from the list of integrals in Appendix E, we find

$$\langle v \rangle = \sqrt{\frac{8k_B T}{\pi m}} \quad \text{(average speed)}. \tag{20-30}$$

Similarly, we can find the average of the square of the speeds $\langle v^2 \rangle$ with

$$\langle v^2 \rangle = \int_0^\infty v^2\, f(v)\, dv. \tag{20-31}$$

Substituting for $f(v)$ from Eq. 20-27 and using generic integral 16 from the list of integrals in Appendix E, we find

$$\langle v^2 \rangle = \frac{3k_B T}{m}. \tag{20-32}$$

The square root of $\langle v^2 \rangle$ is the **root-mean-square speed** v^{rms}. Thus,

$$v^{\text{rms}} = \sqrt{\frac{3k_B T}{m}} \quad \text{(rms speed)}, \tag{20-33}$$

which agrees with Eq. 20-21.

The **most probable speed** v^{prob} is the speed at which $f(v)$ is maximum (see Fig. 20-8*a*). To calculate v^{prob}, we set $df/dv = 0$ (the slope of the curve in Fig. 20-8*a* is zero at the maximum of the curve) and then solve for v. Doing so, we find

$$v^{prob} = \sqrt{\frac{2k_BT}{m}} \quad \text{(most probable speed).} \tag{20-34}$$

What is the relationship between the most probable speed, the average speed, and the rms speed of a molecule? The relationship is fixed.

The most probable speed v^{prob} is always less than the average speed $\langle v \rangle$ which in turn is less than the rms speed v^{rms}. More specifically, $v^{prob} = 0.82\ v^{rms}$ and $\langle v \rangle = 0.92\ v^{rms}$.

This is consistent with the idea that a molecule is more likely to have speed v^{prob} than any other speed, but some molecules will have speeds that are many times v^{prob}. These molecules lie in the *high-speed tail* of a distribution curve like that in Fig. 20-8*a*. We should be thankful for these few, higher speed molecules because they make possible both rain and sunshine (without which we could not exist). We next see why.

Rain: The speed distribution of water molecules in, say, a pond at summertime temperatures can be represented by a curve similar to that of Fig. 20-8*a*. Most of the molecules do not have nearly enough kinetic energy to escape from the water through its surface. However, small numbers of very fast molecules with speeds far out in the tail of the curve can do so. It is these water molecules that evaporate, making clouds and rain a possibility.

As the fast water molecules leave the surface, carrying energy with them, the temperature of the remaining water is maintained by thermal energy transfer from the surroundings. Other fast molecules—produced in particularly favorable collisions—quickly take the place of those that have left, and the speed distribution is maintained.

Sunshine: Let the distribution curve of Fig. 20-8*a* now refer to protons in the core of the Sun. The Sun's energy is supplied by a nuclear fusion process that starts with the merging of two protons. However, protons repel each other because of their electrical charges, and protons of average speed do not have enough kinetic energy to overcome the repulsion and get close enough to merge. Very fast protons with speeds in the tail of the distribution curve can do so, however, and thus the Sun can shine.

20-7 The Molar Specific Heats of an Ideal Gas

Up to now, we have taken the specific heat of a substance as a quantity to be measured. But now, with the kinetic theory of gases, we know something about the structure of matter and where its energy is stored. With this additional information, we can actually calculate and make predictions about what we expect the specific heats of different kinds of gases to be. If we compare our predictions based on kinetic theory to experimental measurements, we get some good agreement and also some surprises. The surprises are among the first hints that the laws of matter at the atomic level are not just Newton's laws scaled down. In other words, we begin to notice that atoms aren't just little billiard balls but something different from any macroscopic object with which we have experience.

To explore this idea, we derive here (from molecular considerations) an expression for the internal energy E^{int} of an ideal gas. In other words, we find an expression for the energy associated with the random motions of the atoms or molecules in the gas. We shall then use that expression to derive the molar specific heats of an ideal gas.

Internal Energy E^{int}

Let us first assume that our ideal gas is a *monatomic gas* (which has individual atoms rather than molecules), such as helium, neon, or argon. Let us also assume that the internal energy E^{int} of our ideal gas is simply the sum of the translational kinetic energies of its atoms.

The average translational kinetic energy of a single atom depends only on the gas temperature and is given by Eq. 20-23 as $\langle K \rangle = \frac{3}{2}k_B T$. A sample of n moles of such a gas contains nN_A atoms. The internal energy E^{int} of the sample is then

$$E^{\text{int}} = (nN_A)\langle K \rangle = (nN_A)(\tfrac{3}{2}k_B T). \qquad (20\text{-}35)$$

Using Eq. 20-6 ($k_B = R/N_A$), we can rewrite this as

$$E^{\text{int}} = \tfrac{3}{2}Nk_B T = \tfrac{3}{2}nRT \qquad \text{(monatomic ideal gas).} \qquad (20\text{-}36)$$

Thus,

> The internal energy E^{int} of an ideal gas is a function of the gas temperature *only;* it does not depend on any other variable.

With Eq. 20-36 in hand, we are now able to derive an expression for the molar specific heat of an ideal gas. Actually, we shall derive two expressions. One is for the case in which the volume of the gas remains constant as thermal energy is transferred to or from it. The other is for the case in which the pressure of the gas remains constant as thermal energy is transferred to or from it. The symbols for these two molar specific heats are C_V and C_P, respectively. (By convention, the capital letter C is used in both cases, even though C_V and C_P represent types of specific heat and not heat capacities.)

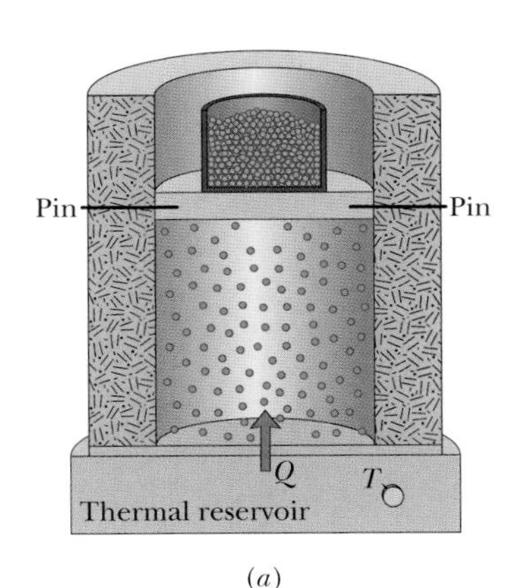

(a)

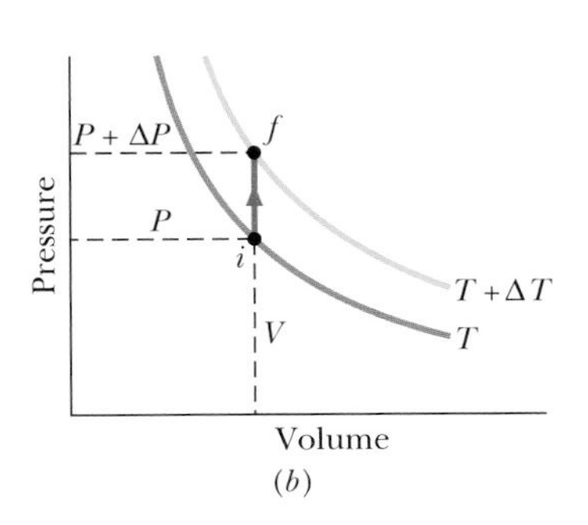

(b)

FIGURE 20-9 (a) The temperature of an ideal gas is raised from T to $T + \Delta T$ in a constant-volume process. Thermal energy is added, but no work is done. (b) The process on a P-V diagram. (The system's insulated lid is not shown.)

Molar Specific Heat at Constant Volume

Figure 20-9*a* shows n moles of an ideal gas at pressure P and temperature T, confined to a cylinder of fixed volume V. This *initial state i* of the gas is marked on the P-V diagram of Fig. 20-9*b*. Suppose that you add a small amount of thermal energy Q to the gas by slowly turning up the temperature of the thermal reservoir. The gas temperature rises a small amount to $T + \Delta T$, and its pressure rises to $P + \Delta P$, bringing the gas to *final state f*.

In such experiments, we would find that the thermal energy transferred Q is related to the temperature change ΔT by

$$Q = nC_V\,\Delta T \qquad \text{(constant volume),} \qquad (20\text{-}37)$$

where C_V is a constant called the **molar specific at constant volume.** Substituting this expression for Q into the first law of thermodynamics as given by Eq. 19-17 ($\Delta E^{\text{int}} = Q - W$) yields

$$\Delta E^{\text{int}} = nC_V\,\Delta T - W. \qquad (20\text{-}38)$$

With the volume held constant, the gas cannot expand and thus cannot do any work. Therefore, $W = 0$, and Eq. 20-38 gives us

$$C_V = \frac{\Delta E^{\text{int}}}{n\,\Delta T}. \tag{20-39}$$

From Eq. 20-36 we know that $E^{\text{int}} = \frac{3}{2}nRT$, so the change in internal energy must be

$$\Delta E^{\text{int}} = \tfrac{3}{2}nR\,\Delta T. \tag{20-40}$$

Substituting this result into Eq. 20-39 yields

$$C_V = \tfrac{3}{2}R = 12.5 \text{ J/mol} \cdot \text{K} \quad \text{(monatomic gas)}. \tag{20-41}$$

As Table 20-2 shows, this prediction that $C_V = \frac{3}{2}R$ based on ideal gas kinetic theory agrees very well with experiment for the real monatomic gases (the case that we have assumed). The experimental values of C_V for *diatomic gases* and *polyatomic gases* (which have molecules with more than two atoms) are greater than the predicted value of $\frac{3}{2}R$. Reasons for this will be discussed in Section 20-8.

TABLE 20-2
Molar Specific Heats

Molecule		Example	C_V (J/mol · K)
	Ideal		$\frac{3}{2}R = 12.5$
Monatomic	Real	He	12.5
(1 atom)	Real	Ar	12.6
	Ideal		$\frac{5}{2}R = 20.8^*$
Diatomic	Real	N_2	20.7
(2 atoms)	Real	O_2	20.8
	Ideal		$3R = 24.9^*$
Polyatomic	Real	NH_4	29.0
(> 2 atoms)	Real	CO_2	29.7

*The presentation of the $\frac{5}{2}R$ and $3R$ will be explained in the next section.

We can now generalize Eq. 20-36 for the internal energy of any ideal gas by substituting C_V for $\frac{3}{2}R$; we get

$$E^{\text{int}} = nC_VT \quad \text{(any ideal gas)}. \tag{20-42}$$

This equation applies not only to an ideal monatomic gas but also to diatomic and polyatomic ideal gases, provided the experimentally determined value of C_V is used. Just as with Eq. 20-37, we see that the internal energy of a gas depends on the temperature of the gas but not on its pressure or density.

When an ideal gas that is confined to a container undergoes a temperature change ΔT, then from either Eq. 20-39 or Eq. 20-42 we can write the resulting change in its internal energy as

$$\Delta E^{\text{int}} = nC_V\,\Delta T \quad \text{(any ideal gas, any process)}. \tag{20-43}$$

This equation tells us:

> A change in the internal energy E^{int} of a confined ideal gas depends on the change in the gas temperature only; it does *not* depend on what type of process produces the change in the temperature.

As examples, consider the three paths between the two isotherms in the *P-V* diagram of Fig. 20-10. Path 1 represents a constant-volume process. Path 2 represents a constant-pressure process (that we are about to examine). Path 3 represents a process in which no thermal energy is exchanged with the system's environment (we discuss this in Section 20-11). Although the values of Q and work W associated with these three paths differ, as do P_f and V_f, the values of ΔE^{int} associated with the three paths are identical and are all given by Eq. 20-43, because they all involve the same temperature change ΔT. Therefore, no matter what path is actually taken between T and $T + \Delta T$, we can *always* use path 1 and Eq. 20-43 to compute ΔE^{int} easily.

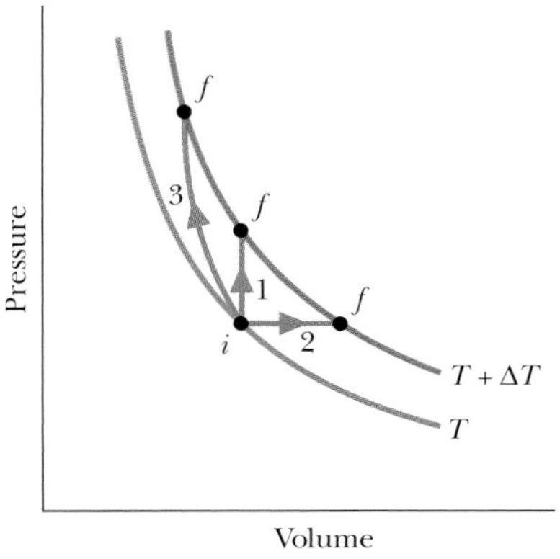

FIGURE 20-10 ■ Three paths representing three different processes that take an ideal gas from an initial state i at temperature T to some final state f at temperature $T + \Delta T$. The change ΔE^{int} in the internal energy of the gas is the same for these three processes and for any others that result in the same change of temperature.

Molar Specific Heat at Constant Pressure

We now assume that the temperature of the ideal gas is increased by the same small amount ΔT as previously, but that the necessary thermal energy (Q) is added with the gas under constant pressure. An experiment for doing this is shown in Fig. 20-11*a*; the *P-V* diagram for the process is plotted in Fig. 20-11*b*. From such experiments we find that the transferred thermal energy Q is related to the temperature change ΔT by

$$Q = nC_P\,\Delta T \qquad \text{(constant pressure)}, \tag{20-44}$$

where C_P is a constant called the **molar specific heat at constant pressure.** This C_P is *greater* than the molar specific heat at constant volume C_V, because energy must now be supplied not only to raise the temperature of the gas but also for the gas to do work—that is, to lift the weighted piston of Fig. 20-11*a*.

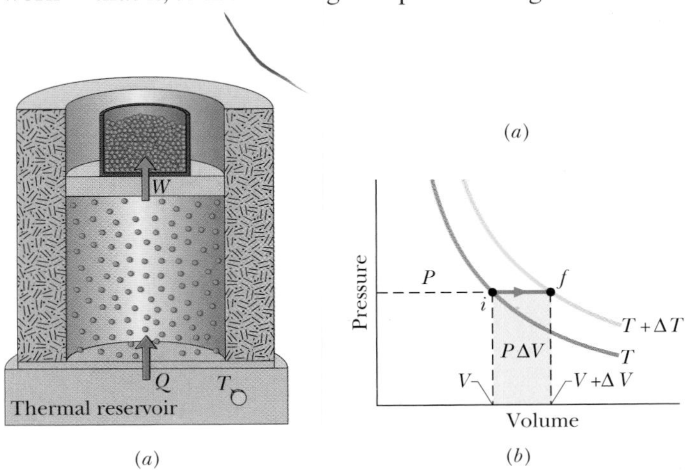

FIGURE 20-11 ■ (*a*) The temperature of an ideal gas is raised from T to $T + \Delta T$ in a constant-pressure process. Thermal energy is added and work is done in lifting the loaded piston. (*b*) The process on a *P-V* diagram. The work $P\Delta V$ is given by the shaded area.

To relate molar specific heats C_P and C_V, we start with the first law of thermodynamics (Eq. 19-17):

$$\Delta E^{\text{int}} = Q - W. \tag{20-45}$$

We next replace each term in Eq. 20-45. For ΔE^{int}, we substitute from Eq. 20-43. For Q, we substitute from Eq. 20-44. To replace W, we first note that since the pressure remains constant, Eq. 20-16 tells us that $W = P\Delta V$. Then we note that, using the ideal gas equation ($PV = nRT$), we can write

$$W = P\,\Delta V = nR\,\Delta T. \qquad (20\text{-}46)$$

Making these substitutions in Eq. 20-45, we find

$$nC_V \Delta T = nC_P \Delta T - nR\Delta T$$

and then dividing through by $n\,\Delta T$,

$$C_V = C_P - R,$$

so

$$C_P = C_V + R \quad \text{(any ideal gas).} \qquad (20\text{-}47)$$

This relationship between C_P and C_V predicted by kinetic theory agrees well with experiment, not only for monatomic gases but for gases in general, as long as their density is low enough so that we may treat them as ideal. As we discuss in Section 19-5, there is very little difference between C_P and C_V for liquids and solids because of their relative incompressibility.

READING EXERCISE 20-4: The figure here shows five paths traversed by a gas on a P-V diagram. Rank the paths according to the change in internal energy of the gas, greatest first.

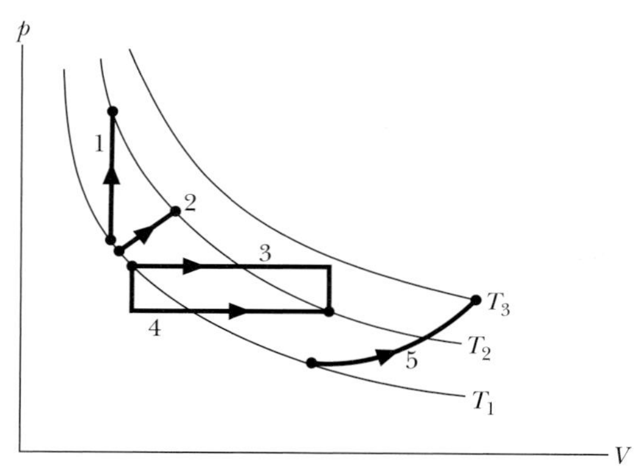

■

TOUCHSTONE EXAMPLE 20-4: Helium Bubble

A bubble of 5.00 mol of helium is submerged at a certain depth in liquid water when the water (and thus the helium) undergoes a temperature increase ΔT of 20.0 C° at constant pressure. As a result, the bubble expands. The helium is monatomic and ideal.

(a) How much thermal energy is added to the helium during the increase and expansion?

SOLUTION ■ One **Key Idea** here is that the thermal energy transferred Q is related to the temperature change ΔT by the molar specific heat of the gas. Because the pressure P is held constant during the addition of energy, we use the molar specific heat at constant pressure C_P and Eq. 20-44,

$$Q = nC_P\,\Delta T, \qquad (20\text{-}48)$$

to find Q. To evaluate C_P we go to Eq. 20-47, which tells us that for any ideal gas, $C_P = C_V + R$. Then from Eq. 20-41, we know that for any *monatomic* gas (like helium), $C_V = \frac{3}{2}R$. Thus, Eq. 20-48 gives us

$$Q = n(C_V + R)\,\Delta T = n(\tfrac{3}{2}R + R)\,\Delta T = n(\tfrac{5}{2}R)\,\Delta T$$
$$= (5.00 \text{ mol})(2.5)(8.31 \text{ J/mol}\cdot\text{K})(20.0 \text{ C}^\circ) \quad \text{(Answer)}$$
$$= 2077.5 \text{ J} \approx 2080 \text{ J}.$$

(b) What is the change ΔE^{int} in the internal energy of the helium during the temperature increase?

SOLUTION ■ Because the bubble expands, this is not a constant-volume process. However, the helium is nonetheless confined (to the bubble). Thus, a **Key Idea** here is that the change ΔE^{int} is the same as *would occur* in a constant-volume process with the same temperature change ΔT. We can easily find the constant-volume change ΔE^{int} with Eq. 20-43:

$$\Delta E^{\text{int}} = nC_V\,\Delta T = n(\tfrac{3}{2}R)\,\Delta T$$
$$= (5.00 \text{ mol})(1.5)(8.31 \text{ J/mol}\cdot\text{K})(20.0 \text{ C}^\circ) \quad \text{(Answer)}$$
$$= 1246.5 \text{ J} \approx 1250 \text{ J}.$$

(c) How much work W is done by the helium as it expands against the pressure of the surrounding water during the temperature increase?

SOLUTION ■ One **Key Idea** here is that the work done by *any* gas expanding against the pressure from its environment is given by Eq. 20-11, which tells us to integrate $P\,dV$. When the pressure is constant (as here), we can simplify that to $W = P\,\Delta V$. When the gas is *ideal* (as here), we can use the ideal gas law (Eq. 20-8) to write $P\,\Delta V = nR\,\Delta T$. We end up with

$$W = nR\,\Delta T$$
$$= (5.00 \text{ mol})(8.31 \text{ J/mol}\cdot\text{K})(20.0 \text{ C}^\circ) \quad \text{(Answer)}$$
$$= 831 \text{ J}.$$

Because we happen to know Q and ΔE^{int}, we can work this problem another way. The **Key Idea** now is that we can account for the energy changes of the gas with the first law of thermodynamics, writing

$$W = Q - \Delta E^{\text{int}} = 2077.5 \text{ J} - 1246.5 \text{ J} \quad \text{(Answer)}$$
$$= 831 \text{ J}.$$

Note that during the temperature increase, only a portion (1250 J) of the thermal energy (2080 J) that is transferred to the helium goes to increasing the internal energy of the helium and thus the temperature of the helium. The rest (831 J) is transferred out of the helium as work that the helium does during the expansion. If the water were frozen, it would not allow that expansion. Then the same temperature increase of 20.0 C° would require only 1250 J of energy, because no work would be done by the helium.

20-8 Degrees of Freedom and Molar Specific Heats

As Table 20-2 shows, the prediction that $C_V = \frac{3}{2}R$ agrees with experiment for monatomic gases. But it fails for diatomic and polyatomic gases. Let us try to explain the discrepancy by considering the possibility that molecules with more than one atom can store internal energy in forms other than *translational* kinetic energy.

Figure 20-12 shows common models of helium (a *monatomic* molecule, containing a single atom), oxygen (a *diatomic* molecule, containing two atoms), and methane (a *polyatomic* molecule). From such models, we would assume that all three types of molecules can have translational motions (say, moving left–right and up–down) and rotational motions (spinning about an axis like a top). However, due to their highly symmetric nature, rotational motions in a monatomic molecule need special consideration. We will return to this point shortly. In addition, we would assume that the diatomic and polyatomic molecules can have oscillatory motions, with the atoms oscillating slightly toward and away from one another, as if attached to opposite ends of a spring.

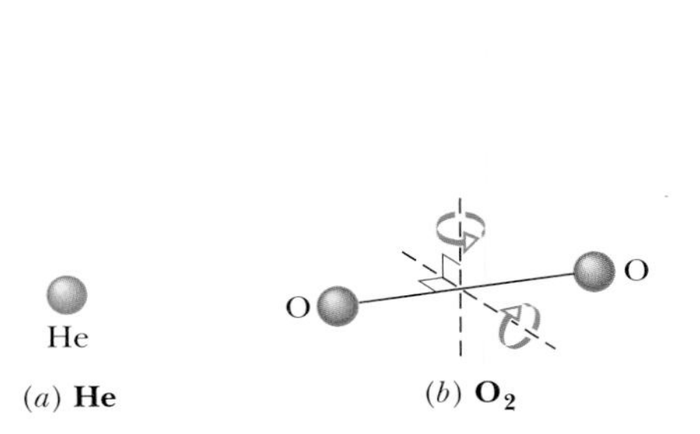

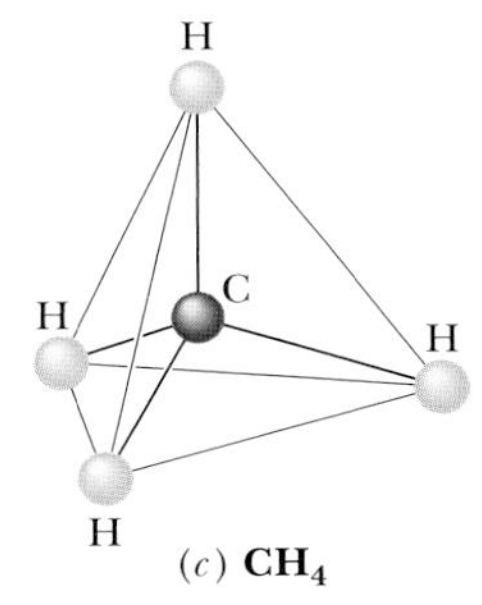

FIGURE 20-12 ■ Models of molecules as used in kinetic theory: (*a*) helium, a typical monatomic molecule; (*b*) oxygen, a typical diatomic molecule; and (*c*) methane, a typical polyatomic molecule. The spheres represent atoms, and the lines between them represent bonds. Two rotation axes are shown for the oxygen molecule.

To keep account of the various ways in which energy can be stored in a gas, James Clerk Maxwell introduced the theorem of the **equipartition of energy:**

> Every kind of molecule has a certain number f of **degrees of freedom,** which are independent ways in which the molecule can store energy. Each such degree of freedom has associated with it—on average—an energy of $\frac{1}{2}k_BT$ per molecule (or $\frac{1}{2}RT$ per mole).

Let us apply the theorem to the translational and rotational motions of the molecules in Fig. 20-12. (We discuss oscillatory motion in the next section.) For the translational motion, superimpose an xyz coordinate system on any gas. The molecules will, in general, have velocity components along all three axes. Thus, gas molecules of all types have three degrees of translational freedom (three ways to move in translation) and, on average, an associated energy of $3(\frac{1}{2}k_BT)$ per molecule.

For the rotational motion, imagine the origin of our xyz coordinate system at the center of each molecule in Fig. 20-12. In a gas, each molecule should be able to rotate with an angular velocity component along each of the three axes, so each gas should have three degrees of rotational freedom and, on average, an additional energy of $3(\frac{1}{2}k_BT)$ per molecule. *However,* experiment shows this is true only for the polyatomic molecules.

A possible solution to this dilemma is that rotations about an axis of symmetry don't count as a degree of freedom. For example, as seen in Fig. 20-12, a single-atom molecule is symmetric about all three (mutually perpendicular) axes through the molecule. Hence, according to our proposed solution, these rotations are not additional degrees of freedom. A diatomic molecule is symmetric about only one axis (the axis through the center of both atoms). Accordingly, a diatomic molecule would have two rather than three degrees of freedom associated with rotation of the molecule.

It appears that modifying our theory in this manner brings us more in alignment with the experimental results. However, one should ask what reasoning (other than experimental evidence) supports this modification of the theory. One thing is clear. If a molecule were rotating about an axis of symmetry, it would be impossible to tell. Unlike a baseball (which has stitches or other marks) molecules have no characteristics that allow us to sense the rotation. Although classical physics gives us no real foundation for ignoring the motion simply because it is indistinguishable from no motion at all, this is what quantum theory would suggest.

So, according to our new model, a monatomic molecule has zero degrees of freedom associated with rotation because any rotation would be about an axis of symmetry. A diatomic molecule has two degrees of freedom associated with rotations about the two axes perpendicular to the line connecting the atoms (the axes are shown in Fig. 20-12*b*) but no degree of freedom for rotation about that line itself. Therefore, a diatomic molecule can have a rotational energy of only $2(\frac{1}{2}k_BT)$ per molecule. A polyatomic molecule has a full three degrees of freedom associated with rotational motion.

To extend our analysis of molar specific heats (C_P and C_V, in Section 20-7) to ideal diatomic and polyatomic gases, it is necessary to retrace the derivations of that analysis in detail. First, we replace Eq. 20-36 ($E^{\text{int}} = \frac{3}{2}nRT$) with $E^{\text{int}} = (f/2)nRT$, where f is the number of degrees of freedom listed in Table 20-3. Doing so leads to the prediction

$$C_V = \left(\frac{f}{2}\right)R = 4.16f \text{ J/mol}\cdot\text{K}, \tag{20-49}$$

which agrees—as it must—with Eq. 20-41 for monatomic gases ($f = 3$). As Table 20-3 shows, this prediction also agrees with experiment for diatomic gases ($f = 5$), but it is too low for polyatomic gases. *Note*: The symbol f used here to denote degrees of freedom should not be confused with $f(v)$ used to describe the velocity distribution function for molecules.

TABLE 20-3
Degrees of Freedom for Various Molecules

		Degrees of Freedom			Predicted Molar Specific Heats	
Molecule	**Example**	**Translational**	**Rotational**	**Total (f)**	**C_V (Eq. 20-47)**	**$C_P = C_V + R$**
Monatomic	He	3	0	3	$\frac{3}{2}R$	$\frac{5}{2}R$
Diatomic	O_2	3	2	5	$\frac{5}{2}R$	$\frac{7}{2}R$
Polyatomic	CH_4	3	3	6	$3R$	$4R$

TOUCHSTONE EXAMPLE 20-5: Internal Energy Change

A cabin of volume V is filled with air (which we consider to be an ideal diatomic gas) at an initial low temperature T_1. After you light a wood stove, the air temperature increases to T_2. What is the resulting change ΔE^{int} in the internal energy of the air in the cabin?

SOLUTION ■ As the air temperature increases, the air pressure P cannot change but must always be equal to the air pressure outside the room. The reason is that, because the room is not airtight, the air is not confined. As the temperature increases, air molecules leave through various openings and thus the number of moles n of air in the room decreases. Thus, one **Key Idea** here is that we *cannot* use Eq. 20-43 ($\Delta E^{\text{int}} = nC_V\,\Delta T$) to find ΔE^{int}, because it requires constant n.

A second **Key Idea** is that we *can* relate the internal energy E^{int} at any instant to n and the temperature T with Eq. 20-42 ($E^{\text{int}} = nC_VT$). From that equation we can then write

$$\Delta E^{\text{int}} = \Delta(nC_VT) = C_V\,\Delta(nT).$$

Next, using Eq. 20-8 ($PV = nRT$), we can replace nT with PV/R, obtaining

$$\Delta E^{\text{int}} = C_V\Delta\left(\frac{PV}{R}\right). \qquad (20\text{-}50)$$

Now, because P, V, and R are all constants, Eq. 20-50 yields

$$\Delta E^{\text{int}} = 0, \qquad \text{(Answer)}$$

even though the temperature changes.

Why does the cabin feel more comfortable at the higher temperature? There are at least two factors involved: (1) You exchange electromagnetic radiation (thermal radiation) with surfaces inside the room, and (2) you exchange energy with air molecules that collide with you. When the room temperature is increased, (1) the amount of thermal radiation emitted by the surfaces and absorbed by you is increased, and (2) the amount of energy you gain through the collisions of air molecules with you is increased.

20-9 A Hint of Quantum Theory

We can improve the agreement of kinetic theory with experiment by including the oscillations of the atoms in a gas of diatomic or polyatomic molecules. For example, the two atoms in the O_2 molecule of Fig. 20-12*b* can oscillate toward and away from each other, with the interconnecting bond acting like a spring. However, experiment shows that such oscillations occur only at relatively high temperatures of the gas—the motion is "turned on" only when the gas molecules have relatively large energies. Rotational motion is also subject to such "turning on," but at a lower temperature.

Figure 20-13 is of help in seeing this turning on of rotational motion and oscillatory motion. The ratio C_V/R for diatomic hydrogen gas (H_2) is plotted there against temperature, with the temperature scale logarithmic to cover several orders of magnitude. Below about 80 K, we find that $C_V/R = 1.5$. This result implies that only the three translational degrees of freedom of hydrogen are involved in the specific heat.

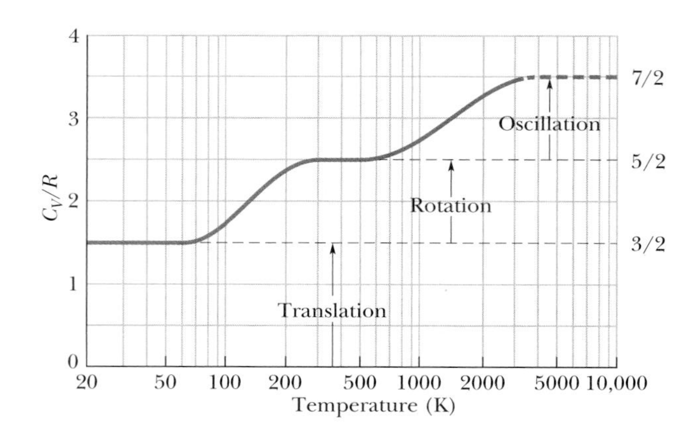

FIGURE 20-13 ■ A plot of C_V/R versus temperature for (diatomic) hydrogen gas. Because rotational and oscillatory motions begin at certain energies, only translation is possible at very low temperatures. As the temperature increases, rotational motion can begin. At still higher temperatures, oscillatory motion can begin.

As the temperature increases, the value of C_V/R gradually increases to 2.5, implying that two additional degrees of freedom have become involved. Quantum theory shows that these two degrees of freedom are associated with the rotational motion of the hydrogen molecules and that this motion requires a certain minimum amount of energy. At very low temperatures (below 80 K), the molecules do not have enough energy to rotate. As the temperature increases from 80 K, first a few molecules and then more and more obtain enough energy to rotate, and C_V/R increases, until all of them are rotating and $C_V/R = 2.5$.

Similarly, quantum theory shows that oscillatory motion of the molecules requires a certain (higher) minimum amount of energy. This minimum amount is not met until the molecules reach a temperature of about 1000 K, as shown in Fig. 20-13. As the temperature increases beyond 1000 K, the number of molecules with enough energy to oscillate increases, and C_V/R increases, until all of them are oscillating and $C_V/R = 3.5$. (In Fig. 20-13, the plotted curve stops at 3200 K because at that temperature, the atoms of a hydrogen molecule oscillate so much that they overwhelm their bond, and the molecule then *dissociates* into two separate atoms.)

The observed fact that rotational degrees of freedom are not excited until sufficiently high temperatures are reached implies that rotational kinetic energy is not a continuous function of angular velocity. Instead, a discrete, quantized energy level must be attained before rotation is excited. This discreteness of energy levels is a hallmark of quantum mechanical behavior. It is interesting to note that some of the issues discussed in this chapter are the first examples (with many more to come) that macroscopic properties of matter, which are easily measured in the laboratory, depend critically on (and provide strong evidence for) the quantum theory we will develop later.

The compatibility between microscopic theory and macroscopic observations when coupled with quantum theory and other phenomena in physics and chemistry provided additional support for the theory that matter is composed of atoms and molecules.

20-10 The Adiabatic Expansion of an Ideal Gas

We saw in Section 18-2 that sound waves are propagated through air and other gases as a series of compressions and expansions; these variations in the transmission medium take place so rapidly that there is no time for thermal energy to be transferred from one part of the medium to another. As we saw in Section 19-8, a process for which $Q = 0$ is an *adiabatic process.* We can ensure that $Q = 0$ either by carrying out the process very quickly (as in sound waves) or by doing it (at any rate) in a well-insulated container. Let us see what the kinetic theory has to say about adiabatic processes.

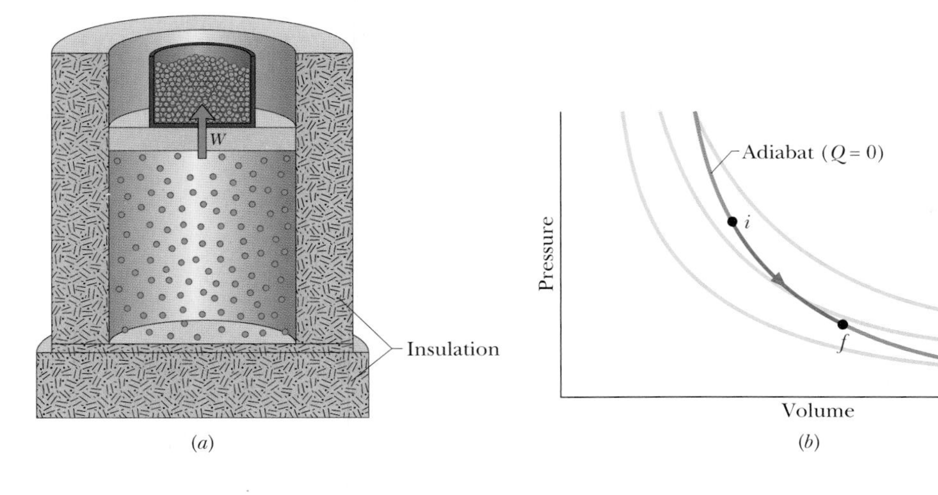

FIGURE 20-14 ■ (*a*) The volume of an ideal gas is increased by removing weight from the piston. The process is adiabatic ($Q = 0$). (*b*) The process proceeds from i to f along an adiabat on a P-V diagram.

Figure 20-14*a* shows our usual insulated cylinder. Its insulating lid is not shown. It now contains an ideal gas and rests on an insulating stand. By removing mass from the piston, we can allow the gas to expand adiabatically (in a slow process rather than a free expansion). As the volume increases, both the pressure and the temperature drop. We shall prove next that the relation between the pressure and the volume during such an adiabatic process is

$$PV^{\gamma} = \text{a constant} \quad \text{(ideal gas adiabatic process)}, \tag{20-51}$$

in which $\gamma = C_P/C_V$, the ratio of the molar specific heats for the gas. On a P-V diagram such as that in Fig. 20-14*b*, the process occurs along a line (called an *adiabat*) that has the equation $P = (\text{a constant})/V^{\gamma}$. Since the gas goes from an initial state i to a final state f, we can rewrite Eq. 20-51 as

$$P_i V_i^{\gamma} = P_f V_f^{\gamma} \quad \text{(ideal gas adiabatic process)}. \tag{20-52}$$

We can also write an equation for an adiabatic process in terms of T and V. To do so, we use the ideal gas equation ($PV = nRT$) to eliminate P from Eq. 20-51, finding

$$\left(\frac{nRT}{V}\right) V^{\gamma} = \text{a constant}.$$

Because n and R are constants, we can rewrite this in the alternative form

$$TV^{\gamma-1} = \text{a constant} \quad \text{(ideal gas adiabatic process)}. \tag{20-53}$$

in which the constant is different from that in Eq. 20-51. When the gas goes from an initial state i to a final state f, we can rewrite Eq. 20-53 as

$$T_i V_i^{\gamma-1} = T_f V_f^{\gamma-1} \quad \text{(adiabatic process)}. \tag{20-54}$$

We can now answer the question that opens this chapter. At the top of an unopened carbonated drink, there is a gas of carbon dioxide and water vapor. Because the pressure of the gas is greater than atmospheric pressure, the gas expands out into the atmosphere when the container is opened. Thus, the gas increases its volume, but that means it must do work to push against the atmosphere. Because the expansion is so rapid, it is adiabatic and the only source of energy for the work is the internal energy of the gas. Because the internal energy decreases, the temperature of the gas must also decrease, which can cause the water vapor in the gas to condense into tiny drops, forming the fog. (Note that Eq. 20-54 also tells us that the temperature must decrease during an adiabatic expansion: Since V_f is greater than V_i, then T_f must be less than T_i.)

Proof of Eq. 20-51

Suppose that you remove some shot from the piston of Fig. 20-14*a*, allowing the ideal gas to push the piston and the remaining shot upward and thus to increase the volume by a differential amount dV. Since the volume change is tiny, we may assume that the pressure P of the gas on the piston is constant during the change. This assumption allows us to say that the work dW done by the gas during the volume increase is equal to $P\,dV$. From Eq. 19-18, the first law of thermodynamics can then be written as

$$dE^{\text{int}} = Q - P\,dV. \qquad (20\text{-}55)$$

Since the gas is thermally insulated (and thus the expansion is adiabatic), we substitute 0 for Q. Then we use Eq. 20-43 to substitute $nC_V\,dT$ for dE^{int}. With these substitutions, and after some rearranging, we have

$$n\,dT = -\left(\frac{P}{C_V}\right)dV. \qquad (20\text{-}56)$$

Now using the ideal gas law ($PV = nRT$) and derivative rule 3 in Appendix E we have

$$P\,dV + V\,dP = nR\,dT. \qquad (20\text{-}57)$$

Replacing R with its equal, $C_P - C_V$, in Eq. 20-57 yields

$$n\,dT = \frac{P\,dV + V\,dP}{C_P - C_V}. \qquad (20\text{-}58)$$

Equating Eqs. 20-56 and 20-58 and rearranging them give

$$\frac{dP}{P} + \left(\frac{C_P}{C_V}\right)\frac{dV}{V} = 0.$$

Replacing the ratio of the molar specific heats with γ and integrating (see integral 5 in Appendix E) yield

$$\ln P + \gamma \ln V = \text{a constant}.$$

Rewriting the left side as $\ln PV^{\gamma}$ and then taking the antilog of both sides, we find

$$PV^{\gamma} = \text{a constant}, \qquad (20\text{-}59)$$

which is what we set out to prove.

Free Expansions

Recall from Section 19-8 that a free expansion of a gas is an adiabatic process that involves no work done on or by the gas, and no change in the internal energy of the gas.

A free expansion is thus quite different from the type of adiabatic process described by Eqs. 20-51 through 20-59, in which work is done and the internal energy changes. Those equations then do *not* apply to a free expansion, even though such an expansion is adiabatic.

Also recall that in a free expansion, a gas is in equilibrium only at its initial and final points; thus, we can plot only those points, but not the expansion itself, on a P-V diagram. In addition, because $\Delta E^{\text{int}} = 0$, the temperature of the final state must be that of the initial state. Thus, the initial and final points on a P-V diagram must be on the same isotherm, and instead of Eq. 20-54 we have

$$T_i = T_f \quad \text{(free expansion)}. \tag{20-60}$$

If we next assume that the gas is ideal (so that $PV = nRT$), because there is no change in temperature, there can be no change in the product PV. Thus, instead of Eq. 20-51 a free expansion involves the relation

$$P_iV_i = P_fV_f \quad \text{(free expansion)}. \tag{20-61}$$

TOUCHSTONE EXAMPLE 20-6: Final Temperature

In Touchstone Example 20-2, 1 mol of oxygen (assumed to be an ideal gas) expands isothermally (at 310 K) from an initial volume of 12 L to a final volume of 19 L.

(a) What would be the final temperature if the gas had expanded adiabatically to this same final volume? Oxygen (O_2) is diatomic and here has rotation but not oscillation.

SOLUTION ■ The **Key Ideas** here are as follows:

1. When a gas expands against the pressure of its environment, it must do work.
2. When the process is adiabatic (no thermal energy is transferred as heat), then the energy required for the work can come only from the internal energy of the gas.
3. Because the internal energy decreases, the temperature T must also decrease.

We can relate the initial and final temperatures and volumes with Eq. 20-54:

$$T_iV_i^{\gamma-1} = T_fV_f^{\gamma-1}. \tag{20-62}$$

Because the molecules are diatomic and have rotation but not oscillation, we can take the molar specific heats from Table 20-3. Thus,

$$\gamma = \frac{C_P}{C_V} = \frac{\frac{7}{2}R}{\frac{5}{2}R} = 1.40.$$

Solving Eq. 20-62 for T_f and inserting known data then yield

$$T_f = \frac{T_iV_i^{\gamma-1}}{V_f^{\gamma-1}} = \frac{(310\text{ K})(12\text{ L})^{1.40-1}}{(19\text{ L})^{1.40-1}}$$
$$= (310\text{ K})\left(\tfrac{12}{19}\right)^{0.40} = 258\text{ K}. \quad \text{(Answer)}$$

(b) What would be the final temperature and pressure if, instead, the gas had expanded freely to the new volume, from an initial pressure of 2.0 Pa?

SOLUTION ■ Here the **Key Idea** is that the temperature does not change in a free expansion:

$$T_f = T_i = 310\text{ K}. \quad \text{(Answer)}$$

We find the new pressure using Eq. 20-61, which gives us

$$P_f = P_i\frac{V_i}{V_f} = (2.0\text{ Pa})\frac{12\text{ L}}{19\text{ L}} = 1.3\text{ Pa}. \quad \text{(Answer)}$$

Problems

SEC. 20-2 ■ THE MACROSCOPIC BEHAVIOR OF GASES

1. Arsenic Find the mass in kilograms of 7.50×10^{24} atoms of arsenic, which has a molar mass of 74.9 g/mol.

2. Gold Gold has a molar mass of 197 g/mol. (a) How many moles of gold are in a 2.50 g sample of pure gold? (b) How many atoms are in the sample?

3. Water If the water molecules in 1.00 g of water were distributed uniformly over the surface of Earth, how many such molecules would there be on 1.00 cm^2 of the surface?

4. It Is Written A distinguished scientist has written: "There are enough molecules in the ink that makes one letter of this sentence to provide not only one for every inhabitant of Earth, but one for every creature if each star of our galaxy had a planet as populous as Earth." Check this statement. Assume the ink sample (molar mass = 18 g/mol) to have a mass of 1 μg, the population of Earth to be 5×10^9, and the number of stars in our galaxy to be 10^{11}.

5. Compute Compute (a) the number of moles and (b) the number of molecules in 1.00 cm^3 of an ideal gas at a pressure of 100 Pa and a temperature of 220 K.

6. Best Vacuum The best laboratory vacuum has a pressure of about 1.00×10^{-18} atm, or 1.01×10^{-13} Pa. How many gas molecules are there per cubic centimeter in such a vacuum at 293 K?

7. Oxygen Gas Oxygen gas having a volume of 1000 cm^3 at 40.0°C and 1.01×10^5 Pa expands until its volume is 1500 cm^3 and its pressure is 1.06×10^5 Pa. Find (a) the number of moles of oxygen present and (b) the final temperature of the sample.

8. Tire An automobile tire has a volume of 1.64×10^{-2} m^3 and contains air at a gauge pressure (pressure above atmospheric pressure) of 165 kPa when the temperature is 0.00°C. What is the gauge pressure of the air in the tires when its temperature rises to 27.0°C and its volume increases to 1.67×10^{-2} m^3? Assume atmospheric pressure is 1.00×10^5 Pa.

9. A Quantity of Ideal Gas A quantity of ideal gas at 10.0°C and 100 kPa occupies a volume of 2.50 m^3. (a) How many moles of the gas are present? (b) If the pressure is now raised to 300 kPa and the temperature is raised to 30.0°C, how much volume does the gas occupy? Assume no leaks.

SEC. 20-3 ■ WORK DONE BY IDEAL GASES

10. Work Done by External Agent Calculate the work done by an external agent during an isothermal compression of 1.00 mol of oxygen from a volume of 22.4 L at 0°C and 1.00 atm pressure to 16.8 L.

11. P, V, T Pressure P, volume V, and temperature T for a certain non-ideal material are related by

$$P = \frac{AT - BT^2}{V},$$

where A and B are constants. Find an expression for the work done by the material if the temperature changes from T_1 to T_2 while the pressure remains constant.

12. A Container Encloses A container encloses two ideal gases. Two moles of the first gas are present, with molar mass M_1. The second gas has molar mass $M_2 = 3M_1$, and 0.5 mol of this gas is present. What fraction of the total pressure on the container wall is attributable to the second gas? (The kinetic theory explanation of pressure leads to the experimentally discovered law of partial pressures for a mixture of gases that do not react chemically: *The total pressure exerted by the mixture is equal to the sum of the pressures that the several gases would exert separately if each were to occupy the vessel alone.*)

13. Air Initially Occupies Air that initially occupies 0.14 m^3 at a gauge pressure of 103.0 kPa is expanded isothermally to a pressure of 101.3 kPa and then cooled at constant pressure until it reaches its initial volume. Compute the work done by the air. (Gauge pressure is the difference between the actual pressure and atmospheric pressure.)

14. A Sample A sample of an ideal gas is taken through the cyclic process *abca* shown in Fig. 20-15; at point a, $T = 200$ K. (a) How many moles of gas are in the sample? What are (b) the temperature of the gas at point b, (c) the temperature of the gas at point c, and (d) the net thermal energy transferred to the gas during the cycle?

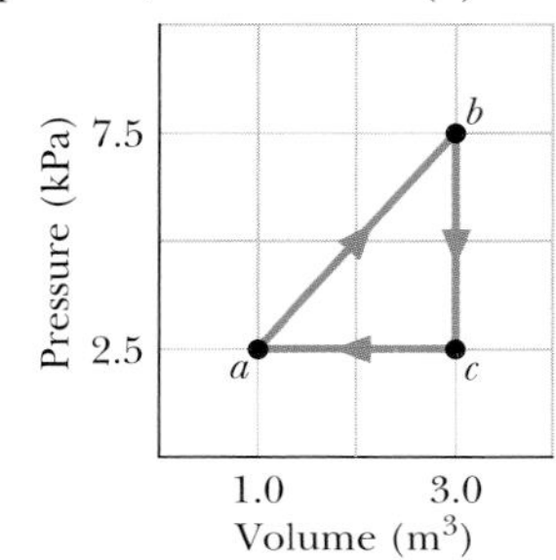

FIGURE 20-15 ■ Problem 14.

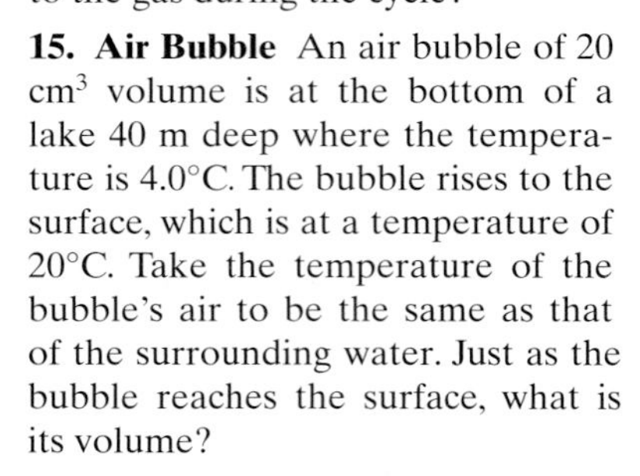

15. Air Bubble An air bubble of 20 cm^3 volume is at the bottom of a lake 40 m deep where the temperature is 4.0°C. The bubble rises to the surface, which is at a temperature of 20°C. Take the temperature of the bubble's air to be the same as that of the surrounding water. Just as the bubble reaches the surface, what is its volume?

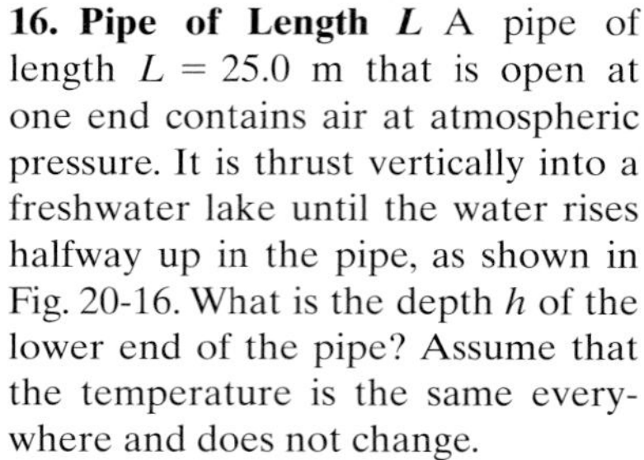

16. Pipe of Length L A pipe of length $L = 25.0$ m that is open at one end contains air at atmospheric pressure. It is thrust vertically into a freshwater lake until the water rises halfway up in the pipe, as shown in Fig. 20-16. What is the depth h of the lower end of the pipe? Assume that the temperature is the same everywhere and does not change.

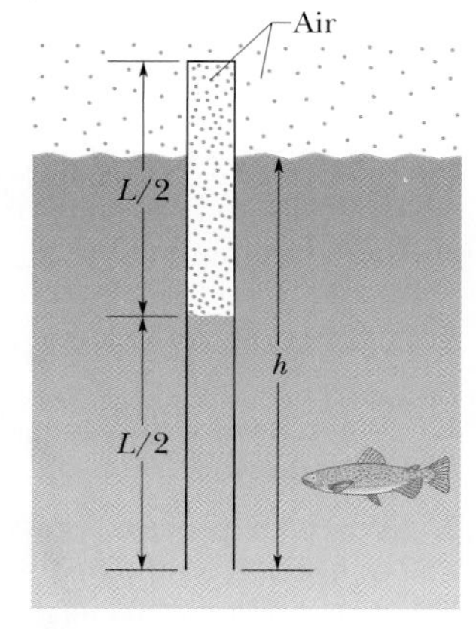

FIGURE 20-16 ■ Problem 16.

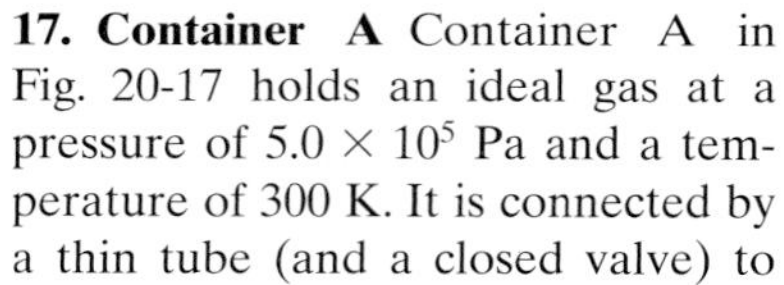

17. Container A Container A in Fig. 20-17 holds an ideal gas at a pressure of 5.0×10^5 Pa and a temperature of 300 K. It is connected by a thin tube (and a closed valve) to container B, with four times the volume of A. Container B holds the same ideal gas at a pressure of 1.0×10^5 Pa and a temperature of 400 K. The valve is opened to allow the pressures to equalize, but the temperature of each container is kept constant at its initial value. What then is the pressure in the two containers?

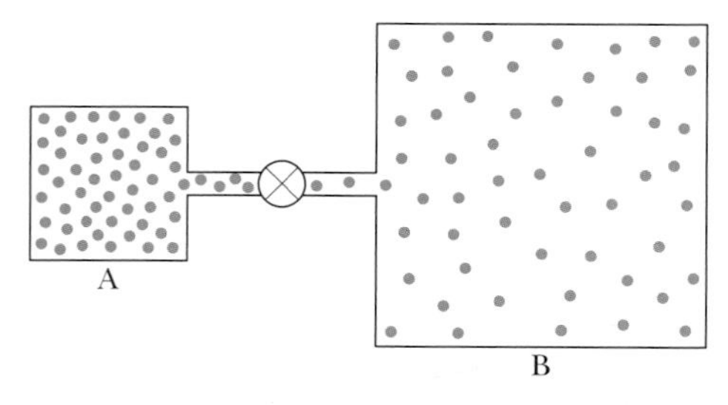

FIGURE 20-17 ■ Problem 17.

SEC. 20-4 ■ PRESSURE, TEMPERATURE, AND MOLECULAR KINETIC ENERGY

18. Helium Atoms Calculate the rms speed of helium atoms at 1000 K. See Appendix F for the molar mass of helium atoms.

19. Lowest Possible The lowest possible temperature in outer space is 2.7 K. What is the root-mean-square speed of hydrogen molecules at this temperature? (The molar mass of hydrogen molecules (H_2) is given in Table 20-1)

20. Speed of Argon Find the rms speed of argon atoms at 313 K. See Appendix F for the molar mass of argon atoms.

21. Sun's Atmosphere The temperature and pressure in the Sun's atmosphere are 2.00×10^6 K and 0.0300 Pa. Calculate the rms speed of free electrons (mass $= 9.11 \times 10^{-31}$ kg) there, assuming they are an ideal gas.

22. Nitrogen Molecule (a) Compute the root-mean-square speed of a nitrogen molecule at 20.0°C. The molar mass of nitrogen molecules (N_2) is given in Table 20-1. At what temperatures will the root-mean-square speed be (b) half that value and (c) twice that value?

23. Hydrogen Molecules A beam of hydrogen molecules (H_2) is directed toward a wall, at an angle of 55° with the normal to the wall. Each molecule in the beam has a speed of 1.0 km/s and a mass of 3.3×10^{-24} g. The beam strikes the wall over an area of 2.0 cm^2, at the rate of 10^{23} molecules per second. What is the beam's pressure on the wall?

24. Density of Gas At 273 K and 1.00×10^{-2} atm, the density of a gas is 1.24×10^{-5} g/cm^3. (a) Find v^{rms} for the gas molecules. (b) Find the molar mass of the gas and identify the gas. (*Hint*: The gas is listed in Table 20-1.)

25. Translational Kinetic Energy What is the average translational kinetic energy of nitrogen molecules at 1600 K?

26. Average Value Determine the average value of the translational kinetic energy of the molecules of an ideal gas at (a) 0.00°C and (b) 100°C. What is the translational kinetic energy per mole of an ideal gas at (c) 0.00°C and (d) 100°C?

27. Evaporating Water Water standing in the open at 32.0°C evaporates because of the escape of some of the surface molecules. The heat of vaporization (539 cal/g) is approximately equal to εn, where ε is the average energy of the escaping molecules and n is the number of molecules per gram. (a) Find ε. (b) What is the ratio of ε to the average kinetic energy of H_2O molecules, assuming the latter is related to temperature in the same way as it is for gases?

28. Alternative Form Show that the ideal gas equation, Eq. 20-8, can be written in the alternative form $P = \rho RT/M$, where ρ is the mass density of the gas and M is the molar mass.

29. Avogadro's Law *Avogadro's law* states that under the same conditions of temperature and pressure, equal volumes of gas contain equal numbers of molecules. Is this law equivalent to the ideal gas law? Explain.

SEC. 20-5 ■ MEAN FREE PATH

30. Nitrogen Molecules The mean free path of nitrogen molecules at 0.0°C and 1.0 atm is 0.80×10^{-5} cm. At this temperature and pressure there are 2.7×10^{19} molecules/cm^3. What is the molecular diameter?

31. Earth's Surface At 2500 km above Earth's surface, the number density of the atmosphere is about 1 molecule/cm^3. (a) What mean free path is predicted by Eq. 20-24 and (b) what is its significance under these conditions? Assume a molecular diameter of 2.0×10^{-8} cm.

32. What Frequency At what frequency would the wavelength of sound in air be equal to the mean free path of oxygen molecules at 1.0 atm pressure and 0.00°C? Take the diameter of an oxygen molecule to be 3.0×10^{-8} cm.

33. Mean Free Path of Jelly Beans Assuming that jelly beans in a bag could behave like ideal gas particles, what is the mean free path for 15 spherical jelly beans in a bag that is vigorously shaken? The volume of the bag is 1.0 L, and the diameter of a jelly bean is 1.0 cm. (Consider bean-bean collisions, not bean-bag collisions.)

34. Mean Free Path for Argon At 20°C and 750 torr pressure, the mean free paths for argon gas (Ar) and nitrogen gas (N_2) are $\lambda_{Ar} = 9.9 \times 10^{-6}$ cm and $\lambda_{N_2} = 27.5 \times 10^{-6}$ cm. (a) Find the ratio of the effective diameter of argon to that of nitrogen. What is the mean free path of argon at (b) 20°C and 150 torr, and (c) −40°C and 750 torr?

35. Particle Accelerator In a certain particle accelerator, protons travel around a circular path of diameter 23.0 m in an evacuated chamber, whose residual gas is at 295 K and 1.00×10^{-6} torr pressure. (a) Calculate the number of gas molecules per cubic centimeter at this pressure. (b) What is the mean free path of the gas molecules if the molecular diameter is 2.00×10^{-8} cm?

SEC. 20-6 ■ THE DISTRIBUTION OF MOLECULAR SPEEDS

36. Twenty-Two Particles Twenty-two particles have speeds as follows (N_i represents the number of particles that have speed v_i):

N_i	2	4	6	8	2
v_i (cm/s)	1.0	2.0	3.0	4.0	5.0

(a) Compute their average speed $\langle v \rangle$. (b) Compute their root-mean-square speed v^{rms}. (c) Of the five speeds shown, which is the most probable speed v^{prob}?

37. Ten Molecules The speeds of 10 molecules are 2.0, 3.0, 4.0 . . . , 11 km/s. (a) What is their average speed? (b) What is their root-mean-square speed?

38. Ten Particles (a) Ten particles are moving with the following speeds: four at 200 m/s, two at 500 m/s, and four at 600 m/s. Calculate their average and root-mean-square speeds. Is $v^{rms} > \langle v \rangle$? (b) Make up your own speed distribution for the 10 particles and show that $v^{rms} \geq \langle v \rangle$ for your distribution. (c) Under what condition (if any) does $v^{rms} = \langle v \rangle$?

39. Compute Temperature (a) Compute the temperatures at which the rms speed for (a) molecular hydrogen and (b) molecular oxygen is equal to the speed of escape from Earth. (c) Do the same for the speed of escape from the Moon, assuming the local gravitational constant on its surface to be 0.16g. (d) The temperature high in Earth's upper atmosphere (in the thermosphere) is about 1000 K. Would you expect to find much hydrogen there? Much oxygen? Explain.

40. Most Probable Speed It is found that the most probable speed of molecules in a gas when it has (uniform) temperature T_2 is the same as the rms speed of the molecules in this gas when it has (uniform) temperature T_1. Calculate T_2/T_1.

41. Hydrogen Molecule A molecule of hydrogen (diameter 1.0×10^{-8} cm), traveling with the rms speed, escapes from a furnace (T = 4000 K) into a chamber containing atoms of *cold* argon (diameter 3.0×10^{-8} cm) at a number density of 4.0×10^{19} atoms/cm³. (a) What is the speed of the hydrogen molecule? (b) If the H_2 molecule collides with an argon atom, what is the closest their centers can be, considering each as spherical? (c) What is the initial number of collisions per second experienced by the hydrogen molecule? (*Hint:* Assume that the cold argon atoms are stationary. Then the mean free path of the hydrogen molecule is given by Eq. 20-25, and not Eq. 20-24.)

42. Two Containers—Same Temperature Two containers are at the same temperature. The first contains gas with pressure P_1, molecular mass m_1, and root-mean-square speed v_1^{rms}. The second contains gas with pressure $2P_1$, molecular mass m_2, and average speed $\langle v_2 \rangle = 2v_1^{\text{rms}}$. Find the mass ratio m_1/m_2.

43. Hypothetical Speeds Figure 20-18 shows a hypothetical speed distribution for a sample of N gas particles (note that $f(v) = 0$ for $v > 2v_0$). (a) Express a in terms of N and v_0. (b) How many of the particles have speeds between $1.5v_0$ and $2.0v_0$? (c) Express the average speed of the particles in terms of v_0. (d) Find v^{rms}.

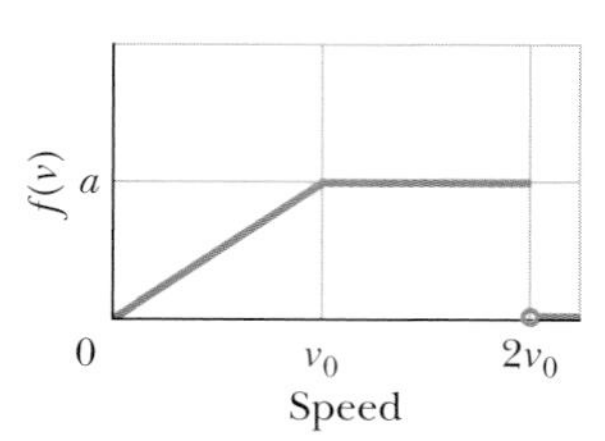

FIGURE 20-18 ■ Problem 43.

SEC. 20-7 ■ THE MOLAR SPECIFIC HEATS OF AN IDEAL GAS

44. Internal Energy What is the internal energy of 1.0 mol of an ideal monatomic gas at 273 K?

45. Isothermal Expansion One mole of an ideal gas undergoes an isothermal expansion. Find the thermal energy Q added to the gas in terms of the initial and final volumes and the temperature. (*Hint*: Use the first law of thermodynamics.)

46. Added as Heat When 20.9 J of thermal energy was added to a particular ideal gas, the volume of the gas changed from 50.0 cm³ to 100 cm³ while the pressure remained constant at 1.00 atm. (a) By how much did the internal energy of the gas change? If the quantity of gas present is 2.00×10^{-3} mol, find the molar specific heat of the gas at (b) constant pressure and (c) constant volume.

47. Three Nonreacting Gases A container holds a mixture of three nonreacting gases: n_1 moles of the first gas with molar specific heat at constant volume C_1, and so on. Find the molar specific heat at constant volume of the mixture, in terms of the molar specific heats and quantities of the separate gases.

48. Ideal Diatomic Gas One mole of an ideal diatomic gas goes from a to c along the diagonal path in Fig. 20-19. During the transition, (a) what is the change in internal energy of the gas, and (b) how much thermal energy is added to the gas? (c) How much thermal energy is required if the gas goes from a to c along the indirect path abc?

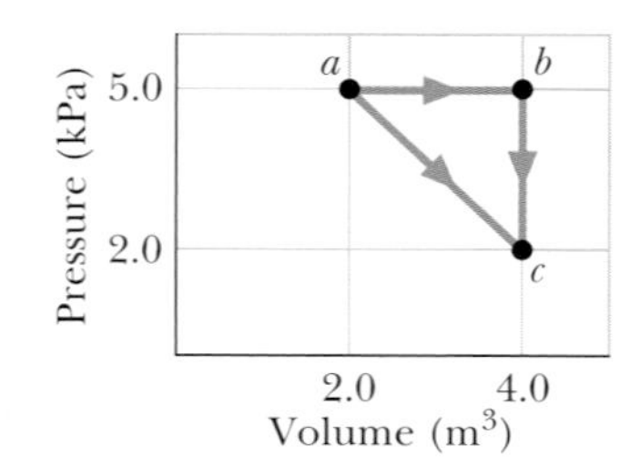

FIGURE 20-19 ■ Problem 48.

49. Gas Molecule The mass of a gas molecule can be computed from its specific heat at constant volume C_V. Take $C_V = 0.075$ cal/g · C° for argon and calculate (a) the mass of an argon atom and (b) the molar mass of argon.

SEC. 20-8 ■ DEGREES OF FREEDOM AND MOLAR SPECIFIC HEATS

50. Heating a Diatomic Gas We give 70 J of thermal energy to a diatomic gas, which then expands at constant pressure. The gas molecules rotate but do not oscillate. By how much does the internal energy of the gas increase?

51. One Mole of Oxygen One mole of oxygen (O_2) is heated at constant pressure starting at 0°C. How much energy Q must be added to the gas to double its volume? (The molecules rotate but do not oscillate.)

52. Oxygen Heating Suppose 12.0 g of oxygen (O_2) is heated at constant atmospheric pressure from 25.0°C to 125°C. (a) How many moles of oxygen are present? (See Table 20-1 for the molar mass.) (b) How much thermal energy is transferred to the oxygen? (The molecules rotate but do not oscillate.) (c) What fraction of the thermal energy absorbed by the oxygen is used to raise the internal energy of the oxygen?

53. Molecular Rotation Suppose 4.00 mol of an ideal diatomic gas, with molecular rotation but not oscillation, experienced a temperature increase of 60.0 K under constant-pressure conditions. (a) How much thermal energy was transferred to the gas? (b) How much did the internal energy of the gas increase? (c) How much work was done by the gas? (d) How much did the translational kinetic energy of the gas increase?

SEC. 20-10 ■ THE ADIABATIC EXPANSION OF AN IDEAL GAS

54. Liter of Gas (a) One liter of a gas with $\gamma = 1.3$ is at 273 K and 1.0 atm pressure. It is suddenly compressed adiabatically to half its original volume. Find its final pressure and temperature. (b) The gas is now cooled back to 273 K at constant pressure. What is its final volume?

55. A Certain Gas A certain gas occupies a volume of 4.3 L at a pressure of 1.2 atm and a temperature of 310 K. It is compressed adiabatically to a volume of 0.76 L. Determine (a) the final pressure and (b) the final temperature, assuming the gas to be an ideal gas for which $\gamma = 1.4$.

56. Adiabatic Process We know that for an adiabatic process PV^γ = a constant. Evaluate the constant for an adiabatic process involving exactly 2.0 mol of an ideal gas passing through the state having exactly $P = 1.0$ atm and $T = 300$ K. Assume a diatomic gas whose molecules have rotation but not oscillation.

57. Let *n* Moles Let n moles of an ideal gas expand adiabatically from an initial temperature T_1 to a final temperature T_2. Prove that the work done by the gas is $nC_V(T_1 - T_2)$, where C_V is the molar specific heat at constant volume. (*Hint*: Use the first law of thermodynamics.)

58. Bulk Modulus For adiabatic processes in an ideal gas, show that (a) the bulk modulus is given by

$$B = -V\frac{dP}{dV} = \gamma P,$$

and therefore (b) the speed of sound in the gas is

$$v_{\text{wave}} = \sqrt{\frac{\gamma P}{\rho}} = \sqrt{\frac{\gamma RT}{M}}.$$

See Eqs. 18-1 and 18-2. Here M is the molar mass and the total mass of the gas is $m = nM$.

59. Molar Specific Heats Air at 0.000°C and 1.00 atm pressure has a density of 1.29×10^{-3} g/cm^3, and the speed of sound in air is 331 m/s at that temperature. Use those data to compute the ratio γ of the molar specific heats of air. (*Hint*: See Problem 58.)

60. Free Expansion (a) An ideal gas initially at pressure P_0 undergoes a free expansion until its volume is 3.00 times its initial volume. What then is its pressure? (b) The gas is next slowly and adiabatically compressed back to its original volume. The pressure after compression is $(3.00)^{1/3}\,P_0$. Is the gas monatomic, diatomic, or polyatomic? (c) How does the average kinetic energy per molecule in this final state compare with that in the initial state?

61. The Cycle One mole of an ideal monatomic gas traverses the cycle of Fig. 20-20. Process $1 \rightarrow 2$ occurs at constant volume, process $2 \rightarrow 3$ is adiabatic, and process $3 \rightarrow 1$ occurs at constant pressure. (a) Compute the thermal energy Q absorbed by the gas, the change in its internal energy ΔE^{int}, and the work done by the gas W, for each of the three processes and for the cycle as a whole. (b) The initial pressure at point 1 is 1.00 atm. Find the pressure and the volume at points 2 and 3. Use 1.00 atm $= 1.013 \times 10^5$ Pa and $R = 8.314$ J/mol · K.

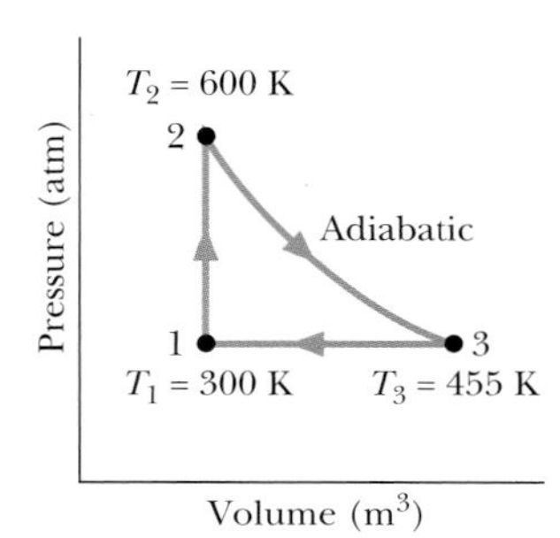

FIGURE 20-20 ■ Problem 61.

Additional Problems

62. Extensive vs. Intensive An intensive variable is one that can be defined locally within a system. Its magnitude does not depend on whether we select the whole system or a part of the system. An extensive variable is one that is defined for the system as a whole; its magnitude does depend on how much of the system we choose to select. Which of the following variables are intensive and which are extensive? Explain your reasoning in each case.
(a) density, (b) pressure, (c) volume, (d) temperature, (e) mass, (f) internal energy, (g) number of moles, and (h) molecular weight.

63. Changes in Gas Molecules An ideal gas is contained in an airtight box. Complete each of the following five statements below to show the quantitative change that will occur. For example, if you want to say that the volume, initially equal to V, quadruples, complete the statement with "$4V$".

(a) If the absolute temperature of the gas is halved, the average speed of a gas molecule, $\langle v \rangle$, becomes_____.

(b) If the average speed of a gas molecule doubles, the pressure, P, on a the wall of the box becomes_____.

(c) If the absolute temperature of the gas is halved, the pressure, P, on a wall of the box becomes________.

(d) If the absolute temperature of the gas is increased by 25%, the total internal energy of the gas, E^{int}, becomes_____.

(e) If the number of gas molecules inside the box is doubled, but the temperature is kept the same, the pressure, P, on a wall of the box becomes______.

64. Scales in a Gas The actual diameter of an atom is about 1 angstrom (10^{-10} m). In order to develop some intuition for the molecular scale of a gas, assume that you are considering a liter of air (mostly N_2 and O_2) at room temperature and a pressure of 10^5 Pa. (a) Calculate the number of molecules in the sample of gas. (b) Estimate the average spacing between the molecules. (c) Estimate the average speed of a molecule using the Maxwell-Boltzmann distribution. (d) Suppose that the gas were rescaled upward so that each atom was the size of a tennis ball (but we don't change the time scale). What would be the average spacing between molecules and the average speed of the molecules in miles/hour?

65. Is the M-B Distribution Wrong? A good student makes the following observation. "If I try to accelerate a small sphere through air, it will be resisted by air drag. If I drop an object, it will eventually reach a terminal velocity where the air is resisting as much as gravity is trying to accelerate (Eq. 6-25). The smaller the ball, the slower is the terminal velocity. But you tell me the Maxwell distribution says that the molecules of air move very rapidly. I estimate that this is much faster than the molecule's terminal velocity, so it can't move that fast. The Maxwell distribution must be wrong." (a) The Newton drag law for a sphere moving through air is calculated by a molecular model to be $D = \rho\pi R^2 v^2$, where D is the magnitude of the drag force, R is the radius of the sphere, ρ is the density of the air, and v is the velocity of the object through the air. Calculate the terminal velocity for a sphere the size and mass of an air molecule falling through a fluid the density of air ($\rho = 1$ kg/m^3). (b) From your estimate in part (a), is this speed greater or less than the average speed the molecule should have given the M-B distribution? What is wrong with the student's argument?

66. Avogadro's Hypothesis Why does Avogadro's hypothesis (that a given volume of gas at a given temperature and pressure has the same number of molecules no matter what kind of gas it is) not hold for liquids and solids?

67. Boiling Molecules When a molecule of a liquid approaches the surface, it experiences a force barrier that tries to keep it in the liquid. Thus it has to do work to escape and loses some of its kinetic energy when it leaves.

(a) Assume that a water molecule can evaporate from the liquid if it hits the surface from the inside with a kinetic energy greater than the thermal energy corresponding to the temperature of boiling water, 100°C. Use this to estimate the numerical value of the work W required to remove a water molecule from the liquid.

(b) Even though the average speed of a molecule in water below the boiling point corresponds to a kinetic energy less than W, some molecules leave anyway and the water evaporates. Explain why this happens.

21 Entropy and the Second Law of Thermodynamics

An anonymous graffito on a wall of the Pecan Street Cafe in Austin, Texas, reads: "Time is God's way of keeping things from happening all at once." Time also has direction—some things happen in a certain sequence and could never happen on their own in a reverse sequence. As an example, an accidentally dropped egg splatters in a cup. The reverse process, a splattered egg re-forming into a whole egg and jumping up to an outstretched hand, will never happen on its own—but why not? Why can't that process be reversed, like a videotape run backward?

What in the world gives direction to time?

The answer is in this chapter.

21-1 Some One-Way Processes

Suppose you come indoors on a very cold day and wrap your cold hands around a warm mug of cocoa. Then your hands get warmer and the mug gets cooler. However, it never happens the other way around. That is, your cold hands never get still colder while the warm mug gets still warmer.

If we assume that the rate of thermal energy transfer from the warm mug and your hands to the room is slow, the system consisting of your hands and the mug is approximately a *closed system,* one that is more or less isolated from (does not interact with) its environment. Here are some other one-way processes that we observe to occur in closed systems: (1) A crate sliding over a horizontal surface eventually stops—but you never see an initially stationary crate on a horizontal surface start to move all by itself. (2) If you drop a glob of putty, it falls to the floor—but an initially motionless glob of putty on the floor never leaps spontaneously into the air. (3) If you puncture a helium-filled balloon in a closed room, the helium gas spreads throughout the room—but the individual helium atoms will never migrate back out of the room and refill the balloon. We call such one-way processes **irreversible,** meaning that they cannot be reversed by means of only small changes in their environment.

Many chemical transformations are also irreversible. For example, when methane gas is burned, each methane molecule mixes with an oxygen molecule. Water vapor and carbon dioxide are given off as shown in the following chemical equation:

$$CH_4 + O_2 \longrightarrow CO_2 + H_2O.$$

This combustion process is irreversible. We don't find water and carbon dioxide spontaneously reacting to produce methane and oxygen gas.

The one-way character of such thermodynamic processes is so pervasive that we take it for granted. If these processes were to occur spontaneously (on their own) in the "wrong" direction, we would be astonished beyond belief. *Yet none of these "wrong-way" events would violate the law of conservation of energy.* In the "cold hands-warm mug" example, energy would be conserved even for a "wrong-way" thermal energy transfer between hands and mug. Conservation of energy would be obeyed even if a stationary crate or a stationary glob of putty suddenly were to transfer internal energy to macroscopic kinetic energy and begin to move. Energy would still be conserved if the helium atoms released from a balloon were, on their own, to clump together again.

Changes in energy within a closed system do not determine the direction of irreversible processes. So, we conclude that the direction must be set by another property that we have not yet considered. We shall discuss this new property quite a bit in this chapter. It is called the *entropy S* of the system. Knowing the *change in entropy* ΔS of a system turns out to be a useful quantity in analyzing thermodynamic processes. It is defined in the next section, but here we can state the central property of entropy change (often called the *entropy postulate*):

> If an irreversible process occurs in a *closed* system, the entropy S of the system always increases; it never decreases.

Entropy, unlike energy, does *not* obey a conservation law. The energy of a closed system is conserved. It always remains constant. For irreversible processes, the *entropy* of a closed system always increases. Because of this property, the change in entropy is sometimes called "the arrow of time." For example, we associate the irreversible breaking of the egg in our opening photograph with the forward direction of time and also with an increase in entropy. The backward direction of time (a videotape run backward) would correspond to the broken egg re-forming into a whole egg and rising into the air. This backward process would result in an entropy decrease and so it never happens.

There are two equivalent ways to define the change in entropy of a system. The first is macroscopic. It is in terms of the system's temperature and any thermal energy transfers between the system and its surroundings. The second is macroscopic or statistical. It is defined by counting the ways in which the atoms or molecules that make up the system can be arranged. We use the first approach in the next section, and the second in Section 21-7.

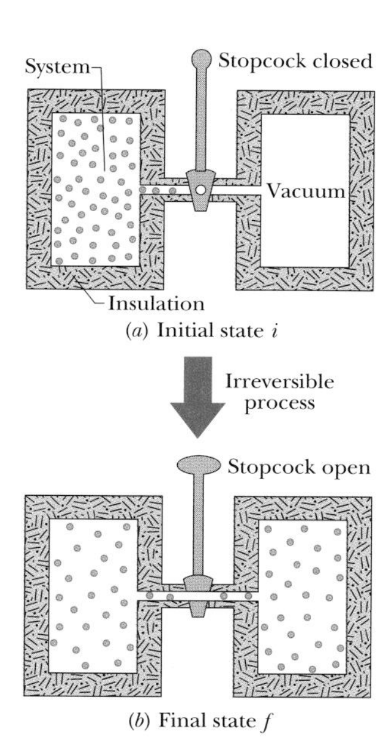

FIGURE 21-1 ■ The free expansion of an ideal gas consisting of a large number of molecules. (*a*) The gas is confined to the left half of an insulated container by a closed stopcock. (*b*) When the stopcock is opened, the gas rushes to fill the entire container. This process is irreversible; that is, it is never observed to occur in reverse, with the gas spontaneously collecting itself in the left half of the container.

READING EXERCISE 21-1: Consider the irreversible process of dropping a glob of putty on the floor. Describe what energy transformations are taking place that allow us to believe that energy is conserved. ■

READING EXERCISE 21-2: List additional everyday phenomena that illustrate irreversibility without violating energy conservation. ■

21-2 Change in Entropy

In this section, we will try to develop a definition for *change in entropy* by looking again at the macroscopic process that we described in Sections 19-8 and 20-10—namely, the free expansion of an ideal gas. Figure 21-1*a* shows the gas in its initial equilibrium state *i*, confined by a closed stopcock to the left half of a thermally insulated container. If we open the stopcock, the gas rushes to fill the entire container, eventually reaching the final equilibrium state *f* shown in Fig. 21-1*b*. Unless the number of gas molecules is small (which is very hard to accomplish), this is an *irreversible* process. Again, what we mean by irreversible is that it is extremely improbable that all the gas particles would return, by themselves, to the left half of the container.

The *P-V* plot of the process in Fig. 21-2 shows the pressure and volume of the gas in its initial state *i* and final state *f*. As we discuss in Section 19-1, the pressure and volume of the gas depend only on the state that the gas is in, and not on the process by which it arrived in that state. Therefore, pressure and volume are examples of *state properties*. **State properties** are properties that depend only on the state of the gas and not on how it reached that state. Other state properties are temperature and internal energy. We now assume that the gas has still another state property—its entropy. Furthermore, we define the **change in entropy** $S_f - S_i$ of a system during a process that takes the system from an initial state *i* to a final state *f* as

$$\Delta S = S_f - S_i \equiv \int_i^f \frac{dQ}{T} \quad \text{(change in entropy defined).} \tag{21-1}$$

Here Q is the *thermal energy transferred* to or from the system during a heating or cooling process, and T is the temperature of the system in kelvin. Thus, an entropy change depends not only on the thermal energy transferred Q, but also on the temperature at which the transfer takes place. Because T is always positive, the sign of ΔS is the same as that of Q (positive if thermal energy is transferred to the system and negative if thermal energy is transferred from the system). We see from this relation (Eq. 21-1) that the SI unit for entropy and entropy change is the joule per kelvin.

There is a problem, however, in applying Eq. 21-1 to the free expansion of Fig. 21-1. As the gas rushes to fill the entire container, the pressure, temperature, and volume of the gas fluctuate unpredictably. In other words, they do not have a sequence of well-defined equilibrium values during the intermediate stages of the change from initial equilibrium state *i* to final equilibrium state *f*. Thus, we cannot trace a pressure–volume path for the free expansion on the *P-V* plot of Fig. 21-2. More importantly, that means that we cannot find a relation between thermal energy transfer Q and temperature T that allows us to integrate as Eq. 21-1 requires.

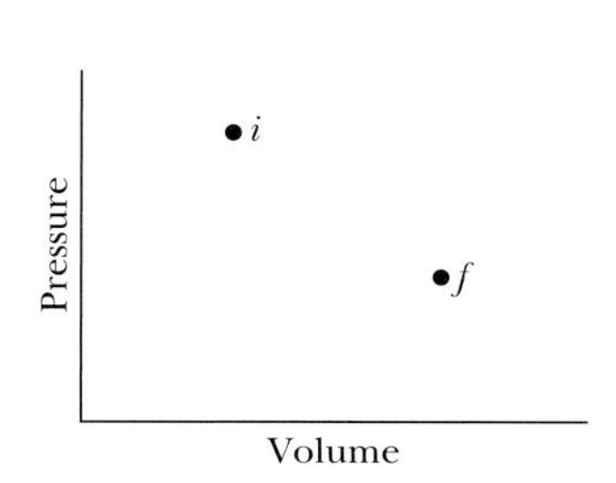

FIGURE 21-2 ■ A *P-V* diagram showing the initial state *i* and the final state *f* of the free expansion of Fig. 21-1. The intermediate states of the gas cannot be shown because they are not equilibrium states.

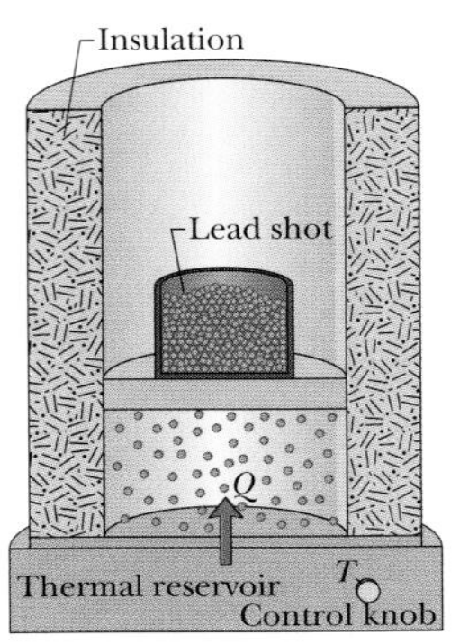

(a) Initial state i

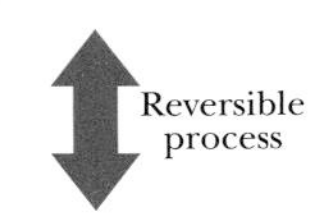

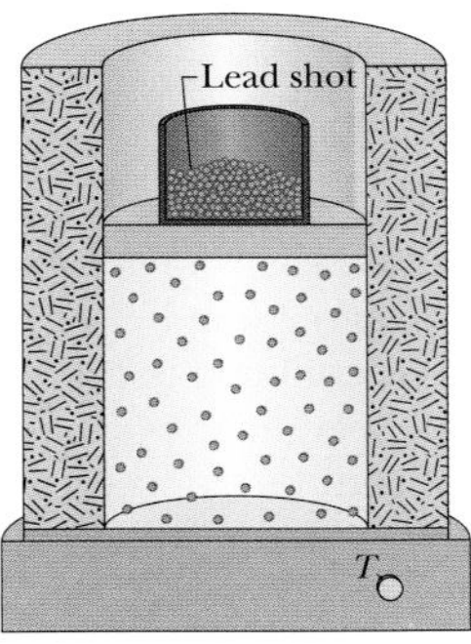

(b) Final state f

FIGURE 21-3 ■ The isothermal expansion of an ideal gas, done in a reversible way. The gas has the same initial state i and same final state f as in the irreversible process of Figs. 21-1 and 21-2.

However, if our assumption is correct and entropy is truly a state property, the difference in entropy between states i and f must depend *only on those states* and not at all on the way the system went from one state to the other. That means that we can replace the irreversible free expansion of Fig. 21-1 with a *reversible* process that connects states i and f. With a reversible process we can trace a pressure–volume path on a P-V plot, and we can find a relation between thermal energy transfer Q and temperature T that allows us to use Eq. 21-1 to obtain the entropy change.

We saw in Section 20-10 that the temperature of an ideal gas does not change during a free expansion. So, $T_i = T_f = T$. Thus, points i and f in Fig. 21-2 must be on the same isotherm. A convenient replacement process is then a reversible isothermal expansion from state i to state f, which actually proceeds *along* that isotherm. Furthermore, because T is constant throughout a reversible isothermal expansion, the integral of Eq. 21-1 is greatly simplified.

Figure 21-3 shows how to produce such a reversible isothermal expansion. We confine the gas to an insulated cylinder that rests on a thermal reservoir maintained at the temperature T. We begin by placing just enough lead shot on the movable piston so that the pressure and volume of the gas are those of the initial state i of Fig. 21-1a. We then remove shot slowly (piece by piece) until the pressure and volume of the gas are those of the final state f of Fig. 21-1b. The temperature of the gas does not change because the gas remains in thermal contact with the reservoir throughout the process.

The reversible isothermal expansion of Fig. 21-3 is physically quite different from the irreversible free expansion of Fig. 21-1. However, *both processes have the same initial state and the same final state. Thus, if entropy is a state property, these two processes must result in the same change in entropy.*

Because we removed the lead shot slowly, the intermediate states of the gas are equilibrium states, so we can plot them on a P-V diagram (Fig. 21-4). To apply Eq. 21-1 to the isothermal expansion, we take the constant temperature T outside the integral, obtaining

$$\Delta S = S_f - S_i = \frac{1}{T}\int_i^f dQ.$$

Because $\int dQ = Q$, where Q is the thermal energy transferred during the process, we have

$$\Delta S = S_f - S_i = \frac{Q}{T} \qquad \text{(change in entropy, isothermal process).} \qquad (21\text{-}2)$$

To keep the temperature T of the gas constant during the isothermal expansion of Fig. 21-3, the thermal energy transferred *from* the reservoir to the gas must have been Q. Thus, Q is positive and the entropy of the gas *increases* during the isothermal process and during the free expansion of Fig. 21-1.

To summarize:

> Assuming entropy is a state property, we can find the entropy change for an irreversible process occurring in a *closed* system by replacing that process with any reversible process that connects the same initial and final states. We can then calculate the entropy change for this reversible process with Eq. 21-1. The change in entropy for an irreversible process connecting the same two states would be the same.

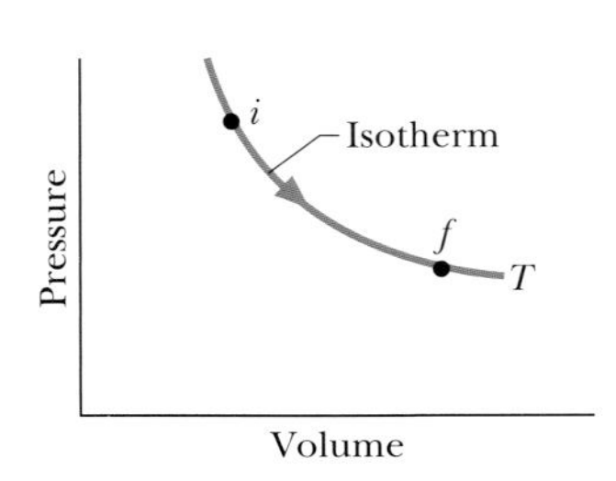

FIGURE 21-4 ■ A P-V diagram for the reversible isothermal expansion of Fig. 21-3. The intermediate states, which are now equilibrium states, are shown.

We can even use this approach if the temperature of the system is not quite constant. That is, if the temperature change ΔT of a system is small relative to the temperature (in kelvin) before and after the process, the entropy change can be approximated as

$$\Delta S = S_f - S_i \approx \frac{Q}{\langle T \rangle} \qquad (21\text{-}3)$$

where $\langle T \rangle$ is the average kelvin temperature of the system during the process.

Entropy as a State Property

In the previous section, we assumed that entropy, like pressure, internal energy, and temperature, is a property of the state of a system and is independent of how that state is reached. The fact that entropy is indeed a state property (or *state function* as state properties are sometimes called) can really only be deduced by careful experiment. However, we will prove entropy is a state property for the special and important case of an ideal gas undergoing a reversible process. This proof will serve two purposes. First, it will verify (for at least this one case) that entropy is a state property (or state function) as we assumed in the section above. Second, it will allow us to develop an expression for the entropy change in an ideal gas as it goes from some initial state i to some final state f via a reversible process.

To make the process reversible, we must make changes slowly in a series of small steps, with the ideal gas in an equilibrium state at the end of each step. For each small step, the thermal energy transfer to or from the gas is dQ, the work done by or on the gas is dW, and the change in internal energy is dE^{int}. These are related by the first law of thermodynamics in differential form (Eq. 19-18):

$$dE^{\text{int}} = dQ - dW.$$

Because the steps are reversible, with the gas in equilibrium states, we can replace dW with $P\,dV$ (Eq. 19-15). Since we are dealing with an ideal gas, we can also replace dE^{int} with $nC_V\,dT$ (Eq. 20-43). Solving for the thermal energy transferred to or from the system in a single small step of the process dQ then leads to

$$dQ = P\,dV + nC_V\,dT.$$

We replace the pressure P in this equation with nRT/V (using the ideal gas law). Then we divide each term in the resulting equation by the temperature T, obtaining

$$\frac{dQ}{T} = nR\frac{dV}{V} + nC_V\frac{dT}{T}.$$

Now let us integrate each term of this equation between an arbitrary initial state i and an arbitrary final state f to get

$$\int_i^f \frac{dQ}{T} = \int_i^f nR\frac{dV}{V} + \int_i^f nC_V\frac{dT}{T}.$$

The quantity on the left is the entropy change $\Delta S(=S_f - S_i)$ as we defined it in Eq. 21-1. Substituting this and integrating the quantities on the right yields an expression for the entropy change in an ideal gas undergoing a reversible process:

$$\Delta S = S_f - S_i = nR\ln\frac{V_f}{V_i} + nC_V\ln\frac{T_f}{T_i}. \tag{21-4}$$

Note that we did not have to specify a particular reversible process when we integrated. Therefore, the integration must hold for all reversible processes that take the gas from state i to state f. Thus, we see that the change in entropy ΔS between the initial and final states of an ideal gas does depend only on properties of the initial state (V_i and T_i) and properties of the final states (V_f and T_f); ΔS does not depend on how the gas changes between the two equilibrium states. Therefore, in at least this one case, we know that entropy must be a state property. In the work that follows in this chapter, we will accept without further proof that entropy is in fact a state property for any system undergoing any process.

READING EXERCISE 21-3: Thermal energy is transferred to water on a stove. Rank the entropy changes of the water as its temperature rises (a) from 20°C to 30°C, (b) from 30°C to 35°C, and (c) from 80°C to 85°C, greatest first. ■

TOUCHSTONE EXAMPLE 21-1: Nitrogen

One mole of nitrogen gas is confined to the left side of the container of Fig. 21-1*a*. You open the stopcock and the volume of the gas doubles. What is the entropy change of the gas for this irreversible process? Treat the gas as ideal.

SOLUTION ■ We need two **Key Ideas** here. One is that we can determine the entropy change for the irreversible process by calculating it for a reversible process that provides the same change in volume. The other is that the temperature of the gas does not change in the free expansion. Thus, the reversible process should be an isothermal expansion—namely, the one of Figs. 21-3 and 21-4.

Since the internal energy of an ideal gas depends only on temperature, $\Delta E^{\text{int}} = 0$ here and so $Q = W$ from the first law. Combining this result with Eq. 20-14 gives

$$Q = W = nRT \ln\left(\frac{V_f}{V_i}\right),$$

in which n is the number of moles of gas present. From Eq. 21-2 the entropy change for this isothermal reversible process is

$$\Delta S^{\text{rev}} = \frac{Q}{T} = \frac{nRT\ln(V_f/V_i)}{T} = nR\ln\frac{V_f}{V_i}.$$

Substituting $n = 1.00$ mol and $V_f/V_i = 2$, we find

$$\Delta S^{\text{rev}} = nR\ln\frac{V_f}{V_i} = (1.00\text{ mol})(8.31\text{ J/mol}\cdot\text{K})(\ln 2)$$

$$= +5.76\text{ J/K}. \qquad \text{(Answer)}$$

Thus, the entropy change for the free expansion (and for all other processes that connect the initial and final states shown in Fig. 21-2) is

$$\Delta S^{\text{irrev}} = \Delta S^{\text{rev}} = +5.76\text{ J/K}.$$

ΔS is positive, so the entropy increases, in accordance with the entropy postulate of Section 21-1.

TOUCHSTONE EXAMPLE 21-2: Copper Blocks

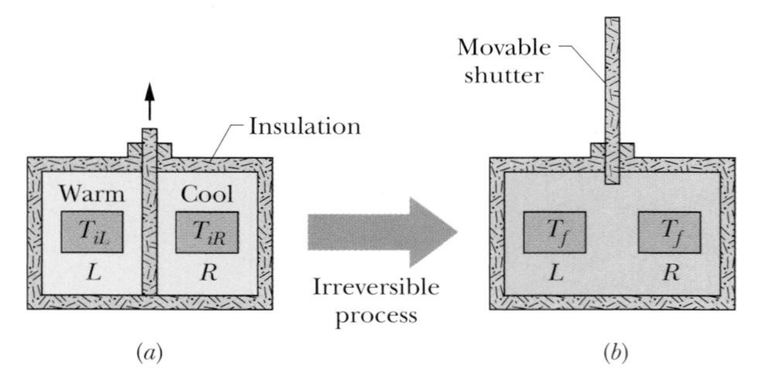

FIGURE 21-5 ■ (*a*) In the initial state, two copper blocks L and R, identical except for their temperatures, are in an insulating box and are separated by an insulating shutter. (*b*) When the shutter is removed, the blocks exchange thermal energy and come to a final state, both with the same temperature T_f. The process is irreversible.

Figure 21-5*a* shows two identical copper blocks of mass $m = 1.5$ kg: block L at temperature $T_{iL} = 60°\text{C}$ and block R at temperature $T_{iR} = 20°\text{C}$. The blocks are in a thermally insulated box and are separated by an insulating shutter. When we lift the shutter, the blocks eventually come to the equilibrium temperature $T_f = 40°\text{C}$ (Fig. 21-5*b*). What is the net entropy change of the two-block system during this irreversible process? The specific heat of copper is 386 J/kg·K.

SOLUTION ■ The **Key Idea** here is that to calculate the entropy change, we must find a reversible process that takes the system from the initial state of Fig. 21-5*a* to the final state of Fig. 21-5*b*. We can calculate the net entropy change ΔS^{rev} of the reversible process using Eq. 21-1, and then the entropy change for the irreversible process is equal to ΔS^{rev}. For such a reversible process we need a thermal reservoir whose temperature can be changed slowly (say, by turning a knob). We then take the blocks through the following two steps, illustrated in Fig. 21-6.

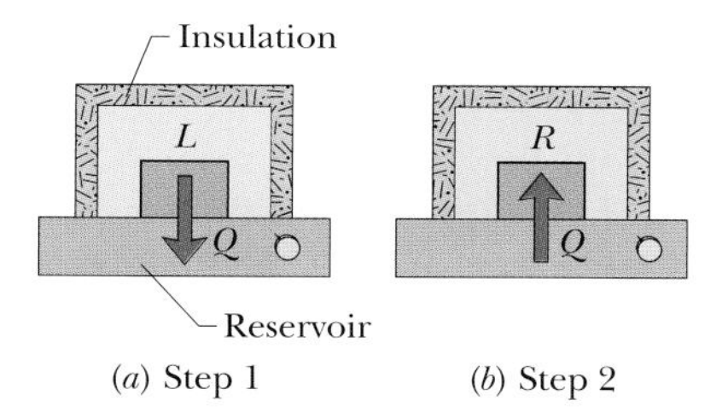

FIGURE 21-6 ■ The blocks of Fig. 21-5 can proceed from their initial state to their final state in a reversible way if we use a reservoir with a controllable temperature (a) to transfer thermal energy reversibly from block L and (b) to transfer thermal energy reversibly to block R.

Step 1. With the reservoir's temperature set at 60°C, put block L on the reservoir. (Since block and reservoir are at the same temperature, they are already in thermal equilibrium.) Then slowly lower the temperature of the reservoir and the block to 40°C. As the block's temperature changes by each increment dT during this process, thermal energy dQ is transferred *from* the block to the reservoir. Using Eq. 19-5, we can write this transferred energy as $dQ = mc\, dT$, where c is the specific heat of copper. According to Eq. 21-1, the entropy change ΔS_L of block L during the full temperature change from initial temperature T_{iL} ($= 60°\text{C} = 333$ K) to final temperature T_f ($= 40°\text{C} = 313$ K) is

$$\Delta S_L = \int_i^f \frac{dQ}{T} = \int_{T_{iL}}^{T_f} \frac{mc\, dT}{T} = mc \int_{T_{iL}}^{T_f} \frac{dT}{T}$$

$$= mc \ln \frac{T_f}{T_{iL}}.$$

Inserting the given data yields

$$\Delta S_L = (1.5\text{ kg})(386\text{ J/[kg}\cdot\text{K]}) \ln \frac{313\text{ K}}{333\text{ K}}$$

$$= -35.86\text{ J/K}.$$

Step 2: With the reservoir's temperature now set at 20°C, put block R on the reservoir. Then slowly raise the temperature of the reservoir and the block to 40°C. With the same reasoning used to find ΔS_L, you can show that the entropy change ΔS_R of block R during this process is

$$\Delta S_R = (1.5\text{ kg})(386\text{ J/[kg}\cdot\text{K]}) \ln \frac{313\text{ K}}{293\text{ K}}$$

$$= +38.23\text{ J/K}.$$

The net entropy change ΔS^{rev} of the two-block system undergoing this two-step reversible process is then

$$\Delta S^{\text{rev}} = \Delta S_L + \Delta S_R$$

$$= -35.85\text{ J/K} + 38.23\text{ J/K} = 2.4\text{ J/K}.$$

Thus, the net entropy change ΔS^{irrev} for the two-block system undergoing the actual irreversible process is $\Delta S^{\text{irrev}} = \Delta S^{\text{rev}} = 2.4$ J/K. This result is positive, in accordance with the entropy postulate of Section 21-1.

21-3 The Second Law of Thermodynamics

Here is a puzzle. We saw in Touchstone Example 21-1 that if we cause the reversible, isothermal process of Fig. 21-3 to proceed from (a) to (b) in that figure, the change in entropy of the gas—which we take as our system—is positive. However, because the process is reversible, we can just as easily make it proceed from (b) to (a), simply by slowly adding lead shot to the piston of Fig. 21-3b until the original volume of the gas is restored. In this reverse isothermal process, the gas must keep its temperature from increasing and so must transfer thermal energy to its surroundings to make up for the work done via the lead shot. Since this thermal energy transfer is *from* our system (the gas), Q is negative. So, from $\Delta S = S_f - S_i = Q/T$ (Eq. 21-2) we find that ΔS must be negative and hence the entropy of the gas must decrease.

Doesn't this decrease in the entropy of the gas violate the entropy postulate of Section 21-1, which states that entropy always increases? No, because that postulate holds only for *irreversible* processes occurring in *closed* systems. The procedure suggested here does not meet these requirements. The process is *not* irreversible and (because there is a heat transfer from the gas to the reservoir) the system—which is the gas alone—is not closed.

However, if we include the reservoir, along with the gas, as part of the system, then we do have a closed system. Let's check the change in entropy of the enlarged system *gas* + *reservoir* for the process that takes it from (*b*) to (*a*) in Fig. 21-3. During this reversible process, thermal energy is transferred from the gas to the reservoir— that is, from one part of the enlarged system to another. Let $|Q|$ represent the amount of energy transferred. With $\Delta S = S_f - S_i = Q/T$ (Eq. 21-2), we can then calculate separately the entropy changes for the gas (which loses $|Q|$ of thermal energy) and the reservoir (which gains $|Q|$ in thermal energy). We get

$$\Delta S_{\text{gas}} = -\frac{|Q|}{T}$$

and

$$\Delta S_{\text{res}} = +\frac{|Q|}{T}.$$

The entropy change of the closed system is the sum of these two quantities, *which* (since the process is isothermal) *is zero*.

With this result, we can modify the entropy postulate of Section 21-1 to include both reversible and irreversible processes:

> If a process occurs in a *closed* system, the entropy of the system increases for irreversible processes and remains constant for reversible processes. It never decreases.

Although entropy may decrease in part of a closed system, there will always be an equal or larger entropy increase in another part of the system, so that the entropy of the system as a whole never decreases. This fact is one form of the **second law of thermodynamics** and can be written as

$$\Delta S \geq 0 \quad \text{(second law of thermodynamics)}, \tag{21-5}$$

where the greater-than sign applies to irreversible processes, and the equals sign to reversible processes. But remember, this relation applies only to closed systems.

In the real world almost all processes are irreversible to some extent because of friction, turbulence, and other factors. So, the entropy of real closed systems undergoing real processes always increases. Processes in which the system's entropy remains constant are idealizations.

21-4 Entropy in the Real World: Engines

Engines, which are fundamentally thermodynamic devices, are everywhere around us and are a big part of what makes modern life possible. However, not all engines are the same. For example, the engine in your car is different from the engine in a typical power plant. Nevertheless, these engines are similar in that they function through the use of a *working substance* that can expand and contract as it exchanges energy with its surroundings. In a power plant, the working substance is often water, in both its vapor and liquid forms. In an automobile engine the working substance is a gasoline-air mixture. If an engine is to do work on a sustained basis, the working substance must operate in a *cycle*. That is, the working substance must pass through a repeating series of thermodynamic processes, called **strokes,** returning again and again to each state in its **cycle.** The fundamental difference between the engine in your car and that in a power plant is that these two engines use different working substances and different types of thermodynamic cycles. That is, the working substances in these two engines undergo different thermodynamic processes.

Most engines that we meet in everyday life are some version of what we will call a *heat engine*. **Heat engines** are devices that take the thermal energy transfers (or heat transfers) Q that result from temperature differences and convert them to useful work W. Internal combustion engines (like in your car) are complicated heat engines that convert chemical energy (from gasoline or diesel fuel) and thermal energy to work. We will not discuss internal combustion engines here. Instead, we will focus on a simpler class of engines in which the working substance simply cycles between two constant temperatures (isotherms) and the engine converts a portion of the resulting thermal energy transfers directly to mechanical work.

The Carnot Engine

We have seen that we can learn much about real gases by analyzing an ideal gas, which obeys the simple law $PV = nRT$. This is a useful plan because, although an ideal gas does not exist, any real gas approaches ideal behavior as closely as you wish if its density is low enough. In much the same spirit we choose to study real engines by analyzing the behavior of an **ideal engine.**

> In an ideal engine, all processes are reversible and no wasteful energy transfers occur due to friction, turbulence, or other processes.

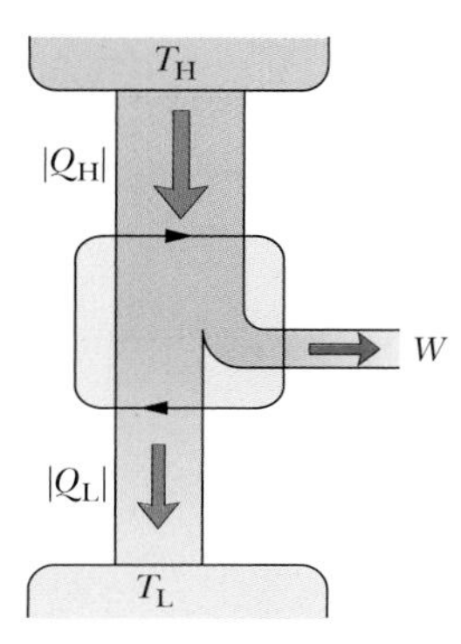

FIGURE 21-7 ■ The elements of a Carnot engine. The two black arrowheads on the central loop suggest the working substance operating in a cycle, as if on a P-V plot. Thermal energy $|Q_H|$ is transferred from the high-temperature reservoir at temperature T_H to the working substance. Thermal energy $|Q_L|$ is transferred from the working substance to the low-temperature reservoir at temperature T_L. Work W is done by the engine (actually by the working substance) on something in the environment.

We shall focus here on a particular ideal engine called an ideal **Carnot engine** named after the French scientist and engineer N. L. Sadi Carnot (pronounced "car-no"), who first proposed the engine's concept in 1824. A Carnot engine is an example of a heat engine. It operates between two constant temperatures and uses the resulting thermal energy transfers directly to do useful work. The ideal Carnot engine is especially important because it turns out to be the best engine of this type.

Figure 21-7 shows the operation of a Carnot engine schematically. During each cycle of the engine, energy $|Q_H|$ is transferred to the working substance from a thermal reservoir at constant temperature T_H and energy $|Q_L|$ is transferred from the working substance to a second thermal reservoir at a constant and lower temperature T_L. The Carnot engine converts the difference between the amount of energy transferred into the system $|Q_H|$ and the amount of energy transferred out of the system $|Q_L|$ into useful work.

The cycle followed by the working substance in a Carnot engine is called a **Carnot cycle.** Figure 21-8*a* shows a P-V plot of the Carnot cycle. As indicated by the arrows, the cycle is traversed in the clockwise direction. Imagine the working substance to be a gas, confined to an insulating cylinder with a weighted, movable piston. Figure 21-8*b* shows how the Carnot cycle might be accomplished. The cylinder may be placed at will on either of the two thermal reservoirs, or on an insulating slab. If we place the cylinder in contact with the high-temperature reservoir at temperature T_H, $|Q_H|$ represents the thermal energy transfer *to* the working substance *from* this reservoir as the gas undergoes an isothermal *expansion* from volume V_a to volume V_b. Similarly, when the working substance is in contact with the low-temperature reservoir at temperature T_L, the gaseous substance undergoes an isothermal *compression* from volume V_c to volume V_d. At the same time energy $|Q_L|$ is transferred *from* the working substance *to* this reservoir. Note that in our engine thermal energy transfers to or from the working substance can take place *only* during the isothermal processes *ab* and *cd* of Fig. 21-8*b*. Thermal energy transfers do not occur in processes *bc* and *da* in that figure, which connect the two isotherms at temperatures T_H and T_L. Therefore, those two processes must be (reversible) adiabatic processes. To ensure this, during processes *bc* and *da*, the cylinder is placed on an insulating slab as the volume of the working substance is changed.

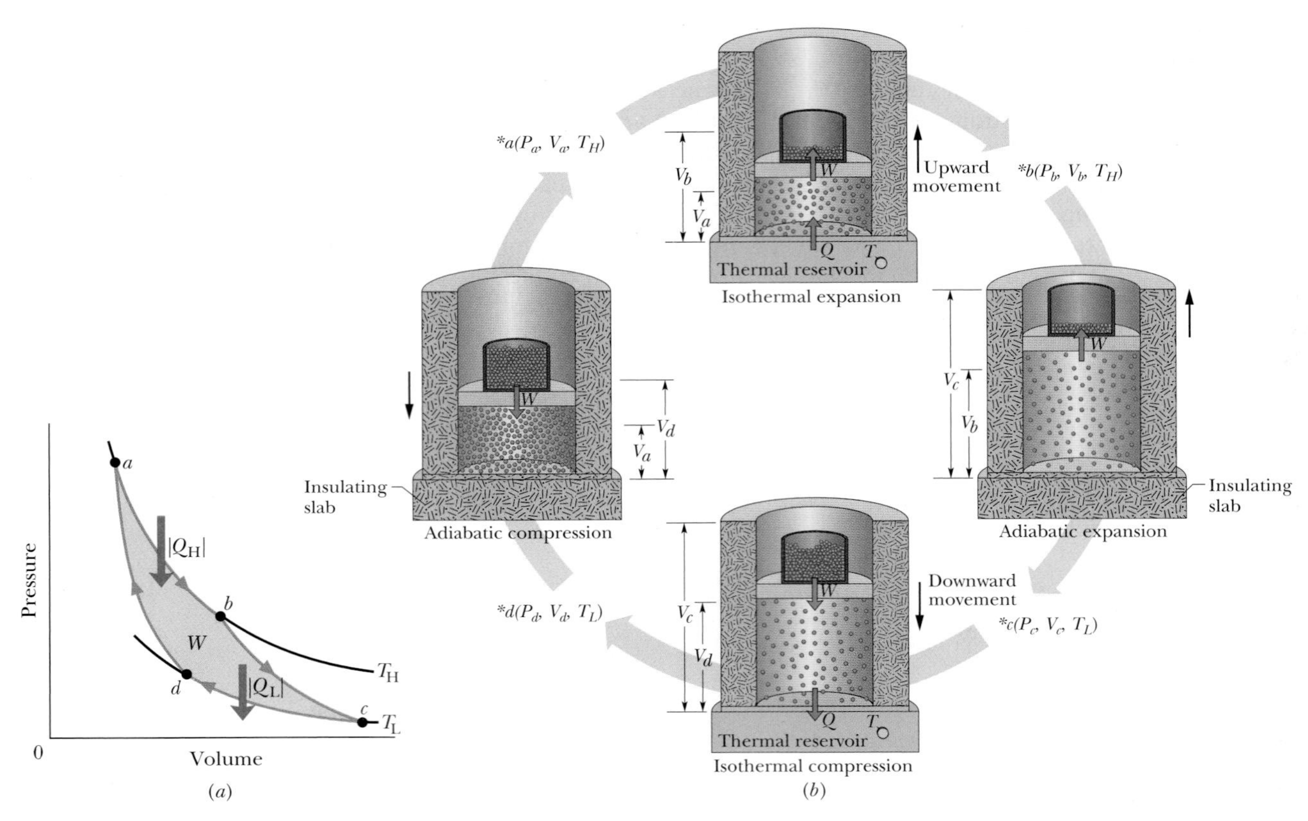

FIGURE 21-8 ■ (*a*) A pressure–volume plot of the cycle followed by the working substance of the Carnot engine in Fig. 21-7. The cycle consists of two isothermal processes (*ab* and *cd*) and two adiabatic processes (*bc* and *da*). The shaded area enclosed by the cycle is equal to the work W per cycle done by the Carnot engine. (*b*) An example of how this set of cycles could be accomplished. The upward motions of a piston during processes *ab* and *bc* are accomplished by slowly removing weight from the piston. The downward motions of the piston during processes *cd* and *da* are accomplished by slowing adding weight to the piston.

Work Done: During the consecutive processes *ab* and *bc* of Fig. 21-8, the working substance is expanding and thus doing positive work as it raises the weighted piston. This work is represented in Fig. 21-8*a* by the area under curve *abc*. During the consecutive processes *cd* and *da*, the working substance is being compressed, which means that it is doing negative work on its environment (the environment is doing positive work on it). This work is represented by the area under curve *cda*. The *net work per cycle*, which is represented by W in Figs. 21-7 and 21-8*a*, is the difference between these two areas. It is a positive quantity equal to the area enclosed by cycle *abcda* in Fig. 21-8*a*. This work W is performed on some outside object. The engine might, for example, be used to lift a weight.

To calculate the net work done by a Carnot engine during a cycle, let us apply the first law of thermodynamics ($\Delta E^{\text{int}} = Q - W$), to the working substance of a Carnot engine. That substance must return again and again to any arbitrarily selected state in that cycle. Thus, if X represents any state property of the working substance, such as pressure, temperature, volume, internal energy, or entropy, we must have $\Delta X = 0$ for every cycle. It follows that $\Delta E^{\text{int}} = 0$ for a complete cycle of the working substance. Recall that Q in Eq. 19-18 is the *net* thermal energy transfer per cycle and W is the net work. We can then write the first law of thermodynamics ($\Delta E^{\text{int}} = Q - W$) for the Carnot cycle as

$$W = |Q_{\text{H}}| - |Q_{\text{L}}|. \tag{21-6}$$

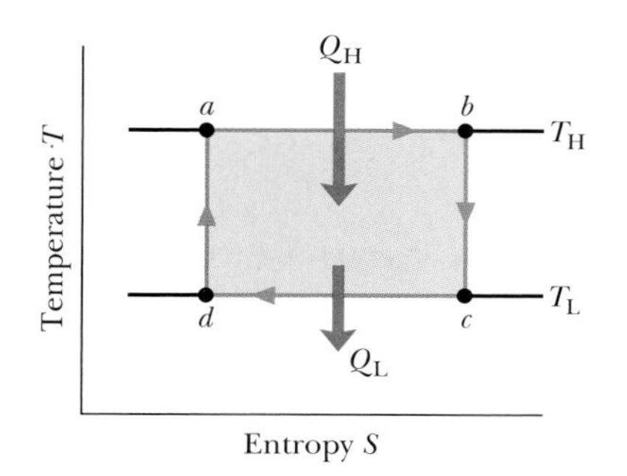

FIGURE 21-9 ■ The Carnot cycle of Fig. 21-8 plotted on a temperature–entropy diagram. During processes ab and cd the temperature remains constant. During processes bc and da the entropy remains constant.

Entropy Changes: Equation 21-1 ($\Delta S = \int dQ/T$) tells us that any thermal energy transfer between a system and its surroundings must involve a change in entropy. To illustrate the entropy changes for a Carnot engine, we can plot the Carnot cycle on a temperature–entropy (T-S) diagram as in Fig. 21-9. The lettered points a, b, c, and d in Fig. 21-9 correspond to the lettered points in the P-V diagram in Fig. 21-8. The two horizontal lines in Fig. 21-9 correspond to the two isothermal processes of the Carnot cycle (because the temperature is constant). Process ab is the isothermal expansion stroke of the cycle. As the high temperature reservoir transfers thermal energy $|Q_H|$ reversibly to the working substance at temperature T_H, its entropy increases. Similarly, during the isothermal compression cd, the working substance transfers thermal energy $|Q_L|$ reversibly to the low temperature reservoir at temperature T_L. In this process the entropy of the working substance decreases.

The two vertical lines in Fig. 21-9 correspond to the two adiabatic processes of the Carnot cycle. Because no thermal energy transfers occur during the adiabatic processes, the entropy of the working substance does not change during either of these processes. So, in a Carnot engine, there are *two* (and only two) reversible thermal energy transfers, and thus two changes in entropy—one at temperature T_H and one at T_L. The net entropy change per cycle is then

$$\Delta S = \Delta S_H + \Delta S_L = \frac{|Q_H|}{T_H} - \frac{|Q_L|}{T_L}. \tag{21-7}$$

Here ΔS_H is positive because energy $|Q_H|$ is *transferred to* the working substance from the surroundings (an increase in entropy) and ΔS_L is negative because energy $|Q_L|$ is *transferred from* the working substance to the surroundings (a decrease in entropy). Because entropy is a state property, we must have $\Delta S = 0$ for a complete cycle. Putting $\Delta S = 0$ in above (Eq. 21-7) requires that

$$\frac{|Q_H|}{T_H} = \frac{|Q_L|}{T_L}. \tag{21-8}$$

Note that, because $T_H > T_L$, we must have $|Q_H| > |Q_L|$. That is, more energy is transferred from the high-temperature reservoir to the engine than the engine transfers to the low-temperature reservoir.

We shall now use our findings on the work done (Eq. 21-6) and entropy change (Eq. 21-8) in an ideal Carnot cycle to derive an expression for the efficiency of an ideal Carnot engine.

Efficiency of an Ideal Carnot Engine

The purpose of any heat engine is to transform as much of the thermal energy, Q_H, transferred to the engine's working medium into useful mechanical work as possible. We measure its success in doing so by its **thermal efficiency** ε, defined as the work the engine does per cycle ("energy we get") divided by the thermal energy transferred to it per cycle ("energy we pay for"):

$$\varepsilon = \frac{\text{energy we get}}{\text{energy we pay for}} = \frac{|W|}{|Q_H|} \quad \text{(efficiency, any engine).} \tag{21-9}$$

For a Carnot engine we can substitute $W = |Q_H| - |Q_L|$ from Eq. 21-6 to write

$$\varepsilon_C = \frac{|Q_H| - |Q_L|}{|Q_H|} = 1 - \frac{|Q_L|}{|Q_H|}. \tag{21-10}$$

Using $|Q_H|/T_H = |Q_L|/T_L$ (Eq. 21-8) we can write this as

$$\varepsilon_C = 1 - \frac{T_L}{T_H} \quad \text{(efficiency, ideal Carnot engine),} \tag{21-11}$$

where the temperatures T_L and T_H are in kelvin. Because $T_L < T_H$, the Carnot engine necessarily has a thermal efficiency less than unity—that is, less than 100%. This is indicated in Fig. 21-7, which shows that only part of the energy transferred to the engine from the high-temperature reservoir causes the engine's working substance to expand and do physical work on the surroundings. The rest of the energy absorbed by the engine provides for the heat transfer to the low-temperature reservoir. We will show in Section 21-6 that no real engine can have a thermal efficiency greater than that calculated for the ideal Carnot engine (Eq. 21-11).

Other Types of Cycles and Real Engines

Efficiency is typically our main concern when designing an engine. For an engine that operates on an ideal Carnot cycle, the efficiency is

$$\varepsilon_C = 1 - \frac{T_L}{T_H}.$$

But remember, an ideal Carnot cycle means that the cycle is composed of the following four processes: a perfectly isothermal (constant temperature) expansion of the working substance, a perfectly adiabatic (zero thermal energy transfer) expansion of the working substance, a perfectly isothermal compression of the working substance, and a perfectly adiabatic compression of the working substance. Perfectly isothermal means the temperature of the working substance cannot change at all during these strokes. Perfectly adiabatic means that there can be no thermal energy transfer at all. These tasks are not easy to accomplish. If you do accomplish them, then you have an *ideal* Carnot cycle and the efficiency of the engine is given by the equation above. Most engines built on the Carnot cycle have efficiencies that are measurably lower than this.

It is important to note that even an ideal Carnot engine cannot have an efficiency of one. That is, it does not do a perfect job of converting thermal energy transferred to it into work. Inspection of the Carnot efficiency expression $\varepsilon_C = 1 - T_L/T_H$ (Eq. 21-11) shows that we can achieve 100% engine efficiency (that is, $\varepsilon = 1$) only if $T_L = 0$ K or $T_H \to \infty$. These requirements are impossible to meet. So, decades of practical engineering experience have led to the following alternative version of the second law of thermodynamics:

> It is impossible to design an engine that converts thermal energy transferred to it from a thermal reservoir to useful work with 100% efficiency.

As we mentioned earlier, Carnot engines are not the only type of heat engine in which the working substance cycles between two constant temperatures and converts some of the associated heat transferred to the engine's working medium to useful work. For example, Fig. 21-10 shows the operating cycle of an ideal **Stirling engine.** Comparison with the Carnot cycle of Fig. 21-8 shows that each cycle includes isothermal energy transfers at temperatures T_H and T_L. However, the two isotherms of the Stirling engine cycle of Fig. 21-10 are connected, not by adiabatic processes (no thermal energy transfer) as for the Carnot engine, but by constant-volume processes. To reversibly increase the temperature of a gas at constant volume from T_L to T_H (as in process *da* of Fig. 21-10) requires a thermal energy transfer to the working substance from a thermal reservoir whose temperature can be varied smoothly between those limits.

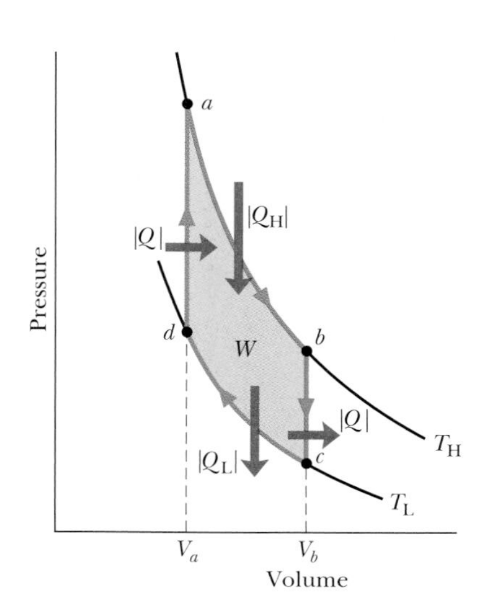

FIGURE 21-10 ■ A P-V plot for the working substance of an ideal Stirling engine, assumed for convenience to be an ideal gas. Processes *ab* and *cd* are isothermal while *bc* and *da* are constant volume.

Note that reversible thermal energy transfers (and corresponding entropy changes) occur in all four of the processes that form the cycle of a Stirling engine, not just two processes as in a Carnot engine. Thus, the derivation that led to the efficiency expression for the Carnot engine (Eq. 21-11) does not apply to an ideal Stirling engine. More important, the efficiency of an ideal Stirling engine, or any other heat engine based on operation between two isotherms, is lower than that of a Carnot engine operating between the same two temperatures. This makes the ideal Carnot engine an ideal version of the ideal type of this class of engine! Of course, common Stirling engines have even lower efficiencies than the ideal Stirling engine discussed here.

Many engines important in our lives operate based on cycles between two isotherms and convert thermal energy transfers to work. For example, consider the nuclear power plant shown in Fig. 21-11. It, like most power plants, is an engine when taken in its entirety. A reactor core (or perhaps a coal-powered furnace) provides the high-temperature reservoir. Thermal energy transfer to the working substance (usually water) is converted to work done on a turbine (which often results in electricity production). The remaining energy $|Q_L|$ is transferred to a low-temperature reservoir, which is usually a nearby river, or the atmosphere (if cooling towers are used). If the power plant shown in Fig. 21-11 operated as an ideal Carnot engine, its efficiency would be about 40%. Its actual efficiency is about 30%.

How does the efficiency of the internal combustion engine compare to that of the ideal Carnot engine? Well, this is a bit like comparing apples and oranges since the internal combustion engine does not operate between two isotherms like the Carnot, Stirling, or power plant engines do. However, we can estimate that if your car could be powered by a Carnot engine, it would have an efficiency of about 55% according to $\varepsilon_C = 1 - T_L/T_H$ (Eq. 21-11). Its actual efficiency (with an internal combustion engine) is probably about 25%.

FIGURE 21-11 ■ The North Anna nuclear power plant near Charlottesville, Virginia, which generates electrical energy at the rate of 900 MW. At the same time, by design, it discards energy into the nearby river at the rate of 2100 MW. This plant—and all others like it—throws away more energy than it delivers in useful form. It is a real counterpart to the ideal engine of Fig. 21-7.

READING EXERCISE 21-4: Three Carnot engines operate between reservoir temperatures of (a) 400 and 500 K, (b) 600 and 800 K, and (c) 400 and 600 K. Rank the engines according to their thermal efficiencies, greatest first. ■

TOUCHSTONE EXAMPLE 21-3: Carnot Engine

Imagine an ideal Carnot engine that operates between the temperatures $T_H = 850$ K and $T_L = 300$ K. The engine performs 1200 J of work each cycle, the duration of each cycle being 0.25 s.

(a) What is the efficiency of this engine?

SOLUTION ■ The **Key Idea** here is that the efficiency ε of an ideal Carnot engine depends only on the ratio T_L/T_H of the temperatures (in kelvins) of the thermal reservoirs to which it is connected. Thus, from Eq. 21-11, we have

$$\varepsilon = 1 - \frac{T_L}{T_H} = 1 - \frac{300\text{ K}}{850\text{ K}} = 0.647 \approx 65\% \qquad \text{(Answer)}$$

(b) What is the average power of this engine?

SOLUTION ■ Here the **Key Idea** is that the average power P of an engine is the ratio of the work W it does per cycle to the time Δt that each cycle takes. For this Carnot engine, we find

$$P = \frac{W}{\Delta t} = \frac{1200\text{ J}}{0.25\text{ s}} = 4800\text{ W} = 4.8\text{ kW}. \qquad \text{(Answer)}$$

(c) How much thermal energy Q_H is extracted from the high-temperature reservoir every cycle?

SOLUTION ■ Now the **Key Idea** is that, for any engine including a Carnot engine, the efficiency ε is the ratio of the work W that is done per cycle to the thermal energy Q_H that is extracted from the high-temperature reservoir per cycle. This relation, $\varepsilon = |W|/|Q_H|$ (Eq. 21-9), gives us

$$Q_H = \frac{W}{\varepsilon} = \frac{1200\text{ J}}{0.647} = 1855\text{ J}. \qquad \text{(Answer)}$$

(d) How much thermal energy Q_L is delivered to the low-temperature reservoir every cycle?

SOLUTION ■ The **Key Idea** here is that for a Carnot engine, the work W done per cycle is equal to the difference in energy transfers $|Q_H| - |Q_L|$. (See Eq. 21-6.) Thus, we have

$$|Q_L| = |Q_H| - W = 1855\text{ J} - 1200\text{ J} = 655\text{ J}. \qquad \text{(Answer)}$$

(e) What entropy change is associated with the energy transfer to the working substance from the high-temperature reservoir? From the working substance to the low-temperature reservoir?

SOLUTION ■ The **Key Idea** here is that the entropy change ΔS during a transfer of thermal energy Q at constant temperature T is given by Eq. 21-2 ($\Delta S = Q/T$). Thus, for the transfer of energy Q_H from the high-temperature reservoir at T_H, we have

$$\Delta S_H = \frac{Q_H}{T_H} = \frac{1855\text{ J}}{850\text{ K}} = +2.18\text{ J/K}.$$

For the transfer of energy Q_L to the low-temperature reservoir at T_L, we have

$$\Delta S_L = \frac{Q_L}{T_L} = \frac{-655\text{ J}}{300\text{ K}} = -2.18\text{ J/K}.$$

Note that the algebraic signs of the two thermal energy transfers are different. Note also that, as Eq. 21-8 requires, the net entropy change of the working substance for one cycle (which is the algebraic sum of the two quantities calculated above) is zero.

TOUCHSTONE EXAMPLE 21-4: Better Than the Ideal?

An inventor claims to have constructed a heat engine that has an efficiency of 75% when operated between the boiling and freezing points of water. Is this possible?

SOLUTION ■ The **Key Idea** here is that the efficiency of a real engine (with its irreversible processes and wasteful energy transfers) must be less than the efficiency of an ideal Carnot engine operating between the same two temperatures. From Eq. 21-11, we find that the efficiency of an ideal Carnot engine operating between the boiling and freezing points of water is

$$\varepsilon = 1 - \frac{T_L}{T_H} = 1 - \frac{(0 + 273)\text{ K}}{(100 + 273)\text{ K}}$$
$$= 0.268 \approx 27\% \qquad \text{(Answer)}$$

Thus, the claimed efficiency of 75% for a real heat engine operating between the given temperatures is impossible.

21-5 Entropy in the Real World: Refrigerators

A heat engine operated in a reverse cycle would require an input of work and transfer thermal energy from a low-temperature reservoir to a high-temperature reservoir as it continuously repeats a set series of thermodynamic processes. We call such a device a **refrigerator.** In a household refrigerator, for example, an electrical compressor does work in order to transfer thermal energy from the food storage compartment (a low-temperature reservoir) to the room (a high-temperature reservoir). Air conditioners and heat pumps are also refrigerators. The differences are only in the nature of the high- and low-temperature reservoirs. For an air conditioner, the low-temperature reservoir is the room that is to be cooled, and the high-temperature reservoir is the (presumably warmer) outdoors. A heat pump is an air conditioner that can also be operated in such a way as to transfer thermal energy to the air in a room from the (presumably cooler) outdoors.

Let us now consider an *ideal refrigerator:*

> In an ideal refrigerator, all processes are reversible and no wasteful energy transfers occur between the refrigerator and its surroundings due to friction, turbulence, or other processes.

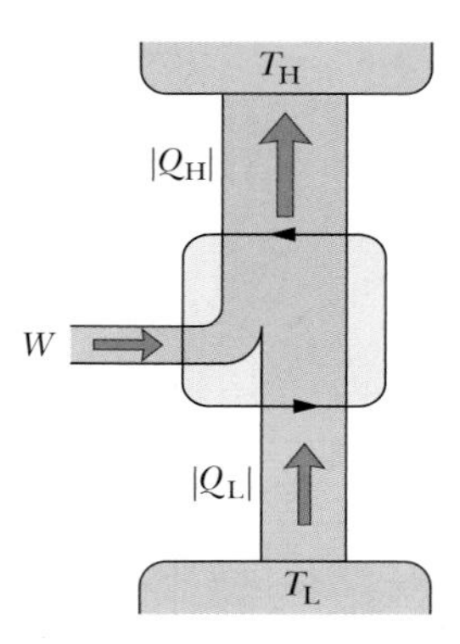

FIGURE 21-12 ■ The elements of a refrigerator. The two black arrowheads on the central loop suggest the working substance operating in a cycle, as if on a P-V plot. Thermal energy Q_L is transferred to the working substance from the low-temperature reservoir. Thermal energy Q_H is transferred to the high-temperature reservoir from the working substance. Work W is done on the refrigerator (on the working substance) by something in the environment.

Figure 21-12 shows the basic elements of an ideal refrigerator that operates based on a Carnot cycle. That is, it is the Carnot engine of Fig. 21-8 operating in reverse. All the energy transfers, either thermal energy or work, are reversed from those of a Carnot engine. Thus, we call such an ideal refrigerator an ideal **Carnot refrigerator.**

The designer of a refrigerator would like to do amount of work W (that we pay for) and cause as large a thermal energy transfer $|Q_L|$ as possible from the low-temperature reservoir (for example, the storage space in a kitchen refrigerator or the room to be cooled by the air conditioner). A measure of the efficiency of a refrigerator, then, is

$$K = \frac{\text{what we want}}{\text{what we pay for}} = \frac{|Q_L|}{|W|} \quad \text{(coefficient of performance, any refrigerator)}, \tag{21-12}$$

where K is called the *coefficient of performance*. For a Carnot refrigerator, the first law of thermodynamics gives $|W| = |Q_H| - |Q_L|$, where $|Q_H|$ is the amount of the thermal energy transfer to the high-temperature reservoir. The coefficient of performance for our ideal Carnot refrigerator then becomes

$$K_C = \frac{|Q_L|}{|Q_H| - |Q_L|}. \tag{21-13}$$

Because an ideal Carnot refrigerator is an ideal Carnot engine operating in reverse, we can again use $|Q_H|/T_H = |Q_L|/T_L$ (Eq. 21-8) and rewrite this expression as

$$K_C = \frac{T_L}{T_H - T_L} \quad \text{(coefficient of performance, Carnot refrigerator)}. \tag{21-14}$$

For typical room air conditioners, $K \approx 2.5$. For household refrigerators, $K \approx 5$. Unfortunately, but logically, the efficiency (and so the value of K) of a given refrigerator is higher the closer the temperatures of the two reservoirs are to each other. For example, a given Carnot air conditioner is more efficient on a warm day than when it is very hot outside.

It would be nice to own a refrigerator that did not require an input of work—that is, one that would run without being plugged in. Figure 21-13 represents an "inventor's dream," of a *perfect refrigerator* that transfers thermal energy Q from a cold reservoir to a warm reservoir without the need for work. Because the unit returns to the same state at the end of each cycle, and entropy is a state property, we know that the change in entropy of the working substance for this imagined refrigerator would be zero for a complete cycle. The entropies of the two reservoirs, however, would change. The entropy change for the low temperature reservoir would be $-|Q|/T_L$, and that for the high temperature reservoir would be $+|Q|/T_H$. Thus, the net entropy change for the entire system is

$$\Delta S = -\frac{|Q|}{T_L} + \frac{|Q|}{T_H}.$$

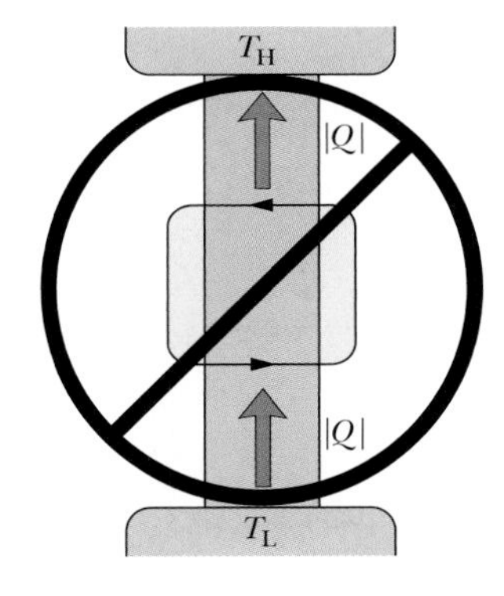

FIGURE 21-13 ■ The elements of a perfect (but *impossible*) refrigerator—that is, one that transfers energy from a low-temperature reservoir to a high-temperature reservoir without any input of work.

Because $T_H > T_L$, the right side of this equation would be negative and thus the net change in entropy per cycle for the closed system *refrigerator + reservoirs* would also be negative. Because such a decrease in entropy violates the second law of thermodynamics $\Delta S \geq 0$ (Eq. 21-5), it must be that a perfect refrigerator cannot exist. That is, if you want your refrigerator to operate, you must plug it in!

This result leads us to another (equivalent) formulation of the second law of thermodynamics:

> It is impossible to design a refrigerator that can cause a thermal energy transfer from a reservoir at a lower temperature to one at a higher temperature without the input of work (that is, with 100% efficiency).

In short, *there are no perfect refrigerators.*

READING EXERCISE 21-5: You wish to increase the coefficient of performance of an ideal Carnot refrigerator. You can do so by (a) running the cold chamber at a slightly higher temperature, (b) running the cold chamber at a slightly lower temperature, (c) moving the unit to a slightly warmer room, or (d) moving it to a slightly cooler room. Assume that the proposed changes in the magnitude of either T_L of T_H are the same in all four cases. List the changes according to the resulting coefficients of performance, greatest first. ■

21-6 Efficiency Limits of Real Engines

As we have just seen, a "perfect" Carnot refrigerator would violate the second law of thermodynamics which states that entropy must always either remain constant or increase. Therefore, we accept that a search for a 100% efficient Carnot refrigerator is futile. They do not exist. But what about Carnot engines? Can we have a "perfect" (that is, 100% efficient) engine?

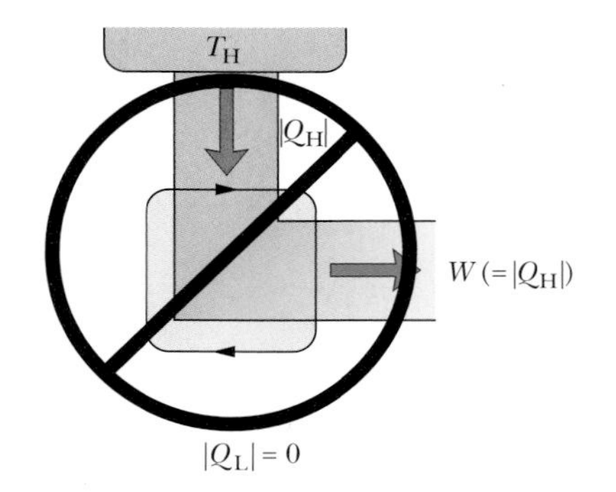

FIGURE 21-14 ■ The elements of a perfect (and *impossible*) engine—that is, one that converts thermal energy transfer Q_H from a high-temperature reservoir directly to work W with 100% efficiency.

Fundamentally, the inefficiency in an ideal Carnot engine is associated with the thermal energy transfer at the low temperature reservoir interface. Naive inventors continually try to improve Carnot engine efficiency by reducing the waste energy $|Q_L|$ transferred to the low-temperature reservoir and, hence, "thrown away" during each cycle. The inventor's dream is to produce the *perfect engine,* diagrammed in Fig. 21-14, in which $|Q_L|$ is reduced to zero and $|Q_H|$ is converted completely into work. For example, if we could do it, a perfect engine on an ocean liner could use thermal energy transferred to it from seawater to drive the propellers, with no fuel cost. An automobile, fitted with such a perfect engine, could use energy transferred from the surrounding air to turn its wheels, again with no fuel cost.

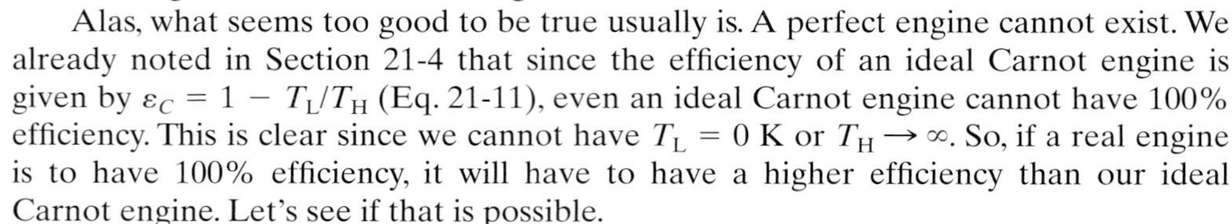

Alas, what seems too good to be true usually is. A perfect engine cannot exist. We already noted in Section 21-4 that since the efficiency of an ideal Carnot engine is given by $\varepsilon_C = 1 - T_L/T_H$ (Eq. 21-11), even an ideal Carnot engine cannot have 100% efficiency. This is clear since we cannot have $T_L = 0$ K or $T_H \rightarrow \infty$. So, if a real engine is to have 100% efficiency, it will have to have a higher efficiency than our ideal Carnot engine. Let's see if that is possible.

Let us assume for a moment that an inventor, working in her garage, has constructed an engine X, which she claims has an efficiency ε_X that is greater than ε_C, where ε_C is the efficiency of an ideal Carnot engine operating between two temperatures. Then,

$$\varepsilon_X > \varepsilon_C \qquad \text{(a claim).} \tag{21-15}$$

Let us connect the inventor's engine X to a Carnot refrigerator, as in Fig. 21-15*a*. We adjust the strokes of the refrigerator so that the work it requires per cycle is just equal to that provided by the engine X. Thus, no (external) work is needed for the operation

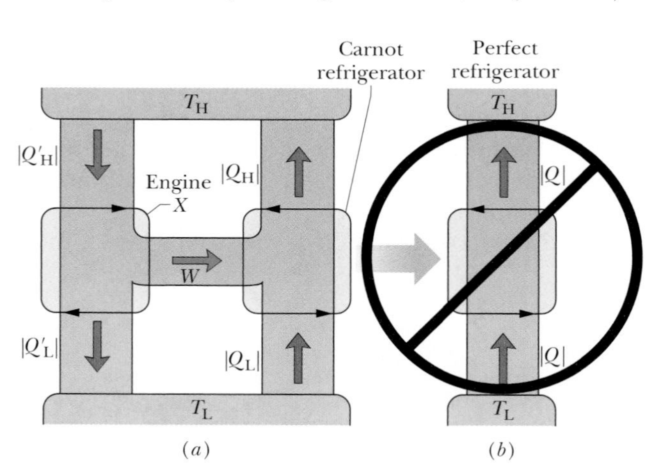

FIGURE 21-15 ■ (*a*) Engine X drives a Carnot refrigerator. (*b*) If, as claimed, engine X is more efficient than a Carnot engine, then the combination shown in (*a*) is equivalent to the perfect refrigerator shown here. This violates the second law of thermodynamics, so we conclude that engine X cannot be more efficient than a Carnot engine.

of the combination *engine* + *refrigerator* of Fig. 21-15*a*, which we take as our system. Since the efficiency of an engine is ε = energy we get/energy we pay for = $|W|/|Q_H|$ (Eq. 21-9), and $\varepsilon_X > \varepsilon_C$, we must have

$$\frac{|W|}{|Q'_H|} > \frac{|W|}{|Q_H|},$$

where Q'_H is the heat transfer at the high-temperature reservoir in engine X. The right side of the inequality is the efficiency of the Carnot refrigerator shown in Fig. 21-15*a* when it operates (in reverse) as an engine. Since $|W|$ is the same in both cases, this inequality requires that

$$|Q_H| > |Q'_H|. \qquad (21\text{-}16)$$

Because the work done by engine X is equal to the work done on the Carnot refrigerator, which is (from Eq. 21-6) $W = |Q_H| - |Q_L|$, we have

$$|Q_H| - |Q_L| = |Q'_H| - |Q'_L|.$$

We can rearrange terms and write this as

$$|Q_H| - |Q'_H| = |Q_L| - |Q'_L| = Q. \qquad (21\text{-}17)$$

Here Q'_L is the thermal energy transfer at the low-temperature reservoir in engine X. Because we found that $|Q_H| > |Q'_H|$ (Eq. 21-16), the quantity Q must be positive. What does Q represent? Careful evaluation of this expression (Eq. 21-17) tells us that Q is the net thermal energy transfer at the high-temperature reservoir as a result of our combined engine plus refrigerator. That is, the net effect of engine X and the Carnot refrigerator, working in combination, is to transfer thermal energy Q from a low-temperature reservoir to a high-temperature reservoir. Notably, this is done with no work input to the combined engine–refrigerator system. Thus, the combination acts like the perfect refrigerator of Fig. 21-13, whose existence is a violation of the second law of thermodynamics.

Thus, we conclude that engine X cannot be more efficient than the ideal Carnot engine. In general, *no real engine can have an efficiency greater than that of a Carnot engine when both engines work between the same two temperatures.* At most, it can have an efficiency equal to that of an ideal Carnot engine. In that case, engine X is an ideal Carnot engine. Since ideal Carnot engines cannot be 100% efficient, this means that *"perfect" (100% efficient) engines are physically impossible.*

21-7 A Statistical View of Entropy

In Chapter 20 we saw that the macroscopic properties of gases can be explained in terms of their microscopic, or molecular, behavior. For one example, recall that we were able to account for the pressure exerted by a gas on the walls of its container in terms of the momentum transferred to those walls by rebounding gas molecules. Such explanations are part of a study called **statistical mechanics.**

Here we shall focus our attention on a single problem, involving the distribution of gas molecules between the two halves of an insulated box. This problem is reasonably simple to analyze, and it allows us to use statistical mechanics to calculate the entropy change for the free expansion of an ideal gas. You will see in Touchstone Example 21-6 that statistical mechanics leads to the same entropy change we obtain in Example 21-2 using thermodynamics.

Figure 21-16 shows a box that contains six identical (and thus indistinguishable) molecules of a gas. At any instant, a given molecule will be in either the left or the

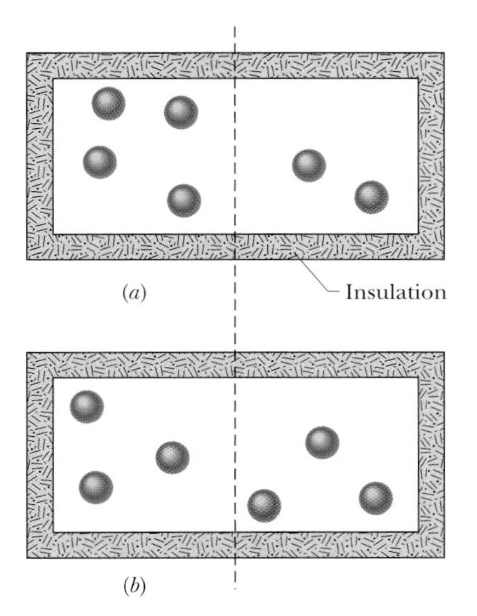

FIGURE 21-16 ■ An insulated box contains six gas molecules. Each molecule has the same probability of being in the left half of the box as in the right half. The arrangement in (*a*) corresponds to configuration III in Table 21-1, and that in (*b*) corresponds to configuration IV.

right half of the box; because the two halves have equal volumes, the molecule has the same likelihood, or probability, of being in either half.

Table 21-1 shows four of the seven possible *configurations* of the six molecules, each configuration labeled with Roman numerals. For example, in configuration I, all six molecules are in the left half of the box ($n_1 = 6$), and none are in the right half ($n_2 = 0$). The three configurations not shown are V with a (2, 4) split, VI with a (1, 5) split, and VII with a (0, 6) split. In configuration II, five molecules are in one half of the box, leaving one molecule in the other half. We see that, in general, a given configuration can be achieved in a number of different ways. We call these different arrangements of the molecules *microstates*. Let us see how to calculate the number of microstates that correspond to a given configuration.

TABLE 21-1
Six Molecules in a Box

Configuration[a] Label	n_1	n_2	Multiplicity W (number of microstates)	Calculation of W (Eq. 21-18)	Entropy 10^{-23} J/K (Eq. 21-19)
I	6	0	1	$6!/(6!\,0!) = 1$	0
II	5	1	6	$6!/(5!\,1!) = 6$	2.47
III	4	2	15	$6!/(4!\,2!) = 15$	3.74
IV	3	3	20	$6!/(3!\,3!) = 20$	4.13
Total number of microstates = 64					

[a]The configurations not listed are $n_1 = 0, n_2 = 6$; $n_1 = 1, n_2 = 5$; and $n_1 = 2, n_2 = 4$. These have the same multiplicities as I, II and III respectively.

Suppose we have N molecules, distributed with n_1 molecules in one half of the box and n_2 in the other half. (Thus $n_1 + n_2 = N$.) Let us imagine that we distribute the molecules "by hand," one at a time. If $N = 6$, we can select the first molecule in six independent ways; that is, we can pick any one of the six molecules. We can pick the second molecule in five ways, by picking any one of the remaining five molecules, and so on. The total number of ways in which we can select all six molecules is the product of these independent ways, or $6 \times 5 \times 4 \times 3 \times 2 \times 1 = 720$. In mathematical shorthand we write this product as $6! = 720$, where 6! is pronounced "six factorial." Your hand-held calculator can probably calculate factorials. For later use you will need to know that $0! = 1$. (Check this on your calculator.)

However, because the molecules are indistinguishable, these 720 arrangements are not all different. In the case that $n_1 = 4$ and $n_2 = 2$ (which is configuration III in Table 21-1), for example, the order in which you put four molecules in one half of the box does not matter, because after you have put all four in, there is no way that you can tell the order in which you did so. The number of ways in which you can order the four molecules is 4! or 24. Similarly, the number of ways in which you can order two molecules for the other half of the box is simply 2! or 2. To get the number of different arrangements that lead to the 4, 2 split of configuration III, we must divide 720 by 24 and also by 2. We call the resulting quantity, which is the number of microstates that correspond to a given configuration, the *multiplicity* W of that configuration. Thus, for configuration III,

$$W_{\text{III}} = \frac{6!}{4!\,2!} = \frac{720}{24 \times 2} = 15.$$

Thus, Table 21-1 tells us there are 15 independent microstates that correspond to configuration III. Note that, as the table also tells us, the total number of microstates for six molecules distributed over four configurations is 42.

Extrapolating from six molecules to the general case of N molecules, we have

$$W = \frac{N!}{n_1!\, n_2!} \quad \text{(multiplicity of configuration).} \tag{21-18}$$

You should verify that Eq. 21-18 gives the multiplicities for all the configurations listed in Table 21-1.

The basic assumption of statistical mechanics is:

> All microstates are equally probable.

In other words, if we were to take a great many snapshots of the six molecules as they jostle around in the box of Fig. 21-16 and then count the number of times each microstate occurred, we would find that all 42 microstates will occur equally often. In other words, the system will spend, on average, the same amount of time in each of the 42 microstates listed in Table 21-1.

Because the microstates are equally probable, but different configurations have different numbers of microstates, the configurations are *not* equally probable. In Table 21-1 configuration IV, with 20 microstates, is the *most probable configuration*, with a probability of 20/64 = 0.313. This means that the system is in configuration IV 31.3% of the time. Configurations I and VII, in which all the molecules are in one half of the box, are the least probable, each with a probability of 1/64 = 0.016 or 1.6%. It is not surprising that the most probable configuration is the one in which the molecules are evenly divided between the two halves of the box, because that is what we expect at thermal equilibrium. However, it *is* surprising that there is any probability, however small, of finding all six molecules clustered in half of the box, with the other half empty. In Touchstone Example 21-5 we show that this state can occur because six molecules is an extremely small number.

For large values of N there are extremely large numbers of microstates, but nearly all the microstates belong to the configuration in which the molecules are divided equally between the two halves of the box, as Fig. 21-17 indicates. Even though the measured temperature and pressure of the gas remain constant, the gas is churning away endlessly as its molecules "visit" all probable microstates with equal probability. However, because so few microstates lie outside the very narrow central configuration peak of Fig. 21-17, we might as well assume that the gas molecules are always divided equally between the two halves of the box. As we shall see, this is the configuration with the greatest entropy.

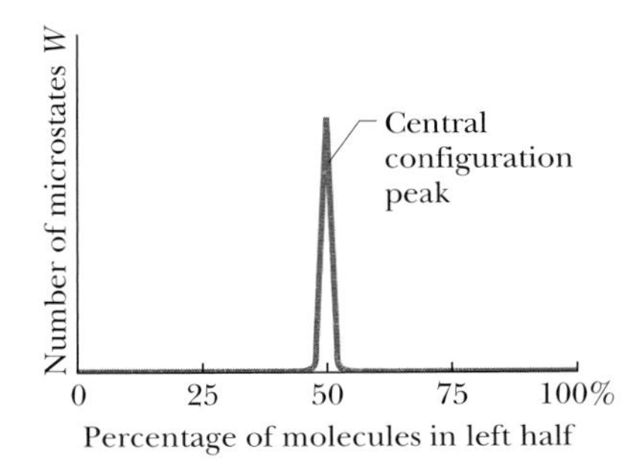

FIGURE 21-17 ■ For a large number of molecules in a box, a plot of the number of microstates that require various percentages of the molecules to be in the left half of the box. Nearly all the microstates correspond to an approximately equal sharing of the molecules between the two halves of the box; those microstates form the central configuration peak on the plot. For $N \approx 10^{22}$ molecules, the central configuration peak is much too narrow to be drawn on this plot.

Probability and Entropy

In 1877, Austrian physicist Ludwig Boltzmann (the Boltzmann of Boltzmann's constant k_B) derived a relationship between the entropy S of a configuration of a gas and the multiplicity W of that configuration. That relationship is

$$S = k_B \ln W \quad \text{(Boltzmann's entropy equation).} \tag{21-19}$$

This famous formula is engraved on Boltzmann's tombstone.

It is natural that S and W should be related by a logarithmic function. The total entropy of two systems is the *sum* of their separate entropies. The probability of occurrence of two independent systems is the *product* of their separate probabilities. Because $\ln ab = \ln a + \ln b$, the logarithm seems the logical way to connect these quantities.

Table 21-1 displays the entropies of the configurations of the six-molecule system of Fig. 21-16, computed using Eq. 21-19. Configuration IV, which has the greatest multiplicity, also has the greatest entropy.

When you use Eq. 21-18 to calculate W, your calculator may signal "OVERFLOW" if you try to find the factorial of a number greater than a few hundred. Fortunately, there is a very good approximation, known as **Stirling's approximation,** not for $N!$ but for $\ln N!$, which as it happens is exactly what is needed in Eq. 21-19. Stirling's approximation is

$$\ln N! \approx N(\ln N) - N \quad \text{(Stirling's approximation).} \tag{21-20}$$

The Stirling of this approximation is not the Stirling of the Stirling engine.

READING EXERCISE 21-6: A box contains one mole of a gas. Consider two configurations: (a) each half of the box contains one-half of the molecules, and (b) each third of the box contains one-third of the molecules. Which configuration has more microstates? ■

TOUCHSTONE EXAMPLE 21-5: Indistinguishable

Suppose that there are 100 indistinguishable molecules in the box of Fig. 21-16. How many microstates are associated with the configuration $n_1 = 50$ and $n_2 = 50$? How many are associated with the configuration $n_1 = 100$ and $n_2 = 0$? Interpret the results in terms of the relative probabilities of the two configurations.

SOLUTION ■ The **Key Idea** here is that the multiplicity W of a configuration of indistinguishable molecules in a closed box is the number of independent microstates with that configuration, as given by Eq. 21-18. For the (n_1, n_2) configuration (50, 50), that equation yields

$$\begin{aligned} W &= \frac{N!}{n_1!\,n_2!} = \frac{100!}{50!\,50!} \\ &= \frac{9.33 \times 10^{157}}{(3.04 \times 10^{64})(3.04 \times 10^{64})} \\ &= 1.01 \times 10^{29}. \end{aligned} \quad \text{(Answer)}$$

Similarly, for the configuration of (100, 0), we have

$$W = \frac{N!}{n_1!\,n_2!} = \frac{100!}{100!\,0!} = \frac{1}{0!} = \frac{1}{1} = 1 \quad \text{(Answer)}$$

Thus, a 50-50 distribution is more likely than a 100-0 distribution by the enormous factor of about 1×10^{29}. If you could count, at one per nanosecond, the number of microstates that correspond to the 50-50 distribution, it would take you about 3×10^{12} years, which is about 750 times longer than the age of the universe. Even 100 molecules is *still* a very small number. Imagine what these calculated probabilities would be like for a mole of molecules—say, about $N = 10^{24}$. You need never worry about suddenly finding all the air molecules clustering in one corner of your room!

TOUCHSTONE EXAMPLE 21-6: Entropy Increase

In Touchstone Example 21-1 we showed that when n moles of an ideal gas doubles its volume in a free expansion, the entropy increase from the initial state i to the final state f is $S_f - S_i = nR \ln 2$. Derive this result with statistical mechanics.

SOLUTION ■ One **Key Idea** here is that we can relate the entropy S of any given configuration of the molecules in the gas to the multiplicity W of microstates for that configuration, using Eq. 21-19 ($S = k_B \ln W$). We are interested in two configurations: the final configuration f (with the molecules occupying the full volume of their container in Fig. 21-1*b*) and the initial configuration i (with the molecules occupying the left half of the container).

A second **Key Idea** is that, because the molecules are in a closed container, we can calculate the multiplicity W of their microstates with Eq. 21-18. Here we have N molecules in the n moles of the gas. Initially, with the molecules all in the left half of the container, their (n_1, n_2) configuration is $(N, 0)$. Then, Eq. 21-18 gives their multiplicity as

$$W_i = \frac{N!}{N!\,0!} = 1.$$

Finally, with the molecules spread through the full volume, their (n_1, n_2) configuration is $(N/2, N/2)$. Then, Eq. 21-18 gives their multiplicity as

$$W_f = \frac{N!}{(N/2)!(N/2)!}.$$

From Eq. 21-19, the initial and final entropies are

$$S_i = k \ln W_i = k \ln 1 = 0$$

and

$$S_f = k_B \ln W_f = k \ln(N!) - 2k_B \ln[(N/2)!]. \tag{21-21}$$

In writing Eq. 21-21, we have used the relation

$$\ln \frac{a}{b^2} = \ln a - 2 \ln b.$$

Now, applying Eq. 21-20 to evaluate Eq. 21-21, we find that

$$\begin{aligned} S_f &= k_B \ln (N!) - 2k_B \ln [(N/2)!] \\ &= k_B[N(\ln N) - N] - 2k_B[(N/2) \ln (N/2) - (N/2)] \\ &= k_B[N(\ln N) - N - N \ln (N/2) + N] \qquad (21\text{-}22) \\ &= k_B[N(\ln N) - N(\ln N - \ln 2)] = Nk_B \ln 2. \end{aligned}$$

From Eq. 20-7 we can substitute nR for Nk_B, where R is the universal gas constant. Equation 21-22 then becomes

$$S_f = nR \ln 2.$$

The change in entropy from the initial state to the final is thus

$$\begin{aligned} S_f - S_i &= nR \ln 2 - 0 \\ &= nR \ln 2, \qquad \text{(Answer)} \end{aligned}$$

which is what we set out to show. In Touchstone Example 21-1 we calculated this entropy increase for a free expansion with thermodynamics by finding an equivalent reversible process and calculating the entropy change for *that* process in terms of temperature and heat transfer. Here we have calculated the same increase with statistical mechanics using the fact that the system consists of molecules.

Problems

SEC. 21-2 ■ CHANGE IN ENTROPY

1. Expands Reversibly A 2.50 mol sample of an ideal gas expands reversibly and isothermally at 360 K until its volume is doubled. What is the increase in entropy of the gas?

2. Reversible Isothermal How much thermal energy must be transferred for a reversible isothermal expansion of an ideal gas at 132 °C if the entropy of the gas increases by 46.0 J/K?

3. Four Moles Four moles of an ideal gas undergo a reversible isothermal expansion from volume V_1 to volume $V_2 = 2V_1$ at temperature $T = 400$ K. Find (a) the work done by the gas and (b) the entropy change of the gas. (c) If the expansion is reversible and adiabatic instead of isothermal, what is the entropy change of the gas?

4. Reversible Isothermal Expansion An ideal gas undergoes a reversible isothermal expansion at 77.0°C, increasing its volume from 1.30 L to 3.40 L. The entropy change of the gas is 22.0 J/K. How many moles of gas are present?

5. Energy Absorbed Find (a) the thermal energy transfer and (b) the change in entropy of a 2.00 kg block of copper whose temperature is increased reversibly from 25°C to 100°C. The specific heat of copper is 386 J/kg · K.

6. Initial Temperature An ideal monatomic gas at initial temperature T_0 (in kelvins) expands from initial volume V_0 to volume $2V_0$ by each of the five processes indicated in the T-V diagram of Fig. 21-18. In which process is the expansion (a) isothermal, (b) isobaric (constant pressure), and (c) adiabatic? Explain your answers. (d) In which processes does the entropy of the gas decrease?

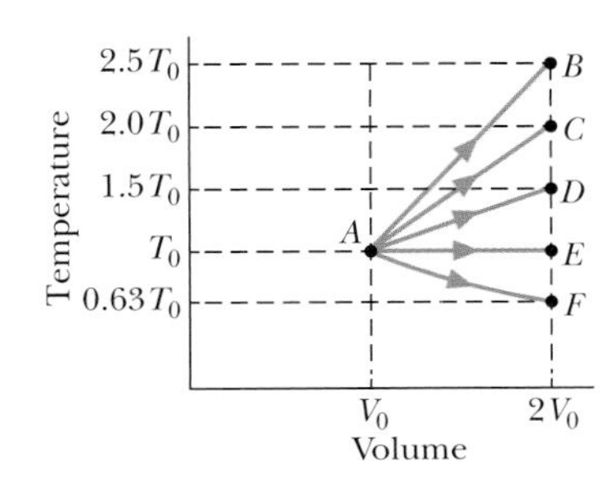

FIGURE 21-18 ■ Problem 6.

7. Entropy Change (a) What is the entropy change of a 12.0 g ice cube that melts completely in a bucket of water whose temperature is just above the freezing point of water? (b) What is the entropy change of a 5.00 g spoonful of water that evaporates completely on a hot plate whose temperature is slightly above the boiling point of water?

8. Ideal Gas Undergoes A 2.0 mol sample of an ideal monatomic gas undergoes the reversible process shown in Fig. 21-19. (a) How much thermal energy is transferred to the gas? (b) What is the change in the internal energy of the gas? (c) How much work is done by the gas?

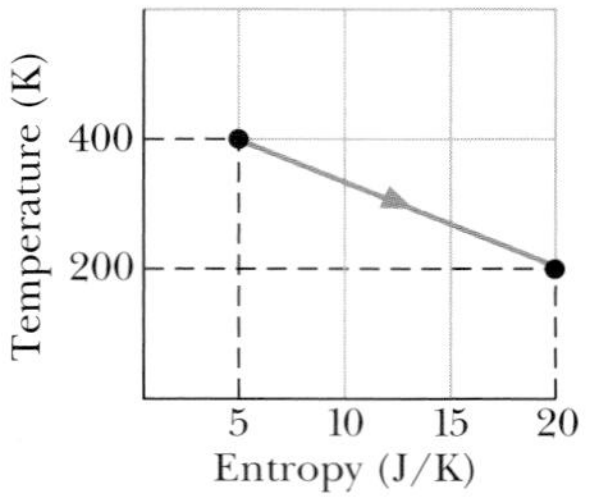

FIGURE 21-19 ■ Problem 8.

9. Aluminum and Water In an experiment, 200 g of aluminum (with a specific heat of 900 J/kg·K) at 100°C is mixed with 50.0 g of water at 20.0°C, with the mixture thermally isolated. (a) What is the equilibrium temperature? What are the entropy changes of (b) the aluminum, (c) the water and (d) the aluminum–water system?

10. Irreversible Process In the irreversible process of Fig. 21-5, let the initial temperatures of identical blocks L and R be 305.5 K and 294.5 K, respectively. Let 215 J be the thermal energy transfer between the blocks required to reach equilibrium. Then for the reversible processes of Fig. 21-6, what are the entropy changes of (a) block L, (b) its reservoir, (c) block R, (d) its reservoir, (e) the two-block system, and (f) the system of the two blocks and the two reservoirs?

11. Reversible Apparatus Use the reversible apparatus of Fig. 21-6 to show that, if the process of Fig. 21-5 happened in reverse, the entropy of the system would decrease, a violation of the second law of thermodynamics.

12. Rotating Not Oscillating An ideal diatomic gas, whose molecules are rotating but not oscillating, is taken through the cycle in Fig. 21-20. Determine for all three processes, in terms of P_1, V_1, T_1, and R: (a) P_2, P_3, and T_3 and (b) W, Q, ΔE^{int}, and ΔS per mole?

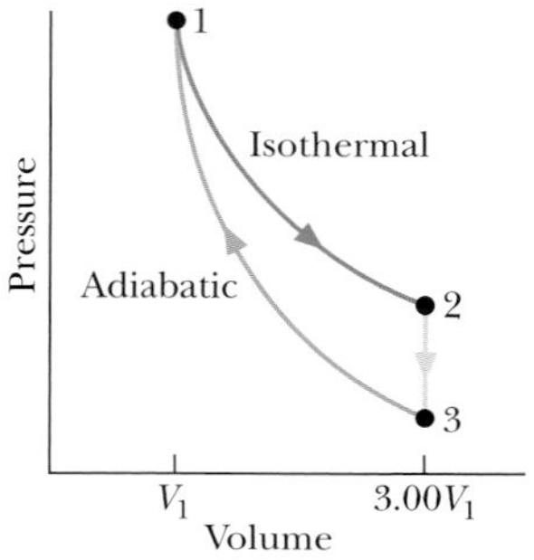

FIGURE 21-20 ■ Problem 12.

13. Copper in Box A 50.0 g block of copper whose temperature is 400 K is placed in an insulating box with a 100 g block of lead whose temperature is 200 K. (a) What is the equilibrium temperature of the two-block system? (b) What is the change in the internal energy of the two-block system between the initial state and the equilibrium state? (c) What is the change in the entropy of the two-block system? (See Table 19-2.)

14. Initial to Final One mole of a monatomic ideal gas is taken from an initial pressure P and volume V to a final pressure $2P$ and volume $2V$ by two different processes: (I) It expands isothermally until its volume is doubled, and then its pressure is increased at constant volume to the final pressure. (II) It is compressed isothermally until its pressure is doubled, and then its volume is increased at constant pressure to the final volume. (a) Show the path of each process on a P-V diagram. For each process calculate, in terms of P and V, (b) the thermal energy absorbed by the gas in each part of the process, (c) the work done by the gas in each part of the process, (d) the change in internal energy of the gas, ΔE^{int}, and (e) the change in entropy of the gas, ΔS.

15. Ice Cube in a Lake A 10 g ice cube at −10°C is placed in a lake whose temperature is 15°C. Calculate the change in entropy of the cube–lake system as the ice cube comes to thermal equilibrium with the lake. The specific heat of ice is 2220 J/kg·K. (*Hint:* Will the ice cube affect the temperature of the lake?)

16. Ice Cube in Thermos An 8.0 g ice cube at −10°C is put into a Thermos flask containing 100 cm³ of water at 20°C. By how much has the entropy of the cube–water system changed when a final equilibrium state is reached? The specific heat of ice is 2220 J/kg·K.

17. Water and Ice A mixture of 1773 g of water and 227 g of ice is in an initial equilibrium state at 0.00°C. The mixture is then, in a reversible process, brought to a second equilibrium state where the water–ice ratio, by mass, is 1:1 at 0.00°C. (a) Calculate the entropy change of the system during this process. (The heat of fusion for water is 333 kJ/kg.) (b) The system is then returned to the initial equilibrium state in an irreversible process (say, by using a Bunsen burner). Calculate the entropy change of the system during this process. (c) Are your answers consistent with the second law of thermodynamics?

18. Cylinder of Gas A cylinder contains n moles of a monatomic ideal gas. If the gas undergoes a reversible isothermal expansion from initial volume V_i to final volume V_f along path I in Fig. 21-21, its change in entropy is $\Delta S = nR\ \ln(V_f/V_i)$. (See Touchstone Example 21-1) Now consider path II in Fig. 21-21, which takes the gas from the same initial state i to state x by a reversible adiabatic expansion, and then from that state x to the same final state f by a reversible constant volume process. (a) Describe how you would carry out the two reversible processes for path II. (b) Show that the temperature of the gas in state x is

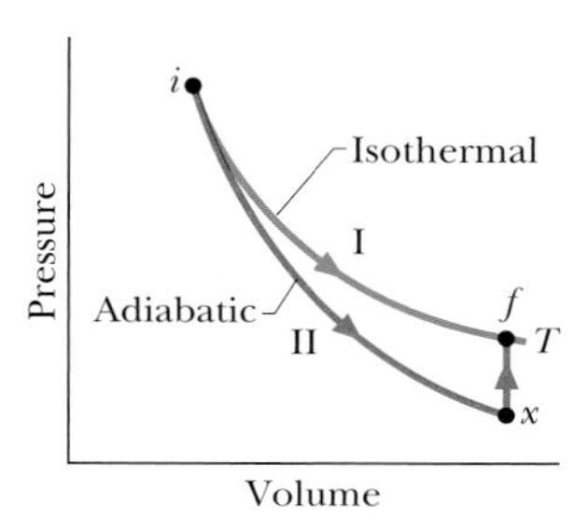

FIGURE 21-21 ■ Problem 18.

$$T_x = T_i(V_i/V_f)^{2/3}.$$

(c) What are the thermal energy transferred along path I (Q_{I}) and the thermal energy transferred along path II (Q_{II})? Are they equal? (d) What is the entropy change ΔS for path II? Is the entropy change for path I equal to it? (e) Evaluate T_x, Q_{I}, Q_{II}, and ΔS for $n = 1$, $T_i = 500$ K, and $V_f/V_i = 2$.

19. Through the Cycle One mole of an ideal monatomic gas is taken through the cycle in Fig. 21-22. (a) How much work is done by the gas in going from state a to state c along path abc? What are the changes in internal energy and entropy in going (b) from b to c and (c) through one complete cycle? Express all answers in terms of the pressure P_0, volume V_0, and temperature T_0 of state a.

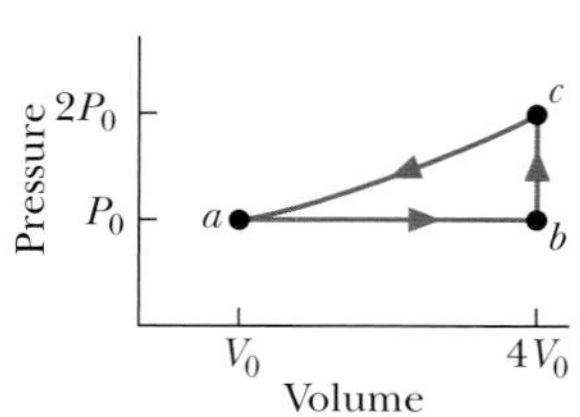

FIGURE 21-22 ■ Problem 19.

20. Initial Pressure and Temperature One mole of an ideal monatomic gas, at an initial pressure of 5.00 kPa and initial temperature of 600 K, expands from initial volume $V_i = 1.00\ \text{m}^3$ to final volume $V_f = 2.00\ \text{m}^3$. During the expansion, the pressure P and volume V of the gas are related by $P = (5.00\ \text{kPa})\exp[(V_i - V)/a]$, where $a = 1.00\ \text{m}^3$. What are (a) the final pressure and (b) the final temperature of the gas? (c) How much work is done by the gas during the expansion? (d) What is the change in entropy of the gas during the expansion? (*Hint*: Use two simple reversible processes to find the entropy change.)

SEC. 21-4 ■ ENTROPY IN THE REAL WORLD: ENGINES

Consider all Carnot engines discussed in these problems to be ideal.

21. Work and Efficiency in a Cycle A Carnot engine absorbs 52 kJ of thermal energy and exhausts 36 kJ as thermal energy in each cycle. Calculate (a) the engine's efficiency and (b) the work done per cycle in kilojoules.

22. Low-Temp Reservoir A Carnot engine whose low-temperature reservoir is at 17°C has an efficiency of 40%. By how much should the temperature of the high-temperature reservoir be increased to increase the efficiency to 50%?

23. Operates Between A Carnot engine operates between 235°C and 115°C, absorbing 6.30×10^4 J per cycle at the higher temperature. (a) What is the efficiency of the engine? (b) How much work per cycle is this engine capable of performing?

24. Reactor In a hypothetical nuclear fusion reactor, the fuel is deuterium gas at temperature of about 7×10^8 K. If this gas could be used to operate a Carnot engine with $T_L = 100°\text{C}$, what would be the engine's efficiency?

25. Carnot Engine A Carnot engine has an efficiency of 22.0%. It operates between constant-temperature reservoirs differing in temperature by 75.0°C. What are the temperatures of the two reservoirs?

26. Engine Has Power A Carnot engine has a power of 500 W. It operates between constant-temperature reservoirs at 100°C and 60.0°C. What are (a) the rate of thermal energy input and (b) the rate of exhaust heat output, in kilojoules per second?

27. Process *bc* One mole of a monatomic ideal gas is taken through the reversible cycle shown in Fig. 21-23. Process bc is an adiabatic expansion, with $P_b = 10.0$ atm and $V_b = 1.00 \times 10^{-3}\ \text{m}^3$. Find (a) the thermal energy transferred to the gas, (b) the thermal energy transferred from the gas, (c) the net work done by the gas, and (d) the efficiency of the cycle.

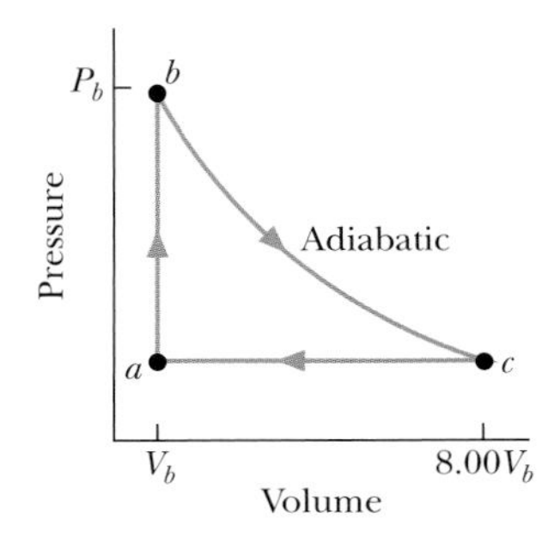

FIGURE 21-23 ■ Problem 27.

28. Enclosed Area Show that the area enclosed by the Carnot cycle on the temperature–entropy plot of Fig. 21-9 represents the net thermal energy transfer per cycle to the working substance.

29. Assume $P = 2P_0$ One mole of an ideal monatomic gas is taken through the cycle shown in Fig. 21-24. Assume that $P = 2P_0$, $V = 2V_0$, $P_0 = 1.01 \times 10^5$ Pa, and $V_0 = 0.0225\ \text{m}^3$. Calculate (a) the work done during the cycle, (b) the thermal energy added during stroke abc, and (c) the efficiency of the cycle. (d) What is the efficiency of a Carnot engine operating between the highest and lowest temperatures that occur in the cycle? How does this compare to the efficiency calculated in (c)?

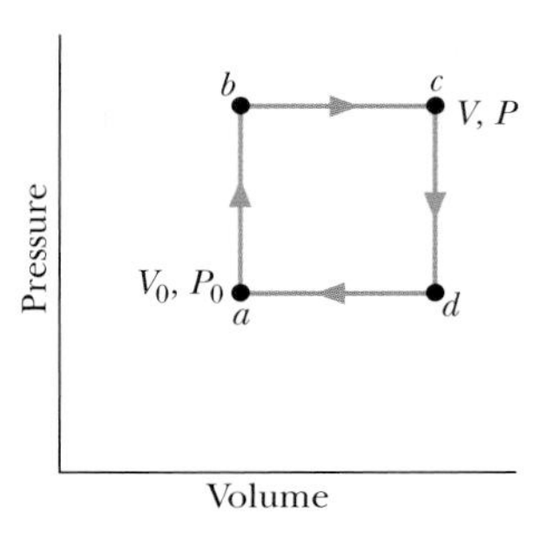

FIGURE 21-24 ■ Problem 29.

30. First Stage In the first stage of a two-stage Carnot engine, thermal energy Q_1 is absorbed at temperature T_1, work W_1 is done, and thermal energy Q_2 is expelled at a lower temperature T_2. The second stage absorbs that energy Q_2, does work W_2, and expels energy Q_3 at a still lower temperature T_3. Prove that the efficiency of the two-stage engine is $(T_1 - T_3)/T_1$.

31. Deep Shaft Suppose that a deep shaft were drilled in Earth's crust near one of the poles, where the surface temperature is −40°C, to a depth where the temperature is 800°C. (a) What is the theoretical limit to the efficiency of an engine operating between these temperatures? (b) If all the thermal energy released as heat into the low-temperature reservoir were used to melt ice that was initially at −40°C, at what rate could liquid water at 0°C be produced by a 100 MW power plant (treat it as an engine)? The specific heat of ice is 2220 J/kg·K; water's heat of fusion is 333 kJ/kg. (Note that the engine can operate only between 0°C and 800°C in this case. Energy exhausted at −40°C cannot be used to raise the temperature of anything above −40°C)

32. Working Substance One mole of an ideal gas is used as the working substance of an engine that operates on the cycle shown in Fig. 21-25. BC and DA are reversible adiabatic processes. (a) Is the gas monatomic, diatomic, or polyatomic? (b) What is the efficiency of the engine?

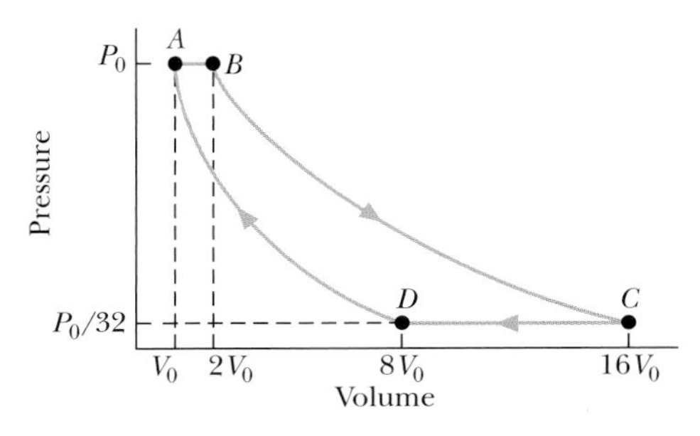

FIGURE 21-25 ■ Problem 32.

33. Gasoline Engine The operation of a gasoline internal combustion engine is represented by the cycle in Fig. 21-26. Assume the gasoline–air intake mixture is an ideal gas and use a compression ratio of 4:1 ($V_4 = 4V_1$). Assume that $P_2 = 3P_1$. (a) Determine the pressure and temperature at each of the vertex points of the P-V diagram in terms of P_1, T_1, and the ratio γ of the molar specific heats of the gas. (b) What is the efficiency of the cycle?

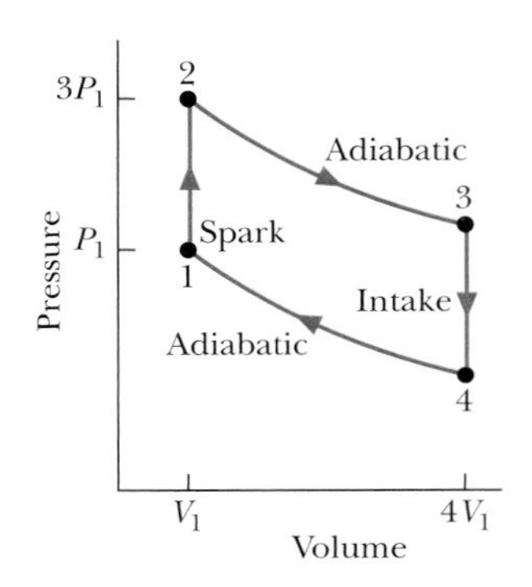

FIGURE 21-26 ■ Problem 33.

SEC. 21-5 ■ ENTROPY IN THE REAL WORLD: REFRIGERATORS

34. Carnot Refrigerator A Carnot refrigerator does 200 J of work to remove 600 J of thermal energy from its cold compartment. (a) What is the refrigerator's coefficient of performance? (b) How much thermal energy per cycle is exhausted to the kitchen?

35. Carnot Air Conditioner A Carnot air conditioner takes energy from the thermal energy of a room at 70°F and transfers it to the outdoors, which is at 96°F. For each joule of electric energy required to operate the air conditioner, how many joules are removed from the room?

36. Heat Pump Transfers The electric motor of a heat pump transfers thermal energy from the outdoors, which is at −5.0°C, to a room, which is at 17°C. If the heat pump were a Carnot heat pump (a Carnot engine working in reverse), how many joules would be transferred to the thermal energy of the room for each joule of electric energy consumed?

37. Heat Pump to Heat Building A heat pump is used to heat a building. The outside temperature is −5.0°C, and the temperature inside the building is to be maintained at 22°C. The pump's coefficient of performance is 3.8, and the heat pump delivers 7.54 MJ of thermal energy to the building each hour. If the heat pump is a Carnot engine working in reverse, at what rate must work be done to run the heat pump?

38. How Much Work How much work must be done by a Carnot refrigerator to transfer 1.0 J of thermal energy (a) from a reservoir at 7.0°C to one at 27°C, (b) from a reservoir at −73°C to one at 27°C, (c) from a reservoir at −173°C to one at 27°C, and (d) from a reservoir at −223°C to one at 27°C?

39. Air Conditioner An air conditioner operating between 93°F and 70°F is rated at 4000 Btu/h cooling capacity. Its coefficient of performance is 27% of that of a Carnot refrigerator operating between the same two temperatures. What horsepower is required of the air conditioner motor?

40. Motor in Refrigerator The motor in a refrigerator has a power of 200 W. If the freezing compartment is at 270 K and the outside air is at 300 K, and assuming the efficiency of a Carnot refrigerator, what is the maximum amount of thermal energy that can be extracted from the freezing compartment in 10.0 min?

41. Engine Driving Refrigerator A Carnot engine works between temperatures T_1 and T_2. It drives a Carnot refrigerator that works between temperatures T_3 and T_4 (Fig. 21-27). Find the ratio $|Q_3|/|Q_1|$ in terms of T_1, T_2, T_3, and T_4.

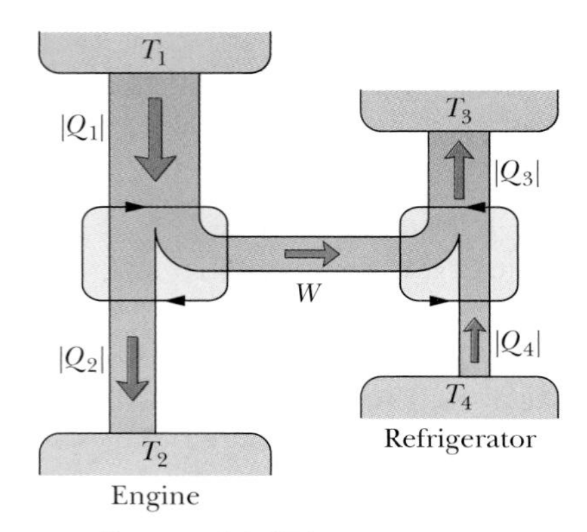

FIGURE 21-27 ■ Problem 41.

SEC. 21-7 ■ A STATISTICAL VIEW OF ENTROPY

42. Construct a Table Construct a table like Table 21-1 for eight molecules.

43. Show for N Molecules Show that for N molecules in a box, the number of possible microstates is 2^N when microstates are defined by whether a given molecule is in the left half of the box or the right half. Check this for the situation of Table 21-1.

44. A Box of N Gas Molecules A box contains N gas molecules, equally divided between its two halves. For $N = 50$: (a) What is the multiplicity of this central configuration? (b) What is the total number of microstates for the system? (*Hint:* See Problem 43.) (c) What percentage of the time does the system spend in its central configuration? (d) Repeat (a) through (c) for $N = 100$. (e) Repeat (a) through (c) for $N = 200$. (f) As N increases, you will find that the system spends *less* time (not more) in its central configuration. Explain why this is so.

45. Three Equal Parts A box contains N gas molecules. Consider the box to be divided into three equal parts. (a) By extension of Eq. 21-18, write an equation for the multiplicity of any given configuration. (b) Consider two configurations: configuration A with equal numbers of molecules in all three thirds of the box, and configuration B with equal numbers of molecules in both halves of the box. What is the ratio W_A/W_B of the multiplicity of configuration A to that of configuration B? (c) Evaluate W_A/W_B for $N = 100$. (Because 100 is not evenly divisible by 3, put 34 molecules into one of the three box parts and 33 in each of the other parts for configuration A.)

46. Four Particles in a Box Four particles are in the insulated box of Fig. 21-16. What are (a) the least multiplicity, (b) the greatest multiplicity, (c) the least entropy, and (d) the greatest entropy of the four-particle system?

Additional Problems

47. Velocity Spread A sample of nitrogen gas (N_2) undergoes a temperature increase at constant volume. As a result, the distribution of molecular speeds increases. That is, the probability distribution function $f(v)$ for the nitrogen molecules spreads to higher values of speed, as suggested in Fig. 20-8*b*. One way to report the spread in $f(v)$ is to measure the difference Δv between the most probable speed v^{prob} and the rms speed v^{rms}. When $f(v)$ spreads to higher speeds, Δv increases. (a) Write an equation relating the change ΔS in the entropy of the nitrogen gas to the initial difference Δv_i and the final difference Δv_f. Assume that the gas is an ideal diatomic gas with rotation but not oscillation of its molecules. Let the number of moles be 1.5 mol, the initial temperature be 250 K, and the final temperature be 500 K. What are (b) the initial difference Δv_i, (c) the final difference Δv_f, and (d) the entropy change ΔS for the gas?

48. Carnot Graph A Carnot engine is set up to produce a certain work W per cycle. In each cycle, thermal energy Q_H is transferred to the working substance of the engine from the higher-temperature thermal reservoir, which is at an adjustable temperature T_H. The lower-temperature thermal reservoir is maintained at temperature $T_L = 250$ K. Figure 21-28 gives Q_H for a range of T_H. if T_H is set at 550 K, what is Q_H?

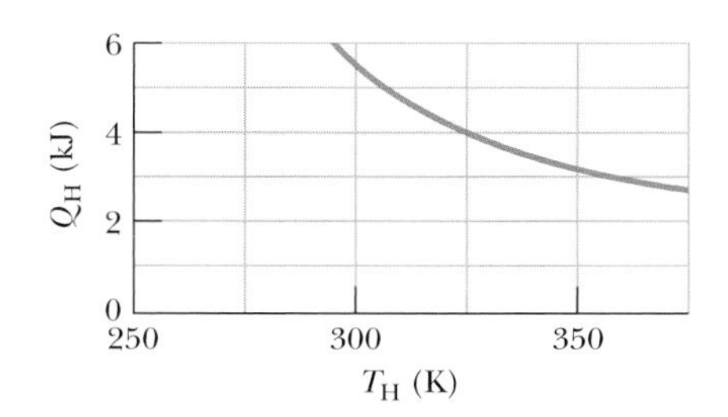

FIGURE 21-28 ■ Problem 48.

49. Gas Sample A gas sample undergoes a reversible isothermal expansion. Figure 21-29 gives the change ΔS in entropy of the gas versus the final volume V_f of the gas. How many moles are in the gas?

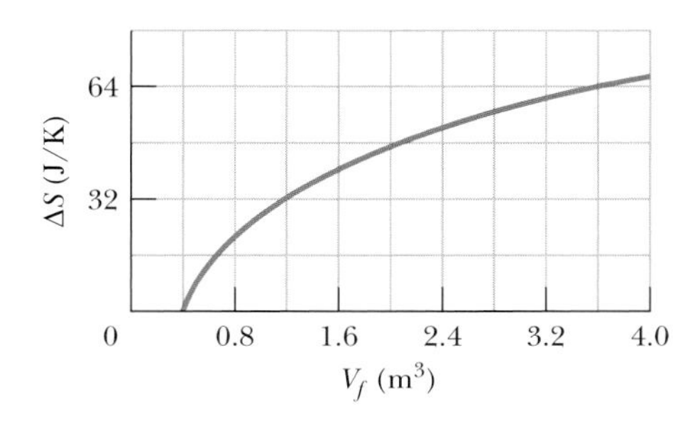

FIGURE 21-29 ■ Problem 49.

50. ΔS Graph A 364 g block is put in contact with a thermal reservoir. The block is initially at a lower temperature than the reservoir. Assume that the consequent heating of the block by the reservoir is reversible. Figure 21-30 gives the change in entropy of the block ΔS until thermal equilbrium is reached. What is the specific heat of the block?

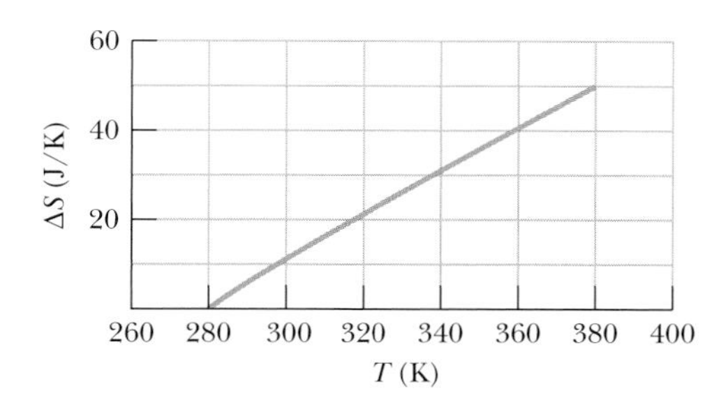

FIGURE 21-30 ■ Problem 50.

51. *C/T* Graph An object of (constant) heat capacity C is heated from an initial temperature T_i to a final temperature T_f by a constant-temperature reservoir at T_f. (a) Represent the process on a graph of C/T versus T, and show graphically that the total change in entropy ΔS of the object–reservoir system is positive. (b) Explain how the use of reservoirs at intermediate temperatures would allow the process to be carried out in a way that makes ΔS as small as desired.

52. Friction A moving block is slowed to a stop by friction. Both the block and the surface along which it slides are in an insulated enclosure. (a) Devise a reversible process to change the system from its initial state (block moving) to its final state (block stationary; temperature of the block and surface slightly increased). Show that this reversible process results in an entropy increase for the closed block–surface system. For the process, you can use an ideal engine and a constant-temperature reservoir. (b) Show, using the same process, but in reverse, that if the temperature of the system were to decrease spontaneously and the block were to start moving again, the entropy of the system would decrease (a violation of the second law of thermodynamics).

53. *T-S* Diagram A diatomic gas of 2 mol is taken reversibly around the cycle shown in the *T-S* diagram of Fig. 21-31. The molecules rotate but do not oscillate. What are the thermal energy transfers Q for (a) the path from point 1 to point 2, (b) the path from point 2 to point 3, and (c) the full cycle? (d) What is the work W for the isothermal process? The volume V_1 at point 1 is 0.200 m^3. What are the volumes at (e) point 2 and (f) point 3?

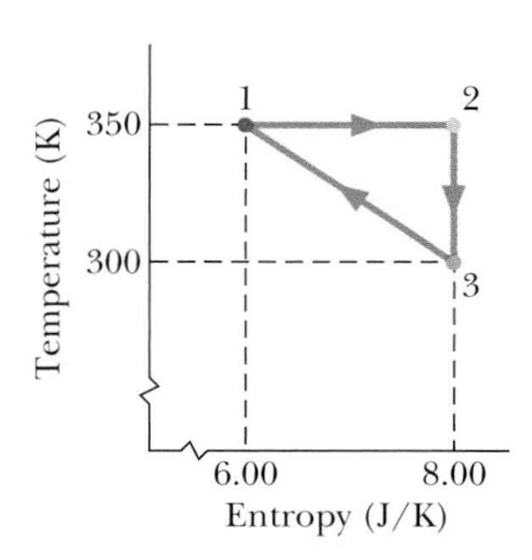

FIGURE 21-31 ■ Problem 53.

What are the changes ΔE^{int} in internal energy for (g) the path from point 1 to point 2, (h) the path from point 2 to point 3, and (i) the full cycle? (*Hint*: Part (h) can be done with one or two lines of calculation using Section 20-8 or with a page of calculation using Section 20-11.) (j) What is the work W for the adiabatic process?

54. Inventor's Engine An inventor has built an engine (engine X) and claims that its efficiency ε_X is greater than the efficiency ε of an ideal engine operating between the same two temperatures. Suppose that you couple engine X to an ideal refrigerator (Fig. 21-32*a*) and adjust the cycle of engine X so that the work per cycle that it provides equals the work per cycle required by the ideal refrigerator. Treat this combination as a single unit and show that if the inventor's claim were true (if $\varepsilon_X > \varepsilon$), the combined unit would act as a perfect refrigerator (Fig. 21-32*b*), transferring thermal energy from the low temperature reservoir to the high-temperature reservoir without the need of work.

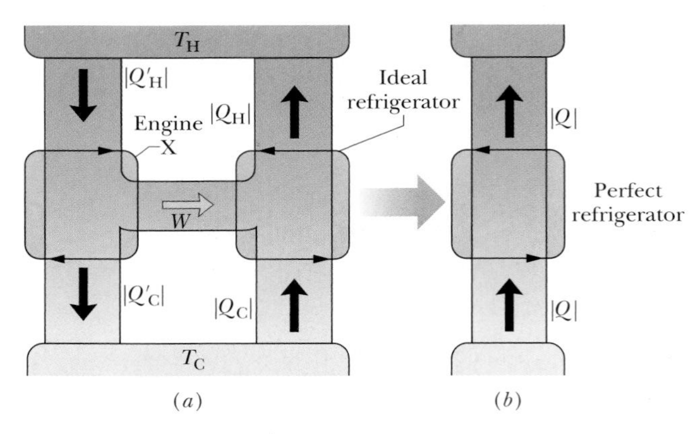

FIGURE 21-32 ■ Problem 54.

55. Ideal Diatomic Gas An ideal diatomic gas with 3.4 mol is taken through a cycle of three processes:

1. its temperature is increased from 200 K to 500 K at constant volume;
2. it is then isothermally expanded to its original pressure;
3. it is then contracted at constant pressure back to its original volume.

Throughout the cycle, the molecules rotate but do not oscillate. What is the efficiency of the cycle?

22 Electric Charge

Nothing happens if you place a plastic comb near tiny scraps of paper, but immediately after you comb your hair or stroke the comb with fur, it will attract the paper scraps. In fact, the attractive force exerted on the paper by the small comb is so strong that it overcomes the opposing gravitational pull of the entire Earth. This phenomenon, commonly called "static cling," occurs between many different objects and is especially easy to observe during cold dry weather.

What causes these pieces of paper to stick to the comb and to one another?

The answer is in this chapter.

22-1 The Importance of Electricity

If you walk across a carpet when it's cold and dry outside, you can produce a spark by bringing your finger close to a metal doorknob. Television advertisements alert us to the problem of "static cling" in clothing. On a grander scale, lightning is familiar to everyone. These phenomena represent a tiny glimpse into the vast number of electric interactions that occur every day.

The phenomenon of electricity plays a major role in modern life. Less than two hundred years ago, fire was almost the only source of heat, the only source of light when the sun or moon was not up, and the only way to cook food. Without electric water pumps, most people did not even have indoor plumbing. It's hard to imagine life without electric lights (not even flashlights), stoves, refrigerators, air conditioners, computers, telephones, radios, televisions, CD players, and a host of other electrical devices. We make extensive use of electricity, but *what is it*? In this chapter we consider this very important question.

So far in our study of the physical world we have learned how the forces acting on objects affect motion. We have also learned about the gravitational force, an action-at-a-distance force, that objects can exert on each other without touching. In this chapter, we will investigate another action-at-a-distance force—the *electrostatic* interaction force. Studying the electrostatic force will provide a foundation for our understanding of the phenomenon of electricity.

We begin our study by looking at the nature of electrical interaction forces between some everyday objects. We then develop the concepts of charging and electric charge as tools for explaining our observations of electrostatic forces on a macroscopic level. However, to obtain a more coherent understanding of electrostatic phenomena, we must turn to the findings of atomic theory.

An understanding of electrostatic interactions will give you insight into the fundamental relationship between electricity and magnetism. In Chapter 30, which is about magnetic fields due to currents, you will discover that although magnetic forces are generated by the interaction between moving charges, they have distinctly different properties than do electrostatic forces. Later, in Chapter 32, you will see how electricity and magnetism are fundamentally related to each other.

READING EXERCISE 22-1: List all the electrical devices that you use in a typical week. ■

22-2 The Discovery of Electric Interactions

FIGURE 22-1 ■ Fossilized resin, known as amber, is popular both for its beauty and for its ability to preserve ancient vegetation and insects, like the bees, wasps, ants, flies and mosquitos seen here. Amber also has electrical properties.

Amber, which is resin that oozed from trees long ago and hardened, has been admired both for its beauty and its ability to preserve early life forms mired in it (Fig. 22-1). Amber has electrical properties of interest to scientists as well. The early Greeks knew that if one rubbed a yellow-brown piece of amber with fur, it would attract bits of straw. The strength of the attraction decreased as the distance between the amber and the straw was increased. The strength of the attraction was also known to fade over time, especially in damp weather.

By the 1600s, this strange force due to amber that was sometimes present and sometimes not, prompted more careful studies. It was subsequently discovered that other materials such as glass can also attract small bits of matter after being rubbed with silk. As was the case for amber, this attractive force diminished with time, especially on humid days, and was not present if the glass had not been rubbed. Additionally, the strength of the attraction decreased as the distance between the glass and small bits of matter increased, just as was the case with the force associated with rubbed amber.

This interaction phenomenon, created by rubbing certain materials with cloth, was named *electrification*. The term is derived from the Greek word for amber, which is *electron*. Any object (not just glass or amber) is defined as becoming *electrified* if:

1. There is an interaction force between this object and another that is present after the objects have been in very close contact, usually through rubbing;
2. The magnitude of this interaction force diminishes with time and is affected by humid weather; and
3. The magnitude of the force decreases with increasing distance between the objects.

Although the similarities between electrified glass and electrified amber were interesting, it was not until 1733 that a French scientist, Charles DuFay, published articles presenting evidence that:

Two amber rods stroked with fur always repel one another.

Two glass rods stroked with silk always repel one another.

A stroked amber rod *attracts* a stroked glass rod (Fig. 22-2).

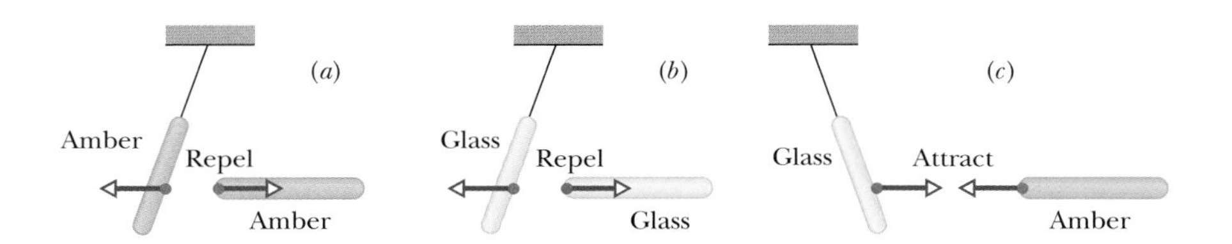

FIGURE 22-2 ■ (*a*) Two amber rods electrified in the same way repel each other. (*b*) Two glass rods electrified in the same way also repel each other. (*c*) An electrified glass rod and an electrified amber rod attract each other.

Provided the weather is not too humid, you may be able to repeat DuFay's observations yourself by replacing the amber and glass rods (which are difficult to find outside of a physics laboratory) with Styrofoam cups and plastic sandwich bags as shown in Fig. 22-3. Place your hand inside a plastic bag and use a rubbing motion to assure that the entire surface of the Styrofoam cup comes in contact with the entire surface of the plastic bag. Then rub another Styrofoam cup with a second sandwich bag in the same manner. If you put one of the cups on its side on a smooth, level, nonmetallic surface and bring the other cup near it, the first cup should roll away as shown in Fig. 22-3*a*. Note that after the two cups have been electrified in a *like manner* they *repel* one another just like DuFay's rods. Now hold the two plastic bags together at the top end. Both plastic bags have also been electrified in a like manner and they repel one another as well as shown in Fig. 22-3*b*. However, an electrified sandwich bag and an electrified Styrofoam cup will be *attracted* to each other just as electrified amber attracts electrified glass. Think about these observations carefully and you must conclude that there are two classes of materials that behave differently when electrified.

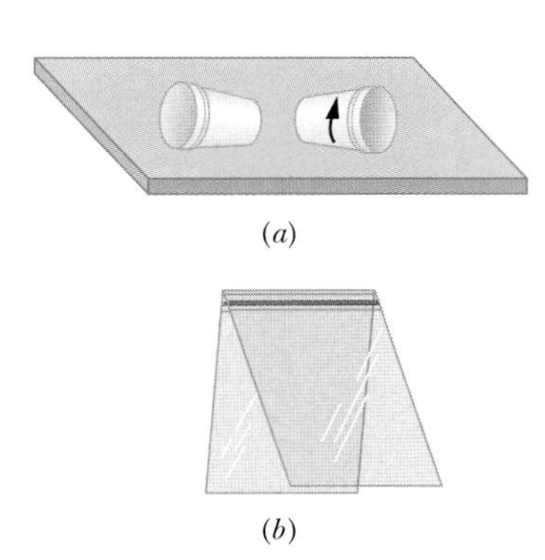

FIGURE 22-3 ■ (*a*) Two Styrofoam cups electrified in the same way repel each other. (*b*) The two sandwich bags used to electrify the cups also repel each other.

Not all types of materials can be electrified. Nevertheless, additional observations with electrified materials lead to the following general statements:

> **OBSERVATION 1.** Two identical objects electrified by the same process always repel one another.
>
> **OBSERVATION 2.** Two different electrified objects will always interact, but they may either repel or attract one another.
>
> **OBSERVATION 3.** Any two objects that have not been electrified will neither repel nor attract one another. (They interact only by means of an imperceptibly small gravitational force.)

Suppose you have electrified two Styrofoam cups so they repel each other. What happens when you give one of the cups extra stroking? The magnitude of the interaction forces between the cups increases. This means that if we think of the first cup (that did not receive extra stroking) as a "standard object," we can determine the degree of electrification of any other object by measuring the magnitude of the electric force exerted on it by the standard object.

One logical way of interpreting our observations regarding electrification is to assume that a substance is added or removed from an object during the stroking process. Extensive experiments done at the end of the 18th century by Benjamin Franklin and others indicate that this is correct. They also found that there are actually two types of the substance involved. Today we call these substances **electrical charge** and say that there are two types of (electrical) charge. When an object contains more of one type of charge than the other, the object is electrified or **electrically charged.** Furthermore, any process of electrification (not just rubbing) is called **charging.** Thus, in the example above, the cup that was stroked for the longer time gained the greater quantity of excess charge. (Quantity of charge is often called **amount** of charge). An object with a greater amount or quantity of charge is observed to experience more force in the presence of a standard electrified object than one with a smaller amount of excess charge.

READING EXERCISE 22-2: The creation of electrified objects can also be done with strips of Scotch™ Magic Tape using a peeling action rather than stroking. In order to charge the tape, cut 2 strips about 10 cm long. (a) If you were to stick the tapes side by side on a table and peel them both off, what do you predict would happen if you then brought the tapes close together? Explain the reasoning for your prediction. (b) Perform the experiment and describe what happens. Is this consistent with your prediction? If not, explain what you think is going on. ■

22-3 The Concept of Charge

Various observations, including our observations using Styrofoam cups and plastic bags, indicate that interaction forces between charged objects can be explained in terms of two (and only two) different kinds of charged matter. The type associated with glass rubbed with silk is one and the type associated with amber rubbed with fur is the other. We cannot prove directly that there are no other types of charge. However, the fact that no one has found a charged object that attracts both charged glass and charged amber leads us to believe that there is no third type of charge.

Today, the terms we associate with these two types of charged matter are *positive* and *negative*. Benjamin Franklin is responsible for assigning these names. He introduced the following definitions:

OBSERVATION 1. An object that is repelled by a glass rod stroked with silk is **positively charged.**

OBSERVATION 2. An object that is repelled by amber (or plastic) stroked with fur is **negatively charged.**

OBSERVATION 3. Any two nonmagnetic objects that do not interact with each other except by gravitational forces are electrically neutral.

The names given to the two varieties of charge are arbitrary. Benjamin Franklin could just as easily have used other words, such as light and dark, to distinguish between the two types of charges. However, we observe that equal amounts of the positive and negative charges combine to produce nonelectrified (i.e., **electrically neutral**) matter. That is, the two types of charge combine algebraically—like positive and negative numbers. So, positive and negative are convenient and appropriate names.

Applying this new terminology to our previous observations leads us to say that if two objects are each repelled by a piece of glass that has been rubbed with silk, then both objects must be positively charged. Furthermore, we hypothesize that these objects repel *because* they contain the same kind of charge or **like charges.** On the other hand, if we find that two objects made of different materials attract after being stroked, we hypothesize that one object has a positive charge while the other has a negative charge. We conclude that objects with **unlike** or **opposite charges** attract.

READING EXERCISE 22-3: Suppose you stroked a smooth wooden rod with a linen cloth and announced that you had created a new type of charge you decided to call *woodolin charge*. (a) If a skeptic asked you to prove that woodolin was really a new type of charge, how would you do it? Specifically what would have to happen if you were to bring two wooden rods together that had both been rubbed with linen? If you were to bring a charged wooden rod near a charged glass rod? Near a charged amber (or plastic) rod? (b) Why do you think most observers agree that there are only two types of known charge? ■

22-4 Using Atomic Theory to Explain Charging

How can we account for the fact that when certain objects are rubbed together they acquire opposite types of charge? One way to make sense of this observed fact is to use a contemporary understanding of the atomic structure of matter. The atomic model that we discuss here has been developed over the past century. We will use it as an explanatory tool without presenting evidence for it.

The Atomic Model

According to modern atomic structure theory, atoms consist of positively charged *protons,* negatively charged *electrons,* and electrically neutral *neutrons*. Electrons and protons have the same amount (although with opposite sign) of charge. We often represent this amount of charge, called the **elementary charge,** with an e (Table 22-1). Hence an electron has a charge of $-1e$ and a proton has a charge of $+1e$.

Protons and neutrons are packed tightly together in a central *nucleus*. They are much more massive than electrons, which lie outside the nucleus as depicted in Fig. 22-4. Most of the atoms that are contained in matter have equal numbers of electrons and protons, so whenever a charged object is at some distance away from the atom, the atom appears to be electrically neutral.

According to contemporary atomic theory, electric charge is an intrinsic characteristic of electrons and protons. You often encounter casual phrases—such as "the

TABLE 22-1
Charges of the Three Fundamental Atomic Particles

Particle	Symbol	Charge
Electron	e or e^-	$-e$
Proton	p	$+e$
Neutron	n	0

Note: The symbols for electron and for electronic charge are the same. This can be confusing.

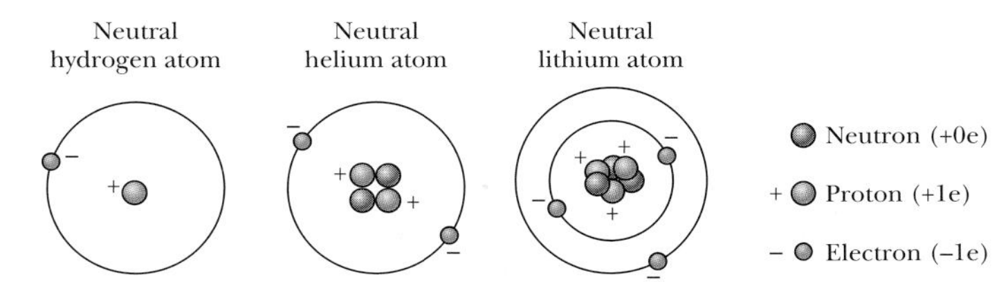

FIGURE 22-4 ■ The structure of the atoms representing the three lightest chemical elements, H, He, and Li. The number of protons that define the element along with the typical number of neutrons in each element's nucleus are shown. The darker circles represent protons, the lighter circles neutrons, and the white circles electrons. The diagram is simplified, as physicists do not actually believe that electrons orbit nuclei in nice neat circles and that the nuclei are much smaller relative to the size of their atoms.

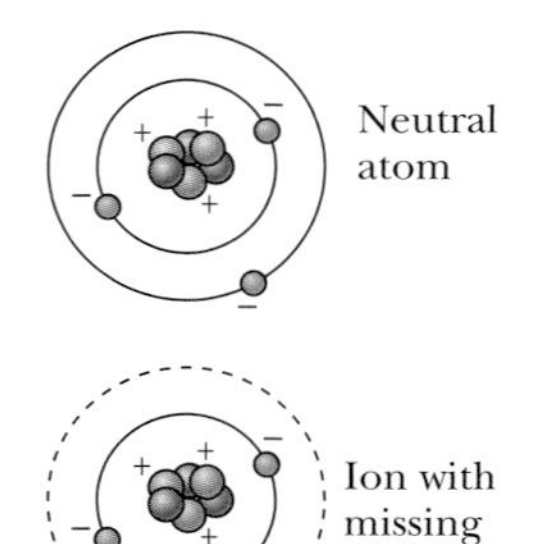

FIGURE 22-5 ■ The upper diagram shows a neutral lithium atom with its full complement of electrons. The bottom atom has lost an outer electron and is now an ion with a net charge of $+1e$ because it has three protons and only two electrons.

charge on a sphere," "the amount of charge transferred," and "the charge carried by the electron." Such phrases can be misleading because they suggest that charge is a substance. You should, however, keep in mind that experiments show electrons and protons are the basic substances. Charge and mass are simply two of their fundamental properties.

The mass of an electron is about 2000 times smaller than that of a neutron or proton. Electrons are attracted to the nucleus because electrons and the protons within the nucleus have opposite charges. However, the electrons that are farthest away from the nucleus are only weakly attracted to the protons within the nucleus and so they don't always remain associated with individual atoms. In many types of materials the electrons are free to wander within the material if they experience forces. If the atom loses an electron it is no longer neutral, but has a net positive charge because there are now more protons than electrons, as seen in Fig. 22-5. Charged atoms are called **ions.** We call mobile electrons **conduction electrons.** If an electric or other force is applied to the atom, only the conduction electrons, with their negative charges, move appreciably. The much more massive positive ions stay fixed in place.

Charge Is Quantized

In Benjamin Franklin's day, electric charge was thought to be a continuous fluid that could "contain" any arbitrary amount of charge. Today we know that fluids, such as air and water, are not continuous but are made up of atoms and molecules. Matter is discrete. In 1909 an American physicist, Robert Millikan, used opposing electric and gravitational forces to balance drops of oil between two electrified metal plates. His famous oil drop experiment and others that followed showed that the "electrical fluid" is not continuous either but is made up of multiples of the elementary charge.

Any positive or negative charge q that has ever been detected as a free particle can be written as

$$q = ne, \quad n = \pm 1, \pm 2, \pm 3, \ldots, \tag{22-1}$$

in which e, the elementary charge, has the value

$$e = 1.60 \times 10^{-19}\ \text{C}. \tag{22-2}$$

The SI unit of charge is the **coulomb** (C), named for Charles Augustin Coulomb, who studied electric forces in the late 1700s. When a physical quantity such as charge can have only discrete values rather than any arbitrary value, we say that the quantity is **quantized.** It is possible, for example, to find a free particle that has no charge at all or a charge of $+10e$ or $-6e$, but not a free particle with a charge of, say, $3.57e$. Modern studies of the structure of neutrons and protons have produced strong evidence that neutrons and protons are made up of tightly bound particles with charges $+2/3e$ and $-1/3e$ that we call quarks, but quarks do not seem to be able to exist as free particles. Hence, extensive experimentation confirms that:

> Charge is quantized. In free particles charge has never been measured to have an amount other than an integer multiple of 1.60×10^{-19} C.

As we noted in the introduction, if you drag your feet as you walk across a carpet, you can produce a spark caused by moving electric charge by bringing your finger close to a metal doorknob. This is a demonstration of a small sample of the vast amount of electric charge that is stored in electrically neutral objects. However, that charge is usually hidden because the object contains equal amounts of positive and negative charge. This was hinted at above when we stated that electrically neutral atoms contain equal numbers of protons and electrons. With such an equality—or *balance*—of the amounts charge there is no *net* charge. If the two types of charge are not in balance, we say an object is *charged* to indicate that it has a charge imbalance, or nonzero net charge.

> Macroscopic objects that are electrically neutral are *not* devoid of charge. Instead, they contain equal numbers of positive protons and negative electrons. This results in a cancellation of their electrical effects.

Charging Is Transferring Electrons

Glass and silk or Styrofoam and plastic become oppositely charged when they are brought into contact and we can use our modern understanding of the atom to explain why. Suppose we observe that Styrofoam becomes positively charged and plastic becomes negatively charged. It is logical to assume that outer electrons associated with atoms in the Styrofoam are attracted to the atoms in the plastic and move over to the plastic. The Styrofoam is now missing electrons so there is a net positive charge on the Styrofoam. The plastic now has excess electrons and has a net negative charge.

In general, experiment shows that an object becomes charged when a very tiny fraction of the mobile electrons with their negative charge are transferred from one object to another. This is why we must rub, stroke, or otherwise make significant contact between two objects for the objects to become charged. Thus, when a Styrofoam cup is stroked with a plastic bag, a very tiny fraction of the electrons near the surface of the Styrofoam cup are transferred to the plastic bag.

Why doesn't the plastic bag get heavier when electrons are transferred to it? Using modern atomic theory, we understand that even if we transfer a lot of electrons to the plastic bag (a typical number might be between 10^9 and 10^{12}) the increase in mass would be less than 10^{-10} kg—not measurable. In ordinary matter, positive charge is much less mobile than negative charge. For this reason, an object becomes positively charged through the *removal of negatively charged electrons* rather than through the addition of positively charged protons.

Charge Is Conserved

Careful measurements reveal that whenever there is excess charge on one of the objects after contact, there are excess charges on the other object too. These charges are equal in amount but opposite in sign. This demonstrates that when electrons are transferred from one object to another, no electrons are destroyed or created in the process. The amount of charge contained in the two objects is constant or conserved.

This hypothesis of **conservation of charge** was first proposed by Benjamin Franklin based on his experiments. It is observed to hold both for large-scale charged bodies and for atoms, nuclei, and elementary particles. No exceptions have ever been found. So, we add electric charge to our list of quantities—including energy, linear momentum, and angular momentum—that are conserved quantities. In summary, extensive experimentation confirms:

> The total amount of electric charge in the universe is conserved. Although particles that carry charge can be transferred from one object to another, the charge associated with particles cannot be created or destroyed.

Force and Quantity of Charge

Because charge is conserved, we can transfer charge from one object to another without changing the total amount of charge in the system. This allows us to perform experiments that indicate how the interaction force between charged objects depends on the amount of charge on each object. These experiments lead to surprisingly simple results when the charged objects are symmetric, made of metal, and are particle-like (so that their dimensions are small compared to the distances between their centers). For example, consider the experiment shown in Fig. 22-6. Two identical uncharged metal spheres are both electrified with a charged plastic rod. We then touch the two spheres together. Since the spheres are identical and the excess electrons repel each other, we expect electrons to travel between the spheres until both spheres have the same number of excess electrons. Next we measure the force exerted by one sphere on the other and record it (Fig. 22-6*a*).

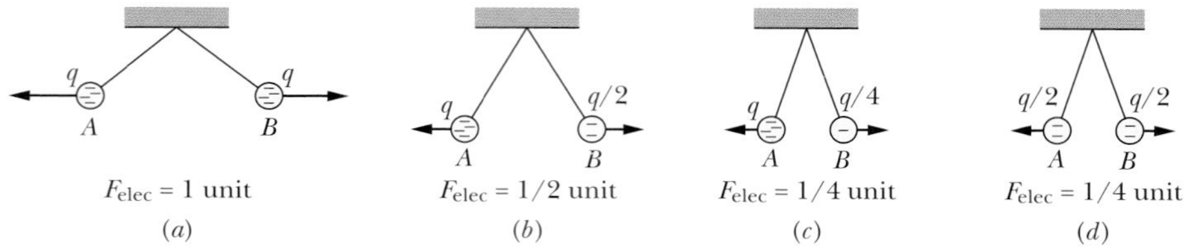

FIGURE 22-6 ■ Depiction of an idealized experiment to measure the forces between small metal spheres that hold different fractions of charge. *Note:* In order to make force measurements for particle-like objects, the distance between the centers of the two balls of identical shape must be more than twice the diameter of a ball.

Then we leave sphere A alone and move sphere B a long distance from sphere A to place it in contact with a third sphere C that is uncharged. The excess electrons on sphere B will now be shared equally between spheres B and C so the number of excess electrons on sphere B will now be half of what it was before. If we return B to its original location and measure the magnitude $F_{\text{elec}} = |\vec{F}_{\text{elec}}|$ of the force between spheres A and B, we find that it is one-half of the force magnitude we first measured (Fig. 22-6*b*). If we repeat this process so we reduce the amount of charge on sphere B to one-fourth of what it was originally, then the magnitude of the interaction force between the spheres is also reduced to one-fourth of what it was originally (Fig. 22-6*c*). In a similar experiment we can reduce the charge on both spheres A and B to half their original values and then the force measures one-fourth the original force between them (Fig. 22-6*d*). These observations, which are summarized in Table 22-2, indicate that the magnitude of the interaction force is proportional to the *product* of the amounts of charge on the two spheres. This relationship is given by

TABLE 22-2

q_A	q_B	$q_A \cdot q_B$	F_{elec}(arbitrary units)
q	q	q^2	$1\,F$
q	$q/2$	$q^2/2$	$1/2\,F$
q	$q/4$	$q^2/4$	$1/4\,F$
$q/2$	$q/2$	$q^2/4$	$1/4\,F$

$$F_{\text{elec}} \propto |q_A||q_B|, \tag{22-3}$$

where the absolute value signs denote charge amounts independent of sign.

> The amount of the charge on a particle-like object can be quantified through measurement of the magnitude of the interaction force between it and a standard charged object that is also particle-like.

The Electroscope

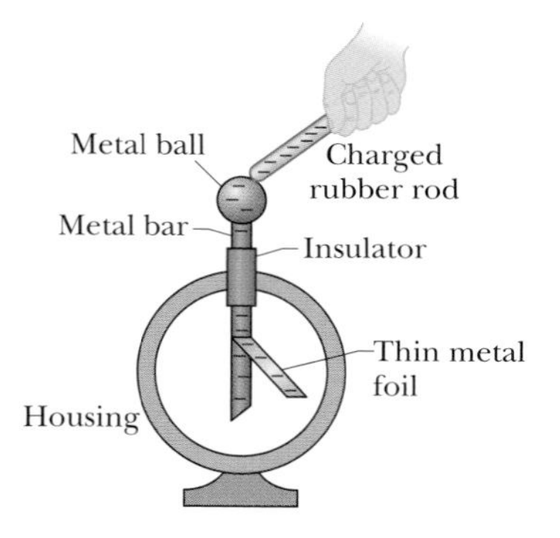

FIGURE 22-7 ■ The electroscope can be used to measure charge. The rise of a metal foil is caused by the repulsion due to an excess of like charged particles distributed on the parts of a metal conducting system. The foil rises in proportion to the net charge contained on the conductor.

The fact that like charges repel has been used in the development of the **electroscope,** a sensitive charge-measuring device, as seen in Fig. 22-7. A net charge can be transferred to an electroscope by stroking the metal ball with a charged rod. If the rod is negatively charged, some of its excess electrons will be transferred to the ball and then they will spread throughout the metal rod and the foil attached to the ball. If a flexible metal leaf is attached to the central conducting bar, the flexible conductor will be repelled from the central charges and rise. As more electrons are transferred to the electroscope, the metal leaf will rise higher. Alternatively if the rod is positively charged it will *attract* electrons from the electroscope, leaving a net positive charge on it. Once again the foil will rise.

READING EXERCISE 22-4: Assuming that solid objects are made up of atoms rather than being continuous, can you think of a plausible way to explain why it is so difficult to pull solids apart or push them together? ■

READING EXERCISE 22-5: Consider the measurements depicted in Fig. 22-6. Suppose you have measured the repulsion force between two identical metal-coated spheres that each have a total negative charge q due to excess electrons. Next, you would like to measure the force on the metal-coated spheres that each have one-fourth of the excess electrons they originally had. Describe how you could use similar uncharged spheres to reduce the excess electrons on each of the original spheres to $q/4$. ■

22-5 Induction

Let's consider some additional observations involving electrical interactions. Typically, bits of straw or paper that have not been rubbed do not attract or repel one another. They are electrically neutral. Thus, it is surprising to find that a plastic comb made negatively charged by rubbing (like the comb shown in the photograph at the beginning of this chapter) can attract bits of electrically neutral paper. It is equally surprising to find that a positively charged glass rod will attract bits of paper.

> **INDUCTION** is the process that causes the attraction we observe between a charged object and an uncharged one.

How can we explain induction? We turn to atomic theory for an explanation. The idea that electrically neutral materials are not devoid of charge, but rather are composed of atoms that have the same number of positive protons and negative electrons, is the first step in developing a viable explanation for induction. The second important idea, mentioned in the last section, is that electrons are more mobile than protons.

Let's begin by considering how induction occurs when a charged object is placed near an uncharged *metal* object. Then in the next section, we will consider induction in a class of nonmetals known as insulators.

Induction in Metals

What happens when we dangle a very small metal rod from a string near a charged object as shown in Fig. 22-8? According to our atomic model, mobile negative electrons in the metal rod are repelled from the negatively charged object. When the mobile electrons in the neutral metal object are repelled they move away and unpaired protons are left behind. The unlike charges at the surfaces of the two objects will now attract as shown in Fig. 22-8*a*. An attraction between a neutral object and a positively charged object occurs as shown in Fig. 22-8*b*. The process of separation of positive protons and negative electrons in the neutral objects is known as **polarization.**

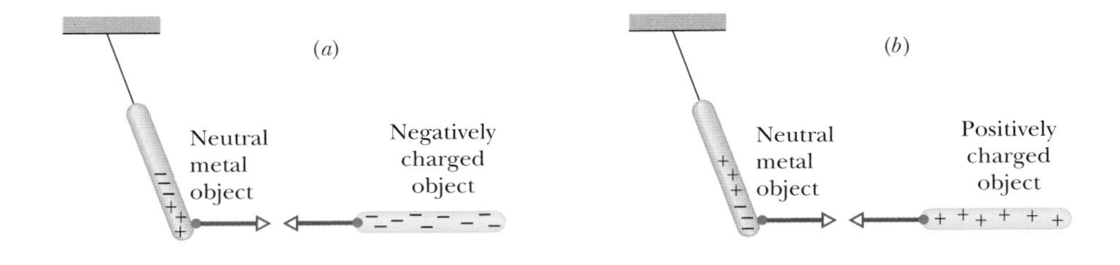

FIGURE 22-8 ■ A tiny neutral metal rod is suspended on a nonconducting thread. The part of the neutral rod that is closest to a charged object will be attracted by either: (*a*) a negatively charged object (such as amber) or (*b*) a positively charged object (such as glass). The extent of the charge separation in the metal rod is exaggerated.

READING EXERCISE 22-6: (a) If we state that two bits of paper are electrically neutral, what observation can you make to verify this is the case? Explain. (b) Can induction be used to determine whether the charged object is positive or negative? Why or why not? ■

READING EXERCISE 22-7: (a) In the figure that follows, objects made of different materials are arranged in six pairs. A is plastic stroked with fur, B is glass stroked with silk, and C is an electrically neutral metal object. State whether the interaction between each pair will be attractive, repulsive, or nonexistent. Explain your reasoning.

(1)	(2)	(3)	(4)	(1)	(1)
A	A	B	B	C	C
B	A	B	C	C	A

(b) According to the sign convention Benjamin Franklin decided to adopt, which of the three objects carry an excess positive charge? An excess negative charge? No excess charge? Explain.

(c) Suppose you were told that A, B, and C are made of new materials but you were not told what those materials are. Would it be possible to determine whether they are negatively charged, positively charged, or electrically neutral by observing how the pairs shown in the diagram above interact? Why or why not? ■

22-6 Conductors and Insulators

Whenever a charged object is near an electrically neutral object, induction and polarization occur. However, the attractive forces are stronger for neutral metal objects than for nonmetal objects. Why? Let's summarize the outcomes of some important observations you can make yourself using metal rods, Styrofoam cups, and plastic bags.

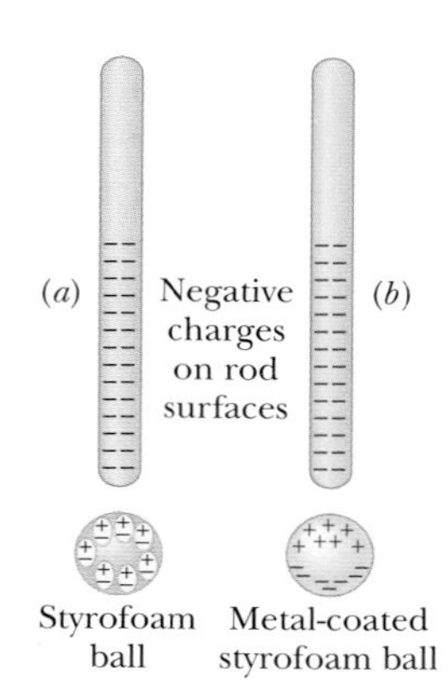

FIGURE 22-9 ■ According to atomic theory: (*a*) polarization induced in an insulator involves very very tiny atomic-scale charge separations as shown in exaggerated form for a Styrofoam ball, whereas (*b*) polarization in a conductor can involve a much larger scale migration of electrons as shown in a metal-coated ball.

OBSERVATION 1. (a) The electrification created on nonmetal objects, like plastic, does not spread out. Instead charge seems to remain in regions where the object is rubbed, and (b) touching charged nonmetal objects removes the electrification only at locations where an object is touched.

OBSERVATION 2. (a) Metal objects can be charged when mounted on nonmetal objects such as glass or plastic but they cannot be charged while being held in someone's hand, and (b) metal objects that are touched anywhere by a person will immediately lose *all* of their charge.

Even though paper is a nonmetal, the ideas of induction and polarization can be used to explain how the charged comb pictured in the puzzler at the beginning of the chapter can be used to pick up a string of paper bits. We assume the first piece of paper is attracted to the charged comb by induction and becomes polarized in the process. Then the excess negative charge at the bottom end of the first paper bit attracts the second paper bit by induction. Since the second paper bit is now polarized, the process continues. (Although experiments show that magnetic forces behave differently than do electrostatic forces, this process of electric polarization looks like a similar process in which a magnet can induce magnetic polarization in a steel paper clip, which then attracts and polarizes a second clip, and so on.)

Atomic Theory and the Behavior of Conductors and Insulators

Once again we can turn to atomic theory to develop a plausible explanation for these new observations. In materials such as metals, tap water, water droplets in air, and the human body, experiments indicate that some of the negative electrons move easily. We call such materials **conductors.** We observe that charge can flow onto or off conductors quite quickly. In other materials, such as glass, chemically pure water, and plastic, the electrons can reposition themselves within an atom but cannot migrate between atoms. We call these materials **nonconductors** or **insulators.** Electrons do not travel from atom to atom very easily in insulators. An exception is that some of the electrons at the surface of an insulator can have a greater affinity for another type of surface. For example, electrons can travel from the surface of a glass rod to the surface of a piece of silk cloth brought into contact with it, leaving the glass rod positively charged.

One implication of this difference in the mobility of charge in insulators and conductors is that the polarization process discussed in Section 22-5 is not as strong in insulators. Neutral conductors and neutral insulators both undergo charge separation (become polarized) when they are brought close to a charged object—but, the electrons in insulators are tightly bound to atoms, and the charge separation (polarization) is only a small fraction of an atomic radius. In contrast, some of the electrons in conductors can move through the material fairly freely and become separated from the atoms to which they were originally associated. This difference is shown in the comparison of the two images in Fig. 22-9. This atomic model provides a plausible explanation for the observed fact that induction is much stronger between a charged object and a conductor than between the same charged object and an insulator.

The difference in mobility of charge carriers in conductors and insulators explains why you cannot charge a metal rod by rubbing if you are holding it. Both you and the rod are conductors. Although the rubbing will cause a charge imbalance on the rod, the excess charge will immediately move from the rod through you to the floor (which is connected to Earth's surface), and the rod will quickly be neutralized. Setting up a pathway for electrons between an object and Earth's surface is called **grounding** the object, and always results in electrically neutralizing the object. If instead of holding the metal rod in your hand, you hold it by an insulating handle, you eliminate the conducting path to Earth, and rubbing can then charge the rod, as long as you do not touch it directly with your hand.

These ideas give us a very functional way to determine whether a material is a conductor or an insulator. If you have two interacting charged objects and you touch one of them, do the objects stop interacting? If so, charge must have been transferred to or from the object. Hence, it must be a conductor. If the transfer of charge does not occur, the object must be an insulator.

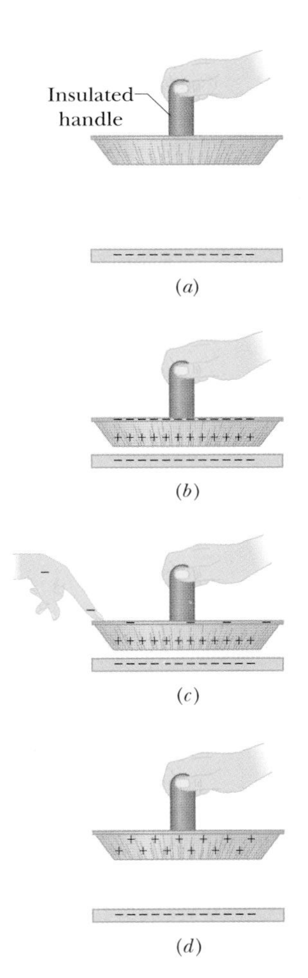

FIGURE 22-10 ■ An electrophorus is an apparatus that can be used to charge a conductor by induction. It consists of an insulated plate such as a slab of Styrofoam and a conducting plate such as an aluminum pie plate. (*a*) The insulating slab is charged negatively after being stroked with fur. (*b*) When a conducting plate is brought near the charged insulator the conductor is polarized so that its free electrons move away from the insulator. (*c*) When the top of the conductor is touched these free electrons move to the hand. (*d*) This leaves a net positive charge on the conductor.

Charging by Induction

The rubbing methods we use to charge insulators such as glass, rubber, and amber do not work well with conductors. Fortunately, you can take advantage of the polarization model (which was used to explain induction) and a process known as **charging by induction** to accomplish this. In the example of charging by induction, an electrically neutral metal plate is brought near a negatively charged insulator shown in Fig. 22-10*b*. The charges in the metal plate are polarized by induction. Since the top of the metal plate now has an excess of electrons, touching it will cause these electrons to flow onto your body. If you now stop touching the rod, the return pathway for electrons is removed. So the metal plate is no longer electrically neutral. If you now move the metal plate away from the charged insulator, the polarization effect disappears and the metal plate remains positively charged due to a deficiency of electrons. Of course, it is also

FIGURE 22-11 ■ This is not a parlor stunt but a serious experiment carried out in 1774 to prove that the human body is a conductor of electricity. The etching shows a person suspended by nonconducting ropes while being charged by a charged rod (which probably touched flesh instead of the trousers). When the person brought his face, left hand, or the conducting ball and rod in his right hand near bits of paper on the plates, charge was induced on the paper, which flew through the intermediate air to him.

possible to develop a similar procedure that will leave an excess of electrons on the metal plate. (Figure 22-11 shows another experiment set up for inducing charge.)

READING EXERCISE 22-8: (a) Make the observation described in Reading Exercise 22-2. Is Scotch™ Magic Tape best described as an insulator or a conductor? Explain your reasoning. (b) Is a balloon an insulator or a conductor? Explain your reasoning. ■

READING EXERCISE 22-9: (a) Can you charge an insulator by induction? Explain your reasoning. (b) Describe the steps you would take to give an object excess negative charge using the process of charging by induction utilizing the electrophorus apparatus shown in Fig. 22-10. ■

22-7 Coulomb's Law

So far all our explanations of electrical phenomena have been qualitative. Can a mathematical law be formulated to quantitatively describe the interaction forces between electric charges?

The observations we depicted in Fig. 22-6 led us to the conclusion that the interaction forces are proportional to the product of the charges on the objects. So the magnitude of the force on either particle is given by $F \propto |q_A q_B|$. But, what observations have been made that would lead to a mathematical relationship that also describes how interaction forces are related to the distance between charged particle-like objects?

Benjamin Franklin observed that a small cork hanging from a silk thread is attracted by induction to the outside of a charged metal can. However, if the cork is dangled inside the can, there are no apparent forces on it. Recall that in Chapter 14 on gravitation we presented a shell theorem Newton derived from the assumption that gravitational forces fall off as the inverse square of the distance between masses. Newton's shell theorem implies that a shell of mass exerts no net gravitational force on a point mass contained within it. Joseph Priestly reasoned that since an analogous shell theorem seems to hold for electric interactions, then the inverse square law ought to hold for electric forces too.

In 1785, Priestley's hypothesis regarding the dependence of electric forces on the inverse square of distance was verified by the experiments of Charles Augustin Coulomb using a sensitive torsion balance to measure the forces between charged spheres.

Coulomb assumed the forces between charged spheres would be the same as if the charge of each object was concentrated at its center. He also found that the forces between the objects lie along a line between their centers. Coulomb used a method like the one we described in Section 22-4 to reduce the charges on the metal spheres in his torsion apparatus by known fractions. Thus, he was able to verify that the interaction forces were proportional to the product of the charges on the interacting objects.

As a result of his careful experiments, Coulomb found that the magnitude of the electric (or Coulomb) force, F^{elec}, between two stationary particle-like charged objects is

$$F^{\text{elec}} = k\frac{|q_A||q_B|}{r^2} \quad \text{(Coulomb's law)}, \tag{22-4}$$

where k is a positive constant of proportionality, r is the distance between the centers of the two objects, q_A is the charge on one of the objects, and q_B is the charge on the other object.

Using modern tools available in many introductory physics laboratories, we can verify the inverse square ($1/r^2$) relationship in Coulomb's law. The experiment is pictured in Fig. 22-12. Two Ping-Pong balls that are covered with conducting paint are stroked with a fur-charged rubber rod so they are negatively charged. One of the balls is hung as a pendulum from a long, nonconducting string. The other ball, which serves as a prod, is attached to a nonconducting rod.

As the prod is moved very slowly toward the hanging ball, the hanging ball is repelled and rises. The hanging ball is displaced further and further from its equilibrium as the prod is brought closer to it. This demonstrates qualitatively that the force exerted by the prod on the hanging ball is greater when the distance, r, between the centers of the two charged balls is smaller. It also indicates that the electrostatic force acts along the line connecting the two charges. We know this because the hanging ball is not pushed off to the side.

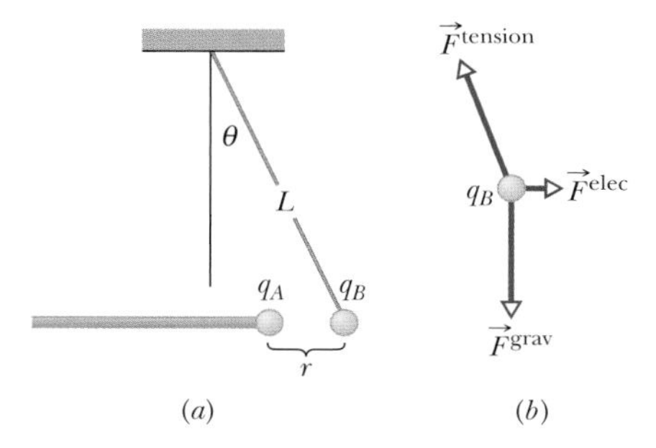

FIGURE 22-12 ▪ A charged metal-coated Ping-Pong ball is repelled from a charged prod. At equilibrium, the vector sum of the gravitational force, the tension in the string, and the Coulomb force on the hanging ball is zero.

Video technology allows us to take this experiment a step further. The motion of the prod inching forward can be captured with a video camera and digitized. Then computer software can be used to perform a frame-by-frame analysis of the angular displacement of the hanging ball and of the distance between the balls. Figure 22-13 is an example of this digital analysis. When the ball is stationary, the net force consists of the vector sum of the gravitational force acting vertically downward, the tension force exerted by the string, and the electric force acting in the horizontal direction. Thus, the magnitude of the electric force on the ball can be calculated from the mass of the ball and its angle of rise, θ, with respect to the vertical.

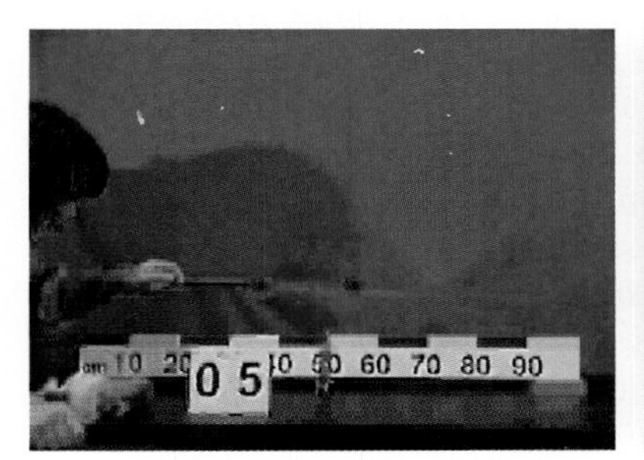
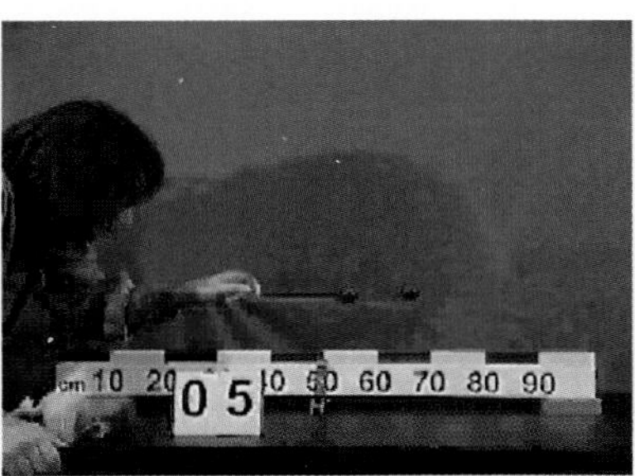
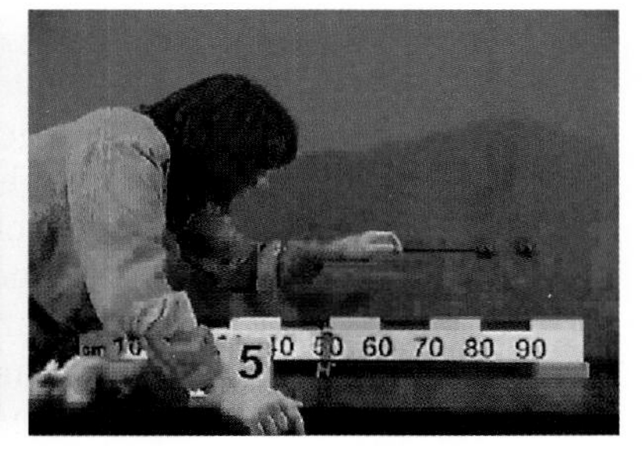

FIGURE 22-13 ▪ Three of twenty-five digitized video frames depicting the forces between two charged balls. The string holding up the hanging ball is too thin to see and its point of attachment is well above the top of the video frames.

A plot of the data is shown in Fig. 22-14. If we try to fit the data, we find that the force between electrical charges falls off with distance as $1/r^2$ just as the gravitational force does. This verifies Coulomb's result summarized in Eq. 22-4.

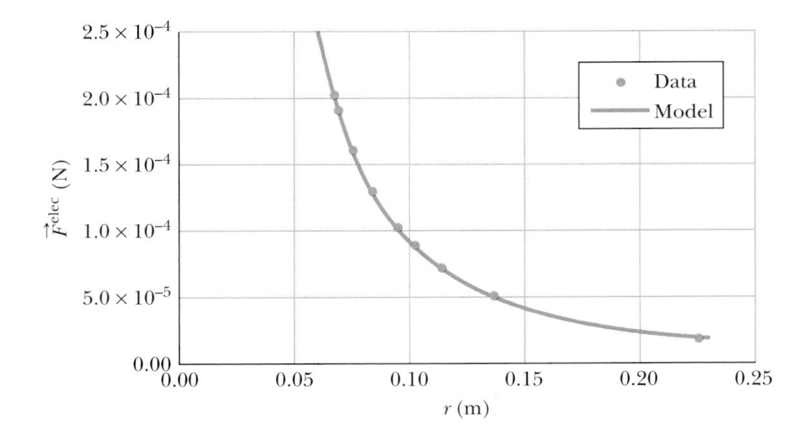

FIGURE 22-14 ▪ A graph of the magnitude of the Coulomb force vs. the distance between two charged Ping-Pong balls each having a mass of 2.40 g. The green line represents an excellent inverse square fit to the red data points. The fit is given by $F^{\text{elec}} = (7.9 \times 10^{-4}\ \text{N} \cdot \text{m}^2)/r^2$. Using a Coulomb constant of $8.99 \times 10^9\ \text{N} \cdot \text{m}^2/\text{C}^2$ (as shown in Eqs. 22-6 and 22-7), it can be shown that each ball carries about 1×10^{-8} coulombs of excess charge. VideoPoint® software was used to obtain the data from video frames (Dson015.mov).

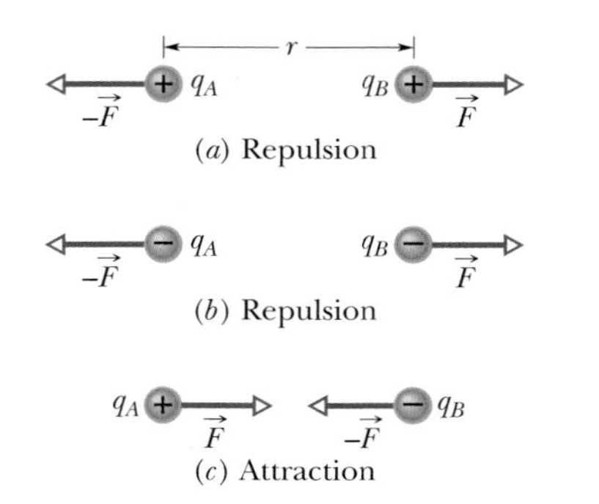

FIGURE 22-15 ■ Two charged particles, separated by distance r, repel each other if their charges are (*a*) both positive or (*b*) both negative. (*c*) They attract each other if their charges are of opposite signs. In each of the three situations, the force acting on one particle is equal in magnitude to the force acting on the other particle but has the opposite direction.

We use the absolute value signs on the force and charges in Eq. 22-4 to remind ourselves that the sign of the force (a vector) indicates the *direction* of the force and not simply whether we are multiplying like or unlike charges. Hence, we should calculate the magnitude of the force between two charged objects using the amounts of the charges. We determine the direction of the force using the attraction and repulsion rules we discussed earlier, remembering that the force always acts along the line connecting the two charges. If the particles *repel* each other, the force on each particle is directed *away from* the other particle (as in Figs. 22-15*a* and *b*). If the particles *attract* each other, the force on each particle is directed *toward* the other particle (as in Fig. 22-15*c*).

As an example of the need for absolute values in the equation

$$F^{\text{elec}}_{A\to B} = k\,\frac{|q_A||q_B|}{r^2},$$

consider the two unlike charges in Fig. 22-15*c*. Inspection of the expression for force above indicates that each particle exerts a force of the same magnitude on the other particle,

$$F_{A\to B} = F_{B\to A}. \tag{22-5}$$

But, electrostatic interactions satisfy Newton's Third Law. The force on the positive charge (in Fig. 22-15*c*) due to the negative charge points to the right. The force on the negative charge due to the positive charge points to the left. If we use explicit positive and negative signs on the charges and don't make use of absolute values, the product $q_A q_B$ (or $q_B q_A$) is always negative and so the force would be negative, regardless of whether we are calculating the force on the positive charge or the force on the negative charge. This cannot be correct, since these two forces point in opposite directions and the sign denotes direction. The force cannot be negative for both charges. In this and every other situation, we avoid this pitfall if we use the absolute values of the charges in our calculations and then determine the sign associated with the force *by thinking* about our coordinate system and the issues of attraction and repulsion.

Curiously, the form of Eq. 22-4 is the same as Newton's law of gravitation presented in Chapter 14 that relates the gravitational force between two particles with masses m_A and m_B to the distance, r, between their centers,

$$F^{\text{grav}}_{A\to B} = G\,\frac{m_A m_B}{r^2}, \tag{Eq. 14-1}$$

in which G is the gravitational constant.

Coulomb's law has survived every experimental test; no exceptions to it have ever been found. It holds even within the atom, correctly describing the force between the positively charged nucleus and each of the negatively charged electrons. This is true even though classical Newtonian mechanics fails in that realm and is replaced there by quantum physics. This simple law also correctly accounts for the forces that bind atoms together to form molecules and for the forces that bind atoms and molecules together to form solids and liquids.

The Coulomb constant k is often replaced by a factor $1/4\pi\varepsilon_0$ where $\varepsilon_0 = 4\pi k$. As you will see later this more complicated expression for the electrostatic constant simplifies many related equations that we have not yet introduced. Substituting the $1/4\pi\varepsilon_0$ term for k gives an alternate form of Coulomb's law as

$$F^{\text{elec}}_{A\to B} = \frac{1}{4\pi\varepsilon_0}\frac{|q_A||q_B|}{r^2} \quad \text{(Coulomb's law)}, \tag{22-6}$$

where k has the value

$$k = \frac{1}{4\pi\varepsilon_0} = 8.99 \times 10^9 \text{ N}\cdot\text{m}^2/\text{C}^2 \qquad \text{(Coulomb constant).} \qquad (22\text{-}7)$$

The quantity ε_0, known as the **electric constant** (sometimes called the *permittivity constant* or simply *epsilon sub zero*), often appears separately in equations and is given by

$$\varepsilon_0 = 8.85 \times 10^{-12} \text{ C}^2/\text{N}\cdot\text{m}^2 \qquad \text{(electric constant).} \qquad (22\text{-}8)$$

READING EXERCISE 22-10: Use the information provided at the end of Section 22-4 and in Fig. 22-6 to explain why the following statements cannot be true: (a) The force between two charged particle-like objects is independent of the charge on the objects. (b) The magnitude of the force between two charged particle-like objects is proportional to $1/|q_A||q_B|$. (c) The force between two charged objects is proportional to $|q_A| + |q_B|$. ■

22-8 Solving Problems Using Coulomb's Law

Coulomb's law can be used to find the forces on particle-like objects having excess charge on them. When solving quantitative (numerical) problems using Coulomb's law, there are several issues to keep in mind. For example, we must be sure to express the charges in coulombs and the distance between the charges in meters. These are the SI units for distance and charge and are required if we are to use the standard value for the Coulomb constant shown in Eq. 22-7. As we discussed in Section 22-7, we must calculate the magnitude of the force using the (positive) absolute value of the charges. We should then make a sketch of the situation, showing the direction of the force and adopting a coordinate system. If the force acts in the negative direction, then we associate a negative sign with the magnitude of the force.

What happens if there are more than two charges interacting as shown in Fig. 22-16?

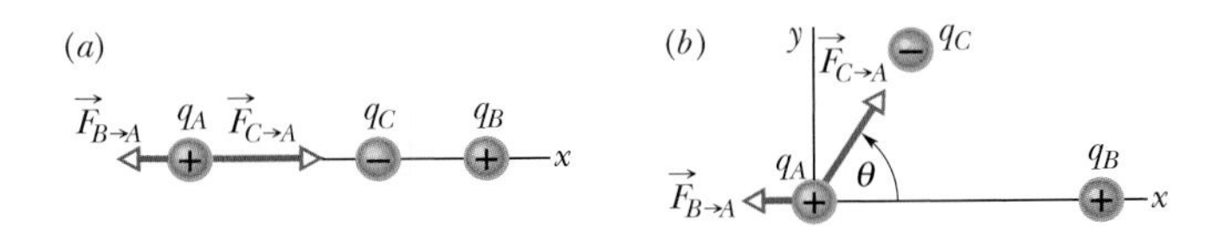

FIGURE 22-16 ■ (*a*) A set of particle-like objects with excess charge on them that lie along a line. The force vectors depict the forces of charges B and C on charge A. (*b*) A similar diagram showing the force vectors on charge A when the other charges do not lie along the same line.

Superposition of Forces

As is the case with all other forces (including the gravitational force), the electrostatic force obeys the principle of superposition. For example, if we have 4 charged particles, they interact *independently in pairs*, and the force on any one of the charges, let us say particle A, is given by the vector sum

$$\vec{F}_A^{\text{net}} = \vec{F}_{B\to A} + \vec{F}_{C\to A} + \vec{F}_{D\to A}, \qquad (22\text{-}9)$$

in which, for example, $\vec{F}_{D\to A}$ is the force acting on particle A due to the presence of particle D.

Often, as is the case for the charges in Fig. 22-16b, the various forces acting on a particle do not all act along the same line. We know how to combine forces such as these, but let's review the process. First we must calculate the magnitudes of the individual forces (in this case, using Coulomb's law), adopt a coordinate system, and determine the directions of the forces. We then calculate the orthogonal (perpendicular) components of each force and determine the direction of these components. For example, this might mean determining the x- and y-components of each of the forces as well as determining whether those components are in the positive or negative direction. We then add or subtract all of the components of forces that act along the same line. We add or subtract depending on whether the components are in the same or opposite directions. This gives us the components of the net (resultant) force. We then use trigonometry to get the magnitude and direction of the resultant force. These steps are presented in brief below.

Steps to Solving Quantitative Problems Using Coulomb's Law

Although the steps that follow are illustrated using the configuration of charges in Fig. 22-17*a*, they should work for any combination of charged particles.

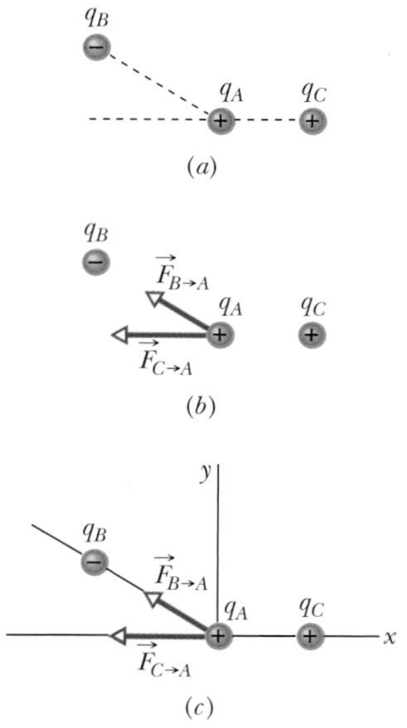

1. Sketch the array of charged objects and draw arrows to represent the anticipated directions of forces on the charged object of interest due to the surrounding charges. An example is shown in Fig. 22-17*b* where q_A is the charge of interest.
2. Use Coulomb's law to calculate the *magnitudes* of the individual forces on the charged particle of interest.
3. Determine the directions of the forces and create a free-body diagram like that shown in Fig. 22-17*b*.
4. Choose a coordinate system and sketch it on your diagram as shown in Fig. 22-17*c*.
5. Calculate the orthogonal (perpendicular) vector components of each force along the coordinate directions using the expressions $F_x = F\cos\theta$ and $F_y = F\sin\theta$, (or equivalent expressions) where F is the magnitude of the force on particle A.
6. Determine the sign of these components based on their directions.
7. Combine all the force components that act along the same line.
8. Combine the resultant components to get the magnitude of the resultant force using the expression

$$[F^{\text{net}}]^2 = [F_x^{\text{net}}]^2 + [F_y^{\text{net}}]^2. \tag{22-10}$$

9. Determine the angle at which the force acts (relative to the positive x axis) using

$$\tan\theta = \frac{F_y^{\text{net}}}{F_x^{\text{net}}}, \tag{22-11}$$

or alternatively express the force in vector notation as $\vec{F}^{\text{net}} = F_x^{\text{net}}\hat{\text{i}} + F_y^{\text{net}}\hat{\text{j}}$.

FIGURE 22-17 ■ Diagrams used to illustrate steps in problem solving using Coulomb's law.

READING EXERCISE 22-11: The figure shows two protons (symbol p) and one electron (symbol e) on an axis. What are the directions of (a) the electrostatic force on the central proton due to the electron, (b) the electrostatic force on the central proton due to the other proton, and (c) the net electrostatic force on the central proton? Are there any points along the line connecting the three charges where the central proton can be moved so that the net force on it is zero? Explain your reasoning and how your answers relate to superposition for forces.

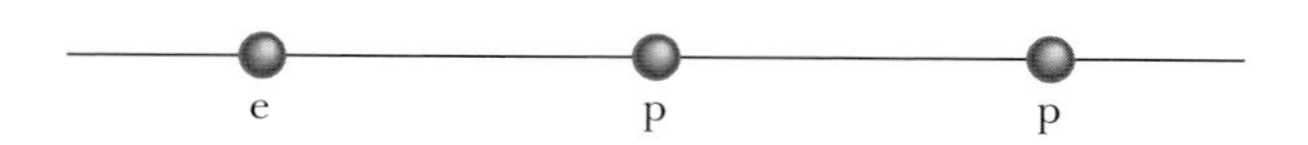

■

TOUCHSTONE EXAMPLE 22-1: Force on a Charge

(a) Figure 22-18*a* shows two positively charged particles fixed in place on an x axis. The charges are $q_A = 1.60 \times 10^{-19}$ C and $q_B = 3.20 \times 10^{-19}$ C, and the particle separation is $R = 0.0200$ m. What are the magnitude and direction of the electrostatic force $\vec{F}_{B \to A}$ on particle A from particle B?

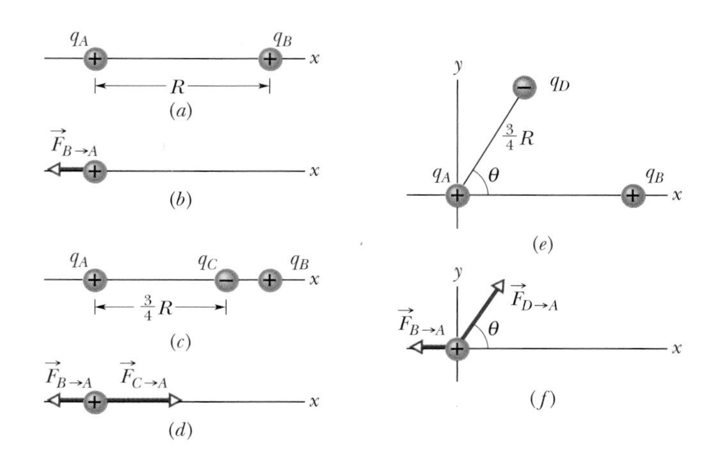

FIGURE 22-18 ■ (*a*) Two charged particles of charges q_A and q_B are fixed in place on an x axis, with separation R. (*b*) The free-body diagram for particle A, showing the electrostatic force on it from particle B. (*c*) Particle C is now fixed in place on the x axis between particles A and B. (*d*) The free-body diagram for particle A due to particles B and C. (*e*) Particle D is fixed in place on a line at angle θ to the x axis with just particles A and B present. (*f*) The new free-body diagram for particle A due to particles B and D.

SOLUTION ■ The **Key Idea** here is that, because both particles are positively charged, particle A is repelled by particle B, with a force magnitude given by Eq. 22-4. Thus, the direction of force $\vec{F}_{B \to A}$ on particle A is *away from* particle B, in the negative direction of the x axis, as indicated in the free-body diagram of Fig. 22-18*b*. Using Eq. 22-4 with separation R substituted for r, we can write the magnitude $F_{B \to A}$ of this force as

$$
\begin{aligned}
F_{B\to A} &= k\frac{|q_A||q_B|}{R^2} \\
&= (8.99 \times 10^9\ \text{N}\cdot\text{m}^2/\text{C}^2) \times \frac{(1.60 \times 10^{-19}\ \text{C})(3.20 \times 10^{-19}\ \text{C})}{(0.0200\ \text{m})^2} \\
&= 1.15 \times 10^{-24}\ \text{N}.
\end{aligned}
$$

Thus, force $\vec{F}_{B \to A}$ has the following magnitude and direction (relative to the positive direction of the x axis):

$$1.15 \times 10^{-24}\ \text{N} \quad \text{and} \quad 180°. \qquad \text{(Answer)}$$

We can also write $\vec{F}_{B \to A}$ in unit-vector notation as

$$\vec{F}_{B\to A} = -(1.15 \times 10^{-24}\ \text{N})\hat{\text{i}}. \qquad \text{(Answer)}$$

(b) Figure 22-18*c* is identical to Fig. 22-18*a* except that particle C now lies on the x axis between particles A and B. Particle C has charge $q_C = -3.20 \times 10^{-19}$ C and is at a distance $\frac{3}{4}R$ from particle A. What is the net electrostatic force $\vec{F}_A^{\text{net}}$ on particle A due to particles B and C?

SOLUTION ■ One **Key Idea** here is that the presence of particle C does not alter the electrostatic force on particle A from particle B. Thus, force $\vec{F}_{B \to A}$ still acts on particle A. Similarly, the force $\vec{F}_{C \to A}$ that acts on particle A due to particle C is not affected by the presence of particle B. Because particles A and C have charge of opposite sign, particle A is attracted to particle C. Thus, force $\vec{F}_{C \to A}$ is directed *toward* particle C, as indicated in the free-body diagram of Fig. 22-18*d*.

To find the magnitude of $\vec{F}_{C \to A}$, we can rewrite Eq. 22-4 as

$$
\begin{aligned}
F_{C\to A} &= k\frac{|q_A||q_C|}{(\frac{3}{4}R)^2} \\
&= (8.99 \times 10^9\ \text{N}\cdot\text{m}^2/\text{C}^2) \times \frac{(1.60 \times 10^{-19}\ \text{C})(3.20 \times 10^{-19}\ \text{C})}{(\frac{3}{4})(0.0200\ \text{m})^2} \\
&= 2.05 \times 10^{-24}\ \text{N}.
\end{aligned}
$$

We can also write $\vec{F}_{C \to A}$ in unit-vector notation:

$$\vec{F}_{C\to A} = +(2.05 \times 10^{-24}\ \text{N})\hat{\text{i}}.$$

A second **Key Idea** here is that the net force $\vec{F}_A^{\text{net}}$ on particle A is the vector sum of $\vec{F}_{B \to A}$ and $\vec{F}_{C \to A}$; that is, from Eq. 22-9, we can write the net force $\vec{F}_A^{\text{net}}$ on particle A in unit-vector notation as

$$
\begin{aligned}
\vec{F}_A^{\text{net}} &= \vec{F}_{B\to A} + \vec{F}_{C\to A} \\
&= -(1.15 \times 10^{-24}\ \text{N})\hat{\text{i}} + (2.05 \times 10^{-24}\ \text{N})\hat{\text{i}} \\
&= (9.00 \times 10^{-25}\ \text{N})\hat{\text{i}}. \qquad \text{(Answer)}
\end{aligned}
$$

Thus, $\vec{F}_A^{\text{net}}$ has the following magnitude and direction (relative to the positive direction of the x axis):

$$9.00 \times 10^{-25}\ \text{N} \quad \text{and} \quad 0°. \qquad \text{(Answer)}$$

(c) Figure 22-18*e* is identical to Fig. 22-18*a* except that particle D is now positioned as shown. Particle D has charge $q_D = -3.20 \times 10^{-19}$ C, is at a distance $\frac{3}{4}R$ from particle A, and lies on a line that makes an angle $\theta = 60°$ with the x axis. What is the net electrostatic force $\vec{F}_A^{\text{net}}$ on particle A due to particles B and D?

SOLUTION ■ The **Key Idea** here is that the net force $\vec{F}_A^{\text{net}}$ is the vector sum of $\vec{F}_{B \to A}$ and a new force $\vec{F}_{D \to A}$ acting on particle A due to particle D. Because particles A and D have charges of opposite sign, particle A is attracted to particle D. Thus, force $\vec{F}_{D \to A}$ on particle A is directed *toward* particle D, at angle $\theta = 60°$, as indicated in the free-body diagram of Fig. 22-18*f*.

To find the magnitude of $\vec{F}_{D\to A}$, we can rewrite Eq. 22-4 as

$$|\vec{F}_{D\to A}| = k\frac{|q_A||q_D|}{(\frac{3}{4}R)^2}$$

$$= (8.99 \times 10^9\ \text{N}\cdot\text{m}^2/\text{C}^2) \times \frac{(1.60 \times 10^{-19}\ \text{C})(3.2 \times 10^{-19}\ \text{C})}{(\frac{3}{4})^2(0.0200\ \text{m})^2}$$

$$= 2.05 \times 10^{-24}\ \text{N}.$$

Then from Eq. 22-9, we can write the net force $\vec{F}_A^{\text{net}}$ on particle A as

$$\vec{F}_A^{\text{net}} = \vec{F}_{B\to A} + \vec{F}_{D\to A}.$$

To evaluate the right side of this equation, we need another **Key Idea**: Because the forces $\vec{F}_{B\to A}$ and $\vec{F}_{D\to A}$ are not directed along the same axis, we *cannot* sum simply by combining their magnitudes. Instead, we must add them as vectors, using one of the following methods.

Method 1. *Summing directly on a vector-capable calculator.* For $\vec{F}_{B\to A}$, we enter the magnitude 1.15×10^{-24} and the angle 180°. For $\vec{F}_{D\to A}$, we enter the magnitude 2.05×10^{-24} and the angle 60°. Then we add the vectors.

Method 2. *Summing in unit-vector notation.* First we rewrite $\vec{F}_{D\to A}$ as

$$\vec{F}_{D\to A} = (F_{D\to A}\cos\theta)\hat{\text{i}} + (F_{D\to A}\sin\theta)\hat{\text{j}}.$$

Substituting 2.05×10^{-24} N for $F_{D\to A}$ and 60° for θ, this becomes

$$\vec{F}_{D\to A} = (1.025 \times 10^{-24}\ \text{N})\hat{\text{i}} + (1.775 \times 10^{-24}\ \text{N})\hat{\text{j}}.$$

Then we sum:

$$\vec{F}_A^{\text{net}} = \vec{F}_{B\to A} + \vec{F}_{D\to A}$$

$$= -(1.15 \times 10^{-24}\ \text{N})\hat{\text{i}} + (1.025 \times 10^{-24}\ \text{N})\hat{\text{i}} + (1.775 \times 10^{-24})\hat{\text{j}}$$

$$\approx (-1.25 \times 10^{-25}\ \text{N})\hat{\text{i}} + (1.78 \times 10^{-24}\ \text{N})\hat{\text{j}}. \quad \text{(Answer)}$$

Method 3. *Summing components axis by axis.* The sum of the x-components gives us

$$F_{Ax}^{\text{net}} = F_{B\to Ax} + F_{D\to Ax} = F_{B\to A} + F_{D\to A}\cos 60°$$

$$= -1.15 \times 10^{-24}\ \text{N} + (2.05 \times 10^{-24}\ \text{N})(\cos 60°)$$

$$= -1.25 \times 10^{-25}\ \text{N}.$$

The sum of the y-components gives us

$$F_{Ay}^{\text{net}} = F_{B\to Ay} + F_{D\to Ay} = 0 + F_{D\to A}\sin 60°$$

$$= (2.05 \times 10^{-24}\ \text{N})(\sin 60°)$$

$$= 1.78 \times 10^{-24}\ \text{N}.$$

The net force $\vec{F}_A^{\text{net}}$ has the magnitude

$$F_A^{\text{net}} = \sqrt{(F_{Ax}^{\text{net}})^2 + (F_{Ay}^{\text{net}})^2} = 1.78 \times 10^{-24}\ \text{N}.$$

To find the direction of $\vec{F}_A^{\text{net}}$, we take

$$\theta = \tan^{-1}\frac{F_{Ay}^{\text{net}}}{F_{Ax}^{\text{net}}} = -86.0°.$$

However, this is an unreasonable result because $\vec{F}_A^{\text{net}}$ must have a direction between the directions of $\vec{F}_{B\to A}$ and $\vec{F}_{D\to A}$. To correct θ, we add 180°, obtaining

$$-86.0° + 180° = 94.0°. \quad \text{(Answer)}$$

TOUCHSTONE EXAMPLE 22-2: Equilibrium Point

Figure 22-19*a* shows two particles fixed in place: a particle of charge $q_A = +8q$ at the origin and a particle of charge $q_B = -2q$ at $x = L$. At what point (other than infinitely far away) can a proton of charge q_p be placed so that it is in *equilibrium* (meaning that the net force on it is zero)? Is that equilibrium *stable* or *unstable*?

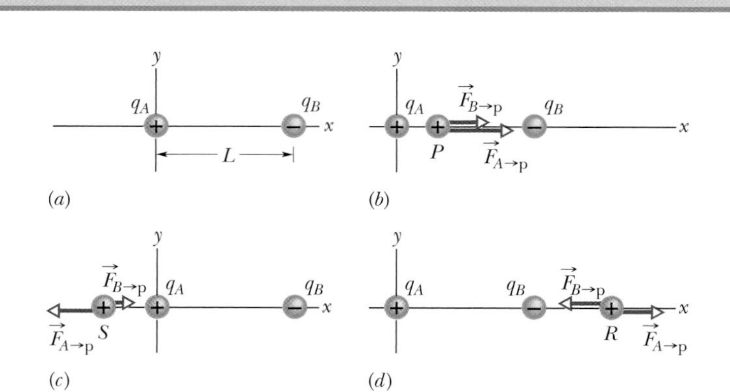

FIGURE 22-19 (*a*) Two particles of charges q_A and q_B are fixed in place on an x axis, with separation L. (*b*)–(*d*) Three possible locations P, S, and R for a proton. At each location, $\vec{F}_{A\to p}$ represents the force on the proton from particle A and $\vec{F}_{B\to p}$ represents the force on the proton from particle B.

SOLUTION ■ The **Key Idea** here is that, if $\vec{F}_{A\to p}$ is the force on the proton due to charge q_A and $\vec{F}_{B\to p}$ is the force on the proton due to charge q_B, then the point we seek is where $\vec{F}_{A\to p} + \vec{F}_{B\to p} = 0$. This condition requires that

$$\vec{F}_{A\to p} = -\vec{F}_{B\to p}. \tag{22-12}$$

This tells us that at the point we seek, the forces acting on the proton due to the other two particles must be of equal magnitudes,

$$|\vec{F}_{A\to p}| = |\vec{F}_{B\to p}|, \tag{22-13}$$

and that the forces must have opposite directions.

A proton has a positive charge. Thus, the proton and the particle of charge q_A are of the same sign, and force $\vec{F}_{A\to p}$ on the proton must point away from q_A. Also, the proton and the particle of charge q_B are of opposite signs, so force $\vec{F}_{B\to p}$ on the proton must point toward q_B. "Away from q_A" and "toward q_B" can be in opposite directions only if the proton is located on the x axis.

If the proton is on the x axis at any point between q_A and q_B, such as P in Fig. 22-19b, then $\vec{F}_{A\to p}$ and $\vec{F}_{B\to p}$ are in the same direction and not in opposite directions as required. If the proton is at any point on the x axis to the left of q_A, such as point S in Fig. 22-9c, then $\vec{F}_{A\to p}$ and $\vec{F}_{B\to p}$ are in opposite directions. However, Eq. 22-4 tells us that $\vec{F}_{A\to p}$ and $\vec{F}_{B\to p}$ cannot have equal magnitudes there: $|\vec{F}_{A\to p}|$ must be greater than $|\vec{F}_{B\to p}|$, because $|\vec{F}_{A\to p}|$ is produced by a closer charge (with lesser r) of greater magnitude ($8q$ versus $2q$).

Finally, if the proton is at any point on the x axis to the right of q_B, such as point R in Fig. 22-19d, then $\vec{F}_{A\to p}$ and $\vec{F}_{B\to p}$ are again in opposite directions. However, because now the charge of greater amount (q_A) is *farther* away from the proton than the charge of lesser amount, there is a point at which $|\vec{F}_{A\to p}|$ is equal to $|\vec{F}_{B\to p}|$. Let x be the coordinate of this point, and let q_p be the charge of the proton. Then with the aid of Eq. 22-4, we can rewrite Eq. 22-13 as

$$k\frac{8|q||q_p|}{x^2} = k\frac{2|q||q_p|}{(x-L)^2}. \tag{22-14}$$

(Note that only the charge amounts appear in Eq. 22-14.) Rearranging Eq. 22-14 gives us

$$\left(\frac{x-L}{x}\right)^2 = \frac{1}{4}.$$

After taking the square roots of both sides, we have

$$\frac{x-L}{x} = \frac{1}{2},$$

which gives us

$$x = 2L. \quad \text{(Answer)}$$

The equilibrium at $x = 2L$ is unstable; that is, if the proton is displaced leftward from point R, then $|\vec{F}_{A\to p}|$ and $|\vec{F}_{B\to p}|$ both increase but $|\vec{F}_{B\to p}|$ increases more (because q_B is closer than q_A), and a net force will drive the proton farther leftward. If the proton is displaced rightward, both $|\vec{F}_{A\to p}|$ and $|\vec{F}_{B\to p}|$ decrease but $|\vec{F}_{B\to p}|$ decreases more, and a net force will then drive the proton farther rightward. In a stable equilibrium, each time the proton was displaced slightly, it would return to the equilibrium position.

22-9 Comparing Electrical and Gravitational Forces

Consider two point-like objects, A and B, separated by a distance r. What are the magnitudes of the electrical and gravitational forces between them?

$$F^{\text{elec}}_{A\to B} = k\frac{|q_A||q_B|}{r^2} \quad \text{(Coulomb's Law)}, \tag{Eq. 22-4}$$

has the same form as that of Newton's equation for the gravitational force between two particles with masses m_A and m_B that are separated by a distance r:

$$F^{\text{grav}}_{A\to B} = G\frac{m_A m_B}{r^2} \quad \text{(Newton's Law of gravitation)}. \tag{14-2}$$

Both of these equations have the distance between the two interacting objects squared and in the denominator of the fraction. That is, they are both "inverse square laws." Both also involve a property of the interacting particles—the mass in one case and the charge in the other. Both the gravitational force and the electrostatic force are conservative forces—the work done by these forces around a closed path is zero. Both forces act along the line connecting the two objects—such forces are called "central" forces.

However, as similar as these forces are, they are not the same force. They are not even different aspects of one force. How do we know this? Electrostatic forces are intrinsically much stronger than gravitational forces. For example, the gravitational attraction between a plastic comb and a small piece of paper is not large enough to overcome the opposing gravitational attraction of Earth on the paper. However, if you rub the comb with fur, the resulting electrostatic force *is* large enough to overcome the gravitational attraction of Earth. Furthermore, the electrostatic force differs from the gravitational forces because the gravitational force is always attractive but the electrostatic force may be *either* attractive or repulsive, depending on the signs of the two charges. This difference arises because although there is only one kind of mass, there are two kinds of charge. That is why absolute value signs are needed in

$$F^{\text{elec}}_{A\to B} = k\,\frac{|q_A||q_B|}{r^2}$$

but not in

$$F^{\text{grav}}_{A\to B} = G\,\frac{m_A m_B}{r^2}.$$

Before concluding our discussion of the electrostatic or Coulomb force, let's compare it to another somewhat similar force—the force associated with magnets.

In addition to amber, the early Greeks knew of another special material that had the ability to attract other objects. They recorded the observation that some naturally occurring "lodestones," known today as the mineral magnetite, would attract iron. Lodestones were the first known magnets. Could the phenomena of amber (electricity) and lodestones (magnetism) be related?

Observation of the interactions between two magnets and two electrified objects shows that the phenomena of electricity and magnetism are *not* the same. Two magnets will either attract or repel one another, depending on their orientation. Two pieces of rubbed amber (or glass) always repel one another, regardless of their orientation.

Hence, the study of electricity and magnetism developed separately for centuries—until 1820, in fact, when Hans Christian Oersted found a connection between them: an electric current in a wire can deflect a magnetic compass needle. The new science of *electromagnetism* (the combination of electrical and magnetic phenomena) was developed further by Michael Faraday, a truly gifted experimenter with a talent for physical intuition and visualization. In fact, Faraday's laboratory notebooks do not contain a single equation. In the mid-19th century, James Clerk Maxwell put Faraday's ideas into mathematical form, introduced many new ideas of his own, and put electromagnetism on a sound theoretical basis.

Table 32-1 shows the basic laws of electromagnetism, now called Maxwell's equations. We plan to work our way through them in the chapters between here and there, but you might want to glance at them now, to see where we are heading.

TOUCHSTONE EXAMPLE 22-3: Nuclear Repulsion

The nucleus in an iron atom has a radius of about 4.0×10^{-15} m and contains 26 protons.

(a) What is the magnitude of the repulsive electrostatic force between two of the protons that are separated by 4.0×10^{-15} m?

SOLUTION ■ The **Key Idea** here is that the protons can be treated as charged particles, so the magnitude of the electrostatic force on one from the other is given by Coulomb's law. Table 22-1 tells us that their charge is $+e$. Thus, Eq. 22-4 gives us

$$\begin{aligned} F^{\text{elec}} &= \frac{ke^2}{r^2} \\ &= \frac{(8.99 \times 10^9\ \text{N}\cdot\text{m}^2/\text{C}^2)(1.60 \times 10^{-19}\ \text{C})^2}{(4.0 \times 10^{-15}\ \text{m})^2} \\ &= 14\ \text{N}. \end{aligned} \qquad \text{(Answer)}$$

This is a small force to be acting on a macroscopic object like a cantaloupe but an enormous force to be acting on a proton. Such forces should blow apart the nucleus of any element but hydrogen (which has only one proton in its nucleus). However, they don't, not even in nuclei with a great many protons. Therefore, there must be some enormous attractive force to counter this enormous repulsive electrostatic force.

(b) What is the magnitude of the gravitational force between those same two protons?

SOLUTION ■ The **Key Idea** here is like that in part (a): Because the protons are particles, the magnitude of the gravitational force on one from the other is given by Newton's equation for the gravitational force (Eq. 14-2). With $m_p (= 1.67 \times 10^{-27}\ \text{kg})$ representing the mass of a proton, Eq. 14-2 gives us

$$F = G\frac{m_p^2}{r^2}$$
$$= \frac{(6.67 \times 10^{-11}\,\mathrm{N \cdot m^2/kg^2})(1.67 \times 10^{-27}\,\mathrm{kg})}{(4.0 \times 10^{-15}\,\mathrm{m})^2}$$
$$= 1.2 \times 10^{-35}\,\mathrm{N}. \qquad \text{(Answer)}$$

This result tells us that the (attractive) gravitational force is far too weak to counter the repulsive electrostatic forces between protons in a nucleus. Instead, the protons are bound together by an enormous force aptly called the *strong nuclear force*—a force that acts between protons (and neutrons) when they are close together, as in a nucleus.

Although the gravitational force is many times weaker than the electrostatic force, it is more important in large-scale situations because it is always attractive. This means that it can collect many small bodies into huge bodies with huge masses, such as planets and stars, that then exert large gravitational forces. The electrostatic force, on the other hand, is repulsive for charges of the same sign, so it is unable to collect either positive charge or negative charge into large concentrations that would then exert large electrostatic forces.

22-10 Many Everyday Forces Are Electrostatic

In Chapter 6 we presented an idealized model of a solid as an array of atoms held together by forces that act like tiny springs that resist both stretching and compression forces. We then used this spring model to help explain the nature of most of the everyday forces encountered in the study of motion including normal forces, friction forces, and tension forces. We made the claim that all of these forces are basically electrical.

Let's look once again at our spring model of solids in light of our new understanding of the nature of the electrostatic forces between protons and electrons in atoms. Since protons and electrons have opposite charges they attract each other. This is what holds individual atoms together and causes a tension force to arise in a string as it resists stretching. Under compression, the outer electrons in the atoms of one object repel the outer electrons in the other object. This is the origin of the normal force. Although we imagine this repulsion starting at the surface, it is happening in other layers of atoms as well. As the electrons from one layer of atoms are being moved closer to those in the next layer, the repulsion forces increase sharply as the electrons are forced closer together.

Thus, we think of a solid as having a delicately balanced equilibrium in which the electron glue holds the atoms together at just the right spacing. Although more detailed analysis of these phenomena requires quantum mechanics, all of the everyday forces we encounter appear to be either gravitational, electrical, or magnetic. We explore the relationship between electrostatic and magnetic interactions further in Chapter 32.

Problems

SEC. 22-4 ■ USING ATOMIC THEORY TO EXPLAIN CHARGING

1. A Large Charge What is the total charge in coulombs of 75.0 kg of electrons?

2. How Many? How many megacoulombs of positive (or negative) charge are in 1.00 mol of neutral molecular-hydrogen gas (H_2)?

3. How Many Electrons How many electrons would have to be removed from a coin to leave it with a charge of $+1.0 \times 10^{-7}$ C?

4. Glass of Water Calculate the number of coulombs of positive charge in 250 $\mathrm{cm^3}$ of (neutral) water (about a glassful).

5. Cosmic Ray Protons Earth's atmosphere is constantly bombarded by *cosmic ray protons* that originate somewhere in space. If the protons all passed through the atmosphere, each square meter of Earth's surface would intercept protons at the average rate of 1500 protons per second. What would be the corresponding rate of charge flow intercepted by the total surface area of the planet?

6. Fibrillation A charge flow of 0.300 C/s through your chest can send your heart into fibrillation, disrupting the flow of blood (and thus oxygen) to your brain. If that current persists for 2.00 min, how many conduction electrons pass through your chest?

7. Beta Decay In *beta decay* a massive fundamental particle changes to another massive particle, and either an electron of charge $-e$ or a positron of charge $+e$ (positive particle with the same amount of charge and mass as an electron) is emitted. (a) If a proton undergoes beta decay to become a neutron, which particle is emitted? (b) If a neutron undergoes beta decay to become a proton, which particle is emitted?

8. Identify X Identify X in the following nuclear reactions (in the first, n represents a neutron): (a) $^{1}\text{H} + {}^{9}\text{Be} \rightarrow \text{X} + \text{n}$; (b) $^{12}\text{C} + {}^{1}\text{H} \rightarrow \text{X}$; (c) $^{15}\text{N} + {}^{1}\text{H} \rightarrow {}^{4}\text{He} + \text{X}$. Appendix F will help.

SEC. 22-8 ■ SOLVING PROBLEMS USING COULOMB'S LAW

9. What Distance At what distance between point charge $q_A = 26.0\ \mu\text{C}$ and point charge $q_B = -47.0\ \mu\text{C}$ will the electrostatic force between them have a magnitude of 5.70 N?

10. Force on Each A point charge of $+3.00 \times 10^{-6}$ C is 12.0 cm from a second point charge of -1.50×10^{-6} C. Calculate the magnitude of the force on each charge.

11. Two Equally Charged Two equally charged particles, held 3.2×10^{-3} m apart, are released from rest. The initial acceleration of the first particle is observed to be 7.0 m/s² and that of the second to be 9.0 m/s². If the mass of the first particle is 6.3×10^{-7} kg, what are (a) the mass of the second particle and (b) the amount of charge on each particle?

12. Isolated Conducting Spheres Identical isolated conducting spheres A and B have the same excess charges and are separated by a distance that is large compared with their diameters (Fig. 22-20*a*). The electrostatic force acting on sphere B due to sphere A is $\vec{F}_{A\rightarrow B}$. Suppose now that a third identical sphere C, having an insulating handle and initially neutral, is touched first to sphere A (Fig. 22-20*b*), then to sphere B (Fig. 22-20*c*), and finally removed (Fig. 22-20*d*). In terms of the force magnitude $F_{A\rightarrow B}$, what is the magnitude of the electrostatic force $\vec{F}'_{A\rightarrow B}$ that now acts on sphere B?

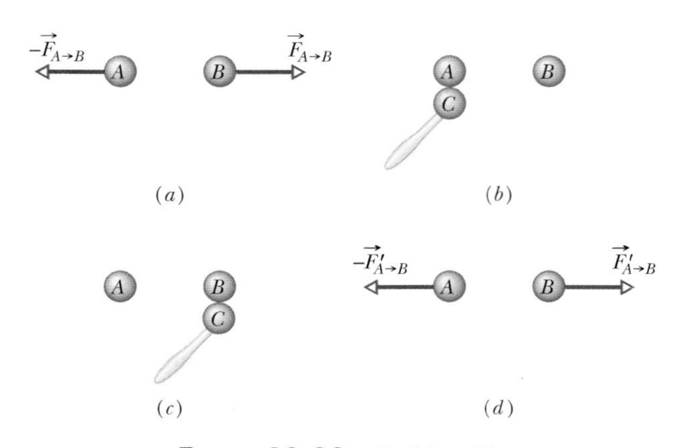

FIGURE 22-20 ■ Problem 12.

13. The Square In Fig. 22-21, what are the (a) horizontal and (b) vertical components of the net electrostatic force on the charged particle in the lower left corner of the square if $q = 1.0 \times 10^{-7}$ C and $a = 5.0$ cm?

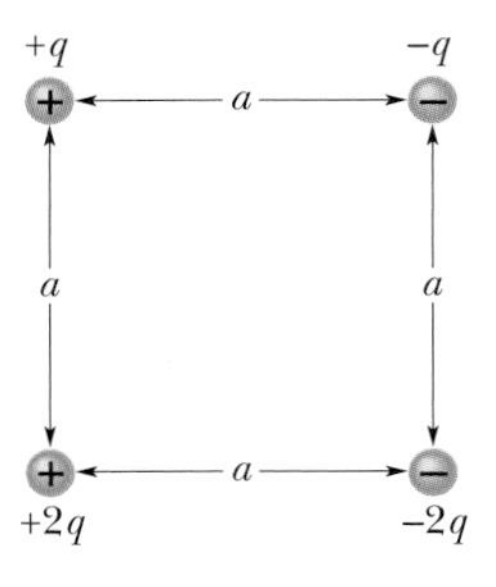

FIGURE 22-21 ■ Problem 13.

14. Where Along the Line Point charges q_A and q_B lie on the x axis at points $x = -d$ and $x = +d$, respectively. (a) How must q_A and q_B be related for the net electrostatic force on point charge $+Q$, placed at $x = +d/2$, to be zero? (b) Repeat (a) but with point charge $+Q$ now placed at $x = +3d/2$.

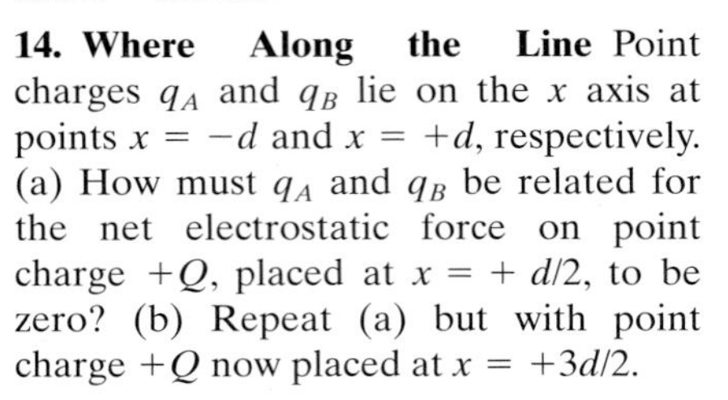

15. Two Identical Spheres Two identical conducting spheres, fixed in place, attract each other with an electrostatic force of 0.108 N when separated by 50.0 cm, center to center. The spheres are then connected by a thin conducting wire. When the wire is removed, the spheres repel each other with an electrostatic force of 0.0360 N. What were the initial charges on the spheres?

16. Three Charges In Fig. 22-22, three charged particles lie on a straight line and are separated by distances d. Charges q_A and q_B are held fixed. Charge q_C is free to move but happens to be in equilibrium (no net electrostatic force acts on it). Find q_A in terms of q_B.

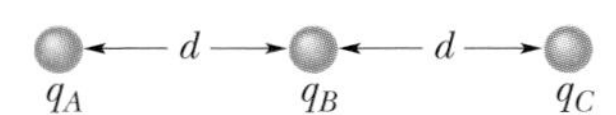

FIGURE 22-22 ■ Problem 16.

17. Two Free Particles Two free particles (that is, free to move) with charges $+q$ and $+4q$ are a distance L apart. A third charge is placed so that the entire system is in equilibrium. (a) Find the location, amount and sign of the third charge. (b) Show that the equilibrium is unstable.

18. Two Fixed Particles Two fixed particles, of charges $q_A = +1.0\ \mu\text{C}$ and $q_B = -3.0\ \mu\text{C}$, are 10 cm apart. How far from each should a third charge be located so that no net electrostatic force acts on it?

19. A Certain Charge Q A certain charge Q is divided into two parts q and $Q - q$, which are then separated by a certain distance. What must q be in terms of Q to maximize the electrostatic repulsion between the two charges?

20. Charges and Coordinates The charges and coordinates of two charged particles held fixed in the xy plane are $q_A = +3.0\ \mu\text{C}$, $x_A = 3.5$ cm, $y_A = 0.50$ cm, and $q_B = -4.0\ \mu\text{C}$, $x_B = -2.0$ cm, $y_B = 1.5$ cm. (a) Find the magnitude and direction of the electrostatic force on q_B. (b) Where could you locate a third charge $q_C = +4.0\ \mu\text{C}$ such that the net electrostatic force on q_B is zero?

21. Identical Ions The magnitude of the electrostatic force between two identical ions that are separated by a distance of 5.0×10^{-10} m is 3.7×10^{-9} N. (a) What is the charge of each ion? (b) How many electrons are "missing" from each ion (thus giving the ion its charge imbalance)?

22. Salt Crystal What is the magnitude of the electrostatic force between a singly charged sodium ion (Na^{+}, of charge $+e$) and an adjacent singly charged chlorine ion (Cl^{-}, of charge $-e$) in a salt crystal if their separation is 2.82×10^{-10} m?

23. Cesium Chloride In the basic CsCl (cesium chloride) crystal structure, Cs^{+} ions form the corners of a cube and a Cl^{-} ion is at the cube's center (Fig. 22-23). The edge length of the cube is 0.40 nm. The Cs^+ ions are each deficient by one electron (and thus each has a charge of $+e$), and the Cl^- ion has one excess electron (and thus has a charge of $-e$). (a) What is the magnitude of the net electrostatic force exerted on the Cl^- ion by the eight Cs^+ ions at the corners of the cube? (b) If one of the Cs^+ ions is missing, the crystal is said to have a *defect;* what is the magnitude of the net electrostatic force exerted on the Cl^- ion by the seven remaining Cs^+ ions?

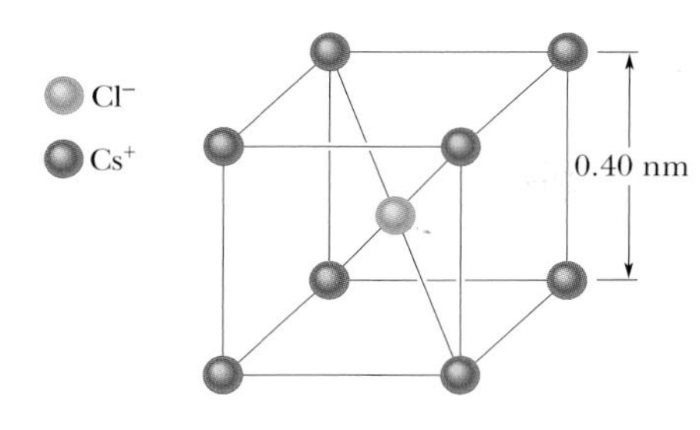

FIGURE 22-23 ■ Problem 23.

24. Water Drops Two tiny, spherical water drops, with identical charges of -1.00×10^{-16} C, have a center-to-center separation of 1.00 cm. (a) What is the magnitude of the electrostatic force acting between them? (b) How many excess electrons are on each drop, giving it its charge imbalance?

25. Beads Figure 22-24 shows four tiny charged beads that can be slid or fixed in place on wires that stretch along x and y axes. A central bead at the crossing point of the wires (the origin) has a charge of $+e$. The other beads each have a charge of $-e$. Initially beads A, B, and C are at distance $d = 10.0$ cm from the central bead, and bead D is at a distance of $d/2$. (a) How far from the central bead must you position bead A so that the direction of the net electrostatic force $\vec{F}^{\,\text{net}}$ on the central bead rotates counterclockwise by 30°? (b) With bead A still in its new position, where must you slide bead C so that the direction of $\vec{F}^{\,\text{net}}$ rotates back by 30°?

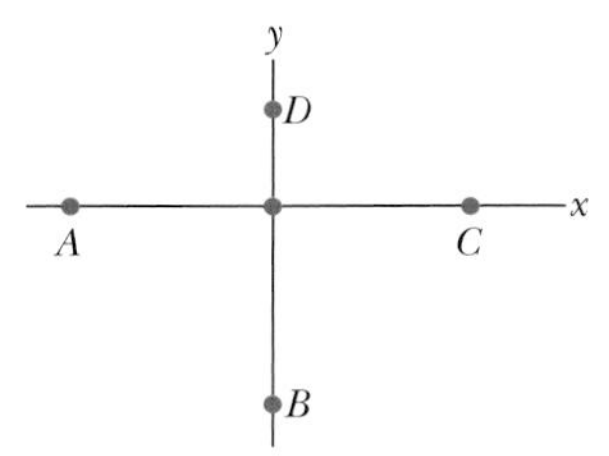

FIGURE 22-24 ■ Problem 25.

26. Two Copper Coins We know that the negative charge on the electron and the positive charge on the proton are equal in amount. Suppose, however, that these amounts differ from each other by 0.00010%. With what force would two copper coins, placed 1.0 m apart, repel each other? Assume that each coin contains 3×10^{22} copper atoms. (*Hint:* A neutral copper atom contains 29 protons and 29 electrons.) What do you conclude?

27. Particles *A* and *B* Figure 22-25a shows charged particles A and B that are fixed in place on an x axis. Particle A has an amount of charge of $|q_A| = 8.00e$. Particle C, with a charge of $q_C = +8.00e$, is initially on the x axis near particle B. Then particle C is gradually moved in the positive direction of the x axis. As a result, the magnitude of the net electrostatic force $\vec{F}_B^{\,\text{net}}$ on particle B due to particles A and C changes. Figure 22-25b gives the x-component of that net force as a function of the position x of particle C. The plot has an asymptote of $F_B^{\text{net}} = 1.5 \times 10^{-25}$ N as $x \to \infty$. As a multiple of e, what is the charge q_B of particle B?

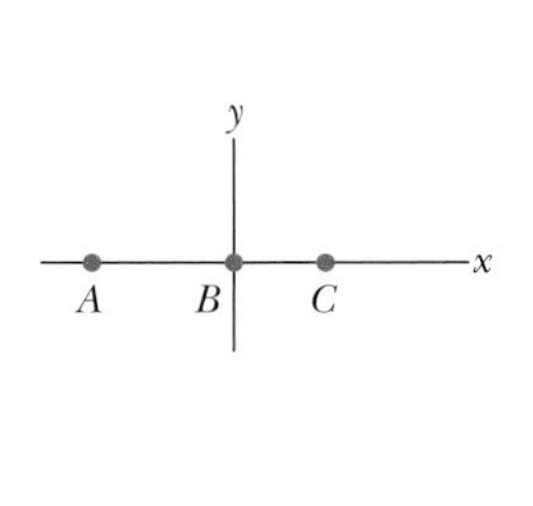

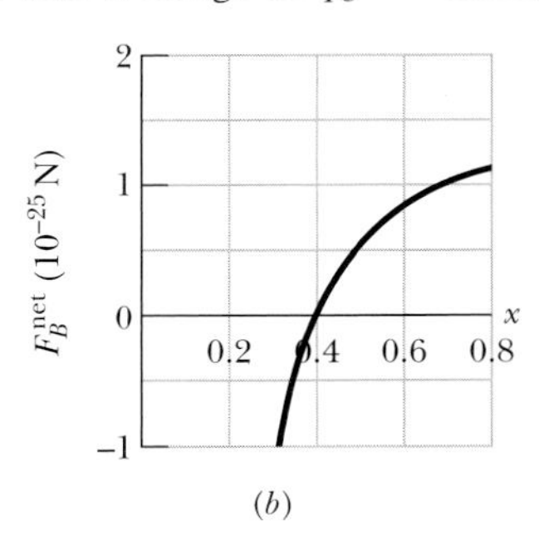

FIGURE 22-25 ■ Problem 27.

28. Above the Floor In Fig. 22-26, a particle of charge $+4e$ is above a floor by distance $d_1 = 2.0$ mm and a particle of charge $+6e$ is on the floor at horizontal distance $d_2 = 6.0$ mm from the first particle. What is the x-component of the electrostatic force on the second particle due to the first particle?

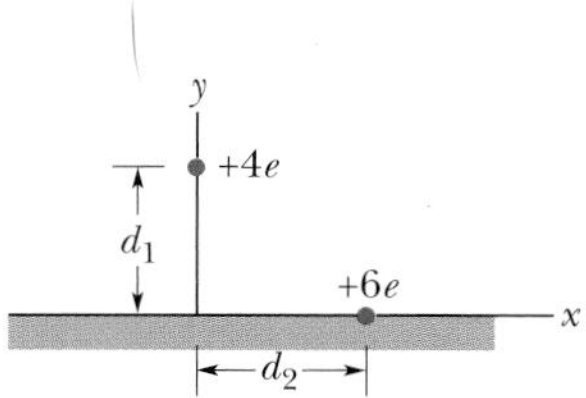

FIGURE 22-26 ■ Problem 28.

29. Fixed on the *x* Axis In Fig. 22-27a, particle A (with charge q_A) and particle B (with charge q_B) are fixed in place on an x axis, 8.00 cm apart. Particle C with a charge $q_C = +5e$ is to be placed on the line between particles A and B, so that they produce a net electrostatic force $\vec{F}_C^{\,\text{net}}$ on it. Figure 22-27b gives the x-component of that force versus the coordinate x at which particle C is placed. What are (a) the sign of charge q_A and (b) the ratio q_B/q_A?

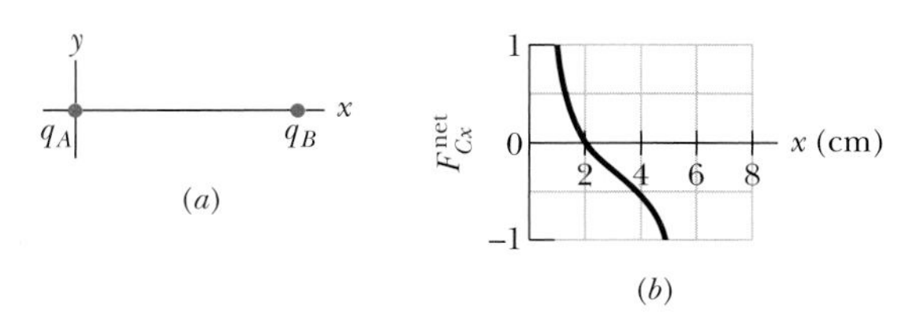

FIGURE 22-27 ■ Problem 29.

30. Four Charged Particles Figure 22-28 shows four charged particles that are fixed along an axis, separated by distance $d = 2.00$ cm. The charges are indicated. Find the magnitude and direction of the net electrostatic force on (a) the particle with charge $+2e$ and (b) the particle with charge $-e$, due to the other particles.

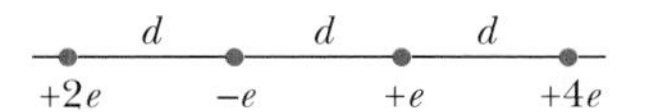

FIGURE 22-28 ■ Problem 30.

31. Split in Two A charge of 6.0 μC is to be split into two parts that are then separated by 3.0 mm. What is the maximum possible magnitude of the electrostatic force between those two parts?

32. Two on the Axis Figure 22-29 shows two particles, each of charge $+2e$, that are fixed on a y axis, each at a distance $d = 17$ cm from the x axis. A third particle, of charge $+4e$, is moved slowly along the x axis, from $x = 0$ to $x = +5.0$ m. At what values of x will the magnitude of the electrostatic force on the third particle from the other two particles be (a) minimum and (b) maximum? What are (c) the minimum magnitude and (d) the maximum magnitude?

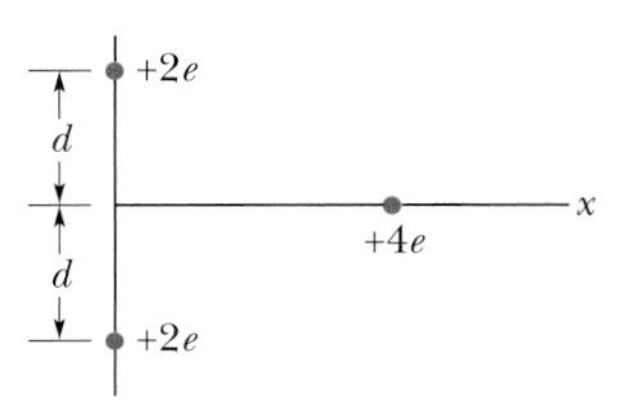

FIGURE 22-29 ■ Problem 32.

33. How Far In Fig. 22-30, how far from the charged particle on the right and in what direction is there a point where a third charged particle will be in balance?

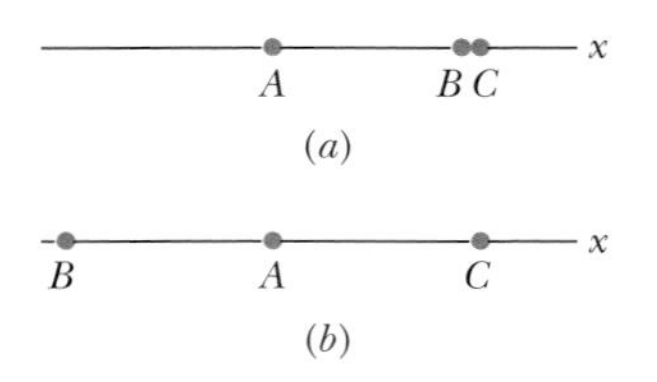

FIGURE 22-30 ■ Problem 33.

34. Three Positive Charges In Fig 22-31*a*, three positively charged particles are fixed on an *x* axis. Particles *B* and *C* are so close to each other that they can be considered to be at the same distance from particle *A*. The net force on particle *A* due to particles *B* and *C* is 2.014×10^{-23} N in the negative direction of the *x* axis. In Fig. 22-31*b*, particle *B* has been moved to the opposite side of *A* but is still at the same distance from it. The net force on *A* is now 2.877×10^{-24} N in the negative direction of the *x* axis. What is the ratio of the charge of particle *C* to that of particle *B*?

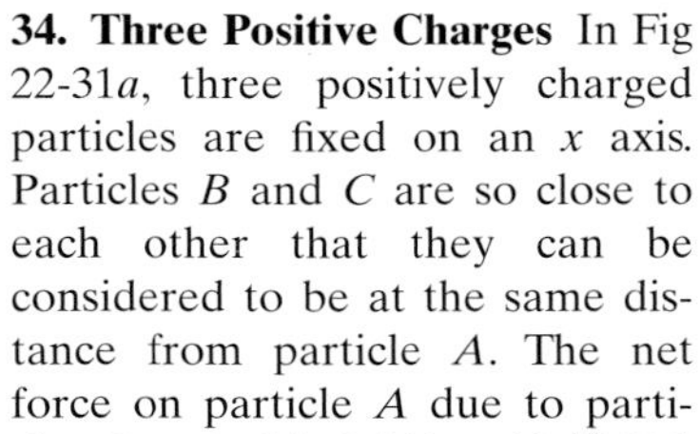

FIGURE 22-31 ■ Problem 34.

35. Fixed at the Origin A particle of charge Q is fixed at the origin of an *xy* coordinate system. At $t = 0$ a particle ($m = 0.800$ g, $q = 4.00\ \mu$C) is located on the *x* axis at $x = 20.0$ cm, moving with a speed of 50.0 m/s in the positive *y* direction. For what value of Q will the moving particle execute circular motion? (Assume that the gravitational force on the particle may be neglected.)

36. Seven Charges Figure 22-32 shows an arrangement of seven positively charged particles that are separated from the central particle by distances of either d ($= 1.0$ cm) or $2d$, as drawn. The charges are indicated. What are the magnitude and direction of the net electrostatic force on the central particle due to the other six particles?

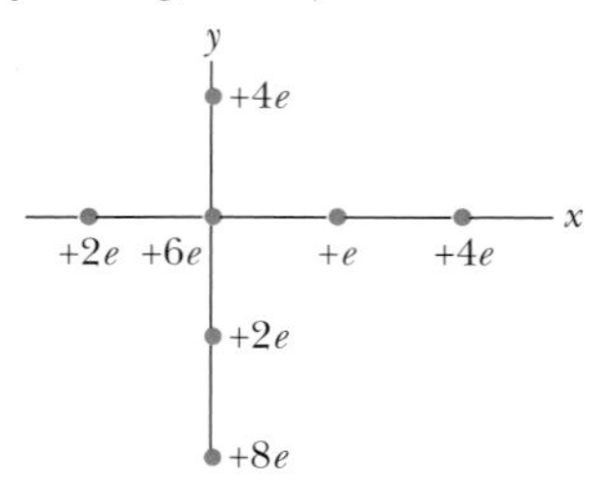

FIGURE 22-32 ■ Problem 36.

37. What is q In Fig. 22-33, what is q in terms of Q if the net electrostatic force on the charged particle at the upper left corner of the square array is to be zero?

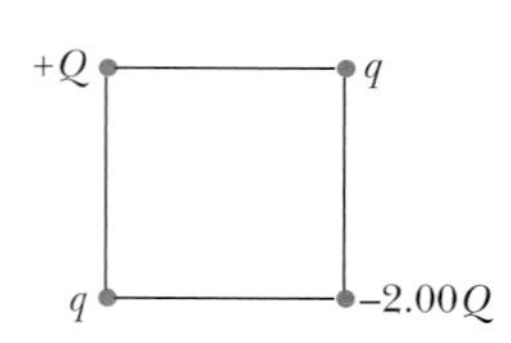

FIGURE 22-33 ■ Problem 37.

38. Charges Figure 22-34*a* shows an arrangement of three charged particles separated by distance d. Particles *A* and *C* are fixed on the *x* axis, but particle *B* can be moved along a circle centered on particle *A*. During the movement, a radial line between *A* and *B* makes an angle θ relative to the positive direction of the *x* axis (Fig. 22-34*b*). The curves in Fig. 22-34*c* give, for two situations, the magnitude F^{net} of the net electrostatic force on particle *A* due to the other particles. That net force magnitude is given as a function of angle θ and as a multiple of a basic force magnitude *F*. For example on curve 1, at $\theta = 180°$, we see that $F^{\text{net}} = 2F$. (a) For the situation corresponding to curve 1, what is the ratio of the charge of particle *C* to that of particle *B* (including sign)? (b) For the situation corresponding to curve 2, what is that ratio?

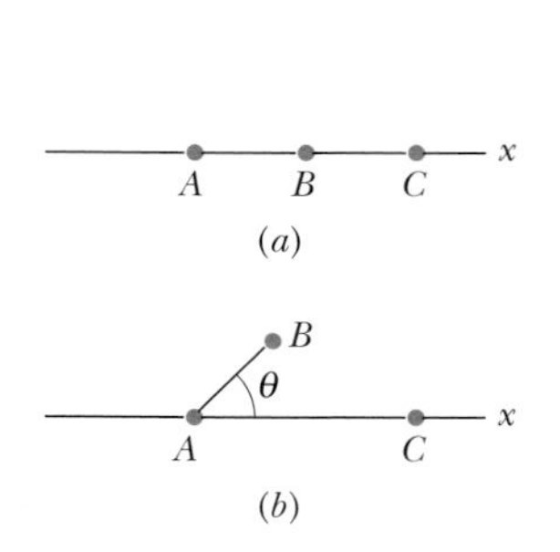

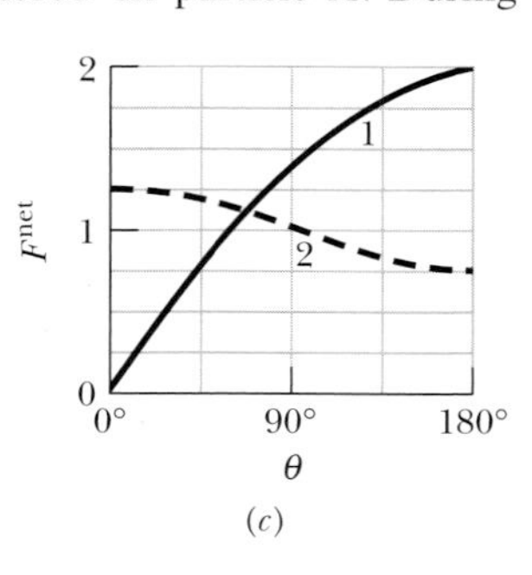

FIGURE 22-34 ■ Problem 38.

39. Two Electrons–Two Ions Figure 22-35 shows two electrons (charge $-e$) on an *x* axis and two negative ions of identical charges $-q$ and at identical angles θ. The central electron is free to move; the other particles are fixed in place at horizontal distances *R* and are intended to hold the free electron in place. (a) Plot the required amount of q versus angle θ if this is to happen. (b) From the plot, determine which values of θ will be needed for physically possible values of $q \leq 5e$.

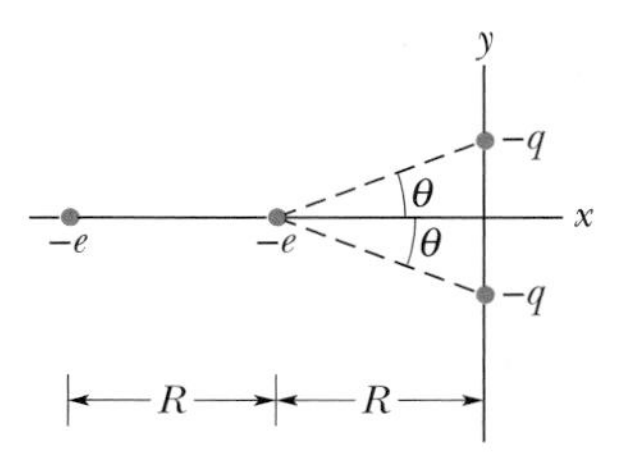

FIGURE 22-35 ■ Problem 39.

40. Diamond Figure 22-36 shows an arrangement of four charged particles, with angle $\theta = 30°$ and distance $d = 2.00$ cm. The two negatively charged particles on the *y* axis are electrons that are fixed in place. The particle at the right has a charge $q_B = +5e$. (a) Find distance *D* such that the net force on q_A, the particle at the left, due to the three other particles, is zero. (b) If the two electrons were moved closer to the *x* axis, would the required value of *D* be greater than, less than, or the same as in part (a)?

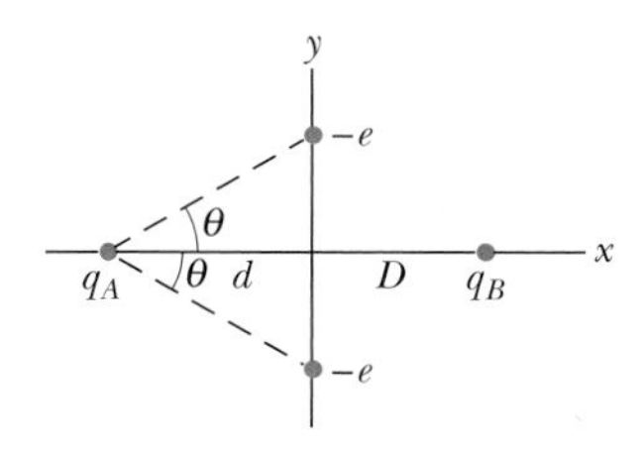

FIGURE 22-36 ■ Problem 40.

41. Each Positive Two particles, each of positive charge q, are fixed in place on an *x* axis, one at $x = 0$ and the other at $x = +d$. A particle of positive charge Q is to be placed along that axis at locations given by $x = \alpha d$. (a) Write expressions, in terms of α, that give the net electrostatic force $\vec{F}^{\text{elec}}$ acting on the third particle when it is in the three regions $x < 0$, $0 < x < d$, and $d < x$. The expressions should give a positive result when $\vec{F}^{\text{elec}}$ acts in the positive direction of the *x* axis and a negative result when $\vec{F}^{\text{elec}}$ acts in the negative direction. (b) Graph the magnitude of $\vec{F}^{\text{elec}}$ versus α for the range $-2 < \alpha < 3$.

42. Particles *A* and *B* In Fig. 22-37, particles *A* and *B* are fixed in place on an *x* axis, at a separation of $L = 8.00$ cm. Their charges are $q_A = +e$ and $q_B = -27e$. Particle *C* with charge $q_C = +4e$ is to be placed on the line between particles *A* and *B*, so that they produce a net electrostatic force $\vec{F}_C^{\text{net}}$ on it. (a) At what coordinate should particle *C* be placed to minimize the magnitude of that force? (b) What is that minimum magnitude?

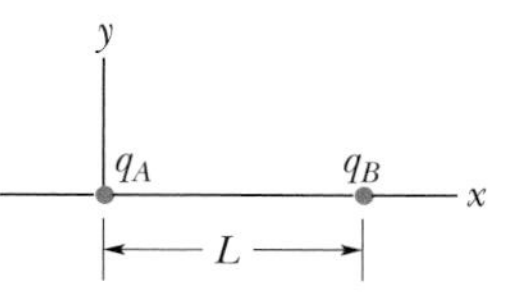

FIGURE 22-37 ■ Problem 42.

Sec. 22-9 ■ Comparing Electrical and Gravitational Forces

43. Earth and Moon (a) What equal positive charges would have to be placed on Earth and on the Moon to neutralize their gravitational attraction? Do you need to know the lunar distance to solve this problem? Why or why not? (b) How many kilograms of hydrogen would be needed to provide the positive charge calculated in (a)?

44. A Particle with Charge Q A particle with charge Q is fixed at each of two opposite corners of a square, and a particle with charge q is placed at each of the other two corners. (a) If the net electrostatic force on each particle with charge Q is zero, what is Q in terms of q? (b) Is there any value of q that makes the net electrostatic force on each of the four particles zero? Explain.

45. Hang from Thread In Fig. 22-38, two tiny conducting balls of identical mass m and identical charge q hang from nonconducting threads of length L. Assume that θ is so small that $\tan\theta$ can be replaced by its approximate equal, $\sin\theta$. (a) Show that, for equilibrium,

$$x = \left(\frac{2kq^2L}{mg}\right)^{1/3},$$

where x is the separation between the balls. (b) If $L = 120$ cm, $m = 10$ g, and $x = 5.0$ cm, what is q?

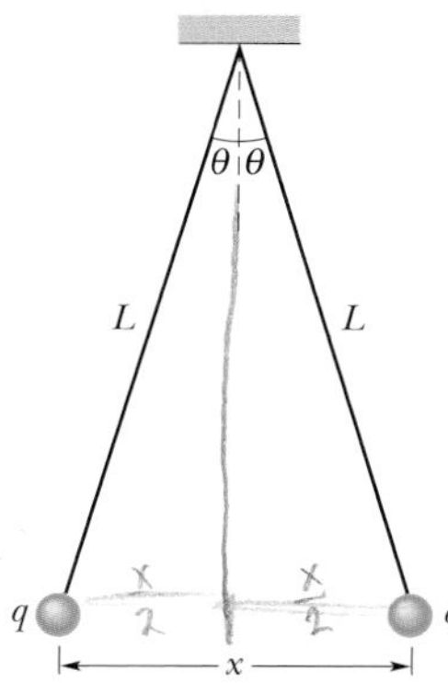

FIGURE 22-38 ■ Problem 45.

46. What Happens? Explain what happens to the balls of Problem 45b if one of them is discharged (loses its charge q to, say, the ground), and find the new equilibrium separation x, using the given values of L and m and the computed value of q.

47. Pivot Figure 22-39 shows a long, nonconducting, massless rod of length L, pivoted at its center and balanced with a block of weight W at a distance x from the left end. At the left and right ends of the rod are attached small conducting spheres with positive charges q and $2q$, respectively. At distance h directly beneath each of these spheres is a fixed sphere with positive charge Q. (a) Find the distance x when the rod is horizontal and balanced. (b) What value should h have so that the rod exerts no vertical force on the bearing when the rod is horizontal and balanced?

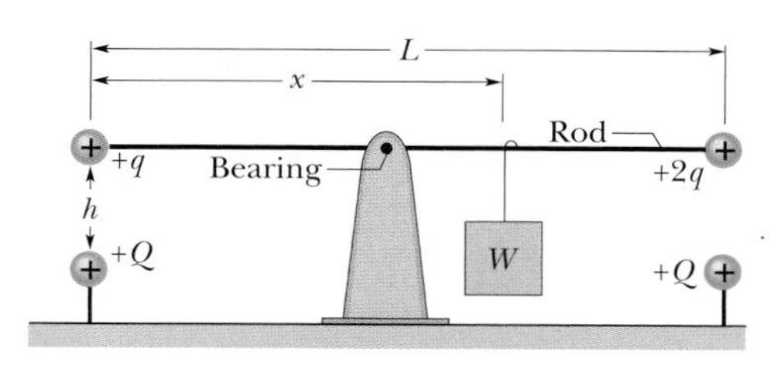

FIGURE 22-39 ■ Problem 47.

48. An Electron in a Vacuum An electron is in a vacuum near the surface of Earth. Where should a second electron be placed so that the electrostatic force it exerts on the first electron balances the gravitational force on the first electron due to Earth?

Additional Problems

49. Opposites Attract It is said that unlike charges attract. You can observe that after the sticky side of a piece of scotch tape is pulled quickly off the smooth side of another piece of tape the tapes attract each other. Perhaps each tape has a like charge and the rule has been stated backwards. Why do you believe the charges on the two tapes are different? *Note:* It is not acceptable to answer "because unlike charges attract and I observed the attraction."

50. Hanging Ball of Foil (a) Explain how a metal conductor such as a hanging ball of aluminum foil can be attracted to a charged insulator *even though the ball of foil has no net charge so that it is electrically neutral.* (b) Can two metal balls with no net charge attract each other? Explain. (c) Can the process of induction cause a neutral conductor to be *repelled* from a charged insulator? Explain.

51. Four Balls Consider four lightweight metal-coated balls suspended on nonconducting threads as shown in Fig. 22-40. Suppose ball A is stroked with a plastic rod that has been rubbed with fur. When you observe interactions between pairs of balls one at a time you find that:

1. B, C, and D are each attracted to A.
2. B and C seem to have no effect on each other.
3. B and C are both attracted to D.

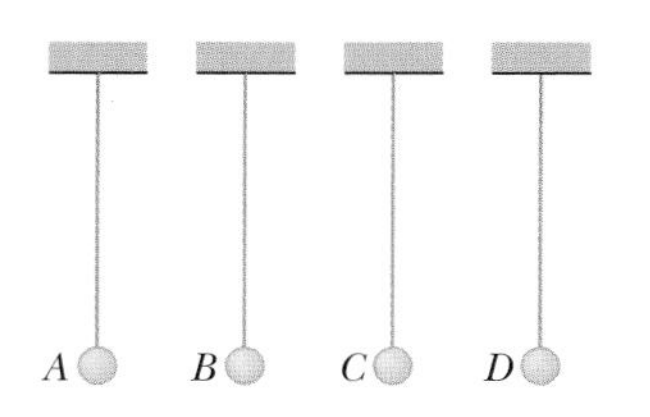

FIGURE 22-40 ■ Problem 51.

Use the concept of electric induction to figure out what type of net charge is on each of the balls: Negative charge? Positive charge? No charge at all? Explain your reasoning. [Based on question 5.9, Arons, *Homework and Test Questions for Introductory Physics Teaching* (Wiley, New York, 1994).]

52. Plastic Rubbed with Fur Suppose you rub a plastic rod with fur that gives it a negative charge. You then bring it close to an uncharged metal coated Styrofoam ball that is suspended from a string. (a) When the rod gets close to the ball, the ball starts moving toward it. Use the concept of induction to explain what happens to the atomic electrons and protons in the ball. Include a sketch of the ball and the rod that shows the excess negative charges on the rod. Also show how the charges are distributed on the ball just before it touches the rod. (b) After the ball touches the rod, it moves away from the rod quickly. Explain why.

53. Small Charged Sphere A small, charged sphere of mass 5.0 g is released 32 cm away from a fixed point charge of $+5.0 \times 10^{-9}$ C. Immediately after release, the sphere is observed to accelerate toward the charge at 2.5 m/s^2. What is the charge on the sphere? *Hint:* The force of gravity can be ignored in your calculation.

54. Lightning Bolt In a lightning bolt electrons travel from a thundercloud to the ground. If there are 1.0×10^{20} electrons in a lightning bolt, how many coulombs of charge are dumped onto the ground?

55. Estimating Charge Two hard rubber spheres of mass ~10 g are rubbed vigorously with fur on a dry day. They are then suspended from a rod with two insulating strings. They are observed to hang at equilibrium as shown in Fig. 22-41, which is drawn approximately to scale. Estimate the amount of charge that is found on each sphere.

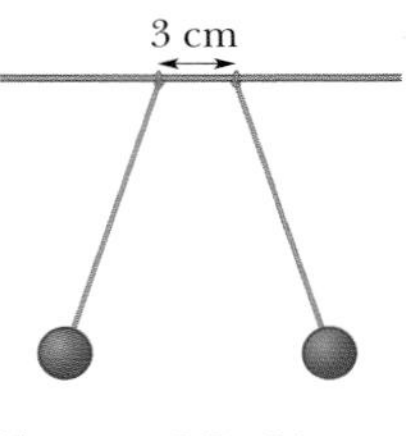

FIGURE 22-41 ■ Problem 55.

56. Various Arrangements Various arrangements of two fixed charges are shown in Fig. 22-42 along with a point labeled *P*. The amount of each charge is the same but the charges are positive or negative as indicated. All the distances between charges and between point *P* and the charge nearest to it are the same. Rank these arrangements in order of the strength of the force (that is, its magnitude) on a tiny positive test charge located at point *P* in each case. Go from greatest to least and indicate when force magnitudes are the same using an equal to sign. For example, if the force magnitudes at (*d*) and (*c*) were the same as each other and were the greatest and (*b*) was less than (*d*) and (*c*) with (*a*) being the least, your answer would be (*d*) = (*b*) > (*c*) > (*a*). Include a diagram and sketch the individual force vectors in each case. [Based on Ranking Task 128 O'Kuma, et. al., *Ranking Task Exercises in Physics* (Prentice Hall, Upper Saddle River NJ, 2000).]

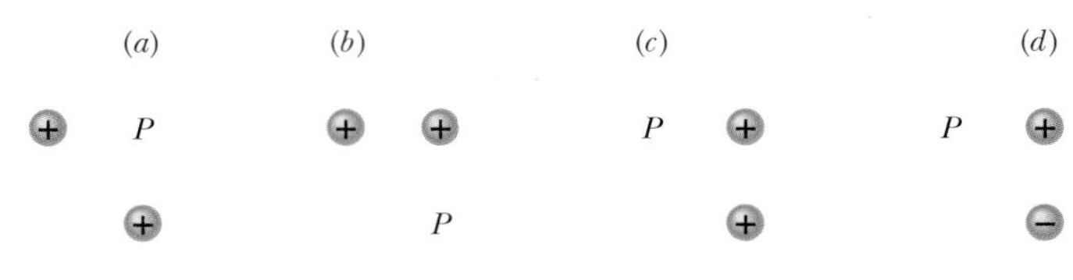

FIGURE 22-42 ■ Problem 56.

23 Electric Fields

Coal-burning power plants account for 56% of the electricity generated in the United States. But coal is dirty, and smokestack emissions containing sulfur dioxide, nitrogen oxides, and fine particles are a health hazard. A demonstration model of a new Advanced Hybrid Particulate Collector (AHPC) was recently installed at South Dakota's Big Stone plant. This new collector virtually eliminates particulate emissions. Although there are filters in the collector, 90% of the smoke particles are removed using another method.

What technology is used to remove most of the pollutants in the AHPC?

The answer is in this chapter.

23-1 Implications of Strong Electric Forces

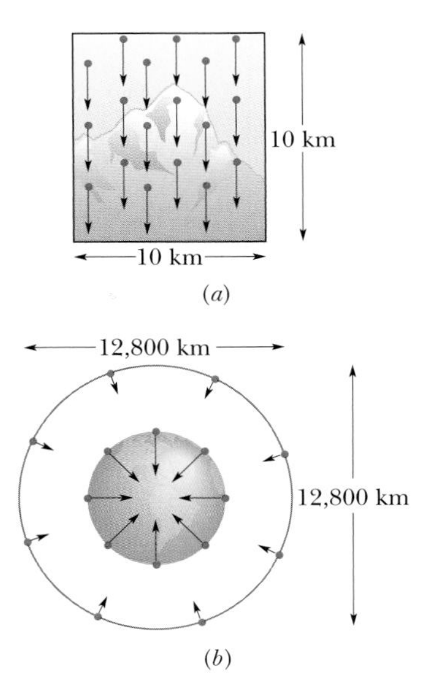

FIGURE 23-1 ■ (*a*) Force field maps for gravitational forces near the Earth are relatively simple: This map shows the uniform nature of the gravitational force on a "test mass" placed at different locations within a vertical and horizontal distance of 10 km. Mt. Everest is in the background. The tails of the force vectors have been placed at various possible test mass locations. (*b*) When distances become much larger than 10 km, the map is not quite so simple. The direction "downward" changes at different points. Also, the gravitational force on a test mass can decrease significantly at altitudes comparable to the Earth's radius.

In Section 3-9 we discuss the experimental fact that the gravitational force the Earth exerts on a mass has essentially the same magnitude and acts in a downward direction for all locations near the surface of the Earth as shown in Fig. 23-1*a*. By comparison the gravitational force exerted on the mass by other objects is negligible. In contrast, we learned in Section 22-2 that the electric forces between small objects are so strong that even Styrofoam cups rubbed with plastic can exert observable forces on each other. Also, it is not easy to calculate the net electric force exerted on a charged object by a complex array of charges in its vicinity. We would have to use the techniques introduced in Section 22-8 that combine Coulomb's law and the principle of superposition to obtain a vector sum of forces. To make matters worse, as our charged object (which we'll call a *test charge*) is moved, we often find the electric forces on it vary in direction and magnitude from point to point. The same can be said for the variation of gravitational forces on a space vehicle when large distances are involved as shown in Fig. 23-1*b*.

How can we describe the effect of the Earth's gravity or the electric force when a small test object is placed at various locations? What is the net effect of a collection of charges on a small test charge located in their vicinity? Answers lie in the concept of a force field.

We begin this chapter by creating a map of the force on a point-like test mass (due to gravitational interactions) or test charge (due to electrostatic interactions) at various locations in space. These maps introduce the concept of a *force field*. We then refine the field concept to define *electric fields* and *gravitational fields*, which are properties of a local space. Knowledge of a gravitational or electric field is useful because it allows us to determine the net force on a small object regardless of its mass or charge.

23-2 Introduction to the Concept of a Field

The temperature at every point in a room can be measured. If the room contains both a good heater and a window open to cold winter air, the temperature at each point in the room might be different. We call the resulting distribution of temperatures around the room a *temperature field*. In much the same way, you can imagine a *pressure field* in the atmosphere. Temperature and pressure are scalar quantities because they have no direction associated with them, so both temperature fields and pressure fields are *scalar fields*. In general,

> A **field** is defined as a representation of any physically measurable quantity that can vary in space.

Vector Fields

We can also have fields associated with vector quantities like force. In contrast to scalar fields like temperature, forces have direction, so it's not good enough to attach just a number to each point in space—the quantity must be identified with a vector. For example, let's ask, "What force would an object feel if it were placed at various locations in space?" and then represent the result pictorially. Consider the simple example of the gravitational force shown in Fig. 23-1*b*. We can calculate (or measure) the gravitational force exerted by the Earth on another object of known mass at several locations. The magnitude of each force vector is directly proportional to both the mass of the Earth and the mass of the object of interest. That is, the gravitational force of the Earth on different objects is different. Objects are pulled more strongly when they have more mass. A similar representation of the electrostatic forces on a test charge at various locations due to a source charge is shown in Fig. 23-2.

Note that there are two separate aspects to creating a map of the fields we discuss above. The first aspect involves the sources of the quantity being measured. For example, the heater and window can be thought of as stimulating thermal energy transfer that can affect the temperature at different locations in the room. In the gravitational case, the Earth and other large astronomical bodies are the sources of gravitational forces. The second aspect involved in creating a field map is a single measurement device that can be moved to different locations. In the case of the pressure field, it is a pressure gauge. In the case of gravitational force, our measurement tool is the motion of the test mass being acted on. Without some kind of sensing tool, we could not "know" the force fields.

In mapping force fields, the fixed objects are called **source objects** because they are the objects that exert forces on another object of interest. We use this other object that the forces are exerted *on* as our measurement tool. We refer to this other object as the **test object.**

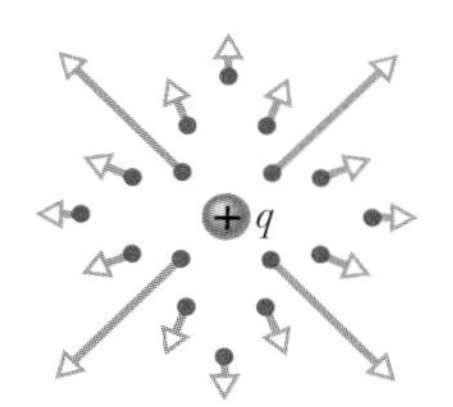

FIGURE 23-2 ■ Vectors representing the direction and relative magnitude of the force that would be exerted *by* a fixed positive source charge shown on a second positive test charge if the second charge was placed at the various locations of the vector tails. The magnitude of the force is given by the length of the vector.

Test Objects Should Not Be Large Enough to Move Source Objects

Since all forces occur in pairs, a test object exerts forces on the source objects, trying to change their locations, and hence to change the nature of the field. This issue makes the determination of the field values very complex unless the amount of a source object is much greater than the amount of a test object or the source object is somehow fixed in space. The fact that the field concept is only useful in cases where the test object does not move the source objects is an example of a universal issue involved in making measurements. Namely, the measurement device or tool should not change the value of the quantity being measured. For example, one should not use a large bathtub thermometer to measure the temperature of water in a tiny cup. The presence of the thermometer itself would affect the temperature of the water. Here are several examples of circumstances in which a field can be mapped:

1. In the Earth's gravitational interactions, the location of a much smaller test mass has a negligible effect on the Earth's location and hence on its force field (Fig. 23-3).
2. A conducting sphere with billions of excess electrons that can act as source charges that can exert a net force on a test charge consisting of a single proton. Each excess electron stays put because the net force from all the other charges in its vicinity is much stronger than that from the single proton (unless the proton gets very close to a particular electron).
3. A few electrons on the surface of an insulator such as a piece of Styrofoam can act as source charges that exert a net force on a single electron. The source charges are trapped on the insulator and do not reconfigure themselves.

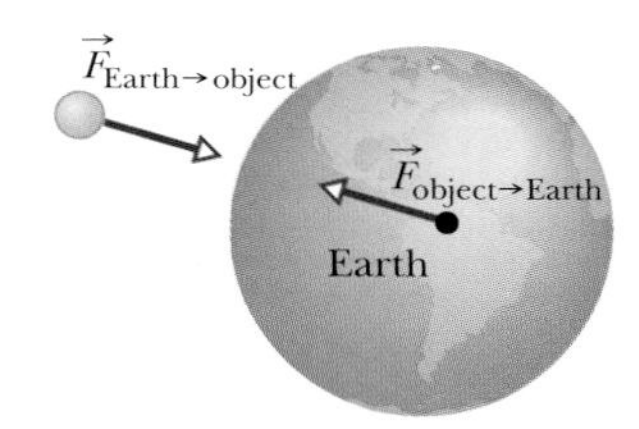

FIGURE 23-3 ■ An Earth–object system in which the interaction forces between a much less massive object and the Earth have, as always, the same magnitude. As the object falls, the movement of the Earth is negligible, so the gravitational force field surrounding the Earth does not change.

> The field concept is useful when the test object is too small to change the location of the source objects or if the source objects are fixed.

Mapping a Vector Field

How can we map the force field for a test object at various locations in space? The procedure that follows can be used to create a field map for any kind of force. The force on the test object could be the result of its interaction with a single source object or a collection of source objects at several locations. The procedure is as follows:

1. Choose a grid or array of points in the vicinity of the sources. The grid should be fine enough to give you a good idea of how the force field looks. However, if the

grid is too fine the procedure will be tedious and the arrows will clutter up the page.

2. Determine the magnitude and direction of the force on the test object at each point on the grid. (Note that we place our test object at one location at a time. If we put down lots of test objects at the same time, they would exert forces on each other and the source object and mess things up.) We must ensure that our test object is not so large that it disturbs the locations of the source objects. We might determine the force at the various points in space experimentally by making direct measurements. Alternatively, if we know the field sources, we can use theoretical relationships such as the law of gravitation or Coulomb's law to make theoretical calculations of the forces on our test object at the different grid locations.

3. Place an arrow representing the force vector at each of the grid locations. Each arrow should point in the direction of the force with a length that is proportional to the force magnitude at that location. It is conventional to locate the tail of the arrow (rather than its tip) at the point for which we have calculated (or measured) the force on the test charge.

   Note: We are free to choose a convenient length for the first arrow we draw on our map. Once the length of the first arrow is chosen, a second arrow can be drawn at another point in space. However, the ratio of the lengths of the two arrows must be the same as the ratio of the magnitudes of the two forces. For example, according to Coulomb's law, if the distance between our test charge and the center of the source charge is doubled, the new force on it has only one-fourth the magnitude. This is shown in Fig. 23-4, where the arrows at twice the distance have one-fourth the length.

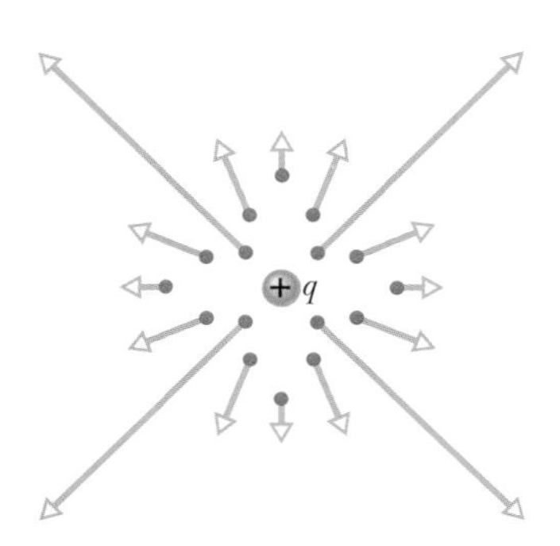

FIGURE 23-4 ■ Here the positive test charge is twice what it was in Fig. 23-2. Coulomb's law tells us that if the test charge is doubled, the electrostatic force it experiences at each location will be doubled.

This type of *vector field plot* is valuable when used to map forces because it immediately tells us important information about characteristics of the forces. For example, Figs. 23-1*b* and 23-4 allow us to infer that the gravitational force exerted by the Earth is everywhere attractive and the electrostatic (or Coulomb) force exerted by one positively charged object on another is everywhere repulsive. These two figures also show us that both of these forces act along the line connecting the centers of the two objects (they are "central forces"). They immediately remind us that these forces are large close to the source (long arrows) and small farther away (short arrows).

READING EXERCISE 23-1: In Fig. 23-4, suppose we chose an arrow that had a length of 36 mm to represent the electrostatic force at a distance of 2 cm from the source charge. What would the length of the arrows representing the magnitude of the force on the same test charge be if it was (a) 4 cm away from the center of the source charge? (b) 6 cm away from the center of the source charge? ■

READING EXERCISE 23-2: In measuring a field that has different values at each point in space, the "test" or measurement device must be small relative to the region of space over which the measurements are to be made. Why? ■

23-3 Gravitational and Electric Fields

Although the lengths of the arrows shown in Figs. 23-1 and 23-4 represent the magnitudes of the gravitational and electrostatic forces experienced by a test object at various locations, we note that the force magnitudes (and hence lengths of the arrows) are different for different test objects.

For example, Figure 23-4 shows the force field of a test object with twice as much charge as the one depicted in Fig. 23-2. Note that every arrow shown in the figure doubles in length. That is, each arrow is scaled by the same factor based on how much larger or smaller the test object's charge is. The force vectors scale as they do because for a given source charge, the electrostatic force is directly proportional to the amount of the test charge. The same is true for the gravitational force due to a given source mass. These changes in the force fields with test object strength are difficult, because we need an infinite number of field plots to represent all different test objects.

Actually, we have already dealt with the problem of test masses for gravitational forces. Recall from Section 14-2 that the magnitude of the gravitational force exerted by a spherical source mass m_s such as the Earth on a test mass m_t (perhaps a ball) is given by

$$F^{\text{grav}}_{s\to t} = |\vec{F}^{\text{grav}}_{s\to t}| = \left(\frac{Gm_s}{r^2}\right)m_t, \tag{23-1}$$

where G is the gravitational constant and r is the distance between the objects. To facilitate calculation of the force *exerted on* various objects by the same source object (typically the Earth) we took advantage of this proportionality to define the *local gravitational strength* g_s as a scalar given by

$$g_s = |\vec{g}_s| \equiv \left(\frac{F^{\text{grav}}_{s\to t}}{m_t}\right) = \left(\frac{Gm_s}{r^2}\right) \quad \text{(spherical source).} \tag{23-2}$$

Using the field concept that we have now developed, we can call the vector $\vec{g}_s$ the **local gravitational field vector.** Combining Eqs. 23-1 and 23-2, we see that the gravitational force exerted by a source mass m_s on a test mass m_t can be determined using the simple expression

$$\vec{F}^{\text{grav}}_{s\to t} = m_t\vec{g}_s. \tag{23-3}$$

In other words, the gravitational force on an object at a certain point in space is the product of its mass and the gravitational field vector at that point. The gravitational field vector is especially convenient because it is solely a property of space. It is completely determined by locations and the masses of the source objects. It is independent of the mass of the test object we might choose to investigate in any given instance.

Similarly, we can take advantage of the direct proportionality involved in electrostatic forces. That is, there is a similar proportionality between the amount of a test charge and the forces exerted on it by a source or a set of sources. Our approach is to define a new field, the *electric field,* which allows us to focus on the influence of the source of the force. To do this, we define the **electric field vector** due to one or more source charges as

$$\vec{E}_s \equiv \frac{\vec{F}^{\text{elec}}_{s\to t}}{q_t} \quad \text{(definition of electric field),} \tag{23-4}$$

where $\vec{F}^{\text{elec}}_{s\to t}$ represents the net electrostatic force a test charge q_t experiences from a set of source charges.

The force in Eq. 23-4 is the vector sum of the electrostatic forces on the test charge q_t from all the source charges. Since the force from each source charge is proportional to q_t, dividing the force by q_t cancels it out in each term, leaving the electric field vector $\vec{E}_s$ *independent* of q_t. Thus, the electric field vector shares the convenient characteristics of the gravitational field vector. It is solely a property of space and source and is independent of the test charge one might choose to probe it. *The electric field depends only on the source charges, not on the test charge.*

TABLE 23-1
Some Electric Field Magnitudes

Field Location or Situation	Value (N/C)
At the surface of a uranium nucleus	$\sim 3 \times 10^{21}$
Within a hydrogen atom, at a radius of 5.29×10^{-11} m	$\sim 5 \times 10^{11}$
Nerve cell membrane	$\sim 1 \times 10^{7}$
Electric breakdown in air (sparking)	$\sim 3 \times 10^{6}$
Near the charged drum of a photocopier	$\sim 1 \times 10^{5}$
Near a charged plastic comb	$\sim 1 \times 10^{3}$
In the lower atmosphere	$\sim 1 \times 10^{2}$
Inside the copper wire of household circuits	$\sim 1 \times 10^{-2}$

According to this definition, the **electric field** is the ratio of the electrostatic force on a test charge to the amount of that test charge. In other words, in SI units the electric field is the force per unit of test charge. This gives an SI unit for electric field as newtons per coulomb (N/C). In comparison, recall that the gravitational field was a measure of force per unit of *mass* and the SI unit for the gravitational field g was the N/kg.

The number given for the electric field can also be thought of as the force exerted on a 1 C test charge. To give you some idea of how much force is exerted on a one coulomb charge in various circumstances, Table 23-1 shows the electric fields that occur in a few physical situations. Remember, though, that one coulomb is a very large amount of charge and not an appropriate test charge. For example, an object with 10 000 more protons than electrons would have only a charge on the order of 10^{-15} C.

Although we use a test charge to determine the electric field associated with a charged object, remember that the electric field's existence is independent of the test charge just as the temperature in a room's existence is independent of whether or not there is a thermometer present to detect it. The test charge is simply the measurement device. The field at point P in Fig. 23-5 exists both before and after the test charge shown in the figure is put there. (We must always assume the test charge does not alter the electric field we are defining.)

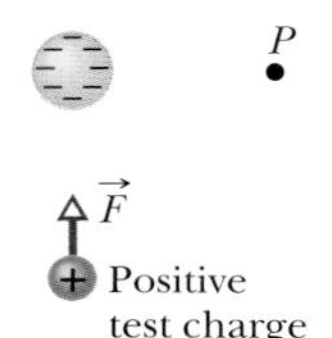

FIGURE 23-5 ■ The electric field associated with a negatively charged source object points inward at all locations outside of the source. Since a legitimate test charge does not noticeably influence the electric field created by the source charges, we think of the electric field as existing whether or not a small test charge is present to experience a force.

The remainder of this chapter is primarily devoted to exploring how to use Coulomb's law and the principle of superposition to find the electric fields associated with a single point charge and with relatively simple arrangements of charged objects. We will also explore the concept of *electric field lines* as an alternative to electric field vectors to represent electric fields visually.

TOUCHSTONE EXAMPLE 23-1: Predicting Forces on Charges

Consider a set of hidden source charges that cannot move. Suppose you are trying to explore the nature of electrical forces in the vicinity of the source charges using a positive test charge given by $q_t = 14$ nC. You discover that at a point A shown in Fig. 23-6, the test charge experiences a force given by $\vec{F}^{\text{elec}}_{s \to tA} = +(2.8\text{ N})\hat{i}$ and at point B it experiences a force given by $\vec{F}^{\text{elec}}_{s \to tB} = -(4.2\text{ N})\hat{j}$.

(*a*) What will the forces at points A and B be on a different point charge given by $q_1 = -15$ nC? Or on another given by $q_2 = +25$ nC? *Note:* nC stands for nanocoulomb, which is 10^{-9} C.

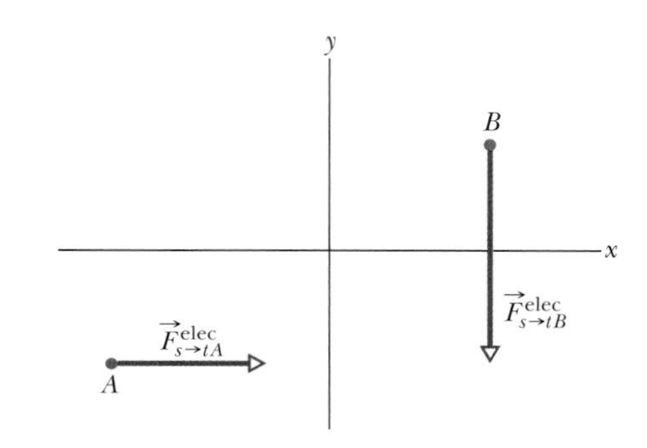

FIGURE 23-6 ■ Two points A and B are in the vicinity of a set of hidden fixed charges. A test charge experiences different electrostatic forces at each location.

SOLUTION ■ The **Key Idea** here is that you can use the information about the electrostatic forces experienced by the test charge to map the electric field in the region of our source charges. Once that is done you can then use your knowledge of the electric field to determine the forces on other charges (as long as the source charges are fixed or are large relative to any additional charges you are bringing into their vicinity). Using the definition of electric field given by Eq. 23-4, we see that the electric field at points A and B are

$$\vec{E}_A \equiv \frac{\vec{F}^{\text{elec}}_{s\to t}}{q_t} = \frac{+(2.8\text{ N})\hat{\text{i}}}{14\times 10^{-9}\text{ C}} = +(2.0\times 10^8\text{ N/C})\hat{\text{i}}$$

$$\vec{E}_B \equiv \frac{\vec{F}^{\text{elec}}_{s\to t}}{q_t} = \frac{-(4.2\text{ N})\hat{\text{i}}}{14\times 10^{-9}\text{ C}} = -(3.0\times 10^8\text{ N/C})\hat{\text{j}}. \qquad (23\text{-}5)$$

Now that we know the values of the electric field at points A and B we can find the forces on other charges by rearranging Eq. 23-5 to solve for electrostatic force and using the values of the new charges q_1 and q_2 in our calculations.

For q_1:

$$\vec{F}^{\text{elec}}_{s\to q_1} = q_1\vec{E}_A = (-15\times 10^{-9}\text{ C})[+(2.0\times 10^8\text{ N/C})\hat{\text{i}}] = (-3.0\text{ N})\hat{\text{i}}$$

$$\vec{F}^{\text{elec}}_{s\to q_1} = q_1\vec{E}_B = (-15\times 10^{-9}\text{ C})[-(3.0\times 10^8\text{ N/C})\hat{\text{j}}] = (+4.5\text{ N})\hat{\text{j}}$$

(Answer)

For q_2:

$$\vec{F}^{\text{elec}}_{s\to q_2} = q_2\vec{E}_A = (+25\times 10^{-9}\text{ C})[+(2.0\times 10^8\text{ N/C})\hat{\text{i}}] = (+5.0\text{ N})\hat{\text{i}}$$

$$\vec{F}^{\text{elec}}_{s\to q_2} = q_2\vec{E}_B = (+25\times 10^{-9}\text{ C})[-(3.0\times 10^8\text{ N/C})\hat{\text{j}}] = (-7.5\text{ N})\hat{\text{j}}$$

(Answer)

(b) Explain how the direction and magnitude of the forces on small charges in an electric field are related to the direction and magnitude of the electric field vector at point A and at point B. *Hint:* Sketching the force vectors will help you visualize the situation.

SOLUTION ■ The **Key Idea** here is that charge is a scalar so when you multiply it by the electric field vector to determine a force, the force vector must either be in exactly the same direction as the electric field vector (for a positive charge) or opposite to it (negative charge). For example, the equations just above show that the force on q_1, which is negative, is in the *opposite* direction from the electric field at both points A and B. On the other hand, the equations just above show that the force on q_2, which is positive, is in the *same* direction as the electric field at points A and B. This is illustrated in Fig. 23-7.

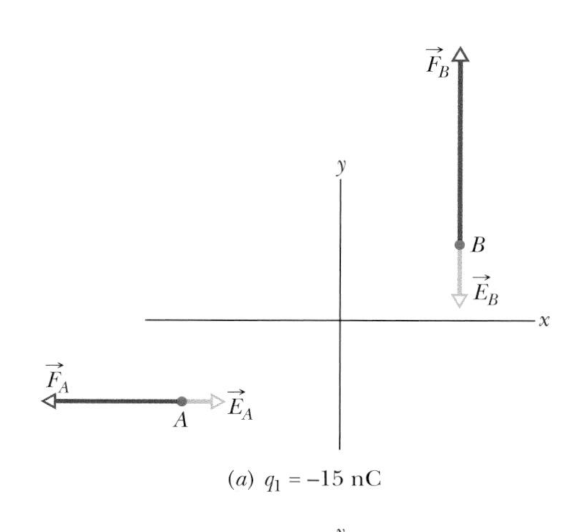

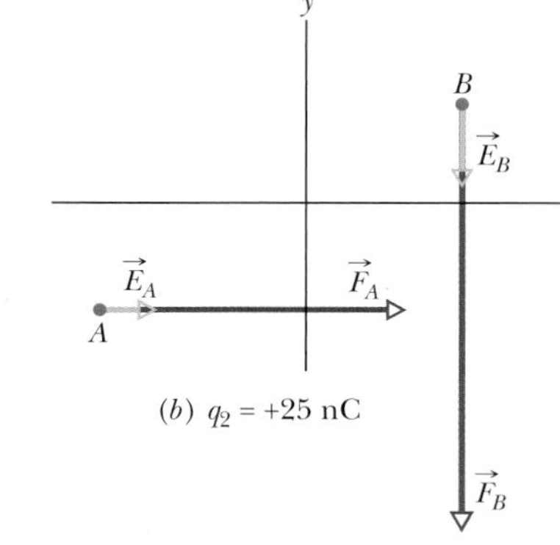

FIGURE 23-7 ■ A diagram showing the force vectors on two charges at points A and B due to an electric field acting on them. (*a*) $q_1 = -15$ nC and (*b*) $q_2 = +25$ nC.

23-4 The Electric Field Due to a Point Charge

The simplest of all possible charge distributions is a charge that can be approximated by a point with zero size. Protons, electrons, nuclei, and ions can all be considered to be point charges. Understanding the interactions of these objects is vital to our understanding of the physical world. In fact, even large charged objects can be viewed as point charges when considered from afar.

In this section, we determine the magnitude of the electric field due to a point-like charge q_s. We do this with the understanding that if we know the mathematical expression for the electric field of the charge, we know the mathematical expression for the force exerted per unit charge on any other charge that we might bring into the region surrounding it.

Here we discuss how to calculate the theoretical value of the electric field created by a positive charge q_s. We can also use another positive charge q_t to probe the region around q_s, testing the magnitude and direction of the force exerted on q_t by q_s. The

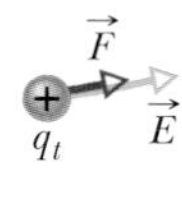

FIGURE 23-8 ■ When a source charge and a point-like test charge are both positive, the force between them is repulsive and both the vector representing the force on the test charge q_t and the electric field vector at its location point radially outward in the same direction. Since the units of $\vec{E}$ and $\vec{F}$ are different, the relative lengths of the vectors are arbitrary.

force vector and electric field vector for two positive point-like charges separated by a distance r are shown in Fig. 23-8. We develop a mathematical expression for the electric field associated with a positive charge below. As we do so, we consider how the situation would be different for negative charges.

We know that q_t experiences a repulsive force caused by q_s because they are like charges. The magnitude of the force $F^{\text{elec}}_{s\to t} = |\vec{F}^{\text{elec}}_{s\to t}|$ on our test charge q_t can be found using Coulomb's law:

$$F^{\text{elec}}_{s\to t} = k\frac{|q_s||q_t|}{r^2}. \tag{23-6}$$

The absolute value signs on q_s and q_t serve as a reminder that the magnitude of the force is independent of the type (sign) of charge we have chosen to use in this development. Using our concept of the electric field vector from Section 23-3 above, we can express the magnitude of the force on the test charge q_t due to the electric field created by the source charge q_s as

$$F^{\text{elec}}_{s\to t} = |q_t|E_s, \tag{23-7}$$

where E_s is the magnitude of the electric field due to the source charge.

By combining Eq. 23-6 with Eq. 23-7, we can express the magnitude of the electric field at a distance r from a point charge of magnitude q_s as

$$E_s = k\frac{|q_s|}{r^2} \quad \text{(magnitude of the electric field due to a point charge).} \tag{23-8}$$

In agreement with our definition of the electric field as the force *per unit* charge, this expression is independent of the amount (and sign) of the test charge q_t we use as the probe. This expression is valid everywhere around the point charge q_s.

The magnitude of the electric field due to a positive or negative point charge is given by the expression above. However, the electric field is a vector. Hence, we must still determine the direction of the electric field associated with our positive point charge q_s. Recall the definition of electric field magnitude in Eq. 23-4,

$$\vec{E}_s \equiv \frac{\vec{F}^{\text{elec}}_{s\to t}}{q_t}. \tag{23-9}$$

For a positive test charge q_t, this means that the direction of the field vector $\vec{E}_s$ is the same as the electrostatic force vector $\vec{F}^{\text{elec}}_{s\to t}$. This force points radially away from q_s as shown in Fig. 23-8. Since the direction of the field is the same as the direction of the force for the positive charge q_t, we know the electric field created by the positive point charge q_s must also point radially away from the charge q_s as shown in Fig. 23-8.

Would the direction of the field change if the charge q_s producing the field was negative rather than positive? The answer is yes. Consider a positive test charge q_t. The vector relationship between force and field (Eq. 23-4) tells us that since q_t is positive, the direction (sign) of the field is still the same as the direction of the force on q_t. However, now q_s is negative and q_t is positive. These unlike charges will attract one another. Hence, as shown in Fig. 23-9, the direction of the force on q_t due to q_s points radially *toward* q_s. Since the directions of the force and the field are the same, so does the electric field. According to the force-field relationship, if we used a negative test charge, the electric field would not change but the force on a negative test charge would act in the opposite direction.

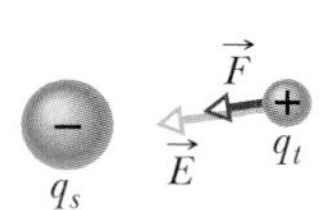

FIGURE 23-9 ■ When the charge on a source charge is negative while a point-like test charge is positive, the forces between them are attractive and both the vector representing the force on the test charge and the electric field vector at its location point radially inward in the same direction.

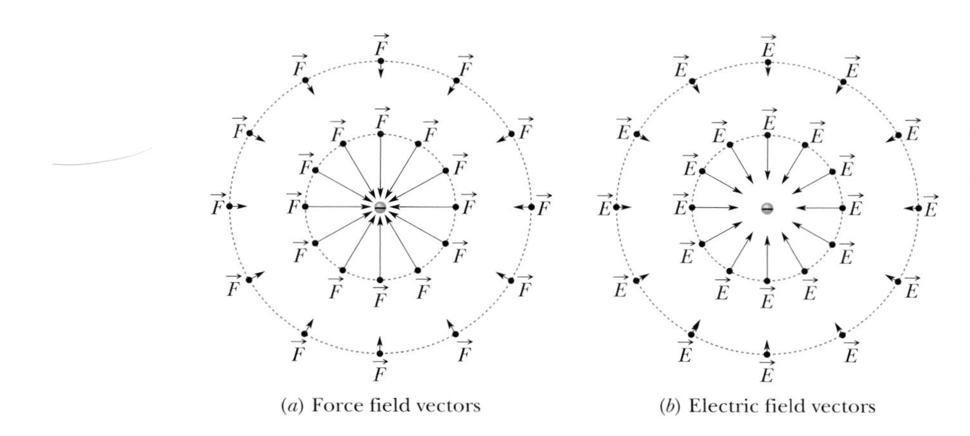

FIGURE 23-10 ■ There are two methods commonly used to depict the pattern of forces a test charge might experience at different locations in space. For the special case of forces associated with a single negative source charge: (*a*) shows a vector force field on a positive test charge, and (*b*) shows a vector electric field. The vector electric field is solely a property of space, independent of the test charge used to generate (*a*).

FORCE AND ELECTRIC FIELD DIRECTIONS: The direction of the force on a positive test charge is always the same as the electric field at the location of the positive test charge. So, at any point in space, the direction of the electric field produced by a positive point charge is radially away from the charge. The direction of the electric field produced by a negative point charge is radially toward the charge. "Radially" means along the line connecting the charge and the point of evaluation.

The Electric Field Vector Representation

The electric field vector representation is extremely useful when we want to determine the force on a charge placed at a given location (Fig. 23-10). We merely have to multiply the electric field vector at the location by the value of the charge. *This method of using the electric field to find force on a charge is valid for any charge, q, that is not large enough to disturb the electric field, regardless of the source of the field.*

READING EXERCISE 23-3: Rewrite the discussion of how we determine the direction of the field using a negative test charge rather than a positive test charge. Does it make any difference which type of charge we decide to use in determining the direction of the field? ■

23-5 The Electric Field Due to Multiple Charges

In the real world, problems are seldom as simple as one charged object exerting a force on another. It is more common for several charges to be present and the force exerted on the test charge to be the net result of the forces due to each of the source charges. As we mention in Section 22-8, experiments involving both gravitational and electrical forces have confirmed that the net force exerted on a test object by a collection of source objects is the vector sum of the forces exerted by each individual source object.

If we place a positive test charge q_t near n point charges $q_A, q_B, \ldots, q_n$, as shown for only three charges in Fig. 23-11, the forces exerted by the individual charges superimpose so that the net force $\vec{F}_t^{\text{net}}$ from the n point charges acting on the test charge is

$$\vec{F}_t^{\text{net}} = \vec{F}_{A\to t} + \vec{F}_{B\to t} + \cdots + \vec{F}_{n\to t}. \tag{23-10}$$

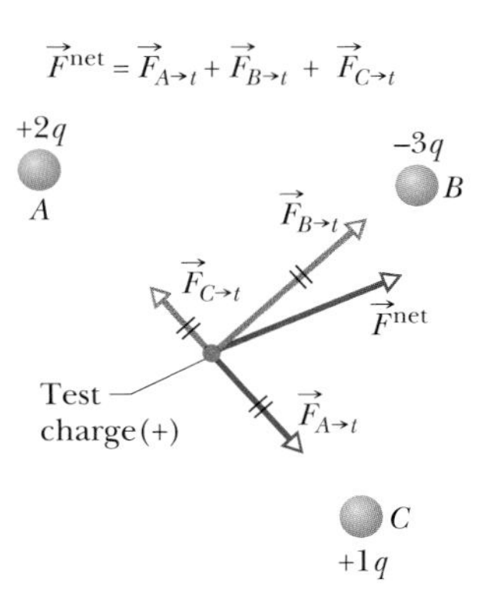

FIGURE 23-11 ■ Three point charges exert forces on a small positive test charge at a point in space. These force vectors superimpose to yield a net force.

For each of the terms in the expression above, we can replace the individual forces with the equivalent expressions based on the definition of the electric field. For example,

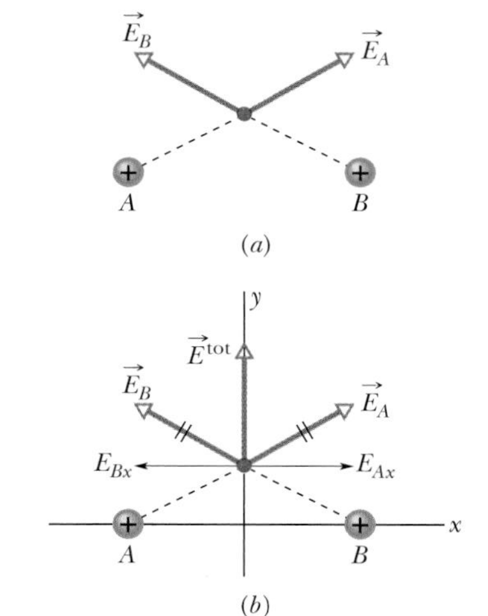

FIGURE 23-12 ■ When charges are arranged symmetrically it is often the case that electric field components cancel each other. In this case the x-components cancel everywhere along a line bisecting the line connecting the charge centers.

$$\vec{F}_{A\to t} = q_t \vec{E}_A, \tag{23-11}$$

where $\vec{E}_A$ is the electric field associated with charge q_A. If we make such replacements for each force term in the expression above we have:

$$\vec{F}_t^{\,net} = q_t \vec{E}^{\,net} = q_t \vec{E}_A + q_t \vec{E}_B + \cdots + q_t \vec{E}_n. \tag{23-12}$$

Here $\vec{E}^{\,net}$ is the resultant electric field associated with the entire group of charges. If we divide both sides of the expression

$$q_t \vec{E}^{\,net} = q_t \vec{E}_A + q_t \vec{E}_B + \cdots + q_t \vec{E}_n \tag{23-13}$$

by q_t, the result is an expression for the net electric field associated with a group of charges. Namely,

$$\vec{E}^{\,net} = \vec{E}_A + \vec{E}_B + \cdots + \vec{E}_n. \tag{23-14}$$

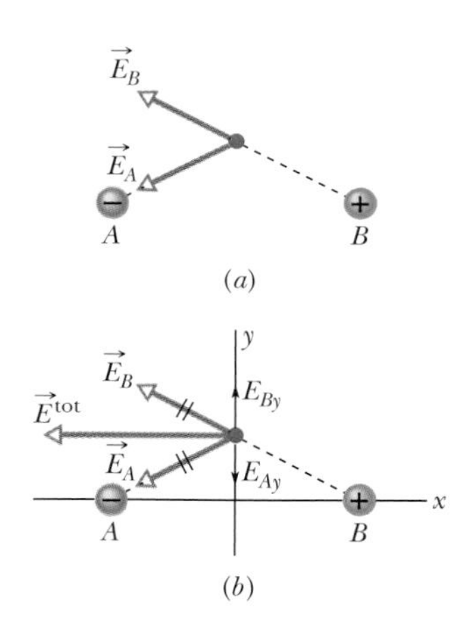

FIGURE 23-13 ■ A symmetrical arrangement for which y-components of the electric field cancel everywhere along a line, bisecting the line connecting the charge centers.

This expression shows us that the principle of superposition applies to electric fields as well as to electrostatic forces. When doing calculations, however, it is important to remember that we are adding vectors here. Hence, the addition is more complex than simply adding numbers together.

If the array of point charges is symmetric, sometimes the addition of vectors at certain points in the vicinity of the charges is simplified. In Figs. 23-12 and 23-13 we show examples of symmetric situations for which the net electric field only has one component everywhere on a line bisecting the line that connects the two charges.

READING EXERCISE 23-4: The figure here shows a proton p and an electron e on an x axis. Draw vectors indicating the direction of the electric field due to the electron and describe the direction in words at (a) point S and (b) point R. Draw vectors indicating the direction of the electric field due to both charges and describe the direction in words at (c) point R and (d) point S.

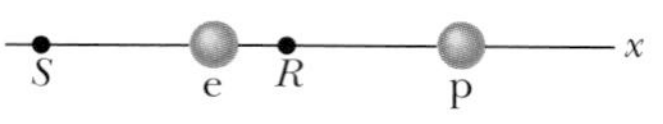

TOUCHSTONE EXAMPLE 23-2: Three Charges

Figure 23-14 shows three particles with charges $q_A = +2Q$, $q_B = -2Q$, and $q_C = -4Q$, each a distance d from the origin. We assume Q is positive. What net electric field $\vec{E}^{\text{net}}$ is produced at the origin?

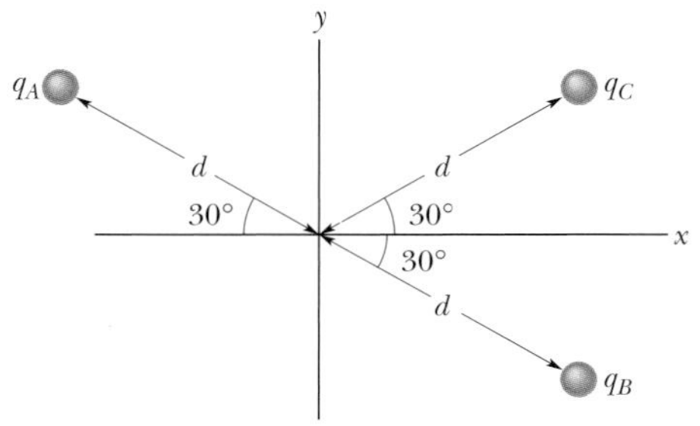

FIGURE 23-14 ■ Three particles with charges q_A, q_B, and q_C are at the same distance d from the origin.

SOLUTION ■ We need to find the electric field vectors $\vec{E}_A$, $\vec{E}_B$, and $\vec{E}_C$ that act at the origin. The **Key Idea** is that we can pick a more convenient coordinate system to describe these electric field vectors. An x'–y' coordinate system that is rotated by 30° in a clockwise direction has q_A and q_B lying along its x' axis, as shown in Fig. 23-15*a*.

Another **Key Idea** is that charges q_A, q_B, and q_C produce electric field vectors $\vec{E}_A$, $\vec{E}_B$, and $\vec{E}_C$, respectively, at the origin, and the net electric field is the vector sum $\vec{E}^{\text{net}} = \vec{E}_A + \vec{E}_B + \vec{E}_C$. To find this sum, we first must find the magnitudes and orientations of the three field vectors. To find the magnitude of $\vec{E}_A$, which is due to q_A, we use Eq. 23-8, substituting d for r and $2Q$ for $|q|$ and obtaining

$$E_A = k\frac{2Q}{d^2}.$$

Similarly, we find the magnitudes of the fields $\vec{E}_B$ and $\vec{E}_C$ to be

$$E_B = k\frac{2Q}{d^2} \quad \text{and} \quad E_C = k\frac{4Q}{d^2}.$$

We next must find the orientations of the three electric field vectors at the origin. Because q_A is a positive charge, the field vector it produces points directly *away* from it, and because q_B and q_C are both negative, the field vectors they produce point directly *toward* each of them. Thus, the three electric fields produced at the origin by the three charged particles are oriented as in Fig. 23-15*b*. (*Caution:* Note that we have placed the tails of the vectors at the point where the fields are to be evaluated; doing so decreases the chance of misinterpretation.)

We can now add the fields vectorially as outlined for forces in Touchstone Example 22-1c. However, here we can use symmetry to simplify the procedure. From Fig. 23-15*b*, we see that $\vec{E}_A$ and $\vec{E}_B$ have the same direction. Hence, their vector sum points along the positive x' axis and has the magnitude

$$\vec{E}_A + \vec{E}_B = k\frac{2Q}{d^2}\hat{\mathrm{i}}' + k\frac{2Q}{d^2}\hat{\mathrm{i}}' = k\frac{4Q}{d^2}\hat{\mathrm{i}}'$$

or

$$E_{Ax'} + E_{Bx'} = k\frac{4Q}{d^2},$$

where $\hat{\mathrm{i}}'$ and $\hat{\mathrm{j}}'$ are unit vectors in the x'–y' coordinate system. This sum happens to equal the magnitude of $\vec{E}_C$.

We must now combine two vectors, $\vec{E}_C$ and the vector sum $\vec{E}_A + \vec{E}_B$, that have the same magnitude. We do this by resolving $\vec{E}_C$ into its x' and y' components.

$$E_{Cx'} = E_C \cos 60° = \frac{1}{2}E_C$$

$$E_{Cy'} = E_C \sin 60° = \frac{\sqrt{3}}{2}E_C.$$

Then we find $E_{x'}^{\text{net}}$ and $E_{y'}^{\text{net}}$ components.

$$E_{x'}^{\text{net}} = E_{Ax'} + E_{Bx'} + E_{Cx'} = k\frac{4Q}{d^2} + \frac{1}{2}k\frac{4Q}{d^2} = \frac{3}{2}\left(k\frac{4Q}{d^2}\right).$$

$$E_{y'}^{\text{net}} = \frac{\sqrt{3}}{2}\left(k\frac{4Q}{d^2}\right).$$

Using vector notation we get

$$\vec{E}^{\text{net}} = \left(k\frac{4Q}{d^2}\right)\left(\frac{3}{2}\hat{\mathrm{i}}' + \frac{\sqrt{3}}{2}\hat{\mathrm{j}}'\right). \qquad \text{(Answer)}$$

The magnitude of $\vec{E}^{\text{net}}$ is given by

$$E^{\text{net}} = |\vec{E}^{\text{net}}| = \sqrt{(E_{x'}^{\text{net}})^2 + (E_{y'}^{\text{net}})^2} = k\frac{6.93Q}{d^2}.$$

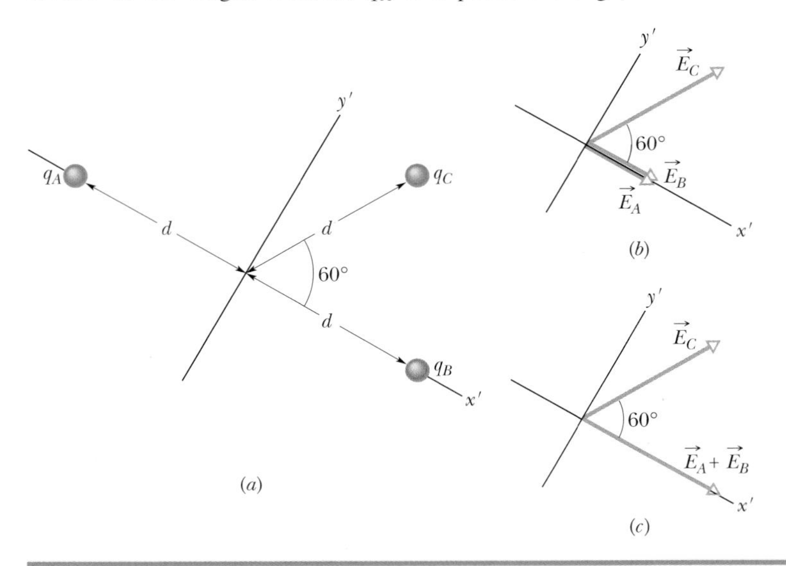

FIGURE 23-15 ■ (*a*) The same three charges in the new $x' - y'$ coordinate system. (*b*) The electric field vectors $\vec{E}_A$, $\vec{E}_B$, and $\vec{E}_C$ at the origin due to the three particles. (*c*) The electric field vector $\vec{E}_C$ and the vector sum $\vec{E}_A + \vec{E}_B$ at the origin.

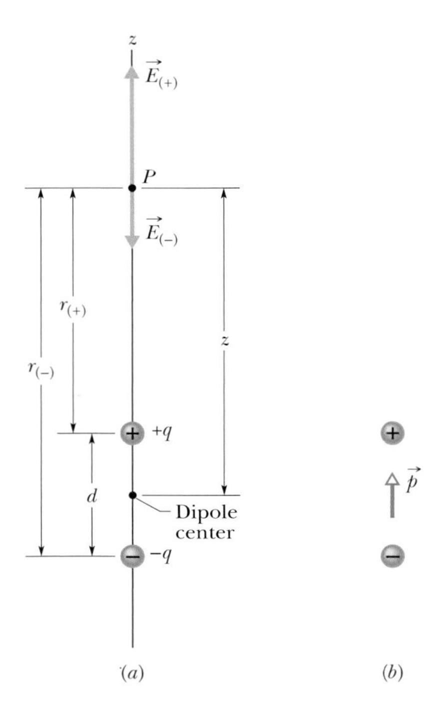

FIGURE 23-16 ■ (*a*) An electric dipole. The electric field vectors $\vec{E}_{(+)}$ and $\vec{E}_{(-)}$ at point P on the dipole axis result from the dipole's two charges. P is at distances $r_{(+)}$ and $r_{(-)}$ from the individual charges that make up the dipole. (*b*) By definition, the dipole moment $\vec{p}$ is a vector that points from the negative to the positive charge of the dipole.

23-6 The Electric Field Due to an Electric Dipole

Figure 23-16*a* shows two charged particles of amount $|q|$ but of opposite sign, separated by a distance d. We call this configuration an **electric dipole.** Separation of positive and negative charge in an electrically neutral object occurs quite naturally. For example, recall the discussion of polarization in Chapter 22. As a result, true electric dipoles and approximations of electric dipoles are reasonably common. Hence, we take some time to develop an expression for the electric field due to a dipole. We start with the idea of superposition of electric fields that we discuss in Section 23-5.

Let us find the electric field due to the dipole of Fig. 23-16*a* at a point P, a distance z from the midpoint of the dipole and on the axis through the particles, which is called the dipole axis. From symmetry, the electric field $\vec{E}$ at point P—and also the fields $\vec{E}_{(+)}$ and $\vec{E}_{(-)}$ due to the separate charges that make up the dipole—must lie along the dipole axis, which we have taken to be a z axis. Applying the superposition principle for electric fields, we find that the magnitude $E = |\vec{E}|$ of the electric field at P is

$$\begin{aligned} E &= |\vec{E}_{(+)}| - |\vec{E}_{(-)}| \\ &= k\frac{|q|}{r_{(+)}^2} - k\frac{|q|}{r_{(-)}^2} \qquad (23\text{-}15) \\ &= \frac{k|q|}{(z - \frac{1}{2}d)^2} - \frac{k|q|}{(z + \frac{1}{2}d)^2}, \end{aligned}$$

where k is the Coulomb constant. Using algebra, we can rewrite this equation as

$$E = \frac{k|q|}{z^2}\left[\left(1 - \frac{d}{2z}\right)^{-2} - \left(1 + \frac{d}{2z}\right)^{-2}\right]. \qquad (23\text{-}16)$$

We are usually interested in the electrical effect of a dipole only at distances that are large compared with the dimensions of the dipole—that is, at distances such that $z >> d$. At such large distances, we have $d/z << 1$ in the expression above. We can then expand the two quantities in the brackets in that equation by the binomial theorem (Appendix E), obtaining for those quantities

$$\left[\left(1 + \frac{2d}{2z(1!)} + \cdots\right) - \left(1 - \frac{2d}{2z(1!)} + \cdots\right)\right].$$

Thus,

$$E = \frac{k|q|}{z^2}\left[\left(1 + \frac{d}{z} + \cdots\right) - \left(1 - \frac{d}{z} + \cdots\right)\right]. \qquad (23\text{-}17)$$

The unwritten terms in these two expansions involve d/z raised to progressively higher powers. Since $d/z << 1$, the contributions of those terms are progressively less, and to approximate the electric field magnitude, E, at large distances, we can neglect them. Then, in our approximation, we can rewrite this expression as

$$E \approx \frac{k|q|}{z^2}\frac{2d}{z} = 2k\frac{|q|d}{z^3} \quad (\text{for } d/z << 1). \qquad (23\text{-}18)$$

The product $|q|d$, which involves the two intrinsic properties of charge q and separation d of the dipole, is the magnitude $|\vec{p}|$ of a vector quantity known as the **electric dipole moment** $\vec{p}$ of the dipole. (The unit of $\vec{p}$ is the coulomb-meter and should not be confused with either momentum or pressure.) Thus, we can rewrite Eq. 23-18 as

$$E = 2k\frac{|\vec{p}|}{z^3} \quad \text{(electric field magnitude for a dipole along axis with } d/z << 1\text{).} \quad (23\text{-}19)$$

The direction of $\vec{p}$ is taken to be from the negative to the positive end of the dipole, as indicated in Fig. 23-16*b*. We can use $\vec{p}$ to specify the orientation of a dipole.

The expression for the electric field due to a dipole shows that if we measure the electric field of a dipole only at distant points, we can never find both $|q|$ and d separately, only their product. The field at distant points would be unchanged if, for example, $|q|$ were doubled and d simultaneously halved. Thus, the dipole moment is a basic property of a dipole.

Although Eq. 23-19 holds only for distant points along the dipole axis, it turns out that $\vec{E}$ for a dipole varies as $1/r^3$ for all distant points, regardless of whether they lie on the dipole axis. Here r is the distance between the point in question and the dipole center.

Inspection of the electric field vectors in Fig. 23-16 shows that the direction of $\vec{E}$ for distant points on the dipole axis is always the direction of the dipole moment vector $\vec{p}$. This is true whether point P in Fig. 23-16*a* is on the upper or the lower part of the dipole axis.

Inspection of Eq. 23-19 shows that if you double the distance of a point from a dipole, the electric field at the point drops by a factor of 8. If you double the distance from a single point charge, however (see Eq. 23-8), the electric field drops only by a factor of 4. Thus the electric field of a dipole decreases more rapidly with distance than does the electric field of a single charge. The physical reason for this rapid decrease in electric field for a dipole is that from distant points a dipole looks like two equal but opposite charges that almost—but not quite—coincide. Thus, their electric fields at distant points almost—but not quite—cancel each other.

23-7 The Electric Field Due to a Ring of Charge

So far we have considered the electric field produced by one or, at most, a few point charges. We now consider charge distributions consisting of a great many closely spaced point charges (perhaps billions) spread along a line, a curve, over a surface, or within a volume. Such distributions can be treated as if they were **continuous** rather than discrete. Since these distributions can include an enormous number of point charges, we find the electric fields that they produce using integral calculus rather than by considering the point charges one by one. In this section we discuss the electric field caused by a ring of charge. In the next chapter, we shall find the field inside a uniformly charged sphere.

When we deal with continuous charge distributions, it is most convenient to express the charge on an object as a *charge density* rather than as a total charge. For a line of charge, for example, we would report the linear charge density (or charge per unit length) λ, whose SI unit is the coulomb per meter. If we have a total amount of charge q distributed uniformly along a line or curve then the **linear charge density** is defined as

$$\lambda \equiv q/L$$

where L is the total length of the path taken by the line or curve. For example, the curve subscribed by the uniformly charged circular ring of radius R shown in Fig. 23-17 has a total length of $L = 2\pi R$.

TABLE 23-2
Some Measures of Charge Density

Name	Symbol	Definition*	SI Unit
Charge	q	q	C
Linear charge density	λ	q/L	C/m
Surface charge density	σ	q/A	C/m^2
Volume charge density	ρ	q/V	C/m^3

*These definitions assume a uniform charge density. Otherwise the charge densities depend on location so that $\lambda(x) = dq/dx$, $\sigma(x, y) = dq/dA(x, y)$, $\lambda(x, y, z) = dq/dV(x, y, z)$.

Table 23-2 summarizes information about the types of charge densities we use in this text.

We may imagine the ring in Fig. 23-17 to be made of plastic or some other insulator, so the charges can be regarded as fixed in place. What is the electric field $\vec{E}$ at point P, a distance z from the plane of the ring along its central axis?

To answer this question, we cannot just use the expression for the electric field set up by a point charge, because the ring is obviously not a point charge. However, we can mentally divide the ring into differential elements of charge. If these charge elements are small they act like point charges, and then we can use Eq. 23-8 to find the electric field magnitude contributed by a single element. This gives us

$$E_q = k\frac{|q|}{r^2},$$

where the value of the coulomb constant k is given in Eq. 22-7. Next, we can add the electric fields set up at location P due to all the differential elements. The vector sum of all those fields gives us the net electric field set up at P by the entire ring.

Let ds be the (arc) length of any differential element of the ring. Since λ is the charge per unit length, the amount of charge in the element is given by

$$|dq| = |\lambda|ds. \tag{23-20}$$

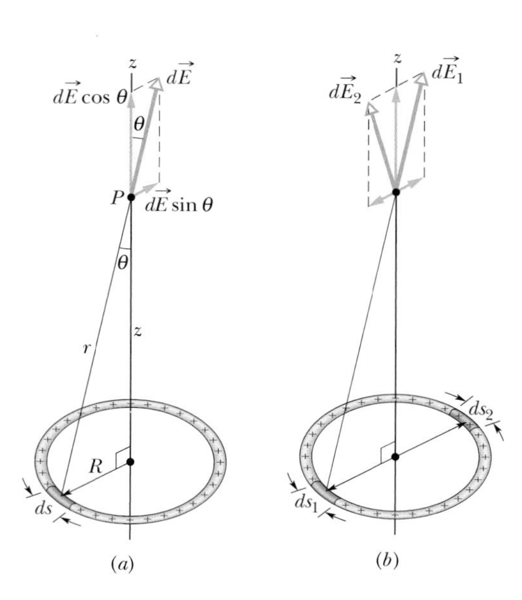

FIGURE 23-17 ■ (a) A ring of uniform positive charge. A differential element of charge occupies a length ds (greatly exaggerated for clarity). This element sets up an electric field $d\vec{E}$ at point P. The magnitude of the component of $d\vec{E}$ along the central axis of the ring is $|d\vec{E}|\cos\theta$. (b) Each ring element ds_1 has an opposite element ds_2. As a result of this symmetry the on-axis $d\vec{E}$ components (along z) add while those perpendicular to the axis cancel (in the x-y plane).

This differential charge sets up a differential electric field $d\vec{E}$ at point P, which is a distance r from the element. Treating the element as a point charge and using the equation above for dq, we can express the magnitude of $d\vec{E}$ as

$$|d\vec{E}| = k\frac{|dq|}{r^2} = k\frac{|\lambda|ds}{r^2}. \tag{23-21}$$

From Fig. 23-17a, we see we can use the Pythagorean theorem to rewrite the equation above as

$$|d\vec{E}| = k\frac{|\lambda|ds}{z^2 + R^2}. \tag{23-22}$$

Figure 23-17a also shows us that the vector $d\vec{E}$ is at an angle θ to the central axis (which we have taken to be a z axis) and has components perpendicular to and parallel to that axis.

Every element of charge in the ring sets up a differential field $d\vec{E}$ at P, with magnitude given by the expression above. All the $d\vec{E}$ vectors have identical components parallel to the central axis. All these $d\vec{E}$ vectors also have components perpendicular to the central axis. However, these perpendicular components are identical in magnitude but point in different directions. In fact, for any perpendicular component that points in a given direction, there is another one that points in the opposite direction as shown in Fig. 23-17b. The sum of this pair of components, like the sum of all other pairs of oppositely-directed components, is zero. Thus, the perpendicular components cancel and we need not consider them further. This leaves only the parallel components. They all have the same direction, so the net electric field at P is just their algebraic sum.

The parallel component of $d\vec{E}$ shown in Fig. 23-17a has magnitude $|d\vec{E}|(\cos\theta)$. The figure also shows us that

$$\cos\theta = \frac{z}{r} = \frac{z}{(z^2 + R^2)^{1/2}}. \tag{23-23}$$

Then combining our expressions for $d\vec{E}$ and $\cos\theta$ gives us the magnitude of the parallel component of $d\vec{E}$,

$$|d\vec{E}_{\parallel}| = |d\vec{E}|\cos\theta = \frac{k|z\lambda|ds}{(z^2 + R^2)^{3/2}}. \tag{23-24}$$

To add the parallel components, $|d\vec{E}|\cos\theta$, produced by all the elements, we integrate this expression around the circumference of the ring, from $s = 0$ to $s = 2\pi R$. Since the only quantity that varies during the integration is s, the other quantities can be moved outside the integral sign. The integration then gives us an electric field magnitude of

$$\begin{aligned} E = |\vec{E}| &= \int |d\vec{E}|\cos\theta = \frac{k|z\lambda|}{(z^2 + R^2)^{3/2}}\int_0^{2\pi R} ds \\ &= \frac{k|z\lambda|(2\pi R)}{(z^2 + R^2)^{3/2}}. \end{aligned} \tag{23-25}$$

Since λ is the charge per unit length of the ring, the term $\lambda(2\pi R)$ is q, the total charge on the ring. We can then rewrite this expression as

$$E = \frac{k|qz|}{(z^2 + R^2)^{3/2}} \quad \text{(electric field magnitude of a charged ring).} \tag{23-26}$$

If the charge on the ring is negative, rather than positive as we have assumed, the magnitude of the field at P is still given by this expression. However, the electric field vector then points toward the ring instead of away from it.

Let us evaluate this equation for the electric field for a point on the central axis so far away that $z >> R$. For such a point, the expression $z^2 + R^2$ can be approximated as z^2, and Eq. 23-26 becomes

$$E = \frac{k|q|}{z^2} \quad \text{(on central axis for } z >> R \text{ at large distance).} \tag{23-27}$$

This is a reasonable result, because from a large distance, the ring simply "looks" like a point charge. So, if we replace z with r then Eq. 23-27 becomes the expression for the electric field due to a point charge.

Let us next check Eq. 23-26 for a point at the center of the ring—that is, for $z = 0$. At that point, this expression tells us that $\vec{E} = 0$. This is a reasonable result, because if we were to place a test charge at the center of the ring, there would be no net electrostatic force acting on it. The force due to any element of the ring would be canceled by the force due to the element on the opposite side of the ring. If the force at the center of the ring is zero, the electric field there also has to be zero.

TOUCHSTONE EXAMPLE 23-3: Charged Arc

Figure 23-18*a* shows a plastic rod having a uniformly distributed charge $-Q$. We assume Q is positive, so $-Q$ is negative. The rod has been bent in a 120° circular arc of radius r. We place coordinate axes such that the axis of symmetry of the rod lies along the x axis and the origin is at the center of curvature P of the rod. In terms of Q and r, what is the electric field $\vec{E}$ due to the rod at point P?

SOLUTION ■ The **Key Idea** here is that, because the rod has a continuous charge distribution, we must find an expression for the electric fields due to differential elements of the rod and then sum those fields via integration. Consider a differential element having arc length ds and located at an angle θ above the x axis (Fig. 23-18*b*). If we let λ represent the linear charge density of the rod, our element ds has a differential charge of magnitude

$$dq = \lambda\, ds. \tag{23-28}$$

Our element produces a differential electric field $d\vec{E}$ at point P, which is a distance r from the element. Treating the element as a point charge, we can rewrite Eq. 23-21 to express the magnitude of $d\vec{E}$ as

$$|d\vec{E}| = k\frac{|dq|}{r^2} = k\frac{|\lambda|ds}{r^2}. \tag{23-29}$$

The direction of $d\vec{E}$ is toward ds, because charge dq is negative.

Our element has a symmetrically located (mirror image) element ds' in the bottom half of the rod. The electric field $d\vec{E}'$ set up at P by ds' also has the magnitude given by Eq. 23-29, but the field vector points toward ds' as shown in Fig. 23-18*b*. If we resolve the electric field vectors due to ds and ds' into x- and y-components as shown in Fig. 23-18*b*, we see that their y-components cancel

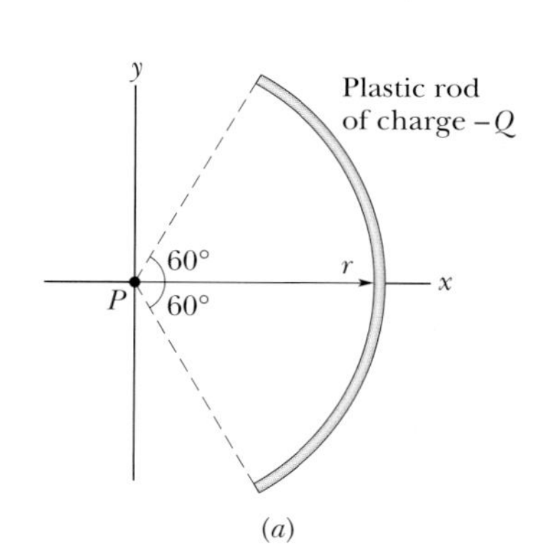

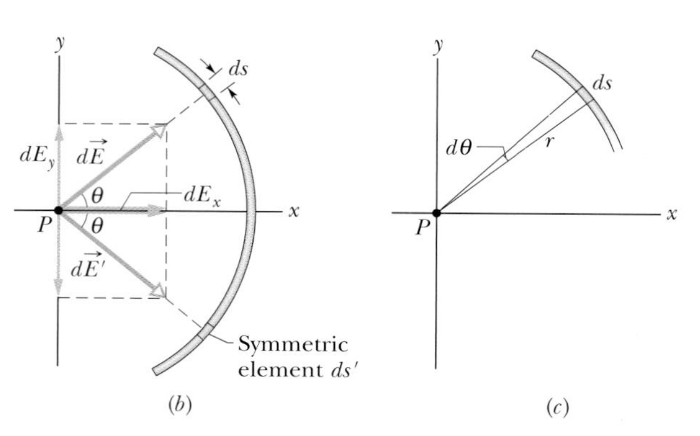

FIGURE 23-18 ■ (*a*) A plastic rod of charge $-Q$ in a circular section of radius r and central angle 120°; point P is the center of curvature of the rod. (*b*) A differential element in the top half of the rod, at an angle θ to the x axis and of arc length ds, sets up a differential electric field $d\vec{E}$ at P. An element ds', symmetric to ds about the x axis, sets up a field $d\vec{E}'$ at P with the same magnitude. (*c*) Arc length ds makes an angle $d\theta$ about point P.

(because they have equal magnitudes and are in opposite directions). We also see that their x-components have equal magnitudes and are in the same direction.

Thus, to find the electric field set up by the rod, we need sum (via integration) only the x-components of the differential electric fields set up by all the differential elements of the rod. From Fig. 23-18*b* and Eq. 23-29, we can write the component dE_x set up by ds as

$$|dE_x| = |d\vec{E}|\cos\theta = k\frac{|\lambda|}{r^2}\cos\theta\, ds. \tag{23-30}$$

Equation 23-30 has two variables, θ and s. Before we can integrate it, we must eliminate one variable. We do so by replacing ds, using the relation

$$ds = r\, d\theta,$$

in which $d\theta$ is the angle at P that includes arc length ds (Fig. 23-18*c*). *Note*: We choose to replace ds here rather than $d\theta$ because we know the angle into which the arc is bent. With this replacement, we can integrate Eq. 23-30 over the angle made by the rod at P, from $\theta = -60°$ to $\theta = 60°$. That will give us E, the magnitude of the electric field at P due to the rod:

$$E = \int |dE_x| = \int_{-60°}^{60°} \frac{k|\lambda|}{r^2}\cos\theta r\, d\theta$$

$$= \frac{k|\lambda|}{r}\int_{-60°}^{60°}\cos\theta\, d\theta = \frac{k|\lambda|}{r}[\sin\theta]_{-60°}^{60°}$$

$$= \frac{k|\lambda|}{r}[\sin 60° - \sin(-60°)]$$

$$= \frac{1.73k|\lambda|}{r}. \tag{23-31}$$

(If we had reversed the limits on the integration, we would have gotten the same result but with a minus sign. Since the integration gives only the magnitude of $\vec{E}$, we would then have discarded the minus sign.)

To evaluate the amount of charge per unit length, $|\lambda|$, we note that the rod has an angle of 120° and so is one-third of a full circle. Its arc length is then $2\pi r/3$, and its linear charge density must be

$$|\lambda| = \frac{\text{charge}}{\text{length}} = \frac{Q}{2\pi r/3} = \frac{0.477Q}{r}.$$

Substituting this into Eq. 23-31 and simplifying give us an electric field magnitude of

$$E = |\vec{E}| = \frac{(1.73)(0.477)kQ}{r^2}$$

$$= \frac{0.83kQ}{r^2}. \qquad \text{(Answer)}$$

The direction of $\vec{E}$ is toward the rod, along the axis of symmetry of the charge distribution. We can write $\vec{E}$ in unit-vector notation as

$$\vec{E} = \frac{0.83kQ}{r^2}\,\hat{\text{i}}.$$

23-8 Motion of Point Charges in an Electric Field

So far we have concentrated on finding electric field values by doing theoretical calculations. However, the electric field concept is especially valuable when we have little or no knowledge about source charges. In such a case, we can use measured values of the forces on our test charge to create an experimentally determined map of the field. In either case, once the electric field is known, we can determine forces on any point-like object with a known quantity of excess charge at any location in the field (provided that our test charge doesn't disturb the source charges). Assuming we know the mass of our charge, knowing the force means we can calculate the magnitude and direction of the particle's acceleration. Knowing the acceleration allows us to accurately predict its subsequent motion.

In the preceding sections we worked at the first of two tasks: given a charge distribution, find the electric field it produces in the surrounding space. Here we begin the second task: to determine what happens to a charged particle when it is in an electric field. When a charged particle is placed in an electric field, an electrostatic force acts on the particle. This force, a vector quantity, is given by

$$\vec{F}_t^{\,\text{elec}} = q_t\vec{E}_s \tag{23-32}$$

in which q_t is the charge of the test particle (including its sign) and $\vec{E}_s$ is the electric field that source charges have produced at the location of the particle. (The field is not the field set up by the test particle itself. A charged particle is not affected by its own electric field.) Equation 23-32 tells us:

> The electrostatic force $\vec{F}^{\text{elec}}$ acting on a charged test particle located in an external electric field $\vec{E}_s$ has the direction of $\vec{E}_s$ if the charge q_t of the test particle is positive and is opposite the direction of $\vec{E}_s$ if q_t is negative.

Knowing the force on the particle allows us to directly calculate the acceleration, but determining the exact motion of the object is more complicated. The electric field determines the force that a charged particle feels. That, in turn, determines its acceleration, *not its velocity*. So at first a charged particle starting from rest follows the direction of the field. This is because without an initial velocity, the direction of the force and acceleration are in the direction of the velocity. However, if the field changes direction, the path of the particle quickly deviates from the direction of the field. If the charged particle is given an initial velocity that is not aligned with the field, it may never follow the direction of the field. A good analogy to this situation is the path of a projectile in Earth's gravitational field. The gravitational field is uniformly directed downward. Yet this is the direction of the force (and acceleration), not necessarily the direction of the velocity at any given moment. The path of a launched projectile may never follow the direction of the gravitational field.

Electrostatic Precipitation

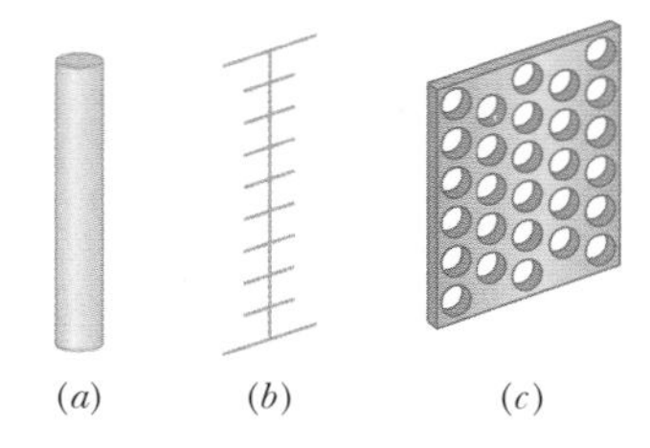

FIGURE 23-19 ■ An end-on side view of AHPC components. (*a*) Tubular filter. (*b*) Discharge electrode. (*c*) Collector plate with holes.

A few years ago, an Advanced Hybrid Particulate Collector (AHPC) was added to the Big Stone coal-fired power plant shown in the chapter's opening photograph. This hybrid collector eliminates essentially all the particulates in the smoke by combining the best features of filtration systems with a new type of **electrostatic precipitation** system. When the smoke from the coal boiler enters the device, more than 90% of the tiny smoke particles become electrically charged and then are attracted to one of the collection plates. The other 10% of particles flow through holes in the collection plates and are trapped by tubular filter bags that are especially efficient at removing extremely small particles (Fig. 23-19*a*).

The use of strong electric fields is a key factor in the effectiveness of electrostatic precipitation. Discharge electrodes (Fig. 23-19*b*) consisting of metal wires are negatively charged. The electric fields surrounding the electrodes are so intense that electrons are discharged. When the electrons fly away from the electrodes and encounter smoke particles, negative ions are formed. These ions are repelled from the electrodes, attracted to neutral metal collector plates (Fig. 23-19*c*), and captured. In short, these electrons and ions act like point charges in the presence of the electrodes' electric field.

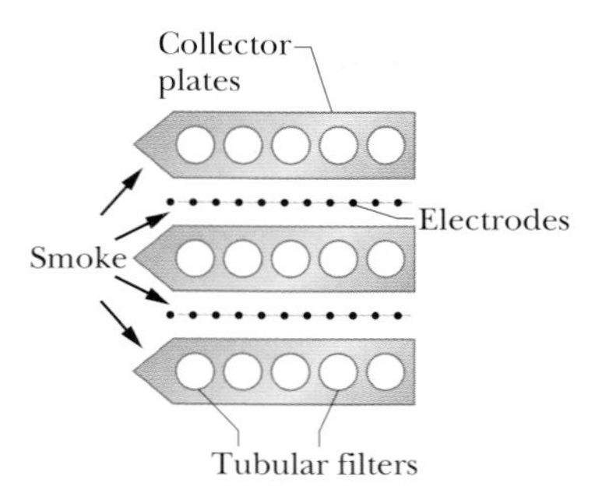

FIGURE 23-20 ■ Simplified AHPC top view showing arrangement of components with collector plates surrounding the filters.

The configuration of components in an AHPC system is shown in Fig. 23-20. A collector plate with holes in it is installed between wire discharge electrodes and filters to protect the filters. On the other side of the electrodes, another collector plate is installed to yield an arrangement in which each row of filters has collector plates on both sides of it. With this arrangement, the collector plates with holes function both as the primary collection surface and as a protective shield for the filters.

READING EXERCISE 23-5: (a) In the figure, what is the direction of the electrostatic force on the electron due to the uniform electric field shown? (b) In which direction does the electron accelerate if it is moving parallel to the y axis before it encounters the electric field? What path does it follow? (c) If, instead, the electron is initially moving rightward, does its speed increase, decrease, or remain constant? What path will it follow in this case? ■

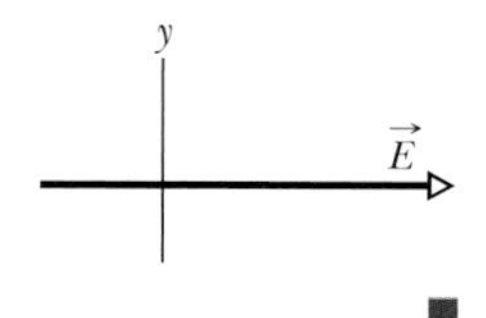

TOUCHSTONE EXAMPLE 23-4: Deflecting an Ink Drop

Figure 23-21 shows the deflecting plates of an ink-jet printer, with superimposed coordinate axes. An ink drop with a mass m of 1.3×10^{-10} kg and an amount of negative charge $|Q| = 1.5 \times 10^{-13}$ C enters the region between the plates, initially moving along the x axis with speed $v_x = 18$ m/s. The length L of the plates is 1.6 cm. The plates are charged and thus produce an electric field at all points between them. Assume that field $\vec{E}$ is downward directed, uniform, and has a magnitude of 1.4×10^6 N/C. What is the vertical deflection of the drop at the right edge of the plates? (The gravitational force on the drop is small relative to the electrostatic force acting on the drop and can be neglected.)

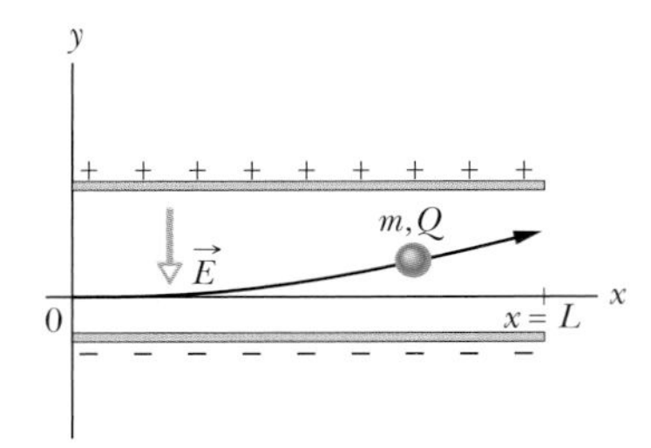

FIGURE 23-21 ■ An ink drop of mass m and an amount of charge $|Q|$ is deflected in the electric field of an ink-jet printer.

SOLUTION ■ The drop is negatively charged and the electric field is directed *downward*. The **Key Idea** here is that, from Eq. 23-4, a constant electrostatic force of magnitude $|Q|E$ acts *upward* on the charged drop. Thus, as the drop travels parallel to the x axis at constant speed v_x, it accelerates upward with some constant acceleration of magnitude a_y where a_y is positive. Applying Newton's Second Law ($F_y^{\text{net}} = ma_y$) for components along the y axis, we find that the y-component of the acceleration is directly proportional to the y-component of the electrostatic force so that

$$a_y = \frac{F_y^{\text{elec}}}{m} = +\frac{|Q|E}{m}. \tag{23-33}$$

Let Δt represent the time required for the drop to pass through the region between the plates. In Chapter 4, we found that during the time Δt the vertical and horizontal displacements of the drop are

$$\Delta y = \tfrac{1}{2}a_y \Delta t^2 \quad \text{and} \quad \Delta x = L = v_x \Delta t, \tag{23-34}$$

respectively. Eliminating Δt between these two equations and substituting Eq. 23-33 for a_y, we find

$$\Delta y = \frac{|Q|EL^2}{2mv_x^2}$$

$$= \frac{(1.5 \times 10^{-13}\ \text{C})(1.4 \times 10^6\ \text{N/C})(1.6 \times 10^{-2}\ \text{m})^2}{(2)(1.3 \times 10^{-10}\ \text{kg})(18\ \text{m/s})^2}$$

$$= 6.4 \times 10^{-4}\ \text{m}$$

$$= 0.64\ \text{mm}. \qquad \text{(Answer)}$$

23-9 A Dipole in an Electric Field

Many electrical effects in matter can be understood by considering matter to be made up of many little electric dipoles. When an electric field is applied to that matter, the dipoles change their orientation in a consistent way. Although each dipole is small, since they all do the same thing, they can produce a substantial electrical effect. In this section, we consider the torque that can be exerted on a dipole that is placed in a uniform electric field.

We have defined the electric dipole moment $\vec{p}$ of an electric dipole to be a vector pointing from the negative to the positive end of the dipole. It turns out that behavior of a dipole in a uniform external electric field $\vec{E}_s$ can be described completely in terms of the two vectors $\vec{E}_s$ and $\vec{p}$, with no need of any details about the dipole's structure.

A molecule of water (H_2O), as shown in Fig. 23-22, is an electric dipole. There the black dots represent the oxygen nucleus (having eight protons) and the two hydrogen nuclei (having one proton each). The colored enclosed areas represent the region in which electrons can be located around the nuclei.

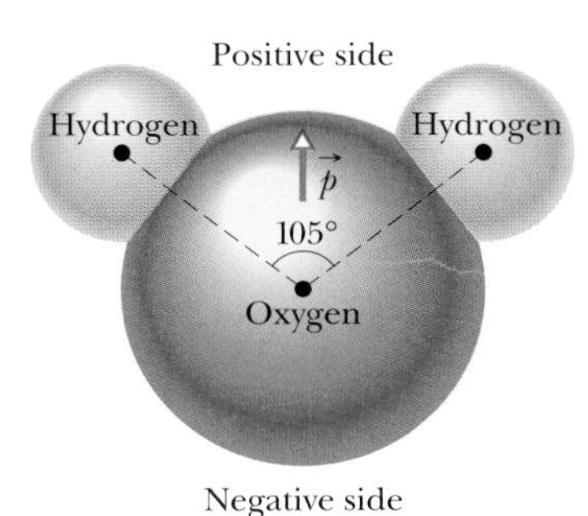

FIGURE 23-22 ■ This H_2O molecule has 3 nuclei (shown as dots). The electrons orbiting the nuclei spend more time near the oxygen nucleus, so the molecule behaves like a dipole. Its moment $\vec{p}$ points from the (negative) oxygen side to the (positive) hydrogen side.

In a water molecule, the two hydrogen atoms and the oxygen atom do not lie on a straight line but form an angle of about 105°. As a result, the molecule has a definite "oxygen side" and "hydrogen side." Moreover, the 10 electrons of the molecule tend to remain closer to the oxygen nucleus than to the hydrogen nuclei. This makes the oxygen side of the molecule slightly more negative than the hydrogen side and creates an electric dipole moment $\vec{p}$ that points along the symmetry axis of the molecule as shown. If the water molecule is placed in an external electric field, it behaves like the idealized electric dipole shown in Fig. 23-16.

To examine this behavior, we now consider what happens to an idealized electric dipole placed in a *uniform* **external electric field** $\vec{E}_s$. This is shown in Fig. 23-23a.

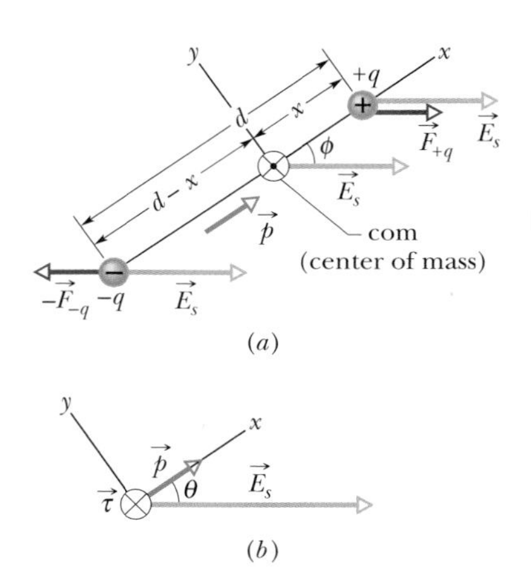

FIGURE 23-23 ■ (*a*) An electric dipole in a uniform electric field $\vec{E}$. Two equal but opposite charges are separated by a distance d. The line between them represents their rigid connection. (*b*) Field $\vec{E}$ causes a torque $\vec{\tau}$ on the dipole. The direction of $\vec{\tau}$ is into the plane of the page, as represented by the symbol ⊗.

Assume that the dipole is a rigid structure that consists of two charges of opposite sign, each having an amount $|q|$, separated by a distance d. Assume that the dipole moment $\vec{p}$ makes an angle ϕ with field $\vec{E}_s$. Recall from our discussion in Section 23-6 above that the electric field far from the dipole depends only on $\vec{p}$ (the product of charge q and separation d). The detailed structure of the water molecule is not important as long as the interaction between it and the electric field isn't strong enough to change the shape of either the molecule or the electric field.

Electrostatic forces act on the charged ends of the dipole. Because the electric field is uniform, those forces act in opposite directions (as shown in Fig. 23-23) and with the same magnitude $|\vec{F}_t^{\text{elec}}| = |q_t \vec{E}_s|$. Thus, because the field is uniform, the net force on the dipole from the field is zero and the center of mass of the dipole does not move. However, the forces on the charged ends do produce a net torque $\vec{\tau}$ on the dipole about its center of mass. Since the charges on a dipole do not necessarily have the same mass, we assume the center of mass lies on the line connecting the charged ends, at some distance x from one end and thus a distance $d - x$ from the other end. From Eq. 11-29 ($|\vec{\tau}| = |\vec{r}||\vec{F}|\sin\phi$), we can express the net torque magnitude $\tau = |\vec{\tau}|$ as

$$\tau = F^{\text{elec}}|x|\sin\phi + F^{\text{elec}}|(d - x)|\sin\phi = F^{\text{elec}}d\sin\phi. \tag{23-35}$$

We can also write the magnitude of the torque in terms of the magnitudes of the electric field $E = |\vec{E}|$ and the dipole moment $p = |\vec{p}| = |q|d$. To do so, we substitute $|q|\,E$ for the magnitude of the electrostatic force, F^{elec}, and $p/|q|$ for the dipole spacing, d, to find an expression for the magnitude of the torque. This magnitude is given by

$$\tau = pE\sin\phi. \tag{23-36}$$

We know the direction of the vector $\vec{\tau}$ is given by the right-hand rule. So, we see the result for both the magnitude and direction can be written in terms of the cross product as

$$\vec{\tau} = \vec{p} \times \vec{E} \quad \text{(torque on a dipole).} \tag{23-37}$$

Vectors $\vec{p}$ and $\vec{E}$ are shown in Fig. 23-23*b*. The torque acting on a dipole tends to rotate $\vec{p}$ (hence the dipole) into the direction of $\vec{E}$, thereby reducing ϕ. In Fig. 23-23, such rotation is clockwise. As we discuss in Chapter 11, we can represent a torque component τ that gives rise to such a rotation as

$$\vec{\tau} = -\tau\hat{k} = -(pE\sin\phi)\hat{k}, \tag{23-38}$$

where $\hat{k}$ is a unit vector pointing along the z axis (not the Coulomb constant).

READING EXERCISE 23-6: The figure shows four orientations of an electric dipole in an external electric field. Rank the orientations according to the magnitude of the torque on the dipole.

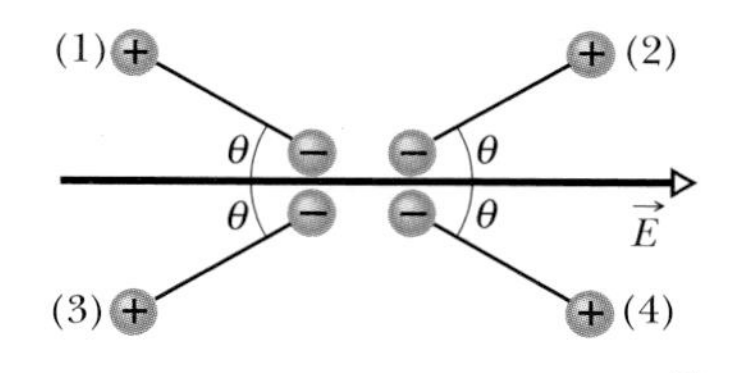

■

23-10 Electric Field Lines

So far in this chapter we have represented electric fields using vector arrows and creating force vector field plots. There is another common method for creating a visual representation of information about electric fields in a region of space. It involves drawing *electric field lines.*

In the electric field line representation, we use continuous lines to convey information about the direction of the field at different points. Since the magnitude and direction of the electric field usually changes smoothly, this turns out to be rather convenient. Michael Faraday, who introduced the idea of electric fields in the 19th century, thought of the space around a charged body as filled with *lines of force.* Although we attach no reality to these lines, they provide a nice way to visualize patterns of changing force. The field line representation is used in Chapter 24 where we introduce Gauss' law. The field line representation is also used in Chapters 29–33 to describe magnetic fields.

Field lines are a good way to visualize the directions of a vector field in a region of space. To draw an **electric field line,** we start at any point and look at the direction of the field at that point. We then draw a short line in that direction. We determine the field direction at the new location and draw another short line in that direction. We continue this process until we reach a charge or get to infinity. Compare Figs. 23-24 and 23-25 for an example. Note that field *lines* shown in Fig. 23-25 and Fig. 23-26 differ from the short straight field *vectors* shown in Fig. 23-24 because they always start or end on the source charge(s). The direction of a straight field line or the direction of the tangent to a curved field line gives the direction of the electric field vector $\vec{E}_s$ at that point. Because the field lines point in the direction of the field, field lines must originate on positive charges and terminate on negative charges.

It is important to note we could draw field lines through every point in space. However, this would not be very helpful since our paper would be totally filled with field lines and we couldn't distinguish one from another. Instead, we choose to draw a few field lines, with the number of lines leaving each positive charge (or ending on each negative charge) being proportional to the amount of each charge. If we choose to have 16 lines originating on a $+4\,\mu\text{C}$ charge, then we should have 8 lines ending on a $-2\,\mu\text{C}$ charge. This scaling of the number of field lines with amount of the charge turns out to be quite convenient since then the field lines are forced to be closely packed together where the field is strong and far apart where it is weak. We can see this in Figs. 23-25 and 23-26. In other words, the average density of field lines (the number of lines crossing through a small area perpendicular to their direction) is proportional to the strength of the field. We then have the following rules:

> At any point along an electric field line, the direction of the corresponding electric field vector is always tangent to the line at that point. **Electric field lines** extend away from positive charge (where they originate) and toward negative charge (where they terminate). The density of field lines is proportional to the strength of the field.

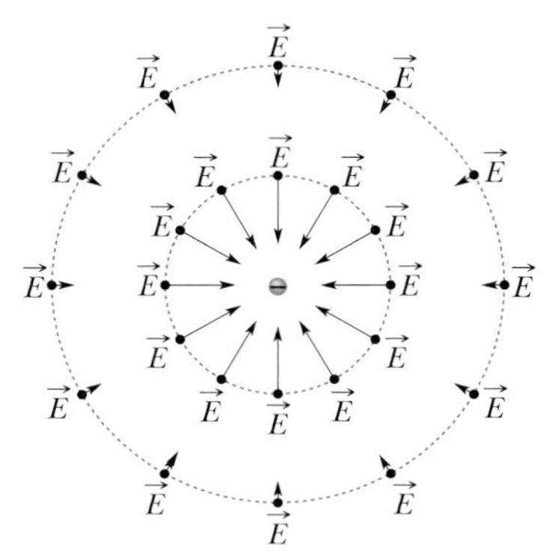

FIGURE 23-24 ■ There are two methods commonly used to depict an electric field. This figure shows a vector electric field map (or plot). See Fig. 23-25 for a comparison method.

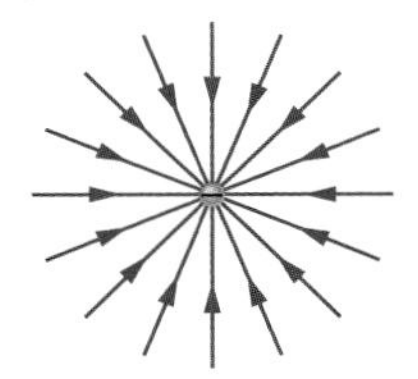

FIGURE 23-25 ■ The second common representation of an electric field is the field line representation shown here. See Fig. 23-24 for a comparison method.

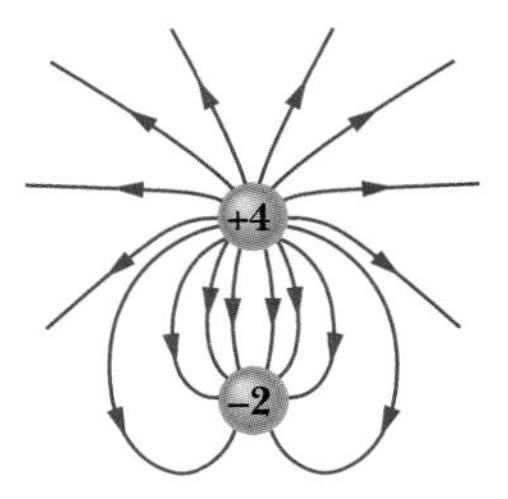

FIGURE 23-26 ■ Electric field lines for a $-2q$ and $+4q$ charge configuration.

Electric Field Near a Nonconducting Sheet

Figure 23-27*a* shows part of an infinitely large, nonconducting *sheet* (or plane) with a uniform distribution of positive charge on its right side. The electric field lines shown in Fig. 23-27*b* are uniformly spaced and always perpendicular to the sheet. Why? If the sheet can be treated as if it is infinitely large, then it looks the same in any direction. The pulls or pushes sideways from any bit of the sheet to one side of the test charge are cancelled by those from a symmetric bit of the sheet on the opposite side of the charge. As a result, the electric force vector at that point must point directly toward the sheet (if the sheet is negative) or directly away from the sheet (if the sheet is positive).

Since the field lines are perpendicular to the sheet and have to start and end on charges, they don't diverge or get closer as you move farther from the sheet. This suggests the surprising result that the field should not get weaker or stronger as you get

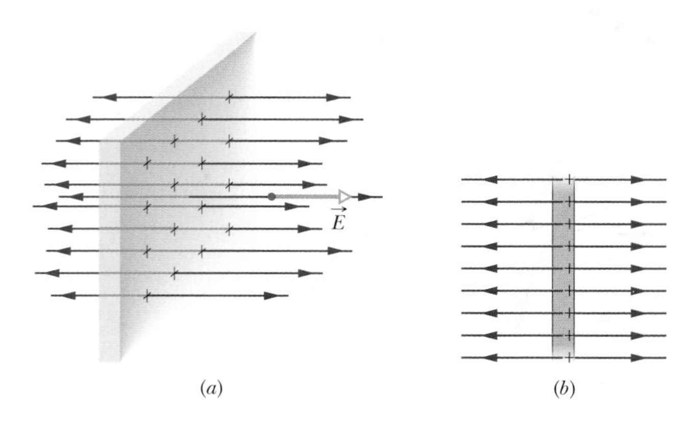

FIGURE 23-27 ■ Depictions of the electric field lines due to a very large, nonconducting sheet with uniformly distributed positive charge on one side. The vector fields shown in both figures are uniform and perpendicular to the charged sheet. (*a*) The electric field vector $\vec{E}$ is shown at the location of a test charge, and (*b*) a side view of *a* showing electric field lines pointing away from the positive charges in the space near the sheet.

farther from the sheet. This is even more strongly suggested by a dimensional analysis argument. An E-field has dimensions that look like those of kq/r^2. For an infinite sheet we can't talk about the total charge (it's infinite) but only the charge density sigma, denoted as σ. Sigma already has units of kq/r^2 so it is not possible to put in an extra distance. Doing the integral (which can be generalized from our integral for the ring) is messy but confirms this result. Because the charge is uniformly distributed along the sheet and all the field vectors have the same magnitude, the sheet creates a **uniform electric field.**

Of course, no real nonconducting sheet (such as a flat expanse of plastic) is infinitely large, but if we consider a region near the middle of a real sheet and not near its edges, the field lines through that region are arranged as in Figs. 23-27*a* and *b*.

Field Lines for Two Positive Charges

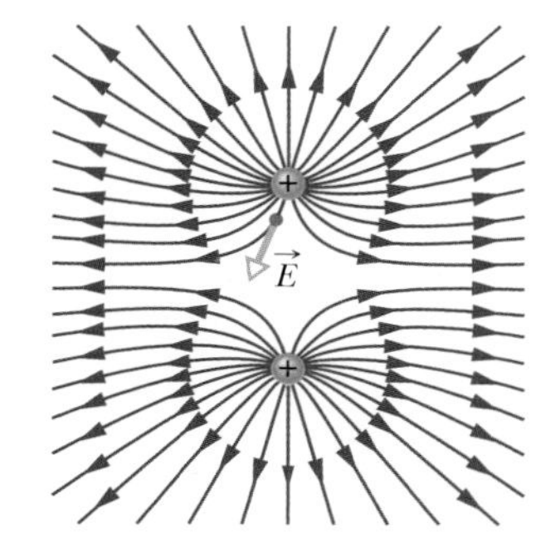

FIGURE 23-28 ■ Field lines for two equal positive point charges. The charges repel each other. (The lines terminate on distant negative charges.) To "see" the actual three-dimensional pattern of field lines, mentally rotate the pattern shown here about an axis passing through both charges in the plane of the page. The three-dimensional pattern and the electric field it represents are said to have rotational symmetry about that axis. The electric field vector at one point is shown; note that it is tangent to the field line through that point.

Figure 23-28 shows the field lines for two equal positive charges. Although we do not often use field lines quantitatively, they are very useful to visualize what is going on. It takes practice to learn to draw electric field lines even for a simple array of point sources. The steps include: (1) creating an electric field map, (2) deciding how detailed the field line representation should be and assigning a certain number of lines per unit charge, (3) placing the assigned number of lines at each point source. The lines should be equally spaced at the source with initial directions that depend on the sign of the source charge. The lines pass radially out from each positive source (and should be marked with an outward arrow). Or the lines pass radially into each negative source (and should be marked with an inward arrow as shown in Figs. 23-25 and 23-26). (4) Each line through the vector field map should be drawn so it is always tangent to the electric field vectors. (5) If the net charge in the region on the map is zero, each line begins on a positive charge and ends on a negative charge. If the net charge is negative, some of the lines come from infinity (off the field line diagram). If the net charge is positive, some of the lines veer off to infinity as shown in Fig. 23-26.

READING EXERCISE 23-7: Explain why the definition that says electric field lines point in the direction of the electric field means electric field lines must originate on positive charges and terminate on negative charges. ■

READING EXERCISE 23-8: Examine Fig. 23-12 showing that the net electric field along a line that bisects the two charges is always perpendicular to the line connecting the charges. How does the construction of this diagram help explain the fact that the net electric field due to a uniformly charged sheet (shown in Fig. 23-27) is always perpendicular to the sheet of charge? ■

Problems

In the following problems, all electric fields referenced are those produced by the source charge(s). That is, $\vec{E} = \vec{E}_s$.

SEC. 23-4 ■ THE ELECTRIC FIELD DUE TO A POINT CHARGE

1. Point Charge What is the amount of charge on a small particle whose electric field 50 cm away has the amount of 2.0 N/C?

2. What Amount? What is the amount of a point charge that would create an electric field of 1.00 N/C at points 1.00 m away?

3. Plutonium-239 A plutonium-239 nucleus of radius 6.64 femptometers has an atomic number $Z = 94$. Assuming that the positive charge is distributed uniformly within the nucleus, what are the magnitude and direction of the electric field at the surface of the nucleus? Assume the influence of the positive charge at the nuclear surface is the same as that of a point charge.

SEC. 23-5 ■ THE ELECTRIC FIELD DUE TO MULTIPLE CHARGES

4. Two Particles Two particles with equal charge amounts 2.0×10^{-7} C but opposite signs are held 15 cm apart. What are the magnitude and direction of $\vec{E}$ at the point midway between the charges?

5. Two Point Charges Two point charges $q_A = 2.1 \times 10^{-8}$ C and $q_B = -4.0q_A$ are fixed in place 50 cm apart. Find the point along the straight line passing through the two charges at which the electric field is zero.

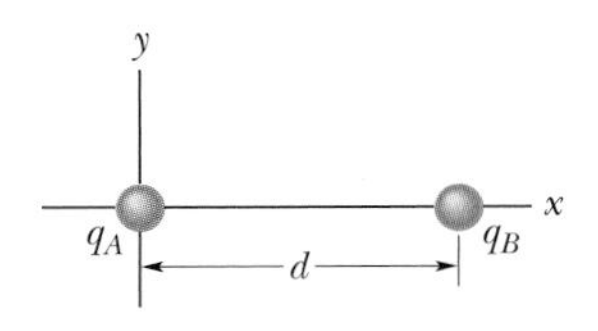

FIGURE 23-29 ■ Problems 6 and 8.

6. Two Fixed Charges In Fig 23-29, two fixed point charges $q_A = +1.0 \times 10^{-6}$ C and $q_B = +3.0 \times 10^{-6}$ C are separated by a distance $d = 10$ cm. Plot their net electric field $\vec{E}(x)$ as a function of x for both positive and negative values of x, taking $\vec{E}$ to be positive when $\vec{E}$ points to the right and negative when $\vec{E}$ points to the left.

7. Four Charges In Fig. 23-30, what is the magnitude of the electric field at point P due to the four point charges shown? The distance d is between charge centers.

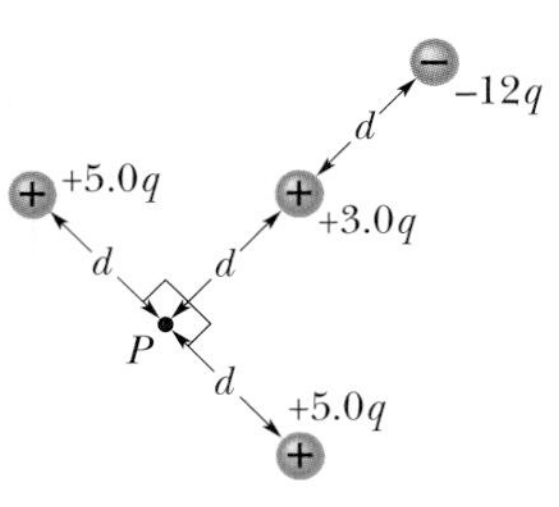

FIGURE 23-30 ■ Problem 7.

8. Separation of d In Fig. 23-29, two fixed point charges $q_A = -5q$ and $q_B = +2q$ are separated by distance d. Locate the point (or points) at which the net electric field due to the two charges is zero.

9. Square What are the magnitude and direction of the electric field at the center of the square of Fig. 23-31 if $q = 1.0 \times 10^{-8}$ C and the distance between charge centers $a = 5.0$ cm?

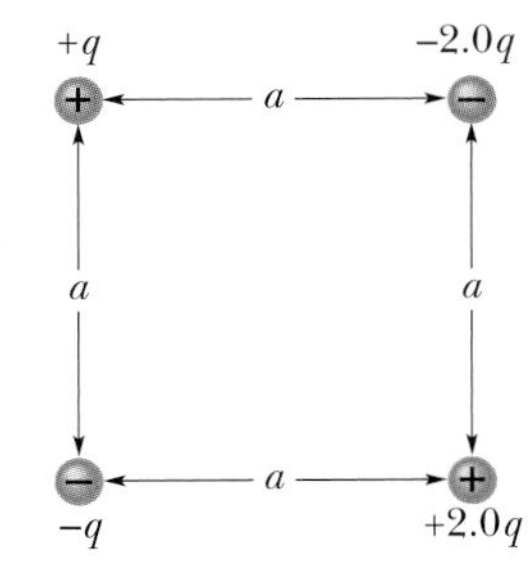

FIGURE 23-31 ■ Problem 9.

10. Three Charges Calculate the direction and magnitude of the electric field at point P in Fig. 23-32, due to the three point charges. The distance between charge centers is a.

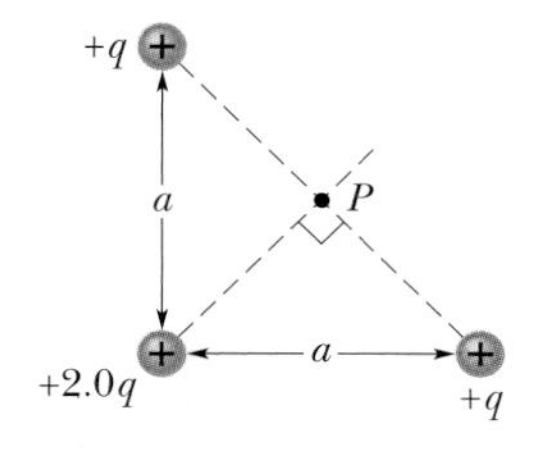

FIGURE 23-32 ■ Problem 10.

11. Equilateral Triangle Two particles, each with an amount of $|q|$ equal to charge of 12 nC, are placed at two of the vertices of an equilateral triangle. The length of each side of the triangle is 2.0 m. What is the magnitude of the electric field at the third vertex of the triangle if (a) both of the charges are positive and (b) one of the charges is positive and the other is negative?

12. Plastic Ring Figure 23-33 shows a plastic ring of radius $R = 50.0$ cm. Two small charged beads are on the ring: Bead 1 of charge $+2.00\ \mu$C is fixed in place at the left side; bead 2 of charge $+6.00\ \mu$C can be moved along the ring. The two beads produce a net electric field of magnitude E at the center of the ring. At what angle θ should bead 2 be positioned such that $E = 2.00 \times 10^5$ N/C?

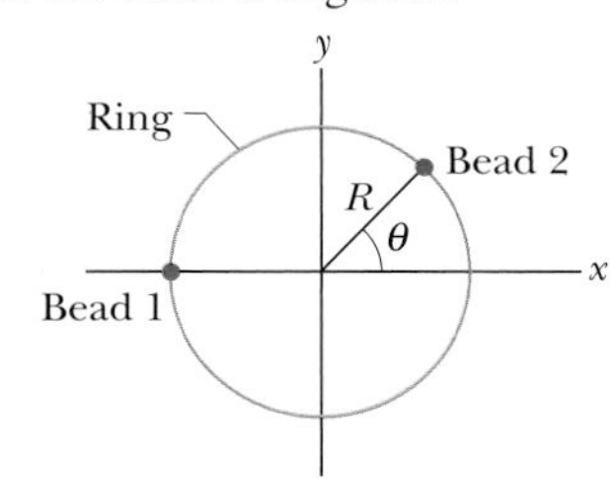

FIGURE 23-33 ■ Problem 12.

13. Three Particles Two Three particles, each with positive charge q, form an equilateral triangle, with each side of length d. What is the magnitude of the electric field produced by the particles at the midpoint of any side?

14. Separation L Figure 23-34a shows two charged particles fixed in place on an x axis with separation L. The ratio q_A/q_B of their charge amounts is 4.00. Figure 23-34b shows the x-component E_x^{net} of their net electric field along the x axis just to the right of particle B. (a) At what value of $x > 0$ is E_x^{net} maximum? (b) If particle B has charge $-q_B = -3e$, what is the value of that maximum?

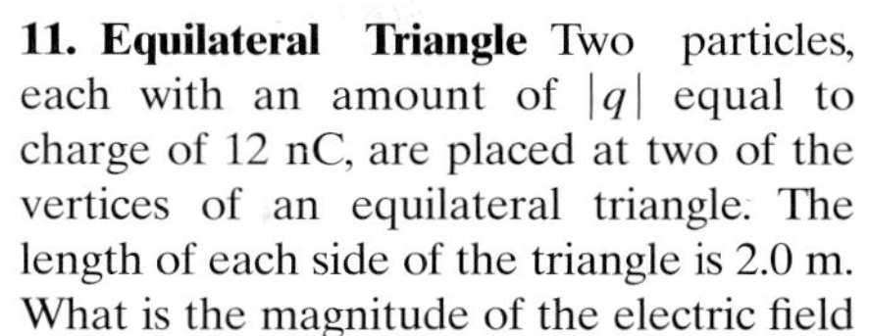

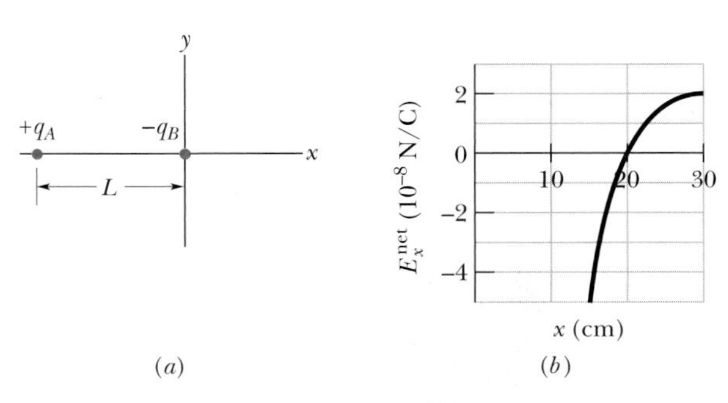

FIGURE 23-34 ■ Problem 14.

15. Two Charges Two Figure 23-35 shows two charged particles on an x axis: $-q = -3.20 \times 10^{-19}$ C at $x = -3.00$ m and $q = 3.20 \times$

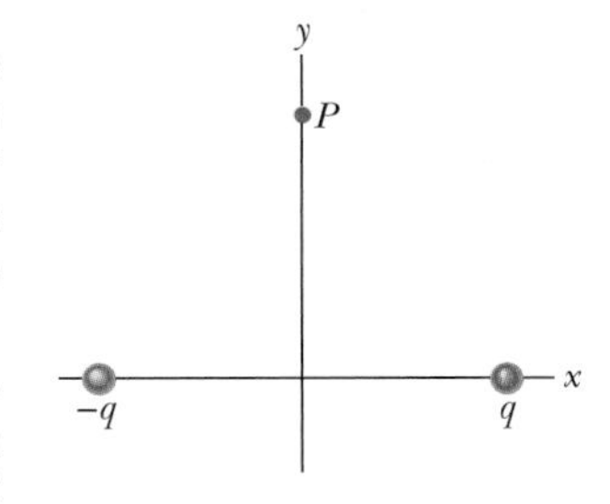

FIGURE 23-35 ■ Problem 15.

10^{-19} C at $x = +3.00$ m. What are the magnitude and direction of the net electric field they produce at point P at $y = 4.00$ m?

16. Eight Charges In Fig. 23-36, eight charged particles form a square array; charge $q = e$ and distance $d = 2.0$ cm. What are the magnitude and direction of the net electric field at the center?

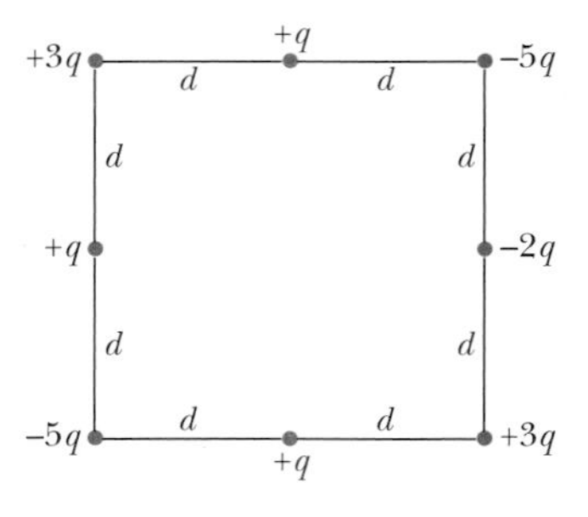

FIGURE 23-36 ■ Problem 16.

SEC. 23-6 ■ THE ELECTRIC FIELD DUE TO AN ELECTRIC DIPOLE

17. Calculate the Moment Calculate the electric dipole moment of an electron and a proton 4.30 nm apart.

18. Field at P In Fig. 23-16, let both charges be positive. Assuming $z \gg d$, show that the magnitude of the vector $\vec{E}$ at point P in that figure is then given by

$$|\vec{E}| = k\frac{2|q|}{z^2}.$$

19. Electric Quadrupole Figure 23-37 shows an electric quadrupole. It consists of two dipoles with dipole moments that are equal in magnitude but opposite in direction. Show that the magnitude of the vector $\vec{E}$ on the axis of the quadrupole for a point P a distance z from its center (assume $z \gg d$) is given by

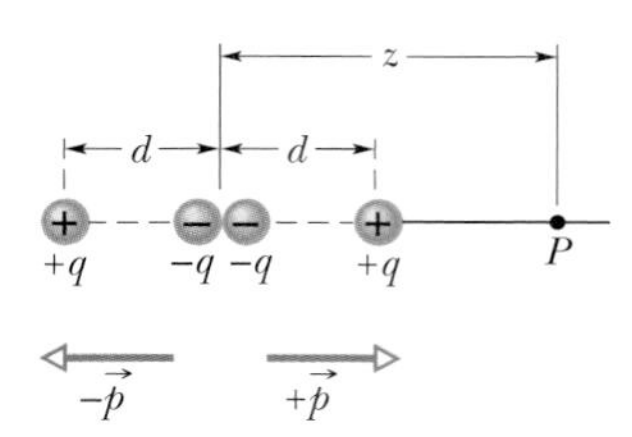

FIGURE 23-37 ■ Problem 19.

$$|\vec{E}| = \frac{3|Q|}{4\pi\varepsilon_0 z^4},$$

in which $Q(= 2qd^2)$ is known as the *quadrupole moment* of the charge distribution.

20. Electric Dipole Find the magnitude and direction of the electric field at point P due to the electric dipole in Fig 23-38. The distance between charge center is d and P is located at a distance $r \gg d$ along the perpendicular bisector of the line joining the charges. Express your answer in terms of the magnitude and direction of the electric dipole moment $\vec{p}$.

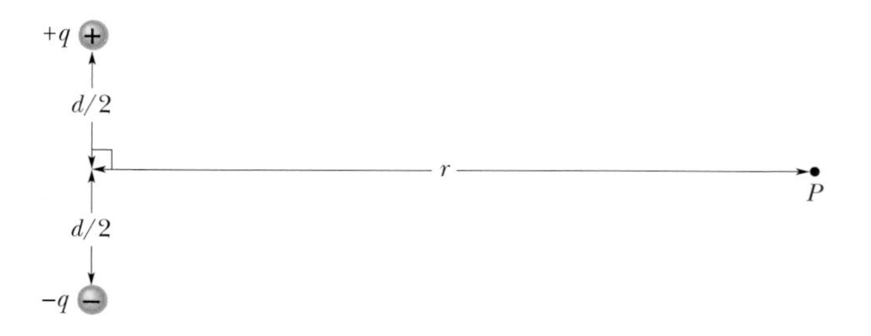

FIGURE 23-38 ■ Problem 20.

SEC. 23-7 ■ THE ELECTRIC FIELD DUE TO A RING OF CHARGE

21. Electron Constrained An electron is constrained to the central axis of the ring of charge of radius R discussed in Section 23-7. Show that the electrostatic force on the electron can cause it to oscillate through the center of the ring with an angular frequency

$$\omega = \sqrt{\frac{eq}{4\pi\varepsilon_0 m R^3}},$$

where q is the ring's charge and m is the electron's mass.

22. Two Rings Figure 23-39 shows two parallel nonconducting rings arranged with their central axes along a common line. Ring A has uniform charge q_A and radius R; ring B has uniform charge q_B and the same radius R. The rings are separated by a distance $3R$. The net electric field at point P on the common line, at distance R from ring A, is zero. What is the ratio q_A/q_B?

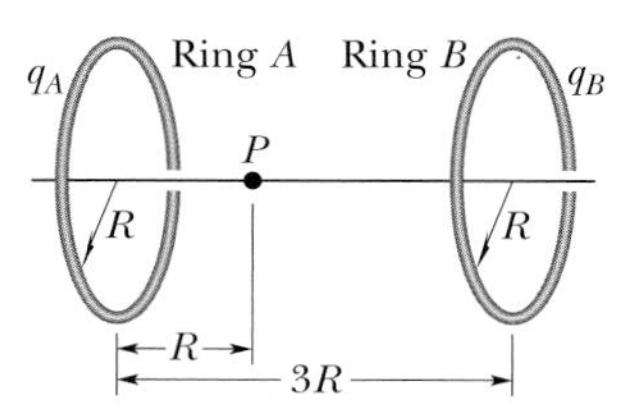

FIGURE 23-39 ■ Problem 22.

23. Thin Glass Rod A thin glass rod is bent into a semicircle of radius r. A charge $+q$ is uniformly distributed along the upper half, and a charge $-q$ is uniformly distributed along the lower half, as shown in Fig. 23-40*a*. Find the magnitude and direction of the electric field $\vec{E}$ at P, the center of the semicircle.

24. Two Curved Plastic Rods In Fig. 23-40*b*, two curved plastic rods, one of charge $+q$ and the other of charge $-q$, form a circle of radius R in an xy plane. The x axis passes through their connecting points, and the charge is distributed uniformaly on both rods. What are the magnitude and direction of the electric field $\vec{E}$ produced at P, the center of the circle?

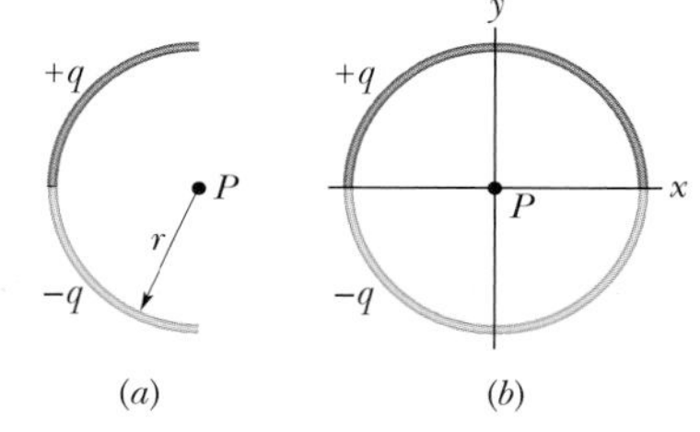

FIGURE 23-40 ■ Problems 23 and 24.

25. Nonconducting Rod In Fig. 23-41, a nonconducting rod of length L has charge $-q$ uniformly distributed along its length. (a) What is the linear charge density of the rod? (b) What is the electric field at point P, a distance a from the end of the rod? (c) If P were very far from the rod compared to L, the rod would look like a point charge. Show that your answer to (b) reduces to the electric field of a point charge for $a \gg L$.

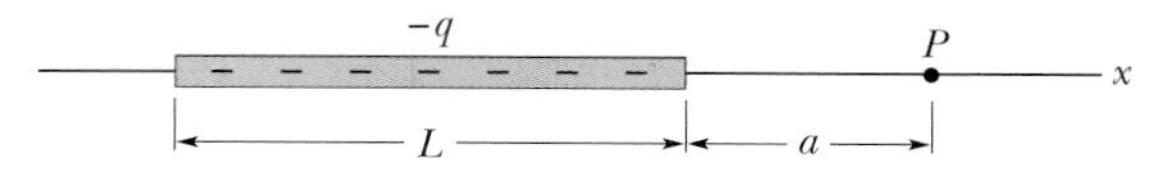

FIGURE 23-41 ■ Problem 25.

26. What Distance? At what distance along the central axis of a ring of radius R and uniform charge is the magnitude of the electric field due to the ring's charge maximum?

27. Semi-Infinite Rod In Fig. 23-42, a "semi-infinite" nonconducting rod (that is, infinite in one direction only) has uniform linear charge density λ. Show that the electric field at point P makes an angle of 45° with the rod and that this result is independent of the distance R. (*Hint:* Separately find the parallel and perpendicular (to the rod) components of the electric field at P, and then compare those components.)

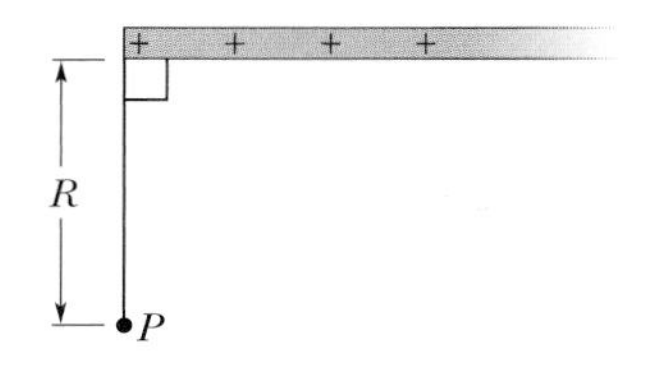

FIGURE 23-42 ■ Problem 27.

28. Length L Rod A thin nonconducting rod of finite length L has a charge q spread uniformly along it. Show that

$$|\vec{E}| = \frac{|q|}{2\pi\varepsilon_0 y} \frac{1}{(L^2 + 4y^2)^{1/2}}$$

gives the magnitude $|\vec{E}|$ of the electric field at point P on the perpendicular bisector of the rod (Fig. 23-43).

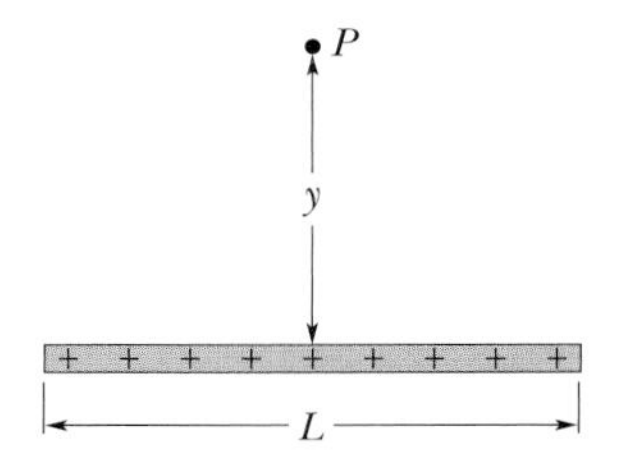

FIGURE 23-43 ■ Problem 28.

29. Density, Density, Density. (a) A charge of $-300e$ is uniformly distributed along a circular arc of radius 4.00 cm, which subtends an angle of 40°. What is the linear charge density along the arc? (b) A charge of $-300e$ is uniformly distributed over one face of a circular disk of radius 2.00 cm. What is the surface charge density over that face? (c) A charge of $-300e$ is uniformly distributed over the surface of a sphere of radius 2.00 cm. What is the surface charge density over that surface? (d) A charge of $-300e$ is uniformly spread through the volume of a sphere of radius 2.00 cm. What is the volume charge density in that sphere?

30. Nonconducting Rod Two A thin nonconducting rod with a uniform distribution of positive charge Q is bent into a circle of radius R (Fig. 23-44). The central axis through the ring is a z axis, with the origin at the center of the ring. What is the magnitude of the electric field due to the rod at (a) $z = 0$ and (b) $z = \pm\infty$? (c) In terms of R, at what values of z is that magnitude maximum? (d) If radius $R = 2.00$ cm and charge $Q = 4.00\ \mu$C, what is the maximum magnitude?

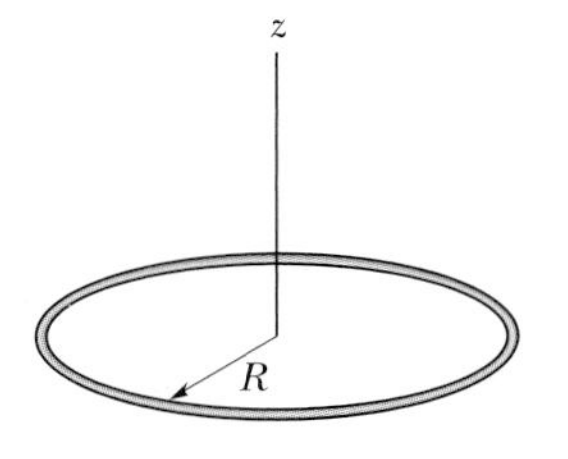

FIGURE 23-44 ■ Problem 30.

31. Circular Rod A circular rod has a radius of curvature R and a uniformly distributed charge Q and it subtends an angle θ (in radians). What is the magnitude of the electric field it produces at the center of curvature?

32. Two Concentric Rings Figure 23-45 shows two concentric rings, of radii R and $R' = 3.00R$, that lie on the same plane. Point P lies on the central z axis, at distance $D = 2.00R$ from the center of the rings. The smaller ring has uniformly distributed charge $+Q$. What must be the uniformly distributed charge on the larger ring if the net electric field at point P due to the two rings is to be zero?

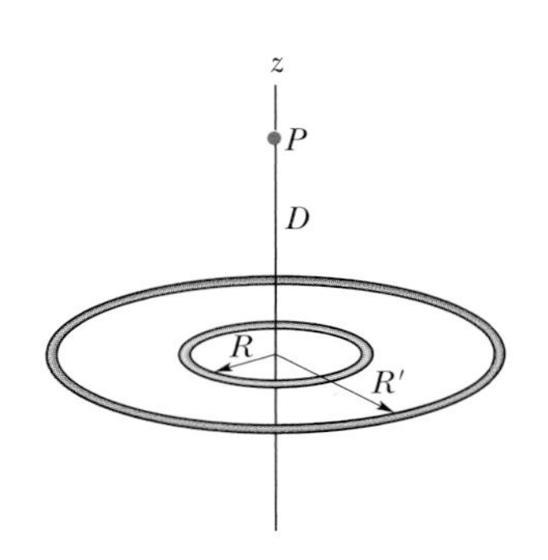

FIGURE 23-45 ■ Problem 32.

33. Charge $+Q$ In Fig. 23-46a, a particle of charge $+Q$ produces an electric field with a magnitude E_{part} at point P, at distance R from it. In Fig. 23-46b, that same amount of charge is spread uniformly along a circular arc that has radius R and subtends an angle θ. The charge on the arc produces an electric field with a magnitude E_{arc} at its center of curvature P. For what value of θ does $E_{\text{arc}} = 0.500$ E_{part}? (*Hint:* You can use a graphical solution.)

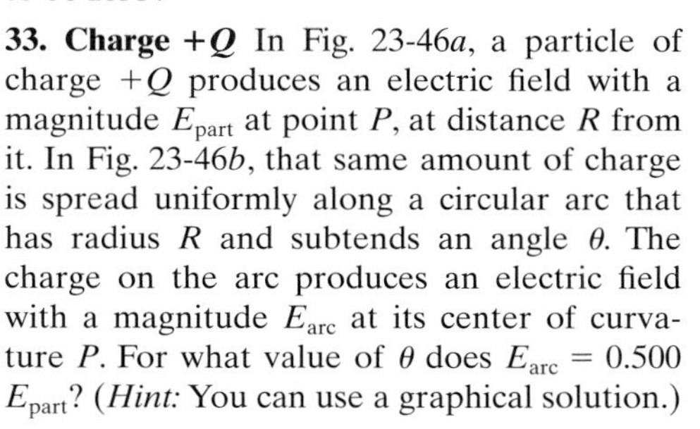

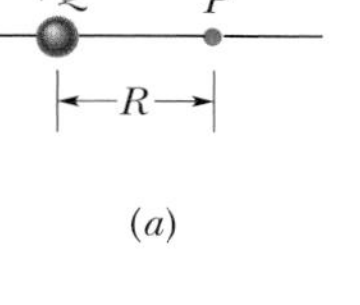

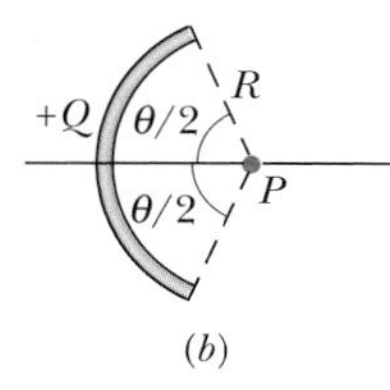

FIGURE 23-46 ■ Problem 33.

34. Half Circle Figure 23-47a shows a nonconducting rod with a uniformly distributed charge $+Q$. The rod forms a half circle with radius R and produces an electric field of magnitude E_{arc} at its center of curvature P. If the arc is collapsed to a point at distance R from P (Fig. 23-47b), by what factor is the magnitude of the electric field at P multiplied?

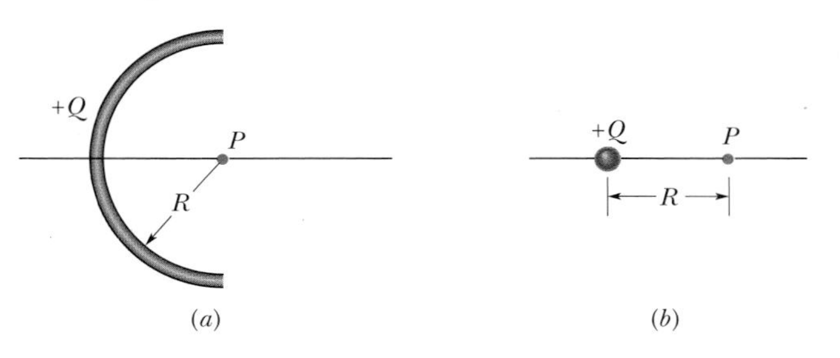

FIGURE 23-47 ■ Problem 34.

SEC. 23-8 ■ MOTION OF POINT CHARGES IN AN ELECTRIC FIELD

35. Electron Released from Rest An electron is released from rest in a uniform electric field of magnitude 2.00×10^4 N/C. Calculate the acceleration of the electron. (Ignore gravitation).

36. Accelerated Electron An electron is accelerated eastward at 1.80×10^9 m/s^2 by an electric field. Determine the magnitude and direction of the electric field.

37. Force Due to Dipole Calculate the magnitude of the force, due to an electric dipole of dipole moment 3.6×10^{-29} C · m, on an electron 25 nm from the center of the dipole, along the dipole axis. Assume that this distance is large relative to the dipole's charge separation.

38. Alpha Particle An alpha particle (the nucleus of a helium atom) has a mass of 6.64×10^{-27} kg and a charge of $+2e$. What are the magnitude and direction of the electric field that will balance the gravitational force on it?

39. Charged Cloud A charged cloud system produces an electric field in the air near Earth's surface. A particle of charge -2.0×10^{-9} C is acted on by a downward electrostatic force of 3.0×10^{-6} N when placed in this field. (a) What is the magnitude of the electric field? (b) What are the magnitude and direction of the electrostatic force exerted on a proton placed in this field? (c) What is the gravitational force on the proton? (d) What is the ratio of the magnitude of the electrostatic force to the magnitude of the gravitational force in this case?

40. Humid Air Humid air breaks down (its molecules become ionized) in an electric field of 3.0×10^6 N/C. In that field, what is the magnitude of the electrostatic force on (a) an electron and (b) an ion with a single electron missing?

41. High-Speed Protons Beams of high-speed protons can be produced in "guns" using electric fields to accelerate the protons. (a) What acceleration would a proton experience if the gun's electric field were 2.00×10^4 N/C? (b) What speed would the proton attain if the field accelerated the proton through a distance of 1.00 cm?

42. Floating a Sulfur Sphere An electric field E with an average magnitude of about 150 N/C points downward in the atmosphere near Earth's surface. We wish to "float" a sulfur sphere weighing 4.4 N in this field by charging the sphere. (a) What charge (both sign and magnitude) must be used? (b) Why is the experiment impractical?

43. Two Oppositely Charged Plates A uniform electric field exists in a region between two oppositely charged plates. An electron is released from rest at the surface of the negatively charged plate and strikes the surface of the opposite plate, 2.0 cm away, in a time 1.5×10^{-8} s. (a) What is the speed of the electron as it strikes the second plate? (b) What is the magnitude of the electric field $\vec{E}$?

44. Field Retards Motion An electron with a speed of 5.00×10^8 cm/s enters an electric field of magnitude 1.00×10^3 N/C, traveling along the field lines in the direction that retards its motion. (a) How far will the electron travel in the field before stopping momentarily and (b) how much time will have elapsed? (c) If the region with the electric field is only 8.00 mm long (too short for the electron to stop within it), what fraction of the electron's initial kinetic energy will be lost in that region?

45. Two Copper Plates Two large parallel copper plates are 5.0 cm apart and have a uniform electric field between them as depicted in Fig. 23-48. An electron is released from the negative plate at the same time that a proton is released from the positive plate. Neglect the force of the particles on each other and find their distance from the positive plate when they pass each other. (Does it surprise you that you need not know the electric field to solve this problem?)

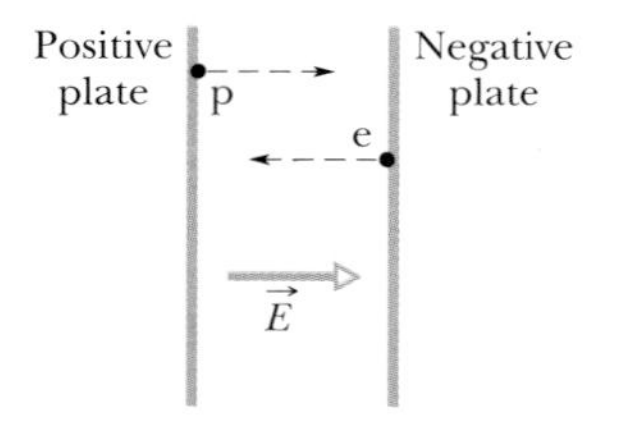

FIGURE 23-48 ■ Problem 45.

46. Velocity Components At some instant the velocity components of an electron moving between two charged parallel plates are $v_x = 1.5 \times 10^5$ m/s and $v_y = 3.0 \times 10^3$ m/s. Suppose that the electric field between the plates is given by $\vec{E} = (120\text{ N/C})\hat{j}$. (a) What is the acceleration of the electron? (b) What will be the velocity of the electron after its x coordinate has changed by 2.0 cm?

47. Uniform Upward Field In Fig. 23-49, a uniform, upward-directed electric field $\vec{E}$ of magnitude 2.00×10^3 N/C has been set up between two horizontal plates by charging the lower plate positively and the upper plate negatively. The plates have length $L = 10.0$ cm and separation $d = 2.00$ cm. An electron is then shot between the plates from the left edge of the lower plate. The initial velocity $\vec{v}_1$ of the electron makes an angle $\theta = 45.0°$ with the lower plate and has a magnitude of 6.00×10^6 m/s. (a) Will the electron strike one of the plates? (b) If so, which plate and how far horizontally from the left edge will the electron strike?

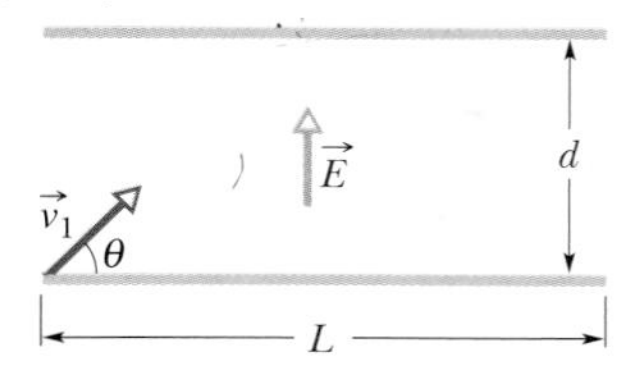

FIGURE 23-49 ■ Problem 47.

48. Charge in an E Field A 10.0 g block with a charge of $+8.00 \times 10^{-5}$ C is placed in electric field $\vec{E} = (3.00 \times 10^3\text{ N/C})\hat{i} - 600\text{ N/C})\hat{j}$. (a) What are the magnitude and direction of the force on the block? (b) If the block is released from rest at the origin at $t = 0.00$ s, what will be its coordinates at $t = 3.00$ s?

49. Entering a Field An electron enters a region of uniform electric field with an initial velocity of 40 km/s in the same direction as the electric field, which has magnitude $E = 50$ N/C. (a) What is the speed of the electron 1.5 ns after entering this region? (b) How far does the electron travel during the 1.5 ns interval?

50. An Electron Is Shot In Fig. 23-50, an electron is shot at an initial speed of $v_1 = 2.00 \times 10^6$ m/s, at angle $\theta_1 = 40°$ from an x axis. It moves in a region with uniform electric field $\vec{E} = (5.00\text{ N/C})\hat{j}$. A screen for detecting electrons is positioned parallel to the y axis, at distance $x = 3.00$ m. In unit-vector notation, what is the velocity of the electron when it hits the screen?

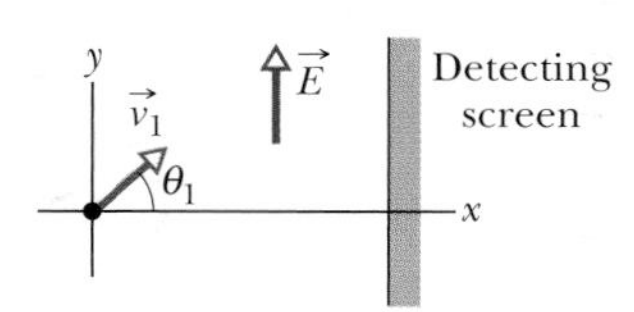

FIGURE 23-50 ■ Problem 50.

51. TV Tube Figure 23-51 shows the deflection-plate system of a conventional TV tube. The length of the plates is 3.0 cm and the electric field between the two plates is 10^6 N/C (vertically up). If the electron enters the plates with a horizontal velocity of 3.9×10^7 m/s, what is the vertical deflection Δy at the end of the plates?

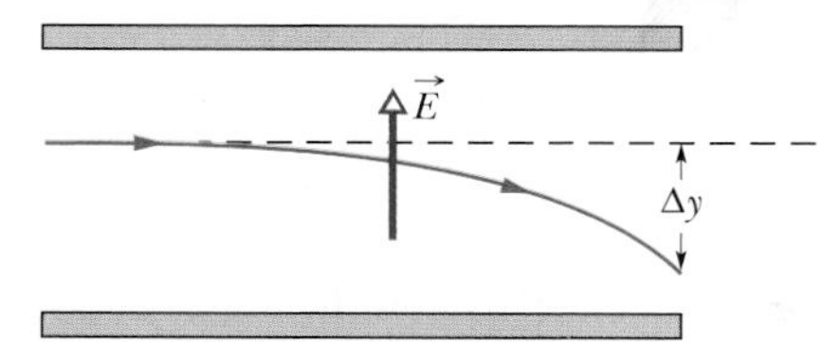

FIGURE 23-51 ■ Problem 51.

SEC. 23-9 ■ A DIPOLE IN AN ELECTRIC FIELD

52. Dipole in a Field An electric dipole, consisting of charges of magnitude 1.50 nC separated by 6.20 μm, is in an electric field of strength 1100 N/C. (a) What is the magnitude of the electric dipole moment? (b) What is the difference between the potential energies corresponding to dipole orientations parallel to and antiparallel to the field?

53. Torque on a Dipole An electric dipole consists of charges $+2e$ and $-2e$ separated by 0.78 nm. It is in an electric field of strength 3.4×10^6 N/C. Calculate the magnitude of the torque on the dipole when the dipole moment is (a) parallel to, (b) perpendicular to, and (c) antiparallel to the electric field.

54. Work Required Find the work required to turn an electric dipole end for end in a uniform electric field $\vec{E}$, in terms of the magnitude $|\vec{p}|$ of the dipole moment, the magnitude $|\vec{E}|$ of the field, and the initial angle θ_1 between $\vec{p}$ and $\vec{E}$.

55. Frequency of Oscillation Find the frequency of oscillation of an electric dipole, of dipole moment $\vec{p}$ and rotational inertia I, for small amplitudes of oscillation about its equilibrium position in a uniform electric field of magnitude $|\vec{E}|$.

56. A Certain Dipole A certain electric dipole is placed in a uniform electric field $\vec{E}$ of magnitude 40 N/C. Figure 23-52 gives the magnitude τ of the torque on the dipole versus the angle θ between $\vec{E}$ and the dipole moment $\vec{p}$. What is the magnitude of $\vec{p}$?

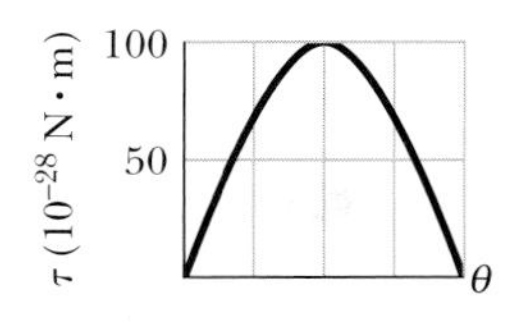

FIGURE 23-52 ■ Problem 56.

57. How Much Energy How much energy is needed to flip an electric dipole from being lined up with a uniform external electric field to being lined up opposite the field? The dipole consists of an electron and a proton at a separation of 2.00 nm, and it is in a uniform field of magnitude 3.00×10^6 N/C.

58. See Graph A certain electric dipole is placed in a uniform electric field $\vec{E}$ of magnitude 20 N/C. Figure 23-53 gives the potential energy U of the dipole versus the angle θ between $\vec{E}$ and the dipole moment $\vec{p}$. What is the magnitude of $\vec{p}$?

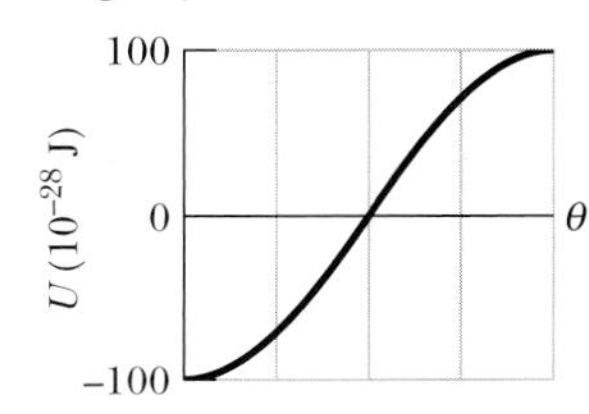

FIGURE 23-53 ■ Problem 58.

SEC. 23-10 ■ ELECTRIC FIELD LINES

59. Twice the Separation In Fig. 23-54, the electric field lines on the left have twice the separation of those on the right. (a) If the magnitude of the field at A is 40 N/C, what force acts on a proton at A? (b) What is the magnitude of the field at B?

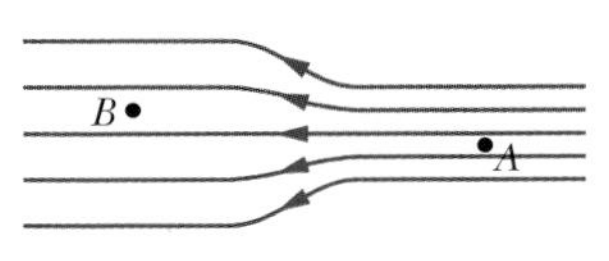

FIGURE 23-54 ■ Problem 59.

60. Sketch Sketch qualitatively the electric field lines both between and outside two concentric conducting spherical shells when a uniform positive charge q_A is on the inner shell and a uniform negative charge $-q_B$ is on the outer. Consider the cases $q_A > q_B$, $q_A = q_B$ and $q_A < q_B$.

61. Thin Circular Disk Sketch qualitatively the electric field lines for a thin, circular, uniformly charged disk of radius R. (*Hint:* Consider as limiting cases points very close to the disk, where the electric field is directed perpendicular to the surface, and points very far from it, where the electric field is like that of a point charge.)

62. Particles and Lines In Fig. 23-55, particles with charges $+1.0q$ and $-2.0q$ are fixed a distance d apart. Find the magnitude and direction of the net electric field at points (a) A, (b) B, and (c) C. (d) Sketch the electric field lines.

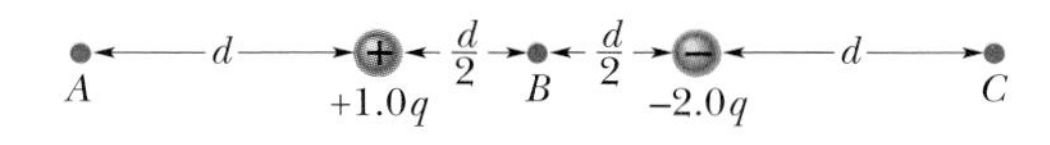

FIGURE 23-55 ■ Problem 62.

63. Three Point Charges Two In Fig. 23-56, three point charges are arranged in an equilateral triangle. (a) Sketch the field lines due to $+Q$ and $-Q$, and from them determine the direction of the force that acts on $+q$ because of the presence of the other two charges. (b) What is the magnitude of that net electric force on $+q$?

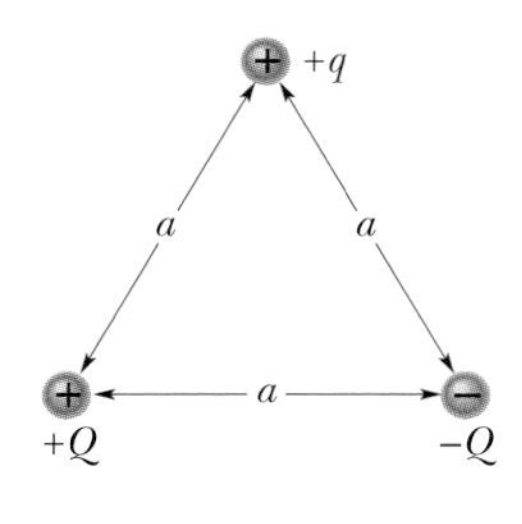

FIGURE 23-56 ■ Problem 63.

Additional Problems

64. Two Electric Charges—Electric Force Figure 23-57 shows seven arrangements of two electric charges. In each figure, a point labeled P is also identified. All of the charges are the same size, 20 nC, but they can be either positive or negative. The charges and point P all lie on a straight line. The distances between adjacent items, either between two charges or between a charge and point P, are all 5 cm. There are no other charges in this region. For this problem, we will place a +5 nC charge at point P.

Rank these arrangements from greatest to least on the basis of the magnitude of the electric force on the +5 nC charge when it is placed at point P. [Based on Ranking Task 126, O'Kuma, et. al., *Ranking Task Exercises* (Prentice Hall, Upper Saddle River, NJ, 2000).]

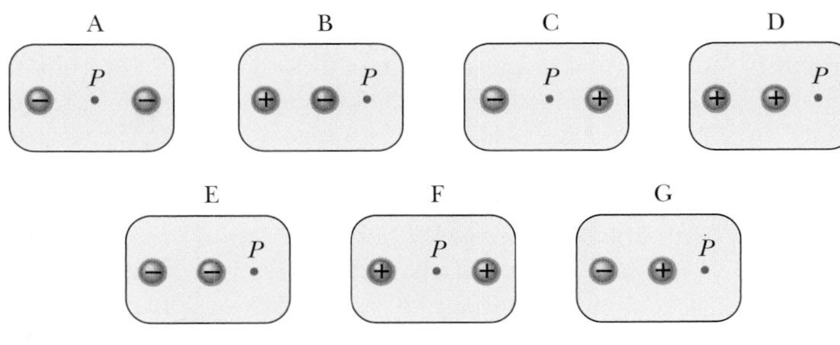

FIGURE 23-57 ■ Problem 64.

65. Fixed and Suspended Charges—Angle Figure 23-58 shows m_A, a stationary sphere with charge q_A. The charge m_B is suspended from the ceiling by a nonconducting string and has charge q_B. The masses m_A and m_B are conducting spheres of the same size. The charges q_A and q_B have the same sign. From the combinations below, rank the angle the string will form with the vertical from highest to lowest value. If any of the angles are the same, state that. Explain your reasoning.

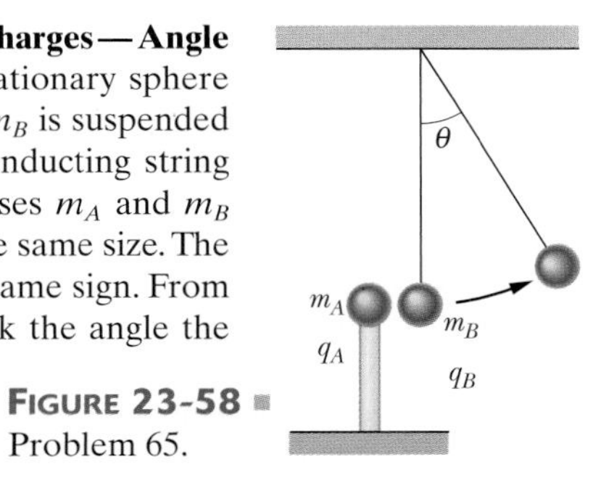

FIGURE 23-58 ■ Problem 65.

(a) $m_A = m$; $q_A = q$; $m_B = m$; $q_B = q$

(b) $m_A = 2m$; $q_A = q$; $m_B = 2m$; $q_B = 2q$

(c) $m_A = m$; $q_A = 2q$; $m_B = m$; $q_B = 2q$

(d) $m_A = 2m$; $q_A = q$; $m_B = 2m$; $q_B = q$

(e) $m_A = m$; $q_A = 2q$; $m_B = 2m$; $q_B = 2q$

(f) $m_A = m$; $q_A = q$; $m_B = m$; $q_B = 2q$

[Based on Ranking Task 138, O'Kuma, et. al., *Ranking Task Exercises* (Prentice Hall, Upper Saddle River, NJ, 2000).]

66. Electric Force on Same Charge Figure 23-59 shows a large region of space that has a uniform electric field in the x direction. At the point (0, 0) m, the electric field is (30 N/C)$\hat{\imath}$. Rank the magnitude of the electric force from greatest to least on a 5 C charge when it is placed at each of the following points: A, (0 m, 0 m); B, (0 m, 3 m); C, (−3 m, 0 m); D, (3 m, 0 m); E, (3 m, 3 m); F, (6 m, 0 m). Explain your reasoning. [Based on Ranking Task 139, O'Kuma, et. al., *Ranking Task Exercises* (Prentice Hall, Upper Saddle River, NJ, 2000).]

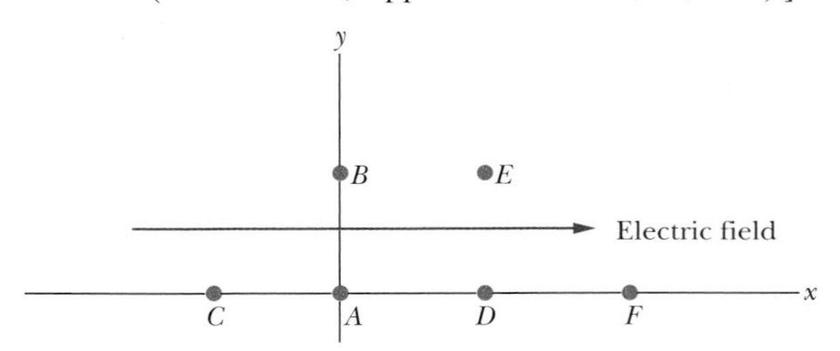

FIGURE 23-59 ■ Problem 66.

67. Dependence of E Figure 23-60 shows a fixed charge (specified by a circle) and a location (specified by the x). A test charge is placed at the x to measure the electric effect of the fixed charge. Complete the following two statements as quantitatively as you can. (For example, if the result is larger by a factor of three, don't say "increases"—say "triples" or "is multiplied by three.") Each statement is meant to be compared with the original situation. (The changes don't accumulate).

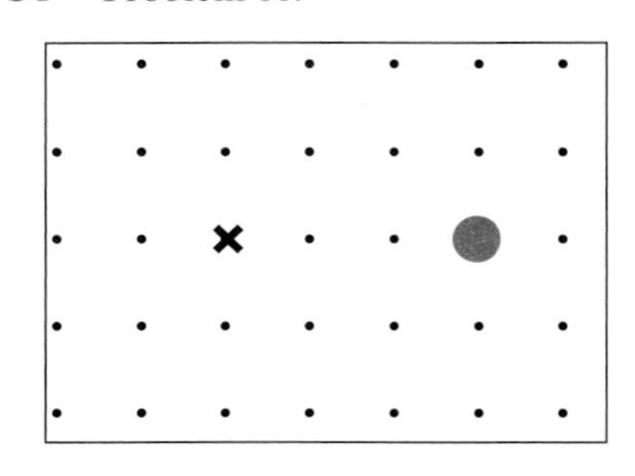

FIGURE 23-60 ■ Problem 67.

(a) If the test charge is replaced by one with half the amount of charge, then the electric field it experiences will ———.
(b) If the fixed charge is replaced by one with twice the amount of charge, then the electric field experiences by the test charge will ———.

68. What Is Going on at P? Figure 23-61 shows two charges of $-q$ arranged symmetrically about the y axis. Each produces an electric field at point P. (a) Are the magnitudes of the fields equal? Why or why not? (b) Does each electric field point toward or away from the charge producing it? Explain. (c) Is the magnitude of the net electric field equal to the sum of the magnitudes of the two field vectors (that is, equal to $2E$)? Why or why not? (d) Do the x-components of the two fields add or cancel? Explain. (e) Do the y-components of the two fields add or cancel? Explain. [Based on question 5.20, Arons, *Homework and Test Questions for Introductory Physics Teaching* (Wiley, New York, 1994).]

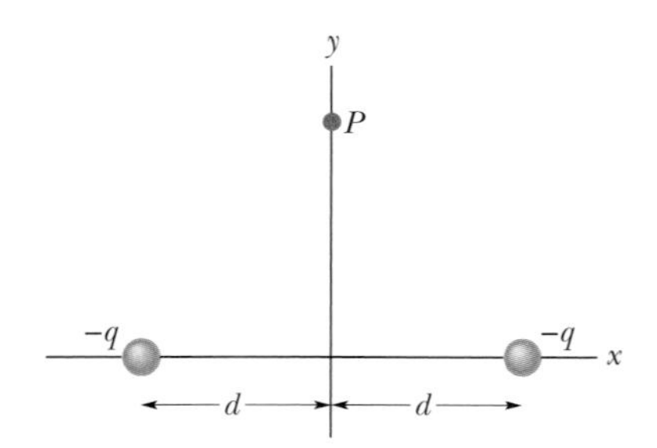

FIGURE 23-61 ■ Problem 68.

69. Field Lines Figure 23-62 shows the region in the neighborhood of a negatively charged conducting sphere and a large positively charged conducting plate extending far beyond the region shown. Someone claims that lines A–F are possible field lines representing the electric field lying in the region between the two conductors. (a) Examine each of the lines and indicate whether it is a correctly drawn field line. If a line is not correct, explain why. (b) Redraw the diagram with a pattern of field lines that is more correct.

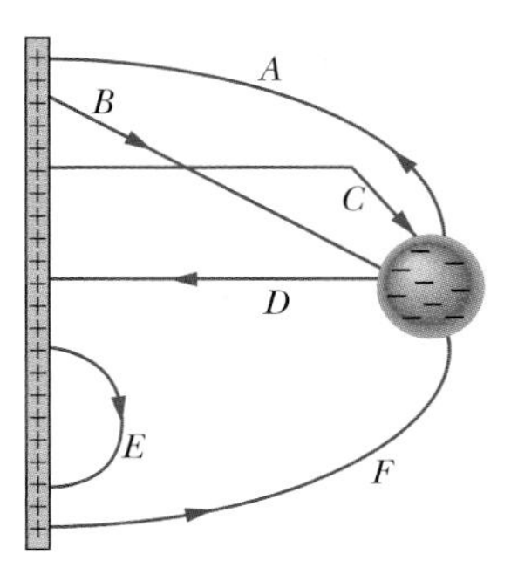

FIGURE 23-62 ■ Problem 69.

70. Field Lines Two Figure 23-63 shows the electric field lines for three point charges that are positive and negative as indicated. (a) Show the *direction* of each of the electric field lines with an arrow, and (b) if the central charge is +1.0 μC what are the values of the outer charges?

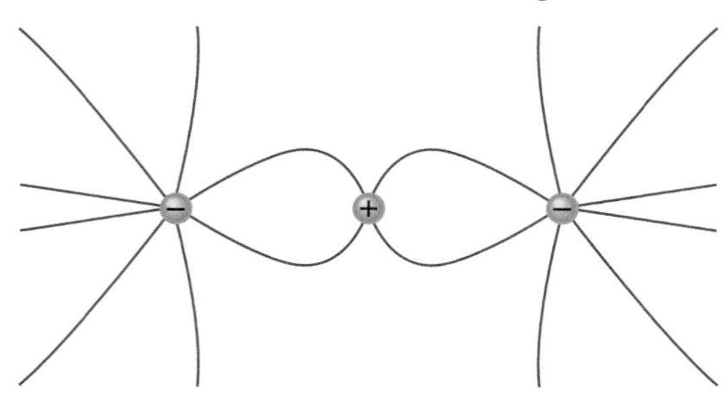

FIGURE 23-63 ■ Problem 70.

71. Field Lines Three Figure 23-64 shows the electric field lines for three point charges separated by a small distance. The two outer charges are identical and the one in the center is different. (a) Determine the ratio, q_A/q_B, of one of the outer charges to the inner one. (b) Determine the signs of q_A and q_B.

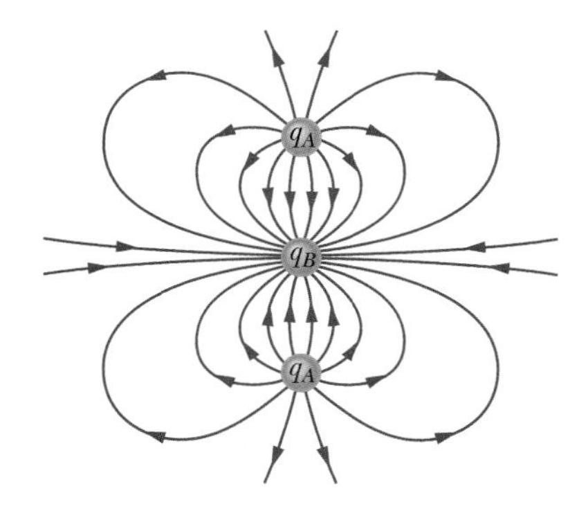

FIGURE 23-64 ■ Problem 71.

72. Functional Dependence and the Electric Field (a) Suppose you want to purchase a sweater in Maryland that has a list price of \$40 for which you pay \$2 in sales tax. Your friend bought the same sweater in Maryland, but it had a list price of \$80 for which she paid \$4 in sales tax. How does the ratio of sales tax to price of the sweater compare for you and your friend [i.e., compare the ratios (sales tax)/(sweater price)]? What does that ratio tell us? As what is that ratio defined? (b) Suppose a charge exerts a repulsive force of 4 N on a test charge of 0.2 μC that is 2 cm from it. However, the charge exerts a repulsive force of 8 N on a test charge of 0.4 μC that is 2 cm from it. How does the ratio of the force on the test charge to the test charge itself compare in each case [i.e., compare (force felt by test charge)/(test charge)]? What does that ratio tell us? What is that ratio defined as? (c) Suppose a charge Q exerts a force F on a test charge q that is placed near it. By how much would the force exerted by Q increase if the test charge increased by a factor of α, where α can be any constant (i.e., $\alpha = -17$ or 5 or 7.812, etc.)? By how much would the ratio of the force on the test charge to the test charge itself increase if the test charge increased by a factor of α?

Explain. (d) When the value of one quantity depends on the value of a second quantity (and perhaps on others), we say that the first quantity is *a function of* the second. *How* the first quantity changes when the second changes is called the *functional dependence.* For example, if $t = As$, we say that t has a linear functional dependence on s. When s doubles, so does t. If s is divided by 10, so is t. As a second example, if we had $y = Bx^2$, we would say that y depends quadratically on x. If x doubles, y quadruples. If x is divided by 10, then y is divided by 100. (Try this with some numbers, picking whatever values of the constants A and B you would like.)

- **(i)** What is the functional dependence on the sales tax paid on the price of the sweater in part (a)? Explain. Write an equation that relates the tax paid (t) to the cost of the sweater (s).
- **(ii)** What is the functional dependence of the sales tax percentage rate on the price of the sweater in part (a)? Explain.
- **(iii)** In part (c), what is the functional dependence of the force magnitude, F, on the amount of the test charge, $|q|$? Explain.
- **(iv)** In part (c), what is the functional dependence of the electric field magnitude established by Q, E_Q, on the test charge, q? Explain.

73. *E*-Field Multiple Representations Figure 23-65*a* displays a grid with coordinates measured in meters. On the grid two charges are placed with their positions indicated as red circles. We call the charge at the position (1 m, 0 m) q_A, and the charge at the position (−1 m, 0 m) q_B. Figure 23-65*b* shows a set of possible vector directions. Below is a list of the components of possible E fields. For each of the following three cases:

I	$q_A = 0$	$q_B = 8\pi\varepsilon_0$	E field at the point $(x, y) = (-1\text{ m}, 1\text{ m})$
II	$q_A = 0$	$q_B = -8\pi\varepsilon_0$	E field at the point $(x, y) = (-1\text{ m}, -1\text{ m})$
III	$q_A = 160\pi\varepsilon_0$	$q_B = -16\pi\varepsilon_0$	E field at the point $(x, y) = (0\text{ m}, -1\text{ m})$

specify an arrow corresponding to the directions of the E field from figure (*b*) and a set of components from the list on the right. Each of your answers should consist of a capital letter and a small letter.

(*Note:* The values of the charges in Coulombs are chosen to make the messy "$4\pi\varepsilon_0$" in Coulomb's law cancel. Don't put in numbers first!)

a. $\sqrt{8}$ N/C)$\hat{\text{i}}$

b. $(-\sqrt{8}$ N/C)$\hat{\text{j}}$

c. (−2 N/C)$\hat{\text{i}}$

d. (2 N/C)$\hat{\text{j}}$

e. $(\frac{1}{2}$ N/C)$\hat{\text{i}}$ + $(\frac{1}{2}\sqrt{5}$ N/C)$\hat{\text{j}}$

f. $(\frac{1}{2}$ N/C)$\hat{\text{i}}$ − $(\sqrt{5}$ N/C)$\hat{\text{j}}$

g. None of the above

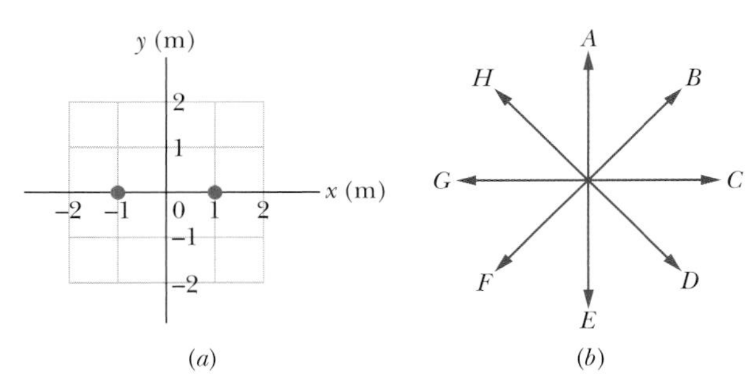

FIGURE 23-65 ■ Problem 73.

74. The Size of an Oil Drop In the Millikan oil drop experiment, an atomizer (a sprayer with a fine nozzle) is used to introduce many tiny droplets of oil between two oppositely charged parallel metal plates. Some of the droplets pick up one or more excess electrons. The charge on the plates is adjusted so that the electric force on the excess electrons exactly balances the weight of the droplet. The idea is to look for a droplet that has the smallest electric force and assume that it has only one excess electron. This lets the observer measure the charge on the electron. Suppose we are using an electric field of 3×10^4 N/C. The charge on one electron is about 1.6×10^{-19} C. Estimate the radius of an oil drop whose weight could be balanced by the electric force of this field on one electron.

75. What's a Field? In this class, we repeatedly refer to an "electric field." Describe what an electric field is. Discuss how you would know a nonzero field was present and how you would measure it.

76. Charge from Field Lines Figure 23-66 shows some representative electric field lines associated with some charges. Both pictures show the same charges, but they are masked in different ways by imaginary closed surfaces drawn for the purpose of hiding the charges from your view.

FIGURE 23-66 ■ Problem 76.

(a) From the field lines in the two pictures, which of the following statements is most likely to be true?

A. There are no charges contained in A.
B. The charge contained in A is positive.
C. The charge contained in A is negative.
D. The total charge contained in A is zero.
E. None of the above can be true.

(b) From the field lines in the two pictures, which of the following statements is most likely to be true?

A. There are no charges contained in B.
B. The charge contained in B is positive.
C. The charge contained in B is negative.
D. The total charge contained in B is zero.
E. None of the above.

(c) From the field lines in the two pictures, which of the following statements is most likely to be true?

A. The charge contained in C is positive and greater in amount than the charge in B.
B. The charge contained in C is positive and smaller in amount than the charge in B.
C. The charged contained in C is negative and greater in amount than the charge in B.
D. The total charge contained in C is negative and smaller in amount than the charge in B.
E. None of the above.

77. Finding the E Field Figure 23-67 shows two charges placed on a coordinate grid. Each of the tic marks on the axes represents 1 m. The amount of the charge is represented by the solid circle is q_A and is at the position (2 m, 0 m), while the charge represented by the open circle is q_B and is at the position (0 m, 2 m). Below is a list of five sets of configurations (labeled a–e) specifying the value of the charges and the positions at which the E field is to be measured. For ease of calculation, these are represented in terms of the Coulomb contant $k = 1/4\pi\varepsilon_0$. On the right is a list of 12 possible electric fields represented as x- and y-components. For each of the five configurations, select the E-field components that represent the field found at that position.

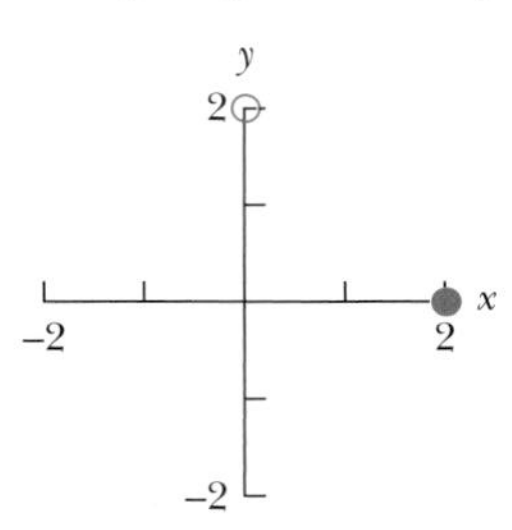

FIGURE 23-67 ■ Problem 77.

Configuration

	q_A	q_B	Position to Test the E Field
(a)	$4/k$	$-4/k$	(0 m, 0 m)
(b)	$4/k$	$-4/k$	(2 m, 2 m)
(c)	$4/k$	0	(0 m, 0 m)
(d)	0	$-8/k$	(−2 m, 2 m)
(e)	0	$-8/k$	(2 m, 0 m)

Possible E-Fields

1. $(1\text{ N/C})\hat{\imath}$
2. $(-1\text{ N/C})\hat{\jmath}$
3. $(1\text{ N/C})\hat{\imath} + (1\text{ N/C})\hat{\jmath}$
4. $(-1\text{ N/C})\hat{\imath}$
5. $(-1\text{ N/C})\hat{\imath} + (1\text{ N/C})\hat{\jmath}$
6. $(1\text{ N/C})\hat{\jmath}$
7. $\left(\frac{1}{\sqrt{2}}\text{ N/C}\right)\hat{\imath} - \left(\frac{1}{\sqrt{2}}\text{ N/C}\right)\hat{\jmath}$
8. $\left(-\frac{1}{\sqrt{2}}\text{ N/C}\right)\hat{\imath} + \left(\frac{1}{\sqrt{2}}\text{ N/C}\right)\hat{\jmath}$
9. $(2\text{ N/C})\hat{\imath}$
10. $(2\text{ N/C})\hat{\jmath}$
11. $(-2\text{ N/C})\hat{\imath}$
12. None of the above

78. Beads on a Ring Two charged beads are on the plastic ring in Fig. 23-68a. Bead 2, which is not shown, is fixed in place on the ring, which has radius $R = 60.0$ cm. Bead 1 is initially at the right side of the ring, at angle $\theta = 0°$. It is then moved to the left side, at angle $\theta = 180°$, through the first and second quadrants of the xy coordinate system. Figure 23-68b gives the x-component of the net electric field produced at the origin by the two beads as a function of θ. Similarly, Fig. 23-68c gives the y-component. (a) At what angle θ is bead 2 located? What are the charges of (b) bead 1 and (c) bead 2?

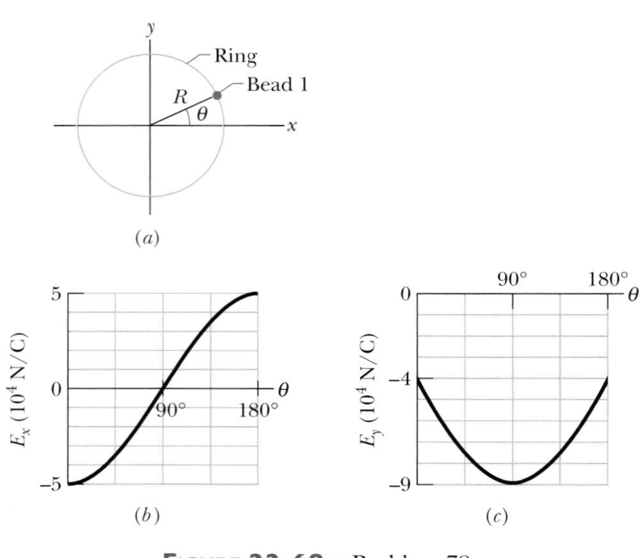

FIGURE 23-68 ■ Problem 78.

24 Gauss' Law

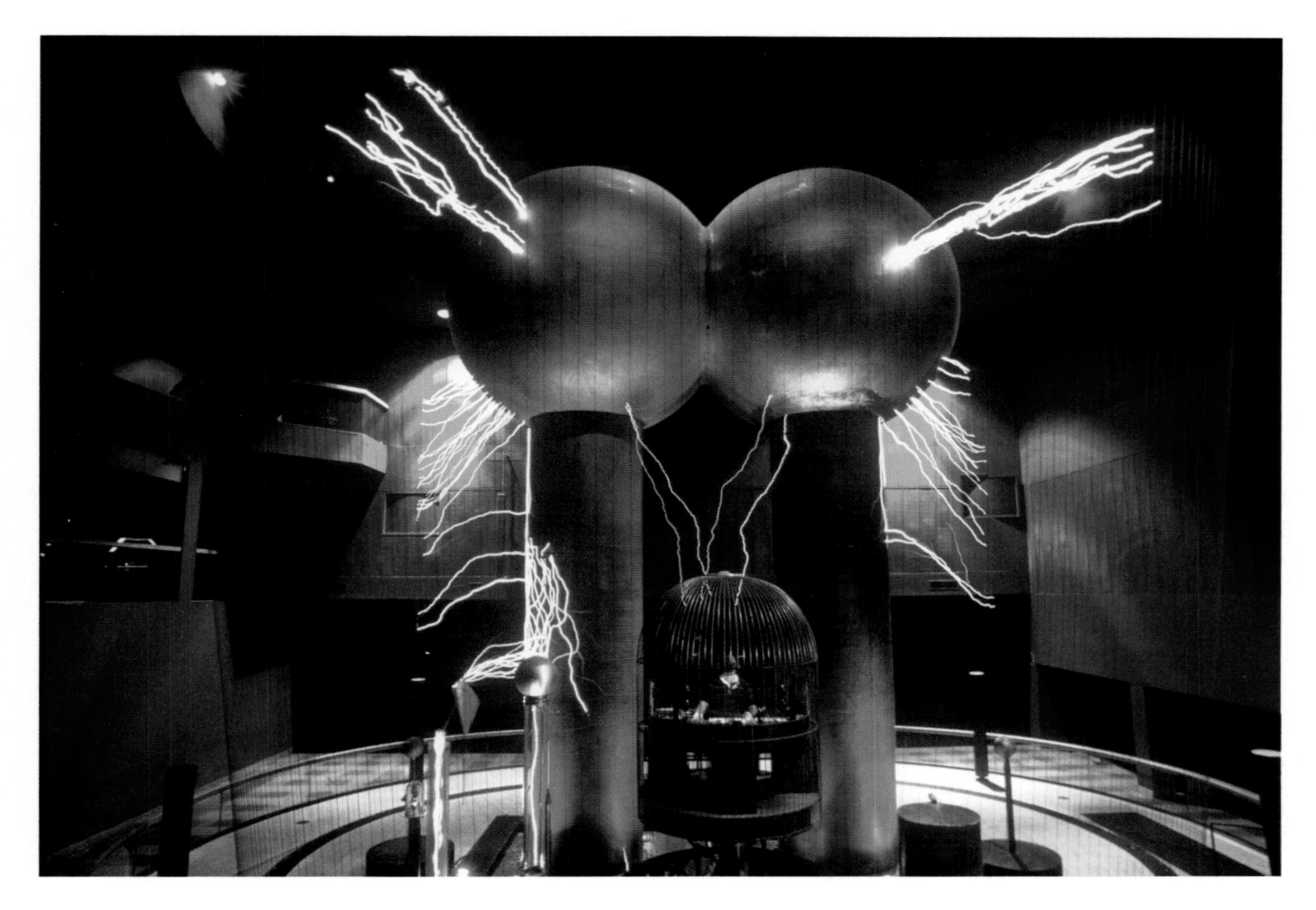

A demonstrator at the Boston Museum of Science is enclosed in a large conducting cage made of wire mesh. An electrical discharge from a giant Van de Graaff generator, like the one discussed in Chapter 25, is charging the metal cage to a dangerously high voltage. Yet the demonstrator cannot detect the fact that the cage is electrically charged even while touching the inside of the cage.

How can a closed conducting surface such as this metal cage or an automobile prevent someone from being harmed by lightning or other high-voltage sources?

The answer is in this chapter.

24-1 An Alternative to Coulomb's Law

We associate a vector electric field with a distribution of charges. The electric field has a vector at every location in space telling us what force a test charge q_t will experience at that location. In Sections 23-5 through 23-10 in the last chapter, we used Coulomb's law and the principle of superposition to calculate the electric field vectors at various points in space due to charges that were distributed in different ways. Although Coulomb's law can be used to calculate the electric force (and hence electric field) exerted on a test charge by any possible arrangement of charges we could imagine, this is usually a very difficult task. For example, even calculating the electric field outside the surface of a hollow, charged, conducting sphere would require us to do a triple integration.

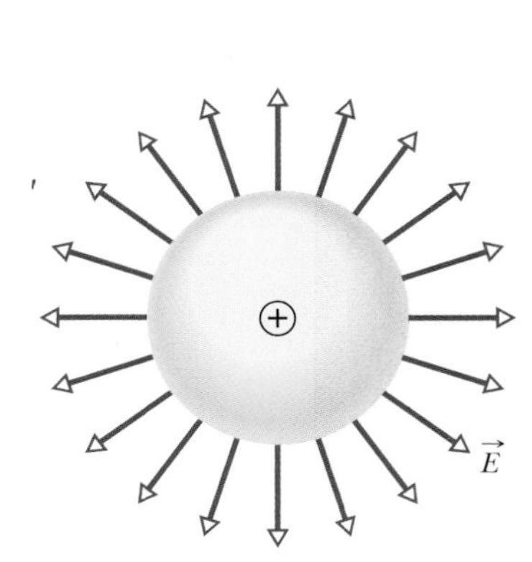

FIGURE 24-1 ■ If a single charge is located at the center of an imaginary sphere, Coulomb's law tells us the magnitudes of the electric field vectors are the same at all points on the surface of the sphere and the direction of each electric field vector is normal (perpendicular) to the surface. Only the field vectors that lie in the plane of the page are shown in this drawing.

In Chapter 23 we used Coulomb's law to find electric fields from charge distributions, but what if we want to turn our calculation around and determine a distribution of charges from an electric field pattern? Unless our distribution of charges is very simple, this reverse calculation is also difficult to perform using Coulomb's law. Thus Coulomb's law appears to be valid but difficult to use in many circumstances. In this chapter we introduce Gauss' law as another method for relating a known electric field to the charge distribution generating it and, conversely, for relating a known charge distribution to its associated electric field. Gauss' law in the integral form discussed in this chapter allows us to find electric fields easily for very symmetrical charge distributions.

To explore how we might find a general relationship between a collection of charges and their electric field, let's consider the electric field associated with the simplest possible charge distribution—a point charge (see Fig. 24-1). By applying Coulomb's law we have already found that the magnitude of the charge's electric field *decreases* as the inverse square of the distance r, as expressed in Eq. 23-8,

$$E = |\vec{E}| = k\frac{|q|}{r^2}.$$

However, if we construct an imaginary spherical surface around our source charge we find that the surface area of the sphere *increases* as the square of the distance of the spherical surface from the source charge. The equation for the surface area is given by $A = 4\pi r^2$. Thus, we see that the product of the electric field magnitude and the surface area of any imaginary spherical boundary is constant no matter how large or small the distance from the charge is, as shown in Eq. 24-1,

$$EA = k\frac{|q|}{r^2}(4\pi r^2) = \frac{1}{4\pi\varepsilon_0}\frac{|q|}{r^2}(4\pi r^2) = \frac{|q|}{\varepsilon_0}. \tag{24-1}$$

Here we use Eq. 22-7 to replace the electrostatic constant k with $1/4\pi\varepsilon_0$ where ε_0 is the electric (or permittivity) constant.

Equation 24-1 is remarkable for two reasons. First, as the electric field magnitude gets smaller, the area over which it can act gets larger by exactly the same factor. Second, the product of the electric field magnitude anywhere on a spherical surface and the area of the spherical surface is *proportional* to the amount of charge $|q|$ enclosed by that surface. Does this proportionality still exist when the closed surface takes on other shapes? These questions were addressed by German mathematician and physicist Carl Friedrich Gauss (1777–1855). We begin our study of Gauss' approach to relating charge distributions, electric fields, and closed surfaces to each other by defining a new quantity called electric flux.

24-2 Electric Flux

For the case of a single point charge at the center of an imaginary sphere, Eq. 24-1 tells us that the product of the electric field magnitude (at the surface of a sphere) and the surface area of the sphere are proportional to the charge. This product EA is known as the **electric flux** through the sphere. In our simple situation the directions of electric field vectors created by the point charge happen to be normal (that is, perpendicular) to the surface of our imaginary sphere at all points along its surface. What if we have a complex array of charges or decide to surround our charge with an imaginary enclosure with a different shape? In that case we need to break our surface into little elements of area and find the component of the electric field vector that is normal to each area element as depicted in Fig. 24-2. We took a similar approach in Section 15-10 in defining *volume flux* for fluids flowing in pipes and streams. If the definitions of volume flux and normal vector for an area are not familiar to you, we suggest you read this earlier section.

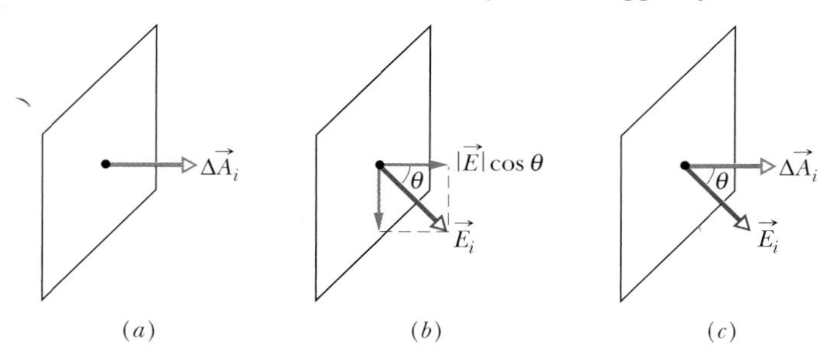

FIGURE 24-2 ■ (*a*) A small area vector element $\Delta\vec{A}$ is perpendicular to the plane of a square loop of area A with a magnitude of A. (*b*) The component of $\vec{E}$ perpendicular to the plane of the loop is $|\vec{E}|\cos\theta$, where θ is the angle between $\vec{E}$ and a normal to the plane. (*c*) The area vector $\Delta\vec{A}$ makes an angle θ with $\vec{E}$.

If we know the nature of the velocity vector field, $\vec{v}$, characterizing the motion of the fluid, we can use the definition of volume flux presented in Chapter 15 to calculate the amount of fluid flowing through any very small element, $\Delta\vec{A}_i$, of a larger surface area*. If we look at the ith element of a larger area, the *volume flux* element, Φ_i, for that small area is defined as the scalar or dot product of the normal vector representing an area element and the velocity vector at the location of the area element as shown in Eq. 15-33,

$$\Phi_i \equiv \vec{v}_i \cdot \Delta\vec{A}_i \quad \text{(volume flux definition for a small area element).}$$

What is a normal vector? Recall that we defined the normal vector to a small flat area to allow us to represent both the magnitude and the orientation of an element of area. If the element of area is part of a closed surface completely surrounding a space, we define the normal vector to be pointing *out* of the surface (Fig. 24-3). The normal vector points at right angles, or normal, to the plane of the area and has a magnitude equal to the area (Fig. 24-4).

Although *electric flux* does not involve the flow of anything, we define it in a way mathematically analogous to volume flux introduced in Chapter 15. An **electric flux element** is defined as the dot product of the normal vector representing an area element and the electric field vector at the location of the area element as shown in Fig. 24-2 and in Eq. 24-2,

$$\Phi_i \equiv (E_i)(\Delta A_i)\cos\theta = \vec{E}_i \cdot \Delta\vec{A}_i \quad \text{(electric flux definition for a small area),} \qquad (24\text{-}2)$$

where E_i and ΔA_i are magnitudes while θ is the angle between the two vectors. If a curved surface like the one in Fig. 24-3 is broken into small area elements, each of the $\Delta\vec{A}_i$ vectors can point in different directions.

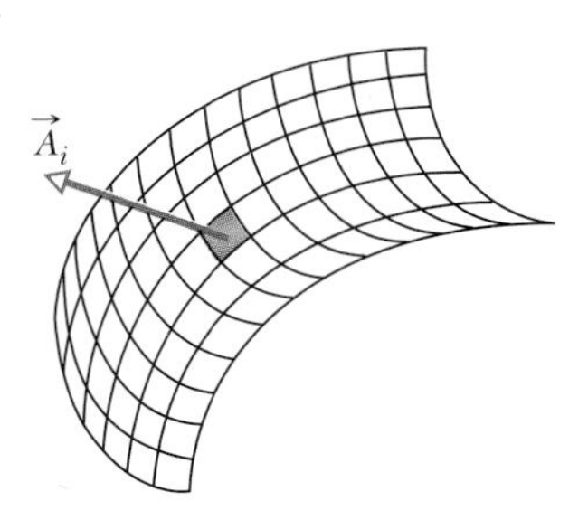

FIGURE 24-3 ■ In order to make net flux calculations, a curved surface area must be divided into N small area elements. Each element must be small enough so it is essentially flat and has electric field vectors that have the same magnitude and direction at every location on a given surface element. The ith area element and its normal vector are shown assuming that an outside piece of a closed surface is being shown here.

*Our use of the symbol ΔA_i instead of just A_i is to signify that the areas are very small. In this context, the delta does not signify change.

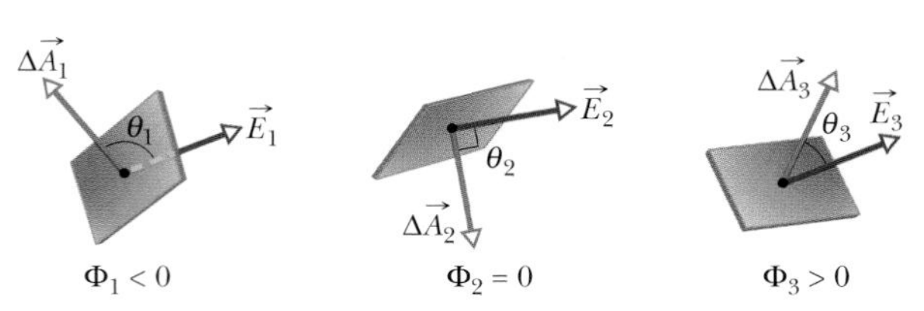

FIGURE 24-4 ■ Three small areas that subtend different angles with respect to various electric field vectors. The first flux element is negative, the second zero, and the third positive. Note that nothing is "flowing" in the case of electric flux to exist.

As is the case for volume flux, if our area is not small enough to be considered as flat or if the electric field vectors are not uniform over the area we choose, then we must break the area into smaller elements that are essentially flat (Fig. 24-4). We can then determine the net electric flux as the sum of individual flux elements. For N flux elements, this is given by

$$\begin{aligned}\Phi^{\text{net}} &= \Phi_1 + \Phi_2 + \cdots + \Phi_N \\ &= \vec{E}_1 \cdot \Delta\vec{A}_1 + \vec{E}_2 \cdot \Delta\vec{A}_2 + \cdots + \vec{E}_N \cdot \Delta\vec{A}_N \quad \text{(net electric flux)}, \\ &= \sum_{n=1}^{N} \vec{E}_n \cdot \Delta\vec{A}_n\end{aligned} \tag{24-3}$$

where $\vec{E}_1$, $\vec{E}_2$, $\vec{E}_3$, and so on represent the electric field vectors at the location of each of the N area elements. The flux associated with an electric field is a scalar, and its SI unit is the newton-meter-squared per coulomb or [$\text{N} \cdot \text{m}^2/\text{C}$].

Some possible orientations for area elements and electric field vectors needed to calculate electric flux elements are shown in Fig. 24-4.

In everyday language the term flux is often used to represent flow or change. This is suggested by expressions such as "an influx of population" or "the economy is in a state of flux." These popular uses of the word flux can be deceptive when applied to electrostatic phenomena that we are dealing with in Chapters 22 through 25. Electric flux can be defined whenever an electric field exists, even when an electric field is static and not changing. Furthermore, even if a redistribution of charges causes an electric field to change over time, the changing flux associated with electric field is not related to the flow of anything.

Instead of representing change or flow, **electric flux** at an area represents the summation over a surface of flux elements. Each flux element represents the product of an essentially flat area element on the surface and the component of the electric field vector that lies along the normal to that area element.

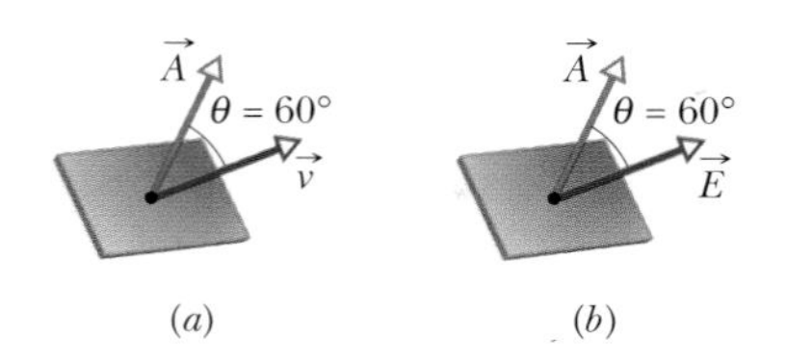

READING EXERCISE 24-1: The figure shows two situations in which the angle between a field vector and the normal vector representing the orientation of the area is $\theta = 60°$. Assume the magnitude of the area in each case is $\Delta A = 2 \times 10^{-4}\ \text{m}^2$. (a) If the imaginary area element is placed at a location in a stream where the magnitude of the stream velocity is $v = 3$ m/s, what is the volume flux through the area? Is anything flowing through the area element? If so, what? (b) Suppose the imaginary area element is placed in an electric field where the magnitude of the field vector is $E = 3$ N/C. What is the electric flux through the area element? Is anything flowing through the area element? If so, what? ■

24-3 Net Flux at a Closed Surface

In the introductory section we posed the question of whether there is a proportionality between an enclosed charge distribution and the flux at a surface that encloses it. To answer this question we need to examine carefully the procedures for determining net electric flux at an imaginary surface that encloses charges. The word "enclose" is important here. In the discussion that follows, we will not be discussing calculations of electric flux at any arbitrary surface. We will limit our discussion to the electric flux at closed surfaces that are continuous and connected. That is, a **closed** surface must be without cuts or edges. Nothing can get into or out of such surfaces without passing through the surface itself.

In order to define the net electric flux at any closed surface, consider Fig. 24-5, which shows an arbitrary (irregularly shaped) imaginary surface immersed in a *nonuniform* electric field. For historical reasons, any imaginary closed surface used in the calculation of a net electric flux is called a **Gaussian surface.** Since the electric field vector might be different at each location on our Gaussian surface, we must divide the entire surface into small area elements and take the sum as shown in Eq. 24-3.

Let's consider the arbitrary closed surface shown in Fig. 24-5. The vectors $\Delta\vec{A}_i$ and $\vec{E}_i$ for each square have some angle θ_i between them. Figure 24-5 shows an enlarged view of three small squares (1, 2, and 3) on the Gaussian surface, and the angle θ_i between $\vec{E}_i$ and $\Delta\vec{A}_i$. Our net flux equation (Eq. 24-3) instructs us to visit each square on the Gaussian surface, to evaluate the scalar product $\vec{E}_i \cdot \Delta\vec{A}_i$ at the location of each, and to sum the results algebraically (that is, with signs included) for all the squares that make up the surface. The sign or a zero resulting from each scalar product determines whether the flux at a square is positive, negative, or zero. Squares like 1, in which $\vec{E}_1$ points inward, make a negative contribution to the sum. Squares like 2, in which $\vec{E}_2$ lies in the surface, make zero contribution. Squares like 3, in which $\vec{E}_3$ points outward, make a positive contribution. (Note that the particular signs for the flux elements discussed above are a consequence of the convention adopted on the previous page; the area vectors point outward for closed surfaces.)

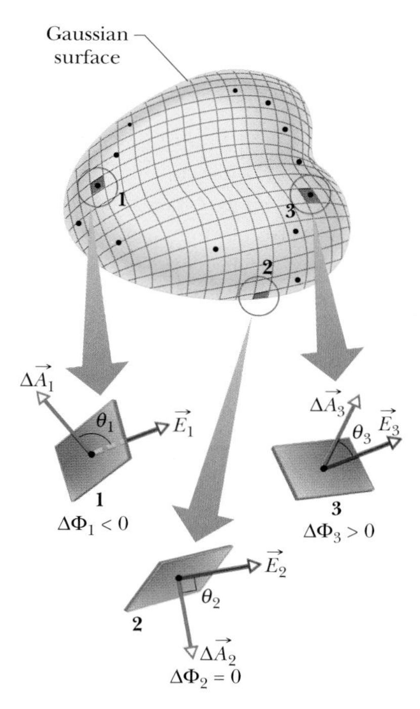

FIGURE 24-5 ■ A Gaussian surface of arbitrary shape is immersed in an electric field. The surface is divided into small area elements. The electric field vectors and the area vectors are shown for three representative area elements marked 1, 2, and 3. The other electric field vectors are not shown.

The exact definition of the flux of the electric field at a surface is found by allowing the area of the squares shown in Fig. 24-5 to become smaller and smaller, approaching a differential limit dA. The normal vectors for each tiny surface area then approach a differential limit $d\vec{A}$. Thus, the electric flux at a closed surface is given by the integral of the electric field components parallel to the normal of each surface area element over the magnitude of each surface area element. In mathematical notation the equation for electric flux becomes

$$\Phi^{\text{net}} \equiv \lim_{\Delta\vec{A}\to 0} \sum_{i=1}^{N} \vec{E}_i \cdot \Delta\vec{A}_i$$

$$= \oint \vec{E} \cdot d\vec{A} \quad \text{(net electric flux at a Gaussian surface).} \quad (24\text{-}4)$$

The circle on the integral sign indicates that the integration is to be taken over the entire closed surface (Gaussian surface).

TOUCHSTONE EXAMPLE 24-1: Net Flux for a Uniform Field

Figure 24-6 shows a Gaussian surface in the form of a cylinder of radius R immersed in a uniform electric field $\vec{E}$, with the cylinder axis parallel to the field. What is the flux Φ^{net} of the electric field through this closed surface?

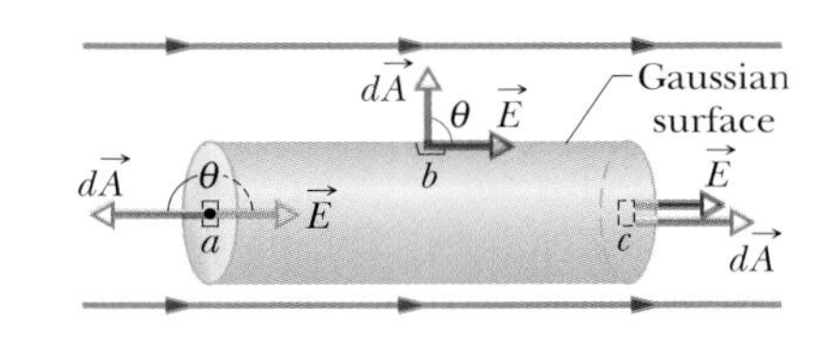

FIGURE 24-6 ■ A cylindrical Gaussian surface, closed by end caps, is immersed in a uniform electric field. The cylinder axis is parallel to the field direction.

SOLUTION ■ The **Key Idea** here is that we can find the flux Φ through the surface by integrating the scalar product $\vec{E} \cdot d\vec{A}$ over the Gaussian surface. We can do this by writing the flux as the sum of three terms: integrals over the left disk cap a, the cylinder surface b, and the right disk cap c. Thus, from Eq. 24-4,

$$\Phi^{\text{net}} = \oint \vec{E} \cdot d\vec{A}$$

$$= \int_a \vec{E} \cdot d\vec{A} + \int_b \vec{E} \cdot d\vec{A} + \int_c \vec{E} \cdot d\vec{A}. \quad (24\text{-}5)$$

For all points on the left cap, the angle θ between $\vec{E}$ and $d\vec{A}$ is $180°$, and the magnitude E of the field is constant. Thus,

$$\int_a \vec{E}\cdot d\vec{A} = \int E(\cos 180°)\,dA = -E\int dA = -EA,$$

where $\int dA$ gives the cap's area, $A(=\pi R^2)$. Similarly, for the right cap, where $\theta = 0$ for all points,

$$\int_c \vec{E}\cdot d\vec{A} = \int E(\cos 0°)\,dA = E\int dA = +EA.$$

Finally, for the cylindrical surface, where the angle θ is $90°$ at all points,

$$\int_b \vec{E}\cdot d\vec{A} = \int E(\cos 90°)\,dA = 0.$$

Substituting these results into Eq. 24-5 leads us to

$$\Phi = -EA + EA = 0. \quad \text{(Answer)}$$

This result is perhaps not surprising because the field lines that represent the electric field all pass entirely through the Gaussian surface, entering through the left end cap, leaving through the right end cap, and giving a net flux of zero.

TOUCHSTONE EXAMPLE 24-2: Flux for a Nonuniform Field

A *nonuniform* electric field given by $\vec{E} = (3.0\ \text{N/C}\cdot\text{m})x\hat{\text{i}} + (4.0\ \text{N/C})\hat{\text{j}}$ pierces the Gaussian cube shown in Fig. 24-7. What is the electric flux through the right face, the left face, and the top face?

SOLUTION ■ The **Key Idea** here is that we can find the flux Φ through the surface by integrating the scalar product $\vec{E}\cdot d\vec{A}$ over each face.

Right face: An area vector $\vec{A}$ is always perpendicular to its surface and always points away from the interior of a Gaussian surface. Thus, the vector $d\vec{A}$ for the right face of the cube must point in the positive x direction. In unit vector notation, then,

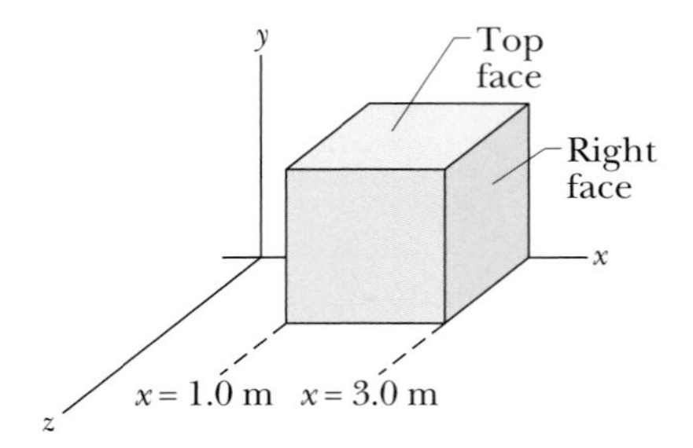

FIGURE 24-7 ■ A Gaussian cube with one edge on the x axis lies within a nonuniform electric field.

$$d\vec{A} = dA\,\hat{\text{i}}.$$

From Eq. 24-4, the flux Φ_r through the right face is then

$$\Phi_r = \int \vec{E}\cdot d\vec{A} = \int [(3.0\ \text{N/C}\cdot\text{m})x\hat{\text{i}} + (4.0\ \text{N/C})\hat{\text{j}}]\cdot(dA\hat{\text{i}})$$

$$= \int [(3.0\ \text{N/C}\cdot\text{m})(x)(dA)\hat{\text{i}}\cdot\hat{\text{i}} + (4.0\ \text{N/C})(dA)\hat{\text{j}}\cdot\hat{\text{i}}]$$

$$= \int (3.0\ \text{N/C}\cdot\text{m})x\,dA + (0.0\ \text{N}\cdot\text{m}^2/\text{C}) = (3.0\ \text{N/C}\cdot\text{m})\int x\,dA.$$

We are about to integrate over the right face, but we note that x has the same value everywhere on that face—namely, $x = 3.0$ m. This means we can substitute that constant value for x. Then

$$\Phi_r = (3.0\ \text{N/C}\cdot\text{m})\int (3.0\ \text{m})\,dA = (9.0\ \text{N/C})\int dA.$$

Now the integral merely gives us the area $A = 4.0\ \text{m}^2$ of the right face, so

$$\Phi_r = (9.0\ \text{N/C})(4.0\ \text{m}^2) = 36\ \text{N}\cdot\text{m}^2/\text{C}. \quad \text{(Answer)}$$

Left face: The procedure for finding the flux through the left face is the same as that for the right face. However, two factors change. (1) The differential area vector $d\vec{A}$ points in the negative x direction and thus $d\vec{A} = -dA\hat{\text{i}}$. (2) The term x again appears in our integration, and it is again constant over the face being considered. However, on the left face, $x = 1.0$ m. With these two changes, we find that the flux Φ_l through the left face is

$$\Phi_l = -12\ \text{N}\cdot\text{m}^2/\text{C}. \quad \text{(Answer)}$$

Top face: The differential area vector $d\vec{A}$ points in the positive y direction and thus $d\vec{A} = dA\hat{\text{j}}$. The flux Φ_t through the top face is then

$$\Phi_t = \int [(3.0\ \text{N/C}\cdot\text{m})x\hat{\text{i}} + (4.0\ \text{N/C})\hat{\text{j}}]\cdot(dA\hat{\text{j}})$$

$$= \int [(3.0\ \text{N/C}\cdot\text{m})(x\,dA)\hat{\text{i}}\cdot\hat{\text{j}} + (4.0\ \text{N/C})(dA)\hat{\text{j}}\cdot\hat{\text{j}}]$$

$$= (0.0\ \text{N}\cdot\text{m}^2/\text{C}) + \int (4.0\ \text{N/C})\,dA) = (4.0\ \text{N/C})\int dA$$

$$= 16\ \text{N}\cdot\text{m}^2/\text{C}. \quad \text{(Answer)}$$

24-4 Gauss' Law

Let's return for a moment to the consequence of Coulomb's law we presented in the first section, where we surrounded a single charge with a spherical Gaussian surface. We found that a flux-like quantity (namely, the product of the magnitude of the electric field at the sphere's surface multiplied by the area of the sphere's surface) is equal to a constant times the enclosed charge. The surprising thing is this is true no matter what the radius of the sphere is, because the amount by which the surface area of the sphere increases just compensates for the amount by which the electric field magnitude decreases. This suggests that the net flux through a Gaussian surface of any shape enclosing a single charge will be proportional to the amount of charge enclosed.

Visualizing Flux through a Gaussian Surface

Since the relationship between flux and charge enclosed by a Gaussian surface is hard to visualize in three dimensions, let's consider the special case of an infinitely long rod that has a uniform charge density. While infinitely long rods do not exist, our result will be valid providing the Gaussian surfaces are far from the ends of the rod. Imagine a Gaussian surface that has the shape of a coin and surrounds a small segment of the rod. This is shown in Fig. 24-8.

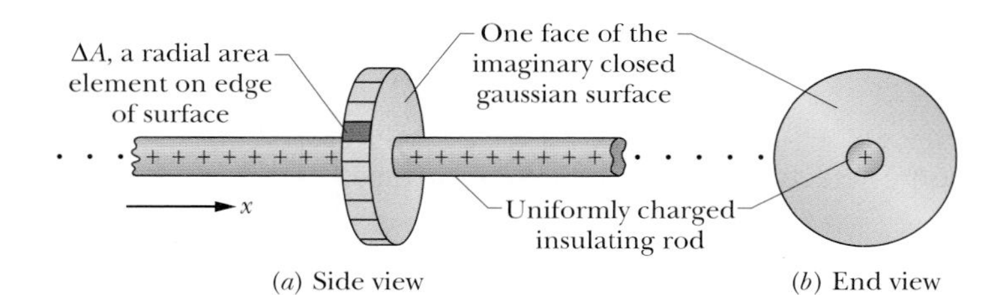

FIGURE 24-8 ■ (*a*)An infinitely long uniformly charged rod has a Gaussian surface that looks like a coin with front and back faces that are perpendicular to the rod and enclose a small charge. (*b*) An end view of the rod and Gaussian surface face can help us visualize flux at the surface's edges.

Because the charged rod is infinitely long it is symmetric about any point on it. As we showed in Section 23-5 (see Fig. 23-12), it turns out the electric field vectors created by a symmetric pair of charges point outward in a radial direction and have no components parallel to the line that the charges lie on (in this case, the line determined by the rod). We can also show that for a thin rod the field magnitude falls off as $1/r$ where r is the radial distance from the center of the rod. (Likewise, a similar negatively charged rod has electric field vectors pointing radially inward). The key factor in surrounding a piece of long rod with a coin-shaped closed surface is that all the flux at the surface will be at the edges and there will be no flux at the faces of the surface. For this reason, we can calculate and depict the "amount" of flux at elements of area on the edges of the surface by looking at an end view of the rod. This is true not only for coin-like closed surfaces that have circular faces but also for any shaped faces so long as the two faces are parallel to each other and perpendicular to the rod. End views depicting flux amounts as green rectangles are shown in Fig. 24-9 for three different imaginary Gaussian surfaces outlined in red.

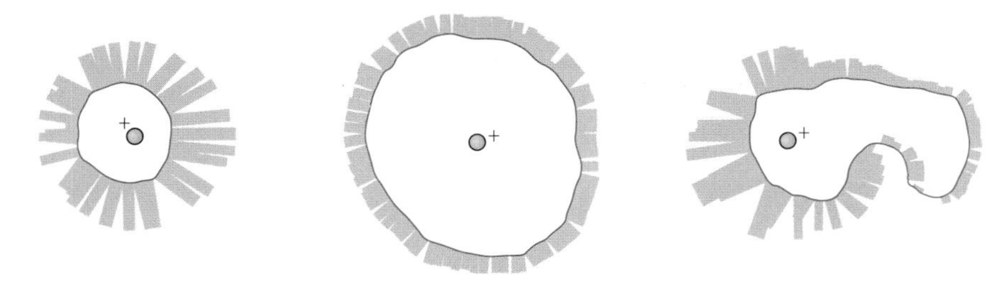

FIGURE 24-9 ■ Three imaginary Gaussian surfaces surround the same point charge. Here the red lines show only a two-dimensional cross section of three-dimensional surfaces. The contribution of electric flux at a series of small area elements is calculated and represented by rectangles. Note that whenever part of a surface is close to the charge, the flux elements are bigger but there are fewer of them. We can see visually that the net flux (which is proportional to the area occupied by all the outgoing flux (shown as green) minus the incoming flux (shown as pink) is approximately the same in the three cases.

A small bundle of enclosed charge yields the same net electric flux at a Gaussian surface no matter what the shape of the surface. By superposition, if there are two charges enclosed by a Gaussian surface, each charge contributes its proportional share to the net flux no matter where each of the charges is located, provided both are *inside* the Gaussian surface. This leads us to a statement of Gauss' law that describes a plausible general relationship between the net flux through a Gaussian surface of any shape and the total enclosed charge no matter how it is distributed.

GAUSS' LAW: The net flux through any imaginary closed surface is directly proportional to the net charge enclosed by that surface.

Based on consideration of SI units, the constant of proportionality must be $1/\varepsilon_0$ where ε_0 is the permittivity constant, so that the mathematical expression of Gauss' law is

$$\Phi^{\text{net}} = \frac{q^{\text{enc}}}{\varepsilon_0} \quad \text{(Gauss' law).} \tag{24-6}$$

By substituting the definition of electric flux at a Gaussian surface, $\Phi^{\text{net}} \equiv \oint \vec{E} \cdot d\vec{A}$, we can also write Gauss' law as

$$\Phi^{\text{net}} = \oint \vec{E} \cdot d\vec{A} = \frac{q^{\text{enc}}}{\varepsilon_0} \quad \text{(Gauss' law).} \tag{24-7}$$

Here, the circle on the integral sign indicates that the surface over which we integrate must be "closed." The use of the permittivity constant for a vacuum, ε_0, in Eqs. 24-6 and 24-7 indicates this form of Gauss' law only holds when the net charge is located in air or some other medium that doesn't polarize easily. In Section 28-6, we modify Gauss' law to include situations in which so-called dielectric materials that can polarize, such as paper, oil, or water, are present. In Fig. 24-10 we show how the net flux can have the same value for two different charge distributions involving the same amount of enclosed charge.

Gauss' law is useful for finding both charge and flux. That is, if we can calculate the net flux through a closed surface, we can deduce the amount of charge enclosed. On the other hand, if we know the amount of charge enclosed, we can use Gauss' law to deduce the net flux through any surface that encloses the charge.

Interpreting Gauss' Law

One use of Gauss' law is to calculate how much net charge is contained inside any closed surface. To make the calculation, you need know only the net electric flux at the surface enclosing the collection of charges. This net flux is related to the strength of the normal components of the electric field at all locations on the surface.

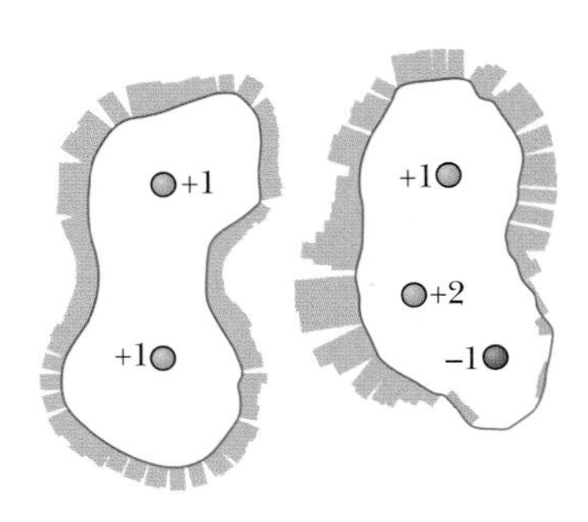

FIGURE 24-10 ■ Each Gaussian surface encloses a different charge distribution but encloses the same net charge. The electric flux calculated at the edges of the surface is represented by green rectangles (outward flux) or pink rectangles (inward flux). The total space covered by all of the green rectangles minus that occupied by the pink rectangles turns out to be the same for the two situations, which is compatible with the predictions of Gauss' law.

In Eqs. 24-6 and 24-7, the net charge q^{enc} is the algebraic sum of all the *enclosed* positive and negative charges, and it can be positive, negative, or zero. We include the sign, rather than just use the amount of enclosed charge, because the sign tells us something about the net flux at the Gaussian surface. Here we continue to use our convention that the normal area vectors representing the area elements of a closed surface point *outward*. If the net charge enclosed, q^{enc}, is positive, its electric field vectors point mostly outward too. This leads to a net flux that is *outward* and positive as shown in Fig. 24-11*a*. If q^{enc} is negative, the area vector still points outward but the electric field vector points inward. This leads to a net flux that is *inward* and negative, as shown in Fig. 24-11. Figure 24-11*c* shows how positive and negative charges inside a Gaussian surface can lead to zero net flux.

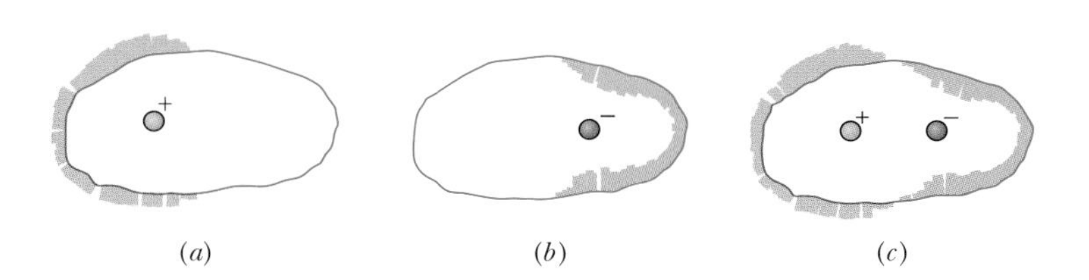

FIGURE 24-11 ■ Each of these Gaussian surfaces has the same shape. (*a*) One unit of enclosed positive charge causes a positive net outward flux shown in green. (*b*) One unit of enclosed negative charge causes a negative net inward flux shown in pink. Note that the amount of negative flux is the same as the amount of positive flux shown in the previous diagram. (*c*) If both the positive and negative charges are enclosed the net charge is zero and so is the net flux.

Charge outside a Gaussian surface, no matter how large or how close it may be, is not included in the term q^{enc} in Gauss' law. We expect this since there is no source of electric field inside the surface, and negative and positive flux elements will cancel each other, as shown in Fig. 24-12. The exact form or location of the charges inside the Gaussian surface is also of no concern; the only things that matter are the amount of the net charge enclosed and its sign. The quantity $\vec{E}$ on the left side of Eq. 24-7, however, is the electric field resulting from *all* charges, both those inside and those outside the Gaussian surface. This may seem to be inconsistent, but keep in mind the electric field due to a charge outside the Gaussian surface contributes zero net flux on the surface (as shown in Fig. 24-12). This is the case even though a charge outside the surface does contribute to the actual values of the electric field at each point on the surface.

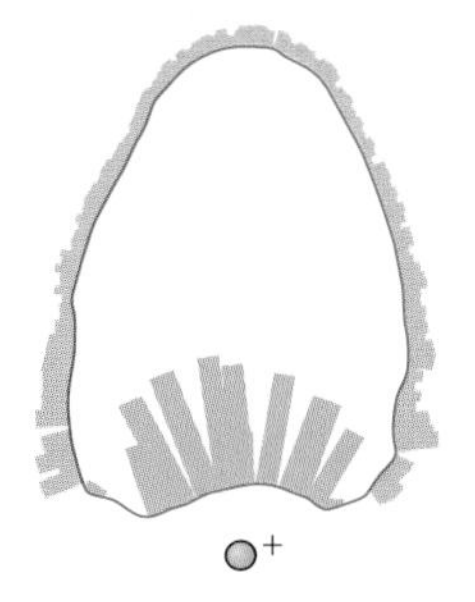

FIGURE 24-12 ■ A charge element along a rod is located *outside* a Gaussian surface. When the electric flux is calculated at each area element using Coulomb's law, its outward values are represented by green rectangles and the inward flux by pink rectangles. The net flux is zero because the negative inward flux at the portion of the surface near the charge just cancels the positive outward flux at the location of the portions of the surface far away from the charge.

Let us apply these ideas to Fig. 24-13, which shows the electric field lines surrounding two point charges, equal in amount but opposite in sign. Four Gaussian surfaces are also shown, in cross section. Let us consider each in turn.

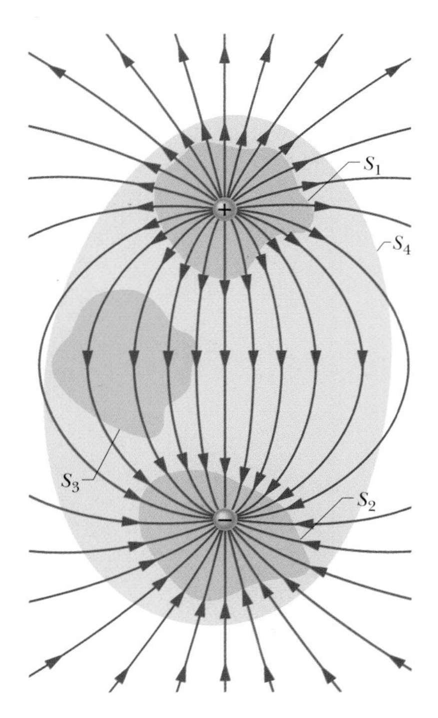

FIGURE 24-13 ■ An idealization showing two point charges of equal amount and opposite sign are shown with the field lines that depict their net electric field as if all lines lie in a plane. The cross sections of four Gaussian surfaces are shown. Surface S_1 encloses the positive charge, S_2 encloses the negative charge, and S_3 encloses no charge. Since S_4 surrounds both charges, it encloses no net charge.

Surface S_1 (encloses only the positive charge): The electric field is dominated by the nearby positive charge and so points outward for the majority of the points on this surface. Thus, the flux of the electric field at this surface is positive, and so is the net charge within the surface, as Gauss' law requires. (That is, if Φ is positive, q^{enc} must be also.)

Surface S_2 (encloses only the negative charge): The electric field is dominated by the nearby negative charge and so points inward for the majority of the points on this surface. Thus, the flux of the electric field is negative and so is the enclosed charge, as Gauss' law requires.

Surface S_3 (encloses no charges): Since $q^{enc} = 0$ and there are comparable contributions to the electric field at points on the surface from both charges, the field on some parts of the surface will point out and on other parts it will point in. Gauss' law (Eq. 24-7) requires the net electric flux through this surface to be zero. That is reasonable because in calculating the net flux, the inward and outward flux elements cancel each other.

Surface S_4 (encloses both charges): This surface encloses no *net* charge, because equal amounts of positive and negative charge are enclosed. Gauss' law requires the net flux of the electric field at this surface be zero. That is reasonable because in this case the field vectors point outward for the portion of the surface nearest to the positive charge (yielding positive flux) and inward for the portion of the surface near the negative charge (yielding negative flux). In calculating the net flux, the positive and negative flux elements cancel each other, even though the field is nonzero along most of the surface.

What would happen if we were to bring an enormous charge Q up close to (but still outside of) surface S_4 in Fig. 24-13? The pattern of the electric field would certainly change, but the net flux for the four Gaussian surfaces would not change. We can understand this because the inward and outward flux elements associated with the added Q at any of the four surfaces would cancel each other, making no contribution to the net flux at any of them. The value of Q would not enter Gauss' law in any way, because Q lies outside all four of the Gaussian surfaces that we are considering.

READING EXERCISE 24-2: The figure shows three situations in which a Gaussian cube sits in an electric field. The arrows indicate the directions of the electric field vectors for the top, front, and right faces of each cube. The flux at the six sides of each cube is listed in the table below. In which situations do the cubes enclose (a) a positive net charge, (b) a negative net charge, and (c) zero net charge?

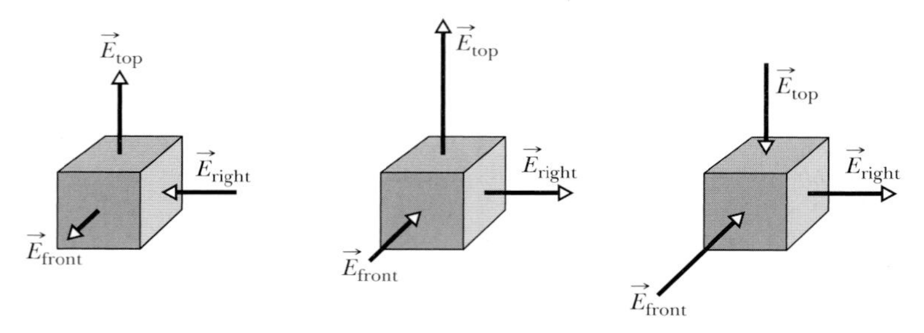

Flux [N · m²/C]

Face	**Cube 1**	**Cube 2**	**Cube 3**
Front	+2	−4	−7
Back	−3	−3	+8
Left	+7	+3	+2
Right	−4	+5	+5
Top	+5	+10	−6
Bottom	−7	−6	−5

■

24-5 Symmetry in Charge Distributions

Why go through all this trouble to develop a method of calculating electric fields that is equivalent to Coulomb's law? We suggested in the introduction to this chapter that it is because Gauss' law makes it possible to calculate the field for highly symmetric charge distributions. What we mean by symmetric charge distributions are arrangements of charges that can be rotated about an axis or reflected in a mirror and still look the same. Figure 24-14 shows several examples of symmetric objects.

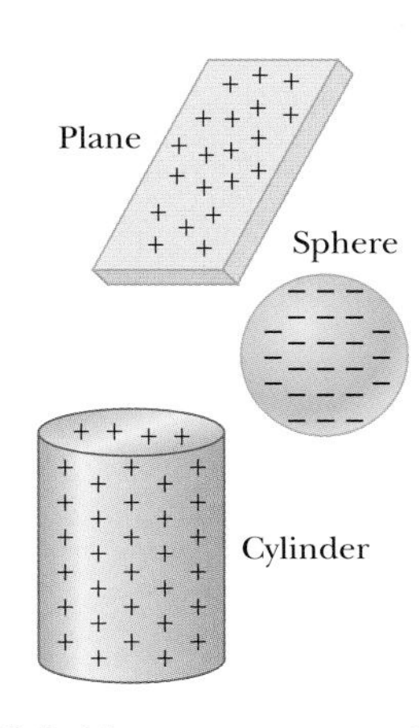

FIGURE 24-14 ■ Some symmetrically charged objects—a plane, a sphere, and a cylinder.

Why do charge distributions need to be symmetric in order for Gauss' law to be helpful in finding an electric field? Because we can use symmetry arguments to find the direction of the electric field and surfaces along which it is constant. This allows us to choose an imaginary Gaussian surface over which the electric field is constant. Then we can take the dot product and turn the vectors into scalar magnitudes. Finally, we know the electric field magnitude is constant at the surface we are integrating over, so we can pull the electric field vector outside of the integral sign. By following the steps we outlined, in some cases Gauss' law can be reduced to

$$\varepsilon_0 \oint E \cos\theta \, dA = \varepsilon_0 E \oint \cos\theta \, dA = |q^{\text{enc}}|.$$

Better still is to be able to find a Gaussian surface over which both the electric field and the angle between the field and area vectors, θ, are constant over the entire area. In that case, both the electric field and the cosine functions can be moved outside the integral and Gauss' law reduces to:

$$(\varepsilon_0 E \cos\theta) \oint dA = |q^{\text{enc}}|.$$

This expression is very easy to evaluate because the integral of dA is simply the magnitude of the total area of the Gaussian surface, which we will denote as A. Hence, if we can find a Gaussian surface over which the field and angle θ are constant, Gauss' law allows us to calculate the electric field of an extended charge distribution without doing an integral. In those cases, Gauss' law tells us that the electric field magnitude is

$$E = \frac{|q^{\text{enc}}|}{\varepsilon_0 A \cos\theta} \qquad \text{(constant } E \text{ and } \theta\text{)}, \qquad (24\text{-}8)$$

where A is the area of the Gaussian surface, θ is the angle between the field and each area vector, and q^{enc} is the net charge *enclosed by the Gaussian surface*. In some cases where the angle, θ, has one value for some parts of a surface and another value for other parts of a surface, we can handle the calculation by breaking the surface integral into parts.

A word of caution: There are only a few charge distributions with sufficient symmetry for Gauss' law to be useful. These include single point charges and spherically symmetric ones. Charge distributions that work with Gauss' law also include the infinitely long cylinder, with cylindrical symmetry, and that of a uniformly charged slab with infinitely long sides with planar symmetry. Fortunately, there are many physical situations for which these geometries are important. Hence, Gauss' law is an extraordinarily useful tool.

However, for many charge distributions, we cannot use Gauss' law to find the field because the flux integral on the left-hand side of the expression

$$\varepsilon_0 \oint \vec{E} \cdot d\vec{A} = q^{\text{enc}}$$

is too complicated to evaluate. In these cases, Gauss' law is still valid but not useful.

24-6 Application of Gauss' Law to Symmetric Charge Distributions

As we determined in the last section, Gauss' law is useful if we already know what the general shape of the vector electric field plot looks like. In some cases we can derive this knowledge from symmetry of the charge distribution without using equations or doing calculations. Only then can we choose an imaginary closed surface and use the mathematical form of Gauss' law to calculate the magnitude of the electric field at points on the surface. In this section we take this approach to determining the electric field for three highly symmetric charge distributions.

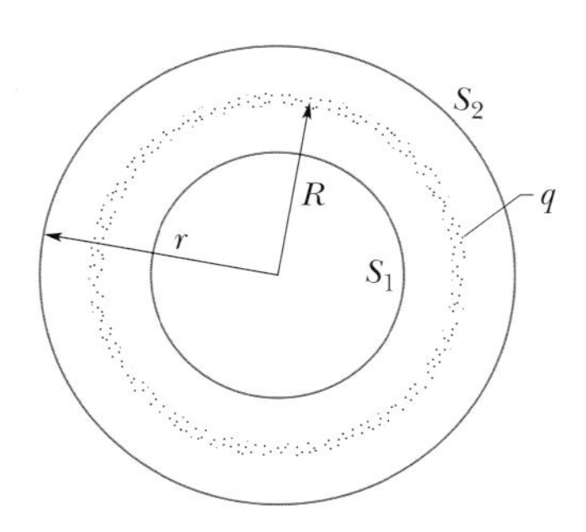

FIGURE 24-15 ■ A thin, charged, spherical shell with total charge q, in cross section. Two Gaussian surfaces S_1 and S_2 are also shown in cross section. Surface S_2 encloses the shell, and S_1 encloses only the empty interior of the shell.

Spherical Symmetry for a Shell of Charge

Figure 24-15 shows a charged spherical shell of total positive charge q and radius R and two concentric spherical Gaussian surfaces, S_1 and S_2. (Note that we chose the shape of the Gaussian surface to mirror the symmetry of the charge distribution.) Because the charge distribution is spherically symmetric no matter how we rotate the spherical shell around its center, the shell looks the same. This means that the electric field must have a spherical symmetry too. Thus, it must have the same magnitude at every point on the spherical Gaussian surface S_2 and it must point in a radial direction. Further, since the area vector points radially outward at all points on S_2, the angle between the electric field $\vec{E}$ and the area $\vec{A}$ is constant. As a result of the spherical symmetry of the distributed charge, we know the electric field also points in a radial direction at all points on S_2. Hence, the angle θ is not only constant but it is also 0° at all points on the surface. Applying Gauss' law to surface S_2 then comes down to evaluating the expression for the electric field magnitude that we derived in Eq. 24-8 for constant E and θ,

$$E = \frac{|q^{\text{enc}}|}{\varepsilon_0 A \cos\theta}.$$

Note that $\cos\theta = \cos 0 = 1$ and the area of a sphere (the Gaussian sphere) of radius r is $4\pi r^2$. Hence, for any $r \geq R$, we find that

$$E = \frac{1}{4\pi\varepsilon_0}\frac{|q^{\text{enc}}|}{r^2} \quad \text{(spherical shell, field at } r \geq R\text{)}. \tag{24-9}$$

What is surprising is that outside the shell the electric field is the same as if the shell of charge were replaced by a single point-like charge, q, provided that the single charge is placed where the center of the shell of charge was. Thus, if the charge on a shell is evenly distributed, a shell of total charge q would produce the same force on a small test charge placed anywhere *outside* the shell as a single point-like charge q would.

> A shell with a uniform charge distribution attracts or repels a charged particle that is outside the shell as if all the shell's charge were concentrated at the center of the shell.

This shell theorem is identical to the one developed by Isaac Newton for gravitation in Section 14-2.

What happens to the electric field inside the shell of charge? Applying Gauss' law to surface S_1, for which $r < R$, leads directly to

$$\vec{E} = 0\ [N/C] \quad \text{(spherical shell, field at } r < R\text{)}, \tag{24-10}$$

because this Gaussian surface encloses no charge. Thus, when a small test charge is enclosed by a shell of uniform charge distribution, the shell exerts no net electrostatic force on it.

> A shell of uniform charge exerts no electrostatic force on a charged particle that is located inside the shell.

A Spherically Symmetric Charge Distribution

Any spherically symmetric charge distribution, such as that of Fig. 24-16, can be constructed with a nest of concentric spherical shells. This is a good starting point for treating a wide variety of charged objects with nearly spherical distribution of charge such as nuclei and atoms. For purposes of applying the two shell theorems stated above, the volume charge density ρ, defined as the charge per unit volume, should have a single value for each shell but need not be the same from shell to shell. Thus, for the charge distribution as a whole, ρ, can vary only with r, the radial distance from the center of the sphere and not with direction. We can then examine the effect of the charge distribution "shell by shell."

In Fig. 24-16*a* the entire charge lies within a Gaussian surface with $r > R$. The charge produces an electric field on the Gaussian surface as if the charge were a point charge located at the center, and Eq. 24-9 holds.

Figure 24-16*b* shows a Gaussian surface with $r < R$. To find the electric field at points on this Gaussian surface, we consider two sets of charged shells—one set inside the Gaussian surface and one set outside. The charge lying *outside* the Gaussian surface does not set up a net electric field on the Gaussian surface. Gauss' law tells us that the charge *enclosed* by the surface sets up an electric field as if that enclosed charge were concentrated at the center. Letting q' represent that enclosed charge, we can then write the electric field magnitude as

$$|\vec{E}| = \frac{1}{4\pi\varepsilon_0}\frac{|q^{\text{enc}}|}{r^2} = \frac{1}{4\pi\varepsilon_0}\frac{|q'(r)|}{r^2} \quad \text{(spherical distribution, field at } r < R\text{)}, \quad (24\text{-}11)$$

where the term $q'(r)$ signifies that q' depends on r. (It is not the product of q' and r.)

Equation 24-11 is valid for any spherically symmetric charge distribution, even one that is not uniform. For example, Fig. 24-16 shows a situation in which the volume charge density is spherically symmetric but larger near the center of the sphere than further out. In other words, Eq. 24-11 is valid whenever $\rho = \rho(r)$ or ρ is a constant. But the equation is not useful unless we know how to use a knowledge of the volume charge density to determine the charge q' enclosed by a sphere of radius r.

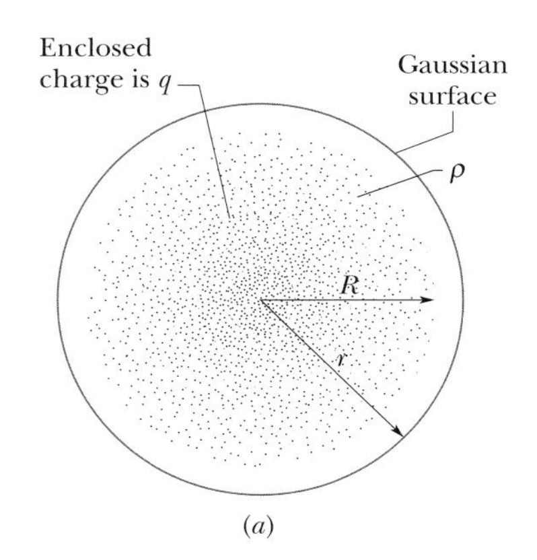

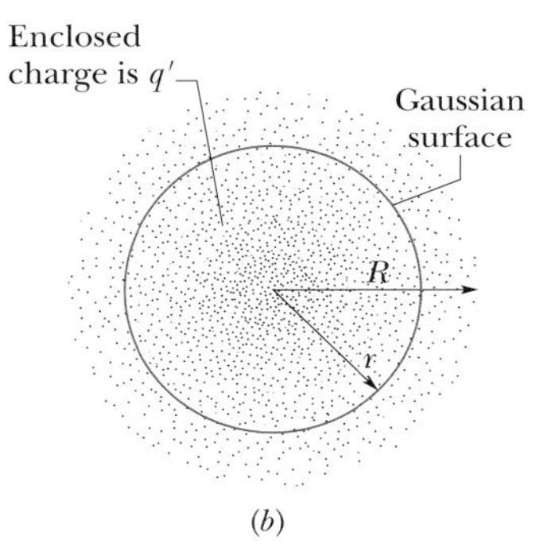

FIGURE 24-16 ■ Spherically symmetric distributions of charge of radius R, whose volume charge density ρ is a function only of distance from the center. The charged object is not a conductor, so the charge is assumed to be fixed in position. A cross-section of concentric spherical Gaussian surface with $r > R$ is shown in (*a*). A similar Gaussian surface with $r < R$ is shown in (*b*).

Spherical Symmetry for a Uniform Volume Charge Distribution

Consider the simple case where the charge is distributed uniformly through the volume of a sphere of radius R containing an excess charge q. In this case it is possible to find the magnitude of the electric field at any location inside the sphere in terms of the total charge in the sphere.

Whenever the total charge q enclosed within a sphere of radius R is distributed uniformly, we can use the definition of volume charge density (presented in Table 23-2) and the knowledge that the volume of a sphere of radius R is given by $\frac{4}{3}\pi R^3$ to write

$$\rho \equiv \frac{q}{V} = \frac{q}{\frac{4}{3}\pi R^3}. \quad (24\text{-}12)$$

Since the charge density is a constant the amount of charge in a smaller sphere of radius r is proportional to its volume. Since its volume is $V' = \frac{4}{3}\pi r^3$ then $q' = \rho V' = \rho\,(\frac{4}{3}\pi r^3)$. Substituting Eq. 24-12 for ρ gives

$$q' = q\frac{r^3}{R^3}. \quad (24\text{-}13)$$

Substituting this into Eq. 24-11 gives us the electric field magnitude in terms of the total charge on the sphere.

$$E = \left(\frac{|q|}{4\pi\varepsilon_0 R^3}\right) r \quad \text{(uniform volume charge density for } r \le R). \tag{24-14}$$

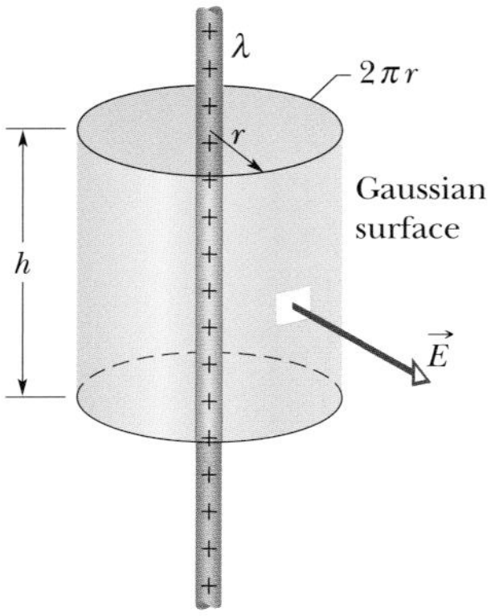

Cylindrical Symmetry for a Uniform Line Charge Distribution

FIGURE 24-17 ■ A Gaussian surface in the form of a closed cylinder surrounds a section of a very long, uniformly charged, cylindrical plastic rod.

Figure 24-17 shows a section of a very long thin cylindrical plastic rod with a uniform distribution of positive charge, so that linear charge density λ (as defined in Table 23-2) is constant. Let us find an expression for the magnitude of the electric field $\vec{E}$ outside of the rod at a distance r from its axis in terms of the linear charge density of the rod. In doing so, we assume that r is small compared to the length of the rod so that we can ignore the effect of the rod's ends.

We start by choosing a Gaussian surface that matches the cylindrical symmetry of the rod. So our imaginary surface is a circular cylinder of radius r and length h, coaxial with the rod. The Gaussian surface must be closed, so we include two end caps as part of the surface. We pick the end caps of the Gaussian surface so they are far from the end of the rod.

Imagine that, while you are not watching, someone rotates the plastic rod around its longitudinal axis or moves it a finite distance along the axis. When you look again at the rod, you will not be able to detect any change in either the appearance of the rod or the behavior of the electric field that surrounds it. Furthermore, when we experiment, we find that if the rod is flipped end for end we still detect no change in the rod's electric field. What does this tell us about the nature of the electric field? If the electric field has only a component that points radially inward or outward from the rod, then the field should be unaffected by the changes in orientation that we have discussed. If however, the field had any component tangent to the rod's surface, pointed toward or away from the rod, we would detect a change in the electric field as we rotated or flipped the rod. Hence, we conclude from these symmetry arguments that at every point on the cylindrical part of the Gaussian surface, the electric field must have the same magnitude $E = |\vec{E}|$ and must be directed radially outward (for a positively charged rod).

Since $2\pi r$ is the circumference of the cylinder and h is its height, the area A of the cylindrical surface is $2\pi rh$. The flux of $\vec{E}$ at this cylindrical surface is then

$$\Phi = EA\cos\theta = E(2\pi rh)\cos 0 = E(2\pi rh).$$

There is no flux at the end caps because $\vec{E}$, being radially directed, is parallel to the end caps at every point, so $\vec{E}$ is perpendicular to the normal and the dot product vanishes. Thus the flux through the cylindrical surface is equal to the net flux ($\Phi^{\text{net}} = \Phi$).

According to Gauss' law, shown in Eq. 24-6,

$$\Phi^{\text{net}} = \frac{q^{\text{enc}}}{\varepsilon_0}.$$

We can find the enclosed charge in terms of the linear charge density, defined as the charge per unit length. If the charge enclosed by the surface that encompasses a length h of the rod has a uniform density λ, then $q^{\text{enc}} = \lambda h$. Thus, the previous two equations reduce to $E(2\pi rh) = |\lambda| h/\varepsilon_0$, so that

$$E = \frac{|\lambda|}{2\pi\varepsilon_0 r} \quad \text{(long line of uniformly distributed charge).} \tag{24-15}$$

This is the expression for the electric field magnitude due to a very long, straight line of uniformly distributed charge, at a point that is a radial distance r from the line. The direction of $\vec{E}$ is radially outward from the line of charge if the charge is positive, and radially inward if it is negative. Equation 24-15 also approximates the field of a *finite* line of charge, at points that are not too near the ends (compared with the distance from the line).

A Sheet of Uniform Charge

Figure 24-18 shows a portion of a thin, very large, sheet with a uniform (positive) surface charge density σ (as defined in Table 23-2). A large sheet of thin plastic wrap, uniformly charged on one side, can serve as a simple example of a nonconducting sheet. A large sheet of aluminum foil serves as an example of a conducting sheet. Let us find the electric field $\vec{E}$ a distance r from the uniformly charged sheet. Here we assume that we are far from the edges of the sheet and that the thickness of the sheet is much less than r.

Even though it doesn't have the same shape as a charged sheet, something called a Gaussian pillbox turns out to make a useful imaginary surface in this case. The pillbox is a closed cylinder with end caps of area A, arranged so that it is perpendicular to the sheet with each end cap located at the same distance from the sheet. This Gaussian pillbox is shown in Fig. 24-18*a*. Using symmetry (considerations like those used earlier in this section or those depicted in Fig. 23-12 and Fig. 23-13 in the previous chapter), $\vec{E}$ must be perpendicular to the sheet and hence to the end caps. Furthermore, since the charge is positive, $\vec{E}$ is directed *away* from the sheet, and thus the electric field vectors point in an outward direction from the two Gaussian end caps. Because the electric field vectors are perpendicular to the normal vectors on the curved surface, there is no flux at this portion of the Gaussian surface. Thus $\vec{E} \cdot d\vec{A}$ is simply EdA—the product of the magnitudes of $\vec{E}$ and $d\vec{A}$. In this case Gauss' law (Eq. 24-7) gives us

$$\oint \vec{E} \cdot d\vec{A} = q^{\text{enc}}/\varepsilon_0.$$

Since there are two caps on our pillbox we need to break the integral into two parts so in terms of the area and electric field magnitudes,

$$EA + EA = \int_{\substack{\text{end}\\ \text{caps}}} EdA = |q^{\text{enc}}|/\varepsilon_0.$$

Next we can find the amount of charge on the sheet enclosed by our Gaussian pillbox in terms of the surface charge density, σ, on the sheet. Since the surface charge is uniform and the surface charge density is defined as the ratio of the charge on a given surface to its area, we know that $\sigma = q^{\text{enc}}/A$. If we replace q^{enc} in the equation above with σA and solve it for the electric field magnitude we get

$$E = \frac{|\sigma|}{2\varepsilon_0} \quad \text{(sheet of uniformly distributed charge).} \qquad (24\text{-}16)$$

The equation holds whether the sheet is conducting or nonconducting as long as the layer of charge on the sheet is thin.

Equation 24-16 tells us that *the electric field has the same value for all locations outside a large uniformly charged sheet and points in a direction that is perpendicular to the sheet.* This result is quite surprising! The fact that the net field is perpendicular to the sheet can be explained using symmetry arguments. But how can it be that as you get farther away from the charged sheet the electric field doesn't decrease? The answer lies in considering the influences of the charges as we move away from the sheet. When a test charge is placed very close to the sheet, the

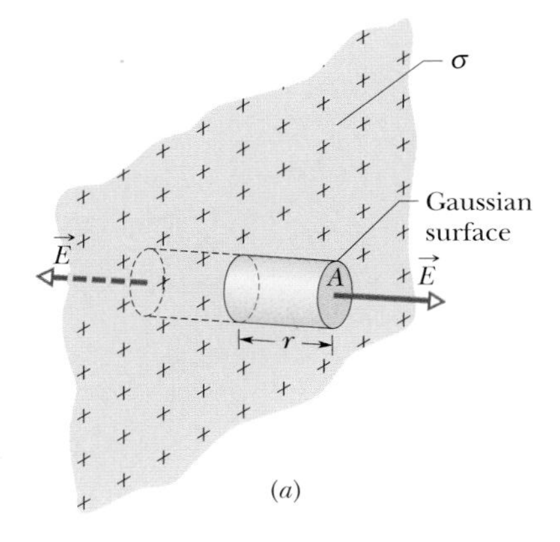

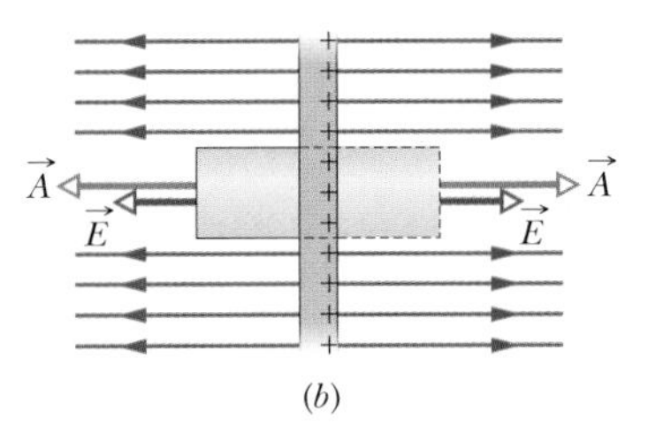

FIGURE 24-18 ■ Perspective view (*a*) and side view (*b*) of a portion of a very large, thin plastic sheet, uniformly charged with surface charge density σ. A closed cylindrical Gaussian surface passes through the sheet and is perpendicular to it.

influence on it by the charge closest to it dominates. If the test charge is moved farther from the sheet the influence of the nearest sheet charge gets weaker, but the normal components of the electric field vectors from neighboring sheet charges start to contribute and compensate for the loss of influence of the nearest sheet charge. If the test charge is moved even farther the influence of the nearest and nearby charges diminish but the components of additional surrounding charges come into play and so on.

Equation 24-16 agrees with what we would have found by integration of the electric field components that are produced by individual charges. That would be a very time-consuming and challenging integration, and note how much more easily we obtain the result using Gauss' law. This is one reason for devoting a whole chapter to Gauss' law. For certain symmetric arrangements of charge, it is much easier to use it than to integrate field components.

READING EXERCISE 24-3: Consider an array of 9 charges evenly distributed on a square insulating sheet as shown in the diagram. Use symmetry arguments to explain why the electric field vector anywhere on a line normal to the central charge and passing through it has no component that is parallel to the sheet.

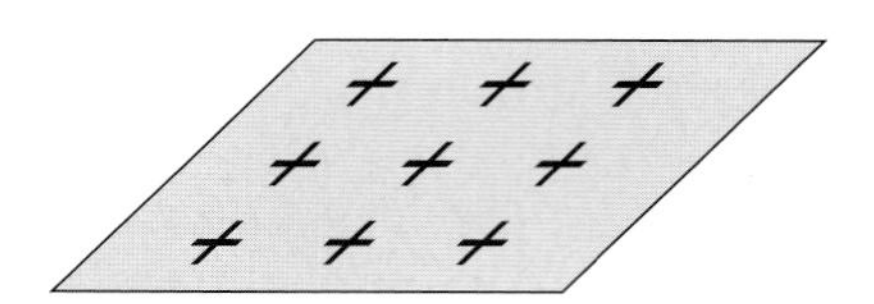

TOUCHSTONE EXAMPLE 24-3: $\vec{E}$ for Two Sheets of Charge

Figure 24-19*a* shows portions of two large, parallel, nonconducting sheets, each with a fixed uniform charge on one side. The amounts of the surface charge densities are $\sigma_{(+)} = 6.8\,\mu\text{C/m}^2$ for the positively charged sheet and $\sigma_{(-)} = 4.3\,\mu\text{C/m}^2$ for the negatively charged sheet.

Find the electric field $\vec{E}$ (a) to the left of the sheets, (b) between the sheets, and (c) to the right of the sheets.

SOLUTION ■ The **Key Idea** here is that with the charges fixed in place, we can find the electric field of the sheets in Fig. 24-19*a* by (1) finding the field of each sheet as if that sheet were isolated and (2) adding the vector fields of the isolated sheets via the superposition principle. (The vector addition is simple here since the fields lie along the same axis. We can add the fields algebraically because they are parallel to each other.) From Eq. 24-16, the magnitude $E_{(+)}$ of the electric field due to the positive sheet at any point is

$$|\vec{E}_{(+)}| = \frac{|\sigma_{(+)}|}{2\varepsilon_0} = \frac{6.8 \times 10^{-6}\ \text{C/m}^2}{(2)(8.85 \times 10^{-12}\ \text{C}^2/\text{N}\cdot\text{m}^2)}$$
$$= 3.84 \times 10^5\ \text{N/C}.$$

Similarly, the magnitude $|\vec{E}_{(-)}|$ of the electric field at any point due to the negative sheet is

$$|\vec{E}_{(-)}| = \frac{|\sigma_{(-)}|}{2\varepsilon_0} = \frac{4.3 \times 10^{-6}\ \text{C/m}^2}{(2)(8.85 \times 10^{-12}\ \text{C}^2/\text{N}\cdot\text{m}^2)}$$
$$= 2.43 \times 10^5\ \text{N/C}.$$

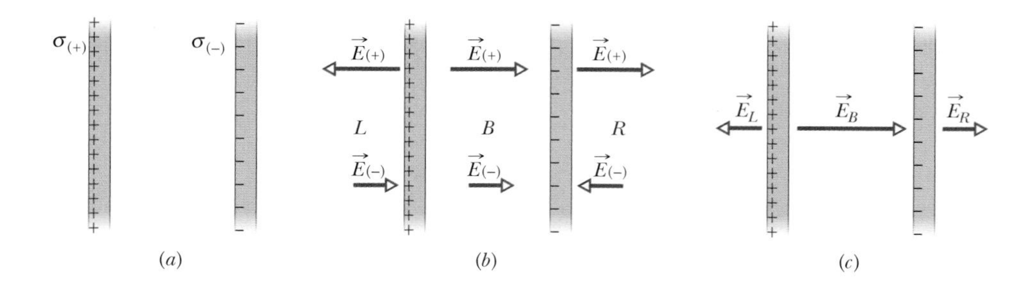

FIGURE 24-19 ■ (*a*) Two large, parallel insulating sheets, uniformly charged on one side. (*b*) The individual electric fields resulting from the two charged sheets. (*c*) The net field due to both charged sheets, found by superposition.

Figure 24-19*b* shows the fields set up by the sheets to the left of the sheets (L), between them (B), and to their right (R).

The resultant fields in these three regions follow from the superposition principle. To the left of the sheets, the field magnitude is

$$|\vec{E}_L| = |\vec{E}_{(+)}| - |\vec{E}_{(-)}|$$
$$= 3.84 \times 10^5 \text{ N/C} - 2.43 \times 10^5 \text{ N/C}$$
$$= 1.4 \times 10^5 \text{ N/C}. \quad \text{(Answer)}$$

Because $|E_{(+)}|$ is larger than $|E_{(-)}|$, the net electric field $\vec{E}_L$ in this region points to the left, as Fig. 24-19*c* shows. To the right of the sheets, the electric field $\vec{E}_R$ has the same magnitude but points to the right, as Fig. 24-19*c* shows.

Between the sheets, the two fields add and we have

$$|\vec{E}_B| = |\vec{E}_{(+)}| + |\vec{E}_{(-)}|$$
$$= 3.84 \times 10^5 \text{ N/C} + 2.43 \times 10^5 \text{ N/C}$$
$$= 6.3 \times 10^5 \text{ N/C}. \quad \text{(Answer)}$$

The electric field $\vec{E}_B$ points to the right.

24-7 Gauss' Law and Coulomb's Law

If Gauss' law and Coulomb's law are equivalent, we should be able to derive each from the other. Here we derive Coulomb's law from Gauss' law and some symmetry considerations.

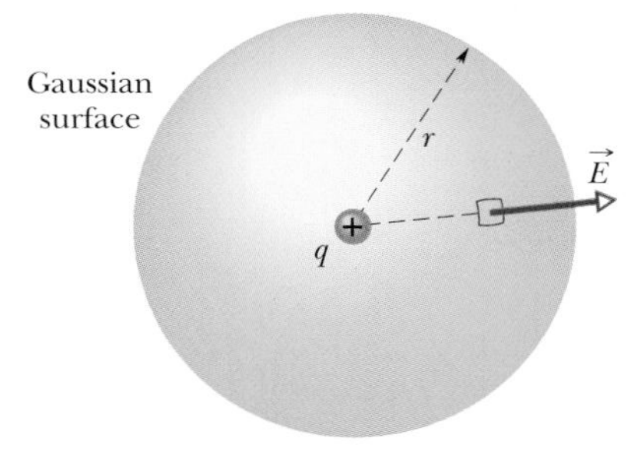

FIGURE 24-20 ■ A spherical Gaussian surface centered on a point charge q.

Figure 24-20 shows a positive point charge q, around which we have drawn a concentric spherical Gaussian surface of radius r. Let us divide this surface into differential areas $d\vec{A}$. By definition, the area vector $d\vec{A}$ at any point is perpendicular to the surface and directed outward from the interior. From the symmetry of the situation, we know at any point the electric field $\vec{E}$ is also perpendicular to the surface and directed outward from the interior. Thus, since the angle θ between $\vec{E}$ and $d\vec{A}$ is zero, we can rewrite Gauss' law expressed in Eq. 24-7 as

$$\oint \vec{E} \cdot d\vec{A} = \oint E dA = q^{\text{enc}}/\varepsilon_0. \quad (24\text{-}17)$$

Here $q^{\text{enc}} = q$. Although the magnitude of the vector $\vec{E}$ varies radially with the distance from q, it has the same value everywhere on the spherical surface. Since the integral in this equation is taken over that surface, the electric field magnitude ($E = |\vec{E}|$) is a constant in the integration and can be brought out in front of the integral sign. That gives us

$$\varepsilon_0 E \oint dA = |q^{\text{enc}}|. \quad (24\text{-}18)$$

The integral is now merely the sum of the magnitudes of all the differential area elements $d\vec{A}$ on the sphere and thus is just the surface area, $4\pi r^2$. Substituting this, we have

$$\varepsilon_0 E(4\pi r^2) = |q^{\text{enc}}|,$$

or since $q = q^{\text{enc}}$

$$E = \frac{1}{4\pi\varepsilon_0} \frac{|q^{\text{enc}}|}{r^2} = k\frac{|q|}{r^2}. \quad (24\text{-}19)$$

This is exactly the electric field due to a point charge (Eq. 23-8), which we found using Coulomb's law. Thus, we have shown that Gauss' law and Coulomb's law give us the same result for the electric field due to a single point-like charge. However, Gauss' law is also valid for complex arrays of charges. It can be shown using the principle of superposition that the information about electric fields obtained by using either Gauss' or Coulomb's law will yield the same results even for charge arrays. The difference between the two laws is this: It is easier to use Coulomb's law if we have an array of a few point-like charges, and it is easier to use Gauss' law if we have certain kinds of highly symmetric charge distributions like those discussed in Section 24-6. In still other situations, it is quite difficult to use either law.

READING EXERCISE 24-4: There is a certain net flux Φ^{net} at a Gaussian sphere of radius r enclosing an isolated charged particle. Suppose the enclosing Gaussian surface is changed to (a) a larger Gaussian sphere, (b) a Gaussian cube with edge length equal to r, and (c) a Gaussian cube with edge length equal to $2r$. In each case, is the net flux at the new Gaussian surface greater than, less than, or equal to Φ^{net}? ■

24-8 A Charged Isolated Conductor

Gauss' law permits us to prove an important theorem about isolated conductors:

> If excess charges are placed on an isolated conductor, that amount of charge will move entirely to the surface of the conductor. Once the charges stop moving, none of the excess charge will be found within the body of the conductor.

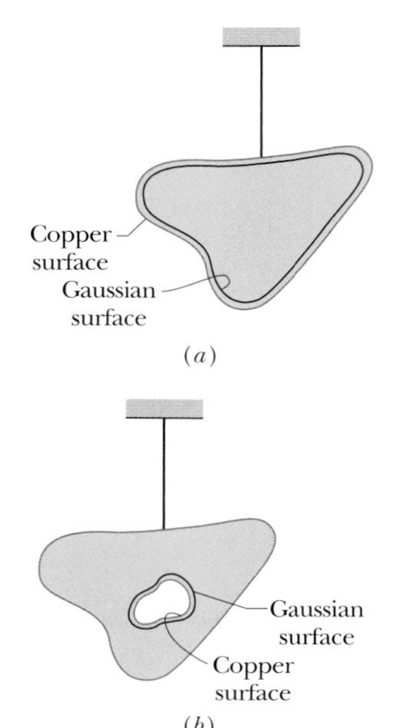

FIGURE 24-21 ■ (*a*) A lump of copper with a charge q hangs from an insulating thread. A Gaussian surface is placed within the metal, just inside the actual surface. (*b*) The lump of copper now has a cavity within it. A Gaussian surface lies within the metal, close to the cavity surface.

This might seem reasonable, considering charges with the same sign repel each other. You might imagine that by moving to the surface, the added charges are getting as far away from each other as they can. We turn to Gauss' law for verification of this speculation.

Figure 24-21*a* shows, in cross section, an isolated lump of copper hanging from an insulating thread and having an excess charge q. We place a Gaussian surface just inside the actual surface of the conductor.

Once the excess charges stop moving, the electric field inside this conductor must be zero. If this were not so, the field would exert forces on the conduction (free) electrons, which are always present in a conductor such as copper, and thus current would always exist within a conductor. (That is, charge would flow from place to place within the conductor.) Of course, there are no such perpetual currents in an isolated conductor, and so we know that the internal electric field is zero.

An internal electric field *does* appear as a conductor is being charged. However, the added charge quickly distributes itself in such a way that the net internal electric field—the vector sum of the electric fields due to all the charges, both inside and outside—is zero. The movement of charge then ceases because there are drag forces known as *resistance* in conductors that dissipate the charges' kinetic energies and eventually bring them to rest. Since the net field is zero, the net force on each charge is zero. So, once the charges are stopped by resistance in the conductor, they remain at rest. Some special materials can be "superconductors" at very low temperatures and allow charges to move without resistance. Therefore, these materials can support long-lasting currents.

If $\vec{E}$ is zero everywhere inside our copper conductor, it must be zero for all points on the Gaussian surface because that surface, though close to the surface of the conductor, is definitely inside the conductor. This means the flux at the Gaussian surface must be zero. Gauss' law then tells us the net charge inside the Gaussian surface must also be zero. Then because the excess charge is not inside the Gaussian surface, it must be outside that surface, which means it must lie on the actual surface of the conductor.

An Isolated Conductor with a Cavity

Figure 24-21*b* shows the same hanging conductor, but now with a cavity totally within the conductor. It is perhaps reasonable to suppose that when we scoop out the electrically neutral material to form the cavity we do not change the distribution of charge or the pattern of the electric field that exists in Fig. 24-21*a*. Again, we can turn to Gauss' law for a quantitative proof.

We draw a Gaussian surface surrounding the cavity, close to its surface but inside the conducting body. Because $\vec{E} = 0$ inside the conductor, there can be no flux at this new Gaussian surface. Therefore, from Gauss' law, that surface can enclose no net charge. We conclude there is no net charge on the cavity walls; all the excess charge remains on the outer surface of the conductor, as in Fig. 24-21*a*.

The Conductor Removed

Consider now an object that has the same shaped surface, but consists of only a conducting shell of charge. This is equivalent to enlarging the cavity of Fig. 24-21*b* until it consumes the entire conductor, leaving only the charges. The electric field would not change at all; it would remain zero inside the thin shell of charge and would remain unchanged for all external points. This reminds us that the electric field is set up by the charges and not by the conductor. The conductor simply provides an initial pathway for the charges to take up their positions.

The External Electric Field

You have seen that the excess charge on an isolated conductor moves entirely to the conductor's surface. However, unless the conductor is spherical, the charge does not distribute itself uniformly. Put another way, the surface charge density σ (charge per unit area) varies over the surface of any nonspherical conductor. Generally, this variation makes the determination of the electric field set up by the surface charges very difficult.

Suppose we know the surface charge density, σ, on a region of a conductor. Then it is easy to use Gauss' law to calculate the electric field just outside the surface of a conductor. To do this, we consider a section of the surface small enough to permit us to neglect any curvature and thus to take the section to be flat. We then imagine a tiny cylindrical Gaussian surface to be embedded in the section as in Fig. 24-22: One end cap is fully inside the conductor, the other is fully outside, and the cylinder is perpendicular to the conductor's surface.

The electric field $\vec{E}$ at and just outside the conductor's surface must also be perpendicular to that surface. If it were not, then it would have a component along the conductor's surface exerting forces on the surface charges, causing them to move. However, such motion would violate our implicit assumption that we are dealing with electrostatic equilibrium. Therefore, $\vec{E}$ is perpendicular to the conductor's surface.

We now sum the flux at the Gaussian surface. There is no flux at the internal end cap, because the electric field within the conductor is zero. There is no flux at the curved surface of the cylinder, because internally (in the conductor) there is no electric field and externally the electric field is parallel to the curved portion of the Gaussian surface. The only flux at the Gaussian surface is at the external end cap, where $\vec{E}$ is perpendicular to the plane of the cap. We assume the cap area A is small enough that the field magnitude $|\vec{E}|$ is constant over the cap. Then the amount of the flux at the cap is $|\vec{E}|A$, and that is the net amount of flux $|\Phi^{net}|$ at the Gaussian surface.

The charge q^{enc} enclosed by the Gaussian surface lies on the conductor's surface in an area A. If σ is the charge per unit area, then q^{enc} is equal to σA. When we substitute σA for q^{enc} and $|\vec{E}|A$ for $|\Phi^{net}|$, Gauss' law, $\varepsilon_0|\Phi^{net}| = |q^{enc}|$, becomes $\varepsilon_0|\vec{E}|A = |\sigma|A$, from which we find

$$|\vec{E}| = \frac{|\sigma|}{\varepsilon_0} \quad \text{(conducting surface).} \qquad (24\text{-}20)$$

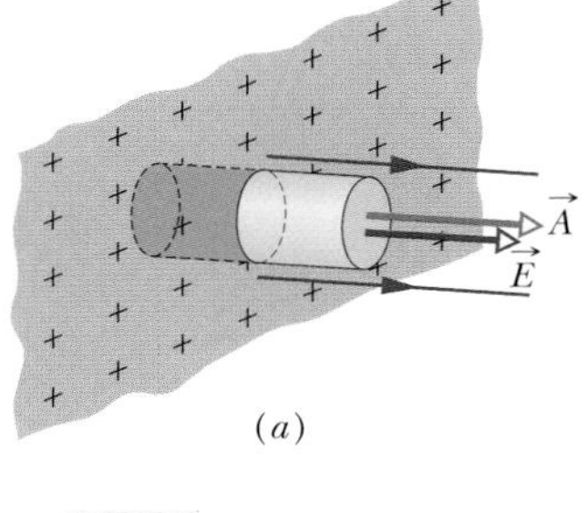

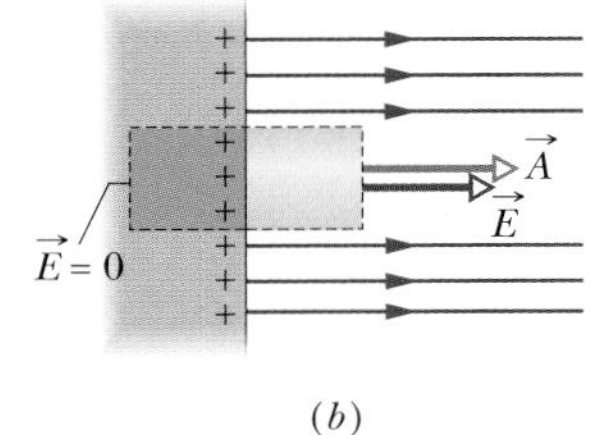

FIGURE 24-22 ■ Perspective view (*a*) and side view (*b*) of a tiny portion of a large, isolated conductor with excess positive charge on its surface. A (closed) cylindrical Gaussian surface, embedded perpendicularly in the conductor, encloses some of the charge. Electric field lines pierce the external end cap of the cylinder, but not the internal end cap. The external end cap has area A and area vector $\vec{A}$.

Thus, the magnitude of the electric field at a location just outside a conductor is proportional to the surface charge density at that location on the conductor. If the charge on the conductor is positive, the electric field is directed away from the conductor as in Fig. 24-22. It is directed toward the conductor if the charge is negative.

The difference between Eq. 24-20 and Eq. 24-16 ($|\vec{E}| = |\sigma|/2\varepsilon_0$) results from the fact that our conductor is no longer thin so that one of our Gaussian pillbox endcaps lies inside the conductor where the electric field is zero. Although the situation in Figs. 24-18 and 24-22 look similar, there is an important difference. There must be other charges in Fig. 24-22 that contribute to making the field zero inside the conductor. Even though these charges are outside the Gaussian surface and therefore do not contribute to the total flux, they change the values of the E field on the surface and therefore change the value we extract.

The field vectors in Fig. 24-22 point toward negative charges somewhere in the environment. If we bring those charges near the conductor, the charge density at any given location on the conductor's surface changes, and so does the magnitude of the electric field. However, the relation between the amount of the surface charge per unit area and the electric field magnitude is still given by Eq. 24-20,

$$E = \frac{|\sigma|}{\varepsilon_0}.$$

The Faraday Cage

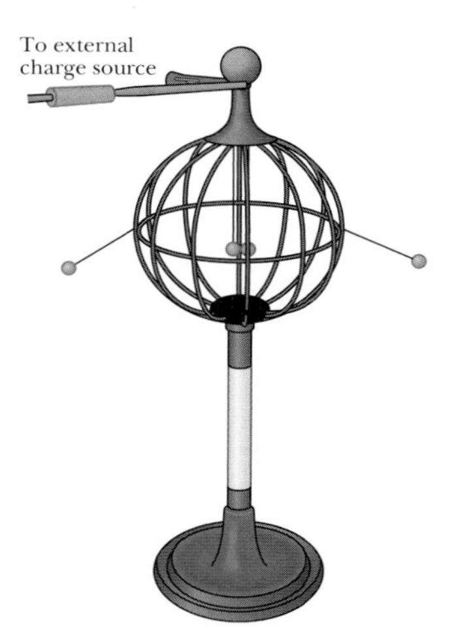

FIGURE 24-23 ■ A charged Faraday cage consisting of a sphere made of curved brass rods. Charges on the outside of the cage travel along conducting strings to the small balls causing them to be repelled from the cage. There is no charge inside the cage so the balls in the cage do not repel.

The fact that an isolated conductor with a cavity has no electric field inside of it has led to the construction of a very valuable electrical device. Many research environments today involve the measurement of very low power electrical signals. This might occur when measuring the electrical signals from the neuron of a live mouse running a maze or while trying to measure the electrical properties of a microscopic device meant as part of a micro-miniaturized computer chip. In our modern world there are numerous electrical signals traveling through space, arising from everything from the 60 Hz power running in our walls to the radio signals from TV stations and cellular phones. These signals can interfere with sensitive electrical measurements.

To prevent these stray electric fields from ruining sensitive measurements, researchers often conduct their experiments inside a thin-walled metal cage known as a Faraday cage. Examples of Faraday cages are shown in the photo on the first page of this chapter as well as in Fig. 24-23. The Faraday cage in Fig. 24-23 is like the object shown in Fig. 24-21*b* except that now the "cavity" takes up almost the whole volume of the material. In addition, the thin metal shell in a Faraday cage is typically made of wire mesh. As long as the mesh is fairly fine, charge can spread out evenly on its surface. This type of cage can prevent even strong electrical signals from producing electric fields inside the cage. How? The external electric field induces charges on the surface of the Faraday cage to move so that the field they produce will precisely cancel the external field at points inside the surface. This rearrangement occurs naturally and is predictable by Gauss' law. This is why a demonstrator in a Faraday cage that is highly charged by a Van de Graaff generator can touch the inside of the cage and survive as shown in the opening photograph. The principle of the Faraday cage is also what makes it safe to be inside an automobile in a lightning storm. Even if lightning strikes your car, the effects inside the conductor are substantially reduced. This would not be the case if you were in a wooden crate, because the lightning could pass right through it. The crate could also catch on fire.

READING EXERCISE 24-5: Suppose a single positive charge is suddenly placed in the cavity shown in Fig. 24-21*b*. What has to happen in the conductor at the cavity walls to ensure that the electric field everywhere inside the conductor remains at zero? ■

TOUCHSTONE EXAMPLE 24-4: Spherical Metal Shell

Figure 24-24*a* shows a cross section of a spherical metal shell of inner radius R. A point charge of $-5.0\,\mu C$ is located at a distance $R/2$ from the center of the shell. If the shell is electrically neutral, what are the (induced) charges on its inner and outer surfaces? Are those charges uniformly distributed? What is the field pattern inside and outside the shell?

SOLUTION ■ Figure 24-24*b* shows a cross section of a spherical Gaussian surface within the metal, just outside the inner wall of the shell. One **Key Idea** here is that the electric field must be zero inside the metal (and thus on the Gaussian surface inside the metal). This means that the electric flux through the Gaussian surface must also be zero. Gauss' law then tells us that the *net* charge enclosed by the Gaussian surface must be zero. With a point charge of $-5.0\,\mu C$ within the shell, a charge of $+5.0\,\mu C$ must lie on the inner wall of the shell.

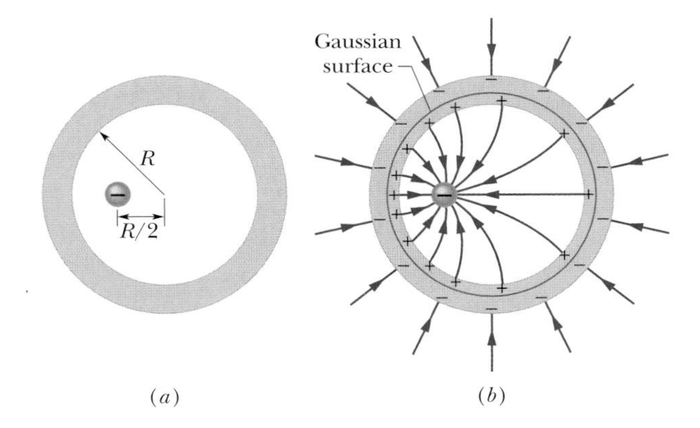

FIGURE 24-24 ■ (*a*) A negative point charge is located within a spherical metal shell that is electrically neutral. (*b*) As a result, positive charge is nonuniformly distributed on the inner wall of the shell, and an equal amount of negative charge is uniformly distributed on the outer wall. The electric field lines are shown.

If the point charge were centered, this positive charge would be uniformly distributed along the inner wall. However, since the point charge is off-center, the distribution of positive charge is skewed, as suggested by Fig. 24-24*b*, because the positive charge tends to collect on the section of the inner wall nearest the (negative) point charge.

A second **Key Idea** is that because the shell is electrically neutral, its inner wall can have a charge of $+5.0\,\mu C$ only if electrons, with a total charge of $-5.0\,\mu C$, leave the inner wall and move to the outer wall. There they spread out uniformly, as is also suggested by Fig. 24-24*b*. This distribution of negative charge is uniform because the shell is spherical and because the skewed distribution of positive charge on the inner wall cannot produce an electric field in the shell to affect the distribution of charge on the outer wall.

The field lines inside and outside the shell are shown approximately in Fig. 24-24*b*. All the field lines intersect the shell and the point charge perpendicularly. Inside the shell the pattern of field lines is skewed owing to the skew of the positive charge distribution. Outside the shell the pattern is the same as if the point charge were centered and the shell were missing. In fact, this would be true no matter where inside the shell the point charge happened to be located.

Problems

SEC. 24-3 ■ NET FLUX AT A CLOSED SURFACE

1. Cube The cube in Fig. 24-25 has edge lengths of 1.40 m and is oriented as shown with its bottom face in the x-y plane at $z = 0.00$ m. Find the electric flux through the right face if the uniform electric field, in newtons per coulomb, is given by (a) $6.00\hat{i}$, (b) $-2.00\hat{j}$, and (c) $-3.00\hat{i} + 4.00\hat{k}$. (d) What is the total flux through the cube for each of these fields?

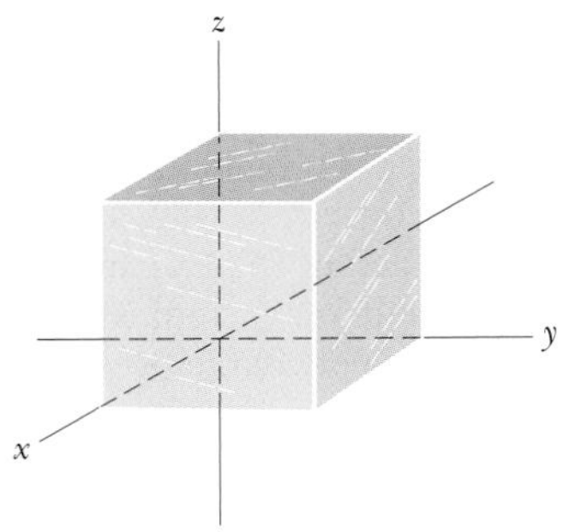

FIGURE 24-25 ■ Problems 1, 5, and 10.

2. Square Surface The square surface shown in Fig. 24-26 measures 3.2 mm on each side. It is immersed in a uniform electric field with magnitude $|\vec{E}| = 1800$ N/C. The field lines make an angle of 35° with a normal to the surface, as shown. Take that normal to be directed "outward," as though the surface were one face of a box. Calculate the electric flux through the surface.

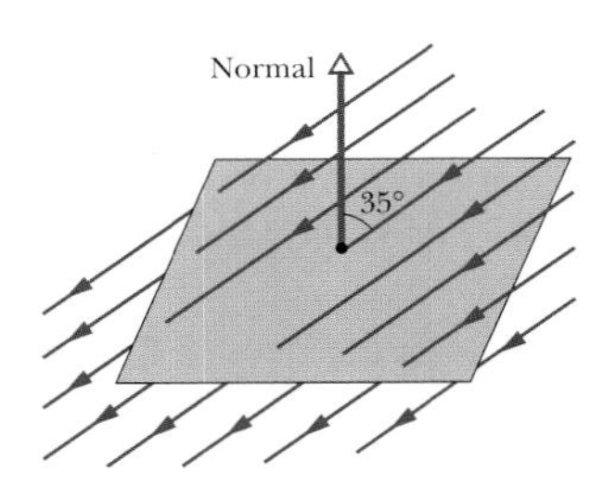

FIGURE 24-26 ■ Problem 2.

SEC. 24-4 ■ GAUSS' LAW

3. Charge at Center of Cube A point charge of $1.8\,\mu C$ is at the center of a cubical Gaussian surface 55 cm on edge. What is the net electric flux through the surface?

4. Four Charges You have four point charges, $2q$, q, $-q$, and $-2q$. If possible describe how you would place a closed surface that encloses at least the charge $2q$ (and perhaps other charges) and through which the net electric flux is (a) 0 (b) $+3q/\varepsilon_0$, and (c) $-2q/\varepsilon_0$.

5. Flux Through Cube Find the net flux through the cube of Problem 1 and Fig. 24-25 if the electric field is given by (a) $\vec{E} = (3.00\,y\,[\text{N/(C}\cdot\text{m)}])\hat{j}$

and (b) $\vec{E} = -(4.00\text{ N/C})\hat{i} + (6.00\text{ N/C} + 3.00\,y\text{ [N/(C}\cdot\text{m)]})\hat{j}$. (c) In each case, how much charge is enclosed by the cube?

6. Butterfly Net In Fig. 24-27, a butterfly net is in a uniform electric field of magnitude $\vec{E}$. The rim, a circle of radius a, is aligned perpendicular to the field. Find the electric flux through the netting.

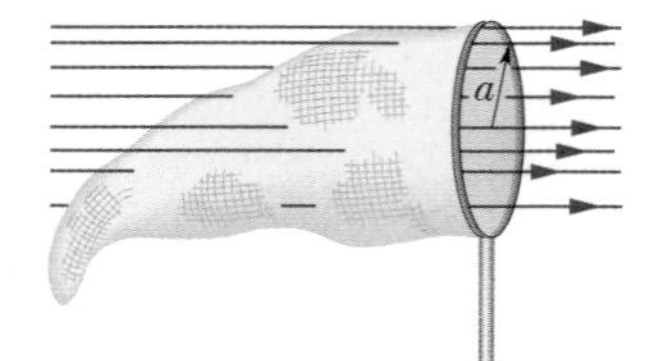

FIGURE 24-27 ■ Problem 6.

7. Earth's Atmosphere It is found experimentally that the electric field in a certain region of Earth's atmosphere is directed vertically down. At an altitude of 300 m the field has magnitude 60.0 N/C; at an altitude of 200 m, the magnitude is 100 N/C. Find the net amount of charge contained in a cube 100 m on edge, with horizontal faces at altitudes of 200 and 300 m. Neglect the curvature of Earth.

8. Shower When a shower is turned on in a closed bathroom, the splashing of the water on the bare tub can fill the room's air with negatively charged ions and produce an electric field in the air as great as 1000 N/C. Consider a bathroom with dimensions of 2.5 m × 3.0 m × 2.0 m. Along the ceiling, floor, and four walls, approximate the electric field in the air as being directed perpendicular to the surface and as having a uniform magnitude of 600 N/C. Also, treat those surfaces as forming a closed Gaussian surface around the room's air. What are (a) the volume charge density ρ and (b) the number of excess elementary charges e per cubic meter in the room's air?

9. Point Charge A point charge q is placed at one corner of a cube of edge a. What is the flux through each of the cube faces? (*Hint*: Use Gauss' law and symmetry arguments.)

10. Surface of Cube At each point on the surface of the cube shown in Fig 24-25, the electric field is along the y-axis. The length of each edge of the cube is 3.0 m. On the right surface of the cube, $\vec{E} = (-34\text{ N/C})\hat{j}$, and on the left face of the cube $\vec{E} = (+20\text{ N/C})\hat{j}$. Determine the net charge contained within the cube.

SEC. 24-6 ■ APPLICATION OF GAUSS' LAW TO SYMMETRIC CHARGE DISTRIBUTIONS

11. Conducting Sphere A conducting sphere of radius 10 cm has an unknown charge. If the electric field 15 cm from the center of the sphere has the magnitude 3.0×10^3 N/C and is directed radially inward, what is the net charge on the sphere?

12. Charge Causes Flux A point charge causes an electric flux of $-750\text{ N}\cdot\text{m}^2/\text{C}$ to pass through a spherical Gaussian surface of 10.0 cm radius centered on the charge. (a) If the radius of the Gaussian surface were doubled, how much flux would pass through the surface? (b) What is the value of the point charge?

13. Rutherford In a 1911 paper, Ernest Rutherford said: "In order to form some idea of the forces required to deflect an α particle through a large angle, consider an atom [as] containing a point positive charge Ze at its center and surrounded by a distribution of negative electricity $-Ze$ uniformly distributed within a sphere of radius R. The electric field E . . . at a distance r from the center for a point inside the atom [is]

$$E = \frac{Ze}{4\pi\varepsilon_0}\left(\frac{1}{r^2} - \frac{r}{R^3}\right)\text{''}.$$

Verify this equation.

14. Concentric Spheres Two charged concentric spheres have radii of 10.0 cm and 15.0 cm. The charge on the inner sphere is 4.00×10^{-8} C, and that on the outer sphere is 2.00×10^{-8} C. Find the electric field (a) at $r = 12.0$ cm and (b) at $r = 20.0$ cm.

15. Proton A proton with speed $v = 3.00 \times 10^5$ m/s orbits just outside a charged sphere of radius $r = 1.00$ cm. What is the charge on the sphere?

16. Charge at Center of Shell A point charge $+q$ is placed at the center of an electrically neutral, spherical conducting shell with inner radius a and outer radius b. What charge appears on (a) the inner surface of the shell and (b) the outer surface? What is the net electric field at a distance r from the center of the shell if (c) $r < a$, (d) $b > r > a$, and (e) $r > b$? Sketch field lines for those three regions. For $r > b$, what is the net electric field due to (f) the central point charge plus the inner surface charge and (g) the outer surface charge? A point charge $-q$ is now placed outside the shell. Does this point charge change the charge distribution on (h) the outer surface and (i) the inner surface? Sketch the field lines now. (j) Is there an electrostatic force on the second point charge? (k) Is there a net electrostatic force on the first point charge? (l) Does this situation violate Newton's Third Law?

17. Solid Nonconducting Sphere A solid nonconducting sphere of radius R has a nonuniform charge distribution of volume charge density $\rho = \rho_s r/R$, where ρ_s is a constant and r is the distance from the center of the sphere. Show (a) that the total charge on the sphere is $Q = \pi\rho_s R^3$ and (b) that

$$|\vec{E}| = k\frac{|Q|}{R^4}r^2$$

gives the magnitude of the electric field inside the sphere.

18. Hydrogen Atom A hydrogen atom can be considered as having a central point-like proton of positive charge $+e$ and an electron of negative charge $-e$ that is distributed about the proton according to the volume charge density $\rho = A\exp(-2r/a_1)$. Here A is a constant, $a_1 = 0.53 \times 10^{-10}$ m is the *Bohr radius*, and r is the distance from the center of the atom. (a) Using the fact that hydrogen is electrically neutral, find A. (b) Then find the electric field produced by the atom at the Bohr radius.

19. Sphere of Radius a In Fig 24-28 an insulating sphere, of radius a and charge $+q$ uniformly distributed throughout its volume, is concentric with a spherical conducting shell of inner radius b and outer radius c. This shell has a net charge of $-q$. Find expressions for the electric field, as a function of the radius r, (a) within the sphere ($r < a$), (b) between the sphere and the shell ($a < r < b$), (c) inside the shell ($b < r < c$), and (d) outside the shell ($r > c$). (e) What are the charges on the inner and outer surfaces of the shell?

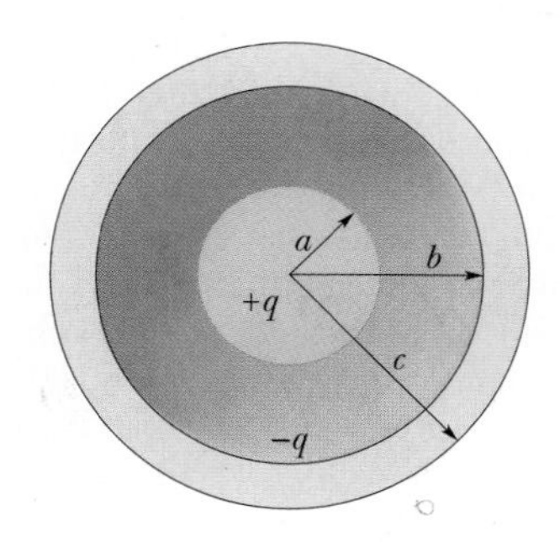

FIGURE 24-28 ■ Problem 19.

20. Uniform Volume Charge Density Figure 24-29a shows a spherical shell of charge with uniform volume charge density ρ. Plot E due to the shell for distances r from the center of the shell ranging from zero to 30 cm. Assume that $\rho = 1.0 \times 10^{-6}$ C/m³, $a = 10$ cm, and $b = 20$ cm.

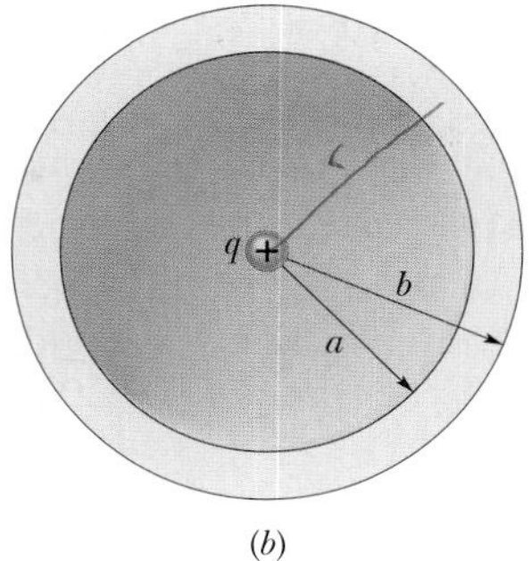

FIGURE 24-29 ■ Problems 20 and 21.

21. Nonconducting Spherical Shell In Fig. 24-29*b*, a nonconducting spherical shell, of inner radius a and outer radius b, has a positive volume charge density $\rho = A/r$ (within its thickness), where A is a constant and r is the distance from the center of the shell. In addition, a positive point charge q is located at that center. What value should A have if the electric field in the shell ($a \le r \le b$) is to be uniform? (*Hint:* The constant A depends on a but not on b.)

22. Show That A nonconducting sphere has a uniform volume charge density ρ. Let $\vec{r}$ be the vector from the center of the sphere to a general point P within the sphere. (a) Show that the electric field at P is given by $\vec{E} = \rho\vec{r}/3\varepsilon_0$. (Note that the result is independent of the radius of the sphere.) (b) A spherical cavity is hollowed out of the sphere, as shown in Fig. 24-30. Using superposition concepts, show that the electric field at all points within the cavity is uniform and equal to $\vec{E} = \rho\vec{a}/3\varepsilon_0$, where $\vec{a}$ is the position vector from the center of the sphere to the center of the cavity. (Note that this result is independent of the radius of the sphere and the radius of the cavity.)

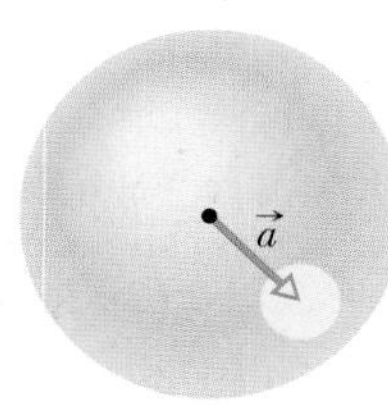

FIGURE 24-30 ■ Problem 22.

23. Spherically Symmetrical A spherically symmetrical but nonuniform volume distribution of charge produces an electric field of magnitude $|\vec{E}| = Kr^4$, directed radially outward from the center of the sphere. Here r is the radial distance from that center, and K is a positive constant. What is the volume density ρ of the charge distribution as a function of r?

24. Long Metal Tube Figure 24-31 shows a section of a long, thin-walled metal tube of radius R, with a positive charge per unit length λ on its surface. Derive expressions for $|\vec{E}|$ in terms of the distance r from the tube axis, considering both (a) $r > R$ and (b) $r < R$. Plot your results for the range $r = 0$ to $r = 5.0$ cm, assuming that $\lambda = 2.0 \times 10^{-8}$ C/m and $R = 3.0$ cm. (*Hint:* Use cylinderical Gaussian surfaces, coaxial with the metal tube.)

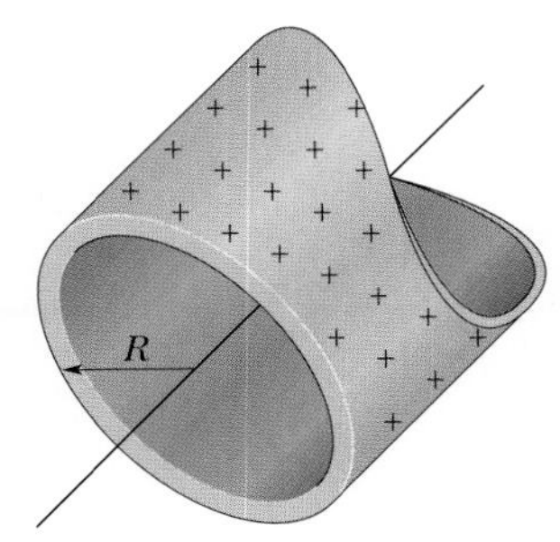

FIGURE 24-31 ■ Problem 24.

25. Infinite Line of Charge An infinite line of charge produces a field magnitude of 4.5×10^4 N/C at a distance of 2.0 m. Calculate the amount of linear charge density $|\lambda|$.

26. Long Straight Wire A long, straight wire has fixed negative charge with a linear charge density of -3.6 nC/m. The wire is to be enclosed by a thin, nonconducting cylinder of outside radius 1.5 cm, coaxial with the wire. The cylinder is to have positive charge on its outside surface with a surface charge density σ such that the net external electric field is zero. Calculate the required σ.

27. Cylindrical Rod A very long conducting cylindrical rod of length L with a total charge $+q$ is surrounded by a conducting cylindrical shell (also of length L) with total charge $-2q$, as shown in Fig. 24-32. Use Gauss' law to find (a) the electric field at points outside the conducting shell, (b) the distribution of charge on the shell, and (c) the electric field in the region between the shell and rod. Neglect end effects.

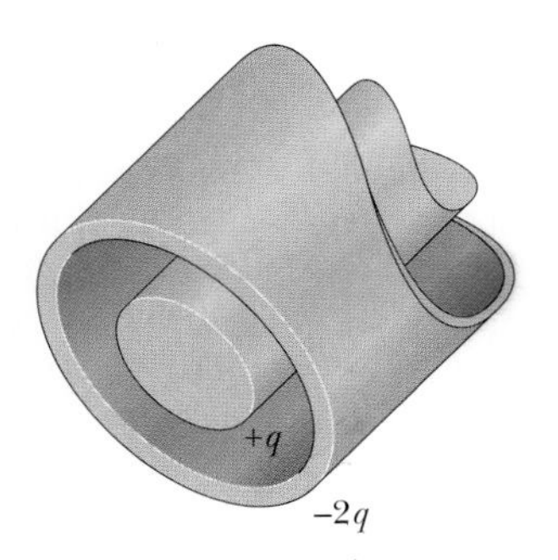

FIGURE 24-32 ■ Problem 27.

28. Solid Cylinder A long, nonconducting, solid cylinder of radius 4.0 cm has a nonuniform volume charge density ρ that is a function of the radial distance r from the axis of the cylinder, as given by $\rho = Ar^2$ with $A = 2.5\ \mu\text{C/m}^5$. What is the magnitude of the electric field at a radial distance of (a) 3.0 cm and (b) 5.0 cm from the axis of the cylinder?

29. Two Concentric Cylinders Two long, charged, concentric cylinders have radii of 3.0 and 6.0 cm. Assume the outer cylinder is hollow. The charge per unit length is 5.0×10^{-6} C/m on the inner cylinder and -7.0×10^{-6} C/m on the outer cylinder. Find the electric field at (a) $r = 4.0$ cm and (b) $r = 8.0$ cm, where r is the radial distance from the common central axis.

30. Geiger Counter Figure 24-33 shows a Geiger counter, a device used to detect ionizing radiation (radiation that causes ionization of atoms). The counter consists of a thin, positively charged central wire surrounded by a concentric, circular, conducting cylinder with an equal negative charge. Thus, a strong radial electric field is set up inside the cylinder. The cylinder contains a low-pressure inert gas. When a particle of radiation enters the device through the cylinder wall, it ionizes a few of the gas atoms. The resulting free electrons (labelled e) are drawn to the positive wire. However, the electric field is so intense that, between collisions with other gas atoms, the free electrons gain energy sufficient to ionize these atoms also. More free electrons are thereby created, and the process is repeated until the electrons reach the wire. The resulting "avalanche" of electrons is collected by the wire generating a signal that is used to record the passage of the original particle of radiation. Suppose that the radius of the central wire is 25 μm, the radius of the cylinder 1.4 cm, and the length of the tube 16 cm. If the electric field component E_r at the cylinder's inner wall is $+2.9 \times 10^4$ N/C, what is the total positive charge on the central wire?

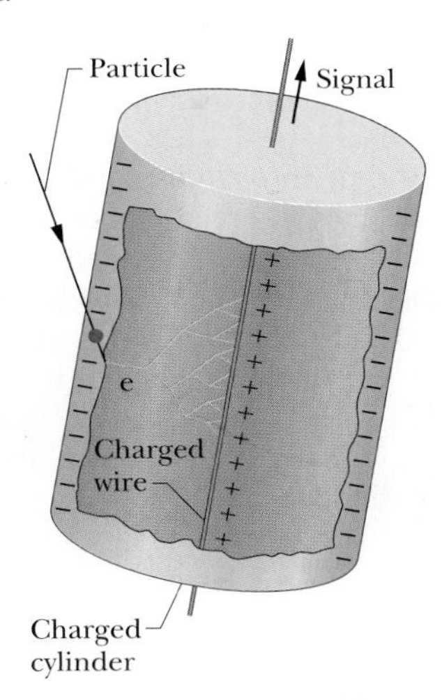

FIGURE 24-33 ■ Problem 30.

31. Charge Is Distributed Uniformly Charge is distributed uniformly throughout the volume of an infinitely long cylinder of radius R. (a) Show that, at a distance r from the cylinder axis (for $r < R$),

$$|\vec{E}| = \frac{|\rho| r}{2\varepsilon_0},$$

where $|\rho|$ is the amount of volume charge density. (b) Write an expression for $|\vec{E}|$ when $r > R$.

32. Parallel Sheets Figure 24-34 shows cross sections through two large, parallel, nonconducting sheets with identical distributions of positive charge with area charge density σ. What is $\vec{E}$ at points (a) above the sheets, (b) between them, and (c) below them?

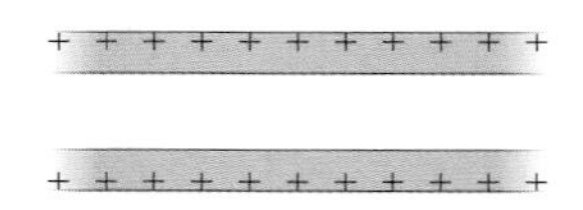

FIGURE 24-34 ■ Problem 32.

33. Square Metal Plate A square metal plate of edge length 8.0 cm and negligible thickness has a total charge of 6.0×10^{-6} C. (a) Estimate the magnitude E of the electric field just off the center of the plate (at, say, a distance of 0.50 mm) by assuming that the charge is spread uniformly over the two faces of the plate. (b) Estimate E at a distance of 30 m (large relative to the plate size) by assuming that the plate is a point charge.

34. Thin Metal Plates Two large, thin metal plates are parallel and close to each other. On their inner faces, the plates have excess surface charge of opposite signs. The amount of charge per unit area is given by $|\sigma| = 7.0 \times 10^{-22}$ C/m², with the negatively charged plate on the left. What are the magnitude and direction of the electric field $\vec{E}$ (a) to the left of the plates, (b) to the right of the plates, and (c) between the plates?

35. Ball on Thread In Fig. 24-35, a small, nonconducting ball of mass $m = 1.0$ mg and charge $q = 2.0 \times 10^{-8}$ C (distributed uniformly through its volume) hangs from an insulating thread that makes an angle $\theta = 30°$ with a vertical, uniformly charged nonconducting sheet (shown in cross section). Considering the gravitational force on the ball and assuming that the sheet extends far vertically and into and out of the page, calculate the surface charge density σ of the sheet.

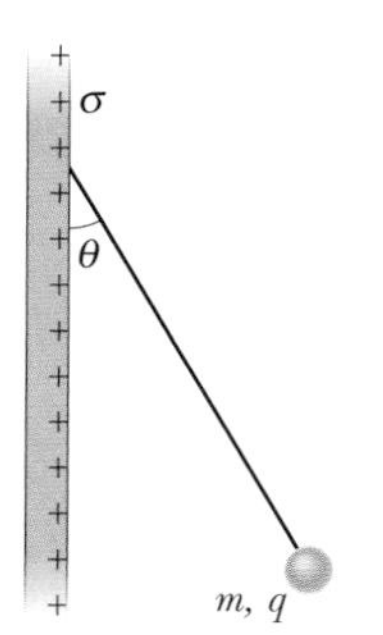

FIGURE 24-35 ■ Problem 35.

36. Large Metal Plates Two large metal plates of area 1.0 m² face each other. They are 5.0 cm apart and have equal but opposite charges on their inner surfaces. If the magnitude $|\vec{E}|$ of the electric field between the plates is 55 N/C, what is the amount of charge on each plate? Neglect edge effects.

37. An Electron Is Shot An electron is shot directly toward the center of a large metal plate that has excess negative charge with surface charge density -2.0×10^{-6} C/m². If the initial kinetic energy of the electron is 1.60×10^{-17} J and if the electron is to stop (owing to electrostatic repulsion from the plate) just as it reaches the plate, how far from the plate must it be shot?

38. Planar Slab A planar slab of thickness d has a uniform volume charge density ρ. Find the magnitude of the electric field at all points in space both (a) within and (b) outside the slab, in terms of x, the distance measured from the central plane of the slab.

SEC. 24-8 ■ A CHARGED ISOLATED CONDUCTOR

39. Photocopying Machine The electric field just above the surface of the charged drum of a photocopying machine has a magnitude $|\vec{E}|$ of 2.3×10^5 N/C. What is the surface charge density on the drum, assuming that the drum is a conductor?

40. Space Vehicles Space vehicles traveling through Earth's radiation belts can intercept a significant number of electrons. The resulting charge buildup can damage electronic components and disrupt operations. Suppose a spherical metallic satellite 1.3 m in diameter accumulates -2.4 μC of charge in one orbital revolution. (a) Find the resulting surface charge density. (b) Calculate the magnitude of the electric field just outside the surface of the satellite due to the surface charge.

41. Charged Sphere A uniformly charged conducting sphere of 1.2 m diameter has a surface charge density of 8.1 μC/m². (a) Find the net charge on the sphere. (b) What is the total electric flux leaving the surface of the sphere?

42. Arbitrary Shape Conductor An isolated conductor of arbitrary shape has a net charge of $+10 \times 10^{-6}$ C. Inside the conductor is a cavity within which is a point charge $q = +3.0 \times 10^{-6}$ C. What is the charge (a) on the cavity wall and (b) on the outer surface of the conductor?

Additional Problems

43. If/Can If the electric field in a region of space is zero, can you conclude there are no electric charges in that region? Explain.

44. If/Than If there are fewer electric field lines leaving a Gaussian surface than there are entering the surface, what can you conclude about the net charge enclosed by that surface?

45. Net Flux What is the net electric flux through each of the closed surfaces in Fig. 24-36 if the value of q is $+1.6 \times 10^{-19}$ C?

46. Net Flux Two What is the net electric flux through each of the closed surfaces in Fig. 24-37 if the value of q is 8.85×10^{-12} C? Explain the reasons for your answers.

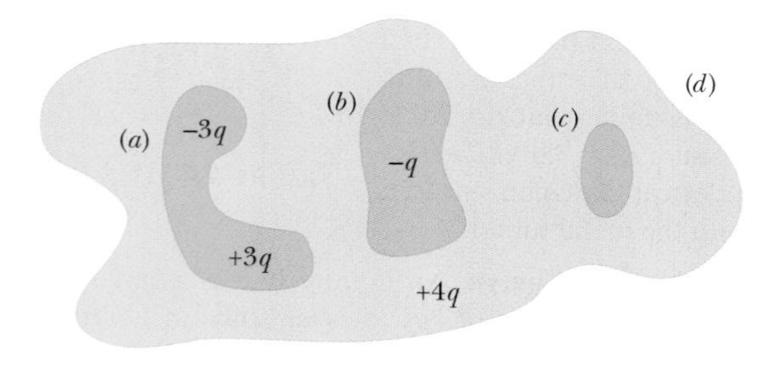

FIGURE 24-36 ■ Problem 45.

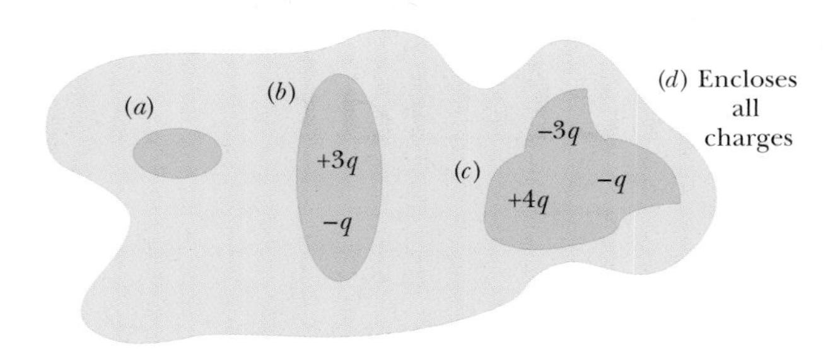

FIGURE 24-37 ■ Problem 46.

47. Fair Weather During fair weather, an electric field of about 100 N/C points vertically downward into Earth's atmosphere. Assuming that this field arises from charge distributed in a spherically symmetric manner over the surface of Earth, determine the *net* charge of Earth and its atmosphere if the radius of Earth and its atmosphere is 6.37×10^6 m.

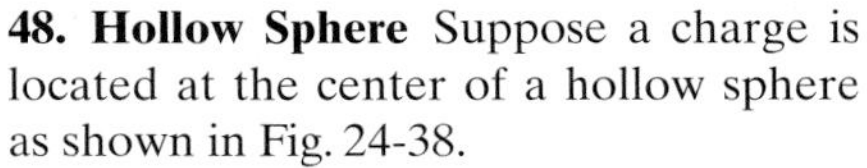

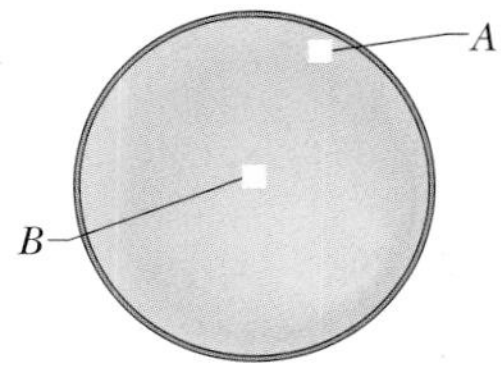

48. Hollow Sphere Suppose a charge is located at the center of a hollow sphere as shown in Fig. 24-38.

FIGURE 24-38 ■ Problem 48.

(a) Are the intersections of the field lines with the surface of the sphere uniformly distributed throughout? In other words, is the density of lines passing through the surface of the sphere uniform? Explain why or why not.

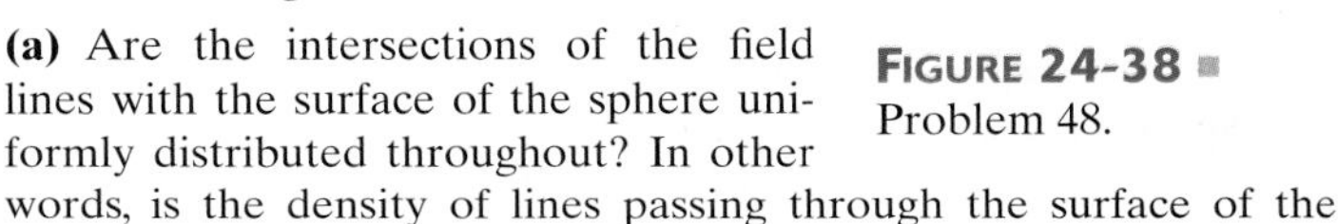

(b) Consider surface elements A and B, which have exactly the same area. Is the number of field lines passing through surface element A greater than, less than, or equal to the number of field lines through surface element B? Explain.
(c) Is the flux through surface element A greater than, less than, or equal to the flux through surface element B? Explain.

49. Center of Cube Suppose a charge is located at the center of the cube shown in Fig. 24-39.

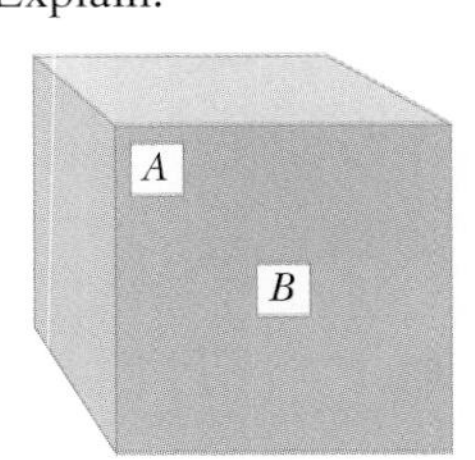

FIGURE 24-39 ■ Problem 49.

(a) Are the intersections of the field lines with a side of the cube uniformly distributed across the side? In other words, is the density of lines passing through the box uniform? Explain why or why not.
(b) Is the number of field lines through surface element A greater than, less than, or equal to the number of field lines through surface element B? Explain.
(c) Is the flux through surface element A greater than, less than, or equal to the flux through surface element B? Explain.

50. Using Gauss' Law Gauss' law is usually written as an equation in the form

$$\oint \vec{E} \cdot d\vec{A} = q^{\text{enc}}/\varepsilon_0.$$

(a) For this equation, specify what each term in this equation means and how it is to be calculated when doing some specific (but arbitrary—not a special case) calculation.

A long thin cylindrical shell like that shown in Fig. 24-40 has length L and radius R with $L \gg R$ and is uniformly covered with a charge Q. If we look for the field near the cylinder somewhere about the middle, we can treat the cylinder as if it were an infinitely long cylinder. Using this assumption, we can calculate the magnitude and direction of the field at a point a distance d from the *axis* of the cylinder (outside the cylindrical shell; i.e., $L \gg d > R$ but d not very close to R) using Gauss's law. Do so by explicitly following the steps below.

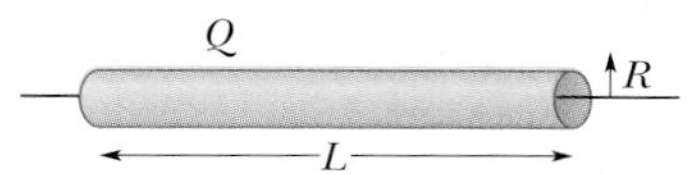

FIGURE 24-40 ■ Problem 50.

(b) Select an appropriate Gaussian surface. Explain why you chose it.
(c) Carry out the integral on the left side of the equation, expressing it in terms of the unknown value of the magnitude of the E field.
(d) What is the relevant value of q for your surface?
(e) Use your results in (c) and (d) in the equation and solve for the magnitude of E.

51. Interpreting Gauss Gauss' law states

$$\oint_A \vec{E} \cdot d\vec{A} = q_A/\varepsilon_0,$$

where A is a surface and q_A is a charge.

(a) Which of the following statements are true about the surface A appearing in Gauss' law for the equation to hold? You may list any number of these statements including all or none.

i. The surface A must be a closed surface (must cover a volume).
ii. The surface A must contain all the charges in the problem.
iii. The surface A must be a highly symmetrical surface like a sphere or a cylinder.
iv. The surface A must be a conductor.
v. The surface A is purely imaginary.
vi. The normals to the surface A must all be in the same direction as the electric field on the surface.

(b) Which of the following statements are true about the charge q_A appearing in Gauss' law? You may list any number of these statements including all or none.

i. The charge q_A must be all the charge lying on the Gaussian surface.
ii. The charge q_A must be the charge lying within the Gaussian surface.
iii. The charge q_A must be all the charge in the problem.
iv. The charge q_A flows onto the Gaussian surface once the surface is established.
v. The electric field E in the integral on the left of Gauss' law is due only to the charge q_A.
vi. The electric field E in the integral on the left on Gauss' law is due to all charges in the problem.

25 Electric Potential

At more than 115 years old, the Eiffel Tower is arguably the world's most famous landmark. Among other things, the tower is an engineering feat, a work of art, a scenic lookout, and a radio tower. Less well known is the tower's ability to protect people, trees, and other buildings from being struck by lightning that might emanate from thunderheads behind it.

How can the Eiffel Tower protect people from lightning?

The answer is in this chapter.

25-1 Introduction

In the last few chapters we have explored the nature of interaction forces between charged particles. We have developed the concept of electric field as a way to represent the forces a point charge would experience at any point in the space surrounding a collection of charges.

In certain situations it is difficult to understand the motions of charges in terms of an electric field. This difficulty is analogous to problems encountered in describing the motion of an object in the presence of gravitational forces. We developed the concepts of work and energy in Chapters 9 and 10 to deal with these problems. We will now investigate the application of the concepts of work and energy to situations in which the forces involved are electrostatic forces. In this chapter we develop the concept of electric potential—commonly referred to as voltage. We then explore some of its properties, including how charges are distributed on a metal conductor placed in an electric field.

Since the concept of potential or voltage is essential to an understanding of electric circuits, we will use the concept of electric potential in the next chapter to help us understand the role that batteries play in maintaining currents.

25-2 Electric Potential Energy

Newton's law for the gravitational force and Coulomb's law for the electrostatic force are mathematically similar. In Section 14-2 we saw that the gravitational force between two particle-like masses depends directly on the product of the masses and inversely with the square of the distance between them (Eq. 14-2). In like manner, the electrostatic forces between two point charges depend directly on the product of the charges and inversely with the square of the distance between them (Eq. 22-4). This similarity gives us a starting point in our search for additional useful concepts related to the interactions between charged objects. In this chapter, we consider whether some of the general features we have established for the gravitational force apply to the electrostatic force as well.

For example, the gravitational force is a *conservative force*. The work done by it is independent of the path along which an object moves. In experimental tests the work done by the electrostatic force has also been found to be *path independent*. If a charged particle moves from point i to point f while an electrostatic force is acting on it, the work W done by the force is the same for all paths between points i and f. Hence, we can infer that the electrostatic force is a conservative force as well.

Definition of Electric Potential Energy

In Chapter 10 we defined potential energy as the energy associated with the configuration of a system of objects that interact and hence exert forces on each other. We then proceeded to define gravitational potential energy as the negative of the amount of gravitational work objects in the system do on each other when their positions relative to one another change. From Eq. 10-5, $\Delta U \equiv -W^{\text{cons}}$ or $\Delta U^{\text{grav}} = -W^{\text{grav}}$ (Eq. 10-6). This general definition of work can be applied to a system of charges that interact by means of electrostatic forces.

Since electrostatic forces, like gravitational forces, are conservative, then it makes sense to assign an **electric potential energy change** ΔU to a system of interacting charges in a similar manner. If we cause or allow a system to change its configuration from an initial potential energy state U_1 to a different final state U_2, the internal electrostatic forces do a total amount of work W^{elec} on the particles in the system. As in

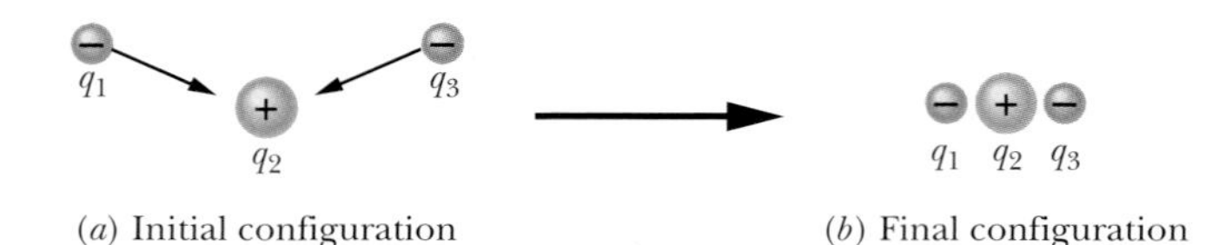

FIGURE 25-1 ■ (*a*) A system of three charges is in an initial configuration in which the charges are separated and have an electric potential energy U_1 associated with them. (*b*) Since both of the negative charges will be attracted to the positive charge, they will coalesce into a final configuration with potential energy U_2. The net electrostatic work the charges do on each other is positive so the system loses potential energy. Thus $U_2 < U_1$ so that $\Delta U < 0$.

Chapter 10, we define the potential energy change ΔU as the negative of the work the system does on itself when it undergoes the reconfiguration. This can be expressed symbolically as

$$\Delta U = U_2 - U_1 \equiv W^{\text{ext}} = -W^{\text{elec}}. \tag{25-1}$$

Figure 25-1 shows a system of charges losing electrostatic potential energy as a result of a natural reconfiguration. Figure 25-2 shows the same system gaining electrostatic potential energy as an external agent (doing positive work) causes the system to reconfigure. This results in the system doing negative work on itself.

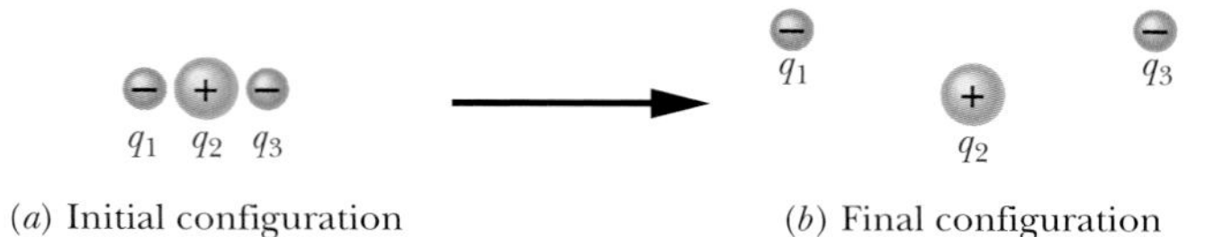

FIGURE 25-2 ■ (*a*) A system of three charges is in an initial configuration in which the charges are close together and have an electric potential energy U_1 associated with them. (*b*) Since q_1 and q_3 are both attracted to q_2, it will take positive external work, W^{ext}, to pull the charges apart. The net electrostatic work the charges do on each other is negative so the system gains potential energy. Thus $U_2 > U_1$ so that $\Delta U > 0$.

As you may recall, we determined that only differences in gravitational potential energy were physically significant. In Chapter 10 the system of masses we considered consisted of the Earth and a single object near its surface. We chose a convenient height at which to set the gravitational potential energy to zero. For example, we may have defined an Earth–object system as having zero potential energy when an object is at floor level or at the level of a tabletop. In doing so, we set the absolute scale for gravitational potential energy differently in different situations. This is legitimate since only potential energy *differences* are meaningful.

Potential energy difference is also of primary importance in keeping track of electric potential energy. Typically, we *define the electric potential energy of a system of charges to be zero when the particles are all infinitely separated from each other*, just as we did in Chapter 14 with the general form of gravitational potential energy. Using this zero of electric potential energy makes sense because the charges making up such a system have no interaction forces in that configuration. Using a standard reference potential (instead of moving it around as we typically do for Earth–object systems) allows us to find unique values of U_1 and U_2. For example suppose several charged particles come together from initially infinite separations (state 1) to form a system of nearby particles (state 2). Then using the conventional reference configuration, the initial potential energy U_1 is zero. If W^{elec} represents the internal work done by the electrostatic forces between particles during the move in from infinity, then from Eq. 25-1,

$$\Delta U = U_2 - U_1 = U_2 = -W^{\text{elec}}. \tag{25-2}$$

Since U_1 is zero, the final potential energy U_2 of the system can simply be denoted as U. Then, in terms of symbols,

$$U \equiv -W^{\text{elec}} \quad \text{(for initial potential energy = 0).} \tag{25-3}$$

As usual, the use of the symbol "≡" signifies that the expression is a definition.

External Forces and Energy Conservation

Since opposite charges attract, they will come together naturally if they are free to move. In these cases the charges "fall together" and the potential energy of the system of charges will be reduced. Similarly, like charges that are free will move apart and their potential energy will also be reduced. However, we can raise the potential energy of a system of charges by using energy from another system. Two common examples of external agents that can raise the potential energy of a system of charges are the Van de Graaff generator and the battery. Van de Graaff generators (see Fig. 25-3) use mechanical energy to force charges of like sign onto metal conductors. Batteries use chemical potential energy (which is actually a combination of electric and quantum effects) to force charges onto an electrode having the same sign charges.

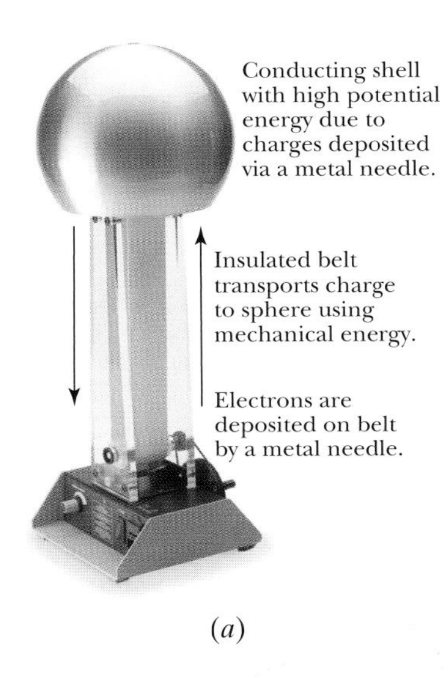
Conducting shell with high potential energy due to charges deposited via a metal needle.
Insulated belt transports charge to sphere using mechanical energy.
Electrons are deposited on belt by a metal needle.
(a)

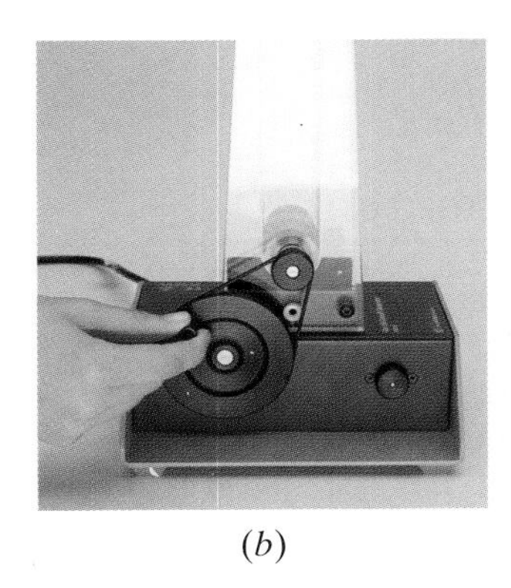
(b)

FIGURE 25-3 ■ A Van de Graaff generator uses mechanical energy from either (*a*) a motor or (*b*) a hand crank to transport charge to a conducting sphere, raising its potential energy. (Photo courtesy of PASCO scientific.)

Suppose an *external force* outside of the system under consideration causes a test particle of charge q to move from an initial location to a final location in the presence of an unchanging electric field generated by the source charges in the system. As the test charge moves, our outside force does work W^{ext} on the charge. At the same time, the electric field does work W^{elec} on it. By the work-kinetic energy theorem, the change ΔK in the kinetic energy of the particle is

$$\Delta K = K_2 - K_1 = W^{\text{ext}} + W^{\text{elec}}.$$

But since $W^{\text{elec}} = -\Delta U$,

$$\Delta K + \Delta U = W^{\text{ext}}. \tag{25-4}$$

Now suppose the particle is stationary before and after the move. Then K_2 and K_1 are both zero, and this reduces to

$$\Delta U = W^{\text{ext}} \quad \text{(for no kinetic energy change).} \tag{25-5}$$

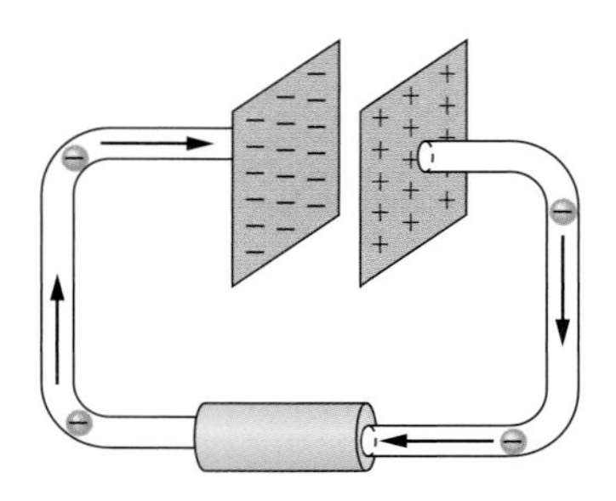

FIGURE 25-4 ■ A 1.5 V D-cell can act as an external agent that does the work needed to move electrons through a wire from a metal plate with excess positive charges to one with excess negative charges.

That is, the work, W^{ext}, done by our external force during the move is equal to the change in electric potential energy—provided there is no change in kinetic energy.

So in what direction will a positive or negative charge move if released? Will the charge move to raise or lower the potential energy of the system? The expression above can be used to determine this. For example, let the external force (perhaps the push or pull of your hand) do positive work. Recall from Section 9-4 the sign convention associated with work in general. If W^{ext} is positive, then ΔU must also be positive (by the equation above) and so we know that $U_2 > U_1$. In other words, the motion of a charge from a lower potential energy to a higher potential energy requires positive work to be done on the system. This motion would not happen if the particle were simply released. Spontaneous or naturally occuring motion is associated with reduced potential energy. This is very similar to the situations encountered in Section 9-8, where we considered an object under the influence of the Earth's attractive gravitational force. Often a battery does this work in a circuit, as in Fig. 25-4.

READING EXERCISE 25-1: Why is a configuration with charges separated by an infinite distance a good choice for our reference (zero) potential energy? Would a zero separation be equally good? Why or why not? ■

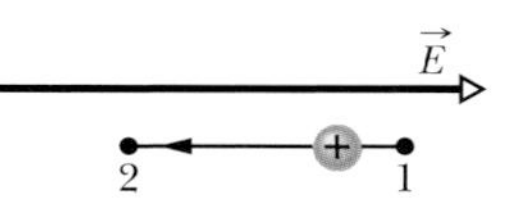

READING EXERCISE 25-2: In the figure, a proton moves from point 1 to point 2 in a uniform electric field directed as shown. (a) Does the electric field do positive or negative work on the proton? (b) Does the electric potential energy of the proton increase or decrease? (c) In this case we don't choose the potential energy to be zero at infinity. Why not? ■

25-3 Electric Potential

When considering gravitational potential energy we dealt primarily with a system consisting of the Earth and a single object much smaller than the Earth. If the object were to fall toward the Earth the interaction forces between them would be equal in magnitude, but as the object moves toward the Earth, the Earth's motion would be negligibly small. Thus the change in the system's gravitational potential energy would simply be the change in potential energy of the falling object. Similarly, as we did in Section 23-2, we can consider systems in which a small "test" charge moves in the presence of an electric field but does not change the electric field significantly. In these systems, the electric potential energy of the system can be calculated as the negative of work done by the electric field on a single test charge as we bring it to a location of interest from infinity.

> In the next several chapters we will focus primarily on systems in which the change in potential energy of a single test charge moving in an electric field is for all practical purposes the same as the change in potential energy of the entire system of charges.

This situation applies if the only charge that moves is our test charge. In this case, the electric field generated by the fixed source charges remains the same, so that the change in the system's potential energy will be proportional to the magnitude of the test charge. (This will not be true if other charges move.)

Defining Electric Potential

Recall that we defined and used the concept of electric field as the *electric force per unit charge* so we could easily analyze the forces experienced by a charge of any sign or magnitude. It is advantageous to develop an analogous concept for the determination of the electric potential energy of a system associated with the change in location of a test charge of any reasonable sign or magnitude. We will do that now, defining **electric potential** as a potential energy *per unit charge.* Once we have chosen a reference configuration with zero energy, our electric potential (potential energy per unit charge) has a unique value at any point in space. For example, suppose we move a test particle of positive charge $q_t = 1.60 \times 10^{-19}$ C from a location at infinity where the electric potential energy is defined as zero to a location in an electric field where the particle has an electric potential energy of 2.40×10^{-17} J. Then the change in electric potential, ΔV, of the system associated with the change in location of a test charge can be calculated as

$$\Delta V = V_2 - V_1 = \frac{2.40 \times 10^{-17}\ \text{J}}{1.60 \times 10^{-19}\ \text{C}} - 0 = 150\ \text{J/C}.$$

Next, suppose we replace that test particle with one having twice as much positive charge, 3.20×10^{-19} C. We would find that, at the same point, the second particle has an electric potential energy of 4.80×10^{-17} J, twice that of the first particle. However, the potential energy per unit charge or electric potential would be the same, still 150 J/C.

Thus, the system potential energy per unit charge, which can be symbolized as U/q_t, is independent of the charge q_t of the test particle we happen to be considering (Fig. 25-5). It is *characteristic only of the electric field* that is present. The potential energy per unit charge at a point in an electric field is defined as the electric potential V (or simply the **potential**) at that point. Thus V is defined as

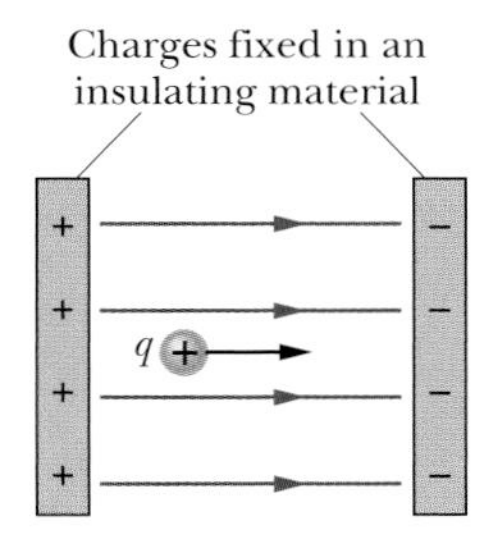

FIGURE 25-5 ■ A test charge moves in an electric field created by a stable configuration of source charges. If the test charge doesn't affect the electric field significantly as it changes location, the change in electric potential, ΔV, of the system (consisting of the source charges and the test charge) is due entirely to the work per unit charge done on the test charge by the electric field.

$$V \equiv \frac{U}{q}. \tag{25-6}$$

Note that potential energy and charge are both scalar quantities, so the electric potential is also a scalar, not a vector.

The *electric potential difference,* ΔV, associated with moving a charge q between any two points 1 and 2 in an electric field is equal to the difference between the potential energy per unit charge at the two points:

$$\Delta V = V_2 - V_1 = \frac{U_2}{q} - \frac{U_1}{q} = \frac{\Delta U}{q}. \tag{25-7}$$

Using $\Delta U = U_2 - U_1 = -W^{\text{elec}}$ (Eq. 25-1) to substitute the work done by electrostatic forces $-W^{\text{elec}}$ for ΔU in the equation above, we can define the potential difference between points 1 and 2 as

$$\Delta V = V_2 - V_1 \equiv -\frac{W^{\text{elec}}}{q} \quad \text{(potential difference defined).} \tag{25-8}$$

That is, the potential difference between two points is the negative of the work done by the electrostatic force to move a unit charge from one point to the other. A potential difference can be positive, negative, or zero, depending on the signs and magnitudes of the charge q and the electrostatic work W^{elec}.

As we already mentioned, we have set $U_1 = 0$ infinitely far from any charges as our reference potential energy. So since $V \equiv U/q$ (Eq. 25-6), the electric potential must also be zero there. Then using Eq. 25-8, we can define the electric potential V (measured relative to infinity) at any point in an electric field to be

$$V = -\frac{W_{\infty}^{\text{elec}}}{q} \quad \text{(potential defined relative to infinity)}, \tag{25-9}$$

where W_{∞}^{elec} is the work done by the electrostatic force on a charged particle as that particle moves in from infinity to point f. As was the case with potential difference ΔV, a potential V can be positive, negative, or zero, depending on the signs and magnitudes of q and W_{∞}^{elec}.

The SI unit for electric potential that follows from Eq. 25-9 is the joule per coulomb. This combination occurs so often that a special unit, the *volt* (abbreviated V) is used to represent it. Thus,

$$1 \text{ volt} \equiv 1 \text{ joule/coulomb}. \tag{25-10}$$

Although the terms electric potential energy and electric potential are very similar, they are not the same thing. This is probably one of the reasons why it is so common to refer to electric potential as **voltage** after its unit—the volt.

This new unit called the volt allows us to adopt a more conventional unit for the electric field $\vec{E}$, which we have measured up to now in newtons per coulomb. With two unit conversions, we obtain

$$1 \text{ N/C} = \left[1\frac{\text{N}}{\text{C}}\right]\left[\frac{1 \text{ V}}{1 \text{ J/C}}\right]\left[\frac{1 \text{ J}}{1 \text{ N}\cdot\text{m}}\right] = 1 \text{ V/m}. \tag{25-11}$$

The conversion factor in the second set of parentheses comes from Eq. 25-10, and that in the third set of parentheses is derived from the definition of the joule. From now on, we shall express values of the electric field in volts per meter rather than in newtons per coulomb.

The Electron Volt

Because we often have situations in which the charges involved are very small (a few times the charge of an electron), we define an energy unit that is a convenient one for energy measurements in the atomic and subatomic domain. One *electron-Volt* (eV) is the energy equal to the work required to move a single positive elementary charge e (the charge magnitude of the electron or the proton) through a potential difference of exactly one volt. Equation 25-8,

$$\Delta V = V_2 - V_1 = -\frac{W^{\text{elec}}}{q} = \frac{-W^{\text{elec}}}{e},$$

tells us that the magnitude of this work is $e\,\Delta V$, so

$$\begin{aligned} 1 \text{ eV} &= e(1 \text{ V}) \\ &= (1.60 \times 10^{-19} \text{ C})(1 \text{ J/C}) = 1.60 \times 10^{-19} \text{ J}, \end{aligned} \tag{25-12}$$

where the units for electron volt are joules because it is actually a unit of energy rather than electric potential.

READING EXERCISE 25-3: In the figure shown in Reading Exercise 25-2, we moved a proton from point i to point f in a uniform electric field directed as shown. (a) Does our external force do positive or negative work? (b) Does the proton move to a point of higher or lower potential? ■

TOUCHSTONE EXAMPLE 25-1: Electron Motion

An electron starts from rest at a point in space at which the electric potential is 9.0 V. If the only force acting on the electron is that associated with the electric potential, how fast will the electron be moving when it passes a second point in space where the electric potential is 10.0 V?

SOLUTION ■ First we need to convince ourselves that this problem describes a physical situation that is even possible. Equation 25-4 tells us that $\Delta K + \Delta U = W^{\text{ext}}$. Since $W^{\text{ext}} = 0$ here and since ΔK must be positive if the electron speeds up, this means that ΔU must be negative. But is it? After all, ΔV is positive here since $V_2 - V_1 = +1.0$ V. However, the charge of the electron is negative, so Eq. 25-7 tells us that:

$$\Delta U = q\Delta V = (-e)\Delta V = (-e)(+1.0\text{ V}) = -1.0\text{ eV},$$

so that
$$\Delta K = -\Delta U = -(-1.0\text{ eV}) = +1.0\text{ eV},$$

which is positive.

The **Key Idea** here is that a negative charge *loses* potential energy and *gains* kinetic energy when it moves from a region of *lower* potential to a region of *higher* potential. This is just the opposite of what would happen to a positive charge!

Now that we know the electron has 1.0 eV of kinetic energy, we need to determine how fast it is going. The **Key Idea** here is that 1 eV $= 1.60 \times 10^{-19}$ J (Eq. 25-12). Then

$$\Delta K = K_2 - K_1 = (\tfrac{1}{2})mv_2^2 - 0,$$

which gives us

$$\begin{aligned} v_2 &= \sqrt{2\Delta K/m} \\ &= \sqrt{2(1.0\text{ eV})(1.60\times10^{-19}\text{ J/eV})/(9.1\times10^{-31}\text{ kg})} \\ &= 5.9\times10^5\text{ m/s.} \end{aligned} \qquad \text{(Answer)}$$

25-4 Equipotential Surfaces

We are interested in what our knowledge of electric potential can tell us about how small test charges might move. We can infer from the discussion above that charged particles will not spontaneously move from one point to another point of equal potential. This is quite analogous to movement of mass in a gravitational field. A skier on a flat surface with no kinetic energy will not spontaneously move from one part of the surface to another. On the other hand, if the skier is on a slope and is free to move, the skier will spontaneously start moving down the slope, from higher to lower potential energy. Thus, it would be useful to know where all the points of equal potential energy are in a given region of space. That way, we can easily infer the directions of the forces on each of the charges. An **equipotential surface** is defined as a surface having the same potential at all points on it. Topographical maps show equipotential surfaces (lines on a two-dimensional map) in regard to gravitational potential energy.

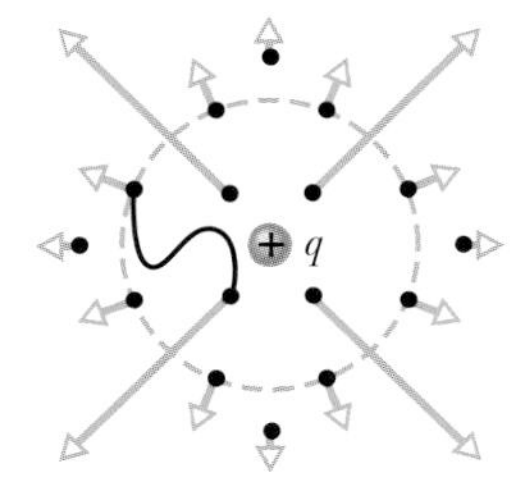

FIGURE 25-6 ■ All of the electric field vectors created by the presence of a single charge point radially outward in three dimensions. If a test charge moves around on a sphere that is centered on the charge (where the dashed circle shows a cross section of the sphere), no work is done on it by the electric field since all the electric field vectors on surface elements of the sphere are normal to the sphere. If the charge is moved from one radius to the other (black squiggly line) it has to move parallel to the field vectors some of the time, and work is done on it.

Let's consider the electric field associated with a source consisting of a single fixed point charge we designate as the source charge. What happens if we place a test charge at a distance r from the source charge and move it around? If we move the charge anywhere on the surface of a sphere of radius r, no electrostatic work is done on the test charge as it is always moving perpendicular to the electric field vectors. However, we cannot move our test charge from one distance from the source charge to another distance without the electric field doing work on it. This is illustrated in Fig. 25-6. Thus, any sphere centered on the source charge is an equipotential surface. If our source charge is positive, then the potential decreases as the distance from the source charge increases. We know this because from $\Delta U = -W^{\text{elec}}$ (Eq. 25-1) a charge naturally moves from high potential energy to low. Thus the equipotential surfaces associated with a positive point charge consist of an infinite family of concentric spheres centered on the source charge. Each sphere has a different potential.

An equipotential surface can be either imaginary, such as a mathematical sphere, or a real, physical surface such as the outside of a wire. The set of all equipotential surfaces fills all of space, since every point in space has some value of electric potential associated with it. We could draw an equipotential surface through any one of these points, just like we can draw a field line through every point in space. However, in order to simplify illustrations and diagrams, we typically show just a few of the surfaces.

No work W^{elec} is done on a charged particle by an electric field when the particle moves between two points i and f on the same equipotential surface. This follows from Eq. 25-8,

$$\Delta V = V_2 - V_1 = -\frac{W^{\text{elec}}}{q},$$

which tells us W^{elec} must be zero if $V_2 = V_1$. Because of the path independence of work (and thus of potential energy and potential), $W^{\text{elec}} = 0$ for *any* path connecting points 1 and 2, regardless of whether that path lies entirely on the equipotential surface. In other words, if the charge moves away from the equipotential surface during the motion, the work done (positive or negative) is exactly canceled by the work done (negative or positive) in moving back onto the surface.

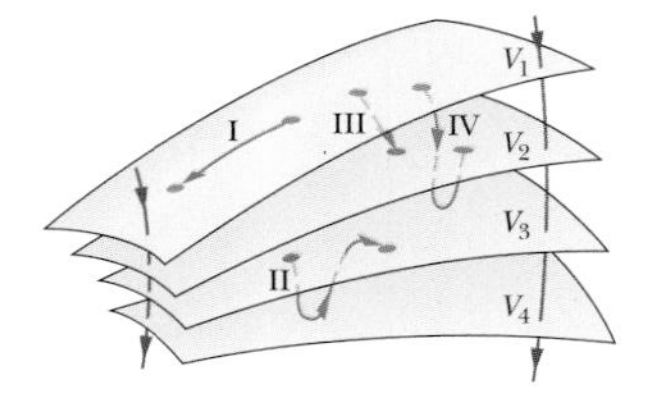

FIGURE 25-7 ■ Portions of four equipotential surfaces at electric potentials $V_1 = 100$ V, $V_2 = 80$ V, $V_3 = 60$ V, and $V_4 = 40$ V. Four paths along which a test charge may move are shown. Two electric field lines are also indicated.

Figure 25-7 shows a *family* of equipotential surfaces associated with the electric field due to some distribution of charges. The work done by the electrostatic force on a charged particle as the particle moves from one end to the other of paths I and II is zero because each of these paths begins and ends on the same equipotential surface. The work done as the charged particle moves from one end to the other of paths III and IV is not zero but has the same value for both these paths because the initial and final potentials are identical for the two paths; that is, paths III and IV connect the same pair of equipotential surfaces.

As we already noted, the equipotential surfaces produced by a point charge or a spherically symmetrical charge distribution are a family of concentric spheres. For a uniform electric field it is not difficult to see that the equipotential surfaces are a family of planes perpendicular to the field lines.

The fact that the value of the potential is constant along an equipotential surface implies that the electric field must always be perpendicular to the equipotential surfaces. Why? Because, if $\vec{E}$ were *not* perpendicular to an equipotential surface, it would have a component lying along that surface. This component would then do work on a

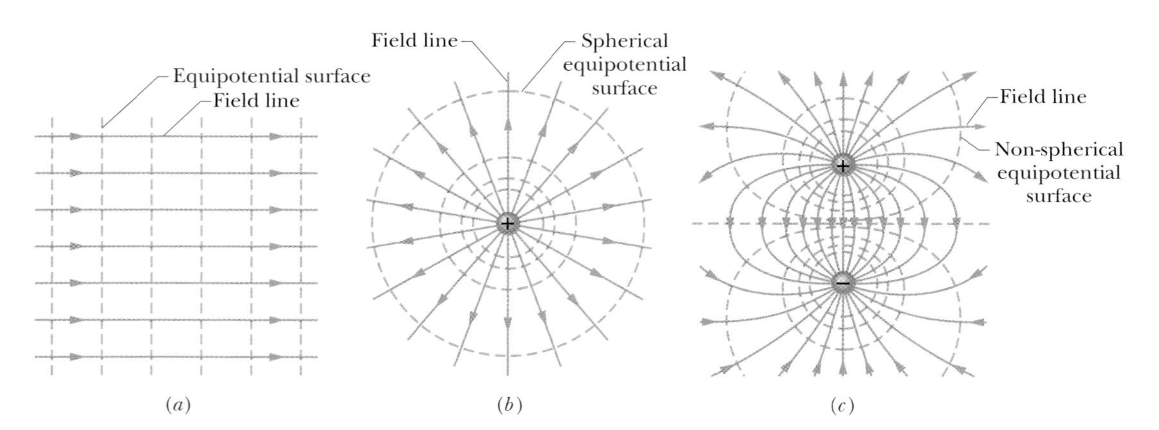

FIGURE 25-8 ■ Electric field lines (solid purple lines with arrows) and cross sections of equipotential surfaces (dashed gold lines) for (a) a uniform field with planar equipotential surfaces, (b) the field of a point charge with spherical equipotential surfaces, and (c) the field of an electric dipole with distorted equipotential surfaces that are not quite spherical.

charged particle as it moved along the surface. However, to prove that work cannot be done if the surface is truly an equipotential surface we use Eq. 25-8 once again,

$$\Delta V = V_2 - V_1 = -\frac{W^{\text{elec}}}{q}.$$

The only possible conclusion is that the electric field lines must be perpendicular to the surface everywhere along it.

If electric field lines are perpendicular to an equipotential surface, then conversely the equipotential surface must be perpendicular to the field lines. Thus, equipotential surfaces are always perpendicular to the direction of the electric field $\vec{E}$, which is tangent to the field lines. Figure 25-8 shows electric field lines and cross sections of the equipotential surfaces for a uniform electric field and for the field associated with a point charge and with an electric dipole.

25-5 Calculating Potential from an *E*-Field

Can we calculate the potential difference between any two points 1 and 2 in an electric field if we know the electric field vector $\vec{E}$ all along any path connecting those points? We can if we can find the work done on a charge by the field as the charge moves from 1 to 2, and then use Eq. 25-8 again,

$$\Delta V = V_2 - V_1 = -\frac{W^{\text{elec}}}{q}.$$

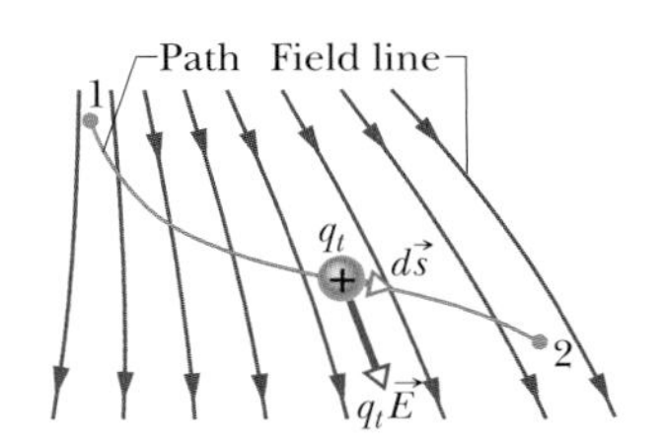

FIGURE 25-9 ■ A test charge q_t moves from point 1 to point 2 along the path shown in a nonuniform electric field represented by curved electric field lines. During a displacement $d\vec{s}$, an electrostatic force $q_t\vec{E}$ acts on the test charge. This force points in the direction of the field line at the location of the test charge.

For example, consider an arbitrary electric field, represented by the field lines in Fig. 25-9, and a positive test charge q_t moving along the path shown from point 1 to point 2. At any point on the path, an electrostatic force $q_t\vec{E}$ acts on the charge as it moves through an infinitesimally small differential displacement $d\vec{s}$. From Chapter 9, we know the differential work dW done on a particle by a force $\vec{F}$ during a displacement $d\vec{s}$ is

$$dW = \vec{F} \cdot d\vec{s}. \tag{25-13}$$

For the situation of Fig. 25-9, $\vec{F} = q_t\vec{E}$, and Eq. 25-13 becomes

$$dW^{\text{elec}} = q_t\vec{E} \cdot d\vec{s}. \tag{25-14}$$

To find the total work W^{elec} done on the particle by the field as the particle moves from point 1 to point 2, we sum—via integration—the differential work done on the charge as it moves through all the differential displacements $d\vec{s}$ along the path:

$$W^{\text{elec}} = q_t\int_1^2 \vec{E} \cdot d\vec{s}. \tag{25-15}$$

If we substitute the total electrical work W^{elec} from Eq. 25-15 into Eq. 25-8, $\Delta V = V_2 - V_1 = -W^{\text{elec}}/q$, we find

$$V_2 - V_1 = -\int_1^2 \vec{E} \cdot d\vec{s}. \tag{25-16}$$

Thus, the potential difference $V_2 - V_1$ between any two points 1 and 2 in an electric field is equal to the negative of the *line integral* (meaning the integral along a particular path) of $\vec{E} \cdot d\vec{s}$ from 1 to 2. However, because the electrostatic force is conservative, all paths (whether easy or difficult to use) yield the same result. So, choose an easy-to-use path.

If the electric field is known throughout a certain region, Eq. 25-16 allows us to calculate the difference in potential between any two points in the field. If we choose the potential V_1 at point 1 to be zero, then Eq. 25-16 becomes

$$V = -\int_1^2 \vec{E} \cdot d\vec{s} \quad \text{(for } V_1 = 0\text{)}, \tag{25-17}$$

where we have dropped the subscript 2 on V_2. Equation 25-17 gives us the potential V at any point 2 in the electric field *relative to the zero potential* at point 1. If we let point 1 be at infinity, then Eq. 25-17 gives us the potential V at any point 2 relative to the zero potential at infinity.

READING EXERCISE 25-4: The figure shows a family of parallel equipotential surfaces (in cross section) and five paths along which we shall move an electron from one surface to another. (a) What is the direction of the electric field associated with the surfaces? (b) For each path, is the work we do positive, negative, or zero? (c) Rank the paths according to the work we do, greatest first.

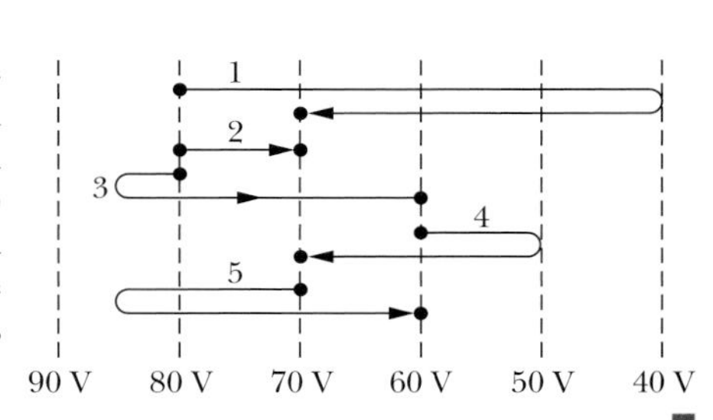

TOUCHSTONE EXAMPLE 25-2: Finding the Potential Difference

(a) Figure 25-10*a* shows two points 1 and 2 in a uniform electric field $\vec{E}$. The points lie on the same electric field line (not shown) and are separated by a distance d. Find the potential difference $V_2 - V_1$ by moving a positive test charge q_t from 1 to 2 along the path shown, which is parallel to the field direction.

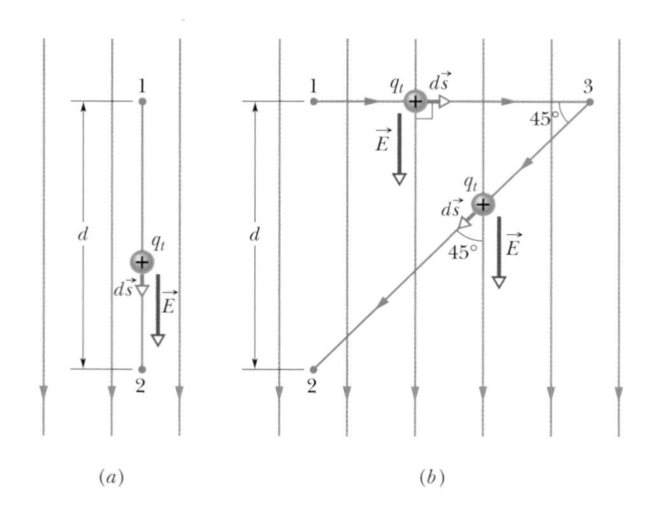

FIGURE 25-10 (*a*) A test charge q_t moves in a straight line from point 1 to point 2, along the direction of a uniform electric field. (*b*) Charge q_t moves along path 1-3-2 in the same electric field.

SOLUTION The **Key Idea** here is that we can find the potential difference between any two points in an electric field by integrating $\vec{E} \cdot d\vec{s}$ along a path connecting those two points according to Eq. 25-16. We do this by mentally moving a test charge q_t along that path, from initial point 1 to final point 2. As we move such a test charge along the path in Fig. 25-10*a*, its differential displacement $d\vec{s}$ always has the same direction as $\vec{E}$. Thus, the angle ϕ between $\vec{E}$ and $d\vec{s}$ is zero and the dot product in Eq. 25-16 is

$$\vec{E} \cdot d\vec{s} = |\vec{E}||d\vec{s}|\cos\phi = |\vec{E}||d\vec{s}|. \tag{25-18}$$

Equations 25-16 and 25-18 then give us

$$V_2 - V_1 = -\int_1^2 \vec{E} \cdot d\vec{s} = -\int_1^2 |\vec{E}||d\vec{s}|. \tag{25-19}$$

Since the field is uniform, E is constant over the path and can be moved outside the integral, giving us

$$V_2 - V_1 = -|\vec{E}|\int_1^2 |d\vec{s}| = -|\vec{E}||\vec{d}|,$$

in which the integral is simply the length d of the path. The minus sign in the result shows that the potential at point 2 in Fig. 25-10*a* is lower than the potential at point 1. This is a general result: The potential always decreases along a path that extends in the direction of the electric field lines.

(b) Now find the potential difference $V_2 - V_1$ by moving the positive test charge q_t from 1 to 2 along the path 1-3-2 shown in Fig. 25-10*b*.

SOLUTION ■ The **Key Idea** of (a) applies here too, except now we move the test charge along a path that consists of two lines: 1-3 and 3-2. At all points along line 1-3, the displacement $\vec{ds}$ of the test charge is perpendicular to $\vec{E}$. Thus, the angle ϕ between $\vec{E}$ and $\vec{ds}$ is 90°, and the dot product $\vec{E}\cdot\vec{ds}$ is 0. Equation 25-16 then tells us that points 1 and 3 are at the same potential: $V_3 - V_1 = 0$.

For line 3-2 we have $\phi = 45°$ and, from Eq. 25-16,

$$V_2 - V_1 = V_2 - V_3 = -\int_3^2 \vec{E}\cdot\vec{ds} = -\int_3^2 |\vec{E}|(\cos 45°)|\vec{ds}|$$

$$= -|\vec{E}|(\cos 45°)\int_3^2 |\vec{ds}|.$$

The integral in this equation is just the length of line 3-2; from Fig. 25-10*b*, that length is d/sin 45°. Thus,

$$V_2 - V_1 = -|\vec{E}|(\cos 45°)\frac{|\vec{d}|}{\sin 45°} = -|\vec{E}||\vec{d}|. \qquad \text{(Answer)}$$

This is the same result we obtained in (a), as it must be; the potential difference between two points does not depend on the path connecting them. Moral: When you want to find the potential difference between two points by moving a test charge between them, you can save time and work by choosing a path that simplifies the use of Eq. 25-16.

25-6 Potential Due to a Point Charge

Imagine a single point charge in space. What would the value of the potential be at a distance of 3 m away from the charge? Consider a point P at a distance R from a fixed particle of positive charge q as in Fig. 25-11. To use Eq. 25-16,

$$V_2 - V_1 = -\int_1^2 \vec{E}\cdot\vec{ds},$$

we imagine that we move a positive test charge q_t from infinity to its final location at point P. We need to bring our test charge from infinity to a point P that is a distance R from the source charge. Because the path we choose will not change our final result, we are free to choose it. Mathematically, the simplest path between infinity and point P involves traveling along the same line that the electric field vectors lie along so no nonradial vector components of the electric field have to be considered.

We must then evaluate the dot product

$$\vec{E}\cdot\vec{ds} = |\vec{E}|\cos\phi|\vec{ds}| = (E)(ds)\cos\phi. \qquad (25\text{-}20)$$

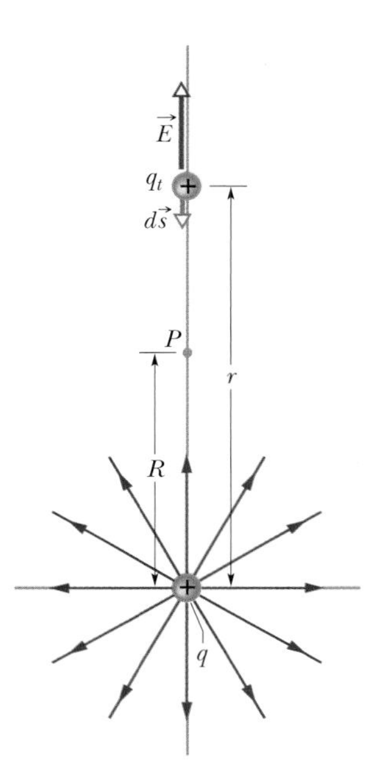

FIGURE 25-11 ■ The positive point charge q produces an electric field $\vec{E}$ and an electric potential ΔV at point P. We find the potential by moving a test charge q_t from its initial location at infinity to a point P. The test charge is shown at distance r from the point charge undergoing differential displacement $\vec{ds}$.

The electric field $\vec{E}$ in Fig. 25-11 is directed radially outward from the fixed particle. So the differential displacement $\vec{ds}$ of the test particle along our chosen path is radially inward and has the opposite direction as $\vec{E}$. That means that the angle $\phi = 180°$ and $\cos\phi = -1$. Because the path is radial, let us write ds as dr. Then, substituting the limits ∞ and R, we can write Eq. 25-16,

$$V_2 - V_1 = -\int_1^2 \vec{E}\cdot\vec{ds},$$

as

$$V_2 - V_1 = -\int_\infty^R E_r\,dr, \qquad (25\text{-}21)$$

where E_r is the component of the electric field in the radial direction. Next we set $V_1 = 0$ (at ∞) and $V_2 = V$ (at R). Then, for the magnitude of the electric field at the site of the test charge, we substitute from Chapter 23:

$$E = k\frac{|q|}{r^2}. \qquad (25\text{-}22)$$

With these changes, Eq. 25-21 then gives us

$$V - 0 = -k|q|\int_{\infty}^{R} \frac{1}{r^2}\,dr = k|q|\left[\frac{1}{r}\right]_{\infty}^{R}$$

$$= k\frac{|q|}{R} \qquad \text{(for positive } q\text{)}. \qquad (25\text{-}23)$$

We want to generalize finding the potential relative to infinity for any distance, not just distance R. So, switching from R to r, we have an expression for potential at a distance r from a source charge of

$$V = k\frac{|q|}{r} \qquad \text{(for positive } q\text{)}.$$

Although we have derived this expression above for a positively charged particle, the derivation also holds for a negatively charged particle as well. However, if q in Fig. 25-11 were a negative charge, the electric field vectors would point in the same direction as the path (radially inward). Thus, the differential displacement $d\vec{s}$ of the test particle along our chosen path has the same direction as $\vec{E}$. That means the angle $\phi = 0°$ and so $\cos\phi = +1$. This introduces a negative sign that remains throughout the derivation and results in a negative final result for the potential. So, we conclude that *the sign of V is the same as the sign of q*. This gives us

$$V = k\frac{q}{r} \qquad \text{(relative to infinity for either sign of charge)}, \qquad (25\text{-}24)$$

as the electric potential V relative to infinity due to a particle of charge q at any radial distance r from the particle.

A positively charged particle produces a positive electric potential. A negatively charged particle produces a negative electric potential.

Figure 25-12 shows a computer-generated plot of Eq. 25-24 for a positively charged particle; the magnitude of V is plotted vertically. Note that the magnitude increases as $r \to 0$. In fact, according to the expression above, V is infinite at $r = 0$, although Fig. 25-12 shows a finite, smoothed-off value there.

Equation 25-24 also gives the electric potential *outside or on the external surface of* a spherically symmetric charge distribution. We can prove this by using an

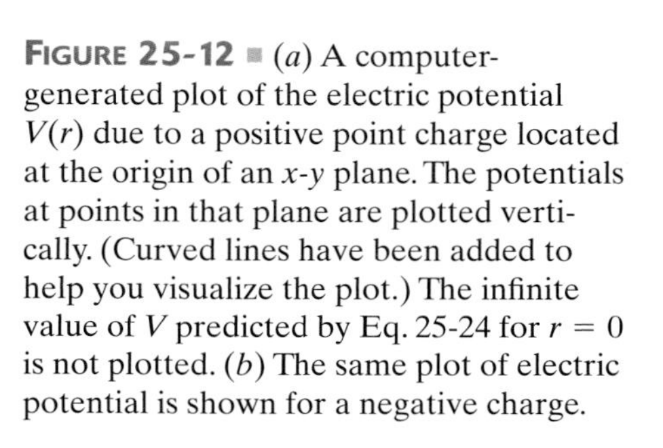

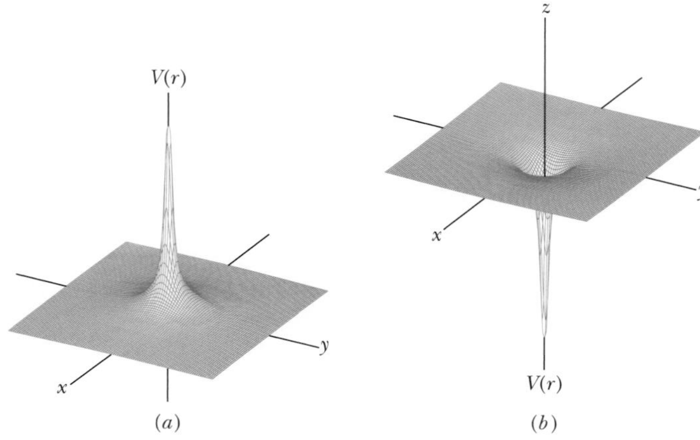

FIGURE 25-12 ■ (*a*) A computer-generated plot of the electric potential $V(r)$ due to a positive point charge located at the origin of an x-y plane. The potentials at points in that plane are plotted vertically. (Curved lines have been added to help you visualize the plot.) The infinite value of V predicted by Eq. 25-24 for $r = 0$ is not plotted. (*b*) The same plot of electric potential is shown for a negative charge.

electrostatic analogy to the shell theorem we found so useful in our study of gravitation (Section 14-2). This theorem allows us to replace the actual spherical charge distribution with an equal charge concentrated at its center. Then the derivation leading to Eq. 25-24 follows, provided we do not consider a point within the actual distribution.

TOUCHSTONE EXAMPLE 25-3: Near a Proton

The nucleus of a hydrogen atom consists of a single proton, which can be treated as a particle (or point charge).

(a) With the electric potential equal to zero at infinite distance, what is the electric potential V due to the proton at a radial distance $r = 2.12 \times 10^{-10}$ m from it?

SOLUTION ■ The **Key Idea** here is that, because we can treat the proton as a particle, the electric potential V it produces at distance r is given by Eq. 25-24,

$$V = k\frac{q}{r}.$$

Here charge q is $e(=1.6 \times 10^{-19}\text{ C})$. Substituting this and the given value for r, we find

$$V = \frac{(8.99 \times 10^9\text{ N}\cdot\text{m}^2/\text{C}^2)(1.60 \times 10^{-19}\text{ C})}{2.12 \times 10^{-10}\text{ m}}$$

$$= 6.78\text{ V}. \qquad \text{(Answer)}$$

(b) What is the electric potential energy U in electron-volts of an electron at the given distance from the nucleus? (The potential energy is actually that of the electron–proton system—the hydrogen atom.)

SOLUTION ■ The **Key Idea** here is that when a particle of charge q is located at a point where the electric potential due to other charges is V, the electric potential energy U is given by Eq. 25-6 ($V = U/q$). Using the electron's charge $-e$, we find

$$U = qV = (-1.60 \times 10^{-19}\text{ C})(6.78\text{ V})$$

$$= -1.0848 \times 10^{-18}\text{ J} = -6.78\text{ eV}. \qquad \text{(Answer)}$$

(c) If the electron moves closer to the proton, does the electric potential energy increase or decrease?

SOLUTION ■ The **Key Ideas** of parts (a) and (b) apply here also. As the electron moves closer to the proton, the electric potential V due to the proton at the electron's position increases because r decreases). Thus, the value of V in part (b) increases. Because the electron is negatively charged, this means that the value of U becomes more negative. Hence, the potential energy U of the electron (that is, of the system or atom) decreases.

25-7 Potential and Potential Energy Due to a Group of Point Charges

Now let's consider what happens when there are lots of charges. First we will look at the case where we only move a small test charge while all the other charges remain fixed. In this case, the changes in the system's potential energy as the test charge moves lead us to the same definition of electric potential, V, as we already developed. At the end of this section we consider the situation in which many charges move and find that the total potential energy of the system changes. In this case, even though the system has a potential energy associated with it, we cannot define an electric potential.

We found in Chapter 23 that the electric field arising from a group of point charges satisfies a superposition principle. That is, the total electric field is the sum of the individual electric fields arising from each individual point charge. Since the potential V is the line integral of the electric field and the integral of a sum of terms is the sum of the integrals, the superposition principle also holds for electrostatic potential.

Hence, we use the principle of superposition to find the electric potential at a particular location due to a group of point charges. We calculate the potential resulting from the influence of each charge in the system one at a time, using Eq. 25-24 with the sign of the charge included. Then we sum the potentials. For n charges, the net potential (measured relative to a zero at infinity) is

$$V = \sum_{i=1}^{n} V_i = k\sum_{i=1}^{n} \frac{q_i}{r_i} \quad \text{(potential due to } n \text{ point charges).} \tag{25-25}$$

Here q_i is the value of the ith charge, and r_i is the radial distance of the given point from the ith charge. The sum in Eq. 25-25 is an *algebraic sum,* not a vector sum like the sum used to calculate the electric field resulting from a group of point charges. Herein lies an important computational advantage of potential over electric field: it is a lot easier to sum several scalar quantities than to sum several vector quantities whose directions and components must be considered.

In Section 25-2, we discussed the electric potential energy of a charged particle as an electrostatic force does work on it. In that section, we assumed that the charges that produced the force were fixed in place, so that neither the force nor the corresponding electric field could be influenced by the presence of the test charge. If we consider a system with charges that move when a test charge moves around, there is no logical way to determine a charge-independent electric potential for it. The electric field will keep changing due to the presence of the test charge. But we can take a broader view and find the electric potential energy of the entire *system* of charges due to the electric field produced *by* those same charges.

We can start simply by pushing two bodies that have charges of like sign into the same vicinity. For example, imagine that we have one excess electron on the conducting shell of the Van de Graaff generator shown in Fig. 25-3 and we want to put a second electron in place. Our second electron is sprayed on the insulated belt and the generator motor does work as it forces the second electron toward the conducting shell in the presence of the first one. The first electron is no doubt relocating and acting on the second electron during the forcing process. Nonetheless, we can keep track of the work the motor does. This work is stored as electric potential energy in the two-body system (provided the kinetic energy of the bodies does not change). As we bring up a third electron we can measure the work we have to bring it up to the shell in the presence of the other two electrons, which are relocating as a result of the interactions of all three electrons. The work needed to bring the third electron to the conducting shell adds to the work needed to bring up the second electron. The total work is stored as the potential energy of the three-body system. This process of doing more work and causing the excess electrons on the shell to relocate goes on until there are billions and billions of electrons on the conducting shell. If you later release the charges but touch the shell with a conductor attached to the ground, you can recover this stored energy, in whole or in part, as the kinetic energy of the charged bodies as they rush away from each other.

We define the electric potential energy *of a system of point charges* in terms of the final locations of all the charges as follows:

> The electric potential energy of a system of point charges that are not moving is equal to the work that must be done by an external agent to assemble the system one charge at a time.

We assume that the charges are stationary both in their initial infinitely distant positions and in their final assembled configuration. In equation form, the total electric potential energy of the system is given by the sum of the potential energies of all the possible pairs in the system so that

$$U = \sum_{\text{all pairs}} k \frac{q_i q_j}{|\vec{r}_i - \vec{r}_j|} \quad \text{(system potential energy of } n \text{ point charges).} \tag{25-26}$$

READING EXERCISE 25-5: So far in this chapter, we have discussed two ways to calculate the electric potential V. Describe how one would calculate the electric potential given information about the charge distribution (the magnitudes of the charges and where they are located). Describe how one would calculate the electric potential given information regarding the electric field $\vec{E}$. ■

READING EXERCISE 25-6: The figure shows three arrangements of two protons. Rank the arrangements according to the net electric potential produced at point P by the protons, greatest first.

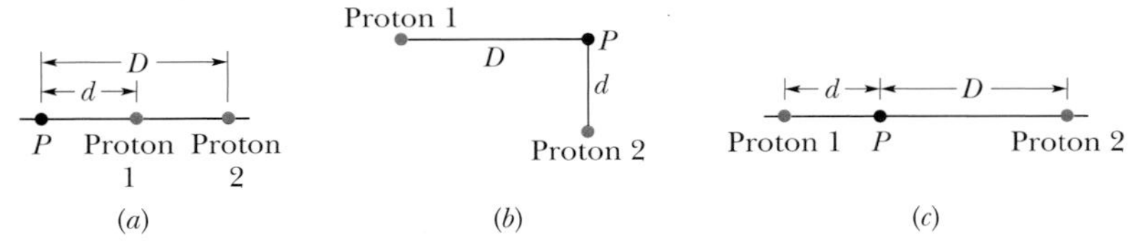

■

TOUCHSTONE EXAMPLE 25-4: A Square of Charges

What is the electric potential at point P, located at the center of the square of point charges shown in Fig. 25-13a? The distance d is 1.3 m, and the charges are

$$q_1 = +12 \text{ nC}, \qquad q_3 = +31 \text{ nC},$$
$$q_2 = -24 \text{ nC}, \qquad q_4 = +17 \text{ nC}.$$

SOLUTION ■ The **Key Idea** here is that the electric potential V at P is the algebraic sum of the electric potentials contributed by the four point charges. (Because electric potential is a scalar, the orientations of the point charges do not matter.) Thus, from Eq. 25-25, we have

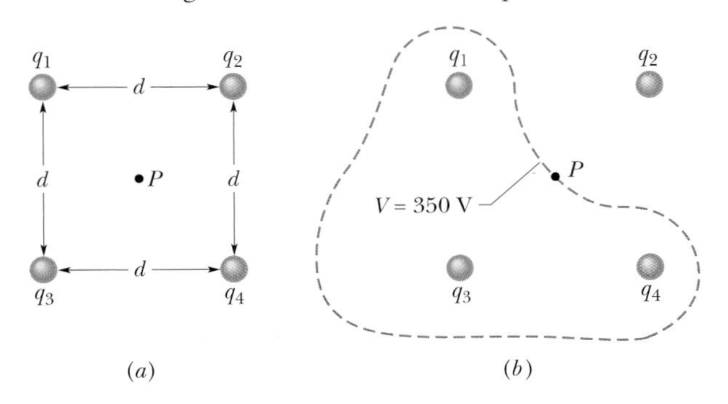

FIGURE 25-13 ■ (a) Four point charges are held fixed at the corners of a square. (b) The closed curve is a cross section, in the plane of the figure, of the equipotential surface that contains point P. (The curve is only roughly drawn.)

$$V = \sum_{i=1}^{4} V_i = k\left(\frac{q_1}{r} + \frac{q_2}{r} + \frac{q_3}{r} + \frac{q_4}{r}\right).$$

The distance r is $d/\sqrt{2}$, which is 0.919 m, and the sum of the charges is

$$q_1 + q_2 + q_3 + q_4 = (12 - 24 + 31 + 17) \times 10^{-9}\text{ C}$$
$$= 36 \times 10^{-9}\text{ C}.$$

Thus,
$$V = \frac{(8.99 \times 10^9 \text{N}\cdot\text{m}^2/\text{C}^2)(36 \times 10^{-9}\text{ C})}{0.919 \text{ m}}$$
$$\approx 350 \text{ V}. \qquad \text{(Answer)}$$

Close to any of the three positive charges in Fig. 25-13a, the potential has very large positive values. Close to the single negative charge, the potential has very large negative values. Therefore, there must be points within the square that have the same intermediate potential as that at point P. The curve in Fig. 25-13b shows the intersection of the plane of the figure with the equipotential surface that contains point P. Any point along that curve has the same potential as point P.

TOUCHSTONE EXAMPLE 25-5: A Dozen Electrons

(a) In Fig. 25-14a, 12 electrons (of charge $-e$) are equally spaced and fixed around a circle of radius R. Relative to $V = 0$ at infinity, what are the electric potential and electric field at the center C of the circle due to these electrons?

SOLUTION ■ The **Key Idea** here is that the electric potential V at C is the algebraic sum of the electric potentials contributed by all the electrons. (Because electric potential is a scalar, the orientations of the electrons do not matter.) Because the electrons all have the same negative charge $-e$ and are all the same distance R from C, Eq. 25-25 gives us

$$\Delta V = -12k\frac{e}{R}. \qquad \text{(Answer)} \qquad (25\text{-}27)$$

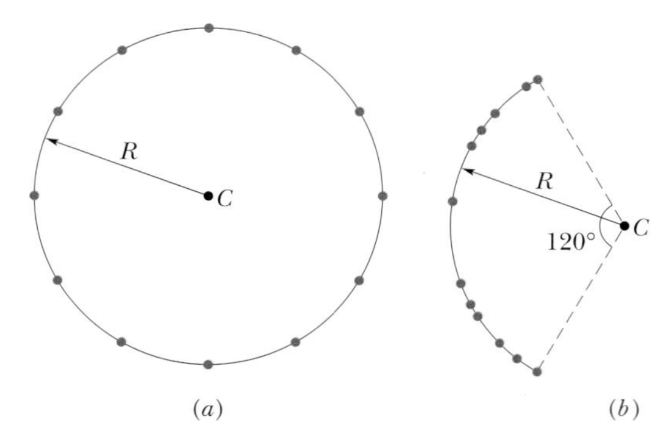

FIGURE 25-14 ■ (*a*) Twelve electrons uniformly spaced around a circle. (*b*) Those electrons are now nonuniformly spaced along an arc of the original circle.

For the electric field at *C*, the **Key Idea** is that electric field is a vector quantity and thus the orientation of the electrons *is* important. Because of the symmetry of the arrangement in Fig. 25-14*a*, the electric field vector at *C* due to any given electron is canceled by the field vector due to the electron that is diametrically opposite it. Thus, at *C*,

$$\vec{E} = 0. \qquad \text{(Answer)}$$

(b) If the electrons are moved along the circle until they are nonuniformly spaced over a 120° arc (Fig. 25-14*b*), what then is the potential at *C*? How does the electric field at *C* change (if at all)?

SOLUTION ■ The potential is still given by Eq. 25-27, because the distance between *C* and each electron is unchanged and orientation is irrelevant. The electric field is no longer zero, because the arrangement is no longer symmetric. There is now a net field that is directed toward the charge distribution.

25-8 Potential Due to an Electric Dipole

Electrically neutral matter is made of equal amounts of positive and negative charges. Electric forces pull in opposite directions on those charges. Thus, an electric field can cause a small separation of the positive and negative charges in matter (called polarization). In addition, many molecules distribute their electrons throughout their volume in a nonuniform way. This results in their having more positive charge on one end and one negative charge on the other end. For example, the water molecule shown in Fig. 23-22 has a nonuniform charge distribution.

A small separation produces an electric field very similar to that of a pair of equal and opposite charges separated by a small distance. If the charges were right on top of each other, their electric fields would cancel and they would appear neutral. But if they are a bit separated, their fields don't cancel perfectly, leaving a field pattern known as an electric dipole. The electric dipole fields produced by molecules play an essential role in a large number of processes in chemistry and biology, as well as in determining the electrical properties of matter such as color and transparency.

Now let us apply Eq. 25-25,

$$V = \sum_{i=1}^{n} V_i = \frac{1}{4\pi\varepsilon_0}\sum_{i=1}^{n}\frac{q_i}{r_i},$$

to an electric dipole to find the potential at an arbitrary point *P* in Fig. 25-15*a*. At *P*, the positive point charge (at distance $r_{(+)}$) sets up potential $V_{(+)}$ and the negative point charge (at distance $r_{(-)}$) sets up potential $V_{(-)}$. Then the net potential at *P* is given by Eq. 25-25 as

$$\begin{aligned} V &= \sum_{i=1}^{2} V_i = V_{(+)} + V_{(-)} = \frac{1}{4\pi\varepsilon_0}\left(\frac{q}{r_{(+)}} + \frac{-q}{r_{(-)}}\right) \\ &= \frac{q}{4\pi\varepsilon_0}\frac{r_{(-)} - r_{(+)}}{r_{(-)}r_{(+)}}. \end{aligned} \tag{25-28}$$

Naturally occurring dipoles—such as those possessed by many molecules—are quite small, so we are usually interested only in points that are relatively far from the

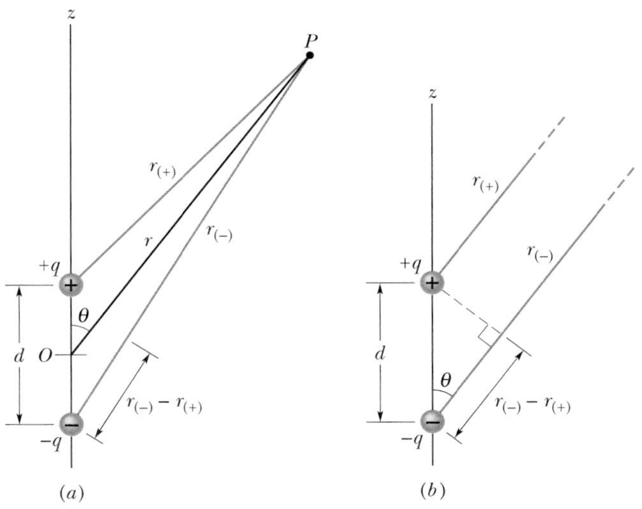

FIGURE 25-15 ■ (*a*) Point P is a distance r from the midpoint O of a dipole. The line OP makes an angle θ with the dipole axis. (*b*) If P is far from the dipole, the lines of lengths $r_{(+)}$ and $r_{(-)}$ are approximately parallel to the line of length r, and the dashed line is approximately perpendicular to the line of length $r_{(-)}$.

dipole, such that $r \gg d$, where d is the distance between the charges. Under those conditions, the approximations that follow from Fig. 25-15*b* are

$$r_{(-)} - r_{(+)} \approx d \cos\theta \quad \text{and} \quad r_{(-)} r_{(+)} \approx r^2.$$

If we substitute these quantities into Eq. 25-28, we can approximate V to be

$$V \approx \frac{q}{4\pi\varepsilon_0} \frac{d\cos\theta}{r^2} \qquad (\text{for } r >> d).$$

Here θ is measured from the dipole axis as shown in Fig. 25-15*a*. We can now write V as

$$V \approx k\frac{p\cos\theta}{r^2} \qquad (\text{electric dipole for } r >> d), \qquad (25\text{-}29)$$

in which $p(=qd)$ is the magnitude of the electric dipole moment $\vec{p}$ defined in Section 23-7. The vector $\vec{p}$ is directed along the dipole axis, from the negative to the positive charge. (Thus, θ is measured from the direction of $\vec{p}$.)

Induced Dipole Moment

Many molecules such as water have *permanent* electric dipole moments. In other molecules (called nonpolar molecules) and in every isolated atom, the centers of the positive and negative charges coincide (Fig. 25-16*a*) and thus no dipole moment is set up. However, if we place an atom or a nonpolar molecule in an external electric field, the field affects the locations of the electrons relative to the nuclei and separates the

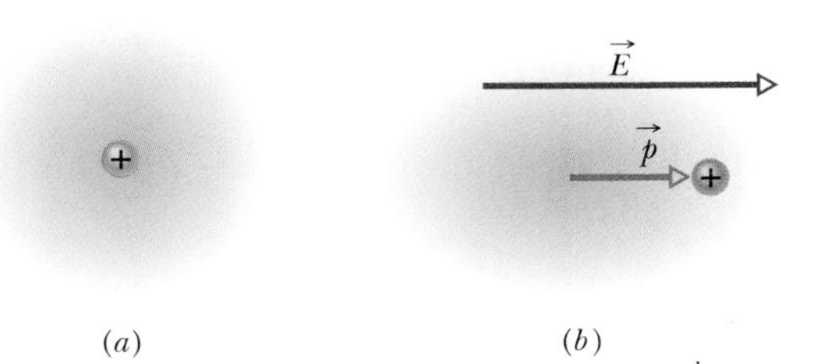

FIGURE 25-16 ■ (*a*) An atom, showing the positively charged nucleus (green) and a cloud of negatively charged electrons (gold shading). The centers of positive and negative charge coincide. (*b*) If the atom is placed in an external electric field $\vec{E}$, the electron orbits are distorted so that the centers of positive and negative charge no longer coincide. An induced dipole moment $\vec{p}$ appears. The distortion is exaggerated here by many orders of magnitude.

centers of positive and negative charge (Fig. 25-16*b*). Because the electrons are negatively charged, they tend to be shifted in a direction opposite the field. This shift sets up a dipole moment $\vec{p}$ pointing in the direction of the field. This dipole moment is said to be induced by the field, and the atom or molecule is then said to be polarized by the field (it has a positive side and a negative side). When the field is removed, the induced dipole moment and the polarization disappear.

READING EXERCISE 25-7: Suppose three points are set at equal (large) distances r from the center of the dipole in Fig. 25-15: Point a is on the dipole axis above the positive charge, point b is on the axis below the negative charge, and point c is on a perpendicular bisector through the line connecting the two charges. Rank the points according to the electric potential of the dipole there, greatest (most positive) first. ■

25-9 Potential Due to a Continuous Charge Distribution

When a charge distribution q is continuous (as on a uniformly charged thin rod or disk), we cannot use a summation to find the potential V at a point P. Instead, we must choose a differential element of charge dq. A differential element of charge is a very small bit of charge, small enough so we can treat it as if it were a point charge. We can then determine the potential dV at P due to dq, and then integrate over the entire charge distribution.

Let us again take the zero of potential to be at infinity. If we treat the element of charge dq as a point charge, then we can use Eq. 25-24,

$$V = k\frac{q}{r},$$

to express the potential dV at point P due to dq:

$$dV = k\frac{dq}{r} \quad \text{(positive or negative } dq\text{)}. \tag{25-30}$$

Here r is the distance between P and dq. To find the total potential V at P, we integrate to sum the potentials due to all the charge elements:

$$V = \int dV = k\int \frac{dq}{r}. \tag{25-31}$$

The integral must be taken over the entire charge distribution. Note that because the electric potential is a scalar, there are *no vector components* to consider in the equation above.

We now examine a continuous charge distribution, a line of charge.

Line of Charge

In Fig. 25-17*a*, a thin, nonconducting rod of length L has a positive charge of uniform linear density λ. Let us determine the electric potential V due to the rod at point P, a perpendicular distance d from the left end of the rod.

We consider a differential element dx of the rod as shown in Fig. 25-17*b*. This (or any other) element of the rod has a differential charge of

$$dq = \lambda\, dx. \tag{25-32}$$

This element produces a potential dV at point P, which is a distance $r = (x^2 + d^2)^{1/2}$ from the element. Treating the element as a point charge, we can use Eq. 25-30,

$$dV = k\frac{dq}{r},$$

to write the potential dV as

$$dV = k\frac{dq}{r} = k\frac{\lambda\,dx}{(x^2 + d^2)^{1/2}}. \tag{25-33}$$

Since the charge on the rod is positive and we have taken $V = 0$ at infinity, we know dV in this expression must be positive.

We now find the total potential V (measured relative to a zero at infinity) produced by the rod at point P by integrating along the length of the rod, from $x = 0$ to $x = L$. We evaluate the integral using an integral table or a symbolic manipulation program like Mathcad or Maple. We then find

$$\begin{aligned} V &= \int dV = \int_0^L k\frac{\lambda}{(x^2 + d^2)^{1/2}}\,dx \\ &= k\lambda\int_0^L \frac{dx}{(x^2 + d^2)^{1/2}} \\ &= k\lambda[\ln(x + (x^2 + d^2)^{1/2})]_0^L \\ &= k\lambda[\ln(L + (L^2 + d^2)^{1/2}) - \ln d]. \end{aligned}$$

We can simplify this result by using the general relation $\ln A - \ln B = \ln(A/B)$. We then find

$$V = k\lambda \ln\left[\frac{L + (L^2 + d^2)^{1/2}}{d}\right]. \tag{25-34}$$

Because V is the sum of positive values of dV, it should be positive—but does this expression give a positive V? Since the argument of the logarithm is greater than one, the logarithm is a positive number and V is indeed positive.

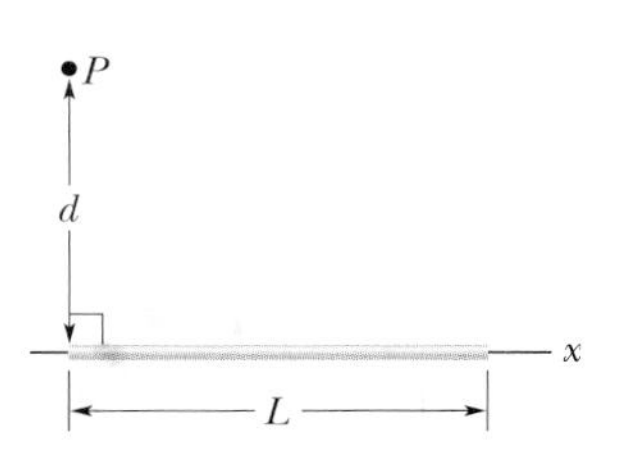

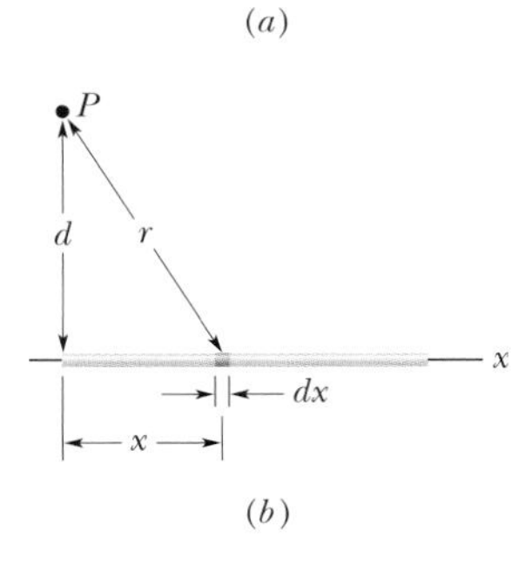

FIGURE 25-17 ■ (a) A thin, uniformly charged rod produces an electric potential V at point P. (b) A differential element of charge produces a differential potential dV at P.

25-10 Calculating the Electric Field from the Potential

In Section 25-5, you saw how to find the potential at a point f if you know the electric field along a path from a reference point to point f. In this section, we propose to go the other way—that is, to find the electric field when we know the potential. As Fig. 25-8 shows, graphically finding the direction of the field is easy: If we know the potential V at all points near an assembly of charges, we can draw in a family of equipotential surfaces. The electric field lines, sketched perpendicular to those surfaces, reveal the direction of $\vec{E}$. What we are seeking here is the mathematical equivalent of this graphical procedure.

Figure 25-18 shows cross sections of a family of closely spaced equipotential surfaces, the potential difference between each pair of adjacent surfaces being dV. As the figure suggests, the field $\vec{E}$ at any point P is perpendicular to the equipotential surface through P.

Suppose a positive test charge q_t moves through a displacement $d\vec{s}$ from one equipotential surface to the adjacent surface. From Eq. 25-8, we can relate the change

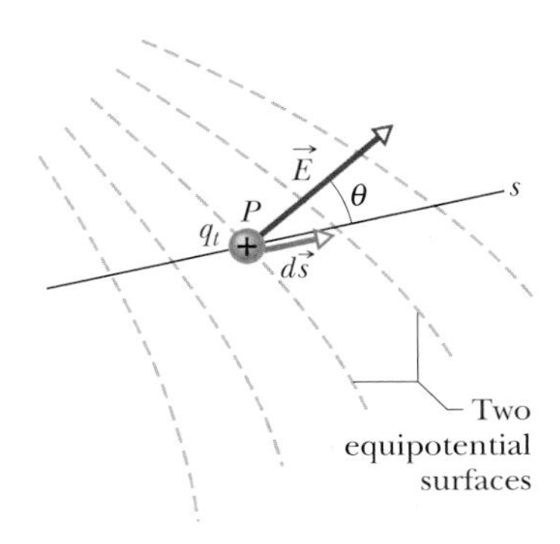

FIGURE 25-18 ■ A test charge q_t undergoes a displacement $d\vec{s}$ from one equipotential surface to another. (The separation between the surfaces has been exaggerated for clarity.) The displacement $d\vec{s}$ makes an angle θ with the direction of the electric field $\vec{E}$.

in electric potential to the work done by the electric field on our test charge

$$\Delta V = V_2 - V_1 = -\frac{W^{\text{elec}}}{q_t}.$$

Let's consider the potential difference associated with an infinitesimally small displacement denoted by $d\vec{s}$. We see that the electric field does an infinitesimal amount of work on the test charge during the move. Using Eq. 25-8, we can denote this as $-q_t dV$. From Eq. 25-14, $dW = q_t\vec{E}\cdot d\vec{s}$, and Fig. 25-18, we see that the infinitesimal work done by the force may also be written as $(q_t\vec{E})\cdot d\vec{s}$ or $q_t|\vec{E}|(\cos\phi)|d\vec{s}|$, where ϕ is the angle between the electric field and displacement vectors as shown in Fig. 25-18. Equating these two expressions for the work yields

$$-q_t\, dV = q_t|\vec{E}|(\cos\phi)|d\vec{s}|, \tag{25-35}$$

or

$$\vec{E}(\cos\phi) = -\frac{dV}{ds}. \tag{25-36}$$

Since $E_s = |\vec{E}|\cos\phi$ is the component of $\vec{E}$ in the direction of $d\vec{s}$, the equation above becomes

$$E_s = -\frac{\partial V}{\partial s}. \tag{25-37}$$

We have added a subscript to the component of $\vec{E}$ and switched to the partial derivative symbols to emphasize that this expression involves only the variation of ΔV along a specified axis (here called the s axis) and only the component of $\vec{E}$ along that axis. In words, Eq. 25-37 is essentially the inverse of Eq. 25-16,

$$V_2 - V_1 = -\int_1^2 \vec{E}\cdot d\vec{s},$$

and states:

> The component of $\vec{E}$ in any direction is the negative of the rate of change of the electric potential with distance in that direction. Hence, $\vec{E}$ points in the direction of decreasing electric potential V.

If we take the s axis to be, in turn, the x, y, and z axes, we find that the x-, y-, and z-components of $\vec{E}$ at any point are

$$E_x = -\frac{\partial V}{\partial x}; \qquad E_y = -\frac{\partial V}{\partial y}; \qquad E_z = -\frac{\partial V}{\partial z}. \tag{25-38}$$

Thus, if we know V for all points in the region around a charge distribution—that is, if we know the function $V(x,y,z)$—we can find the components of $\vec{E}$, and thus $\vec{E}$ itself, at any point by taking partial derivatives. Each component of the electric field is simply the negative of the slope of the curve representing the electric potential vs. distance along each chosen axis.

For the simple situation in which the electric field $\vec{E}$ is uniform, the equipotential surfaces are a set of parallel planes that lie perpendicularly to the direction of the electric field. In addition, for a given potential difference, the distance between any two equipotential planes is the same. So, when the component of the electric field along the direction of $d\vec{s}$ is uniform, we can rewrite Eq. 25-37 ($E_s = -\partial V/\partial s$) in terms of the magnitude of the electric field $E = |\vec{E}|$ as

$$E = \left|\frac{\Delta V}{\Delta s}\right|, \tag{25-39}$$

where Δs is the component of displacement perpendicular to the equipotential surfaces. Equation 25-36 tells us that whenever the potential is constant along a surface so that $\Delta V = 0$, the electric field is zero. The component of the electric field is zero in any direction parallel to the equipotential surfaces. Thus, for a given potential difference ΔV, the magnitude of the electric field is given by the magnitude of the potential difference divided by the distance between any two equipotential surfaces.

READING EXERCISE 25-8: The figure shows three pairs of parallel plates with the same separation, and the electric potential of each plate. The electric field between the plates is uniform and perpendicular to the plates. (a) Rank the pairs according to the magnitude of the electric field between the plates, greatest first. (b) For which pair is the electric field pointing rightward? (c) If an electron is released midway between the third pair of plates, does it remain there, move rightward at constant speed, move leftward at constant speed, accelerate rightward, or accelerate leftward?

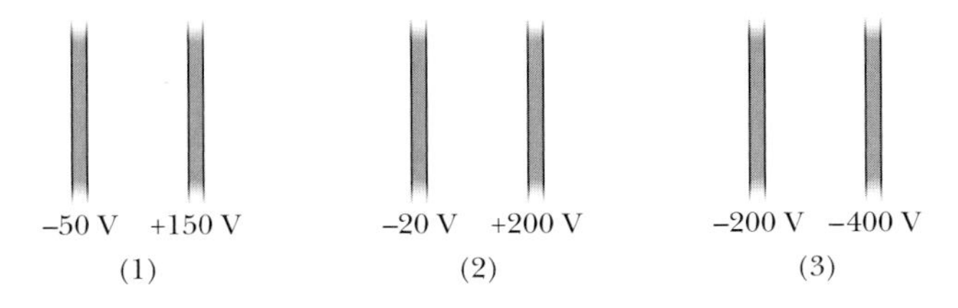

■

READING EXERCISE 25-9: In what ways is the superposition principle for energy discussed above the same as, and different from, the superposition principle for electric field? ■

TOUCHSTONE EXAMPLE 25-6: Obtaining $\vec{E}$ from V

The electric potential at any point on the axis of a uniformly charged disk is given by

$$V = \frac{\sigma}{2\varepsilon_0}(\sqrt{z^2 + R^2} - z).$$

Starting with this expression, derive an expression for the electric field at any point on the axis of the disk.

SOLUTION ■ We want the electric field $\vec{E}$ as a function of distance z along the axis of the disk. For any value of z, the direction of $\vec{E}$ must be along that axis because the disk has circular symmetry about that axis. Thus, we want the component E_z of $\vec{E}$ in the direction of z. Then the **Key Idea** is that this component is the negative of the rate of change of the electric potential with distance z. Thus, from the last of Eqs. 25-38, we can write

$$E_z = -\frac{\partial V}{\partial z} = -\frac{\sigma}{2\varepsilon_0}\frac{d}{dz}(\sqrt{z^2 + R^2} - z)$$

$$= \frac{\sigma}{2\varepsilon_0}\left(1 - \frac{z}{\sqrt{z^2 + R^2}}\right). \qquad \text{(Answer)}$$

25-11 Potential of a Charged Isolated Conductor

In Section 24-8, we concluded $\vec{E} = 0$ for all points inside an electrically isolated conductor. We then used Gauss' law to prove that an excess charge placed on an isolated conductor lies entirely on its surface. (This is true even if the conductor has an empty internal cavity.) Here we use the first of these facts to prove an extension of the second:

An excess charge placed on an isolated conductor will distribute itself on the surface of that conductor so that all points of the conductor—whether on the surface or inside—come to the same potential. This is true even if the conductor has an internal cavity and even if that cavity contains a net charge.

This fact is rather obvious since any potential difference inside a conductor requires an electric field inside it. The nonzero electric field would, in turn, cause the free conduction electrons to redistribute themselves until the potential difference disappears.

The mathematical proof that an electrically isolated conductor is an equipotential region follows directly from Eq. 25-16,

$$V_2 - V_1 = -\int_1^2 \vec{E} \cdot d\vec{s}.$$

Since $\vec{E} = 0$ for all points within a conductor, it follows directly that $V_2 = V_1$ for all possible pairs of points i and f in the conductor.

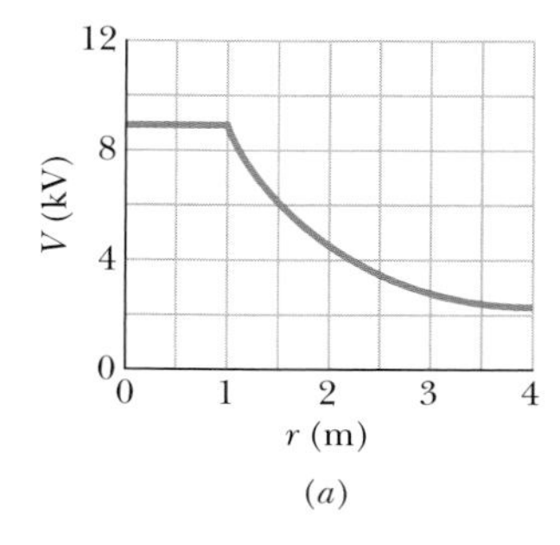

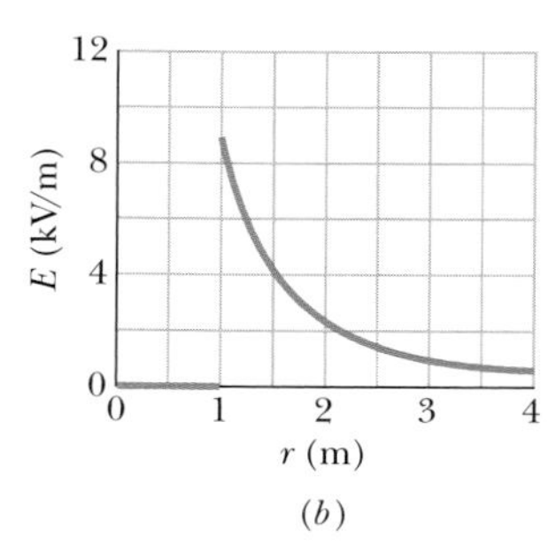

FIGURE 25-19 ■ (*a*) A plot of $V(r)$ both inside and outside a charged spherical shell of radius 1.0 m. (*b*) A plot of the electric field magnitude, $E(r)$, for the same shell.

A Spherical Shell with No External Electric Field

Figure 25-19*a* shows a plot of potential against radial distance r from the center for an isolated spherical conducting shell of 1.0 m radius, having a net excess charge of 1.0 μC. In the absence of an external field, we know by symmetry the surface charges will be uniformly distributed over the surface of the shell. For points outside the shell, we can calculate $V(r)$, the electric potential. Obviously this potential also has a spherical symmetry and can be given by Eq. 25-24,

$$V = k\frac{q}{r},$$

because the total charge on the shell, denoted as q, behaves for external points as if it were concentrated at the center of the shell. That equation holds right up to the surface of the shell. Now let us push a small test charge through the shell—assuming a small hole exists—to its center. No extra work is needed to do this because no net electric force acts on the test charge once it is inside the shell. Thus, the potential at all points inside the shell has the same value as on the surface, as shown in the Fig. 25-19*a* graph.

The Fig. 25-19*b* graph shows the variation of electric field with radial distance for the same shell. Note that $\vec{E} = 0$ everywhere inside the shell. The curves of Fig. 25-19*b* can be derived from the curve of Fig. 25-19*a* by differentiating with respect to r, using Eq. 25-37 (the derivative of a constant, recall, is zero). The curve of Fig. 25-19*a* can be derived from the curves of Fig. 25-19*b* by integrating

$$E_s = -\frac{\partial V}{\partial s}.$$

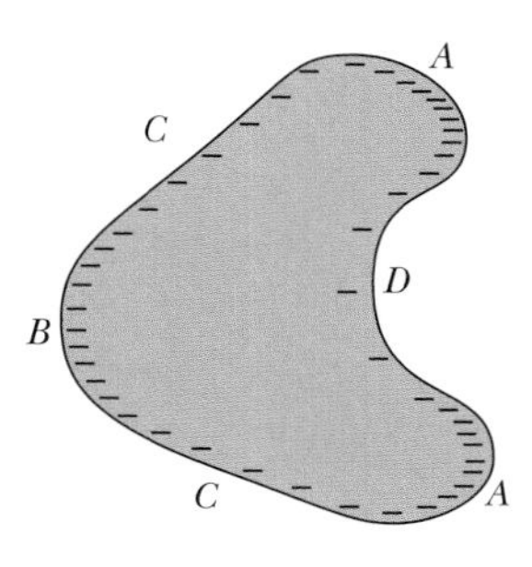

FIGURE 25-20 ■ The magnitude of the charge density on a conductor is greatest on a convex surface with a small radius of curvature (A) and least on a concave surface having small radius of curvature (D). The ranking of the magnitude of the charge density is $A > B > C > D$.

The Charge Distribution on a Nonspherical Conductor

Consider a nonspherical charged conductor. Assume the conductor is electrically isolated and there is no external electric field in its vicinity. It turns out its surface charges do not distribute themselves uniformly. When compared to the uniform density of excess charge on a spherical conductor, the charges redistribute themselves so there is a higher charge density when the radius of curvature is convex and small and a lower charge density where the radius of curvature is concave and small (Fig. 25-20). Why? We can use the characteristics of equipotential surfaces to develop a qualitative explanation for this phenomenon.

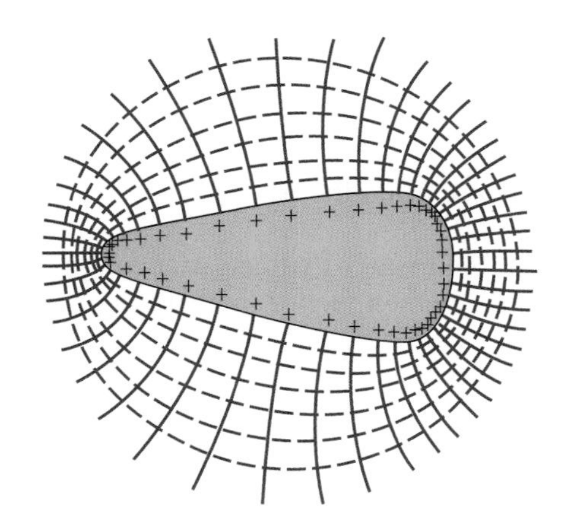

FIGURE 25-21 ■ The net positive charge on an odd-shaped isolated conductor distributes itself on the conductor's surface so the electric field generated by it is zero inside and normal to the surface elements of the conductor. This requires the equipotential surfaces (shown with dotted lines) to be closest together on the left where the conductor's convex radius of curvature is smallest. The electric field lines (shown with solid lines) and the excess charges also have the greatest density on the left where the curvature of the conductor's surface is smallest.

The explanation is as follows: There is no electric field inside the conductor, and the electric field at each point on the surface of the conductor must be normal (in other words perpendicular) to the surface. This requirement is obvious since any component of electric field parallel to the surface would cause free electrons to reconfigure themselves until all tangential components along the surface disappear. This also means the entire surface of our conductor is an equipotential surface no matter what its shape is. However, if we are far away from our charged conductor, the equipotential surfaces look more and more like those of a point charge. Thus, the family of equipotential surfaces that are each ΔV apart from the previous one become more and more spherical in shape. As the successive equipotential surfaces morph (change shape) slowly from that of our odd-shaped surface to that of a sphere, the parts of the equipotential surfaces near small-radius convex surface elements must be closer together than those elements having large radii of curvature. This is shown in Fig. 25-21. Now, equipotential surfaces more closely spaced occur where the electric field is the strongest and can do the most work on test charges, but the electric field is largest where the charge density that is its source is largest. The implication is that:

> On an isolated conductor the concentration of charges and hence the strength of the electric field is greater near sharp points where the curvature is large.

An Isolated Conductor in an External Electric Field

Suppose an *uncharged* isolated conductor is placed in an *external electric field*, as in Fig. 25-22. The electric field at the conductor's surface must have the same characteristics as it does when no external field is present. However, this doesn't mean its charges will be distributed in the same way as if no external electric field were present. All points of the conductor still come to a single potential regardless of whether the conductor is electrically neutral or has an excess charge. The free conduction electrons distribute themselves on the surface in such a way that the electric field they produce at interior points cancels the external electric field that would otherwise be there. Furthermore, the electron distribution causes the net electric field at all points on the surface to be normal to the surface. If the conductor in Fig. 25-22 could somehow be removed, leaving the surface charges frozen in place, the pattern of the electric field would remain absolutely unchanged for both exterior and interior points.

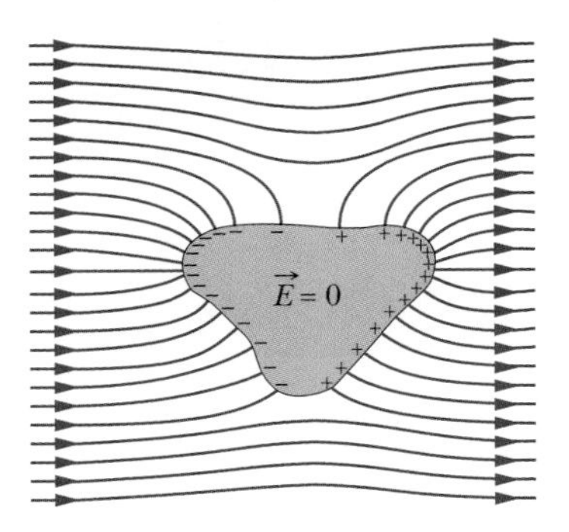

FIGURE 25-22 ■ An uncharged conductor is suspended in an external electric field. The free electrons in the conductor distribute themselves on the surface as shown, so as to reduce the net electric field inside the conductor to zero and make the net field outside normal to each surface element.

One common natural source of an external electric field that can affect isolated metal objects are excess negative charges at the bases of clouds contributing to the onset of thunderstorms. Such an external electric field can cause charge separation in conducting objects at the Earth's surface such as golf clubs and rock hammers. Since

FIGURE 25-23 ■ This enhanced photograph shows the result of an overhead cloud system creating a strong electric field $\vec{E}$ near a woman's head. Many of the hair strands extended along the field, which was perpendicular to the equipotential surfaces and greatest where those surfaces were closest, near the top of her head.

these objects have points where the curvature is high, the surface charge density—and thus the external electric field, which is proportional to it—may reach very high values. The air around sharp points may become ionized, producing the corona discharge that golfers and mountaineers see on their tools when thunderstorms threaten. Such corona discharges, like hair that stands on end, are often the precursors of lightning strikes.

The cells and blood inside a human body contain salt water that acts as a conductor. The natural oil found on hair is also conductive. A person placed in a strong electric field can act like an uncharged conductor. For example, the woman shown in Fig. 25-23 was standing on a platform connected to the mountainside, and was at about the same potential as the mountainside. Overhead, a cloud system that had a high degree of charge separation with excess negative charges at its base moved in and created a strong electric field around her and the mountainside. Electrostatic forces due to this field drove some of the conduction electrons in the woman downward through her body, leaving her head and strands of her hair positively charged. The magnitude of this electric field was apparently large, but less than the value of about 3×10^6 V/m needed to cause electrical breakdown of the air molecules. (That value was exceeded when lightning struck the platform shortly after the picture was taken.)

As we just discussed, the surface charges on a nonspherical conductor concentrate in regions where the curvature is greatest. Thus, we expect the electric field to be greatest near the top of the woman's head—an equipotential surface. This suspicion is confirmed because the strands of her hair, containing excess positive charge, are pulled out most strongly where her head has the most curvature. Also, the strands of hair are extended along the direction of $\vec{E}$ perpendicular to her head. Since the magnitude of $\vec{E}$ was greatest just above her head, this is where the equipotential surfaces were most closely spaced. A sketch showing this close spacing is shown in Fig. 25-23.

The lesson here is simple. If an electric field causes the hairs on your head to stand up, you'd better run for shelter rather than pose for a snapshot.

What If Lightning Might Strike?

FIGURE 25-24 ■ A large spark jumps to a car's body and then exits by moving across the insulating left front tire, leaving the person inside unharmed because the electric potential difference remains zero inside the car.

Speaking of lightning, what is the best way to protect yourself if lightning strikes? There are two ways to protect yourself using your knowledge of how conductors behave in electric fields. One is to enclose yourself in a relatively spherical conducting shell. The other is to use a lightning rod.

Using a Spherical Shell: If you enclose yourself inside a more or less spherical cavity, the electric field inside the cavity is guaranteed to be zero. A car (unless it is a convertible) is almost ideal (Fig. 25-24) because it protects the passengers from the effects of lightning for the same reason that the Faraday cage shown in Chapter 24 protects the demonstrator from the high voltage caused by the transfer of charge to the cage by a Van de Graaff generator.

Using a Lightning Rod: If you live in an area where thunderstorms are common, you can embed the base of a tall metal lightning rod in the ground. Recall that the bottoms of thunderclouds have an excess of negative charge that creates strong electric fields at the Earth's surface. What happens if a conducting rod, like the Eiffel Tower, has a sharp point *and* is taller than its immediate surroundings? A couple of factors come into play. First, the distance $|\Delta s|$ between the cloud bases and the top of a lightning rod is smaller than the distance to the ground, even though the electric field strength near the top of a tall rod is not really uniform. Equation 25-39 ($E = |\Delta V/\Delta s|$) tells us that the magnitude of the electric field between the cloud bases and the top of the rod is greater than that between the clouds and the ground. Second, as we discussed earlier in this section the magnitude of the electric field near a conductor that has a sharp point is quite strong compared to that on level ground. This means that the free electrons at the top of a lightning rod (such as the Eiffel Tower) will move toward the ground leaving a large accumulation of positive metal ions at the sharp point at the top of the tower. The tip of the rod will attract electrons from the atmosphere to it and down to the ground in a corona discharge process that can serve to prevent a major discharge or lightning strike in the vicinity of the tower. Lightning is shown hitting the Eiffel Tower in Fig. 25-25.

READING EXERCISE 25-10: The figure below shows the region in the neighborhood of a negatively charged conducting sphere and a large positively charged conducting plate extending far beyond the region shown. Someone claims lines A through F are possible field lines describing the electric field lying in the region between the two conductors. (a) Examine each of the lines and indicate whether it is a correctly drawn field line. If a line is not correct, explain why. (b) Redraw the diagram with a pattern of field lines that is more nearly correct. (Based on Arnold Arons' *Homework and Test Questions,* Wiley, New York, 1994.) ■

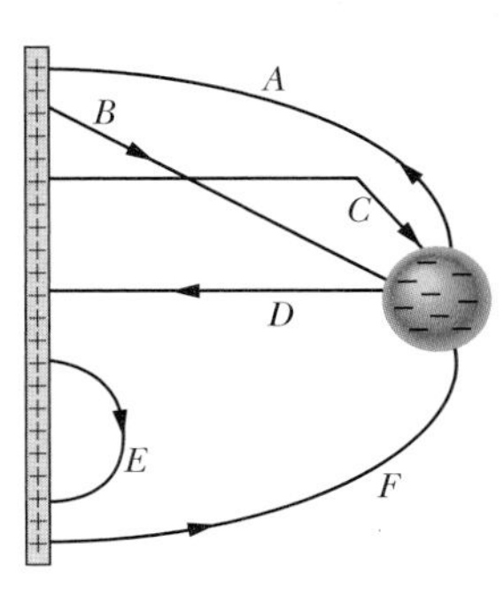

READING EXERCISE 25-11: Why are the equipotential surfaces shown in Fig. 25-23 closer together just above the woman's head than they are at the side of her head? ■

FIGURE 25-25 ■ In this historic 1902 postcard photo, bolts of lightning are shown converging at the top of the Eiffel Tower. The tower is acting as a "lightning rod" protecting people, trees, and other buildings from being struck by lightning.

Problems

SEC. 25-3 ■ ELECTRIC POTENTIAL

1. Car Battery A particular 12 V car battery can send a total charge of 3.0×10^5 C through a circuit, from one terminal to the other. (a) How many coulombs of charge does this represent? (b) If this entire charge undergoes a potential difference of 12 V, how much energy is involved?

2. Ground and Cloud The electric potential difference between the ground and a cloud in a particular thunderstorm is 1.2×10^9 V. What is the magnitude of the change in the electric potential energy (in multiples of the electron-volt) of an electron that moves between the ground and the cloud?

3. Lightning Flash In a given lightning flash, the potential difference between a cloud and the ground is 1.0×10^9 V and the quantity of charge transferred is 30 C. (a) What is the decrease in energy of that transferred charge. (b) If all that energy could be used to accelerate a 1000 kg automobile from rest, what would be the automobile's final speed? (c) If the energy could be used to melt ice, how much ice would it melt at 0°C? The heat of fusion of ice is 3.33×10^5 J/kg.

SEC. 25-5 ■ CALCULATING THE POTENTIAL FROM AN *E*-FIELD

4. From *A* to *B* When an electron moves from A to B along an electric field line in Fig. 25-26, the electric field does 3.94×10^{-19} J of work on it. What are the electric potential differences (a) $V_B - V_A$, (b) $V_C - V_A$, and (c) $V_C - V_B$?

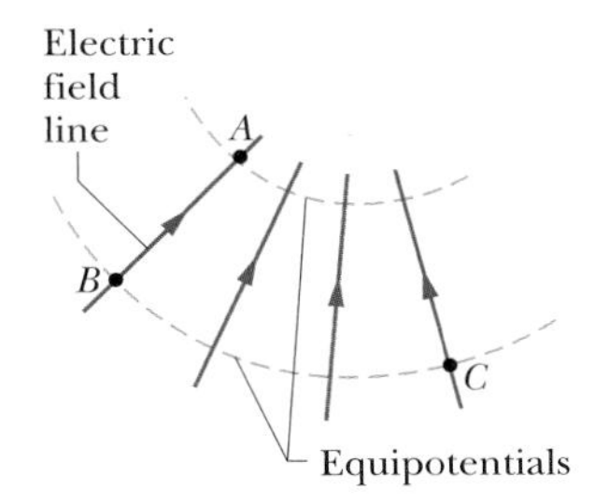

FIGURE 25-26 ■ Problem 4.

5. Infinite Sheet An infinite nonconducting sheet has a surface charge density $\sigma = 0.10\ \mu\text{C/m}^2$ on one side. How far apart are equipotential surfaces whose potentials differ by 50 V?

6. Parallel Plates Two large, parallel, conducting plates are 12 cm apart and have charges of equal magnitude and opposite sign on their facing surfaces. An electrostatic force of 3.9×10^{-15} N acts on an electron placed anywhere between the two plates. (Neglect fringing.) (a) Find the electric field at the position of the electron. (b) What is the potential difference between the plates?

7. Geiger Counter A Geiger counter has a metal cylinder 2.00 cm in diameter along whose axis is stretched a wire 1.30×10^{-4} cm in diameter. If the potential difference between the wire and the cylinder is 850 V, what is the electric field at the surface of (a) the wire and (b) the cylinder? (*Hint:* Use the result of Problem 30 of Chapter 24.)

8. Field Inside The electric field inside a nonconducting sphere of radius R, with charge spread uniformly throughout its volume, is radially directed and has magnitude

$$E(r) = |\vec{E}(r)| = \frac{|q|r}{4\pi\varepsilon_0 R^3}.$$

Here q (positive or negative) is the total charge within the sphere, and r is the distance from the sphere's center. (a) Taking $V = 0$ at the center of the sphere, find the electric potential $V(r)$ inside the sphere. (b) What is the difference in electric potential between a point on the surface and the sphere's center? (c) If q is positive, which of those two points is at the higher potential?

9. Uniformly Distributed A charge q is distributed uniformly throughout a spherical volume of radius R. (a) Setting $V = 0$ at infinity, show that the potential at a distance r from the center, where $r < R$, is given by

$$V = \frac{q(3R^2 - r^2)}{8\pi\varepsilon_0 R^3}.$$

(*Hint:* See Section 24-6.) (b) Why does this result differ from that in (a) of Problem 8? (c) What is the potential difference between a point on the surface and the sphere's center? (d) Why doesn't this result differ from that of (b) of Problem 8?

10. Infinite Sheet Two Figure 25-27 shows, edge-on, an infinite nonconducting sheet with positive surface charge density σ on one side. (a) Use Eq. 25-16 and Eq. 24-16 to show that the electric potential of an infinite sheet of charge can be written $V = V_0 - (\sigma/2\varepsilon_0)z$, where V_0 is the electric potential at the surface of the sheet and z is the perpendicular distance from the sheet. (b) How much work is done by the electric field of the sheet as a small positive test charge q_0 is moved from an initial position on the sheet to a final position located a distance z from the sheet?

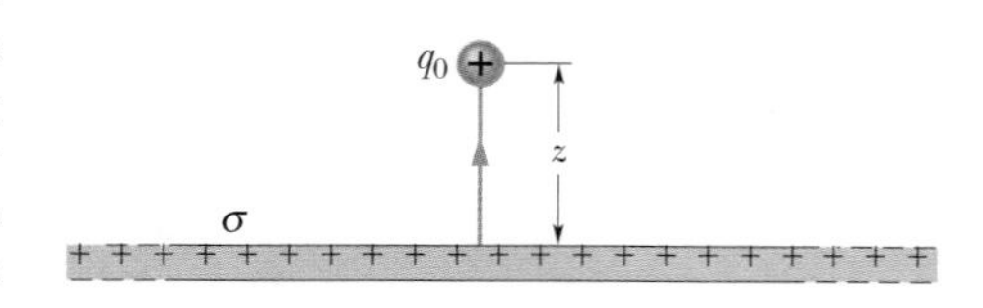

FIGURE 25-27 ■ Problem 10.

11. Thick Spherical Shell A thick spherical shell of charge Q and uniform volume charge density ρ is bounded by radii r_1 and r_2, where $r_2 > r_1$. With $V = 0$ at infinity, find the electric potential ΔV as a function of the distance r from the center of the distribution, considering the regions (a) $r > r_2$, (b) $r_2 > r > r_1$, and (c) $r < r_1$. (d) Do these solutions agree at $r = r_2$ and $r = r_1$? (*Hint:* See Section 24-6.)

SEC. 25-7 ■ POTENTIAL AND POTENTIAL ENERGY DUE TO A GROUP OF POINT CHARGES

12. Space Shuttle As a space shuttle moves through the dilute ionized gas of Earth's ionosphere, its potential is typically changed by -1.0 V during one revolution. By assuming that the shuttle is a sphere of radius 10 m, estimate the amount of charge it collects.

13. Diametrically Opposite Consider a point charge $q = 1.0\ \mu$C, point A at distance $d_1 = 2.0$ m from q, and point B at distance $d_2 = 1.0$ m. (a) If these points are diametrically opposite each other, as in Fig. 25-28a, what is the electric potential difference $V_A - V_B$? (b) What is that electric potential difference if points A and B are located as in Fig. 25-28b?

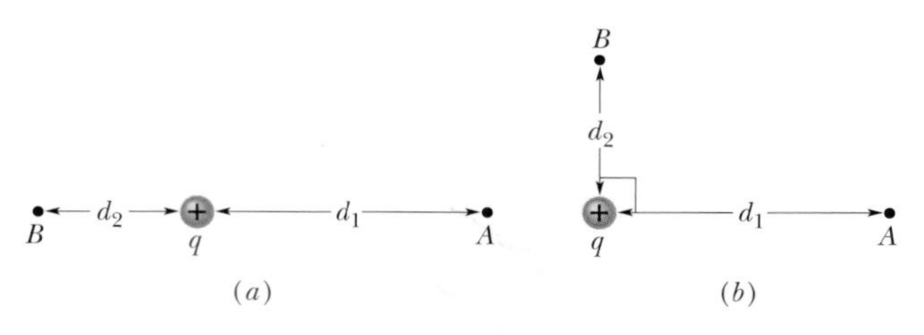

FIGURE 25-28 ■ Problem 13.

14. Field Lines and Equipotentials Figure 25-29 shows two charged particles on an axis. Sketch the electric field lines and the equipotential surfaces in the plane of the page for (a) $q_1 = +q$ and $q_2 = +2q$ and (b) $q_1 = +q$ and $q_2 = -3q$.

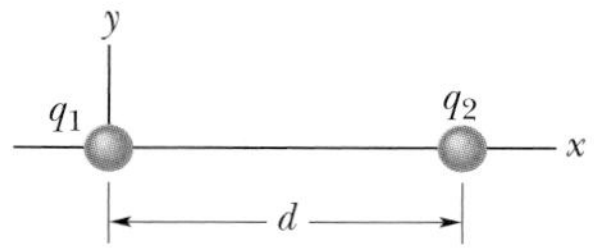

FIGURE 25-29 ■ Problems 14, 15, 16.

15. In Terms of d In Fig. 25-29, set $V = 0$ at infinity and let the particles have charges $q_1 = +q$ and $q_2 = -3q$. Then locate (in terms of the separation distance d) any point on the x axis (other than at infinity) at which the net potential due to the two particles is zero.

16. E-Field Is Zero Two particles, of charges q_1 and q_2, are separated by distance d in Fig. 25-29. The net electric field of the particles is zero at $x = d/4$. With $V = 0$ at infinity, locate (in terms of d) any point on the x axis (other than at infinity) at which the electric potential due to the two particles is zero.

17. Spherical Drop of Water A spherical drop of water carrying a charge of 30 pC has a potential of 500 V at its surface (with $V = 0$ at infinity). (a) What is the radius of the drop? (b) If two such drops of the same charge and radius combine to form a single spherical drop, what is the potential at the surface of the new drop?

18. Charge and Charge Density What are (a) the charge and (b) the charge density on the surface of a conducting sphere of radius 0.15 m whose potential is 200 V (with $V = 0$ at infinity)?

19. Field Near Earth An electric field of approximately 100 V/m is often observed near the surface of Earth. If this were the field over the entire surface, what would be the electric potential of a point on the surface? (Set $V = 0$ at infinity.)

20. Center of Rectangle In Fig. 25-30, point P is at the center of the rectangle. With $V = 0$ at infinity, what is the net electric potential at P due to the six charged particles?

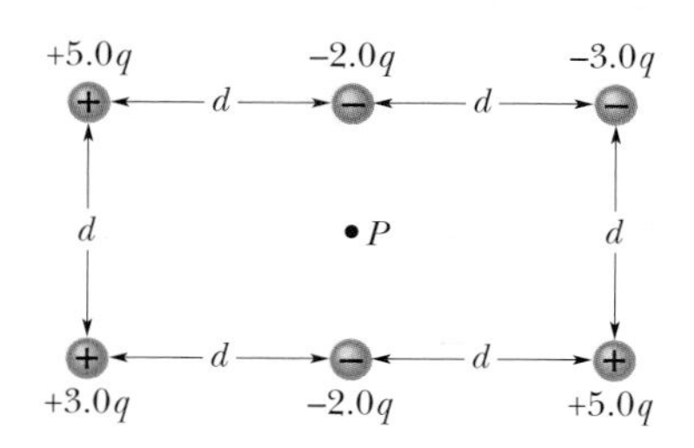

FIGURE 25-30 ■ Problem 20.

21. Potential at P In Fig. 25-31, what is the net potential at point P due to the four point charges, if $V = 0$ at infinity?

22. Potential Energy (a) What is the electric potential energy of two electrons separated by 2.00 nm? (b) If the separation increases, does the potential energy increase or decrease?

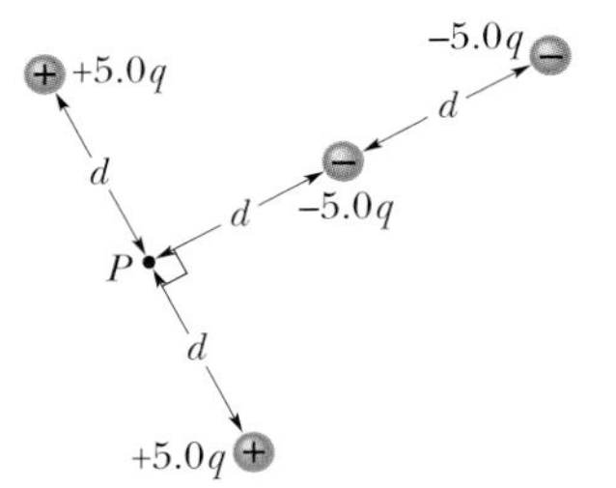

+5.0q, −5.0q, −5.0q, +5.0q, d, P

FIGURE 25-31 ■ Problem 21.

23. Work Required Derive an expression for the work required to set up the four-charge configuration of Fig. 25-32, assuming the charges are initially infinitely far apart.

24. Electric Potential Energy What is the electric potential energy of the charge configuration of Fig. 25-13a? Use the numerical values provided in Touchstone Example 25-4.

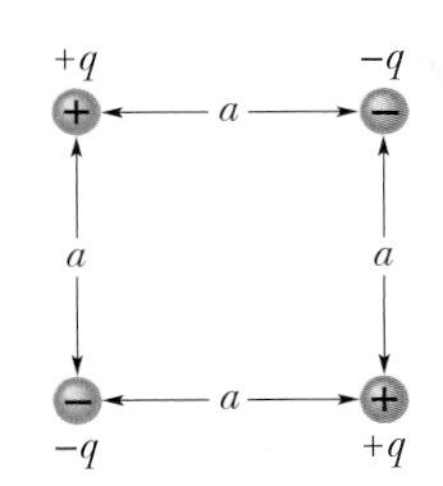

FIGURE 25-32 ■ Problem 23.

25. The Rectangle In the rectangle of Fig. 25-33, the sides have lengths 5.0 cm and 15 cm, $q_1 = -5.0\ \mu C$, and $q_2 = +2.0\ \mu C$. With $V = 0$ at infinity, what are the electric potentials (a) at corner A and (b) at corner B? (c) How much work is required to move a third charge $q_3 = +3.0\ \mu C$ from B to A along a diagonal of the rectangle? (d) Does this work increase or decrease the electric energy of the three-charge system? Is more, less, or the same work required if q_3 is moved along paths that are (e) inside the rectangle but not on a diagonal and (f) outside the rectangle?

FIGURE 25-33 ■ Problem 25.

26. How Much Work In Fig. 25-34, how much work is required to bring the charge of $+5q$ in from infinity along the dashed line and place it as shown near the two fixed charges $+4q$ and $-2q$? Take distance $d =$ 1.40 cm and charge $q = 1.6 \times 10^{-19}$ C.

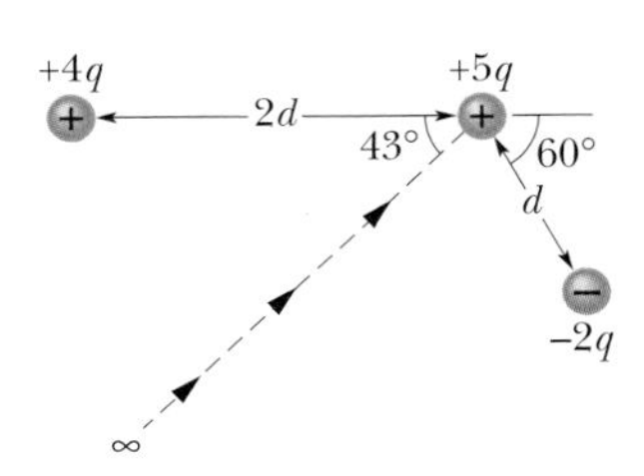

FIGURE 25-34 ■ Problem 26.

27. A Particle of Positive Charge A particle of positive charge Q is fixed at point P. A second particle of mass m and negative charge $-q$ moves at constant speed in a circle of radius r_1, centered at P. Derive an expression for the work W that must be done by an external agent on the second particle to increase the radius of the circle of motion to r_2.

28. How Much Energy Calculate (a) the electric potential established by the nucleus of a hydrogen atom at the average distance ($r = 5.29 \times 10^{-11}$ m) of the atom's electron (take $V = 0$ at infinite distance), (b) the electric potential energy of the atom when the electron is at this radius, and (c) the kinetic energy of the electron, assuming it to be moving in a circular orbit of this radius centered on the nucleus. (d) How much energy is required to ionize the hydrogen atom (that is, to remove the electron from the nucleus so that the separation is effectively infinite)? Express all energies in electron-volts.

29. Fixed at Point P A particle of charge q is fixed at point P, and a second particle of mass m and the same charge q is initially held a distance r_1 from P. The second particle is then released. Determine its speed when it is distance r_2 from P. Let $q = 3.1\ \mu C$, $m = 20$ mg, $r_1 = 0.90$ mm, and $r_2 = 2.5$ mm.

30. Thin Plastic Ring A charge of -9.0 nC is uniformly distributed around a thin plastic ring of radius 1.5 m that lies in the yz plane with its center at the origin. A point charge of -6.0 pC is located on the x axis at $x = 3.0$ m. Calculate the work done on the point charge by an external force to move the point charge to the origin.

31. Tiny Metal Spheres Two tiny metal spheres A and B of mass $m_A = 5.00$ g and $m_B = 10.0$ g have equal positive charges $q = 5.00\ \mu C$. The spheres are connected by a massless nonconducting string of length $d = 1.00$ m, which is much greater than the radii of the spheres. (a) What is the electric potential energy of the system? (b) Suppose you cut the string. At that instant, what is the acceleration of each sphere? (c) A long time after you cut the string, what is the speed of each sphere?

32. Conducting Shell on Support A thin, spherical, conducting shell of radius R is mounted on an isolating support and charged to a potential of $-V$. An electron is then fired from point P at distance r from the center of the shell ($r \gg R$) with initial speed v_1 and directly toward the shell's center. What value of v_1 is needed for the electron to just reach the shell before reversing direction?

33. Two Electrons Two electrons are fixed 2.0 cm apart. Another electron is shot from infinity and stops midway between the two. What is its initial speed?

34. Charged, Parallel Surfaces Two charged, parallel, flat conducting surfaces are spaced $d = 1.00$ cm apart and produce a potential difference $\Delta V = 625$ V between them. An electron is projected from one surface directly toward the second. What is the initial speed of the electron if it stops just at the second surface?

35. An Electron Is Projected An electron is projected with an initial speed of 3.2×10^5 m/s directly toward a proton that is fixed in place. If the electron is initially a great distance from the proton, at what distance from the proton is the speed of the electron instantaneously equal to twice the initial value?

SEC. 25-8 ■ POTENTIAL DUE TO AN ELECTRIC DIPOLE

36. Ammonia The ammonia molecule NH_3 has a permanent electric dipole moment equal to 1.47 D, where 1 D = 1 debye unit = 3.34×10^{-30} C·m. Calculate the electric potential due to an ammonia molecule at a point 52.0 nm away along the axis of the dipole. (Set $V = 0$ at infinity.)

37. Three Particles Figure 25-35 shows three charged particles located on a horizontal axis. For points (such as P) on the axis with $r \gg d$, show that the electric potential $V(r)$ is given by

$$V(r) = \frac{kq}{r}\left(1 + \frac{2d}{r}\right).$$

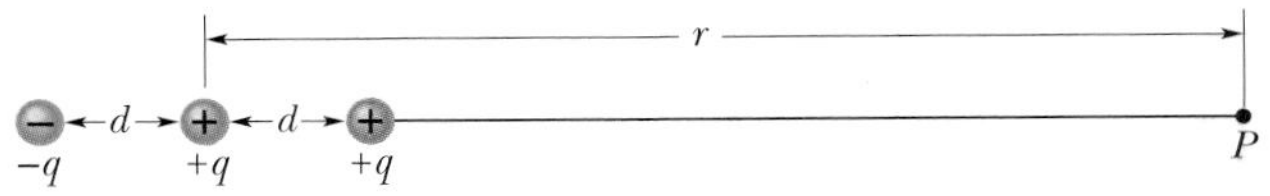

−q, d, +q, d, +q, r, P

FIGURE 25-35 ■ Problem 37.

(*Hint:* The charge configuration can be viewed as the sum of an isolated charge and a dipole.)

SEC. 25-9 ■ POTENTIAL DUE TO A CONTINUOUS CHARGE DISTRIBUTION

38. Plastic Rod (*a*) Figure 25-36*a* shows a positively charged plastic rod of length L and uniform linear charge density λ. Setting $V = 0$ at infinity and considering Fig. 25-17 and Eq. 25-34, find the electric potential at point P without written calculation. (b) Figure 25-36*b* shows an identical rod, except that it is split in half and the right half is negatively charged; the left and right halves have the same magnitude λ of uniform linear charge density. With V still zero at infinity, what is the electric potential at point P in Fig. 25-36*b*?

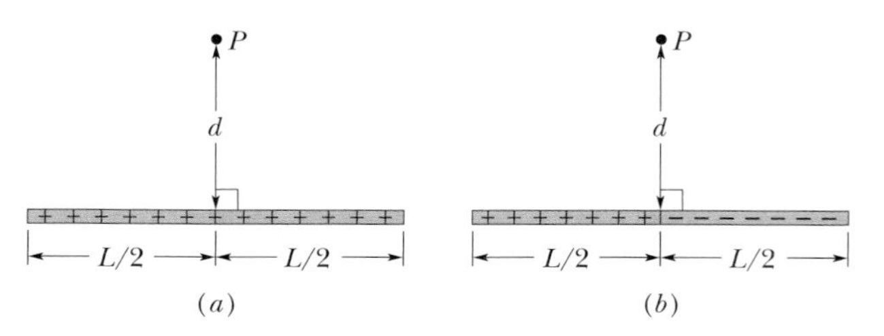

FIGURE 25-36 ■ Problem 38.

39. Nonlinear Charge Density The plastic rod shown in Fig. 25-37 has length L and a nonuniform linear charge density $\lambda = cx$, where c is a positive constant. With $V = 0$ at infinity, find the electric potential at point P_1 on the axis, at distance d from one end.

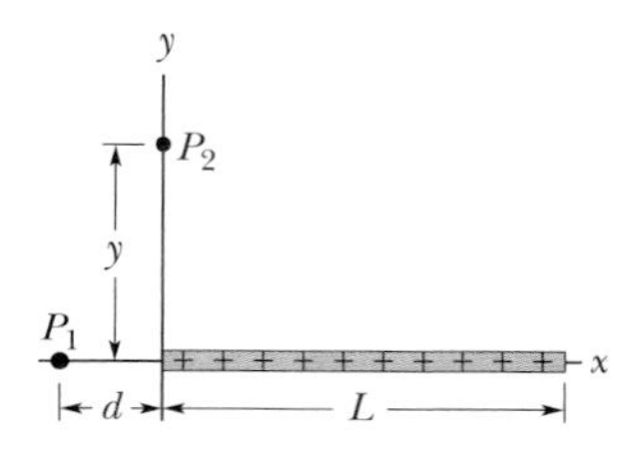

FIGURE 25-37 ■ Problems 39, 40, 44, 45.

40. Rod of Length *L* Figure 25-37 shows a plastic rod of length L and uniform positive charge Q lying on an x axis. With $V = 0$ at infinity, find the electric potential at point P_1 on the axis, at distance d from one end of the rod.

SEC. 25-10 ■ CALCULATING THE ELECTRIC FIELD FROM THE POTENTIAL

41. Points in the *xy* Plane The electric potential at points in an xy plane is given by $V = (2.0\ \text{V/m}^2)x^2 - (3.0\ \text{V/m}^2)y^2$. What are the magnitude and direction of the electric field at the point (3.0 m, 2.0 m)?

42. Parallel Metal Plates Two large parallel metal plates are 1.5 cm apart and have equal but opposite charges on their facing surfaces. Take the potential of the negative plate to be zero. If the potential halfway between the plates is then +5.0 V, what is the electric field in the region between the plates?

43. Show That (a) Using Eq. 25-31, show that the electric potential at a point on the central axis of a thin ring of charge of radius R and a distance z from the ring is

$$V = \frac{kq}{\sqrt{z^2 + R^2}}.$$

(b) From this result, derive an expression for the E-field magnitude $|\vec{E}| = E$ at points on the ring's axis; compare your result with the calculation of E in Section 23-7

44. Why Not The plastic rod of length L in Fig. 25-37 has the nonuniform linear charge density $\lambda = cx$, where c is a positive constant. (a) With $V = 0$ at infinity, find the electric potential at point P_2 on the y axis, a distance y from one end of the rod. (b) From that result, find the electric field component E_y at P_2. (c) Why cannot the field component E_x at P_2 be found using the result of (a)?

45. Find Component (a) Use the result of Problem 39 to find the electric field component E_x at point P_1 in Fig. 25-37 (*Hint:* First substitute the variable x for the distance d in the result.) (b) Use symmetry to determine the electric field component E_y at P_1.

SEC. 25-11 ■ POTENTIAL OF A CHARGED ISOLATED CONDUCTOR

46. Hollow Metal Sphere An empty hollow metal sphere has a potential of +400 V with respect to ground (defined to be at $V = 0$) and has a charge of 5.0×10^{-9} C. Find the electric potential at the center of the sphere.

47. Excess Charge What is the excess charge on a conducting sphere of radius $r = 0.15$ m if the potential of the sphere is 1500 V and $V = 0$ at infinity?

48. Widely Separated Consider two widely separated conducting spheres, 1 and 2, the second having twice the diameter of the first. The smaller sphere initially has a positive charge q, and the larger one is initially uncharged. You now connect the spheres with a long thin wire. (a) How are the final potentials V_1 and V_2 of the spheres related? (b)What are the final charges q_1 and q_2 on the spheres, in terms of q? (c) What is the ratio of the final surface charge density of sphere 1 to that of sphere 2?

49. Two Metal Spheres Two metal spheres, each of radius 3.0 cm, have a center-to-center separation of 2.0 m. One has a charge of $+1.0 \times 10^{-8}$ C; the other has a charge of -3.0×10^{-8} C. Assume that the separation is large enough relative to the size of the spheres to permit us to consider the charge on each to be uniformly distributed (the spheres do not affect each other). With $V = 0$ at infinity, calculate (a) the potential at the point halfway between their centers and (b) the potential of each sphere.

50. Charged Metal Sphere A charged metal sphere of radius 15 cm has a net charge of 3.0×10^{-8} C. (a) What is the electric field at the sphere's surface? (b) If $V = 0$ at infinity, what is the electric potential at the sphere's surface? (c) At what distance from the sphere's surface has the electric potential decreased by 500 V?

51. Surface Charge Density (a) If Earth had a net surface charge density of 1.0 electron per square meter (a very artificial assumption), what would its potential be? (Set $V = 0$ at infinity.) (b) What would be the electric field due to the Earth just outside its surface?

52. Concentric Spheres Two thin, isolated, concentric conducting spheres of radii R_1 and R_2 (with $R_1 < R_2$) have charges q_1 and q_2. With $V = 0$ at infinity, derive expressions for the electric field magnitude $E(r)$ and the electric potential $V(r)$, where r is the distance from the center of the spheres. Plot $E(r)$ and $V(r)$ from $r = 0$ to $r = 4.0$ m for $R_1 = 0.50$ m, $R_2 = 1.0$ m, $q_1 = +2.0\ \mu$C, and $q_2 = +1.0\ \mu$C.

Additional Problems

53. Work Done Consider a charge $q = -2.0\ \mu C$ that moves from A to B or C to D along the paths shown in Fig. 25-38. This charge is moving in the presence of a uniform electric field of magnitude $E = 100$ N/C.

(a) What is the total work done on the charge if the distance between A and B is 0.62 m?

(b) What is the total work done on the charge if the distance between C and D is 0.58 m?

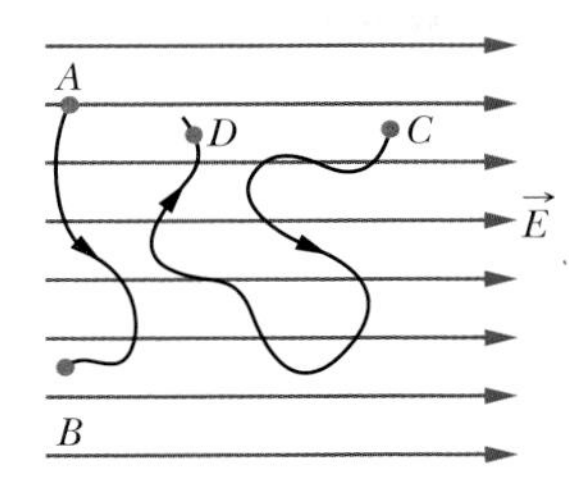

FIGURE 25-38 ■ Problem 53.

54. Orienteering an Electric Potential. (a) Figure 25-39 shows a contour plot of part of a range of hills in Virginia. The outer part of the figure is at sea level (marked 0). Each contour line from the region marked 0 shows a level 10 m higher than the previous line. The maximum height is 70 m and is shown by the number 70.

Answer the following questions by giving the pair of grid markers (a letter and a number) closest to the point being requested.

i. Where is there a steep cliff?
ii. Where is there a pass between two hills?
iii. Where is the easiest climb up the hill?

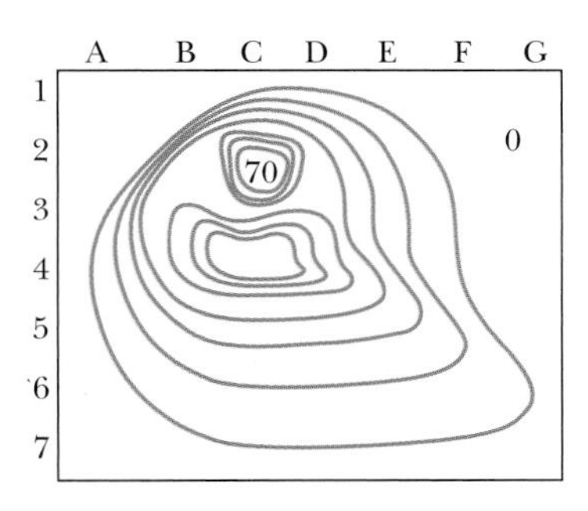

FIGURE 25-39 ■ Problem 54.

(b) Now suppose the figure represents a plot of the electric equipotentials for the surface of a glass plate, and the numbers now represent voltage. The maximum is 70 V and each contour line from the region marked 0 shows a level 10 V higher than the previous line.

i. Where would a test charge placed on the glass feel the strongest electric force? In what direction would the force point?
ii. Is there a place on the glass where a charge could be placed so it feels no electric force? Where?

26 Current and Resistance

When the zeppelin *Hindenburg* was built, it was the pride of Germany. Almost three football fields long, it was the largest flying machine ever built. Although the zeppelin was kept aloft by 16 cells of highly flammable hydrogen gas, it made many uneventful trans-Atlantic trips. However, on May 6, 1937, the *Hindenburg* burst into flames while landing at a U.S. naval air station in New Jersey during a rainstorm. While its handling ropes were being let down to a ground crew, ripples were sighted on the outer fabric near the rear of the ship. Seconds later, flames erupted from that region and 32 seconds after that the Hindenburg fell to the ground.

After so many successful flights of hydrogen zeppelins, why did this one burst into flames?

The answer is in this chapter.

26-1 Introduction

The interpretation of electrostatics experiments (described in Chapters 22 through 25) is that matter consists of two kinds of electrical charges, positive and negative. At least some negative charge can be moved from one object to another, leaving the first positively charged (with a deficit of negative charge) and the second negatively charged (with an excess of negative charge). Once the charges stopped moving we explored the electrostatic forces between them.

It turns out that the electrical devices we encounter most often in modern life such as computers, lights, and telephones are not purely electrostatic but involve moving charges which we will come to call *electric currents*. In addition, natural phenomena such as lightning, the flow of protons between the Earth's magnetic poles, and cosmic ray currents involve electric currents.

In this chapter we explore electric currents, or charge flow, with a primary focus on how current passes through conductors in electric circuits. We will see that the critical idea is to understand that a potential difference across a conductor causes a flow of charge (a current) through that conductor.

26-2 Batteries and Charge Flow

By the end of the 18th century, Alessandro Volta had discovered that when two metal plates were placed in contact with a moist piece of metal, they seemed to have electrical properties like those of rubbed amber and glass. To magnify this effect, Volta piled up pairs of unlike metals. When he grasped the plates (terminals) at each end of the pile with his hands, he claimed to feel electric charges move through his body on a continuous basis. Volta had invented the battery, and his experience with early batteries is an indication that there is a connection between electric charges, as discussed in Chapters 22–25, and the continuous flow of electricity created by batteries and other power sources. However, it is not obvious without further investigation that there is actually a connection between the sensation of electric flow that Volta experienced and the electric charges we believe exist based on the electrostatic observations discussed in Chapter 22.

In order to investigate this further, let's examine the results of several experiments involving a metal wire connected to oppositely charged conducting plates as shown in Fig. 26-1*a* and the same wire connected to a battery instead as shown in Fig. 26-1*b*.

Experiment 1 (Electrostatic Discharge): Suppose we use glass and amber rods that have been rubbed to transfer electrons to or from conducting plates. We can use a hanging amber or glass rod shown in Fig. 22-2 to verify that we have excess electric charge on each plate. Since it is easier to add excess electrons to a conductor than remove electrons from a conductor, the negatively charged plate will tend to have a

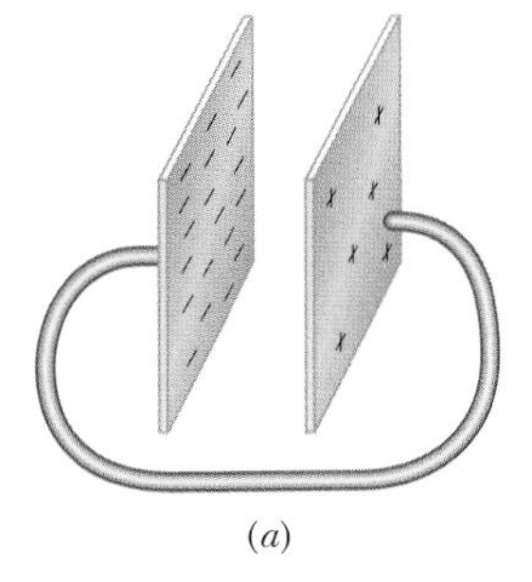
(*a*)

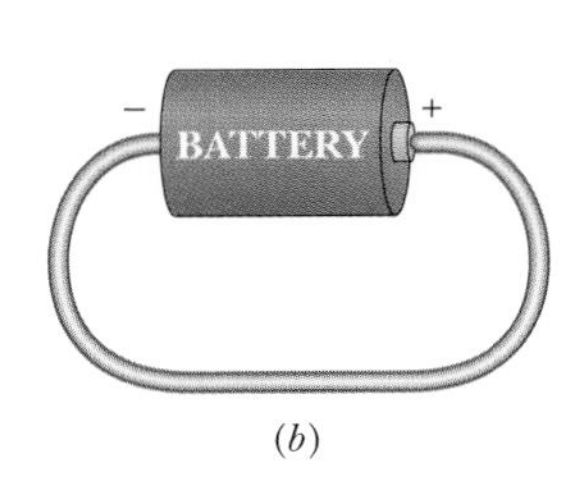

(*b*)

FIGURE 26-1 ■ (*a*) When a conducting wire connects two oppositely charged plates, charge flows from the negatively charged plate to the positively charged plate until both plates have the same number of excess electrons. As a result the wire becomes hot. (*b*) When a battery is placed between the ends of the wire instead, the wire also becomes hot, indicating that charge is also flowing.

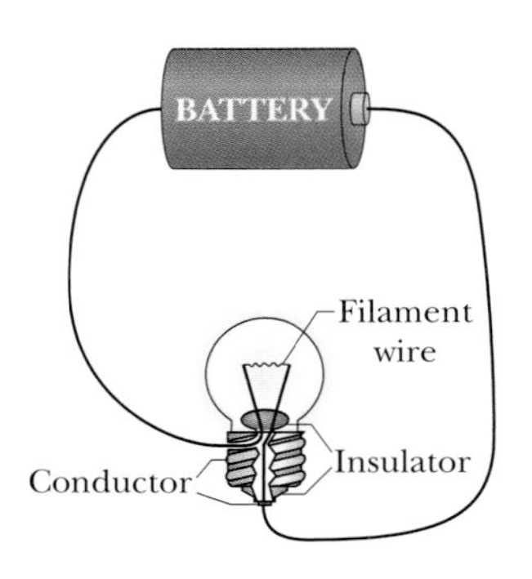

FIGURE 26-2 ■ If there is a complete conducting loop between the two terminals of a battery, a bulb will stay lit until the battery runs down.

greater magnitude of charge. Initially the negative electrons on the left plate repel each other and spread out but cannot leave the plate. Since a positive test charge placed between the plates will be repelled from the positive plate (on the right) and attracted to the negative plate (on the left), we know there is an electric field between the plates. So, Eq. 25-17 tells us there will be a potential difference between the plates.

If we connect the two plates with a piece of thread nothing happens. But when a conducting wire is connected across the two plates, (Fig. 26-1*a*) we observe that excess electrons on the left-hand plate will flow to the right-hand plate until both plates have the same number of excess electrons on them. This is not surprising since we expect the repulsive forces between the electrons on the left plate to push the charges through the wire while the attractive forces on the right plate pull on the charges. If we have enough excess charge on the plates, the wire will feel hot just after the discharge and then cool down again. If the wire has a properly connected small bulb in the middle of it, the bulb will light up briefly and then go out. We conclude from these observations that charge is flowing through the wire for a short time.

Experiment 2 (Battery Current): As we mentioned in Chapter 25, a battery is capable of doing work on electric charges and increasing their potential energy. So there must be a potential difference across its terminals. If we connect a piece of thread between the terminals of a battery nothing happens. On the other hand, if we connect a wire between the terminals of a battery, we observe that the wire gets very hot and stays that way for a long time as shown in Fig. 26-1*b*. If we also properly connect a bulb to the middle of the wire, as shown in Fig. 26-2, the bulb stays continuously lit until the battery eventually runs down. (In the next section we discuss how to connect a bulb to a battery properly, so that it lights.)

Because, at first, the electrostatic charging in Experiment 1 has the same result as the battery in Experiment 2, we infer that the underlying electric effects are the same in both cases. The hot wires and the lighting of bulbs lead us to conclude that charge is flowing through the wires. We call this flow of charge **electric current.**

READING EXERCISE 26-1: Although you were not provided with any details, what sensations might Volta have felt that led him to believe that electric charge was flowing through his body? ■

26-3 Batteries and Electric Current

There are some additional observations that help us understand the nature of electric current. Suppose we want to use a battery and perhaps a wire to light a flashlight bulb. By fiddling around we discover that many of the possible arrangements for lighting a bulb *do not work*. For example, none of the arrangements shown in Fig. 26-3 work.

To understand why these arrangements do not work, we need to examine a flashlight bulb much more carefully. The flashlight bulb consists of a piece of thin conducting "filament" wire encapsulated in glass that has no air inside. This wire glows and so gives off light when electric current passes through it. One end of the filament wire is in contact with a conductor that surrounds the bottom part of the bulb. The other end is connected to another conductor at the bulb's base. These conductors are separated by an insulator. A cutaway diagram of the bulb is shown in Fig. 26-4.

FIGURE 26-3 ■ Three of many arrangements of a battery and bulb and wire that do not cause the bulb to light.

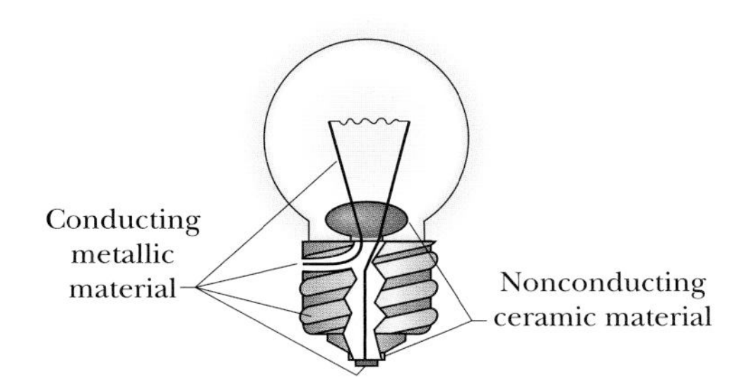

FIGURE 26-4 ■ A cutaway diagram of a flashlight bulb.

After some more fiddling we discover that all of the arrangements of wires, bulb, and battery that cause the bulb to light have one thing in common. They all have a continuous, complete loop or **circuit** for current to pass from one terminal of a battery through conductors back to other terminal of the battery. In addition to the arrangement shown in Fig. 26-2, another of the many arrangements that forms a complete loop and causes a bulb to light is shown in Fig. 26-5.

When bulb filaments get old they sometimes break. In this case the circuit is incomplete and our "burned out" bulb does not light. Another requirement is that the battery must have a potential difference between its terminals. When a battery loses its potential difference after much use we refer to it as a "dead battery."

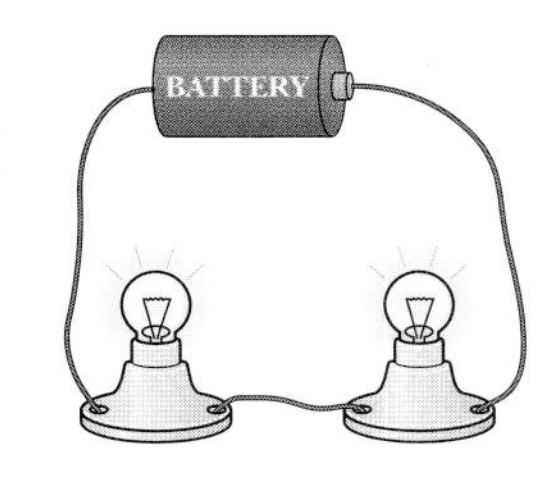

FIGURE 26-5 ■ When two identical bulbs in holders are connected in a row to a battery, they have the same brightness as each other. We conclude that the same current is passing through both bulbs. This indicates that the battery is not a source of excess charge used up by the bulbs.

What Is Stored in a Battery?

It is commonly (and wrongly) believed that batteries store excess charge that can be "used up" in a circuit, and that a battery is "dead" when this excess charge is used up. The fact that people often refer to "charging" and "discharging" batteries is evidence of this belief. Careful observation tells us that this idea is wrong. The excess charge a fresh alkaline flashlight battery would have to store to keep a flashlight bulb lit as long as it does is more than 20 000 coulombs. This is a hundred million times the amount of charge we can typically place on a light metal-coated ball on a string. Yet, we observe no forces between such a charged ball and a fresh battery. There are also no forces between a charged ball and the wires carrying current in a circuit.

> Observations indicate that both batteries and any current-carrying wires connected to them are *electrically neutral.*

We conclude from these observations that batteries do not store charge. Batteries store energy. The energy in the battery is transformed to mechanical energy, light energy, and thermal energy as it pushes charges through wires and bulbs. Thus our observations support the idea that a battery acts as a pump that absorbs electrons at the negative terminal and releases higher potential energy electrons from the positive terminal. We discuss how chemical reactions can create a charge pump in more detail in Section 27-6.

If we connect two identical bulbs to the same battery as shown in Fig. 26-5, they shine with the same brightness. Based on this observation we conclude that the same current is passing through both bulbs.

> When wires and other conducting elements such as bulbs are placed between the battery terminals to make a continuous loop or circuit, the battery acts as a pump that pushes charge carriers already available in the wires around the loop. The battery is not a source of charge and electrical elements like bulbs do not use up charge.

Figure 26-6*a* shows a very simplified representation of a small segment of wire made up of electrically neutral atoms. In Fig. 26-6*b* the ends of the wire segment are

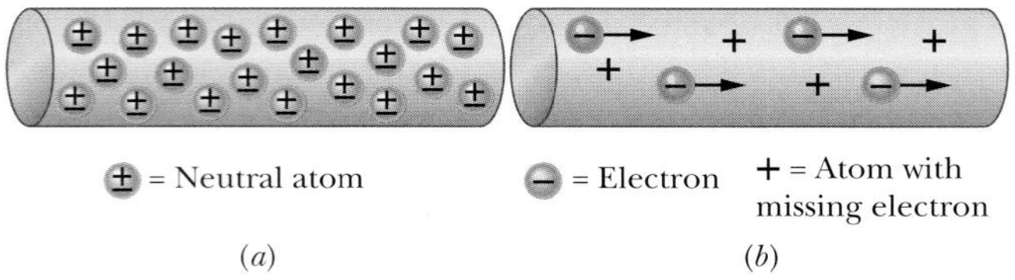

FIGURE 26-6 ■ (*a*) A representation of many electrically neutral atoms in a wire. (*b*) A diagram that shows a potential difference across the ends of the wire so a very small fraction of the electrons surrounding atoms start moving and a few ions with missing conduction electrons are present. These ions have excess positive charge. *Note:* The neutral atoms are still present but are not shown.

± = Neutral atom (*a*)
− = Electron + = Atom with missing electron (*b*)

connected to a battery (not shown). A few of the conduction electrons in the metal start moving, but the stationary charges, consisting of neutral atoms and ions (with missing conduction electrons) still exist in the wire. The stationary ions neutralize the moving conduction electrons. To reduce clutter, Fig. 26-6*b* shows the stationary ions but not the neutral atoms. In other figures in this chapter we just show moving electrons and not the stationary ions. This type of depiction can give the false impression that there is excess charge in the wire. This is not so. Conducting wires are electrically neutral.

Defining Current Mathematically

Figure 26-2 shows a complete circuit with a battery (or other power source) that maintains a constant potential difference across its terminals. In this case, charge pushed through the circuit by the battery flows through a conducting wire, and then through the filament of a bulb, which is usually a very thin wire. In order to think more carefully about the current, we need to develop a mathematical definition for current.

Figure 26-7 shows a section of a conducting loop with different cross-sectional areas in which a current has been established. If net charge dq passes through a hypothetical plane (such as a) in time dt, then the current through that plane is defined as

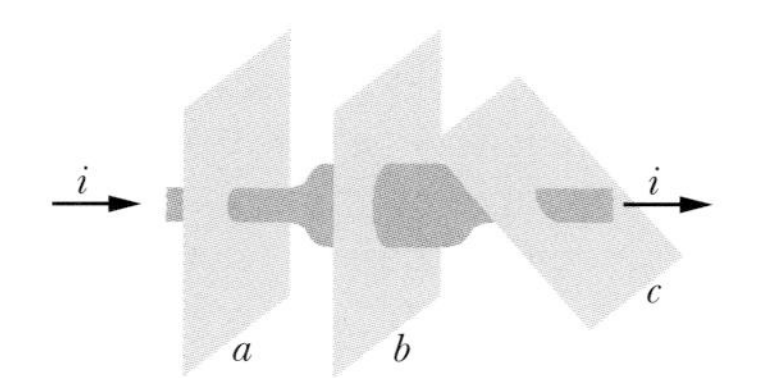

FIGURE 26-7 ■ The current i or charge per unit time through the conductor has the same value at imaginary planes a, b, and c as long as the planes cut through the entire conductor at the points of intersection.

$$i \equiv \frac{dq}{dt} \quad \text{(definition of current).} \tag{26-1}$$

Regardless of the details of the geometry of the charge flow, we can find the net charge passing through any plane in a time interval extending from 0 to t by integration:

$$q = \int dq = \int_0^t i\,dt. \tag{26-2}$$

Measurements of current through various locations in a single loop circuit show that the current is the same in all parts of a circuit where there are no junctions or alternate paths for the current to take. The current or rate of charge flow is the same passing through the imaginary planes a, b, and c shown in Fig. 26-7. Indeed, the current is the same for any plane that passes completely through the conducting elements in a continuous circuit with no branches, no matter what their locations or orientations. That is, a charge carrier must pass through plane a for every charge carrier that passes through plane c.

The unit for current is called the *ampere* (A), and it can be related to the coulomb by the expression

$$1 \text{ ampere} = 1 \text{ A} = 1 \text{ coulomb/second} = 1 \text{C/s}.$$

The Directions of Currents

How can we tell whether there are positive or negative charges moving when a current is established in electrically neutral conductors? When we place a conducting wire between the plates shown in Fig. 26-8, charge carriers flow until the plates are neutralized.

It is not possible for us to design an experiment based on macroscopic observations that will allow us to tell whether the charge carriers are positive or negative because the end result (neutralized plates) will be the same in either case. Early experimenters with electricity had no knowledge of atomic structure and could only use macroscopic observations of electrical effects to guide them. They assumed that charge carriers were positive. Even though we now know that negatively charged electrons are the charge carriers in conductors, for historical reasons we will stick with the assumption that the charge carriers are positive. This historical assumption makes it easier to use traditional references on electricity, and all the characteristics of circuits we will study on a macroscopic level will be exactly the same. Furthermore, this early assumption would have been correct if Benjamin Franklin had decided to designate the excess charges on rubber rods as positive and those on glass as negative instead of the other way around!

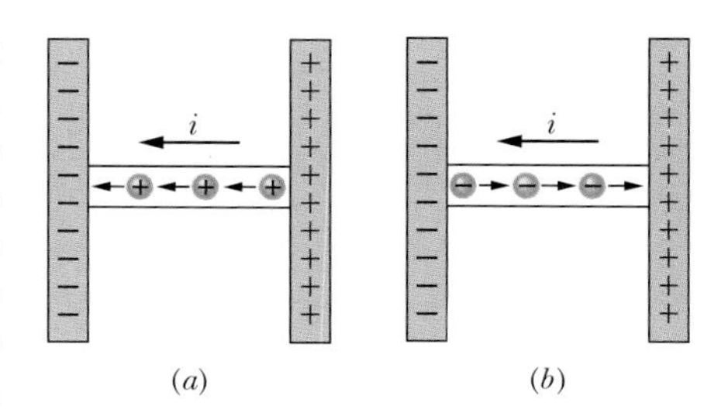

FIGURE 26-8 ■ No macroscopically-oriented experiment will allow us to detect whether the charge carriers in a conducting wire are (*a*) positive or (*b*) negative. So we define the current flow to be from right to left in both cases. Although stationary charges are not depicted, the conducting wires are neutral.

Although the charge carriers in conductors are negative, other currents, for example, protons streaming out of our Sun, create positive currents. Also charge carriers in fluids can be either positive ions (atoms with missing electrons) or negative electrons or ions (atoms with extra electrons). In fact, the movement of charge within most batteries is due to the migration of positive ions that undergo chemical reactions. Also, currents in biological systems are carried by sodium and potassium ions, which are positive charge carriers.

Current arrows show only a direction (or sense) of flow of charge carriers along the connected conductors as they bend and turn between battery terminals, not a fixed direction in space. Since current is actually a flux, which is a scalar quantity, *these current arrows do not represent vectors with magnitude and direction.*

> A current arrow, although not a mathematical vector, is drawn in the direction in which positive charge carriers would move through wires and circuit elements from a higher potential to a lower (more negative) potential, even though the actual charge carriers are usually negative and move in the opposite direction.

Charge Conservation at Junctions

So far we have only considered circuits like the one shown in Fig. 26-9*a*, ones for which there is only one path for charge carriers to follow. Such circuits are called **series** circuits. However, it is also common to find circuits or portions of circuits in which charge carriers encounter a junction where they can take either of two (or more) paths as shown in Fig. 26-9*b*. We call this type of circuit a **parallel** circuit. Although we introduce the terms "series" and "parallel" here, we will focus on the quantitative evaluation of series and parallel circuits in Chapter 27.

Figure 26-9*b* shows the moving charge carriers splitting up at a junction and then moving in parallel. If the bulbs are *identical,* how do the currents split at junction 1?

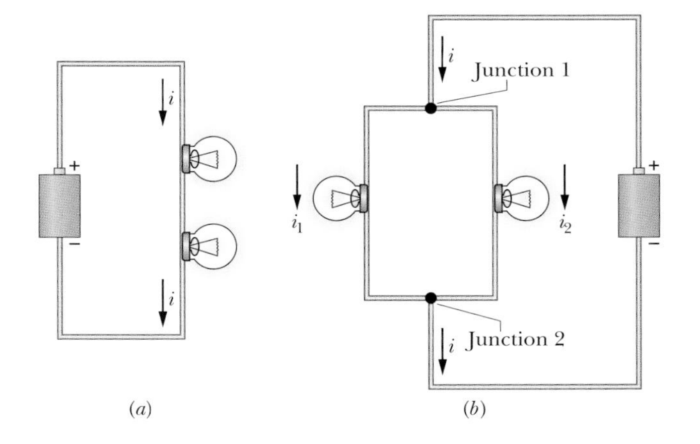

FIGURE 26-9 ■ We use a lightbulb as an example of a circuit element. (*a*) A series connection involves two or more circuit elements that are connected together so that the same current that passes through one element must pass through the other element. The potential differences across the elements is the sum of the drops across each element. (*b*) A parallel connection requires that one terminal of each two or more elements are connected together at one point and then the other terminal of each of the elements is connected together at another point. These points of connection are called junctions. Because of the connections at the junctions the potential difference across each element is the same when the parallel network is placed in a circuit.

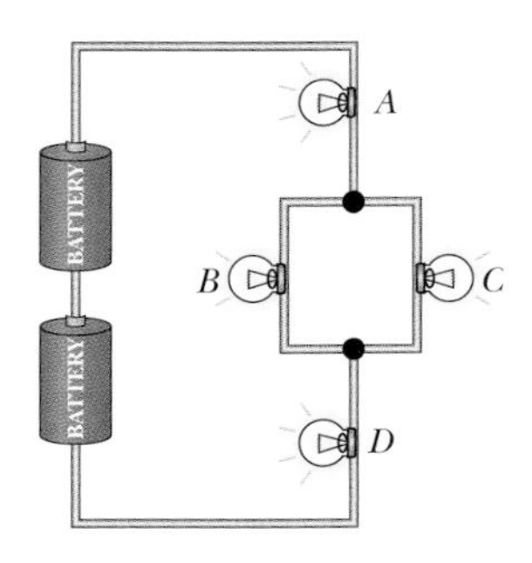

FIGURE 26-10 ■ To verify what we think would happen to current at the junctions, four identical lightbulbs and a battery are connected in a circuit having both series and parallel elements. Observations tell us that the brightness of bulb A is the same as that in bulb D. Bulbs B and C have the same brightness as each other but share the battery's current, so they are much dimmer than A and D. Note that even though they are not adjacent, bulbs A and D are in series.

What happens when they come back together at junction 2? Since the bulbs are identical, we expect that the current coming into junction 1 will divide equally so half takes the left path and half the right. If charge is conserved, when the currents combine at junction 2, they should add up to the original current. We can verify this by making a very simple observation with a couple of flashlight batteries in series and four bulbs as shown in Fig. 26-10. Bulbs A and D have the same brightness as each other. This fact indicates that the amount of current through bulb A is the same amount of current passing through bulb D and back through the battery. The fact that bulbs B and C have the same brightness as each other but are dimmer than A and D suggests that the current is splitting in half at the junction. We conclude

$$i = i_1 + i_2 = \frac{i}{2} + \frac{i}{2} \quad \text{(special case with identical parallel elements)}, \tag{26-3}$$

where i is the total current and i_1 and i_2 are the currents in the two branches.

If charge is conserved, the magnitudes of the currents in two parallel branches *must* add to yield the magnitude of the current in the original conductor even when the branches have different circuit elements. This statement, called Kirchhoff's current law, states that in general

$$i = i_1 + i_2 \quad \text{(three-way junction of Fig. 26-11}a\text{)}. \tag{26-4}$$

We treat Kirchhoff's circuit laws in more detail in Chapter 27.

Experiments indicate that bending or reorienting the wires in space does not change the validity of Eq. 26-4. The fact that current is not affected by wire orientation can be explained by the accumulation of static surface charges. These charges keep the electric field associated with a potential difference pointing along a wire, regardless of how the wire twists. The lack of influence of bending is depicted in Fig. 26-11.

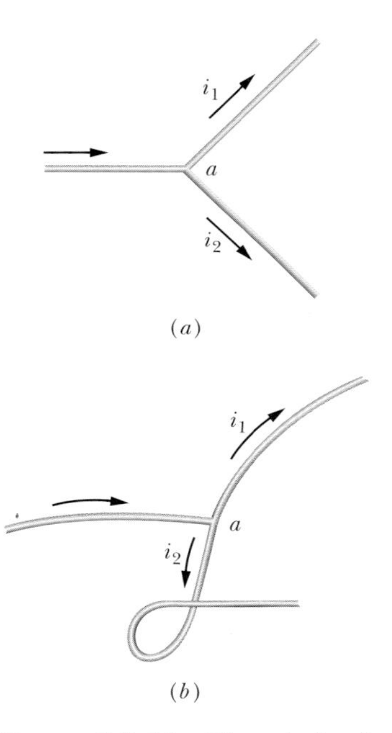

FIGURE 26-11 ■ The relation $i = i_1 + i_2$ is true at junction a no matter what the orientation of the three wire segments is.

READING EXERCISE 26-2: Explain why the word "circuit," as used in everyday speech, is an appropriate term for application to electrical situations. ■

READING EXERCISE 26-3: Suppose a battery sets up a flow of charges through wires and a bulb. (a) Will the overall circuit, consisting of the battery, bulb, and wires, remain electrically neutral, become positive, or become negative? Explain. (b) Will the wires remain electrically neutral, become positive, or become negative? (c) What ordinary observations would support your answers? ■

READING EXERCISE 26-4: Apply your understanding of the concept of flux, the nature of electrical charge, and the definition of current presented to explain why a flow of equal amounts of opposite charge in the same direction would not be considered a current. That is, explain why there must be a net flow of charge through a surface for there to be a current. ■

READING EXERCISE 26-5: The figure below shows a portion of a circuit. What are the magnitude and direction of the current i in the lower right-hand wire?

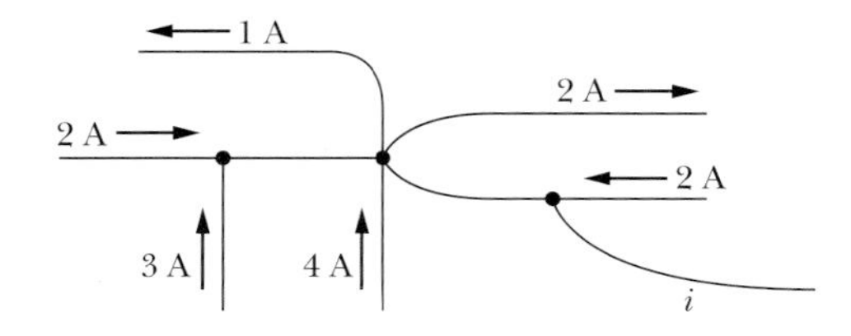

■

TOUCHSTONE EXAMPLE 26-1: Charged Fuel

If you've ever gone to a gas station to fill a gas can with fuel for your lawn mower, you may have noticed the sign that tells you to take the gas can out of your car and place it on the ground before you fill it with gasoline. Why is this important? As fuel is pumped from its underground storage tank, it can acquire a net electrical charge. If so, as you pump fuel into a container, the can will build up a net electrical charge if it is electrically isolated from its surroundings. If this charge builds up to a sufficient level, it can create a spark, igniting the fumes around your container with very unfortunate consequences.

Suppose the maximum safe charge that can be deposited on your 5.0 gal gas can is 1.0 μC.

(a) What is the maximum safe charge per liter that the fuel you are pumping can have?

SOLUTION ■ The **Key Idea** here is simply that the maximum safe "charge density" is

$$(1.0\ \mu\text{C})/(5.0\ \text{gal}) = (0.20\ \mu\text{C/gal})(264\ \text{gal/m}^3)(1\ \text{m}^3/1000\ \text{L})$$
$$= 0.0528\ \mu\text{C/L}. \qquad \text{(Answer)}$$

(b) If the pump delivers fuel at a rate of 8.0 gallons per minute, what is the maximum safe electrical current associated with the flow of the fuel into the can?

SOLUTION ■ The **Key Idea** here is that the fuel delivery rate is

$$(8.0\ \text{gal/min})(1\ \text{L}/0.264\ \text{gal})(1\ \text{min}/60\ \text{s}) = 0.50505\ \text{L/s}.$$

Since each liter of fuel can deliver no more than 0.0528 μC safely, the maximum safe electrical current is just (0.0528 μC/L) (0.50505 L/s) = 0.027 μC/s = 27 nA.

26-4 Circuit Diagrams and Meters

As we move into the remaining sections in this chapter and the next, we will be drawing electric circuits with elements such as batteries, bulbs, wires, and switches. We will also be introducing new elements such as resistors and meters for measuring current and voltage.

Symbols for Basic Circuit Elements

Before proceeding with our study of current and resistance, we pause and introduce a few of the symbols scientists and engineers have created to represent circuit elements. Figure 26-12 shows the common symbols used to make the circuits we discuss in this chapter easier to draw.

Using these symbols, the circuit shown in Fig. 26-2 with a switch added can be represented as shown in Fig. 26-13.

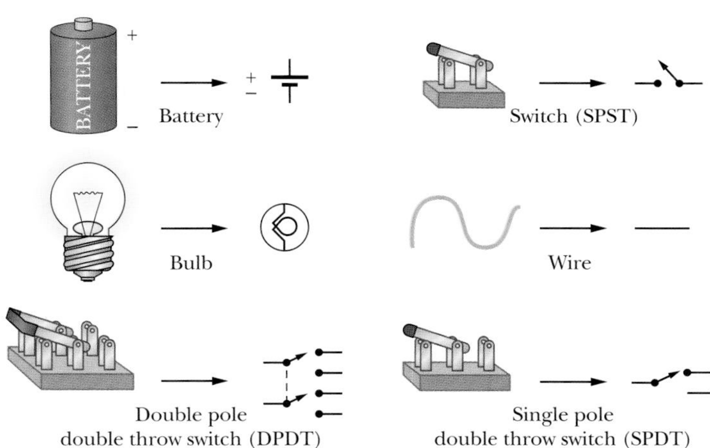

FIGURE 26-12 ■ Some circuit symbols.

Meters

Current and potential differences are very important properties of electrical circuits, and we have well-established convenient

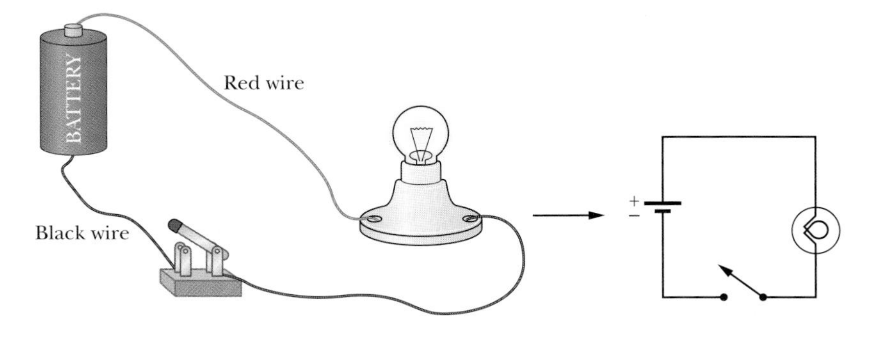

FIGURE 26-13 ■ A circuit sketch and corresponding diagram.

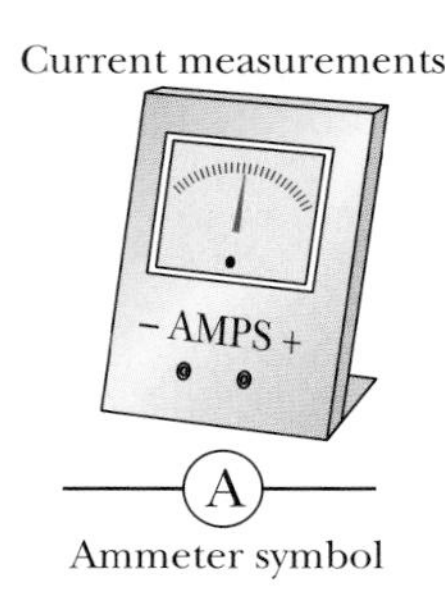

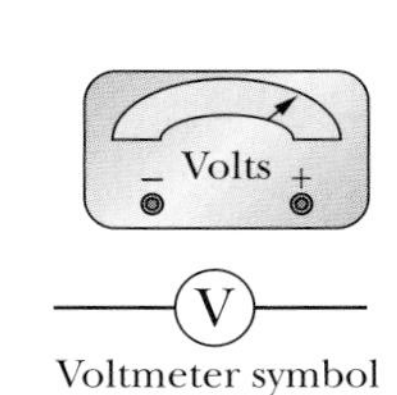

FIGURE 26-14 ■ An analog ammeter for measuring current and an analog voltmeter for measuring potential difference (or "voltage"), along with their circuit symbols.

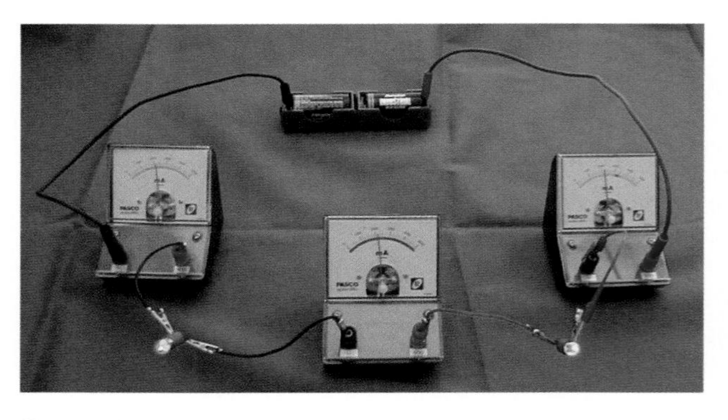

FIGURE 26-15 ■ Three analog ammeters measure the same current flowing through three locations in a series circuit consisting of two #14 flashlight bulbs.

ways to measure these quantities using meters. The device with which one measures current is called an **ammeter.** Potential difference is measured with a device called a **voltmeter.** An ammeter and voltmeter along with their circuit symbols are depicted in Fig. 26-14.

Since an ammeter measures current *through* a circuit (or a branch of a more complex circuit), it is placed in *series* with circuit elements. A voltmeter measures the potential difference between two locations (or points) in a circuit, so a voltmeter is placed across or in *parallel* with the two points of interest. This is shown in Fig. 26-15.

Often ammeters and voltmeters are combined in a device used to measure either potential difference or current. When the two or more meters are combined, the meter is typically called a **multimeter.** A digital multimeter is shown in Fig. 26-16. Many modern digital multimeters are also capable of measuring other quantities we will discuss, such as resistance and capacitance.

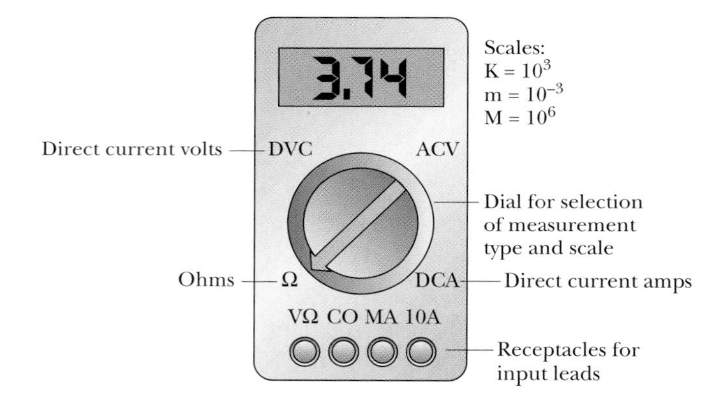

FIGURE 26-16 ■ The digital multimeter pictured can be configured to act as an ammeter to measure current *through* a given part of a circuit, a voltmeter to measure potential difference *across* any two points in a circuit, or the resistance of any circuit element.

READING EXERCISE 26-6: In Fig. 26-17, the voltmeter is attached across the bulb and the ammeter is inserted into the circuit. Why are these devices connected this way? How would the ammeter reading change if it were inserted in the circuit before the bulb instead of after it? ■

FIGURE 26-17 ■ A basic circuit for measuring the current flowing *through* a circuit element as a function of potential difference *across* it.

26-5 Resistance and Ohm's Law

In professional applications of physics like designing electronic devices, we often need to know what effect adding more circuit elements will have on the flow of current. Given devices like ammeters and voltmeters, with which we can measure current and potential difference, we can do quantitative studies of the relationship between current and potential difference. For example, what will happen to the current in a circuit element, such as a bulb, that is part of a circuit if we add more batteries in series with our original battery? What will happen to the current in a conducting wire as voltage increases? The experimental setup for this investigation is shown in Fig. 26-17. The results are presented in Fig. 26-18 as graphs of applied potential difference ΔV and the resulting current i in two different circuit elements.

We can draw several interesting conclusions from looking at the two graphs in Fig. 26-18. First, we see in both graphs that as the potential difference increases, the amount of current through a given device increases. Second, it is not possible to tell how much current exists just by knowing the potential difference across a circuit element. For instance, when 1.0 V is placed across the lightbulb, the current through it is greater than the current in the Nichrome wire with the same potential difference across it. Third, for the length of Nichrome wire, the current is directly proportional to the potential difference, ΔV, across it. Thus, if we know the slope of the line, we can predict the current associated with any value of ΔV. Because of this direct proportionality, we refer to the Nichrome wire as a **linear** device. For the lightbulb, there is no convenient direct proportionality, so it is called a **nonlinear** device.

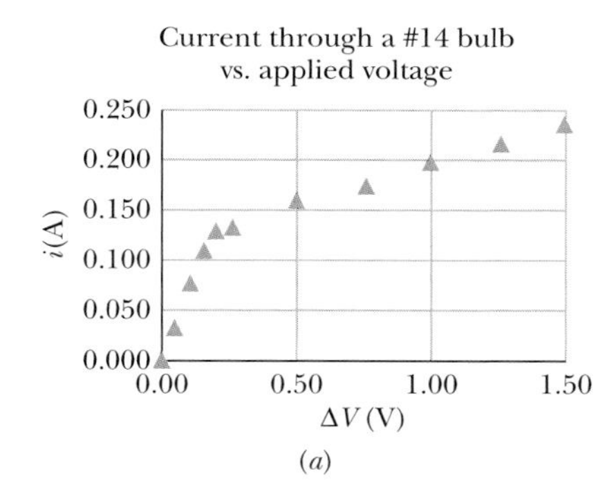

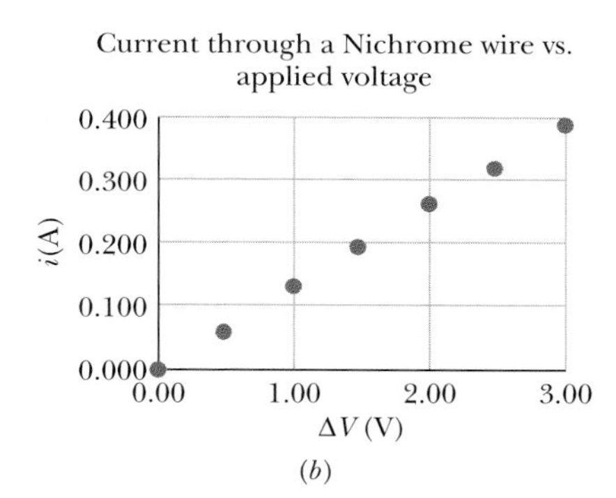

FIGURE 26-18 ■ Graph (a) shows ammeter data for current passing through a #14 lightbulb as a function of potential difference between the terminals of the lightbulb. Graph (b) shows ammeter data for the current through a length of cylindrical Nichrome wire as a function of potential difference between the ends of the wire.

Definition of Resistance

In both the small bulb and the Nichrome wire, once we measure a specific potential difference, ΔV, across a circuit element and the corresponding current through it we have a measure of the *resistance* of the element to current but only at that ΔV. The **resistance** of a given circuit element is defined as the ratio of the potential difference across the element to the current through the element. When a small potential difference causes a relatively large current, the circuit element has a small resistance to flow of charge. Conversely, when the same potential difference produces a current that is small, we say the resistance is large. For example, in the data presented in Fig. 26-18, a potential difference of 1V across the bulb causes a current of 0.19A to flow, while the same potential difference across that Nichrome wire causes only 0.13A to flow. So we say that at the specific potential difference of 0.25V, the Nichrome wire has more resistance than the bulb.

We define resistance as the ratio of potential difference applied to the current that results:

$$R \equiv \frac{\Delta V}{i} \quad \text{(definition of } R\text{)}. \tag{26-5}$$

Here we use the notation ΔV to emphasize we are dealing with the *difference* in potential between two locations in a circuit, which changes the potential energy of the charges as they flow. When discussing circuits, potential difference is often referred to by an alternate name of **voltage.**

The SI unit for resistance that follows from Eq. 26-5 is the volt per ampere. This combination occurs so often that we give it a special name, the **ohm** (symbol Ω); that is,

$$1\,\text{ohm} = 1\,\Omega = 1\,\text{volt/ampere} = 1\,\text{V/A}. \quad (26\text{-}6)$$

If we rewrite Eq. 26-5 as

$$i = \frac{\Delta V}{R},$$

it emphasizes the fact that the potential difference across a device with resistance R *produces* an electric current. The most common way to express the definition of resistance in Eq. 26-5 is

$$\Delta V = iR. \quad (26\text{-}7)$$

For a linear device like Nichrome wire we will get the same value for R no matter what potential difference we impress across the device. However, we must be *careful* in the case of a nonlinear device like a light bulb to specify at what potential difference we are measuring the current, i, in order to determine its resistance.

Ohm's Law

As we just pointed out, our Nichrome wire has the same resistance no matter what the value of the applied potential difference (as shown in Fig. 26-18*b*). Other conducting devices, such as lightbulbs, have resistances that change with the applied potential difference (as shown in Fig. 26-18*a*). Although both the Nichrome wire and the bulb contain metallic conductors, the wire in the bulb is so thin that its temperature rises noticeably as the potential difference increases, and the bulb's resistance increases.

In 1827, George Simm Ohm, a Bavarian, reported that he had observed a linear relationship between current and potential difference for metallic conductors kept at a fairly constant temperature. Because of this, linear devices such as the length of Nichrome wire are sometimes referred to as **ohmic.**

> A device is said to obey Ohm's law whenever the current through it is *always* directly proportional to the potential difference applied. That is, the device's resistance is constant in the $\Delta V = iR$ relation.

Many elements used in electric circuits, whether they are conductors like copper or semiconductors like pure silicon or silicon containing special impurities, obey Ohm's law within some range of values of potential difference. If the current in a resistive device is large enough to cause significant temperature changes in it, then Ohm's law often breaks down.

It is sometimes contended that $R = \Delta V/i$ (or $\Delta V = iR$) is a statement of Ohm's law. That is not true! This equation is the defining equation for resistance, and it applies to all conducting devices, whether they obey Ohm's law or not. If we measure the potential difference ΔV across and the current i through any device, even a bulb or other non-ohmic device, we can find its resistance *at that value of* ΔV as $R \equiv \Delta V/i$. The essence of Ohm's law, however, is a plot of i versus ΔV that is a straight line, so that the value of R is independent of the value of ΔV.

Resistors

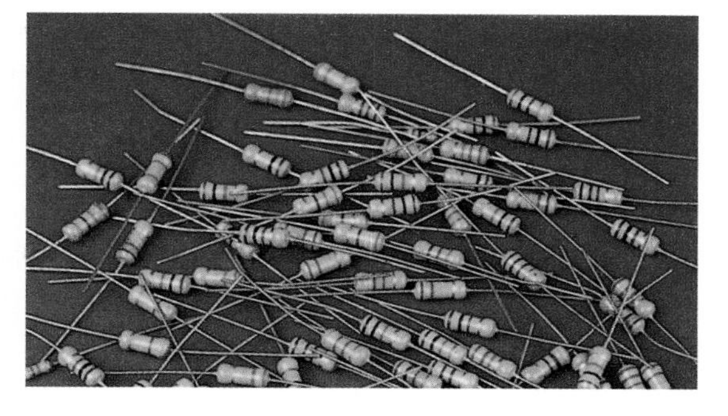

FIGURE 26-19 ■ An assortment of carbon resistors. The circular bands are color-coding marks that identify the value of the resistance.

A conductor whose function in a circuit is to obey Ohm's law so that it provides a specified resistance to the flow of charge independent of the potential difference impressed across it is called a **resistor** (see Fig. 26-19). Carbon resistors are the most standard sources of ohmic resistance used in electrical circuits for several reasons. Unlike a lightbulb, a resistor has a resistance that remains constant as current changes. Carbon resistors are inexpensive to manufacture, and they can be produced with a large range of resistances. The circuit diagram symbol for a resistor is shown in Fig. 26-20.

A typical carbon resistor contains graphite, a form of carbon, suspended in a hard glue binder. It usually is surrounded by a plastic case with a color code painted on it as shown in Fig. 26-21.

READING EXERCISE 26-7: The following table gives the current i (in amperes) through three devices for several values of potential difference ΔV (in volts). From these data, determine which devices, if any, obey Ohm's law.

Device 1		Device 2		Device 3	
ΔV	i	ΔV	i	ΔV	i
2.00	4.50	2.00	1.50	2.00	6.50
3.00	6.75	3.00	2.50	3.00	8.75
4.00	9.00	4.00	3.00	4.00	11.00

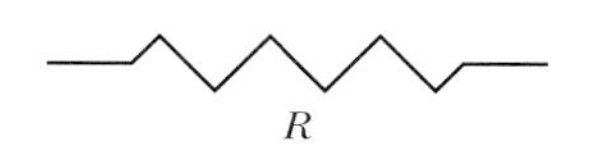

FIGURE 26-20 ■ Circuit diagram symbol for an ohmic resistor.

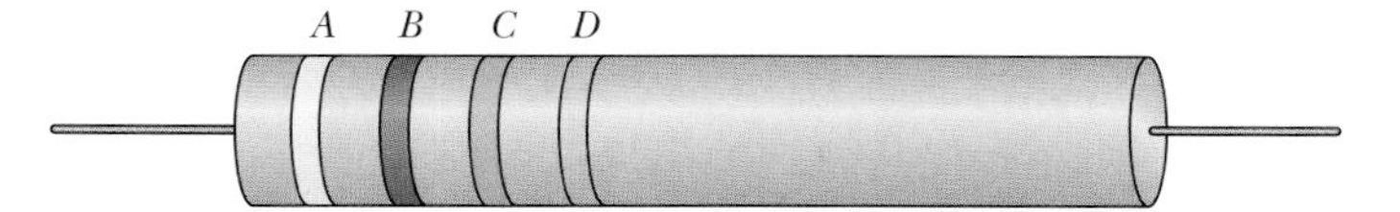

FIGURE 26-21 ■ Depiction of the four color bands on a color-coded resistor with $R = 47\ \text{K}\Omega \pm 10\%$. See Table 26-1 for details

26-6 Resistance and Resistivity

Next we consider how the resistance of ohmic circuit elements such as metal wires or carbon resistors depends on their geometries. That is, how does the resistance of a short, broad object change if we stretch it so it is long and thin? To determine this, we fix our investigation on a single material. For example, we might experiment with copper wire. Relatively thick copper wire is commonly used in electric circuits because it has a very low resistance compared to other circuit elements. Thus, it can be used to connect circuit elements without adding much resistance to a circuit.

TABLE 26-1
The Resistor Code[a]

Black	= 0	Blue	= 6
Brown	= 1	Violet	= 7
Red	= 2	Gray	= 8
Orange	= 3	White	= 9
Yellow	= 4	Silver	= ±10%
Green	= 5	Gold	= ±5%

[a]The value in ohms $= AB \times 10^C \pm D$. (AB means the A band digit placed beside the B band digit, not A times B). The colors on bands A, B, and C represent the digits shown in Table 26-1. The D band represents the "tolerance" of the resistor. No band denotes ±20%, a silver band denotes ±10%, and a gold band denotes ±5%. For example, a resistor with bands of Blue-Gray-Red-Silver has a value: $AB \times 10^C \pm D = 68 \times 10^2\,\Omega \pm 10\%$ or $(6800 \pm 680)\,\Omega$, since $A = 6$, $B = 8$, $C = 2$, $D =$ silver, (±10%).

Observations

Consider a conducting wire with a potential difference across its ends as shown in Fig. 26-22. To start with, we will keep the thickness of the wire fixed and just decrease its length. If we apply a potential difference across the ends of the wire and use current and potential difference measurements, we can determine its resistance as a function of length. We find that its resistance is proportional to its length L. Thus, we can write

$$R = kL.$$

If instead we fix the length of the wire and decrease its thickness or cross-sectional area A, then the measured resistance of the wire increases as its cross-sectional area

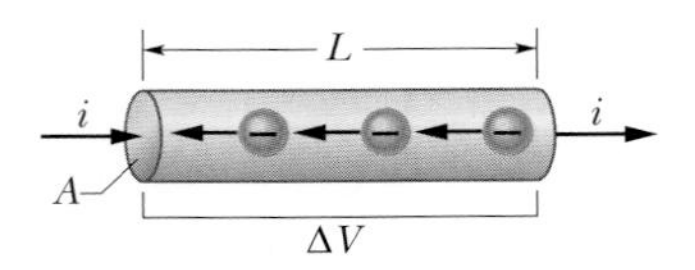

FIGURE 26-22 ■ A potential difference ΔV is applied between the ends of a conducting wire of length L and cross section A, establishing a current i. Although the stationary ions that neutralize the conduction electrons that make up the current are not shown, the wire is, as always, essentially neutral electrically.

decreases. In fact we get an inverse relationship so that

$$R = k'\frac{1}{A}.$$

To combine these two results, we write that R is proportional to L and inversely proportional to A with a new proportionality constant, ρ, which we define as the **resistivity** of the wire. Thus,

$$R = \rho\frac{L}{A}. \tag{26-8}$$

The results of these resistivity observations are important for two reasons. First, the fact that resistance varies inversely with cross-sectional area implies that current passes through the volume of the conductor, and not just along the surface. This knowledge will be useful as we continue to think about how charge moves through wires and other circuit elements.

Second, we know that every conducting material has a resistivity ρ. Is it the same for all materials? The answer is no. Is it the same if the length (or area) of a wire is changed? The answer is yes. What we observe is that if we apply the same potential difference between the ends of geometrically similar (same L and same A) rods of copper and of glass, very different currents result. This investigation reveals that resistivity varies with material. That is, it is a property of the *material* from which the object is fashioned.

We have just made an important distinction:

> Resistance is a property of an object. Resistivity is a property of a material.

It is important to note that resistivity is analogous in many ways to the concept of density. Density depends only on the kind of material being used (such as lead or Styrofoam). The density can be used to calculate the mass of a certain volume of a substance. Similarly, resistivity depends only on the material being used in the wire and not on the length or cross-sectional area of the wire. If you know the resistivity of a material then the resistance of a given wire can be calculated using Eq. 26-8 once its length and cross-sectional area are known.

Variation of Resistivity with Temperature

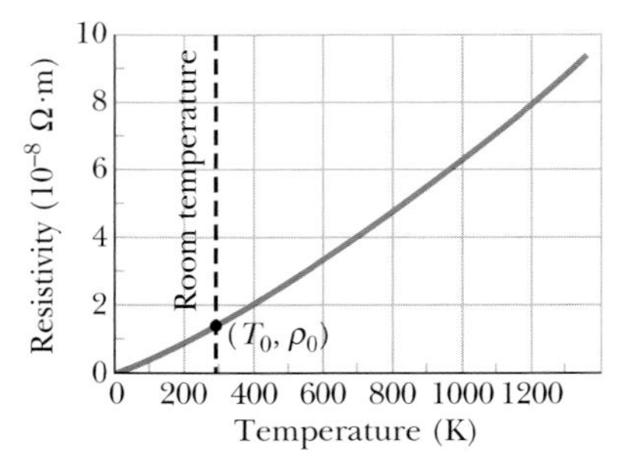

FIGURE 26-23 ■ The resistivity of copper as a function of temperature. The dot on the curve marks a convenient reference point ($T_0 = 293$ K and $\rho_0 = 1.69 \times 10^{-8}\ \Omega\cdot\text{m}$).

The values of most physical properties vary with temperature, and resistivity is no exception. Figure 26-23, for example, shows the variation of this property for copper over a wide temperature range. The relation between temperature and resistivity for copper—and for metals in general—is fairly linear over the temperature range commonly found in circuits. For such linear relations we can write an approximation based on the results of measurements as

$$\rho - \rho_0 \approx \rho_0\alpha(T - T_0) \quad \text{(approx. temperature dependence of } \rho\text{)}. \tag{26-9}$$

Here T_0 is a selected reference temperature and ρ_0 is the resistivity at that temperature. Usually $T_0 = 293$ K (room temperature), for which $\rho_0 = 1.69 \times 10^{-8}\,\Omega \cdot \text{m}$ for copper. This approximate relationship is good enough for most engineering purposes.

Because temperature enters into this expression only as a difference, it does not matter whether you use the Celsius or Kelvin scale in that equation because the sizes of degrees on these scales are identical. The quantity α, called the **temperature coefficient of resistivity,** is chosen so that the equation gives good agreement with

TABLE 26-2
Resistivities of Some Materials at Room Temperature (20°C)

Material	Resistivity, ρ ($\Omega \cdot m$)	Temperature Coefficient of Resistivity, α (K^{-1})
	Typical Metals	
Silver	1.62×10^{-8}	4.1×10^{-3}
Copper	1.69×10^{-8}	4.3×10^{-3}
Aluminum	2.75×10^{-8}	4.4×10^{-3}
Tungsten	5.25×10^{-8}	4.5×10^{-3}
Iron	9.68×10^{-8}	6.5×10^{-3}
Platinum	10.6×10^{-8}	3.9×10^{-3}
Manganin[a]	48.2×10^{-8}	0.002×10^{-3}
	Typical Semiconductors	
Silicon, pure	2.5×10^{3}	-70×10^{-3}
Silicon, *n*-type[b]	8.7×10^{-4}	
Silicon, *p*-type[c]	2.8×10^{-3}	
	Typical Insulators	
Glass	$10^{10} - 10^{14}$	
Fused quartz	$\sim 10^{16}$	

[a]An alloy specifically designed to have a small value of α.
[b]Pure silicon doped with phosphorus impurities to a charge carrier density of $10^{23}\ m^{-3}$.
[c]Pure silicon doped with aluminum impurities to a charge carrier density of $10^{23}\ m^{-3}$.

experimental values for temperatures in the chosen range. Some values of α for metals are listed in Table 26-2.

The *Hindenburg*

When the zeppelin *Hindenburg* was preparing to land on May 6th, 1937, the handling ropes were let down to the ground crew. Exposed to the rain, the ropes became wet (and thus were able to conduct a current). In this condition, the ropes "grounded" the metal framework of the zeppelin to which they were attached; that is, the wet ropes formed a conducting path between the framework and the ground, making the electric potential of the framework the same as the ground's. This should have also grounded the outer fabric of the zeppelin. The *Hindenburg,* however, was the first zeppelin to have its outer fabric painted with a sealant of large electrical resistivity. The fabric remained at the electric potential of the atmosphere at the zeppelin's altitude of about 43 m. Due to the rainstorm, that potential was large relative to the potential at ground level.

The handling of the ropes apparently ruptured one of the hydrogen cells and released hydrogen between that cell and the zeppelin's outer fabric, causing the reported rippling of the fabric. There was then a dangerous situation: the fabric was wet with conducting rainwater and was at a potential much different from the framework of the zeppelin. Apparently, charge flowed along the wet fabric and then sparked through the released hydrogen to reach the metal framework of the zeppelin, igniting the hydrogen in the process. The burning rapidly ignited the cells of hydrogen in the zeppelin and brought the ship down. If the sealant on the outer fabric of the *Hindenburg* had been of less resistivity (like that of other zeppelins), the *Hindenburg* disaster probably would not have occurred.

READING EXERCISE 26-8: Sketch a graph of i vs ΔV for a Nichrome wire like that in Fig. 26-22 but with the diameter of the wire cut in half. ■

READING EXERCISE 26-9: In the section above, we cited the fact the resistance of a wire to current was inversely proportional to the cross-sectional area of the wire as evidence that the current passes through the volume of the wire rather than along the surface of the wire. (a) Justify this assertion. (b) What expression would you expect to replace Eq. 26-8 if the current was along the surface of the wire instead? ■

READING EXERCISE 26-10: The figure shows three cylindrical copper conductors along with their face areas and lengths. Rank them according to the current through them, greatest first, when the same potential difference ΔV is placed across their lengths.

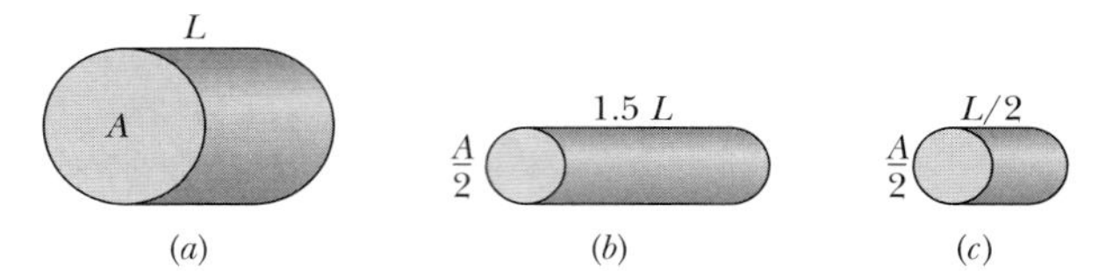

■

26-7 Power in Electric Circuits

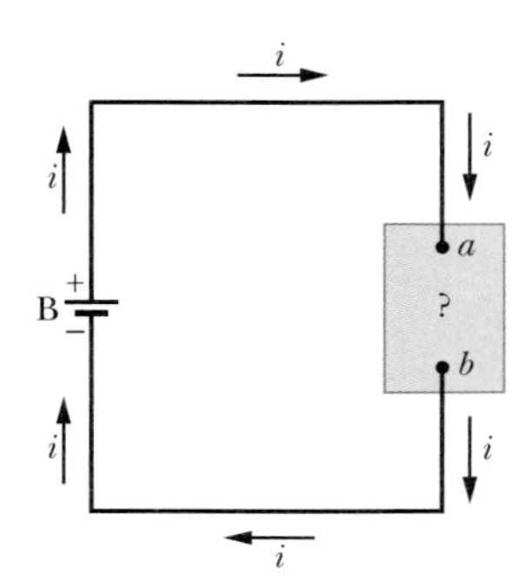

FIGURE 26-24 ■ A battery B sets up a current i in a circuit containing an unspecified conducting device.

Batteries store a certain amount of chemical energy. This chemical energy is transformed to electrical and other forms of energy as current flows through various circuit elements. At times we are interested in the rate at which a battery's energy is used up by a circuit. Just as we did in Section 9-10 where power is defined as the rate at which work is done by a force, we also use the term power to describe the rate at which electrical energy is delivered to a circuit.

We start our consideration of power by examining the energy delivered to an electrical device that is connected to a battery by ideal wires. Figure 26-24 shows a circuit consisting of a battery B that is connected by wires to an unspecified conducting device. The device might be a resistor, a storage battery (a rechargeable battery), a motor, or some other electrical device. If the wires in the circuit are thick enough they are ideal because they have essentially no resistance. When current is present in a wire with no resistance the entire wire is at the same potential. In other words, there is no potential difference between one end of an ideal wire and the other end. In this case, a battery maintains a potential difference of magnitude ΔV across its own terminals, and thus across the terminals of the unspecified device, with a greater potential at terminal a of the device than at terminal b.

Since there is an external conducting path between the two terminals of the battery, and since the battery maintains a fixed potential difference, the battery produces a steady current i in the circuit. This current is directed from terminal a to terminal b. The amount of charge dq moving between those terminals in time interval dt is equal to $i\,dt$. This charge dq moves through a decrease in potential difference across the terminals of the device of magnitude ΔV, and thus its electric potential energy U decreases in magnitude by the amount

$$dU = -dq\,\Delta V = -i\,dt(\Delta V).$$

The wire coils within a toaster have appreciable resistance. When there is a current through them, electrical energy is transferred to thermal energy of the coils, increasing their temperature. The coils then emit infrared radiation and visible light that can toast bread.

The principle of conservation of energy tells us that the decrease in electric potential energy from a to b is accompanied by a transfer of energy to some other form. Since $P = dW/dt$ (Eq. 9-48), the power P associated with that transfer is the rate at which the battery does work. Since $dW = -dU = i\,dt(\Delta V)$, we get

$$P = i\,\Delta V \quad \text{(rate of electric energy transfer).} \tag{26-10}$$

Moreover, this power P is also the rate at which energy is transferred from the battery to the unspecified device. If that device is a motor connected to a mechanical load, the energy is transferred as work done on the load. If the device is a storage battery being charged, the energy is transferred to stored chemical energy in the storage battery. We know from observations that if the device is a resistor, the energy is transferred to internal thermal energy, tending to increase the resistor's temperature.

The unit of power following from the equation above is the volt-ampere (V · A). We can write it as

$$1\ \text{V} \cdot \text{A} = \left(1\frac{\text{J}}{\text{C}}\right)\left(1\frac{\text{C}}{\text{s}}\right) = 1\frac{\text{J}}{\text{s}} = 1\ \text{W}.$$

The course of an electron moving through a resistor at constant speed is much like that of a stone falling through syrup at constant terminal speed. The average kinetic energy of the electron remains constant, and its lost electric potential energy appears as thermal energy in the resistor and its surroundings. On a microscopic scale this energy transfer is due to collisions between the electron and the molecules of the resistor, which leads to an increase in the temperature of the resistor lattice. The mechanical energy thus transferred to thermal energy is *lost* because the transfer cannot be reversed. This energy transfer due to atomic collisions is discussed in more detail in Sections 26-10 and 26-11.

For a resistor or some other device with resistance R, we can combine Eqs. 26-5 ($R = \Delta V/i$) and 26-10 to obtain, for the rate of electric energy loss (or dissipation) due to a resistance, either

$$P = i^2R \quad \text{(resistive dissipation)} \tag{26-11}$$

or

$$P = \frac{(\Delta V)^2}{R} \quad \text{(resistive dissipation).} \tag{26-12}$$

Caution: We must be careful to distinguish these two new equations from Eq. 26-10: $P = i\,\Delta V$ applies to electric energy transfers of all kinds; $P = i^2R$ and $P = (\Delta V)^2/R$ apply only to the transfer of electric potential energy to thermal energy in a device with resistance.

READING EXERCISE 26-11: A potential difference ΔV is connected across a device with resistance R, causing current i through the device. Rank the following variations according to the change in the rate at which electrical energy is converted to thermal energy due to the resistance, greatest change first: (a) ΔV is doubled with R unchanged, (b) i is doubled with R unchanged, (c) R is doubled with ΔV unchanged, (d) R is doubled with i unchanged. ■

TOUCHSTONE EXAMPLE 26-2: Heating Wire

You are given a length of uniform heating wire made of a nickel-chromium-iron alloy called Nichrome; it has a resistance R of 72 Ω. At what rate is energy dissipated in each of the following situations? (1) A potential difference of 120 V is applied across the full length of the wire. (2) The wire is cut in half, and a potential difference of 120 V is applied across the length of each half.

SOLUTION ■ The **Key Idea** is that a current in a resistive material produces a transfer of electrical energy to thermal energy; the rate of transfer (dissipation) is given by Eqs. 26-10 to 26-12. Because we know the potential ΔV and resistance R, we use Eq. 26-12, which yields, for situation 1,

$$P = \frac{(\Delta V)^2}{R} = \frac{(120\ \text{V})^2}{72\ \Omega} = 200\ \text{W}. \quad \text{(Answer)}$$

In situation 2, the resistance of each half of the wire is (72 Ω)/2, or 36 Ω. Thus, the dissipation rate for each half is

$$P' = \frac{(120\ \text{V})^2}{36\ \Omega} = 400\ \text{W},$$

and that for the two halves is

$$P = 2P' = 800 \text{ W}. \qquad \text{(Answer)}$$

This is four times the dissipation rate of the full length of wire. Thus, you might conclude that you could buy a heating coil, cut it in half, and reconnect it to obtain four times the heat output. Why is this unwise? (What would happen to the amount of current in the coil?)

26-8 Current Density in a Conductor

We defined current so that it was a scalar—basically a "count" of the amount of charge crossing a surface per second with a sign to tell us in which direction the charge is crossing the surface—in the direction we choose as positive or opposite to it. Since in a current, charges are actually moving and have a velocity associated with them, there is a vector "hidden" in the concept of current. We can make it explicit by defining a new concept, the **current density.** If we have a volume that contains a set of moving charged particles, let the charge on each particle be e, let the density of the charges be n (number per unit volume), and let their average velocity be $\langle\vec{v}\rangle$. We then define the current density (or current percentage of cross-sectional area) as

$$\vec{J} \equiv ne\langle\vec{v}\rangle \quad \text{(definition of current density).} \tag{26-13}$$

As is the case for the volume flux of water described in Eq. 15-33, the total amount of charge flowing through a given element of area can be defined as the dot product of the current density and an area element. If the area element is infinitesimal we can write the amount of current through it as $\vec{J} \cdot d\vec{A}$, where $d\vec{A}$ is the area vector of the element, perpendicular to the plane of the area element. The total *conventional current* through the surface of a cross section of wire is then

$$i = \int \vec{J} \cdot d\vec{A}. \tag{26-14}$$

In most electrical conductors the charge carriers are negative. As we mentioned earlier, the term "conventional current" refers to the direction of flow of positive charge carriers. For a typical conductor such as copper, the electrons are moving in the opposite direction to the direction of the conventional current.

In Section 26-6 we concluded that in steady current through a conductor the charges must be flowing throughout the volume of the conductor. The key evidence for this is the inverse proportionality between resistance and the cross-sectional area of a conductor.* If we further assume that the direction of the current is parallel to $d\vec{A}$, then $\vec{J}$ is also uniform and parallel to $d\vec{A}$. In this case Eq. 26-14 can be rewritten in terms of the magnitudes of the current density and area.

$$|i| = \int J dA = J \int dA = JA,$$

so

$$J = \frac{|i|}{A}, \tag{26-15}$$

*However, steady charge flow throughout a conductor is not true for the high-frequency alternating currents we treat in Chapter 33.

where A is the total area of the surface. From these equations, we see that the SI unit for current density is the ampere per square meter (A/m^2).

In Chapter 23 we represented an electric field with electric field lines. Figure 26-25 shows how current density can be represented with a similar set of lines, which we can call *streamlines*. The current, which is toward the right in Fig. 26-25, makes a transition from the wider conductor at the left to the narrower conductor at the right. Because charge is conserved during the transition, the current or rate at which the charges flow through the wire cannot change. However, the current density (or rate of charge flow per unit of cross-sectional area) does change—it is greater in the narrower conductor. The spacing of the streamlines suggests this increase in current density; streamlines that are closer together imply greater current density.

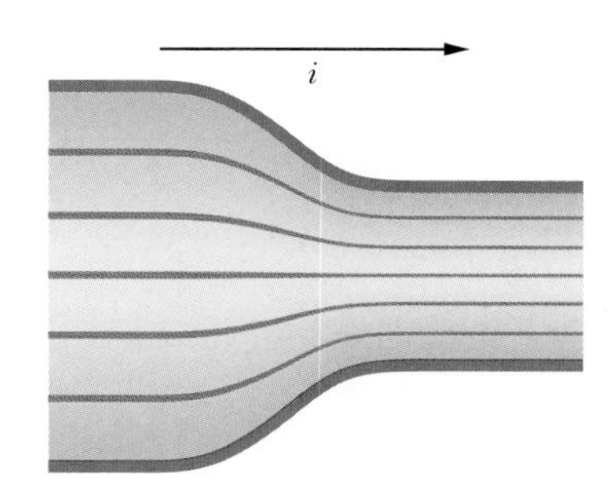

FIGURE 26-25 ■ Streamlines representing current density in the flow of charge through a constricted conductor.

READING EXERCISE 26-12: The sketches below show several copper wires with the same potential difference across them. Rank the current density magnitude from largest to smallest.

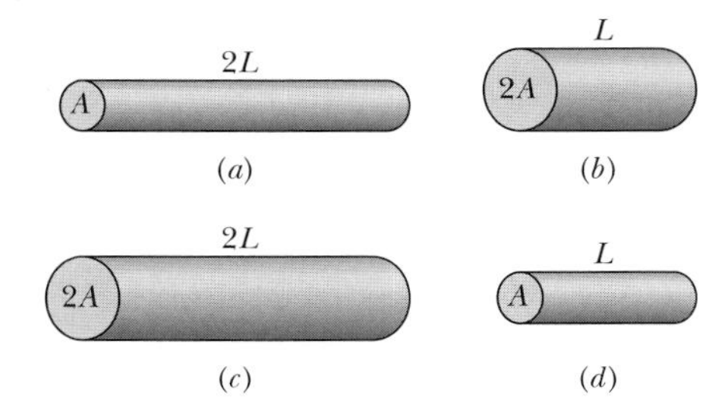

■

26-9 Resistivity and Current Density

Although i, ΔV, and R are the quantities that are directly measurable in electrical circuits, if we want to think more explicitly about what is happening in terms of the motion of charges it makes sense to reframe our Ohm's law relation in terms of the forces (or the electric field) and the current density. This gives us a generic relation that describes how forces affect the motion of charges without relying in any way on the properties of specific circuit elements in the way that Ohm's law does.

Recall that, for materials that obey Ohm's law, the resistance of a segment of a conductor R is related to the potential difference ΔV across it as well as the conventional current i passing through it. This relationship is given by

$$\Delta V = iR. \qquad \text{(Eq. 26-7)}$$

We can write this expression in an alternate form if we replace the potential difference ΔV with an expression involving the electric field $\vec{E}$. From Chapter 25, we know that the relationship between the electric field and the potential difference between two locations a and b is

$$V_a - V_b = \int_a^b \vec{E} \cdot d\vec{s}.$$

For a wire of length L with one end at location a and the other at location b, (Fig. 26-26), the electric field $\vec{E}$ set up within the wire is constant. As a result, the expression above can be expressed in terms of the electric field magnitude E and the length of the wire L as

$$V_a - V_b = \pm EL,$$

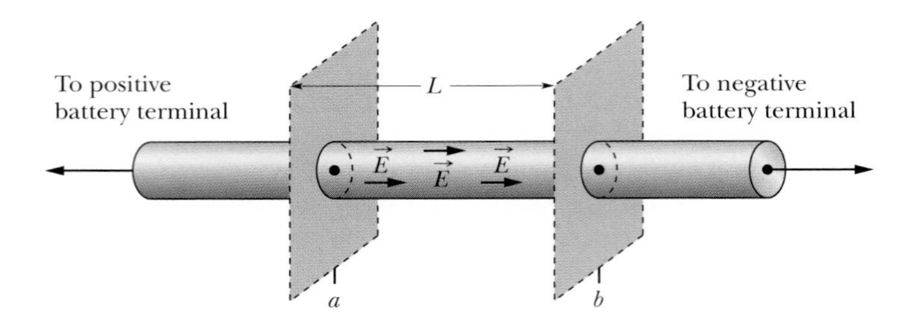

FIGURE 26-26 ■ A length L between points a and b along a current-carrying conductor.

where we use the plus sign if $\vec{E}$ and $d\vec{s}$ are in the same direction and the minus sign if $\vec{E}$ and $d\vec{s}$ point in opposite directions. Combining the expression above with

$$R = \frac{V_a - V_b}{i}$$

and ignoring signs gives us
$$R = \frac{EL}{i} = \frac{EL}{JA}.$$

The substitution for i comes from the relationship between current i, current density $\vec{J}$, and the cross section of the wire A. We compare this relation with that presented earlier when we introduced ρ as the resistivity of the material in Eq. 26-8:

$$R = \rho\frac{L}{A}.$$

By combining the previous two equations, we see that resistivity can be defined in terms of the magnitudes of the microscopic quantities $\vec{E}$ and $\vec{J}$ as

$$\rho \equiv \frac{E}{J} \quad \text{(definition of } \rho\text{).} \tag{26-16}$$

If we combine the SI units of $\vec{E}$ and $\vec{J}$ we get, for the unit of ρ, the ohm-meter ($\Omega \cdot$ m):

$$\text{units of } \rho = \frac{\text{units of } E}{\text{units of } J} = \frac{\text{V/m}}{\text{A/m}^2} = \frac{\text{V}}{\text{A}}\text{m} = \Omega \cdot \text{m}.$$

(Do not confuse the *ohm-meter*, the unit of resistivity, with the *ohmmeter*, which is an instrument that measures resistance.)

Since $\vec{E}$ and $\vec{J}$ always point in the same direction, we can rewrite this expression in vector form as

$$\vec{E} = \rho\vec{J}. \tag{26-17}$$

However, be aware that these two relations hold only for *isotropic* materials—materials whose electrical properties are the same in all directions (like the metals used to make wires).

26-10 A Microscopic View of Current and Resistance

Our macroscopic studies tell us that there is a current in a conductor whenever there is a potential difference across it. Whenever Ohm's law holds, the current is directly proportional to the potential difference that causes it. Let's consider a length L of thin conducting wire with a potential difference of ΔV between its ends. What happens microscopically to the charge carriers in this situation?

We already know that the conduction electrons in a metal serve as charge carriers, and that when there is a steady current, we can represent the density of electrons as n and the charge on each electron as e. What does Ohm's law tell us about the average velocity $\langle \vec{v} \rangle$ of these electrons? When Ohm's law holds so that $\Delta V = iR$ (Eq. 26-7), then according to Eq. 26-13, the current density is proportional to the average velocity of the charge carriers by definition,

$$\vec{J} \equiv ne\langle \vec{v} \rangle. \qquad \text{(Eq. 26-13)}$$

Since $\vec{E} = \rho \vec{J}$ (Eq. 26-17), we find that the electric field $\vec{E}$ across the wire (associated with potential difference ΔV across the wire) is also proportional to the average velocity of the charge carriers,

$$\vec{E} = \rho ne\langle \vec{v} \rangle. \qquad (26\text{-}18)$$

However, the electrostatic force on a charge carrier is given by $\vec{F}^{\text{elec}} = e\vec{E}$ (Eq. 23-4), so that

$$\vec{F}^{\text{elec}} = e\vec{E} = \rho ne^2\langle \vec{v} \rangle. \qquad (26\text{-}19)$$

This is a dramatic and interesting result. It tells us that the average velocity, $\langle \vec{v} \rangle$, of a charge carrier is proportional to the electrostatic force on it! However, if the electrostatic force is the only force acting on the electron, then Newton's Second Law tells us that the electron should accelerate and not maintain a constant average velocity. To maintain a constant velocity, the *net force* on the charge carrier must be zero. Thus, there must be a second force. This situation is very similar to that associated with air drag where an object falling in the presence of a gravitational force reaches a terminal velocity as a result of an air drag acting in the opposite direction. Using Eq. 6-24 we see that

$$\vec{F}^{\text{net}} = \vec{F}^{\text{elec}} + \vec{D} = \rho ne^2\langle \vec{v} \rangle + \vec{D} = 0$$

so that

$$\vec{D} = -e\vec{E} = -\rho ne^2\langle \vec{v} \rangle. \qquad (26\text{-}20)$$

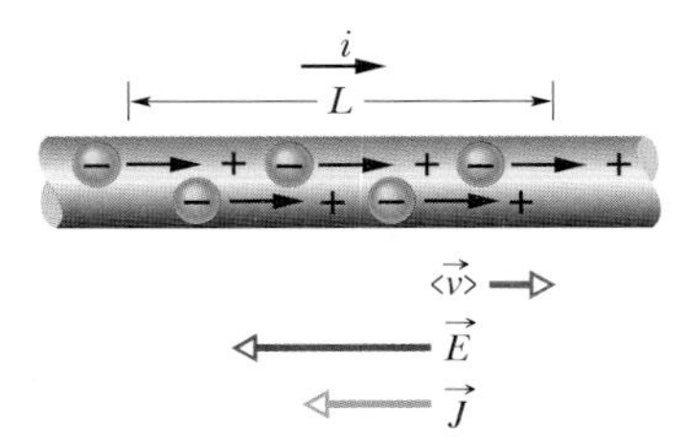

FIGURE 26-27 ■ Conduction electrons which are negative charge carriers drift at an average velocity $\langle \vec{v} \rangle$ in the opposite direction of the applied electric field $\vec{E}$. Their size is greatly exaggerated. By convention, the direction of the current density $\vec{J}$ and the sense of the arrow representing the flow of conventional current are drawn in that same direction.

This leads us to conclude that there must be a drag force that is proportional to the average velocity of the charge carriers. The air drag force on a falling object is attributed to the action of many small air molecules hitting the falling object as it moves. Similarly, we can imagine that a charge carrier is being slowed down by hitting many stationary atoms and ions as it passes through the conductor. The interactions between charge carriers and the atoms in a conductor can only be described properly using quantum mechanics. Nonetheless, we attempt to picture the flow of charge past positive ions in Fig. 26-27.

What Is a Typical Average Charge Carrier Speed?

Solving $\vec{J} \equiv ne\langle \vec{v} \rangle$ (Eq. 26-13) for the average velocity and recalling Eq. 26-15 ($J = |i|/A$), we obtain the following expression for the average speed of the charge carrier,

$$|\langle \vec{v} \rangle| = \frac{|i|}{nAe} = \frac{J}{ne}. \qquad (26\text{-}21)$$

The product ne, whose SI unit is the coulomb per cubic meter (C/m^3), is the *carrier charge density*.

At this point we can use Eq. 26-21 to find a typical value for the average speed for electrons flowing in a copper wire. Since copper has one conduction electron per atom we can use measurements for the density of copper atoms of $n = 8.5 \times 10^{28}$ atoms/m^3. Assume that our wire carries a current of 1.0 A and has a diameter of 2 mm so its cross-sectional area is 3×10^{-6} m^2. Then, according to Eq. 26-21, the average speed of the electrons is about

$$|\langle\vec{v}\rangle| = \frac{|i|}{nAe} = \frac{1 \text{ C/s}}{(8.5 \times 10^{28} \text{ atoms/m}^3)(3 \times 10^{-6} \text{ m}^2)(1.6 \times 10^{-19} \text{ C/atom})}$$
$$\approx 2.5 \times 10^{-5} \text{ m/s}.$$

This typical average speed is extremely small compared to very high speed random thermal motion of the electrons. It would take an electron about 11 hours to move across a 10 cm stretch of wire. Although the conduction electrons move along a wire very slowly like tired snails, there are so many of them that the current can actually be relatively large.

A Microscopic View of Resistivity

We can carry our microscopic analysis further, by relating the resistivity of a conductor to the properties of its charge carriers and the average time between electron collisions. If an electron of mass m is placed in an electric field of magnitude E, the electron will experience an acceleration given by Newton's Second Law:

$$\vec{a} = \frac{\vec{F}}{m} = \frac{e\vec{E}}{m}. \tag{26-22}$$

The nature of the collisions experienced by conduction electrons is such that, after a typical collision, each electron will—so to speak—completely lose its memory of its previous average velocity. Between collisions a conduction electron will have a mean free path λ like that derived in Section 20-5 for molecules traveling in a gas. However, it moves with a typical random speed $v^{\text{eff}} = \lambda/\tau$ where τ is the average time between collisions. Each electron will then start off fresh after every encounter, moving off in a random direction. In the average time τ between collisions, a typical electron will undergo an acceleration $\vec{a}$ in a direction opposite to that of the electric field as shown in Fig. 26-27. Thus, the average speed (often called the drift speed), the electron acquires in that direction is given by $|\langle\vec{v}\rangle| = a\tau$. Using Eq. 26-22 we get

$$|\langle\vec{v}\rangle| = a\tau = \frac{eE\tau}{m}. \tag{26-23}$$

Combining this result with $\vec{J} = ne\langle\vec{v}\rangle$ yields the average velocity of

$$\langle\vec{v}\rangle = \pm\frac{\vec{J}}{ne} = \pm\frac{e\vec{E}\tau}{m},$$

where we use the plus (+) sign for positive charge carriers and the minus (−) sign for negative charge carriers. We can combine the last two terms in the previous equation and solve for $\vec{E}$ to get

$$\vec{E} = \left(\frac{m}{e^2 n\tau}\right)\vec{J}.$$

This equation shows a proportionality between the electric field in a wire and the amount of current. Note that the magnitude of the electric field in a wire is in turn proportional to the potential difference across the wire. Thus, our microscopic picture of resistivity for metallic conductors is consistent with our macroscopic measurements, and it predicts a proportionality between potential difference and current.

Comparing the equation above with Eq. 26-17 ($\vec{E} = \rho\vec{J}$) leads to an expression for the resistivity in terms of the mass and charge of the carriers, the charge density n, and the average time between collisions

$$\rho = \frac{m}{e^2 n\tau}. \qquad (26\text{-}24)$$

Conductivity

As well as referring to the resistivity of a material, we often speak of the conductivity σ of a material. This is simply the reciprocal of its resistivity, so

$$\sigma \equiv \frac{1}{\rho} \quad \text{(definition of } \sigma\text{)}. \qquad (26\text{-}25)$$

The SI unit of conductivity is the reciprocal ohm-meter $(\Omega \cdot \text{m})^{-1}$. The unit name mhos per meter is sometimes used (mho is ohm backward). The definition of conductivity, σ, allows us to write Eq. 26-17($\vec{E} = \rho\vec{J}$) in the alternative form

$$\vec{J} = \sigma\vec{E}. \qquad (26\text{-}26)$$

READING EXERCISE 26-13: The figure shows positive charge carriers moving leftward through a wire. Are the following leftward or rightward: (a) the conventional current i, (b) the current density $\vec{J}$, (c) the electric field $\vec{E}$ in the wire? *Hint*: You may want to review the discussion of conventional current in Section 26-8. ■

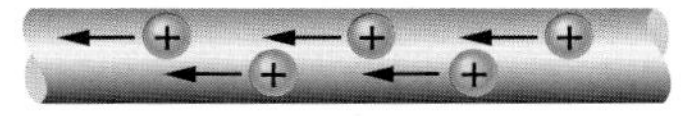

TOUCHSTONE EXAMPLE 26-3: Mean Free Time

What is the mean free time τ between collisions for the conduction electrons in copper?

SOLUTION ■ The **Key Idea** here is that the mean free time τ of copper is approximately constant, and in particular does not depend on any electric field that might be applied to a sample of the copper. Thus, we need not consider any particular value of applied electric field. However, because the resistivity ρ displayed by copper under an electric field depends on τ, we can find τ from Eq. 26-24 ($\rho = m/e^2 n\tau$). That equation gives us

$$\tau = \frac{m}{ne^2\rho}.$$

Taking the value of n, the number of conduction electrons per unit volume in copper, to be $8.5 \times 10^{28}\ \text{m}^{-3}$, and taking the value of ρ from Table 26-2, the denominator then becomes

$$(8.5 \times 10^{28}\ \text{m}^{-3})(1.6 \times 10^{-19}\ \text{C})^2(1.69 \times 10^{-8}\ \Omega\cdot\text{m})$$
$$= 3.67 \times 10^{-17}\ \text{C}^2\cdot\Omega/\text{m}^2$$
$$= 3.67 \times 10^{-17}\ \text{kg/s},$$

where we converted units as

$$\frac{\text{C}^2\cdot\Omega}{\text{m}^2} = \frac{\text{C}^2\cdot\text{V}}{\text{m}^2\cdot\text{A}} = \frac{\text{C}^2\cdot\text{J/C}}{\text{m}^2\cdot\text{C/s}} = \frac{\text{kg}\cdot\text{m}^2/\text{s}^2}{\text{m}^2/\text{s}} = \frac{\text{kg}}{\text{s}}.$$

Using these results and substituting for the electron mass m, we then have

$$\tau = \frac{9.1 \times 10^{-31}\ \text{kg}}{3.67 \times 10^{-17}\ \text{kg/s}} = 2.5 \times 10^{-14}\ \text{s}. \qquad \text{(Answer)}$$

(b) The mean free path λ of the conduction electrons in a conductor is the average distance traveled by an electron between collisions. (This definition parallels that in Section 20-5 for the mean free path of molecules in a gas.) What is λ for the conduction electrons in copper?

SOLUTION ■ The **Key Idea** here is that the distance d any particle travels in a certain time t at a constant speed v is $d = vt$. To estimate v^{eff}, the speed at which the electrons typically move between collisions, we can think of the electrons as a "gas" of particles in thermal equilibrium with their surroundings inside the metal wire. Equation 20-21 then tells us that a typical electron has a kinetic energy related to the Kelvin temperature of its environment by $(\frac{1}{2})m\langle v^2\rangle = (\frac{3}{2})k_BT$ (where k_B is the Boltzmann constant). Taking the electron's effective speed in a room temperature (300 K) environment to be $v^{\rm rms} = \sqrt{\langle v^2\rangle}$ gives

$$v^{\rm eff} = \sqrt{3k_BT/m} = \sqrt{(3(1.38 \times 10^{-23}\ {\rm J/K})(300\ {\rm K})/(9.11 \times 10^{-31}\ {\rm kg}))}$$
$$= 1.168 \times 10^5\ {\rm m/s}$$

and

$$\lambda = v^{\rm eff}\tau = (1.168 \times 10^5\ {\rm m/s})(2.5 \times 10^{-14}\ {\rm s})$$
$$= 2.9 \times 10^{-9}\ {\rm m} = 2.9\ {\rm nm}. \qquad \text{(Answer)}$$

This is about 10 times the distance between nearest-neighbor atoms in a copper lattice. While this is a reasonable sounding result, it turns out that the actual value of λ is about 10 times larger than this due to quantum effects.

26-11 Other Types of Conductors

In the last few chapters we have assumed that the conductors under consideration are metallic like copper or nichrome. As you can see from Table 26-2, one of the distinctive properties of metallic conductors is that they have positive temperature coefficients indicating that their resistivities *increase* with temperature. This property seems reasonable since the thermal energy in the metal lattice causes the atoms in the metal to vibrate more, which further impedes the flow of conduction electrons. In addition, Eq. 26-9 indicates that this increase of resistivity with temperature is approximately *linear*.

There are other types of conductors with resistivities that do not simply increase linearly with temperature. The most important of these are semiconductors, which lie at the heart of the microelectronic revolution. The resistivity of semiconductors decreases more or less linearly with temperature. Superconductors are another class of conductors that do not have the same temperature behavior as conductors. Although the resistivity of superconductors increases with temperature, it does so in a very nonlinear fashion.

Because of the importance of semiconductors and superconductors we describe some of their properties here. Both of these nonmetallic conductors have some amazing properties that we describe briefly in this section. However, in the next few chapters we return to the study—within the framework of classical physics—of *steady* currents of *conduction electrons* moving through *metallic conductors*.

Semiconductors

The basic element found in virtually all semiconductors is either silicon or germanium. Table 26-3 compares the properties of silicon—a typical semiconductor—and copper—a typical metallic conductor. We see that silicon has significantly fewer charge carriers, a much higher resistivity, and a temperature coefficient of resistivity that is both large and negative. Thus, although the resistivity of copper increases with temperature, that of pure silicon decreases.

Pure silicon has such a high resistivity that it is effectively an insulator and not of much direct use in microelectronic circuits. However, its resistivity can be greatly

TABLE 26-3
Some Electrical Properties of Copper and Silicon[a]

Property	Copper	Silicon
Type of material	Metal	Semiconductor
Charge carrier density, m^{-3}	9×10^{28}	1×10^{16}
Resistivity, $\Omega \cdot \text{m}$	2×10^{-8}	3×10^{3}
Temperature coefficient of resistivity, K^{-1}	$+4 \times 10^{-3}$	-70×10^{-3}

[a]Rounded to one significant figure for easy comparison.

reduced in a controlled way by adding minute amounts of specific "impurity" atoms in a process called *doping*. Table 26-2 gives typical values of resistivity for silicon before and after doping with two different impurities, phosphorus and aluminum. Most semiconducting devices, such as transistors and junction diodes, are fabricated by the selective doping of different regions of the silicon with impurity atoms of different kinds.

A full explanation of the difference in resistivity between semiconductors and metallic conductors requires an understanding of quantum theory developed to explain atomic behavior. However, the difference has to do with the probability that electrons in a material can be made mobile. As we discuss in Section 22-6, in a metallic conductor some of the outermost electrons associated with an atom can move from one atom to the next without any additional energy. Thus, the electric field set up in the wire when a potential difference is applied drives current through a conductor.

In an insulator, considerable energy is required to free electrons so they can move through the material. Thermal energy cannot supply enough energy, and neither can any reasonable electric field applied to the insulator. Thus, no electrons are available to move through the insulator, and hence no current occurs even with an applied electric field. A semiconductor is like an insulator *except* that the energy required to free some electrons can be adjusted through doping. Doping can supply either electrons or positive charge carriers held very loosely within the material that are easy to get moving.*

In a semiconductor, the density of charge carriers is small but increases very rapidly with temperature as the increased thermal agitation makes more charge carriers available. This causes a *decrease* of resistivity with increasing temperature, as indicated by the negative temperature coefficient of resistivity for silicon in Table 26-3. The same increase in collision rate we noted for metals also occurs for semiconductors, but its effect is swamped by the rapid increase in the number of charge carriers.

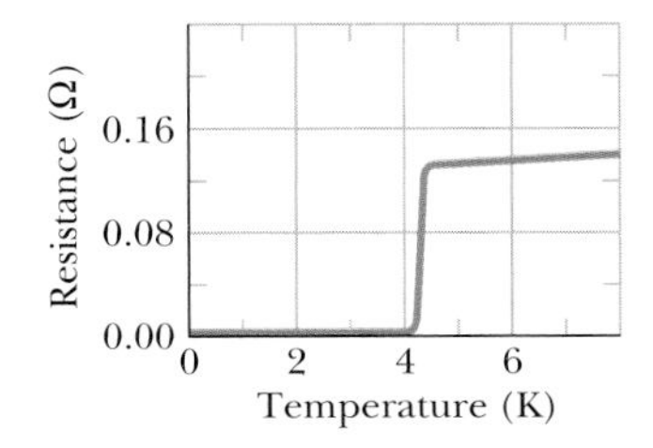

FIGURE 26-28 ■ The resistance of mercury drops to zero at a temperature of about 4 K.

Superconductors

In 1911, Dutch physicist Kamerlingh Onnes discovered that the resistivity of mercury absolutely disappears at temperatures below about 4 K (Fig. 26-28). This phenomenon of **superconductivity** is of vast potential importance in technology because it means charge can flow through a superconducting conductor without producing thermal energy losses. Currents created in a superconducting ring, for example, have persisted for several years without any measurable decrease; the electrons making up the current require a force and a source of energy at start-up time, but not thereafter.

Prior to 1986, the technological development of superconductivity was throttled by the cost of producing the extremely low temperatures that were required to achieve the effect. In 1986, however, new ceramic materials were discovered that become superconducting at considerably higher (and thus cheaper to produce) temperatures. Practical application of superconducting devices at room temperature may eventually become feasible.

Superconductivity is a much different phenomenon from conductivity. In fact, the best of the normal conductors, such as silver and copper, cannot become superconducting at any temperature, and the new ceramic superconductors are actually insulators when they are not at low enough temperatures to be in a superconducting state.

One explanation for superconductivity is that the electrons making up the current move in coordinated pairs. One of the electrons in a pair may electrically distort the molecular structure of the superconducting material as it moves through, creating a short-lived concentration of positive charge nearby. The other electron in the pair may then be attracted toward this positive charge. According to the theory, such coor-

A disk-shaped magnet is levitated above a superconducting material that has been cooled by liquid nitrogen. The goldfish is along for the ride.

*Explaining what positive charge carriers are and how they move is complex. For now just consider the charge carriers as negative (that is, electrons).

dination between electrons would prevent them from colliding with the molecules of the material and thus would eliminate electrical resistance. The theory worked well to explain the pre-1986, lower temperature superconductors, but new theories appear to be needed for the newer, higher temperature superconductors.

Problems

SEC. 26-3 ■ BATTERIES AND ELECTRIC CURRENT

1. Coulombs and Electrons A current of 5.0 A exists in a 10 Ω resistor for 4.0 min. How many (a) coulombs and (b) electrons pass through any cross section of the resistor in this time?

2. Charged Belt A charged belt, 50 cm wide, travels at 30 m/s between a source of charge and a sphere. The belt carries charge into the sphere at a rate corresponding to 100 μA. Compute the surface charge density on the belt.

3. Isolated Sphere An isolated conducting sphere has a 10 cm radius. One wire carries a current of 1.000 002 0 A into it. Another wire carries a current of 1.000 000 0 A out of it. How long would it take for the sphere to increase in potential by 1000 V?

SEC. 26-5 ■ RESISTANCE AND OHM'S LAW

4. Electrical Cable An electrical cable consists of 125 strands of fine wire, each having 2.65 $\mu\Omega$ resistance. The same potential difference is applied between the ends of all the strands and results in a total current of 0.750 A. (a) What is the current in each strand? (b) What is the applied potential difference? (c) What is the resistance of the cable?

5. Electrocution A human being can be electrocuted if a current as small as 50 mA passes near the heart. An electrician working with sweaty hands makes good contact with the two conductors he is holding, one in each hand. If his resistance is 2000 Ω, what might the fatal voltage be?

SEC. 26-6 ■ RESISTANCE AND RESISTIVITY

6. Trolley Car A steel trolley-car rail has a cross-sectional area of 56.0 cm^2. What is the resistance of 10.0 km of rail? The resistivity of the steel is $3.00 \times 10^{-7}\ \Omega \cdot$ m.

7. Conducting Wire A conducting wire has a 1.0 mm diameter, a 2.0 m length, and a 50 mΩ resistance. What is the resistivity of the material?

8. A Wire A wire 4.00 m long and 6.00 mm in diameter has a resistance of 15.0 mΩ. A potential difference of 23.0 V is applied between the ends. (a) What is the current in the wire? (b) Calculate the resistivity of the wire material. Identify the material. (Use Table 26-2.)

9. A Coil A coil is formed by winding 250 turns of insulated 16-gauge copper wire (diameter = 1.3 mm) in a single layer on a cylindrical form of radius 12 cm. What is the resistance of the coil? Neglect the thickness of the insulation (Use Table 26-2.)

10. What Temperature (a) At what temperature would the resistance of a copper conductor be double its resistance at 20.0°C? (Use 20.0°C as the reference point in Eq. 26-9; compare your answer with Fig. 26-23.) (b) Does this same "doubling temperature" hold for all copper conductors regardless of shape or size?

11. Longer Wire A wire with a resistance of 6.0 Ω is drawn out through a die so that its new length is three times its original length. Find the resistance of the longer wire, assuming that the resistivity and density of the material are unchanged.

12. A Certain Wire A certain wire has a resistance R. What is the resistance of a second wire, made of the same material, that is half as long and has half the diameter?

13. Two Conductors Two conductors are made of the same material and have the same length. Conductor A is a solid wire of diameter 1.0 mm. Conductor B is a hollow tube of outside diameter 2.0 mm and inside diameter 1.0 mm. What is the resistance ratio R_A/R_B, measured between their ends?

14. Flashlight Bulb A common flashlight bulb is rated at 0.30 A and 2.9 V (the values of the current and voltage under operating conditions). If the resistance of the bulb filament at room temperature (20°C) is 1.1 Ω, what is the temperature of the filament when the bulb is on? The filament is made of tungsten.

15. Metal Rod When a metal rod is heated, not only its resistance but also its length and its cross-sectional area change. The relation $R = \rho L/A$ suggests that all three factors should be taken into account in measuring ρ at various temperatures. (a) If the temperature changes by 1.0 C°, what percentage changes in R, L, and A occur for a copper conductor? (b) The coefficient of linear expansion for copper is 1.7×10^{-5}/K. What conclusion do you draw?

16. Gauge Number If the gauge number of a wire is increased by 6, the diameter is halved; if a gauge number is increased by 1, the diameter decreases by the factor $2^{1/6}$ (see the table in Problem 32). Knowing this, and knowing that 1000 ft of 10-gauge copper wire has a resistance of approximately 1.00 Ω, estimate the resistance of 25 ft of 22-gauge copper wire.

SEC. 26-7 ■ POWER IN ELECTRIC CIRCUITS

17. X-Ray Tube A certain x-ray tube operates at a current of 7.0 mA and a potential difference of 80 kV. What is its power in watts?

18. A Student A student kept his 9.0 V, 7.0 W radio turned on at full volume from 9:00 P.M. until 2:00 A.M. How much charge went through it?

19. Space Heater A 120 V potential difference is applied to a space heater whose resistance is 14 Ω when hot. (a) At what rate is electric energy transferred to heat? (b) At 5.0¢/kW·h, what does it cost to operate the device for 5.0 h?

20. Thermal Energy Thermal energy is produced in a resistor at a rate of 100 W when the current is 3.00 A What is the resistance?

21. Energy Is Dissipated An unknown resistor is connected between the terminals of a 3.00 V battery. Energy is dissipated in the resistor at the rate of 0.540 W. The same resistor is then connected between the terminals of a 1.50 V battery. At what rate is energy now dissipated?

22. Space Heater Two A 120 V potential difference is applied to a space heater that dissipates 500 W during operation. (a) What is its resistance during operation? (b) At what rate do electrons flow through any cross section of the heater element?

23. Radiant Heater A 1250 W radiant heater is constructed to operate at 115 V. (a) What will be the current in the heater? (b) What is the resistance of the heating coil? (c) How much thermal energy is produced in 1.0 h by the heater?

24. Heating Element A heating element is made by maintaining a potential difference of 75.0 V across the length of a Nichrome wire that has a 2.60×10^{-6} m^2 cross section. Nichrome has a resistivity of 5.00×10^{-7} $\Omega \cdot$m. (a) If the element dissipates 5000 W, what is its length? (b) If a potential difference of 100 V is used to obtain the same dissipation rate, what should the length be?

25. Nichrome Heater A Nichrome heater dissipates 500 W when the applied potential difference is 110 V and the wire temperature is 800°C. What would be the dissipation rate if the wire temperature were held at 200°C by immersing the wire in a bath of cooling oil? The applied potential difference remains the same, and α for Nichrome at 800°C is 4.0×10^{-4}/K.

26. 100 W Lightbulb A 100 W lightbulb is plugged into a standard 120 V outlet. (a) How much does it cost per month to leave the light turned on continuously? Assume electric energy costs 12¢/kW·h. (b) What is the resistance of the bulb? (c) What is the current in the bulb? (d) Is the resistance different when the bulb is turned off?

27. Linear Accelerator A linear accelerator produces a pulsed beam of electrons. The pulse current is 0.50 A, and each pulse has a duration of 0.10 μs. (a) How many electrons are accelerated per pulse? (b) What is the average current for an accelerator operating at 500 pulses/s? (c) If the electrons are accelerated to an energy of 50 MeV, what are the average and peak powers of the accelerator?

28. Cylindrical Resistor A cylindrical resistor of radius 5.0 mm and length 2.0 cm is made of material that has a resistivity of 3.5×10^{-5} $\Omega \cdot$m. What is the potential difference when the energy dissipation rate in the resistor is 1.0 W?

29. Copper Wire A copper wire of cross-sectional area 2.0×10^{-6} m^2 and length 4.0 m has a current of 2.0 A uniformly distributed across that area. How much electric energy is transferred to thermal energy in 30 min?

SEC. 26-8 ■ CURRENT DENSITY IN A CONDUCTOR

30. Small But Measurable A small but measurable current of 1.2×10^{-10} A exists in a copper wire whose diameter is 2.5 mm. Assuming the current is uniform, calculate (a) the current density and (b) the average electron speed.

31. A Beam A beam contains 2.0×10^8 doubly charged positive ions per cubic centimeter, all of which are moving north with a speed of 1.0×10^5 m/s. (a) What are the magnitude and direction of the current density $\vec{J}$? (b) Can you calculate the total current i in this ion beam? If not what additional information is needed?

32. The U.S. Electric Code The (United States) National Electric Code, which sets maximum safe currents for insulated copper wires of various diameters, is given (in part) in the table. Plot the safe current density as a function of diameter. Which wire gauge has the maximum safe current density? ("Gauge" is a way of identifying wire diameters, and 1 mil = 10^{-3} in.)

Gauge	4	6	8	10	12	14	16	18
Diameter, mils	204	162	129	102	81	64	51	40
Safe current, A	70	50	35	25	20	15	6	3

33. A Fuse A fuse in an electric circuit is a wire that is designed to melt, and thereby open the circuit, if the current exceeds a predetermined value. Suppose that the material to be used in a fuse melts when the current density rises to 440 A/cm^2. What diameter of cylindrical wire should be used to make a fuse that will limit the current to 0.50 A?

34. Near Earth Near the Earth, the density of protons in the solar wind (a stream of particles from the Sun) is 8.70 cm^{-3}, and their speed is 470 km/s. (a) Find the current density of these protons. (b) If the Earth's magnetic field did not deflect them, the protons would strike the planet. What total current would the Earth then receive?

35. Steady Beam A steady beam of alpha particles ($q = +2e$) traveling with constant kinetic energy 20 MeV carries a current of 0.25 μA. (a) If the beam is directed perpendicular to a plane surface, how many alpha particles strike the surface in 3.0 s? (b) At any instant, how many alpha particles are there in a given 20 cm length of the beam? (c) Through what potential difference is it necessary to accelerate each alpha particle from rest to bring it to an energy of 20 MeV?

36. Current Density (a) The current density across a cylindrical conductor of radius R varies in magnitude according to the equation

$$J = J_0\left(1 - \frac{r}{R}\right),$$

where r is the distance from the central axis. Thus, the current density has a maximum magnitude of $J_0 = |\vec{J}_0|$ at that axis ($r = 0$) and decreases linearly to zero at the surface ($r = R$). Calculate the current in terms of J_0 and the conductor's cross-sectional area $A = \pi R^2$. (b) Suppose that, instead, the current density is a maximum J_0 at the cylinder's surface and decreases linearly to zero at the axis: $J = J_0\, r/R$. Calculate the magnitude of the current. Why is the result different from that in (a)?

37. How Long How long does it take electrons to get from a car battery to the starting motor? Assume the current is 300 A and the electrons travel through a copper wire with cross-sectional area 0.21 cm^2 and length 0.85 m. (*Hint:* Assume one conduction electron per atom and take the number density of copper atoms to be 8.5×10^{28} atoms/m^2.)

38. Nichrome A wire of Nichrome (a nickel-chromium-iron alloy commonly used in heating elements) is 1.0 m long and 1.0 mm^2 in cross-sectional area. It carries a current of 4.0 A when a 2.0 V potential difference is applied between its ends. Calculate the conductivity σ of Nichrome.

39. When Applied When 115 V is applied across a wire that is 10 m long and has a 0.30 mm radius, the current density is 1.4×10^4 A/m^2. Find the resistivity of the wire.

40. Truncated Right-Circular Cone A resistor has the shape of a truncated right-circular cone (Fig. 26-29). The end radii are a and b, and the altitude is L. If the taper is small, we may assume that the current density is uniform across any cross section. (a) Calculate the resistance of this object. (b) Show that your answer reduces to $\rho(L/A)$ for the special case of zero taper (that is, for $a = b$).

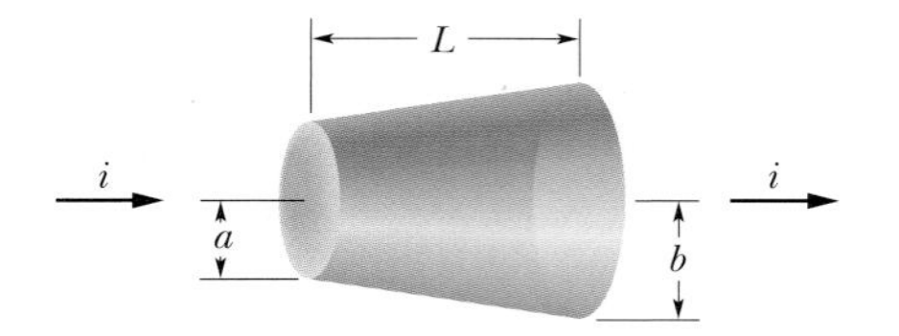

FIGURE 26-29 ■ Problem 40.

SEC. 26-10 ■ A MICROSCOPIC VIEW OF CURRENT AND RESISTANCE

41. Gas Discharge Tube A current is established in a gas discharge tube when a sufficiently high potential difference is applied across the two electrodes in the tube. The gas ionizes; electrons move toward the positive terminal and singly charged positive ions toward the negative terminal. (a) What is the magnitude of the current in a hydrogen discharge tube in which 3.1×10^{18} electrons and 1.1×10^{18} protons move past a cross-sectional area of the tube each second? (b) What is the direction of the current density $\vec{J}$?

42. A Block A block in the shape of a rectangular solid has a cross-sectional area of 3.50 cm^2 across its width, a front-to-rear length of 15.8 cm, and a resistance of 935 Ω. The material of which the block is made has 5.33×10^{22} conduction electrons/m^3. A potential difference of 35.8 V is maintained between its front and rear faces. (a) What is the current in the block? (b) If the current density is uniform, what is its value? (c) What is the average or drift speed of the conduction electrons? (d) What is the magnitude of the electric field in the block?

43. Earth's Lower Atmosphere Earth's lower atmosphere contains negative and positive ions that are produced by radioactive elements in the soil and cosmic rays from space. In a certain region, the atmospheric electric field strength is 120 V/m, directed vertically down. This field causes singly charged positive ions, at a density of 620/cm^3, to drift downward and singly charged negative ions, at a density of 550/cm^3, to drift upward (Fig. 26-30). The measured conductivity of the air in that region is $2.70 \times 10^{-14}(1/\Omega \cdot \text{m})$. Calculate (a) the average ion speed, assumed to be the same for positive and negative ions, and (b) the current density.

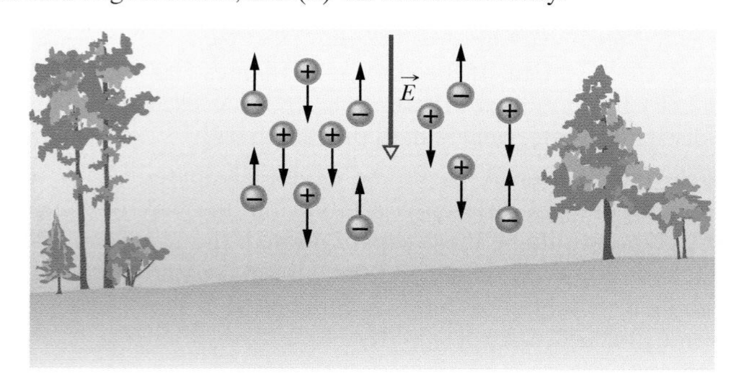

FIGURE 26-30 ■ Problem 43.

Additional Problems

44. Saving on Your Electric Bill Fluorescent bulbs deliver the same amount of light using much less power. If one kW-hr costs 12¢, estimate the amount of money you would save each month by replacing all the 75 W incandescent bulbs in your house by 10 W fluorescent ones than incandescent ones. *Be sure to clearly state your assumptions.*

45. Building a Water Heater The nickel-chromium alloy Nichrome has a resistivity of about 10^{-6} Ω-m. Suppose you want to build a small heater out of a coil of Nichrome wire and a 6 V battery in order to heat 30 ml of water from a temperature of 20 C to 40 C in 1 min. Assume the battery has negligible internal resistance.
(a) How much heat energy (in joules) do you need to do this?
(b) How much power (in watts) do you need to do it in the time indicated?
(c) What resistance should your Nichrome coil have in order to produce this much power in heat?
(d) Can you create a coil having these properties? (*Hint:* Can you find a plausible length and cross-sectional area for your wire that will give you the resistance you need?)
(e) If the internal resistance of the battery were 1/3 Ω, how would it affect your calculation? (Only explain what you would have to do; don't recalculate the size of your coil.)

46. A Confusing Thing One of the most confusing things about wiring circuits and figuring out what you've done is that many arrangements are electrically equivalent. Unless you have unusual powers of visualization it is often hard to recognize this. For example, three of the circuits shown in Fig. 26-31 are electrically equivalent and one is not. Answer questions (a) through (d) that follow.
(a) Which circuit is not like the others? Explain why it's different.
(b) Draw circuit diagrams for each of the arrangements and label each diagram as A, B, C, or D. (c) Examine your diagrams. Is it possible for neat circuit diagrams that look superficially different to represent the same set of electrical connections?

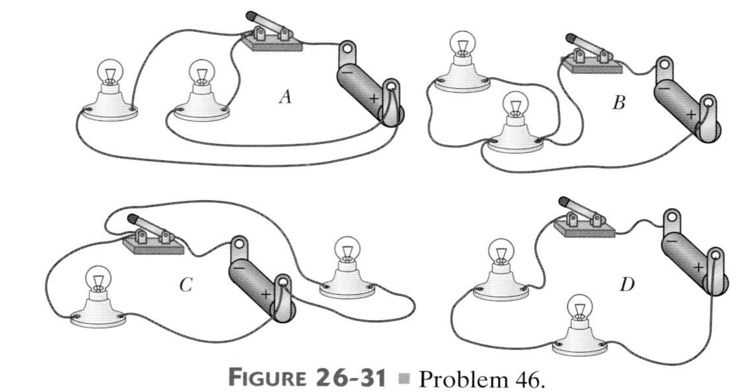

FIGURE 26-31 ■ Problem 46.

47. Draw the Circuit Diagram Draw a neat circuit diagram for each of the two circuits shown in Fig. 26-32 using the standard symbols for bulbs, batteries, and switches.

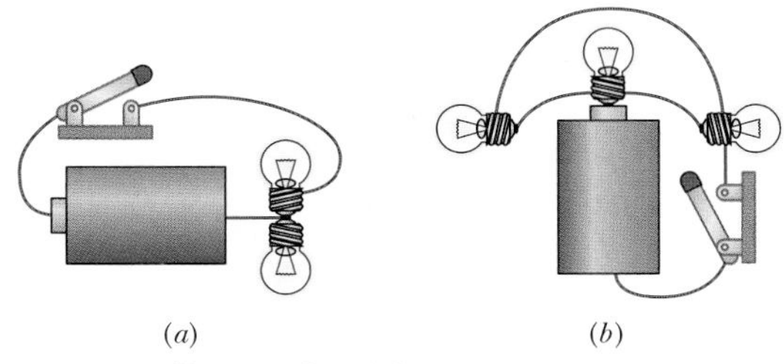

FIGURE 26-32 ■ Problem 47.

48. Charge Through Conductor The charge passing through a conductor increases over time as $q(t) = (1.6\ \text{C/s}^2)t^2 + (2.2\ \text{C/s})t$, where t is in seconds. (a) What equation describes the current in the circuit as a function of time? (b) What is the current in the conductor at $t = 0.0$ s and at $t = 2.0$ s?

49. Increases Over Time The charge passing through a conductor increases over time as $q(t) = (1.5\ \text{C/s}^3)t^3 - (4.5\ \text{C/s}^2)t^2 + (2\ \text{C/s})t$, where t is in seconds. (a) What equation describes the current in the circuit as a function of time? (b) What is the current in the conductor at $t = 0.0$ s and at $t = 1.0$ s?

50. 1994 Honda Accord Consider a 1994 Honda Accord with a battery that is rated at 52 ampere-hours. This battery is supposed to be able to deliver 1 ampere of current to electrical devices in a car for at least 52 hours or 2 amperes for 26 hours, and so on. Suppose you leave the car lights turned on when you park the car and the car lights draw 20 amperes of current. How long will it be before your battery is dead?

51. The Resistance of a Pocket Calculator A typical AAA battery delivers a nearly constant voltage of 1.5 V and stores about 3 kJ of energy. From the time it takes you to use up the batteries in your calculator, estimate the resistance of your calculator. (If you don't have a calculator of this type, make a plausible estimate of how long it might take to use up the batteries. Give some reason for your estimate.)

27 Circuits

The electric eel (*Electrophorus*) lurks in rivers of South America, killing the fish on which it preys with pulses of current. It does so by producing a potential difference of several hundred volts along its length; the resulting current in the surrounding water, from near the eel's head to the tail region, can be as much as one ampere. If you were to brush up against this eel while swimming, you might wonder (after recovering from the very painful stun):

How can the electric eel manage to produce a current that large without shocking itself?

The answer is in this chapter.

27-1 Electric Currents and Circuits

Knowing how to analyze circuits by predicting the currents through their elements and the potential differences across them is a valuable skill. Such knowledge enables engineers and scientists to design electrical devices and helps them make productive use of existing devices. Our goal in this chapter is to understand the behavior of relatively simple electric circuits by applying concepts such as current, potential difference, and resistors developed in the previous chapter. We will start by considering very simple ideal circuits and then go on to consider circuits with multiple loops and batteries such as those shown in Fig. 27-1. Toward the end of the chapter we will introduce the concept of emf or electromotive force associated with batteries and other power sources. In particular, we will consider how to extend our analysis to the behavior of circuits powered by nonideal batteries that have internal resistance.

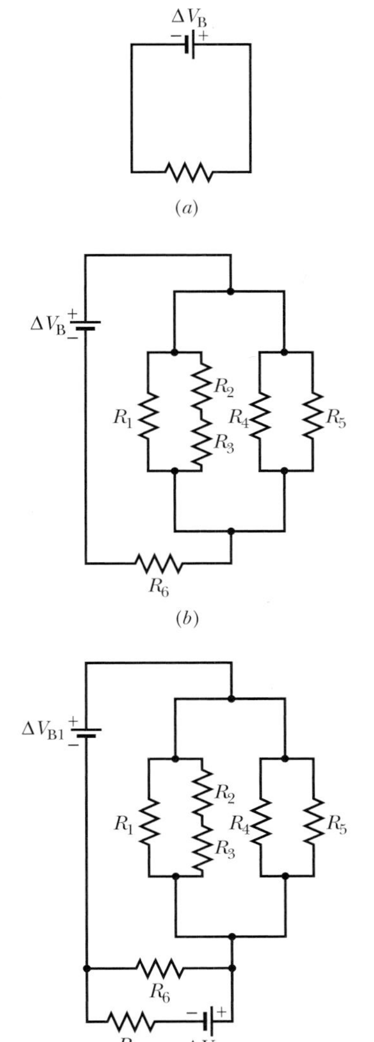

FIGURE 27-1 ■ Several types of ideal circuits we will learn to analyze in this chapter consist of ideal batteries, conducting wires with negligible resistance and ohmic resistors. (*a*) A single-loop circuit. (*b*) A single-battery, multiple-loop circuit. (*c*) A multiple-loop circuit with multiple batteries.

Ideal Circuits

As we so often do in developing physical ideas, we start by analyzing how a system behaves under ideal conditions. Only then do we introduce real-world complexities that require us to modify our methods of analysis. The ideal circuits we consider first have three characteristics:

1. **They are powered by ideal batteries.** As stated in Section 26-3, an ideal battery "maintains a constant potential difference across its terminals." This means there is a negligible amount of "electric friction" and the potential difference, ΔV_B, across the terminals of an ideal battery stays the same, regardless of the amount of charge flowing through it. But as the chemical potential energy of a real battery decreases, it develops some *internal resistance,* and the potential difference across its terminals decreases if its current increases.
2. **All circuit elements, other than the battery and connecting wires, are ohmic devices having a significant resistance.** As discussed in Section 26-5, an *ohmic device has a constant value of resistance, R, that is not a function of the amount of current passing through it.* Although lightbulbs and some other circuit elements are not ohmic, standard carbon resistors obey Ohm's law and have a constant resistance over a large current range. We make use of the fact that the potential difference across the terminals of an ohmic device is directly proportional to the current, i, flowing through it and is given by $\Delta V = iR$ (Eq. 26-7).
3. **Ideal conducting wires connect the battery to circuit elements.** Copper wiring is used in most circuits found in consumer devices, households, and industries. We can use Eq. 26-7 and data from Table 26-2 to determine that the resistance of a 30 cm length of common 22 gauge copper wire is about 0.1 Ω. If this wire was connected to a 10 Ω resistor, the additional resistance of the wire would add 1% to the overall resistance. In connecting larger resistors, the influence of the resistance of the wire is even smaller. Because the resistance in the wire is so small, the potential difference between the ends of even a relatively long continuous connecting wire is for all practical purposes negligible. In ideal circuits, we assume there is no potential drop across connecting wires.

READING EXERCISE 27-1: Show that the resistance of a 30 cm (≈ 12 inch) length of 22 gauge copper wire of diameter 0.024 cm has a resistance of about 0.1 Ω. *Hint:* You will need to use information from Table 26-2 along with Eq. 26-7. ■

27-2 Current and Potential Difference in Single-Loop Circuits

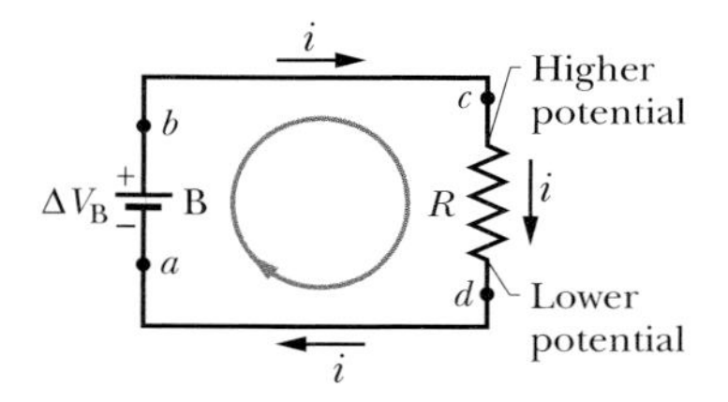

FIGURE 27-2 ■ A single-loop circuit in which a resistor R is connected across an ideal battery B with potential difference ΔV_B. The resulting current i is the same throughout the circuit.

Suppose we want to design or operate an electrical device such as a CD player or refrigerator. The operation of the given device will require a certain minimum current or potential difference. How would we calculate the amount of current in a circuit or the potential difference between two points within the device? That is the topic of this section.

We start out our discussion of current in circuits by focusing on the part of the circuit outside of the battery. That is, we will focus on current that passes from one battery terminal, through the circuit, and back to the other terminal. At the end of the chapter we will review and extend our previous discussions about what goes on inside devices like batteries and generators.

Consider the simple *single-loop* circuit of Fig. 27-2 consisting of an ideal battery, a resistor, R, and two ideal connecting wires. Unless otherwise indicated, we assume that wires in circuits have negligible resistance. Their function, then, is merely to provide pathways along which charge carriers can move. Through use of stored chemical energy (a form of internal potential energy), the battery keeps one of its terminals (called the positive terminal and often labeled +) at a higher electric potential than the other terminal (called the negative terminal and labeled −).

The mobile negative charge carriers in the circuit wires move preferentially toward the positive terminal and away from the negative terminal. As a result, for the circuit shown in Fig. 27-2, we have a net flow of negative charge in a counterclockwise direction. In Chapter 26, we discussed the fact that a flow of negative electrons in one direction is macroscopically indistinguishable from a flow of positive charges in the other direction. For historical reasons we continue the practice established in that chapter of working with current as if the charge carriers are positive.

The direction of the conventional current in the circuit shown in Fig. 27-2 is noted with arrows that are labeled i. Unless otherwise noted, we will continue the practice of using conventional (positive) current in our analysis of electric circuits. We will reach the same conclusions about the fundamental behavior of circuits as we would if we had used electron currents.

To begin learning how to calculate currents in circuits, let's start with the ideal circuit depicted in Fig. 27-2. We have marked the points just before and after each element with the letters a, b, c, and d. Let's start at point a and proceed around the circuit in either direction, adding any changes in potential we encounter. Once we return to our starting point, we must also have returned to our starting potential. In words, the potential energy change per unit of charge traveling through the battery plus the potential energy change of the charge traveling through the wires and the resistors must be zero. This can be denoted as

$$\Delta V_{a\to b} + \Delta V_{b\to c} + \Delta V_{c\to d} + \Delta V_{d\to a} = \Delta V_{a\to a} = 0\text{ V}.$$

For our simple circuit in Fig. 27-2 the charges gain potential while traveling from a to b due to the energy boost from the battery so that $\Delta V_{a\to b} = V_b - V_a = \Delta V_B$. The charges then flow freely from b to c through the first segment of the ideal conductor with no potential loss since the wire has a negligible resistance. Then the charges flow through the resistor, R. Finally, they flow back to point a, through another length of ideal wire.

$$\Delta V_B + \Delta V_{c\to d} = 0\text{ V},$$

where ΔV_B represents a positive change in potential per unit charge as charges proceed from point a to point b by moving through the battery. Recall that if our ohmic resistor has a fixed value R, then we noted in Eq. 26-7 that $\Delta V = iR$ where i is the current passing through the circuit. However, Eq. 26-7 didn't specify whether the ΔV refers to $\Delta V_{c \to d}$ or $\Delta V_{d \to c} = -\Delta V_{c \to d}$. It is clear from the context that if we proceed through the loop from c to d, $\Delta V_{c \to d}$ must be negative so it will cancel the ΔV_B, which we know is positive. This tells us the following about the mathematics of finding the potential difference across a resistor:

$$\Delta V_{d \to c} = iR \quad \text{and} \quad \Delta V_{c \to d} = -iR.$$

In other words, charges lose potential as they travel through a resistor. This makes sense physically because resistors give off energy in the form of heat and light. So our battery acts as a pump to increase the potential energy of a charge and the charge loses potential energy in passing through a resistive device.

This can be summarized as the loop rule.

LOOP RULE: The algebraic sum of the changes in potential encountered in a complete traversal of any loop of a circuit must be zero.

This is often referred to as *Kirchhoff's loop rule* (or *Kirchhoff's voltage law*), after German physicist Gustav Robert Kirchhoff. This rule is analogous to what happens when you hike around a mountain. If you start from any point on a mountain and return to the same point after walking around it, the algebraic sum of the changes in elevation you encounter must be zero. Thus, you end up at the same gravitational potential as you had before you started. Although we developed this rule through consideration of a single-loop circuit, it also holds for any complete loop in a *multiloop* circuit, no matter how complicated.

In Fig. 27-2, we will start at point a, whose potential is V_a, and mentally walk clockwise around the circuit until we are back at a, keeping track of potential changes as we move. (Our starting point is at the low-potential terminal of the battery—the negative terminal.) The potential difference between the battery terminals is equal to ΔV_B. When we pass through the battery from the low to high-potential terminal, the change in potential is positive.

As we walk along the top wire to the top end of the resistor, there is no potential change because the wire has negligible resistance; it is at the same potential as the high-potential terminal of the battery. So too is the top end of the resistor. When we pass through the resistor in the direction of the current flow, the potential decreases by an amount equal to $-iR$. We know the potential decreases because we are moving from the higher potential terminal of the resistor to the lower potential terminal.

For a walk around a single-loop circuit of total resistance R in *the direction of the current* our loop rule gives us

$$\Delta V_B - iR = 0\text{ V}.$$

Solving this equation for i gives us

$$i = \frac{\Delta V_B}{R} \quad \text{(single-loop circuit).} \tag{27-1}$$

If we apply the loop rule to a complete walk around a single-loop circuit of total resistance R *against the direction of current*, the rule gives us

$$-\Delta V_B + iR = 0\text{ V},$$

and we again find that

$$i = \frac{\Delta V_B}{R} \quad \text{(single-loop circuit).}$$

Thus, you may mentally circle a loop in either direction to apply the loop rule.

To prepare for circuits more complex than Fig. 27-2, let us summarize two rules for finding potential differences as we move around a chosen loop:

> **RESISTANCE RULE:** For a move through a resistor in the direction of the conventional current, the change in potential is $-iR$; in the opposite direction of current flow it is $+iR$.

> **POTENTIAL RULE:** For a move through a source of potential difference from low potential (for example, the negative terminal on a battery denoted a) to high potential (for example, the positive terminal on a battery denoted b) the change in potential is positive and given by $V_b - V_a = \Delta V_B$; in the opposite direction it is negative and given by $V_a - V_b = -\Delta V_B$.

What happens to the amount of current as it passes through a resistor? Is the current going into the resistor the same as the current coming out of the resistor? Or does a resistor (for example, a lightbulb) "use up" current? Recall that in Fig. 26-5 we depicted observations involving batteries and bulbs that clearly showed current is constant throughout a single loop circuit when resistors are connected in series. You can easily replicate these observations using fresh flashlight batteries, copper wires, and 1.5 V bulbs.

READING EXERCISE 27-2: It is asserted above that we can infer that the current flow into and out of a resistor is the same because three lightbulbs connected in series glow equally brightly. Suppose the resistors shown in Fig. 27-3a are lightbulbs. Describe the brightness of the third bulb relative to the first and second bulbs under the following assumptions: (a) All the current is used up by the first bulb; (b) most of the current is used up by the first bulb; (c) a small amount of the current was used up by the first bulb. ■

READING EXERCISE 27-3: The figure to the right shows the conventional current i in a single-loop circuit with a battery B and a resistor R (and wires of negligible resistance). At points a, b, and c, rank (a) the amount of the current and (b) the electric potential, greatest first. ■

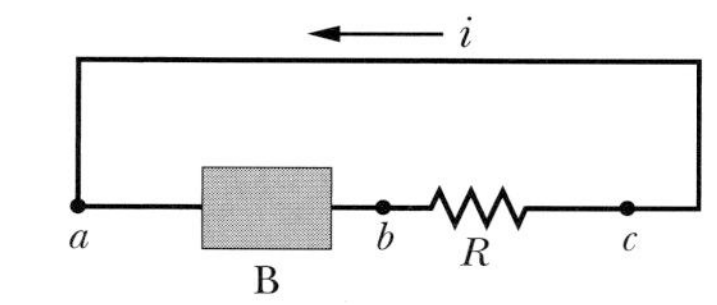

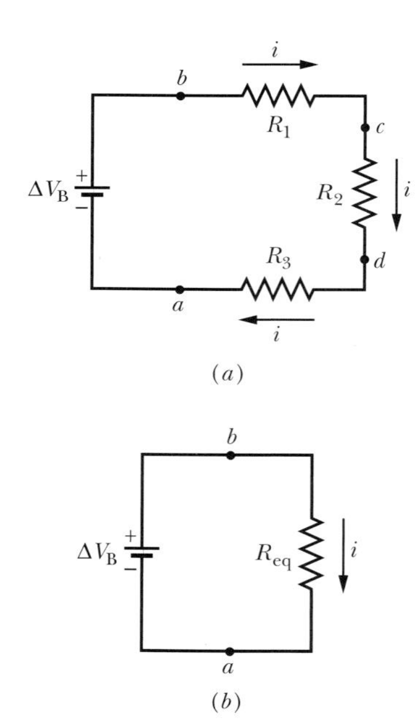

FIGURE 27-3 ■ (a) Three resistors are connected in series between points a and b. (b) An equivalent circuit, with the three resistors replaced with their equivalent resistance R_{eq}.

27-3 Series Resistance

We now turn our attention to more complicated single-loop circuits. Figure 27-3a shows three resistors connected in series to an ideal battery with potential difference ΔV_B between its terminals. Note that the three resistors are connected one after another between b and c, c and d, and d and a. Also an ideal battery maintains a potential difference across the series of resistors (between points a and b). If we apply the loop rule for charges moving in the direction of conventional current from point a at the negative terminal of the battery and proceeding through the loop until we encounter point a, again we get

$$\Delta V_{a\to b} + \Delta V_{b\to c} + \Delta V_{c\to d} + \Delta V_{d\to a} = 0\ V. \qquad (27\text{-}2)$$

Because we know that current is not used up by a resistor, we know the current flowing through the loop is the same everywhere, and so the current through each resistor must be the same. We also assume there is no potential difference along any segment of wire. If we consider the three resistors separately, applying the loop rule in the same manner (starting at the positive terminal of the battery and proceeding through the loop in the direction of conventional current) gives

$$\Delta V_B + (-iR_1) + (-iR_2) + (-iR_3) = 0\text{ V}.$$

By rearranging terms in the equation above we get

$$\Delta V_B - i(R_1 + R_2 + R_3) = 0\text{ V}, \tag{27-3}$$

and defining an equivalent resistance as $R_{eq} = R_1 + R_2 + R_3$ we find that Eq. 27-3 reduces to the same form as Eq. 27-1 with the equivalent resistance playing the role of the resistance in a circuit that has only one resistance. This is illustrated in Fig. 27-3.

Equating these two expressions tells us two things. First, the potential difference across the whole series of resistors is equal to the sum of the potential differences across the three resistors. Second, the potential difference across the whole series of resistors is equal to the potential difference across our ideal battery. Figure 27-3*b* shows the equivalent resistance, with a new resistor R_{eq}, that can replace the three resistors of Fig. 27-3*a*.

The result $R_{eq} = R_1 + R_2 + R_3$ is not surprising because it is compatible with the experimental findings we presented in Section 26-6: the resistance of a length of wire is directly proportional to its length (Eq. 26-8). Imagine three different carbon resistors like those depicted in the previous chapter (Fig. 26-21). Suppose these resistors are connected by ideal conductors (with almost no resistivity) having the same graphite material in their centers each with the same cross-sectional area. Giving the resistors different values of resistance would involve having the centers of the resistors be three different lengths. We would then expect the total resistance to be proportional to the sum of the three lengths of the resistors' graphite centers.

Obviously, we can extend our method of finding the equivalent resistance from 3 to N resistors by expanding Eq. 27-3 into the equation

$$R_{eq} = R_1 + R_2 + R_3 + \cdots + R_N = \sum_{j=1}^{N} R_j \quad \text{(N resistors in series).} \tag{27-4}$$

Note that when resistors are in series, their equivalent resistance is always *greater* than that of any of the individual resistors. Also, the current moving through resistors wired in series can move along only a single route. If there are additional routes so the currents in different resistors are different, the resistors are not connected in series.

In general:

> If N resistors in series were covered by a box, the resistors could be replaced by a single equivalent resistor with a value. $R_{eq} = R_1 + R_2 + R_3 + \cdots + R_N$. Someone making measurements outside the box could not tell whether there is a single equivalent resistor or a series of individual resistors.

In short, we conclude that if we replace a series of resistors with a single equivalent resistor, the new circuit will have the same overall potential differences and currents as the original one (so long as we don't measure potential drops between the resistors wired in series).

More on Ammeters

Analog ammeters work by measuring the torque exerted by magnetic forces on a current-carrying wire. We discuss more about their operation in Chapter 29 on magnetic

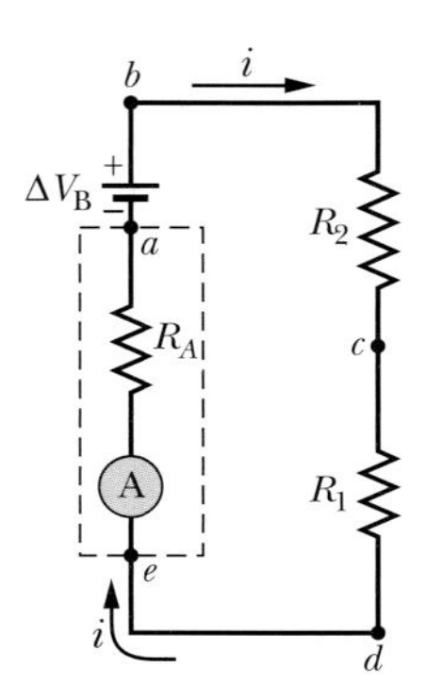

FIGURE 27-4 ■ This depicts how an ammeter can be inserted into a series circuit to measure the current. The third resistor represents the small resistance R_A of the ammeter itself.

fields. However, we continue our discussion of these devices from Chapter 26 and consider some important attributes the ammeter must have.

Recall from Chapter 26 that to measure the current in a wire, you are to break or cut the wire and insert the ammeter in series with an arm of the circuit so the current to be measured passes through the meter. (In Fig. 27-4, ammeter A is set up to measure current i).

When measuring the current in a circuit (or anything else for that matter) it is imperative that the measurement tool does not significantly change the quantity you are trying to measure. Hence, it is essential that the resistance R_A of the ammeter be very small compared to other resistances in the circuit. Otherwise, the presence of the meter will significantly change the current flow in the circuit, and measured current will be an inaccurate representation of the true current.

READING EXERCISE 27-4: In Fig. 27-3*a*, if $R_1 > R_2 > R_3$, rank the three resistances according to (a) the current through them and (b) the potential difference across them, greatest first. ■

READING EXERCISE 27-5: Consider an ammeter inserted into the circuit shown in Fig. 27-4. Compare the amount of current flowing through R_1 under the following three conditions: (a) without the ammeter inserted, (b) when the ammeter has a resistance much less than the equivalent resistance of $R_1 + R_2$, and (c) when the ammeter has a resistance equal to the equivalent resistance of $R_1 + R_2$. Explain your reasoning. Discuss the implications of your result on designing an ammeter. ■

27-4 Multiloop Circuits

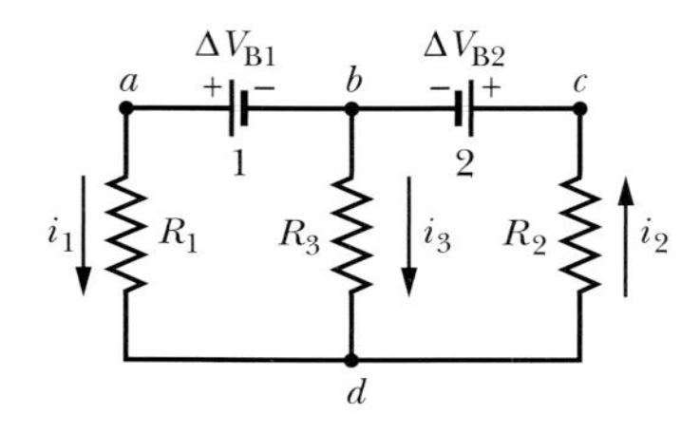

FIGURE 27-5 ■ A multiloop circuit consisting of three branches: left-hand branch *bad*, right-hand branch *bcd*, and central branch *bd*. The circuit has three loops we could choose to follow: left-hand loop *badb*, right-hand loop *bcdb*, and big loop *badcb*.

Figure 27-5 shows a circuit containing more than one loop. There are two points (b and d) at which the current branches split off or come together. We call such branching points **junctions.** For the circuit shown in Fig. 27-5, we would say there are two junctions, at b and d, and there are three *branches* connecting these junctions. The branches are the left branch (*bad*), the right branch (*bcd*), and the central branch (*bd*).

What are the currents in the three branches? We arbitrarily label the currents, using a different subscript for each branch. Because current is not used up and there are no additional branching points, current i_1 has the same value everywhere in branch *bad*, i_2 has the same value everywhere in branch *bcd*, and i_3 is the current through branch *bd*. The directions of the currents are assigned arbitrarily.

Consider junction d for a moment: charge comes into that junction via incoming currents i_1 and i_3, and it leaves via outgoing current i_2. Because charged particles neither accumulate nor disperse at the junction, the total incoming charge must be equal to the total outgoing charge. Hence, through conservation of charge arguments, we conclude that the total current coming into junction d must equal the total current leaving junction d,

$$i_{in} = i_{out},$$

or

$$i_1 + i_3 = i_2. \tag{27-5}$$

You can easily check that application of this condition to junction b leads to exactly the same equation. This expression for the current in branch 2 thus suggests a general principle:

> **JUNCTION RULE:** The sum of the currents entering any junction must be equal to the sum of the currents leaving that junction.

This rule is often called *Kirchhoff's junction rule* (or *Kirchhoff's current law*). It is simply a statement of the conservation of charge for a steady flow of charge—there is neither a buildup nor a depletion of charge at a junction. Thus, our basic tools for solving complex circuits are the *loop rule* (based on the conservation of energy) and the *junction rule* (based on the conservation of charge).

The relationship between i_1, i_2, and i_3 above is a single equation involving three unknowns. To solve the circuit completely (that is, to find all three currents), we need two more equations involving those same unknowns. We obtain them by applying the loop rule twice. In the circuit of Fig. 27-5, we have three loops from which to choose: the left-hand loop (*badb*), the right-hand loop (*bcdb*), and the big loop (*badcb*). Which two loops we choose turns out not to matter so long as we manage to pass through all the circuit elements at least once. For now, let's choose the left-hand loop and the right-hand loop.

If we traverse the left-hand loop in a counterclockwise direction from point b, the loop rule gives us

$$\Delta V_{B1} - i_1 R_1 + i_3 R_3 = 0\ \text{V}, \tag{27-6}$$

where ΔV_{B1} is the difference in potential between the terminals of battery 1. If we traverse the right-hand loop in a counterclockwise direction from point b, the loop rule gives us an equation involving battery 2,

$$-i_3 R_3 - i_2 R_2 - \Delta V_{B2} = 0\ \text{V}. \tag{27-7}$$

We now have three equations (Eqs. 27-5, 27-6, and 27-7) containing the three unknown currents, and they can be solved by a variety of mathematical techniques.

If we had applied the loop rule to the big loop, we would have obtained (moving counterclockwise from b) the equation

$$\Delta V_{B1} - i_1 R_1 - i_2 R_2 - \Delta V_{B2} = 0\ \text{V}.$$

This equation may look like fresh information, but in fact it is only the sum of Eqs. 27-6 and 27-7. (It would, however, yield the proper results when used with Eq. 27-5 and either 27-6 or 27-7.)

It is important to note that the assumed direction of the currents in a branch of the circuit do not have to be correct to get a correct solution. We must only keep track of the assumptions we have made. If in solving the resulting algebraic expressions we find that one of our currents turns out to have a negative value, then (because of the negative value) we know we made a wrong assumption about the direction of the current in that branch of the circuit.

In general, the total number of equations needed will be equal to the total number of independent loops in the circuit. The number of independent loops is simply the minimum number of loops needed to cover every branch in the circuit. Although some branches could be covered twice, every circuit element would be "covered" at least once. For example, we need at least two equations to cover all the loops in the circuit in Fig. 27-5 and at least three equations to cover all the loops in the more complex circuit in Fig. 27-6.

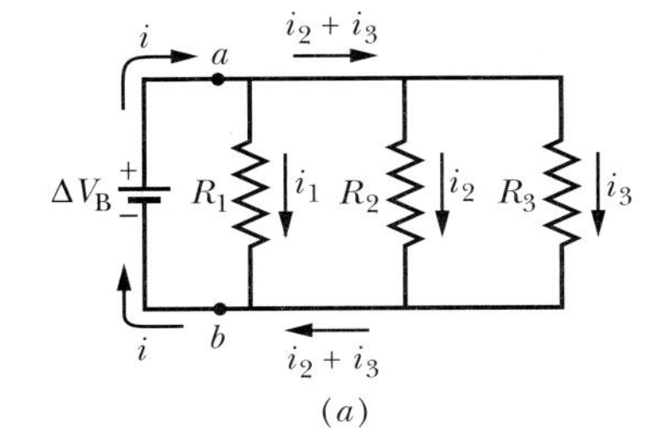

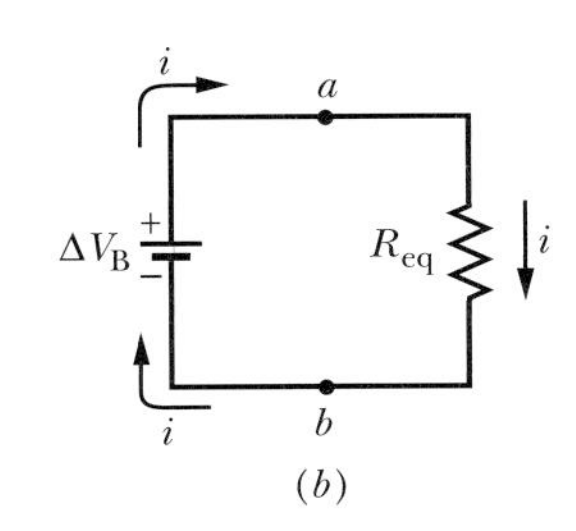

FIGURE 27-6 ■ (*a*) Three resistors connected in parallel across points a and b. (*b*) An equivalent circuit, with the three resistors replaced with their equivalent resistance R_{eq}.

27-5 Parallel Resistance

Figure 27-6*a* shows three resistances connected by branching junctions. Resistances that are parts of separate loops like those in Fig. 27-6*a* are said to be connected *in parallel* to the battery. Resistors connected "in parallel" are directly wired together on one side and directly wired together on the other side, and a potential difference ΔV

is applied across the pair of connected sides. Thus, the resistances have the same potential difference ΔV across them, producing a current through each. Because we are assuming ideal wires, there is no potential difference across the wires. Therefore, the potential across the top branch of the circuit is constant everywhere equal to the potential at the positive pole of the battery, and the potential across the bottom branch of the circuit is constant everywhere equal to the potential at the negative pole of the battery. In general,

> When a potential difference ΔV is applied across resistances connected in parallel, each resistor has the same potential difference ΔV across it.

Notice that we have again labeled the currents in each of the branches i_1, i_2, and i_3. We have discussed the way in which the current into a junction is equal to the current out of the junction. We have not yet discussed in what proportions currents divide when there is a branch (a choice of path) in a circuit. Are all three currents i_1, i_2, and i_3 equal? If not, which of these currents is largest? The answer to this question becomes clear when we write out the expressions for current through each of the resistors in Fig. 27-6 using the potential rule for loops. For the case pictured here, we have

$$i_1 = \frac{\Delta V}{R_1}, \quad i_2 = \frac{\Delta V}{R_2}, \quad \text{and} \quad i_3 = \frac{\Delta V}{R_3}. \tag{27-8}$$

Since each resistor is connected so it has the same potential difference across it, it is straightforward to see how the sizes of the currents compare to each other. If the resistances are all equal, the current through each is the same. However, if the three resistances are not equal, more current flows through the smaller resistances. This outcome is consistent with what we might predict based solely on an understanding that a resistor is just a device that resists the flow of current.

If we want to simplify how we think about a circuit that has resistors wired in parallel (like that shown in Fig. 27-6*a*), we can treat the three resistors in parallel as if they have been replaced by a single equivalent resistor R_{eq}. Figure 27-6*b* shows the three parallel resistances replaced with an equivalent resistance R_{eq}. The applied potential difference ΔV_B is maintained by a battery. We can see from this figure that the potential difference across the equivalent resistance would have to be the same as the potential difference applied across each of the original resistors. Furthermore, the equivalent resistor would have to have the same total current ($i_1 + i_2 + i_3$) through it as the original three resistors.

> Resistances connected in parallel can be replaced with an equivalent resistance R_{eq}. If the equivalent resistance has the same potential difference applied across it, then the current through it will equal the sum of currents flowing through the original resistors.

To derive an expression for R_{eq} in Fig. 27-6*b*, we first write the current in each of the resistors in Fig. 27-6*a* as

$$i_1 = \frac{\Delta V}{R_1}, \quad i_2 = \frac{\Delta V}{R_2}, \quad \text{and} \quad i_3 = \frac{\Delta V}{R_3}, \tag{Eq. 27-8}$$

where ΔV is the potential difference between *a* and *b*. If we apply the junction rule at point *a* in Fig. 27-6*a* and then substitute these values, we find

$$i = i_1 + i_2 + i_3 = \Delta V\left(\frac{1}{R_1} + \frac{1}{R_2} + \frac{1}{R_3}\right). \tag{27-9}$$

If we instead consider the parallel combination with the equivalent resistance R_{eq} (Fig. 27-6*b*), we have

$$i = \frac{\Delta V}{R_{eq}} = \Delta V\left(\frac{1}{R_{eq}}\right). \qquad (27\text{-}10)$$

Comparing the two equations above leads to

$$\frac{1}{R_{eq}} = \frac{1}{R_1} + \frac{1}{R_2} + \frac{1}{R_3}. \qquad (27\text{-}11)$$

The result $1/R_{eq} = 1/R_1 + 1/R_2 + 1/R_3$ is not surprising because it is compatible with the experimental findings we presented in Section 26-6: the resistance of a length of wire is inversely proportional to its cross-sectional area (Eq. 26-8). To see this connection, imagine three different carbon resistors like those depicted in the last chapter (Fig. 26-21) connected in parallel. Then giving them different values of resistance would involve having the centers of the resistors have three different cross-sectional areas. Because the resistors are connected in parallel, we would then expect the total cross-sectional area to be the sum of the three cross-sectional areas of the resistors' graphite centers so $A_{eq} = A_1 + A_2 + A_3$. Since the cross-sectional area and resistance are inversely proportional, we get $1/R_{eq} = 1/R_1 + 1/R_2 + 1/R_3$.

Extending Eq. 27-11 to the case of n resistors, we have

$$\frac{1}{R_{eq}} = \sum_{j=1}^{n} \frac{1}{R_j} \quad (n \text{ resistors in parallel}). \qquad (27\text{-}12)$$

Since we often deal with the case of two resistors in parallel, it is worth it for us to consider this case a bit more. For the case of two resistors, the equivalent resistance is

$$\frac{1}{R_{eq}} = \frac{1}{R_1} + \frac{1}{R_2}.$$

With a bit of algebra, this becomes

$$R_{eq} = \frac{R_1 R_2}{R_1 + R_2} \quad (2 \text{ resistors in parallel}). \qquad (27\text{-}13)$$

If you accidentally took the equivalent resistance to be the sum divided by the product, you would notice at once that this result would be dimensionally incorrect.

Note that when two or more resistors are connected in parallel, the equivalent resistance is smaller than any of the combining resistances.

More on the Voltmeter

Recall from our discussion in Chapter 26 that a meter used to measure potential differences is called a *voltmeter*. To measure the potential difference between any two points in the circuit, the voltmeter terminals are connected across those points, without breaking or cutting the wire. In Fig. 27-7, voltmeter V is set up to measure the potential difference across a resistor R_1. The voltmeter is inserted in parallel to R_1 by connecting its terminals to points d and e in the circuit.

To prevent the voltmeter from affecting a measurement, it is essential that the resistance R_V of a voltmeter be *very large* compared to the resistance of the circuit element across which the voltmeter is connected. Otherwise, the meter becomes an important circuit element by drawing a significant current through itself. This change

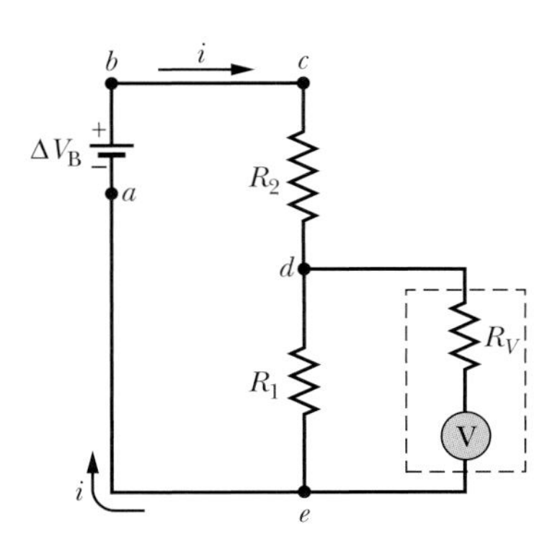

FIGURE 27-7 ■ A single-loop circuit, showing how to connect a voltmeter (V). The third resistor R_V represents the resistance of the voltmeter itself. We assume that R_V is very large compared to R_1 and R_2.

in current flow can alter the potential difference to be measured. On the other hand, even if the potential difference across the voltmeter is large, if a very small current flows through the voltmeter, the flow of current through R_1 will not change very much.

READING EXERCISE 27-6: A battery, with potential ΔV_B across it, is connected to a combination of two identical resistors and a current i flows through the battery. What is the potential difference across and the current through either resistor if the resistors are (a) in series, and (b) in parallel? ■

READING EXERCISE 27-7: Consider the voltmeter inserted into the circuit shown in Fig. 27-7. Describe what would happen if the voltmeter has a resistance $R_V \ll R_1$. How would this affect the potential difference measured across the resistor R_1? Describe what would happen if the voltmeter has a resistance $R_V \gg R_1$. How would this affect the potential difference measured across the resistor R_1? Which case would give the most "accurate" measure of the potential difference across the resistor when the voltmeter is not a part of the circuit? ■

READING EXERCISE 27-8: Suppose the resistors in Fig. 27-6*a* are all identical lightbulbs. Rank the brightness of the three bulbs. Compare the brightness of each of the bulbs to the brightness of one of the bulbs alone connected to the same battery. ■

TOUCHSTONE EXAMPLE 27-1: One Battery and Four Resistances

Figure 27-8*a* shows a multiloop circuit containing one ideal battery and four resistances with the following values:

$$R_1 = 20\ \Omega, \quad R_2 = 20\ \Omega, \quad \Delta V_B = 12\ \text{V},$$
$$R_3 = 30\ \Omega, \quad \text{and} \quad R_4 = 8.0\ \Omega.$$

(a) What is the current through the battery?

SOLUTION ■ First note that the current through the battery must also be the current through R_1. Thus, one **Key Idea** here is that we might find that current by applying the loop rule to a loop that includes R_1 because the current would be included in the potential difference across R_1. Either the left-hand loop or the big loop will do. Noting that the potential difference arrow of the battery points upward so the current the battery supplies is clockwise, we might apply the loop rule to the left-hand loop, clockwise from point a. With i being the current through the battery, we would get

$$+\Delta V_B - iR_1 - iR_2 - iR_4 = 0\ \text{V}. \quad \text{(incorrect)}$$

However, this equation is incorrect because it assumes that R_1, R_2, and R_4 all have the same current i. Resistances R_1 and R_4 do have the same current, because the current passing through R_4 must pass through the battery and then through R_1 with no change in value. However, that current splits at junction point b—only part passes through R_2, and the rest through R_3.

To distinguish the several currents in the circuit, we must label them individually as in Fig. 27-8*b*. Then, circling clockwise from a, we can write the loop rule for the left-hand loop as

$$+\Delta V_B - i_1R_1 - i_2R_2 - i_1R_4 = 0\ \text{V}.$$

Unfortunately, this equation contains two unknowns, i_1 and i_2; we need at least one more equation to find them.

A second **Key Idea** is that an easier option is to simplify the circuit of Fig. 27-8*b* by finding equivalent resistances. Note carefully that R_1 and R_2 are *not* in series and thus cannot be replaced with an equivalent resistance. However, R_2 and R_3 are in parallel, so we can use either Eq. 27-12 or Eq. 27-13 to find their equivalent resistance R_{23}. From the latter,

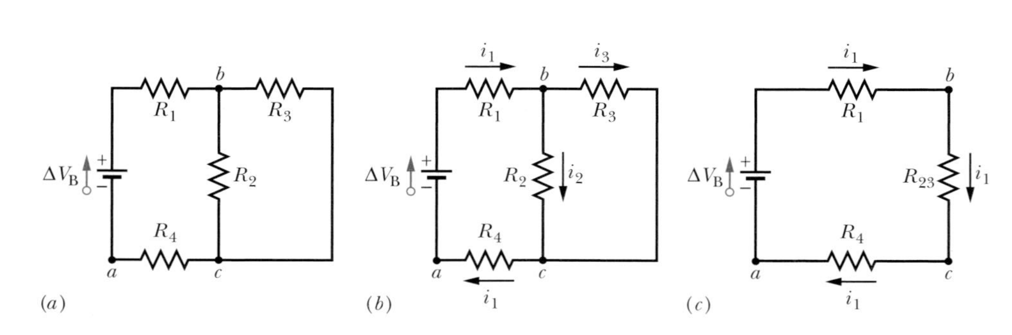

FIGURE 27-8 ■ (*a*) A multiloop circuit with an ideal battery of potential difference ΔV_B and four resistances. (*b*) Assumed currents through the resistances. (*c*) A simplification of the circuit, with resistances R_2 and R_3 replaced with their equivalent resistance R_{23}. The current through R_{23} is equal to that through R_1 and R_4.

$$R_{23} = \frac{R_2 R_3}{R_2 + R_3} = \frac{(20\ \Omega)(30\ \Omega)}{50\ \Omega} = 12\ \Omega.$$

We can now redraw the circuit as in Fig. 27-8*c*; note that the current through R_{23} must be i_1 because charge that moves through R_1 and R_4 must also move through R_{23}. For this simple one-loop circuit, the loop rule (applied clockwise from point *a*) yields

$$+\Delta V_B - i_1 R_1 - i_1 R_{23} - i_1 R_4 = 0\ \text{V}.$$

Substituting the given data, we find

$$12\ \text{V} - i_1(20\ \Omega) - i_1(12\ \Omega) - i_1(8.0\ \Omega) = 0\ \text{V},$$

which gives us

$$i_1 = \frac{12\ \text{V}}{40\ \Omega} = 0.30\ \text{A}. \quad \text{(Answer)}$$

(b) What is the current i_2 through R_2?

SOLUTION ■ One **Key Idea** here is that we must work backward from the equivalent circuit of Fig. 27-8*c*, where R_{23} has replaced the parallel resistances R_2 and R_3. A second **Key Idea** is that, because R_2 and R_3 are in parallel, they both have the same potential difference across them as their equivalent R_{23}. We know the current through R_{23} is $i_1 = 0.30$ A. Thus, we can use Eq. 26-5 $R = \Delta V/i$ to find the potential difference ΔV_{23} across R_{23}:

$$\Delta V_{23} = i_1 R_{23} = (0.30\ \text{A})(12\ \Omega) = 3.6\ \text{V}.$$

The potential difference across R_2 is thus 3.6 V, so the current i_2 in R_2 must be, by Eq. 26-5,

$$i_2 = \frac{\Delta V_2}{R_2} = \frac{3.6\ \text{V}}{20\ \Omega} = 0.18\ \text{A}. \quad \text{(Answer)}$$

(c) What is the current i_3 through R_3?

SOLUTION ■ We can answer by using the same technique as in (b), or we can use this **Key Idea**: The junction rule tells us that at point *b* in Fig. 27-8*b*, the incoming current i_1 and the outgoing currents i_2 and i_3 are related by

$$i_1 = i_2 + i_3.$$

This gives us

$$i_3 = i_1 - i_2 = 0.30\ \text{A} - 0.18\ \text{A} = 0.12\ \text{A}. \quad \text{(Answer)}$$

TOUCHSTONE EXAMPLE 27-2: Three Batteries and Five Resistances

Figure 27-9 shows a circuit with three ideal batteries in it. Two of these batteries labeled ΔV_{B2} are identical. The circuit elements have the following values:

$$\Delta V_{B1} = 3.0\ \text{V}, \quad \Delta V_{B2} = 6.0\ \text{V}, \quad R_1 = 2.0\ \Omega, \quad R_2 = 4.0\ \Omega.$$

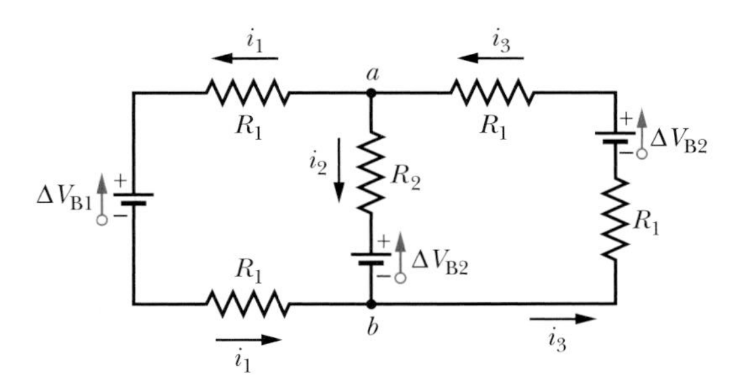

FIGURE 27-9 ■ A multiloop circuit with three ideal batteries and five resistances.

Find the amount and direction of the current in each of the three branches.

SOLUTION ■ It is not worthwhile to try to simplify this circuit, because no two resistors are in parallel, and the resistors that are in series (those in the right branch or those in the left branch) present no problem. So our **Key Idea** is to apply the junction and loop rules to this circuit.

Using arbitrarily chosen directions for the currents as shown in Fig. 27-9, we apply the junction rule at point *a* by writing

$$i_3 = i_1 + i_2. \quad (27\text{-}14)$$

An application of the junction rule at junction *b* gives only the same equation, so we next apply the loop rule to any two of the three loops of the circuit. We first arbitrarily choose the left-hand loop, arbitrarily start at point *a*, and arbitrarily traverse the loop in the counterclockwise direction, obtaining

$$-i_1 R_1 - \Delta V_{B1} - i_1 R_1 + \Delta V_{B2} + i_2 R_2 = 0\,\text{V}.$$

Substituting the given data and simplifying yield

$$i_1(4.0\ \Omega) - i_2(4.0\ \Omega) = 3.0\,\text{V}. \quad (27\text{-}15)$$

For our second application of the loop rule, we arbitrarily choose to traverse the right-hand loop clockwise from point *a*, finding

$$+i_3 R_1 - \Delta V_{B2} + i_3 R_1 + \Delta V_{B2} + i_2 R_2 = 0\,\text{V}.$$

Substituting the given data and simplifying yield

$$i_2(4.0\ \Omega) + i_3(4.0\ \Omega) = 0\,\text{V}. \quad (27\text{-}16)$$

Using Eq. 27-14 to eliminate i_3 from Eq. 27-16 and simplifying give us

$$i_1(4.0\ \Omega) + i_2(8.0\ \Omega) = 0\,\text{V}. \quad (27\text{-}17)$$

We now have a system of two equations (Eqs. 27-15 and 27-17) in two unknowns (i_1 and i_2) to solve either by hand (which is easy enough here) or with a math computer software package. (One solution technique is Cramer's rule, given in Appendix E.) We find

$$i_2 = -0.25 \text{ A}.$$

(The minus sign signals that our arbitrary choice of direction for i_2 in Fig. 27-9 is wrong; i_2 should point up through ΔV_{B2} and R_2.) Substituting $i_2 = -0.25$ A into Eq. 27-17 and solving for i_1 then give us

$$i_1 = 0.50 \text{ A}. \quad \text{(Answer)}$$

With Eq. 27-14 we then find that

$$i_3 = i_1 + i_2 = 0.25 \text{ A}. \quad \text{(Answer)}$$

The positive answers we obtained for i_1 and i_3 signal that our choices of directions for these currents are correct. We can now correct the direction for i_2 and write its amount as

$$i_2 = 0.25 \text{ A}. \quad \text{(Answer)}$$

27-6 Batteries and Energy

So far we have discussed ideal batteries that can be characterized as maintaining a constant potential difference between their terminals no matter what current is flowing through them. Also, we have concentrated on analyzing what happens in the part of the circuit that lies outside the battery. In this section we consider more about what goes on inside batteries and how real, not so ideal, batteries behave.

The amazing thing about a battery is that positive charge carriers enter with a low potential energy and other carriers emerge from the battery at a higher potential. Energy transformations inside a battery enable charges to overcome the forces exerted on them by the electric field inside the battery. Positive carriers seem to move opposite to the battery's electric field, whereas negative charge carriers move with it. There must be some other force present inside an energy-providing device enabling charges to swim upstream against electrical forces. The outdated term given to this "force" is electromotive force. Its abbreviation, which we still use today, is emf. How is this "force" defined? Where does it come from in a typical battery?

We define the emf, $\mathscr{E}$, of a battery in terms of the work done per unit charge on charges flowing into it:

$$\mathscr{E} \equiv \frac{dW}{dq} \quad \text{(definition of electromotive force).} \qquad (27\text{-}18)$$

In words, the battery emf is the work per unit charge it does to move charge from one terminal to the other. The SI unit for emf is the joule/coulomb. In Chapter 25 we defined one joule/coulomb as the *volt*. There must be some source of energy within a battery, enabling it to do work on the charges. The energy source may be chemical, as in a battery (or a fuel cell). Temperature differences may supply the energy, as in a thermopile; or the Sun may supply it, as in a solar cell. As you can see, the term electromotive force is very misleading since it is not a force at all, but has the same units as electrostatic potential (energy per unit charge). Furthermore emf is a scalar quantity and is not a vector quantity like a force is.

When a battery is connected to a circuit, it transfers energy to the charge carriers passing through it. Let's look at one example of how chemical action can do this. For this purpose we will consider the chemical reactions that take place inside one cell of a lead acid battery used in most automobiles. A lead acid battery consists of several cells wired together in series. Each cell has two metal plates surrounded by a liquid bath of chemicals. In a lead-acid cell, the negative plate is made of pure lead, and the positive plate is made of lead-oxide. These plates are immersed in sulfuric acid mixed with water. The acid dissociates in the water into hydronium ions (H_3O^+) and bisulfate ions (HSO_4^-). This is shown in Fig. 27-10. Both the lead and lead oxide can react

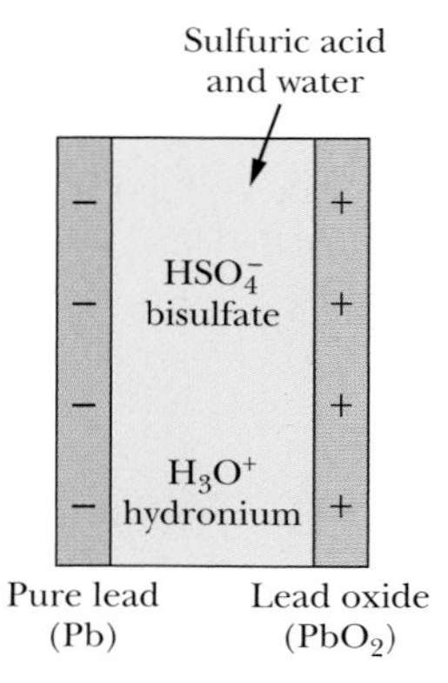

FIGURE 27-10 ■ The chemical constituents of the lead acid battery.

with the bisulfate ions as follows:

$$Pb + HSO_4^- + H_2O \rightarrow PbSO_4 + H_3O^+ + 2e^-$$

$$PbO_2 + HSO_4^- + 3H_3O^+ + 2e^- \rightarrow PbSO_4 + 4H_2O$$

The two electrons produced on the pure lead plate pile up on it. The second reaction removes the two electrons it needs from the lead oxide plate. Thus, each time the pair of reactions occur, electrons are added to the negative plate and removed from the positive plate. If the cell were not connected to a circuit, the reactions would stop when the charge difference gets so large that the energy needed to put more charges on the plates is greater than the energy released by the reactions. If the battery is connected to an external circuit, then as the charges flow through the circuit, they are removed from one plate and put back on the other; the process can keep going until all the sulfuric acid (HSO_4) is consumed.

Note that when we talk about a battery as a charge pump, this is somewhat misleading because the electrons removed by the chemical reaction at one battery terminal (plate) are not the same electrons released at the other terminal.

There are hundreds of different types of chemical batteries. The lead-acid battery action described here simply serves as an example of how chemical reactions can cause charge separation in a battery.

27-7 Internal Resistance and Power

In our evaluation of circuits up to this point, we have assumed the current passes through the battery (or other emf source) without encountering any resistance within it. We call such a battery or other emf device "ideal."

An **ideal emf device** is one that lacks any resistance to the movement of charge through it. The potential difference between the terminals of an *ideal* emf device is equal to the emf of the device. For example, an ideal battery with an emf of 12.0 V has a potential difference of 12.0 V between its terminals. Very fresh alkaline batteries are nearly ideal.

A **real emf device** has internal resistance to the movement of charge through it. For a real emf device (for example, a real battery), the only situation for which the potential difference between its terminals is equal to its emf is when the device is not connected to a circuit, and thus does not have current through it. However, when the device has current through it, the potential difference between its terminals differs from its emf.

Figure 27-11*a* shows circuit elements that describe the behavior of a real battery, with internal resistance r, wired to an external resistor of resistance R. The internal

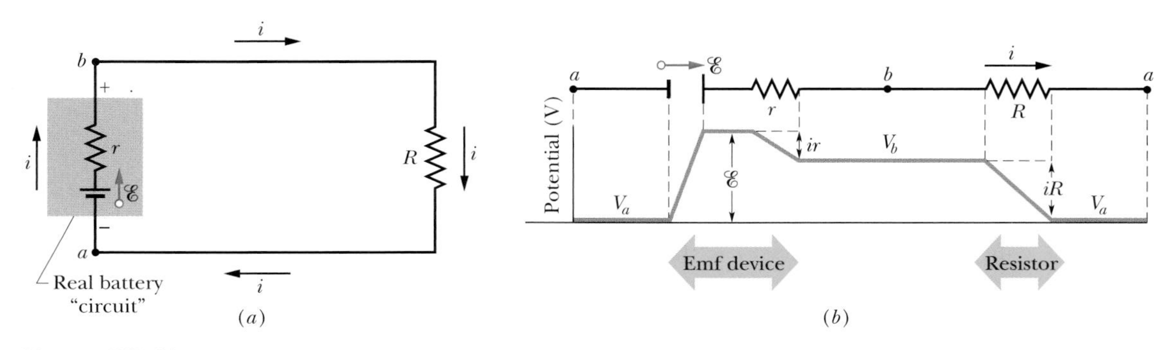

FIGURE 27-11 ■ (*a*) A single-loop circuit containing a real battery having internal resistance r and emf $\mathscr{E}$. (*b*) The same circuit, now spread out in a line. The potentials encountered in traversing the circuit clockwise from a are also shown. The potential V_a is arbitrarily assigned a value of zero, and other potentials in the circuit are graphed relative to V_a.

resistance of the battery is the electrical resistance of the conducting materials of the battery and thus is an unavoidable feature of any real battery. However, as an illustration, a real battery is depicted in Fig. 27-11*b* as if it could be separated into an ideal battery with potential difference $\mathscr{E}$ between its terminals and a resistor of resistance r. The order in which the symbols for these separated parts are drawn does not matter.

If we apply the potential (loop) rule, proceeding clockwise and beginning at point a, the *changes* in potential give us

$$\Delta V_{a\to b} + \Delta V_R = 0\text{ V},$$

or

$$\mathscr{E} + \Delta V_{\text{internal resistance}} + \Delta V_R = 0\text{ V}. \tag{27-19}$$

It is customary to keep track of potential differences as if the charge carriers are positive. Thus, we go through both resistances in the direction of the *conventional* current (defined in the previous chapter as the direction of flow we would find if the charge carriers were positive instead of negative):

$$\mathscr{E} - ir - iR = 0\text{ V}. \tag{27-20}$$

Solving for the current, we find

$$i = \frac{\mathscr{E}}{R + r}. \tag{27-21}$$

Note that this equation reduces to Eq. 27-1 if the battery is ideal so that $r = 0\ \Omega$.

Figure 27-11*b* shows graphically the changes in electric potential around the circuit. (To better link Fig. 27-11*b* with the *closed circuit* in Fig. 27-11*a*, imagine curling the graph into a cylinder with point a at the left overlapping point a at the right.) Note how traversing the circuit is like walking up and down a (potential) mountain and returning to your starting point—you also return to the starting elevation.

In this book, if a battery is not described as real or if no internal resistance is indicated, you can assume for simplicity that it is ideal.

Implications of Internal Resistance in Real EMF Devices

To understand the implications of internal resistance in emf devices for real circuits, let's try to make our understanding a bit more quantitative. To start with, let's see how $\Delta V_B = \Delta V_{a\to b} = V_b - V_a$, the potential difference across the battery terminals in Fig. 27-11, is affected by the existence of an internal resistance in the battery. To calculate $V_b - V_a$, we start at point a and follow the shorter path around to b, which takes us clockwise through the battery. We then have

$$V_a + \mathscr{E} - ir = V_b,$$

or

$$V_b - V_a = \Delta V_B = \mathscr{E} - ir, \tag{27-22}$$

where r is the internal resistance of the battery and $\mathscr{E}$ is the emf of the battery. This expression tells us the potential difference of the battery is equal to the emf minus the drop in potential associated with internal resistance.

Furthermore, if we refer back to Eq. 27-21,

$$i = \frac{\mathscr{E}}{R + r},$$

and substitute this expression for current (in the circuit shown in Fig. 27-11) into our expression for the potential difference across the battery terminals, we get

$$\Delta V_B = \mathscr{E} - \left(\frac{\mathscr{E}r}{R + r}\right).$$

With some algebra, we get the following generally applicable expression:

$$\Delta V_B = \mathscr{E}\frac{R}{R + r}. \qquad (27\text{-}23)$$

For example, suppose that in Fig. 27-11, $\mathscr{E} = 12$ V, $R = 10\ \Omega$, and $r = 2.0\ \Omega$. Then the equation above tells us the potential across the battery's terminals is

$$\Delta V_B = (12\text{ V})\frac{10\ \Omega}{10\ \Omega + 2.0\ \Omega} = 10\text{ V}.$$

In "pumping" charge through itself, the battery (via electrochemical reactions) does work per unit charge of $\mathscr{E} = 12$ J/C, or 12 V. However, because of the internal resistance of the battery, it produces a potential difference of only 10 J/C, or 10 V, across its terminals.

If the internal resistance becomes large compared to the overall resistance in the circuit, the available potential difference of the battery, electrical generator, or other emf device will drop significantly. This drop in available potential difference results in a reduction in the amount of current in the circuit. This is especially important to consider when circuits are designed with a low resistance so they will carry a large current.

For example, consider the circuit shown in Fig. 27-6 (three resistors in parallel with a battery) and let $R = 3\ \Omega$ for each resistor. The equivalent resistance in the circuit is $R_{eq} = 1\ \Omega$. If the potential difference source is taken to be an ideal battery (internal resistance $r = 0$), the current in the circuit is

$$i = \frac{\Delta V_B}{R_{eq}} = \frac{12\text{ V}}{1\ \Omega} = 12\text{ A}.$$

The 12 amps are split evenly between each branch (because the resistances are all equal), so each resistor has 4 amps of current flowing through it.

However, if the potential difference source is a real battery with $\mathscr{E} = 12$ V and internal resistance $r = 2.0\ \Omega$, then the available potential difference from the battery is

$$\Delta V_B = (12\text{ V})\frac{1\ \Omega}{1\ \Omega + 2.0\ \Omega} = 4\text{ V}.$$

The total current in the circuit is then

$$i = \frac{\Delta V_B}{R_{eq}} = \frac{4\text{ V}}{1\ \Omega} = 4\text{ A}.$$

This current is still split between each of the branches of the circuit, so for the case of the real battery, the current flowing through each resistor is now only 4/3 amp. In comparison to the 4 amps produced by the ideal battery, one can see how the internal resistance of an emf device can play a significant role in the functioning of real circuits.

Power

When a battery or some other type of emf device does work on the charge carriers to establish a current i, it transfers energy from its source of energy (such as the chemical source in a battery) to the charge carriers. Because a real emf device has an internal resistance r, it also transfers energy to internal thermal energy via resistive dissipation, discussed in Chapter 26. Let us relate these transfers.

The net rate P of energy transfer from the emf device to the charge carriers is given by

$$P = i\,\Delta V, \tag{27-24}$$

where ΔV is the potential across the terminals of the emf device. (Note that this is the power associated with the transfer). If we apply this expression to the circuit shown in Fig. 27-11 (from Eq. 27-24 above), we can substitute $\Delta V_B = \mathscr{E} - ir$ into Eq. 27-24 to find

$$P = i(\mathscr{E} - ir) = i\mathscr{E} - i^2 r. \tag{27-25}$$

We see that the term $i^2 r$ in Eq. 27-25 is the rate P_r of energy transfer to thermal energy within the emf device:

$$P_r = i^2 r \quad \text{(internal dissipation rate).} \tag{27-26}$$

Then the term $i\mathscr{E}$ in Eq. 27-25 must be the rate P_{emf} at which the emf device transfers energy to *both* the charge carriers and to internal thermal energy. Thus,

$$P_{emf} = i\mathscr{E} \quad \text{(power of emf device).} \tag{27-27}$$

If a battery is being *recharged*, with a "wrong way" current through it, the energy transfer is then from the charge carriers to the battery—both to the battery's chemical energy and to the energy dissipated in the internal resistance r. The rate of change of the chemical energy is given by Eq. 27-27, the rate of dissipation is given by Eq. 27-26, and the rate at which the carriers supply energy is given by Eq. 27-24.

As is the case for mechanics, the accepted SI unit for electrical power is the watt. One watt is equal to one joule-sec.

TOUCHSTONE EXAMPLE 27-3: Two Real Batteries

Let's consider a circuit with two *nonideal* batteries that have internal resistances. Since the potential differences across the terminals of these batteries are not constant, we characterize each battery in terms of its emf ($\mathscr{E}_1$ or $\mathscr{E}_2$) and internal resistances (r_1 or r_2). The emfs and resistances in the circuit of Fig. 27-12*a* have the following values:

$$\mathscr{E}_1 = 4.4\text{ V},\quad \mathscr{E}_2 = 2.1\text{ V},\quad r_1 = 2.3\ \Omega,\quad r_2 = 1.8\ \Omega,\quad R = 5.5\ \Omega.$$

(a) What is the current i in the circuit?

SOLUTION ■ The **Key Idea** here is that we can get an expression involving the current i in this single-loop circuit by applying the loop rule. Although knowing the direction of i is not necessary, we can easily determine it from the emfs of the two batteries. Because $\mathscr{E}_1$ is greater than $\mathscr{E}_2$, battery 1 controls the direction of i, so that direction is clockwise. Let us then apply the loop rule by going counterclockwise—against the current—and starting at point a. We find

$$-\mathscr{E}_1 + ir_1 + iR + ir_2 + \mathscr{E}_2 = 0\text{ V}.$$

Check that this equation also results if we apply the loop rule clockwise or start at some point other than a. Also, take the time to compare this equation term by term with Fig. 27-12*b*, which shows the potential changes graphically (with the potential at point a arbitrarily taken to be zero).

Solving the above loop equation for the current i, we obtain

$$i = \frac{\mathscr{E}_1 - \mathscr{E}_2}{R + r_1 + r_2} = \frac{4.4\text{ V} - 2.1\text{ V}}{5.5\ \Omega + 2.3\ \Omega + 1.8\ \Omega}$$

$$= 0.2396\text{ A} \approx 240\text{ mA}. \qquad \text{(Answer)}$$

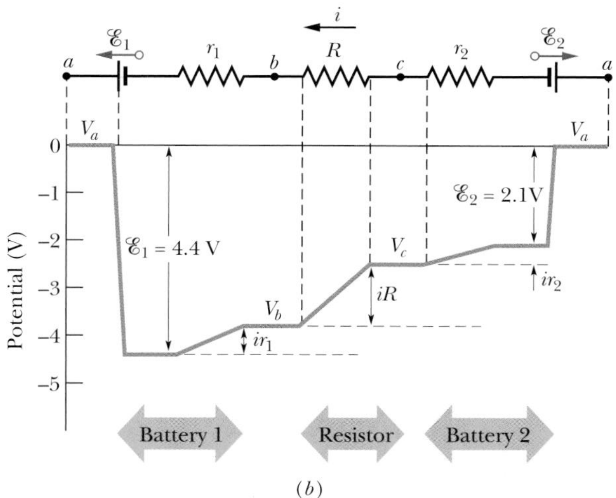

FIGURE 27-12 ■ (a) A single-loop circuit containing two real batteries and a resistor. The batteries oppose each other; that is, they tend to send current in opposite directions through the resistor. (b) A graph of the potentials encountered in traversing this circuit counterclockwise from point a, with the potential at a arbitrarily taken to be zero. (To better link the circuit with the graph, mentally cut the circuit at a and then unfold the left side of the circuit toward the left and the right side of the circuit toward the right.)

(b) What is the potential difference between the terminals of battery 1 in Fig. 27-12a?

SOLUTION ■ The **Key Idea** is to sum the potential differences between points a and b. Let us start at point b (effectively the negative terminal of battery 1) and travel clockwise through battery 1 to point a (effectively the positive terminal), keeping track of potential changes. We find that

$$V_b - ir_1 + \mathscr{E}_1 = V_a,$$

which gives us

$$\begin{aligned} V_a - V_b &= -ir_1 + \mathscr{E}_1 \\ &= -(0.2396\ \text{A})(2.3\ \Omega) + 4.4\ V \\ &= +3.84\ \text{V} \approx 3.8\ \text{V}, \end{aligned}$$

which is less than the emf of the battery. You can verify this result by starting at point b in Fig. 27-12a and traversing the circuit counterclockwise to point a.

TOUCHSTONE EXAMPLE 27-4: Electric Eel

Electric fish generate current with biological cells called *electroplaques,* which are physiological emf devices. The electroplaques in the South American eel shown in the photograph that opens this chapter are arranged in 140 rows, each row stretching horizontally along the body and each containing 5000 electroplaques. The arrangement is suggested in Fig. 27-13a; each electroplaque has an emf $\mathscr{E}$ of 0.15 V and an internal resistance r of 0.25 Ω. The water surrounding the eel completes a circuit between the two ends of the electroplaque array, one end at the animal's head and the other near its tail.

(a) If the water surrounding the eel has resistance $R_{\text{water}} = 800\ \Omega$, how much current can the eel produce in the water?

SOLUTION ■ The **Key Idea** here is that we can simplify the circuit of Fig. 27-13a by replacing combinations of emfs and internal resistances with equivalent emfs and resistances. We first consider a single row. The total emf $\mathscr{E}_{\text{row}}$ along a row of 5000 electroplaques is the sum of the emfs:

$$\mathscr{E}_{\text{row}} = 5000\mathscr{E} = (5000)(0.15\ \text{V}) = 750\ \text{V}.$$

The total resistance R_{row} along a row is the sum of the internal resistances of the 5000 electroplaques:

$$R_{\text{row}} = 5000r = (5000)(0.25\ \Omega) = 1250\ \Omega.$$

We can now represent each of the 140 identical rows as having a single emf $\mathscr{E}_{\text{row}}$ and a single resistance R_{row}, as shown in Fig. 27-13b.

In Fig. 27-13b, the emf between point a and point b on any row is $\mathscr{E}_{\text{row}} = 750$ V. Because the rows are identical and because they are all connected together at the left in Fig. 27-13b, all points b in

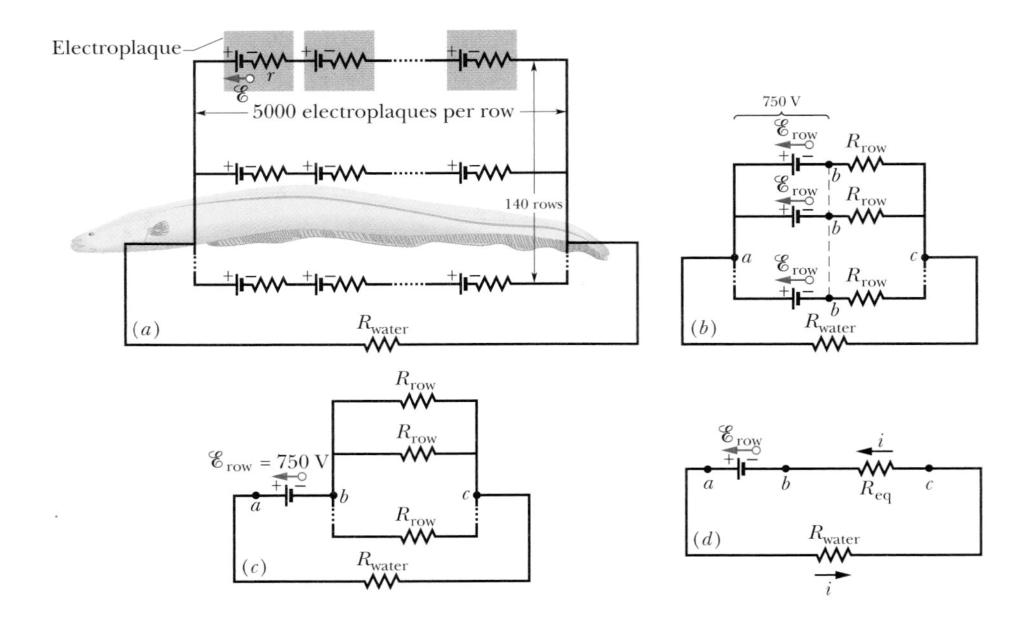

FIGURE 27-13 ■ (a) A model of the electric circuit of an eel in water. Each electroplaque of the eel has an emf $\mathscr{E}$ and internal resistance r. Along each of 140 rows extending from the head to the tail of the eel, there are 5000 electroplaques. The surrounding water has resistance R_{water}. (b) The emf $\mathscr{E}_{\text{row}}$ and resistance R_{row} of each row. (c) The emf between points a and b is $\mathscr{E}_{\text{row}}$. Between points b and c are 140 parallel resistances R_{row}. (d) The simplified circuit, with R_{eq} replacing the parallel combination.

that figure are at the same electric potential. Thus, we can consider them to be connected so that there is only a single point b. The emf between point a and this single point b is $\mathscr{E}_{\text{row}} = 750$ V, so we can draw the circuit as shown in Fig. 27-13c.

Between points b and c in Fig. 27-13c are 140 resistances $R_{\text{row}} = 1250\ \Omega$, all in parallel. The equivalent resistance R_{eq} of this combination is given by Eq. 27-12 as

$$\frac{1}{R_{\text{eq}}} = \sum_{j=1}^{140} \frac{1}{R_j} = 140\frac{1}{R_{\text{row}}},$$

or

$$R_{\text{eq}} = \frac{R_{\text{row}}}{140} = \frac{1250\ \Omega}{140} = 8.93\ \Omega.$$

Replacing the parallel combination with R_{eq}, we obtain the simplified circuit of Fig. 27-13d. Applying the loop rule to this circuit counterclockwise from point b, we have

$$\mathscr{E}_{\text{row}} - iR_{\text{water}} - iR_{\text{eq}} = 0\ \text{V}.$$

Solving for i and substituting the known data, we find

$$i = \frac{\mathscr{E}_{\text{row}}}{R_{\text{water}} + R_{\text{eq}}} = \frac{750\ \text{V}}{800\ \Omega + 8.93\ \Omega}$$
$$= 0.927\ \text{A} \approx 0.93\ \text{A}. \qquad \text{(Answer)}$$

If the head or tail of the eel is near a fish, much of this current could pass along a narrow path through the fish, stunning or killing it.

(b) How much current i_{row} travels through each row of Fig. 27-13a?

SOLUTION ■ The **Key Idea** here is that since the rows are identical, the current into and out of the eel is evenly divided among them:

$$i_{\text{row}} = \frac{i}{140} = \frac{0.927\ \text{A}}{140} = 6.6 \times 10^{-3}\ \text{A}. \qquad \text{(Answer)}$$

Thus, the current through each row is small, about two orders of magnitude smaller than the current through the water. This tends to spread the current through the eel's body, so that it need not stun or kill itself when it stuns or kills a fish.

Problems

SECS. 27-2 AND 27-3 ■ CURRENT AND POTENTIAL DIFFERENCE IN SINGLE LOOP CIRCUITS, SERIES RESISTANCE

1. Three Resistors In Fig. 27-14, take $R_1 = R_2 = R_3 = 10\ \Omega$. If the potential difference across the ideal battery is $\Delta V_B = 12$ V, find: (a) the equivalent resistance of the circuit and (b) the direction the current flows in the circuit. (c) Which point, A or B, is at higher potential?

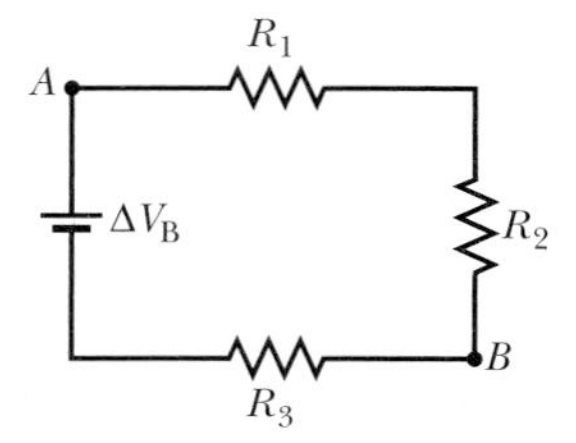

FIGURE 27-14 ■ Problems 1, 3, and 5.

2. Two Ideal Batteries Figure 27-15 shows two ideal batteries with $\Delta V_{B1} = 12$ V and $\Delta V_{B2} = 8$ V. (a) What is the direction of the current in the resistor? (b) Which battery is doing positive work? (c) Which point, A or B, is at the higher potential?

FIGURE 27-15 ■ Problem 2.

3. Total Current In Fig. 27-14, take $R_1 = 10\ \Omega$, $R_2 = 15\ \Omega$, and $R_3 = 20\ \Omega$. If the potential difference across the ideal battery is $\Delta V_B = 15$ V, find: (a) the equivalent resistance of the circuit, (b) the current through each of the resistors, and (c) the total current in the circuit.

4. If Potential at P Is In Fig. 27-16, if the potential at point P is 100 V, what is the potential at point Q?

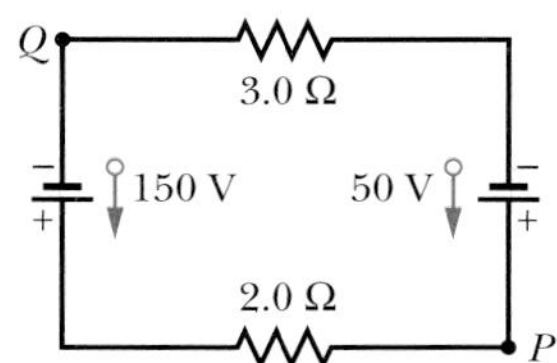

FIGURE 27-16 ■ Problem 4.

5. Voltages In Fig. 27-14, take $R_1 = 12\ \Omega$, $R_2 = 15\ \Omega$, and $R_3 = 25\ \Omega$. If the potential difference across the ideal battery is $\Delta V_B = 15$ V, find the potential differences across each of the resistors.

6. Neglecting Wires Figure 27-17 shows a 6.00 Ω resistor connected to a 12.0 V battery by means of two copper wires. The wires each have length 20.0 cm and radius 1.00 mm. In such circuits we generally neglect the potential differences along wires and the transfer of energy to thermal energy in them. Check the validity of this neglect for the circuit of Fig. 27-17: What are the potential differences across (a) the resistor and (b) each of the two sections of wire? At what rate is energy lost to thermal energy in (c) the resistor and (d) each of the two sections of wire?

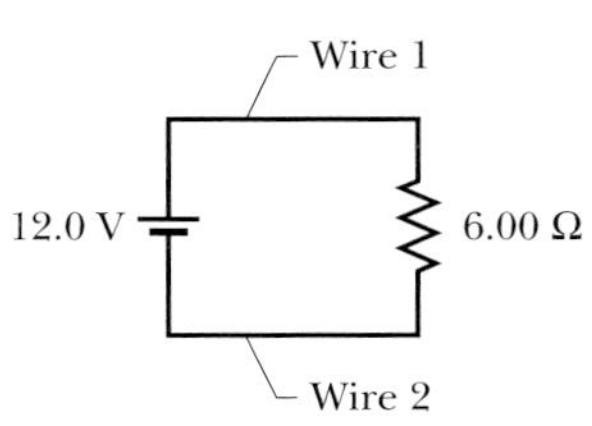

FIGURE 27-17 ■ Problem 6.

7. Single Loop The current in a single-loop circuit with one resistance R is 5.0 A. When an additional resistance of 2.0 Ω is inserted in series with R, the current drops to 4.0 A. What is R?

8. Ohmmeter A simple ohmmeter is made by connecting an ideal 1.50 V flashlight battery in series with a resistance R and an ammeter that reads from 0 to 1.00 mA, as shown in Fig. 27-18. Resistance R is adjusted so that when the clip leads are shorted together, the meter deflects to its full-scale value of 1.00 mA. What external resistance across the leads results in a deflection of (a) 10%, (b) 50%, and (c) 90% of full scale? (d) If the ammeter has a resistance of 20.0 Ω and the internal resistance of the battery is negligible, what is the value of R?

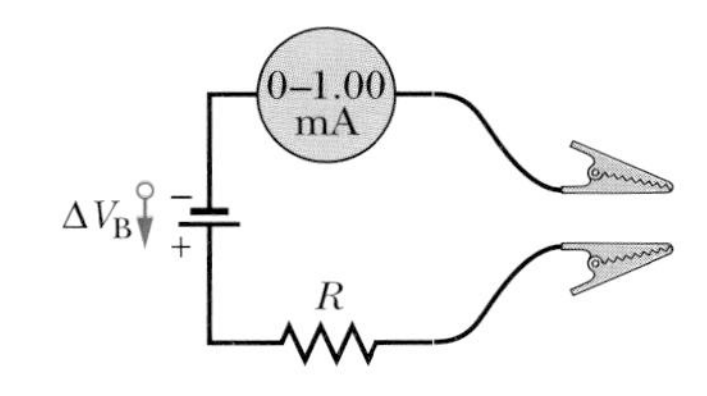

FIGURE 27-18 ■ Problem 8.

SECS. 27-4 AND 27-5 ■ MULTILOOP CIRCUITS AND PARALLEL RESISTANCE

9. Sizes and Directions What are the sizes and directions of the currents through resistors (a) R_2 and (b) R_3 in Fig. 27-19, where each of the three resistances is 4.0 Ω?

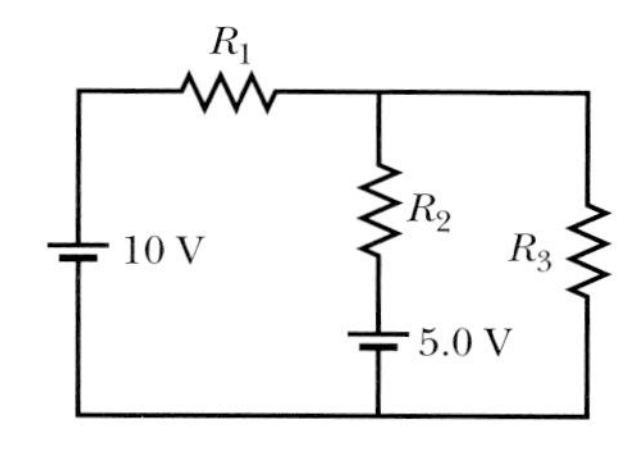

FIGURE 27-19 ■ Problem 9.

10. Changes The resistances in Figs. 27-20a and b are all 6.0 Ω, and the batteries are ideal 12 V batteries. (a) When switch S in Fig. 27-20a is closed, what is the change in the electric potential difference ΔV_{R_1} across resistor 1, or does ΔV_{R_1} remain the same? (b) When switch S in Fig. 27-20b is closed what is the change in the electric potential difference ΔV_{R_1} across resistor 1, or does ΔV_{R_1} remain the same?

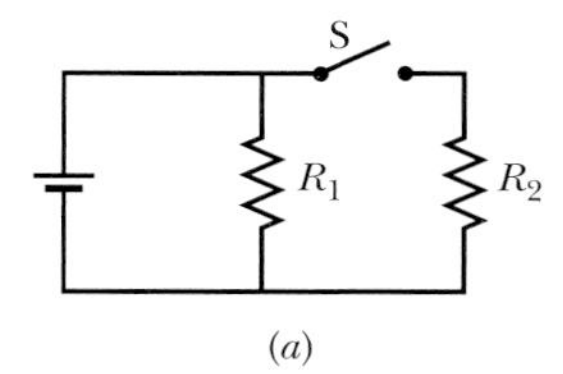

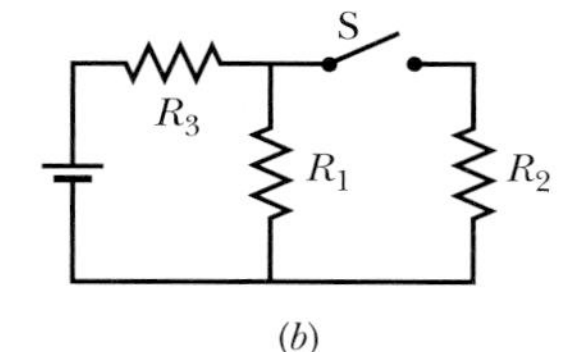

FIGURE 27-20 ■ Problem 10.

11. Equivalent (a) In Fig. 27-21, what is the equivalent resistance of the network shown? (b) What is the current in each resistor? Put $R_1 = 100\ \Omega$, $R_2 = R_3 = 50\ \Omega$, $R_4 = 75\ \Omega$, and $\Delta V_B = 6.0$ V; assume the battery is ideal.

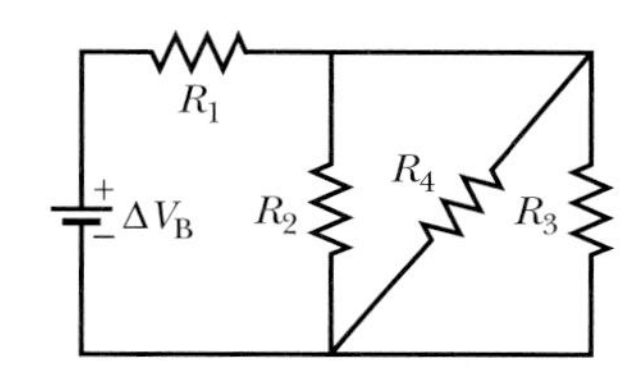

FIGURE 27-21 ■ Problem 11.

12. Plots Plot 1 in Fig. 27-22a gives the electric potential difference ΔV_{R_1} set up across R_1 versus the current i that can appear in resistor 1. Plots 2 and 3 are similar plots for resistors 2 and 3, respectively. Figure 27-22b shows a circuit with those three resistors and a 6.0 V battery. What is the current in resistor 2 in that circuit?

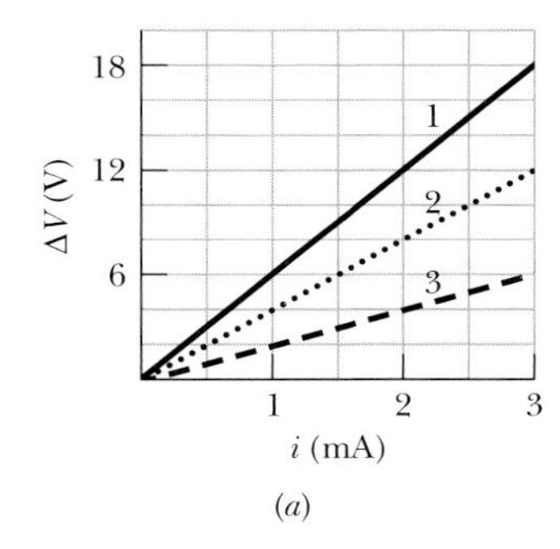

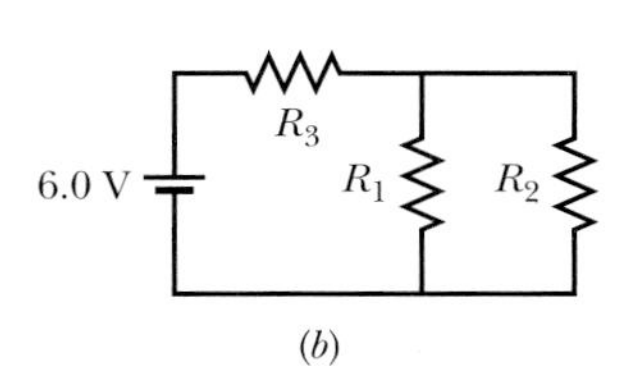

FIGURE 27-22 ■ Problem 12.

13. Equivalent Resistance Two In Fig. 27-23, $R = 10\ \Omega$. What is the equivalent resistance between points A and B? (*Hint:* This circuit section might look simpler if you first assume that points A and B are connected to a battery.)

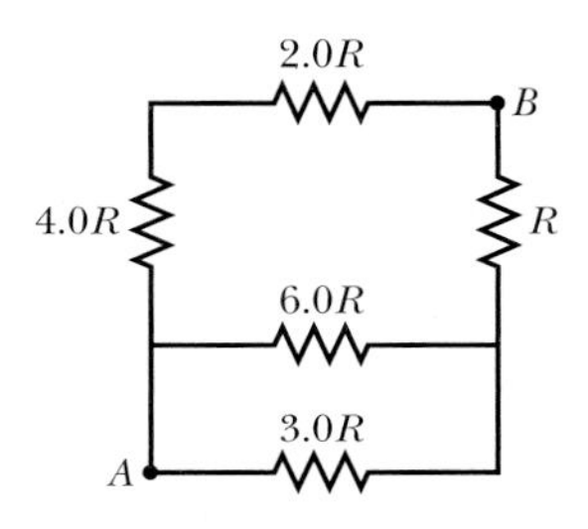

FIGURE 27-23 ■ Problem 13.

14. Three Switches Figure 27-24 shows a circuit containing three switches, labeled S_1, S_2, and S_3. Find the current at a for all possible combinations of switch settings. Put $\Delta V_B = 120$ V, $R_1 = 20.0\ \Omega$, and $R_2 = 10.0\ \Omega$. Assume that the battery has no resistance.

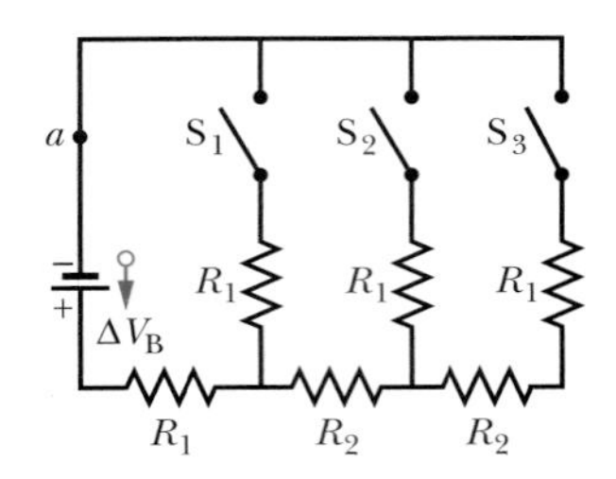

FIGURE 27-24 ■ Problem 14.

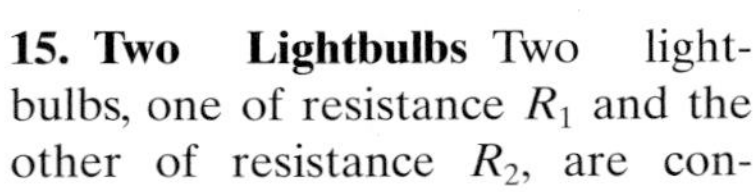

15. Two Lightbulbs Two lightbulbs, one of resistance R_1 and the other of resistance R_2, are connected to a battery (a) in parallel and (b) in series. Which bulb is brighter in each case if $R_1 = R_2$? How is your answer different if $R_1 > R_2$?

16. Calculate Potential In Fig. 27-5, calculate the potential difference between points c and d by as many paths as possible. Assume that $\Delta V_{B1} = 4.0$ V, $\Delta V_{B2} = 1.0$ V, $R_1 = R_2 = 10\ \Omega$, and $R_3 = 5.0\ \Omega$.

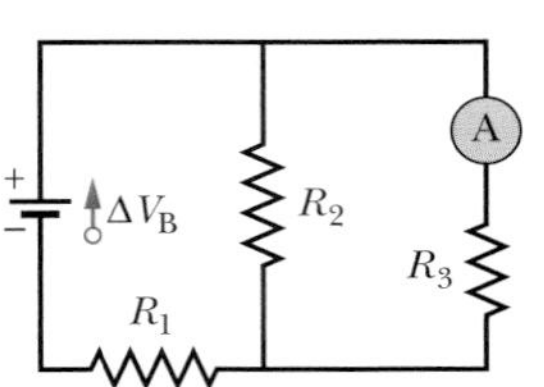

FIGURE 27-25 ■ Problem 17.

17. Ammeter (a) In Fig. 27-25, determine what the ammeter will read, assuming $\Delta V_B = 5.0$ V (for the ideal battery), $R_1 = 2.0\ \Omega$, $R_2 = 4.0\ \Omega$, and $R_3 = 6.0\ \Omega$. (b) The ammeter and the source of emf are now physically interchanged. Show that the ammeter reading remains unchanged.

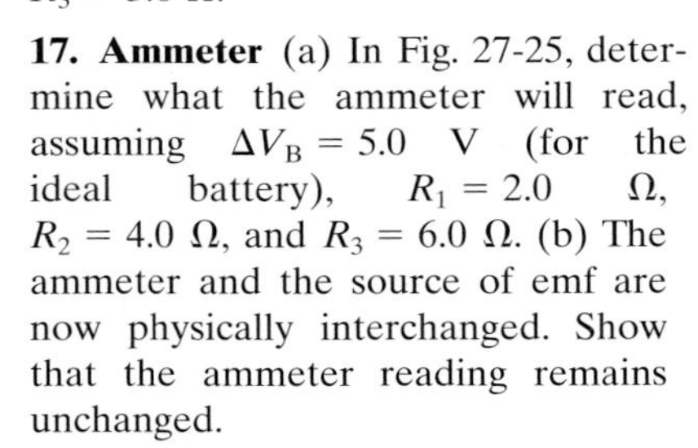

18. Equivalent Resistance In Fig. 27-26, find the equivalent resistance between points (a) F and H and (b) F and G. (*Hint*: for each pair of points, imagine that a battery is connected across the pair).

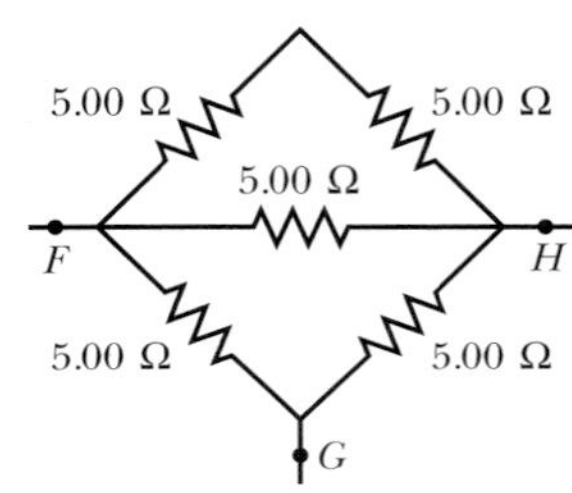

FIGURE 27-26 ■ Problem 18.

19. Current in Each In Fig. 27-27 find the current in each resistor and the potential difference between points a and b. Put $\Delta V_{B1} = 6.0$ V, $\Delta V_{B2} = 5.0$ V, $\Delta V_{B3} = 4.0$ V, $R_1 = 100\ \Omega$, and $R_2 = 50\ \Omega$.

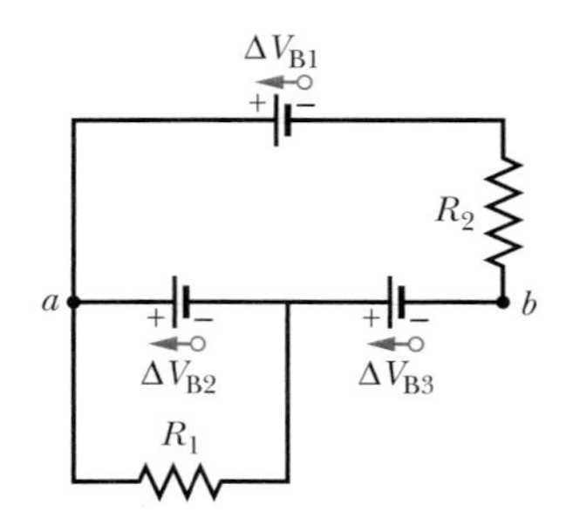

FIGURE 27-27 ■ Problem 19.

20. Two Resistors By using only two resistors—singly, in series, or in parallel—you are able to obtain resistances of 3.0, 4.0, 12, and 16 Ω. What are the two resistances?

21. Wire of Radius A copper wire of radius $a = 0.250$ mm has an aluminum jacket of outer radius $b = 0.380$ mm. (a) There is a current $i = 2.00$ A in the composite wire. Using Table 26-2, calculate the current in each material. (b) If a potential difference $V = 12.0$ V between the ends maintains the current, what is the length of the composite wire?

FIGURE 27-28 ■ Problem 22.

22. Between D and E In Fig. 27-28, find the equivalent resistance between points D and E. (*Hint*: Imagine that a battery is connected between points D and E.)

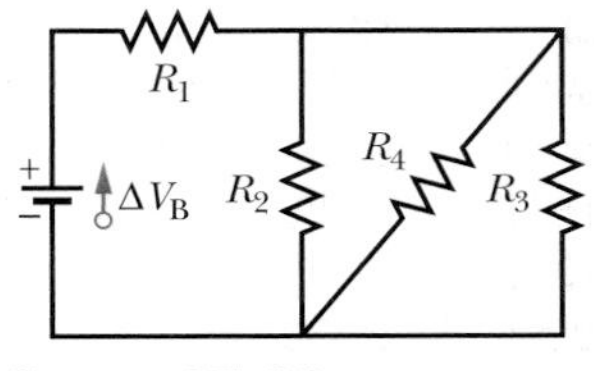

FIGURE 27-29 ■ Problem 24.

23. Four Resistors Four 18.0 Ω resistors are connected in parallel across a 25.0 V battery. What is the current through the battery?

24. Network Shown (a) In Fig. 27-29, what is the equivalent resistance of the network shown? (b) What is the current in each resistor? Put $R_1 = 100\ \Omega$, $R_2 = R_3 = 50\ \Omega$, $R_4 = 75\ \Omega$, and $\Delta V_B = 6.0$ V; assume the battery is ideal.

25. Nine Copper Wires Nine copper wires of length l and diameter d are connected in parallel to form a single composite conductor of resistance R. What must be the diameter D of a single copper wire of length l if it is to have the same resistance?

26. Voltmeter A voltmeter (of resistance $R_{\Delta V}$) and an ammeter (of resistance R_A) are connected to measure a resistance R, as in Fig. 27-30*a*. The resistance is given by $R = \Delta V/i$, where ΔV is the voltmeter reading and i is the current in the resistance R. Some of the current i' registered by the ammeter goes through the voltmeter, so that the ratio of the meter readings ($=\Delta V/i'$) gives only an *apparent* resistance reading R'. Show that R and R' are related by

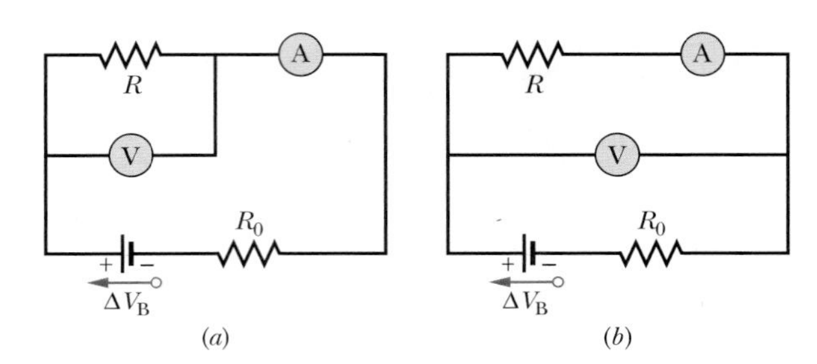

FIGURE 27-30 ■ Problems 26 to 28.

$$\frac{1}{R} = \frac{1}{R'} - \frac{1}{R_{\Delta V}}.$$

Note that as $R_{\Delta V} \to \infty$, $R' \to R$. Ignore R_0 for now.

27. Ammeter and Voltmeter (See Problem 26.) If an ammeter and a voltmeter are used to measure resistance, they may also be connected as in Fig. 27-30*b*. Again the ratio of the meter readings gives only an apparent resistance R'. Show that now R' is related to R by

$$R = R' - R_A,$$

in which R_A is the ammeter resistance. Note that as $R_A \to 0\ \Omega$, $R' \to R$. Ignore R_0 for now.

28. What Will the Meters Read (See Problems 26 and 27.) In Fig. 27-30, the ammeter and voltmeter resistances are 3.00 Ω and 3.00 Ω, respectively. Take $\Delta V_B = 12.0$ V for the ideal battery and $R_0 = 100\ \Omega$. If $R = 85.0\ \Omega$, (a) what will the meters read for the two different connections (Figs. 27-30*a* and *b*)? (b) What apparent resistance R' will be computed in each case?

29. Given a Number You are given a number of 10 Ω resistors, each capable of dissipating only 1.0 W without being destroyed. What is the minimum number of such resistors that you need to combine in series or in parallel to make a 10 Ω resistance that is capable of dissipating at least 5.0 W?

30. Asymptote In Fig. 27-31*a*, resistor 3 is a variable resistor and the battery is an ideal 12 V battery. Figure 27-31*b* gives the current *i* through the battery as a function of R_3. The curve has an asymptote of 2.0 mA as $R_3 \to \infty$. What are (a) resistance R_1 and (b) resistance R_2?

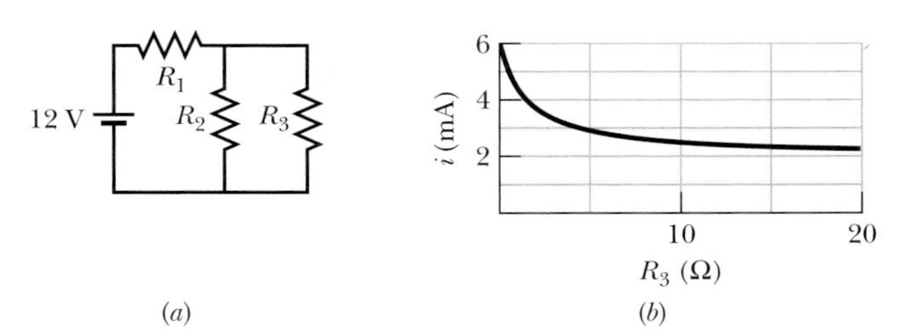

FIGURE 27-31 ■ Problem 30.

31. Box Figure 27-32 shows a section of a circuit. The electric potential difference between points *A* and *B* that connect the section to the rest of the circuit is $V_A - V_B = 78$ V, and the current through the 6.0 Ω resistor is 6.0 A. Is the device represented by "Box" absorbing or providing energy to the circuit and at what rate?

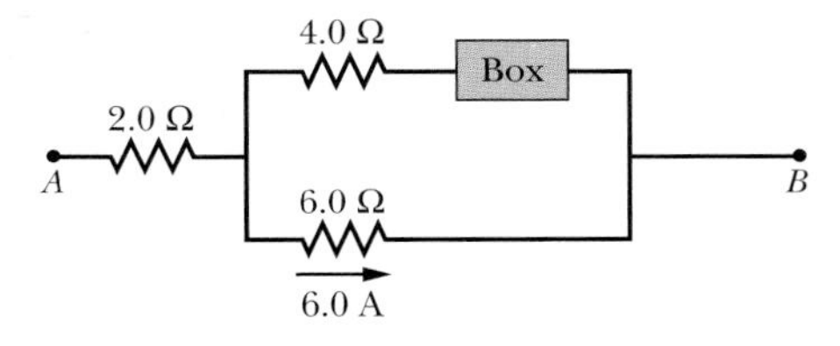

FIGURE 27-32 ■ Problem 31.

32. Arrangement of *N* Resistors In Fig. 27-33, a resistor and an arrangement of *n* resistors in parallel are connected in series with an ideal battery. All the resistors have the same resistance. If one more identical resistor were added in parallel to the *n* resistors already in parallel, the current through the battery would change by 1.25%. What is the value of *n*?

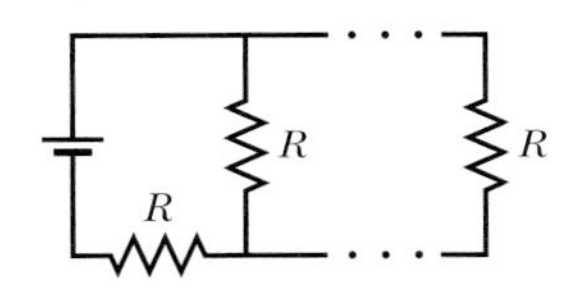

FIGURE 27-33 ■ Problem 32.

33. Rate of Energy Transfer In Fig. 27-34, where each resistance is 4.00 Ω, what are the sizes and directions of currents (a) i_1 and (b) i_2? At what rates is energy being transferred at (c) the 4.00 V battery and (d) the 12.0 V battery, and for each, is the battery supplying or absorbing energy?

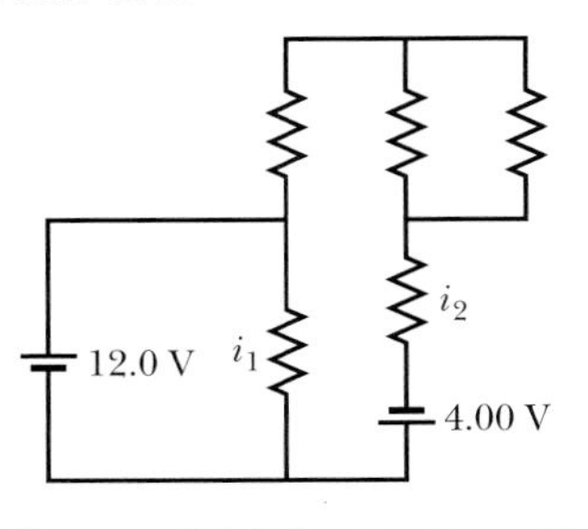

FIGURE 27-34 ■ Problem 33.

34. Both Batteries Are Ideal Both batteries in Fig. 27-35*a* are ideal. ΔV_{B1} of battery 1 has a fixed value but ΔV_{B2} of battery 2 can be varied between 1.0 V and 10 V. The plots in Fig. 27-35*b* give the currents through the two batteries as a function of ΔV_{B2}. You must decide which plot corresponds to which battery, but for both plots, a negative current occurs when the direction of the current through the battery is opposite the direction of that battery's potential difference. What are (a) ΔV_{B1} (b) resistance R_1, and (c) resistance R_2?

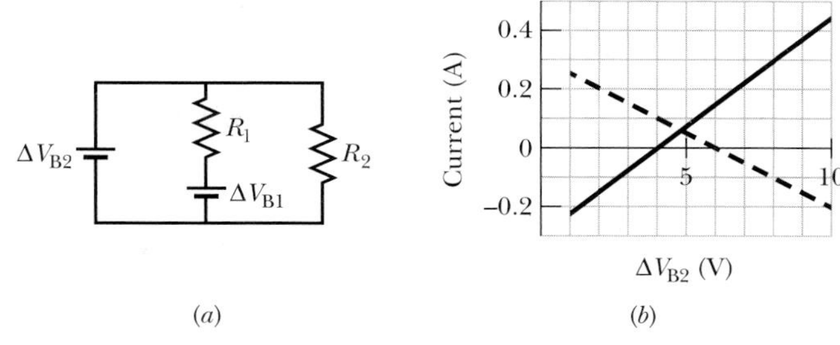

FIGURE 27-35 ■ Problem 34.

35. Work Done by Ideal Battery (a) How much work does an ideal battery with $\Delta V_B = 12.0$ V do on an electron that passes through the battery from the positive to the negative terminal? (b) If 3.4×10^{18} electrons pass through each second, what is the power of the battery?

36. Portion of a Circuit Figure 27-36 shows a portion of a circuit. The rest of the circuit draws current *i* at the connections *A* and *B*, as indicated. Take $\Delta V_{B1} = 10$ V, $\Delta V_{B2} = 15$ V, $R_1 = R_2 = 5.0$ Ω, $R_3 = R_4 = 8.0$ Ω, and $R_5 = 12$ Ω. For each of four values of *i*—0, 4.0, 8.0, and 12 A—find the current through each ideal battery and state whether the battery is charging or discharging. Also find the potential difference ΔV_{AB} between points *A* and *B*.

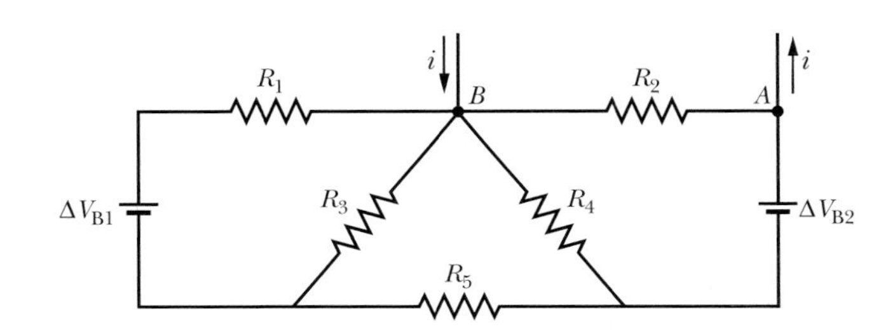

FIGURE 27-36 ■ Problem 36.

37. Adjusted Value In Fig. 27-37, R_s is to be adjusted in value by moving the sliding contact across it until points *a* and *b* are brought to the same potential. (One tests for this condition by momentarily connecting a sensitive ammeter between *a* and *b*; if these points are at the same potential, the ammeter will not deflect.) Show that when this adjustment is made, the following relation holds:

$$R_x = R_s\left(\frac{R_2}{R_1}\right).$$

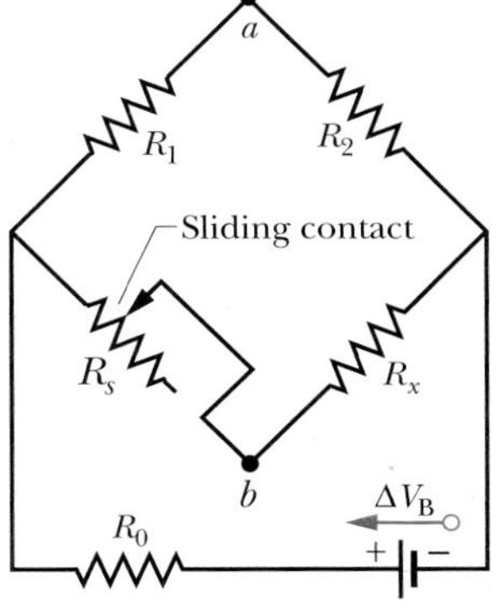

FIGURE 27-37 ■ Problem 37.

An unknown resistance (R_x) can be measured in terms of a standard (R_s) using this device, which is called a Wheatstone bridge.

38. What Are the Currents In Fig. 27-38, what are currents (a) i_2, (b) i_4, (c) i_1, (d) i_3, and (e) i_5?

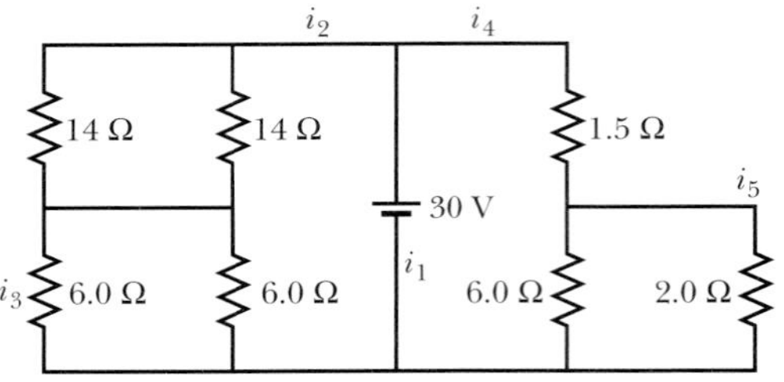

FIGURE 27-38 ■ Problem 38.

39. Sizes and Directions Two What are the sizes and directions of (a) current i_1 and (b) current i_2 in Fig. 27-39, where each resistance is 2.00 Ω? (Can you answer this making only mental calculations?) (c) At what rate is energy being transferred in the 5.00 V battery at the left, and is the energy being supplied or absorbed by the battery?

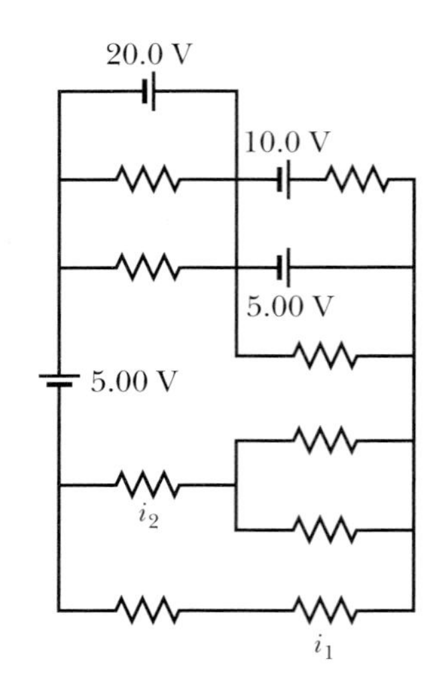

FIGURE 27-39 ■ Problem 39.

40. Size and Direction Three (a) What are the size and direction of current i_1 in Fig. 27-40, where each resistance is 2.0 Ω? What are the powers of (b) the 20 V battery, (c) the 10 V battery, and (d) the 5.0 V battery, and for each, is energy being supplied or absorbed?

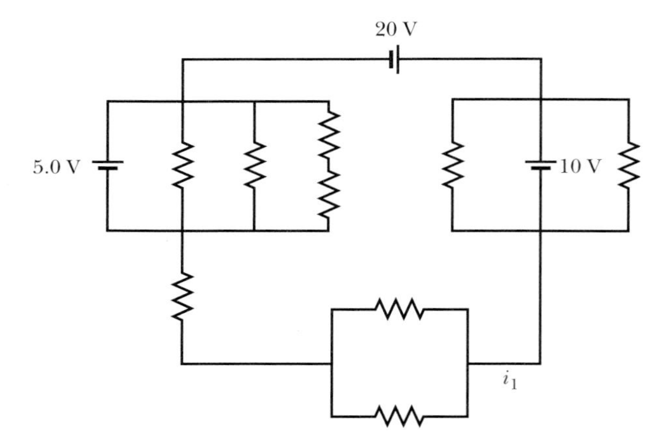

FIGURE 27-40 ■ Problem 40.

41. Size and Direction Four (a) What are the size and direction of current i_1 in Fig. 27-41? (b) How much energy is dissipated by all four resistors in 1.0 min?

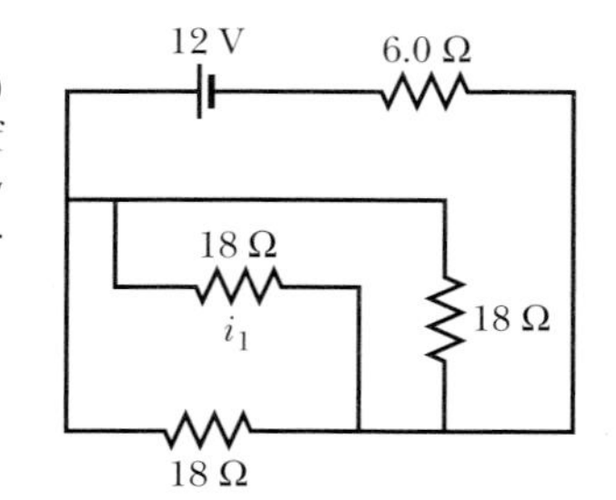

FIGURE 27-41 ■ Problem 41.

SEC. 27-7 ■ INTERNAL RESISTANCE AND POWER

42. Chemical Energy A 5.0 A current is set up in a circuit for 6.0 min by a rechargeable battery with a 6.0 V emf. By how much is the chemical energy of the battery reduced?

43. Flashlight Battery A standard flashlight battery can deliver about 2.0 W·h of energy before it runs down. (a) If a battery costs 80¢, what is the cost of operating a 100 W lamp for 8.0 h using batteries? (b) What is the cost if energy is provided at 12¢ per kilowatt-hour?

44. Power Supplied Power is supplied by a device of emf $\mathscr{E}$ to a transmission line with resistance R. Find the ratio of the power dissipated in the line for $\mathscr{E} = 110\ 000$ V to that dissipated for $\mathscr{E} = 110$ V, assuming the power supplied is the same for the two cases.

45. Car Battery A certain car battery with a 12 V emf has an initial charge of 120 A·h. Assuming that the potential across the terminals stays constant until the battery is completely discharged, for how long can it deliver energy at the rate of 100 W?

46. Energy Transferred A wire of resistance 5.0 Ω is connected to a battery whose emf $\mathscr{E}$ is 2.0 V and whose internal resistance is 1.0 Ω. In 2.0 min, (a) how much energy is transferred from chemical to electrical form? (b) How much energy appears in the wire as thermal energy? (c) Account for the difference between (a) and (b).

47. Assume the Batteries Assume that the batteries in Fig. 27-42 have negligible internal resistance. Find (a) the current in the circuit, (b) the power dissipated in each resistor, and (c) the power of each battery, stating whether energy is supplied by or absorbed by it.

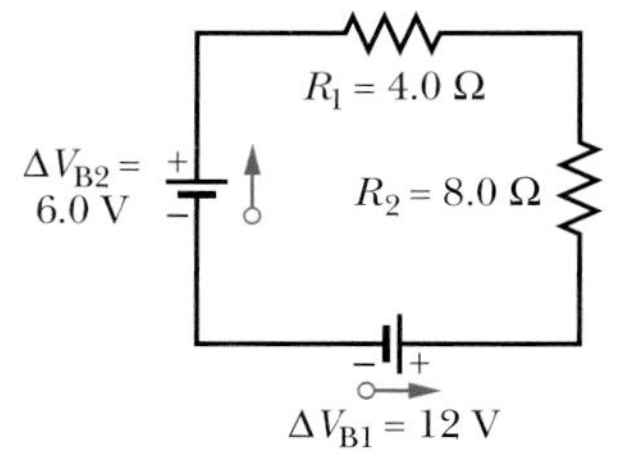

FIGURE 27-42 ■ Problem 47.

48. Both Batteries In Fig. 27-43*a*, both batteries have emf $\mathscr{E} = 1.20$ V and the external resistance R is a variable resistor. Figure 27-43*b* gives the electric potentials ΔV_T between the terminals of each battery as functions of R: Curve 1 corresponds to battery 1 and curve 2 corresponds to battery 2. What are the internal resistances of (a) battery 1 and (b) battery 2?

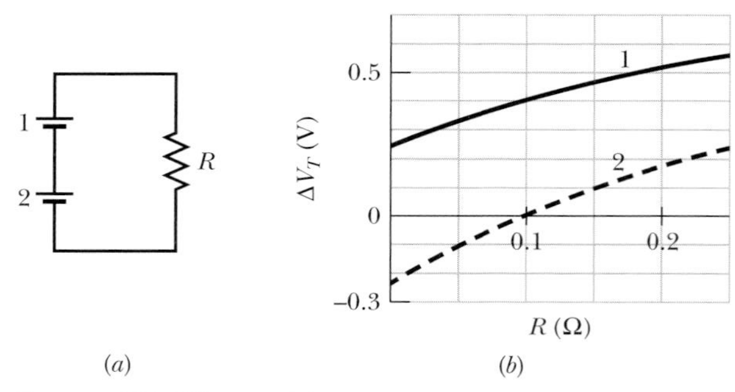

FIGURE 27-43 ■ Problem 48.

49. Find Internal Resistance The following table gives the electric potential difference ΔV_T across the terminals of a battery as a function of current i being drawn from the battery. (a) Write an equation that represents the relationship between the terminal potential difference ΔV_T and the current i. Enter the data into your graphing calculator and perform a linear regression fit of ΔV_T versus i. From the parameters of the fit, find (b) the battery's emf and (c) its internal resistance.

i (A):	50	75	100	125	150	175	200
ΔV_T (V):	10.7	9.0	7.7	6.0	4.8	3.0	1.7

50. Make Plots In Fig. 27-11*a*, put $\mathscr{E} = 2.0$ V and $r = 100\ \Omega$. Plot (a) the current and (b) the potential difference across R, as functions of R over the range 0 to 500 Ω. Make both plots on the same graph. (c) Make a third plot by multiplying together, for various values of R, the corresponding values on the two plotted curves. What is the physical significance of this third plot?

51. Energy Converted A car battery with a 12 V emf and an internal resistance of 0.040 Ω is being charged with a current of 50 A. (a) What is the potential difference across its terminals? (b) At what rate is energy being dissipated as thermal energy in the battery? (c) At what rate is electric energy being converted to chemical energy? (d) What are the answers to (a) and (b) when the battery is used to supply 50 A to the starter motor?

52. What Value of *R* (a) In Fig. 27-44, what value must R have if the current in the circuit is to be 1.0 mA? Take $\mathscr{E}_1 = 2.0$ V, $\mathscr{E}_2 = 3.0$ V, and $r_1 = r_2 = 3.0\ \Omega$. (b) What is the rate at which thermal energy appears in R?

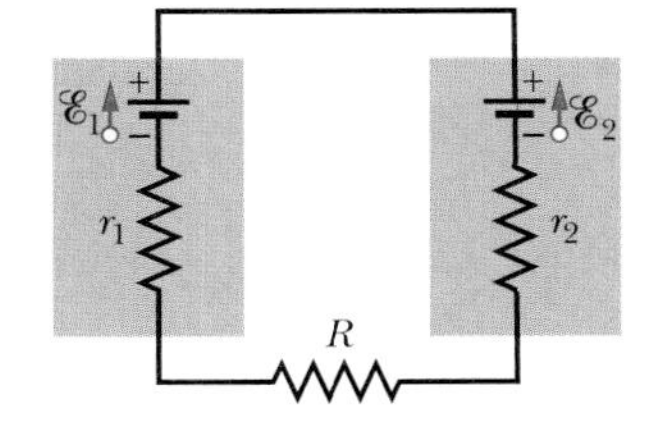

FIGURE 27-44 ■ Problem 52.

53. Circuit Section In Fig. 27-45, circuit section AB absorbs energy at a rate of 50 W when a current $i = 1.0$ A passes through it in the indicated direction. (a) What is the potential difference between A and B? (b) emf device X does not have internal resistance. What is its emf? (c) What is its *polarity* (the orientation of its positive and negative terminals)?

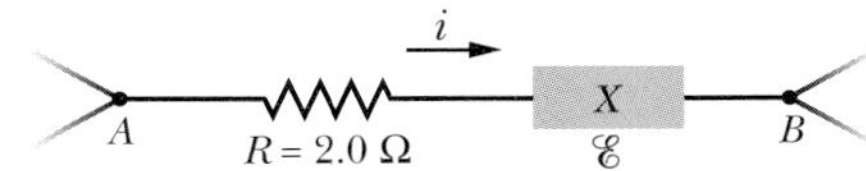

FIGURE 27-45 ■ Problem 53.

54. Lights of an Auto When the lights of an automobile are switched on, an ammeter in series with them reads 10 A and a voltmeter connected across them reads 12 V. See Fig. 27-46. When the electric starting motor is turned on, the ammeter reading drops to 8.0 A and the lights dim somewhat. If the internal resistance of the battery is 0.050 Ω and that of the ammeter is negligible, what are (a) the emf of the battery and (b) the current through the starting motor when the lights are on?

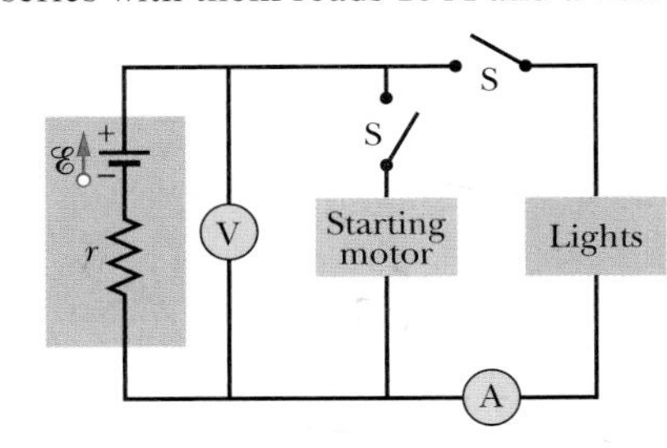

FIGURE 27-46 ■ Problem 54.

55. Same EMF Two batteries having the same emf $\mathscr{E}$ but different internal resistances r_1 and r_2 ($r_1 > r_2$) are connected in series to an external resistance R. (a) Find the value of R that makes the potential difference zero between the terminals of one battery. (b) Which battery is it?

56. Starting Motor The starting motor of an automobile is turning too slowly, and the mechanic has to decide whether to replace the motor, the cable, or the battery. The manufacturer's manual says that the 12 V battery should have no more than 0.020 Ω internal resistance, the motor no more than 0.200 Ω resistance, and the cable no more than 0.040 Ω resistance. The mechanic turns on the motor and measures 11.4 V across the battery, 3.0 V across the cable, and a current of 50 A. Which part is defective?

57. Maximum Power (a) In Fig. 27-11*a*, show that the rate at which energy is dissipated in R as thermal energy is a maximum when $R = r$. (b) Show that this maximum power is $P = \mathscr{E}^2/4r$.

58. Solar Cell A solar cell generates a potential difference of 0.10 V when a 500 Ω resistor is connected across it, and a potential difference of 0.15 V when a 1000 Ω resistor is substituted. What are (a) the internal resistance and (b) the emf of the solar cell? (c) The area of the cell is 5.0 cm², and the rate per unit area at which it receives energy from light is 2.0 mW/cm². What is the efficiency of the cell for converting light energy to thermal energy in the 1000 Ω external resistor?

59. Maximum Energy Two batteries of emf $\mathscr{E}$ and internal resistance r are connected in parallel across a resistor R, as in Fig. 27-47*a*. (a) For what value of R is the rate of electrical energy dissipation by the resistor a maximum? (b) What is the maximum energy dissipation rate?

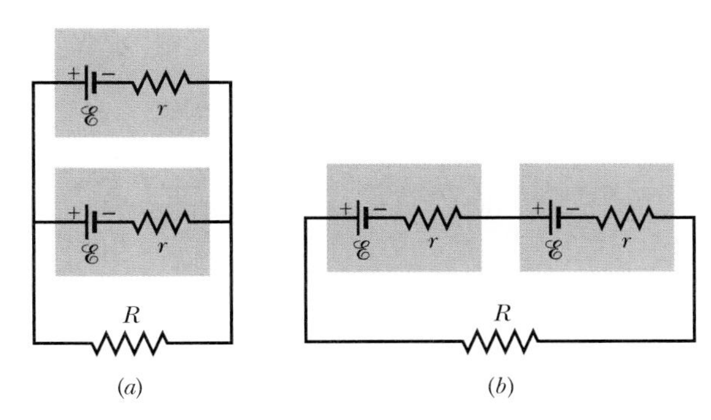

FIGURE 27-47 ■ Problems 59 and 60.

60. Either Parallel or Series You are given two batteries of emf $\mathscr{E}$ and internal resistance r. They may be connected either in parallel (Fig. 27-47*a*) or in series (Fig. 27-47*b*) and are to be used to establish a current in a resistor R. (a) Derive expressions for the current in R for both arrangements. Which will yield the larger current (b) when $R > r$ and (c) when $R < r$?

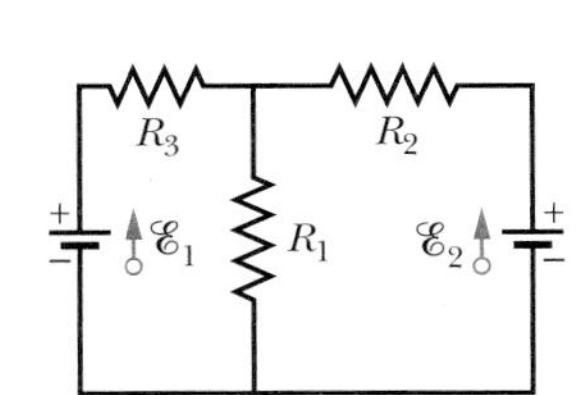

FIGURE 27-48 ■ Problem 61.

61. Batteries Are Ideal In Fig. 27-48, $\mathscr{E}_1 = 3.00$ V, $\mathscr{E}_2 = 1.00$ V, $R_1 = 5.00\ \Omega$, $R_2 = 2.00\ \Omega$, $R_3 = 4.00\ \Omega$, and both batteries are ideal. What is the rate at which energy is dissipated in (a) R_1, (b) R_2, and (c) R_3? What is the power of (d) battery 1 and (e) battery 2?

62. For What Value of *R* In the circuit of Fig. 27-49, for what value of R will the ideal battery transfer energy to the resistors (a) at a rate of 60.0 W, (b) at the maximum possible rate, and (c) at the minimum possible rate? (d) What are those rates?

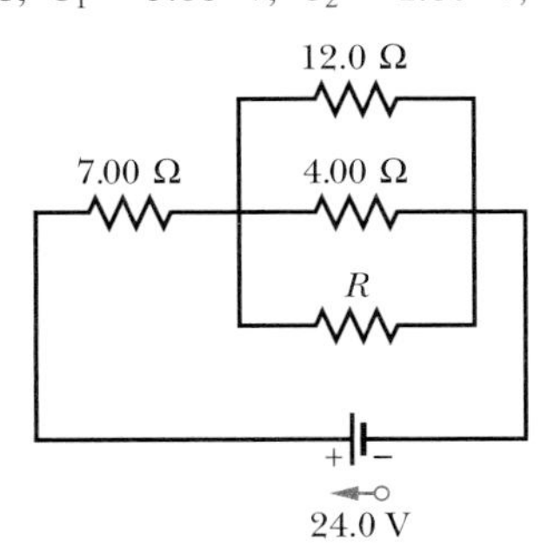

FIGURE 27-49 ■ Problem 62.

63. Calculate Current (a) Calculate the current through each ideal battery in Fig. 27-50. Since the batteries are ideal $\mathscr{E} = \Delta V_B$ in each case. Assume that $R_1 = 1.0\ \Omega$, $R_2 = 2.0\ \Omega$, $\mathscr{E}_1 = 2.0$ V and $\mathscr{E}_2 = \mathscr{E}_3 = 4.0$ V. (b) Calculate $V_a - V_b$.

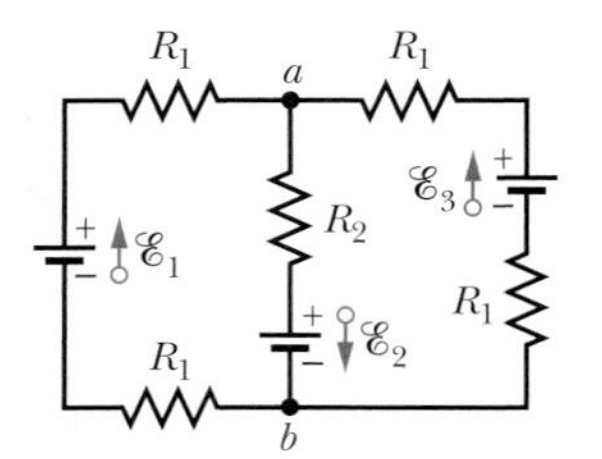

FIGURE 27-50 ■ Problem 63.

64. Constant Value In the circuit of Fig. 27-51, $\mathscr{E}$ has a constant value but R can be varied. Find the value of R that results in the maximum heating in that resistor. The battery is ideal.

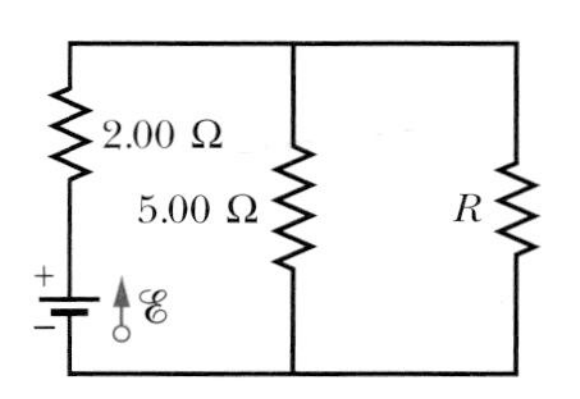

FIGURE 27-51 ■ Problem 64.

Additional Problems

65. True or False For the circuit in Fig. 27-52, indicate whether the statements are true or false. If a statement is false, give a correct statement.

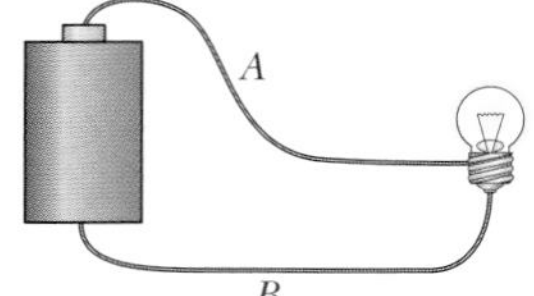

FIGURE 27-52 ■ Problem 65.

(a) Some of the current is used up when the bulb is lit; the current in wire B is smaller than the current in wire A.
(b) A current probe will have the same readings if connected to read the current in wire A or wire B. The current flows from the battery, through wire A, through the bulb, and then back to the battery through wire B.
(c) The current flows toward the bulb in both wires A and B.
(d) The (positive) current flows from the battery, through wire A, and then back to the battery through wire B.
(e) If wire A is left connected but wire B is disconnected, the bulb will still light.

66. Use the Model (a) Use our model for electric current to rank the networks shown in Fig. 27-53 in order by resistance. Explain your reasoning. (b) If a battery were connected to each of the circuits, in which case would the current through the battery be the largest? The smallest? Explain your reasoning.

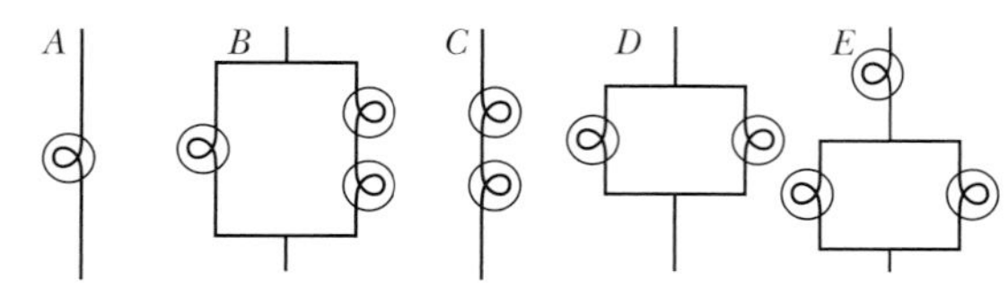

FIGURE 27-53 ■ Problem 66.

67. Examine the Circuits Examine the circuits shown in Fig. 27-54 and indicate whether you think each of the following two statements are true or false. Please explain your reasoning.

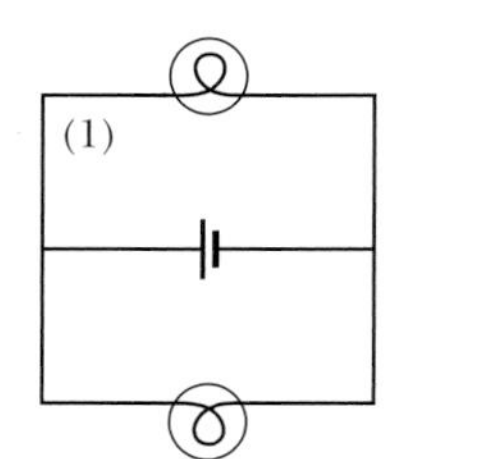

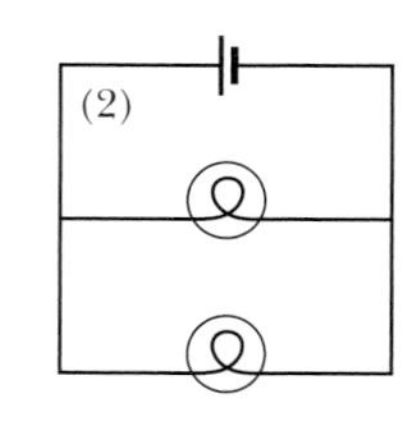

FIGURE 27-54 ■ Problem 67.

(a) Circuits 1 and 2 are different. The brightness of the two bulbs in circuit 1 are the same, but in circuit 2 the bulb closest to the battery in brighter than the bulb that is further away.
(b) Circuit diagrams only show electrical connections, so the drawings in circuits 1 and 2 are electrically equivalent and the brightness of the two bulbs is the same in both circuits 1 and 2.

68. Which Diagram (a) Identify which of the nice, neat circuit diagrams (A, B, C, or D) in Fig. 27-55c corresponds to the messy circuit drawing in Fig. 27-55a. Explain the reasons for your answer.

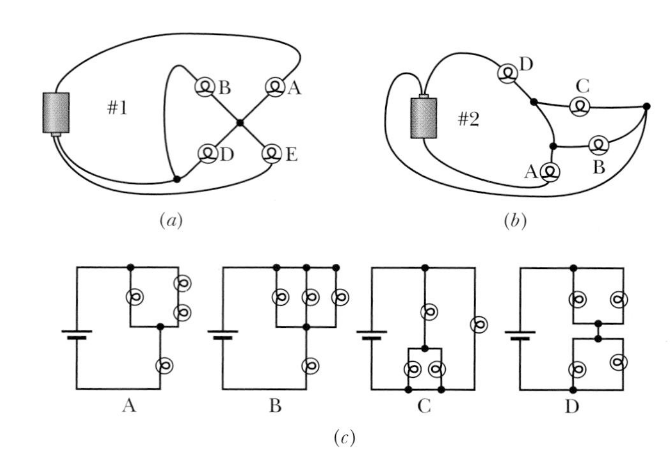

FIGURE 27-55 ■ Problem 68.

(b) Which neat circuit diagram corresponds to the messy circuit drawing in Fig. 27-55b. Explain the reasons for your answer.

69. At Which Point (a) For the circuit in Fig. 27-56, at which point A, B, C, D or E is the voltage the lowest? Explain. (b) At which point is the potential energy of a positive charge the highest? Explain. (c) At which point is the current the largest? Explain.

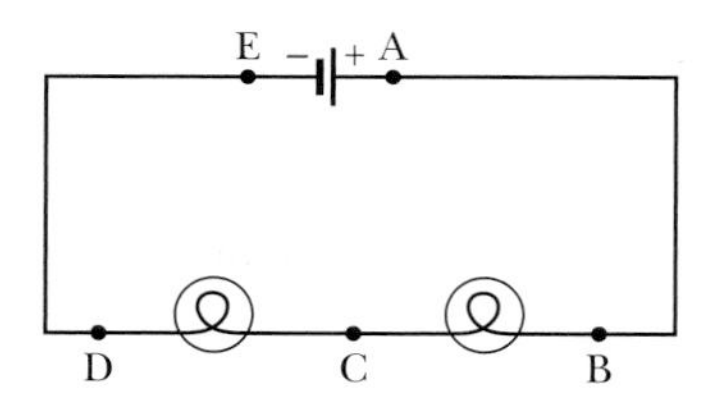

FIGURE 27-56 ■ Problem 69.

70. Bulbs 1 Through 6 (a) For the circuit shown in Fig. 27-57, rank bulbs 1 through 6 in order of descending brightness. Explain the reasoning for your ranking. (b) Now assume that the filament of lightbulb 6 breaks. Again rank the bulbs in order of descending brightness. Explain the reasoning for your ranking.

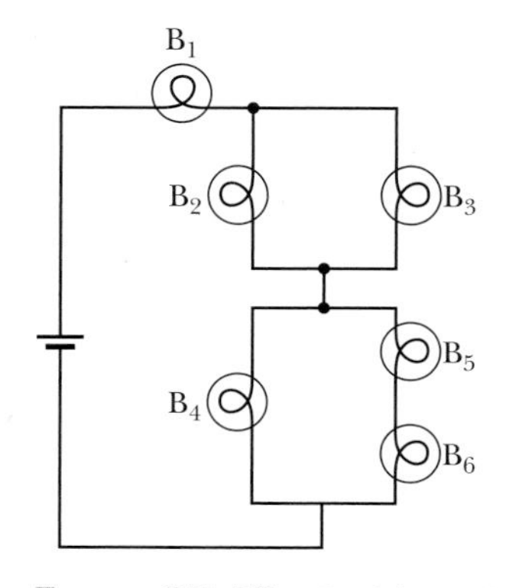

FIGURE 27-57 ■ Problem 70.

71. The Circuit Diagram The circuit diagram in Fig. 27-58 shows two unlabeled resistors attached to identical bulbs. Explain how you would interpret the brightness of bulbs A and B to decide which resistor is larger.

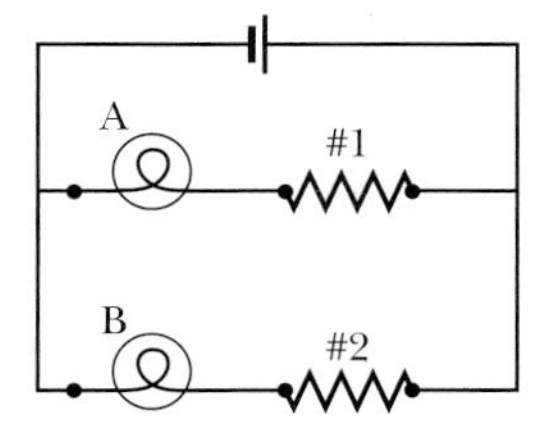

FIGURE 27-58 ■ Problem 71.

72. Three Circuits Which of the three circuits shown in Fig. 27-59, if any, are electrically identical? Which are different? Explain your answers.

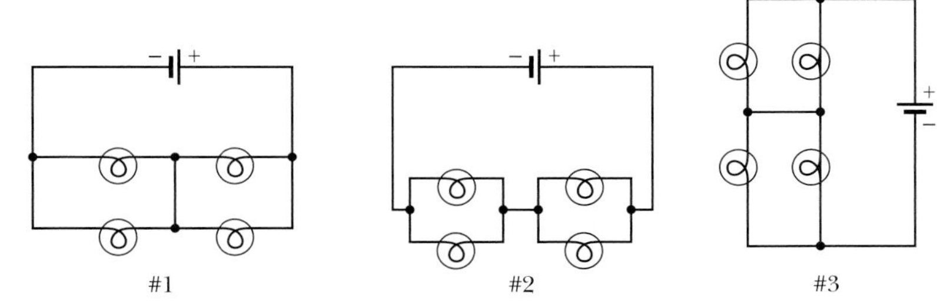

FIGURE 27-59 ■ Problem 72.

73. An Unscrewed Bulb Examine the circuit shown in Fig. 27-60. (a) Rank the bulbs according to brightness and explain your reasoning. (b) How will the brightness of bulbs 1 and 3 change if bulb 4 is unscrewed? Explain. (c) How will the brightness of bulbs 1, 3, 5, and 6 change if a conducting wire is connected between points A and F? Explain.

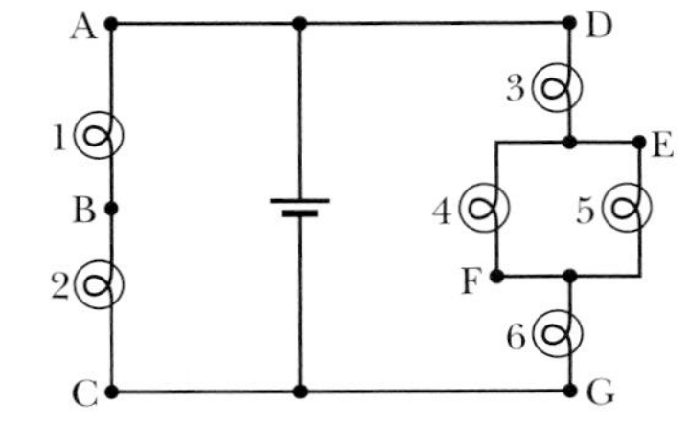

FIGURE 27-60 ■ Problem 73.

74. Examine the Circuit Examine the circuit shown in Fig. 27-61. (a) Assume that the switch is *open*. State which bulbs or combination of bulbs are in series, and in parallel. (b) Assume that the switch is *closed*. State whether the bulbs in the circuit are arranged in series or parallel.

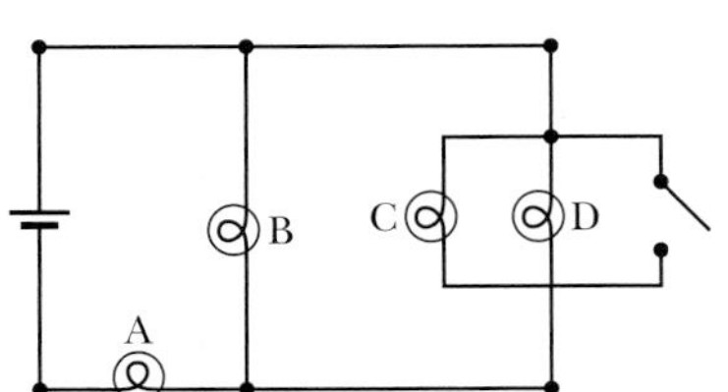

FIGURE 27-61 ■ Problem 74.

75. Examine the Circuit Two Examine the circuit shown in Fig. 27-62. (a) Assume that the switch is *open*. Rank the bulbs according to brightness and explain your reasoning. (b) Assume that the switch is *closed*. Rank the bulbs according to brightness and explain your reasoning.

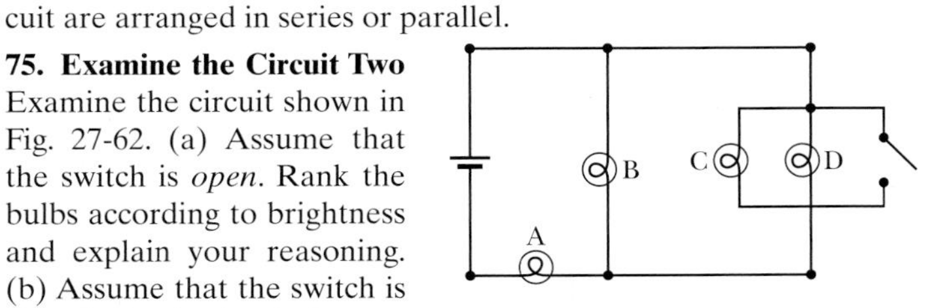

FIGURE 27-62 ■ Problem 75.

76. More Current If the batteries in Fig. 27-63 are identical, which circuit draws more current? Circuit A? Circuit B? Neither? Show your calculations and reasoning.

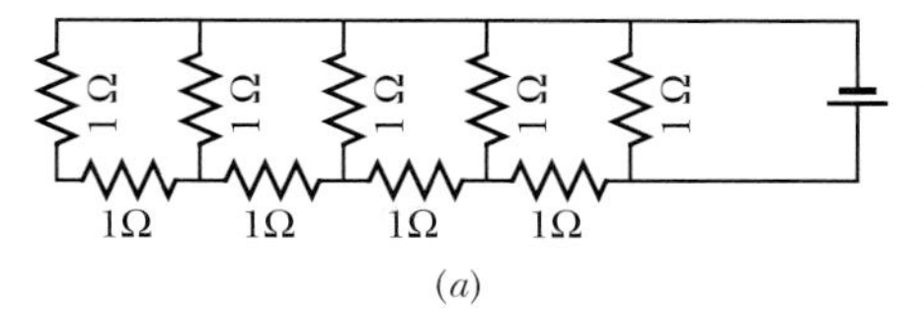

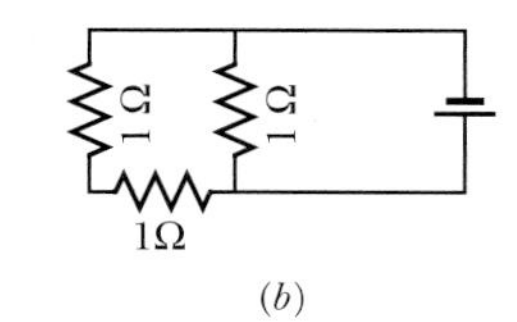

FIGURE 27-63 ■ Problem 76.

77. Which Are Connected In the circuits shown in Fig. 27-64, state which resistors are connected in series with which other resistors, which are connected in parallel with which other resistors, and which are neither in series nor parallel.

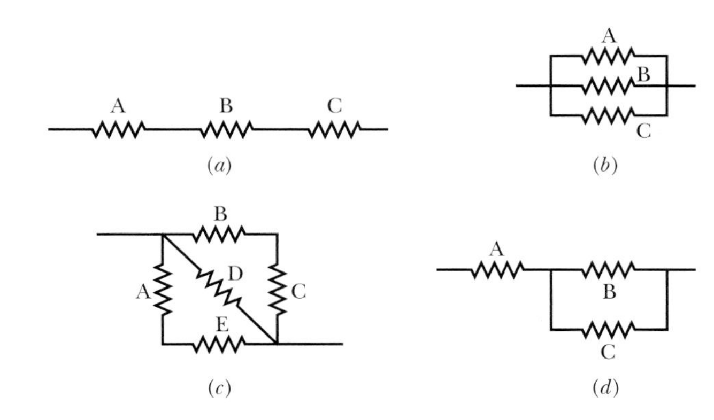

FIGURE 27-64 ■ Problem 77.

78. Lots of Batteries and a Bulb Figure 27-65 shows identical batteries connected in different arrangements to the same lightbulb. Assume the batteries have negligible internal resistances. The positive terminal of each battery is marked with a plus. Rank these arrangements on the basis of bulb brightness from the highest to the lowest. Please explain your reasoning.

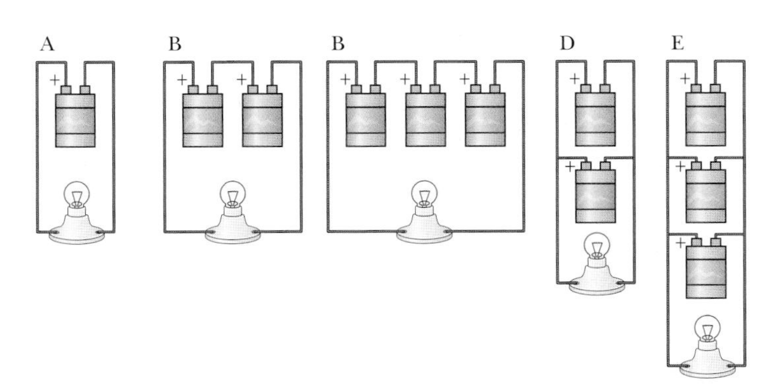

FIGURE 27-65 ■ Problem 78.

79. Constant Current Source We have studied batteries that provide a fixed voltage across their terminals. In that case, we had to examine our circuit and use our physical principles in order to calculate the current through the battery. In neuroscience, it is sometimes useful to use a constant current source (CCS), which instead provides a fixed amount of current through itself. In this case, we have to use our physical principles in order to calculate the voltage drop across the source.

Suppose we have a constant current source (denoted CSS) that always provides a current of $i_c = 10^{-6}$ amps. For the three circuits shown in Fig. 27-66, find the voltage drop across the current source. Each resistor has a resistance $R = 2000\ \Omega$. (If you prefer, you may leave your answer in terms of the symbols i_c and R.)

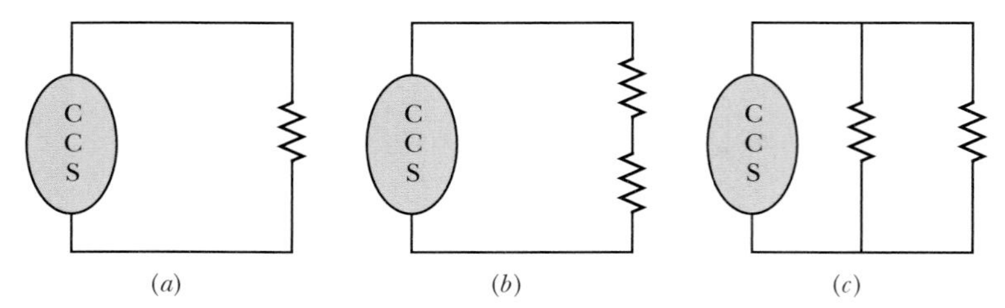

FIGURE 27-66 ■ Problem 79.

80. Tracking Around a Circuit The circuit shown in Fig. 27-67 contains an ideal battery and three resistors. The battery has an emf of 1.5 V, $R_1 = 2\ \Omega$, $R_2 = 3\ \Omega$, and $R_3 = 5\ \Omega$. Also shown in Fig. 27-67 is a graph tracking some quantity around the circuit. Make three copies of this graph. On the first, plot the voltage a test charge would experience as it moved through the circuit. On the second, plot the electric field a test charge would experience as it moved through the circuit. On the third, plot the current one would measure crossing a plane perpendicular to the wire of the circuit as one goes through the circuit.

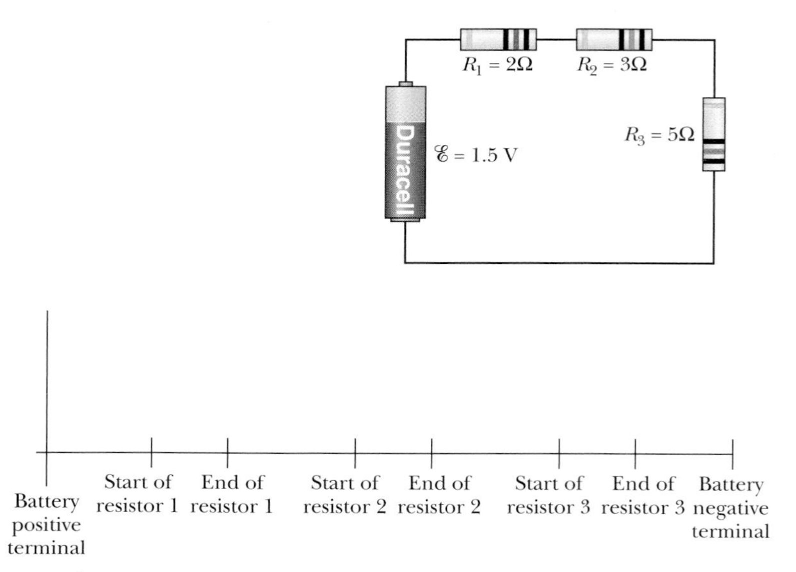

FIGURE 27-67 ■ Problem 80.

81. Modeling a Nerve Membrane (From a homework set in a graduate course in synaptic physiology) As a result of a complex set of biochemical reactions, the cell membrane of a nerve cell pumps ions (Na^+ and K^+) back and forth across itself, thereby maintaining an electrostatic potential difference from the inside to the outside of the membrane. Modifications on the conditions can result in changes in those potentials.

Part of the process can be modeled by treating the membrane as if it were a simple electric circuit consisting of batteries, resistors, and a switch. A simple model of the membrane of a nerve cell is shown in Fig. 27-68. It consists of two batteries (ion pumps) with voltages $\Delta V_1 = 100$ mV and $V_2 = 50$ mV. The resistance to flow across the membrane is represented by two resistors with resistances $R_1 = 10$ K Ω and $R_2 = 90$ K Ω. The variability is represented by a switch, S_1.

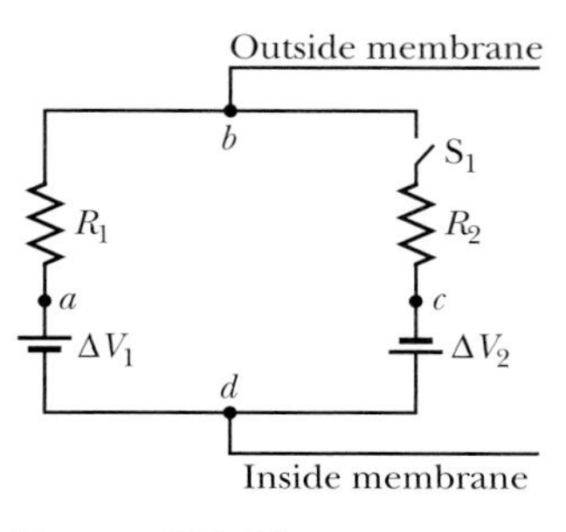

FIGURE 27-68 ■ Problem 81.

Four points on the circuit are labeled by the letters *a–d*. The point *b* represents the outside of the membrane and the point *d* the inside of the membrane.

(a) What is the voltage difference across the membrane (i.e., between *d* and *b*) when the switch is open?
(b) What is the current flowing around the loop when the switch is closed?
(c) What is the voltage drop across the resistor R_1 when the switch is open? Closed?
(d) What is the voltage drop across the resistor R_2 when the switch is open? Closed?
(e) What is the potential difference across the membrane (i.e., between *d* and *b*) when the switch is closed?
(f) If the locations of resistances R_1 and R_2 were reversed, would the voltages across the cell membrane be different?

82. Find the Five Currents Consider the circuit in Fig. 27-69. (a) Apply the junction rule to junctions *d* and *a* and the loop rule to the three loops to produce five simultaneous, linearly independent equations. (b) Represent the five linear equations by the matrix equation $[A][B] = [C]$, where

$$[B] = \begin{bmatrix} i_1 \\ i_2 \\ i_3 \\ i_4 \\ i_5 \end{bmatrix}.$$

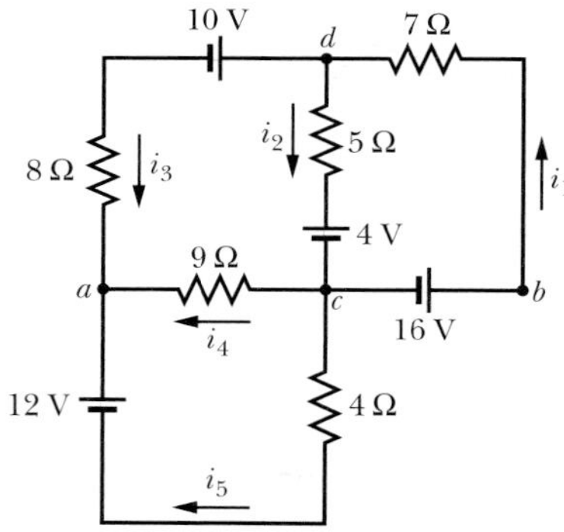

FIGURE 27-69 ■ Problem 82 and 83.

What are the matrices $[A]$ and $[C]$? (c) Have the calculator perform $[A]^{-1}[C]$ to find the values of i_1, i_2, i_3, i_4, and i_5.

83. Knowing the Currents For the same situation as in Problem 82 and having already solved for the five unknown currents, do the following. (a) Find the electric potential difference across the 9 Ω resistor. (b) Find the rate at which work is being done on the 7 Ω resistor. (c) Find the rate at which the 12 V battery is doing work on the circuit. (d) Find the rate at which the 4 V battery is doing work on the circuit. (e) Of the points in the circuit labeled *a* and *c*, which is at the higher electric potential?

28 Capacitance

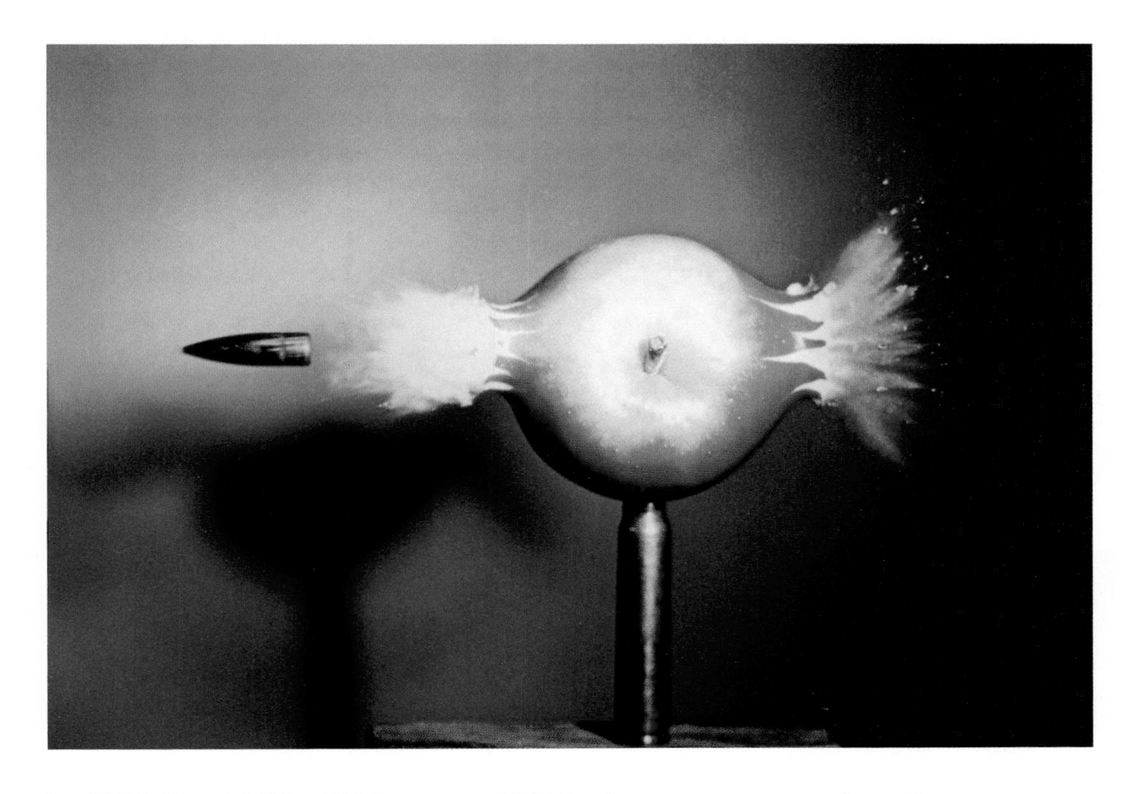

In 1964 Harold "Doc" Edgerton of MIT, who was renowned for his ability to take high-quality stop-action photos, captured this image of a bullet penetrating an apple. This stop-action photo was made by leaving the camera shutter open and tripping a high-speed electronic flash device at just the right time to illuminate the apple and bullet. Since the bullet was moving at 900 m/s, Edgerton used a flash with a duration of only 0.3 μs. This meant that the bullet only moved 0.3 mm during the flash. If we were doing ordinary photography we would probably illuminate the apple with a 100 W lightbulb using an exposure time of 1/20 s to provide about 5 J of energy for illumination. But providing the necessary 5 J of electrical energy for the illumination in a time period of 0.3 μs requires 15 MW of power.

How is it possible to provide the energy needed to stop a bullet's action when it is only illuminated for a tiny fraction of a second?

The answer is in this chapter.

28-1 The Uses of Capacitors

In Chapter 26 we discussed transferring excess charge to a pair of metal plates as shown in Fig. 26-1. The pair of metal plates is an example of the basic component of a **capacitor.** A capacitor can be constructed using any two conductors separated by an insulator. If we connect each conductor making up a capacitor to one of the terminals of a source of potential difference such as a battery, one conductor acquires a net positive charge while the other conductor acquires the same amount of net negative charge. The conductors can be any shape. Figure 28-1 shows some possible capacitor

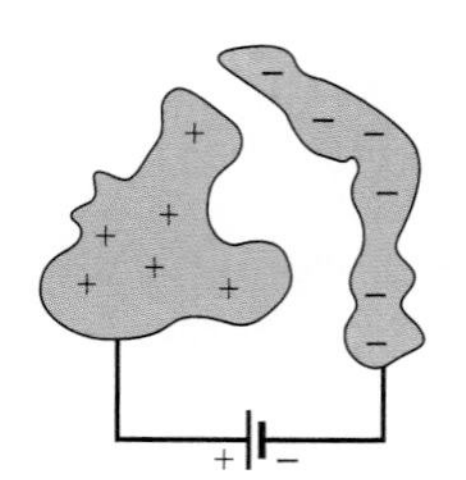

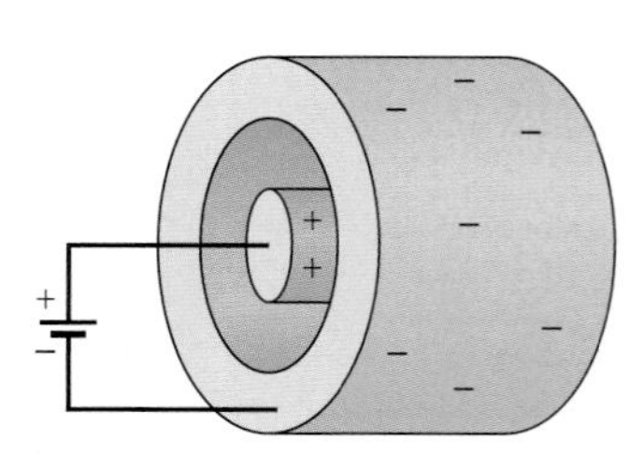

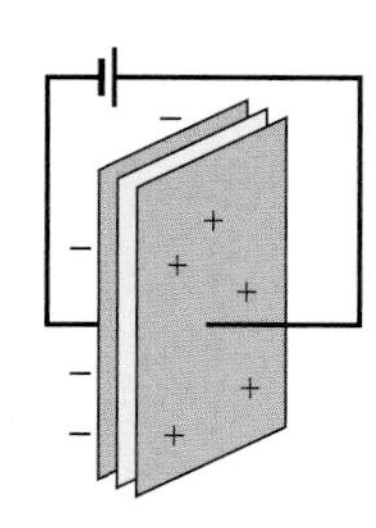

Parallel plate capacitor
with paper and air
as an insulator

FIGURE 28-1 ■ Three capacitors of different sizes and shapes have been connected to a battery. They each consist of a pair of conductors separated by an insulator. In each case the battery removes electrons from one of the two conductors, leaving it with excess positive charge and forces the same number of electrons to the opposite conductor.

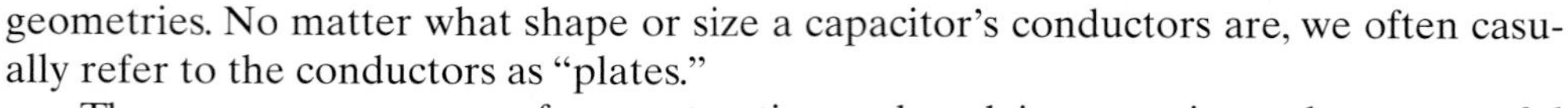

geometries. No matter what shape or size a capacitor's conductors are, we often casually refer to the conductors as "plates."

There are many reasons for constructing and studying capacitors: they are useful circuit elements and they can store energy.

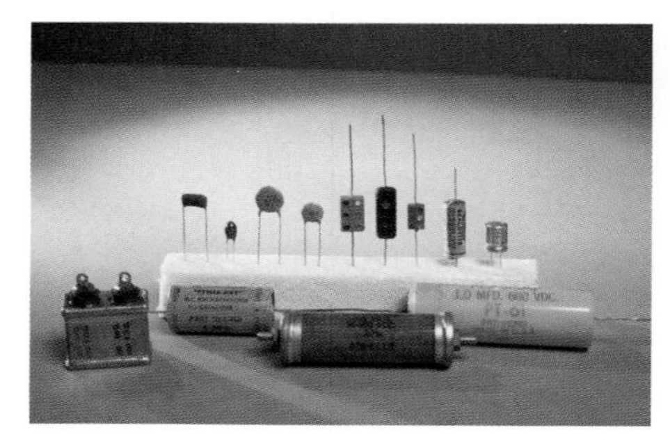

FIGURE 28-2 ■ An assortment of capacitors commonly found in electrical circuits. The structures of these devices are hidden.

Capacitors in Electrical Circuits

Since a capacitor consists of conductors separated by an insulator, no current can flow *through* it. So at first glance, it doesn't seem to make sense to use a capacitor as a circuit element. Surprisingly, capacitors have very interesting and useful properties in circuits with changing currents through their other components. For example, variable capacitors are vital elements that enable us to tune radio and television receivers. They are found in most household electrical devices. Capacitors are used to control the frequency of the flashing lights used for warning signals at construction sites. The coaxial cables used to carry high-frequency microwave and radio signals are cylindrical capacitors. Microscopic capacitors are used in communications and computers to shape the timing and strength of time-varying signal transmissions. Figure 28-2 shows some of the many sizes and shapes of capacitors commonly found in electric circuits.

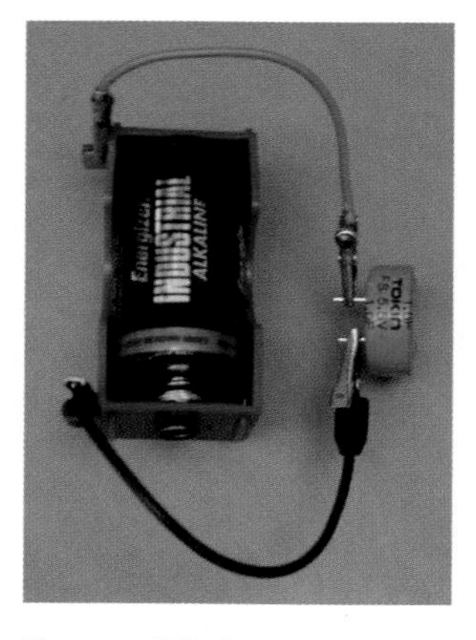

FIGURE 28-3 ■ When a battery is connected across the terminals of a capacitor, the capacitor stores electrical energy.

Capacitors as Energy Storage Devices

Just as you can store potential energy by pulling a bowstring, stretching a spring, compressing a gas, or lifting a book, you can also store electrical energy in the electric field found inside a "charged" capacitor as shown in Figs. 28-3 and 28-4. For example, energy storage in microscopic capacitors enables them to function as memory devices in modern digital computers and in the charge-coupled devices (CCDs) used in video cameras. Energy stored in capacitors can also be used to keep computer circuits running smoothly during brief power outages. A much larger capacitor lies at the heart of a battery-powered photoflash unit. This capacitor accumulates electrical energy relatively slowly during the time between flashes, building up an electric field as it does so.

The electric field across the capacitor plates stores energy that can be released rapidly to create an intensive flash of light. (It is important to note that because capacitors are storehouses for electrical energy, some electrical devices can give you a nasty shock if you open them and accidentally touch both terminals of a capacitor—even when the device is turned off.)

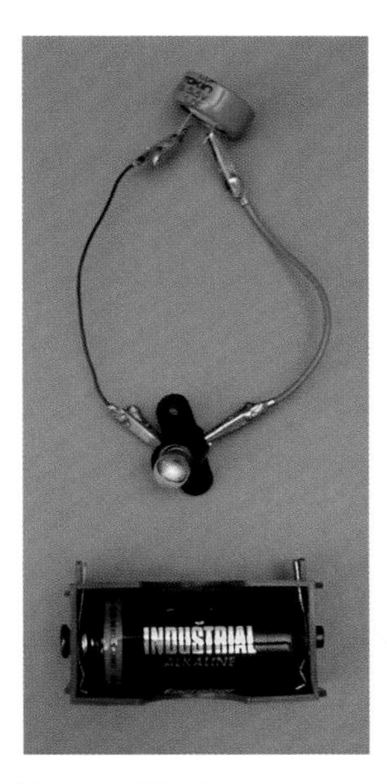

FIGURE 28-4 ■ When a "charged" capacitor is disconnected from its battery and wired in series with a bulb, the energy stored in it can light the bulb for a short period of time.

28-2 Capacitance

Figure 28-5 shows a capacitor made from a conventional arrangement of a pair of metal plates. A device consisting of two parallel conducting plates of area A separated by a distance d is called a *parallel-plate capacitor*. The circuit symbol we use to represent a capacitor (⊣⊢) is based on the structure of a parallel-plate capacitor but is used for capacitors of all shapes. For the purpose of defining capacitance in a simple manner, we will consider an ideal capacitor as two flat parallel conductors (or **plates**) with a perfect insulator between its plates. This perfect insulator allows absolutely no current to pass between them. For simplicity, at first we choose to consider the situation where there is no matter (such as air, glass, or plastic) between the capacitor plates. We just have a vacuum between the plates. We further assume we will charge our capacitor with an ideal battery. Recall that an ideal battery has no internal resistance, so its emf and the potential difference across its terminals are always the same. In Section 28-6 and those following we will relax some of these idealized restrictions.

Equal and Opposite Excess Charge on Plates

When capacitor plates of any shape are connected to a battery or some other voltage source, electrons flow from the negative terminal of the battery through the connecting wire and onto one plate of the capacitor. Meanwhile, the positive terminal of the battery attracts electrons from the other plate. These electrons are pulled through the wires of the circuit, away from the capacitor plate, and leave behind an excess of positive metal atoms with missing electrons. During this process, we cannot find an electric field outside of the capacitor so the overall capacitor seems to be electrically neutral. Hence, we must conclude that at any given time one plate has net or excess charge of $+q$ while the other has a net charge of $-q$. The chemical reactions taking place in the battery are complex, so the electrons pulled off one plate are not necessarily the same ones being pushed through the wires of the circuit onto the other plate. However, the battery does deposit one electron on the negative plate for every one it pulls off the positive plate. We will call this process *charge separation*. Sometimes the process is called *charging*.

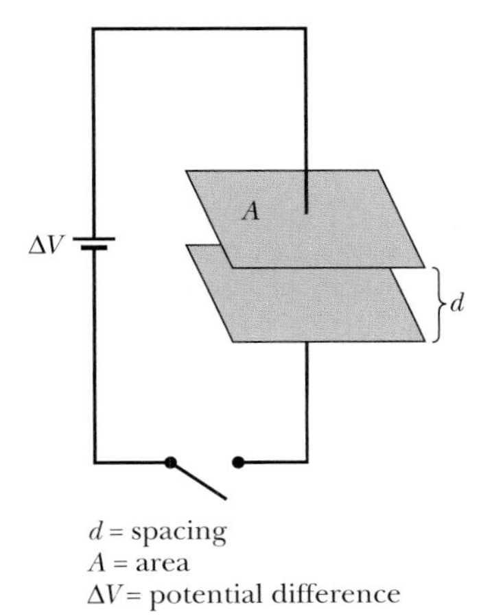

FIGURE 28-5 ■ A parallel-plate capacitor with identical plates of area A and spacing d is connected to a battery with potential difference ΔV. The plates have equal and opposite excess charges of amount $|q|$ on their facing surfaces.

> To understand how a capacitor works, it is important to note that charge separation occurs as a result of charge flow in the wires of the circuit. Charges are not transferred from one plate to the other inside an ideal capacitor.

Why Do Capacitor Plates Stop Accumulating Charge?

Observations show that the battery eventually stops pulling electrons off the positively charged plate and depositing electrons on the negatively charged plate. This is because as electrons build up on the negative plate, they oppose the battery's action and start repelling the flow of additional electrons. Similarly, it becomes harder and harder for the battery to pull electrons off the positive plate as the atoms carrying positive net charge pull back on them. When enough charge has accumulated on the plates, the force exerted on an electron by the battery and the oppositely directed forces exerted on it by the other charges on a plate cancel each other. No more electrons can flow from one plate to the other. We can use a high-quality voltmeter to measure the potential difference across a capacitor just disconnected from a battery. This measurement shows that *charge separation stops when the potential difference across a capacitor is the same as the potential difference across the battery*.

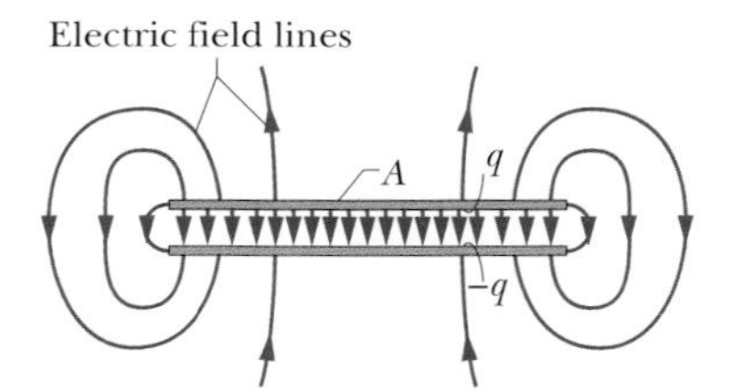

FIGURE 28-6 ■ As the field lines show, the electric field due to the charged plates is uniform in the central region between the plates. The field is not uniform at the edges of the plates, as indicated by the "fringing" of the field lines there.

Factors Affecting Charge Separation Capacity

By convention we refer to the *charge on a capacitor* as $|q|$, the absolute value of the net charge on each plate. Although we refer to a capacitor with charges q and $-q$ on its plates as "charged," a capacitor is electrically neutral so we are actually describing its charge separation created by a voltage source. What factors might affect the capacity for charge separation in a parallel-plate capacitor? We can use our knowledge of electrostatics to explore the effects of several factors. In particular, we will explore how we expect charge to depend on the potential difference across the battery terminals and on geometric factors such as the area of the plates and their spacing (Fig. 28-6):

1. **Potential Difference, ΔV:** For a given capacitor of any shape, we would expect the charge separation to be larger when the potential difference the battery places across the capacitor plates is larger. How much larger? Consider a group of n charges distributed on the plates of a capacitor. Since the plates are conductors, each one is an equipotential surface. According to Eq. 25-25 we can find the electric potential at a given point on a plate relative to infinity. We just need to know the locations of the group of n charges distributed on the capacitor plates. The potential is given by

$$V = k\sum_{i=1}^{n} \frac{q_i}{r_i} \qquad \text{(Eq. 25-25)}$$

where r_i represents the radial distance between the point where the potential is being calculated and the location of the ith charge. By examining this equation we can see that if the potential is to be doubled, there needs to be twice as much charge at each location on the capacitor plates. We expect the amount of the charge separation on a capacitor to be proportional to the potential difference across its plates. We predict

$$|q| \propto |\Delta V|.$$

As you will see in the next subsection, the constant of proportionality between the amount of excess charge on each plate and the potential difference across the plates for a given capacitor is known as its *capacitance*. We will deal more formally with the definition of capacitance and its units in the next section.

2. **Influence of Plate Area, A:** Consider a parallel-plate capacitor. For a given potential difference and plate spacing, d, how do we expect the charge separation capacity to depend on the area of the plates? When the plates have a large area, the electrons the battery is trying to push on the negative plate have more room to spread out. Likewise, the unneutralized atoms left behind when electrons are pulled off the positive plate can be distributed further apart. We expect that as the area of the plates increases, it will be easier to remove or deposit electrons on them.

   A simple experiment can be done to show that the charge separation capacity is in fact directly proportional to area. In this experiment, two sheets of aluminum

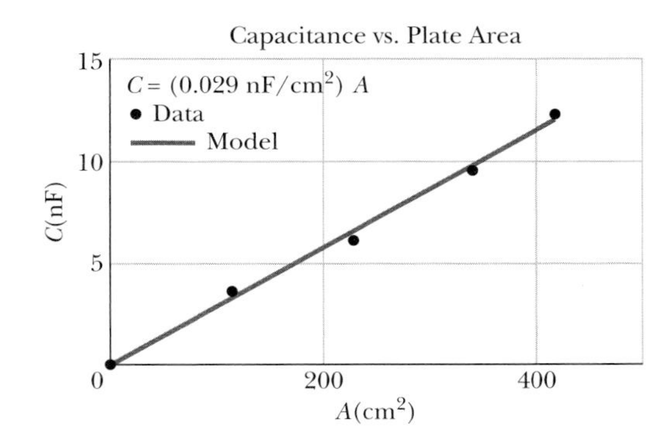

FIGURE 28-7 ■ Two rectangular pieces of aluminum foil are wedged between the insulating pages of a book. A multimeter is used to measure the capacitance of the system. The result shows capacitance increasing in direct proportion to the area of the conducting aluminum plates.

foil are placed opposite each other and separated by the insulating pages of a book. A multimeter like that described in Section 26-4 can be used to measure capacitance. Measurements are taken for different areas of foil. The results are shown in Fig. 28-7. We derive this relationship theoretically in the next section.

3. **Influence of Plate Spacing, d:** Once again we consider a parallel-plate capacitor. For a given potential difference and plate area, A, how do we expect the charge separation capacity to depend on the spacing between the plates? When the plates have a small spacing, the excess positive charges on one plate are quite close to the excess negative charges on the other plate. Since opposite charges attract each other, these charges pull on each other across the insulating gap even though they cannot cross the gap. This attraction helps to counterbalance the repulsion between the like charges on each plate. As the spacing between plates becomes smaller, we expect the overall capacity for the charge separation caused by the battery action to become larger.

   A simple experiment can be done to show that the charge separation capacity does in fact increase as the spacing between plates decreases. In this experiment, two sheets of aluminum foil are placed opposite each other and separated by the insulating pages of a book. A multimeter is used to measure capacitance as different numbers of pages are inserted between the foil plates. The results are shown in Fig. 28-8. This graph shows that the capacitance of the foil plate system is inversely proportional to the spacing, d, between the plates. We derive this relationship theoretically in the next section.

Defining Capacitance

As we just discussed in the last subsection, the amount of the excess charge on each plate of a capacitor, $|q|$, and the size of the potential difference, $|\Delta V|$, across it should be proportional to each other, so

$$|q| = C|\Delta V|. \tag{28-1}$$

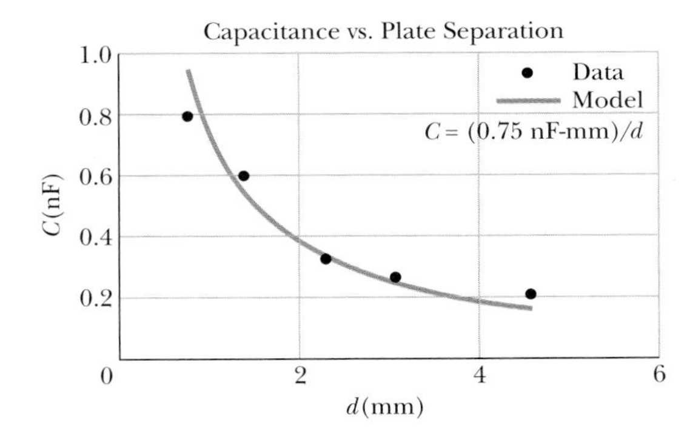

FIGURE 28-8 ■ Two rectangular pieces of aluminum foil are wedged between the insulating pages of a book. The capacitance of the system is measured as a function of the spacing between the plates. The result shows that capacitance is inversely proportional to the spacing.

The proportionality constant C is defined as the **capacitance.** The capacitance is a measure of how much excess charge must be put on each of the plates to produce a certain potential difference between them: the greater the capacitance, the larger the charge separation created by a given potential difference.

For a parallel-plate capacitor, experimental results have shown us its capacitance depends directly on the plate areas and inversely on the spacing between plates. We will see in Sections 28-6 and 28-7 that capacitance will also depend on the nature of the insulating material inserted between the plates. Capacitors having different shapes will not have the same simple relationships between plate area and spacing. In the next section, we will use the definition of electric potential and Gauss' law to identify the theoretical geometric factors for several different types of capacitors including parallel-plate, cylindrical, and spherical capacitors.

Capacitance Units

The SI unit of capacitance following from this expression is the coulomb per volt. This unit occurs so often that it is given a special name, the *farad* (F):

$$1 \text{ farad} = 1 \text{ F} = 1 \text{ coulomb per volt} = 1 \text{ C/V}. \tag{28-2}$$

As you will see, the farad is a very large unit. Fractions of the farad, such as the microfarad ($1\ \mu\text{F} = 10^{-6}$ F) and the picofarad ($1 \text{ pF} = 10^{-12}$ F), are more convenient units in practice. A summary of units and their common notations is shown in Table 28-1.

TABLE 28-1
Units of Capacitance

microfarad: $10^{-6}\text{ F} = 1\ \mu\text{F}$
nanofarad: $10^{-9}\text{ F} = 1\text{ nF} = 1000\ \mu\mu\text{F}$
picofarad: $10^{-12}\text{ F} = 1\text{ pF} = 1\ \mu\mu\text{F}$

READING EXERCISE 28-1: Does the capacitance C of a capacitor increase, decrease, or remain the same (a) when the excess charge of amount $|q|$ on its plates is doubled and (b) when the potential difference ΔV_c across it is tripled? ■

28-3 Calculating the Capacitance

Our task here is to calculate the capacitance of a capacitor once we know its geometry. Because we will consider a number of different geometries, it seems wise to develop a general plan to simplify the work. In brief, our plan is as follows:

1. Assume a charge of amount $|q|$ on each of the "plates."
2. Calculate the electric field $\vec{E}$ between the plates in terms of this amount of charge, using Gauss' law.
3. Knowing $\vec{E}$, calculate the potential difference ΔV between the plates from

$$V_2 - V_1 = -\int_1^2 \vec{E} \cdot d\vec{s}. \quad \text{(Eq. 25-16)}$$

4. Calculate C from $|q| = C|\Delta V|$ (Eq. 28-1).

Before we start, we can simplify the calculation of both the electric field and the potential difference by making certain assumptions. We discuss each in turn.

Calculating the Electric Field

To relate the electric field $\vec{E}$ between the plates of a capacitor to the amount of excess charge $|q|$ on either plate, we shall use Gauss' law:

$$\varepsilon_0 \oint \vec{E} \cdot d\vec{A} = q^{\text{net}} = q. \tag{28-3}$$

Here q is the net charge enclosed by a Gaussian surface, and $\oint \vec{E} \cdot d\vec{A}$ is the net electric flux through that surface. In all cases we shall consider, the Gaussian surface will be such whenever electric flux passes through it, $\vec{E}$ will have a uniform magnitude $E = |\vec{E}|$, and the vectors $\vec{E}$ and $d\vec{A}$ will be parallel. This equation will then reduce to

$$|q| = \varepsilon_0 EA \quad \text{(special case of Eq. 28-3)}, \tag{28-4}$$

in which A is the area of the part of the Gaussian surface through which flux passes. For convenience, we shall always draw the Gaussian surface in such a way it completely encloses the charge on the positive plate; see Fig. 28-9 for an example.

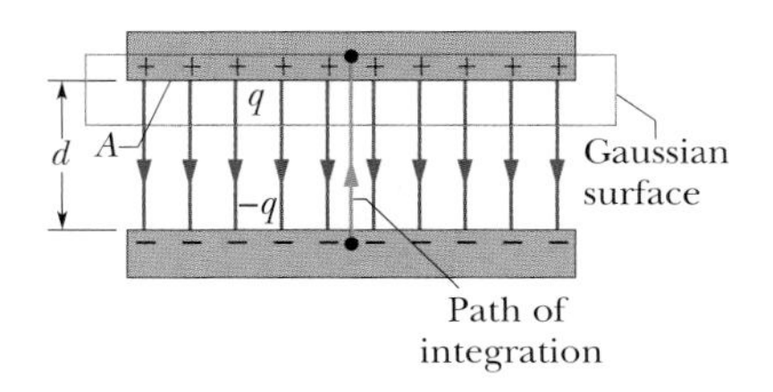

FIGURE 28-9 ■ A charged parallel-plate capacitor. A Gaussian surface encloses the charge on the positive plate. The integration of Eq. 28-6 is taken along a path extending directly from the negative plate to the positive plate.

Calculating the Potential Difference

In the notation of Chapter 25 (Eq. 25-16), the potential difference between the plates of a capacitor is related to the field $\vec{E}$ by

$$V_2 - V_1 = -\int_1^2 \vec{E} \cdot d\vec{s}, \tag{28-5}$$

in which the integral is to be evaluated along any path starting on one plate and ending on the other. We shall always choose a path following an electric field line, from the negative plate to the positive plate. For this path, the vectors $\vec{E}$ and $d\vec{s}$ will have opposite directions, so the dot product $\vec{E} \cdot d\vec{s}$ will be equal to $-|\vec{E}||d\vec{s}|$. The right side of this equation will then be positive. Letting ΔV represent the difference, $V_2 - V_1$, we can then recast the relationship as

$$\Delta V = -\int_-^+ |\vec{E}||d\vec{s}| \quad \text{(special case of Eq. 28-5)}, \tag{28-6}$$

in which the " $-$ " and " $+$ " remind us that our path of integration starts on the negative plate and ends on the positive plate.

We are now ready to apply $|q| = \varepsilon_0 EA$ (Eq. 28-4) and $\Delta V = -\int_-^+ |\vec{E}||d\vec{s}|$ (Eq. 28-6) to some particular cases.

A Parallel-Plate Capacitor

We assume, as Fig. 28-9 suggests, that the plates of our parallel-plate capacitor are so large and so close together we can neglect the fringing of the electric field at the edges of the plates, taking $\vec{E}$ to be constant throughout the region between the plates. This configuration was used in old-time radios. As we will see in Chapter 33, the frequency of an oscillating circuit depends on the capacitance. In old radios (those built before the time that tiny transistors became ubiquitous), the dial was connected to a set of nested metal plates. When the dial was turned, some of the plates rotated while others stayed fixed. By turning the dial, the overlap of the plates changed, changing the capacitance and thereby the frequency of the signal selected.

We draw a Gaussian surface enclosing just the excess charge q on the positive plate, as in Fig. 28-9. Recall from above that

$$|q| = \varepsilon_0 EA, \tag{28-7}$$

where A is the area of each of the plates.

Equation 28-6 yields

$$\Delta V = \int_-^+ |\vec{E}||d\vec{s}| = |\vec{E}| \int_0^d ds = Ed. \tag{28-8}$$

Here, $|\vec{E}| = E$ can be placed outside the integral because it is a constant; the second integral then is simply the plate separation d.

Combining these two expressions with the relation $|q| = C|\Delta V|$ (Eq. 28-1), we find

$$C = \frac{\varepsilon_0 A}{d} \quad \text{(parallel-plate capacitor).} \tag{28-9}$$

This theoretical relationship matches the results of the experiments we presented in the last section. The capacitance does indeed depend only on geometrical factors—namely, the plate area A and the plate separation d. Note that C increases as we increase the plate area A or decrease the separation d.

As an aside, we point out that this expression suggests one of our reasons for writing the electrostatic constant in Coulomb's law in the form $1/4\pi\varepsilon_0$. If we had not done so, the expression for the capacitance of a parallel-plate capacitor above—which is used more often in engineering practice than Coulomb's law—would have been less simple in form. We note further that it permits us to express the permittivity constant ε_0 in a unit more appropriate for use in problems involving capacitors; namely,

$$\varepsilon_0 = 8.85 \times 10^{-12}\ \text{F/m} = 8.85\ \text{pF/m}. \tag{28-10}$$

We have previously expressed this constant as

$$\varepsilon_0 = 8.85 \times 10^{-12}\ \text{C}^2/\text{N}\cdot\text{m}^2. \tag{28-11}$$

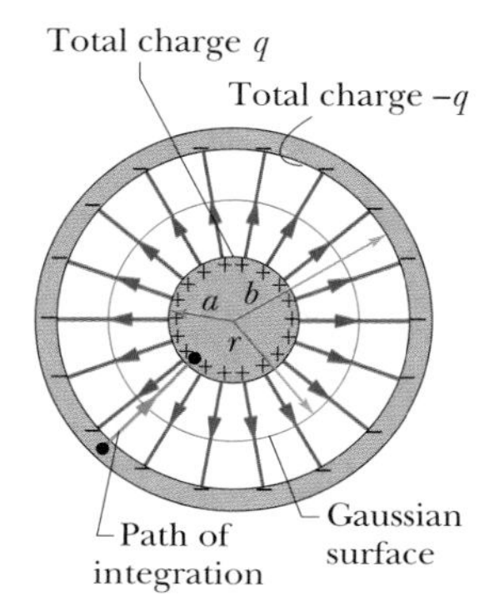

FIGURE 28-10 ■ A cross section of a long cylindrical capacitor, showing a cylindrical Gaussian surface of radius r (that encloses the positive "plate") and the radial path of integration along which Eq. 28-6 is to be applied. If we visualize the central conductor as the cross section of a sphere rather than that of a long cylindrical wire then this figure also illustrates a spherical capacitor.

A Cylindrical Capacitor

Fig. 28-10 shows, in cross section, a cylindrical capacitor of length L formed by two coaxial cylinders of radii a and b. We assume $L \gg b$ so we can neglect the fringing of the electric field occurring at the ends of the cylinders. Each plate contains an amount of excess charge $|q|$. This configuration is important because coaxial cables are used in the communications industry for the long distance transmission of electrical signals (Fig 28-11).

The electric field inside the cylinder is highly symmetrical, so we can use Gauss's law to determine its values. As a Gaussian surface, we choose a cylinder

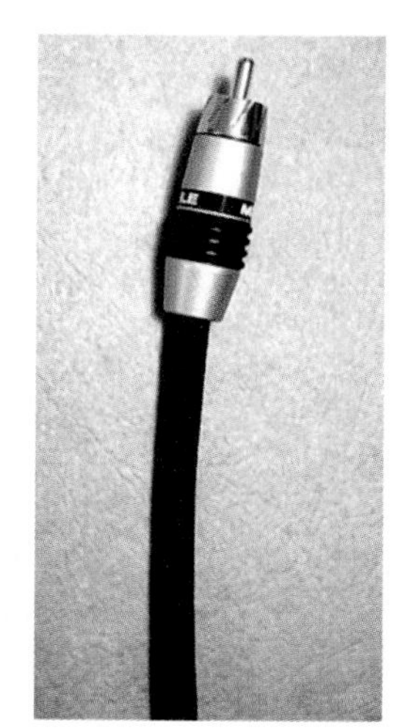

FIGURE 28-11 ■ Coaxial cables and connectors are used for long-distance transmission of television and radio signals. The cable consists of a central conducting wire surrounded by a layer of insulation and then a cylindrical conductor. All three elements are centered on the same axis. Coaxial cables are good examples of cylindrical capacitors.

of length L and radius r, closed by end caps and placed as is shown in Fig. 28-10. Then

$$|q| = \varepsilon_0|\vec{E}|A = \varepsilon_0|\vec{E}|(2\pi rL),$$

in which $2\pi rL$ is the area of the curved part of the Gaussian surface. There is no flux through the end caps. Solving for $|\vec{E}|$ yields

$$|\vec{E}| = \frac{|q|}{2\pi\varepsilon_0 Lr}. \tag{28-12}$$

Substitution of this result into our general expression for potential difference yields

$$\Delta V = \int_{-}^{+} \vec{E}\cdot d\vec{s} = -\frac{q}{2\pi\varepsilon_0 L}\int_b^a \frac{dr}{r} = \frac{q}{2\pi\varepsilon_0 L}\ln\left(\frac{b}{a}\right), \tag{28-13}$$

where here $ds = -dr$ (we integrated radially inward). From the relation $C = |q/\Delta V|$, we then have

$$C = 2\pi\varepsilon_0 \frac{L}{\ln(b/a)} \quad \text{(cylindrical capacitor).} \tag{28-14}$$

We see that the capacitance of a cylindrical capacitor, like that of a parallel-plate capacitor, depends only on geometrical factors, in this case L, b, and a.

A Spherical Capacitor

Fig. 28-10 can also serve as a central cross section of a capacitor consisting of two concentric spherical shells, of radii a and b. As a Gaussian surface we draw a sphere of radius r concentric with the two shells; then

$$|q| = \varepsilon_0 EA = \varepsilon_0 E(4\pi r^2),$$

in which $4\pi r^2$ is the area of the spherical Gaussian surface. We solve this equation for $|\vec{E}|$, obtaining

$$E = |\vec{E}| = k\frac{|q|}{r^2} = \frac{1}{4\pi\varepsilon_0}\frac{|q|}{r^2}, \tag{28-15}$$

which we recognize as the expression for the electric field due to a uniform spherical charge distribution from Chapter 24.

If we substitute this expression into Eq. 28-6, we find

$$\Delta V = \int_{-}^{+} \vec{E}\cdot d\vec{s} = -\frac{|q|}{4\pi\varepsilon_0}\int_b^a \frac{dr}{r^2} = \frac{|q|}{4\pi\varepsilon_0}\left(\frac{1}{a} - \frac{1}{b}\right) = \frac{|q|}{4\pi\varepsilon_0}\frac{b-a}{ab}, \tag{28-16}$$

where again we have substituted $-dr$ for ds. If we now substitute this into $|q| = C|\Delta V|$ (Eq. 28-1) and solve for C, we find

$$C = 4\pi\varepsilon_0 \frac{ab}{b-a} \quad \text{(spherical capacitor).} \tag{28-17}$$

An Isolated Sphere

We can assign a capacitance to a *single* isolated spherical conductor of radius R by assuming that the "missing plate" is a conducting sphere of infinite radius. After all, the field lines leaving the surface of a positively charged isolated conductor must end somewhere; the walls of the room in which the conductor is housed can serve effectively as our sphere of infinite radius.

To find the capacitance of the isolated conductor, we first rewrite the expression for a spherical capacitor above as

$$C = 4\pi\varepsilon_0 \frac{a}{1 - a/b}.$$

If we then let $b \rightarrow \infty$ and substitute R for a, we find

$$C = 4\pi\varepsilon_0 R \quad \text{(isolated sphere)}. \tag{28-18}$$

Note that this formula and the others we have derived for capacitance (Eqs. 28-9, 28-14, and 28-17) involve the constant ε_0 multiplied by a quantity having the dimensions of a length.

READING EXERCISE 28-2: Consider capacitors charged by and then removed from the same battery. Does the charge on the capacitor plates increase, decrease, or remain the same in each of the following situations? (a) The plate separation of a parallel-plate capacitor is increased. (b) The radius of the inner cylinder of a cylindrical capacitor is increased. (c) The radius of the outer spherical shell of a spherical capacitor is increased. ■

READING EXERCISE 28-3: Consider capacitors charged by identical batteries. If the capacitors stay connected to the batteries, does the amount of excess charge on the capacitor plates increase, decrease, or remain the same in each of the following situations? (a) The plate separation of a parallel-plate capacitor is increased. (b) The radius of the inner cylinder of a cylindrical capacitor is increased. (c) The radius of the outer spherical shell of a spherical capacitor is increased. ■

28-4 Capacitors in Parallel and in Series

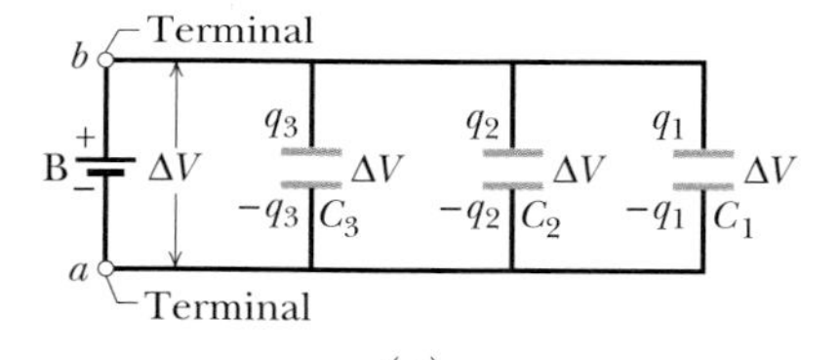

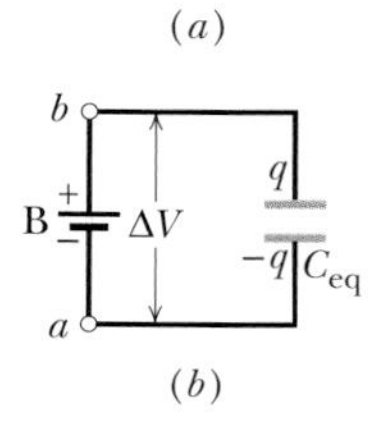

FIGURE 28-12 ■ (*a*) Three capacitors connected in parallel to battery B. The battery maintains a positive potential difference $\Delta V = V_b - V_a$ across its terminals and thus across each fully charged capacitor. (*b*) The equivalent capacitor, with capacitance C_{eq}, replaces the parallel combination.

When there is a combination of capacitors in a circuit, we can often replace that combination with an **equivalent capacitor**—that is, a single capacitor having the same behavior as the actual combination of capacitors. With such a replacement, we can simplify circuits This is similar to the approach we took with resistors in Chapter 27. In addition, circuits often have what is termed *stray capacitance* due to the presence of conductors and insulators in other types of circuit elements. Knowing how the effective capacitance of such elements might combine with each other and other capacitors in the vicinity is vital to the design of high-performance circuits. In this section we discuss the behavior of two basic types of capacitor combinations—parallel and series.

Capacitors in Parallel

Figure 28-12*a* shows an electric circuit in which three capacitors are connected *in parallel* with battery B. This description has little to do with where the capacitor plates appear in the diagram. Rather, "in parallel" means that one plate of each capacitor is wired directly to one plate of the other capacitors. The opposite plates of the capacitors are also wired to each other. When the parallel combination is connected to a battery, the battery's potential difference ΔV_B is applied across all three capacitors as shown in Fig. 28-12*a*.

We can anticipate how the parallel combination will behave by considering the special case in which all three capacitors are parallel-plate capacitors with the same spacing. What happens in this case is that the effective area of the plates of the combined network of capacitors is equal to the sum of the three areas. Using Eq. 28-9 we see

$$C_{eq} = \frac{\varepsilon_0 A}{d} = \frac{\varepsilon_0 (A_1 + A_2 + A_3)}{d} = \frac{\varepsilon_0 A_1}{d} + \frac{\varepsilon_0 A_2}{d} + \frac{\varepsilon_0 A_3}{d} = C_1 + C_2 + C_3.$$

Even if the three capacitors are of different types with each having a different geometry, we expect the effective area of the combination will be increased. The proof of the pudding is in the experiment. It turns out that a multimeter set to measure capacitance can be used to verify

$$C_{eq} = C_1 + C_2 + C_3,$$

for parallel combinations of three capacitors of all sorts of different types. Since the potential difference across a parallel combination of capacitors connected to a voltage source is the same, we can use the expression $|q| = C|\Delta V|$ (Eq. 28-1) to show that if $C_{eq} = C_1 + C_2 + C_3$, then

$$\frac{|q_{eq}|}{|\Delta V|} = \frac{|q_1|}{|\Delta V|} + \frac{|q_2|}{|\Delta V|} + \frac{|q_3|}{|\Delta V|},$$

so that

$$|q_{eq}| = |q_1| + |q_2| + |q_3|.$$

In general,

> When a potential difference ΔV is applied across several capacitors connected in parallel, that potential difference ΔV is applied across each capacitor. The total amount of the excess charge $|q|$ found on each plate of the equivalent capacitor is equal to the sum of the excess charge amounts found on each of the capacitors.

When we analyze a circuit of capacitors in parallel, we can simplify it with this mental replacement:

> Capacitors connected in parallel can be replaced with an equivalent capacitor that has the same total charge $|q|$ and the same potential difference ΔV as the actual capacitors.

We can easily extend our method for finding the equivalent capacitance for three capacitors to any number of capacitors. For n capacitors wired in parallel,

$$C_{eq} = \sum_{j=1}^{n} C_j \quad (n \text{ capacitors in parallel}). \tag{28-19}$$

To find the equivalent capacitance of a parallel combination, we simply add the individual capacitances.

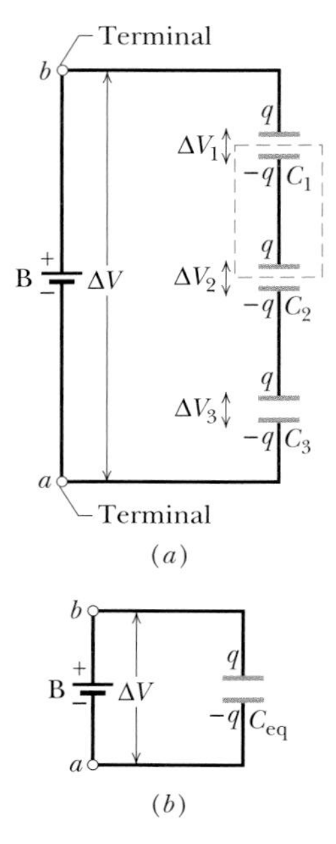

FIGURE 28-13 ■ (*a*) Three capacitors connected in series to battery B. The battery maintains a positive potential difference ΔV between the top and bottom plates of the series combination. (*b*) The equivalent capacitor, with capacitance C_{eq}, replaces the series combination.

Capacitors in Series

Figure 28-13*a* shows three capacitors connected *in series* to battery B. This description has little to do with where the capacitors are located on the drawing. Rather, "in series" means the capacitors are wired serially, one after the other, so a battery can set up a potential difference ΔV across the two ends of the series as shown in Fig. 28-13*a*.

Let's consider what goes on with the charges on the capacitor plates by following a *chain reaction* of events, in which the charging of each capacitor causes the charging of the next capacitor. We start with capacitor 3 and work upward to capacitor 1. When the battery is first connected to the series of capacitors, it produces a net charge $-q$ on the bottom plate of capacitor 3. That charge then repels negative charge from the top plate of capacitor 3 (leaving it with a net or excess charge $+q$). The repelled negative charge moves to the bottom plate of capacitor 2 (giving it charge $-q$). That excess negative charge on the bottom plate of capacitor 2 then repels negative charge from the top plate of capacitor 2 (leaving it with charge $+q$) to the bottom plate of capacitor 1 (giving it a net charge $-q$). Finally the excess charge on the bottom plate of capacitor 1 helps move negative charge from the top plate of capacitor 1 to the battery, leaving that top plate with net charge $+q$. We see then that the potential differences existing across the capacitors in the series produce identical amounts of excess charge $|q|$ on their plates.

Since the amounts of excess charge on each pair of plates in a series connection are the same, we can use Eq. 28-1, $|q| = C|\Delta V|$, to summarize our reasoning in equation form:

$$|q_1| = |q_2| = |q_3| = |q|,$$

and so

$$|\Delta V_1| = \frac{|q|}{C_1}, \quad |\Delta V_2| = \frac{|q|}{C_2}, \quad \text{and} \quad |\Delta V_3| = \frac{|q|}{C_3}.$$

The total potential difference ΔV due to the battery is the sum of these three potential differences. Thus,

$$|\Delta V| = |\Delta V_1| + |\Delta V_2| + |\Delta V_3|,$$

so that

$$\frac{|q|}{C_{eq}} = \frac{|q|}{C_1} + \frac{|q|}{C_2} + \frac{|q|}{C_3}.$$

The equivalent capacitance is then

$$C_{eq} = \frac{|q|}{|\Delta V|} \quad \text{and also} \quad \frac{1}{C_{eq}} = \frac{1}{C_1} + \frac{1}{C_2} + \frac{1}{C_3}.$$

> When a potential difference of size $|\Delta V|$ is applied across several capacitors connected in series, each of the capacitors has the same amount of excess charge $|q|$ on its plates. The sum of the potential differences across the entire network of capacitors is equal to the size of the applied potential difference $|\Delta V|$.

Here is an important point about capacitors in series: When charge is shifted from one capacitor to another in a series of capacitors, it can move along only one route, such as from capacitor 3 to capacitor 2 in Fig. 28-13*a*. If there are additional routes, the capacitors are not in series. Hence, when we analyze a circuit of capacitors in series, we can simplify it with this mental replacement:

TABLE 28-2
Series and Parallel Resistors and Capacitors

Resistors		Capacitors	
Series	**Parallel**	**Series**	**Parallel**
$R_{eq} = \sum_{j=1}^{n} R_j$	$\frac{1}{R_{eq}} = \sum_{j=1}^{n} \frac{1}{R_j}$	$\frac{1}{C_{eq}} = \sum_{j=1}^{n} \frac{1}{C_j}$	$C_{eq} = \sum_{j=1}^{n} C_j$
Eq. 27-4	Eq. 27-12	Eq. 28-20	Eq. 28-19
1. Same current through all resistors	1. Same potential difference across all resistors	1. Same excess charge on all capacitors	1. Same potential difference across all capacitors
2. Potential differences across each resistor add	2. Currents through each resistor add	2. Potential differences across each capacitor add	2. Excess charges on attached plates add

> Capacitors connected in series can be replaced with an equivalent capacitor having the same amount of excess charge $|q|$ on each plate and the same size of potential difference $|\Delta V|$ as the size of the total potential differences across the individual capacitors.

We can easily extend our method of determining the equivalent capacitance of a set of capacitors wired in series from three capacitors to n capacitors by using the expression

$$\frac{1}{C_{eq}} = \sum_{j=1}^{n} \frac{1}{C_j} \quad (n \text{ capacitors in series}). \qquad (28\text{-}20)$$

Using this expression, you can show that the equivalent of a series of capacitances is always less than the least capacitance in the series. This can also be predicted qualitatively since the effective insulated separation between the top and bottom plate increases since $d = d_1 + d_2 + d_3$. According to Eq. 28-9, capacitance is inversely proportional to plate separation.

Table 28-2 summarizes the equivalence relations for resistors and capacitors in series and in parallel. It also presents the information about potential differences and charges on the combinations we determined by thinking about the physics of how the charges move and distribute themselves in these different geometrical configurations.

READING EXERCISE 28-4: A battery with a potential difference ΔV is used to store an amount of excess charge $|q|$ on each of two identical capacitors and is then disconnected. The two capacitors are then connected to each other. What is the potential difference across each capacitor and the amount of excess charge on each capacitor plate when the capacitors are wired (a) in parallel and (b) in series? ■

TOUCHSTONE EXAMPLE 28-1: Equivalent Capacitance

(a) Find the equivalent capacitance for the combination of capacitances shown in Fig. 28-14a, across which potential difference ΔV is applied. Assume

$$C_1 = 12.0\ \mu\text{F}, \quad C_2 = 5.30\ \mu\text{F}, \quad \text{and} \quad C_3 = 4.50\ \mu\text{F}.$$

SOLUTION ■ The **Key Idea** here is that any capacitors connected in series can be replaced with their equivalent capacitor, and any capacitors connected in parallel can be replaced with their equivalent capacitor. Therefore, we should first check whether any of the capacitors in Fig. 28-14a are in parallel or series.

Capacitors 1 and 3 are connected one after the other, but are they in series? No. The potential ΔV that is applied to the capacitors forces excess charge on the bottom plate of capacitor 3. That charge causes charge to shift from the top plate of capacitor 3. However, note that the shifting charge can move to the bottom

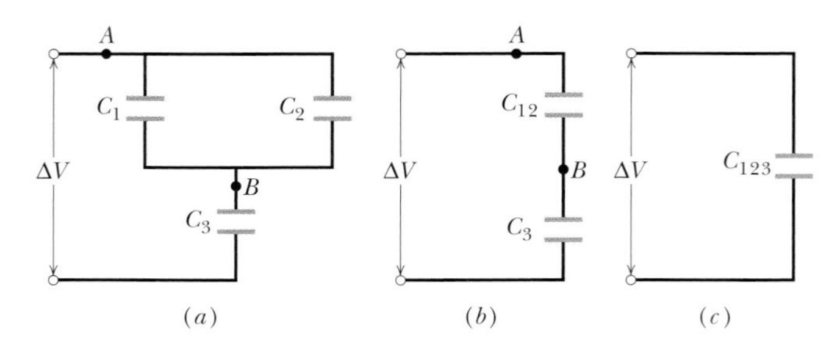

FIGURE 28-14 ■ (*a*) Three capacitors. (*b*) C_1 and C_2, a parallel combination, are replaced by C_{12}. (*c*) C_{12} and C_3, a series combination, are replaced by the equivalent capacitance C_{123}.

plates of both capacitor 1 and capacitor 2. Because there is more than one route for the shifting charge, capacitor 3 is *not* in series with capacitor 1 (or capacitor 2).

Are capacitor 1 and capacitor 2 in parallel? Yes. Their top plates are directly wired together and their bottom plates are directly wired together, and electric potential is applied between the top-plate pair and the bottom-plate pair. Thus, capacitor 1 and capacitor 2 are in parallel, and Eq. 28-19 tells us that their equivalent capacitance C_{12} is

$$C_{12} = C_1 + C_2 = 12.0\ \mu\text{F} + 5.30\ \mu\text{F} = 17.3\ \mu\text{F}.$$

In Fig. 28-14*b*, we have replaced capacitors 1 and 2 with their equivalent capacitor, call it capacitor 12 (say "one two"). (The connections at points *A* and *B* are exactly the same in Figs. 28-14*a* and *b*.)

Is capacitor 12 in series with capacitor 3? Again applying the test for series capacitances, we see that the charge that shifts from the top plate of capacitor 3 must entirely go to the bottom plate of capacitor 12. Thus, capacitor 12 and capacitor 3 are in series, and we can replace them with their equivalent C_{123}, as shown in Fig. 28-14*c*.

From Eq. 28-20, we have

$$\frac{1}{C_{123}} = \frac{1}{C_{12}} + \frac{1}{C_3} = \frac{1}{17.3\ \mu\text{F}} + \frac{1}{4.50\ \mu\text{F}} = 0.280\ \mu\text{F}^{-1},$$

from which

$$C_{123} = \frac{1}{0.280\ \mu\text{F}^{-1}} = 3.57\ \mu\text{F}. \qquad \text{(Answer)}$$

(b) The potential difference that is applied to the input terminals in Fig. 28-14*a* is $V = 12.5$ V. What is the excess charge on each plate of C_1?

SOLUTION ■ One **Key Idea** here is that, to get the excess charge q_1 on each plate of capacitor 1, we now have to work backward to that capacitor, starting with the equivalent capacitor 123. Since the given potential difference $\Delta V = 12.5$ V is applied across the actual combination of three capacitors in Fig. 28-14*a*, it is also applied across capacitor 123 in Fig. 28-14*c*. Thus, Eq. 28-1 ($|q| = C|\Delta V|$) gives us

$$|q_{123}| = C_{123}|\Delta V| = (3.57\ \mu\text{F})(12.5\ \text{V}) = 44.6\ \mu\text{C}.$$

A second **Key Idea** is that the series capacitors 12 and 3 in Fig. 28-1*b* have the same charge as their equivalent capacitor 123. Thus, capacitor 12 has charge $q_{12} = q_{123} = 44.6\ \mu\text{C}$. From Eq. 28-1, the potential difference across capacitor 12 must be

$$|\Delta V_{12}| = \frac{|q_{12}|}{C_{12}} = \frac{44.6\ \mu\text{C}}{17.3\ \mu\text{F}} = 2.58\ \text{V}.$$

A third **Key Idea** is that the parallel capacitors 1 and 2 both have the same potential difference as their equivalent capacitor 12. Thus, capacitor 1 has the potential difference $\Delta V_1 = \Delta V_{12} = 2.58$ V. Thus, from Eq. 28-1, the excess charge on each plate of capacitor 1 must be

$$|q_1| = C_1|\Delta V_1| = (12.0\ \mu\text{F})(2.58\ \text{V})$$
$$= 31.0\ \mu\text{C}. \qquad \text{(Answer)}$$

28-5 Energy Stored in an Electric Field

Work must be done by an external agent to charge a capacitor. Starting with an uncharged capacitor, for example, imagine—using "magic tweezers"—that you remove electrons from one plate and deposit them one at a time to the other plate. The electric field building up in the space between the plates has a direction that tends to oppose further separation of charge. As excess charge accumulates on the capacitor plates, you have to do increasingly larger amounts of work to transfer additional electrons. In practice, this work is done not by "magic tweezers" but by a battery, at the expense of its store of chemical energy.

We visualize the work required to charge a capacitor as being stored in the form of **electric potential energy** *U* in the electric field between the plates. You can recover this energy at will, by discharging the capacitor in a circuit, just as you can recover the potential energy stored in a stretched bow by releasing the bowstring to transfer the energy to the kinetic energy of an arrow. Another example is carrying rocks up a hill against gravity. Energy is stored because of the hill's height and can be recovered by letting the rocks fall down again. In a capacitor, we can recover the stored energy by connecting wires to the ends.

Suppose that at a given instant, a charge $|q'|$ has been moved from one plate of a capacitor, through the wires in the circuit, to the other plate. The amount of the potential difference $|\Delta V'|$ between the plates at that instant will be $|q'|/C$. If an extra increment of charge $|dq'|$ is then removed from one plate and deposited on the other, the amount of the increment of work required will be (from Chapter 25)

$$|dW| = |\Delta V'||dq'| = \frac{|q'|}{C}|dq'|.$$

The work required to bring the total capacitor charge separation up to a final value $|q|$ is

$$W = \int dW = \frac{1}{C}\int_0^q q'|dq'| = \frac{|q|^2}{2C}.$$

This work is stored as potential energy U in the capacitor, and since $q^2 = |q|^2$

$$U = \frac{q^2}{2C} \quad \text{(potential energy).} \tag{28-21}$$

From $|q| = C|\Delta V|$, we can also write this as

$$U = \tfrac{1}{2}C(\Delta V)^2 = \tfrac{1}{2}q\Delta V \quad \text{(potential energy).} \tag{28-22}$$

These relations hold no matter what the geometry of the capacitor is.

To gain some physical insight into energy storage, consider two parallel-plate capacitors identical except that capacitor 1 has twice the plate separation of capacitor 2. Then capacitor 1 has twice the volume between its plates and also, from Eq. 28-9, half the capacitance of capacitor 2. Equation 28-4 tells us that if both capacitors have the same amount of charge $|q|$, the electric fields between their plates are identical. Equation 28-21 tells us capacitor 1 has twice the stored potential energy of capacitor 2. Of two otherwise identical capacitors with the same charge and same electric field, the one with twice the volume between its plates has twice the stored potential energy. Arguments like this tend to verify our earlier assumption:

> The potential energy of a charged capacitor may be viewed as being stored in the electric field between its plates.

A High-Speed Electronic Flash Unit

The ability of a capacitor to store potential energy is the basis of *high-speed electronic flash* devices, like those used in stop-action photography. In an electronic flash unit, a battery charges a capacitor relatively slowly to a high potential difference, storing a large amount of energy in the capacitor. The battery maintains only a modest potential difference; an electronic circuit repeatedly uses that potential difference to greatly increase the potential difference of the capacitor. The power, or rate of energy transfer, during this process is also modest.

When a high-speed flash unit fires, the capacitor releases its stored energy by sending a burst of electric current through a Xenon gas discharge tube that gives off a brief flash of white light. As an example, when a 200 μF capacitor in a high-speed flash unit is charged to 300 V, Eq. 28-22 gives the energy stored in the capacitor as

$$U = \tfrac{1}{2}C(\Delta V)^2 = \tfrac{1}{2}(200 \times 10^{-6}\ \text{F})(300\ \text{V})^2 = 9\ \text{J}.$$

As mentioned in the puzzler at the beginning of this chapter, this should be more than enough energy to provide the illumination needed to take a photograph with ordinary film. Suppose the flashtube in the high-speed flash unit Edgerton used to take the photo of the bullet passing through the apple has a very rapid discharge rate. If the Xenon tube takes only one-third of a microsecond to discharge, then the power associated with the discharge is

$$P = \frac{U}{t} = \frac{9\text{ J}}{0.33 \times 10^{-6}\text{ s}} = 27 \times 10^{6}\text{ W} = 27\text{ MW}.$$

Energy Density

In a parallel-plate capacitor, neglecting fringing, the electric field has the same value at all points between the plates. The **energy density** u—that is, the potential energy per unit volume between the plates—should also be uniform. We can find u by dividing the total potential energy by the volume Ad of the space between the plates. Using Eq. 28-22, we obtain

$$u = \frac{U}{Ad} = \frac{C(\Delta V)^2}{2Ad}.$$

With Eq. 28-9 ($C = \varepsilon_0 A/d$), this result becomes

$$u = \tfrac{1}{2}\varepsilon_0\left(\frac{\Delta V}{d}\right)^2.$$

However, from Eq. 25-39, $\Delta V/d$ equals the electric field magnitude $|\vec{E}| = E$, so

$$u = \tfrac{1}{2}\varepsilon_0 E^2 \quad \text{(energy density)}. \tag{28-23}$$

Although we derived this result for the special case of a parallel-plate capacitor, it holds generally, whatever may be the source of the electric field. If an electric field $\vec{E}$ exists at any point in space, we can think of that point as a site of electric potential energy whose amount per unit volume is given by Eq. 28-23.

TOUCHSTONE EXAMPLE 28-2: Redistributing Charge

(a) Capacitor 1, with $C_1 = 3.55\ \mu$F, is charged to a potential difference $\Delta V_0 = 6.30$ V, using a 6.30 V battery. The battery is then removed and the capacitor is connected as in Fig. 28-15 to an uncharged capacitor 2, with $C_2 = 8.95\ \mu$F. When switch S is closed, charge flows between the capacitors until they have the same potential difference ΔV. Find ΔV.

SOLUTION ■ The situation here differs from Touchstone Example 28-1 because an applied electric potential is *not* maintained across a combination of capacitors by a battery or some other source. Here, just after switch S is closed, the only applied electric potential is that of capacitor 1 on capacitor 2, and that potential is decreasing. Thus, although the capacitors in Fig. 28-15 are connected end to end, in this situation they are not *in series;* and although they are drawn parallel, in this situation they are not *in parallel.*

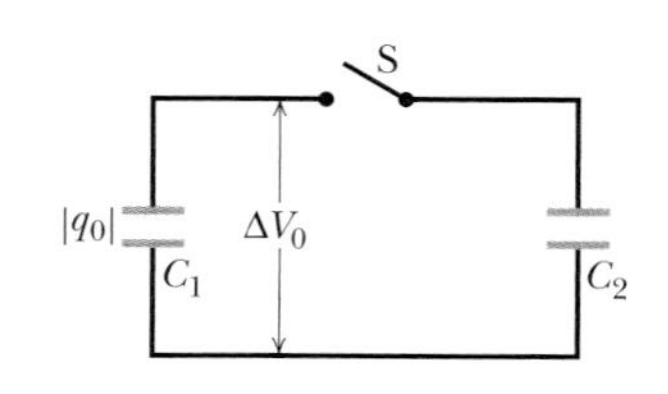

FIGURE 28-15 ■ A potential difference ΔV_0 is applied to capacitor 1 and the charging battery is removed. Switch S is then closed so that the charge on capacitor 1 is shared with capacitor 2.

To find the final electric potential (when the system comes to equilibrium and charge stops flowing), we use this **Key Idea**: After the switch is closed, the original excess charge $|q_0|$ on each plate of capacitor 1 is redistributed (shared) between capacitor 1 and capacitor 2. When equilibrium is reached, we can relate the original charge $|q_0|$ with the final charges $|q_1|$ and q_2 by writing

$$|q_0| = |q_1| + |q_2|.$$

Applying the relation $|q| = C|\Delta V|$ (Eq. 28-1) to each term of this equation yields

$$C_1|\Delta V_0| = C_1|\Delta V| + C_2|\Delta V|,$$

from which

$$|\Delta V| = |\Delta V_0|\frac{C_1}{C_1 + C_2} = \frac{(6.30\text{ V})(3.55\ \mu\text{F})}{3.55\ \mu\text{F} + 8.95\ \mu\text{F}}$$

$$= 1.79\text{ V}. \qquad \text{(Answer)}$$

When the capacitors reach this steady value of electric potential difference, the charge flow stops.

(b) How much energy is stored in the original capacitor when it is first charged up?

SOLUTION ■ The **Key Idea** here is that the potential energy stored in a capacitor, given by Eq. 28-22, is just

$$U = (\tfrac{1}{2})C(\Delta V)^2$$

$$= (\tfrac{1}{2})(3.55\ \mu\text{F})(6.30\text{ V})^2$$

$$= 70.4\ \mu\text{J}. \qquad \text{(Answer)}$$

(c) How much energy is stored in the two capacitors after they are connected together?

SOLUTION ■ The **Key Idea** here is that the potential energy stored in *each* capacitor, given by Eq. 28-22, so that

$$U^{\text{total}} = U_1 + U_2 = (\tfrac{1}{2})C_1(\Delta V_1)^2 + (\tfrac{1}{2})C_2(\Delta V_2)^2$$

$$= (\tfrac{1}{2})((3.55\ \mu\text{F}) + (8.95\ \mu\text{F}))(1.79\text{ V})^2$$

$$= 20.0\ \mu\text{J}. \qquad \text{(Answer)}$$

But how can this be? Before the second capacitor was placed across the first one, there was over 70 μJ of energy stored in the system. What happened to the 50 μJ of energy that seems to have vanished when the second capacitor was charged from the first one? You might argue that the "lost" energy must have been dissipated as heat in the resistance of the wires connecting the two capacitors. But suppose we used superconducting wires with zero resistance? Then where does the missing energy go? The answer, as you will learn in Chapters 33 and 34, is that the charge would oscillate back and forth between the two capacitors until the 50 μJ of "excess" energy was radiated away in the form of electromagnetic waves.

28-6 Capacitor with a Dielectric

If you fill the space between the plates of a capacitor with a *dielectric,* which is usually an insulating material such as mineral oil or plastic, what happens to the capacitance? Michael Faraday—to whom the whole concept of capacitance is largely due and for whom the SI unit of capacitance is named—first looked into this matter in 1837. Using simple equipment much like that shown in Fig. 28-16, he found that the capacitance *increased* by a numerical factor κ, which he called the dielectric constant of the insulating material. Table 28-3 shows some dielectric materials and their dielectric constants. The dielectric constant of a vacuum is unity by definition. Because air is mostly empty space, its measured dielectric constant is only slightly greater than unity.

FIGURE 28-16 ■ The simple electrostatic apparatus used by Faraday. An assembled apparatus (second from left) forms a spherical capacitor consisting of a central brass ball and a concentric brass shell. Faraday placed dielectric materials in the space between the ball and the shell.

Another effect of the introduction of a dielectric is to limit the potential difference that can be applied between the plates to a certain value ΔV^{max}, called the *breakdown potential*. If this value is substantially exceeded, the dielectric material will break down and form a conducting path between the plates. That is, when the capacitor is filled with a dielectric, the charge separation you can maintain with a given potential difference increases. Every dielectric material has a characteristic *dielectric strength,* which is the maximum value of the electric field that it can tolerate without breakdown. A few such values are listed in Table 28-3.

As we discussed in connection with Eq. 28-18, the capacitance of any capacitor can be written in the form

$$C = \varepsilon_0 L, \qquad (28\text{-}24)$$

in which L has the dimensions of a length. For example, $L = A/d$ for a parallel-plate capacitor. Faraday's discovery was, with a dielectric *completely* filling the space between the plates, Eq. 28-24 becomes

$$C = \kappa\varepsilon_0 L = \kappa C_{\text{air}}, \qquad (28\text{-}25)$$

where C_{air} is the value of the capacitance with only air between the plates.

TABLE 28-3
Some Properties of Dielectrics[a]

Material	Dielectric Constant κ	Dielectric Strength (kV/mm)
Air (1 atm)	1.00054	3
Polystyrene	2.6	24
Paper	3.5	16
Transformer oil	4.5	
Pyrex	4.7	14
Ruby mica	5.4	
Porcelain	6.5	
Tantalum oxide	11.6	
Silicon	12	
Germanium	16	
Ethanol	25	
Water (20°C)[b]	80.4	
Water (25°C)[b]	78.5	
Titania ceramic	130	
Strontium titanate	310	8
For a vacuum, κ = unity.		

[a]Measured at room temperature, except for the water.
[b]Note that water is not an insulating material. It is listed because it has dielectric properties.

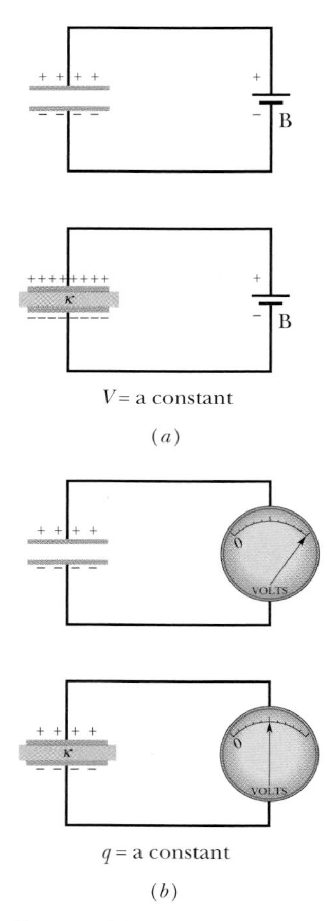

FIGURE 28-17 ■ (*a*) If the potential difference between the plates of a capacitor is maintained, as by battery B, the effect of a dielectric is to increase the excess charge on each plate. (*b*) If the charge on the capacitor plates is maintained, as in this case, the effect of a dielectric is to reduce the potential difference between the plates. The scale shown is that of a *potentiometer,* a device used to measure potential difference (here, between the plates). A capacitor cannot discharge through a potentiometer.

Figure 28-17 provides some insight into Faraday's experiments. In Fig. 28-17*a* the battery ensures that the potential difference ΔV between the plates will remain constant. When a dielectric slab is inserted between the plates, the excess amount of charge $|q|$ on the plates increases by a factor of κ, where κ is always greater than 1; the additional charge is delivered to the capacitor plates by the battery. In Fig. 28-17*b* there is no battery and therefore the amount of excess charge $|q|$ must remain constant when the dielectric slab is inserted; then the potential difference ΔV between the plates decreases by a factor of κ. Both these observations are consistent (through the relation $|q| = C|\Delta V|$) with the increase in capacitance caused by the dielectric.

Comparison of Eqs. 28-24 and 28-25 suggests that the effect of a dielectric can be summed up in more general terms:

> In a region completely filled by a dielectric material of dielectric constant κ, all electrostatic equations containing the permittivity constant ε_0 are to be modified by replacing ε_0 with $\kappa\varepsilon_0$.

A point charge inside a dielectric produces an electric field that, by Coulomb's law, has the magnitude

$$|\vec{E}| = \frac{1}{4\pi\kappa\varepsilon_0}\frac{|q|}{r^2}. \tag{28-26}$$

Also, the expression for the electric field just outside an isolated conductor immersed in a dielectric (see Eq. 24-20) becomes

$$|\vec{E}| = \frac{|\sigma|}{\kappa\varepsilon_0}. \tag{28-27}$$

Both these equations show that *for a fixed distribution of charges, the effect of a dielectric is to weaken the magnitude of the electric field that would otherwise be present.* In addition, the amount of energy stored is reduced because work must be done by the field to pull in the dielectric.

TOUCHSTONE EXAMPLE 28-3: A Dielectric's Energetics

A parallel-plate capacitor whose capacitance C is 13.5 pF is charged by a battery to a potential difference $\Delta V = 12.5$ V between its plates. The charging battery is now disconnected and a porcelain slab ($\kappa = 6.50$) is slipped between the plates. What is the potential energy of the device, both before and after the slab is put into place?

SOLUTION The **Key Idea** here is that we can relate the potential energy U of the capacitor to the capacitance C and either the potential ΔV (with Eq. 28-22) or the capacitor charge $|q|$ (with Eq. 28-21):

$$U_1 = \tfrac{1}{2}C\Delta V^2 = \frac{q^2}{2C}.$$

Because we are given the initial potential $\Delta V(=12.5\text{V})$, we use Eq. 28-22 to find the initial stored energy:

$$U_1 = \tfrac{1}{2}CV^2 = \tfrac{1}{2}(13.5 \times 10^{-12}\text{ F})(12.5\text{ V})^2$$
$$= 1.055 \times 10^{-9}\text{ J} = 1055\text{ pJ} \approx 1100\text{ pJ}. \qquad \text{(Answer)}$$

To find the final potential energy U_2 (after the slab is introduced), we need another **Key Idea**: Because the battery has been disconnected, the amount of excess charge on each capacitor plate cannot change when the dielectric is inserted. However, the potential *does* change. Thus, we must now use Eq. 28-21 (based on q) to write the final potential energy U_2, but now that the slab is within the capacitor, the capacitance is κC. We then have

$$U_2 = \frac{q^2}{2\kappa C} = \frac{U_1}{\kappa} = \frac{1055\text{ pJ}}{6.50} = 162\text{ pJ} \approx 160\text{ pJ}. \qquad \text{(Answer)}$$

When the slab is introduced, the potential energy decreases by a factor of κ.

The "missing" energy, in principle, would be apparent to the person who introduced the slab. The capacitor would exert a tiny tug on the slab and would do work on it, in amount

$$W = U_1 - U_2 = (1055 - 162)\text{ pJ} = 893\text{ pJ}.$$

If the slab were allowed to slide between the plates with no restraint and if there were no friction, the slab would oscillate back and forth between the plates with a (constant) mechanical energy of 893 pJ, and this system energy would transfer back and forth between kinetic energy of the moving slab and potential energy stored in the electric field.

28-7 Dielectrics: An Atomic View

What happens, in atomic and molecular terms, when we put a dielectric in an electric field? There are two possibilities, depending on the nature of the molecules:

1. *Polar dielectrics.* The molecules of some dielectrics, like water, have permanent electric dipole moments. In such materials (called *polar dielectrics*), the electric dipoles tend to line up with an external electric field as in Fig. 28-18. Because the molecules are continuously jostling each other as a result of their random thermal motion, this alignment is not complete, but it becomes more complete as the magnitude of the applied field is increased (or as the temperature, and thus the jostling, is decreased). The alignment of the electric dipoles produces an electric field directed opposite the applied field and smaller in magnitude.
2. *Nonpolar dielectrics.* Regardless of whether they have permanent electric dipole moments, molecules acquire dipole moments by induction when placed in an external electric field. In Section 25-9 (see Fig. 25-16), we saw that this occurs because the external field tends to "stretch" the molecules, slightly separating the centers of negative and positive charge.

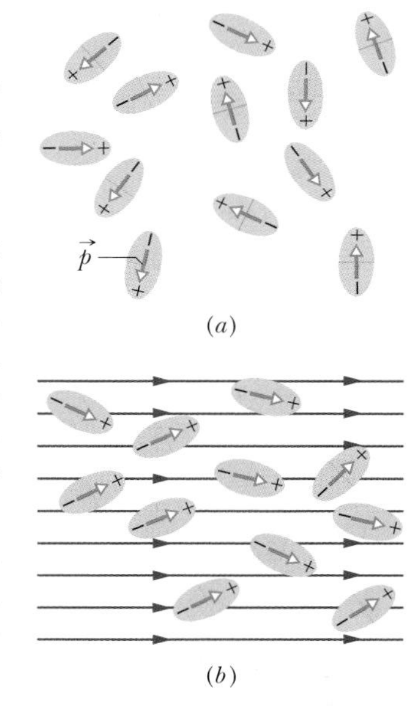

FIGURE 28-18 (*a*) Molecules with a permanent electric dipole moment, showing their random orientation in the absence of an external electric field. (*b*) An electric field is applied, producing partial alignment of the dipoles. Thermal agitation prevents complete alignment.

Figure 28-19*a* shows a nonpolar dielectric slab with no external electric field applied. An electric field $\vec{E}_0$ is present due to the excess charges shown on the capacitor plates in Fig. 28-19*a*. The result is a slight separation of the centers of the positive and negative charge distributions within the slab, producing positive charge on one face of the slab (due to the positive ends of dipoles there) and negative charge on the opposite

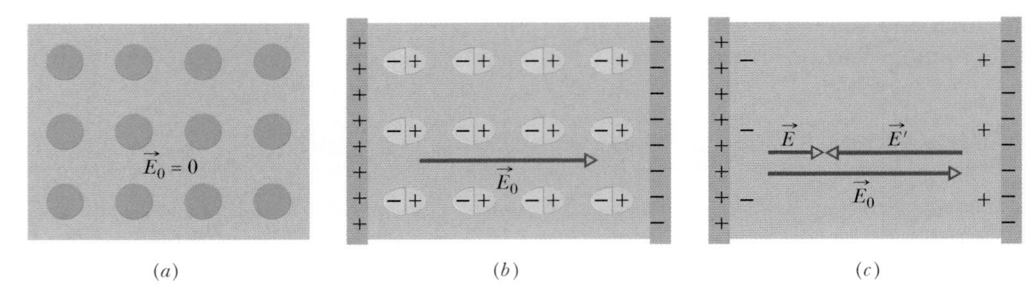

FIGURE 28-19 ■ (*a*) A nonpolar dielectric slab. The circles represent the electrically neutral atoms within the slab. (*b*) An electric field is applied via charged capacitor plates; the field slightly stretches the atoms, separating the centers of positive and negative charge. (c) The separation produces surface charges on the slab faces. These charges set up a field $\vec{E}'$, which opposes the applied field $\vec{E}_0$. The resultant field $\vec{E}$ inside the dielectric (the vector sum of $\vec{E}_0$ and $\vec{E}'$) has the same direction as $\vec{E}_0$ but smaller magnitude.

face (due to the negative ends of dipoles there). The slab as a whole remains electrically neutral and—within the slab—there is no excess charge in any volume element.

Figure 28-19*c* shows that the induced surface charges on the faces produce an electric field $\vec{E}'$, in the direction opposite the applied electric field $\vec{E}_0$. The resultant field $\vec{E}$ inside the dielectric (the vector sum of fields $\vec{E}_0$ and $\vec{E}'$) has the direction of $\vec{E}_0$ but is smaller in magnitude.

Both the field $\vec{E}'$ produced by the surface charges in Fig. 28-19*c* and the electric field produced by the permanent electric dipoles in Fig. 28-18 act in the same way—they oppose the applied field $\vec{E}$. (Inside the material, the $\vec{E}$ field fluctuates wildly, depending on whether you are close to one side of a molecule or another. The effects we are looking at are the average effects of the molecules.) Thus, the effect of both polar and nonpolar dielectrics is to weaken any applied field within them, as between the plates of a capacitor. As a result, a given charge separation can be maintained at a lower potential difference, ΔV, with a dielectric than with a vacuum. This means that a capacitor with a dielectric added has a higher capacitance.

We can now see why the dielectric porcelain slab in Touchstone Example 28-3 is pulled into the capacitor: As it enters the space between the plates, the excess surface charge appearing on each slab face has a sign that is opposite to that of the excess charge on the nearby capacitor plate. Thus, slab and plates attract each other.

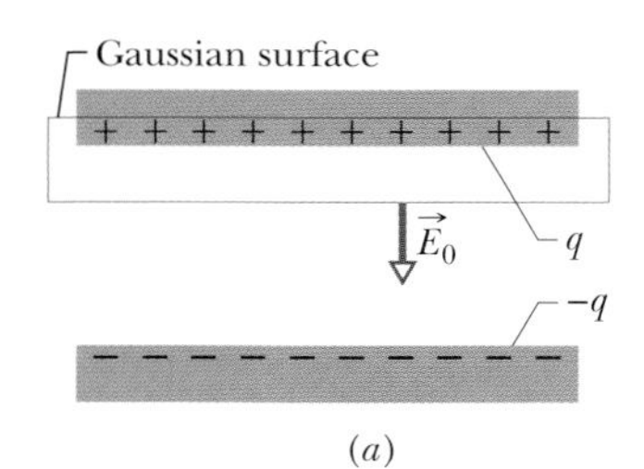

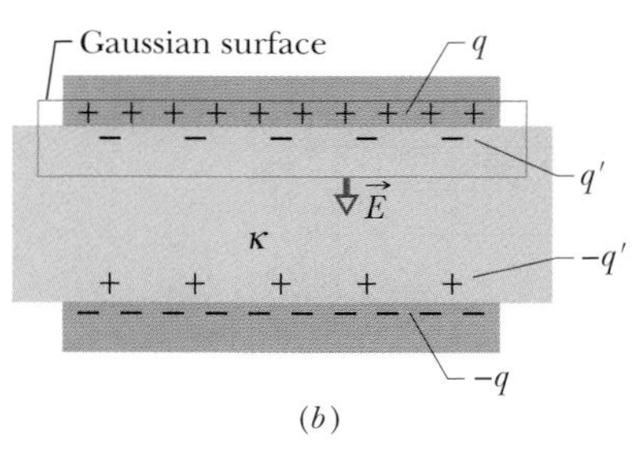

FIGURE 28-20 ■ A parallel-plate capacitor (*a*) without and (*b*) with a dielectric slab inserted. The excess charge q on the plates is assumed to be the same in both cases.

28-8 Dielectrics and Gauss' Law

In our discussion of Gauss' law in Chapter 24, we assumed that the charges existed in a vacuum. Here we shall see how to modify and generalize that law if dielectric materials, such as those listed in Table 28-3, are present. Figure 28-20 shows a parallel-plate capacitor of plate area A, both with and without a dielectric. We assume the amount of excess charge $|q|$ on the plates is the same in both situations. Note the field between the plates induces charge buildup on the faces of the dielectric by one of the methods discussed in Section 28-7.

For the situation of Fig. 28-20*a*, without a dielectric, we can find the electric field $\vec{E}_0$ between the plates as we did in Fig. 28-9: We enclose the excess charge q on the top plate with a Gaussian surface and then apply Gauss' law. Letting $E_0 = |\vec{E}_0|$ represent the magnitude of the field, we find

$$\left|\varepsilon_0 \oint \vec{E} \cdot d\vec{A}\right| = \varepsilon_0 E_0 A = |q^{\text{net}}| = |q|, \tag{28-28}$$

or

$$E_0 = \frac{|q|}{\varepsilon_0 A}. \tag{28-29}$$

In Fig. 28-20*b*, with the dielectric in place, we can find the electric field between the plates (and within the dielectric) by using the same Gaussian surface. However, now the surface encloses two types of charge: it still encloses a net charge q on the top plate but it now also encloses the induced charge q' on the top face of the dielectric. The excess charge on each conducting plate is said to be *free charge* because it can move through the circuit if we change the electric potential of the plate. The induced charge on the surfaces of the dielectric is bound charge. It's stuck to the molecules of an insulator. It can only be displaced from its original position by microscopic amounts and cannot move from the surface.

The amount of net charge enclosed by the Gaussian surface in Fig. 28-20*b* is $|q + q'|$, so Gauss' law now gives

$$\left|\varepsilon_0 \oint \vec{E} \cdot d\vec{A}\right| = \varepsilon_0 EA = |q + q'|, \tag{28-30}$$

or

$$E = \frac{|q + q'|}{\varepsilon_0 A}. \tag{28-31}$$

Since q' and q have different signs, this means that the effect of the dielectric is to weaken the original field E_0 by a factor of κ, so we may write

$$E = \frac{E_0}{\kappa} = \frac{|q|}{\kappa \varepsilon_0 A}. \tag{28-32}$$

Comparison of Eqs. 28-31 and 28-32 shows

$$|q^{\text{net}}| = |q + q'| = \frac{|q|}{\kappa}. \tag{28-33}$$

Equation 28-33 shows correctly that the amount of induced surface charge is less than that of the excess free charge and is zero if no dielectric is present (then, $\kappa = 1$ in Eq. 28-33).

By substituting for $|q + q'|$ from Eq. 28-33 in Eq. 28-30, we can write Gauss' law in the form

$$\varepsilon_0 \oint \kappa \vec{E} \cdot d\vec{A} = q \quad \text{(Gauss' law with dielectric)}, \tag{28-34}$$

where q is the net free charge on the plate of interest. Here we drop the absolute value sign to account for the fact that the excess charge on a plate of interest, q, can be either positive or negative.

This important equation, although derived for a parallel-plate capacitor, is true generally and is the most general form in which Gauss' law can be written. Note the following:

1. The flux integral now involves $\kappa\vec{E}$, not just $\vec{E}$. (The vector $\varepsilon_0 \kappa \vec{E}$ is sometimes called the electric displacement $\vec{D}$, so Eq. 28-34 can be written in the form $\oint \vec{D} \cdot d\vec{A} = q$).
2. The amount of excess charge $|q|$ enclosed by the Gaussian surface is now taken to be the free charge only. The induced surface charge is deliberately ignored on the right side of Eq. 28-34, having been taken fully into account by introducing the dielectric constant κ on the left side.
3. Equation 28-34 differs from Eq. 24-7, our original statement of Gauss' law, only in that ε_0 in the latter equation has been replaced by $\kappa\varepsilon_0$. We keep κ inside the integral of Eq. 28-34 to allow for cases in which κ is not constant over the entire Gaussian surface.

Gauss's law still holds when charged molecules are present, but it's hard to use, since we don't know where those molecular charges are. We only know their average effect, which is summarized by the measured constant κ. Here, we saw how to create a form of Gauss's law including the effect of the molecules automatically, and this allows us to work only with the charges we control directly—the "free" charges.

TOUCHSTONE EXAMPLE 28-4: Adding a Dielectric

Figure 28-21 shows a parallel-plate capacitor of plate area A and plate separation d. A potential difference ΔV_0 is applied between the plates. The battery is then disconnected, and a dielectric slab of thickness b and dielectric constant κ is placed between the plates as shown. Assume

$$A = 115\text{ cm}^2,\quad d = 1.24\text{ cm},\quad \Delta V_0 = 85.5\text{ V},$$
$$b = 0.780\text{ cm},\quad\text{and}\quad \kappa = 2.61.$$

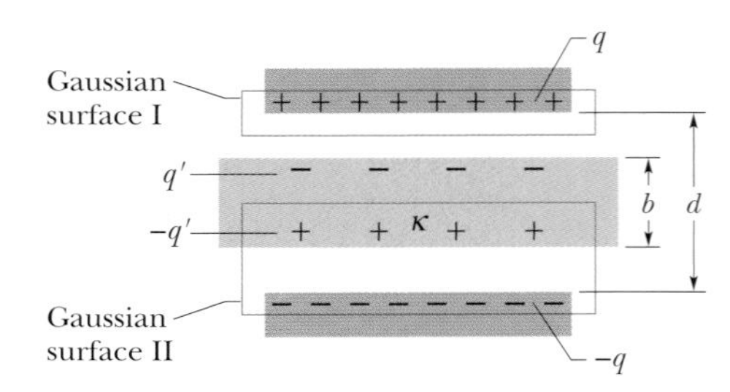

FIGURE 28-21 ■ A parallel-plate capacitor containing a dielectric slab that only partially fills the space between the plates.

(a) What is the capacitance C_0 before the dielectric slab is inserted?

SOLUTION ■ From Eq. 28-9 we have

$$C_0 = \frac{\varepsilon_0 A}{d} = \frac{(8.85 \times 10^{-12}\text{ F/m})(115 \times 10^{-4}\text{ m}^2)}{1.24 \times 10^{-2}\text{ m}}$$
$$= 8.21 \times 10^{-12}\text{ F} = 8.21\text{ pF}. \qquad \text{(Answer)}$$

(b) What is the amount of free excess charge that appears on each plate?

SOLUTION ■ From Eq. 28-1,

$$|q| = C_0|\Delta V_0| = (8.21 \times 10^{-12}\text{ F})(85.5\text{ V})$$
$$= 7.02 \times 10^{-10}\text{ C} = 702\text{ pC}. \qquad \text{(Answer)}$$

Because the charging battery was disconnected before the slab was introduced, the free charge remains unchanged as the slab is put into place.

(c) What is the magnitude of the electric field E_0 in the gaps between the plates and the dielectric slab?

SOLUTION ■ A **Key Idea** here is to apply Gauss' law, in the form of Eq. 28-34, to Gaussian surface I in Fig. 28-21—that surface passes through the gap, and so it encloses *only* the free charge on the upper capacitor plate. Because the area vector $d\vec{A}$ and the field vector $\vec{E}_0$ are both directed downward, the dot product in Eq. 28-34 becomes

$$\vec{E}_0 \cdot d\vec{A} = |\vec{E}_0| dA \cos 0° = E_0 dA.$$

Equation 28-34 then becomes

$$\varepsilon_0 \kappa E_0 \oint dA = q.$$

The integration now simply gives the surface area A of the plate. Thus, we obtain

$$\varepsilon_0 \kappa |\vec{E}_0| \oint dA = q,$$

or

$$E_0 = \frac{q}{\varepsilon_0 \kappa A}.$$

One more **Key Idea** is needed before we evaluate E_0; that is, we must put $\kappa = 1$ here because Gaussian surface I does not pass through the dielectric. Since the charge q on the upper plate is positive, we have

$$E_0 = \frac{q}{\varepsilon_0 \kappa A} = \frac{7.02 \times 10^{-10}\text{ C}}{(8.85 \times 10^{-12}\text{ F/m})(1)(115 \times 10^{-4}\text{ m}^2)}$$
$$= 6900\text{ V/m} = 6.90\text{ kV/m}. \qquad \text{(Answer)}$$

Note that the value of E_0 does not change when the slab is introduced because the amount of charge enclosed by Gaussian surface I in Fig. 28-21 does not change.

(d) What is the magnitude of the electric field E_1 in the dielectric slab?

SOLUTION ■ The **Key Idea** here is to apply Eq. 28-34 to Gaussian surface II in Fig. 28-21. That surface encloses free charge $-q$ and induced charge $-q'$, but we ignore the latter when we use Eq. 28-34. We find

$$\varepsilon_0 \oint \kappa \vec{E}_1 \cdot d\vec{A} = -\varepsilon_0 \kappa E_1 A = -q. \qquad (28\text{-}35)$$

(The first minus sign in this equation comes from the dot product $\vec{E}_1 \cdot d\vec{A}$, because now the field vector $\vec{E}_1$ is directed downward and the area vector dA' is directed upward.) Equation 28-35 gives us

$$E_1 = \frac{q}{\varepsilon_0 \kappa A} = \frac{E_0}{\kappa} = \frac{6.90\text{kV/m}}{2.61} = 2.64\text{ kV/m}. \qquad \text{(Answer)}$$

(e) What is the potential difference ΔV between the plates after the slab has been introduced?

SOLUTION ■ The **Key Idea** here is to find ΔV by integrating along a straight-line path extending directly from the bottom plate to the top plate. Within the dielectric, the path length is b and the electric field is E_1. Within the two gaps above and below the dielectric, the total path length is $d - b$ and the electric field is E_0. Equation 28-6 then yields

$$\Delta V = \int_{-}^{+} |\vec{E}||d\vec{s}| = E_0(d - b) + E_1 b$$
$$= (6900 \text{ V/m})(0.0124 \text{ m} - 0.00780 \text{ m}) + (2640 \text{ V/m})(0.00780 \text{ m})$$
$$= 52.3 \text{ V}. \quad \text{(Answer)}$$

This is less than the original potential difference of 85.5 V.

(f) What is the capacitance with the slab in place?

SOLUTION ■ The **Key Idea** now is that the capacitance C is related to the free charge q and the potential difference ΔV via Eq. 28-1, just as when a dielectric is not in place. Taking q from (b) and ΔV from (e), we have

$$C = \frac{|q|}{|\Delta V|} = \frac{7.02 \times 10^{-10} \text{ C}}{52.3 \text{ } V}$$
$$= 1.34 \times 10^{-11} \text{ F} = 13.4 \text{ pF}. \quad \text{(Answer)}$$

This is greater than the original capacitance of 8.21 pF.

28-9 *RC* Circuits

In preceding sections we dealt only with circuits in which the currents did not vary with time. Here we begin a discussion of time-varying currents.

Charging a Capacitor

The capacitor of capacitance C in Fig. 28-22 is initially uncharged. To charge it, we close switch S on point a. This completes an *RC series circuit* consisting of the capacitor, an ideal battery of emf ε, and a resistance R. Since an ideal battery has no internal resistance, its emf is the same as the potential difference across the battery, ΔV_B.

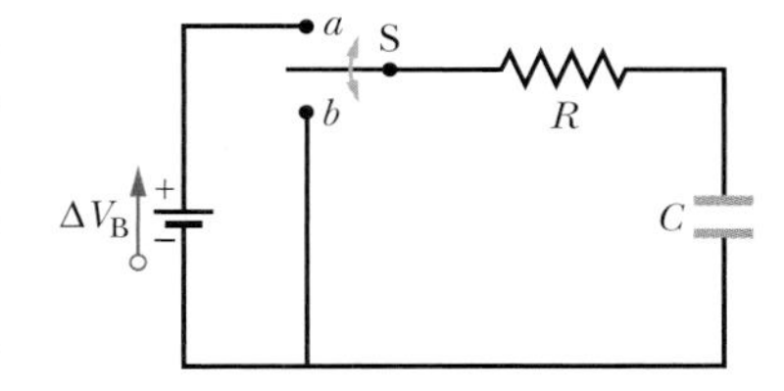

FIGURE 28-22 ■ When switch S is closed on a, the capacitor is *charged* through the resistor. When the switch is afterward closed on b, the capacitor *discharges* through the resistor.

From Section 28-2, we already know that as soon as the circuit is complete, charge begins to flow (current exists) between a capacitor plate and a battery terminal on each side of the capacitor. This current increases the amount of excess charge on the plates, q and the size of the potential difference $|\Delta V_C| = |q|/C$ across the capacitor. When that potential difference across the capacitor equals the potential difference across the battery (which here is equal to the emf of the battery, ΔV_B), the current is zero. From Eq. 28-1 ($|q| = C|\Delta V_C|$), the *equilibrium* (final) amount of excess *charge* on each plate of the fully charged capacitor is equal to $C|\Delta V_B|$.

Here we want to examine the charging process. In particular we want to know how the amount of excess charge $|q(t)|$ on each capacitor plate, the potential difference $\Delta V_C(t)$ across the capacitor, and the current $i(t)$ in the circuit vary with time during the charging process. We begin by applying the loop rule to the circuit, traversing it clockwise from the negative terminal of the battery. We find

$$\Delta V_B - iR - \frac{q}{C} = 0, \quad (28\text{-}36)$$

where q represents the excess charge on the top plate of the capacitor, which is positive in this case.

The last term on the left side represents the potential difference across the capacitor. The term is negative because the capacitor's top plate, which is connected to the battery's positive terminal, is at a higher potential than the lower plate. Thus, there is a drop in potential as we move down through the capacitor.

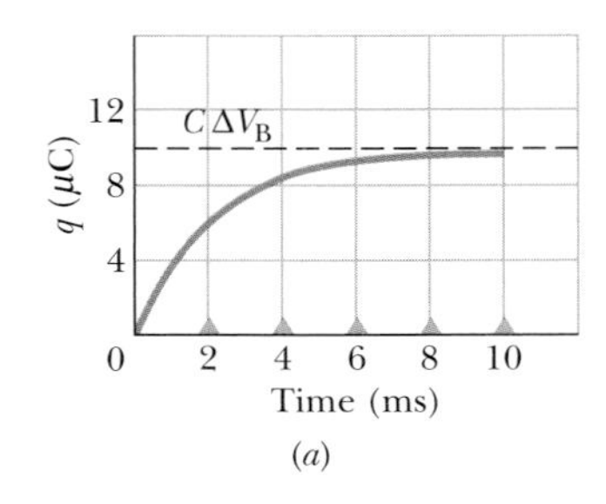

(*a*)

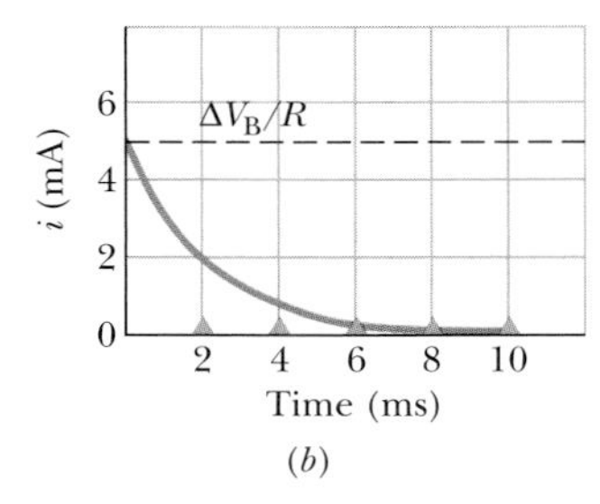

(*b*)

FIGURE 28-23 ■ (*a*) A plot of Eq. 28-39, which shows the buildup of excess charge on the capacitor plates of Fig. 28-22. (*b*) A plot of Eq. 28-40. The charging current in the circuit of Fig. 28-22 declines as the capacitor becomes more fully charged. The curves are plotted for $R = 2000\ \Omega$, $C = 1\ \mu$F, and $\Delta V_B = 10$ V. The small triangles represent successive intervals of one time constant τ.

We cannot immediately solve Eq. 28-36 because it contains two variables, i and q. However, those variables are not independent but are related by

$$i = \frac{dq}{dt}. \tag{28-37}$$

Substituting this for i and rearranging, we find

$$R\frac{dq}{dt} + \frac{q}{C} = \Delta V_B \quad \text{(charging equation).} \tag{28-38}$$

This differential equation describes the time variation of the excess positive charge q on the top plate of the capacitor shown in Fig. 28-23. To solve it, we need to find the function $q(t)$ that satisfies this equation and also satisfies the condition the capacitor be initially uncharged: $q = 0$ C at $t = 0$ s.

The solution to Eq. 28-38 is

$$q = C\Delta V_B(1 - e^{-t/RC}) \quad \text{(charging a capacitor).} \tag{28-39}$$

(Here e is the exponential base, 2.718 . . . , and not the elementary charge.) You can verify by substitution that Eq. 28-39 is indeed a solution to Eq. 28-38. We can see that this expression does indeed satisfy our required initial condition, because at $t = 0$ the term $e^{-t/RC}$ is unity, so the equation gives $q = 0$. Note also that as t goes to ∞ (that is, a long time later), the term $e^{-t/RC}$ goes to zero; so the equation gives the proper value for the full (equilibrium) excess charge on the positive plate of the capacitor—namely, $q = C\Delta V_B$. A plot of $q(t)$ for the charging process is given in Fig. 28-23*a*.

The derivative of $q(t)$ is the positive current $i(t)$ charging the capacitor:

$$i = \frac{dq}{dt} = \left(\frac{\Delta V_B}{R}\right)e^{-t/RC} \quad \text{(charging a capacitor).} \tag{28-40}$$

A plot of $i(t)$ for the charging process is given in Fig. 28-23*b*. Note that the current has the initial value $\Delta V_B/R$ and it decreases to zero as the capacitor becomes fully charged.

> A capacitor being charged initially acts like ordinary connecting wire relative to the charging current. A long time later, it acts like a broken wire.

By combining $|q| = C|\Delta V_C|$ (Eq. 28-1) and $q = C\Delta V_B(1 - e^{-t/RC})$ (Eq. 28-39), we find the potential difference $\Delta V_C(t)$ across the capacitor during the charging process is

$$|\Delta V_C| = \frac{q}{C} = |\Delta V_B(1 - e^{-t/RC})| \quad \text{(charging a capacitor).} \tag{28-41}$$

This tells us $\Delta V_C = 0$ at $t = 0$ and $\Delta V_C = \Delta V_B$ when the capacitor is fully charged as the time approaches infinity ($t \to \infty$).

The Time Constant

The product RC appearing in the equations above has the dimensions of time (both because the argument of an exponential must be dimensionless and because, in fact, $1.0\ \Omega \times 1.0\ \text{F} = 1.0$ s). RC is called the **capacitive time constant** of the circuit and is represented with the symbol τ:

$$\tau = RC \quad \text{(time constant).} \tag{28-42}$$

From the expression for the excess charge as a function of time on one plate of a charging capacitor $q = C\Delta V_B(1 - e^{-t/RC})$ (Eq. 28-39), we can now see that at time $t = \tau\,(=RC)$, the excess charge on the top plate of the initially uncharged capacitor of Fig. 28-22 has increased from zero to

$$q = C\Delta V_B(1 - e^{-1}) = 0.63C\Delta V_B. \qquad (28\text{-}43)$$

In words, after the first time constant, τ, the amount of excess charge has increased from zero to 63% of its final value, $C\Delta V_B$. In Fig. 28-22, the small triangles along the time axes mark successive intervals of one time constant during the charging of the capacitor. The charging times for RC circuits are often stated in terms of τ. The greater τ is, the greater is the charging time.

Discharging a Capacitor

Assume that now the capacitor of Fig. 28-22 is fully charged to a potential ΔV_0 equal to the potential difference, ΔV_B, of the battery. At a new time $t = 0$, switch S is thrown from a to b so the capacitor can *discharge* through resistance R. How do the excess charge $q(t)$ on the top plate of the capacitor and the current $i(t)$ through the discharge loop of capacitor and resistance now vary with time?

The differential equation describing $q(t)$ in this case is similar to the one we worked with for the case of charging Eq. 28-38, except now there is no battery in the discharge loop and so $\Delta V_B = 0$. Thus,

$$R\frac{dq}{dt} + \frac{q}{C} = 0 \quad \text{(discharging equation)}, \qquad (28\text{-}44)$$

where the current term, dq/dt, and the voltage across the capacitor, q/C, can be positive or negative. The solution to this differential equation is

$$q = q_0 e^{-t/RC} \quad \text{(discharging a capacitor)}, \qquad (28\text{-}45)$$

where $|q_0|(=C|\Delta V_0|)$ is the initial amount of excess charge on the capacitor plates. You can verify by substitution that Eq. 28-45 is indeed a solution of Eq. 28-44.

Equation 28-45 tells us that the amount of excess charge on each capacitor plate decreases exponentially with time, at a rate set by the capacitive time constant $\tau = RC$. At time $t = \tau$, the capacitor's excess charge has been reduced to $|q_0|e^{-1}$, or about 37% of the initial value. That is, the amount of excess charge on the plates has decreased by 63%. Note that a greater τ means a greater discharge time.

Differentiating Eq. 28-45 gives us the current $i(t)$:

$$i = \frac{dq}{dt} = -\left(\frac{q_0}{RC}\right)e^{-t/RC} \quad \text{(discharging a capacitor)}. \qquad (28\text{-}46)$$

This tells us the current also decreases exponentially with time, at a rate set by τ. The initial current i_0 is equal to q_0/RC. Note that you can find i_0 by simply applying the loop rule to the circuit at $t = 0$ the moment when the capacitor's initial potential ΔV_0 is connected across the resistance R. So the current must be

$$i_0 = \frac{\Delta V_0}{R} = \frac{(q_0/C)}{R} = \frac{q_0}{RC}.$$

The minus sign in the discharging capacitor expression (Eq. 28-46) can be ignored; it merely means the amount of excess charge on the plate is decreasing.

READING EXERCISE 28-5: The table gives four sets of values for the circuit elements in Fig. 28-22. Rank the sets according to (a) the initial current (as the switch is closed on a) and (b) the time required for the current to decrease to half its initial value, greatest first.

	1	2	3	4
ΔV_B (V)	12.0	12.0	10.0	10.0
R (Ω)	2.0	3.0	10.0	5.0
C (μF)	3.0	2.0	0.5	2.0

TOUCHSTONE EXAMPLE 28-5: Discharging a Capacitor

A capacitor of capacitance C is discharging through a resistor of resistance R.

(a) In terms of the time constant $\tau = RC$, when will the excess charge on each plate of the capacitor be half its initial value?

SOLUTION ■ The **Key Idea** here is that the excess charge on each plate of the capacitor varies according to Eq. 28-45,

$$q = q_0 e^{-t/RC},$$

in which q_0 is the initial charge. We are asked to find the time t at which $q = \frac{1}{2}q_0$ or at which

$$\tfrac{1}{2}q_0 = q_0 e^{-t/RC}. \quad (28\text{-}47)$$

After canceling q_0, we realize that the time t we seek is "buried" inside an exponential function. To expose the symbol t in Eq. 28-47, we take the natural logarithms of both sides of the equation. (The natural logarithm is the inverse function of the exponential function.) We find

$$\ln\tfrac{1}{2} = \ln(e^{-t/RC}) = -\frac{t}{RC},$$

or

$$t = (-\ln\tfrac{1}{2})RC = 0.69RC = 0.69\tau. \quad \text{(Answer)}$$

(b) When will the energy stored in the capacitor be half its initial value?

SOLUTION ■ There are two **Key Ideas** here. First, the energy U stored in a capacitor is related to the charge $|q|$ on the each plate according to Eq. 28-21 ($U = q^2/2C$). Second, that charge is decreasing according to Eq. 28-45. Combining these two ideas gives us

$$U = \frac{q^2}{2C} = \frac{q_0^2}{2C} e^{-2t/RC} = U_0 e^{-2t/RC},$$

in which U_0 is the initial stored energy. We are asked to find the time at which $U = \frac{1}{2}U_0$, or at which

$$\tfrac{1}{2}U_0 = U_0 e^{-2t/RC}.$$

Canceling U_0 and taking the natural logarithms of both sides, we obtain

$$\ln\tfrac{1}{2} = -\frac{2t}{RC},$$

or

$$t = -RC\frac{\ln\frac{1}{2}}{2} = 0.35RC = 0.35\tau. \quad \text{(Answer)}$$

It takes longer (0.69τ versus 0.35τ) for the *charge* to fall to half its initial value than for the *stored energy* to fall to half its initial value. Does this result surprise you?

Problems

SEC. 28-2 ■ CAPACITANCE

1. Electrometer An electrometer is a device used to measure static charge—an unknown excess charge is placed on the plates of the meter's capacitor, and the potential difference is measured. What minimum charge can be measured by an electrometer with a capacitance of 50 pF and a voltage sensitivity of 0.15 V?

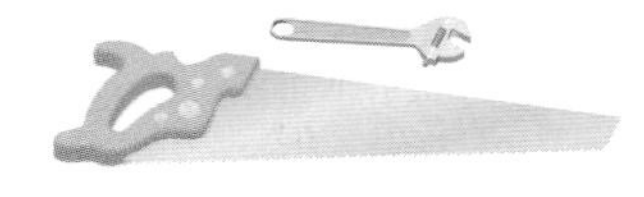

FIGURE 28-24 ■ Problem 2.

2. Two Metal Objects The two metal objects in Fig. 28-24 have net (or excess) charges of +70 pC and −70 pC, which result in a 20 V potential difference between them. (a) What is the capacitance of the system? (b) If the excess charges are changed to +200 pC and −200 pC, what does the capacitance become? (c) What does the potential difference become?

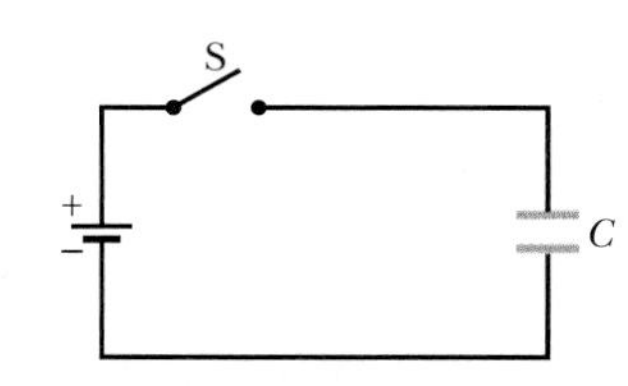

FIGURE 28-25 ■ Problem 3.

3. Initially Uncharged The capacitor in Fig. 28-25 has a capacitance of

25 μF and is initially uncharged. The battery provides a potential difference of 120 V. After switch S is closed, how much charge will pass through it?

SEC. 28-3 ■ CALCULATING THE CAPACITANCE

4. Show That If we solve Eq. 28-9 for ε_0 we see that its SI unit is the farad per meter. Show that this unit is equivalent to that obtained earlier for ε_0—namely, the coulomb squared per newton-meter squared ($C^2/N \cdot m^2$).

5. Circular Plates A parallel-plate capacitor has circular plates of 8.2 cm radius and 1.3 mm separation. (a) Calculate the capacitance. (b) What excess charge will appear on each of the plates if a potential difference of 120 V is applied?

6. Two Flat Metal Plates You have two flat metal plates, each of area 1.00 m^2, with which to construct a parallel-plate capacitor. If the capacitance of the device is to be 1.00 F, what must be the separation between the plates? Could this capacitor actually be constructed?

7. Spherical Drop of Mercury A spherical drop of mercury of radius R has a capacitance given by $C = 4\pi\varepsilon_0 R$. If two such drops combine to form a single larger drop what is its capacitance?

8. Spherical Capacitor The plates of a spherical capacitor have radii 38.0 mm and 40.0 mm. (a) Calculate the capacitance. (b) What must be the plate area of a parallel-plate capacitor with the same plate separation and capacitance?

9. Two Spherical Shells Suppose that the two spherical shells of a spherical capacitor have approximately equal radii. Under these conditions the device approximates a parallel-plate capacitor with $b - a = d$. Show that Eq. 28-17 does indeed reduce to Eq. 28-9 in this case.

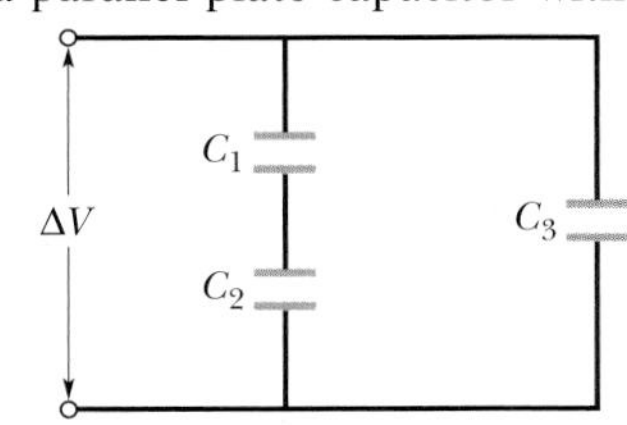

FIGURE 28-26 ■ Problems 10 and 30.

SEC. 28-4 ■ CAPACITORS IN PARALLEL AND IN SERIES

10. Equivalent In Fig. 28-26, find the equivalent capacitance of the combination. Assume that $C_1 = 10.0$ μF, $C_2 = 5.00$ μF, and $C_3 = 4.00$ μF.

11. How Many How many 1.00 μF capacitors must be connected in parallel to store an excess charge of 1.00 C with a potential of 110 V across the capacitors?

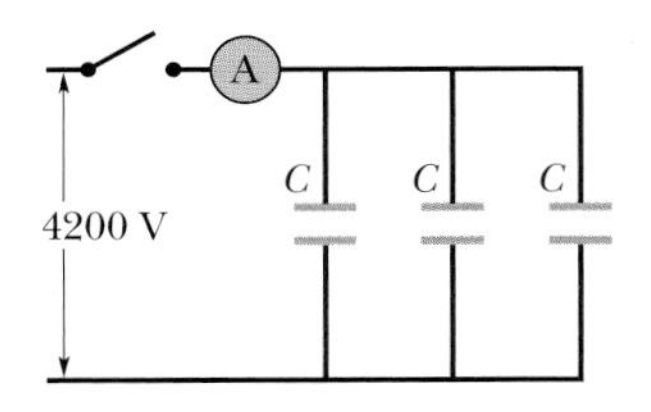

FIGURE 28-27 ■ Problem 12.

12. Each Uncharged Each of the uncharged capacitors in Fig. 28-27 has a capacitance of 25.0 μF. A potential difference of 4200 V is established when the switch is closed. How many coulombs of charge then pass through meter A?

13. Combo In Fig. 28-28 find the equivalent capacitance of the combination. Assume that $C_1 = 10.0$ μF, $C_2 = 5.00$ μF, and $C_3 = 4.00$ μF.

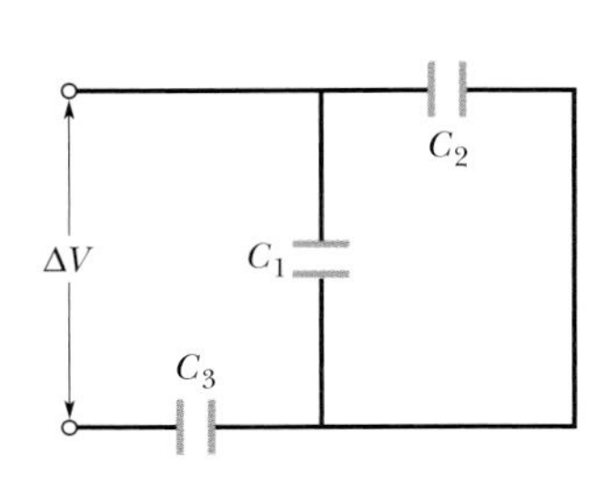

FIGURE 28-28 ■ Problems 13, 14, and 28.

14. Breaks Down In Fig. 28-28 suppose that capacitor 3 breaks down electrically, becoming equivalent to a conducting path. What *changes* in (a) the amount of excess charge and (b) the potential difference occur for capacitor 1? Assume that $\Delta V = 100$ V.

15. Two in Series Figure 28-29 shows two capacitors in series; the center section of length b is movable vertically. Show that the equivalent capacitance of this series combination is independent of the position of the center section and is given by $C = \varepsilon_0 A/(a - b)$, where A is the plate area.

FIGURE 28-29 ■ Problem 15.

16. Battery Potential In Fig. 28-30, the battery has a potential difference of 10 V and the five capacitors each have a capacitance of 10 μF. What is the excess charge on (a) capacitor 1 and (b) capacitor 2?

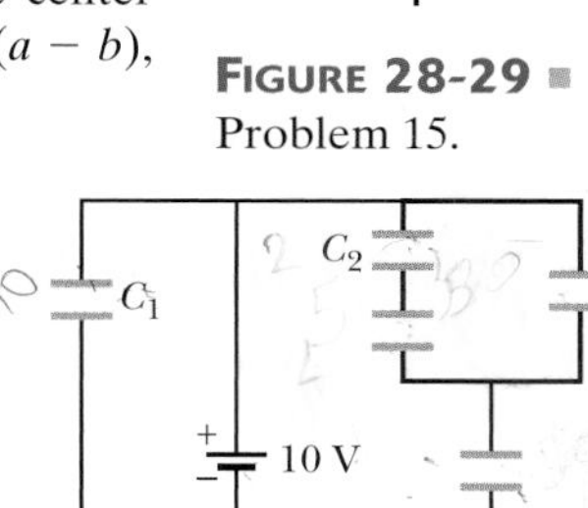

FIGURE 28-30 ■ Problem 16.

17. Parallel with Second 100 pF capacitor is charged to a potential difference of 50 V, and the charging battery is disconnected. The capacitor is then connected in parallel with a second (initially uncharged) capacitor. If the potential difference across the first capacitor drops to 35 V, what is the capacitance of this second capacitor?

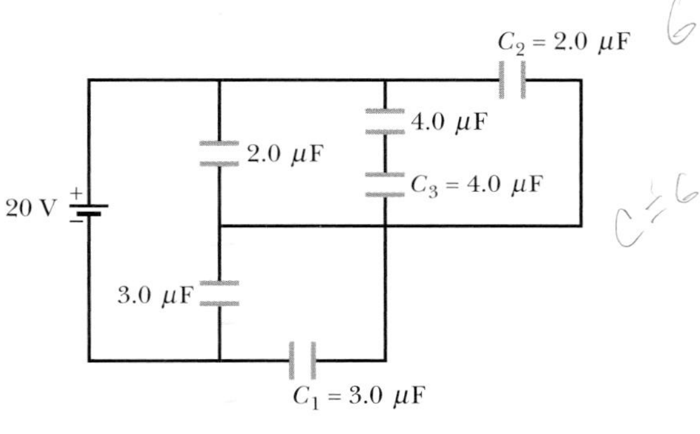

FIGURE 28-31 ■ Problem 18.

18. Charge Stored In Fig. 28-31, the battery has a potential difference of 20 V. Find (a) the equivalent capacitance of all the capacitors and (b) the excess charge stored by that equivalent capacitance. Find the potential across and charge on (c) capacitor 1, (d) capacitor 2, and (e) capacitor 3.

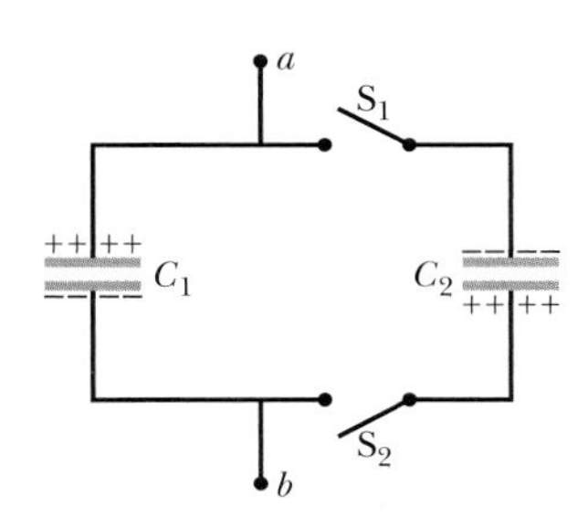

FIGURE 28-32 ■ Problem 19.

19. Opposite Polarity In Fig. 28-32, the capacitances are $C_1 = 1.0$ μF and $C_2 = 3.0$ μF and both capacitors are charged to a potential difference of $\Delta V = 100$ V but with opposite polarity as shown. Switches S_1, and S_2 are now closed. (a) What is now the potential difference between points a and b? What are now the amounts of excess charge on capacitors (b) 1 and (c) 2?

20. Battery Supplies In Fig. 28-33, battery B supplies 12 V. Find the excess charge on each capacitor (a) first when only switch S_1

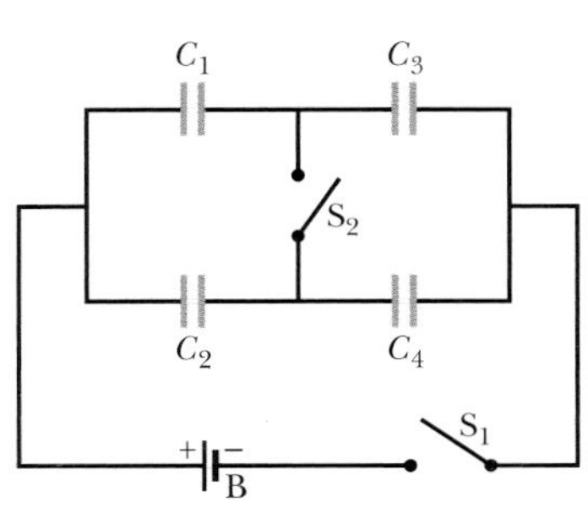

FIGURE 28-33 ■ Problem 20.

switch S_2 is also closed. Take $C_1 = 1.0$ μF, $C_2 = 2.0$ μF, $C_3 = 3.0$ μF, and $C_4 = 4.0$ μF.

21. Switch Is Thrown When switch S is thrown to the left in Fig. 28-34, the plates of capacitor 1 acquire a potential difference ΔV_0. Capacitors 2 and 3 are initially uncharged. The switch is now thrown to the right. What are the final amounts of excess charge $|q_1|, |q_2|$, and $|q_3|$ on the capacitors?

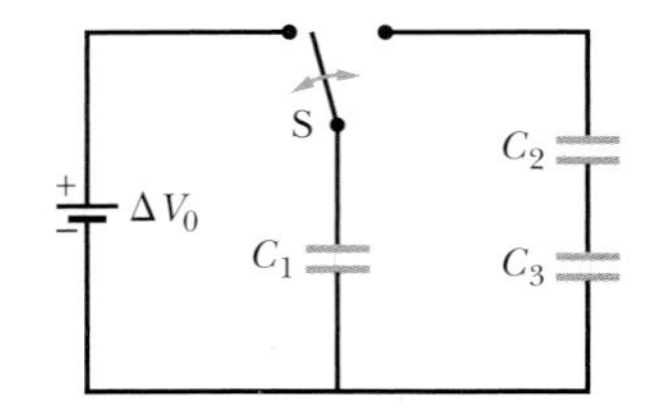

FIGURE 28-34 ■ Problem 21.

SEC. 28-5 ■ ENERGY STORED IN AN ELECTRIC FIELD

22. Air How much energy is stored in one cubic meter of air due to the "fair weather" electric field of magnitude 150 V/m?

23. Capacitance Required What capacitance is required to store an energy of 10 kW · h at a potential difference of 1000 V?

24. Air-Filled Capacitor A parallel-plate air-filled capacitor having area 40 cm^2 and plate spacing 1.0 mm is charged to a potential difference of 600 V. Find (a) the capacitance, (b) the amount of excess charge on each plate, (c) the stored energy, (d) the electric field between the plates, and (e) the energy density between the plates.

25. Two Capacitors Two capacitors, of 2.0 and 4.0 μF capacitance, are connected in parallel across a 300 V potential difference. Calculate the total energy stored in the capacitors.

26. Connected Bank A parallel-connected bank of 5.00 μF capacitors is used to store electric energy. What does it cost to charge the 2000 capacitors of the bank to 50,000 V assuming 12.0¢/kW · h?

27. One Capacitor One capacitor is charged until its stored energy is 4.0 J. A second uncharged capacitor is then connected to it in parallel. (a) If the charge distributes equally, what is now the total energy stored in the electric fields? (b) Where did the excess energy go?

28. Find In Fig. 28-28 find (a) the excess charge, (b) the potential difference, and (c) the stored energy for each capacitor. Assume the numerical values of Problem 13, with $\Delta V = 100$ V.

29. Plates of Area *A* A parallel-plate capacitor has plates of area A and separation d and is charged to a potential difference ΔV. The charging battery is then disconnected, and the plates are pulled apart until their separation is $2d$. Derive expressions in terms of A, d, and ΔV for (a) the new potential difference; (b) the initial and final stored energies, U_i and U_f and (c) the work required to separate the plates.

30. Find the Charge In Fig. 28-26, find (a) the excess charge, (b) the potential difference, and (c) the stored energy for each capacitor. Assume the numerical values of Problem 10, with $\Delta V = 100$ V.

31. Cylindrical Capacitor A cylindrical capacitor has radii a and b as in Fig. 28-10. Show that half the stored electric potential energy lies within a cylinder whose radius is $r = \sqrt{ab}$.

32. Metal Sphere A charged isolated metal sphere of diameter 10 cm has a potential of 8000 V relative to $V = 0$ at infinity. Calculate the energy density in the electric field near the surface of the sphere.

33. Force of Magnitude (a) Show that the plates of a parallel-plate capacitor attract each other with a force of magnitude given by $F = q^2/2\varepsilon_0 A$. Do so by calculating the work needed to increase the plate separation from x to $x + dx$, with the excess charge $|q|$ remaining constant. (b) Next show that the magnitude of the force per unit area (the *electrostatic stress*) acting on either capacitor plate is given by $\frac{1}{2}\varepsilon_0 E^2$. (Actually, this is the force per unit area on *any* conductor of *any* shape with an electric field $\vec{E}$ at its surface.)

SEC. 28-6 ■ CAPACITOR WITH A DIELECTRIC

34. Wax An air-filled parallel-plate capacitor has a capacitance of 1.3 pF. The separation of the plates is doubled and wax is inserted between them. The new capacitance is 2.6 pF. Find the dielectric constant of the wax.

35. Convert It Given a 7.4 pF air-filled capacitor, you are asked to convert it to a capacitor that can store up to 7.4 μJ with a maximum potential difference of 652 V. What dielectric in Table 28-3 should you use to fill the gap in the air capacitor if you do not allow for a margin of error?

36. Separation A parallel-plate air-filled capacitor has a capacitance of 50 pF. (a) If each of its plates has an area of 0.35 m^2, what is the separation? (b) If the region between the plates is now filled with material having $\kappa = 5.6$, what is the capacitance?

37. Coaxial Cable A coaxial cable used in a transmission line has an inner radius of 0.10 mm and an outer radius of 0.60 mm. Calculate the capacitance per meter for the cable. Assume that the space between the conductors is filled with polystyrene.

38. Construct a Capacitor You are asked to construct a capacitor having a capacitance near 1 nF and a breakdown potential in excess of 10 000 V. You think of using the sides of a tall Pyrex drinking glass as a dielectric, lining the inside and outside curved surfaces with aluminum foil to act as the plates. The glass is 15 cm tall with an inner radius of 3.6 cm and an outer radius of 3.8 cm. What are the (a) capacitance and (b) breakdown potential of this capacitor?

39. Certain Substance A certain substance has a dielectric constant of 2.8 and a dielectric strength of 18 MV/m. If it is used as the dielectric material in a parallel-plate capacitor, what minimum area should the plates of the capacitor have to obtain a capacitance of 7.0×10^{-2} μF and to ensure that the capacitor will be able to withstand a potential difference of 4.0 kV?

40. Two Dielectrics A parallel-plate capacitor of plate area A is filled with two dieletrics as in Fig. 28-35*a*. Show that the capacitance is

$$C = \frac{\varepsilon_0 A}{d} \frac{\kappa_1 + \kappa_2}{2}.$$

Check this formula for limiting cases. (*Hint:* Can you justify this arrangement as being two capacitors in parallel?)

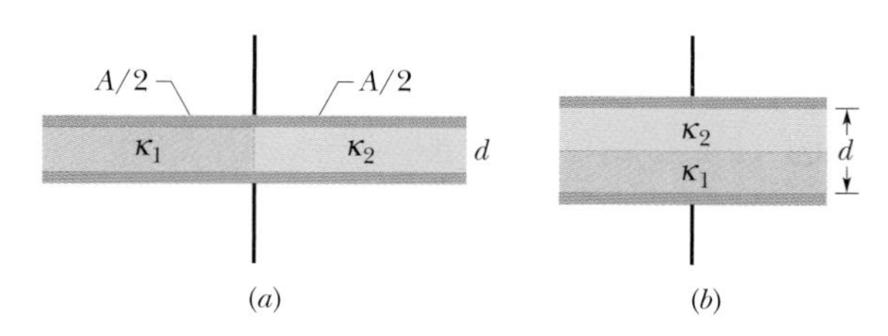

FIGURE 28-35 ■ Problems 40 and 41.

41. Limiting Cases A parallel-plate capacitor of plate area A is filled with two dielectrics as in Fig. 28-35b. Show that the capacitance is

$$C = \frac{2\varepsilon_0 A}{d}\frac{\kappa_1 \kappa_2}{\kappa_1 + \kappa_2}.$$

Check this formula for limiting cases. (*Hint:* Can you justify this arrangement as being two capacitors in series?)

42. What is Capacitance What is the capacitance of the capacitor, of plate area A, shown in Fig. 28-36? (*Hint:* See Problems 40 and 41.)

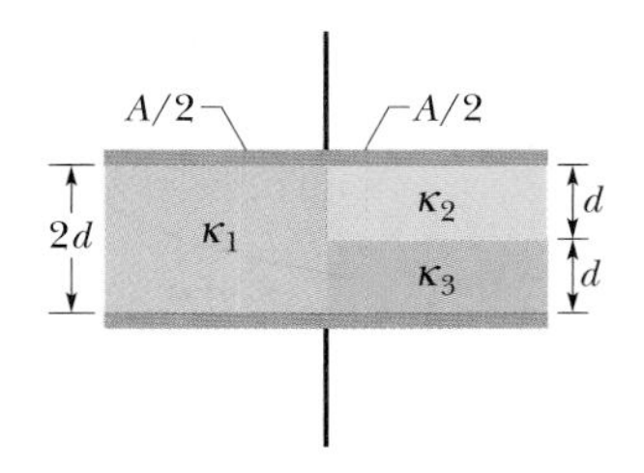

FIGURE 28-36 ■ Problem 42.

SEC. 28-8 ■ DIELECTRICS AND GAUSS' LAW

43. Mica A parallel-plate capacitor has a capacitance of 100 pF, a plate area of 100 cm^2, and a mica dielectric ($\kappa = 5.4$) completely filling the space between the plates. At 50 V potential difference, calculate (a) the electric field magnitude E in the mica, (b) the amount of excess free charge on each plate, and (c) the amount of induced surface charge on the mica.

44. Electric Field Two parallel plates of area 100 cm^2 are given excess charges of equal amounts 8.9×10^{-7} C but opposite signs. The electric field within the dielectric material filling the space between the plates is 1.4×10^6 V/m. (a) Calculate the dielectric constant of the material. (b) Determine the amount of bound charge induced on each dielectric surface.

45. Concentric Conducting Shells The space between two concentric conducting spherical shells of radii b and a (where $b > a$) is filled with a substance of dielectric constant κ. A potential difference ΔV exists between the inner and outer shells. Determine (a) the capacitance of the device, (b) the excess free charge q on the inner shell, and (c) the charge q' induced along the surface of the inner shell.

SEC. 28-9 ■ *RC* CIRCUITS

46. Initial Charge A capacitor with initial excess charge of amount $|q_0|$ is discharged through a resistor. In terms of the time constant τ, how long is required for the capacitor to lose (a) the first one-third of its charge and (b) two-thirds of its charge?

47. How Many Time Constants How many time constants must elapse for an initially uncharged capacitor in an RC series circuit to be charged to 99.0% of its equilibrium charge?

48. Leaky Capacitor The potential difference between the plates of a leaky (meaning that charges leak directly across the "insulated" space between the plates) 2.0 μF capacitor drops to one-fourth its initial value in 2.0 s. What is the equivalent resistance between the capacitor plates?

49. Time Constant A 15.0 kΩ resistor and a capacitor are connected in series and then a 12.0 V potential difference is suddenly applied across them. The potential difference across the capacitor rises to 5.00 V in 1.30 μs. (a) Calculate the time constant of the circuit. (b) Find the capacitance of the capacitor.

50. Flashing Lamp Figure 28-37 shows the circuit of a flashing lamp, like those attached to barrels at highway construction sites. The fluorescent lamp L (of negligible capacitance) is connected in parallel across the capacitor C of an RC circuit. There is a current through the lamp only when the potential difference across it reaches the breakdown voltage V_L; in this event, the capacitor discharges completely through the lamp and the lamp flashes briefly. Suppose that two flashes per second are needed. For a lamp with breakdown voltage $\Delta V_L = 72.0$ V, wired to a 95.0 V ideal battery and a 0.150 μF capacitor, what should be the resistance R?

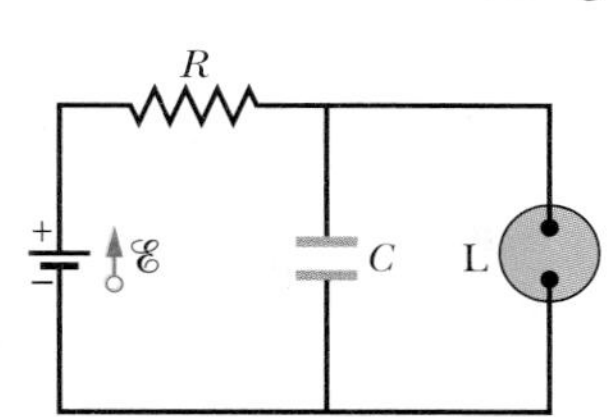

FIGURE 28-37 ■ Problem 50.

51. Initial Potential Difference A capacitor with an initial potential difference of 100 V is discharged through a resistor when a switch between them is closed at $t = 0$. At $t = 10.0$ s, the potential difference across the capacitor is 1.00 V. (a) What is the time constant of the circuit? (b) What is the potential difference across the capacitor at $t = 17.0$ s?

52. Electronic Arcade Game A controller on an electronics arcade games consists of a variable resistor connected across the plates of a 0.220 μF capacitor. The capacitor is charged to 5.00 V, then discharged through the resistor. The time for the potential difference across the plates to decrease to 0.800 V is measured by a clock inside the game. If the range of discharge times that can be handled effectively is from 10.0 μs to 6.00 ms, what should be the resistance range of the resistor?

53. Initial Stored Energy A 1.0 μF capacitor with an initial stored energy of 0.50 J is discharged through a 1.0 MΩ resistor. (a) What is the initial amount of excess charge on the capacitor plates? (b) What is the current through the resistor when the discharge starts? (c) Determine ΔV_C, the potential difference across the capacitor, and ΔV_R, the potential difference across the resistor, as functions of time. (d) Express the production rate of thermal energy in the resistor as a function of time.

Additional Problems

54. Capacitance (a) What is the physical definition and description of a capacitor? (b) What is the mathematical definition of capacitance? (c) Based on the physical description of a capacitor, why would you expect it to hold more excess charge on each of its conducting surfaces when the voltage difference between the two pieces of conductor increases?

55. Net Charge What is the net charge on a capacitor in a circuit? Is it ever possible for the amount of excess charge on one conductor to be different from the amount of excess charge on the other conductor? Explain.

56. Attraction and Repulsion Consider the attraction and repulsion of different types of charge. (a) Explain why you expect to find that the amount of excess charge a battery can pump onto a parallel-plate capacitor will double if the area of each plate doubles. (b) Explain why you expect to find that the amount of excess charge a battery can pump onto a parallel-plate capacitor will be cut in half if the distance between each plate doubles.

57. Three Parallel-Plate Capacitors Suppose you have three parallel-plate capacitors as follows:

Capacitor 1: Area A, spacing d
Capacitor 2: Area A, spacing $2d$
Capacitor 3: Area $2A$, spacing d

The three graph lines (labeled a, b, and c) in Fig. 28-38 represent data for the amounts of excess of charge on the plates of each capacitor as a function of the potential difference across it. Which capacitor {l, 2, or 3} belongs to which line {a, b, and c}? Explain your reasoning carefully.

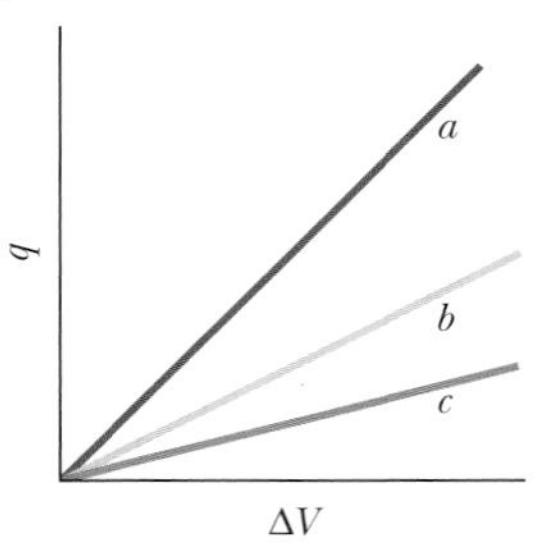

FIGURE 28-38 ■ Problem 57.

58. Capacitors in Series Give as clear an explanation as possible as to why it is physically reasonable to expect that two identical parallel-plate capacitors that are placed in series ought to have half the capacitance as one capacitor. *Hints:* What happens to the effective spacing between the first plate of capacitor 1 and the second plate of capacitor 2 when they are wired in series? What does the fact that like charges repel and opposites attract have to do with anything?

59. Capacitors in Parallel Give as clear an explanation as possible as to why it is physically reasonable to expect that two identical parallel-plate capacitors placed in parallel ought to have twice the capacitance as one capacitor. *Hints:* What happens to the effective area of capacitors wired in parallel? What does the fact that like charges repel and opposites attract have to do with anything?

60. *Charge Ratios on Capacitors* (Adapted from a TYC WS Project ranking task by D. Takahashi). Eight capacitor circuits are shown in Fig. 28-39. All of the capacitors are identical and all are fully charged. The batteries are also identical. In each circuit, one capacitor is labeled C_1 and another is labeled C_2. Assuming $|q_1|$ denotes the amount of excess charge on C_1, $|q_2|$ denotes the amount of excess charge on C_2, and the value of the ratio is denoted $|q_1/q_2|$, rank the circuit in which the value of the ratio $|q_1/q_2|$ is largest *first,* and rank the circuit in which the value of the ratio is the smallest *last*. If two or more circuits result in identical values for the ratio, give these circuits equal ranking. Express your ranking symbolically. (For example, suppose the ratio was highest for D and G and lowest for A and E with the in-between ratios being equal, then the symbolic ranking would be

$$D = G > B = C = H = F > A = E$$

(*Beware:* This is only a sample, not a correct answer!)

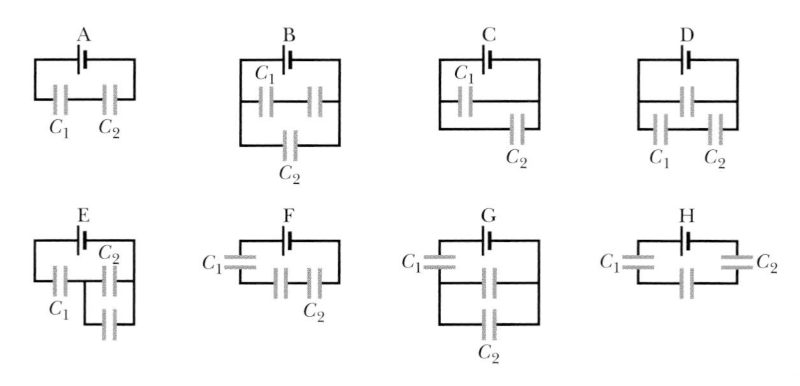

FIGURE 28-39 ■ Problem 60.

61. Physicists Claim Physicists claim that charge never flows *through* an ideal capacitor. Yet when an uncharged capacitor is first placed in series with a resistor and a battery, current flows through the battery and the resistor. Explain how this is possible.

62. Voltage Graphs Figure 28-40 shows plots of voltage across the capacitor as a function of time for three different capacitors that have each been separately discharged through the same resistor. Rank the plots according to the capacitances, the greatest first. Explain the reasons for your rankings.

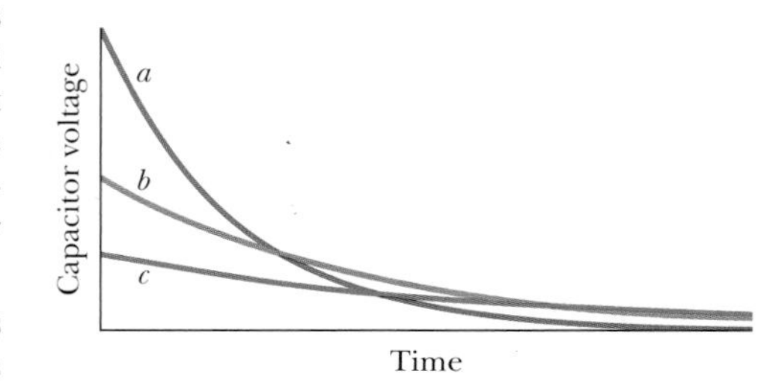

FIGURE 28-40 ■ Problem 62.

63. A Cell Membrane The inner and outer surfaces of a cell membrane carry excess negative and positive charge, respectively. Because of these charges, a potential difference of about 70 mV exists across the membrane. The thickness of the membrane is 8 nm.

(a) If the membrane were empty (filled with air), what would be the magnitude of the electric field inside the membrane?

(b) If the dielectric constant of the membrane were $\kappa = 3$ what would the field be inside the membrane?

(c) Cells can carry ions across a membrane *against the field* ("uphill") using a variety of active transport mechanisms. One mechanism does so by using up some of the cell's stored energy converting ATP to ADP. How much work does it take to carry one sodium ion (charge = +e) across the membrane against the field? Calculate your answer in eV, joules, and kcal/mole (the last for 1 mole of sodium ions).

29 Magnetic Fields

Ocean water contains huge quantities of the light atomic nuclei found in "heavy water" needed to produce fusion power. If we could produce a cost-effective fusion reactor, the world's power problems could be solved. We have known this for over 50 years and still not produced fusion power. Why? A key problem is that it takes a temperature of at least 100 million degrees Celsius to force two light nuclei to fuse together. At this temperature, any material we tried to squeeze together to fuse would be so hot that it would vaporize any material it touches. The torus-shaped chamber of the large Tokomak reactor in this photo was built in an attempt to contain fusion reactions.

How can this Tokomak contain matter at 100 million degrees Celsius?

The answer is in this chapter.

FIGURE 29-1 ■ A large electromagnet is used to collect and transport scrap metal at a steel mill.

29-1 A New Kind of Force?

In Chapter 14 we studied gravitational interaction forces that we experience on an everyday basis. Gravitational forces are so weak that it takes a source the size of a planet or star to produce a noticeable effect. This made the study of the effects of gravity near the Earth's surface relatively simple. In most cases, we treat the gravitational force on an object as a constant.

Then, in Chapter 22, we studied the electrostatic force—a long-range force that is much stronger than the gravitational force. If you run a comb through your hair, a bit of paper near the comb hops up and sticks to the comb. The electrostatic force exerted on the paper by the comb is somewhat larger than the gravitational force that the whole Earth exerts on the paper.

Are there any other long-range (or, action-at-a-distance) forces, or are we done? If you think about your personal experiences, you probably have had the opportunity to play with small disk-shaped refrigerator magnets or pairs of bar magnets. On a larger scale, **electromagnets** are used for sorting scrap metal (Fig. 29-1) and many other things. Magnets are fun because they behave in such an unusual way. You can use one magnet to chase a second magnet around a table without even touching it. But if you come at the magnet from a slightly different direction, it will suddenly seem to change what it's doing and will be pulled toward the other. A refrigerator magnet will seem to leap to the door of the refrigerator, being drawn to it from a distance. Clearly a long-range force is at work here. But is it a new kind of force? Or is it merely a form of gravitational or electrical force?

29-2 Probing Magnetic Interactions

We know from our everyday experiences with small bar magnets that we can *feel* a force on one bar magnet as it interacts with another. This means we can use a bar magnet as a test object for investigating the nature of magnetic interactions. In order to answer the question of whether magnetic interactions are really gravitational or electrostatic forces, let's investigate what happens when a small bar magnet or disk-shaped refrigerator magnet experiences a significant force.

Is the Magnetic Force a Type of Gravitational Force?

The force on our test magnet near the Earth's surface is clearly *in addition* to the gravitational attraction of the Earth. The fact that a refrigerator magnet can stick to the refrigerator and not fall means that it is experiencing a force that is stronger than the gravitational force exerted on it by the entire Earth.

What happens if we replace our test magnet with another *nonmagnetic* object of equal mass and the same shape? We find that the magnetic force disappears. Hence, we must surmise that the force we detected with the bar magnet is not a gravitational force associated with the presence of another object. It is too strong and exists only for certain probe objects. Furthermore, we know from playing with magnets that the force can be attractive or repulsive. As we know, this is not true for the gravitational force.

Is the Magnetic Force a Type of Electrostatic Force?

Could the magnetic force be the electrostatic force we have learned about? After all, the magnetic force, like the electrostatic force, is sometimes attractive and sometimes repulsive. To test this idea, we replace our test magnet with a test charge (such as a tiny Styrofoam ball charged by a rubber rod) at the former location of our test magnet. Again, we find that our new probe (the charge) is only weakly attracted—as is any charged object to a neutral object. So, we must also surmise that *the force the bar magnet detects is not a type of electrostatic force.*

The Magnetic Force and a Moving Charge

We have just described observations that show that forces between magnets are fundamentally different from either electrostatic or gravitational forces. So it appears that we have a new action-at-a-distance force to learn about. This force can be either attractive or repulsive. We can detect this force with a magnet, and so we will refer to it as a **magnetic force.**

Having completed our investigations of electric force in earlier chapters, we now take the electric charge we had been using as a probe and move it rapidly away from a magnet. When we do this, we find something strange. When we *move* the charge, we do detect a force!

OBSERVATION: A magnet exerts a force on a moving charged object, but not on a stationary charged object.

Furthermore, when we try moving the charge at different velocities, we find that the larger the magnitude of the velocity, the larger is the force exerted on the charge. Is the same true for *uncharged,* nonmagnetic masses? Experimentation shows the same is *not* true for uncharged masses. No magnetic force is detected when an uncharged, nonmagnetic mass is used as a probe—regardless of whether the probe is moving or stationary.

In the early 19th century, both Oersted and Ampère discovered that magnets interact with moving charges. In fact, these two scientists showed that current-carrying wires both exert forces on and feel forces from bar magnets. Their observations provide us with important information in our quest to understand the magnetic force. We have found that magnetic forces are not just exerted on other magnets. Magnetic forces are also exerted on a nonmagnetic small charged particle in rough proportion to the degree to which the particle is *both* charged and moving. What is the simplest relationship between magnetic force, charge, and velocity that is consistent with our observations? Mathematically stated, it is a proportional relationship given by

$$|\vec{F}^{\text{mag}}| \propto |q| v,$$

where $|q|$ represents the amount of electric charge on the particle and v is the particle's speed.

Is this relationship correct? Well, if it is, we should see a doubling of the force when we double the velocity of the charged particle we are using as a probe. Experimentally, this does turn out to be the case. Furthermore, we also find that doubling the charge on the probe doubles the force detected. Hence, the linear relationship expressed above is a good start toward a more precise mathematical description of the magnetic force on a moving charged particle. We will return to experimentation as a means for developing a precise expression for the magnetic force in just a moment.

29-3 Defining a Magnetic Field $\vec{B}$

When we play with two bar magnets, we quickly see that the magnetic force can be attractive or repulsive. Furthermore, if we observe more carefully, we find that the strength of the force decreases as the distance between the two magnets is increased. These observations are distinctly reminiscent of our observations of the electrostatic force between two charges. So our first guess in developing a model of the magnetic force might turn out to be somewhat similar to our model of the electrostatic force.

In order to develop a model of magnetism that parallels our model of electrostatics, we should have two different kinds of "magnetic charges." These conceptual objects are referred to as **magnetic monopoles.** We can model our bar magnet as containing a south and a north pole with at least some separation between them. If we assume that like poles repel and unlike poles attract, then this model allows us to correctly predict all our observations. Playing with bar magnets informs us that poles of the same kind repel one another and poles of different kinds attract one another. This is just as we found for electric charges. However, careful observation of the interaction between bar magnets shows that their behavior is similar to that of electric dipoles. Recall that an electric dipole consists of two charges of opposite sign with a small spacing between them. If two electric dipoles that are placed with all their charges lying on the same line are brought together, they will attract. Why? Because a negative charge from the end of one dipole will be closest to the positive charge of the other dipole. However, if we turn one of the electric dipoles around so the dipoles are anti-aligned, then the two like charges will be closest together. Now the dipoles will repel.

Two bar magnets when aligned and then anti-aligned will behave just like electric dipoles. For this reason, we often refer to magnets as *magnetic dipoles*. That is, one end appears to be one kind of magnetic charge and the other end appears to be the other kind of magnetic charge. By convention, we can assign names to the poles of a bar magnet as follows. If we suspend a bar magnet by a string placed halfway between its ends and take other magnetic sources away from its vicinity, one pole of the magnet will point more or less north and the other more or less south. We can call the north-pointing end the north pole of the magnet and the other end the south pole of the magnet.

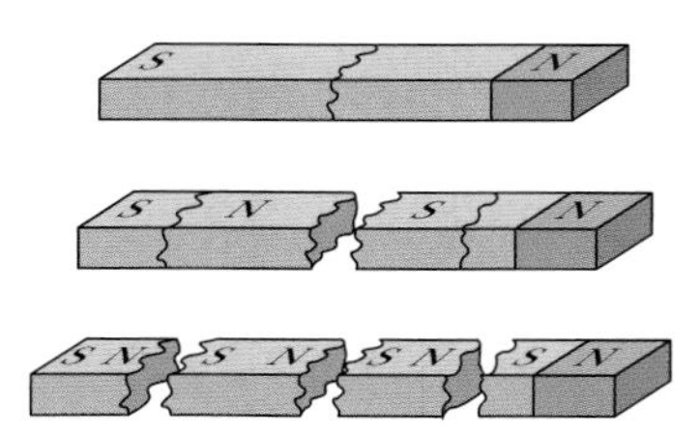

FIGURE 29-2 ■ Whenever a magnet is broken into pieces between its poles, the pieces behave like smaller, weaker magnets.

This idea that a bar magnet is a magnetic dipole with a north charge at one end and a south charge at the other end provides us with a start in describing magnetic interactions. However, to continue with the analogy between the magnetism and electricity, we would like to isolate a magnetic charge. After all, we can separate a negative charge from a positive charge. So we need to be able to separate the north pole of a bar magnet from the south pole of a bar magnet. To do this, we take our bar magnet and cut it in half. But, when we do this we find a surprising thing. The result of breaking the bar magnet in half is simply that we have two weaker half-sized bar magnets. Each one still behaves as a dipole with both a north and a south pole. If we again try to break the magnet in half, we find we have a still smaller magnet, but still with a north and south pole (Fig. 29-2). In fact, if we break the magnet down into subatomic parts, we find that even the electrons, protons, and neutrons within atoms behave as magnetic dipoles (that is, *very* little bar magnets).

As it turns out, the magnetic effect of a bar magnet arises from the combination of the effects of the little bar magnets in the electrons in iron, nickel, and cobalt aligning with each other and producing a strong effect. Each electron's magnet is small, but when you turn them in the same direction and add them all up, the total effect is strong—the full magnetic effect of the bar magnet. So, in short, although the existence of separate magnetic charges (or magnetic monopoles) have been predicted by some physicists, they have never actually been found.

Does the fact that we cannot find an isolated magnetic monopole mean that we must abandon our effort to find parallels between magnetic and electrostatic forces? Not at all. In Chapter 23, we found that the concept of an electric field was quite useful. With so many different possible sources of significant electrostatic forces, it was helpful to think about the force field associated with a given charge (the source of electrostatic force)—without having to decide on what object the force will be exerted on. That is, we wanted to separate the discussion of the source of the force from the discussion of the object the force is exerted on. So we defined the electric field $\vec{E}$ as

$$\vec{E} = \frac{\vec{F}^{\text{elec}}}{q}. \quad \text{(Eq. 23-4)}$$

We determined the electric field $\vec{E}$ at a point by putting a test particle of charge q at rest at that point and measuring the electrostatic force $\vec{F}^{\text{elec}}$ acting on the particle. We saw that *electric charges* set up an electric field that can then affect other electric charges.

Perhaps the same idea could be useful to us in describing magnetic forces. If we could develop a parallel concept of a magnetic field, we could separate the issue of sources of magnetic forces from discussions of the objects that magnetic forces are exerted on. This would be helpful since the concept of a magnetic monopole is so problematic. If a magnetic monopole were available, we could define the magnetic field $\vec{B}$ in a way similar to that used for electric fields. However, because such particles have not been found, we must using another method to define a magnetic field $\vec{B}$.

For nonmagnetic particles, we have already observed that the magnetic force is proportional to the charge and the magnitude of the velocity of the particle being acted on (the probe). We can use this information and define the magnetic field in terms of the force $\vec{F}^{\text{mag}}$ exerted on a moving, electrically charged test particle. The magnitude of the force seems to depend on the direction of the particle's velocity $\vec{v}$ as well. We will examine this effect in more detail in the next section, but for now we define the magnetic field $\vec{B}$ in terms of the *maximum* force magnitude we measure after trying all different directions for $\vec{v}$. So we can express the *magnitude* of the magnetic field $\vec{B}$ in terms of this maximum force magnitude as:

$$B = \frac{F^{\text{mag}}_{\text{max}}}{|q|v}, \quad \text{(29-1)}$$

where q is the particle's charge and v is its speed.

Having defined the magnitude of the magnetic field is a big step forward. It is a concept that will turn out to be extremely useful. Right now, it is helpful because we have not identified the source of the force exerted on our probe. But, having defined the magnitude of the magnetic field in this way, we can at least say that we know that there is a vector magnetic field in the region of space we have been probing. We make extensive use of the concept of a magnetic field in this chapter. Next we turn our attention to this issue of how to define the direction of the magnetic field.

29-4 Relating Magnetic Force and Field

In order to determine the direction of the magnetic field, we can fire a charged particle through a region of space where a magnetic field $\vec{B}$ is known to exist. If we shoot the charged particle in various directions, we find something surprising—the direction of $\vec{F}^{\text{mag}}$ is always perpendicular to the direction of $\vec{v}$ (Fig. 29-3). After many such trials we find that when the particle's velocity $\vec{v}$ is along a particular axis through the region of space, force $\vec{F}^{\text{mag}}$ is zero. Furthermore, we find that for all other directions of $\vec{v}$, the *magnitude* of $\vec{F}^{\text{mag}}$ depends on the direction of $\vec{v}$. In fact, it is proportional to $|\vec{v}|\sin\phi$ where ϕ is the angle between the zero-force axis and the direction of $\vec{v}$. Thinking back to our work on torque and angular momentum, these observations suggest that a cross product is involved. But a cross product of what two vectors?

Clearly, one of the two vectors involved in the cross product is the velocity vector. Our observation that the force is zero when the velocity is along a certain axis implies

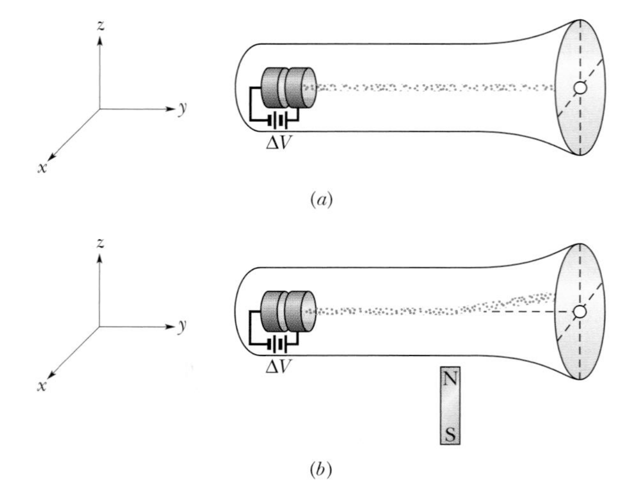

FIGURE 29-3 ■ (*a*) An electron beam is accelerated by a voltage source and travels through an evacuated glass tube to the center of a phosphorescent screen. (*b*) If a magnet is oriented vertically and placed just below the beam (along the $+z$ axis), the electrons are deflected horizontally along the $-x$ axis.

that the other vector must be aligned with this "zero magnetic force" axis. Referring back to our definition of the magnetic field magnitude, B, in Eq. 29-1, we note that the magnitude of the observed magnetic force is given by

$$F_{\text{max}}^{\text{mag}} = |q|vB,$$

where v is the particle speed and $|q|$ is the amount of charge the particle has. Suppose the direction of the magnetic field is taken to be along the "zero magnetic force" axis. We could then represent all of our observations with the following vector equation, known as the *magnetic force law* or ***Lorentz force law*:**

$$\vec{F}^{\text{mag}} = q\vec{v} \times \vec{B} \qquad \text{(magnetic force law).} \qquad (29\text{-}2)$$

That is, the force $\vec{F}^{\text{mag}}$ on the particle is equal to the charge q times the cross product of its velocity $\vec{v}$ and the magnetic field $\vec{B}$. If this expression is correct, the force on a negatively charged particle should be opposite in direction from the force on a positively charged particle. This does in fact turn out to be the case.

Furthermore, expressing the magnetic force on a charged particle moving through a magnetic field as $\vec{F}^{\text{mag}} = q\vec{v} \times \vec{B}$ requires that we adopt a standard convention for the *direction* of the magnetic field. That is,

> The direction of a magnetic field is defined to be related to the direction of the force on and the velocity of a positively charged particle by $\vec{F}^{\text{mag}} = q\vec{v} \times \vec{B}$.

Although this is not a very intuitive statement of how one goes about finding the direction of a magnetic field, we are forced to use it if we want to use $\vec{F}^{\text{mag}} = q\vec{v} \times \vec{B}$ to determine the magnitude and direction of the magnetic force on a moving charged particle.

Using the mathematical definition of a cross product to evaluate this expression, we see that we can write the magnitude of the magnetic force as

$$F^{\text{mag}} = |q\vec{v}||\vec{B}|\sin\phi = |q|vB\sin\phi, \qquad (29\text{-}3)$$

where ϕ is the smaller angle (the one whose value lies between 0° and 180°) between the directions of velocity $\vec{v}$ and magnetic field $\vec{B}$.

We have seen that magnetic force and electric force are not the same. However, a magnetic force *is* exerted on a moving charged particle as well as on bar magnets. This suggests that there is a profound connection between electricity and magnetism—even though they are *not* the same thing. As it turns out, the theory of relativity, treated in Chapter 38, reveals a deep underlying connection between $\vec{E}$ and $\vec{B}$. Furthermore, much of the technology that makes our lives more comfortable today results from an understanding of this relationship. In Chapter 30, we show how moving electrical charges can create magnetic fields and in Chapter 31 we show an even deeper and more surprising link between electricity and magnetism (called Faraday's law). What we find is that a magnetic field can, if it changes in time, create an electric field without any electric charge present!

Finding the Magnetic Force on a Moving Charged Particle

Equation 29-3 reveals that the magnitude of the force $\vec{F}^{\text{mag}}$ acting on a particle in a magnetic field is proportional to the amount of charge $|q|$ and speed v of the particle. Thus, the force is equal to zero if the charge is zero or if the particle is stationary. Equation 29-3 also tells us that the magnitude of the force is zero if $\vec{v}$ and $\vec{B}$ are either parallel ($\phi = 0°$) or antiparallel ($\phi = 180°$), and the force is a maximum when $\vec{v}$ and $\vec{B}$ are perpendicular to each other.

Equation 29-2 tells us all this and the direction of $\vec{F}^{\text{mag}}$. From Section 12-4, we know that the cross product $\vec{v} \times \vec{B}$ in Eq. 29-2 is a vector that is perpendicular to the two vectors $\vec{v}$ and $\vec{B}$. The right-hand rule (Fig. 29-4*a*) specifies that the thumb of the right hand points in the direction of $\vec{v} \times \vec{B}$ when the fingers sweep $\vec{v}$ into $\vec{B}$. If q is positive, then (by Eq. 29-2) the force $\vec{F}^{\text{mag}}$ has the same sign as $\vec{v} \times \vec{B}$ and thus must be in the same direction. That is, for positive q, $\vec{F}^{\text{mag}}$ is directed along the thumb as in Fig. 29-4*b*. If q is negative, then the force $\vec{F}^{\text{mag}}$ and the cross product $\vec{v} \times \vec{B}$ have opposite signs and thus must be in opposite directions. For negative q, $\vec{F}^{\text{mag}}$ is directed opposite the thumb as in Fig. 29-4*c*.

Regardless of the sign of the charge, however,

> The force $\vec{F}^{\text{mag}}$ acting on a charged particle moving with velocity $\vec{v}$ through a magnetic field $\vec{B}$ is *always* perpendicular to $\vec{v}$ and $\vec{B}$.

Thus, $\vec{F}^{\text{mag}}$ *never* has a component parallel to $\vec{v}$. This means that $\vec{F}^{\text{mag}}$ cannot change the particle's speed $v = |\vec{v}|$ (and thus it cannot change the particle's kinetic energy). The force can change only the direction of $\vec{v}$ (and thus the direction of travel); only in this sense can $\vec{F}^{\text{mag}}$ accelerate the particle. If there are no other forces acting on the charged particle and the velocity of the particle is perpendicular to the direction of the magnetic field, this means that the particle will move in a circle. If the particle has a component perpendicular to the magnetic field *and* a component of velocity parallel

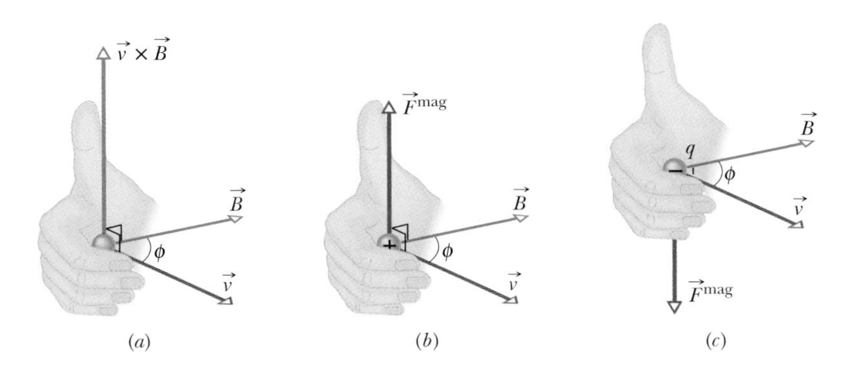

FIGURE 29-4 ■ (*a*) The right-hand rule (in which $\vec{v}$ is swept into $\vec{B}$ through the smaller angle ϕ between them) gives the direction of $\vec{v} \times \vec{B}$ as the direction of the thumb. (*b*) If q is positive, then the direction of $\vec{F}^{\text{mag}} = q\vec{v} \times \vec{B}$ is in the direction of $\vec{v} \times \vec{B}$. (*c*) If q is negative, then the direction of $\vec{F}^{\text{mag}}$ is opposite that of $\vec{v} \times \vec{B}$.

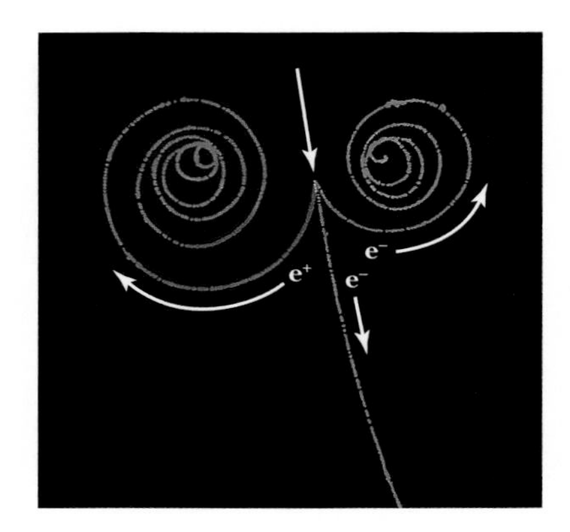

FIGURE 29-5 ■ Color enhanced tracks showing two electrons (e^-) and a positron (e^+) in a bubble chamber that is immersed in a uniform magnetic field that is directed out of the plane of the page.

to the magnetic field, the particle will move along a *helix* of constant radius. These paths are discussed in more detail in Section 29-5.

To develop a feeling for the relationship between the magnetic force on a moving charged particle and the magnetic field, $\vec{F}^{\,\text{mag}} = q\vec{v} \times \vec{B}$, consider Fig. 29-5. This figure shows some tracks left by charged particles moving rapidly through a *bubble chamber* at the Lawrence Berkeley Laboratory. The chamber, which is filled with liquid hydrogen, is immersed in a strong uniform magnetic field that is directed out of the plane of the figure. An incoming gamma ray particle—which leaves no track because it is uncharged—transforms into an electron (spiral track marked e^-) and a positron (track marked e^+) while it knocks an electron out of a hydrogen atom (long track marked e^-). At first these newly created charged particles are moving in the same direction as the gamma ray. As they move, they each experience a magnetic force of magnitude $F^{\text{mag}} = |q|vB$ and begin to move in a circular path given by $F^{\text{mag}} = mv^2/r$. Since $qvB = mv^2/r$, a particle has a path of radius $r = mv/|q|B$. You can use Eq. 29-2 and Fig. 29-4 to confirm that the three tracks made by these two negative particles and one positive particle curve in the proper directions. It is interesting to note that the electrons and positron do not move in a pure circle. Instead, they move in a shrinking spiral because they are slowed down through their interaction with the gas in the bubble chamber. This makes sense because $r = mv/|q|B$ and as each particle's speed, v, becomes smaller, so does its radius r. When this happens, the magnetic force, which is proportional to the particle's velocity, decreases and so the radius of the particle's path decreases.

What Produces a Magnetic Field?

We have discussed how a charged plastic rod produces a vector field—the electric field $\vec{E}$—at all points in the space around it. Similarly, a magnet produces a vector field—the **magnetic field** $\vec{B}$—at all points in the space around it. You get a hint of that magnetic field whenever you attach a note to a refrigerator door with a small magnet, or accidentally erase a computer disk by bringing it near a strong magnet. The magnet acts on the door or disk *by means of* its magnetic field.

In a common type of magnet, a wire coil is wound around an iron core and a current is sent through the coil; the strength of the magnetic field is determined by the size of the current. In industry, such **electromagnets** are used for sorting scrap metal (Fig. 29-1) among many other things. You are probably more familiar with **permanent magnets**—magnets, like the refrigerator-door type, that do not need current to have a magnetic field.

How then are magnetic fields set up? We know about two ways to create magnetic fields. (1) We observe that moving electrically charged particles, such as the current in a wire or charged beams of cosmic rays create magnetic fields. (2) We find that elementary particles such as protons, neutrons, and electrons have *intrinsic* magnetic moments that create magnetic fields. In Chapter 30 we discuss how moving charges create magnetic fields, and in Chapter 32 we consider the role of intrinsic magnetic moments in the creation of magnetic fields. In this chapter we stay focused on how to represent magnetic fields and how they influence charged particles that are moving.

The SI unit for $\vec{B}$ that follows from Eqs. 29-2 and 29-3 is the newton per coulomb-meter per second. For convenience, the SI unit for magnetic field is called the tesla (T):

$$1 \text{ tesla} = 1 \text{ T} = 1 \frac{\text{newton}}{(\text{coulomb})(\text{meter/second})}.$$

Recalling that a coulomb per second is an ampere, we have

$$1 \text{ T} = 1 \frac{\text{newton}}{(\text{coulomb/second})(\text{meter})} = 1 \frac{\text{N}}{\text{A} \cdot \text{m}}. \tag{29-4}$$

TABLE 29-1
Some Approximate Magnetic Fields

At the surface of a neutron star	10^8 T
Near a big electromagnet	1.5 T
Near a small bar magnet	10^{-2} T
At Earth's surface	10^{-4} T
In interstellar space	10^{-10} T
Smallest value in a magnetically shielded room	10^{-14} T

An earlier (non-SI) unit for $\vec{B}$, that is still in common use is the *gauss* (G), and

$$1 \text{ tesla} = 10^4 \text{ gauss}. \tag{29-5}$$

Table 29-1 lists the magnetic fields that occur in a few situations. Note that Earth's magnetic field near the planet's surface is about 10^{-4} T (= 100 μT or 1 gauss).

READING EXERCISE 29-1: The figure shows three situations in which a charged particle with velocity $\vec{v}$ travels through a uniform magnetic field $\vec{B}$. In each situation, what is the direction of the magnetic force $\vec{F}^{\,mag}$ on the particle?

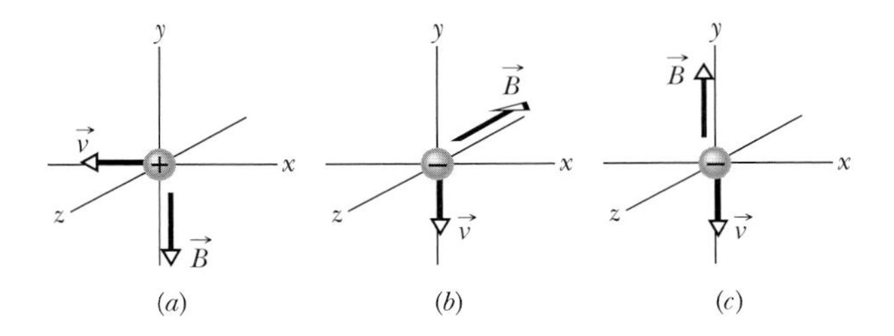

Magnetic Field Lines

We can represent magnetic fields with field lines, as we did for electric fields. Similar rules apply; that is, (1) the direction of the tangent to a magnetic field line at any point gives the direction of $\vec{B}$ at that point, and (2) the spacing of the lines represents the magnitude of $\vec{B}$—the magnetic field is stronger where the lines are closer together, and conversely.

Figure 29-6*a* shows how the magnetic field near a *bar magnet* (a permanent magnet in the shape of a bar) can be represented by magnetic field lines. The lines all pass through the magnet, and they all form closed and continuous loops (even those that are not shown closed in the figure). They don't start or end anywhere. Since electric field lines begin and end on electric charges, this is consistent with our assumption that there are no magnetic charges (monopoles). As shown with field lines, the external magnetic effects of a bar magnet are strongest near its ends, where the field lines are most closely spaced. Thus, the magnetic field of the bar magnet in Fig. 29-6*b* collects the iron filings mainly near the two ends of the magnet. Overall, outside of the bar magnet the field lines look just like they would for an electric dipole, but inside the magnet they point in the opposite direction.

The (closed) field lines enter one end of a magnet and exit the other end. The end of a magnet from which the field lines emerge is called the *north pole* of the magnet; the other end, where field lines enter the magnet, is called the *south pole*. (Remember that the direction of the field line is related to the direction of the force on a moving positively charged particle.) Some of the magnets we use to fix notes on refrigerators

FIGURE 29-6 ■ (*a*) The magnetic field lines for a bar magnet. (*b*) A "cow magnet"—a bar magnet that is intended to be slipped down into the rumen (first stomach) of a cow to prevent accidentally ingested bits of scrap iron from reaching the cow's intestines. The iron filings at its ends reveal the directions of the magnetic field lines in the vicinity of the magnet.

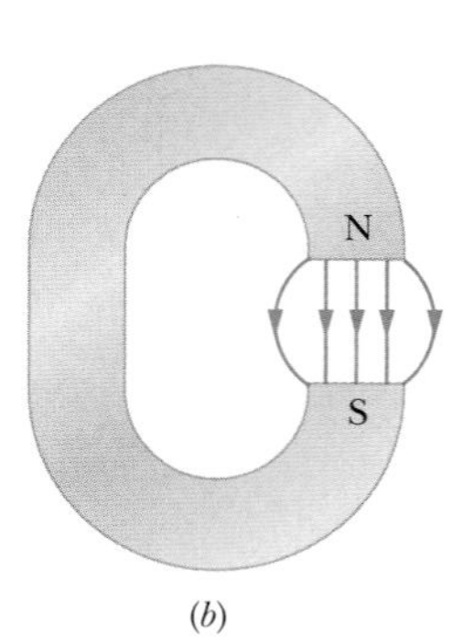

FIGURE 29-7 ■ (*a*) A horseshoe magnet and (*b*) a C-shaped magnet. (Only a few of the possible of the external field lines are shown.)

are short bar magnets. Figure 29-7 shows two other common shapes for magnets: a *horseshoe magnet* and a magnet that has been bent around into the shape of a C so that the *pole faces* are facing each other. (The magnetic field between the pole faces can then be approximately uniform.) Regardless of the shape of the magnets, if we place two of them near each other we find:

> Opposite magnetic poles attract each other, and like magnetic poles repel each other.

Earth has a magnetic field that is produced in its core. We discuss current theories about the nature and origin of the Earth's magnetic field in Section 32-9. On Earth's surface, we can detect this magnetic field with a compass, which is essentially a slender bar magnet on a low-friction pivot. This bar magnet, or this needle, turns because its north pole end is attracted toward the Arctic region, or North Pole, of Earth. Thus, the *south* pole of Earth's magnetic field must be located toward the North Pole. Logically, we then should call the pole there a south pole. However, because we call that direction north, we are trapped into the statement that Earth has a *geomagnetic north pole* in that direction.

With more careful measurement we would find that in the northern hemisphere, the magnetic field lines of Earth generally point down into Earth and toward the Arctic. In the southern hemisphere, they generally point up out of Earth and away from the Antarctic—that is, away from Earth's *geomagnetic south pole*.

TOUCHSTONE EXAMPLE 29-1: Proton in a Magnetic Field

A uniform magnetic field $\vec{B}$, with magnitude 1.2 mT, is directed vertically upward throughout the volume of a laboratory chamber. A proton with kinetic energy 5.3 MeV enters the chamber, moving horizontally from south to north. What is the magnitude of the magnetic deflecting force acting on the proton as it enters the chamber? The proton mass is 1.67×10^{-27} kg. (Neglect Earth's magnetic field.)

SOLUTION ■ Because the proton is charged and moving through a magnetic field, a magnetic force $\vec{F}^{\text{mag}}$ can act on it. The **Key Idea** here is that, because the initial direction of the proton's velocity is not along a magnetic field line, $\vec{F}^{\text{mag}}$ is not simply zero. To find the magnitude of $\vec{F}^{\text{mag}}$, we can use Eq. 29-3 provided we first find the proton's speed $|\vec{v}| = v$. We can find v from the given kinetic energy, since $K = \frac{1}{2}mv^2$. Solving for $|\vec{v}|$, we find

$$v = \sqrt{\frac{2K}{m}} = \sqrt{\frac{(2)(5.3\ \text{MeV})(1.60 \times 10^{-13}\ \text{J/MeV})}{1.67 \times 10^{-27}\ \text{kg}}}$$

$$= 3.2 \times 10^{7}\ \text{m/s}.$$

Equation 29-3 then yields

$$\begin{aligned} F^{\text{mag}} &= |q|vB_{\sin}\phi \\ &= (1.60 \times 10^{-19}\ \text{C})(3.2 \times 10^{7}\ \text{m/s}) \\ &\quad \times (1.2 \times 10^{-3}\ \text{T})(\sin 90°) \\ &= 6.1 \times 10^{-15}\ \text{N}. \end{aligned}$$ (Answer)

This may seem like a small force, but it acts on a particle of small mass, producing a large magnitude of acceleration; namely,

$$a = \frac{F^{mag}}{m} = \frac{6.1 \times 10^{-15}\text{ N}}{1.67 \times 10^{-27}\text{ kg}} = 3.7 \times 10^{12}\text{ m/s}^2.$$

To find the direction of $\vec{F}^{mag}$, we use the **Key Idea** that $\vec{F}^{mag}$ has the direction of the cross product $q\vec{v} \times \vec{B}$. Because the charge q is positive, $\vec{F}^{mag}$ must have the same direction as $\vec{v} \times \vec{B}$, which can be determined with the right-hand rule for cross products (as in Fig. 29-4*b*). We know that $\vec{v}$ is directed horizontally from south to north and $\vec{B}$ is directed vertically up. The right-hand rule shows us that the deflecting force $\vec{F}^{mag}$ must be directed horizontally from west to east, as Fig. 29-8 shows. (The array of dots in the figure represents a magnetic field directed out of the plane of the figure. An array of Xs would have represented a magnetic field directed into that plane.)

If the charge of the particle were negative, the magnetic deflecting force would be directed in the opposite direction—that is, horizontally from east to west. This is predicted automatically by Eq. 29-2, if we substitute a negative value for q.

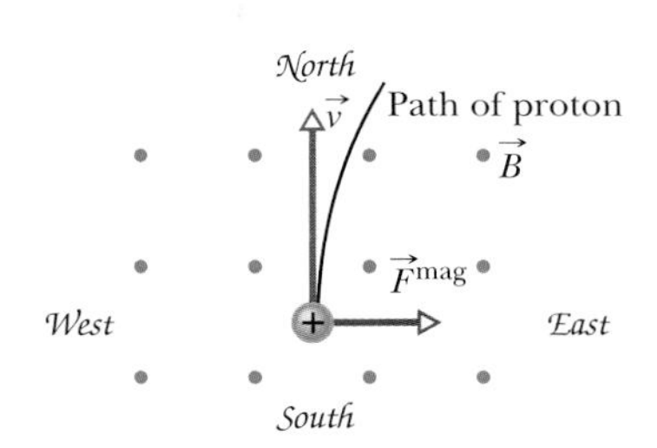

FIGURE 29-8 ■ An overhead view of a proton moving from south to north with velocity $\vec{v}$ in a chamber. A magnetic field is directed vertically upward in the chamber, as represented by the array of dots (which resemble the tips of arrows). The proton is deflected toward the east.

29-5 A Circulating Charged Particle

Remember that when we studied projectile motion we found that the (vertical) gravitational acceleration had no effect on the horizontal velocity of the projectile. Furthermore, when we studied uniform circular motion, we found that the (radial) centripetal acceleration only changed the direction of the object's velocity (keeping it moving in a circle), but did not speed it up or slow it down. This is a general relationship: The component of acceleration that is perpendicular to the direction of velocity only changes the direction of the velocity, not the magnitude.

We have a similar situation here. If we have a charged particle whose size is small enough to ignore, the magnetic force the particle feels is always perpendicular to its velocity and not its magnitude. As we established earlier, if the velocity and magnetic field are perpendicular (and there are no other forces on the particle), the particle will move in a circle.

If a particle moves in a circle at constant speed, we can be sure that the net force acting on the particle is constant in magnitude and is centripetal. That is, the force points toward the center of the circle, always perpendicular to the particle's velocity. Think of a stone tied to a string and whirled in a circle on a smooth horizontal surface, or of a satellite moving in a circular orbit around the Earth. In the first case, the tension in the string provides the necessary force and centripetal acceleration. In the second case, Earth's gravitational attraction provides the force and acceleration.

Figure 29-9 shows another example of a centripetal magnetic force: A beam of electrons is projected into a chamber by an *electron gun* G. The electrons enter in the plane of the page with speed v and move in a region of uniform magnetic field $\vec{B}$ directed out of the plane of the figure. As a result, a magnetic force $\vec{F}^{mag} = q\vec{v} \times \vec{B}$ continually deflects the electrons, and because the particle's velocity, $\vec{v}$, and the magnetic field it passes through, $\vec{B}$, are always perpendicular to each other, this deflection causes the electrons to follow a circular path. The path is visible in the photo because atoms of gas in the chamber emit light when some of the circulating electrons collide with them.

We would like to determine the parameters that characterize the circular motion of these electrons, or of any particle having an amount of charge $|q|$ and mass m moving perpendicular to a uniform magnetic field $\vec{B}$ at speed v. From Eq. 29-3, the force acting on the particle has a magnitude of $|q|vB$. From Newton's Second Law $(\vec{F} = m\vec{a})$ applied to uniform circular motion (Eq. 5-34),

$$F^{mag} = m\frac{v^2}{r}, \tag{29-6}$$

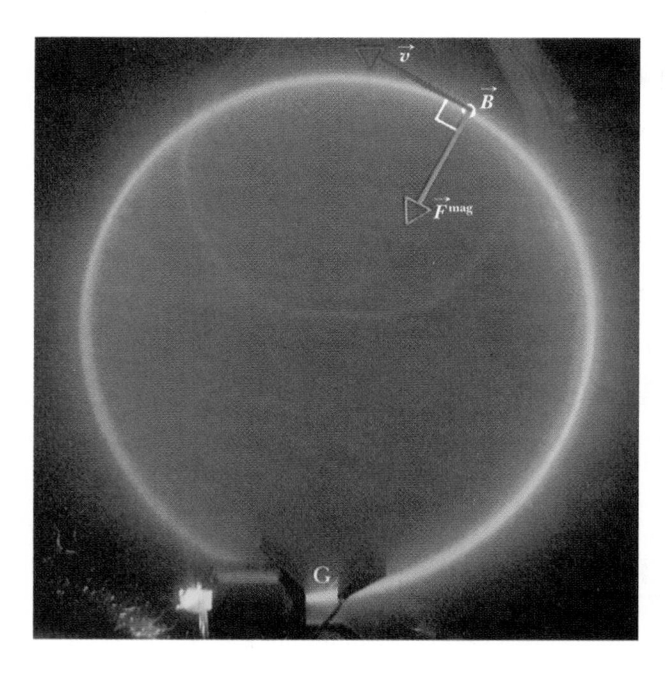

FIGURE 29-9 ■ Electrons circulating in a chamber containing gas at low pressure (their path is the glowing circle). A uniform magnetic field $\vec{B}$, pointing directly out of the plane of the page, fills the chamber. Note the radially directed magnetic force $\vec{F}^{mag}$; for circular motion to occur, $\vec{F}^{mag}$ must point toward the center of the circle. Use the right-hand rule for cross products to confirm that $\vec{F}^{mag} = q\vec{v} \times \vec{B}$ gives $\vec{F}^{mag}$ the proper direction. (Don't forget to incude the sign of q.)

we have

$$|q|vB = \frac{mv^2}{r}. \tag{29-7}$$

Solving for r, we find the radius of the circular path as

$$r = \frac{mv}{|q|B} \quad \text{(radius of circular path).} \tag{29-8}$$

The period T (the time for one full revolution) is equal to the circumference divided by the speed:

$$T = \frac{2\pi r}{v} = \frac{2\pi}{v}\frac{mv}{|q|B} = \frac{2\pi m}{|q|B} \quad \text{(period).} \tag{29-9}$$

The frequency f (the number of revolutions per unit time) is

$$f = \frac{1}{T} = \frac{|q|B}{2\pi m} \quad \text{(frequency).} \tag{29-10}$$

The angular frequency ω of the motion is then

$$\omega = 2\pi f = \frac{|q|B}{m} \quad \text{(angular or cyclotron frequency).} \tag{29-11}$$

The quantities T, f, and ω do not depend on the speed of the particle (provided that speed is much less than the speed of light). Fast particles move in large circles and slow ones in small circles, but all particles with the same charge-to-mass ratio q/m take the same time T (the period) to complete one round trip. A bigger velocity makes the particle travel in a larger circle. The increase in speed is exactly compensated by the increase in distance, so the time it takes to go around the circle is the same. We see later that this plays an important role in the construction of a charged particle accelerator known as a **cyclotron.** Using Eq. 29-2, you can show that if you are looking in the direction of $\vec{B}$, the direction of rotation for a positive particle is always counterclockwise; the direction for a negative particle is always clockwise.

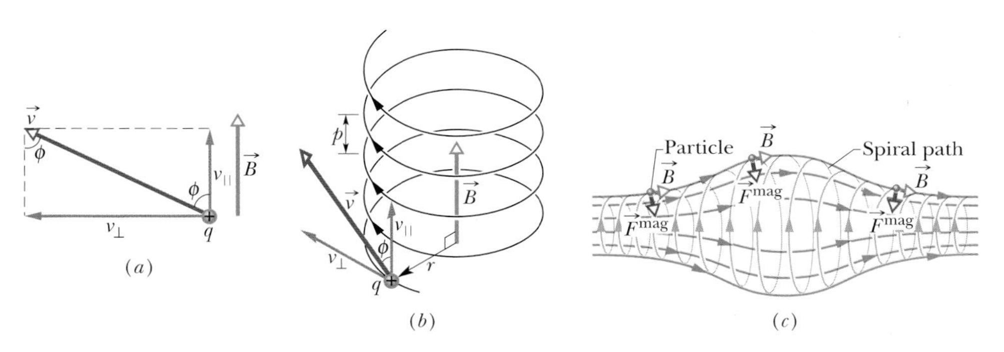

FIGURE 29-10 ■ (a) A charged particle moves in a uniform magnetic field $\vec{B}$, its velocity $\vec{v}$ making an angle ϕ with the field direction. (b) The particle follows a helical path, of radius r and pitch p. (c) A charged particle spiraling in a nonuniform magnetic field. (The particle can become trapped, spiraling back and forth between the strong field regions at either end.) Note that the magnetic force vectors at the left and right sides have a component pointing toward the center of the figure.

Helical Paths

As we discussed in regard to the electrons and positron in the bubble chamber of Figure 29-5, if the velocity of a charged particle moving through a magnetic field is changing, the particle will move in a shrinking spiral, rather than a circle. One way this can happen is for the particle to be slowed by frictional or other forces. Furthermore, if the velocity of a charged particle has a component parallel to the (uniform) magnetic field, the particle will move in a helical path about the direction of the field vector. Figure 29-10a, for example, shows the velocity vector $\vec{v}$ of such a particle resolved into two components, one parallel to $\vec{B}$ and one perpendicular to it:

$$v_{\parallel} = |\vec{v}|\cos\phi \quad \text{and} \quad v_{\perp} = |\vec{v}|\sin\phi. \tag{29-12}$$

The parallel component determines the *pitch* p of the helix—that is, the distance between adjacent turns (Fig. 29-10b). The perpendicular component determines the radius of the helix and is the quantity to be substituted for $|\vec{v}|$ in Eq. 29-8.

Figure 29-10c shows a charged particle spiraling in a nonuniform magnetic field. The more closely spaced field lines at the left and right sides indicate that the magnetic field is stronger there. When the field at an end is strong enough, the particle "reflects" from that end. If the particle reflects from both ends, it is said to be trapped in a *magnetic bottle*.

Confining Particles in a Tokomak Reactor

In the chapter opener we explained that in order to induce fusion reactions capable of releasing large amounts of energy, we must fuse light atoms together. To do this we need to confine ions having very high energy, and hence high temperature. Magnetic fields are ideal for containing the ions because both the ions and the electrons are charged and will spiral along magnetic field lines instead of hitting the walls of a containment vessel.

Scientists have not yet been able to confine charged particles at high enough temperatures to achieve controlled fusion. However, experiments reveal that one of the most effective configurations of magnetic field lines for containing the light atomic ions is shaped like a torus. A torus is basically a donut shape. The containment vessel of the Joint European Torus, commonly known as a tokomak, is shown at the beginning of this chapter. In a tokomak reactor, the magnetic field is produced by a series of magnetic coils that are evenly spaced around the torus-shaped containment vessel as shown in Fig. 29-11. The magnetic field lines form continuous loops inside the ring of the torus. In theory, when a tokomak is working properly, the high temperature ions and electrons should revolve in helical paths around the field lines. An ion can then travel in a continuous loop until it undergoes a fusion reaction with another ion.

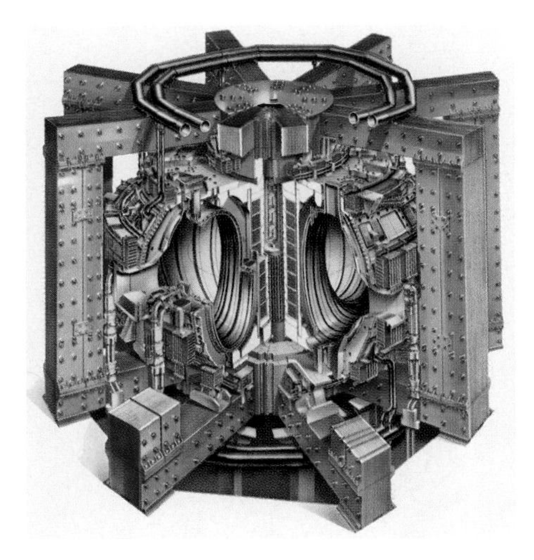

FIGURE 29-11 ■ A cutaway drawing of the JET tokomak showing the donut shaped containment vessel and surrounding magnetic coils.

Particles Trapped in the Earth's Magnetic Field

The terrestrial magnetic field acts as a magnetic bottle, trapping electrons and protons; the trapped particles form the *Van Allen radiation belts,* which loop well above the Earth's atmosphere between Earth's north and south geomagnetic poles. These particles bounce back and forth, from one end of this magnetic bottle to the other, within a few seconds.

When a large solar flare shoots additional energetic electrons and protons into the radiation belts, an electric field is produced in the region where electrons normally reflect. This field eliminates the reflection and instead drives electrons down into the atmosphere, where they collide with atoms and molecules of air, causing that air to emit light. This light forms the aurora—a curtain of light that hangs down to an altitude of about 100 km. Green light is emitted by oxygen atoms, and pink light is emitted by nitrogen molecules, but often the light is so dim that we perceive only white light.

READING EXERCISE 29-2: The figure shows the circular paths of two particles that travel at the same speed in a uniform magnetic field $\vec{B}$, which is directed into the page. One particle is a proton; the other is an electron (which is less massive). The relative sizes of the circles are not to scale. (a) Which particle follows the smaller circle, and (b) does that particle travel clockwise or counterclockwise? ■

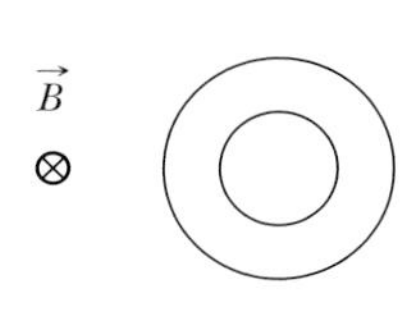

TOUCHSTONE EXAMPLE 29-2: Mass Spectrometer

Figure 29-12 shows the essentials of a *mass spectrometer,* which can be used to measure the mass of an ion; an ion of mass m (to be measured) and charge q is produced in source S. The initially stationary ion is accelerated by the electric field due to a potential difference ΔV. The ion leaves S and enters a separator chamber in which a uniform magnetic field $\vec{B}$ is perpendicular to the path of the ion. The magnetic field causes the ion to move in a semicircle, striking (and thus altering) a photographic plate at distance x from the entry slit. Suppose that in a certain trial $B = 80.000$ mT and $\Delta V = 1000.0$ V, and ions of charge $q = +1.6022 \times 10^{-19}$ C strike the plate at $x = 1.6254$ m. What is the mass m of the individual ions, in unified atomic mass units ($1 \text{ u} = 1.6605 \times 10^{-27}$ kg)?

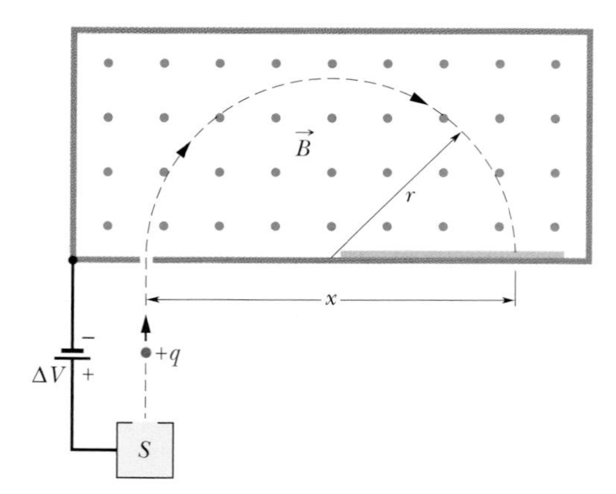

FIGURE 29-12 ■ Essentials of an early model of a mass spectrometer. A positive ion, after being accelerated from its source S by potential difference ΔV, enters a chamber of uniform magnetic field $\vec{B}$. There it travels through a semicircle of radius r and strikes a photographic plate at a distance x from where it entered the chamber.

SOLUTION ■ One **Key Idea** here is that, because the (uniform) magnetic field causes the (charged) ion to follow a circular path, we can relate the ion's mass m to the path's radius r with Eq. 29-8 ($r = m|\vec{v}|/|q\vec{B}|$). From Fig. 29-12 we see that $r = x/2$, and we are given the magnitude $|\vec{B}|$ of the magnetic field. However, we don't know the ion's speed v in the magnetic field, after it has been accelerated due to the potential difference ΔV.

To relate v and ΔV, we use the **Key Idea** that mechanical energy ($E^{\text{mec}} = K + U$) of the mass spectrometer system is conserved during the acceleration. When the ion emerges from the source, its kinetic energy is approximately zero. At the end of the acceleration, its kinetic energy is $\frac{1}{2}mv^2$. Also, during the acceleration, the positive ion moves through a change in potential of $-\Delta V$. Thus, because the ion has positive charge q, its potential energy changes by $-q\Delta V$. If we now write the conservation of the system's mechanical energy as

$$\Delta K + \Delta U = 0,$$

we get

$$\tfrac{1}{2}mv^2 - q\Delta V = 0$$

or

$$v = \sqrt{\frac{2|q\Delta V|}{m}}. \tag{29-13}$$

Substituting this into Eq. 29-8 gives us

$$r = \frac{mv}{|q\vec{B}|} = \frac{m}{|q|B}\sqrt{\frac{2|q\Delta V|}{m}} = \frac{1}{B}\sqrt{\frac{2m|\Delta V|}{|q|}}.$$

Thus,

$$x = 2r = \frac{2}{B}\sqrt{\frac{2m|\Delta V|}{|q|}}.$$

Solving this for m and substituting the given data yield

$$m = \frac{B^2|q|x^2}{8|\Delta V|}$$

$$= \frac{(0.080000\text{ T})^2(1.6022 \times 10^{-19}\text{ C})(1.6254\text{ m})^2}{8(1000.0\text{ V})}$$

$$= 3.3863 \times 10^{-25}\text{ kg} = 203.93\text{ u}. \quad \text{(Answer)}$$

29-6 Crossed Fields: Discovery of the Electron

As we have seen, both an electric field $\vec{E}$ and a magnetic field $\vec{B}$ can produce a force on a charged particle. When the two fields are perpendicular to each other, they are said to be *crossed fields*. Here we shall examine what happens to charged particles—namely, electrons—as they move through crossed fields. We use as our example the experiment that led to the discovery of the electron in 1897 by J. J. Thomson at Cambridge University.

Figure 29-13 shows a modern, simplified version of Thomson's experimental apparatus—a *cathode ray tube* (which is like the picture tube in a standard television set). Charged particles (which we now know as electrons) are emitted by a hot filament at the rear of the evacuated tube and are accelerated by an applied potential difference ΔV. After the electrons pass through a slit in screen C, they form a narrow beam. They then pass through a region of crossed $\vec{E}$ and $\vec{B}$ fields, headed toward a fluorescent screen S, where they produce a spot of light (on a television screen the spot is part of the picture). The forces on the charged particles in the crossed-fields region can deflect them from the center of the screen. By controlling the magnitudes and directions of the fields, Thomson could thus control where the spot of light appeared on the screen. Recall that the force on a negatively charged particle due to an electric field is directed opposite the field. Thus, for the particular field arrangement of Fig. 29-13, electrons are forced up the page by the electric field $\vec{E}$ and down the page by the magnetic field $\vec{B}$; that is, the forces are *in opposition*. Thomson's procedure was equivalent to the following series of steps:

1. Set $\vec{E} = 0$ N/C and $\vec{B} = 0$ T and note the position of the spot on screen S due to the undeflected beam.
2. Turn on $\vec{E}$ and measure the resulting beam deflection.
3. Maintaining $\vec{E}$, now turn on $\vec{B}$ and adjust its value until the beam returns to the undeflected position. (With the forces in opposition, they can be made to cancel.)

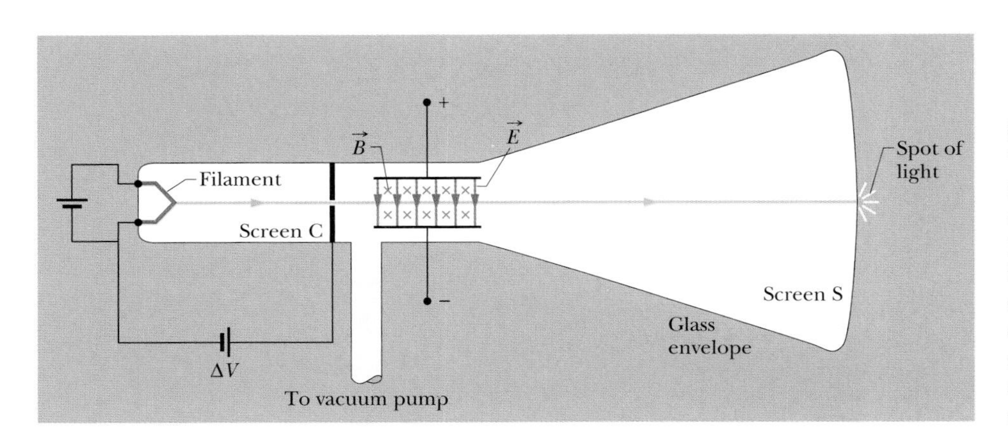

FIGURE 29-13 ■ A modern version of J. J. Thomson's apparatus for measuring the ratio of mass to the amount of charge of an electron. The electric field $\vec{E}$ is established by connecting a battery across the deflecting-plate terminals. The magnetic field $\vec{B}$ is set up by means of a current in a system of coils (not shown). The magnetic field shown is into the plane of the figure, as represented by the array of Xs (which resemble the feathered ends of arrows).

We discussed the deflection of a charged particle moving perpendicular to an electric field $\vec{E}$ between two plates (step 2 here) in Touchstone Example 23-4. We found that the magnitude of the deflection of the particle at the far end of the plates is

$$|\Delta y| = \frac{|q|EL^2}{2mv^2}, \tag{29-14}$$

where v is the particle's initial speed (which was v_x in Touchstone Example 23-4), m its mass, and q its charge, and L is the length of the plates. So long as the particle's deflection is small, we can apply this same equation to the beam of electrons in Fig. 29-13; if necessary, we can calculate the deflection by measuring the deflection of the beam on screen S and then working back to calculate the deflection y at the end of the plates. (Because the direction of the deflection is set by the sign of the particle's charge, Thomson was able to show that the particles lighting up his screen were negatively charged.)

When the two fields in Fig. 29-13 are adjusted so that the two deflecting forces cancel (step 3), we have from Eqs. 29-1 and 29-3,

$$|q|E = |q|vB\sin(90°) = |q|vB,$$

so the particle speed v is given by the ratio of the field magnitudes

$$v = \frac{E}{B}. \tag{29-15}$$

Thus, the crossed fields allow us to measure the speed of the charged particles passing through them. Substituting Eq. 29-15 for $|\vec{v}|$ in Eq. 29-14 and rearranging yield

$$\frac{m}{|q|} = \frac{B^2L^2}{2|\Delta y|E}, \tag{29-16}$$

in which all quantities on the right can be measured. Thus, the crossed fields allow us to measure the mass-charge amount ratio $m/|q|$ of the particles moving through Thomson's apparatus.

Thomson claimed that these particles are found in all matter. He also claimed that they are lighter than the lightest known atom (hydrogen) by a factor of more than 1000. (The exact ratio proved later to be 1836.15.) His $m/|q|$ measurement, coupled with the boldness of his two claims, is considered to be the moment of "discovery of the electron."

READING EXERCISE 29-3: The figure shows four directions for the velocity vector $\vec{v}$ of a positively charged particle moving through a uniform electric field $\vec{E}$ (directed out of the page and represented by an encircled dot) and a uniform magnetic field $\vec{B}$ (pointing to the left). (a) Rank directions 1, 2, 3, and 4 according to the magnitude of the net force on the particle, greatest first. (b) Of all four directions, which might result in a net force of zero?

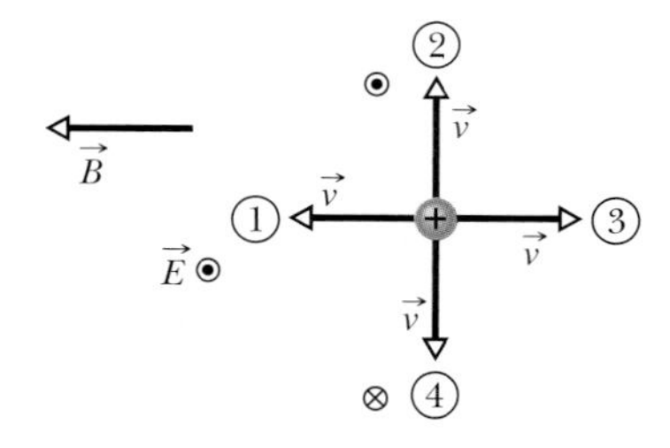

29-7 The Hall Effect

In Chapters 22 and 26 we claimed that currents in solid conductors are due to moving electrons, and that the positive nuclei are at rest. What evidence do we have for this claim? In the late 1870s, Edwin H. Hall, a 24-year-old graduate student at the

Johns Hopkins University, investigated the deflection of electric current passing through copper wire when the wire is placed in a magnetic field. The result of his work, which is called the **Hall effect** after him, allows us to answer important questions about the nature of charge carriers. For example, Hall's findings allowed him to determine whether charge carriers in a conductor are positive or negative. In addition, Hall's measurements enabled him to deduce the number of charge carriers per unit volume contained in a given conductor.

What happens to a current-carrying metal wire in a magnetic field if the charge carriers are positive and negative charges are at rest? Figure 29-14*a* shows a copper strip of width d, carrying a current i that is assumed to be made up of positive charge carriers (the convention at the time) moving from the top of the figure to the bottom. The charge carriers drift (with an average speed $|\langle\vec{v}\rangle|$) in the direction of the current, from top to bottom. At the instant shown in Fig. 29-14*a*, an external magnetic field $\vec{B}$, pointing into the plane of the figure, has just been turned on. From Eq. 29-2 we see that a deflecting magnetic force $\vec{F}^{\text{mag}}$ will act on each drifting positive charge, pushing it toward the right edge of the strip.

As time goes on, positive charges pile up on the right edge of the strip, leaving uncompensated negative charges in fixed positions at the left edge. The separation of positive and negative charges produces a constant electric field $\vec{E}$ within the strip, pointing from right to left. This field exerts an average electrostatic force $\langle\vec{F}^{\text{elec}}\rangle$ on a typical positive charge, tending to push it back toward the left.

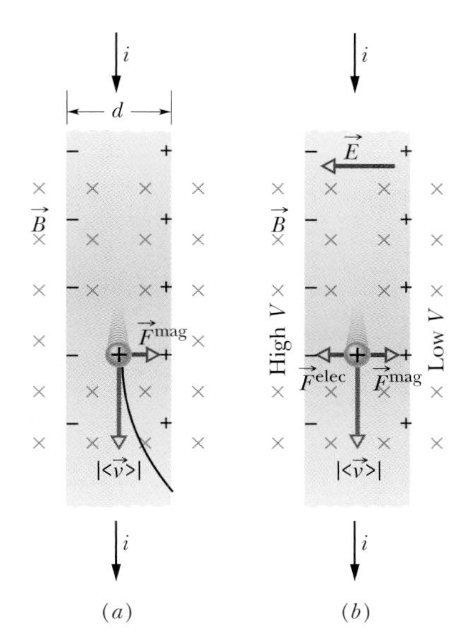

FIGURE 29-14 ■ What would happen if a positive current were to flow through a strip of copper immersed in a magnetic field $\vec{B}$? (*a*) As soon as the magnetic field is turned on, the positive charges follow a curved path as shown. (*b*) A short time later positive charges pile up on the right side of the strip. Thus, the right side of the strip has a higher potential than the left side. Since the higher potential is observed on the left not the right, we conclude that the *charge carriers are not positive.*

An equilibrium quickly develops in which the electric force on each positive charge (pushing left) builds up until it just cancels the magnetic force (pushing right). When this happens, as Fig. 29-14*b* shows, the force due to $\vec{B}$ and the force due to $\vec{E}$ are in balance. The drifting positive charges then move along the strip toward the bottom of the page at an average velocity $\langle\vec{v}\rangle$, with no further collection of positive charge on the right edge of the strip and thus no further increase in the electric field $\vec{E}$.

A *Hall potential difference* ΔV is associated with the electric field across strip width d. Because the field is constant, we use Eq. 25-39 to get

$$|\Delta V| = Ed. \tag{29-17}$$

By connecting a voltmeter across the width, we can measure the potential difference between the two edges of the strip. Moreover, the voltmeter can tell us which edge is at higher potential. This information, in turn, tells us whether our charge carriers are positive or negative.

So what do we find? For the situation of Fig. 29-14*a*, we find that the *left* edge is at *higher* potential, meaning we have a buildup of positive charge there. This result is inconsistent with our assumption that the charge carriers are positive.

Suppose we make the opposite assumption, that the charge carriers in current i are negative, as shown in Fig. 29-15. The negative charge carriers drift (with an average speed $|\langle\vec{v}\rangle|$) in the *opposite* direction of the conventional current, from bottom to top. You can use the magnetic force law (Eq. 29-2) to convince yourself that as these charge carriers move from bottom to top in the strip, they are pushed to the right edge by $\vec{F}^{\text{mag}}$ and thus that the *left* edge is at higher potential. Because that last statement is in fact what we actually observe with a voltmeter, we conclude that the charge carriers must be negative.

Now for the quantitative part. When the electric and magnetic forces are in balance (Fig. 29-14*b*), Eqs. 29-1 and 29-3 give us a relationship between the magnitudes of the electric and magnetic fields:

$$eE = e|\langle\vec{v}\rangle|B, \tag{29-18}$$

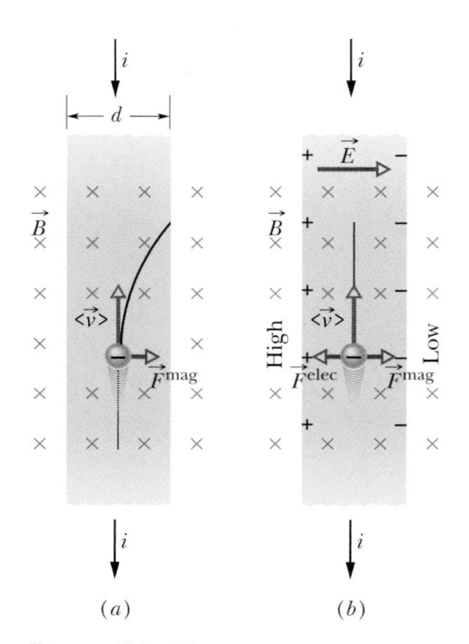

FIGURE 29-15 ■ What should happen when the conventional current, i, actually consists of a negative electron current flowing in the opposite direction? (a) As soon as the magnetic field is turned on, electrons follow the curved path shown. (b) A short time later negative charges pile up on the right side of the strip so that a higher potential develops on the left. Since this prediction matches experimental findings we must conclude that the *charge carriers are negative*.

where e is the amount of the charge on the electron. From Eq. 26-21, the average or drift speed $|\langle\vec{v}\rangle|$ is

$$|\langle\vec{v}\rangle| = \frac{|\vec{J}|}{ne} = \frac{|i|}{neA}, \tag{29-19}$$

in which $|\vec{J}|(=|i|/A)$ is the current density in the strip, A is the cross-sectional area of the strip, and n is the *number density* of charge carriers (their number per unit volume).

In Eq. 29-18, substituting $|\Delta V|/d$ for E (Eq. 29-17) and substituting for $|\langle\vec{v}\rangle|$ with the rightmost term in Eq. 29-19, we obtain

$$n = \frac{|i|B}{e\ell|\Delta V|}, \tag{29-20}$$

in which $\ell = A/d$ is the thickness of the strip. With this equation we can find n from measurable quantities.

It is also possible to use the Hall effect to measure directly the average or drift speed $|\langle\vec{v}\rangle|$ of the charge carriers, which you may recall is of the order of centimeters per hour. In this clever experiment, the metal strip is moved mechanically through the magnetic field in a direction opposite that of the drift velocity of the charge carriers. The speed of the moving strip is then adjusted until the Hall potential difference vanishes. At this condition, with no Hall effect, the velocity of the charge carriers *with respect to the laboratory frame* must be zero, so the velocity of the strip must be equal in magnitude but opposite in direction to the velocity of the negative charge carriers.

TOUCHSTONE EXAMPLE 29-3: Motional Potential Difference

Figure 29-16 shows a solid metal cube, of edge length $d = 1.5$ cm, moving in the positive y direction at a constant velocity $\vec{v}$ of magnitude 4.0 m/s. The cube moves through a uniform magnetic field $\vec{B}$ of magnitude 0.050 T directed toward positive z.

(a) Which cube face is at a lower electric potential and which is at a higher electric potential because of the motion through the field?

SOLUTION ■ One **Key Idea** here is that, because the cube is moving through a magnetic field $\vec{B}$, a magnetic force $\vec{F}^{\,\text{mag}}$ acts on its charged particles, including its conduction electrons. A second **Key Idea** is how $\vec{F}^{\,\text{mag}}$ causes an electric potential difference between certain faces of the cube. When the cube first begins to move through the magnetic field, its electrons do also. Because each electron has charge $q = -e$ and is moving through a magnetic field with velocity $\vec{v}$, the magnetic force $\vec{F}^{\,\text{mag}}$ acting on it is given by Eq. 29-2. Because q is negative, the direction of $\vec{F}^{\,\text{mag}}$ is opposite the cross product $\vec{v} \times \vec{B}$, which is in the positive direction of the x axis in Fig. 29-16. Thus, $\vec{F}^{\,\text{mag}}$ acts in the negative direction of the x axis, toward the left face of the cube (which is hidden from view in Fig. 29-16).

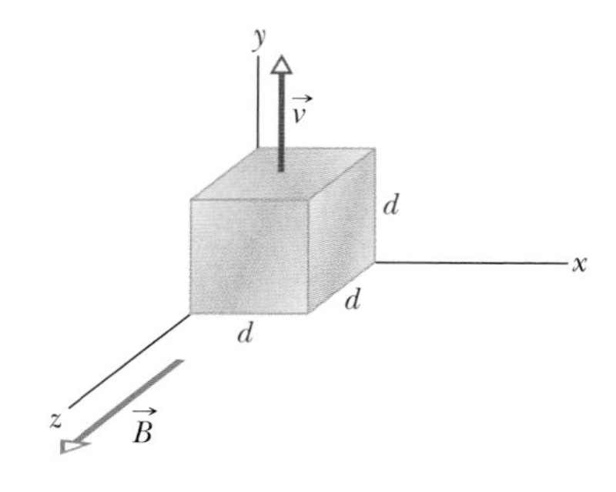

FIGURE 29-16 ■ A solid metal cube of edge length d moves at constant velocity $\vec{v}$ through a uniform magnetic field $\vec{B}$.

Most of the electrons are fixed in place in the molecules of the cube. However, because the cube is a metal, it contains conduction electrons that are free to move. Some of those conduction electrons are deflected by $\vec{F}^{\,\text{mag}}$ to the left cube face, making that face negatively charged and leaving the right face positively charged. This charge separation produces an electric field $\vec{E}$ directed from the positively charged right face to the negatively charged left face. Thus, the left face is at a lower electric potential, and the right face is at a higher electric potential.

(b) What is the potential difference between the faces of higher and lower electric potential?

SOLUTION ■ The **Key Ideas** here are these:

1. The electric field $\vec{E}$ created by the charge separation produces an electric force $\vec{F}^{\,\text{elec}} = q\vec{E}$ on each electron. Because q is

negative, this force is directed opposite the field $\vec{E}$—that is, toward the right. Thus on each electron, $\vec{F}^{\,elec}$ acts toward the right and $\vec{F}^{\,mag}$ acts toward the left.

2. When the cube had just begun to move through the magnetic field and the charge separation had just begun, the magnitude of $\vec{E}$ began to increase from zero. Thus, the magnitude of $\vec{F}^{\,elec}$ also began to increase from zero and was initially smaller than the magnitude $\vec{F}^{\,mag}$. During this early stage, the net force on any electron was dominated by $\vec{F}^{\,mag}$, which continuously moved additional electrons to the left cube face, increasing the charge separation.

3. However, as the charge separation increased, eventually magnitude $|\vec{F}^{\,elec}|$ became equal to magnitude $|\vec{F}^{\,mag}|$. The net force on any electron was then zero, and no additional electrons were moved to the left cube face. Thus, the magnitude of $\vec{F}^{\,elec}$ could not increase further, and the electrons were then in equilibrium.

We seek the potential difference ΔV between the left and right cube faces after equilibrium was reached (which occurred quickly). We can obtain the magnitude of ΔV with Eq. 29-17 ($|\Delta V| = Ed$) provided we first find the magnitude $|\vec{E}| = E$ of the electric field at equilibrium. We can do so with the equation for the balance of force magnitudes ($|\vec{F}^{\,elec}| = |\vec{F}^{\,mag}|$).

For F^{elec}, we substitute $|q|E$. For F^{mag}, we substitute $|q|vB \sin\phi$ from Eq. 29-3. From Fig. 29-16, we see that the angle ϕ between v and B is 90°; so $\sin\phi = 1$. We can now write ($F^{elec} = F^{mag}$) as

$$|q|E = |q|vB\sin 90° = |q|vB.$$

This gives us $E = vB$, so Eq. 29-17 ($|\Delta V| = Ed$) becomes

$$|\Delta V| = |V_{left} - V_{right}| = vBd. \tag{29-21}$$

Substituting known values gives us

$$|\Delta V| = (4.0\text{ m/s})(0.050\text{ T})(0.015\text{ m})$$
$$= 0.0030\text{ V} = 3.0\text{ mV}.$$

Since the left face of the cube has excess negative charges, the right face is at a higher potential than the left face by 3.0 mV. (Answer)

29-8 Magnetic Force on a Current-Carrying Wire

We have just seen that a magnetic field exerts a sideways force on electrons moving in a wire. This force must then be transmitted to the wire itself, because the conduction electrons cannot escape sideways out of the wire.

In Fig. 29-17*a*, a vertical wire, carrying no current and fixed in place at both ends, extends through the gap between the vertical pole faces of a magnet represented by the shaded circle. The magnetic field between the faces is directed outward from the page. In Fig. 29-17*b*, a current is sent upward through the wire; the wire deflects to the right. In Fig. 29-17*c*, we reverse the direction of the current and the wire deflects to the left.

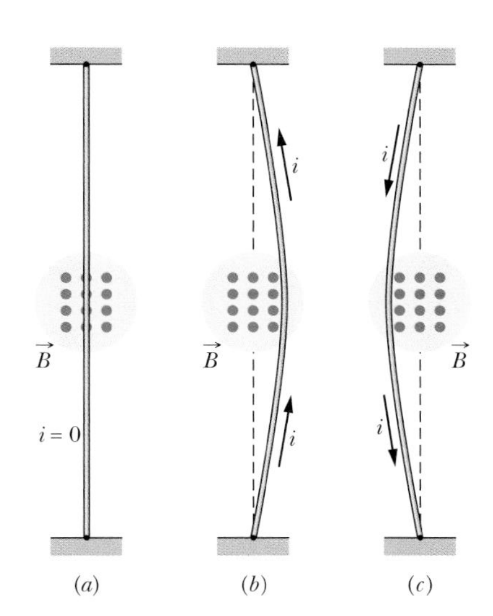

FIGURE 29-17 ■ A flexible wire passes between the pole faces of a magnet (only the farther pole face is shown). (*a*) Without current in the wire, the wire is straight. (*b*) With upward current, the wire is deflected rightward. (*c*) With downward current, the deflection is leftward. Connections for getting the current into one end of the wire and out of the other are not shown.

Figure 29-18 shows what happens inside the wire of Fig. 29-17. We see one of the conduction electrons, drifting downward with an assumed average (drift) speed $|\langle\vec{v}\rangle|$. Equation 29-3, in which we must put $\phi = 90°$, tells us that a force of magnitude $F^{mag} = e|\langle\vec{v}\rangle|B$ must act on a typical electron. From Eq. 29-2 we see that this force must be directed to the right. We expect then that the wire as a whole will experience a force to the right, in agreement with Fig. 29-17*b*.

If, in Fig. 29-18, we were to reverse *either* the direction of the magnetic field *or* the direction of the current, the force on the wire would reverse, being directed now to the left. Note too that it does not matter whether we consider negative charges drifting downward in the wire (the actual case) or positive charges drifting upward. The direction of the deflecting force on the wire is the same. We are safe then in dealing with a current of positive charge.

Consider a length L of the wire in Fig. 29-18. All the conduction electrons in this section of wire will drift past a plane that is parallel to xx' (shown in Fig. 29-18) in a time $\Delta t = L/|\langle\vec{v}\rangle|$. Thus, in that time the charge that will pass through the plane is given by

$$q = i\Delta t = i\frac{L}{|\langle\vec{v}\rangle|}.$$

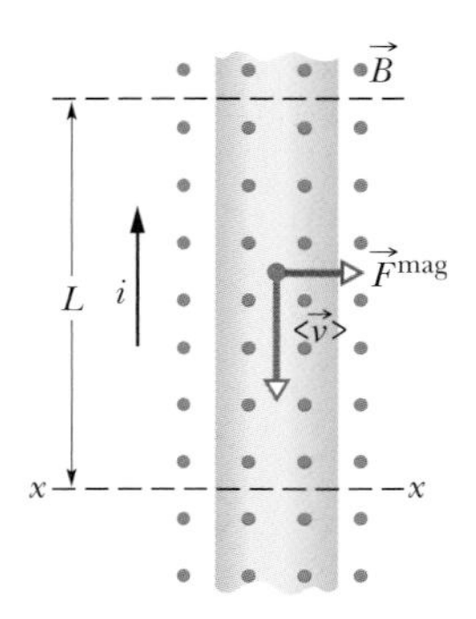

FIGURE 29-18 ■ A close-up view of a section of the wire of Fig. 29-17*b*. The current direction is upward, which means that electrons drift downward. A magnetic field that emerges from the plane of the page causes the electrons and the wire to be deflected to the right.

Substituting this into Eq. 29-3 yields the following expressions for the magnitude of the magnetic force

$$F^{mag} = |q||\langle\vec{v}\rangle|B\sin\phi = \frac{|i|L|\langle\vec{v}\rangle|B}{|\langle\vec{v}\rangle|}\sin 90^\circ$$

or

$$F^{mag} = |i|LB. \tag{29-22}$$

This equation gives the magnetic force that acts on a length L of straight wire carrying a current i and immersed in a magnetic field $\vec{B}$ that is perpendicular to the wire.

If the magnetic field is *not* perpendicular to the wire, as in Fig. 29-19, the magnetic force is given by a generalization of Eq. 29-22:

$$\vec{F}^{mag} = |i|\vec{L}\times\vec{B} \quad \text{(force on a current).} \tag{29-23}$$

Here $\vec{L}$ is a *length vector* that has magnitude $|\vec{L}|$ and is directed along the wire segment in the direction of the (conventional) current. The magnitude of the magnetic field is

$$F^{mag} = |i|LB\sin\phi, \tag{29-24}$$

where ϕ is the smaller angle between the directions of $\vec{L}$ and $\vec{B}$. The direction of $\vec{F}^{mag}$ is that of the cross product $\vec{L}\times\vec{B}$, because we take current i to be a positive quantity. Equation 29-23 tells us that $\vec{F}^{mag}$ is always perpendicular to the plane defined by $\vec{L}$ and $\vec{B}$, as indicated in Fig. 29-19.

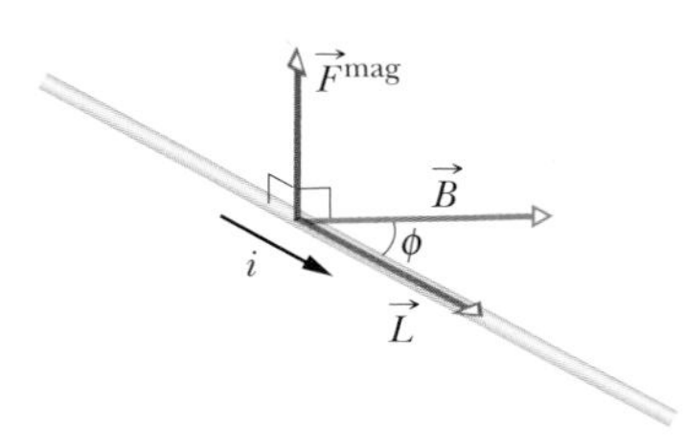

FIGURE 29-19 ■ A wire carrying current i makes an angle ϕ with magnetic field $\vec{B}$. The wire has length L in the field and length vector $\vec{L}$ (in the direction of the current). A magnetic force $\vec{F}^{mag} = i\vec{L}\times\vec{B}$ acts on the wire.

Equation 29-23 is equivalent to Eq. 29-2 in that either can be taken as the defining equation for $\vec{B}$. In practice, we define $\vec{B}$ from Eq. 29-23. It is much easier to measure the magnetic force acting on a wire than that on a single moving charge.

If a wire is not straight or the field is not uniform, we can imagine it broken up into small straight segments and apply Eq. 29-23 to each short segment $d\vec{L}$. The force on the wire as a whole is then the vector sum of all the forces on the segments that make it up. In the differential limit, we can write

$$d\vec{F}^{mag} = i\,d\vec{L}\times\vec{B}, \tag{29-25}$$

and we can find the resultant force on any given arrangement of currents by integrating Eq. 29-25 over that arrangement.

In using Eq. 29-25, bear in mind that there is no such thing as an isolated current-carrying wire segment of length $d\vec{L}$. There must always be a way to introduce the current into the segment at one end and take it out at the other end.

READING EXERCISE 29-4: The figure shows a current i through a wire in a uniform magnetic field $\vec{B}$, as well as the magnetic force $\vec{F}^{mag}$ acting on the wire. The field is oriented so that the magnitude force is a maximum. In what direction is the field?

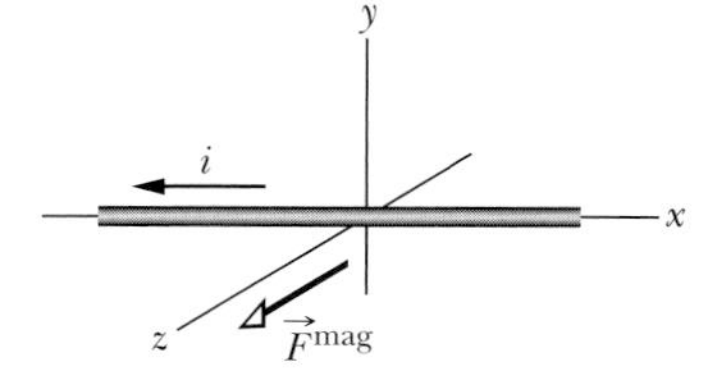

■

TOUCHSTONE EXAMPLE 29-4: Levitating a Wire

A straight, horizontal length of copper wire has a current $i = 28$ A through it. If this current is directed out of the page as shown in Fig. 29-20, what are the magnitude and direction of the minimum magnetic field $\vec{B}$ needed to suspend the wire—that is, to balance the gravitational force on it? The linear density (mass per unit length) is 46.6 g/m.

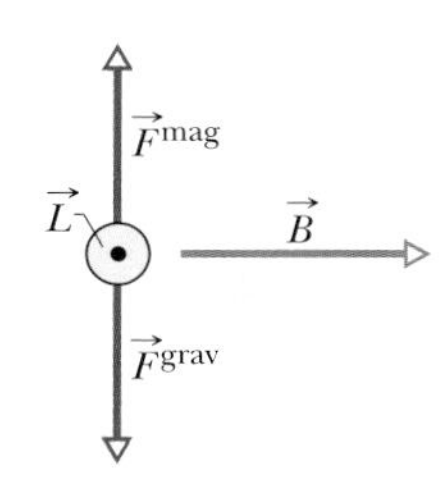

FIGURE 29-20 ■ A current-carrying wire (shown in cross section) can be made to "float" in a magnetic field. The current in the wire emerges from the plane of the page, and the magnetic field is directed to the right.

SOLUTION ■ One **Key Idea** is that, because the wire carries a current, a magnetic force $\vec{F}^{\,mag}$ can act on the wire if we place it in a magnetic field $\vec{B}$. To balance the downward gravitational force $\vec{F}^{\,grav}$ on the wire, we want $\vec{F}^{\,mag}$ to be directed upward (Fig. 29-20).

A second **Key Idea** is that the direction of $\vec{F}^{\,mag}$ is related to the directions of $\vec{B}$ and the wire's length vector $\vec{L}$ by Eq. 29-23. Because $\vec{L}$ is directed horizontally (and the current is taken to be positive), Eq. 29-23 and the right-hand rule for cross products tell us that $\vec{B}$ must be horizontal and rightward (in Fig. 29-20) to give the required upward $\vec{F}^{\,mag}$.

The magnitude of $\vec{F}^{\,mag}$ is given by Eq. 29-24 ($|\vec{F}^{\,mag}| = |i\vec{L}||\vec{B}|\sin\phi$). Because we want $\vec{F}^{\,mag}$ to balance $\vec{F}^{\,grav}$, we want

$$|i|LB\sin\phi = mg, \qquad (29\text{-}26)$$

where mg is the magnitude of $\vec{F}^{\,grav}$ and m is the mass of the wire. We also want the minimal field magnitude B for $\vec{F}^{\,mag}$ to balance $\vec{F}^{\,grav}$. Thus, we need to maximize $\sin\phi$ in Eq. 29-26. To do so, we set $\phi = 90°$, thereby arranging for $\vec{B}$ to be perpendicular to the wire. We then have $\sin\phi = 1$, so Eq. 29-26 yields a magnetic field magnitude of

$$|\vec{B}| = B = \frac{mg}{|i|L\sin\phi} = \frac{(m/L)g}{|i|}. \qquad (29\text{-}27)$$

We write the result this way because we know m/L, the linear density of the wire. Substituting known data then gives us a magnitude of

$$B = \frac{(46.6 \times 10^{-3}\ \text{kg/m})(9.8\ \text{m/s}^2)}{28\ \text{A}}$$
$$= 1.6 \times 10^{-2}\ \text{T}. \qquad \text{(Answer)}$$

This is about 160 times the strength of Earth's magnetic field. As stated in the second paragraph of this solution, the right-hand rule tells us that B must point to the right.

29-9 Torque on a Current Loop

Much of the world's work is done by electric motors. The forces that do this work are magnetic. In principle a direct current motor can be constructed from a single loop of current-carrying wire that is immersed in a magnetic field and is attached to a battery. If the current were to flow through the loop in the same direction all the time, the magnetic field would push on this loop in one direction at one instant of time, but would reverse the direction of the force when the loop was rotated halfway around. We would get a vibration that would quickly damp out. We can, however, get a continuous rotation if we use a connection, called a commutator, that reverses the current direction when the loop has gone halfway around (Fig. 29-21). Then, the force will continue to push the loop in the same direction and the motor will spin. Although many essential details have been omitted, the figure does suggest how the action of a magnetic field on a current loop produces rotary motion. To understand how the dc motor works in detail, we need to understand how a magnetic field can cause a current-carrying wire loop to rotate by exerting a torque on it.

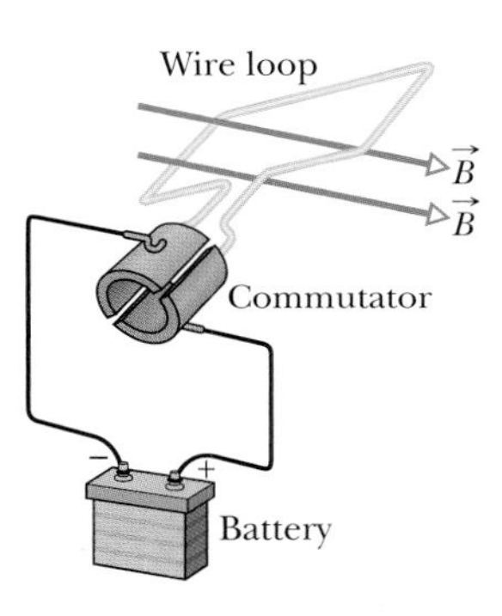

FIGURE 29-21 ■ The elements of an electric motor. A rectangular loop of wire, carrying a current and free to rotate about a fixed axis, is placed in a magnetic field. Magnetic forces on the wire produce a torque that rotates it. A commutator reverses the direction of the current every half-revolution so that the torque always acts in the same direction.

How a Current Loop Can Experience a Torque

Figure 29-22a shows a front view of a rectangular loop of sides a and b. The loop is carrying a current i and is immersed in a uniform magnetic field $\vec{B}$. We start our consideration of the torque on the loop with a special case in which the plane of the loop is parallel to the magnetic field as shown in Fig. 29-22a.

Let's use Eq. 29-24 to find the forces on each side of the loop for our special case. For sides 1 and 3 the vector $\vec{L}$ points in the direction of the current and has magnitude a. The angle between $\vec{L}$ and $\vec{B}$ for these is $\phi = 0°$. Thus, the magnitude of the forces acting on this side is

$$F_1^{mag} = F_3^{mag} = |i|aB\sin 0° = 0\ \text{N}.$$

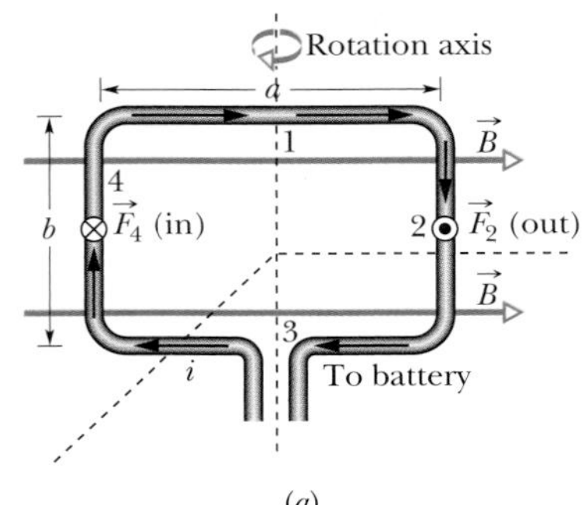

(a)
Front view (maximum torque)

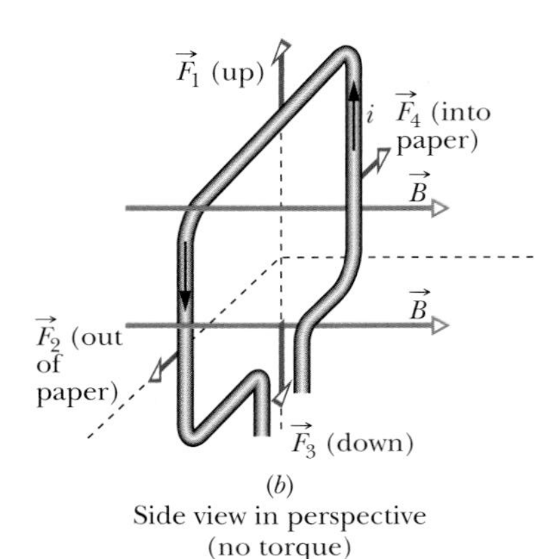

(b)
Side view in perspective (no torque)

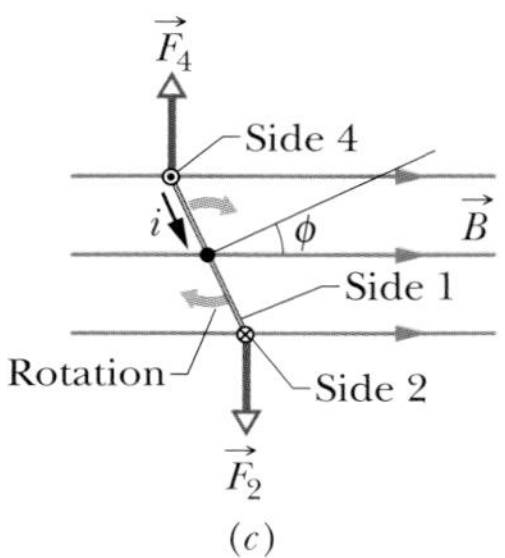

(c)
Top view (intermediate torque)

FIGURE 29-22 ■ A rectangular loop, of length a and width b and carrying a current i, is located in a uniform magnetic field. A torque $\vec{\tau}$ that is perpendicular to the magnetic field acts on the loop. The angle, ϕ, perpendicular or normal to the plane of the loop and the B-field varies. (*a*) The plane of the loop is aligned with the magnetic field so that $\phi = 90°$. (*b*) A perspective drawing of the loop after it has rotated to $\phi = 0°$ due to the torque exerted on it by the magnetic field. (*c*) A top view of the loop when it is part way between $\phi = 90°$ (part *a*) and $\phi = 0°$ (part *b*).

The situation is different for sides 2 and 4. For them, $\vec{L}$, which has magnitude b, is perpendicular to $\vec{B}$ so $\phi = 90°$. Thus, the forces $\vec{F}_2$ and $\vec{F}_4$ have the common magnitude given by

$$F_2^{\text{mag}} = F_4^{\text{mag}} = |i|bB\sin 90° = |i|bB. \tag{29-28}$$

However, since the direction of the current is different on each of these sides, the right-hand rule tells us that these two forces point in opposite directions. The vector $\vec{F}_2$ points out of the page while the vector $\vec{F}_4$ points into the page. However, as Fig. 29-22*a* shows, these two forces do *not* share the same line of action so they *do* produce a net torque. The torque tends to rotate the loop toward an orientation for which the plane of the loop is perpendicular to the direction of the magnetic field $\vec{B}$. At $\phi = 90°$ that torque has a moment arm of magnitude $a/2$ about the central axis of the loop. The magnitude of the torque due to forces $\vec{F}_2$ and $\vec{F}_4$ is then (see Fig. 29-22*a*),

$$\tau_{90} = \frac{a}{2}F_2 + \frac{a}{2}F_4 = |i|\frac{a}{2}(bB) + |i|\frac{a}{2}(bB) = |i|abB. \tag{29-29}$$

As the coil in Fig. 29-22*a* starts to rotate, the moment arm between sides 2 and 4 decreases, and it reaches zero when the loop is in the position shown in Fig. 29-22*b*. In general, the torque on the loop is given by

$$\tau' = |i|abB\sin\phi, \tag{29-30}$$

where ϕ is the smaller angle normal to the area subtended by the loop and the external magnetic field (Fig. 22-22*c*).

Suppose we replace the single loop of current with a *coil* of N loops, or *turns*. Further, suppose that the turns are wound tightly enough that they can be approximated as all having the same dimensions and lying in a plane. Then the turns form a *flat coil* and a torque $\vec{\tau}'$ with the magnitude found in Eq. 29-29 acts on each of the turns. The total torque on the coil then has magnitude

$$\tau = N\tau' = N|i|AB\sin\phi, \tag{29-31}$$

in which $A(= ab)$ is the area enclosed by the coil. Equation 29-31 holds for all flat coils, no matter what their shape, provided the magnetic field is uniform.

How a DC Motor Works

Consider the operation of a motor like that shown in Fig. 29-21. When the coil is at the point where the plane of the coil is perpendicular to the field direction so $\phi = 0°$, the polarity of the battery is suddenly reversed. Since the coil is accelerated by the initial torque on it, it sails past the point where $\phi = 0°$, and a new torque takes over and continues to rotate the coil in the same direction. This automatic reversal of the current occurs every half cycle and is accomplished with a commutator that electrically connects the rotating coil with the stationary contacts connected to the battery (or other power source).

29-10 The Magnetic Dipole Moment

We can describe the current-carrying coil of the preceding section with a single vector $\vec{\mu}$, its magnetic dipole moment. The direction of the magnetic dipole $\vec{\mu}$ is determined by another right hand rule similar to the one shown in Fig. 29-4. If you wrap your right hand around the coil in the direction of the positive current, your thumb points in the direction of the magnetic dipole $\vec{\mu}$.

We define the magnitude of $\mu = |\vec{\mu}|$ as

$$\mu = N|i|A \quad \text{(magnetic moment magnitude)}, \tag{29-32}$$

in which N is the number of turns in the coil, $|i|$ is the magnitude current through the coil, and A is the area enclosed by each turn of the coil. (Equation 29-32 tells us that the unit of $\vec{\mu}$ is the ampere-square meter.) Using $\vec{\mu}$, we can rewrite Eq. 29-31 for the magnitude of the torque on the coil due to a magnetic field as

$$\tau = |\vec{\mu}||\vec{B}|\sin\phi = \mu B \sin\phi, \tag{29-33}$$

in which ϕ is the smallest angle between the vectors $\vec{\mu}$ and $\vec{B}$.

We can generalize this to the vector relation

$$\vec{\tau} = \vec{\mu} \times \vec{B}, \tag{29-34}$$

which reminds us very much of the corresponding equation for the torque exerted by an *electric* field on an *electric* dipole—namely, Eq. 23-37:

$$\vec{\tau} = \vec{p} \times \vec{E}.$$

In each case the torque exerted by the external field—either magnetic or electric—is equal to the vector product of the corresponding dipole moment and the field vector.

A magnetic dipole in an external magnetic field has a **magnetic potential energy** that depends on the dipole's orientation in the field. For electric dipoles,

$$U(\theta) = -\vec{p} \cdot \vec{E}.$$

In strict analogy, we can write for the magnetic case

$$U(\theta) = -\vec{\mu} \cdot \vec{B}. \tag{29-35}$$

A magnetic dipole has its lowest energy ($-|\vec{\mu}||\vec{B}|\cos\theta = -\mu B$) when its dipole moment $\vec{\mu}$ is lined up with the magnetic field (Fig. 29-23). It has its highest energy ($-\mu B\cos 180° = +\mu B$) when the vector $\vec{\mu}$ is directed opposite the field.

When a magnetic dipole rotates in the presence of a magnetic field from an initial orientation θ_1 to another orientation θ_2, the work $W_{B\to\mu}$ done on the dipole by the magnetic field is

$$W_{B\to\mu} = -\Delta U = -(U_2 - U_1), \tag{29-36}$$

where U_2 and U_1 are calculated with Eq. 29-35. If an external torque acts on the dipole during the change in its orientation, then work $W_{\text{ext}\to\mu}$ is done on the dipole by the external torque. *If the dipole is stationary* before and after the change in its orientation, then work $W_{\text{ext}\to\mu}$ is the negative of the work done on the dipole by the field. Thus,

$$W_{\text{ext}\to\mu} = -W_{B\to\mu} = U_2 - U_1. \tag{29-37}$$

So far, we have identified only a current-carrying coil as a magnetic dipole. However, a simple bar magnet is also a magnetic dipole, as is a rotating sphere of charge. Earth itself is (approximately) a magnetic dipole. And, most subatomic particles, including the electron, the proton, and the neutron, have magnetic dipole moments. As you will see in Chapter 32, all these quantities can be viewed as current loops. For comparison, some approximate magnetic dipole moments are shown in Table 29-2.

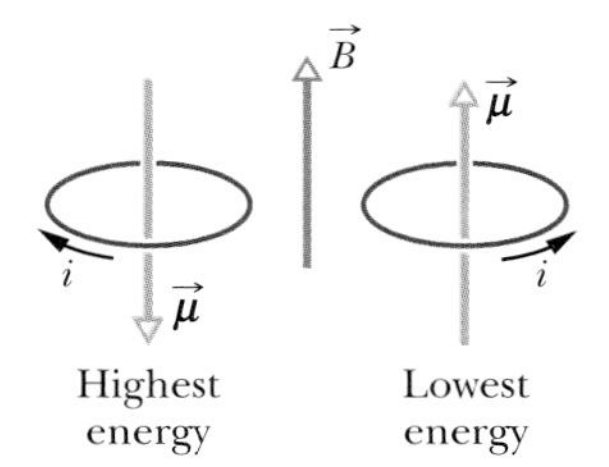

FIGURE 29-23 ■ The orientations of highest and lowest energy of a magnetic dipole in an external magnetic field $\vec{B}$. In each case, the direction of the current i determines the direction of the magnetic dipole moment $\vec{\mu}$ shown in Fig. 29-23 via the right-hand rule.

TABLE 29-2
Some Magnetic Dipole Moments

A small bar magnet	5 J/T
Earth	8.0×10^{22} J/T
A proton	1.4×10^{-26} J/T
An electron	9.3×10^{-24} J/T

READING EXERCISE 29-5: The figure shows four orientations, at angle θ, of a magnetic dipole moment $\vec{\mu}$ in a magnetic field. Rank the orientations according to (a) the magnitude of the torque on the dipole and (b) the potential energy of the dipole, greatest first.

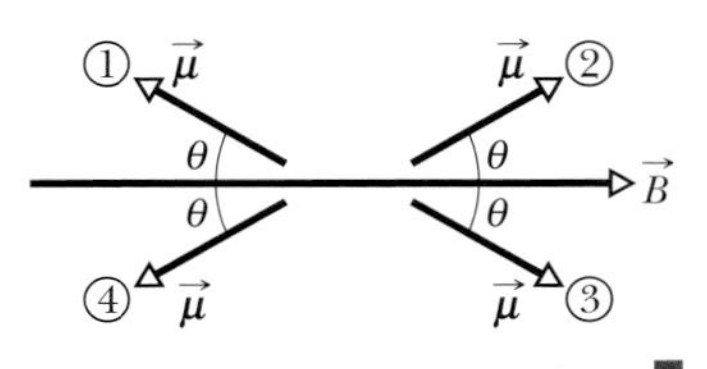

TOUCHSTONE EXAMPLE 29-5: Coil in an External Magnetic Field

Figure 29-24 shows a circular coil with 250 turns, an area A of $2.52 \times 10^{-4}\ \text{m}^2$, and a current of 100 μA. The coil is at rest in a uniform magnetic field of magnitude $|\vec{B}| = 0.85$ T, with its magnetic dipole moment $\vec{\mu}$ initially aligned with $\vec{B}$.

(a) In Fig. 29-24, what is the direction of the current in the coil?

SOLUTION ■ The **Key Idea** here is to apply the right-hand rule to the coil by curling your fingers around the current in the coil so your right thumb points in the $\vec{\mu}$ direction. Thus, in the wires on the near side of the coil—those we see in Fig. 29-24—the current is from top to bottom.

(b) How much work would the torque applied by an external agent have to do on the coil to rotate it 90° from its initial orientation, so that $\vec{\mu}$ is perpendicular to $\vec{B}$ and the coil is again at rest?

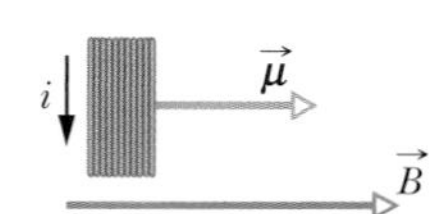

FIGURE 29-24 ■ A side view of a circular coil carrying a current and oriented so that its magnetic dipole moment $\vec{\mu}$ is aligned with magnetic field $\vec{B}$.

SOLUTION ■ The **Key Idea** here is that the work $W_{\text{ext}\to\mu}$ done by the applied torque would be equal to the change in the coil's potential energy due to its change in orientation. From Eq. 29-37 ($W_{\text{ext}\to\mu} = U_2 - U_1$), we find

$$\begin{aligned} W_{\text{ext}\to\mu} &= U(90°) - U(0°) \\ &= -\mu B\cos 90° - (-\mu B\cos 0°) = 0 + \mu B \\ &= \mu B. \end{aligned}$$

Substituting for $\vec{\mu}$ from Eq. 29-32 ($\mu = N|i|A$), we find that

$$\begin{aligned} W_{\text{ext}\to\mu} &= (N|i|AB) \\ &= (250)(100 \times 10^{-6}\ \text{A})(2.52 \times 10^{-4}\ \text{m}^2)(0.85\ \text{T}) \\ &= 5.356 \times 10^{-6}\ \text{J} \approx 5.4\ \mu\text{J}. \qquad \text{(Answer)} \end{aligned}$$

29-11 The Cyclotron

Physicists have been able to use their understanding of how charged particles behave in magnetic fields to develop devices that can accelerate protons to high speeds. These high-energy protons are extremely useful to scientists for several reasons. Collisions between energetic protons and matter allow them to learn about the nature of atomic and subatomic particles. High-energy protons and ions can also be used to create new radioactive elements. In addition, physicians can use high-energy protons to destroy tumors in cancer patients. In 1939, E. O. Lawrence was awarded a Nobel Prize in physics for the development of the cyclotron—the first of many magnetic accelerators capable of accelerating protons, ions, and electrons.

The principles that govern the operation of the cyclotron are quite simple. Figure 29-9 showed experimental evidence that a charged particle projected into an evacuated chamber perpendicular to a uniform magnetic field moves in a circular orbit. We used the magnetic force law (Eq. 29-2) to derive the frequency of revolution of the orbit. In Eq. 29-10 we found that $f = |q|B/2\pi m$. This is known as the cyclotron frequency, and its derivation had a rather surprising outcome. The frequency, f, with which a charged particle moves in its circular orbit depends only on its charge, its mass, and the magnetic field strength. So f is independent of speed. This is because a particle with low speed moves in a small circle whereas one with a higher speed moves in a larger circle. The particle speeds and orbital sizes are related in such a way that all charged particles take the same amount of time to make a revolution in a uniform magnetic field. (At least this is true for all speeds that are well below the speed of light.) Lawrence used the fact that the orbital frequency of a charged particle does not depend on its speed in the design of the cyclotron.

The original cyclotron was first used to accelerate protons. It consisted of two hollow semicircular disks shaped more or less like a capital D as shown in Fig. 29-25. In early cyclotrons, the dees, as they are called, were made of copper sheeting. The diameter of a dee was only about one meter. The dees were then placed in a vacuum chamber and oriented perpendicular to a large uniform magnetic field having a strength of a few teslas. There was a small gap between them. The dees were connected to an electrical oscillator that can alternate the potential difference across the gap between them at exactly the same frequency as an orbiting proton would have in the magnetic field. This arrangement is shown in Fig. 29-26.

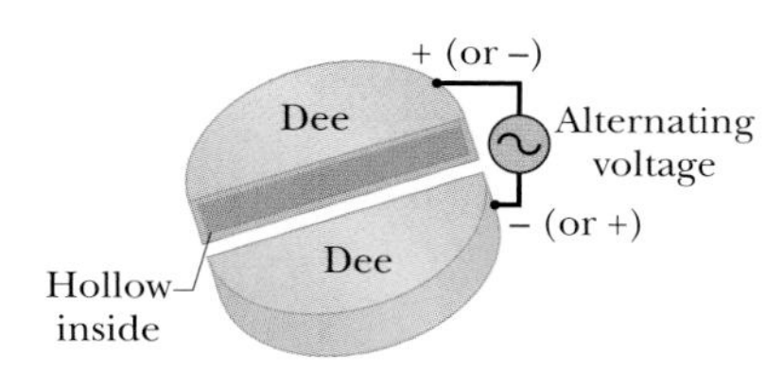

FIGURE 29-25 ■ Cyclotron dees are hollow semicircular metal containers that are open along their diameters.

To begin the operation of the original cyclotron, an oscillator was set at the cyclotron frequency. Then hydrogen gas was leaked into the vacuum chamber. Next a beam of high-energy electrons was injected into the center of the chamber so that other electrons were knocked out of hydrogen atoms. This ionization process produced protons. At a time when the oscillator caused the left dee to be at a lower potential than the right dee, the proton received a kick in the direction of the right dee. It moved into the right dee where the electric field was zero. However, the magnetic field penetrated the dee and caused the proton to start into a small, low-speed orbit. Only half a cycle later the proton reached the gap again. Since the oscillator was tuned to the cyclotron frequency, the potential of the right dee was now lower than that of the left dee and the proton got another kick as it crossed the gap. The proton then proceeded into another circular orbit that involved a larger speed and radius, given by Eq. 29-8,

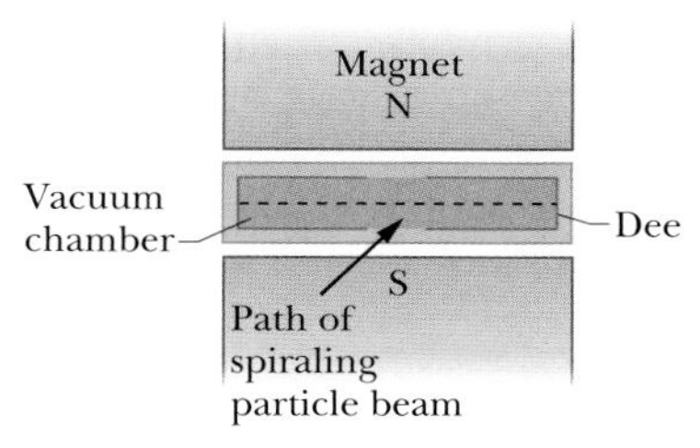

FIGURE 29-26 ■ Cutaway view of dees placed between the poles of a large electromagnet. The dotted line showns the plane in which the paths of the particles orbit.

$$r = \frac{mv}{|q|B}.$$

When the proton reached the gap again it completed one full cycle but so had the alternating voltage oscillator. Thus the proton got another kick. This process continued, with the circulating proton always being in step with the oscillations of the dee potential. When the proton finally spiraled out to the edge of the dee system, a deflector plate sent it out through a portal. The path of such a proton is shown in Fig. 29-27.

Recall that the key to the operation of the cyclotron is that the frequency f at which the proton circulates in the field (and that does not depend on its speed) must be equal to the fixed frequency f_{osc} of the electrical oscillator, or

$$f = f_{osc} \quad \text{(resonance condition).} \tag{29-38}$$

This *resonance condition* says that, if the energy of the circulating proton is to increase, energy must be fed to it at a frequency f_{osc} that is equal to the natural frequency f at which the proton circulates in the magnetic field.

Combining Eqs. 29-10 and 29-38 allows us to write the resonance condition as

$$|q|B = 2\pi m f_{osc}. \tag{29-39}$$

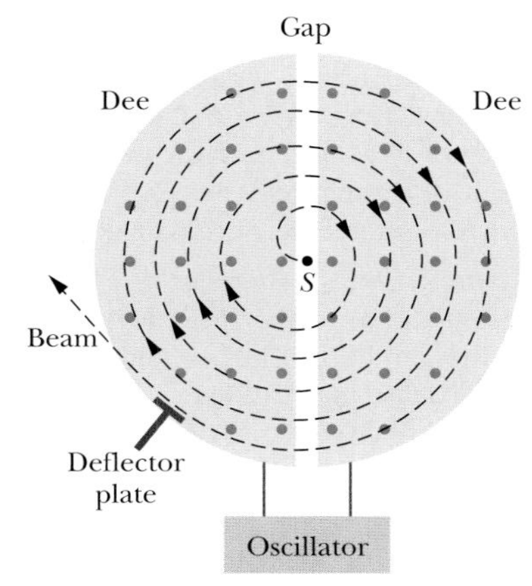

FIGURE 29-27 ■ Top view of dees showing the path of a charged particle beam in a cyclotron. Each time the particle passes through the gap three things happen: (1) the particle gets a kick and is accelerated to a higher speed, (2) the oscillator changes the sign of the gap's potential difference, and (3) the particle goes into a new semicircular orbit with a larger radius than before.

For the proton, q and m are fixed. The oscillator (we assume) is designed to work at a single fixed frequency f_{osc}. We can then either "tune" the cyclotron by varying either $\vec{B}$ or f_{osc} until Eq. 29-39 is satisfied. Then many protons can circulate through the magnetic field and emerge as a beam.

If the cyclotron is powerful enough to accelerate protons, electrons, or ions to speeds close to that of light, relativistic effects come into play. In such cases the simple resonance condition between orbital and oscillator frequencies no longer hold. More sophisticated magnetic field-based high-energy accelerators called synchrotrons and betatrons have been designed. We introduce relativistic effects in Chapter 38 where we discuss special relativity.

TOUCHSTONE EXAMPLE 29-6: Cyclotron

Suppose a cyclotron is operated at an oscillator frequency of 12 MHz and has a dee radius $R = 53$ cm.

(a) What is the magnitude of the magnetic field needed for deuterons to be accelerated in the cyclotron? A deuteron is the nucleus of deuterium, an isotope of hydrogen. It consists of a proton and a neutron and thus has the same charge as a proton. Its mass is $m = 3.34 \times 10^{-27}$ kg.

SOLUTION ■ The **Key Idea** here is that, for a given oscillator frequency f_{osc}, the magnetic field magnitude B required to accelerate any particle in a cyclotron depends on the ratio m/q of mass to charge for the particle, according to Eq. 29-39. For deuterons and the oscillator frequency $f_{osc} = 12$ MHz, we find

$$B = \frac{2\pi m f_{osc}}{q} = \frac{(2\pi)(3.34 \times 10^{-27}\ \text{kg})(12 \times 10^6\ \text{s}^{-1})}{1.60 \times 10^{-19}\ \text{C}}$$

$$= 1.57\ \text{T} \approx 1.6\ \text{T}. \qquad \text{(Answer)}$$

Note that, to accelerate protons, B would have to be reduced by a factor of 2, providing the oscillator frequency remained fixed at 12 MHz.

(b) What is the resulting kinetic energy of the deuterons?

SOLUTION ■ One **Key Idea** here is that the kinetic energy $\frac{1}{2}mv^2$ of a deuteron exiting the cyclotron is equal to the kinetic energy it had just before exiting, when it was traveling in a circular path with a radius approximately equal to the radius R of the cyclotron dees. A second **Key Idea** is that we can find the speed v of the deuteron in that circular path with Eq. 29-8 ($r = mv/|q|B$). Solving that equation for v, substituting R for r, and then substituting known data, we find

$$v = \frac{RqB}{m} = \frac{(0.53\ \text{m})(1.60 \times 10^{-19}\ \text{C})(1.57\ \text{T})}{3.34 \times 10^{-27}\ \text{kg}}$$

$$= 3.99 \times 10^7\ \text{m/s}.$$

This speed corresponds to a kinetic energy of

$$K = \tfrac{1}{2}mv^2$$

$$= \tfrac{1}{2}(3.34 \times 10^{-27}\ \text{kg})(3.99 \times 10^7\ \text{m/s})^2 \qquad \text{(Answer)}$$

$$= 2.7 \times 10^{-12}\ \text{J},$$

or about 17 MeV.

Problems

SEC. 29-3 ■ DEFINING A MAGNETIC FIELD $\vec{B}$

1. Alpha Particle An alpha particle travels at a velocity $\vec{v}$ of magnitude 550 m/s through a uniform magnetic field $\vec{B}$ of magnitude 0.045 T. (An alpha particle has a charge of $+3.2 \times 10^{-19}$ C and a mass of 6.6×10^{-27} kg.) The angle between $\vec{v}$ and $\vec{B}$ is 52°. What are the magnitudes of (a) the force $\vec{F}^{\,mag}$ acting on the particle due to the field and (b) the acceleration of the particle due to $\vec{F}^{\,mag}$? (c) Does the speed of the particle increase, decrease, or remain equal to 550 m/s?

2. TV Camera An electron in a TV camera tube is moving at 7.20×10^6 m/s in a magnetic field of strength 83.0 mT. (a) Without knowing the direction of the field, what can you say about the greatest and least magnitudes of the force acting on the electron due to the field? (b) At one point the electron has an acceleration of magnitude 4.90×10^{14} m/s^2. What is the angle between the electron's velocity and the magnetic field?

3. Proton Traveling A proton traveling at 23.0° with respect to the direction of a magnetic field of strength 2.60 mT experiences a magnetic force of 6.50×10^{-17} N. Calculate (a) the proton's speed and (b) its kinetic energy in electron-volts.

4. Force on Charges An electron that has velocity

$$\vec{v} = (2.0 \times 10^6\ \text{m/s})\hat{\text{i}} + (3.0 \times 10^6\ \text{m/s})\hat{\text{j}}$$

moves through the magnetic field $\vec{B} = (0.030\ \text{T})\hat{\text{i}} - (0.15\ \text{T})\hat{\text{j}}$. (a) Find the force on the electron. (b) Repeat your calculation for a proton having the same velocity.

5. Television Tube Each of the electrons in the beam of a television tube has a kinetic energy of 12.0 keV. The tube is oriented so that the electrons move horizontally from geomagnetic south to geomagnetic north. The vertical component of Earth's magnetic field points down and has a magnitude of 55.0 μT. (a) In what direction will the beam deflect? (b) What is the magnitude of the acceleration of a single electron due to the magnetic field? (c) How far will the beam deflect in moving 20.0 cm through the television tube?

SEC. 29-5 ■ A CIRCULATING CHARGED PARTICLE

6. Accelerated from Rest An electron is accelerated from rest by a potential difference of 350 V. It then enters a uniform magnetic field of magnitude 200 mT with its velocity perpendicular to the field. Calculate (a) the speed of the electron and (b) the radius of its path in the magnetic field.

7. Field Perpendicular to Beam A uniform magnetic field is applied perpendicular to a beam of electrons moving at 1.3×10^6 m/s. What is the magnitude of the field if the electrons travel in a circular arc of radius 0.35 m?

8. Heavy Ions Physicist S. A. Goudsmit devised a method for measuring the masses of heavy ions by timing their periods of revolution in a known magnetic field. A singly charged ion of iodine makes 7.00 rev in a field of 45.0 mT in 1.29 ms. Calculate its mass, in atomic mass units. (Actually, the method allows mass measurements to be carried out to much greater accuracy than these approximate data suggest.)

9. Kinetic Energy An electron with kinetic energy 1.20 keV circles in a plane perpendicular to a uniform magnetic field. The orbit radius is 25.0 cm. Find (a) the speed of the electron, (b) the magnetic field, (c) the frequency, and (d) the period of the motion.

10. Circular Path An alpha particle ($q = +2e, m = 4.00$ u) travels in a circular path of radius 4.50 cm in a uniform magnetic field with magnitude $B = 1.20$ T. Calculate (a) its speed, (b) its period of revolution, (c) its kinetic energy in electron-volts, and (d) the potential difference through which it would have to be accelerated to achieve this energy.

11. Frequency of Revolution (a) Find the frequency of revolution of an electron with an energy of 100 eV in a uniform magnetic field of magnitude 35.0 μT. (b) Calculate the radius of the path of this electron if its velocity is perpendicular to the magnetic field.

12. Source of Electrons A source injects an electron of speed $v = 1.5 \times 10^7$ m/s into a uniform magnetic field of magnitude $B = 1.0 \times 10^{-3}$ T. The velocity of the electron makes an angle $\theta = 10°$ with the direction of the magnetic field. Find the distance d from the point of injection at which the electron next crosses the field line that passes through the injection point.

13. Beam of Electrons A beam of electrons whose kinetic energy is K emerges from a thin-foil "window" at the end of an accelerator tube. There is a metal plate a distance d from this window and perpendicular to the direction of the emerging beam (Fig. 29-28). Show that we can prevent the beam from hitting the plate if we apply a uniform magnetic field $\vec{B}$ such that its magnitude is

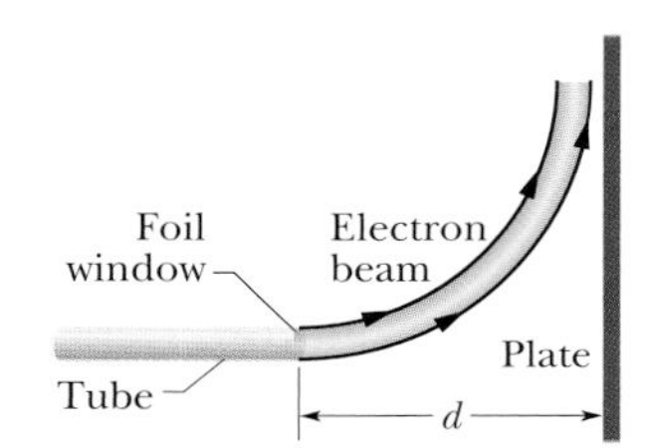

FIGURE 29-28 ■ Problem 13.

$$B \geq \sqrt{\frac{2mK}{e^2d^2}},$$

in which m and e are the electron mass and charge. How should $\vec{B}$ be oriented?

14. Proton, Deuteron, Alpha A proton, a deuteron ($q = +e, m = 2.0$ u), and an alpha particle ($q = +2e, m = 4.0$ u) with the same kinetic energies enter a region of uniform magnetic field $\vec{B}$, moving perpendicular to $\vec{B}$. Compare the radii of their circular paths.

15. Nuclear Experiment In a nuclear experiment a proton with kinetic energy 1.0 MeV moves in a circular path in a uniform magnetic field. What energy must (a) an alpha particle ($q = +2e, m = 4.0$ u) and (b) a deuteron ($q = +e, m = 2.0$ u) have if they are to circulate in the same circular path?

16. Uniform Magnetic Field A proton of charge $+e$ and mass m enters a uniform magnetic field $\vec{B} = B\hat{i}$ with an initial velocity $\vec{v} = v_{1x}\hat{i} + v_{1y}\hat{j}$. Find an expression in unit-vector notation for its velocity $\vec{v}$ at any later time t.

17. Mass Spectrometer A certain commercial mass spectrometer (see Touchstone Example 29-2) is used to separate uranium ions of mass 3.92×10^{-25} kg and charge 3.20×10^{-19} C from related species. The ions are accelerated through a potential difference of 100 kV and then pass into a uniform magnetic field, where they are bent in a path of radius 1.00 m. After traveling through 180° and passing through a slit of width 1.00 mm and height 1.00 cm, they are collected in a cup. (a) What is the magnitude of the (perpendicular) magnetic field in the separator? If the machine is used to separate out 100 mg of material per hour, calculate (b) the current of the desired ions in the machine and (c) the thermal energy produced in the cup in 1.00 h.

18. Half Circle In Fig 29-29, a charged particle moves into a region of uniform magnetic field B, goes through half a circle, and then exits that region. The particle is either a proton or an electron (you must decide which). It spends 130 ns within the region. (a) What is the magnitude $|\vec{B}|$? (b) If the particle is sent back through the magnetic field (along the same initial path) but with 2.00 times its previous kinetic energy, how much time does it spend within the field?

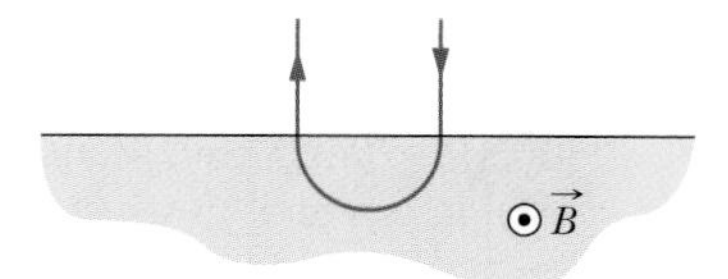

FIGURE 29-29 ■ Problem 18.

19. Positron A positron with kinetic energy 2.0 keV is projected into a uniform magnetic field $\vec{B}$ of magnitude 0.10 T, with its velocity vector making an angle of 89° with $\vec{B}$. Find (a) the period, (b) the pitch p, and (c) the radius r of its helical path.

20. Neutral Particle A neutral particle is at rest in a uniform magnetic field $\vec{B}$. At time $t = 0$ it decays into two charged particles, each of mass m. (a) If the charge of one of the particles is $+q$, what is the charge of the other? (b) The two particles move off in separate paths, both of which lie in the plane perpendicular to $\vec{B}$. At a later time the particles collide. Express the time from decay until collision in terms of m, $|\vec{B}|$, and $|q|$.

SEC. 29-6 ■ CROSSED FIELDS: DISCOVERY OF THE ELECTRON

21. Horizontal Motion An electron with kinetic energy 2.5 keV moves horizontally into a region of space in which there is a downward-directed uniform electric field of magnitude 10 kV/m. (a) What are the magnitude and direction of the (smallest) uniform magnetic field that will cause the electron to continue to move horizontally? Ignore the gravitational force, which is small. (b) Is it possible for a proton to pass through the combination of fields undeflected? If so, under what circumstances?

22. At One Instant A proton travels through uniform magnetic and electric fields. The magnetic field is $\vec{B} = (-2.5 \text{ mT})\hat{i}$. At one instant the velocity of the proton is $\vec{v} = (2000 \text{ m/s})\hat{j}$. At that instant, what is the magnitude of the net force acting on the proton if the electric field is (a) $(4.0 \text{ V/m})\hat{k}$ and (b) $(4.0 \text{ V/m})\hat{i}$?

23. Potential Difference An electron is accelerated through a potential difference of 1.0 kV and directed into a region between two parallel plates separated by 20 mm with a potential difference of 100 V between them. The electron is moving perpendicular to the electric field of the plates when it enters the region between the plates. What magnitude of uniform magnetic field, applied perpendicular to both the electron path and the electric field, will allow the electron to travel in a straight line?

24. Electric and Magnetic Field An electric field of magnitude 1.50 kV/m and a magnetic field of 0.400 T act on a moving electron to produce no net force. (a) Calculate the minimum speed $|\vec{v}|$ of the electron. (b) Draw a set of vectors $\vec{E}$, $\vec{B}$, and $\vec{v}$ that could yield the net force.

25. Ion Source An ion source is producing ions of ^{6}Li (mass = 6.0 u), each with a charge of $+e$. The ions are accelerated by a potential difference of 10 kV and pass horizontally into a region in which there is a uniform vertical magnetic field of magnitude $|\vec{B}| =$ 1.2 T. Calculate the strength of the smallest electric field, to be set up over the same region, that will allow the ^{6}Li ions to pass through undeflected.

26. Initial Velocity An electron has an initial velocity of $(12.0 \text{ km/s})\hat{j} + (15.0 \text{ km/s})\hat{k}$ and a constant acceleration of $(2.00 \times 10^{12} \text{ m/s}^2)\hat{i}$ in a region in which uniform electric and magnetic fields are present. If $\vec{B} = (400\ \mu\text{T})\hat{i}$, find the electric field $\vec{E}$.

SEC. 29-7 ■ THE HALL EFFECT

27. Field Ratio (a) In Fig 29-14, show that the ratio of the magnitudes of the Hall electric field $\vec{E}$ to the electric field $\vec{E}^{\text{curr}}$ responsible for moving charge (the current) along the length of the strip is

$$\frac{E}{E^{\text{curr}}} = \frac{B}{ne\rho}$$

where ρ is the resistivity of the material and n is the number density of the charge carriers and e is the amount of charge on the electron. (b) Compute this ratio numerically for Problem 28. (See Table 26-2.)

28. Strip of Copper A strip of copper 150 μm wide is placed in a uniform magnetic field $\vec{B}$ of magnitude 0.65 T, with $\vec{B}$ perpendicular to the strip. A current $i = 23$ A is then sent through the strip such that a Hall potential difference ΔV appears across the width of the strip. Calculate ΔV. (The number of charge carries per unit volume for copper is 8.47×10^{28} electrons/m^3.)

29. Metal Strip A metal strip 6.50 cm long, 0.850 cm wide, and 0.760 mm thick moves with constant velocity $\vec{v}$ through a uniform magnetic field of magnitude $|\vec{B}| = 1.20$ mT directed perpendicular to the strip, as shown in Fig. 29-30. A potential difference of 3.90 μV is measured between points x and y across the strip. Calculate the speed $|\vec{v}|$.

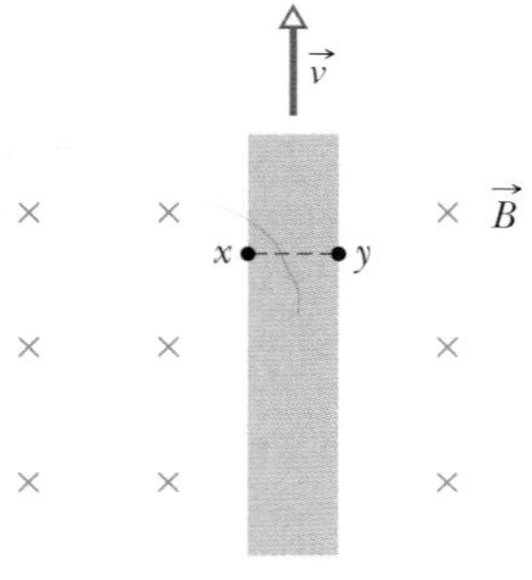

FIGURE 29-30 ■ Problem 29.

SEC. 29-8 ■ MAGNETIC FORCE ON A CURRENT-CARRYING WIRE

30. A Wire Carries a Current A wire 1.80 m long carries a current of 13.0 A and makes an angle of 35.0° with a uniform magnetic field of magnitude $B = 1.50$ T. Calculate the magnitude of the magnetic force on the wire.

31. Horizontal Conductor A horizontal conductor that is part of a power line carries a current of 5000 A from south to north. The magnitude of the Earth's magnetic field is 60.0 μT. The field is directed toward the north and is inclined downward at 70° to the horizontal. Find the magnitude and direction of the magnetic force on 100 m of the conductor due to Earth's field.

32. Along the x Axis A wire 50 cm long lying along the x axis carries a current of 0.50 A in the positive x direction. It passes through a magnetic field $\vec{B} = (0.0030 \text{ T})\hat{j} + (0.0100 \text{ T})\hat{k}$. Find the magnetic force on the wire.

33. A Wire of Length A wire of 62.0 cm length and 13.0 g mass is suspended by a pair of flexible leads in a uniform magnetic field of magnitude 0.440 T (Fig. 29-31). What are the magnitude and direction of the current required to remove the tension in the supporting leads?

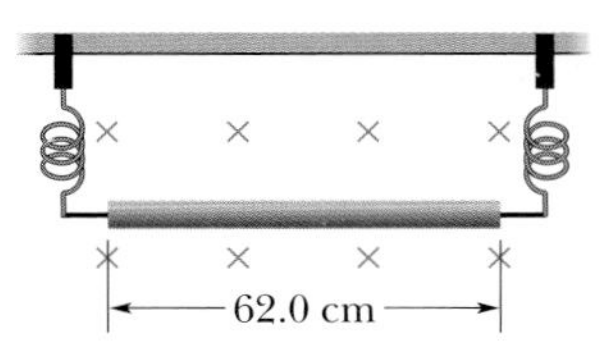

FIGURE 29-31 ■ Problem 33.

34. Electric Train Consider the possibility of a new design for an electric train. The engine is driven by the force on a conducting axle due to the vertical component of Earth's magnetic field. To produce the force, current is maintained down one rail, through a conducting wheel, through the axle, through another conducting wheel, and then back to the source via the other rail. (a) What amount of current is needed to provide a modest force of magnitude 10kN? Take the vertical component of Earth's field to be 10 μT and the length of the axle to be 3.0 m. (b) At what rate would electric energy be lost for each ohm of resistance in the rails? (c) Is such a train totally or just marginally unrealistic?

35. Copper Rod A 1.0 kg copper rod rests on two horizontal rails 1.0 m apart and carries a current of 50 A from one rail to the other. The coefficient of static friction between rod and rails is 0.60. What is the magnitude of the smallest magnetic field (not necessarily vertical) that would cause the rod to slide?

SEC. 29-9 ■ TORQUE ON A CURRENT LOOP

36. Current Loop A single-turn current loop, carrying a current of 4.00 A, is in the shape of a right triangle with sides 50.0, 120, and 130 cm. The loop is in a uniform magnetic field of magnitude 75.0 mT whose direction is parallel to the current in the 130 cm side of the loop. (a) Find the magnitude of the magnetic force on each of the three sides of the loop. (b) Show that the total magnetic force on the loop is zero.

37. Rectangular Coil Figure 29-32 shows a rectangular 20-turn coil of wire, of dimensions 10 cm by 5.0 cm. It carries a current of 0.10 A and is hinged along one long side. It is mounted in the xy plane, at 30° to the direction of a uniform magnetic field of magnitude 0.50 T. Find the magnitude and direction of the torque acting on the coil about the hinge line.

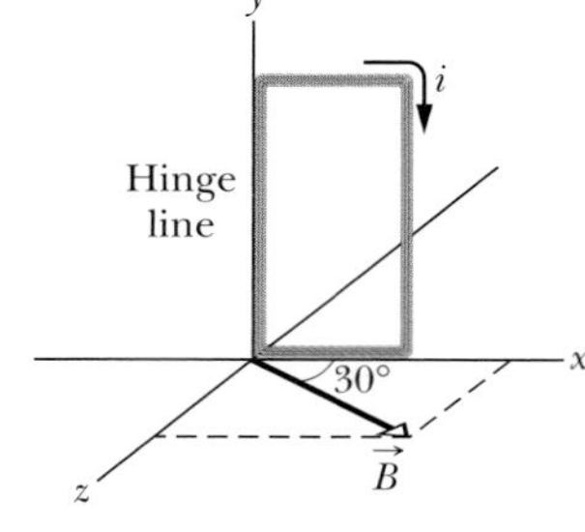

FIGURE 29-32 ■ Problem 37.

38. Arbitrarily Shaped Coil Prove that the relation $\tau = N|i|AB \sin \phi$ (Eq. 29-31) holds for closed loops of arbitary shape and not only for rectangular loops as in Fig. 29-22. (*Hint:* Replace the loop of arbitrary shape with an assembly of adjacent long, thin, approximately rectangular loops that are nearly equivalent to the loop of arbitrary shape as far as the distribution of current is concerned.)

39. Show That A length L of wire carries a current i. Show that if the wire is formed into a circular coil, then the magnitude of the maximum torque in a given magnetic field is developed when the coil has one turn only. Also show that maximum torque has the magnitude $\tau = L^2 iB/4\pi$.

40. Zero Total Force A closed wire loop with current i is in a uniform magnetic field $\vec{B}$, with the plane of the loop at angle θ to the direction of $\vec{B}$. Show that the total magnetic force on the loop is zero. Does your proof also hold for a nonuniform magnetic field?

41. Wire Ring Figure 29-33 shows a wire ring of radius a that is perpendicular to the general direction of a radially symmetric, diverging magnetic field. The magnetic field at the ring is everywhere of the same magnitude $|\vec{B}|$, and its direction at the ring everywhere makes an angle θ with a normal to the plane of the ring. The twisted lead wires have no effect on the problem. Find the magnitude and direction of the force the field exerts on the ring if the ring carries a positive current i.

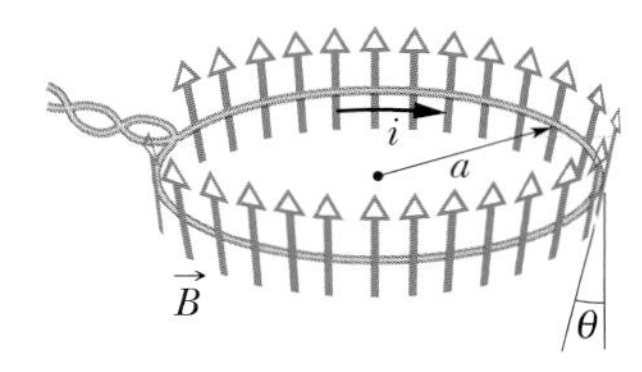

FIGURE 29-33 ■ Problem 41.

42. Maximum Torque A particle of charge q moves in a circular wire loop of radius a with speed $|\vec{v}|$. Find the maximum torque exerted on the loop by a uniform magnetic field of magnitude $|\vec{B}|$.

43. Wooden Cylinder Figure 29-34 shows a wooden cylinder with mass $m = 0.250$ kg and length $L = 0.100$ m, with $N = 10.0$ turns of wire wrapped around it longitudinally, so that the plane of the wire coil contains the axis of the cylinder. Also the plane of the coil is parallel to the inclined plane. There is a vertical, uniform magnetic field of magnitude 0.500 T. What is the least amount of current $|i|$ through the coil that will prevent the cylinder from rolling down a plane inclined at an angle θ to the horizontal?

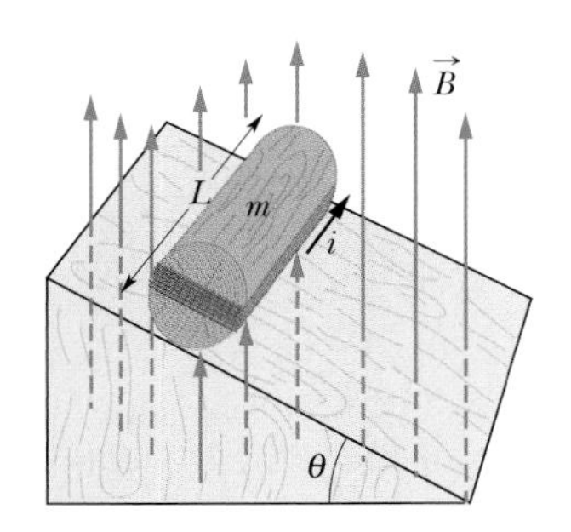

FIGURE 29-34 ■ Problem 43.

SEC. 29-10 ■ THE MAGNETIC DIPOLE MOMENT

44. Earth's Moment The magnitude of magnetic dipole moment of Earth is 8.00×10^{22} J/T. Assume that this is produced by charges flowing in Earth's molten outer core. If the radius of their circular path is 3500 km, calculate the amount of current associated with each moving charge.

45. Calculate the Current A circular coil of 160 turns has a radius of 1.90 cm. (a) Calculate the current that results in a magnetic dipole moment of $2.30 \text{ A} \cdot \text{m}^2$. (b) Find the maximum magnitude of torque that the coil, carrying this current, can experience in a uniform 35.0 mT magnetic field.

46. Moment and Torque A circular wire loop whose radius is 15.0 cm carries an amount of current of 2.60 A. It is placed so that the normal to its plane makes an angle of 41.0° with a uniform magnetic field of magnitude 12.0 T. (a) Calculate the magnitude of the magnetic dipole moment of the loop. (b) What is the magnitude of torque that acts on the loop?

47. Right Triangle A current loop, carrying an amount of current of 5.0 A, is in the shape of a right triangle with sides 30, 40, and 50 cm. The loop is in a uniform magnetic field of magnitude 80 mT whose direction is parallel to the current in the 50 cm side of the loop. Find the magnitude of (a) the magnetic dipole moment of the loop and (b) the torque on the loop.

48. Wall Clock A stationary circular wall clock has a face with a radius of 15 cm. Six turns of wire are wound around its perimeter; the wire carries a current of 2.0 A in the clockwise direction. The clock is located where there is a constant, uniform external magnetic field of magnitude 70 mT (but the clock still keeps perfect time). At exactly 1:00 P.M., the hour hand of the clock points in the direction of the external magnetic field. (a) After how many minutes will the minute hand point in the direction of the torque on the winding due to the magnetic field? (b) Find the torque magnitude.

49. Concentric Loops Two concentric, circular wire loops, of radii 20.0 and 30.0 cm, are located in the xy plane; each carries a clockwise current of 7.00 A (Fig. 29-35). (a) Find the magnitude of the net magnetic dipole moment of this system. (b) Repeat for reversed current in the inner loop.

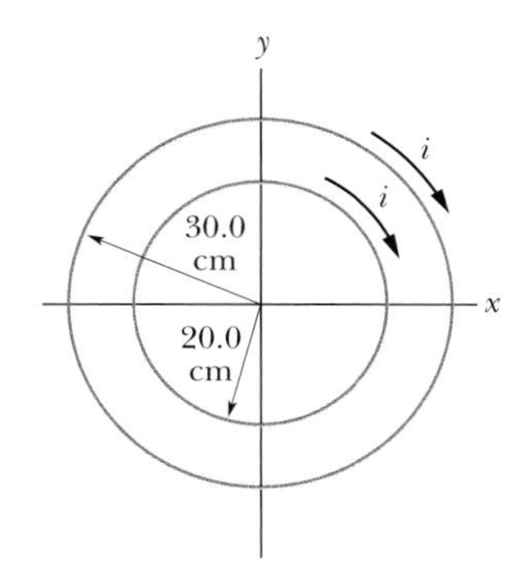

FIGURE 29-35 ■ Problem 49.

50. *ABCDEFA* Figure 29-36 shows a current loop $ABCDEFA$ carrying a current $i = 5.00$ A. The sides of the loop are parallel to the coordinate axes, with $AB = 20.0$ cm, $BC = 30.0$ cm, and $FA = 10.0$ cm. Calculate the magnitude and direction of the magnetic dipole moment of this loop. (*Hint:* Imagine equal and opposite currents i in the line segment AD; then treat the two rectangular loops $ABCDA$ and $ADEFA$.)

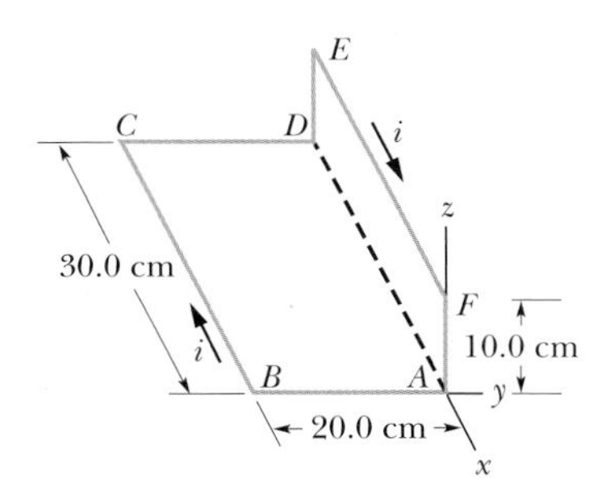

FIGURE 29-36 ■ Problem 50.

51. Circular Loop A circular loop of wire having a radius of 8.0 cm carries a current of 0.20 A. A vector of unit length and parallel to the dipole moment $\vec{\mu}$ of the loop is given by $0.60\hat{\text{i}} - 0.80\hat{\text{j}}$. If the loop is located in a uniform magnetic field given by $\vec{B} = (0.25 \text{ T})\hat{\text{i}} + (0.30 \text{ T})\hat{\text{k}}$, find (a) the torque on the loop (in unit-vector notation) and (b) the magnetic potential energy of the loop.

Additional Problems

52. Permanent Magnet You can observe that a permanent magnet can exert forces on moving charges or currents. (a) If a magnet exerts a force on a moving charge, would the magnet experience any forces? Explain. (b) In the case of the gravitational or electrostatic interaction between two objects, each object has a common property, such as mass in the case of gravitational interaction or excess charge in the case of the electrostatic interaction. A permanent magnet and a moving electron seem very different. Can you think of any way that they might have a common property? Explain.

53. U-Shaped Magnet An electron having a velocity of magnitude v enters a region between the poles of a U-shaped magnet. This region has a uniform magnetic field, $\vec{B}$, pointing out of the paper in the positive z direction as shown in Fig. 29-37.

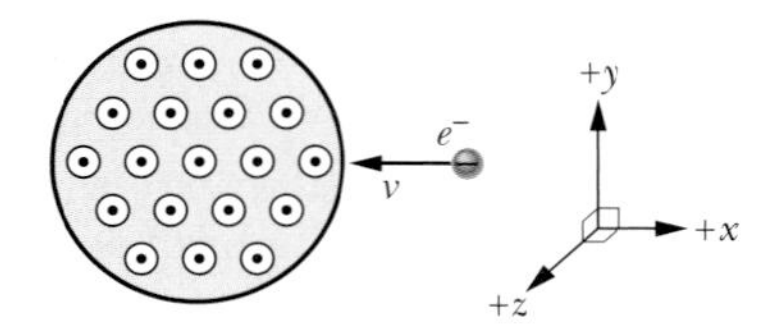

FIGURE 29-37 ■ Problem 53.

(a) If the magnetic field points out of the paper, where is the north pole of the magnet—in front of or behind the image shown? (b) Use the right-hand rule to find the direction of force on the electron as it passes into the region where the magnetic field is uniform. (c) Sketch the path of the electron, assuming that the magnetic field is relatively weak. (d) If the speed of the electron is 4.79×10^6 m/s and the magnitude of the magnetic field is 0.234 T, what is the magnitude of the force on the electron?

54. A Velocity Selector A group of physicists at Argonne National Laboratory in Illinois wants to bombard metals with monoenergetic beams of alpha particles to study radiation damage. (Alpha particles are helium nuclei, which consist of two neutrons and two protons and thus have a net charge of $+2e$ where e is the amount of the charge on the electron.) They have managed to create a beam of alpha particles from the decay of radioactive elements, but some of the alpha particles lose energy as they collide with other atoms in the source. As a new physicist assigned to the group you have been asked to use a velocity selector to select only the alpha particles in the beam that are close to one velocity and get rid of the others. The velocity selector consists of: (1) a power supply capable of delivering large potential differences between capacitor plates and (2) a large permanent magnet that has a uniform magnetic field perpendicular to the beam. The setup for the velocity selector is shown in Fig. 29-38. The direction of the B-field is out of the paper. Your magnet has a field of 0.22 T and the capacitor plates have a spacing of 2.5 cm. You are asked to figure out how the velocity selector works and then tell your group what voltage to put across the capacitor plates to select a velocity of 4.2×10^6 m/s.

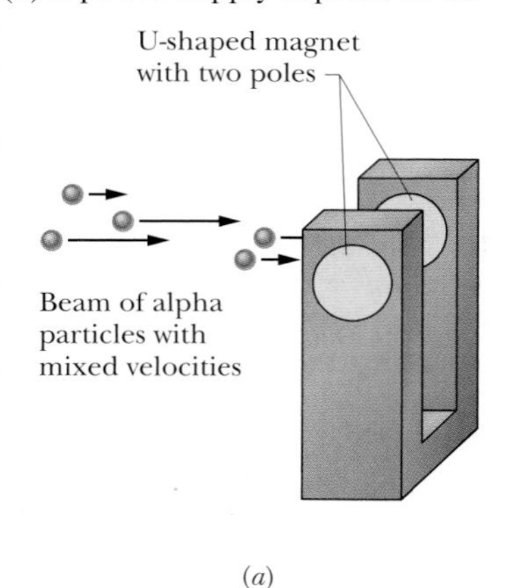

FIGURE 29-38a ■ Problem 54.

This is your first job and you feel overwhelmed by the assignment, but you calm down and begin to analyze the situation one step at a time. You come up with the following:

(a) The magnet is oriented so its magnetic field is out of the paper in the diagram you are given, so you use the right-hand rule to determine the direction of the magnetic force on an alpha particle passing from left to right into the magnetic field. What direction did you come up with for the force?

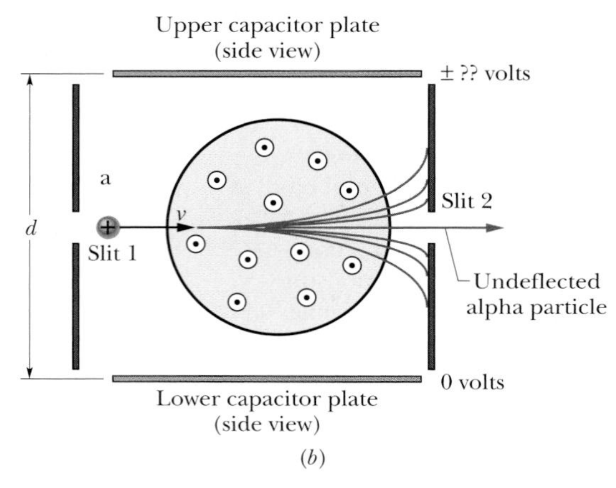

FIGURE 29-38b ■ Enlarged view of the region between the poles of the magnet showing a single alpha particle having a speed v entering the region of uniform magnetic field.

(b) You realize that by using $\vec{F}^{\,\text{mag}} = q\vec{v} \times \vec{B}$, you can calculate the magnitude of force on an alpha particle moving at speed v just as it enters the uniform magnetic field as a function of the charge on the alpha particle and the magnitude of the magnetic field B. What is the expression for the magnitude of the force in terms of e, v, and B?

(c) You realize that you might be able to put just the right voltage across the two capacitor plates so that the electrical force on a given alpha particle will be equal in magnitude and opposite in direction to the Lorentz magnetic force. Then any alpha particles with just the right velocity will pass straight through the poles of the magnet without being deflected. First you think about whether the voltage on the upper capacitor plate should be positive or negative to give a canceling force. What do you decide?

(d) Next you realize that if you know the electric field between the plates and the charge on the alpha particle then you can compute the electrical force on it. What is the relationship between the electrical force $\vec{F}^{\,\text{elec}}$, charge, q, and electric field $\vec{E}$?

(e) Finally, you use the fact that the magnitude of the electric field between capacitor plates is given by $E = |\Delta V|/d$ where d is the spacing between the plates. Show that the voltage needed to have the electrical force and the magnetic force be "equal and opposite" can be calculated using the equation $|\Delta V| = vBd$. Calculate the voltage needed.

55. Region A—Region B Figure 29-39 shows a charged particle that is moving in the positive x direction when it encounters region A with a uniform magnetic field. Its path is bent in a half-circle and then moves into region B also with a uniform magnetic field. The particle undergoes another half revolution. Finally it passes between two charged capacitor plates and is deflected downward in the negative y direction.

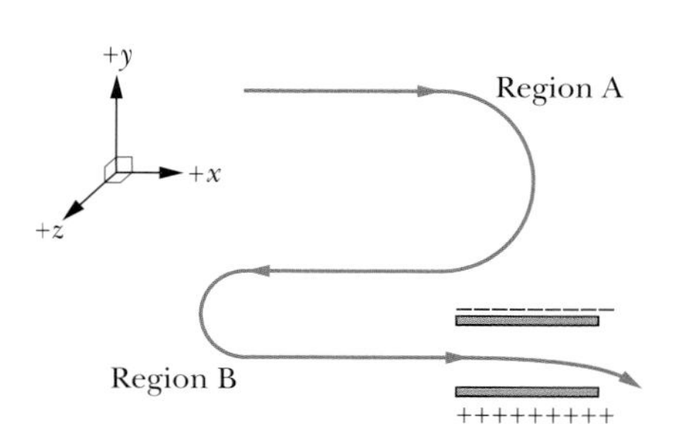

FIGURE 29-39 ■ Problem 55.

(a) Is the charge positive or negative? Explain.
(b) What is the direction of the magnetic field in region A? Explain.
(c) What is the direction of the magnetic field in region B? Explain.
(d) Which region has the larger magnetic field, A or B? Explain.

56. A Mass Spectrometer It is possible to accelerate ions to a known kinetic energy in an electric field. Sometimes chemists and physicists do this as part of a method to identify the chemical elements present in a beam of ions by determining the mass of each ion. This can be done by bending the ion beam in a uniform magnetic field and measuring the radius of the semicircular path each ion takes. A device that does this is called a mass spectrometer. A schematic of a mass spectrometer is shown in Fig. 29-40.

Boron is the fifth element in the periodic table so it always has 5 protons. However, different isotopes of boron have 3, 5, 6, 7, or 8 neutrons in addition to the 5 protons to make up boron-8, boron-10, boron-11 and so on. As a research chemist for the Borax Company you have been asked to use a mass spectrometer to determine the relative abundance of different isotopes of boron in a sample of boron obtained from a mine near Death Valley in California. You decide to accelerate a beam of singly charged boron ions (i.e., those that have lost one of their orbital electrons). You use an accelerating potential difference of -2.68×10^3 volts. The boron beam then enters a uniform magnetic field you set up to have a magnetic flux density of 0.182 T in a direction perpendicular to the direction of the boron beam. You observe two bright spots on your photographic plate with the spot corresponding to a radius of 13.0 cm having four times the intensity of the one corresponding to a radius of 13.6 cm. There are very faint spots at 11.6 cm, 14.2 cm, and 14.8 cm. Which isotope of boron has approximately 80% abundance? Which one has about 20% abundance? Which ones are present in only trace amounts? Please show all your reasoning and calculations. *Hints:* (1) An atomic mass unit is given by 1.66×10^{-27} kg, which is close to the mass of the proton and neutron. (2) Find the velocity of each isotope of boron in meters per second just after it has been accelerated by the potential difference of -2.68×10^3 volts. (3) It is helpful to do the calculations for each of the five isotopes on a spreadsheet.

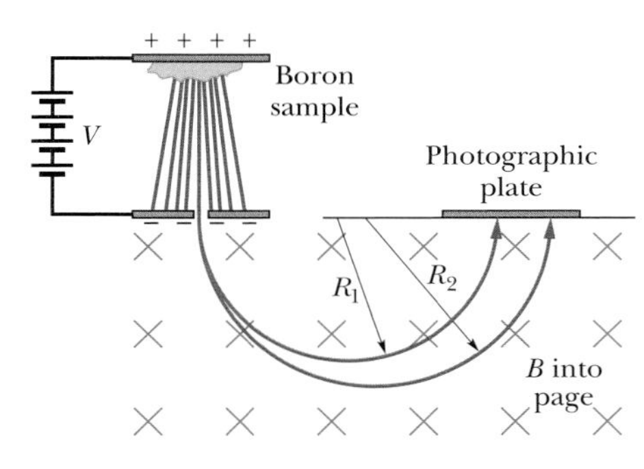

FIGURE 29-40 ■ Problem 56.

57. Bubble Chamber Tracks Energetic gamma rays like those coming from outer space can disappear near a heavy nucleus producing a rapidly moving pair of particles consisting of an electron and a positron. (A positron is a small positively charged particle that has the same mass and amount of charge as an electron). This process is called pair production. A device called a bubble chamber allows one to observe the path taken by electron–positron pairs produced by gamma rays. The study of bubble chamber tracks in the presence of magnetic fields has revealed a great deal about high-energy gamma rays, the processes of pair production, and the loss of energy by electrons and positrons. A sample bubble chamber track is shown in Fig. 29-41.

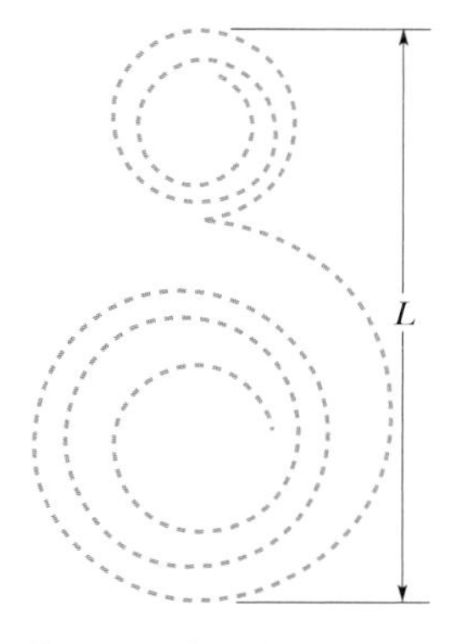

FIGURE 29-41 ■ Problem 57.

(a) If the magnetic field is uniform pointing into the paper, which trajectory (the upper one or the lower one) shows the motion of the positron? Explain your reasoning. (b) In which part of the spiral does the positron have the greatest energy—the large radius part or the small radius part? Explain the reasons for your answer. (c) Is the electron moving faster, slower, or at the same speed as the positron at the point in time when the two particles are created? Cite the evidence for your answer. (d) Suppose the bubble chamber photograph in Fig. 29-41 is an enlargement of the actual event so that the length L is actually only 2.4×10^{-3} m. Show that the radius of curvature of the electron path just after the electron is created is approximately 0.8×10^{-3} m. *Hint:* Measure L in picture units to find a scale factor and then measure the appropriate feature of the electron path in picture units and use the scale factor to find R in meters. (e) Use the Lorentz force law and the expression for centripetal force to find the equation relating the speed of the electron to B, R, e, and m. (f) Suppose the magnitude of the magnetic field in the bubble chamber is $B = 0.54$ T. Calculate the approximate speed of the electron when it is first created in the bubble chamber.

58. Three Force Fields We have studied three long-range forces: gravity, electricity, and magnetism. Compare and contrast these three forces giving at least one feature that all three forces have in common, and at least one feature that distinguishes each force.

59. Comparing $\vec{E}$ and $\vec{B}$ Fields We have studied two fields: electric and magnetic. Explain why we introduce the idea of field, and compare and contrast the electric and magnetic fields. In your comparison, be certain to discuss at least one similarity and one difference.

60. Anti-matter Ion Cosmic Rays An international consortium is presently building a device to look for anti-matter nuclei in cosmic rays to help us decide whether there are galaxies made of anti-matter. Anti-matter is just like ordinary matter except the basic particles (anti-protons and anti-electrons) have opposite charge from ordinary matter counterparts. Anti-protons are negative, and anti-electrons (positrons) are positive.

A schematic of the device is shown in Fig. 29-42. A cosmic ray—say, a carbon nucleus or an anti-carbon nucleus—enters the device at the left where its position and velocity are measured. It then passes through a (reasonably uniform) magnetic field. Its path is bent in one direction if its charge is positive and in the opposite direction it its charge is negative. Its deflection is measured as it goes out of the device.

(a) In Fig. 29-42, what is the direction of the magnetic field? How do you know?
(b) Which path is followed by each particle in the device? How do you know?
(c) If you were given the magnetic field, B, the size of the device, D, the amount of charge on the incoming particle, q, and the mass of the incoming particle, M, would this be enough to calculate the displacement of the charge, d? If so, describe briefly how you would

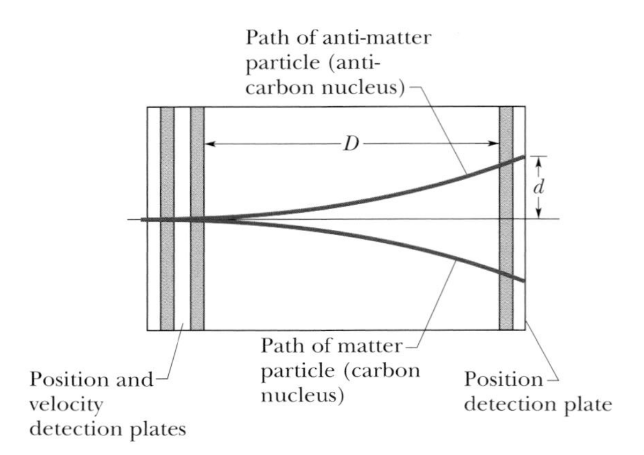

FIGURE 29-42 ■ Problem 60.

do it (but don't do it). If not, explain what additional information you would need (but don't estimate it).

61. Magnets and Charge A bar magnet is hung from a string through its center as shown in Fig. 29-43. A charged rod is brought up slowly into the position shown. In what direction will the magnet tend to rotate? Suppose the charged rod is replaced by a bar magnet with the north pole on top. In what direction will the magnet tend to rotate? Is there a difference between what happens to the hanging magnet in the two situations? Explain why you either do or do not think so.

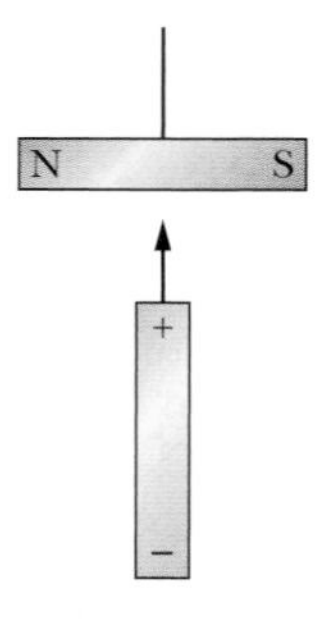

FIGURE 29-43 ■ Problem 61.

* C14 is a radioactive isotope of carbon that behaves chemically almost identically to its more common but slightly lighter sibling, C12. The amount of C14 in the atmosphere stays about constant since it is being produced continually by cosmic rays. Once carbon from the air is bound into an organic substance, the C14 will decay with half of them vanishing every 5730 years. The ratio of C14 to C12 in an organic substance therefore tells how long ago it died.

62. Buying a Mass Spectrometer* You are assigned the task of working with a desktop-sized magnetic spectrometer for the purpose of measuring the ratio of C^{12} to C^{14} atoms in a sample in order to determine the sample's age. For this problem, let's concentrate on the magnet that will perform the separation of masses. Suppose you have burned and vaporized the sample so that the carbon atoms are in a gas. You now pass this gas through an "ionizer" that on the average strips one electron from each atom. You then accelerate the ions by putting them through an electrostatic accelerator—two capacitor plates with small holes that permit the ions to enter and leave. (From the University of Washington Physics Education Group)

The two plates are charged so that they are at a voltage difference of ΔV volts. The electric field produced by charges on the capacitor plates accelerates the ions to an energy of $q\Delta V$. These are then introduced into a nearly constant, vertical magnetic field. If we ignore gravity, the magnetic field will cause the charged particles to follow a circular path in a horizontal plane. The radius of the circle will depend on the atom's mass. (Assume the whole device will be placed inside a vacuum chamber.)

Answer these three questions about how the device works.

(a) We want to keep the voltage at a moderate level. If ΔV is 1000 volts, how big of a magnetic field would we require to have a plausible tabletop-sized instrument? Is this a reasonable magnetic field to have with a tabletop-sized magnet?
(b) Do the C^{12} and C^{14} atoms hit the collection plate far enough apart? (If they are not separated by at least a few millimeters at the end of their path, we will have trouble collecting the atoms in separate bins.)
(c) Can we get away with ignoring gravity? (*Hint:* Calculate the time it would take the atom to travel its semicircle and calculate how far it would fall in that time.)

30 Magnetic Fields Due to Currents

This is the way we presently launch materials into space. However, when we begin mining the Moon and the asteroids, where we will not have a source of fuel for such conventional rockets, we shall need a more effective way. Electromagnetic launchers may be the answer. A small prototype, the *electromagnetic rail gun*, can accelerate a projectile from rest to a speed of 10 km/s (36 000 km/h) within 1 ms.

How can such rapid acceleration possibly be accomplished?

The answer is in this chapter.

30-1 Introduction

When people first began to study magnetism scientifically (say, starting from Gilbert's *Treatise de Magnete* in 1600), they focused on the properties of magnets. For example, they studied lodestones (pieces of iron naturally magnetized by the Earth's magnetic field) and found that magnets interact with other magnets through an "action-at-a-distance" force that we now call magnetism. Magnetism was found to be a third distinct noncontact force to add to the list of the two already known: gravity and electricity.

As we learned in the previous chapter, *stationary* electric charges and magnets do not interact (except for the polarization effects that stationary charges can induce in all objects). However, *moving* electric charges do experience a force in the presence of a magnet. Since magnets can exert forces on other magnets, could it be that moving charges behave like magnets?

We have postulated the existence of an entity called the magnetic field in order to introduce a magnetic force law that provides a mathematical description of the force that a permanent magnet can exert on moving electrical charges. Newton's Third Law states that whenever one object exerts a force on another object, the latter object exerts an equal and opposite force on the former. So, if a magnet exerts a force on a current-carrying wire, shouldn't the wire exert an equal and opposite force on the magnet? The symmetry demanded by Newton's Third Law leads us to predict that if moving charges feel forces as they pass through magnetic fields, then they should be capable of exerting forces on the sources of these magnetic fields. In the early 19th century the Danish physicist Hans Christian Oersted demonstrated that an electric current does indeed exert forces on a magnet in its vicinity.

In this chapter we describe how to determine the magnetic fields associated with current-carrying wires and the forces they exert on other wires and magnets. We begin with a summary of Oersted's observations of magnetic phenomena associated with current-carrying wires. We also discuss the work of Biot and Savart, two French scientists. Biot and Savart made a series of careful observations to formulate a mathematical expression describing the magnetic field from a short segment of current-carrying wire, doing for magnetism what Coulomb did for electricity. Next we show how the Biot–Savart law and an alternative law known as Ampère's law (much as Gauss' law was an alternative to Coulomb's) can be used to calculate the magnetic fields and forces associated with various configurations of current-carrying wires. The ability to make such calculations has had a tremendous impact on the design of devices ranging from electric toothbrushes to gigantic particle accelerators.

30-2 Magnetic Effects of Currents—Oersted's Observations

The Earth has a relatively weak magnetic field that interacts with magnets. This phenomenon was exploited for navigational purposes through the development of the compass—a small bar magnet suspended so it pivots freely. Hence a compass is a sensitive magnetic field detector. Oersted and other scientists used the orientation of a compass to detect magnetic fields and determine their directions. By convention, the north-seeking pole of a magnet points in the direction of the magnetic field at its location.

In 1820, H. C. Oersted reported on a famous experiment connecting magnetism with electric currents. He placed a conducting wire along the north–south line of the Earth's magnetic field and laid a compass on top of the wire. The needle pointed along the wire (and the Earth's north–south line). When Oersted connected the ends of the wire to the terminals of a battery, the compass needle swung *perpendicular* to the wire as shown in Fig. 30-1, demonstrating that moving charges in a wire affect a compass in the same way a magnet does. Oersted also noticed that when the direction of the current is reversed, the compass needle flips so it points in the opposite direction.

Oersted found that moving charged particles, such as a current in a wire, create magnetic fields. Oersted's observation was especially surprising because this was the first known instance in which the force on an object (in this case the compass) was not observed to act along a line connecting it with the source of the force (in this case the wire). Within a week of the time that Oersted announced his observations, a French physicist, André Marie Ampère, began to refine them. Ampère noted that the magnetic field lines lay in concentric circles around the wire. His careful observations revealed that a long current-carrying wire sets up a magnetic field that orients small compass magnets so they are tangent to a circle centered on the wire that lies in a plane perpendicular to the wire. The alignment of iron filings, which act like small compasses, is shown in Fig. 30-2. Drawing the direction of the compass needle alignments at many different points that completely surround the wire results in an image of concentric circles like those shown in Fig. 30-3.

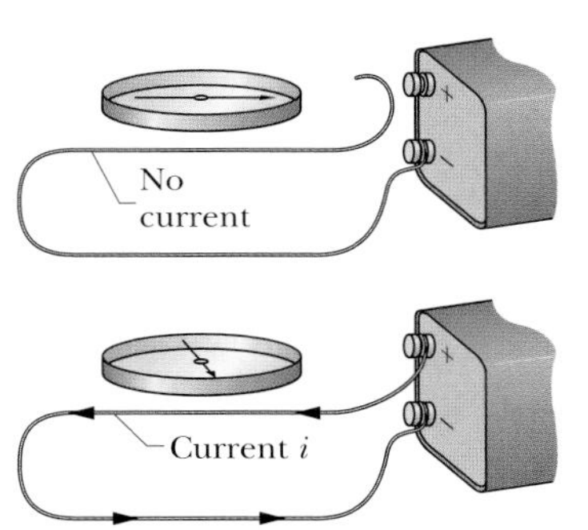

FIGURE 30-1 ■ Oersted's experiment showing how a compass needle becomes aligned in a direction that is perpendicular to the direction of the current in a length of wire.

Ampère also developed a graphic way of relating the direction of conventional current (that is, traveling from the positive to the negative terminal of a battery) and the orientation of the magnetic field, which is indicated by the direction of the north pole of a compass needle. Ampère stated his **right-hand rule** as follows:

> Encircle the wire with the fingers of the right hand, thumb extended in the direction of positive current. The fingers then point in the direction of deflection of the north pole.

FIGURE 30-2 ■ Iron filing slivers that have been sprinkled onto cardboard collect in concentric circles when a strong current is sent through the central wire. The filings are magnetized and align themselves like tiny compasses in the direction of the magnetic field produced by the current.

This right-hand rule is shown graphically in Fig. 30-4.

You can easily replicate the following observations made by Oersted, Ampère, and many others in the early 19th century using a battery, wire, a piece of cardboard, and one or more small compasses:

- The compass needles are more strongly deflected when they are close to the wire than when they are far from the wire.
- For a given current, the amount of needle deflection depends only on the needle's radial distance from the wire.
- At a given radial distance from the wire, increasing the current in the wire increases the needle deflection.
- The direction of the needle deflection flips (change by 180°) if you reverse the direction of the current flow.
- Drawing the directions of the needle orientations at many different points that completely surround the wire results in an image of concentric circles like those shown in Fig. 30-3.

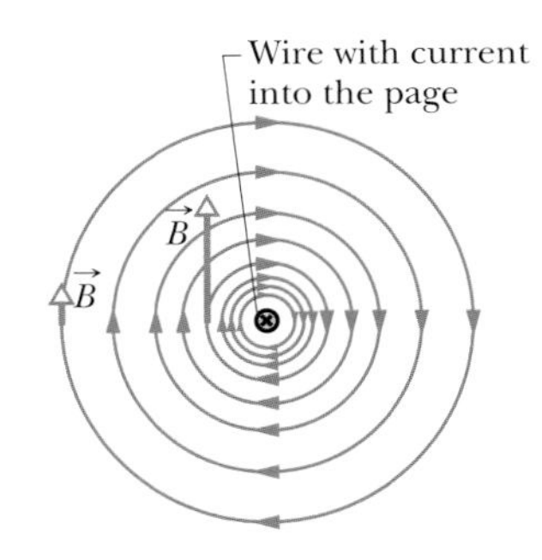

FIGURE 30-3 ■ The magnetic field lines produced by a current in a long straight wire form concentric circles around the wire. Here the current is into the page, as indicated by the ×. The field lines are farther apart as the distance from the wire increases, signifying a decrease in the magnitude of the field with distance.

READING EXERCISE 30-1: In each of the following situations, assume that the magnetic field associated with a current-carrying wire can point up, down, left, right, into the page, or out of the page. (a) If the direction of the conventional current in the wire is out of the page, what is the direction of the magnetic field it generates at point 1? (b) At point 2? (c) If the direction of the conventional current in the wire is into the page, what is the direction of the magnetic field it generates at point 1? (d) At point 2? ■

•1 (*a*) ⊙ Wire •2 (*b*)   •1 (*c*) ⊗ Wire •2 (*d*)

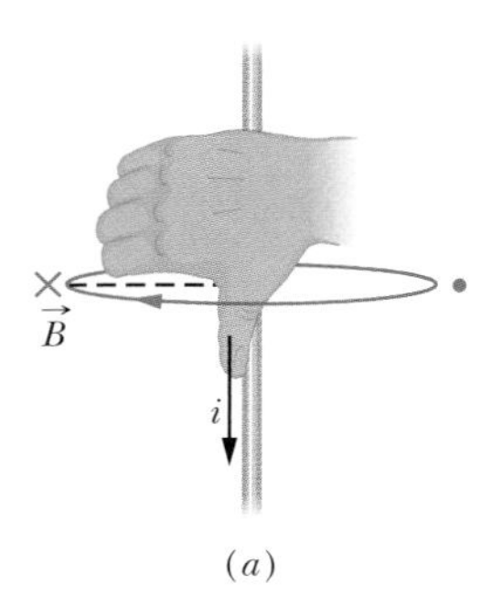

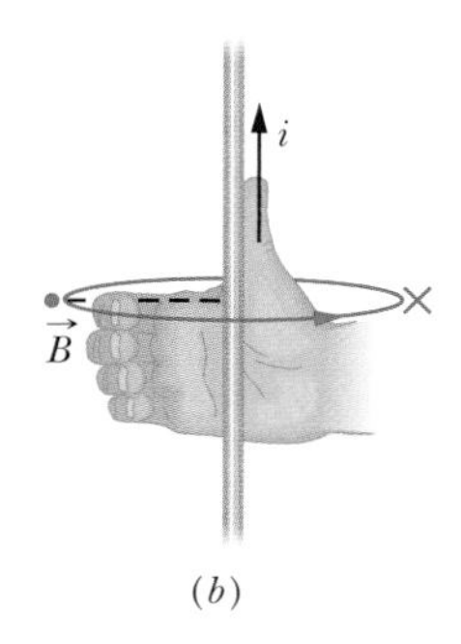

FIGURE 30-4 ■ Ampère's right-hand rule gives the direction of the magnetic field relative to the conventional current in a wire. (*a*) The situation of Fig. 30-3, seen from the side. The magnetic field $\vec{B}$ at any point to the left of the wire is perpendicular to the dashed radial line and directed into the page, in the direction of the fingertips, as indicated by the $\times$. (*b*) If the current is reversed, $\vec{B}$ at any point to the left is still perpendicular to the dashed radial line but now is directed out of the page, as indicated by the dot.

30-3 Calculating the Magnetic Field Due to a Current

It is very useful to be able to compute the net magnetic field created by a current-carrying wire. We would also like to be able to do this either for long straight wires or for any wire no matter how it bends around.

Two French physicists named Biot and Savart (rhymes with "Leo and bazaar") were able to develop a mathematical description of the magnetic field in the vicinity of a short segment of current-carrying wire. To do this, these investigators made a set of very clever experimental measurements:

- First, the two investigators positioned magnets around their experimental setup in order to cancel out the local magnetic field of the Earth.
- Next they placed sharp bends in a current-carrying wire so they could observe the approximate effect that an "isolated" short element of wire would have.
- Then they ran a known current through the wire and measured the direction of the magnetic field produced by the small wire segment at various locations using the final orientation of the suspended compass needles.
- Finally, they measured the relative magnitude of the torque on the suspended compass needles before they reached their final orientation and thus the relative force applied to the needles. In doing this, they were actually making measurements of the strength of the field at various locations.

Given what we know of the observations summarized in the previous section, it is not surprising that Biot and Savart found that the magnitude of the magnetic field contribution $|d\vec{B}| = dB$ is directly proportional to the amount of the current $|i|$ and the length of the small segment of wire. They also found that the magnitude of the magnetic field at a point P in space decreases as the inverse square of the distance between the segment of wire and point P. The two investigators proposed that the magnitude of the field contribution $d\vec{B}$ produced at a point P by a segment of wire $d\vec{s}$ carrying a current i is

$$dB = \frac{\mu_0}{4\pi}\frac{|i\,ds|\sin\phi}{r^2}, \tag{30-1}$$

where $d\vec{s}$ is a vector of magnitude ds equal to the length of the piece of wire and direction given by the direction of the current. ϕ is the angle between the directions of $d\vec{s}$ and $\vec{r}$, where $\vec{r}$ is the vector that extends from $d\vec{s}$ to point P. (See Figure 30-5*b*.) The symbol μ_0 is called the *magnetic constant* (or permeability). By definition its value in SI units is exactly

$$\mu_0 \equiv 4\pi \times 10^{-7}\,\mathrm{T\cdot m/A} \approx 1.26 \times 10^{-6}\,\mathrm{T\cdot m/A} \quad \text{(magnetic constant).} \tag{30-2}$$

Equation 30-1 is similar in many ways to that found for the differential electric field from a small segment of wire holding static charge described by Eq. 23-21. However, the perpendicular relationship between the direction of a segment of wire and the magnetic field it produces is a new phenomenon. Fortunately, it turns out that a vector crossproduct can be used to find the direction of the magnetic field contribution. The direction of $d\vec{B}$, shown as being into the page in Fig. 30-5*b*, is the same as that given by the cross product $d\vec{s} \times \vec{r}$. We can therefore recast Eq. 30-1 in vector form as

$$d\vec{B} = \frac{\mu_0}{4\pi}\frac{i\,d\vec{s} \times \vec{r}}{r^3} \quad \text{(Biot–Savart law).} \tag{30-3}$$

This vector equation is known as the **Biot–Savart law.** The law, which was experimentally deduced, is an inverse-square law (the exponent in the denominator of Eq. 30-3 is 3 only because of the factor $\vec{r}$ in the numerator). How can we use this law to calculate the net magnetic field $\vec{B}$ produced at a point by various distributions of current?

If our goal is to calculate the magnetic field that is produced by a given current *distribution* based on the field produced by *segments* of the distribution, perhaps we should use the same basic procedure we used in Chapter 23 to calculate the electric field produced by a given distribution of charged particles. Let us quickly review that basic procedure. We first mentally divide the charge distribution into charge elements dq, as is done for a charge distribution of arbitrary shape in Fig. 30-5*a*. We then calculate the field $d\vec{E}$ produced at some point P by a single charge element. Because the electric fields contributed by different elements can be superimposed, we calculate the net field $\vec{E}$ at P by summing, via integration, the contributions $d\vec{E}$ from all the elements.

Recall that we express the magnitude of $d\vec{E}$ as

$$|dE| = k\frac{|dq|}{r^2}, \tag{30-4}$$

in which r is the distance between the charge element dq and point P. For a positively charged element, the direction of $d\vec{E}$ is that of $\vec{r}$, where $\vec{r}$ is the vector that extends from the charge element dq to the point P. Using $\vec{r}$, we can rewrite Eq. 30-4 in vector form as

$$d\vec{E} = k\frac{dq}{r^3}\vec{r}, \tag{30-5}$$

which indicates that the direction of the vector $d\vec{E}$ produced by a positively charged element is the direction of the vector $\vec{r}$. Note that just as is the case for the Biot-Savart law this is an inverse-square law ($d\vec{E}$ depends on inverse r^2) in spite of the r^3 term in the denominator. This is because the $\vec{r}$ term in the numerator cancels one of the r's in the denominator.

We can use the same basic procedure to calculate the magnetic field due to a current. Figure 30-5*b* shows a wire of arbitrary shape carrying a current i. We want to find the magnetic field $\vec{B}$ at a nearby point P. We first mentally divide the wire into differential elements ds and then define for each element a length vector $d\vec{s}$ that has length ds and whose direction is the direction of the current in ds. We can then define a differential *current-length element* to be $i\,d\vec{s}$; we wish to calculate the field $d\vec{B}$ produced at P by a single current-length element. From experiment we find that magnetic fields, like electric fields, can be superimposed to find a net field. Thus, we can calculate the net field $\vec{B}$ at P by summing contributions for discrete sources or by integrating the contributions $d\vec{B}$ from all the current-length elements in a continuous source. However, this summation (or integration) is more challenging than the process associated with electric fields because of a complexity. The charge element dq that produces an electric field is a scalar, but a current-length element $i\,d\vec{s}$ that produces a magnetic field is the product of a scalar and a vector.

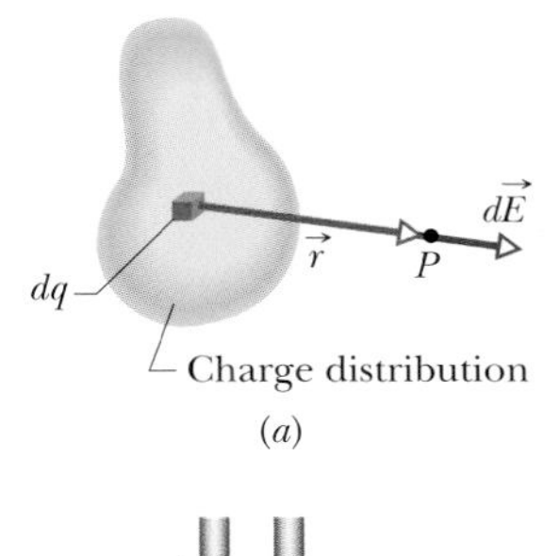

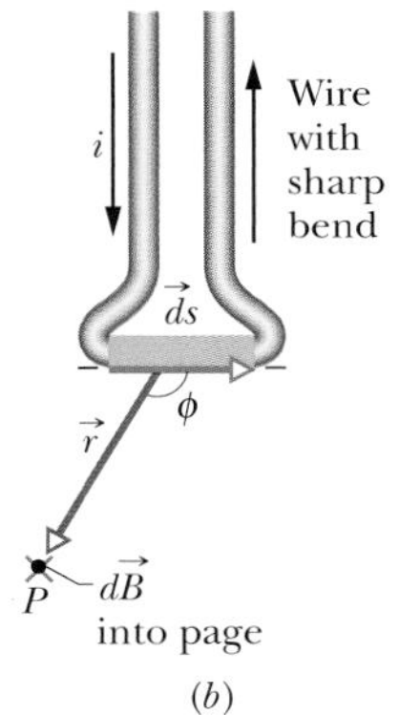

FIGURE 30-5 ■ (*a*) A charge element dq produces a differential electric field $d\vec{E}$ at point P. (*b*) A current-length element $i\,d\vec{s}$, isolated by sharp bends in the wire, produces a differential magnetic field $d\vec{B}$ at point P. The $\times$ (the tail of an arrow) at the dot for point P indicates that $d\vec{B}$ is directed into the page there for the special case where i and $d\vec{s}$ are parallel. If a small magnetic compass needle is used to detect the magnetic field, then its north pole points into the page.

Magnetic Field Due to a Current in a Long Straight Wire

Shortly we shall use the law of Biot and Savart to prove that the magnitude of the magnetic field at a perpendicular distance R from a long (infinite) straight wire carrying a current i is given by

$$B = \frac{\mu_0 |i|}{2\pi R} \quad \text{(long straight wire).} \tag{30-6}$$

The field magnitude $B = |\vec{B}|$ in Eq. 30-6 depends only on the amount of current and the perpendicular distance R of the point from the wire. We shall show in our derivation that the field lines of $\vec{B}$ form concentric circles around the wire, as Fig. 30-3 shows and as the iron filings in Fig. 30-2 suggest. The increase in the spacing of the lines in Fig. 30-3 with increasing distance from the wire represents the $1/R$ decrease in the magnitude of $\vec{B}$ predicted by $B = \mu_0|i|/2\pi R$ (Eq. 30-6). The lengths of the two vectors $\vec{B}$ in Fig. 30-3 also show the $1/R$ decrease when we use Ampère's right-hand rule for finding the direction of the magnetic field set up by a current-length element, such as a section of a long wire. What we are really doing is describing the orientation of concentric circles centered on the wire. A careful review of Fig. 30-3 yields two additional points that are often quite useful in solving magnetic field problems. Namely, the magnetic field $\vec{B}$ due to a current-carrying wire at any point is *tangent to a magnetic field line* and it is *perpendicular to a dashed radial line connecting the point and the current.*

Proof of Equation 30-6

Figure 30-6, which is just like Fig. 30-5*b* except that now the wire is straight and of infinite length, illustrates the task at hand; we seek the field $\vec{B}$ at point P, a perpendicular distance R from the wire. The magnitude of the differential magnetic field produced at P by the current-length element $|i d\vec{s}|$ located a distance r from P is given by Eq. 30-1:

$$|d\vec{B}| = \frac{\mu_0}{4\pi}\frac{|i d\vec{s}|\sin\phi}{r^2}.$$

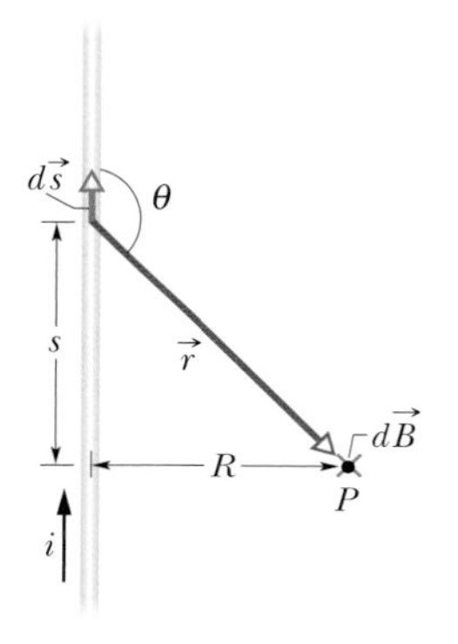

FIGURE 30-6 ■ Calculating the magnetic field produced by a current i in a long straight wire. Using either Ampère's right-hand rule or $d\vec{s} \times \vec{r}$, we find $d\vec{B}$ at P is directed into the page as shown.

Since the direction of $d\vec{s}$ is always in the direction of the current, we find that the direction of $d\vec{B}$ in Fig. 30-6 (given by $d\vec{s} \times \vec{r}$) is into the page.

Note that $d\vec{B}$ at point P has this same direction (into the page) for all the current-length elements into which the wire can be divided. Thus, we can find the magnitude of the magnetic field produced at P by the current-length elements in the upper half of the infinitely long wire by integrating dB in Eq. 30-1 from 0 to ∞.

Now consider a current-length element in the lower half of the wire, one that is as far below P as $d\vec{s}$ is above P. By Eq. 30-6, the magnetic field produced at P by this current-length element has the same magnitude and direction as that from $i\,d\vec{s}$ in Fig. 30-6. Further, the magnetic field produced by the lower half of the wire is exactly the same as that produced by the upper half. To find the magnitude of the total magnetic field $\vec{B}$ at P, we need only multiply the result of our integration by 2. We get

$$B = 2\int_0^\infty dB = \frac{\mu_0|i|}{2\pi}\int_0^\infty \frac{|\sin\phi\, ds|}{r^2}. \tag{30-7}$$

The variables ϕ, s, and r in this equation are not independent but (see Fig. 30-6) are related by

$$r = \sqrt{s^2 + R^2}$$

and

$$\sin\phi = \sin(\pi - \phi) = \frac{R}{\sqrt{s^2 + R^2}}.$$

Using these substitutions along with the solution to integral 19 in Appendix E, Eq. 30-7 describing the magnitude of the magnetic field becomes

$$B = \frac{\mu_0|i|}{2\pi}\int_0^\infty \frac{R\,ds}{(s^2 + R^2)^{3/2}} = \frac{\mu_0|i|}{2\pi R}\left[\frac{s}{(s^2 + R^2)^{1/2}}\right]_0^\infty.$$

Substituting the limits in the expression above gives a B-field magnitude of

$$B = \frac{\mu_0|i|}{2\pi R}, \quad \text{(infinite straight wire)}, \qquad (30\text{-}8)$$

which is the relation we set out to prove. Note that the magnitude of the magnetic field at P due to either the lower half or the upper half of the infinite wire in Fig. 30-6 is half this value; that is,

$$B = \frac{\mu_0|i|}{4\pi R} \quad \text{(semi-infinite straight wire)}. \qquad (30\text{-}9)$$

Magnetic Field Due to a Current in a Circular Arc of Wire

To find the magnetic field produced at a point by a current in a curved wire, we would again use Eq. 30-1 to write the magnitude of the field produced by a single current-length element, and we would again integrate to find the net field produced by all the current-length elements. That integration can be difficult, depending on the shape of the wire; it is fairly straightforward, however, when the wire is a circular arc and the point is the center of curvature.

Figure 30-7a shows such an arc-shaped wire with central angle ϕ_C, radius R, and center C, carrying current i. At C, each current-length element $i\,d\vec{s}$ of the wire produces a magnetic field element of magnitude dB given by Eq. 30-1. Moreover, as Fig. 30-7b shows, no matter where the element is located on the wire, the angle ϕ between the vectors $d\vec{s}$ and $\vec{r}$ is 90°; also, $r = R$. Thus, by substituting R for r and 90° for ϕ, we obtain from Eq. 30-1,

$$dB = \frac{\mu_0}{4\pi}\frac{|i|\,ds\sin 90°}{R^2} = \frac{\mu_0}{4\pi}\frac{|i|\,ds}{R^2}. \qquad (30\text{-}10)$$

The field at C due to each current-length element in the circular arc has this same magnitude.

An application of the right-hand rule anywhere along the wire (as in Fig. 30-7c) will show that all the differential fields $d\vec{B}$ have the same direction at C—directly out of the page. Thus, the total field at C is simply the sum (via integration) of all the fields $d\vec{B}$. We use the identity $ds = R\,d\phi$ to change the variable of integration from ds to $d\phi$ and obtain, from Eq. 30-10, a magnitude of

$$B = \int dB = \int_0^{\phi_C} \frac{\mu_0}{4\pi}\frac{|i|R\,d\phi}{R^2} = \frac{\mu_0|i|}{4\pi R}\int_0^{\phi_C} d\phi.$$

Integrating, we find that

$$B = \frac{\mu_0|i|\phi_C}{4\pi R} \quad \text{(at center of circular arc)}. \qquad (30\text{-}11)$$

Note that this equation gives us the magnitude of the magnetic field *only* at the center of curvature of a circular arc of current. When you insert data into the equation, you must be careful to express ϕ_C in radians rather than degrees. For example, to find the magnitude of the magnetic field at the center of a full circle of current,

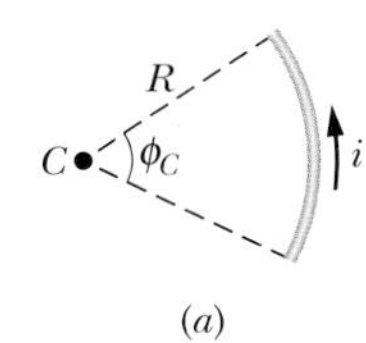

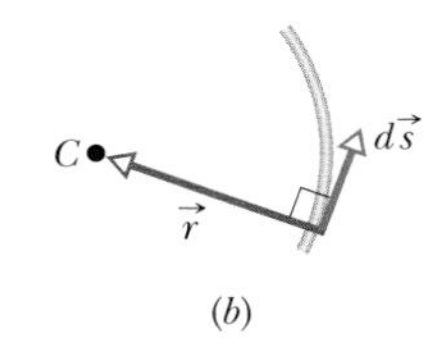

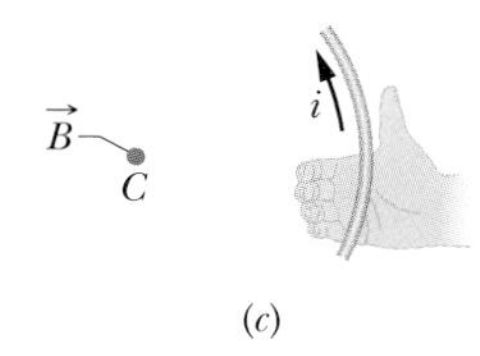

FIGURE 30-7 ■ (a) A wire in the shape of a circular arc with center C carries current i. (b) For any element of wire along the arc, the angle between the directions of $d\vec{s}$ and $\vec{r}$ is 90°. (c) Determining the direction of the magnetic field at the center C due to the current in the wire; the field is out of the page, in the direction of the fingertips, as indicated by the colored dot at C.

you would substitute 2π for ϕ_C in Eq. 30-11, finding

$$B = \frac{\mu_0|i|(2\pi)}{4\pi R} = \frac{\mu_0|i|}{2R} \quad \text{(at center of full circle).} \tag{30-12}$$

READING EXERCISE 30-2: A uniform magnetic field is directed toward the right in the plane of the paper as shown in the diagram that follows. A wire oriented perpendicular to the plane of the paper carries a current i. Suppose that the resultant magnetic field at point 1 due to a superposition of the uniform magnetic field of magnitude $|\vec{B}|$ and the magnetic field of the wire at point 1 is zero. (a) Is the direction of the current in the wire into or out of the paper? Explain how you arrived at your conclusion. (b) Assume that point 2 lies at the same distance from the center of the wire as point 1 and that the length of the vector assigned to represent the magnitude of the uniform external magnetic field is that shown to the left. Construct a vector arrow showing the length and direction of the resultant magnetic field vector at point 2. Explain how you deduced what the vector should be. (Adapted from A. Arons, *Homework and Test Questions for Introductory Physics Teaching,* John Wiley and Sons, 1947.)

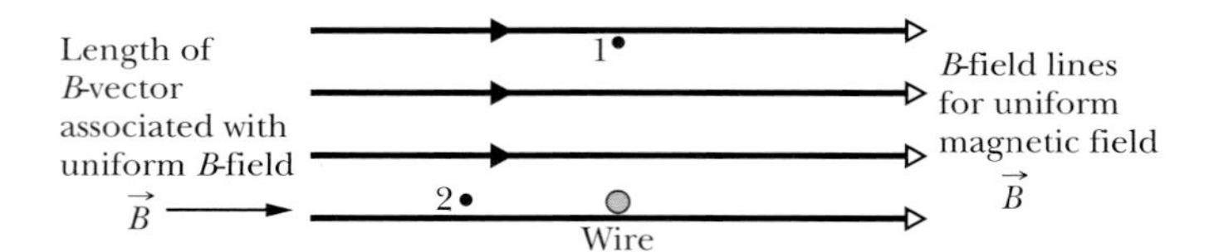

TOUCHSTONE EXAMPLE 30-1: An Arc and Two Straight Lines

The wire in Fig. 30-8*a* carries a current i and consists of a circular arc of radius R and central angle $\pi/2$ rad, and two straight sections whose extensions intersect the center C of the arc. What magnetic field $\vec{B}$ does the current produce at C?

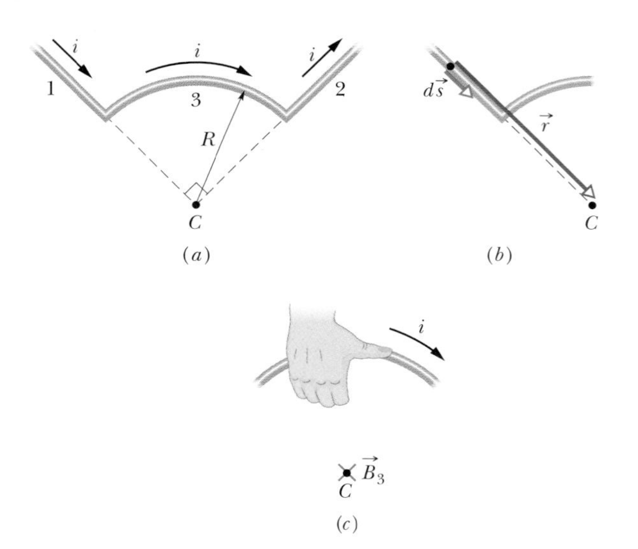

FIGURE 30-8 (*a*) A wire consists of two straight sections (1 and 2) and a circular arc (3), and carries current i. (*b*) For a current-length element in section 1, the angle between $d\vec{s}$ and $\vec{r}$ is zero. (*c*) Determining the direction of magnetic field $\vec{B}_3$ at C due to the current in the circular arc; the field is into the page there.

SOLUTION One **Key Idea** here is that we can find the magnetic field $\vec{B}$ at point C by applying the Biot–Savart law of Eq. 30-3 to the wire. A second **Key Idea** is that the application of Eq. 30-3 can be simplified by evaluating $\vec{B}$ separately for the three distinguishable sections of the wire—namely, (1) the straight section at the left, (2) the straight section at the right, and (3) the circular arc.

Straight sections. For any current-length element in section 1, the angle ϕ between $d\vec{s}$ and $\vec{r}$ is zero (Fig. 30-8*b*), so Eq. 30-1 gives us

$$|d\vec{B}_1| = \frac{\mu_0}{4\pi}\frac{|i\,d\vec{s}\,|\sin\phi}{r^2} = \frac{\mu_0}{4\pi}\frac{|i\,d\vec{s}\,|\sin 0}{r^2} = 0\text{ T}.$$

Thus, the current along the entire length of wire in straight section 1 contributes no magnetic field at C:

$$\vec{B}_1 = 0\text{ T}.$$

The same situation prevails in straight section 2, where the angle ϕ between $d\vec{s}$ and $\vec{r}$ for any current-length element is 180°. Thus,

$$\vec{B}_2 = 0\text{ T}.$$

Circular arc. The **Key Idea** here is that application of the Biot–Savart law to evaluate the magnetic field at the center of a circular arc leads to Eq. 30-11 ($|B| = \mu_0|i|\phi/4\pi R$). Here the central angle ϕ of the arc is $\pi/2$ rad. Thus from Eq. 30-11, the magnitude of the magnetic field $\vec{B}_3$ at the arc's center C is

$$|\vec{B}_3| = \frac{\mu_0|i|(\pi/2)}{4\pi R} = \frac{\mu_0|i|}{8R}.$$

To find the direction of $\vec{B}_3$, we apply the right-hand rule displayed in Fig. 30-4. Mentally grasp the circular arc with your right hand as suggested in Fig. 30-8*c*, with your thumb in the direction of the current. The direction in which your fingers curl around the wire indicates the direction of the magnetic field lines around the wire. In the region of point C (inside the circular arc), your fingertips point *into the plane* of the page. Thus, $\vec{B}_3$ is directed into that plane.

Net field. Generally, when we must combine two or more magnetic fields to find the net magnetic field, we must combine the fields as vectors and not simply add their magnitudes. Here, however, only the circular arc produces a magnetic field at point C. Thus, we can write the magnitude of the net field $\vec{B}$ as

$$|\vec{B}| = |\vec{B}_1 + \vec{B}_2 + \vec{B}_3| = 0 + 0 + \left|\frac{\mu_0 i}{8R}\right| = \left|\frac{\mu_0 i}{8R}\right|. \quad \text{(Answer)}$$

The direction of $\vec{B}$ is the direction of $\vec{B}_3$—namely, into the plane of Fig. 30-8.

TOUCHSTONE EXAMPLE 30-2: Two Long Parallel Wires

Figure 30-9*a* shows two long parallel wires carrying currents i_1 and i_2 in opposite directions. What are the magnitude and direction of the net magnetic field at point P? Assume the following values: $i_1 = 15$ A, $i_2 = 32$ A, and $d = 5.3$ cm.

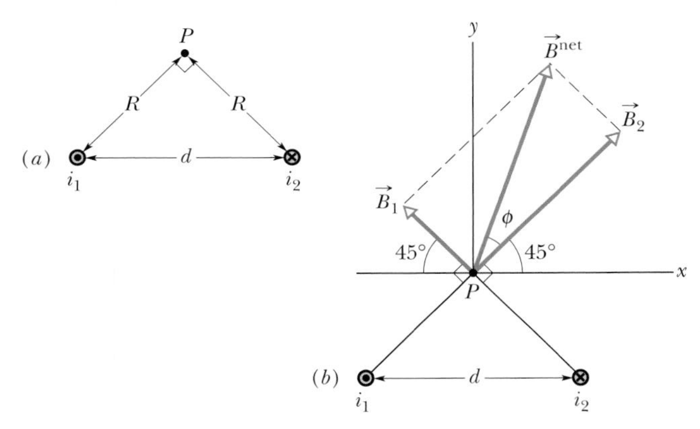

FIGURE 30-9 ■ (*a*) Two wires carry currents i_1 and i_2 in opposite directions (out of and into the page). Note the right angle at P. (*b*) The separate fields $\vec{B}_1$ and $\vec{B}_2$ are combined vectorially to yield the net field $\vec{B}^{net}$.

SOLUTION ■ One **Key Idea** here is that the net magnetic field $\vec{B}$ at point P is the vector sum of the magnetic fields due to the currents in the two wires. A second **Key Idea** is that we can find the magnetic field due to any current by applying the Biot–Savart law to the current. For points near the current in a long straight wire, that law leads to Eq. 30-6.

In Fig. 30-9*a*, point P is distance R from both currents i_1 and i_2. Thus, Eq. 30-6 tells us that at point P those currents produce magnetic fields $\vec{B}_1$ and $\vec{B}_2$ with magnitudes

$$B_1 = \frac{\mu_0 |i_1|}{2\pi R} \quad \text{and} \quad B_2 = \frac{\mu_0 |i_2|}{2\pi R}.$$

In the right triangle of Fig. 30-9*a*, note that the base angles (between sides R and d) are both 45°. Thus, we may write $\cos 45° = R/d$ and replace R with $d \cos 45°$. Then the field magnitudes B_1 and B_2 become

$$B_1 = \frac{\mu_0 |i_1|}{2\pi d \cos 45°} \quad \text{and} \quad B_2 = \frac{\mu_0 |i_2|}{2\pi d \cos 45°}.$$

We want to combine $\vec{B}_1$ and $\vec{B}_2$ to find their vector sum, which is the net field $\vec{B}^{net}$ at P. To find the directions of $\vec{B}_1$ and $\vec{B}_2$, we apply the right-hand rule of Fig. 30-4 to each current in Fig. 30-9*a*. For wire 1, with current out of the page, we mentally grasp the wire with the right hand, with the thumb pointing out of the page. Then the curled fingers indicate that the field lines run counterclockwise. In particular, in the region of point P, they are directed upward to the left. Recall that the magnetic field at a point near a long, straight current-carrying wire must be directed perpendicular to a radial line between the point and the current. Thus, $\vec{B}_1$ must be directed upward to the left as drawn in Fig. 30-9*b*. (Note carefully the perpendicular symbol between vector $\vec{B}_1$ and the line connecting point P and wire 1.)

Repeating this analysis for the current in wire 2, we find that $\vec{B}_2$ is directed upward to the right as drawn in Fig. 30-9*b*. (Note the perpendicular symbol between vector $\vec{B}_2$ and the line connecting point P and wire 2.)

We can now vectorially add $\vec{B}_1$ and $\vec{B}_2$ to find the net magnetic field $\vec{B}^{net}$ at point P, either by using a vector-capable calculator or by resolving the vectors into components and then combining the components of $\vec{B}^{net}$. However, in Fig. 30-9*b*, there is a third method: Because $\vec{B}_1$ and $\vec{B}_2$ are perpendicular to each other, they form the legs of a right triangle, with $\vec{B}^{net}$ as the hypotenuse. The Pythagorean theorem then gives us

$$B^{net} = \sqrt{B_1^2 + B_2^2} = \frac{\mu_0}{2\pi d(\cos 45°)}\sqrt{i_1^2 + i_2^2}$$

$$= \frac{(4\pi \times 10^{-7}\,\text{T}\cdot\text{m/A})\sqrt{(15\,\text{A})^2 + (32\,\text{A})^2}}{(2\pi)(5.3 \times 10^{-2}\,\text{m})(\cos 45°)}$$

$$= 1.89 \times 10^{-4}\,\text{T} \approx 190\,\mu\text{T}. \quad \text{(Answer)}$$

The angle ϕ between the directions of $\vec{B}^{net}$ and $\vec{B}_2$ in Fig. 30-9*b* follows from

$$\phi = \tan^{-1}\frac{B_1}{B_2},$$

which, with B_1 and B_2 as given above, yields

$$\phi = \tan^{-1}\frac{i_1}{i_2} = \tan^{-1}\frac{15\,\text{A}}{32\,\text{A}} = 25°.$$

The angle between the direction of $\vec{B}^{net}$ and the x axis shown in Fig. 30-9*b* is then

$$\phi + 45° = 25° + 45° = 70°. \quad \text{(Answer)}$$

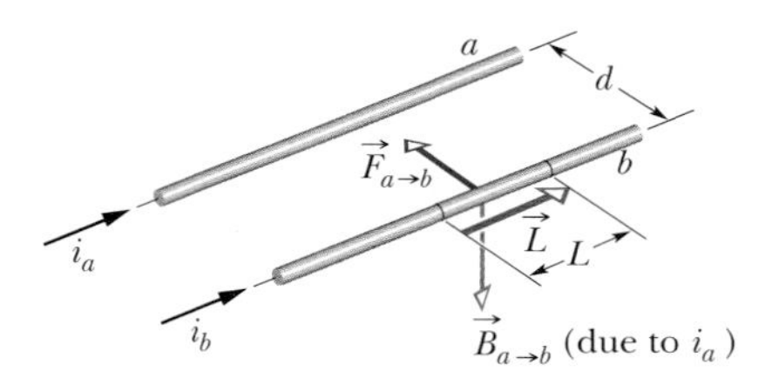

FIGURE 30-10 ■ Two parallel wires carrying currents in the same direction attract each other. $\vec{B}_{a\to b}$ is the magnetic field at wire b produced by the current in wire a. $\vec{F}_{a\to b}$ is the resulting force acting on wire b because it carries current in field $\vec{B}_{a\to b}$.

30-4 Force Between Parallel Currents

Back in 1820, when Ampère was first replicating Oersted's observations, he predicted that two current-carrying wires in parallel would exert forces on each other. This is a logical consequence of the Biot–Savart law, which quantifies the magnetic field surrounding a current-carrying wire, and the magnetic force law, which describes the force on a current in the presence of a magnetic field. Indeed, Ampère observed that there is a mutual interaction between the two wires. In other words, each wire exerts a force on the other. As shown in Fig. 30-10, the application of the right-hand rules that accompany the Biot–Savart law (Eq. 30-3) and the expression for the magnetic force on a current (Eq. 29-22) lead us to predict that wires that carry currents in the same direction will attract, whereas wires that carry currents in opposite directions will repel. It is interesting that the attractions and repulsions are opposite to the electrostatic and magnetic relationships, where unlike charges or poles attract and like charges or poles repel.

We can use the two equations just mentioned to derive a third equation that describes the forces between two parallel current-carrying wires. Why do we want to determine these interaction forces? Three reasons come to mind. First, we can compare the measurement of these forces to the forces predicted by our third equation to verify the Biot–Savart law. Second, these mutual interaction forces enable us to define the ampere as the SI unit of current. Finally, by understanding the nature of these forces we can design an electromagnetic launcher (like that mentioned in the "puzzler" on the first page of this chapter).

Figure 30-10 shows two parallel wires, separated by a distance d and carrying currents i_a and i_b. The first step in analyzing the forces between these wires is to find an expression for the force on wire b due to the current in wire a. The current in wire a produces a magnetic field $\vec{B}_{a\to b}$ at the location of wire b, and it is this magnetic field produced by wire a that actually causes wire b to experience a force denoted as $\vec{F}_{a\to b}$. According to Eq. 30-6, the magnitude of $B_{a\to b}$ at every point along wire b is

$$B_{a\to b} = \frac{\mu_0 |i_a|}{2\pi d}. \tag{30-13}$$

The right-hand rule tells us that the direction of $\vec{B}_{a\to b}$ at wire b is down, as shown in Fig. 30-10.

Now that we have determined the magnetic field vector, we can find the force that wire a produces on wire b. The expression for the force on a length of current-carrying wire (Eq. 29-22) tells us that the force on wire b is

$$\vec{F}_{a\to b} = i_b\vec{L} \times \vec{B}_{a\to b}, \tag{30-14}$$

where $\vec{L}$ is the length vector (direction given by the direction of current i) of the wire. In Fig. 30-10 the vectors $\vec{L}$ and $\vec{B}_{a\to b}$ are perpendicular, so using Eqs. 30-13 and 30-14, we can express the magnitude of the force on wire b due to the current in wire a as

$$F_{a\to b} = |i_b| L B_{a\to b} \sin 90° = \frac{\mu_0 L |i_a i_b|}{2\pi d}. \tag{30-15}$$

The direction of $\vec{F}_{a\to b}$ is the direction of the cross product $\vec{L} \times \vec{B}_{a\to b}$. Applying the right-hand rule for cross products to $\vec{L}$ and $\vec{B}_{a\to b}$ in Fig. 30-10, we find that $\vec{F}_{a\to b}$ points directly toward wire a, as shown.

The general procedure for finding the force on a current-carrying wire is this:

> To find the force on a current-carrying wire due to a second current-carrying wire, first find the field due to the second wire at the site of the first wire. Then find the force on the first wire due to that field.

We could now use this procedure to compute the force on wire a due to the current in wire b. We would find that the force has the same magnitude but is in the opposite direction. This is true regardless of whether the currents are the same or in opposite directions. Once again, Newton's Third Law holds:

> Parallel currents attract, and antiparallel currents repel.

The forces acting between currents in parallel wires provide us with the basis for defining the ampere, which is one of the seven SI base units. It is appropriately named after André Marie Ampère, who was the first to demonstrate the forces acting between parallel currents. The official SI definition, adopted in 1946, is:

> The **ampere** is that constant current which, if maintained in two straight, parallel conductors of infinite length, of negligible circular cross section, and placed 1 m apart in a vacuum, would produce between each of these conductors a force equal to 2×10^{-7} newton per meter of length.

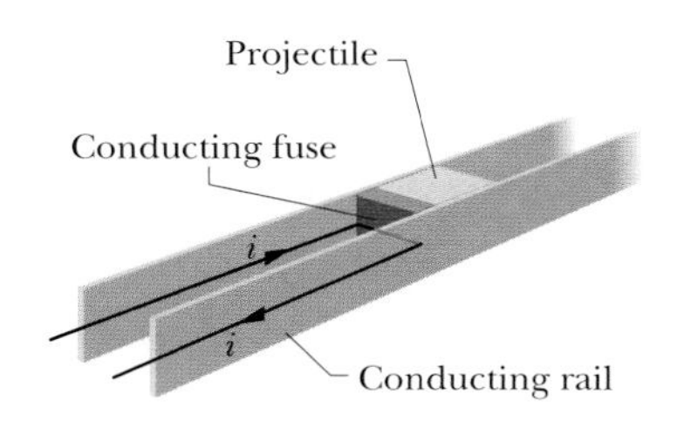

(a)

(b)

FIGURE 30-11 ■ (a) A rail gun, as a current i is set up in it. The current rapidly causes the conducting fuse to vaporize. (b) The current produces a magnetic field $\vec{B}$ between the rails, and the field causes a force $\vec{F}$ to act on the conducting gas, which is part of the current path. The gas propels the projectile along the rails, launching it.

Rail Gun

A rail gun is a device in which a magnetic force can accelerate a projectile to a high speed in a short time. The basics of a rail gun are shown in Fig. 30-11a. A large current flows in a circuit consisting of two conducting rails joined by a conducting "fuse" (such as a narrow piece of copper) between the rails, and then back to the current source along the second rail. The projectile to be fired lies on the far side of the fuse and fits loosely between the rails. Immediately after the current is established, the fuse element melts and vaporizes, creating a conducting gas between the rails where the fuse had been.

The right-hand rule of Fig. 30-4 shows that the current in the rails of Fig. 30-11a produces a magnetic field that is directed downward between the rails. The net magnetic field $\vec{B}$ exerts a force $\vec{F}$ on the gas due to the current i through the gas (Fig. 30-11b). Using Eq. 30-14 and the right-hand rule for cross products, we find that $\vec{F}$ points outward along the rails. As the gas is forced outward along the rails, it pushes the projectile, accelerating it by as much as $5 \times 10^7\,\text{m/s}^2$ or $(5 \times 10^6\,g)$, and then launches it with a speed of 10 km/s, all within less than one millisecond.

READING EXERCISE 30-3: The figure shows three long, straight, parallel, equally spaced wires with identical amounts of current either into or out of the page. Rank the wires according to the magnitude of the force on each due to the currents in the other two wires, greatest first. ■

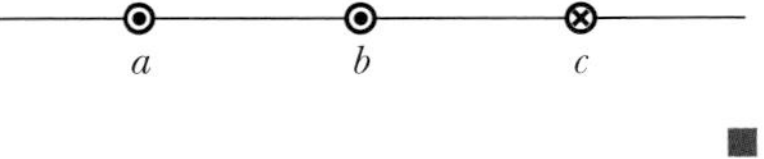

30-5 Ampère's Law

We can find the net electric field due to *any* distribution of charges with the inverse-square law for the differential field $d\vec{E}$ (Eq. 30-5), but if the distribution is complicated, we may have to use a computer. Recall, however, that if the distribution has planar, cylindrical, or spherical symmetry, we can apply Gauss' law to find the net electric field with considerably less effort.

Similarly, we can find the net magnetic field due to any distribution of currents with the inverse-square law for the differential field $d\vec{B}$ (Eq. 30-3), but again we may have to use a computer for a complicated distribution. However, if the distribution has enough symmetry, we can apply *Ampère's law* to find the magnetic field with considerably less effort. This law, which can be derived from the Biot–Savart law, has traditionally been credited to André Marie Ampère (1775–1836), for whom the SI unit of current is named. However, the law actually was advanced by English physicist James Clerk Maxwell.

Ampère's law is

$$\oint \vec{B} \cdot d\vec{s} = \mu_0 i^{\text{enc}} \quad \text{(Ampère's law).} \tag{30-16}$$

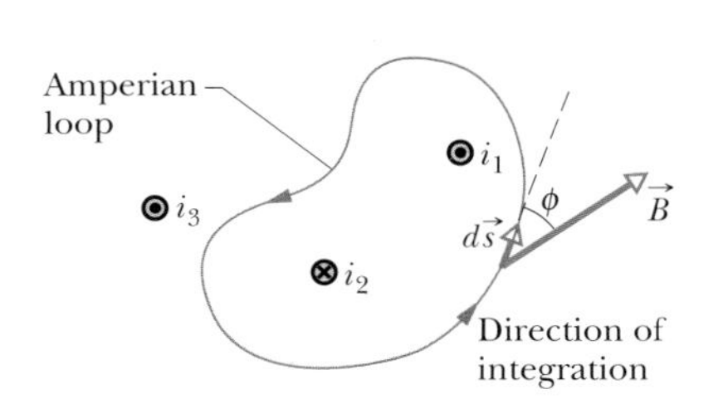

FIGURE 30-12 ■ Ampère's law applied to an arbitrary Ampèrian loop that encircles two long straight wires but excludes a third wire. Note the directions of the currents.

The circle on the integral sign means that the scalar (or dot) product $\vec{B} \cdot d\vec{s}$ is to be integrated around an imaginary *closed* loop, called an *Ampèrian loop*. The current i^{enc} on the right is the *net* current encircled by that loop.

In Gauss' law we choose a closed surface on which to evaluate the integral. The integral flux is proportional to the net charge enclosed by the surface. In Ampère's law, we choose a closed loop on which to evaluate the integral. The integral is proportional to the net current passing through the loop.

To see the meaning of the scalar product $\vec{B} \cdot d\vec{s}$ and its integral, let us first apply Ampère's law to the general situation shown in Fig. 30-12. This figure depicts the cross sections of three long straight wires that carry currents i_1, i_2, and i_3 either directly into or directly out of the page. An arbitrary Ampèrian loop lying in the plane of the page encircles two of the currents but not the third. The counterclockwise direction marked on the loop indicates the arbitrarily chosen direction of integration for Eq. 30-16.

To apply Ampère's law, we mentally divide our imaginary loop into short, nearly straight, directed pieces, $d\vec{s}$. The direction of each of these pieces is tangent to the loop along the direction of integration. Assume that at the location of the element $d\vec{s}$ shown in Fig. 30-12, the net magnetic field due to the three currents is $\vec{B}$. Because the wires are perpendicular to the page, we know that the magnetic field at $d\vec{s}$ due to each current is in the plane of Fig. 30-12; thus, the net magnetic field $\vec{B}$ at $d\vec{s}$ must also be in that plane. However, we do not know the orientation of $\vec{B}$ within the plane. In Fig. 30-12, $\vec{B}$ is arbitrarily drawn at an angle ϕ to the direction of $d\vec{s}$.

The scalar product $\vec{B} \cdot d\vec{s}$ on the left side of Eq. 30-16 is then equal to $(B \cos\phi)ds$. Thus, Ampère's law can be written as

$$\oint \vec{B} \cdot d\vec{s} = \oint (B \cos \phi)\, ds = \mu_0 i^{\text{enc}}. \tag{30-17}$$

We can now interpret the scalar product $\vec{B} \cdot d\vec{s}$ as being the product of a length ds of the Ampèrian loop and the field component $B \cos \phi$ that is tangent to the loop. Then we can interpret the integration as being the summation of all such products around the entire loop.

When we can actually perform this integration, we do not need to know the direction of $\vec{B}$ before integrating. Instead, we arbitrarily assume $\vec{B}$ to be generally in the direction of integration (as in Fig. 30-12). Then we use the following curled fingers-straight thumb right-hand rule to assign a plus sign or a minus sign to each of the currents that make up the net encircled current i^{enc}:

> **CURLED-STRAIGHT RIGHT-HAND RULE FOR AMPÈRE'S LAW:** Curl your right hand around the Ampèrian loop, with the fingers pointing in the direction of integration. A current through the loop in the general direction of your outstretched thumb is assigned a plus sign, and a current generally in the opposite direction is assigned a minus sign.

Finally, we solve Eq. 30-17 for the magnitude of $\vec{B}$. Once we have chosen a coordinate system to describe the system, we can use Ampère's law right-hand rule to decide whether $\vec{B}$ is positive or negative.

In Fig. 30-13 we apply the curled-straight rule for Ampère's law to the situation of Fig. 30-12. With the indicated counterclockwise direction of integration, the net current encircled by the loop is

$$i^{\text{enc}} = i_1 - i_2.$$

(Current i_3 is not encircled by the loop.) We can then rewrite Eq. 30-17 as

$$\left|\oint(B\cos\phi)ds\right| = \mu_0|(i_2 - i_1)|. \quad (30\text{-}18)$$

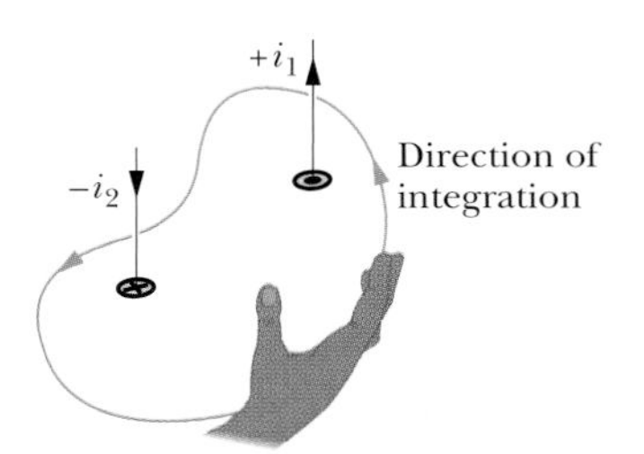

FIGURE 30-13 ■ A right-hand rule for Ampère's law, to determine the signs for currents encircled by an Ampèrian loop. The situation is that of Fig. 30-12.

You might wonder why, since current i_3 contributes to the magnetic-field magnitude B on the left side of Eq. 30-18, it is not needed on the right side. The answer is that the contributions of current i_3 to the magnetic field cancel out because the integration in Eq. 30-18 is made around the full loop. In contrast, the contributions of an encircled current to the magnetic field do not cancel out.

We cannot solve Eq. 30-18 for the magnitude B of the magnetic field, because for the situation of Fig. 30-12 we do not have enough information to solve the integral. However, we do know the magnitude of the integral; it must be equal to the value of $\mu_0|(i_1 - i_2)|$, which is set by the net current passing through the loop. Next we apply Ampère's law to two situations in which symmetry does allow us to solve the integrals and determine the magnetic fields.

The Magnetic Field Outside a Long Straight Wire with Current

Figure 30-14 shows a long straight wire that carries current i (assumed to be uniformly distributed) that points directly out of the page. The equation for the magnetic field magnitude, B, produced by a long straight wire (Eq. 30-6) tells us that B depends only on the radial distance from the wire. That is, the field $\vec{B}$ has cylindrical symmetry about the wire. We can take advantage of that symmetry to simplify the integral in Ampère's law (Eqs. 30-16 and 30-17) if we encircle the wire with a concentric circular Ampèrian loop of radius r, as in Fig. 30-14. The magnetic field $\vec{B}$ then has the same magnitude B at every point on the loop. We shall integrate counterclockwise, so that $d\vec{s}$ has the direction shown in Fig. 30-14.

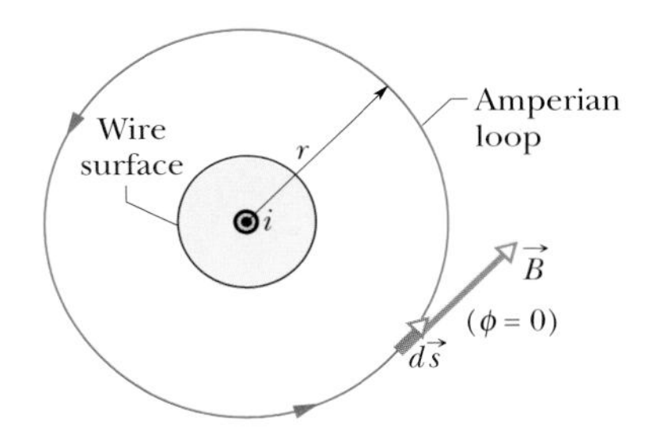

FIGURE 30-14 ■ Using Ampère's law to find the magnetic field produced by a current i in a long straight wire. The Ampèrian loop is a concentric circle that lies outside the wire.

We can further simplify the quantity $B\cos\phi$ in Eq. 30-17 by noting that $\vec{B}$ is tangent to the loop at every point along the loop, as is $d\vec{s}$. Thus, $\vec{B}$ and $d\vec{s}$ are parallel at each point on the loop. Then at every point the angle ϕ between $d\vec{s}$ and $\vec{B}$ is 0° (so $\cos\phi = +1$). The magnitude of the integral in Eq. 30-17 then becomes

$$\left|\oint \vec{B}\cdot d\vec{s}\right| = \oint(B\cos\phi)\,ds = B\left|\oint ds\right| = B(2\pi r).$$

Note that $\oint ds$ above is the summation of all the line segment lengths ds around the circular loop; that is, it simply gives the circumference $2\pi r$ of the loop.

The right side of Ampère's law becomes $+\mu_0|i|$ and we then have

$$B(2\pi r) = \mu_0 i$$

or

$$B = \frac{\mu_0|i|}{2\pi r}. \quad (30\text{-}19)$$

With a slight change in notation, this is Eq. 30-6, which we derived earlier—with considerably more effort—using the Biot-Savart law. We know that the correct direction of $\vec{B}$ must be the counterclockwise one shown in Fig. 30-14 when i is positive. When i is negative, the correct direction for $\vec{B}$ is clockwise.

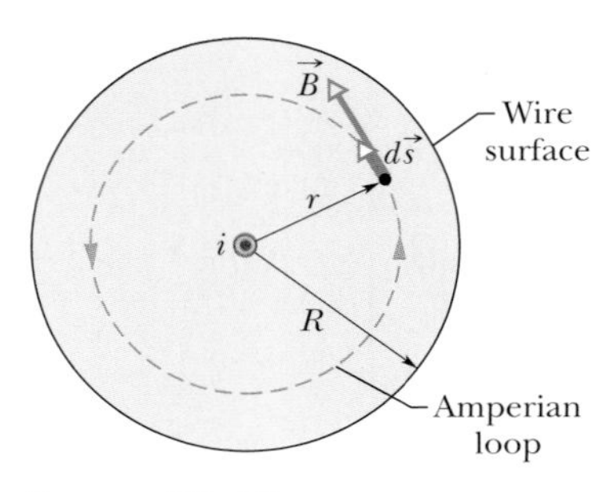

FIGURE 30-15 ■ Using Ampère's law to find the magnetic field that a current i produces inside a long straight wire of circular cross section. The current is uniformly distributed over the cross section of the wire and emerges from the page. An Ampèrian loop is drawn inside the wire.

The Magnetic Field Inside a Long Straight Wire with Current

Figure 30-15 shows the cross section of a long straight wire of radius R that carries a uniformly distributed current i either directly out of the page or directly into the page. Because the current is uniformly distributed over a cross section of the wire, the magnetic field $\vec{B}$ that it produces must have cylindrical symmetry. Thus, to find the magnetic field at points inside the wire, we can again use an Amperian loop of radius r, as shown in Fig. 30-15, where now $r < R$. Symmetry again requires that $\vec{B}$ is tangent to the loop, as shown, so the left side of Ampère's law again yields

$$\oint \vec{B} \cdot d\vec{s} = B \left| \oint d\vec{s} \right| = B(2\pi r). \tag{30-20}$$

To find the right side of Ampère's law, we note that because the current is uniformly distributed, the current i^{enc} encircled by the loop is proportional to the area encircled by the loop; that is,

$$i^{\text{enc}} = i \frac{\pi r^2}{\pi R^2} = i \frac{r^2}{R^2}. \tag{30-21}$$

Then Ampère's law gives us

$$B(2\pi r)\mu_0 |\, i^{\text{enc}} | = \mu_0 |\, i \,| \frac{r^2}{R^2}$$

or

$$B = \left(\frac{\mu_0 |\, i \,|}{2\pi R^2} \right) r. \tag{30-22}$$

Thus, inside the wire, the magnitude B of the magnetic field is proportional to r; that magnitude is zero at the center and a maximum at the surface, where $r = R$. Note that Eqs. 30-19 and 30-22 give the same value for B at $r = R$; that is, the expressions for the magnetic field outside the wire and inside the wire yield the same result at the surface of the wire.

READING EXERCISE 30-4: The figure shows three equal currents i (two parallel and one antiparallel) and four Amperian loops. Rank the loops according to the magnitude of $\oint \vec{B} \cdot d\vec{s}$ along each, greatest first. ■

TOUCHSTONE EXAMPLE 30-3: Hollow Conducting Cylinder

Figure 30-16a shows the cross section of a long hollow conducting cylinder with inner radius a = 2.0 cm and outer radius b = 4.0 cm. The cylinder carries a current out of the page, and the current density in the cross section is given by $|\vec{J}| = cr^2$, with $c = 3.0 \times 10^6$ A/m^4 and r in meters. What is the magnitude of the magnetic field $\vec{B}$ at a point that is 3.0 cm from the central axis of the cylinder?

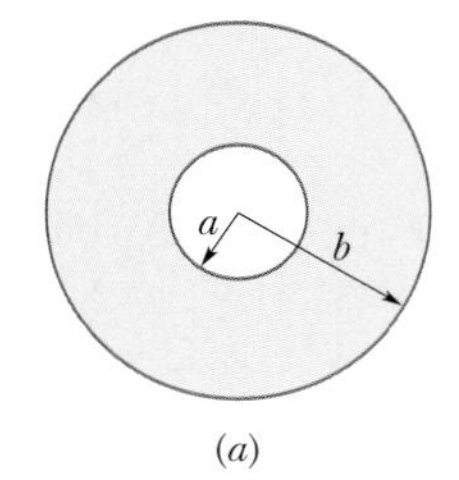

(a)

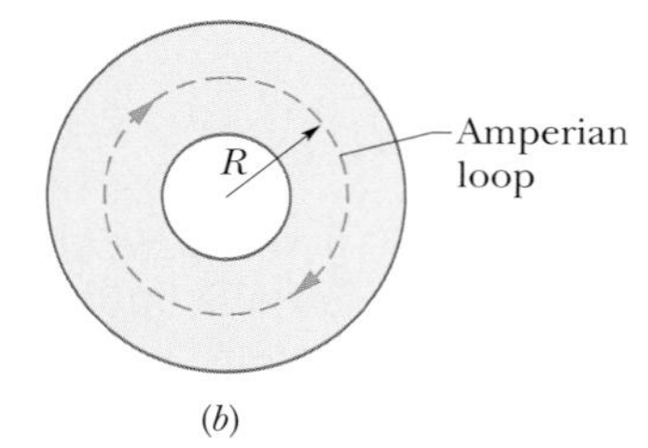

(b)

FIGURE 30-16 ■ (a) Cross section of a conducting cylinder of inner radius a and outer radius b. (b) An Ampèrian loop of radius R is added to compute the magnetic field at points that are a distance R from the central axis.

SOLUTION ■ The point at which we want to evaluate $\vec{B}$ is inside the material of the conducting cylinder, between its inner and outer radii. We note that the current distribution has cylindrical symmetry (it is the same all around the cross section for any given radius). Thus, the **Key Idea** here is that the symmetry allows us to use Ampère's law to find $\vec{B}$ at the point. We first draw the Ampèrian loop shown in Fig. 30-16b. The loop is concentric with the cylinder and has radius R = 3.0 cm, because we want to evaluate $\vec{B}$ at that distance from the cylinder's central axis.

Next, we must compute the current i^{enc} that is encircled by the Ampèrian loop. However, a second **Key Idea** is that we *cannot* set up a proportionality as in Eq. 30-21, because here the current is not uniformly distributed. Instead, we must integrate the current density from the cylinder's inner radius a to the loop radius r. Since $\vec{J}$ and $d\vec{A}$ are parallel, $\vec{J} \cdot d\vec{A} = J dA$, so

$$|i^{\text{enc}}| = \left|\int \vec{J} \cdot d\vec{A}\right| = \left|\int_a^R cr^2(2\pi r\, dr)\right|$$

$$= \left|2\pi c \int_a^R r^3\, dr\right| = \left|2\pi c \left[\frac{r^4}{4}\right]_a^R\right|$$

since $$|i^{\text{enc}}| = \frac{\pi c(R^4 - a^4)}{2}, \quad \text{since } R > a.$$

The direction of integration indicated in Fig. 30-16*b* is (arbitrarily) clockwise. Applying the right-hand rule for Ampère's law to that loop, we find that we should take i^{enc} as negative because the current is directed out of the page but our thumb is directed into the page.

We next evaluate the left side of Ampère's law exactly as we did in Fig. 30-15, and we again obtain Eq. 30-20. Then Ampère's law,

$$\oint \vec{B} \cdot d\vec{s} = \mu_0 i^{\text{enc}},$$

gives us

$$(B\cos\phi)(2\pi R) = -\frac{\mu_0 \pi c}{2}(R^4 - a^4),$$

where $\cos\phi = \cos 0° = +1$ if $\vec{B}$ is parallel to $d\vec{s}$ and $\cos\phi = \cos 180° = -1$ if $\vec{B}$ is antiparallel to $d\vec{s}$. Solving for $(B\cos\phi)$ for $\phi = 180°$and substituting known data yield

$$(B\cos\phi) = -\frac{\mu_0 c}{4R}(R^4 - a^4)$$

$$-B = -\frac{(4\pi \times 10^{-7}\ \text{T}\cdot\text{m/A})(3.0 \times 10^6\ \text{A/m}^4)}{4(0.030\ \text{m})} \times [(0.030\ \text{m})^4 - (0.020\ \text{m})^4]$$

$$-B = -2.0 \times 10^{-5}\ \text{T}.$$

Thus, the magnetic field $\vec{B}$ at a point 3.0 cm from the central axis is

$$B = 2.0 \times 10^{-5}\ \text{T} \qquad \text{(Answer)}$$

and forms magnetic field lines that are directed opposite our direction of integration, hence counterclockwise in Fig. 30-16*b*.

30-6 Solenoids and Toroids

Magnetic Field of a Solenoid

We now turn our attention to another situation in which Ampère's law proves useful. It concerns the magnetic field produced by the current in a long, tightly wound helical coil of wire. Such a coil is called a **solenoid** (Fig. 30-17). Solenoids are very common electrical devices that are important in many technological applications.

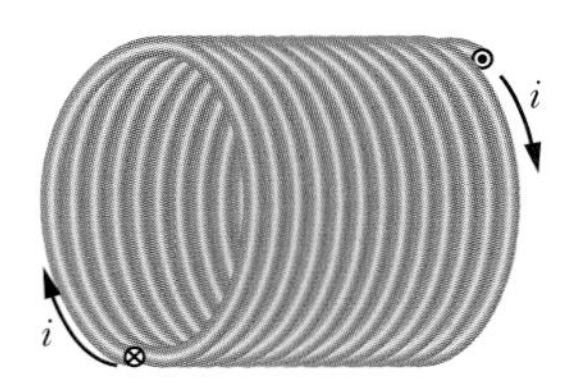

FIGURE 30-17 ■ A solenoid carrying current i.

To make the calculation simpler here, we will assume that the length of the solenoid is much greater than the diameter. Figure 30-18 shows a section through a portion of a "stretched-out" solenoid. The solenoid's magnetic field is the vector sum of the fields produced by the individual turns (loops) that make up the solenoid. For points very close to each turn, the wire behaves magnetically almost like a long straight wire, and the lines of $\vec{B}$ there are almost concentric circles. Figure 30-18 suggests that the field tends to cancel between adjacent turns. It also suggests that, at points inside the solenoid and reasonably far from the wire, $\vec{B}$ is approximately

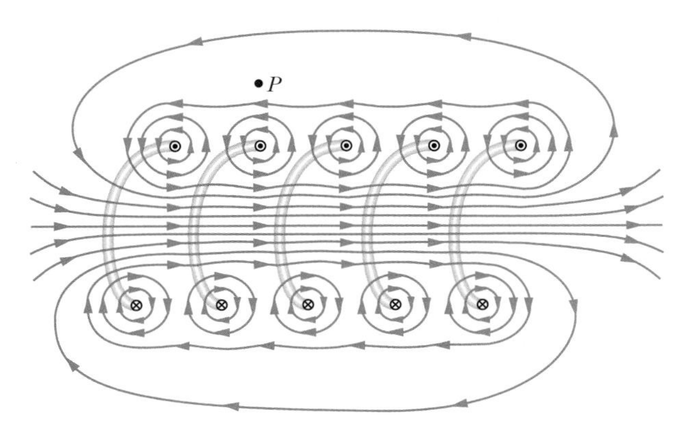

FIGURE 30-18 ■ A vertical cross section through the central axis of a "stretched-out" solenoid. The back portions of five turns are shown, as are the magnetic field lines due to a current through the solenoid. Each turn produces circular magnetic field lines near it. Near the solenoid's axis, the field lines combine into a net magnetic field that is directed along the axis. The closely spaced field lines there indicate a strong magnetic field. Outside the solenoid the field lines are widely spaced; the field there is very weak.

parallel to the (central) solenoid axis. In the limiting case of an *ideal solenoid,* which is infinitely long and consists of tightly packed (*close-packed*) turns of square wire, the field inside the coil is uniform and parallel to the solenoid axis.

At points above the solenoid, such as P in Fig. 30-18, the field set up by the upper parts of the solenoid turns (marked ⊙) is directed to the left (as drawn near P) and tends to cancel the field set up by the lower parts of the turns (marked ⊗), which is directed to the right (not drawn). In the limiting case of an ideal solenoid, the magnetic field outside the solenoid is zero. Taking the external field to be zero is an excellent assumption for a real solenoid if its length is much greater than its diameter and if we consider external points such as point P that are not near either end of the solenoid. The direction of the magnetic field along the solenoid axis is given by a curled-straight right-hand rule: Grasp the solenoid with your right hand so that your fingers follow the direction of the current in the windings; your extended right thumb then points in the direction of the axial magnetic field.

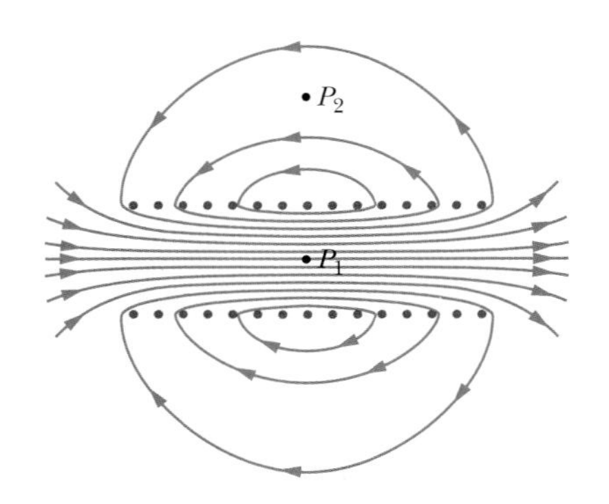

FIGURE 30-19 ■ Magnetic field lines for a real solenoid of finite length. The field is strong and uniform at interior points such as P_1 but relatively weak at external points such as P_2.

Figure 30-19 shows the lines of $\vec{B}$ for a real solenoid. The spacing of the lines of $\vec{B}$ in the central region shows that the field inside the coil is fairly strong and uniform over the cross section of the coil. The external field, however, is relatively weak.

Let us now apply Ampère's law,

$$\oint \vec{B} \cdot d\vec{s} = \mu_0 i^{\text{enc}}, \tag{30-23}$$

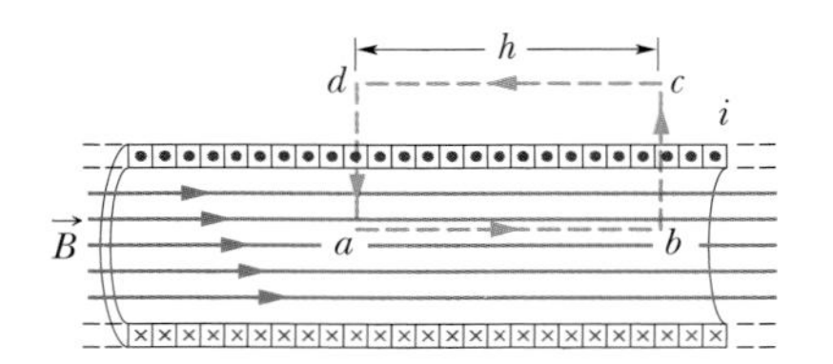

FIGURE 30-20 ■ Application of Ampère's law to a section of a long ideal solenoid carrying a current i. The Ampèrian loop is the rectangle $abcd$.

to the ideal solenoid of Fig. 30-20, where $\vec{B}$ is uniform within the solenoid and zero outside it, using the rectangular Amperian loop $abcda$. We write $\oint \vec{B} \cdot d\vec{s}$ as the sum of four integrals, one for each loop segment:

$$\oint \vec{B} \cdot d\vec{s} = \int_a^b \vec{B} \cdot d\vec{s} + \int_b^c \vec{B} \cdot d\vec{s} + \int_c^d \vec{B} \cdot d\vec{s} + \int_d^a \vec{B} \cdot d\vec{s}. \tag{30-24}$$

The first integral on the right of Eq. 30-24 is Bh, where B is the magnitude of the uniform field $\vec{B}$ inside the solenoid and h is the (arbitrary) length of the segment from a to b. The second and fourth integrals are zero because for every element ds of these segments, $\vec{B}$ either is perpendicular to ds or is zero, and thus $\vec{B} \cdot d\vec{s}$ is zero. The third integral, which is taken along a segment that lies outside the solenoid, is zero because $\vec{B} = 0$ at all external points. Thus, $\oint \vec{B} \cdot d\vec{s}$ for the entire rectangular loop has the value Bh.

The net current i^{enc} encircled by the rectangular Ampèrian loop in Fig. 30-20 is not the same as the current i in the solenoid windings because the windings pass more than once through this loop. Let n be the number of turns per unit length of the solenoid; then the loop encloses nh turns, so

$$i^{\text{enc}} = i(nh).$$

Ampère's law then gives us

$$Bh = \mu_0 |i| nh,$$

or

$$B = n\mu_0 |i| \quad \text{(inside ideal solenoid).} \tag{30-25}$$

Although we derived Eq. 30-25 for an infinitely long ideal solenoid, it holds quite well for actual solenoids if we apply it only at interior points, well away from the solenoid ends. Equation 30-25 is consistent with the experimental fact that the magnetic field magnitude $|\vec{B}| = B$ within a solenoid does not depend on the diameter or the length of the solenoid and that B is uniform over the solenoidal cross section. A solenoid thus

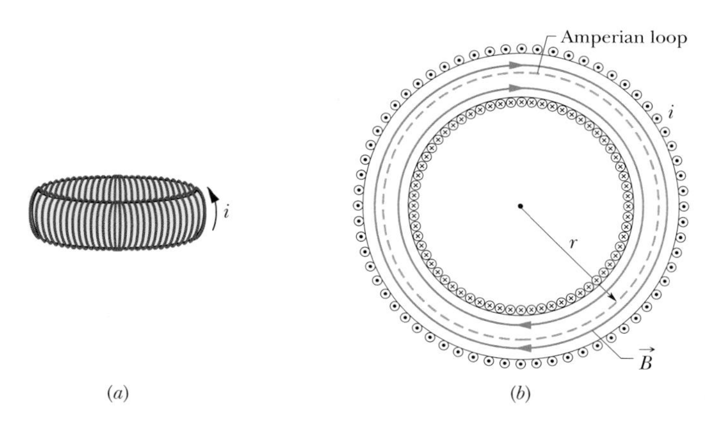

FIGURE 30-21 ■ (*a*) A toroid carrying a current i. (*b*) A horizontal cross section of the toroid. The interior magnetic field (inside the doughnut-shaped tube) can be found by applying Ampère's law with the Ampèrian loop shown.

provides a practical way to set up a known uniform magnetic field for experimentation, just as a parallel-plate capacitor provides a practical way to set up a known uniform electric field.

Magnetic Field of a Toroid

Figure 30-21*a* shows a **toroid,** which may be described as a solenoid bent into the shape of a hollow doughnut. What magnetic field $\vec{B}$ is set up at its interior points (within the hollow of the doughnut)? We can find out from Ampère's law and the symmetry of the toroid.

From the symmetry, we see that the lines of $\vec{B}$ form concentric circles inside the toroid, directed as shown in Fig. 30-21*b*. Let us choose a concentric circle of radius r as an Ampèrian loop and traverse it in the clockwise direction. Ampère's law (Eq. 30-16) yields

$$B(2\pi r) = N\mu_0|i|,$$

where i is the current in the toroid windings (and is positive for those windings enclosed by the Ampèrian loop) and N is the total number of turns. This gives

$$B = \frac{N\mu_0|i|}{2\pi}\frac{1}{r} \quad \text{(toroid).} \tag{30-26}$$

In contrast to the situation for a solenoid, $\vec{B}$ is not constant over the cross section of a toroid. With Ampère's law, it is easy to show that $\vec{B} = 0$ for points outside an ideal toroid (as if the toroid were made from an ideal solenoid).

The direction of the magnetic field within a toroid follows from our curled-straight right-hand rule: Grasp the toroid with the fingers of your right hand curled in the direction of the current in the windings; your extended right thumb points in the direction of the magnetic field.

30-7 A Current-Carrying Coil as a Magnetic Dipole

So far we have examined the magnetic fields produced by current in a long straight wire, a solenoid, and a toroid. We turn our attention here to the field produced by a coil carrying a current. You saw in Section 29-10 that such a coil behaves as a magnetic dipole in that, if we place it in an external magnetic field $\vec{B}$, a torque $\vec{\tau}$ given by

$$\vec{\tau} = \vec{\mu} \times \vec{B} \tag{30-27}$$

acts on it. Here $\vec{\mu}$ is the magnetic dipole moment of the coil and has the magnitude NiA, where N is the number of turns (or loops), i is the current in each turn, and A is the area enclosed by each turn.

Recall that the direction of $\vec{\mu}$ is given by a curled-straight right-hand rule: Grasp the coil so that the fingers of your right hand curl around it in the direction of the current; your extended thumb then points in the direction of the dipole moment $\vec{\mu}$.

Magnetic Field of a Coil

We turn now to the other aspect of a current-carrying coil as a magnetic dipole. What magnetic field does *it* produce at a point in the surrounding space? The problem does not have enough symmetry to make Ampère's law useful, so we must turn to the Biot–Savart law. For simplicity, we first consider only a coil with a single circular loop and only points on its central axis, which we take to be a z axis. We shall show that the magnetic field at such points only has a z-component, B_z which is given by

$$\vec{B} = B_z\hat{\mathrm{k}} = \frac{\mu_0 i R^2}{2(R^2 + z^2)^{3/2}}\hat{\mathrm{k}}, \tag{30-28}$$

where R is the radius of the circular loop and z is the distance of the point in question from the center of the loop. Furthermore, the direction of the magnetic field $\vec{B}$ is the same as the direction of the magnetic dipole moment $\vec{\mu}$ of the loop.

For axial points far from the loop, we have $z \gg R$ in Eq. 30-28. With that approximation, the equation for the z-component of $\vec{B}$, which is a function of z only, reduces to

$$B_z \approx \frac{\mu_0 i R^2}{2z^3}.$$

Recalling that πR^2 is the area A of the loop and extending our result to include a coil of N turns, we can write this equation as

$$B_z = \frac{\mu_0}{2\pi}\frac{NiA}{z^3}.$$

Further, since $\vec{B}$ and $\vec{\mu}$ have the same direction, we can write the equation in vector form, substituting from $\mu = NiA$ (Eq. 29-32):

$$\vec{B} = B_z\hat{\mathrm{k}} = \frac{\mu_0}{2\pi}\frac{\vec{\mu}}{z^3} \quad \text{(current-carrying coil).} \tag{30-29}$$

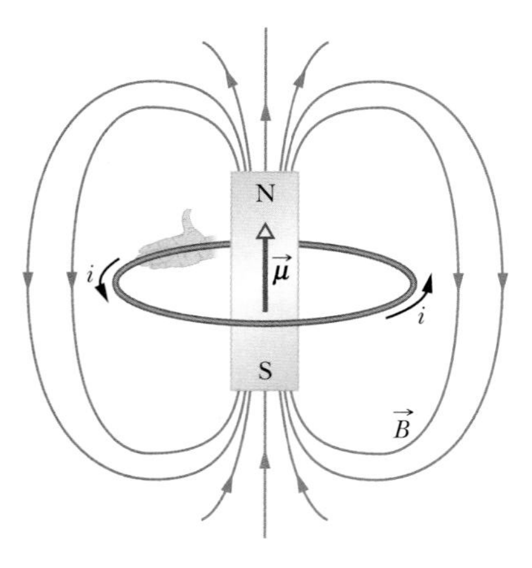

FIGURE 30-22 ■ A current loop produces a magnetic field like that of a bar magnet and thus has associated north and south poles. The magnetic dipole moment $\vec{\mu}$ of the loop, given by a curled-straight right-hand rule, points from the south pole to the north pole, in the direction of the field $\vec{B}$ within the loop.

Note that the magnetic constant μ_0 and the magnetic moment vector $\vec{\mu}$ are completely different quantities with different units. The choice of the symbol μ to represent both quantities is unfortunate.

In summary, we have two ways in which we can regard a current-carrying coil as a magnetic dipole: (1) it experiences a torque when we place it in an external magnetic field; (2) it generates its own intrinsic magnetic field, given by Eq. 30-29 for distant points along its axis. Figure 30-22 shows some magnetic field lines for a current loop; one side of the loop acts as a north pole (in the direction of $\vec{\mu}$) and the other side as a south pole, as suggested by the lightly drawn magnet in the figure.

Proof of Equation 30-28

Figure 30-23 shows the back half of a circular loop of radius R carrying a current i. Consider a point P on the axis of the loop, a distance z from its plane. Let us apply the Biot–Savart law to a differential element $d\vec{s}$ of the loop, located at the left side of the loop. The length vector $d\vec{s}$ for this element points perpendicularly out of the page. The angle θ between $d\vec{s}$ and $\vec{r}$ in Fig. 30-23 is 90°; the plane formed by these two vectors is perpendicular to the plane of the figure and contains both $\vec{r}$ and $d\vec{s}$. Using the Biot–Savart law and the right-hand rule, we see that the differential field $d\vec{B}$ produced at point P by the current in this element is perpendicular to this plane. Thus $d\vec{B}$ lies in the plane of the figure, perpendicular to $\vec{r}$ (as indicated in Fig. 30-23).

Let us resolve $d\vec{B}$ into two components: $d\vec{B}_{\parallel}$ along the axis of the loop and $d\vec{B}_{\perp}$ perpendicular to this axis. From the symmetry, the vector sum of all the perpendicular components $d\vec{B}_{\perp}$ due to all the loop elements ds is zero. This leaves only the axial components $d\vec{B}_{\parallel}$ and we have the magnitude of the axial component given by

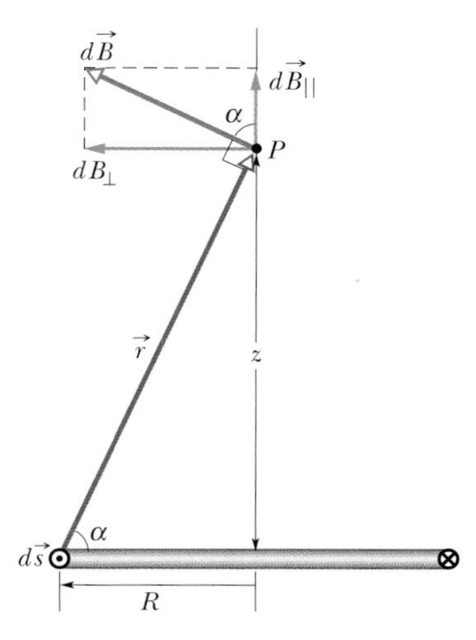

FIGURE 30-23 ■ A current loop of radius R. The plane of the loop is perpendicular to the page and only the back half of the loop is shown. We use the law of Biot and Savart to find the magnetic field at point P on the central axis of the loop.

$$B_{\parallel} = \int dB_{\parallel}.$$

For the element $d\vec{s}$ in Fig. 30-23, the Biot–Savart law (Eq. 30-1) tells us that the magnitude of the axial magnetic field component at distance r is

$$dB_{\parallel} = \frac{\mu_0}{4\pi}\frac{i\,ds\sin 90°}{r^2}.$$

We also have

$$dB_{\parallel} = dB\cos\alpha.$$

Combining these two relations, we obtain

$$dB_{\parallel} = \frac{\mu_0 i\cos\alpha\, ds}{4\pi r^2}. \qquad (30\text{-}30)$$

Figure 30-23 shows that r and α are not independent but are related to each other. Let us express each in terms of the variable z, the distance between point P and the center of the loop. The relations are

$$r = \sqrt{R^2 + z^2} \qquad (30\text{-}31)$$

and

$$\cos\alpha = \frac{R}{r} = \frac{R}{\sqrt{R^2 + z^2}}. \qquad (30\text{-}32)$$

Substituting Eqs. 30-31 and 30-32 into Eq. 30-30, we find

$$dB_{\parallel} = \frac{\mu_0 iR}{4\pi(R^2 + z^2)^{3/2}}\,ds.$$

Note that i, R, and z have the same values for all elements $d\vec{s}$ around the loop, so when we integrate this equation, we find that the magnitude of the axial field component is given as

$$\begin{aligned} B_{\parallel} &= \oint dB_{\parallel} \\ &= \frac{\mu_0 iR}{4\pi(R^2 + z^2)^{3/2}}\oint ds, \end{aligned}$$

or, since $\oint ds$ is simply the circumference $2\pi R$ of the loop, the axial or z-component of the magnetic field is

$$\vec{B} = B_z\hat{k} = \frac{\mu_0 i R^2}{2(R^2 + z^2)^{3/2}}\hat{k},$$

which is Eq. 30-28, the relation we sought to prove.

READING EXERCISE 30-5: The figure here shows four arrangements of circular loops of radius r or $2r$, centered on vertical axes (perpendicular to the loops) and carrying identical currents in the directions indicated. Assume the sizes of the loops are exaggerated and that $z \gg R$. Rank the arrangements according to the magnitude of the net magnetic field at the dot, midway between the loops on the central axis, greatest first.

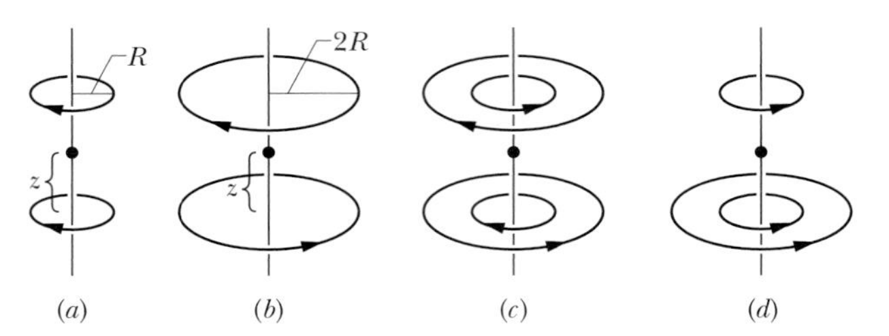

Problems

SEC. 30-3 ■ CALCULATING THE MAGNETIC FIELD DUE TO A CURRENT

1. Surveyor A surveyor is using a magnetic compass 6.1 m below a power line in which there is a steady current of 100 A. (a) What is the magnitude of the magnetic field at the site of the compass due to the power line? (b) Will this interfere seriously with the compass reading? The horizontal component of the Earth's magnetic field at the site is 20 μT.

2. Electron Gun The electron gun in a traditional television tube fires electrons of kinetic energy 25 keV at the screen in a circular beam 0.22 mm in diameter; 5.6×10^{14} electrons arrive each second. Calculate the magnitude of the magnetic field produced by the beam at a point 1.5 mm from the beam axis.

3. Philippines At a certain position in the Philippines, the magnitude of the Earth's magnetic field of 39 μT is horizontal and directed due north. Suppose the net field is zero exactly 8.0 cm above a long, straight, horizontal wire that carries a constant current. What are (a) the size and (b) the direction of the current?

4. Locate Points A long wire carrying a current of 100 A is placed in a uniform external magnetic field of 5.0 mT. The wire is perpendicular to this magnetic field. Locate the points at which the net magnetic field is zero.

5. Particle with Positive Charge A particle with positive charge q is a distance d from a long straight wire that carries a current i; the particle is traveling with speed $|\vec{v}|$ perpendicular to the wire. What are the direction and magnitude of the force on the particle if it is moving (a) toward and (b) away from the wire?

6. Semicircular Arcs A straight conductor carrying a current i splits into identical semicircular arcs as shown in Fig. 30-24. What is the magnitude of the magnetic field at the center C of the resulting circular loop?

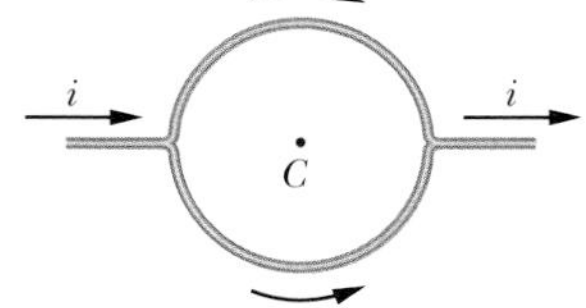

FIGURE 30-24 ■ Problem 6.

7. Two Semi-Infinite A wire carrying current i has the configuration shown in Fig. 30-25. Two semi-infinite straight sections, both tangent to the same circle, are connected by a circular arc, of central angle ϕ, along the circumference of the circle, with all sections lying in the same plane. What must ϕ be in order for $|\vec{B}|$ to be zero at the center of the circle?

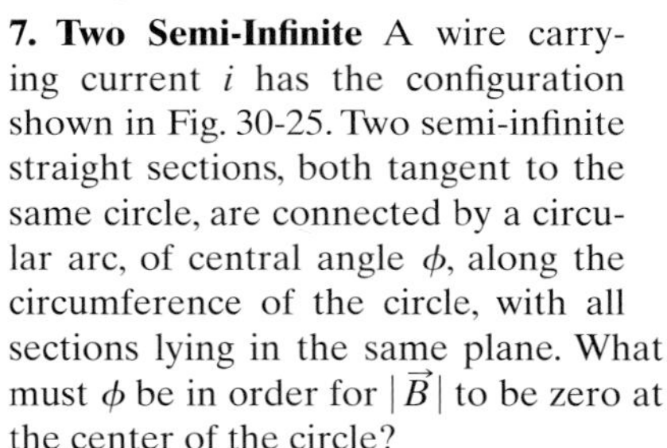

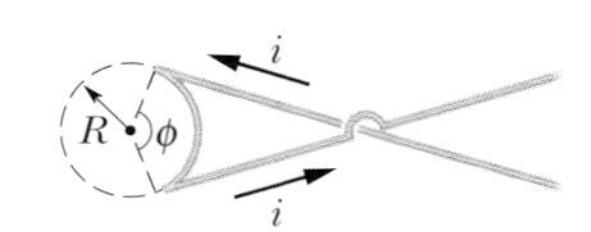

FIGURE 30-25 ■ Problem 7.

8. Use Biot–Savart Use the Biot–Savart law to calculate the magnitude and direction of the magnetic field $\vec{B}$ at C, the common center of the semicircular arcs AD and HJ in Fig. 30-26a. The two arcs, of radii R_2 and R_1, respectively, form part of the circuit $ADJHA$ carrying current i.

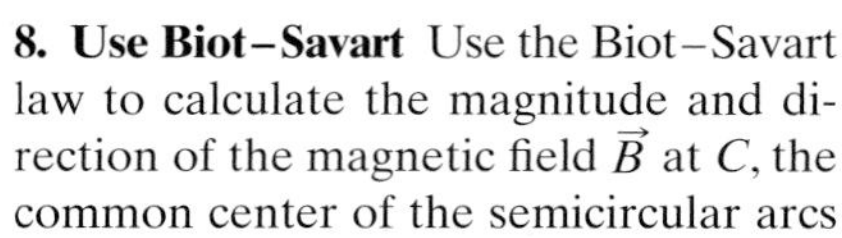

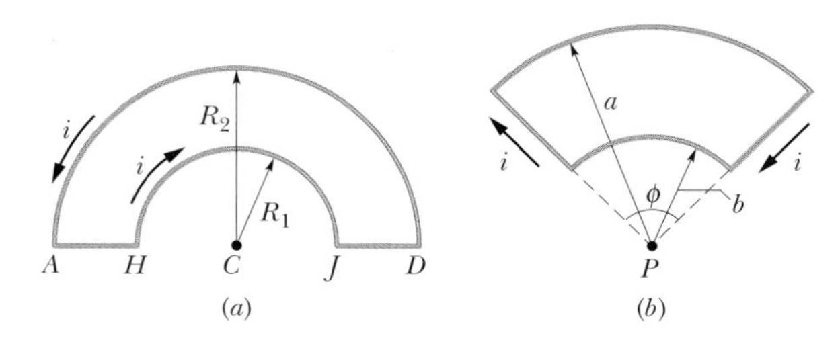

FIGURE 30-26 ■ Problems 8 and 9.

9. Curved Segments In the circuit of Fig. 30-26*b*, the curved segments are arcs of circles of radii a and b with common center P. The straight segments are along radii. Find the magnitude and direction of the magnetic field $\vec{B}$ at point P, assuming a current i in the circuit.

10. Magnitude and Directions The wire shown in Fig. 30-27 carries current i. What are the magnitude and direction of the magnetic field $\vec{B}$ produced at the center C of the semicircle by (a) each straight segment of length L, (b) the semicircular segment of radius R, and (c) the entire wire?

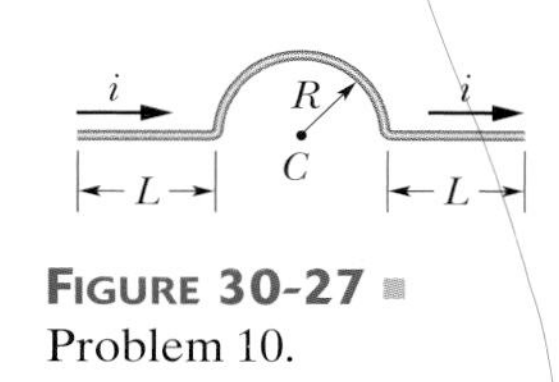

FIGURE 30-27 ■ Problem 10.

11. Straight Wire In Fig. 30-28, a straight wire of length L carries current i. Show that the magnitude of the magnetic field $\vec{B}$ produced by this segment at P_1, a distance R from the segment along a perpendicular bisector, is

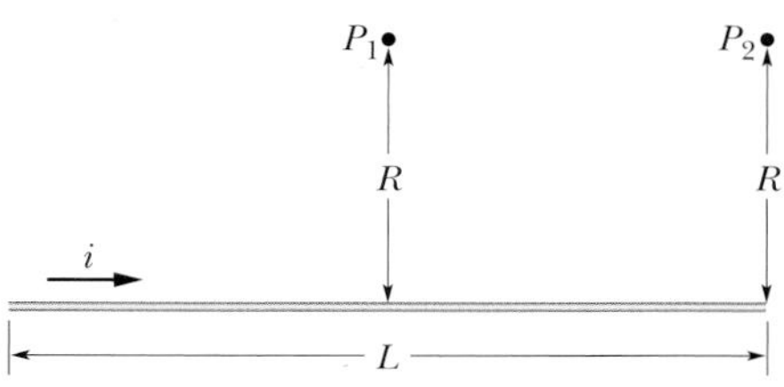

FIGURE 30-28 ■ Problems 11 and 13.

$$B = \frac{\mu_0 |i|}{2\pi R} \frac{L}{(L^2 + 4R^2)^{1/2}}.$$

Show that this expression for $|\vec{B}|$ reduces to an expected result as $L \to \infty$.

12. Square Loop A square loop of wire of edge length a carries current i. Using the results of Problem 11, show that, at the center of the loop, the magnitude of the magnetic field produced by the current is

$$B = \frac{2\sqrt{2}\mu_0 |i|}{\pi a}.$$

13. Length L In Fig. 30-28, a straight wire of length L carries current i. Show that

$$B = \frac{\mu_0 |i|}{4\pi R} \frac{L}{(L^2 + R^2)^{1/2}}$$

gives the magnitude of the magnetic field $\vec{B}$ produced by the wire at P_2, a perpendicular distance R from one end of the wire.

14. Rectangular Loop Using the results of Problem 11, show that the magnitude of the magnetic field produced at the center of a rectangular loop of wire of length L and width W, carrying a current i, is

$$B = \frac{2\mu_0 |i|}{\pi} \frac{(L^2 + W^2)^{1/2}}{LW}.$$

15. Square Loop Two A square loop of wire of edge length a carries current i. Using the results of Problem 11, show that the magnitude of the magnetic field produced at a point on the axis of the loop and a distance x from its center is

$$B(x) = |\vec{B}(x)| = \frac{4\mu_0 |i| a^2}{\pi(4x^2 + a^2)(4x^2 + 2a^2)^{1/2}}.$$

Prove that this result is consistent with the result of Problem 12.

16. Length a In Fig. 30-29, a straight wire of length a carries a current i. Show that the magnitude of the magnetic field produced by the current at point P is $B = \sqrt{2}\mu_0 |i| / 8\pi a$.

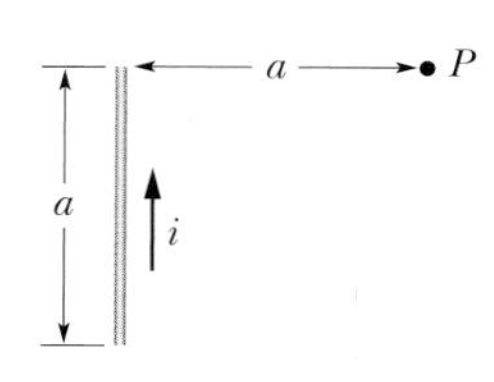

FIGURE 30-29 ■ Problem 16.

17. Two Wires Two wires, both of length L, are formed into a circle and a square, and each carries current i. Show that the square produces a greater magnetic field at its center than the circle produces at its center. (See Problem 12.)

18. Magnetic Field Find the magnitude and direction of the magnetic field $\vec{B}$ at point P in Fig. 30-30. (See Problem 16.)

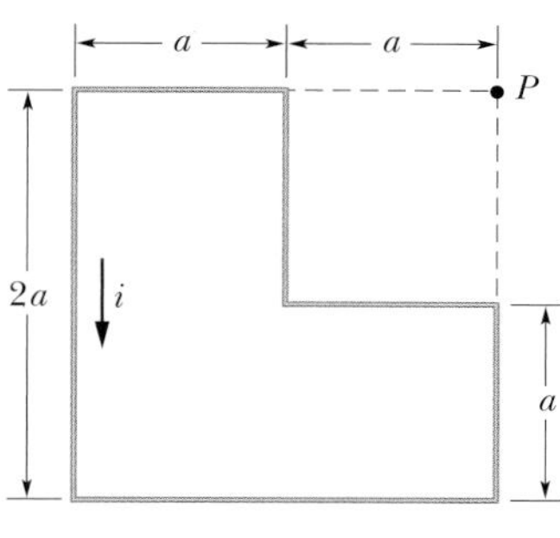

FIGURE 30-30 ■ Problem 18.

19. Long Thin Ribbon Figure 30-31 shows a cross section of a long thin ribbon of width w that is carrying a uniformly distributed total current i into the page. Calculate the magnitude and direction of the magnetic field $\vec{B}$ at a point P in the plane of the ribbon at a distance d from its edge. (*Hint:* Imagine the ribbon to be constructed from many long, thin, parallel wires.)

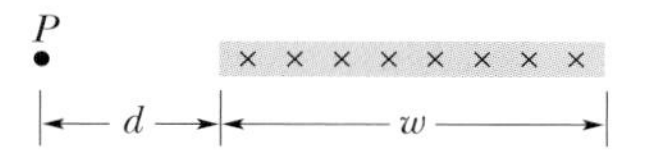

FIGURE 30-31 ■ Problem 19.

20. Find Magnitude and Direction Find the magnitude and direction of the magnetic field $\vec{B}$ at point P in Fig. 30-32, for $|i| = 10$ A and $a = 8.0$ cm. (See Problems 13 and 16.)

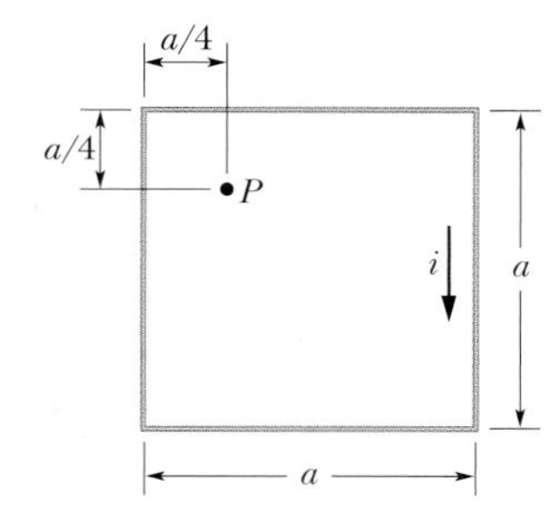

FIGURE 30-32 ■ Problem 20.

21. Perpendicular Bisector Figure 30-33 shows two very long straight wires (in cross section) that each carry currents of 4.00 A directly out of the page. Distance $d_1 = 6.00$ m and distance $d_2 = 4.00$ m. What is the magnitude of the net magnetic field at point P, which lies on a perpendicular bisector to the wires?

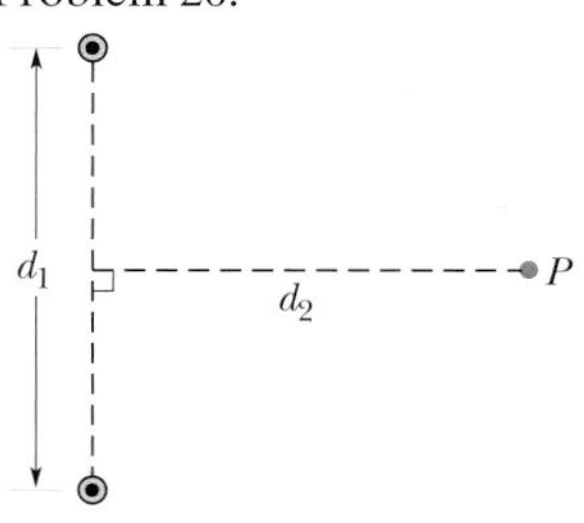

FIGURE 30-33 ■ Problem 21.

22. Greatest and 10% In Fig. 30-34, point P is at perpendicular distance $R = 2.00$ cm from a very long straight wire carrying a current. The magnetic field $\vec{B}$ set up at point P is due to contributions from all the identical current-length elements $i\,d\vec{s}$ along the wire. What is the distance s to the current-length element that makes (a) the greatest contribution to field $\vec{B}$ and (b) 10% of the greatest contribution?

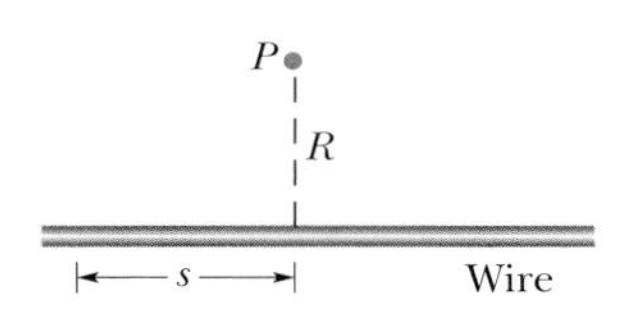

FIGURE 30-34 ■ Problem 22.

SEC. 30-4 ■ FORCE BETWEEN TWO PARALLEL CURRENTS

23. Two Parallel Wires Two long parallel wires are 8.0 cm apart. What equal currents must be in the wires if the magnetic field halfway between them is to have a magnitude of 300 μT? Answer for both (a) parallel and (b) antiparallel currents.

24. i and $3i$ Two long parallel wires a distance d apart carry currents of i and $3i$ in the same direction. Locate the point or points at which their magnetic fields cancel.

25. Two Parallel Wires Two Two long, straight, parallel wires, separated by 0.75 cm, are perpendicular to the plane of the page as shown in Fig. 30-35. Wire 1 carries a current of 6.5 A into the page. What must be the current (magnitude and direction) in wire 2 for the resultant magnetic field at point P to be zero?

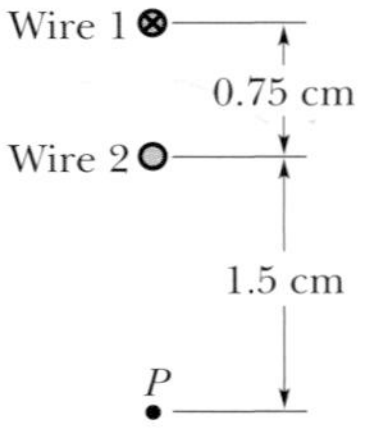

FIGURE 30-35 ■ Problem 25.

26. Five Parallel Wires Figure 30-36 shows five long parallel wires in the xy plane. Each wire carries a current $i = 3.00$ A in the positive x direction. The separation between adjacent wires is $d = 8.00$ cm. In unit-vector notation, what are the magnitude and direction of the magnetic force per meter exerted on each of these five wires by the other wires?

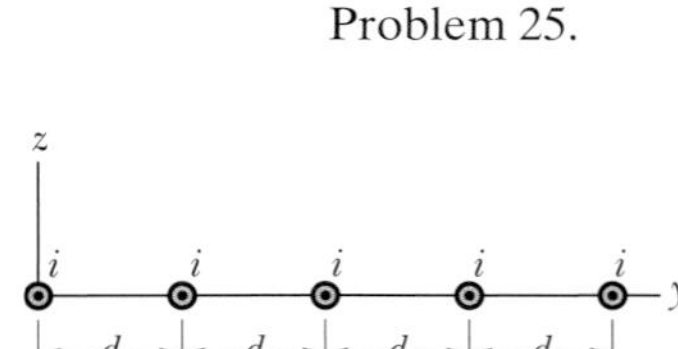

FIGURE 30-36 ■ Problem 26.

27. Four Long Wires Four long copper wires are parallel to each other, their cross sections forming the corners of a square with sides $a = 20$ cm. A 20 A current exists in each wire in the direction shown in Fig. 30-37. What are the magnitude and direction of $\vec{B}$ at the center of the square?

28. Four Currents Form a Square Four identical parallel currents i are arranged to form a square of edge length a as in Fig. 30-37, *except* that they are *all* out of the page. What is the force per unit length (magnitude and direction) on any one wire?

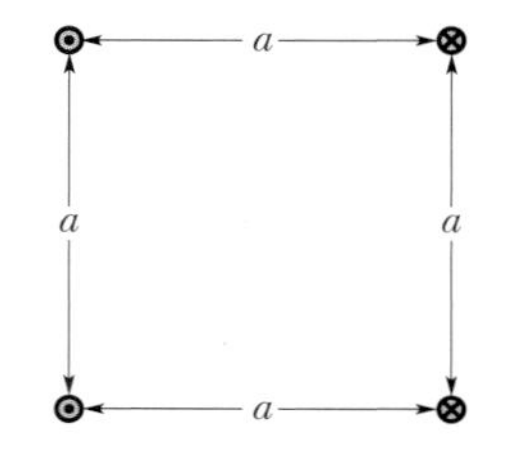

FIGURE 30-37 ■ Problems 27, 28, and 29.

29. Force per Unit Length In Fig. 30-37, what is the force per unit length acting on the lower left wire, in magnitude and direction, with the current directions as shown? The currents are i.

30. Idealized Schematic Figure 30-38 is an idealized schematic drawing of a rail gun. Projectile P sits between two wide rails of circular cross section; a source of current sends current through the rails and through the (conducting) projectile itself (a fuse is not used). (a) Let w be the distance between the rails, R the radius of the rails, and i the current. Show that the magnitude of the force on the projectile is directed to the right along the rails and is given approximately by

$$F = |\vec{F}| = \frac{i^2\mu_0}{\pi} \ln \frac{w + R}{R}.$$

(b) If the projectile starts from the left end of the rails at rest, find the speed v at which it is expelled at the right. Assume that $|i| = 450$ kA, $w = 12$ mm, $R = 6.7$ cm, $L = 4.0$ m, and the mass of the projectile is $m = 10$ g.

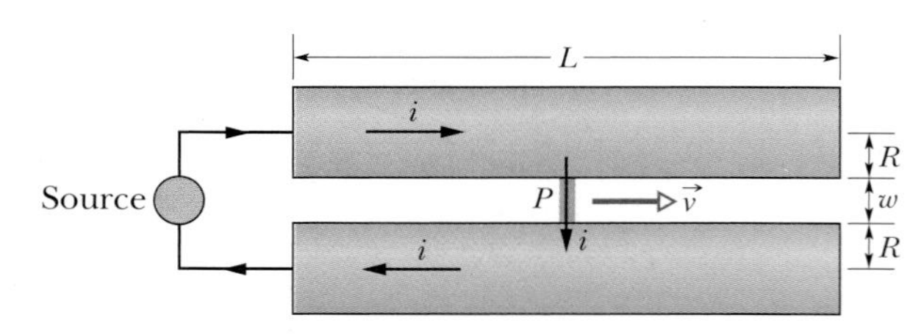

FIGURE 30-38 ■ Problem 30.

31. Rectangular Loop Two In Fig. 30-39, the long straight wire carries a current of 30 A and the rectangular loop carries a current of 20 A. Calculate the resultant force acting on the loop. Assume that $a = 1.0$ cm, $b = 8.0$ cm, and $L = 30$ cm.

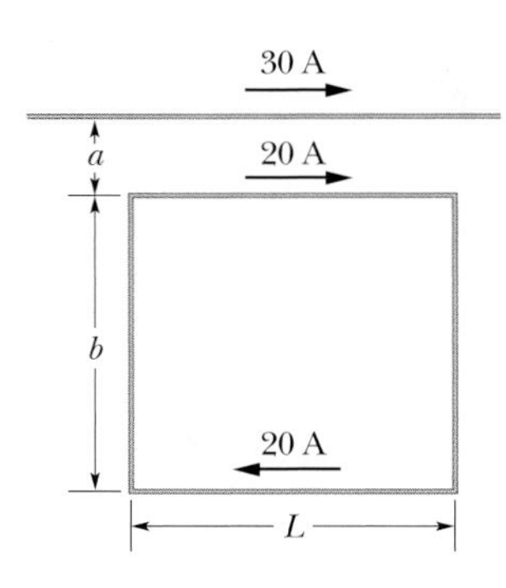

FIGURE 30-39 ■ Problem 31.

SEC. 30-5 ■ AMPÈRE'S LAW

32. Eight Wires Eight wires cut the page perpendicularly at the points shown in Fig. 30-40. A wire labeled with the integer k ($k = 1, 2, \ldots, 8$) carries the current ki. For those with odd k, the current is out of the page; for those with even k, it is into the page. Evaluate $\oint \vec{B} \cdot d\vec{s}$ along the closed path in the direction shown.

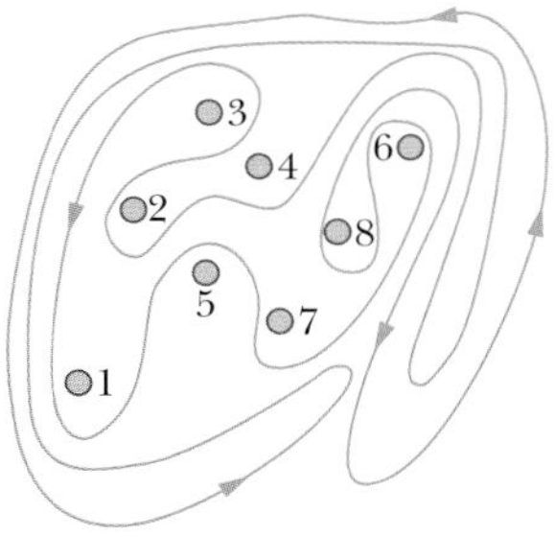

FIGURE 30-40 ■ Problem 32.

33. Eight Conductors Each of the eight conductors in Fig. 30-41 carries 2.0 A of current into or out of the page. Two paths are indicated for the line integral $\oint \vec{B} \cdot d\vec{s}$. What is the value of the integral for the path (a) at the left and (b) at the right?

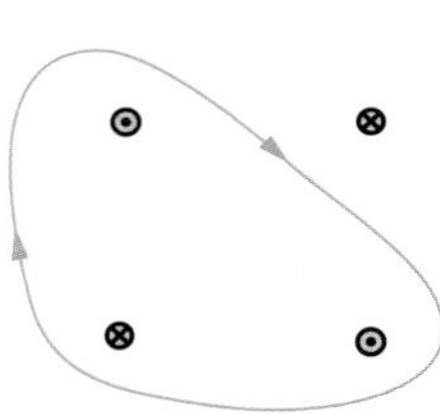
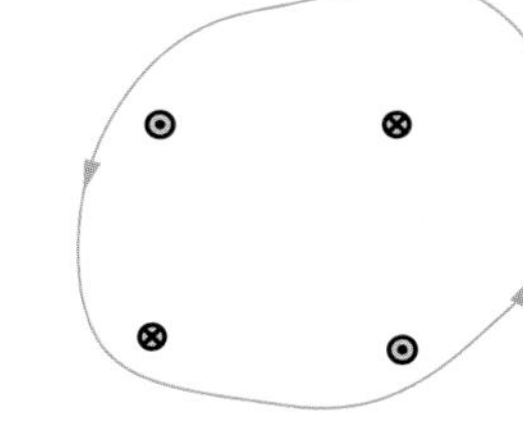

FIGURE 30-41 ■ Problem 33.

34. Cross Section of a Cylindrical Conductor Figure 30-42 shows a cross section of a long cylindrical conductor of radius a, carrying a uniformly distributed current i. Assume that $a = 2.0$ cm and $i = 100$ A, and plot the magnitude of the magnetic field $|\vec{B}(r)| = B(r)$ over the range $0 < r < 6.0$ cm.

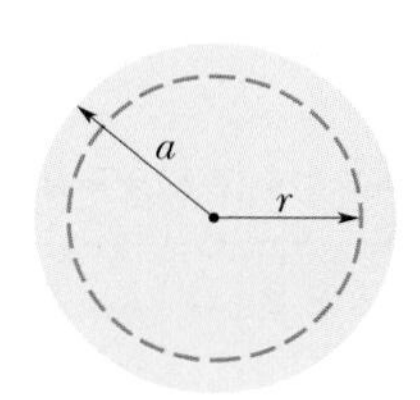

FIGURE 30-42 ■ Problem 34.

35. Cannot Drop to Zero Show that a uniform magnetic field $\vec{B}$ cannot drop abruptly to zero (as is suggested by the lack of field lines to the right of point a in Fig. 30-43) as one moves perpendicular to $\vec{B}$, say along the horizontal arrow in the figure. (*Hint:* Apply Ampère's law to the rectangular path shown by the dashed lines.) In actual magnets "fringing" of the magnetic field lines always occurs, which means that $\vec{B}$ approaches zero in a gradual manner. Modify the field lines in the figure to indicate a more realistic situation.

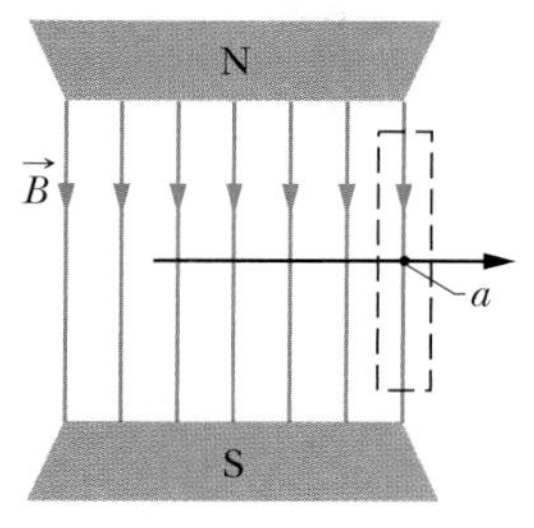

FIGURE 30-43 ■ Problem 35.

36. Two Square Conducting Loops Two square conducting loops carry currents of 5.0 and 3.0 A as shown in Fig. 30-44. What is the value of $\oint \vec{B} \cdot d\vec{s}$ for each of the two closed paths shown?

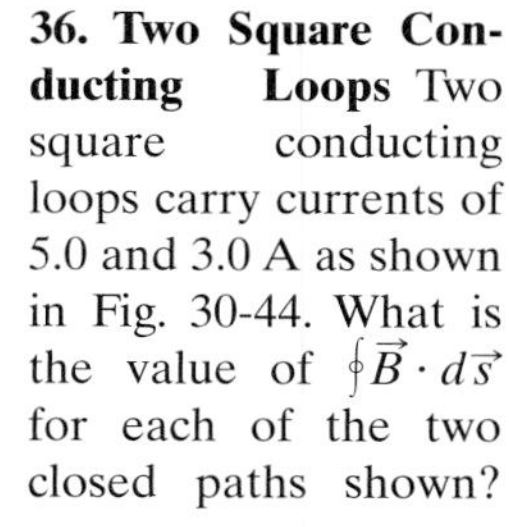

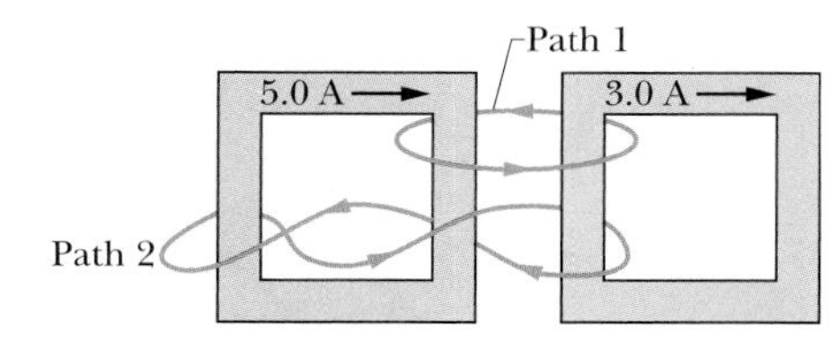

FIGURE 30-44 ■ Problem 36.

37. Current Density The current density inside a long, solid, cylindrical wire of radius a is in the direction of the central axis and varies linearly with radial distance r from the axis according to $|\vec{J}| = |J_0| r/a$. Find the magnitude and direction of the magnetic field inside the wire.

38. Uniformly Distributed Current A long straight wire (radius = 3.0 mm) carries a constant current distributed uniformly over a cross section perpendicular to the axis of the wire. If the magnitude of the current density is 100 A/m^2, what are the magnitudes of the magnetic fields (a) 2.0 mm from the axis of the wire and (b) 4.0 mm from the axis of the wire?

39. Cylindrical Hole Figure 30-45 shows a cross section of a long cylindrical conductor of radius a containing a long cylindrical hole of radius b. The axes of the cylinder and hole are parallel and are a distance d apart; a current i is uniformly distributed over the tinted area. (a) Use superposition to show that the magnitude of the magnetic field at the center of the hole is

$$B = \frac{\mu_0 |i| d}{2\pi(a^2 - b^2)}.$$

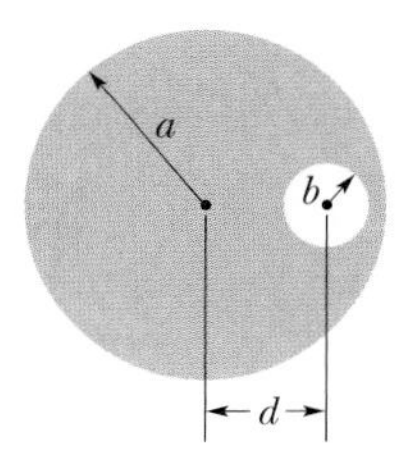

FIGURE 30-45 ■ Problem 39.

(b) Discuss the two special cases $b = 0$ and $d = 0$. (*Hint:* Regard the cylindrical hole as resulting from the superposition of a complete cylinder (no hole) carrying a current in one direction and a cylinder of radius b carrying a current in the opposite direction, both cylinders having the same current density.)

40. Circular Pipe A long circular pipe with outside radius R carries a (uniformly distributed) current i into the page as shown in Fig. 30-46. A wire runs parallel to the pipe at a distance of $3R$ from center to center. Find the amount and direction of the current in the wire such that the net magnetic field at point P has the same magnitude as the net magnetic field at the center of the pipe but is in the opposite direction.

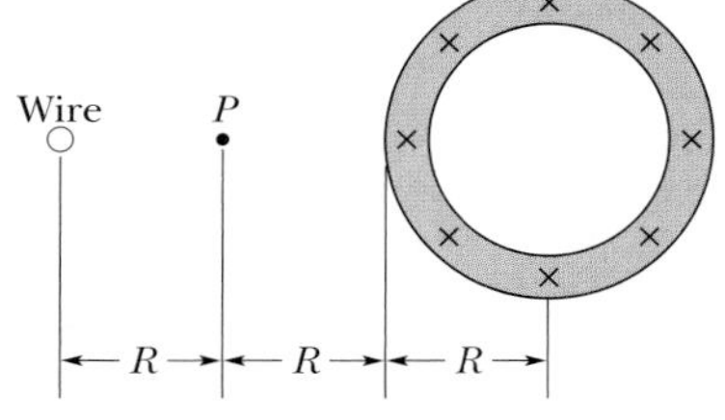

FIGURE 30-46 ■ Problem 40.

41. Conducting Sheet Figure 30-47 shows a cross section of an infinite conducting sheet lying in the x-y plane, carrying a current per unit x-length of λ; the current emerges perpendicularly out of the page. (a) Use the Biot–Savart law and symmetry to show that for all points P above the sheet, and all points P' below it, the magnetic field $\vec{B}$ is parallel to the sheet and directed as shown. (b) Use Ampère's law to prove that $B = \frac{1}{2}\mu_0|\lambda|$ at all points P and P'.

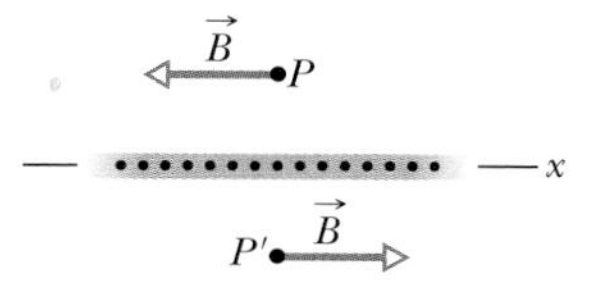

FIGURE 30-47 ■ Problems 41 and 48.

42. Field at P is Zero Figure 30-48 shows, in cross section, two long straight wires; the 3.0 A current in the right-hand wire is out of the page. What are the size and direction of the current in the left-hand wire if the net magnetic field at point P is to be zero?

FIGURE 30-48 ■ Problem 42.

SEC. 30-6 ■ SOLENOIDS AND TOROIDS

43. Field Inside Solenoid A 200-turn solenoid having a length of 25 cm and a diameter of 10 cm carries a current of 0.30 A. Calculate the magnitude of the magnetic field $\vec{B}$ inside the solenoid.

44. Field Inside Solenoid Two A solenoid that is 95.0 cm long has a radius of 2.00 cm and a winding of 1200 turns; it carries a current of 3.60 A. Calculate the magnitude of the magnetic field inside the solenoid.

45. Toroid A toroid having a square cross section, 5.00 cm on a side, and an inner radius of 15.0 cm has 500 turns and carries a current of magnitude 0.800 A. (It is made up of a square solenoid—instead of a round one as in Fig. 30-21—bent into a doughnut shape.) What is the magnitude of the magnetic field inside the toroid at (a) the inner radius and (b) the outer radius of the toroid?

46. Length of Wire A solenoid 1.30 m long and 2.60 cm in diameter carries a current of 18.0 A. The magnitude of the magnetic field inside the solenoid is 23.0 mT. Find the length of the wire forming the solenoid.

47. Field Inside Toroid In Section 30-6, we showed that the magnitude of the magnetic field at any radius r *inside* a toroid is given by

$$B = \frac{\mu_0 |i| N}{2\pi r}.$$

Show that as you move from any point just inside a toroid to a point just outside, the magnitude of the *change* in $\vec{B}$ that you encounter is just $\mu_0|\lambda|$. Here $|\lambda|$ is the amount of current per unit length along a circumference of radius r within the toroid. Compare this with the similar result found in Problem 48. Isn't the equality surprising?

48. Solenoid as Cylindrical Conductor Treat an ideal solenoid as a thin cylindrical conductor whose current per unit length, measured parallel to the cylinder axis, is λ. (a) By doing so, show that the magnitude of the magnetic field inside an ideal solenoid can be written as $B = \mu_0|\lambda|$. This is the value of the *change* in $\vec{B}$ that you encounter as you move from inside the solenoid to outside, through the solenoid wall. (b) Show that the same change occurs as you move through an infinite flat current sheet such as that of Fig. 30-47 (see Problem 41). Does this equality surprise you?

49. Direction of Field A long solenoid with 10.0 turns/cm and a radius of 7.00 cm carries a current of 20.0 mA. A current of 6.00 A exists in a straight conductor located along the central axis of the solenoid. (a) At what radial distance from the axis will the direction of the resulting magnetic field be at 45.0° to the axial direction? (b) What is the magnitude of the magnetic field there?

50. Find Current in Solenoid A long solenoid has 100 turns/cm and carries current i. An electron moves within the solenoid in a circle of radius 2.30 cm perpendicular to the solenoid axis. The speed of the electron is 0.0460c (c = speed of light). Find the amount of current $|i|$ in the solenoid.

SEC. 30-7 ■ A CURRENT-CARRYING COIL AS A MAGNETIC DIPOLE

51. Magnetic Dipole What is the magnetic dipole moment $\vec{\mu}$ of the solenoid described in Problem 43?

52. One Turn Coil Figure 30-49a shows a length of wire carrying a current i and bent into a circular coil of one turn. In Fig. 30-49b the same length of wire has been bent more sharply, to give a coil of two turns, each of half the original radius. (a) If B_a and B_b are the magnitudes of the magnetic fields at the centers of the two coils, what is the ratio B_b/B_a? (b) What is the ratio of the magnitude of the dipole moments, μ_b/μ_a of the coils?

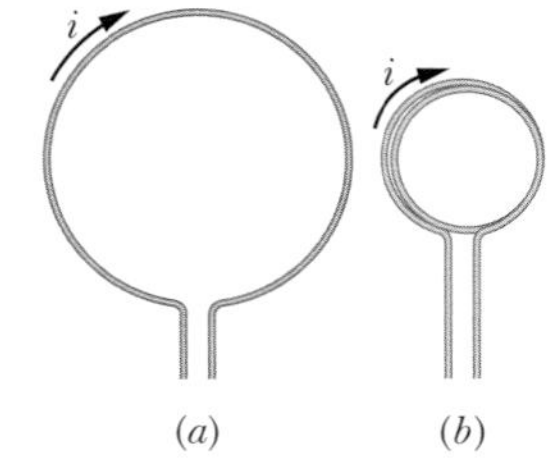

FIGURE 30-49 ■ Problem 52.

53. Student's Electromagnet A student makes a short electromagnet by winding 300 turns of wire around a wooden cylinder of diameter d = 5.0 cm. The coil is connected to a battery producing a current of 4.0 A in the wire. (a) What is the magnetic moment of this device? (b) At what axial distance $z \gg d$ will the magnetic field of this dipole have the magnitude 5.0 μT (approximately one-tenth that of the Earth's magnetic field)?

54. Helmholtz Figure 30-50 shows an arrangement known as a Helmholtz coil. It consists of two circular coaxial coils, each of N turns and radius R, separated by a distance R. The two coils carry equal currents i in the same direction. Find the magnitude of the net magnetic field at P, midway between the coils.

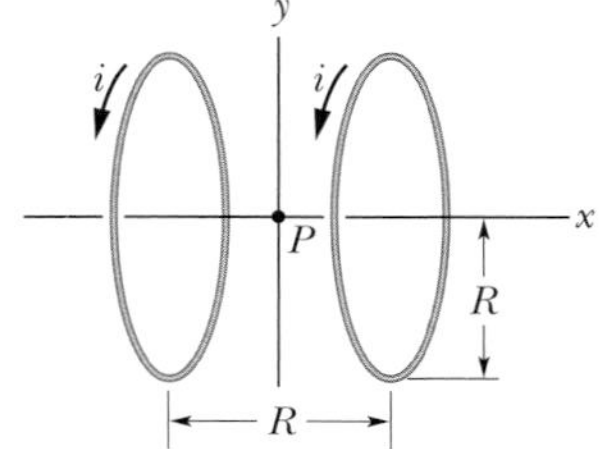

FIGURE 30-50 ■ Problems 54, 55, and 57.

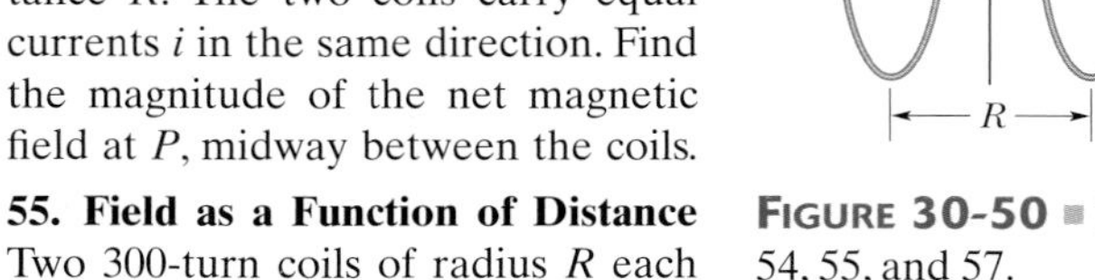

55. Field as a Function of Distance Two 300-turn coils of radius R each carry a current i. They are arranged a distance R apart, as in Fig. 30-50. For R = 5.0 cm and i = 50 A, plot the magnitude $|B(x)| = B(x)$ of the net magnetic field as a function of distance x along the common x axis over the range $x = -5$ cm to $x = +5$ cm, taking $x = 0$ at the midpoint P. (Such coils provide an especially uniform field $\vec{B}$ near point P.) (*Hint:* See Eq. 30-28).

56. Square Current Loop The magnitude $B(x)$ of the magnetic field at points on the axis of a square current loop of side a is given in Problem 15. (a) Show that the axial magnetic field of this loop, for $x \gg a$, is that of a magnetic dipole (see Eq. 30-29). (b) What is the magnitude of the magnetic dipole moment of this loop?

57. Let the Separation Be In Problem 54 (Fig. 30-50), let the separation of the coils be a variable s (not necessarily equal to the coil radius R). (a) Show that the first derivative of the magnitude of the net magnetic field of the coils (dB/dx) vanishes at the midpoint P regardless of the value of s. Why would you expect this to be true from symmetry? (b) Show that the second derivative (d^2B/dx^2) also vanishes at P, provided $s = R$. This accounts for the uniformity of B near P for this particular coil separation.

58. *abcdefgha* A conductor carries a current of 6.0 A along the closed path *abcdefgha* involving 8 of the 12 edges of a cube of side 10 cm as shown in Fig. 30-51. (a) Why can one regard this as the superposition of three square loops: *bcfgb, abgha,* and *cdefc*? (*Hint:* Draw currents around those square loops.) (b) Use this superposition to find the magnetic dipole moment $\vec{\mu}$ (magnitude and direction) of the closed path. (c) Calculate the magnitude and direction of the magnetic field $\vec{B}$ at the points (x, y, z) = (0.0 m, 5.0 m, 0.0 m) and (5.0 m, 0.0 m, 0.0 m).

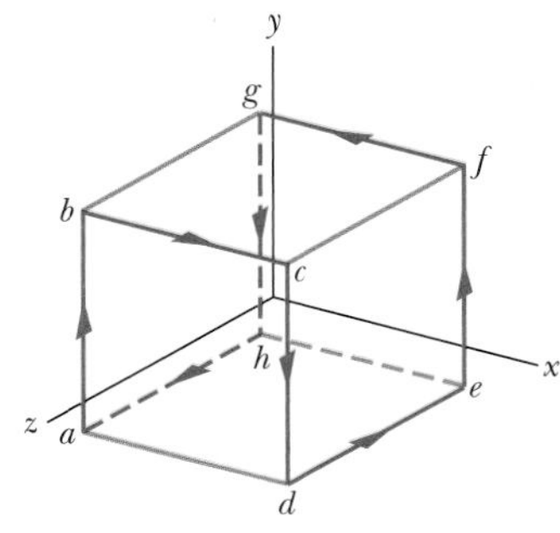

FIGURE 30-51 ■ Problem 58.

59. What Torque A circular loop of radius 12 cm carries a current of 15 A. A flat coil of radius 0.82 cm, having 50 turns and a current of 1.3 A, is concentric with the loop. (a) What magnetic field $\vec{B}$ (magnitude and direction) does the loop produce at its center? (b) What torque acts on the coil? Assume that the planes of the loop and coil are perpendicular and that the magnetic field due to the loop is essentially uniform throughout the volume occupied by the coil.

60. Two Different Arcs A length of wire is formed into a closed circuit with radii a and b, as shown in Fig. 30-52 and carries a current i. (a) What are the magnitude and direction of $\vec{B}$ at point P? (b) Find the magnetic dipole moment of the circuit.

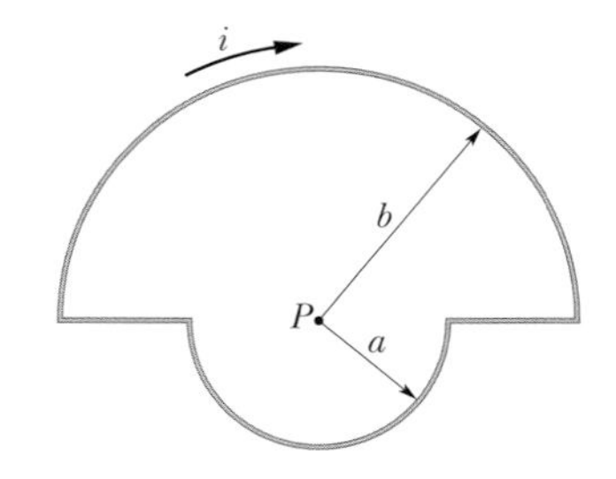

FIGURE 30-52 ■ Problem 60.

Additional Problems

61. Cross Section of a Wire Figure 30-53 shows the cross section of a wire that is perpendicular to the plane of the paper. Suppose a compass is placed at location A, which is a distance r from the wire. The compass points in the direction shown in the diagram. (a) Resketch the diagram and draw arrows to show what direction you expect the compass to point if it were moved to locations B and C. *Note:* Use the symbol ⊙ if the flow is out of the page and the symbol ⊗ if the flow is into the page. (b) Indicate in what direction *positive current* is flowing through the wire and describe the rule you are using to deduce the direction of current in the wire. (c) What is the direction of the flow of *electrons* through the wire?

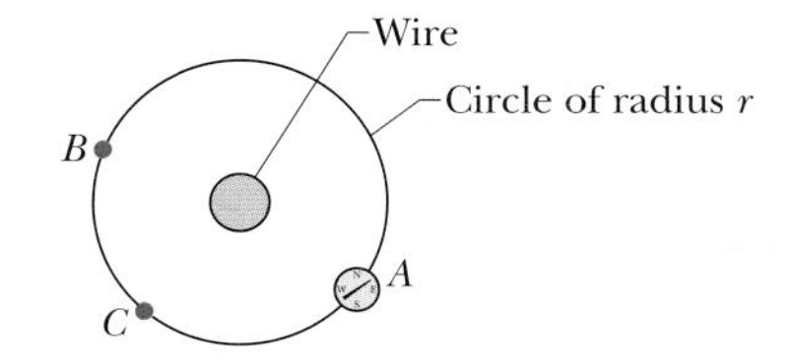

FIGURE 30-53 ■ Problem 61.

62. Wires in a *B*-Field A uniform magnetic field is directed toward the right in the plane of the paper as shown in Fig. 30-54. A wire lying perpendicular to the plane of the paper at location A carries

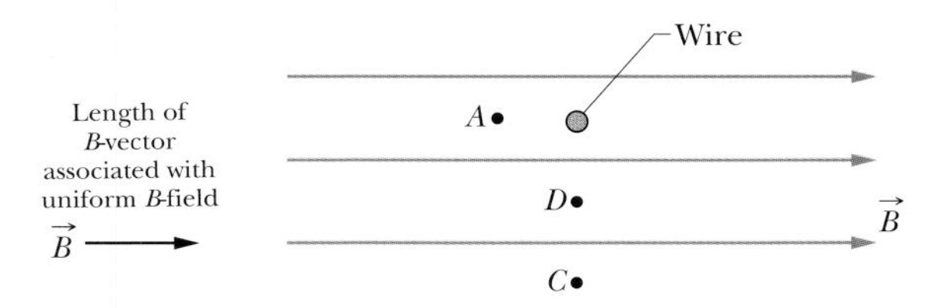

FIGURE 30-54 ■ Problem 62.

a current i. Suppose that the resultant magnetic field at point D due to a superposition of the uniform magnetic field of magnitude B and the magnetic field of the wire of magnitude B_w is zero. (a) Is the direction of the current in the wire into or out of the paper? Explain how you arrived at your conclusion. (b) Assume that point A lies at the same distance from the center of the wire as point D and that the length of the vector assigned to represent the magnitude of the uniform external magnetic field is that shown on the right. Construct a vector diagram showing the net magnetic field vector B_A^{net} at point A. (c) Assume that point C is twice the distance from the center of the wire as point D. Construct a vector diagram showing the net magnetic field vector, B_C^{net}, at point C. (Adapted from A. Arons, Homework and Test Questions for Introductory Physics Teaching, John Wiley and Sons, 1994.)

63. Earth's Field The magnitude of the Earth's magnetic field, B, at either geomagnetic pole, is about 7×10^{-5} T. Using a model in which you assume that this field is produced by a single current loop at the equator, determine the current that would generate such a field ($R_e = 6.37 \times 10^6$ m). *Hint:* The magnitudes of the magnetic field due to a single current loop of radius R at a distance R from its center and perpendicular to the plane of the loop is given by the equation

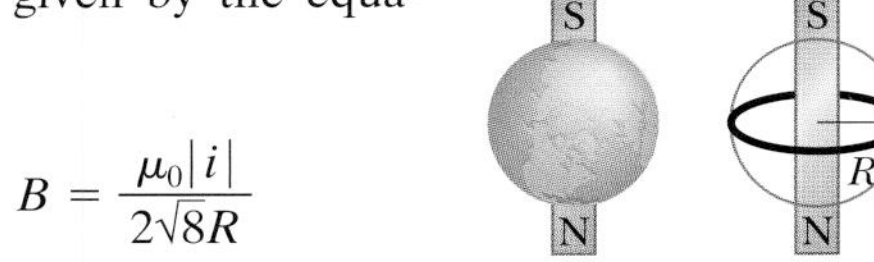

$$B = \frac{\mu_0 |i|}{2\sqrt{8}R}$$

and see Fig. 30-55.

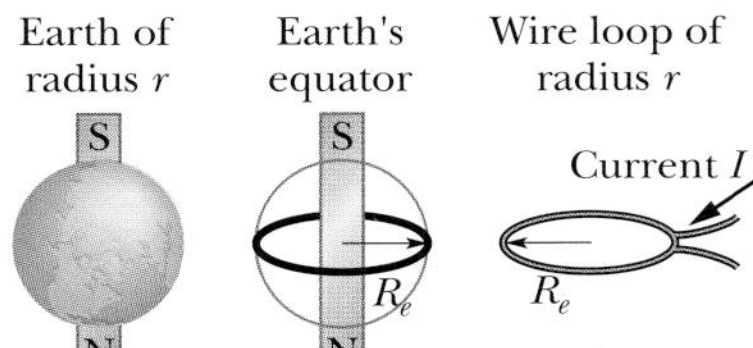

FIGURE 30-55 ■ Problem 63.

64. Comparing Electric and Magnetic Forces One In this problem we consider situations corresponding to three different long thin lines of matter containing charges: 1. A copper wire carrying an electric current from left to right, 2. A long amber rod that has been rubbed with fur and has a uniform excess of negative charge, and 3.

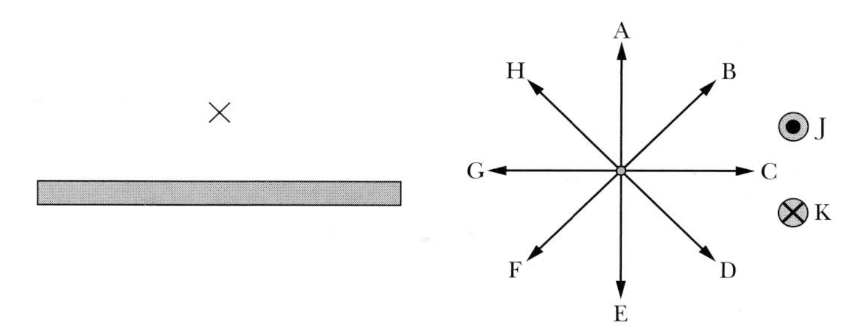

FIGURE 30-56 ■ Problem 64.

A beam of electrons passing from left to right through a vacuum inside a cathode ray tube. The direction of the electric current and of the electron flow are from left to right. Figure 30–56 shows a location marked x and a set of directions with labels on the right.

(a) For each of the three lines of matter, indicate in what direction the electric and magnetic fields at the location x would point. To indicate the direction, use one of the letters associated with a directional arrow on the "compass" in Fig. 30-56. If any of the fields are zero, write 0.

(b) Now consider placing a positive charge at the location x. In one case it is stationary, while in a second case it is moving in the direction C (to the right). Indicate the direction nearest to the total force the charge would feel. (Ignore gravity and air resistance.) Do this for all three lines of matter and for both cases.

65. Comparing Electric and Magnetic Forces Two Figure 30-57 shows a long wire carrying a current i to the right and a long amber rod with a charge density (charge/unit length) of λ. Assume that i and λ are both positive.

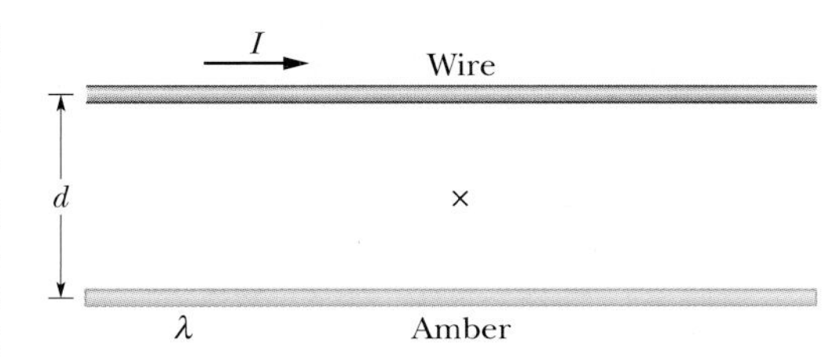

FIGURE 30-57 ■ Problem 65.

(a) The two are separated by a distance d. The point marked x is halfway between them. Copy this figure onto your paper and draw arrows to represent the following. (Be sure to label your arrows clearly to show which one is which.)

i. the direction of the magnetic field at the point marked x
ii. the direction of the electric field at the point marked x
iii. the direction of the electric force that a positive charge q placed at x would feel
iv. the direction of the magnetic force that a positive charge q placed at x would feel if it were moving to the right.

(b) The current, i, is +10 A, the charge density, λ, is -1 nC/m ($= 10^{-9}$ C/m) (note that it is negative), and the distance between the wires is 40 cm. At the instant shown, a proton with charge $q = 1.6 \times 10^{-19}$ C is moving into the page with a speed $v = 10^6$ m/s. Ignoring gravity, what is the magnitude and direction of the net force the proton feels at that time?

66. Direction of Magnetic Forces Figure 30-58 shows a cross section of four long parallel wires (labeled A through D) taken in a plane perpendicular to the wires. One or more of the wires may be carrying a current. If a wire carries a current, i_0, it is in the direction indicated and has strength $|i_0|$.

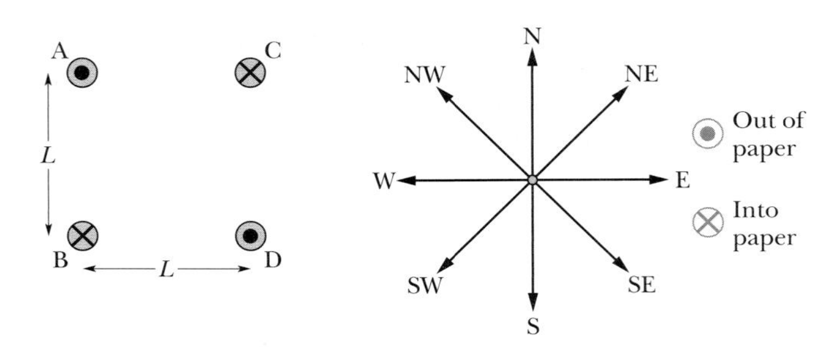

FIGURE 30-58 ■ Problem 66.

For each of the four vector quantities listed in (i) through (iv) below give the direction of the quantity. To indicate the direction, use one of the directions on the "compass" in Fig. 30-58. If the magnitude of the quantity is zero, write "0." If it is nonzero but in none of the indicated directions, write "Other."

i. Only wires B and D are carrying current. The direction of the force on wire D is____.
ii. Only wires B and D are carrying current. The direction of the force felt by an electron traveling in the E direction (on the compass) is____.
iii. Only wires B and D are carrying current. The direction of the force felt by an electron traveling in the N direction (on the compass) is____.
iv. All four wires are carrying current. The direction of the net force felt by wire A is____.

67. Magnetic Forces and Fields Figure 30-59 shows a cross section of three long parallel wires (labeled A through C) taken in a plane perpendicular to the wires. One or more of the wires may be carrying a current. If a wire carries a current, i_0, it is in the direction indicated and has strength $|i_0|$. For each of the five vector quantities (1) through (5) shown, indicate the direction of the quantity on the compass in Fig. 30-59. If the magnitude of the quantity is zero, write "0." If the result is not zero but points in a direction other than one of those indicated, write "other."

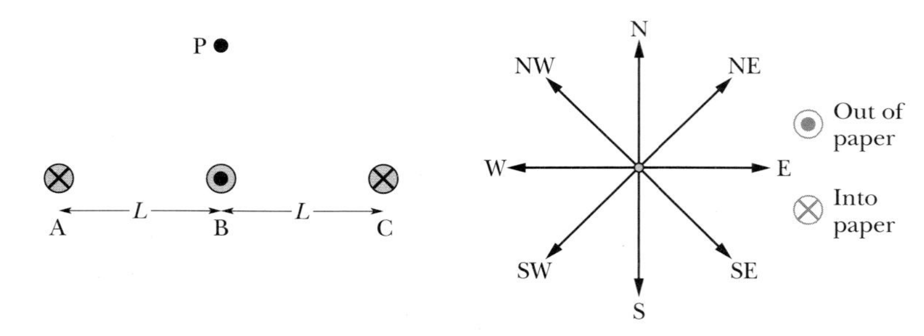

FIGURE 30-59 ■ Problem 67.

1. The magnetic field at point P if only wire A is carrying a current
2. The magnetic field at wire C if only wire A is carrying a current
3. The magnetic force on wire C if only wires A and C carry currents
4. The magnetic force on wire C if only wire A is carrying a current
5. The magnetic force on a proton at P traveling to the right (i.e., in direction E) if only wire B is carrying a current.

68. Right-Hand Rules During our discussions of magnetism and rotation we have encountered a number of different right-hand rules for obtaining the direction or sign of various quantities. Describe three right-hand rules. In your discussion of each one, include a statement of the equation or law in which the rule is applied, and whether the rule is "fundamental" or derived from a more basic principle.

69. Magnetic Forces Figure 30-60 shows parts of two long, current-carrying wires labeled 1 and 2. The wires lie in the same plane and cross at right angles at the point indicated. When carrying a current, each wire carries the same amount of current in the direction shown. At the right is shown a set of coordinate directions for describing the direction of vectors.

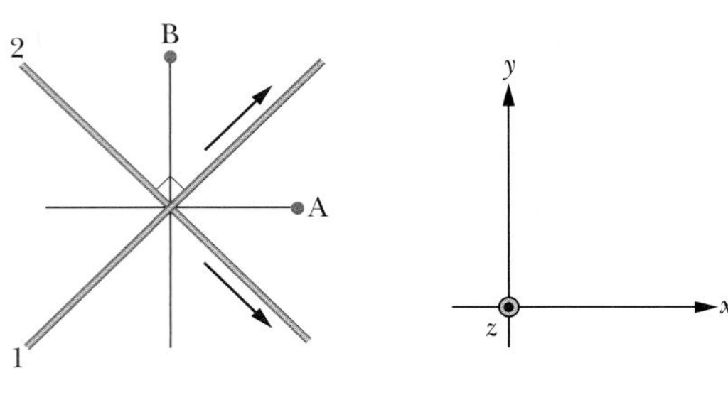

FIGURE 30-60 ■ Problem 69.

For each of the vectors discussed, indicate the direction of the vector using the coordinate system shown. For example, you might specify "the $+x$ direction" or "the $-z$ direction" or "in the x-y plane at 45° between the $+x$ and $+y$ directions." If the magnitude of the vector requested is zero, write "0."

(a) The direction of the force on a positively charged ion at the point B moving in the $+y$ direction if only wire 1 carries current
(b) The direction of the force on a positively charged ion at the point B moving in the $-z$ direction if both wires carry current
(c) The direction of the force on a positively charged ion at the point A moving in the $+x$ direction if only wire 2 carries current

For the next two parts of the problem, select which answer is correct if both wires carry current.

(d) The magnetic force on wire 1 will

i. push it in the $-z$ direction
ii. push it in the $+z$ direction
iii. tend to rotate it clockwise about the joining point
iv. tend to rotate it counterclockwise about the joining point
v. none of the above

(e) The magnetic force on wire 2 will

i. push it in the $-z$ direction
ii. push it in the $+z$ direction
iii. tend to rotate it clockwise about the joining point
iv. tend to rotate it counterclockwise about the joining point
v. none of the above

70. Constrained to a Circle Figure 30-61 shows, in cross section, two long straight wires held against a plastic cylinder of radius 20.0 cm. Wire 1 carries current $i_1 = 60.0$ mA out of the page and is fixed in place at the left side of the cylinder. Wire 2 carries current $i_2 = 40.0$ mA out of the page and can be moved around the cylinder. At what angle θ_2 should wire 2 be positioned such that the net magnetic field at the origin from the two currents has a magnitude of 80.0 nT?

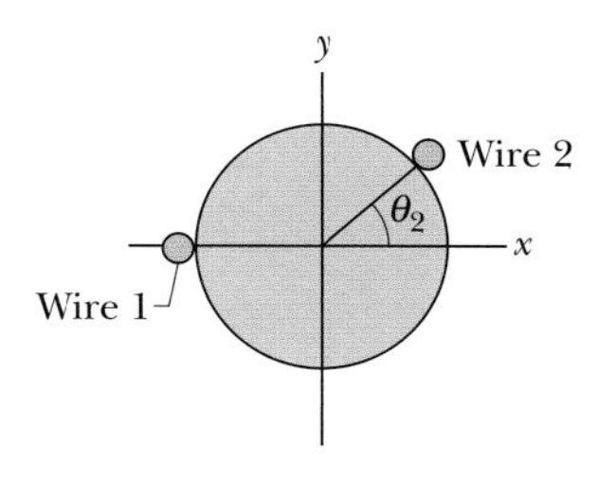

FIGURE 30-61 ■ Problem 70.

71. Element Length Figure 30-62a shows an element of length $ds = 1.00\ \mu m$ in a very long straight wire carrying current. The current in that element sets up a differential magnetic field $d\vec{B}$ at points in the surrounding space. Figure 30-62b gives the magnitude dB of the field in pico-Teslas (10^{-12} T) for points 2.5 cm from the element, as a function of angle θ between the wire and a straight line to the point. What is the magnitude of the magnetic field set up by the entire wire at perpendicular distance 2.5 cm from the wire?

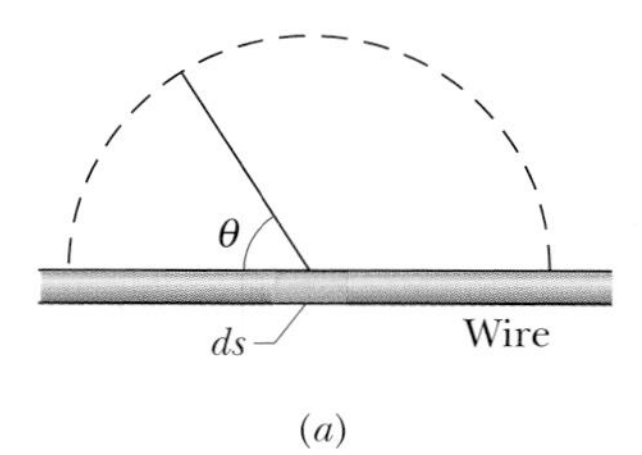

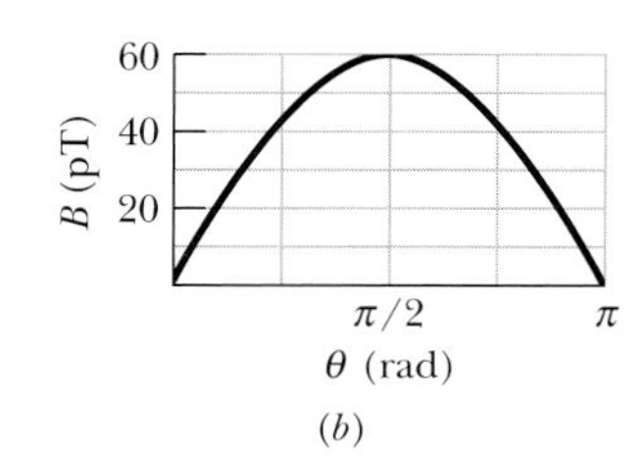

FIGURE 30-62 ■ Problem 71.

72. Where Is Wire 2 Two long straight thin wires with current lie against an equally long plastic cylinder, at radius $R = 20.0$ cm from the cylinder's central axis. Figure 30-63a shows, in cross section, the cylinder and wire 1 but not wire 2. With wire 2 fixed in place, wire 1 is moved around the cylinder, from angle $\theta_1 = 0°$ to angle $\theta_1 = 180°$, through the first and second quadrants of the xy coordinate system. The net magnetic field $\vec{B}$ at the center of the cylinder is measured as a function of θ_1. Figure 30-63b gives the x-component B_x of that field in micro-Teslas (10^{-6} T) and Fig. 30-63c gives the y-component B_y, both as functions of θ_1. (a) At what angle θ_2 is wire 2 located? What are the size and direction of the currents in (b) wire 1 and (c) wire 2?

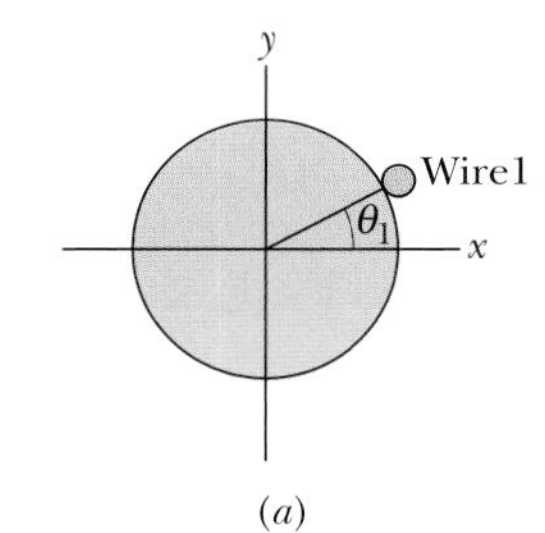

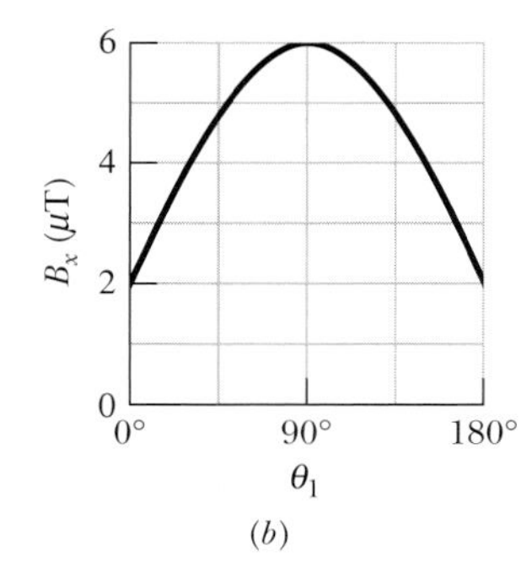

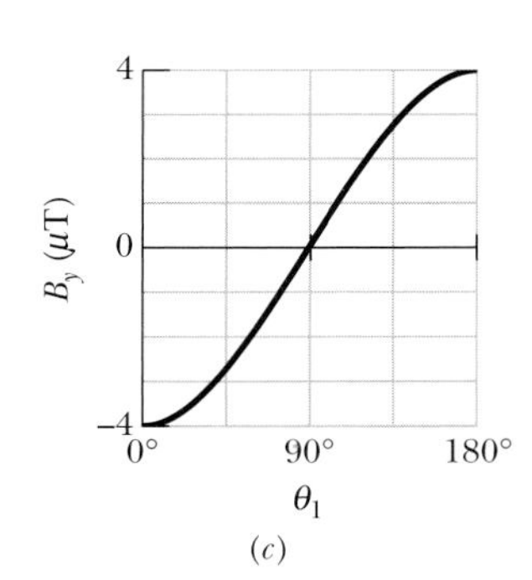

FIGURE 30-63 ■ Problem 72.

73. The Ratio of Currents Figure 30-64a shows, in cross section, two long, parallel wires carrying current and separated by distance L. The ratio $|i_1/i_2|$ of their current amounts is 4.00; the directions of the currents are not indicated. Figure 30-64b shows the y-component B_y in nano-Teslas (10^{-9} T) of their net magnetic field along the x axis to the right of wire 2. (a) At what value of $x > 0$ is B_y maximum? (b) If $|i_2| = 3$ mA, what is the value of the maximum? What are the directions of (c) current i_1 and (d) current i_2?

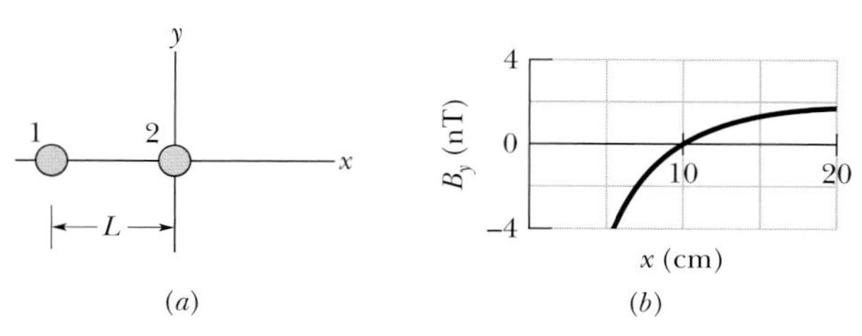

FIGURE 30-64 ■ Problem 73.

74. Same Radius Different Current In Fig. 30-65a two circular loops, with different currents but the same radius of 4.0 cm, are centered on a y axis. They are initially separated by distance $L = 3.0$ cm, with loop 2 positioned at the origin of the axis. The currents in the two loops produce a net magnetic field at the origin, with y-component B_y. That component is to be measured as loop 2 is gradually moved in the positive direction of the y axis. Figure 30-65b gives B_y in micro-Teslas (10^{-6} T) as a function of the position y of loop 2. The curve approaches an asymptote of $B_y = 7.20\ \mu T$ as $y \to \infty$. What are (a) current i_1 in loop 1 and (b) current i_2 in loop 2?

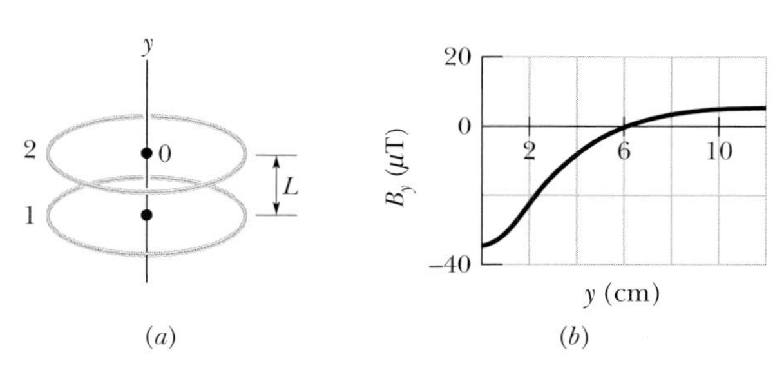

FIGURE 30-65 ■ Problem 74.

75. How Many Revolutions An electron is shot into one end of a solenoid, as it enters the uniform magnetic field within the solenoid, its speed is 800 m/s and its velocity vector makes an angle of 30° with the central axis of the solenoid. The solenoid carries 4.0 A and has 8000 turns along its length. How many revolutions does the electron make along its helical path within the solenoid by the time it emerges from the solenoid's opposite end? (In a real solenoid, where the field is not uniform at the two ends, the number of revolutions would be slightly less than the answer here.)

76. Force per Unit Length Two Figure 30-66 shows wire 1 in cross section; the wire is long and straight, carries a current of 4.00 mA out of the page, and is at distance $d_1 = 2.4$ cm from a surface. Wire 2, which is parallel to wire 1 and also long, is at horizontal distance $d_2 = 5.0$ cm from wire 1 and carries a current of 6.80 mA into the page. What is the x component of the magnetic force per unit length on wire 2 due to the current in wire 1?

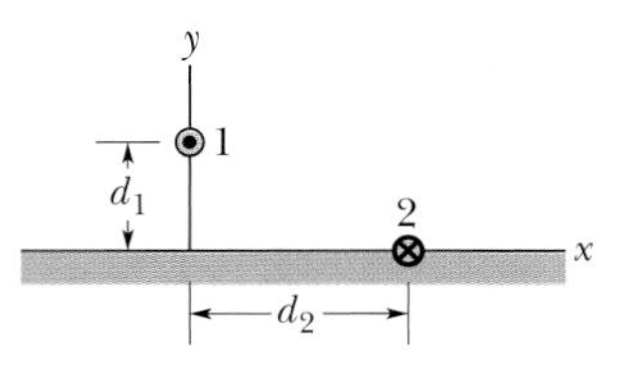

FIGURE 30-66 ■ Problem 76.

31 Induction and Maxwell's Equations

The General Motors EV1 electric car was marketed in the southwest with two generations of vehicles in 1997 and 1999. Production was completed in 1999 and all leases have been assigned. The EV1 had no engine, tailpipe, valves, pistons, timing belts, or crankshaft. The EV1 came with an inductive charging system in which there was no metal-to-metal connection. The charger, which plugged into a 220-volt outlet, had a paddle which, when inserted into the charge port at the front of the car, provided the electricity to re-charge the batteries.

How can electric car batteries be charged without making electrical contact with the power source?

The answer is in this chapter.

31-1 Introduction

FIGURE 31-1 ■ Michael Faraday, a famous English scientist, is credited with the discovery of electromagnetic induction.

In the previous chapter we discovered that the moving charges that make up electric currents create magnetic fields. We also learned that both permanent magnets and moving charges can exert forces on each other. These discoveries have powerful practical consequences. They allow us to build electromagnets to create large magnetic fields. More significantly, they enable us to harness the forces these large magnetic fields can exert on moving charges to create electric motors capable of moving massive objects.

In 1820, when Oersted observed that electric currents create magnetic fields, a number of prominent scientists began to look for ways to use magnetic fields to create currents. For more than a decade, scientists searched for current induced by static magnetic fields and failed to find it. By 1831, both Michael Faraday (Fig 31-1) and an American physicist, Joseph Henry, had discovered that a *changing* magnetic field is required to induce electric current. This phenomenon is called **electromagnetic induction.**

The discovery of electromagnetic induction, usually credited to Faraday, was of tremendous technological importance. Induction made it possible to create electric power from motion. Indeed, by the end of the 19th century, systems had been developed for the generation and transmission of electric power. Applications of Faraday's and Henry's discoveries are found in the design of thousands of electrical devices including transformers, high-speed trains, inductive battery chargers, and electric guitar pickups.

Although the practical benefits of the discovery of induction are tremendous, so is its impact on science. Many scientists view Faraday's law of induction as one of the most profound laws in all of classical physics because it "closed the loop" between magnetism and electricity. By combining Faraday's law with Ampère's law, we can understand how electricity and magnetism can be treated as complimentary aspects of the same phenomenon. By the middle of the 19th century, James Clerk Maxwell incorporated the ideas of Faraday and others into a famous set of four equations describing electromagnetic phenomena. In this chapter you will learn about the characteristics of electromagnetic induction and about Maxwell's synthesis of electromagnetic interactions.

READING EXERCISE 31-1: Why did it take so long for scientists working in the early 19th century to actually observe magnetic induction? ■

31-2 Induction by Motion in a Magnetic Field

Let us start our treatment of electromagnetic induction by considering what happens if we move a coil of conducting wire at a constant velocity through a uniform magnetic field and then out of the field as shown in Fig. 31-2. This is not the observation made by Faraday. We will describe that later. Notice that the diagram shows the plane of the coil is always perpendicular to the direction of the magnetic field. Under what conditions can a current be induced? We will consider this situation from both an experimental and a theoretical perspective.

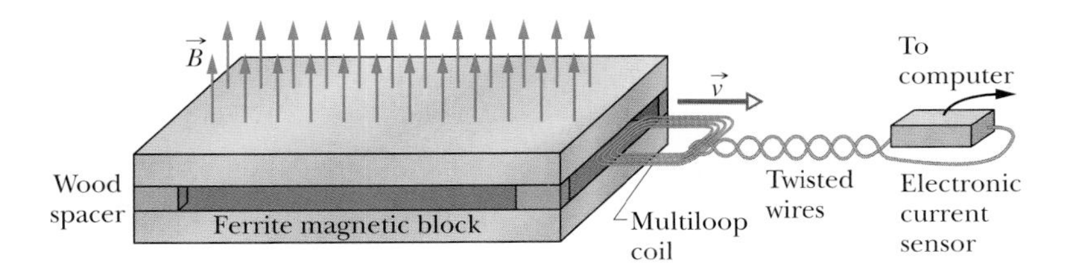

FIGURE 31-2 ■ It is not difficult to measure the current induced in a coil of wire while it is being pulled out of the gap between a pair of ferrite blocks separated by wooden spacers. The magnetic field in the central area between the magnetic blocks is essentially uniform. The ends of a multiloop coil are connected to an electronic current sensor.

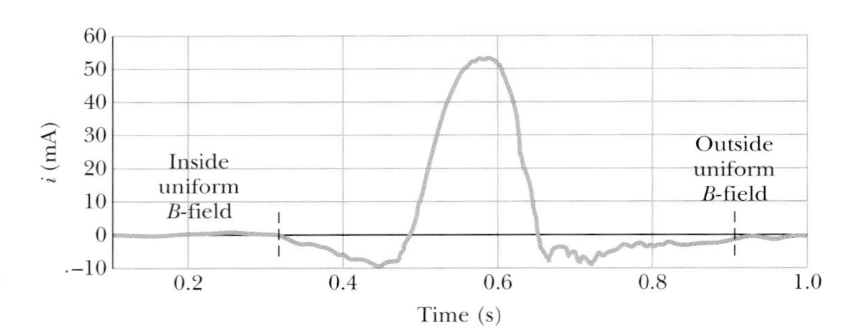

FIGURE 31-3 ■ A computer data acquisition system is used to measure the induced current 200 times a second as the coil shown in Fig. 31-2 is pulled steadily out of the uniform magnetic field in the central part of the gap between the two magnetic ferrite blocks. From 0.0 s to 0.3 s the entire coil is in the uniform magnetic field. After 0.9 s the coil is entirely outside the magnetic field. Between 0.3 s and 0.9 s part of the coil is in the *B*-field and part is outside of it.

A Conductor Moving Through a Magnetic Field—Observation

If we connect the ends of the coil to an ammeter, we see the needle jump back and forth a bit erratically during the time that the coil is passing out of the gap between the magnets. When the whole area of the coil is still in the central part of the gap between the magnets, the ammeter needle points to zero. When the coil has completely emerged from the region of space influenced by the magnets, the ammeter needle points to zero once again. This current jump can be seen in more detail using an electronic current sensor as shown in Fig. 31-3.

Both casual observation with a sensitive ammeter and the data gathered using an electronic current sensor show that for this situation:

- When the coil is not moving, there is no induced current no matter what the steady magnetic field is like at its location.
- When the coil is moving through a region where the magnetic field is entirely uniform or zero, there is no induced current.
- When the coil is moving through a region where the steady magnetic field is not uniform, a current is induced.

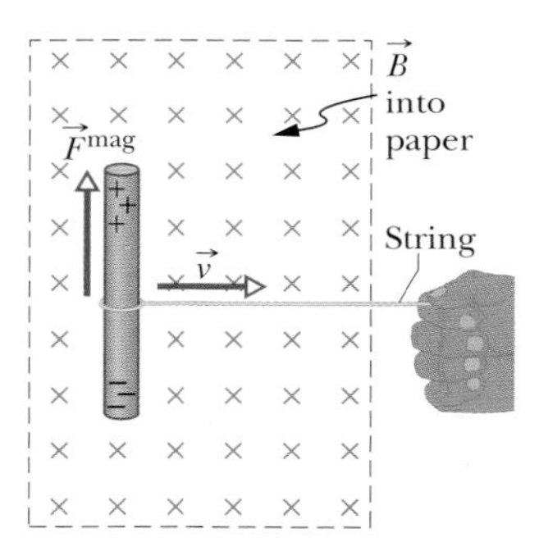

FIGURE 31-4 ■ A piece of wire is pulled through a uniform magnetic field at a constant velocity and becomes polarized.

We can draw the following conclusion from these observations:

> **OBSERVATION:** When a conducting loop moves perpendicular to a magnetic field, a current will be induced whenever the coil experiences a *changing* magnetic field.

A Conductor Moving Through a Magnetic Field—Theory

Although it's not obvious without reflection, you are capable of predicting that a *changing* magnetic field is required to induce an electric current in a moving coil. This induction is a natural consequence of the magnetic force laws described in Eqs. 29-2 and 29-23.

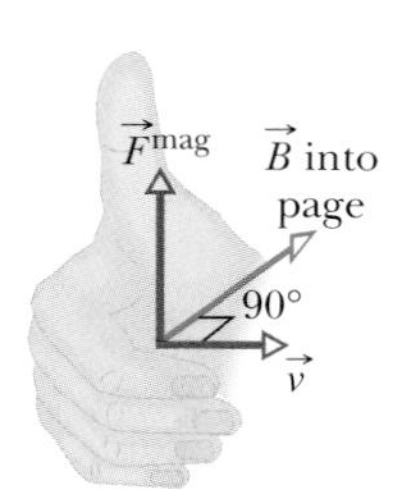

FIGURE 31-5 ■ The right-hand rule for the magnetic force law provides an "upward" force on positive changes and a "downward" force on mobile negative charges present in the wire shown in Fig. 31-4.

Straight Conductor Moving in a Uniform $\vec{B}$-Field: Let's start by using the force law to predict what happens to a straight piece of conducting wire if we pull it at a constant velocity in a direction perpendicular to a uniform magnetic field (Fig. 31-4). Each of the charges in the conductor experiences a force given by the magnetic force law, $\vec{F}^{mag} = q\vec{v} \times \vec{B}$ (Eq. 29-2). The direction of the force is given by the right-hand rule for cross products as shown in Fig. 31-5. Since there are mobile electrons in metals, these electrons will move toward the bottom of the wire, exposing fixed excess positive charge at the top of the wire. Thus, a current will flow in the wire for an instant until the electric field created by the charge separation opposes any further electron flow.

Loop Moving in a Uniform $\vec{B}$-Field: Perhaps we can induce a current by forming a closed loop instead of a single length of wire. This doesn't help because both the left and right segments (a and c in Fig. 31-6) are perpendicular to the motion, so they become polarized in the same manner. This merely results in excess charge piling up on the top and bottom segments (b and d as shown in Fig. 31-6).

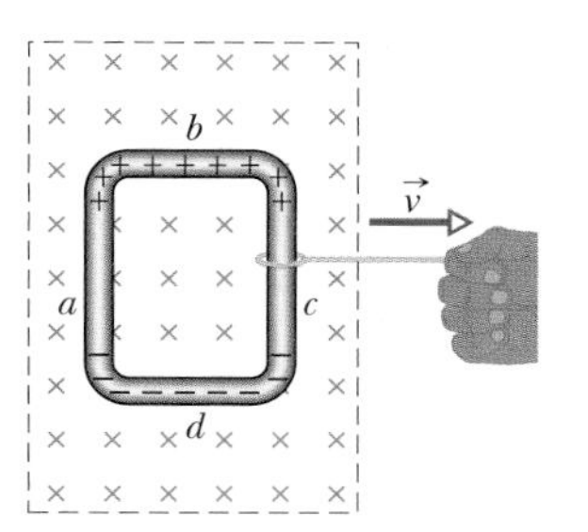

FIGURE 31-6 ■ A wire loop is pulled by a string through a uniform magnetic field at a constant velocity. Although excess charge accumulates on the top and bottom segments (b and d), no current is induced in the loop.

Loop Moving from a Uniform $\vec{B}$-Field to No Field: The easiest way to create a current using the polarization caused by the magnetic force law is to pull our loop in such a way that segment a is inside the magnetic field and segment c is not. Now what happens? We can see from Fig. 31-7 that the electrons in segment a, which is the trailing loop segment, continue to have magnetic forces exerted on them. But there are no forces on electrons in segment c of the loop because these electrons are not in the magnetic field. The only force that contributes to the current flow in the loop is the force on the left segment of wire, so the electrons in this segment of wire are pushed downward. The result is a net flow of electrons in a counterclockwise direction. Since "conventional current" as defined in Chapter 26 represents the flow of positive charge carriers, conventional current flow would be clockwise as shown in Fig. 31-7.

Nonuniform $\vec{B}$-Field: Theoretically we still expect to be able to induce a current in our loop in any nonuniform magnetic field. For example, suppose the magnetic field in Fig. 31-7 is weaker (but not necessarily zero) on the right side of the loop (near segment c) than it is at the left (near segment a). In this case, the magnetic forces on electrons in the left and right segments of the loop will no longer be equal and the forces on the charges in one of the segments will overpower those on the charges in the other segments. This will cause a net current to be induced.

Our theoretical considerations enable us to conclude that by applying the magnetic force law, we can predict the results of the observations presented in the first part of this section: When a conducting loop moves perpendicular to a magnetic field, then a current will be induced whenever the coil experiences a *changing* magnetic field through it.

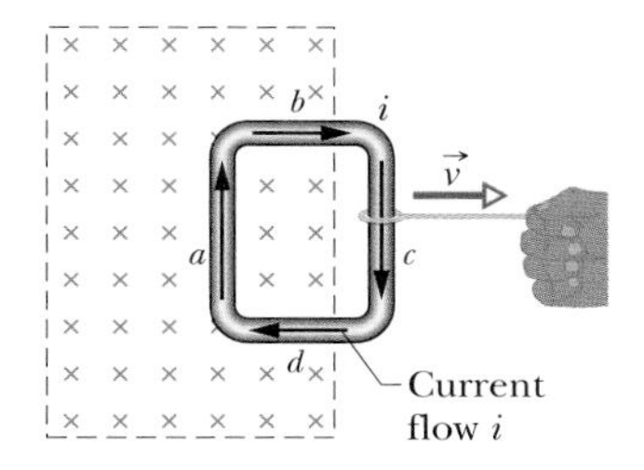

FIGURE 31-7 ■ A wire loop is pulled by a string through a region where the magnetic field is uniform on one side and zero on the other. Electrons from segment a are allowed to flow counterclockwise around the loop. The conventional current flow is clockwise.

READING EXERCISE 31-2: In the discussion above, we determined that the forces on the electrons in the top and bottom segments of the wire loop shown in Fig. 31-6 did not contribute to the current flow. Why is this the case? ■

READING EXERCISE 31-3: Suppose the magnetic field shown in Fig. 31-6 varies continuously in such a way that it is always stronger on the right than it is on the left. What will be the direction of the resulting (conventional) current in the loop? Explain. ■

31-3 Induction by a Changing Magnetic Field

Michael Faraday made a significant contribution to physics when he asked: What happens if instead of moving the wire loop in a magnetic field we keep the wire loop *stationary* and move a magnet toward or away from the loop to create a "moving" or changing magnetic field? One might argue that since the electrons in the wire are not moving in this case, the velocity of the loop segments used in the magnetic force law expression is zero and so there should be no force on the electrons and therefore no current. On the other hand, in many ways these two situations are the same. In order to answer his question, Faraday made observations similar to those discussed below.

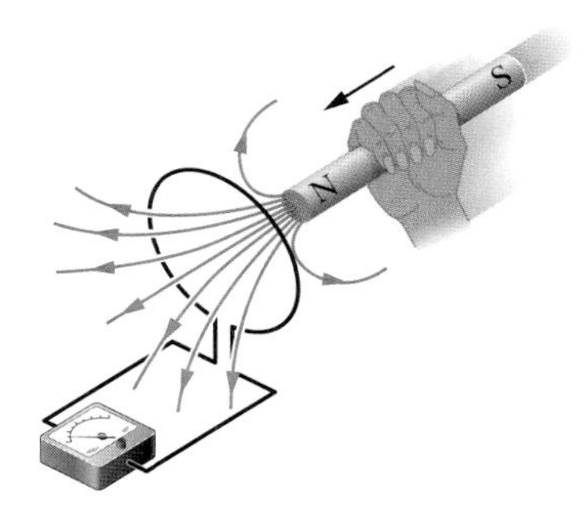

FIGURE 31-8 ■ An idealized setup showing a current meter registering nonzero currents in a stationary wire loop when a magnet is moving near the loop. (Typically a multiturn loop is needed to generate a detectable current.)

Observation 1, with a magnet: Figure 31-8 shows a conducting loop connected to a sensitive ammeter. Since there is no battery or other source of emf included, there is no current in the circuit. What happens if we move a bar magnet toward the loop? We observe that a current suddenly appears in the circuit! But the current disappears as soon as the magnet stops moving. If we then move the magnet away, a current again

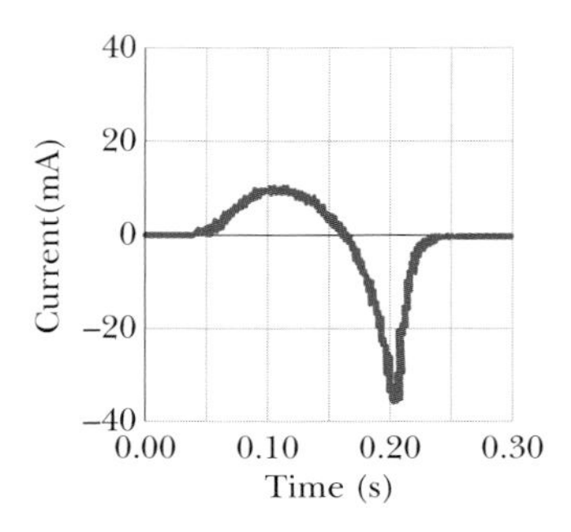

FIGURE 31-9 ■ The current induced as a magnet is dropped through a stationary multiturn coil (like that shown in Fig. 31-8). A computer data acquisition system is used to record current data at 2000 points/second.

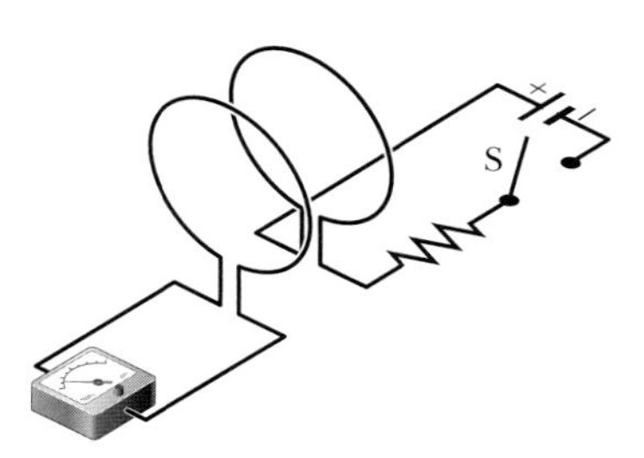

FIGURE 31-10 ■ An idealized setup showing an ammeter registering a current in the left-hand wire loop while switch S is being closed or opened (to turn the current in the right-hand loop on and off). No motion of the coils is involved. Faraday made essentially the same observation using multiloop coils.

suddenly appears, but now in the opposite direction. If we experimented for a while, we would observe the following:

1. A current appears only if there is relative motion between the loop and the magnet (one must move relative to the other, but it doesn't matter which one); the current disappears when the relative motion between them ceases. See Fig. 31-9 for a graph of the current induced by a magnet dropped through a stationary coil.
2. Faster motion produces a greater current.
3. If moving the magnet's north pole toward the loop causes, say, clockwise current, then moving the north pole away causes counterclockwise current. Moving the south pole toward or away from the loop also causes current, but in the reversed direction.

We call the current produced in the loop an **induced current;** the work done per unit charge to produce that current (to move the conduction electrons that constitute the current) is called an **induced emf,** and the process of producing the current and emf is called induction. Currents that are caused by batteries in a circuit and those caused by induction in a wire loop are the same—mobile electrons are flowing through wires.

Observation 2, replacing the magnet with a current-carrying coil: Let us now perform a second observation. For this observation we use the apparatus of Fig. 31-10, with the two conducting loops close to each other but not touching. If we close switch S to turn on a current in the right-hand loop, the meter suddenly and briefly registers a current—an induced current—in the left-hand loop. If we then open the switch, another sudden and brief induced current appears in the left-hand loop, but in the opposite direction. We get an induced current (and thus an induced emf) only when the current in the right-hand loop is changing (either when turning on or off) and not when it is constant (even if the current is large). The outcome of this second observation is not surprising. We know from Ampère's law (Chapter 30) that the magnitude of the magnetic field surrounding a current-carrying wire increases as the current increases and its direction changes when the direction of current changes.

Faraday also noticed that the actual amount of magnetic field present at the area enclosed by the loop does not matter. Instead, the values of the induced emf and induced current are determined by the *rate* at which the amount changes.

When we pull all of these observations together, the way Faraday did, we conclude that

> Induced emf and current are present whenever the magnetic field present in the area subtended by the conducting loop *changes* for any reason.
>
> The amount of induced emf and current increases as a function of the rate of change of the magnetic field present at the area subtended by the loop.

Charging an Electric Car by Induction

Suppose we replace the switch in Fig. 31-10 with a source of current in the right-hand loop that varies over time sinusoidally. Then we create a magnetic field at the location of the left-hand loop that is also changing sinusoidally in time. This time-varying magnetic field then induces current in the left-hand loop that varies with time as well. By using some circuitry to filter out the negative current, we can use this induced current to charge a battery even though there is *no electrical contact* between the right- and left-hand loops. This type of noncontact charging is also used for charging familiar devices such as electric toothbrushes. Although an actual charger for an electric toothbrush or car like that discussed in the chapter-opening puzzler has more loops of wire and electrical circuits in it, it works on the same principle of electromagnetic induction discovered by Henry and Faraday in the early 19th century.

One drawback of inductive charging is that it is slower than direct charging. This is not a problem for electric toothbrush charging, but it is for electric car charging. This is probably one reason why the inductively charged General Motors EV1 cars like the one shown in the chapter puzzler have been taken off the market. You will learn more about the practical applications of induction in Chapter 32.

READING EXERCISE 31-4: Can the magnetic force law be used to explain why a current appears in a stationary loop when a bar magnet is brought close to it? If so, use your understanding of this force law to explain how this happens. If not, justify why not. ■

READING EXERCISE 31-5: Consider the induced current data shown in Fig. 31-9. The magnet is accelerating as it falls through the stationary coil. The magnet is dropped in free fall. The extrema of currents are about +8mA and −35mA. Why is the negative extremum larger? ■

31-4 Faraday's Law

We can enhance the predictive power of Faraday's qualitative observations by developing a mathematical formulation of electromagnetic induction. The mathematical expression that describes electromagnetic induction is commonly known as **Faraday's law.** Although we derive Faraday's law for a simplified situation using concepts and laws that we have already introduced, it can be applied to virtually any situation.

Magnetic Flux

To begin we use the concept of magnetic flux to quantify the amount of magnetic field at the area enclosed by a loop. In Chapter 24, in a similar situation, we needed to calculate the amount of an electric field present on a surface. There we determined electric flux for a small element of essentially flat area in Eq. 24-2 as $\Phi^{\text{elec}} = \vec{E}\cdot\Delta\vec{A}$ (the dot product of the normal vector representing a small area and the electric field vector at the location of the area). By analogy, the *magnetic flux* at the surface of a small area element $\Delta\vec{A}$ that is located in a magnetic field $\vec{B}$ is defined as

$$\Phi^{\text{mag}} = \vec{B}\cdot\Delta\vec{A} \qquad \text{(magnetic flux at an area } \Delta\vec{A}). \tag{31-1}$$

Simply put, the flux of magnetic field Φ^{mag} at an area element A is the product of the area element and the component of the field *perpendicular* to it for a uniform magnetic field. The validity of this basic definition depends on the assumption that the magnetic field $\vec{B}$ is uniform over the surface element $\Delta\vec{A}$. If the field varies over the area, we must break the area up into little pieces in such a way that the field will be about constant for each piece. We then calculate the flux in each little piece and perform an integration to add up all the little contributions in analogy to the more general definition of electric flux.

From Eq. 31-1, we see that the SI unit for magnetic flux is the tesla-square meter, which is called the weber (abbreviated Wb):

$$1 \text{ weber} = 1 \text{ Wb} = 1 \text{ T}\cdot\text{m}^2. \tag{31-2}$$

Using our simplified formulation of magnetic flux, we are now ready to derive Faraday's law.

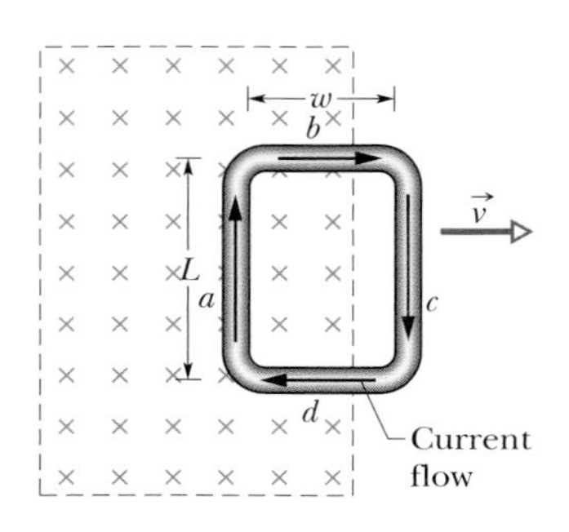

FIGURE 31-11 ■ A wire loop is moving at a constant velocity through a region where the magnetic field is uniform on one side and zero on the other. While this is happening, the magnetic flux at the area subtended by the coil is decreasing at a constant rate.

A Simplified Derivation of Faraday's Law

Consider the simple situation depicted in Fig. 31-7 in which a wire loop is being pulled out of a uniform magnetic field at a constant velocity. Next we derive the relationship between the emf induced in the loop and the rate of change of the magnetic flux enclosed by the loop. To help us with the derivation we have redrawn the situation and introduced symbols for the dimensions of the loop and the axis along which it moves in Fig. 31-11.

According to the magnetic force law, each charge in the left part of the loop (segment a) will experience a force of magnitude $F^{\text{mag}} = qvB$. As the positive and negative charges separate, an electric field of magnitude

$$E = \frac{F^{\text{mag}}}{|q|} = vB \tag{31-3}$$

will be generated. If segment a has a length L, then the potential difference of induced emf across it is given by

$$\mathscr{E} = EL = vBL. \tag{31-4}$$

Next we need to relate the right side of Eq. 31-4 to the rate at which the magnetic flux at the area subtended by the loop is decreasing as it moves out of the uniform B-field. If we designate the loop as being pulled in the x direction, then its velocity component can be expressed as $v_x = dx/dt$. Note that the area of the moving loop is decreasing at a rate given by $dA/dt = -L\,dx/dt = -Lv_x$. Since the magnetic field that subtends the left part of the area enclosed by the loop is constant, the rate of change of the magnetic flux at the loop can be expressed as

$$\frac{d\Phi^{\text{mag}}}{dt} = \frac{d(BA)}{dt} = B\frac{dA}{dt} = -v_x BL. \tag{31-5}$$

Combining Eqs. 31-4 with 31-5, we get an expression for Faraday's law for a single loop or coil,

$$\mathscr{E} = -\frac{d\Phi^{\text{mag}}}{dt} \quad \text{(Faraday's law for a single-turn coil).} \tag{31-6}$$

As you will see in the next section, the induced emf $\mathscr{E}$ tends to oppose the flux change, and the minus sign indicates that opposition. Faraday's law can also be expressed in words:

> The amount of the emf $\mathscr{E}$ induced in a conducting loop is equal to the rate at which the magnetic flux Φ^{mag} at the area enclosed by the loop changes with time.

FIGURE 31-12 ■ It is quite easy to verify Faraday's law with modern apparatus and computer data acquisition systems. Here a student holds a small multiturn pickup coil inside a larger field coil that is generating a "sawtooth" magnetic field that increases and then decreases continuously. The B-field is shown on the jagged dark red trace on the computer screen. The induced current in the pickup coil is shown by the squarish lighter green trace. (Photo courtesy of PASCO scientific.)

If we change the magnetic flux at a coil of N turns, an induced emf appears in every turn and the total emf induced in the coil is the sum of these individual induced emfs. If the coil is tightly wound (closely packed), so that the same magnetic flux Φ^{mag} is present in each turn, the total emf induced in the coil is

$$\mathscr{E} = -N\frac{d\Phi^{\text{mag}}}{dt} \quad \text{(Faraday's law for an } N\text{-turn coil).} \tag{31-7}$$

Although we have used simple geometry to derive Faraday's law (Eq. 31-7), experiments (such as the one shown in Fig. 31-12) have verified that the mathematical expression we have derived is true for any situation where the flux enclosed by a set of conducting loops or coils is changing. In fact, there are many ways to change the magnetic flux at a coil and thus induce emfs and currents:

1. Change the magnitude B of the magnetic field within the coil.
2. Change the area of the coil, or the portion of that area that happens to lie within the magnetic field (for example, by expanding the coil or sliding it out of the field).
3. Change the angle between the direction of the magnetic field $\vec{B}$ and the area of the coil (for example, by rotating the coil so that $\vec{B}$ is first perpendicular to the plane of the coil and then is along that plane).

Later on in the chapter we will derive a more general form of Faraday's law that relates flux change to electric field induction even when no charges or conducting loops are present.

READING EXERCISE 31-6: The graph gives the magnitude $B(t)$ of a magnetic field that exists throughout the area subtended by a conducting loop, perpendicular to the plane of the loop. Although it changes with time, at any particular instant the magnetic field is uniform over the area of the loop. (a) Rank the five time intervals (a, b, c, d, and e) shown on the graph according to the amount of the emf $|\mathscr{E}|$ induced in the loop, greatest first. (b) Explain your reasoning.

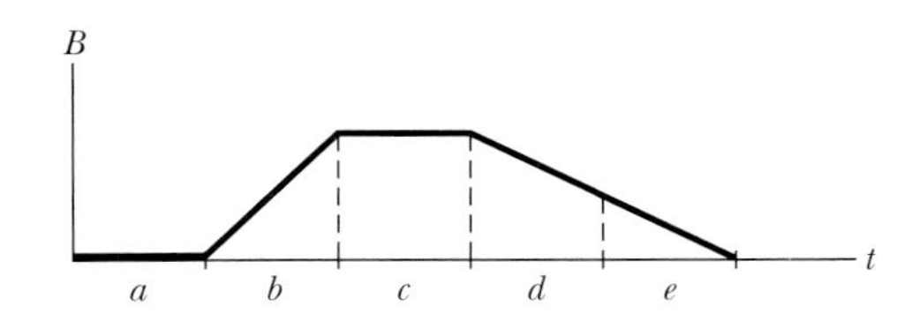

TOUCHSTONE EXAMPLE 31-1: Coil in a Long Solenoid

The long solenoid S shown (in cross section) in Fig. 31-13 has 220 turns/cm and carries a current $i = 1.5$ A; its diameter D is 3.2 cm. At its center we place a 130-turn, closely packed coil C of diameter $d = 2.1$ cm. The current in the solenoid is reduced to zero at a steady rate in 25 ms. What is the size of the emf $|\mathscr{E}|$ that is induced in coil C while the current in the solenoid is changing?

SOLUTION ■ The **Key Ideas** here are these:

1. Because coil C is located in the interior of the solenoid, it lies within the magnetic field produced by current i in the solenoid; thus, there is a magnetic flux Φ^{mag} present in coil C.
2. Because current i decreases, flux Φ^{mag} also decreases.
3. As Φ^{mag} decreases, emf $\mathscr{E}$ is induced in coil C, according to Faraday's law.

Because coil C consists of more than one turn, we apply Faraday's law in the form of Eq. 31-7 ($\mathscr{E} = -N d\Phi^{\text{mag}}/dt$), where the number

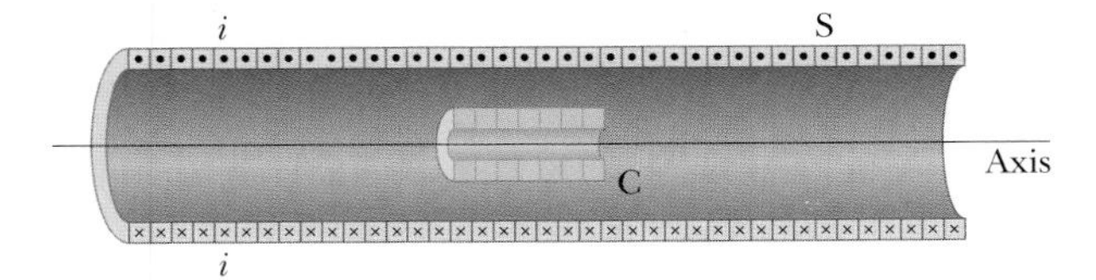

FIGURE 31-13 ■ A coil C is located inside solenoid S, which carries current i.

of turns N is 130 and $d\Phi^{\text{mag}}/dt$ is the rate at which the flux in each turn changes.

Because the current in the solenoid decreases at a steady rate, flux Φ^{mag} also decreases at a steady rate and we can write $d\Phi^{\text{mag}}/dt$ as $\Delta\Phi^{\text{mag}}/\Delta t$. Then, to evaluate $\Delta\Phi^{\text{mag}}$, we need the final and initial flux. The final flux Φ_f^{mag} is zero because the final current in the solenoid is zero. To find the initial flux Φ_i^{mag}, we need two more **Key Ideas:**

4. The flux at the area enclosed by each turn of coil C depends on the area A and orientation of that turn in the solenoid's magnetic field $\vec{B}$. Because $\vec{B}$ is uniform and directed perpendicular to area A, the flux is given by Eq. 31-1 ($\Phi^{\text{mag}} = BA$).
5. The magnitude B of the magnetic field in the interior of a solenoid depends on the solenoid's current i and its number n of turns per unit length, according to Eq. 30-25 ($B = n\mu_0|i|$).

For the situation of Fig. 31-13, A is $\frac{1}{4}\pi d^2$ ($=3.46 \times 10^{-4}$ m^2) and n is 220 turns/cm, or 22 000 turns/m. Substituting Eq. 30-25 into Eq. 31-1 then leads to

$$\begin{aligned}\Phi_i^{\text{mag}} &= BA = (n\mu_0|i|)A \\ &= (22\,000 \text{ turns/m})(4\pi \times 10^{-7}\,\text{T}\cdot\text{m/A})(1.5\text{ A})(3.46 \times 10^{-4}\text{ m}^2) \\ &= 1.44 \times 10^{-5}\,\text{Wb}.\end{aligned}$$

Now we can write

$$\frac{d\Phi^{\text{mag}}}{dt} = \frac{\Delta\Phi^{\text{mag}}}{\Delta t} = \frac{\Phi_f^{\text{mag}} - \Phi_i^{\text{mag}}}{\Delta t}$$

$$= \frac{(0 - 1.44 \times 10^{-5}\ \text{Wb})}{25 \times 10^{-3}\ \text{s}}$$

$$= -5.76 \times 10^{-4}\ \text{Wb/s} = -5.76 \times 10^{-4}\ \text{V}.$$

We are interested only in the size of the emf, so we ignore the minus signs here and in Eq. 31-7, writing

$$|\mathscr{E}| = \left|N\frac{d\Phi^{\text{mag}}}{dt}\right| = (130\ \text{turns})(5.76 \times 10^{-4}\ \text{V})$$

$$= 7.5 \times 10^{-2}\ \text{V} = 75\ \text{mV}. \qquad \text{(Answer)}$$

31-5 Lenz's Law

Soon after Faraday proposed his law of induction, Heinrich Friedrich Lenz devised a rule—now known as Lenz's law—for determining the direction of an induced current in a loop:

> An induced current has a direction such that the magnetic field due to the current opposes the change in the magnetic flux that has induced the current.

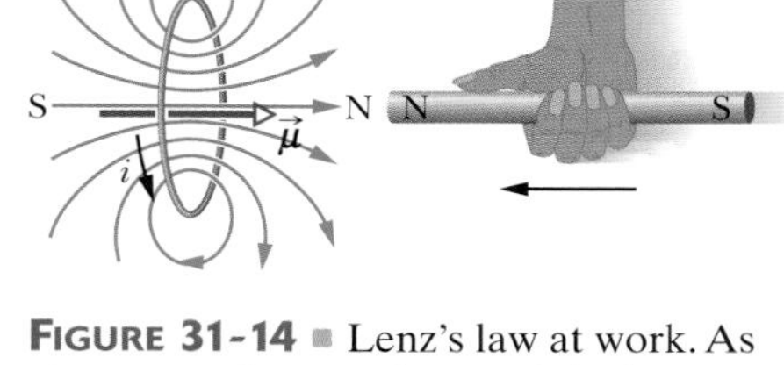

FIGURE 31-14 ■ Lenz's law at work. As the magnet is moved toward the loop, a current is induced in the loop. The current produces its own magnetic field, with magnetic dipole moment $\vec{\mu}$ oriented so as to oppose the motion of the magnet. Thus, the induced current must be counterclockwise as shown.

It is important to notice that it is the *change* in the flux that determines the direction of the induced current rather than the direction of the magnetic field or motion. Furthermore, the direction of an induced emf is that of the induced current. To get a feel for Lenz's law, let us apply it in two different but equivalent ways to Fig. 31-14, where the north pole of a magnet is being moved toward a conducting loop.

1. **Opposition to Flux Change.** In Fig. 31-14, with the magnet initially distant, there is no magnetic flux at the area encircled by the loop. As the north pole of the magnet then nears the loop with its magnetic field $\vec{B}$ directed *toward the left,* the flux at the loop increases. To oppose this increase in flux, the induced current i must set up its own field $\vec{B}_i$ *directed toward the right* inside the loop, as shown in Fig. 31-15*a*; then the rightward flux of field $\vec{B}_i$ opposes the increasing leftward flux of field $\vec{B}$. The right-hand rule of Fig. 30-19 then tells us that i must be counterclockwise in Fig. 31-15*a*.

2. **Opposition to Pole Movement.** The approach of the magnet's north pole in Fig. 31-14 increases the magnetic flux in the loop and thereby induces a current in the loop. From Fig. 30-22, we know that the loop then acts as a magnetic dipole with a south pole and a north pole, and that its magnetic dipole moment $\vec{\mu}$ is directed from south to north. To oppose the magnetic flux increase being caused by the approaching magnet, the loop's north pole (and thus $\vec{\mu}$) must face *toward* the approaching north pole so as

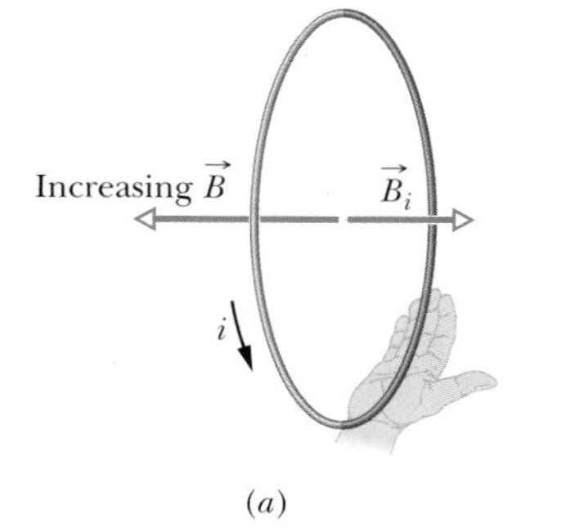

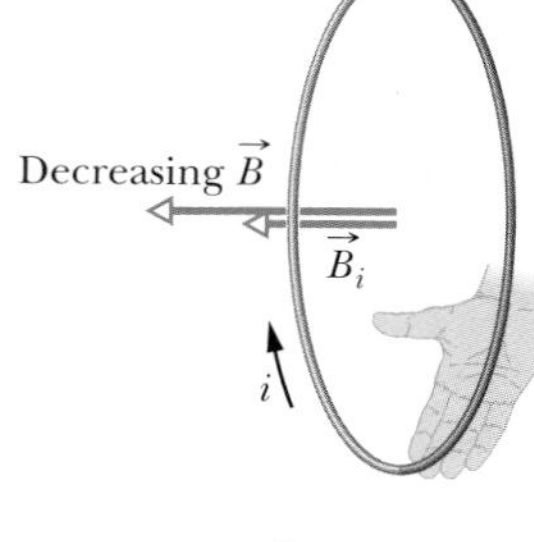

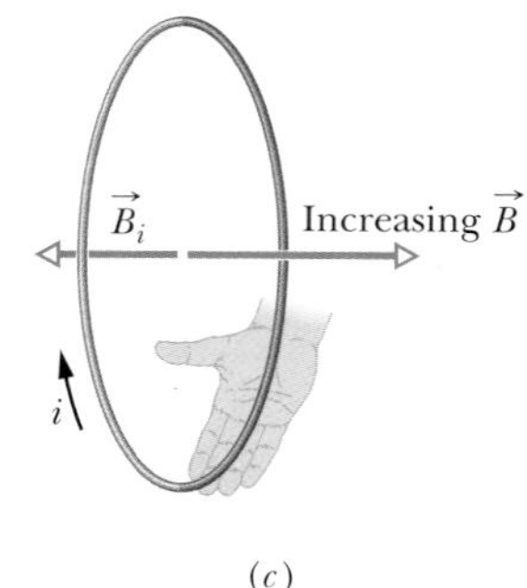

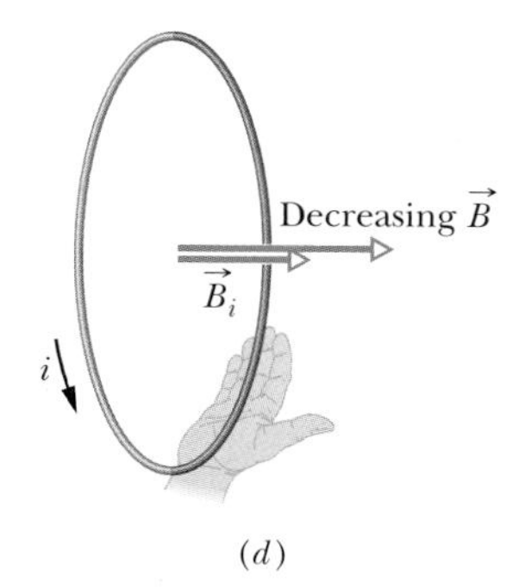

FIGURE 31-15 ■ The current i induced in a loop has the direction such that the current's magnetic field $\vec{B}_i$ opposes the change in the magnetic field $\vec{B}$ inducing i. The field $\vec{B}_i$ is always directed opposite an increasing field $\vec{B}$ shown in (*a*) and (*c*) and in the same direction as a decreasing field $\vec{B}$ shown in (*b*) and (*d*). The curled-straight right-hand rule gives the direction of the induced current based on the direction of the induced field.

to repel it (Fig. 31-14). Then the curled-straight right-hand rule for $\vec{\mu}$ (Fig. 30-22) tells us that the current induced in the loop must be counterclockwise in Fig. 31-14.

If we next pull the magnet away from the loop, a current will again be induced. Now, however, the loop will have a south pole facing the retreating north pole of the magnet, so as to oppose the retreat. Thus, the induced current will be clockwise.

As we noted above, be careful to remember that the flux of $\vec{B}_i$ always opposes the *change* in the flux of $\vec{B}$, but that does not always mean that $\vec{B}_i$ points opposite $\vec{B}$. For example, if we pull the magnet away from the loop in Fig. 31-14, the flux Φ^{mag} from the magnet is still directed to the left at the area subtended by the loop, but it is now decreasing. The flux of $\vec{B}_i$ must now be to the left inside the loop, to oppose the *decrease* in Φ^{mag}, as shown in Fig. 31-15*b*. Thus, $\vec{B}_i$ and $\vec{B}$ are now in the same direction.

Figures 31-15*c* and *d* show the situations in which the south pole of the magnet approaches and retreats from the loop, respectively. Figure 31-16 is a photo of a demonstration of Lenz's law in action.

FIGURE 31-16 ■ This demonstration of Lenz's law occurs when an electromagnet is switched on suddenly. The current induced in a metal ring opposes the electromagnet's current. The repulsive forces between the magnet and the ring cause the ring to jump more than a meter. (Photo courtesy of PASCO scientific.)

Electric Guitars

Soon after rock began in the mid-1950s, guitarists switched from acoustic guitars to electric guitars—but it was Jimi Hendrix who first used the electric guitar as an electronic instrument. He was able to create new sounds that continue to influence rock music today. What is it about an electric guitar that enabled Hendrix to make different sounds?

Whereas an acoustic guitar depends for its sound on the acoustic resonance produced in the hollow body of the instrument by the oscillations of the strings, an electric guitar like that being played by Hendrix in Fig. 31-17 is a solid instrument, so there is no body resonance. Instead, the oscillations of the metal strings are sensed by electric "pickups" that send signals to an amplifier and a set of speakers.

The basic construction of a pickup is shown in Fig. 31-18. Wire connecting the instrument to the amplifier is coiled around a small magnet. The magnetic field of the magnet produces a north and south pole in the section of the metal string just above the magnet. That section of string then has its own magnetic field. When the string is plucked and thus made to oscillate, its motion relative to the coil changes the flux of its magnetic field at the area encircled by the coil, inducing a current in the coil. As the string oscillates toward and away from the coil, the induced current changes direction at the same frequency as the string's oscillations, thus relaying the frequency of oscillation to the amplifier and speaker.

On a Stratocaster©, there are three groups of pickups, placed near the bridge at the end of the wide part of the guitar body. The group closest to the bridge better detects the high-frequency oscillations of the strings; the group farthest from the near end better detects the low-frequency oscillations. By throwing a toggle switch on the guitar, the musician can select which group or which pair of groups will send signals to the amplifier and speakers.

To gain further control over his music, Hendrix sometimes rewrapped the wire in the pickup coils of his guitar to change the number of turns. In this way, he altered the amount of emf induced in the coils and thus their relative sensitivity to string oscillations. Even without this additional measure, you can see that the electric guitar offers far more control over the sound that is produced than can be obtained with an acoustic guitar.

FIGURE 31-17 ■ Jimi Hendrix playing his Fender Stratocaster©. This guitar has three groups of six electric pickups each (within the wide part of the body). A toggle switch (at the bottom of the guitar) allows the musician to determine which group of pickups sends signals to an amplifier and thus to a speaker system.

READING EXERCISE 31-7: Lenz's law states: "An induced current has a direction such that the magnetic field due to the current opposes the change in the magnetic flux that induces the current." (a) Suppose there is a magnetic field directed into the plane of this page and that the strength of the field is decreasing. Would a magnetic field that opposes this change in magnetic flux be directed into the page, out of the page, or in some other direction? Explain your reasoning. (b) Suppose that there is a magnetic field directed into the plane of this page that is increasing in strength. Would a magnetic field that opposes this change in magnetic flux be directed into the page, out of the page, or in some other direction? Explain your reasoning. ■

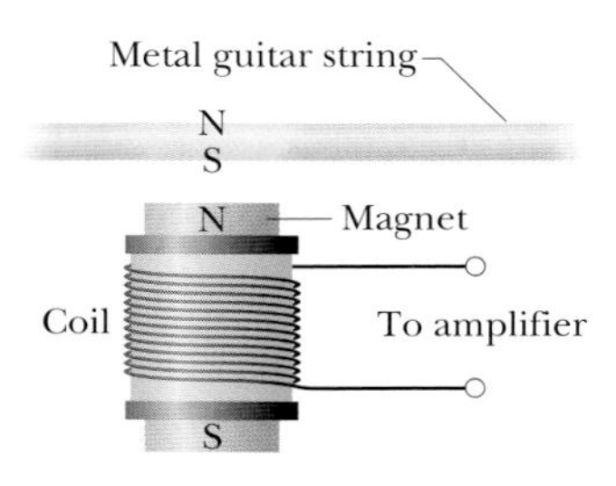

FIGURE 31-18 ■ A side view of an electric guitar pickup. When the metal string (which acts like a magnet) oscillates, it causes a variation in magnetic flux that induces a current in the coil.

READING EXERCISE 31-8: The figure shows three situations in which identical circular conducting loops are in uniform magnetic fields that are either increasing (Inc) or decreasing (Dec) in magnitude at identical rates. In each, the dashed line coincides with a diameter. (a) Rank the situations according to the amount of the current induced in the loops, greatest first. (b) Explain your reasoning.

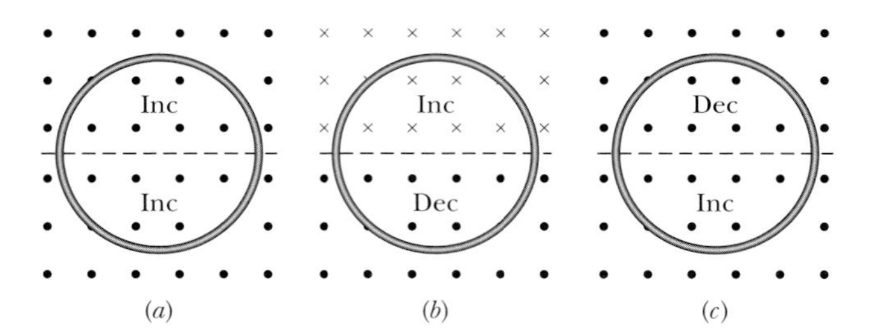

■

TOUCHSTONE EXAMPLE 31-2: Induced Emf

Figure 31-19 shows a conducting loop consisting of a half-circle of radius $r = 0.20$ m and three straight sections. The half-circle lies in a uniform magnetic field $\vec{B}$ that is directed out of the page; the field magnitude is given by $B = (4.0\ \text{T/s}^2)t^2 + (2.0\ \text{T/s})t + 3.0\ \text{T}$. An ideal battery with $\mathscr{E}_{\text{bat}} = 2.0$ V is connected to the loop. The resistance of the loop is 2.0 Ω.

(a) What are the amount and direction of the emf $\mathscr{E}^{\text{ind}}$ induced around the loop by field $\vec{B}$ at $t = 10$ s?

SOLUTION ■ One **Key Idea** here is that, according to Faraday's law, $\mathscr{E}^{\text{ind}}$ is equal to the negative rate $d\Phi^{\text{mag}}/dt$ at which the magnetic flux at the area encircled by the loop changes. A second **Key Idea** is that the flux at the loop depends on the loop's area A and its orientation in the magnetic field $\vec{B}$. Because $\vec{B}$ is uniform and is perpendicular to the plane of the loop, the flux is given by Eq. 31-1 ($\Phi^{\text{mag}} = BA$). Using this equation and realizing that only the field magnitude B changes in time (not the area A), we rewrite Faraday's law, Eq. 31-6, as

$$|\mathscr{E}^{\text{ind}}| = \left|\frac{d\Phi^{\text{mag}}}{dt}\right| = \left|\frac{d(BA)}{dt}\right| = A\left|\frac{dB}{dt}\right|.$$

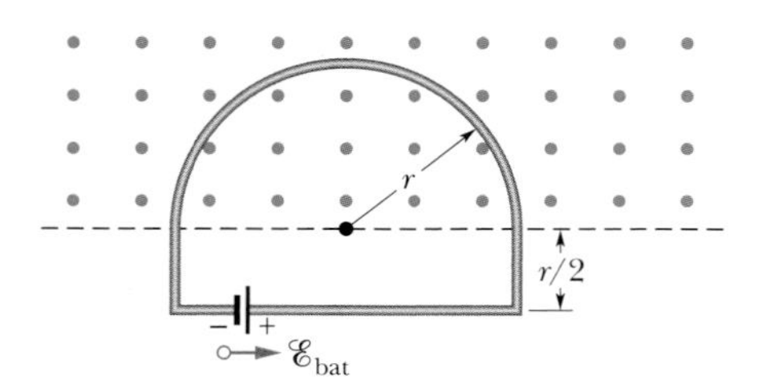

FIGURE 31-19 ■ A battery is connected to a conducting loop consisting of a half-circle of radius r that lies in a uniform magnetic field. The field is directed out of the page; its magnitude is changing.

A third **Key Idea** is that, because the flux penetrates the loop only within the half-circle, the area A in this equation is $\frac{1}{2}\pi r^2$. Substituting this and the given expression for B yields

$$|\mathscr{E}^{\text{ind}}| = A\left|\frac{dB}{dt}\right| = \frac{\pi r^2}{2}\frac{d}{dt}[(4.0\ \text{T/s})t^2 + (2.0\ \text{T/s})t + 3.0\ \text{T}]$$

$$= \frac{\pi r^2}{2}[(8.0\ \text{T/s}^2)t + (2.0\ \text{T/s})].$$

At $t = 10$ s, then,

$$|\mathscr{E}^{\text{ind}}| = \frac{\pi(0.20\ \text{m})^2}{2}[(8.0\ \text{T/s})(10\ \text{s}) + (2.0\ \text{T/s})]$$

$$= 5.152\ \text{V} \approx 5.2\ \text{V}. \qquad \text{(Answer)}$$

To find the direction of $\mathscr{E}^{\text{ind}}$, we first note that in Fig. 31-19 the flux at the loop is out of the page and increasing. Then the **Key Idea** here is that the induced field B^{ind} (due to the induced current) must oppose that increase, and thus be into the page. Using the curled-straight right-hand rule (Fig. 30-8c), we find that the induced current *contribution* must be clockwise around the loop. The induced emf $\mathscr{E}^{\text{ind}}$ must then also be clockwise.

(b) What is the current in the loop at $t = 10$ s?

SOLUTION ■ The **Key Idea** here is that two emfs tend to move charges around the loop. The induced $\mathscr{E}^{\text{ind}}$ tends to drive a current clockwise around the loop; the battery's $\mathscr{E}_{\text{bat}}$ tends to drive a current counterclockwise. Because $\mathscr{E}^{\text{ind}}$ is greater than $\mathscr{E}_{\text{bat}}$, the net emf $\mathscr{E}^{\text{net}}$ is clockwise, and thus so is the current. To find the current at $t = 10$ s, we use $i = \mathscr{E}/R$:

$$i = \frac{\mathscr{E}^{\text{net}}}{R} = \frac{\mathscr{E}^{\text{ind}} - \mathscr{E}_{\text{bat}}}{R}$$

$$= \frac{5.152\ \text{V} - 2.0\ \text{V}}{2.0\ \Omega} = 1.58\ \text{A} \approx 1.6\ \text{A}. \qquad \text{(Answer)}$$

31-6 Induction and Energy Transfers

Let us return to the simple situation we considered in Fig. 31-7. What are the consequences of the fact that a clockwise current is induced when the loop is pulled to the right and a counterclockwise current is induced when the loop is pushed to the left? If one pushes the loop back and forth (right and left), the result is an alternating current in the loop. This is current just like the current in our household electric system. It is a current that could run a motor, light a bulb, or provide heating through the resistive dissipation. If it took no effort on our part to push the loop back and forth, we could solve the energy crisis. Of course, it does take effort (work) on our part to push and pull the loop back and forth.

If you want to drag a metal loop out of a magnetic field at a constant velocity, you have to exert a force on the loop to balance the magnetic force associated with the charges moving in the magnetic field. This requires you to do work on the loop, but doing work adds energy to a system. We certainly cannot violate the principle of conservation of energy. So, where does this energy go? One place the energy could go is into an increase in the internal energy of the loop's wires. Since we observe a temperature rise in the wires, we conclude that the work done has been transformed into thermal energy—one form of internal energy. This makes sense. There is a current i in the loop that has some resistance R, and we learned in Section 26-7 that the electric power dissipation (or rate of thermal energy increase in the wires) is given by

$$P = i^2R \qquad \text{(resistive dissipation).} \qquad \text{(Eq. 26-11)}$$

How does this rate of energy loss compare to the rate we are doing work? Perhaps they are the same. In that case, we might conclude that the work we do in moving the loop is transformed into thermal energy in the loop. Let's work out the details.

Figure 31-11 shows a situation involving induced current. A rectangular loop of wire of width L has one end in a uniform external magnetic field that is directed perpendicularly into the plane of the loop. This field may be produced, for example, by a large electromagnet. The dashed lines in the figure show the assumed limits of the magnetic field; the fringing of the field at its edges is neglected. You are asked to pull this loop to the right at a constant velocity $\vec{v}$.

In the situation of Fig. 31-11, the flux of the field at the loop is changing with time. Let us now calculate the rate at which you do mechanical work as you pull steadily on the loop. The amount of work done by a force $\vec{F}$ in moving a loop a small distance $d\vec{x}$ in a time dt is

$$dW = \vec{F} \cdot d\vec{x}.$$

For simplicity, let us consider a force $\vec{F}$, which is completely in the direction of the displacement $d\vec{x}$. Then

$$dW = \vec{F} \cdot d\vec{x} = F dx.$$

The rate of doing work (which is called the *power* P) is

$$P = \frac{dW}{dt} = F\frac{dx}{dt}.$$

So

$$P = Fv, \qquad (31\text{-}8)$$

where v is the speed at which we move the loop.

Suppose that we wish to find an expression for the power, P, in terms of the magnitude B of the magnetic field and the characteristics of the loop—namely, its resistance

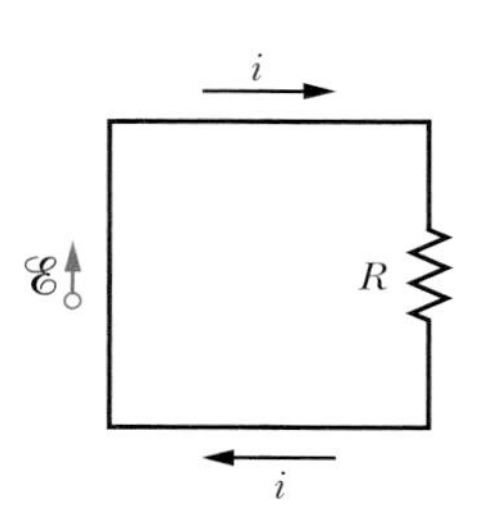

FIGURE 31-20 ■ A circuit diagram for the loop of Fig. 31-7 while it is moving.

R to current and its dimension L. As you move the loop to the right in Fig. 31-11, the portion of its area within the magnetic field decreases. Thus, the flux at the loop also decreases and, according to Lenz's law, a current is produced in the loop. It is the presence of this current that causes the force that opposes your pull.

To find the amount of the current, we first apply Faraday's law for a single loop in conjunction with Eq. 31-4. We can write the amount of this emf as

$$|\mathscr{E}| = \left|\frac{d\Phi^{\text{mag}}}{dt}\right| = BvL. \tag{31-9}$$

Figure 31-20 shows the loop depicted in Fig. 31-7, as a circuit. The induced emf, $\mathscr{E}$, is represented on the left, and the collective resistance R of the loop is represented on the right. The direction of the induced current i is shown as in Fig. 31-7, and we have already established that $\mathscr{E}$ must have the same direction as the conventional current, i.

To find the amount of the induced current, we cannot apply the loop rule for potential differences in a circuit because, as you will see in Section 31-7, we cannot define a potential difference for an induced emf. However, we can apply the equation $i = \mathscr{E}/R$. With Eq. 31-9, the current amount becomes

$$|i| = \frac{BvL}{R}. \tag{31-10}$$

Because three segments of the loop in Fig. 31-7 carry this current through the magnetic field, sideways deflecting forces act on those segments. From Chapter 29, we know that the magnitude of such a deflecting force is given in general notation by

$$F_d = |i\vec{L} \times \vec{B}|. \tag{31-11}$$

The deflecting forces acting on segments a, b, and d of the loop shown in Fig. 31-7 can be denoted as $\vec{F}_a$, $\vec{F}_b$, and $\vec{F}_d$. Application of the right-hand rule to each of these segments shows that the forces are perpendicular to each segment and point outward from the loop. Note, however, that from the symmetry, $\vec{F}_b$ and $\vec{F}_d$ are oppositely directed and equal in magnitude, so they cancel. This leaves only $\vec{F}_a$, which is directed opposite the force $\vec{F}$ you apply to the loop. Therefore, $\vec{F} = -\vec{F}_a$.

Using Eq. 31-11 to obtain the magnitude of $\vec{F}_a$ and noting that the angle between $\vec{B}$ and the length vector $\vec{L}$ for the left segment is 90°, we can write

$$F = F_a = |i|BL \sin 90° = |i|BL. \tag{31-12}$$

Substituting Eq. 31-10 for i in Eq. 31-12 then gives us

$$F = \frac{B^2vL^2}{R}. \tag{31-13}$$

Since B, L, and R are constants, the speed v at which you move the loop is constant if the magnitude F of the force you apply to the loop is also constant.

By substituting Eq. 31-13 into Eq. 31-8, we find the rate at which you do work on the loop as you pull it out of the magnetic field:

$$P = Fv = \frac{B^2v^2L^2}{R} \quad \text{(rate of doing work)}. \tag{31-14}$$

To complete our analysis, let us find the rate at which internal energy appears in the loop as you pull it along at constant speed. We calculate it from Eq. 26-11,

$$P = i^2R. \tag{31-15}$$

Substituting for i from Eq. 31-10, we find

$$P = \left(\frac{BvL}{R}\right)^2 R = \frac{B^2v^2L^2}{R} \quad \text{(rate of internal energy gain)}, \tag{31-16}$$

which is exactly equal to the rate at which you are doing work on the loop (Eq. 31-14). Thus, the work that you do in pulling the loop through the magnetic field is transferred to thermal energy in the loop, manifesting itself as a small increase in the loop's temperature.

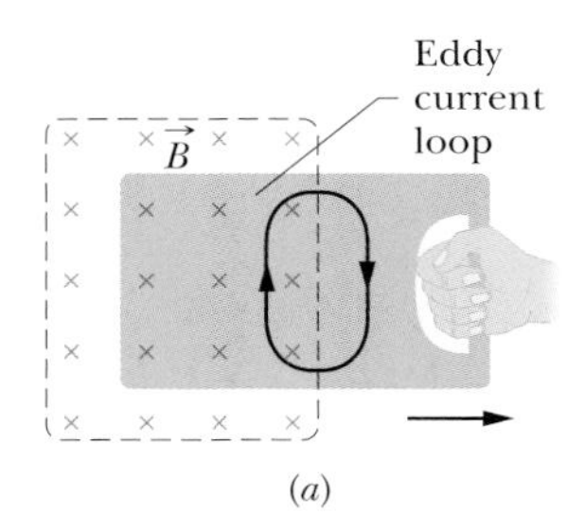

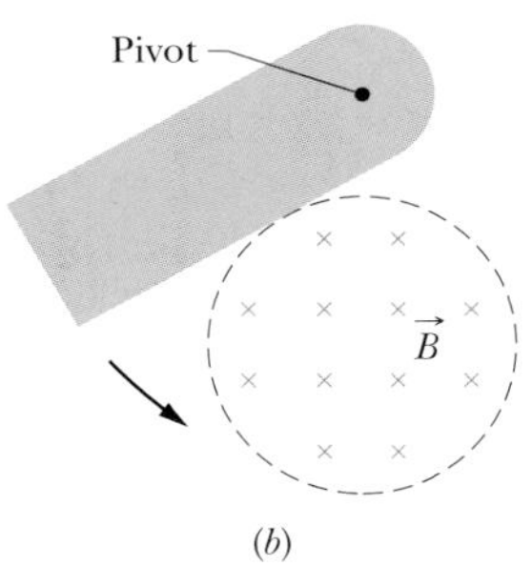

FIGURE 31-21 ■ (*a*) As you pull a solid conducting plate out of a magnetic field, *eddy currents* are induced in the plate. A typical loop of eddy current is shown; it has the same clockwise sense of circulation as the current in the conducting loop of Fig. 31-7. (*b*) A conducting plate is allowed to swing like a pendulum about a pivot and into a region of magnetic field. As it enters and leaves the field, eddy currents are induced in the plate.

Eddy Currents

Suppose we replace the conducting loop of Fig. 31-7 with a solid conducting plate as shown in Fig. 31-21*a*. If we then move the plate out of the magnetic field, the relative motion of the field and the conductor again induces a current in the conductor. Thus, we again encounter an opposing force and must do work because of the induced current. With the plate, however, the conduction electrons making up the induced current do not follow one path as they do with the loop. Instead, the electrons swirl about within the plate as if they were caught in an eddy (or whirlpool) of water. Such a current is called an *eddy current* and can be represented as in Fig. 31-21*a as if* it followed a single path.

Eddy currents are used to cook food on an induction stove. To do this an oscillating current is sent through a conducting coil that lies just below the cooking surface. The magnetic field produced by that current oscillates and induces an oscillating current in the conducting cooking pan. Because the pan has some resistance to that current, the electrical energy of the current is continuously transformed to the pan's energy, resulting in a temperature increase of the pan and the food in it. What's amazing is that the stove itself might not get hot at all—only the pan.

As with the conducting loop of Fig. 31-7, the current induced in the plate results in mechanical energy being dissipated as it increases the pan's thermal energy. The dissipation is more apparent in the arrangement of Fig. 31-21*b*; a conducting plate, free to rotate about a pivot, is allowed to swing down through a magnetic field like a pendulum. Each time the plate enters and leaves the field, a portion of its mechanical energy is transferred to its thermal energy. After several swings, no mechanical energy remains and the warmed-up plate just hangs from its pivot.

READING EXERCISE 31-9: The figure shows four wire loops, with edge lengths of either L or $2L$. All four loops will move through a region of uniform magnetic field $\vec{B}$ (directed out of the page) at the same constant velocity. (a) Rank the four loops according to the maximum amount of the emf induced as they move through the field, greatest first. (b) Explain your reasoning.

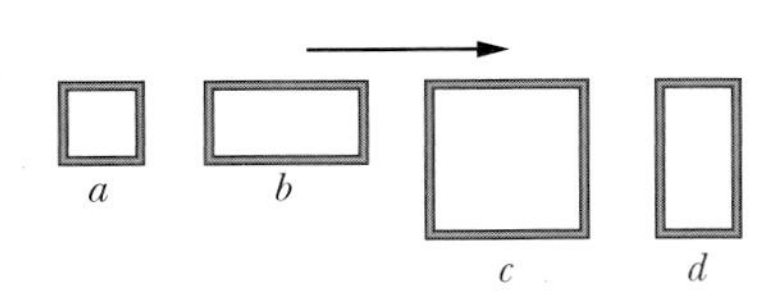

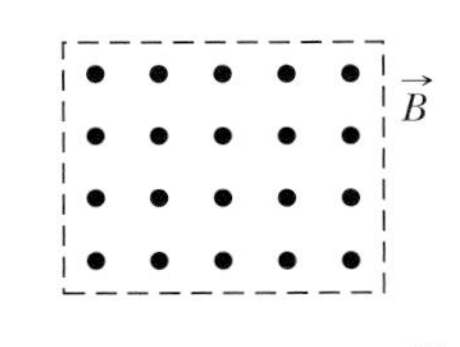

■

31-7 Induced Electric Fields

Let us place a copper ring of radius r in a uniform external magnetic field, as in Fig. 31-22*a*. The field—neglecting fringing—fills a cylindrical volume of radius R. Suppose that we increase the strength of this field at a steady rate, perhaps by increasing—in an appropriate way—the current in the windings of the electromagnet that produces the field. The magnetic flux at the ring will then change at a steady rate and—by Faraday's law—an induced emf and thus an induced current will appear in the ring. From Lenz's law we can deduce that the direction of the induced current is counterclockwise in Fig. 31-22*a*.

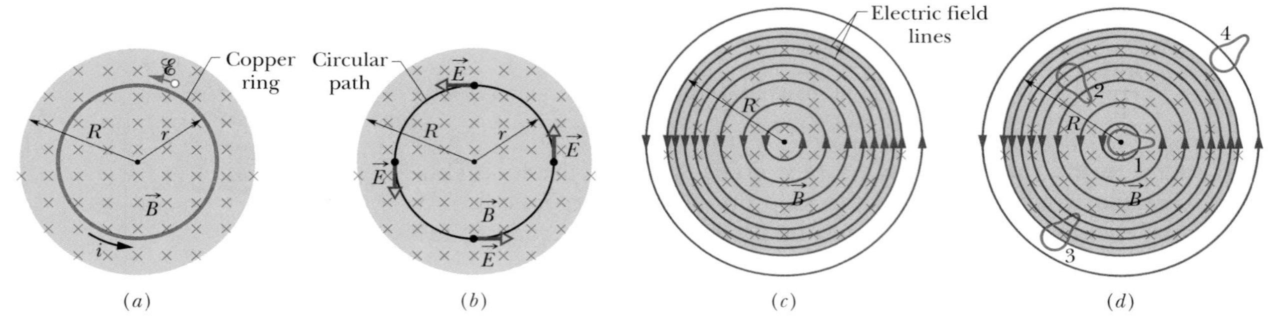

FIGURE 31-22 ■ (*a*) If the magnetic field increases at a steady rate, a constant induced current appears, as shown, in the copper ring of radius r. (*b*) An induced electric field exists even when the ring is removed; the electric field is shown at four points. (*c*) The complete picture of the induced electric field, displayed as field lines. (*d*) Four similar closed paths that enclose identical areas. Equal emfs are induced around paths 1 and 2, which lie entirely within the region of the changing magnetic field. A smaller emf is induced around path 3, which only partially lies in that region. No emf is induced around path 4, which lies entirely outside the magnetic field.

If there is a current in the copper ring, an electric field must be present along the ring; an electric field is needed to do the work of moving the conduction electrons. Moreover, the electric field must have been produced by the changing magnetic flux. This **induced electric field** $\vec{E}$ is just as real as an electric field produced by static charges; either field will exert a force $q\vec{E}$ on a particle of charge q.

By this line of reasoning, we are led to a more general and informative restatement of Faraday's law of induction:

> A changing magnetic field produces an electric field.

The striking feature of this statement is that the electric field is induced even if there is no copper ring.

To fix these ideas, consider Fig. 31-22*b*, which is just like Fig. 31-22*a* except the copper ring has been replaced by a hypothetical circular path of radius r. We assume, as previously, that the magnetic field $\vec{B}$ is increasing in magnitude at a constant rate dB/dt. The electric field induced at various points around the circular path must—from the symmetry—be tangent to the circle, as Fig. 31-22*b* shows.* Hence, the circular path is an electric field line. There is nothing special about the circle of radius r, so the electric field lines produced by the changing magnetic field must be a set of concentric circles, as in Fig. 31-22*c*.

As long as the magnetic field is increasing with time, the electric field represented by the circular field lines in Fig. 31-22*c* will be present. If the magnetic field remains constant with time, there will be no induced electric field and thus no electric field lines. If the magnetic field is decreasing with time (at a constant rate), the electric field lines will still be concentric circles as in Fig. 31-22*c*, but they will now have the opposite direction. All this is what we have in mind when we say: A changing magnetic field produces an electric field.

A Reformulation of Faraday's Law

Consider a particle of charge q moving around the circular path of Fig. 31-22*b*. The work W done on it in one revolution by the induced electric field is $q\mathscr{E}$, where $\mathscr{E}$ is the

---

* Arguments of symmetry would also permit the lines of $\vec{E}$ around the circular path to be radial, rather than tangential. However, such radial lines would imply that there are free charges, distributed symmetrically about the axis of symmetry, on which the electric field lines could begin or end; there are no such charges.

induced emf—that is, the work done per unit charge in moving the test charge around the path. From another point of view, the work is

$$\int \vec{F} \cdot d\vec{s} = qE(2\pi r), \tag{31-17}$$

where $|q|E$ is the magnitude of the force acting on the test charge and $2\pi r$ is the distance over which that force acts. Setting these two expressions for W equal to each other and canceling q, we find that

$$|\mathscr{E}| = 2\pi rE. \tag{31-18}$$

More generally, we can rewrite Eq. 31-17 to give the work done on a particle of charge q moving along any closed path:

$$W = \oint \vec{F} \cdot d\vec{s} = q\oint \vec{E} \cdot d\vec{s}. \tag{31-19}$$

(The circle indicates that the integral is to be taken around the closed path.) Substituting $q\mathscr{E}$ for W, we find that

$$\mathscr{E} = \oint \vec{E} \cdot d\vec{s}. \tag{31-20}$$

This integral reduces at once to Eq. 31-18 if we evaluate it for the special case of Fig. 31-22*b*.

With Eq. 31-20, we can expand the meaning of induced emf. Previously, induced emf meant the work per unit charge done in maintaining current due to a changing magnetic flux, or it meant the work done per unit charge on a charged particle that moves around a closed path in a changing magnetic flux. However, we can see in Fig. 31-22*b* and Eq. 31-20 that an induced emf can exist without the need of a current or particle: An induced emf is the sum—via integration—of quantities $\vec{E} \cdot d\vec{s}$ around a closed path, where $\vec{E}$ is the electric field induced by a changing magnetic flux and $d\vec{s}$ is a differential length vector along the closed path.

If we combine Eq. 31-20 with Faraday's law in Eq. 31-6 ($\mathscr{E} = -d\Phi^{mag}/dt$), we can rewrite Faraday's law as

$$\oint \vec{E} \cdot d\vec{s} = -\frac{d\Phi^{mag}}{dt} \quad \text{(Faraday's law, general formula).} \tag{31-21}$$

This equation says simply that a changing magnetic field induces an electric field. The changing magnetic field appears on the right side of this equation, the electric field on the left.

Faraday's law in the form of Eq. 31-21 can be applied to *any* closed path that can be drawn in a changing magnetic field. But $\oint \vec{E} \cdot d\vec{s}$ can only be evaluated for symmetrical situations. Figure 31-22*d*, for example, shows four such paths, all having the same shape and area but located in different positions in the changing field. For paths 1 and 2, the induced emfs $\mathscr{E}(=\oint \vec{E} \cdot d\vec{s})$ are equal because these paths lie entirely in the magnetic field and thus have the same value of $d\Phi^{mag}/dt$. This is true even though the electric field vectors at points along these paths are different, as indicated by the patterns of electric field lines in the figure. For path 3 the induced emf is smaller because the enclosed flux Φ^{mag} (hence, $d\Phi^{mag}/dt$) is smaller, and for path 4 the induced emf is zero, even though the electric field is not zero at any point on the path.

A New Look at Electric Potential

Induced electric fields are produced not by static charges but by a changing magnetic flux. Although electric fields produced in either way exert forces on charged particles, there is an important difference between them. The difference is not in the way they affect charges at a given point (the electric force on a charge q in this field is still qE), but in their global properties. Their field lines behave differently and there is a problem defining the electric potential associated with induced electric fields. The simplest evidence of this difference is that the field lines of induced electric fields form closed loops, as in Fig. 31-22*c*. Field lines produced by static charges never do so but rather must start on positive charges and end on negative charges. Since the induced fields are not caused by charges, there is no place for the field lines to start or end. Instead, they form closed loops, similar to those of magnetic fields. (But these are still electric fields! They act on stationary charges whereas magnetic fields don't.)

So, a varying *magnetic field* is accompanied by circular *electric field lines*. An electric current is known to be accompanied by circular magnetic field lines. But is an electric *current* the only source of circular magnetic field lines? Might it be possible that a varying *electric field* is accompanied by a circulating *magnetic field?* This is a question we will consider in the next chapter.

What we are immediately concerned with is that the electric field lines make closed loops, which has a powerful implication for trying to define an electrostatic potential. Since the potential difference equals the work per unit charge, if we carry a charge around a loop of electric field line, the $\vec{E}$ field always acts in the direction of motion, so every small step we make makes a positive contribution to the work. But since the field follows a loop, we can come back to our starting point after having only done positive work! The implication is:

> Electric potential has meaning only for electric fields that are produced by static charges; it has no meaning for electric fields that are produced by induction.

You can understand this statement quantitatively by considering what happens to a charged particle that makes a single journey around the circular path in Fig. 31-22*b*. It starts at a certain point and, on its return to that same point, has experienced an emf $\mathscr{E}$ of, let us say, 5 V; that is, work of 5 J/C has been done on the particle, and thus the particle should then be at a point that is 5 V greater in potential. However, that is impossible because the particle is back at the same point, which cannot have two different values of potential. We must conclude that potential has no meaning for electric fields that are set up by changing magnetic fields.

We can take a more formal look by recalling Eq. 25-16, which defines the potential difference between two points 1 and 2 in an electric field $\vec{E}$:

$$V_2 - V_1 = -\int_1^2 \vec{E} \cdot d\vec{s}. \tag{31-22}$$

In Chapter 25 we had not yet encountered Faraday's law of induction, so the electric fields involved in the derivation of Eq. 25-16 were those due to static charges. If 1 and 2 in Eq. 31-22 are the same point, the path connecting them is a closed loop, V_1 and V_2 are identical, and Eq. 31-22 reduces to

$$\oint \vec{E} \cdot d\vec{s} = 0\ \text{V}. \tag{31-23}$$

However, when a changing magnetic flux is present, this integral is *not* zero but is $-d\Phi^{\text{mag}}/dt$, as Eq. 31-21 asserts. Thus, assigning electric potential to an induced electric field leads us to a contradiction. We must conclude that electric potential difference is path dependent for the electric fields associated with induction.

READING EXERCISE 31-10: The figure shows five lettered regions in which a uniform magnetic field extends either directly out of the page (as in region *a*) or into the page. The field is increasing in magnitude at the same steady rate in all five regions; the regions are identical in area. Also shown are four numbered paths along which $\oint \vec{E} \cdot d\vec{s}$ has the magnitudes given below in terms of a unit "mag." Determine whether the magnetic fields in regions *b* through *e* are directed into or out of the page.

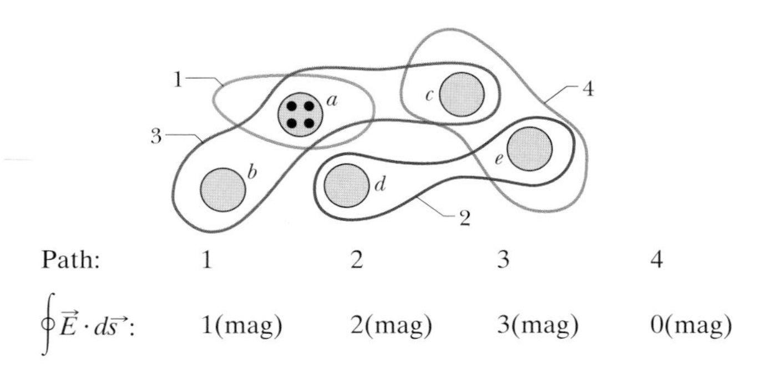

Path:	1	2	3	4
$\oint \vec{E} \cdot d\vec{s}$:	1(mag)	2(mag)	3(mag)	0(mag)

TOUCHSTONE EXAMPLE 31-3: Inducing an Electric Field

In Fig. 31-22*b*, take $R = 8.5$ cm and $dB/dt = 0.13$ T/s.

(a) Find an expression for the magnitude E of the induced electric field at points within the magnetic field, at radius r from the center of the magnetic field. Evaluate the expression for $r = 5.2$ cm.

SOLUTION ■ The **Key Idea** here is that an electric field is induced by the changing magnetic field, according to Faraday's law. To calculate the field magnitude E, we apply Faraday's law in the form of Eq. 31-21. We use a circular path of integration with radius $r \leq R$ because we want E for points within the magnetic field. We assume from the symmetry that $\vec{E}$ in Fig. 31-22*b* is tangent to the circular path at all points. The path vector $d\vec{s}$ is also always tangent to the circular path, so the dot product $\vec{E} \cdot d\vec{s}$ in Eq. 31-21 must have the magnitude $E\,ds$ at all points on the path. We can also assume from the symmetry that E has the same value at all points along the circular path. Then the left side of Eq. 31-21 becomes

$$\oint \vec{E} \cdot d\vec{s} = \oint E\,ds = E \oint ds = E(2\pi r). \qquad (31\text{-}24)$$

(The integral $\oint ds$ is the circumference $2\pi r$ of the circular path.)

Next, we need to evaluate the right side of Eq. 31-21. Because $\vec{B}$ is uniform over the area A encircled by the path of integration and is directed perpendicular to that area, the magnetic flux is given by Eq. 31-1:

$$\Phi^{\text{mag}} = BA = B(\pi r^2). \qquad (31\text{-}25)$$

Substituting this and Eq. 31-24 into Eq. 31-21 and dropping the minus sign, we find that the magnitude of the electric field is

$$E(2\pi r) = (\pi r^2)\frac{dB}{dt}$$

or

$$E = \frac{r}{2}\frac{dB}{dt}. \qquad \text{(Answer) } (31\text{-}26)$$

Equation 31-26 gives the magnitude of the electric field at any point for which $r \leq R$ (that is, within the magnetic field). Substituting given values yields, for the magnitude of $\vec{E}$ at $r = 5.2$ cm,

$$E = \frac{(5.2 \times 10^{-2}\text{ m})}{2}(0.13\text{ T/s})$$
$$= 0.0034\text{ V/m} = 3.4\text{ mV/m}. \qquad \text{(Answer)}$$

(b) Find an expression for the magnitude E of the induced electric field at points that are outside the magnetic field, at radius r. Evaluate the expression for $r = 12.5$ cm.

SOLUTION ■ The **Key Idea** of part (a) applies here also, except that we use a circular path of integration with radius $r = R$, because we want to evaluate E for points outside the magnetic field. Proceeding as in (a), we again obtain Eq. 31-24. However, we do not then obtain Eq. 31-25, because the new path of integration is now outside the magnetic field, and we need this **Key Idea**: The magnetic flux encircled by the new path is only that in the area πR^2 of the magnetic field region. Therefore,

$$\Phi^{\text{mag}} = BA = B(\pi R^2). \qquad (31\text{-}27)$$

Substituting this and Eq. 31-24 into Eq. 31-21 (without the minus sign) and solving for the magnitude of $\vec{E}$ yield

$$E = \frac{R^2}{2r}\frac{dB}{dt}. \qquad \text{(Answer) } (31\text{-}28)$$

Since E is not zero here, we know that an electric field is induced even at points that are outside the changing magnetic field, an

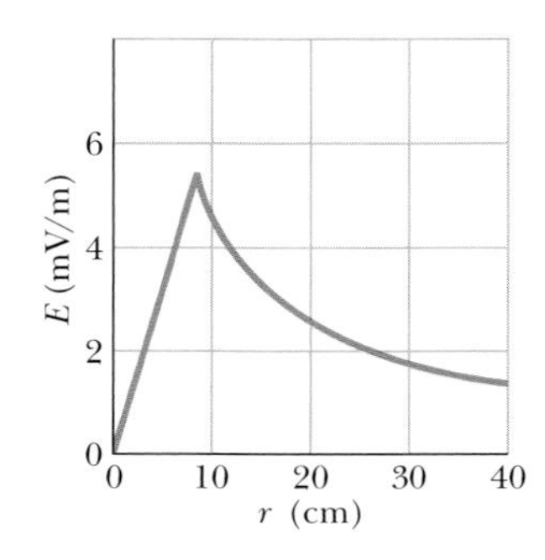

FIGURE 31-23 ■ A plot of the induced electric field $E(r)$ for the conditions given in Touchstone Example 31-3.

important result that (as you shall see in Section 32-5) makes transformers possible. With the given data, Eq. 31-28 yields the magnitude of $\vec{E}$ at $r = 12.5$ cm:

$$E = \frac{(8.5 \times 10^{-2}\text{ m})^2}{(2)(12.5 \times 10^{-2}\text{ m})}(0.13\text{ T/s})$$

$$= 3.8 \times 10^{-3}\text{ V/m} = 3.8\text{ mV/m}. \qquad \text{(Answer)}$$

Equations 31-26 and 31-28 give the same result, as they must, for $r = R$. Figure 31-23 shows a plot of $E(r)$ based on these two equations.

31-8 Induced Magnetic Fields

Let's consider a region in space where no electric currents are present. As we have seen, a changing magnetic flux induces an electric field, and we end up with Faraday's law of induction in the form

$$\oint \vec{E} \cdot d\vec{s} = -\frac{d\Phi^{\text{mag}}}{dt} \quad \text{(Faraday's law of induction).} \qquad (31\text{-}29)$$

Here $\vec{E}$ is the electric field induced along a closed loop by the changing magnetic flux Φ^{mag} encircled by that loop. Because symmetry is often so powerful in physics, we should be tempted to ask whether induction can occur in the opposite sense; that is, can a changing electric flux induce a magnetic field?

The answer is that it can; furthermore, the equation governing the induction of a magnetic field is almost symmetric with Eq. 31-21. We often call it Maxwell's law of induction after James Clerk Maxwell, and we write it as

$$\oint \vec{B} \cdot d\vec{s} = \mu_0\varepsilon_0\frac{d\Phi^{\text{elec}}}{dt} \quad \text{(Maxwell's law of induction—no currents).} \qquad (31\text{-}30)$$

Here $\vec{B}$ is the magnetic field induced along a closed loop by the changing electric flux Φ^{elec} in the region encircled by that loop.

As an example of this sort of induction, we consider the charging of a parallel-plate capacitor with circular plates, as shown in Fig. 31-24*a*. (Although we shall focus on this particular arrangement, a changing electric flux will always induce a magnetic field whenever it occurs.) We assume that the charge on the capacitor is being increased at a steady rate by a constant current i in the connecting wires. Then the amount of the electric field between the plates must also be increasing at a steady rate.

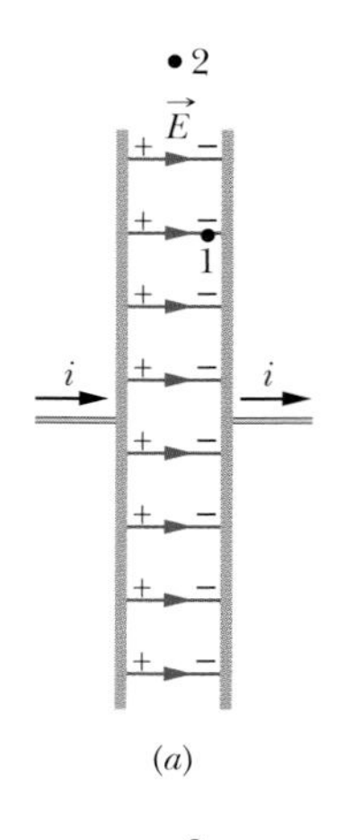

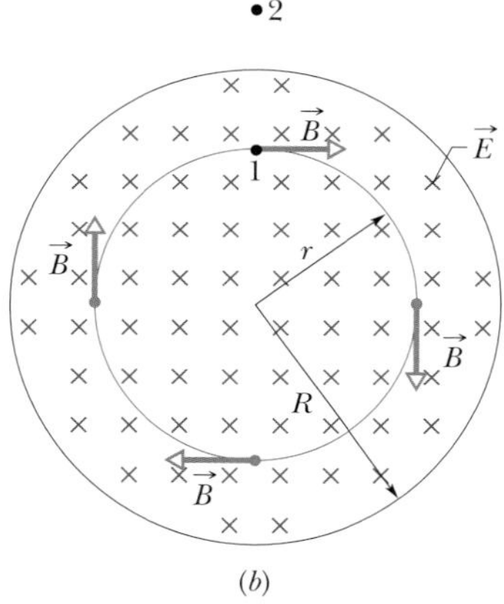

FIGURE 31-24 ■ (*a*) A circular parallel-plate capacitor, shown in side view, is being charged by a constant current i. (*b*) A view from within the capacitor, toward the plate at the right. The electric field $\vec{E}$ is uniform, is directed into the page (toward the plate), and grows in magnitude as the charge on the capacitor increases. The magnetic field $\vec{B}$ induced by this changing electric field is shown at four points on a circle with a radius r less than the plate radius R.

Figure 31-24*b* is a view of the right-hand plate of Fig. 31-24*a* from between the plates. The electric field is directed into the page. Let us consider a circular loop through point 1 in Figs. 31-24*a* and *b*, concentric with the capacitor plates and with a radius smaller than that of the plates. Because the electric field at the area subtended by the loop is changing, the electric flux at the loop must also be changing. According to Eq. 31-22, this changing electric flux induces a magnetic field around the loop.

Experiment proves that a magnetic field $\vec{B}$ *is* indeed induced around such a loop, directed as shown. This magnetic field has the same magnitude at every point around the loop and thus has circular symmetry about the central axis of the capacitor plates.

If we now consider a larger loop—say, through point 2 outside the plates in Figs. 31-24*a* and *b*—we find that a magnetic field is induced around that loop as well. Thus, while the electric field is changing, magnetic fields are induced between the plates, both inside and outside the gap. When the electric field stops changing, these induced magnetic fields disappear.

Although Eq. 31-30 is similar to Eq. 31-29, the equations differ in two ways. First, Eq. 31-30 has the two extra symbols, μ_0 and ε_0, but they appear only because we employ SI units. Second, Eq. 31-30 lacks the minus sign of Eq. 31-29. That difference in sign means that the induced electric field $\vec{E}$ and the induced magnetic field $\vec{B}$ have opposite directions when they are produced in otherwise similar situations.

To see this opposition of directions, examine Fig. 31-25, in which an increasing magnetic field $\vec{B}$, directed into the page, induces an electric field $\vec{E}$. The induced field $\vec{E}$ is counterclockwise, whereas the induced magnetic field $\vec{B}$ in Fig. 31-24*b* is clockwise.

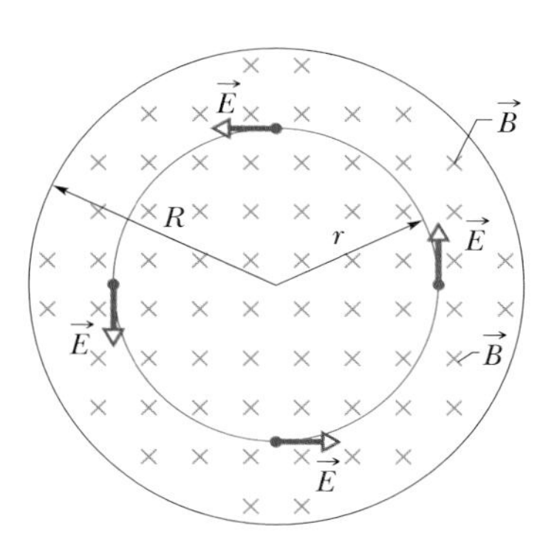

FIGURE 31-25 ■ A uniform magnetic field $\vec{B}$ in a circular region. The field, directed into the page, is increasing in magnitude. The electric field $\vec{E}$ induced by the changing magnetic field is shown at four points on a circle concentric with the circular region. Compare this situation with that of Fig. 31-24*b*.

Ampère–Maxwell Law

Now recall that the left side of Eq. 31-30, the integral of the dot product $\vec{B}\cdot d\vec{s}$ around a closed loop, appears in another equation—namely, Ampère's law:

$$\oint \vec{B}\cdot d\vec{s} = \mu_0 i^{\text{enc}} \qquad \text{(Ampère's law)}, \qquad (31\text{-}31)$$

where i^{enc} is the current encircled by the closed loop. Thus, our two equations that specify the magnetic field $\vec{B}$ produced by means other than a magnetic material (that is, by a current and by a changing electric field) give the field in exactly the same form. We can combine the two equations into the single equation

$$\oint \vec{B}\cdot d\vec{s} = \mu_0\varepsilon_0\frac{d\Phi^{\text{elec}}}{dt} + \mu_0 i^{\text{enc}} \qquad \text{(Ampère–Maxwell law)}. \qquad (31\text{-}32)$$

When there is a current but no change in electric flux (such as with a wire carrying a constant current), the first term on the right side of Eq. 31-32 is zero, and Eq. 31-32 reduces to Eq. 31-31, Ampère's law. When there is a change in electric flux but no current (such as inside or outside the gap of a charging capacitor), the second term on the right side of Eq. 31-32 is zero, and Eq. 31-32 reduces to Eq. 31-30, Maxwell's law of induction.

READING EXERCISE 31-11: Referring back to Chapter 30, where we first studied Ampère's law, describe how we found the direction of the magnetic field produced by a current. What did the magnetic field lines look like for a long, straight, current-carrying wire? Discuss any connections or similarities between the case of the current-carrying wire and the case shown in Fig. 31-24. ■

TOUCHSTONE EXAMPLE 31-4: Inducing a Magnetic Field

A parallel-plate capacitor with circular plates of radius R is being charged as in Fig. 31-24*a*.

(a) Derive an expression for the magnitude of the magnetic field at radii r for the case $r \le R$.

SOLUTION ■ The **Key Idea** here is that a magnetic field can be set up by a current and by induction due to a changing electric flux; both effects are included in Eq. 31-32. There is no current between the capacitor plates of Fig. 31-24, but the electric flux there is changing. Thus, Eq. 31-32 reduces to

$$\oint \vec{B}\cdot d\vec{s} = \mu_0\varepsilon_0\frac{d\Phi^{\text{elec}}}{dt}. \qquad (31\text{-}33)$$

We shall separately evaluate the left and right sides of this equation.

Left side of Eq. 31-33: We choose a circular Ampèrian loop with a radius $r \le R$ as shown in Fig. 31-24, because we want to evaluate the magnetic field for $r \le R$—that is, inside the capacitor. The magnetic field $\vec{B}$ at all points along the loop is tangent to the loop, as is the path element $d\vec{s}$. Thus, $\vec{B}$ and $d\vec{s}$ are either parallel or antiparallel at each point of the loop. For simplicity, assume they are parallel (the choice does not alter our outcome here). Then

$$\oint \vec{B}\cdot d\vec{s} = \oint B \cdot ds \cos 0° = \oint B\, ds.$$

Due to the circular symmetry of the plates, we can also assume that $\vec{B}$ has the same magnitude at every point around the loop. Thus, B can be taken outside the integral on the right side of the above equation. The integral that remains is $\oint ds$, which simply gives the circumference $2\pi r$ of the loop. The left side of Eq. 31-33 is then $(B)(2\pi r)$.

Right side of Eq. 31-33: We assume that the electric field $\vec{E}$ is uniform between the capacitor plates and directed perpendicular to the plates. Then the electric flux Φ^{elec} encircled by the Ampèrian loop is EA, where A is the area encircled by the loop within the electric field. Thus, the right side of Eq. 31-33 is $\mu_0\varepsilon_0 d(EA)/dt$.

Substituting our results for the left and right sides into Eq. 31-33, we get

$$B(2\pi r) = \mu_0\varepsilon_0 \frac{d(EA)}{dt}.$$

Because A is a constant, we write $d(EA)$ as $A\, dE$, so we have

$$B(2\pi r) = \mu_0\varepsilon_0 A \frac{dE}{dt}. \qquad (31\text{-}34)$$

We next use this **Key Idea:** The area A that is encircled by the Ampèrian loop within the electric field is the full area πr^2 of the loop, because the loop's radius r is less than (or equal to) the plate radius R. Substituting πr^2 for A in Eq. 31-34 and solving the result for B give us, for $r \leq R$,

$$B = \frac{\mu_0\varepsilon_0 r}{2}\frac{dE}{dt}. \qquad \text{(Answer) (31-35)}$$

This equation tells us that, inside the capacitor, B increases linearly with increased radial distance r, from zero at the center of the plates to a maximum value at the plate edges (where $r = R$).

(b) Evaluate the field magnitude B for $r = R/5 = 11.0$ mm and $dE/dt = 1.50 \times 10^{12}$ V/m·s.

SOLUTION From the answer to (a), we have

$$\begin{aligned} B &= \tfrac{1}{2}\mu_0\varepsilon_0 r \frac{dE}{dt} \\ &= \tfrac{1}{2}(4\pi \times 10^{-7}\ \text{T}\cdot\text{m/A})(8.85 \times 10^{-12}\ \text{C}^2/\text{N}\cdot\text{m}^2) \\ &\quad \times (11.0 \times 10^{-3}\ \text{m})(1.50 \times 10^{12}\ \text{V/m}\cdot\text{s}) \\ &= 9.18 \times 10^{-8}\ \text{T}. \end{aligned} \qquad \text{(Answer)}$$

(c) Derive an expression for the induced magnetic field for the case $r \geq R$.

SOLUTION Our procedure is the same as in (a) except we now use an Ampèrian loop with a radius r that is greater than the plate radius R, to evaluate B outside the capacitor. Evaluating the left and right sides of Eq. 31-33 again leads to Eq. 31-34. However, we then need this subtle **Key Idea:** The electric field exists only between the plates, *not* outside the plates. Thus, the area A that is encircled by the Ampèrian loop in the electric field is *not* the full area πr^2 of the loop. Rather, A is only the plate area πR^2.

Substituting πR^2 for A in Eq. 31-34 and solving the result for B give us, for $r \geq R$,

$$B = \frac{\mu_0\varepsilon_0 R^2}{2r}\frac{dE}{dt}. \qquad \text{(Answer) (31-36)}$$

This equation tells us that, outside the capacitor, B decreases with increased radial distance r, from a maximum value at the plate edges (where $r = R$). By substituting $r = R$ into Eqs. 31-35 and 31-36, you can show that these equations are consistent; that is, they give the same maximum value of B at the plate radius.

The magnitude of the induced magnetic field calculated in (b) is so small that it can scarcely be measured with simple apparatus. This is in sharp contrast to the magnitudes of induced electric fields (Faraday's law), which can be measured easily. This experimental difference exists partly because induced emfs can easily be multiplied by using a coil of many turns. No technique of comparable simplicity exists for multiplying induced magnetic fields. In any case, the experiment suggested by this sample problem has been done, and the presence of the induced magnetic fields has been verified quantitatively.

31-9 Displacement Current

If you compare the two terms on the right side of Eq. 31-32, you will see that the product $\varepsilon_0(d\Phi^{elec}/dt)$ in the first term must have the units associated with a current. Since no charge actually flows, historically, that product has been treated as being a fictitious current called the **displacement current** i^{dis}:

$$i^{dis} = \varepsilon_0 \frac{d\Phi^{elec}}{dt} \quad \text{(displacement current).} \qquad (31\text{-}37)$$

"Displacement" is a poorly chosen term in that nothing is being displaced, but we are stuck with the word. Nevertheless, we can now rewrite Eq. 31-32 as

$$\oint \vec{B} \cdot d\vec{s} = \mu_0 i_{dis}^{enc} + \mu_0 i^{enc} \quad \text{(Ampère–Maxwell law)}, \tag{31-38}$$

in which i_{dis}^{enc} is the displacement current that is encircled by the integration loop.

Let us again focus on a charging capacitor with circular plates, as in Fig. 31-26*a*. The real current i that is charging the plates changes the electric field $\vec{E}$ between the plates. The fictitious displacement current i^{dis} between the plates is associated with that changing field $\vec{E}$. Let us relate these two currents.

The amount of excess charge $|q|$ on each of the plates at any time is related to the magnitude $|\vec{E}| = E$ of the field between the plates at that time by Eq. 28-4:

$$|q| = \varepsilon_0 A E, \tag{31-39}$$

in which A is the plate area. To get the real current i, we differentiate Eq. 31-39 with respect to time, finding

$$\frac{d|q|}{dt} = |i| = \varepsilon_0 A \frac{dE}{dt}. \tag{31-40}$$

To get the displacement current i^{dis}, we can use Eq. 31-37. Assuming that the electric field $\vec{E}$ between the two plates is uniform (we neglect any fringing), we can replace the electric flux Φ^{elec} in that equation with EA. Then Eq. 31-37 becomes

$$|i^{dis}| = \varepsilon_0 \left| \frac{d\Phi^{elec}}{dt} \right| = \varepsilon_0 \left| \frac{d(EA)}{dt} \right| = \varepsilon_0 A \left| \frac{dE}{dt} \right|. \tag{31-41}$$

Comparing Eqs. 31-40 and 31-41, we see that the real current i charging the capacitor and the fictitious displacement current i^{dis} between the plates have the same value:

$$i^{dis} = i \quad \text{(displacement current in a capacitor)}. \tag{31-42}$$

Thus, we can consider the fictitious displacement current i^{dis} to be simply a continuation of the real current i from one plate, across the capacitor gap, to the other plate. Because the electric field is uniformly spread over the plates, the same is true of this fictitious displacement current i^{dis}, as suggested by the spread of current arrows in Fig. 31-26*a*. Although no charge actually moves across the gap between the plates, the idea of the fictitious current i^{dis} can help us to quickly find the direction and magnitude of an induced magnetic field, as follows.

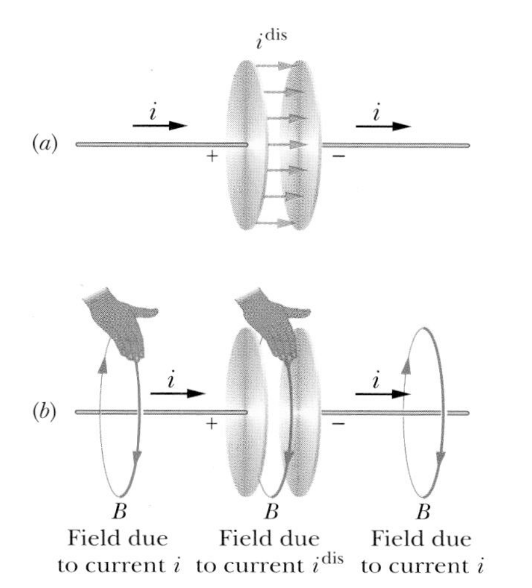

FIGURE 31-26 ■ (*a*) The displacement current i^{dis} between the plates of a capacitor that is being charged by a current i. (*b*) The right-hand rule for finding the direction of the magnetic field around a wire with a real current (as at the left) also gives the magnetic field direction around a displacement current (as in the center).

Finding the Induced Magnetic Field

In Chapter 30 we found the direction of the magnetic field produced by a real current i by using the right-hand rule of Fig. 30-4. We can apply the same rule to find the direction of an induced magnetic field produced by a fictitious displacement current i^{dis}, as shown in the center of Fig. 31-26*b* for a capacitor.

We can also use i^{dis} to find the magnitude of the magnetic field induced by a charging capacitor with parallel circular plates of radius R. We simply consider the space between the plates to be an imaginary circular wire of radius R carrying the imaginary current i^{dis}. Then, from Eq. 30-22, the magnitude of the magnetic field at a point inside the capacitor at radius r from the center is

$$B = \left(\frac{\mu_0 |i^{dis}|}{2\pi R^2} \right) r \quad \text{(inside a circular capacitor)}. \tag{31-43}$$

Similarly, from Eq. 30-19, the magnitude of the magnetic field at a point outside the capacitor at radius r is

$$B = \frac{\mu_0 |i^{\text{dis}}|}{2\pi r} \quad \text{(outside a circular capacitor).} \tag{31-44}$$

READING EXERCISE 31-12: Discuss the ways in which it is useful for us to think of the quantity $\varepsilon_0\, d\Phi^{\text{elec}}/dt$ as a current. ■

TOUCHSTONE EXAMPLE 31-5: Displacement Current

The circular parallel-plate capacitor in Touchstone Example 31-4 is being charged with a current i.

(a) Between the plates, what is the magnitude of $\oint \vec{B}\cdot d\vec{s}$, in terms of μ_0 and i, at a radius $r = R/5$ from their center?

SOLUTION ■ The first **Key Idea** of Touchstone Example 31-4a holds here too. However, now we can replace the product $\varepsilon_0 d\Phi^{\text{elec}}/dt$ in Eq. 31-32 with a fictitious displacement current i^{dis}. Then integral $\oint \vec{B}\cdot d\vec{s}$ is given by Eq. 31-38, but because there is no real current i between the capacitor plates, the equation reduces to

$$\oint \vec{B}\cdot d\vec{s} = \mu_0 i^{\text{enc}}_{\text{dis}}. \tag{31-45}$$

Because we want to evaluate $\oint \vec{B}\cdot d\vec{s}$ at radius $r = R/5$ (within the capacitor), the integration loop encircles only a portion $i^{\text{enc}}_{\text{dis}}$ of the total displacement current i^{dis}. A second **Key Idea** is to assume that i^{dis} is uniformly spread over the full plate area. Then the portion of the displacement current encircled by the loop is proportional to the area encircled by the loop:

$$\frac{(\text{encircled displacement current } i^{\text{enc}}_{\text{dis}})}{(\text{total displacement current } i^{\text{dis}})} = \frac{\text{encircled area } \pi r^2}{\text{full plate area } \pi R^2}.$$

This gives us a current magnitude of

$$i^{\text{enc}}_{\text{dis}} = i^{\text{dis}}\frac{\pi r^2}{\pi R^2}.$$

Substituting this into Eq. 31-45, we obtain

$$\oint \vec{B}\cdot d\vec{s} = \mu_0 i^{\text{dis}}\frac{\pi r^2}{\pi R^2}. \tag{31-46}$$

Now substituting $i^{\text{dis}} = i$ (from Eq. 31-42) and $r = R/5$ into Eq. 31-46 leads to

$$\oint \vec{B}\cdot d\vec{s} = \mu_0 i\frac{(R/5)^2}{R^2} = \frac{\mu_0 i}{25}. \quad \text{(Answer)}$$

(b) In terms of the maximum induced magnetic field, what is the magnitude of the magnetic field induced at $r = R/5$, inside the capacitor?

SOLUTION ■ The **Key Idea** here is that, because the capacitor has parallel circular plates, we can treat the space between the plates as an imaginary wire of radius R carrying the imaginary current i^{dis}. Then we can use Eq. 31-43 to find the induced magnetic field magnitude B at any point inside the capacitor. At $r = R/5$, that equation yields

$$B = \left(\frac{\mu_0 |i^{\text{dis}}|}{2\pi R^2}\right) r = \frac{\mu_0 |i^{\text{dis}}|(R/5)}{2\pi R^2} = \frac{\mu_0 |i^{\text{dis}}|}{10\pi R}. \tag{31-47}$$

The maximum field magnitude B^{max} within the capacitor occurs at $r = R$. It is

$$B^{\text{max}} = \left(\frac{\mu_0 |i^{\text{dis}}|}{2\pi R^2}\right) R = \frac{\mu_0 |i^{\text{dis}}|}{2\pi R}. \tag{31-48}$$

Dividing Eq. 31-47 by Eq. 31-48 and rearranging the result, we find

$$B = \frac{B^{\text{max}}}{5}, \quad \text{(Answer)}$$

We should be able to obtain this result with a little reasoning and less work. Equation 31-43 tells us that inside the capacitor, B increases linearly with r. Therefore, a point $\frac{1}{5}$ the distance out to the full radius R of the plates, where B^{max} occurs, should have a field B that is $\frac{1}{5}B^{\text{max}}$.

31-10 Gauss' Law for Magnetic Fields

In this chapter and the two that precede it, we have investigated several fundamental aspects of electricity and magnetism. Furthermore, we have seen many ways in which magnetism and electricity are connected. When combined as a set of laws, these ideas provide a framework from which we can understand all of the electromagnetic phenomena that fill our world, much like Newton's laws do in regard to forces and motion.

However, there remains one last idea that we must discuss before our view of electromagnetism is complete. This idea is contained in an idea known as *Gauss' law for magnetic fields*. Gauss' law for magnetic fields is a formal way of saying that magnetic monopoles do not exist. The law asserts that the net magnetic flux Φ^{mag} at any closed Gaussian surface is zero:

$$\Phi^{mag} = \oint \vec{B} \cdot d\vec{A} = 0 \quad \text{(Gauss' law for magnetic fields).} \tag{31-49}$$

Contrast this with Gauss' law for electric fields,

$$\Phi^{elec} = \oint \vec{E} \cdot d\vec{A} = \frac{q^{enc}}{\varepsilon_0} \quad \text{(Gauss' law for electric fields).}$$

In both equations, the integral is taken over a *closed* Gaussian surface. Gauss' law for electric fields says that this integral (the net electric flux at the surface) is proportional to the net electric charge q^{enc} enclosed by the surface. Gauss' law for magnetic fields says that there can be no net magnetic flux at the surface because there can be no net "magnetic charge" (individual magnetic poles) enclosed by the surface. The simplest magnetic structure that can exist and thus be enclosed by a Gaussian surface is a dipole, which consists of both a source and a sink for the field lines. Thus, there must always be as much magnetic flux into the surface as out of it, and the net magnetic flux must always be zero.

Gauss' law for magnetic fields holds for more complicated structures than a magnetic dipole, and it holds even if the Gaussian surface does not enclose the entire structure. Gaussian surface II near the bar magnet of Fig. 31-27 encloses no poles, and we can easily conclude that the net magnetic flux at it is zero. Gaussian surface I is more difficult to understand. It may seem to enclose only the north pole of the magnet because it encloses the label N and not the label S. However, a south pole must be associated with the lower boundary of the surface, because magnetic field lines enter the surface there. (The enclosed section is like one piece of the broken cylindrical magnet in Fig. 31-28.) Thus, Gaussian surface I encloses a magnetic dipole and the net flux at the surface is zero.

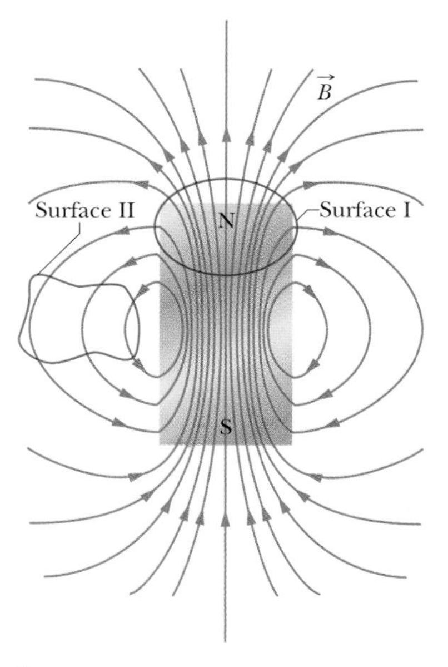

FIGURE 31-27 ■ The field lines for the magnetic field $\vec{B}$ of a short bar magnet. The red curves represent cross sections of closed, three-dimensional Gaussian surfaces.

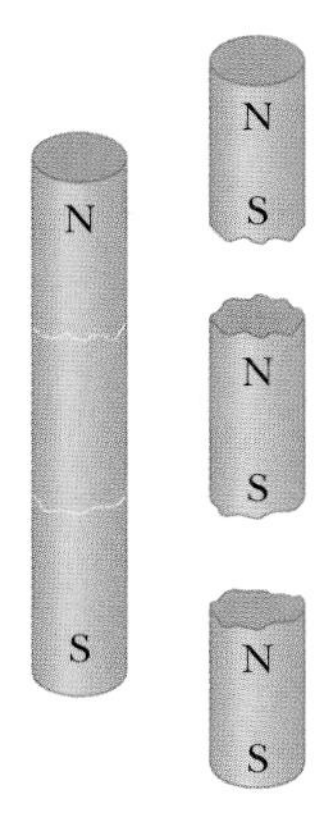

FIGURE 31-28 ■ If you break a magnet, each fragment becomes a separate magnet, with its own north and south poles.

READING EXERCISE 31-13: The figure below shows four closed surfaces with flat top and bottom faces and curved sides. The table gives the areas A of the faces and the magnitudes B of the uniform and perpendicular magnetic fields at those faces; the units of A and B are arbitrary but consistent. (a) Rank the surfaces according to the magnitudes of the magnetic flux at their curved sides, greatest first. (b) Explain your reasoning.

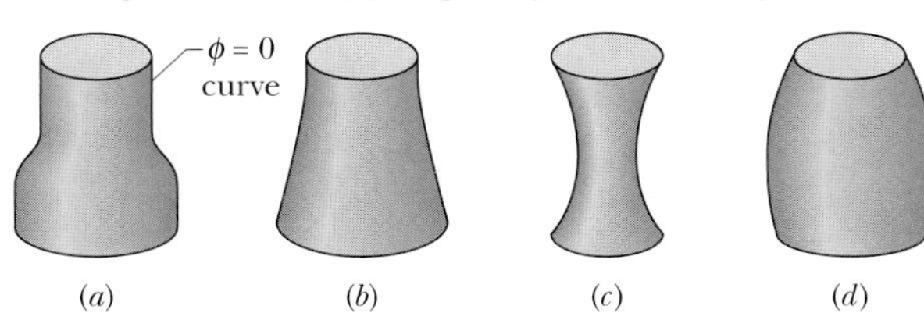

Surface	A_{top}	B_{top}, direction	A_{bot}	B_{bot}, direction
a	2	6, outward	4	3, inward
b	2	1, inward	4	2, inward
c	2	6, inward	2	8, outward
d	2	3, outward	3	2, outward

■

31-11 Maxwell's Equations in a Vacuum

Many 18th and 19th century scientists contributed to our understanding of electricity and magnetism including Franklin, Coulomb, Gauss, Oersted, Biot, Savart, Lorentz, Ampère, Henry, Faraday, and Maxwell. But it was James Clerk Maxwell who reformulated many of the basic equations describing electric and magnetic effects we have already presented. A special case of Maxwell's equations are shown in Table 31-1 for situations in which no dielectric or magnetic materials are present.

It is amazing that these four rather compact equations can be used to *derive a complete description of all electromagnetic interactions that were understood by the end of the 19th century*. Taken together they describe a diverse range of phenomena, from how a compass needle points north to how a car starts when you turn the ignition key. They have been used to design electric motors, cyclotrons, television transmitters and receivers, telephones, fax machines, radar, and microwave ovens.

In addition, many of the equations you have seen since Chapter 22 can be derived from Maxwell's equations. Perhaps the most exciting intellectual outcome of Maxwell's equations is their prediction of electromagnetic waves and our eventual understanding of the self-propagating nature of these waves that will be introduced in Chapter 34. Maxwell's picture of electromagnetic wave propagation was not fully appreciated until scientists abandoned the idea that all waves had to propagate through an elastic medium and accepted Einstein's theory of special relativity formulated in the early part of the 20th century.

Because we now know that visible light is a form of electromagnetic radiation, these equations provide the basis for many of the equations you will see in Chapters 34 through 37, which introduce you to optics and optical devices such as telescopes and eyeglasses.

The significance of Maxwell's equations should not be underestimated. Richard Feynman, a leading famous 20th-century physicist, recognized this when he stated:

> Now we realize that the phenomena of chemical interaction and ultimately of life itself are to be understood in terms of electromagnetism . . . . The electrical forces, enormous as they are, can also be very tiny, and we can control them and use them in many ways . . . From a long view of the history of mankind—seen from, say, ten thousand years from now—there can be little doubt that the most significant event of the nineteenth century will be judged as Maxwell's discovery of the laws of electrodynamics.

TABLE 31-1
Maxwell's Equations for Vacuum[a]

Name	Equation	
Gauss' law for electricity (Eq. 24-7)	$\oint \vec{E} \cdot d\vec{A} = q/\varepsilon_0$	Relates net electric flux to net enclosed electric charge
Gauss' law for magnetism (Eq. 31-49)	$\oint \vec{B} \cdot d\vec{A} = 0$	Relates net magnetic flux to net enclosed magnetic charge
Faraday's law (Eq. 31-7)	$\oint \vec{E} \cdot d\vec{s} = -\frac{d\Phi^{\text{mag}}}{dt}$	Relates induced electric field to changing magnetic flux
Ampère–Maxwell law (Eq. 31-32)	$\oint \vec{B} \cdot d\vec{s} = \mu_0 \varepsilon_0 \frac{d\Phi^{\text{elec}}}{dt} + \mu_0 i$	Relates induced magnetic field to changing electric flux and to current

[a]Written on the assumption that no dielectric or magnetic materials are present.

READING EXERCISE 31-14: Discuss several ways in which Gauss' law for electricity and Gauss' law for magnetism are similar. Discuss several ways in which they are different. ■

READING EXERCISE 31-15: Discuss several ways in which Faraday's law and the Ampère–Maxwell law are similar. Discuss several ways in which they are different. ■

Problems

SEC. 31-4 ■ FARADAY'S LAW

1. UHF Antenna A UHF television loop antenna has a diameter of 11 cm. The magnetic field of a TV signal is normal to the plane of the loop and, at one instant of time, its magnitude is changing at the rate 0.16 T/s. The magnetic field is uniform. What emf is induced in the antenna?

2. Small Loop A small loop of area A is inside of, and has its axis in the same direction as, a long solenoid of n turns per unit length and current i. If $i = I^{max} \sin \omega t$, find the magnitude of the emf induced in the loop.

3. Magnetic Flux The magnetic flux encircled by the loop shown in Fig. 31-29 increases according to the relation $\Phi^{mag} = (6.0 \text{ mWb/s}^2)t^2 + (3.7 \text{ mWb/s})t$. (a) What is the magnitude of the emf induced in the loop when $t = 2.0$ s? (b) What is the direction of the current through R?

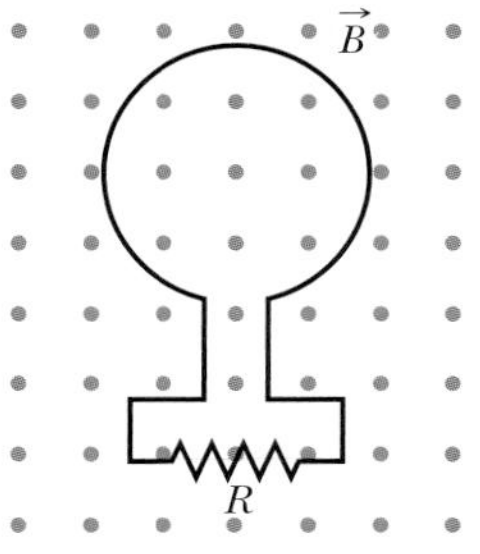

FIGURE 31-29 ■ Problems 3 and 13.

4. Calculate emf The magnitude of the magnetic field encircled by a single loop of wire, 12 cm in radius and of 8.5 Ω resistance, changes with time as shown in Fig. 31-30. Calculate the magnitude of the emf in the loop as a function of time. Consider the time intervals (a) $t_1 = 0.0$ s to $t_2 = 2.0$ s, (b) $t_2 = 2.0$ s to $t_3 = 4.0$ s, (c) $t_3 = 4.0$ s to $t_4 = 6.0$ s. The (uniform) magnetic field is perpendicular to the plane of the loop.

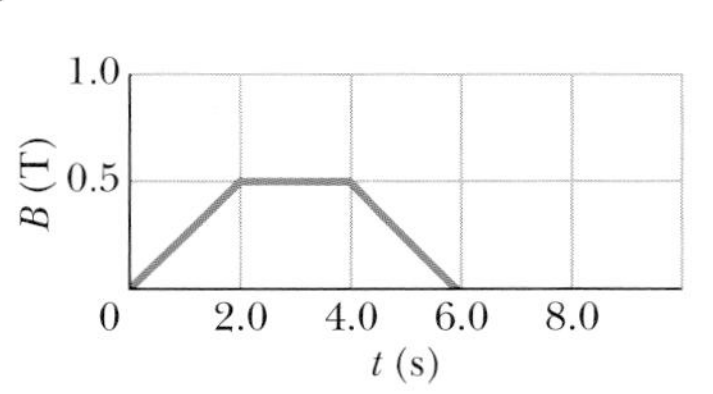

FIGURE 31-30 ■ Problem 4.

5. Uniform Magnetic Field A uniform magnetic field is normal to the plane of a circular loop 10 cm in diameter and made of copper wire (of diameter 2.5 mm). (a) Calculate the resistance of the wire. (See Table 26-2.) (b) At what rate must the magnetic field change with time if an induced current of 10 A is to appear in the loop?

6. Current in Solenoid The current in the solenoid of Touchstone Example 31-1 changes, not as stated there, but according to $i = (3.0 \text{ A/s})t + (1.0 \text{A/s}^2)t^2$. (a) Plot the induced emf in the coil from $t_1 = 0.0$ s to $t_2 = 4.0$ s. (b) The resistance of the coil is 0.15 Ω. What is the current in the coil at $t = 2.0$ s?

7. Coil Outside Solenoid In Fig. 31-31 a 120-turn coil of radius 1.8 cm and resistance 5.3 Ω is placed *outside* a solenoid like that of Touchstone Example 31-1. If the current in the solenoid is changed as in that sample problem, what current appears in the coil while the solenoid current is being changed?

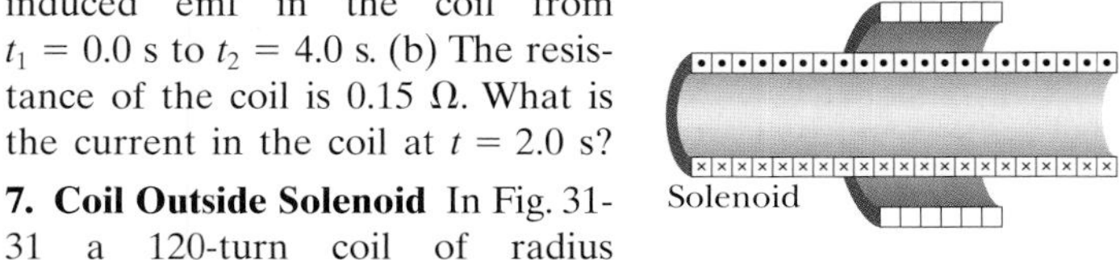

FIGURE 31-31 ■ Problem 7.

8. Elastic Conducting Material An elastic conducting material is stretched into a circular loop of 12.0 cm radius. It is placed with its plane perpendicular to a uniform 0.800 T magnetic field. When released, the radius of the loop starts to shrink at an instantaneous rate of 75.0 cm/s. What magnitude of emf is induced in the loop at that instant?

9. Square Loop A square loop of wire is held in a uniform, magnetic field 0.24 T directed perpendicularly to the plane of the loop. The length of each side of the square is decreasing at a constant rate of 5.0 cm/s. What emf is induced in the loop when the length is 12 cm?

10. Rectangular Loop A rectangular loop (area = 0.15 m²) turns in a uniform magnetic field, $B = 0.20$ T. When the angle between the field and the normal to the plane of the loop is $\pi/2$ rad and increasing at 0.60 rad/s, what emf is induced in the loop?

SEC. 31-5 ■ LENZ'S LAW

11. Two Parallel Loops Though not to scale, Fig. 31-32 shows two parallel loops of wire with a common axis. The smaller loop (radius r) is above the larger loop (radius R) by a distance $x \gg R$. Consequently, the magnetic field due to the current i in the larger loop is nearly constant throughout the smaller loop. Suppose that x is increasing at the constant rate of $dx/dt = v$. (a) Determine the magnetic flux at the area bounded by the smaller loop as a function of x. (*Hint:* See Eq. 30-29.) In the smaller loop, find (b) the induced emf and (c) the direction of the induced current.

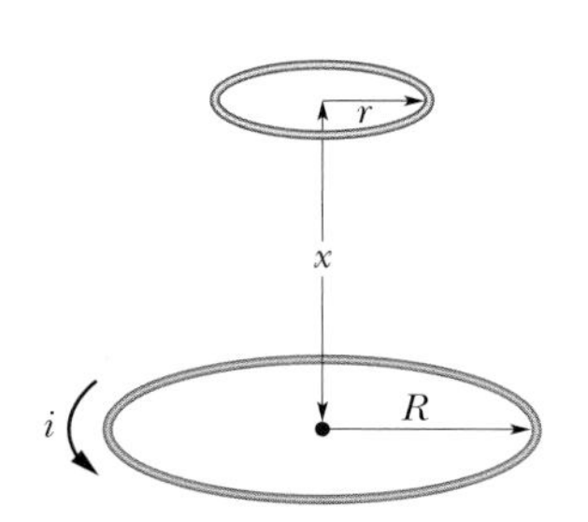

FIGURE 31-32 ■ Problem 11.

12. Circular Loop In Fig. 31-33, a circular loop of wire 10 cm in diameter (seen edge-on) is placed with its normal at an angle $\theta = 30°$ with the direction of a uniform magnetic field $\vec{B}$ of magnitude 0.50 T. The loop is then rotated such that the normal rotates in a cone about the field direction at the constant rate of 100 rev/min; the angle θ remains unchanged during the process. What is the emf induced in the loop?

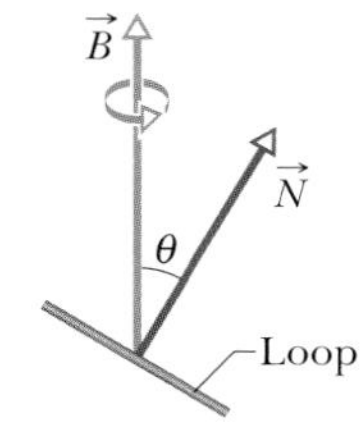

FIGURE 31-33 ■ Problem 12.

13. Flux At Loop In Fig. 31-29 let the flux encircled by the loop be $\Phi^{\text{mag}}(0)$ at time $t_1 = 0$. Then let the magnetic field $\vec{B}$ vary in a continuous but unspecified way, in both magnitude and direction, so that at time t_2 the flux is represented by $\Phi^{\text{mag}}(t_2)$. (a) Show that the net charge $q(t_2)$ that has passed through resistor R in time t_2 is

$$q(t_2) = \frac{1}{R}\,[\Phi^{\text{mag}}(0) - \Phi^{\text{mag}}(t_2)]$$

and is independent of the way $\vec{B}$ has changed. (b) If $\Phi^{\text{mag}}(t_2) = \Phi^{\text{mag}}(0)$ in a particular case, we have $q(t_2) = 0$. Is the induced current necessarily zero throughout the interval from 0 to t_2?

14. Big Loop, Little Loop A small circular loop of area 2.00 cm^2 is placed in the plane of, and concentric with, a large circular loop of radius 1.00 m. The current in the large loop is changed uniformly from 200 A to −200 A (a change in direction) in a time of 1.00 s, beginning at $t_1 = 0$. (a) What is the magnitude of the magnetic field at the center of the small circular loop due to the current in the large loop at $t_1 = 0$ s, $t_2 = 0.500$ s, and $t_3 = 1.00$ s? (b) What is the magnitude of the emf induced in the small loop at $t_2 = 0.500$ s? (Since the inner loop is small, assume the field $\vec{B}$ due to the outer loop is uniform over the area of the smaller loop.)

15. Copper Wire on Wooden Core One hundred turns of insulated copper wire are wrapped around a wooden cylindrical core of cross-sectional area $1.20 \times 10^{-3}\,\text{m}^2$. The two ends of the wire are connected to a resistor. The total resistance in the circuit is 13.0 Ω. If an externally applied uniform longitudinal magnetic field in the core changes from 1.60 T in one direction to 1.60 T in the opposite direction, how much charge flows through the circuit? (*Hint:* See Problem 13.)

16. Earth's Field At a certain place, Earth's magnetic field has magnitude $|\vec{B}| = 0.590$ gauss and is inclined downward at an angle of 70.0° to the horizontal. A flat horizontal circular coil of wire with a radius of 10.0 cm has 1000 turns and a total resistance of 85.0 Ω. It is connected to a meter with 140 Ω resistance. The coil is flipped through a half-revolution about a diameter, so that it is again horizontal. How much charge flows through the meter during the flip? (*Hint:* See Problem 13.)

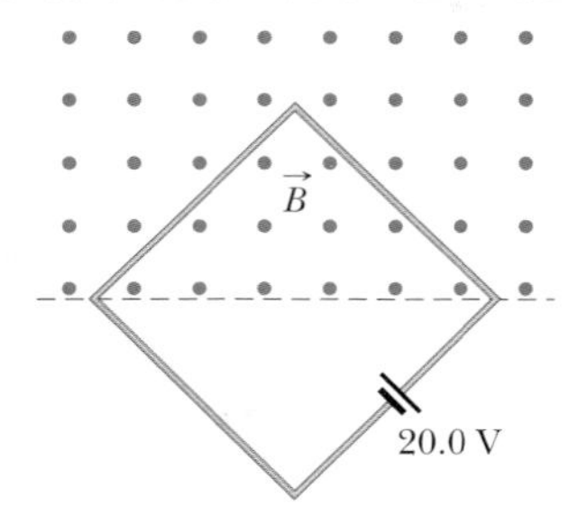

FIGURE 31-34 ■ Problem 17.

17. Square Loop A square wire loop with 2.00 m sides is perpendicular to a uniform magnetic field, with half the area of the loop in the field as shown in Fig. 31-34. The loop contains a 20.0 V battery with negligible internal resistance. If the magnitude of the field varies with time according to $B = (0.0420\ \text{T}) - (0.870\ \text{T/s})t$, what are (a) the magnitude of the net emf in the circuit and (b) the direction of the current through the battery?

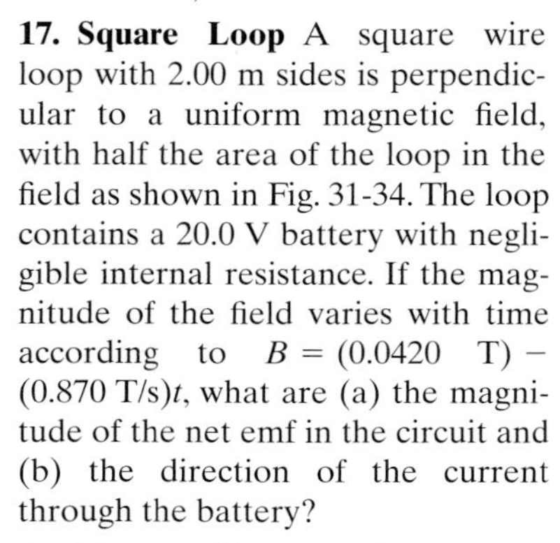

18. Three Circular Segments A wire is bent into three circular segments, each of radius $r = 10$ cm, as shown in Fig. 31-35. Each segment is a quadrant of a circle, ab lying in the xy plane, bc lying in the yz plane, and ca lying in the zx plane. (a) If a uniform magnetic field $\vec{B}$ points in the positive x direction, what is the magnitude of the emf developed in the wire when $\vec{B}$ increases at the rate of 3.0 mT/s in the x direction? (b) What is the direction of the current in segment bc?

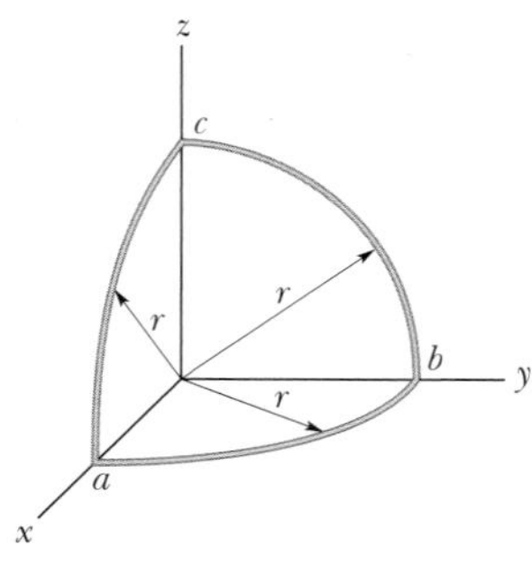

FIGURE 31-35 ■ Problem 18.

19. Rectangular Coil A rectangular coil of N turns and of length a and width b is rotated at frequency f in a uniform magnetic field $\vec{B}$, as indicated in Fig. 31-36. The coil is connected to co-rotating cylinders, against which metal brushes slide to make contact. If we arbitrarily define emf as being positive during the first quarter-turn, (a) show that the emf induced in the coil is given (as a function of time t) by

$$\mathcal{E} = 2\pi f NabB \sin(2\pi ft) = \mathcal{E}_0 \sin(2\pi ft).$$

This is the principle of the commercial alternating-current generator. (b) Design a loop that will produce an emf with $\mathcal{E}_0 = 150$ V when rotated at 60.0 rev/s in a uniform magnetic field of 0.500 T.

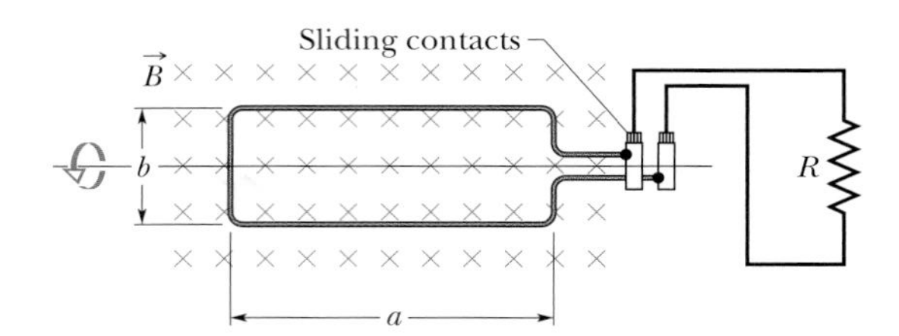

FIGURE 31-36 ■ Problem 19.

20. Semicircle A stiff wire bent into a semicircle of radius a is rotated with frequency f in a uniform magnetic field, as suggested in Fig. 31-37. What are (a) the frequency and (b) the amplitude of the varying emf induced in the loop?

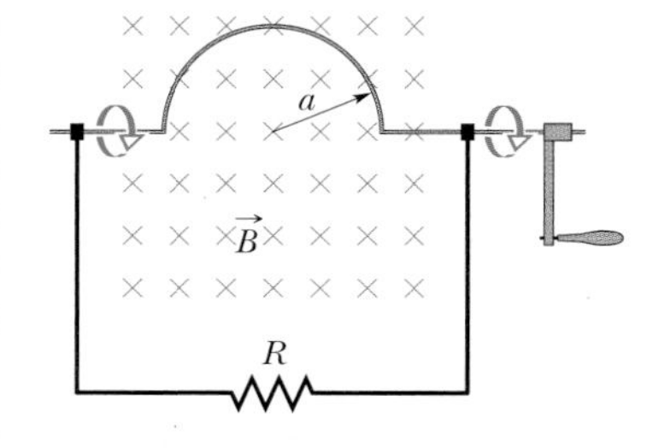

FIGURE 31-37 ■ Problem 20.

21. Electric Generator An electric generator consists of 100 turns of wire formed into a rectangular loop 50.0 cm by 30.0 cm, placed entirely in a uniform magnetic field with magnitude $B = 3.50$ T. What is the maximum value of the emf produced when the loop is spun at 1000 rev/min about an axis perpendicular to $\vec{B}$?

22. Closed Circular Loop In Fig. 31-38, a wire forms a closed circular loop, with radius $R = 2.0$ m and resistance 4.0 Ω. The circle is centered on a long straight wire; at time $t = 0$, the current in the long straight wire is 5.0 A rightward. Thereafter, the current changes according to $i = 5.0\ \text{A} - (2.0\ \text{A/s}^2)t^2$. (The straight wire is insulated, so there is no electrical contact between it and the wire of the loop.) What are the magnitude and direction of the current induced in the loop at times $t > 0$?

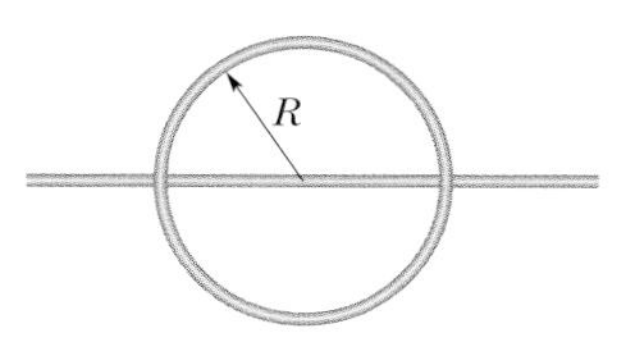

FIGURE 31-38 ■ Problem 22.

23. Square Loop Two In Fig. 31-39, the square loop of wire has sides of length 2.0 cm. A magnetic field is directed out of the page; its magnitude is given by $B = (4.0\ \text{T/m} \cdot \text{s}^2)\ t^2 y$, where B is in teslas, t is in seconds, and y is in meters. Determine the emf around the square at $t = 2.5\ \text{s}$ and indicate whether its direction is clockwise or counterclockwise.

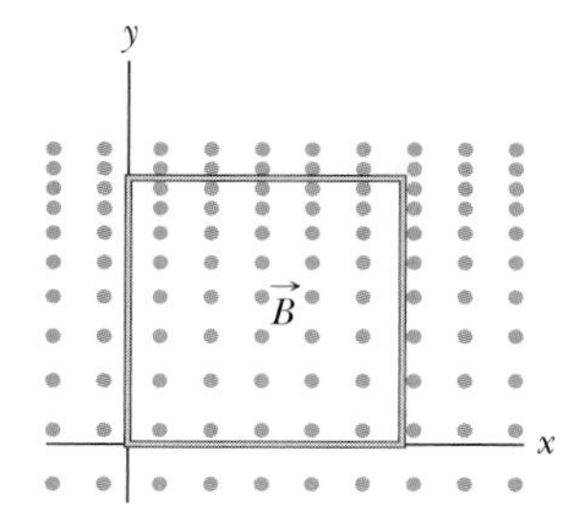

FIGURE 31-39 ■ Problem 23.

24. Square Loop Three For the situation shown in Fig. 31-40, $a = 12.0$ cm and $b = 16.0$ cm. The current in the long straight wire is given by $i = (4.50\ \text{A/s}^2)t^2 - (10.0\ \text{A/s})t$, where i is in amperes and t is in seconds. (a) Find the magnitude of the emf in the square loop at $t = 3.00$ s. (b) Indicate whether the direction of the induced current in the loop is clockwise or counterclockwise at $t = 3.00$ s.

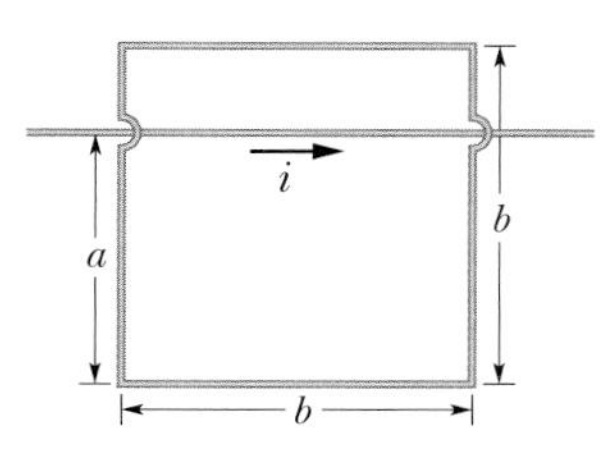

FIGURE 31-40 ■ Problem 24.

25. Parallel Copper Wires Two long, parallel copper wires of diameter 2.5 mm carry currents of 10 A in opposite directions. (a) Assuming that their central axes are 20 mm apart, calculate the magnetic flux per meter of wire that exists in the space between those axes. (b) What fraction of this flux lies inside the wires? (c) Repeat part (a) for parallel currents.

26. Rectangular Wire Loop A rectangular loop of wire with length a, width b, and resistance R is placed near an infinitely long wire carrying current i, as shown in Fig. 31-41. The distance from the long wire to the center of the loop is r. Find (a) the magnitude of the magnetic flux encircled by the loop and (b) the amount of induced current in the loop $|i^{\text{ind}}|$ as it moves away from the long wire with velocity $\vec{v}$. (c) Indicate whether the induced current is clockwise or counterclockwise.

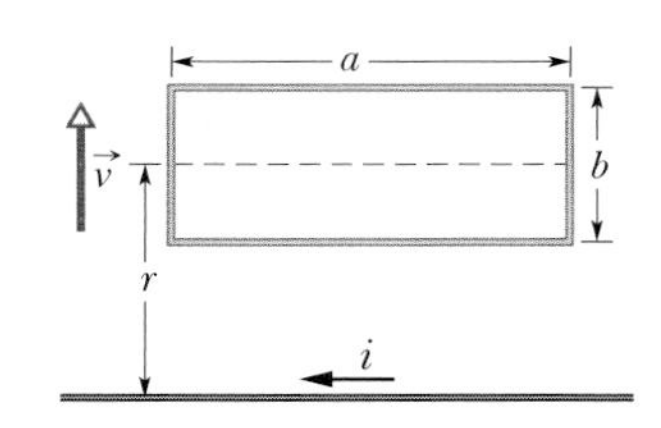

FIGURE 31-41 ■ Problem 26.

SEC 31-6 ■ INDUCTION AND ENERGY TRANSFERS

27. Internal Energy If 50.0 cm of copper wire (diameter = 1.00 mm) is formed into a circular loop and placed perpendicular to a uniform magnetic field that is increasing at the constant rate of 10.0 mT/s, at what rate does internal energy increase in the loop?

28. Loop Antenna A loop antenna of area A and resistance R is perpendicular to a uniform magnetic field $\vec{B}$. The field drops linearly to zero in a time interval Δt. Find an expression for the total internal energy added to the loop.

29. Rod on Rails A metal rod is forced to move with constant velocity $\vec{v}$ along two parallel metal rails, connected with a strip of metal at one end, as shown in Fig. 31-42. A magnetic field of magnitude $|\vec{B}| = 0.350$ T points out of the page. (a) If the rails are separated by 25.0 cm and the speed of the rod is 55.0 cm/s, what emf is generated? (b) If the rod has a resistance of 18.0 Ω and the rails and connector have negligible resistance, what is the current in the rod? (c) At what rate is mechanical energy being transformed to thermal energy?

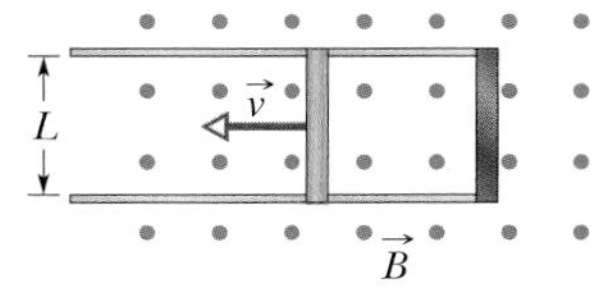

FIGURE 31-42 ■ Problem 29 and Problem 31.

30. Find Terminal Speed In Fig. 31-43, a long rectangular conducting loop, of width L, resistance R, and mass m, is hung in a horizontal, uniform magnetic field $\vec{B}$ that is directed into the page and that exists only above line aa. The loop is then dropped; during its fall, it accelerates until it reaches a certain terminal speed v_t. Ignoring air drag, find that terminal speed.

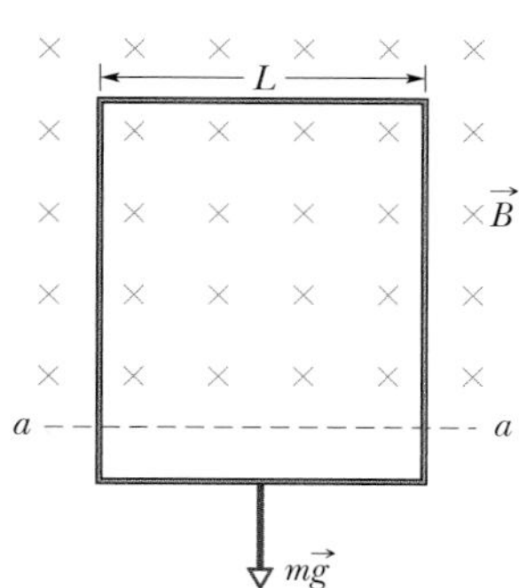

FIGURE 31-43 ■ Problem 30.

31. Rod on Rails Two The conducting rod shown in Fig. 31-42 has length L and is being pulled along horizontal, frictionless conducting rails at a constant velocity $\vec{v}$. The rails are connected at one end with a metal strip. A uniform magnetic field $\vec{B}$, directed out of the page, fills the region in which the rod moves. Assume that $L = 10$ cm, $v = 5.0$ m/s, and $B = 1.2$ T. (a) What is the magnitude of the emf induced in the rod? (b) What is the magnitude and direction (clockwise or counterclockwise) of the current in the conducting loop? Assume that the resistance of the rod is 0.40 Ω and that the resistance of the rails and metal strip is negligibly small. (c) At what rate is thermal energy added to the rod? (d) What magnitude of force must be applied to the rod by an external agent to maintain its motion? (e) At what rate does this external agent do work on the rod? Compare this answer with the answer to (c).

32. Rods Bent into V Two straight conducting rails form a right angle where their ends are joined. A conducting bar in contact with the rails starts at the vertex at time $t = 0$ and moves with a constant velocity of magnitude 5.20 m/s along them, as shown in Fig. 31-44. A magnetic field of magnitude $B = 0.350$ T is directed out of the page. Calculate (a) the flux through the triangle formed by the rails and bar at $t = 3.00$ s and (b) the magnitude of emf around the triangle at that time. (c) If we write the emf as $\mathscr{E} = at^n$, where a and n are constants, what is the value of n?

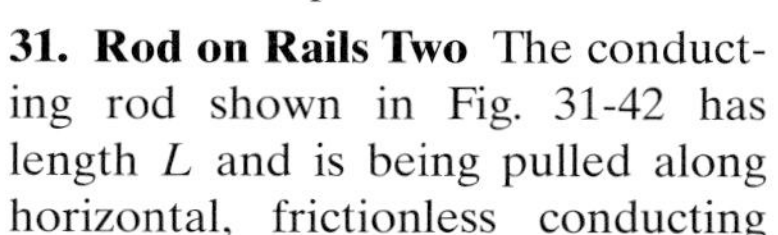
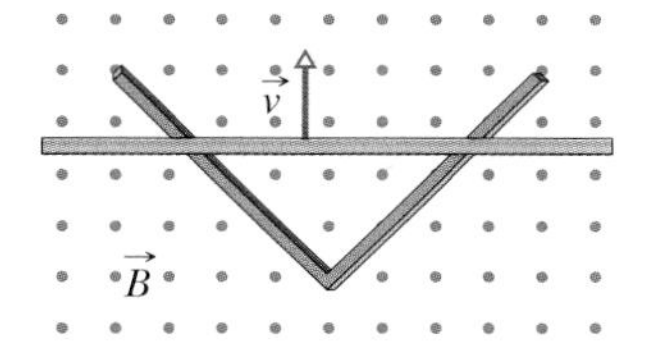

FIGURE 31-44 ■ Problem 32.

33. Rod on Conducting Rails Two Figure 31-45 shows a rod of length L caused to move at constant speed v along horizontal conducting rails. The magnetic field in which the rod moves is *not uniform* but is provided by a current i in a long wire parallel to the rails. Assume that $v = 5.00$ m/s, $a = 10.0$ mm, $L = 10.0$ cm, and $i = 100$ A. (a) Calculate the magnitude of the emf induced in the rod. (b) What is the magnitude of the current in the conducting loop?

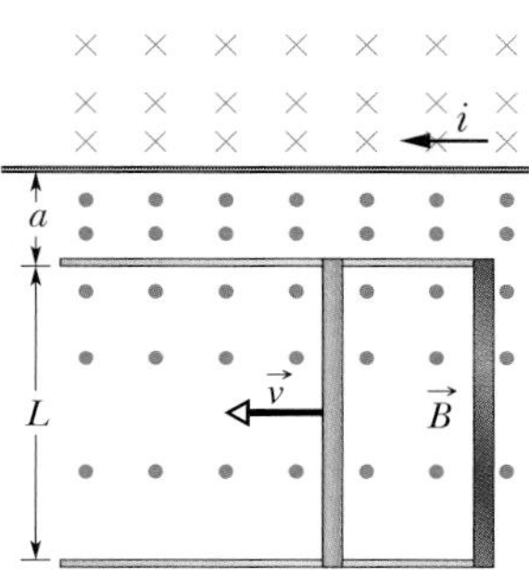

FIGURE 31-45 ■ Problem 33.

Assume that the resistance of the rod is 0.400 Ω and that the resistance of the rails and the strip that connects them at the right is negligible. (c) At what rate is internal energy added to the rod? (d) What magnitude of force must be applied to the rod by an external agent to maintain its motion? (e) At what rate does this external agent do work on the rod? Compare this answer to that for (c).

SEC. 31-7 ■ INDUCED ELECTRIC FIELDS

34. Two Circular Regions Figure 31-46 shows two circular regions R_1 and R_2 with radii $r_1 = 20.0$ cm and $r_2 = 30.0$ cm. In R_1 there is a uniform magnetic field of magnitude $B_1 = 50.0$ mT into the page, and in R_2 there is a uniform magnetic field of magnitude $B_2 = 75.0$ mT out of the page (ignore any fringing of these fields). Both fields are decreasing at the rate of 8.50 mT/s. Calculate the integral $\oint \vec{E} \cdot d\vec{s}$ for each of the three dashed paths.

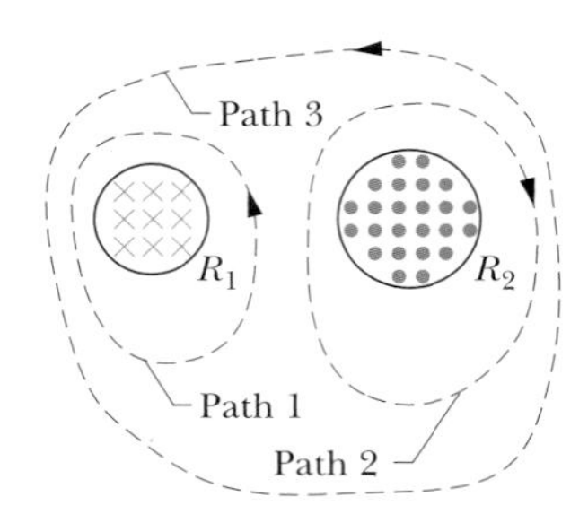

FIGURE 31-46 ■ Problem 34.

35. Long Solenoid A long solenoid has a diameter of 12.0 cm. When a current i exists in its windings, a uniform magnetic field of magnitude $B = 30.0$ mT is produced in its interior. By decreasing i, the field is caused to decrease at the rate of 6.50 mT/s. Calculate the magnitude of the induced electric field (a) 2.20 cm and (b) 8.20 cm from the axis of the solenoid.

36. Magnet Lab Early in 1981 the Francis Bitter National Magnet Laboratory at M.I.T. commenced operation of a 3.3-cm-diameter cylindrical magnet that produces a 30 T field, then the world's largest steady-state field. The field magnitude can be varied sinusoidally between the limits of 29.6 and 30.9 T at a frequency of 15 Hz. When this is done, what is the maximum value of the magnitude of the induced electric field at a radial distance of 1.6 cm from the axis? (*Hint:* See Touchstone Example 31-3.)

37. Drop to Zero Prove that the electric field $\vec{E}$ in a charged parallel-plate capacitor cannot drop abruptly to zero (as is suggested at point a in Fig. 31-47), as one moves perpendicular to the field, say, along the horizontal arrow in the figure. Fringing of the field lines always occurs in actual capacitors, which means that $\vec{E}$ approaches zero in a continuous and gradual way (see Problem 35 in Chapter 30). (*Hint:* Apply Faraday's law to the rectangular path shown by the dashed lines).

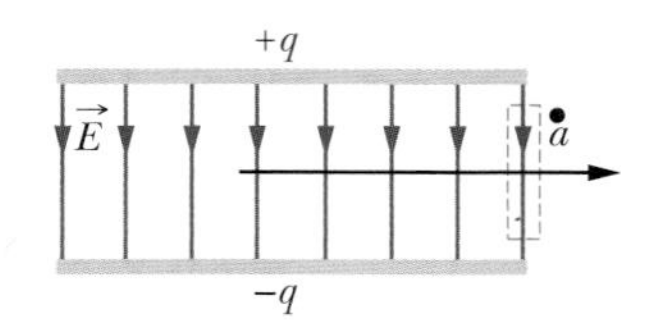

FIGURE 31-47 ■ Problem 37.

SEC 31-8 ■ INDUCED MAGNETIC FIELDS

38. Charging Capacitor Touchstone Example 31-4 describes the charging of a parallel-plate capacitor with circular plates of radius 55.0 mm. At what two radii r from the central axis of the capacitor is the magnitude of the induced magnetic field equal to 50% of its maximum value?

39. Induced Magnetic Field The induced magnetic field 6.0 mm from the central axis of a circular parallel-plate capacitor and between the plates has magnitude of 2.0×10^{-7} T. The plates have radius 3.0 mm. At what rate $|d\vec{E}/dt|$ is the electric field magnitude between the plates changing?

40. Parallel-Plate Capacitor Suppose that a parallel-plate capacitor has circular plates with radius $R = 30$ mm and a plate separation of 5.0 mm. Suppose also that a sinusoidal potential difference with a maximum value of 150 V and a frequency of 60 Hz is applied across the plates. That is,

$$\Delta V = (150 \text{ V}) \sin[2\pi(60 \text{ Hz})t].$$

(a) Find $B^{\max}(R)$, the maximum value of the magnitude of the induced magnetic field that occurs at $r = R$. (b) Plot $B^{\max}(r)$ for $0 < r < 10$ cm.

41. Uniform Electric Flux Figure 31-48 shows a circular region of radius $R = 3.00$ cm in which a uniform electric flux is directed out of the page. The total electric flux enclosed by the region is given by $\Phi^{\text{elec}} = (3.00 \text{ mV} \cdot \text{m/s})t$, where t is time. What is the magnitude of the magnetic field that is induced at radial distances (a) 2.00 cm and (b) 5.00 cm?

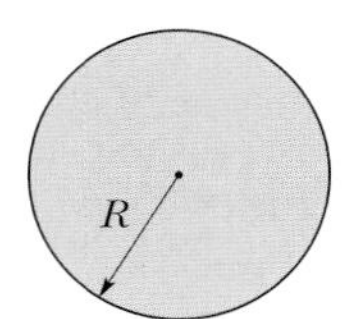

FIGURE 31-48 ■ Problems 41 through 44, and 57, 59, and 60.

42. Nonuniform Electric Flux Figure 31-48 shows a circular region of radius $R = 3.00$ cm in which an electric flux is directed out of the page. The flux encircled by a concentric circle of radius r is given by $\Phi^{\text{elec}} = (0.600 \text{ V} \cdot \text{m/s})(r/R)t$, where $r \leq R$ and t is time. What is the magnitude of the induced magnetic field at radial distances (a) 2.00 cm and (b) 5.00 cm?

43. Uniform Electric Field In Fig. 31-48, a uniform electric field is directed out of the page within a circular region of radius $R = 3.00$ cm. The magnitude of the electric field is given by $E = (4.5 \times 10^{-3} \text{ V/m} \cdot \text{s})t$, where t is time. What is the magnitude of the induced magnetic field at radial distances (a) 2.00 cm and (b) 5.00 cm?

44. Nonuniform Electric Field In Fig. 31-48, an electric field is directed out of the page within a circular region of radius $R = 3.00$ cm. The magnitude of the electric field is given by $E = (0.500 \text{ V/m} \cdot \text{s})(1 - r/R)t$, where t is the time and r is the radial distance ($r \leq R$). What is the magnitude of the induced magnetic field at radial distances (a) 2.00 cm and (b) 5.00 cm?

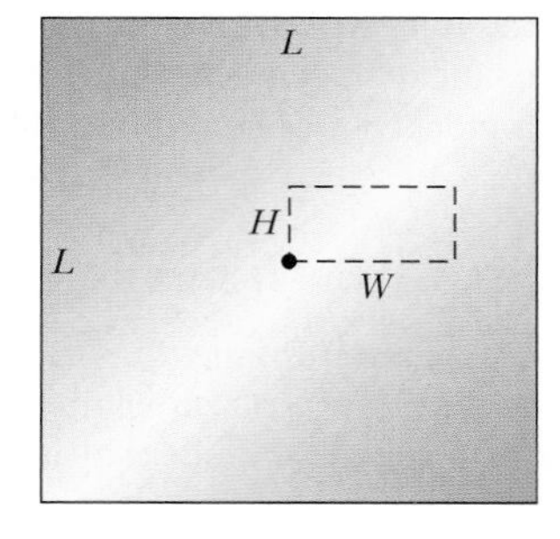

FIGURE 31-49 ■ Problem 45.

45. Discharging Capacitor A capacitor with square plates of edge length L is being discharged by a current of 0.75 A. Figure 31-49 is a head-on view of one of the plates from inside the capacitor. A dashed rectangular path is shown. If $L = 12$ cm, $W = 4.0$ cm, and $H = 2.0$ cm, what is the value of $\oint \vec{B} \cdot d\vec{s}$ around the dashed path?

46. Charging Capacitor The circuit in Fig. 31-50 consists of switch S, a 12.0 V ideal battery, a 20.0 MΩ resistor, and an air-filled capacitor. The capacitor has parallel circular plates of radius 5.00 cm,

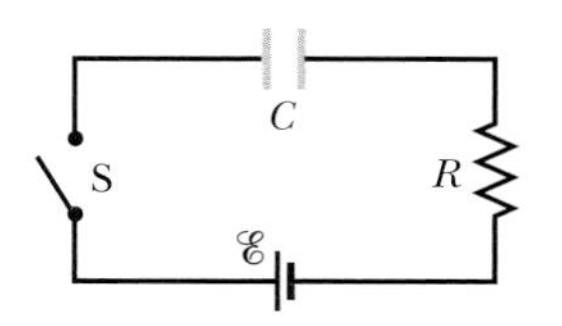

FIGURE 31-50 ■ Problem 46.

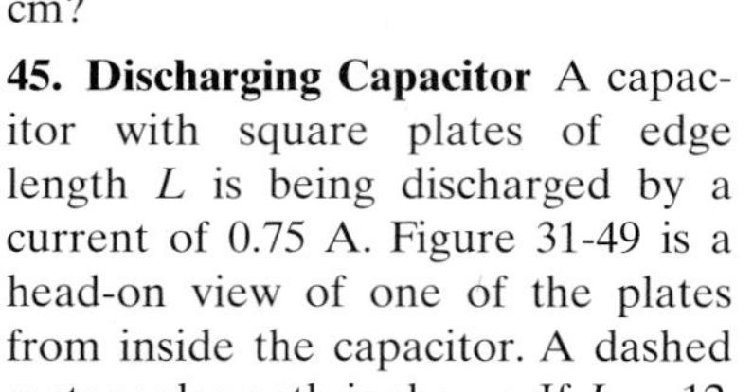

separated by 3.00 mm. At time $t = 0$ s, switch S is closed to begin charging the capacitor. The electric field between the plates is uniform. At $t = 250\ \mu s$, what is the magnitude of the magnetic field within the capacitor, at radial distance 3.00 cm?

SEC. 31-9 ■ DISPLACEMENT CURRENT

47. Prove That Displacement Prove that the displacement current in a parallel-plate capacitor of capacitance C can be written as $i^{dis} = C(d\Delta V/dt)$, where ΔV is the potential difference between the plates.

48. At What Rate At what rate must the potential difference between the plates of a parallel-plate capacitor with a 2.0 μF capacitance be changed to produce a displacement current of 1.5 A?

49. Current Density For the situation of Touchstone Example 31-4, show that the magnitude of the current density of the displacement current is $J^{dis} = \varepsilon_0 (dE/dt)$ for $r \leq R$.

50. Being Discharged A parallel-plate capacitor with circular plates of radius 0.10 m is being discharged. A circular loop of radius 0.20 m is concentric with the capacitor and halfway between the plates. The displacement current through the loop is 2.0 A. At what rate is the magnitude of the electric field between the plates changing?

51. Displacement Current As a parallel-plate capacitor with circular plates 20 cm in diameter is being charged, the current density of the displacement current in the region between the plates is uniform and has a magnitude of 20 A/m². (a) Calculate the magnitude B of the magnetic field at a distance $r = 50$ mm from the axis of symmetry of this region. (b) Calculate dE/dt in this region.

52. Electric Field The magnitude of the electric field between the two circular parallel plates in Fig. 31-51 is $E = (4.0 \times 10^5\ \text{V} \cdot \text{m}) - (6.0 \times 10^4\ \text{V} \cdot \text{m/s})t$, with E in volts per meter and t in seconds. At $t = 0$ s, the field is upward as shown. The plate area is 4.0×10^{-2} m². For $t \geq 0$ s, (a) what are the magnitude and direction of the displacement current between the plates and (b) is the direction of the induced magnetic field clockwise or counterclockwise around the plates?

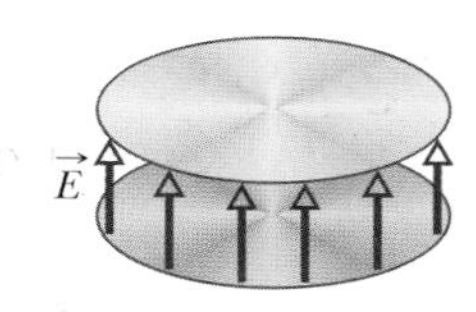

FIGURE 31-51 ■ Problem 52.

53. Magnitude of Electric Field The magnitude of a uniform electric field collapses to zero from an initial strength of 6.0×10^5 N/C in a time of 15 μs in the manner shown in Fig. 31-52. Calculate the amount of displacement current, $|i|$, through a 1.6 m² area perpendicular to the field, during each of the time intervals, a, b, and c shown on the graph. (Ignore the behavior at the ends of the intervals.)

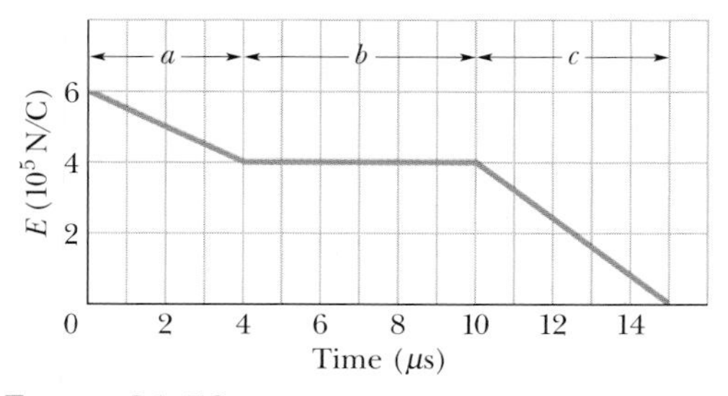

FIGURE 31-52 ■ Problem 53.

54. Displacement Current Two A parallel-plate capacitor with circular plates is being charged. Consider a circular loop centered on the central axis between the plates. The loop radius is 0.20 m, the plate radius is 0.10 m, and the displacement current through the loop is 2.0 A. What is the rate at which the magnitude of the electric field between the plates is changing?

55. Square Plates A parallel-plate capacitor has square plates 1.0 m on a side as shown in Fig. 31-53. A current of 2.0 A charges the capacitor, producing a uniform electric field $\vec{E}$ between the plates, with $\vec{E}$ perpendicular to the plates. (a) What is the displacement current i^{dis} through the region between the plates? (b) What is dE/dt in this region? (c) What is the displacement current through the square dashed path between the plates? (d) What is $\oint \vec{B} \cdot d\vec{s}$ around this square dashed path?

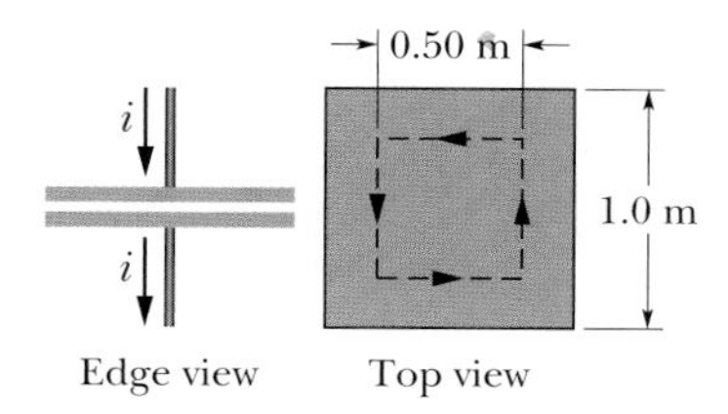

FIGURE 31-53 ■ Problem 55.

56. Consider a Loop A capacitor with parallel circular plates of radius R is discharging via a current of 12.0 A. Consider a loop of radius $R/3$ that is centered on the central axis between the plates. (a) How much displacement current is encircled by the loop? The maximum induced magnetic field has a magnitude of 12.0 mT. (b) At what radial distance from the central axis of the plate is the magnitude of the induced magnetic field 3.00 mT?

57. Uniform Displacement-Current Density. Figure 31-48 shows a circular region of radius $R = 3.00$ cm in which a displacement current is directed out of the page. The magnitude of the displacement current has a uniform density $J^{dis} = 6.00$ A/m². What is the magnitude of the magnetic field due to the displacement current at radial distances (a) 2.00 cm and (b) 5.00 cm?

58. Actual and Displacement Figure 31-54a shows current i that is produced in a wire of resistivity $1.62 \times 10^{-8}\ \Omega \cdot \text{m}$ in the direction indicated. The magnitude of the current versus time t is shown in Fig. 31-54b. Point P is at radius 9.00 mm from the wire's center. Determine the magnitude of the magnetic field at point P due to the real current i in the wire at (a) $t_1 = 20$ ms, (b) $t_2 = 40$ ms, (c) $t_3 = 60$ ms, and (d) $t_4 = 70$ ms. Next, assume that the electric field driving the current is confined to the wire. Then determine the magnitude of the magnetic field at point P due to the displacement current i^{dis} in the wire at (e) $t_1 = 20$ ms, (f) $t_2 = 40$ ms, (g) $t_3 = 60$ ms, and (h) $t_4 = 70$ ms. (i) When both magnetic fields are present at point P, what are their directions in Fig. 31-54a?

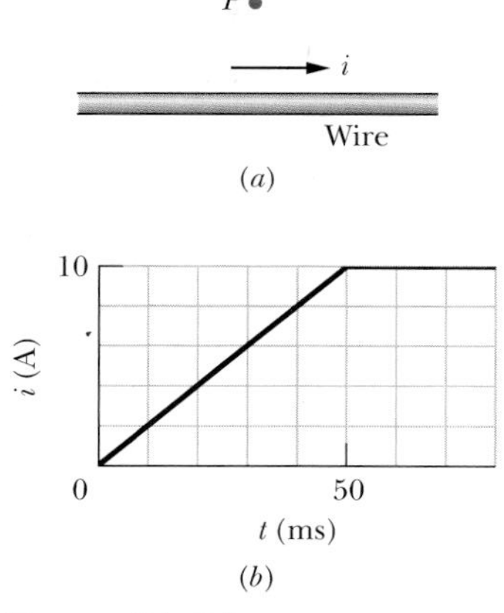

FIGURE 31-54 ■ Problem 58.

59. Nonuniform Displacement-Current Density. Figure 31-48 shows a circular region of radius $R = 3.00$ cm in which a displacement current is directed out of the page. The displacement current has a density of magnitude $J^{dis} = (4.00\ \text{A/m}^2)(1 - r/R)$, where r is the radial distance $r \leq R$. What is the magnitude of the magnetic field due to the displacement current at radial distances (a) 2.00 cm and (b) 5.00 cm?

60. Uniform Displacement Current. Figure 35-48 shows a circular region of radius $R = 3.00$ cm in which a uniform displacement current $i^{dis} = 0.500$ A is directed out of the page. What is the magnitude of the magnetic field due to the displacement current at radial distances (a) 2.00 cm and (b) 5.00 cm?

SEC. 31-10 ■ GAUSS' LAW FOR MAGNETIC FIELDS

61. Rolling a Sheet of Paper Imagine rolling a sheet of paper into a cylinder and placing a bar magnet near its end as shown in Fig. 31-55. (a) Sketch the magnetic field lines that pass through the surface of the cylinder. (b) What can you say about the sign of $\vec{B} \cdot d\vec{A}$ for every area $d\vec{A}$ on the surface? (c) Does this result contradict Gauss' law for magnetism? Explain.

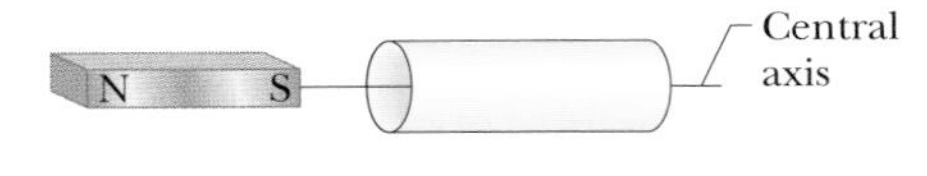

FIGURE 31-55 ■ Problem 61.

62. Die Suppose the magnetic flux at each of five faces of a die (singular of "dice") is given by $\Phi^{mag} = \pm N$ Wb, where $N(= 1$ to $5)$ is the number of spots on the face. The flux is positive (outward) for N even and negative (inward) for N odd. What is the flux at the sixth face of the die? Is it directed in or out?

63. Right Circular Cylinder A Gaussian surface in the shape of a right circular cylinder with end caps has a radius of 12.0 cm and a length of 80.0 cm. One end encircles an inward magnetic flux of 25.0 μWb. At the other end there is a uniform magnetic field of 1.60 mT, normal to the surface and directed outward. What is the net magnetic flux at the curved surface?

64. Weird Shape Figure 31-56 shows a closed surface. Along the flat top face, which has a radius of 2.0 cm, a magnetic field $\vec{B}$ of magnitude 0.30 T is directed outward. Along the flat bottom face, a magnetic flux of 0.70 mWb is directed outward. What are (a) the magnitude and (b) the net magnetic flux at the curved part of the surface?

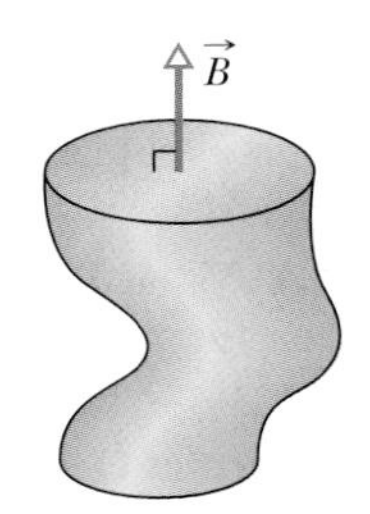

FIGURE 31-56 ■ Problem 64.

Additional Problems

65. Power from a Tether A few years ago, the space shuttle *Columbia* tried an experiment with a tethered satellite. The satellite was released from the shuttle and slowly reeled out on a long conducting cable as shown in Fig. 31-57 (not to scale). For this problem we will make the following approximations:

The shuttle is moving at a constant velocity.

The Earth's magnetic field is constant and uniform.

The line of the tether, the velocity of the system, and the magnetic field are all perpendicular to each other.

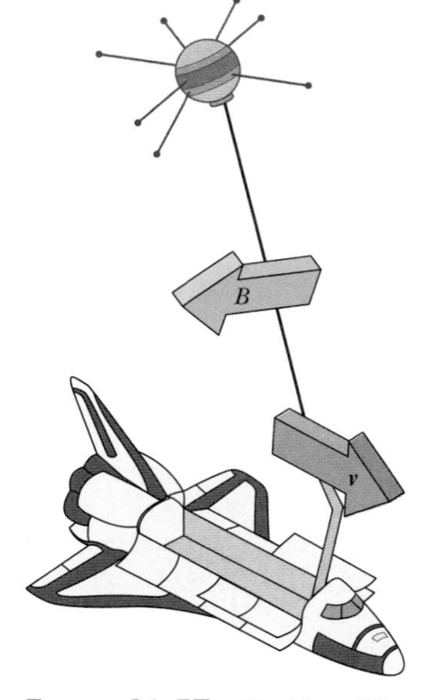

FIGURE 31-57 ■ Problem 65.

The Earth's field produces an emf from one end of the cable to the other. The idea is to use a system like this to generate electric power in space more efficiently than with solar panels.

(a) Explain why a voltage difference is produced.
(b) If the Earth's magnetic field is given by a magnitude $\vec{B}$, the shuttle– satellite system is moving with a velocity $\vec{v}$, and the tether has a length L, calculate the magnitude of the emf $\mathscr{E}$ from one end of the tether to the other.
(c) At the shuttle's altitude, the Earth's field is about 0.3 gauss and the shuttle's speed is about 7.5 km/s. The tether is 20 km long (!). What is the expected potential difference in volts?
(d) At the altitude of the shuttle, the thin atmosphere is lightly ionized, allowing a current of about 0.5 amps to flow from the satellite back to the shuttle through the thin air. What is the resistance of the 20 km of ionized air?

66. Building a Generator The apparatus shown in Fig. 31-58 can be used to build a motor. This device can also be used to build a generator that will produce a voltage. (a) Explain the setup that one would use to make a motor and explain how it works. Do the same for the generator. (b) Estimate the maximum voltage that would be produced if you cranked the generator by hand. (*Hint:* As a comparison for estimating the strength of the bar magnet, the Earth's magnetic field at our location is about 0.4 gauss.)

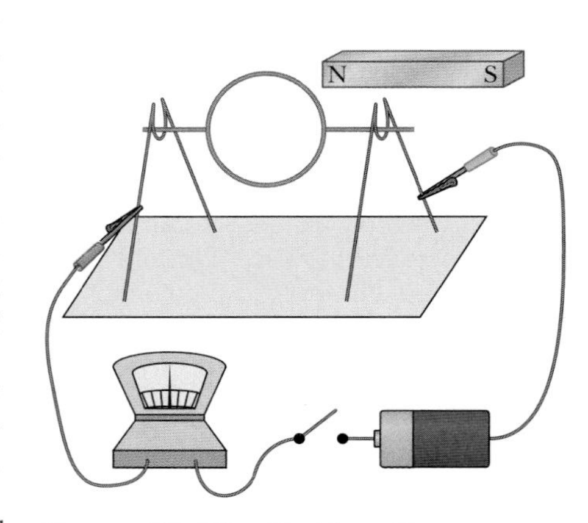

FIGURE 31-58 ■ Problem 66.

67. Faraday's Law Faraday's law describes the emf produced by magnetic fields in a variety of circumstances. State and discuss Faraday's law, being careful to include a discussion of different physical situations that may be described by the statement of the law.

68. Magnetic Field, Force, and Torque Figure 31-59 shows two long, current-carrying wires and a bar magnet. At the right is shown a compass specifying set of direction labels. For each of the vectors (a)–(e) below, select the direction label that best gives the direction of the item. If the magnitude of the item is zero, write 0. If none of the directions are correct, write N.

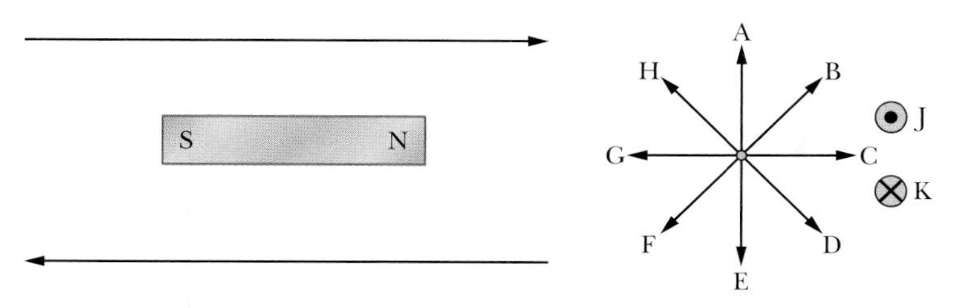

FIGURE 31-59 ■ Problem 68.

(a) The magnetic field due to the lower wire at the center of the upper wire
(b) The force on the lower wire due to the magnetic field from the upper wire
(c) The net torque acting on the upper wire
(d) The magnetic field due to the currents at the center of the magnet
(e) The net force acting on the lower wire due to the bar magnet

69. *B* Increases in Time In Fig. 31-60*a*, a uniform magnetic field $\vec{B}$ increases in magnitude with time t as given by Fig. 31-60*b*. A circular conducting loop of area 8.0×10^{-4} m^2 lies in the field, in the plane of the page. The amount of charge q that has passed point A on the loop is given in Fig. 31-60*c* as a function of t. What is the loop's resistance?

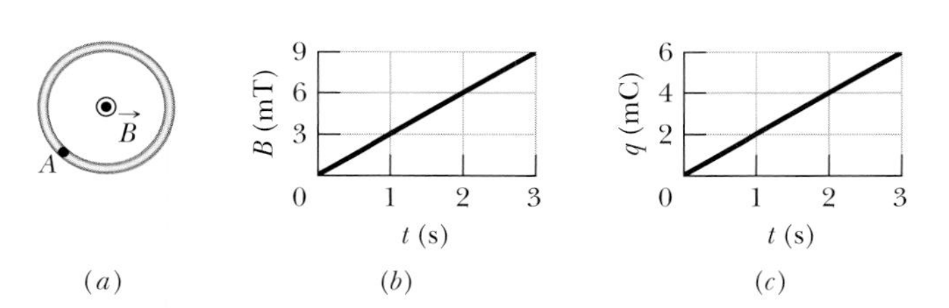

FIGURE 31-60 ■ Problem 69.

70. Circular Loop Around a Solenoid In Fig. 31-61*a*, a circular loop of wire is concentric with a solenoid and lies in a plane that is perpendicular to the solenoid's central axis. The loop has radius 6.00 cm. The solenoid has radius 2.00 cm, consists of 8000 turns per meter, and has a current i_{sol} that varies with time t as given in Fig. 31-61*b*. Figure 31-61*c* shows, as a function of time, the energy $E^{thermal}$ that is transformed to thermal energy in the loop. What is the loop's resistance?

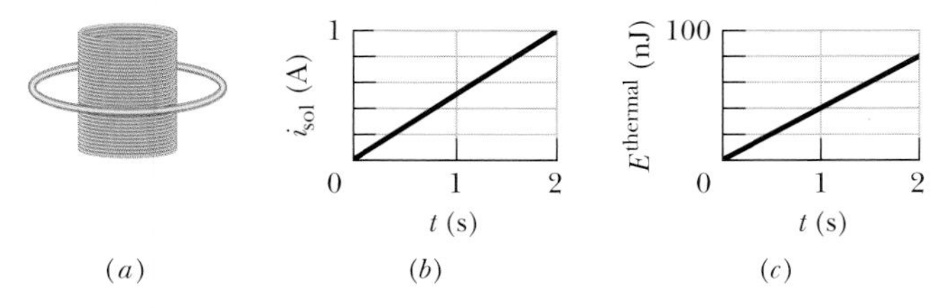

FIGURE 31-61 ■ Problem 70.

71. Magnitudes and Direction Figure 31-62*a* shows a wire that forms a rectangle and has a resistance of 5.0 mΩ. Its interior is split into three equal areas with different magnetic fields $\vec{B}_1$, $\vec{B}_2$, and $\vec{B}_3$ that are either directly out of or into the page, as indicated. The fields are uniform within each region. Figure 31-62*b* gives the change in the z components B_z of the three fields with time t. What are the magnitude and direction of the current induced in the wire?

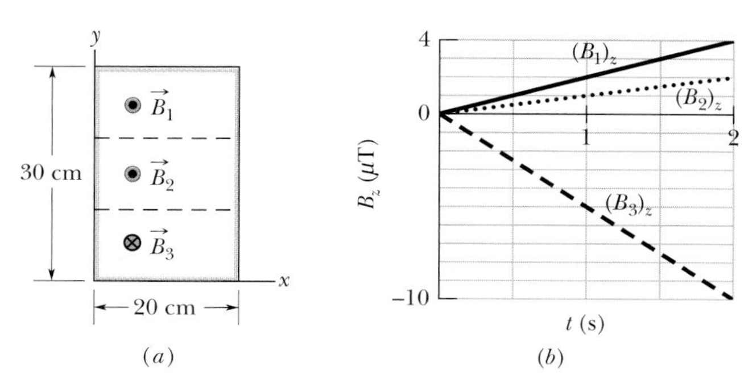

FIGURE 31-62 ■ Problem 71.

72. Two Concentric Regions Figure 31-63*a* shows two concentric circular regions in which uniform magnetic fields can change. Region 1, with radius $r_1 = 1.0$ cm, has an outward magnetic field $\vec{B}_1$ that is increasing in magnitude. Region 2, with radius $r_2 = 2.0$ cm, has an outward magnetic field $\vec{B}_2$ that may also be changing. Imagine that a conducting ring of radius R is centered on the two regions and then the emf $\mathscr{E}$ around the ring is determined. Figure 31-63*b* gives emf $\mathscr{E}$ as a function of the square of the ring's radius, R^2, to the outer edge of region 2. What are the rates of B-field magnitude change (a) dB_1/dt and (b) dB_2/dt? (c) Is the magnitude of $\vec{B}_2$ increasing, decreasing, or remaining constant?

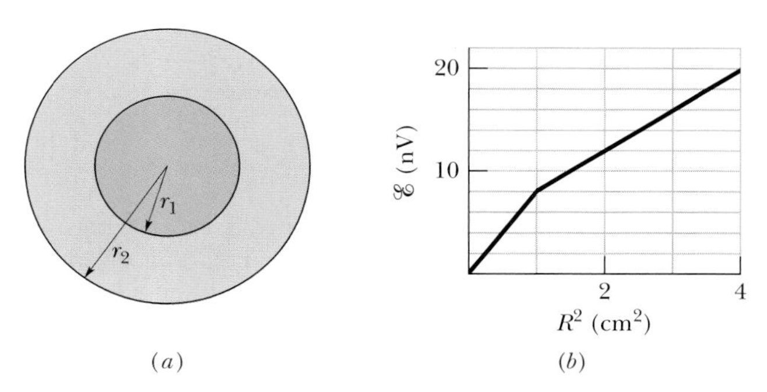

FIGURE 31-63 ■ Problem 72.

73. Pulled at Constant Speed Figure 31-64*a* shows a rectangular conducting loop of resistance $R = 0.020$ Ω, height $H = 1.5$ cm, and length $D = 2.5$ cm being pulled at constant speed $v = 40$ cm/s through two regions of uniform magnetic field. Figure 31-64*b* gives the current i induced in the loop as a function of the position x of the right side of the loop. For example, a current of 3.0 μA is induced clockwise as the loop enters region 1. What are the magnitudes and directions of the magnetic field in (a) region 1 and (b) region 2?

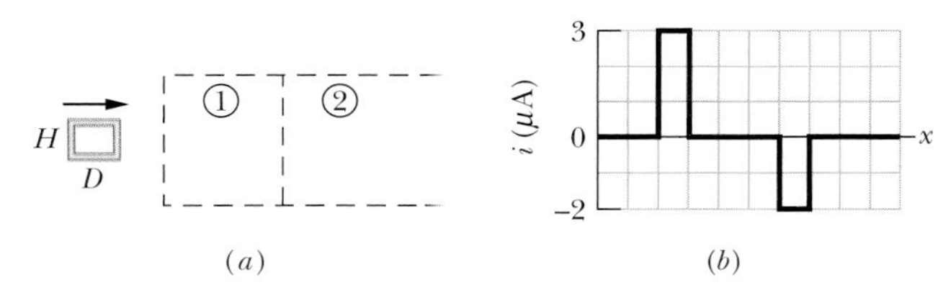

FIGURE 31-64 ■ Problem 73.

74. Plane Loop A plane loop of wire consisting of a single turn of area 8.0 cm^2 is perpendicular to a magnetic field that increases uniformly in magnitude from 0.50 T to 2.5 T in a time of 1.0 s. What is the resulting induced current if the coil has a total resistance of 2.0 Ω?

75. At What Rate Must *B* Change The plane of a rectangular coil of dimensions 5.0 cm by 8.0 cm is perpendicular to the direction of magnetic field B. If the coil has 75 turns and a total resistance of 8.0 Ω, at what rate must the magnitude of B change in order to induce a current of 0.10 A in the windings of the coil?

76. Rod on Rails 3 In the arrangement shown in Fig. 31-65, a conducting rod rolls to the right along parallel conducting rails connected on one end by a 6.0 Ω resistor. A 2.5 T magnetic field is directed *into* the paper. Let $L = 1.2$ m. Neglect the mass of the bar and friction. (a) Calculate the applied force required to move the bar to the right at a *constant* speed of 2.0 m/s. (b) At what rate is energy dissipated in the resistor?

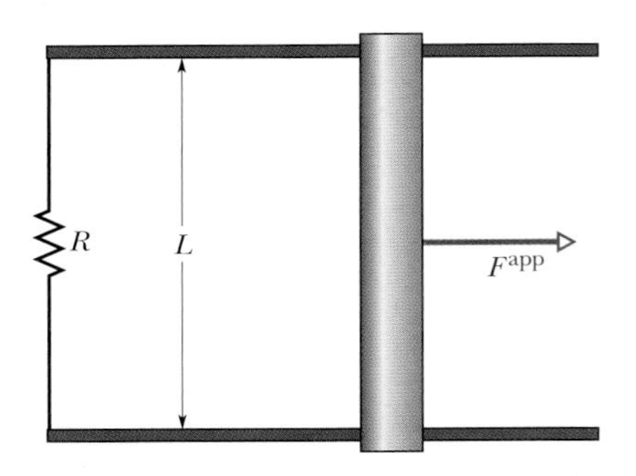

FIGURE 31-65 ■ Problem 76.

77. An Engineer An engineer has designed a setup with a small pickup coil placed in the center of a large field coil as shown in Fig. 31-66. Both coils have many turns of conducting wire. The field coil produces a magnetic field that is proportional in magnitude to the amount of current flowing through its wires. The pickup coil is smaller and its many turns can sense or "pick up" the changing magnetic field in the field coil. The pickup coil produces an emf that is proportional in magnitude to the rate of change of the magnetic field and the angle ϕ. Here ϕ is the angle between the normal to the field coil and the normal to the pickup coil. You have been hired as a consultant to check on the reliability of the engineer's work. You figure out how to use Faraday's law along with proportional reasoning to check on the validity of the results that have been reported without doing any formal calculations or measurements. Sketches from the engineer's notebook are shown in Fig. 31-66*b*. (a) Look at the graph pair in Fig. 31-66*b*. Sketch the measured emf induced in the pickup coil if the engineer has adjusted the scope so the maximum emf is the first positive grid line and the minimum emf is on the first negative grid line. Assume that the normal to each of the coils is pointing in the same direction. (b) According to the engineer's notebook, she fed exactly the same pattern of current to the field coil but she turned the pickup coil so its normal makes an angle of +45° with respect to the normal to the plane of the field coil. Carefully sketch the pattern of emf observed in the pickup coil. What is the maximum and minimum amplitude of the emf in "grid" units? (c) What happens when she flips the pickup coil over around so its normal is 180° from the normal to the field coil? Sketch the emf and use the correct signs for the values of the induced emf for this situation. Explain the reasons for the shape and magnitude of your sketch in each case.

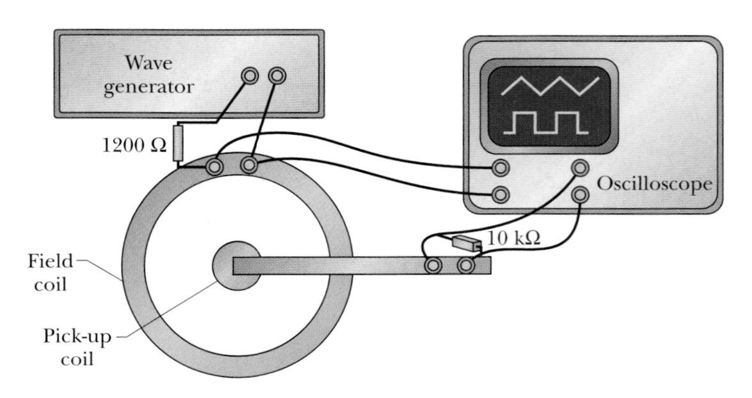

FIGURE 31-66*a* ■ Problem 77.

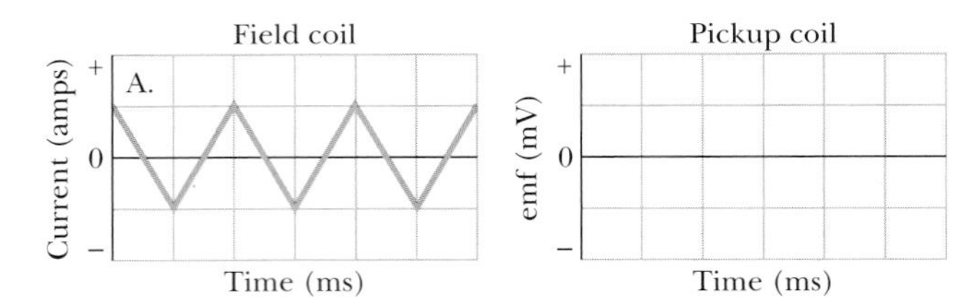

FIGURE 31-66*b* ■ Problem 77.

78. Engineer Task 2 You are still double-checking the work of the engineer from Problem 77. Consider the graph shown in Fig. 31-67. (a) What should our honest and competent engineer have reported for the pattern of emf values as a function of time? Assume that once again the normal to the pickup coil is in the same direction as the normal to the field coil. Please take care to sketch not only the shape of the emf graph but also its proper magnitude using the same gain setting on the oscilloscope as you did in Problem 77. Use a solid line for your sketch. (b) Suppose the engineer reduced the number of turns in the pickup coil by a factor of 2 and redid the measurements. Sketch a new graph showing the shape and proper magnitudes for the expected pickup coil emf using a dashed line. Explain the reasons for the shape and magnitude of your sketch in each case.

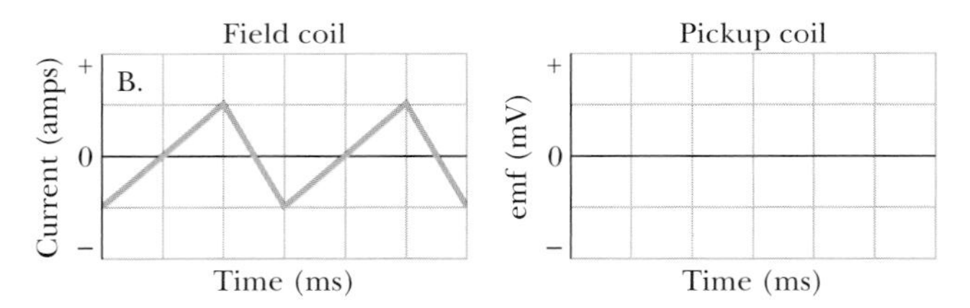

FIGURE 31-67 ■ Problem 78.

79. Engineer Task 3 You are still double-checking the work of the engineer from Problem 31-77. Assume that the number of turns in both the field and pickup coils is the same as in that problem, as is the oscilloscope setting. Consider the graph shown in Fig. 31-68. (a) What should our honest and competent engineer have reported for the pattern of current fed into the field coil as a function of time? Assume that once again the normal to the pickup coil is in the same

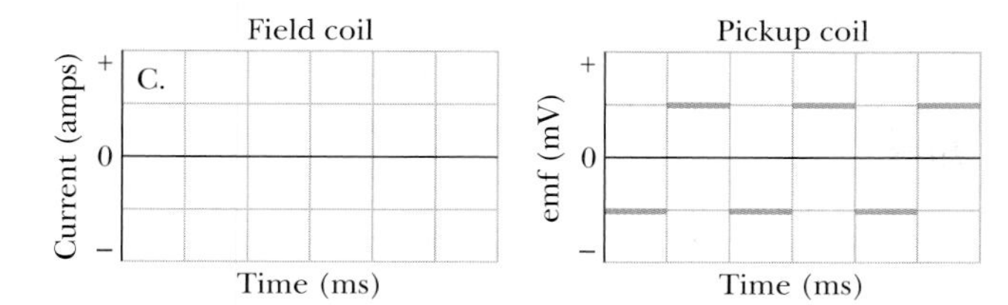

FIGURE 31-68 ■ Problem 79.

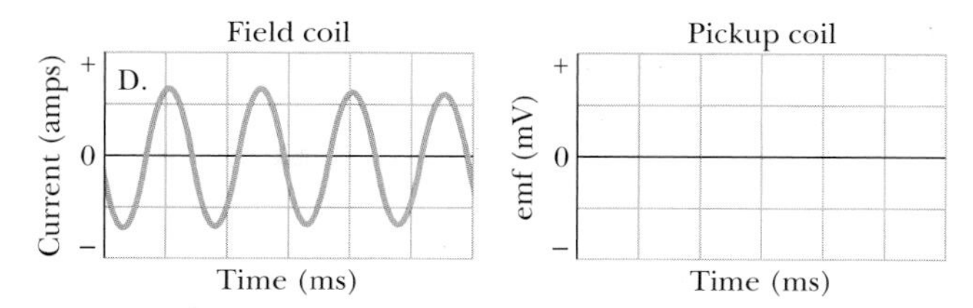

FIGURE 31-69 ■ Problem 80.

direction as the normal to the field coil. Please take care to sketch not only the shape of the emf graph but also its proper magnitude using the same gain setting on the oscilloscope as you did in Problem 31-77. Use a solid line for your sketch. (b) Suppose the engineer reduced the number of turns in the pickup coil by a factor of 2 and redid the measurements. Sketch a new graph showing the shape and proper magnitudes for emf in the field coil using a dashed line. Explain the reasons for the shape and magnitude of your sketch in each case.

80. Engineer Task 4 You are still double-checking the work of the engineer from Problem 31-77. Assume that the number of turns in both the field and pickup coils is the same as in that problem. Consider the graph shown in Fig. 31-69. What should our honest and competent engineer have reported for the pattern of emf induced in the pickup coil if the oscilloscope gain is adjusted to give a maximum value of emf of +2 oscilloscope grid units and a minimum value of −2 oscilloscope units? (*Hint:* What is the derivative of the sine function?) Explain the reason for the shape and magnitude of your sketch in each case.

81. Ring of Copper Figure 31-70 shows a ring of copper with its plane perpendicular to the axis of the nearby rod-shaped magnet. In which of the following situations will a current be induced in the ring? Choose all correct answers.

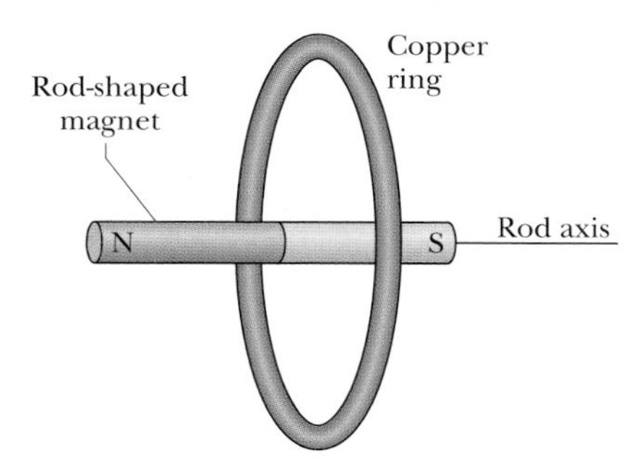

FIGURE 31-70 ■ Problem 81.

(a) The magnet is moved horizontally toward the left.
(b) The ring is moved away from the magnet.
(c) The ring is rotated around any of its diameters.
(d) The magnet is moved up or down.
(e) The ring is rotated around its center in the plane in which it lies.

32 Inductors and Magnetic Materials

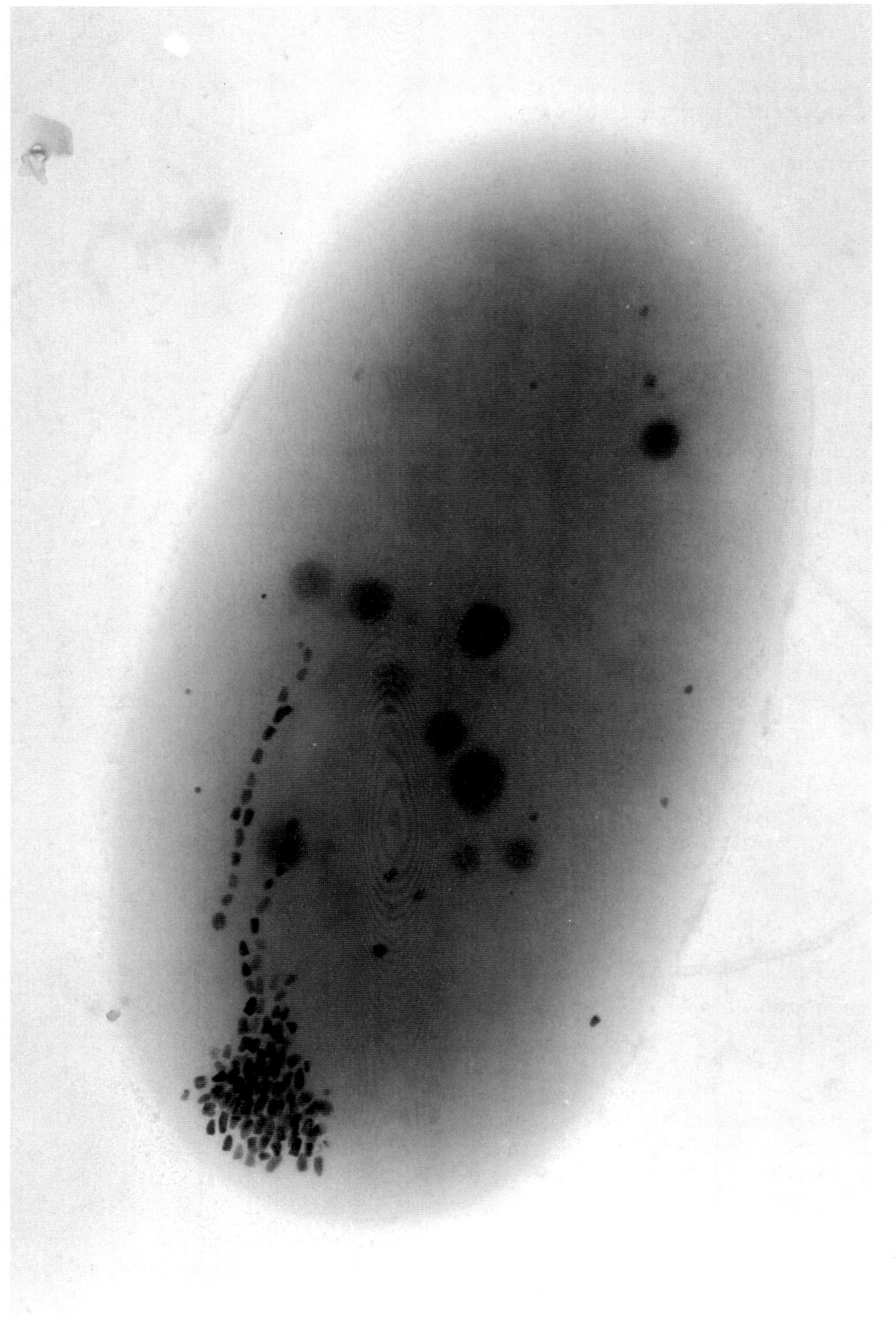

This is a microscopic view of a bacterium found in Australia that will swim to the muddy bottom of a pond to escape oxygen in its environment and find the nutrients it needs to survive. But if this bacterium were transported to a pond in the United States, it would swim to the top of the pond and die.

How does this bacterium know how to swim down in Australia but not in the U.S.?

The answer is in this chapter.

32-1 Introduction

In the previous chapter we described how an electric car or toothbrush could be charged without electrical contacts. Likewise the guitar pickup described in Section 31-5 amplifies sound. These devices make practical use of inductance. In this chapter we consider some additional practical uses of inductance phenomena in common electric circuit elements known as inductors and transformers. You will consider the basic behaviors of these elements in circuits where the voltage changes in time. Then you will move on to what appears to be an unrelated topic—the behavior of magnetic materials.

The simplest magnetic structure contained in magnetic materials is a magnetic dipole. We will trace the origin of magnetic dipoles, and the associated magnetic properties of materials back to atoms and electrons. You will then reconsider inductors and transformers and learn how magnetic materials can be used to enhance their performance. Some of the first inductors are pictured in Fig. 32-1.

Finally, we will discuss recent theories that enable us explain why the Earth behaves like a huge magnetic dipole, and we will consider the possible role induction plays in explaining the characteristics and changing nature of the Earth's magnetic field.

FIGURE 32-1 ■ The crude inductors with which Michael Faraday discovered the law of induction. In those days amenities such as insulated wire were not commercially available. It is said that Faraday insulated his wires by wrapping them with strips cut from one of his wife's petticoats.

32-2 Self-Inductance

Let's explore how the phenomenon of inductance introduced in the previous chapter can be useful in the design of electric circuits with changing currents. In Section 31-3 we saw that when two coils are near each other, a changing current in one of the coils can induce an emf in the other according to Faraday's law ($\mathscr{E} = -N d\,\Phi^{\text{mag}}/dt$). But if the second coil is part of an electric circuit, the current induced in it can also induce an emf in the first coil. This phenomenon, known as **mutual induction,** is used in the design of *inductive chargers*—noncontact charging systems like those used for electric toothbrushes and other devices. In multiple-loop coils the emfs produced by mutual induction are proportional to the number of loops in the coil. For this reason mutual induction is also used in the design of **transformers**—devices that can transform time-varying voltages to larger or smaller time-varying voltages.

In addition, when current in a single coil with one or more loops changes, this induces an emf in the *same* coil. This emf is produced as the result to the changing flux the coil produces in the area it encloses. This process is known as **self induction.** In general,

> A self-induced emf $\mathscr{E}_L$ appears in any coil whenever its current is changing.

According to Lenz's law the self-induced emf acts to oppose the change of current in the coil. For this reason, coils of wire called **inductors** (sometimes called "chokes") are useful in circuits whenever it is desirable to stabilize currents. In a circuit diagram, an inductor is denoted by a symbol that looks like helical loops of wire (ℓℓℓ). See Figs. 32-2 and 32-3. In addition, inductors can be combined with resistors and capacitors to modify the characteristics in circuits driven by oscillating voltage sources. In this chapter we consider the role of inductors in stabilizing currents. In the next chapter we will study the behavior of inductors in circuits with oscillating voltages.

A typical **inductor** consists of a wire that is coiled into a very large number of loops wrapped around a piece of hollow cardboard or perhaps a magnetic rod. Inductors come in many shapes. A common shape is a **solenoid,** which consists of a tightly

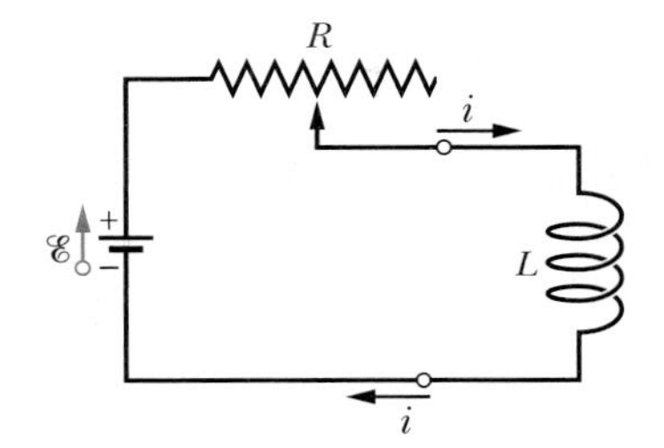

FIGURE 32-2 ■ If the current in a coil is changed by varying the contact position on a variable resistor, a self-induced emf $\mathscr{E}_L$ will appear in the coil while the current is changing.

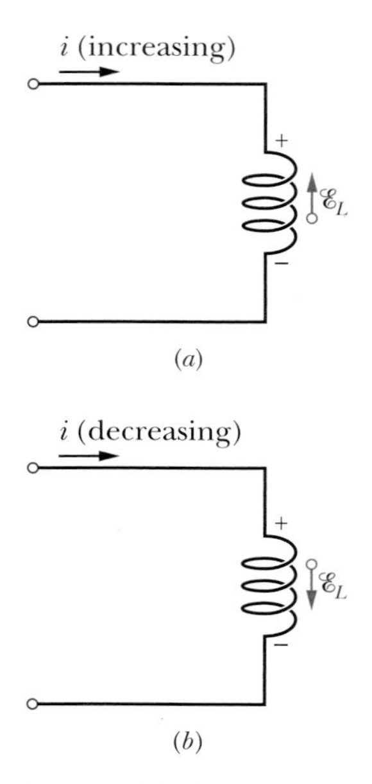

FIGURE 32-3 ■ The arrow and the + and − signs on either side of the inductor indicate the direction the emf $\mathscr{E}_L$ acts in relative to the direction of the current in the circuit alongside the coil. (*a*) The current i is increasing and the self-induced emf $\mathscr{E}_L$ appears along the coil in a direction such that it opposes the increase. (*b*) The current i is decreasing and the self-induced emf appears in a direction such that it opposes the decrease.

wound helical coil of wire—like the one Faraday wound shown in the lower part of Fig. 32-1. Because the magnetic field inside a solenoid is very uniform, it is not difficult to calculate the emfs created by current changes in solenoids. For this reason, we shall consider a solenoid as our basic type of inductor. Also, at first we assume that all inductors are air-core inductors that have no magnetic materials such as iron in their vicinity to distort their magnetic fields.

The Mathematics of Self-Inductance

We start our mathematical treatment of self-inductance with a solenoid-shaped inductor of length l and total number of loops N. When a charge flows through an inductor, the coil produces a magnetic field inside its coils whose strength is *directly proportional* to the current. For an ideal solenoid, the magnitude of the magnetic field is given by Eq. 30-25,

$$B = \mu_0 n|i| \quad \text{(inside an ideal solenoid),} \qquad \text{(Eq. 30-25)}$$

where μ_0 is the magnetic constant and n the number of turns per unit length.

This magnetic field yields an amount of flux over the area A enclosed by the coil of $|\Phi^{\text{mag}}| = BA = n(\mu_0 A|i|)$. Now, if we try to change the current by changing the resistance in the circuit shown in Fig. 32-2, then the magnetic field and hence the flux at the center of the coil changes. According to Faraday's law this change in flux will produce an emf in the coil given by Eq. 31-7 ($\mathscr{E} = -N d\Phi^{\text{mag}}/dt$). According to Lenz's law this emf will act to oppose the change in the current. Thus, if you close a switch that connects a voltage source to an inductor, the induced "back" emf will retard the rise in current through the circuit. An emf that acts to oppose a change in current is known as a **back emf.** Alternately, if a current already exists in a circuit then opening a switch will slow the rate of reduction of the current (Fig. 32-3). Applying Faraday's law and noting that the total number of turns N is the product of the turns per unit length, n, and the length, l, of the solenoid gives us

$$\mathscr{E}_L = -N\frac{d\Phi^{\text{mag}}}{dt} = -nl\frac{d\Phi^{\text{mag}}}{dt} = -\mu_0 A n^2 l\frac{di}{dt} \quad \text{(solenoidal air-core inductor),} \qquad (32\text{-}1)$$

where $\mathscr{E}_L$ is the self-induced emf in the solenoid.

If the solenoid is very much longer than its radius, then Eq. 32-1 expresses its inductance to a good approximation. However, we have neglected the spreading of the magnetic field lines near the ends of the solenoid, just as the parallel-plate capacitor formula $C = \varepsilon_0 A/d$ neglects the fringing of the electric field lines near the edges of the capacitor plates.

Equation 32-1 tells us that the amount of self-induced back emf is directly proportional to the rate of change of the current through the coil. The minus sign tells us that $\mathscr{E}_L$ is a back emf. It is customary to combine the product of constants (which for a solenoid is $\mu_0 A n^2 l$) and write this proportionality between the self-induced emf and the rate of current change as

$$\mathscr{E}_L = -L\frac{di}{dt}, \qquad (32\text{-}2)$$

where L is known as the self-inductance of the coil. As we learned in Chapter 31, the minus sign in the equation indicates that the emf acts to oppose the change in current. From Eq. 32-2 we see that when the inductance L is large, a large emf will be produced for a given rate of current change.

This combination of terms (such as area, length, and so on) that makes up the constant of proportionality, L, is only valid for a long solenoid. The terms will be different if the coil has a flat shape or if the inductor wire is wrapped around an iron core. In addition, since any electric circuit is basically a loop of some sort, all circuits have a certain amount of self-inductance even when no inductor is present. Self-inductance is usually negligible, but it can be significant when high-voltage circuits are switched on or off or when the circuit current oscillates at high frequencies. If we have a complicated geometry and cannot calculate inductance simply, the inductance L can be determined experimentally by measuring both the emf and the rate of change of current and taking the ratio of these quantities. Thus, for any geometry the **self-inductance** of an inductor or a circuit can be defined as the ratio of the induced emf to the rate of current change or

$$L \equiv -\frac{\mathscr{E}_L}{di/dt} \quad \text{(self-inductance defined).} \tag{32-3}$$

For any inductor having a self-inductance L, Eqs. 32-1 and 32-2 tell us that $N(d\phi^{mag}/dt) = L\, di/dt$. Thus we conclude that $\mathscr{E}_L = -Nd\ \Phi^{mag}/dt = -\ L\ di/dt$, so $Li = N\Phi^{mag}$, where N is the number of turns in the coil producing flux and i is the current in the coil producing the flux. The windings of the inductor are said to be *linked* by the shared flux, and the product $N\Phi^{mag}$ is called the *magnetic flux linkage*. This leads us to an alternate definition of inductance (which is equivalent to that given in Eq. 32-3):

$$L \equiv \frac{N\Phi^{mag}}{i} \quad \text{(alternative definition of self-inductance).} \tag{32-4}$$

The inductance L is thus a measure of the flux linkage produced by the inductor per unit of current.

Because the SI unit of magnetic flux is the tesla-square meter, the SI unit of inductance is the tesla square-meter per ampere ($\text{T}\cdot\text{m}^2/\text{A}$). We call this the **henry** (H), after American physicist Joseph Henry, the co-discoverer, with Faraday, of the law of induction. Thus,

$$1 \text{ henry} = 1 \text{ H} = 1 \text{T}\cdot\text{m}^2/\text{A}. \tag{32-5}$$

In any inductor (such as a flat coil, a solenoid, or a toroid) a self-induced emf appears whenever the current changes with time. The amount of the current has no influence on the amount of induced emf. Only the rate of change of the current matters.

You can find the *direction* of a self-induced emf from Lenz's law. The minus signs in Eqs. 32-2 and 32-3 indicate that—as the law states and Fig. 32-2 shows—the self-induced emf $\mathscr{E}_L$ has an orientation such that it opposes the change in current i.

Ideal Inductors

In Section 31-7 we saw that we cannot define an electric potential for an emf that is induced by a changing magnetic flux. This means that when a self-induced emf is produced, we cannot define an electric potential within the inductor itself. However, electric potentials can still be defined at points in a circuit that are not within the inductor—points where the electric fields are due to charge distributions.

Moreover, we can define a self-induced potential difference ΔV_L *across an inductor* (between its terminals, which we assume to be outside the region of changing flux). If the inductor is ideal so that its wire has negligible resistance, the amount of the measured voltage change ΔV_L is equal to the amount of the self-induced emf $\mathscr{E}_L$.

If, instead, the wire in the inductor has resistance R_L, we mentally separate the inductor into a resistance R_L (which we take to be outside the region of changing flux) and an ideal inductor of self-induced emf $\mathscr{E}_L$. As with a real battery of emf $\mathscr{E}$ and internal resistance R, the potential difference across the terminals of a real inductor then differs from the emf. Unless otherwise indicated, we assume here that inductors are ideal.

READING EXERCISE 32-1: (a) What happens to the inductance of a solenoid if: (a) the number of turns per unit length doubles, (b) the cross-sectional area enclosed by the windings doubles? ■

READING EXERCISE 32-2: The figure shows an emf $\mathscr{E}_L$ induced in a coil. Which of the following can describe the current through the coil: (a) constant and rightward, (b) constant and leftward, (c) increasing and rightward, (d) decreasing and rightward, (e) increasing and leftward, (f) decreasing and leftward? ■

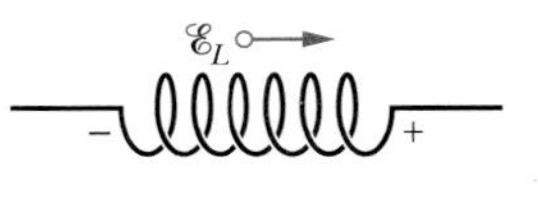

32-3 Mutual Induction

In this section we return to the case of two interacting coils, which we started discussing in the previous section. We saw earlier that if two coils are close together as in Fig. 32-4 (or Fig. 31-10), a steady current i in one coil will set up a magnetic flux Φ^{mag} at the other coil (*linking* the other coil). If we change the current, i, in the first coil with time, an emf $\mathscr{E}$ given by Faraday's law ($\mathscr{E} = -N d\,\Phi^{\text{mag}}/dt$) will be induced in the second coil. We called this process **mutual induction,** to suggest the mutual interaction of the two coils and to distinguish it from *self-induction,* in which only one coil is involved.

Let us look at mutual induction quantitatively. For any inductor having a self-inductance L, Eq. 32-3 tells us that

$$L \equiv -\frac{\mathscr{E}_L}{di/dt}$$

where i is the current in the coil producing the flux. Figure 32-4a shows two circular coils near each other that share a common central axis. Assume there is a steady current i_1 in coil 1, produced by the battery in the external circuit. This current creates a magnetic field represented by the lines of $\vec{B}_1$ in the figure. Coil 2 is connected to a

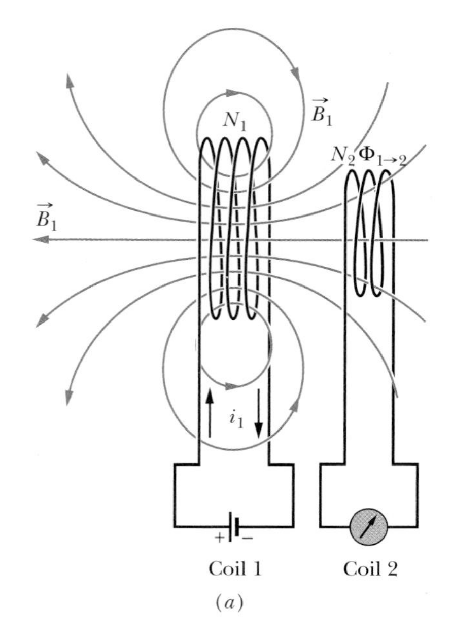

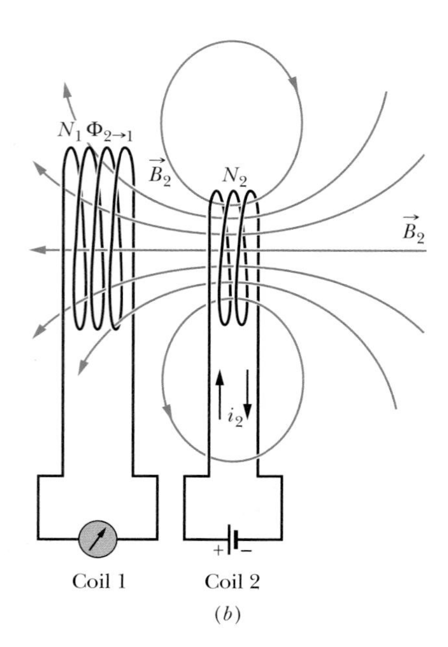

FIGURE 32-4 ■ Mutual induction. (a) If the current in coil 1 changes, an emf will be induced in coil 2. (b) If the current in coil 2 changes, an emf will be induced in coil 1.

sensitive meter but contains no battery. A magnetic flux $\Phi_{1\to 2}$ (the flux associated with the current in coil 1 that passes through coil 2) links the N_2 turns of coil 2.

Suppose that by external means we cause i_1 to vary with time. Then by analogy to the definition of self-inductance, we can write a mutual induction equation that is analogous to Eq. 32-2,

$$\mathscr{E}_2 = -M_{1\to 2}\frac{di_1}{dt}.$$

This leads us to define the mutual inductance $M_{1\to 2}$ of coil 2 due to coil 1 as

$$M_{1\to 2} \equiv -\frac{\mathscr{E}_2}{di_1/dt} \quad \text{(mutual inductance defined).} \tag{32-6}$$

Once again we can formulate an alternate definition of mutual induction using the relationship between flux linkage in coil 2 and the current in coil 1, which is $M_{1\to 2}i_1 = N_2\Phi_{1\to 2}$. The factor N_2 is the number of turns in coil 2 and the factor $\Phi_{1\to 2}$ is the magnetic flux present inside coil 2 due to coil 1. This allows us to define mutual inductance as

$$M_{1\to 2} \equiv \frac{N_2\Phi_{1\to 2}}{i_1} \quad \text{(alternate definition of mutual inductance).} \tag{32-7}$$

If we take the time derivative of all terms in the expression $M_{1\to 2}i_1 = N_2\Phi_{1\to 2}$ we can write

$$\mathscr{E}_2 = -M_{1\to 2}\frac{di_1}{dt} = -N_2\frac{d\Phi_{1\to 2}}{dt}. \tag{32-8}$$

According to Faraday's law, the right side of this equation is just the amount of the emf $\mathscr{E}_2$ appearing in coil 2 due to the changing current in coil 1. As usual, the minus sign reminds us that induced emf acts to oppose the change in current.

Let us now interchange the roles of coils 1 and 2, as in Fig. 32-4*b*; that is, we set up a current i_2 in coil 2 by means of a battery, and this produces a magnetic flux $\Phi_{2\to 1}$ that links coil 1. If we change i_2 with time, we have, by the arguments given above,

$$\mathscr{E}_1 = -M_{2\to 1}\frac{di_2}{dt} = -N_1\frac{d\Phi_{2\to 1}}{dt}. \tag{32-9}$$

Thus, we see that the emf induced in either coil is proportional to the rate of change of current in the other coil. The proportionality constants $M_{1\to 2}$ and $M_{2\to 1}$ seem to be different. We assert, without proof, that they are in fact the same so that no subscripts are needed. (This conclusion is true but is not obvious.) Thus, we have

$$M_{1\to 2} = M_{2\to 1} = M, \tag{32-10}$$

and we can rewrite Eqs. 32-9 and 32-10 as

$$\mathscr{E}_2 = -M\frac{di_1}{dt} \tag{32-11}$$

and

$$\mathscr{E}_1 = -M\frac{di_2}{dt}. \tag{32-12}$$

The induction is indeed mutual. The SI unit for M (as for L) is the henry.

TOUCHSTONE EXAMPLE 32-1: Two Coupled Coils

Figure 32-5 shows two circular close-packed coils, the smaller (radius R_2, with N_2 turns) being coaxial with the larger (radius R_1, with N_1 turns) and in the same plane.

(a) Derive an expression for the mutual inductance M for this arrangement of these two coils, assuming that $R_1 \gg R_2$.

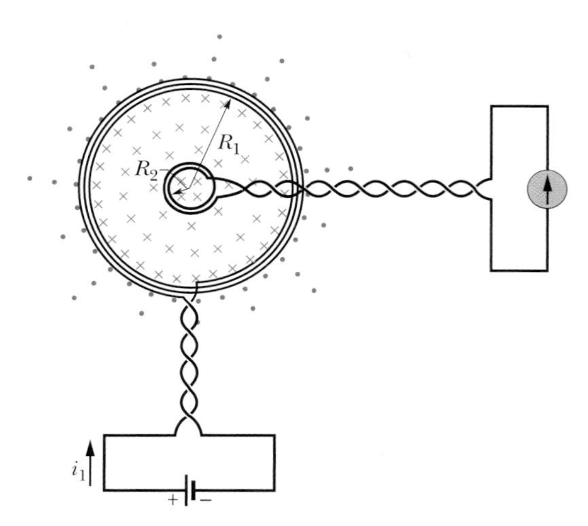

FIGURE 32-5 ■ A small coil is located at the center of a large coil. The mutual inductance of the coils can be determined by sending current i_1 through the large coil.

SOLUTION ■ The **Key Idea** here is that the mutual inductance M for these coils is the ratio of the flux linkage ($N\Phi$) through one coil to the current i in the other coil, which produces that flux linkage. Thus, we need to assume that currents exist in the coils; then we need to calculate the flux linkage in one of the coils.

The magnetic field through the larger coil due to the smaller coil is nonuniform in both magnitude and direction, so the flux in the larger coil due to the smaller coil is nonuniform and difficult to calculate. However, the smaller coil is small enough for us to assume that the magnetic field through it due to the larger coil is approximately uniform. Thus, the flux in it due to the larger coil is also approximately uniform. Hence, to find M we shall assume a current i_1 in the larger coil and calculate the flux linkage $N_2\Phi_{1\to2}$ in the smaller coil:

$$M_{1\to2} = \frac{N_2\Phi_{1\to2}}{i_1}. \tag{32-13}$$

A second **Key Idea** is that the flux $\Phi_{1\to2}$ through each turn of the smaller coil is, from Eq. 31-1,

$$\Phi_{1\to2} = B_1A_2,$$

where B_1 is the magnitude of the magnetic field at points within the small coil due to the larger coil, and $A_2(=\pi R_2^2)$ is the area enclosed by the coil. Thus, the flux linkage in the smaller coil (with its N_2 turns) is

$$N_2\Phi_{1\to2} = N_2B_1A_2. \tag{32-14}$$

A third **Key Idea** is that to find B_1 at points within the smaller coil, we can use Eq. 30-28, with z set to 0 because the smaller coil is in the plane of the larger coil. That equation tells us that each turn of the larger coil produces a magnetic field of magnitude $\mu_0 i_1/2R_1$ at points within the smaller coil. Thus, the larger coil (with its N_1 turns) produces a total magnetic field of magnitude

$$B_1 = N_1\frac{\mu_0 i_1}{2R_1} \tag{32-15}$$

at points within the smaller coil.

Substituting Eq. 32-15 for B_1 and πR_2^2 for A_2 in Eq. 32-14 yields

$$N_2\Phi_{1\to2} = \frac{\pi\mu_0 N_1 N_2 R_2^2 i_1}{2R_1}.$$

Substituting this result into Eq. 32-7, and using Eq. 32-10, we find

$$M = M_{1\to2} = \frac{N_2\Phi_{1\to2}}{i_1} = \frac{\pi\mu_0 N_1 N_2 R_2^2}{2R_1}. \qquad \text{(Answer)} \tag{32-16}$$

Just as capacitance does not depend on the amount of charge on capacitor plates, mutual inductance, M, does not depend on the current in the coils.

(b) What is the value of M for $N_1 = N_2 = 1200$ turns, $R_2 = 1.1$ cm, and $R_1 = 15$ cm?

SOLUTION ■ Equation 32-16 yields

$$M = \frac{(\pi)(4\pi \times 10^{-7}\ \text{H/m})(1200)(1200)(0.011\ \text{m})^2}{(2)(0.15\ \text{m})}$$
$$= 2.29 \times 10^{-3}\ \text{H} \approx 2.3\ \text{mH}. \qquad \text{(Answer)}$$

Consider the situation if we reverse the roles of the two coils—that is, if we produce a current i_2 in the smaller coil and try to calculate M from Eq. 32-7 in the form

$$M_{2\to1} = \frac{N_2\Phi_{2\to1}}{i_2}.$$

The calculation of $\Phi_{2\to1}$ (the nonuniform flux of the smaller coil's magnetic field encompassed by the larger coil) is not simple. If we were to do the calculation numerically using a computer, we would find M to be 2.3 mH, as above! This emphasizes that Eq. 32-10 ($M_{1\to2} = M_{2\to1} = M$) is not obvious.

32-4 *RL* Circuits (With Ideal Inductors)

In Section 28-9 we saw that if we suddenly switch an emf $\mathscr{E}$ on in a series circuit containing a resistor R and a capacitor C, the charge on the capacitor q does not build up immediately to its final equilibrium value $C\mathscr{E}$ but approaches it in an exponential fashion:

$$q = C\mathscr{E}(1 - e^{t/\tau_C}). \tag{32-17}$$

The rate at which the charge builds up is determined by the capacitive time constant τ_C, defined in Eq. 28-42 as

$$\tau_C = RC. \tag{32-18}$$

If we suddenly remove the emf from this same circuit, the charge does not immediately fall to zero but approaches zero in an exponential fashion:

$$q = q_0 e^{-t/\tau_C}. \tag{32-19}$$

The time constant τ_C describes the fall of the charge as well as its rise and q_0 is the initial charge on the capacitor.

An analogous slowing of the rise (or fall) of the current occurs if we introduce an emf $\mathscr{E}$ into (or remove it from) a single-loop circuit containing a resistor R and an inductor L. We assume the inductor is ideal and has a resistance R_L that is much less than R. When the switch S in Fig. 32-6 is closed on a, for example, the current in the resistor starts to rise. If the inductor were not present, the current would rise rapidly to a steady value $\mathscr{E}/R$. Because of the inductor, however, a self-induced emf $\mathscr{E}_L$ appears in the circuit. As predicted from Lenz's law, this emf opposes the rise of the current. This means that it opposes the battery emf $\mathscr{E}$ in polarity. Thus the current in the resistor responds to the *difference* between two emfs, a constant one $\mathscr{E}$ due to the battery, and a variable one $\mathscr{E}_L(= -L\, di/dt)$ due to self-induction. As long as $\mathscr{E}_L$ is present, the current in the resistor will be less than $\mathscr{E}/R$. As time goes on, the rate at which the current increases becomes less rapid and the amount of the self-induced emf, which is proportional to di/dt, becomes smaller. Thus, the current in the circuit approaches $\mathscr{E}/R$ asymptotically.

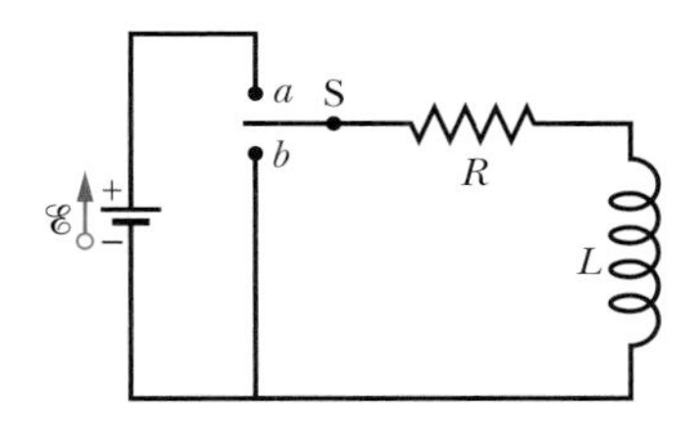

FIGURE 32-6 ■ An *RL* circuit. When switch S is closed on a, the current rises and approaches a limiting value of $\mathscr{E}/R$.

We can generalize these results as follows: When a switch is opened or closed in a dc circuit, an inductor initially acts to oppose changes in the current through it. A long time later, it acts like ordinary connecting wire that has some resistance R_L.

Now let us analyze the situation quantitatively. With the switch S in Fig. 32-6 thrown to a, the circuit is equivalent to that of Fig. 32-7. Let us apply the loop rule, starting at point x in this figure and moving clockwise around the loop along with current i.

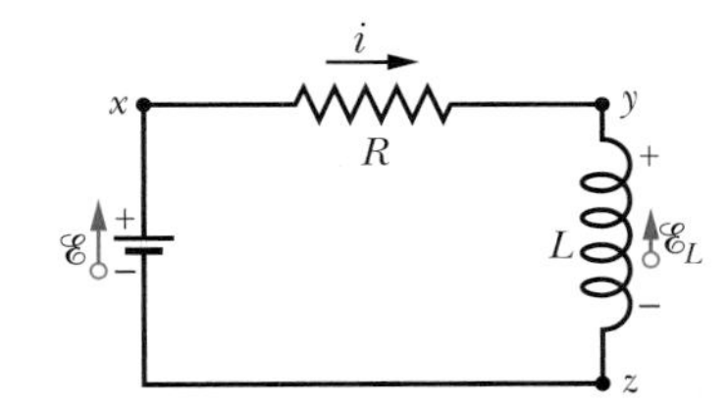

FIGURE 32-7 ■ The circuit of Fig. 32-6 with the switch closed on a. We apply the loop rule for circuits clockwise, starting at x.

1. *Resistor*. Because we move through the resistor in the direction of current i, the electric potential decreases by iR. Thus, as we move from point x to point y where these points lie *outside* the inductor, we encounter a potential change of $-iR$.

2. *Inductor*. Because current i is changing, there is a self-induced emf $\mathscr{E}_L$ in the inductor. The amount of $\mathscr{E}_L$ is given by Eq. 32-2 as $L\, di/dt$. The direction of $\mathscr{E}_L$ is upward in Fig. 32-7 because current i is downward through the inductor and increasing. Thus, as we move from point y to point z, opposite the direction of $\mathscr{E}_L$, we encounter a potential change of $-L\, di/dt$.

3. *Battery*. As we move from point z back to starting point x, we encounter a potential change of $+\mathscr{E}$ due to the battery's emf.

Thus, the loop rule gives us

$$-iR - L\frac{di}{dt} + \mathscr{E} = 0$$

or

$$L\frac{di}{dt} + Ri = \mathscr{E} \quad (RL \text{ circuit}). \tag{32-20}$$

Equation 32-20 is a differential equation involving the variable i and its first derivative di/dt. To solve it, we seek the function $i(t)$ such that when $i(t)$ and its first derivative are substituted in Eq. 32-20, the equation is satisfied and the initial condition $i(0) = 0$ A is satisfied.

Equation 32-20 and its initial condition are of exactly the form of Eq. 28-38 for an RC circuit, with i replacing q, L replacing R, and R replacing $1/C$. The solution of Eq. 32-20 must then be of exactly the form of Eq. 28-39 with the same replacements. That solution is

$$i = \frac{\mathscr{E}}{R}(1 - e^{-(R/L)t}), \tag{32-21}$$

which we can rewrite as

$$i = \frac{\mathscr{E}}{R}(1 - e^{-t/\tau_L}) \quad \text{(rise of current)}. \tag{32-22}$$

Here τ_L, the inductive time constant, is given by

$$\tau_L = \frac{L}{R} \quad \text{(time constant)}. \tag{32-23}$$

What happens to the current described in Eq. 32-22 between the time the switch is closed (at time $t = 0$ s) and a later time ($t \to \infty$)? If we substitute $t = 0$ s into Eq. 32-22, the exponential becomes $e^{-0} = 1$. Thus, Eq. 32-22 tells us that the current is initially $i = 0$ A, as expected. Next, if we let t go to infinity, then the exponential goes to $e^{-\infty} = 0$. Thus, Eq. 32-22 tells us that the current goes to its equilibrium value of $\mathscr{E}/R$.

We can also examine the potential differences in the circuit. The graphs of Fig. 32-8 show experimental data describing how the potential differences $|\Delta V_R| = iR$ across a resistor and $|\Delta V_L| = L\,di/dt$ across an inductor vary with time for particular

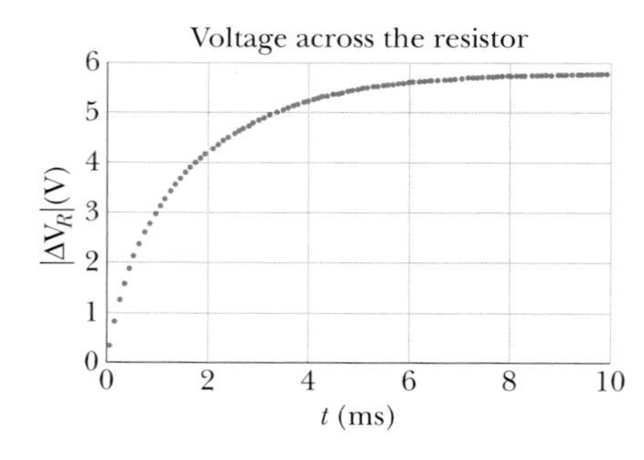

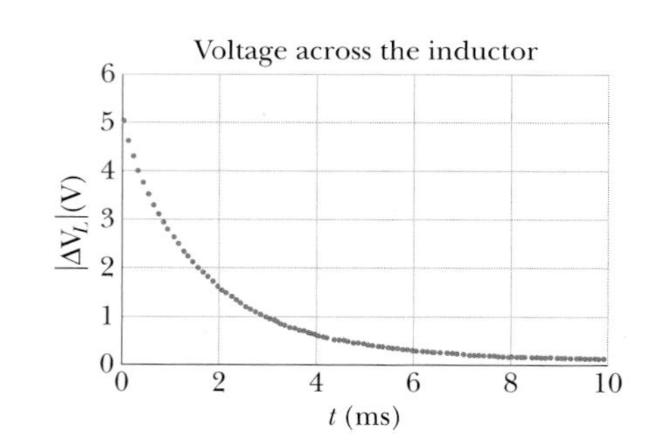

FIGURE 32-8 ■ A computer data acquisition system is used to record the time variation of potential differences (a) ΔV_R across the resistor in Fig. 32-7 and (b) ΔV_L across the inductor in that circuit. The data were obtained at 10 000 samples per second for $R = 9830\ \Omega$, $L \approx 20$ H, and $\mathscr{E} = 5.88$ V. The inductor has a direct current resistance of 167 Ω, so it is not ideal. The data show some different characteristics than those predicted by Eqs. 32-22 and 32-23.

values of $\mathscr{E}$, L, and R. Compare this figure carefully with the corresponding figure for an RC circuit (Fig. 28-23).

To show that the quantity $\tau_L(=L/R)$ has the dimension of time, we convert from henries per ohm as follows:

$$1\frac{\text{H}}{\Omega} = 1\frac{\text{H}}{\Omega}\left(\frac{1\ \text{V}\cdot\text{s}}{1\ \text{H}\cdot\text{A}}\right)\left(\frac{1\ \Omega\cdot\text{A}}{1\ \text{V}}\right) = 1\ \text{s}.$$

The first quantity in parentheses is a conversion factor based on Eq. 32-20, and the second one is a conversion factor based on the relation $\Delta V = iR$.

The physical significance of the time constant follows from Eq. 32-21. If we put $t = \tau_L = L/R$ in this equation, it reduces to

$$i = \frac{\mathscr{E}}{R}(1 - e^{-1}) = 0.63\frac{\mathscr{E}}{R}. \tag{32-24}$$

Thus, the time constant τ_L is the time it takes the current in the circuit to reach about 63% of its final equilibrium value $\mathscr{E}/R$. Since the potential difference ΔV_R across the resistor is proportional to the current i, a graph of the increasing current versus time has the same shape as that of ΔV_R in Fig. 32-8*a*.

If the switch S in Fig. 32-6 is closed on *a* long enough for the equilibrium current $\mathscr{E}/R$ to be established and then is thrown to *b*, the effect will be to remove the battery from the circuit. (The connection to *b* must actually be made an instant before the connection to *a* is broken. A switch that does this is called a *make-before-break* switch.)

With the battery gone, the current through the resistor will decrease. However, because of the inductor it cannot drop immediately to zero but must decay to zero over time. The differential equation that governs the decay can be found by putting $\mathscr{E} = 0$ in the RL circuit voltage loop equation (Eq. 32-20):

$$L\frac{di}{dt} + iR = 0. \tag{32-25}$$

By analogy with Eqs. 28-44 and 28-45, the solution of this differential equation that satisfies the initial condition $i(0) = i_0 = \mathscr{E}/R$ is

$$i = \frac{\mathscr{E}}{R}e^{-t/\tau_L} = i_0 e^{-t/\tau_L} \quad \text{(decay of current).} \tag{32-26}$$

We see that both current rise (Eq. 32-21) and current decay (Eq. 32-26) in an RL circuit are governed by the same inductive time constant, τ_L.

We have used i_0 in Eq. 32-26 to represent the current at time $t = 0$. In our case that happened to be $\mathscr{E}/R$, but it could be any other initial value.

TOUCHSTONE EXAMPLE 32-2: Two Inductors and Three Resistors

Figure 32-9*a* shows a circuit that contains three identical resistors with resistance $R = 9.0\ \Omega$, two identical ideal inductors with inductance $L = 2.0$ mH, and an ideal battery with emf $\mathscr{E} = 18$ V.

(a) What is the current i through the battery just after the switch is closed?

SOLUTION ■ The **Key Idea** here is that just after the switch is closed, the inductor acts to oppose a change in the current through it. Because the current through each inductor is zero before the switch is closed, it will also be zero just afterward. Thus, immediately after the switch is closed, the inductors act as broken wires, as indicated in Fig. 32-9*b*. We then have a single-loop circuit

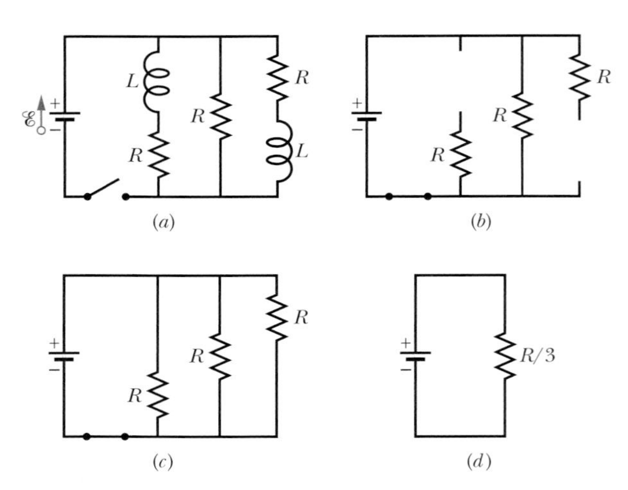

FIGURE 32-9 ■ (*a*) A multiloop *RL* circuit with an open switch. (*b*) The equivalent circuit just after the switch has been closed. (*c*) The equivalent circuit a long time later. (*d*) The single-loop circuit that is equivalent to circuit (*c*).

for which the loop rule gives us

$$\mathscr{E} - iR = 0.$$

Substituting given data, we find that

$$i = \frac{\mathscr{E}}{R} = \frac{18\text{ V}}{9.0\ \Omega} = 2.0\text{ A}. \qquad \text{(Answer)}$$

(b) What is the current i through the battery long after the switch has been closed?

SOLUTION ■ The **Key Idea** here is that long after the switch has been closed, the currents in the circuit have reached their equilibrium values, and the inductors act as simple connecting wires, as indicated in Fig. 32-9*c*. We then have a circuit with three identical resistors in parallel; from Eq. 27-12, their equivalent resistance is $R^{eq} = R/3 = (9.0\ \Omega)/3 = 3.0\ \Omega$. The equivalent circuit shown in Fig. 32-9*d* then yields the loop equation $\mathscr{E} - iR^{eq} = 0.0$ V, or

$$i = \frac{\mathscr{E}}{R^{eq}} = \frac{18\text{ V}}{3.0\ \Omega} = 6.0\text{ A}. \qquad \text{(Answer)}$$

32-5 Inductors, Transformers, and Electric Power

In most countries the electrical power used in homes and industries involves voltages and currents that change over time periodically, often sinusoidally. Such power is usually referred to as alternating current or ac electricity. Alternating electrical power is usually generated using induction. An **ac generator** simply consists of a magnet or electromagnet rotating inside an inductor coil or, alternatively, an inductor coil rotating in a magnetic field like that shown in Fig. 32-10.

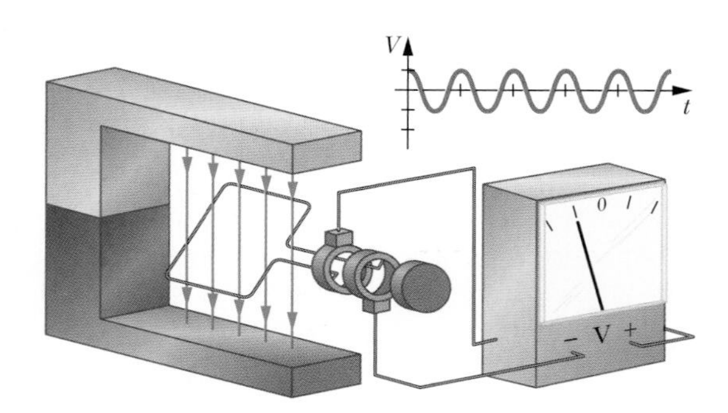

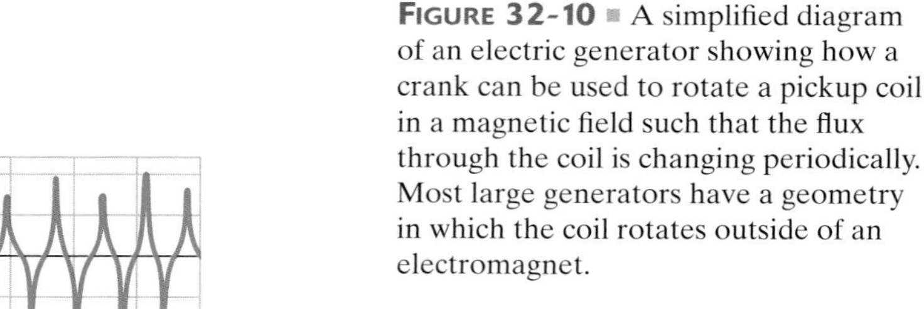

FIGURE 32-10 ■ A simplified diagram of an electric generator showing how a crank can be used to rotate a pickup coil in a magnetic field such that the flux through the coil is changing periodically. Most large generators have a geometry in which the coil rotates outside of an electromagnet.

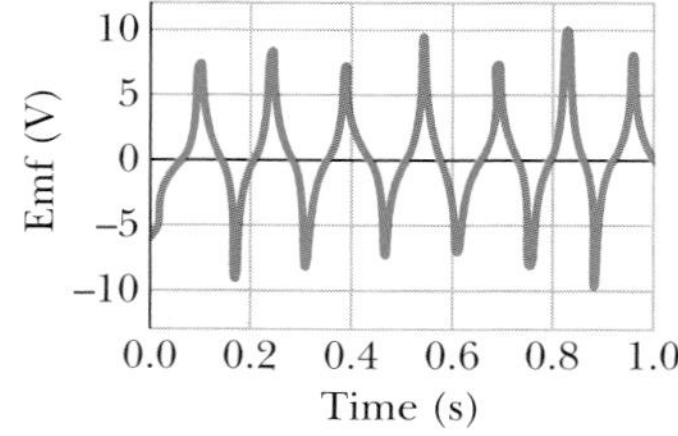

FIGURE 32-11 ■ A periodic emf is induced in a coil that is being turned by a hand crank in the presence of a magnet. The generator is similar to that shown in Fig. 32-10. If the coil were less bulky and the $\vec{B}$-field were more uniform, the emf would vary sinusoidally when the crank is turned steadily.

Generators don't care what form of energy is used to cause the rotation (Fig. 32-11). The shaft can turn when steam produced by a coal-fired or nuclear power plant pushes on propeller-like blades. In hydroelectric plants, falling water can provide the rotational energy. Since the potential difference and current in a generator vary sinusoidally, the voltages and currents are reported as root mean square (or rms) values. The use of rms values is explained in the next chapter, where we deal with alternating-current circuits in more detail.

The Role of Transformers

Generators typically produce power at low voltage, but it is important to transmit this power from generation stations to consumers with minimum energy loss. It turns out that the losses are minimized when ac power is transmitted at high voltages. The reason has to do with how power loss is related to current and voltage. The total power generated is given by Eq. 26-10 as $P^{gen} = i^{gen}\Delta V^{gen}$. If this power is transmitted to consumers with an rms current i^{gen} flowing over long distances, then the power lost in heating transmission lines is given by

$$P^{lost} = (i^{gen})^2 R \quad \text{(power lost in transmission)}, \tag{32-27}$$

where R is the total resistance of the wires that make up the transmission lines. The power available to consumers is then $P^{gen} - P^{lost}$. Although we can't get something for nothing, it is obvious from Eq. 32-27 that reorganizing the generated power so that it is transmitted at high voltage and low current would greatly reduce the transmission losses. In other words, we would like to achieve

$$P^{gen} = i^{gen}\Delta V^{gen} = i^{trans}\Delta V^{trans}, \tag{32-28}$$

where $\Delta V^{trans} \gg \Delta V^{gen}$ so that $i^{trans} \ll i^{gen}$.

As an example, consider the 735 kV line used to transmit electric energy from the La Grande 2 hydroelectric plant in Quebec to Montreal, 1000 km away. Suppose that the current is 500 A. Then from Eq. 32-28, energy is supplied at the average rate

$$P^{gen} = i^{gen}\Delta V^{gen} = (7.35 \times 10^5\,\text{V})(500\,\text{A}) = 368\,\text{MW}.$$

The resistance of the transmission line is about 0.220 Ω/km. Thus, there is a total resistance of about 220 Ω for the 1000 km stretch. Energy is dissipated due to that resistance at a rate of about

$$P^{lost} = (i^{gen})^2 R = (500\,\text{A})^2(220\,\Omega) = 55.0\,\text{MW},$$

which is nearly 15% of the supply rate.

Imagine what would happen if we could halve the current and double the voltage. Energy would be supplied by the plant at the same average rate of 368 MW as before, but now energy would be dissipated at the much lower rate of about

$$P^{lost} = (i^{gen})^2 R = (250\,\text{A})^2(220\,\Omega) = 13.8\,\text{MW},$$

This rate of energy loss is *only 4% of the supply rate*. Hence the general energy transmission rule: Transmit at the highest possible voltage and the lowest possible current. There is an upper limit to the voltage that can be used. If the voltage gets too high, the power line insulation and the surrounding air will not be able to prevent the current from passing through them and leaking to the ground.

The Ideal Transformer

The *ideal transformer* in Fig. 32-12 consists of two coils, a *primary* and a *secondary*. These coils have different numbers of turns and are wound around the same iron core. The coils experience mutual induction. The iron core concentrates the flux so that it is the same in both coils. (We will discuss the role iron plays in Section 32-7 on ferromagnetism.) In use, the **primary coil,** of N_p turns, is connected to an alternating-current generator whose emf $\mathscr{E}$ at any time t is given by

$$\mathscr{E} = \mathscr{E}^{max}\sin\omega t. \tag{32-29}$$

The **secondary coil,** of N_s turns, is connected to load resistance R, but its circuit is an open circuit as long as switch S is open (which we assume for the present). Thus, there

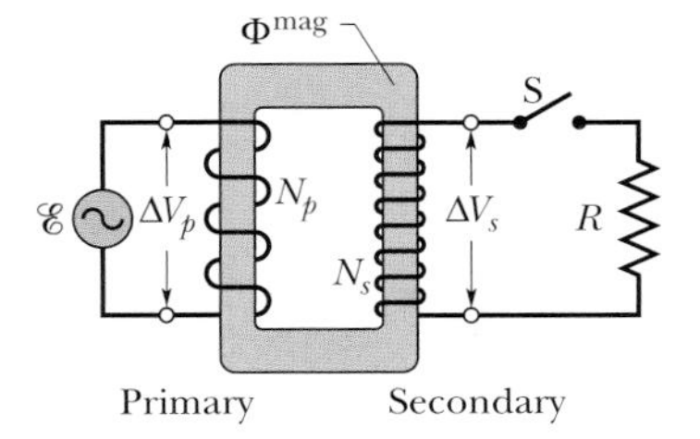

FIGURE 32-12 ■ An ideal transformer (two coils wound on an iron core) in a basic transformer circuit. An ac generator produces current in the coil at the left (the *primary*). The coil at the right (the *secondary*) is connected to the resistive load R when switch S is closed.

can be no current through the secondary coil. We assume further for this ideal transformer that the resistances of the primary and secondary coils (or **windings**) are negligible as are energy losses in the iron core. Well-designed, high-capacity transformers can have energy losses as low as 1%, so our assumptions are reasonable.

For the assumed conditions, the primary winding (or *primary*) is a pure inductance that carries a small alternating primary current i_p. This current induces an alternating magnetic flux Φ^{mag} in the iron core. Because the core extends through the secondary winding (or *secondary*), this induced flux also extends through the turns of the secondary. At any given time the flux in the primary and secondary coils are the same. Therefore, Faraday's law of induction (Eq. 31-7) tells us that the amount of the induced emf per turn, denoted as $\mathscr{E}_{\text{turn}}$ is the same for both the primary and the secondary coils. Also, the voltage ΔV_p across the primary is equal to the emf induced in the primary, and the voltage ΔV_s across the secondary is equal to the emf induced in the secondary. Thus, we can write

$$\mathscr{E}_{\text{turn}} = \frac{d\Phi^{\text{mag}}}{dt} = \frac{\Delta V_p}{N_p} = \frac{\Delta V_s}{N_s},$$

and thus,

$$\Delta V_s = \Delta V_p \frac{N_s}{N_p} \quad \text{(transformation of voltage).} \tag{32-30}$$

If $N_s > N_p$, the transformer is called a **step-up transformer** because it steps the primary's voltage ΔV_p up to a higher voltage ΔV_s. Alternatively, if $N_s < N_p$, the device is a **step-down transformer.**

So far, with switch S open, no energy is transferred from the generator to the rest of the circuit. Now let us close S to connect the secondary to the resistive load R. (In general, the load would also contain inductive and capacitive elements, but here we neglect the capacitance.) We find that now energy is transferred from the generator. To see why, let's explore what happens when we close switch S.

1. An alternating current i_s appears in the secondary circuit, with corresponding energy dissipation rate $i_s^2 R = (\Delta V_s^2)/R$ in the resistive load. Since the emf produced in the secondary coil is a back emf that opposes the direction of the change in current in the primary, the secondary current is out of phase with the primary current.
2. This current produces its own alternating magnetic flux in the iron core, and this flux induces (from Faraday's law and Lenz's law) an opposing emf in the primary windings.
3. The voltage ΔV_p of the primary, however, cannot change in response to this opposing emf because it must always be equal to the emf $\mathscr{E}$ that is provided by the generator; closing switch S cannot change this fact.

In order to relate i_s to i_p, we can apply the principle of conservation of energy. For the ideal transformer without losses in the magnetic core, the power drawn from the primary is equal to the power transferred to the secondary (via the alternating magnetic field linking the two coils). Conservation of energy requires that

$$i_p \Delta V_p = i_s \Delta V_s. \tag{32-31}$$

Substituting for ΔV_s from Eq. 32-30, we find that

$$i_s = i_p \frac{N_p}{N_s} \quad \text{(transformation of currents).} \tag{32-32}$$

This equation tells us that the amount of the current i_s in the secondary can be greater than, less than, or the same as the amount of current i_p in the primary, depending on the *ratio of turns (or loops) in the coils given by* N_p/N_s.

Current i_p appears in the primary circuit because of the resistive load R in the secondary circuit. To find i_p, we substitute $i_s = \Delta V_s/R$ into Eq. 32-32 and then we substitute for ΔV_s from Eq. 32-30. We find

$$i_p = \frac{1}{R}\left(\frac{N_s}{N_p}\right)^2 \Delta V_p. \tag{32-33}$$

This equation has the form $i_p = \Delta V_p/R_{\text{eq}}$, where equivalent resistance R_{eq} is

$$R_{\text{eq}} = \left(\frac{N_p}{N_s}\right)^2 R. \tag{32-34}$$

Here R is the actual resistance in the secondary circuit and R_{eq} is the value of the load resistance as "seen" by the generator. The generator produces the current i_p and voltage ΔV_p as if it were connected to a resistance R_{eq}.

Impedance Matching

Equation 32-34 suggests still another function for the transformer. For maximum transfer of energy from an emf device to a resistive load, the resistance of the emf device and the resistance of the load must be equal. The same relation holds for ac circuits (discussed in Chapter 33) except that the *impedance* (rather than just the resistance) of the generator must be matched to that of the load. Often this condition is not met. For example, in a music-playing system, the amplifier can have high impedance and the speaker set have low impedance. We can match the impedances of the two devices by coupling them through a transformer with a suitable turns ratio N_p/N_s.

READING EXERCISE 32-3: An alternating-current emf device has a smaller resistance than that of the resistive load; to increase the transfer of energy from the device to the load, a transformer will be connected between the two. (a) Should N_s be greater than or less than N_p? (b) Will that make it a step-up or step-down transformer? ■

32-6 Magnetic Materials—An Introduction

Today, magnets and magnetic materials are ubiquitous. In addition to naturally magnetic lodestones, magnets are also in VCRs, audiocassettes, credit cards, electronic speakers, audio headsets, and even the inks in paper money. In fact, some breakfast cereals that are "iron fortified" contain small bits of magnetic materials (you can collect them from a slurry of cereal and water with a magnet). In this section we are interested in understanding more about why so-called bulk matter, made of billions upon billions of individual atoms, has magnetic properties.

Characteristics of Magnetic Materials

When we speak of magnetism in everyday conversation, we usually have a mental picture of a bar magnet, a disk magnet (probably clinging to a refrigerator door), or even a tiny compass needle. That is, we picture a *ferromagnetic* material made of iron having strong, permanent magnetism. Although most bulk matter does not behave like the familiar iron bar magnets, it turns out that almost all bulk materials have some magnetic behaviors. There are three general types of magnetism: ferromagnetism, paramagnetism, and diamagnetism.

1. ***Ferromagnetism*** is present if a material produces a strong magnetic field of its own in the presence of an external field, and if its magnetic field partially persists after the external field is removed. We usually use the term *ferromagnetic material,* and also the common term *magnetic material*, to refer to materials that exhibit primarily ferromagnetism. Iron, nickel, and cobalt (and compounds and alloys of these elements) are ferromagnetic.
2. ***Paramagnetism*** is present if a material that is placed in an external magnetic field is attracted to the region of greater magnetic field and produces a magnetic field of its own—but only while it is in the presence of the external field. The term *paramagnetic material* usually refers to materials that exhibit primarily paramagnetism. This type of magnetism is exhibited by materials such as liquid oxygen and aluminum as well as transition elements, rare earth elements, and actinide elements (see Appendix G).
3. ***Diamagnetism*** is present if a material that is placed near a magnet is repelled from the region of greater magnetic field. This is opposite to the behavior of the other two types of magnetism. Diamagnetism is exhibited by all common materials, but it is so weak that it is masked if the material exhibits magnetism of either of the other two types. Thus, the term *diamagnetic material* refers to materials that only exhibit diamagnetism. Metals such as bismuth, copper, gold, silver, and lead, as well as many nonmetals such as water and most organic compounds, are diamagnetic. Because people and other animals are made largely of water and organic compounds, they are diamagnetic too.

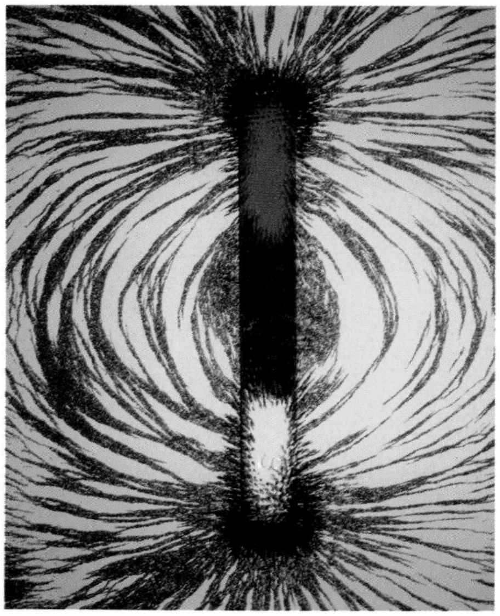

What causes magnetism? Why are there three types of magnetism? We now believe that magnetism is caused by tiny magnetic dipoles that are intrinsic to the atoms contained in all materials. For this reason, understanding the characteristics of magnetic dipoles is essential to understanding the behavior of magnetic materials. We will conclude this section with a discussion of magnetic dipoles, and then in the next two sections we will explore how the characteristics of atomic magnetic dipoles help us understand the three types of magnetism.

FIGURE 32-13 ■ A bar magnet is a magnetic dipole. The orientations of the iron filings suggest the direction of magnetic field lines.

Characteristics of Magnetic Dipoles

Both the bar magnets with which we are familiar and small coils of wire carrying current are magnetic dipoles. Let's review some of the characteristics of magnetic dipoles that we have already discussed. A magnetic dipole:

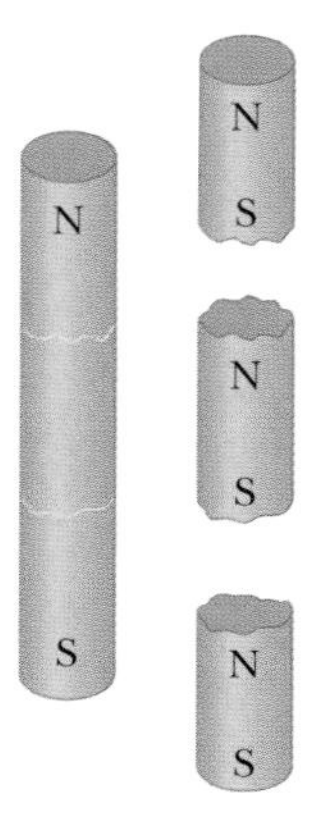

FIGURE 32-14 ■ If you break a bar magnet, each fragment becomes a smaller magnet, with its own north and south poles. It is impossible to break a fragment into separate north and south poles.

- Has a magnetic field pattern associated with it similar to that of an electric dipole (like that shown in Fig. 32-13 or that described by the equations derived in Sections 23-6 and 30-7).
- *Always has two poles,* which we have chosen to call north (seeking) and south (seeking) because of the way they behave when placed in the Earth's magnetic field, as shown in Fig. 32-14 (see Section 29-3 for a review).
- Can be described by a magnetic dipole moment $\vec{\mu}$, which is a vector quantity whose *magnitude* tells us how *strong* the magnetic field associated with the dipole is and whose direction tells how the field pattern is oriented. The orientation is along the axis of a bar magnet pointing from its south pole and to its north pole or perpendicular to the plane of a current-carrying coil with a direction determined by the right-hand rule (Section 29-10). *Note:* Our use of conventional notation is unfortunate here. The magnetic moment $\vec{\mu}$ should not be

confused with the permeability constant μ_0 or μ that sometimes appears in the same equation.

- Will attempt to align its magnetic dipole moment with an external magnetic field, $\vec{B}$, because the dipole experiences a torque, $\vec{\tau}$, given by $\vec{\tau} = \vec{\mu} \times \vec{B}$ (Section 30-7).
- Has a potential energy U in an external magnetic field given by $U = -\vec{\mu} \cdot \vec{B}$, so that the dipole's potential energy is a minimum when its dipole moment is aligned with an external magnetic field.

Magnetism in Atoms

We believe that the combined effect of tiny magnetic dipole moments in atoms are responsible for all magnetic interactions in bulk matter. Before we attempt to explain why different materials exhibit certain types of magnetism, we need to discuss what is known about atomic magnetism.

The focus of this book is on classical physics. However, understanding atomic phenomena requires some familiarity with quantum physics, which is in general beyond the scope of this book. So, we will present some basic ideas of quantum physics that apply to atomic magnetism without discussing the existing body of experimental evidence.

So far in our classical treatment of magnetism we have already identified two sources of magnetic dipole fields: (1) electric charges that create a current if they move in a loop and (2) magnetic dipoles consisting of a bar or rod of magnetized iron. Also, some effects of atomic magnetism can be explained using a classical model that identifies two types of atomic magnetic dipoles—orbital and spin. First, if we think of electrons as "orbiting" around a nucleus, then an orbit is a current loop with an orbital magnetic moment. Second, we think of the electron as having an intrinsic magnetic dipole moment that we call spin. This model is quite comfortable because it is rather like the familiar picture of the Earth spinning about its own axis as it orbits the Sun. But when we try to predict the magnetic behavior of various types of materials using this classical model, its usefulness is limited and it is completely wrong in many ways.

The bad classical predictions are not surprising since quantum physics, devised to explain atomic behavior, tells us that: (1) We cannot think of electrons as having distinct orbits. Instead we visualize them as swarming about in the vicinity of a nucleus without having distinct paths. So all we can know is something about the probability of finding the electron at various locations in the vicinity of the nucleus and that these probabilities are different for each type of atom or molecule. (2) The spin magnetic moments are a fundamental property of electrons and should not be thought of as being produced by an electron spinning about an internal axis. (3) The spin and orbital magnetic moments associated with atomic electrons are quantized. This means they can only have certain values.

Next let's examine the characteristics of these two types of atomic magnetic moments in more detail.

Spin Magnetic Dipole Moment

An electron has an intrinsic **spin magnetic dipole moment** $\vec{\mu}^{\,\text{spin}}$. (By *intrinsic,* we mean that $\vec{\mu}^{\,\text{spin}}$ is a basic characteristic of an electron, like its mass and electric charge.) According to quantum theory,

1. $\vec{\mu}^{\,\text{spin}}$ itself cannot be measured directly. Only its component along a single axis can be well-defined (and therefore measured) at any one time.
2. A measured component of $\vec{\mu}^{\,\text{spin}}$ is *quantized*, which is a general term that means it is restricted to certain values.

Let us assume that the component of the spin magnetic moment $\vec{\mu}^{\,\text{spin}}$ is measured along the z axis of a coordinate system you have chosen. Then the measured component μ_z^{spin} can have only the two values given by

$$\mu_z^{\text{spin}} = +\frac{eh}{4\pi m} \quad \text{or} \quad \mu_z^{\text{spin}} = -\frac{eh}{4\pi m}, \tag{32-35}$$

where $h = 6.63 \times 10^{-34}\ \text{J}\cdot\text{s}$ and is the well-known Planck constant used often in quantum physics. The constants e and m represent the charge and mass of the electron, respectively. The plus and minus signs given in Eq. 32-35 describe the direction of μ_z^{spin} along the chosen z axis. The plus sign indicates that μ_z^{spin} is parallel to the z axis, and the electron is said to be "spin up." When μ_z^{spin} is antiparallel to the z axis, the minus sign is used and the electron is said to be "spin down."

The combination of constants in Eq. 32-35 is called the *Bohr magneton* μ_B, which can be calculated from the known values of Planck's constant and the electron charge and mass:

$$\mu_B = \frac{eh}{4\pi m} = 9.27 \times 10^{-24}\ \text{J/T} \quad \text{(Bohr magneton value for an electron).} \tag{32-36}$$

Spin magnetic dipole moments of electrons and other elementary particles can be expressed in terms of μ_B. In terms of the Bohr magneton, we can substitute in to Eq. 32-35 to rewrite the expression for the two possible values of μ_z^{spin} as

$$\mu_z^{\text{spin}} = +\mu_B \quad \text{or} \quad \mu_z^{\text{spin}} = -\mu_B. \tag{32-37}$$

When an electron is placed in an external magnetic field $\vec{B}^{\text{ext}}$, a potential energy U can be associated with the orientation of the electron's spin magnetic dipole moment $\vec{\mu}^{\,\text{spin}}$ just as a potential energy can be associated with the orientation of the magnetic dipole moment $\vec{\mu}$ of a current loop placed in an external magnetic field $\vec{B}^{\text{ext}}$. From Eq. 29-35, the potential energy for the electron due to its spin orientation has only two possible values

$$U^{\text{spin}} = -\vec{\mu}^{\,\text{spin}} \cdot \vec{B}^{\text{ext}} = -\mu_z^{\text{spin}} B^{\text{ext}} = \pm\, \mu_B B^{\text{ext}}, \tag{32-38}$$

where the z axis is taken to be in the direction of $\vec{B}^{\text{ext}}$.

Again, although we use the word "spin" here, according to quantum theory the fact that electrons have intrinsic magnetic moments does not mean that they spin like tops.

Protons and neutrons also have intrinsic magnetic dipole moments. In fact, these nuclear magnetic moments are a critical element in the development of magnetic resonance imaging—a valuable diagnostic tool in medicine. The masses of protons and neutrons are almost 2000 times that of the electron, so the magnetic moment for these particles is much smaller than that of the electron. For this reason, the contributions of nuclear dipole moments to the magnetic fields of atoms are negligible.

Orbital Magnetic Dipole Moment

An electron that is part of an atom has an additional dipole magnetic moment. This is called its **orbital magnetic moment** $\vec{\mu}^{\,\text{orb}}$. Again, although we use the "orbital" here, electrons do not orbit the nucleus of an atom like planets orbiting the Sun. According to quantum physics, an "orbit" roughly defines a region in space where the electron is most likely to be found. The orientation of this region specifies the direction of the electron's orbital angular momentum. The so-called "outer electrons" in an atom with many electrons will tend to be found further from its nucleus. An outer electron has a larger orbital angular momentum and, hence, a larger magnetic moment. It turns out that in any given atom there are typically more than two possible quantized values for the z-components of orbital magnetic moments. We can express these possible components along a chosen z axis in terms of the Bohr magneton as

$$\mu_z^{orb} = -m_l^{max}\mu_B, \ldots, -3\mu_B, -2\mu_B, -1\mu_B, 0\mu_B, 1\mu_B, 2\mu_B, 3\mu_B, \ldots, m_l^{max}\mu_B, \quad (32\text{-}39)$$

where m_l^{max} is an integer that designates the magnitude of the orbital magnetic moment component an electron can have.

When an atom is placed in an external magnetic field $\vec{B}^{ext}$, an orbital potential energy U^{orb} can be associated with the orientation of the orbital magnetic dipole moment of each electron in the atom. Its value is

$$U^{orb} = -\vec{\mu}^{orb} \cdot \vec{B}^{ext} = -\mu_z^{orb} B^{ext}, \quad (32\text{-}40)$$

where the z axis is taken in the direction of $\vec{B}^{ext}$ so that $\vec{B}^{ext} = \vec{B}^{ext}\hat{k}$.

The Magnetic Dipole Moment of an Atom

Each electron in an atom has an orbital magnetic dipole moment and a spin magnetic dipole moment that combine vectorially. The resultant of these two vector quantities combines vectorially with similar resultants for all other electrons in the atom, and the resultant for each atom combines with those for all the other atoms in a bulk sample of matter. If the combination of all these magnetic dipole moments produces a magnetic field, then the material is magnetic.

We have one more step in preparing to explain the magnetic behavior of bulk material on the basis of the magnetic behavior of its atoms. We need to consider how the spin and orbital magnetic moments associated with all the electrons in a single atom of a certain element (such as iron, arsenic, and so on) could combine to determine its total magnetic moment.

If the spin and orbital magnetic moments of all the electrons in a given atom lined up with each other and then if all the individually aligned atoms lined up with each other in a solid or liquid, we would have an incredibly strong magnet. Most materials are not strongly magnetic because whenever possible it is natural for the electron magnetic moments in an atom to cancel out. Atomic electrons are located in regions around the nucleus called shells. The number of electrons in a shell is governed by a quantum mechanical rule known as the *Pauli exclusion principle*. This exclusion principle requires that no two electrons in the same shell can have both the same components of orbital magnetic moment and the same components of spin magnetic moment. When all different combinations of orbital and spin states have occurred in a shell it is full. Once a shell is full the pairing of electrons in their locations with respect to the nucleus cancel each other. Each successive shell has more possible orbital magnetic moment components than its electrons can have. In addition, the electrons in a given shell have more energy than the ones in the previous shell. If the atom is in its lowest possible energy state, the shells usually fill in order.

In an atom with many electrons, the number of electrons in a full shell is $4n + 2$, where n is 0, 1, 2, 3, and so on. So the first shell can have 2 electrons, the second 6 electrons, the third 10, the fourth 14, and so on. Typically it is unpaired outermost electrons in an atom that determine the magnetic behavior of bulk matter. Magnetism depends critically on how atoms combine with each other as a result of the sharing and interaction of the outermost electrons.

READING EXERCISE 32-4: In this section, we discuss three types of magnetism in materials. Which one is associated with a refrigerator magnet? A standard paper clip? A piece of silver wire? Explain your reasoning. ■

READING EXERCISE 32-5: The figure shows the spin orientations of two particles in an external magnetic field $\vec{B}^{ext}$. (a) If the particles are electrons, which spin orientation is at lower potential energy? (b) If, instead, the particles are protons, which spin orientation is at lower potential energy? Explain your reasoning.

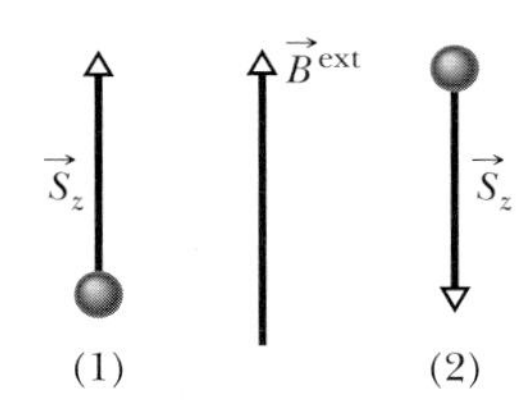

■

32-7 Ferromagnetism

Iron, cobalt, nickel, gadolinium, dysprosium, and alloys of these become strongly magnetized in the presence of an external magnetic field. Because they retain this magnetism when the external field is removed, we call them ferromagnetic.

Atomic Magnetic Moments in Ferromagnetism

Although most heavy elements are not ferromagnetic, all ferromagnetic materials are relatively heavy elements with complex electronic structures. The lightest of these is iron, which has 26 electrons. The best explanation to date for iron's ferromagnetism involves the complex behavior of its electrons. The 20 innermost electrons are paired in such a way that their spin and orbital magnetic moments cancel each other. The other 6 electrons behave in a manner that is unusual for most materials. Instead of piling into the third shell that has plenty of room for them, 2 of the 6 electrons move out into the fourth shell. These 2 electrons form outer conduction electrons. The key to the ferromagnetism of iron is that the third shell is unfilled, which allows 4 of 14 electrons in that shell to have spin magnetic moments that end up being aligned. However, these aligned electrons do not participate in chemical bonding with other atoms. The detailed quantum mechanical calculations reveal that this unusual arrangement of electrons gives an individual iron *atom* both a net magnetic moment and a lower energy—a situation that is similar for cobalt and nickel.

Even though individual iron atoms have permanent magnetic dipole moments, we might assume that their orientations relative to each other are random, leaving a bulk sample of iron with no net magnetic moment. We know this is not the case. Although various explanations have been put forth to explain why individual atoms line up in ferromagnetic materials, the situation is not well understood. It is currently believed that the spin-aligned electrons in the third shell, which causes the magnetism, influence the outermost conduction electrons, which are wandering through the material. Because of the Pauli exclusion principle, the spin magnetic moment of a conduction electron will have a tendency to be aligned in a direction opposite to that of the third-shell electrons. This anti-aligned conduction electron could, in turn, influence the alignment of the third shell electrons in a neighboring atom. This interaction could align the third-shell spin magnetic moments in the two neighboring atoms, and so on. The jargon for this quantum physical effect, in which spins of the electrons in one atom interact with those of neighboring atoms via conduction electrons, is called **exchange coupling.** The result is an alignment of the magnetic dipole moments of the atoms, in spite of the randomizing effects of thermal energy that causes atomic collisions. We currently believe that this type of coupling is what gives ferromagnetic materials their permanent magnetism.

Magnetic Domains

Exchange coupling in which spins of the electrons in one atom interact with those of neighboring atoms via conduction electrons, produces strong alignment of adjacent atomic dipoles in a ferromagnetic *material.* So we might expect that all the atoms in a sample of iron would align themselves into a permanent magnet even in the absence of an external magnetic field. This doesn't happen. Instead, a piece of iron, nickel, or cobalt is always made up of a number of *magnetic domains*. Each domain is a region in which the alignment of the atomic dipoles is essentially perfect. The domains, however, are not all aligned. For the sample as a whole, the domains are so oriented that they largely cancel each other as far as their external magnetic effects are concerned.

Two reasons are often given for the existence of domains in ferromagnetic materials. First, calculations reveal that a pure sample with perfectly aligned atoms (known as a *single crystal*) has a lower energy state when there are distinct domains with boundaries between them. Second, most real samples have impurities that can cause even more boundaries between domains to form.

A photograph of a single crystal of nickel is shown in Fig. 32-15. A suspension of powdered iron oxide was sprinkled on the crystal surface. The domain boundaries, which are thin regions in which the alignment of the elementary dipoles changes from a certain orientation in one domain to a different orientation in the other, are the sites of intense, but highly localized and nonuniform, magnetic fields. The suspended iron oxide particles are attracted to some of the more prominent boundaries and show up as the white lines. Although the atomic dipoles in each domain are completely aligned as shown by the arrows, the crystal as a whole has a very small resultant magnetic moment.

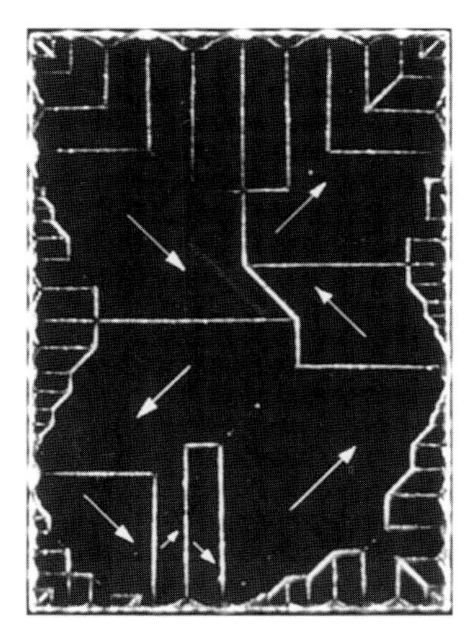

FIGURE 32-15 ■ A photograph of domain patterns within a single crystal of nickel; white lines reveal the boundaries of the domains. The white arrows superimposed on the photograph show the orientations of the magnetic dipoles within the domains and thus the orientations of the net magnetic dipoles of the domains. The crystal as a whole is unmagnetized if the net magnetic field (the vector sum over all the domains) is zero.

Actually, a piece of iron as we ordinarily find it is not a single crystal but an assembly of many tiny crystals, randomly arranged; we call it a polycrystalline solid. Each tiny crystal, however, has its array of variously oriented domains, just as in Fig. 32-15. We can magnetize such a specimen by placing it in an external magnetic field B_z^{ext} of gradually increasing strength, and measuring the magnetization B_z^{M} of the iron. (The measurement process is explained in the next subsection on Bulk Properties.) A common way to display the results is to plot a magnetization curve. If the piece of iron had all of its magnetic dipoles aligned perfectly with the external field, its magnetization would be a maximum represented by $B_{\text{max}}^{\text{M}}$. The magnetization curve consists of a plot of the ratio $B_z^{\text{M}}/B_{\text{max}}^{\text{M}}$ as a function of the external field (shown in Fig. 32-16). Note that $B^{\text{M}}/B_{\text{max}}^{\text{M}}$ is always less than one, so the iron does not become perfectly magnetized.

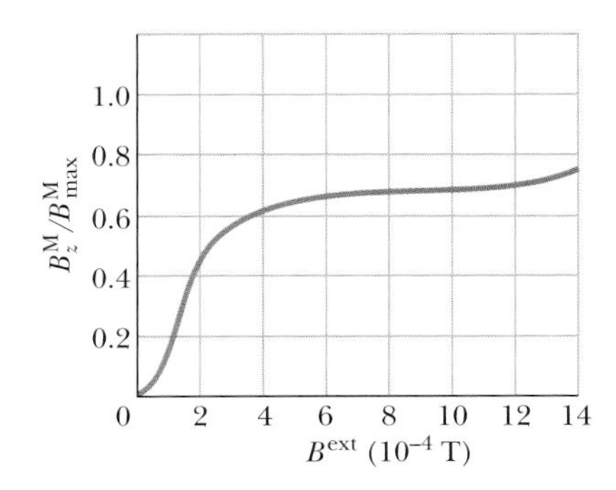

FIGURE 32-16 ■ A magnetization curve for a ferromagnetic core material in the Rowland ring of Fig. 32-17. On the vertical axis, 1.0 corresponds to complete alignment (saturation) of the atomic dipoles within the material.

By photographing domain patterns as in Fig. 32-15, we see two microscopic effects that serve to explain the shape of the magnetization curve: One effect is a growth in size of the domains that are oriented along the external field at the expense of those that are not. The second effect is a shift of the orientation of the dipoles within a domain, as a unit, to become closer to the field direction.

Exchange coupling and domain shifting give us the following result:

> A ferromagnetic material placed in an external magnetic field $\vec{B}^{\text{ext}}$ develops a strong magnetic dipole moment in the direction of $\vec{B}^{\text{ext}}$. If the field is nonuniform, the ferromagnetic material is attracted toward a region of greater magnetic field from a region of lesser field.

Bulk Properties of Ferromagnetic Materials

If the temperature of a ferromagnetic material is raised above a certain critical value, called the **Curie temperature,** the exchange coupling ceases to be effective. Most such materials then become simply paramagnetic. That is, the dipoles still tend to align with an external field but much more weakly, and thermal agitation can now more easily disrupt the alignment. The Curie temperature for iron is 1043 K (= 770°C).

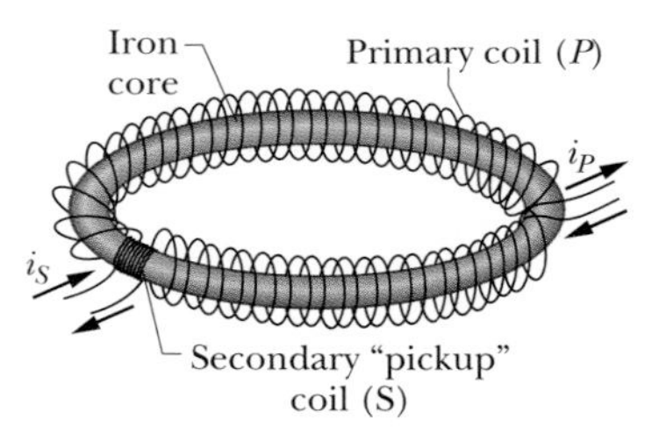

FIGURE 32-17 ■ A toroidal Rowland ring coil in which a current i_P is sent through a primary coil P. This current is used to study the behavior of the ferromagnetic material of the iron core inside the windings. The extent of magnetization of the core determines the total magnetic field $\vec{B}$ within coil P. Field $\vec{B}$ can be measured by means of a secondary or "pickup" coil.

We can express the extent to which a given paramagnetic sample is magnetized by finding the ratio of its magnetic dipole moment to its volume V. This vector quantity, the magnetic dipole moment per unit volume, is called the **magnetization** $\vec{M}$ of the sample, and its magnitude is

$$\vec{M} = \frac{\text{measured magnetic moment}}{V}. \tag{32-41}$$

The unit of $\vec{M}$ is the ampere-square meter per cubic meter, or ampere per meter (A/m). Complete alignment of the atomic dipole moments, called **saturation** of the sample, corresponds to the maximum magnetization of magnitude $M^{\text{max}} = N\mu/V$ where N is the number of atoms in the volume V.

The magnetization of a ferromagnetic material such as iron can be studied using a toroidal coil called a *Rowland ring* (Fig. 32-17). A Rowland ring is basically a long solenoid with an iron cylinder at its core, except the whole thing is bent into the shape of a donut. Assume that the ring's primary coil P has n turns per unit length and carries current i_P. If the iron core were not present, the magnitude of the magnetic field inside the coil caused by the "external" solenoid windings (as distinct from the magnetization of a core material inside the windings) would be given by Eq. 30-25,

$$B^{\text{ext}} = n\mu_0|i_P| \qquad \text{(no iron core)}. \tag{32-42}$$

Here μ_0 represents the magnetic constant (or permeability) of air (and is not a magnetic moment).

If an iron core is present, the magnitude of the magnetic field B inside the coil is proportional to B^{ext} but is on the order of 1000 to 10 000 times greater due to the magnetization of the iron core. This magnetization results from the alignment of the atomic dipole moments within the iron. The field $\vec{B}$ inside the coil should be the vector sum of the field $\vec{B}^{\text{ext}}$ contributed by the coil without the core and the field $\vec{B}^{\text{M}}$ contributed by the magnetization of the core. Since the magnitude of $\vec{B}^{\text{ext}}$ field is much smaller than that produced by the core magnetization

$$\vec{B} = \vec{B}^{ext} + \vec{B}^{\text{M}} \approx \vec{B}^{\text{M}}. \tag{32-43}$$

To determine $\vec{B}^{\text{M}}$ we use a secondary coil S to measure $\vec{B}$ and hence $\vec{B}^{\text{M}}$. If needed, we compute $\vec{B}^{\text{ext}}$ using Eq. 32-42.

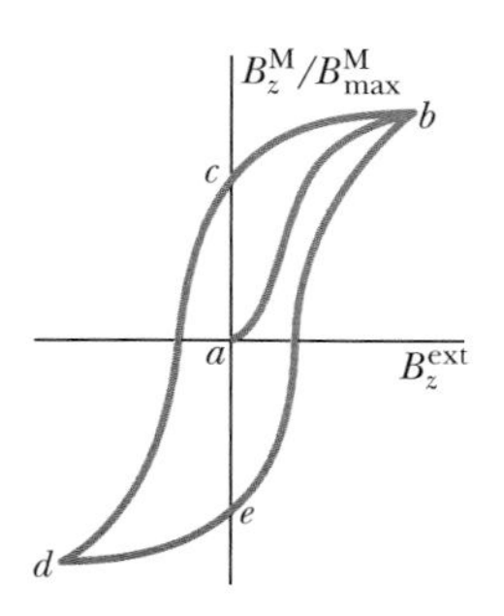

FIGURE 32-18 ■ A magnetization curve (ab) for a ferromagnetic specimen and an associated hysteresis loop ($bcdeb$).

Figure 32-18 shows a magnetization curve for a ferromagnetic material in a Rowland ring: the ratio of magnitudes $B^{\text{M}}/B^{\text{M}}_{\text{max}}$ is plotted as a function of B^{ext} (where $B^{\text{M}}_{\text{max}}$ is the maximum possible value of B^{M}, corresponding to saturation). The curve is similar to that for the magnetization curve for a paramagnetic substance shown in Fig. 32-19. Both curves are measures of the extent to which an applied magnetic field can align the atomic dipole moments of a material.

For the ferromagnetic core described by the graph in Fig. 32-16, the alignment of the dipole moments is about 70% complete for $B^{\text{ext}} \approx 1 \times 10^{-3}$ T. If B^{ext} were increased to 1 T, the alignment would be almost complete (but $B^{\text{ext}} = 1$ T, and thus almost complete saturation, is quite difficult to achieve).

Hysteresis

Magnetization curves for ferromagnetic materials are not retraced as we increase and then decrease and then reverse the external magnetic field $\vec{B}^{\text{ext}}$. Let's assume that we choose the z axis to be along the direction of the external magnetic field. Figure 32-18 is a plot of the z-component of the magnetization field B_z^{M} versus the z-component of the external field B_z^{ext} during the following operations with a Rowland ring: (1) Starting with the iron unmagnetized (point a), increase the current in the toroid until $B_z^{\text{ext}} = n\mu_0|i|$ has the value corresponding to point b; (2) reduce the current in the toroid winding (and thus B^{ext}) back to zero (point c); (3) reverse the toroid current and increase it in amount until B^{ext} has the value corresponding to point d; (4) reduce the current to zero again (point e); (5) reverse the current once more until point b is reached again.

The lack of retraceability shown in Fig. 32-18 is called **hysteresis,** and the curve *bcdeb* is called a *hysteresis loop*. Note that at points c and e the iron core is magnetized, even though there is no current in the toroid windings; this is the familiar phenomenon of permanent magnetism. In fact when engineers are designing permanent magnets, they look for materials that have a high degree of hysteresis.

Hysteresis can be understood through the concept of magnetic domains. When the magnetic field in the coil due to the current in the solenoid windings, $\vec{B}^{ext}$, is increased and then decreased back to its initial value, the domains do not return completely to their original configuration but retain some "memory" of their alignment after the initial increase. This memory of magnetic materials is essential for the magnetic storage of information, as on cassette tapes and computer disks.

This memory of the alignment of domains can also occur naturally. When lightning sends currents along multiple tortuous paths through the ground, the currents produce intense magnetic fields that can suddenly magnetize any ferromagnetic material in nearby rock. Because of hysteresis, such rock material retains some of that magnetization after the lightning strike (after the currents disappear) then becomes lodestones.

Inductors and Transformers with Iron Cores

Based on our discussion above of the Rowland ring, it is clear that the use of iron and iron alloys in inductors and transformers can literally increase the performance of these devices by a thousandfold or more.

A great deal of engineering has gone into the design of cores for large inductors and high-performance transformers. For example, these cores should not behave like permanent magnets with large hysteresis. Instead, they should have small hysteresis so that the magnetization of the core can change rapidly in the presence of alternating currents. In addition, transformer cores are not single hunks of iron. Rather, they are built up in layers to prevent eddy currents from being induced in the cores that could reduce the efficiency of the power transfer from the primary to secondary coils in a transformer.

READING EXERCISE 32-6: Iron is a ferromagnetic material. Why then isn't every piece of iron—for example, an iron nail—a naturally strong magnet? ■

READING EXERCISE 32-7: What is hysteresis and why does it occur? ■

32-8 Other Magnetic Materials

Paramagnetism

In paramagnetic materials, the spin and orbital magnetic dipole moments of the electrons in individual atoms do not cancel but add vectorially to give each *atom* a net (and permanent) magnetic dipole moment $\vec{\mu}$. In the absence of an external magnetic field, these atomic dipole moments are randomly oriented, and the net magnetic dipole moment of the *material* is zero. However, if a sample of the material is placed in an external magnetic field $\vec{B}^{ext}$, the magnetic dipole moments tend to line up with the field, which gives the sample a net magnetic dipole moment not unlike that found in a ferromagnetic sample. However, paramagnetic materials lack the exchange coupling needed to set up permanent magnetic domains. Paramagnetism is fairly weak compared to ferromagnetism because the forces of alignment from external magnetic

Liquid oxygen is suspended between the two pole faces of a magnet because the liquid is paramagnetic and is magnetically attracted to the magnet.

fields are smaller than the randomizing forces due to thermal motions. Also, paramagnetic materials do not retain their magnetism once an external magnetic field is turned off.

> A paramagnetic material placed in an external magnetic field $\vec{B}^{\text{ext}}$ develops a magnetic dipole moment in the direction of $\vec{B}^{\text{ext}}$. If the field is not uniform, the paramagnetic material is attracted toward a region of greater magnetic field from a region of lesser field.

As is the case for ferromagnetism, we can express the extent to which a given paramagnetic sample is magnetized by measuring the magnetization $\vec{M}$ (defined in Eq. 32-41). In 1895, Pierre Curie discovered that the magnitude of the magnetization of a paramagnetic sample is directly proportional to the external magnetic field magnitude B^{ext} and inversely proportional to the temperature T in kelvins; that is,

$$M = C\frac{B^{\text{ext}}}{T}. \tag{32-44}$$

Equation 32-44 is known as **Curie's law,** and C is called the **Curie constant.** Curie's law is reasonable in that increasing $\vec{B}^{\text{ext}}$ tends to align the atomic dipole moments in a sample and thus to increase $\vec{M}$, whereas increasing T tends to disrupt the alignment via thermal agitation and thus to decrease $\vec{M}$. However, the law is actually an approximation that is valid only when the ratio B^{ext}/T is not too large.

Figure 32-19 shows the ratio M/M^{max} as a function of B^{ext}/T for a sample of the salt potassium chromium sulfate, in which chromium ions are the paramagnetic substance. The plot is called a **magnetization curve.** The straight line for Curie's law fits the experimental data at the left, for B^{ext}/T below about 0.5 T/K. The curve that fits all the data points is based on quantum physics. The data on the right side, near saturation, are very difficult to obtain because they require very strong magnetic fields (about 100 000 times Earth's field), even at the very low temperatures noted in Fig. 32-19.

Diamagnetism

The *atoms* in diamagnetic materials have no net magnetic dipole moments. However, diamagnetic *materials* do undergo a very weak nonpermanent alignment in the presence of an external magnetic field. The strength of the alignments is still proportional to the strength of the external magnetic field (as is the case for both ferro- and paramagnetism). However, the behavior of diamagnetic materials is not very temperature dependent.

The most interesting characteristic of diamagnetism is that in the presence of an external magnetic field that is nonuniform, each atom experiences a net force that is directed *away* from the region of greater magnetic field. Thus, in diamagnetism the

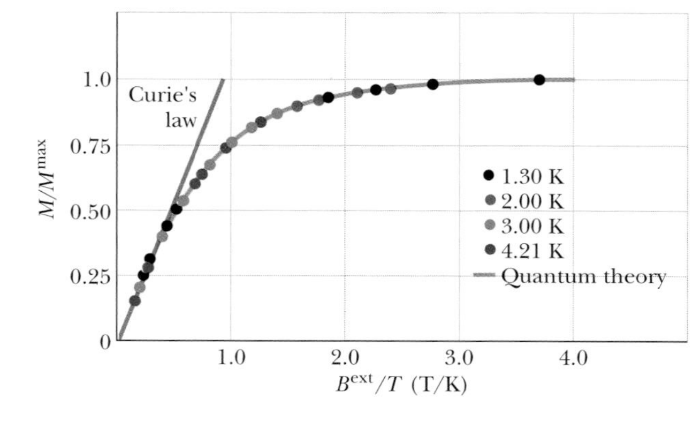

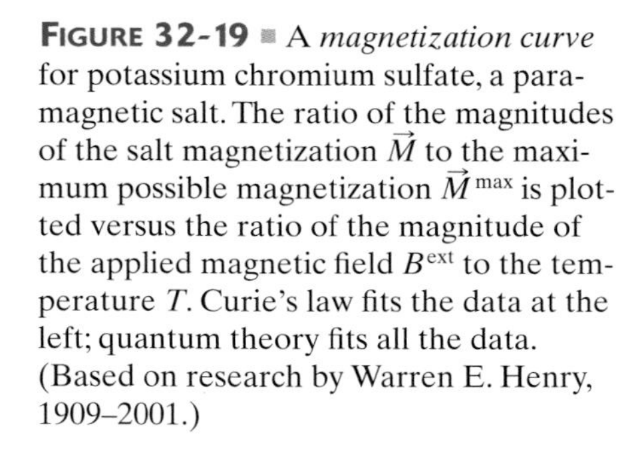
FIGURE 32-19 ■ A *magnetization curve* for potassium chromium sulfate, a paramagnetic salt. The ratio of the magnitudes of the salt magnetization $\vec{M}$ to the maximum possible magnetization $\vec{M}^{\text{max}}$ is plotted versus the ratio of the magnitude of the applied magnetic field B^{ext} to the temperature T. Curie's law fits the data at the left; quantum theory fits all the data. (Based on research by Warren E. Henry, 1909–2001.)

alignment of atomic magnetic moments with an external magnetic field is opposite to that associated with ferromagnetic and paramagnetic materials. In general,

> A diamagnetic material placed in an external magnetic field $\vec{B}^{\text{ext}}$ develops a magnetic dipole moment directed opposite $\vec{B}^{\text{ext}}$. If the field is nonuniform, the diamagnetic material is repelled from a region of greater magnetic field toward a region of lesser field.

Animals like the frog shown in Fig. 32-20 are diamagnetic. This frog has been placed in the diverging magnetic field near the top end of a vertical current-carrying solenoid; every atom in the frog was repelled upward, away from the region of stronger magnetic field at that end of the solenoid. The frog moved upward into weaker and weaker magnetic field until the upward magnetic force balanced the gravitational force on it, and there it hung in midair. People are also diamagnetic, so if we built a large enough solenoid, we could also suspend a person in midair.

FIGURE 32-20 ■ An overhead view of a diamagnetic frog that is being levitated in a magnetic field. The $\vec{B}$-field is produced by current in a vertical solenoid below the frog. The solenoid's upward magnetic force on the frog balances the downward gravitational force on the frog. (The frog is not in discomfort; the sensation is like floating in water, which frogs don't seem to mind.)

READING EXERCISE 32-8: The figure here shows two paramagnetic spheres located near the south pole of a bar magnet. Are (a) the magnetic forces on the spheres and (b) the magnetic dipole moments of the spheres directed toward or away from the bar magnet? (c) Is the magnetic force on sphere 1 greater than, less than, or equal to that on sphere 2? ■

READING EXERCISE 32-9: The figure shows two diamagnetic spheres located near the south pole of a bar magnet. Are (a) the magnetic forces on the spheres and (b) the magnetic dipole moments of the spheres directed toward or away from the bar magnet? (c) Is the magnetic force on sphere 1 greater than, less than, or equal to that on sphere 2? ■

32-9 The Earth's Magnetism

The Earth has a magnetic field associated with it that behaves approximately like that of a magnetic dipole. In other words, the Earth's magnetic field can be thought of as being produced by a bar magnet that straddles the center of the planet with its axis more or less aligned with the Earth's rotation axis. Figure 32-21 is an idealized depiction of the Earth's dipole field that ignores the distortion of field lines caused by charged particles streaming out of the Sun and other factors.

Characteristics of the Earth's Magnetic Field

For the idealized magnetic field shown in Fig. 32-21, the Earth's magnetic dipole moment $\vec{\mu}$ has a magnitude of 8.0×10^{22} J/T. The point where the Earth's rotation axis intersects the surface is known at the *geographic north pole*. In 2001, the geological survey of Canada placed the direction of the Earth's dipole moment at an angle of $\theta = 8.7°$ from the rotation axis (RR) of the Earth.* The *dipole axis* (MM in Fig. 32-21) lies along $\vec{\mu}$ and intersects the Earth's surface at the *geomagnetic north and south poles*, These days the magnetic north pole is estimated to be somewhere in the Arctic Ocean north of Canada and the south pole is in the Antarctic Ocean. Since the poles are currently moving at about 40 km/yr, the possibility exists that the magnetic north pole could pass north of Alaska and in about fifty years end up in Siberia, although this outcome is not certain.

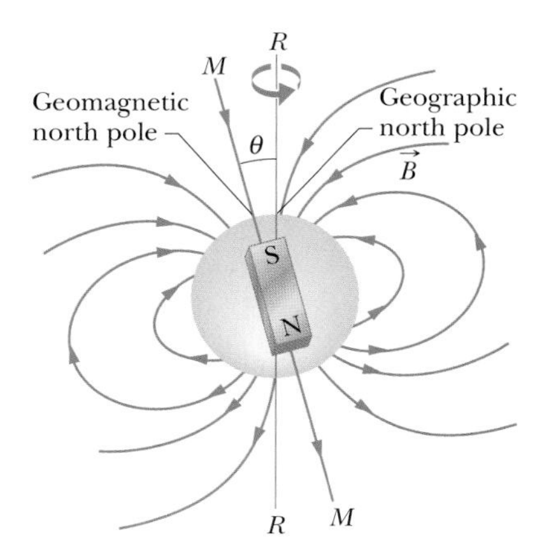

FIGURE 32-21 ■ An idealized view of the Earth's magnetic field as a dipole field. At present, dipole axis MM makes an angle of 8.7° with Earth's rotational axis RR. The "south pole" of the dipole is in Earth's northern hemisphere.

*http://www.geolab.nrcan.gc.ca/geomag/northpole_e.shtml

The lines of the magnetic field $\vec{B}$ generally emerge in the southern hemisphere and reenter Earth in the northern hemisphere. Thus, the magnetic pole that is in the Earth's northern hemisphere and known as a "north magnetic pole" *is really the south pole of the Earth's magnetic dipole*. This means that the north pole of a compass is attracted to the Earth's geographic north pole.

The direction of the magnetic field varies from location to location on the Earth. The field direction at any location on the Earth's surface is commonly specified in terms of two angles. The **field declination** is the angle (left or right) between geographic north (which is toward 90° latitude) and the horizontal component of the field. The **field inclination** is the angle (up or down) between a horizontal plane and the field's direction.

The field's inclination and declination at a given location can be measured with a *compass* and a *dip meter*. A **compass** is simply a needle-shaped magnet that is mounted so it can rotate freely about a vertical axis. When it is held in a horizontal plane, the north-pole end of the needle points, generally, toward the geomagnetic north pole (really a south magnetic pole). The angle between the compass needle and geographic north is the field declination.* A dip meter, used to measure inclination, is simply another needle-shaped magnet mounted so it can rotate freely about a *horizontal* axis. If the plane of the dip meter is aligned with the direction of the compass needle used to measure the declination, then the angle the dip meter needle makes with the horizontal is defined as the inclination angle. The magnetic north pole is defined as the location in the northern hemisphere for which the dip angle is 90°.

Causes of the Earth's Magnetism

The mechanisms that produce the Earth's magnetic field are not completely understood. However, it is helpful to begin our discussion of the latest models with a consideration of what is known about the Earth's formation and structure.

The Earth's Structure: Measurements of the spread of seismic waves tell us that the structure of the Earth is rather like that of a chocolate-covered cherry with gooey liquid between the cherry and the chocolate. This structure makes sense when we consider the currently accepted theory that the Earth was formed five billion years ago as a conglomeration of colliding meteorites and comets. Iron and other dense elements from meteorites were pulled by gravitational forces toward the center of the Earth. Compounds made of lighter elements, as well as the water contained in comets, migrated toward the surface. In between the solid core at the center of the Earth and the solid crust at the Earth's surface there is the gooey liquid consisting of molten lava (Fig. 32-22).

FIGURE 32-22 ■ Seismic data reveal that the Earth has an **inner core** (white) of solid iron with a radius of about 1200 km, an **outer core** (yellow) of iron rich molten lava about 2200 km thick, a more or less solid **mantle** (orange and red—not to scale) of less dense matter about 2600 km thick, and a very thin **crust** of rocks and soils at the surface with an average thickness of 20 km.

Continuous Molten Lava Currents: Many scientists believe that most of the Earth's magnetic field is produced by electromagnetic interactions that depend on the molten lava acting like a moving electrical conductor. We know from our study of Faraday's law that if even a small magnetic field is present in a region of the core, the electrical currents can be induced in the conducting fluid that travels through it. These induced currents can, in turn, produce magnetic fields that can act on other parts of the liquid core that are also moving. Thus a continuous cycle of induction and magnetic field production can take place as long as the material in the liquid core keeps flowing. In principle, this process is rather like that described for the generator shown in Fig. 32-10.

Two mechanisms have been proposed that explain the flow of molten lava in the liquid core. One possible mechanism is thermal convection produced by the temperature

*Inclination is the angle that a magnetic needle makes with the plane of the horizon. It is also called the angle of dip. Declination is the angle between magnetic north and geographic north.

difference between the hot solid inner core and the much cooler mantle. A second proposed mechanism for the flow of lava involves condensation of the heavier elements onto a growing inner core. This causes lighter, less dense, elements to flow toward the Earth's surface. In either case liquid convection currents are produced that are not unlike those in a pot of boiling water.

The Earth's magnetic field depends critically on the existence of *continuous* convection currents that requires the solid core to remain very hot for billions of years. Some scientists believe that nuclear energy in the core is being transformed to thermal energy through the decay of heavy radioactive elements. Other scientists have suggested that thermal energy can be released if the inner core expands by condensing material from the liquid core.

Changes in the Earth's Magnetic Field Over Time: Some mysterious characteristics of the Earth's magnetic field have been gleaned from fossil records and other geomagnetic measurements. The strength of the field and the location of the magnetic poles are constantly changing. For example, in recent years the geographic location of the magnetic poles has changed by an average of about 100 meters a day. These relatively small day-to-day changes are not obvious to someone a long distance from a magnetic pole who uses a compass and dip meter to measure a local field direction. It's another story when longer time scales are involved. We can use simple instruments to detect changes over a time period of a year or more. When even longer time periods are considered, the changes have been dramatic. In fact, the orientations of magnetized minerals imbedded in ancient rocks indicate that the Earth's magnetic field has completely reversed itself many times in the Earth's five billion year history, though reversals seem to take 1000 years or more.

FIGURE 32-23 ▪ This image shows one of many configurations of the Earth's magnetic field lines created by the model developed by Glatzmaier and Roberts.

The Glatzmaier/Roberts Model: A few years ago, two scientists, Gary Glatzmaier and Paul Roberts, developed a comprehensive numerical model of the electromagnetic and fluid dynamic processes in the Earth's interior. When this model was run on a CRAY supercomputer for thousands of hours these investigators were able to simulate over 300,000 years of magnetic field conditions. Their results showed many of the key features revealed by geological data, including the existence of a dipole field outside the Earth, a preference for approximate alignment between the Earth's dipole moment and its rotation axis, field strength variations, migration of the magnetic poles over the Earth's surface, and several field reversals. One such configuration of magnetic field lines is seen in Fig. 32-23.

There is still a great deal to be learned about the actual mechanisms responsible for the continual changes in the Earth's magnetic field, but scientists expect to resolve many of their uncertainties within the next few decades.

Magnetic Bacteria

The survival of many organisms depends on their ability to sense the Earth's magnetic field. For example, it is believed that the Earth's dipole field is critical to the navigation of migrating birds and fish as well as certain types of bacteria.

Magnetotactic bacteria are one-celled organisms that can be found almost anywhere in the world where there are ponds, marshes, or muddy lake bottoms. Many species of these bacteria are anaerobic or microanaerobic and must burrow in mud both to get away from oxygen and to feed on nutrients. Notice that on the lower left side of the bacterium shown in the photo at the beginning of this chapter there is a string of tiny 100-nanometer-long particles. These particles, known as magnetosomes, are oriented along the bacterium's long axis. An enlarged view of a set of magnetosomes is shown in Fig. 32-24.

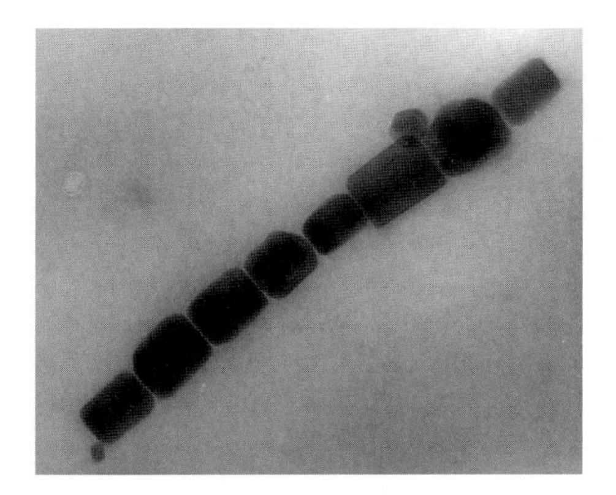

FIGURE 32-24 ▪ The type of bacterium shown in the puzzler at the beginning of the chapter is magnetotactic because it contains a chain of dense iron-rich magnetosomes each having a length of about 100 nm. The chain shown in this transmission electron micrograph has a net magnetic moment and tends to align itself with the Earth's magnetic field.

Magnetotactic bacteria synthesize these magnetic particles out of iron-oxygen or iron sulfur compounds. Each magnetosome is just big enough to have a permanent magnetic dipole moment and just small enough to be a single ferromagnetic domain. When strung together like a set of microscopic refrigerator magnets, the array has a net dipole moment. So instead of bumbling around randomly, these bacteria align with the Earth's magnetic field. This allows them to swim naturally along field lines.

Through natural selection, the bacteria that have their magnetosome strings oriented so they swim down along magnetic field lines to the mud at the bottom of a pond or lake will survive and multiply. Those that don't will swim up and die. An examination of the pattern of the Earth's magnetic field lines shown in Fig. 32-21 reveals that "down" is opposite to the direction of the field lines in the southern hemisphere and in the same direction as the field lines in the northern hemisphere. Thus, an Australian bacterium evolved to swim down in its normal habitat would swim up if transported to the United States. Alternatively, a healthy bacterium that evolves in the United States would be preset by evolutionary processes to have its magnetosomes oriented in the opposite direction, so it will swim down.

It is interesting to note that the orientations of bacterial magnetosome strings in fossils have helped scientists piece together evidence for past changes in the Earth's magnetic field.

READING EXERCISE 32-10: Describe the ways in which the Earth's magnetic field varies over its surface. Does the Earth's magnetic field vary in time as well? ■

Problems

SEC. 32-2 ■ SELF-INDUCTANCE

1. Close-Packed Coil The inductance of a close-packed coil of 400 turns is 8.0 mH. Calculate the magnetic flux through the coil when the current is 5.0 mA.

2. Circular Coils and Flux A circular coil has a 10.0 cm radius and consists of 30.0 closely wound turns of wire. An externally produced magnetic field of magnitude 2.60 mT is perpendicular to the coil. (a) If no current is in the coil, what is the magnitude of the magnetic flux that links its turns? (b) When the current in the coil is 3.80 A in a certain direction, the net flux through the coil is found to vanish. What is the inductance of the coil?

3. Equal Currents, Opposite Directions Two long parallel wires, both of radius a and whose centers are a distance d apart, carry equal currents in opposite directions. Show that, neglecting the flux within the wires, the inductance of a length l of such a pair of wires is given by

$$L = \frac{\mu_0 l}{\pi} \ln \frac{d - a}{a}$$

(*Hint:* Calculate the flux through a rectangle of which the wires form two opposite sides.)

4. Wide Copper Strip A wide copper strip of width W is bent to form a tube of radius R with two parallel planar extensions, as shown in Fig. 32-25. There is a current i through the strip, distributed uniformly over its width. In this way a "one-turn solenoid" is formed. (a) Derive an expression for the magnitude of the magnetic field $\vec{B}$ in the tubular part (far away from the edges). (*Hint:* Assume that the magnetic field outside this one-turn solenoid is negligibly small.) (b) Find the inductance of this one-turn solenoid, neglecting the two planar extensions.

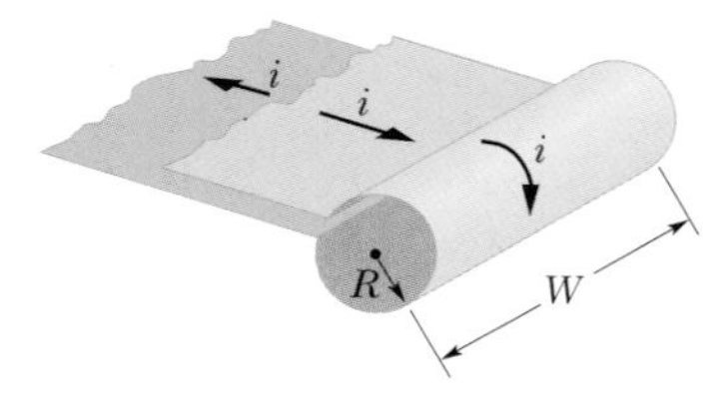

FIGURE 32-25 ■ Problem 4.

5. Inductor Carries Steady Current A 12 H inductor carries a steady current of 2.0 A. How can a 60 V self-induced emf be made to appear in the inductor?

6. At a Given Instant At a given instant the current and self-induced emf in an inductor are directed as indicated in Fig. 32-26. (a) Is the current increasing or decreasing? (b) The induced emf is 17 V and the rate of change of the current is 25 kA/s; find the inductance.

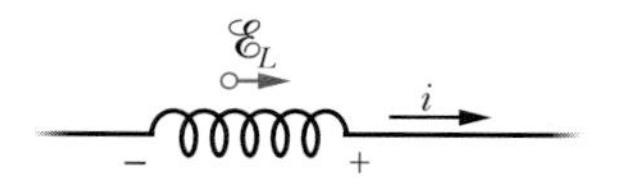

FIGURE 32-26 ■ Problem 6.

7. Inductors in Series Two inductors L_1 and L_2 are connected in series and are separated by a large distance. (a) Show that the equivalent inductance is given by

$$L_{eq} = L_1 + L_2.$$

(*Hint:* Review the derivations for resistors in series and capacitors in series. Which is similar here?) (b) Why must their separation be large for this relationship to hold? (c) What is the generalization of (a) for N inductors in series?

8. Current Varies with Time The current i through a 4.6 H inductor varies with time t as shown by the graph of Fig. 32-27. The inductor has a resistance of 12 Ω. Find the magnitude of the induced emf $\mathscr{E}$ during the time intervals (a) $t_1 = 0$ to $t_2 = 2$ ms, (b) $t_2 = 2$ ms to $t_3 = 5$ ms, (c) $t_3 = 5$ ms to $t_4 = 6$ ms. (Ignore the behavior at the ends of the intervals.)

FIGURE 32-27 ■ Problem 8.

9. At What Rate At time $t = 0$ ms, a 45 V potential difference is suddenly applied to the leads of a coil with inductance $L = 50$ mH and resistance $R = 180\ \Omega$. At what rate is the current through the coil increasing at $t = 1.2$ ms?

10. Inductors in Parallel Two inductors L_1 and L_2 are connected in parallel and separated by a large distance. (a) Show that the equivalent inductance is given by

$$\frac{1}{L_{eq}} = \frac{1}{L_1} + \frac{1}{L_2}.$$

(*Hint:* Review the derivations for resistors in parallel and capacitors in parallel. Which is similar here?) (b) Why must their separation be large for this relationship to hold? (c) What is the generalization of (a) for N inductors in parallel?

11. What Is L The inductance of a closely wound coil is such that an emf of 3.0 mV is induced when the current changes at the rate of 5.0 A/s. A steady current of 8.0 A produces a magnetic flux of 40 μWb through each turn. (a) Calculate the inductance of the coil. (b) How many turns does the coil have?

SEC. 32-3 ■ MUTUAL INDUCTION

12. Coil 1, Coil 2 Coil 1 in Fig. 32-4 has $L_1 = 25$ mH and $N_1 = 100$ turns. Coil 2 has $L_2 = 40$ mH and $N_2 = 200$ turns. The coils are rigidly positioned with respect to each other; their mutual inductance M is 3.0 mH. A 6.0 mA current in coil 1 is changing at the rate of 4.0 A/s. (a) What magnetic flux $\Phi_{1\to 2}$ links coil 2, and what self-induced emf appears there? (b) What magnetic flux $\Phi_{2\to 1}$ links coil 1, and what mutually induced emf appears there?

13. Two Coils at Fixed Locations Two coils are at fixed locations. When coil 1 has no current and the current in coil 2 increases at the rate 15.0 A/s, the emf in coil 1 is 25.0 mV. (a) What is their mutual inductance? (b) When coil 2 has no current and coil 1 has a current of 3.60 A, what is the flux linkage in coil 2?

14. Two Solenoids Two solenoids are part of the spark coil of an automobile. When the current in one solenoid falls from 6.0 A to zero in 2.5 ms, an emf of 30 kV is induced in the other solenoid. What is the mutual inductance M of the solenoids?

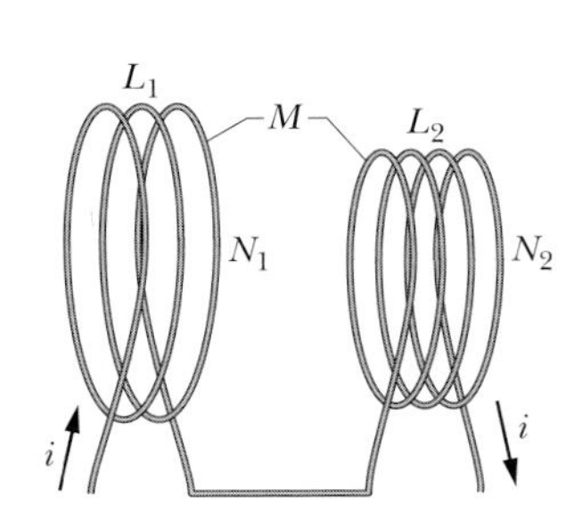

FIGURE 32-28 ■ Problem 15.

15. Two Connected Coils Two coils, connected as shown in Fig. 32-28, separately have inductances L_1 and L_2. Their mutual inductance is M. (a) Show that this combination can be replaced by a single coil of equivalent inductance given by

$$L_{eq} = L_1 + L_2 + 2M.$$

(b) How could the coils in Fig. 32-28 be reconnected to yield an equivalent inductance of

$$L_{eq} = L_1 + L_2 - 2M?$$

(This problem is an extension of Problem 7, but the requirement that the coils be far apart has been removed.)

16. Coil Around Solenoid A coil C of N turns is placed around a long solenoid S of radius R and n turns per unit length as in Fig. 32-29. Show that the mutual inductance for the coil–solenoid combination is given by $M = \mu_0 \pi R^2 n N$. Explain why M does not depend on the shape, size, or possible lack of close-packing of the coil.

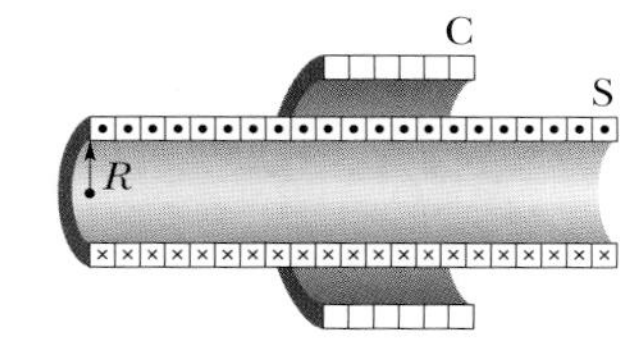

FIGURE 32-29 ■ Problem 16.

17. Coaxial Solenoid Figure 32-30 shows, in cross section, two coaxial solenoids. Show that the mutual inductance M for a length l of this solenoid–solenoid combination is given by $M = \pi R_1^2 l \mu_0 n_1 n_2$, in which n_1 and n_2 are the respective numbers of turns per unit length and R_1 is the radius of the inner solenoid. Why does M depend on R_1 and not on R_2?

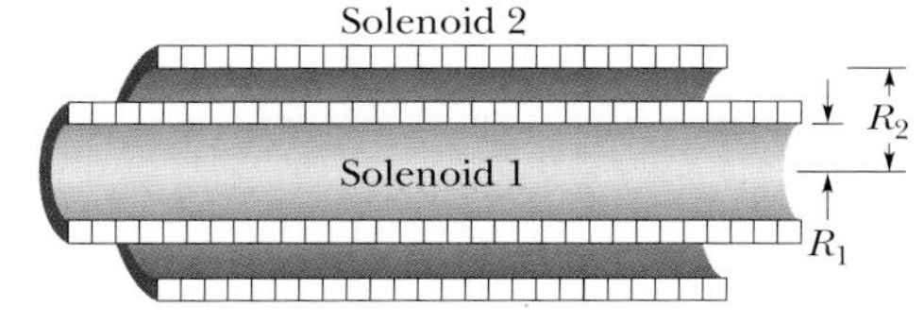

FIGURE 32-30 ■ Problem 17.

18. Coils Over a Toroid Figure 32-31 shows a coil of N_2 turns wound as shown around part of a toroid of N_1 turns. The toroid's inner radius is a, its outer radius is b, and its height is h. Show that the mutual inductance M for the toroid–coil combination is

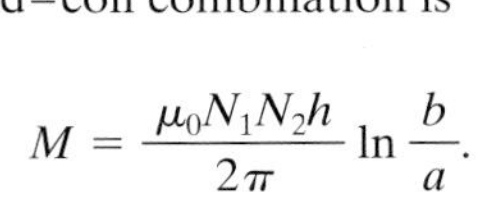

$$M = \frac{\mu_0 N_1 N_2 h}{2\pi} \ln \frac{b}{a}.$$

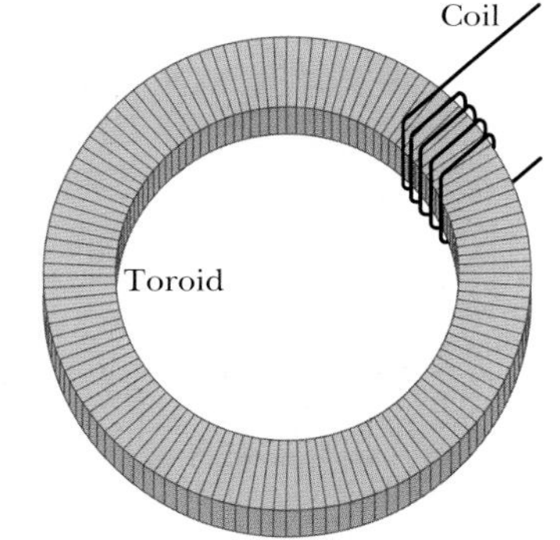

FIGURE 32-31 ■ Problem 18.

19. Rectangular Loop A rectangular loop of N close-packed turns is positioned near a long straight wire as shown in Fig. 32-32. (a) What is the mutual inductance M for the loop–wire combination? (b) Evaluate M for $N = 100$, $a = 1.0$ cm, $b = 8.0$ cm, and $l = 30$ cm.

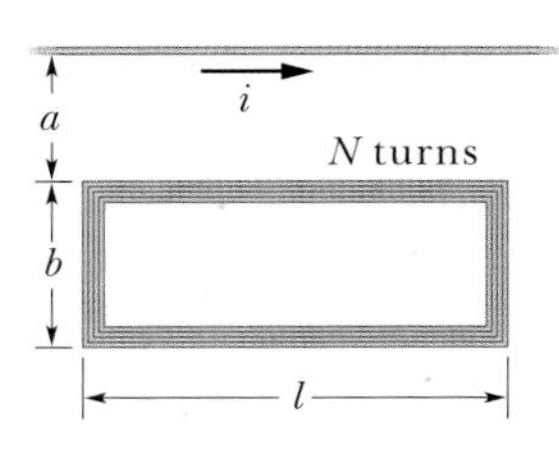

FIGURE 32-32 ■ Problem 19.

SEC. 32-4 ■ *RL* CIRCUITS (WITH IDEAL INDUCTORS)

20. Inductive Time Constant The current in an RL circuit builds up to one third of its steady-state value in 5.00 s. Find the inductive time constant.

21. How Long Must We Wait In terms of τ_L, how long must we wait for the current in an RL circuit to build up to within 0.100% of its equilibrium value?

22. In Terms of the emf Consider the RL circuit of Fig. 32-6. In terms of the battery emf $\mathscr{E}$, (a) what is the self-induced emf ΔV_2 when the switch has just been closed on a, and (b) what is ΔV_2 when $t = 2.0\tau_L$? (c) In terms of τ_L, when will ΔV_2 be just one-half the battery emf $\mathscr{E}$?

23. First Second The current in an RL circuit drops from 1.0 A to 10 mA in the first second following removal of the battery from the circuit. If L is 10 H, find the resistance R in the circuit.

24. emf Varies with Time Suppose the emf of the battery in the circuit of Fig. 32-7 varies with time t so that the current is given by $i(t) = 3.0\text{ A} + (5.0\text{ A/s})t$, where i is in amperes and t is in seconds. Take $R = 4.0\ \Omega$ and $L = 6.0$ H, and find an expression for the battery emf as function of time. (*Hint:* Apply the loop rule.)

25. Solenoid A solenoid having an inductance of 6.30 μH is connected in series with a 1.20 kΩ resistor. (a) If a 14.0 V battery is inserted into the circuit, how long will it take for the current through the resistor to reach 80.0% of its final value? (b) What is the current through the resistor at time $t = 1.0\tau_L$?

26. Wooden Toroidal Core A wooden torodial core with a square cross section has an inner radius of 10 cm and an outer radius of 12 cm. It is wound with one layer of wire (of diameter 1.0 mm and resistance per meter 0.020 Ω/m). What are (a) the inductance and (b) the inductive time constant of the resulting toroid? Ignore the thickness of the insulation on the wire.

27. Suddenly Applied At time $t = 0$ ms, a 45.0 V potential difference is suddenly applied to a coil with $L = 50.0$ mH and $R = 180\ \Omega$. At what rate is the current increasing at $t = 1.20$ ms?

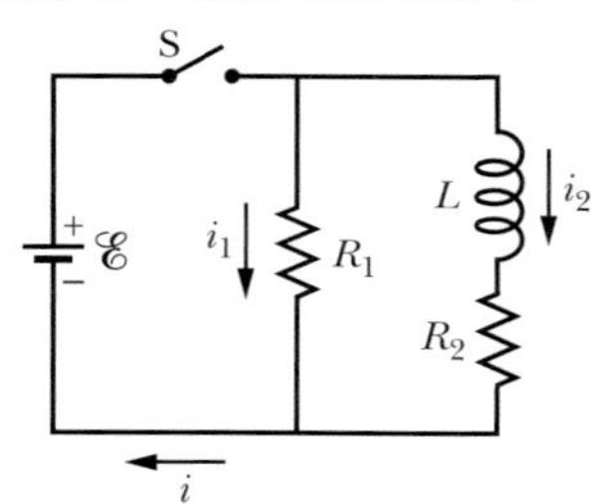

FIGURE 32-33 ■ Problem 28.

28. In the Circuit In the circuit of Fig. 32-33, $\mathscr{E} = 10$ V, $R_1 = 5.0\ \Omega$, $R_2 = 10\ \Omega$, and $L = 5.0$ H. For the two separate conditions (I) switch S just closed and (II) switch S closed for a long time, calculate (a) the current i_1 through R_1, (b) the current i_2 through R_2, (c) the current i through the switch, (d) the potential difference across R_2, (e) the potential difference across L, and (f) the rate of change di_2/dt.

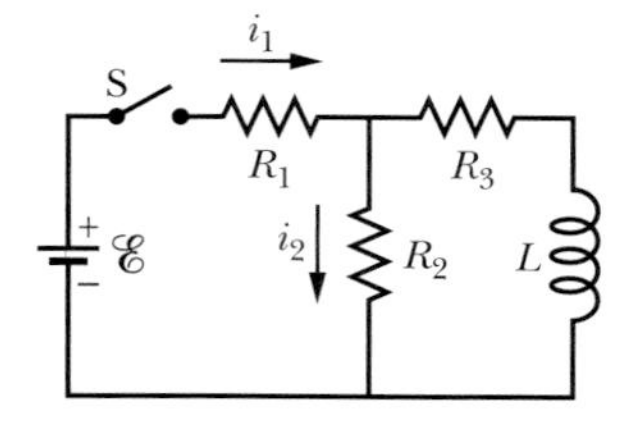

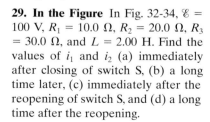

FIGURE 32-34 ■ Problem 29.

29. In the Figure In Fig. 32-34, $\mathscr{E} = 100$ V, $R_1 = 10.0\ \Omega$, $R_2 = 20.0\ \Omega$, $R_3 = 30.0\ \Omega$, and $L = 2.00$ H. Find the values of i_1 and i_2 (a) immediately after closing of switch S, (b) a long time later, (c) immediately after the reopening of switch S, and (d) a long time after the reopening.

30. What Is the Constant Figure 32-35a shows a circuit consisting of an ideal battery with emf $\mathscr{E} = 6.00\ \mu$V, a resistance R, and a small wire loop of area 5.0 cm^2. For the time interval $t_1 = 10$ to $t_2 = 20$ s, an external magnetic field is set up throughout the loop. The field is uniform, its direction is into the page in Fig. 32-35a, and the field magnitude is given by $B = at$, where B is in teslas, a is a

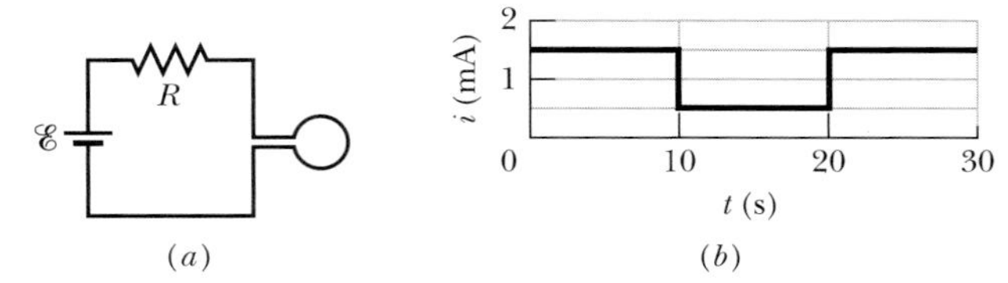

FIGURE 32-35 ■ Problem 30.

constant with units of teslas per second, and t is in seconds. Figure 32-35b gives the current i in the circuit before, during, and after the external field is set up. Find a.

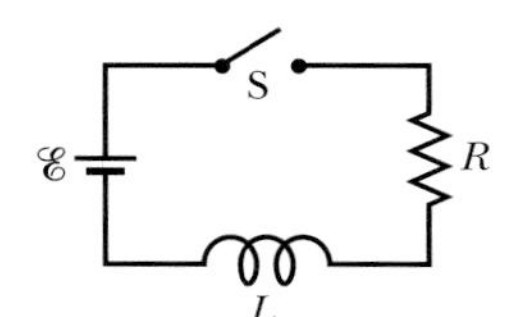

FIGURE 32-36 ■ Problems 31, 34, and 61.

31. Once the Switch Is Closed Once the switch S is closed in Fig. 32-36 the time required for the current to reach any obtainable value depends, in part, on the value of resistance R. Suppose the emf $\mathscr{E}$ of the ideal battery is 12 V and the inductance of the ideal (resistanceless) inductor is 18 mH. How much time is needed for the current to reach 2.00 A if R is (a) 1.00 Ω, (b) 5.00 Ω, and (c) 6.00 Ω? (d) Why is there a huge jump between the answers to (b) and (c)? (e) For what value of R is the time required for the current to reach 2.00 A least? (f)What is that least time? (*Hint:* Rethink Eq. 32-21.)

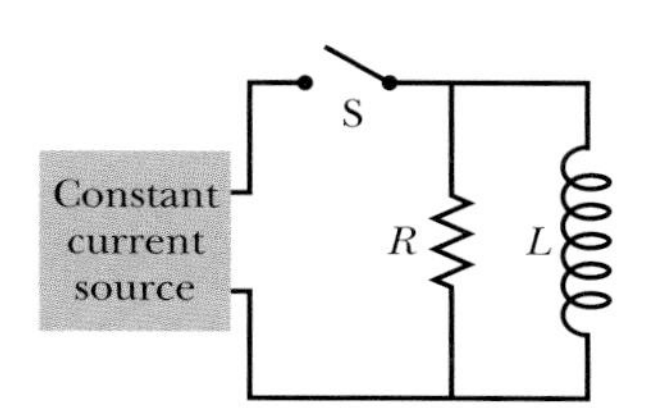

FIGURE 32-37 ■ Problem 32.

32. Circuit Shown In the circuit shown in Fig. 32-37, switch S is closed at time $t = 0$. Thereafter, the constant current source, by varying its emf, maintains a constant current i out of its upper terminal. (a) Derive an expression for the current through the inductor as a function of time. (b) Show that the current through the resistor equals the current through the inductor at time $t = (L/R)\ln 2$.

33. When Is the Flux Equal In Fig. 32-38a, switch S has been closed on A long enough to establish a steady current in the inductor of inductance $L_1 = 5.00$ mH and the resistor of resistance $R_1 = 25\ \Omega$. Similarly, in Fig. 32-38b, switch S has been closed on A long enough to establish a steady current in the inductor of inductance $L_2 = 3.00$ mH and the resistor of resistance $R_2 = 30\ \Omega$. The ratio Φ_{02}/Φ_{01} of the magnetic flux through a turn in inductor 2 to that in inductor 1 is 1.5. At time $t = 0$, the two switches are closed on B. At what time t is the flux through a turn in the two inductors equal?

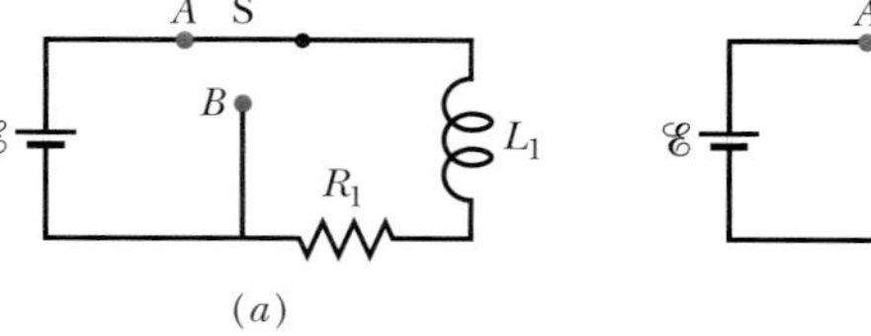

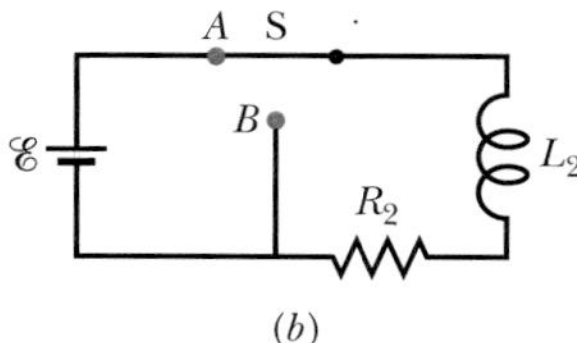

FIGURE 32-38 ■ Problem 33.

34. When Is emf Equal Switch S in Fig. 32-36 is closed at time $t = 0$, initiating the buildup of current in the 15.0 mH inductor and the 20.0 Ω resistor. At what time is the emf across the inductor equal to the potential difference across the resistor?

SEC. 32-5 ■ INDUCTORS, TRANSFORMERS, AND ELECTRIC POWER

35. A Transformer A transformer has 500 primary turns and 10 secondary turns. (a) If ΔV_p is 120 V (rms), what is ΔV_s with an open circuit? (b) If the secondary now has a resistive load of 15 Ω, what are the currents in the primary and secondary?

36. A Generator A generator supplies 100 V to the primary coil of a transformer of 50 turns. If the secondary coil has 500 turns, what is the secondary voltage?

37. Audio Amplifier In Fig. 32-39 let the rectangular box on the left represent the (high-impedance) output of an audio amplifier, with $r = 1000\ \Omega$. Let $R = 10\ \Omega$ represent the (low-impedance) coil of a loudspeaker. For maximum transfer of energy to the load R we must have $R = r$, and that is not true in this case. However, a transformer can be used to "transform" resistances, making them behave electrically as if they were larger or smaller than they actually are. Sketch the primary and secondary coils of a transformer that can be introduced between the amplifier and the speaker in Fig. 32-39 to match the impedances. What must be the turns ratio?

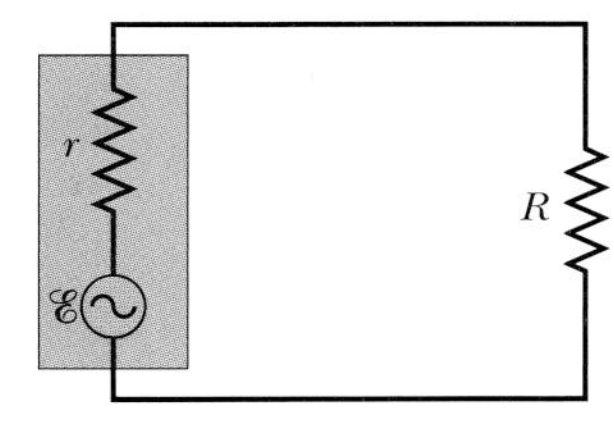

FIGURE 32-39 ■ Problem 37.

38. Autotransformer Figure 32-40 shows an "autotransformer." It consists of a single coil (with an iron core). Three taps T_N are provided. Between taps T_1 and T_2 there are 200 turns, and between taps T_2 and T_3 there are 800 turns. Any two taps can be considered the "primary terminals" and any two taps can be considered the "secondary terminals." List all the ratios by which the primary voltage may be changed to a secondary voltage.

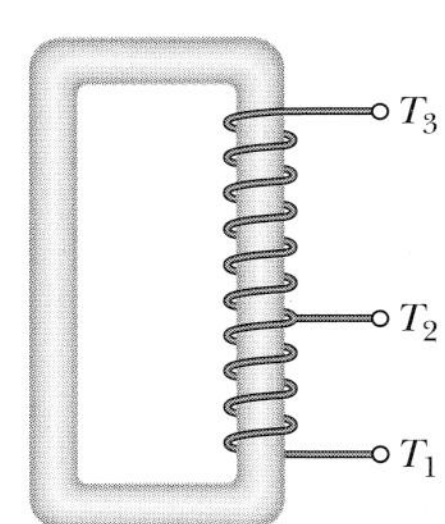

FIGURE 32-40 ■ Problem 38.

SEC. 32-6 ■ MAGNETIC MATERIALS—AN INTRODUCTION

39. Orbital Magnetic Dipole What is the measured component of the orbital magnetic dipole moment of an electron with (a) $m_l = 1$ and (b) $m_l = -2$?

40. Energy Difference What is the energy difference between parallel and antiparallel alignment of the z-component of an electron's spin magnetic dipole moment with an external magnetic field of magnitude 0.25 T, directed parallel to the z axis?

41. Electron in an Atom If an electron in an atom has an orbital angular momentum with $m_l = 0$, (a) what is the component μ_z^{orb}? If the atom is in an external magnetic field $\vec{B}$ of magnitude 35 mT and directed along z axis, what are the potential energies associated with the orientations of (b) the electron's orbital magnetic dipole moment and (c) the electron's spin magnetic dipole moment? (d) Repeat (a) through (c) for $m_l = -3$.

42. Spin Magnetic Moment An electron is placed in a magnetic field $\vec{B}$ that is directed along a z axis. The energy difference between parallel and antiparallel alignments of the z-component of the electron's spin magnetic moment with $\vec{B}$ is 6.00×10^{-25} J. What is the magnitude of $\vec{B}$?

43. How Many Suppose that ±4 are the limits to the values of m_l for an electron in an atom. (a) How many different values of the z-component μ_z^{orb} of the electron's orbital magnetic dipole moment are possible? (b) What is the greatest magnitude of those possible values? Next, suppose that the atom is in a magnetic field of magnitude 0.250 T, in the positive direction of the z axis. What are (c) the maximum potential energy and (d) the minimum potential energy associated with those possible values of μ_z^{orb}?

44. NMR and MRI Nuclear Magnetic Resonance (NMR) and Magnetic Resonance Imaging (MRI) exploit the interactions between charged particles and very strong magnetic fields in order to produce images (including images of soft tissue). The magnetic field in a certain MRI machine is 0.5 Tesla. What is the maximum difference in energy that one might measure for a single electron placed in this field?

SEC. 32-7 ■ FERROMAGNETISM

45. Saturation Magnetization The saturation magnetization M^{max} of the ferromagnetic metal nickel is 4.70×10^5 A/m. Calculate the magnetic moment of a single nickel atom. (The density of nickel is 8.90 g/cm³ and its molar mass is 58.71 g/mol.)

46. Iron The dipole moment associated with an atom of iron in an iron bar has magnitude 2.1×10^{-23} J/T. Assume that all the atoms in the bar, which is 5.0 cm long and has a cross-sectional area of 1.0 cm², have their dipole moments aligned. (a) What is the magnitude of the dipole moment of the bar? (b) What is the magnitude of the torque that must be exerted to hold this magnet perpendicular to an external field of 1.5 T? (The density of iron is 7.9 g/cm³.)

47. Earth's Magnetic Moment The magnetic dipole moment of Earth has magnitude 8.0×10^{22} J/T. (a) If the origin of this magnetism were a magnetized iron sphere at the center of the Earth, what would be its radius? (b) What fraction of the volume of the Earth would such a sphere occupy? Assume complete alignment of the dipoles. The density of the Earth's inner core is 14 g/cm³. The magnetic dipole moment of an iron atom is 2.1×10^{-23} J/T. (*Note:* The Earth's inner core is in fact thought to be in both liquid and solid forms and partly iron, but a permanent magnet as the source of the Earth's magnetism has been ruled out by several considerations. For one, the temperature is certainly above the Curie point.)

48. Mines and Boreholes Measurements in mines and boreholes indicate that the Earth's interior temperature increases with depth at the average rate of 30 C°/km. Assuming a surface temperature of 10°C, at what depth does iron cease to be ferromagnetic? (The Curie temperature of iron varies very little with pressure.)

SEC. 32-8 ■ OTHER MAGNETIC MATERIALS

49. Electron Assume that an electron of mass m and charge magnitude e moves in a circular orbit of radius r about a nucleus. A uniform magnetic field $\vec{B}$ is then established perpendicular to the plane of the orbit. Assuming also that the radius of the orbit does not change and that the change in the speed of the electron due to field $\vec{B}$ is small, find an expression for the change in the orbital magnetic dipole moment of the electron due to the field.

50. Loop Model Figure 32-41 shows a loop model (loop L) for a diamagnetic material. (a) Sketch the magnetic field lines through and about the material due to the bar magnet. (b) What are the directions of the loop's net magnetic dipole moment $\vec{\mu}$ and the conventional current i in the loop? (c) What is the direction of the magnetic force on the loop?

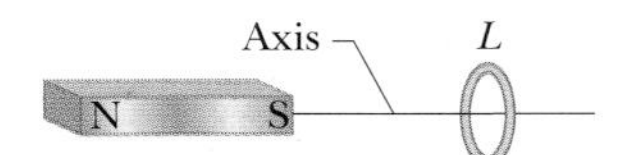

FIGURE 32-41 ■ Problems 50 and 54.

51. Cylindrical Magnet A magnet in the form of a cylindrical rod has a length of 5.00 cm and a diameter of 1.00 cm. It has a uniform magnetization of 5.30×10^3 A/m. What is the magnitude of its magnetic dipole moment?

52. Paramagnetic Gas A magnetic field of magnitude 0.50 T is applied to a paramagnetic gas whose atoms have an intrinsic magnetic dipole moment of magnitude 1.0×10^{-23} J/T. At what temperature will the mean kinetic energy of translation of the gas atoms be equal to the energy required to reverse such a dipole end for end in this magnetic field?

53. Paramagnetic Salt A sample of the paramagnetic salt to which the magnetization curve of Fig. 32-19 applies is to be tested to see whether it obeys Curie's law. The sample is placed in a uniform 0.50 T magnetic field that remains constant throughout the experiment. The magnetization M is then measured at temperatures ranging from 10 to 300 K. Will Curie's law be valid under these conditions?

54. Paramagnetic Material Repeat Problem 50 for the case in which loop L is the model for a paramagnetic material.

55. Electron's Kinetic Energy An electron with kinetic energy K travels in a circular path that is perpendicular to a uniform magnetic field, the electron's motion is subject only to the force due to the field. (a) Show that the magnetic dipole moment of the electron due to its orbital motion has magnitude $\mu = K/|\vec{B}|$ and that it is in the direction opposite that of $\vec{B}$. (b) What are the magnitude and direction of the magnetic dipole moment of a positive ion with kinetic energy K_{ion} under the same circumstances? (c) An ionized gas consists of 5.3×10^{21} electrons/m^3 and the same number density of ions. Take the average electron kinetic energy to be 6.2×10^{-20} J and the average ion kinetic energy to be 7.6×10^{-21} J. Calculate the magnetization of the gas when it is in a magnetic field of 1.2 T.

56. Magnetization Curve A sample of the paramagnetic salt to which the magnetization curve of Fig. 32-17 applies is held at room temperature (300 K). At what applied magnetic field will the degree of magnetic saturation of the sample be (a) 50% and (b) 90%? (c) Are these fields attainable in the laboratory?

SEC. 32-9 ■ THE EARTH'S MAGNETISM

57. New Hampshire In New Hampshire the average horizontal component of the Earth's magnetic field in 1912 was 16 μT and the average inclination or "dip" was 73°. What was the corresponding magnitude of the Earth's magnetic field?

58. Earth's Field Assume the average value of the vertical component of the Earth's magnetic field is 43 μT (downward) for all of Arizona, which has an area of 2.95×10^5 km^2, and calculate the net magnetic flux through the rest of the Earth's surface (the entire surface excluding Arizona). Is that net magnetic flux outward or inward?

59. Earth's Field Two Use the results of Problem 60 to predict the Earth's magnetic field (both magnitude and inclination) at (a) the geomagnetic equator, (b) a point at geomagnetic latitude 60°, and (c) the north geomagnetic pole.

60. Magnetic Field of Earth The magnetic field of the Earth can be approximated as the magnetic field of a dipole, with horizontal and vertical components, at a point a distance r from the Earth's center, given by

$$B_h = \frac{\mu_0 \mu}{4\pi r^3} \cos \lambda_m, \qquad B_v = \frac{\mu_0 \mu}{2\pi r^3} \sin \lambda_m,$$

where λ_m is the *magnetic latitude* (this type of latitude is measured from the geomagnetic equator toward the north or south geomagnetic pole). Assume that the Earth's magnetic dipole moment is $\mu = 8.00 \times 10^{22}$ A · m^2. (a) Show that the magnitude of the Earth's field at latitude λ_m is given by

$$B = \frac{\mu_0 \mu}{4\pi r^3} \sqrt{1 + 3 \sin^2 \lambda_m}\,.$$

(b) Show that the inclination ϕ_i of the magnetic field is related to the magnetic latitude λ_m by

$$\tan \phi_i = 2 \tan \lambda_m.$$

Additional Problems

61. Rate of Energy Transfer In Fig. 32-36, a 12.0 V ideal battery, a 20 Ω resistor, and an ideal inductor are connected by a switch at time $t = 0$ s. At what rate is the battery transferring energy to the inductor's field at $t = 1.61\tau_L$?

62. Compass Needle You place a magnetic compass on a horizontal surface, allow the needle to settle into equilibrium position, and then give the compass a gentle wiggle to cause the needle to oscillate about that equilibrium position. The frequency of oscillation is 0.312 Hz. The Earth's magnetic field at the location of the compass has a horizontal component of 18.0 μT. The needle has a magnetic moment of 0.680 mJ/T. What is the needle's rotational inertia about its (vertical) axis of rotation?

63. Induced Current in a Coil A long narrow coil is surrounded by a short wide coil as shown in Fig. 32-42. Both coils have negligible resistance. The short wide coil has a diameter d_S, n_S turns per unit length, and a length S. Its ends are connected through a resistor of resistance R. The long narrow inner coil has a diameter d_L, n_L turns per unit length, and a length L. Its ends are connected across a variable power source.

For each of the partial sentences below, indicate whether they are correctly completed by the phrase greater than (>), less than (<), or the same as (=). If you cannot determine which is the case from the information given, indicate not sufficient information (NSI).

The current through an inner coil is increased from 0.0 amps to 0.1 amps over a period of 10 seconds in a smooth fashion according to the rule

$$i_L(t) = (0.01 \text{ A/s})\, t.$$

(a) The magnitude of the current in the long narrow coil at time $t = 1$ s is _________ the current in that coil at time $t = 5$ s.

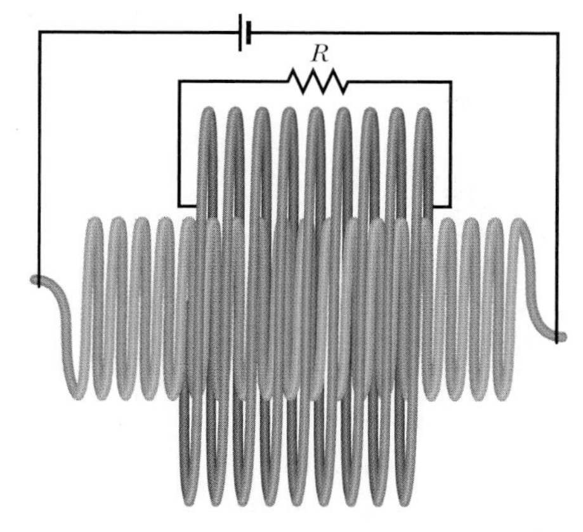

FIGURE 32-42 ■ Problem 63.

(b) The magnitude of the current in the short wide coil at time $t = 1$ s is_________ the current in that coil at time $t = 5$ s.
(c) The magnitude of the current in the long narrow coil at time $t = 1$ s is _________ the current in the short wide coil at that same time.
(d) If the long narrow coil was compressed to half its length (without changing its diameter) before the current was turned on, the current in the short wide coil would be _________ it was without the compression.

64. Inducing Current Figure 32-43 shows a solenoid and two hoops. When the switch is closed, the solenoid carries a current in the direction indicated. The planes of the small loops are parallel to the planes of the hoops of the solenoid.

Hoop 1 consists of a single turn of resistive wire that has a resistance per unit length of λ. Hoop 2 consists of N turns of the same wire. Each hoop is a circle of radius r.

(a) The switch is closed and remains closed for a few seconds. Hoop 1 is then moved to the right. Is there a current flow induced? If there is a current, indicate the direction and explain how you figured it out.

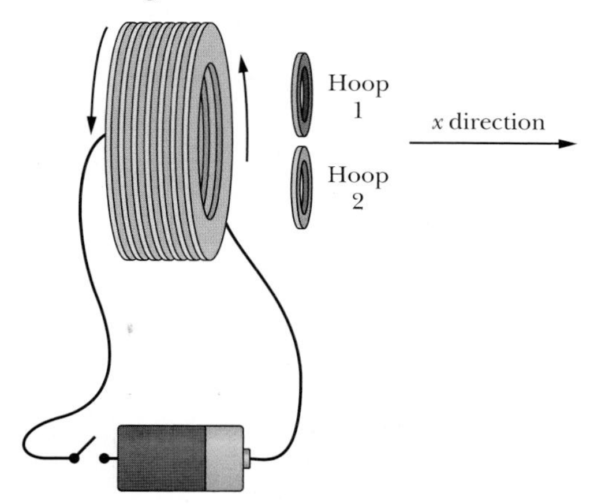

FIGURE 32-43 ■ Problem 64.

(b) The hoops are now returned to their original locations and held fixed. The switch is opened. For a short time, the magnetic field at the hoops decreases like

$$B_x(t) = B_x(0) - \gamma t,$$

where γ is a constant with units of gauss per second. Is there a current flow induced in the hoops? If there is a current, indicate the direction.
(c) Calculate the current flow in each hoop for situation (b).
(d) If $B_x(0) = 10$ gauss, $\gamma = 2.5$ gauss/s, the resistivity of the wire is 1 $\mu\Omega$/m, and the hoops have a radius of 2 cm, calculate the current induced in hoop 1 as the B-field from the solenoid begins to fall.

33 Electromagnetic Oscillations and Alternating Current

When a high-voltage power transmission line requires repair, a utility company cannot just shut it down, perhaps blacking out an entire city. Repairs must be made while the lines are electrically "hot." The man outside the helicopter in this photograph has just replaced a spacer between 500 kV lines *by hand*, a procedure that requires considerable expertise.

How does he manage this repair without being electrocuted?

The answer is in this chapter.

33-1 Advantages of Alternating Current

So far we have confined our study of electric circuits to **direct current** or **dc** circuits in which the direction of current does not change over time. In reality the vast majority of electric power systems and electrical devices involve **alternating current** or *ac* circuits where the current direction is continuously oscillating back and forth. Why did ac power become so popular? By 1879 the famous American inventor, Thomas Edison, refined the electric lightbulb invented by Humphry Davy in England. Almost overnight, there was high demand in Europe and the United States for the creation of systems for the generation and distribution of electric power.

Edison's quest for a practical lightbulb was apparently motivated by his desire to promote the use of the dc power system that he and his colleagues were developing. Indeed, his dc power station in lower Manhattan quickly become a monopoly, but only temporarily. By 1888 the Serbian immigrant Nikola Tesla (Fig. 33-1) had patented a complete system of alternating current generators, transformers, transmission lines, and induction motors. Shortly thereafter, entrepreneur-inventor George Westinghouse (Fig. 33-2) purchased Tesla's patents. After the Westinghouse Company's ac system was featured at the 1893 Chicago World Fair, more than 80% of all electrical devices were powered by ac circuits.

FIGURE 33-1 ■ Nikola Tesla, an eccentric Serbian-American scientist and electrical engineer, invented the first successful ac power generation system.

You already know some of the key factors that render ac power superior to dc. In Section 32-5 we discussed how electricity generated by induction naturally produces alternating current. We discussed how transformers can be used to step up voltages so that power can be transmitted more efficiently over long distances.* There were other factors that favored the Westinghouse system. Alternating current power transmission requires far less copper wire than Edison's dc system. Furthermore, the ac induction motors invented by Tesla were so efficient and easy to manufacture that they quickly became the heart of almost all labor-saving household devices, including water pumps, washing machines, dryers, electric drills, blenders, dishwashers, and garbage disposals.

There are other more recent inventions that we now take for granted that operate on ac circuits. Examples include radio and television transmission (treated in Chapter 34) and reception, computer monitors, and even the graphic equalizers in hi-fi equipment. Thus, without an understanding of ac circuits, it is impossible to understand how modern electrical systems and devices work. So, the major focus of this chapter is to use what you already know about induction and dc circuits to help you understand ac circuits. Resistors, capacitors, and inductors are the basic building blocks of both ac and dc circuits. We have already studied the independent functioning of each. In addition, we learned about dc resistor-capacitor (RC) combinations in Chapter 28 and dc resistor-inductor (RL) combinations in Chapter 32.

We begin this chapter by deriving equations that quantify the energy and energy density stored in the magnetic field created by current flowing through an inductor. We also review what we learned in Section 28-5 about the energy and energy density stored in a capacitor's electric field due to its charge. This will prepare you to study *electromagnetic oscillations* in several types of ac circuits where energy shuttles back and forth between the magnetic field in an inductor and the electric field in a capacitor.

FIGURE 33-2 ■ George Westinghouse developed the first ac power distribution system in the United States. It was based on Tesla's design. The system went online at Niagara Falls in 1896. After only a few years it was found to be superior to existing dc systems.

*The method of repairing high-voltage lines shown in the opening photograph is patented by Scott H. Yenzer and is licensed exclusively to Haverfield Corporation of Gettysburg, Pennsylvania. As the lineman approaches a hot line, the electric field surrounding the line brings his body to nearly the potential of the line. To match the two potentials, he then extends a conducting "wand" to the line. To avoid being electrocuted, he must be isolated from anything electrically connected to the ground. To ensure that his body is always at a single potential—that of the line he is working on—he wears a conducting suit, hood, and gloves, all of which are electrically connected to the line via the wand.

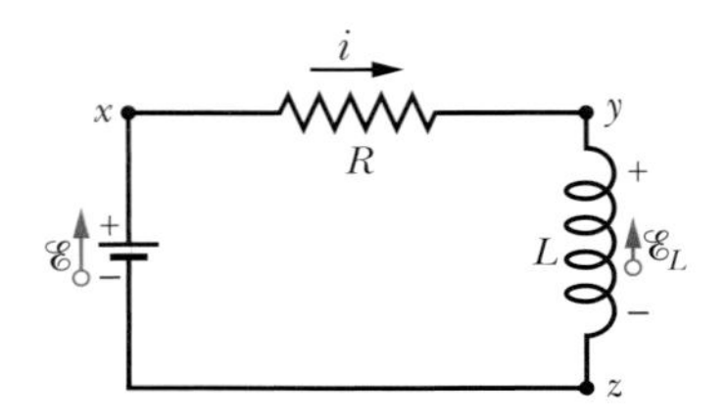

FIGURE 33-3 ■ The circuit of Fig. 32-6 with the switch closed on a. We apply the loop rule for circuits clockwise, starting at x.

33-2 Energy Stored in a $\vec{B}$-Field

When we pull two particles with opposite signs of charge away from each other, the resulting electric potential energy is stored in the electric field of the particles. We get this energy back from the field by letting the particles move toward each other again. In the same way we can consider energy to be stored in a magnetic field.

To derive a quantitative expression for that stored energy, consider Fig. 33-3, which shows a source of emf $\mathscr{E}$ connected to a resistor R and an inductor L. After a switch is closed, the growth of current can be described by Eq. 32-20, which is restated here for convenience,

$$\mathscr{E} = L\frac{di}{dt} + iR. \tag{33-1}$$

This differential equation follows immediately from the loop rule for potential differences in single-loop circuits. If we multiply each side of this expression by the current i we obtain

$$\mathscr{E}i = L\,i\frac{di}{dt} + i^2R, \tag{33-2}$$

which has the following physical interpretation in terms of work and energy:

1. If a charge dq passes through the battery of emf $\mathscr{E}$ in Fig. 33-3 in time dt, the battery does work on it in the amount $\mathscr{E}\,dq$. The rate at which the battery does work is $(\mathscr{E}\,dq)/dt$, or $\mathscr{E}i$. Thus, the left side of Eq. 33-2 represents the rate at which the emf device delivers energy to the rest of the circuit.
2. The rightmost term in Eq. 33-2 represents the rate at which energy is transformed to thermal energy in the resistor.
3. Energy that is delivered to the circuit but does not appear as thermal energy must, by the conservation-of-energy hypothesis, be stored in the magnetic field of the inductor. Since Eq. 33-2 represents conservation of energy for RL circuits, the middle term must represent the rate dU^{mag}/dt at which energy is stored in the magnetic field.

Thus

$$\frac{dU^{\text{mag}}}{dt} = Li\frac{di}{dt}. \tag{33-3}$$

We can write this as

$$dU^{\text{mag}} = Li\,di.$$

Integrating yields

$$\int_0^{U^{\text{mag}}} dU^{\text{mag}} = \int_0^i Li\,di$$

or

$$U^{\text{mag}} = \tfrac{1}{2}Li^2 \quad \text{(magnetic energy)}, \tag{33-4}$$

which represents the total energy stored in the magnetic field of an inductor L carrying a current i. Note the similarity in form between this expression and the expression for the energy a capacitor stores in its electric field due to its capacitance

C and charge q. That equation is given by Eq. 28-21 and restated here for convenience as

$$U^{\text{elec}} = \tfrac{1}{2}\left(\frac{1}{C}\right)q^2. \tag{33-5}$$

(The variable i corresponds to q, and the constant L corresponds to $1/C$.)

33-3 Energy Density of a $\vec{B}$-Field

Since a typical inductor has the shape of either a solenoid or a toroid (a solenoid bent into a donut shape) it is often useful to know the magnetic field energy per unit volume stored in the magnetic field of this type of inductor. Consider a length l near the middle of a long solenoid of cross-sectional area A carrying current i. The volume associated with this length is Al. The energy U^{mag} stored by the length l of the solenoid must lie entirely within this volume because the magnetic field outside such a solenoid is approximately zero. Moreover, the stored energy must be uniformly distributed within the solenoid because the magnetic field inside a solenoid is also essentially uniform.

Thus, the energy u^{mag} stored per unit of magnetic field volume is given by

$$u^{\text{mag}} = \frac{U^{\text{mag}}}{Al}.$$

But since

$$U^{\text{mag}} = \tfrac{1}{2}L\,i^2,$$

we have

$$u^{\text{mag}} = \frac{Li^2}{2Al} = \left(\frac{L}{l}\right)\frac{i^2}{2A}.$$

Here L/l is the inductance of length l of the solenoid. Since the self-induced emf $\mathscr{E}_L = -L\,di/dt$ (Eq. 32-2) and $\mathscr{E}_L = -\mu_0 An^2l\,di/dt$ (Eq. 32-1) for an air-filled solenoid, we can replace L/l in the expression above to find

$$u^{\text{mag}} = \tfrac{1}{2}\mu_0 n^2 i^2, \tag{33-6}$$

where n is the number of turns per unit length. By using Eq. 30-25 ($B = n\mu_0|i|$) we can write this *energy density* as

$$u^{\text{mag}} = \frac{1}{2}\frac{B^2}{\mu_0} \quad \text{(magnetic energy density).} \tag{33-7}$$

The **magnetic energy density,** u^{mag}, is the density of stored energy at any point where the magnetic field is $\vec{B}$. Even though we derived it by considering a special case, the solenoid, it turns out that Eq. 33-7 holds for all magnetic fields, no matter how they are generated. Equation 33-7 is comparable to Eq. 28-23; namely,

$$u^{\text{elec}} = \tfrac{1}{2}\varepsilon_0 E^2, \tag{33-8}$$

which gives the energy density (in a vacuum) at any point in an electric field. Note that both u^{mag} and u^{elec} are proportional to the square of the appropriate field, $\vec{B}$ or $\vec{E}$.

READING EXERCISE 33-1: The table lists the number of turns per unit length, current, and cross-sectional area for three solenoids. Rank the solenoids according to the magnetic energy density within them, greatest first.

Solenoid	Turns per Unit Length	Current	Area
a	$2n_1$	i_1	$2A_1$
b	n_1	$2i_1$	A_1
c	n_1	i_1	$6A_1$

■

33-4 *LC* Oscillations, Qualitatively

We now turn our attention to how electromagnetic oscillations can occur in various types of circuits. We begin with the consideration of a dc circuit. Of the three circuit elements, resistance R, capacitance C, and inductance L, we have so far discussed dc circuits with the series combinations RC (Section 28-9) and RL (Section 32-4). In these two kinds of circuits we found that under certain circumstances, the charge, current, and potential differences across circuit elements can grow or decay exponentially. The exponential nature of the growth and decay curves is the result of energy losses in the resistor. The time constant associated with the exponential growth or decay is denoted by τ, which is either called capacitive or inductive depending on which circuit element is present.

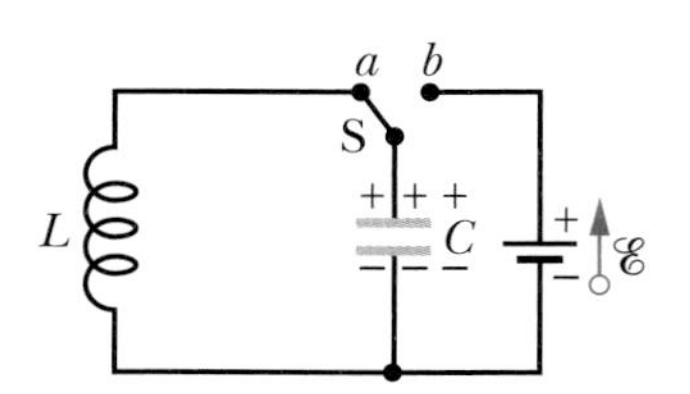

FIGURE 33-4 ■ A series LC circuit. where a switch is thrown from b to a so that an ideal inductor and a charged capacitor are in series.

What if there is almost no resistance in a circuit to dissipate energy? We now examine qualitatively the two-element combination LC in a series circuit. Then in Section 33-6 we will derive equations that describe the behavior of the circuit.

We assume that our LC circuit shown in Fig. 33-4 has a negligible resistance. You will see that in this case, the potential difference across the circuit elements is alternately associated with the inductor and capacitor. Why? The sequence of events is as follows:

- This initial state of the circuit at $t = 0$ as shown in Fig. 33-5*a*. The bar graphs for energy included there indicate that at this instant, with zero current through the inductor and maximum charge on the capacitor, the energy U^{mag} of the magnetic field is zero and the energy U^{elec} of the electric field is a maximum.
- As soon as the switch is thrown from b to a, the potential difference across the capacitor will start a flow of charge from one capacitor plate through the inductor to the other capacitor plate. The back emf generated by the inductor will slow the rate of capacitor discharge. The energy stored in the capacitor's electric field will decrease while the current through the inductor begins to increase its magnetic field energy as shown in Fig. 33-5*b*. Eventually as the capacitor is fully discharged, all the energy in its electric field will be transformed into energy stored in the magnetic field. At this point there will be no potential difference across the capacitor as shown in Fig. 33-5*c*.
- Without the potential difference, the current flowing through the inductor will start to decrease. However, in response to this changing current, a self-induced current will be set up through the inductor: this current will be in the same direction as the original current. Thus excess positive charge will begin to build up on the lower capacitor plate while the upper plate will begin to accumulate negative charge as shown in Fig. 33-5*d*. This will continue until no current flows through the inductor and the charges on the capacitor plates will be opposite in sign to

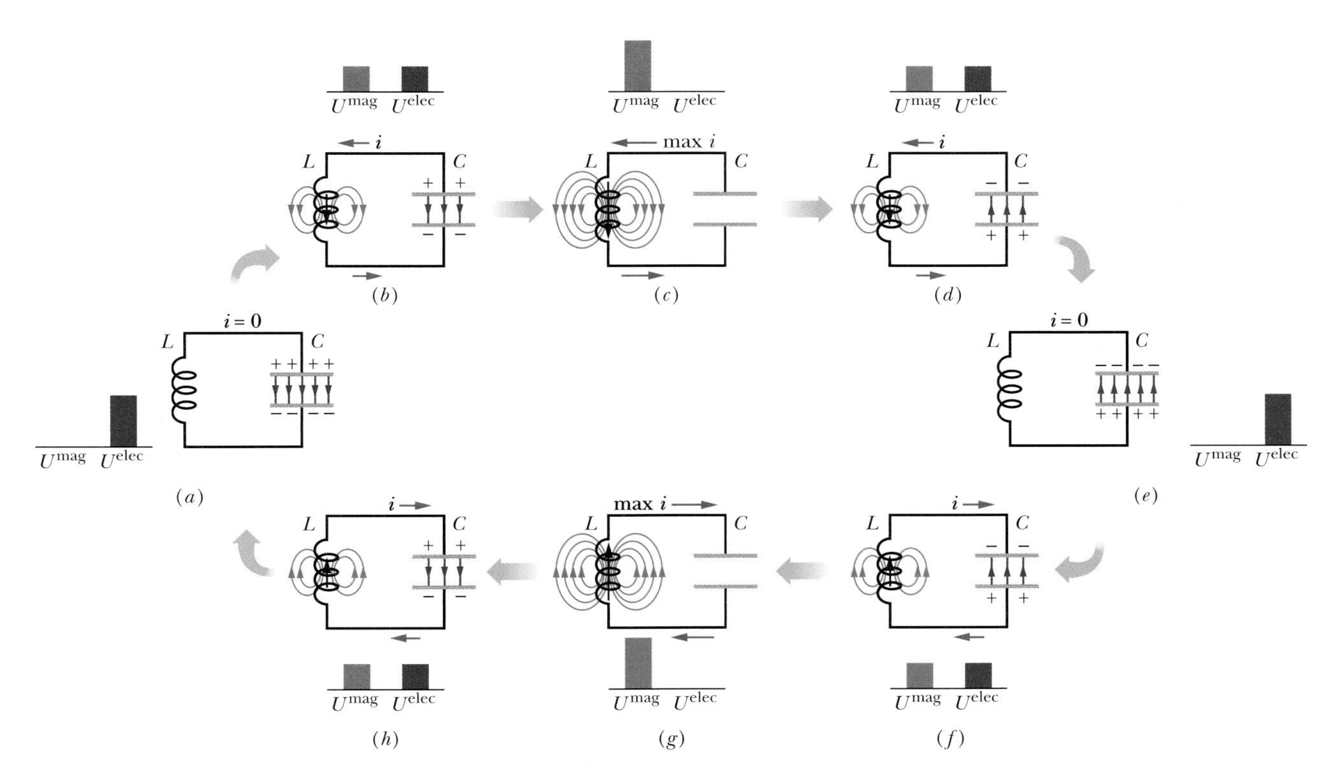

FIGURE 33-5 ■ Stages in a cycle of oscillation of an ideal *LC* circuit. The small bar graphs show levels of stored magnetic and electric energies. The inductor magnetic field lines and the capacitor electric field lines are shown. (*a*) Capacitor with maximum charge, no current. (*b*) Capacitor discharging, current increasing. (*c*) Capacitor fully discharged, current maximum. (*d*) Capacitor charging with opposite polarity to that in (*a*), current decreasing. (*e*) Capacitor with maximum charge with opposite polarity to that in (*a*), no current. (*f*) Capacitor discharging, current increasing with direction opposite that in (*b*). (*g*) Capacitor fully discharged, current maximum. (*h*) Capacitor charging, current decreasing.

that just after the switch was thrown. At this time all the circuit energy is in the capacitor's electric field once again as shown in Fig. 33-5*e*.

- Events 1 and 2 happen again but with the current flowing in the opposite direction until the capacitor is back to its original state (at $t = 0$) as shown in Fig. 33-5*f*, *g*, *h*, and *a*. Without a resistor in the circuit to dissipate energy, the maximum current through the inductor and the maximum amount of charge on the capacitor plates do not decay with time. In theory, for a perfectly ideal inductor, these oscillations can continue forever.

In the next section we do a mathematical analysis that fortunately agrees with observations that the current in the inductor i and the charge on the upper capacitor plate q vary *sinusoidally* with time as shown in Fig. 33-6. The resulting oscillations of the capacitor's electric field and the inductor's magnetic field are said to be **electromagnetic oscillations.**

Parts *a* through *h* of Fig. 33-5 show succeeding stages of the oscillations in a simple *LC* circuit. We know that the energy stored in the electric field of the capacitor at any time is given by $U^{elec} = \frac{1}{2}(q^2/C)$ (Eq. 33-5) where q is the charge on the capacitor at that time. The energy stored in the magnetic field of the inductor at any time is given by $U^{mag} = \frac{1}{2}L\,i^2$ (Eq. 33-4) where i is the current through the inductor at that time.

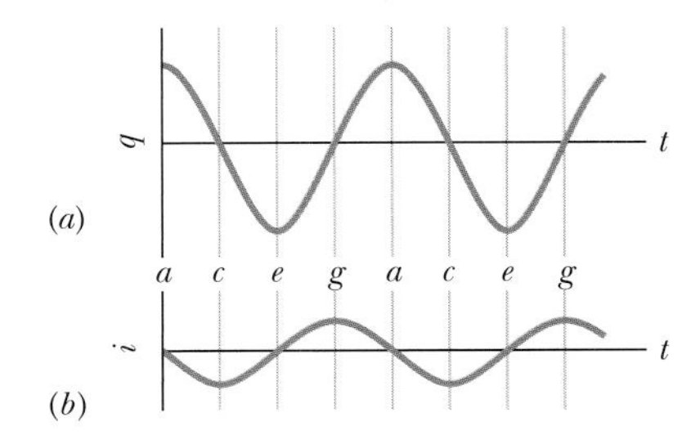

FIGURE 33-6 ■ (*a*) The charge q on the upper plate of the capacitor in an ideal *LC* circuit with almost no resistance (Fig. 33-5) as a function of time. (*b*) The current i in the circuit of Fig. 33-5 (q and i are determined from measurements of ΔV_C and ΔV_R across a very small resistor added to the circuit). The letters refer to the correspondingly labeled oscillation stages in Fig. 33-5.

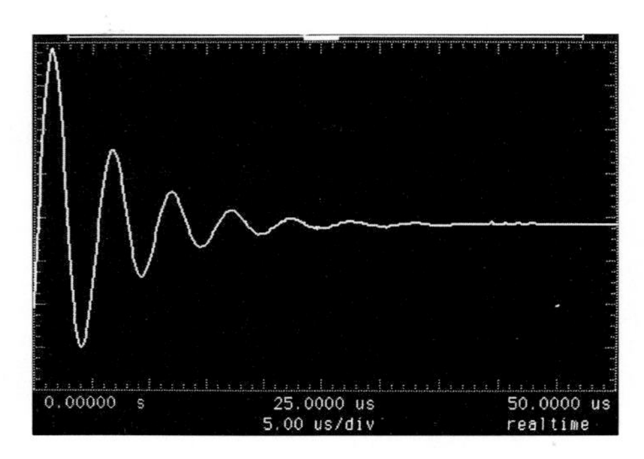

FIGURE 33-7 ■ An oscilloscope trace showing how the oscillations in an RLC circuit actually die away because energy is dissipated in the resistor as thermal energy.

In an actual LC circuit, the oscillations will not continue indefinitely because there is always some resistance present that will drain energy from the electric and magnetic fields and dissipate it as thermal energy (the circuit becomes warmer). The oscillations, once started, will die away as suggested in Fig. 33-7 (which displays a potential difference vs. time for a similar LC circuit with a resistor added to it). Compare this figure with Fig. 16-27, which shows the decay of mechanical oscillations caused by damping forces acting on a physical pendulum.

READING EXERCISE 33-2: A charged capacitor and an inductor are connected in series at time $t = 0$. In terms of the period T of the resulting sinusoidal oscillations shown in Fig. 33-6, determine how much later the following reach their maximums: (a) the charge on the capacitor; (b) the voltage across the capacitor, with its original polarity; (c) the energy stored in the electric field; and (d) the current. ■

TOUCHSTONE EXAMPLE 33-1: *LC* Oscillation

A 1.5 μF capacitor is charged to 57 V. The charging battery is then disconnected, and a 12 mH coil is connected in series with the capacitor so that LC oscillations occur. What is the maximum current in the coil? Assume that the circuit contains no resistance.

SOLUTION ■ The **Key Ideas** here are these:

1. Because the circuit contains no resistance, the electromagnetic energy of the circuit is conserved as the energy is transferred back and forth between the electric field of the capacitor and the magnetic field of the coil (inductor).
2. At any time t, the energy $U^{mag}(t)$ of the magnetic field is related to the current $i(t)$ through the coil by Eq. 33-4 ($U^{mag} = Li^2/2$). When all the energy is stored as magnetic energy, the current is at its maximum value I and that energy is $U^{max}_{mag} = LI^2/2$.
3. At any time t, the energy $U^{elec}(t)$ of the electric field is related to the charge $q(t)$ on the capacitor by Eq. 33-5 ($U^{elec} = q^2/2C$). When all the energy is stored as electric energy, the charge is at its maximum value Q and that energy is $U^{max}_{elec} = Q^2/2C$.

With these ideas, we can now write the conservation of energy as

$$U^{max}_{mag} = U^{max}_{elec}$$

or

$$\frac{LI^2}{2} = \frac{Q^2}{2C}.$$

Solving for I gives us

$$I = \sqrt{\frac{Q^2}{LC}}.$$

We know L and C, but not Q. However, with Eq. 28-1 ($|q| = C|\Delta V|$) we can relate Q to the maximum potential difference ΔV across the capacitor, which is the initial potential difference of 57 V. Thus, substituting $|Q| = C|\Delta V|$ leads to a maximum current magnitude of

$$I = \Delta V\sqrt{\frac{C}{L}} = (57\text{ V})\sqrt{\frac{1.5 \times 10^{-6}\text{ F}}{12 \times 10^{-3}\text{ H}}}$$

$$= 0.637\text{ A} \approx 640\text{ mA}. \qquad \text{(Answer)}$$

33-5 The Electrical–Mechanical Analogy

Let us look a little at an analogy between the ideal oscillating LC system like that shown in Fig. 33-5 and an oscillating block–spring system that experiences no friction forces. Two kinds of energy are involved in the block-spring system. One is potential energy of the compressed or extended spring; the other is kinetic energy of the moving block. These two energies are given by the familiar equations in the left energy column in Table 33-1.

The table also shows, in the right energy column, the two kinds of energy involved in LC oscillations. By looking across the table, we can see an analogy between the forms of the two pairs of energies—the mechanical energies of the block–spring system and the electromagnetic energies of the LC oscillator. The equations for velocity and current at the bottom of the table help us see the details of the analogy. They tell us that

TABLE 33-1
Comparison of the Energy in Two Oscillating Systems

Block–Spring System		LC Oscillator	
Element	**Energy**	**Element**	**Energy**
Spring	Potential, $\frac{1}{2}kx^2$	Capacitor	Electric, $\frac{1}{2}(1/C)q^2$
Block	Kinetic, $\frac{1}{2}mv^2$	Inductor	Magnetic, $\frac{1}{2}Li^2$
Block velocity, $v = dx/dt$		Circuit current, $i = dq/dt$	

the charge q corresponds to the displacement x and the current i corresponds to the block velocity v (in both equations, the former is differentiated to obtain the latter).

These correspondences suggest that in the energy expressions for an LC oscillator, the inverse of the capacitance is mathematically like the spring constant in a block–spring system. It is easy to place charge on the capacitor when $1/C$ is small, just as it's easy to displace a spring when k is small. The inductor is like the block mass. The inductor resists a change of the current in the circuit, and the mass on a spring resists a change in velocity.

In summary,

q corresponds to x, 1/C corresponds to k, i corresponds to v, L corresponds to m.

In Section 16-3 we saw that the angular frequency of oscillation of a (frictionless) block–spring system is

$$\omega = \sqrt{\frac{k}{m}} \quad \text{(block–spring system).} \tag{33-9}$$

The correspondences listed above suggest that to find the angular frequency of oscillation for a (resistanceless) LC circuit, k should be replaced by 1/C and m by L, yielding

$$\omega = \frac{1}{\sqrt{LC}} \quad (LC \text{ circuit}). \tag{33-10}$$

Experimentally, we find this expression is correct. We derive it more formally in the next section.

READING EXERCISE 33-3: What is the standard unit for angular frequency in Eq. 33-10 above? Show that this expression does in fact yield the correct unit. (*Hint:* The unit of ampere can be written as a coulomb per second. How would one write the units of henry and farad in the fundamental units of kilograms, meters, seconds, and coulombs?) ■

33-6 *LC* Oscillations, Quantitatively

Here we want to show explicitly that Eq. 33-10 for the angular frequency of LC oscillations is theoretically valid and that the oscillations should be sinusoidal. At the same time, we want to examine even more closely the analogy between LC oscillations and block–spring oscillations. We start by extending somewhat our earlier treatment of the mechanical block–spring oscillator.

The Block–Spring Oscillator—A Review

We analyzed block-spring oscillations in Chapter 16 in terms of energy transfers and did not—at that early stage—derive the fundamental differential equation that governs those oscillations. We do so now.

We can write, for the total energy U of a block–spring oscillator with a massless spring at any instant,

$$U = U^{\text{blk}} + U^{\text{spr}} = \tfrac{1}{2}mv^2 + \tfrac{1}{2}kx^2, \tag{33-11}$$

where U^{blk} and U^{spr} are, respectively, the kinetic energy of the moving block and the potential energy of the stretched or compressed spring. If there is no friction—which we assume—the total energy U remains constant with time, even though the values of velocity v and displacement x vary. In more formal language, $dU/dt = 0$. This leads to

$$\frac{dU}{dt} = \frac{d}{dt}(\tfrac{1}{2}mv^2 + \tfrac{1}{2}kx^2) = mv\frac{dv}{dt} + kx\frac{dx}{dt} = 0. \tag{33-12}$$

However, by definition, $v \equiv dx/dt$ and $dv/dt \equiv d^2x/dt^2$. With these substitutions, Eq. 33-12 becomes

$$m\frac{d^2x}{dt^2} + kx = 0 \qquad \text{(block–spring oscillations).} \tag{33-13}$$

Equation 33-13 is the fundamental differential equation that governs the frictionless block–spring oscillations. It involves the displacement x and its second derivative with respect to time.

The general solution to Eq. 33-13—that is, the function $x(t)$ that describes the block–spring oscillations—is (as we saw in Eq. 16-5)

$$x(t) = X\cos(\omega t + \phi) \qquad \text{(displacement),} \tag{33-14}$$

where X is the amplitude (or maximum displacement) of the mechanical oscillations undergoing simple harmonic motion, ω is the angular frequency of the oscillations, and ϕ is a phase constant.

The *LC* Oscillator

Now let us analyze the oscillations of an ideal LC circuit with no resistance. We proceed exactly as we just did for the block–spring oscillator. The total energy U present at any instant in an oscillating LC circuit is given by

$$U = U^{\text{mag}} + U^{\text{elec}} = \tfrac{1}{2}L\,i^2 + \frac{1}{2}\left(\frac{1}{C}\right)q^2 \tag{33-15}$$

where U^{mag} is the energy stored in the magnetic field of the inductor and U^{elec} is the energy stored in the electric field of the capacitor. Since we have assumed the circuit resistance to be zero, no energy is transferred to thermal energy and U remains constant with time. In more formal language, dU/dt must be zero. This leads to

$$\frac{dU}{dt} = \frac{d}{dt}\left(\frac{Li^2}{2} + \frac{q^2}{2C}\right) = Li\frac{di}{dt} + \frac{q}{C}\frac{dq}{dt} = 0. \tag{33-16}$$

However, $i = dq/dt$ and $di/dt = d^2q/dt^2$. With these substitutions, Eq. 33-16 becomes

$$L\frac{d^2q}{dt^2} + \frac{1}{C}q = 0 \qquad (LC \text{ oscillations).} \tag{33-17}$$

This is the differential equation that describes the oscillations of a resistanceless LC circuit. Careful comparison shows that Eqs. 33-17 and 33-13 have exactly the same mathematical form, differing only in the symbols used.

Charge and Current Oscillations

Since the differential equations are mathematically identical, their solutions must also be mathematically identical. Because q corresponds to x, we can write the general solution of Eq. 33-17, giving $q(t)$ as a function of time, by analogy to Eq. 33-14 as

$$q(t) = Q\cos(\omega t + \phi) \qquad \text{(charge)}, \tag{33-18}$$

where Q is the amplitude or maximum amount of charge on the capacitor during the charge variations, while ω represents the angular frequency of the electromagnetic oscillations, and ϕ is the phase constant.

Taking the first derivative of Eq. 33-18 with respect to time gives us the time-varying current $i(t)$ of the LC oscillator:

$$i(t) = \frac{dq}{dt} = -\omega Q\sin(\omega t + \phi) \qquad \text{(current)}. \tag{33-19}$$

The amplitude I of this sinusoidally varying current is

$$I = \omega Q, \tag{33-20}$$

so we can rewrite Eq. 33-19 as

$$i(t) = -I\sin(\omega t + \phi). \tag{33-21}$$

Angular Frequencies

We can test whether Eq. 33-18 is a solution of Eq. 33-17 by substituting it and its second derivative with respect to time into Eq. 33-17. The first derivative of Eq. 33-18 is Eq. 33-19. The second derivative is then

$$\frac{d^2q}{dt^2} = -\omega^2 Q\cos(\omega t + \phi).$$

Substituting for q and d^2q/dt^2 in Eq. 33-17, we obtain

$$-L\omega^2 Q\cos(\omega t + \phi) + \left(\frac{1}{C}\right)Q\cos(\omega t + \phi) = 0.$$

Canceling $Q\cos(\omega t + \phi)$ and rearranging lead to

$$\omega = \frac{1}{\sqrt{LC}}.$$

Thus, Eq. 33-18 is indeed a solution of Eq. 33-17 if ω has the constant value $1/\sqrt{LC}$. Note that this expression for ω is exactly that given by Eq. 33-10, which we arrived at by examining correspondences.

The phase constant ϕ in Eq. 33-18 is determined by the conditions that prevail at any certain time—say, $t = 0$. If the conditions yield $\phi = 0$ at $t = 0$, Eq. 33-18 requires that $q = Q$ and Eq. 33-19 requires that $i = 0$; these are the initial conditions represented in Fig. 33-5.

Electric and Magnetic Energy Oscillations

The electric energy stored in the LC circuit at any time t is, from Eqs. 33-5 and 33-18,

$$U^{\text{elec}} = \frac{q^2}{2C} = \frac{Q^2}{2C}\cos^2(\omega t + \phi). \tag{33-22}$$

The magnetic energy is, from Eqs. 33-4 and 33-19,

$$U^{\text{mag}} = \tfrac{1}{2}Li^2 = \tfrac{1}{2}L\omega^2 Q^2 \sin^2(\omega t + \phi).$$

Substituting for ω from Eq. 33-10 then gives us

$$U^{\text{mag}} = \frac{Q^2}{2C}\sin^2(\omega t + \phi). \tag{33-23}$$

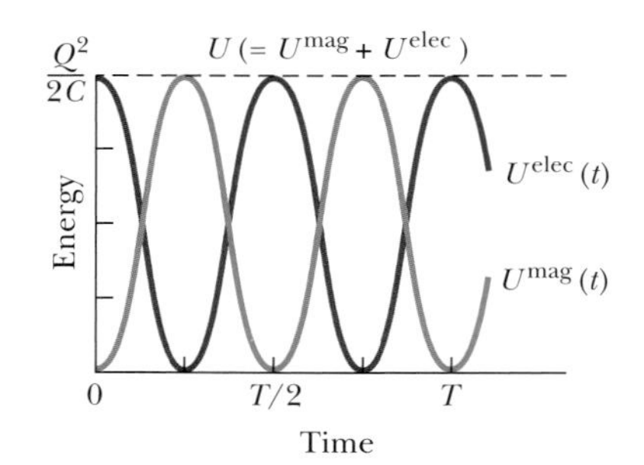

FIGURE 33-8 ■ The stored magnetic energy and electric energy in the circuit of Fig. 33-5 as a function of time. Note that their sum remains constant. T is the period of oscillation.

Figure 33-8 shows plots of $U^{\text{elec}}(t)$ and $U^{\text{mag}}(t)$ for the case of $\phi = 0$. Note that

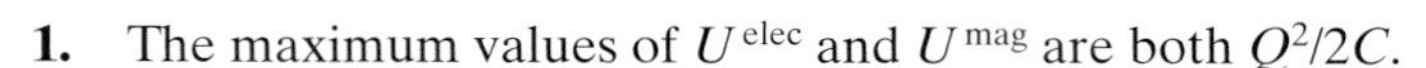

1. The maximum values of U^{elec} and U^{mag} are both $Q^2/2C$.
2. At any instant the sum of U^{elec} and U^{mag} is equal to $Q^2/2C$, a constant.
3. When U^{elec} is maximum, U^{mag} is zero, and conversely.

READING EXERCISE 33-4: A capacitor in an LC oscillator has a maximum potential difference of 20 V and a maximum energy of 160 μJ. When the capacitor has a potential difference of 5 V and an energy of 10 μJ, what are (a) the emf across the inductor and (b) the energy stored in the magnetic field? ■

TOUCHSTONE EXAMPLE 33-2: *LC* Oscillation Continued

For the situation described in Touchstone Example 33-1, let the coil (inductor) be connected to the charged capacitor at time $t = 0$. The result is an LC circuit like that in Fig. 33-4.

(a) What is the potential difference $\Delta v_L(t)$ across the inductor as a function of time?

SOLUTION ■ One **Key Idea** here is that the current and potential differences of the circuit undergo sinusoidal oscillations. Another **Key Idea** is that we can still apply the loop rule to this oscillating circuit—just as we did for the nonoscillating circuits of Chapter 27. At any time t during the oscillations, the loop rule and Fig. 33-4 give us

$$\Delta v_L(t) = \Delta v_C(t); \tag{33-24}$$

that is, the potential difference Δv_L across the inductor must always be equal to the potential difference Δv_C across the capacitor, so that the net potential difference around the circuit is zero. Thus, we will find $\Delta v_L(t)$ if we can find $\Delta v_C(t)$, and we can find $\Delta v_C(t)$ from $q(t)$ with Eq. 28-1 $|q| = C|\Delta V|$.

Because the potential difference $\Delta v_C(t)$ is maximum when the oscillations begin at time $t = 0$, the charge q on the capacitor must also be maximum then. Thus, phase constant ϕ must be zero, so that Eq. 33-18 gives us

$$q = Q\cos\omega t. \tag{33-25}$$

(Note that this cosine function does indeed yield maximum q ($= Q$) when $t = 0$.) To get the potential difference $\Delta v_C(t)$, we divide both sides of Eq. 33-25 by C to write

$$\frac{q}{C} = \frac{Q}{C}\cos\omega t,$$

and then use Eq. 28-1 to write

$$\Delta v_C = \Delta V_C \cos\omega t. \tag{33-26}$$

Here, ΔV_C is the amplitude of the oscillations in the potential difference Δv_C across the capacitor.

Next, substituting $\Delta v_C = \Delta v_L$ from Eq. 33-24, we find

$$\Delta v_L = \Delta V_C \cos \omega t. \qquad (33\text{-}27)$$

We can evaluate the right side of this equation by first noting that the amplitude ΔV_C is equal to the initial (maximum) potential difference of 57 V across the capacitor. Then, using the values of L and C from Touchstone Example 33-1, we find ω with Eq. 33-10:

$$\omega = \frac{1}{\sqrt{LC}} = \frac{1}{[(0.012\text{ H})(1.5 \times 10^{-6}\text{ F})]^{0.5}}$$
$$= 7454\text{ rad/s} \approx 7500\text{ rad/s}.$$

Thus, Eq. 33-27 becomes

$$\Delta v_L = (57\text{ V})\cos(7500\text{ rad/s})t. \qquad \text{(Answer)}$$

(b) What is the maximum rate $(di/dt)^{\max}$ at which the current i changes in the circuit?

SOLUTION ■ The **Key Idea** here is that, with the charge on the capacitor oscillating as in Eq. 33-18, the current is in the form of Eq. 33-19. Because $\phi = 0$, that equation gives us

$$i = -\omega Q \sin \omega t.$$

Then
$$\frac{di}{dt} = \frac{d}{dt}(-\omega Q \sin\omega t) = -\omega^2 Q \cos \omega t.$$

We can simplify this equation by substituting $C\Delta V_C$ for Q (because we know C and ΔV_C but not Q) and $1/\sqrt{LC}$ for ω according to Eq. 33-10. We get

$$\frac{di}{dt} = -\frac{1}{LC} C\Delta V_C \cos \omega t = -\frac{\Delta V_C}{L}\cos \omega t.$$

This tells us that the current changes at a varying (sinusoidal) rate, with its maximum rate of change being

$$\frac{\Delta V_C}{L} = \frac{57\text{ V}}{0.012\text{ H}} = 4750\text{ A/s} \approx 4800\text{ A/s}. \qquad \text{(Answer)}$$

33-7 Damped Oscillations in an *RLC* Circuit

A circuit containing resistance, inductance, and capacitance is called an *RLC circuit*. We shall here discuss only *series RLC circuits* like that shown in Fig. 33-9. With a resistance R present, the total *electromagnetic energy U* of the circuit (the sum of the electric energy and magnetic energy) is no longer constant; instead, it decreases with time as energy is transferred to thermal energy in the resistance. Because of this loss of energy, the oscillations of charge, current, and potential difference continuously decrease in amplitude, and the oscillations are referred to as *damped*. As you will see, they are damped in exactly the same way as those of the damped block–spring oscillator of Section 16-8.

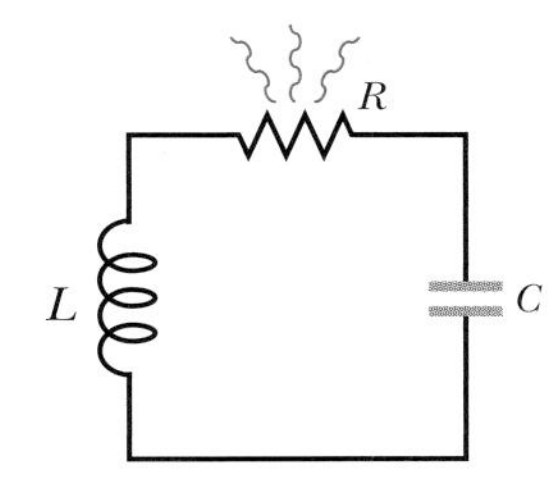

FIGURE 33-9 ■ A series *RLC* circuit. As the charge contained in the circuit oscillates back and forth through the resistance, electromagnetic energy is dissipated as thermal energy, damping (decreasing the amplitude of) the oscillations.

To analyze the oscillations of our *RLC* circuit, we write an equation for the total electromagnetic energy U in the circuit at any instant. Because the resistance does not store electromagnetic energy, we can use Eq. 33-15:

$$U = U^{\text{mag}} + U^{\text{elec}} = \frac{Li^2}{2} + \frac{q^2}{2C}. \qquad (33\text{-}28)$$

Now, however, this total energy decreases as energy is transferred to thermal energy. The rate of that transfer is, from Eq. 26-11,

$$\frac{dU}{dt} = -i^2 R, \qquad (33\text{-}29)$$

where the minus sign indicates that U decreases. By differentiating Eq. 33-28 with respect to time and then substituting the result in Eq. 33-29, we obtain

$$\frac{dU}{dt} = Li\frac{di}{dt} + \frac{q}{C}\frac{dq}{dt} = -i^2 R.$$

Substituting dq/dt for i and d^2q/dt^2 for di/dt, we obtain

$$L\frac{d^2q}{dt^2} + R\frac{dq}{dt} + \frac{1}{C}q = 0 \quad (RLC\text{ circuit}), \qquad (33\text{-}30)$$

which is the differential equation that describes damped oscillations in an RLC circuit. The solution to Eq. 33-30 is

$$q = Qe^{-Rt/2L}\cos(\omega' t + \phi), \tag{33-31}$$

where

$$\omega' = \sqrt{\omega^2 - (R/2L)^2}, \tag{33-32}$$

with $\omega = 1/\sqrt{LC}$, as with an undamped oscillator. Equation 33-31 tells us how the charge on the capacitor oscillates in a damped RLC circuit. That equation is the electromagnetic counterpart of Eq. 16-37, which gives the displacement of a damped block–spring oscillator.

Equation 33-31 describes a sinusoidal oscillation (the cosine function) with an *exponentially decaying amplitude* $Qe^{-Rt/2L}$ (the factor that multiplies the cosine) as shown in Fig. 33-7. The angular frequency ω' of the damped oscillations is always less than the angular frequency ω of the undamped oscillations; however, we shall here consider only situations for which R is small enough for us to replace ω' with ω.

Let us next find an expression for the energy of the electric field in the capacitor, which is given by Eq. 33-5 ($U^{\text{elec}} = q^2/2C$). By substituting Eq. 33-31 into Eq. 33-5, we obtain

$$U^{\text{elec}} = \frac{q^2}{2C} = \frac{[Qe^{-Rt/2L}\cos(\omega' t + \phi)]^2}{2C} = \frac{Q^2}{2C}e^{-Rt/L}\cos^2(\omega' t + \phi). \tag{33-33}$$

Thus, the energy of the electric field oscillates according to a cosine-squared term and the amplitude of that oscillation decreases exponentially with time.

If we do a similar derivation for the energy of the magnetic field in the inductor we find that it too oscillates in such a way that its amplitude decreases exponentially in time. Since energy is being traded back and forth between the inductor and the capacitor, the total electromagnetic energy (which is the sum of the electric and magnetic energies) does not oscillate. Instead it just decays exponentially as the total energy is transformed to thermal energy by the total resistance in the circuit.

TOUCHSTONE EXAMPLE 33-3: Decaying Oscillation

A series RLC circuit has inductance $L = 12$ mH, capacitance $C = 1.6\ \mu$F, and resistance $R = 1.5$ V.

(a) At what time t will the amplitude of the charge oscillations in the circuit be 50% of its initial value?

SOLUTION ■ The **Key Idea** here is that the amplitude of the charge oscillations decreases exponentially with time t: According to Eq. 33-31, the charge amplitude at any time t is $Qe^{-Rt/2L}$, in which Q is the amplitude at time $t = 0$. We want the time when the charge amplitude has decreased to $0.50Q$—that is, when

$$Qe^{-Rt/2L} = 0.50Q.$$

Canceling Q and taking the natural logarithms of both sides, we have

$$-\frac{Rt}{2L} = \ln 0.50.$$

Solving for t and then substituting given data yield

$$t = -\frac{2L}{R}\ln 0.50 = -\frac{(2)(12 \times 10^{-3}\ \text{H})(\ln 0.50)}{1.5\ \Omega}$$

$$= 0.0111\ \text{s} \approx 11\ \text{ms}. \qquad \text{(Answer)}$$

(b) How many oscillations are completed within this time?

SOLUTION ■ The **Key Idea** here is that the time for one complete oscillation is the period $T' = 2\pi/\omega'$, where the angular frequency for decaying LC oscillations is given by Eq. 33-32 ($\omega' = \sqrt{\omega^2 - (R/2L)^2}$) where

$$\omega^2 = 1/(LC) = 1/[(0.012\ \text{H})(1.6 \times 10^{-6}\ \text{F})]$$

$$= 52.1 \times 10^6\ (\text{rad/s})^2,$$

while

$$(R/2L)^2 = [(1.5\ \Omega)/(2)(0.012\ \text{H})]^2$$
$$= 3.91 \times 10^3 (\text{rad/s})^2.$$

Since $(R/2L)^2 \ll \omega^2$, we can neglect $(R/2L)^2$ compared with ω^2, so here $\omega' \cong \omega = \sqrt{52.1 \times 10^6\ (\text{rad/s}^2)} = 7.22 \times 10^3$ rad/s.

The time for one period of the decaying oscillation is then

$$T' = \frac{2\pi}{\omega'} \cong \frac{2\pi}{\omega} = 2\pi/(7.22 \times 10^3\ \text{rad/s})$$
$$= 0.871 \times 10^{-3}\ \text{s}.$$

Thus, in the time interval $\Delta t = 0.0111$ s, the number of complete oscillations is

$$\frac{\Delta t}{T'} = \frac{0.0111\ \text{s}}{0.871 \times 10^{-3}\ \text{s}} \cong 13. \qquad \text{(Answer)}$$

Thus, the amplitude decays by 50% in about 13 complete oscillations. This damping is less severe than that shown in Fig. 33-7, where the amplitude decays by a little more than 50% in one oscillation.

33-8 More About Alternating Current

The oscillations in an *RLC* circuit will not damp out if an external emf device supplies enough energy to make up for the energy dissipated as thermal energy in the resistance R. As we discussed at the beginning of the chapter, the United States and most other countries deliver alternating current or ac electricity. These oscillating emfs and currents vary sinusoidally with time, reversing direction (in North America) 120 times per second and thus having frequency $f = 60$ Hz.

At first sight this may seem to be a strange arrangement. We have seen that the drift speed of the conduction electrons in household wiring may typically be 4×10^{-5} m/s. If we now reverse their direction every 1/120th of a second, such electrons can move only about 3×10^{-7} m in a half-cycle. At this rate, a typical electron can drift past no more than about 10 atoms in the wiring before it is required to reverse its direction. How, you may wonder, can the electron ever get anywhere?

Although this question may be worrisome, it is a needless concern. The conduction electrons do not have to "get anywhere." This is similar to the idea that the molecules in a spring do not have to move far longitudinally to transmit energy a long distance. Here, the electrons don't have to move far to have a long-range effect. When we say that the current in a wire is one ampere, we mean that charge passes through any plane cutting across that wire at the rate of one coulomb per second. The speed at which the charge carriers cross that plane does not matter directly; one ampere may correspond to many charge carriers moving very slowly or to a few moving very rapidly.

Furthermore, the signal to the electrons to reverse directions—which originates in the alternating emf provided by the power company's generator—is propagated along the conductor at a speed close to that of light. All electrons, no matter where they are located, get their reversal instructions at about the same instant. Finally, we note that for many devices, such as lightbulbs and toasters, the direction of motion is unimportant as long as the electrons do move so as to transfer energy to the device via collisions with atoms in the device.

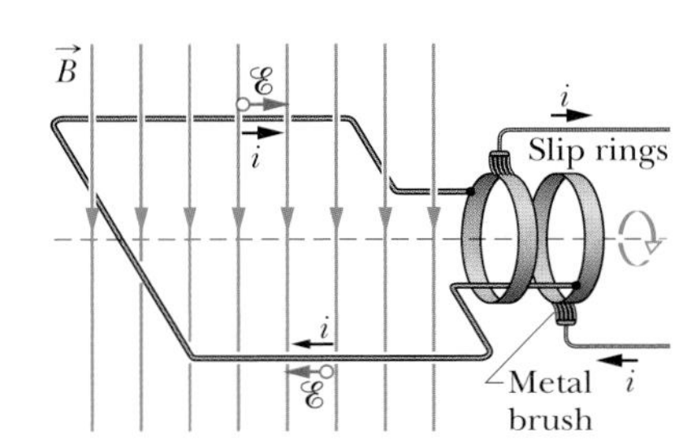

FIGURE 33-10 ■ The basic mechanism of an alternating-current generator is a conducting loop rotated in an external magnetic field. In practice, the alternating emf induced in a coil of many turns of wire is made accessible by means of slip rings attached to the rotating loop. Each ring is connected to one end of the loop wire and is electrically connected to the rest of the generator circuit by a conducting brush against which it slips as the loop (and it) rotates.

Generator Equations

Figure 33-10 shows a simplified model of an ac generator like that shown in Fig. 32-12. As the conducting loop is forced to rotate through the external magnetic field $\vec{B}$, a sinusoidally oscillating emf $\mathscr{E}$ is induced in the loop:

$$\mathscr{E} = \mathscr{E}^{\max} \sin \omega^{\text{dr}} t. \qquad (33\text{-}34)$$

The *angular frequency* ω^{dr} of the emf is equal to the angular speed with which the loop rotates in the magnetic field, the *phase* of the emf is $\omega^{\text{dr}} t$, and the *amplitude* of the emf is $\mathscr{E}^{\max}$(where the superscript stands for maximum). When the rotating loop is part of a closed conducting path, this emf produces (*drives*) a sinusoidal (alternating) current along the path with the same angular frequency ω^{dr}, which then is called the **driving angular frequency.** Following Eq. 33-21, we can write the current as

$$i = -I\sin(\omega^{\text{dr}}t + \phi) = I\sin(\omega^{\text{dr}}t - \phi'), \tag{33-35}$$

where I is the amplitude or maximum value of the driven current. (The phase $\omega^{\text{dr}}t - \phi'$ of the current is traditionally written with a minus sign instead of as $\omega^{\text{dr}}t + \phi$.) We include the phase constant ϕ' in Eq. 33-35 to emphasize that the current i may not be in phase with the emf $\mathscr{E}$. (As you will see, the phase constant depends on the circuit to which the generator is connected.) We can also write the current i in terms of the **driving frequency** f^{dr} of the emf, by substituting $2\pi f^{\text{dr}}$ for ω^{dr} in Eq. 33-35.

33-9 Forced Oscillations

We have seen that once started, the charge, potential difference, and current in both undamped LC circuits and damped RLC circuits (with small enough R) oscillate at angular frequency $\omega = 1/\sqrt{LC}$. Such oscillations are said to be *free oscillations* (free of any external emf), and the angular frequency ω is said to be the circuit's **natural angular frequency.**

When the external alternating emf of Eq. 33-34 is connected to an RLC circuit, the oscillations of charge, potential difference, and current are said to be *driven oscillations* or *forced oscillations*. These oscillations always occur at the driving angular frequency ω^{dr}:

> No matter what the natural angular frequency ω of a circuit is, forced oscillations of charge, current, and potential difference in the circuit always occur at the driving angular frequency ω^{dr}.

However, as you will see in Section 33-11, the amplitudes of the oscillations very much depend on how close ω^{dr} is to ω. When the two angular frequencies match—a condition known as resonance—the amplitude I of the current in the circuit is maximum.

33-10 Representing Oscillations with Phasors: Three Simple Circuits

Later in this chapter, we shall connect an external alternating emf device to a series RLC circuit as in Fig. 33-11. We shall then find expressions for the amplitude I and phase constant ϕ of the sinusoidally oscillating current in terms of the amplitude $\mathscr{E}^{\max}$ and angular frequency ω^{dr} of the external emf. First, however, let us consider three simpler circuits, each having an external emf and only one other circuit element: R, C, or L. We start with a resistive element (a purely *resistive load*). We continue to use the convention here that uppercase letters such as Q, I, and V represent constants, while lowercase letters represent time-varying quantities such as q, i, and v.

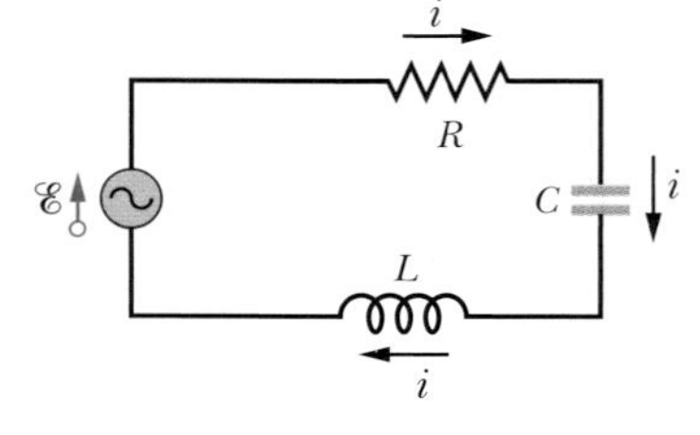

FIGURE 33-11 ■ A single-loop circuit containing a resistor, a capacitor, and an inductor. A generator, represented by a sine wave in a circle, produces an alternating emf that establishes an alternating current; the directions of the emf and current are indicated here at only one instant.

A Resistive Load

Figure 33-12*a* shows a circuit containing a resistance element of value R and an ac generator with the alternating emf of Eq. 33-34. By the loop rule, we have

$$\mathscr{E} - \Delta v_R = 0.$$

Note that in this context, Δv_R represents a potential difference across the resistance element and not a velocity change. With Eq. 33-34, this gives us

$$\Delta v_R = \mathscr{E}^{\max} \sin \omega^{dr} t.$$

Because the amplitude ΔV_R of the alternating potential difference (or voltage) across the resistance is equal to the amplitude $\mathscr{E}^{\max}$ of the alternating emf, we can write this as

$$\mathscr{E} = \Delta v_R = \Delta V_R \sin \omega^{dr} t. \tag{33-36}$$

From the definition of resistance ($R = \Delta v_R/i$), we can now write the current i_R in the resistor as

$$i_R = \frac{\Delta v_R}{R} = \frac{\Delta V_R}{R} \sin \omega^{dr} t. \tag{33-37}$$

From Eq. 33-35, we can also write this current as

$$i_R = I_R \sin(\omega^{dr} t - \phi), \tag{33-38}$$

where I_R is the amplitude of the current i_R passing through the resistance. Comparing Eqs. 33-37 and 33-38, we see that for a purely resistive load the phase constant $\phi = 0°$. We also see that the voltage amplitude and current amplitude are related by

$$\Delta V_R = I_R R \quad \text{(resistor).} \tag{33-39}$$

Although we found this relation for the circuit of Fig. 33-12*a*, it applies to any resistance in any ac circuit.

By comparing Eqs. 33-36 and 33-37, we see that the time-varying quantities Δv_R and i_R are both functions of $\sin \omega^{dr} t$ with $\phi = 0°$. Thus, these two quantities are *in phase,* which means that their corresponding maxima (and minima) occur at the same times. Figure 33-12*b*, which is a plot of $\Delta v_R(t)$ and $i_R(t)$, illustrates this fact. Note that Δv_R and i_R do not decay here, because the generator supplies energy to the circuit to make up for the energy dissipated in R.

Since we have chosen to drive our circuits with a voltage that varies as $\sin(\omega t)$, and constant speed motion around a circle has x- and y-components that vary sinusoidally, it is convenient to use a component of a rotating vector to represent our oscillations. Such a rotating vector representation is called a *phasor*. Recall from Section 17-12 that phasors are vectors that rotate around an origin. Those that represent the voltage across and current in the resistor of Fig. 33-12*a* are shown in Fig. 33-12*c* at an arbitrary time t. Such phasors have the following properties:

Angular speed: Both phasors rotate counterclockwise about the origin with an angular speed equal to the angular frequency ω^{dr} of both Δv_R and i_R.

Length: The length of each phasor represents the amplitude of the alternating quantity: ΔV_R for the voltage and i_R for the current.

Projection: The projection of each phasor on the *vertical* axis represents the value of the alternating quantity at time t: Δv_R for the voltage and I_R for the current.

Rotation angle: The rotation angle of each phasor is equal to the phase of the alternating quantity at time t. In Fig. 33-12*c*, the voltage and current are in phase, so their phasors always have the same phase $\omega^{dr} t$ and the same rotation angle, and thus they rotate together.

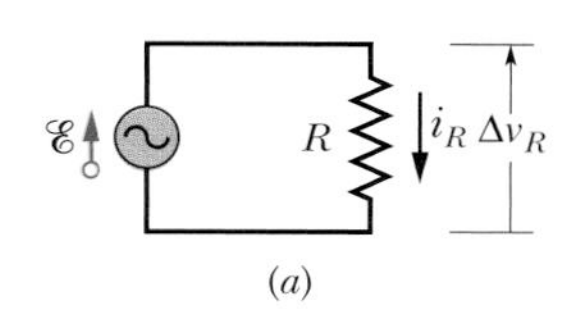

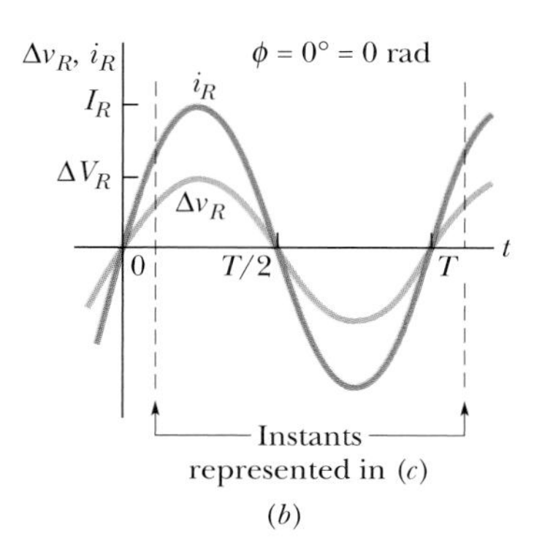

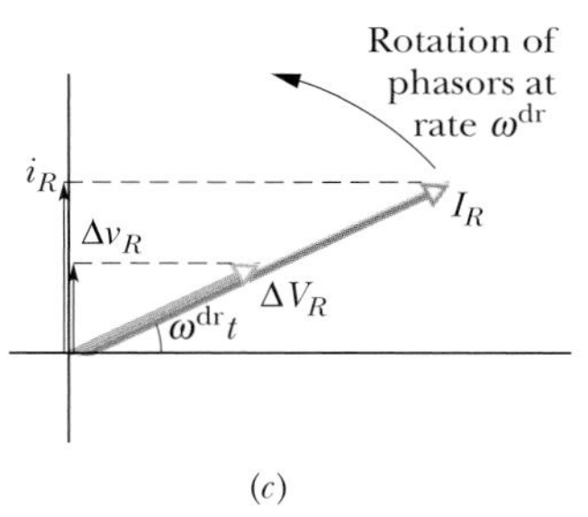

FIGURE 33-12 ■ (*a*) A resistor is connected across an alternating-current generator. (*b*) The current i_R and the potential difference Δv_R across the resistor are plotted on the same graph, both versus time t. They are in phase and complete one cycle in one period T. (*c*) A phasor diagram shows the same thing as (*b*).

Mentally follow the rotation. Can you see that when the phasors have rotated so that $\omega^{dr}t = 90°$ (so they point vertically upward), then $\Delta v_R = \Delta V_R$? Equations 33-36 and 33-38 give the same results. For this case, the voltage and the current oscillate together. Hence, it is not especially useful to introduce the phasor representation. However, the value of this representation becomes clearer with more complex situations such as the capacitor and inductor discussed below.

A Capacitive Load

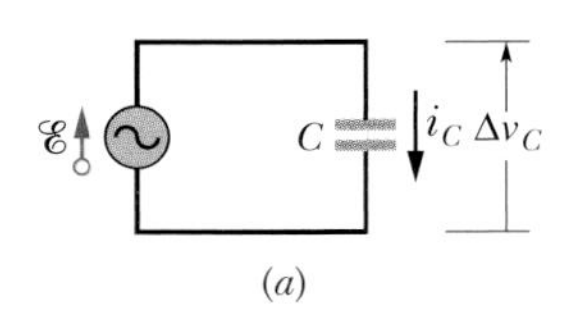

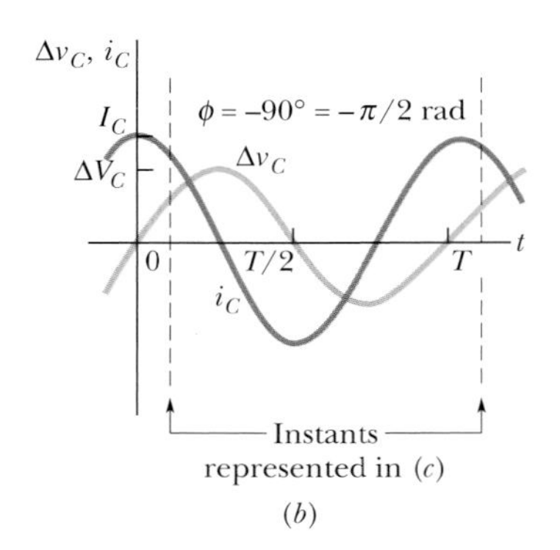

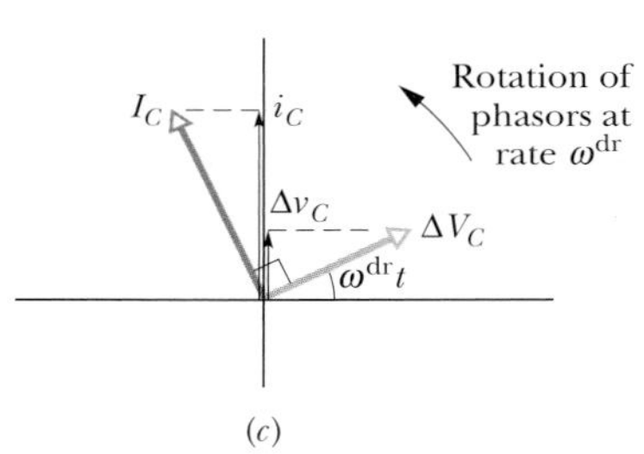

FIGURE 33-13 ■ (*a*) A capacitor is connected across an alternating-current generator. (*b*) The current in the capacitor leads the voltage by 90°(= $\pi/2$ rad). (*c*) A phasor diagram shows the same thing.

Figure 33-13*a* shows a circuit containing a capacitance and a generator with the alternating emf of Eq. 33-34. Using the loop rule and proceeding as we did when we obtained Eq. 33-36, we find that the potential difference across the capacitor is

$$\Delta v_C = \Delta V_C \sin \omega^{dr} t, \tag{33-40}$$

where ΔV_C is the amplitude of the alternating voltage across the capacitor. From the definition of capacitance we can also write

$$q_C = C\Delta v_C = C\Delta V_C \sin \omega^{dr} t. \tag{33-41}$$

Our concern, however, is with the current rather than the charge. Thus, we differentiate Eq. 33-41 to find

$$i_C = \frac{dq_C}{dt} = \omega^{dr} C\Delta V_C \cos \omega^{dr} t. \tag{33-42}$$

We now modify Eq. 33-42 in two ways. First, for reasons of symmetry of notation, we introduce the quantity X_C, called the *capacitive reactance* of a capacitor, defined as

$$X_C = \frac{1}{\omega^{dr} C} \quad \text{(capacitive reactance).} \tag{33-43}$$

Its value depends not only on the capacitance but also on the driving angular frequency ω^{dr}. We know from the definition of the capacitive time constant ($\tau = RC$) that the SI unit for C can be expressed as seconds per ohm. Applying this to Eq. 33-43 shows that the SI unit of X_C is the *ohm*, just as for resistance R.

Second, we replace $\cos \omega^{dr} t$ with a phase-shifted sine:

$$\cos \omega^{dr} t = \sin(\omega^{dr} t - (-90°)).$$

You can verify this identity by shifting a sine curve in the negative direction by 90°.

With these two modifications, Eq. 33-42 becomes

$$i_C = \left(\frac{\Delta V_C}{X_C}\right) \sin(\omega^{dr} t + 90°). \tag{33-44}$$

From Eq. 33-35, we can also write the current i_C in C as

$$i_C = I_C \sin(\omega^{dr} t - \phi), \tag{33-45}$$

where I_C is the amplitude of i_C. Comparing Eqs. 33-44 and 33-45, we see that for a purely capacitive load the phase constant ϕ for the current is $-90°$. We also see that the voltage amplitude and current amplitude are related by

$$\Delta V_C = I_C X_C \quad \text{(capacitor).} \tag{33-46}$$

Although we found this relation for the circuit of Fig. 33-13*a*, it applies to any capacitance in any ac circuit. Note that using phasors allowed us to write the equation associated with a capacitor in such a way that it looks just like Ohm's law for a resistor. Although this expression is true for the amplitude of the phasor, the full voltage and current don't look like Ohm's law because of the shift in phase. We'll see in the next case that a similar thing happens for the inductor.

Comparison of Eqs. 33-40 and 33-44, or inspection of Fig. 33-13*b*, shows that the quantities Δv_C and i_C are 90°, or one-quarter cycle, out of phase. Furthermore, we see that i_C leads Δv_C, which means that, if you monitored the current i_C and the potential difference Δv_C in the circuit of Fig. 33-13*a*, you would find that i_C reaches its maximum before Δv_C does, by one-quarter cycle.

This relation between i_C and Δv_C is illustrated by the phasor diagram of Fig. 33-13*c*. As the phasors representing these two quantities rotate counterclockwise together, the phasor labeled I_C does indeed lead that labeled ΔV_C, and by an angle of 90°; that is, the phasor I_C coincides with the vertical axis one-quarter cycle before the phasor ΔV_C does. Be sure to convince yourself that the phasor diagram of Fig. 33-13*c* is consistent with Eqs. 33-40 and 33-44.

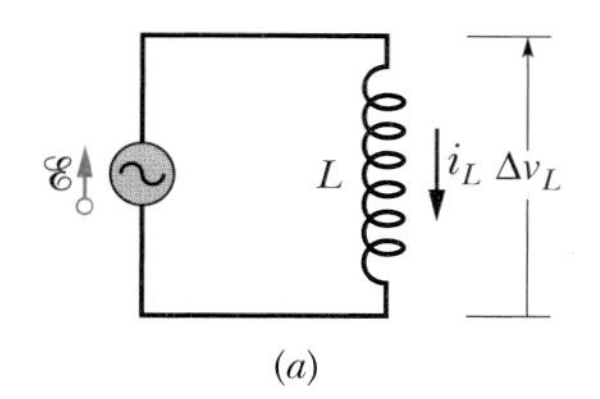

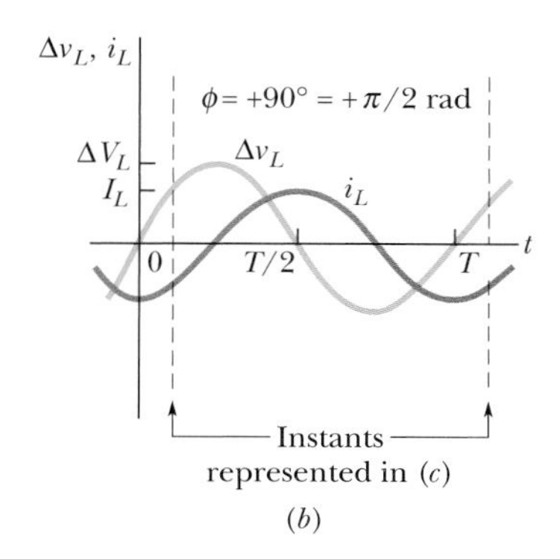

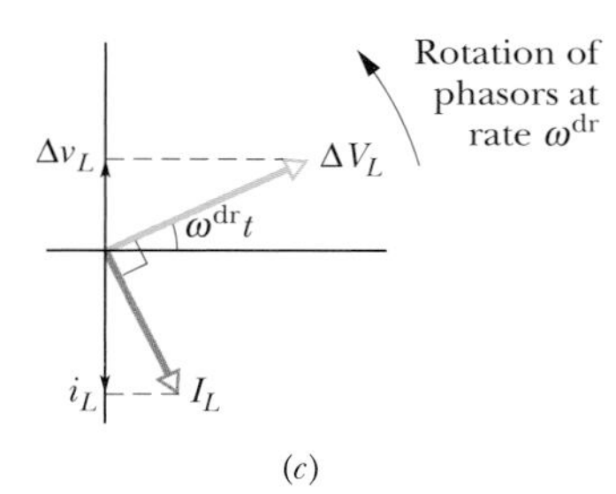

FIGURE 33-14 ■ (*a*) An inductor is connected across an alternating-current generator. (*b*) The current in the inductor lags the voltage by 90°(= π/2 rad). (*c*) A phasor diagram shows the same thing.

An Inductive Load

Now let's consider a third situation in which we connect an external alternating emf to a circuit containing just one of our basic circuit elements. Figure 33-14*a* shows a circuit containing an inductance and a generator with the alternating emf of Eq. 33-34. Using the loop rule and proceeding as we did to obtain Eq. 33-36, we find that the potential difference across the inductance is

$$\Delta v_L = \Delta V_L \sin \omega^{dr} t, \tag{33-47}$$

where ΔV_L is the amplitude of Δv_L. From Eq. 32-2, we can write the potential difference across an inductance L, in which the current is changing at the rate di_L/dt, as

$$\Delta v_L = L \frac{di_L}{dt}. \tag{33-48}$$

If we combine Eqs. 33-47 and 33-48 we have

$$\frac{di_L}{dt} = \frac{\Delta V_L}{L} \sin \omega^{dr} t. \tag{33-49}$$

Our concern, however, is with the current rather than with its time derivative. We find the former by integrating Eq. 33-49, obtaining

$$i_L = \int di_L = \frac{\Delta V_L}{L} \int \sin \omega^{dr} dt = -\left(\frac{\Delta V_L}{\omega^{dr} L}\right) \cos \omega^{dr} t. \tag{33-50}$$

We now modify this equation in two ways. First, for reasons of symmetry of notation, we introduce the quantity X_L, called the **inductive reactance** of an inductor, which is defined as

$$X_L = \omega^{dr} L \quad \text{(inductive reactance)}. \tag{33-51}$$

The value of X_L depends on the driving angular frequency ω^{dr}. The unit of the inductive time constant τ_L indicates that the SI unit of X_L is the *ohm*, just as it is for X_C and for R.

Second, we replace the function $-\cos\omega^{\rm dr}t$ in Eq. 33-50 with a phase-shifted sine—namely,

$$-\cos\omega^{\rm dr}t = \sin(\omega^{\rm dr}t - (+90°)).$$

You can verify this identity by shifting a sine curve in the positive direction by 90°.

With these two changes, Eq. 33-50 becomes

$$i_L = \left(\frac{\Delta V_L}{X_L}\right)\sin(\omega^{\rm dr}t - 90°). \tag{33-52}$$

From Eq. 33-35, we can also write this current in the inductance as

$$i_L = I_L \sin(\omega^{\rm dr}t - \phi), \tag{33-53}$$

where I_L is the amplitude of the current i_L. Comparing Eqs. 33-52 and 33-53, we see that for a purely inductive load the phase constant ϕ for the current is $+90°$. We also see that the voltage amplitude and current amplitude are related by

$$\Delta V_L = I_L X_L \quad \text{(inductor)}. \tag{33-54}$$

Although we found this relation for the circuit of Fig. 33-14*a*, it applies to any inductance in any ac circuit.

Comparison of Eqs. 33-47 and 33-52, or inspection of Fig. 33-14*b*, shows that the quantities i_L and Δv_L are 90° out of phase. In this case, however, i_L *lags* Δv_L; that is, if you monitored the current i_L and the potential difference Δv_L in the circuit of Fig. 33-14*a*, you would find that i_L reaches its maximum value after Δv_L does, by one-quarter cycle.

The phasor diagram of Fig. 33-14*c* also contains this information. As the phasors rotate counterclockwise in the figure, the phasor labeled I_L does indeed lag that labeled ΔV_L, and by an angle of 90°. Be sure to convince yourself that Fig. 33-14*c* represents Eqs. 33-47 and 33-52.

READING EXERCISE 33-5: The figure shows, in (*a*), a sine curve $S(t) = \sin(\omega^{\rm dr}t)$ and three other sinusoidal curves $A(t)$, $B(t)$, and $C(t)$, each of the form $\sin(\omega^{\rm dr}t - \phi)$. (a) Rank the three other curves according to the value of ϕ, most positive first and most negative last. (b) Which curve corresponds to which phasor in (*b*) of the figure? (c) Which curve leads the others?

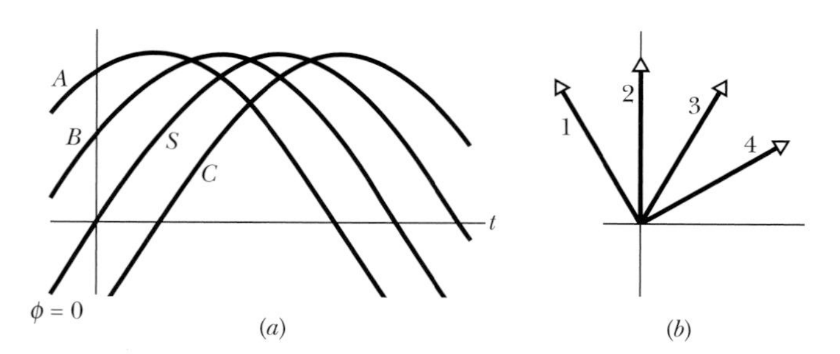

■

33-11 The Series *RLC* Circuit

We are now ready to apply the alternating emf of Eq. 33-34,

$$\mathscr{E} = \mathscr{E}^{\max}\sin\omega^{\rm dr}t \quad \text{(applied emf)}, \tag{33-55}$$

TABLE 33-2
Phase and Amplitude Relations for Alternating Currents and Voltages

Circuit Element	Symbol	Resistance or Reactance	Phase of the Current	Phase Constant (or Angle ϕ)	Amplitude Relation
Resistor	R	R	In phase with v_R	0° (= 0 rad)	$\Delta V_R = I_R R$
Capacitor	C	$X_C = 1/\omega^{dr}C$	Leads Δv_C by 90° (= $\pi/2$ rad)	−90° (= $-\pi/2$ rad)	$\Delta V_C = I_C X_C$
Inductor	L	$X_L = \omega^{dr} L$	Lags Δv_L by 90° (= $\pi/2$ rad)	+90° (= $+\pi/2$ rad)	$\Delta V_L = I_L X_L$

to the full *RLC* circuit of Fig. 33-11. Because R, L, and C are in series, the same current

$$i = I\sin(\omega^{dr}t - \phi) \qquad (33\text{-}56)$$

is driven in all three of them. We wish to find the current amplitude I and the phase constant ϕ. The solution is simplified by the use of phasor diagrams.

Table 33-2 summarizes the relations between the current i and the voltage V for each of the three kinds of circuit elements we have considered. When an applied alternating voltage produces an alternating current in them, the current is in phase with the voltage across a resistor, leads the voltage across a capacitor, and lags the voltage across an inductor.

The Current Amplitude

We start with Fig. 33-15*a*, which shows the phasor representing the current of Eq. 33-56 at an arbitrary time t. The length of the phasor is the current amplitude I, the projection of the phasor on the vertical axis is the current i at time t, and the angle of rotation of the phasor is the phase $\omega^{dr}t - \phi$ of the current at time t.

Figure 33-15*b* shows the phasors representing the voltages across R, L, and C at the same time t. Each phasor is oriented relative to the angle of rotation of current phasor I in Fig. 33-15*a*, based on the information in Table 33-2:

Resistor: Here current and voltage are in phase, so the angle of rotation of voltage phasor ΔV_R is the same as that of phasor I.

Capacitor: Here current leads voltage by 90°, so the angle of rotation of voltage phasor ΔV_C is 90° less than that of phasor I.

Inductor: Here current lags voltage by 90°, so the angle of rotation of voltage phasor ΔV_L is 90° greater than that of phasor I.

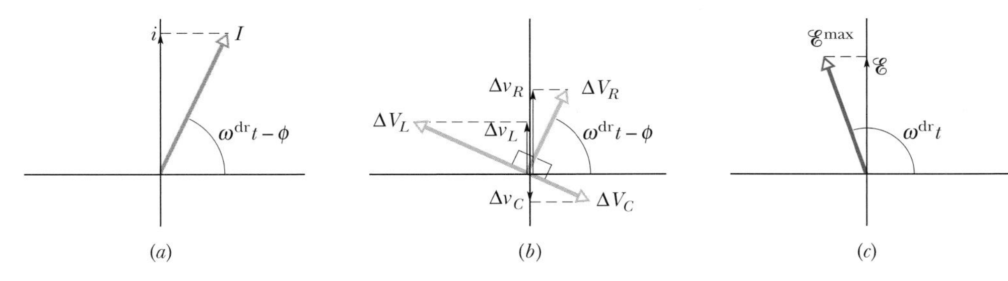

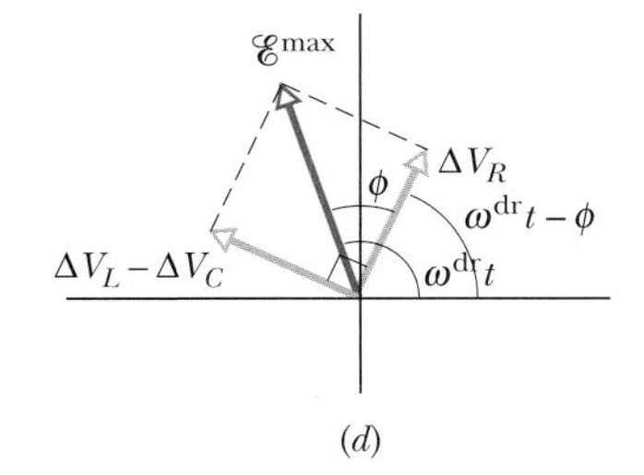

FIGURE 33-15 ■ (*a*) A phasor representing the alternating current in the driven *RLC* circuit of Fig. 33-11 at time t. The amplitude I, the instantaneous value i, and the phase $\omega^{dr}t - \phi$ are shown. (*b*) Phasors representing the voltages across the inductor, resistor, and capacitor, oriented with respect to the current phasor in (*a*). (*c*) A phasor representing the alternating emf that drives the current of (*a*). (*d*) The emf phasor is equal to the vector sum of the three voltage phasors of (*b*). Here, voltage phasors ΔV_L and ΔV_C have been added to yield their net phasor $(\Delta V_L - \Delta V_C)$.

Figure 33-15*b* also shows the instantaneous voltages Δv_R, Δv_C, and Δv_L across R, C, and L at time t; those voltages are the projections of the corresponding phasors on the vertical axis of the figure.

Figure 33-15*c* shows the phasor representing the applied emf of Eq. 33-55. The length of the phasor is the emf amplitude $\mathcal{E}^{\max}$; the projection of the phasor on the vertical axis is the emf $\mathcal{E}$ at time t, and the angle of rotation of the phasor is the phase $\omega_d t$ of the emf at time t.

From the loop rule we know that at any instant the sum of the voltages Δv_R, Δv_C, and Δv_L is equal to the applied emf $\mathcal{E}$:

$$\mathcal{E} = \Delta v_R + \Delta v_C + \Delta v_L. \tag{33-57}$$

Thus, at time t the projection $\mathcal{E}$ in Fig. 33-15*c* is equal to the algebraic sum of the projections Δv_R, Δv_C, and Δv_L in Fig. 33-15*b*. In fact, as the phasors rotate together, this equality always holds. This means that phasor $\mathcal{E}^{\max}$ in Fig. 33-15*c* must be equal to the vector sum of the three voltage phasors ΔV_R, ΔV_C, and ΔV_L in Fig. 33-15*b*.

That requirement is indicated in Fig. 33-15*d*, where phasor $\mathcal{E}^{\max}$ is drawn as the sum of phasors ΔV_R, ΔV_L, and ΔV_C. Because phasors ΔV_L and ΔV_C have opposite directions in the figure, we simplify the vector sum by first combining ΔV_L and ΔV_C to form the single phasor $\Delta V_L - \Delta V_C$. Then we combine that single phasor with ΔV_R to find the net phasor. Again, the net phasor must coincide with phasor $\mathcal{E}^{\max}$, as shown.

Both triangles in Fig. 33-15*d* are right triangles. Applying the Pythagorean theorem to either one yields

$$(\mathcal{E}^{\max})^2 = \Delta V_R^2 + (\Delta V_L - \Delta V_C)^2. \tag{33-58}$$

From the amplitude information displayed in Table 33-2 we can rewrite this as

$$(\mathcal{E}^{\max})^2 = (IR)^2 + (IX_L - IX_C)^2, \tag{33-59}$$

and then rearrange it to the form

$$I = \frac{\mathcal{E}^{\max}}{\sqrt{R^2 + (X_L - X_C)^2}}. \tag{33-60}$$

The denominator in Eq. 33-60 is called the **impedance** Z of the circuit for the driving angular frequency ω^{dr}:

$$Z = \sqrt{R^2 + (X_L - X_C)^2} \quad \text{(impedance defined).} \tag{33-61}$$

We can then write Eq. 33-60 as

$$I = \frac{\mathcal{E}^{\max}}{Z}. \tag{33-62}$$

If we substitute for X_C and X_L from Eqs. 33-43 and 33-51, we can write Eq. 33-60 more explicitly as

$$I = \frac{\mathcal{E}^{\max}}{\sqrt{R^2 + (\omega^{\text{dr}}L - 1/\omega^{\text{dr}}C)^2}} \quad \text{(current amplitude)} \tag{33-63}$$

We have now accomplished half our goal: We have obtained an expression for the current amplitude I in terms of the sinusoidal driving emf and the circuit elements in a series RLC circuit.

The value of I depends on the difference between $\omega^{dr}L$ and $1/\omega^{dr}C$ in Eq. 33-63 or, equivalently, the difference between X_L and X_C in Eq. 33-60. In either equation, it does not matter which of the two quantities is greater because the difference is always squared.

The current that we have been describing in this section is the *steady-state current* that occurs after the alternating emf has been applied for some time. When the emf is first applied to a circuit, a brief *transient current* occurs. Its duration (before settling down into the steady-state current) is determined by the time constants $\tau_L = L/R$ and $\tau_C = RC$ as the inductive and capacitive elements "turn on." This transient current can be large and can, for example, destroy a motor on startup if it is not properly taken into account in the motor's circuit design.

The Phase Constant

From the right-hand phasor triangle in Fig. 33-15*d* and from Table 33-2 we can write

$$\tan\phi = \frac{\Delta V_L - \Delta V_C}{\Delta V_R} = \frac{IX_L - IX_C}{IR}, \tag{33-64}$$

which gives us

$$\tan\phi = \frac{X_L - X_C}{R} \quad \text{(phase constant).} \tag{33-65}$$

This is the other half of our goal: an equation for the phase constant ϕ in a sinusoidally driven series *RLC* circuit. In essence, it gives us three different results for the phase constant, depending on the relative values of X_L and X_C:

$\boldsymbol{X_L > X_C}$**:** The circuit is said to be *more inductive than capacitive*. Equation 33-65 tells us that ϕ is positive for such a circuit, which means that phasor I rotates behind phasor $\mathscr{E}^{max}$ (Fig. 33-16*a*). A plot of $\mathscr{E}$ and i versus time is like that in Fig. 33-16*b*. (The phasors in Figs. 33-16*c* and *d* were drawn assuming $X_L > X_C$.)

$\boldsymbol{X_C > X_L}$**:** The circuit is said to be *more capacitive than inductive*. Equation 33-65 tells us that ϕ is negative for such a circuit, which means that phasor I rotates ahead of phasor $\mathscr{E}^{max}$ (Fig. 33-16*c*). A plot of $\mathscr{E}$ and i versus time is like that in Fig. 33-16*d*.

$\boldsymbol{X_C = X_L}$**:** The circuit is said to be in *resonance*, a state that is discussed next. Equation 33-65 tells us that $\phi = 0°$ for such a circuit, which means that phasors $\mathscr{E}^{max}$ and I rotate together (Fig. 33-16*e*). A plot of $\mathscr{E}$ and i versus time is like that in Fig. 33-16*f*.

As an illustration, let us reconsider two extreme circuits. In the *purely inductive circuit* of Fig. 33-14*a*, where X_L is nonzero and $X_C = R = 0$, Eq. 33-65 tells us that $\phi = +90°$ (the greatest value of ϕ), consistent with Fig. 33-14*c*. In the *purely capacitive circuit* of Fig. 33-13*a*, where X_C is nonzero and $X_L = R = 0$, Eq. 33-65 tells us that $\phi = -90°$ (the least value of ϕ), consistent with Fig. 33-13*c*.

Resonance

Equation 33-63 gives the current amplitude I in an *RLC* circuit as a function of the driving angular frequency ω^{dr} of the external alternating emf. For a given resistance R, that amplitude is a maximum when the quantity $\omega^{dr}L - 1/\omega^{dr}C$ in the denominator is zero—that is, when

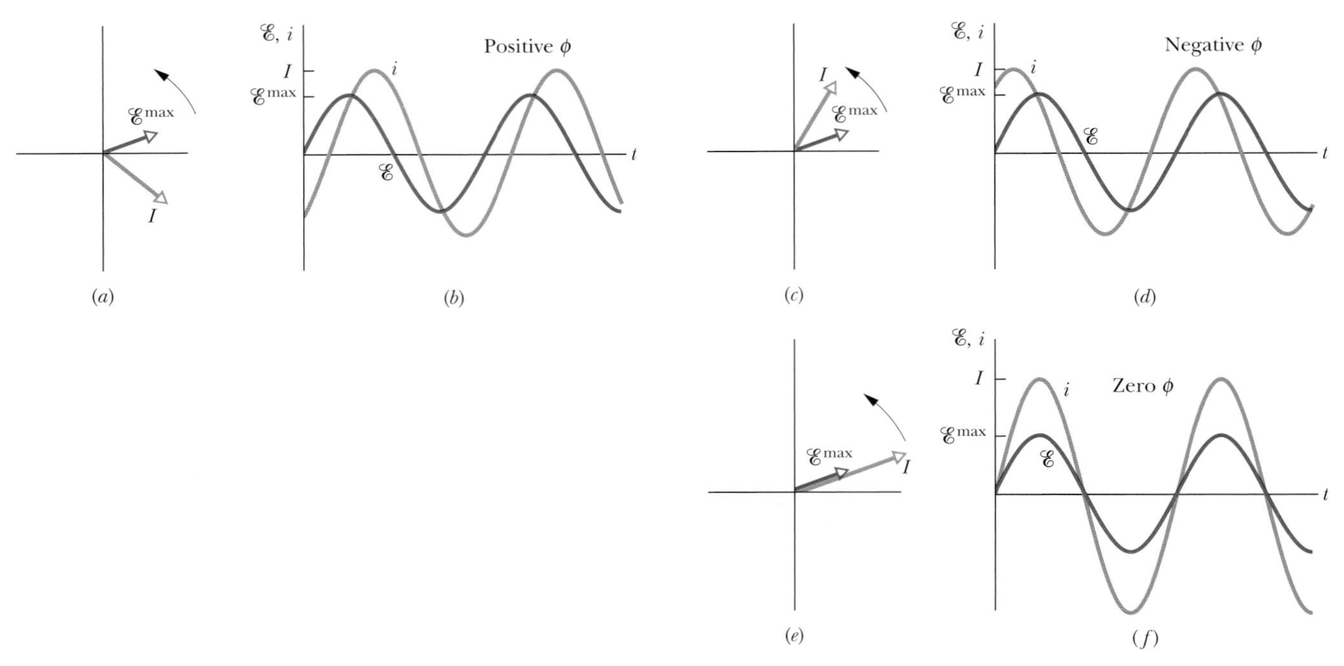

FIGURE 33-16 ■ Phasor diagrams and graphs of the alternating emf and current for the driven *RLC* circuit of Fig. 33-11. In the phasor diagram of (*a*) and the graph of (*b*), the current *i* lags the driving emf $\mathscr{E}$ and the current's phase constant ϕ is positive. In (*c*) and (*d*), the current *i* leads the driving emf $\mathscr{E}$ and its phase constant ϕ is negative. In (*e*) and (*f*), the current *i* is in phase with the driving emf $\mathscr{E}$ and its phase constant ϕ is zero.

$$\omega^{dr} L = \frac{1}{\omega^{dr} C}$$

or

$$\omega^{dr} = \frac{1}{\sqrt{LC}} \quad \text{(for maximum } I\text{).} \tag{33-66}$$

Because the natural angular frequency ω of the *RLC* circuit is also equal to $1/\sqrt{LC}$, the maximum value of I occurs when the driving angular frequency matches the natural angular frequency—that is, at resonance. Thus, in an *RLC* circuit, resonance and maximum current amplitude I occur when

$$\omega^{dr} = \omega = \frac{1}{\sqrt{LC}} \quad \text{(resonance).} \tag{33-67}$$

Figure 33-17 shows three *resonance curves* for sinusoidally driven oscillations in three series *RLC* circuits differing only in R. Each curve peaks at its maximum current amplitude I when the ratio ω^{dr}/ω is 1.00, but the maximum value of I decreases with increasing R. (The maximum I is always $\mathscr{E}^{max}/R$; to see why, combine Eqs. 33-61 and 33-62.) In addition, the curves increase in width (measured in Fig. 33-17 at half the maximum value of I) with increasing R.

To make physical sense of Fig. 33-17, consider how the reactances X_L and X_C change as we increase the driving angular frequency ω^{dr}, starting with a value much less than the natural frequency ω. For small ω^{dr}, reactance $X_L (= \omega^{dr} L)$ is small and reactance $X_C (1/\omega^{dr} C)$ is large. Thus, the circuit is mainly capacitive and the impedance is dominated by the large X_C, which keeps the current low.

As we increase ω^{dr}, reactance X_C remains dominant but decreases while reactance X_L increases. The decrease in X_C decreases the impedance, allowing the current

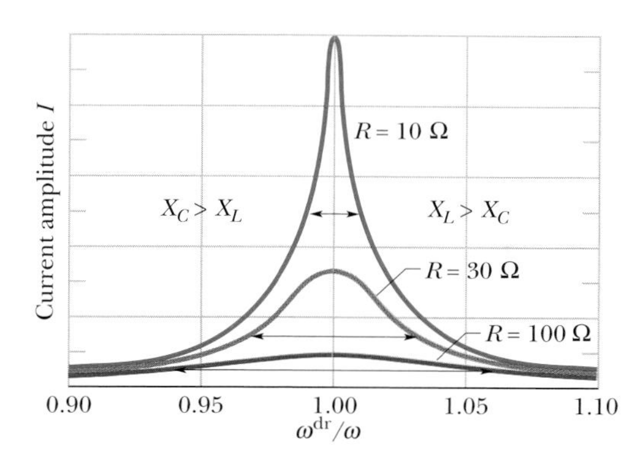

FIGURE 33-17 ■ *Resonance curves* for the driven *RLC* circuit of Fig. 33-11 with $L = 100\ \mu\text{H}$, $C = 100$ pF, and three values of R. The current amplitude I of the alternating current depends on how close the driving angular frequency ω^{dr} is to the natural angular frequency ω. The horizontal arrow on each curve measures the curve's width at the half-maximum level, a measure of the sharpness of the resonance. To the left of $\omega^{dr}/\omega = 1.00$, the circuit is mainly capacitive, with $X_C > X_L$ to the right, it is mainly inductive, with $X_L > X_C$.

to increase, as we see on the left side of any resonance curve in Fig. 33-17. When the increasing X_L and the decreasing X_C reach equal values, the current is greatest and the circuit is in resonance, with $\omega^{dr} = \omega$.

As we continue to increase ω^{dr}, the increasing reactance X_L becomes progressively more dominant over the decreasing reactance X_C. The impedance increases because of X_L and the current decreases, as on the right side of any resonance curve in Fig. 33-17. In summary, then: The low-angular-frequency side of a resonance curve is dominated by the capacitor's reactance, the high-angular-frequency side is dominated by the inductor's reactance, and resonance occurs between the two regions.

READING EXERCISE 33-6: Here are the capacitive reactance and inductive reactance, respectively, for three sinusoidally driven series *RLC* circuits: (1) 50 Ω, 100 Ω; (2) 100 Ω, 50 Ω; (3) 50 Ω, 50 Ω. (a) For each, does the current lead or lag the applied emf, or are the two in phase? (b) Which circuit is in resonance? ■

TOUCHSTONE EXAMPLE 33-4: Series *RLC* Circuit

In Fig. 33-11 let $R = 200\ \Omega$, $C = 15.0\ \mu\text{F}$, $L = 230$ mH, $f^{dr} = 60.0$ Hz, and $\mathscr{E}^{max} = 36.0$ V.

(a) What is the current amplitude I?

SOLUTION ■ The **Key Idea** here is that current amplitude I depends on the amplitude $\mathscr{E}^{max}$ of the driving emf and on the impedance Z of the circuit, according to Eq. 33-62 ($I = \mathscr{E}^{max}/Z$). Thus, we need to find Z, which depends on the circuit's resistance R, capacitive reactance X_C, and inductive reactance X_L.

The circuit's only resistance is the given resistance R. Its only capacitive reactance is due to the given capacitance. From Table 33-2, ($X_C = 1/\omega^{dr}C$), with $\omega^{dr} = 2\pi f^{dr}$, we can write

$$X_C = \frac{1}{2\pi f^{dr}C} = \frac{1}{(2\pi)(60.0\ \text{Hz})(15.0 \times 10^{-6}\ \text{F})} = 177\ \Omega.$$

From Table 33-2 ($X_L = \omega^{dr}L$), with $\omega^{dr} = 2\pi f^{dr}$, we can write

$$X_L = 2\pi f^{dr}L = (2\pi)(60.0\ \text{Hz})(230 \times 10^{-3}\ \text{H}) = 86.7\ \Omega.$$

Thus, the circuit's impedance is

$$Z = \sqrt{R^2 + (X_L - X_C)^2} = \sqrt{(200\ \Omega)^2 + (86.7\ \Omega - 177\ \Omega)^2} = 219\ \Omega.$$

We then find

$$|I| = \frac{|\mathscr{E}^{max}|}{Z} = \frac{36.0\ \text{V}}{219\ \Omega} = 0.164\ \text{A}. \qquad \text{(Answer)}$$

(b) What is the phase constant ϕ of the current in the circuit relative to the driving emf?

SOLUTION ■ The **Key Idea** here is that the phase constant depends on the inductive reactance, the capacitive reactance, and the resistance of the circuit, according to Eq. 33-65. Solving that equation for ϕ leads to

$$\phi = \tan^{-1}\frac{X_L - X_C}{R} = \tan^{-1}\frac{86.7\ \Omega - 177\ \Omega}{200\ \Omega} = -24.3° = -0.424\ \text{rad}. \qquad \text{(Answer)}$$

The negative phase constant is consistent with the fact that the load is mainly capacitive; that is, $X_C > X_L$.

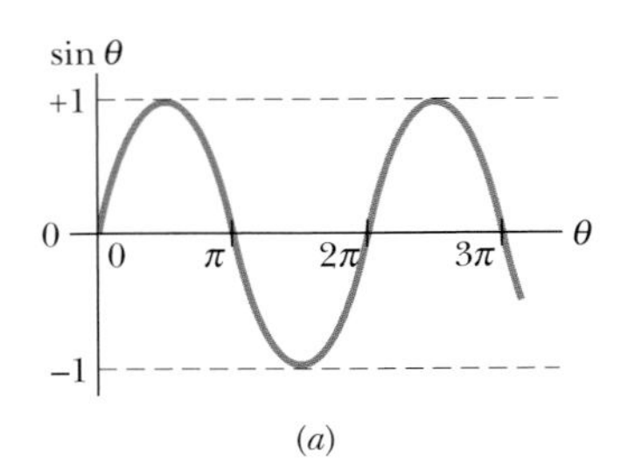

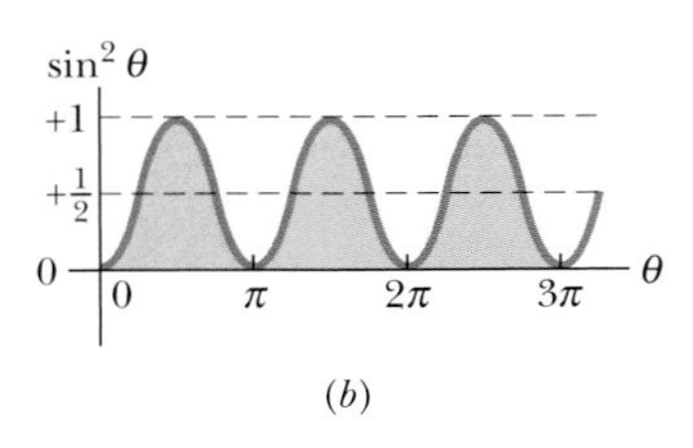

FIGURE 33-18 ■ (*a*) A plot of sin θ versus θ. The average value over one cycle is zero. (*b*) A plot of $\sin^2\theta$ versus θ. The average value over one cycle is $\frac{1}{2}$.

33-12 Power in Alternating-Current Circuits

In the *RLC* circuit of Fig. 33-11, the source of energy is the alternating-current generator. Some of the energy that it provides is stored in the electric field in the capacitor, some is stored in the magnetic field in the inductor, and some is dissipated as thermal energy in the resistor. In steady-state operation—which we assume—the average energy stored in the capacitor and inductor together remains constant. The net transfer of energy is thus from the generator to the resistor, where electromagnetic energy is dissipated as thermal energy.

The instantaneous rate at which energy is dissipated in the resistor can be written, with the help of Eqs. 27-26 and 33-35, as

$$P = i^2R = [I\sin(\omega^{\text{dr}}t - \phi)]^2 R = I^2R\sin^2(\omega^{\text{dr}}t - \phi), \tag{33-68}$$

where I is the maximum value of the current.

The *average* rate at which energy is dissipated in the resistor, however, is the average of Eq. 33-68 over time. Although the average value of sin θ, where θ is any variable, is zero (Fig. 33-18*a*), the average value of $\sin^2\theta$ over one complete cycle is 1/2 (Fig. 33-18*b*). (Note in Fig. 33-18*b* how the shaded areas under the curve but above the horizontal line marked +1/2 exactly fill in the unshaded spaces below that line.) Thus, we can write, from Eq. 33-68,

$$\langle P\rangle = \frac{I^2R}{2} = \left(\frac{I}{\sqrt{2}}\right)^2 R. \tag{33-69}$$

The quantity $I/\sqrt{2}$ is defined as the **root-mean-square,** or **rms,** value of the current *i in a cycle.* The square of the rms current is the average of the squares of the instantaneous currents in a cycle. It represents an effective dc current that would produce the same heating effect as an ac current with a maximum value of I. So,

$$I^{\text{rms}} \equiv \frac{I}{\sqrt{2}} \quad \text{(definition of rms current).} \tag{33-70}$$

We can now rewrite Eq. 33-69 as

$$\langle P\rangle = (I^{\text{rms}})^2R \quad \text{(average power).} \tag{33-71}$$

Equation 33-71 looks much like Eq. 26-11 for dc currents ($P = i^2R$); the purpose of defining rms current is that we can use it to compute the average rate of energy dissipation for alternating-current circuits using essentially the same equation we use for dc circuits.

We can also define rms values of voltages and emfs for alternating-current circuits in terms of the maximum values of those quantities in a cycle. So

$$\Delta V^{\text{rms}} = \frac{\Delta V}{\sqrt{2}} \quad \text{(rms voltage)} \quad \text{and} \quad \mathscr{E}^{\text{rms}} = \frac{\mathscr{E}^{\text{max}}}{\sqrt{2}} \quad \text{(rms emf).} \tag{33-72}$$

Alternating-current instruments, such as ammeters and voltmeters, are usually calibrated to read I^{rms}, ΔV^{rms}, and $\mathscr{E}^{\text{rms}}$. Thus, if you plug an alternating-current voltmeter into a household electric outlet and it reads 120 V, it represents an rms voltage. The maximum value of the potential difference at the outlet is $\sqrt{2} \times (120\text{ V})$, or 170 V.

Because the proportionality factor $1/\sqrt{2}$ in Eqs. 33-70 and 33-72 is the same for all three variables, we can write Eqs. 33-62 and 33-60 as

$$I^{\rm rms} = \frac{\mathscr{E}^{\rm rms}}{Z} = \frac{\mathscr{E}^{\rm rms}}{\sqrt{R^2 + (X_L - X_C)^2}}, \qquad (33\text{-}73)$$

and, indeed, this is the form that we almost always use.

We can use the relationship $I^{\rm rms} = \mathscr{E}^{\rm rms}/Z$ to recast Eq. 33-71 in a useful equivalent way. We write

$$\langle P \rangle = \frac{\mathscr{E}^{\rm rms}}{Z} I^{\rm rms} R = \mathscr{E}^{\rm rms} I^{\rm rms} \frac{R}{Z}. \qquad (33\text{-}74)$$

From Fig. 33-15*d*, Table 33-2, and Eq. 33-62, however, we see that R/Z is just the cosine of the phase constant ϕ:

$$\cos\phi = \frac{\Delta V_R}{\mathscr{E}^{\max}} = \frac{IR}{IZ} = \frac{R}{Z}. \qquad (33\text{-}75)$$

Equation 33-74 then becomes

$$\langle P \rangle = \mathscr{E}^{\rm rms} I^{\rm rms} \cos\phi \qquad \text{(average power)}, \qquad (33\text{-}76)$$

in which the term $\cos\phi$ is called the **power factor.** Because $\cos\phi = \cos(-\phi)$, Eq. 33-76 is independent of the sign of the phase constant ϕ.

To maximize the rate at which energy is supplied to a resistive load in an *RLC* circuit, we should keep the power factor $\cos\phi$ as close to unity as possible. This is equivalent to keeping the phase constant ϕ in Eq. 33-35 as close to zero as possible. If, for example, the circuit is highly inductive, it can be made less so by putting more capacitance in the circuit, connected in series. Adding capacitive reactance counters the excess inductive reactance in the circuit. This makes Z closer in value to R and so reduces the phase constant and increases the power factor in Eq. 33-76. Power companies place series-connected capacitors throughout their transmission systems to get these results.

READING EXERCISE 33-7: (a) If the current in a sinusoidally driven series *RLC* circuit leads the emf, would we increase or decrease the capacitance to increase the rate at which energy is supplied to the resistance? (b) Will this change bring the resonant angular frequency of the circuit closer to the angular frequency of the emf or move it further away? ■

TOUCHSTONE EXAMPLE 33-5: Power Factor

A series RLC circuit, driven with $\mathscr{E}^{\rm rms} = 120$ V at frequency $f^{\rm dr} = 60.0$ Hz, contains a resistance $R = 200\ \Omega$, an inductance with $X_L = 80.0\ \Omega$, and a capacitance with $X_C = 150\ \Omega$.

(a) What are the power factor $\cos\phi$ and phase constant ϕ of the circuit?

SOLUTION ■ The **Key Idea** here is that the power factor $\cos\phi$ can be found from the resistance R and impedance Z via Eq. 33-75 ($\cos\phi = R/Z$). To calculate Z, we use Eq. 33-61:

$$Z = \sqrt{R^2 + (X_L - X_C)^2}$$
$$= \sqrt{(200\ \Omega)^2 + (80.0\ \Omega - 150\ \Omega)^2} = 211.90\ \Omega.$$

Equation 33-75 then gives us

$$\cos\phi = \frac{R}{Z} = \frac{200\ \Omega}{211.90\ \Omega} = 0.9438 \approx 0.944. \qquad \text{(Answer)}$$

Taking the inverse cosine then yields

$$\phi = \cos^{-1} 0.944 = \pm 19.3°.$$

Both $+19.3°$ and $-19.3°$ have a cosine of 0.944. To determine which sign is correct, we must consider whether the current leads or lags the driving emf. Because $X_C > X_L$, this circuit is mainly capacitive, with the current leading the emf. Thus, ϕ must be negative:

$$\phi = -19.3°. \qquad \text{(Answer)}$$

We could, instead, have found ϕ with Eq. 33-65. A calculator would then have given us the complete answer, with the minus sign.

(b) What is the average rate $\langle P \rangle$ at which energy is dissipated in the resistance?

SOLUTION ■ One way to answer this question is to use this **Key Idea:** Because the circuit is assumed to be in steady-state operation, the rate at which energy is dissipated in the resistance is equal to the rate at which energy is supplied to the circuit, as given by Eq. 33-76 ($\langle P \rangle = \mathscr{E}^{\text{rms}} I^{\text{rms}} \cos\phi$).

We are given the rms driving emf $\mathscr{E}^{\text{rms}}$ and we know $\cos\phi$ from part (a). To find I^{rms} we use the **Key Idea** that the rms current is determined by the rms value of the driving emf and the circuit's impedance Z (which we know), according to Eq. 33-73:

$$I^{\text{rms}} = \frac{\mathscr{E}^{\text{rms}}}{Z}.$$

Substituting this into Eq. 33-76 then leads to

$$\langle P \rangle = \mathscr{E}^{\text{rms}} I^{\text{rms}} \cos\phi = \frac{(\mathscr{E}^{\text{rms}})^2}{Z} \cos\phi$$

$$= \frac{(120\text{ V})^2}{211.90\ \Omega}(0.9438) = 64.1\text{ W}. \qquad \text{(Answer)}$$

A second way to answer the question is to use the **Key Idea** that the rate at which energy is dissipated in a resistance R depends on the square of the rms current I^{rms} through it, according to Eq. 33-69. We then find

$$\langle P \rangle = (I^{\text{rms}})^2 R = \frac{(\mathscr{E}^{\text{rms}})^2}{Z^2} R$$

$$= \frac{(120\text{ V})^2}{(211.90\ \Omega)^2}(200\ \Omega) = 64.1\text{ W}. \qquad \text{(Answer)}$$

(c) What new capacitance C^{new} is needed to maximize $\langle P \rangle$ if the other parameters of the circuit are not changed?

SOLUTION ■ One **Key Idea** here is that the average rate $\langle P \rangle$ at which energy is supplied and dissipated is maximized if the circuit is brought into resonance with the driving emf. A second **Key Idea** is that resonance occurs when $X_C = X_L$. From the given data, we have $X_C > X_L$. Thus, we must decrease X_C to reach resonance. From Eq. 33-43 ($X_C = 1/\omega^{\text{dr}} C$), we see that this means we must increase C to the new value C^{new}.

Using Eq. 33-43, we can write the condition $X_C = X_L$ as

$$\frac{1}{\omega^{\text{dr}} C^{\text{new}}} = X_L.$$

Substituting $2\pi f^{\text{dr}}$ for ω^{dr} (because we are given f^{dr} and not ω^{dr}) and then solving for C^{new}, we find

$$C^{\text{new}} = \frac{1}{2\pi f^{\text{dr}} X_L} = \frac{1}{(2\pi)(60\text{ Hz})(80.0\ \Omega)}$$

$$= 3.32 \times 10^{-5}\text{ F} = 33.2\ \mu\text{F}. \qquad \text{(Answer)}$$

Following the procedure of part (b), you can show that with C^{new}, $\langle P \rangle$ would then be at its maximum value of 72.0 W.

Problems

SEC. 33-2 ■ ENERGY STORED IN A $\vec{B}$-FIELD

1. Current Is Zero Suppose that the inductive time constant for the circuit of Fig. 33-19 is 37.0 ms and the current in the circuit is zero at time $t = 0$ s. At what time does the rate at which energy is dissipated in the resistor equal the rate at which energy is being stored in the inductor?

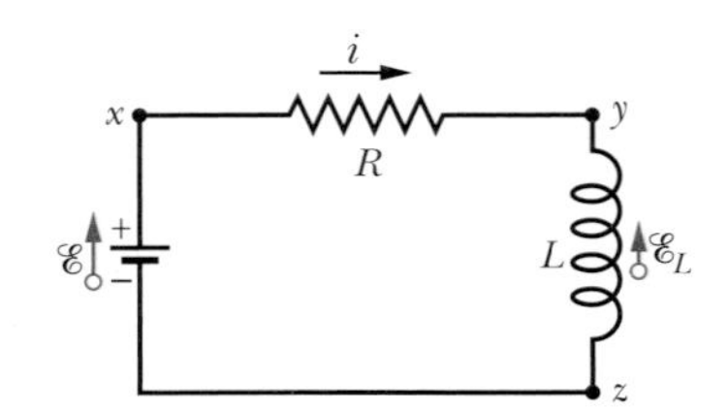

FIGURE 33-19 ■ Problems 1, 2, and 6.

2. Consider the Circuit Consider the circuit of Fig. 33-19. In terms of the inductive time constant, at what instant after the battery is connected will the energy stored in the magnetic field of the inductor be half its steady-state value?

3. Coil Connected in Series A coil is connected in series with a 10.0 k Ω resistor. A 50.0 V battery is applied across the two devices, and the current reaches a value of 2.00 mA after 5.00 ms. (a) Find the inductance of the coil. (b) How much energy is stored in the coil at this same moment?

4. Rates A coil with an inductance of 2.0 H and a resistance of 10 Ω is suddenly connected to a resistanceless battery with $\mathscr{E} =$ 100 V. At 0.10 s after the connection is made, what are the rates at which (a) energy is being stored in the magnetic field, (b) thermal energy is appearing in the resistance, and (c) energy is being delivered by the battery?

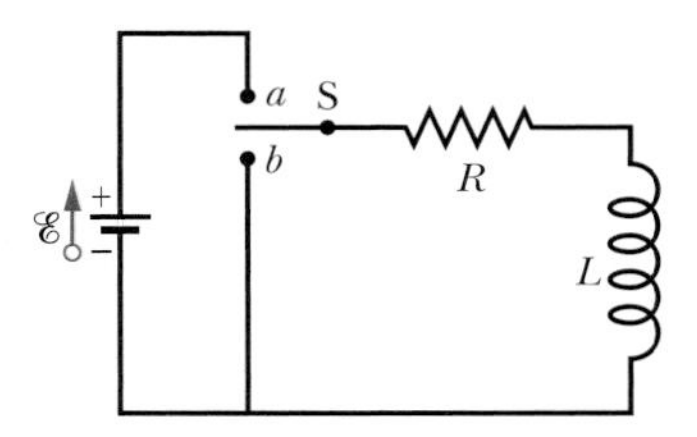

FIGURE 33-20 ■ Problem 5.

5. Prove That Prove that, after switch S in Fig. 33-20 has been thrown from a to b, all the energy stored in the inductor will ultimately appear as thermal energy in the resistor.

6. Energy Delivered For the circuit of Fig. 33-19, assume that $\mathscr{E} =$ 10.0 V, $R = 6.70\ \Omega$, and $L = 5.50$ H. The battery is connected at time $t = 0$ s. (a) How much energy is delivered by the battery during the first 2.00 s? (b) How much of this energy is stored in the magnetic field of the inductor? (c) How much of this energy is dissipated in the resistor?

SEC. 33-3 ■ ENERGY DENSITY OF A $\vec{B}$-FIELD

7. Energy Density A solenoid that is 85.0 cm long has a cross-sectional area of 17.0 cm^2. There are 950 turns of wire carrying a current of 6.60 A. (a) Calculate the energy density of the magnetic field inside the solenoid. (b) Find the total energy stored in the magnetic field there (neglect end effects).

8. Toroidal Inductor A toroidal inductor with an inductance of 90.0 mH encloses a volume of 0.0200 m^3. If the average energy density in the toroid is 70.0 J/m^3, what is the current through the inductor?

9. Magnitude of *E*-Field What must be the magnitude of a uniform electric field if it is to have the same energy density as that possessed by a 0.50 T magnetic field?

10. Interstellar Space The magnetic field in the interstellar space of our galaxy has a magnitude of about 10^{-10} T. How much energy is stored in this field in a cube 10 light-years on edge? (For scale, note that the nearest star is 4.3 light-years distant and the radius of our galaxy is about 8×10^4 light-years.)

11. Length of Copper Wire A length of copper wire carries a current of 10 A, uniformly distributed through its cross section. Calculate the energy density of (a) the magnetic field and (b) the electric field at the surface of the wire. The wire diameter is 2.5 mm, and its resistance per unit length is 3.3 Ω/km.

12. Circular Loop A circular loop of wire 50 mm in radius carries a current of 100 A. (a) Find the magnetic field strength at the center of the loop. (b) Calculate the energy density at the center of the loop.

SEC. 33-4 ■ *LC* OSCILLATIONS, QUALITATIVELY

13. What Is the Capacitance What is the capacitance of an oscillating LC circuit if the maximum charge on the capacitor is 1.60 μC and the total energy is 140 μJ?

14. Maximum Charge In an oscillating LC circuit, $L = 1.10$ mH and $C = 4.00$ μF. The maximum charge on the capacitor is 3.00 μC. Find the maximum current.

15. Total Energy An oscillating LC circuit consists of a 75.0 mH inductor and a 3.60 μF capacitor. If the maximum charge on the capacitor is 2.90 μC, (a) what is the total energy in the circuit and (b) what is the maximum current?

16. Electric to Magnetic Energy In a certain oscillating LC circuit the total energy is converted from electric energy in the capacitor to magnetic energy in the inductor in 1.50 μs. (a) What is the period of oscillation? (b) What is the frequency of oscillation? (c) How long after the magnetic energy is a maximum will it be a maximum again?

17. Maximum Positive Charge The frequency of oscillation of a certain LC circuit is 200 kHz. At time $t = 0$ s, plate A of the capacitor has maximum positive charge. At what times $t > 0$ s will (a) plate A again have maximum positive charge, (b) the other plate of the capacitor have maximum positive charge, and (c) the inductor have maximum magnetic field?

SEC. 33-5 ■ THE ELECTRICAL–MECHANICAL ANALOGY

18. SHM A 0.50 kg body oscillates in simple harmonic motion on a spring that, when extended 2.0 mm from its equilibrium, has an 8.0 N restoring force. (a) What is the angular frequency of oscillation? (b) What is the period of oscillation? (c) What is the capacitance of an LC circuit with the same period if L is chosen to be 5.0 H?

19. Energy The energy in an oscillating LC circuit containing a 1.25 H inductor is 5.70 μJ. The maximum charge on the capacitor is 175 μC. Find (a) the mass, (b) the spring constant, (c) the maximum displacement, and (d) the maximum speed for a mechanical system with the same period.

SEC. 33-6 ■ *LC* OSCILLATIONS, QUANTITATIVELY

20. Loudspeakers LC oscillators have been used in circuits connected to loudspeakers to create some of the sounds of electronic music. What inductance must be used with a 6.7 μF capacitor to produce a frequency of 10 kHz, which is near the middle of the audible range of frequencies?

21. Initially a Maximum In an oscillating LC circuit with $L = 50$ mH and $C = 4.0$ μF, the current is initially a maximum. How long will it take before the capacitor is fully charged for the first time?

22. Single Loop A single loop consists of inductors ($L_1, L_2, \ldots$), capacitors ($C_1, C_2, \ldots$), and resistors ($R_1, R_2, \ldots$) connected in series as shown, for example, in Fig. 33-21*a*. Show that regardless of the sequence of these circuit elements in the loop, the bahavior of this circuit is identical to that of the simple LC circuit shown in Fig. 33-21*b*. (*Hint:* Consider the loop rule and see Problem 7 in Chapter 32.)

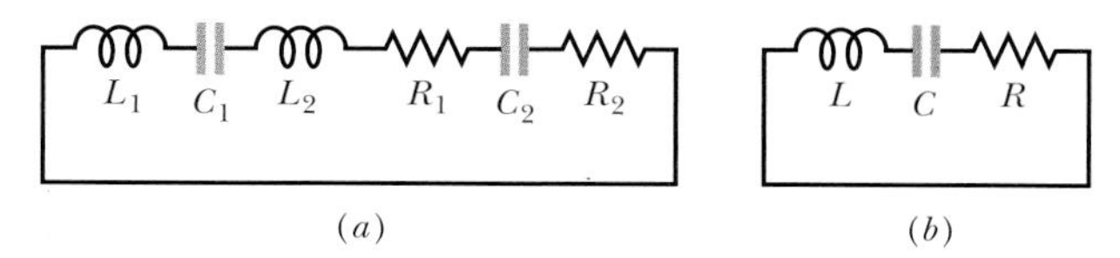

FIGURE 33-21 ■ Problem 22.

23. Maximum Voltage An oscillating LC circuit consisting of a 1.0 nF capacitor and a 3.0 mH coil has a maximum voltage of 3.0 V. (a) What is the maximum charge on the capacitor? (b) What is the maximum current through the circuit? (c) What is the maximum energy stored in the magnetic field of the coil?

24. Maximum Potential Difference In an oscillating LC circuit in which $C = 4.00$ μF, the maximum potential difference across the capacitor during the oscillations is 1.50 V and the maximum current through the inductor is 50.0 mA. (a) What is the inductance L? (b) What is the frequency of the oscillations? (c) How much time is required for the charge on the capacitor to rise from zero to its maximum value?

25. Switch Is Thrown In the circuit shown in Fig. 33-22 the switch is kept in position a for a long time. It is then thrown to position b. (a) Calculate the frequency of the resulting oscillating current. (b) What is the amplitude of the current oscillations?

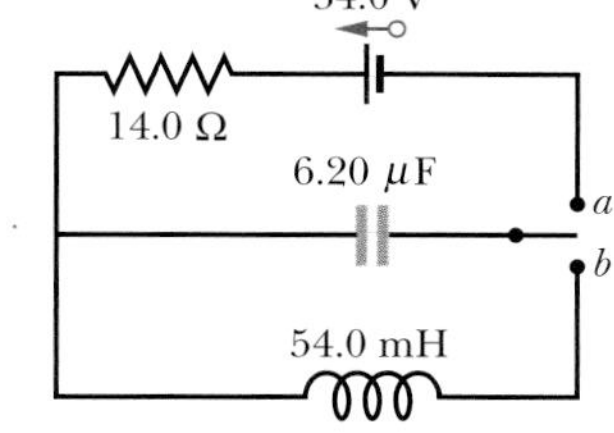

FIGURE 33-22 ■ Problem 25.

26. One Inductor, Two Capacitors You are given a 10 mH inductor and two capacitors, of 5.0 μF and 2.0 μF capacitance. List the oscillation frequencies that can be generated by connecting these elements in various combinations.

27. Variable Capacitor A variable capacitor with a range from 10 to 365 pF is used with a coil to form a variable-frequency LC

circuit to tune the input to a radio. (a) What ratio of maximum to minimum frequencies may be obtained with such a capacitor? (b) If this circuit is to obtain frequencies from 0.54 MHz to 1.60 MHz, the ratio computed in (a) is too large. By adding a capacitor in parallel to the variable capacitor, this range may be adjusted. What should be the capacitance of this added capacitor, and what inductance should be used to obtain the desired range of frequencies?

28. Energy Stored in Magnetic Field In an oscillating LC circuit, 75.0% of the total energy is stored in the magnetic field of the inductor at a certain instant. (a) In terms of the maximum charge on the capacitor, what is the charge there at that instant? (b) In terms of the maximum current in the inductor, what is the current there at that instant?

29. Capacitor Is Charging In an oscillating LC circuit, $L = 25.0$ mH and $C = 7.80$ μF. At time $t = 0$ s the current is 9.20 mA, the charge on the capacitor is 3.80 μC, and the capacitor is charging. (a) What is the total energy in the circuit? (b) What is the maximum charge on the capacitor? (c) What is the maximum current? (d) If the charge on the capacitor is given by $q = |Q| \cos(\omega t + \phi)$, what is the phase angle ϕ? (e) Suppose the data are the same, except that the capacitor is discharging at $t = 0$ s. What then is ϕ?

30. Varied by a Knob An inductor is connected across a capacitor whose capacitance can be varied by turning a knob. We wish to make the frequency of oscillation of this LC circuit vary linearly with the angle of rotation of the knob, going from 2×10^5 to 4×10^5 Hz as the knob turns through 180°. If $L = 1.0$ mH, plot the required capacitance C as a function of the angle of rotation of the knob.

31. Oscillating LC Circuit In an oscillating LC circuit, $L = 3.00$ mH and $C = 2.70$ μF. At $t = 0$ s the charge on the capacitor is zero and the current is 2.00 A. (a) What is the maximum charge that will appear on the capacitor? (b) In terms of the period T of oscillation, how much time will elapse after $t = 0$ until the energy stored in the capacitor will be increasing at its greatest rate? (c) What is this greatest rate at which energy is transferred to the capacitor?

32. Angular Frequency A series circuit containing inductance L_1 and capacitance C_1 oscillates at angular frequency ω. A second series circuit, containing inductance L_2 and capacitance C_2, oscillates at the same angular frequency. In terms of ω, what is the angular frequency of oscillation of a series circuit containing all four of these elements? Neglect resistance. (*Hint:* Use the formulas for equivalent capacitance and equivalent inductance; see Section 28-4 and Problem 7 in Chapter 32.)

33. Current as Function of Time In an oscillating LC circuit with $C = 64.0$ μF, the current as a function of time is given by $i = (1.60 \text{ A}) \sin[(2500 \text{ rad/s})\ t + 0.680 \text{ rad}]$, where t is in seconds. (a) How soon after $t = 0$ s will the current reach its maximum value? What are (b) the inductance L and (c) the total energy?

34. Three Identical Inductors Three identical inductors L and two identical capacitors C are connected in a two-loop circuit as shown in Fig. 33-23. (a) Suppose the currents are as shown in Fig. 33-23a. What is the current in the middle inductor? Write the loop equations and show that they are satisfied if the current oscillates with angular frequency $\omega = 1/\sqrt{LC}$. (b) Now suppose the currents are as shown in Fig. 33-23b. What is the current in the middle inductor? Write the loop equations and show that they are satisfied if the current oscillates with angular frequency $\omega = 1/\sqrt{3LC}$. Because the circuit can oscillate at two different frequencies, we cannot find an equivalent single-loop LC circuit to replace it.

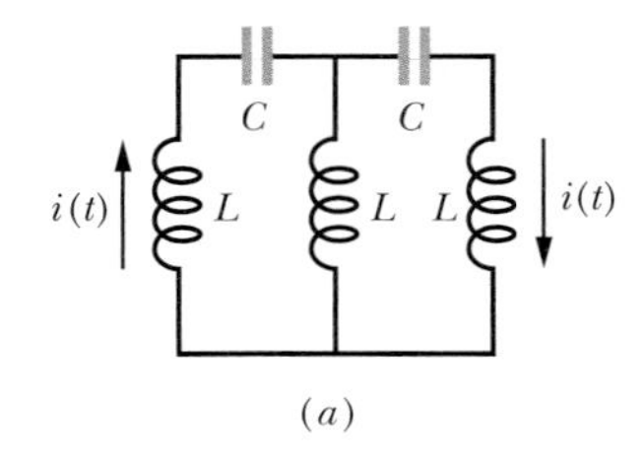

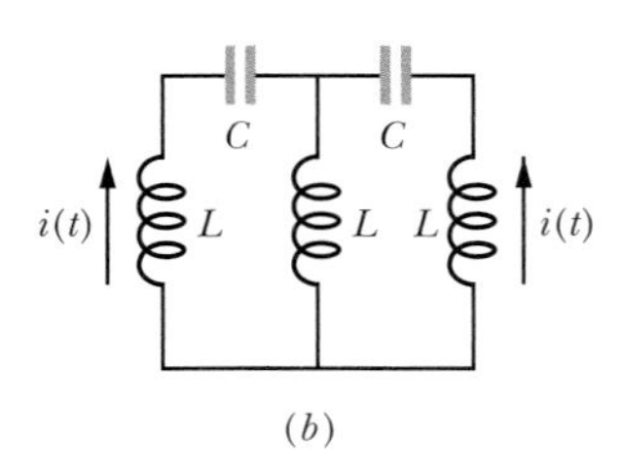

FIGURE 33-23 ■ Problem 34.

35. Capacitor One, Capacitor Two In Fig. 33-24, capacitor 1 with $C_1 = 900$ μF is initially charged to 100 V and capacitor 2 with $C_2 = 100$ μF is uncharged. The inductor has an inductance of 10.0 H. Describe in detail how one might charge capacitor 2 to 300 V by manipulating switches S_1 and S_2.

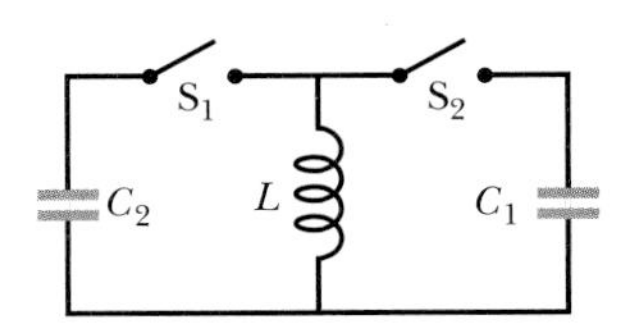

FIGURE 33-24 ■ Problem 35.

SEC. 33-7 ■ DAMPED OSCILLATIONS IN AN *RLC* CIRCUIT

36. Damped LC Consider a damped LC circuit. (a) Show that the damping term $e^{-Rt/2L}$ (which involves L but not C) can be rewritten in a more symmetric manner (involving L and C) as $e^{-\pi R(\sqrt{C/L})t/T}$. Here T is the period of oscillation (neglecting resistance). (b) Using (a), show that the SI unit of $\sqrt{L/C}$ is the ohm. (c) Using (a), show that the condition that the fractional energy loss per cycle be small is $R \ll \sqrt{L/C}$.

37. What Resistance What resistance R should be connected in series with an inductance $L = 220$ mH and capacitance $C = 12.0$ μF for the maximum charge on the capacitor to decay to 99.0% of its initial value in 50.0 cycles? (Assume $\omega' \approx \omega$.)

38. Single-Loop Circuit A single-loop circuit consists of a 7.20 Ω resistor, a 12.0 H inductor, and a 3.20 μF capacitor. Initially the capacitor has a charge of 6.20 μC and the current is zero. Calculate the charge on the capacitor N complete cycles later for $N = 5$, 10, and 100.

39. Oscillating Series RLC In an oscillating series RLC circuit, find the time required for the maximum energy present in the capacitor during an oscillation to fall to half its initial value. Assume $q = Q$ at $t = 0$.

40. No Charge on Capacitor At time $t = 0$ s there is no charge on the capacitor of a series RLC circuit but there is current I through the inductor. (a) Find the phase constant ϕ in Eq. 33-31 for the circuit. (b) Write an expression for the charge q on the capacitor as a function of time t and in terms of the current amplitude and angular frequency ω' of the oscillations.

41. Fraction of Energy Lost In an oscillating series RLC circuit, show that the fraction of the energy lost per cycle of oscillation, $\Delta U/U$, is given to a close approximation by $2\pi R/\omega L$. The quantity $\omega L/R$ is often called the Q of the circuit (for *quality*). A high-Q circuit has low resistance and a low fractional energy loss ($= 2\pi/Q$) per cycle.

SEC. 33-10 ■ REPRESENTING OSCILLATIONS WITH PHASORS: THREE SIMPLE CIRCUITS

42. Amplitude A 1.50 μF capacitor is connected as in Fig. 33-13a to an ac generator with $|\mathscr{E}^{\max}| = 30.0$ V. What is the amplitude of the resulting alternating current if the frequency of the emf is (a) 1.00 kHz and (b) 8.00 kHz?

43. AC Generator A 50.0 mH inductor is connected as in Fig. 33-14*a* to an ac generator with $|\mathscr{E}^{max}| = 30.0$ V. What is the amplitude of the resulting alternating current if the frequency of the emf is (a) 1.00 kHz and (b) 8.00 kHz?

44. Frequency of emf Is A 50 Ω resistor is connected as in Fig. 33-12*a* to an ac generator with $|\mathscr{E}^{max}| = 30.0$ V. What is the amplitude of the resulting alternating current if the frequency of the emf is (a) 1.00 kHz and (b) 8.00 kHz?

45. At What Frequency (a) At what frequency would a 6.0 mH inductor and a 10 μF capacitor have the same reactance? (b) What would the reactance be? (c) Show that this frequency would be the natural frequency of an oscillating circuit with the same L and C.

46. When the Current Is Maximum An ac generator has emf $\mathscr{E} = \mathscr{E}^{max} \sin \omega^{dr} t$, with $\mathscr{E}^{max} = 25.0$ V and $\omega^{dr} = 377$ rad/s. It is connected to a 12.7 H inductor. (a) What is the maximum value of the current? (b) When the current is a maximum, what is the emf of the generator? (c) When the emf of the generator is −12.5 V and increasing in magnitude, what is the current?

47. At What Time An ac generator has emf $\mathscr{E} = \mathscr{E}^{max} \sin(\omega^{dr} t - \pi/4)$, where $\mathscr{E}^{max} = 30.0$ V and $\omega^{dr} = 350$ rad/s. The current produced in a connected circuit is $i(t) = I \sin(\omega^{dr} t - 3\pi/4)$, where $I =$ 620 mA. (a) At what time after $t = 0$ does the generator emf first reach a maximum? (b) At what time after $t = 0$ does the current first reach a maximum? (c) The circuit contains a single element other than the generator. Is it a capacitor, an inductor, or a resistor? Justify your answer. (d) What is the value of the capacitance, inductance, or resistance, as the case may be?

48. Generator from Above The ac generator of Problem 46 is connected to a 4.15 μF capacitor. (a) What is the maximum value of the current? (b) When the current is a maximum, what is the emf of the generator? (c) When the emf of the generator is −12.5 V and increasing in magnitude, what is the current?

SEC. 33-11 ■ THE SERIES *RLC* CIRCUIT

49. Find Z, ϕ, and I (a) Find Z, ϕ, and I for the situation of Touchstone Example 33-4 with the capacitor removed from the circuit, all other parameters remaining unchanged. (b) Draw to scale a phasor diagram like that of Fig. 33-15*d* for this new situation.

50. Find Z, ϕ, and I Two (a) Find Z, ϕ, and I for the situation of Touchstone Example 33-4 with the inductor removed from the circuit, all other parameters remaining unchanged. (b) Draw to scale a phasor diagram like that of Fig. 33-15*d* for this new situation.

51. Find Z, ϕ, and I Three (a) Find Z, ϕ, and I for the situation of Touchstone Example 33-4 with $C = 70.0$ μF, the other parameters remaining unchanged. (b) Draw a phasor diagram like that of Fig. 33-15*d* for this new situation and compare the two diagrams closely.

52. Adjustable Frequency In Fig 33-25, a generator with an adjustable frequency of oscillation is connected to a variable resistance R, a capacitor of $C = 5.50$ μF, and an inductor of inductance L. The amplitude of the current produced in the circuit by the generator is at half-maximum level when the generator's frequency is 1.30 or 1.50 kHz. (a) What is L? (b) If R is increased, what happens to the frequencies at which the current amplitude is at half-maximum level?

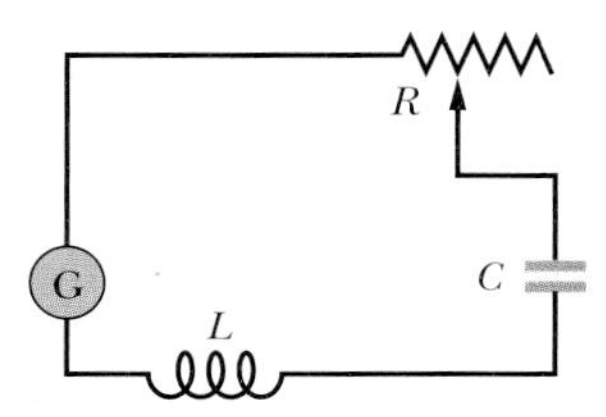

FIGURE 33-25 ■ Problem 52.

53. At Resonance In an RLC circuit, can the amplitude of the voltage across an inductor be greater than the amplitude of the generator emf? Consider an RLC circuit with $\mathscr{E}^{max} = 10$ V, $R = 10$ Ω, $L =$ 1.0 H, and $C = 1.0$ μF. Find the amplitude of the voltage across the inductor at resonance.

54. Emf Is Maximum When the generator emf in Touchstone Example 33-4 is a maximum, what is the voltage across (a) the generator, (b) the resistance, (c) the capacitance, and (d) the inductance? (e) By summing these with appropriate signs, verify that the loop rule is satisfied.

55. Unknown Resistance A coil of inductance 88 mH and unknown resistance and a 0.94 μF capacitor are connected in series with an alternating emf of frequency 930 Hz. If the phase constant between the applied voltage and the current is 75°, what is the resistance of the coil?

56. Capacitive Reactance An ac generator with $\mathscr{E}^{max} = 220$ V and operating at 400 Hz causes oscillations in a series RLC circuit having $R = 220$ Ω, $L = 150$ mH, and $C = 24.0$ μF. Find (a) the capacitive reactance X_C, (b) the impedance Z, and (c) the current amplitude I. A second capacitor of the same capacitance is then connected in series with the other components. Determine whether the values of (d) X_C, (e) Z, and (f) I increase, decrease, or remain the same.

57. Half-Width An RLC circuit such as that of Fig. 33-11 has $R = 5.00$ Ω, $C = 20.0$ μF, $L = 1.00$ H, and $\mathscr{E}^{max} = 30.0$ V. (a) At what angular frequency ω^{dr} will the current amplitude have its maximum value, as in the resonance curves of Fig. 33-17? (b) What is this maximum value? (c) At what two angular frequencies ω_1^{dr} and ω_2^{dr} will the current amplitude be half this maximum value? (d) What is the fractional half-width $[= (\omega_1^{dr} - \omega_2^{dr})/\omega]$ of the resonance curve for this circuit?

58. Generator in Series An ac generator is to be connected in series with an inductor of $L = 2.00$ mH and a capacitance C. You are to produce C by using capacitors of capacitances $C_1 = 4.00$ μF and $C_2 = 6.00$ μF, either singly or together. What resonant frequencies can the circuit have, depending on how you use C_1 and C_2?

59. Fractional Half-Width Show that the fractional half-width (see Problem 57) of a resonance curve is given by

$$\frac{\Delta\omega^{dr}}{\omega} = \sqrt{\frac{3C}{L}}\, R,$$

in which ω is the angular frequency at resonance and $\Delta\omega^{dr}$ is the width of the resonance curve at half-amplitude. Note that $\Delta\omega^{dr}/\omega$ increases with R, as Fig. 33-17 shows. Use this formula to check the answer to Problem 57d.

60. Adjustable Frequency Two In Fig. 33-26, a generator with an adjustable frequency of oscillation is connected to resistance $R = 100$ Ω, inductances $L_1 =$ 1.70 mH and $L_2 = 2.30$ mH, and capacitances $C_1 = 4.00$ μF,

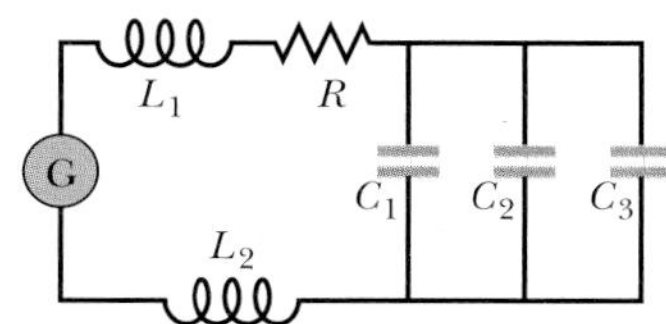

FIGURE 33-26 ■ Problem 60.

$C_2 = 2.50\ \mu\text{F}$, and $C_3 = 3.50\ \mu\text{F}$. (a) What is the resonant frequency of the circuit? (*Hint:* See Problem 7 in Chapter 32.) What happens to the resonant frequency if (b) the value of R is increased, (c) the value of L_1 is increased, and (d) capacitance C_3 is removed from the circuit?

SEC. 33-12 ■ POWER IN ALTERNATING-CURRENT CIRCUITS

61. Thermal Energy What direct current will produce the same amount of thermal energy, in a particular resistor, as an alternating current that has a maximum value of 2.60 A?

62. AC Voltmeter An ac voltmeter with large impedance is connected in turn across the inductor, the capacitor, and the resistor in a series circuit having an alternating emf of 100 V(rms); it gives the same reading in volts in each case. What is this reading?

63. AC Voltage What is the maximum value of an ac voltage whose rms value is 100 V?

64. Give or Take (a) For the conditions in Problem 46c, is the generator supplying energy to or taking energy from the rest of the circuit? (b) Repeat for the conditions of Problem 48c.

65. Average Rate of Dissipation Calculate the average rate of energy dissipation in the circuits of Problems 43, 44, 49, and 50.

66. Energy Is Supplied Show that the average rate at which energy is supplied to the circuit of Fig. 33-11 can also be written as $\langle P \rangle = (\mathscr{E}^{\text{rms}})^2 R/Z^2$. Show that this expression for average power gives reasonable results for a purely resistive circuit, for an RLC circuit at resonance, for a purely capacitive circuit, and for a purely inductive circuit.

67. Air Conditioner An air conditioner connected to a 120 V rms ac line is equivalent to a 12.0 Ω resistance and a 1.30 Ω inductive reactance in series. (a) Calculate the impedance of the air conditioner. (b) Find the average rate at which energy is supplied to the appliance.

68. Oscillating *RLC* In a series oscillating RLC circuit, $R = 16.0\ \Omega$, $C = 31.2\ \mu\text{F}$, $L = 9.20$ mH, and $\mathscr{E} = |\mathscr{E}^{\text{max}}| \sin \omega^{\text{dr}} t$ with $|\mathscr{E}^{\text{max}}| = 45.0$ V and $\omega^{\text{dr}} = 3000$ rad/s. For time $t = 0.442$ ms find (a) the rate at which energy is being supplied by the generator, (b) the rate at which the energy in the capacitor is changing, (c) the rate at which the energy in the inductor is changing, and (d) the rate at which energy is being dissipated in the resistor. (e) What is the meaning of a negative result for any of (a), (b), and (c)? (f) Show that the results of (b), (c), and (d) sum to the result of (a).

69. Black Box Figure 33-27 shows an ac generator connected to a "black box" through a pair of terminals. The box contains an RLC circuit, possibly even a multiloop circuit, whose elements and connections we do not know. Measurements outside the box reveal that

$$\mathscr{E}(t) = (75.0\ \text{V}) \sin \omega^{\text{dr}}$$

and
$$i(t) = (1.20\ \text{A}) \sin (\omega^{\text{dr}} t + 42.0°).$$

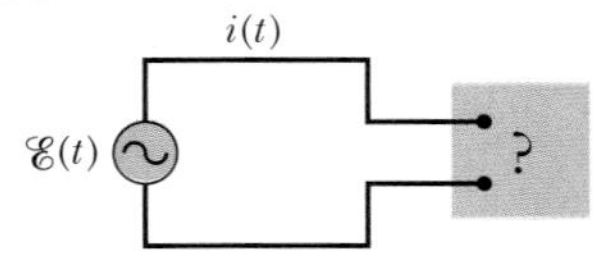

FIGURE 33-27 ■ Problem 69.

(a) What is the power factor? (b) Does the current lead or lag the emf? (c) Is the circuit in the box largely inductive or largely capacitive? (d) Is the circuit in the box in resonance? (e) Must there be a capacitor in the box? An inductor? A resistor ? (f) At what average rate is energy delivered to the box by the generator? (g) Why don't you need to know the angular frequency ω^{dr} to answer all these questions?

70. Average Rate In Fig. 33-28 show that the average rate at which energy is dissipated in resistance R is a maximum when R is equal to the internal resistance r of the ac generator. (In the text discussion we have tacitly assumed that $r = 0$.)

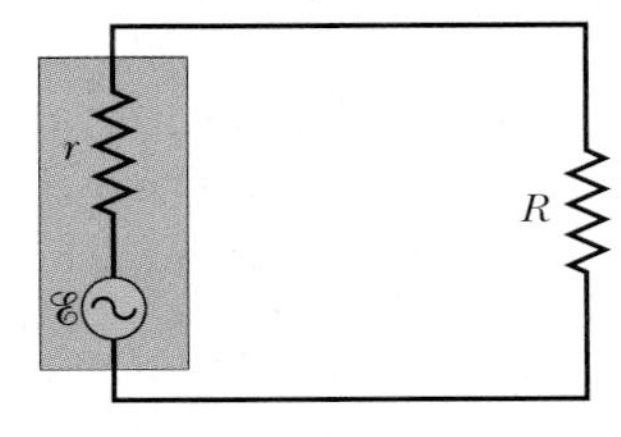

FIGURE 33-28 ■ Problem 70.

71. Energy Is Dissipated In an RLC circuit such as that of Fig. 33.11 assume that $R = 5.00\ \Omega$, $L = 60.0$ mH $f^{\text{dr}} = 60.0$ Hz, and $|\mathscr{E}^{\text{max}}| = 30.0$ V. For what values of the capacitor would the average rate at which energy is dissipated in the resistance be (a) a maximum and (b) a minimum? (c) What are these maximum and minimum energy dissipation rates? What are (d) the corresponding phase angles and (e) the corresponding power factors?

72. Light Dimmer A typical "light dimmer" used to dim the stage lights in a theater consists of a variable inductor L (whose inductance is adjustable between zero and L^{max}) connected in series with the lightbulb B as shown in Fig. 33-29. The electrical supply is 120 V (rms) at 60.0 Hz; the lightbulb is rated as "120 V, 1000 W." (a) What L^{max} is required if the rate of energy dissipation in the lightbulb is to be varied by a factor of 5 from its upper limit of 1000 W? Assume that the resistance of the lightbulb is independent of its temperature. (b) Could one use a variable resistor (adjustable between zero and R^{max}) instead of an inductor? If so, what R^{max} is required? Why isn't this done?

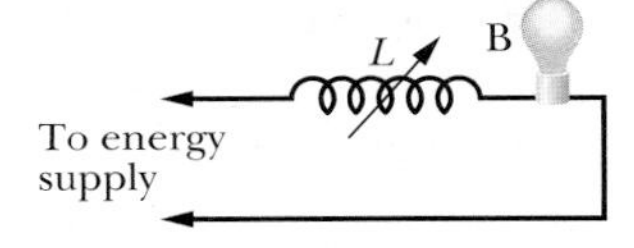

FIGURE 33-29 ■ Problem 72.

73. Sinusoidal Voltage In Fig. 33-30, $R = 15.0\ \Omega$, $C = 4.70\ \mu\text{F}$, and $L = 25.0$ mH. The generator provides a sinusoidal voltage of 75.0 V (rms) and frequency $f = 550$ Hz.
(a) Calculate the rms current.
(b) Find the rms voltages ΔV_{ab}, ΔV_{bc}, ΔV_{cd}, ΔV_{bd}, ΔV_{ad}. (c) At what average rate is energy dissipated by each of the three circuit elements?

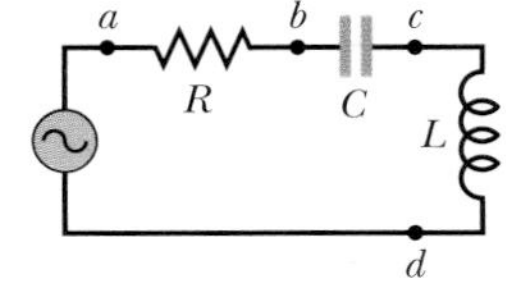

FIGURE 33-30 ■ Problem 73.

34 Electromagnetic Waves

A comet is often referred to as a dirty snowball. As a comet swings around the Sun, ice on its surface vaporizes, releasing trapped dust and charged particles. The "solar wind," mostly consisting of protons streaming away from the Sun, forces the charged particles released by the comet into a straight "tail" that points radially away from the Sun. The dust continues to travel in the comet's orbit.

Why does most of the dust released by the comet remain in the tail of dust on the right in the photograph?

The answer is in this chapter.

FIGURE 34-1 ■ James Clerk Maxwell (1831–1879).

34-1 Introduction

At the end of Chapter 31 we presented James Clerk Maxwell's (Fig. 34-1) four equations that synthesized 50 years of research on electricity and magnetism. Maxwell's equations are compact and elegant in the way that they reveal the inherent symmetry between electrical and magnetic phenomena. But their importance also stems from the fact that they led Maxwell and others to predict a wide range of new phenomena. The design of the communication systems that support radio and television broadcasting, cellular phones, and the Internet are all informed by Maxwell's equations. So is much of our understanding of the nature of light. For these reasons, Maxwell's equations are considered to be the crowning achievement of 19th-century theoretical physics.

One of the most incredible outcomes of Maxwell's work was his prediction of electromagnetic waves in 1864, long before investigators were able to generate and detect them. In fact, Maxwell's hypothesis was not taken seriously until almost 25 years later, when Heinrich Hertz first generated and detected electromagnetic waves. We begin this chapter by reviewing the remarkable chain of reasoning that led Maxwell to postulate the existence of this yet unknown type of wave. We also consider why it took Hertz until 1887 to confirm Maxwell's prediction. Next we describe how the electromagnetic wave pulses and continuous waves used for radio transmission are generated. We also examine how and why the orientation of radio and TV antennas is related to an idea called "polarization" of electromagnetic waves.

Maxwell predicted that all electromagnetic waves would move at a speed that was quite close to the measured value for the speed of visible light in air. For this reason he correctly asserted from the beginning that visible light was an electromagnetic wave. We can use what we learn about electromagnetic waves in this chapter to build a foundation for our study of optics in Chapter 35, where we will learn about how light waves interact with lenses and mirrors to form images.

34-2 Maxwell's Prediction of Electromagnetism

Maxwell's prediction of the electromagnetic wave was a result of the way in which he rewrote and reinterpreted the mathematical expressions for experimentally determined laws named after Faraday and Ampère. We start with a review of these two laws.

Generating Fields in the Absence of Conductors

Faraday's law

$$\mathscr{E} = -\frac{d\Phi^{\text{mag}}}{dt} \qquad \text{(Eq. 31-6)}$$

was originally based on his measurements of the electric currents induced *within a conducting loop* when the magnetic flux enclosed by it changes. The fact that an electric field is needed to produce the current in the conductor led Maxwell to a bold conjecture. He predicted that an electric field would *always* be induced in the region of varying magnetic flux regardless of whether or not a conducting loop was present. Thus, Faraday's law in the more general form shown below should probably be called the Faraday–Maxwell law:

$$\mathscr{E} = \oint \vec{E} \cdot d\vec{s} = -\frac{d\Phi^{\text{mag}}}{dt} \quad \text{(Faraday's law).} \qquad \text{(Eq. 31-21)}$$

We owe to Faraday and Maxwell the discovery that a changing magnetic field produces an electric field. We owe to Maxwell alone the symmetric prediction that a changing electric field ought to produce a magnetic field. Maxwell made this prediction by noting a mathematical similarity between Ampère's law, which describes the nature of the magnetic field generated by current flow in a conductor

$$\oint \vec{B} \cdot d\vec{s} = \mu_0 i^{\text{enc}} \quad \text{(Ampère's law)}, \qquad \text{(Eq. 30-16)}$$

and the general formulation of Faraday's law

$$\mathscr{E} = \oint \vec{E} \cdot d\vec{s} = -\frac{d\Phi^{\text{mag}}}{dt} \quad \text{(Faraday's law)}. \qquad \text{(Eq. 31-21)}$$

At this point Maxwell speculated that the magnetic field induced along a closed loop (Ampère's law) was more generally a function of the rate of change of the electric flux enclosed by the loop. He predicted that any region in space with changing electric flux (like that found between plates when a capacitor is charged or discharged) could induce a magnetic field in the region—even without the presence of a conductor. This notion prompted Maxwell in 1861 to invent the concept of displacement current (discussed in Section 31-9). This fictitious current was devised to describe the possible, but as yet unobserved, magnetic effects of changing electric flux. Maxwell incorporated his displacement current concept into the reformulation of Ampère's law that we presented in Section 31-8,

$$\oint \vec{B} \cdot d\vec{s} = \mu_0 \varepsilon_0 \frac{d\Phi^{\text{elec}}}{dt} + \mu_0 i^{\text{enc}} \quad \text{(Ampère–Maxwell law)}. \qquad \text{(Eq. 31-32)}$$

Since there was no experimental evidence that changing electric flux could generate a magnetic field in the absence of real current, the proposed Ampère–Maxwell law was perhaps the most remarkable of Maxwell's many predictions.

Note that in free space, with no conductors present, the $\mu_0 i$ term in the expression above (Eq. 31-32) disappears and it becomes symmetric to Faraday's law (Eq. 31-21). The symmetry between these equations provided the basis for Maxwell's belief in the existence of electromagnetic waves almost 25 years before Hertz was able to generate and detect them. To find out why, read on.

Electromagnetic Wave Propagation

Maxwell used the symmetric relationship between Faraday's law and the Ampère–Maxwell law to predict a phenomenon that at first glance seems quite bizarre. He noted that if a changing magnetic field could create a changing electric field, then the changing electric field could, in turn, create another changing magnetic field. These changing fields could continuously generate each other and propagate, carrying "electromagnetic" energy with them.

In his 1861 paper (where he also introduced the displacement current concept) Maxwell rewrote the Faraday and Ampère–Maxwell laws as differential equations and solved them to describe $\vec{E}$ and $\vec{B}$ fields separately as functions of time. This produced differential equations that have the same algebraic form as those that describe the propagation of pressure variations in air (sound waves), ripples on the surface of a pond (water waves), and transverse displacements of a stretched string. Hence, the concept of an "electromagnetic wave" was born.

A rigorous treatment of Maxwell's mathematical description of electromagnetic waves would require us to solve differential equations and interpret the results. These methods are beyond the scope of this text. Instead we present some new results that arise from Maxwell's equations and show that they are consistent with experimental results. This will provide us with insight into the generation and properties of electromagnetic radiation. For example, one outcome of solving Maxwell's wave equations was the revelation that an electromagnetic wave ought to travel in a vacuum at a speed given by

$$c = \frac{1}{\sqrt{\mu_0 \varepsilon_0}} \quad \text{(predicted wave speed in a vacuum).} \tag{34-1}$$

Maxwell was quite surprised to find that the wave speed depends on the familiar electric and magnetic constants, ε_0 and μ_0, previously measured in static, electric, and magnetic experiments. He was equally surprised to find that a calculation of the speed in a vacuum agreed well with what was known at the time to be the speed of light. This led Maxwell to make the additional, rather bold, hypothesis that light was an electromagnetic wave.

34-3 The Generation of Electromagnetic Waves

In this section we give a more detailed description of how electromagnetic waves are generated. Observations summarized in Maxwell's equations reveal that electromagnetic waves (sometimes called "radiation") should be generated by accelerating charges. We begin by considering the generation of radio-frequency waves used for radio and TV transmission. These waves (wavelength $\lambda \approx 1\text{m}$) provide a source of radiation (the emitted waves) that is both macroscopic and of manageable dimensions so that classical physics rules. Some electromagnetic waves, including x-rays, gamma rays, and visible light, are *radiated* (emitted) from sources that are of atomic or nuclear size, where quantum physics rules.

At this point, you may find it helpful to quickly review mechanical waves discussed in Chapter 17. It was there that we first introduced important and pertinent wave-related concepts such as wavelength and frequency. We begin this section by considering the analogy between the generation and propagation of an electromagnetic *wave pulse* and a pulse traveling on a stretched string (Section 17-3) or along the surface of a pond (Section 18-1). We end the section with a discussion of how sinusoidal radio-frequency electromagnetic waves are generated.

An Electromagnetic Wave Pulse

How does a wave pulse propagate? Let's consider a more familiar situation in which a small bucket of water is dumped onto the surface of a pond. The water that is dumped on the pond's surface will undergo a rapid oscillation as it falls and rises again. However, the water that is dumped on the surface cannot undergo this oscillation without causing the ring of water that surrounds it to begin oscillating. This oscillation is passed along to the next ring of surrounding water, and so on. We see a two-dimensional wave crest like that shown in Fig. 18-3 traveling along the pond's surface at a constant speed.

In 1842, prior to Maxwell's prediction of electromagnetic waves, Joseph Henry observed electrical oscillations produced by a spark discharge from a capacitor. Because the theoretical basis for explaining electrical oscillations was not yet developed, the significance of these oscillations was not appreciated. Hertz generated electrical oscillations once again in 1888 using an induction coil to create sparks.

In order to better understand the connection between electromagnetic waves, electric oscillations, and the spark discharges with which Henry and Hertz experimented, we need to pull together ideas from other chapters. First, we know from

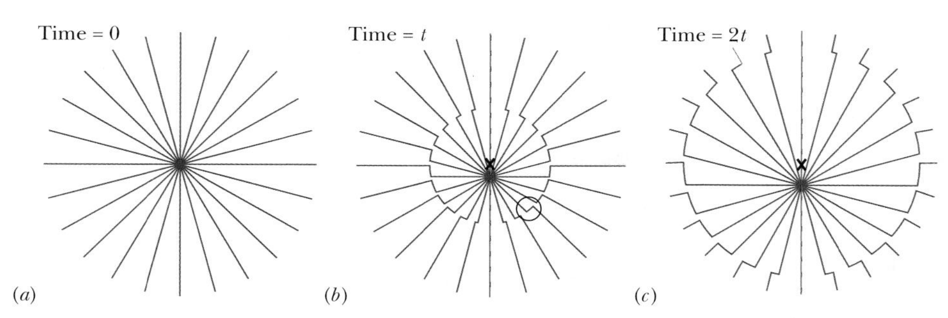

FIGURE 34-2 ■ (*a*) Electric field lines in the plane of a resting point charge at the exact moment the jerk starts. (*b*) After an elapsed time t the kink at the interface between the old set of lines and the set associated with the jerk has traveled a distance ct. The original location of the charge is marked with an X and the line of motion is shown with a red dashed line. The region within the small circle is shown enlarged in Fig. 34-3. (*c*) After a time $2t$, the kink is at a distance $2ct$. (Diagrams adapted from *Electric & Magnetic Interactions: The Movies,* © 1996 Ruth Chabay and Bruce Sherwood, Carnegie Mellon University.)

Section 23-10 that electric field arrows indicate the direction of the electric field. Second, although we do not discuss this until Section 38-3, Einstein's principle of relativity establishes that nothing can travel faster than the speed of light. This includes information about the position of a charge. Keeping these ideas in mind, we can visualize how the field lines from a charge must readjust when the charge undergoes a sudden and brief acceleration like the spark discharge with which Henry and Hertz experimented. Specifically, suppose that at a time $t_1 = 0$ s a point charge initially at rest is suddenly accelerated straight downward and then stopped, all in an extremely short time period Δt. This is shown in Fig. 34-2. The original location of the charge is marked with an "X" in Fig. 34-2*b* and *c*.

What happens? As discussed below Maxwell's equations would lead us to predict that the accelerating charge will generate a three-dimensional electromagnetic wave crest. Furthermore, if you consider how this situation is like the momentary acceleration of the bit of mass at the end of a taut string when it is given a quick downward jerk, the generation of a pulse makes some intuitive sense as well. Let's look at the situation more carefully.

Before the acceleration, the electric field lines associated with the charge look just like those depicted Fig. 34-2*a*. Electromagnetic information travels at the speed of light, so after a time period $\Delta t = t$ the field lines will point to the charge's new position at the end of the acceleration—but only up to a distance ct. This is shown in Fig. 34-2*b*. Beyond that distance, the field lines will point to the charge's old position because the "news" of the charge's changed position will not have had time to spread that far.

At the distance ct, the field lines will have a kink joining the new set and the old set. This kink is increasingly close to perpendicular to the direction of motion of the wave front the farther you get from the accelerating charge. If we assume that field lines are continuous and that the field vectors are tangent to the line at each point, then the $\vec{E}$ field vectors in the kink are also increasingly perpendicular to the direction of motion of the wavefronts. The kink travels outward at the speed of light so that it has moved to a distance of $2ct$ after a time of $\Delta t = 2t$. This pulse continues to move out at speed c from the original location of the charge. Figure 34-3 shows a magnified view of a circled piece of Fig. 34-2*b* with extra field vectors drawn in.

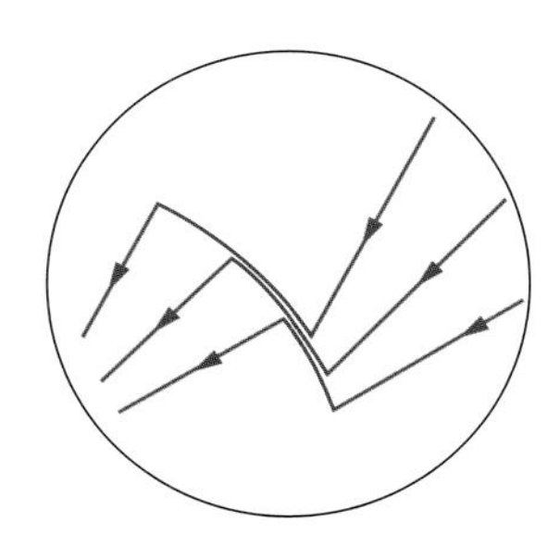

FIGURE 34-3 ■ Here more $\vec{E}$ field vectors are assigned to the charge and shown in a magnified view of a circled piece of Fig. 34-2*b*. The crest of the wave pulse at time t show that the density of electric field lines is higher than the density of lines created when the charge is resting. Note that the $\vec{E}$ field in the kink points in a direction that is nearly perpendicular to the propagating electromagnetic wave pulse.

In the brief moment that the charge is being accelerated, the "current" created by it is increasing ($di/dt = d^2q/dt > 0$). We can use Ampère's law to determine the direction of the increasing magnetic field associated with this increasing "current." Taking the line of motion to be the direction of our "current" (shown with a red dashed line in Fig. 34-2*b*) and using Ampère's law, we see that the magnetic field points into the page to the left of the line of motion of the charge and out of the page to the right. The direction of these $\vec{B}$ field vectors (into or out of the page) is always perpendicular to that of the $\vec{E}$ field vectors. So, we surmise that the $\vec{B}$ field vectors associated with the kink will point perpendicular to the direction of motion of the kink. There are no propagating $\vec{E}$ or $\vec{B}$ fields along the line of motion of the charge. We define the wave disturbance as the propagation of the $\vec{E}$ field and $\vec{B}$ field vectors that are each perpendicular to the direction of motion of the wave. So, we have an electromagnetic wave pulse that is *transverse*. (See Chapter 17 for a review of wave properties if this last comment is not clear.)

Although this informal consideration of a single wave pulse created by a sudden acceleration of a charge is not rigorous or complete, it does give us a useful qualitative picture of how a pulsed electromagnetic wave might be generated.

Continuous Electromagnetic Wave Generation

How can we generate a continuous electromagnetic wave? What corresponds to the generation of the sinusoidal oscillation in a stretched string as shown in Fig. 17-6? Suppose we cause charges to oscillate back and forth along a line with sinusoidal or simple harmonic motion (SHM). We know from Section 16-4 that these charges will also have a sinusoidal acceleration, and so by the discussion above, we know that they will generate continuous electromagnetic (EM) waves.

Figure 34-4 is a schematic of an apparatus that can be used to generate radio frequency waves. The apparatus is an LC oscillator like that described in Section 33-6, but with a broadcast antenna coupled to it. This oscillator, mentioned in the previous sentence, establishes a sinusoidal current with an angular frequency given by $\omega = 1/\sqrt{LC}$. An ac generator provides a source of energy to compensate for thermal losses in the oscillator circuit and also for the energy carried away by the radiated electromagnetic wave.

The LC oscillator is coupled by a transformer and a transmission line to an *antenna,* which consists essentially of two thin, solid, conducting rods. Through this coupling, the sinusoidally varying current in the oscillator causes charge to oscillate sinusoidally along the antenna rods. The antenna behaves like an electric dipole described in Section 25-8, except that its electric dipole moment along the antenna changes sinusoidally in magnitude and direction over time. Since the charges are oscillating, they are continually accelerating and thereby produce electromagnetic radiation.

Because the dipole moment varies in the antenna in magnitude and direction, the electric and magnetic fields produced by the dipole vary in magnitude and direction. However, the changes in the electric and magnetic fields do not happen everywhere instantaneously. Rather, the changes travel outward from the antenna at the speed of light c. Together the changing fields form an electromagnetic wave that travels away from the antenna at speed c. The angular frequency of this wave, ω, is the same as that of the LC oscillator.

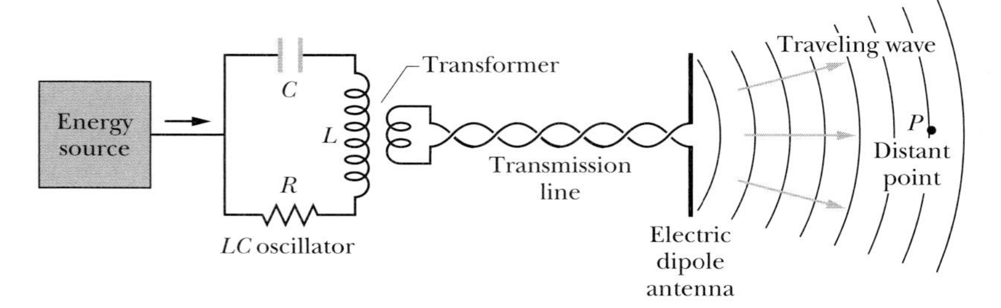

FIGURE 34-4 ■ Apparatus for generating a traveling electromagnetic wave at a "shortwave" radio frequency. An LC oscillator produces a sinusoidal current in an antenna. This generates an oscillating magnetic field and thus an EM wave. P is a distant point at which a detector (consisting of a dipole receiving antenna) could be placed to monitor the wave traveling past it.

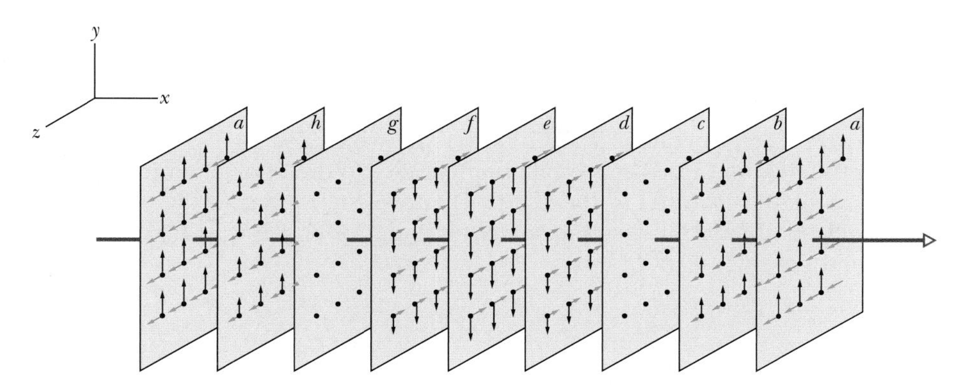

FIGURE 34-5 ■ A "snapshot" of some equally spaced planes of a sinusoidal electromagnetic wave traveling in the $+x$ direction. In this particular wave, the darker, vertically oriented $\vec{E}$ field vectors always point along the y axis. The lighter, horizontally oriented $\vec{B}$ field vectors always point along the z axis. The planes are labeled a through h to correspond to the $\vec{E}$ and $\vec{B}$ field vector configurations shown in Fig. 34-6.

The wave moves out in all directions except along the line of motion of the charges. In a direction perpendicular to the antenna the wavefronts will be approximately spherical in shape. If we go to a point that is far from the charge, the spherical wavefronts seem almost flat (just as the Earth seems flat to us because we are so far from its center). If we block all but a small piece of this wavefront, we get what looks like a series of planes marching forward at the speed of light. Hence, we often refer to electromagnetic waves as "plane waves." Although the plane wave approximation is a simplification of the real situation, it describes the nature of waves quite well when the distance from the source, in this case the antenna, is large compared to the wavelength of electromagnetic wave. The value to us in looking at electromagnetic waves as **"plane waves"** is that they are simpler to deal with mathematically than spherical waves.

There are several ways to depict a traveling sinusoidal electromagnetic plane wave. One of these is shown in Fig. 34-5. Here we show a "snapshot" of a few equally spaced planes of a sinusoidal electromagnetic wave traveling in the $+x$ direction. In this particular wave, the darker, vertically oriented $\vec{E}$ field vectors always point along the y axis. The lighter, horizontally oriented $\vec{B}$ field vectors always point along the z axis. This diagram emphasizes the fact that, at a given instant, field vectors are the same everywhere in a given y-z plane. The figure also shows that as the wave passes a point in space, the field vector values will sinusoidally vary from a maximum to a minimum and back again. This is shown by the dashed lines representing curves through the tips of the $\vec{E}$ vectors and the $\vec{B}$ vectors at the bottom of each plane. The planes are labeled a through h to correspond to the $\vec{E}$ and $\vec{B}$ field vector configurations shown in Fig. 34-6.

Figure 34-6 shows how the electric field $\vec{E}$ and the magnetic field $\vec{B}$ change with time as one wavelength of the wave sweeps past the distant point P of Fig. 34-4. In each part of Fig. 34-5, the wave is traveling directly along the x axis. There are several key features of any sinusoidal plane electromagnetic wave that are shown in Fig. 34-5 and Fig. 34-6 that are present whenever a plane electromagnetic wave is created with charges that oscillate sinusoidally along a line:

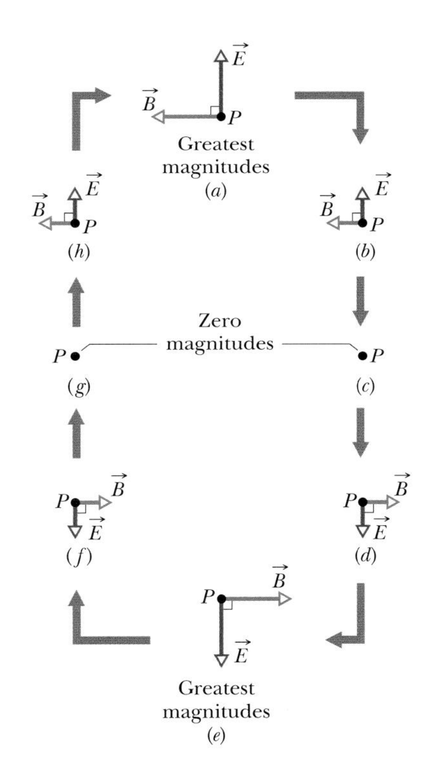

FIGURE 34-6 ■ (a)–(h) The variation in the electric field $\vec{E}$ and the magnetic field $\vec{B}$ at the distant point P of Fig. 34-4 as one wavelength of the electromagnetic wave shown in Fig. 34-5 travels past it. In this perspective, the wave is traveling directly out of the page. The two fields vary sinusoidally in magnitude and direction. Each of the planes that correspond to parts of this diagram (a)–(h) are also shown in Fig. 34-5.

1. The electric and magnetic fields $\vec{E}$ and $\vec{B}$ are always perpendicular to the direction of travel of the wave. Thus, the wave is a *transverse wave,* as discussed in Chapter 17.
2. The electric field is always perpendicular to the magnetic field.
3. The cross product $\vec{E} \times \vec{B}$ always points in the direction of travel of the wave.
4. For a single simple plane wave, the fields always vary sinusoidally over time, just like the transverse waves discussed in Chapter 17. Moreover, the fields vary with the same frequency and are *in phase* (in step) with each other. More complex fields can be described mathematically as a superposition of plane waves.

34-4 Describing Electromagnetic Wave Properties Mathematically

Suppose an electromagnetic wave like that shown in Fig. 34-5 is traveling toward P, the electric field in Fig. 34-5 is oscillating parallel to the y axis, and the magnetic field is then oscillating parallel to the z axis. Then we can represent the electric and magnetic fields mathematically as sinusoidal functions of position x (along the path of the wave) and time t:

$$\vec{E} = \vec{E}^{\,\text{max}} \sin(kx - \omega t), \tag{34-2}$$

and

$$\vec{B} = \vec{B}^{\,\text{max}} \sin(kx - \omega t), \tag{34-3}$$

where $\vec{E}^{\,\text{max}}$ and $\vec{B}^{\,\text{max}}$ represent the amplitudes of the fields and, as in Chapter 17, ω and k are the angular frequency and wave number, respectively. From these equations, we note that each type of field forms its own wave. Equation 34-2 gives the *electric wave component* of the electromagnetic wave, and Eq. 34-3 gives the *magnetic wave component*. As we already realize from considering Maxwell's formulation, these two wave components cannot exist independently. The wave propagates because a changing electric field creates a changing magnetic field, which generates another changing electric field, and so on.

At this point it is useful to devise a second "snapshot" representation of the plane electromagnetic wave. This is shown in Fig. 34-7*b*. Instead of emphasizing the planar nature of a selected sample of wave fronts as we did in Fig. 34-5, we show just one vector that represents the length of electric and magnetic field vectors in each sample plane. The curves through the tips of the vectors display the sinusoidal nature of the oscillations described by $\vec{E} = \vec{E}^{\,\text{max}} \sin(kx - \omega t)$, and $\vec{B} = \vec{B}^{\,\text{max}} \sin(kx - \omega t)$, (Eqs. 34-2 and 34-3) above. The wave components $\vec{E}$ and $\vec{B}$ are depicted as in phase, perpendicular to each other, and perpendicular to the wave's direction of travel.

Figure 34-7*a* shows how the new representation is tied to the old one. Only three sample planes with $\vec{E}$ vectors along one line, separated by a half wavelength $\lambda/2(= \pi/k)$ of the wave, are shown.

In some cases waves are traveling in approximately the same direction and form a beam, such as a laser beam or a beam of radio waves. A beam can be represented with a "ray," which is just a line showing the direction of motion of the beam. This is also

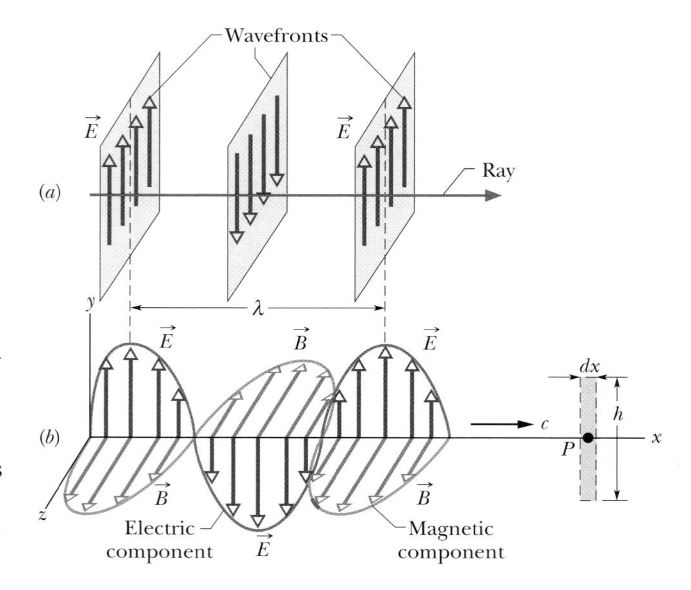

FIGURE 34-7 ■ (*a*) A plane electromagnetic wave represented with a ray and three wavefronts separated by a half wavelength $\lambda/2$. (*b*) The same wave represented in a "snapshot" of its electric field $\vec{E}$ and magnetic field $\vec{B}$ at points on the x axis, along which the wave travels at speed c. As it travels past point P, the fields vary as shown in Fig. 34-5. The dashed rectangle at P is used in Fig. 34-8*a*.

shown in Fig. 34-7*a*. "Rays" of light will become increasing prominent in our discussions of image formations in Chapters 35 and 36.

Interpretation of Fig. 34-7*b* requires some care. The similar drawings for a transverse wave on a taut string that we discussed in Chapter 17 represented the up and down displacement of sections of the string as the wave passed (*something actually moved*). Figure 34-7*b* is more abstract. At the instant shown, the electric and magnetic fields each have a certain magnitude and direction (but always perpendicular to the x axis) at each point along the x axis. We choose to represent these vector quantities with a pair of arrows for each point, so we must draw arrows of different lengths for different points, all directed away from the x axis, like thorns on a rose stem. For each line parallel to the x axis there is a similar picture. In viewing figures such as this, it is important to remember that the length of the arrows represents the field values along the line chosen as the x axis. Neither the arrows nor the sinusoidal curves represent a sideways displacement of anything.

A Most Curious Wave

From our previous work with waves in Chapters 17 and 18, we know that the speed of the wave is ω/k (Eq. 17-12). However, it is customary to use the symbol c (rather than v) to denote an electromagnetic wave speed in a vacuum (or air):

> All electromagnetic waves, including visible light, have the same speed c in a vacuum. Hence, c is called "the speed of light."*

The waves we discussed in Chapters 17 and 18 require a *medium* (some material) through which or along which to travel. We had waves traveling along a string, through the Earth, and through the air. Maxwell and other 19th-century investigators assumed there was a medium through which electromagnetic waves traveled. However, we now believe that electromagnetic waves are curiously different in that they require no medium for travel. They can, indeed, travel through a medium such as air or glass, but they can also travel through the vacuum of space between a distant star and the Earth.

Once Albert Einstein proposed the special theory of relativity in 1905, scientists realized that visible light waves and other electromagnetic waves were special entities. The reason is that light has the same speed in any inertial frame of reference. If you send a beam of light along an axis and ask several observers to measure its speed while they move at different speeds along that axis, either in the direction of the light or opposite it, the observers will all measure the *same speed* for the light. This result is an amazing one and quite different from what would have been found if those observers had measured the speed of any other type of wave. For other waves, the speed of the observers relative to the wave would have affected their measurements. The implications of this are striking and include some seemingly bizarre effects that we will learn about in Chapter 38 on special relativity.

As we saw in Chapter 1, the speed of light (any electromagnetic wave) in vacuum has the exact value

$$c = 299\,792\,458 \text{ m/s.}^{\dagger}$$

The Ratio of $\vec{E}$ to $\vec{B}$ and the Induced Electric Field

We can use Faraday's law

$$\mathscr{E} = \oint \vec{E} \cdot d\vec{s} = -\frac{d\Phi^{\text{mag}}}{dt}$$

* The letter c comes from the Latin word *celar,* which means "fast".

† The value is "exact" so the meter could be redefined as a length of path traveled by light in a specified time. (See Section 1-6 for details.)

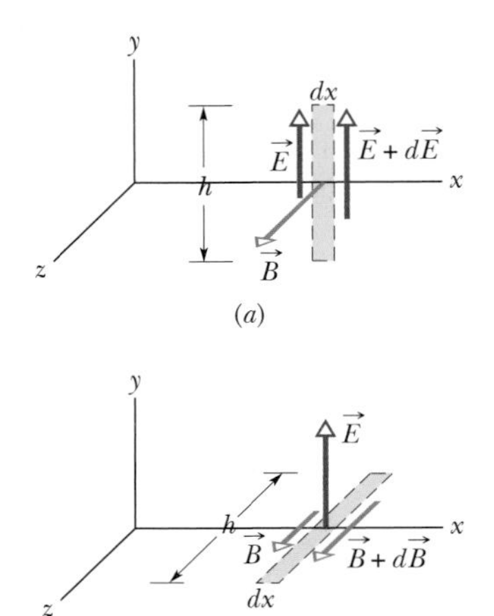

FIGURE 34-8 ■ (*a*) As the electromagnetic wave travels rightward past point P in Fig. 34-7, the sinusoidal variation of the magnetic field $\vec{B}$ through a rectangle centered at P induces electric fields along the rectangle. At the instant shown, $\vec{B}$ is decreasing in magnitude and the induced electric field is therefore greater in magnitude on the right side of the rectangle than on the left. (*b*) The sinusoidal variation of the electric field through this rectangle, located (but not shown) at point P in Fig. 34-8*a*, induces magnetic fields along the rectangle. The instant shown is that of Fig. 34-8*a*: $\vec{E}$ is decreasing in magnitude, and the induced magnetic field is greater in magnitude on the right side of the rectangle than on the left.

to find the ratio of the electric and magnetic fields at any location along an electromagnetic wave. We start by considering the dashed rectangle (of dimensions dx and h in Fig. 34-7*b*) that is fixed at point P on the x axis and in the xy plane. As the electromagnetic wave moves to the right past the rectangle, the magnetic flux Φ^{mag} through the rectangle changes and—according to Faraday's law of induction—induced electric fields appear throughout the region of the rectangle. We take $\vec{E}$ and $\vec{E} + d\vec{E}$ to be the induced fields along the two long sides of the rectangle. These induced electric fields are, in fact, the electric component of the electromagnetic wave at those points.

Let us consider these fields at the instant when the magnetic wave component passing through the rectangle is the small section marked with red on the line marked "a magnetic field component" in Fig. 34-7*b*. Just then, the magnetic field through the rectangle points in the positive z direction and is decreasing in magnitude (the magnitude was greater just before the red section arrived). Because the magnetic field is out of the page and decreasing, the magnetic flux Φ^{mag} through the rectangle is also decreasing. According to Faraday's law, this change in flux is opposed by induced electric fields. This implies that a counterclockwise induced current would appear along the rectangle if it were a conductor (which it is not). This in turn implies that a counterclockwise induced electric field would have to appear along the rectangle. So, the induced electric field vectors $\vec{E}$ and $\vec{E} + d\vec{E}$ are indeed oriented as shown in Fig. 34-8*a*, with the magnitude of $\vec{E} + d\vec{E}$ greater than that of $\vec{E}$. Otherwise, the net induced electric field would not act counterclockwise around the rectangle.

Let us now apply Faraday's law of induction,

$$\oint \vec{E} \cdot d\vec{s} = -\frac{d\Phi^{\text{mag}}}{dt}, \tag{34-4}$$

proceeding counterclockwise around the rectangle of Fig. 34-8*a*. There is no contribution to the integral from the top or bottom of the rectangle because $\vec{E}$ and $d\vec{s}$ are perpendicular there. The integral then has the value

$$\oint \vec{E} \cdot d\vec{s} = (|\vec{E}| + d|\vec{E}|)h - |\vec{E}|h = h\, d|\vec{E}|. \tag{34-5}$$

The flux Φ^{mag} through this rectangle is

$$\Phi^{\text{mag}} = |\vec{B}|h\, dx, \tag{34-6}$$

where $|\vec{B}|$ is the magnitude of $\vec{B}$ within the rectangle and $h\,dx$ is the area of the rectangle. Differentiating this expression (Eq. 34-6) with respect to t gives

$$\frac{d\Phi^{\text{mag}}}{dt} = h\, dx \frac{d|\vec{B}|}{dt}. \tag{34-7}$$

If we substitute this result (Eq. 34-7) and $\oint \vec{E} \cdot d\vec{s} = h\, d|\vec{E}|$ (Eq. 34-5) into Faraday's law $\oint \vec{E} \cdot d\vec{s} = -d\Phi^{\text{mag}}/dt$ (Eq. 34-4), we find

$$h\, d|\vec{E}| = -h\, dx \frac{d|\vec{B}|}{dt}$$

or

$$\frac{d|\vec{E}|}{dx} = -\frac{d|\vec{B}|}{dt}. \tag{34-8}$$

We gather from our electromagnetic wave representations and from $\vec{E} = \vec{E}^{\text{max}} \sin(kx - \omega t)$, and $\vec{B} = \vec{B}^{\text{max}} \sin(kx - \omega t)$ (Eqs. 34-2 and 34-3) that both $|\vec{B}|$

and $|\vec{E}|$ are functions of two variables, x and t. However, in evaluating $d|\vec{E}|/dx$, we can assume that t is constant because we consider only an "instantaneous snapshot." Also, in evaluating $d|\vec{B}|/dt$ we can assume that x is constant because we are dealing with the time rate of change of $|\vec{B}|$ at a particular place, the point P in Fig. 34-7b. The derivatives under these circumstances are partial derivatives, and Eq. 34-8 must be written

$$\frac{\partial|\vec{E}|}{\partial x} = -\frac{\partial|\vec{B}|}{\partial t}. \tag{34-9}$$

The minus sign in this equation is appropriate and necessary because, although $|\vec{E}|$ is increasing with x at the site of the rectangle in Fig. 34-8, $|\vec{B}|$ is decreasing with t.

From $\vec{E} = \vec{E}^{\max}\sin(kx - \omega t)$ (Eq. 34-2) we have

$$\frac{\partial \vec{E}}{\partial x} = k\vec{E}^{\max}\cos(kx - \omega t)$$

and from $\vec{B} = \vec{B}^{\max}\sin(kx - \omega t)$ (Eq. 34-3)

$$\frac{\partial \vec{B}}{\partial t} = -\omega\vec{B}^{\max}\cos(kx - \omega t).$$

Then $\partial|\vec{E}|/\partial x = -\partial|\vec{B}|/\partial t$ (Eq. 34-9) reduces to

$$k|\vec{E}^{\max}|\cos(kx - \omega t) = \omega|\vec{B}^{\max}|\cos(kx - \omega t). \tag{34-10}$$

The ratio ω/k for a traveling wave is its speed, which we are calling c. Hence, we see that

$$\frac{|\vec{E}^{\max}|}{|\vec{B}^{\max}|} = c \quad \text{(amplitude ratio).} \tag{34-11}$$

If we divide $\vec{E} = \vec{E}^{\max}\sin(kx - \omega t)$ by $\vec{B} = \vec{B}^{\max}\sin(kx - \omega t)$ (Eqs. 34-2 and 34-3) and then substitute into Eq. 34-10, we find that the ratio of magnitudes of the fields at every instant is given by

$$\frac{|\vec{E}|}{|\vec{B}|} = c \quad \text{(magnitude ratio).} \tag{34-12}$$

Induced Magnetic Field and the Equation for Wave Speed

If we use the Ampère–Maxwell law

$$\oint \vec{B}\cdot d\vec{s} = \mu_0\varepsilon_0\frac{d\Phi^{\text{elec}}}{dt} + \mu_0 i^{\text{enc}},$$

we can find an alternative expression for the wave speed in the case where no real current is present (so $i^{\text{enc}} = 0$ A). We start with Fig. 34-8b, which shows another dashed rectangle at point P of Fig. 34-7; this one is in the xz plane. As the electromagnetic wave moves rightward past this new rectangle, the electric flux Φ^{elec} through the rectangle changes and—according to the Ampère–Maxwell law of induction—induced magnetic fields appear throughout the region of the rectangle. These induced magnetic fields are, in fact, the magnetic field components of the electromagnetic wave.

We see from Fig. 34-7 that at the instant chosen for the magnetic field in Fig. 34-8, the electric field through the rectangle of Fig. 34-8*b* is directed as shown. Recall that at the chosen instant, the magnetic field in Fig. 34-8*a* is decreasing. Because the two fields are in phase, the electric field in Fig. 34-8*b* must also be decreasing, and so must the electric flux Φ^{elec} through the rectangle. By applying the same reasoning we applied to Fig. 34-8*a*, we see that the changing flux Φ^{elec} will induce a magnetic field with vectors $\vec{B}$ and $\vec{B} + d\vec{B}$ oriented as shown in Fig. 34-8*b*, where $\vec{B} + d\vec{B}$ is greater than $\vec{B}$.

Let us also apply Maxwell's law of induction with no real current present,

$$\oint \vec{B} \cdot d\vec{s} = \mu_0 \varepsilon_0 \frac{d\Phi^{elec}}{dt}, \qquad (34\text{-}13)$$

by proceeding counterclockwise around the dashed rectangle of Fig. 34-8*b*. Only the long sides of the rectangle contribute to the integral, whose value is

$$\oint \vec{B} \cdot d\vec{s} = -(|\vec{B}| + d|\vec{B}|)h + |\vec{B}|h = -h\, d|\vec{B}|. \qquad (34\text{-}14)$$

The flux Φ^{elec} through the rectangle is

$$\Phi^{elec} = (|\vec{E}|)(h\, dx), \qquad (34\text{-}15)$$

where $|\vec{E}|$ is the average magnitude of $\vec{E}$ within the rectangle. Differentiating this expression with respect to t gives

$$\frac{d\Phi^{elec}}{dt} = h\, dx \frac{d|\vec{E}|}{dt}.$$

If we substitute this and $\oint \vec{B} \cdot d\vec{s} = -h\, d|\vec{B}|$ (Eq. 34-14 from above) into Maxwell's law of induction we find that

$$-h\, d|\vec{B}| = \mu_0 \varepsilon_0 \left(h dx \frac{d|\vec{E}|}{dt} \right).$$

Changing to partial-derivative notation as we did before (Eq. 34-9),

$$-\frac{\partial |\vec{B}|}{\partial x} = \mu_0 \varepsilon_0 \frac{\partial |\vec{E}|}{\partial t}. \qquad (34\text{-}16)$$

Again, the minus sign in this equation makes sense because, although $\vec{B}$ is increasing with x at point P in the rectangle in Fig. 34-8*b*, $\vec{E}$ is decreasing with t.

Evaluating this expression by using $\vec{E} = \vec{E}^{max} \sin(kx - \omega t)$, and $\vec{B} = \vec{B}^{max} \sin(kx - \omega t)$ (Eqs. 34-2 and 34-3) leads to

$$-k|\vec{B}^{max}|\cos(kx - \omega t) = -\mu_0 \varepsilon_0 \omega |\vec{E}^{max}| \cos(kx - \omega t),$$

which we can write as

$$\frac{|\vec{E}^{max}|}{|\vec{B}^{max}|} = \frac{1}{\mu_0 \varepsilon_0 (\omega/k)} = \frac{1}{\mu_0 \varepsilon_0 c}.$$

Combining this with $|\vec{E}^{max}|/|\vec{B}^{max}| = c$ (Eq. 34-11) leads at once to

$$c = \frac{1}{\sqrt{\mu_0 \varepsilon_0}} \quad \text{(wave speed)}, \qquad (34\text{-}17)$$

which is exactly Eq. 34-1.

READING EXERCISE 34-1: The magnetic field $\vec{B}$ through the rectangle of Fig. 34-8*a* is shown at a different instant in part 1 of the accompanying figure; $\vec{B}$ is directed in the xz plane, parallel to the z axis, and its magnitude is increasing. (a) Complete part 1 by drawing the induced electric fields, indicating both directions and relative magnitudes (as in Fig. 34-8*a*). (b) For the same instant, complete part 2 of the figure by drawing the electric field of the electromagnetic wave. Also draw the induced magnetic fields, indicating both directions and relative magnitudes (as in Fig. 34-8*b*).

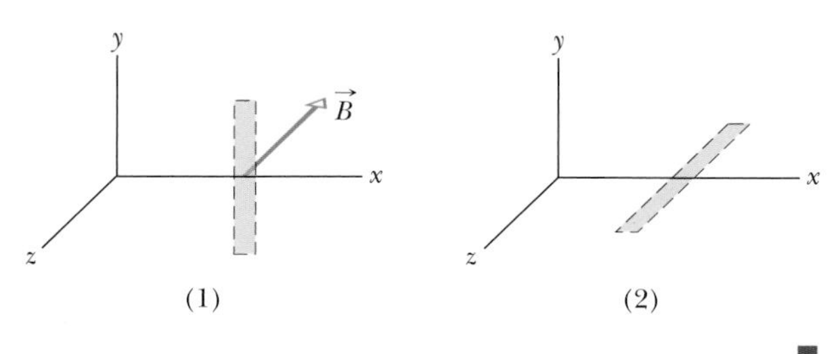

34-5 Transporting Energy with Electromagnetic Waves

From our experience with capacitors, we know that energy is stored in an electric field. Likewise, from our experience with inductors we know that energy is stored in a magnetic field. So, it makes sense that as an electromagnetic wave moves through space, it carries energy with it. Sunbathers will confirm this hunch. An electromagnetic wave can transport energy and deliver it to a body on which it falls. In this section, we develop an expression that will allow us to quantify the rate per unit area at which energy is transported by an electromagnetic wave. This quantity is a measure of the **intensity** of the electromagnetic wave.

From Chapter 28 on capacitors, we know that the energy per unit area or *energy density* within an electric field is

$$u^{\text{elec}} = \tfrac{1}{2}\varepsilon_0 E^2.$$

Because you also know (from this chapter) that $\vec{E} = c\vec{B}$ and c is such a very large number, you might conclude that the energy associated with the electric field in an electromagnetic wave is much greater than that associated with the magnetic field. That conclusion is incorrect; the densities of the two energies are exactly equal. To show this, we substitute $c\vec{B}$ for $\vec{E}$; then we can write

$$u^{\text{elec}} = \tfrac{1}{2}\varepsilon_0 E^2 = \tfrac{1}{2}\varepsilon_0 c^2 B^2.$$

If we now substitute for c with (Eq. 34-1)

$$c = \frac{1}{\sqrt{\varepsilon_0 \mu_0}},$$

we get

$$u^{\text{elec}} = \tfrac{1}{2}\varepsilon_0 \frac{1}{\mu_0 \varepsilon_0} B^2 = \frac{B^2}{2\mu_0}.$$

In Section 33-3, we found that $B^2/2\mu_0$ is the energy density u^{mag} of a magnetic field $\vec{B}$. So, we see that $u^{\text{elec}} = u^{\text{mag}}$ everywhere along an electromagnetic wave.

> The energy density of the electric field is equal to the energy density of the magnetic field at every instant and for every point along an electromagnetic wave.

The total energy density for the electromagnetic wave is the sum of the energy density associated with the magnet field and the energy density associated with the electric field:

$$u^{\text{tot}} = u^{\text{elec}} + u^{\text{mag}} = \tfrac{1}{2}\varepsilon_0 E^2 + \frac{1}{2\mu_0}B^2.$$

However, since these values are equal to one another, we can also write the total energy density as twice either value:

$$u^{\text{tot}} = 2u^{\text{elec}} = \varepsilon_0 E^2$$

or

$$u^{\text{tot}} = 2u^{\text{mag}} = \frac{1}{\mu_0}B^2.$$

This result is quite helpful in developing an expression for the rate of energy (or power) transport across a unit of area perpendicular to the direction of propagation of the electromagnetic wave. At any instant, the **rate of energy transport per unit area** is

$$S = \left(\frac{\text{energy/time}}{\text{area}}\right)^{\text{inst}} = \left(\frac{\text{power}}{\text{area}}\right)^{\text{inst}} \quad \text{(instantaneous).} \qquad (34\text{-}18)$$

Note that from this we can see that the SI unit for S must be the watt per square meter (W/m^2). Since an electromagnetic wave moves with a speed c, in a time period Δt, the wave travels a distance $c\,\Delta t$. During that motion, if it passes through a surface of some area A, the volume of space through which the wave passes is $c\,\Delta tA$. The total energy transported by the wave is the total energy density (electric and magnetic) multiplied by this volume. That is, the total energy (U) transported by the wave to an area A in a time period Δt is

$$U = u^{\text{tot}}c\,\Delta tA,$$

where

$$u^{\text{tot}} = \varepsilon_0 E^2.$$

Hence,

$$U = \varepsilon_0 E^2 c\,\Delta tA.$$

As expressed by Eq. 34-18 above, the rate of energy transport is this total energy transported divided by the area through which the wave travels and the time period. That is,

$$S = \frac{\varepsilon_0 E^2 c\,\Delta tA}{\Delta tA} = \varepsilon_0 E^2 c.$$

Since $c^2 = 1/\varepsilon_0\mu_0$, we can multiply this expression by one in the form of $c^2\varepsilon_0\mu_0$. This gives an alternative expression for the instantaneous energy flow rate:

$$S = \frac{1}{c\mu_0}E^2 = \varepsilon_0 c\,E^2 \quad \text{(instantaneous energy flow rate).} \qquad (34\text{-}19)$$

By substituting $\vec{E} = \vec{E}^{\text{max}}\sin(kx - \omega t)$ into Eq. 34-19, we could obtain an equation for the energy transport rate as a function of time. More useful in practice, however, is the average energy transported over time. For that, we need to find the time-averaged value of S, written $\langle S\rangle$. We call this quantity the **intensity** I of the wave. Thus the intensity I is

$$I = \langle S \rangle = \langle \frac{\text{energy/time}}{\text{area}} \rangle = \langle \frac{\text{power}}{\text{area}} \rangle. \tag{34-20}$$

With $S = E^2/c\mu_0$ (Eq. 34-19), we find

$$I = \langle S \rangle = \frac{1}{c\mu_0}\langle E^2 \rangle = \frac{1}{c\mu_0}\langle (E^{\text{max}})^2 \sin^2(kx - \omega t) \rangle. \tag{34-21}$$

Over a full cycle, the average value of $\sin^2\theta$, for any angular variable θ, is $\frac{1}{2}$. In addition, we define a new quantity E^{rms}, the root-mean-square value of the electric field magnitude as

$$E^{\text{rms}} \equiv \frac{E^{\text{max}}}{\sqrt{2}}. \tag{34-22}$$

We can then rewrite Eq. 34-21 as

$$I = \frac{1}{c\mu_0}(E^{\text{rms}})^2. \tag{34-23}$$

If we combine the ideas that we have developed above for the rate of energy transported by an electromagnetic wave per unit area with knowledge of the direction in which the wave is traveling, we can define a vector $\vec{S}$ that describes both the energy transport rate S *and* the direction in which the transfer in occurring. This vector is an important quantity, so we give it a name. It is called the **Poynting vector** after John Henry Poynting (1852–1914), who first discussed its properties.

The direction of the Poynting vector $\vec{S}$ of an electromagnetic wave at any point gives the wave's direction of travel and so the direction of energy transport at that point.

We can combine everything that we have developed in this section into a single expression defining the Poynting vector $\vec{S}$ as

$$\vec{S} = \frac{1}{\mu_0}\vec{E} \times \vec{B} \quad \text{(instantaneous Poynting vector)}, \tag{34-24}$$

where magnitude $|\vec{S}|$ of the Poynting vector is the instantaneous intensity of the electromagnetic wave. $\vec{E}$ and $\vec{B}$ are perpendicular to each other in an electromagnetic wave. Hence, the magnitude of $\vec{E} \times \vec{B}$ is $|\vec{E}||\vec{B}|$. Since $|\vec{E}| = c|\vec{B}|$, this is consistent with the expression that we developed above,

$$S = |\vec{S}| = \frac{1}{c\mu_0}E^2,$$

for the magnitude of energy transport rate. You can confirm for yourself that the cross product gives the correct direction for the wave propagation, and hence for the energy transport.

Variation of Intensity with Distance

The intensity variation of electromagnetic radiation with distance is often complex—especially when the source beams the radiation in a particular direction (like a searchlight at a movie premiere). We know that as a wavefront spreads out over a wider surface, its energy density must diminish, but how? Let's consider the simplest case we can imagine. Assume that the source is a *point source* that emits the light **isotropically**—

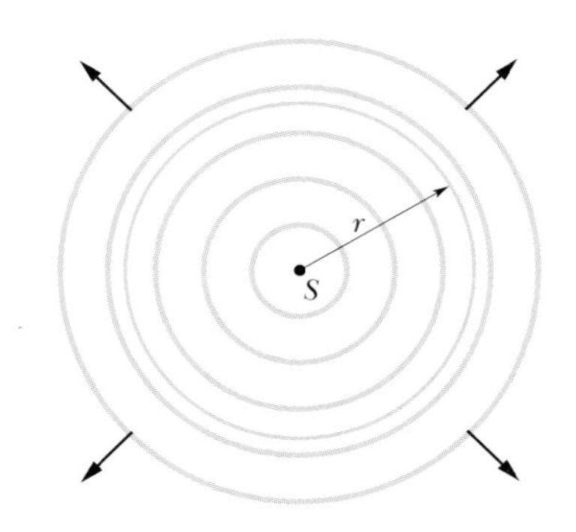

FIGURE 34-9 ■ A point source S emits electromagnetic waves uniformly in all directions. The spherical wavefronts pass through an imaginary sphere of radius r that is centered on S.

that is, with equal intensity in all directions. The spherical wavefronts spreading from such an isotropic point source S at a particular instant are shown in cross section in Fig. 34-9.

In a vacuum there is no mechanism for dissipation of energy, so it is conserved. Let us center an imaginary sphere of radius r on the source, as shown in Fig. 34-9. All the energy emitted by the source must pass through the sphere. Thus, the rate at which energy is transferred through the sphere by the radiation must equal the rate at which energy is emitted by the source—that is, the power P_s of the source. The intensity I (= power/area) at the sphere must then be

$$I = \frac{P_s}{4\pi r^2}, \quad (34\text{-}25)$$

where $4\pi r^2$ is the area of the sphere. This expression tells us that the intensity of the electromagnetic radiation from an isotropic point source decreases with the square of the distance r from the source.

Since from Eq. 34-23 we also know

$$I = \frac{1}{c\mu_0}(E^{\text{rms}})^2,$$

we can equate these two expressions to give

$$\frac{P_s}{4\pi r^2} = \frac{(E^{\text{rms}})^2}{c\mu_0},$$

or

$$(E^{\text{rms}})^2 = \frac{c\mu_0 P_s}{4\pi r^2}.$$

Simplification gives the relationship between the average electric field E^{rms} for a radiating charge, the power of the source P_s and the distance from the source r:

$$E^{\text{rms}} = \sqrt{\frac{c\mu_0 P_s}{4\pi}}\,\frac{1}{r}. \quad (34\text{-}26)$$

This equation tells us that the electric field associated with an isotropic radiation point source falls off as $1/r$, rather than as $1/r^2$ as it does for a static electric field.

READING EXERCISE 34-2: The figure gives the electric field of an electromagnetic wave at a certain point and a certain instant. The wave is transporting energy in the negative z direction. What is the direction of the magnetic field of the wave at that point and instant?

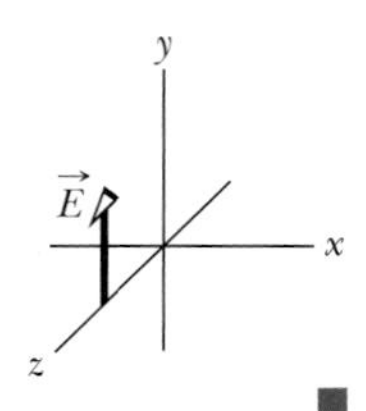

■

TOUCHSTONE EXAMPLE 34-1: Isotropic Light Source

As we stated in the introduction, visible light is now known to consist of electromagnetic waves. An observer is 1.8 m from an isotropic point light source whose power P_s is 250 W. Calculate the rms values of the electric and magnetic fields due to the source at the position of the observer.

SOLUTION ■ The first two **Key Ideas** here are these:

1. The rms value E^{rms} of the electric field in light is related to the intensity I of the light by $I = (E^{\text{rms}})^2/c\mu_0$.

2. Because the source is a point source emitting light with equal intensity in all directions, the intensity I at any distance r from the source is related to the source's power P_s via Eq. 34-25 ($I = P_s/4\pi r^2$).

Putting these two ideas together gives us

$$I = \frac{P_s}{4\pi r^2} = \frac{(E^{\text{rms}})^2}{c\mu_0},$$

which leads to

$$E^{\text{rms}} = \sqrt{\frac{P_s c\mu_0}{4\pi r^2}}$$

$$= \sqrt{\frac{(250\ \text{W})(3.00 \times 10^8\ \text{m/s})(4\pi \times 10^{-7}\ \text{H/m})}{(4\pi)(1.8\ \text{m})^2}}$$

$$= 48.1\ \text{V/m} \approx 48\ \text{V/m}. \qquad \text{(Answer)}$$

The third **Key Idea** here is that magnitudes of the electric field and magnetic field of an electromagnetic wave at any instant and at any point in the wave are related by the speed of light c according to Eq. 34-12 ($|\vec{E}|/|\vec{B}| = c$). Thus, the rms values of those fields are also related by Eq. 34-12 and we can write

$$B^{\text{rms}} = \frac{E^{\text{rms}}}{c}$$

$$= \frac{48.1\ \text{V/m}}{3.00 \times 10^8\ \text{m/s}}$$

$$= 1.6 \times 10^{-7}\ \text{T}. \qquad \text{(Answer)}$$

Note that E^{rms} (= 48 V/m) is appreciable as judged by ordinary laboratory standards, but B^{rms}(=1.6 × 10^{-7} T) is quite small. This difference helps to explain why most instruments used for the detection and measurement of electromagnetic waves are designed to respond to the electric component of the wave. It is wrong, however, to say that the electric component of an electromagnetic wave is "stronger" than the magnetic component. You cannot compare quantities that are measured in different units. As we have seen, the electric and magnetic components are on an equal basis as far as the propagation of the wave is concerned, because their average energies, which can be compared, are exactly equal.

34-6 Radiation Pressure

Electromagnetic waves carry linear momentum as well as energy. This means that we can exert a pressure—a **radiation pressure**—on an object by shining light on it. However, the pressure must be very small because, for example, you do not feel a camera flash pushing on you when it is used to take your photograph.

To see how Maxwell related radiation pressure to light intensity, let us shine a beam of electromagnetic radiation—visible light, for example—on an object for a time interval Δt. Further, let us assume that the object is free to move and that the radiation is entirely **absorbed** (taken up) by the object. This means that during the interval Δt, the object gains an energy ΔU from the radiation. Maxwell showed that the object also gains linear momentum. As usual we can represent momentum with a lowercase p and an object's momentum change as $\Delta\vec{p} = \vec{p}_2 - \vec{p}_1$. The magnitude $|\Delta\vec{p}\,|$ of the momentum change of the object is related to the energy change ΔU by

$$|\Delta\vec{p}\,| = \frac{\Delta U}{c} \quad \text{(total absorption)}, \qquad (34\text{-}27)$$

where c is the speed of light. The direction of the momentum change of the object is the direction of the incident (incoming) beam that the object absorbs.

Instead of being absorbed, the radiation can be reflected by the object; that is, the radiation can be sent off in a new direction as if it bounced off the object. If the radiation is entirely reflected back along its original path, the magnitude of the momentum change of the object is twice that given above, or

$$|\Delta\vec{p}\,| = \frac{2\,\Delta U}{c} \quad \text{(total reflection back along path)}. \qquad (34\text{-}28)$$

In the same way, an object undergoes twice as much momentum change when a perfectly elastic tennis ball is bounced from it as when it is struck by a perfectly inelastic ball (a lump of wet putty, say) of the same mass and velocity. If the incident radiation is partly absorbed and partly reflected, the momentum change of the object is between $\Delta U/c$ and $2\,\Delta U/c$.

From Newton's Second Law, we know that a change in momentum is related to a force by

$$\vec{F} = \frac{|\Delta \vec{p}\,|}{\Delta t}. \tag{34-29}$$

To find expressions for the force exerted by radiation in terms of the intensity I of the radiation, suppose that a flat surface of area A, perpendicular to the path of the radiation, intercepts the radiation. In time interval Δt, the energy intercepted by area A is

$$\Delta U = IA\,\Delta t. \tag{34-30}$$

If the energy is completely absorbed, then Eq. 34-27 tells us that $|\Delta \vec{p}\,| = IA\,\Delta t/c$. Then with $\vec{F} = |\Delta \vec{p}\,|\Delta t$ (Eq. 34-29), the magnitude of the force on the area A is

$$|\vec{F}| = \frac{IA}{c} \quad \text{(total absorption)}. \tag{34-31}$$

Similarly, if the radiation is totally reflected back along its original path, Eq. 34-28 tells us that $|\Delta \vec{p}\,| = 2IA\,\Delta t/c$ and, from Eq. 34-29,

$$|\vec{F}| = \frac{2IA}{c} \quad \text{(total reflection back along path)}. \tag{34-32}$$

If the radiation is partly absorbed and partly reflected, the magnitude of the force on area A is between the values of IA/c and $2IA/c$.

The force per unit area on an object due to radiation is the radiation pressure P. (Note that we represent pressure, as usual, with a capital P.) We can find the pressure for total absorption and total reflection by dividing both sides of each equation (34-31 and 34-32) by A. We obtain

$$P^{\text{absorp}} = \frac{I}{c} \quad \text{(pressure for total wave absorption)}, \tag{34-33}$$

and

$$P^{\text{refl}} = \frac{2I}{c} \quad \text{(pressure for total wave reflection back along path)}. \tag{34-34}$$

Just as with fluid pressure in Chapter 15, the SI unit of radiation pressure is the pascal (Pa) which equals a newton per square meter (N/m^2).

The development of laser technology has permitted researchers to achieve radiation pressures much greater than, say, that due to a camera flashlamp. This comes about because a beam of laser light—unlike a beam of light from a small lamp filament—can be focused to a tiny spot only a few wavelengths in diameter. This permits the delivery of great amounts of energy and momentum to small objects placed at that spot.

READING EXERCISE 34-3: Light of uniform intensity shines perpendicularly on a totally absorbing surface, fully illuminating the surface. If the area of the surface is decreased, do (a) the radiation pressure and (b) the radiation force on the surface increase, decrease, or stay the same? ■

TOUCHSTONE EXAMPLE 34-2: Comet Dust

When dust is released by a comet, it does not continue along the comet's orbit because radiation pressure from sunlight pushes it radially outward from the Sun. Assume that a dust particle is spherical with radius R, has density $\rho = 3.5 \times 10^3$ kg/m^3, and totally absorbs the sunlight it intercepts. For what value of R does the gravitational force $\vec{F}^{\,\text{grav}}$ on the dust particle due to the Sun just balance the radiation force $\vec{F}^{\,\text{rad}}$ on it from the sunlight?

SOLUTION ■ We can assume that the Sun is far enough from the particle to act as an isotropic point source of light. Then because we are told that the radiation pressure pushes the particle radially outward from the Sun, we know that the radiation force $\vec{F}^{\,\text{rad}}$ on the particle is directed radially outward from the center of the Sun. At the same time, the gravitational force $\vec{F}^{\,\text{grav}}$ on the particle is directed radially inward toward the center of the Sun. Thus, we can write the balance of these two forces as

$$|\vec{F}^{\,\text{rad}}| = |\vec{F}^{\,\text{grav}}|. \tag{34-35}$$

Let us consider these forces separately.

Radiation force: To evaluate the left side of Eq. 34-35, we use these three **Key Ideas.**

1. Because the particle is totally absorbing, the force magnitude $|\vec{F}^{\,\text{rad}}|$ can be found from the intensity I of sunlight at the particle's location and the particle's cross-sectional area A, via Eq. 34-31 ($|\vec{F}| = IA/c$).
2. Because we assume that the Sun is an isotropic point source of light, we can use Eq. 34-25 ($I = P_{\text{sun}}/4\pi r^2$) to relate the Sun's power P_{sun} to the intensity I of the sunlight at the particle's distance r from the Sun.
3. Because the particle is spherical, its cross-sectional area A is πR^2 (*not* half its surface area).

Putting these three ideas together gives us

$$|\vec{F}^{\,\text{rad}}| = \frac{IA}{c} = \frac{P_{\text{sun}}\pi R^2}{4\pi r^2 c} = \frac{P_{\text{sun}}R^2}{4r^2 c}. \tag{34-36}$$

Gravitational force: The **Key Idea** here is Newton's law of gravitation (Eq. 14-2), which gives us the magnitude of the gravitational force on the particle as

$$|\vec{F}^{\,\text{grav}}| = \frac{GM_{\text{sun}}m}{r^2}, \tag{34-37}$$

where M_{sun} is the Sun's mass and m is the particle's mass. Next, the particle's mass is related to its density ρ and volume $V(=\frac{4}{3}\pi R^3$, for a sphere) by

$$\rho = \frac{m}{V} = \frac{m}{\frac{4}{3}\pi R^3}.$$

Solving this for m and substituting the result into Eq. 34-37 give us

$$|\vec{F}^{\,\text{grav}}| = \frac{GM_{\text{sun}}\rho\left(\frac{4}{3}\pi R^3\right)}{r^2}. \tag{34-38}$$

Then substituting Eqs. 34-36 and 34-38 into Eq. 34-35 and solving for R yield

$$R = \frac{3P_{\text{sun}}}{16\pi c\rho GM_{\text{sun}}}.$$

Using the given value of ρ and the known values of G (Appendix B) and M_{sun} (Appendix C), we can evaluate the denominator:

$$\begin{aligned}&(16\pi)(3 \times 10^8 \text{ m/s})(3.5 \times 10^3 \text{ kg/m}^3)\\ &\times (6.67 \times 10^{-11}\,\text{N}\cdot\text{m}^2/\text{kg}^2)(1.99 \times 10^{30}\,\text{kg})\\ &= 7.0 \times 10^{33} \text{ N/s}.\end{aligned}$$

Using P_{sun} from Appendix C, we then have

$$R = \frac{(3)(3.9 \times 10^{26}\text{ W})}{7.0 \times 10^{33}\text{ N/s}} = 1.7 \times 10^{-7}\text{ m}. \quad \text{(Answer)}$$

Note that this result is independent of the particle's distance r from the Sun.

Dust particles with radius $R \approx 1.7 \times 10^{-7}$ m follow an approximately straight path like path b in Fig. 34-10. For larger values of R, comparison of Eqs. 34-36 and 34-38 shows that, because $\vec{F}^{\,\text{grav}}$ varies with R^3 and $\vec{F}^{\,\text{rad}}$ varies with R^2, the gravitational force $\vec{F}^{\,\text{grav}}$ dominates the radiation force $\vec{F}^{\,\text{rad}}$. Thus, such particles follow a path that is curved toward the Sun like path c in Fig. 34-10. Similarly, for smaller values of R, the radiation force dominates, and the dust follows a path that is curved away from the Sun like path a. The composite of these dust particles is the dust tail of the comet.

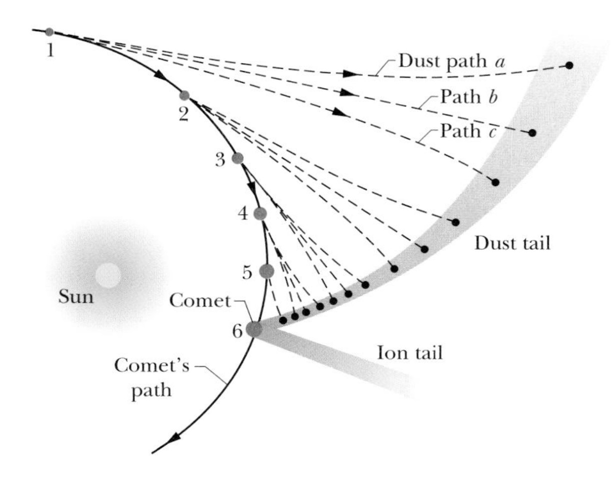

FIGURE 34-10 ■ A comet is now at position 6. Dust it has released at five previous positions has been pushed outward by radiation pressure from sunlight, has taken the dashed paths, and now forms the comet's curved dust tail. Its ion trail points directly away from the Sun.

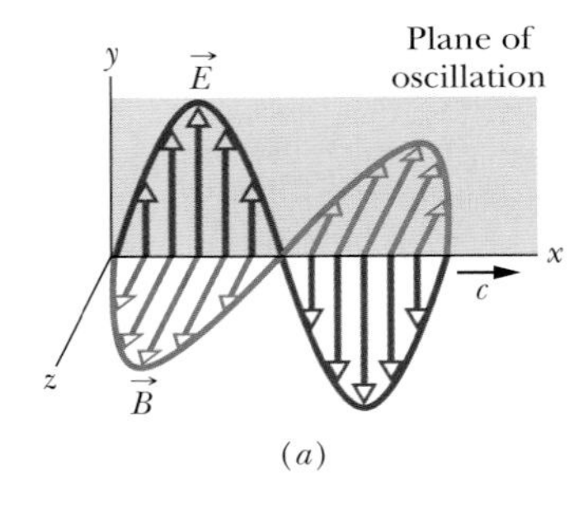

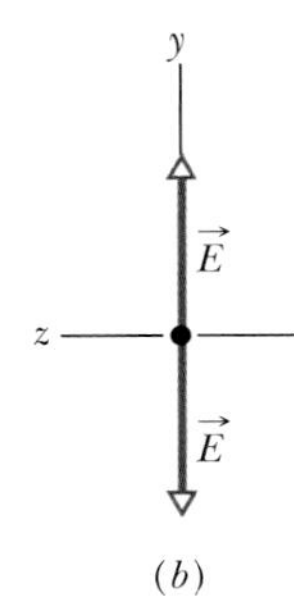

FIGURE 34-11 ■ (a) The plane of oscillation of a polarized electromagnetic wave. (b) To represent the polarization, we view the plane of oscillation "head-on" and indicate the possible directions of the oscillating electric field with two arrows, which we refer to as a "double arrow."

34-7 Polarization

At the beginning of the chapter we talked about a charge oscillating along a line. We observed that the charge's oscillation produced kinks in its field lines that are interpreted as wavefronts. We noted that the electric field in the kink was perpendicular to the direction of the propagation and pointed along the line of oscillation of the charge. This means that the electric field in the outgoing wave points back and forth in a single direction. It doesn't wander around pointing in all directions as it would if the electromagnetic wave was the result of many charges oscillating in many different directions. We call such a wave **polarized.** This is an important concept since many of our sources of radiation—such as antennas—impart this property.

Figure 34-11a shows an electromagnetic wave with its electric field oscillating parallel to the vertical y axis. The plane containing the $\vec{E}$ vectors is called the **plane of oscillation** of the wave (hence, the wave is said to be *plane-polarized* parallel to the y axis). We can represent the wave's polarization (state of being polarized) by showing the direction of the electric field oscillations in a "head-on" view of the plane of oscillation, as in Fig. 34-11b. The two vertical arrows in that figure indicate that as the wave travels past us, its electric field oscillates vertically, continuously changing between being directed up and down along the y axis. VHF (very high frequency) television antennas in England are oriented vertically, but those in North America are horizontal. The difference is due to the direction of oscillation of the electromagnetic waves carrying the TV signal. In England, the transmitting equipment is designed to produce waves that are polarized vertically; that is, their electric field oscillates vertically. Thus, for the electric field of the incident television waves to drive a current along an antenna (and provide a signal to a television set), the antenna must be vertical. In North America, the antenna must be horizontal.

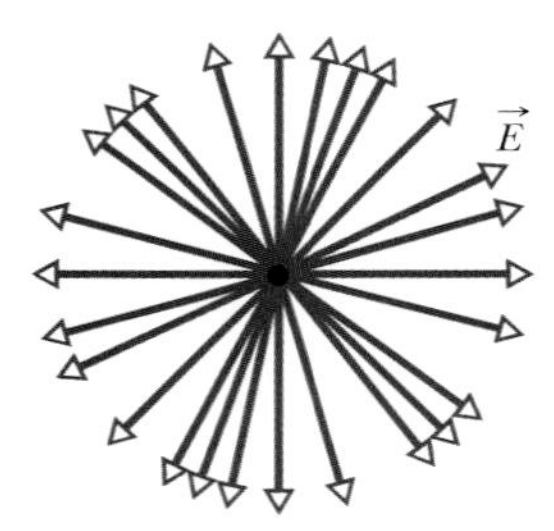

FIGURE 34-12 ■ Unpolarized light consists of waves with randomly directed electric fields. Here the waves are all traveling along the same axis, directly out of the page, and all have the same amplitude $|\vec{E}|$.

Polarized Light

The electromagnetic waves emitted by a television station all have the same polarization because they are generated by electrons moving up and down (or right and left) along a transmission antenna. On the other hand, electromagnetic waves emitted by any common source of *light* (such as the Sun or a lightbulb) are generated by the individual atoms or molecules that comprise the light source. The fact that light is generated by individual atoms or molecules within an object means that there is no preferred *orientation* associated with the electromagnetic waves that make up the light emerging from a source, even in cases where there is a preferred direction of travel.

We call electromagnetic waves with random orientations (like the light produced by atoms and molecules in the sun) **randomly polarized** or **unpolarized.** This is because the direction of the electric field at a given point in space changes direction quickly and randomly. It is still perpendicular to both the direction of travel of the wave and the magnetic field vector on a moment by moment basis, but its orientation in space changes continuously. Figure 34-12 shows an unpolarized electromagnetic wave traveling into or out of the page. If we try to represent a head-on view of the oscillations over some time period, we do not have a simple drawing with a single double arrow like that of Fig. 34-11b; instead we have a mess of double arrows like that in Fig. 34-12.

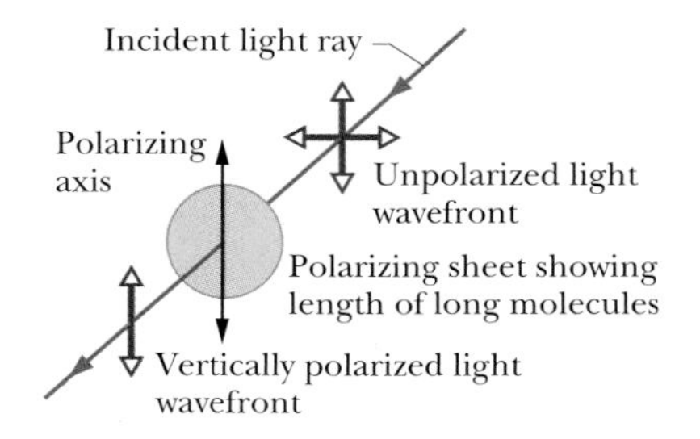

FIGURE 34-13 ■ Unpolarized light becomes polarized when it is sent through a polarizing sheet. The polarization is then parallel to the polarizing axis of the sheet.

We can (and often do) transform randomly polarized (or unpolarized) visible light into polarized light by sending it through a *polarizing sheet,* as is shown in Fig. 34-13. Such sheets, commercially known as Polaroids or Polaroid filters, were invented in 1932 by Edwin Land while he was an undergraduate student. A polarizing sheet consists of certain long molecules embedded in plastic. When the sheet is manufactured, it is stretched to align the long molecules in parallel rows, like rows in a plowed field. When light is then sent through the sheet, electric field components perpendicular to the long molecules pass through the sheet, while components parallel to the long molecules are absorbed and disappear.

This is not surprising. The electrons surrounding long molecules are more free to move up and down along the molecular axis and absorb the radiation. Those perpendicular to the long axis are not as free to oscillate. We shall not dwell on the orientation of the molecules but, instead, shall assign a *polarizing axis* to the sheet, along which electric field components are passed:

> An electric field component parallel to the polarizing axis is passed (*transmitted*) by a polarizing sheet; a component perpendicular to it is absorbed.

Thus, the electric field of the light emerging from the sheet consists of only the components that are parallel to the polarizing axis of the sheet. Hence, the light is polarized in that direction. In Fig. 34-13, the vertical electric field components are transmitted by the sheet; the horizontal components are absorbed. The transmitted waves are then vertically polarized.

In some situations, light is **partially polarized** (its field oscillations are not completely random as in Fig. 34-12 nor are they parallel to a single axis as in Fig. 34-11*b*). Partially polarized light can be viewed as a superposition of plane polarized and unpolarized light waves.

Intensity of Transmitted Polarized Light

We now consider the intensity of light transmitted by a polarizing sheet. We start with unpolarized light, whose electric field oscillations we can resolve into *y*- and *z*-components as represented in Fig. 34-12*b*. Further, we can arrange for the *y* axis to be parallel to the polarizing direction of the sheet. Then only the *y*-components of the light's electric field are passed by the sheet; the *z*-components are absorbed. As suggested by Fig. 34-12*b*, if the original waves are randomly oriented, the sum of the *y*-components and the sum of the *z*-components are equal. When the *z*-components are absorbed, half the original intensity I_0 of the light is lost. The intensity I of the emerging polarized light is then

$$I = \tfrac{1}{2} I_0. \tag{34-39}$$

Let us call this the *one-half rule;* we can use it *only* when the light reaching a polarizing sheet is unpolarized.

Suppose now that the light reaching a polarizing sheet is already polarized. Figure 34-14 shows a polarizing sheet in the plane of the page and the electric field $\vec{E}$ of such a polarized light wave traveling toward the sheet (and thus prior to any absorption). We can resolve $\vec{E}$ into two components relative to the polarizing axis of the sheet: parallel component E_y is transmitted by the sheet, and perpendicular component E_z is absorbed. Since θ is the angle between $\vec{E}$ and the polarizing axis of the sheet, the transmitted parallel component is

$$E_y = |\vec{E}|\cos\theta. \tag{34-40}$$

Recall that the intensity of an electromagnetic wave (such as our light wave) is proportional to the square of the electric field's magnitude $I = (E^{\text{rms}})^2/c\mu_0$ (Eq. 34-23). In our present case then, the intensity I of the emerging wave is proportional to E_y^2 and

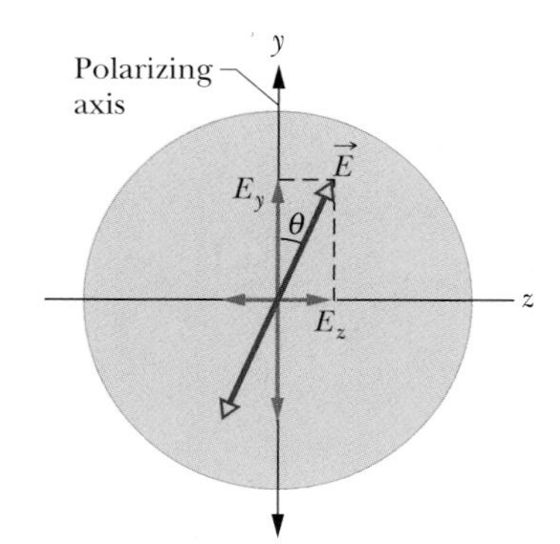

FIGURE 34-14 ■ Polarized light approaching a polarizing sheet. The electric field $\vec{E}$ of the light can be resolved into components E_y (parallel to the polarizing axis of the sheet) and E_z (perpendicular to that axis). Component E_y will be transmitted by the sheet; component E_z will be absorbed. *Note:* The long molecules are oriented perpendicular to the polarizing axis.

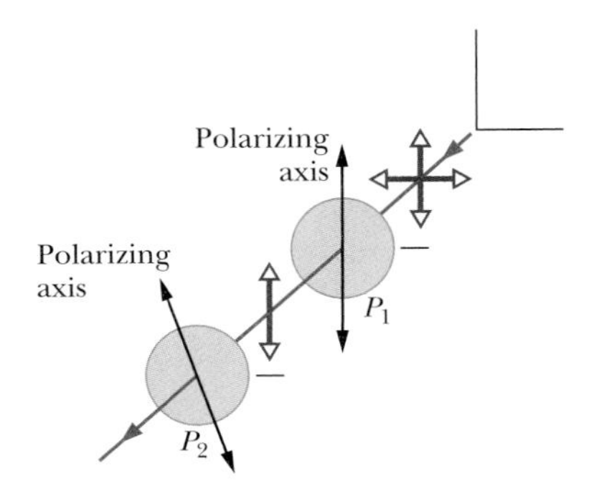

FIGURE 34-15 ■ The light transmitted by polarizing sheet P_1 is vertically polarized, as represented by the vertical double arrow. The amount of that light that is then transmitted by polarizing sheet P_2 depends on the angle between the polarization axis of that light and the polarizing axis of P_2.

the intensity I_0 of the original wave is proportional to $\vec{E}^2$. Hence, from $E_y = |\vec{E}|\cos\theta$ (Eq. 34-40) we can write $I/I_0 = \cos^2\theta$, or

$$I = I_0 \cos^2\theta \qquad \text{(Malus' law).} \qquad (34\text{-}41)$$

This expression was first introduced in the nineteenth century by Ètienne Malus, a French mathematician who studied polarized light. So, Eq. 34-41 is called Malus' law or the *cosine-squared rule.* We can use it *only* when the light reaching a polarizing sheet is already polarized. Then the transmitted intensity I is a maximum and is equal to the original intensity I_0 when the original wave is polarized parallel to the polarizing axis of the sheet (when θ in Eq. 34-41 is 0° or 180°). I is zero when the original wave is polarized perpendicular to the polarizing axis of the sheet (when θ is 90°).

Figure 34-15 shows an arrangement in which initially unpolarized light is sent through two polarizing sheets P_1 and P_2. (Often, the first sheet is called the *polarizer,* and the second the *analyzer.*) Because the polarizing axis of P_1 is vertical, the light transmitted by P_1 to P_2 is polarized vertically. If the polarizing axis of P_2 is also vertical, then all the light transmitted by P_1 is transmitted by P_2. If the polarizing axis of P_2 is horizontal, none of the light transmitted by P_1 is transmitted by P_2. We reach the same conclusions by considering only the *relative* orientations of the two sheets: If their polarizing axes are parallel, all the light passed by the first sheet is passed by the second sheet. If those axes are perpendicular (the sheets are said to be *crossed*), no light is passed by the second sheet. These two extremes are displayed with polarized sunglasses in Fig. 34-16.

Finally, if the two polarizing axes of Fig. 34-15 make an angle between 0° and 90°, some of the light transmitted by P_1 will be transmitted by P_2. The intensity of that light is determined by $I = I_0 \cos^2\theta$ (Eq. 34-41).

Light can be polarized by means other than polarizing sheets, such as by reflection (discussed in Section 34-10) and by scattering from atoms or molecules. In *scattering,* light that is intercepted by an object, such as a molecule, is sent off in many, perhaps random, directions. An example is the scattering of sunlight by molecules in the atmosphere, which gives the sky its general glow.

Although direct sunlight is unpolarized, light from much of the sky is at least partially polarized by such scattering. Bees use the polarization of sky light in navigating to and from their hives. Similarly, the Vikings used it to navigate across the North Sea when the daytime Sun was below the horizon (because of the high latitude of the North Sea). These early seafarers had discovered certain crystals (now called cordierite) that changed color when rotated in polarized light. By looking at the sky through such a crystal while rotating it about their line of sight, they could locate the hidden Sun and thus determine which way was south.

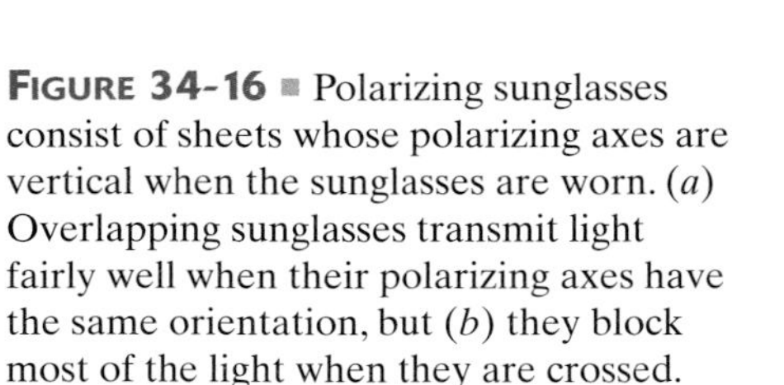

FIGURE 34-16 ■ Polarizing sunglasses consist of sheets whose polarizing axes are vertical when the sunglasses are worn. (*a*) Overlapping sunglasses transmit light fairly well when their polarizing axes have the same orientation, but (*b*) they block most of the light when they are crossed.

TOUCHSTONE EXAMPLE 34-3: Polarizing Sheets

Figure 34-17 shows a system of three polarizing sheets in the path of initially unpolarized light. The polarizing axis of the first sheet is parallel to the y axis, that of the second sheet is 60° counterclockwise from the y axis, and that of the third sheet is parallel to the x axis. What fraction of the initial intensity I_0 of the light emerges from the system, and how is that light polarized?

SOLUTION ■ The **Key Ideas** here are these:

1. We work through the system sheet by sheet, from the first one encountered by the light to the last one.
2. To find the intensity transmitted by any sheet, we apply either the one-half rule or the cosine-squared rule, depending on whether the light reaching the sheet is unpolarized or already polarized.
3. The light that is transmitted by a polarizing sheet is always polarized parallel to the polarizing axis of the sheet.

First sheet: The original light wave is represented in Fig. 34-17*b*, using the head-on, double-arrow representation of Fig. 34-11*b*. Because the light is initially unpolarized, the intensity I_1 of the light transmitted by the first sheet is given by the one-half rule (Eq. 34-39):

$$I_1 = \tfrac{1}{2} I_0.$$

Because the polarizing axis of the first sheet is parallel to the y axis, the polarization of the light transmitted by it is also, as shown in the head-on view of Fig. 34-17*c*.

Second sheet: Since the light reaching the second sheet is polarized, the intensity I_2 of the light transmitted by that sheet is given by the cosine-squared rule (Eq. 34-41). The angle θ in the rule is the angle between the polarization axis of the entering light (parallel to the y axis) and the polarizing axis of the second sheet (60° counterclockwise from the y axis), and so θ is 60°. Then

$$I_2 = I_1 \cos^2 60°.$$

The polarization of this transmitted light is parallel to the polarizing axis of the sheet transmitting it—that is, 60° counterclockwise from the y axis, as shown in the head-on view of Fig. 34-17*d*.

Third sheet: Because the light reaching the third sheet is polarized, the intensity I_3 of the light transmitted by that sheet is given by the cosine-squared rule. The angle θ is now the angle between the polarization axis of the entering light (Fig. 34-17*d*) and the polarizing axis of the third sheet (parallel to the x axis), and so $\theta = 30°$. Thus,

$$I_3 = I_2 \cos^2 30°.$$

This final transmitted light is polarized parallel to the x axis (Fig. 34-17*e*). We find its intensity by substituting first for I_2 and then for I_1 in the equation above:

$$I_3 = I_2 \cos^2 30° = (I_1 \cos^2 60°)\cos^2 30°$$
$$= (\tfrac{1}{2} I_0)\cos^2 60° \cos^2 30° = 0.094 I_0.$$

Thus,

$$\frac{I_3}{I_0} = 0.094. \quad \text{(Answer)}$$

That is to say, 9.4% of the initial intensity emerges from the three-sheet system.

If we now remove the second sheet, what fraction of the initial intensity emerges from the system?

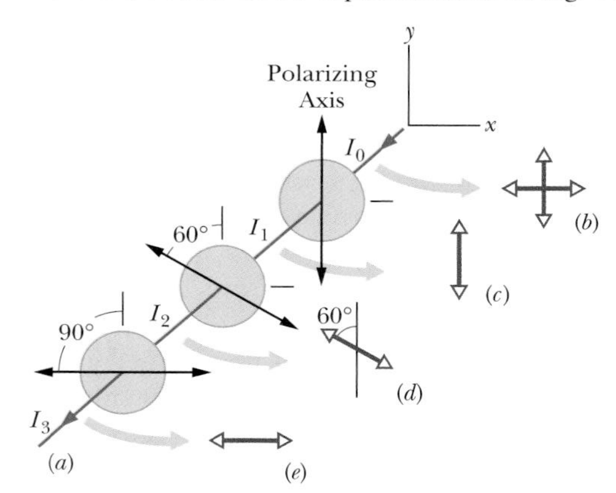

FIGURE 34-17 ■ (*a*) Initially unpolarized light of intensity I_0 is sent into a system of three polarizing sheets. The intensities I_1, I_2, I_3 of the light transmitted by the sheets are labeled. Shown also are the polzarizations, from head-on views, of (*b*) the initial light and the light transmitted by (*c*) the first sheet, (*d*) the second sheet, and (*e*) the third sheet.

34-8 Maxwell's Rainbow

In Maxwell's time (the mid-1800s), the visible, infrared, and ultraviolet forms of light were the only electromagnetic waves known. Spurred on by Maxwell's work, however, Heinrich Hertz discovered what we now call radio waves and verified that they move through the laboratory at the same speed as visible light.

As Fig. 34-18 shows, we now know a wide *spectrum* (or range) of electromagnetic waves, referred to by one imaginative writer as "Maxwell's rainbow" but generally referred to as "light" or "electromagnetic radiation" by physicists.

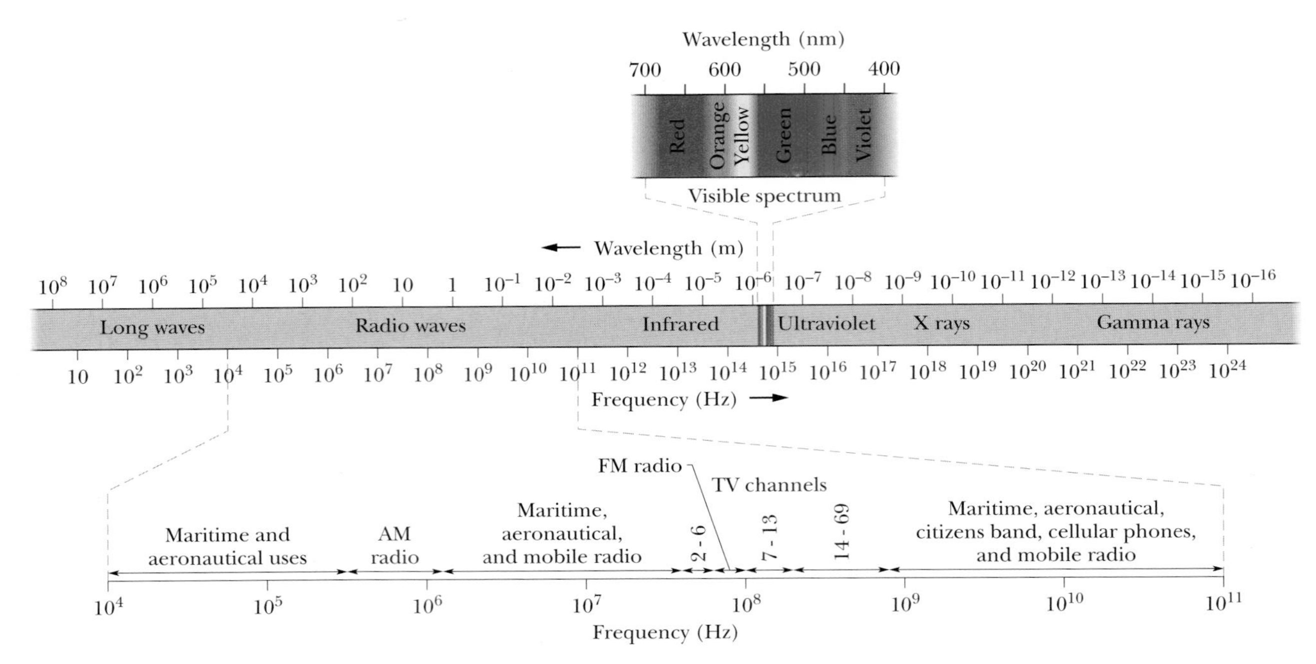

FIGURE 34-18 ■ The electromagnetic spectrum.

Consider the extent to which we are bathed in electromagnetic waves throughout this spectrum. The Sun, whose radiations define the environment in which we as a species have evolved and adapted, is the dominant source. We are also crisscrossed by radio, television and cellular phone signals. Microwaves from radar systems and from telephone relay systems may reach us. There are electromagnetic waves from lightbulbs, from the heated engine blocks of automobiles, from x-ray machines, from lightning flashes, and from buried radioactive materials. Beyond this, radiation reaches us from stars and other objects in our galaxy and from other galaxies. Electromagnetic waves also travel in the other direction. Television signals, transmitted from the Earth since about 1950, have now transmitted news about us (along with episodes of *I Love Lucy*) to whatever technically sophisticated inhabitants there may be on the planets that encircle the nearest 400 or so stars.

In the wavelength scale in Fig. 34-18 (and similarly the corresponding frequency scale), each scale marker represents a change in wavelength (and correspondingly in frequency) by a factor of 10. The scale is open-ended; the wavelengths of electromagnetic waves have no inherent upper or lower bounds.

There are many regions of the electromagnetic spectrum in Fig. 34-18, including radio waves, infrared light produced by the thermal motion of charged particles, visible and ultraviolet light emitted by energetic atoms, x rays that are generated when charged particles collide with solid matter, and gamma rays that originate inside atomic nuclei. These regions denote roughly defined wavelength ranges within which certain kinds of sources and detectors of electromagnetic waves are in common use. Other regions of Fig. 34-18, such as those labeled television and AM radio, represent specific wavelength bands assigned by law for certain commercial or other purposes. There are no gaps in the electromagnetic spectrum—and all electromagnetic waves, no matter where they lie in the spectrum, travel through *free space* (vacuum) with the same speed c.

The visible region of the spectrum is of course of particular interest to us. Figure 34-19 shows the relative sensitivity of the human eye to visible light of various wavelengths. The center of the visible region is about 555 nm, which produces the sensation that we call yellow-green.

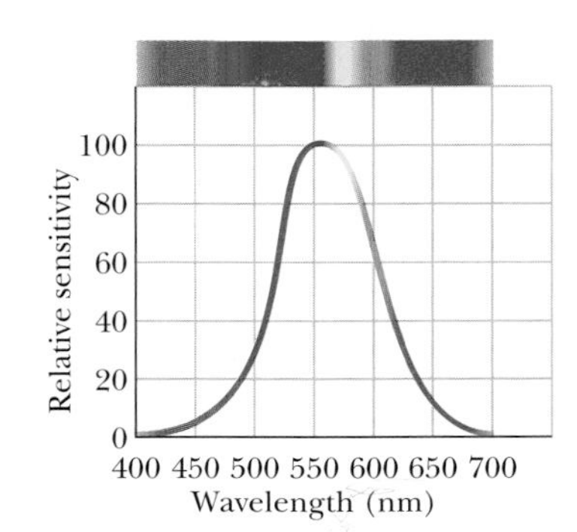

FIGURE 34-19 ■ The relative sensitivity of the average human eye to electromagnetic waves at different wavelengths. This portion of the electromagnetic spectrum to which the eye is sensitive is called visible light.

The limits of this visible spectrum are not well defined because the eye sensitivity curve approaches the zero-sensitivity line asymptotically at both long and short wavelengths. If we take the limits, arbitrarily, as the wavelengths at which eye sensitivity has dropped to 1% of its maximum value, these limits are about 430 and 690 nm. However, the eye can detect electromagnetic waves somewhat beyond these limits if they are intense enough.

Problems

SEC. 34-3 ■ THE GENERATION OF ELECTROMAGNETIC WAVES

1. What Inductance What inductance must be connected to a 17 pF capacitor in an oscillator capable of generating 550 nm (i.e., visible) electromagnetic waves? Comment on your answer.

2. Wavelength What is the wavelength of the electromagnetic wave emitted by the oscillator–antenna system of Fig. 34-4 if $L = 0.253\ \mu\text{H}$ and $C = 25.0$ pF?

SEC. 34-4 ■ DESCRIBING ELECTROMAGNETIC WAVE PROPERTIES MATHEMATICALLY

3. Electric Field The electric field of a certain plane electromagnetic wave is given by $E_x = 0$; $E_y = 0$; $E_z = (2.0\ \text{V/m})\cos[(\pi \times 10^{15}\ \text{s}^{-1})(t - x/c)]$, with $c = 3.0 \times 10^8$ m/s. The wave is propagating in the positive x direction. Write expressions for the components of the magnetic field of the wave.

4. Plane Wave A plane electromagnetic wave has a maximum electric field of 3.20×10^{-4} V/m. Find the maximum magnetic field.

SEC. 34-5 ■ TRANSPORTING ENERGY WITH ELECTROMAGNETIC WAVES

5. Neodymium–Glass Lasers Some neodymium–glass lasers can provide 100 terawatts of power in 1.0 ns pulses at a wavelength of 0.26 μm. How much energy is contained in a single pulse?

6. Poynting Vector Show, by finding the direction of the Poynting vector $\vec{S}$, that the directions of the electric and magnetic fields at all points in Figs. 34-6 to 34-8 are consistent at all times with the assumed directions of propagation.

7. Radiation Emitted The radiation emitted by a laser spreads out in the form of a narrow cone with circular cross section. The angle θ of the cone (see Fig. 34-20) is called the *full-angle beam divergence.* An argon laser, radiating at 514.5 nm, is aimed at the Moon in a ranging experiment. If the beam has a full-angle beam divergence of 0.880 μrad, what area on the Moon's surface is illuminated by the laser?

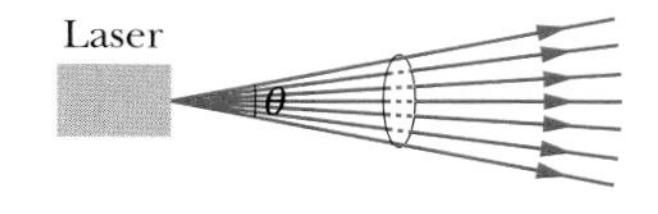

FIGURE 34-20 ■ Problem 7.

8. Closest Neighbor Our closest stellar neighbor, Proxima Centauri, is 4.3 lightyears away. It has been suggested that TV programs from our planet have reached this star and may have been viewed by the hypothetical inhabitants of a hypothetical planet orbiting it. Suppose a television station on Earth has a power of 1.0 MW. What is the intensity of its signal at Proxima Centauri?

9. Plane Radio Wave In a plane radio wave the maximum value of the electric field component is 5.00 V/m. Calculate (a) the maximum value of the magnetic field component and (b) the wave intensity.

10. Intensity What is the intensity of a plane traveling electromagnetic wave if $|\vec{B}^{\max}|$ is 1.0×10^{-4} T?

11. Maximum Electric Field The maximum electric field at a distance of 10 m from anisotropic point light source is 2.0 V/m. What are (a) the maximum value of the magnetic field and (b) the average intensity of the light there? (c) What is the power of the source?

12. Sunlight Sunlight just outside the Earth's atmosphere has an intensity of 1.40 kW/m^2. Calculate $|\vec{E}^{\max}|$ and $|\vec{B}^{\max}|$ for sunlight there, assuming it to be a plane wave.

13. An Airplane An airplane flying at a distance of 10 km from a radio transmitter receives a signal of intensity 10 μW/m^2. Calculate (a) the amplitude of the electric field at the airplane due to this signal, (b) the amplitude of the magnetic field at the airplane, and (c) the total power of the transmitter, assuming the transmitter to radiate uniformly in all directions.

14. Frank Drake Frank D. Drake, an investigator in the SETI (Search for Extra-Terrestrial Intelligence) program, once said that the large radio telescope in Arecibo, Puerto Rico, "can detect a signal which lays down on the entire surface of the earth a power of only one picowatt." (a) What is the power that would be received by the Arecibo antenna for such a signal? The antenna diameter is 300 m. (b) What would be the power of a source at the center of our galaxy that could provide such a signal? The galactic center is 2.2×10^4 ly away. Take the source as radiating uniformly in all directions.

15. Isotropic Point Source An isotropic point source emits light at wavelength 500 nm, at the rate of 200 W. A light detector is positioned 400 m from the source. What is the maximum rate $\partial B/\partial t$ at which the magnetic component of the light changes with time at the detector's location?

16. Magnetic Component The magnetic component of an electromagnetic wave in vacuum has an amplitude of 85.8 nT and an angular wave number of 4.00 m^{-1}. What are (a) the frequency of the wave, (b) the rms value of the electric component, and (c) the intensity of the light?

17. Magnetic Component Two The magnetic component of a polarized wave of light is

$$B_x = (4.0 \times 10^{-6}\ \text{T})\sin[(1.57 \times 10^7\ \text{m}^{-1})y + \omega t]$$

(a) Parallel to which axis is the light polarized? What are the (b) frequency and (c) intensity of the light?

18. Rms Value of Electric Component An electromagnetic wave with a wavelength of 450 nm travels through vacuum in the negative direction of a y axis with its electric component directed parallel to the x axis. The rms value of the electric component is 5.31×10^{-6} V/m. Write an equation for the magnetic component in the form of Eq. 34-3, but complete with numbers.

19. Direct Solar Radiation The intensity of direct solar radiation that is not absorbed by the atmosphere on a particular summer day is 100 W/m^2. How close would you have to stand to a 1.0 kW electric heater to feel the same intensity? Assume that the heater radiates uniformly in all directions.

20. Isotropic Point Source Two The intensity I of light from an isotropic point light source is determined as a function of the distance r from the source. Figure 34-21 gives intensity I versus the inverse square r^{-2} of that distance. What is the power of the source?

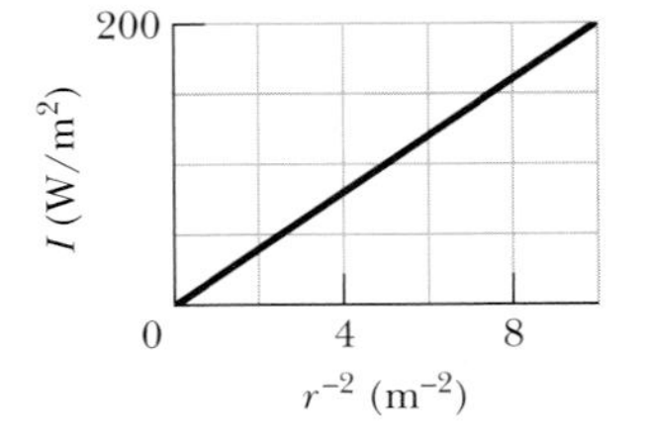

FIGURE 34-21 ■ Problem 20.

21. What Is the Power During a test, a NATO surveillance radar system, operating at 12 GHz and 180 kW of power, attempts to detect an incoming stealth aircraft at 90 km. Assume that the radar beam is emitted uniformly over a hemisphere. (a) What is the intensity of the beam when it reaches the aircraft's location? The aircraft reflects radar waves as though it has a cross-sectional area of only 0.22 m^2. (b) What is the power of the aircraft's reflection? Assume that the beam is reflected uniformly over a hemisphere. Back at the radar site, what are the (c) intensity, (d) maximum value of the electric field vector, and (e) rms value of the magnetic field of the reflected (and now detected) radar beam?

22. Average Energy Transport Show that in a plane traveling electromagnetic wave the intensity—that is, the average rate of energy transport per unit area—is given by

$$\langle S \rangle = \frac{(E^{\max})^2}{2\mu_0 c} = \frac{(B^{\max})^2}{2\mu_0}.$$

SEC. 34-6 ■ RADIATION PRESSURE

23. High-Power Laser High-power lasers are used to compress a plasma (a gas of charged particles) by radiation pressure. A laser generating pulses of radiation of peak power 1.5 GW is focused onto 1.0 mm^2 of high-electron-density plasma. Find the pressure exerted on the plasma if the plasma reflects all the light pulses directly back along their paths.

24. Black Cardboard A black, totally absorbing piece of cardboard of area $A = 2.0$ cm^2 intercepts light with an intensity of 10 W/m^2 from a camera strobe light. What radiation pressure is produced on the cardboard by the light?

25. Radiation Pressure What is the radiation pressure 1.5 m away from a 500 W lightbulb? Assume that the surface on which the pressure is exerted faces the bulb and is perfectly absorbing and that the bulb radiates uniformly in all directions.

26. Radiation from the Sun Radiation from the Sun reaching the Earth (just outside the atmosphere) has an intensity of 1.4 kW/m^2. (a) Assuming that the Earth (and its atmosphere) behaves like a flat disk perpendicular to the Sun's rays and that all the incident energy is absorbed, calculate the force on the Earth due to radiation pressure. (b) Compare it with the force due to the Sun's gravitational attraction.

27. Electromagnetic Wave A plane electromagnetic wave, with wavelength 3.0 m, travels in vacuum in the positive x direction with its electric field $\vec{E}$, of amplitude 300 V/m, directed along the y axis. (a) What is the frequency f of the wave? (b) What are the direction and amplitude of the magnetic field associated with the wave? (c) What are the values of k and ω if $\vec{E} = \vec{E}^{\max} \sin(kx - \omega t)$? (d) What is the time-averaged rate of energy flow in watts per square meter associated with this wave? (e) If the wave falls on a perfectly absorbing sheet of area 2.0 m^2, at what rate is momentum delivered to the sheet and what is the radiation pressure exerted on the sheet?

28. Helium–Neon Laser A helium–neon laser of the type often found in physics laboratories has a beam power of 5.00 mW at a wavelength of 633 nm. The beam is focused by a lens to a circular spot whose effective diameter may be taken to be equal to 2.00 wavelengths. Calculate (a) the intensity of the focused beam, (b) the radiation pressure exerted on a tiny perfectly absorbing sphere whose diameter is that of the focal spot, (c) the force exerted on this sphere, and (d) the magnitude of the acceleration imparted to it. Assume a sphere density of 5.00×10^3 kg/m^3.

29. Normally Incident Prove, for a plane electromagnetic wave that is normally incident on a plane surface, that the radiation pressure on the surface is equal to the energy density in the incident beam. (This relation between pressure and energy density holds no matter what fraction of the incident energy is reflected.)

30. Laser Beam In Fig. 34-22, a laser beam of power 4.60 W and diameter 2.60 mm is directed upward at one circular face (of diameter $d < 2.60$ mm) of a perfectly reflecting cylinder, which is made to "hover" by the beam's radiation pressure. The cylinder's density is 1.20 g/cm^3. What is the cylinder's height H?

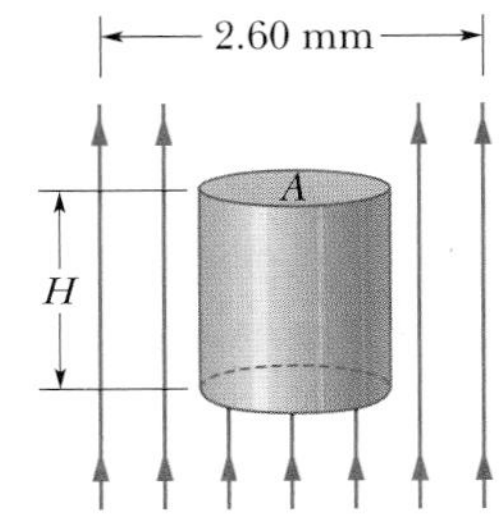

FIGURE 34-22 ■ Problem 30.

31. Small Spaceship A small spaceship whose mass is 1.5×10^3 kg (including an astronaut) is drifting in outer space with negligible gravitational forces acting on it. If the astronaut turns on a 10 kW laser beam, what speed will the ship attain in 1.0 day because of the momentum carried away by the beam?

32. Average Pressure Prove that the average pressure of a stream of bullets striking a plane surface perpendicularly is twice the kinetic energy density in the stream outside the surface. Assume that the bullets are completely absorbed by the surface. Contrast this with Problem 29.

33. Particle in Solar System A particle in the solar system in under the combined influence of the Sun's gravitational attraction and the radiation force due to the Sun's rays. Assume that the particle is a sphere of density 1.0×10^3 kg/m^3 and that all the incident light is absorbed. (a) Show that, if its radius is less than some critical radius R, the particle will be blown out of the solar system. (b) Calculate the critical radius.

34. Radiation Propelled It has been proposed that a spaceship might be propelled in the solar system by radiation pressure, using a large sail made of foil. How large must the sail be if the radiation force is to be equal in magnitude in the Sun's gravitational attraction? Assume that the mass of the ship + sail is 1500 kg, that the sail is perfectly reflecting, and that the sail is oriented perpendicular to the Sun's rays. See Appendix C for needed data. (With a larger sail, the ship is continually driven away from the Sun.)

35. Totally Absorbing Someone plans to float a small, totally absorbing sphere 0.500 m above an isotropic point source of light, so that the upward radiation force from the light matches the downward gravitational force on the sphere. The sphere's density is 19.0 g/cm^3 and its radius is 2.00 mm. (a) What power would be required of the light source? (b) Even if such a source were made, why would the support of the sphere be unstable?

36. *frac* Radiation of intensity I is normally incident on an object that absorbs a fraction *frac* of it and reflects the rest back along the original path. What is the radiation pressure on the object?

SEC. 34-7 ■ POLARIZATION

37. Unpolarized Light A beam of unpolarized light of intensity 10 mW/m^2 is sent through a polarizing sheet as in Fig. 34-13. (a) Find the maximum value of the electric field of the transmitted beam. (b) What radiation pressure is exerted on the polarizing sheet?

38. Magnetic Field Equations The magnetic field equations for an electromagnetic wave in vacuum are $B_x = B \sin(ky + \omega t)$, $B_y = B_z = 0$. (a) What is the direction of propagation? (b) Write the electric field equations. (c) Is the wave polarized? If so, in what direction?

39. Three Polarizing Sheets In Fig. 34-23, initially unpolarized light is sent through three polarizing sheets whose polarizing axes make angles of $\theta_1 = 40°$, $\theta_2 = 20°$, and $\theta_3 = 40°$ with the direction of the y axis. What percentage of the light's initial intensity is transmitted by the system? (*Hint:* Be careful with the angles.)

40. Initially Unpolarized In Fig. 34-23, initially unpolarized light is sent through three polarizing sheets whose polarizing axes make angles of $\theta_1 = \theta_2 = \theta_3 = 50°$ with the direction of the y axis. What percentage of the initial intensity is transmitted by the system of the three sheets? (*Hint:* Be careful with the angles.)

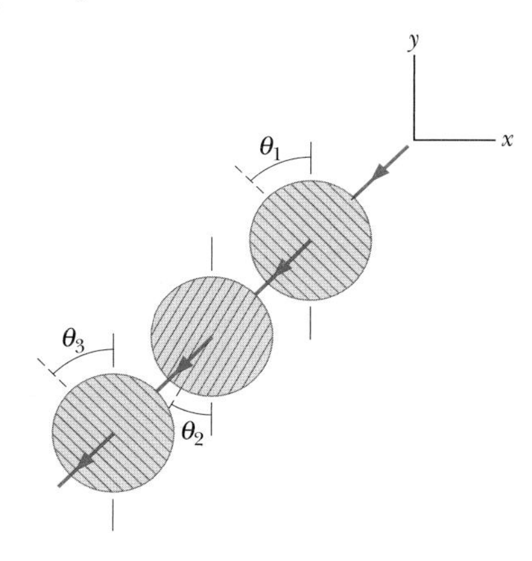

FIGURE 34-23 ■ Problems 39 and 40.

41. Vertically Polarized A horizontal beam of vertically polarized light of intensity 43 W/m^2 is sent through two polarizing sheets. The polarizing axis of the first is at 70° to the vertical, and that of the second is horizontal. What is the intensity of the light transmitted by the pair of sheets?

42. Two Polarizing Sheets A beam of polarized light is sent through a system of two polarizing sheets. Relative to the polarization axis of that incident light, the polarizing axes of the sheets are at angles θ for the first sheet and 90° for the second sheet. If 0.10 of the incident intensity is transmitted by the two sheets, what is θ?

43. Partially Polarized A beam of partially polarized light can be considered to be a mixture of polarized and unpolarized light. Suppose we send such a beam through a polarizing filter and then rotate the filter through 360° while keeping it perpendicular to the beam. If the transmitted intensity varies by a factor of 5.0 during the rotation, what fraction of the intensity of the original beam is associated with the beam's polarized light?

44. What Is the Intensity Suppose that in Problem 41 the initial beam is unpolarized. What then is the intensity of the transmitted light?

45. Rotate the Polarization We want to rotate the direction of polarization of a beam of polarized light through 90° by sending the beam through one or more polarizing sheets. (a) What is the minimum number of sheets required? (b) What is the minimum number of sheets required if the transmitted intensity is to be more than 60% of the original intensity?

46. At a Beach At a beach the light is generally partially polarized due to reflections off sand and water. At a particular beach on a particular day near sundown, the horizontal component of the electric field vector is 2.3 times the vertical component. A standing sunbather puts on polarizing sunglasses; the glasses eliminate the horizontal field component. (a) What fraction of the light intensity received before the glasses were put on now reaches the sunbather's eyes? (b) The sunbather, still wearing the glasses, lies on his side. What fraction of the light intensity received before the glasses were put on now reaches his eyes?

47. Four Polarizing Sheets An unpolarized beam of light is sent through a stack of four polarizing sheets, oriented so that the angle between the polarizing directions of adjacent sheets is 30°. What fraction of the incident intensity is transmitted by the system?

48. Four Polarizing Sheets Two In Fig. 34-24, unpolarized light with an intensity of 25 W/m^2 is sent into a system of four polarizing sheets. What is the intensity of the light that emerges from the system?

49. Two Polarizing Sheets Two A beam of unpolarized light is sent through two polarizing sheets placed one on top of the other. What must be the angle between the polarizing directions of the sheets if the intensity of the transmitted light is to be one-third the incident intensity?

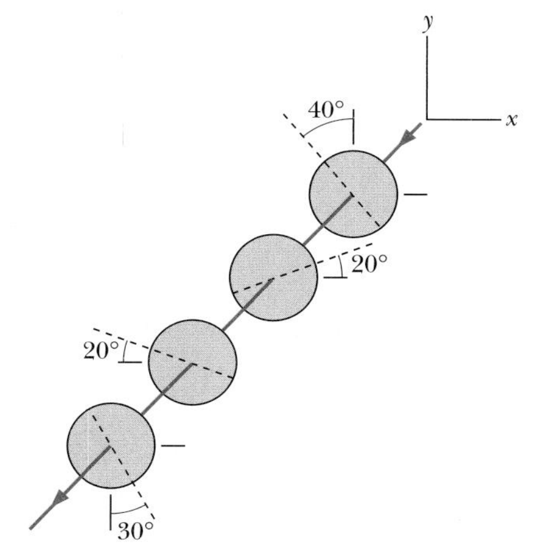

FIGURE 34-24 ■ Problem 48.

50. Two Polarizing Sheets Three In Fig. 34-25*a*, unpolarized light is

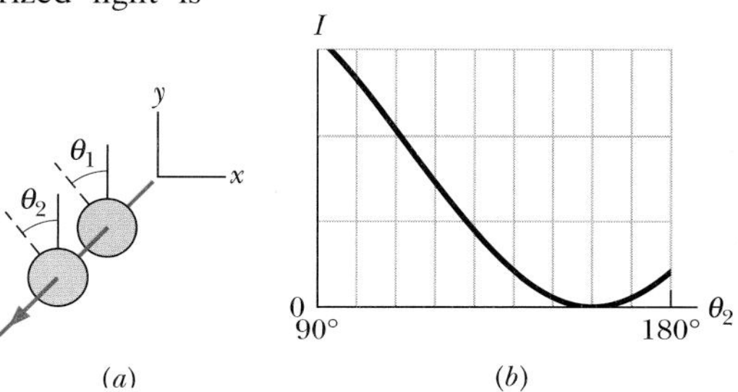

FIGURE 34-25 ■ Problem 50.

sent through a system of two polarizing sheets. The angles θ_1 and θ_2 of the polarizing axes of the sheets are measured counterclockwise from the positive direction of the y axis (they are not drawn to scale in the figure). Angle θ_1 is fixed but angle θ_2 can be varied. Figure 34-25b gives the intensity of the light emerging from sheet 2 as a function of θ_2. (The scale of the intensity axis is not indicated.) What percentage of the light's initial intensity is transmitted by the two-sheet system when $\theta_2 = 90°$?

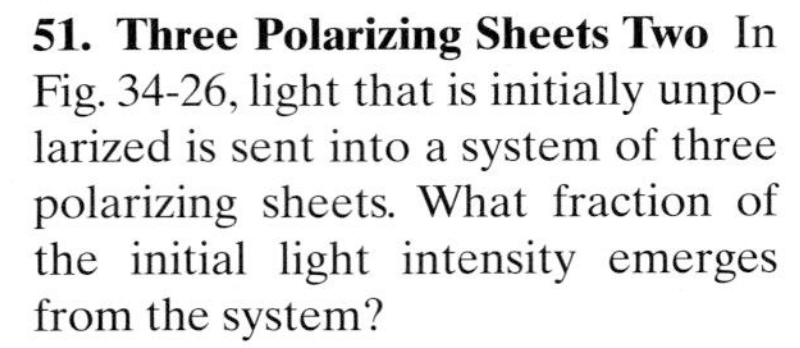

FIGURE 34-26 ■ Problem 51.

51. Three Polarizing Sheets Two In Fig. 34-26, light that is initially unpolarized is sent into a system of three polarizing sheets. What fraction of the initial light intensity emerges from the system?

52. Three Polarizing Sheets Three In Fig. 34-27a, unpolarized light is sent through a system of three polarizing sheets. The angles θ_1, θ_2, and θ_3 of the polarizing axes of the sheets are measured counterclockwise from the positive direction of the y axis (they are not drawn to scale). Angles θ_1 and θ_3 are fixed but angle θ_2 can be varied. Figure 34-28 gives the intensity of the light emerging from sheet 3 as a function of θ_2. (The scale of the intensity axis is not indicated.) What percentage of the light's initial intensity is transmitted by the three-sheet system when $\theta_2 = 90°$?

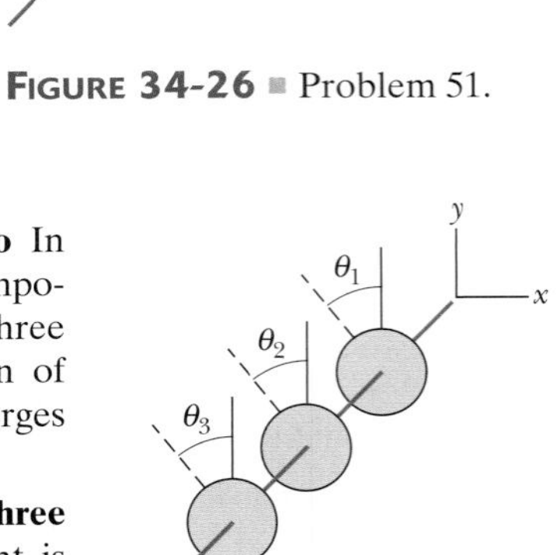

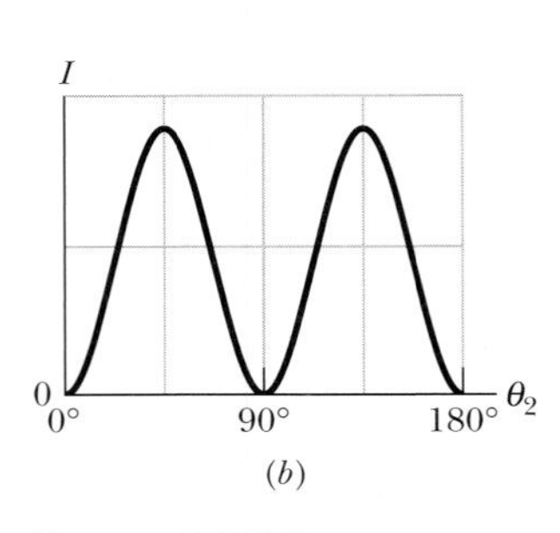

FIGURE 34-27 ■ Problems 52 and 54.

53. Three Polarizing Sheets Four A system of three polarizing sheets is shown in Fig. 34-29. When initially unpolarized light is sent into the system, the intensity of the transmitted light is 5.0% of the initial intensity. What is the value of θ?

54. Three Polarizing Sheets Five In Fig. 34-27a, unpolarized light is sent through a system of three polarizing sheets. The angles θ_1, θ_2, and θ_3 of the polarizing axes of the sheets are measured counterclockwise from the positive direction of the y axis (they are not drawn to scale). Angles θ_1 and θ_3 are fixed but angle θ_2 can be varied. Figure 34-27b gives the intensity of the light emerging from sheet 3 as a function of θ_2. (The scale of the intensity axis is not indicated.) What percentage of the light's initial intensity is transmitted by the three-sheet system when $\theta_2 = 30°$?

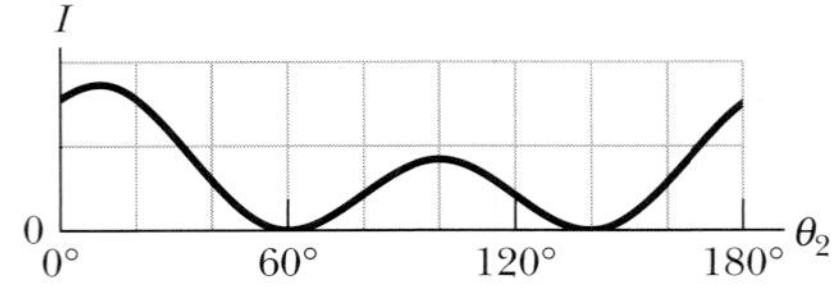

FIGURE 34-28 ■ Problem 52.

SEC. 34-8 ■ MAXWELL'S RAINBOW

55. How Long (a) How long does it take a radio signal to travel 150 km from a transmitter to a receiving antenna? (b) We see a full Moon by reflected sunlight. How much earlier did the light that enters our eye leave the Sun? The Earth–Moon and Earth–Sun distances are 3.8×10^5 km and 1.5×10^8 km. (c) What is the round-trip travel time for light between the Earth and a spaceship orbiting Saturn, 1.3×10^9 km distant? (d) The Crab nebula, which is about 6500 light-years (ly) distant, is thought to be the result of a supernova explosion recorded by Chinese astronomers in A.D. 1054. In approximately what year did the explosion actually occur?

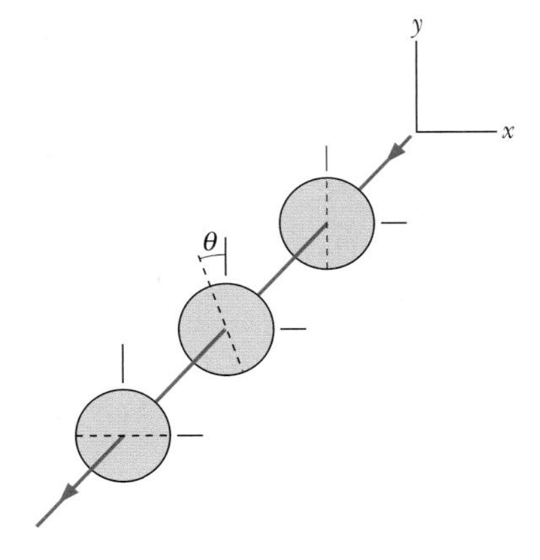

FIGURE 34-29 ■ Problem 53.

56. Project Seafarer Project Seafarer was an ambitious proposal to construct an enormous antenna, buried underground on a site about 10 000 km^2 in area. Its purpose was to transmit signals to submarines while they were deeply submerged. If the effective wavelength were 1.0×10^4 Earth radii, what would be (a) the frequency and (b) the period of the radiations emitted? Ordinarily, electromagnetic radiations do not penetrate very far into conductors such as seawater.

57. At What Wavelengths (a) At what wavelengths does the eye of a standard observer have half its maximum sensitivity? (b) What are the wavelength, frequency, and period of the light for which the eye is the most sensitive?

58. Helium–Neon Laser Two A certain helium–neon laser emits red light in a narrow band of wavelengths centered at 632.8 nm and with a "wavelength width" (such as on the scale of Fig. 34-18) of 0.0100 nm. What is the corresponding "frequency width" for the emission?

59. Speed of Light One method for measuring the speed of light, based on observations by Roemer in 1676, consisted of observing the apparent times of revolution of one of the moons of Jupiter. The true period of revolution is 42.5 h. (a) Taking into account the finite speed of light, how would you expect the apparent time for one revolution to change as the Earth moves in its orbit from point x to point y in Fig. 34-30? (b) What observations would be needed to compute the speed of light? Neglect the motion of Jupiter in its orbit. Figure 34-30 is not drawn to scale.

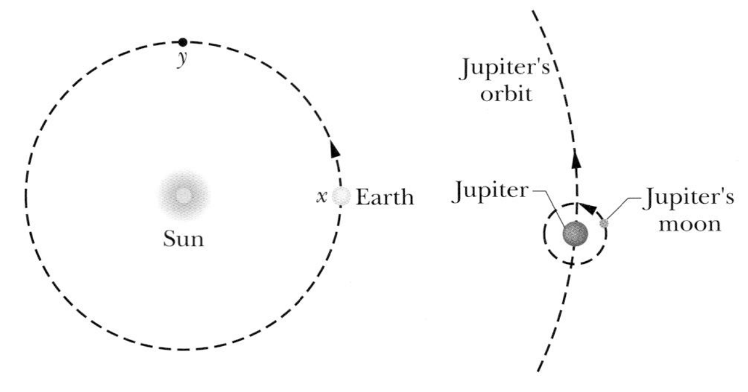

FIGURE 34-30 ■ Problem 59.

Additional Problems

60. Wave of Frequency An electromagnetic wave with frequency 400 terahertz travels through vacuum in the positive direction of an x axis. The wave is polarized, with its electric field directed parallel to the y axis, with amplitude $E^{\max}$. At time $t = 0$, the electric field at point P on the x axis has a value of $+E^{\max}/4$ and is decreasing with time. What is the distance along the x axis from point P to the first point with $E = 0$ if we search in (a) the negative direction and (b) the positive direction of the x axis?

61. Earth's Surface At the Earth's surface, what intensity of light is needed to suspend a totally absorbing spherical particle against its own weight if the mass of the particle is 2.0×10^{-13} kg and its radius is 2.0 μm?

62. Tracking a Plane Wave in a Box An oscillating current in an antenna is producing an electromagnetic wave. The region shown in Fig. 34-31 enclosed by a dashed box (not to scale) is far from the antenna. In it, the field produced is well approximated by a plane wave traveling in the z direction and having its E-field pointed along the x direction (using the coordinate system shown).

(a) You perform a series of measurements of the electric field at the origin of your coordinate system and obtain a result that points in the y direction and is well represented by the function

$$E(t) = E_0 \cos(\omega t).$$

What result would you find if instead of at the origin, you repeated the experiment at a point with coordinates {0, 0, z}? Explain how you know.
(b) What result would you get if you made your measurements at a point in the box with coordinates {2, 3, z} cm? (The point is still well within the dashed box.)
(c) For what values of z would you find exactly the same result as you found at the origin?

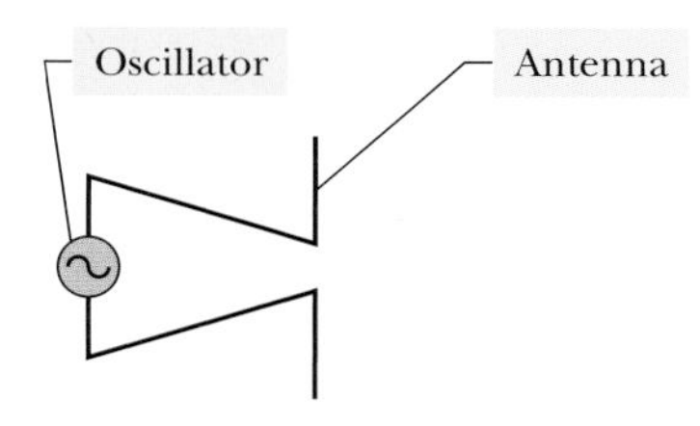

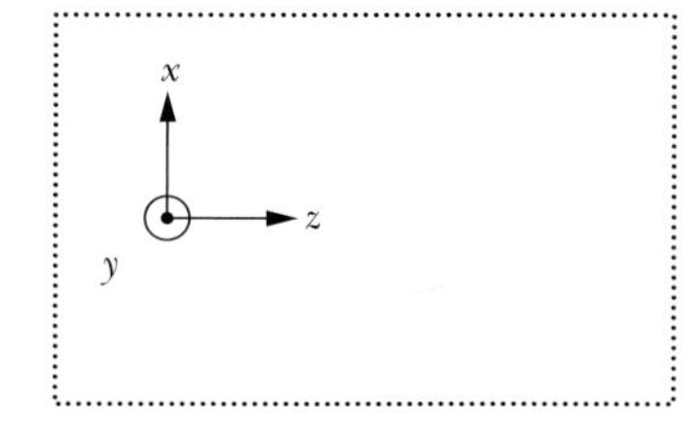

FIGURE 34-31 ■ Problem 62.

63. Electromagnetic Light After completing the construction of his equations for electromagnetism, Maxwell proposed that visible light was actually an electromagnetic wave. Discuss whether or not this is plausible and what evidence there is for his hypothesis.

64. *E*-Field and String Pulses Compare and contrast the propagation of a pulse on a string and an electromagnetic pulse shown in Fig. 34-32. In particular, address the similarities and differences for **(a)** how the pulse "knows" to move from one position to the next; **(b)** what will happen if the wave is passed through a slit. (See the figure.)

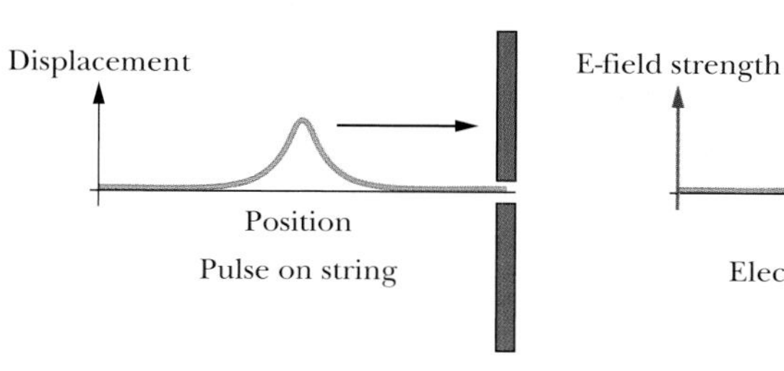

FIGURE 34-32 ■ Problem 64.

65. Measuring the Speed of Light à la Galileo Galileo tried to measure the speed of light by having two people stand on hills about 5 km apart. Each would hold a shuttered lantern. The first would open his lantern and when the second saw the light, he would open his lantern. The first person would then measure how much time it took between the time he first opened his lantern and the time he saw the light returning.

(a) How much time would it take the light to travel between the two hills?
(b) Is this a good way to measure the speed of light? Support your argument with a brief explanation that includes some quantitative discussion of the uncertainty in the measurement.

66. Solar Power for Your House The amount of energy from the sun that reaches the ground is on the order of 1 kW/m^2. Use this information to estimate the area you would need for a solar energy collector to provide all the electricity in your house. Explain carefully your assumptions and reasoning.

67. Boiling Water in a Microwave Most of you have had the experience of using a microwave oven to boil a cup of water. [If you have not, ask a friend or roommate to help you estimate the time in part (a)]. According to a Pyrex measuring cup that is marked in both English and SI units, one cup contains about 230 ml.

(a) From the amount of time it takes to heat one cup of water from room temperature to boiling in a microwave oven, estimate the power that the oven delivers to the water in watts (joules/second).
(b) Assuming that electromagnetic radiation is flowing into the cup in the microwave from all sides, estimate the electromagnetic energy flux, S, in W/m^2.
(c) From the flux you calculated in (b), estimate the strength of the electric and magnetic fields in a microwave oven.

68. Speed of Light and the GPS System Although light appears to travel at a speed that is for all practical purposes infinite, for some modern purposes the time delay due to light travel time is of great importance. The Global Positioning System (GPS) allows you to determine your position from comparison of the time delays between radio signals from 4 satellites at a height of 20,000 km above the surface of the earth. (There are actually 24 of these satellites. Your GPS picks out the closest 4 to your current position.) In order to get some idea of how important the speed of light is in establishing your position with one of these gadgets, make some simple assumptions. Assume that a satellite is almost directly overhead. Then figure out how far the satellite will move in the time it takes light (the radio signal) to get from the satellite to your GPS receiver. This estimates how far off the reading of your position would be if your device didn't include the speed of light in its calculations. To do this:

(a) Figure out what speed the satellite must be traveling to be in a circular orbit.
(b) Estimate the time it would take for a radio signal to get from the satellite to your receiver.
(c) Estimate how far the satellite would move in that time. If you ignore light travel time, this tells about how wrong you would get the satellite's position (and therefore how wrong you would get your position).

69. Laser Eye Surgery Laser eye surgery is carried out by delivering highly intense bursts of energy using electromagnetic waves. A typical laser used in such surgery has a wavelength of 190 nm (ultraviolet light) and produces bursts of light that last for 1 ms. The laser delivers an energy of 0.5 mJ to a circular spot on the cornea with a diameter of 1 mm. (The light is well approximated by a plane wave for the short distance between the laser and the cornea.)

(a) Assuming that the energy of a single pulse is delivered to a volume of the cornea about 1 mm^3, and assuming that the pulses are delivered so quickly that the energy deposited has no time to flow out of that volume, how many pulses are required to raise the temperature of that volume from 20°C to 100°C? (Assume that the cornea has a heat capacity similar to that of water.)
(b) Estimate the maximum strength of the electric field in one of these pulses.

70. Insolation of the Earth The power radiated by the sun is 3.9×10^{26} W. The Earth orbits the Sun in a nearly circular orbit of radius 1.5×10^{11} m. The Earth's axis of rotation is tilted by 23° relative to the plane of the orbit (see Fig. 34-33) so sunlight does not strike the equator perpendicularly.

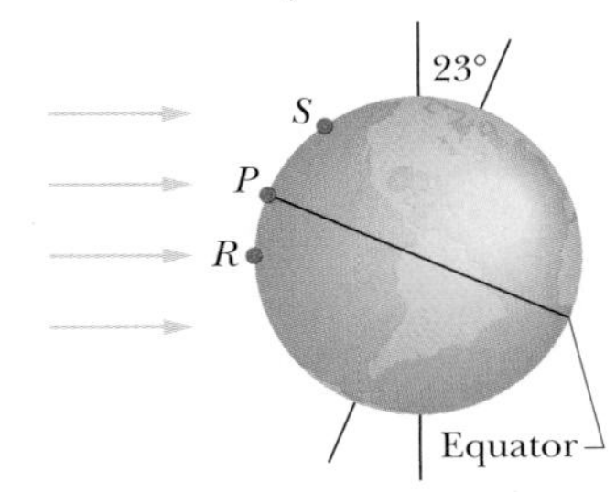

FIGURE 34-33 ■ Problem 70.

(a) At the time of year depicted in Fig. 34-33, what power strikes a 1 m^2 patch of horizontal flat land at the equator at the point P?
(b) Will a 1 m^2 patch of horizontal flat land at the point R or S receive more radiation?
(c) Explain how your answer to part (b) tells you at which of the points R or S it is summer or winter.

35 Images

Edouard Manet's *A Bar at the Folies-Bergère* has enchanted viewers ever since it was painted in 1882. Part of its appeal lies in the contrast between an audience ready for entertainment and a bartender whose eyes betray her fatigue. Its appeal also depends on subtle distortions of reality that Manet hid in the painting—distortions that give an eerie feel to the scene even before you recognize what is "wrong."

Can you find those subtle distortions of reality?

The answer is in this chapter.

35-1 Introduction

In Chapter 34 we began our study of electromagnetic waves. In this chapter, our focus is on electromagnetic image formation. Although most of us are familiar with optical images of visible light created by lenses and mirrors, image formation occurs in many other regions of the electromagnetic spectrum, including x rays, ultraviolet and infrared light, microwaves, and radio waves. In this chapter we will concentrate on images formed by visible light.

Electromagnetic waves with wavelengths that are within or near the visually detectable range (typically 400 to 700 nanometers) are commonly referred to as **light.** Visible light and its interaction with materials in our everyday world is of obvious importance. For many species, the ability to **see** (the visual system of an organism's formation of a mental image that is based on the detection of light) is often a necessary condition for survival. Light-based (or **optical**) instruments including eyeglasses, microscopes, and mirrors are important to most people on a daily basis. Whether these instruments have allowed you or someone around you to read the words on a page, diagnose a bacterial infection, or detect the car behind you as you back up in the parking lot, an optical instrument has inevitably impacted your life today. The human eye is an optical instrument.

In Chapter 34, we focused on the wave nature of electromagnetic radiation. In this chapter we will use a simplified wave model call the *ray model* of light. Although a light wave spreads as it moves away from its source, we can often approximate its travel as being in a straight line. For example, we did so in Chapter 34 for the light wave in Fig. 34-7*a*. This straight-line approximation is the basis of the ray model of light in which light waves are represented as lines called **rays** or **beams.** The study of the properties of light waves under this approximation is called *geometrical optics*. It is a perfectly productive (and simpler) approach to understanding how optical instruments function. We will use it extensively in this chapter.

In order to understand how the optical instruments that are so important in our lives work, we first need to understand some fundamental concepts related to the way light interacts with objects around us. For example,

1. Each small area on the surface of most ordinary objects scatters incident light rays in many directions. So anyone who has a line of sight to parts of an object's surface that are illuminated can see them.
2. Objects with smooth surfaces act as mirrors that reflect a light ray in a single direction rather than scattering it.
3. In transparent materials that are uniform like glass, plastic, air, and water, a ray of light travels in a straight line. However, light rays that cross an interface between two different transparent materials at an angle change direction at that interface.
4. When your eye receives neighboring rays of light that are diverging slightly, your brain assumes that the rays are coming from a common point. This is how your brain constructs a visual model of your surroundings.

In the first few sections of this chapter, we discuss "reflection" and "refraction." These are the two fundamental scattering processes that can occur when light strikes an object with a smooth surface such as a mirror or polished glass. This discussion will provide the foundation required to understand image formation by simple mirrors and lenses, which is covered in Sections 35-6 to 35-10. Finally, we will use our understanding of mirrors and lenses to consider more complex optical devises like microscopes and telescopes. We will continue to use both the wave model and the ray model of light in Chapters 36 and 37.

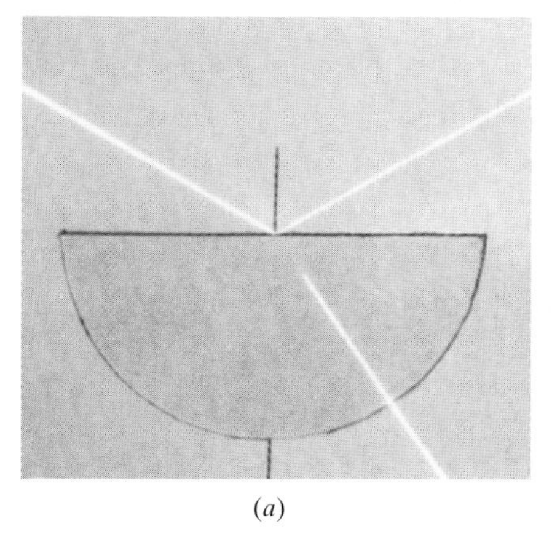

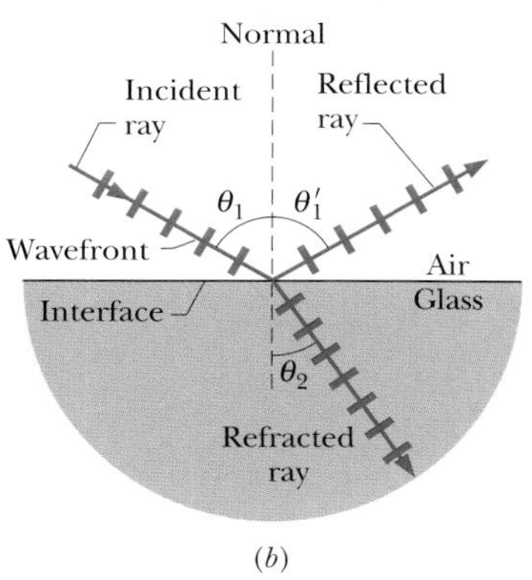

(*a*) (*b*)

FIGURE 35-1 ■ (*a*) A black-and-white photograph showing the reflection and refraction of an incident beam of light by a horizontal plane glass surface. At the bottom surface, which is curved, the beam is perpendicular to the surface, so the refraction there does not bend the beam. (Note: It is not possible to view a beam of light from the side as it passes through air or a lens. The path of the beams are visible because they are skimming along a piece of paper that is underneath the lens.) (*b*) A representation of (*a*) using rays. The angles of incidence (θ_1), of reflection (θ_1'), and of refraction (θ_2) are marked.

35-2 Reflection and Refraction

The black-and-white photograph in Fig. 35-1*a* shows an example of light waves traveling in approximately straight lines. A narrow beam of incoming light (the **incident beam**), angled downward from the left and traveling through air, encounters a *plane* (flat) glass surface. Part of the light seems to bounce off the smooth glass surface, forming a smooth glass **reflected beam** directed upward toward the right. The rest of the light travels through the surface and into the glass, forming a **refracted beam** directed downward to the right. Since the surface of the glass is smooth and the glass is uniform, the refracted light forms a beam rather than being scattered in many directions. Because light can travel through the glass like this, the glass is said to be *transparent;* that is, we can see through it. (In this chapter we shall consider only transparent materials with smooth surfaces.)

The passage of light from one homogeneous surface to another with a smooth surface (for example, from air to glass) is called **refraction,** and the light is said to be refracted. Unless an incident beam of light is perpendicular to a surface, refraction at the surface (or **interface**) changes the light's direction of travel. For this reason, the beam is said to be "bent" by the refraction. Note that as shown in Fig. 35-1*a*, the bending occurs only at the surface.

In Figure 35-1*b*, the beams of light in the photograph are represented with an *incident ray,* a *reflected ray,* and a *refracted ray* (and wave fronts). Each ray is oriented with respect to a line, called the normal, that is perpendicular to the surface at the point of reflection and refraction. In Fig. 35-1*b*, the **angle of incidence** is θ_1, the **angle of reflection** is θ_1', and the **angle of refraction** is θ_2. These are all measured *relative to the normal* as a line perpendicular to the surface. The plane containing the incident ray and the normal is the plane of incidence, which is in the plane of the page in Fig. 35-1*b*.

Experiment shows that reflection and refraction from smooth transparent surfaces are governed by two laws:

Law of reflection: A reflected ray lies in the plane of incidence and has an angle of reflection equal to the angle of incidence. In Fig. 35-1*b*, this means that

$$\theta_1' = \theta_1 \quad \text{(reflection from a smooth surface).} \tag{35-1}$$

(We shall now usually drop the prime on the angle of reflection.) As we will see in the sections that follow, the law of reflection is the fundamental basis for understanding image formation from any kind of mirror. Hence, this simple statement is really quite important.

TABLE 35-1
Some Indices of Refraction[a]

Medium	Index	Medium	Index
Vacuum	Exactly 1	Typical crown glass	1.52
Air (STP)[b]	1.00029	Sodium chloride	1.54
Water (20°C)	1.33	Polystyrene	1.55
Acetone	1.36	Carbon disulfide	1.63
Ethyl alcohol	1.36	Heavy flint glass	1.65
Sugar solution (30%)	1.38	Sapphire	1.77
Fused quartz	1.46	Heaviest flint glass	1.89
Sugar solution (80%)	1.49	Diamond	2.42

[a] For a wavelength of 589 nm (yellow sodium light).
[b] STP means "standard temperature (0°C) and pressure (1 atm)."

Law of refraction (or Snell's law): A refracted ray lies in the plane of incidence and has an angle of refraction θ_2 that is related to the angle of incidence θ_1 by

$$n_2 \sin\theta_2 = n_1 \sin\theta_1 \quad \text{(refraction in a transparent medium)}. \tag{35-2}$$

Here each of the symbols n_1 and n_2 is a dimensionless constant, called the index of refraction, that is associated with a medium involved in the refraction. We derive this equation, called Snell's law, in Chapter 36. As we shall discuss there, the index of refraction, n, of a medium is equal to c/v, where v is the speed of light in that medium and c is its speed in vacuum.

Table 35-1 gives the indices of refraction for visible light of vacuum and some common substances. For vacuum, n is defined to be exactly 1; for air, n is very close to 1.0 (an approximation we shall often make). No material used in basic optical devices has an index of refraction less than 1.

We can rearrange Eq. 35-2 as

$$\sin\theta_2 = \frac{n_1}{n_2}\sin\theta_1 \tag{35-3}$$

to compare the angle of refraction θ_2 with the angle of incidence θ_1. We can then see that the relative value of θ_2 depends on the relative values of n_2 and n_1. In fact, we can have three basic results:

1. If n_2 is equal to n_1, then θ_2 is equal to θ_1. In this case, refraction does not bend the light beam, which continues in the undeflected direction, as in Fig. 35-2*a*.
2. If n_2 is greater than n_1, then θ_2 is less than θ_1. In this case, refraction bends the light beam away from the undeflected direction and toward the normal, as in Fig. 35-2*b*.

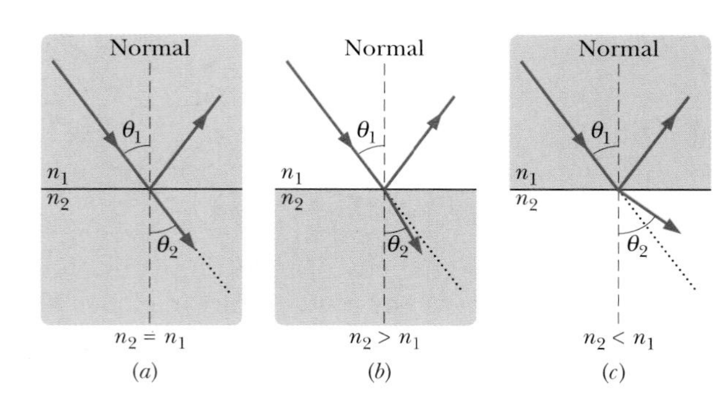

FIGURE 35-2 ■ Light refracting from a medium with an index of refraction n_1 and into a medium with an index of refraction n_2. (*a*) The beam does not bend when $n_2 = n_1$; the refracted light then travels in the *undeflected direction* (the dotted line), which is the same as the direction of the incident beam. The beam bends (*b*) toward the normal when $n_2 > n_1$ and (*c*) away from the normal when $n_2 < n_1$.

3. If n_2 is less than n_1, then θ_2 is greater than θ_1. In this case, refraction bends the light beam away from the undeflected direction and away from the normal, as in Fig. 35-2*c*.

Refraction cannot bend a beam so much that the refracted ray is on the same side of the normal as the incident ray.

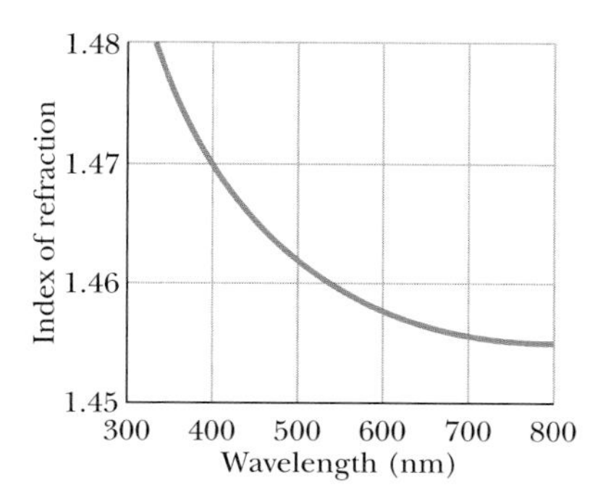

FIGURE 35-3 ■ The index of refraction as a function of wavelength for fused quartz. The graph indicates that a beam of short-wavelength light, for which the index of refraction is higher, is bent more upon entering or leaving quartz than a beam of long-wavelength light.

Chromatic Dispersion

The index of refraction n encountered by light in any medium except a vacuum depends on the wavelength of the light. The dependence of n on wavelength implies that when a light beam consists of rays of different wavelengths, the rays will be refracted at different angles by a surface; that is, the light will be spread out by the refraction. This spreading of light is called **chromatic dispersion,** in which "chromatic" refers to the colors associated with the individual wavelengths (as discussed in Section 34-8) and "dispersion" refers to the spreading of the light according to its wavelengths. The refractions of Figs. 35-1 and 35-2 do not show chromatic dispersion because the beams are *monochromatic* (of a single wavelength or color).

Generally, the index of refraction of a given medium is *greater* for a shorter wavelength (corresponding to, say, blue light) than for a longer wavelength (say, red light). As an example, Fig. 35-3 shows how the index of refraction of fused quartz depends on the wavelength of light. Such dependence means that when a beam with waves of both blue and red light is refracted through a surface, such as from air into quartz or vice versa, the blue *component* (the ray corresponding to the wave of blue light) bends more than the red component.

A beam of *white light* consists of components of all (or nearly all) the colors in the visible spectrum with approximately uniform intensities. When you see such a beam, you perceive white rather than the individual colors. In Fig. 35-4*a*, a beam of white light in air is incident on a glass surface. (Because the pages of this book are white, a beam of white light is represented with a gray ray here. Also, a beam of monochromatic light is generally represented with a red ray.) Of the refracted light in Fig. 35-4*a*, only the red and blue components are shown. Because the blue component is bent more than the red component, the angle of refraction θ_{2b} for the blue component is *smaller* than the angle of refraction θ_{2r} for the red component. (Remember, angles are measured relative to the normal.) In Fig. 35-4*b*, a ray of white light in glass is incident on a glass–air interface. Again, the blue component is bent more than the red component, but now θ_{2b} is greater than θ_{2r}.

To increase the color separation, we can use a solid glass prism with a triangular cross section, as in Fig. 35-5*a*. The dispersion at the first surface (on the left in Fig. 35-5*a, b*) is then enhanced by that at the second surface.

The most charming example of chromatic dispersion is a rainbow. As shown in Fig. 35-6, when white sunlight is intercepted by a falling raindrop, some of the light re-

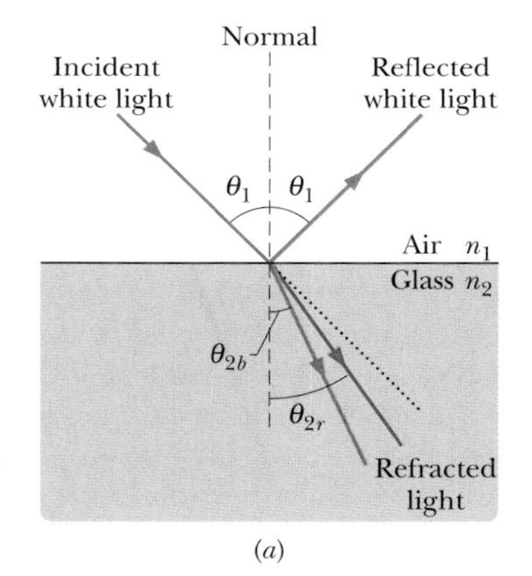

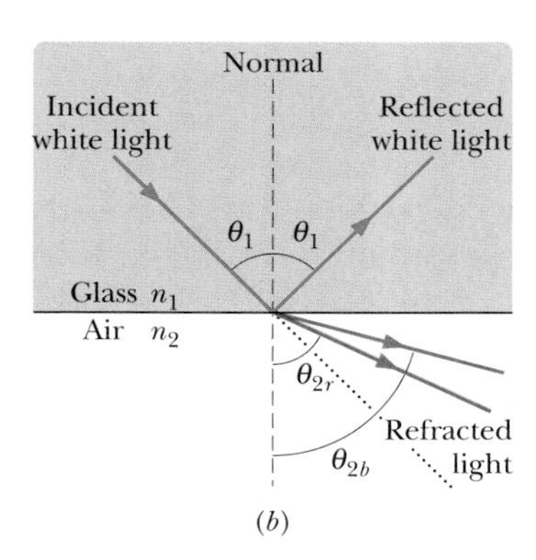

FIGURE 35-4 ■ Chromatic dispersion of white light. The blue component is bent more than the red component. (*a*) Passing from air to glass, the blue component ends up with the smaller angle of refraction. (*b*) Passing from glass to air, the blue component ends up with the greater angle of refraction.

FIGURE 35-5 ■ (*a*) A triangular prism separating white light into its component colors. (*b*) Chromatic dispersion occurs at the first surface and is increased at the second surface.

fracts into the drop and then reflects from the drop's inner surface, via total internal reflection (discussed in the next section). Finally it refracts out of the drop (Fig. 35-6). As with a prism, the first refraction separates the sunlight into its component colors, and the second refraction increases the separation.

The rainbow you see is formed by light that emerges from many drops; the red comes from drops angled slightly higher in the sky, the blue from drops angled slightly lower, and the intermediate colors from drops at intermediate angles. All the drops sending separated colors to you are angled at about 42° from a point that is directly opposite the Sun in your view. If the rainfall is extensive and brightly lit, you see a circular arc of color, with red on top and blue on bottom. Your rainbow is a personal one, because another observer intercepts light from other drops.

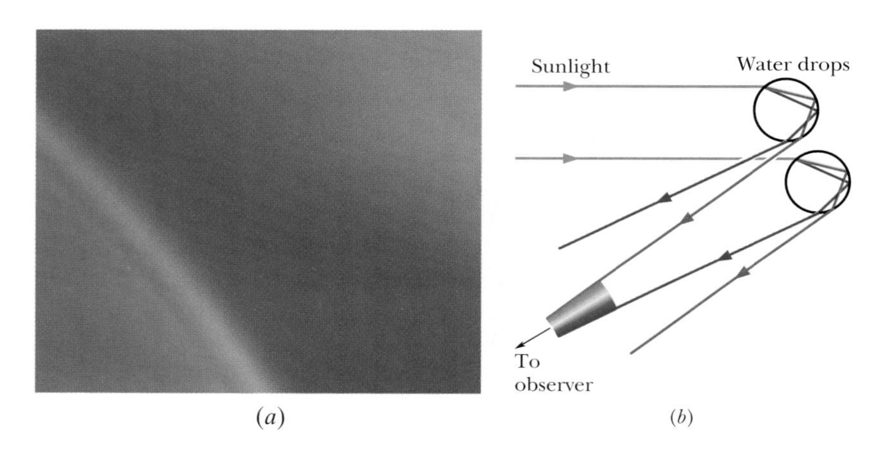

FIGURE 35-6 ■ (*a*) A rainbow is always a circular arc that is centered on the direction you would look if you looked directly away from the Sun. Under normal conditions, you are lucky if you see a long arc, but if you are looking downward from an elevated position, you might actually see a full circle. (*b*) The separation of colors when sunlight refracts into and out of falling raindrops leads to a rainbow. The figure shows the situation for the Sun on the horizon (the rays of sunlight are then horizontal). The paths of red and blue rays from two drops are indicated. Many other drops also contribute red and blue rays, as well as the intermediate colors of the visible spectrum.

READING EXERCISE 35-1: Which of the three drawings (if any) show physically possible refraction?

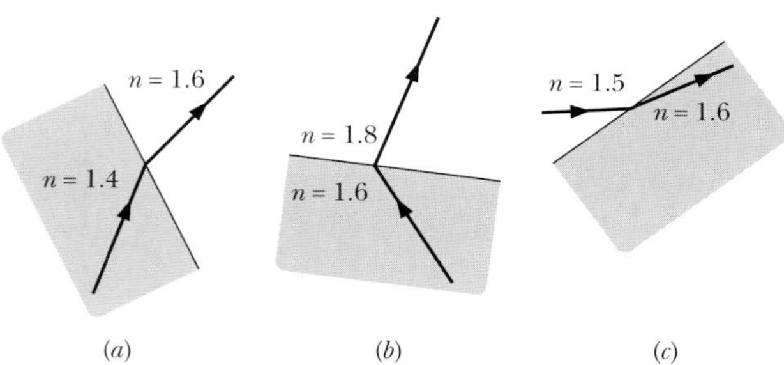

■

TOUCHSTONE EXAMPLE 35-1: Reflection and Refraction

(a) In Fig. 35-7*a*, a beam of monochromatic light reflects and refracts at point A on the interface between material 1 with index of refraction $n_1 = 1.33$ and material 2 with index of refraction $n_2 = 1.77$. The incident beam makes an angle of 50° with the interface. What is the angle of reflection at point A? What is the angle of refraction there?

SOLUTION ■ The **Key Idea** of any reflection is that the angle of reflection is equal to the angle of incidence. Further, both angles are measured between the corresponding light ray and a normal to the interface at the point of reflection. In Fig. 35-7*a*, the normal at point A is drawn as a dashed line through the point. Note that the angle of incidence θ_1 is not the given 50° but rather is $90° - 50° = 40°$. Thus, the angle of reflection is

$$\theta_1' = \theta_1 = 40°. \quad \text{(Answer)}$$

The light that passes from material 1 into material 2 undergoes refraction at point A on the interface between the two materials. The **Key Idea** of any refraction is that we can relate the angle of incidence, the angle of refraction, and the indexes of refraction of the two materials via Eq. 35-2:

$$n_2 \sin\theta_2 = n_1 \sin\theta_1. \tag{35-4}$$

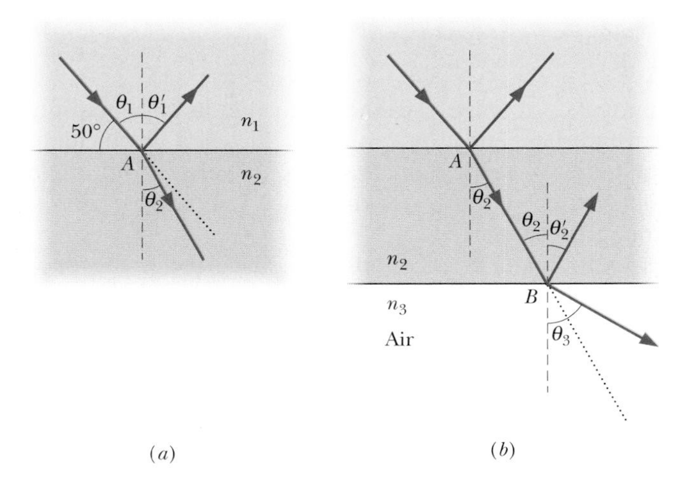

FIGURE 35-7 ■ (*a*) Light reflects and refracts at point A on the interface between materials 1 and 2. (*b*) The light that passes through material 2 reflects and refracts and point B on the interface between materials 2 and 3 (air).

Again we measure angles between light rays and a normal, here at the point of refraction. Thus, in Fig. 35-7*a*, the angle of refraction is the angle marked θ_2. Solving Eq. 35-4 for θ_2 gives us

$$\theta_2 = \sin^{-1}\left(\frac{n_1}{n_2}\sin\theta_1\right) = \sin^{-1}\left(\frac{1.33}{1.77}\sin 40°\right)$$
$$= 28.88° \approx 29°. \quad \text{(Answer)}$$

This result means that the beam swings toward the normal (it was at 40° to the normal and is now at 29°). The reason is that when the light travels across the interface, it moves into a material with a greater index of refraction.

(b) The light that enters material 2 at point A then reaches point B on the interface between material 2 and material 3, which is air, as shown in Fig. 35-7*b*. The interface through B is parallel to that through A. At B, some of the light reflects and the rest enters the air. What is the angle of reflection? What is the angle of refraction into the air?

SOLUTION ■ We first need to relate one of the angles at point B with a known angle at point A. Because the interface through point B is parallel to that through point A, the incident angle at B must be equal to the angle of refraction θ_2, as shown in Fig. 35-7*b*. Then for reflection, we use the same **Key Idea** as in (a): the law of reflection. Thus, the angle of reflection at B is

$$\theta_2' = \theta_2 = 28.88° \approx 29°. \quad \text{(Answer)}$$

Next, the light that passes from material 2 into the air undergoes refraction at point B, with refraction angle θ_3. Thus, the **Key Idea** here is again to apply the law of refraction, but this time by writing Eq. 35-4 as

$$n_3 \sin\theta_3 = n_2 \sin\theta_2. \tag{35-5}$$

Solving for θ_3 then leads to

$$\theta_3 = \sin^{-1}\left(\frac{n_2}{n_3}\sin\theta_2\right) = \sin^{-1}\left(\frac{1.77}{1.00}\sin 28.88°\right)$$
$$= 58.75° \approx 59°. \quad \text{(Answer)}$$

This result means that the beam swings away from the normal (it was at 29° to the normal and is now at 59°). The reason is that when the light travels across the interface, it moves into a material (air) with a lower index of refraction.

35-3 Total Internal Reflection

Figure 35-8 shows rays of monochromatic light from a point source S in glass incident on the interface between the glass and air. For ray a, which is perpendicular to the interface, part of the light reflects at the interface and the rest travels through it with no change in direction.

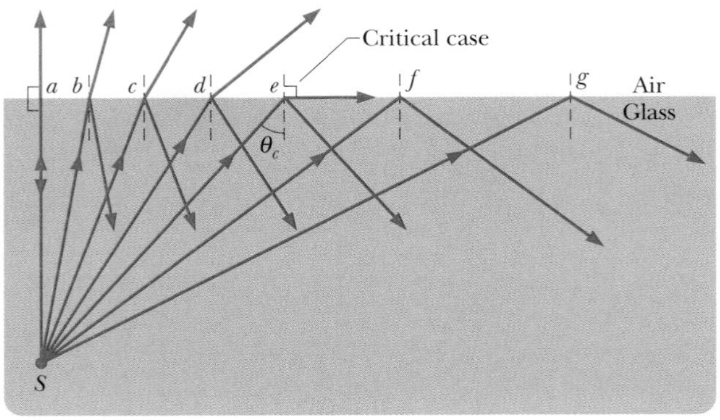

FIGURE 35-8 ■ Total internal reflection of light from a point source S in glass occurs for all angles of incidence greater than the critical angle θ_c. At the critical angle, shown at point e, the refracted ray points along the air–glass interface.

For rays b through e, which have progressively larger angles of incidence at the interface, there are also both reflection and refraction at the interface. As the angle of incidence increases, the angle of refraction increases; for ray e it is 90°, which means that the refracted ray points directly along the interface. The angle of incidence that gives this situation is called the critical angle θ_c. For angles of incidence larger than θ_c, such as for rays f and g, there is no refracted ray and all the light is reflected; this effect is called total internal reflection.

To find θ_c, we use Eq. 35-2; we arbitrarily associate subscript 1 with the glass and subscript 2 with the air, and then we substitute θ_c for θ_1 and 90° for θ_2, finding

$$n_1 \sin \theta_c = n_2 \sin 90°,$$

which gives us

$$\theta_c = \sin^{-1}\frac{n_2}{n_1} \quad \text{(critical angle).} \tag{35-6}$$

FIGURE 35-9 ■ Light sent into one end of an optical fiber like those shown here is transmitted to the opposite end with little loss of light through the sides of the fiber.

Because the sine of an angle cannot exceed unity, n_2 cannot exceed n_1 in this equation. This restriction tells us that total internal reflection cannot occur when the incident light is in the medium of lower index of refraction. If source S were in the air in Fig. 35-8, all its rays that are incident on the air–glass interface (including f and g) would be both reflected and refracted at the interface.

Total internal reflection has found many applications in medical technology. For example, a physician can search for an ulcer in the stomach of a patient by running two thin bundles of *optical fibers* (Fig. 35-9) down the patient's throat. Light introduced at the outer end of one bundle undergoes repeated total internal reflection within the fibers so that, even though the bundle provides a curved path, most of the light ends up exiting the other end and illuminating the interior of the stomach. Some of the light reflected from the interior then comes back up the second bundle in a similar way, to be detected and converted to an image on a monitor's screen for the physician to view.

TOUCHSTONE EXAMPLE 35-2: Triangular Prism

Figure 35-10 shows a triangular prism of glass in air; an incident ray enters the glass perpendicular to one face and is totally reflected at the far glass–air interface as indicated. If θ_1 is 45°, what can you say about the index of refraction n of the glass?

SOLUTION ■ One **Key Idea** here is that because the light ray is totally reflected at the interface, the critical angle θ_c for that interface must be less than the incident angle of 45°. A second **Key Idea** is that we can relate the index of refraction n of the glass to θ_c with the law of refraction, Eq. 35-2. Substituting $n_2 = 1$ (for the

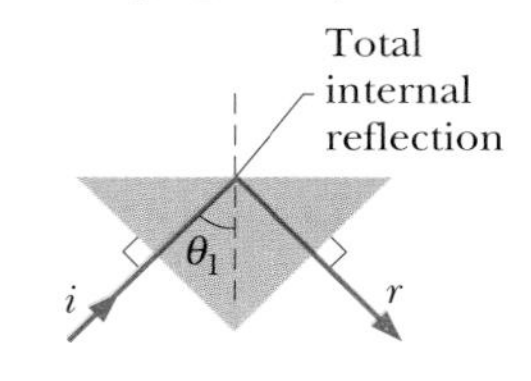

FIGURE 35-10 ■ The incident ray i is totally internally reflected at the glass–air interface, becoming the reflected ray r.

air) and $n_1 = n$ (for the glass) into Eq. 35-2 yields

$$\theta_c = \sin^{-1}\frac{n_2}{n_1} = \sin^{-1}\frac{1}{n}.$$

Because θ_c must be less than the incident angle of 45°, and the sine function is increasing between 0° and 90°,

$$\sin^{-1}\frac{1}{n} < 45°,$$

which gives us

$$\frac{1}{n} < \sin 45°$$

or

$$n > \frac{1}{\sin 45°} = 1.4. \quad \text{(Answer)}$$

The index of refraction of the glass must be greater than 1.4; otherwise, total internal reflection would not occur for the incident ray shown.

35-4 Polarization by Reflection

As we discuss in Chapter 34, you can increase and decrease the glare you see in sunlight that has been reflected from, say, water by looking through a polarizing sheet (such as a polarizing sunglass lens) and then rotating the sheet's polarizing axis around your line of sight. You can do so because reflected light is fully or partially polarized by the reflection from a surface.

Figure 35-11 shows a ray of unpolarized light incident on a glass surface. Let us resolve the electric field vectors of the light into two components. The *perpendicular components* are perpendicular to the plane of incidence and thus also to the page in Fig. 35-11; these components are represented with dots (as if we see the tips of the vectors). The *parallel components* are parallel to the plane of incidence and the page; they are represented with double-headed arrows. Because the light is unpolarized, these two components are of equal magnitude.

In general, the reflected light also has both components but with unequal magnitudes. This means that the reflected light is partially polarized—the electric fields oscillating along one direction have greater amplitudes than those oscillating along other directions. However, when the light is incident at a particular incident angle, called the *Brewster angle* θ_B, the reflected light has only perpendicular components, as shown in Fig. 35-11. The reflected light is then fully polarized perpendicular to the plane of incidence. The parallel components of the incident light do not disappear; they and perpendicular components form the light that is refracted through the glass surface.

Glass, water, and the other dielectric materials discussed in Section 28-6 can partially and fully polarize light by reflection. When you intercept sunlight reflected from such a surface, you see a bright spot (the glare) on the surface where the reflection takes place. If the surface is horizontal as in Fig. 35-11, the reflected light is partially or fully polarized horizontally. To eliminate such glare from horizontal surfaces, the lenses in polarizing sunglasses are mounted with their polarizing direction vertical.

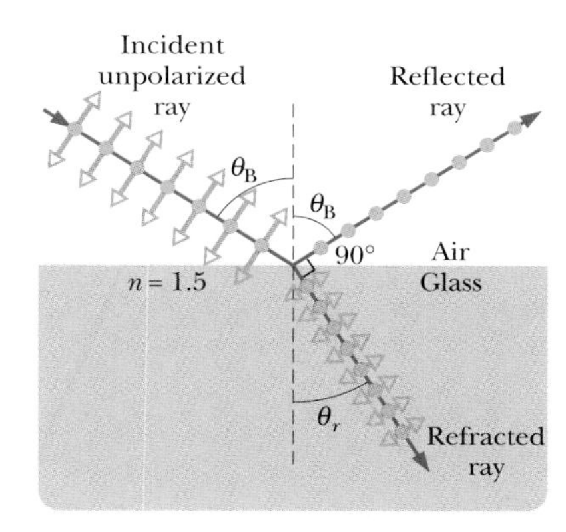

FIGURE 35-11 A ray of unpolarized light in air is incident on a glass surface at the Brewster angle θ_B. The electric fields along that ray have been resolved into components perpendicular to the page (the plane of incidence) and components parallel to the page. The reflected light consists only of components perpendicular to the page and is thus polarized in that direction. The refracted light consists of the original components parallel to the page and weaker components perpendicular to the page; this light is partially polarized.

Brewster's Law

For light incident at the Brewster angle θ_B, we find experimentally that the reflected and refracted rays are perpendicular to each other. Because the reflected ray is reflected at the angle θ_B in Fig. 35-11 and the refracted ray is at an angle $\theta_2 = \theta_r$, we have

$$\theta_B + \theta_r = 90°. \quad (35\text{-}7)$$

These two angles can also be related with Eq. 35-2. Arbitrarily assigning subscript 1 in Eq. 35-2 to the material through which the incident and reflected rays travel, we have, from that equation,

$$n_1 \sin\theta_B = n_2 \sin\theta_r.$$

Combining these equations leads to

$$n_1 \sin \theta_B = n_2 \sin(90° - \theta_B) = n_2 \cos \theta_B,$$

which gives us

$$\theta_B = \tan^{-1} \frac{n_2}{n_1} \quad \text{(Brewster angle).} \tag{35-8}$$

(Note carefully that the subscripts in Eq. 35-8 are not arbitrary because of our decision as to their meaning.) If the incident and reflected rays travel *in air,* we can approximate n_1 as unity and let n represent n_2 in order to write Eq. 35-8 as

$$\theta_B = \tan^{-1} n \quad \text{(only if } n_1 = 1\text{).} \tag{35-9}$$

This simplified version of Eq. 35-8 is also named after Sir David Brewster, who verified both equations experimentally in 1812.

35-5 Two Types of Image

Up to this point, most of our study of physics in general and optics in particular has been focused on understanding physical phenomena. We have not had to pay particular attention to the role of the observer in determining information about the physical system. (Even though there always is such an observer assumed, we also assume the measurements can be made gently enough so as not to disturb the system so the observer can be ignored.) That will need to change now. In studying image formation, we cannot be concerned only with what happens to the light. We must also be concerned with what it looks like to the observer. For this reason, we have to think about how your eyes and brain interpret the signals they receive. This new concern with the observer continues as we move into the study of relativity in Chapter 38 and as you move into quantum physics in your later studies.

For you to see an object, your eye must intercept some of the light rays spreading from the object and then redirect them onto the retina at the rear of the eye. Your visual system, starting with the retina and ending with the visual cortex at the rear of your brain, automatically and subconsciously processes the information provided by the light. That system identifies edges, orientations, textures, shapes, and colors and then rapidly brings to your consciousness an **image** (a reproduction derived from light) of the object; you perceive and recognize the object as being in the direction from which the light rays came and at the proper distance.

Your visual system goes through this processing and recognition even if the light rays do not come directly from the object, but instead reflect toward you from a mirror or refract through the lenses in a pair of binoculars. However, independent of whether the light rays come directly from an object or indirectly from a reflection or refraction event, the visual system in the human brain always forms an image as follows:

> The apparent location of an object is the common point from which the diverging straight line light rays seem to have come (even if the light rays have actually been bent). See Figs. 35-13 and 35-14 for examples.

For example, if the light rays have been reflected toward you from a standard flat mirror, the object appears to be behind the mirror because the rays you intercept come from that direction. Of course, the object is not back there. This type of image, which is called a **virtual image,** truly exists only on your retina but nevertheless is *said* to exist at the perceived location.

A **real image** differs in that it can be formed on a surface, such as a card or a movie screen. You can see a real image (otherwise movie theaters would be empty), but the existence of the image does not depend on your seeing it and it is present even if you are not.

In this chapter we explore several ways in which virtual and real images are formed by reflection (as with mirrors) and refraction (as with lenses). We also distinguish between the two types of image more carefully. We start by considering an example of a natural virtual image.

A Common Mirage

A common example of a virtual image is a pool of water that appears to lie on the road some distance ahead of you on a sunny day, but that you can never reach. The pool is a *mirage* (a type of illusion), formed by light rays coming from the low section of the sky in front of you (Fig. 35-12*a*). As the rays approach a road that has been heated by the Sun, they travel through progressively warmer air that has been heated by the road. With an increase in air temperature, the speed of light in air increases slightly and, correspondingly, the index of refraction of the air closer to the road decreases continuously. Thus, as the rays descend, encountering progressively smaller indices of refraction, they gradually bend more and more toward the horizontal (Fig. 35-12*b*).

Once a ray is horizontal, somewhat above the road's surface, it still bends because the lower portion of each associated wave front is in slightly warmer air and is moving slightly faster than the upper portion of the wavefront (Fig. 35-12*c*). This nonuniform motion of the wavefronts bends the ray upward. As the ray then ascends, it continues to bend upward through progressively greater indexes of refraction (Fig. 35-12*d*).

If some of this light enters your eyes, your visual system automatically infers that it originated along a backward extension of the rays you have intercepted and, to make sense of the light, assumes that it came from the road surface. If the light happens to be bluish from blue sky, the mirage appears bluish, like water. Because the air is probably turbulent due to the heating, the mirage shimmies, as if water waves were present. The bluish coloring and the shimmy enhance the illusion of a pool of water, but you are actually seeing a virtual image of a low section of the sky.

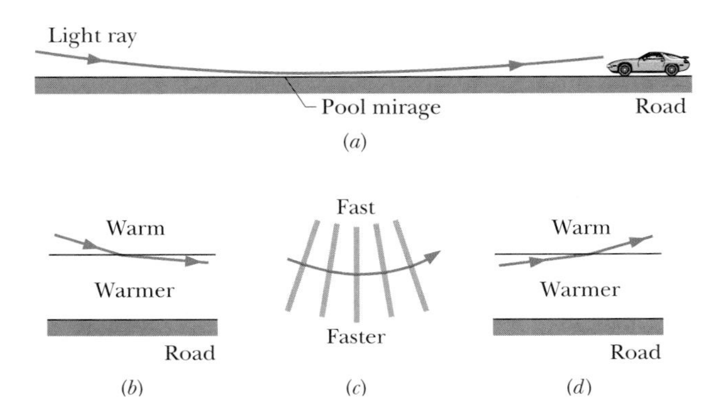

FIGURE 35-12 ■ (*a*) A ray from a low section of the sky refracts through air that is heated by a road (without reaching the road). An observer who intercepts the light perceives it to be from a pool of water on the road. (*b*) Bending (exaggerated) of a light ray descending across an imaginary boundary from warm air to warmer air. (*c*) Shifting of wavefronts and associated bending of a ray, which occurs because the lower ends of wavefronts move faster in warmer air. (*d*) Bending of a ray ascending across an imaginary boundary to warm air from warmer air.

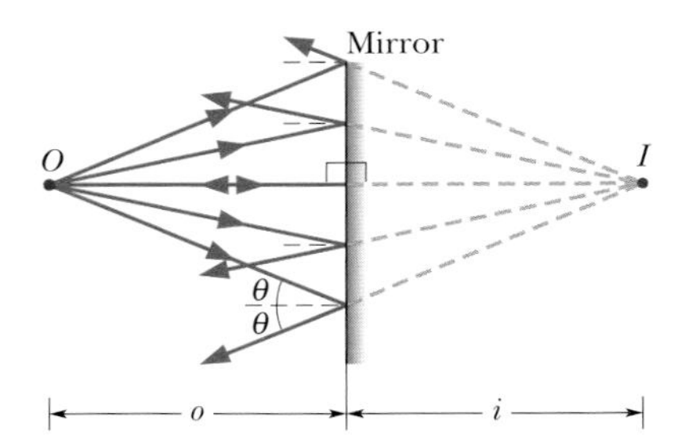

FIGURE 35-13 ■ A point source of light O, called the object, is a perpendicular distance o in front of a plane mirror. Light rays reaching the mirror from O reflect from the mirror. If your eye intercepts some of the reflected rays, you perceive a point source of light I to be behind the mirror, at a perpendicular distance i. The perceived source I is a virtual image of object O.

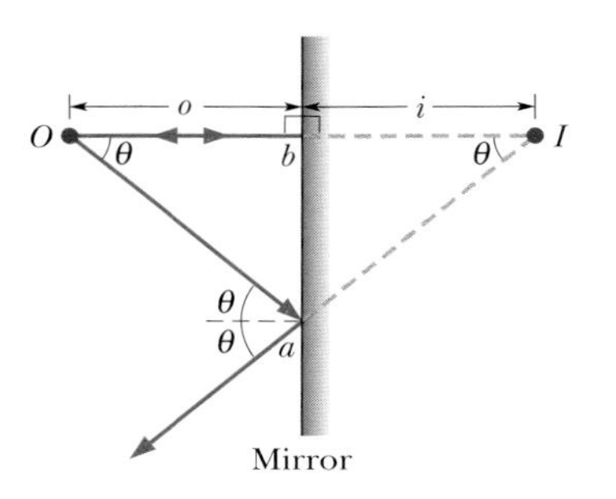

FIGURE 35-14 ■ Two rays from Fig. 35-13. Ray Oa makes an arbitrary angle θ with the normal to the mirror surface. Ray Ob is perpendicular to the mirror.

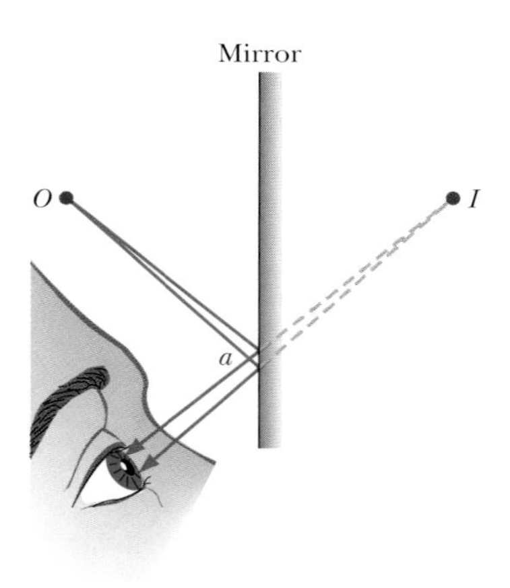

FIGURE 35-15 ■ A "pencil" of rays from O enters the eye after reflection at the mirror. Only a small portion of the mirror near a is involved in this reflection. The light appears to originate at point I behind the mirror.

35-6 Plane Mirrors

A **mirror** is a very smooth surface that can reflect a beam of light (or other electromagnetic radiation) in one direction instead of either scattering it widely in many directions or absorbing it. A shiny metal surface acts as a mirror; a concrete wall does not. In this section we examine the images that a **plane mirror** (a flat reflecting surface) can produce.

Figure 35-13 shows a point source of light O, which we shall call the *object,* at a perpendicular distance o in front of a plane mirror. The light that is incident on the mirror is represented with rays spreading from O.

> *Caution:* Since O is a point while o is a distance, it is important to distinguish between the two, not to confuse them with zero, and to write each one differently.

The reflection of that light is represented with reflected rays spreading from the mirror. If we extend the reflected rays backward (behind the mirror), we find that the extensions intersect at a point that is a perpendicular distance i behind the mirror.

If you look into the mirror of Fig. 35-13, your eyes intercept some of the reflected light. To make sense of what you see, you perceive a point source of light located at the point of intersection of the extensions. This point source is the image I of object O. It is called a point image because it is a point, and it is a virtual image because the rays do not actually pass through it. (As you will see, rays do pass through a point of intersection for a real image.) Your eyes (and brain) trace the rays back to their apparent intersection point and are fooled into thinking that's where the object is.

Figure 35-14 shows two rays selected from the many rays in Fig. 35-13. One reaches the mirror at point b, perpendicularly. The other reaches it at an arbitrary point a, with an angle of incidence θ. The extensions of the two reflected rays are also shown. The right triangles $aOba$ and $aIba$ have a common side and three equal angles and are thus congruent, so their horizontal sides are congruent. That is,

$$Ib = Ob, \tag{35-10}$$

where Ib and Ob are the distances from the mirror to the image and the object, respectively. Equation 35-10 tells us that the image is as far behind the mirror as the object is in front of it. By convention (that is, to get our equations to work out), *object distances o are taken to be positive quantities, and image distances i for virtual images* (as here) are taken to be negative quantities. Thus, Eq. 35-10 can be written as $|i| = o$, or as

$$i = -o \quad \text{(plane mirror).} \tag{35-11}$$

Only rays that are fairly close together can enter the eye after reflection at a mirror. For the eye position shown in Fig. 35-15, only a small portion of the mirror near point a (a portion smaller than the pupil of the eye) is useful in forming the image. To find this portion, close one eye and look at the mirror image of a small object such as the tip of a pencil. Then move your fingertip over the mirror surface until you cannot see the image. Only light coming from that small portion of the mirror under your fingertip produced the image.

Extended Objects

In Fig. 35-16, an extended object O, represented by an upright arrow, is at perpendicular distance o in front of a plane mirror. Each small portion of the object that faces the mirror acts like the point source O of Figs. 35-13 and 35-14. If you intercept the light reflected by the mirror, you perceive a virtual image I that is a composite of the virtual point images of all those portions of the object and seems to be at distance i behind the mirror. Distances i and o are related by Eq. 35-11.

We can also locate the image of an extended object as we did for a point object in Fig. 35-13: we draw some of the rays that reach the mirror from the top of the object, draw the corresponding reflected rays, and then extend those reflected rays behind the mirror until they intersect to form an image of the top of the object. We then do the same for rays from the bottom of the object. As shown in Fig. 35-16, we find that virtual image I has the same orientation and *height* (measured parallel to the mirror) as object O.

FIGURE 35-16 ■ An extended object O and its virtual image I in a plane mirror.

Manet's *A Bar at the Folies-Bergère*

In *A Bar at the Folies-Bergère* you see the barroom via reflection by a large mirror on the wall behind the woman tending bar, but the reflection is subtly wrong in three ways. First note the bottles at the left. Manet painted their reflections in the mirror but misplaced them, painting them farther toward the front of the bar than they should be.

Now note the reflection of the woman. Since your view is from directly in front of the woman, her reflection should be behind her, with only a little of it (if any) visible to you; yet Manet painted her reflection well off to the right. Finally, note the reflection of the man facing her. He must be you, because the reflection shows that he is directly in front of the woman, and thus he must be the viewer of the painting. You are looking into Manet's work and seeing your reflection well off to your right. The effect is eerie because it is not what we expect from either a painting or a mirror.

READING EXERCISE 35-2: In the figure you look into a system of two vertical parallel mirrors A and B separated by distance d. A grinning gargoyle is perched at point O, a distance $0.2d$ from mirror A. Each mirror produces a *first* (least deep) image of the gargoyle. Then each mirror produces a *second* image with the object being the first image in the opposite mirror. Then each mirror produces a *third* image with the object being the second image in the opposite mirror, and so on—you might see hundreds of grinning gargoyle images. How deep behind mirror A are the first, second, and third images in mirror A? ■

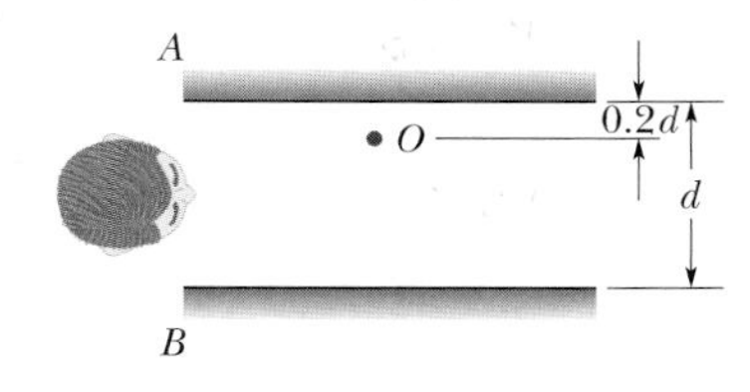

READING EXERCISE 35-3: Is an object in a mirror reversed left to right? How does this happen? Why isn't it also upside down? If you look at your image in a flat mirror, your left hand becomes your right hand, but your head and feet are not interchanged. Can you explain why? ■

35-7 Spherical Mirrors

We turn now from images produced by plane mirrors to images produced by mirrors with curved surfaces. In particular, we shall consider spherical mirrors, which are simply mirrors in the shape of a small section of the surface of a sphere. A plane mirror is in fact a spherical mirror with an infinitely large *radius of curvature*. This is like treating a small piece of the Earth as if it were flat. If the Earth has a large enough radius, we can't tell the difference.

A plane mirror fooled your visual system into thinking an object was in a different place than it really was by bending the rays so they didn't come directly from the

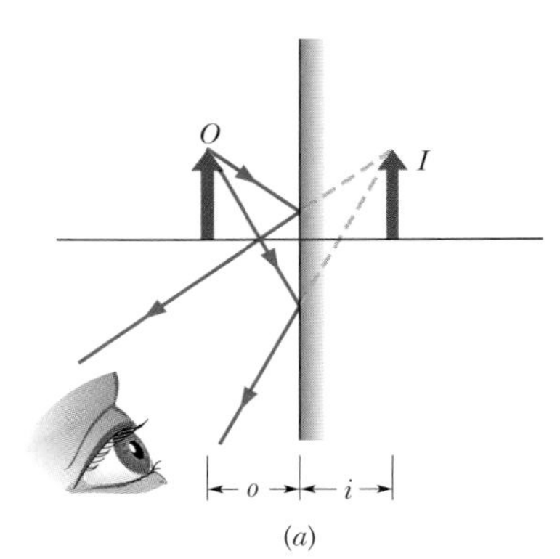

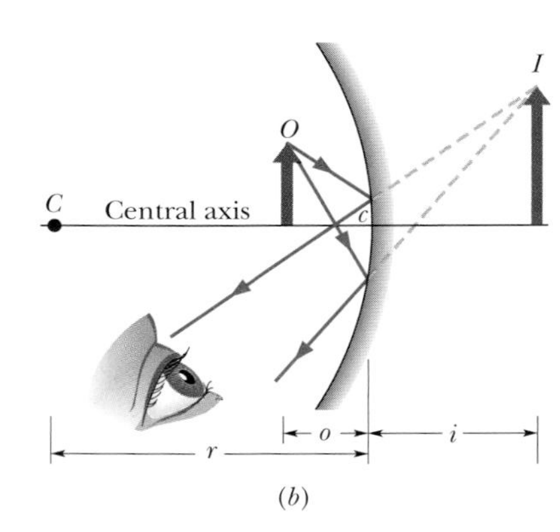

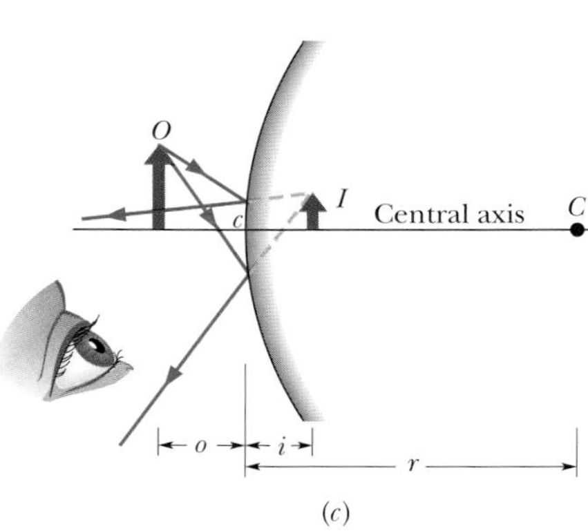

FIGURE 35-17 ■ (*a*) An object *O* forms a virtual image *I* in a plane mirror. (*b*) If the mirror is bent so that it becomes *concave*, the image moves farther away and becomes larger. (*c*) If the plane mirror is bent so that it becomes *convex*, the image moves closer and becomes smaller.

object in straight lines. A curved mirror does even more interesting things by bending the rays in a different way.

Making a Spherical Mirror

We start with the plane mirror of Fig. 35-17*a*, which faces leftward toward an object *O* and an observer. We make a **concave mirror** by curving the mirror's surface so it is concave ("caved in") as in Fig. 35-17*b*. Curving the surface in this way changes several characteristics of the mirror and the image it produces of the object:

1. *The center of curvature C* (the center of the sphere of which the mirror's surface is part) was infinitely far from the plane mirror; it is now closer and in front of the concave mirror.
2. *The field of view*—the extent of the scene that is reflected to the observer—was wide for the plane mirror; it is now smaller.
3. The image of the object was as far behind the plane mirror as the object was in front; the image is farther behind the concave mirror; that is, $|i|$ is greater.
4. The height or size of the image was equal to the height or size of the object for the plane mirror. The height of the image is now greater than the height of the object. This feature is why many makeup mirrors and shaving mirrors are concave—they produce a larger image of a face.

We can make a **convex mirror** by curving a plane mirror so its surface is *convex* ("flexed out") as in Fig. 35-17*c*. Curving the surface in this way:

1. Moves the center of curvature *C* to *behind* the mirror.
2. *Increases* the field of view. It is wider with a convex mirror than with a plane mirror.
3. Moves the image of the object *closer* to the mirror as compared to the plane mirror.
4. *Shrinks* the size of the image. It is now smaller than the actual size of the object.

Side view mirrors on cars and store surveillance mirrors are usually convex to take advantage of the increase in the field of view—more of the store can then be monitored with a single mirror.

In looking at Fig. 35-17*b* and *c*, you should note that when both surfaces of a curved mirror are reflective, the side the observer is on determines whether the mirror is convex or concave.

Focal Points of Spherical Mirrors

In order to figure out how a ray of light reflects from a curved mirror, we will look at a very small region of mirror around the point that the ray we are considering strikes. For the small enough region, the mirror looks flat (like a bit of the Earth looks flat although it is actually curved). Then, we use our plane mirror principle from above: angle of incidence is equal to angle of reflection (Eq. 35-1). To do this, first we need to draw a normal to the surface of the mirror at the point that the ray strikes. Then,

> For a spherical mirror, the reflected ray is in the plane determined by the incident ray and the normal to the surface. The angle between the normal and the reflected ray is equal to the angle between the normal and the incident ray.

For a plane mirror, the magnitude of the image distance i is always equal to the object distance o. Before we can determine how these two distances are related for a spherical mirror, we find it convenient to consider the reflection of light from an object O located an effectively infinite distance in front of a spherical mirror, on the mirror's *central axis*. That axis extends through the center of curvature C and the center c of the mirror. Because of the great distance between the object and the mirror, the light waves spreading from the object are nearly plane waves when they reach the mirror along the central axis. This means that the rays representing the light waves are all parallel to the central axis when they reach the mirror.

When these parallel rays reach a concave mirror like that of Fig. 35-18*a*, those near the central axis are reflected through a common point F; two of these reflected rays are shown in the figure. If we placed a (small) card at F, a point image of the infinitely distant object O would appear on the card since rays actually converge at that point. (This would occur for any infinitely distant object.) Point F is called the **focal point** (or **focus**) of the mirror, and its distance from the center of the mirror is the **focal length** f of the mirror.

If we now substitute a convex mirror for the concave mirror, we find that the parallel rays are no longer reflected through a common point. Instead, they diverge as shown in Fig. 35-18*b*. However, if your eye or a camera lens intercepts some of the reflected light, you perceive the light as originating from a point source behind the mirror. Although no rays actually converge behind the convex mirror, this perceived source is located where extensions of the reflected rays pass through a common point (F in Fig. 35-18*b*). That point is the focal point (or focus) F of the convex mirror, and its distance from the mirror surface is the focal length f of the mirror. If we placed a card at this focal point, an image of object O would not appear on the card, so this focal point is a virtual focal point and is *not* like that of a concave mirror.

To distinguish the actual focal point of a concave mirror from the perceived focal point of a convex mirror, the former is said to be a *real focal point* and the latter is said to be a *virtual focal point*. Moreover, the focal length f of a concave mirror is taken to be a positive quantity, and that of a convex mirror a negative quantity. For mirrors of both types, the focal length f is related to the radius of curvature r of the mirror by

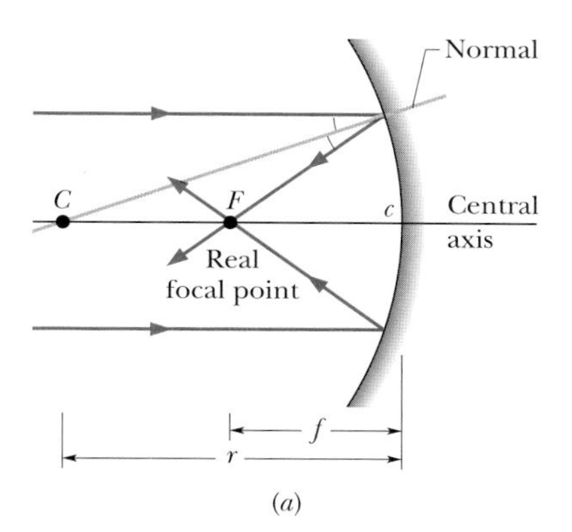

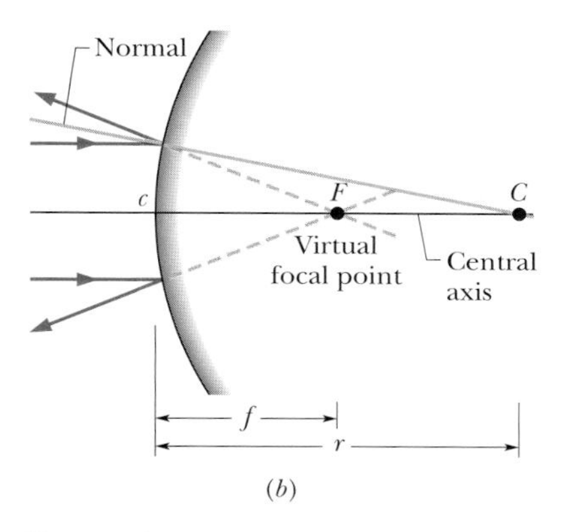

FIGURE 35-18 ■ (*a*) In a concave mirror, incident parallel light rays are brought to a real focus at F, on the same side of the mirror as the light rays. (*b*) In a convex mirror, incident parallel light rays seem to diverge from a virtual focus at F, on the side of the mirror opposite the light rays.

$$f = \tfrac{1}{2}r \quad \text{(spherical mirror)}, \tag{35-12}$$

where, consistent with the signs for the focal length, r is a positive quantity for a concave mirror and a negative quantity for a convex mirror.

35-8 Images from Spherical Mirrors

With the focal point of a spherical mirror defined, we can find the relation between image distance i and object distance o for concave and convex spherical mirrors. We discussed the law of reflection (Eq. 35-1) in Section 35-2. This law states that the angle of reflection is equal to the angle of incidence. It alone is the foundation required to understand the change in direction of light rays at the surface of a mirror. Ultimately, it is the direction of the reflected light rays that determines where the image is perceived to be located.

We begin by placing the object O *inside the focal point* of the concave mirror—that is, between the mirror and its focal point F (Fig. 35-19*a*). An observer can then see a virtual image of O in the mirror: The image appears to be behind the mirror, and it has the same orientation as the object.

If we now move the object away from the mirror until it is at the focal point, the image moves farther back from the mirror until it is at infinity (Fig. 35-19*b*).

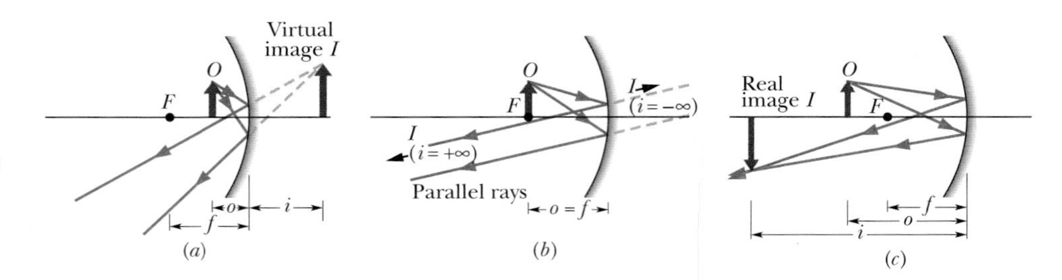

FIGURE 35-19 ■ (*a*) An object O inside the focal point of a concave mirror, and its virtual image I. (*b*) The object at the focal point F. (*c*) The object outside the focal point, and its real image I.

The image is then ambiguous and imperceptible because neither the rays reflected by the mirror nor the ray extensions behind the mirror cross to form an image of O.

If we next move the object *outside the focal point*—that is, farther away from the mirror than the focal point—the rays reflected by the mirror converge to form an *inverted* image of object O (Fig. 35-19*c*) in front of the mirror. That image moves in from infinity as we move the object farther outside F. If you were to hold a card at the position of the image, the rays converging at that point would scatter in all directions, when those scattered rays entered your eyes, your brain would see them as coming from the card, and so an image of the object would appear on the card. In this case, the image is said to be *focused* on the card by the mirror. (The verb "focus," which in this context means to produce an image, differs from the noun "focus," which is another name for the focal point.) Because this image can actually appear on a surface, it is a real image—the rays actually intersect to create the image, regardless of whether an observer is present. The image distance i of a real image is a positive quantity, in contrast to that for a virtual image. We also see that

> Real images form on the side of a mirror where the object is, and virtual images form on the opposite side.

As we shall prove in Section 35-8, when light rays from an object make only small angles with the central axis of a spherical mirror and the proper sign conventions are chosen, a simple equation relates the object distance o, the image distance i, and the focal length f:

$$\frac{1}{o}+\frac{1}{i}=\frac{1}{f} \quad \text{(spherical mirror).} \tag{35-13}$$

We assume such small angles in figures such as Fig. 35-19, but for clarity the rays are drawn with exaggerated angles. With that assumption, Eq. 35-13 applies to any concave, convex, or plane mirror. For a convex or plane mirror, only a virtual image can be formed, regardless of the object's location on the central axis. As shown in the example of a convex mirror in Fig. 35-17*c*, the image is always on the opposite side of the mirror from the object and has the same orientation as the object.

The size of an object or image, as measured *perpendicular* to the mirror's central axis, is called the object or image *height*. Let h represent the height of the object and h' the height of the image. Then the ratio h'/h is called the **lateral magnification** m produced by the mirror. However, by convention, the lateral magnification always includes a plus sign when the image orientation is the same as that of the object and a minus sign when the image orientation is opposite that of the object (that is, upside down). For this reason, we write the formula for m as

$$|m|=\frac{h'}{h} \quad \text{(lateral magnification).} \tag{35-14}$$

TABLE 35-2
Organizing Table for Mirrors

Mirror Type	Object Location	Image Location	Image Type	Image Orientation	Sign of f	Sign of r	Sign of i	Sign of m
Plane	Anywhere							
Concave	Inside F							
	Outside F							
Convex	Anywhere							

We shall soon prove that the lateral magnification can also be written as

$$m = -\frac{i}{o} \quad \text{(lateral magnification).} \tag{35-15}$$

For a plane mirror for which $i = -o$, we have $m = +1$. The magnification of 1 means that the image is the same size as the object. The plus sign means that the image and the object have the same orientation. For the concave mirror of Fig. 35-19*c*, $m \approx -1.5$.

Equations 35-12 through 35-15 hold for all plane mirrors, concave spherical mirrors, and convex spherical mirrors. In addition to those equations, you have been asked to absorb a lot of information about these mirrors, and you should organize it for yourself by filling in Table 35-2. Under Image Location, note whether the image is on the *same* side of the mirror as the object or on the opposite side. Under Image Type, note whether the image is *real* or *virtual*. Under Image Orientation, note whether the image has the *same* orientation as the object or is *inverted*. Under Sign, give the sign of the quantity or fill in ± if the sign is ambiguous. You will need this organization to tackle homework or examinations.

Locating Images by Drawing Rays

Figures 35-20*a* and *b* show an object O in front of a concave mirror. In general, a point on an object puts out a spray of rays in all directions. For most of these rays, we need a protractor to calculate where a ray that hits our mirror (and later in the

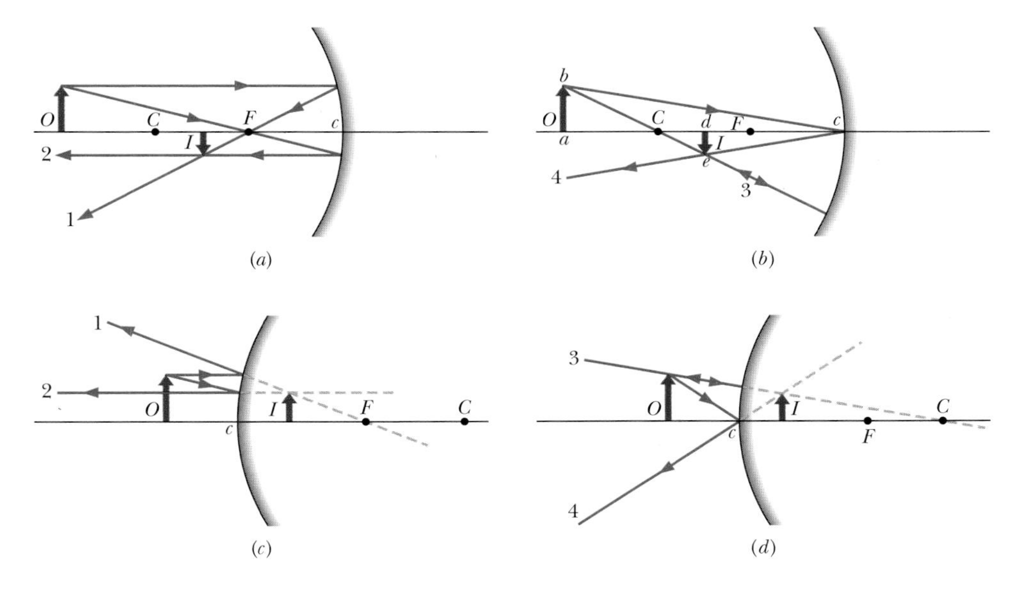

FIGURE 35-20 ■ (*a*, *b*) Four rays that can easily be drawn to find the image of an object in a concave mirror. For the object position shown, the image is real, inverted, and smaller than the object. (*c*, *d*) Four similar rays for the case of a convex mirror. For a convex mirror, the image is always virtual, oriented like the object, and smaller than the object. [In (*c*), ray 2 is initially directed toward focal point F. In (*d*), ray 3 is initially directed toward center of curvature C.]

chapter, our lens) will go. But for four special rays, we can easily draw where they are going to go, using the focal point and symmetry. We can graphically locate the image of any off-axis point of the object by drawing a *ray diagram* with any two of four special rays through the point:

1. A ray that is initially parallel to the central axis reflects through the focal point F (ray 1 in Fig. 35-20*a*).
2. A ray that reflects from the mirror after passing through the focal point emerges parallel to the central axis (ray 2 in Fig. 35-20*a*).
3. A ray that reflects from the mirror after passing through the center of curvature C returns along itself (ray 3 in Fig. 35-20*b*).
4. A ray that reflects from the mirror at its intersection c with the central axis is reflected symmetrically about that axis (ray 4 in Fig. 35-20*b*).

The image of the point is at the intersection of the two special rays you choose. The image of the object can then be found by locating the images of two or more of its off-axis points. You need to modify the descriptions of the rays slightly to apply them to convex mirrors, as in Figs. 35-20*c* and *d*. By referring to Figs. 35-20*c* and *d*, you can easily write a description of what happens to rays 1, 2, 3, and 4 as they are reflected from a convex mirror.

Proof of Equation 35-15

We are now in a position to derive Eq. 35-15 ($m = -i/o$), the equation for the lateral magnification of an object reflected in a mirror. Consider ray 4 in Fig. 35-20*b*. It is reflected at point c so that the incident and reflected rays make equal angles with the axis of the mirror at that point.

The two right triangles abc and cde in the figure are similar, so we can write

$$\frac{de}{ab} = \frac{cd}{ca}.$$

The quantity on the left (apart from the question of sign) is the lateral magnification m produced by the mirror. Since we indicate an inverted image as a *negative* magnification, we symbolize this as $-m$. However, $cd = i$ and $ca = o$, so we have at once

$$m = -\frac{i}{o} \quad \text{(magnification)}, \tag{35-16}$$

which is the relation we set out to prove.

READING EXERCISE 35-4: Use Figs. 35-20*c* and *d* to modify the rays developed for a concave mirror to describe what happens to rays 1, 2, 3, and 4 when they are incident on a convex mirror. ■

READING EXERCISE 35-5: A Central American vampire bat, dozing on the central axis of a spherical mirror, is magnified by $m = -4$. Is its image (a) real or virtual, (b) inverted or of the same orientation as the bat, and (c) on the same side of the mirror as the bat or on the opposite side? ■

TOUCHSTONE EXAMPLE 35-3: Tarantula

A tarantula of height h sits cautiously before a spherical mirror whose focal length has absolute value $|f| = 40$ cm. The image of the tarantula produced by the mirror has the same orientation as the tarantula and has height $h' = 0.20h$.

(a) Is the image real or virtual, and is it on the same side of the mirror as the tarantula or the opposite side?

SOLUTION ■ The **Key Idea** here is that because the image has the same orientation as the tarantula (the object), it must be virtual and on the opposite side of the mirror. (You can easily see this result if you have filled out Table 35-2).

(b) Is the mirror concave or convex, and what is its focal length f, sign included?

SOLUTION ■ We *cannot* tell the type of mirror from the type of image, because both types of mirror can produce virtual images. Similarly, we cannot tell the type of mirror from the sign of the focal length f, as obtained from Eq. 35-12 or 35-13, because we lack enough information to use either equation. However—and this is the **Key Idea** here—we can make use of the magnification information. We know that the ratio of image height h' to object height h is 0.20. Thus, from Eq. 35-14 we have

$$|m| = \frac{h'}{h} = 0.20.$$

Because the object and image have the same orientation, we know that m must be positive: $m = +0.20$. Substituting this into Eq. 35-15 and solving for, say, i gives us

$$i = -0.20o,$$

which does not appear to be of help in finding f. However, it is helpful if we substitute it into Eq. 35-13. That equation gives us

$$\frac{1}{f} = \frac{1}{i} + \frac{1}{o} = \frac{1}{-0.20o} + \frac{1}{o} = \frac{1}{o}(-5 + 1),$$

from which we find

$$f = -o/4.$$

Now we have it: Because o is positive, f must be negative, which means that the mirror is convex with

$$f = -40 \text{ cm}. \qquad \text{(Answer)}$$

35-9 Spherical Refracting Surfaces

We now turn from images formed by reflections to images formed by refraction through smooth surfaces of transparent materials, such as glass. We've seen that curved mirrors scatter light rays so that our eyes see them in new and different ways—in different places, of different sizes, and perhaps upside down. But the most powerful applications of the bending of light rays come when we consider the effect of the refraction of rays passing through transparent materials. We can then understand how the human eye works and construct optical devices with which we can look at objects and bring them into focus (eye glasses), make them bigger (microscopes and telescopes), or create images for storing (cameras). We shall consider only spherical surfaces, with radius of curvature r and center of curvature C. The light will be emitted by a point object O in a medium with index of refraction n_1; it will refract through a spherical surface into a medium of index of refraction n_2.

To determine where the image forms (that is, whether it is real or virtual), we need to once again consider the change in direction of the rays of light that strike the refracting surface. However, as opposed to image formation by mirrors where we are concerned with reflected rays, we are concerned here with the ray that enters the refracting material and is bent (refracted). The final answer to whether the image is virtual (assuming that an observer intercepts the rays) or real (no observer necessary) depends on the relative values of n_1 and n_2 and on the geometry of the situation. Specifically, to understand image formation by spherical refracting surfaces, we draw a normal to the surface. Then,

This insect has been entombed in amber for about 25 million years. Because we view the insect through a curved refracting surface, the image we see does not coincide with the insect.

> The refracted ray is in the plane determined by the incident ray and the normal. The relationship between the angle of refraction and the angle of incidence is represented in the law of refraction (or Snell's law): $n_1 \sin\theta_1 = n_2 \sin\theta_2$ (Eq. 35-2).

Six possible results are shown in Fig. 35-21. In each part of the figure, the medium with the greater index of refraction is shaded, and an object O is located on a central axis passing through the center of curvature of the refracting surface. O is always in

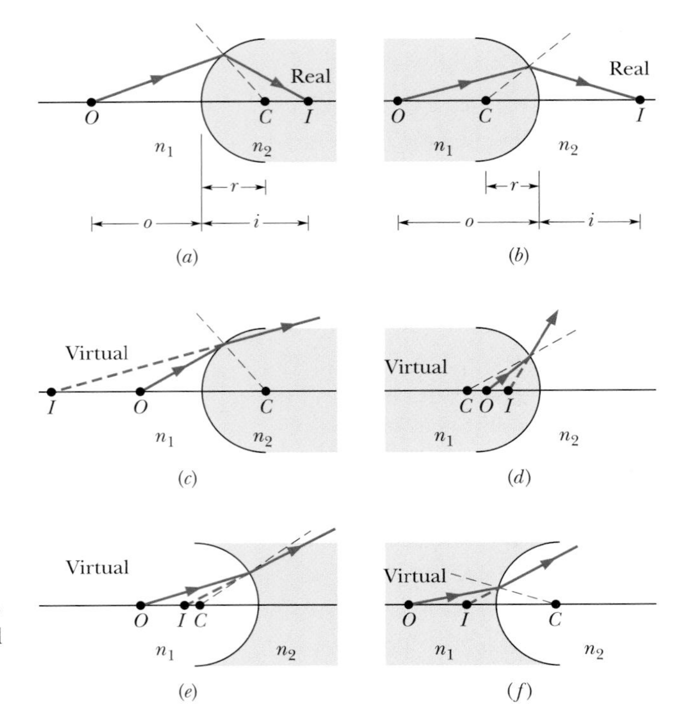

FIGURE 35-21 ■ Six possible ways in which an image can be formed by refraction through a spherical surface of radius r and center of curvature C. The surface separates a medium with index of refraction n_1 from a medium with index of refraction n_2. The point object O is always in the medium with n_1, to the left of the surface. The material with the lesser index of refraction is unshaded (think of it as being air, and the other material as being glass). Real images are formed in (a) and (b); virtual images are formed in the other four situations.

the medium with index of refraction n_1, to the left of the refracting surface. In each part, a representative ray is shown refracting through the surface. Another ray along the axis has $\theta_1 = 0$, which means $\theta_2 = 0$, so it is undeviated. The undeviated ray on the central axis and the ray refracting from an off-axis surface point along the central axis and suffice to determine the position of the image in each case.

At the point of refraction of each ray, the normal to the refracting surface is a radial line through the center of curvature C. Because of the refraction, the ray bends toward the normal if it is entering a medium of greater index of refraction, and away from the normal if it is entering a medium of lesser index of refraction. If the refracted ray is then directed toward the central axis, it and other (undrawn) rays will form a real image on that axis. If it is directed away from the central axis, it cannot form a real image; however, backward extensions of it and other refracted rays can form a virtual image, provided (as with mirrors) some of those rays are intercepted by an observer.

Real images I are formed (at image distance i) in parts a and b of Fig. 35-21, where the refraction directs the ray *toward* the central axis. Virtual images are formed in parts c and d, where the refraction directs the ray *away* from the central axis. Note that, in these four parts, real images are formed when the object is relatively far from the refracting surface, and virtual images are formed when the object is nearer the refracting surface. In the final situations (Figs. 35-21e and f), refraction always directs the ray away from the central axis and virtual images are always formed, regardless of the object distance.

Note the following major difference from reflected images:

> For a single spherical refracting surface, real images form on the side of a refracting surface that is opposite the object, and virtual images form on the same side as the object.

In Section 35-12, we shall show that (for light rays making only small angles with the central axis)

$$\frac{n_1}{o} + \frac{n_2}{i} = \frac{n_2 - n_1}{r}. \tag{35-17}$$

Just as with mirrors, the object distance o is positive, and the image distance i is positive for a real image and negative for a virtual image. However, to keep all the signs correct in Eq. 35-17, we must use the following rule for the sign of the radius of curvature r:

> When the object faces a convex refracting surface, the radius of curvature r is positive. When it faces a concave surface, r is negative.

Be careful: This is just the reverse of the sign convention we have for mirrors. Figure 35-20*a* shows the case of a image formation by a mirror in which the values of o, i, r, and f are all positive. Figure 35-21*a* shows the case of image formation by a lens in which these values are all positive. Also, don't forget that we write o, which is a distance, differently than O, which is a point, and 0, which is zero. Be careful not to confuse these quantities.

READING EXERCISE 35-6: A bee is hovering in front of the concave spherical refracting surface of a glass sculpture. (a) Which of the general situations of Fig. 35-21 is like this situation? (b) Is the image produced by the surface real or virtual, and is it on the same side as the bee or the opposite side? ■

TOUCHSTONE EXAMPLE 35-4: Jurassic Mosquito

A Jurassic mosquito is discovered embedded in a chunk of amber, which has index of refraction 1.6. One surface of the amber is spherically convex with radius of curvature 3.0 mm (Fig. 35-22). The mosquito head happens to be on the central axis of that surface and, when viewed along the axis, appears to be buried 5.0 mm into the amber. How deep is it really?

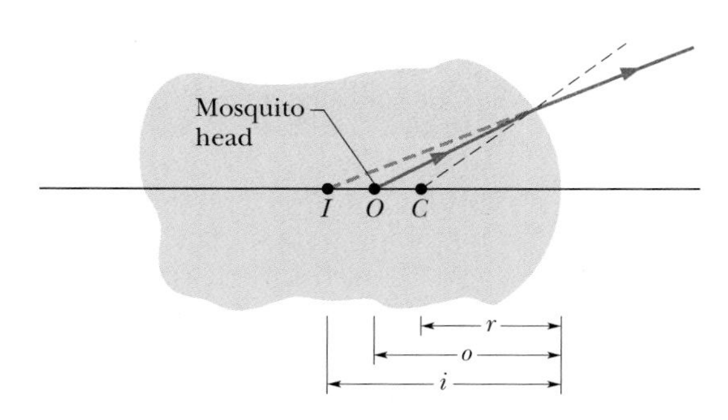

FIGURE 35-22 ■ A piece of amber with a mosquito from the Jurassic period, with the head buried at point O. The spherical refracting surface at the right end, with center of curvature C, provides an image I to an observer intercepting rays from the object at O.

SOLUTION ■ The **Key Idea** here is that the head only appears to be 5.0 mm into the amber because the light rays that the observer intercepts are bent by refraction at the convex amber surface. The image distance i differs from the actual object distance o according to Eq. 35-17. To use that equation to find the actual object distance, we first note:

1. Because the object (the head) and its image are on the same side of the refracting surface, the image must be virtual and so $i = -5.0$ mm.
2. Because the object is always taken to be in the medium of index of refraction n_1, we must have $n_1 = 1.6$ and $n_2 = 1.0$.
3. Because the object faces a concave refracting surface, the radius of curvature r is negative and so $r = -3.0$ mm.

Making these substitutions in Eq. 35-17,

$$\frac{n_1}{o} + \frac{n_2}{i} = \frac{n_2 - n_1}{r},$$

yields

$$\frac{1.6}{o} + \frac{1.0}{-5.0\text{ mm}} = \frac{1.0 - 1.6}{-3.0\text{ mm}}$$

and

$$o = 4.0\text{ mm}. \qquad \text{(Answer)}$$

35-10 Thin Lenses

A **lens** is a transparent object with two refracting surfaces whose central axes coincide. The common central axis is the central axis of the lens. When a lens is surrounded by air, light refracts from the air into the lens, crosses through the lens, and then refracts back into the air. Each refraction can change the direction of travel of the light.

A lens that causes light rays initially parallel to the central axis to converge is (reasonably) called a **converging lens.** If, instead, it causes such rays to diverge, the lens is a **diverging lens.** When an object is placed in front of a lens of either type, refraction by the lens's surface of light rays from the object can produce an image of the object.

We shall consider only the special case of a **thin lens**—that is, a lens in which the thickest part is thin compared to the object distance o, the image distance i, and the radii of curvature r_1 and r_2 of the two surfaces of the lens. We shall also consider only light rays that make small angles with the central axis (they are exaggerated in the figures here). In Section 35-12 we shall prove that for such rays, a thin lens has a focal length f. Moreover, i and o are related to each other by

$$\frac{1}{f} = \frac{1}{o} + \frac{1}{i} \quad \text{(thin lens)}, \qquad (35\text{-}18)$$

which is the same form of equation we had for mirrors. We shall also prove that when a thin lens with index of refraction n is surrounded by air, this focal length f is given by

$$\frac{1}{f} = (n - 1)\left(\frac{1}{r_1} - \frac{1}{r_2}\right) \quad \text{(thin lens in air)}, \qquad (35\text{-}19)$$

which is often called the *lens maker's equation*. Here r_1 is the radius of curvature of the lens surface nearer the object, and r_2 is that of the other surface. The signs of these radii are found with the rules in Section 35-9 for the radii of spherical refracting surfaces. If the lens is surrounded by some medium other than air (say, corn oil) with index of refraction n_{medium}, we replace n in Eq. 35-19 with n/n_{medium}. Keep in mind the basis of Eqs. 35-18 and 35-19:

> A lens can produce an image of an object only if it bends light rays, but it can bend light rays only if its index of refraction differs from that of the surrounding medium.

Figure 35-23a shows a thin lens with convex outer surfaces, or *sides*. Once again, we figure out how the rays of light will be bent by taking a small part of the lens, near where the ray hits, and treating it as flat surface. Then, using the law of refraction (Snell's law, Eq. 35-2) which tells us $n_1 \sin \theta_1 = n_2 \sin \theta_2$, we know how the rays will be bent.

When rays that are parallel to the central axis of the lens are sent through the lens, they refract twice, as is shown enlarged in Fig. 35-23b. This double refraction causes the rays to converge and pass through a common point F_2 at a distance f from the center of the lens. Hence, this lens is a converging lens; further, a *real* focal point (or focus) exists at F_2 (because the rays really do pass through it), and the associated focal length is f. When rays parallel to the central axis are sent in the opposite direction through the lens, we find another real focal point at F_1 on the other side of the lens. For a thin lens, these two focal points are equidistant from the lens.

Because the focal points of a converging lens are real, we take the associated focal lengths f to be positive, just as we do with a real focus of a concave mirror. However, signs in optics can be tricky, so we had better check this in Eq. 35-19. The left side of that equation is positive if f is positive; how about the right side? We examine it term by term. Because the index of refraction n of glass or any other material is greater than 1, the term $(n - 1)$ must be positive. Because the source of the light (which is the object) is at the left and faces the convex left side of the lens, the

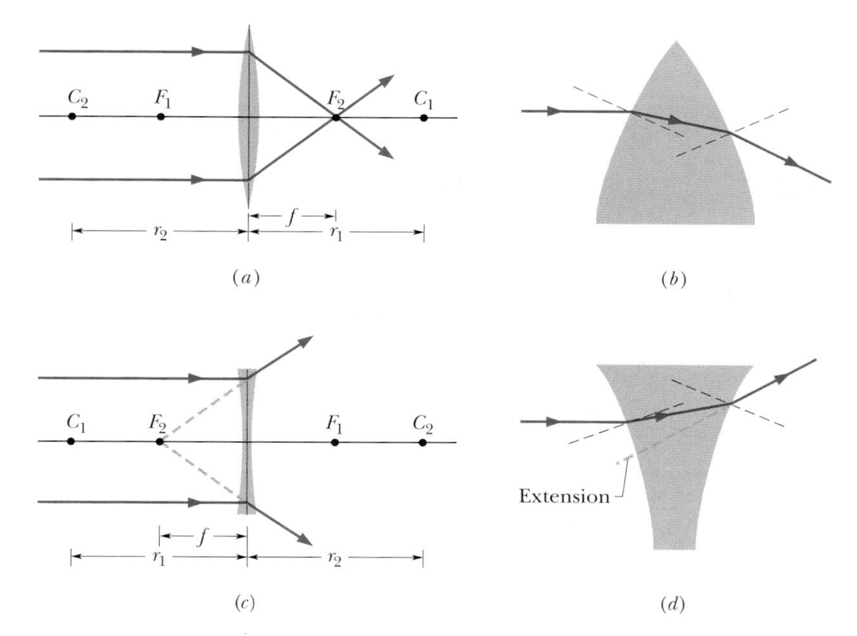

FIGURE 35-23 ■ (*a*) Rays initially parallel to the central axis of a converging lens are made to converge to a real focal point F_2 by the lens. The lens is thinner than drawn, with a width like that of the vertical line through it, where we shall consider all the bending of rays to occur. (*b*) An enlargement of the top part of the lens of (*a*); normals to the surfaces are shown dashed. Note that both refractions of the ray at the surfaces bend the ray downward, toward the central axis. (*c*) The same initially parallel rays are made to diverge by a diverging lens. Extensions of the diverging rays pass through a virtual focal point F_2. (*d*) An enlargement of the top part of the lens of (*c*). Note that both refractions of the ray at the surfaces bend the ray upward, away from the central axis.

radius of curvature r_1 of that side must be positive according to the sign rule for refracting surfaces. Similarly, because the object faces a concave right side of the lens, the radius of curvature r_2 of that side must be negative. Thus, the term $(1/r_1 - 1/r_2)$ is positive, the whole right side of Eq. 35-19 is positive, and all the signs are consistent.

Figure 35-23*c* shows a thin lens with concave outer surfaces. When rays that are parallel to the central axis of the lens are sent through this lens, they refract twice, as is shown enlarged in Fig. 35-23*d*; these rays *diverge,* never passing through any common point, and so this lens is a diverging lens. However, extensions of the rays do pass through a common point F_2 at a distance f from the center of the lens. Hence, the lens has a *virtual* focal point at F_2. (If your eye intercepts some of the diverging rays, you perceive a bright spot to be at F_2, as if it is the source of the light.) Another virtual focus exists on the opposite side of the lens at F_1, symmetrically placed if the lens is thin. Because the focal points of a diverging lens are virtual, we take the focal length f to be negative. *Note:* For a thin lens, the two focal points are at the same distance from the lens on either side even if the curvatures of the two sides are not equal and opposite.

A fire is being started by focusing sunlight onto newspaper by means of a converging lens made of clear ice. The lens was made by freezing water in the shallow vessel (which has a curved bottom).

Images from Thin Lenses

We now consider the types of image formed by converging and diverging lenses. In thinking about the image formed of an object by a lens, it is important to keep in mind that the object scatters light falling on it in all directions. However, only those rays of light that fall on the lens are refracted and have their directions changed. Thus, all of the rays that are incident on the lens contribute to the image that forms.*

Figure 35-24*a* shows an object O outside the focal point F_1 of a converging lens. The two rays drawn in the figure show that the lens forms a real, inverted image I of the object on the side of the lens opposite the object.

* In Section 35-8, we use Eq. 35-12 to construct diagrams (Figs. 35-19 and 35-20) for spherical mirrors in which the center curvature of the mirror is replaced by its focal point. We can construct similar diagrams for thin lenses using Eq. 35-19 to determine the locations of F based on the value of f.

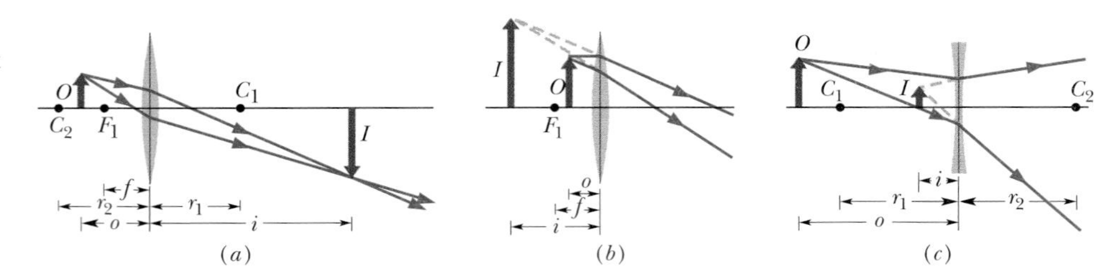

FIGURE 35-24 ■ (*a*) A real, inverted image I is formed by a converging lens when the object O is outside the focal point F_1. (*b*) The image I is virtual and has the same orientation as O when O is inside the focal point. (*c*) A diverging lens forms a virtual image I, with the same orientation as the object O, whether O is inside or outside the focal point of the lens.

When the object is placed inside the focal point F_1, as in Fig. 35-24*b*, the lens forms a virtual image I on the same side of the lens as the object and with the same orientation but larger in size. In this situation, the lens acts as a magnifying glass. Hence, a converging lens can form either a real image or a virtual image, depending on whether the object is outside or inside the focal point, respectively.

Figure 35-24*c* shows an object O in front of a diverging lens. Regardless of the object distance (regardless of whether O is inside or outside the virtual focal point), this lens produces a virtual image that is on the same side of the lens as the object and has the same orientation.

As with mirrors, we take the image distance i to be positive when the image is real and negative when the image is virtual. However, the locations of real and virtual images from lenses are the reverse of those from mirrors:

> Real images form on the side of a lens that is opposite the object, and virtual images form on the side where the object is.

The lateral magnification m produced by converging and diverging lenses is given by Eqs. 35-14 and 35-15, the same as for mirrors.

You have been asked to absorb a lot of information in this section, and you should organize it for yourself by filling in Table 35-3 for thin lenses. Under Image Location note whether the image is on the *same* side of the lens as the object or on the *opposite* side. Under Image Type note whether the image is *real* or *virtual*. Under Image Orientation note whether the image has the same orientation as the object or is inverted.

TABLE 35-3
Organizing Table for Lenses

Lens Type	Object Location	Image			Sign		
		Location	Type	Orientation	of f	of i	of m
Converging	Inside F						
	Outside F						
Diverging	Anywhere						

Locating Images of Extended Objects with Principal Rays

Converging Lens: Object Outside of F_1 Figure 35-25*a* shows an object O outside focal point F_1 of a converging lens. We can graphically locate the image of any off-axis point on such an object (such as the tip of the arrow in Fig. 35-25*a*) by drawing a ray diagram with any two of three easy-to-draw principal rays through the point. These principal rays are chosen for convenience from the infinite number of rays that pass through the lens to form the image:

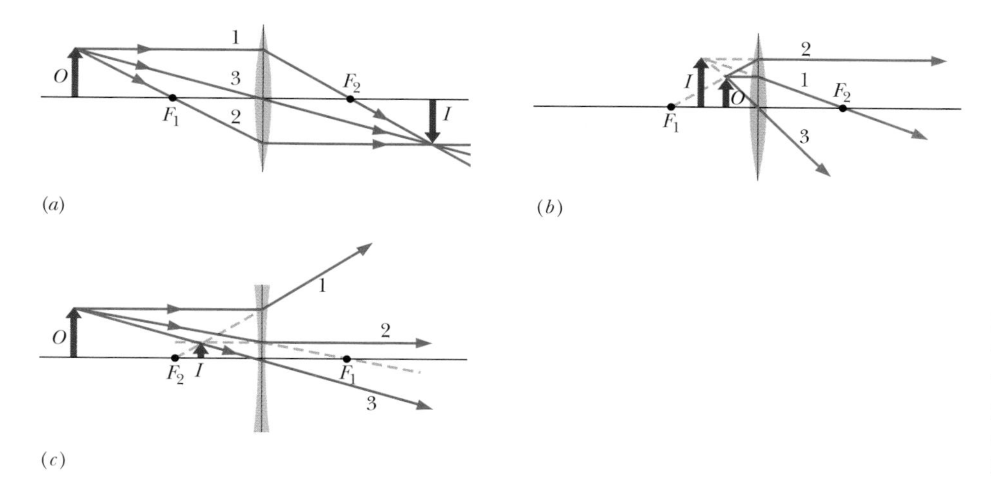

FIGURE 35-25 ■ Three special rays allow us to locate an image formed by a thin lens whether the object O is (*a*) outside or (*b*) inside the focal point of a converging lens, or (*c*) anywhere in front of a diverging lens.

1. A ray (from outside the focal point) that is initially parallel to the central axis of a converging lens will pass through focal point F_2 (ray 1 in Fig. 35-25*a*).
2. A ray that initially passes through focal point F_1 will emerge from a converging lens parallel to the central axis (ray 2 in Fig. 35-25*a*).
3. A ray that is initially directed toward the center of a converging lens will emerge from the lens with no change in its direction (ray 3 in Fig. 35-25*a*) because the ray encounters the two sides of the lens where they are almost parallel.

The image of the point is located where the rays intersect on the far side of the converging lens. The image of the object is found by locating the images of two or more of its points.

Rules for Other Situations Figure 35-25*b* shows how the extensions of the three special rays can be used to locate the image of an object placed inside focal point F_1 of a converging lens. Notice that although ray 2 is determined by the focal point, no part of the ray goes through it. It is the reversal of ray 1—what ray 1 would do if it came from the other side of the lens. We suggest you develop rules like these three for Fig. 35-25*a* and 35-25*b*.

You need to modify the descriptions of rays 1 and 2 to use them to locate an image placed (anywhere) in front of a diverging lens. In Fig. 35-25*c*, for example, we find the intersection of ray 3 and the backward extensions of rays 1 and 2.

Two-Lens Systems

When an object O is placed in front of a system of two lenses whose central axes coincide, we can locate the final image of the system (that is, the image produced by the lens farther from the object) by working in steps. Let lens 1 be the nearer lens and lens 2 the farther lens.

Step 1. We let o_1 represent the distance of object O from lens 1. We then find the distance i_1 of the image produced by lens 1, by use of Eq. 35-18. The image could be real or virtual.

Step 2. Now, ignoring the presence of lens 1, we treat the image found in step 1 *as the object* for lens 2. If this new object is located beyond lens 2, the object distance o_2 for lens 2 is taken to be negative. (Note this exception to the rule that says the object distance is positive; the exception occurs because the object here is on the side opposite the source of light.) Otherwise, o_2 is taken to be positive as usual.

We then find the distance i_2 of the (final) image produced by lens 2 by use of Eq. 35-18.

A similar step-by-step solution can be used for any number of lenses or if a mirror is substituted for lens 2.

The overall lateral magnification M produced by a system of two lenses is the product of the lateral magnifications m_1 and m_2 produced by the two lenses:

$$M = m_1 m_2. \tag{35-20}$$

TOUCHSTONE EXAMPLE 35-5: Praying Mantis

A praying mantis preys along the central axis of a thin symmetric lens, 20 cm from the lens. The lateral magnification of the mantis provided by the lens is $m = -0.25$, and the index of refraction of the lens material is 1.65.

(a) Determine the type of image produced by the lens, the type of lens, whether the object (mantis) is inside or outside the focal point, on which side of the lens the image appears, and whether the image is inverted.

SOLUTION ■ The **Key Idea** here is that we can tell a lot about the lens and the image from the given value of m. From it and Eq. 35-16 ($m = -i/o$), we see that

$$i = -(m)(o) = 0.25o.$$

Even without finishing the calculation, we can answer the questions. Because o is positive, i here must be positive. That means we have a real image, which means we have a converging lens (the only lens that can by itself produce a real image). The object must be outside the focal point (the only way a real image can be produced). Also, the image is inverted and on the side of the lens opposite the object. (That is how a converging lens makes a real image.)

(b) What are the two radii of curvature of the lens?

SOLUTION ■ The **Key Ideas** here are these:

1. Because the lens is symmetric, r_1 (for the surface nearer the object) and r_2 have the same magnitude r.
2. Because the lens is a converging lens, the object faces a convex surface on the nearer side and so $r_1 = +r$. Similarly, it faces a concave surface on the farther side and so $r_2 = -r$.
3. We can relate these radii of curvature to the focal length f via the lens maker's equation, Eq. 35-19 (our only equation involving the radii of curvature of a lens).
4. We can relate f to the object distance o and image distance i via Eq. 35-18.

We know o but we do not know i. Thus, our starting point is to finish the calculation for i in part (a); we obtain

$$i = (0.25)(20\text{ cm}) = 5.0\text{ cm}.$$

Now Eq. 35-18 gives us

$$\frac{1}{f} = \frac{1}{o} + \frac{1}{i} = \frac{1}{20\text{ cm}} + \frac{1}{5.0\text{ cm}},$$

from which we find $f = 4.0$ cm.

Equation 35-19 then gives us

$$\frac{1}{f} = (n-1)\left(\frac{1}{r_1} - \frac{1}{r_2}\right) = (n-1)\left(\frac{1}{+r} - \frac{1}{-r}\right)$$

or, with known values inserted,

$$\frac{1}{4.0\text{ cm}} = (1.65 - 1)\frac{2}{r},$$

which yields

$$r = (0.65)(2)(4.0\text{ cm}) = 5.2\text{ cm}. \quad \text{(Answer)}$$

TOUCHSTONE EXAMPLE 35-6: Jalapeño Seed

Figure 35-26a shows a jalapeño seed O_1 that is placed in front of two thin symmetrical coaxial lenses 1 and 2, with focal lengths $f_1 = +24$ cm and $f_2 = +9.0$ cm, respectively, and with lens separation $L = 10$ cm. The seed is 6.0 cm from lens 1. Where does the system of two lenses produce an image of the seed?

SOLUTION ■ We could locate the image produced by the system of lenses by tracing light rays from the seed through the two lenses. However, the **Key Idea** here is that we can, instead, calculate the location of that image by working through the system in steps, lens by lens. We begin with the lens closer to the

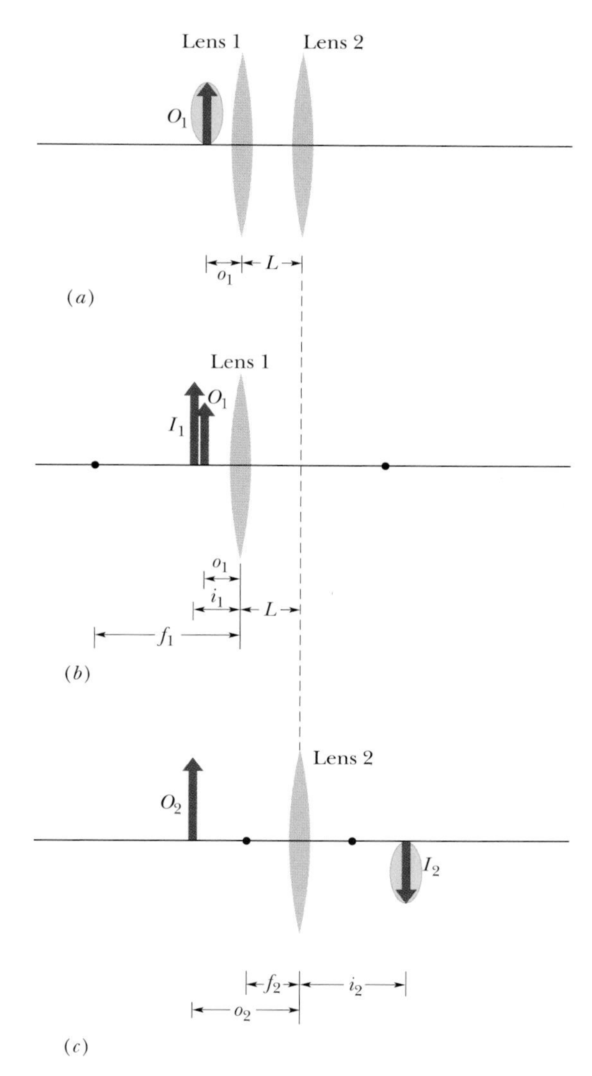

FIGURE 35-26 ■ (*a*) Seed O_1 is distance o_1 from a two-lens system with lens separation L. We use the arrow to orient the seed. (*b*) The image I_1 produced by lens 1 alone. (*c*) Image I_1 acts as object O_2 for lens 2 alone, which produces the final image I_2.

seed. The image we seek is the final one—that is, image I_2 produced by lens 2.

Lens 1. Ignoring lens 2, we locate the image I_1 produced by lens 1 by applying Eq. 35-18 to lens 1 alone:

$$\frac{1}{o_1} + \frac{1}{i_1} = \frac{1}{f_1}.$$

The object O_1 for lens 1 is the seed, which is 6.0 cm from the lens; thus, we substitute $o_1 = +6.0$ cm. Also substituting the given value of f_1, we then have

$$\frac{1}{+6.0 \text{ cm}} + \frac{1}{i_1} = \frac{1}{+24 \text{ cm}},$$

which yields $i_1 = -8.0$ cm.

This tells us that image I_1 is 8.0 cm to the left of lens 1 and virtual. (We could have guessed that it is virtual by noting that the seed is inside the focal point of lens 1.) Since I_1 is virtual, it is on the same side of the lens as object O_1 and has the same orientation as the seed, as shown in Fig. 35-26*b*.

Lens 2. In the second step of our solution, the **Key Idea** is that we can treat image I_1 as an object O_2 for the second lens and now ignore lens 1. We first note that this object O_2 is outside the focal point of lens 2. So the image I_2 produced by lens 2 must be real, inverted, and on the side of the lens opposite O_2. Let us see.

The distance o_2 between this object O_2 and lens 2 is, from Fig. 35-26*c*,

$$o_2 = L + |i_1| = 10 \text{ cm} + 8.0 \text{ cm} = 18 \text{ cm}.$$

Then Eq. 35-18, now written for lens 2, yields

$$\frac{1}{+18 \text{ cm}} + \frac{1}{i_2} = \frac{1}{+9.0 \text{ cm}}.$$

Hence,

$$i_2 = +18 \text{ cm}. \quad \text{(Answer)}$$

The plus sign confirms our guess: Image I_2 produced by lens 2 is real, inverted, and on the side of lens 2 opposite O_2, as shown in Fig. 35-26*c*.

35-11 Optical Instruments

The human eye is a remarkably effective organ, but its range can be extended in many ways by optical instruments such as eyeglasses, simple magnifying lenses, motion picture projectors, cameras (including TV cameras), microscopes, and telescopes. Many such devices extend the scope of our vision beyond the visible range; satellite-borne infrared cameras, x-ray microscopes, and radio telescopes are examples. Furthermore, electron microscopes and magnets in particle accelerators can focus electron and proton beams.

The mirror and thin-lens formulas can be applied only as approximations to most sophisticated optical instruments. The lenses in typical laboratory microscopes are by no means "thin." In most optical instruments the lenses are compound lenses; that is, they are made of several components, the interfaces rarely being exactly spherical. A more complex treatment is needed for these systems. Now we discuss three optical instruments, assuming for simplicity, that the thin-lens formulas apply.

Simple Magnifying Lens

The normal human eye can focus a sharp image of an object on the retina (at the rear of the eye) if the object is located anywhere from infinity to a certain point called the *near point* P_n. If you move the object closer to the eye than the near point, the perceived retinal image becomes fuzzy. The location of the near point normally varies with age. We have all heard about people who claim not to need glasses but read their newspapers at arm's length; their near points are receding. To find your own near point, remove your glasses or contacts if you wear any, close one eye, and then bring this page closer to your open eye until it becomes indistinct. In what follows, we take the near point to be 25 cm from the eye, a bit more than the typical value for 20-year-olds.

Figure 35-27*a* shows an object O placed at the near point P_n of an eye. The size of the image of the object produced on the retina depends on the angle θ that the object occupies in the field of view from that eye. By moving the object closer to the eye, as in Fig. 35-27*b*, you can increase the angle and, hence, the possibility of distinguishing details of the object. However, because the object is then closer than the near point, it is no longer in focus; that is, the image is no longer clear.

You can restore the clarity by looking at O through a converging lens, placed so that O is just inside the focal point F_1 of the lens, which is at focal length f (Fig. 35-27*c*). What you then see is the virtual image of O produced by the lens. That image is farther away than the near point; thus, the eye can see it clearly.

Moreover, the angle θ' occupied by the virtual image is larger than the largest angle θ that the object alone can occupy and still be seen clearly. The *angular magnification* m_θ (not to be confused with lateral magnification m) of what is seen is

$$m_\theta = \theta'/\theta.$$

In words, the angular magnification of a simple magnifying lens is a comparison of the angle occupied by the image the lens produces with the angle occupied by the object when the object is moved to the near point of the viewer.

From Fig. 35-27, assuming that O is at the focal point of the lens, and approximating $\tan\theta$ as θ and $\tan\theta'$ as θ' for small angles, we have

$$\theta \approx h/25 \text{ cm} \quad \text{and} \quad \theta' \approx h/f.$$

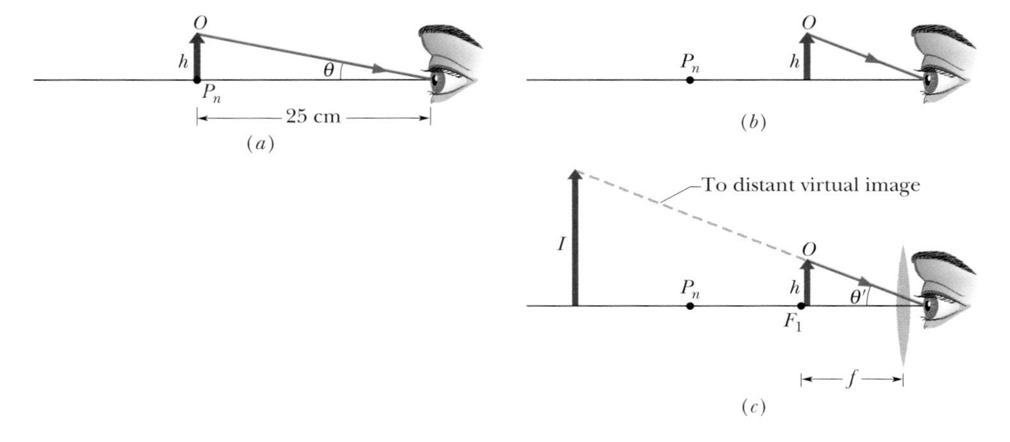

FIGURE 35-27 ■ (*a*) An object O of height h, placed at the near point of a human eye, occupies angle θ in the eye's view. (*b*) The object is moved closer to increase the angle, but now the observer cannot bring the object into focus. (*c*) A converging lens is placed between the object and the eye, with the object just inside the focal point F_1 of the lens. The image produced by the lens is then far enough away to be focused by the eye, and the image occupies a larger angle θ' than object O does in (*a*).

We then find that

$$m_\theta \approx \frac{25\text{ cm}}{f} \quad \text{(maximum magnification).} \tag{35-21}$$

Note that this is an equation for a person with a near point of 25 cm. It is an example of how the study of optics depends on the observer. That is, the value of the magnification depends not only on the lens but also on the person using the lens.

Compound Microscope

Figure 35-28 shows a thin-lens version of a compound microscope. The instrument consists of an *objective* (the front lens) of focal length f_{obj} and an *eyepiece* (the lens near the eye) of focal length f_{eye}. It is used for viewing small objects that are very close to the objective.

The object O to be viewed is placed just outside the first focal point F_{obj} of the objective lens, close enough to F_{obj} that we can approximate its distance o from the lens as being f_{obj}. Combining this with a consideration of the thin-lens equation

$$\frac{1}{f} = \frac{1}{o} + \frac{1}{i} \quad \text{(Eq. 35-18)}$$

when an object under a microscope is very close to the focal point, we see that object distance o is approximately equal to focal length of the objective lens f_{obj}, so $1/f_{\text{obj}} - 1/o$ is very close to zero. As a result, the image distance i is very sensitive to *exactly how close* the object is to the focal point.

The separation between the lenses is then adjusted so that the enlarged, inverted, real image I produced by the objective is located just inside the first focal point F_1' of the eyepiece. The *tube length* s shown in Fig. 35-28 is actually large relative to f_{obj}, and we can approximate the distance i between the objective and the image I as being length s.

From Eq. 35-15, and using our approximations for o and i, we can write the lateral magnification produced by the objective as

$$m_{\text{obj}} = -\frac{i}{o} = -\frac{s}{f_{\text{obj}}}. \tag{35-22}$$

Since the image I is located just inside the focal point F_1' of the eyepiece, the eyepiece acts as a simple magnifying lens, and an observer sees a final (virtual, inverted) image I' through it. The overall magnification of the instrument is the product of the lateral magnification m_{obj} produced by the objective (Eq. 35-22) and the angular magnification produced by the eyepiece $m_{\text{eye}} = m_\theta$ (Eq. 35-21) so that

$$M = m_{\text{obj}} m_{\text{eye}} = -\frac{s}{f_{\text{obj}}}\frac{25\text{ cm}}{f_{\text{eye}}} \quad \text{(microscope magnification).} \tag{35-23}$$

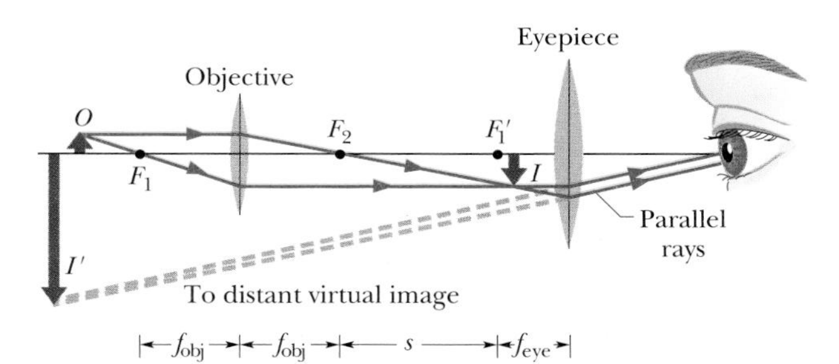

FIGURE 35-28 ■ A thin-lens representation of a compound microscope (not to scale). The objective produces a real image I of object O just inside the focal point F_1' of the eyepiece. Image I then acts as an object for the eyepiece, which produces a virtual final image I' that is seen by the observer. The objective has focal length f_{obj}; the eyepiece has focal length f_{eye}; and s is the tube length.

The microscope designer must also take into account the difference between real lenses and the ideal thin lenses we have discussed. A real lens with spherical surfaces does not form sharp images, a flaw called **spherical aberration.** Also, because refraction by the two surfaces of a real lens depends on wavelength, a real lens does not focus light of different wavelengths to the same point, a flaw called **chromatic aberration.**

Refracting Telescope

Telescopes come in a variety of forms. The form we describe here is the simple refracting telescope that consists of an objective and an eyepiece; both are represented in Fig. 35-29 with simple lenses, although in practice, as is also true for most microscopes, each lens is usually a compound-lens system to reduce distortions.

The lens arrangements for telescopes and for microscopes are similar, but telescopes are designed to view large objects, such as galaxies, stars, and planets, at large distances, whereas microscopes are designed for just the opposite purpose. This difference requires that in the telescope of Fig. 35-29 the second focal point of the objective F_2 coincide with the first focal point of the eyepiece F_1', whereas in the microscope of Fig. 35-28 these points are separated by the tube length s.

In Fig. 35-29*a*, parallel rays from a distant object strike the objective, making an angle θ_{obj} with the telescope axis and forming a real, inverted image at the common focal point F_2, F_1'. This image I acts as an object for the eyepiece, through which an observer sees a distant (still inverted) virtual image I'. The rays defining the image make an angle θ_{eye} with the telescope axis.

The angular magnification m_θ of the telescope is $\theta_{eye}/\theta_{obj}$. From Fig. 35-29*b*, for rays close to the central axis, we can write $\theta_{obj} = h'/f_{obj}$ and $\theta_{eye} \approx h'/f_{eye}$, which gives us

$$m_\theta = -\frac{f_{obj}}{f_{eye}} \quad \text{(telescope)}, \tag{35-24}$$

where the minus sign indicates that I' is inverted. In words, the angular magnification of a telescope is a comparison of the angle occupied by the image the telescope produces with the angle occupied by the distant object as seen without the telescope.

Magnification is only one of the design factors for an astronomical telescope and is indeed easily achieved. A good telescope needs light-gathering power, which determines how bright the image is. This is important for viewing faint objects such as distant galaxies and is accomplished by making the objective diameter as large as possible. A telescope also needs resolving power, which is the ability to distinguish

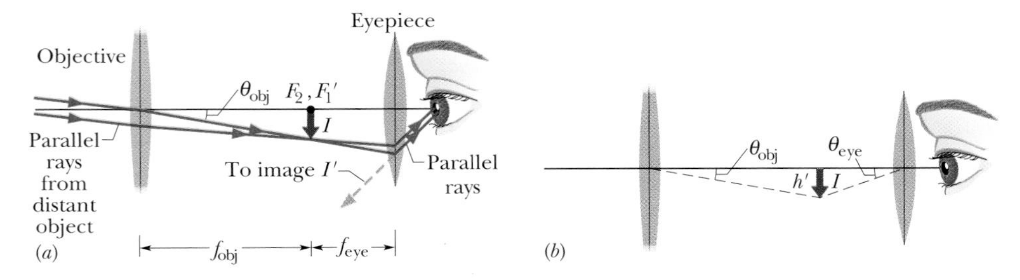

FIGURE 35-29 ■ (*a*) A thin-lens representation of a refracting telescope. The objective produces a real image I of a distant source of light (the object), with approximately parallel light rays at the objective. (One end of the object is assumed to lie on the central axis.) Image I, formed at the common focal points F_2 and F_1', acts as an object for the eyepiece, which produces a virtual final image I' at a great distance from the observer. The objective has focal length f_{obj}; the eyepiece has focal length f_{eye}. (*b*) Image I has height h' and takes up angle θ_{obj} measured from the objective and angle θ_{eye} measured from the eyepiece.

between two distant objects (stars, say) whose angular separation is small. Field of view is another important design parameter. A telescope designed to look at galaxies (which occupy a tiny field of view) is much different from one designed to track meteors (which move over a wide field of view).

The telescope designer must also take into account the differences between real lenses and the ideal thin lenses we have discussed. Designers use compound lens systems to minimize spherical and chromatic aberrations.

This brief discussion by no means exhausts the design parameters of astronomical telescopes—many others are involved. We could make a similar listing for any other high-performance optical instrument.

35-12 Three Proofs

The Spherical Mirror Formula (Eq. 35-13)

Let us prove that Eq. 35-13, $1/i + 1/o = 1/f$ is true for a spherical mirror. Figure 35-30 shows a point object O placed on the central axis of a concave spherical mirror, outside its center of curvature C. Here, we use the reflection principle that comes from treating the mirror as approximately flat near where the ray hits. A ray from O that makes an angle α with the axis intersects the axis at I after reflection from the mirror at a. A ray that leaves O along the axis is reflected back along itself at c and also passes through I. Thus, I is the image of O; it is a real image because light actually passes through it. Let us find the image distance i.

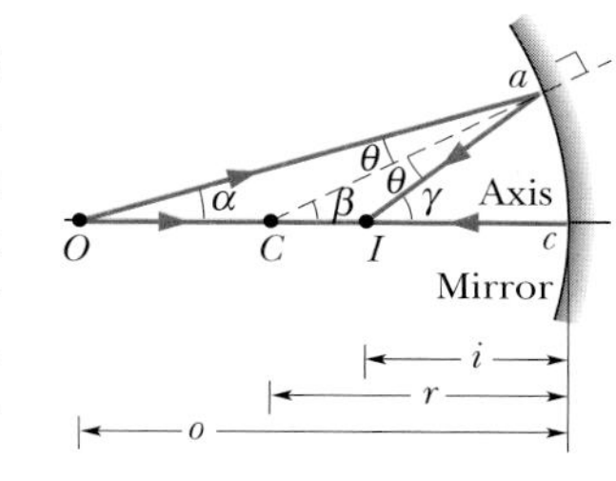

FIGURE 35-30 ■ A concave spherical mirror forms a real point image I by reflecting light rays from a point object O.

A trigonometry theorem that is useful here tells us that an exterior angle of a triangle is equal to the sum of the two opposite interior angles. Applying this to triangles OaC and OaI in Fig. 35-30 yields

$$\beta = \alpha + \theta \quad \text{and} \quad \gamma = \alpha + 2\theta.$$

If we eliminate θ between these two equations, we find

$$\alpha + \gamma = 2\beta. \tag{35-25}$$

We can write angles α, β, and γ, in radian measure, as

$$\alpha \approx \frac{\widehat{ac}}{cO} = \frac{\widehat{ac}}{o}, \qquad \beta = \frac{\widehat{ac}}{cC} = \frac{\widehat{ac}}{r},$$

and

$$\gamma \approx \frac{\widehat{ac}}{cI} = \frac{\widehat{ac}}{i}. \tag{35-26}$$

Only the equation for β is exact, because the center of curvature of arc ac is at C. Here, $\widehat{ac}$ is the arc extending from the point that the ray reflects from the mirror to the central axis. However, the equations for α and γ are approximately correct if these angles are small enough (that is, for rays close to the central axis). Substituting Eqs. 35-26 into Eq. 35-25, using Eq. 35-12 to replace r with $2f$, and canceling $\widehat{ac}$ lead exactly to Eq. 35-13, the relation that we set out to prove.

The Refracting Surface Formula (Eq. 35-17)

Next we prove that when light from an object passes through a medium having an index of refraction n_1 and encounters a smooth spherical refracting surface of a

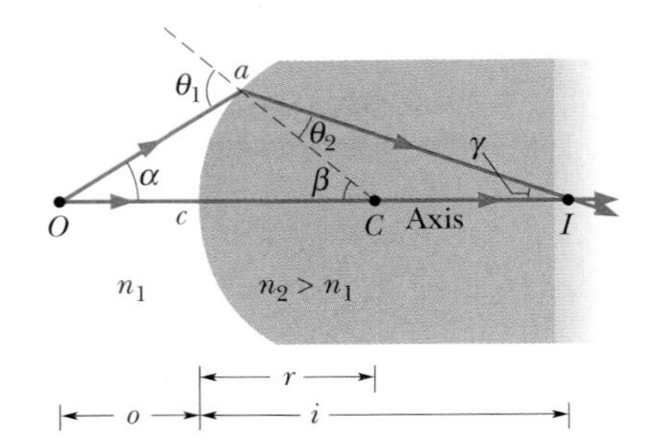

FIGURE 35-31 ■ A real point image I of a point object O is formed by refraction at a spherical convex surface between two media.

medium with an index of refraction n_2, the image and object distances are related by Eq. 35-17,

$$\frac{n_1}{o} + \frac{n_2}{i} = \frac{n_1 - n_2}{r}.$$

Let's start by considering how an incident ray from point object O in Fig. 35-31 is refracted. According to Eq. 35-2, $n_1 \sin\theta_1 = n_2 \sin\theta_2$.

If α is small, θ_1 and θ_2 will also be small and we can replace the sines of these angles with the angles themselves. Thus, the equation above becomes

$$n_1\theta_1 \approx n_2\theta_2. \tag{35-27}$$

We again use the fact that an exterior angle of a triangle is equal to the sum of the two opposite interior angles. Applying this to triangles COa and ICa yields

$$\theta_1 = \alpha + \beta \quad \text{and} \quad \beta = \theta_2 + \gamma. \tag{35-28}$$

If we use Eqs. 35-28 to eliminate θ_1 and θ_2 from Eq. 35-27, we find

$$n_1\alpha + n_2\gamma = (n_2 - n_1)\beta. \tag{35-29}$$

In radian measure the angles α, β, and γ, are

$$\alpha \approx \frac{\widehat{ac}}{o}; \qquad \beta = \frac{\widehat{ac}}{r}; \qquad \gamma \approx \frac{\widehat{ac}}{i}. \tag{35-30}$$

Only the second of these equations is exact. The other two are approximate because I and O are not the centers of circles of which ac is a part. However, for α small enough (for rays close to the axis), the inaccuracies in Eqs. 35-30 are small. Substituting Eqs. 35-30 into Eq. 35-29 leads directly to Eq. 35-17, the relation we set out to prove.

The Thin-Lens Formulas (Eqs. 35-18 and 35-19)

Finally, we set out to show that for a thin lens, the object and image distances are related to the focal length of the lens by Eq. 35-18, $1/o + 1/i = 1/f$. Our plan is to consider each lens surface as a separate refracting surface, and to use the image formed by the first surface as the object for the second.

We start with the thick glass "lens" of length L in Fig. 35-32a whose left and right refracting surfaces are ground to radii r' and r''. A point object O' is placed near the left surface as shown. A ray leaving O' along the central axis is not deflected on entering or leaving the lens.

A second ray leaving O' at an angle α with the central axis intersects the left surface at point a', is refracted, and intersects the second (right) surface at point a''. The ray is again refracted and crosses the axis at I'', which, being the intersection of two rays from O', is the image of point O', formed after refraction at two surfaces.

Figure 35-32b shows that the first (left) surface also forms a virtual image of O' at I'. To locate I', we use Eq. 35-17,

$$\frac{n_1}{o'} + \frac{n_2}{i'} = \frac{n_2 - n_1}{r'}.$$

Putting $n_1 = 1$ for air and $n_2 = n$ for lens glass and bearing in mind that the image distance is negative (that is, $i = -i'$ in Fig. 35-32b), we obtain

$$\frac{1}{o'} - \frac{n}{|i'|} = \frac{n-1}{r'}. \tag{35-31}$$

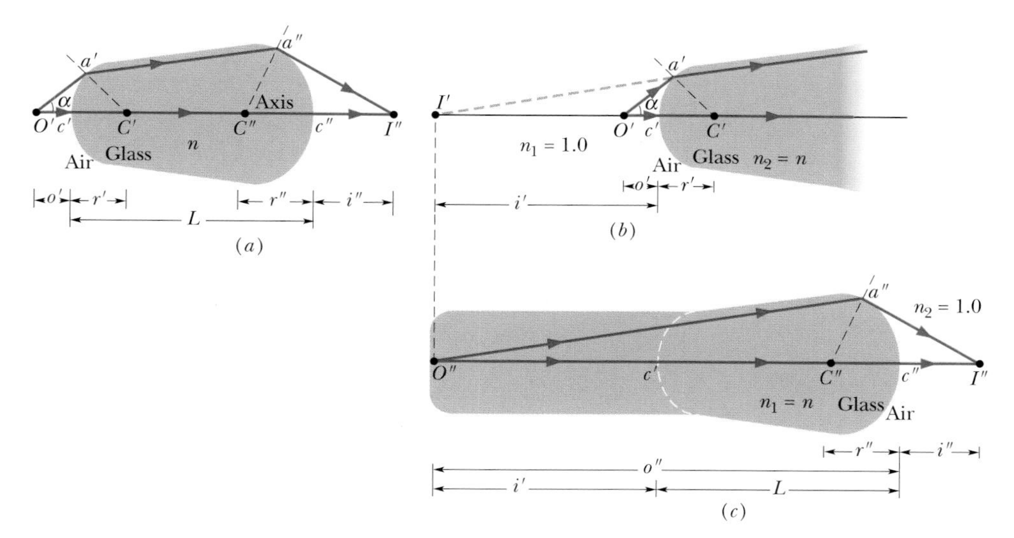

FIGURE 35-32 ■ (a) Two rays from point object O' form a real image I'' after refracting through two spherical surfaces of a "lens." The object faces a convex surface at the left side of the lens and a concave surface at the right side. The ray traveling through points a' and a'' is actually close to the central axis through the lens. (b) The left side and (c) the right side of the "lens" in (a), shown separately.

Figure 35-32c shows the second surface again. Unless an observer at point a'' were aware of the existence of the first surface, the observer would think that the light striking that point originated at point I' in Fig. 35-32b and that the region to the left of the surface was filled with glass as indicated. Thus, the (virtual) image I' formed by the first surface serves as a real object O'' for the second surface. The distance of this object from the second surface is

$$o'' = |i'| + L. \tag{35-32}$$

To apply Eq. 35-17 to the second surface, we must insert $n_1 = n$ and $n_2 = 1$ because the object now is effectively imbedded in glass. If we substitute with Eq. 35-31, then Eq. 35-17 becomes

$$\frac{n}{|i'| + L} + \frac{1}{i''} = \frac{1 - n}{r''}. \tag{35-33}$$

Let us now assume that the thickness L of the "lens" in Fig. 35-32a is so small that we can neglect it in comparison with our other linear quantities (such as o', i', o'', i'', r', and r''). In all that follows we make this *thin-lens approximation*. Putting $L = 0$ in Eq. 35-33 and rearranging the right side lead to

$$\frac{n}{i'} + \frac{1}{i''} = -\frac{n - 1}{r''}. \tag{35-34}$$

Adding Eqs. 35-31 and 35-34 leads to

$$\frac{1}{o'} + \frac{1}{i''} = (n - 1)\left(\frac{1}{r'} - \frac{1}{r''}\right).$$

Finally, calling the original object distance simply o and the final image distance simply i leads to

$$\frac{1}{o} + \frac{1}{i} = (n - 1)\left(\frac{1}{r'} - \frac{1}{r''}\right), \tag{35-35}$$

which, with $r' \equiv r_1$ and $r'' = r_2$, reduces to Eqs. 35-18 and 35-19, the relations we set out to prove.

Problems

SEC. 35-2 ■ REFLECTION AND REFRACTION

1. Light in a Vacuum Light in vacuum is incident on the surface of a glass slab. In the vacuum the beam makes an angle of 32.0° with the normal to the surface, while in the glass it makes an angle of 21.0° with the normal. What is the index of refraction of the glass?

2. Two Perpendicular Surfaces Figure 35-33 shows light reflecting from two perpendicular reflecting surfaces A and B. Find the angle between the incoming ray i and the outgoing ray r'.

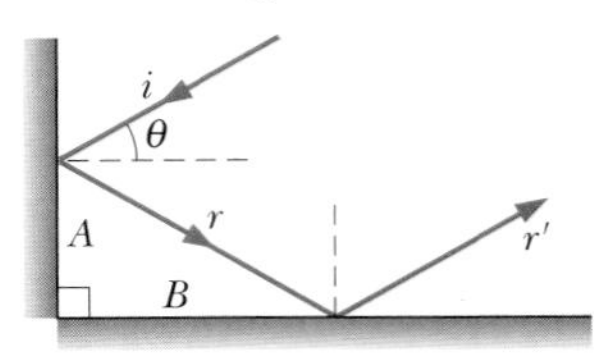

FIGURE 35-33 ■ Problem 2.

3. Rectangular Metal Tank When the rectangular metal tank in Fig. 35-34 is filled to the top with an unknown liquid, an observer with eyes level with the top of the tank can just see the corner E; a ray that refracts toward the observer at the top surface of the liquid is shown. Find the index of refraction of the liquid.

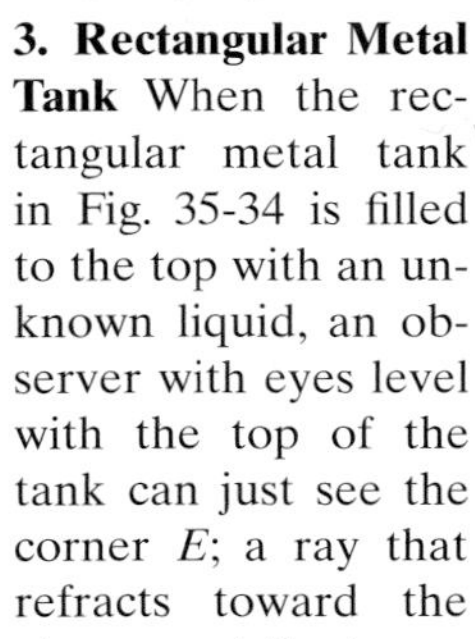

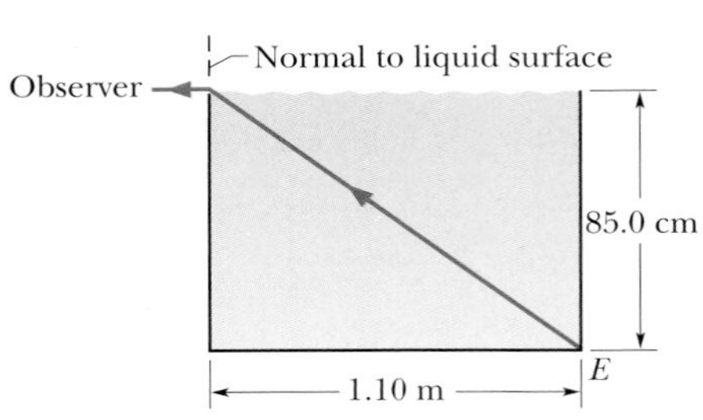

FIGURE 35-34 ■ Problem 3.

4. Claudius Ptolemy In about A.D. 150, Claudius Ptolemy gave the following measured values for the angle of incidence θ_1 and the angle of refraction θ_2 for a light beam passing from air to water:

θ_1	θ_2	θ_1	θ_2
10°	8°00′	50°	35°00′
20°	15°30′	60°	45°30′
30°	22°30′	70°	45°30′
40°	29°00′	80°	50°00′

(a) Are these data consistent with the law of refraction? (b) If so, what index of refraction results? These data are interesting as perhaps the oldest recorded physical measurements.

5. Vertical Pole In Fig. 35-35, a 2.00-m-long vertical pole extends from the bottom of a swimming pool to a point 50.0 cm above the water. Sunlight is incident at 55.0° above the horizon. What is the length of the shadow of the pole on the level bottom of the pool?

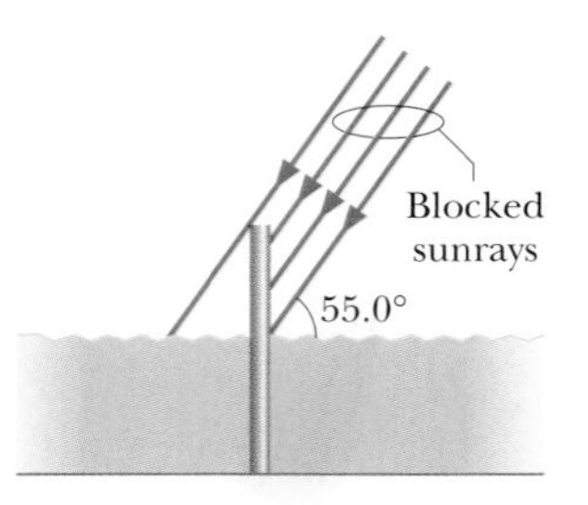

FIGURE 35-35 ■ Problem 5.

6. Four Transparent Materials In Fig. 35-36, light is incident at angle $\theta_1 = 40.1°$ on a boundary between two transparent materials. Some of the light then travels down through the next three layers of transparent materials, while some of it reflects upward and then escapes into the air. What are the values of (a) θ_5 and (b) θ_4?

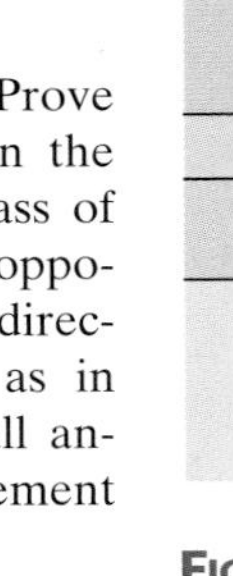

FIGURE 35-36 ■ Problem 6.

7. Sideways Displacement Prove that a ray of light incident on the surface of a sheet of plate glass of thickness t emerges from the opposite face parallel to its initial direction but displaced sideways, as in Fig. 35-37. Show that, for small angles of incidence θ, this displacement is given by

$$x = t\theta\frac{n-1}{n},$$

where n is the index of refraction of the glass and θ is measured in radians.

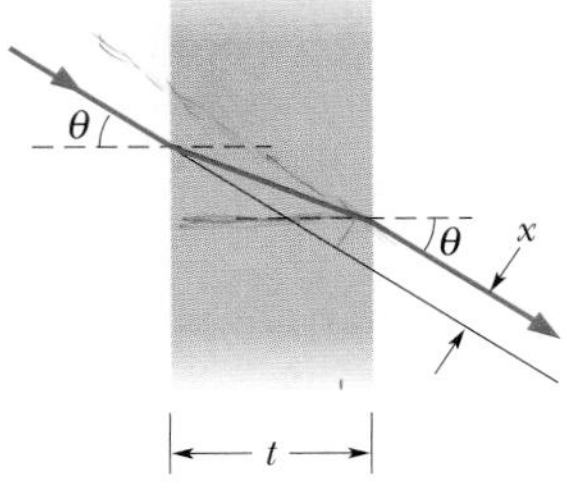

FIGURE 35-37 ■ Problem 7.

8. White Light A ray of white light makes an angle of incidence of 35° on one face of a prism of fused quartz; the prism's cross section is an equilateral triangle. Sketch the light as it passes through the prism, showing the paths traveled by rays representing (a) blue light, (b) yellow-green light, and (c) red light.

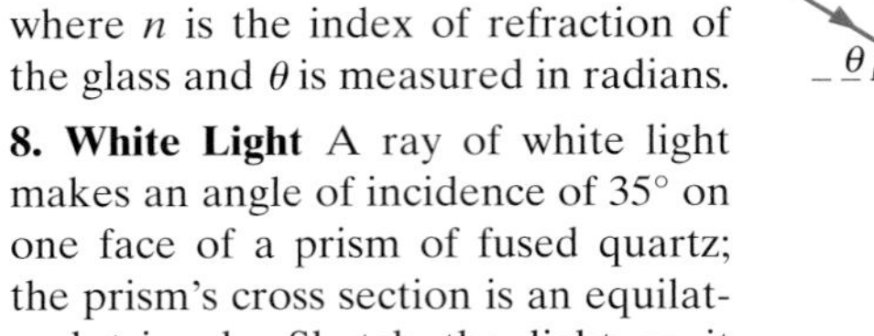

9. Triangular Prism In Fig. 35-38, a ray is incident on one face of a triangular glass prism in air. The angle of incidence θ is chosen so that the emerging ray also makes the same angle θ with the normal to the other face. Show that the index of refraction n of the glass prism is given by

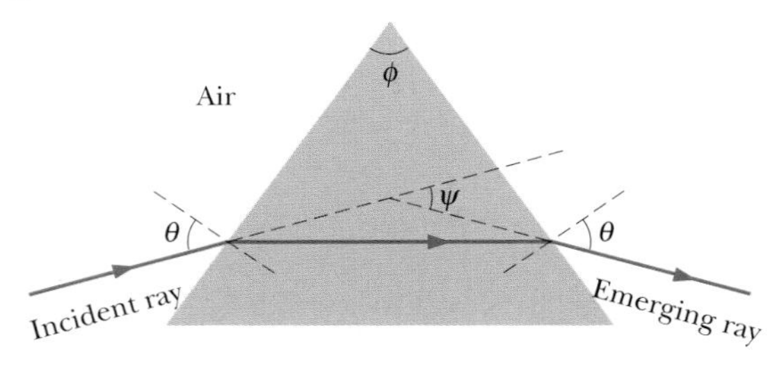

FIGURE 35-38 ■ Problems 9 and 18.

$$n = \frac{\sin\frac{1}{2}(\psi + \phi)}{\sin\frac{1}{2}\phi},$$

where ϕ is the vertex angle of the prism and ψ is the *deviation angle,* the total angle through which the beam is turned in passing through the prism. (Under these conditions the deviation angle ψ has the smallest possible value, which is called the *angle of minimum deviation.*)

10. Perpendicular Mirrors In Fig. 35-39 two perpendicular mirrors form the sides of a vessel filled with

FIGURE 35-39 ■ Problem 10.

water. (a) A light ray is incident from above, normal to the water surface. Show that the emerging ray is parallel to the incident ray. Assume that there are reflections at both mirror surfaces. (b) Repeat the analysis for the case of oblique incidence with the incident ray in the plane of the figure.

SEC. 35-3 ■ TOTAL INTERNAL REFLECTION

11. Glass Slab In Fig. 35-40 a light ray enters a glass slab at point A and then undergoes total internal reflection at point B. What minimum value for the index of refraction of the glass can be inferred from this information?

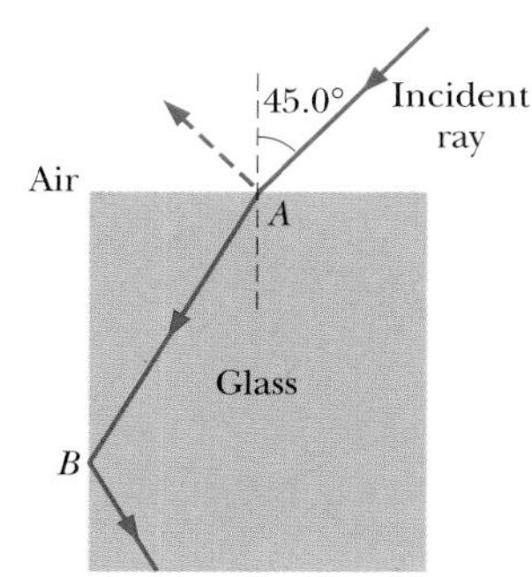

FIGURE 35-40 ■ Problem 11.

12. Benzene The index of refraction of benzene is 1.8. What is the critical angle for a light ray traveling in benzene toward a plane layer of air above the benzene?

13. Perpendicular to Face In Fig. 35-41, a ray of light is perpendicular to the face ab of a glass prism ($n = 1.52$). Find the largest value of the angle ϕ so that the ray is totally reflected at face ac if the prism is immersed (a) in air and (b) in water.

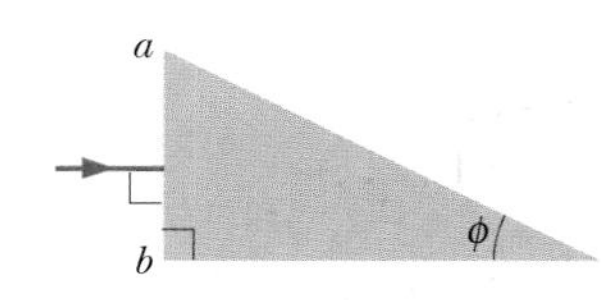

FIGURE 35-41 ■ Problem 13.

14. Point Source A point source of light is 80.0 cm below the surface of a body of water. Find the diameter of the circle at the surface through which light emerges from the water.

15. Solid Glass Cube A solid glass cube, of edge length 10 mm and index of refraction 1.5, has a small spot at its center. (a) What parts of each cube face must be covered to prevent the spot from being seen, no matter what the direction of viewing? (Neglect light that reflects inside the cube and then refracts out into the air.) (b) What fraction of the cube surface must be so covered?

16. Fused Quartz A ray of white light travels through fused quartz that is surrounded by air. If all the color components of the light undergo total internal reflection at the surface, then the reflected light forms a reflected ray of white light. However, if the color component at one end of the visible range (either blue or red) partially refracts through the surface into the air, there is less of that component in the reflected light. Then the reflected light is not white but has the tint of the opposite end of the visible range. (If blue were partially lost to refraction, then the reflected beam would be reddish, and vice versa.) Is it possible for the reflected light to be (a) bluish of (b) reddish? (c) If so, what must be the angle of incidence of the original white light on the quartz surface? (See Fig. 35-31.)

17. 90° Prism In Fig. 35-42, light enters a 90° trianglular prism at point P with incident angle θ and then some of it refracts at point Q with an angle of refraction of 90°. (a) What is the index of refraction of the prism in terms of θ? (b) What, numerically, is the maximum value that the index of refraction can have? Explain what happens to the light at Q if the incident angle at Q is (c) increased slightly and (d) decreased slightly.

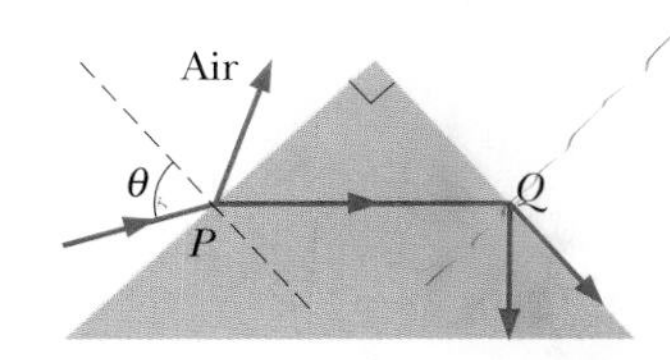

FIGURE 35-42 ■ Problem 17.

18. Apex Angle Given Suppose the prism of Fig. 35-38 has apex angle $\phi = 60.0°$ and index of refraction $n = 1.60$. (a) What is smallest angle of incidence θ for which a ray can enter the left face of the prism and exit the right face? (b) What angle of incidence θ is required for the ray to exit the prism with an identical angle θ for its refraction, as it does in Fig. 35-38? (See Problem 9.)

SEC. 35-4 ■ POLARIZATION BY REFLECTION

19. Light in Water Light traveling in water of refractive index 1.33 is incident on a plate of glass with index of refraction 1.53. At what angle of incidence is the reflected light fully polarized?

20. Completely Polarized (a) At what angle of incidence will the light reflected from water be completely polarized? (b) Does this angle depend on the wavelength of the light?

SEC. 35-6 ■ PLANE MIRRORS

21. Moth A moth at about eye level is 10 cm in front of a plane mirror; you are behind the moth, 30 cm from the mirror. What is the distance between your eyes and the apparent position of the moth's image in the mirror?

22. Hummingbird You look through a camera toward an image of a hummingbird in a plane mirror. The camera is 4.30 m in front of the mirror. The bird is at camera level, 5.00 m to your right and 3.30 m from the mirror. What is the distance between the camera and the apparent position of the bird's image in the mirror?

23. Two Vertical Mirrors Figure 35-43a is an overhead view of two vertical plane mirrors with an object O placed between them. If you look into the mirrors, you see multiple images of O. You can find them by drawing the reflection in each mirror of the angular region between the mirrors, as is done for the left-hand mirror in Fig. 35-43b. Then draw the reflection of the reflection. Continue this on the left and on the right until the reflections meet or overlap at the rear of the mirrors. Then you can count the number of images of O. (a) If $\theta = 90°$, how many images of O would you see? (b) Draw their locations and orientations (as in Fig. 35-43b).

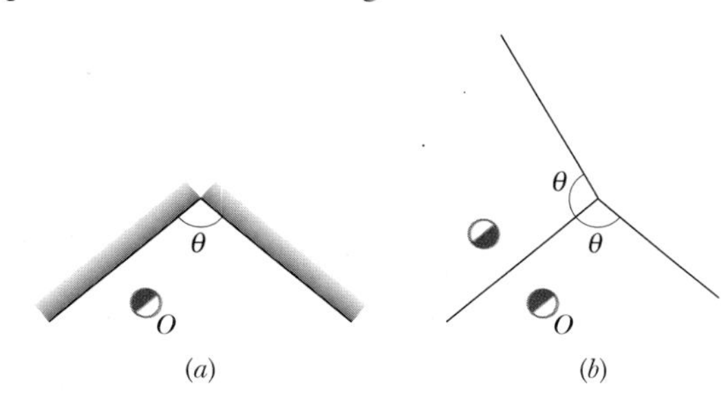

FIGURE 35-43 ■ Problems 23 and 24.

24. Repeat Repeat Problem 23 for the mirror angle θ equal to (a) 45°, (b) 60°, and (c) 120°. (d) Explain why there are several possible answers for (c).

25. Prove That Prove that if a plane mirror is rotated through an angle α, the reflected beam is rotated through an angle 2α. Show that this result is reasonable for $\alpha = 45°$.

26. Corridor Figure 35-44 shows an overhead view of a corridor with a plane mirror M mounted at one end. A burglar B sneaks along the corridor directly toward the center of the mirror. If $d = 3.0$ m, how far from the mirror will she be when the security guard S can first see her in the mirror?

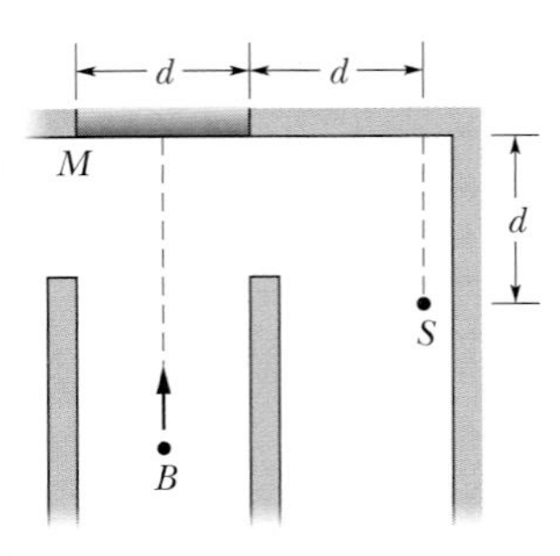

FIGURE 35-44 ■ Problem 26.

27. S and d You put a point source of light S a distance d in front of a screen A. How is the light intensity at the center of the screen changed if you put a completely reflecting mirror M a distance d behind the source, as in Fig. 35-45? (*Hint:* Use Eq. 34-25.)

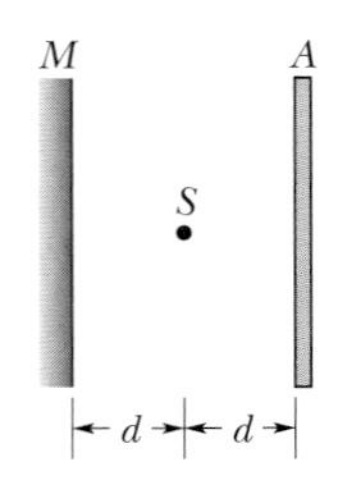

FIGURE 35-45 ■ Problem 27.

28. Small Lightbulb Figure 35-46 shows a small lightbulb suspended above the surface of the water in a swimming pool. The bottom of the pool is a large mirror. How far below the mirror's surface is the image of the bulb? (*Hint:* Construct a diagram of two rays like that of Fig. 35-14, but take into account the bending of light rays by refraction. Assume that the rays are close to a vertical axis through the bulb, and use the small-angle approximation that $\sin\theta \approx \tan\theta$.)

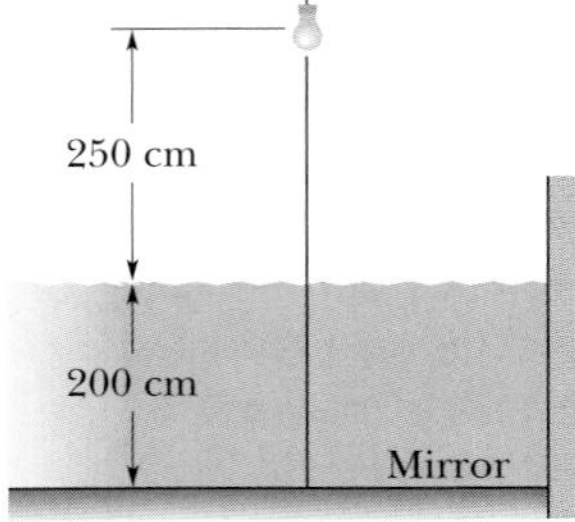

FIGURE 35-46 ■ Problem 28.

SEC. 35-8 ■ IMAGES FROM SPHERICAL MIRRORS

29. Concave Shaving Mirror A concave shaving mirror has a radius of curvature of 35.0 cm. It is positioned so that the (upright) image of a man's face is 2.50 times the size of the face. How far is the mirror from the face?

30. Fill in Table Fill in Table 35-4, each row of which refers to a different combination of an object and either a plane mirror, a spherical convex mirror, or a spherical concave mirror. Distances are in centimeters. If a number lacks a sign, find the sign. Sketch each combination and draw in enough rays to locate the object and its image.

31. Short Straight Object A short straight object of length L lies along the central axis of a spherical mirror of focal length f, a distance o from the mirror. (a) Show that its image in the mirror has a length L' where

$$L' = L\left(\frac{f}{o - f}\right)^2.$$

(*Hint:* Locate the two ends of the object.) (b) Show that the *longitudinal magnification* m' ($= L'/L$) is equal to m^2, where m is the lateral magnification.

32. Luminous Point (a) A luminous point is moving at speed v_O toward a spherical mirror with radius of curvature r, along the central axis of the mirror. Show that the image of this point is moving at speed

$$v_I = -\left(\frac{r}{2o - r}\right)^2 v_O,$$

where o is the distance of the luminous point from the mirror at any given time. (*Hint:* Start with Eq. 35-13.) Now assume that the mirror is concave, with $r = 15$ cm, and let $v_O = 5.0$ cm/s. Find the speed of the image when (b) $o = 30$ cm (far outside the focal point), (c) $o = 8.0$ cm (just outside the focal point), and (d) $o = 10$ mm (very near the mirror).

SEC. 35-9 ■ SPHERICAL REFRACTING SURFACES

33. Parallel Light Rays A beam of parallel light rays from a laser is incident on a solid transparent sphere of index of refraction n (Fig. 35-47). (a) If a point image is produced at the back of the sphere, what is the index of refraction of the sphere? (b) What index of refraction, if any, will produce a point image at the center of the sphere?

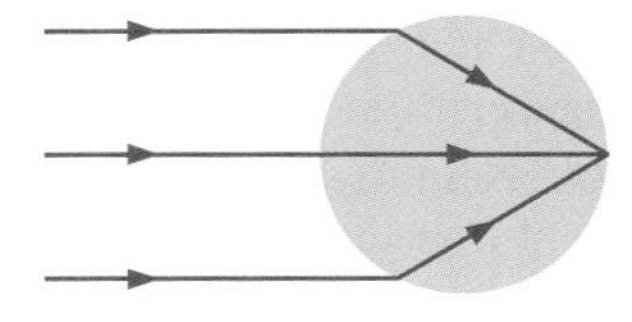
FIGURE 35-47 ■ Problem 33.

34. Fill in Table Two Fill in Table 35-5, each row of which refers to a different combination of a point object and a spherical refracting surface separating two media with different indexes of refraction. Distances are in centimeters. If a number lacks a sign, find the sign. Sketch each combination and draw in enough rays to locate the object and image.

TABLE 35-4
Problem 30: Mirrors

	Type	f	r	i	o	m	Real Image?	Inverted Image?
(a)	Concave	20			+10			
(b)					+10	+1.0	No	
(c)		+20			+30			
(d)					+60	−0.50		
(e)			−40	−10				
(f)		20				+0.10		
(g)	Convex		40	4.0				
(h)					+24	0.50		Yes

TABLE 35-5
Problem 34: Spherical Refracting Surfaces

	n_1	n_2	o	i	r	Inverted Image?
(a)	1.0	1.5	+10		+30	
(b)	1.0	1.5	+10	−13		
(c)	1.0	1.5		+600	+30	
(d)	1.0		+20	−20	−20	
(e)	1.5	1.0	+10	−6.0		
(f)	1.5	1.0		−7.5	−30	
(g)	1.5	1.0	+70		+30	
(h)	1.5		+100	+600	−30	

35. Coin in a Pool You look downward at a coin that lies at the bottom of a pool of liquid with depth d and index of refraction n (Fig. 35-48). Because you view with two eyes, which intercept different rays of light from the coin, you perceive the coin to be where extensions of the intercepted rays cross, at depth d_a instead of d. Assuming that the intercepted rays in Fig. 35-48 are close to a vertical axis through the coin, show that $d_a = d/n$. (*Hint:* Use the small-angle approximation that $\sin\theta \approx \tan\theta \approx \theta$.)

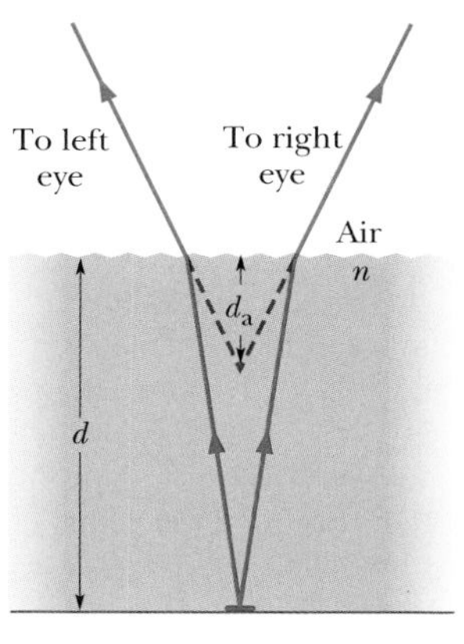

FIGURE 35-48 ■ Problem 35.

36. Carbon Tetrachloride A 20-mm-thick layer of water ($n = 1.33$) floats on a 40-mm-thick layer of carbon tetrachloride ($n = 1.46$) in a tank. A coin lies at the bottom of the tank. At what depth below the top water surface do you perceive the coin? (*Hint:* Use the result and assumptions of Problem 35 and work with a ray diagram of the situation.)

SEC. 35-10 ■ THIN LENSES

37. Thin Diverging Lens An object is 20 cm to the left of a thin diverging lens having a 30 cm focal length. What is the image distance i? Find the image position with a ray diagram.

38. Image of Sun You produce an image of the Sun on a screen, using a thin lens whose focal length is 20.0 cm. What is the diameter of the image? (See Appendix C for needed data on the Sun.)

39. Double-Convex A double-convex lens is to be made of glass with an index of refraction of 1.5. One surface is to have twice the radius of curvature of the other and the focal length is to be 60 mm. What are the radii?

40. One Side Is Flat A lens is made of glass having an index of refraction of 1.5. One side of the lens is flat, and the other is convex with a radius of curvature of 20 cm. (a) Find the focal length of the lens. (b) If an object is placed 40 cm in front of the lens, where will the image be located?

41. Newtonian Form The formula

$$\frac{1}{o} + \frac{1}{i} = \frac{1}{f}$$

is called the *Gaussian* form of the thin-lens formula. Another form of this formula, the *Newtonian* form, is obtained by considering the distance x from the object to the first focal point and the distance x' from the second focal point to the image. Show that

$$xx' = f^2$$

is the Newtonian form of the thin-lens formula.

42. Movie Camera A movie camera with a (single) lens of focal length 75 mm takes a picture of a 180-cm-high person standing 27 m away. What is the height of the image of the person on the film?

43. Illuminated Slide An illuminated slide is held 44 cm from a screen. How far from the slide must a lens of focal length 11 cm be placed to form an image of the slide's picture on the screen?

44. Fill in Table Three To the extent possible, fill in Table 35-6, each row of which refers to a different combination of an object and a thin lens. Distances are in centimeters. For the type of lens, use C for converging and D for diverging. If a number (except for the index of refraction) lacks a sign, find the sign. Sketch each combination and draw in enough rays to locate the object and image.

TABLE 35-6
Problem 44: Thin Lenses

	Type	f	r_1	r_2	i	o	n	m	Real Image?	Inverted Image?
(a)	C	10				+20				
(b)		+10				+5.0				
(c)		10				+5.0		> 1.0		
(d)		10				+5.0		< 1.0		
(e)			+30	−30		+10	1.5			
(f)			−30	+30		+10	1.5			
(g)			−30	−60		+10	1.5			
(h)						+10		0.50		No
(i)						+10		−0.50		

45. Show That Show that the distance between an object and its real image formed by a thin converging lens is always greater than or equal to four times the focal length of the lens.

46. Diverging and Converging A diverging lens with a focal length of −15 cm and a converging lens with a focal length of 12 cm have a common central axis. Their separation is 12 cm. An object of height 1.0 cm is 10 cm in front of the diverging lens, on the common central axis. (a) Where does the lens combination produce the final image of the object (the one produced by the second, converging lens)? (b) What is the height of that image? (c) Is the image real or virtual? (d) Does the image have the same orientation as the object or is it inverted?

47. Final Image A converging lens with a focal length of +20 cm is located 10 cm to the left of a diverging lens having a focal length of −15 cm. If an object is located 40 cm to the left of the converging lens, locate and describe completely the final image formed by the diverging lens.

48. Location and Size An object is 20 cm to the left of a lens with a focal length of +10 cm. A second lens of focal length +12.5 cm is 30 cm to the right of the first lens. (a) Find the location and relative size of the final image. (b) Verify your conclusions by drawing the lens system to scale and constructing a ray diagram. (c) Is the final image real of virtual? (d) Is it inverted?

49. Two Thin Lenses Two thin lenses of focal lengths f_1 and f_2 are in contact. Show that they are equivalent to a single thin lens with

$$f = \frac{f_1 f_2}{f_1 + f_2}$$

as its focal length.

50. Real Inverted In Fig 35-49, a real inverted image I of an object O is formed by a certain lens (not shown); the object–image separation is $d = 40.0$ cm, measured along the central axis of the lens. The image is just half the size of the object. (a) What kind of lens must be used to produce this image? (b) How far from the object must the lens be placed? (c) What is the focal length of the lens?

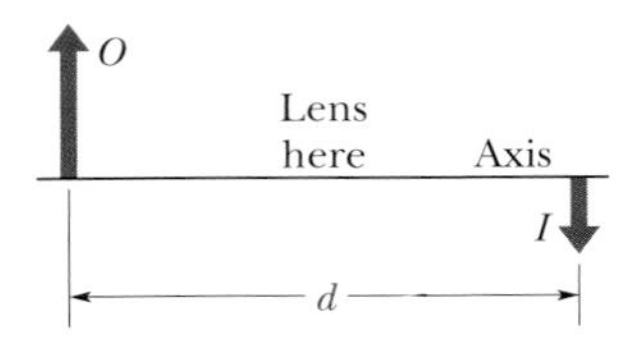

FIGURE 35-49 ■ Problem 50.

51. Object–Screen Distance A luminous object and a screen are a fixed distance D apart. (a) Show that a converging lens of focal length f, placed between object and screen, will form a real image on the screen for two lens positions that are separated by a distance

$$d = \sqrt{D(D - 4f)}.$$

(b) Show that

$$\left(\frac{D - d}{D + d}\right)^2$$

gives the ratio of the two image sizes for these two positions of the lens.

SEC. 35-11 ■ OPTICAL INSTRUMENTS

52. Astronomical Telescope If an angular magnification of an astronomical telescope is 36 and the diameter of the objective is 75 mm, what is the minimum diameter of the eyepiece required to collect all the light entering the objective from a distant point source on the telescope axis?

53. Microscope In a microscope of the type shown in Fig. 35-28, the focal length of the objective is 4.00 cm, and that of the eyepiece is 8.00 cm. The distance between the lenses is 25.0 cm. (a) What is the tube length s? (b) If image I in Fig. 35-28 is to be just inside focal point F_1', how far from the objective should the object be? What then are (c) the lateral magnification m of the objective, (d) the angular magnification m_θ of the eyepiece, and (e) the overall magnification M of the microscope?

54. Magnifying Lens A simple magnifying lens of focal length f is placed near the eye of someone whose near point P_n is 25 cm from the eye. An object is positioned so that its image in the magnifying lens appears at P_n. (a) What is the lens's angular magnification? (b) What is the angular magnification if the object is moved so that its image appears at infinity? (c) Evaluate the angular magnifications of (a) and (b) for $f = 10$ cm. (Viewing an image at P_n requires effort by muscles in the eye, whereas for many people viewing an image at infinity requires no effort.)

55. Human Eye Figure 35-50a shows the basic structure of a human eye. Light refracts into the eye through the cornea and is then further redirected by a lens whose shape (and thus ability to focus the light) is controlled by muscles. We can treat the cornea and eye lens as a single effective thin lens (Fig. 35-50b). A "normal" eye can focus parallel light rays from a distance object O to a point on the retina at the back of the eye, where processing of the visual information begins. As an object is brought close to the eye, however, the muscles must change the shape of the lens so that rays form an inverted real image on the retina (Fig. 35-50c). (a) Suppose that for the parallel rays of Figs. 35-50a and b, the focal length f of the effective thin lens of the eye is 2.50 cm. For an object at distance $o = 40.0$ cm, what focal length f' of the effective lens is required for it to be seen clearly? (b) Must the eye muscles increase or decrease the radii of curvature of the eye lens to give focal length f'?

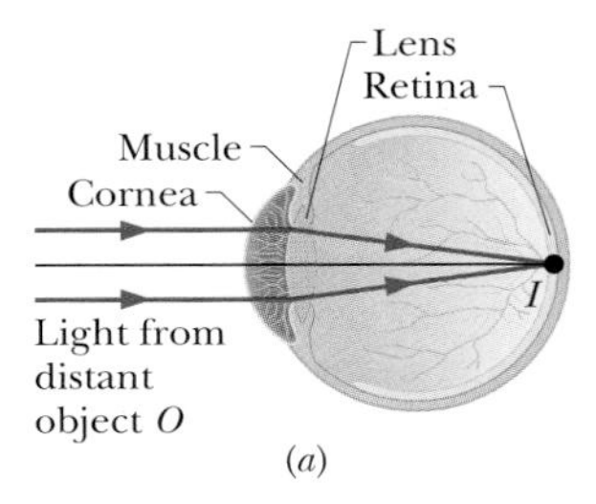

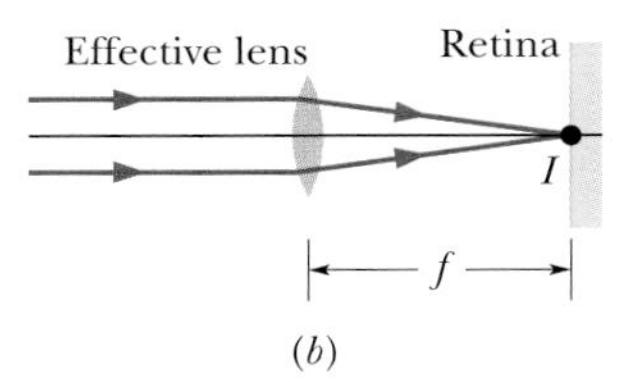

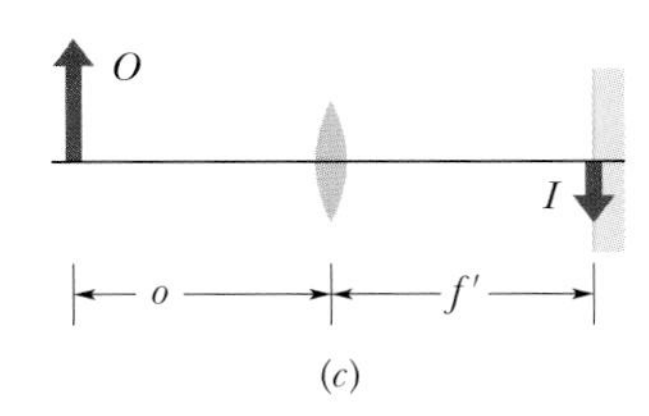

FIGURE 35-50 ■ Problem 55.

56. Compound Microscope An object is 10.0 mm from the objective of a certain compound microscope. The lenses are 300 mm apart and the intermediate image is 50.0 mm from the eyepiece. What overall magnification is produced by the instrument?

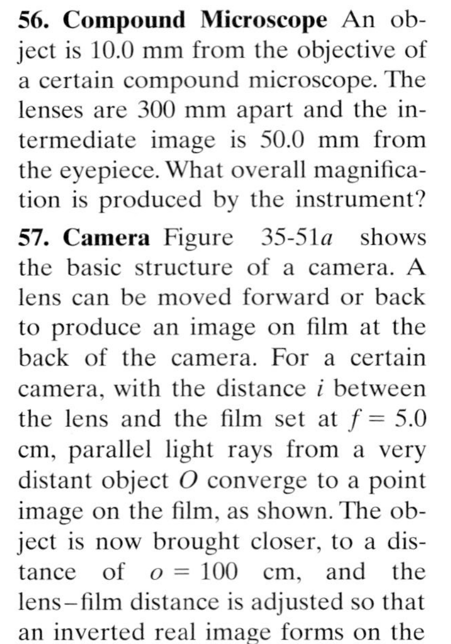

57. Camera Figure 35-51a shows the basic structure of a camera. A lens can be moved forward or back to produce an image on film at the back of the camera. For a certain camera, with the distance i between the lens and the film set at $f = 5.0$ cm, parallel light rays from a very distant object O converge to a point image on the film, as shown. The object is now brought closer, to a distance of $o = 100$ cm, and the lens–film distance is adjusted so that an inverted real image forms on the film (Fig. 35-51b). (a) What is the lens–film distance i now? (b) By how much was i changed?

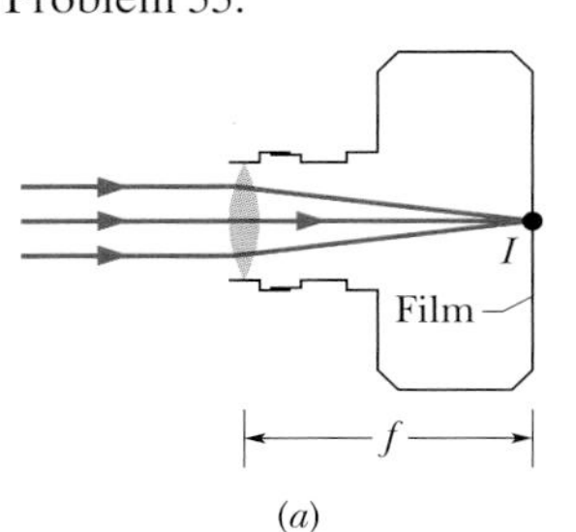

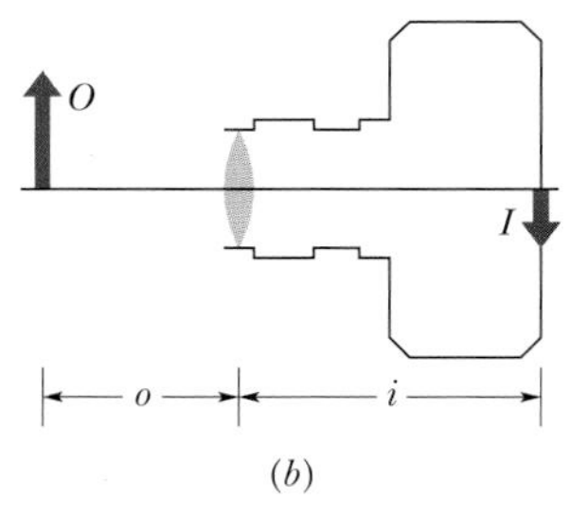

FIGURE 35-51 ■ Problem 57.

Additional Problems

58. Bizarre Behavior with Light You may have observed people with cameras behaving strangely:

(a) At a conference in North Carolina, one of the physics graduate students (who should have known better!) tried to take a picture of the overheads projected on a white screen. Since the room was darkened, he used a flash. Explain why this is a bad idea and what his pictures are likely to show.

(b) Someone mentioned to the student that he probably should not be using his flash, so he turned it off. He then proceeded to try and take pictures of the participants in the darkened room! Explain why this is a bad idea and what his pictures are likely to show.

(c) A woman on an airplane at night with a camera was impressed with the view of the city lights in the dark as the plane flew over Washington, D.C. She stood back in the aisle with her camera and tried to take a picture through the window using her flash. Explain why this is a bad idea and what her pictures are likely to show.

59. Closer Than They May Appear When a *T. rex* pursues a jeep in the movie *Jurassic Park,* we see a reflected image of the (very large) *T. rex* via a side-view mirror, on which is printed the (then darkly humorous) warning: "Objects in mirror are closer than they appear." Is the mirror flat, convex, or concave? Why do you think so?

60. Where Can You See the Bulb? In Fig. 35-52, M is a plane mirror; B is a very small bright lightbulb that can be treated as a point source of light; and H is an opaque housing that does not transmit light. An observer can stand anywhere along a line O to try to see the image of the lightbulb in the mirror. By using relevant rays of light, determine those locations along the line O from which the image of B is visible and those locations from which it is not visible. Mark the regions along line O accordingly, and explain the reasoning you used in drawing the rays. (Arons, Arnold, *A Guide to Introductory Physics Teaching,* John Wiley and Sons, New York, 1990.)

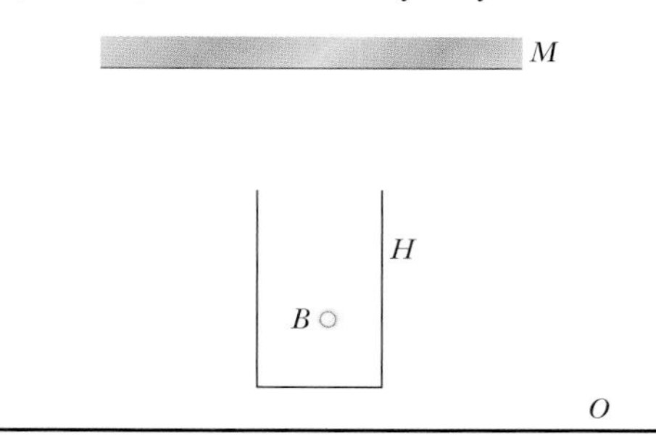

FIGURE 35-52 ■ Problem 60.

61. Who Sees What? Figure 35-53 shows a small object (represented by an arrow) in front of a curved mirror. At the tip of the arrow is a black dot. The mirror is a piece of a sphere. *The center of the sphere is marked in the picture with an x.* Eyes corresponding to three different observers are shown. For each question, explain how you got your result. Be sure to include a ray diagram as part of your explanation.

(a) How many black dots will the observer at position A see? Where will the dots appear to be? Specify quantitatively how far from the mirror the dots will appear to be and how far off the axis they will be.

(b) How many black dots will the observer at position B see? Explain how you know.

(c) How many black dots will the observer at position C see? Explain how you know.

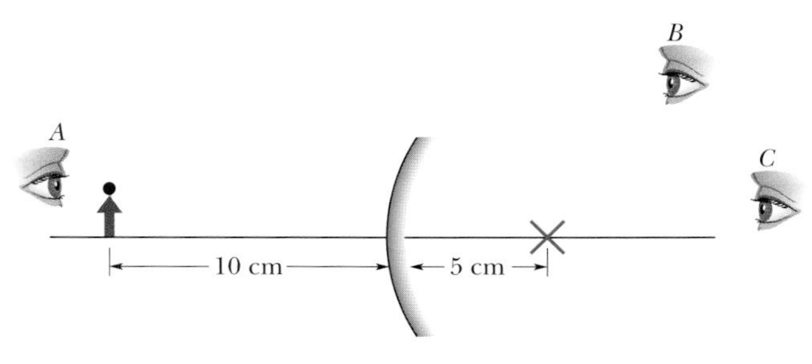

FIGURE 35-53 ■ Problem 61.

62. The Camera and the Slide Projector Address each part of this question in two ways: (1) by drawing and interpreting appropriate geometrical diagrams and (2) by appealing to the lens equation and the expression for lateral magnification and demonstrating your result mathematically. If your two approaches do not agree, explain which one is correct and why the other is wrong.

(a) Suppose you are using a camera and wish to have a larger image of a distant object than you are obtaining with the lens currently in use. Would you change to a lens with a longer or a shorter focal length? Explain your reasoning. (*Hint:* Note that the object distance is essentially fixed.)

(b) Suppose you are using a slide projector and wish to obtain a larger image on the screen. You cannot achieve this by moving the screen farther from the projector because you are already using the entire length of the room. Would you change to a lens with a longer or a shorter focal length than the one you are using? Explain your reasoning. (*Hint:* Note that the image distance is essentially fixed.)

63. Mirrors and Lenses Each of the parts of this problem has a description of an object and an optical device (lens or mirror). A sketch is shown in Fig. 35-54. For each case, specify whether

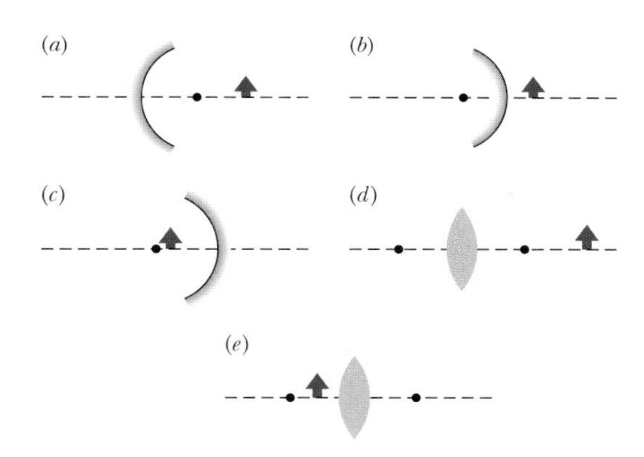

FIGURE 35-54 ■ Problem 63.

- The image is real (R), virtual (V), or no image is formed (N).
- The image is on the same side of the device as the object (S) or the opposite side (O). If there is no image put a null mark (∅).
- If an image is formed, on which side of the system must the observer be in order to see it, left (−) or right (+)?

For each problem you should therefore give three answers (for example, VO +). For the mirrors, the center is shown. For the lenses, the focal points are shown. The radius of curvature of the mirrors is R, and the focal length of the lenses is f.

(a) An object on the right side of a spherical mirror, a distance $s > R$ from the mirror. The mirror is concave toward the object.

(b) An object on the right side of a spherical mirror, a distance $s < R/2$ from the mirror. The mirror is convex toward the object.

(c) An object on the left side of a spherical mirror, a distance $R > s > R/2$ from the mirror. The mirror is concave toward the object.
(d) An object on the right side of a convex lens, a distance $s > f$ from the lens.
(e) An object on the left side of a convex lens, a distance $s < f$ from the lens.

64. The Diverging Lens In Fig. 35-55, point A (marked by a circle) is the top of a small object (indicated as an arrow). Near it is a concave lens, as shown. The focal points of the lens are marked with black dots.

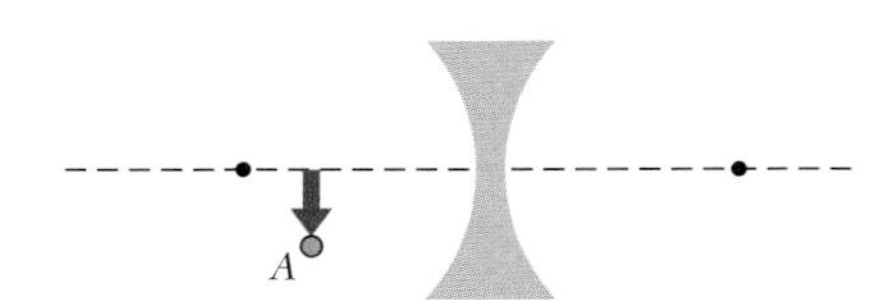

FIGURE 35-55 ■ Problem 64.

(a) Using a ray diagram, show where an image of point A would be formed.
(b) If the focal length of the lens is 8 cm and the object is 6 cm from the lens, where will the image be?
(c) If the object is 1 cm tall, how tall will the image be?
(d) Will the image created by the lens be real or virtual?
(e) Where will you have to be to see the image?

65. The Half Lens A projector has an arrangement of lenses as shown in Fig. 35-56. A bulb illuminates an object (a slide) and the light then passes through a lens that creates an image on a distant screen as shown.

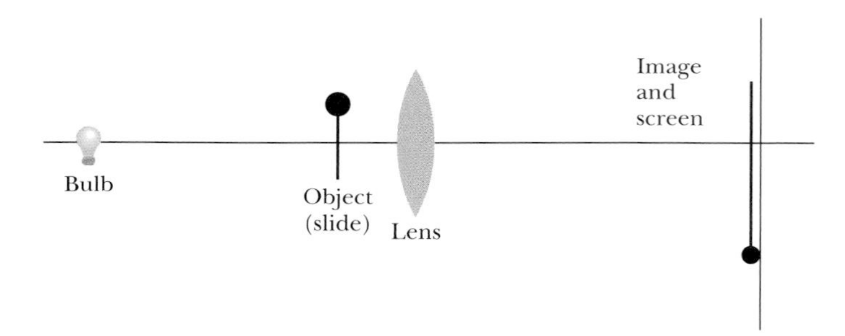

FIGURE 35-56 ■ Problem 65.

When a sheet of a cardboard is brought up to cover the lower half of the lens, what happens to the image on the screen?

(a) The top half of the image disappears.
(b) The bottom half of the image disappears.
(c) The image remains but is weaker (not as bright).
(d) The image remains unchanged.
(e) The bottom half of the image becomes weaker, the top is unchanged.
(f) The top half of the image becomes weaker, the bottom is unchanged.
(g) Something else happens. (Tell what it is.)

Explain your reasoning, drawing whatever rays are needed to make your point clear.

66. Alice and the Looking Glass Alice faces a looking glass (mirror) and is standing at a level so that her eyes appear to her to be right at the top of the mirror as shown in Fig. 35-57. At the position she is standing, she can just see her hands at the bottom of the mirror. If she steps back far enough,

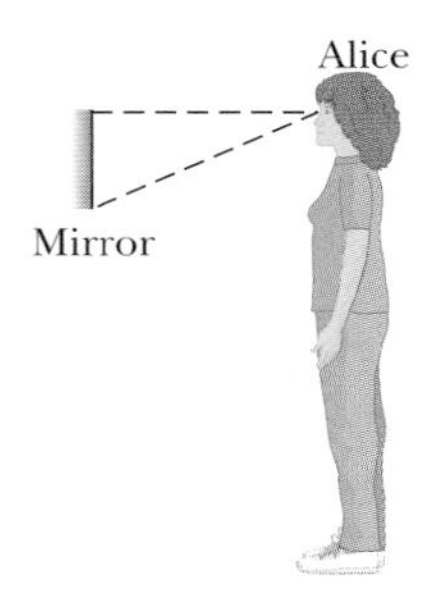

FIGURE 35-57 ■ Problem 66.

(a) She will eventually be able to see all of herself in the mirror at the same time.
(b) There will be no change in how much of herself she can see.
(c) She will see less of herself as she steps back.
(d) Some other result (explain).

Choose the letter of the choice that completes the sentence correctly and explain why you think so with a few sentences and some rays on the diagram.

67. A Bigger Lens Point O in Fig. 35-58 is a source of light. Two rays from O are shown passing through a thin converging lens and crossing each other at the point marked C.

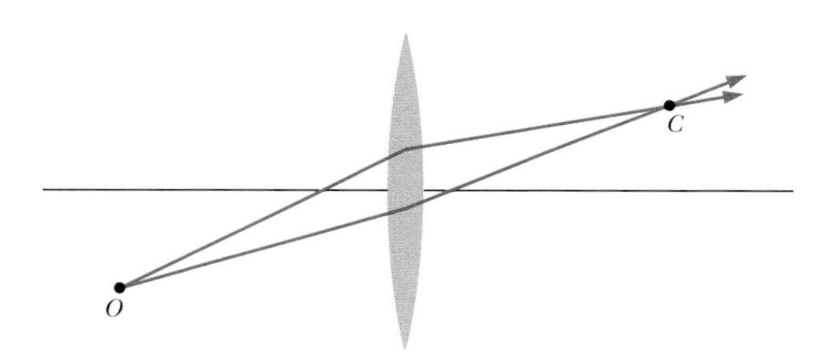

FIGURE 35-58 ■ Problem 67.

(a) On a copy of the figure on your answer sheet, find the two principal foci of the lens by drawing appropriate rays. Label the foci F1 and F2. Explain your reasoning.
(b) Suppose the lens is replaced by another lens having the same focal length but a larger diameter. Indicate whether each of the following partial sentences is correctly completed by the phrase greater than (>), less than (<), or the same as (=).

- The distance of the image from the principal axis is ______ it was with the smaller lens.
- The brightness of the image with the large lens is ______ it was with the smaller lens.

68. Where's the Image? Figure 35-59 shows a thin lens (indicated by a gray rectangle) and a coordinate system. The x axis passes

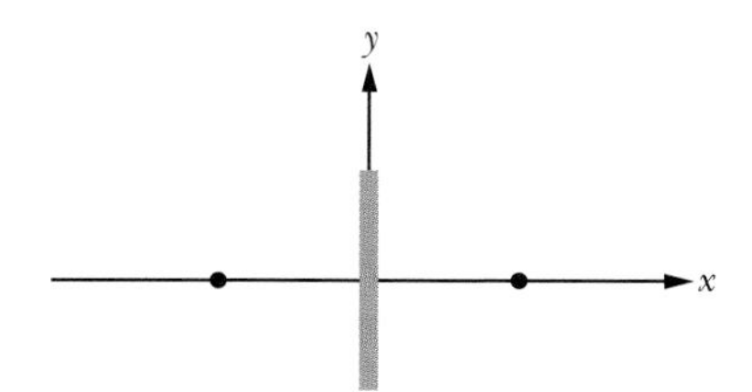

FIGURE 35-59 ■ Problem 68.

through the center of the lens and runs along its axis of symmetry, with the positive x direction indicated by the arrowhead. The lens may be treated as being of negligible thickness and has a focal length f. The points $(x, y) = (f, 0)$ and $(-f, 0)$ are marked by black dots.

A small object is placed at the position $(x, 0)$. For each of the four cases (i)–(iv) below, indicate whether the location of the image formed $(= x')$ is on the positive or negative side of the axis, and closer to the lens than the focal point or farther away.

i. $-f < x < 0$ iii. $x < f < 0$

ii. $f < 0 < x$ iv. $0 < f < x$

Hint: Your answer should take a form such as $(x' > f > 0)$ to indicate that the image is on the positive side of the axis and farther away than the focal point or $(0 > x' > f)$ to indicate that the image is on the negative side of the axis between the lens and the focal point. Note that in some cases the focal length specified is negative and in some cases it is positive.

36 Interference

At first glance, the top surface of the *Morpho* butterfly's wing is simply a beautiful blue-green. There is something strange about the color, however, for it almost glimmers, unlike the colors of most objects—and if you change your perspective, or if the wing moves, the tint of the color changes. The wing is said to be iridescent, and the blue-green we see hides the wing's "true" dull brown color that appears on the bottom surface.

What is so different about the top surface that gives us this arresting display?

The answer is in this chapter.

36-1 Interference

Sunlight, as the rainbow shows us, is a composite of all the colors of the visible spectrum. The colors reveal themselves in the rainbow because the incident wavelengths refract and so are bent through different angles as they pass through raindrops that produce the bow. However, soap bubbles and oil slicks can also show striking colors produced not by refraction but by constructive and destructive **interference** of light waves. The interfering waves combine either to enhance or to suppress certain colors in the spectrum of the incident sunlight. Interference of light waves is thus a superposition phenomenon. Hence, this chapter is a significant point of connection between much of what we have just learned in Chapters 34 and 35 about electromagnetic waves in general and light in particular and what we learned earlier regarding the interference of waves on a string and sound waves in Chapters 17 and 18.

Interference, which can lead to the selective enhancement or suppression of wavelengths, has many applications. When light encounters an ordinary glass surface, for example, about 4% of the incident energy is reflected, thus weakening the transmitted beam by that amount. This unwanted loss of light can be a real problem in optical systems with many components. A thin, transparent "interference film," deposited on the glass surface, can reduce the amount of reflected light (and thus enhance the transmitted light) by destructive interference. The bluish cast of a camera lens reveals the presence of such a coating. Interference coatings can also be used to enhance—rather than reduce—the ability of a surface to reflect light.

To understand interference, we must go beyond the restrictions of geometrical optics and employ the full power of wave optics. In fact, as you will see, the existence of interference phenomena is perhaps our most convincing evidence that light is a wave—because interference cannot be explained other than with waves.

36-2 Light as a Wave

The first person to advance a convincing wave theory for light was Dutch physicist Christian Huygens, in 1678. Although much less comprehensive than the later electromagnetic theory of Maxwell, Huygens' theory was simpler mathematically and remains useful today. Its great advantages are that it accounts for the laws of reflection and refraction in terms of waves and gives physical meaning to the index of refraction.

Huygens' wave theory is based on a geometrical construction that allows us to tell where a given wavefront will be at any time in the future if we know its present position. This construction is based on **Huygens' principle,** which is:

> All points on a wavefront serve as point sources of spherical secondary wavelets. After a time Δt, the new position of the wavefront will be that of a surface tangent to these secondary wavelets.*

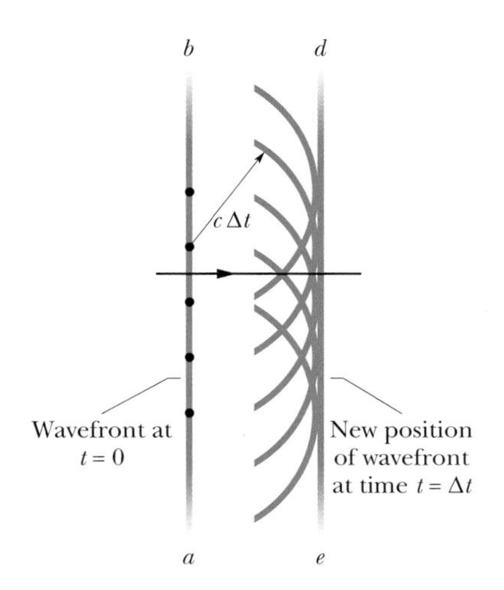

FIGURE 36-1 ■ The propagation of a plane wave in vacuum, as portrayed by Huygens' principle.

Here is a simple example. At the left in Fig. 36-1, the present location of a wavefront of a plane wave traveling to the right in vacuum is represented by plane *ab*, perpendicular to the page. Where will the wavefront be at time Δt later? We let several points on plane *ab* (the dots) serve as sources of spherical secondary wavelets that are emitted at $t = 0$. At time Δt, the radius of all these spherical wavelets will have grown to

* When using this principle in calculations, there is a factor which gives a greater wave amplitude in the direction of propagation of the original wavefront. This prevents the back wavelets from combining to create a backwards wavefront.

$c\Delta t$, where c is the speed of light in vacuum. We draw plane *de* tangent to these wavelets at time Δt. This plane represents the wavefront of the plane wave at time Δt; it is parallel to plane *ab* and a perpendicular distance $c\Delta t$ from it.

The Law of Refraction

We now use Huygens' principle to derive the law of refraction or Snell's law, $n_1 \sin\theta_1 = n_2 \sin\theta_2$ (Eq. 35-2). Recall that here n_1 is the index of refraction in the medium from which the light is incident, n_2 is the index of refraction in the refracting medium, and θ_1 and θ_2 are the angles of incidence and refraction, respectively. There are two key ideas behind this derivation:

1. The oscillating electromagnetic wave (with its oscillating E- and B-fields) hitting the surface of a material drives the electrons in the surface to oscillate and hence to reradiate. As a result, the outgoing wave the electrons produce will have the same frequency as the incoming wave.
2. The wavelength is determined by how far the wave propagates into the media while the electrons at the surface are undergoing one full oscillation. Since the relationship between wavelength λ, period T, and speed of the wave v is $v = \lambda/T$, and the wave speed in a denser medium is slower, the wavelength is smaller in a denser medium.

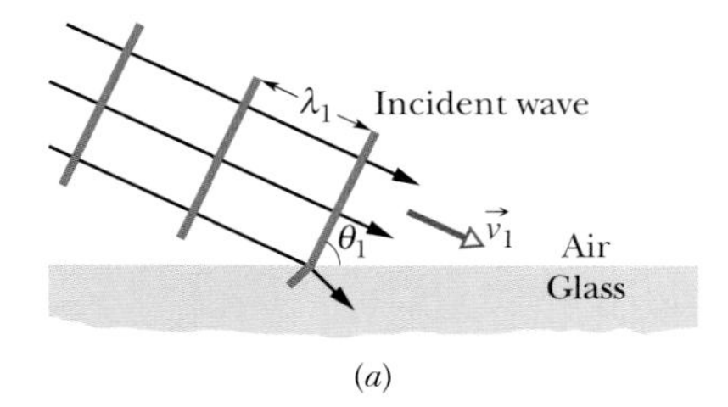

(*a*)

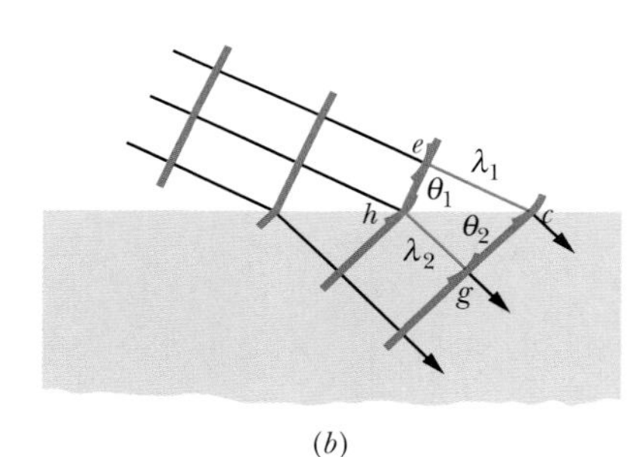

(*b*)

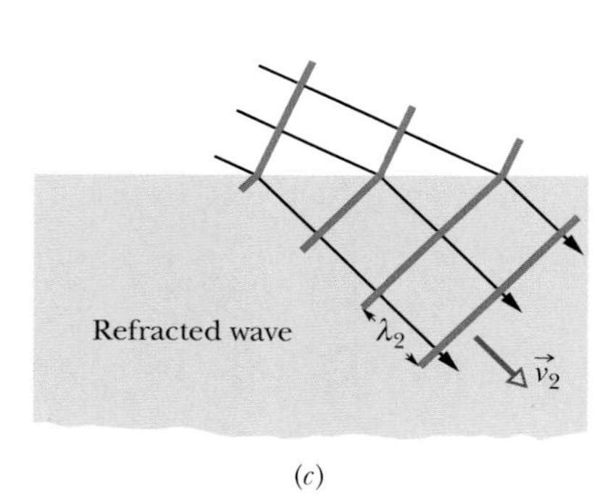

(*c*)

FIGURE 36-2 ■ The refraction of a plane wave at an air–glass interface, as portrayed by Huygens' principle. The wavelength in glass is smaller than that in air. For simplicity, the reflected wave is not shown. Parts (*a*) through (*c*) represent three successive stages of the refraction.

Figure 36-2 shows three stages in the refraction of several wavefronts at a plane interface between air (medium 1) and glass (medium 2). By convention, we choose the wavefronts in the incident beam to be separated by λ_1, the wavelength in medium 1. Let the speed of light in air be v_1 and that in glass be v_2. We assume that $v_2 < v_1$, which happens to be true. (Since in this chapter we do not use vector components, we run no risk of confusing the magnitude of a velocity vector with its components. Therefore, we will simplify notation and write the speed as simply v, rather than $|\vec{v}|$.)

Angle θ_1 in Fig. 36-2*a* is the angle between the wavefront and the interface; it has the same value as the angle between the *normal* to the wavefront (that is, the incident ray) and the *normal* to the interface. Thus, θ_1 is the angle of incidence.

As the incident light wave moves into the glass, a wavefront at point *e* will travel a distance of λ_1 to point *c*. The time period required for the wave to travel this distance is the distance divided by the speed of the wavelet, or λ_1/v_1. In this same time period, a Huygens wavelet (a new wave created by the oscillating electrons in the material) at point *h* will travel to point *g*, at the reduced speed v_2 and with wavelength λ_2. Thus, this time period must also be equal to λ_2/v_2. By equating these times, we obtain the relation

$$\frac{\lambda_1}{\lambda_2} = \frac{v_1}{v_2}, \tag{36-1}$$

which shows that the wavelengths of light in two media are proportional to the speeds of light in those media.

By Huygens' principle, the refracted wavefront must be tangent to an arc of radius λ_2 centered on *h*, say, at point *g*. The refracted wavefront must also be tangent to an arc of radius λ_1 centered on *e*, say, at *c*. Then the refracted wavefront must be oriented as shown. Note that θ_2, the angle between the refracted wavefront and the interface, is actually the angle of refraction.

For the right triangles *hce* and *hcg* in Fig. 36-2*b* we may write

$$\sin\theta_1 = \frac{\lambda_1}{hc} \quad \text{(for triangle } hce\text{)}$$

and

$$\sin\theta_2 = \frac{\lambda_2}{hc} \quad \text{(for triangle } hcg\text{)}.$$

Dividing the first of these two equations by the second and using Eq. 36-1, we find

$$\frac{\sin\theta_1}{\sin\theta_2} = \frac{\lambda_1}{\lambda_2} = \frac{v_1}{v_2}. \tag{36-2}$$

We can define an **index of refraction** n for each medium as the ratio of the speed of light in vacuum to the speed of light v in the medium. Thus,

$$n \equiv \frac{c}{v} \quad \text{(definition of index of refraction)}. \tag{36-3}$$

In particular, for our two media, we have

$$n_1 = \frac{c}{v_1} \quad \text{and} \quad n_2 = \frac{c}{v_2}. \tag{36-4}$$

If we combine Eqs. 36-2 and 36-4 we find

$$\frac{\sin\theta_1}{\sin\theta_2} = \frac{c/n_1}{c/n_2} = \frac{n_2}{n_1}, \tag{36-5}$$

or

$$n_1 \sin\theta_1 = n_2 \sin\theta_2 \quad \text{(law of refraction)}, \tag{36-6}$$

as introduced in Chapter 35. This result demonstrates that Huygen's Principle is a construct that can be used to explain the law of refraction.

Wavelength and Index of Refraction

We have now seen that the wavelength of light changes when the speed of the light changes, as happens when light crosses an interface from one medium into another. Further, the speed of light in any medium depends on the index of refraction of the medium, according to Eq. 36-3. Thus, the wavelength of light in any medium depends on the index of refraction of the medium. Let a certain monochromatic light have wavelength λ and speed c in vacuum and wavelength λ_n and speed v in a medium with an index of refraction n. Now we can rewrite Eq. 36-1 as

$$\lambda_n = \lambda \frac{v}{c}. \tag{36-7}$$

Using Eq. 36-3 to substitute $1/n$ for v/c then yields

$$\lambda_n = \frac{\lambda}{n}. \tag{36-8}$$

This equation relates the wavelength of light in any medium to its wavelength in vacuum. It tells us that the greater the index of refraction of a medium, the smaller is the wavelength of light in that medium.

What about the frequency of the light? Let f_n represent the frequency of the light in a medium with index of refraction n. Then from the general relation of Eq. 17-12 ($v = \lambda f$), we can write

$$f_n = \frac{v}{\lambda_n}.$$

Substituting Eqs. 36-3 and 36-8 then gives us

$$f_n = \frac{c/n}{\lambda/n} = \frac{c}{\lambda} = f,$$

where f is the frequency of the light in vacuum. Thus, although the speed and wavelength of light are different in the medium than in vacuum, the frequency of the light in the medium is the same as it is in vacuum. Since we started this discussion with the assumption that the periods of the waves were the same, this result indicates that our equations are consistent.

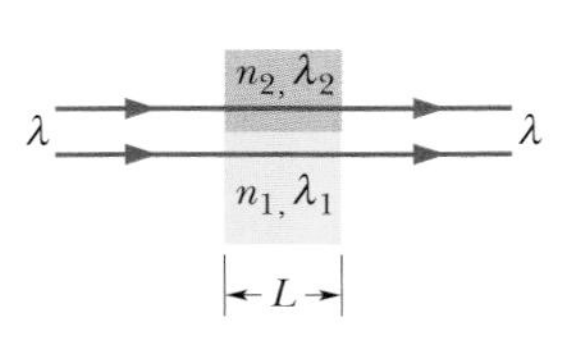

FIGURE 36-3 ■ Two light rays with the same initial wavelength λ in air travel through two media having different indexes of refraction. During that time, $\lambda_1 \neq \lambda_2$.

The fact that the wavelength of light depends on the index of refraction via Eq. 36-8 is important in certain situations involving the interference of light waves. For example, in Fig. 36-3, the *waves of the rays* (that is, the waves associated with the rays) have identical wavelengths λ and are initially in phase in air ($n \approx 1$). One of the waves travels through medium 1 of index of refraction n_1 and length L. The other travels through medium 2 of index of refraction n_2 and the same length L. Each ray acquires a different wavelength when traveling through its medium. When the waves leave the two media, they will have the same wavelength once again—their wavelength λ in air. However, because their wavelengths differed in the two media, the two waves may no longer be in phase.

> The phase difference between two light waves can change if the waves travel through different materials having different indexes of refraction.

As we shall discuss soon, this phase difference change can determine how the light waves will interfere if they reach some common point.

To find their new phase difference in terms of wavelengths, we first count the number N_1 of wavelengths there are in the length L of medium 1. From Eq. 36-8, the wavelength in medium 1 is $\lambda_{n1} = \lambda/n_1$, so

$$N_1 = \frac{L}{\lambda_{n1}} = \frac{Ln_1}{\lambda}. \tag{36-9}$$

In general, the term Ln is known as the **optical path difference**. Similarly, we count the number N_2 of wavelengths there are in the length L of medium 2, where the wavelength is $\lambda_{n2} = \lambda/n_2$:

$$N_2 = \frac{L}{\lambda_{n2}} = \frac{Ln_2}{\lambda}. \tag{36-10}$$

To find the new phase difference between the waves, we subtract the smaller of N_1 and N_2 from the larger. Assuming $n_2 > n_1$, we obtain

$$N_2 - N_1 = \frac{Ln_2}{\lambda} - \frac{Ln_1}{\lambda} = \frac{L}{\lambda}(n_2 - n_1). \tag{36-11}$$

Thus the phase difference is simply the optical path length difference divided by the wavelength of the light in a vacuum.

Suppose Eq. 36-11 tells us that the waves now have a phase difference of 45.6 wavelengths. That is equivalent to taking the initially in-phase waves and shifting one of them by 45.6 wavelengths. However, a shift of an integer number of wavelengths (such as 45) would put the waves back in phase, so it is only the decimal fraction (here, 0.6) that is important. A phase difference of 45.6 wavelengths is equivalent to an *effective phase difference* of 0.6 wavelength.

A phase difference of 0.5 wavelength puts two waves exactly out of phase. If the waves had equal amplitudes and were to reach some common point, they would then undergo fully destructive interference, producing darkness at that point. With an effective phase difference of 0.0 wavelength, they would, instead, undergo fully constructive interference, resulting in brightness at the common point. Our effective phase difference of 0.6 wavelength is an intermediate situation, but closer to destructive interference, and the waves would produce a dimly illuminated common point.

We can also express phase difference in terms of radians and degrees, as we have done already.

> A phase difference of one wavelength is equivalent to phase differences of 2π rad or 360°.

READING EXERCISE 36-1: The figure shows a monochromatic ray of light traveling across parallel interfaces, from an original material *a*, through layers of material *b* and *c*, and then back into material *a*. Rank the materials according to the speed of light in them, greatest first.

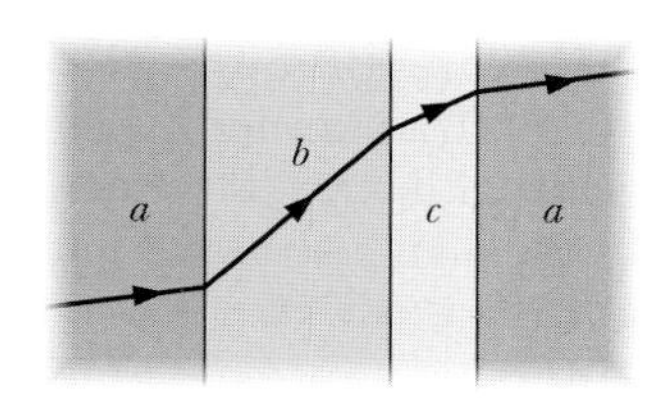

TOUCHSTONE EXAMPLE 36-1: Phase Difference and Interference

In Fig. 36-3, the two light waves that are represented by the rays have wavelength 550.0 nm before entering media 1 and 2. They also have equal amplitudes and are in phase. Medium 1 is now just air, and medium 2 is a transparent plastic layer of index of refraction 1.600 and thickness 2.600 μm.

(a) What is the phase difference of the emerging waves in wavelengths, radians, and degrees? What is their effective phase difference (in wavelengths)?

SOLUTION ■ One **Key Idea** here is that the phase difference of two light waves can change if they travel through different media, with different indexes of refraction. The reason is that their wavelengths are different in the different media. We can calculate the change in phase difference by counting the number of wavelengths that fits into each medium and then subtracting those numbers. When the path lengths of the waves in the two media are identical, Eq. 36-11 gives the result. Here we have $n_1 = 1.000$ (for the air), $n_2 = 1.600$, $L = 2.600$ μm, and $\lambda = 550.0$ nm. Thus, Eq. 36-11 yields

$$N_2 - N_1 = \frac{L}{\lambda}(n_2 - n_1) = \frac{2.600 \times 10^{-6}\text{ m}}{5.500 \times 10^{-7}\text{ m}}(1.600 - 1.000) = 2.84. \quad \text{(Answer)}$$

Thus, the phase difference of the emerging waves is 2.84 wavelengths. Because 1.0 wavelength is equivalent to 2π rad and 360°, you can show that this phase difference is equivalent to

$$\text{phase difference} = 17.8 \text{ rad} \approx 1020°. \quad \text{(Answer)}$$

A second **Key Idea** is that the effective phase difference is the decimal part of the actual phase difference *expressed in wavelengths*. Thus, we have

$$\text{effective phase difference} = 0.84 \text{ wavelength}. \quad \text{(Answer)}$$

You can show that this is equivalent to 5.3 rad and about 300°. *Caution:* We do *not* find the effective phase difference by taking the decimal part of the actual phase difference as expressed in radians or degrees. For example, we do *not* take 0.8 rad from the actual phase difference of 17.8 rad.

(b) If the rays of the waves were angled slightly so that the waves reached the same point on a distant viewing screen, what type of interference would the waves produce at that point?

SOLUTION ■ The **Key Idea** here is to compare the effective phase difference of the waves with the phase differences that give the extreme types of interference. Here the effective phase difference of 0.84 wavelength is between 0.5 wavelength (for fully destructive interference, or the darkest possible result) and 1.0 wavelength (for fully constructive interference, or the brightest possible result), but closer to 1.0 wavelength. Thus, the waves would produce intermediate interference that is closer to fully constructive interference—they would produce a relatively bright spot.

36-3 Diffraction

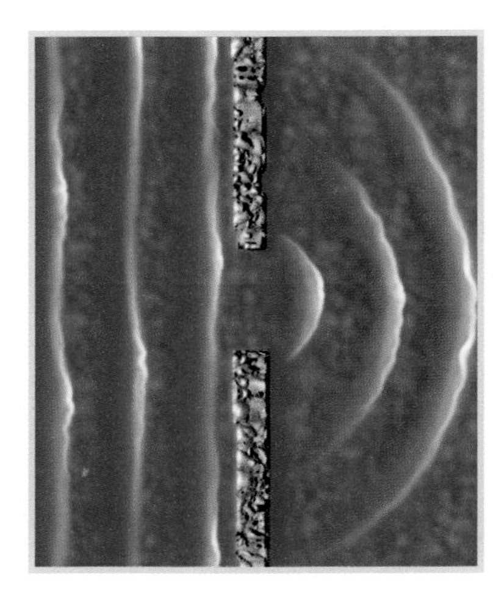

FIGURE 36-4 ■ The diffraction of water waves in a ripple tank. The waves are produced by an oscillating paddle at the left. As they move from left to right, they flare out through an opening in a barrier along the water surface.

In the next section we shall discuss the experiment that first proved that light is a wave. To prepare for that discussion, we must introduce the idea of **diffraction** of waves, a phenomenon that we explore much more fully in Chapter 37. Its essence is this: If a wave encounters a barrier that has an opening of dimensions similar to the wavelength, the part of the wave that passes through the opening will flare (spread) out—will *diffract*—into the region beyond the barrier. The flaring out is consistent with the spreading of the wavelets in the Huygens construction of Fig. 36-1. Diffraction occurs for waves of all types, not just light waves. Figure 36-4 shows the diffraction of water waves traveling across the surface of water in a shallow tank.

Figure 36-5*a* shows the situation schematically for an incident plane wave of wavelength λ encountering a slit that has width $a = 6.0\lambda$ and extends into and out of the page. The wave flares out on the far side of the slit. Figures 36-5*b* (with $a = 3.0\lambda$) and 36-5*c* ($a = 1.5\lambda$) illustrate the main feature of diffraction: the narrower the slit, the greater the diffraction.

Diffraction limits geometrical optics, in which we represent an electromagnetic wave with a ray. If we actually try to form a ray by sending light through a narrow slit, or through a series of narrow slits, diffraction will always defeat our effort because it always causes the light to spread. Indeed, the narrower we make the slits (in the hope of producing a narrower beam), the greater the spreading is. Thus, geometrical optics holds only when slits or other apertures that might be located in the path of light have dimensions that are much larger than the wavelength of the light.

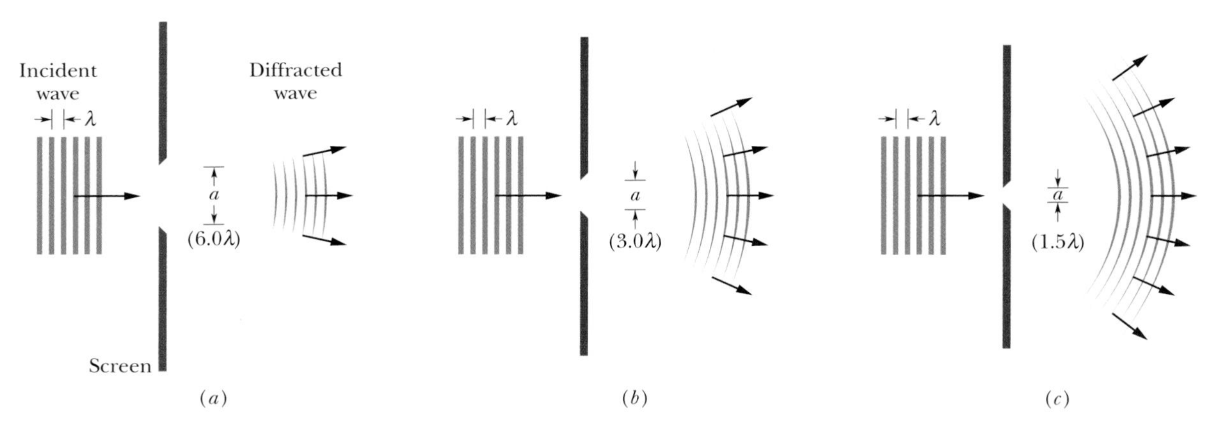

FIGURE 36-5 ■ Diffraction represented schematically. For a given wavelength λ, the diffraction is more pronounced the smaller the slit width a. The figures show the cases for (*a*) slit width $a = 6.0\lambda$, (*b*) slit width $a = 3.0\lambda$, and (*c*) slit width $a = 1.5\lambda$. In all three cases, the screen and the length of the slit extend well into and out of the page, perpendicular to it.

36-4 Young's Interference Experiment

In 1801, Thomas Young experimentally proved that light is a wave, contrary to what most other scientists then thought. He did so by demonstrating that light undergoes interference, as do water waves, sound waves, and waves of all other types. In addition, he was able to measure the average wavelength of sunlight; his value, 570 nm, is

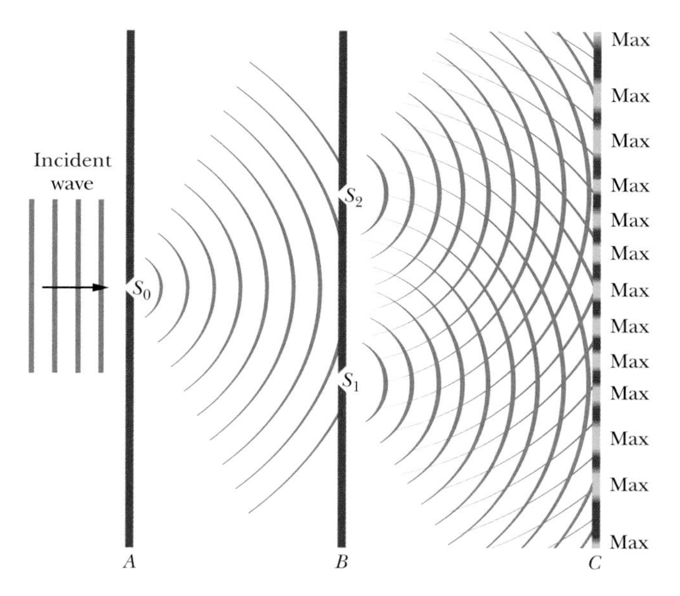

FIGURE 36-6 ■ In Young's interference experiment, incident monochromatic light is diffracted by slit S_0, which then acts as a point source of light that emits semicircular wavefronts. As that light reaches screen B, it is diffracted by slits S_1 and S_2, which then act as two point sources of light. The light waves traveling from slits S_1 and S_2 overlap and undergo interference, forming an interference pattern of maxima and minima on viewing screen C. This figure is a cross section; the screens, slits, and interference pattern extend into and out of the page. Between screens B and C, the semicircular wavefronts centered on S_2 depict the waves that would be there if only S_2 were open. Similarly, those centered on S_1 depict waves that would be there if only S_1 were open.

impressively close to the modern accepted value of 555 nm. We shall here examine Young's historic experiment as an example of the interference of light waves.

Figure 36-6 gives the basic arrangement of Young's experiment. Light from a distant monochromatic source illuminates slit S_0 in screen A. The emerging light then spreads via diffraction to illuminate two slits S_1 and S_2 in screen B. Diffraction of the light by these two slits sends overlapping circular waves into the region beyond screen B, where the waves from one slit interfere with the waves from the other slit.

The "snapshot" of Fig. 36-6 depicts the interference of the overlapping waves from very small slits. However, we cannot see evidence for the interference except where a viewing screen C intercepts the light. Where it does so, points of interference maxima form visible bright rows—called *bright bands, bright fringes,* or (loosely speaking) *maxima*—that extend across the screen (into and out of the page in Fig. 36-6). Dark regions—called *dark bands, dark fringes,* or (loosely speaking) *minima*—result from fully destructive interference and are visible between adjacent pairs of bright fringes. (*Maxima* and *minima* more properly refer to the center of a band.) The pattern of bright and dark fringes on the screen is called an **interference pattern.** Figure 36-7 is a photograph of part of the interference pattern as seen from the left in Fig. 36-6. The fringes that appear on a flat screen get further apart and dimmer as the distance from the center of the screen increases.

Note: These are very small slits, so we assume diffraction acts to spread out the waves passing through each slit. This explains why the waves from the two slits can overlap at a distant point. However, for now we ignore the fact that a diffracted wave has more intensity in the direction of the wavefront incident on the slit. In Chapter 37 we will analyze diffraction mathematically, which is the cause of the weakening intensity away from the center of the pattern seen in Fig. 36-7.

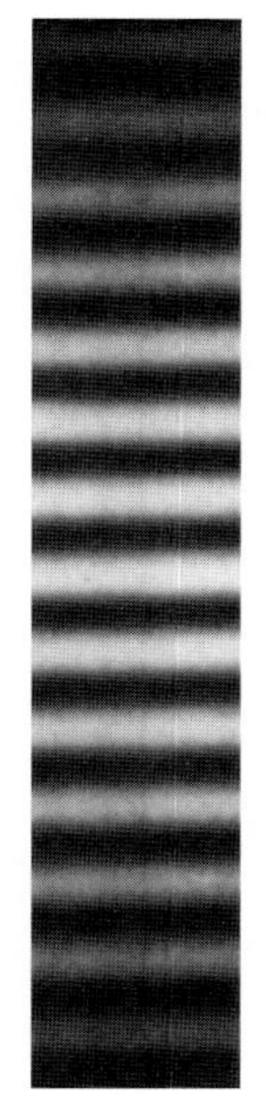

FIGURE 36-7 ■ A photograph of the interference pattern produced by the arrangement shown in Fig 36-6. (The photograph is a front view of part of screen C.) The alternating maxima and minima are called *interference fringes* (because they resemble the decorative fringe sometimes used on clothing and rugs).

Locating the Fringes

Light waves produce fringes in a *Young's double-slit interference experiment,* as it is called, but what exactly determines the locations of the fringes? To answer, we shall use the arrangement in Fig. 36-8*a*. There, a plane wave of monochromatic light is incident on two slits S_1 and S_2 in screen B; the light diffracts through the slits and produces an interference pattern on screen C. We draw a central axis from the point halfway between the slits to screen C as a reference. We then pick, for discussion, an arbitrary point P on the screen, at angle θ to the central axis. This point intercepts the wave of ray r_1 from the bottom slit and the wave of ray r_2 from the top slit.

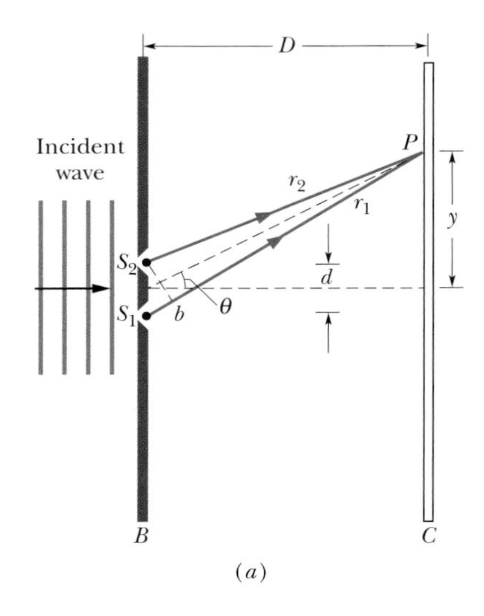

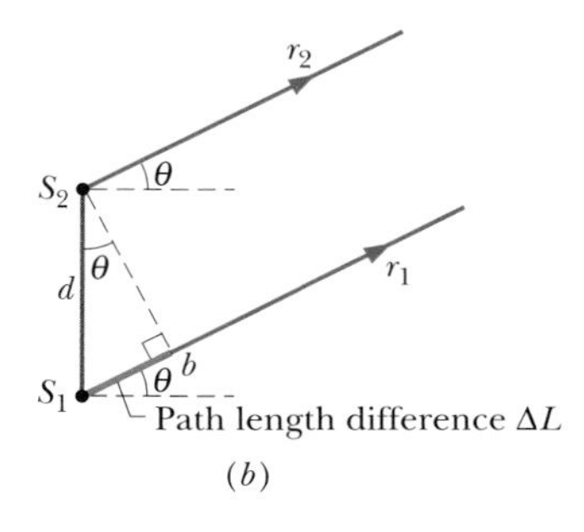

FIGURE 36-8 ■ (*a*) Waves from slits S_1 and S_2 (which extend into and out of the page) with spacing d combine at P, an arbitrary point on screen C at distance y from the central axis. The angle θ serves as a convenient locator for P. (*b*) For a slit screen distance of $D \gg d$, we can approximate rays r_1 and r_2 as being parallel, at angle θ to the central axis. *Note:* In a typical demonstration of these effects d might be on the order of 1 mm or some fraction thereof and D might be 1–2 meters or more.

These waves are in phase when they pass through the two slits because there they are just portions of the same incident wave. However, once they have passed the slits, the two waves must travel different distances to reach P. We saw a similar situation in Section 18-4 with sound waves and concluded that

> The phase difference between two waves can change if the waves travel paths of different lengths.

The change in phase difference is due to the *path length difference* ΔL in the paths taken by the waves. Consider two waves initially exactly in phase, traveling along paths with a path length difference ΔL, and then passing through some common point. When ΔL is zero or an integer number of wavelengths, the waves arrive at the common point exactly in phase and they interfere fully constructively there. If that is true for the waves of rays r_1 and r_2 in Fig. 36-8, then point P is part of a bright fringe. When, instead, ΔL is an odd multiple of half a wavelength, the waves arrive at the common point exactly out of phase and they interfere fully destructively there. If that is true for the waves of rays r_1 and r_2, then point P is part of a dark fringe (and, of course, we can have intermediate situations of interference and thus intermediate illumination at P.) Thus,

> What appears at each point on the viewing screen in a Young's double-slit interference experiment is determined by the path length difference ΔL of the rays reaching that point.

We can specify where each bright or dark fringe is located on the screen by giving the angle θ from the central axis to that fringe. To find θ, we must relate it to ΔL. We start with Fig. 36-8*a* by finding a point b along ray r_1 such that the path length from points b to P equals the path length from S_2 to P. Then the path length difference ΔL between the two rays is the distance from S_1 to b.

The relation between this S_1-to-b distance and θ is complicated, but we can simplify it considerably if we arrange for the distance D from the slits to the screen to be much greater than the slit separation d. Then we can approximate rays r_1 and r_2 as being parallel to each other and at angle θ to the central axis (Fig. 36-8*b*). We can also approximate the triangle formed by S_1, S_2, and b as being a right triangle, and approximate the angle inside that triangle at S_2 as being θ. Then, for that triangle, $\sin\theta = \Delta L/d$ and thus

$$\Delta L = d\sin\theta \quad \text{(path length difference).} \tag{36-12}$$

For a bright fringe, we saw that ΔL must be zero or an integer number of wavelengths. Using Eq. 36-12, we can write this requirement as

$$\Delta L = d\sin\theta = (\text{integer})(\lambda), \tag{36-13}$$

or as
$$d\sin\theta = m\lambda, \quad \text{for } m = 0, 1, 2, \ldots \quad \text{(maxima—bright fringes).} \tag{36-14}$$

For a dark fringe, ΔL must be an odd multiple of half a wavelength. Again using Eq. 36-12, we can write this requirement as

$$\Delta L = d\sin\theta = (\text{odd number})(\tfrac{1}{2}\lambda), \tag{36-15}$$

or as
$$d\sin\theta = (m + \tfrac{1}{2})\lambda, \quad \text{for } m = 0, 1, 2, \ldots \quad \text{(minima—dark fringes).} \tag{36-16}$$

With Eqs. 36-14 and 36-16, we can find the angle θ to any fringe and thus locate that fringe; further, we can use the values of m to label the fringes. For $m = 0$, Eq. 36-14 tells us that a bright fringe is at $\theta = 0$—that is, on the central axis. This *central maximum* is the point at which waves arriving from the two slits have a path length difference $\Delta L = 0$, hence zero phase difference.

For, say, $m = 2$, Eq. 36-14 tells us that *bright* fringes are at

$$\theta = \sin^{-1}\left(\frac{2\lambda}{d}\right)$$

above and below the central axis. Waves from the two slits arrive at these two fringes with $\Delta L = 2\lambda$ and with a phase difference of two wavelengths. These fringes are said to be the *second-order fringes* (meaning $m = 2$) or the *second side maxima* (the second maxima to the side of the central maximum), or they are described as being the second fringes from the central maximum.

For $m = 1$, Eq. 36-16 tells us that *dark* fringes are at

$$\theta = \sin^{-1}\left(\frac{1.5\lambda}{d}\right)$$

above and below the central axis. Waves from the two slits arrive at these two fringes with $\Delta L = 1.5\lambda$ and with a phase difference, in wavelengths, of 1.5. These fringes are called the *second dark fringes* or *second minima* because they are the second dark fringes from the central axis. (The first dark fringes, or first minima, are at locations for which $m = 0$ in Eq. 36-16.)

We derived Eqs. 36-14 and 36-16 for the situation $D >> d$. However, they also apply if we place a converging lens between the slits and the viewing screen and then move the viewing screen closer to the slits, to the focal point of the lens. (The screen is then said to be in the *focal plane* of the lens; that is, it is in the plane perpendicular to the central axis at the focal point.) One property of a converging lens is that it focuses all rays that are parallel to one another to the same point on its focal plane. Thus, the rays that now arrive at any point on the screen (in the focal plane) were exactly parallel (rather than approximately) when they left the slits. They are like the initially parallel rays in Fig. 35-23*a* that are directed to a point (the focal point) by a lens.

READING EXERCISE 36-2: In Fig. 36-8, what are ΔL (as a multiple of the wavelength) and the phase difference (in wavelengths) for the two rays if point P is (a) a third side maximum and (b) a third minimum? ■

TOUCHSTONE EXAMPLE 36-2: Distance Between Adjacent Maxima

What is the distance on screen C in Fig. 36-8*a* between adjacent maxima near the center of the interference pattern? The wavelength λ of the light is 546 nm, the slit separation d is 0.12 mm, and the slit–screen separation D is 55 cm. Assume that θ in Fig. 36-8 is small enough to permit use of the approximations $\sin\theta \approx \tan\theta \approx \theta$, in which θ is expressed in radian measure.

SOLUTION ■ First, let us pick a maximum with a low value of m to ensure that it is near the center of the pattern. Then one **Key Idea** is that, from the geometry of Fig. 36-8*a*, the maximum's vertical distance y_m from the center of the pattern is related to its angle θ from the central axis by

$$\tan\theta \approx \theta = \frac{y_m}{D}.$$

A second **Key Idea** is that, from Eq. 36-14, this angle θ for the mth maximum is given by

$$\sin\theta \approx \theta = \frac{m\lambda}{d}.$$

If we equate these two expressions for θ and solve for y_m, we find

$$y_m = \frac{m\lambda D}{d}. \quad (36\text{-}17)$$

For the next farther out maximum, we have

$$y_{m+1} = \frac{(m+1)\lambda D}{d}. \quad (36\text{-}18)$$

We find the distance between these adjacent maxima by subtracting Eq. 36-17 from Eq. 36-18:

$$\Delta y = y_{m+1} - y_m = \frac{\lambda D}{d}$$

$$= \frac{(546 \times 10^{-9}\text{ m})(55 \times 10^{-2}\text{ m})}{0.12 \times 10^{-3}\text{ m}}$$

$$= 2.50 \times 10^{-3}\text{ m} \approx 2.5\text{ mm}. \quad \text{(Answer)}$$

As long as d and θ in Fig. 36-8*a* are small, the separation of the interference fringes is independent of m; that is, the fringes are evenly spaced.

36-5 Coherence

For the interference pattern to appear on viewing screen C in Fig. 36-6, the light waves reaching any point P on the screen must have a phase difference that does not vary in time. That is the case in Fig. 36-6, because the waves passing through slits S_1 and S_2 are portions of the single light wave that illuminates the slits. Because the phase difference remains constant, the light from slits S_1 and S_2 is said to be completely **coherent.**

Direct sunlight is partially coherent; that is, sunlight waves intercepted at two points have a constant phase difference only if the points are very close. If you look closely at your fingernail in bright sunlight, you can see a faint interference pattern called *speckle* that causes the nail to appear to be covered with specks. You see this effect because light waves scattering from very close points on the nail are sufficiently coherent to interfere with one another at your eye. The slits in a double-slit experiment, however, are not close enough, and in direct sunlight, the light at the slits would be **incoherent.** To get coherent light, we would have to send the sunlight through a single slit as in Fig. 36-6; because that single slit is small, light that passes through it is coherent. In addition, the smallness of the slit causes the coherent light to spread sufficiently via diffraction to illuminate both slits in the double-slit experiment.

If we replace the double slits with two similar but independent monochromatic light sources, such as two fine incandescent wires, the phase difference between the waves emitted by the sources varies rapidly and randomly. (This occurs because the light is emitted by vast numbers of atoms in the wires, acting randomly and independently for extremely short times—of the order of nanoseconds.) As a result, at any given point on the viewing screen, the interference between the waves from the two sources varies rapidly and randomly between fully constructive and fully destructive. The eye (and most common optical detectors) cannot follow such changes, and no interference pattern can be seen. The fringes disappear, and the screen is seen as being uniformly illuminated.

A *laser* differs from common light sources in that its atoms emit light in a cooperative manner, thereby making the light coherent. Moreover, the light is almost monochromatic, is emitted in a thin beam with little spreading, and can be focused to a width that almost matches the wavelength of the light.

36-6 Intensity in Double-Slit Interference

Equations 36-14 and 36-16 tell us how to locate the maxima and minima of the double-slit interference pattern on screen C of Fig. 36-8 as a function of the angle θ in that figure. Here we wish to derive an expression for the intensity I of the fringes as a function of θ.

The light leaving the slits is in phase. However, let us assume that the light waves from the two slits are not in phase when they arrive at point P. Instead, the electric field components of those waves at point P are not in phase and vary with time as

$$E_1 = E_0 \sin \omega t \tag{36-19}$$

and

$$E_2 = E_0 \sin(\omega t + \phi), \tag{36-20}$$

where ω is the angular frequency of the waves and ϕ is the phase constant of wave E_2. Note that when θ is small the two waves have approximately the same amplitude E_0 and a phase difference of ϕ. Because that phase difference does not vary, the waves are coherent. We shall show that these two waves will combine at P to produce an illumination of intensity I given by

$$I = 4I_0 \cos^2 \tfrac{1}{2}\phi, \tag{36-21}$$

where

$$\phi = \frac{2\pi d}{\lambda} \sin\theta. \tag{36-22}$$

In Eq. 36-21, I_0 is the intensity of the light that arrives on the screen from one slit when the other slit is temporarily covered. We assume that the slits are so narrow in comparison to the wavelength that this single-slit intensity is essentially uniform over the central region of the screen in which we wish to examine the fringes.

Equations 36-21 and 36-22, which together tell us how the intensity I of the fringe pattern varies with the angle θ in Fig. 36-8, necessarily contain information about the location of the maxima and minima. Let us see if we can extract it.

Study of Eq. 36-21 shows that intensity maxima will occur when

$$\tfrac{1}{2}\phi = m\pi \qquad \text{for } m = 0, 1, 2, \ldots. \tag{36-23}$$

If we put this result into Eq. 36-22, we find

$$2m\pi = \frac{2\pi d}{\lambda} \sin\theta \qquad \text{for } m = 0, 1, 2, \ldots,$$

or

$$\Delta L = d \sin\theta = m\lambda \qquad \text{for } m = 0, 1, 2, \ldots \quad \text{(maxima)}, \tag{36-24}$$

which is exactly Eq. 36-14, the expression that we derived earlier for the locations of the maxima.

The minima in the fringe pattern occur when

$$\tfrac{1}{2}\phi = (m + \tfrac{1}{2})\pi \qquad \text{for } m = 0, 1, 2, \ldots.$$

If we combine this relation with Eq. 36-22 we are led at once to

$$d \sin\theta = (m + \tfrac{1}{2})\lambda \qquad \text{for } m = 0, 1, 2, \ldots \quad \text{(minima)}, \tag{36-25}$$

which is just Eq. 36-16, the expression we derived earlier for the locations of the fringe minima.

Figure 36-9, which is a plot of Eq. 36-21, shows the intensity of double-slit interference pattern fringes near the central maxima as a function of the phase difference ϕ

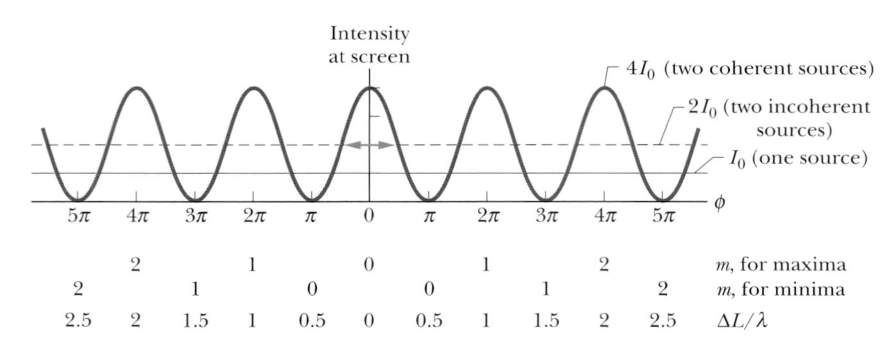

FIGURE 36-9 ■ A plot of Eq. 36-21, showing the intensity of a double-slit interference pattern as a function of the phase difference between the waves when they arrive from the two slits. I_0 is the (uniform) intensity that would appear on the screen if one slit were covered. The average intensity of the fringe pattern is $2I_0$, and the *maximum* intensity (for coherent light) is $4I_0$.

between the waves at the screen. The horizontal solid line is I_0, the (uniform) intensity on the screen when one of the slits is covered up. Note in Eq. 36-21 and the graph that the intensity I (which is always positive) varies from zero at the fringe minima to $4I_0$ at the fringe maxima.

If the waves from the two sources (slits) were *incoherent,* so that no enduring phase relation existed between them, there would be no fringe pattern and the intensity would have the uniform value $2I_0$ for all points on the screen; the horizontal dashed line in Fig. 36-9 shows this uniform value.

Interference cannot create or destroy energy but merely redistributes it over the screen. Thus, the *average* intensity on the screen must be the same $2I_0$ regardless of whether the sources are coherent. This follows at once from Eq. 36-21; if we substitute $\frac{1}{2}$, the average value of the cosine-squared function, this equation reduces to $\langle I \rangle = 2I_0$.

Proof of Eqs. 36-21 and 36-22

We shall combine the electric field components E_1 and E_2, given by Eqs. 36-19 and 36-20, respectively, by the method of phasors discussed in Section 17-12. In Fig. 36-10*a*, the waves with components E_1 and E_2 are represented by phasors of magnitude E_0 that rotate around the origin at angular speed ω. The values of E_1 and E_2 at any time are the projections of the corresponding phasors on the vertical axis. Figure 36-10*a* shows the phasors and their projections at an arbitrary time t. Consistent with Eqs. 36-19 and 36-20, the phasor for E_1 has a rotation angle ωt and the phasor for E_2 has a rotation angle $\omega t + \phi$.

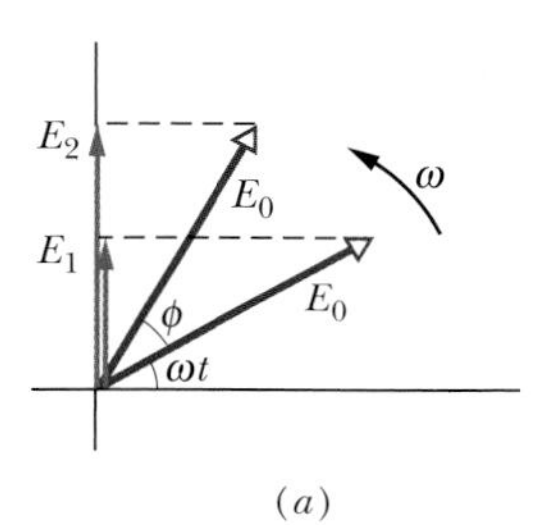

To combine the field components E_1 and E_2 at any point P in Fig. 36-8, we add their phasors as if they were vectors (which they are not), as shown in Fig. 36-10*b*. The magnitude of the phasor sum is the amplitude E of the resultant wave at point P, and that wave has a certain phase constant β. To find the amplitude E in Fig. 36-10*b*, we first note that the two angles marked β are equal because they are opposite equal-length sides of a triangle. From the theorem (for triangles) that an exterior angle (ϕ) is equal to the sum of the two opposite interior angles ($\beta + \beta$), we see that $\beta = \frac{1}{2}\phi$. Thus, we have

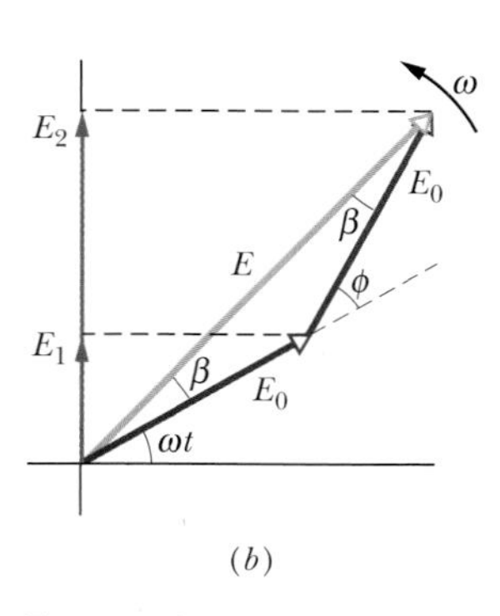

FIGURE 36-10 ■ (*a*) Phasors representing, at time t, the electric field components given by Eqs. 36-19 and 36-20. Both phasors have magnitude E_0 and rotate with angular speed ω. Their phase difference is ϕ. (*b*) Addition of the two phasors gives the phasor representing the resultant wave, with amplitude E and phase constant β.

$$E = 2(E_0 \cos \beta) = 2E_0 \cos \tfrac{1}{2}\phi. \tag{36-26}$$

If we square each side of this relation we obtain

$$E^2 = 4E_0^2 \cos^2 \tfrac{1}{2}\phi. \tag{36-27}$$

Now, from Eq. 34-24, we know that the intensity of an electromagnetic wave is proportional to the square of its amplitude. Therefore, the waves we are combining in Fig. 36-10*b*, whose amplitudes are E_0, each have an intensity I_0 that is proportional to

E_0^2, and the resultant wave, with amplitude E, has an intensity I that is proportional to E_2. Thus,

$$\frac{I}{I_0} = \frac{E^2}{E_0^2}.$$

Substituting Eq. 36-27 into this equation and rearranging then yield

$$I = 4I_0 \cos^2 \tfrac{1}{2}\phi,$$

which is Eq. 36-21, which we set out to prove.

It remains to prove Eq. 36-22, which relates the phase difference ϕ between the waves arriving at any point P on the screen of Fig. 36-8 to the angle θ that serves as a locator of that point.

The phase difference ϕ in Eq. 36-20 is associated with the path difference $\Delta L = S_1b$ in Fig. 36-8*b*. If ΔL is $\frac{1}{2}\lambda$, then ϕ is π; if ΔL is λ, then ϕ is 2π, and so on. This suggests

$$\text{(phase difference)} = \frac{2\pi}{\lambda}\text{(path length difference)}. \tag{36-28}$$

The path difference ΔL in Fig. 36-8*b* is $d \sin \theta$, so Eq. 36-28 becomes

$$\phi = \frac{2\pi d}{\lambda} \sin\theta,$$

which is Eq. 36-22, the other equation that we set out to prove.

Combining More Than Two Waves

In a more general case, we might want to find the resultant of more than two sinusoidally varying waves at a point. The general procedure is this:

1. Construct a series of phasors representing the waves to be combined. Draw them end to end, maintaining the proper phase relations between adjacent phasors.
2. Construct the vector-like phasor sum of this array. The length of this sum gives the amplitude of the resultant phasor. The angle between the phasor sum and the first phasor is the phase of the resultant with respect to this first phasor. The projection of this resultant-sum phasor on the vertical axis gives the time variation of the resultant wave.

TOUCHSTONE EXAMPLE 36-3: Three Light Waves

Three light waves combine at a certain point where their electric field components are

$$E_1 = E_0 \sin \omega t,$$
$$E_2 = E_0 \sin(\omega t + 60°),$$
$$E_3 = E_0 \sin(\omega t - 30°).$$

Find their resultant component $E(t)$ at that point.

SOLUTION ■ The resultant wave is

$$E(t) = E_1(t) + E_2(t) + E_3(t).$$

The **Key Idea** here is two-fold: We can use the method of phasors to find this sum, and we are free to evaluate the phasors at any time t. To simplify the solution we choose $t = 0$, for which the phasors representing the three waves are shown in Fig. 36-11. We can add these three phasors either directly on a vector-capable calcula-

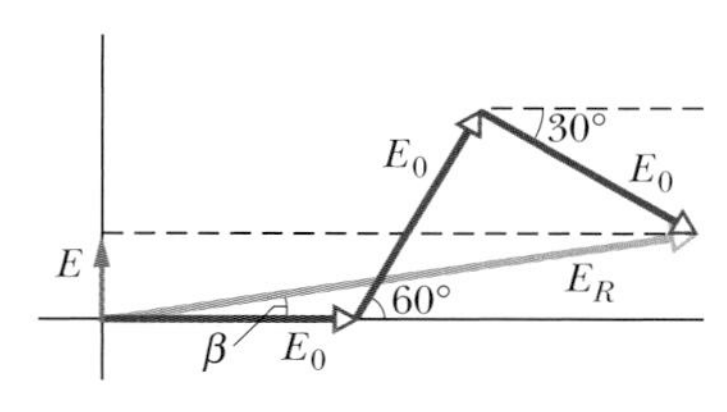

FIGURE 36-11 ■ Three phasors, representing waves with equal amplitudes E_0 and with phase constants 0°, 60°, and −30°, shown at time $t = 0$. The phasors combine to give a resultant phasor with magnitude E_R, at angle β.

tor or by components. For the component approach, we first write the sum of their horizontal components as

$$\sum E_h = E_0 \cos 0 + E_0 \cos 60° + E_0 \cos(-30°) = 2.37E_0.$$

The sum of their vertical components, which is the value of E at $t = 0$, is

$$\sum E_v = E_0 \sin 0 + E_0 \sin 60° + E_0 \sin(-30°) = 0.366E_0.$$

The resultant wave $E(t)$ thus has an amplitude E_R of

$$E_R = \sqrt{(2.37E_0)^2 + (0.366E_0)^2} = 2.4E_0,$$

and a phase angle β relative to the phasor representing E_1 of

$$\beta = \tan^{-1}\left(\frac{0.366E_0}{2.37E_0}\right) = 8.8°.$$

We can now write, for the resultant wave $E(t)$,

$$E = E_R \sin(\omega t + \beta)$$
$$= 2.4\, E_0 \sin(\omega t + 8.8°). \quad \text{(Answer)}$$

Be careful to interpret the angle β correctly in Fig. 36-11: It is the constant angle between E_R and the phasor representing E_1 as the four phasors rotate as a single unit around the origin. The angle between E_R and the horizontal axis in Fig. 36-11 does not remain equal to β.

36-7 Interference from Thin Films

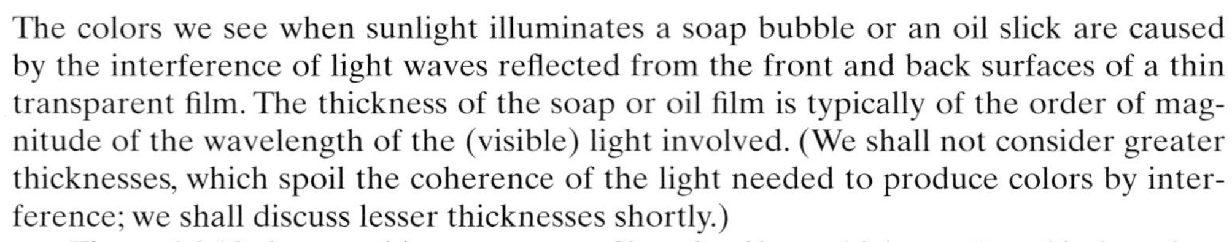

The colors we see when sunlight illuminates a soap bubble or an oil slick are caused by the interference of light waves reflected from the front and back surfaces of a thin transparent film. The thickness of the soap or oil film is typically of the order of magnitude of the wavelength of the (visible) light involved. (We shall not consider greater thicknesses, which spoil the coherence of the light needed to produce colors by interference; we shall discuss lesser thicknesses shortly.)

Figure 36-12 shows a thin transparent film of uniform thickness L and index of refraction n_2, illuminated by bright light of wavelength λ from a distant point source. For now, we assume that air lies on both sides of the film and thus that $n_1 = n_3$ in Fig. 36-12. For simplicity, we also assume that the light rays are almost perpendicular to the film ($\theta \approx 0$). We are interested in whether the film is bright or dark to an observer viewing it almost perpendicularly. (Since the film is brightly illuminated, how could it possibly be dark? You will see.)

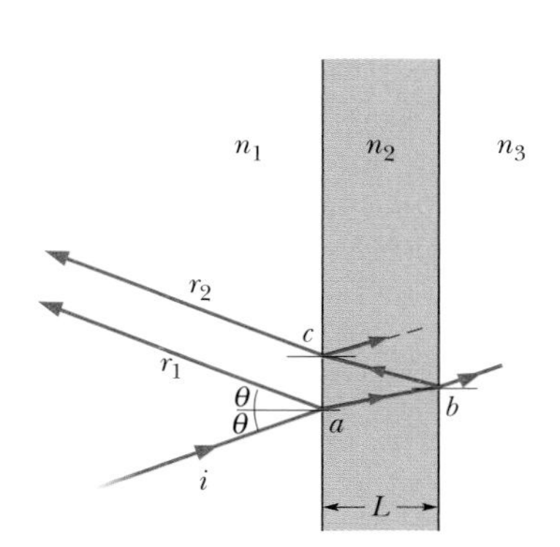

FIGURE 36-12 ■ Light waves, represented with ray i, are incident on a thin film of thickness L and index of refraction n_2. Rays r_1 and r_2 represent light waves that have been reflected by the front and back surfaces of the film, respectively. (All three rays are actually nearly perpendicular to the film.) The interference of the waves of r_1 and r_2 with each other depends on their phase difference. The index of refraction n_1 of the medium at the left can differ from the index of refraction n_3 of the medium at the right, but for now we assume that both media are air, with $n_1 = n_3 = 1.0$, which is less than n_2.

The incident light, represented by ray i, hits the front (left) surface of the film at point a and undergoes both reflection and refraction there. The reflected ray r_1 enters the observer's eye. The refracted light crosses the film to point b on the back surface, where it undergoes both reflection and refraction. The light reflected at b crosses back through the film to point c, where it undergoes both reflection and refraction. The light refracted at c, represented by ray r_2, also enters the observer's eye.

If the light waves of rays r_1 and r_2 are exactly in phase at the eye, they produce an interference maximum, and region ac on the film is bright to the observer. If they are exactly out of phase, they produce an interference minimum, and region ac is dark to the observer, *even though it is illuminated.* If there is some intermediate phase difference, there are intermediate interference and intermediate brightness.

Thus, the key to what the observer sees is the phase difference between the waves of rays r_1 and r_2. Both rays are derived from the same ray i, but the path involved in producing r_2 involves light traveling twice across the film (a to b, and then b to c), whereas the path involved in producing r_1 involves no travel through the film. Because θ is close to zero, we approximate the path length difference between the waves of r_1 and r_2 as $2L$. However, to find the phase difference between the waves, we cannot just find the number of wavelengths λ that is equivalent to a path length difference of $2L$. This simple approach is impossible for two reasons: (1) the path length difference occurs in a medium other than air, and (2) the reflections involved can change the phase.

> The phase difference between two waves can change if one or both are reflected.

Before we continue our discussion of interference from thin films, we must discuss changes in phase that are caused by reflections.

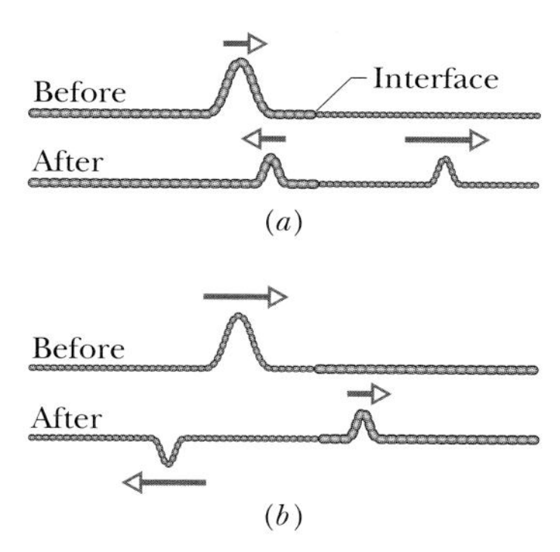

FIGURE 36-13 ■ Phase changes when a pulse is reflected at the interface between two stretched strings of different linear densities. The wave speed is greater in the lighter string. (*a*) The incident pulse is in the denser string. (*b*) The incident pulse is in the lighter string. Only here is there a phase change, and only in the reflected wave.

Reflection Phase Shifts

Refraction at an interface never causes a phase change—but reflection can, depending on the indices of refraction on the two sides of the interface. Figure 36-13 shows what happens when reflection of light waves causes a phase change, using as an example mechanical wave pulses on a denser string (along which pulse travel is relatively slow) and a lighter string (along which pulse travel is relatively fast). This effect is just like the one we discussed in Section 17-10. Recall that a pulse traveling down a string will reflect differently from an end of the string that is tied to a post (a fixed end) than from an end that is tied to a ring that can slide on the post (an open end) (see Fig. 17-25). Reflecting off a heavy string is like reflecting off a fixed end. Reflecting off a light string is like reflecting off an open end.

When a pulse traveling relatively slowly along the denser string in Fig. 36-13*a* reaches the interface with the lighter string, the pulse is partially transmitted and partially reflected, with no change in orientation. For light, this situation corresponds to the incident wave traveling in the medium of greater index of refraction n (recall that greater n means slower speed). In that case, the wave that is reflected at the interface does not undergo a change in phase; that is, the *reflection phase shift* is zero.

When a pulse traveling more quickly along the lighter string in Fig. 36-13*b* reaches the interface with the denser string, the pulse is again partially transmitted and partially reflected. The transmitted pulse again has the same orientation as the incident pulse, but now the reflected pulse is inverted. For a sinusoidal wave, such an inversion involves a phase change of π rad, or half a wavelength. For light, this situation corresponds to the incident wave traveling in the medium of lesser index of refraction (with greater speed). In that case, the wave that is reflected at the interface undergoes a phase shift of π rad, or half a wavelength.

We can summarize these results for light in terms of the index of refraction of the medium off which (or from which) the light reflects:

Reflection	Reflection phase shift	Phase change (rad)
Off lower index	$0.0\ \lambda$	0
Off higher index	$0.5\ \lambda$	π

An aid to remembering this is "reflection off high, phase change is pi."

Equations for Thin-Film Interference

In this chapter we have now seen three ways in which the phase difference between two light waves can change:

1. by reflection,
2. by the waves traveling along paths of different lengths,
3. by the waves traveling through media of different indexes of refraction.

When light reflects from a thin film, producing the waves of rays r_1 and r_2 in Fig. 36-12, all three ways are involved. Let us consider them one by one.

We first reexamine the two reflections in Fig. 36-12. At point a on the front interface, the incident wave (in air) reflects from the medium having the higher of the two indexes of refraction, so the wave of reflected ray r_1 has its phase shifted by 0.5 wavelength. At point b on the back interface, the incident wave reflects from the medium

TABLE 36-1
An Organizing Table for Thin-Film Interference in Air[a]

	r_1	r_2
Reflection phase shifts	0.5 wavelength	0
Path length difference	$2L$	
Index in which path length difference occurs	n_2	

[a]Valid for $n_2 > n_1$ and $n_2 > n_3$

(air) having the lower of the two indexes of refraction, so the wave reflected there is not shifted in phase by the reflection, and thus neither is the portion of it that exits the film as ray r_2. We can organize this information with the first line in Table 36-1. It tells us that, so far, as a result of the reflection phase shifts, the waves of r_1 and r_2 have a phase difference of 0.5 wavelength and thus are exactly out of phase.

Based on the information presented in Table 36-1, we have the following rules for predicting constructive interference (in phase) and destructive interference (out of phase):

For a film thickness L, two waves of wavelength λ traveling through a film with index of refraction n_2 will:

constructively interfere (in phase) if $\quad 2L = \dfrac{\text{odd number}}{2} \times \dfrac{\lambda}{n_2},$

destructively interfere (out of phase) if $\quad 2L = \text{integer} \times \dfrac{\lambda}{n_2}.$

These equations are valid for $n_2 > n_1$ and $n_2 > n_3$.

Now we must consider the path length difference $2L$ that occurs because the wave of ray r_2 crosses the film twice. (This difference $2L$ is shown in the expressions above. If the waves of r_1 and r_2 are to be exactly in phase so that they produce fully constructive interference, the path length $2L$ must cause an additional phase difference of 0.5, 1.5, 2.5, . . . wavelengths. Only then will the net phase difference be an integer number of wavelengths. Thus, for a bright film, we must have

$$2L = \frac{\text{odd number}}{2} \times \text{wavelength} \quad \text{(in-phase waves).} \tag{36-29}$$

The wavelength we need here is the wavelength λ_{n2} of the light in the medium containing path length $2L$—that is, in the medium with index of refraction n_2. Thus, we can rewrite Eq. 36-29 as

$$2L = \frac{\text{odd number}}{2} \times \lambda_{n2} \quad \text{(in-phase waves).} \tag{36-30}$$

If, instead, the waves are to be exactly out of phase so that there is fully destructive interference, the path length $2L$ must cause either no additional phase difference or a phase difference of 1, 2, 3, . . . , wavelengths. Only then will the net phase difference be an odd number of half-wavelengths. For a dark film, we must have

$$2L = \text{integer} \times \text{wavelength}, \tag{36-31}$$

where, again, the wavelength is the wavelength λ_{n2} in the medium containing $2L$. Thus, this time we have

$$2L = \text{integer} \times \lambda_{n2} \quad \text{(out-of-phase waves).} \tag{36-32}$$

Now we can use Eq. 36-8 ($\lambda_n = \lambda/n$) to write the wavelength of the wave of ray r_2 inside the film as

$$\lambda_{n2} = \frac{\lambda}{n_2}, \tag{36-33}$$

where λ is the wavelength of the incident light in vacuum (and approximately also in air). Substituting Eq. 36-33 into Eq. 36-30 and replacing "odd number/2" with $(m + \frac{1}{2})$ give us

$$2L = (m + \tfrac{1}{2})\frac{\lambda}{n_2} \quad \text{for } m = 0, 1, 2, \ldots \quad \text{(maxima—bright film in air).} \tag{36-34}$$

Similarly, with m replacing "integer," Eq. 36-32 yields

$$2L = m\frac{\lambda}{n_2} \quad \text{for } m = 0, 1, 2, \ldots \quad \text{(minima—dark film in air).} \tag{36-35}$$

For a given film thickness L, Eqs. 36-34 and 36-35 tell us the wavelengths of light for which the film appears bright and dark, respectively, one wavelength for each value of m. Intermediate wavelengths give intermediate brightnesses. For a given wavelength λ, Eqs. 36-34 and 36-35 tell us the thicknesses of the films that appear bright and dark in that light, respectively, one thickness for each value of m. Intermediate thicknesses give intermediate brightnesses.

A special situation arises when a film is so thin that L is much less than λ, say, $L < 0.1\lambda$. Then the path length difference $2L$ can be neglected, and the phase difference between r_1 and r_2 is due *only* to reflection phase shifts. If the film of Fig. 36-12, where the reflections cause a phase difference of 0.5 wavelength, has thickness $L < 0.1\lambda$, then r_1 and r_2 are exactly out of phase, and thus the film is dark, regardless of the wavelength and even the intensity of the light that illuminates it. This special situation corresponds to $m = 0$ in Eq. 36-35. We shall count any thickness $L < 0.1\lambda$ as being the least thickness specified by Eq. 36-35 to make the film of Fig. 36-12 dark. (Every such thickness will correspond to $m = 0$.) The next greater thickness that will make the film dark is that corresponding to $m = 1$.

Figure 36-14 shows a vertical soap film whose thickness increases from top to bottom because the weight of the film has caused it to slump. Bright white light illuminates the film. However, the top portion is so thin that it is dark. In the (somewhat thicker) middle we see fringes, or bands, whose color depends primarily on the wavelength at which reflected light undergoes fully constructive interference for a particular thickness. Toward the (thickest) bottom of the film the fringes become progressively narrower and the colors begin to overlap and fade.

FIGURE 36-14 ■ The reflection of light from a soapy water film spanning a vertical loop. The top portion is so thin that the light reflected there undergoes destructive interference, making that portion dark. Colored interference fringes, or bands, decorate the rest of the film but are marred by circulation of liquid within the film as the liquid is gradually pulled downward by gravitation.

Iridescence of a *Morpho* Butterfly Wing

A surface that displays colors due to thin-film interference is said to be *iridescent* because the tints of the colors change as you change your view of the surface. The iridescence of the top surface of a *Morpho* butterfly wing is due to thin-film interference of light reflected by thin terraces of transparent cuticle-like material on the wing. These terraces are arranged like wide, flat branches on a tree-like structure that extends perpendicular to the wing.

Suppose you look directly down on these terraces as white light shines directly down on the wing. Then the light reflected back up to you from the terraces undergoes fully constructive interference in the blue-green region of the visible spectrum. Light in the yellow and red regions, at the opposite end of the spectrum, is weaker because it undergoes only intermediate interference. Thus, the top surface of the wing looks blue-green to you.

If you intercept light that reflects from the wing in some other direction, the light has traveled along a slanted path through the terraces. Then the wavelength at which there is fully constructive interference is somewhat different from that for light reflected directly upward. Thus, if the wing moves in your view so that the angle at which you view it changes, the color at which the wing is brightest changes somewhat, producing the iridescence or brilliant rainbow-like colors of the wing.

TOUCHSTONE EXAMPLE 36-4: Brightest Reflected Light

White light, with a uniform intensity across the visible wavelength range of 400 to 690 nm, is perpendicularly incident on a water film, of index of refraction $n_2 = 1.33$ and thickness $L = 320$ nm, that is suspended in air. At what wavelength λ is the light reflected by the film brightest to an observer?

SOLUTION ■ The **Key Idea** here is that the reflected light from the film is brightest at the wavelengths λ for which the reflected rays are in phase with one another. The equation relating these wavelengths λ to the given film thickness L and film index of refraction n_2 is either Eq. 36-34 or Eq. 36-35, depending on the reflection phase shifts for this particular film.

To determine which equation is needed, we should fill out an organizing table like Table 36-1. However, because there is air on both sides of the water film, the situation here is exactly like that in Fig. 36-12, and thus the table would be exactly like Table 36-1. Then from Table 36-1, we see that the reflected rays are in phase (and thus the film is brightest) when

$$2L = \frac{\text{odd number}}{2} \times \frac{\lambda}{n_2},$$

which leads to Eq. 36-34:

$$2L = (m + \tfrac{1}{2})\frac{\lambda}{n_2}.$$

Solving for λ and substituting for L and n_2, we find

$$\lambda = \frac{2n_2L}{m + \frac{1}{2}} = \frac{(2)(1.33)(320\text{ nm})}{m + \frac{1}{2}} = \frac{851\text{ nm}}{m + \frac{1}{2}}.$$

For $m = 0$, this gives us $\lambda = 1700$ nm, which is in the infrared region. For $m = 1$, we find $\lambda = 567$ nm, which is yellow-green light, near the middle of the visible spectrum. For $m = 2$, $\lambda = 340$ nm, which is in the ultraviolet region. Thus, the wavelength at which the light seen by the observer is brightest is

$$\lambda = 567\text{ nm}. \qquad \text{(Answer)}$$

TOUCHSTONE EXAMPLE 36-5: Magnesium Fluoride Film

In Fig. 36-15, a glass lens is coated on one side with a thin film of magnesium fluoride (MgF_2) to reduce reflection from the lens surface. The index of refraction of MgF_2 is 1.38; that of the glass is 1.50. What is the least coating thickness that eliminates (via interference) the reflections at the middle of the visible spectrum ($\lambda = 550$ nm)? Assume that the light is approximately perpendicular to the lens surface.

SOLUTION ■ The **Key Idea** here is that reflection is eliminated if the film thickness L is such that light waves reflected from the two film interfaces are exactly out of phase. The equation relating L to the given wavelength λ and the index of refraction n_2 of the thin film is either Eq. 36-34 or Eq. 36-35, depending on the reflection phase shifts at the interfaces.

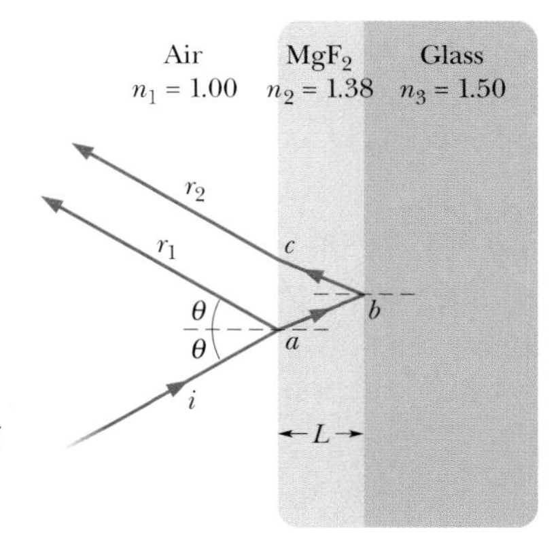

FIGURE 36-15 ■ Unwanted reflections from glass can be suppressed (at a chosen wavelength) by coating the glass with a thin transparent film of magnesium fluoride of the properly chosen thickness.

To determine which equation is needed, we fill out an organizing table like Table 36-1. At the first interface, the incident light is in air, which has a lesser index of refraction than the MgF_2 (the thin film). Thus, we fill in 0.5 wavelength under r_1 in our organizing table (meaning that the waves of ray r_1 are shifted by 0.5λ at the first interface). At the second interface, the incident light is in the MgF_2, which has a lesser index of refraction than the glass on the other side of the interface. Thus, we fill in 0.5 wavelength under r_2 in our table.

Because both reflections cause the same phase shift, they tend to put the waves of r_1 and r_2 in phase. Since we want those waves to be *out of phase,* their path length difference $2L$ must be an odd number of half-wavelengths:

$$2L = \frac{\text{odd number}}{2} \times \frac{\lambda}{n_2}.$$

This leads to Eq. 36-34. Solving that equation for L then gives us the film thicknesses that will eliminate reflection from the lens and coating:

$$L = (m + \tfrac{1}{2})\frac{\lambda}{2n_2}, \qquad \text{for } m = 0, 1, 2, \ldots . \qquad (36\text{-}36)$$

We want the least thickness for the coating—that is, the least L. Thus, we choose $m = 0$, the least possible value of m. Substituting it and the given data in Eq. 36-36, we obtain

$$L = \frac{\lambda}{4n_2} = \frac{550 \text{ nm}}{(4)(1.38)} = 99.6 \text{ nm}. \qquad \text{(Answer)}$$

TOUCHSTONE EXAMPLE 36-6: Red Light

Figure 36-16*a* shows a transparent plastic block with a thin wedge of air at the right. (The wedge thickness is exaggerated in the figure.) A broad beam of red light, with wavelength $\lambda = 632.8$ nm, is directed downward through the top of the block (at an incidence angle of 0°). Some of the light is reflected back up from the top and bottom surfaces of the wedge, which acts as a thin film (of air) with a thickness that varies uniformly and gradually from L_L at the left-hand end to L_R at the right-hand end. (The plastic layers above and below the wedge of air are too thick to act as thin films.) An observer looking down on the block sees an interference pattern consisting of six dark fringes and five bright red fringes along the wedge. What is the change in thickness $\Delta L = (L_R - L_L)$ along the wedge?

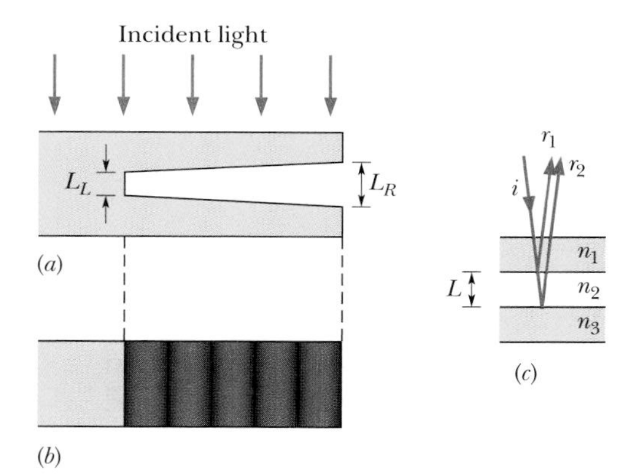

FIGURE 36-16 ■ (*a*) Red light is incident on a thin, air-filled wedge in the side of a transparent plastic block. The thickness of the wedge is L_L at the left end and L_R at the right end. (*b*) The view from above the block: an interference pattern of six dark fringes and five bright red fringes lies over the region of the wedge. (*c*) A representation of the incident ray i, reflected rays r_1 and r_2, and thickness L of the wedge anywhere along the length of the wedge.

SOLUTION ■ One **Key Idea** here is that the brightness at any point along the left–right length of the air wedge is due to the interference of the waves reflected at the top and bottom interfaces of the wedge. A second **Key Idea** is that the variation of brightness in the pattern of bright and dark fringes is due to the variation in the thickness of the wedge. In some regions, the thickness puts the reflected waves in phase and thus produces a bright reflection (a bright red fringe). In other regions, the thickness puts the reflected waves out of phase and thus produces no reflection (a dark fringe).

Because the observer sees more dark fringes than bright fringes, we can assume that a dark fringe is produced at both the left and right ends of the wedge. Thus, the interference pattern is that shown in Fig. 36-16*b*, which we can use to determine the change in thickness ΔL of the wedge.

Another **Key Idea** is that we can represent the reflection of light at the top and bottom interfaces of the wedge, at any point along its length, with Fig. 36-16*c*, in which L is the wedge thickness at that point. Let us apply this figure to the left end of the wedge, where the reflections give a dark fringe.

We know that, for a dark fringe, the waves of rays r_1 and r_2 in Fig. 36-16*c* must be out of phase. We also know that the equation relating the film thickness L to the light's wavelength λ and the film's index of refraction n_2 is either Eq. 36-34 or Eq. 36-35, depending on the reflection phase shifts. To determine which equation gives a dark fringe at the left end of the wedge, we should fill out an organizing table like Table 36-1.

At the top interface of the wedge, the incident light is in the plastic, which has a greater index of refraction than the air beneath that interface. Thus, we fill in 0 under r_1 in our organizing table. At the bottom interface of the wedge, the incident light is in air, which has a lesser index of refraction than the plastic beneath that interface. Thus, we fill in 0.5 wavelength under r_2 in our organizing table. Therefore, the reflections alone tend to put the waves of r_1 and r_2 out of phase.

Since the waves are, in fact, out of phase at the left end of the air wedge, the path length difference $2L$ at that end of the wedge must be given by

$$2L = \text{integer} \times \frac{\lambda}{n_2},$$

which leads to Eq. 36-35:

$$2L = m\frac{\lambda}{n_2}, \quad \text{for } m = 0, 1, 2, \ldots. \quad (36\text{-}37)$$

Here is another **Key Idea:** Eq. 36-37 holds not only for the left end of the wedge but also at any point along the wedge where a dark fringe is observed, including the right end—with a different integer value of m for each fringe. The least value of m is associated with the least thickness of the wedge where a dark fringe is observed. Progressively greater values of m are associated with progressively greater thicknesses of the wedge where a dark fringe is observed. Let m_L be the value at the left end. Then the value at the right end must be $m_L + 5$ because, from Fig. 36-16*b*, the right end is located at the fifth dark fringe from the left end.

We want the change ΔL in thickness, from the left end to the right end of the wedge. To find it we first solve Eq. 36-37 twice—once for the thickness L_L at the left end and once for the thickness L_R at the right end:

$$L_L = (m_L)\frac{\lambda}{2n_2}, \qquad L_R = (m_L + 5)\frac{\lambda}{2n_2}. \quad (36\text{-}38)$$

To find the change in thickness ΔL, we can now subtract L_L from L_R and substitute known data, including $n_2 = 1.00$ for the air within the wedge:

$$\Delta L = L_R - L_L = \frac{(m_L + 5)\lambda}{2n_2} - \frac{m_L\lambda}{2n_2} = \frac{5}{2}\frac{\lambda}{n_2}$$

$$= \frac{5}{2}\frac{632.8 \times 10^{-9}\text{ m}}{1.00}$$

$$= 1.58 \times 10^{-6}\text{ m}. \quad \text{(Answer)}$$

36-8 Michelson's Interferometer

An **interferometer** is a device that can be used to measure lengths or changes in length with great accuracy by means of interference fringes. We describe the form originally devised and built by A. A. Michelson in 1881.

Consider light that leaves point P on extended source S in Fig. 36-17 and encounters *beam splitter* M. A beam splitter is a mirror that transmits half the incident light and reflects the other half. In the figure we have assumed, for convenience, that this mirror possesses negligible thickness. At M the light thus divides into two waves. One proceeds by transmission toward mirror M_1; the other proceeds by reflection toward mirror M_2. The waves are entirely reflected at these mirrors and are sent back along their directions of incidence, each wave eventually entering telescope T. What the observer sees is a pattern of curved or approximately straight interference fringes; in the latter case the fringes resemble the stripes on a zebra.

The path length difference for the two waves when they recombine at the telescope is $2d_2 - 2d_1$, and anything that changes this path length difference will cause a change in the phase difference between these two waves at the eye. As an example, if mirror M_2 is moved by a distance $\frac{1}{2}\lambda$, the path length difference is changed by λ and the fringe pattern is shifted by one fringe (as if each dark stripe on a zebra had moved to where the adjacent dark stripe had been). Similarly, moving mirror M_2 by $\frac{1}{4}\lambda$ causes a shift by half a fringe (each dark zebra stripe shifts to where the adjacent white stripe was).

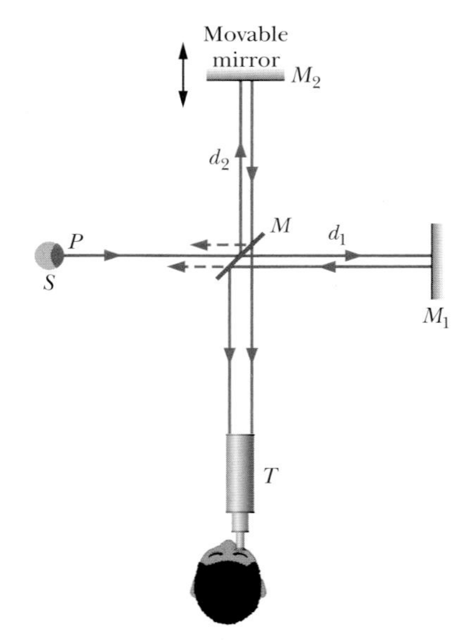

FIGURE 36-17 ■ Michelson's interferometer, showing the path of light originating at point P of an extended source S. Mirror M splits the light into two beams, which reflect from mirrors M_1 and M_2 back to M and then to telescope T. In the telescope an observer sees a pattern of interference fringes.

A shift in the fringe pattern can also be caused by the insertion of a thin transparent material into the optical path of one of the mirrors—say, M_1. If the material has thickness L and index of refraction n, then the number of wavelengths along the light's to-and-fro path through the material is, from Eq. 36-9,

$$N_m = \frac{2L}{\lambda_n} = \frac{2Ln}{\lambda}. \quad (36\text{-}39)$$

The number of wavelengths in the same thickness $2L$ of air before the insertion of the material is

$$N_a = \frac{2L}{\lambda}. \quad (36\text{-}40)$$

When the material is inserted, the light returned by mirror M_1 undergoes a phase change (in terms of wavelengths) of

$$N_m - N_a = \frac{2Ln}{\lambda} - \frac{2L}{\lambda} = \frac{2L}{\lambda}(n - 1). \tag{36-41}$$

For each phase change of one wavelength, the fringe pattern is shifted by one fringe. Thus, by counting the number of fringes through which the material causes the pattern to shift, and substituting that number for $N_m - N_a$ in Eq. 36-41, you can determine the thickness L of the material in terms of λ.

By such techniques the lengths of objects can be expressed in terms of the wavelengths of light. In Michelson's day, the standard of length—the meter—was chosen by international agreement to be the distance between two fine scratches on a certain metal bar preserved at Sèvres, near Paris. Michelson was able to show, using his interferometer, that the standard meter was equivalent to 1 553 163.5 wavelengths of a certain monochromatic red light emitted from a light source containing cadmium. For this careful measurement, Michelson received the 1907 Nobel Prize in physics. His work laid the foundation for the eventual abandonment (in 1961) of the meter bar as a standard of length and for the redefinition of the meter in terms of the wavelength of light. By 1983, as we have seen, even this wavelength standard was not precise enough to meet the growing requirements of science and technology, and it was replaced with a new standard based on a defined value for the speed of light as discussed in Section 1-6.

Problems

SEC. 36-2 ■ LIGHT AS A WAVE

1. Yellow Sodium Light The wavelength of yellow sodium light in air is 589 nm. (a) What is its frequency? (b) What is its wavelength in glass whose index of refraction is 1.52? (c) From the results of (a) and (b) find its speed in this glass.

2. Sapphire vs. Diamond How much faster, in meters per second, does light travel in sapphire than in diamond? See Table 35-1.

3. Yellow Light The speed of yellow light (from a sodium lamp) in a certain liquid is measured to be 1.92×10^8 m/s. What is the index of refraction of this liquid for the light at this wavelength?

4. Fused Quartz What is the speed in fused quartz of light of wavelength 550 nm? (See Fig. 35-3.)

5. Ocean Wave Ocean waves moving at a speed of 4.0 m/s are approaching a beach at an angle of 30° to the normal, as shown from above in Fig. 36-18. Suppose the water depth changes abruptly at a certain distance from the beach and the wave speed there drops to 3.0 m/s. Close to the beach, what is the angle θ between the direction of wave motion and the normal? (Assume the same law of refraction as for light.) Explain why most waves come in normal to a shore even though at large distances they approach at a variety of angles.

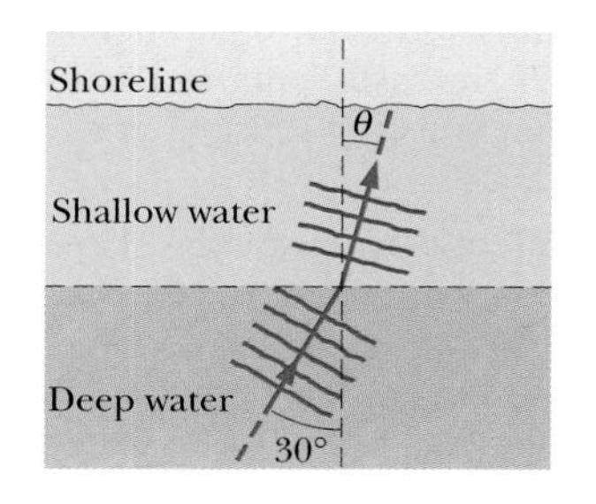

FIGURE 36-18 ■ Problem 5.

6. Two Pulses In Fig. 36-19. Two pulses of light are sent through layers of plastic with the indexes of refraction indicated and with thicknesses of either L or $2L$ as shown. (a) Which pulse travels through the plastic in less time? (b) In terms of L/c, what is the difference in the traversal times of the pulses?

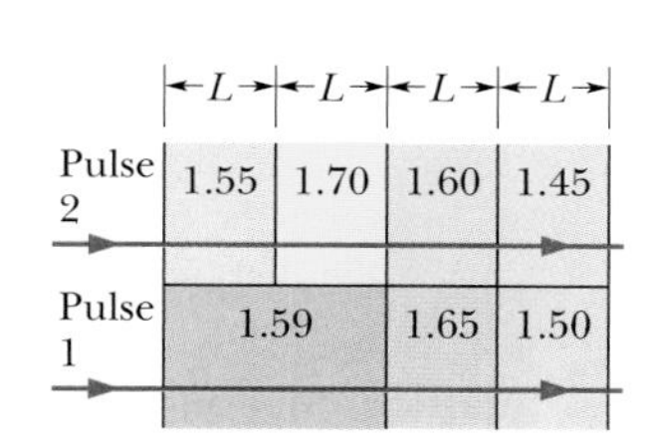

FIGURE 36-19 ■ Problem 6.

7. Two Waves In Fig. 36-3, assume that two waves of light in air, of wave length 400 nm, are initially in phase. One travels through a glass layer of index of refraction $n_1 = 1.60$ and thickness L. The other travels through an equally thick plastic layer of index of refraction $n_2 = 1.50$. (a) What is the least value L should have if the waves are to end up with a phase difference of 5.65 rad? (b) If the waves arrive at some common point after emerging, what type of interference do they undergo?

8. Two Media Suppose that the two waves in Fig. 36-3 have wavelength 500 nm in air. In wavelengths, what is their phase difference after traversing media 1 and 2 if (a) $n_1 = 1.50$, $n_2 = 1.60$, and $L = 8.50\ \mu$m; (b) $n_1 = 1.62$, $n_2 = 1.72$, and $L = 8.50\ \mu$m; and (c) $n_1 = 1.59$, $n_2 = 1.79$, and $L = 3.25\ \mu$m? (d) Suppose that in each of these three situations the waves arrive at a common point after emerging. Rank the situations according to the brightness the waves produce at the common point.

9. Initially in Phase Two waves of light in air, of wavelength 600.0 nm, are initially in phase. They then travel through plastic layers as shown in Fig. 36-20, with $L_1 = 4.00\ \mu$m, $L_2 = 3.50\ \mu$m,

$n_1 = 1.40$, and $n_2 = 1.60$. (a) In wavelengths, what is their phase difference after they both have emerged from the layers? (b) If the waves later arrive at some common point, what type of interference do they undergo?

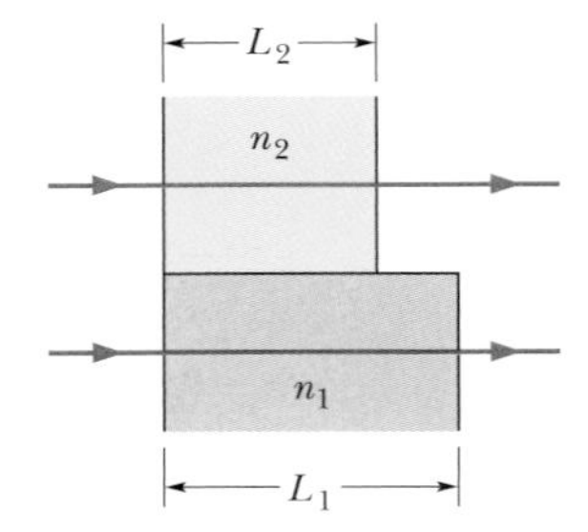

FIGURE 36-20 ■ Problem 9.

10. Two Light Waves In Fig. 36-3, assume that the two light waves, of wavelength 620 nm in air, are initially out of phase by π rad. The indexes of refraction of the media are $n_1 = 1.45$ and $n_2 = 1.65$. (a) What is the least thickness L that will put the waves exactly in phase once they pass through the two media? (b) What is the next greater L that will do this?

SEC. 36-4 ■ YOUNG'S INTERFERENCE EXPERIMENT

11. Green Light Monochromatic green light, of wavelength 550 nm, illuminates two parallel narrow slits 7.70 μm apart. Calculate the angular deviation (θ in Fig. 36-8) of the third-order (for $m = 3$) bright fringe (a) in radians and (b) in degrees.

12. Phase Difference What is the phase difference of the waves from the two slits when they arrive at the mth dark fringe in a Young's double-slit experiment?

13. Blue-Green Light Suppose that Young's experiment is performed with blue-green light of wavelength 500 nm. The slits are 1.20 mm apart, and the viewing screen is 5.40 m from the slits. How far apart are the bright fringes.

14. Angular Separation In a double-slit arrangement the slits are separated by a distance equal to 100 times the wavelength of the light passing through the slits. (a) What is the angular separation in radians between the central maximum and an adjacent maximum? (b) What is the distance between these maxima on a screen 50.0 cm from the slits?

15. Interference Fringes A double-slit arrangement produces interference fringes for sodium light ($\lambda = 589$ nm) that have an angular separation of 3.50×10^{-3} rad. For what wavelength would the angular separation be 10.0% greater?

16. Immersed in Water A double-slit arrangement produces interference fringes for sodium light ($\lambda = 589$ nm) that are 0.20° apart. What is the angular fringe separation if the entire arrangement is immersed in water ($n = 1.33$)?

17. Radio Frequency Sources Two radio-frequency point sources separated by 2.0 m are radiating in phase with $\lambda = 0.50$ m. A detector moves in a circular path around the two sources in a plane containing them. Without written calculation, find how many maxima it detects.

18. Long-Range Radio Waves Sources A and B emit long-range radio waves of wavelength 400 m, with the phase of the emission from A ahead of that from source B by 90°. The distance r_A from A to a detector is greater than the corresponding distance r_B by 100 m. What is the phase difference at the detector?

19. Two Interference Patterns In a double-slit experiment the distance between slits is 5.0 mm and the slits are 1.0 m from the screen. Two interference patterns can be seen on the screen: one due to light with wavelength 480 nm, and the other due to light with wavelength 600 nm. What is the separation on the screen between the third-order ($m = 3$) bright fringes of the two interference patterns?

FIGURE 36-21 ■ Problems 20 and 27.

20. Identical Radiators In Fig. 36-21, S_1 and S_2 are identical radiators of waves that are in phase and of the same wavelength λ. The radiators are separated by distance $d = 3.00\lambda$. Find the greatest distance from S_1, along the x axis, for which fully destructive interference occurs. Express this distance in wavelengths.

21. Mica Flake A thin flake of mica ($n = 1.58$) is used to cover one slit of a double-slit interference arrangement. The central point on the viewing screen is now occupied by what had been the seventh bright side fringe ($m = 7$) before the mica was used. If $\lambda = 550$ nm, what is the thickness of the mica? (*Hint:* Consider the wavelength of the light within the mica.)

22. Laser Light Laser light of wavelength 632.8 nm passes through a double-slit arrangement at the front of a lecture room, reflects off a mirror 20.0 m away at the back of the room, and then produces an interference pattern on a screen at the front of the room. The distance between adjacent bright fringes is 10.0 cm. (a) What is the slit separation? (b) What happens to the pattern when the lecturer places a thin cellophane sheet over one slit, thereby increasing by 2.50 the number of wavelengths along the path that includes the cellophane?

SEC. 36-6 ■ INTENSITY IN DOUBLE-SLIT INTERFERENCE

23. Same Frequency Two waves of the same frequency have amplitudes 1.00 and 2.00. They interfere at a point where their phase difference is 60.0°. What is the resultant amplitude?

24. Find Sum Find the sum y of the following quantities:

$$y_1 = 10 \sin \omega t \quad \text{and} \quad y_2 = 8.0 \sin(\omega t + 30°).$$

25. Use Phasors Add the quantities

$$y_1 = 10 \sin \omega t$$
$$y_2 = 15 \sin(\omega t + 30°)$$
$$y_3 = 5.0 \sin(\omega t - 45°)$$

using the phasor method.

26. Sketch Intensity Light of wavelength 600 nm is incident normally on two parallel narrow slits separated by 0.60 mm. Sketch the intensity pattern observed on a distant screen as a function of angle θ from the pattern's center for the range of values $0 \leq \theta \leq 0.0040$ rad.

27. Electromagnetic Waves S_1 and S_2 in Fig. 36-21 are point sources of electromagnetic waves of wavelength 1.00 m. They are in phase and separated by $d = 4.00$ m, and they emit at the same power. (a) If a detector is moved to the right along the x axis from source S_1, at what distances from S_1 are the first three interference maxima detected? (b) Is the intensity of the nearest minimum exactly zero? (*Hint:* Does the intensity of a wave from a point source remain constant with an increase in distance from the source?).

28. Horizontal Arrow The double horizontal arrow in Fig. 36-9 marks the points on the intensity curve where the intensity of the central fringe is half the maximum intensity. Show that the angular separation $\Delta\theta$ between the corresponding points on the viewing screen is

$$\Delta\theta = \frac{\lambda}{2d}$$

if θ in Fig. 36-8 is small enough so that $\sin\theta \approx \theta$.

29. Wider Slit Suppose that one of the slits of a double-slit interference experiment is wider than the other, so the amplitude of the light reaching the central part of the screen from one slit, acting alone, is twice that from the other slit, acting alone. Derive an expression for the light intensity I at the screen as a function of θ, corresponding to Eqs. 36-21 and 36-22.

SEC. 36-7 ■ INTERFERENCE FROM THIN FILMS

30. Reflections In Fig. 36-22, light wave W_1 reflects once from a reflecting surface while light wave W_2 reflects twice from that surface and once from a reflecting sliver at distance L from the mirror. The waves are initially in phase and have a wavelength of 620 nm. Neglect the slight tilt of the rays. (a) For what least value of L are the reflected waves exactly out of phase? (b) How far must the sliver be moved to put the waves exactly out of phase again?

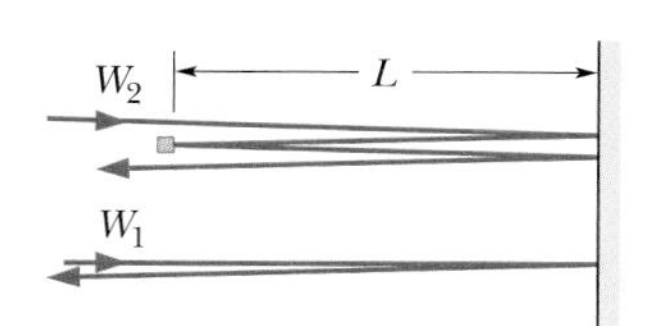

FIGURE 36-22 ■ Problems 30 and 32.

31. Bright Light Bright light of wavelength 585 nm is incident perpendicularly on a soap film ($n = 1.33$) of thickness 1.21 μm, suspended in air. Is the light reflected by the two surfaces of the film closer to interfering fully destructively or fully constructively?

32. Exactly Out of Phase Suppose the light waves of Problem 30 are initially exactly out of phase. Find an expression for the values of L (in terms of the wavelength λ) that put the reflected waves exactly in phase.

33. Soap Film Light of wavelength 624 nm is incident perpendicularly on a soap film (with $n = 1.33$) suspended in air. What are the least two thicknesses of the film for which the reflections from the film undergo fully constructive interference?

34. Camera Lens A camera lens with index of refraction greater than 1.30 is coated with a thin transparent film of index of refraction 1.25 to eliminate by interference the reflection of light at wavelength λ that is incident perpendicularly on the lens. In terms of λ, what minimum film thickness is needed?

35. Rhinestones The rhinestones in costume jewelry are glass with index of refraction 1.50. To make them more reflective, they are often coated with a layer of silicon monoxide of index of refraction 2.00. What is the minimum coating thickness needed to ensure that light of wavelength 560 nm and of perpendicular incidence will be reflected from the two surfaces of the coating with fully constructive interference?

36. Five Sections In Fig. 36-23, light of wavelength 600 nm is incident perpendicularly on five sections of a transparent structure suspended in air. The structure has index of refraction 1.50. The thickness of each section is given in terms of $L = 4.00$ μm. For which sections will the light that is reflected from the top and bottom surfaces of that section undergo fully constructive interference?

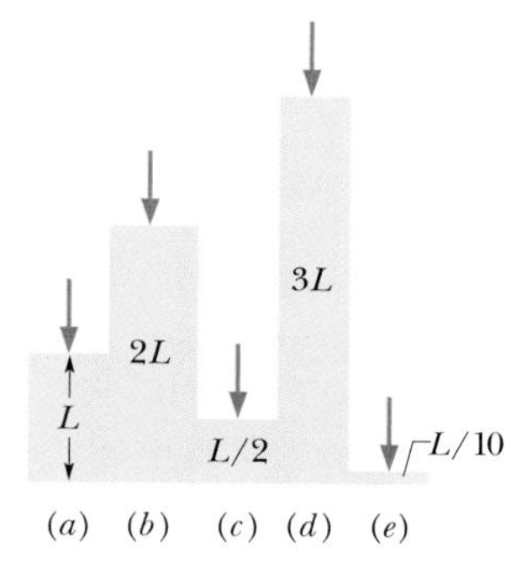

FIGURE 36-23 ■ Problem 36.

37. Coat Glass We wish to coat flat glass ($n = 1.50$) with a transparent material ($n = 1.25$) so that reflection of light at wavelength 600 nm is eliminated by interference. What minimum thickness can the coating have to do this?

38. Four Thin Layers In Fig. 36-24, light is incident perpendicularly on four thin layers of thickness L. The indexes of refraction of the thin layers and of the media above and below these layers are given. Let λ represent the wavelength of the light in air and n_2 represent the index of refraction of the thin layer in each situation. Consider only the transmission of light that undergoes no reflection or two reflections, as in Fig. 36-24a. For which of the situations does the expression

$$\lambda = \frac{2Ln_2}{m}, \qquad \text{for } m = 1, 2, 3, \ldots,$$

give the wavelengths of the transmitted light that undergoes fully constructive interference?

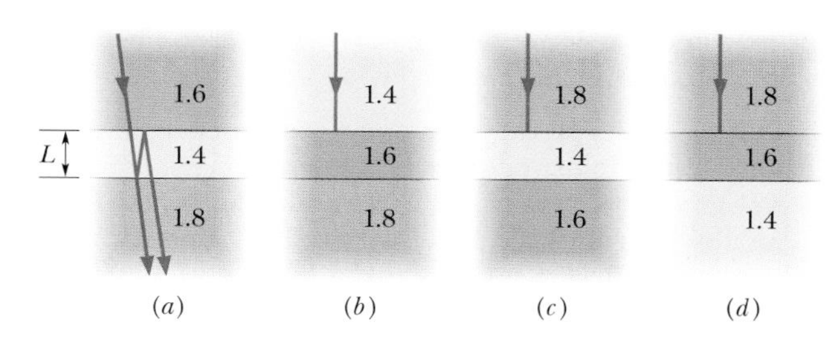

FIGURE 36-24 ■ Problems 38 and 39.

39. Leaking Tanker A disabled tanker leaks kerosene ($n = 1.20$) into the Persian Gulf creating a large slick on top of the water ($n = 1.30$). (a) If you are looking straight down from an airplane, while the Sun is overhead, at a region of the slick where its thickness is 460 nm, for which wavelength(s) of visible light is the reflection brightest because of constructive interference? (b) If you are scuba diving directly under this same region of the slick, for which wavelength(s) of visible light is the transmitted intensity strongest? (*Hint:* Use Fig. 36-24a with appropriate indexes of refraction.)

40. Plane Wave A plane wave of monochromatic light is incident normally on a uniform thin film of oil that covers a glass plate. The wavelength of the source can be varied continuously. Fully destructive interference of the reflected light is observed for wavelengths of 500 and 700 nm and for no wavelengths in between. If the index of refraction of the oil is 1.30 and that of the glass is 1.50, find the thickness of the oil film.

41. Monochromatic Light A plane monochromatic light wave in air is perpendicularly incident on a thin film of oil that covers a glass plate. The wavelength of the source may be varied continuously. Fully destructive interference of the reflected light is observed for wavelengths of 500 and 700 nm and for no wavelength in between. The index of refraction of the glass is 1.50. Show that the index of refraction of the oil must be less than 1.50.

42. Soap Film Two The reflection of perpendicularly incident white light by a soap film in air has an interference maximum at 600 nm and a minimum at 450 nm, with no minimum in between. If $n = 1.33$ for the film, what is the film thickness, assumed uniform?

43. Glass Plates In Fig. 36-25, a broad beam of light of wavelength 683 nm is sent directly downward through the top plate of a pair of glass plates. The plates are 120 mm long, touch at the left end, and are separated by a wire of diameter 0.048 mm at the right end. The air between the plates acts as a thin film. How many bright fringes will be seen by an observer looking down through the top plate?

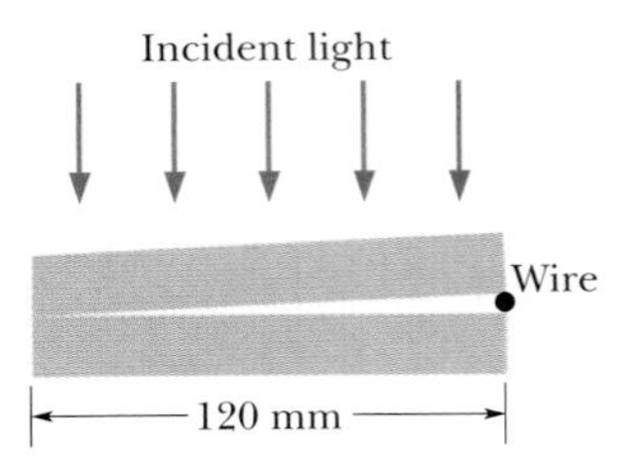

FIGURE 36-25 ■ Problems 43 and 44.

44. Directly Downward In Fig. 36-25, white light is sent directly downward through the top plate of a pair of glass plates. The plates touch at the left end and are separated by a wire of diameter 0.048 mm at the right end; the air between the plates acts as a thin film. An observer looking down through the top plate sees bright and dark fringes due to that film. (a) Is a dark fringe or a bright fringe seen at the left end? (b) To the right of that end, fully destructive interference occurs at different locations for different wavelengths of the light. Does it occur first for the red end or the blue end of the visible spectrum?

45. Wedge-Shaped A broad beam of light of wavelength 630 nm is incident at 90° on a thin, wedge-shaped film with index of refraction 1.50. An observer intercepting the light transmitted by the film sees 10 bright and 9 dark fringes along the length of the film. By how much does the film thickness change over this length?

46. Acetone A thin film of acetone ($n = 1.25$) coats a thick glass plate ($n = 1.50$). White light is incident normal to the film. In the reflections, fully destructive interference occurs at 600 nm and fully constructive interference at 700 nm. Calculate the thickness of the acetone film.

47. Two Glass Plates Two glass plates are held together at one end to form a wedge of air that acts as a thin film. A broad beam of light of wavelength 480 nm is directed through the plates, perpendicular to the first plate. An observer intercepting light reflected from the plates sees on the plates an interference pattern that is due to the wedge of air. How much thicker is the wedge at the sixteenth bright fringe than it is at the sixth bright fringe, counting from where the plates touch?

48. Broad Beam A broad beam of monochromatic light is directed perpendicularly through two glass plates that are held together at one end to create a wedge of air between them. An observer intercepting light reflected from the wedge of air, which acts as a thin film, sees 4001 dark fringes along the length of the wedge. When the air between the plates is evacuated, only 4000 dark fringes are seen. Calculate the index of refraction of air from these data.

49. Radius of Curvature Figure 36-26*a* shows a lens with radius of curvature R lying on plane glass plate and illuminated from above by light with wavelength λ. Figure 36-26*b* (a photograph taken from above the lens) shows that circular interference fringes (called *Newton's rings*) appear, associated with the variable thickness d of the air film between the lens and the plate. Find the radii r of the interference maxima assuming $r/R \ll 1$.

50. Newtons's Rings One In a Newton's rings experiment (see Problem 49), the radius of curvature R of the lens is 5.0 m and the lens diameter is 20 mm. (a) How many bright rings are produced? Assume that $\lambda = 589$ nm. (b) How many bright rings would be produced if the arrangement were immersed in water ($n = 1.33$)?

51. Newton's Rings Two A Newton's rings apparatus is to be used to determine the radius of curvature of a lens (see Fig. 36-26 and Problem 49). The radii of the nth and $(n + 20)$th bright rings are measured and found to be 0.162 and 0.368 cm, respectively, in light of wavelength 546 nm. Calculate the radius of curvature of the lower surface of the lens.

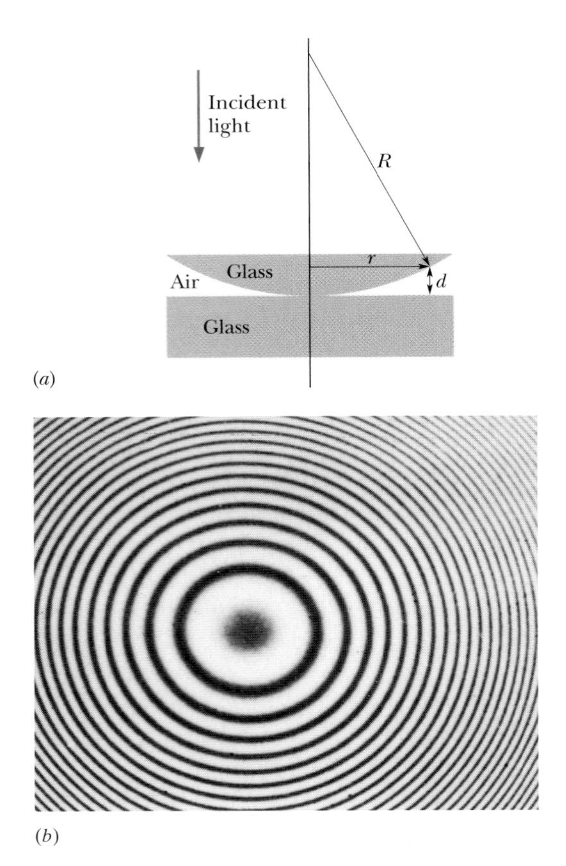

FIGURE 36-26 ■ Problems 49 through 52.

52. Newton's Rings Three (a) Use the result of Problem 49 to show that, in a Newton's rings experiment, the difference in radius between adjacent bright rings (maxima) is given by

$$\Delta r = r_{m+1} - r_m \approx \tfrac{1}{2}\sqrt{\lambda R/m},$$

assuming $m \gg 1$. (b) Now show that the *area* between adjacent bright rings is given by

$$A = \pi\lambda R,$$

assuming $m \gg 1$. Note that this area is independent of m.

53. Microwave Transmitter In Fig. 36-27, a microwave transmitter at height a above the water level of a wide lake transmits microwaves of wavelength λ toward a receiver on the opposite shore, a distance x above the water level. The microwaves reflecting from the water interfere with the microwaves arriving directly from the transmitter. Assuming that the lake width D is much greater than a and x, and that $\lambda \geq a$, at what values of x is the signal at the receiver maximum? (*Hint:* Does the reflection cause a phase change?)

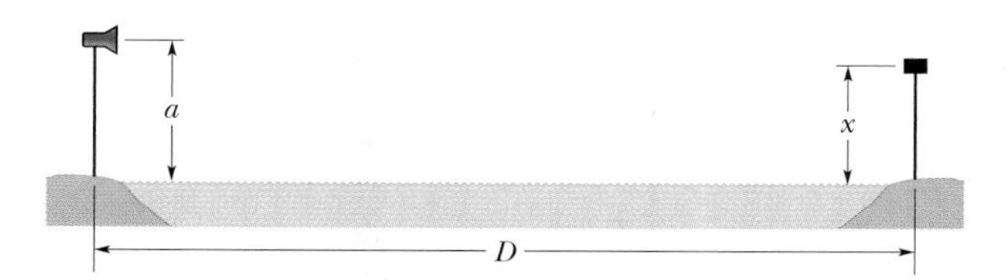

FIGURE 36-27 ■ Problem 53.

SEC. 36-8 ■ MICHELSON'S INTERFEROMETER

54. Thin Film A thin film with index of refraction $n = 1.40$ is placed in one arm of a Michelson interferometer, perpendicular to the optical path. If this causes a shift of 7.0 fringes of the pattern produced by light of wavelength 589 nm, what is the film thickness?

55. Move the Mirror If mirror M_2 in a Michelson interferometer (Fig. 36-17) is moved through 0.233 mm, a shift of 792 fringes occurs. What is the wavelength of the light producing the fringe pattern?

56. Light at Two Wavelengths The element sodium can emit light at two wavelengths, $\lambda_1 = 589.10$ nm and $\lambda_2 = 589.59$ nm. Light from sodium is being used in a Michelson interferometer (Fig. 36-17). Through what distance must mirror M_2 be moved to shift the fringe pattern for one wavelength by 1.00 fringe more than the fringe pattern for the other wavelength?

57. Airtight Chamber In Fig. 36-28, an airtight chamber 5.0 cm long with glass windows is placed in one arm of a Michelson interferometer. Light of wavelength $\lambda = 500$ nm is used. Evacuating the air from the chamber causes a shift of 60 fringes. From these data, find the index of refraction of air at atmospheric pressure.

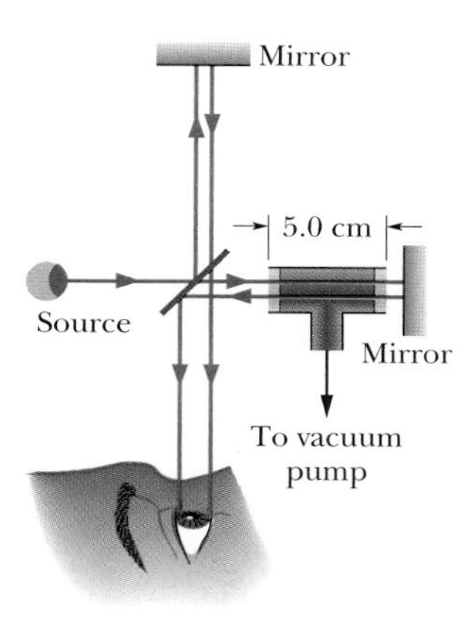

FIGURE 36-28 ■ Problem 57.

58. Observed Intensity Write an expression for the intensity observed in a Michelson interferometer (Fig. 36-17) as a function of the position of the moveable mirror. Measure the position of the mirror from the point at which $d_2 = d_1$.

Additional Problems

59. Arranging the Patio Speakers You have set up two stereo speakers on your patio as shown in the top view diagram in Fig. 36-29. You are worried that at certain positions you will lose frequencies as a result of interference. The coordinate grid on the edge of the picture has its large tick marks separated by 1 m. For ease of calculation, make the following assumptions:

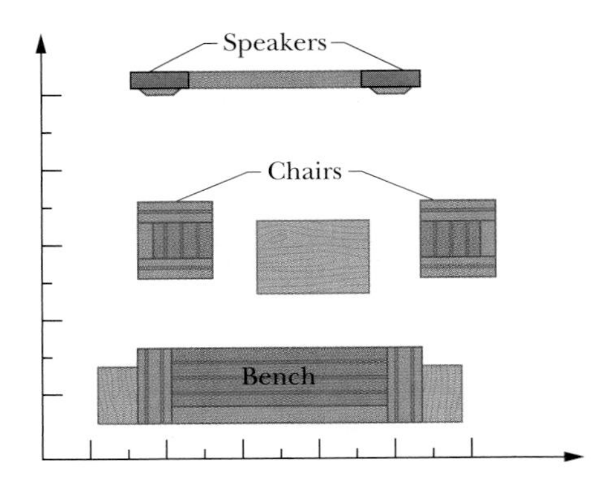

FIGURE 36-29 ■ Problem 59.

- Assume that the relevant objects lie on integer or half-integer grid points of the coordinate system.
- Take the speed of sound to be 343 m/s.
- Ignore the reflection of sound from the house, trees, etc.
- Assume that the speakers are in phase.

(a) What will happen if you are sitting in the middle of the bench?
(b) If you are sitting in the lawn chair on the left, what will be the lowest frequency you will lose to destructive interference?
(c) Can you restore the frequency lost in part (b) by switching the leads to one of the speakers, thereby reversing the phase of that source?
(d) With the leads reversed, what will happen to the sound for a person sitting at the center of the bench?

60. What Happens If a Double Slit Winks? When a laser beam is incident on a double slit, a closeup of the center of the pattern looks like that shown in Fig. 36-30. If one of the slits is covered (the left one) but the other slit remains open, what will the pattern look like? Explain how you know.

FIGURE 36-30 ■ Problem 60.

37 Diffraction

Georges Seurat painted *Sunday Afternoon on the Island of La Grande Jatte* using not brush strokes in the usual sense, but rather a myriad of small colored dots, in a style of painting now known as pointillism. You can see the dots if you stand close enough to the painting, but as you move away from it, they eventually blend and cannot be distinguished. Moreover, the color that you see at any given place on the painting changes as you move away—which is why Seurat painted with the dots.

What causes this change in color?

The answer is in this chapter.

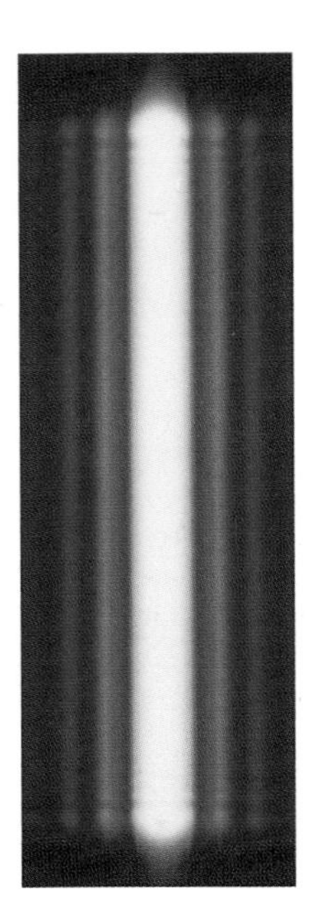

FIGURE 37-1 ■ This diffraction pattern appeared on a viewing screen when light that had passed through a narrow but tall vertical slit reached the screen. Diffraction causes light to flare out perpendicular to the long sides of the slit. That produces an interference pattern consisting of a broad central maximum less intense and narrower secondary (or side) maxima, with minima between them.

37-1 Diffraction and the Wave Theory of Light

In Chapter 36 we defined diffraction rather loosely as the flaring of light as it emerges from a narrow slit. More than just flaring occurs, however, because the light produces an interference pattern called a **diffraction pattern.** For example, when monochromatic light from a distant source (or a laser) passes through a narrow slit and is then intercepted by a viewing screen, the light produces on the screen a diffraction pattern like that in Fig. 37-1. This pattern consists of a broad and intense (very bright) central maximum and a number of narrower and less intense maxima (called **secondary** or **side** maxima) to both sides. In between the maxima are minima.

Such a pattern would be totally unexpected in geometrical optics: If light traveled in straight lines as rays, then the slit would allow some of those rays through and they would form a sharp, bright rendition of the slit on the viewing screen. As in Chapter 36, we again must conclude that geometrical optics is only an approximation.

Diffraction of light is not limited to situations of light passing through a narrow opening (such as a slit or pinhole). It also occurs when light passes an edge, such as the edges of the razor blade whose diffraction pattern is shown in Fig. 37-2. Note the lines of maxima and minima that run approximately parallel to the edges, at both the inside edges of the blade and the outside edges. As the light passes, say, the vertical edge at the left, it flares left and right and undergoes interference, producing the pattern along the left edge. The rightmost portion of that pattern actually lies within what would have been the shadow of the blade if geometrical optics prevailed.

You encounter a common example of diffraction when you look at a clear blue sky and see tiny specks and hair-like structures floating in your view. These *floaters,* as they are called, are produced when light passes the edges of tiny deposits in the vitreous humor, the transparent material filling most of your eyeball. What you are seeing when a floater is in your field of vision is the diffraction pattern produced on the retina by one of these deposits. If you sight through a pinhole in an otherwise opaque sheet so as to make the light entering your eye approximately a plane wave, you can distinguish individual maxima and minima in the patterns.

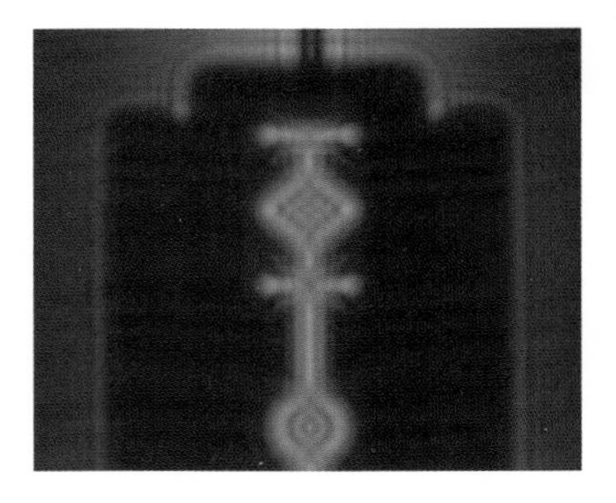

FIGURE 37-2 ■ The diffraction pattern produced by a razor blade in monochromatic light. Note the lines of alternating maximum and minimum intensity.

The Fresnel Bright Spot

Diffraction finds a ready explanation in the wave theory of light. However, this theory, originally advanced in the late 1600s by Huygens and used 123 years later by Young to explain double-slit interference, was very slow in being adopted, largely because it ran counter to Newton's theory that light was a stream of particles.

Newton's view was the prevailing view in French scientific circles of the early 19th century, when Augustin Fresnel was a young military engineer. Fresnel, who believed in the wave theory of light, submitted a paper to the French Academy of Sciences describing his experiments with light and his wave-theory explanations of them.

In 1819, the Academy, dominated by supporters of Newton and thinking to challenge the wave point of view, organized a prize competition for an essay on the subject of diffraction. Fresnel won. The Newtonians, however, were neither converted nor silenced. One of them, S. D. Poisson, pointed out the "strange result" that if Fresnel's theories were correct, then light waves should flare into the shadow region of a sphere as they pass the edge of the sphere, producing a bright spot at the center of the shadow. The prize committee arranged to have Dominique Argo test the famous mathematician's prediction. He discovered (see Fig. 37-3) that the predicted *Fresnel bright spot,* as we call it today, was indeed there!* Nothing builds confidence in a

* Since Poisson predicted the spot and Argo discovered it, an alternate name is the Poisson-Argo bright spot.

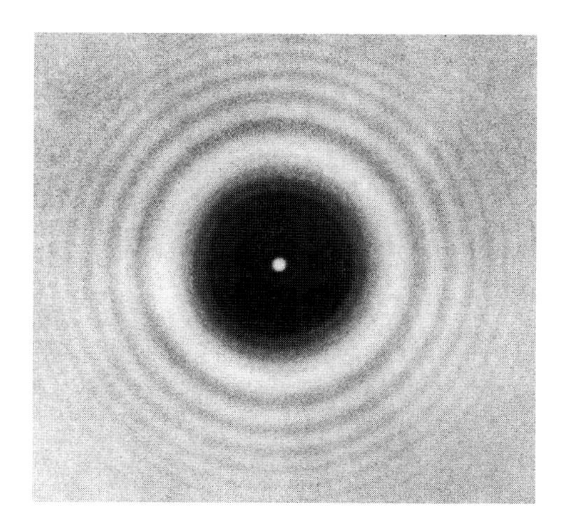

FIGURE 37-3 ■ A photograph of the diffraction pattern of a disk. Note the concentric diffraction rings and the Fresnel bright spot at the center of the pattern. This experiment is essentially identical to that arranged by the committee testing Fresnel's theories, because both the sphere they used and the disk used here have a cross section with a circular edge.

theory so much as having one of its unexpected and counterintuitive predictions verified by experiment.

37-2 Diffraction by a Single Slit: Locating the Minima

Let us now examine the diffraction pattern of plane waves of light of wavelength λ that are diffracted by a single, long, narrow slit of width a in an otherwise opaque screen B, as shown in cross section in Fig. 37-4a. (In that figure, the slit's length extends into and out of the page, and the incoming wavefronts are parallel to screen B.) When the diffracted light reaches viewing screen C, waves from different points within the slit undergo interference and produce a diffraction pattern of bright and dark fringes (interference maxima and minima) on the screen. To locate the fringes, we shall use a procedure somewhat similar to the one we used to locate the fringes in a two-slit interference pattern. However, diffraction is more mathematically challenging, and here we shall be able to find equations for only the dark fringes.

Before we do that, however, we can justify the central bright fringe seen in Fig. 37-1 by noting that the Huygens wavelets from all points in the slit travel about the same distance to reach the center of the pattern and thus are in phase there. As for the other bright fringes, we can say only that they are approximately halfway between adjacent dark fringes.

To find the dark fringes, we shall use a clever (and simplifying) strategy that involves pairing up all the rays coming through the slit and then finding what conditions cause the wavelets of the rays in each pair to cancel each other. Figure 37-4a shows how we apply this strategy to locate the first dark fringe, at point P_1. First, we mentally divide the slit into two zones of equal widths $a/2$. Then we extend to P_1 a light ray r_1 from the top point of the top zone and a light ray r_2 from the top point of the bottom zone. A central axis is drawn from the center of the slit to screen C, and P_1 is located at an angle θ to that axis.

The wavelets of the pair of rays r_1 and r_2 are in phase within the slit because they originate from the same wavefront passing through the slit, along the width of the slit. However, to produce the first dark fringe they must be out of phase by $\lambda/2$ when they reach P_1; this phase difference is due to their path length difference, with the wavelet of r_2 traveling a longer path to reach P_1 than the wavelet of r_1. To display this path length difference, we find a point b on ray r_2 such that the path length from b to P_1 matches the path length of ray r_1. Then the path length difference between the two rays is the distance from the center of the slit to b.

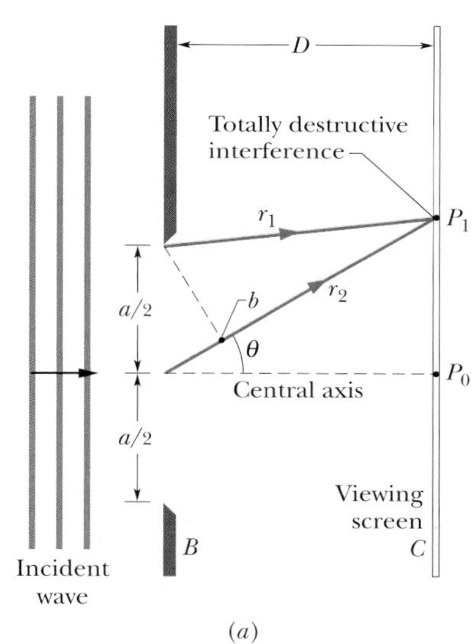

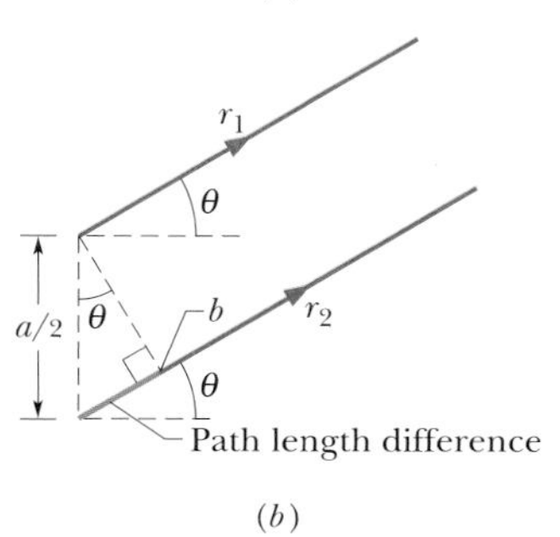

FIGURE 37-4 ■ (a) Waves from the top points of two zones of width $a/2$ undergo totally destructive interference at point P_1 on viewing screen C. (b) For $D \gg a$, we can approximate rays r_1 and r_2 as being parallel, at angle θ to the central axis.

When viewing screen C is near screen B, as in Fig. 37-4a, the diffraction pattern on C is difficult to describe mathematically. However, we can simplify the mathematics considerably if we arrange for the distance between the slit and screen D to be much larger than the slit width a. Then we can approximate rays r_1 and r_2 as being parallel, at angle θ to the central axis (Fig. 37-4b). We can also approximate the triangle formed by point b, the top point of the slit, and the center point of the slit as being a right triangle, and one of the angles inside that triangle as being θ. The path length difference between rays r_1 and r_2 (which is still the distance from the center of the slit to point b) is then equal to $(a/2)\sin\theta$.

We can repeat this analysis for any other pair of rays originating at corresponding points in the two zones (say, at the midpoints of the zones) and extending to point P_1. Each such pair of rays has the same path length difference $(a/2)\sin\theta$. Setting this common path length difference equal to $\lambda/2$ (our condition for the first dark fringe), we have

$$\frac{a}{2}\sin\theta = \frac{\lambda}{2},$$

which gives us

$$a\sin\theta = \lambda \qquad \text{(first minimum for } D \gg a\text{)}. \tag{37-1}$$

Given slit width a and wavelength λ, Eq. 37-1 tells us the angle θ of the first dark fringe above and (by symmetry) below the central axis.

Note that if we begin with $a > \lambda$ and then narrow the slit while holding the wavelength constant, we increase the angle at which the first dark fringes appear; that is, the extent of the diffraction (the extent of the flaring and the width of the pattern) is *greater* for a *narrower* slit. When we have reduced the slit width to the wavelength (that is, $a = \lambda$), the angle of the first dark fringes is 90°. Since the first dark fringes mark the two edges of the central bright fringe, that bright fringe must then cover the entire viewing screen.

We find the second dark fringes above and below the central axis as we found the first dark fringes, except that we now divide the slit into *four* zones of equal widths $a/4$, as shown in Fig. 37-5a. We then extend rays r_1, r_2, r_3, and r_4 from the top points of the zones to point P_2, the location of the second dark fringe above the central axis. To produce that fringe, the path length difference between r_1 and r_2, that between r_2 and r_3, and that between r_3 and r_4 must all be equal to $\lambda/2$.

For $D >> a$, we can approximate these four rays as being parallel, at angle θ to the central axis. To display their path length differences, we extend a perpendicular line through each adjacent pair of rays, as shown in Fig. 37-5b, to form a series of right triangles, each of which has a path length difference as one side. We see from the top triangle that the path length difference between r_1 and r_2 is $(a/4)\sin\theta$. Similarly, from the bottom triangle, the path length difference between r_3 and r_4 is also $(a/4)\sin\theta$. In fact, the path length difference for any two rays that originate at corresponding points in two adjacent zones is $(a/4)\sin\theta$. Since in each such case the path length difference is equal to $\lambda/2$, we have

$$\frac{a}{4}\sin\theta = \frac{\lambda}{2},$$

which gives us

$$a\sin\theta = 2\lambda \qquad \text{(second minimum for } D \gg a\text{)}. \tag{37-2}$$

We could now continue to locate dark fringes in the diffraction pattern by splitting up the slit into more zones of equal width. We would always choose an even num-

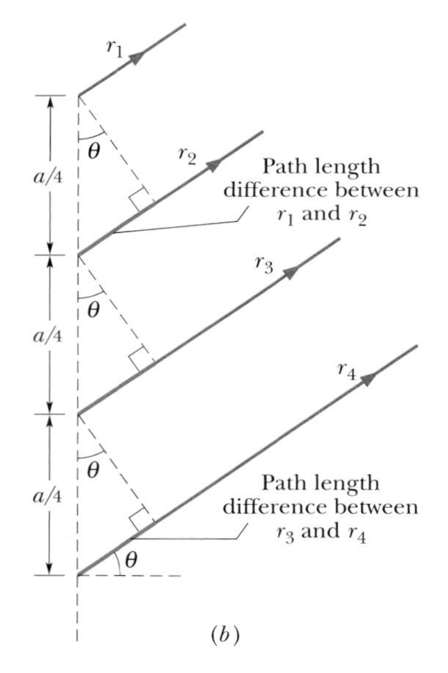

FIGURE 37-5 ■ (a) Waves from the top points of four zones of width $a/4$ undergo totally destructive interference at point P_2. (b) For $D >> a$, we can approximate rays r_1, r_2, r_3, and r_4 as being parallel, at angle θ to the central axis.

ber of zones so that the zones (and their waves) could be paired as we have been doing. We would find that the dark fringes above and below the central axis can be located with the following general equation:

$$a \sin\theta = m\lambda, \qquad \text{for } m = 1, 2, 3, \ldots \qquad \text{(single slit minima—dark fringes).} \qquad (37\text{-}3)$$

You can remember this result in the following way. Draw a triangle like the one in Fig. 37-4*b*, but for the full slit width a, and note that the path length difference between the top and bottom rays from the slit equals $a \sin\theta$. Thus, Eq. 37-3 says:

> In a single-slit diffraction experiment, dark fringes are produced where the path length differences ($a \sin\theta$) between the top and bottom rays are equal to $\lambda, 2\lambda, 3\lambda \ldots$.

This may seem to be wrong, because the waves of those two particular rays will be exactly in phase with each other when their path length difference is an integer number of wavelengths. However, they each will still be part of a pair of waves that are exactly out of phase with each other; thus, *each* will be canceled by some other wave.

READING EXERCISE 37-1: We produce a diffraction pattern on a viewing screen by means of a long narrow slit illuminated by blue light. Does the pattern expand away from the bright center (the maxima and minima shift away from the center) or contract toward it if we (a) switch to yellow light or (b) decrease the slit width? ■

TOUCHSTONE EXAMPLE 37-1: White Light, Red Light

A slit of width a is illuminated by white light (which consists of all the wavelengths in the visible range).

(a) For what value of a will the first minimum for red light of wavelength $\lambda = 650$ nm appear at $\theta = 15°$?

SOLUTION ■ The **Key Idea** here is that diffraction occurs separately for each wavelength in the range of wavelengths passing through the slit, with the locations of the minima for each wavelength given by Eq. 37-3 ($a \sin\theta = m\lambda$). When we set $m = 1$ (for the first minimum) and substitute the given values of θ and λ, Eq. 37-3 yields

$$a = \frac{m\lambda}{\sin\theta} = \frac{(1)(650 \text{ nm})}{\sin 15°}$$
$$= 2511 \text{ nm} \approx 2.5\ \mu\text{m}. \qquad \text{(Answer)}$$

For the incident light to flare out that much (±15° to the first minima) the slit has to be very fine indeed—about four times the wavelength. For comparison, note that a fine human hair may be about 100 μm in diameter.

(b) What is the wavelength λ' of the light whose first side diffraction maximum is at 15°, thus coinciding with the first minimum for the red light?

SOLUTION ■ The **Key Idea** here is that the first side maximum for any wavelength is about halfway between the first and second minima for that wavelength. Those first and second minima can be located with Eq. 37-3 by setting $m = 1$ and $m = 2$, respectively. Thus, the first side maximum can be located *approximately* by setting $m = 1.5$. Then Eq. 37-3 becomes

$$a \sin\theta = 1.5\lambda'.$$

Solving for λ' and substituting known data yield

$$\lambda' = \frac{a \sin\theta}{1.5} = \frac{(2511 \text{ nm})(\sin 15°)}{1.5}$$
$$= 430 \text{ nm}. \qquad \text{(Answer)}$$

Light of this wavelength is violet. The first side maximum for light of wavelength 430 nm will always coincide with the first minimum for light of wavelength 650 nm, no matter what the slit width is. If the slit is relatively narrow, the angle θ at which this overlap occurs will be relatively large, and conversely for a wide slit the angle is small.

37-3 Intensity in Single-Slit Diffraction, Qualitatively

In Section 37-2 we saw how to find the positions of the minima and the maxima in a single-slit diffraction pattern. Now we turn to a more general problem: Find an expression for the intensity I of the pattern as a function of θ, the angular position of a point on a viewing screen.

To do this, we divide the slit of Fig. 37-4*a* into N zones of equal widths Δx small enough that we can assume each zone acts as a source of Huygens wavelets. We wish to superimpose the wavelets arriving at an arbitrary point P on the viewing screen, at angle θ to the central axis, so that we can determine the amplitude E_θ of the magnitude of the electric field of the resultant wave at P. The intensity of the light at P is then proportional to the square of that amplitude.

To find E_θ, we need the phase relationships among the arriving wavelets. The phase difference between wavelets from adjacent zones is given by

$$\text{(phase difference)} = \left(\frac{2\pi}{\lambda}\right)\text{(path length difference)}.$$

For point P at angle θ, the path length difference between wavelets from adjacent zones is $\Delta x \sin\theta$, so the phase difference $\Delta\phi$ between wavelets from adjacent zones is

$$\Delta\phi = \left(\frac{2\pi}{\lambda}\right)(\Delta x \sin\theta). \tag{37-4}$$

We assume that the wavelets arriving at P all have the same amplitude ΔE. To find the amplitude E_θ of the resultant wave at P, we add the amplitudes ΔE via phasors. To do this, we construct a diagram of N phasors, one corresponding to the wavelet from each zone in the slit.

For point P_0 at $\theta = 0$ on the central axis of Fig. 37-4*a*, Eq. 37-4 tells us that the phase difference $\Delta\phi$ between the wavelets is zero; that is, the wavelets all arrive in phase. Figure 37-6*a* is the corresponding phasor diagram; adjacent phasors represent wavelets from adjacent zones and are arranged head to tail. Because there is zero phase difference between the wavelets, there is zero angle between each pair of adjacent phasors. The amplitude E_θ of the net wave at P_θ is the vector-like sum of these phasors. This arrangement of the phasors turns out to be the one that gives the greatest value for the amplitude E_θ. We call this value $E^{\max}$; that is, $E^{\max}$ is the value of E_θ for $\theta = 0$.

We next consider a point P that is at a small angle θ to the central axis. Equation 37-4 now tells us that the phase difference $\Delta\phi$ between wavelets from adjacent zones is no longer zero. Figure 37-6*b* shows the corresponding phasor diagram; as before,

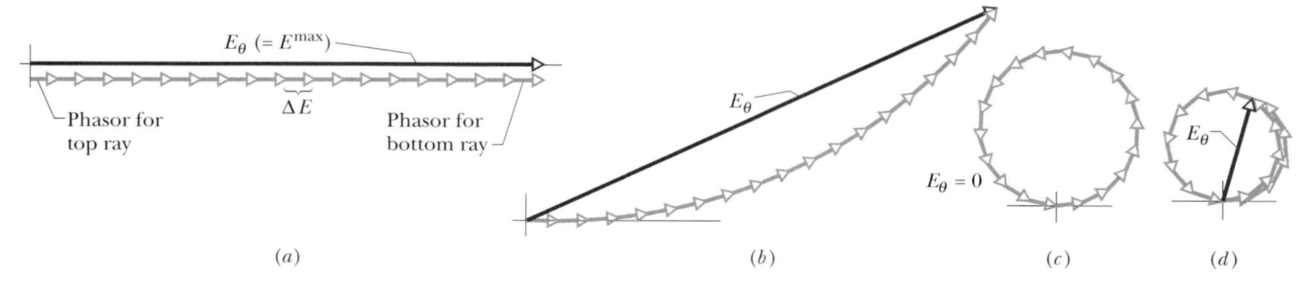

FIGURE 37-6 ■ Phasor diagrams for $N = 18$ phasors, corresponding to the division of a single slit into 18 zones. Resultant amplitudes E_θ are shown for (*a*) the central maximum at $\theta = 0$, (*b*) a point on the screen lying at a small angle θ to the central axis, (*c*) the first minimum, and (*d*) the first side maximum.

the phasors are arranged head to tail, but now there is an angle $\Delta\phi$ between adjacent phasors. The amplitude E_θ at this new point is still the vector sum of the phasors, but it is smaller than the amplitude in Fig. 37-6*a*, which means that the intensity of the light is less at this new point P than at P_0.

If we continue to increase θ, the angle $\Delta\phi$ between adjacent phasors increases, and eventually the chain of phasors curls completely around so that the head of the last phasor just reaches the tail of the first phasor (Fig. 37-6*c*). The amplitude E_θ is now zero, which means that the intensity of the light is also zero. We have reached the first minimum, or dark fringe, in the diffraction pattern. The first and last phasors now have a phase difference of 2π rad, which means that the path length difference between the top and bottom rays through the slit equals one wavelength. Recall that this is the condition we determined for the first diffraction minimum.

As we continue to increase θ, the angle $\Delta\phi$ between adjacent phasors continues to increase, the chain of phasors begins to wrap back on itself, and the resulting coil begins to shrink. Amplitude E_θ now increases until it reaches a maximum value in the arrangement shown in Fig. 37-6*d*. This arrangement corresponds to the first side maximum in the diffraction pattern.

If we increase θ a bit more, the resulting shrinkage of the coil decreases E_θ, which means that the intensity also decreases. When θ is increased enough, the head of the last phasor again meets the tail of the first phasor. We have then reached the second minimum.

We could continue this qualitative method of determining the maxima and minima of the diffraction pattern but, instead, we shall now turn to a quantitative method.

READING EXERCISE 37-2: The figures represent, in smoother form (with more phasors) than Fig. 37-6, the phasor diagrams for two points of a diffraction pattern that are on opposite sides of a certain diffraction maximum. (a) Which maximum is it? (b) What is the approximate value of m (in Eq. 37-3) that corresponds to this maximum?

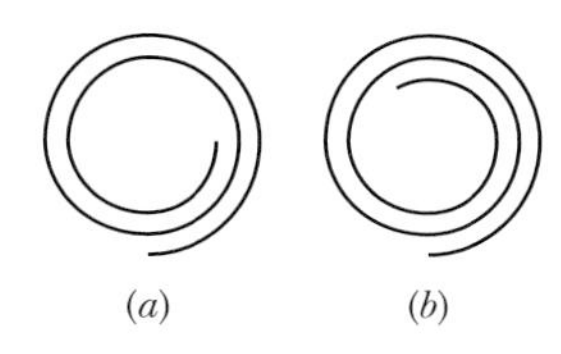

■

37-4 Intensity in Single-Slit Diffraction, Quantitatively

Equation 37-3 tells us how to locate the minima of the single-slit diffraction pattern on screen C of Fig. 37-4*a* as a function of the angle θ in that figure. Here we wish to derive an expression for the intensity I_θ of the pattern as a function of θ. We state, and shall prove below, that the intensity is given by

$$I_\theta = I^{\max}\left(\frac{\sin\alpha}{\alpha}\right)^2, \tag{37-5}$$

where

$$\alpha = \frac{1}{2}\Delta\phi = \frac{\pi a}{\lambda}\sin\theta. \tag{37-6}$$

The symbol α is just a convenient connection between the angle θ that locates a point on the viewing screen and the light intensity I_θ at that point. $I^{\max}$ is the greatest value of the intensity I_θ in the pattern and occurs at the central maximum (where $\theta = 0$), and $\Delta\phi$ is the phase difference (in radians) between the top and bottom rays from the slit width a.

Study of Eq. 37-5 shows that intensity minima will occur where

$$\alpha = m\pi, \qquad \text{for } m = 1, 2, 3, \ldots. \tag{37-7}$$

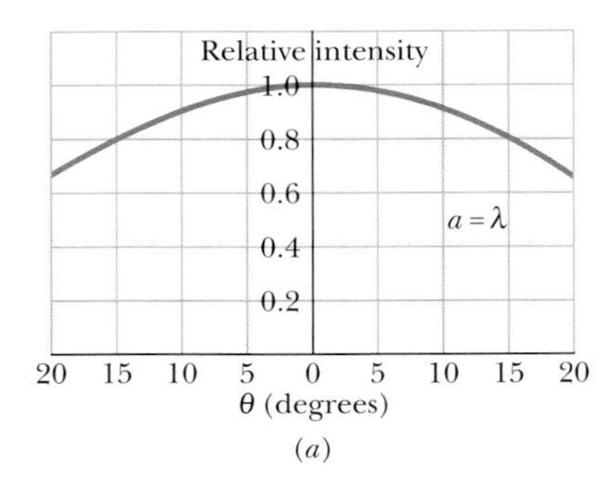

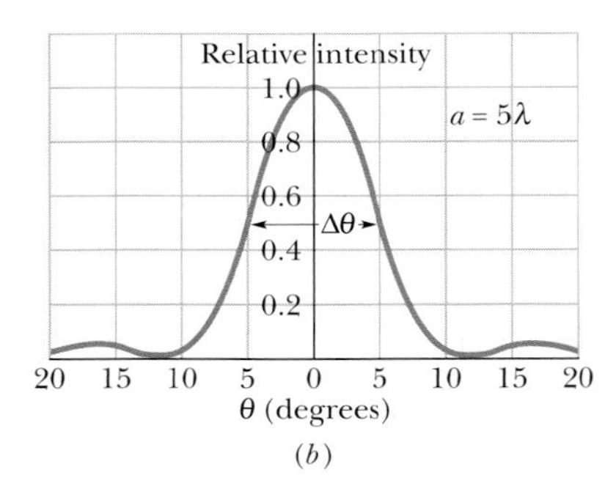

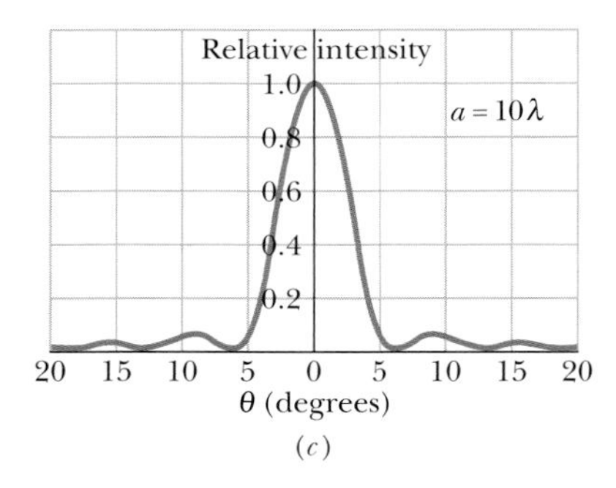

FIGURE 37-7 ■ The relative intensity in single-slit diffraction for three values of the ratio a/λ. The wider the slit is, the narrower is the central diffraction maximum.

If we put this result into Eq. 37-6 we find

$$m\pi = \frac{\pi a}{\lambda}\sin\theta, \qquad \text{for } m = 1, 2, 3, \ldots,$$

or

$$a\sin\theta = m\lambda, \qquad \text{for } m = 1, 2, 3, \ldots \quad \text{(minima—dark fringes)}, \tag{37-8}$$

which is exactly Eq. 37-3, the expression that we derived earlier for the location of the minima.

Figure 37-7 shows plots of the intensity of a single-slit diffraction pattern, calculated with Eqs. 37-5 and 37-6 for three slit widths: $a = \lambda$, $a = 5\lambda$, and $a = 10\lambda$. Note that as the slit width increases (relative to the wavelength), the width of the *central diffraction maximum* (the central hill-like region of the graphs) decreases; that is, the light undergoes less flaring by the slit. The secondary maxima also decrease in width (and become weaker). In the limit of slit width a being much greater than wavelength λ, the secondary maxima due to the slit disappear; we then no longer have single-slit diffraction (but we still have diffraction due to the edges of the wide slit, like that produced by the edges of the razor blade in Fig. 37-2).

Proof of Eqs. 37-5 and 37-6

The arc of phasors in Fig. 37-8 represents the wavelets that reach an arbitrary point P on the viewing screen of Fig. 37-4, corresponding to a particular small angle θ. The amplitude E_θ of the resultant wave at P is the vector sum of these phasors. If we divide the slit of Fig. 37-4 into infinitesimal zones of width Δx, the arc of phasors in Fig. 37-8 approaches the arc of a circle; we call its radius R as indicated in that figure. The length of the arc must be E^{max}, the amplitude at the center of the diffraction pattern, because if we straightened out the arc we would have the phasor arrangement of Fig. 37-6*a* (shown lightly in Fig. 37-8).

The angle $\Delta\phi$ in the lower part of Fig. 37-8 is the difference in phase between the infinitesimal vectors at the left and right ends of arc E^{max}. From the geometry, $\Delta\phi$ is also the angle between the two radii marked R in Fig. 37-8. The dashed line in that figure, which bisects $\Delta\phi$, then forms two congruent right triangles. From either triangle we can write

$$\sin\tfrac{1}{2}\Delta\phi = \frac{E_\theta}{2R}. \tag{37-9}$$

In radian measure, $\Delta\phi$ is (with E^{max} considered to be a circular arc)

$$\Delta\phi = \frac{E^{\text{max}}}{R}.$$

Solving this equation for R, substituting the result into Eq. 37-9 and re-arranging terms yields

$$E_\theta = \frac{E^{\text{max}}}{\frac{1}{2}\Delta\phi}\sin\tfrac{1}{2}\Delta\phi. \tag{37-10}$$

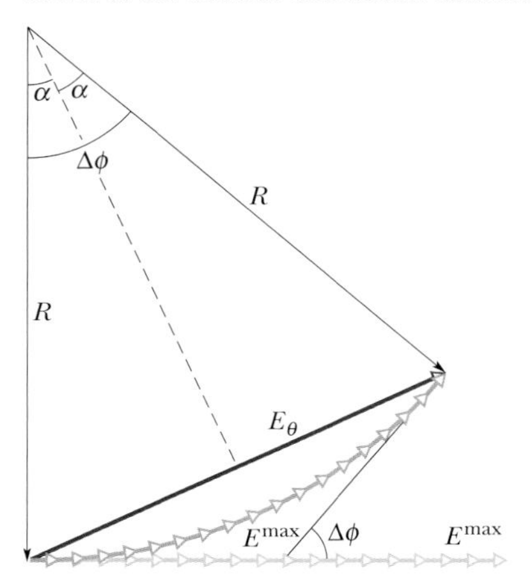

FIGURE 37-8 ■ A construction used to calculate the intensity in single-slit diffraction. The situation shown corresponds to that of Fig. 37-6*b*.

In Section 34-4 we saw that the intensity of an electromagnetic wave is proportional to the square of the amplitude of its electric field. Here, this means that the maximum intensity I^{max} (which occurs at the center of the diffraction pattern) is proportional to $(E^{\text{max}})^2$ and the intensity I_θ at angle θ is proportional to E_θ^2. Thus, we may write

$$\frac{I_\theta}{I^{\max}} = \frac{E_\theta^2}{(E^{\max})^2}. \qquad (37\text{-}11)$$

Substituting for E_θ with Eq. 37-10 and then substituting $\alpha = \frac{1}{2}\Delta\phi$, we are led to the following expression for the intensity as a function of θ:

$$I_\theta = I^{\max}\left(\frac{\sin\alpha}{\alpha}\right)^2.$$

This is exactly Eq. 37-5, one of the two equations we set out to prove.

The second equation we wish to prove relates α to θ: The phase difference $\Delta\phi$ between the rays from the top and bottom of the entire slit may be related to a path length difference with Eq. 37-4; it tells us that

$$\Delta\phi = \left(\frac{2\pi}{\lambda}\right)(a\sin\theta),$$

where a is the sum of the widths Δx of the infinitesimal zones. However, $\Delta\phi = 2\alpha$, so this equation reduces to Eq. 37-6.

READING EXERCISE 37-3: Two wavelengths, 650 and 430 nm, are used separately in a single-slit diffraction experiment. The figure shows the results as graphs of intensity I versus angle θ for the two diffraction patterns. If both wavelengths are then used simultaneously, what color will be seen in the combined diffraction pattern at (a) angle A and (b) angle B?

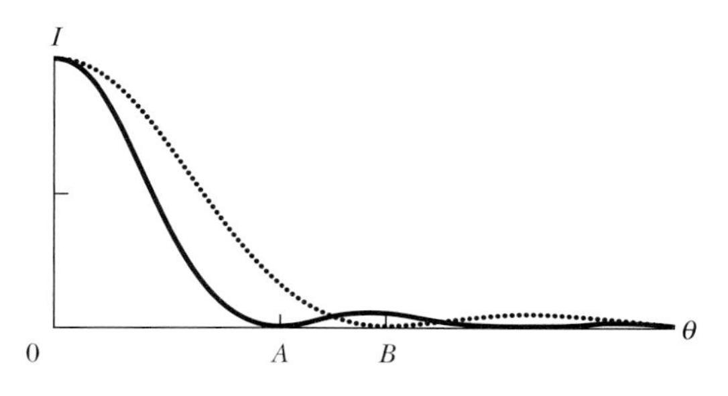

TOUCHSTONE EXAMPLE 37-2: Maxima Intensities

Find the intensities of the first three secondary maxima (side maxima) in the single-slit diffraction pattern of Fig. 37-1, measured relative to the intensity of the central maximum.

SOLUTION ■ One **Key Idea** here is that the secondary maxima lie approximately halfway between the minima, whose angular locations are given by Eq. 37-7 ($\alpha = m\pi$). The locations of the secondary maxima are then given (approximately) by

$$\alpha = (m + \tfrac{1}{2})\pi, \qquad \text{for } m = 1, 2, 3, \ldots,$$

with α in radian measure.

A second **Key Idea** is that we can relate the intensity I at any point in the diffraction pattern to the intensity $I^{\max}$ of the central maximum via Eq. 37-5. Thus, we can substitute the approximate values of α for the secondary maxima into Eq. 37-5 to obtain the relative intensities at those maxima. We get

$$\frac{I}{I^{\max}} = \left(\frac{\sin\alpha}{\alpha}\right)^2 = \left(\frac{\sin(m+\frac{1}{2})\pi}{(m+\frac{1}{2})\pi}\right)^2, \qquad \text{for } m = 1, 2, 3, \ldots.$$

The first of the secondary maxima occurs for $m = 1$, and its relative intensity is

$$\frac{I_1}{I^{\max}} = \left(\frac{\sin(1+\frac{1}{2})\pi}{(1+\frac{1}{2})\pi}\right)^2 = \left(\frac{\sin 1.5\pi}{1.5\pi}\right)^2$$
$$= 4.50 \times 10^{-2} \approx 4.5\%. \qquad \text{(Answer)}$$

For $m = 2$ and $m = 3$ we find that

$$\frac{I_2}{I^{\max}} = 1.6\% \quad \text{and} \quad \frac{I_3}{I^{\max}} = 0.83\%. \qquad \text{(Answer)}$$

Successive secondary maxima decrease rapidly in intensity. Figure 37-1 was deliberately overexposed to reveal them.

37-5 Diffraction by a Circular Aperture

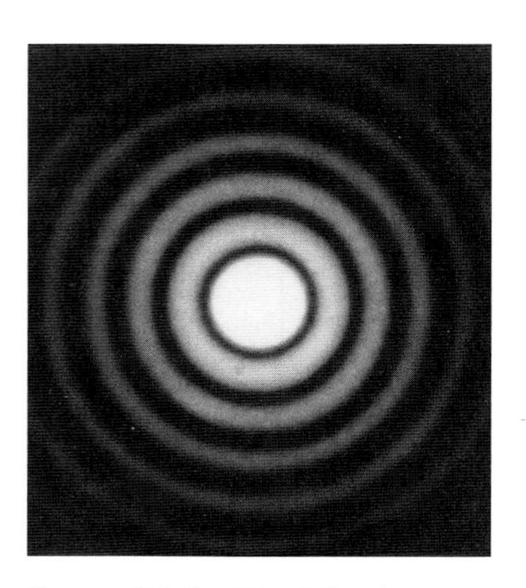

FIGURE 37-9 ■ The diffraction pattern of a circular aperture. Note the central maximum and the circular secondary maxima. The figure has been overexposed to bring out these secondary maxima, which are much less intense than the central maximum.

Here we consider diffraction by a circular aperture—that is, a circular opening such as a circular lens, through which light can pass. Figure 37-9 shows the image of a distant point source of light (a star, for instance) formed on photographic film placed in the focal plane of a converging lens. This image is not a point, as geometrical optics would suggest, but a circular disk surrounded by several progressively fainter secondary rings. Comparison with Fig. 37-1 leaves little doubt that we are dealing with a diffraction phenomenon. Here, however, the aperture is a circle of diameter d rather than a rectangular slit.

The analysis of such patterns is complex. It shows, however, that the first minimum for the diffraction pattern of a circular aperture of diameter d is located by

$$\sin\theta = 1.22\frac{\lambda}{d} \quad \text{(first minimum—circular aperture).} \tag{37-12}$$

The angle θ here is the angle from the central axis to any point on that (circular) minimum. Compare this with Eq. 37-1,

$$\sin\theta = \frac{\lambda}{a} \quad \text{(first minimum—single slit),} \tag{37-13}$$

which locates the first minimum for a long narrow slit of width a. The main difference is the factor 1.22, which enters because of the circular shape of the aperture.

Resolvability

The fact that lens images are diffraction patterns is important when we wish to *resolve* (distinguish) two distant point objects whose angular separation is small. Figure 37-10 shows, in three different cases, the visual appearance and corresponding intensity pattern for two distant point objects (stars, say) with small angular separation. In Figure 37-10a, the objects are not resolved because of diffraction; that is, their diffraction patterns (mainly their central maxima) overlap so much that the two objects cannot be distinguished from a single point object. In Fig. 37-10b the objects are barely resolved, and in Fig. 37-10c they are fully resolved.

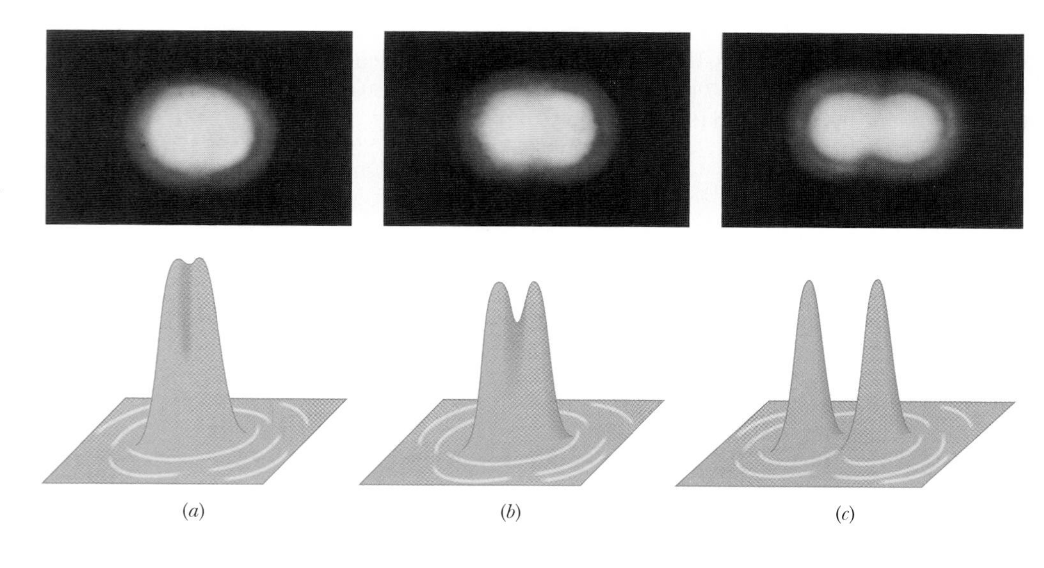

FIGURE 37-10 ■ At the top, the images of two point sources (stars), formed by a converging lens. At the bottom, representations of the image intensities. In (a) the angular separation of the sources is too small for them to be distinguished; in (b) they can be marginally distinguished, and in (c) they are clearly distinguished. Rayleigh's criterion is just satisfied in (b), with the central maximum of one diffraction pattern coinciding with the first minimum of the other.

In Fig. 37-10*b* the angular separation of the two point sources is such that the central maximum of the diffraction pattern of one source is centered on the first minimum of the diffraction pattern of the other, a condition called **Rayleigh's criterion** for resolvability. From Eq. 37-12, two objects that are barely resolvable by this criterion must have an angular separation θ_R of

$$\theta_R = \sin^{-1}\frac{1.22\lambda}{d}.$$

Since the angles involved are small, we can replace $\sin \theta_R$ with θ_R expressed in radians:

$$\theta_R = 1.22\frac{\lambda}{d} \quad \text{(Rayleigh's criterion—circular aperture).} \quad (37\text{-}14)$$

Rayleigh's criterion for resolvability is only an approximation, because resolvability depends on many factors, such as the relative brightness of the sources and their surroundings, turbulence in the air between the sources and the observer, and the functioning of the observer's visual system. Experimental results show that the least angular separation that can actually be resolved by a person is generally somewhat greater than the value given by Eq. 37-14. However, for the sake of calculations here, we shall take Eq. 37-14 as being a precise criterion: If the angular separation θ between the sources is greater than θ_R, we can resolve the sources; if it is less, we cannot.

Rayleigh's criterion can explain the colors in Seurat's *Sunday Afternoon on the Island of La Grande Jatte* (or any other pointillistic painting). When you stand close enough to the painting, the angular separations θ of adjacent dots are greater than θ_R and thus the dots can be seen individually. Their colors are the colors of the paints Seurat used. However, when you stand far enough from the painting, the angular separations θ are less than θ_R and the dots cannot be seen individually. The resulting blend of colors coming into your eye from any group of dots can then cause your brain to "make up" a color for that group—a color that may not actually exist in the group. In this way, Seurat uses your visual system to create the colors of his art.

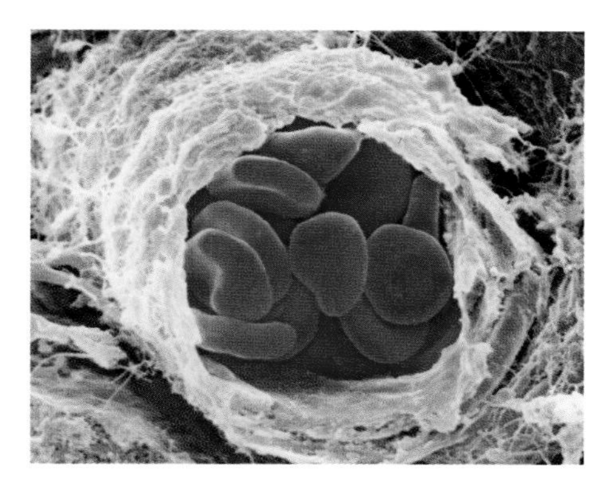

FIGURE 37-11 ■ A false-color scanning electron micrograph of red blood cells traveling through an arterial branch.

When we wish to use a lens instead of our visual system to resolve objects of small angular separation, it is desirable to make the diffraction pattern as small as possible. According to Eq. 37-14, this can be done either by increasing the lens diameter or by using light of a shorter wavelength.

For this reason ultraviolet light is often used with microscopes; because of its shorter wavelength, it permits finer detail to be examined than would be possible for the same microscope operated with visible light. It turns out that under certain circumstances, a beam of electrons behaves like a wave. In an *electron microscope* such beams may have an effective wavelength that is 10^{-5} of the wavelength of visible light. They permit the detailed examination of tiny structures, like that in Fig. 37-11, that would be blurred by diffraction if viewed with an optical microscope.

READING EXERCISE 37-4: Suppose you can barely resolve two red dots, due to diffraction by the pupil of your eye. If we increase the general illumination around you so that the pupil decreases in diameter, does the resolvability of the dots improve or diminish? Consider only diffraction. (You might experiment to check your answer.) ■

TOUCHSTONE EXAMPLE 37-3: Circular Converging Lens

A circular converging lens, with diameter $d = 32$ mm and focal length $f = 24$ cm, forms images of distant point objects in the focal plane of the lens. Light of wavelength $\lambda = 550$ nm is used.

(a) Considering diffraction by the lens, what angular separation must two distant point objects have to satisfy Rayleigh's criterion?

SOLUTION ■ Figure 37-12 shows two distant point objects P_1 and P_2, the lens, and a viewing screen in the focal plane of the lens. It also shows, on the right, plots of light intensity I versus position on the screen for the central maxima of the images formed by the lens. Note that the angular separation θ_o of the objects equals the angular separation θ_i of the images. Thus, the **Key Idea** here is that if the images are to satisfy Rayleigh's criterion for resolvability, the angular separations on both sides of the lens must be given by Eq. 37-14 (assuming small angles). Substituting the given data, we obtain from Eq. 37-14

$$\theta_o = \theta_i = \theta_R = 1.22\frac{\lambda}{d}$$
$$= \frac{(1.22)(550 \times 10^{-9}\text{ m})}{32 \times 10^{-3}\text{ m}} = 2.1 \times 10^{-5}\text{ rad.} \quad \text{(Answer)}$$

At this angular separation, each central maximum in the two intensity curves of Fig. 37-12 is centered on the first minimum of the other curve.

(b) What is the separation Δx of the centers of the *images* in the focal plane? (That is, what is the separation of the *central* peaks in the two curves?)

SOLUTION ■ The **Key Idea** here is to relate the separation Δx to the angle θ_i, which we now know. From either triangle between the lens and the screen in Fig. 37-12, we see that tan $\theta_1/2 = \Delta x/2f$. Rearranging this and making the approximation tan $\theta<\theta$, we find

$$\Delta x = f\theta_i, \qquad (37\text{-}15)$$

where θ_i is in radian measure. Substituting known data then yields

$$\Delta x = (0.24\text{ m})(2.1 \times 10^{-5}\text{ rad}) = 5.0\ \mu\text{m}. \quad \text{(Answer)}$$

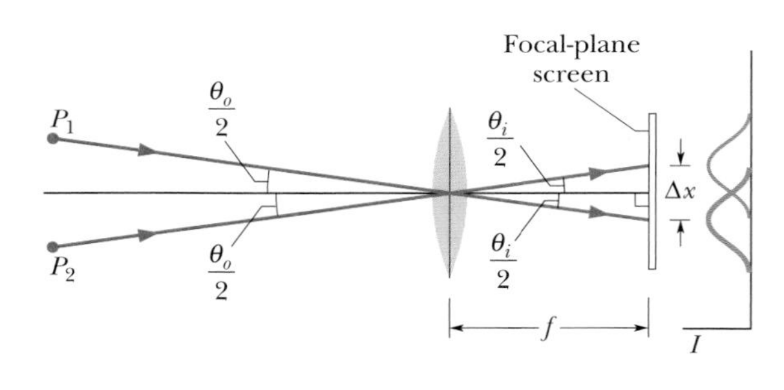

FIGURE 37-12 ■ Light from two distant point objects P_1 and P_2 passes through a converging lens and forms images on a viewing screen in the focal plane of the lens. Only one representative ray from each object is shown. The images are not points but diffraction patterns, with intensities approximately as plotted at the right. The angular separation of the objects is θ_o and that of the images is θ_i; the central maxima of the images have a separation Δx.

37-6 Diffraction by a Double Slit

In the double-slit experiments of Chapter 36, we implicitly assumed that the slits were narrow compared to the wavelength of the light illuminating them; that is, $a \ll \lambda$. For such narrow slits, the central maximum of the diffraction pattern of either slit covers the entire viewing screen. Moreover, the interference of light from the two slits produces bright fringes that all have approximately the same intensity (Fig. 36-9).

In practice with visible light, however, the condition $a \ll \lambda$ is rarely met. For relatively wide slits, the interference of light from two slits produces bright fringes that do not all have the same intensity. That is, the intensities of the fringes produced by double-slit interference (as discussed in Chapter 36) are modified by diffraction of the light passing through each slit (as discussed in this chapter).

As an example, the intensity plot of Fig. 37-13*a* (like that in Fig. 36-9) suggests the double-slit interference pattern that would occur if the slits were infinitely narrow (for $a \ll \lambda$); all the bright interference fringes would have the same intensity. The intensity plot of Fig. 37-13*b* is that for diffraction by a single actual slit; the diffraction pattern has a broad central maximum and weaker secondary maxima at $\pm 1.7°$. The plot of Fig. 37-13*c* suggests the interference pattern for two actual slits. That plot was constructed by using the curve of Fig. 37-13*b* as an *envelope* on the intensity plot in Fig. 37-13*a*. The positions of the fringes are not changed; only the intensities are affected.

Figure 37-14*a* shows an actual pattern in which both double-slit interference and diffraction are evident. If one slit is covered, the single-slit diffraction pattern of Fig. 37-14*b* results. Note the correspondence between Figs. 37-14*a* and 37-13*c* and between Figs. 37-14*b* and 37-13*b*. In comparing these figures, bear in mind that 37-14 has been deliberately overexposed to bring out the faint secondary maxima and that two secondary maxima (rather than one) are shown.

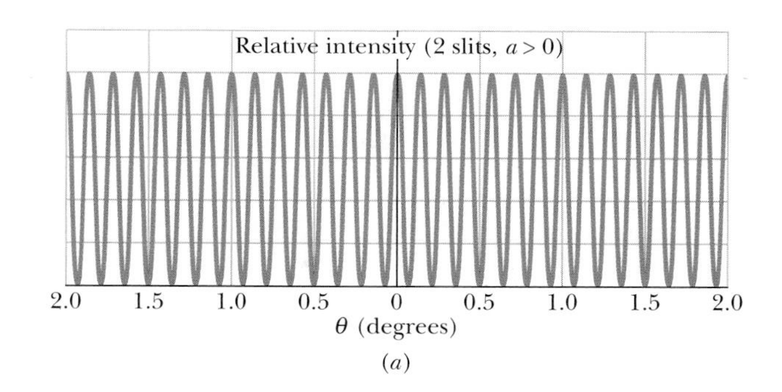

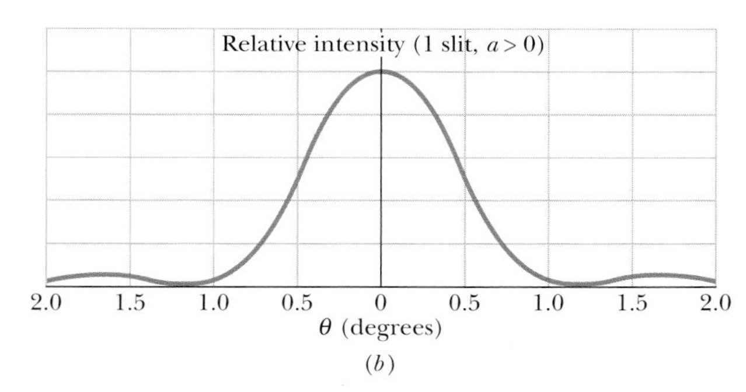

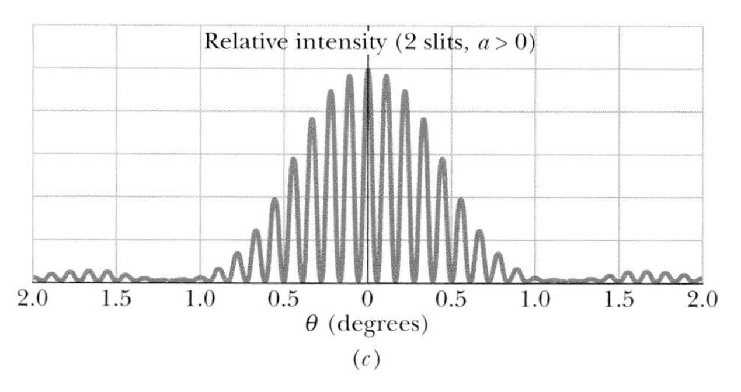

FIGURE 37-13 ■ (*a*) The intensity plot to be expected in a double-slit interference experiment with vanishingly narrow slits (here the distance between the center of the slits is $d = 25$ mm and the incident light is reddish-orange with $\lambda = 623$ mm). (*b*) The intensity plot for diffraction by a typical slit of width $a = 0.031$ mm (not vanishingly narrow). (*c*) The intensity plot to be expected for two slits of width $a = 0.031$ mm. The curve of (*b*) acts as an envelope, limiting the intensity of the double-slit fringes in (*a*). Note that the first minima of the diffraction pattern of (*b*) eliminate the double-slit fringes that would occur near 1.2° in (*c*).

With diffraction effects taken into account, the intensity of a double-slit interference pattern is given by

$$I(\theta) = I^{\max}(\cos^2 \beta)\left(\frac{\sin \alpha}{\alpha}\right)^2 \quad \text{(double slit)}, \tag{37-16}$$

in which

$$\beta = \frac{\pi d}{\lambda} \sin\theta \tag{37-17}$$

and

$$\alpha = \frac{\pi a}{\lambda} \sin\theta. \tag{37-18}$$

Here d is the distance between the centers of the slits, and a is the slit width. Note carefully that the right side of Eq. 37-16 is the product of $I^{\max}$ and two factors. (1) The

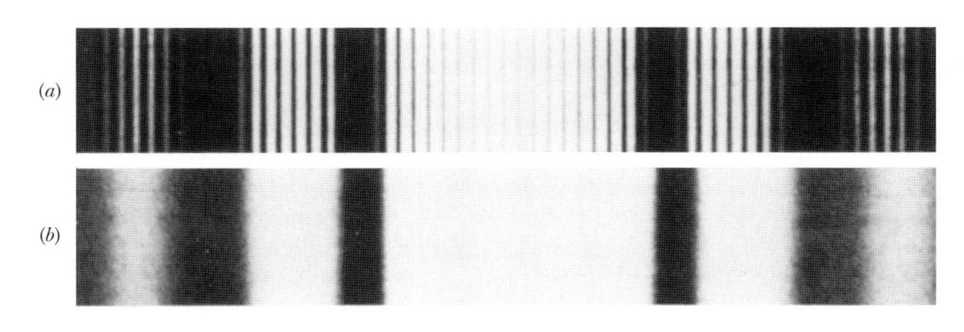

FIGURE 37-14 ■ (*a*) Interference fringes for an actual double-slit system; compare with Fig. 37-13*c*. (*b*) The diffraction pattern of a single slit; compare with Fig. 37-13*b*.

interference factor $\cos^2 \beta$ is due to the interference between two slits with slit separation d (as given by Eqs. 36-17 and 36-18). (2) The *diffraction factor* $[(\sin \alpha)/\alpha]^2$ is due to diffraction by a single slit of width a (as given by Eqs. 37-5 and 37-6).

Let us check these factors. If we let $a \to 0$ in Eq. 37-18, for example, then $\alpha \to 0$ and using L'Hopital's rule, we find that $(\sin \alpha)/\alpha \to 1$. Equation 37-16 then reduces, as it must, to an equation describing the interference pattern for a pair of vanishingly narrow slits with slit separation d. Similarly, putting $d = 0$ in Eq. 37-17 is equivalent physically to causing the two slits to merge into a single slit of width a. Then Eq. 37-17 yields $\beta = 0$ and $\cos^2 \beta = 1$. In this case Eq. 37-16 reduces, as it must, to an equation describing the diffraction pattern for a single slit of width a.

The double-slit pattern described by Eq. 37-16 and displayed in Fig. 37-14a combines interference and diffraction in an intimate way. Both are superposition effects, in that they result from the combining of waves with different phases at a given point. If the combining waves originate from a small number of elementary coherent sources—as in a double-slit experiment with $a \ll \lambda$—we call the process *interference*. If the combining waves originate in a single wavefront—as in a single-slit experiment—we call the process *diffraction*. This distinction between interference and diffraction (which is somewhat arbitrary and not always adhered to) is a convenient one, but we should not forget that both are superposition effects and usually both are present simultaneously (as in Fig. 37-14a).

TOUCHSTONE EXAMPLE 37-4: Bright Fringes

Let's consider a double slit with an unusually small spacing. Suppose the wavelength λ of the light source is 405 nm, the slit separation d is 19.44 μm, and the slit width a is 4.050 μm. Consider the interference of the light from the two slits and also the diffraction of the light through each slit.

(a) How many bright interference fringes are within the central peak of the diffraction envelope?

SOLUTION ■ Let us first analyze the two basic mechanisms responsible for the optical pattern produced in the experiment:

Single-slit diffraction: The **Key Idea** here is that the limits of the central peak are the first minima in the diffraction pattern due to either slit, individually. (See Fig. 37-13.) The angular locations of those minima are given by Eq. 37-3 ($a \sin\theta = m\lambda$). Let us write this equation as $a \sin\theta = m_1\lambda$, with the subscript 1 referring to the one-slit diffraction. For the first minima in the diffraction pattern, we substitute $m_1 = 1$, obtaining

$$a \sin\theta = \lambda. \tag{37-19}$$

Double-slit interference: The **Key Idea** here is that the angular locations of the bright fringes of the double-slit interference pattern are given by Eq. 36-14, which we can write as

$$d \sin\theta = m_2\lambda, \qquad \text{for } m_2 = 1, 2, 3, \ldots. \tag{37-20}$$

Here the subscript 2 refers to the double-slit interference.

We can locate the first diffraction minimum within the double-slit fringe pattern by dividing Eq. 37-20 by Eq. 37-19 and solving for m_2. By doing so and then substituting the given data, we obtain

$$m_2 = \frac{d}{a} = \frac{19.44\ \mu\text{m}}{4.050\ \mu\text{m}} = 4.8.$$

This tells us that the bright interference fringe for $m_2 = 4$ fits into the central peak of the one-slit diffraction pattern, but the fringe for

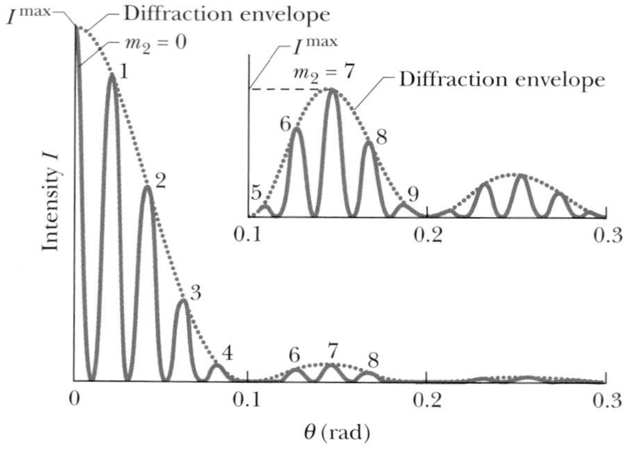

FIGURE 37-15 ■ One side of the intensity plot for a two-slit interference experiment; the diffraction envelope is indicated by the dotted curve. The smaller inset shows (vertically expanded) the intensity plot within the first and second side peaks of the diffraction envelope.

$m_2 = 5$ does not fit. Within the central diffraction peak we have the central bright fringe ($m_2 = 0$), and four bright fringes (up to $m_2 = 4$) on each side of it. Thus, a total of nine bright fringes of the double-slit interference pattern are within the central peak of the diffraction envelope. The bright fringes to one side of the central bright fringe are shown in Fig. 37-15.

(b) How many bright fringes are within either of the first side peaks of the diffraction envelope?

SOLUTION ■ The **Key Idea** here is that the outer limits of the first side diffraction peaks are the second diffraction minima, each of which is at the angle θ given by $a \sin \theta = m_1 \lambda$ with $m_1 = 2$:

$$a \sin\theta = 2\lambda \tag{37-21}$$

Dividing Eq. 37-20 by Eq. 37-21, we find

$$m_2 = \frac{2d}{a} = \frac{(2)(19.44\ \mu\text{m})}{4.050\ \mu\text{m}} = 9.6.$$

This tells us that the second diffraction minimum occurs just before the bright interference fringe for $m_2 = 10$ in Eq. 37-20. Within either first side diffraction peak we have the fringes from $m_2 = 5$ to $m_2 = 9$ for a total of five bright fringes of the double-slit interference pattern (shown in the inset of Fig. 37-15). However, if the $m_2 = 5$ bright fringe, which is almost eliminated by the first diffraction minimum, is considered too dim to count, then only four bright fringes are in the first side diffraction peak.

37-7 Diffraction Gratings

One of the most useful tools in the study of light and of objects that emit and absorb light is the **diffraction grating.** A diffraction grating is a device that uses **interference** phenomena to seperate a beam of light by wavelength. A diffraction grating is a more elaborate form of the double-slit arrangement of Fig. 36-8. This device has a much greater number N of slits, often called *rulings,* perhaps as many as several thousand per millimeter. An idealized grating consisting of only five slits is represented in Fig. 37-16. When monochromatic light is sent through the slits, it forms narrow interference fringes that can be analyzed to determine the wavelength of the light. (Diffraction gratings can also be opaque surfaces with narrow parallel grooves arranged like the slits in Fig. 37-16. Light then scatters back from the grooves to form interference fringes rather than being transmitted through open slits.)

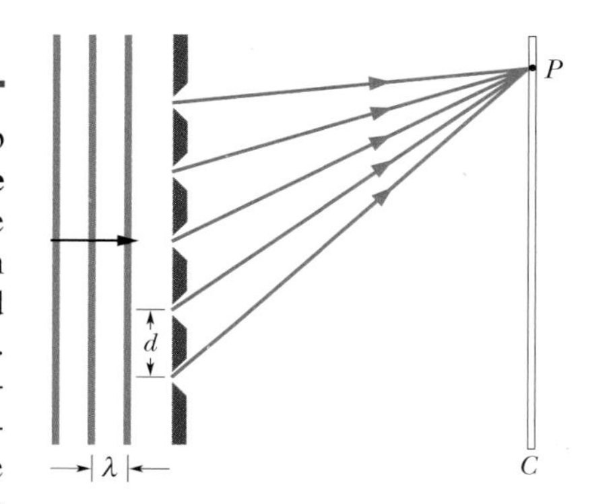

FIGURE 37-16 ■ An idealized diffraction grating, consisting of only five rulings, that produces an interference pattern on a distant viewing screen C.

With monochromatic light incident on a diffraction grating, if we gradually increase the number of slits from two to a large number N, the intensity plot changes from the typical double-slit plot of Fig. 37-13*c* to a much more complicated one and then eventually to a simple graph like that shown in Fig. 37-17*a*. The pattern you would see on a viewing screen using monochromatic red light from, say, a helium-neon laser, is shown in Fig. 37-17*b*. The maxima are now very narrow (and so are called *lines*); they are separated by relatively wide dark regions.

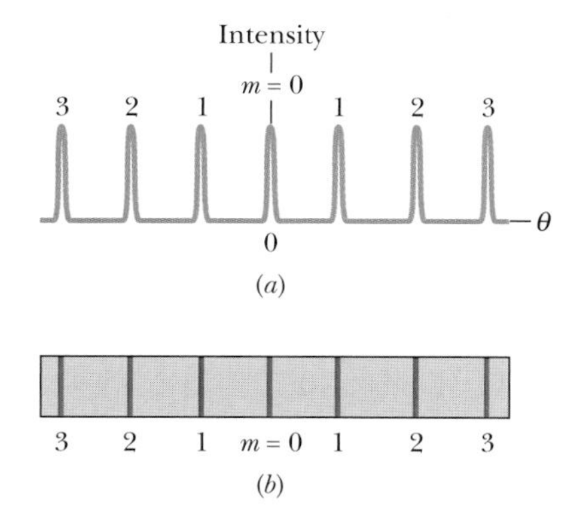

FIGURE 37-17 ■ A diffraction grating illuminated with a single wavelength of light. (*a*) The intensity plot produced by a diffraction grating with a great many rulings consists of narrow peaks, here labeled with their order numbers m. (*b*) The corresponding bright fringes seen on the screen are called lines and are here also labeled with order numbers m. Lines of the zeroth, first, second, and third orders are shown.

We use a familiar procedure to find the locations of the bright lines on the viewing screen. We first assume that the screen is far enough from the grating so that the rays reaching a particular point P on the screen are approximately parallel when they leave the grating (Fig. 37-18). Then we apply to each pair of adjacent rulings the same reasoning we used for double-slit interference. The separation d between rulings is called the *grating spacing.* (If N rulings occupy a total width w, then $d = w/N$.) The path length difference between adjacent rays is again $d \sin\theta$ (Fig. 37-18), where θ is the angle from the central axis of the grating (and of the diffraction pattern) to point P. A line will be located at P if the path length difference between adjacent rays is an integer number of wavelengths—that is, if

$$d \sin\theta = m\lambda, \quad \text{for } m = 0, 1, 2, \ldots \quad \text{(maxima—lines)}, \tag{37-22}$$

where λ is the wavelength of the light. Each integer m represents a different line; hence these integers can be used to label the lines, as in Fig. 37-17. The integers are then called the *order numbers,* and the lines are called the zeroth-order line (the central line, with $m = 0$), the first-order line, the second-order line, and so on.

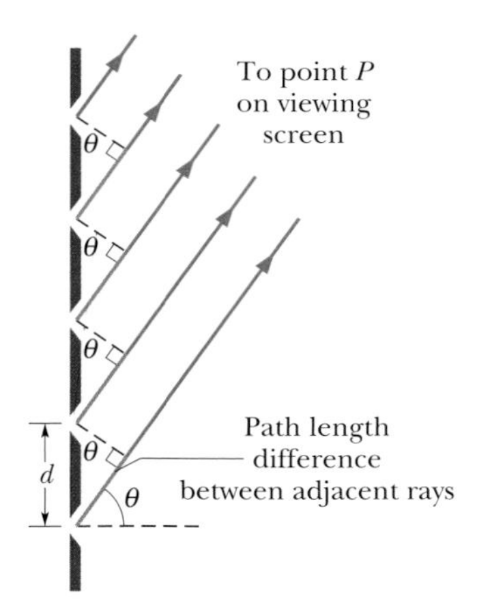

FIGURE 37-18 ■ The rays from the rulings in a diffraction grating to a distant point P are approximately parallel. The path length difference between each two adjacent rays is $d \sin\theta$, where θ is measured as shown. (The rulings extend into and out of the page.)

If we rewrite Eq. 37-22 as $\theta = \sin^{-1}(m\lambda/d)$ we see that, for a given diffraction grating, the angle from the central axis to any line (say, the third-order line) depends on the wavelength of the light being used. Thus, when light of an unknown wavelength is sent through a diffraction grating, measurements of the angles to the higher-order lines can be used in Eq. 37-22 to determine the wavelength. Even light of several unknown wavelengths can be distinguished and identified in this way. We cannot do that with the double-slit arrangement of Section 36-4, even though the same equation and wavelength dependence apply there. In double-slit interference, the bright fringes due to different wavelengths overlap too much to be distinguished.

Width of the Lines

A grating's ability to resolve (separate) lines of different wavelengths depends on the width of the lines. We shall here derive an expression for the *half-width* of the central line (the line for which $m = 0$) and then state an expression for the half-widths of the higher-order lines. We measure the half-width of the central line as the angle $\Delta\theta_{hw}$ from the center of the line at $\theta = 0$ outward to where the line effectively ends and darkness effectively begins with the first minimum (Fig. 37-19). At such a minimum, the N rays from the N slits of the grating cancel one another. (The actual width of the central line is, of course $2(\Delta\theta_{hw})$, but line widths are usually compared via half-widths.)

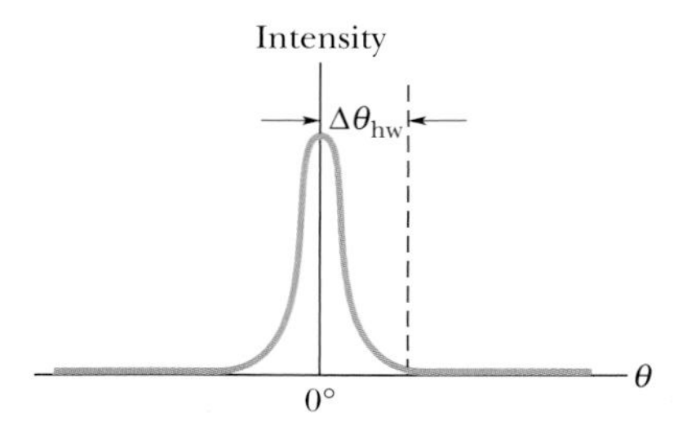

FIGURE 37-19 ■ The half-width $\Delta\theta_{hw}$ of the central line is measured from the center of that line to the adjacent minimum on a plot of I versus θ like Fig. 37-17*a*.

In Section 37-2 we were also concerned with the cancellation of a great many rays, there due to diffraction through a single slit. We obtained Eq. 37-3, which, because of the similarity of the two situations, we can use to find the first minimum here. It tells us that the first minimum occurs where the path length difference between the top and bottom rays equals λ. For single-slit diffraction, this difference is $a \sin\theta$. For a grating of N rulings, each separated from the next by distance d, the distance between the top and bottom rulings is Nd (Fig. 37-20), so the path length difference between the top and bottom rays here is $Nd \sin\Delta\theta_{hw}$. Thus, the first minimum occurs where

$$Nd \sin \Delta\theta_{hw} = \lambda. \tag{37-23}$$

Because $\Delta\theta_{hw}$ is small, $\sin\Delta\theta_{hw} \approx \Delta\theta_{hw}$ (in radian measure). Substituting this in Eq. 37-23 gives the half-width of the central line as

$$\Delta\theta_{hw} = \frac{\lambda}{Nd} \quad \text{(half-width of central line).} \tag{37-24}$$

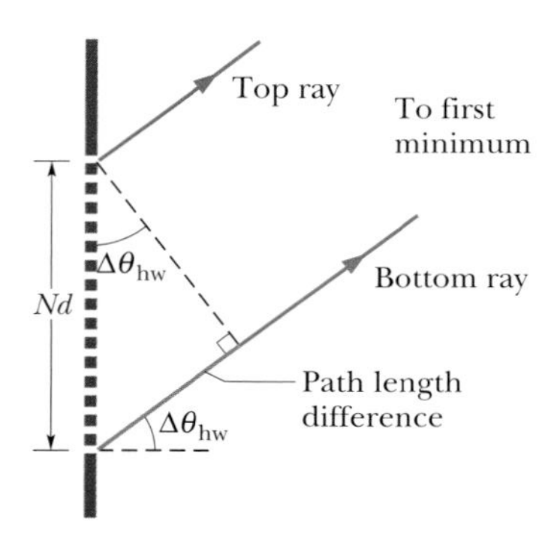

FIGURE 37-20 ■ The top and bottom rulings of a diffraction grating of N rulings are separated by distance Nd. The top and bottom rays passing through these rulings have a path length difference of $Nd \sin \Delta\theta_{hw}$, where $\Delta\theta_{hw}$ is the angle to the first minimum. (The angle is here greatly exaggerated for clarity.)

We state without proof that the half-width of any other line depends on its location relative to the central axis and is

$$\Delta\theta_{hw} = \frac{\lambda}{Nd\cos\theta} \quad \text{(half-width of line at } \theta\text{).} \tag{37-25}$$

Note that for light of a given wavelength λ and a given ruling separation d, the widths of the lines decrease with an increase in the number N of rulings. Thus, of two diffraction gratings, the grating with the larger value of N is better able to distinguish between wavelengths because its diffraction lines are narrower and so produce less overlap. But the line width of a monochromatic light beam is determined by the number of slits that the beam encounters. In a diffraction grating spectrometer, a collimating telescope can be used to illuminate all N slits of the grating.

The Diffraction Grating Spectrometer

Diffraction gratings are widely used to determine the wavelengths that are emitted by sources of light ranging from lamps to stars. Figure 37-21 shows a simple *grating spectroscope* in which a grating is used for this purpose. Light from source S is focused by lens L_1 on a vertical slit S_1 placed in the focal plane of lens L_2. The light emerging from tube C (called a *collimator*) is a plane wave and is incident perpendicularly on grating G, where it is diffracted into a diffraction pattern, with the $m = 0$ order diffracted at angle $\theta = 0$ along the central axis of the grating.

We can view the diffraction pattern that would appear on a viewing screen at any angle θ simply by orienting telescope T in Fig. 37-21 to that angle. Lens L_3 of the telescope then focuses the light diffracted at angle θ (and at slightly smaller and larger angles) onto a focal plane FF' within the telescope. When we look through eyepiece E, we see a magnified view of this focused image.

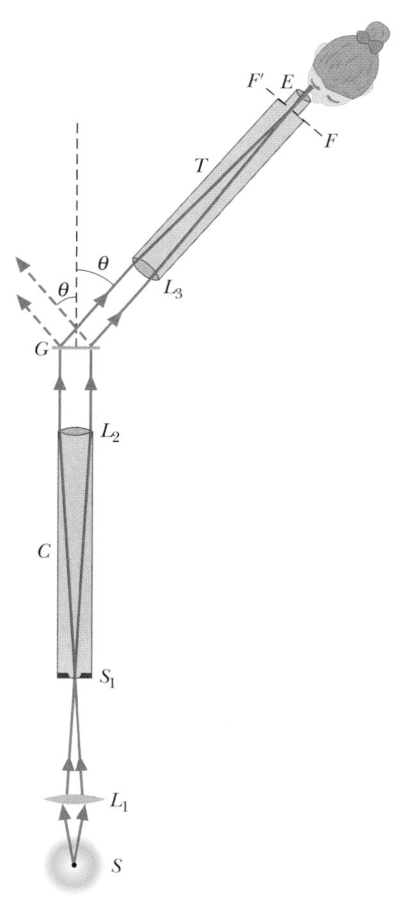

FIGURE 37-21 ■ A simple type of grating spectroscope used to analyze the wavelengths of light emitted by source S.

By changing the angle θ of the telescope, we can examine the entire diffraction pattern. For any order number other than $m = 0$, the original light is spread out according to wavelength (or color) so that we can determine, with Eq. 37-22, just what wavelengths are being emitted by the source. If the source emits a number of discrete wavelengths, what we see as we rotate the telescope horizontally through the angles corresponding to an order m is a vertical line of color for each wavelength, with the shorter-wavelength line at a smaller angle $m = 0$ than the longer-wavelength line.

For example, the light emitted by a hydrogen lamp, which contains hydrogen gas, has four discrete wavelengths in the visible range. If our eyes intercept this light directly, it appears to be white. If, instead, we view it through a grating spectroscope, we can distinguish, in several orders, the lines of the four colors corresponding to these visible wavelengths. (Such lines are called *emission lines.*) Four orders are represented in Fig. 37-22. In the central order ($m = 0$), the lines corresponding to all four wavelengths are superimposed, giving a single white line at $\theta = 0$. The colors are separated in the higher orders.

The third order is not shown in Fig. 37-22 for the sake of clarity; it actually overlaps the second and fourth orders. The fourth-order red line is missing because it is not formed by the grating used here. That is, when we attempt to solve Eq. 37-22 for

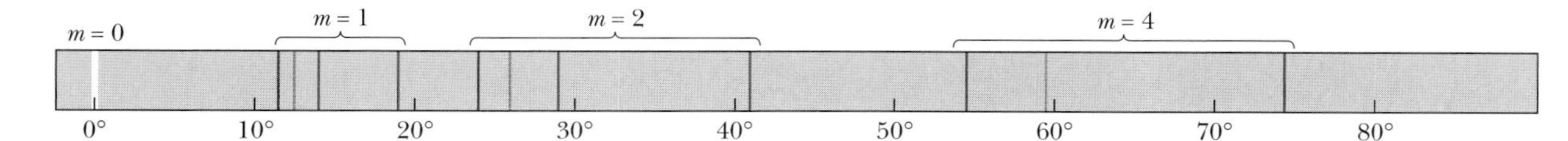

FIGURE 37-22 ■ The zeroth, first, second, and fourth orders of the visible emission lines from hydrogen. Note that the lines are farther apart at greater angles. (The lines are also dimmer and wider, although that is not shown here. Also, the third order line is eliminated for clarity.)

the angle θ for the red wavelength when $m = 4$, we find that sin θ is greater than unity, which is not possible. The fourth order is then said to be *incomplete* for this grating; it might not be incomplete for a grating with greater spacing d, which will spread the lines less than in Fig. 37-22. Figure 37-23 is a photograph of the visible emission lines produced by cadmium.

FIGURE 37-23 ■ The visible emission lines of cadmium, as seen through a grating spectroscope.

READING EXERCISE 37-5: The figure shows lines of different orders produced by a diffraction grating in monochromatic red light. (a) Is the center of the pattern to the left or right? (b) If we switch to monochromatic green light, will the half-widths of the lines then produced in the same orders be greater than, less than, or the same as the half-widths of the lines shown? ■

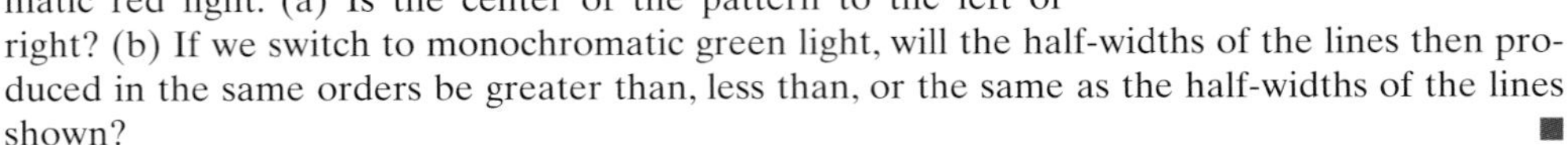

37-8 Gratings: Dispersion and Resolving Power

The fine rulings, each 0.5 μm wide, on a compact disc function as a diffraction grating. When a small source of white light illuminates a disc, the diffracted light forms colored "lanes" that are the composite of the diffraction patterns from the rulings.

There are two characteristics that are important in the design of a diffraction grating spectrometer. First, the different wavelengths of light in a beam should be spread out. This characteristic is called **dispersion.** The second characteristic is the **resolving power** of the spectrometer. It should have a narrow line width for each wavelength so the lines are sharp.

Dispersion

To be useful in distinguishing wavelengths that are close to each other (as in a grating spectroscope), a grating must spread apart the diffraction lines associated with the various wavelengths. This spreading, called **dispersion,** is defined as

$$D = \frac{\Delta\theta}{\Delta\lambda} \quad \text{(dispersion defined).} \tag{37-26}$$

Here $\Delta\theta$ is the angular separation of two lines whose wavelengths differ by $\Delta\lambda$. The greater D is, the greater is the distance between two emission lines whose wavelengths differ by $\Delta\lambda$. We show below that the dispersion of a grating at angle θ is given by

$$D = \frac{m}{d\cos\theta} \quad \text{(dispersion of a grating).} \tag{37-27}$$

Thus, to achieve higher dispersion we must use a grating of smaller grating spacing d and work in a higher order m. Note that the dispersion does not depend on the number of rulings. The SI unit for D is the degree per meter or the radian per meter.

Proof of Eq. 37-27

Let us start with Eq. 37-22, the expression for the locations of the lines in the diffraction pattern of a grating:

$$d\sin\theta = m\lambda.$$

Let us regard θ and λ as variables and take differentials of this equation. We find

$$d\cos\theta\,(d\theta) = m\,(d\lambda),$$

where the differentials $d\theta$ and $d\lambda$ are placed in parentheses to distinguish them from the product of the center to center slit spacing d and the angle θ or wavelength λ.

For small enough angles, we can write these differentials as small differences, obtaining

$$d\cos\theta\,(\Delta\theta) = m(\Delta\lambda), \tag{37-30}$$

or

$$\frac{(\Delta\theta)}{(\Delta\lambda)} = \frac{m}{d\cos\theta}.$$

The ratio on the left is simply D (see Eq. 37-26), so we have indeed derived Eq. 37-27.

Resolving Power

To *resolve* lines whose wavelengths are close together (that is, to make the lines distinguishable), the line should also be as narrow as possible. Expressed otherwise, the grating should have a high **resolving power *R*,** defined as

$$R = \frac{\langle\lambda\rangle}{\Delta\lambda} \quad \text{(resolving power defined).} \tag{37-28}$$

Here $\langle\lambda\rangle$ is the mean wavelength of two emission lines that can barely be recognized as separate, and $\Delta\lambda$ is the wavelength difference between them. The greater R is, the closer two emission lines can be and still be resolved. We shall show below that the resolving power of a grating is given by the simple expression

$$R = Nm \quad \text{(resolving power of a grating).} \tag{37-29}$$

To achieve high resolving power, we must spread out the light beam so it is incident on many rulings (large N in Eq. 37-29).

Proof of Eq. 37-29

We start with Eq. 37-30, which was derived from Eq. 37-22, the expression for the locations of the lines in the diffraction pattern formed by a grating. Here $\Delta\lambda$ is the small wavelength difference between two waves that are diffracted by the grating, and $\Delta\theta$ is the angular separation between them in the diffraction pattern. If $\Delta\theta$ is to be the smallest angle that will permit the two lines to be resolved, it must (by Rayleigh's criterion) be equal to the half-width of each line, which is given by Eq. 37-25:

$$\Delta\theta_{\text{hw}} = \frac{\lambda}{Nd\cos\theta}.$$

If we substitute $\Delta\theta_{\text{hw}}$ as given here for $\Delta\theta$ in Eq. 37-30, we find that

$$\frac{\lambda}{N} = m\,\Delta\lambda,$$

from which it readily follows that

$$R = \frac{\lambda}{\Delta\lambda} = Nm.$$

This is Eq. 37-29, which we set out to derive.

Dispersion and Resolving Power Compared

The resolving power of a grating must not be confused with its dispersion. Table 37-1 shows the characteristics of three gratings, all illuminated with light of wavelength $\lambda = 589$ nm, whose diffracted light is viewed in the first order ($m = 1$ in Eq. 37-22). You should verify that the values of D and R as given in the table can be calculated with Eqs. 37-27 and 37-29, respectively. (In the calculations for D, you will need to convert radians per meter to degrees per micrometer.)

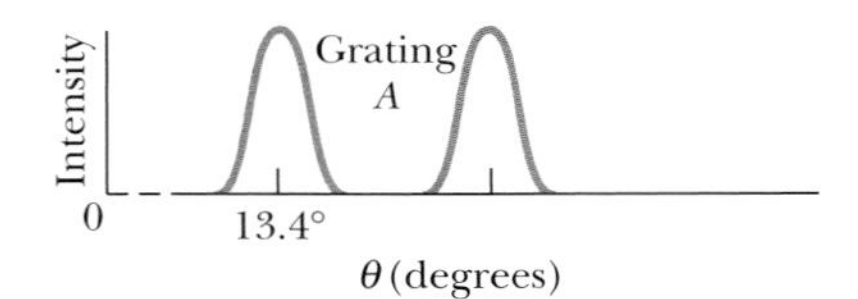

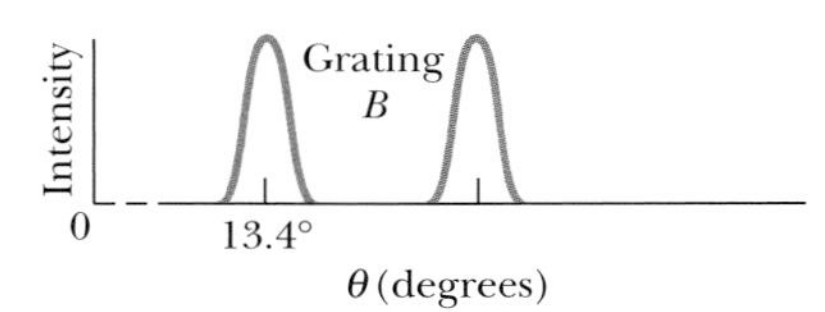

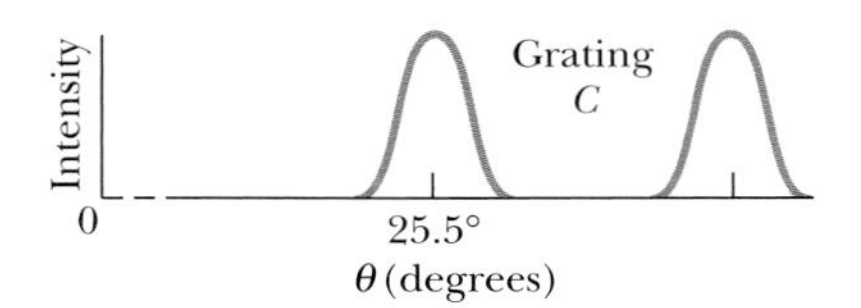

FIGURE 37-24: The intensity patterns for light of two wavelengths sent through the gratings of Table 37-1. Grating B has the highest resolving power and grating C the highest dispersion.

TABLE 37-1
Three Gratings[a]

	Specifications		Calculated Values		
Grating	N	d (nm)	θ	D (°/μm)	R
A	10 000	2540	13.4°	23.2	10 000
B	20 000	2540	13.4°	23.2	20 000
C	10 000	1370	25.5°	46.3	10 000

[a]Data are for $\lambda = 589$ nm and $m = 1$.

For the conditions noted in Table 37-1, gratings A and B have the same *dispersion* and A and C have the same *resolving power.*

Figure 37-24 shows the intensity patterns (also called *line shapes*) that would be produced by these gratings for two lines of wavelengths λ_1 and λ_2, in the vicinity of $\lambda = 589$ nm. Grating B, with the higher resolving power, produces narrower lines and thus is capable of distinguishing lines that are much closer together in wavelength than those in the figure. Grating C, with the higher dispersion, produces the greater angular separation between the lines.

TOUCHSTONE EXAMPLE 37-5: Diffraction Grating

A diffraction grating has 1.26×10^4 rulings uniformly spaced over width $w = 25.4$ mm (so that it has 496 lines/mm). It is illuminated at normal incidence by yellow light from a sodium vapor lamp. This light contains two closely spaced emission lines (known as the sodium doublet) of wavelengths 589.00 nm and 589.59 nm.

(a) At what angle does the first-order maximum occur (on either side of the center of the diffraction pattern) for the wavelength of 589.00 nm?

SOLUTION ■ The **Key Idea** here is that the maxima produced by the diffraction grating can be located with Eq. 37-22 ($d \sin \theta = m\lambda$). The grating spacing d for this diffraction grating is

$$d = \frac{w}{N} = \frac{25.4 \times 10^{-3}\text{ m}}{1.26 \times 10^4}$$
$$= 2.016 \times 10^{-6}\text{ m} = 2016\text{ nm}.$$

The first-order maximum corresponds to $m = 1$. Substituting these values for d and m into Eq. 37-22 leads to

$$\theta = \sin^{-1}\frac{m\lambda}{d} = \sin^{-1}\frac{(1)(589.00\text{ nm})}{2016\text{ nm}}$$
$$= 16.99° \approx 17.0°. \qquad \text{(Answer)}$$

(b) Using the dispersion of the grating, calculate the angular separation between the two lines in the first order.

SOLUTION ■ One **Key Idea** here is that the angular separation $\Delta\theta$ between the two lines in the first order depends on their wavelength difference $\Delta\lambda$ and the dispersion D of the grating, according to Eq. 37-26 ($D = \Delta\theta/\Delta\lambda$). A second **Key Idea** is that the dispersion D depends on the angle θ at which it is to be evaluated. We can assume that, in the first order, the two sodium lines occur close enough to each other for us to evaluate D at the angle $\theta = 16.99°$ we found in part (a) for one of those lines. Then Eq. 37-27 gives the dispersion as

$$D = \frac{m}{d\cos\theta} = \frac{1}{(2016\text{ nm})(\cos 16.99°)}$$
$$= 5.187 \times 10^{-4}\text{ rad/nm}.$$

From Eq. 37-26, we then have

$$\Delta\theta = D\,\Delta\lambda = (5.187 \times 10^{-4}\text{ rad/nm})(589.59\text{ nm} - 589.00\text{ nm})$$
$$= 3.06 \times 10^{-4}\text{ rad} = 0.0175°. \qquad \text{(Answer)}$$

You can show that this result depends on the grating spacing d but not on the number of rulings there are in the grating.

(c) What is the least number of rulings a grating can have and still be able to resolve the sodium doublet in the first order?

SOLUTION ■ One **Key Idea** here is that the resolving power of a grating in any order m is physically set by the number of rulings N in the grating according to Eq. 37-29 ($R = Nm$). A second **Key Idea** is that the least wavelength difference $\Delta\lambda$ that can be resolved depends on the average wavelength involved and the resolving power R of the grating, according to Eq. 37-28 ($R = \langle\lambda\rangle/\Delta\lambda$).

For the sodium doublet to be barely resolved, $\Delta\lambda$ must be their wavelength separation of 0.59 nm, and $\langle\lambda\rangle$ must be their average wavelength of 589.30 nm.

Putting these ideas together, we find that the least number of rulings for a grating to resolve the sodium doublet is

$$N = \frac{R}{m} = \frac{\langle\lambda\rangle}{m\Delta\lambda}$$

$$= \frac{589.30 \text{ nm}}{(1)(0.59 \text{ nm})} = 999 \text{ rulings.} \qquad \text{(Answer)}$$

37-9 X-Ray Diffraction

X rays are electromagnetic radiation whose wavelengths are of the order of 1 Å (= 0.1 nm = 10^{-10} m). Compare this with a wavelength of 550 nm (= 5.5×10^{-7} m) at the center of the visible spectrum. Figure 37-25 shows that x rays are produced when electrons escaping from a heated filament F are accelerated by a potential difference V and strike a metal target T.

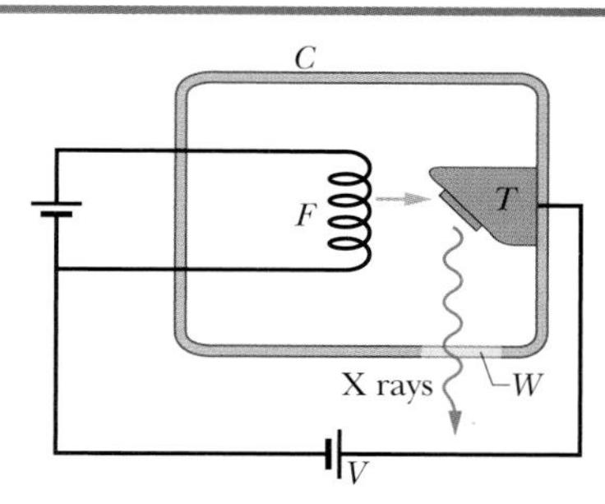

FIGURE 37-25: X rays are generated when electrons leaving heated filament F are accelerated through a potential difference V and strike a metal target T. The "window" W in the evacuated chamber C is transparent to x rays.

A standard optical diffraction grating cannot be used to discriminate between different wavelengths in the x-ray wavelength range. For λ = 1 Å (= 0.1 nm) and d = 3000 nm, for example, Eq. 37-22 shows that the first-order maximum occurs at

$$\theta = \sin^{-1}\frac{m\lambda}{d} = \sin^{-1}\frac{(1)(0.1 \text{ nm})}{3000 \text{ nm}} = 0.0019°.$$

This is too close to the central maximum to be practical. A grating with $d \approx \lambda$ is desirable, but, since x-ray wavelengths are about equal to atomic diameters, such gratings cannot be constructed mechanically.

In 1912, it occurred to German physicist Max von Laue that a crystalline solid, which consists of a regular array of atoms, might form a natural three-dimensional "diffraction grating" for x rays. The idea is that, in a crystal such as sodium chloride (NaCl), a basic unit of atoms (called the *unit cell*) repeats itself throughout the array. In NaCl four sodium ions and four chlorine ions are associated with each unit cell. Figure 37-26*a* represents a section through a crystal of NaCl and identifies this basic unit. The unit cell is a cube measuring a_0 on each side.

When an x-ray beam enters a crystal such as NaCl, x rays are *scattered*—that is, redirected—in all directions by the crystal structure. In some directions the scattered

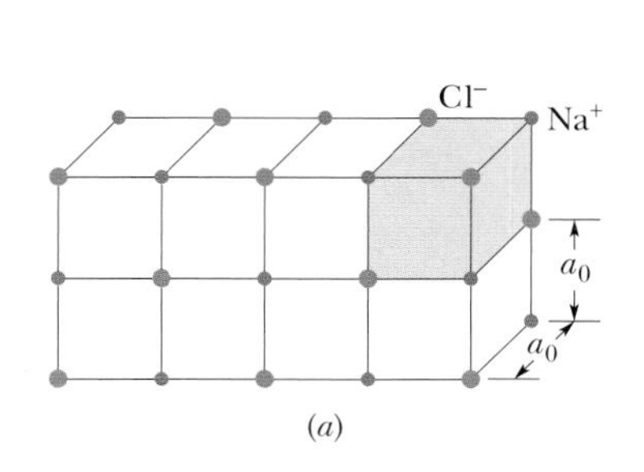

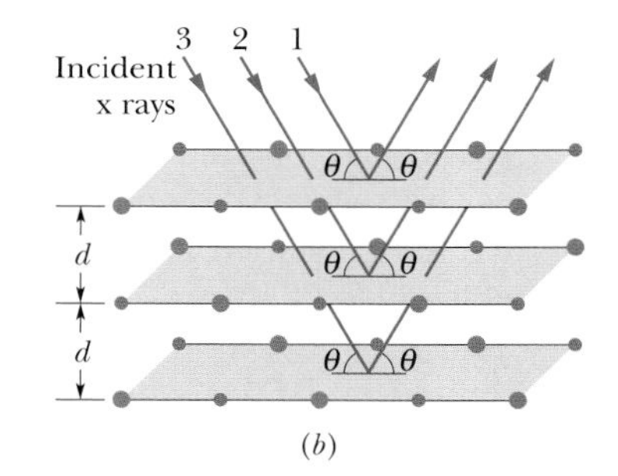

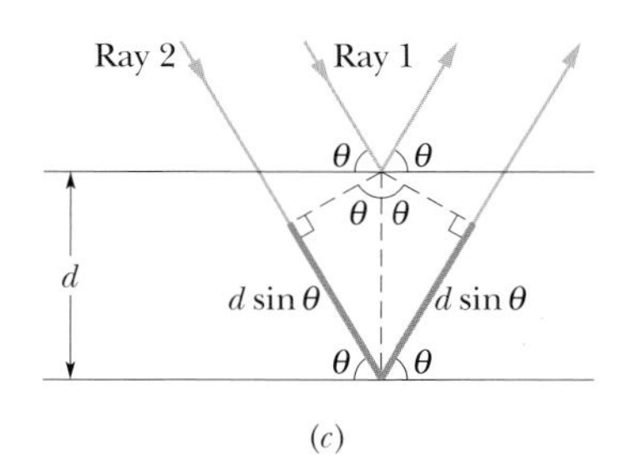

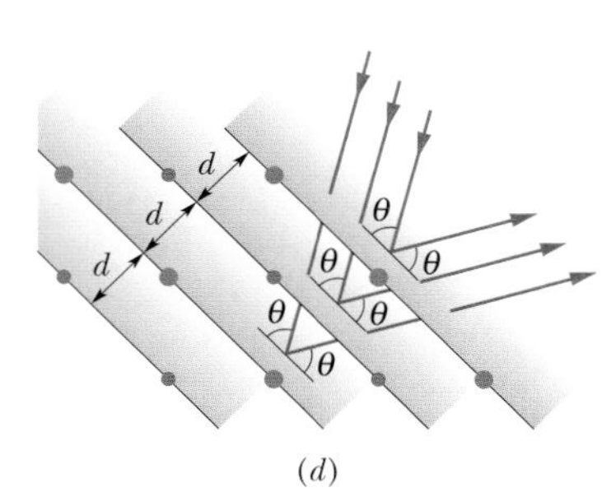

FIGURE 37-26: (*a*) The cubic structure of NaCl, showing the sodium and chlorine ions and a unit cell (shaded). (*b*) Incident x rays undergo diffraction by the structure of (*a*). The x rays are diffracted as if they were reflected by a family of parallel planes, with the angle of reflection equal to the angle of incidence, both angles measured relative to the planes (not relative to a normal as in optics). (*c*) The path length difference between waves effectively reflected by two adjacent planes is $2d \sin\theta$. (*d*) A different orientation of the incident x rays relative to the structure. A different family of parallel planes now effectively reflects the x rays.

waves undergo destructive interference, resulting in intensity minima; in other directions the interference is constructive, resulting in intensity maxima. This process of scattering and interference is a form of diffraction, although it is unlike the diffraction of light traveling through a slit or past an edge as we discussed earlier.

Although the process of diffraction of x rays by a crystal is complicated, the maxima turn out to be in directions as if the x rays were reflected by a family of parallel *reflecting planes* (or *crystal planes*) that extend through the atoms within the crystal and that contain regular arrays of the atoms. (The x rays are not actually reflected; we use these fictional planes only to simplify the analysis of the actual diffraction process.)

Figure 37-26*b* shows three of the family of planes, with *interplanar spacing* d, from which the incident rays shown are said to reflect. Rays 1, 2, and 3 reflect from the first, second, and third planes, respectively. At each reflection the angle of incidence and the angle of reflection are represented with θ. Contrary to the custom in optics, these angles are defined relative to the *surface* of the reflecting plane rather than a normal to that surface. For the situation of Fig. 37-26*b*, the interplanar spacing happens to be equal to the unit cell dimension a_0.

Figure 37-26*c* shows an edge-on view of reflection from an adjacent pair of planes. The waves of rays 1 and 2 arrive at the crystal in phase. After they are reflected, they must again be in phase, because the reflections and the reflecting planes have been defined solely to explain the intensity maxima in the diffraction of x rays by a crystal. Unlike light rays, the x rays have negligible refraction when entering the crystal; moreover, we do not define an index of refraction for this situation. Thus, the relative phase between the waves of rays 1 and 2 as they leave the crystal is set solely by their path length difference. For these rays to be in phase, the path length difference must be equal to an integer multiple of the wavelength λ of the x rays.

By drawing the dashed perpendiculars in Fig. 37-26*c*, we find that the path length difference is $2d \sin \theta$. In fact, this is true for any pair of adjacent planes in the family of planes represented in Fig. 37-26*b*. Thus, we have, as the criterion for intensity maxima for x-ray diffraction,

$$2d \sin\theta = m\lambda, \qquad \text{for } m = 1, 2, 3, \ldots \quad \text{(Bragg's law)}, \tag{37-31}$$

where m is the order number of an intensity maximum. Equation 37-31 is called **Bragg's law** after British physicist W. L. Bragg, who first derived it. (He and his father shared the 1915 Nobel Prize for their use of x rays to study the structures of crystals.) The angle of incidence and reflection in Eq. 37-31 is called a *Bragg angle*.

Regardless of the angle at which x rays enter a crystal, there is always a family of planes from which they can be said to reflect so that we can apply Bragg's law. In Fig. 37-26*d*, the crystal structure has the same orientation as it does in Fig. 37-26*a*, but the angle at which the beam enters the structure differs from that shown in Fig. 37-26*b*. This new angle requires a new family of reflecting planes, with a different interplanar spacing d and different Bragg angle θ, in order to explain the x-ray diffraction via Bragg's law.

Figure 37-27 shows how the interplanar spacing d can be related to the unit cell dimension a_0. For the particular family of planes shown there, the Pythagorean theorem gives

$$5d = \sqrt{5}a_0,$$

or

$$d = \frac{a_0}{\sqrt{5}}. \tag{37-32}$$

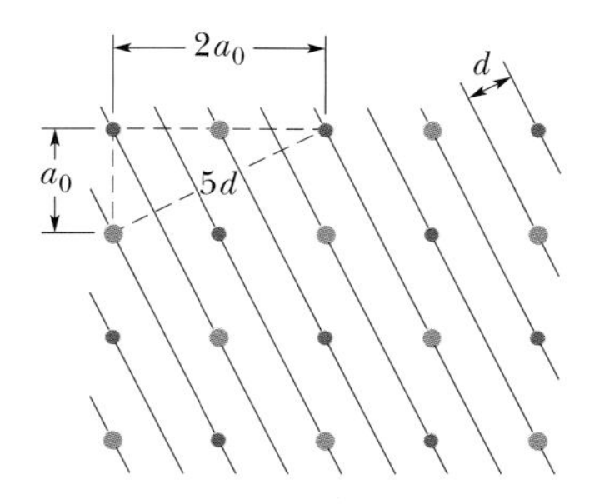

FIGURE 37-27: A family of planes through the structure of Fig. 37-26*a*, and a way to relate the edge length a_0 of a unit cell to the interplanar spacing d.

Figure 37-27 suggests how the dimensions of the unit cell can be found once the interplanar spacing has been measured by means of x-ray diffraction.

X-ray diffraction is a powerful tool for studying both x-ray spectra and the arrangement of atoms in crystals. To study spectra, a particular set of crystal planes, having a known spacing d, is chosen. These planes effectively reflect different wavelengths at different angles. A detector that can discriminate one angle from another can then be used to determine the wavelength of radiation reaching it. The crystal itself can be studied with a monochromatic x-ray beam, to determine not only the spacing of various crystal planes but also the structure of the unit cell.

Problems

SEC. 37-2 ■ DIFFRACTION BY A SINGLE SLIT: LOCATING THE MINIMA

1. Narrow Slit Light of wavelength 633 nm is incident on a narrow slit. The angle between the first diffraction minimum on one side of the central maximum and the first minimum on the other side is 1.20°. What is the width of the slit?

2. Distance Between Monochromatic light of wavelength 441 nm is incident on a narrow slit. On a screen 2.00 m away, the distance between the second diffraction minimum and the central maximum is 1.50 cm. (a) Calculate the angle of diffraction θ of the second minimum. (b) Find the width of the slit.

3. Single Slit A single slit is illuminated by light of wavelengths λ_a and λ_b, chosen so the first diffraction minimum of the λ_a component coincides with the second minimum of the λ_b component. (a) What relationship exists between the two wavelengths? (b) Do any other minima in the two diffraction patterns coincide?

4. First and Fifth The distance between the first and fifth minima of a single-slit diffraction pattern is 0.35 mm with the screen 40 cm away from the slit, when light of wavelength 550 nm is used. (a) Find the slit width. (b) Calculate the angle θ of the first diffraction minimum.

5. Plane Wave A plane wave of wavelength 590 nm is incident on a slit with a width of $a = 0.40$ nm. A thin converging lens of focal length +70 cm is placed between the slit and a viewing screen and focuses the light on the screen. (a) How far is the screen from the lens? (b) What is the distance on the screen from the center of the diffraction pattern to the first minimum?

6. Sound Waves Sound waves with frequency 3000 Hz and speed 343 m/s diffract through the rectangular opening of a speaker cabinet and into a large auditorium. The opening, which has a horizontal width of 30.0 cm, faces a wall 100 m away (Fig. 37-28). Where along that wall will a listener be at the first diffraction minimum and thus have difficulty hearing the sound? (Neglect reflections).

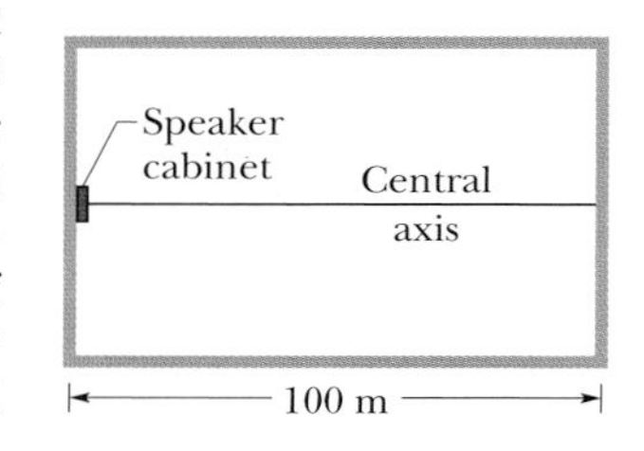

FIGURE 37-28 ■ Problem 6.

7. Central Maximum A slit 1.00 mm wide is illuminated by light of wavelength 589 nm. We see a diffraction pattern on a screen 3.00 m away. What is the distance between the first two diffraction minima on the same side of the central diffraction maximum?

SEC. 37-4 ■ INTENSITY IN SINGLE-SLIT DIFFRACTION, QUANTITATIVELY

8. Off Central Axis A 0.10-mm-wide slit is illuminated by light of wavelength 589 nm. Consider a point P on a viewing screen on which the diffraction pattern of the slit is viewed; the point is at 30° from the central axis of the slit. What is the phase difference between the Huygens wavelets arriving at point P from the top and midpoint of the slit? (*Hint:* See Eq. 37-4.)

9. Explain Quantitatively If you double the width of a single slit, the intensity of the central maximum of the diffraction pattern increases by a factor of 4, even though the energy passing through the slit only doubles. Explain this quantitatively.

10. Monochromatic Monochromatic light with wavelength 538 nm is incident on a slit with width 0.025 mm. The distance from the slit to a screen is 3.5 m. Consider a point on the screen 1.1 cm from the central maximum. (a) Calculate θ for that point. (b) Calculate α. (c) Calculate the ratio of the intensity at this point to the intensity at the central maximum.

11. FWHM The full width at half-maximum (FWHM) of a central diffraction maximum is defined as the angle between the two points in the pattern where the intensity is one-half that at the center of the pattern. (See Fig. 37-7*b*.) (a) Show that the intensity drops to one-half the maximum value when $\sin^2 \alpha = \alpha^2/2$. (b) Verify that $\alpha = 1.39$ rad (about 80°) is a solution to the transcendental equation of (a). (c) Show that the FWHM is $\Delta\theta = 2\sin^{-1}(0.443\lambda/a)$, where a is the slit width. (d) Calculate the FWHM of the central maximum for slits whose widths are 1.0, 5.0, and 10 wavelengths.

12. Babinet's Principle A monochromatic beam of parallel light is incident on a "collimating" hole of diameter $x \gg \lambda$. Point P lies in the geometrical shadow region on a *distant* screen (Fig. 37-29*a*). Two diffracting objects, shown in Fig. 37-29*b*, are placed in turn

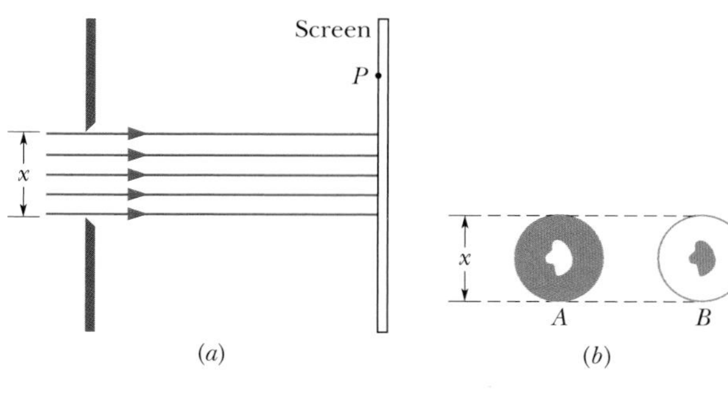

FIGURE 37-29 ■ Problem 12.

over the collimating hole. A is an opaque circle with a hole in it and B is the "photographic negative" of A. Using superposition concepts, show that the intensity at P is identical for the two diffracting objects A and B.

13. Values of α (a) Show that the values of α at which intensity maxima for single-slit diffraction occur can be found exactly by differentiating Eq. 37-5 with respect to α and equating the result to zero, obtaining the condition $\tan\alpha = \alpha$. (b) Find the values of α satisfying this relation by plotting the curve $y = \tan\alpha$ and the straight line $y = \alpha$ and finding their intersections or by using a calculator with an equation solver to find an appropriate value of α (or by using trial and error). (c) Find the (noninteger) values of m corresponding to successive maxima in the single-slit pattern. Note that the secondary maxima do not lie exactly halfway between minima.

SEC. 37-5 ■ DIFFRACTION BY A CIRCULAR APERTURE

14. Entopic Halos At night many people see rings (called *entopic halos*) surrounding bright outdoor lamps in otherwise dark surroundings. The rings are the first of the side maxima in diffraction patterns produced by structures that are thought to be within the cornea (or possibly the lens) of the observer's eye. (The central maxima of such patterns overlap the lamp.) (a) Would a particular ring become smaller or larger if the lamp were switched from blue to red light? (b) If a lamp emits white light, is blue or red on the outside edge of the ring? (c) Assume that the lamp emits light at wavelength 550 nm. If a ring has an angular diameter of 2.5°, approximately what is the (linear) diameter of the structure in the eye that causes the ring?

15. Headlights The two headlights of an approaching automobile are 1.4 m apart. At what (a) angular separation and (b) maximum distance will the eye resolve them? Assume that the pupil diameter is 5.0 mm, and use a wavelength of 550 nm for the light. Also assume that diffraction effects alone limit the resolution so that Rayleigh's criterion can be applied.

16. An Astronaut An astronaut in a space shuttle claims she can just barely resolve two point sources on the Earth's surface, 160 km below. Calculate their (a) angular and (b) linear separation, assuming ideal conditions. Take $\lambda = 540$ nm and the pupil diameter of the astronaut's eye to be 5.0 mm.

17. Moon's Surface Find the separation of two points on the Moon's surface that can just be resolved by the 200 in. (= 5.1 m) telescope at Mount Palomar, assuming that this separation is determined by diffraction effects. The distance from the Earth to the Moon is 3.8×10^5 km. Assume a wavelength of 550 nm for the light.

18. Large Room The wall of a large room is covered with acoustic tile in which small holes are drilled 5.0 mm from center to center. How far can a person be from such a tile and still distinguish the individual holes, assuming ideal conditions, the pupil diameter of the observer's eye to be 4.0 mm, and the wavelength of the room light to be 550 nm?

19. Estimate Linear Separation Estimate the linear separation of two objects on the planet Mars that can just be resolved under ideal conditions by an observer on Earth (a) using the naked eye and (b) using the 200 in. (= 5.1 m) Mount Palomar telescope. Use the following data: distance to Mars = 8.0×10^7 km, diameter of pupil = 5.0 mm, wavelength of light = 550 nm.

20. Radar System The radar system of a navy cruiser transmits at a wavelength of 1.6 cm, from a circular antenna with a diameter of 2.3 m. At a range of 6.2 km, what is the smallest distance that two speedboats can be from each other and still be resolved as two separate objects by the radar system?

21. Tiger Beetles The wings of tiger beetles (Fig. 37-30) are colored by interference due to thin cuticle-like layers. In addition, these layers are arranged in patches that are 60 μm across and produce different colors. The color you see is a pointillistic mixture of thin-film interference colors that varies with perspective. Approximately what viewing distance from a wing puts you at the limit of resolving the different colored patches according to Rayleigh's criterion? Use 550 nm as the wavelength of light and 3.00 mm as the diameter of your pupil.

FIGURE 37-30 ■ Problem 21. Tiger beetles are colored by pointillistic mixtures of thin-film interference colors.

22. Discovery In June 1985, a laser beam was sent out from the Air Force Optical Station on Maui, Hawaii, and reflected back from the shuttle *Discovery* as it sped by, 354 km overhead. The diameter of the central maximum of the beam at the shuttle position was said to be 9.1 m, and the beam wavelength was 500 nm. What is the effective diameter of the laser aperture at the Maui ground station? (*Hint:* A laser beam spreads only because of diffraction; assume a circular exit aperture.)

23. Millimeter-Wave Radar Millimeter-wave radar generates a narrower beam than conventional microwave radar, making it less vulnerable to antiradar missiles. (a) Calculate the angular width of the central maximum, from first minimum to first minimum, produced by a 220 GHz radar beam emitted by a 55.0-cm-diameter circular antenna. (The frequency is chosen to coincide with a low-absorption atmospheric "window.") (b) Calculate the same quantity for the ship's radar described in Problem 20.

24. Circular Obstacle A circular obstacle produces the same diffraction pattern as a circular hole of the same diameter (except very near $\theta = 0$). Airborne water drops are examples of such obstacles. When you see the Moon through suspended water drops, such as in a fog, you intercept the diffraction pattern from many drops.

FIGURE 37-31 ■ Problem 24. The corona around the Moon is a composite of the diffraction patterns of airborne water drops.

The composite of the central diffraction maxima of those drops forms a white region that surrounds the Moon and may obscure it. Figure 37-31 is a photograph in which the Moon is obscured. There are two, faint, colored rings around the Moon (the larger one may be too faint to be seen in your copy of the photograph). The smaller ring is on the outer edge of the central maxima from the drops; the somewhat larger ring is on the outer edge of the smallest of the secondary maxima from the drops (see Fig. 37-3). The color is visible because the rings are adjacent to the diffraction minima (dark rings) in the patterns. (Colors in other parts of the pattern overlap too much to be visible.)

(a) What is the color of these rings on the outer edges of the diffraction maxima? (b) The colored ring around the central maxima in Fig. 37-31 has an angular diameter that is 1.35 times the angular diameter of the Moon, which is 0.50°. Assume that the drops all have about the same diameter. Approximately what is that diameter?

25. Allegheny Observatory (a) What is the angular separation of two stars if their images are barely resolved by the Thaw refracting telescope at the Allegheny Observatory in Pittsburgh? The lens diameter is 76 cm and its focal length is 14 m. Assume $\lambda = 550$ nm. (b) Find the distance between these barely resolved stars if each of them is 10 light-years distant from Earth. (c) For the image of a single star in this telescope, find the diameter of the first dark ring in the diffraction pattern, as measured on a photographic plate placed at the focal plane of the telescope lens. Assume that the structure of the image is associated entirely with diffraction at the lens aperture and not with lens "errors".

26. Soviet–French Experiment In a joint Soviet–French experiment to monitor the Moon's surface with a light beam, pulsed radiation from a ruby laser ($\lambda = 0.69\ \mu$m) was directed to the Moon through a reflecting telescope with a mirror radius of 1.3 m. A reflector on the Moon behaved like a circular plane mirror with radius 10 cm, reflecting the light directly back toward the telescope on the Earth. The reflected light was then detected after being brought to a focus by this telescope. What fraction of the original light energy was picked up by the detector? Assume that for each direction of travel all the energy is in the central diffraction peak.

SEC. 37-6 ■ DIFFRACTION BY A DOUBLE SLIT

27. Bright Fringes Suppose that the central diffraction envelope of a double-slit diffraction pattern contains 11 bright fringes and the first diffraction minima eliminate (are coincident with) bright fringes. How many bright fringes lie between the first and second minima of the diffraction envelope?

28. Slit Separation In a double-slit experiment, the slit separation d is 2.00 times the slit width w. How many bright interference fringes are in the central diffraction envelope?

29. Eliminate Bright Fringes (a) In a double-slit experiment, what ratio of d to a causes diffraction to eliminate the fourth bright side fringe? (b) What other bright fringes are also eliminated?

30. Two Slits Two slits of width a and separation d are illuminated by a coherent beam of light of wavelength λ. What is the linear separation of the bright interference fringes observed on a screen that is at a distance D away?

31. How Many (a) How many bright fringes appear between the first diffraction-envelope minima to either side of the central maximum in a double-slit pattern if $\lambda = 550$ nm, $d = 0.150$ mm, and $a = 30.0\ \mu$m? (b) What is the ratio of the intensity of the third bright fringe to the intensity of the central fringe?

32. Intensity Vs. Position Light of wavelength 440 nm passes through a double slit, yielding a diffraction pattern whose graph of intensity I versus angular position θ is shown in Fig. 37-32. Calculate the (a) slit width and (b) slit separation. (c) Verify the displayed intensities of the $m = 1$ and $m = 2$ interference fringes.

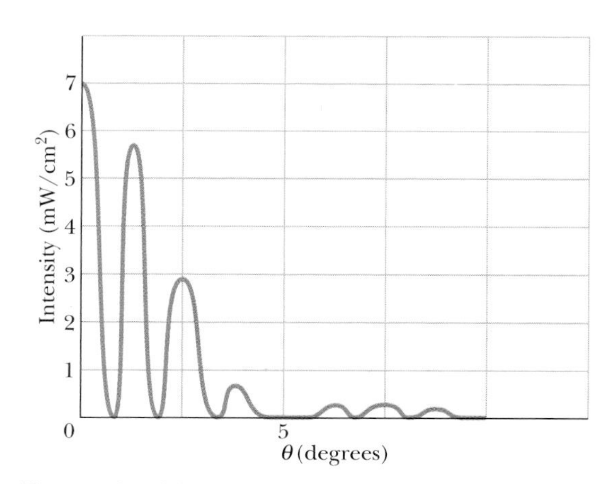

FIGURE 37-32 ■ Problem 32.

SEC. 37-7 ■ DIFFRACTION GRATINGS

33. Calculate d A diffraction grating 20.0 mm wide has 6000 rulings. (a) Calculate the distance d between adjacent rulings. (b) At what angles θ will intensity maxima occur on a viewing screen if the radiation incident on the grating has a wavelength of 589 nm?

34. Visible Spectrum A grating has 315 rulings/mm. For what wavelengths in the visible spectrum can fifth-order diffraction be observed when this grating is used in a diffraction experiment?

35. How Many Orders A grating has 400 lines/mm. How many orders of the entire visible spectrum (400–700 nm) can it produce in a diffraction experiment, in addition to the $m = 0$ order?

36. Confuse a Predator Perhaps to confuse a predator, some tropical gyrinid beetles (whirligig beetles) are colored by optical interference that is due to scales whose alignment forms a diffraction grating (which scatters light instead of transmiting it). When the incident light rays are perpendicular to the grating, the angle between the first-order maxima (on opposite sides of the zeroth-order maximum) is about 26° in light with a wavelength of 550 nm. What is the grating spacing of the beetle?

37. Two Adjacent Maxima Light of wavelength 600 nm is incident normally on a diffraction grating. Two adjacent maxima occur at angles given by $\sin\theta = 0.2$ and $\sin\theta = 0.3$. The fourth-order maxima are missing. (a) What is the separation between adjacent slits? (b) What is the smallest slit width this grating can have? (c) Which orders of intensity maxima are produced by the grating, assuming the values derived in (a) and (b)?

38. Normal Incidence A diffraction grating is made up of slits of width 300 nm with separation 900 nm. The grating is illuminated by monochromatic plane waves of wavelength $\lambda = 600$ nm at normal incidence. (a) How many maxima are there in the full diffraction pattern? (b) What is the width of a spectral line observed in the first order if the grating has 1000 slits?

39. Visible Spectrum Assume that the limits of the visible spectrum are arbitrarily chosen as 430 and 680 nm. Calculate the number of rulings per millimeter of a grating that will spread the first-order spectrum through an angle of 20°.

40. Gaseous Discharge Tube With light from a gaseous discharge tube incident normally on a grating with slit separation 1.73 μm, sharp maxima of green light are produced at angles $\theta = \pm 17.6°$, 37.3°, −37.1°, 65.2°, and −65.0°. Compute the wavelength of the green light that best fits these data.

41. Show That Light is incident on a grating at an angle ψ as shown in Fig. 37-33. Show that bright fringes occur at angles θ that satisfy the equation

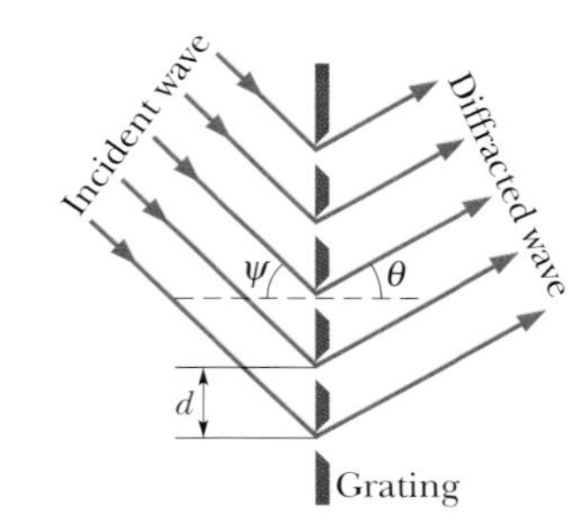

FIGURE 37-33 ■ Problem 41.

$$d(\sin\psi + \sin\theta) = m\lambda, \quad \text{for } m = 0, 1, 2, \ldots.$$

(Compare this equation with Eq. 37-22.) Only the special case $\psi = 0$ has been treated in this chapter.

42. Plot A grating with $d = 1.50$ μm is illuminated at various angles of incidence by light of wavelength 600 nm. Plot, as a function of the angle of incidence (0 to 90°), the angular deviation of the first-order maximum from the incident direction. (See Problem 41.)

43. Derive Derive Eq. 37-25, the expression for the half-widths of lines in a grating's diffraction pattern.

44. Spectrum Is Formed A grating has 350 rulings per millimeter and is illuminated at normal incidence by white light. A spectrum is formed on a screen 30 cm from the grating. If a hole 10 mm square is cut in the screen, its inner edge being 50 mm from the central maximum and parallel to it, what is the range in the wavelengths of the light that passes through the hole?

45. Derive Two Derive this expression for the intensity pattern for a three-slit grating (ignore diffraction effects);

$$I_\theta = \tfrac{1}{9} I^{\max}(1 + 4\cos\phi + 4\cos^2\phi),$$

where $\phi = (2\pi d \sin\theta)/\lambda$. Assume that $a \ll \lambda$; be guided by the derivation of the corresponding double-slit formula (Eq. 36-21).

SEC. 37-8 ■ GRATINGS: DISPERSION AND RESOLVING POWER

46. D Line The D line in the spectrum of sodium is a doublet with wave-lengths 589.0 and 589.6 nm. Calculate the minimum number of lines needed in a grating that will resolve this doublet in the second-order spectrum. See Touchstone Example 37-5.

47. Hydrogen–Deuterium Mix A source containing a mixture of hydrogen and deuterium atoms emits red light at two wavelengths whose mean is 656.3 nm and whose separation is 0.180 nm. Find the minimum number of lines needed in a diffraction grating that can resolve these lines in the first order.

48. Smallest Wavelength A grating has 600 rulings/mm and is 5.0 mm wide. (a) What is the smallest wavelength interval it can resolve in the third order at $\lambda = 500$ nm? (b) How many higher orders of maxima can be seen?

49. Dispersion Show that the dispersion of a grating is $D = (\tan\theta)/\lambda$.

50. Sodium Doublet With a particular grating the sodium doublet (see Touchstone Example 37-5) is viewed in the third order at 10° to the normal and is barely resolved. Find (a) the grating spacing and (b) the total width of the rulings.

51. Resolving Power A diffraction grating has resolving power $R = \langle\lambda\rangle/\Delta\lambda = Nm$. (a) Show that the corresponding frequency range Δf that can just be resolved is given by $\Delta f = c/Nm\lambda$. (b) From Fig. 37-18, show that the times required for light to travel along the ray at the bottom of the figure and the ray at the top differ by an amount $\Delta t = (Nd/c)\sin\theta$. (c) Show that $(\Delta f)(\Delta t) = 1$, this relation being independent of the various grating parameters. Assume $N \gg 1$.

52. Product (a) In terms of the angle θ locating a line produced by a grating, find the product of that line's half-width and the resolving power of grating. (b) Evaluate that product for the grating of Problem 38, for the first order.

SEC. 37-9 ■ X-RAY DIFFRACTION

53. Second-Order Reflection X rays of wavelength 0.12 nm are found to undergo second-order reflection at a Bragg angle of 28° from a lithium fluoride crystal. What is the interplanar spacing of the reflecting planes in the crystal?

54. Diffraction by Crystal Figure 37-34 is a graph of intensity versus angular position θ for the diffraction of an x-ray beam by a crystal. The beam consists of two wavelengths, and the spacing between the reflecting planes is 0.94 nm. What are the two wavelengths?

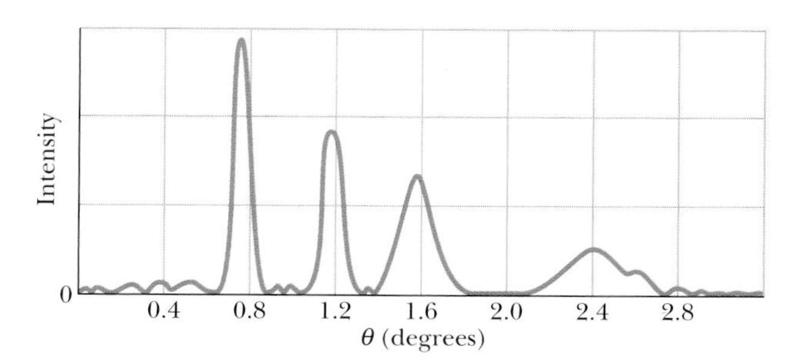

FIGURE 37-34 ■ Problem 54.

55. NaCl Crystal An x-ray beam of a certain wavelength is incident on a NaCl crystal, at 30.0° to a certain family of reflecting planes of spacing 39.8 pm. If the reflection from those planes is of the first order, what is the wavelength of the x rays?

56. Two Beams An x-ray beam of wavelength A undergoes first-order reflection from a crystal when its angle of incidence to a crystal face is 23°, and an x-ray beam of wavelength 97 pm undergoes third-order reflection when its angle of incidence to that face is 60°. Assuming that the two beams reflect from the same family of reflecting planes, find the (a) interplanar spacing and (b) wavelength A.

57. Not Possible Prove that it is not possible to determine both wavelength of incident radiation and spacing of reflecting planes in a crystal by measuring the Bragg angles for several orders.

58. Reflection Planes In Fig. 37-35, first-order reflection from the reflection planes shown occurs when an x-ray beam of wavelength 0.260 nm makes an angle of 63.8° with the top face of the crystal. What is the unit cell size a_0?

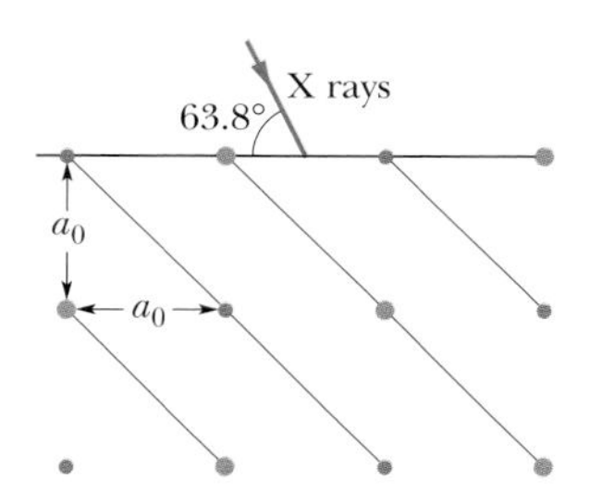

FIGURE 37-35 ■ Problem 58.

59. Square Crystal Consider a two-dimensional square crystal structure, such as one side of the structure shown in Fig. 37-26*a*. One interplanar spacing of reflecting planes is the unit cell size a_0. (a) Calculate and sketch the next five smaller interplanar spacings. (b) Show that your results in (a) are consistent with the general formula

$$d = \frac{a_0}{h^2 + k^2},$$

where h and k are relatively prime integers (they have no common factor other than unity).

60. X-Ray Beam In Fig. 37-36, an x-ray beam of wavelengths from 95.0 pm to 140 pm is incident at 45° to a family of reflecting planes with spacing $d = 275$ pm. At which wavelengths will these planes produce intensity maxima in their reflections?

61. NaCl In Fig. 37-36, let a beam of x-rays of wavelength 0.125 nm be incident on an NaCl crystal at an angle of 45.0° to the top face of the crystal and a family of reflecting planes. Let the reflecting planes have separation $d = 0.252$ nm. Through what angles must the crystal be turned about an axis that is perpendicular to the plane of the page for these reflecting planes to give intensity maxima in their reflections?

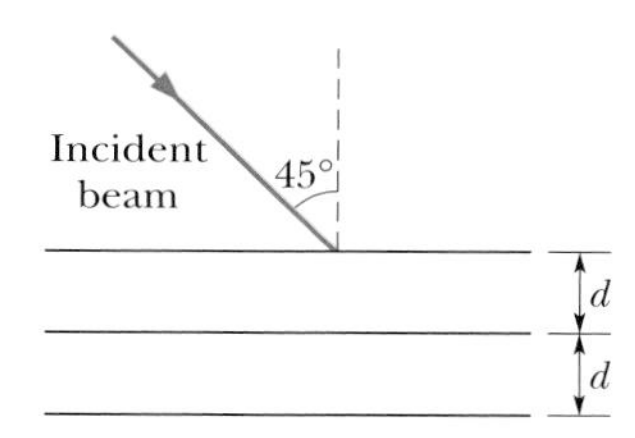

FIGURE 37-36 ■ Problems 60 and 61.

Additional Problems

62. Changing Interference Consider a plane wave of monochromatic green light, $\lambda = 500$ nm, that is incident normally upon two identical narrow slits (the widths of the individual slits are much less than λ). The slits are separated by a distance $d = 30\ \mu$m. An interference pattern is observed on a screen located a distance L away from the slits. On the screen, the location nearest the central maximum where the intensity is zero (i.e., the first dark fringe) is found to be 1.5 cm from this central point. Let this particular position on the screen be referred to as P_1. (a) Calculate the distance, L, to the screen. Show all work. (b) In each of the parts below, one change has been made to the problem above (in each case, all parameters not explicitly mentioned have the value or characteristics stated above). For each case, explain briefly whether the light intensity at location P_1 remains zero or not. If not, does P_1 become the location of a maximum constructive interference (bright) fringe? In each case, explain your reasoning.

(1) One of the two slits is made slightly narrower, so that the amount of light passing through it is less than that through the other.

(2) The wavelength is doubled so that $\lambda = 1000$ nm.

(3) The two slits are replaced by a single slit whose width is exactly 60 μm.

63. Hearing and Seeing Around a Corner We can make the observation that we can hear around corners (somewhat) but not see around corners. Estimate why this is so by considering a doorway and two kinds of waves passing through it: (1) a beam of red light ($\lambda = 660$ nm), and (2) a sound wave playing an "A" ($f = 440$ Hz). (See Fig. 37-37.) Treat these two waves as plane waves passing through a slit whose width equals the width of the door. (a) Find the angle that gives the position of the first dark diffraction fringe. (b) From that, assuming you are 2 m back from the door, estimate how far outside the door you could be and still detect the wave. (See the picture for a clarification. The distance x is desired.)

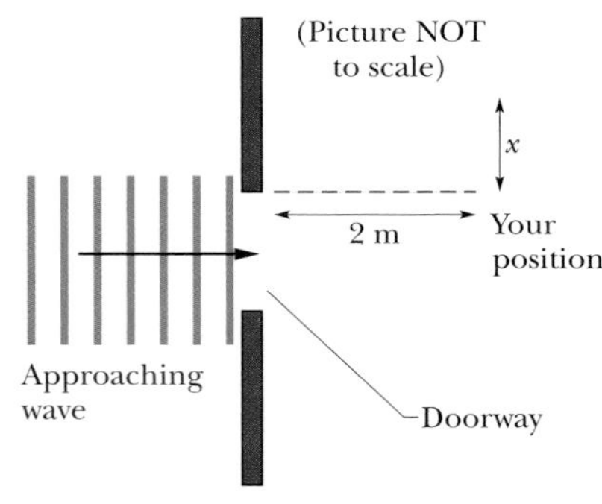

FIGURE 37-37 ■ Problem 63.

38 Special Relativity

Guest Author: Edwin F. Taylor *Massachusetts Institute of Technology*

Billions of dollars have been spent constructing gigantic particle accelerators, such as this one 4 miles in circumference at Fermi National Accelerator Laboratory. More and more advanced accelerators give more and more energy and momentum to particles being accelerated. Decades of experimentation have verified that every particle, however great its energy and momentum, moves slower than the speed of light in a vacuum.

How can the energy of a particle increase without limit while its speed remains slower than the speed of light?

The answer is in this chapter.

38-1 Introduction

Special relativity and *general* relativity both describe the behavior of radiation and matter moving at or near the speed of light. However, special relativity is limited to situations in which gravitational effects can be neglected. Both special and general relativity are called *classical* theories because they do not describe atomic or molecular effects, for which quantum theory is needed.

Unfortunately, special relativity has a reputation for being difficult and mathematically complex. But if you understand basic algebra and square roots, you have the necessary mathematical tools to comprehend it. What makes special relativity seem difficult is that we have no direct experience with objects moving anywhere near the speed of light. It's no wonder that our idea of space and time, molded by everyday experience, is limited. As a result, the predictions of special relativity—fully verified by experiment—strike us as outlandish and outrageous. But these outlandish predictions not only make special relativity fascinating, they also provide us with deep insights into the nature of space and time—the arena in which we all live and in which science operates.

In this chapter we will show how the outlandish predictions of special relativity can be deduced logically from a single principle proposed by Albert Einstein at the beginning of the 20th century.

FIGURE 38-1 ■ While waiting at a stoplight, you find it hard to tell whether the car next to you is rolling forward or you are drifting backward—unless you are looking at a fixed object such as the speed limit sign.

38-2 Origins of Special Relativity

While waiting at a stoplight, you notice that the car next to yours appears to be moving forward slowly (Fig. 38-1). Instead you suddenly realize that you are drifting backward, so you slam on the brakes to avoid bumping another car behind you. Before you step on the brakes, which car is standing still? Which is moving? Without seeing a "stationary" object such as a sign post, you cannot tell! Are such observations about relative motion trivial or profound? Can we cover the windows of our car and carry out some experiment inside—any experiment at all—to detect whether we are in motion or at rest?

Special relativity grew out of questions raised in the late 1800s and early 1900s about relative motions of material objects and waves. Some of these questions involved comparisons between everyday phenomena involving boats and ocean waves. Other questions were raised about the relative motions of objects and light waves. For example, consider ocean waves that move slowly past a swimmer moving in the same direction as the waves. The same waves will move rapidly past a second swimmer traveling in the opposite direction. Will the same thing happen when someone moves toward or away from light waves? Will observers traveling in opposite directions measure different speeds for the same light wave? Do light waves move in a *medium* the way ocean waves move in the medium of water?

FIGURE 38-2 ■ Albert Einstein in the early 1900s at the patent office in Bern, Switzerland, where he was employed when he published his article on special relativity. In later life he was known to dress much more informally.

From the age of 16 Albert Einstein (Fig. 38-2) puzzled over a thought experiment: Suppose you run very fast while looking at yourself in a mirror that you hold up in front of you. What happens as your running speed approaches the speed of light? Will the light waves move more and more slowly past you? In modern terms, *can you surf light waves?*

While Einstein was growing up, other people were trying to answer such questions with experiments. Some scientists hypothesized that light moves in a medium they called *ether*. In the late 1800s, Albert W. Michelson and Edward W. Morley carried out experiments with light trying to measure the motion of the Earth through this ether, under the assumption that the ether was at rest with respect to the Sun or some other location (such as the center of our galaxy). They used the fastest-moving object available to them: the Earth itself. The Earth moves around the sun at approximately 30 kilometers per second in one direction in January and in the opposite direction past the sun in July. Michelson and Morley could detect no motion of the Earth through the hypothesized ether. These negative results caused great puzzlement.

In 1905 Einstein, a 26-year-old patent examiner in Bern, Switzerland, published a paper that changed the face of science.

READING EXERCISE 38-1: You are sitting in a train that stopped at a station ten minutes ago. Suddenly you notice that a second train on the track next to you is gliding past you. You feel a slight vibration that tells you your train is rolling slowly along the track. Is the second train in motion or at rest? ■

38-3 The Principle of Relativity

Einstein's special relativity theory does not assume that light moves through a medium. Even so, it appears that Einstein did not base his ideas on Michelson and Morley's earlier failure to detect ether. Instead, Einstein treasured simplicity, logic, physical intuition, and his now famous thought experiments. He started from a clean assertion that he called the *Principle of Relativity*. Think of an automobile or train either at rest or moving at constant velocity. Define each of these enclosures as a *reference frame*. Then the *Principle of Relativity* says:

> All the laws of physics are the same in every reference frame.

In other words: Pull down the shades in your room or vehicle. Then carry out as many experiments as you need to create the laws of physics. Someone who carries out the same experiments inside another vehicle will discover the same laws, as long as this new vehicle moves at a constant velocity relative to yours.

The laws of physics contain fundamental numerical constants, such as the charge e on the electron, Planck's constant h, and the speed of light c in a vacuum. According to the Principle of Relativity, each of these constants must have the same numerical value when measured in any reference frame. In particular, all observers measure the speed of light in a vacuum to have the value presented back in Chapter 1 of $c =$ 299 792 458 m/s ($\sim 3 \times 10^8$ m/s). The equality of the speed of light in all reference frames eliminates the need to postulate the existence of *ether* through which light propagates. The predictions of special relativity about space, time, mass, and motion all spring from the single *Principle of Relativity,* including the postulate of the "universal speed" of light.

The Principle of Relativity solved Einstein's puzzler about running fast while holding a mirror in front of you. You will not observe light waves to slow down as you move faster. Why not? Because, says the Principle of Relativity, light always moves past you with the same speed c, no matter how fast you run along the ground. *You cannot surf light waves!*

"Relativity theory" is a misleading term that Einstein avoided for years. What we call the special theory of relativity is based on the Principle of Relativity, which tells us that the laws of nature are the *same* for observers in different reference frames. These laws are *not* relative. *General* relativity employs an even more radical version than special relativity, of the Principle of Relativity—that the laws of nature are independent of the observer's viewpoint.

READING EXERCISE 38-2: While standing beside a railroad track, we are startled by a boxcar traveling past us at half the speed of light. A passenger (shown in the figure) standing at the front of the boxcar fires a laser pulse toward the rear of the boxcar. The pulse is absorbed at the back of the boxcar. While standing beside the track we measure the speed of the pulse through the open side door. (a) Is our measured value of the speed of the pulse greater than, equal to, or less than its speed measured by the rider? (b) Is our measurement of the distance between emission and absorption of the light pulse greater than, equal to, or less than the distance between emission and absorption measured by the rider? (c) What conclusion can you draw about the relation between the times of flight of the light pulse as measured in the two reference frames? ■

TOUCHSTONE EXAMPLE 38-1: Communications Storm!

A sunspot emits a tremendous burst of particles that travels toward the Earth. An astronomer on the Earth sees the emission through a solar telescope and issues a warning. The astronomer knows that when the particle pulse arrives it will wreak havoc with broadcast radio transmission. Communications systems require ten minutes to switch from over-the-air broadcast to underground cable transmission. What is the maximum speed of the particle pulse emitted by the Sun such that the switch can occur in time, between warning and arrival of the pulse? Take the sun to be 500 light-seconds distant from the Earth.

SOLUTION ■ It takes 500 seconds for the warning light flash to travel the distance of 500 light-seconds between the Sun and the Earth and enter the astronomer's telescope. If the particle pulse moves at half the speed of light, it will take twice as long as light to reach the Earth. If the pulse moves at one-quarter the speed of light, it will take four times as long to make the trip. We generalize this by saying that if the pulse moves with speed v/c, it will take time Δt_{pulse} to make the trip given by the expression:

$$\Delta t_{\text{pulse}} = \frac{500 \text{ s}}{v_{\text{pulse}}/c}.$$

How long a warning time does the Earth astronomer have between arrival of the light flash carrying information about the pulse and the arrival of the pulse itself? It takes 500 seconds for the light to arrive. Therefore the warning time is the difference between the pulse transit time and the transit time of light:

$$\Delta t_{\text{warning}} = \Delta t_{\text{pulse}} - 500 \text{ s}.$$

But we know that the minimum possible warning time is 10 min = 600 s.

Therefore we have

$$600 \text{ s} = \frac{500 \text{ s}}{v_{\text{pulse}}/c} - 500 \text{ s},$$

which gives the maximum value for v_{pulse} if there is to be sufficient time for warning:

$$v_{\text{pulse}} = 0.455\,c. \qquad \text{(Answer)}$$

Observation reveals that pulses of particles emitted from the sun travel much slower than this maximum value. So we would have a much longer warning time than calculated here.

38-4 Locating Events with an Intelligent Observer

In devising special relativity, Einstein stripped science to its bare essentials. The essence of science is the description of *events*—occurrences in space and time. Science has a simple task: to tell us how one event is related to another event. One of the most important outcomes of special relativity is the ability to predict how events observed in one reference frame will look to an observer in another frame. We need to start by carefully defining what events are and how to observe them intelligently.

> An event is an occurrence that happens at a unique place and time.

Examples of events include a collision, an explosion, the emission of a light flash, and the fleeting touch of a friend's hand. When can an occurrence be called an event? When an observer finds it sufficiently localized in space and time to serve her purposes. Your birth was an event unique in both time and place for a genealogist who studies family trees. Your birth mother, however, experienced the process as a *series* of events, from first contraction (maybe at home) to delivery (perhaps in a hospital). Since your birth mother's experiences spanned both time and space, she might not call your birth an event (at least while it is taking place!).

Locating an event in space and time is not always as simple as it might seem at first because of the time delay between event and observation. Think about observing a lightning flash in the night sky. We count the seconds, "one-thousand-one, one-thousand-two, one-thousand-three." Then we hear a crash of thunder. "Wow, lightning struck only one kilometer away from us!" We know this from our knowledge that it takes sound about three seconds to travel one kilometer in air. In making our calculation, we *assume* that the time it takes the lightning flash to reach us is negligible. This means that the lapse between receiving the flash and hearing the thunder is entirely due to the travel time of sound. In this case the signal travels with the speed of sound.

A pulse of high-energy particles may move at nearly the speed of light. How do we determine the time of events that occur along its path? Suppose a pulse of high-energy protons emerges from a particle accelerator and passes through detector A, where we are standing. The pulse continues its flight to arrive at detector B that lies 30 meters away. *When* did the pulse arrive at detector B? We arrange in advance for detector B to send us a light flash when the pulse arrives there. We time the arrival of this light flash at detector A and from this arrival time we *subtract* the known time delay that results when the light flash travels 30 meters. This *difference* gives us the time at which the pulse arrived at detector B.

To account for the delay due to the speed of light, we define the **intelligent observer** to be someone who takes into account the time delays required to locate distant events in space and time. Standing by detector A in the example above, we acted as intelligent observers in determining the time at which the pulse reached detector B.

READING EXERCISE 38-3: The Minute Waltz by Friedrich Chopin takes more than a minute for most pianists to perform. Halfway through playing the Minute Waltz at a recital, will you think of your performance as a single event? Is your performance a single event for those who printed the program for the recital? Looking back ten years later, will you think of it as a single event? ■

READING EXERCISE 38-4: When the pulse of protons passes through detector A (next to us), we start our clock from the time $t = 0$ microseconds. The light flash from detector B arrives back at detector A at a time $t = 0.225$ microsecond (0.225×10^{-6} second) later. (a) At what time did the pulse arrive at detector B? (b) Use the result from part (a) to find the speed at which the proton pulse moved, as a fraction of the speed of light. ■

TOUCHSTONE EXAMPLE 38-2: Simultaneous?

You are an intelligent observer standing next to beacon A, which emits a flash of light every 10 s. 100 km distant from you is a second beacon, beacon B, stationary with respect to you, that also emits a light flash every 10 s. You want to know whether or not each flash is emitted from remote beacon B simultaneous with (at the same time as) the flash from your own beacon A. Explain how to do this without leaving your position next to beacon A. Be specific and use numerical values. Assume that light travels 3×10^8 m/s.

SOLUTION ■ You are an intelligent observer, which means that you know how to take into account the speed of light in determining the time of a remote event, in this case the time of emission of a flash by the distant beacon B. You measure the time lapse between emission of a flash by your beacon A and your reception of the flash from beacon B. If this time lapse is just that required for light to move from beacon B to beacon A, then the two emissions occur at the same time. The two beacons are 100 km $= 10^5$ m apart. Call this distance L. Then the time Δt for a light flash to move from B to A is:

$$t = \frac{L}{c} = \frac{10^5\ \text{m}}{3 \times 10^8\ \text{m/s}} = 3.33 \times 10^{-4}\ \text{s}, \qquad \text{(Answer)}$$

or 0.333 ms. If this is the time you record between the flash of nearby beacon A and reception of the flash from distant beacon B, then you are justified in saying that the two beacons emit their flashes simultaneously in your frame.

38-5 Laboratory and Rocket Latticeworks of Clocks

There are difficulties with the procedure used by our intelligent observer. First, she needs to make a separate calculation for each remote event. This is bothersome. Second, and more fundamental, she cannot calculate the time delay in reporting remote events unless she *already knows the location of every event she wants to measure.* Sometimes information about event location is easily available, sometimes not. We need a general, conceptually simple way to observe both the location and time of events.

In principle, one way to do this is to assemble a cubical lattice of meter sticks with a recording clock at each intersection (Fig. 38-3). Using this latticework, we say that the *position* of an event is that of the recording clock nearest to the event. The *time* of the event is the time recorded on that nearby clock. *Observing an event* then reduces to recording the position of the clock nearest to the event and the time for the event recorded on that clock. Now there is no delay in recording the position and the time of any event that occurs in the lattice.

Synchronizing Latticework Clocks

Before we can actually observe events with our latticework of meter sticks and clocks, we need to set all the clocks in the lattice to read the *same time*. But how can we *synchronize* all the clocks in the latticework? One method would be to carry a traveling clock around the lattice and synchronize each lattice clock with it. This approach is not only time-consuming but incorrect. You will see in the next section that a clock traveling through the lattice runs at a different rate than a resting clock as recorded by clocks in the lattice. In fact, if you set a lattice clock to the time of the traveling clock and then later bring it back after it has traveled to other clocks, you will find that the traveling clock no longer agrees with that lattice clock!

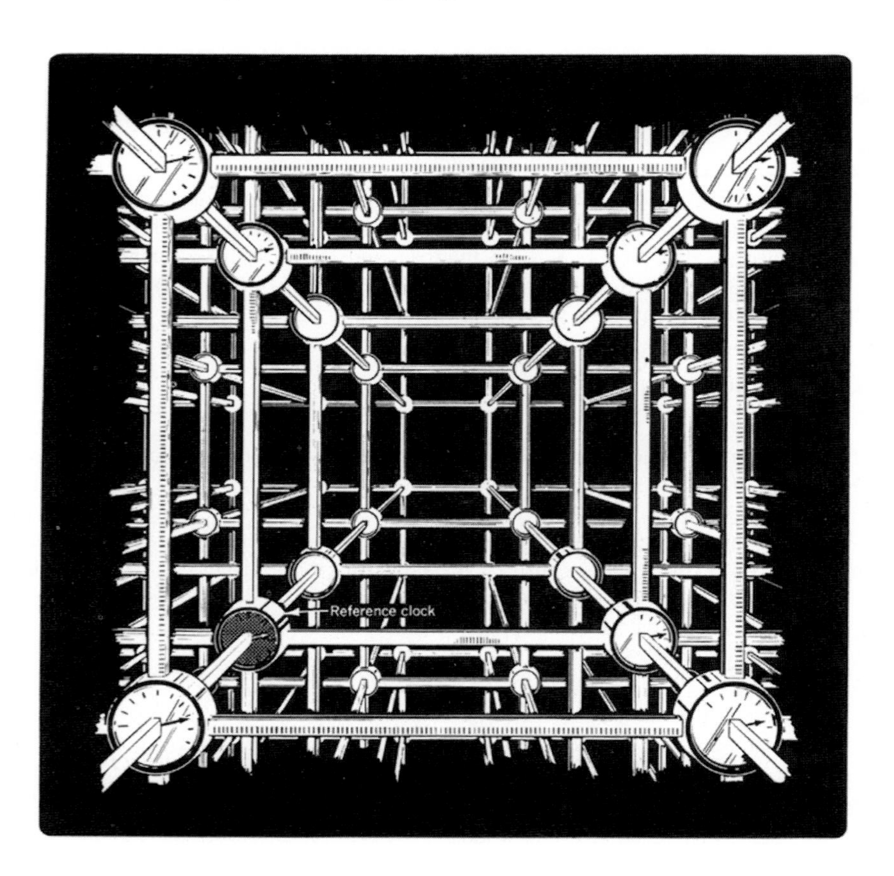

FIGURE 38-3 ■ Latticework of meter sticks and clocks.

Instead of a traveling clock, we use the speed of light to synchronize all the lattice clocks. Our procedure starts by picking one clock in the lattice as the standard or **reference clock.** We know the distance between the reference clock and every other clock in the lattice. At midnight the reference clock sends out a **synchronizing flash** of light. When an observer at any one of the distant clocks receives the flash, she quickly sets the time on her clock to midnight *plus* the time it took for the light to reach her over the known distance from the reference clock at the known speed of light. We say that after this procedure is complete the clocks in the lattice read the *same time* as one another—they are *synchronized* with respect to this lattice.

Now our latticework of synchronized clocks is ready to record the position and time of events that occur during any experiment. Analyzing the results of that experiment means relating events by collecting event data from all recording clocks in the lattice and analyzing these data at some central location.

Laboratory and Rocket Frames as Inertial Reference Frames

We often hear talk in special relativity about the *laboratory frame* and the *rocket frame.* Envision *each* of these frames as having a latticework of rods and clocks (Fig. 38-4). The rocket coasts at constant velocity in unpowered flight. By convention we assign the positive x direction to be the direction of motion of the rocket with respect to the laboratory lattice.

AN IMPORTANT ASIDE: Strictly speaking, reference frames used in special relativity must be inertial frames, frames with respect to which Newton's First Law of motion holds: *A free particle at rest remains at rest and a free particle in motion continues that motion in a straight line at constant speed* (see Section 3-2). Obviously the surface of the Earth is not an inertial reference frame; a stone released from rest accelerates downward! However, for a particle moving at a substantial fraction of the speed of light with respect to the Earth, the acceleration of gravity can usually be ignored. In this chapter we make no distinction between inertial frames and those at rest or moving at constant velocity with respect to the Earth's surface.

We can detect and record a single event using overlapping rocket and laboratory lattice works. If the right rear tire of your car hits a nail, it goes flat with a *bang*. For you as the driver (in the "rocket frame") the bang occurs at the right rear of your car. For the observer on the road (the "laboratory frame"), the bang takes place where the nail sticks up at one end of the bridge that your car has just crossed. Neither you nor the road observer "owns" the event. You both have equal status in observing and recording the bang. The bang exists, *and all other events exist,* independent of reference frames. Events are the nails on which all of science hangs.

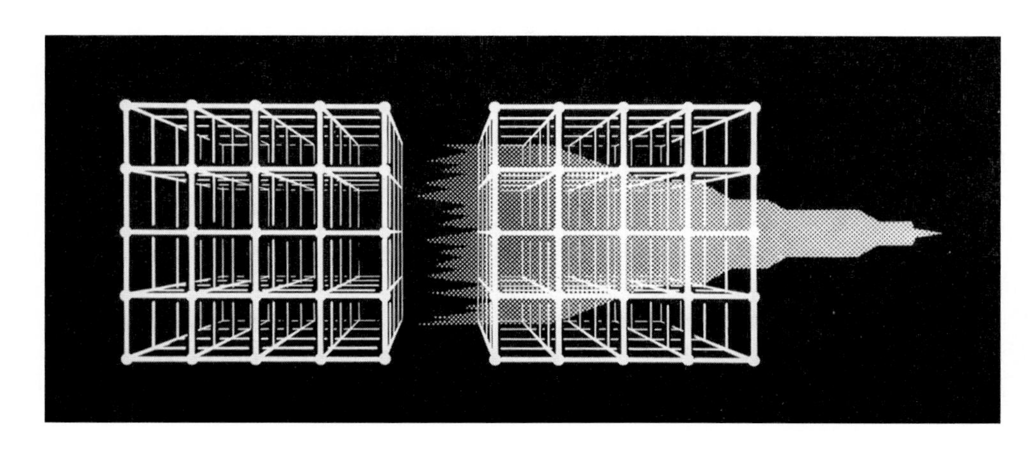

FIGURE 38-4 ■ Laboratory and rocket frames. A moment ago the two latticeworks were intermeshed. By convention, the rocket frame moves in the positive x direction of the laboratory frame.

TOUCHSTONE EXAMPLE 38-3: Synchronizing Clocks

You are stationed at a latticework clock with the coordinates $x = 3 \times 10^8$ m, $y = 4 \times 10^8$ m, and $z = 0$ m. The reference clock at coordinates $x = y = z = 0$ emits a reference flash at exactly midnight on its clock. You want your clock to be synchronized with (set to the same time as) the reference clock. To what time do you immediately set your clock when you receive the reference flash?

SOLUTION ■ Your distance D from the reference clock is

$$D = [(3 \times 10^8 \text{ m})^2 + (4 \times 10^8 \text{ m})^2 + 0 \text{ m}]^{1/2} = [25 \text{ m}]^{1/2} \times 10^8$$
$$= 5 \times 10^8 \text{ m}.$$

The time Δt that it takes the reference flash to reach you is therefore

$$\Delta t = \frac{D}{c} = \frac{5 \times 10^8 \text{ m}}{3 \times 10^8 \text{ m/s}} = 1.66 \text{ s}. \qquad \text{(Answer)}$$

So when you receive the reference flash, you quickly set your clock to 1.66 seconds after midnight.

38-6 Time Stretching

Every year hundreds of email messages, letters, and articles "disproving" relativity are sent to textbook authors and scientific journal editors. Many of these papers are extremely ingenious, showing considerable insight and sometimes representing years of labor. (Indeed, fighting a new idea often helps us to understand it and make it our own. As you continue reading this chapter, you may want to make a note of the ideas that seem paradoxical or outrageous. By the time you finish the chapter, see if you can refute or defend some of your initial objections to relativity.)

A primary target of writers who object to special relativity is **time stretching,** the conclusion that the time between two events can have different values as measured in laboratory and rocket frames in relative motion. The clearest case of time stretching is this: *The time between two ticks measured on a clock at rest is always less than the time between the same two ticks measured in a reference frame in which the clock is moving.* Many people remember this result by using a not-quite-exact motto: *Moving clocks run slow.* However we express it, this conclusion is so obviously ridiculous that it stimulates dozens of skeptics to write letters and articles.

Verification of Time Stretching

Time stretching is verified experimentally every day as part of the ongoing enterprise of experimental physics. Here are two examples of time stretching in action.

Time stretching with atomic clocks: In October 1971, J. C. Hafele and R. E. Keating of the U.S. Naval Observatory sent atomic clocks (like the one described in Section 1-5) around the Earth on regularly scheduled commercial airliners. One clock circled the globe traveling eastward, the other clock traveled westward. When the clocks were finally brought together, they did not read the same time. Also, the reading on both clocks was different from that of a third atomic clock, which stayed at home in one place on the Earth's surface. Why the different readings? Think of the center of the Earth as at rest. (Actually, the center of the Earth is in free fall around the sun.) With respect to the Earth's center, the speed of the eastward-moving clock is *added* to the speed of the Earth's rotation; it is the "faster-moving" clock. In contrast, the speed of the westward-moving clock is *subtracted* from the eastward motion of the Earth's surface; this is the "slower-moving" clock. The stay-at-home clock moves with the Earth's surface at a speed that is intermediate between that of the other two clocks. The result? The "faster-moving" eastward-going clock runs slow compared with the stay-at-home clock of intermediate speed. And the "slower-moving" westward-going clock runs fast compared with the stay-at-home clock. In the Hafele–Keating experiment, the magnitudes of the different readings corresponded to the predictions of relativity. There was, however, at least one complication: The airplanes changed altitude as they took off, flew their courses, and landed. General relativity, the theory that includes gravitational effects, predicts that changes in altitude, as well as relative speeds, affect the relative rates at which clocks run. The results of the Hafele–Keating experiment were actually consistent with the predictions of general relativity too, but that is a story for another day.

Time stretching with pions: Our second example of time stretching is more technical than the Hafele–Keating experiment, but a lot more convincing. It involves measuring the lifetimes of **pions,** also called pi-mesons or π^+ mesons. These short-lived particles can be created during cosmic ray interactions or when a beam of protons energized by a particle accelerator strikes a target. On average, half the pions in a beam will decay into other particles in 18 nanoseconds (18×10^{-9} seconds) as measured by a clock carried with the pions. In this pion frame, half of the remainder will decay in the next 18 nanoseconds, and so on. We call this time the pion **half-life** ($t_{1/2}$). If pions are moving at nearly the speed of light, how far can a pion beam travel before half the pions decay? If the time were the same in our laboratory as it is in the rest frame of the speeding pions, the maximum distance would be approximately equal to

$$c \times t_{1/2} = (3 \times 10^8\,\text{m/s}) \times (18 \times 10^{-9}\,\text{s}) = 5.4\,\text{m}.$$

However, experiment shows that the flying pions travel tens of meters before half of them decay. We conclude that in our laboratory frame the time for half of the pions to decay is much greater than it is in the rest frame of the pions. Time stretching!

Why Time Stretching Makes Sense

Objections to time stretching have always failed because they attack a result based on an utterly simple idea: All the laws of nature are the same in every reference frame (the Principle of Relativity). In particular, the speed of light is invariant (that is, it has the same value) in every reference frame. The invariance of the speed of light leads directly to the difference in time between two events as measured in laboratory and rocket frames. To illustrate this, let's consider the ticking of a "light clock" diagrammed in Fig. 38-5.

While riding in a transparent unpowered rocket ship, you fire a flash of light upward toward a mirror that you hold on a stick 3 meters directly above you (the left-hand panel in Fig. 38-5). The flash reflects at the mirror and returns to you. Call the emission of light event A and its reception upon return event B. For you, events A and B occur at the same place. Between the events, the light moves first straight up

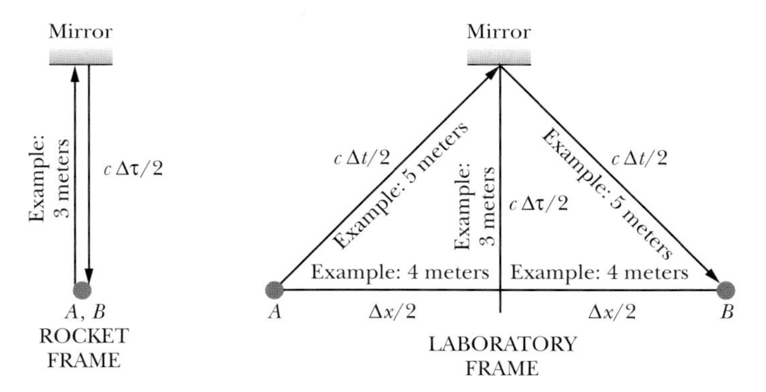

FIGURE 38-5 ■ A flash of light emitted at event A reflects from a mirror and returns to the source, arriving as event B. Events A and B are recorded in both the rocket and laboratory frames. Einstein tells us that each observer measures the same speed of light. Therefore different path lengths for the light flash in rocket and laboratory frames mean that different times between events A and B are measured in the two frames. (The meanings of the symbols Δx, $c\,\Delta t$, and $c\,\Delta\tau$ are discussed in the text.)

3 meters then straight down 3 meters. *For you the total time between events A and B equals the time that it takes light to travel a total of 6 meters.*

The unpowered spaceship in which you carry out these experiments moves from left to right past the rest of us, who stand in another transparent container arbitrarily labeled "laboratory" (the right-hand panel in Fig. 38-5). We also observe the same flash of light emitted at event A, reflected at your mirror, and received again at event B. But for us in the laboratory, you and your mirror move together to the right. Thus for us the path of the light slants upward from A along the 5-meter-long (for example) hypotenuse of the first right triangle, reflects from the speeding mirror, then slants back down along the 5-meter hypotenuse of the second right triangle to meet you again at event B. Therefore *for us events A and B are separated by the time it takes light to travel a total of 10 meters.* For us a longer time lapses between events A and B than you measured in your rocket frame.

That's it! Longer path length for light, longer time for light to travel that path at its "universal speed," therefore longer time between events as measured in that reference frame. No one has ever found an acceptable way around this simple and powerful result. The light clock demonstrates the longer time between two events in one frame than in another frame. Hence the name for this effect is **time stretching** or **time dilation** (*dilation* is a medical term for *stretching*).

The light flash and mirror make a kind of clock that we define as a **light clock.** The Principle of Relativity assures us that *all kinds of clocks* at rest in a frame, once calibrated, must run at the same rate as one another as observed in every frame in uniform relative motion with respect to the first frame. Otherwise we could tell which frame we are in by detecting different rates of different clocks all at rest in our frame. In any given frame, *properly calibrated clocks of every kind run at the same rate as one another,* including the "clock" of your body—namely, the aging process. Suppose the mirror was so high above your head that it took 6 years in your rocket for the light to return to you—and 10 years by our laboratory clocks. Then you would age 6 years between these new events A and B because your body's "aging clock" and your "light clock" ride together in your rocket frame. In contrast, between events A and B we in the laboratory frame would age 10 years, and our light clock would also advance 10 years. Between events A and B you would age less than we do!

How strange it is that the speed of light is *invariant* for all observers, no matter what their relative velocity! But experiment continually verifies this result. Consequences of the invariance of the speed of light include the fact that clocks run at different rates for observers in relative motion. Experiment continually verifies this result as well. More than one hundred years of the most rigorous testing have validated beyond reasonable doubt that the speed of light is the same in all reference frames and that clocks tick at different rates when observed in frames that are in motion with respect to each other.

READING EXERCISE 38-5: Suppose that a beam of pions moves so fast that at 25 meters from the target in the laboratory frame exactly half of the original number remain undecayed. As an experimenter, you want to put more distance between the target and your detectors. You are satisfied to have one-eighth of the initial number of pions remaining when they reach your detectors. How far can you place your detectors from the target? ■

READING EXERCISE 38-6: A set of clocks is assembled in a stationary boxcar. They include a quartz wristwatch, a balance wheel alarm clock, a pendulum grandfather clock, a cesium atomic clock, fruit flies with average individual lifetimes of 2.3 days, a clock based on radioactive decay of nuclei, and a clock timed by marbles rolling down a track. The clocks are adjusted to run at the same rate as one another. The boxcar is then gently accelerated along a smooth horizontal track to a final velocity of 300 km/hr. At this constant final speed, which clocks will run at a different rate from the others as measured in that moving boxcar? ■

38-7 The Metric Equation

Time stretching occurs whenever the rocket has a velocity with respect to the laboratory frame. We can prove this in general using symbols in Fig. 38-5. In the laboratory frame we measure Δx as the distance between events A and B and we measure Δt as the time between the events (that is, the time it takes the light flash to slant upward along one hypotenuse at the speed c and then slant downward to the second event). Here delta (Δ) indicates the *lapse* of time and the letter t tells us that the elapsed time refers to the time between two events in our laboratory frame. One hypotenuse has a length given by

$$\text{length} = \text{velocity} \times \text{time} = c\ \Delta t/2.$$

In your rocket frame the flash moves vertically upward to the mirror and back down again in the time between events A and B. To describe your rocket frame time lapse between events we use the notation $\Delta\tau$. Here delta (Δ) indicates the *lapse* of time and the Greek letter tau (τ) tells us that the elapsed time refers to the time between two events in your unpowered rocket, in which they occur at the same place. So for you the upward distance covered in time $\Delta\tau/2$ is equal to $c\ \Delta\tau/2$. This vertical span is the same as the vertical leg shared by the right triangles in the diagram at the right. Hence we have expressions for the lengths of all sides of both of these right triangles. If we use the Pythagorean Theorem for right triangles we get

$$(\Delta x)^2 + (c\ \Delta\tau)^2 = (c\ \Delta t)^2 \qquad (38\text{-}1)$$

where each term has dimensions of length squared. If we rearrange the terms in Eq. 38-1 so all the terms that refer to the laboratory frame are on the right, we get a squared time-like interval called the **metric equation:**

$$(c\ \Delta\tau)^2 = (c\ \Delta t)^2 - (\Delta x)^2 \quad \text{(squared time-like interval).} \qquad (38\text{-}2)$$

Now suppose two events occur in the same place ($\Delta x = 0$), such as two sequential ticks of a clock in its rest frame. The time lapse $\Delta\tau$ between the events measured on the clock at rest is called the **proper time** or **wristwatch time.** The German term for proper time is *Eigenzeit,* meaning "one's own time." The square root of the difference of squares on the right side of Eq. 38-2 has the formal name **invariant time-like interval.** The interval is *invariant* because it has the same value as calculated by all observers. It is *time-like* because the magnitude of the time part $c\ \Delta t$ is greater than the magnitude of the space part Δx. (Both sides of Eq. 38-2 are necessarily positive).

The metric equation 38-2 is one of the most amazing equations in all of physics. Look at its outrageous implications:

- First, the metric equation relates *two different* measures of the time between the *same* two events. These are: (1) the time recorded on clocks in the reference frame in which the events occur at different places, and also (2) the wristwatch time read on the clock carried by a traveler who records the two events as occurring at the same place. The ability to relate these two times is one of the greatest scientific innovations in history.
- Second, the metric equation reveals an even deeper insight—space and time combine in a single expression on the right side. We no longer speak of space and time separately, but as a unity: **space-time**!

A wealth of other insights can be gleaned from Eq. 38-2. For example:

1. The time between events A and B as measured in the two frames *cannot have the same value* if the laboratory and rocket frames are in relative motion.

2. Laboratory observers can correctly predict the proper time observers in the rocket frame measure between events A and B on their rocket-frame wristwatches, in spite of the fact that this time is not the same as the time measured in the laboratory frame. (Observers in the laboratory simply put their values for Δt and Δx into Eq. 38-2 and calculate the value of the wristwatch time $\Delta\tau$ that the rocket observer measures.)

3. If the rocket speed relative to the laboratory frame is reduced, then both $\Delta x/2$ (the length of the horizontal leg in Fig. 38-5) and $c\,\Delta t/2$ (the hypotenuse) will be smaller than before. But these terms become smaller in such a way that the difference between their squares, which represents the vertical distance between the rocket observer and her mirror $c\,\Delta\tau/2$, will remain the same. So the metric equation (Eq. 38-2) will still hold. No matter how fast or slow the rocket is, the value of the proper time $\Delta\tau$ (also known as the invariant time-like interval) remains the same. Hence we call this interval between two events **invariant,** meaning that it has the same value as measured in *all* reference frames in uniform relative motion.

If a rocket passes by our laboratory frame at a speed v, we can derive an equation that relates $\Delta\tau$ and Δt directly by setting $\Delta x = v\,\Delta t$. Substituting this expression into Eq. 38-2 and dividing through by c^2 gives us

$$(\Delta\tau)^2 = (\Delta t)^2 - \left(\frac{\Delta x}{c}\right)^2 = (\Delta t)^2 - \left(\frac{v\,\Delta t}{c}\right)^2 = \left(1 - \frac{v^2}{c^2}\right)(\Delta t)^2.$$

Taking the square root of both sides gives us an expression known as the *time-stretching* or *time dilation equation,*

$$\Delta\tau = \sqrt{1 - v^2/c^2}\,\Delta t \qquad \text{(time-stretching equation).} \qquad (38\text{-}3)$$

The time-stretching equation (Eq. 38-3) gives us the value of wristwatch time $\Delta\tau$ between two events that occur a time Δt apart in some reference frame. In this equation, v is the speed required for an observer in the rocket frame to move directly from one event to the other event. The equation encompasses all possible values of speed v from the very slow to the very fast.

What Happens at High and Low Speeds?

The speeds we observe in everyday life are so much smaller than the speed of light c that the value of v/c is extremely small compared to 1. Thus, for low relative speeds, the expression $(1 - v^2/c^2)$ is approximately equal to 1. The time-stretching equation (Eq. 38-3) tells us that in this case Δt and $\Delta\tau$ are essentially equal; the time between events A and B is the same for you in a passing airplane as it is for us standing on Earth. So at very low relative speeds special relativity is consistent with our everyday assumption that time is a universal quantity, that everyone measures the time between two events to have the same value. This is the approximating assumption used in Newtonian mechanics.

In contrast, at a high relative speed v the outcome is quite different from what happens at everyday speeds. Imagine that you start from Earth (event A: departure from Earth) and travel to the star Alpha Centauri, about 4 light-years from Earth (event B: arrival at Alpha Centauri). Both events A and B (departure and arrival) occur at the position of your cockpit. Equation 38-3 tells us that by making v/c closer and closer to the value unity, your trip can take place in shorter and shorter wristwatch time $\Delta\tau$ *as measured in your spaceship*. (This is true even though the time Δt measured in the Earth frame can never be less than the time it takes light to move from Earth to Alpha Centauri—4 years.) By extension of this argument, we arrive at a result that frees the human spirit, if not yet the human body. Given sufficient rocket speed, we can go anywhere in the universe in the lifetime of a single astronaut! At least this is the prediction of special relativity.

Nature's Speed Limit

The time-stretching equation (Eq. 38-3) also gives evidence that the natural speed limit of the universe is the speed of light c. Imagine that we in the laboratory measured your rocket speed v to be greater than the speed of light c. Then v/c (and also v^2/c^2) would have a value greater than unity, and the expression on the right side of Eq. 38-3 would include the square root of a negative number. This would mean that the time measurement $\Delta\tau$ would be proportional to the square root of a negative number. But this is impossible: No real time can be proportional to the square root of a negative number. Careful study and experiment have led to the conclusion, consistent with this formula, that no object can be accelerated to a speed v greater than the speed of light c in a vacuum. Experiment verifies this: Many nations together have spent billions of dollars to build and operate huge particle accelerators that use electric and magnetic fields to urge protons or electrons to ever-higher energies. At higher and higher energies, these particles approach closer and closer to the speed of light but have never been observed to exceed this speed.

READING EXERCISE 38-7: Find the rocket speed v at which the time $\Delta\tau$ between ticks on the rocket clock is recorded by the laboratory clock as $\Delta t = 1.01\ \Delta\tau$. ■

TOUCHSTONE EXAMPLE 38-4: Satellite Clock Runs Slow?

An Earth satellite in circular orbit just above the atmosphere circles the Earth once every $T = 90$ min. Take the radius of this orbit to be $r = 6500$ kilometers from the center of the Earth. How long a time will elapse before the reading on the satellite clock and the reading on a clock on the Earth's surface differ by one microsecond? For purposes of this approximate analysis, assume that the Earth does not rotate and ignore gravitational effects due the difference in altitude between the two clocks (gravitational effects described by general relativity).

SOLUTION ■ First we need to know the speed of the satellite in orbit. From the radius of the orbit we compute the circumference and divide by the time needed to cover that circumference:

$$v = \frac{2\pi r}{T} = \frac{2\pi \times 6500\text{ km}}{90 \times 60\text{ s}} = 7.56\text{ km/s}. \tag{38-4}$$

Light speed is almost exactly $c = 3 \times 10^5$ km/s, so the satellite moves at the fraction of the speed of light given by

$$\frac{v}{c} = \frac{7.56\text{ km/s}}{3 \times 10^5\text{ km/s}} = 2.52 \times 10^{-5}. \tag{38-5}$$

or

$$v^2/c^2 = (2.52 \times 10^{-5})^2 = 6.35 \times 10^{-10}. \tag{38-6}$$

The relation between the time lapse $\Delta\tau$ recorded on the satellite clock and the time lapse Δt on the clock on Earth (ignoring the Earth's rotation and gravitational effects) is given by Eq. 38-3. Square both sides of that equation to obtain:

$$(\Delta\tau)^2 = (1 - v^2/c^2)(\Delta t)^2. \tag{38-7}$$

We want to know the difference between Δt and $\Delta\tau$. Rearrange this equation to give the difference of squares:

$$v^2/c^2(\Delta t)^2 = (\Delta t)^2 - (\Delta\tau)^2 \equiv (\Delta t - \Delta\tau)(\Delta t + \Delta\tau). \tag{38-8}$$

Substituting the numerical result of Eq. 38-6 into Eq. 38-7, we see that $\Delta\tau$ and Δt have very nearly the same value. Therefore we can set

$$\Delta t + \Delta\tau \approx 2\Delta t. \tag{38-9}$$

With this substitution, Eq. 38-8 becomes

$$v^2/c^2(\Delta t/2) \approx \Delta t - \Delta\tau. \tag{38-10}$$

Substitute from Eq. 38-6:

$$\Delta t - \Delta\tau \approx 3.18 \times 10^{-10}\Delta t. \tag{38-11}$$

We are asked to find the elapsed Δt for which the satellite clock and the Earth clock differ in their reading by one microsecond = 10^{-6} second. Rearrange Eq. 38-11 to read

$$\Delta t \approx \frac{\Delta t - \Delta\tau}{3.18 \times 10^{-10}} = \frac{10^{-6}\text{ s}}{3.18 \times 10^{-10}} = 3.14 \times 10^3\text{ s}. \quad \text{(Answer)}$$

This is approximately equal to 52 minutes, or a little less than one hour. A difference of one microsecond between atomic clocks is easily detectable.

38-8 Cause and Effect

The analysis thus far has omitted from our consideration a large number of possible pairs of events. Suppose two events occur at the same time but not at the same place in a reference frame. For example, what if two firecracker explosions occur simultaneously, one in New York City, the other in San Francisco? Since $\Delta\tau = 0$ for this pair of events, Eq. 38-2 for the space-time interval becomes

$$(c\,\Delta\tau)^2 = (c\,\Delta t)^2 - (\Delta x)^2 \rightarrow -(\Delta x)^2 \qquad (\text{for } \Delta t = 0).$$

What can this expression possibly mean? The left side contains the square of a time, obviously a positive quantity. Yet on the right is a negative quantity. No clock records a time lapse $\Delta\tau$ whose square is a negative quantity! We have a contradiction here, and a contradiction that applies in a similar way to all possible pairs of events simultaneous in some frame.

The problem is not with physics but with mathematical notation. Pairs of simultaneous events were not envisioned in the derivation of the proper time equation (Eq. 38-2) based on Fig. 38-5. For this new class of event-pairs we need a new formalism. To achieve this, reverse the order of squared quantities on the right side of Eq. 38-2 and give the result a different name. Earlier we used the notation $\Delta\tau$ (involving the Greek letter tau—denoted τ) to represent the elapsed *time* between two events measured on a clock for which the events occur at the same *place*. For our new expression we use the notation $\Delta\sigma$ (involving the Greek letter sigma—denoted σ) to represent the *distance* between two events measured in the frame in which they occur at the same time. This new equation has the form

$$(\Delta\sigma)^2 = (\Delta x)^2 - (c\,\Delta t)^2 \qquad \text{(squared space-like interval).} \qquad (38\text{-}12)$$

The distance $\Delta\sigma$ between two events, measured in a frame in which the events occur at the same time ($\Delta t = 0$), is called the **proper distance.** This square root of the difference of squares on the right side of Eq. 38-12 also has the formal name **invariant space-like interval**—space-like because the space part Δx is greater than the time part $c\,\Delta t$.

The right-hand side of Eq. 38-12 also describes the space and time separations between these two events as measured in a second frame that moves past the first; in the second frame Δx and Δt are both different from zero. The proper distance $\Delta\sigma$, like the wristwatch time $\Delta\tau$, is an *invariant* in the following sense: Observers in relative motion may measure different values of Δx and different values of Δt between these two events. However, when each observer substitutes these values into Eq. 38-12, he will obtain the same numerical value for the proper distance $\Delta\sigma$. And the value of $\Delta\sigma$ is just the distance between the two events as measured in that particular reference frame in which they occur at the same time.

Some important consequences for events separated by a space-like interval can be read from Eq. 38-12:

- If Δx is greater in magnitude than $c\,\Delta t$ in one frame, then Δx is greater in magnitude than $c\,\Delta t$ in all frames. Why? Because $\Delta\sigma$ is an invariant, so $(\Delta\sigma)^2$ has the same value whatever the values of Δx and $c\,\Delta t$ in a particular frame. Both sides of Eq. 38-12 must remain positive. But $c\,\Delta t$ is the distance that light can travel in the time available between these events. Equation 38-12 says that Δx is greater than this distance $c\,\Delta t$ between the two events in that frame. Nothing, not even a light flash, can move fast enough to travel from one event to the other in the elapsed time Δt between them. Therefore, for events connected by a space-like interval, one event *cannot* cause the other event as observed in *any* frame.

- By definition, $\Delta\sigma$ is the separation between two events in a reference frame in which the time between these events is zero so the events are simultaneous. But the right side of Eq. 38-12 contains *both* Δt *and* Δx, implying that for another frame in relative motion Δt is *not* zero. That is, in this other frame the two events are *not* simultaneous. As an example, for observers in a rocket streaking across the continent from New York toward San Francisco, the two firecracker explosions in New York and San Francisco will *not* occur at the same time. This leads to a major result of special relativity: *Two events simultaneous in one frame are not necessarily simultaneous in other frames in motion relative to the first.* For many people this is the most difficult concept in special relativity, harder to believe even than the difference in clock rates described by the time-stretching equation (Eq. 38-3). In Section 38-9 we elaborate on this result, which is called the **relativity of simultaneity.**

Suppose that two events are separated in space Δx and time Δt so that a flash of light moving directly between them can *just* make it from one event to the other event in the time Δt. Then the distance between them is given by $\Delta x = c\,\Delta t$. In this case both the proper distance $\Delta\sigma$ and the proper time $\Delta\tau$ are zero:

$$(c\,\Delta\tau)^2 = (\Delta\sigma)^2 = 0 \quad \text{(squared light-like interval).} \tag{38-13}$$

Two events that can be connected by a direct light flash are said to be related by an **invariant light-like interval** or **null interval**—null because the space part Δx is equal in magnitude to the time part $c\,\Delta t$, so the difference between their squares is zero, or null.

Equations 38-2, 38-12, and 38-13 embrace all possible cause-and-effect relations between pairs of events that occur along the x direction as described by special relativity. Equation 38-2 describes two events separated by a time-like interval. Something moving more slowly than light, a rocket for example, can travel directly between these two events in the time between them, so it is possible for the earlier event to cause the later event. This possible cause-and-effect relation between an earlier and a later event is preserved in every reference frame. In contrast, not even light can travel between the two events separated by a space-like interval described in Eq. 38-12, so that neither one of these two events can cause the other event. This *lack* of possible cause-and-effect relation is preserved in every frame. Equation 38-13 provides the boundary between these two cases: the relation between two events that can be connected only by a direct light flash. The earlier event in this pair can cause the other event only through a directly connecting light flash. This cause-and-effect relation between the earlier and later events is also maintained in every reference frame.

In brief, the three-fold categories of time-like, space-like, and light-like intervals between pairs of events preserve the possible cause-and-effect relation between these events in *all* reference frames. Special relativity may be weird, but at least it reaffirms the fact that cause comes before effect for all observers—a statement that most of us consider to be a central requirement of science.

READING EXERCISE 38-8: Points on the surfaces of the Earth and the Moon that face each other are separated by a distance of 3.76×10^8 meters. How long does it take light to travel between these points? A firecracker explodes at each of these two points; the time between these explosions is one second. Is it possible that one of these explosions caused the other explosion? ■

38-9 Relativity of Simultaneity

In the previous section we obtained the following result from the space-like form of the metric equation:

> Two events that are simultaneous in one frame are not necessarily simultaneous in a second frame in uniform relative motion.

This result becomes clear when we consider what observers on the ground and the train see as each measures the time between the same two events, as shown in Fig. 38-6. Suppose lightning bolts strike both ends of the train, emitting flashes and leaving char marks on both the train and the track (top image in the figure). Assume that flashes from the front and back of the train reach the observer on the ground at the same time (bottom image in the figure). This ground observer measures his distance from the two char marks on the track and finds these distances to be equal. He concludes that, for him, the two lightning bolts struck *simultaneously*. In contrast, the rider at the middle of the train sees the flash from the front of the train first (because in Fig. 38-6 she moves toward the light flash coming from the front of the train and away from the light flash coming from the back). She measures her distance from the char marks on the two ends of the train and finds these distances equal. Following the Principle of Relativity, she assumes that the speed of light has the same value in her train frame as in every other frame. She concludes that, for her, the lightning struck the front end of the train first. Her reasoning is explained in the caption to Fig. 38-6.

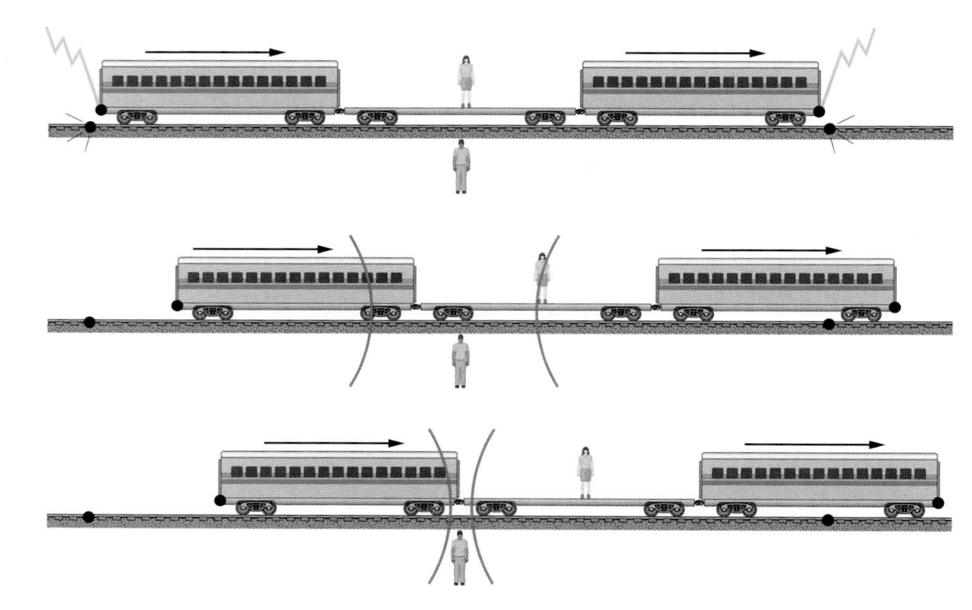

FIGURE 38-6 ■ Einstein's Train Paradox illustrating the relativity of simultaneity. *Top:* Lightning strikes the front and back ends of a moving train, leaving char marks on both track and train. Each emitted flash spreads out in all directions. *Center:* Observer riding in the middle of the train concludes that the two strokes are *not* simultaneous. Her argument: "(1) I am equidistant from the front and back char marks on the train. (2) Light has the standard speed in my frame, and equal speed in both directions. (3) The flash arrived from the front of the train first. (4) Therefore, the flash must have left the front of the train first; the front lightning bolt fell before the rear lightning bolt fell. I conclude that the lightning strokes were not simultaneous." *Bottom:* Observer standing by the tracks halfway between the char marks on the tracks concludes that the two lightning strokes were simultaneous, since the flashes from the strokes reach him at the same time and he is equidistant from the char marks on the track. *Conclusion:* Two events that are simultaneous in one frame may *not* be simultaneous in another frame.

READING EXERCISE 38-9: Susan, the rider on the train pictured in Fig. 38-6, is carrying an audio tape player. When she receives the light flash from the front of the train she switches on the tape player, which plays *very loud* music. When she receives the light flash from the back end of the train, Susan switches off the tape player. Will Sam, the observer on the ground, be able to hear this music? Later Susan and Sam meet for coffee and examine the tape player. Will they agree that some tape has been wound from one spool to the other? ■

TOUCHSTONE EXAMPLE 38-5: Principle of Relativity Applied

Divide the following items into two lists. On one list, labeled SAME, place items that name properties and laws that are always the *same* in every frame. On the second list, labeled MAY BE DIFFERENT, place items that name properties that can be *different* in different frames:

a. the time between two given events
b. the distance between two given events
c. the numerical value of Planck's constant h
d. the numerical value of the speed of light c
e. the numerical value of the charge e on the electron
f. the mass m of an electron (measured at rest)
g. the elapsed time on the wristwatch of a person moving between two given events
h. the order of elements in the periodic table
i. Newton's First Law of Motion ("A particle initially at rest remains at rest, and . . .")
j. Maxwell's equations that describe electromagnetic fields in a vacuum
k. the distance between two simultaneous events

SOLUTION ■ The Principle of Relativity says that the laws of physics are the same in every frame. So items (i) and (j) should go on the SAME list, along with item (h). The Principle of Relativity extends to the values of fundamental constants, so items (c), (d), (e), and (f) should also go on the SAME list.

In contrast, as we have seen in this chapter, the time between a pair of events (item a) may be different in different frames. The same is true for the distance between two events (item b). So these go in the DIFFERENT list.

This leaves two items, (g) and (k). Item (g), the time on the wristwatch of a person moving between two given events (the so-called "wristwatch time") is an invariant, the same as calculated using space and time separations measured in any frame (Eq. 38-2). So this goes on the SAME list. The same is true of item (k), the "proper distance" between two events. This is also an invariant and goes on the SAME list.

In summary, here are the two lists requested: (Answer)

THE SAME IN ALL FRAMES	MAY BE DIFFERENT IN DIFFERENT FRAMES
c. numerical value of h	a. time between two given events
d. numerical value of c	b. distance between two given events
e. numerical value of e	
f. mass of electron (at rest)	
g. wristwatch time between two events	
h. order of elements in the periodic table	
i. Newton's First Law of Motion	
j. Maxwell's equations	
k. distance between two simultaneous events	

38-10 Momentum and Energy

Shortly after his first paper on special relativity was published, Einstein submitted a paper that added the most famous equation of all time, $E = mc^2$, to his theory. This equation tells us that every particle in the universe with mass is a storehouse of energy, useful to us provided we can find ways to transform this mass into other forms of energy. The explosion of a nuclear weapon and burning of a star provide spectacular examples of transformations of mass to energy, but every single energy-emitting reaction—down to the burning of a match—carries with it a conversion of mass to a significant amount of energy. For example, the wood in a kitchen match contains about 30,000 calories (or 30 food calories). Because of the huge magnitude of the conversion factor c^2, the corresponding predicted change in mass of the combustion products is less than 2 billionths of a gram.

Where Does $E = mc^2$ Come From?

How does the famous $E = mc^2$ equation grow out of the special theory of relativity discussed so far in this chapter? The connection is not direct. In this section we shall present arguments for the development of $E = mc^2$ using equations we have already introduced in this and earlier chapters. We shall also explore some of the consequences of the equivalence of mass and energy. Please be patient and follow the logic. It will be rewarding.

Imagine that a moving particle emits two flashes a time $\Delta\tau$ apart as recorded on its own wristwatch. We use these two emissions to track the motion of the particle. These two flashes can be related using the metric equation (38-2):

$$(c\,\Delta\tau)^2 = (c\,\Delta t)^2 - (\Delta x)^2$$

where the values of Δx and Δt are measured with respect to the laboratory frame. Starting with this equation, we can extract some important information about the momentum and energy of the particle. We start by multiplying both sides of the equation by $m^2 c^2/(\Delta\tau)^2$, where m is the mass of the particle. This gives us

$$(mc^2)^2 = \left(mc^2 \frac{\Delta t}{\Delta\tau}\right)^2 - \left(mc \frac{\Delta x}{\Delta\tau}\right)^2. \tag{38-14}$$

Note that the famous expression mc^2 appears on the left. The second term on the right contains the fraction $\Delta x/\Delta\tau$—namely, the distance Δx traveled by the particle as measured by our laboratory observer, divided by the time $\Delta\tau$ it takes to move this distance as recorded on the wristwatch carried by the particle. This measures a kind of velocity. Mass times velocity yields the formula for momentum; call it p. The laboratory observer reckons the momentum to have the value

$$p \equiv m \frac{\Delta x}{\Delta\tau} \quad \text{(lab observer's definition of particle momentum).} \tag{38-15}$$

But, why does the lab observer use $\Delta\tau$ in Eq. 38-15 rather than Δt to define momentum? Newtonian mechanics assumes that the time Δt between two events is a universal quantity, with the same value as measured in all reference frames. But relativity shows us (Eq. 38-3) that the time between the two flashes emitted by the particle has a different value when measured in different frames. We have chosen to use the invariant proper time $\Delta\tau$ (as recorded on the wristwatch carried by the particle) to be the time to use in reckoning the particle's momentum. So Eq. 38-15 results from a decision about time that Newton did not have to make.

What about the first term on the right side of Eq. 38-14, the one containing the squared ratio of time lapses $(\Delta t/\Delta\tau)^2$? According to Eq. 38-3, this squared ratio is related to the ratio of the particle velocity and the speed of light by the equation

$$\left(\frac{\Delta t}{\Delta\tau}\right)^2 = \frac{1}{1 - v^2/c^2}. \tag{38-16}$$

The $(mc^2)^2$ term on the left in Eq. 38-14 has the units of energy squared. Some powerful results follow if we assume that the first term on the right of that equation is the square of the total energy, E, of the particle. Then, using Eqs. 38-14 and 38-16, energy can be written in two ways:

$$E = mc^2 \frac{\Delta t}{\Delta\tau} = \frac{mc^2}{(1 - v^2/c^2)^{1/2}}. \tag{38-17}$$

We can substitute the definition of the momentum, p, from Eq. 38-15 and the definition of energy, E, from Eq. 38-17 into Eq. 38-14 to get

$$(mc^2)^2 = E^2 - (pc)^2. \tag{38-18}$$

When the particle is at rest, the momentum $p = 0$ and this equation takes the famous form

$$E_{\text{rest}} = mc^2 \qquad \text{(rest energy of a particle of mass } m\text{).} \tag{38-19}$$

Note that Eq. 38-19 describes only a particle that is *at rest* in a given frame. For a particle in motion, observers in that frame must use Eq. 38-17 to predict its energy.

Is there experimental evidence that the expressions for energy and momentum derived above have a useful reality? Yes, overwhelming evidence. In analyzing decades of experiments with high-speed particles, conservation of energy and momentum continue to be valid in special relativity *provided* that one uses the *relativistic* expressions for energy and momentum. In analyzing high-speed particle collisions in an isolated system, one adds up the total energy of particles before a collision, using Eq. 38-17 for each particle (being sure to include the rest energy of any particles at rest). This number will be equal to the total energy of the system of particles after the collision, no matter how many particles are destroyed or created in the process. A similar conservation law holds true for *each* spatial component of the total momentum of the particles, using Eq. 38-15 for the x-component and similar equations for other directions (substituting Δy or Δz for Δx).

Relativistic Kinetic Energy

When a given particle is *not* at rest in a given reference frame, then its momentum p is *not* zero as measured in that frame. In this case the total particle energy, E, must be greater than its rest value to keep the right side of Eq. 38-18 a constant, equal to the left side. The *increase in energy of a particle due to its motion* is called **kinetic energy.** In special relativity the kinetic energy is defined as the difference between the total energy and the rest energy:

$$K \equiv E - E_{\text{rest}} = \frac{mc^2}{(1 - v^2/c^2)^{1/2}} - mc^2. \tag{38-20}$$

Equations 38-17 and 38-20 show that the total energy—and therefore also the kinetic energy—increases without limit as the particle speed v approaches the speed of light c (that is, as v/c approaches 1). And indeed we can add as much kinetic energy as we want to a moving particle in order to increase the energy of collision with other particles, as is done in ever more powerful and ingeniously designed particle accelerators. Yet even the highest-energy particle never moves faster than light as measured in any frame (Fig. 38-7). This result provides the answer to the question asked at the beginning of this chapter: *How can the energy of a particle increase without limit while its speed remains slower than the speed of light?*

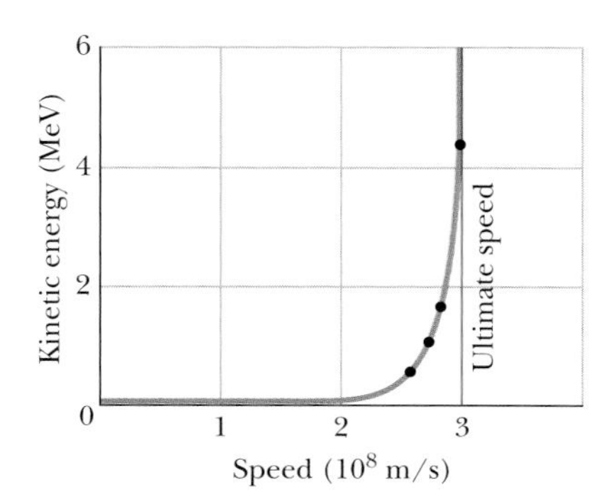

FIGURE 38-7 ■ The dots show measured values of the kinetic energy of an electron plotted against its measured speed. No matter how much energy is given to an electron (or to any other particle having mass), it cannot be accelerated to a speed that equals or exceeds the ultimate limiting speed c. (The curve drawn through the dots shows the predictions of Einstein's special theory of relativity.)

Kinetic Energy at Everyday Speeds

In our everyday world the fastest speed we encounter is probably that of a fighter plane moving above the speed of sound at Mach 3 (three times the speed of sound or about 1000 m/s). This speed is not even close to the speed of light. In fact, $v/c = (1000)/(3 \times 10^8\,\text{m/s}) \approx 3 \times 10^{-6}$.

Equation 38-20 looks complicated. However, for speeds much less than the speed of light (that is, for $v \ll c$) the equation reduces to the Newtonian expression for kinetic energy. To see this, we can use the following approximation:

$$(1 + d)^n \approx 1 + nd \qquad \text{for } |d| \ll 1 \text{ and } |nd| \ll 1. \tag{38-21}$$

Apply this approximation to the first term on the right of Eq. 38-20. Then in the limit of small values for v^2/c^2, Eq. 38-20 reduces to

$$K = mc^2(1 - v^2/c^2)^{-1/2} - mc^2$$
$$\approx mc^2(1 + v^2/2c^2) - mc^2 \qquad \text{(for } v \ll c\text{)}. \qquad (38\text{-}22)$$
$$\approx \frac{1}{2}mv^2$$

In brief, the expression for kinetic energy reduces to the Newtonian form for speeds very much less than the speed of light. This limiting case helps to justify our assignment of the name *energy* to the relativistic expressions shown in Eqs. 38-17 and 38-20. The total energy in these equations includes *both* relativistic kinetic energy and rest energy.

Proper Time and Proper Distance in Three Space Dimensions

So far, for simplicity, we have used one space dimension x in the equations of this chapter. Of course, there are three space dimensions, the additional two dimensions often labeled with the symbols y and z. The more general expressions for the time-like and space-like intervals, Eqs. 38-2 and 38-12, are

$$(c\,\Delta\tau)^2 = (c\,\Delta t)^2 - [(\Delta x)^2 + (\Delta y)^2 + (\Delta z)^2] \qquad \text{(squared time-like interval)}, \qquad (38\text{-}23)$$

and $$(\Delta\sigma)^2 = [(\Delta x)^2 + (\Delta y)^2 + (\Delta z)^2] - (c\,\Delta t)^2 \qquad \text{(squared space-like interval)}. \qquad (38\text{-}24)$$

READING EXERCISE 38-10: Find the speed v at which the energy E of a particle is equal to twice its rest energy. ■

TOUCHSTONE EXAMPLE 38-6: Energy of a Fast Particle

A particle of mass m moves so fast that its total energy is equal to 1.1 times its rest energy.

(a) What is the speed v of the particle?

SOLUTION ■ From Eq. 38-19, the rest energy is equal to

$$E_{\text{rest}} = mc^2.$$

The statement of the example says we are looking for a speed such that the energy is 1.1 times the rest energy. From Eq. 38-17, we have

$$E = \frac{mc^2}{(1 - v^2/c^2)^{1/2}} = 1.1E_{\text{rest}} = 1.1mc^2.$$

Cancel mc^2 from the second and fourth of these equal quantities and equate the results:

$$\frac{1}{(1 - v^2/c^2)^{1/2}} = 1.1.$$

Square both sides and solve the resulting equation for v^2/c^2:

$$v^2/c^2 = 0.1735$$

from which

$$v = 0.416\,c. \qquad \text{(Answer)}$$

(b) What is the kinetic energy of the particle?

SOLUTION ■ From Eq. 38-20, the kinetic energy is just the total energy minus the rest energy. In our case,

$$K = E - E_{\text{rest}} = 1.1E_{\text{rest}} - E_{\text{rest}} = 0.1E_{\text{rest}}. \qquad \text{(Answer)}$$

TOUCHSTONE EXAMPLE 38-7: Decay of K° Meson

A K° meson at rest decays into a π^+ meson and a π^- meson. As is often done in particle physics, we express the masses of these particles in terms of energy divided by c^2. The masses of the two π mesons are identical, so call both m_π. Then

$$m_K = 497.7 \text{ MeV}/c^2$$
$$m_\pi = 139.6 \text{ MeV}/c^2.$$

What is the speed of each of the π mesons after the decay?

SOLUTION ■ This is assumed to be an isolated system, in which both the total relativistic energy and the total relativistic momentum are conserved. The initial K° meson is at rest, so there is zero momentum before and therefore also after the collision. From this we conclude that the momenta labeled p_1 and p_2 in Fig. 38-8 have equal magnitudes but opposite directions. From Eq. 38-18, this guarantees that the two resulting particles also have equal energies after the decay. The conservation of energy equates *total* relativistic energy before and after the collision, and this total energy includes rest energy. Before the decay there is only the rest energy of the K° meson. After the decay, there are two π mesons of equal mass and equal energy. From Eq. 38-17,

$$m_K c^2 = 2E_\pi = \frac{2m_\pi c^2}{(1 - v^2/c^2)^{1/2}}.$$

Into the first and last expressions in this equation, substitute the values for the masses from the earlier equations. The factors c^2 cancel, and the units MeV cancel, to yield the equation

$$497.7 = \frac{2 \times 139.6}{(1 - v^2/c^2)^{1/2}}.$$

Square both sides of this equation, solve for v^2/c^2, and take the square root. The answer is

$$v = 0.828c. \qquad \text{(Answer)}$$

This is the speed of each π meson as the two move in opposite directions after the decay.

BEFORE

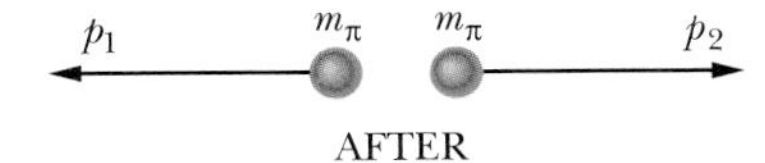

AFTER

FIGURE 38-8 ■

38-11 The Lorentz Transformation

Most textbooks on special relativity do not emphasize *invariant* quantities that have the same value for all observers. Instead, they focus on the so-called **Lorentz transformation** equations. These equations connect the distinct space and time separations between two events as measured in one frame with those separations as measured in another frame moving past the first. We display the Lorentz transformation equations here without deriving them. (The derivation is not difficult and depends only on the Principle of Relativity and some symmetry arguments.)

In writing the Lorentz transformations it is customary to let unprimed coordinates represent measurements made in the laboratory frame and primed coordinates represent corresponding measurements made in a rocket frame that moves past the laboratory with relative speed v^{rel} along the positive x direction. Then the Lorentz equations that transform the laboratory space and time separations to rocket space and time separations are

$$\begin{aligned} \Delta x' &= \frac{\Delta x - v^{\text{rel}}\Delta t}{(1 - (v^{\text{rel}})^2/c^2)^{1/2}} \\ \Delta t' &= \frac{\Delta t - (v^{\text{rel}}\Delta x/c^2)}{(1 - (v^{\text{rel}})^2/c^2)^{1/2}} \quad \text{(Lorentz transformation).} \qquad (38\text{-}25) \\ \Delta y' &= \Delta y \\ \Delta z' &= \Delta z \end{aligned}$$

What happens to these equations when they describe our everyday life in which typical moving objects are trains, airplanes, and automobiles? These vehicles move at speeds very much less than the speed of light. Assume that v^{rel} is much less than c, so that $v^{rel}/c \ll 1$ in Eqs. 38-25. Then the first two of these equations become

$$\begin{aligned} \Delta x' &= \Delta x - v^{rel}\Delta t \\ \Delta t' &= \Delta t \end{aligned} \qquad (v^{rel}/c \ll 1, \text{ Galilean transformation}). \qquad (38\text{-}26)$$

These equations are called the **Galilean transformation equations** because they lay out the consequences of relative motion first described by Galileo Galilei in the 1630s. The second equation tells us that for small relative velocities the time between two events has the same value for the moving observer as for the stationary observer. This is certainly typical of our experience; we do not need to reset our watches after an automobile ride! The first equation makes everyday sense as well. A race begins with a starting gun at the starting line and ends with the firing of an "ending gun" at the finish line when the winner crosses it. For observers in the stands the two firings occur a distance Δx and a time Δt apart. Running as fast as we can, we come in second, behind the winner. For us the ending gun goes off a distance ahead of us given by the distance of the racecourse, Δx , minus the distance we have run ($v^{rel}\Delta t$) at speed v^{rel} during the time Δt between the starting and ending guns. This is just what the first Galilean transformation (Eq. 38-26) tells us.

On the other hand, the Lorentz transformation (Eq. 38-25) gives us the rocket (primed) coordinates of an event if we know the laboratory (unprimed) coordinates of that event. But "laboratory" is just a label; it could represent simply another unpowered spaceship. Then the only difference between laboratory and rocket frames is the artificial difference we have given them. The rocket moves in the *positive x* direction with respect to the laboratory, so the laboratory moves in the *negative x* direction with respect to the rocket. It follows that the inverse transformation—the one that gives unprimed laboratory space and time separations in terms of primed rocket space-time separations—can be derived from the Lorentz transformation (Eq. 38-25) merely by interchanging primed and unprimed coordinates and reversing the sign of v^{rel}, making all v-terms positive in the numerators. (Reversing the sign of v^{rel} does not change the sign of $(v^{rel})^2$ in the denominators.) This leads to the equations

$$\begin{aligned} \Delta x &= \frac{\Delta x' + v^{rel}\Delta t'}{(1 - (v^{rel})^2/c^2)^{1/2}} \\ \Delta t &= \frac{\Delta t' + v^{rel}\Delta x'/c^2}{(1 - (v^{rel})^2/c^2)^{1/2}} \qquad \text{(the inverse Lorentz transformation)}. \qquad (38\text{-}27) \\ \Delta y &= \Delta y' \\ \Delta z &= \Delta z' \end{aligned}$$

38-12 Lorentz Contraction

The Lorentz transformation equations predict a relativistic effect important in the history of the subject—namely, that we as observers will measure an object moving past us at high speed to be shortened—contracted—along its direction of relative motion (but not changed in dimension perpendicular to this direction).

The Lorentz transformation describes the space and time separations between a pair of events. What events can we use to measure the length of a moving object? One choice is the explosions of two firecrackers, one at each end of the object *at the same time* ($\Delta t = 0$) in our frame. Then we can define the length to be the distance Δx between the explosions as measured in our frame. By setting $\Delta t = 0$ in the first term of Eqs. 38-25 and multiplying through by the square-root quantity, we get

$$\sqrt{1 - v^2/c^2}\,\Delta x' = \Delta x. \qquad (38\text{-}28)$$

This equation tells us that the length Δx we measure for the object in the laboratory frame—the distance between simultaneous firecracker explosions—is less than the distance $\Delta x'$ between the two ends as measured in the rocket frame in which the object is at rest.

The Lorentz contraction is a curiosity, and is not used very often to analyze experiments. However, it is the consequence of a deeper principle, the relativity of simultaneity, discussed following Eq. 38-12 in Section 38-8 and in Section 38-9. The two firecrackers may explode at the same time in our frame ($\Delta t = 0$) but not in the frame of the rocket ($\Delta t' \neq 0$), as you can see by substituting $\Delta t = 0$ into the second of Eqs. 38-25. The same-time explosions at the two ends of the moving rod in our frame yield a measure of the length of the moving object in our frame. In contrast, the *lack* of simultaneity in the rocket frame does not change the rocket measurement of $\Delta x'$ because the object is at rest in the rocket frame; the distance between the explosions at the two ends is the same whether or not these explosions occur at the same time. However, this lack of simultaneity allows observers in the two frames to account for the difference in measured length of the object.

READING EXERCISE 38-11: What is the speed v of a passing rocket in the case that we measure the length of the rocket to be half its length as measured in a frame in which the rocket is at rest? ■

38-13 Relativity of Velocities

The speed of light c is the ultimate speed according to special relativity. No object can be accelerated from rest to a speed greater than c. Yet some have found in this statement a paradox that challenges the validity of special relativity. Someone who objects to relativity says, "I ride in a rocket moving at 3/4 the speed of light with respect to the laboratory. From my rocket I launch a stone forward at 3/4 the speed of light as measured in my rocket frame. The result should be an object moving at $3/4 + 3/4 = 1.5$ times the speed of light in the laboratory frame. But relativity says that nothing can move faster than light. So my thought experiment shows special relativity to be illogical—and disproves it!"

Since the Lorentz transformation deals with space and time separation between *events*, we can use it to investigate the validity of this thought experiment. Let the stone that you launch forward from your rocket emit two flashes close together. Call the separation between these flash emissions $\Delta x'$ and $\Delta t'$ in your rocket frame and Δx and Δt in our laboratory frame. Then we can derive the velocities of the stone in the two frames from the differential limits of $\Delta x'/\Delta t'$ (velocity in rocket frame) and $\Delta x/\Delta t$ (velocity in laboratory frame).

Next we can use the first two entries in the Lorentz transformation (38-27):

$$\Delta x = \frac{\Delta x' + v^{\text{rel}}\Delta t'}{(1 - (v^{\text{rel}})^2/c^2)^{1/2}}$$

$$\Delta t = \frac{\Delta t' + v^{\text{rel}}\Delta x'/c^2}{(1 - (v^{\text{rel}})^2/c^2)^{1/2}}.$$

Then we can divide corresponding sides of these two equations into each other. The square root expressions in the denominators cancel and we have

$$\frac{\Delta x}{\Delta t} = \frac{\Delta x' + v^{\text{rel}}\Delta t'}{\Delta t' + v^{\text{rel}}\Delta x'/c^2}. \tag{38-29}$$

On the right side of this equation, divide both the numerator and denominator by $\Delta t'$:

$$\frac{\Delta x}{\Delta t} = \frac{\Delta x'/\Delta t' + v^{rel}}{1 + v^{rel}(\Delta x'/\Delta t')/c^2}. \tag{38-30}$$

Finally, we can take the differential limit and define u as the velocity of the stone in the laboratory frame and u' as the velocity of the stone in the rocket frame. Then Eq. 38-30 becomes

$$u = \frac{u' + v^{rel}}{1 + u'v^{rel}/c^2} \quad \text{(law of addition of velocities).} \tag{38-31}$$

Although Eq. 38-31 is called the **law of addition of velocities,** this is not a good name, because the addition is not simple.

How does the law of addition of velocities support the conclusion that nothing can move faster than the speed of light? Let's use this law to find the value of the stone's velocity u in our laboratory frame when the rocket moves away from us at 3/4 the speed of light ($v^{rel} = 0.75c$) while the stone moves away from the rocket at 3/4 the speed of light ($u' = 0.75c$) as measured in the rocket frame. Substituting these values into Eq. 38-31 yields

$$u = \frac{0.75c + 0.75c}{1 + (0.75c)^2/c^2} = \frac{1.5c}{1 + 0.5625} = 0.96c. \tag{38-32}$$

The result is that we observe the stone to move at the speed $0.96c$ in our laboratory frame. Its speed does not exceed the speed of light. Once again relativity saves itself from disproof!

READING EXERCISE 38-12: A rocket moves with speed $0.9c$ in our laboratory frame. A flash of light is sent forward from the front end of the rocket. Is the speed of that flash equal to $1.9c$ as measured in our laboratory frame? If not, what is the speed of the light flash in our frame? Verify your answer using Eq. 38-31. ■

TOUCHSTONE EXAMPLE 38-8: Relative Speed of Light

A rocket moves with speed $0.9c$ in our laboratory frame. A flash of light is fired *backward* from the rear of the rocket.

(a) What is the *speed* of that light flash in our laboratory frame?

SOLUTION ■ Using the Principle of Relativity, we can give an answer immediately, without doing any calculations. The speed of light is the same—invariant—in all reference frames. Therefore, the speed of the light flash in our laboratory frame will be c as usual, and in the backward direction. The equations should verify this result.

Use Eq. 38-31:

$$u = \frac{u' + v^{rel}}{1 + u'v^{rel}/c^2}.$$

In the example, $v^{rel} = 0.9c$ and $u' = -c$, with a minus since the light is fired out the back of the rocket:

$$u = \frac{-c + 0.9c}{1 - 0.9c^2/c^2} = \frac{-0.1c}{0.1} = -c. \quad \text{(Answer)}$$

Therefore the speed of the light flash is equal to c in our laboratory frame, as it has to be and as predicted at the beginning of this solution.

(b) What is the *direction* of the light flash in our laboratory frame?

SOLUTION ■ The minus sign in the most recent equation tells us that in the laboratory frame the light flash moves in a direction opposite to that of the rocket in the laboratory frame.

38-14 Doppler Shift

A car approaches us and passes while sounding its horn. We hear the pitch of the horn decrease as it passes, a change called the **Doppler shift** or **Doppler effect.** In Chapter 18 we found that the Doppler effect depends on two velocities—namely, the velocity of the source and the velocity of the detector with respect to the air, the medium that transmits the waves.

Light and other electromagnetic waves follow a different rule for the Doppler shift, because they require no transmitting medium and can travel through a vacuum. For this reason, the Doppler effect for light depends on only one velocity, the relative velocity between source and detector.

Suppose that a source of light moves away from us with speed v^{rel}, while sending light backward toward us. Let the frequency of the light as measured by the source be denoted f_0. We detect a smaller frequency f than the frequency f_0 and for two reasons: (1) Each wave crest is emitted at a greater distance from us than the previous wave crest, so it has to travel farther to reach us, and (2) for us the traveling clock runs slow (Eq. 38-3). These two effects combine to change the frequency of the light that we receive according to the equation

$$f = f_0\left[\frac{1 - (v^{rel}/c)}{1 + (v^{rel}/c)}\right]^{1/2} \quad \text{(source moving away).} \qquad (38\text{-}33)$$

In the case that the source moves *toward* us with speed v^{rel}, we simply reverse the sign of v^{rel}, which occurs twice in Eq. 38-33.

We can measure the frequencies emitted by luminous elements, such as hydrogen excited by an electric discharge in the laboratory. Looking out at stars in nearby galaxies, we can identify hydrogen from the pattern of emitted frequencies. We notice that galaxies farther from us have light shifted downward in frequency compared with their laboratory value and conclude that these galaxies are moving away from us; we call this the **red shift.**

Solving Eq. 38-33 for v^{rel}, we can determine the velocity with which a galaxy recedes from us. This analysis is approximately correct for nearby galaxies. However, the red shift due to the most remote galaxies is not due to the Doppler shift described by special relativity. Rather, the stretching out of the light waves heading toward us occurs because space itself is stretching as the universe expands over time. Thus, general relativity is required to describe this stretching of space with time.

READING EXERCISE 38-13: A not-too-distant galaxy is moving directly away from the Earth. Light from this galaxy includes a pattern of frequencies recognized as those emitted by hydrogen gas. We detect one of these frequencies to have the value $f = 0.9\,f_0$, where f_0 is the corresponding frequency for light from hydrogen gas at rest in the laboratory. How fast is the distant galaxy moving away from the Earth? ■

TOUCHSTONE EXAMPLE 38-9: Colliding with Andromeda

According to some predictions, the Andromeda galaxy, currently two million light-years away from us, is moving toward our galaxy and the two will collide in three billion years. Light of a particular frequency f_0 is emitted from hydrogen gas in the stars of the Andromeda galaxy. What is the frequency f of that light measured on Earth?

SOLUTION ■ Equation 38-33 describes the case in which the source is moving directly *away* from us. But in this example, the Andromeda galaxy is moving directly *toward* us. For a source moving directly toward us, reverse the sign of v^{rel} in Eq. 38-33:

$$f = f_0\left[\frac{1 + (v^{rel}/c)}{1 - (v^{rel}/c)}\right]^{1/2} \quad \text{(source approaching).}$$

Andromeda is two million light-years distant and is predicted to reach our galaxy in three billion years. Therefore it must be moving at the following fraction of the speed of light,

$$\frac{v^{\rm rel}}{c} = \frac{2 \times 10^6\ \text{(light) years}}{3 \times 10^9\ \text{years}} = 6.667 \times 10^{-4}.$$

This has a magnitude very much less than one, so we can apply approximation Eq. 38-21 to the equation above:

$$\begin{aligned} f &= f_0\left[\frac{1 + (v^{\rm rel}/c)}{1 - (v^{\rm rel}/c)}\right]^{1/2} \\ &= f_0\{1 + (v^{\rm rel}/c)\}^{1/2}\{1 - (v^{\rm rel}/c)\}^{-1/2} \\ &\approx f_0\{1 + (v^{\rm rel}/2c)\}\{1 + (v^{\rm rel}/2c)\} \\ &\approx f_0\{1 + (v^{\rm rel}/c) + (v^{\rm rel}/2c)^2\}. \end{aligned}$$

Now, the second term in the last parenthesis has the approximate value 10^{-3}, while the third term has the approximate value 10^{-7}. Therefore we neglect the third term and reach our result:

$$f \approx f_0\{1 + (v^{\rm rel}/c)\} = f_0\{1 + 6.667 \times 10^{-4}\} \approx f_0\{1 + 7 \times 10^{-4}\}.$$

(Answer)

This expression tells us how much higher is the frequency of light from Andromeda than the frequency of the light from a source of the same atoms viewed at rest in a laboratory on Earth.

Bibliography and Acknowledgments

For a fuller treatment of special relativity that follows the outline presented in this chapter, see *Spacetime Physics,* 2nd Edition, by Edwin F. Taylor and John Archibald Wheeler, W. H. Freeman, New York, 1992, ISBN 0-7167-2327-1. For the relativity of simultaneity, see pages 62–63 and 128–131. For a derivation of the Lorentz transformation equations, see pages 99–103. For a derivation of the Doppler equations, see pages 114 and 263. Some exercises from the Taylor–Wheeler text were adapted for exercises in the present chapter with permission of one author. The present chapter also adapts some excerpts from the entry "Special Relativity" by Edwin F. Taylor in Volume 3 of *The Macmillan Encyclopedia of Physics,* John S. Rigden, Editor in Chief, Simon & Schuster Macmillan, New York, 1996, ISBN 0-02-864588-X.

For a treatment of special relativity that pays more attention to the experimental foundations, see *Special Relativity,* by A. P. French, W. W. Norton, New York, 1968, ISBN 0-393-09793-5. For the relativity of simultaneity, see page 74. For a derivation of the Lorentz transformation equations, see pages 76–82. For a derivation of the Doppler equations, see pages 134–146. Several exercises at the end of Chapter 38 were adapted from the French text with permission of the author.

Albert Einstein's original publication on special relativity is "Zur Elektrodynamik bewegter Körper" ("On the Electrodynamics of Moving Bodies") in *Annalen der Physik,* Volume 17, pages 891–921 (1905). This is reprinted, along with many other original articles on the subject, in the book *The Principle of Relativity* by Albert Einstein, H. A. Lorentz, H. Weyl, H. Minkowski, and others, Dover Books, New York 1952, ISBN 0-486-60081-5.

Arthur I. Miller presents a modern English translation of Einstein's original article, together with a wealth of information on the historical and scientific background and immediate consequences, in his book *Albert Einstein's Special Theory of Relativity: Emergence (1905) and Early Interpretation (1905–1911),* Addison Wesley, Reading, Massachusetts, 1981, ISBN 0-201-04680-6.

The example of the loud tape player in Reading Exercise 38-9 was devised by Rachel Scherr, Stamatis Vokos, Peter Shaffer, and Andrew Boudreaux.

The author thankfully acknowledges both strategic and detailed suggestions and advice from Patrick J. Cooney, Priscilla W. Laws, and Edward F. Redish. Their help has greatly improved both the physics content and the pedagogical effectiveness of the text, though they are not responsible for any remaining errors.

Problems

SEC. 38-2 ■ ORIGINS OF SPECIAL RELATIVITY

1. Chasing Light. What fraction of the speed of light does each of the following speeds v represent? That is, what is the value of the ratio v/c? (a) A typical rate of continental drift, 3 cm/y. (b) A highway speed limit of 100 km/h. (c) A supersonic plane flying at Mach 2.5 = 3100 km/h. (d) The Earth in orbit around the Sun at 30 km/s. (e) What conclusion(s) do you draw about the need for special relativity to describe and analyze most everyday phenomena? (*Note: Some* everyday phenomena can be derived from relativity. For example, magnetism can be described as arising from electrostatics plus special relativity applied to the slow-moving charges in wires.)

SEC. 38-3 ■ THE PRINCIPLE OF RELATIVITY

2. Fast Computation. A "serial computer," one that carries out one instruction at a time, executes an instruction by transmitting data from the memory to the processor (where computation takes place) and then transmitting the result back to the memory. Estimate the maximum size of a serial "teraflop" computer, one that carries out 10^{12} instructions per second.

3. Examples of the Principle of Relativity. Identical experiments are carried out (1) in a high-speed train moving at constant speed along a horizontal track with the shades drawn and (2) in a closed freight container on the platform as the train passes. Copy the following list and mark with a "yes" quantities that will necessarily be the same as measured in the two frames. Mark with a "no" quantities that are not necessarily the same as measured in the two frames. (a) The time it takes for light to travel one meter in a vacuum; (b) the kinetic energy of an electron accelerated from rest through a voltage difference of one million volts; (c) the time for half the number of radioactive particles at rest to decay; (d) the mass of a proton; (e) the structure of DNA for an amoeba; (f) Newton's Second Law of Motion: $F = ma$; (g) the value of the downward acceleration of gravity g.

4. Riding to Alpha Centauri. You are taking a trip from the solar system to our nearest visible neighbor, Alpha Centauri, approximately 4 light-years distant. At launch you experienced a period of acceleration that increased your speed with respect to Earth from zero to nearly half the speed of light. Now your spaceship is coasting in unpowered flight. Compare and contrast the observations you make now with those you made before the rocket took off from the Earth's surface. Be as specific and detailed as possible. Distinguish between observations made inside the cabin with the windows covered and those made looking out of uncovered windows at the front, side, and back of the cabin.

SEC. 38-4 ■ LOCATING EVENTS WITH AN INTELLIGENT OBSERVER

5. Deducing a Speed. A pulse of protons arrives at detector D, where you are standing. Prior to this, the pulse passed through detector C, which lies 60 meters upstream. Detector C sent a light flash in your direction at the same instant that the pulse passed through it. At detector D you receive the light flash and the proton pulse separated by a time of 2 nanoseconds (2×10^{-9} s). What is the speed of the proton pulse?

6. Eruption from the Sun. You see a sudden eruption on the surface of the Sun. From solar theory you predict that the eruption emitted a pulse of particles that is moving toward the Earth at one-eighth the speed of light. How long do you have to seek shelter from the radiation that will be emitted when the particle pulse hits the Earth? Take the light-travel time from the Sun to the Earth to be 8 minutes.

SEC. 38-5 ■ LABORATORY AND ROCKET LATTICEWORKS OF CLOCKS

7. Synchronizing a Clock. In a vast latticework of meter sticks and clocks, you stand next to a lattice clock whose coordinates are x = 8 km, y = 40 km, z = 44 km. When you receive the synchronizing flash, to what time do you quickly set your clock?

8. Earth's Surface Inertial? Quite apart from effects due to the Earth's rotational and orbital motion, a laboratory reference frame on the Earth is not an inertial frame, as required by a strict interpretation of special relativity. It is not inertial because a particle released from rest at the Earth's surface does not remain at rest; it falls! Often, however, the events in an experiment for which one needs special relativity happen so quickly that we can ignore effects due to gravitational acceleration. Consider, for example, a proton moving horizontally at speed $v = 0.992c$ through a 10-m-wide detector in a laboratory test chamber. (a) How long will the transit through that detector take? (b) How far does the proton fall vertically during this time lapse? (c) What do you conclude about the suitability of the laboratory as an inertial frame in this case?

SEC. 38-6 ■ TIME STRETCHING

9. Light Clock for a Faster Rocket. Redo Fig. 38-5 with a vertical distance $c\ \Delta\tau/2 = 7$ m and horizontal distance in the lab frame $\Delta x/2 = 24$ m. Find the ratio of the times $\Delta t/\Delta\tau$ between events A and B recorded on laboratory and rocket clocks.

SEC. 38-7 ■ THE METRIC EQUATION

10. Where and When? Two firecrackers explode at the same place in the laboratory and are separated by a time of 12 years. (a) What is the spatial distance between these two events in a rocket in which the events are separated in time by 13 years? (b) What is the relative speed of the rocket and laboratory frames? Express your answer as a fraction of the speed of light.

11. Traveling to Vega. Jocelyn DeGuia takes off from Earth and moves toward the star Vega, which is 26 ly distant from Earth. Assume that Earth and Vega are relatively at rest and Jocelyn moves at $v = 0.99c$ in the Earth–Vega frame. How much time will have elapsed on Earth (a) when Jocelyn reaches Vega and (b) when Earth observers receive a radio signal reporting that Jocelyn has arrived? (c) How much will Jocelyn age during her outward trip?

12. Travel to the Dog Star. In the 24th century the fastest available interstellar rocket moves at $v = 0.75c$. Mya Allen is sent in this rocket at full (constant) speed to Sirius, the Dog Star, the brightest star in the heavens as seen from Earth, which is a distance 8.7 ly as measured in the Earth frame. Assume Sirius is at rest with respect to Earth. Mya stays near Sirius, slowly orbiting around that Dog Star, for 7 years as recorded on her wristwatch while making observations and recording data, then returns to Earth with the same speed $v = 0.75c$. According to Earth-linked observers: (a) When does Mya arrive at Sirius? (b) When does Mya leave Sirius? (c) When does Mya arrive back at Earth? According to Mya's wristwatch: (d) When does she arrive at Sirius? (e) When does she leave Sirius? (f) When does she arrive back on Earth?

13. Fast-Moving Muons. The half-life of stationary muons is measured to be 1.6 microseconds. Half of any initial number of stationary muons decays in one half-life. Cosmic rays colliding with atoms in the upper atmosphere of the Earth create muons, some of which move downward toward the Earth's surface. The mean lifetime of high-speed muons in one such burst is measured to be 16 microseconds. (a) Find the speed of these muons relative to the Earth. (b) Moving at this speed, how far will the muons move in one half-life? (c) How far would this pulse move in one half-life if there were no relativistic time stretching? (d) In the relativistic case, how far will the pulse move in 10 half-lives? (e) An initial pulse consisting of 10^8 muons is created at a distance above the Earth's surface given in part (d). How many will remain at the Earth's surface? Assume that the pulse moves vertically downward and none are lost to collisions. (Ninety-nine percent of the Earth's atmosphere lies below 40 km altitude.)

14. Lifetime of a Fast Particle. An unstable high-energy particle is created in a collision inside a detector and leaves a track 1.05 mm long before it decays while still in the detector. Its speed relative to the detector was $0.992c$. How long did the particle live as recorded in its rest frame?

15. Living a Thousand Years in One Year. You wish to make a round trip from Earth in a spaceship, traveling at constant speed in a straight line for 6 months on your watch and then returning at the same constant speed. You wish, further, to find Earth to be 1000 years older on your return. (a) What is the value of your constant speed with respect to Earth? (b) How much do you age during the trip? (c) Does it matter whether or not you travel in a straight line? For example, could you travel in a huge circle that loops back to Earth?

16. Birthdays. An astronaut traveling in an unpowered spaceship celebrates his 18th, 19th, 20th, and 21st birthdays. Five Earth-years elapse between the 18th and 21st birthday parties. Find (a) the spatial separation between the 18th and 21st birthday parties in the Earth frame and (b) the speed of his spaceship with respect to Earth.

SEC. 38-8 ■ CAUSE AND EFFECT

17. Relations Between Events. The table shows the t and x coordinates of three events as observed in the laboratory frame.

Laboratory Coordinates of Three Events

Event	t years	x light-years
Event 1	2	1
Event 2	7	4
Event 3	5	6

On a piece of paper list vertically every pair of these events: (1, 2), (1, 3), (2, 3). (a) Next to each pair write "time-like," "light-like," or "space-like" for the relationship between those two events. (b) Next to each pair, write "Yes" if it is possible for one of the events to cause the other event and "No" if a cause and effect relation between them is not possible. (For full benefit of this exercise, construct and analyze your own tables.)

18. Proper Distance and Proper Time. Use the equations in Chapter 38 to show the following general results: (a) Given that two events P and Q have a space-like separation, show that in all such cases a reference frame can be found in which the two events occur at the same time. Also show that with respect to this frame the distance between the two events is equal to the proper distance between them. (b) Given that two events P and R have a time-like separation, show that in all such cases a reference frame can be found in which the two events occur at the same place. Also show that in this frame the time lapse between the two events is equal to the proper time between them. (c) Given that two events R and W have a light-like separation, show that in all such cases a light flash can be found that moves from R to W. Also show that the proper time and proper distance between R and W are both equal to zero.

SEC. 38-9 ■ RELATIVITY OF SIMULTANEITY

19. Symmetric Relativity of Simultaneity. In the thought experiment pictured in Fig. 38-6, we arbitrarily chose events so that the two light flashes from the lightning strikes arrived simultaneously at the ground observer. Analyze a new version of this experiment in which a completely different pair of lightning strikes fall at the two ends of the train such that the resulting light flashes arrive simultaneously at the position of the rider at the center of the train. View the experiment in the rest frame of the train. In this new version of the experiment, which lightning bolt falls first according to the observer on the ground?

SEC. 38-10 ■ MOMENTUM AND ENERGY

20. Boosting the Speed. How much work must be done to increase the speed of an electron (a) from $0.08c$ to $0.09c$? (b) from $0.98c$ to $0.99c$? Note that the increase in speed is the same in both cases.

21. Lightbulb Radiating Mass. How much mass does a 100 W lightbulb dissipate (in heat and light) when it burns for one full year?

22. Proton Crosses Galaxy. Find the energy of a proton that crosses our galaxy (diameter 100 000 light-years) in one minute of its own time.

23. Converting Mass to Energy. The values of the masses in the reaction

$$p + {}^{19}F \rightarrow \alpha + {}^{16}O$$

have been determined by a mass spectrometer to have the values:

$$m(p) = 1.007825u,$$
$$m(F) = 18.998405u,$$
$$m(\alpha) = 4.002603u,$$
$$m(O) = 15.994915u.$$

Here u is the atomic mass unit (Section 1.7). How much energy is released in this reaction? Express your answer in both kilograms and MeV.

24. Aspirin-Powered Automobile. An aspirin tablet contains 5 grains of aspirin (medicinal unit), which is equal to 325 mg. For how many kilometers would the energy equivalent of this mass power an automobile? Assume 12.75 km/L and a heat of combustion of 3.65×10^7 J/L for the gasoline used in the automobile.

25. Converting Energy to Mass. Two freight trains, each of mass 6×10^6 kg (6 000 metric tons) travel in opposite directions on the same track with equal speeds of 150 km/hr. They collide head-on and come to rest. (a) Calculate in joules the kinetic energy $(1/2)mv^2$ for each train before the collision. (Newtonian expression OK for everyday speeds!) (b) After the collision, the mass of the trains plus the mass of the track plus the mass of the roadbed plus the mass of the surrounding air plus the mass of emitted sound and light has increased by what number of milligrams?

26. Electrically Accelerated Electron. Through what voltage must an electron be accelerated from rest in order to increase its energy to 101% of its rest energy?

27. Powerful Proton. A proton exits an accelerator with a kinetic energy equal to N times its rest energy. Find expressions for its (a) speed and (b) momentum.

28. Relativistic Chemistry. One kilogram of hydrogen combines chemically with 8 kilograms of oxygen to form water; about 10^8 J of energy is released. Ten metric tons (10^4 kg) of hydrogen combines with oxygen to produce water. (a) Does the resulting water have a greater or less mass than the original hydrogen plus oxygen? (b) What is the numerical magnitude of this difference in mass? (c) A smaller amount of hydrogen and oxygen is weighed, then combined to form water, which is weighed again. A very good chemical balance is able to detect a fractional change in mass of 1 part in 10^8. By what factor is this sensitivity more than enough—or insufficient—to detect the fractional change in mass in this reaction?

29. Finding the Mass. (a) Find an equation for the unknown mass m of a particle if you know its momentum p and its kinetic energy K. Show that this expression reduces to an expected result for nonrelativistic particle speeds. (b) Find the mass of a particle whose kinetic energy is $K = 55.0$ MeV and whose momentum is $p = 121$ MeV/c. Express your answer as a decimal fraction or multiple of the mass m_e of the electron.

30. A Box of Light. Estimate the power in kilowatts used to light a city of 8 million inhabitants. If all this light generated during one hour in the evening could be captured and put in a box, how much would the mass of the box increase?

31. Creating a Proton–Antiproton Pair. Two protons, each of mass m, are fired toward one another with equal energy (see Fig. 38-9). They collide and create an additional proton–antiproton pair, each with the proton mass m. (a) Show that the lowest total energy E of the incident protons for this creation to take place leaves the resulting four particles at rest with respect to one another. The value of this minimum energy for each incident particle is called the **threshold energy.** (b) What is the threshold *kinetic* energy K of each incident particle for this creation to occur? Express your answer in terms of the rest energy of the proton. (c) Given that the mass of a proton is approximately equal to 1 GeV/c^2, what is the value of the threshold kinetic energy of each incident proton? Explain why this result is reasonable.

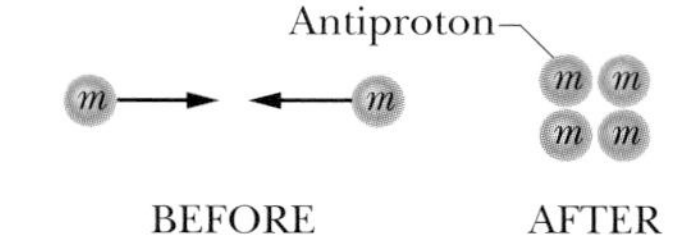

FIGURE 38-9 ■ Problem 31.

SEC. 38-11 ■ THE LORENTZ TRANSFORMATION

32. Really Simultaneous? (a) Two events occur at the same time in the laboratory frame and at the laboratory coordinates ($x_1 = 10$ km, $y_1 = 4$ km, $z_1 = 6$ km) and ($x_2 = 10$ km, $y_2 = 7$ km, $z_2 = -10$ km). Will these two events be simultaneous in a rocket frame moving with speed $v^{\text{rel}} = 0.8c$ in the x direction in the laboratory frame? Explain your answer. (b) Three events occur at the same time in the laboratory frame and at the laboratory coordinates (x_0, y_1, z_1), (x_0, y_2, z_2), and (x_0, y_3, z_3), where x_0 has the same value for all three events. Will these three events be simultaneous in a rocket frame moving with speed v^{rel} in the laboratory x direction? Explain your answer. (c) Use your results of parts (a) and (b) to make a general statement about simultaneity of events in laboratory and rocket frames.

33. Transformation of y-velocity. A particle moves with uniform speed $v_y' = \Delta y'/\Delta t'$ in the y' direction with respect to a rocket frame that moves along the x axis of a laboratory frame. Find expressions for the x-component and for the y-component of the particle's velocity in the laboratory frame.

34. Transformation of Velocity Direction. A particle moves with speed v' in the $x'y'$ plane of the rocket frame and in a direction that makes an angle ϕ' with the x' axis. Find the angle ϕ that the velocity vector of this particle makes with the x axis of the laboratory frame. (*Hint:* Transform space and time displacements rather than velocities.)

35. The Headlight Effect. A flash of light is emitted at an angle ϕ' with respect to the x' axis of the rocket frame. (a) Show that the angle ϕ the direction of motion of this flash makes with respect to the x axis of the laboratory frame is given by the equation

$$\cos\phi = \frac{\cos\phi' + v^{\text{rel}}/c}{1 + (v^{\text{rel}}/c)\cos\phi'}.$$

Optional: Show that your answer to Problem 34 gives the same result when the velocity v' is given the value c. (b) A light source at rest in the rocket frame emits light uniformly in all directions. In the rocket frame 50% of this light goes into the forward hemisphere of a sphere surrounding the source. Show that in the laboratory frame this 50% of the light is concentrated in a narrow forward cone of half-angle ϕ_0 whose axis lies along the direction of motion of the particle. Derive the following expression for the half-angle ϕ_0:

$$\cos\phi_0 = v^{\text{rel}}/c.$$

This result is called the **headlight effect.** (c) What is the half-angle ϕ_0 in degrees for a light source moving at 99% of the speed of light?

SEC. 38-12 ■ LORENTZ CONTRACTION

36. Electron Shrinks Distance. An evacuated tube at rest in the laboratory has a length 3.00 m as measured in the laboratory. An electron moves at speed $v = 0.999\,987c$ in the laboratory along the axis of this evacuated tube. What is the length of the tube measured in the rest frame of the electron?

37. Passing Time. A spaceship of rest length 100 m passes a laboratory timing station in 0.2 microseconds measured on the timing station clock. (a) What is the speed of the spaceship in the laboratory frame? (b) What is the Lorentz-contracted length of the spaceship in the laboratory frame?

38. Transformation of Angles. A meter stick lies at rest in the rocket frame and makes an angle ϕ' with the x' axis as measured by the rocket observer. The laboratory observer measures the x- and y-components of the meter stick as it streaks past. From these components the laboratory observer computes the angle ϕ that the stick makes with his x axis. (a) Find an expression for the angle ϕ in terms of the angle ϕ' and the relative speed v^{rel} between rocket and laboratory frames. (b) What is the length of the "meter" stick measured by the laboratory observer? (c) *Optional:* Why is your expression in part (a) different from equations derived in Problems 34 and 35?

39. Traveling to the Galactic Center. (a) Can a person, in principle, travel from Earth to the center of our galaxy, which is 23 000 ly distant, in one lifetime? Explain using either length contraction or time dilation arguments. (b) What constant speed with respect to the galaxy is required to make the trip in 30 y of the traveler's lifetime?

40. Limo in the Garage. Carman has just purchased the world's longest stretch limo, which has proper length $L_c = 30.0$ m. Part (a) of Figure 38-10 shows the limo parked at rest in front of a garage of proper length $L_g = 6.00$ m, which has front and back doors. Looking at the limo parked in front of the garage, Carman says there is no way that the limo can fit into the garage. "*Au contraire!*" shouts Garageman, "Under the right circumstances the limo can fit into the garage with both garage doors closed and room to spare!" Garageman envisions a fast-moving limo that takes up exactly one-third of the proper length of the garage. Part (b) of Figure 38-10 shows the speeding limo just as the front garage door closes behind it as recorded in the garage frame. Part (c) of Figure 38-10 shows the limo just as the back garage door opens in front of it as recorded in the garage frame. Find the speed of the limo with respect to the garage required for this scenario to take place.

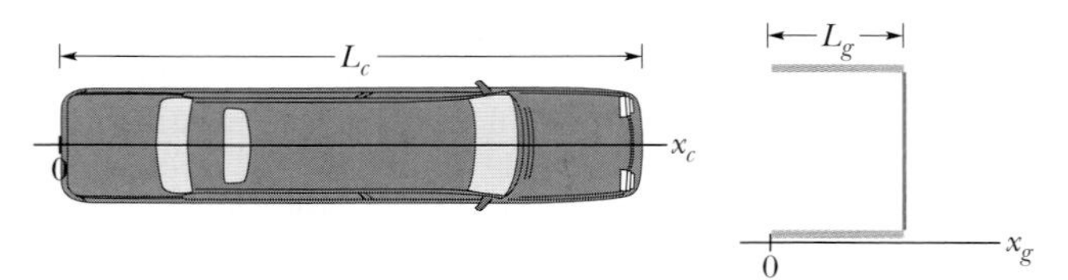

(*a*) Stretch limo and garage, both at rest in frame of garage.

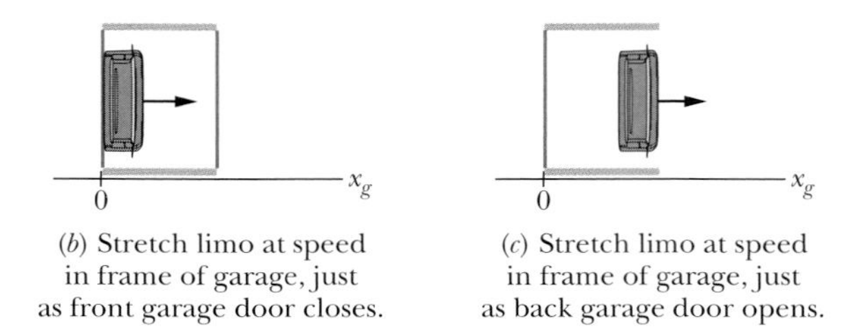

(*b*) Stretch limo at speed in frame of garage, just as front garage door closes.

(*c*) Stretch limo at speed in frame of garage, just as back garage door opens.

FIGURE 38-10 ■ Problem 40.

SEC. 38-13 ■ RELATIVITY OF VELOCITIES

41. Backfire. An unpowered rocket moves past you in the positive x direction at speed $v^{rel} = 0.9c$. This rocket fires a bullet out the back that you measure to be moving at speed $v_{bullet} = 0.3c$ in the positive x direction. With what speed relative to the rocket did the rocket observer fire the bullet out the back of her ship?

42. Separating Galaxies. Galaxy A is measured to be receding from us on Earth with a speed of $0.3c$. Galaxy B, located in precisely the opposite direction, is also receding from us at the same speed. What recessional velocity will an observer on galaxy A measure (a) for our galaxy, and (b) for galaxy B?

43. Decaying K° Meson. Touchstone Example 38-7 concluded that when a K° meson at rest decays into two daughter π mesons, they move in opposite directions in the rest frame of the original K° meson, each with a speed of $0.828c$. Now suppose that the initial K° meson moves with speed $v^{rel} = 0.9c$ as measured in the laboratory frame. What are the maximum and minimum speeds of the daughter π mesons with respect to the laboratory?

44. Transit Time. An unpowered spaceship whose rest length is 350 meters has a speed $0.82c$ with respect to Earth. A micrometeorite, also with speed of $0.82c$ with respect to Earth, passes the spaceship on an antiparallel track that is moving in the opposite direction. How long does it take the micrometeorite to pass the spaceship as measured on the ship?

SEC. 38-14 ■ DOPPLER SHIFT

45. Listening to the Traveler. A spaceship moving away from Earth at a speed $0.900c$ radios its reports back to Earth using a frequency of 100 MHz measured in the spaceship frame. To what frequency must Earth's receivers be tuned in order to receive the reports?

46. Speed Trap. How fast would you have to approach a red traffic light in order that it appears green to you?

47. Redshift Factor z. Astrophysicists describe the redshift of receding astronomical objects using the **redshift factor z,** defined implicitly in the following equation:

$$\lambda_{observed} \equiv (1 + z)\lambda_{emitted}.$$

Here $\lambda_{observed}$ is the wavelength of light observed from Earth, while $\lambda_{emitted}$ is the wavelength of the light emitted from the source as measured in the rest frame of the source. The emitted wavelength is known if one knows the emitting atom, identified from the pattern of different wavelengths characteristic of that atom. Astrophysicists measuring the redshifts of light from extremely remote quasars calculate a z-factor in the neighborhood of $z \approx 6$. Use the Doppler shift equations of special relativity to determine how fast such quasars are moving away from Earth. *Note:* Actually, for such distant objects the unmodified Doppler shift formula of special relativity does not apply. Instead, one thinks of the space between Earth and the source expanding as the universe expands; the wavelength of the light expands with this expansion of the universe as it travels from the source quasar to us.

48. Receding Galaxy. Figure 38-11 shows a graph of intensity versus wavelength for light reaching Earth from galaxy NGC 7319, which is about 3×10^8 light-years away. The most intense light is emitted by the oxygen in that galaxy. In a laboratory, that emission is at wavelength $\lambda = 513$ nm, but in the light from NGC 7319 it has been shifted to 525 nm due to the Doppler effect. (Indeed, all the emissions from that galaxy have been shifted.) (a) According to special relativity Doppler shift theory, what is the radial speed of galaxy NGC 7319 relative to Earth? (b) Is the relative motion toward or away from Earth?

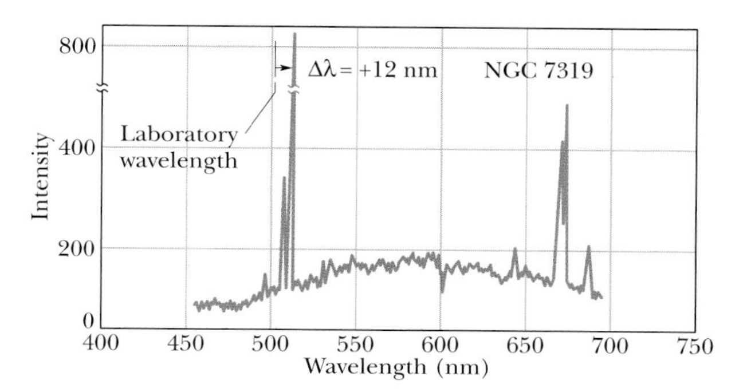

FIGURE 38-11 ■ Problem 48

Additional Problems

49. Exodus from Earth. A billion years from now our Sun will increase its heat, destroying life on Earth. Still later the sun will expand as a red giant, swallowing the Earth and annihilating any remaining life on all planets in the solar system. In anticipation of these catastrophes, an advanced Earth civilization a million years from now develops a transporter mechanism that reduces living beings to data and sends the data by radio to planets orbiting younger stars. The living beings on Earth are destroyed by this process but are reconstituted and restored to life on the distant planets. Your descendent Rasmia Kirmani leaves Earth as data at a time we will take to be zero and is quickly reconstituted after arrival of her data set on the planet Zircon, 100 ly distant from Earth. Assume that Earth and Zircon are relatively at rest.

(a) How much does Rasmia age during her outward trip to Zircon?
(b) How much older is Earth and its civilization when Rasmia is resurrected on Zircon?
(c) Rasmia has a productive and happy life on Zircon and dies as a pioneer hero after 150 years living on that planet. How soon after her departure from Earth can Rasmia's obituary be received on Earth?
(d) Over the millennia between our time and then, specialists whom we now call geneticists discover that there is no such thing as a superperson (man or woman), but rather that a minimum variety of genetic types must be maintained and continually recombined (by whatever method is then current) in order to sustain a healthy population. To this end, several dozen healthy individuals are deconstructed on Earth and transported to Zircon, where each individual is quickly reproduced in thousands of copies (using the same data set over and over) in order to populate the planet rapidly. It takes 5 full generations from birth to death, each generation an average of 200 years, to determine whether or not the new population has been successfully established. How soon after transmission of the dozens of original data sets from Earth can Earth's people learn whether or not this project has been successful?

50. Electron in Orbit. Use Newtonian mechanics to calculate the speed of an electron in the lowest Bohr orbit, which has one quantum of angular momentum:

$$mvr = \hbar = \frac{h}{2\pi}.$$

Carry out this calculation for (a) hydrogen ($Z = 1$) and (b) uranium ($Z = 92$). Insofar as the Bohr model of the atom can be trusted, is relativity required to find the correct answer for (c) hydrogen, (d) uranium?

51. Super Cosmic Rays. The Giant Shower Array detector, spread over 100 square kilometers in Japan, detects pulses of particles from cosmic rays. Each detected pulse is assumed to originate in a single high-energy cosmic proton that strikes the top of the Earth's atmosphere. The highest energy of a single cosmic ray proton inferred from the data is 10^{20} eV. How long would it take that proton to cross our galaxy (10^5 light-years in diameter) as recorded on the wristwatch of the proton? (The answer is not zero!)

52. Synchronization by a Traveling Clock. Evelyn Brown does not approve of our latticework of rods and clocks and the use of a light flash to synchronize them.

(a) "I can synchronize my clocks in any way I choose!" she exclaims. Is she right?

(b) Evelyn wants to synchronize two identical clocks, called Big Ben and Little Ben, which are at rest with respect to one another and separated by one million kilometers in their rest frame. She uses a third clock, identical in construction with the first two, that travels with constant velocity between them. As her moving clock passes Big Ben, it is set to read the same time as Big Ben. When the moving clock passes Little Ben, that outpost clock is set to read the same time as the traveling clock. "Now Big Ben and Little Ben are synchronized," says Evelyn Brown. Is Evelyn's method correct?

(c) After Evelyn completes her synchronization of Little Ben by her method, how does the reading of Little Ben compare with the reading of a nearby clock on a latticework at rest with respect to Big Ben (and Little Ben) and synchronized by our standard method using a light flash? Evaluate in milliseconds any difference between the reading on Little Ben and the nearby lattice clock in the case that Evelyn's traveling clock moved at a constant velocity of 500 000 kilometers per hour from Big Ben to Little Ben.

(d) Evaluate the difference in the reading between the Evelyn-Brown-synchronized Little Ben and the nearby lattice clock when Evelyn's synchronizing traveling clock moves 1000 times as fast as the speed given in part (c).

53. Down with Relativity! Sara Settlemyer is an intelligent layperson who carefully reads articles about science in the public press. She has the objections to relativity listed below. Respond to each of Sara's objections clearly, decisively, and politely—without criticizing her!

(a) "Observer A says that observer B's clock runs slow, while B says that A's clock runs slow. This is a logical contradiction. Therefore relativity should be abandoned."

(b) "Observer A says that B's meter sticks are contracted along their direction of relative motion. B says that A's meter sticks are contracted. This is a logical contradiction. Therefore relativity should be abandoned."

(c) "Anybody with common sense knows that travel at high speed in the direction of a receding light pulse decreases the speed with which the pulse recedes. Hence a flash of light cannot have the same speed for observers in relative motion. With this disproof of the Principle of Relativity, all of relativity collapses."

(d) "Relativity is preoccupied with how we *observe* things, not with what is *really* happening. Therefore relativity is not a scientific theory, since science deals with *reality*."

(e) "Relativity offers no way to describe an event without coordinates, and no way to speak about coordinates without referring to one or another particular reference frame. However, physical events have an existence independent of all choice of coordinates and reference frames. Therefore the special relativity you talk about in this chapter cannot be the most fundamental theory of events and the relation between events."

54. The Photon as a Zero-Mass Particle. A **photon,** the quantum of light, can be considered to be a zero-mass particle.

(a) Using this definition and Eq. 38-18, show that the relation between energy and momentum for the photon is $E = |pc|$, where the "absolute value" vertical lines ensure that energy is positive.

(b) A π^o meson decays rapidly into two gamma rays (high-energy photons). In the rest frame of the original π^o meson, what are the relative directions of the two outgoing photons?

(c) If the mass of the π^o meson is 135 MeV/c^2, what is the energy of each outgoing gamma ray?

55. Pair Production with Gamma Rays. Two gamma rays of equal energy E_p and equal and opposite momenta are incident on a nucleus. (See Figure 38-12.) The collision leads to annihilation of the gamma rays and creation of an electron–positron pair. The lowest energy (the "threshold energy") of incident photons for this production leaves the resulting electron and positron at rest with respect to the nucleus. (The nucleus acts as midwife to this birth and is not changed by the interaction.)

(a) What is the threshold energy E_p of each photon for this creation to take place?

(b) Generalize Eq. 38-18 to *define* the mass M_s of a system of particles, given the total energy E_s and net momentum p_s of the system:

$$M_s^2c^4 \equiv E_s^2 - p_s^2c^2.$$

What is the mass M_s of the system of particles after the collision? Before the collision?

(c) *Mass without mass?* Now let the "nuclear" mass m become less and less. In the limit $m \to 0$, what is the mass of the system after the collision? Before the collision? Before the collision, you apparently have a system with mass composed of "particles," each of which has zero mass. Does this make sense?

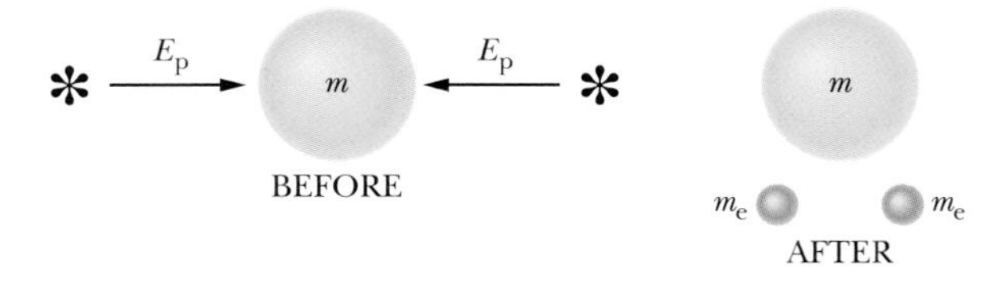

FIGURE 38-12 ■ Problem 55.

56. Resonant Absorption of a Gamma Ray. A gamma ray (an energetic photon) falls on a nucleus of initial mass m, initially at rest. The energy E_p of the incoming gamma ray matches the energy separation between the lowest energy of the nucleus and its first excited state, so the incident photon is absorbed. We want to know the mass m^* of the excited nucleus. (see Fig. 38-13.)

(a) Show that the conservation of energy and momentum equations are, in an obvious notation:

$$E_p + mc^2 = E_{m^*}$$

and
$$\frac{E_p}{c} = p_{m^*} = \frac{(E_{m^*}^2 - m^{*2}c^4)^{1/2}}{c}.$$

(b) Combine the two conservation equations to find an expression for m^* as a function of E_p, m, and c.
(c) Show that for very small values of E_p the limiting result is $m^* = m$. Explain why this limiting result is reasonable.

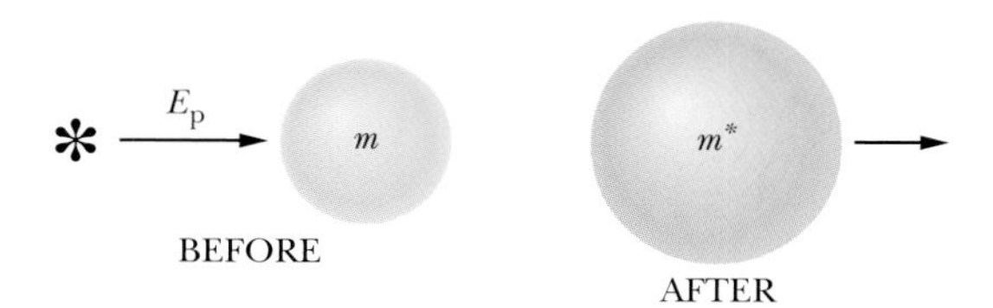

FIGURE 38-13 ■ Problem 56.

57. Photon Braking. A radioactive nucleus of known initial mass M and known initial total energy E_M emits a gamma ray (high-energy photon) in the direction of its motion, drops to its stable nonradioactive state of known mass m, and comes to rest. (see Fig. 38-14). Find an expression for the total energy E_M of the incoming nucleus. The unknown energy E_p of the outgoing gamma ray should not appear in your expression.

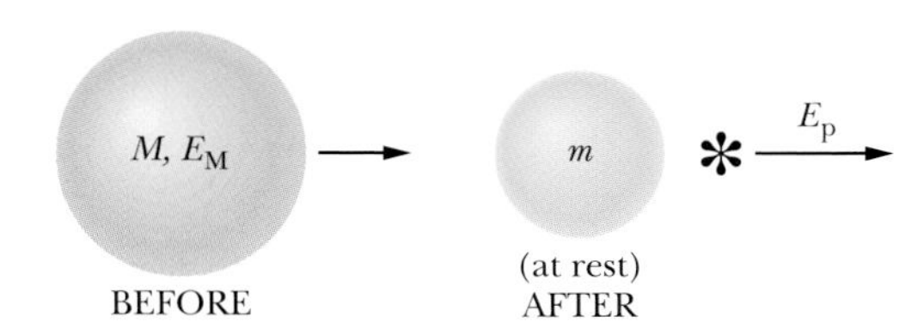

FIGURE 38-14 ■ Problem 57.

58. Limo and Garage Paradox. Review Problem 40, in which we concluded that a limo of proper length 30 m can fit into a garage of proper length 6 m with room to spare. This result is possible because the speeding limo is observed by Garageman to be Lorentz-contracted. Carman protests that in the rest frame of the limo (in which the limo is its full proper length) it is the *garage* that is Lorentz-contracted. As a result, he claims, there is no possibility whatever that the limo can fit into the garage. What could be the possible basis for resolving this paradox? (*Hint:* Think about the space and time locations of two events: event A, front garage door closes and event B, rear garage door opens.)

59. Twin Paradox. The famous **twin paradox** is often introduced as follows: Two identical twins grow up together on Earth. When they reach adulthood, one twin zooms to a distant star and returns to find her stay-at-home sister much older than she is. Thus far no paradox. But Alexis Allen formulates the Twin Paradox for us:

> "The theory of special relativity tells us that all motion is relative. With respect to the traveling twin, the Earth-bound twin moves away and then returns. Therefore it is the Earth-bound twin who should be younger than the 'traveling' twin. But when they meet again at the same place, it cannot possibly be that each twin is younger than the other twin. This Twin Paradox disproves relativity."
>
> The paradox is usually resolved by realizing that the traveling twin turns around. Everyone agrees which twin turns around, since the reversal of direction slams the poor traveler against the bulkhead of the decelerating starship, breaking her collarbone. The turnaround, evidenced by the broken collarbone, destroys the symmetry required for the paradox to hold. Good-bye Twin Paradox!
>
> Still, Alexis's father Cyril Allen has his doubts about this resolution of the paradox. "Your solution is extremely unsatisfying. It forces me to ask: What if the retro-rockets malfunction and will not fire at all to slow me down as I approach a distant star a thousand light-years from Earth? Then I cannot even stop at that star, much less turn around and head back to Earth. Instead, I continue moving away from Earth forever at the original constant speed. Does this mean that as I pass the distant star, one thousand light-years from Earth, it is no longer possible to say that I have aged less than my Earth-bound twin? But if not, then I would never have even gotten to the distant star at all during my hundred-year lifetime! Your resolution of the Twin Paradox is insufficient and unsatisfying."
>
> Write a half-page response to Cyril Allen, answering his objections politely but decisively.

60. The Runner on the Train Paradox. A train moves at 10 km/h along the track. A passenger sprints toward the rear of the train at 10 km/h with respect to the train. Our knee-jerk motto says that the train clocks "run slow" with respect to clocks on the track, and the runner's watch "runs slow" with respect to train clocks. Therefore the runner's watch should "run doubly slow" with respect to clocks on the track. But the runner is at rest with respect to the track. What gives? (This example illustrates the danger of the simple knee-jerk motto "Moving clocks run slow.")

61. A Summer Evening's Fantasy. You and a group of female and male friends stand outdoors at dusk watching the Sun set and noticing the planet Venus in the same direction as the Sun. An alien ship lands beside you at the same instant that you see the Sun explode. The aliens admit that earlier they shot a laser flash at the Sun, which caused the explosion. They warn that the Sun's explosion emitted an immense pulse of particles that will blow away Earth's atmosphere. In confirmation, a short time after the aliens land you notice Venus suddenly change color. You and your friends plead with the aliens to take your group away from Earth in order to establish the human gene pool elsewhere. They agree. Describe the conditions under which your escape plan will succeed. Be specific and use numbers. Assume that the Sun is 8 light-minutes from Earth and Venus is 2 light-minutes from Earth.

APPENDIX A The International System of Units (SI)*

1 SI Base Units

1. The SI Base Units

Quantity	Name	Symbol	Definition
length	meter	m	"... the length of the path traveled by light in vacuum in 1/299 792 458 of a second." (1983)
mass	kilogram	kg	"... this prototype [a certain platinum–iridium cylinder] shall henceforth be considered to be the unit of mass." (1889)
time	second	s	"... the duration of 9 192 631 770 periods of the radiation corresponding to the transition between the two hyperfine levels of the ground state of the cesium-133 atom." (1967)
electric current	ampere	A	"... that constant current which, if maintained in two straight parallel conductors of infinite length, of negligible circular cross section, and placed 1 meter apart in vacuum, would produce between these conductors a force equal to 2×10^{-7} newton per meter of length." (1946)
thermodynamic temperature	kelvin	K	"... the fraction 1/273.16 of the thermodynamic temperature of the triple point of water." (1967)
amount of substance	mole	mol	"... the amount of substance of a system which contains as many elementary entities as there are atoms in 0.012 kilogram of carbon-12." (1971)
luminous intensity	candela	cd	"... the luminous intensity, in a given direction, of a source that emits monochromatic radiation of frequency 540×10^{12} hertz and that has a radiant intensity in that direction of 1/683 watt per steradian." (1979)

2 The SI Supplementary Units

2. The SI Supplementary Units

Quantity	Name of Unit	Symbol
plane angle	radian	rad
solid angle	steradian	sr

*Adapted from "The International System of Units (SI)," National Bureau of Standards Special Publication 330, 2001 edition. The definitions above were adopted by the General Conference of Weights and Measures, an international body, on the dates shown. In this book we do not use the candela.

3 Some SI Derivations

3. Some SI Derived Units

Quantity	Name of Unit	Symbol	In Terms of other SI Units
area	square meter	m^2	
volume	cubic meter	m^3	
frequency	hertz	Hz	s^{-1}
mass density (density)	kilogram per cubic meter	kg/m^3	
speed, velocity	meter per second	m/s	
rotational velocity	radian per second	rad/s	
acceleration	meter per second per second	m/s^2	
rotational acceleration	radian per second per second	rad/s^2	
force	newton	N	$kg \cdot m/s^2$
pressure	pascal	Pa	N/m^2
work, energy, quantity of heat	joule	J	$N \cdot m$
power	watt	W	J/s
quantity of electric charge	coulomb	C	$A \cdot s$
potential difference, electromotive force	volt	V	W/A
electric field strength	volt per meter (or newton per coulomb)	V/m	N/C
electric resistance	ohm	Ω	V/A
capacitance	farad	F	$A \cdot s/V$
magnetic flux	weber	Wb	$V \cdot s$
inductance	henry	H	$V \cdot s/A$
magnetic flux density	tesla	T	Wb/m^2
magnetic field strength	ampere per meter	A/m	
entropy	joule per kelvin	J/K	
specific heat	joule per kilogram kelvin	$J/(kg \cdot K)$	
thermal conductivity	watt per meter kelvin	$W/(m \cdot K)$	
radiant intensity	watt per steradian	W/sr	

4 Mathematical Notation

Poorly chosen mathematical notation can be a source of considerable confusion to those trying to learn and to do physics. For example, ambiguity in the meaning of a mathematical symbol can prevent a reader from understanding the meaning of a crucial relationship. It is also difficult to solve problems when the symbols used ot represent different quantities are not distinctive. In this text we have taken special care to use mathematical notation in ways that allow important distinctions to be easily visible both on the printed page and in handwritten work.

An excellent starting point for clear mathematical notation is the U.S. National Institute of Standard and Technology's Special Publication 811 (SP 811), *Guide for the Use of the International System of Units (SI)*, available at http://physics.nist.gov/cuu/Units/bibliography.html. In addition to following the National Institute guidelines, we have made a number of systematic choices to facilitate the translation of printed notation into handwritten mathematics. For example:

- Instead of making vectors bold, vector quantities (even in one dimension) are denoted by an arrow above the symbol. So printed equations look like handwritten equations. Example: $\vec{v}$ rather than $\boldsymbol{v}$ is used to denote an instantaneous velocity.

- In general, each vector component has an explicit subscript denoting that it represents the component along a chosen coordinate axis. The one exception is the position vector, $\vec{r}$, whose components are simply written as x, y, and z. For example, $\vec{r} = x\hat{\imath} + y\hat{\jmath} + z\hat{k}$, whereas, $\vec{v} = v_x\hat{\imath} + v_y\hat{\jmath} + v_z\hat{k}$.
- To emphasize the distinction between a vector's components and its magnitude, we write the magnitude of a vector, such as $\vec{F}$, as $|\vec{F}|$. However, when it is obvious that a magnitude is being described, we use the plain symbol (such as F with no coordinate subscript) to denote a vector's magnitude.
- We often choose to spell out the names of objects that are associated with mathematical variables—writing, for example, $\vec{v}_{\text{ball}}$ and not $\vec{v}_{\text{b}}$ for the velocity of a ball.
- Numerical subscripts most commonly denote sequential times, positions, velocities, and so on. For example, x_1 is the x-component of the position of some object at time t_1, whereas x_2 is the value of that parameter at some later time t_2. We have avoided using the subscript zero to denote initial values, as in x_0 to denote "the initial position along the x axis," to emphasize that *any* time can be chosen as the initial time for consideration of the subsequent time evolution of a system.
- To avoid confusing the numerical time sequence labels with object labels, we prefer to use capital letters as object labels. For example, we would label two particles A and B rather than 1 and 2. Thus, $\vec{p}_{\text{A}\,1}$ and $\vec{p}_{\text{B}\,1}$ would represent the translational momenta of two particles before a collision whereas $\vec{p}_{\text{A}\,2}$ and $\vec{p}_{\text{B}\,2}$ would be their momenta after a collision.
- To avoid excessively long strings of subscripts, we have made the unconventional choice to write all adjectival labels as *super*scripts. Thus, Newton's Second Law is written $\vec{F}^{\text{net}} = m\vec{a}$ whereas the sum of the forces acting on a certain object might be written as $\vec{F}^{\text{net}} = \vec{F}^{\text{grav}} + \vec{F}^{\text{app}}$. To avoid confusion with mathematical exponents, an adjectival label is never a single letter.
- Following a usage common in contemporary physics, the time average of a variable $\vec{v}$ is denoted as $\langle \vec{v} \rangle$ and not as $\vec{v}_{\text{avg}}$.
- Physical constants such as e, c, g, G, are all **positive** scalar quantities.

5 Significant Figures and the Precision of Numerical Results

Quoting the result of a calculation or a measurement to the correct number of significant figures is merely a way of telling your reader roughly how precise you believe the result to be. Quoting too many significant figures overstates the precision of your result and quoting too few implies less precision than the result may actually possess. So how many significant figures should you quote when reporting your result.

Determining Significant Figures

Before answering the question of how many significant figures to quote, we need to have a clear method for determining how many significant figures a reported number has. The standard method is quite simple:

> **METHOD FOR COUNTING SIGNIFICANT FIGURES:** Read the number from left to right, and count the first nonzero digit and all the digits (zero or not) to the right of it as significant.

Using this rule, 350 mm, 0.000350 km, and 0.350 m each has *three* significant figures. In fact, each of these numbers merely represents the same distance, expressed in different units. As you can see from this example, the number of *decimal places* that a number has is *not* the same as its number of *significant figures*. The first of these distances has zero decimal places, the second has six decimal places, and the third has three, yet all three of these numbers have three significant figures.

One consequence of this method is especially worth noting. Trailing zeros count as significant figures. For example, 2700 m/s has four significant figures. If you really meant it to have only three significant figures, you would have to write it either as 2.70 km/s (changing the unit) or 2.70×10^3 m/s (using scientific notation.)

A Simple Rule for Reporting Significant Figures in a Calculated Result

Now that you know how to count significant figures, how many should the result of a calculation have? A simple rule that will work in most calculations is:

SIGNIFICANT FIGURES IN A CALCULATED RESULT: The common practice is to quote the result of a calculation to the number of significant figures of the *least* precise number used in the calculation.

Although this simple rule will often either understate or (less frequently) overstate the precision of a result, it still serves as a good rule-of-thumb for everyday numerical work. In introductory physics you will only rarely encounter data that are known to better than two, three, or four significant figures. This simple rule then tells you that you can't go very far wrong if you round off all your final results to three significant figures.

There are two situations in which the simple rule should *not* be applied to a calculation. One is when an exact number is involved in the calculation and another is when a calculation is done in parts so that intermediate results are used.

1. ***Using Exact Data*** There are some obvious situations in which a number used in a calculation is exact. Numbers based on counting items are exact. For example, if you are told that there are 5 people on an elevator, there are exactly 5 people, not 4.7 or 5.1. Another situation arises when a number is exact by definition. For example, the conversion factor 2.54 cm/inch does *not* have three significant figures because the inch is *defined* to be exactly 2.5400000 . . . cm. *Data that are known exactly should not be included when deciding which of the original data has the fewest significant figures.*
2. ***Significant Figures in Intermediate Results*** Only the final result at the end of your calculation should be rounded using the simple rule. Intermediate results should never be rounded. Spreadsheet software takes care of this for you, as does your calculator if you store your intermediate results in its memory rather than writing them down and then rekeying them. If you must write down intermediate results, keep a few more significant figures than your final result will have.

Understanding and Refining the Simple Significant Figure Rule

Quoting the result of a calculation or measurement to the correct number of significant figures is a way of indicating its precision. You need to understand what limits the precision of data before you fully understand how to use the simple rule or its exceptions.

Absolute Precision There are two ways of talking about precision. First there is *absolute precision*, which tells you explicitly the smallest scale division of the measurement. It's always quoted in the same units as the measured quantity. For example, saying "I measured the length of the table to the nearest centimeter" states the absolute precision of the measurement. The absolute precision tells you how many *decimal places* the measurement has; it alone does not determine the number of significant figures. Example: if a table is 235 cm long, then 1 cm of absolute precision translates into three significant figures. On the other hand, if a table is for a doll's house and is only 8 cm long, then the same 1 cm of absolute precision has only one significant figure.

Relative Precision Because of this problem with absolute precision, scientists often prefer to describe the precision of data *relative* to the size of the quantity being measured. To use the previous examples, the *relative precision* of the length of the real table in the previous example is 1 cm out of 235 cm. This is usually stated as a ratio (1 part in 235) or as a percentage ($1/235 = 0.004255 \approx 0.4\%$). In the case of the toy table, the same 1 cm of absolute precision yields a relative precision of only 1 part in 8 or $1/8 = 0.125 = 12.5\%$.

Inconsistencies between Significant Figures and Relative Precision There is an inconsistency that goes with using a certain number of significant figures to express relative precision. Quoted to the same number of significant figures, the relative precision of results can be quite different. For example, 13 cm and 94 cm both have two significant figures. Yet the first is specified to only 1 part in 13 or $1/13 \approx 10\%$, whereas the second is known to 1 part in 94 or $1/94 \approx 1\%$. This bias toward greater relative precision for results with larger first significant figures is one weakness of using significant figures to track the precision of calculated results. You can partially address this problem, by including one more significant figure than the simple rule suggests, when the final result of a calculation has a 1 as its first significant figure.

Multiplying and Dividing When multiplying or dividing numbers, the *relative* precision of the result cannot exceed that of the least precise number used. Since the number of significant figures in the result tells us its relative precision, the simple rule is all that you need when you multiply or divide. For example, the area of a strip of paper of measured size is 280 cm by 2.5 cm would be correctly reported, according to the simple rule, as 7.0×10^2 cm^2. This result has only two significant figures since the less precise measurement, 2.5 cm, that went into the calculation had only two significant figures. Reporting this result as 700 cm^2 would not be correct since this result has three significant figures, exceeding the relative precision of the 2.5 cm measurement.

Addition and Subtraction When adding or subtracting, you line up the decimal points before you add or subtract. This means that it's the *absolute* precision of the least precise number that limits the precision of the sum or the difference. This can lead to some exceptions to the simple rule. For example, adding 957 cm and 878 cm yields 1835 cm. Here the result is reliable to an absolute precision of about 1 cm since both of the original distances had this reliability. But the result then has four significant figures whereas each of the original numbers had only three. If, on the other hand, you take the difference between these two distances you get 79 cm. The difference is still reliable to about 1 cm, but that absolute precision now translates into only two significant figures worth of relative precision. So, you should be careful when adding or subtracting, since addition can actually increase the relative precision of your result and, more important, subtraction can reduce it.

Evaluating Functions What about the evaluation of functions? For example, how many significant figures does the sin(88.2°) have? You can use your calculator to answer this question. First use your calculator to note that $\sin(88.2°) = 0.999506$. Now add 1 to the least significant decimal place of the argument of the function and evaluate it again. Here this gives $\sin(88.3°) = 0.999559$. Take the last significant figure in the result to be *the first one from the left that changed* when you repeated the calculation. In this example the first digit that changed was the 0; it became a 5 (the second 5) in the recalculation. So, using the empirical approach gives you five significant figures.

APPENDIX B

Some Fundamental Constants of Physics*

Constant	Symbol	Computational Value	Best (1998) Value	
			Value[a]	Uncertainty[b]
Speed of light in a vacuum	c	3.00×10^{8} m/s	2.997 924 58	exact
Elementary charge	e	1.60×10^{-19} C	1.602 176 462	0.039
Gravitational constant	G	6.67×10^{-11} $m^3/s^2 \cdot kg$	6.673	1500
Universal gas constant	R	8.31 $J/mol \cdot K$	8.314 472	1.7
Avogadro constant	N_A	6.02×10^{23} mol^{-1}	6.022 141 99	0.079
Boltzmann constant	k_B	1.38×10^{-23} J/K	1.380 650 3	1.7
Stefan–Boltzmann constant	σ	5.67×10^{-8} $W/m^2 \cdot K^4$	5.670 400	7.0
Molar volume of ideal gas at STP[d]	V_m	2.27×10^{-2} m^3/mol	2.271 098 1	1.7
Electric constant (permittivity)	ϵ_0	8.85×10^{-12} $C^2/N \cdot m^2$	8.854 187 817 62	exact
Coulomb constant	$k = 1/4\pi\epsilon_0$	8.99×10^{9} $N \cdot m^2/C^2$	8.987 551 787	5×10^{-10}
Magnetic constant (permeability)	μ_0	1.26×10^{-6} N/A^2	1.256 637 061 43	exact
Planck constant	h	6.63×10^{-34} $J \cdot s$	6.626 068 76	0.078
Electron mass[c]	m_e	9.11×10^{-31} kg	9.109 381 88	0.079
		5.49×10^{-4} u	5.485 799 110	0.0021
Proton mass[c]	m_p	1.67×10^{-27} kg	1.672 621 58	0.079
		1.0073 u	1.007 276 466 88	$1.3 \times .10^{-4}$
Ratio of proton mass to electron mass	m_p/m_e	1840	1836.152 667 5	0.0021
Electron charge-to-mass ratio	e/m_e	1.76×10^{11} C/kg	1.758 820 174	0.040
Neutron mass[c]	m_n	1.68×10^{-27} kg	1.674 927 16	0.079
		1.0087 u	1.008 664 915 78	5.4×10^{-4}
Hydrogen atom mass[c]	m_{1H}	1.0078 u	1.007 825 031 6	0.0005
Deuterium atom mass[c]	m_{2H}	2.0141 u	2.014 101 777 9	0.0005
Helium atom mass[c]	m_{4He}	4.0026 u	4.002 603 2	0.067
Muon mass	m_μ	1.88×10^{-28} kg	1.883 531 09	0.084
Electron magnetic moment	μ_e	9.28×10^{-24} J/T	9.284 763 62	0.040
Proton magnetic moment	μ_p	1.41×10^{-26} J/T	1.410 606 663	0.041
Bohr magneton	μ_B	9.27×10^{-24} J/T	9.274 008 99	0.040
Nuclear magneton	μ_N	5.05×10^{-27} J/T	5.050 783 17	0.040
Bohr radius	r_B	5.29×10^{-11} m	5.291 772 083	0.0037
Rydberg constant	R	1.10×10^{7} m^{-1}	1.097 373 156 854 8	7.6×10^{-6}
Electron Compton wavelength	λ_C	2.43×10^{-12} m	2.426 310 215	0.0073

[a]Values given in this column should be given the same unit and power of 10 as the computational value.
[b]Parts per million.
[c]Masses given in u are in unified atomic mass units, where $1\ u = 1.660\ 538\ 73 \times 10^{-27}$ kg.
[d]STP means standard temperature and pressure: 0°C and 1.0 atm (0.1 MPa).

*The values in this table were selected from the 1998 CODATA recommended values (www.physics.nist.gov).

APPENDIX C Some Astronomical Data

Some Distances from Earth			
To the Moon*	3.82×10^8 m	To the center of our galaxy	2.2×10^{20} m
To the Sun*	1.50×10^{11} m	To the Andromeda Galaxy	2.1×10^{22} m
To the nearest star (Proxima Centauri)	4.04×10^{16} m	To the edge of the observable universe	$\sim 10^{26}$ m

* Mean distance.

The Sun, Earth, and the Moon

Property	Unit	Sun		Earth	Moon
Mass	kg	1.99×10^{30}		5.98×10^{24}	7.36×10^{22}
Mean radius	m	6.96×10^8		6.37×10^6	1.74×10^6
Mean density	kg/m^3	1410		5520	3340
Free-fall acceleration at the surface	m/s^2	274		9.81	1.67
Escape velocity	km/s	618		11.2	2.38
Period of rotation[a]	—	37 d at poles[b]	26 d at equator[b]	23 h 56 min	27.3 d
Radiation power[c]	W	3.90×10^{26}			

[a] Measured with respect to the distant stars, [b] The Sun, a ball of gas, does not rotate as a rigid body; [c] Just outside Earth's atmosphere solar energy is received, assuming normal incidence, at the rate of 1340 W/m^2.

Some Properties of the Planets

	Mercury	Venus	Earth	Mars	Jupiter	Saturn	Uranus	Neptune	Pluto
Mean distance from Sun, 10^6 km	57.9	108	150	228	778	1430	2870	4500	5900
Period of revolution, y	0.241	0.615	1.00	1.88	11.9	29.5	84.0	165	248
Period of rotation,[a] d	58.7	−243[b]	0.997	1.03	0.409	0.426	−0.451[b]	0.658	6.39
Orbital speed, km/s	47.9	35.0	29.8	24.1	13.1	9.64	6.81	5.43	4.74
Equatorial diameter, km	4880	12 100	12 800	6790	143 000	120 000	51 800	49 500	2300
Mass (Earth = 1)	0.0558	0.815	1.000	0.107	318	95.1	14.5	17.2	0.002
Surface value of g,[c] m/s^2	3.78	8.60	9.78	3.72	22.9	9.05	7.77	11.0	0.5
Escape velocity,[c] km/s	4.3	10.3	11.2	5.0	59.5	35.6	21.2	23.6	1.1

[a] Measured with respect to the distant stars.
[b] Venus and Uranus rotate opposite their orbital motion.
[c] Gravitational acceleration measured at the planet's equator.

APPENDIX D Conversion Factors

Conversion factors may be read directly from these tables. For example, 1 degree = 2.778×10^{-3} revolutions, so $16.7° = 16.7 \times 2.778 \times 10^{-3}$ rev. The SI units are fully capitalized. Adapted in part from G. Shortley and D. Williams, *Elements of Physics*, 1971, Prentice-Hall, Englewood Cliffs, N.J.

Solid Angle

1 sphere = 4π steradians = 12.57 steradians

Plane Angle

	°	′	″	RADIAN	rev
1 degree =	1	60	3600	1.745×10^{-2}	2.778×10^{-3}
1 minute =	1.667×10^{-2}	1	60	2.909×10^{-4}	4.630×10^{-5}
1 second =	2.778×10^{-4}	1.667×10^{-2}	1	4.848×10^{-6}	7.716×10^{-7}
1 RADIAN =	57.30	3438	2.063×10^{5}	1	0.1592
1 revolution =	360	2.16×10^{4}	1.296×10^{6}	6.283	1

Length

	cm	METER	km	in.	ft	mi
1 centimeter =	1	10^{-2}	10^{-5}	0.3937	3.281×10^{-2}	6.214×10^{-6}
1 METER =	100	1	10^{-3}	39.37	3.281	6.214×10^{-4}
1 kilometer =	10^{5}	1000	1	3.937×10^{4}	3281	0.6214
1 inch =	2.540	2.540×10^{-2}	2.540×10^{-5}	1	8.333×10^{-2}	1.578×10^{-5}
1 foot =	30.48	0.3048	3.048×10^{-4}	12	1	1.894×10^{-4}
1 mile =	1.609×10^{5}	1609	1.609	6.336×10^{4}	5280	1

1 angström = 10^{-10} m
1 nautical mile = 1852 m = 1.151 miles = 6076 ft
1 fermi = 10^{-15} m
1 light-year = 9.460×10^{12} km
1 parsec = 3.084×10^{13} km
1 fathom = 6 ft
1 Bohr radius = 5.292×10^{-11} m
1 yard = 3 ft
1 rod = 16.5 ft
1 mil = 10^{-3} in.
1 nm = 10^{-9} m

Area

	METER^2	cm^2	ft^2	in.^{2}
1 SQUARE METER =	1	10^{4}	10.76	1550
1 square centimeter =	10^{-4}	1	1.076×10^{-3}	0.1550
1 square foot =	9.290×10^{-2}	929.0	1	144
1 square inch =	6.452×10^{-4}	6.452	6.944×10^{-3}	1

key: 1 square mile = 2.788×10^{7} ft^2 = 640 acres; 1 barn = 10^{-28} m^2; 1 acre = 43 560 ft^2; 1 hectare = 10^{4} m^2 = 2.471 acres.

Volume

	METER^3	cm^3	L	ft^3	in.^{3}
1 CUBIC METER =	1	10^{6}	1000	35.31	6.102×10^{4}
1 cubic centimeter =	10^{-6}	1	1.000×10^{-3}	3.531×10^{-5}	6.102×10^{-2}
1 liter =	1.000×10^{-3}	1000	1	3.531×10^{-2}	61.02
1 cubic foot =	2.832×10^{-2}	2.832×10^{4}	28.32	1	1728
1 cubic inch =	1.639×10^{-5}	16.39	1.639×10^{-2}	5.787×10^{-4}	1

key: 1 U.S. fluid gallon = 4 U.S. fluid quarts = 8 U.S. pints = 128 U.S. fluid ounces = 231 in.^{3} 1 British imperial gallon = 277.4 in.^{3} = 1.201 U.S. fluid gallons.

Mass

Quantities in the colored areas are not mass units but are often used as such. When we write, for example, 1 kg "=" 2.205 lb, this means that a kilogram is a *mass* that *weighs* 2.205 pounds at a location where g has the standard value of 9.80665 m/s^2.

g	KILOGRAM	slug	u	oz	lb	ton
1 gram = 1	0.001	6.852×10^{-5}	6.022×10^{23}	3.527×10^{-2}	2.205×10^{-3}	1.102×10^{-6}
1 KILOGRAM = 1000	1	6.852×10^{-2}	6.022×10^{26}	35.27	2.205	1.102×10^{-3}
1 slug = 1.459×10^{4}	14.59	1	8.786×10^{27}	514.8	32.17	1.609×10^{-2}
1 atomic mass unit = 1.661×10^{-24}	1.661×10^{-27}	1.138×10^{-28}	1	5.857×10^{-26}	3.662×10^{-27}	1.830×10^{-30}
1 ounce = 28.35	2.835×10^{-2}	1.943×10^{-3}	1.718×10^{25}	1	6.250×10^{-2}	3.125×10^{-5}
1 pound = 453.6	0.4536	3.108×10^{-2}	2.732×10^{26}	16	1	0.0005
1 ton = 9.072×10^{5}	907.2	62.16	5.463×10^{29}	3.2×10^{4}	2000	1

1 metric ton = 1000 kg

Time

y	d	h	min	SECOND
1 year = 1	365.25	8.766×10^{3}	5.259×10^{5}	3.156×10^{7}
1 day = 2.738×10^{-3}	1	24	1440	8.640×10^{4}
1 hour = 1.141×10^{-4}	4.167×10^{-2}	1	60	3600
1 minute = 1.901×10^{-6}	6.944×10^{-4}	1.667×10^{-2}	1	60
1 SECOND = 3.169×10^{-8}	1.157×10^{-5}	2.778×10^{-4}	1.667×10^{-2}	1

Speed

ft/s	km/h	METER/SECOND	mi/h	cm/s
1 foot per second = 1	1.097	0.3048	0.6818	30.48
1 kilometer per hour = 0.9113	1	0.2778	0.6214	27.78
1 METER per SECOND = 3.281	3.6	1	2.237	100
1 mile per hour = 1.467	1.609	0.4470	1	44.70
1 centimeter per second = 3.281×10^{-2}	3.6×10^{-2}	0.01	2.237×10^{-2}	1

1 knot = 1 nautical mi/h = 1.688 ft/s 1 mi/min = 88.00 ft/s = 60.00 mi/h

Force

dyne	NEWTON	lb	pdl
1 dyne = 1	10^{-5}	2.248×10^{-6}	7.233×10^{-5}
1 NEWTON = 10^{5}	1	0.2248	7.233
1 pound = 4.448×10^{5}	4.448	1	32.17
1 poundal = 1.383×10^{4}	0.1383	3.108×10^{-2}	1

1 ton = 2000 lb

Pressure

atm	dyne/cm^2	inch of water	cm Hg	PASCAL	lb/in.^{2}	lb/ft^2
1 atmosphere = 1	1.013×10^6	406.8	76	1.013×10^5	14.70	2116
1 dyne per centimeter^2 = 9.869×10^{-7}	1	4.015×10^{-4}	7.501×10^{-5}	0.1	1.405×10^{-5}	2.089×10^{-3}
1 inch of water[a] at 4°C = 2.458×10^{-3}	2491	1	0.1868	249.1	3.613×10^{-2}	5.202
1 centimeter of mercury[a] at 0°C = 1.316×10^{-2}	1.333×10^4	5.353	1	1333	0.1934	27.85
1 PASCAL = 9.869×10^{-6}	10	4.015×10^{-3}	7.501×10^{-4}	1	1.450×10^{-4}	2.089×10^{-2}
1 pound per inch^2 = 6.805×10^{-2}	6.895×10^4	27.68	5.171	6.895×10^3	1	144
1 pound per foot^2 = 4.725×10^{-4}	478.8	0.1922	3.591×10^{-2}	47.88	6.944×10^{-3}	1

[a] Where the acceleration of gravity has the standard value of 9.80665 m/s^2.

1 bar = 10^6 dyne/cm^2 = 0.1 MPa   1 millibar = 10^3 dyne/cm^2 = 10^2 Pa   1 torr = 1 mm Hg

Energy, Work, Heat

Btu	erg	ft·lb	hp·h	JOULE	cal	kW·h	eV	MeV
1 British thermal unit = 1	1.055×10^{10}	777.9	3.929×10^{-4}	1055	252.0	2.930×10^{-4}	6.585×10^{21}	6.585×10^{15}
1 erg = 9.481×10^{-11}	1	7.376×10^{-8}	3.725×10^{-14}	10^{-7}	2.389×10^{-8}	2.778×10^{-14}	6.242×10^{11}	6.242×10^5
1 foot-pound = 1.285×10^{-3}	1.356×10^7	1	5.051×10^{-7}	1.356	0.3238	3.766×10^{-7}	8.464×10^{18}	8.464×10^{12}
1 horsepower-hour = 2545	2.685×10^{13}	1.980×10^6	1	2.685×10^6	6.413×10^5	0.7457	1.676×10^{25}	1.676×10^{19}
1 JOULE = 9.481×10^{-4}	10^7	0.7376	3.725×10^{-7}	1	0.2389	2.778×10^{-7}	6.242×10^{18}	6.242×10^{12}
1 calorie = 3.969×10^{-3}	4.186×10^7	3.088	1.560×10^{-6}	4.186	1	1.163×10^{-6}	2.613×10^{19}	2.613×10^{13}
1 kilowatt hour = 3413	3.600×10^{13}	2.655×10^6	1.341	3.600×10^6	8.600×10^5	1	2.247×10^{25}	2.247×10^{19}
1 electron-volt = 1.519×10^{-22}	1.602×10^{-12}	1.182×10^{-19}	5.967×10^{-26}	1.602×10^{-19}	3.827×10^{-20}	4.450×10^{-26}	1	10^{-6}
1 million electron-volts = 1.519×10^{-16}	1.602×10^{-6}	1.182×10^{-13}	5.967×10^{-20}	1.602×10^{-13}	3.827×10^{-14}	4.450×10^{-20}	10^{-6}	1

Power

Btu/h	ft·lb/s	hp	cal/s	kW	WATT
1 British thermal unit per hour = 1	0.2161	3.929×10^{-4}	6.998×10^{-2}	2.930×10^{-4}	0.2930
1 foot-pound per second = 4.628	1	1.818×10^{-3}	0.3239	1.356×10^{-3}	1.356
1 horsepower = 2545	550	1	178.1	0.7457	745.7
1 calorie per second = 14.29	3.088	5.615×10^{-3}	1	4.186×10^{-3}	4.186
1 kilowatt = 3413	737.6	1.341	238.9	1	1000
1 WATT = 3.413	0.7376	1.341×10^{-3}	0.2389	0.001	1

Magnetic Field

gauss	TESLA	milligauss
1 gauss = 1	10^{-4}	1000
1 TESLA = 10^4	1	10^7
1 milligauss = 0.001	10^{-7}	1

1 tesla = 1 weber/meter^2

Magnetic Flux

maxwell	WEBER
1 maxwell = 1	10^{-8}
1 WEBER = 10^8	1

APPENDIX E Mathematical Formulas

Geometry

Circle of radius r: circumference $= 2\pi r$; area $= \pi r^2$.
Sphere of radius r: area $= 4\pi r^2$; volume $= \frac{4}{3}\pi r^3$.
Right circular cylinder of radius r and height h:
area $= 2\pi r^2 + 2\pi rh$; volume $= \pi r^2 h$.
Triangle of base a and altitude h: area $= \frac{1}{2}ah$.

Quadratic Formula

If $ax^2 + bx + c = 0$, then $x = \dfrac{-b \pm \sqrt{b^2 - 4ac}}{2a}$.

Trigonometric Functions of Angle θ

$\sin\theta = \dfrac{y}{r}$ $\cos\theta = \dfrac{x}{r}$

$\tan\theta = \dfrac{y}{x}$ $\cot\theta = \dfrac{x}{y}$

$\sec\theta = \dfrac{r}{x}$ $\csc\theta = \dfrac{r}{y}$

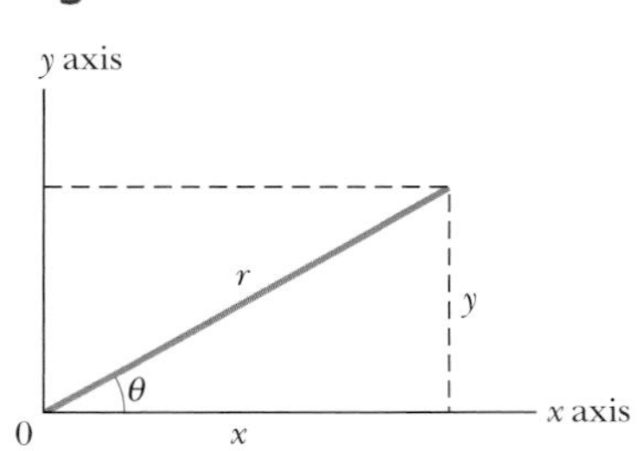

Pythagorean Theorem

In this right triangle,
$a^2 + b^2 = c^2$

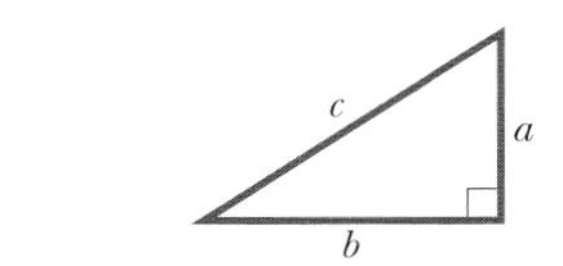

Triangles

Angles are A, B, C
Opposite sides are a, b, c
Angles $A + B + C = 180°$
$\dfrac{\sin A}{a} = \dfrac{\sin B}{b} = \dfrac{\sin C}{c}$
$c^2 = a^2 + b^2 - 2ab\cos C$
Exterior angle $D = A + C$

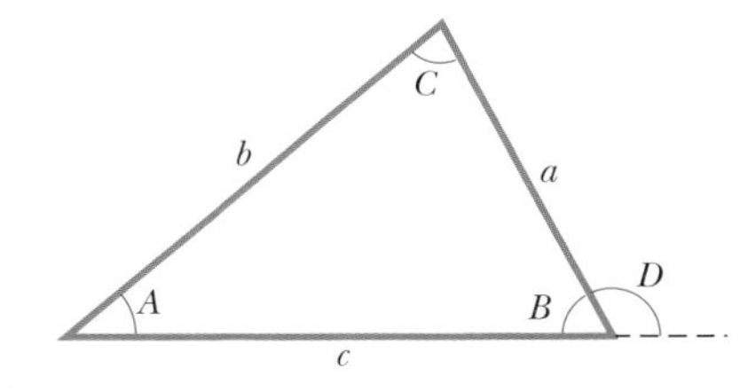

Mathematical Signs and Symbols

$=$ equals
$\approx$ equals approximately
$\sim$ is the order of magnitude of
$\neq$ is not equal to
$\equiv$ is identical to, is defined as
$>$ is greater than ($\gg$ is much greater than)
$<$ is less than ($\ll$ is much less than)
$\geq$ is greater than or equal to (or, is no less than)
$\leq$ is less than or equal to (or, is no more than)
$\pm$ plus or minus
$\propto$ is proportional to
Σ the sum of
$\langle x \rangle$ the average value of x

Trigonometric Identities

$\sin(90° - \theta) = \cos\theta$
$\cos(90° - \theta) = \sin\theta$
$\sin\theta/\cos\theta = \tan\theta$
$\sin^2\theta + \cos^2\theta = 1$
$\sec^2\theta - \tan^2\theta = 1$
$\csc^2\theta - \cot^2\theta = 1$
$\sin 2\theta = 2\sin\theta\cos\theta$
$\cos 2\theta = \cos^2\theta - \sin^2\theta = 2\cos^2\theta - 1 = 1 - 2\sin^2\theta$
$\sin(\alpha \pm \beta) = \sin\alpha\cos\beta \pm \cos\alpha\sin\beta$
$\cos(\alpha \pm \beta) = \cos\alpha\cos\beta \mp \sin\alpha\sin\beta$
$\tan(\alpha \pm \beta) = \dfrac{\tan\alpha \pm \tan\beta}{1 \mp \tan\alpha\tan\beta}$
$\sin\alpha \pm \sin\beta = 2\sin\frac{1}{2}(\alpha \pm \beta)\cos\frac{1}{2}(\alpha \mp \beta)$
$\cos\alpha + \cos\beta = 2\cos\frac{1}{2}(\alpha + \beta)\cos\frac{1}{2}(\alpha - \beta)$
$\cos\alpha - \cos\beta = -2\sin\frac{1}{2}(\alpha + \beta)\sin\frac{1}{2}(\alpha - \beta)$

Binomial Theorem

$$(1 + x)^n = 1 + \frac{nx}{1!} + \frac{n(n-1)x^2}{2!} + \cdots \qquad (x^2 < 1)$$

Exponential Expansion

$$e^x = 1 + x + \frac{x^2}{2!} + \frac{x^3}{3!} + \cdots$$

Logarithmic Expansion

$$\ln(1 + x) = x - \tfrac{1}{2}x^2 + \tfrac{1}{3}x^3 - \cdots \qquad (|x| < 1)$$

Trigonometric Expansions (θ in radians)

$$\sin\theta = \theta - \frac{\theta^3}{3!} + \frac{\theta^5}{5!} - \cdots$$

$$\cos\theta = 1 - \frac{\theta^2}{2!} + \frac{\theta^4}{4!} - \cdots$$

$$\tan\theta = \theta + \frac{\theta^3}{3} + \frac{2\theta^5}{15} + \cdots$$

Cramer's Rule

Two simultaneous equations in unknowns x and y,

$$a_1x + b_1y = c_1 \quad \text{and} \quad a_2x + b_2y = c_2,$$

have the solutions

$$x = \frac{\begin{vmatrix} c_1 & b_1 \\ c_2 & b_2 \end{vmatrix}}{\begin{vmatrix} a_1 & b_1 \\ a_2 & b_2 \end{vmatrix}} = \frac{c_1b_2 - c_2b_1}{a_1b_2 - a_2b_1}$$

and

$$y = \frac{\begin{vmatrix} a_1 & c_1 \\ a_2 & c_2 \end{vmatrix}}{\begin{vmatrix} a_1 & b_1 \\ a_2 & b_2 \end{vmatrix}} = \frac{a_1c_2 - a_2c_1}{a_1b_2 - a_2b_1}.$$

Products of Vectors

Let $\hat{\mathrm{i}}$, $\hat{\mathrm{j}}$, and $\hat{\mathrm{k}}$ and be unit vectors in the x, y, and z directions. Then

$$\hat{\mathrm{i}} \cdot \hat{\mathrm{i}} = \hat{\mathrm{j}} \cdot \hat{\mathrm{j}} = \hat{\mathrm{k}} \cdot \hat{\mathrm{k}} = 1, \qquad \hat{\mathrm{i}} \cdot \hat{\mathrm{j}} = \hat{\mathrm{j}} \cdot \hat{\mathrm{k}} = \hat{\mathrm{k}} \cdot \hat{\mathrm{i}} = 0,$$

$$\hat{\mathrm{i}} \times \hat{\mathrm{i}} = \hat{\mathrm{j}} \times \hat{\mathrm{j}} = \hat{\mathrm{k}} \times \hat{\mathrm{k}} = 0,$$

$$\hat{\mathrm{i}} \times \hat{\mathrm{j}} = \hat{\mathrm{k}}, \qquad \hat{\mathrm{j}} \times \hat{\mathrm{k}} = \hat{\mathrm{i}}, \qquad \hat{\mathrm{k}} \times \hat{\mathrm{i}} = \hat{\mathrm{j}}.$$

Any vector $\vec{a}$ with components a_x, a_y, and a_z along the x, y, and z axes can be written as

$$\vec{a} = a_x\hat{\mathrm{i}} + a_y\hat{\mathrm{j}} + a_z\hat{\mathrm{k}}.$$

Let $\vec{a}$, $\vec{b}$, and $\vec{c}$ be arbitrary vectors with magnitudes a, b, and c. Then

$$\vec{a} \times (\vec{b} + \vec{c}) = (\vec{a} \times \vec{b}) + (\vec{a} \times \vec{c})$$

$$(s\vec{a}) \times \vec{b} = \vec{a} \times (s\vec{b}) = s(\vec{a} \times \vec{b}) \quad (s = \text{a scalar}).$$

Let θ be the smaller of the two angles between $\vec{a}$ and $\vec{b}$. Then

$$\vec{a} \cdot \vec{b} = \vec{b} \cdot \vec{a} = a_xb_x + a_yb_y + a_zb_z = ab\cos\theta$$

$$\vec{a} \times \vec{b} = -\vec{b} \times \vec{a} = \begin{vmatrix} \hat{\mathrm{i}} & \hat{\mathrm{j}} & \hat{\mathrm{k}} \\ a_x & a_y & a_z \\ b_x & b_y & b_z \end{vmatrix}$$

$$= \hat{\mathrm{i}}\begin{vmatrix} a_y & a_z \\ b_y & b_z \end{vmatrix} - \hat{\mathrm{j}}\begin{vmatrix} a_x & a_z \\ b_x & b_z \end{vmatrix} + \hat{\mathrm{k}}\begin{vmatrix} a_x & a_y \\ b_x & b_y \end{vmatrix}$$

$$= (a_yb_z - b_ya_z)\hat{\mathrm{i}} + (a_zb_x - b_za_x)\hat{\mathrm{j}} + (a_xb_y - b_xa_y)\hat{\mathrm{k}}$$

$$|\vec{a} \times \vec{b}| = ab\sin\theta$$

$$\vec{a} \cdot (\vec{b} \times \vec{c}) = \vec{b} \cdot (\vec{c} \times \vec{a}) = \vec{c} \cdot (\vec{a} \times \vec{b})$$

$$\vec{a} \times (\vec{b} \times \vec{c}) = (\vec{a} \cdot \vec{c})\vec{b} - (\vec{a} \cdot \vec{b})\vec{c}$$

Derivatives and Integrals

In what follows, the letters u and v stand for any functions of x, and a and m are constants. To each of the indefinite integrals should be added an arbitrary constant of integration. The *Handbook of Chemistry and Physics* (CRC Press Inc.) gives a more extensive tabulation.

Derivatives

1. $\dfrac{dx}{dx} = 1$

2. $\dfrac{d}{dx}(au) = a\dfrac{du}{dx}$

3. $\dfrac{d}{dx}(u + v) = \dfrac{du}{dx} + \dfrac{dv}{dx}$

4. $\dfrac{d}{dx}x^m = mx^{m-1}$

5. $\dfrac{d}{dx}\ln x = \dfrac{1}{x}$

6. $\dfrac{d}{dx}(uv) = u\dfrac{dv}{dx} + v\dfrac{du}{dx}$

7. $\dfrac{d}{dx}e^x = e^x$

8. $\dfrac{d}{dx}\sin x = \cos x$

9. $\dfrac{d}{dx}\cos x = -\sin x$

10. $\dfrac{d}{dx}\tan x = \sec^2 x$

11. $\dfrac{d}{dx}\cot x = -\csc^2 x$

12. $\dfrac{d}{dx}\sec x = \tan x \sec x$

13. $\dfrac{d}{dx}\csc x = -\cot x \csc x$

14. $\dfrac{d}{dx}e^u = e^u\dfrac{du}{dx}$

15. $\dfrac{d}{dx}\sin u = \cos u\dfrac{du}{dx}$

16. $\dfrac{d}{dx}\cos u = -\sin u\dfrac{du}{dx}$

Integrals

1. $\int dx = x$

2. $\int au\,dx = a\int u\,dx$

3. $\int (u + v)\,dx = \int u\,dx + \int v\,dx$

4. $\int x^m dx = \dfrac{x^{m+1}}{m+1} \quad (m \neq -1)$

5. $\int \dfrac{dx}{x} = \ln|x|$

6. $\int u\dfrac{dv}{dx}\,dx = uv - \int v\dfrac{du}{dx}\,dx$

7. $\int e^x\,dx = e^x$

8. $\int \sin x\,dx = -\cos x$

9. $\int \cos x\,dx = \sin x$

10. $\int \tan x\,dx = \ln|\sec x|$

11. $\int \sin^2 x\,dx = \frac{1}{2}x - \frac{1}{4}\sin 2x$

12. $\int e^{-ax}\,dx = -\dfrac{1}{a}e^{-ax}$

13. $\int xe^{-ax}\,dx = -\dfrac{1}{a^2}(ax + 1)e^{-ax}$

14. $\int x^2e^{-ax}\,dx = -\dfrac{1}{a^3}(a^2x^2 + 2ax + 2)e^{-ax}$

15. $\int_0^\infty x^n e^{-ax}\,dx = \dfrac{n!}{a^{n+1}}$

16. $\int_0^\infty x^{2n}e^{-ax^2}\,dx = \dfrac{1\cdot 3\cdot 5\cdots(2n-1)}{2^{n+1}a^n}\sqrt{\dfrac{\pi}{a}}$

17. $\int \dfrac{dx}{\sqrt{x^2 + a^2}} = \ln(x + \sqrt{x^2 + a^2})$

18. $\int \dfrac{x\,dx}{(x^2 + a^2)^{3/2}} = -\dfrac{1}{(x^2 + a^2)^{1/2}}$

19. $\int \dfrac{dx}{(x^2 + a^2)^{3/2}} = \dfrac{x}{a^2(x^2 + a^2)^{1/2}}$

20. $\int_0^\infty x^{2n+1}\,e^{-ax^2}\,dx = \dfrac{n!}{2a^{n+1}} \quad (a > 0)$

21. $\int \dfrac{x\,dx}{x + d} = x - d\ln(x + d)$

APPENDIX F Properties of Common Elements

All physical properties are for a pressure of 1 atm unless otherwise specified.

Element	Symbol	Atomic Number Z	Molar Mass, g/mol	Density, g/cm^3 at 20°C	Melting Point, °C	Boiling Point, °C	Specific Heat, J/(g·°C) at 25°C
Aluminum	Al	13	26.9815	2.699	660	2450	0.900
Antimony	Sb	51	121.75	6.691	630.5	1380	0.205
Argon	Ar	18	39.948	1.6626×10^{-3}	−189.4	−185.8	0.523
Arsenic	As	33	74.9216	5.78	817 (28 atm)	613	0.331
Barium	Ba	56	137.34	3.594	729	1640	0.205
Beryllium	Be	4	9.0122	1.848	1287	2770	1.83
Bismuth	Bi	83	208.980	9.747	271.37	1560	0.122
Boron	B	5	10.811	2.34	2030	—	1.11
Bromine	Br	35	79.909	3.12 (liquid)	−7.2	58	0.293
Cadmium	Cd	48	112.40	8.65	321.03	765	0.226
Calcium	Ca	20	40.08	1.55	838	1440	0.624
Carbon	C	6	12.01115	2.26	3727	4830	0.691
Cesium	Cs	55	132.905	1.873	28.40	690	0.243
Chlorine	Cl	17	35.453	3.214×10^{-3} (0°C)	−101	−34.7	0.486
Chromium	Cr	24	51.996	7.19	1857	2665	0.448
Cobalt	Co	27	58.9332	8.85	1495	2900	0.423
Copper	Cu	29	63.54	8.96	1083.40	2595	0.385
Fluorine	F	9	18.9984	1.696×10^{-3}(0°C)	−219.6	−188.2	0.753
Gadolinium	Gd	64	157.25	7.90	1312	2730	0.234
Gallium	Ga	31	69.72	5.907	29.75	2237	0.377
Germanium	Ge	32	72.59	5.323	937.25	2830	0.322
Gold	Au	79	196.967	19.32	1064.43	2970	0.131
Hafnium	Hf	72	178.49	13.31	2227	5400	0.144
Helium	He	2	4.0026	0.1664×10^{-3}	−269.7	−268.9	5.23
Hydrogen	H	1	1.00797	0.08375×10^{-3}	−259.19	−252.7	14.4
Indium	In	49	114.82	7.31	156.634	2000	0.233
Iodine	I	53	126.9044	4.93	113.7	183	0.218
Iridium	Ir	77	192.2	22.5	2447	(5300)	0.130
Iron	Fe	26	55.847	7.874	1536.5	3000	0.447
Krypton	Kr	36	83.80	3.488×10^{-3}	−157.37	−152	0.247
Lanthanum	La	57	138.91	6.189	920	3470	0.195
Lead	Pb	82	207.19	11.35	327.45	1725	0.129
Lithium	Li	3	6.939	0.534	180.55	1300	3.58
Magnesium	Mg	12	24.312	1.738	650	1107	1.03
Manganese	Mn	25	54.9380	7.44	1244	2150	0.481
Mercury	Hg	80	200.59	13.55	−38.87	357	0.138
Molybdenum	Mo	42	95.94	10.22	2617	5560	0.251
Neodymium	Nd	60	144.24	7.007	1016	3180	0.188

Element	Symbol	Atomic Number Z	Molar Mass, g/mol	Density, g/cm^3 at 20°C	Melting Point, °C	Boiling Point, °C	Specific Heat, J/(g·°C) at 25°C
Neon	Ne	10	20.183	0.8387×10^{-3}	−248.597	−246.0	1.03
Nickel	Ni	28	58.71	8.902	1453	2730	0.444
Niobium	Nb	41	92.906	8.57	2468	4927	0.264
Nitrogen	N	7	14.0067	1.1649×10^{-3}	−210	−195.8	1.03
Osmium	Os	76	190.2	22.59	3027	5500	0.130
Oxygen	O	8	15.9994	1.3318×10^{-3}	−218.80	−183.0	0.913
Palladium	Pd	46	106.4	12.02	1552	3980	0.243
Phosphorus	P	15	30.9738	1.83	44.25	280	0.741
Platinum	Pt	78	195.09	21.45	1769	4530	0.134
Plutonium	Pu	94	(244)	19.8	640	3235	0.130
Polonium	Po	84	(210)	9.32	254	—	—
Potassium	K	19	39.102	0.862	63.20	760	0.758
Radium	Ra	88	(226)	5.0	700	—	—
Radon	Rn	86	(222)	9.96×10^{-3} (0°C)	(−71)	−61.8	0.092
Rhenium	Re	75	186.2	21.02	3180	5900	0.134
Rubidium	Rb	37	85.47	1.532	39.49	688	0.364
Scandium	Sc	21	44.956	2.99	1539	2730	0.569
Selenium	Se	34	78.96	4.79	221	685	0.318
Silicon	Si	14	28.086	2.33	1412	2680	0.712
Silver	Ag	47	107.870	10.49	960.8	2210	0.234
Sodium	Na	11	22.9898	0.9712	97.85	892	1.23
Strontium	Sr	38	87.62	2.54	768	1380	0.737
Sulfur	S	16	32.064	2.07	119.0	444.6	0.707
Tantalum	Ta	73	180.948	16.6	3014	5425	0.138
Tellurium	Te	52	127.60	6.24	449.5	990	0.201
Thallium	Tl	81	204.37	11.85	304	1457	0.130
Thorium	Th	90	(232)	11.72	1755	(3850)	0.117
Tin	Sn	50	118.69	7.2984	231.868	2270	0.226
Titanium	Ti	22	47.90	4.54	1670	3260	0.523
Tungsten	W	74	183.85	19.3	3380	5930	0.134
Uranium	U	92	(238)	18.95	1132	3818	0.117
Vanadium	V	23	50.942	6.11	1902	3400	0.490
Xenon	Xe	54	131.30	5.495×10^{-3}	−111.79	−108	0.159
Ytterbium	Yb	70	173.04	6.965	824	1530	0.155
Yttrium	Y	39	88.905	4.469	1526	3030	0.297
Zinc	Zn	30	65.37	7.133	419.58	906	0.389
Zirconium	Zr	40	91.22	6.506	1852	3580	0.276

The values in parentheses in the column of molar masses are the mass numbers of the longest-lived isotopes of those elements that are radioactive. Melting points and boiling points in parentheses are uncertain. The data for gases are valid only when these are in their usual molecular state, such as H_2, He, O_2, Ne, etc. The specific heats of the gases are the values at constant pressure. *Primary source*: Adapted fron J. Emsley, *The Elements*, 3rd ed., 1998, Clarendon Press, Oxford (www.webelements.com). Data on newest elements are current.

APPENDIX G

Periodic Table of the Elements

Legend: Metals; Metalloids; Nonmetals

Period	IA (Alkali metals)	IIA	IIIB	IVB	VB	VIB	VIIB	VIIIB	VIIIB	VIIIB	IB	IIB	IIIA	IVA	VA	VIA	VIIA	0 (Noble gases)
1	1 H																	2 He
2	3 Li	4 Be											5 B	6 C	7 N	8 O	9 F	10 Ne
3	11 Na	12 Mg											13 Al	14 Si	15 P	16 S	17 Cl	18 Ar
4	19 K	20 Ca	21 Sc	22 Ti	23 V	24 Cr	25 Mn	26 Fe	27 Co	28 Ni	29 Cu	30 Zn	31 Ga	32 Ge	33 As	34 Se	35 Br	36 Kr
5	37 Rb	38 Sr	39 Y	40 Zr	41 Nb	42 Mo	43 Tc	44 Ru	45 Rh	46 Pd	47 Ag	48 Cd	49 In	50 Sn	51 Sb	52 Te	53 I	54 Xe
6	55 Cs	56 Ba	57-71 *	72 Hf	73 Ta	74 W	75 Re	76 Os	77 Ir	78 Pt	79 Au	80 Hg	81 Tl	82 Pb	83 Bi	84 Po	85 At	86 Rn
7	87 Fr	88 Ra	89-103 †	104 Rf	105 Db	106 Sg	107 Bh	108 Hs	109 Mt	110 Ds	111 Uua	112 Uub	113	114 Uuq	115	116	117	118

Transition metals (groups IIIB–IIB). THE HORIZONTAL PERIODS.

Inner transition metals

Lanthanide series *	57 La	58 Ce	59 Pr	60 Nd	61 Pm	62 Sm	63 Eu	64 Gd	65 Tb	66 Dy	67 Ho	68 Er	69 Tm	70 Yb	71 Lu
Actinide series †	89 Ac	90 Th	91 Pa	92 U	93 Np	94 Pu	95 Am	96 Cm	97 Bk	98 Cf	99 Es	100 Fm	101 Md	102 No	103 Lr

The names of elements 104 through 109 (Rutherfordium, Dubnium, Seaborgium, Bohrium, Hassium, and Meitnerium, respectively) were adopted by the International Union of Pure and Applied Chemistry (IUPAC) in 1997. As of early 2004, the discovery of elements 110 through 115 have been reported in scientific journals. See www.webelements.com for the latest information and newest elements.

Answers to Reading Exercises and Odd-Numbered Problems

(Answers that involve a proof, graph, or otherwise lengthy solution are not included.)

Chapter 1

RE 1-1: Examples include second or hour, meter or inch, and gram or kilogram.

RE 1-2: A 12-inch ruler would more likely change less over time than your foot, especially if you are still growing.

RE 1-3: The length of one day or the time it takes for the earth to rotate 360° about its own axis is not constant, because the speed of the earth's rotation is slowly decreasing with time.

RE 1-4: (a) Since 24 h of time occurs for each 360° of rotation or 4 min for each degree of longitude or 240 s for each degree of longitude, 20 min and 13 s will relate to a rotation or longitude change of (1213 s)/(240 s/deg) = 5.05 degrees of longitude change. (b) If the clock is off by 2 min or 120 s, the longitude will be off by (120 s)/(240 s/degree) = 0.5 degrees of longitude. (c) 360° or one revolution relates to one circumference of length. Therefore $0.5°/360° = x/(24\,000$ nautical miles), or $x = 33.3$ nautical miles off course. Sailor beware!

RE 1-5: (a) If your watches are synchronized, you should measure the same time for the flash. For the same duration of time between the flash and thunder you both should have accurate watches and be located close to one another. (b) No, the 12 h (smaller) clock shows a time of 7:44 or a total elapsed time of 464 min since 12 o'clock. This is 464 min/(1440 min/day) = .322 day elapsed. The 10 h (larger) clock shows a time of 8.23 hours elapsed since 10 o'clock (12 o'clock on the other scale) or 8.23/(20 hr/day) = .412 day elapsed.

RE 1-6: One of many possible procedures would be to use the balance to determine the amount of clay equal to 1 kg. Divide the clay into 1000 equal volume pieces. Assuming the density of the clay is uniform, each clay piece now has a mass of 1 gram. Use these pieces with the balance and the object whose mass is to be determined to find its mass.

RE 1-7: (a) It is correct to write 1 min/60 s = 1 because 1 minute and 60 seconds are the same *length* of time. It is meaningless to say 1/60 = 1 when no units are specified. These numbers are not the same in the absence of the context of the units. (b) In terms of conversion factors and chain-link conversions, the number of minutes in a day is given by

$$1\text{ d} = (1\text{ d})\left(\frac{24\text{ h}}{1\text{ d}}\right)\left(\frac{60\text{ min}}{1\text{ h}}\right) = 1440\text{ m}.$$

RE 1-8: (a) 2. (b) Exact, if the cows were counted. (c) 6. Remember that the leading zeros don't count. (d) 7. Trailing zeros do count. (e) Exact, by definition.

RE 1-9: (a) 11. (b) Probably 3, we can't be sure. (c) 2.09×10^{10} ft. (d) 10^{10} ft (ten to the tenth feet).

RE 1-10: (a) You should keep all digits for intermediate results; thus you should use $A = 1.96$ cm^2 for calculating V. (b) 2.7 cm^3; in this situation the answer can be to no more significant figures than the original data. (c) 2.8 cm^3.

RE 1-11: (a) 27; (b) 198.0; (c) 0.6; (d) 0.9986, see *Evaluating Functions* in Appendix A, Section 5. (e) Since five is an exact number, the four significant numbers in the average length limits the answer to 10.67 m.

RE 1-12: (a) 0.01 s; (b) .01 s out of 1.78 s or .01/1.78 = 0.00562, or about 0.6%.

Problems

1. (a) 0.98 ft/ns; (b) 0.30 mm/ps. **3.** C, D, A, B, E; the important criterion is the constancy of the daily variation, not its magnitude. **5.** 0.12 AU/min **7.** 2.1 h. **9.** 1.21×10^{12} μs. **11.** (a) 160 rods; (b) 40 chains. **13.** (a) 4.00×10^4 km; (b) 5.10×10^8 km^2; (c) 1.08×10^{12} km^3. **15.** 1.9×10^{22} cm^3. **17.** 1.1×10^3 acre-feet. **19.** 9.0×10^{49}. **21.** (a) 10^3 kg; (b) 158 kg/s. **23.** (a) 1.18×10^{-29} m^3. **25.** 3.8 mg/s. **27.** 8×10^2 km. **29.** 6.0×10^{26}. **31.** (a) 60.8 W; (b) 43.3 Z. **33.** 89 km. **35.** $\approx 1 \times 10^{36}$. **37.** 700 to 1500. **39.** (a) 293 U.S. bushels; (b) 3.81×10^3 U.S. bushels. **41.** 9.4×10^{-3}. **43.** 5.95 km. **45.** 1.9×10^5 kg. **47.** 2×10^4 to 4×10^4. **49.** 10.7. **59.** (a) 13 597 kg; (b) 4917 L; (c) 6172 kg; (d) 20 075 L; (e) 45%

Chapter 2

RE 2-1: (b), (c), and (d).

RE 2-2: Correct order: (c), (b), and (a).

RE 2-3: Yes, the displacement can be positive as long as the particle moves to a less negative position.

RE 2-4: (a) Average velocity is the displacement divided by the total time $\langle v_x \rangle = 10$ mi/30 min $= 0.33$ mi/min due east. (b) Average speed is the total distance traveled divided by the total time $\langle s \rangle = 30$ mi/30 min $= 1$ mi/min. (c) The answers are different because the displacement is different from the total distance traveled in the 30 minute time period.

RE 2-5: Instantaneous speed. The speedometer only tells you the speed at which you are currently driving, not your acceleration or direction.

RE 2-6: (a) Remember that the velocity is the time derivative of the position equation. The velocity will be constant if it has no time dependence. Position equations 1 and 4 give a constant velocity. (b) The velocity is negative in equations 2 and 3.

RE 2-7: In returning to x_1 the total displacement $\Delta x = x_1 - x_1$ is zero. Since $\langle v_x \rangle = \Delta x/\Delta t$, the average velocity is also zero.

RE 2-8: (a) +, (b) −, (c) −, (d) +; remember that $\vec{a}$ will have the same direction as $\Delta\vec{v}$ or $\vec{v}_2 - \vec{v}_1$.

RE 2-9: The equations of Table 2-1 apply when a_x is constant. Take the second derivative of x with respect to t to find a_x. Only equations 1, 3 and 4 give a constant a_x ($a_x = 0$ is a constant).

Problems

1. 414 ms. **3.** (a) +40 km/h; (b) 40 km/h. **5.** (a) 73 km/h; (b) 68 km/h; (c) 70 km/h; (d) 0. **7.** (a) 0, −2, 0, 12 m; (b) +12 m; (c) +7 m/s. **9.** 1.4 m. **11.** (a) −6 m/s; (b) negative x direction; (c) 6 m/s; (d) first smaller, then zero, and then larger; (e) yes (t = 2s); (f) no. **13.** 100 m. **15.** (a) velocity squared; (b) acceleration; (c) m^2/s^2, m/s^2. **17.** 20 m/s^2, in the direction opposite to its initial velocity. **19.** (a) m/s^2, m/s^3; (b) 1.0 s; (c) 82 m; (d) −80 m; (e) 0, −12, −36, −72 m/s; (f) −6, −18, −30, −42 m/s^2. **21.** 0.10 m. **23.** (a) 1.6 m/s; (b) 18 m/s. **25.** (a) 3.1×10^6 s = 1.2 months; (b) 4.6×10^{13} m. **27.** 1.62×10^{15} m/s^2. **29.** 2.5 s. **31.** (a) 3.56 m/s^2; (b) 8.43 m/s. **33.** (a) 5.00 m/s; (b) 1.67 m/s^2; (c) 7.50 m. **35.** (a) 0.74 s; (b) −6.2 m/s^2. **37.** (a) 10.6 m; (b) 41.5 s. **39.** (a) 30 s; (b) 300 m. **41.** (a) 54 m, 18 m/s, −12 m/s^2; (b) 64 m at t = 4.0 s; (c) 24 m/s at t = 2.0 s; (d) −24 m/s^2; (e) 18 m/s. **49.** (a) 0.75 s; (b) 50 m. **57.** Since there is some latitude in what might be considered "the right answer" here, we have elected to mention some Web sites (current as of May 2002) where graphs for model rocket kinematics are shown: http://www.rocket-roar.com/rap/alt.html; http://mks.niobrara.com/altitude.html; http://www.boilerbay.com/rockets/; **59.** 40 m.

Chapter 3

RE 3-1: (a) The velocity of the cart on the carpet goes to zero at t = 1.1 s. (b) The velocity of the cart on the track at t = 1.1 s is approximately 0.65 m/s, so it still has (0.65 m/s/0.80 m/s) or 81% of its initial speed.

RE 3-2: (a) An elevator or car starting or stopping, or a merry-go-round moving at a constant speed. (b) The person feels heavy during startup and light during stopping. Objects, such as a marble, start to move with no apparent reason on the merry-go-round floor.

RE 3-3: (a) No acceleration: Sliding a block along a table with a small steady force or shoving on a huge object like a desk or car, etc., can result in either constant velocity motion or an inability to move the object (desk or car). (b) Acceleration: Pushing hard on a sliding block, pushing on a rolling ball, pushing or pulling someone on a vehicle with wheels, etc.

RE 3-4: You would attach one end of the rubber band to a post and hook the other end of the rubber band to a calibrated spring scale. Then you would record the unstretched length of the rubber band and the fact that the force on it is 0 N. Next you would pull on the rubber band with the spring scale until it reads 1 N and record the new length of the rubber band. Then you would repeat the process as the spring scale reads 2 N, 3 N, etc., recording the rubber-band length each time. In that way you can generate either a look-up table or a graph of force vs. rubber-band length. If greater precision is needed, you could take data for many more force-scale readings.

RE 3-5: (a) $\vec{F} = (-26\text{ N})\hat{\text{i}}$, $\vec{a} = (-0.42\text{ m/s}^2)\hat{\text{i}}$; (b) $m = F/a = 62$ kg; (c) 62 kg

RE 3-6: (a) The mass measurement in part (b) above uses the ratio of the force to acceleration and hence is the inertial mass. (b) We assumed that the student is on the surface of planet Earth and that the bathroom scale was calibrated for the same planet.

RE 3-7: In both cases (a) and (b) the acceleration is zero, therefore the net force must also be zero. This will require all three forces to add to zero as vectors. (a) This requires $\vec{F}_C$ to point to the left in the diagram with a magnitude of 2 N so $\vec{F}_C = (-2\text{ N})\hat{\text{i}}$. (b) Since the acceleration is also zero in this case, we still have $\vec{F}_C = (-2\text{ N})\hat{\text{i}}$.

RE 3-8: (a) Bottom right cart has a net force of −5 N, top left has +4 N, top right has −1 N, and bottom left has a net force of zero. (b) Since the acceleration and net force are directly proportional, the accelerations rank in the same order.

RE 3-9: In the chosen coordinate system, all the accelerations in the v vs. t graphs shown is Fig. 3-2 are negative since the slopes are negative. (a) The box on carpet acceleration is about −3.9 m/s^2 as determined by calculating the slope of the v vs. t graph. Slope = (0.00 − 0.90)(m/s)/(0.23 − 0.00)(s). (b) The cart on track acceleration is about −0.15 m/s^2 as determined by calculating the slope of the v vs. t graph. Slope = (0.62 − 0.80)(m/s)/(1.2 − 0.0)(s).

RE 3-10: (a) There appear to be no other horizontal forces on the moving objects except friction. Thus, we can assume that the net force on each object is due to a friction force. This friction force seems to be constant since the acceleration is constant and we assume that $F_x^{\text{net}} = ma_x$. (b) Box on carpet $F_x^{\text{fric}} = ma_x$ = 0.5 kg × (− 3.9 m/s^2) = −2 N. It points to the left. (c) Cart on track $F_x^{\text{fric}} = ma_x$ = 0.5 kg × (−0.15 m/s^2) = −0.08 N. It also points to the left.

RE 3-11: (a) A tossed object is changing its velocity at all times. Just before it reaches the top of its flight it has a positive velocity and just after it has a negative velocity. Since acceleration is rate of change of velocity over time, even the instantaneous acceleration doesn't go to zero over an infinitesimal time interval. (b) The Fig. 3-22 graph of velocity vs. time is linear with a constant negative slope. Since slope of a v_y vs. t graph represents the acceleration component a_y, then a_y = constant so $\vec{a} = a_y\hat{\text{j}}$ is constant.

RE 3-12: Change every x in the two equations in Table 2-1 to a y. Then replace a_y (previously a_x) with $-g$.

RE 3-13: (a) The unmagnetized paperclip will be attracted to the magnet and, in turn, the magnet will be attracted toward the paperclip. Newton's Third Law tells us that these attractive forces will be equal in magnitude to one another but opposite in direction; the force on the magnet will be to the left and the force on the paperclip will be to the right. (b) Newton's Third Law applies to all forces of interaction of which this is just one example.

Problems

1. 16 N. **3.** (a) 0.02 m/s^2; (b) 8×10^4 km; (c) 2×10^3 m/s. **5.** 1.2×10^5 N. **7.** (a) 4.9×10^5 N; (b) 1.5×10^6 N. **9.** (a) 245 m/s^2; (b) 20.4 kN. **11.** (a) 8.0 m/s; (b) $+x$ direction. **13.** 8.0 cm/s^2. **15.** 1.8×10^4 N. **17.** (a) 31.3 kN; (b) 24.4 kN. **19.** $2Ma/(a + g)$. **21.** 2.4 N. **23.** (a) 1.23 N; (b) 2.46 N; (c) 3.69 N; (d) 4.92 N; (e) 6.15 N; (f) 0.25 N. **25.** (a) 3.2 s; (b) 1.3 s. **27.** (a) 3.70 m/s; (b) 1.74 m/s; (c) 0.154 m. **29.**

4.0 m/s. **31.** 22 cm and 89 cm below the nozzle. **33.** (a) 5.4 s; (b) 41 m/s. **35.** (a) 1.23 cm; (b) 4 times, 9 times, 16 times, 25 times. **37.** (a) 29.4 m; (b) 2.45 s. **39.** (a) 3260 N (b) 2.7×10^3 kg; (c) 1.2 m/s **41.** (a) 17 s; (b) 290 m. **43.** (a) 11 N; (b) 2.2 kg; (c) 0; (d) 2.2 kg. **45.** (a) 494 N, up; (b) 494 N, down. **47.** (a) 1.1 N. **49.** 5.1 m/s. **51.** (a) 466 N; (b) 527 N.

Chapter 4

RE 4-1: Displacement (1) is identical as the ball ends up going a net distance of 6 meters north and 3 meters west. Displacement (2) is different. It actually has an equal magnitude but the ball has moved in the opposite direction. *Note*: Displacement does not depend on where something starts or ends, but only on how much and in what direction its position has changed relative to where it started.

RE 4-2: (a) The maximum magnitude occurs when the two vectors point in the same direction. This gives a magnitude for vector $\vec{c}$ of 3 m + 4 m = 7 m. (This answer is not correct without a unit attached.) (b) The minimum magnitude occurs when the two vectors point in the opposite directions. This gives a magnitude for vector $\vec{c}$ of 4 m − 3 m = 1 m.

RE 4-3: Methods (c), (d), and (f) work since the parallelogram methods (c) and (d) show that the same correct resultant can be obtained regardless of the order in which components are added. Method (f) shows an equivalent construction using components. All the other vectors point in the wrong directions.

RE 4-4: The vectors in figures (b) and (d) have the same components as the standard vector.

RE 4-5: Compare Figs 4-12 and 4-13.

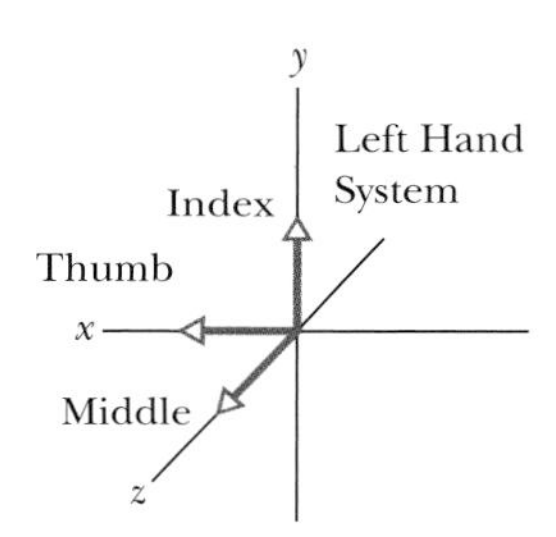

RE 4-6: (a & b). The x- and y-components of $\vec{d}_1$ are both positive. The x-component of $\vec{d}_2$ is positive but the y-component points down in a negative direction. (c) Using the parallelogram method to get the vector sum of $\vec{d}_1$ and $\vec{d}_2$ results in a vector that has both x- and y-components that are positive.

RE 4-7: This is a kind of artificial question since units of force and acceleration are different as are units of displacement and velocity. However, if the scalars (mass and time respectively) act as compressors or stretchers, then the simplistic answers would be (a) The force vector would point in the same direction as the acceleration vector but be three times as long. (b) The velocity vector would point off in the same direction as the displacement vector and be twice as long since the displacement was divided by 0.5 s.

RE 4-8:

(a) $\vec{F} = m\vec{a} = 3.0\text{ kg}[(1.8\text{ m/s}^2)\hat{\text{i}} + (1.0\text{ m/s}^2)\hat{\text{j}}] = (5.4\text{ N})\hat{\text{i}} + (3.0\text{ N})\hat{\text{j}}.$

(b) $\langle\vec{v}\rangle = \dfrac{\Delta\vec{r}}{\Delta t} = \dfrac{(3.2\text{ m})\hat{\text{i}} + (-0.8\text{ m})\hat{\text{j}}}{0.5\text{ s}} = (6.4\text{ m/s})\hat{\text{i}} + (-1.6\text{ m/s})\hat{\text{j}}.$

Problems

1. The displacements should be (a) parallel, (b) antiparallel, (c) perpendicular. **3.** (a) 5; (b) 1; (c) 7. **5.** (a) −2.5 m; (b) −6.9 m. **7.** (a) 47.2 m; (b) 122°. **9.** (a) 168 cm; (b) 32.5° above the floor. **11.** (a) 6.42 m; (b) no; (c) yes; (d) yes; (e) a possible answer: $(4.30\text{ m})\hat{\text{i}} + (3.70\text{ m})\hat{\text{j}} + (3.00\text{ m})\hat{\text{k}}$; (f) 7.96 m. **13.** (a) 370 m; (b) 36° north of east; (c) 425 m; (d) the distance. **15.** (a) $(-9\text{ m})\hat{\text{i}} + (10\text{ m})\hat{\text{j}}$; (b) 13 m; (c) + 132°. **17.** (a) 4.2 m; (b) 40° east of north; (c) 8.0 m; (d) 24° north of west. **19.** (a) $(3.0\text{ m})\hat{\text{i}} - (2.0\text{ m})\hat{\text{j}} + (5.0\text{ m})\hat{\text{k}}$; (b) $(5.0\text{ m})\hat{\text{i}} - (4.0\text{ m})\hat{\text{j}} - (3.0\text{ m})\hat{\text{k}}$; (c) $(-5.0\text{ m})\hat{\text{i}} + (4.0\text{ m})\hat{\text{j}} + (3.0\text{ m})\hat{\text{k}}$. **21.** (a) 38 m; (b) 320°; (c) 130 m; (d) 1.2°; (e) 62 m; (f) 130°. **23.** (a) 1.59 m; (b) 12.1 m; (c) 12.2 m; (d) 82.5°. **29.** (a) Put axes along cube edges, with the origin at one corner. Diagonals are $a\hat{\text{i}} + a\hat{\text{j}} + a\hat{\text{k}}$, $a\hat{\text{i}} + a\hat{\text{j}} - a\hat{\text{k}}$, $a\hat{\text{i}} - a\hat{\text{j}} - a\hat{\text{k}}$, $a\hat{\text{i}} - a\hat{\text{j}} + a\hat{\text{k}}$; (b) 54.7°; (c) $\sqrt{3}\,a$. **31.** 4.1. **33.** (a) 103 km; (b) 60.9° north of due west. **35.** (a) 15 m; (b) south; (c) 6.0 m; (d) north. **37.** 5.0 km, 4.3° south of due west. **39.** 5.39 m at 21.8° left of forward. **41.** (a) 4.28 m; (b) 11.7 m. **43.** (a) −80 m; (b) 110 m; (c) 143 m; (d) +168° (counterclockwise). **45.** 3.6 m. **47.** (a) 1.84 m; (b) 69° north of east. **49.** (a) 9.51 m; (b) 14.1 m; (c) 13.4 m; (d) 10.5 m. **51.** (a) $9.19\hat{\text{i}} + 7.71\hat{\text{j}}$; (b) $14.0\hat{\text{i}} + 3.41\hat{\text{j}}$

Chapter 5

RE 5-1: (a) No, because in Fig. 5-5 the vertical positions of the ball on the right are the same as those of the ball on the left. (b) No. The horizontal positions of the ball on the right are equally spaced, indicating that horizontal velocity of the ball is constant and unaffected by the falling.

RE 5-2: The skateboarder's vertical motion is independent of his horizontal velocity. This is why the skateboarder lands back on his skateboard after his jump.

RE 5-3: (a) At each of the three points, the force vector points straight down and has a constant magnitude and (b) the same is true for the three acceleration vectors. (c) The horizontal component of each of the three velocity vectors points to the right and has a constant size. The vertical component of the velocity at the left point is directed straight upward and is slightly larger than the common size of the horizontal velocity components. The vertical component of the velocity at the center point is zero, while at the right point it is directed downward and is smaller in size than the horizontal velocity component.

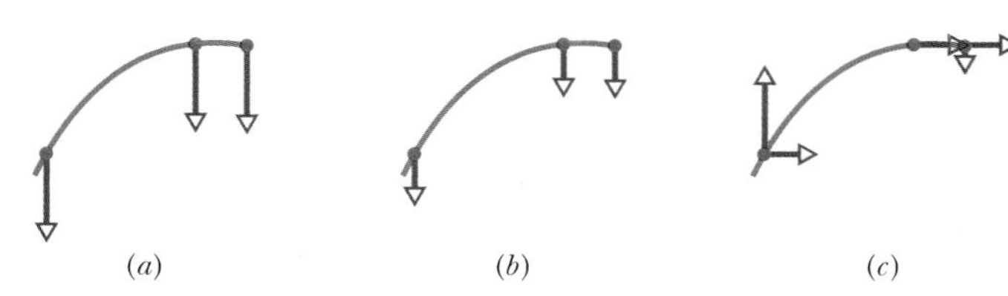

RE 5-4: (a) The x-component of velocity is not changing and is the slope of Fig. 5-9. From the data in the figures, the slope is about 2.3 m/s. The initial y-component of velocity is the initial slope of Fig 5-10, which is about 3.5 m/s. The launch angle will be the inverse tangent of 3.5/2.3 or about 57°. (b) Using a protractor about 57°, too.

RE 5-5: (a) The horizontal component of velocity remains constant. (b) The vertical component of velocity is changing constantly as there is a vertical acceleration. (c) The horizontal component of its acceleration is zero. The only force (gravity) acting is in the vertical direction. (d) The vertical component of its acceleration is constant (9.8 m/s² downward).

RE 5-6: (a) Using Eq. 5-15 and noting that $\Delta x = 8$ m and $\Delta y = -6$ m gives a displacement of $\Delta\vec{r} = (8\text{ m})\hat{\text{i}} + (-6\text{ m})\hat{\text{j}}$. (b) No, since it has components along both axes.

RE 5-7: (a) When traveling clockwise, the x-component of the particle's velocity is positive when it is in the I and II quadrant, and its y-component is negative in the I quadrant, so the particle is now in the I quadrant. (b) When traveling counterclockwise, the x-component of the particle's velocity is positive when it is in the III and IV quadrant, and it's y-component is negative in the III quadrant, so the particle is then in the III quadrant.

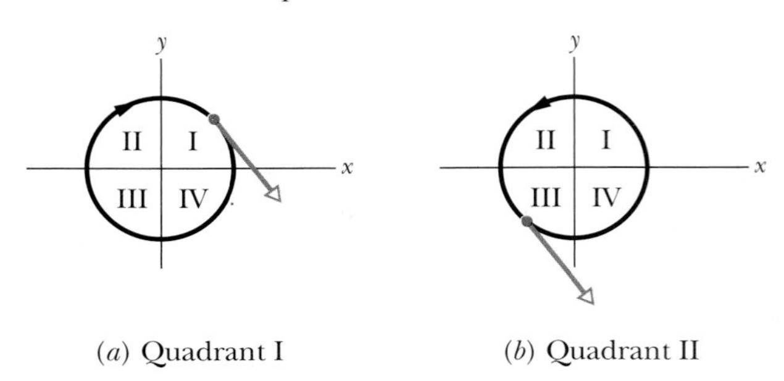

(a) Quadrant I (b) Quadrant II

RE 5-8: Remember that the x-component of acceleration will be in the direction of the change in the x-component of velocity and the y-component of acceleration will be in the direction of the change in the y-component of velocity. Just knowing the trajectory or path of the particle does not give you the direction of the acceleration. You also need to know how the velocity is changing as the particle travels along its trajectory. Therefore, if the change in the velocity vector is in the direction of the path of the particle, $\vec{a}$ will be tangent to the trajectory. However, you will study other situations (Section 5-7) where $\vec{a}$ is actually perpendicular to the trajectory, that is, the change in the velocity is perpendicular to the trajectory.

RE 5-9: The centripetal force is always inward toward the center of the curve. According to Newton's First Law, the passenger wants to travel in a straight line unless acted upon by a force. The centripetal force acts on the passenger through the friction between the passenger and the car seat. If that frictional force is not strong enough, the passenger tends to travel in a straight line and slides to the outside edge of the seat, where both the seat and the side of the car can provide the centripetal force needed to move your body in a curved path.

Problems

1. (a) 62 ms; (b) 480 m/s. **3.** (a) 0.205 s; (b) 0.205 s; (c) 20.5 cm; (d) 61.5 cm. **5.** (a) 2.00 ns; (b) 2.00 mm; (c) 1.00×10^7 m/s; (d) 2.00×10^6 m/s. **7.** (a) 16.9 m; (b) 8.21 m; (c) 27.6 m; (d) 7.26 m; (e) 40.2 m; (f) 0. **9.** 4.8 cm. **13.** (a) 11 m; (b) 23 m; (c). 17 m/s; (d) 63° below the horizontal. **15.** (a) 24 m/s; (b) 65° above the horizontal. **17.** (a) 10 s; (b) 897 m. **19.** the third. **21.** (a) 202 m/s; (b) 806 m; (c) 161 m/s; (d) −171 m/s. **23.** (a) yes; (b) 2.56 m. **25.** between the angles 31° and 63° above the horizontal. **27.** (a) $(-5.0\text{ m})\hat{\text{i}} + (8.0\text{ m})\hat{\text{j}}$; (b) 9.4 m; (c) 122°; (e) $(8\text{ m})\hat{\text{i}} + (-8\text{ m})\hat{\text{j}}$; (f) 11 m; (g) −45°. **29.** (a) $(-7.0\text{ m})\hat{\text{i}} + (12\text{ m})\hat{\text{j}}$; (b) x axis. **31.** 8.43 m at −129°. **33.** 7.59 km/h, 22.5° east of north; **35.** (a) $(3.00\text{ m/s})\hat{\text{i}} + (-8.00\text{ m/s}^2)t\,\hat{\text{j}}$; (b) $(3.00\text{ m/s})\hat{\text{i}} + (-16.00\text{ m/s})\hat{\text{j}}$; (c) 16.3 m/s; (d) −79.4° **37.** 0.421 m/s at 3.1° west of due north. **39.** (a) $(6.00\text{ m})\hat{\text{i}} + (-106\text{ m})\hat{\text{j}}$; (b) $(19.0\text{ m/s})\hat{\text{i}} + (-224\text{ m/s})\hat{\text{j}}$; (c) $(24.0\text{ m/s}^2)\hat{\text{i}} + (-336\text{ m/s}^2)\hat{\text{j}}$; (d) −85.2° to $+x$. **41.** (a) $(-1.5\text{ m/s})\hat{\text{j}}$; (b) $(4.5\text{ m})\hat{\text{i}} + (-2.25\text{ m})\hat{\text{j}}$. **43.** (a). 45 m; (b) 22 m/s. **45.** (a) $(8\text{ m/s}^2)t\,\hat{\text{j}}$; (b) $(8\text{ m/s}^2)\hat{\text{j}}$. **47.** (a) 22 m; (b) 15 s. **49.** (a) 7.49 km/s; (b) 8.00 m/s². **51.** (a) 19 m/s; (b) 35 rev/min; (c) 1.7 s. **53.** (a) 0.034 m/s²; (b) 84 min. **55.** (a) 12 s; (b) 4.1 m/s², down; (c) 4.1 m/s², up. **57.** 160 m/s². **59.** (4.00 m, 6.00 m)

Chapter 6

RE 6-1: If you gather the tails of the three vectors shown in the helicopter diagram, you get the free-body diagram shown in (c).

RE 6-2: Use the balance in Fig. 3-9 and place one object on the left pan and the other object on the right pan. If the two objects have the same mass they will balance one another. They would have the same weight if they both gave the same reading on the spring scale. Also you could realize that if they have the same mass, they have the same weight since $W = mg$ and g is a constant. The weight and mass are not the same. The weight is a force, and the mass is mass. Yes, since the weight and mass are proportional, the ratios are the same.

RE 6-3: It's true that the planet is yanking down on the patient but this is a force equal to his weight. However, since the normal force from the floor is equal and opposite, there is no net force and hence no acceleration.

RE 6-4: (a) In this case, at constant speed a equals zero and thus the net force must equal zero, requiring $\vec{N}$ and $\vec{F}^{\text{grav}}$ to be equal in magnitude and opposite in direction. (b) Since the only two forces acting on the block are $\vec{N}$ and $\vec{F}^{\text{grav}}$, to have an upward acceleration we must have a net upward force, meaning that the magnitude of $\vec{N}$ is now larger than that of $\vec{F}^{\text{grav}}$. (c) Slowing down means an acceleration or net force in the downward direction, requiring the magnitude of $\vec{F}^{\text{grav}}$ to be larger than that of $\vec{N}$. What do you think would happen to $\vec{N}$ if the elevator cable broke and the block fell freely with $a = g$?

RE 6-5: In both answers to follow we are assuming the only forces acting on the block in the horizontal direction are the friction force and the pull of the cord, which is what the force sensor is measuring. Since in both cases there is no acceleration, these two forces must be equal and opposite, allowing us to equate the force sensor reading to the frictional force. (a) From the graph it looks like the block breaks free when the force is about 9.5 N. The total mass is 0.7956 kg, and the normal force that equals the weight is mg; therefore using Eq. 6-11, we find $\mu^{\text{stat}} = 9.5/(0.7956 \times 9.8) = 1.22$. Notice that the coefficient of friction has no units. (b) From the graph, the force needed to keep the block moving at a constant speed is about 3.0 N. Using Eq. 6-10, $\mu^{\text{kin}} = 3.0/(0.7956 \times 9.8) = 0.38$.

RE 6-6: (a) Zero; (b) 5 N; (c) No; (d) Yes, there is now a net force of 2 N on the block causing it to accelerate; (e) 8 N.

RE 6-7: It is true that friction has both a bad side and a good side. Friction always tries to retard motion. If you desire that motion then friction is bad—for example, the pistons in your car engine—and we do everything we can (lubricants) to eliminate it. However, there are other times when we don't want motion (slippage) to occur, as when we are walking or riding a bike, and the force of friction allows us to do these activities.

RE 6-8: Think of the cord as an object with a mass you are trying to accelerate with only two forces—the one at one end from the hand and the other at the other end from the block. We will assume that the length of the cord hanging down on each side is the same so we can ignore the force of gravity on the cord. (a) If the cord is not accelerating then the magnitudes of the two forces are equal and cancel. (b) If the block is accelerating then so is the cord and the force of the hand on the cord is greater than that of the block. (c). In this case the acceleration is opposite to b and the pull force of the hand is less than the pull force due to the block.

RE 6-9: Look at Eq. 6-25. The only things in this equation that will change with the size of the drops are the mass, m, and the cross-sectional area A. So for this exercise v_t^2 is proportional to m/A. How will this ratio change with the size of the drops? A changes as r^2 and m changes as (ρ_{water})(volume) and since volume goes as r^3, we finally determine that m/A and hence v_t^2 goes as r. Therefore, large drops have greater speeds than small drops.

Problems

1. (a) $F_x = 1.88$ N; (b) $F_y = 0.684$ N; (c) $(1.88\text{ N})\hat{\text{i}} + (0.684\text{ N})\hat{\text{j}}$. **3.** 2.9 m/s². **5.** $(3\text{ N})\hat{\text{i}} + (-11\text{ N})\hat{\text{j}}$. **7.** (a) $(-32\text{ N})\hat{\text{i}} + (-21\text{ N})\hat{\text{j}}$; (b) 38 N; (c) 213° from $+x$. **9.** (a) 108 N; (b) 108 N; (c) 108 N. **11.** (a) 200 N; (b) 120 N. **13.** 0.61. **15.** (a) 190 N; (b) 0.56 m/s². **17.** (a) 0.13 N; (b) 0.12. **19.** (a) no; (b) $(-12\text{ N})\hat{\text{i}} + (5\text{ N})\hat{\text{j}}$. **23.** (a) 300 N; (b) 1.3 m/s². **25.** (a) 66 N; (b) 2.3 m/s². **27.** (b) 3.0×10^7 N. **29.** 100 N. **31.** (a) 0; (b) 3.9 m/s² down the incline; (c) 1.0 m/s² down the incline. **33.** (a) 3.5 m/s²; (b) 0.21 N; (c) blocks move independently. **35.** 490 N **37.** (a) 6.1 m/s², leftward; (b) 0.98 m/s², leftward. **39.** $g(\sin\theta - \sqrt{2}\,\mu^{kin}\cos\theta)$. **41.** 9.9 s. **43.** 6200 N. **45.** 2.3. **47.** 1.5 mm. **49.** (a) 68 N (b) 73 N. **51.** (a) 2.2×10^{-3} N; (b) 3.7×10^{-3} N. **53.** (a) 4.6×10^3 N for each bolt; (b) 5.8×10^3 N. **55.** (a) 180 N; (b) 640 N. **57.** (a) 3.1 N; (b) 14.7 N. **59.** (a) 6.8×10^3 N (b) − 21° or 159°. **61.** (b) $F/(m + M)$; (c) $MF/(m + M)$; (d) $F(m + 2M)/2\ (m + M)$. **63.** 1.8×10^4 N. **65.** about 48 km/h. **67.** 21 m. **69.** $\sqrt{Mgr/m}$. **71.** (a) light; (b) 778 N; (c) 223 N. **73.** 2.2 km. **75.** (b) 8.74 N; (c) 37.9 N, radially inward; (d) 6.45 m/s. **77.** (a) $\sqrt{Rg\tan(\theta + \tan^{-1}(\mu^{stat}))}$; (b) graph; (c) 41.3 m/s; (d) 21.2 m/s. **81.** (a) 3.0 N, up the incline; (b) 3.0 N, up the incline; (c) 1.6 N, up the incline; (d) 4.4 N, up the incline; (e) 1.0 N, down the incline. **83.** 0.54

Chapter 7

RE 7-1: (a) The 60 s encounter between the *Titanic* and an iceberg was a collision. (b) A tennis ball encountering a racket for 2 s is not a collision.

RE 7-2: (a) $|\vec{F}_1| > |\vec{F}_3| > |\vec{F}_2| = |\vec{F}_4| = 0$. Since the slopes represent $\Delta\vec{p}/\Delta t$ the magnitude is greatest where the slope is steepest. Thus, ranking is by steepness of slope. (b) Since the momentum is initially positive, the particle speeds up in region 1, drifts in region 2, and *slows down in region 3,* where its momentum is becoming less positive (and hence more negative).

RE 7-3: The change in the egg's momentum is $m\vec{v}_2 - m\vec{v}_1$, and since $\vec{v}_2$ is zero the change is just $m\vec{v}_1$. The time you take in catching the egg does not affect the momentum change since the initial and final velocities are still the same. However the time taken in the catch will affect the average force the egg experiences. Since the change in momentum equals the impulse, which equals the average force times the time the force acts, making the time of the catch longer makes the average force on the egg less and hence a greater likelihood of a successful catch. In order to make Δt as large as possible, you move your hands and body backwards once the catch is made in order to bring the egg to zero speed over the largest time interval possible.

RE 7-4: (a) p_{1x} is to the right and $+$, p_{2x} is to the left and $-$, therefore Δp_x is $-$. Remember that Δ is always final minus initial, and here we have a negative number minus a positive number giving a negative result. (b) Δp_y is zero since the y component of the momentum does not change in the bounce. (c) The direction of $\Delta\vec{p}$ is left. To see this, draw the two momentum vectors and subtract the initial from the final. Remember: To subtract vectors add the negative of the second to the first.

RE 7-5: (1) Assuming the carts are frictionless, the system consisting of the firecracker and the two carts is an isolated system and momentum should be conserved. In fact, if the firecracker is initially at rest and explodes symmetrically, then the carts should move off at the same speed in opposite directions. (2) Assuming the carts are not frictionless, then the track and the table and the Earth become part of the system. We might not see the carts come off with the same speeds in opposite directions. Instead the Earth might move (imperceptibly) to make up the difference. However, momentum is always conserved, so it should be so for our new system.

RE 7-6: (a) Zero, since no external forces are acting and hence the total momentum is conserved. (b) No, since the y-component of momentum must also be conserved. (c) The second piece must be moving in the negative direction on the x axis so that the total momentum after the explosion is zero.

RE 7-7: We need a mass for the grapefruit—let's say 1.0 kg. The grapefruit's momentum starts at zero and goes to (1 kg)(2 m/s) = 2 kg · m/s, therefore $\Delta p = 2$ kg · m/s. The change in the Earth's momentum will be equal and opposite, therefore the change in the Earth's speed will be 2 kg · m/s divided by the mass of the Earth. If you look at the inside front cover of this text, you find $m_{Earth} = 5.98 \times 10^{24}$ kg. Dividing, you get $v_{Earth} = 3.3 \times 10^{-25}$ m/s. Did you feel the Earth move?

Problems

1. 24 km/h. **3.** (a) $(-4.0 \times 10^4\text{ kg}\cdot\text{m/s})\,\hat{\text{i}}$; (b) west. **5.** (a) 30°; (b) $(-0.572\text{ kg}\cdot\text{m/s})\,\hat{\text{j}}$. **7.** 2.5 m/s. **9.** 3000 N. **11.** 67 m/s, in opposite direction. **13.** (a) 42 N · s; (b) 2100 N. **15.** (a) $(7.4 \times 10^3\text{ N}\cdot\text{s})\hat{\text{i}} + (-7.4 \times 10^3\text{ N}\cdot\text{s})\hat{\text{j}}$; (b) $(-7.4 \times 10^3\text{ N}\cdot\text{s})\hat{\text{i}}$; (c) 2.3×10^3 N; (d) 2.1×10^4 N; (e) −45°. **17.** 10 m/s. **19.** (a) 1.0 kg · m/s; (b) 10 N; (c) 1700 N; (d) the answer for (b) includes time between pellet collisions. **21.** 41.7 cm/s. **23.** (a) 46 N; (b) none. **25.** ≈2 mm/y. **27.** 3.0 mm/s, away from the stone. **29.** (a) 4.6 m/s; (b) 3.9 m/s; (c) 7.5 m/s. **31.** increases by 4.4 m/s. **33.** 190 m/s. **35.** (a) $\{m_A/(m_A + m_B)\}v_{A\,1}$. **37.** (a) 7290 m/s; (b) 8200 m/s. **39.** 4400 km/h. **41.** 8.1 m/s at 38° south of east. **43.** (a) 11.4 m/s; (b) 95.1° clockwise from $+x$. **45.** (a) 61.7 km/h; (b) 63.4° south of west. **47.** (a) 2.5 m/s. **49.** 1.0 m/s north. **51.** (a) 1.4×10^{-22} kg · m/s; (b) 150°; (c) 120°. **53.** 14 m/s, 135° from the other pieces. **55.** 3.0 m/s. **57.** 120°. **59.** (a) 4.15×10^5 m/s; (b) 4.84×10^5 m/s. **61.** (a) 41°; (b) 4.76 m/s; (c) no. **63.** 2.0 m/s, $-x$ direction. **65.** 108 m/s. **67.** (a) 1.57×10^6 N; (b) 1.35×10^5 kg; (c) 2.08 km/s. **69.** 2.2×10^{-3}

Chapter 8

RE 8-1: (a) At the center; (b) in the lower right quadrant; (c) on the negative y axis; (d) at the center; (e) in the lower left quadrant; (f) at the center

RE 8-2: (a) The spacing between successive halfway points is the same, which suggests that the velocity represented by these points is constant.

(b) $v = |\Delta\vec{r}|/\Delta t = 0.41\text{ m}/[(12/15)\text{s}] = 0.51$ m/s

RE 8-3: Since there are no outside forces on the system, the center of mass of the system will not change. Thus, the skaters will end up meeting at the origin of the original coordinate system in all three situations (a), (b), and (c). The only difference is that in case (a) Ethel will be holding one end of the "massless" pole at the end, in case (b) Fred will be holding an end of the "massless" pole, and in case (c) one-third of the "massless" pole will be sticking out behind Fred and two-thirds will be sticking out behind Ethel.

Problems

1. (a) -4.5 m; (b) -5.5 m. **3.** (a) 4600 km; (b) $0.73R_e$. **5.** (a) 1.1 m; (b) 1.3 m; (c) shifts toward topmost particle. **7.** (a) -0.25 m; (b) 0. **9.** 6.8×10^{-12} m from the nitrogen atom, along axis of symmetry. **11.** (a) $H/2$; (b) $H/2$; (c) descends to lowest point and then ascends to $H/2$; (d) $\frac{HM}{m}\left(\sqrt{1 + \frac{m}{M}} - 1\right)$. **13.** $x_{\text{com}} = B/2$ and $y_{\text{com}} = H/3$. **15.** $x_{\text{com}} = B/2$ and $y_{\text{com}} = 4R/(3\pi)$. **17.** (a) 0,0; (b) 0. **19.** (-1.50 m, -1.43 m). **21.** 29 m. **23.** 72 km/h. **25.** (a) 28 cm; (b) 2.3 m/s. **27.** 53 m. **29.** (a) halfway between the containers; (b) 26 mm toward the heavier container; (c) down; (d) -1.6×10^{-2} m/s^2. **31.** 4.2 m. **33.** 1.2 m/s, 132° counterclockwise from east. **37.** (a) 33 m/s; (b) 8.7 m/s. **39.** (a) 540 m/s; (b) 40.4°. **41.** (a) $0.2000v^{\text{rel}}$; (b) $0.2103v^{\text{rel}}$; (c) $0.2095v^{\text{rel}}$. **43.** (a) 1.0 m/s north; (b) 3 m north

Chapter 9

RE 9-1: (a) Decreases. (b) Remains the same. Remember that the kinetic energy is a scalar and depends on the velocity squared, so -2 m/s and 2 m/s give the same kinetic energy. (c) Negative for situation (a) and zero for situation (b). Situation (b) is interesting. How can the net work done be zero? Try breaking the velocity change into two changes: first from -2 m/s to zero, then from zero to 2 m/s. For the first change the work is negative and for the second change the work is positive. When we add the two works together, we get zero for the total.

RE 9-2: $c > a > b = d$

RE 9-3: Use Eq. 9-19: (a) positive; (b) negative; (c) zero. Think through your calculated answers. Do they make sense? For example, in (a) as the block moves from -3 cm to the origin, the spring force and displacement are in the same direction giving a positive work; from the origin to 2 cm the spring force and displacement are in opposite directions giving a negative work, but the positive work is larger because the displacement is larger giving a net positive work.

RE 9-4: $d > c > b > a$

RE 9-5: The power is zero at all times since $\vec{F}$ and $\vec{v}$ are always perpendicular in uniform circular motion.

Problems

1. 1.2×10^6 m/s. **3.** (a) 3610 J; (b) 1900 J; (c) 1.1×10^{10} J. **5.** (a) 2.9×10^7 m/s; (b) 2.1×10^{-13} J. **7.** (a) 7.5×10^4 J; (b) 3.8×10^4 kg · m/s; (c) 38° south of east. **9.** 1.18×10^4 kg. **11.** (a) 3.7 m/s; (b) 1.3 N·s; (c) 1.8×10^2 N. **13.** (a) 42 J; (b) 30 J; (c) 12 J; (d) 6.48 m/s, positive direction of x axis; (e) 5.48 m/s, positive direction of x axis; (f) 3.46 m/s, positive direction of x axis. **15.** AB: +, BC: 0, CD: −, DE: +. **17.** (a) 170 N; (b) 340 m; (c) -5.8×10^4 J; (d) 340 N; (e) 170 m; (f) -5.8×10^4 J. **19.** 800 J. **21.** (a) 98 N; (b) 4.0 cm; (c) 3.9 J; (d) -3.9 J. **23.** 0, by both methods. **25.** (a) -0.043 J; (b) -0.13 J. **27.** (a) 6.0 N; (b) $-$ 2.5 N; (c) 15 N **29.** 15.3 J. **31.** (a) 590 J; (b) 0; (c) 0; (d) 590 J. **33.** 6.8 J. **35.** (a) 1.20 J; (b) 1.10 m/s. **37.** (a) 1.50 J; (b) increases. **39.** (a) 1.2×10^4 J; (b) $-$ 1.1×10^4 J; (c) 1100 J; (d) 5.4 m/s. **41.** (a) $-3Mgd/4$; (b) Mgd; (c) $Mgd/4$; (d) $\sqrt{gd/2}$. **43.** 20 J. **45.** (a) 8.84×10^3 ; (b) 7.84×10^3 J; (c) 6.84×10^3 J. **47.** (a) 2.3 J; (b) 2.6 J. **49.** 490 W. **51.** (a) 0.83 J; (b) 2.5 J; (c) 4.2 J; (d) 5.0 W. **53.** 740 W. **55.** 68 kW. **57.** (a) 1.8×10^5 ft · lb; (b) 0.55 hp. **59.** (a) 8.8 m/s; (b) 2600 J; (c) 1.6 kW. **61.** 24 W **63.** (a) 2.1×10^6 kg; (b) $\sqrt{100 + 1.5t}$ m/s; (c) $(1.5 \times 10^6)/\sqrt{100 + 1.5t}$ N; (d) 6.7 km **65.** (a) $\approx 1 \times 10^5$ megatons; (b) $\approx$ ten million bombs

Chapter 10

RE 10-1: No, for the force to be conservative the work done in going between two points must not depend on the path taken. Also, if you go from 2 to 1 instead of 1 to 2 the work will change sign. Therefore, for the force in the exercise to be conservative the work for the bottom path should have a negative sign.

RE 10-2: A Hot Wheels® car that traverses path b should lose more kinetic energy than one that traverses path a. This is because path b is longer so the friction forces have more distance to act on path b.

RE 10-3: The kinetic energy of the barbell is zero before the lift and zero after the lift, as evidenced by the fact that y vs. t is a constant at $t = 0.0$ s and at $t = 2.0$ s. Since the kinetic energy change $\Delta K = 0.0$ J, then the net work on the barbells should be zero. An examination of graph 10-10b shows that the positive work is approximately given by the area under the F^{net} vs. y curve. $W^+ =$ area under the positive portion of the curve $= (0.5)(116\text{ N})(.15\text{ m}) = +8.7$ J and $W^- =$ area under the negative portion of the curve $(0.5)(58\text{ J})(.45 - .15)\text{ m} = -8.7$ J. So $W^{\text{net}} = W^+ + W^- = 0.0$ J.

RE 10-4: Use Eq. 10-13. Note that the change in the potential energy is the negative of the area under the curves in the figure. The most positive will be (3) and the least positive (2).

RE 10-5: Without friction, the decrease in the potential energy will equal the increase in the kinetic energy. (a) Therefore, since all four blocks are losing the same amount of potential energy, they will all have the same kinetic energy at point B. (b) Since the kinetic energies are the same, the speeds are the same.

RE 10-6: Use the equation $F_x^{\text{int}}(x) = -dU(x)/dx$. The force is the negative of the slope of the U vs. x curve. (a) Ranking *magnitudes* with the greatest first: CD, AB, BC. (b) The slope is negative, hence the force is in the positive x direction.

RE 10-7: $b > a > c$ as determined by the equation $\Delta E^{\text{thermal}} = f_x^{\text{kin}}\Delta x$.

RE 10-8: (a) 4 kg · m/s; (b) 8 kg · m/s; (c) assuming an elastic collision, 3 J.

RE 10-9: (a) 2 kg · m/s. (b) Since the initial y-component is zero, the final must be zero. Therefore, the final y-component of momentum for the target is 3 kg · m/s.

Problems

1. 89 N/cm. **3.** (a) 4.31 mJ; (b) -4.31 mJ; (c) 4.31 mJ; (d) -4.31 mJ; (e) all increase. **5.** (a) mgL; (b) $-mgL$; (c) 0; (d) $-mgL$; (e) mgL; (f) 0; (g) same. **7.** (a) 184 J; (b) -184 J; (c) -184 J. **9.** -320 J **11.** (a) 2.08 m/s; (b) 2.08 m/s; (c) increase. **13.** (a) $\sqrt{2gL}$; (b) $2\sqrt{gL}$; (c) $\sqrt{2gL}$; (d) all the same. **15.** (a) 260 m; (b) same; (c) decrease. **17.** (a) 21.0 m/s; (b) 21.0 m/s; (c) 21.0 m/s. **19.** (a) 0.98 J; (b) -0.98 J; (c) 3.1 N/cm. **21.** (a) 39.2 J; (b) 39.2 J; (c) 4.00 m. **23.** (a) 35 cm; (b) 1.7 m/s. **25.** 10 cm. **27.** 1.25 cm. **31.** (a) $2\sqrt{gL}$; (b) $5mg$; (c) 71°. **33.** $mgL/32$. **37.** (a) $1.12(A/B)^{1/6}$; (b) repulsive; (c) attractive. **39.** (a) -3.7 J; (c) 1.29 m; (d) 9.12 m; (e) 2.16 J ; (f) 4.0 m; (g) $(4 - x)e^{-x/4}$ N; (h) 4 m. **41.** (a) 30.1 J; (b) 30.1 J; (c) 0.22. **43.** (a) 5.6 J; (b) 3.5 J. **45.** 11 kJ. **47.** 20 ft · lb. **49.** (a) 1.5 MJ; (b) 0.51 MJ; (c) 1.0 MJ; (d) 63 m/s. **51.** (a) 67 J;

(b) 67 J; (c) 46 cm. **53.** (a) 31.0 J; (b) 5.35 m/s; (c) conservative. **55.** (a) 44 m/s; (b) 0.036. **57.** (a) −0.90 J; (b) 0.46 J; (c) 1.0 m/s. **59.** 1.2 m. **63.** in the center of the flat part. **65.** (a) 216 J; (b) 1180 N; (c) 432 J; (d) motor also supplies thermal energy to crate and belt. **67.** (a) 0.2 to 0.3 MJ; (b) same amount. **69.** (a) 860 N; (b) 2.4 m/s. **71.** (a) $mR(\sqrt{2gh} + gt)$; (b) 5.06 kg. **73.** (a) $mv_1/(m + M)$; (b) $M/(m + M)$. **75.** 25 cm. **79.** (a) 41°; (b) 4.76 m/s; (c) no. **81.** (a) 6.9 m/s, 30° to $+x$ direction; (b) 6.9 m/s, −30° to $+x$ direction; (c) 2.0 m/s, $-x$ direction. **83.** (a) 99 g; (b) 1.9 m/s; (c) 0.93 m/s. **85.** 7.8%. **87.** (a) 1.2 kg; (b) 2.5 m/s. **89.** (a) 100 g; (b) 1.0 m/s. **91.** (a) 1.9 m/s, to the right; (b) yes; (c) no, total kinetic energy would have increased. **93.** (a) 1/3; (b) $4h$. **95.** 1.0 kg. **97.** (c) 11%; (d) 10%; (e) 79%

Chapter 11

RE 11-1: (a) Positive, since θ is increasing. (b) Negative, since θ is decreasing.

RE 11-2: (a) Positive; (b) negative; (c) negative; (d) positive

RE 11-3: Find the angular acceleration, α, by taking the second derivative of θ with respect to t. The accelerations for (a) and (d) do not depend on t and are therefore constant, and hence the equations of Table 11-1 apply.

RE 11-4: Since the speeds are being squared, v^2 and ω^2 will always be positive quantities.

RE 11-5: (a) Yes, the centripetal acceleration; (b) no, since α is zero; (c) yes; (d) yes, since α is no longer zero.

RE 11-6: Calculate mr^2 for each, and you'll find they are all the same.

RE 11-7: (1) > (2) > (4) > (3). Remember that I depends not only on the mass but also on how far that mass is from the chosen axis.

RE 11-8: $I_a = I_d = mr^2$, $I_b = \frac{1}{2}mr^2$, $I_c = \frac{5}{8}mr^2$, so $a = d > c > b$.

RE 11-9: $A = C > D > B = E =$ zero. For A and C, ϕ is 90°; for D, ϕ is between zero and 90°; for E, ϕ is zero; and for C, r is zero.

RE 11-10: (a) Same direction. (b) Less.

Problems

1. (a) $a + 3bt^2 - 4ct^3$; (b) $6bt - 12ct^2$. **3.** (a) 5.5×10^{15} s; (b) 26. **5.** (a) 2 rad; (b) 0; (c) 130 rad/s; (d) 32 rad/s²; (e) no. **7.** 11 rad/s. **9.** (a) −67 rev/min²; (b) 8.3 rev. **11.** 200 rev/min. **13.** 8.0 s. **15.** (a) 44 rad; (b) 5.5 s, 32 s; (c) −2.1 s, 40 s. **17.** (a) 340 s; (b) -4.5×10^{-3} rad/s²; (c) 98 s. **19.** 1.8 m/s², toward the center. **21.** 0.13 rad/s. **23.** (a) 3.0 rad/s; (b) 30 m/s; (c) 6.0 m/s²; (d) 90 m/s². **25.** (a) 3.8×10^3 rad/s; (b) 190 m/s. **27.** (a) 7.3×10^{-5} rad/s; (b) 350 m/s; (c) 7.3×10^{-5} rad/s; (d) 460 m/s. **29.** 16 s. **31.** (a) -2.3×10^{-9} rad/s²; (b) 2600 y; (c) 24 ms. **33.** 12.3 kg · m². **35.** (a) 1100 J; (b) 9700 J. **37.** (a) $5md^2 + 8/3Md^2$; (b) $(5/2m + 4/3M)d^2\omega^2$. **39.** 0.097 kg · m². **41.** $\frac{1}{3}M(a^2 + b^2)$. **45.** 4.6 N · m. **47.** (a) $r_1F_A \sin\theta_1 - r_2F_B \sin\theta_2$; (b) −3.8 N · m. **49.** (a) 28.2 rad/s²; (b) 338 N · m. **51.** (a) 155 kg · m²; (b) 64.4 kg. **53.** 130 N. **55.** (a) 6.00 cm/s²; (b) 4.87 N; (c) 4.54 N; (d) 1.20 rad/s²; (e) 0.0138 kg · m². **57.** (a) 1.73 m/s²; (b) 6.92 m/s². **59.** 396 N · m. **61.** (a) $mL^2\omega^2/6$; (b) $L^2\omega^2/6g$. **63.** 5.42 m/s **65.** $\frac{3}{2}\sqrt{\frac{g}{L}}$. **67.** (a) $[(3g/H)(1 - \cos\theta)]^{0.5}$; (b) $3g(1 - \cos\theta)$; (c) $3/2g \sin\theta$; (d) 41.8°. **69.** (a) $0.083519ML^2 \approx 0.084ML^2$; (b) low by (only) 0.22%

Chapter 12

RE 12-1: (a) When is the sin of the angle between the vectors zero? Sin is zero for 0° and 180°. (b) Here the sin needs to equal ±1. This occurs at 90° and 270°. (c) Here $|\vec{c}||\vec{d}| \sin\phi = 3 \cdot 4 \sin\phi = 6$ so $\phi = \sin^{-1}(6/12)$ so $\phi = 30°$ or 150°.

RE 12-2: The time rate of change of the rotational momentum is equal to the net torque. 3 > 1 > 2 = 4 = zero.

RE 12-3: (a) 1 = 3 > 2 = 4 > 5 = zero, since $r_\perp$ is 4 m for both 1 and 3 and 2 m for both 2 and 4 and zero for 5. (b) Particles 2 and 3 have negative rotational momentum about o, since $\vec{\ell} = \vec{r} \times \vec{p}$ points into the page for each of them.

RE 12-4: (a) Since the rate of change of the rotational momentum is equal to the applied torque, which is the same for all three cases, all three objects increase their rotational momentum at the same rate; and assuming all three started from rest, they will all have the same rotational momentum at any given time. (b) Look at Table 11-2 (Some Rotational Inertias). Note that $I_{\text{hoop}} > I_{\text{disk}} > I_{\text{sphere}}$. Since $L = I\omega$ and they all have the same L, the object with the biggest I will have the smallest ω; $\omega_{\text{sphere}} > \omega_{\text{disk}} > \omega_{\text{hoop}}$.

RE 12-5: (a) Decrease, since although the total mass of the system has not changed, it is distributed closer to the axis of rotation. (b) Remain the same, since there is no net external torque. (c) If I decreases and L is constant, then ω must increase.

Problems

1. (a) 59.3 rad/s; (b) 9.31 rad/s²; (c) 70.7 m. **3.** −3.15 J. **5.** 1/50 **7.** (a) 8.0°; (b) more. **9.** (a) 13 cm/s²; (b) 4.4 s; (c) 55 cm/s; (d) 1.8×10^{-2} J; (e) 1.4 J; (f) 27 rev/s. **11.** (a) 10 s; (b) 897 m. **13.** the third. **17.** (a) 10 N · m, parallel to yz plane, at 53° to $+y$; (b) 22 N · m, $-x$. **19.** (a) (50 N · m)$\hat{k}$; (b) 90°. **21.** (a) (−170 kg · m²/s)$\hat{k}$; (b) (+56 N · m)$\hat{k}$; (c) (+56 kg · m²/s²)$\hat{k}$. **23.** (a) 0; (b) $8t$ N · m, in $-z$ direction; (c) $2/\sqrt{t}$ N · m, $-z$; (d) $8/t^3$ N · m, $+z$. **25.** 9.8 kg · m²/s. **27.** (a) 0; (b) (8.0 N · m)$\hat{i}$ + (8.0 N · m)$\hat{k}$. **29.** (a) mvd; (b) no; (c) 0, yes. **31.** (a) −1.47 N · m; (b) 20.4 rad; (c) −29.9 J; (d) 19.9 W. **33.** (a) $14md^2$; (b) $4md^2\omega$; (c) $14md^2\omega$. **35.** $\omega_1 R_A R_B I_A/(I_A R_B^2 + I_B R_A^2)$. **37.** (a) 3.6 rev/s; (b) 3.0; (c) in moving the bricks in, the forces on them from the man transferred energy from internal energy of the man to kinetic energy. **39.** (a) 267 rev/min; (b) ²/₃. **41.** (a) 149 kg · m²; (b) 158 kg · m²/s; (c) 0.746 rad/s **43.** $\frac{m}{M + m}\left(\frac{v}{R}\right)$ **45.** (a) $(mRv - I\omega_1)/(I + mR^2)$; (b) no, energy transferred to internal energy of cockroach. **47.** 3.4 rad/s. **49.** (a) 0.148 rad/s; (b) 0.0123; (c) 181°. **51.** The day would be longer by about 0.8 s. **53.** (a) 18 rad/s; (b) 0.92 **55.** (a) 0.24 kg · m²; (b) 1800 m/s **57.** $\theta = \cos^{-1}\left[1 - \frac{6m^2h}{d(2m + M)(3m + M)}\right]$. **59.** 11.0 m/s **61.** (a) 0.180 m ; (b) clockwise

Chapter 13

RE 13-1: Situations (c), (e), and (f) can yield static equilibrium, since in each case both the net force and the net torque can be zero. In (a), (b), and (d) the net force can be zero but the net torque cannot.

RE 13-2: The apple's center of gravity will end up directly below the rod, since only in that position is the net torque on the apple *stably* zero. The net torque on the apple is also zero when the apple's center of gravity is directly *above* the rod, but this is an *unstable* equilibrium point and the slightest rotation will cause the apple to rotate away from this position.

RE 13-3: You are better off if there is no friction between the ladder and the wall. With no friction between the ladder and the ground, the ground cannot exert any *horizontal* force to counter the horizontal force that the wall must exert on the ladder to keep it in place.

RE 13-4: In each of these three cases, the net horizontal force is zero independent of the magnitudes of the three unknown forces. This leaves only two independent equations for equilibrium—namely, net vertical force equals zero and net torque equals zero. But we have three unknowns to solve for. Since we can't do this, each of these three situations is indeterminate.

RE 13-5: Equation 13-29 tells us that, for elastic stretching, Young's modulus is just the stress (F/A) divided by the strain ($\Delta L/L$). Relative to rod 1, rod 2 has the same stress and twice the strain, and so its Young's modulus is half that of rod 1. By the same reasoning, rod 3 also has half the Young's modulus of rod 1, and rod 4 has a Young's modulus that is four times larger than that of rod 1. So, from higher to lower Young's modulus, rod 4 is the largest, rod 1 is next, and rods 2 and 3 tie for smallest.

RE 13-6: During bending, the particles on the inside of the bend are pushed closer together while those on the outside of the bend are pulled farther apart. During a shear deformation, adjacent planes of particles shift laterally with respect to one another. While the planes remain the same distance from one another, the "springs" (bonds) between adjacent planes are each stretched by the same amount.

Problems

1. (a) 2; (b) 7 **3.** (a) $(-27\text{ N})\hat{\text{i}} + (2\text{ N})\hat{\text{j}}$; (b) 176° counterclockwise from $+x$ direction **5.** 7920 N **7.** (a) $(mg/L)\,\sqrt{L^2 + r^2}$; (b) mgr/L **9.** (a) 1160 N, down; (b) 1740 N, up; (c) left; (d) right **11.** 74 g **13.** (a) 280 N; (b) 880 N, 71° above the horizontal **15.** (a) 8010 N; (b) 3.65 kN; (c) 5.66 kN **17.** 71.7 N **19.** (a) 5.0 N; (b) 30 N; (c) 1.3 m **21.** $mg\sqrt{\dfrac{2rh - h^2}{r - h}}$ **23.** (a) 192 N; (b) 96.1 N; (c) 55.5 N **25.** (a) 6630 N; (b) 5740 N; (c) 5960 N **27.** 2.20 m **29.** 0.34 **31.** (a) 211 N; (b) 534 N; (c) 320 N **33.** (a) 445 N; (b)0.50; (c) 315 N **35.** (a) slides at 31°; (b) tips at 34° **37.** (a) 6.5×10^6 N/m²; (b) 1.1×10^{-5} m **39.** (a) 867 N; (b) 143 N; (c) 0.165 **41.** 44 N

Chapter 14

RE 14-1: The ratio (relative amount) of the magnitudes of these two forces depends only on the square of the ratio of the two center-to-center distances between the Earth and the other mass.

RE 14-2: Equation 14-2 tells us that g (the acceleration of a freely falling body) is just (Gm^{Earth}/r^2) where r is the distance to the center of the Earth to the point where g is measured. This is consistent with the model that assumes that the Moon stays in its orbit simply because it is in free fall. Although the observations are consistent with this model, this does not "prove" that the model is "true." It only establishes that this model is "good enough" to account for the data at hand.

RE 14-3: Since the location of the particle lies outside each of the spheres at the same distance from the center of the sphere in each case, each of the spheres will exert exactly the *same* magnitude force on the particle.

RE 14-4: $\vec{F}^{\text{grav}}$ due to the Earth *always* points directly toward the center of the Earth. However, the object's apparent weight associated with $\vec{N}$, (the "normal" force exerted on an object "at rest" on the surface of the rotating Earth), is *not* always directed exactly *away* from the center of the Earth! In fact, $\vec{N}$ points directly away from the center of the Earth only at the Earth's poles and at its equator. Why?

RE 14-5: In each case the direction of $\vec{F}^{\text{grav}}$ would be toward the center of the Earth. Case A: The magnitude of $\vec{F}^{\text{grav}}$ would decrease as $1/r^2$ where r is the distance to the center of the Earth. Case B: The magnitude of $\vec{F}^{\text{grav}}$ would be proportional to r, *and hence decrease.* Case C: Because of the considerably higher density of the Earth's core compared with its surface crust, the magnitude of $\vec{F}^{\text{grav}}$ would increase at first but then decrease to zero as we moved toward the center of the Earth.

RE 14-6: (a) The gravitational potential energy of the ball–sphere system increases. (b) The gravitational force between the ball and the sphere is attractive (inward), and the displacement is outward. Since the force and displacement are in opposite directions, the work done by the gravitational force is negative.

Problems

1. 19 m **3.** 29 pN **5.** 1/2 **7.** 2.60×10^5 km **9.** 0.017 N, toward the 300 kg sphere **11.** 3.2×10^{-7} N **13.** $\dfrac{GmM}{d^2}\left[1 - \dfrac{1}{8(1 - R/2d)^2}\right]$ **15.** 2.6×10^6 m **17.** (b) 1.9 h **21.** (a) $0.414R$; (b) $0.5R$ **23.** (a) $(3.0 \times 10^{-7}\text{ N/kg})m$; (b) $(3.3 \times 10^{-7}\text{ N/kg})m$; (c) $(6.7 \times 10^{-7}\text{ N/kg·m})mr$ **25.** (a) 9.83 m/s²; (b) 9.84 m/s²; (c) 9.79 m/s² **27.** (a) -1.3×10^{-4} J; (b) less; (c) positive; (d) negative **29.** (a) 0.74; (b) 3.7 m/s²; (c) 5.0 km/s **31.** (a) 5.0×10^{-11} J; (b) -5.0×10^{-11} J **35.** (a) 1700 m/s; (b) 250 km; (c) 1400 m/s **37.** (a) 82 km/s; (b) 1.8×10^4 km/s **39.** 2.5×10^4 km

Chapter 15

RE 15-1: Half the weight of the woman, (125 lb/2)(9.8 N/2.2 lb) ≈ 300 N, is supported by her two spike heels. Let's say that each heel makes contact with 1 cm² = 10^{-4} m² of the floor. Then the pressure of her heels on the floor is $P = F/A = (300\text{ N}/10^{-4}\text{ m}^2) = 3 \times 10^6$ Pa. This estimate is close to that presented in the table. This pressure is high because of the small contact area over which this otherwise modest force is applied. An automobile has a much larger contact area.

RE 15-2: If air and water are made up of molecules that are about the same size and mass, then the average distance between the molecules in air at sea level must be about $1000^{1/3} = 10$ times larger than those of the water. This suggests that there is significantly more empty space around each air molecule, allowing them to be compressed closer together by quite a bit before they fill all of the available volume.

RE 15-3: The force the air exerts on the book is about (10^5 N/m²) (2.54×10^{-2} m/in)² (2.2 lb/9.8 N) (8 in)(10 in) ≅ 1200 lb. The close fit and the flexibility of the rubber mat prevents air from leaking into the space between the mat and the smooth tabletop, holding the mat down against the table with close to the full 1200 lb of force the air exerts on the top surface of the mat. The rougher surface of the book, as well as its rigidity, let air readily leak into the space between the book and the table when you start to pick up the book. This "equalizes" the pressure on each side of the book, reducing the net force that the air exerts on the book to a negligible amount.

RE 15-4: The density of air is only about one-thousandth that of water.

RE 15-5: The pressure at a depth Δy is the *same* in each container of oil. The shape of the container does not matter.

RE 15-6: Compressible fluids, like compressible springs, can "absorb" work and store it as elastic potential energy. Increasing the pressure of the compressible fluid in the hydraulic jack will thus store some of the work done on the fluid as elastic potential energy and slightly reduce the amount of work that the fluid does on the output by that amount.

RE 15-7: The pressure at the bottom of this container is determined solely by the depth of the fluid above the bottom and the pressure that the air exerts on the surface of that fluid. In particular, the weight of the "extra" fluid that lies outside the central column is *not* carried by the horizontal bottom of the container and does *not* increase the pressure there.

RE 15-8: (a) Since the penguin floats in each of the three fluids, each fluid supplies a buoyant force exactly equal to the penguin's weight, so each fluid supplies the *same* buoyant force ($A = B = C$). (b) The penguin must displace the amount of fluid that matches her weight. Thus she *displaces* more of the least dense fluid B than of A, and even less of the most dense fluid C ($B > A > C$).

RE 15-9: You need to make sure that the weight of the canoe and its load is less than that of the water it displaces before the water starts coming in over the top edge of the hull. Although a chunk of concrete cannot displace its weight with water, a thin concrete canoe and its riders can.

RE 15-10: The net flow into (+) and out of (−) the entire system must be zero. So: $+x + (4 + 8 + 4 - 6 + 5 - 2)\ \text{cm}^3/\text{s} = 0\ \text{cm}^3/\text{s}$ or $x = -13\ \text{cm}^3/\text{s}$, so fluid flows out of the unlabeled pipe at a rate of $13\ \text{cm}^3/\text{s}$.

RE 15-11: (a) The area of face 1 is $4.0\ \text{cm}^2 = 4.0 \times 10^{-4}\ \text{m}^2$. Face 2 has an area of $5.7\ \text{cm}^2 = 5.7 \times 10^{-4}\ \text{m}^2$. Face 3 has an area of $8.0\ \text{cm}^2 = 8.0 \times 10^{-4}\ \text{m}^2$. (b) The total surface area of all 6 faces is $69.7\ \text{cm}^2 = 69.7 \times 10^{-4}\ \text{m}^2$.

RE 15-12: (a) The flux through any face is $v\ \Delta A \cos(\theta)$. Thus the flux through face 1 is (0.5 m/s) $(4.0 \times 10^{-4}\ \text{m}^2)$ (cos (0)) $= +2.0 \times 10^{-4}\ \text{m}^3/\text{s}$. The flux through face 2 is (0.5 m/s) $(5.7 \times 10^{-4}\ \text{m}^2)$ (cos (45°)) $= +2.0 \times 10^{-4}\ \text{m}^3/\text{s}$. The flux through face 3 is (0.5 m/s) $(8.0 \times 10^{-4}\ \text{m}^2)$ (cos (180°)) $= -4.0 \times 10^{-4}\ \text{m}^3/\text{s}$. (b) The flux through the front, back, and bottom faces is zero because $\theta = 90°$ for each of these faces and so the cos (θ) term in the expression for the flux is zero. (c) Adding the contributions from all six faces yields zero net flux through this closed surface, as expected.

RE 15-13: (a) The volume flow rate is the *same* through each of the four sections. (b) The flow speed is largest in section 1, followed by section 2 and section 3, where it will be the same, and finally section 4 has the smallest flow speed. Recall that the flow speed is inversely proportional to the local cross-sectional area of the pipe. (c) The pressure will be greatest in section 4, less in section 3, still less in section 2, and least in section 1. The pressure difference between sections 2 and 3 is due to their difference in altitude. The pressure differences between sections at the same altitude are due to differences in the flow speed.

Problems

1. 1.1×10^5 Pa or 1.1 atm **3.** 2.9×10^4 N **5.** 0.074 **7.** (b) 26 kN **9.** 5.4×10^4 Pa **11.** (a) 5.3×10^6 N; (b) 2.8×10^5N; (c) 7.4×10^5 N; (d) no **13.** 7.2×10^5 N **15.** $\frac{1}{4}\rho g A(h_2 - h_1)^2$ **17.** 1.7 km **19.** (a) $\rho g W D^2/2$; (b) $\rho g W D^3/6$; (c) $D/3$ **21.** (a) 7.9 km; (b) 16 km **23.** 4.4 mm **25.** (a) $2.04 \times 10^{-2}\ \text{m}^3$; (b) 1570 N **27.** (a) 670 kg/m^3; (b) 740 kg/m^3 **29.** (a) 1.2 kg; (b) 1300 kg/m^3 **31.** 57.3 cm **33.** 0.126 m^3 **35.** (a) 45 m^2; (b) car should be over center of slab if slab is to be level **37.** (a) 9.4 N; (b) 1.6 N **39.** 8.1 m/s **41.** 66 W **43.** (a) 2.5 m/s; (b) 2.6×10^5 Pa **45.** (a) 3.9 m/s; (b) 88 kPa **47.** (a) $1.6 \times 10^{-3}\ \text{m}^3/\text{s}$; (b) 0.90 m **49.** 116 m/s **51.** (a) 6.4 m^3; (b) 5.4 m/s; (c) 9.8×10^4 Pa **53.** (a) 74 N; (b) 150 m^3 **55.** (b) $2.0 \times 10^{-2}\ \text{m}^3/\text{s}$ **57.** (b) 63.3 m/s

Chapter 16

RE 16-1: The amplitude and the angular frequency will stay the same. The initial phase will differ from ϕ_0 by 90° or $\pi/2$ since you can think of a cosine as a sine that has been shifted 90° to the left.

RE 16-2: (a) When $t = 2.00\,T$ the particle will have moved through two full oscillations and will be back where it started from—namely, at $x = -X$. (b) When $t = 3.50T$ the particle will have moved through three full oscillations and an additional half oscillation and so will be at $x = +X$. (c) When $t = 5.25T$ the particle will have moved through five full oscillations and an additional quarter oscillation and so will be at $x = 0$.

RE 16-3: Equation 16-12 tells us that the period of a mass on a given spring increases as the amount of oscillating mass increases. The fact that the mass of the spring itself oscillates along with the mass on its end suggests that some of the spring's mass should be included in the mass that appears in Eq. 16-12. Since the spring oscillates with a progressively smaller amplitude as we go from its moving end to its fixed end, only some fraction of the spring's mass needs to be included in this corrected total oscillating mass.

RE 16-4: Only (a) implies simple harmonic motion. Although (b) is a restoring type of force, it is quadratic, not linear in x. Force (c) is repulsive rather than attractive, driving the particle away from $x = 0$ rather than back toward it. Force (d) is both repulsive and nonlinear.

RE 16-5: The particle's velocity component is zero at $t = t_2$ and t_4. The particle is moving to the left at its greatest speed at $t = t_1$ and it is moving to the right at its greatest speed at $t = t_3$. Considering v_x as a mathematical function of t, we can indeed say that v_x is a minimum at t_1 and a maximum at t_3, but do remember that it is actually moving at its fastest speed when the velocity component is both a minimum and a maximum.

RE 16-6: The vertical component a_x of the acceleration is *increasing* in regions 1 and 2 and it is *decreasing* in regions 3 and 4. Note, however, that the *magnitude* of this acceleration is actually *decreasing* in regions 1 and 3 while the magnitude is *increasing* in regions 2 and 4. Pause and reflect on this!

RE 16-7: In each of these cases, the net force acting on the pendulum mass is proportional to the mass itself. Since Newton's Second Law tells us that acceleration is net force divided by mass, the mass cancels out here and so acceleration will be independent of the mass in these cases.

RE 16-8: "Same shape and size" for these three pendula means that the rotational inertia of each is simply proportional to its mass *with the same constant of proportionality in each case*. "Suspended at the same point" means the same distance from the point of suspension to the center of mass in each case. Since I/m and h are the same for each of the three, Eq. 16-26 tells us that each will have the same period.

RE 16-9: Since $K = 3$ J and $U = 2$ J at *a* given point, then the total mechanical energy of this system is $E = K + U = 5$ J *at every point* in its motion. Conservation of mechanical energy rules! (a) In particular, when the block is at $x = 0$, the system's potential energy is zero and so its kinetic energy must be 5 J. (b) At $x = -X$, the system's kinetic energy is zero so then $U = 5$ J.

RE 16-10: From Eq. 16-39 the time it takes for the mechanical energy of a damped oscillator to fall to one-fourth (or to any given fraction, for that matter) of its initial value is proportional to m/b. The ratio of m/b for set 2 is 4/6 = 2/3 that of set 1, and for set 3 it is 1/3 of that for set 1. Thus set 1 takes the longest time to lose one-fourth of its mechanical energy, followed by set 2, then by set 3. (set 1 > set 2 > set 3)

Problems

1. (a) 0.50 s; (b) 2.0 Hz; (c) 18 cm **3.** (a) 0.500 s; (b) 2.00 Hz; (c) 12.6 rad/s; (d) 79.0 N/m; (e) 4.40 m/s; (f) 27.6 N **5.** $f > 500$ Hz **7.** (a) 6.28×10^5 rad/s; (b) 1.59 mm **9.** (a) 1.0 mm; (b) 0.75 m/s; (c) 570 m/s^2 **11.** (a) 1.29×10^5 N/m; (b) 2.68 Hz **13.** 7.2 m/s **15.** 2.08 h **17.** 3.1 cm **19.** (a) 5.58 Hz; (b) 0.325 kg; (c) 0.400 m **21.** (a) 2.2 Hz; (b) 56 cm/s; (c) 0.10 kg; (d) 20.0 cm below y_i 23. (a) $0.183A$; (b) same direction **29.** (a) $(n+1)k/n$; (b) $(n+1)k$; (c) $\sqrt{(n+1)/n}f$; (d) $\sqrt{n+1}f$ **31.** (a) 39.5 rad/s; (b) 34.2 rad/s; (c) 124 rad/s^2 **33.** 99 cm **35.** 5.6 cm **37.** (a) $2\pi\sqrt{\dfrac{L^2 + 12d^2}{12gd}}$; (b) increases for $d < L/\sqrt{12}$, decreases for $d > L/\sqrt{12}$; (c) increases; (d) no change **39.** (a) 0.205 kg·m^2; (b) 47.7 cm; (c) 1.50 s **41.** $2\pi\sqrt{m/3k}$ **43.** (a) 0.35 Hz; (b) 0.39 Hz; (c) 0 **45.** (b) smaller **47.** (a) $(r/R)\sqrt{k/m}$; (b) $\sqrt{k/m}$; (c) no oscillation **49.** 37 mJ **51.** (a) 2.25 Hz; (b) 125 J; (c) 250 J; (d) 86.6 cm **53.** (a) $\frac{3}{4}$; (b) $\frac{1}{4}$; (c) $x^{max}/\sqrt{2}$ **55.** (a) 16.7 cm; (b) 1.23% **57.** 0.39 **59.** (a) 14.3 s; (b) 5.27 **61.** (a) $F^{max}/b\omega$; (b) F^{max}/b

Chapter 17

RE 17-1: (a) None of these graphs correctly shows the displacement of the rope versus position along the rope at $t = 0$ s. Graph (b) is the closest to correct of the four but fails to show the considerable length of undisturbed rope that lies between $x = 0$ and the trailing (left) edge of the pulse at $t = 0$ s. (b) Graph (a) correctly shows the displacement of the rope versus time at $x = x_1$ as the pulse passes by.

RE 17-2: Realizing that each of these phase expressions is of the form $(kx - \omega t)$ and that the wavelength $\lambda = 2\pi/k$, we see that wave 1 has the smallest wavelength and thus the largest k so it must correspond to case (c). Wave 2 has the smallest k and so must go with case (a), and wave 3 has the middle value of k and so matches case (b).

RE 17-3: The velocity v_y^{string} describes the up and down (transverse) motion of a particular small segment of the string as the wave passes by that location. Typically the velocity varies rapidly between positive and negative values as the wave goes by. The magnitude of the maximum of this velocity increases with increasing amplitude of passing wave having the same wavelength and frequency. The other velocity, v_x^{wave} tells us how fast and in what direction any given *crest* of the wave itself moves along the rope. For a uniform rope its time does not vary at all and it does not depend on the amplitude of the wave. In the derivation in this section, v_x^{wave} is used to obtain the mass of the segment of string under study and v_y^{string} is used to obtain the momentum change of the segment.

RE 17-4: If you increase the frequency of the oscillations driving waves in a string, holding the tension constant, then (a) the speed of the waves remains the same and (b) the wavelength decreases. If you instead increase the tension keeping the driving frequency constant, then (c) the wave speed increases and (d) the wavelength also increases.

RE 17-5: (a) Equation (1) represents the interference of a pair of waves traveling in the positive x direction. (b) Equation (3) represents the interference of a pair of waves traveling in the negative x direction. (c) Equation (2) represents the interference of a pair of waves traveling in opposite directions.

RE 17-6: (a) The missing frequency is 75 Hz. (b) The seventh harmonic has a frequency of 7×75 Hz = 525 Hz.

Problems

1. (a) 3.49 m^{-1}; (b) 31.5 m/s **3.** (a) 0.68 s; (b) 1.47 Hz; (c) 2.06 m/s **7.** (a) $y(x,t) = 2.0 \sin 2\pi(0.10x - 400t)$, with x and y in cm and t in s; (b) 50 m/s; (c) 40 m/s **9.** (a) 11.7 cm; (b) π rad **11.** 129 m/s **13.** (a) 15 m/s; (b) 0.036 N **15.** $y(x,t) = 0.12 \sin(141x + 628t)$, with y in mm, x in m, and t in s **17.** (a) $2\pi y^{max}/\lambda$; (b) no **19.** (a) 5.0 cm; (b) 40 cm; (c) 12 m/s; (d) 0.033 s; (e) 9.4 m/s; (f) $5.0 \sin(16x + 190t + 0.93)$, with x in m, y in cm, and t in s **21.** 2.63 m from the end of the wire from which the later pulse originates **25.** $1.4y^{max}$ **27.** (a) 0.31 m; (b) 1.64 rad; (c) 2.2 mm **29.** (a) 140 m/s; (b) 60 cm; (c) 240 Hz **31.** (a) 82.0 m/s; (b) 16.8 m; (c) 4.88 Hz **33.** 7.91 Hz, 15.8 Hz, 23.7 Hz **35.** (a) 105 Hz; (b) 158 m/s **37.** (a) 0.25 cm; (b) 120 cm/s; (c) 3.0 cm; (d) zero **39.** (a) 50 Hz; (b) $y = 0.50 \sin[\pi(x \pm 100t)]$, with x in m, y in cm, and t in s **41.** (a) 1.3 m; (b) $y = 0.002 \sin(9.4x)\cos(3800t)$, with x and y in m and t in s **43.** (a) 2.0 Hz; (b) 200 cm; (c) 400 cm/s; (d) 50 cm, 150 cm, 250 cm, etc.; (e) 0, 100 cm, 200 cm, etc. **47.** (a) 323 Hz; (b) eight **49.** 5.0 cm

Chapter 18

RE 18-1: We express units in terms of very basic elements of length [L], mass [M] and time [T]. Since B is a force per unit area, its units are $B \sim \dfrac{[M][L]/[T^2]}{[L^2]} = [M]/[L][T^2]$. ρ is a mass per unit volume so $\rho \sim [M]/[L^3]$. $B/\rho \sim [L^2]/[T^2]$ and $\sqrt{B/\rho} \sim [L]/[T]$ or a "velocity" given by [m]/[s] in SI units.

RE 18-2: The measured wave speed for the round trip is $v^{wave} = (2)(2.4\ \text{m})/(.0133 - .0002)(\text{s}) = 366$ m/s, which is in reasonable agreement with the stated 343 m/s in room-temperature air.

RE 18-3: Since energy per unit time passing through a surface that faces a source of sound is just the product of the sound intensity there and the area of the surface, and since sound intensity falls off with distance from the source, (a) the intensity of the sound is the same at surfaces 1 and 2 and is smaller at surface 3, and (b) the areas of surfaces 1 and 2 are equal while that of surface 3 is larger.

RE 18-4: The second harmonic of the longer pipe B has the same frequency as the first harmonic of the shorter pipe A.

RE 18-5: (a) and (e) have greater detected frequency than emitted frequency. (b) and (f) have reduced detected frequencies. (c) and (d) are indeterminate.

Problems

1. divide the time by 3 **3.** (a) 79 m, 41 m; (b) 89 m **5.** 1900 km **7.** 40.7 m **9.** (a) 0.0762 mm; (b) 0.333 mm **11.** (a) 343 $(1 + 2m)$ Hz, with m being an integer from 0 to 28; (b) $686m$ Hz, with m being an integer from 1 to 29 **13.** (a) 143 Hz, 429 Hz, 715 Hz; (b) 286 Hz, 572 Hz, 858 Hz **15.** 17.5 cm **17.** 15.0 mW **19.** (a) 1000; (b) 32 **21.** (a) 59.7; (b) 2.81×10^{-4} **23.** (a) 5000; (b) 71; (c) 71 **25.** (a) 5200 Hz; (b) amplitude$_{SAD}$/amplitude$_{SBD}$ = 2 **27.** (a) 57.2 cm; (b) 42.9 cm **29.** (a) 405 m/s; (b) 596 N; (c) 44.0 cm; (d) 37.3 cm **31.** (a) 1129, 1506, and 1882 Hz **33.** 12.4 m **35.** (a) node; (c) 22 s **37.** 45.3 N **39.** 387 Hz **41.** 0.02 **43.** 17.5 kHz **45.** (a) 526 Hz; (b) 555 Hz **47.** (a) 1.02 kHz; (b) 1.04 kHz **49.** 155 Hz **51.** (a) 485.8 Hz; (b) 500.0 Hz; (c) 486.2 Hz; (d) 500.0 Hz **53.** (a) 598 Hz; (b) 608 Hz; (c) 589 Hz **55.** (a) 42°; (b) 11 s

Chapter 19

RE 19-1: Some properties that are measurable include mass, volume, hardness, elasticity, and breaking strength. Flavor and color, for example, are less easily quantified.

RE 19-2: For comfort we often want to maintain the temperature inside our homes significantly higher (in winter) or lower (in summer) than that of the environment outside. Thermal insulation inside the walls of our homes reduces the amount of heat energy that would otherwise flow out of the house in winter or into the house in summer, reducing the expenditure of energy needed to maintain a comfortable interior temperature.

RE 19-3: Equation 19-5 tells us that, for the same amount of heat energy added to the same mass, the temperature increase is inversely proportional to the specific heat of the material being heated. Thus object A has a greater specific heat than object B.

RE 19-4: The good news for the firefighter is that each kilogram of water sprayed on the fire can remove a relatively large amount of heat from the burning object. The bad news is that this heat can easily be transferred to the firefighter's body if the steam condenses on her skin. One gram of steam at 100 °C condensing on one gram of (water-like) flesh at 37 °C will yield two grams of water-like material at a temperature of (100 °C + 37 °C)/2 = 69 °C.

RE 19-5: For the net work done by the gas on its environment to be positive, the top curve must go from left (lower pressure) to right (higher pressure.) For maximum positive work each cycle that area on the P-V diagram enclosed by the cycle must be as large as possible. So curves c and e yield the maximum possible positive work here.

RE 19-6: (a) The change in the internal energy of the gas is the same in each case. (b) The work done by the gas is greatest for path 4, then path 3, path 2, and finally path 1. (c) The thermal energy added to the gas is also greatest for path 4, then path 3, path 2, and finally path 1.

RE 19-7: (a) For any cyclic process, $Q - W = 0$ or $Q = W$. (b) Because the net work that the gas does on its environment is negative here, that means that the net thermal energy Q transfer to the system is also negative and has the same value as the work. Thus thermal energy equal in magnitude to the work done by the system must be transferred *from the gas to the environment* each cycle.

RE 19-8: (a) Plates 2 and 3 will be tied for the largest increase in their vertical heights, followed by plate 1 and then plate 4. (b) Plate 3 will have the greatest increase in area, followed by plate 2, with plates 1 and 4 tied for last place.

RE 19-9: The hole gets larger as the plate's temperature increases, as would a circle drawn on the plate.

RE 19-10: The pressure at the base of the rod decreases, since the weight of the rod remains constant while the area of the base supporting that weight increases a bit.

RE 19-11: The greater the thermal conductivity, the smaller is the temperature difference between the two faces of samples of the same thickness. Since the temperature differences here, going from left to right, are 10 C°, 5 C°, 15 C°, and 5 C°, slabs b and d tie for greatest thermal conductivity, followed by slab a, with slab c having the smallest thermal conductivity.

Problems

1. (a) 320 °F; (b) −12.3 °F **3.** (a) Dimensions are inverse time. **5.** (a) 523 J/kg·K; (b) 26.2 J/mol·K; (c) 0.600 mole **7.** 42.7 kJ **9.** 1.9 times as great **11.** (a) 33.9 Btu; (b) 172 F° **13.** 160 s **15.** 2.8 days **17.** 742 kJ **19.** 73 kW **21.** 33 g **23.** (a) 0°C; (b) 2.5°C **25.** A: 120 J, B: 75 J, C: 30 J **27.** −30 J **29.** (a) 6.0 cal; (b) −43 cal; (c) 40 cal; (d) 18 cal, 18 cal **31.** 348 K **33.** (a) −40°; (b) 575°; (c) Celsius and Kelvin cannot give the same reading **35.** 960 μm **37.** 2.731 cm **39.** 29 cm^3 **41.** 0.26 cm^3 **43.** 360°C **47.** 0.68 s/h, fast **49.** 7.5 cm **51.** (a) 0.13 m; (b) 2.3 km **53.** 1660 J/s **55.** (a) 16 J/s; (b) 0.048 g/s **57.** 0.50 min **59.** (a) 17 kW/m^2; (b) 18 W/m^2 **61.** 0.40 cm/h

Chapter 20

RE 20-1: Processes (a), (b), (d), and (e) start and end on the same isotherm because each has PV = 12 units.

RE 20-2: (a) The average translational kinetic energy doubles when the temperature in kelvins of the gas doubles. (b) The average translational kinetic energy would be zero if the temperature of the gas were 0 K. However, all real gases condense into liquids before reaching 0 K.

RE 20-3: (a) The average kinetic energy of each of the three types of molecules is the same. (b) Since that is true, the rms speed of each is inversely related to its molecular mass, so type 3 has the greatest rms speed, followed by type 2, with type 1 the smallest.

RE 20-4: Since the internal energy of an ideal gas depends only on its temperature, path 5 has the greatest change in E^{int}, followed by the other four paths, all of which tie for second place.

Problems

1. 0.933 kg **3.** 6560 **5.** (a) 5.47×10^{-8} mol; (b) 3.29×10^{16} **7.** (a) 0.0388 mol; (b) 220°C **9.** (a) 106; (b) 0.892 m^3 **11.** $A(T_2 - T_1) - B(T_2^2 - T_1^2)$ **13.** 5600 J **15.** 100 cm^3 **17.** 2.0×10^5 Pa **19.** 180 m/s **21.** 9.53×10^6 m/s **23.** 1.9 kPa **25.** 3.3×10^{-20} J **27.** (a) 6.75×10^{-20} J; (b) 10.7 **31.** (a) 6×10^9 km **33.** 15 cm **35.** (a) 3.27×10^{10}; (b) 172 m **37.** (a) 6.5 km/s; (b) 7.1 km/s **39.** (a) 1.0×10^4 K; (b) 1.6×10^5 K; (c) 440 K, 7000 K; (d) hydrogen, no; oxygen, yes **41.** (a) 7.0 km/s; (b) 2.0×10^{-8} cm; (c) 3.5×10^{10} collisions/s **43.** (a) $\frac{2}{3}v_0$; (b) $N/3$; (c) $122v_0$; (d) $1.31v_0$ **45.** $RT\ \ln(V_f/V_i)$ **47.** $(n_1C_1 + n_2C_2 + n_3C_3)/(n_1 + n_2 + n_3)$ **49.** (a) 6.6×10^{-26} kg; (b) 40 g/mol **51.** 8000 J **53.** (a) 6980 J; (b) 4990 J; (c) 1990 J; (d) 2990 J **55.** (a) 14 atm; (b) 620 K **59.** 1.40 **61.** (a) In joules, in the order Q, ΔE^{int}, W: $1 \rightarrow 2$: 3740, 3740, 0; $2 \rightarrow 3$: 0, −1810, 1810; $3 \rightarrow 1$: −3220, −1930, −1290; cycle: 520, 0, 520; (b) V_2 = 0.0246 m^3, p_2 = 2.00 atm, V_3 = 0.0373 m^3, p_3 = 1.00 atm

Chapter 21

RE 21-1: As the putty falls to the floor, gravitational potential energy is converted into translational kinetic energy. When the putty hits the floor it looks at first as if all the mechanical energy is somehow destroyed. But a closer look at the putty after the fall reveals that the putty and the floor beneath it have warmed up a bit. How so? As the putty collides with the floor, the floor does work on the putty. Since the putty doesn't bounce back, the work done on the putty system serves to raise its internal energy. Now the putty's temperature rises and it begins transferring microscopic thermal energy to its surroundings (air and floor) until thermal equilibrium is achieved. Even more careful observations show, in fact, that all of the mechanical energy present in the system before the fall is still there after the putty hits the floor, just in other forms, primarily as thermal energy.

RE 21-2: Driving a nail into a board, letting your hot coffee cool, and saying hello to your friend are all examples of irreversible processes in the sense that if you saw a movie of them running backward you would know something was wrong.

RE 21-3: Process (c) and (b) involve the same amount of heat energy transfer to the water, while (a) adds twice as much heat to the water. Process (a) also happens at the lowest average temperature, so it involves the greatest entropy change of the water, followed by (b) and then (c).

RE 21-4: Equation 21-11 relates the efficiency of a Carnot engine to the two thermodynamic temperatures between which it operates. Applying this to these three cases yields Carnot efficiencies of (a) 0.20, (b) 0.25, and (c) 0.33, so ranking the efficiencies, greatest first, yields (c), then (b), then (a).

RE 21-5: (a) Raising the lower temperature T_L by (δT) increases the numerator of Eq. 21-14 by (δT) and simultaneously decreases the denominator by the same amount. This yields the *greatest increase* in the coefficient of performance of the refrigerator. (b) Lowering the lower temperature T_L by (δT) decreases the numerator of Eq. 21-14 by (δT) and simultaneously decreases the denominator by the same amount. This yields the *greatest decrease* in the coefficient of performance of the refrigerator. (c) Increasing the higher temperature T_H by (δT) makes the denominator bigger by (δT) with no change in the numerator, decreasing the coefficient of performance of the refrigerator, but not as much as in (b). (d) Decreasing the higher temperature T_H by (δT) makes the denominator smaller by (δT) with no change in the numerator, increasing the coefficient of performance of the refrigerator, but not as much as in (a). So, from greatest to least, the changes in the coefficient of performance of the refrigerator are (a), (d), (c), and finally (b).

RE 21-6: If we had, say, 6 molecules, then the number of microstates corresponding to 3 molecules in each half of the box would be $6!/(3!)^2 = 20$. Generalizing Eq. 21-18 to three bins in the box with 2 molecules in each bin would have $6!/(2!)^3 = 90$. In this case a greater number of microstates is associated with dividing the box up into a larger number of equally populated equal subvolumes. This remains true as the number of molecules is increased, so (b) has more microstates than (a).

Problems

1. 14.4 J/K **3.** (a) 9220 J; (b) 23.0 J/K; (c) 0 **5.** (a) 5.79×10^4 J; (b) 173 J/K **7.** (a) 14.6 J/K; (b) 30.2 J/K **9.** (a) 57.0°C; (b) −22.1 J/K; (c) +24.9 J/K; (d) +2.8 J/K **13.** (a) 320 K; (b) 0; (c) +1.72 J/K **15.** +0.75 J/K **17.** (a) −943 J/K; (b) +943 J/K; (c) yes **19.** (a) $3p_0V_0$; (b) $\Delta E^{\text{int}} = 6RT_0$, $\Delta S = \frac{3}{2}R \ln 2$; (c) both are zero **21.** (a) 31%; (b) 16 kJ **23.** (a) 23.6%; (b) 1.49×10^4 J **25.** 266 K and 341 K **27.** (a) 1470 J; (b) 554 J; (c) 918 J; (d) 62.4% **29.** (a) 2270 J; (b) 14800 J; (c) 15.4% (d) 75.0%, greater **31.** (a) 78%; (b) 81 kg/s **33.** (a) $T_2 = 3T_1$, $T_3 = 3T_1/4^{\gamma-1}$, $T_4 = T_1/4^{\gamma-1}$, $p_2 = 3p_1$, $p_3 = 3p_1/4^{\gamma}$, $p_4 = p_1/4^{\gamma}$; (b) $1 - 4^{1-\gamma}$ **35.** 21 J **37.** 440 W **39.** 0.25 hp **41.** $[1 - (T_2/T_1)]/[1 - (T_4/T_3)]$ **45.** (a) $W = N!/(n_1!n_2!n_3!)$; (b) $[(N/2)!\,(N/2)!]/[(N/3)!\,(N/3)!\,(N/3)!]$; (c) 4.2×10^{16}

Chapter 22

RE 22-1: Electric stove, microwave, lights, car starter motor, toothbrush, computer, tape recorder, CD player, FM radio, amplifier, etc.

RE 22-2: (a) Since the two tapes have identical histories, they should have like charges and repel. (b) The observations were consistent with my predictions. The two tapes repelled each other.

RE 22-3: (a) If woodolin was a new type of charge then two wooden rods charged with linen would repel each other. A wooden rod would have to attract *both* the charged amber (or plastic) rod *and* the charged glass rod. (b) According to the text statements, this observation has never been made. It has always been the case that a suspected new type of charge (such as woodolin) always repels either a charged amber rod or a charged glass rod and attracts the other type rod. This makes it the same as one of the existing charges.

RE 22-4: A very simple explanation is that in a solid, all parts are stiff. But since one can melt ice into water and then boil water into a gas (water vapor) the atomic explanation seems quite plausible.

RE 22-5: I would discharge one of the spheres by touching it. Then I would allow the two spheres to touch each other. They should share the charge q equally so each sphere has charge $q/2$. If I *repeat* the process, then each sphere will have charge $(q/2)/2 = q/4$.

RE 22-6: (a) If the paper bits are uncharged, then there is no mutual attraction or repulsion. (b) "Induction" always causes the neutral object to be *attracted* toward the charged object, independent of the sign of the charge on the charged object. So, no, you can't tell the sign of the charge on the charged object in this way.

RE 22-7: (1) *A*, *B* is attractive (unlike charges), (2) *A*, *A* is repulsive (like charges), (3) *B*, *B* is repulsive (like charges), (4) *B*, *C* attract (by induction), (5) *C*, *C* nonexistent forces (both neutral), and (6) *C*, *A* attract (by induction).

RE 22-8: (a) Scotch tape acts like an insulator since charge doesn't draw away as you handle the tape at its ends. (b) A balloon behaves like an insulator, because when you charge it, it can stick by induction to a wall rather than touch and pull away.

RE 22-9: (a) No, since charges on it are not mobile. (b) If we start with a positively charged glass plate as the bottom plate in Fig. 22-10 and perform all the same steps, the aluminum pie plate will be negatively charged.

RE 22-10: All of these assertions are inconsistent with the experimental results in the text.

RE 22-11: (a) The central proton is attracted toward the electron, so this force is to the left. (b) The central proton is repelled by the other proton, so this force is also to the left. (c) Thus the net force on the central proton is to the left. (d) There are no locations along the line connecting the charges where the force on the former central proton can be zero. Since the magnitudes of the charges on the proton and electron are the same, the only location where the force magnitudes on the other proton are zero is halfway between the first two, but we know the forces don't cancel there.

Problems

1. -1.32×10^{13} C **3.** 6.3×10^{11} **5.** 122 mA **7.** (a) positron; (b) electron **9.** 1.38 m **11.** (a) 4.9×10^{-7} kg; (b) 7.1×10^{-11} C **13.** (a) 0.17 N; (b) -0.046 N **15.** either $-1.00\ \mu$C and $+3.00\ \mu$C or $+1.00\ \mu$C and $-3.00\ \mu$C **17.** (a) charge $-4q/9$ must be located on the line joining the two positive charges, a distance $L/3$ from charge $+q$. **19.** $q = Q/2$ **21.** (a) 3.2×10^{-19} C; (b) two **23.** (a) 0; (b) 1.9×10^{-9} N **25.** (a) 6.05 cm; (b) 6.05 cm from central bead **27.** $+13e$ **29.** (a) positive; (b) +9 **31.** 9.0 kN **33.** $1.72a$, directly rightward **35.** $-11.1\ \mu$C **37.** $q = 0.71Q$ **39.** (b) $1e$, 0.654 rad; $2e$, 0.889 rad; $3e$, 0.988 rad; $4e$, 1.047 rad; $5e$, 1.088 rad **41.** (a) Let $J = qQ/4\pi\varepsilon_0 d^2$. For $\alpha < 0$, $F = -J[\alpha^{-2} + (1 + |\alpha|)^{-2}]$; for $0 < \alpha < 1$, $F = J[\alpha^{-2} - (1-\alpha)^{-2}]$; for $1 < \alpha$, $F = J[\alpha^{-2} + (\alpha - 1)^{-2}]$ **43.** (a) 5.7×10^{13} C, no; (b) 6.0×10^5 kg **45.** (b) $\pm 2.4 \times 10^{-8}$ C **47.** (a) $\frac{L}{2}\left(1 + \frac{1}{4\pi\varepsilon_0}\frac{qQ}{Wh^2}\right)$; (b) $\sqrt{3qQ/4\pi\varepsilon_0 W}$

Chapter 23

RE 23-1: $F^{elec} \propto 1/r^2$. Thus at 4 cm, F^{elec} would be $(1/2)^2 = 1/4$ of its value at 2 cm, or 9 mm. At 6 cm, F^{elec} would be $(1/3)^2 = 1/9$ of its value at 2 cm, or 4 mm.

RE 23-2: Since the force on the test object to the sources, $\vec{F}_{s\to t}$, varies from point to point in space, the test object must be small enough spatially to test the "local" value rather than the average value over too large a volume of space.

RE 23-3: The type of test charge makes no difference! For a negative test charge we would still use Eq. 23-9 to determine the electric field vector. But, the new $(\vec{F}^{elec})' = -\vec{F}^{elec}$ and the new negative charge $q_t' = -q_t$. So $\vec{E}_s'$ will equal E_s' (no charge).

RE 23-4: (a) Rightward, (b) leftward, (c) leftward, (d) rightward (p and e have the same charge magnitude and p is farther).

RE 23-5: (a) To the left, (b) to the left in a parabolic path, (c) its speed decreases at first, then increases. It will move in a straight line first rightward, then leftward.

RE 23-6: All four experience the same magnitude torque.

RE 23-7: Near a positive charge, $\vec{E}$ points always *away* from the charge; near a negative charge, $\vec{E}$ points always *toward* the charge.

RE 23-8: Just as for the two equidistant point charges in Fig. 23-10, we can "pair up" equal patches of charge equidistant from the point at which we are calculating $\vec{E}$ for all such patches of charge on the sheet, canceling the contributions to $\vec{E}$ parallel to the sheet.

Problems

1. 56 pC **3.** 3.07×10^{21} N/C, radially outward **5.** 50 cm from q_A and 100 cm from q_B **7.** 0 **9.** 1.02×10^5 N/C, upward **11.** (a) 47 N/C; (b) 27 N/C **13.** $4kQ/3d^2$ or $Q/3\pi\varepsilon_0 d^2$ **15.** 1.38×10^{-10} N/C, 180° from $+x$ **17.** 6.88×10^{-28} C · m **23.** $q/\pi^2\varepsilon_0 r^2$, vertically downward **25.** (a) $-q/L$; (b) $q/4\pi\varepsilon_0 a(L + a)$ **29.** (a) -1.72×10^{-15} C/m; (b) -3.82×10^{-14} C/m²; (c) -9.56×10^{-15} C/m²; (d) -1.43×10^{-12} C/m³ **31.** $E = 2k|Q|(\sin\theta/2)/\theta R^2$ **33.** 217° **35.** 3.51×10^{15} m/s² **37.** 6.6×10^{-15} N **39.** (a) 1.5×10^3 N/C; (b) 2.4×10^{-16} N, up; (c) 1.6×10^{-26} N; (d) 1.5×10^{10} **41.** (a) 1.92×10^{12} m/s²; (b) 1.96×10^5 m/s **43.** (a) 2.7×10^6 m/s; (b) 1000 N/C **45.** 27 μm **47.** (a) yes; (b) upper plate, 2.73 cm **49.** (a) 27 km/s; (b) 50 μm **51.** 5.2 cm **53.** (a) 0; (b) 8.5×10^{-22} N · m; (c) 0 **55.** $(1/2\pi)\sqrt{pE/I}$ **57.** 1.92×10^{-21} J **59.** (a) 6.4×10^{-18} N; (b) 20 N/C **63.** (a) to the right in the figure; (b) $(2kqQ \cos 60°)/a^2$

Chapter 24

RE 24-1: (a) $\phi = \vec{v} \cdot \Delta\vec{A} = (3 \text{ m/s})(2 \times 10^{-4} \text{ m}^2) \cos 60° = (3 \times 10^{-4} \text{ m}^3/\text{s})$. Whatever fluid that is represented by this vector velocity field is flowing through this surface area dA. (b) $\phi = \vec{E} \cdot \Delta\vec{A} = (3 \text{ N/C})(2 \times 10^{-4} \text{ m}^2) \cos 60° = 3 \times 10^{-4}$ N · m²/C. Nothing is flowing through the small area. Instead, the flux represents the product of the E-field component normal to the area.

RE 24-2: To find the answers we simply sum the flux through all six faces. We get $\phi^{net}_{cube\ 1} = 0$ N · m²/C, $\phi^{net}_{cube\ 2} = +5$ N · m²/C, and $\phi^{net}_{cube\ 3} = -3$ N · m²/C. (a) Cube 2, (b) cube 3, and (c) cube 1.

RE 24-3: The central charge always acts along the central line. For each noncentral charge (for example, the one to the left) that acts at a point on this central line, there is a conjugate charge (in this example, the one to the right of center) that is exactly the same distance from the point as the original point. The E-field vectors have the same magnitude. The E-components perpendicular to the plane act in the same direction and add vectorially. The parallel components act in opposite directions and cancel.

RE 24-4: Since Gauss' law states that $\phi^{net} = q^{enc}/\varepsilon_0$ as long as the same charge is enclosed by the new Gaussian surfaces, ϕ^{net} is unchanged.

RE 24-5: Negative charges would be induced on the inside surface of the cavity so that $q^{enc} = q^{induced} + q^{center} = 0$. Thus, the net flux at the cavity's Gaussian surface would be zero.

Problems

1. (a) 0; (b) -3.92 N · m²/C; (c) 0; (d) 0 for each field **3.** 2.0×10^5 N · m²/C **5.** (a) 8.23 N · m²/C; (b) 8.23 N · m²/C; (c) 72.8 pC in each case **7.** 3.54 μC **9.** 0 through each of the three faces meeting at q, $q/24\varepsilon_0$ through each of the other faces **11.** -7.5 nC **15.** -1.04 nC **19.** (a) $E = (q/4\pi\varepsilon_0 a^3)r$; (b) $E = q/4\pi\varepsilon_0 r^2$; (c) 0; (d) 0; (e) inner, $-q$; outer, 0 **21.** $q/2\pi a^2$ **23.** $6K\varepsilon_0 r^3$ **25.** 5.0 μC/m **27.** (a) $E = q/2\pi\epsilon_0 LR$, radially inward; (b) $-q$ on both inner and outer surfaces; (c) $E = q/2\mu\epsilon_0 Lr$, radially outward **29.** (a) 2.3×10^6 N/C, radially out; (b) 4.5×10^5 N/C, radially in **31.** (b) $\rho R^2/2\varepsilon_0 r$ **33.** (a) 5.3×10^7 N/C; (b) 60 N/C **35.** 5.0 nC/m² **37.** 0.44 mm **39.** 2.0 μC/m² **41.** (a) 37 μC; (b) 4.1×10^6 N · m²/C

Chapter 25

RE 25-1: Question 1: Because charges that are infinitely far apart exert no forces on each other. Question 2: Zero separation between particles would involve infinite attractive or repulsive forces.

RE 25-2: (a) If we assume the E-field does not change as a result of the reconfiguration of the charge then the positive charge displacement is opposite to the direction of the E-field, so the E-field does negative work. (b) It takes external work to move the charge against the field so ΔU increases, and (c) because we are interested in the *change* of electric potential between points 1 and 2.

RE 25-3: (a) The external force does positive work. (b) The proton moves to a higher potential so $V_2 > V_1$.

RE 25-4: (a) The E-field acts from left to right. (b) Positive external work is done on the electron in paths 1, 2, 3, and 5. Negative work is done on Path 4. (c) $\Delta V_3 > \Delta V_1 = \Delta V_2 = \Delta V_5 > \Delta V_4$.

RE 25-5: Given the charge distribution, we can simply add the contribution to the potential at a point P due to each of the charges, taken separately, using Eq. 25-25. If all we know is $\vec{E}(\vec{r})$ then we must calculate $V(\vec{r})$ using Eq. 25-17.

RE 25-6: V at P is the same for all three of these configurations. The potential at P due to each proton only depends on how far away that proton is from P and not on the direction.

RE 25-7: Using Eq. 25-29 for case (a) $\theta = 0$ and so $\cos\theta = +1$, for case (b) $\theta = 180°$ and $\cos\theta = -1$, and for case (c) $\theta = 90°$ so $\cos\theta = 0$. All other terms remain constant, so ranked from most to least positive, $V_a > V_c > V_b$.

RE 25-8: (a) $E_2 > E_1 = E_3$, (b) Pair 3. (c) It accelerates leftward.

RE 25-9: Since potential energy is a scalar quantity, its superposition involves only scalar addition while the superposition of electric fields requires adding vectors.

RE 25-10: (a) A is wrong since it originates on $-$ and terminates on $+$. B is wrong since it is not perpendicular to the plate near the plate. C is wrong since it has a kink. D is wrong for the same reason as A. E is wrong since it both originates and terminates on a $+$ charge. F is ok. (b) A correct drawing would have curves like A, D, and F but with arrows pointing toward the negatively-charged sphere.

RE 25-11: Because her skin is a conductor and thus an equipotential surface. Charges will redistribute so they have a higher density near the top of her head, which has more curvature than the sides of her head. The strength of the electric field is higher where the charges bunch so the equipotential surfaces are closer together than they were.

Problems

1. (a) 3.0×10^5 C; (b) 3.6×10^6 J **3.** (a) 3.0×10^{10} J; (b) 7.7 km/s; (c) 9.0×10^4 kg **5.** 8.8 mm **7.** (a) 136 MV/m; (b) 8.82 kV/m **9.** (b) because $V = 0$ point is chosen differently; (c) $q/(8\pi\varepsilon_0 R)$; (d) potential differences are independent of the choice for the $V = 0$ point **11.** (a) $Q/4\pi\varepsilon_0 r$; (b) $\dfrac{\rho}{3\varepsilon_0}\left(\dfrac{3}{2}r_2^2 - \dfrac{1}{2}r^2 - \dfrac{r_1^3}{r}\right)$, $\rho = \dfrac{Q}{\frac{4\pi}{3}(r_2^3 - r_1^3)}$; (c) $\dfrac{\rho}{2\varepsilon_0}(r_2^2 - r_1^2)$, with ρ as in (b); (d) yes **13.** (a) -4.5 kV; (b) -4.5 kV **15.** $x = d/4$ and $x = -d/2$ **17.** (a) 0.54 mm; (b) 790 V **19.** 6.4×10^8 V **21.** $2.5q/4\pi\varepsilon_0 d$ **23.** $-0.21q^2/\varepsilon_0 a$ **25.** (a) $+6.0 \times 10^4$ V; (b) -7.8×10^5 V; (c) 2.5 J; (d) increase; (e) same; (f) same **27.** $W = \dfrac{qQ}{8\pi\varepsilon_0}\left(\dfrac{1}{r_1} - \dfrac{1}{r_2}\right)$ **29.** 2.5 km/s **31.** (a) 0.225 J; (b) A, 45.0 m/s²; B, 22.5 m/s²; (c) A, 7.75 m/s, B, 3.87 m/s **33.** 0.32 km/s **35.** 1.6×10^{-9} m **39.** $(c/4\pi\varepsilon_0)[L - d\ln(1 + L/d)]$ **41.** 17 V/m at 135° counterclockwise from $+x$ **45.** (a) $\dfrac{Q}{4\pi\varepsilon_0 d(d + L)}$, leftward; (b) 0 **47.** 2.5×10^{-8} C **49.** (a) -180 V; (b) 2700 V, -8900 V **51.** (a) -0.12 V; (b) 1.8×10^{-8} N/C, radially inward

Chapter 26

RE 26-1: Volta probably felt a tingling sensation or perhaps a shock or jolt that would cause him to let go of the terminals.

RE 26-2: "Circuit" means a full round trip around some route. This is just what the electric charge does.

RE 26-3: (a) If the overall circuit had $q^{net} = 0$ before the switch was closed, it will remain charge neutral after the switch is closed since the circuit is a closed system and charge is neither created nor destroyed, but merely flows around the circuit. (b) Individual wires in the circuit can and do acquire a (small) net positive or negative charge, but this charge must come from other parts of the circuit.

RE 26-4: Electrical current *is* the net transport of charge past a given point in a circuit in a given time. If equal amounts of positive charge moving, say, right, and negative charge moving right go past the same point, there is no net transport of charge past that point, so $i = 0$ A.

RE 26-5: Let's assume that the unknown current i flows from right to left. Then the net current flowing into (+) or out of (−) the *middle* node is $(+2 + 3 + 4 - 1 + 2 - 2 + i)$ A. But currents must all add to zero at this node. Thus $i = -8$ A, meaning our assumption was wrong and that $i = 8$ A flowing from left to right.

RE 26-6: A voltmeter is attached *across* a circuit element because it is designed to measure the potential difference *between* the ends of the circuit. An ammeter is inserted in a branch of a circuit because it is designed to measure the current *through* that part of the circuit. In a series circuit where there are no branches or alternate paths for current to flow, it doesn't matter whether the ammeter is placed before or after a series circuit element.

RE 26-7: Device 1 is ohmic since $(\Delta V/i) = 2.25\ \Omega$ = constant and $i = 0$ A when $\Delta V = 0$ V. Device 2 is nonohmic since $(\Delta V/i) \neq$ constant. Device 3 is nonohmic. Although a plot of ΔV vs. i is a straight line, i is nonzero at $\Delta V = 0$, so i is not proportional to ΔV.

RE 26-8: If the cross-sectional area of the Nichrome wire is cut in half, its resistance will double, so the slope of the i vs. ΔV graph which is $1/R$ will be cut in half.

RE 26-9: (a) $R \propto 1/r^2$ for most wires, suggesting the current flows through the whole cross-sectional area of the wire, not just on its surface as indicated in Eq. 26-8. (b) If the current flowed only in a thin layer near the surface then I'd expect $R \propto 1/r$ since the circumference is $2\pi r$ for a wire with a circular cross section.

RE 26-10: Since $R = \rho L/A$, $(a) = (c) > (b)$.

RE 26-11: $(a) = (b) > (d) > (c)$.

RE 26-12: Only the cross-sectional area A matters in comparing current densities, so $(a) = (d) > (b) = (c)$.

RE 26-13: Since the current density is $(I/A) = (\Delta V/(RA))$ and $RA = \rho L$, we see here that the current density is just inversely proportional to the length of each wire. So $(b) = (d) > (a) = (c)$.

Problems

1. (a) 1200 C; (b) 7.5×10^{21} **3.** 5.6 ms **5.** 100 V **7.** $2.0 \times 10^{-8}\ \Omega \cdot$ m **9.** 2.4 Ω **11.** 54 Ω **13.** 3.0 **15.** (a) 0.43%, 0.0017%, 0.0034% **17.** 560 W **19.** (a) 1.0 kW; (b) 25 ¢ **21.** 0.135 W **23.** (a) 10.9 A; (b) 10.6 Ω; (c) 4.5 MJ **25.** 660 W **27.** (a) 3.1×10^{11}; (b) 25 μA; (c) 1300 W, 25 MW **29.** (a) 17 mV/m; (b) 243 J **31.** (a) 6.4 A/m², north; (b) no, cross-sectional area **33.** 0.38 mm **35.** (a) 2×10^{12}; (b) 5000; (c) 10 MV **37.** 13 min **39.** $8.2 \times 10^{-4}\ \Omega \cdot$ m **41.** (a) 0.67A; (b) toward the negative terminal **43.** (a) 1.73 cm/s; (b) 3.24 pA/m²

Chapter 27

RE 27-1: $R = \rho L/A$; $A = \pi r^2 = \frac{1}{4}\pi d^2$

$\rho_{Cu} = 1.7 \times 10^{-8}\,\Omega\cdot\text{m}$; $d = 2.4 \times 10^{-4}$ m; $L = 0.30$ m

$\therefore R = (1.7 \times 10^{-8}\,\Omega\cdot\text{m})(0.30\text{ m})/(\frac{1}{4}\pi(2.4 \times 10^{-4}\text{ m})^2)$

$= 0.113\,\Omega$.

RE 27-2: (a) If all the current were "used up" in the first bulb, the second and third bulbs would be dark. (b) If most of the current were "used up" in the first bulb, the second bulb would glow more dimly than the first and the third bulb would glow more dimly than the second. (c) If only a small amount were "used up" in the first bulb, the third would be dimmer than the second, and the second would be a bit dimmer than the first.

RE 27-3: $i_a = i_b = i_c$ and $V_b > V_c > V_a$.

RE 27-4: (a) $i_1 = i_2 = i_3$. (b) $\Delta V_1 > \Delta V_2 > \Delta V_3$.

RE 27-5: Since the ammeter is wired *in series* with the resistors, its resistance *adds* to theirs. (a) With no ammeter, the current will be largest. (b) with $R_A \ll R_1 + R_2$ the current will be reduced, but only a little. (c) With $R_A = R_1 + R_2$ the current would be cut in half. Thus a good ammeter should have as *small* a resistance as possible.

RE 27-6: (a) $R_1 = R_2$ in series so $i_1 = i_2$ and $\Delta V_1 = \Delta V_2 = \frac{1}{2}\Delta V_B$ and so $i = i_1 = i_2 = \Delta V_B/(R_1 + R_2)$. (b) $R_1 = R_2$ in parallel so $i_1 = i_2$ and $\Delta V_1 = \Delta V_2 = \Delta V_B$ so now $i = i_1 + i_2 = 2\Delta V_B/R_1 = 2\Delta V_B/R_2$.

RE 27-7: Note that R_V is in parallel with R_1. Thus if $R_V << R_1$, the effective resistance between d and e in Fig. 27-7 would be dramatically decreased from R_1 to less than R_V. This would "pull down" ΔV_{de} to a smaller value that it had before I installed the voltmeter. However, if $R_V >> R_1$, then the effective resistance between d and e remains just about R_1 and the value of ΔV_{de} is about what it was without the meter present. Thus $R_V >> R$ gives more accurate measurements of potential differences.

RE 27-8: Since the bulbs are identical and wired in parallel, $i_1 = i_2 = i_3$. If only one bulb were connected to the battery its brightness would be the same as before, since the potential difference across it is still ΔV_B.

Problems

1. (a) 30 Ω; (b) clockwise; (c) A **3.** (a) 45 Ω; (b) 0.33 A each; (c) 0.33 A **5.** $V_1 = 3.5$ V; $V_2 = 4.3$ V; $V_3 = 7.2$ V **7.** 8.0 Ω **9.** (a) 0; (b) 1.25 A, downward **11.** (a) 120 Ω; (b) $i_1 = 51$ mA, $i_2 = i_3 = 19$ mA, $i_4 = 13$ mA **13.** 20 Ω **15.** (a) bulb 2; (b) bulb 1 **17.** 0.45 A **19.** $i_1 = -50$ mA, $i_2 = 60$ mA, $V_{ab} = 9.0$ V **21.** (a) Cu: 1.11 A, A1: 0.893 A; (b) 126 m **23.** 5.56 A **25.** $3d$ **29.** nine **31.** providing energy, 360 W **33.** (a) 3.0 A, downward; (b) 1.6 A, downward; (c) 6.4 W, supplying; (d) 55.2 W, supplying **35.** (a) 12 eV (1.9×10^{-18} J); (b) 6.5 W **39.** (a) 7.50 A, leftward; (b) 10.0 A, leftward; (c) 87.5 W, supplied **41.** (a) 0.33 A, rightward; (b) 720 J **43.** (a) \$320; (b) 4.8 cents **45.** 14 h 24 min **47.** (a) 0.50 A; (b) $P_1 = 1.0$ W, $P_2 = 2.0$ W; (c) $P_1 = 6.0$ W supplied, $P_2 = 3.0$ W absorbed **49.** (a) $V_T = -ir + \mathscr{E}$; (b) 13.6 V; (c) 0.060 Ω **51.** (a) 14 V; (b) 100 W; (c) 600 W; (d) 10 V, 100 W **53.** (a) 50 V; (b) 48 V; (c) B is connected to the negative terminal **55.** (a) $r_1 - r_2$; (b) battery with r_1 **59.** (a) $R = r/2$; (b) $P^{\max} = \varepsilon^2/2r$ **61.** (a) 0.346 W; (b) 0.050 W; (c) 0.709 W; (d) 1.26 W; (e) −0.158 W **63.** (a) battery 1, 0.67 A down; battery 2, 0.33 A up; battery 3, 0.33 A up; (b) 3.3 V

Chapter 28

RE 28-1: The capacitance of a capacitor remains the same, whatever the amount of excess charge on its plates and whatever potential difference is applied across it. Doubling $|q|$ doubles ΔV_c while tripling ΔV_c triples $|q|$.

RE 28-2: Each of these three types of capacitors becomes electrically isolated when removed from a battery so the excess charge on each of the "plates" does not change.

RE 28-3: In these cases ΔV is constant and C and hence $|q|$ must change when spacings change, so $|q|$ (a) decreases, (b) increases, and (c) decreases.

RE 28-4: Each capacitor initially has the same $|q|$ and the same $|\Delta V|$. (a) Wiring them in parallel, positive plate to positive and negative to negative, leaves $|q|$ and $|\Delta V|$ on each unchanged. Wiring them in parallel, positive to negative, makes $|\Delta V| = 0$ and so $|q| = 0$ on each. (b) Wiring them in series leaves these quantities unchanged.

RE 28-5: (a) Since $i_0 = |\Delta V_B|/R$, $(i_0)_1 > (i_0)_2 > (i_0)_4 > (i_0)_3$. (b) Since $t_{(1/2)}$ is proportional to $\tau = RC$, $(t_{(1/2)})_4 > (t_{(1/2)})_1 = (t_{(1/2)})_2 > (t_{(1/2)})_3$.

Probl.ems

1. 7.5 pC **3.** 3.0 mC **5.** (a) 140 pF; (b) 17 nC **7.** $5.04\pi\varepsilon_0 R$ **11.** 9090 **13.** 3.16 μF **17.** 43 pF **19.** (a) 50 V; (b) 5.0×10^{-5} C; (c) 1.5×10^{-4} C

21. $q_1 = \dfrac{C_1C_2 + C_1C_3}{C_1C_2 + C_1C_3 + C_2C_3}C_1\Delta V_0$,

$q_2 = q_3 = \dfrac{C_2C_3}{C_1C_2 + C_1C_3 + C_2C_3}C_1\Delta V_0$

23. 72 F **25.** 0.27 J **27.** (a) 2.0 J **29.** (a) $2\Delta V$; (b) $U_i = \varepsilon_0 A\Delta V^2/2d$, $U_f = 2U_i$; (c) $\varepsilon_0 A\Delta V^2/2d$ **35.** Pyrex **37.** 81 pF/m **39.** 0.63 m² **43.** (a) 10 kV/m; (b) 5.0 nC; (c) 4.1 nC

45. (a) $C = 4\pi\varepsilon_0\kappa\left(\dfrac{ab}{b-a}\right)$; (b) $q = 4\pi\varepsilon_0\kappa\Delta V\left(\dfrac{ab}{b-a}\right)$; (c) $q' = q(1 - 1/\kappa)$ **47.** 4.6 **49.** (a) 2.41 μs; (b) 161 pF **51.** (a) 2.17 s; (b) 39.6 mV **53.** (a) 1.0×10^{-3} C; (b) 1.0×10^{-3} A; (c) $\Delta V_C = 1.0 \times 10^3\,e^{-t}$ V, $\Delta V_R = 1.0 \times 10^3\,e^{-t}$ V; (d) $P = e^{-2t}$ W

Chapter 29

RE 29-1: (a) z axis, (b) $-x$ axis, (c) no direction since $\vec{F} = 0$ N.

RE 29-2: (a) The electron, because it's less massive and "bends" more easily in the presence of a perpendicular force, (b) the electron travels clockwise.

RE 29-3: $\vec{F}^{\text{net}} = \vec{F}^{\text{elec}} + \vec{F}^{\text{mag}}$. The force exerted on the charge by the E-field is the same in all 4 cases and points out of the page. In cases 1 and 3, $\vec{B}$ and $\vec{v}$ are parallel so there is no magnetic force on the charged particle. In cases 2 and 4, $\vec{B}$ and $\vec{v}$ are perpendicular with magnetic forces out of and into the page respectively. (a) In terms of force magnitude $|\vec{F}_2^{\text{net}}| > |\vec{F}_1^{\text{net}}| = |\vec{F}_3^{\text{net}}|$. $|\vec{F}_4^{\text{net}}|$ can take on any value from zero to larger than $|\vec{F}_4^{\text{net}}|$ and so can not be ranked. (b) A zero net force is only possible for case 4.

RE 29-4: The equation $|\vec{F}^{\text{mag}}| = |i\vec{L} \times \vec{B}|$ is a maximum for a given $|\vec{B}|$ when $\vec{B}$ is perpendicular to both $\vec{F}^{\text{mag}}$ and $\vec{L}$. This is true whenever $\vec{B} = \pm|\vec{B}|\hat{\text{j}}$. Trying each direction, the right-hand rule yields $\vec{B}$ pointing along the $-y$ axis.

RE 29-5: (a) $\tau = |\vec{\mu}||\vec{B}|\sin\phi$ where $\phi = \theta$ for cases 2 and 3 and $\phi = \pi - \theta$ for cases 1 and 4. But $\sin\theta = \sin(\pi - \theta)$ so τ is the same for all 4 cases. (b) $U(\phi) = -\vec{\mu}\cdot\vec{B} = -|\vec{\mu}||\vec{B}|\cos\phi$. Now for cases 2 and 3, $\phi = \theta < \pi/2$ so $\cos\theta > 0$, and for cases 1 and 4 $\phi = \pi - \theta > \pi/2$ $\cos\phi = -\cos\theta < 0$, thus $U_1 = U_4 > U_3 = U_2$.

Problems

1. (a) 6.2×10^{-18} N; (b) 9.5×10^8 m/s²; (c) remains equal to 550 m/s **3.** (a) 400 km/s; (b) 835 eV **5.** (a) east; (b) 6.28×10^{14} m/s²; (c) 2.98 mm **7.** 21 μT **9.** (a) 2.05×10^7 m/s; (b) 467 μT; (c) 13.1 MHz; (d) 76.3 ns **11.** (a) 0.978 MHz; (b) 96.4 cm **15.** (a) 1.0 MeV; (b) 0.5 MeV **17.** (a) 495 mT; (b) 22.7 mA; (c) 8.17 MJ **19.** (a) 0.36 ns; (b) 0.17 nm; (c) 1.5 mm **21.** (a) 3.4×10^{-4} T, horizontal and to the left as viewed along $\vec{v}_1$; (b) yes, if its velocity is the same as the electron's velocity **23.** 0.27 mT **25.** 680 kV/m **27.** (b) 2.84×10^{-3} **29.** 38.2 cm/s **31.** 28.2N, horizontally west **33.** 467 mA, from left to right **35.** 0.10 T, at 31° from the vertical **37.** 4.3×10^{-3} N·m, negative y **41.** $2\pi aiB \sin\theta$, normal to the plane of the loop (up) **43.** 2.45 A **45.** (a) 12.7 A; (b) 0.0805 N·m **47.** (a) 0.30 J/T; (b) 0.024 N·m **49.** (a) 2.86 A·m²; (b) 1.10 A·m² **51.** (a) $(8.0 \times 10^{-4}\ \text{N}\cdot\text{m})(-1.2\hat{\imath} - 0.90\hat{\jmath} + 1.0\hat{k})$; (b) -6.0×10^{-4} J

Chapter 30

RE 30-1: (a) $\vec{B}$ is to the left at point 1, (b) $\vec{B}$ is up at point 2, (c) $\vec{B}$ is to the right at point 1, (d) $\vec{B}$ is down at point 2.

RE 30-2: (a) If $\vec{B}^{\text{net}} = 0$ at point 1 then the current in the wire is coming *out* of the page. (b) Since $\vec{B}^{\text{net}} = \vec{B}^{\text{ext}} + \vec{B}^{\text{wire}}$, and since $\vec{B}^{\text{wire}}$ at point 2 points straight down and has the same magnitude as $\vec{B}^{\text{ext}}$, $\vec{B}^{\text{net}}$ is directed 45 degrees down and toward the right at point 2 and its magnitude is $\sqrt{2}\, B^{\text{ext}}$.

RE 30-3: $F_b > F_c > F_a$.

RE 30-4: $\oint \vec{B}\cdot d\vec{s} = \mu_0 i^{\text{enc}}$ where i^{enc} is the *net* current flowing *through* the loop. Therefore, $\left|\frac{1}{\mu_0}\oint \vec{B}\cdot d\vec{s}\right| = i$ for case (a)

$= 0$ for case (b)
$= i$ for case (c)
$= 2i$ for case (d).

(d) > (a) = (c) > (b).

RE 30-5: For $z \gg R$, $|\vec{B}|$ due to any *one* loop is proportional to $|\vec{\mu}| = iA$. Since all the i's are equal, $|\vec{B}| \propto A$ for *each* loop. Taking the directions of the currents into account and calling B_1 the magnetic field magnitude for one *small* loop, and $B_2 = 4B_1$, the magnetic field magnitude for one *large* loop, $B_a = 2B_1$; $B_b = 0$; $B_c = 0$; $B_d = 2B_1 + B_2 = 2B_1 + 4B_1 \therefore |\vec{B}_d| > |\vec{B}_a| > |\vec{B}_b| = |\vec{B}_c| = 0$.

Problems

1. (a) 3.3 μT; (b) yes **3.** (a) 16 A; (b) west to east **5.** $\mu_0 qvi/2\pi d$, antiparallel to i; (b) same magnitude, parallel to i **7.** 2 rad **9.** $\frac{\mu_0 i\theta}{4\pi}\left(\frac{1}{b} - \frac{1}{a}\right)$, out of page. **19.** $(\mu_0 i/2\pi w)\ \ln(1 + w/d)$, up **21.** 256 nT **23.** (a) it is impossible to have other than $B = 0$ midway between them; (b) 30 A **25.** 4.3 A, out of page **27.** 80 μT, up the page **29.** $0.791\mu_0 i^2/\pi a$, 162° counterclockwise from the horizontal **31.** 3.2 mN, toward the wire **33.** (a) $(-2.0\ \text{A})\mu_0$; (b) 0 **37.** $\mu_0 J_0 r^2/3a$ **43.** 0.30 mT **45.** (a) 533 μT; (b) 400 μT **49.** (a) 4.77 cm; (b) 35.5 μT **51.** 0.47 A·m² **53.** (a) 2.4 A·m²; (b) 46 cm **59.** (a) 79 μT; (b) 1.1×10^{-6} N·m

Chapter 31

RE 31-1: They were trying to relate induction to the presence of a magnetic field rather than to a changing field.

RE 31-2: Since the magnetic field is uniform, the left and right segments are polarized symmetrically as shown in the diagram. There is no favored direction in which current can flow.

RE 31-3: This case is similar to the one shown in Fig. 31-7. However, now the polarization will always be stronger on the *right* side of the coil than it is on the left side, so the current will flow continuously in a *counter* clockwise direction.

RE 31-4: Yes, since observations show that the $\vec{v}$ in the magnetic force law ($\vec{F} = q\vec{v} \times \vec{B}$) turns out to be the relative velocity between the object producing the B-field and the charge.

RE 31-5: The magnet is accelerating downward as it falls at $\vec{a} = (-9.8\ \text{m/s}^2)\hat{\jmath}$. By the time its rear end is passing through the area subtended by the loop, it is traveling faster than the front pole was as it passed by, so the rate of change of the B-field is greater at $t = 0.20$ s than it was at $t = 0.10$ s and the amount of induced current is also greater.

RE 31-6: (a) $b > d = e > a = c$. (b) The magnitude $|dB/dt|$ determines that the amount of induced emf is greatest when the slope is greatest.

RE 31-7: (a) into the page to add to the decreasing field, (b) out of the page to subtract from the decreasing field.

RE 31-8: In each semicircular area $|d\Phi^{\text{mag}}/dt|$ is identical. The only issue is the "sense" of the induced emf contributed by each semicircle. Using Lenz's law, loop (a) has a nonzero, clockwise (CW) induced current. Loop (b) has a counterclockwise (CCW) current in both the upper and lower halves, so $|i_a| = |i_b|$. In loop (c), the induced emfs in the upper and lower half circles cancel one another out so $|i_c| = 0$ so $|i_a| = |i_b| > |i_c|$.

RE 31-9: As each loop enters or leaves the region where $B \neq 0$, $|\mathscr{E}| = |d\Phi^{\text{mag}}/dt| \propto (h)(v)$ where h = height of the loop and v is its speed. Since v = constant for each, $|\mathscr{E}_c| = |\mathscr{E}_d| = 2|\mathscr{E}_a| = 2|\mathscr{E}_b|$.

RE 31-10: (a) Out (given), (b) out since path 3 has $|\mathscr{E}|$ = 3(mag), (c) out since path 3 has $|\mathscr{E}|$ = 3(mag), (e) in, since path 4 has $|\mathscr{E}| = 0$, (d) in since path 2 has $|\mathscr{E}|$ = 2(mag).

RE 31-11: When we pointed a right thumb in the direction of the current our fingers wrapped around the wire in the direction of the magnetic field. This is consistent with the direction of the magnetic field shown in Fig. 31-24.

RE 31-12: The quantity $i^{\text{dis}} = \varepsilon_0 d\Phi^{\text{elec}}/dt$ has the units of current. We can use the right hand rule to find the direction of $\vec{B}$ and we can use it to find the magnitude of $\vec{B}$ induced by a capacitor.

RE 31-13: (a) $|\Phi_d| > |\Phi_b| > |\Phi_c| > |\Phi_a|$. Since $\oint \vec{B}\cdot d\vec{A} = 0$ (Eq. 31-49),

$$\Phi^{\text{net}} = \oint_{\text{ends}} \vec{B}\cdot d\vec{A} + \oint_{\text{curve}} \vec{B}\cdot d\vec{A} \text{ so } \Phi_{\text{curve}} = -\oint_{\text{ends}} \vec{B}\cdot d\vec{A}.$$

RE 31-14: They both involve the integration of a field vector over a closed Gaussian surface. Each integral determines a net flux at the closed surface that is proportional to the net electric or magnetic charge enclosed by the surface. The major difference between the electric and magnetic situation is that the net magnetic charge enclosed is always zero (that is, north and south poles always appear together), and the net electric charge enclosed can be positive, negative, or zero.

RE 31-15: A statement of Faraday's law is that a changing magnetic field produces an electric field. The Ampère-Maxwell law states that a changing electric field produces a magnetic field. So there is a mathematical symmetry between the two fields.

Problems

1. 1.5 mV **3.** (a) 31 mV; (b) right to left **5.** (a) 1.1×10^{-3} Ω; (b) 1.4 T/s **7.** 30 mA **9.** 2.9 mV **11.** (a) $\mu_0 i R^2 \pi r^2/2x^3$; (b) $3\mu_0 i \pi R^2 r^2 v/2x^4$; (c) in the same direction as the current in the large loop **13.** (b) no **15.** 29.5 mC **17.** (a) 21.7 V; (b) counterclockwise **19.** (b) design it so that $Nab = (5/2\pi)$ m^2 **21.** 5.50 kV **23.** 80 μV, clockwise **25.** (a) 13 μWb/m; (b) 17%; (c) 0 **27.** 3.66 μW **29.** (a) 48.1 mV; (b) 2.67 mA; (c) 0.128 mW **31.** (a) 600 mV, up the page; (b) 1.5 A, clockwise; (c) 0.90 W; (d) 0.18 N; (e) same as (c) **33.** (a) 240 μV; (b) 0.600 mA; (c) 0.144 μW; (d) 2.88 $\times 10^{-8}$ N; (e) same as (c) **35.** (a) 71.5 μV/m; (b) 143 μV/m **39.** 2.4 $\times 10^{13}$ V/m·s **41.** (a) 1.18×10^{-19} T; (b) 1.06×10^{-19} T **43.** (a) 5.01 $\times 10^{-22}$ T; (b) 4.51×10^{-22} T **45.** 52 nT · m **51.** (a) 0.63 μT; (b) 2.3 $\times 10^{12}$ V/m·s **53.** (a) 710 mA; (b) 0; (c) 1.1 A **55.** (A) 2.0 A; (b) 2.3 $\times 10^{11}$ V/m·s; (c) 0.50 A; (d) 0.63 μT·m **57.** (a) 75.4 nT; (b) 67.9 nT **59.** (a) 27.9 nT; (b) 15.1 nT **61.** (b) sign is minus; (c) no, there is compensating positive flux through open end near magnet **63.** 47.4 μWb, inward

Chapter 32

RE 32-1: Combine Eqs. 32-1 and 32-2 to get $L = \mu_0 A n^2 l$. (a) If n doubles $L \rightarrow 4L$. (b) If l doubles $A \rightarrow 2A$.

RE 32-2: (d) decreasing rightward or (e) increasing and leftward.

RE 32-3: (a) $R_{eq} = (N_p/N_s)^2 R$ we want R_{eq} seen by the generator to be smaller. So N_s must be greater than N_p. (b) This would be a step up transformer.

RE 32-4: A refrigerator magnet is ferromagnetic; a standard paper clip is also ferromagnetic, since it is made of steel, a ferromagnetic material; a silver wire is diamagnetic (the book says so).

RE 32-5: (a) Spin down or (2). (b) Since the proton has the opposite sign of charge, spin up or (1).

RE 32-6: A ferromagnetic material must have well more than 50% of its domains aligned with each other to act like a strong magnet. If no one alignment of the domains dominates, then it is not a permanent magnet.

RE 32-7: Hysteresis is a lack of retraceability of a magnetization curve. It occurs because the reorientation of domains are not completely reversible.

RE 32-8: (a) $\vec{F}^{mag}$ is directed *toward* the magnet. (b) The dipole moments are also directed *toward*. (c) The force on sphere 1 is *less*.

RE 32-9: (a) $\vec{F}^{mag}$ is directed *away* from the magnet. (b) The dipole moments are also directed *away*. (c) The force on sphere 1 is *less*.

RE 32-10: The Earth's B-field has a different declination and inclination at different locations at any one time. But, it also varies in time. Currently the geographic poles are moving daily. They can also reverse themselves in time periods on the scale of 1000 years.

Problems

1. 0.10 μWb **5.** let the current change at 5.0 A/s **7.** (b) so that the changing magnetic field of one does not induce current in the other; (c) $L_{eq} = \sum_{j=1}^{N} L_j$ **9.** 12 A/s **11.** (a) 0.60 mH; (b) 120 **13.** (a) 1.67 mH; (b) 6.00 mWb **15.** (b) have the turns of the two solenoids wrapped in opposite directions **17.** magnetic field exists only within the cross section of solenoid 1 **19.** (a) $\frac{\mu_0 N l}{2\pi} \ln\left(1 + \frac{b}{a}\right)$; (b) 13 μH **21.** $6.91\tau_L$ **23.** 46 Ω **25.** (a) 8.45 ns; (b) 7.37 mA **27.** 10.6 A/s **29.** (a) $i_1 = i_2 = 3.33$ A; (b) $i_1 = 4.55$ A; $i_2 = 2.73$ A; (c) $i_1 = 0$, $i_2 = 1.82$ A (reversed); (d) $i_1 = i_2 = 0$ **31.** (a) 3.28 ms; (b) 6.45 ms; (c) infinite time; (d) for R = 6.0 Ω, the current of the 2.00 A is the equilibrium current, given by $\mathscr{E}$/R = (12 V)/(6.0 Ω); it takes an infinite time to reach. For R = 5.00 Ω, the current of 2.00 A is less than the equilibrium current and requires a finite time to reach. (e) 0; (f) 3 ms **33.** 81.1 μs **35.** (a) 2.4 V; (b) 3.2 mA, 0.16 A **37.** 10 **39.** (a) -9.3×10^{-24} J/T; (b) 1.9×10^{-23} J/T **41.** (a) 0; (b)0; (c) 0; (d) $\pm 3.2 \times 10^{-25}$ J; (e) -3.2×10^{-34} J·s, 2.8×10^{-23} J/T, $+9.7 \times 10^{-25}$ J, $\pm 3.2 \times 10^{-25}$ J **43.** (a) nine; (b) 4 $\mu_B = 3.71 \times 10^{-23}$ J/T; (c) $+9.27 \times 10^{-24}$ J; (d) -9.27×10^{-24} J **45.** 5.15×10^{-24} A·m^2 **47.** (a) 180 km; (b) 2.3×10^{-5} **49.** $\Delta\mu = e^2 r^2 B/4m$ **51.** 20.8 mJ/T **53.** yes **55.** (b) K_i/B, opposite to the field; (c) 310 A/m **57.** 55 μT **59.** (a) 31.0 μT, 0°; (b) 55.9μT, 73.9°; (c) 62.0 μT, 90°

Chapter 33

RE 33-1: Using Eq. 33-6, $a = b > c$. (Note that coil area doesn't matter here.)

RE 33-2: At $t = 0$ s, U^{elec} = max and $U^{mag} = 0$. T = period = $1/f$. (a) $|q(t)|$ is a maximum again at $t = T/2$. (b) Δv_C is next the same at $t = T$. (c) U^{elec} is next a maximum at $t = T/2$. (d) i is next a maximum at $t = T/4$.

RE 33-3: The unit for ω is [rad/s]. Since $L = \mathscr{E}_L/(di/dt)$, we get [H] = [V/(A/s)]. Since $C = q/\Delta V$ we get [F] = [Q/V]. $\omega = 1/\sqrt{LC}$ and the units of $1/\sqrt{LC}$ are $[1/(\text{V}\cdot\text{s/A})(Q/\text{V})]^{1/2}$ but [A] = [Q/s] so $[1/\text{s}^2]^{1/2}$ or [1/s]. This matches the ω unit of [rad/s].

RE 33-4: (a) According to the loop rule, $\Delta v_C + \Delta v_L = 0$. Since $\mathscr{E}_L = \Delta v_L$, $\mathscr{E}_L = -5$ V. (b) $U^{mag} = U - U^{elec} = 160$ μJ − 10 μJ = 150 μJ.

RE 33-5: (a) C > B > A. (b) 1 & A, 2 & B, 3 & S, 4 & C. (c) A.

RE 33-6: (a) (1) lags, (2) leads, (3) in phase. (b) (3) ($\omega^{dr} = \omega$ when $X_L = X_C$).

RE 33-7: (a) Increase since the circuit is mainly capacitive; increase C to decrease X_C to be closer to resonance for maximum ⟨P⟩. (b) Closer.

Problems

1. 25.6 ms **3.** (a) 97.9 H; (b) 0.196 mJ **7.** (a) 34.2 J/m^3; (b) 49.4 mJ **9.** 1.5×10^8 V/m **11.** (a) 1.0 J/m^3; (b) 4.8×10^{-15} J/m^3 **13.** 9.14 nF **15.** (a) 1.17 μJ; (b) 5.58 mA **17.** with n a positive integer: (a) $t = n(5.00$ μs); (b) $t = (2n - 1)(2.50$ μs); (c) $t = (2n - 1)(1.25$ μs) **19.** (a) 1.25 kg; (b) 372 N/m; (c) 1.75×10^{-4} m; (d) 3.02 mm/s **21.** 7.0×10^{-4} s **23.** (a) 3.0 nC; (b) 1.7 mA; (c) 4.5 nJ **25.** (a) 275 Hz; (b) 364 mA **27.** (a) 6.0:1; (b) 36 pF, 0.22 mH **29.** (a) 1.98 μJ; (b) 5.56 μC; (c) 12.6 mA; (d) −46.9°; (e) +46.9° **31.** (a) 0.180 mC; (b) $T/8$; (c) 66.7 W **33.** (a) 356 μs; (b) 2.50 mH; (c) 3.20 mJ **35.** Let T_2 (= 0.596 s) be the period of the inductor plus the 900 μF capacitor and let T_1 (= 0.199 s) be the period of the inductor plus the 100 μF capacitor. Close S_2, wait $T_2/4$; quickly close S_1, then open S_2; wait $T_1/4$ and then open S_1. **37.** 8.66 mΩ **39.** (L/R) ln 2 **43.** (a) 0.0955 A; (b) 0.0119 A **45.** (a) 0.65 kHz; (b) 24 Ω **47.** (a) 6.73 ms; (b) 11.2 ms; (c) inductor; (d) 138 mH **49.** (a) $X_C = 0$, $X_L = 86.7$ Ω, $Z = 218$ Ω, $I = 165$ mA, $\phi = 23.4°$ **51.** (a) $X_C = 37.9$ Ω, $X_L = 86.7$ Ω, $Z = 206$ Ω, $I = 175$ mA, $\phi = 13.7°$ **53.** 1000 V **55.** 89 Ω **57.** (a) 224 rad/s; (b) 6.00 A; (c)

228 rad/s, 219 rad/s; (d) 0.040 **61.** 1.84 A **63.** 141 V **65.** 0, 9.00 W, 2.73 W, 1.82 W **67.** (a) 12.1 Ω; (b) 1.19 kW **69.** (a) 0.743; (b) leads; (c) capacitive; (d) no; (e) yes, no, yes; (f) 33.4 W **71.** (a) 117 μF; (b) 0; (c) 90.0 W, 0; (d) 0°, 90°; (e) 1, 0 **73.** (a) 2.59 A; (b) 38.8 V, 159 V, 224 V, 64.2 V, 75.0 V; (c) 100 W for R, 0 for L and C.

Chapter 34

RE 34-1: (a) Since the induced emf around the dotted loop must oppose the increase in $\vec{B}$. $\vec{E}$ on the right of the rectangle points down in the negative y direction. $\vec{E} + d\vec{E}$ on the left has a greater magnitude and points in the same direction. (b) Since $\vec{E} \times \vec{B}$ must be in the positive x direction, $\vec{B}$ on the right points into the paper in the negative z direction. $\vec{B} + d\vec{B}$ on the left points in the same direction as $\vec{B}$ but has a greater magnitude.

RE 34-2: In the positive x direction.

RE 34-3: For total absorption, $P_r = I/c$ independent of area, but $F_r = P_r A$ so it decreases as the area decreases.

Problems

1. 5.0×10^{-21} H **3.** $B_x = 0$, $B_y = -6.7 \times 10^{-9} \cos[\pi \times 10^{15}(t - x/c)]$, $B_z = 0$ in SI units **5.** 0.10 MJ **7.** 8.88×10^4 m^2 **9.** (a) 16.7 nT; (b) 33.1 mW/m^2 **11.** (a) 6.7 nT; (b) 5.3 mW/m^2; (c) 6.7 W **13.** (a) 87 mV/m; (b) 0.30 nT; (c) 13 kW **15.** 3.44×10^6 T/s **17.** (a) z axis; (b) 7.5×10^{14} Hz; (c) 1.9 kW/m^2 **19.** 89 cm **21.** (a) 3.5 μW/m^2; (b) 0.078 μW; (c) 1.5×10^{-17} W/m^2; (d) 110 nV/m; (e) 0.25 fT **23.** 1.0×10^7 Pa **25.** 5.9×10^{-8} Pa **27.** (a) 100 MHz; (b) 1.0 μT along the z axis; (c) 2.1 m^{-1}, 6.3×10^8 rad/s; (d) 120 W/m^2; (e) 8.0×10^{-7} N, 4.0×10^{-7} Pa **31.** 1.9 mm/s **33.** (b) 580 nm **35.** (a) 4.68×10^{11} W; (b) any chance disturbance could move the sphere from being directly above the source, and then the two force vectors would no longer be along the same axis **37.** (a) 1.9 V/m; (b) 1.7×10^{-11} Pa **39.** 3.1% **41.** 4.4 W/m^2 **43.** 2/3 **45.** (a) 2 sheets; (b) 5 sheets **47.** 0.21 **49.** 35° **51.** 0.031 **53** 19.6° or 70.4° (= 90° − 19.6°) **55.** (a) 0.50 ms; (b) 8.4 min; (c) 2.4 h; (d) 5500 B.C. **57.** (a) 515 nm, 610 nm; (b) 555 nm, 5.41×10^{14} Hz, 1.85×10^{-15} s **59.** it would steadily increase; (b) the summed discrepancies between the apparent time of eclipse and those observed from x; the radius of Earth's orbit

Chapter 35

RE 35-1: a

RE 35-2: $0.2d, 1.8d, 2.2d$.

RE 35-3: When you look into a flat mirror, you see the portion of light scattering off your face that bounces off the mirror and travels straight back into your eyes. But you assume that the light entering your eyes has traveled in a straight line to reach you, so you see an image of your face behind the mirror. The image of your face is right side up. The light from your hair hits the mirror at a slight angle and then bounces into your eyes from above which is why you see your hair on top. Left and right are a different story. If you are standing face to face with another person and your right ear points toward the east, her left ear will point toward the east. If, instead, you face a flat mirror, the light from your right ear will bounce off the mirror and enter your eyes from the east. Even though your east ear is the east ear of the image, your right ear has become the left ear of the image.

RE 35-4: Ray 1: A ray that is initially parallel to the central axis reflects as if it came originally from the focal point *behind* the mirror. Ray 2: A ray that comes from the object and is traveling toward the focal point behind the mirror emerges parallel to the central axis. Ray 3: A ray that comes from the object and is traveling toward the center of curvature C of the mirror returns along itself. Ray 4: A ray that comes from the object and reflects from the mirror at its intersection c from the central axis is reflected symmetrically from the central axis.

RE 35-5: (a) Real; (b) inverted; (c) same.

RE 35-6: (a) e; (b) virtual, same.

Problems

1. 1.48 **3.** 1.26 **5.** 1.07 m **11.** 1.22 **13.** (a) 49°; (b) 29° **15.** (a) cover the center of each face with an opaque disk of radius 4.5 mm; (b) about 0.63 **17.** (a) $\sqrt{1 + \sin^2\theta}$; (b) $\sqrt{2}$; (c) light emerges at the right; (d) no light emerges at the right **19.** 49.0° **21.** 40 cm **23.** (a) 3 **27.** new illumination is 10/9 of the old **29.** 10.5 cm **33.** (a) 2.00; (b) none **37.** $i = -12$ cm **39.** 45 mm, 90 mm **43.** 22 cm **47.** same orientation, virtual, 30 cm to the left of the second lens; $m = 1$ **53.** (a) 13.0 cm; (b) 5.23 cm; (c) −3.25; (d) 3.13; (e) − 10.2 **55.** (a) 2.35 cm; (b) decrease **57.** (a) 5.3 cm; (b) 3.0 mm

Chapter 36

RE 36-1: b (least n), c, a.

RE 36-2: (a) 3λ, 3; (b) 2.5λ, 2.5.

Problems

1. (a) 5.09×10^{14} Hz; (b) 388 nm; (c) 1.97×10^8 m/s **3.** 1.56 **5.** 22°, refraction reduces θ **7.** (a) 3.60 μm; (b) intermediate, closer to fully constructive interference **9.** (a) 0.833; (b) intermediate, closer to fully constructive interference **11.** (a) 0.216 rad; (b) 12.4° **13.** 2.25 mm **15.** 648 nm **17.** 16 **19.** 0.072 mm **21.** 6.64 μm **23.** 2.65 **25.** $y = 27\sin(\omega t + 8.5°)$ **27.** (a) 1.17 m, 3.00 m, 7.50 m; (b) no **29.** $I = \frac{1}{9}I_m[1 + 8\cos^2(\pi d \sin\theta/\lambda)]$, I_m = intensity of central maximum **31.** Fully constructively **33.** 0.117μm, 0.352 μm **35.** 70.0 nm **37.** 120 nm **39.** (a) 552 nm; (b) 442 nm **43.** 140 **45.** 1.89μm **47.** 2.4 μm **49.** $\sqrt{(m + \frac{1}{2})\lambda R}$, for $m = 0, 1, 2, \ldots$ **51.** 1.00 m **53.** $x = (D/2a)(m + \frac{1}{2})\lambda$, for $m = 0, 1, 2, \ldots$ **55.** 588 nm **57.** 1.00030

Chapter 37

RE 37-1: (a) expand, (b) expand

RE 37-2: (a) second side maximum, (b) 2.5

RE 37-3: (a) red, (b) violet

RE 37-4: Diminish

RE 37-5: (a) left, (b) less.

Problems

1. 60.4 μm **3.** (a) $\lambda_a = 2\lambda_b$; (b) coincidences occur when $m_b = 2m_a$ **5.** (a) 70 cm; (b) 1.0 mm **7.** 1.77 mm **11.** (d) 53°, 10°, 5.1° **13.** (b) 0 rad, 4.493 rad, etc.; (c) −0.50, 0.93, etc. **15.** (a) 1.3×10^{-4} rad; (b) 10 km **17.** 50 m **19.** (a) 1.1×10^4 km; (b) 11 km **21.** 27 cm **23.** (a) 0.347°; (b) 0.97° **25.** (a) 8.7×10^{-7} rad; (b) 8.4×10^7 km; (c) 0.025 mm **27.** five **29.** (a) 4; (b) every fourth bright fringe is missing **31.** (a) nine; (b) 0.255 **33.** (a) 3.33 μm; (b) 0.0°, ±10.2°, ±20.7°, ±32.0°, ±45.0°, ±62.2° **35.** three **37.** (a) 6.0 μm; (b) 1.5 μm; (c) $m = 0, 1, 2, 3, 5, 6, 7, 9,$ **39.** 1100 **47.** 3650 **53.** 0.26 nm **55.** 39.8 pm **59.** (a) $a_0/\sqrt{2}$, $a_0/\sqrt{5}$, $a_0/\sqrt{10}$, $a_0/\sqrt{13}$, $a_0/\sqrt{17}$ **61.** 30.6°, 15.3° (clockwise); 3.08°, 37.8° (counterclockwise)

Chapter 38

RE 38-1: We observe that the second train is moving with respect to our train. The "slight vibration" we feel is evidence that our own train is moving along the tracks, but this does not tell us either the speed or the direction of that motion. Without this information on our own motion, we cannot determine whether or not the second train is at rest with respect to the tracks.

RE 38-2: (a) Our measured value of the speed of light is equal to its value measured by the rider. (b) With respect to our frame, it takes some time for the light to move from one end of the boxcar to the other. During that time the boxcar moves in a direction opposite to that of the light. As a result, we measure the distance between emission and absorption of the light to be smaller than the length of the boxcar. (c) Part (b) shows that the distance between emission and absorption is shorter in our frame than in the frame of the rider on the boxcar. The speed of light is the same for both of us. Therefore, the time between emission and detection is shorter as measured in our frame is shorter than the time measured in the boxcar frame. (You should revisit this analysis after reading Section 38-12 Lorentz Contraction. Will this re-analysis lead to the same conclusion or a different one?)

RE 38-3: These questions concern individual impressions, so there are no objective answers. Here are mine: Halfway through the performance I would experience it as a whole series of events: hard parts, easy parts, mistakes! Those who printed the program probably listed the Minute Waltz as one event in the concert. Looking back ten years later, I will probably (but not necessarily) remember it as a single event.

RE 38-4: (a) Recall that, in general, distance = velocity*time. We know the velocity (c) and the distance (30 meters) of the returning light pulse. Therefore the time taken for this return is (30 m)/(3×10^8 m/s) = 10^{-7} second = 0.1 microsecond. Therefore the pulse arrived at detector B 0.225 − 0.1 = 0.125 microsecond after it passed us at detector A. (b) The proton pulse left detector A at $t = 0$ and, according to part (a) arrived at detector B at $t = 0.125$ microseconds. Therefore its speed from A to B is (30 m)/(0.125×10^{-6} s) = 2.4×10^8 m/sec, or 2.4/3 = 0.8 of the speed of light.

RE 38-5: Decay reduces the remaining number of pions by a factor of two for every 25 meters of distance they travel (at that particular speed, whatever it is). So there will be one-quarter remaining after 50 meters of travel and one-eighth at a distance of 75 meters from the target.

RE 38-6: All the clocks will run at the rate of every other clock. If this were not so, you could use the difference between rates of different clocks to detect which inertial reference frame you are in, contrary to the principle of relativity.

RE 38-7: Rearrange Eq. 38-3 to read $\Delta\tau/\Delta t = \sqrt{1 - v^2/c^2}$. Square both sides of this equation, solve for v^2/c^2, and substitute the values given in the statement of the exercise, $v^2/c^2 = 1 - \Delta\tau/\Delta t = 1 - 1/1.01 = 0.0099$. Take the square root of both sides to obtain approximately $v/c = 0.1$. This is a rough-and-ready criterion for the speed above which relativistic effects become significant in reasonably accurate experiments.

RE 38-8: The time a light pulse takes to travel one way from Earth's surface to the moon's surface is 3.76×10^8 m/3.00×10^8 m = 1.25 second. The two firecrackers, one on each surface explode one second apart in the earth-moon frame. Nothing, not even light can travel from the first explosion to the second explosion. Therefore one explosion cannot have caused the other one.

RE 38-9: Music has been emitted from the tape player. There are vibrations in the air. This is a fact that must be true in both frames of reference. (For example, it might be arranged to have the noise set off a firecracker, whose explosion must be acknowledged by all.) Air currents and distance permitting, Sam on the ground will be able to hear the music sometime (with what distortions we do not bother to analyze here). When Sam and Susan meet over coffee, they will both verify that some tape has been wound from one spool to the other in the tape recorder.

RE 38-10: Rearrange Eq. 38-17 to read $E/mc^2 = (1 - v^2/c^2)^{-1/2}$. Take the reciprocal of both sides, then square both sides and substitute values for the ratio of energy to rest energy given in the statement of the exercise. The result is $(mc^2/E)^2 = 1/4 = 1 - v^2/c^2$. Rearrange and take a square root to obtain $v = \sqrt{3/4}\,c = 0.866c$.

RE 38-11: The algebraic equations for this solution are essentially identical to those for the solution to the preceding reading exercise 38-10. Rearrange Eq. 38-28 to read $\Delta x'/\Delta x = (1 - v^2/c^2)^{-1/2}$. Take the reciprocal of both sides, then square both sides and substitute values for the ratio of measured lengths given in the statement of the exercise. The result is $(\Delta x'/\Delta x)^2 = 1/4 = 1 - v^2/c^2$. Rearrange and take a square root to obtain $v = \sqrt{3/4}\,c = 0.866c$.

RE 38-12: The light flash will move with speed c in our frame; this is a basic assumption of special relativity (Section 38-3). Verify this result by substituting the values $u' = c$ and $v^{rel} = 0.9c$ into Eq. 38-31.

$$u = \frac{c + v^{rel}}{1 + cv^{rel}/c^2} = \frac{c + 0.9c}{1 + 0.9c^2/c^2} = \frac{1.9c}{1.9} = c \text{ as we predicted.}$$

RE 38-13: Square both sides of Eq. 38-33 and multiply through by the resulting denominator: $(f/f_0)^2(1 + v^{rel}/c) = (1 - v^{rel}/c)$. Solve for v^{rel}

$$v^{rel} = \frac{1 - (f/f_0)^2}{1 + (f/f_0)^2}c = \frac{1 - 0.81}{1 + 0.81}c = 0.1c.$$

Problems

1. (a) $v/c = 3.16 \times 10^{-18}$ (b) $v/c = 9.26 \times 10^{-8}$ (c) $v/c = 2.87 \times 10^{-6}$ (d) $v/c = 10^{-4}$ **3.** EACH of the identical experiments should give the same result in the uniformly moving train as in the closed freight container. **5.** $v/c = 0.990$ or $v = 2.97 \times 10^8$ m/s **7.** You set your clock to the time 2×10^{-4} s. **9.** $\Delta\tau = 4.7 \times 10^{-8}$ s and $\Delta t = 17 \times 10^{-8}$ s. Therefore $\Delta t/\Delta\tau = 3.6$ **11.** (a) 26.3 y (b) 52.3 y (c) 3.71 y **13.** (a) $v/c = 0.995$ (b) 4.8×10^3 m (c) 480 m (d) 48 km (e) 9.8×10^4 particles will survive. **15.** (a) $v/c = 0.9999995$ (b) one year (c) It does not matter as long as the acceleration is small. **17.** (1, 2) timelike, yes; (1, 3) spacelike, no; (2,3) lightlike, yes **21.** 3.51×10^{-8} kg/y or about 35 micrograms/year **23.** 1.4467×10^{-29} kg, or 8.127 MeV **25.** (a) 1.04×10^{10} J (b) 0.116 mg **27.** (a) $v/c = [N(N+2)]^{1/2}/(N+1)(b)p = [N(N+2)]^{1/2}\,mc$ **29.** (a) $m[p^2/(2K)] - [K/(2c^2)]$. For slow particle speed this reduces to the first term, which becomes m, as expected. (b) $m/m_e = 206$ **31.** (a) The lowest total energy after the collision (equal to the total energy before the collision) leaves the products at rest. (b) Kinetic energy of each incident proton is equal to the rest energy (the mass) of one proton. (c) This incident kinetic energy is equal to 1 GeV, which is reasonable since in the zero-total-momentum frame all the incident kinetic energy goes into the creation of mass, provided that the products remain at rest. **33.** $v_x = v^{rel}$ and $v_y = v_y'[1 - (v^{rel})^2/c^2]^{1/2}$ **35.** (a) $\cos\phi = [\cos\phi' + v^{rel}/c]/[1 +$

$(v^{rel}/c)\cos\phi']$ (b) $\cos\phi_o = v^{rel}/c$ (c) $\phi_o = 8.1°$ **37.** (a) $v = 2.6 \times 10^8$ m/s (b) $L = 50$ m. **39.** (a) Yes, at an appropriate speed, proper time between two timelike events can be made as small as desired. (b) $v = 0.999\,999\,15c$ **41.** velocity with respect to the rocket $= -0.82c$ **43.** Minimum and maximum values occur when daughter particles move along direction of relative motion. $u_+ = 0.990\ c$ and $u_- = 0.282\ c$ **45.** $f = 22.9$ MHz **47.** $v^{rel} = 0.96\ c$ **49.** (a) She does not age at all. (b) Both earth and Zircon age 100 *y*. (c) 350 *y* (d) 1200 *y* on earth **51.** 31.6 s **55.** (a) 0.511 MeV (b) $M_{sys} = m + 2m_e$ (c) Mass of the system is $2m_e$ both before and after the collision. **57.** $E_M = (M^2 + m^2)c^2/(2m)$ **61.** Partial answer: Let T be the time lapse between the instant we see the sun explode and the instant we see Venus change color. Then we have time $T/3$ to escape earth after we see Venus change color. This assumes that the alien ship moves faster than the pulse emitted by the sun.

Photo Credits

Dedication Photo courtesy Jean Arons.

Introduction Opener: Courtesy of the Archives, California Institute of Technology. Reproduced with permission of the Estate of Richard Feynman. Figure I-1: FOXTROT ©1999 Bill Amend. Reprinted with permission of UNIVERSAL PRESS SYNDICATE. All rights reserved.

Preface Figures P-1, P-2 and P-3: Courtesy Priscilla Laws.

Chapter 1 Opener: Larry Bray/Taxi/Getty Images. Figure 1-2: Detlev van Ravenswaay/Photo Researchers. Figure 1-3: Courtesy Randall Feenstra, Carnegie Mellon University. Figure 1-4: ©Steven Pitkin. Figure 1-5: Courtesy National Institute of Standards and Technology. Figure 1-7: Courtesy Bureau International des Poids et Mesures, France. Figure 1-8: Courtesy Fisher Scientific. Figures 1-9, 1-10 and 1-11 (left): Courtesy Vernier Software and Technology. Figure 1-11 (right): Courtesy Pasco Scientific. Figure 1-12: Courtesy Ron Thornton. Figures 1-13 and 1-14: Courtesy Priscilla Laws. Figure 1-15: Getty Images News and Sport Services. Figure 1-20: Worldscapes/Age Fotostock America, Inc. Figure 1-21: Lynda Richardson/Corbis Images. Figure 1-22: Corbis Digital Stock.

Chapter 2 Opener: Niagara Gazette/Corbis Sygma. Figures 2-6, 2-7 and 2-9: Courtesy Pat Cooney. Figure 2-10: Courtesy Priscilla Laws. Figure 2-13: Courtesy U.S. Air Force. Figure 2-16: Courtesy Ron Thornton. Figures 2-18, 2-19, 2-20 and 2-21b: Courtesy Priscilla Laws. Figure 2-30: Courtesy Vernier Software and Technology. Figure 2-42: Courtesy Priscilla Laws.

Chapter 3 Opener: Nicole Duplaix/Corbis Images. Figure 3-1: American Institute of Physics/Photo Researchers. Figures 3-2, 3-3 and 3-5: Courtesy Priscilla Laws. Figure 3-8: Courtesy Vernier Software and Technology. Figure 3-10: Courtesy Ohaus Corporation. Figures 3-12 and 3-13: Courtesy Priscilla Laws. Page 67: ©AP/Wide World Photos. Figure 3-19a: Courtesy Priscilla Laws. Figure 3-20: James A. Sugar/Corbis Images. Figures 3-22 and 3-28: Courtesy Priscilla Laws. Figure 3-34: Clive Newton/Corbis Images. Figures 3-45, 3-55 and 3-56: Courtesy Priscilla Laws.

Chapter 4 Opener: Courtesy David des Marais, Cave Research Foundation. Figure 4-2: FOXTROT ©1999 Bill Amend. Reprinted with permission of UNIVERSAL PRESS SYNDICATE. All rights reserved.

Chapter 5 Opener: ©AP/Wide World Photos. Figure 5-1: Photo by Andrew Davidhazy/RIT. Figure 5-2: Bettmann/Corbis Images. Figure 5-4: Courtesy Priscilla Laws. Figure 5-5: Richard Megna/Fundamental Photographs. Figure 5-6: Courtesy Priscilla Laws. Figure 5-7: Jammie Budge/Gamma-Presse, Inc. Figures 5-9 and 5-10: Courtesy Priscilla Laws. Figure 5-23: Adam Woolfitt/Corbis Images. Figure 5-30, Problem 6: Allsport/Getty Images.

Chapter 6 Opener: © Natural History Photographic Agency. Page 145: ©2003 Tom Thaves. Reprinted with permission. Figure 6-14: Courtesy Priscilla Laws. Figure 6-16: Courtesy Pasco Scientific. Figure 6-17: Courtesy Priscilla Laws. Figure 6-26: Jean Y. Ruszniewki/Stone/Getty Images. Figure 6-28: Joe McBride/Stone/Getty Images. Figure 6-29: Courtesy Priscilla Laws. Figure 6-41: Jerry Schad/Photo Researchers. Figure 6-47: Susan Copen Oken/Dot, Inc. Figure 6-73: Peter Turnley/Corbis Images. Figure 6-84: Courtesy Joe Redish. Figure 6-92a: Courtesy Priscilla Laws.

Chapter 7 Opener: Terje Rakke/The Image Bank/Getty Images. Figure 7-1a: Charles and Josette Lenars/Corbis Images. Figure 7-1b: Science Photo Library/Photo Researchers. Figure 7-1c: Photo by Andrew Davidhazy/RIT. Figure 7-5: Courtesy Robert Teese. Figure 7-9: Courtesy Priscilla Laws. Figure 7-10: Courtesy PASCO scientific. Figures 7-11, 7-12, 7-13 and 7-14: Courtesy Priscilla Laws. Figure 7-17: Courtesy NASA. Page 198: FOXTROT ©2000 Bill Amend. Reprinted with permission of UNIVERSAL PRESS SYNDICATE.. All rights reserved. Figure 7-19: Courtesy NASA. Figure 7-21: George Long/Sports Illustrated/Time, Inc. Picture Collection. Figure 7-22: Superman #48, October 1990. ©D.C. Comics. All rights reserved. Reprinted with permission. Figures 7-33 and 7-38: Courtesy Priscilla Laws.

Chapter 8 Opener: Lois Greenfield. Figure 8-1a: Richard Megna/Fundamental Photographs. Figures 8-6, 8-10, 8-11 and 8-12: Courtesy Priscilla Laws. Figure 8-23: Adam Crowley/PhotoDisc, Inc./Getty Images.

Chapter 9 Opener and Figure 9-22: ©AP/Wide World Photos. Figure 9-21: ©Photri.

Chapter 10 Opener: Malcolm S. Kirk/Peter Arnold, Inc. Figure 10-1: Dimitri Lundt/Corbis Images. Figure 10-7: Photo provided courtesy of Mattel, Inc. Figure 10-9a: ©AP/Wide World Photos. Figure 10-14: Courtesy Priscilla Laws. Figure 10-19: Courtesy Mercedes-Benz of North America.

Chapter 11 Opener: Arthur Tilley/Stone/Getty Images. Figure 11-1a: Doug Pensinger/Getty Images News and Sport Services. Figure 11-1b: Duomo/Corbis Images. Figures 11-2: Courtesy PASCO scientific and Priscilla Laws. Figure 11-9: Courtesy Priscilla Laws. Page 308: Calvin and Hobbes ©1990 Bill Watterson. Reprinted with permission of UNIVERSAL PRESS SYNDICATE. All rights reserved. Figure 11-13: Roger Ressmeyer/Corbis Images. Figure 11-16: Courtesy Test Devices, Inc. Figure 11-26: Courtesy Lick Observatory. Figure 11-34: Courtesy Lawrence Livermore Laboratory, University of California. Figure 11-50: Courtesy Mark Luetzelschwab.

Chapter 12 Opener: Image courtesy Ringling Brothers and Barnum & Bailey®, THE GREATEST SHOW ON EARTH. Figure 12-1: Richard Megna/Fundamental Photographs. Figure 12-2: Courtesy PASCO Scientific. Figure 12-15: From *Shepp's World's Fair* Photographed by James W. Shepp and Daniel P. Shepp, Globe Publishing Co., Chicago and Philadelphia, 1893. Photo provided courtesy of Jeffery Howe.

Chapter 13 Opener: Greg Epperson/Age Fotostock America, Inc. Page 363 (top): David Noton/Age Fotostock America, Inc. Page 363 (bottom): Andy Levin/Photo Researchers. Page 375 (right): Courtesy PASCO Scientific. Page 375 (bottom): Courtesy Vishay Micro-Measurements Group, Raleigh, NC. Page 378: Worldscapes/Age Fotostock America, Inc.

Chapter 14 Opener: Courtesy NASA. Page 386: Courtesy Jon Lomberg. Page 394: Mark Simons, California Institute of Technology/Photo Researchers. Page 406 (center): Courtesy National Radio Astronomy Observatory. Page 406 (right): Courtesy NASA.

Chapter 15 Opener: Peter Atkinson/The Image Bank/Getty Images. Page 419 (left): Courtesy Vernier Software and Technology. Page 419 (right): Courtesy PASCO Scientific. Page 425 (top and bottom): Corbis Sygma. Page 427: Courtesy Carol Everett. Page 428: Will McIntyre/Photo Researchers. Page 429 (left): Courtesy D. H. Peregrine,University of Bristol. Page 429 (right): Courtesy Volvo North America Corporation. Page 443: David Parker/Science Photo Library/Photo Researchers.

Chapter 16 Opener: Owen Franken/Corbis Images. Page 445 (left): PhotoDisc, Inc./Getty Images. Page 445 (right): Digital Vision/Getty Images. Pages 451 and 456: Courtesy Priscilla Laws. Page 474: David Wall/Age Fotostock America, Inc.

Chapter 17 Opener: John Visser/Bruce Coleman, Inc. Page 493: Courtesy Education Development Center. Page 500: Richard Megna/Fundamental Photographs. Page 501: Courtesy Thomas D. Rossing, Northern Illinois University.

Chapter 18 Opener: Stephen Dalton/Animals Animals. Page 513 (top): Courtesy Virginia Jackson. Page 513 (bottom): Courtesy Kerry Browne. Page 514: Courtesy Sara Settlemyer. Page 521: Ben Rose/The Image Bank/Getty Images. Page 524 (top): Bob Gruen/Star File. Page 524 (bottom): Jaroslav Kubec/HAGA/The Image Works. Page 534: U.S. Navy photo by Ensign John Gay.

Chapter 19 Opener: Courtesy Dr. Masato Ono, Tamagawa University. Page 541 (top): PhotoDisc, Inc./Getty Images. Page 541 (center): Adam Hart-Davis/Photo Researchers. Page 541 (bottom): Damien Lovegrove/Photo Researchers. Page 546: Courtesy Pat Cooney. Page 548: Courtesy Priscilla Laws. Page 563: ©AP/Wide World Photos. Page 568: Alfred Pasieka/Photo Researchers. Page 569: Courtesy Dr. Masato Ono, Tamagawa University.

Chapter 20 Opener: Tom Branch/Photo Researchers.

Chapter 21 Opener: Stephen Dalton/Photo Researchers. Page 619:Tim Wright/-Corbis Images.

Chapter 22 Opener: Fundamental Photographs. Page 634: Vaughan Fleming/Photo Researchers. Page 644: Johann Gabriel Doppelmayr, Neuentdeckte Phaenenomena von Bewünderswurdigen Würckungen der Natur, Nuremberg 1744. Page 645: Courtesy Priscilla Laws.

Chapter 23 Opener: Courtesy Paula Brakke.

Chapter 24 Opener: Peter Menzel.

Chapter 25 Opener: Larry Lee/Corbis Images. Page 717: Courtesy PASCO scientific. Pages 738 (top) and 739: Courtesy NOAA. Page 738 (bottom): Courtesy Westinghouse Corporation.

Chapter 26 Opener: Corbis-Bettmann. Page 752: Courtesy Priscilla Laws. Page 755: The Image Works. Page 758: Tim Flach/Stone/Getty Images. Page 767: Courtesy Shoji Tonaka/International Superconductivity Technology Center, Tokyo, Japan.

Chapter 27 Opener: George Grall/National Geographic Society.

Chapter 28 Opener: Photo by Harold E. Edgerton. ©The Harold and Esther Edgerton Family Trust, courtesy of Palm Press, Inc. Page 800 (top): Lester V. Bergman/Corbis Images. Pages 800 (bottom) and 801: Courtesy Priscilla Laws. Page 806: Courtesy Timothy Settlemyer. Page 815: The Royal Institution, England/Bridgeman Art Library/NY.

Chapter 29 Opener: EFDA-JET/Photo Researchers. Page 830: Jeremy Walker/Photo Researchers. Page 836: Lawrence Berkeley Laboratory/Photo Researchers. Page 837: Courtesy Dr. Richard Cannon, Southeast Missouri State University, Cape Girardeau. Page 840: Courtesy John Le P. Webb, Sussex University, England. Page 841: Courtesy EFDA-JET, www.jet.efda.org.

Chapter 30 Opener: NASDA/Gamma-Presse, Inc. Page 863: Courtesy Education Development Center.

Chapter 31 Opener: Copyright General Motors Corporation. Page 889: Science Photo Library/Photo Researchers. Pages 894 and 897 (top): Courtesy PASCO scientific. Page 897 (bottom): Joseph Sia/Archive Photos/Hulton Archive/Getty Images.

Chapter 32 Opener: Courtesy Dr. Timothy St. Pierre, University of Western Australia. Page 923: The Royal Institution, England/Bridgeman Art Library/NY. Pages 936 and 943: Yoav Levy/Phototake. Page 941: Courtesy Ralph W. DeBlois. Page 945: Courtesy Andre Geim, University of Manchester, U.K. Page 946: Mehau Kulyk/Photo Researchers. Page 947 (top): Courtesy Greg Foss, Pittsburgh Supercomputing Center; research and data: Gary Glatzmaier, USC; Earth map provided by NOAA/NOS. Page 947 (bottom): Courtesy Dr. Timothy St. Pierre, University of Western Australia.

Chapter 33 Opener: Photo by Rick Diaz, provided courtesy Haverfield Helicopter Co. Page 955 (top): Corbis Images. Page 955 (bottom): Bettmann/Corbis Images. Page 960: Courtesy Agilent Technologies.

Chapter 34 Opener: Chris Madeley/Science Photo Library/Photo Researchers. Page 986: Baldwin H. Ward/Corbis Images. Page 1006: Diane Schiumo/Fundamental Photographs.

Chapter 35 Opener: Courtesy Courtauld Institute Galleries, London. Page 1017: From *PSSC Physics,* 2nd edition; ©1975 D.C. Heath and Co. with Education Development Center, Newton, MA. Page 1020 (top): Courtesy Bausch & Lomb. Pages 1020 (bottom) and 1022: PhotoDisc, Inc./Getty Images. Page 1033: Dr. Paul A. Zahl/Photo Researchers. Page 1037: Courtesy Matthew J. Wheeler.

Chapter 36 Opener: Gail Shumway/Taxi/Getty Images. Pages 1063 and 1081: From Michel Cagnet, Maurice Franzon, and Jean Claude Thierr, *Atlas of Optical Phenomena,* Springer-Verlag, New York, 1962. Reproduced with permission. Page 1073: Richard Megna/Fundamental Photographs. Page 1080: Courtesy Bausch & Lomb.

Chapter 37 Opener: Georges Seurat, *A Sunday on La Grande Jatte,* 1884; oil on canvas, 207.5 × 308 cm, Helen Birch Bartlett Memorial Collection; photograph ©2003, The Art Institute of Chicago. All rights reserved. Page 1084: Ken Kay/Fundamental Photographs. Pages 1085, 1092 (top) and 1095: From Michel Cagnet, Maurice Franzon, and Jean Claude Thierr, *Atlas of Optical Phenomena,* Springer-Verlag, New York, 1962. Reproduced with permission. Page 1093: P.M. Motta & S. Correr/Photo Researchers. Page 1099: Department of Physics, Imperial College/Science Photo Library/Photo Researchers. Page 1100: Steve Percival/Science Photo Library/Photo Researchers. Page 1106: Kjell B. Sandved/Bruce Coleman, Inc. Page 1107: Pekka Parvianen/Photo Researchers.

Chapter 38 Opener: Courtesy Fermi National Accelerator Laboratory. Page 1112: Corbis Images.

Data Credits

Chapter 15 Page 417(top): Data obtained by David Vernier. Page 417 (bottom): Data obtained by Priscilla Laws.

Chapter 16 Pages 446, 457, 464, and 465: Data obtained by Priscilla Laws.

Chapter 18 Pages 516 and 517: Data obtained by Priscilla Laws. Pages 526-527: Data obtained by Priscilla Laws and David and Ginger Hildebrand.

Chapter 19 Pages 544-545: Data obtained by Priscilla Laws.

Chapter 33 Pages 645, 803, and 892: Courtesy Priscilla Laws.

Index

Page references followed by italic *table* indicate material in tables.
Page references followed by italic *n* indicate material in footnotes.

C

F

G

H

I

N

O

P

Q

R

S

T

U

V

Mathematical Formulas*

Quadratic Formula

If $ax^2 + bx + c = 0$, then $x = \dfrac{-b \pm \sqrt{b^2 - 4ac}}{2a}$

Binomial Theorem

$$(1 + x)^n = 1 + \frac{nx}{1!} + \frac{n(n-1)x^2}{2!} + \cdots \qquad (x^2 < 1)$$

Products of Vectors

Let θ be the smaller of the two angles between $\vec{a}$ and $\vec{b}$. Then

$$\vec{a} \cdot \vec{b} = \vec{b} \cdot \vec{a} = a_x b_x + a_y b_y + a_z b_z = |\vec{a}||\vec{b}| \cos \theta$$

$$\vec{a} \times \vec{b} = -\vec{b} \times \vec{a} = \begin{vmatrix} \hat{\mathrm{i}} & \hat{\mathrm{j}} & \hat{\mathrm{k}} \\ a_x & a_y & a_z \\ b_x & b_y & b_z \end{vmatrix}$$

$$= \hat{\mathrm{i}} \begin{vmatrix} a_y & a_z \\ b_y & b_z \end{vmatrix} - \hat{\mathrm{j}} \begin{vmatrix} a_x & a_z \\ b_x & b_z \end{vmatrix} + \hat{\mathrm{k}} \begin{vmatrix} a_x & a_y \\ b_x & b_y \end{vmatrix}$$

$$= (a_y b_z - b_y a_z)\hat{\mathrm{i}} + (a_z b_x - b_z a_x)\hat{\mathrm{j}} + (a_x b_y - b_x a_y)\hat{\mathrm{k}}$$

$$|\vec{a} \times \vec{b}| = |\vec{a}||\vec{b}| \sin \theta$$

Trigonometric Identities

$$\sin \alpha \pm \sin \beta = 2 \sin \tfrac{1}{2}(\alpha \pm \beta) \cos \tfrac{1}{2}(\alpha \mp \beta)$$

$$\cos \alpha + \cos \beta = 2 \cos \tfrac{1}{2}(\alpha + \beta) \cos \tfrac{1}{2}(\alpha - \beta)$$

Derivatives and Integrals

$$\frac{d}{dx} \sin x = \cos x \qquad \int \sin x \, dx = -\cos x$$

$$\frac{d}{dx} \cos x = -\sin x \qquad \int \cos x \, dx = \sin x$$

$$\frac{d}{dx} e^x = e^x \qquad \int e^x \, dx = e^x$$

$$\int \frac{dx}{\sqrt{x^2 + a^2}} = \ln(x + \sqrt{x^2 + a^2})$$

$$\int \frac{x \, dx}{(x^2 + a^2)^{3/2}} = -\frac{1}{(x^2 + a^2)^{1/2}}$$

$$\int \frac{dx}{(x^2 + a^2)^{3/2}} = \frac{x}{a^2(x^2 + a^2)^{1/2}}$$

Cramer's Rule

Two simultaneous equations in unknowns x and y,

$$a_1 x + b_1 y = c_1 \qquad \text{and} \qquad a_2 x + b_2 y = c_2,$$

have the solutions

$$x = \frac{\begin{vmatrix} c_1 & b_1 \\ c_2 & b_2 \end{vmatrix}}{\begin{vmatrix} a_1 & b_1 \\ a_2 & b_2 \end{vmatrix}} = \frac{c_1 b_2 - c_2 b_1}{a_1 b_2 - a_2 b_1}$$

and

$$y = \frac{\begin{vmatrix} a_1 & c_1 \\ a_2 & c_2 \end{vmatrix}}{\begin{vmatrix} a_1 & b_1 \\ a_2 & b_2 \end{vmatrix}} = \frac{a_1 c_2 - a_2 c_1}{a_1 b_2 - a_2 b_1}.$$

* See Appendix E for a more complete list.

The Greek Alphabet

Alpha	A	α	Iota	I	ι	Rho	P	ρ
Beta	B	β	Kappa	K	κ	Sigma	Σ	σ
Gamma	Γ	γ	Lambda	Λ	λ	Tau	T	τ
Delta	Δ	δ	Mu	M	μ	Upsilon	Y	υ
Epsilon	E	ϵ	Nu	N	ν	Phi	Φ	ϕ, φ
Zeta	Z	ζ	Xi	Ξ	ξ	Chi	X	χ
Eta	H	η	Omicron	O	o	Psi	Ψ	ψ
Theta	Θ	θ	Pi	Π	π	Omega	Ω	ω